DOWNSTREAM INDUSTRIAL BIOTECHNOLOGY

DOWNSTREAM INDUSTRIAL BIOTECHNOLOGY

Recovery and Purification

Edited By

MICHAEL C. FLICKINGER

Golden LEAF Biomanufacturing Training and Education Center (BTEC)
Department of Chemical and Biomolecular Engineering
North Carolina State University, Raleigh
North Carolina, USA

WILEY

A JOHN WILEY & SONS, INC., PUBLICATION

Published by John Wiley & Sons, Inc., Hoboken, New Jersey
Published simultaneously in Canada

Library of Congress Cataloging-in-Publication Data:

Encyclopedia of industrial biotechnology. Selections.
Downstream industrial biotechnology : recovery and purification / edited by Michael C. Flickinger.
pages cm
Includes bibliographical references and index.
ISBN 978-1-118-13124-4 (hardback)
1. Biotechnology–Encyclopedias. I. Flickinger, Michael C., editor of compilation. II. Title.
TP248.16.E533 2013
660.6–dc23

2012030526

10 9 8 7 6 5 4 3 2 1

CONTENTS

PREFACE

Downstream Industrial Biotechnology is a compilation of essential in depth articles, organized topically and listed in alphabetical format, for biopharmaceutical, bioprocess and biologics process scientists, engineers and regulatory professionals from the comprehensive seven volumes of the *Encyclopedia of Industrial Biotechnology*. Process development for the manufacture of complex biomolecules involves solving many scientific, compliance and technical problems quickly in order to support pilot, preclinical and clinical development, technology transfer and manufacturing start-up. Every organization develops new processes from accumulated process knowledge. Accumulated process knowledge has a very significant impact on accelerating the time to market (and reducing the financial resources required) of products manufactured using recombinant DNA and living microbes, cells, transgenic plants or transgenic mammals. However, when an entirely new upstream platform or downstream unit operation is needed, there are few books that will quickly provide the depth of industry-relevant background. *Downstream Industrial Biotechnology* can fill this void as an advanced desk reference. This volume includes relevant biology, protein purification and engineering literature with abundant process examples provide by industry subject matter experts (SMEs) and academic scholars. This desk reference will also be useful for advanced biomanufacturing students and professionals to quickly gain in depth knowledge on how to design processes (and facilities) capable of being licensed to manufacture enzymes, biopharmaceutical intermediates, human and veterinary biopharmaceuticals or vaccines. The opportunity is yours to leverage the combined knowledge from scores of industry professionals from around the world who have contributed to *Downstream Industrial Biotechnology* to reduce the time and cost to deliver engineered proteins, biomolecules and cost-effective biologics to the market and especially to millions of patients worldwide.

Professor Michael C. Flickinger, Editor

Golden LEAF Biomanufacturing Training and Education Center (BTEC)
Department of Chemical and Biomolecular Engineering
North Carolina State University
Raleigh, North Carolina, 27695-7928, USA

Contributors

Muhammad Aasim, Downstream Bioprocessing Laboratory, School of Engineering and Science, Jacobs University, Bremen, Germany

Oscar Aguilar, Centro de Biotecnología Tecnológico de Monterrey, Monterrey, México

Mattias Ahnfelt, GE Healthcare Bio-Sciences AB, Uppsala, Sweden

Hazel Aranha, GAEA Resources Inc., Northport, New York, USA

Claude Artois, University of Surrey, Guildford, Surrey, United Kingdom; SmithKline Beecham Biologicals, Rixensart, Belgium

Hans Axelsson, Alfa Laval AB, Tumba, Sweden

Diana C.S. Azevedo, Federal University of Ceará, Fortaleza-CE, Brazil

H.S.C. Barbosa, Center of Chemistry, University of Minho, Campus de Gualtar, Braga, Portugal

Sonja Berensmeier, Technische Universität München, Institute of Biochemical Engineering, Garching, Germany

Joseph Bertolini, CSL Bioplasma, Broadmeadows, Victoria, Australia

Darcy Birse, Fast Trak Biopharma Services, GE Healthcare, Piscataway, New Jersey, USA

Eggert Brekkan, GE Healthcare Bio-Sciences AB, Uppsala, Sweden

Phil J. Bremer, University of Otago, Dunedin, New Zealand

Kurt Brorson, Office of Biotech Products, Center for Drug Evaluation and Research, Food and Drug Administration, Silver Spring, Maryland, USA

Kurt Brorson, Office of Pharmaceutical Science, Center for Drug Evaluation and Research, United States Food and Drug Administration

Thierry Burnouf, Human Protein Process Sciences, Lille, France

Trent Carrier, Invitrogen, part of Life Technologies, Grand Island, New York, USA

David Clark, Centocor R&D, Spring House, Pennsylvania, USA

Efrem Curcio, University of Calabria, Arcavacata di Rende (CS), Italy

Jean Didelez, University of Surrey, Guildford, Surrey, United Kingdom; SmithKline Beecham Biologicals, Rixensart, Belgium

Gianluca Di Profio, Institute on Membrane Technology (ITM-CNR), c/o University of Calabria, Arcavacata di Rende (CS), Italy; University of Calabria, Arcavacata di Rende (CS), Italy

Dennis Dobie, Fluor Daniel, Marlton, New Jersey, USA

Ed Domanico, Tri-Clover, Valencia, California, USA

Enrico Drioli, Institute on Membrane Technology ITM-CNR, At University of Calabria, Rende, Italy

Zhiwu Fang, Amgen Inc., Systems Informatics, Thousand Oaks, California, USA

Patrick Florent, University of Surrey, Guildford, Surrey, United Kingdom; SmithKline Beecham Biologicals, Rixensart, Belgium

Matthias Franzreb, Karlsruhe Institute of Technology, Institute for Functional Interfaces, Eggenstein-Leopoldshafen, Germany

Pete Gagnon, Validated Biosystems, San Clemente, California, USA

F.A.P. Garcia, University of Coimbra, Coimbra, Portugal

Tom Gervais, Centocor R&D Spring House, Pennsylvania, USA

Iraj Ghazi, The Ohio State University, Columbus, Ohio, USA

Siddartha Ghose, Aston University, Birmingham, United Kingdom

Guy Godeau, University of Surrey, Guildford, Surrey, United Kingdom; SmithKline Beecham Biologicals, Rixensart, Belgium

Susanne Gräslund, Structural Genomics Consortium, Karolinska Institutet, Stockholm, Sweden

Tingyue Gu, Ohio University, Athens, Ohio, USA

Martin Hammarström, Structural Genomics Consortium, Karolinska Institutet, Stockholm, Sweden

Richard Hassett, Invitrogen, part of Life Technologies, Grand Island, New York, USA

Eva Heldin, GE Healthcare Bio-Sciences AB, Uppsala, Sweden

Nathaniel G. Hentz, PhD, North Carolina State University, Golden LEAF Biomanufacturing Training and Education Center, Raleigh, North Carolina, USA

Birgit Hickstein, Clausthal University of Technology, Institute of Chemical Process Engineering, Clausthal-Zellerfeld, Germany

Timothy John Hobley, Technical University of Denmark, Systems of Biology, Lyngby, Denmark

Tony Hunt, Advanced Minerals Corporation, Santa Barbara, California, USA

Omkar Joshi, Bayer HealthCare LLC, Berkeley, California, USA

Varsha S. Joshi, Chemical Engineering Department, Indian Institute of Technology Delhi, Hauz Khas, New Delhi, India

Adalberto Pessoa Jr., School of Pharmaceutical Sciences, University of São Paulo, Brazil

Amaro G. Barreto Jr., Escola de Química, Universidade Federal do Rio de Janeiro, Rio de Janeiro-RJ, Brazil

Ivanildo J. Silva Jr., Federal University of Ceará, Fortaleza-CE, Brazil

Beth H. Junker, Bioprocess R&D Merck Research Laboratories, Rahway, New Jersey, USA

Manohar Kalyanpur, Consultant, Bioseparations & Pharmaceutical Validation, Plaisir, France

Ingo Kampen, Technische Universität, Institute for Particle Technology, Braunschweig, Germany

Mansoor A. Khan, Office of Pharmaceutical Science, Center for Drug Evaluation and Research, United States Food and Drug Administration

Alexandros Koulouris, Intelligen Europe, Thermi, Greece

Maria-Regina Kula, Heinrich Heine University Düsseldorf, Jülich, Germany

Ingo Kampen Arno Kwade, Technische Universität, Institute for Particle Technology, Braunschweig, Germany

Per Kårsnäs, Institute of Biology and Chemical Engineering, Mälardalens högskola, Eskilstuna, Sweden

Marcelo Fernández Lahore, Downstream Bioprocessing Laboratory, School of Engineering and Science, Jacobs University, Bremen, Germany

Philippe Lam, Pharmaceutical Development Genentech, Inc., South San Francisco, California, USA

Chung Lim Law, The University of Nottingham, Malaysia Campus, Selangor, Malaysia

Jinsong Liu, Product Development, Abraxis BioScience, Melrose Park, Illinois, USA

Scott Lute, Office of Biotech Products, Center for Drug Evaluation and Research, Food and Drug Administration, Silver Spring, Maryland, USA

Pérola O. Magalhães, University of Brasília, Brasília, DF, Brazil

Robert Z. Maigetter, Centocor R&D, Spring House, Pennsylvania, USA

J.C. Marcos, Center of Chemistry, University of Minho, Campus de Gualtar, Braga, Portugal

Joseph McGuire, Oregon State University, Corvallis, Oregon, USA

George Miesegaes, Office of Biotech Products, Center for Drug Evaluation and Research, Food and Drug Administration, Silver Spring, Maryland, USA

Jamie Moore, Pharmaceutical Development Genentech, Inc., South San Francisco, California, USA

Manuel Mota, IBB, Centro de Eng. Biológica, University of Minho, Portugal

Arun S. Mujumdar, National University of Singapore, Singapore

P.T. Noble, Fluor Daniel GmbH, Wiesbaden, Germany

Jeffery N. Odum, CPIP Biotech Sector Lead & Director of Operations Integrated Project Services

Jeffery Odum, IPS, Morrisville (RTP), North Carolina, USA

Victor Papavasileiou, Intelligen Europe, Leiden, The Netherlands

Jun T. Park, Office of Pharmaceutical Science, Center for Drug Evaluation and Research, United States Food and Drug Administration

Steve Peppers, Invitrogen, part of Life Technologies, Grand Island, New York, USA

Demetri Petrides, Intelligen, Inc., Scotch Plains, New Jersey, USA

Urs Alexander Peuker, TU Bergakademie Freiberg, Institute for Mechanical Process Engineering and Mineral Processing, Freiberg, Germany

John Pieracci, Biogen Idec, San Diego, California, USA

Tom Piombino, Integrated Project Services, Inc., Lafayette Hill, Pennsylvania, USA

Tina Pitarresi, Fast Trak Biopharma Services, GE Healthcare, Piscataway, New Jersey, USA

Mirjana Radosevich, Human Protein Process Sciences, Lille, France

Anurag S. Rathore, Chemical Engineering Department, Indian Institute of Technology Delhi, Hauz Khas, New Delhi, India

Erik K. Read, Office of Pharmaceutical Science, Center for Drug Evaluation and Research, United States Food and Drug Administration

James J. Reilly, Laureate Pharma, Inc., Princeton, New Jersey, USA

Robert van Reis, Genentech, Inc., South San Francisco, California, USA

Craig Robinson, GE Healthcare, Westborough, Massachusetts, USA

Carl A. Rockburne, The Rockburne Group, Atlanta, Georgia, USA

Gustav Rodrigo, GE Healthcare Bio-Sciences AB, Uppsala, Sweden

Cesar C. Santana, School of Chemical Engineering, State University of Campinas, Campinas-SP, Brazil

Maria Schäfer, TU Bergakademie Freiberg, Institute for Mechanical Process Engineering and Mineral Processing, Freiberg, Germany

Catherine H. Schein, Sealy Center for Structural Biology and Molecular Biophysics, Sealy Center for Vaccine Development, University of Texas Medical Branch, Galveston, Texas, USA

Richard Brent Seale, University of Otago, Dunedin, New Zealand

Klaus Selber, Heinrich Heine University Düsseldorf, Jülich, Germany

Rakhi B. Shah, Office of Pharmaceutical Science, Center for Drug Evaluation and Research, United States Food and Drug Administration

Bryan Shingle, Centocor R&D Spring House, Pennsylvania, USA

Charles Siletti, Intelligen, Inc., Mt. Laurel, New Jersey, USA

Eduardo V. Soares, Bioengineering Laboratory, Superior Institute of Engineering from Porto Polytechnic Institute, Porto, Portugal; IBB-Institute for Biotechnology and Bioengineering, Centre for Biological Engineering, Universidade do Minho, Braga, Portugal

Gail Sofer, GE Healthcare, Piscataway, New Jersey, USA

Bob Stover, Tri-Clover, Valencia, California, USA

Jörg Thömmes, Biogen Idec, San Diego, California, USA

Owen Thomas, University of Birmingham, Biochemical Engineering, Birmingham, United Kingdom

Claudio Thomasin, Centocor R&D, Spring House, Pennsylvania, USA

Michiel E. Ultee, Laureate Pharma, Inc., Princeton, New Jersey, USA

Greg Van Slyke, Invitrogen, part of Life Technologies, Grand Island, New York, USA

Rami Reddy Vennapusa, Downstream Bioprocessing Laboratory, School of Engineering and Science, Jacobs University, Bremen, Germany

Gary Walsh, Industrial Biochemistry and Materials Surface Sciences Institute, University of Limerick, Limerick City, Ireland

Dave A. Wareheim, Centocor R&D Spring House, Pennsylvania, USA

Sandy Weinberg, Clayton State University, Atlanta, Georgia, USA

Christian Wood, Centocor R&D Spring House, Pennsylvania, USA

David W. Wood, The Ohio State University, Columbus, Ohio, USA

Alexander Yelshin, Polotsk State University, Novopolotsk, Belarus

Inna Yelshina, Polotsk State University, Novopolotsk, Belarus

Jonathan Yourkin, GE Instruments, Boulder, Colorado

David (Xiaojian) Zhao, Invitrogen, part of Life Technologies, Grand Island, New York, USA

Andrew L. Zydney, The Pennsylvania State University, University Park, Pennsylvania, USA

PART I

INTRODUCTION

INTRODUCTION

Downstream biomanufacturing processes increase product concentration and purity, while decreasing process volume. Therefore, *decreasing process volume without loss of product* is essential to increase product purity, while at the same time eliminating product contaminants. The biochemistry of different products (peptides, proteins, hormones, low-molecular-weight metabolic intermediates, complex antigens etc.), all of which are liable to degradation, dictates that different separation methods be used to isolate and purify these products from contaminating biomolecules produced by the upstream process. Optimal downstream product yield is the yield of recovered product in the appropriate final biologically active form and purity. Purified but inactive product is a contaminant, reduces overall process yield, and may have serious consequences on clinical safety and efficacy. That is why downstream process design has the greatest impact on the overall biomanufacturing cost.

As product purity increases, more product can be lost to inactivation, nonspecific binding to equipment surfaces, binding to membranes, and chromatography media or by precipitation, thus decreasing the recovery of product. Because of these potential losses, each additional separation step may reduce overall yield. Therefore, downstream separation scientists and engineers are continually seeking to eliminate or combine unit operations to minimize the number of process steps in order to maximize product recovery at a specified concentration and purity.

Section II of *Downstream Industrial Biotechnology* includes detailed methods used for the initial steps of cell separation, cell disruption (for intracellular products), filter aids and adsorbents for rapid protein capture and initial volume reduction. Each of these steps is critically affected by upstream process design (volume, product concentration, and contaminants derived from the growth media or host cells), which impacts every subsequent step of downstream product recovery and purification. In particular, cell separation and cell disruption methods can have a dramatic effect on contributing (or minimizing) contaminants such as nucleic acids, host cell proteins, cell membrane fragments or pyrogenic lipopolysaccharides that need to be removed from the final product in subsequent separation steps.

Although each upstream process decision impacts downstream product recovery and purification, not all contaminants come from upstream operations. In some cases contaminants can also be generated by downstream operations, as inactivated product (due to heating, proteolysis, photoinactivation or precipitation), bioburden or microbial contamination introduced during downstream operations (from the environment, water, operations staff etc.) or contaminants derived from materials in direct contact with the product (extractable, leachable contaminants).

The downstream steps described in Section III are optimized by absorbent surface area, selectivity, binding capacity, and degree of volume reduction to purify product in the concentration range needed for each subsequent step to meet overall criteria of scale, stability, purity, and potency. Therefore, close integration of the characteristics of the upstream biological system that produces the product with the engineering and optimal performance of the downstream product separation, concentration. and purification operations are essential. This means that separation engineers, bioseparation and bioanalytical scientists, and manufacturing operations staff with broad expertise in working with labile biological molecules all need to work and communicate effectively as a team to design a downstream process that can be scaled from the laboratory bench and transferred to the manufacturing scale. It also means that downstream process scientists must *continually provide feedback information to upstream* process engineers and scientists to minimize the impact of upstream changes (cell line changes, media composition changes, the addition of antifoam, degradation of product during in-process storage or holds) on downstream separation operations. Therefore, the companion volumes of *Upstream Industrial Biotechnology* should also be consulted when designing a downstream process.

Each downstream step requires process development and optimization (for purity, overall yield) because of the complexity of the structure of the biological molecules being purified and the complexity of contaminants. Section III also includes approaches for scale down of purification operations. Each downstream step is expensive to optimize at the pilot or manufacturing scale. This expense is not only due to the scale of the equipment and expense of the separation media, but also because of the large quantity of valuable product needed to carry out optimization studies at scale.

Downstream operations require specialized equipment designed for separation of proteins, peptides, virus, particulate antigens or low-molecular-weight biomolecules while minimizing product degradation. Sections IV and V focus on large scale equipment design and fluid transfer systems, and describe in detail many types of industrial bioseparation equipment. Of particular concern for products derived from mammalian cell lines are effective methods for virus inactivation and viral filtration that can be validated with model virus challenge. These methods are described in section V.

Not only do the upstream and downstream processes need to be designed to meet cGMPs and be capable of being licensed, but the facility used to carry out the process also must be designed so that it can be licensed. Section VI and VII of *Downstream* address facility design, facility validation, clean-in-place (CIP) and sterilization-in-place (SIP) methods. A major advance in facility design for downstream processes is the growing impact of single use (SU) disposable downstream materials and this is described in Section VI.

The overall goal of all downstream operations is not only to purify bulk product for formulation, but to achieve regulatory compliance and licensure so that final formulated and filled product can be released to consumers, physicians or patients. Section VII describes how Process Analytical Technology (PAT), bioburden testing and Quality by Design (QbD) impact downstream process design and contribute to regulatory compliance both for the USFDA and European regulatory agencies.

1

BIOPROCESS DESIGN, COMPUTER-AIDED

Victor Papavasileiou
Intelligen Europe, Leiden, The Netherlands
Charles Siletti
Intelligen, Inc., Mt. Laurel, New Jersey
Alexandros Koulouris
Intelligen Europe, Thermi, Greece
Demetri Petrides
Intelligen, Inc., Scotch Plains, New Jersey

1.1 INTRODUCTION

Bioprocess design is the conceptual work done prior to commercialization of a biological product. Given information on the potential market demand for a new product, bioprocess design endeavors to answer the following questions: What are the required amounts of raw materials and utilities for manufacturing a certain amount of product per year? What is the required size of process equipment and supporting utilities? Can the product be manufactured in an existing facility or is a new plant required? What is the total capital investment for a new facility? What is the manufacturing cost? How long does a single batch take? What is the minimum time between consecutive batches? During the course of a batch, what is the demand for various resources (e.g. raw materials, labor, and utilities)? Which process steps or resources are the likely production bottlenecks? What process and equipment changes can increase throughput? What is the environmental impact of the process? Which design is the "best" among several plausible alternatives?

Bioprocess design and project economic evaluation require the integration of knowledge from many different scientific and engineering disciplines. Design and evaluation are also carried out at various levels of detail. Table 1.1 presents a common classification of design and cost estimates and typical engineering costs for a $50 million capital investment project (1).

Order-of-magnitude estimates are usually practiced by experienced engineers who have worked on similar projects in the past. They take minutes or hours to complete, but the error in the estimate can be as high as 50%. Table 1.2 provides a good example of information typically employed for order-of-magnitude estimates of the capital investment for cell culture facilities. It lists capital investment for cell culture facilities of various sizes built in the last 10 years. The last column displays unit cost of capital investment expressed in millions of US dollars per cubic meter of production bioreactor capacity. The numbers range between 2.5 and 6.2 and for the more recent facilities the numbers are in the 5–6.2 range. Consequently, using the data of Table 1.2, one can safely estimate the capital investment for a new cell culture facility with production bioreactor capacity of 100 m^3 to be in the range of $500–650 million.

Engineers employed by operating companies usually perform level 2 and 3 studies. Such studies take days or weeks to complete using appropriate computer aids. The main objective of such a study is to evaluate alternatives and pinpoint areas of high cost and low yield. The results

Downstream Industrial Biotechnology: Recovery and Purification, First Edition. Edited by Michael C. Flickinger.

TABLE 1.1. Types of Design Estimates

Level	Type of Estimate	Accuracy	Cost ($1000)
1	Order-of-magnitude estimate (ratio estimate) based on similar previous cost data	≤50%	—
2	Project planning estimate (budget estimation) based on knowledge of major equipment items	≤30%	20–40
3	Preliminary engineering (scope estimate) based on sufficient data to permit the estimate to be budgeted	≤25%	50–100
4	Detailed engineering (capital approval stage) based on almost complete process data	≤15%	100–200
5	Procurement and construction (contractor's estimate) based on complete engineering drawings, specifications, and site surveys	≤10%	3000–7000

TABLE 1.2. Capital Investment for Cell Culture Facilities

Company	Capacity (m^3)	Completion Year	Investment ($ million)	Unit Cost ($ million/m^3)
Genentech	$8 \times 15 = 120$	2001	300	2.5
Amgen	$8 \times 8 = 64$	2002	300	4.7
Wyeth	$6 \times 15 = 90$	2003	325	3.6
Biogen Idec	$6 \times 15 = 90$	2005	450	5.0
BMS	$6 \times 20 = 120$	2009	750	6.2

are used to plan future research and development and to generate project budgets.

Level 4 and 5 studies are usually performed by engineering and construction companies that are hired to build new plants for promising new products that are at an advanced stage of development. These detailed estimates are beyond the scope of this chapter. Instead, the rest of this chapter will focus on level 2 and 3 studies. It should also be noted that opportunities for creative process design work are usually limited to preliminary studies. By the time detailed engineering work is initiated, a process is more than 80% fixed. Furthermore, most of the important decisions for capital expenditure and product commercialization are based on results of preliminary process design and cost analysis. This is why it is so important for a new engineer to master the skills of preliminary process design and cost analysis.

1.2 BENEFITS FROM THE USE OF COMPUTER AIDS

Process design calculations are greatly facilitated by the use of computer aids, such as spreadsheets, process simulators, finite capacity scheduling (FCS), and other specialized tools. Use of appropriate computer aids allows the process design team to quickly and accurately redo the entire series of calculations with a different set of assumptions and other input data. The benefits from the use of such tools depend on the type of product, the stage of development, and the size of the investment. For commodity biological products such as biofuels, minimization of capital and operating costs are the primary benefits. For high-value biopharmaceuticals, systematic process development that shortens the time to commercialization is the primary motivation. Figure 1.1 shows a pictorial representation of the benefits from the use of computer aids at the various stages of the commercialization process.

1.2.1 Idea Generation

When product and process ideas are first conceived, process modeling tools are used for project screening, selection, and strategic planning on the basis of preliminary economic analyses.

1.2.2 Process Development

During this phase, the company's process development groups look into the various options available for synthesizing, purifying, characterizing, and formulating the final product. At this stage, the process undergoes constant change. Typically, a large number of scientists and engineers are involved in the improvement and optimization of individual processing steps. The use of process simulation tools at this stage can introduce a common language of communication and facilitate team interaction. A computer model of the entire process can provide a common reference and evaluation framework to facilitate process development. The impact of process changes can be readily evaluated and documented in a

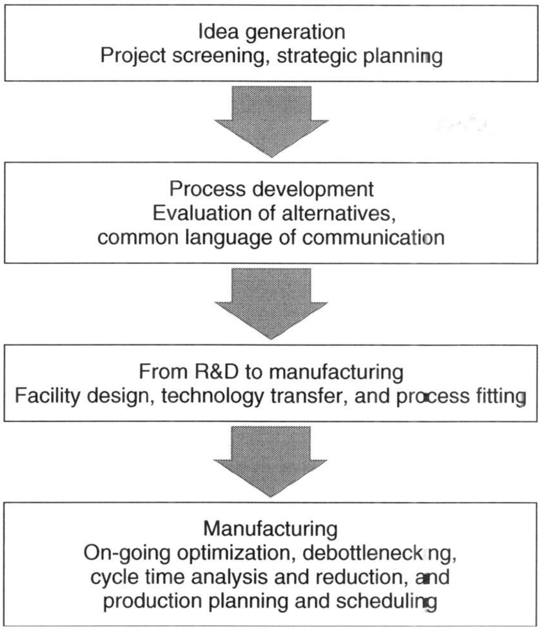

Figure 1.1. Benefits from the use of computer aids

systematic way. Once a reliable model is available, it can be used to pinpoint the cost-sensitive areas of a complex process. These are usually steps of high capital and operating cost or low yield and production throughput. The findings from such analyses can be used to focus further lab and pilot plant studies to optimize those portions of the process. The ability to experiment on the computer with alternative process setups and operating conditions reduces the costly and time-consuming laboratory and pilot plant effort.

The environmental impact of a process is another issue that can be readily evaluated with computer models. Material balances calculated for the projected large-scale manufacturing reveal the environmental hot spots. These are usually process steps that use organic solvents and other regulated materials of high disposal costs. Environmental issues not addressed during process development may lead to serious headaches during manufacturing. This is especially true for biopharmaceuticals because after a process has been approved by the regulatory agencies, it is extremely costly and time consuming to implement process changes.

1.2.3 Facility Design and/or Selection

With process development near completion at the pilot plant level, simulation tools are used to systematically design and optimize the process for commercial production. Availability of a good computer model can greatly facilitate the transfer of a new process from the pilot plant to the large-scale facility. If a new facility needs to be built, process simulators can be used to size process equipment and supporting utilities, and estimate the required capital investment. In transferring production to existing manufacturing sites, process simulators can be used to evaluate the various sites from a capacity and cost point of view and select the most appropriate one. The same can apply to outsourcing of manufacturing to contract manufacturers.

1.2.4 Manufacturing

In large-scale manufacturing, simulation tools are mainly used for on-going process optimization and debottlenecking studies. Other computer aids that play an important role in manufacturing include FCS, manufacturing resource planning (MRP), and enterprise resource planning (ERP) tools. FCS tools play an important role in batch chemical manufacturing. They are used to generate production schedules on an on-going basis in a way that does not violate constraints related to the limited availability of equipment, labor resources, utilities, inventories of materials, and so on. FCS tools close the gap between ERP/MRP tools and the plant floor (2). Production schedules generated by ERP/MRP tools are typically based on coarse process representations and approximate plant capacities and, as a result, solutions generated by those tools may not be feasible, especially for multiproduct facilities that operate at high capacity utilization. This can often lead to late orders that require expediting and/or to large inventories in order to maintain customer responsiveness. "Lean manufacturing" principles, such as just-in-time production, low work-in-progress (WIP), and low product inventories cannot be implemented without good production scheduling tools that can accurately estimate capacity (3,4).

1.3 COMMERCIALLY AVAILABLE TOOLS

Process simulation programs, also known as process simulators, have been in use in the chemical and petrochemical industries since the early 1960s. Established simulators for those industries include: Aspen Plus and HYSYS from Aspen Technology, Inc. (Cambridge, MA), ChemCAD from Chemstations, Inc. (Houston, TX), and PRO/II from SimSci-Esscor, Inc. (Lake Forest, CA).

The above simulators have been designed to model primarily continuous processes and their transient behavior. Most biological products, however, are produced in batch and semicontinuous mode (5,6). Such processes are best modeled with batch process simulators that account for time-dependency and sequencing of events. Batches from Batch Process Technologies, Inc. (West Lafayette, IN) was the first simulator specific to batch processes. It was commercialized in the mid-1980s. All of its operation models are dynamic and simulation always involves integration of differential equations over a period of time. In the mid-1990s, Aspen Technology (Cambridge, MA) introduced Batch Plus, a recipe-driven simulator that targeted batch pharmaceutical processes. Around the same time, Intelligen, Inc. (Scotch Plains, NJ) introduced SuperPro Designer. A unique feature of SuperPro is its ability to model batch as well as continuous processes (7).

Discrete-event simulators have also found applications in the bioprocessing industries. Established tools of this type include ProModel from ProModel Corporation (Orem, UT), Arena and Witness from Rockwell Automation, Inc. (Milwaukee, WI), Extend from Imagine That, Inc. (San Jose, CA), and FlexSim from FlexSim Software Products, Inc. (Orem, UT). The focus of models developed with such tools is usually on the minute-by-minute time-dependency of events and the animation of the process. Material balances, equipment sizing, and cost analysis tasks are usually out of the scope of such models. Some of these tools are quite customizable and third-party companies occasionally use them as platforms to create industry-specific modules. For instance, BioPharm Services, Ltd. (Bucks, UK) have created a module that runs on top of Extend and focuses on biopharmaceuticals.

MS Excel from Microsoft is another common platform for creating models for integrated processes that focus on material balances, equipment sizing, and cost analysis. Some companies have even developed models in Excel that capture the time-dependency of batch processes. This is typically done by writing extensive code (in the form of macros and subroutines) in VBA (Visual Basic for Applications) that comes with Excel. K-TOPS from Alfa Laval Biokinetics, Inc. (Philadelphia, PA) belongs to this category.

In terms of production scheduling, established tools include Infor SCM from Infor Global Solutions (Alpharetta, GA), Optiflex from i2 Technologies, Inc. (Irving, TX), SAP APO from SAP AG (Walldorf, Germany), ILOG Plant PowerOps from ILOG SA (Gentilly, France), Aspen SCM (formerly Aspen MIMI) from Aspen Technology, Inc. (Cambridge, MA), and so on. Their success in the biochemical industries, however, has been rather limited so far. Their primary focus on discrete manufacturing (as opposed to batch chemical manufacturing) and their approach to scheduling from a mathematical optimization viewpoint are some of the reasons for the limited market penetration.

SchedulePro from Intelligen, Inc. (Scotch Plains, NJ) is a new FCS tool that focuses on scheduling of batch and semicontinuous biochemical and related processes. It is a recipe-driven tool with emphasis on generation of feasible solutions that can be readily improved by the user in an interactive manner.

The rest of this chapter will address, through an illustrative example, the use of simulation and scheduling tools for evaluating and optimizing integrated biochemical processes. Analysis and assessment of additional bioprocesses can be found in the literature (8).

1.4 MONOCLONAL ANTIBODY EXAMPLE

Monoclonal antibodies (Mabs) are the fastest growing segment within the biopharmaceutical industry (9). More than 20 Mabs and Fc fusion proteins are approved for sale in the United States and Europe and approximately 200 Mabs are in clinical trials for a wide variety of indications (2). The market is predicted to grow by around 20% per year and reach $17 billion in 2008 (10).

The high-dose demand for several Mabs translates into annual production requirement for purified product in the metric ton range. Such a process is modeled and analyzed with SuperPro Designer in the rest of this chapter. Figure 1.2 displays the flow sheet of the overall process. The generation of the flow sheet was based on information available in the patent and technical literature combined with our engineering judgment and experience with such processes. The computer files for this example are available as part of the evaluation version of SuperPro Designer at the website www.intelligen.com/literature. Additional examples dealing with other biopharmaceuticals as well as commodity biological products are available at the same website.

To model an integrated process on the computer using SuperPro Designer, the user starts by developing a flow sheet that represents the overall process. The flow sheet is developed by putting together the required unit procedures (see the next paragraph for an explanation), and joining them with material flow streams. Next, the user initializes the flow sheet by registering the various materials that are used in the process and specifying operating conditions and performance parameters for the various operations.

Most biopharmaceutical processes operate in batch mode. This is in contrast to petrochemical and other high-throughput industries that use continuous processes. In continuous production, a piece of equipment performs the same action all the time. In batch processing, on the other hand, a piece of equipment goes through a cycle of operations. For instance, an inoculum preparation step (P-5 in SBR1) includes the following operations (Fig. 1.3): *SIP, SET UP, TRANSFER IN-1(media), TRANSFER IN-2 (inoculum), FERMENT (fermentation operation), TRANSFER OUT (emptying vessel), CIP (cleaning in place)*. In SuperPro, the set of operations that compose a processing step is called a *unit procedure* (as opposed to a unit operation). The individual tasks contained in a procedure (e.g. transfer in, Ferment, and CIP) are called *operations*.

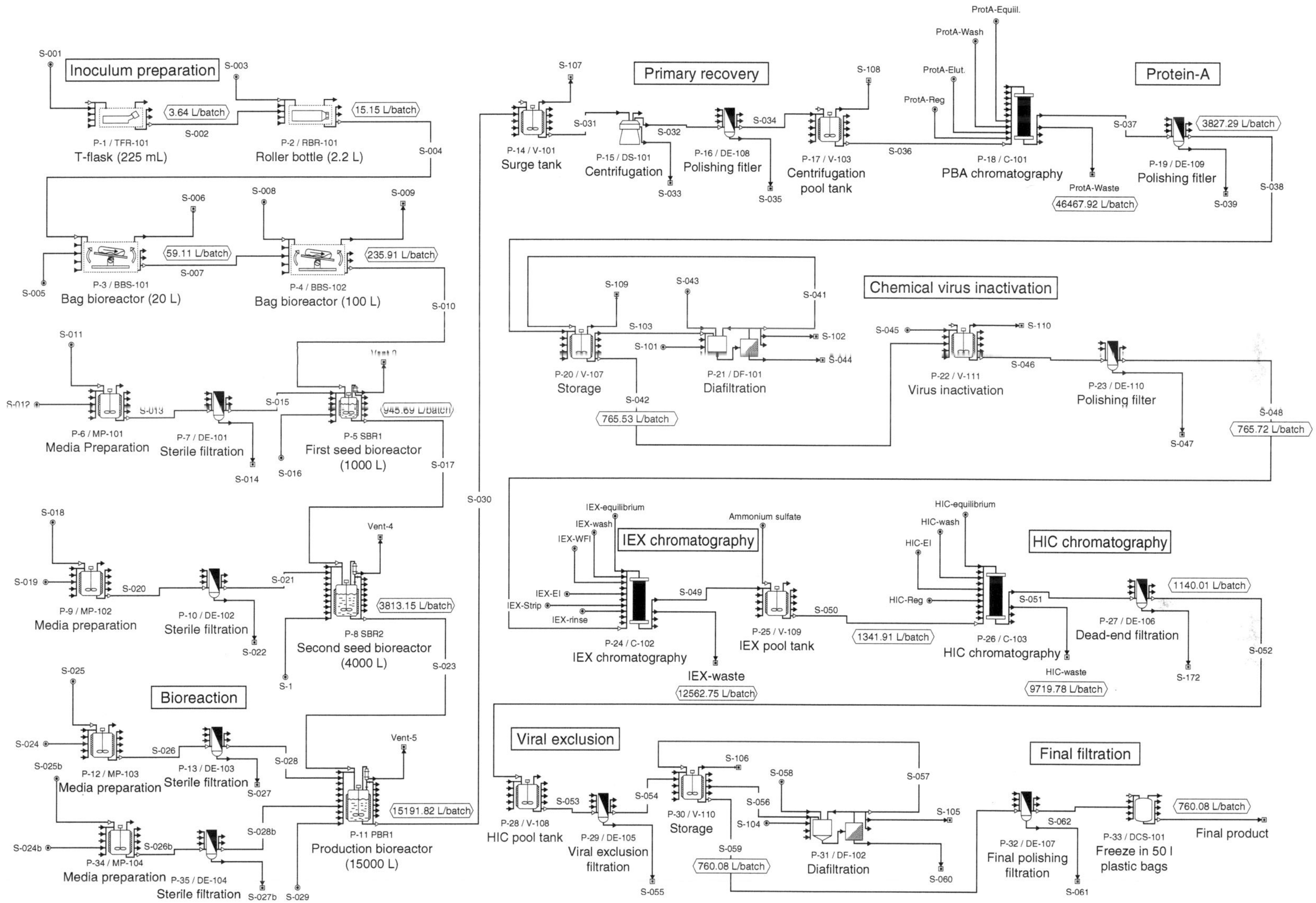

Figure 1.2. Monoclonal antibody production flow sheet.

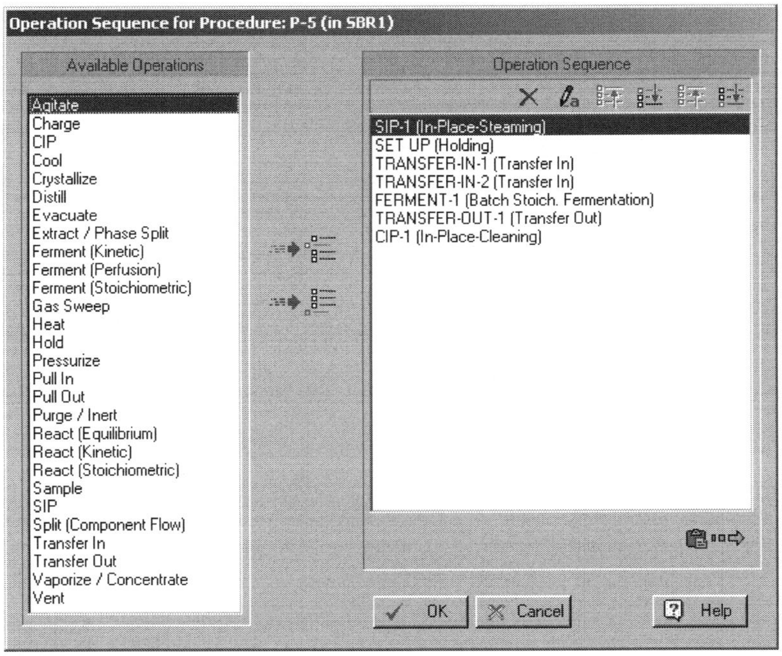

Figure 1.3. The operations associated with the P-5 unit procedure of Fig. 1.2. (This figure is available in full color at http://onlinelibrary.wiley.com/book/10.1002/9780470054581.)

A unit procedure is represented on the screen with a single equipment icon. In essence, a unit procedure is the recipe that describes the sequence of actions required to complete a single processing step. Figure 1.3 displays the dialog through which the recipe of a vessel unit procedure is specified. On the left-hand side of that dialog, the program displays the operations that are available in a vessel procedure; on the right-hand side, it displays the registered operations. The hierarchical representation of batch processes (also known as recipes) using unit procedures and operations is an approach that is recommended by the Instrument Society of America (ISA) because it facilitates modeling, control, and scheduling of batch operations (11).

For every operation within a unit procedure, the simulator includes a mathematical model that performs material and energy balance calculations. On the basis of the material balances, it performs equipment-sizing calculations. If multiple operations within a unit procedure dictate different sizes for a certain piece of equipment, the software reconciles the different demands and selects an equipment size that is appropriate for all operations. The equipment is sized so that it is large enough that it will not be overfilled during any operation, but it is no larger than necessary (to minimize capital costs). If the equipment size is specified by the user, the simulator checks to make sure that the vessel is not overfilled. In addition, the tool checks to ensure that the vessel contents will not fall below a user-specified minimum volume (e.g. a minimum stir volume) for applicable operations.

1.4.1 Process Description

1.4.1.1 Upstream. The upstream part is split in two sections: the inoculum preparation section and the bioreaction section. The inoculum is initially prepared in 225-mL T-flasks. The material is first moved to 2.2-L roller bottles, then to 20-L and subsequently to 100-L disposable bag bioreactors. Sterilized media is fed at the appropriate amount in all of these four initial steps (3.6, 11.4, 43.6, 175.4 kg/batch, respectively). The broth is then moved to the first (1000 L) and second (4000 L) seed bioreactor. For the seed bioreactors the media powder is dissolved in water for injection (WFI) in two prep tanks (MP-101 & MP-102) and then sterilized/fed to the reactors through 0.2-μm dead-end filters (DE-101 and DE-102). In the bioreaction section, serum-free low-protein media powder is dissolved in WFI in a stainless-steel tank (MP-103). The solution is sterilized using a 0.2-μm dead-end polishing filter (DE-103). A stirred-tank bioreactor (production bioreactor, PBR1) is used to grow the cells, which produce the therapeutic Mab. The production bioreactor operates under a fed batch mode. High media concentrations are inhibitory to the cells, so half of the media is added at the start of the process and the rest is fed at a variable rate during fermentation. The concentration of media powder in the initial feed solution is 17 g/L. The fermentation time is 12 days. The volume of broth generated per bioreactor batch is approximately 15,000 L, which contains roughly 22.6 kg of product (the product titer is approximately 1.5 g/L).

1.4.1.2 Downstream. Between the downstream unit procedures there are 0.2-μm dead-end filters to ensure sterility. The generated biomass and other suspended compounds are removed using a Disc-Stack centrifuge (DS-101). During this step, roughly 2% of Mab is lost in the solids waste stream resulting in a product yield of 98%. The bulk of the contaminant proteins are removed using a Protein-A affinity chromatography column (C-101). The following operating assumptions were made: (i) resin binding capacity is 15 g of product per liter of resin, (ii) the eluant or elution buffer is a 0.6% w/w solution of acetic acid and its volume is equal to 5 column volumes (CVs), (iii) the product is recovered in 2 CVs of eluant with a recovery yield of 90%, and (iv) the total volume of the solution for column equilibration, wash, and regeneration is 14 CVs. The entire procedure takes approximately 27 h and requires a resin volume of 362 L. The protein solution is then concentrated fivefold and diafiltered 2 times (in P-21/DF-101) using WFI as diluent. This step takes approximately 5 h and requires a membrane of 15 m^2. The product yield is 97%. The concentrated protein solution is then chemically treated for 1.5 h with Polysorbate 80 to inactivate viruses (in P-22/V-111). An ion exchange chromatography step follows (P-24/C-102). The following operating assumptions were made: (i) the resin's binding capacity is 40 g of product per liter of resin, (ii) a gradient elution step is used with a sodium chloride concentration ranging from 0.0 to 0.1 M and a volume of 5 CVs, (iii) the product is recovered in 2 CVs of eluant buffer with a yield on Mab of 90%, and (iv) the total volume of the solutions for column equilibration, wash, regeneration and rinse is 16 CVs. The step takes approximately 22.3 h and requires a resin volume of 158 L. Ammonium sulfate is then added to the ion exchange (IEX) eluate (in P-25/V-109) to a concentration of 0.75 M to increase the ionic strength in preparation for the hydrophobic interaction chromatography (HIC; P-26/C-103) that follows. The following operating assumptions were made for the HIC step: (i) the resin binding capacity is 40 g of product per liter of resin, (ii) the eluant is a sodium chloride (4% w/w) sodium di-hydrophosphate (0.3% w/w) solution and its volume is equal to 5 CVs, (iii) the product is recovered in 2 CVs of eluant buffer with a recovery yield of 90%, and (iv) the total volume of the solution for column equilibration, wash, and regeneration is 12 CVs. The step takes approximately 22 h and requires a resin volume of 142 L. A viral exclusion step (DE-105) follows. It is a dead-end type of filter with a pore size of 0.02 μm. This step takes approximately 2.3 h and requires a membrane of 1.45 m^2. Finally, the HIC elution buffer is exchanged for the product bulk storage (PBS) buffer and concentrated 1.5-fold (in DF-102). This step takes approximately 4 h and requires a membrane of 7 m^2. The approximately 580 L of final protein solution is stored in fifteen 50-L disposable storage bags (DCS-101). Approximately, 14.6 kg of Mab are produced per batch. The overall yield of the downstream operations is approximately 64.5%.

1.4.2 Material Balance

Table 1.3 provides a summary of the overall material balance of the process. Note the large amount of WFI utilized per batch. A major part of WFI is consumed for cleaning and buffer preparation. Approximately, 14.6 kg of Mab are produced per batch.

TABLE 1.3. Raw Material Requirements

	Requirement		
Raw Material	(kg/yr)	(kg/batch)	(kg/kg MP)
Inoculation Media	374	4.68	0.32
WFI	9,403,568	117,545	8,058
Phosphoric acid	44,113	551.41	37.80
Sodium hydroxide	34,164	427.05	29.28
Serum-free media	35,882	448.52	30.75
EDTA, sodium	2,544	31.80	2.18
Sodium chloride	53,600	670.00	45.93
TRIS base	1,272	15.90	1.09
TRIS HCl	3,815	47.69	3.27
Acetic acid	3,457	43.21	2.96
Sodium citrate	623	7.78	0.53
KCl	1	0.01	0.001
KH_2PO_4	1	0.01	0.001
Na_2HPO_4	1,817	22.72	1.56
NaH_2PO_4	105	1.31	0.09
Ammonium sulfate	8,104	101.31	6.94
Polysorbate 80	5	0.06	0.01
Total	9,593,440	119,918	8,221

MP, purified Mab.

1.4.3 Scheduling and Cycle Time Reduction

Figure 1.4 displays the Gantt chart of the process for four consecutive batches. The schedule represents a plant that has a single production train. The cleaning-in-place (CIP) skids can be seen at the top of the graph. The batch time is approximately 50 days. This is the time required from the start of inoculum preparation to the final product purification of a single batch. A new batch is initiated every 2 weeks (14 days). The production bioreactor (PBR1) is the time (scheduling) bottleneck. On an annual basis the plant processes 20 batches and produces approximately 292 kg of purified Mab. It is clear from the chart that under these conditions the downstream train is underutilized and the cycle time of the process—the time between consecutive batch starts—is relatively long. The cycle time of the process can be reduced and the plant throughput increased by installing multiple bioreactor trains that operate in staggered mode (out of phase) and feed the same purification train. Figure 1.5 represents a case where four bioreactor trains feed the same purification train. The new cycle time is 3.5 days, which is one-fourth of the original. Under these conditions, the plant processes 80 batches per year and produces approximately 1167 kg of Mab per year. Some biopharmaceutical companies have installed more than four bioreactor trains per purification train aiming at cycle times as low as 2 days.

1.4.4 Sizing of Batch Utilities

Another characteristic of batch processing is the variable demand of resources such as labor, utilities and raw materials as a function of time. Sizing of WFI systems is a common challenge during the design of new facilities and the retrofit of existing ones. WFI is used for preparing media and buffer solutions, for cleaning equipment, for generating clean steam, and so on. A WFI system consists of a distillation unit that generates the distilled water, a surge tank, and a circulation loop for delivering the material around the plant. The capacity may be limited by any of the following:

- The process can not, on average, consume more water than the still can generate.
- The process peak demand can not exceed the capacity of the circulation system.

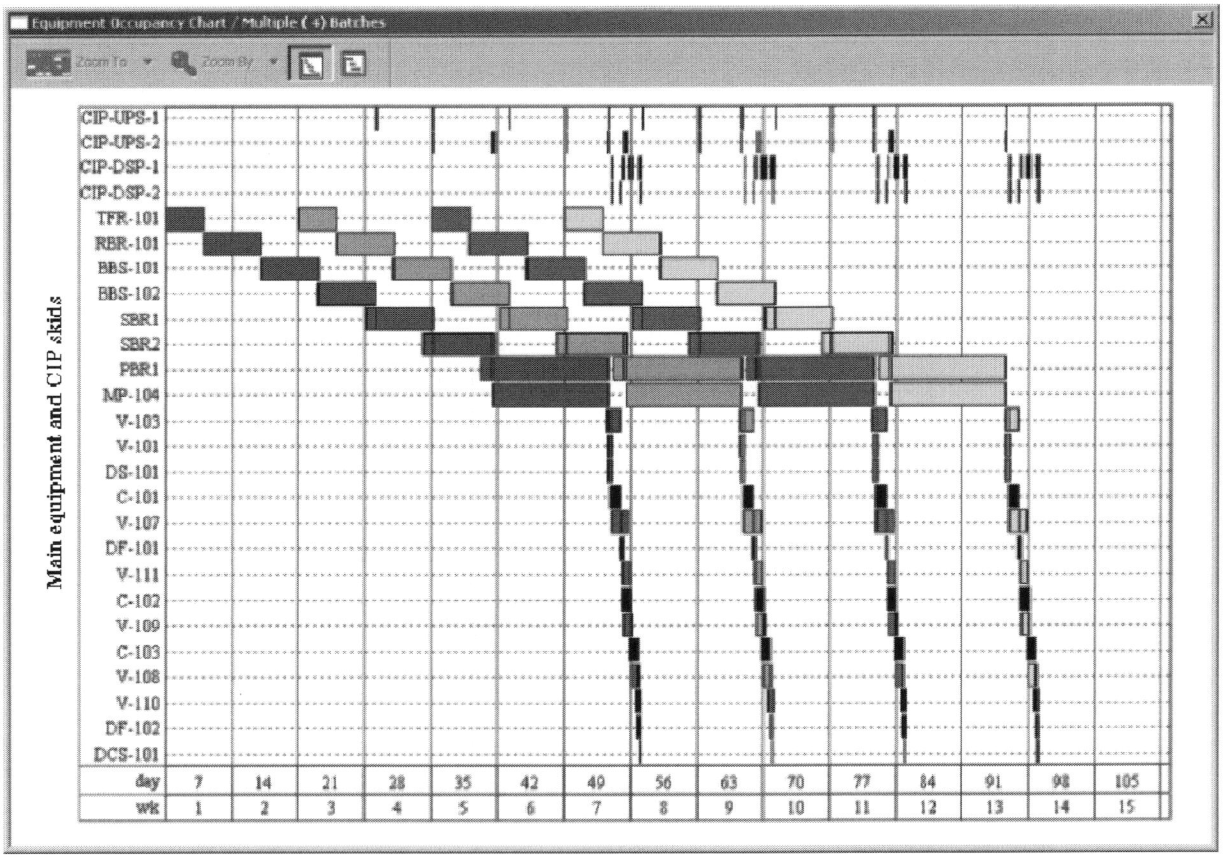

Figure 1.4. One bioreactor train feeding one purification train. (This figure is available in full color at http://onlinelibrary.wiley.com/book/10.1002/9780470054581.)

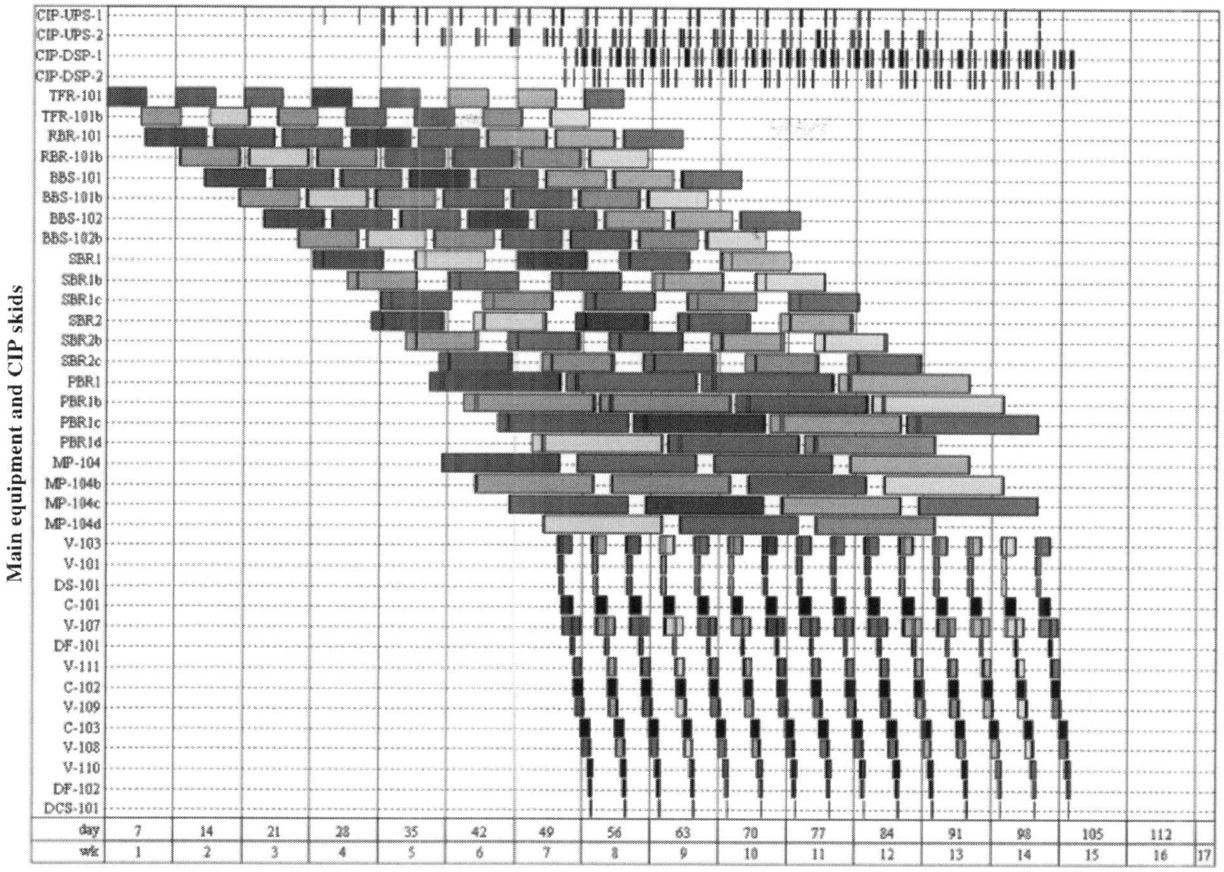

Figure 1.5. Four bioreactor trains feeding one purification train. (This figure is available in full color at http://onlinelibrary.wiley.com/book/10.1002/9780470054581.)

- The surge vessel must be large enough to maintain capacity during peak operation.
- In some plants, periodic sanitization cycles may interrupt all purified water draws.

Process modeling can provide reasonable estimates for the sizes of the still, the surge tank, and the pumping capacity of the circulation loop. Figure 1.6 displays the demand for WFI of the Mab process over time. The plots show the instantaneous and the 12-h average (heavy-line) demands. The chart also shows the 12-h cumulative amount that corresponds to the y-axis on the right. The peak instantaneous demand indicates the minimum pumping capacity for the system (11,500 kg/h or 50.7 gpm). The peak 12-h average rate provides an estimate for the capacity of still (1800 kg/h or 8 gpm), and the corresponding peak 12-h accumulation is an estimate of the surge tank capacity of 25,000 L. The trade-off between still rate and surge capacity can be examined by changing the averaging time. Selecting a longer period predicts a larger surge tank and a lower still rate.

Figure 1.7 displays the inventory profile of WFI in the surge tank for a tank size of 25,000 L and a still rate of 3500 L/h. The generation still is turned on when the level in the tank falls below 30% and it remains on until the tank is full. The operation rate of the still is depicted by the blue step-function lines. (The reader is requested to refer to the online version of this chapter for color indication.)

1.4.5 Economic Evaluation

Cost analysis and project economic evaluation are important for a number of reasons. For a new product, if the company lacks a suitable manufacturing facility with available capacity, it must decide whether to build a new plant or outsource the production. Building a new plant is a major capital expenditure (Table 1.2) and a lengthy process. To make a decision, management must have information on capital investment required and time to complete the facility. When production is outsourced, a cost-of-goods analysis serves as a basis for negotiation with contract manufacturers. A sufficiently detailed computer model can be used as the basis for the discussion and negotiation of the terms. Contract manufacturers usually base their estimates on requirements of facility/equipment utilization and labor per batch, which

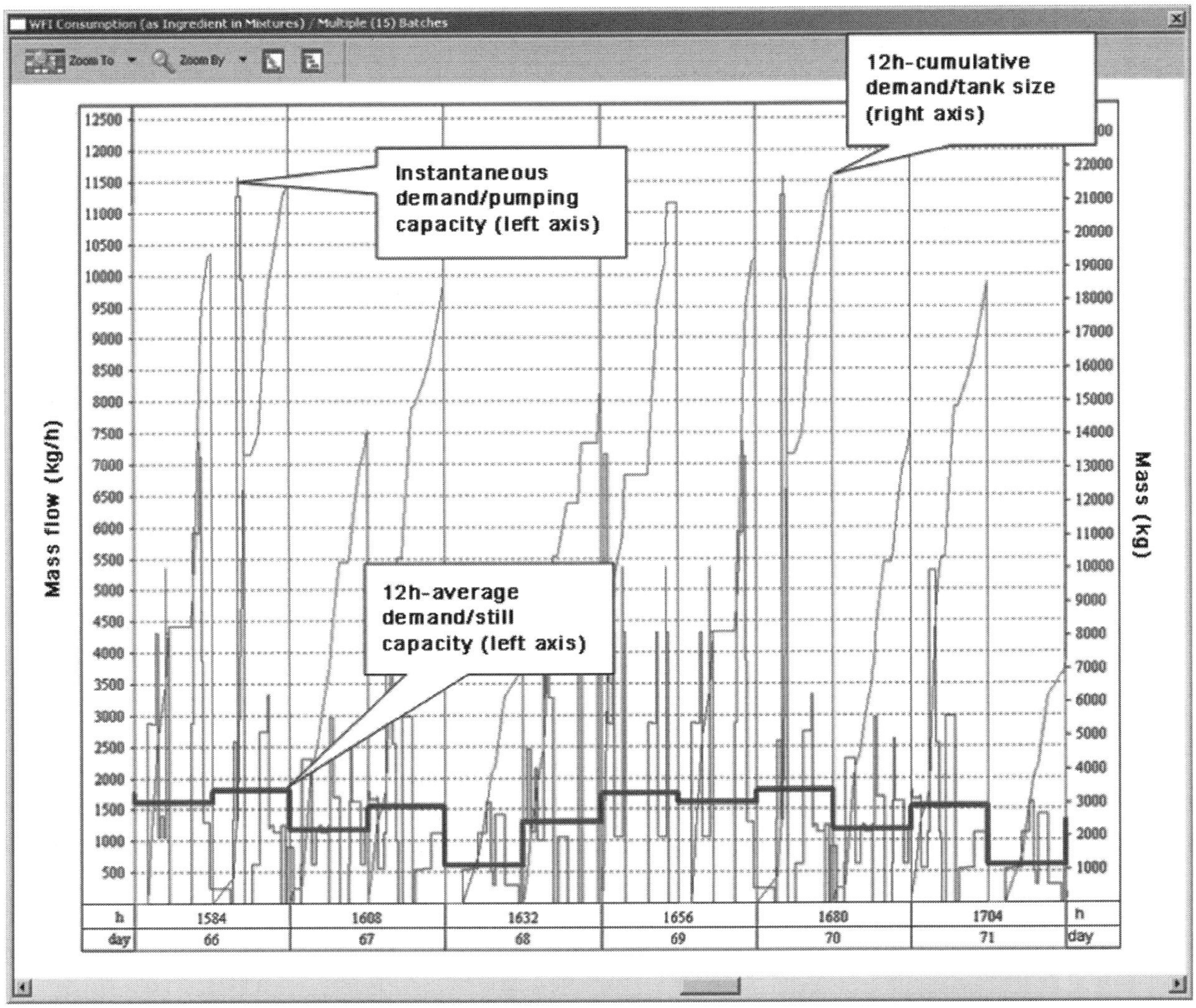

Figure 1.6. WFI demand as a function of time. (This figure is available in full color at http://onlinelibrary.wiley.com/book/10.1002/9780470054581.)

is information that is provided by a good model. SuperPro performs thorough cost analysis and project economic evaluation calculations. It estimates capital as well as operating cost. The cost of equipment is estimated using built-in cost correlations that are based on data derived from a number of vendors and literature sources. The fixed capital investment is estimated based on equipment cost and using various multipliers, some of which are equipment specific (e.g. installation cost) while others are process specific (e.g. cost of piping and buildings). The approach is described in detail in the literature (12–14). The rest of this section provides a summary of the cost analysis results for this example process.

Table 1.4 provides a list of major equipment items along with their purchase costs (generated by SuperPro Designer). The total equipment cost for a plant of this capacity (four production bioreactors each having a working volume of 15,000 L) is around $24 million. Approximately, a quarter of the equipment cost is associated with the four production bioreactors. The cost of vessels and filters that are seen in Fig. 1.2 but are missing from the table are accounted for under the "Cost of Unlisted Equipment" item. The economic evaluation also takes into account the vessels required for buffer preparation and holding that are not included in Fig. 1.2. A full model that includes all buffer preparation and holding activities and other advanced process modeling features can be downloaded from www.intelligen.com/literature.

Table 1.5 displays the various items included in the direct fixed capital (DFC) investment. The total DFC for a plant of this capacity is around $240 million or approximately 10 times the total equipment cost. The total capital investment that includes the cost of start-up and validation is around $300 million.

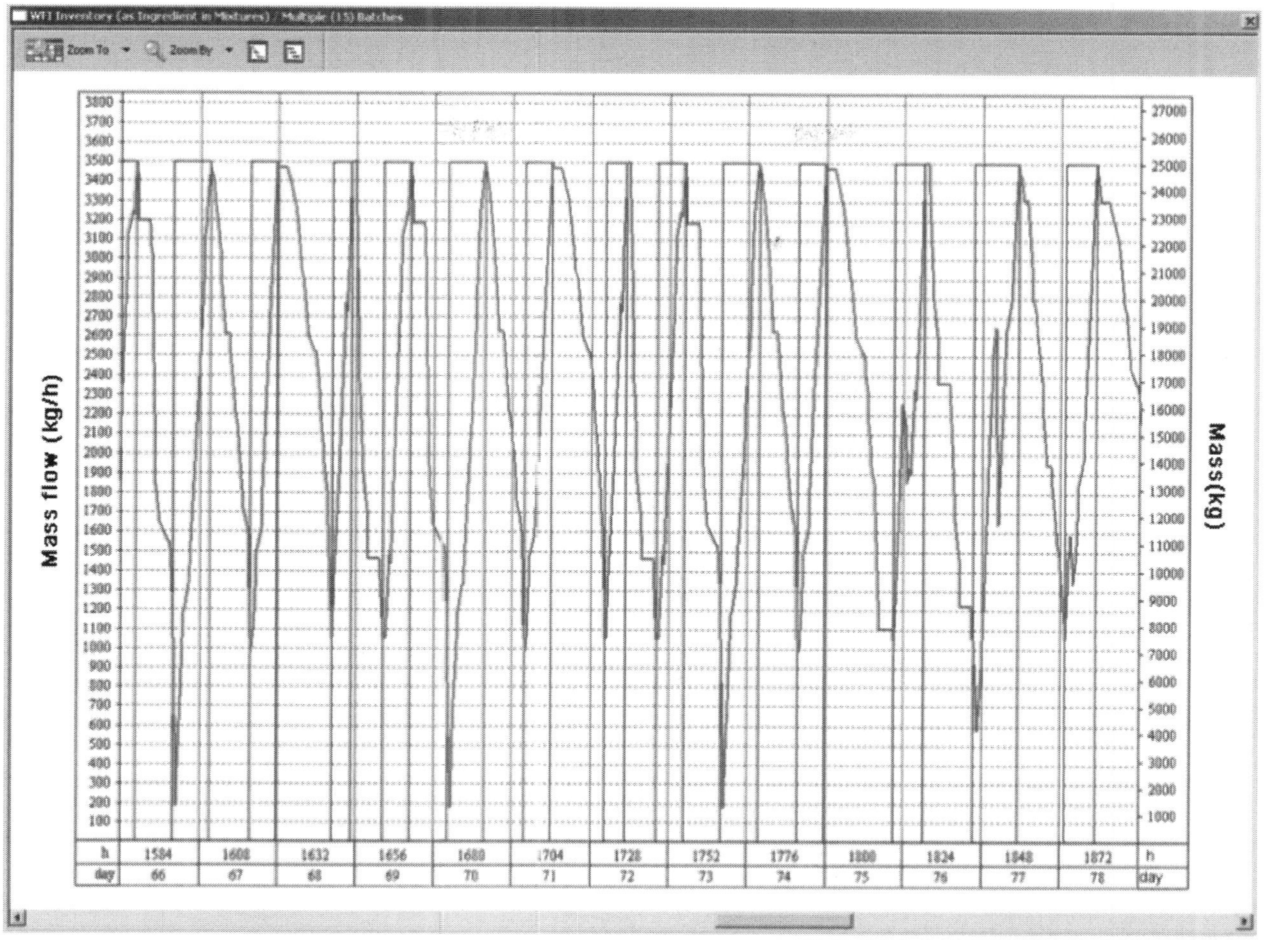

Figure 1.7. WFI inventory profile in the surge tank. (This figure is available in full color at http://onlinelibrary.wiley.com/book/10.1002/9780470054581.)

Table 1.6 provides a summary of the operating cost. The total annual operating cost is $111.6 million resulting in a unit production cost of around $95.6/g (1167 kg of purified product are produced annually). The facility-dependent cost is the most important item, accounting for around 40% of the overall operating cost. This is common for high-value biopharmaceuticals. Depreciation of the fixed capital investment and maintenance of the facility are the main contributors to this cost. Raw materials account for around 18% of the overall cost. Serum-free media accounts for 90% of the raw materials cost (Table 1.7). A price of $500/kg was assumed for serum-free media in dry powder form. Labor and consumables come third and fourth, respectively, each accounting for around 17% of the overall manufacturing cost. Consumables include the cost of chromatography resins and membrane filters that need to be replaced on a regular basis. The replacement of the Protein-A resin accounts for 62.3% of the total consumables cost (Table 1.8). A unit cost of $6000/L and a replacement frequency of 60 cycles were assumed for the Protein-A resin. Approximately 63% of the manufacturing cost is associated with the upstream section (inoculum preparation and fermentation) and 37% with the downstream section (product recovery and purification).

1.4.6 Sensitivity Analysis

After a model of the entire process has been developed on the computer, tools like SuperPro Designer can be used to ask and readily answer "what if" questions and to carry out sensitivity analysis with respect to key design variables. In this example, we looked at the impact of the number of bioreactor trains, product titer, and bioreactor volume on unit production cost.

Figure 1.8 displays the impact of bioreactor trains on the unit production cost. The cost analysis calculations of the previous section correspond to the case of four production bioreactors (each having a working volume of 15,000 L) feeding a single purification train and resulting in a unit cost of around $96/g. If just a single bioreactor train feeds the purification train, then, the manufacturing cost increases by 50%. For more than four

TABLE 1.4. Major Equipment Specification and Purchase Costs (Year 2007, Prices in USD)

Quantity	Name	Description	Unit Cost ($)	Cost ($)
4	SBR1	Seed reactor Vessel volume = 1,200 L	500,000	2,000,000
4	MP-101	Blending tank Vessel volume = 800 L	148,000	296,000
4	MP-102	Blending tank Vessel volume = 3,200 L	180,000	360,000
3	SBR2	Stirred reactor Vessel volume = 4,800 L	640,000	1,920,000
4	PBR1	Stirred reactor Vessel volume = 19,000 L	1,468,000	5,872,000
4	MP-103	Blending tank Vessel volume = 11,500 L	205,000	410,000
4	V-101	Blending tank Vessel volume = 17,000 L	223,000	446,000
4	V-103	Blending tank Vessel volume = 16,100 L	221,000	442,000
4	V-107	Blending tank Vessel volume = 3,250 L	181,000	362,000
4	DF-101	Diafilter Membrane area = 15.2 m^2	44,000	88,000
1	C-102	PBA chromatography column Column volume = 158 L	195,000	195,000
4	V-109	Blending tank Vessel volume = 1,150 L	156,000	312,000
1	C-103	PBA chromatography column Column volume = 143 L	188,000	188,000
4	V-110	Blending tank Vessel volume = 1,000 L	153,000	306,000
1	DF-102	Diafilter Membrane area = 7.3 m^2	28,000	28,000
2	DS-101	Disk-stack centrifuge Throughput = 2,000 L/h	400,000	800,000
1	C-101	PBA chromatography column Column volume = 362 L	291,000	291,000
2	V-108	Blending tank Vessel volume = 1,000 L	153,000	306,000
2	V-111	Blending tank Vessel volume = 650 L	144,000	288,000
5	MP-104	Blending tank Vessel volume = 1,250 L	157,000	785,000
		Cost of unlisted equipment		8,390,000
Total				24,085,000

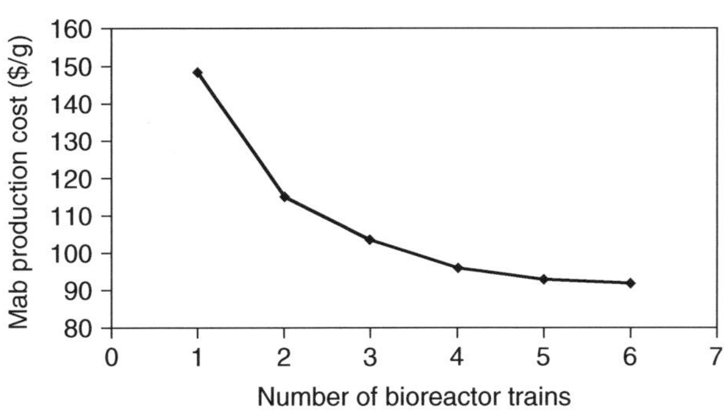

Figure 1.8. Production cost of Mab as a function of the number of bioreaction trains.

production bioreactors per purification train, the unit cost drops a bit and asymptotically approaches a value of around $90/g. Multiple-production bioreactors that feed a single purification train lead to reduced manufacturing cost because the plant throughput is increased (it is proportional to the number of bioreactors) without the need for additional capital investments in the purification train. Four to six bioreactor trains per purification train is probably the optimum number for cell culture processes that have a fermentation time of around 12 days. Such processes operate with cycle times ranging between 3.5 and 2.5 days.

TABLE 1.5. Fixed Capital Estimate Summary (Year 2007, Prices in USD)

Total plant direct cost (TPDC) (physical cost)		
Equipment purchase cost	24,085,000	
Installation	10,877,000	
Process piping	8,430,000	
Instrumentation	9,634,000	
Insulation	723,000	
Electrical	2,409,000	
Buildings	60,213,000	
Yard improvement	3,613,000	
Auxiliary facilities	9,634,000	
TPDC		129,617,000
Total plant indirect cost (TPIC)		
10. Engineering	32,404,000	
11. Construction	45,366,000	
TPIC		77,770,000
Total plant cost (TPC = TPDC + TPIC)		
TPC		207,387,000
Contractor's fee and contingency (CFC)		
12. Contractor's fee	10,369,000	
13. Contingency	20,739,000	
CFC = 12+13		31,108,000
Direct fixed capital cost (DFC = TPC + CFC)		
DFC		238,495,000

TABLE 1.6. Operating Cost Summary (Year 2007, Prices in USD)

Cost Item	Annual Cost ($/yr)	Proportion of Total (%)
Raw materials	19,925,000	17.85
Labor-dependent	18,538,000	16.61
Facility-dependent	45,077,000	40.39
Laboratory/QC/QA	9,269,000	8.30
Consumables	18,597,000	16.66
Waste treatment/ disposal	167,000	0.15
Utilities	37,000	0.03
Total	111,611,000	100.00

Figure 1.9 displays the impact of product titer and bioreactor volume on the unit production cost. All points correspond to four production bioreactors feeding a single purification train. For low product titers, the bioreactor volume has a considerable effect on the unit production cost. For instance, for a bioreactor product titer of 0.5 g/L, going from 10,000 L to 15,000 L and finally to 20,000 L of production bioreactor volume, the unit cost is reduced from approximately $330/g to $250/g and $210/g, respectively. For high product titers (e.g. around 2.5 g/L), on the other hand, the impact of bioreactor scale is not as important. This can be explained by the fact that at high product titers, a major part of the manufacturing cost is associated with the purification train. It is therefore wise to shift R&D efforts from cell culture to product purification as the product titer in the bioreactor increases. A key assumption for the results of the sensitivity analysis is that the composition and cost of the cell culture media is independent of product titer.

1.4.7 Variability and Uncertainty Analysis

Process simulation tools typically used for batch process design, cycle time reduction, and cost estimation employ

TABLE 1.7. Raw Materials Cost Breakdown (Year 2007, Prices in USD)

Bulk Raw Material	Unit Cost ($/kg)	Annual Amount (kg)	Annual Cost ($)	%
Inoculation media	6.147	18,720	115,072	0.58
H3PO4 (5% w/w)	0.143	882,259	125,722	0.63
NaOH (0.5 M)	0.245	745,872	182,783	0.92
WFI	0.150	3,914,847	587,227	2.95
Serum-free media	500.0	35,882	17,940,824	90.04
Protein-A equilibrium	0.153	1,271,787	194,520	0.98
Protein-A elution	0.153	576,131	88,425	0.44
Protein-A regulating buffer	0.168	345,873	58,013	0.29
NaOH (0.1 *M*)	0.243	517,328	125,773	0.63
IEX-equilibrium-buffer	0.186	189,473	35,268	0.18
IEX-wash-buffer	0.221	190,310	42,017	0.21
IEX-El-buffer	0.347	10,727	3,726	0.02
NaCI (1 *M*)	0.368	116,341	42,848	0.22
Ammonium sulfate	8.000	8,104	64,835	0.33
HIC-equilibrium-buffer	0.909	75,133	68,279	0.34
HIC-wash-buffer	0.538	178,584	96,078	0.48
HIC-El-buffer	0.305	173,202	52,763	0.26
NaOH (1 *M*)	0.336	250,320	84,168	0.42
PBS	0.182	92,548	16,870	0.08
Polysorbate 80	1.833	5	9	0.00
Total		9,593,446	19,925,219	100.00

TABLE 1.8. Consumables Cost Breakdown (Year 2007, Prices in USD)

Consumable	Units Cost ($)	Annual Amount	UOM	Annual Cost ($)	%
2.2-L Roller bottle	6	640	item	3,840	0.02
Dft DEF cartridge	1,000	720	item	720,000	3.87
225-mL T-Flask	2	1,440	item	2,880	0.02
Dft membrane	400	89	m2	35,799	0.19
50-L Bag	5	1,200	item	6,000	0.03
100-L Cell Bag	300	400	item	120,000	0.65
Viral exclusion membrane	13,356	160	item	2,136,960	11.49
20-L cell bag	100	480	item	48,000	0.26
Protein-A resin	6,000	1,930	L	11,580,800	62.27
SP-Seph HP resin	2,500	758	L	1,895,632	10.19
HIC Butyl Seph HP resin	3,000	682	L	2,047,283	11.01
Total				18,597,194	100.00

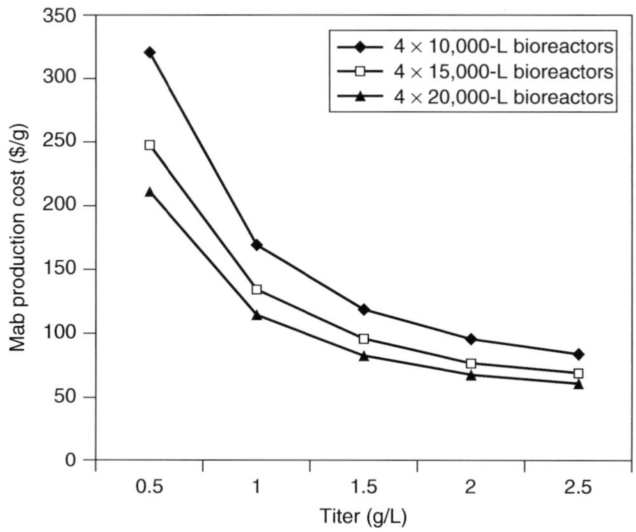

Figure 1.9. Production cost of Mab as a function of product titer and production bioreactor volume.

deterministic (cause and effect) models. They model the "average" or "expected" situation commonly referred to as the *base case* or *most likely scenario*. However, variability occurs in all bioprocesses despite best efforts to ameliorate their effects. Modeling many cases can help determine the range of performance with respect to key process parameters. However, such an approach does not account for the relative likelihood of the various cases. Monte Carlo simulation is a practical means of quantifying the variability and uncertainty in process parameters (15). In Monte Carlo simulation, uncertain input variables are represented with probability distributions. A simulation calculates numerous scenarios of a model by repeatedly picking values from a user-defined probability distribution for the uncertain variables and using those values for the model to calculate and analyze the outputs in a statistical way to quantify risk. For models developed in SuperPro, Monte Carlo simulation can be performed by combining SuperPro with Crystal Ball from Decissioneering, Inc. (Denver, Colorado). Crystal Ball is an Excel add-in application that facilitates Monte Carlo simulation. It enables the user to designate the uncertain input variables, specify their probability distributions, and select the output (decision) variables whose values are recorded and analyzed during the simulation. For each simulation trial (scenario), Crystal Ball generates random values for the uncertain input variables selected in frequency dictated by their probability distributions using the Monte Carlo method. Crystal Ball also calculates the uncertainty involved in the outputs in terms of their statistical properties, mean, median, mode, variance, standard deviation, and frequency distribution.

For the base case of this example we estimated an annual throughput of 1167 kg of purified Mab and a unit production cost of $95.6/g. Let us assume that our objective is to reliably manufacture at least 1100 kg/year of purified product at a cost of no more than $100/g. If there is variability and uncertainty in some key process and market parameters, how confident are we about the values of our objective?

To illustrate the benefits of such an exercise, we assigned probability distributions to some parameters that affect the annual throughput of the plant and the unit production cost. Since the production bioreactors are the time bottleneck of the process (i.e. have the longest cycle time), delays in the harvesting of production bioreactors will impact the annual number of batches and, consequently, the annual plant throughput and the unit production cost. For the base case scenario, a fermentation time of 12 days was assumed. To investigate the impact of variability in the harvesting of production bioreactors, a Weibull distribution was assumed for the fermentation time as part of this exercise (Fig. 1.10a) with values varying from 11.5 to 14 days and a mean value of approximately 12 days. The variability in the fermentation time accounts for the combined variability of the entire inoculum preparation and fermentation line because any delays in inoculum preparation affect the start and, consequently, the harvesting of production bioreactor batches.

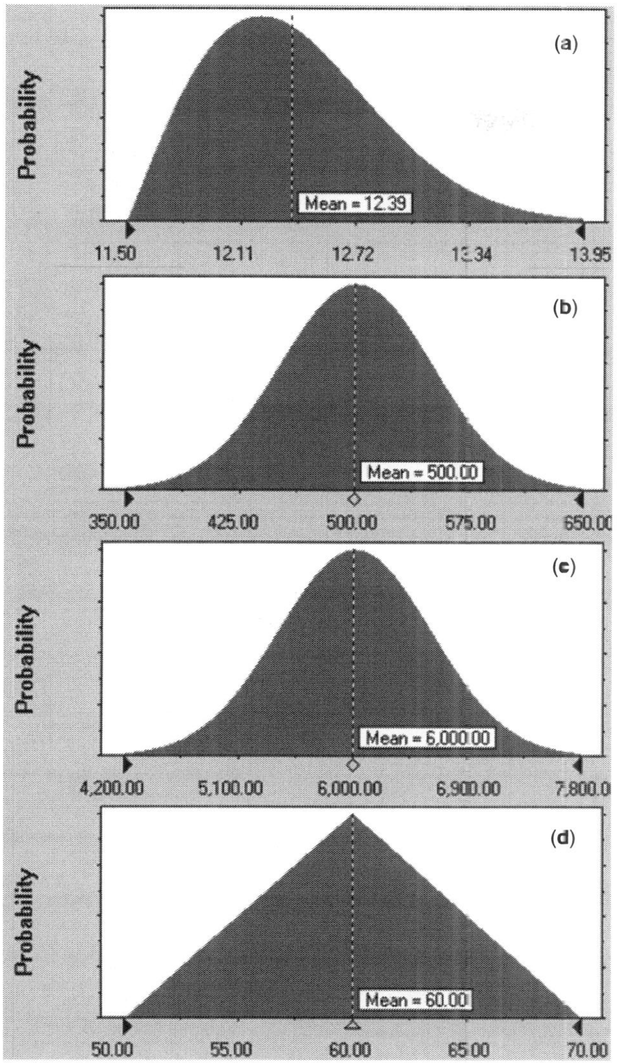

Figure 1.10. Assumed probability distribution for (a) fermentation time in days, (b) serum-free media price in dollars per kilogram, (c) Protein-A resin price in dollars per liter, and (d) Protein-A resin replacement frequency in cycles. (This figure is available in full color at http://onlinelibrary.wiley.com/book/10.1002/9780470054581.)

The unit cost of the serum-free media, which accounts for 90% of the raw materials cost, was the second parameter that was investigated. As part of this exercise, its cost was represented with a normal distribution (Fig. 1.10b) whose mean is $500/kg (equal to the base case value) and the standard deviation is 50.

Finally, the Protein-A resin, which dominates the cost of consumables, was also investigated. Figure 1.10c represents the probability distribution for the unit cost of Protein-A. The mean of $6000/L corresponds to the value of the base case. Figure 1.10d represents the probability distribution for the replacement of the Protein-A resin. The mean of 60 cycles corresponds to the base case value. If this type of analysis is done for an existing facility, historical data should be used to derive the probability distributions. Crystal Ball has the capability to fit experimental data.

The decision (output) variables considered in this example are the production cost and the annual throughput of the facility. Figures 1.11 and 1.12 display the results of the Monte Carlo simulation. The analysis reveals that the production cost will be less that $100/g (or $100,000/kg) with a certainty of 91% (dark area of Fig. 1.11). Similarly, the annual throughput of the facility will be higher than 1100 kg of purified product with a certainty of 87% (Fig. 1.12). Such findings constitute a quantification of the risk associated with a process and can assist the management of a company in making decisions on whether to proceed or not with a project idea. Additional information on Monte Carlo simulation and risk assessment can be found in the literature (16,17).

1.5 DESIGN AND OPERATION OF MULTIPRODUCT FACILITIES

Many biopharmaceutical facilities produce more than one product in parallel. The multiple production lines in such facilities typically share utilities (e.g. WFI and clean steam) and labor resources. They may also share auxiliary equipment (e.g. CIP skids and transfer panels), buffer preparation and holding tanks, and even main processing equipment. However, the sharing of resources across multiple lines renders the design and operation of such facilities more challenging. Computer models developed for such environments must capture the interaction among production lines at the facility level. During the design of such facilities, appropriate computer models assist engineers in sizing the shared utilities and figuring out equipment requirements. During operation, similar models are used for generating feasible production schedules that respect all the major constraints. Owing to the inherent variability of biological processes, scheduling tools employed in the biopharmaceutical industry must be able to handle rescheduling easily. Production scheduling results are communicated through Gantt charts and reports that provide information on tasks that need to be executed during a certain time period. The simple example that follows illustrates some of the challenges associated with the design and operation of multiproduct facilities.

Figure 1.13 displays a simplified flow diagram of a two-line Mab facility. Each line has its own bioreactor suites and purification train. However, the two lines share a centrifugal separator for biomass removal (broth clarification). They also share a buffer preparation and holding area for the first two chromatography steps. A single CIP skid, CIP-2, is available in that area for cleaning all the buffer preparation and holding tanks.

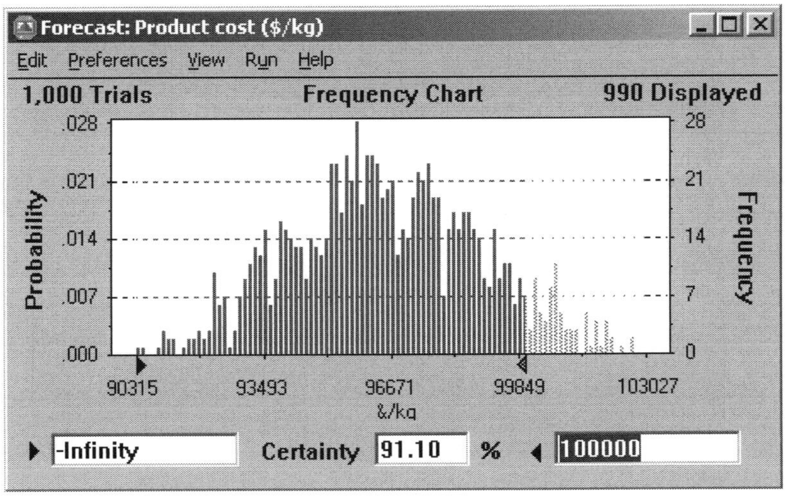

Figure 1.11. Calculated probability distribution for the production cost (in $/kg). (This figure is available in full color at http://onlinelibrary.wiley.com/book/10.1002/9780470054581.)

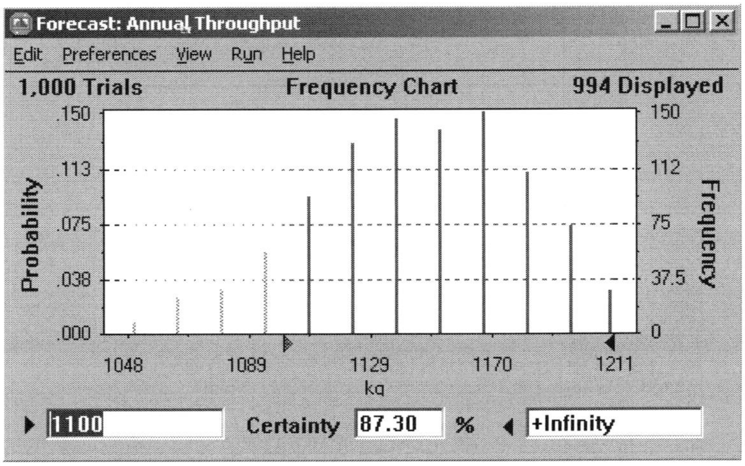

Figure 1.12. Calculated probability distribution for the annual throughput of the process (in kg/year). (This figure is available in full color at http://onlinelibrary.wiley.com/book/10.1002/9780470054581.)

The facility and the two recipes were modeled using SchedulePro (Intelligen, Scotch Plains, NJ). Figure 1.14 displays a typical production schedule for this facility with a 3.5-day cycle time for each product line (represented by the different colors). The top line of the chart displays the occupancy of CIP-2, the single CIP skid used for cleaning the tanks of the shared buffer preparation and holding area. Under normal conditions, CIP-2 can clean all the tanks. However, its utilization is quite high and that may lead to conflicts in cases where the schedule deviates from its normal one.

For instance, what if there was a 24-h delay in the harvesting of the bioreactor of the fourth batch? That would lead to the delay of the start of the downstream operations. When such a delay is introduced to the model, the algorithm warns the user about the conflicts that it will create and offers to reschedule. Rescheduling typically involves delaying of activities that have not started yet. If the user does not authorize rescheduling, the tool simply displays the conflicts created by the delay (Fig. 1.15). SchedulePro creates multiple lines for conflicting equipment, draws a red frame around the conflicting activities, and displays an "exclamation" mark on the y axis. (The reader is requested to refer to the online version of this chapter for color indication.) The user can manually resolve the conflicts through drag-and-drop of tasks and local rescheduling and evaluate whether a solution with the current available resources is possible.

Elimination of conflicts related to CIP-2 through the installation of a second CIP skid can be readily evaluated

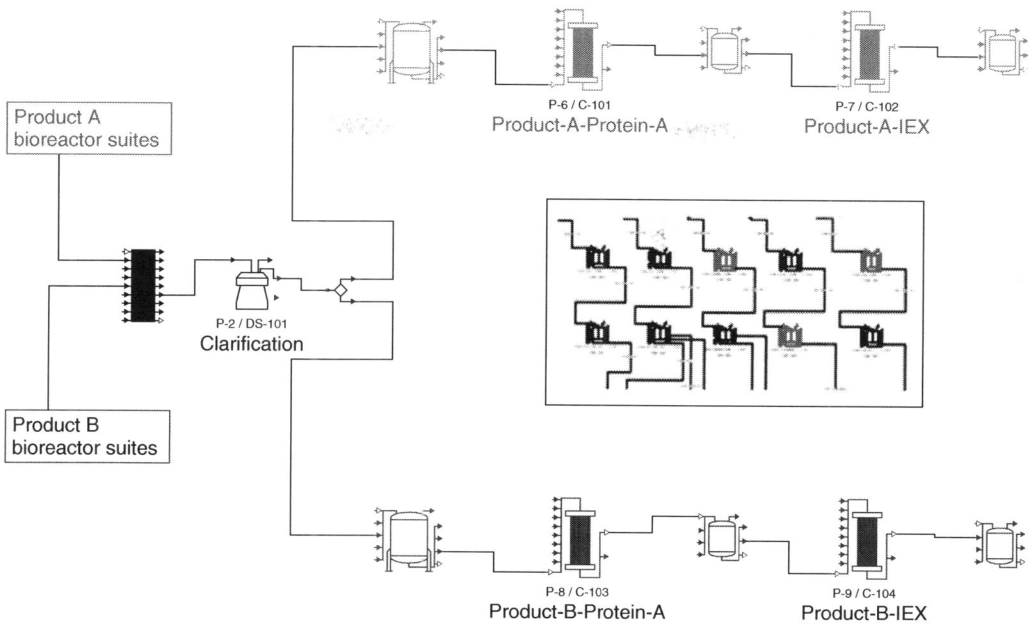

Figure 1.13. Two-product Mab facility with common buffer preparation and holding area. (This figure is available in full color at http://onlinelibrary.wiley.com/book/10.1002/9780470054581.)

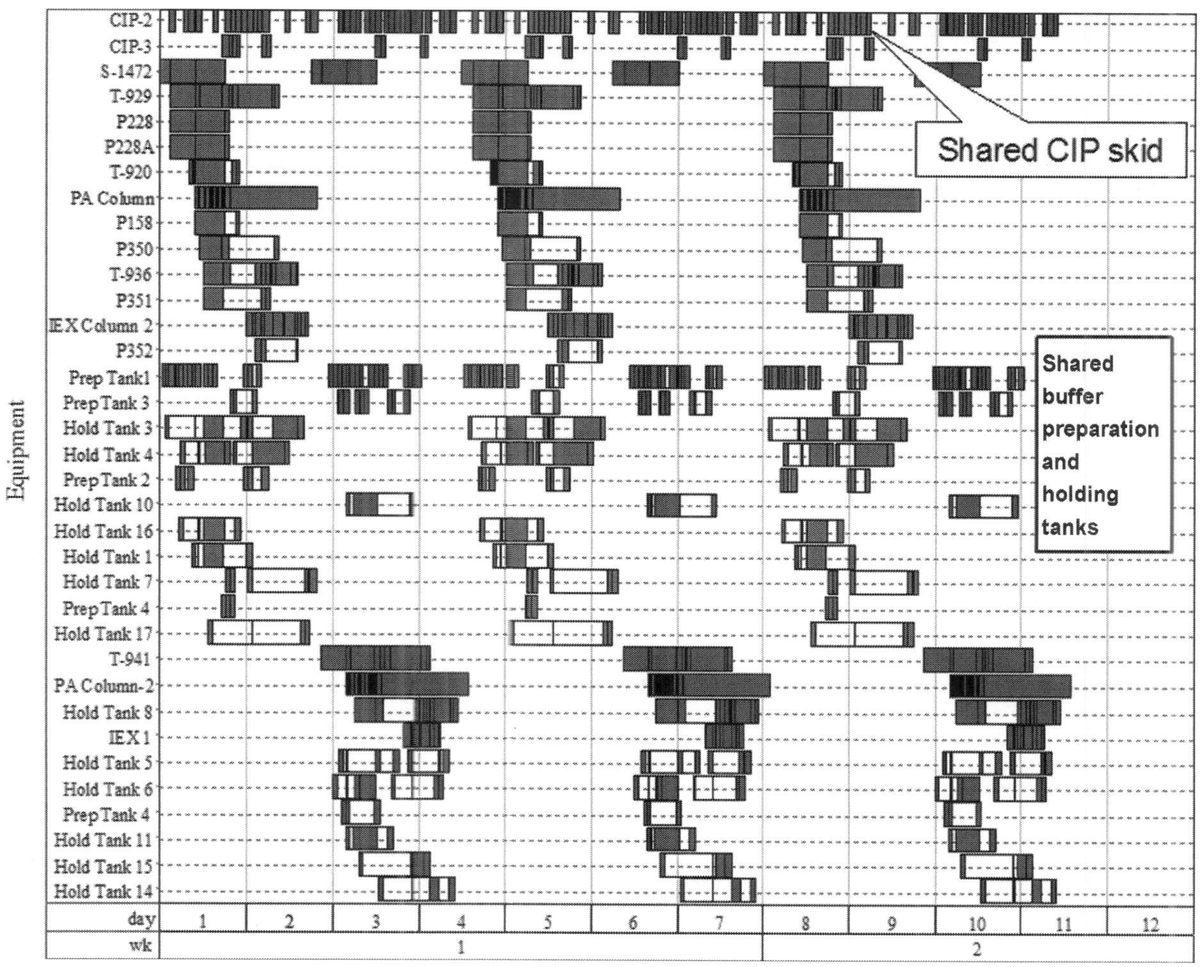

Figure 1.14. Production schedule for the two-line Mab facility. (This figure is available in full color at http://onlinelibrary.wiley.com/book/10.1002/9780470054581.)

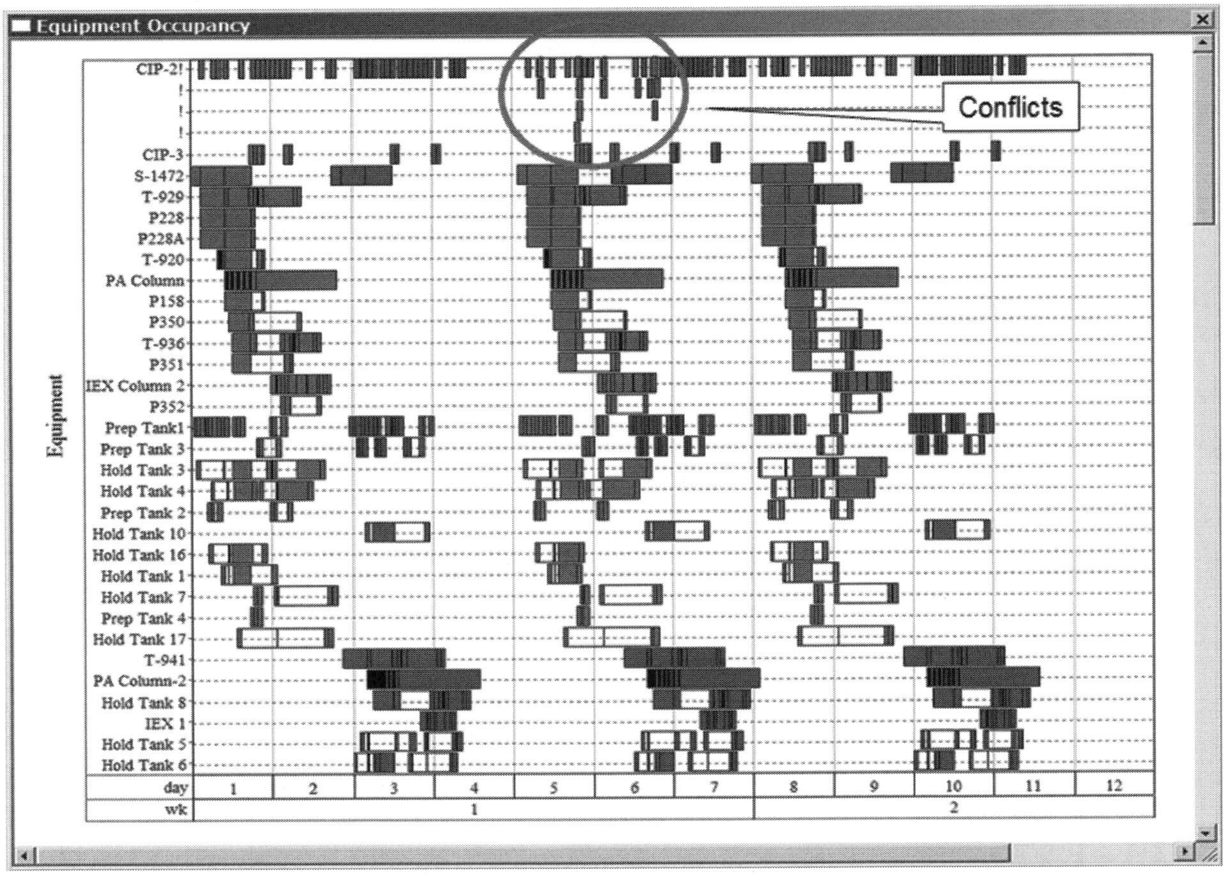

Figure 1.15. Introducing a delay and viewing conflicts. (This figure is available in full color at http://onlinelibrary.wiley.com/book/10.1002/9780470054581.)

with tools like SchedulePro. Other issues that can be easily investigated include potential conflicts related to the limited availability of labor, utilities, inventories of raw materials, storage and treatment capacity for waste materials, and so on.

1.6 SUMMARY AND CONCLUSIONS

Process simulation and production scheduling tools can play an important role throughout the life cycle of process development and product commercialization. In process development, simulation tools are becoming increasingly useful as a means to analyze and evaluate process alternatives. During the transition from development to manufacturing, they facilitate technology transfer and process fitting. Production scheduling tools play a valuable role in manufacturing. They are used to generate feasible production schedules and enable manufacturing personnel to efficiently handle process delays and equipment failures. Such tools also facilitate capacity analysis and debottlenecking tasks.

The biopharmaceutical industry has only recently begun making significant use of process simulation and scheduling tools. Increasingly, universities are incorporating the use of such tools in their curricula. In the future, we can expect to see increased use of these technologies and tighter integration with other enabling information technologies, such as supply chain tools, manufacturing execution systems (MES), batch process control systems, and process analytics tools (PAT). The result will be more robust processes and efficient manufacturing leading to more affordable biological products.

REFERENCES

1. Douglas JM. Conceptual design of chemical processes. New York: McGraw-Hill; 1988.
2. Pavlou AK, Belsey MJ. Eur J Pharm Biopharm 2005; 59: 389–396.
3. Plenert G, Kirchmier B. Finite capacity scheduling–management, selection, and implementation. New York: John Wiley & Sons; 2000.
4. Pinedo ML. Planning and scheduling in manufacturing and services. New York: Springer Science; 2005.

5. Korovessi E, Linningerr AA. Batch processes. Boca Raton, FL: Taylor & Francis; 2006.
6. Hwang F. Pharm Eng 1997; January/February 28–43.
7. Petrides DP, Koulouris A, Lagonikos PT. Pharm Eng 2002; 22: 56–64.
8. Heinzle E, Biwer A, Cooney C. Development of sustainable bioprocesses. West Sussex: John Wiley & Sons; 2006.
9. Walsh G. Nat Biotechnol 2006; 24: 769–775.
10. Langer ES, Junker B, editors. Advances in large scale biopharmaceutical manufacturing and scale-up production. Rockville, IN: ASM Press & ISTM; 2004. pp. 152–190.
11. Parshall J, Lamb L. Applying S88–batch control from a user's perspective. Research Triangle Park, NC: ISA; 2000.
12. Harrison RG, Todd P, Rudge SR, Petrides DP. Bioseparations science and engineering. New York: Oxford University Press; 2003.
13. Peters MS, Timmerhaus KD. Plant design and economics for chemical engineers. 4th ed. New York; McGraw-Hill; 1991.
14. Valle-Riestra JF. Project evaluation in the chemical process industries. New York: McGraw-Hill; 1983.
15. Mun J. Applied risk analysis. Hoboken, NJ: John Wiley & Sons; 2004.
16. Achilleos EC, Calandranis JC, Petrides DP. Pharm Eng 2006; 26(4): 34–40.
17. Papavasileiou V, Koulouris A, Siletti C, Petrides D. Pharm TechnolInnovations 2006; 22(1): s28–s38.

PART II

DOWNSTREAM RECOVERY OF CELLS AND PROTEIN CAPTURE

2

CELL SEPARATION, CENTRIFUGATION

HANS AXELSSON

AlfaLaval AB, Tumba, Sweden

2.1 INTRODUCTION

The first application of centrifugal separation outside milk processing was harvesting of cells in baker's yeast production; centrifugation is still the only choice for cell separation and protein recovery in many fermentation processes. In many such processes, the most optimal phase separation is obtained by combining centrifugation with some type of fine filtration.

2.2 CENTRIFUGAL SEPARATION

Centrifugal separation is a sedimentation operation accelerated by centrifugal force. Thus, a prerequisite for the separation is a difference in density between the phases. This applies to both solid–liquid and liquid–liquid separation.

The settling velocity V_g of a solids particle (or droplet) under the influence of gravity alone is given by Stokes' law (refer to section on nomenclature for abbreviations and symbols):

$$V_g = \frac{\rho_p - \rho_f}{18 \cdot \eta} \cdot d_p^2 \cdot g \tag{2.1}$$

In the centrifuge field, the settling velocity becomes:

$$V_c = V_g \cdot Z \tag{2.2}$$

where

$$Z = \frac{\omega^2 \cdot r}{g} \tag{2.3}$$

is called relative centrifugal force (RCF) or G-number.

It is however, not the RCF alone that determines the performance of a given centrifugal separator. The residence time of the particle or droplet in the centrifugal force field is also important, so that a larger volume of the rotor will increase the possible flow rate. In the most common centrifugal separators, the disk bowl machines, a given bowl material can cope with a certain peripheral speed; the stronger the bowl material, the higher the peripheral speed. This leads to the fact that the maximal RCF is inversely proportional to the bowl diameter, but not necessarily that the lower RCF machine has a lower performance.

2.3 TYPES OF CENTRIFUGAL SEPARATORS

2.3.1 General

What is determining the design of a centrifuge is the method by which the solids phase in the process material is handled. The first two types developed were applied to duties where the resulting products were in liquid form, cream and skimmed milk, and yeast cream and cell-free wort. In the beginning of this century, centrifuges were applied to processes with solids, amorphous or crystalline. In the biotechnology industry, the solids phase is often recovered in a diluted suspension or is, even at high solids concentrations, aslurry with good flow properties.

Types of machines are described in the work by Rushton *et al.* (1). The most comprehensive description of centrifuges can be found in the book by Sokolov (2). Brochures and technical publications from sedimenting centrifuge manufacturers such as Alfa Laval AB, also provide information on machine types and sizes.

Downstream Industrial Biotechnology: Recovery and Purification, First Edition. Edited by Michael C. Flickinger.

In this chapter, centrifugal filtration will not be dealt with because of the limited use in this industry, with the exception of crystal recovery in antibiotic production. For a description of various centrifugal filters, see Rushton *et al*. (1).

2.3.2 Disk Bowl Machines

2.3.2.1 Solid Bowl. A solid bowl machine is shown in Fig. 2.1a. The feed enters from above through a stationary pipe that leads into a volume formed by the distributor, the component that carries a multitude of thin conical disks, less than 1-mm thick, separated by spacers, usually 0.3–1.5 mm thick. On the inside of the distributor are 4–16 radial wings; however, a large bowl needs more wings. The purpose of the wings is to start the rotation of the incoming liquid. This design principle is the same for all disk bowls with feed from above. Like all other disk bowl machines, it may have one or two liquid outlets. The outlet(s) may be open or

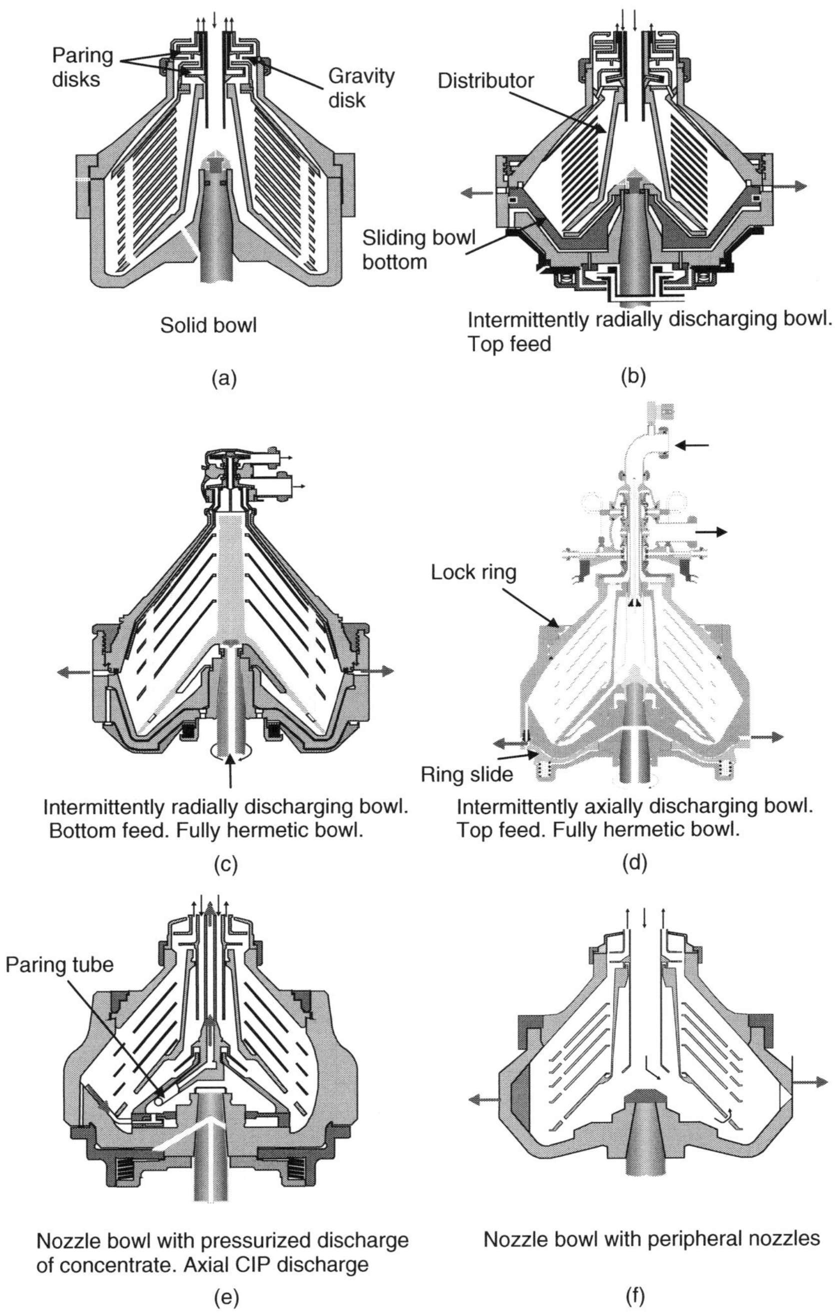

Figure 2.1. Vertical cuts of various disc centrifuge bowls. (This figure is available in full color at http://onlinelibrary.wiley.com/book/10.1002/9780470054581.)

equipped with paring disks. The paring disk is a stationary device, and has the shape of a pump wheel. Liquid fills the chamber around the paring disk. Part of the inner layer of the rotating liquid in the chamber is pared off, enters the pump wheel and is forced out by the static pressure created when the kinetic energy of the rotating liquid is converted to static pressure. The version in Fig. 2.1a has two outlets, each equipped with a paring disk. The solids accumulate at the periphery and must be removed manually. In the separation of two nonmiscible liquids, a coalescence of the dispersed droplets takes place at an internal interface in the bowl. The positioning of this interface, its diameter D_i is very important. The nearer it is placed to the light-liquid outlet, the more contaminated the light liquid will be with droplets of the heavy phase. The position is determined by a gravity disk (Fig. 2.1a), over which the heavy liquid flows. With every machine a set of gravity disks with different inner diameters is supplied. The gravity disk's inner diameter is given by the relation:

$$D_h = \sqrt{D_i^2\left(1 - \frac{\rho_{fl}}{\rho_{fh}}\right) + D_l^2\frac{\rho_{fl}}{\rho_{fh}}} \qquad (2.4)$$

where D_i and D_l are the diameter of the interface and the outflow diameter of the light phase, respectively.

If properly designed, and if the feed flow rate is high enough to prevent sedimentation in the bowl, this type of machine can be used to recover a dilute cell concentrate in the heavy-liquid outlet. This is also valid for three-phase solids-ejecting machines. Solid bowl machines are made in a large number of sizes with bowl diameters ranging from 140 to 750 mm.

2.3.2.2 Solids-Ejecting Separators, Radial Discharge. Of all types of disk bowl machines, the solids-ejecting separators with radial discharge (Fig. 2.1b, c) are the most common. The feed zone in the machine in Fig 2.1c resembles that of the machine in Fig. 2.1a. With an automated, periodic partial discharge of the sediment, it is possible to obtain a considerably higher concentration of the solids than in the peripheral nozzle machine (Fig. 2.1f). In its upper position, the sliding bowl bottom (cf. Fig. 2.1b) keeps the bowl closed. At discharge it is pressed downwards during about 0.1 sec, making it possible for the solids to escape through large ports in the outer bowl wall. This is illustrated in a video clip (the reader is requested to refer to the online version for Videoclip1.mov). To make the discharge at the right moment, the discharge can be initiated by a self-triggering mechanism that hydraulically senses the solids level in the solids space (3). Three-phase separation units are available. The solids-ejecting machines can be equipped with hermetic seals and a hermetic feed from below in a hollow spindle (Fig. 2.1c). In these true hermetic machines the split between the exiting light and heavy liquids can be made by controlling the back pressure in the outlet pipes. The heavy liquid can contain large amounts of solids (cells or proteins) if the machine is properly designed. The solids-ejecting separators are today available in about 10 sizes with bowl diameters between 180 and 1000 mm. Several small- and medium-sized units are available in versions equipped for BL2-LS requirements. A pilot-scale sterilizable separator module is seen in Fig. 2.2. A large-scale unit is seen in Fig. 2.3. A machine with radial intermittent discharge has been developed primarily for recovery of protein precipitates (4). When the solids space is full of solids, the liquid in the bowl is decanted and a discharge takes place by a pneumatic mechanism.

2.3.2.3 Solids-Ejecting Separators, Axial Discharge. When comparing axially and radially discharging separators (Fig. 2.1b, d) one finds that the former has a different geometry. The lock ring that holds together the upper and lower parts of the rotor (Fig. 2.1d) can be placed at a smaller diameter, because no sliding bowl bottom needs to be fitted. This reduces the stresses in the thread of the lock ring. Also, the solids discharge takes place through a few small axial channels, which do not induce large stresses in the bowl shell. These two factors make it possible to increase the bowl speed considerably above that of the radially discharging machines, so that the RCF is at least doubled. The discharge system is operated by pressurized air, which forces a ring slide, equipped with valve seats, downward, opening the axial channels. To facilitate the transport of solids toward the channels, the

Figure 2.2. Solids-ejecting disc bowl centrifuge of pilot size for sterilizable contained installation. Type Culturefuge™ 100. The height of the separrator only is 1.3 m. Courtesy Alfa Laval AB. (This figure is available in full color at http://onlinelibrary.wiley.com/book/10.1002/9780470054581.)

Figure 2.3. Radially solids-ejecting disk bowl centrifuge in production size. Capacity in breweries up to 90 m^3/h. Type BREW 3000. The height of the separator is 2.2 m. Courtesy Alfa Laval AB. (This figure is available in full color at http://onlinelibrary.wiley.com/book/10.1002/9780470054581.)

bowl is machined to have a star-shaped inner contour. This type of machine exists at present in three sizes with bowl diameters ranging from 500 to 900 mm. BL-2LS versions are available. A medium-sized unit is seen in Fig. 2.4.

2.3.2.4 Nozzle Machines with Pressurized Discharge of Concentrate. The apparent viscosity of bacteria and yeast suspensions as well as some protein precipitates decreases with increasing velocity gradient. Therefore, they are able to flow even at high concentrations in channels or tubes from the periphery of the bowl in nozzle machines with pressurized discharge of concentrate (Fig. 2.1e). The tubes empty into a central chamber, where a centripetal pump, a paring tube, picks up the concentrate and pumps it out of the bowl. The paring tube is a stationary radial tube acting in the same way as a paring disk (Fig. 2.1a). The nozzles are placed at the end of the concentrate tubes, just in front of the chamber. This machine type is suitable only for applications where the solids have the specific flow properties mentioned above, and in the absence of other type of solids. In some machines of this type, a vortex chamber is placed in front of the nozzle (5). Liquid enters the chamber tangentially, creating a viscosity-dependent whirl, and leaves through the nozzle in the center of the chamber. The viscosity effect gives a self-regulating function so that varying solids flow rates (within limits) to the machine will result in a constant concentration of the concentrate,

Figure 2.4. Axially solids-ejecting disk bowl centrifuge in production size for bacteria and mammal cell recovery. Capacity up to 12 m^3/h. Type BTAX 215S. The height of the separator including piping is 1.8 m. Courtesy Alfa Laval AB. (This figure is available in full color at http://onlinelibrary.wiley.com/book/10.1002/9780470054581.)

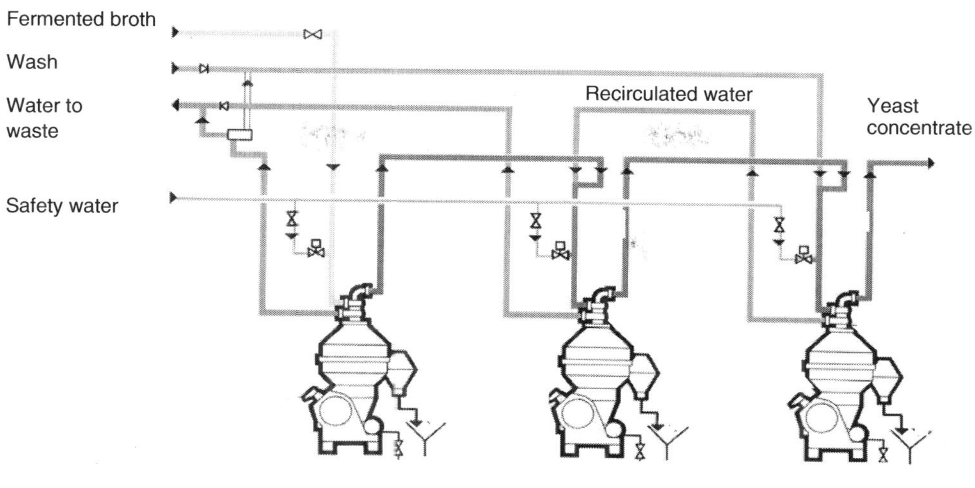

(a)

(b)

Figure 2.5. (a) Flow sheet for a countercurrent washing system with two washing stages for yeast. Centrifuges with pressurized yeast outlet. (b) Nozzle machines with pressurized solids discharge for yeast separation, installed for countercurrent washing. Capacity on bakers' yeast up to 100 m^3/h. Type FEUX 2 4. The height of the separator including piping is 2.3 m. [Courtesy of Alfa Laval AB.] (This figure is available in full color at http://onlinelibrary.wiley.com/book/10.1002/9780470054581.)

thereby reducing the risk of clogging the machine. The machine type was originally developed for bakers' yeast production. The yeast, being grown on molasses as the carbon source, needs to be washed clean from substrate residues by water. A three-stage countercurrent yeast harvest and washing system is shown in Fig. 2.5. One of the units of this type (Fig. 2.6) is available in a sterilizable, contained and aseptic version, conforming to BL2-LS. In order to make this possible, the cleaning-in-place (CIP) has to be efficient. Therefore, the sterilizable machine uses the discharge system of the machine in Fig. 2.1d during CIP. This system is also used in noncontained versions. Bowl diameters between 500 and 900 mm are manufactured.

2.3.2.5 Nozzle Machines with Peripheral Nozzles. In machines with peripheral nozzles (Fig. 2.1f), nozzles with a diameter of 0.5–3 mm are situated at the periphery of the bowl, which has sloping walls toward the nozzles. The number of nozzles varies between 4 in small bowls and 20 in the largest bowls. It can be equipped with a recirculation device for the sediment built as follows: Inside the feed tube is a separate central tube for recirculated liquid. The central tube enters into a chamber at the bottom of the

Figure 2.6. Nozzle machine with pressurized solids discharge for yeast and bacteria separation. Capacity on *E. coli* about 3000 l/h. Type BTUX 510. The height of the separator is 1.8 m. Sterilizable and contained installation. Courtesy Alfa Laval AB. (This figure is available in full color at http://onlinelibrary.wiley.com/book/10.1002/9780470054581.)

bowl, from which tubes lead the recirculated liquid along the bowl bottom. Each tube ends just in front of the nozzle. This makes it possible to increase the concentration of the solids phase without decreasing the nozzle size which leads to increased risk of nozzle clogging. The nozzle flow rate is proportional to number of nozzles, distance from center of rotation, bowl speed, and square of the nozzle diameter. The peripheral nozzle machines are used for the largest flow rates; the bowl diameters are up to 1050 mm.

2.3.3 Decanter Centrifuges

In the 1940s, the solid bowl separator with scroll discharge of solids, the decanter centrifuge (Fig. 2.7), was developed into the design it has today. It is intended for use with process liquids containing a high percentage of suspended solids. It is equipped with a screw conveyor that rotates at a speed slightly higher or lower than the bowl speed (the reader is requested to refer to the online version for Videoclip2.mov).

The differential speed between the conveyor and bowl can be varied. Careful control of this parameter, torque control, and some newer design features (e.g. the baffle disk), have recently made the decanter interesting to use for biological sludges as well. However, the conventional mechanical design of these machines does not allow more than 5000 G for a medium-diameter machine (6).

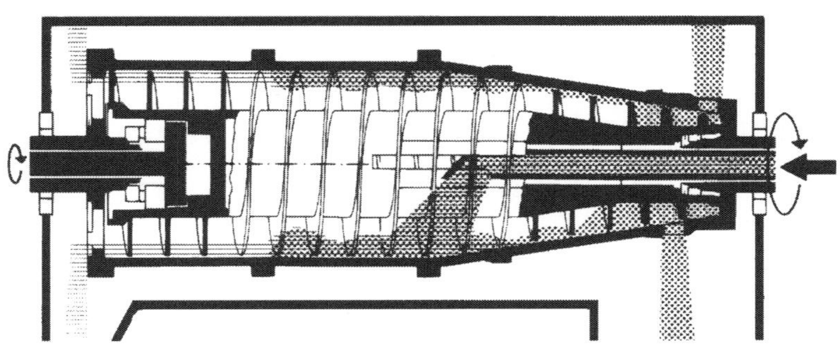

Figure 2.7. Decanter centrifuge. Vertical cut.

For three-phase decanters the dimensioning of the gravity disk (Fig. 2.1a) follows in principle Equation 2.4. It is, however, very important to take the weir head into account. The weir head is the thickness of the liquid layer on the edges, over which the liquid flows. This thickness varies with the flow rate. A variation of such parameters as the light–heavy phase ratio gives a very unstable interface position $D_i(7)$, especially in combination with a density ratio near 1 (Eq. 2.4).

The range of bowl diameters is very wide, from 150 to 1200 mm; the bowl length is often 4–5 times the diameter.

2.3.4 Other Bowl Designs

2.3.4.1 Tubular Bowl. In the tubular bowl (Fig. 2.8) the rotor, driven from above, consists of a long cylinder, into which liquid is pumped from below. The design shown in Fig. 2.8 is equipped for separation of two liquid phases. Most tubular centrifuges are used for separating very valuable solids from a liquid. The liquid is pumped out by a paring disk (Fig. 2.1a), avoiding aerosol and foam formation. To remove the solids accumulated at the periphery, the machine must be stopped and dismantled. The bowl diameters are up to 130 mm with a solids volume of up to 6 L. In the industrial models, 20,000 G are generated. Sterilizable and contained units are available.

In a recent design (8), the separated solids are removed intermittently by a piston. The unit is available in four sizes, can reach a G-number of 20,000 and is designed with CIP and sterilization-in-place (SIP) in mind.

In a design for liquid–liquid separation in multistage extraction plants (9), bowl diameters up to 500 mm are available. The bowl speed is very low, so that the G-number is well below 1000. The mixing of the two phases takes place in the annulus between the rotor and the housing.

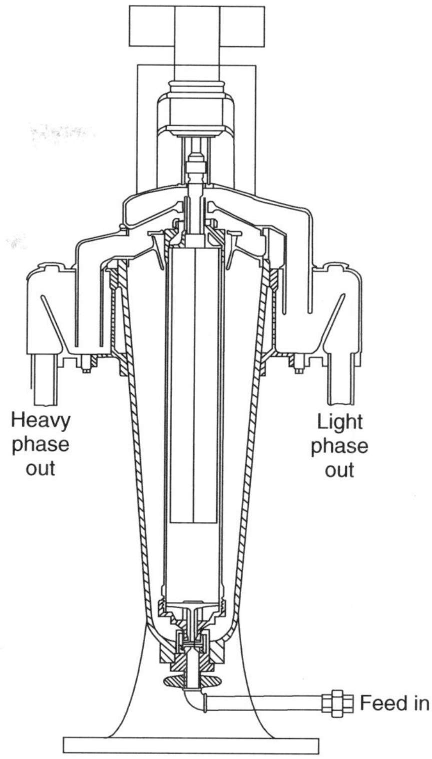

Figure 2.8. Tubular bowl centrifuge. Vertical cut.

2.3.4.2 Multichamber Bowl. In the multichamber bowl machine (Fig. 2.9), the process liquid flows in the annuli between concentric cylinders.

Having a relatively low separation capability, this type of machine is used for recovery of valuable solids, when the solids volume of the tubular bowl is too small. Some designs include cooling channels in the bowl wall, a cooling jacket around the bowl, and a cooled centripetal pump (paring disk). The maximum bowl diameter is about 500 mm.

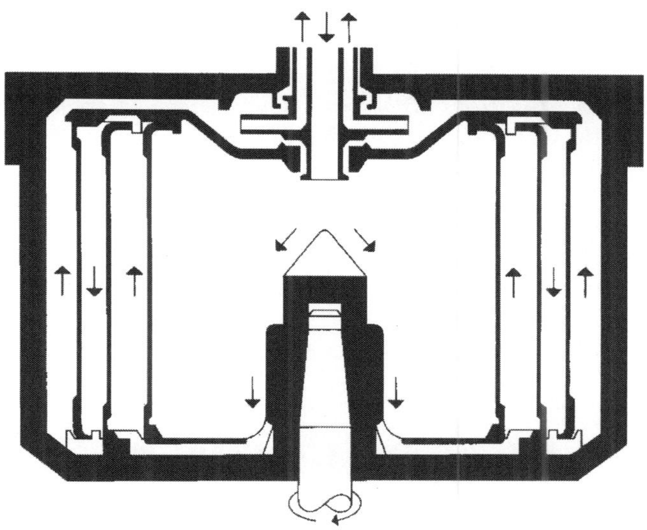

Figure 2.9. Multichamber bowl. Vertical cut.

2.3.4.3 Centritech® Machines. In the Centritech machine (Fig. 2.10), a disposable separation insert of plastic is placed in a rotor, forming an annular separation chamber. The feed liquid enters at the top rim of the insert and the main liquid outflow is at the opposite end of the top rim of the insert. The heavy-liquid phase (cell concentrate) is withdrawn intermittently through the bottom outlet. Through the principle of the inverted comma, no seals between rotating and nonrotating parts are needed. The G-number is at most 300. The mechanics are further described in Apelman *et al.* (10). Its only described application is separation of mammalian cells from culture fluid. Two sizes of the machine are available, and the diameter of the insert is about 300 mm in the larger unit.

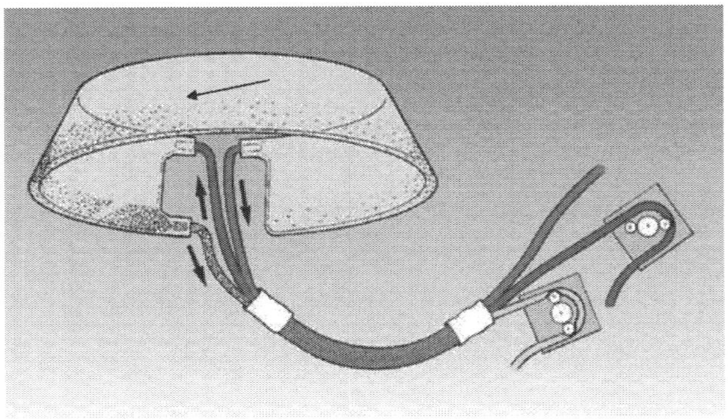

Figure 2.10. Centritech® Cell separation insert. The arrow in the separation chamber indicates the direction of the solids flow. Courtesy PneumaticScaleAngelus div. of Barry–Wehmiller Companies, Inc. (This figure is available in full color at http://onlinelibrary.wiley.com/book/10.1002/9780470054581.)

2.3.4.4 Inverted Chamber Bowl with Intermittently Operated Scraper. The inverted chamber bowl machine (Fig. 2.11) is sold under the name Powerfuge™ (11). Its bowl has its opening downward, where the separated liquid leaves under gravity and is collected in a chute. The feed is from the top. When the bowl is full of solids, it is decelerated. At low speeds, the asymmetrically positioned scraper is engaged and removes the solids. Three sizes are available with bowl diameters from 150 to 400 mm. They are designed for CIP and SIP. The smallest unit can develop 20,000 G.

2.3.4.5 Inverted Chamber Bowl with Intermittent Drainage of Solids. This recent design (Fig. 2.12) is primarily intended for separation of shear-sensitive cells. The separation takes place in the annulus between the core and the bowl wall. The cells are collected at the bowl wall, and the clarified liquid flows over the bottom edge of the bowl into the stationary outer annular bottom chamber from where the liquid is drawn off continuously. When the bowl is full of solids, the feed is shut off, and the bowl decelerated. Once the bowl stops, the accumulated solids flow down into the stationary inner annular chamber from where they are drawn off; then a new cycle can start.

2.3.5 Summary

In Table 2.1, the characteristics of various separator types are found. It has to be remembered that the hydraulic capacities given by centrifuge manufacturers may indicate the maximum possible feed flow rates that often are far above the flow rates that will give satisfactory separation results for difficult separations, such as bacteria and proteins.

2.4 FLUID AND PARTICLE DYNAMICS

2.4.1 Introduction

A successful scale-up method should be based on a mathematical model which accurately describes the separation process, the hydrodynamics, and particle dynamics inside the bowl. The phenomena going on inside the centrifuge bowl are, however, so complicated that no mathematical model is able to predict in absolute terms the separation efficiency as a function of machine and process media parameters. Axelsson (12) reviewed some of the flow phenomena, which are briefly discussed in this section, along with examples of more recent findings.

2.4.2 Centrifuges with Conical Disks

2.4.2.1 Fluid and Particle Dynamics in the Disk Stack. The disks in the disk bowl split the flow in thin layers. The Reynolds number for the flow between the discs is

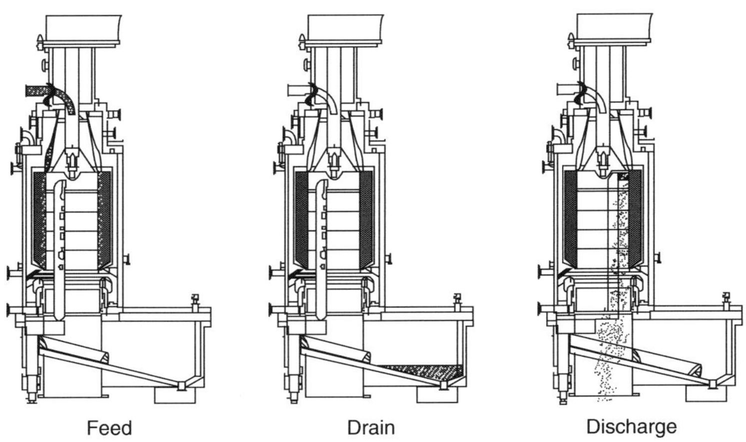

Figure 2.11. Vertical cut of a Powerfuge®, one-chamber bowl with scraper. Cycle of operation. Courtesy PneumaticScaleAngelus div. of Barry–Wehmiller Companies, Inc. (This figure is available in full color at http://onlinelibrary.wiley.com/book/10.1002/9780470054581.)

TABLE 2.1. Characteristics of Separator Types

Type	Mode of Solids Discharge	Feed Flow Rate (L/h)	Feed Solids Content, % by Volume	Solids Flow Rate (L/h)	ΣValue (m^2)	Maximum G-Force Developed	Consistency of Solids
Disk solid bowl	Manual	20–100,000	<1	0	1,000–300,000	10,000	Firm paste
Solids ejecting, radial	Intermittent	20–100,000	<25	<3,000	1,000–500,000	14,500	Thick flowing slurry
Solids ejecting, axial	Intermittent	1,000–150,000	<15	<1,000	110,000–220,000	15,000	Thick flowing slurry
Nozzle, pressurized discharge[a]	Continuous	1,000–180,000	4–30	>150, <40,000	69,000–180,000	15,000	Thick flowing slurry
—	—	—	—		—	—	—
Peripheral nozzle	Continuous	300–500,000	2–30	>3000, <140,000	35,000–180,000	11,000	Thin slurry
—	—	—	—		—	—	—
Decanter centrifuge	Continuous with scroll	300–200,000	5–50	<~50,000	400–25,000	10,000	Thick slurry to semisolid
Tubular bowl, liquid/solid	Manual	20–7,000	<1	0	1400–4500	31,000	Firm paste
Tubular bowl, liquid/liquid/solid	Flushing	<40,000	—	0	—	100–1000	Gel
Tubular bowl	Intermittent with piston	<11,200	—	—	1700–43,000	<20,000	Firm paste
Multichamber bowl	Manual	100—20,000	<5	0	—	9,000	Firm paste
Centritech machines[b]	Intermittent	5–100	<1	<15	Not applicable	100	Very thin slurry
Inverted solid bowl	Intermittent with scraper	10–1700	1–30	<~300	800–~11,200	20,000	Firm paste

[a]Only for slurry of single cells (cf text).
[b]Only for mammalian cells.

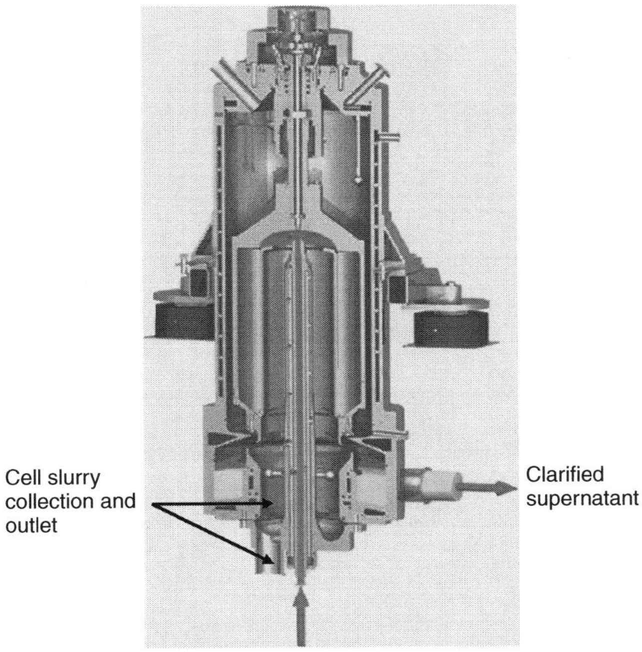

Figure 2.12. Vertical cut of a Viafuge® V12, one-chamber bowl with intermittent solids drainage. System height is 2.3 m. Courtesy PneumaticScaleAngelus div. of Barry–Wehmiller Companies, Inc. (This figure is available in full color at http://onlinelibrary.wiley.com/book/10.1002/9780470054581.)

usually low, and the flow is laminar, resulting in an easy particle settling. The critical Reynolds number for transition to turbulent flow increases with decreasing disk spacing and increasing bowl speed (2), which is beneficial for the separation efficiency because the settling distance at a small disk spacing will be short, and the high bowl speed gives a large settling velocity. Also, a small disk spacing keeps the liquid rotating at or near the bowl speed, thus reducing the pressure drop in the disk stack.

In Fig. 2.13 is a vertical cut of two discs in the disk stack, where the flow is driven toward the center by a radial pressure gradient. The character of the velocity profile between the disks depend on a dimensionless number, λ, which is defined as:

$$\lambda = t \cdot \sqrt{\frac{\omega \cdot \sin\alpha}{\nu}} \tag{2.5}$$

When λ is small, the velocity profile is parabolic, but for λ values above 5, normally encountered in industrial centrifuges, the radial velocity component splits up in two thin layers, the Ekman layers (Fig. 2.13). In these layers, all the net transport of liquid toward the center takes place. Because separated particles (heavy phase) are collected at the underside of the disk, they are subjected to a strong shear force from the Ekman layer that drags the particle toward the liquid outlet. The velocity in these layers

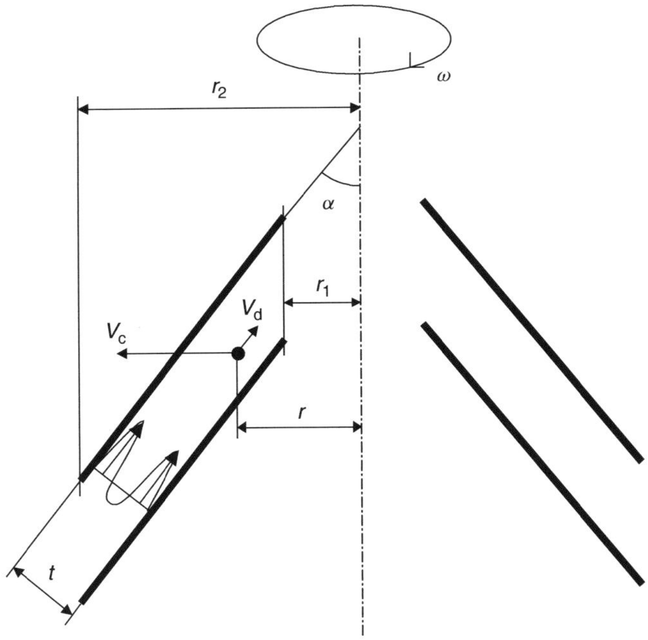

Figure 2.13. Flow profile between conical discs at high λ value. Particle separation.

increases with increasing λ, such as with increasing bowl speed. Therefore, a higher bowl speed cannot always be used fully because of the counteracting shear force.

In between the Ekman layers is the geostrophic layer. Figure 2.14a shows a picture of a laboratory model of the space between two discs of which the upper one is transparent. The speed of rotation is 100 times lower than in a commercial centrifuge, and to retain the important λ value, the disk spacing t has been increased by a factor of 10. The picture is taken with a co-rotating camera. The streamlines in the geostrophic layer have been visualized by dye injection and in Fig. 2.14b they can be compared with the calculated streamlines (13). Particle separation has also been studied in the same geometry (14). It has been found that the drag force on the particles collected at the underside of the disk depends on the position on the disk and that some particles are easily separated, whereas others are forced toward the center. The drag force can be computed (15) and the calculations of separation efficiency including the drag force verify these findings.

There is also an accumulation of particles at the radial rib, along which the solids slide toward the periphery (14). At the end of the rib there is a wake forcing the particles back toward the center until the bulk weight of the solids becomes large enough to cause an intermittent discharge of particles into the solids space outside the perimeter of the disk.

2.4.2.2 Inlet. All bowls in Fig. 2.1–2.12 except the hermetic machines (Fig. 2.1c, d) and the Centritech machine (Fig. 2.10) have an open chamber in the center of

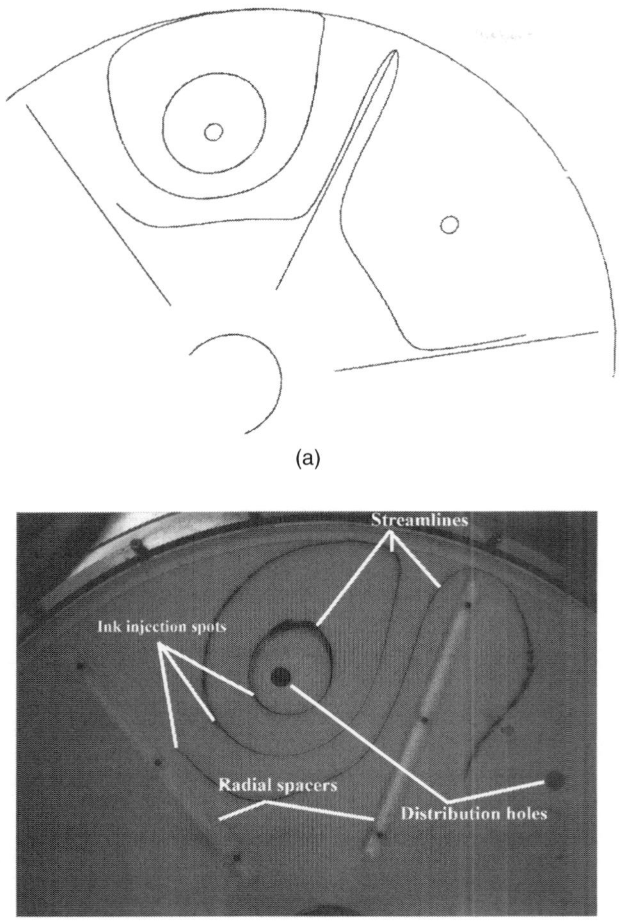

Figure 2.14. Comparison of (a) calculated and (b) actual flow patterns in a disk bowl. (This figure is available in full color at http://onlinelibrary.wiley.com/book/10.1002/9780470054581.)

the bowl, where the nonrotating feed liquid is received. In most inlets, the liquid forms a cylindrical interface with the gas (usually air) in the receiving chamber, usually called distributor (Fig. 2.1b) On the inside of the distributor is mounted a number of radial wings (not shown in Fig. 2.1b) to bring the nonrotating incoming liquid up to the rotational speed. When the feed liquid hits the rotating wings, the result is a turbulence that whips air into the feed and splits particles that are susceptible to shear; the further out from the center the interface, the larger the turbulence. If it is not practical to use a hermetic inlet, the interface in the inlet must have as small an inner radius as possible. Also, the unavoidable energy dissipation must be distributed over a large volume thereby diminishing the shear force. One design that allows smoother inlet conditions and a reduced air entrainment is the disk inlet (16).

A mathematical model for feed zone breakage of shear-sensitive particles has been verified for a conventional pilot-scale solids-ejecting separator (17). The extent of breakage was found to be independent of flow rate. This agrees with the finding that temperature increase in the separated liquid after the separator is also independent of flow rate.

By means of computational fluid dynamics (CFD) Boychyn *et al.* (18) simulated the flow field in the feed zone of a top-fed centrifuge for two cases, one where there air was present in the feed zone, and one with no air present (flooded conditions). They found that in the former case with air present, the maximum energy dissipation rate expressed as watt per kilogram, was 2 times higher than in the flooded case. This had the effect of a considerably higher particle breakage in the separated protein precipitate suspension and consequently a worse clarification. In a later paper (19), the CFD was used to model the flow forces in the feed zone of a multichamber bowl. The forces were then reproduced in a small rotating-disk device that was used to estimate the performance of three centrifuge types, separating a protein precipitate, and good predictions of centrifuge performance were found. A comparative study by the same research group (20) on a mammalian cell suspension separated in two commercial centrifuges of a small size, one with an inlet with air present as in Fig. 2.1b, and the other a fully hermetic inlet with no air, no wings and a large volume for gently accelerating the feed as in Fig. 2.1c (refer to the description of the solid bowl in section "Disk Bowl Machines"). In other respects, the machines were identical. A 2.5-fold increase in throughput for the same clarification performance was achieved by the change to the hermetic centrifuge. The result had been predicted by a combination of CFD and mimicking of the feed shearing conditions that was performed in a combination of pretreatment of the fluid in a rotating-disk device with subsequent centrifugation in a lab-scale test tube centrifuge.

2.4.2.3 Outlet. The liquid streams leaving the separator have picked up energy in the form of kinetic energy and heat due to inner friction. When the liquid is decelerated outside the separator, the kinetic energy is converted to heat. The temperature increase is proportional to the square of the bowl speed and to the square of the radius at which the liquid leaves the separator. It is therefore not surprising that the discharge in a peripheral nozzle or a solids-ejecting machine causes breakage of cells because of the high kinetic energy of the fluid. The consequence of this is that if the particles to be separated are susceptible to shear and cell integrity is important, the cell concentrate should leave the centrifuge as near to the center as possible, for example, in a machine according to Fig. 2.1c or Fig. 2.1e. It has been reported (21) that in such machines the breakage of *Escerichia coli* cells is low. The breakage of these cells in a jet stream and upon impact has been demonstrated in small-scale experiments (22). The results are important for the design of centrifugal separators with intermittent discharge of cells from the bowl periphery.

The outlet from the centrifuge is often of the paring disk type (Fig. 2.1a). For a good function, this outlet must be designed to avoid air entrainment and large shear forces. This is especially important for chemically sensitive substances, such as proteins. A hermetic design (Fig. 2.1c) can be used in some applications.

2.4.3 Decanter Centrifuges

The objective of a decanter is usually to concentrate the solids present in concentrated suspensions. The hydrodynamics of a decanter are complicated. The conveyor adds turbulence and forces the liquid into a helical path (2). Below the top layer, vortices are formed. The transport of solids with low inner friction will influence the sedimentation process. The turbulence at the inlet zone, with possible resuspension of separated solids, and the lower solids concentration toward the liquid outlet contribute to the fact that a slender decanter will separate better than a short one with the same theoretical size (23).

2.5 THE THEORETICAL SIZE OF A CENTRIFUGAL SEPARATOR

2.5.1 Introduction

From the descriptions in the previous section it is evident that mathematical models describing the flow phenomena in centrifuges are very difficult to build. However, the attempts made are successful when machines of different sizes but with identical geometry are compared, for example, for scale-up purposes (refer to section titled "Pilot Machine Testing").

2.5.2 The Σ Theory

2.5.2.1 The Generalized Σ Formula. The most commonly used quantity to characterize centrifuges, the area equivalent (Σ) concept, was presented by Ambler (24,25). In its derivation he considered particles with a critical diameter d_c that were separated to 50%. Today however, the most commonly used definition of the critical particle is the particle that is separated to 100%. This does not influence the formulae for the Σ value, but the value for the feed flow rate Q will be halved.

Σ is defined as:

$$Q_{\text{theor}} = V_g \cdot \Sigma \tag{2.6}$$

where V_g is the Stokes settling velocity (Eq. 2.1). Σ has the general expression

$$\Sigma = \frac{V \cdot \omega^2 \cdot r_e}{g \cdot s_e} \tag{2.7}$$

The equation for the critical diameter becomes:

$$d_c = \left(\frac{18 \cdot \eta \cdot Q_{\text{theor}}}{(\rho_p - \rho_f)} \cdot \frac{s_e}{V \cdot \omega^2 \cdot r_e} \right)^{1/2} = \left(\frac{18 \cdot \eta \cdot Q_{\text{theor}}}{\Sigma \cdot (\rho_p - \rho_f) \cdot g} \right)^{1/2} \tag{2.8}$$

The derivation of Equation 2.8 is based on the following assumptions:

- viscous drag is determining the particle movement;
- the flow in disk bowls between the disks is laminar and symmetrical;
- the liquid rotates at the same speed as the bowl;
- The particle concentration is low (no hindered settling).
- the particle always moves at its final settling velocity;
- this settling velocity (V_c) is proportional to the G-force.

Ambler uses the critical particle in his formula; therefore his analysis is a special case of Svarovsky's grade efficiency function (26), where for each particle size the ratio of sedimented mass to the mass in the feed material is determined and integrated over all particle sizes to get a more realistic measure of the separation efficiency.

2.5.2.2 Σ for Disk Bowl Centrifuges. The flow system in between two conical disks is shown in Fig. 2.13. The particle has a radial velocity component because of the centrifugal force (V_c), which increases with increasing r. It has also a drag-dependent velocity component V_d, which can have any size or direction depending on the flow situation round the particle. In the Σ theory it is assumed that V_d depends on uniform radial plug flow parallel to the disks. The derivation of the equation for Σ is found in the paper by Ambler (25) and the result is shown in Table 2.2.

2.5.3 Summary

In Table 2.2, the resulting expressions of the Σ value for four different bowl geometries in continuous centrifuges are shown. These have been obtained by inserting actual values of V, r_e, and s_e for the various geometries.

2.5.4 Separation Efficiency

The assumptions in the Σ theory are not fulfilled in reality, as mentioned previously. Therefore Equation 2.6 should include an efficiency factor, so that in practice the flow rate is reduced at the requested separation performance. The efficiency factors have been estimated to have values

TABLE 2.2. Σ Equations for Various Separator Types

Machine Type	Equation for Σ	Symbols Used	References
Disk bowl	$\frac{\pi \cdot \omega^2}{g} \cdot \frac{2}{3} \cdot N \cdot (r_2^3 - r_1^3) \cdot \cot\alpha$	r_2 = Maximum radius of disk r_1 = Minimum radius of disk N = Number of disks α = Half cone angle of disk	24,25
Decanter	$\frac{\pi \cdot \omega^2}{g}\left[L_1\left(\frac{3}{2} \cdot r_2^2 + \frac{1}{2} \cdot r_1^2\right) + L_2\left(\frac{r_2^2 + 3r_2r_1 + 4r_1^2}{4}\right)\right]$	L_1 = Length of cylindrical part L_2 = Length of conical part r_1 = Inner radius of liquid r_2 = Inner radius of bowl	25,26
Tubular bowl	$\frac{\pi \cdot \omega^2}{g} L \frac{r_2^2 - r_1^2}{\ln\left(\frac{2r_2^2}{r_2^2 + r_1^2}\right)}$	r_1 = Inner radius of liquid r_2 = Inner radius of bowl L = Inner length of bowl	24

according to Table 2.3 (12). This table was first compiled in 1984 from data in literature and has been amended with data from Boychyn *et al.* (27). By careful measurement of the density difference and particle dimensions, it has been shown that in one case an efficiency factor of maximum of 16% was found to fit the experimental data in a pilot-size disk bowl machine (17). However, this probably also included effects of hindered settling (refer to section titled "Hindered Settling").

In the absence of hindered settling, the equation describing the relationship between the feed flow rate, the Σ value, and the settling rate, will thus be:

$$Q_{\text{act}} = \mu \cdot V_g \cdot \Sigma \qquad (2.9)$$

2.5.5 The KQ Formula for Disk Bowl Centrifuges

In 1961 a semiempirical equation for the separation area was published (28). It is based on the finding that separation data fitted the relation:

$$\frac{V_c}{V_g} = \left(\frac{\omega^2 \cdot r}{g}\right)^k \qquad (2.10)$$

with $k = 0.75$ better than the classical analysis where $k = 1$ (Eqs 2.2, 2.3 and 2.6). This fact is a consequence of the increasing shear forces in the Ekman layers (section titled "Fluid and Particle Dynamics in the Disk Stack"). For practical purposes the following relation is used:

$$KQ = 280 \cdot \left(\frac{n}{1000}\right)^{1.5} N \cdot \cot\alpha \left(r_2^{2.75} - r_1^{2.75}\right) \qquad (2.11)$$

where n = bowl speed in rpm and r_2 and r_1 are the radii of the disk in centimeters. This form of the *KQ* formula is dimensionally incorrect, but it is mathematically possible to make it correct, expressing the *KQ* value as an entity with the dimension area. In practice the difference between the *KQ* and the Σ for scale-up purposes is negligible, compared to other inaccuracies.

TABLE 2.3. Efficiency Factors for Various Separator Types (μ, %)

Disk bowl machines	40–73
Decanter centrifuges	54–67
Tubular bowls	90–98
Multichamber bowls	88

2.5.6 Hindered Settling

An early experimental investigation into hindered settling of cells (29) shows that complex phenomena occur at high concentrations. A model with several experimental constants was developed. It showed that the settling rates drop very fast with increasing solids content, by a factor of about 2 at 20% by volume, and 15–20 at 50% by volume. The classical model used in the work by Clarkson *et al.* (17) gave similar results. In the tests by Aiba *et al.* (29), it was also found that cells may settle flocculated at higher concentrations. Hindered settling is one further example of the difficulty to calculate theoretically the performance of centrifuges.

2.6 SELECTION OF TYPE AND SIZE OF CENTRIFUGE

2.6.1 Introduction

The choice of type and size of centrifuges is a combined theoretical and practical exercise. In most cases the final choice for a new centrifuge application cannot be made without a test of the type of centrifuge preliminary selected, at between 25% and 100% of the final plant flow rate. There at least two reasons for this. One is the elimination of the technical risk. The other is most common in the pharmaceutical industry. During the preapproval inspection, those in the inspecting and license-giving institutions are concerned with the comparability between batches of product from the development phase and from the production phase (30). Because the validation procedures are costly and time-consuming, it is very important to select at least the right type of centrifuge already in pilot scale. It is important that alternative process and plant designs are discussed before the centrifuge type and size are selected, (Table 2.4). By selecting a number of important plant design parameters, Saite *et al*. (31) devised a selection method, a Window of Operation, for each of three different types of centrifuge that are known to be suitable for the type of solids (*Pichia pastoris yeast*). The parameters chosen were shift length, maximum flow rate capacity, and minimum yield. It is important to include the right types of centrifuge and economics in the analysis as well as a sensitivity analysis when varying the parameter values.

2.6.2 Main Process Parameters

It is important to define at least the characteristics of the duty listed in Table 2.5. The feedflow rate is often determined by the available or necessary time to process one batch. The separability can be estimated on a bench scale by test tube centrifugation (spin test), see following section titled "Test Tube Centrifugation (Spin Test)." The flow of solids is the feed flow rate multiplied by the percent by volume of suspended solids. It is the solids flow rather than the suspended solid content alone that influences the choice of machine type. Different machine types differ in their ability to produce high concentration of the separated solids (Table 2.1). The required degree of separation has a strong influence on the possible flow rate in a given machine. To increase separation efficiency of *E. coli* from 90% to 99%, the flow rate must be decreased by 30% to 35% (32). To reach 99.9% a further reduction of about 30% is necessary. Many cell suspensions, including yeast, have the same narrow particle size distribution and the numbers are the same. With the wider distribution of cell debris, for example, the flow rate reduction percentages are much higher. Solids rheology, which is the reaction of the solids to pressure and shear forces, is very important for the choice of machine type. Pseudoplastic (shear thinning) solids (such as unflocculated yeast, rod-shaped bacteria, and cocci, as well as some protein precipitates) will flow out of machine types in Fig. 2.1c and Fig. 2.1e at 80% by volume of suspended solids, which often has positive consequences (reduced cell breakage). Shear thinning solids may have an angle of repose of close to 0 degrees (refer to section "Important Properties of the Process Streams"), but this is not always the case, such as with flocculated or mycelium-forming bacteria. On the other hand, dilatant solids (starch or calcium carbonate) will not be discharged in concentrated form without manual handling from any other machines than those in Fig. 2.1f and Fig. 2.7. The rheology of inclusion body concentrates is not much studied, but they are preferably discharged from the machine in Fig. 2.1b. Because of the content of a valuable dissolved substance in the entrained liquid, washing of solids may be necessary.

Another reason for washing solids is, of course, to get rid of impurities from the cell mass, such as bakers' yeast. This is usually done by resuspension and further separation steps (Fig. 2.5).

TABLE 2.4. Process and Plant Considerations

Possibility of increasing particle size
Possibility of reducing viscosity
Chemical stability of product
Containment requirements
Aseptic requirements
Sterilization requirements
Mode of operation (hours/day, days/year)
Integration with unit operations upstream and downstream
Scale of pilot- and full-scale operation
Handling of solids from the centrifuge

TABLE 2.5. Main Process Parameters

Feed flow rate
Flow of solids
Required concentration of separated solids
Solids washing requirements
Separability (V_g)
Required degree of separation
Solids rheology

2.6.3 Test Tube Centrifugation (Spin Test)

2.6.3.1 Determination of Separability. The most important task for a spin test is the determination of separability, that is, the settling velocity. In practice, it is difficult to calculate, because often the information needed in Stokes' law is not complete. The spin test is done by spinning a sample of the feed stream for different periods of time or different G-numbers or both. For each of these spins, the supernatant is analyzed with regard to the important process parameter, such as residual solids, turbidity, cell count, and, in the case of light solvent, water content. Thereafter, the time of the spin with the acceptable supernatant quality is recorded, and the necessary Σ is calculated according to Equation 2.16.

2.6.3.2 The Spin Test as an Analytical Tool. The spin test is also a valuable tool for measuring the volume percentages of the different phases. The solids content in percentage by volume from the spin test will be a very important parameter for the next step, which is to make a crude mass balance round the centrifuge. The spin test performed under different time and G-force conditions also gives an indication of solids compressibility and rheology. It is useful to prepare a sample of the solids phase after spinning, to get a good indication of its dry solids content after the centrifugation. In cases where the fermentation substrate contains suspended solids, residues of these may be present after fermentation. A spin test will make it possible to identify the nature and amount of these, which is very important since they may obstruct the solids phase flow in nozzle and solids-ejecting machines.

2.6.3.3 ***Theory of the Test Tube Centrifuge.*** The swing-out test tube (bottle) centrifuge is shown schematically in Fig. 2.15. In order for a particle to be separated out with 100% efficiency it has to move from radius r_1 to radius r_2 in the spin time (T). The shaded volume with height (h) is the volume taken out for sampling of the result. The following equations describe the particle movement:

$$V_C = \frac{dr}{dt} \tag{2.12}$$

$$V_c = V_g \cdot \frac{r \cdot \omega^2}{g} \tag{2.13}$$

Insertion of Equation 2.13 in Equation 2.12 and rearranging and integration gives:

$$\int_0^T dt = \frac{g}{V_g \cdot \omega^2} \int_{r_1}^{r_2} \frac{dr}{r} \tag{2.14}$$

$$T = \frac{g}{V_g \cdot \omega^2} \cdot \ln \frac{r_2}{r_1} \tag{2.15}$$

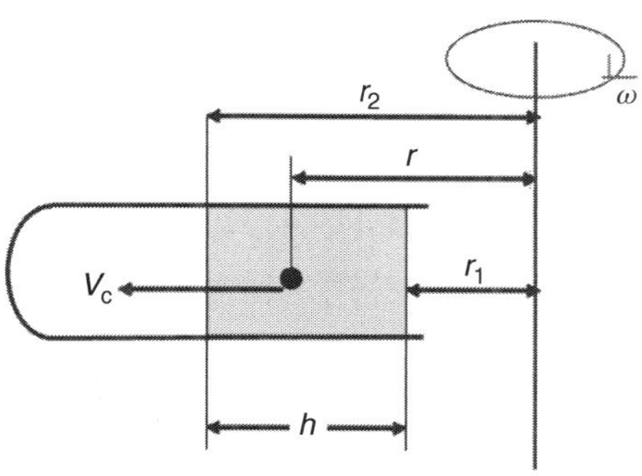

Figure 2.15. Separation of particle in a swing-out test tube centrifuge.

In the test tube centrifuge, the efficiency factor μ in Equation 2.12 is 100% because the settling process is considered ideal. Therefore, Equation 2.6 is valid and:

$$\Sigma = \frac{T \cdot Q_{\text{theor}} \cdot \omega^2}{g \cdot \ln \frac{r_2}{r_1}} \tag{2.16}$$

and

$$\frac{Q_{\text{theor}}}{\Sigma} = \frac{g \cdot \ln \frac{r_2}{r_1}}{T \cdot \omega^2} = V_g \tag{2.17}$$

For a test tube centrifuge with angle head, $\ln(r_2/r_1)$ is exchanged for $\ln(1 + D/r_1 \cdot \cos\beta)$.

For short spin times, it is necessary to include the contribution to the settling that the acceleration and deceleration phases give. An equation for this is presented in (33). In those systems where the particles may be susceptible to shear, it is advisable to pretreat the sample by subjecting it to shear in a rotating-disk device (20).

2.6.4 Important Properties of the Process Streams

In Table 2.6 some solids phase data, such as particle size and density difference, have been collected. These data show that the settling velocity of the solids, and therefore the separability, of many important biological processes is very low and belong to the most demanding centrifuge applications. The separability varies from very high for bakers' yeast to very low for bacterial cell debris, a factor of 500. The data given in Table 2.6 are very approximate. Cell size will vary with fermentation conditions. Cell debris size distribution is very wide and depends on the method of cell breakage. The density difference between cell and medium depends on the osmotic pressure, and also of the possible presence of inclusion bodies. Apparent density and the density difference of inclusion bodies vary with size, so that larger ones have lower density (34).

As many as possible of the items in Table 2.7 should be defined before the selection of a centrifuge is made. Some of the items need further comments. Apart from Stokes' law the density is important for the stress in the bowl material. Low pH and high temperature may, together with even a moderate chloride concentration, lead to difficulties with corrosion. In the biological process industrial steels 316 and 316L are often preferred (45). These steels have, however, such a low tensile strength that they are less suitable for high speed centrifuges. Instead, duplex steel has been developed, it has high tensile strength coupled with a good corrosion resistance (46). The corrosive properties of process materials can be very serious but it is in many cases possible to reduce the corrosion rate by intermittent washing with caustic solution every 4 to 5 h of running to reestablish a passivating layer on the metal surfaces. Erosion may

TABLE 2.6. Size, Density Difference, and Relative Feed Flow Rate for Some Solid or Liquid Systems

Type of Solids	Particle Size (μm)	Density Difference (kg/m^3)	Relative Throughput[a] (Relative to Bakers' Yeast)
Cell debris	0.2–0.5 (Mean, bacteria) 2 (median, yeast)	—	0.21
Streptococcae	1 × 11 (chain)	—	1–23–6
Clostridia	1 to 3	—	5
E.coli	1 × 2	70–85	2–6
Actinomyces spp.	10 × 10–143 × 143	1 (for large particles)	1–7
Inclusion bodies	0.7 × 0.7–1.3 × 1.3	29–154	5–8
Corynebacterium	—	—	6
Bacillus spp.	1 × 3	—	6–10
Aspergillus spp.	—	—	15
Mammalian cells	20 to 40	70	20
Flocculated material (e.g. debris)	—	—	20
Chlorella	4 × 4	—	30
Saccharomyces (alcohol yeast)	4 × 6	90	50–75
Saccharomyces (baker's yeast)	6 × 8	90	100

[a]Proportional to Q_{act}/Σ.
Source: Refs 12,29,34–44.

TABLE 2.7. Important Process Liquid Properties

Chemical composition of liquid and solid phases
Densities of liquid and solid phases
Viscosities of liquid and solid phases
Particle size distribution
pH
Temperature
Vapor pressure
Solubility of one phase in the other
Corrosivity
Erosivity
Shear sensitivity
Angle of repose
Toxicity
Explosion hazard

Source: Adapted from Ref. 47.

be a problem with silt-containing process water or other raw materials such as molasses. This may necessitate the use of hydrocyclones to remove these particles. The angle of repose (refer to section titled "Main Process Parameters") is another useful characteristic of the solids phase. It is the angle of a cone that solids will form if allowed to settle freely such as sand being dumped from a lorry bed. In centrifuges, cones are formed between the nozzles in a nozzle bowl. If the angle of repose is 0, or close to 0, degrees, the inner friction is low and the flow properties are extremely good. The shear sensitivity was mentioned in the section titled "Centrifuges with Conical Disks.". The most shear-sensitive cells are mammalian cells, for which the shear forces must be minimized in the case the cells are used in perfusion systems.

In the Centritech machines the RCF is limited to 300, more than one order of magnitude lower than for other centrifuges. The cells may be so vulnerable that the residence time in the centrifugal force field must be minimized (48). Some protein precipitates are also sensitive to shear (17,27), including those formed in the maturation tanks in breweries. Many bacteria are sensitive, especially starter culture bacteria, where the activity of the cultures depends on intact cells. Yeast cells are the most robust cells, but they are also sheared to some extent at discharge in radial solids-ejecting centrifuges.

2.6.5 Improvement of Separability

According to Stokes' law the settling velocity improves when the density difference and the particle density increase and when the viscosity decreases. Bowden (49) lists physical, chemical, and biological techniques for broth conditioning. The most commonly used ones are coagulation and flocculation, but also methods that induce cell lysis can improve aggregation of cells, such as aging, change of pH, and addition of enzymes, or other chemicals. In some cases, it is possible to flocculate submicron suspensions, so that a decanter centrifuge can recover the precipitate (cell debris) with good efficiency and a dry cake (42). Another example of this is given by Riske *et al*. (50), who used chitosan to flocculate NS0 cells and cell debris and found a 6 to 7-fold improvement in throughput capacity in a centrifuge or filter system without adversely affecting antibody recovery and purity. An example of product loss (extracellular enzyme) after flocculation of a *Bacillus subtilis* fermented culture with a polyelectrolyte is given by Pearson *et al.* (51). Any addition of chemicals for flocculation or other purposes have to be monitored for possible adverse effects on the process performance.

2.6.6 Pilot Machine Testing

After maximizing the separability of the feed stream, the preliminary Q_{theor}/Σ is determined by spin tests. The machine type and size are tentatively chosen based on the preliminary mass balance and on the requested Q_{act}. With the separation efficiency μ, Q_{act}/Σ can be calculated from the equation:

$$\frac{Q_{\text{act}}}{\Sigma} = \mu \cdot V_g \tag{2.18}$$

It is often advisable to make a test with the chosen type of machine. The purpose of such a test is not only to obtain a more reliable Q_{act}/Σ value, but to also assess solids flow properties, emulsification and particle break-up tendencies, (52) and clogging risks. In order to reduce the consumption of feed material, Maybury *et al.* (33) blocked off part of a disk stack of the centrifuge and reduced the volume of material by 75%. They also found that steady state in terms of clarification was reached after feeding the centrifuge with 10 bowl volumes of liquid.

2.6.7 Scale-Up

For all types of centrifugal separators, the Q_{act}/Σ value can be used for scale-up of the sedimentation performance if some general considerations are taken into account:

- The type of centrifuge should be similar.
- The geometry should be similar; for decanters, feed inlet positions, ratio of length to diameter, and beach angle, and for disk machines, cone angle, disk spacing, and number of feed holes.
- The bowl speed should give similar linear velocity of, for example, wings when treating shear-sensitive process feeds.
- The λ value (Eq. 2.5) should be almost the same in all the machines.
- The larger machine should have no hydraulic limitations in relation to the calculated flows. The limitations could be inlet and outlet pressure drops, upper and lower limits of paring disk capacity (Fig. 2.1a), and the upper and lower limit of solids flow.
- There should be no mechanical limitations, for example power transmission and gear-box torque in decanters.

The equation to use in scale-up is :

$$\left(\frac{Q_{\text{act}}}{\Sigma}\right)_1 = \left(\frac{Q_{\text{act}}}{\Sigma}\right)_2 (= \mu \cdot V_g) \tag{2.19}$$

where indexes 1 and 2 denote the two machines. Direct calculations of V_g and μ are thus avoided.

2.7 DESCRIPTION OF SOME APPLICATIONS

2.7.1 Introduction

The most important centrifuge application in biotechnology is cell harvesting where either the liquid phase or the solid phase contains (or is) the product. Downstream of the harvesting stage, the number of applications for centrifuges is growing. The operations performed are many: classification, dewatering, liquid–liquid extraction, and precipitate recovery. The applications discussed in this section are examples only; there is an infinite variation in the downstream process design.

2.7.2 Cell Harvesting

2.7.2.1 ***Introduction.*** As can be seen from Table 2.5, the throughput capacity of a given centrifuge that is, Q_{act}/Σ may vary by a factor of 100 depending on the microorganism used.

2.7.2.2 ***Yeasts.*** The second application of disk bowl centrifuges developed after milk skimming is concentration of baker's yeast. Today, nozzle machines with pressurized discharge (Fig. 2.1e) are very common in the industry. The flow rates are the highest in the biotechnology industry, up to 150 m^3/h of fermented broth. Cell concentrations in the fermenters have increased over the years, and have now reached above 25% by volume. The bakers' yeast is washed, often twice and countercurrently to increase shelf life (Fig. 2.5). The pressurized discharge of the solids concentrate makes it easy to use an inline mixer to mix the wash water with the cells.

The Q_{act}/Σ is about 2/10^7 m/s for a 99.9% reduction of the cell count.

In distilleries, nozzle bowl centrifuges have been used for a long time. The centrifuge has been the prerequisite for the development of new alcohol fermentation systems (12). In the continuous fermentation from spent sulfite liquor, centrifuges are used for recycling of the yeast. This is also the case in the batchwise Melle–Boinot process, where the recovered yeast inoculates the next batch. In the continuous Biostil process, in which the distillery slop is returned to the fermenter, the centrifuge recovers the yeast from the feed to the distillation column and returns it to the fermenter.

In breweries, centrifuges are used in several process steps. In the separation step after fermentation, green beer separation, the cell count can be as high as 80×10^6 cells per mL (2% by volume); however, during the emptying of the fermentation tank the solids concentration decreases. Radial solids-ejecting machines are used, equipped with self-triggering and special outlet devices to prevent air contact. At a cell reduction of 99% and 20×10^6 cells/mL, the Q_{act}/Σ is about $2/10^7$ m/s, increasing somewhat with higher cell count. The flow rates are up to 90 m^3/h. In this step, the same machine type as for bakers' yeast could be used if the solids flow to the machine could be controlled. Instruments (53–55) measuring the dielectric constant or the inductive permittivity may be used.

For the separation step after lagering and before final filtration, preclarification, the cell count is much smaller, and less solids-handling capacity is needed. Therefore, the bowl of the solids-ejecting machine can have the shape of the bowl in Fig. 2.1c. The hermetic inlet is an advantage since the separation of the shear-sensitive protein particles is helped by the gentle acceleration. The bowl has only one liquid outlet. The solids are ejected at the periphery. Because the emphasis is on cell reduction the Q_{act}/Σ is about $5/10^8$ m/s. Flow rates are up to 90 m^3/h.

The corresponding separation step is also present in wine making. Because the same machine unit is used also in must separation before fermentation, radial solids-ejecting machines are used.

In rDNA yeast processes, the same type of machine as for baker's yeast can be used. Such machines are available for BL-2LS containment.

2.7.2.3 Bacteria. In modern biotechnology, solids ejecting machines, nozzle machines with pressurized discharge, and decanters are used to harvest bacteria. For microorganisms with sizes about 1 μm, axially discharging machines (Fig. 2.1d) are used. If the solids concentration is so high that the solids discharge frequency has to be more than 1 per 60 s, a radially discharging machine (Fig. 2.1b) has to be employed, but this is less cost-effective. The vortex nozzle machine (5) (Fig. 2.1e) has the disadvantage of accepting only clear substrate media. Residues of substrate may clog the concentrate tubes. If a clear substrate is used, this type of machine is as cost-effective as the axially discharging machine but without its solids-handling limitation. A further possibility for clear media is the hermetic machine in Fig. 2.1c, where the cell concentrates leaves the machine continuously and under pressure. The intermittent discharge mechanism is used for CIP.

A typical Q_{act}/Σ value for 99% recovery of *E. coli* K 12 in a disk-stack machine is $1/10^8$ m/s. But the variation between various types of *E. coli* can be very large indeed; between 30% and 300% of the above value has been encountered by the author.

Vornefeld (56) describes an unusual process with the type of tube centrifuge presented above (9). *E. coli* cells were dissolved in buffer solution, a solvent was added to recover the intracellular product, and the mixture was separated. After the treatment, the cell mass was gelatinous, and accumulated at the interface between the solvent and the water phase, from where it was periodically flushed out at CIP.

Corynebacterium, somewhat larger cells used in amino acid production, are harvested in nozzle machines with pressurized discharge often with washing of the bacteria (repulping) and a second separation to improve the yield. A typical Q_{act}/Σ is 5 times of the *E. coli*. *Actinomyces* mycelia can be harvested by radial discharging machines: as in nozzle machines, the cells are washed (often in two steps) to improve the yield.

The high solids-handling capability of decanters is also used to concentrate solids from nozzle machines or from ultrafiltration units, working on fermenter broths in production of amino acids and antibiotics, respectively. The improvements are yield and less costly waste disposal (e.g. drying costs).

Single-cell protein production from hydrocarbons or their simple derivatives is today a not-so-profitable business but the technology exists. Mayer and Woernle (57) presented a flow sheet for biomass downstream processing. Bacteria are preconcentrated in large nozzle machines to about 6% by weight Dry Solids (DS) and then flocculated. This biomass is then dewatered to about 30% by weight DS in decanters and goes after granulating to drying. The liquid phase from the decanter is clarified in high speed separators and sent to sterilization together with the clarified liquid from the preconcentration step. From sterilization it is recycled back to the fermenter.

2.7.2.4 Mammalian Cells. It is possible to use high speed centrifuges also in mammalian cell processes (58) in which the cells are discarded after removal. Mammalian cells are more sensitive to shear than most bacterial and yeast cells. The unavoidable shear in the inlet will cause some rupture of the cells, but the cell breakage can be minimized by special inlet designs, such as the air-free bottom-feed ((20,59); Fig. 2.1c) or the disk inlet (60), or by designing for a filled inlet without air (18,44,52). Very often the separation takes place at reduced bowl speed to minimize shear. Special paring discs (Fig. 2.1a) and paring disk chambers can be adopted to minimize air entrapment, a phenomenon that has proven to be detrimental to sensitive proteins. Formerly, filtration in several steps had to be performed after centrifugation, but the number of steps has decreased (61–64). Care must be taken that air contact in the separated liquid is avoided, since some protein substances may be destroyed.

The centrifugation of intact cells for perfusion must be performed at very low pressure and shear forces (65). Special machines have been developed recently for this purpose (10,48,66,67).Cruz *et al.* (68) compared perfusion with other types of culture and found that the chemostat culture was more stable. A hydrocyclone has been used with good result, as cell retention device (69).

2.7.3 Cell Debris

In those cases where the intracellular products do not form inclusion bodies, the cell debris often has to be removed after cell breakage. The separation is difficult because of the very fine particles produced (Table 2.6). Moreover, the viscosity may be high. Dilution of the suspension reduces the viscosity but the Q_{act}/Σ may still be 5%–10% of that of the whole cell. It is necessary to use high speed solids-ejecting machines for this step. In cases where it is possible to flocculate the debris, the Q_{act}/Σ increases by a factor of at least 10 (21). It has been reported (39), that borax flocculated yeast cell debris can be removed to about 80% in a pilot-size decanter centrifuge at a Q_{act}/Σ of $4/10^4$ m/s. Yeast hydrolysate is an important food product. For the phase separation step, Garcia *et al.* (70) preferred a centrifuge to decantation and filtration.

2.7.4 Inclusion Body Recovery and Washing

In cases where the substance forms inclusion bodies, it is advantageous to remove cell debris and other cell constituents before the next processing step (71,72). The inclusion body has a higher settling velocity than the cell debris, so that a classifying centrifugation will recover the inclusion bodies while leaving the cell debris in the centrifugate. It is found that there is a strong interaction between the homogenization method (i.e. the number of passes) and the resulting particle size distribution on the one hand, and the ease of washing the inclusion bodies on the other hand,(73). There exists an optimum number of passes. The classification split is not 100% sharp, making it necessary to repeat the separation a number of times. This is done either by resuspending the recovered inclusion bodies batchwise, or by a continuous simultaneous feeding of the inclusion body concentrate and water into a well-mixed flow-through feed tank continuing until the required purity has been reached. Axial or preferably radial discharging solids-ejecting machines are applied at a Q_{act}/Σ of 1 to $2/10^8$ m/s. To reduce the consumption of chemicals in the next stage, the final separation is often a dewatering step. This can be performed in a high speed decanter at a Q_{act}/Σ of $2/10^6$ m/s.

Alternatives to the homogenization/centrifugation route for inclusion body processing have been proposed, such as chemical extraction. In a theoretical simulation (74), this route, where the phase separation is made in decanter centrifuges, may have an advantage over the other route because of a lower prerefolding cost, at least on a small scale.

2.7.5 Extraction of Cell-Free Broth

2.7.5.1 Solvent Extraction. Since 50 years liquid–liquid extraction has been used to concentrate and recover penicillin from filtered broth. This was often performed in multistage centrifugal extractors. Today the trend for filtered broth is to use a two-stage, countercurrent system with mixers in between disk bowl machines, which are of the three-phase solids-ejecting type. A hermetic inlet is to be preferred to reduce the risk of emulsification. Many antibiotics need only a one-stage extraction, for example, erythromycin and tetracycline. Often the antibiotic-loaded solvent needs to be washed with water to reduce the content of impurities, such as proteins. The same type of separator is used for this step.

2.7.5.2 Two-Phase Aqueous Systems. The poly ethylene glycol (PEG)/dextran or PEG/salt systems used in protein purification are characterized by a very small density difference between the phases, from 100 down to 12 kg/m^3. The viscosity of the polymer phase could be high, perhaps 100 mPas. The interfacial tension is often small, leading to easy splitting of drops. Therefore, it is not surprising that the Q_{act}/Σ may be rather low, from $5/10^8$ down to 1 to $2/10^9$ m/s. The equipment used for the systems with low density difference is radial solids-ejecting machines. Decanters may be used when the density difference is larger. It is important for optimum performance that the interface between the light and the heavy phase in the bowl is in the right position. Equation 2.4 shows the relation between the three liquid levels in the bowl. The interface should usually be placed just outside of the disk stack in the disk bowl machines.

The two-phase aqueous systems can also be used for partitioning cell debris and other impurities from a dissolved protein product (75). A PEG 4000/potassium phosphate system was used. No density difference data are given. The Q_{act}/Σ was $2/10^7$ m/s.

2.7.5.3 Precipitation. Precipitation is often used to isolate and concentrate antibiotics (e.g. rifamycin, erythromycin, and tetracycline). The suitable machine is the radial solids-ejecting disk bowl type. The methods and parameters involved in protein precipitation have been reviewed (76).

2.7.6 Whole-Broth Extraction

2.7.6.1 Solvent Extraction. Whole-broth extraction of penicillin was the first development of this technique (77). The solids flow is so high that decanter centrifuges must be applied. Many of the constituents of the broth (e.g. macromolecules and the mycelium) have emulsion stabilizing properties, so that de-emulsifying agents must be added (78). After extraction, the rich solvent is separated to reduce the water content before further processing. The solvent-contaminated mycelium must be further treated before disposal.

2.7.6.2 Two-Phase Aqueous Systems. The two-phase aqueous systems can be applied for extraction also in the presence of cell debris. The phase separation is made in small solids-ejecting or nozzle disk bowl machines. Hustedt *et al*. reported on three enzyme recovery processes that use cross-current extraction (79). In (80), Kepka reports on pilot-scale extraction of an intracellular enzyme from an *E.coli* homogenate in a thermoseparating thermopolymer/starch system. It was shown that the phase separation in a pilot solids-ejecting centrifuge (Fig. 2.1b) could be made if the density difference was at least 12 kg/m^3. Provided that the density difference between phases is large enough, also extraction decanter centrifuges can be used. Among the commercial uses of this technology on a large scale is the recovery of enzyme in a polymer–salt system, where the separation is performed in a three-phase radial solids-ejecting centrifuge.

2.8 INSTALLATION AND OPERATION

2.8.1 Plant Design

In modern biotechnology plants, the design philosophy must be documented as part of the validation procedure. These procedures are becoming common also in the food industry. Guidelines with more stringent demands for safe design of plants are being issued. In Europe, the standard for performance criteria for centrifuges in biotechnology deals with leak tightness, cleanability, and sterilizability. Centrifuge installations designed according to BL-2LS are seen in Figs. 2.2 and 2.6.

2.8.2 Cleaning-in-Place

Automated CIP is today regarded as a quality guarantee. The validation of CIP procedures is one of the necessary protocols to obtain approval to manufacture bioproducts. An automated and logged CIP procedure is repeatable and traceable, less costly than manual cleaning and safer for the plant personnel than manual cleaning is. Centrifuges and their peripherals are nowadays designed with greater cleanability in mind. Continuous efforts are being devoted to developing designs without dead pockets and with polished surfaces. With use of programmable logic controllers (PLCs) for valve control, the CIP of a disk-stack centrifuge (or decanter) has been greatly simplified. A CIP cycle is comprised of several steps of rinsing with different cleaning liquids. The sequence could be: cold water, warm (60–70°C) 4% caustic, 3% nitric acid if necessary, hot water, and finally cold water. It is often necessary to use some type of detergent to improve the cleaning effect, mixed with the caustic or the water. In the radial solids-ejecting machines and the modern pressurized discharge nozzle machines, the discharge mechanism is activated especially during the initial stages.

2.8.3 Containment, Asepsis, and Explosion Protection

These three areas are related in the sense that the installation designer must define a space round the centrifuge, including tanks and piping, with controlled interfaces with the environment. In the case of containment, the task is to protect process operators from harmful process substances. The National Institute of Health's concept of Biosafety level is an attempt to classify the risks involved (81). The interface with the environment is sterile High Efficiency Particulate Air (HEPA) filters. This is a case of "no leaking out." The reverse case "no leaking in" applies for asepsis and explosion protection. The barriers are formed by HEPA filters and liquid seal pots, respectively.

In both these cases, it is necessary to keep the centrifuge system under a slight, positive pressure relative to the environment, either by lowering the pressure in the room (asepsis) or increasing it in the centrifuge system. For explosion protection, this is done by filling the system by inert gas, such as nitrogen, and bubbling the inert gas through a water seal pot. Other measures taken with explosion hazards have been reviewed (82). Centrifuges in closed systems have to be modified in a number of ways: mechanical instead of labyrinth seals, cooling jacket around the bowl cover, and devices that can handle the pressure wave at discharge.

2.8.4 Sterilization

Sterilization with saturated steam at temperatures above 100°C is considered by many to be the reliable way for avoiding contamination, and it is prescribed for BL2-LS. For the centrifuge manufacturer this means that in addition to the modifications described in the previous section, the bowl cover must be designed as a pressure vessel. But the most complicated changes concern the installation, because the bowl's internal and external surfaces, the cover, the service liquid lines, and so on must all be sterilized. To reach the desired temperature (above at least 121°C), condensate must be drained away. Krook *et al*. (5) discuss the sterilizable system for a medium-sized machine with pressurized nozzle discharge in some detail. The sterilization efficiency for a radially solids-ejecting centrifuge of pilot scale has been investigated (83,84). It was found that it was not possible to sterilize the whole system at once. By trying different combinations of open and closed valves, regimes were developed for heating different sections, and these were combined to sterilize the whole system. In the end, sterilization was validated with different chemical and biological indicators. In (83), a short overview of different methods used as monitoring systems is given.

2.9 CENTRIFUGATION VERSUS MICROFILTRATION

After the development of sterilizable microfilters, there is an alternative available to centrifugation in separating solids and liquids. A general rule is that with increasing particle size and increasing scale of operation, centrifugation becomes more interesting than microfiltration and vice versa. It is conceivable that *E. coli* on a 200 L fermenter scale is a microfiltration application. Many factors must be considered in the choice. What is becoming more apparent is that the choice done in pilot scale becomes the most difficult one, since the scale-up and validation procedures makes it easier to keep the same technology on the larger scale, even if it is not the optimal choice. In many cases a combination of a filter and a centrifuge is the best answer.

NOMENCLATURE

d_p	Particle diameter m
D	Inner diameter of test tube M
$D_{h,l}$	Diameter of heavy- and light-liquid outlet m
D_i	Diameter of interface m
g	equal to 9.81 m/s^2
h	Distance between inner and outer radius in test tube m
k	Constant in Equation (2.10)
KQ	Semiempirical measure of centrifuge size, Equation (2.11)
n	Bowl speed rpm
N	Number of discs
$Q_{act,\ theor}$	Actual and theoretical feed flow rate m^3/s
$r_{e,1,2}$	Equivalent, minimum and maximum radius m
S_e	Equivalent settling distance m
t	Distance between disks m
T	Spin time s
V	Bowl volume m^3
$V_{g,c}$	Settling velocity at 1 g and in centrifugal field m/s
V_d	Velocity component caused by drag forces m/s
Z	Relative centrifugal force (RCF) or G-number
α	Half cone angle of disk
β	Angle of tube in test tube centrifuge
η	Dynamic viscosity of fluid mPas
λ	Dimensionless number, defined in Equation 2.5
μ	Separation efficiency factor
ν	Kinematic viscosity of fluid m^2/s
ρ_p	Particle density kg/m^3
$\rho_{f,fh,fl}$	Fluid density, general, heavy liquid, light liquid kg/m^3
ω	Velocity of rotation rad/s
Σ	Area equivalent according to Ambler m^2

REFERENCES

1. Rushton A, Ward AS, Holdich RG. Solid liquid filtration and separation technology. Weinheim: VCH Verlagsgesellschaft mbH; 1996.
2. Sokolov VI. Moderne industriezentrifugen. Berlin: VEB Verlag Technik; 1971.
3. Barker TA. Filtr Sep 1975; 12: 33–36.
4. Mackel NW, Bersch S. CIT Plus 2001; 4(3): 52–53.
5. Krook G, Axelsson H, Thorsson C. In: Blazej A, Privarova V, editors. Environmental biotechnology. Amsterdam: Elsevier; 1991.
6. Bergjohann G. Chem Technol 1988; 17: 85–86.
7. Litvine F. Inf Chim 1989; 310: 327–334.
8. Celeros, Inc. brochure, Novi, Michigan; 2007.
9. Meikrantz DL, Meikrantz SB, Macaluso LL. Chem Eng Commun 2001; 188: 115–127.
10. Apelman S, Björling T. Biotech Forum Eur 1991; 8: 356–358.
11. Simpson TW. Adv Filtr Sep Technol 2002; 15: 233–245.
12. Axelsson H, Cooney CL, Humphrey AE, editors. Comprehensive biotechnology. Volume 2, The principles of biotechnology: engineering considerations. Oxford: Pergamon Press Ltd.; 1985. pp. 325–346, Chapter 21.
13. Lagerstedt T, Nåbo O. In: Veret C, editor. Flow visualization IV. Washington (DC): Hemisphere; 1986.
14. Nåbo O, Lagerstedt T. In: Reznicek R, editor. Flow visualization V. Washington (DC): Hemisphere; 1990.
15. Nåbo O, Carlsson C-G. Proc Filtech Conf 1987; 1: 1–7.
16. Borgström L, Carlsson C-G, Inge C, Lagerstedt T, Moberg H. Appl Sci Res 1994; 53: 35–50.
17. Clarkson AI, Bulmer M, Titchener-Hooker NJ. Bioprocess Biosyst Eng 1996; 14: 81–89.
18. Boychyn BM, Yim SSS, Ayazi Shamlou P, Bulmer M, More J, Hoare M. Chem Eng Sci 2001; 56: 4759–4770.
19. Boychyn M, Yim SSS, Bulmer M, More J, Bracewell DG, Hoare M. Bioprocess Biosyst Eng 2004; 26(6): 385–391.
20. Hutchinson N, Bingham N, Murrell N, Farid S, Hoare M. Biotechnol Bioeng 2006; 95(3): 483–491.
21. Lindman B. Meded Fac Landbouwwet Rijksuniv Gent 1987; 52: 1383–1387.
22. Chan G, Booth AJ, Mannweiler K, Hoare M. Biotechnol Bioeng 2006; 95(4): 671–683.
23. Madsen B. Institution of Chemical Engineers Symposium Series No. 113; 1989, 301–317.
24. Ambler CM. Chem Eng Prog 1952; 48: 150–158.
25. Ambler CM. J Biochem Microbiol Technol Eng 1959; 1: 185–205.
26. Svarovsky L. In: Svarovsky L, editor. Solid–liquid separation. London: Butterworths; 1977. pp. 124–147.

27. Boychyn M, Doyle W, Bulmer M, More J, Hoare M. Biotechnol Bioeng 2000; 69(1): 1–10.
28. Sullivan FE, Erikson RA. Ind Eng Chem 1961; 53: 434–438.
29. Aiba S, Kitai S, Heima H. J Gen Appl Microbiol 1964: 243–256.
30. Avallone H, D'Eramo P. Pharm Eng 1992; 12: 36–39.
31. Saite H, King JMP, Baganz F, Hoare M, Titchener-Hooker NJ. Biotechnol Bioeng 2006; 95(6): 1218–1227.
32. Higgins JJ, Lewis DJ, Daly WH, Mosqueira FG, Dunnill P, Lilly MD. Biotechnol Bioeng 1978; 20: 159–182.
33. Maybury JP, Mannweiler K, Titchener-Hooker NJ, Hoare M, Dunnill P. Bioprocess Eng 1998; 18: 191–199.
34. Taylor G, Hoare M, Gray DR, Marston FAO. Biotechnology (N Y) 1986; 4: 553–557.
35. Datar RV, Rosén C-G. In: Stephanopoulos G, editor. Biotechnology. Volume 3, Bioprocessing. 2nd ed. Weinheim, Germany: VCH Verlagsgesellschaft mbH; 1993. pp. 469–503.
36. Datar R. Filtr Sep 1984; 21: 402–406.
37. Titchener-Hooker NJ, Gritsis D, Olbrich R, Mannweiler K, Gardiner SAM, Fish NM, Hoare M. Pharm Technol Int 1991; 3: 42–48.
38. Hwang SO. Biotechnol Tech 1996; 10: 157–160.
39. Clarkson AI, Lefevre P, Titchener-Hooker NJ. Biotechnol Prog 1993; 9: 462–467.
40. Siddiqi SF, Titchener-Hooker NJ, Ayazi Shamlou P. Biotechnol Bioeng 1996; 50: 145–150.
41. Agerkvist I, Enfors S-O. Biotechnol Bioeng 1990; 36: 1083–1089.
42. Bentham AC, Bonnerjea J, Orsborn CB, Ward PN, Hoare M. Biotechnol Bioeng 1990; 36: 397–401.
43. Datar R, Rosén C-G. Chem Eng J 1987; 34: B49–B56.
44. Kempken R, Preissman A, Bertold W. Biotechnol Bioeng 1995; 46: 132–138.
45. Curiel GJ, Hauser G, Peschel P, Timperley DA. Trends Food Sci Technol 1993; 4: 225–229.
46. Weibull I. Mater Des 1987; 8: 35–40, 82–88.
47. Wakeman RJ. Filtr Sep 1995; 32: 337–341.
48. Johnson M, Lanthier S, Massie B, Lefebvre G, Kamen AA. Biotechnol Prog 1996; 12: 855–864.
49. Bowden C. Chem Eng (Rugby) 1985; 415: 50–54.
50. Riske F, Schroeder J, Belliveau J, Kang X, Kutzko J, Menon MK. J Biotechnol 2007; 128: 813–823.
51. Pearson CR, Heng M, Gebert M, Glatz CE. Biotechnol Bioeng 2004; 87(1): 61–68.
52. Maybury JP, Hoare M, Dunnill P. Biotechnol Bioeng 2000; 67(3): 265–273.
53. Austin GD, Watson RWJ, D'Amore T. Biotechnol Bioeng 1994; 43: 337–341.
54. Siano SA. Biotechnol Bioeng 1997; 55: 289–304.
55. Siems G. Brauwelt Int 1997;(2): 132–137.
56. Vornefeld M. Chem Anal Verfaren 2002;(04): 114.
57. Mayer M, Woernle R. Chem Ing Tech 1985; 57: 152–153.
58. Backer MP, Metzger LS, Slaber PL, Nevitt KK, Boder GB. Biotechnol Bioeng 1988; 32: 993–1000.
59. Lander R, Daniels C, Meacle F. Bioprocess Int 2005; 3(10): 32–40.
60. Axelsson H. Filtr Sep 2000; 37(4): 20–23.
61. Yavorsky D, Blanck R, Lambalot C, Brunkow R. Pharm Biotechnol 2003: 2–8.
62. Winter C. Paper no BIOT 156, Presented at the 227th ACS National Meeting; 2004 Mar 31; Anaheim (CA).
63. Bux R, Gaus W, Kempken R, Hannapehl M, Bisschoph T, Regel J. Chem Ing Tech 2006; 78(9): 1402.
64. Pailhes M, Lambalot C, Barloga R. Bioprocess J 2004: 55–58.
65. Castilho LR, Medronho RA. Adv Biochem Eng Biotechnol 2002; 74: 129–169.
66. Jäger V. In: Spier RE, Griffiths JB, Berthold W, editors. Animal cell technology: developments, processes and products. Oxford: Butterworth-Heinemann; 1992.
67. Chatzisavido N, Björling T, Fenge C, Boork S, Lindner-Olsson E, Apelman S. In: Kobayashi T, Kitagawa Y, Okamura K, editors. Animal cell technology: basic and applied aspects. Dordrect: Kluwer Academic Publishers; 1993. pp. 463–468.
68. Cruz HJ, Conradt HS, Dunker R, Peixoto CM, Cunha AE, Thomaz M, Burger C, Dias EM, Clemente J, Moreira JL, Rieke E, Carrondo MJT. J Biotechnol 2002; 96: 169–183.
69. Pinto RCV, Medronho RA, Castilho LR. Cytotechnology 2007, DOI 10.1007/s10616-007-9108-x.
70. Garcia T, Villar J, González M. Filtr Sep 2005; 42(8): 36–37.
71. Graumann K, Premstaller A. Biotechnol J 2006; 1: 164–186.
72. van Hee P, Middelberg APJ, van der Lans RGJM, van der Wielen LAM. Biotechnol Bioeng 2004; 88(1): 100–110.
73. Middelberg PJ, O'Neill BK, Bogle IDL. Trans Inst Chem Eng 1992; 70: 8–12.
74. Lee GH, Cooney D, Middelberg APJ, Choe WS. Bioprocess Biosyst Eng 2006; 29: 73–90.
75. Datar R, Rosén C-G. J Biotechnol 1986; 3: 207–219.
76. Bell DJ, Hoare M, Dunnill P. In: Fiechter A, editor. Advances in biochemical engineering/biotechnology. Berlin: Springer-Verlag; 1983. pp. 1–72.
77. Katinger H, Wibbelt F, Scherfler H. Verfahrenstechnik 1981; 15: 179–182.
78. Szabo C. Biotechnol Bioeng 1992; 40: 247–251.
79. Hustedt H, Kroner K-H, Papamichael N. Process Biochem 1988; 23(5): 129–137.
80. Kepka C, Collet E, Persson J, Ståhl Å, Lagerstedt T, Tjerneld F, Veide A. J Biotechnol 2003; 103: 165–181.
81. Hambleton P, Melling J, Salusbury TT, editors. Biosafety in industrial biotechnology. London; Chapman & Hall: 1994.
82. Lindley J. Chem Eng (Rugby) 1985; (411): 41–44.
83. Frude MJ, Simpson MT. Process Biochem 1993; 28: 297–303.
84. Schmidt M, Krützfeldt R, Ross A. Process Biochem 1999; 34: 769–776.

3

CELL DISRUPTION, MICROMECHANICAL PROPERTIES

INGO KAMPEN AND ARNO KWADE
Technische Universität, Institute for Particle Technology, Braunschweig, Germany

3.1 INTRODUCTION

Microbial micromechanical properties play an important role in several biotechnological processes. In addition to the release of intracellular products through cell disruption, the mechanical properties influence the maximum shear stress, which can be applied to the cells, the resistance to osmotic pressure, and the uptake of extracellular compounds. The cell wall is considered to be the main factor for the mechanical strength in all microbial cells.

Mechanical properties can be measured in several ways: locally by an atomic force microscope (AFM), globally by squeezing single cells between two flat surfaces and integrally by applying shear stress to a bacterial suspension or using cell disruption devices. These investigations show that microbial cells have a tremendous resistance against mechanical stress.

On account of this durability, combined with the small size of microbial cells, only a few devices are capable of cell disruption. Industrially the most commonly used devices are the high pressure homogenizer, the stirred media mill, and the ultrasonic homogenizer. At the laboratory scale, a few more devices are used, such as the French Press, X-Press, and vibratory mill.

The next pages will give a short overview of the composition of microbial cells, a review of single cell manipulation, a description of micromechanical properties, and the operation methods of high pressure homogenizer, stirred media mill, and ultrasonic homogenizer. Finally, the behavior of different cells in different devices will be discussed.

3.2 MICROORGANISMS: COMPOSITION AND MORPHOLOGY

Cells used in biotechnological production can have significant differences. Apart from single-celled bacteria of different origins, fungal, mammalian, and plant cells are also used. They all have different properties, such as size, composition, stability, and growth characteristics. This section describes the composition and characteristics of bacteria as well as fungi.

Bacteria have a size between 0.5–5 μm. The morphology of the different genera is quite variable. The most common shapes are spherical cells, rods, spirals, and comma-shaped rods. Bacteria consist of 70–80% water. Polymers are the main part of the dry weight: proteins 50%, cell wall components 10–20%, RNA 10–20%, DNA 3–4%, and lipids 10% (1). Bacteria are prokaryotes and in contrast to eukaryotes, have a simpler composition. The cytoplasm membrane prevents the cell components from diffusing into the medium, but provides only low resistance against mechanical stress. The main obstacle for cell disruption is the cell wall, which surrounds the cytoplasm. It is also responsible for the shape of the microorganism. After enzymatic degradation of the cell wall, independent from the primary shape of the cell, spherical protoplasts are generated. They are surrounded by the cytoplasm membrane alone. The fact that protoplasts burst at low salt concentrations shows that the cell wall is also responsible for resistance against osmotic pressure.

The stability of the cell wall depends on many factors. Apart from the genera of the microbe, the cell wall is

Downstream Industrial Biotechnology: Recovery and Purification, First Edition. Edited by Michael C. Flickinger.

influenced by the conditions of fermentation. Hull and Middelberg (2) found that in contrast to cells in the stationary phase, cells in the growth phase are easier to disrupt in a high pressure homogenizer.

3.2.1 Cell Wall Structure of Gram Positive and Gram Negative Bacteria

Depending on the structure of their cell wall, bacteria are divided into Gram positive and Gram negative bacteria. This classification is based on a staining procedure introduced by Gram (1884), which reveals the different abilities of the cell wall types to be shown through staining with a special crystal violet/iodide dye. Figure 3.1 gives a schematic overview of the different structures of these two cell wall types.

The inner part of the cell is surrounded by the cytoplasmic membrane. It consists of a lipid bilayer with the hydrophilic heads pointing to the outside. The cytoplasmic membrane contains specific proteins (integral membrane proteins) which control the in- and outflow of substances and therefore enable a specialized transport of compounds in and out of the cell. But it is easily disrupted by mechanical stress. The main protection against mechanical stress is given by the cell wall. It is composed of a heteropolymer, which is linked by peptide bounds. The heteropolymer consists of 50–500 alternating *N*-acetylglucosamine and *N*-acetylmuraminic acid residues. In the cell wall, the peptide bonds are arranged parallel to the main axis of the bacterium, and the hetero polymer chains are aligned perpendicular to the main axis of the bacterium (4). This arrangement leads to the anisotropic mechanical properties of the bacterial cell wall.

The cell wall of Gram negative cells is only composed of one or two layers of this heteropolymer, while the cell walls of Gram positive cells consists of up to 25 layers. Thus, Gram negative cells are much easier to disrupt. The murein layer of Gram negative cells is covered by the outer membrane. Some of the lipoproteins, lipopolysaccharides (e.g.endotoxins) and other lipids are covalently linked to the murein layer. The outer membrane represents up to 80% of the dry weight of the cell wall of Gram negative bacteria, but it only has a small effect on the mechanical resistance of the bacterium.

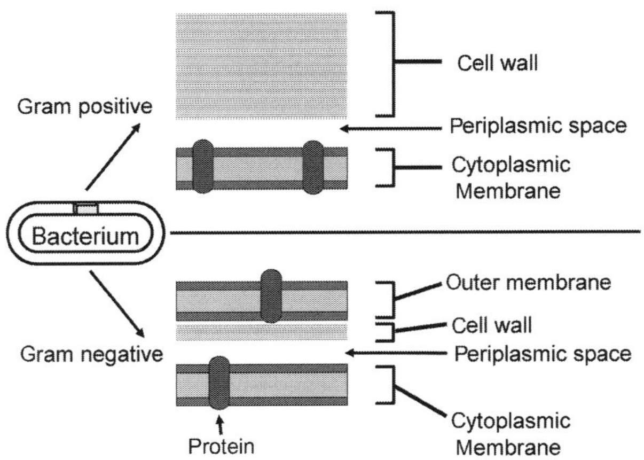

Figure 3.1. Schematic of the cell wall composition from Gram positive and Gram negative bacteria (3). (This figure is available in full color at http://onlinelibrary.wiley.com/book/10.1002/9780470054581.)

3.2.2 Cell Wall Structure of Fungi

The cell wall structure of fungi differs greatly from the bacterial cell wall structure. The phospholipid bilayer is surrounded by various layers of polysaccharides. The exact composition of these polysaccharides (e.g. glucane, mannane, chitin, cellulose) depends on the taxonomic group. In the yeast *Saccharomyces cerevisiae* the cell wall is mainly composed of β-1,3 glucan (containing 1–2% chitin). A layer of β-1,3 glucan is connected to the outer mannoproteins of the cell wall. The thickness of the cell wall is around 100–200 nm.

Another fact that has to be taken into account, is the shape of the fungi. While yeast usually grows in single cells, many fungi, like *Aspergillus niger*, form hyphae. These long-branched filaments can cause trouble in the cell disruption device, especially when growing as a pellet, in millimeter range.

3.3 MICROMECHANICAL PROPERTIES OF MICROORGANISMS

The measurement and interpretation of these values is difficult and therefore only a few sources are found in literature. Most of these investigations were carried out with the fungus *S. cerevisiae*. The investigations can be divided into three categories: local, global, and integral.

3.3.1 Local Measurement of Micromechanical Properties

Measurements of mechanical properties from distinct parts of a cell are usually done with an AFM. In the first step, the cells have to be imaged with the cantilever tip (Fig. 3.2a). This can be done in liquid, but requires a fixation of the microorganisms. Using the AFM image a specific part of the cell can be chosen and measured. This is done by force–distance spectroscopy, where the tip of the cantilever is pressed into the cell while a force distance curve is recorded. For the date processing, three parameters are calibrated: the z-piezo movement, the spring constant of the cantilever, and the correlation between the detector signal and the deflection of the cantilever. The calibration procedures are described elsewhere (5–8).

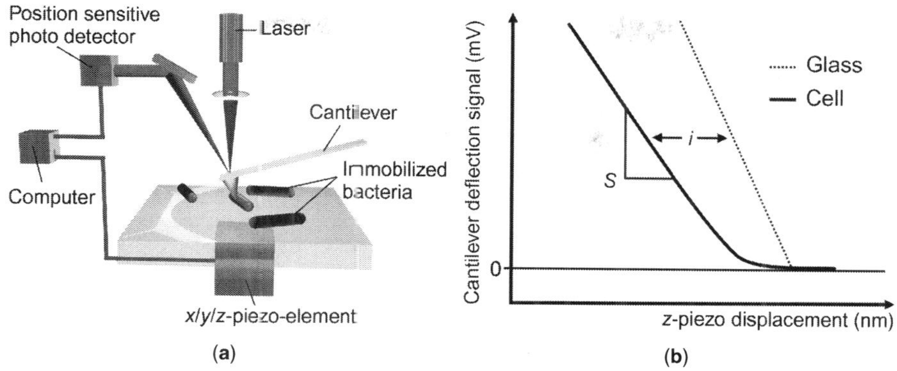

Figure 3.2. (a) Schematic of an atomic force microscope; (b) force distance curve, from the linear part the slope S is used for the calculation of a cellular spring constant k_Z. (This figure is available in full color at http://onlinelibrary.wiley.com/book/10.1002/9780470054581.)

For measurements only small indents can be performed. Otherwise the tip, which usually has a tip radius of 10–20 nm, will damage the cell.

From the linear part of the force/displacement curve the elastic behavior of the cell wall can be derived (Fig. 3.2b). In the literature two models for data analysis can be found: the herzian contact model, to calculate the cellular E-modulus E_Z, and the two spring model that allows the calculation of a cellular spring constant k_Z.

The indentation of a spherically shaped indenter into an elastic material can be described by Equation 3.1, given by Herz (9).

$$F = \frac{4}{3} E_r \sqrt{R} \cdot i^{1.5} \tag{3.1}$$

In this equation, F is the Force, i the indentation depth, R the radius of the indenter (usually the tip radius) and E_r the reduced Young's modulus, which is a combination of the Young's modules from the sample, the indenter, and the Poisson's ratio ν of both materials (Eq. 3.2).

$$\frac{1}{E_r} = \left(\frac{1-\nu^2}{E}\right)_{\text{indenter}} + \left(\frac{1-\nu^2}{E}\right)_{\text{material}} \tag{3.2}$$

The indentation depth can easily be extracted from the force distance curve by comparison with an indent onto glass as a nondeformable material.

The most frequently used model for data analysis in the measurement of local microbial material properties is the two spring model. Because of the elastic material behavior of the cell, it can be used for the analysis of the force distance curves. In this model, two springs, one representing the cantilever (k_C), the other one representing the cell (k_Z) are matched in a row. Taking the stiffness S (the slope of the force distance curve) into account, the cell spring constant k_Z can be calculated as shown by Equation 3.3.

$$k_Z = k_C \frac{S}{1-S} \tag{3.3}$$

For comparison of cellular spring constants or E-moduli, it is important to measure under equal environmental conditions because the answer is not only derived from the cell wall, but from the whole cell. This means the measured values depend on several parameters like temperature, pH (10), and osmotic pressure. Also, the angle between surface and tip are important. If the cell is not indented at a right angle, the torsion of the cantilever will affect the results. Despite these parameters the indentation speed and number does not influence the data (11). Probing a statistically sufficient number of yeast cells, the results for the cellular spring constants show a log-normal distribution. For the yeast *S. cerevisiae*, the distribution is shown.

Table 3.1 shows a summary of local measurements of cells described in literature.

The influence of extracellular osmotic pressure on the cellular spring constant was investigated by Arfsten on *S. cerevisiae* (11) and Arnoldi *et al*. on *Magnetospirillum gryphiswaldense* (6). While Arnoldi *et al*. measured no effect of extracellular osmotic pressure, Arfsten *et al*. identified an increase of the cellular spring constant k_Z at low glucose concentrations. These results showed that the osmotic pressure always has to be taken into account when measuring the mechanical properties of cells by AFM.

The relationship between the cellular spring constant and the intrinsic mechanical parameters of the cell wall is complicated. Deformation, banding, and stretching of the cell wall are to be taken into account. Current works use finite element approaches for the extraction of these parameters, but further investigations have to be done to buildup a realistic model for the tests.

TABLE 3.1. Organisms Tested by Local Measurements with AFM[a]

Organism	Results	References
Aspergillus nidulans (fungus)	$k_Z = 0.17–0.29$ N/m	12
A. nidulans (spores)	$k_Z = 110–300$ N/m	13
Enterococcus hirae (Gram +)	$k_Z = 0.5–0.6$ N/m	14
Escherichia coli (Gram −)	$k_Z = 0.044–0.13$ N/m	8
E. coli (Gram −)	$k_Z = 0.194$ N/m	15
E. coli (Gram −)	$E_Z = 12.8$ MPa	16
Lactobacillus johnsonii (Gram +)	$k_Z = 0.016–0.02$ N/m	17
Lactobacillus crispatus (Gram +)	$k_Z = 0.053$ N/m	17
Lactobacillus helveticus (Gram +)	$k_Z = 0.02–0.031$ N/m	17
Magnetospirillum gryphiswaldense (Gram −)	$k_Z = 0.042$ N/m	6
Pseudomonas aeroginosa (Gram −)	$k_Z = 0.02–0.03$ N/m	14
P. aeroginosa (Gram −)	$k_Z = 0.044$ N/m	18
Pseudomonas putida (Gram −)	$k_Z = 0.014–0.114$ N/m	19
Saccharomyces cerevisiae (fungus)	$E_Z = 0.6$ MPa N/m	20
S. cerevisiae (budding scarf)	$E_Z = 6.1$ MPa N/m	20
S. cerevisiae (fungus)	$k_Z = 0.07–3.2$ N/m	11
Shewanella putrefaciens (Gram −)	$k_Z = 0.01–0.022$ N/m $E_Z = 0.037–0.21$ MPa	10
S. putrefaciens (Gram −)	$k_Z = 0.065$ N/m, $E_Z = 0.098$ MPa	21

[a]Derived from Ref. 11.

3.3.2 Global Measurement of Micromechanical Properties

Global measurement stands for squeezing single cells between two parallel surfaces. In the past, several experiments were carried out using an optical glass fiber with a flattened end to compress cells (22). Currently a nanoindenter (23) is used, which is a very accurate device adapted to the field of material sciences. The measurement of single cells requires an immobilization of the cells to the bottom, preventing the cells to move because of the flow caused from the approaching probe. This can be done by the use of adhesive glass slides. Another important step is the degree to which both surfaces are parallel. Otherwise the cells will slip out of the gap during loading and cannot be stressed. The motion of the probe has to be very smooth and accurate while a force transducer measures the applied forces at the loading and unloading cycle. From the obtained force/distance curves the deformation ratio (d), the applied force (F), and the deformation energy, (W) can be calculated. For cell disruption the most interesting data is the bursting force F_B, the bursting energy E_B, and the relative deformation at the bursting point d_B.

Figure 3.3a shows the extraction of this data from a force distance curve. Figure 3.3b shows an example for a bursted yeast cell.

Most of these investigations were carried out using the yeast *S. cerevisiae*. Only few reports exist for other microorganisms (Table 3.2). The data points out that yeast cells are much harder to disrupt than the Gram negative *Escherichia coli* cells and even more difficult than Gram positive *Staphylococcus epidermis* cells. Investigations on

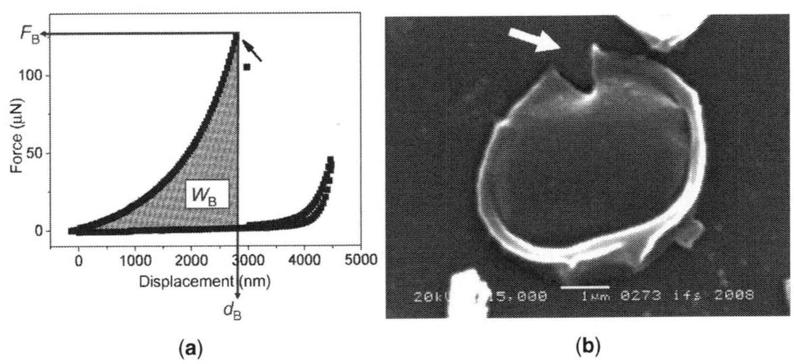

Figure 3.3. (a) Force/distance curve of a compression test and the derived parameters; (b) SEM image of a *S. cerevisiae* cell after compression. The arrow shows the cell failure (24). (This figure is available in full color at http://onlinelibrary.wiley.com/book/10.1002/9780470054581.)

TABLE 3.2. Published Works on Global Measurements of Microbial Cells

Organism	Bursting force	References
Saccharomyces cerevisiae	101 ± 24 μN	25
Saccharomyces cerevisiae	74–96 μN	26
Saccharomyces cerevisiae	65–105 μN	27
Saccharomyces cerevisiae	55–100 μN	28
Saccharomyces cerevisiae	120–220 μN	24
Escherichia coli	1–9 μN	22
Staphylococcus epidermis	3–34 μN	22

spores from the fungus *A. niger* show breaking forces around 250–550 μN (not published). Table 3.2 gives a short summary of reported values.

The variations in bursting forces (Table 3.2) are the result of biological diversity as well as from different environmental conditions. Arfsten *et al*. investigated several effects (osmotic pressure, compression speed, and the effect of multiple stress events) on the mechanical characteristics (W_B, F_B, d_B) of *S. cerevisiae* (24). They found that the osmotic pressure has a very large effect on these parameters (Fig. 3.4). At high osmolarities, water flows out of the cells, which leads to smaller cells with lower internal pressure. This pressure reduction has to be compensated for, by the application of a higher bursting force F_B, and therefore a higher bursting energy W_B. Stressing shrunken cells, the cell wall is able to expand more, leading to higher deformation ratios d_B.

Not only bursting, but also smaller deformations of a cell can give information about the mechanical behavior. Arfsten *et al* (23) showed that a small deformation ($d_B = 0.1$) has no effect on *S. cerevisiae* cells. At higher deformation ratios ($d_B = 0.3$) the cells could not fully recover their previous shape. This effect could also be seen in the mechanical parameters when breaking the cells after predeformation. The cells were squeezed five times to the desired d_B and finally broken. While a predeformation of $d_B = 0.1$ showed no significant effect, a deformation of $d_B = 0.3$ causes an increase in F_B and W_B around a quarter (24).

From these results the following facts can be derived for cell disruption experiments:

- For mechanical cell disruption, it is good to work at low osmolarities (as long as the product can handle this).
- Cells should be disrupted by one stress event. Multiple stress events aggravate the process.
- The necessary stress energies in a device can be estimated by compression tests (this is true only for compressive stress events, shear stress has not been investigated yet).

The derivation of intrinsic parameters of the cell wall from compression testing results has been reported in literature (24,29–31). All articles investigated the yeast *S. cerevisiae* and calculated the Young's modulus of the cell wall. The results are between 98 and 172 MPa. While the glucan of the yeast cell wall seemed to have isometric failure behavior, the organisation of the bacterial murein layer makes it possible that it has a favored breaking direction. Electron microscope pictures revealed ring-shaped cell debris after cell disruption by a high pressure homogenizer (3). With this, it is evident that bacterial cell walls have a nonisometric mechanical behavior. This makes the determination of intrinsic parameters more difficult. For the bacterial cell wall at least two Young's modules have to be determined.

3.3.3 Integral Measurement of Micromechanical Properties

Only few integral measurements have been reported to test mechanical stability regarding cell disruption. Most work investigated the effect of shear stress on growing cells in a bioreactor. It was shown that shear stress may inactivate enzymes (32,33) and have a great impact

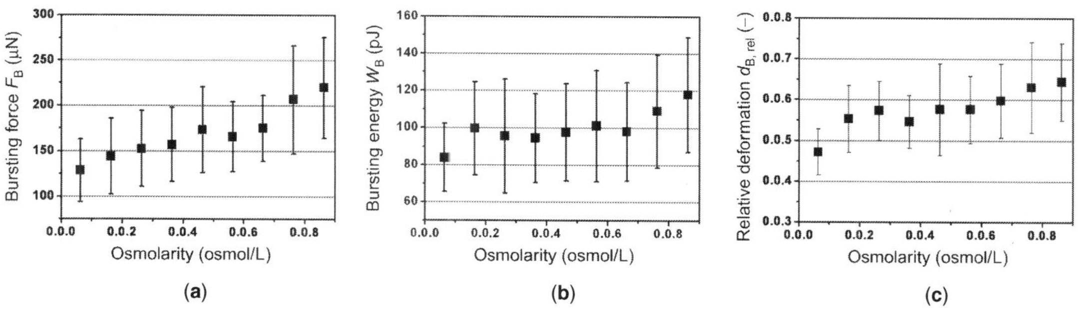

Figure 3.4. Effect of osmolarity on the bursting force F_B, the bursting Energy W_B and the deformation at the bursting point d_3 (24).

on the productivity of several organisms (34,35). Also, morphological changes on growing fungi could be seen (36,37). Several investigations were carried out with a clay–polymer floc system as a model material for biological cells (38–41). Interestingly the highest shear stress is derived by the gassing process, not by the stirrer.

In the realm of cell disruption, Tonnius (42), showed a very high resistance of *S. cerevisiae* towards shear stress. He stressed the yeast in a cuette flow at a shear stress of 17.000 Pa without any significant cell disruption. Also, investigations with a rotor stator device on sewage sludge showed only small changes in cell disruption (43). Further work was done using cell disruption devices; this is discussed in the next section.

3.4 CELL DISRUPTION

In order to purify intracellular products, they have to be released from the cell into the mostly aqueous medium first. In some cases it is enough to slightly damage the cell wall in order to enable the product of interest to diffuse from the cytoplasm or the periplasmatic space into the medium. In other cases it is necessary to mechanically open the cells. Only the mechanical cell disruption devices will be discussed here as they are industrially the most applied techniques for cell disruption.

All methods for cell disruption have to meet certain criteria. Depending on the process, they can be weighted differently:

- No product damage or denaturation
  Proteins especially, tend to denature when exposed to high temperatures, resulting in a conformational change which leads to a loss of function. Even if a rise in temperature does not directly damage the enzyme, proteases can be significantly more active at higher temperatures and affect the function of the protein. Furthermore, the adsorption of the protein to the walls of the cell disruption device can lower the enzyme activity in the medium.
- High degree of disruption
- Sterility of the cell disruption device
  The production of biotechnological products is usually done with genetically modified organisms. Therefore it is necessary to sterilize the equipment after use.
- No contamination of the product
  This is especially important for pharmaceutical products. Some cell disruption methods cannot be applied in the pharmaceutical industry due to the fact that harmful chemical agents are used. Their later removal usually requires a significant big effort and a loss of the product.
- Easy cell debris removal
  The cell debris is removed by filtration, centrifugation, or expanded bed chromatography. For all these techniques, large cell debris is beneficial.
- Low energy consumption
  This is not only important to keep the costs low but also because nearly all energy contributed to the system is converted into heat.
- Low time consumption
- Low investment costs.

Cell disruption methods can roughly be divided into chemical, biological, and physical methods. Figure 3.5 shows an overview of the most common methods.

The majority of the cell disruption methods are only used on a laboratory scale. On an industrial scale mainly autolysis, high pressure homogenizers, and stirred media mills are used.

3.4.1 Measurement of Cell Disruption Results

The results of cell disruption measurement is usually done through the measurement of the concentration of released cell compounds like proteins, nitrogen, or the activity of a

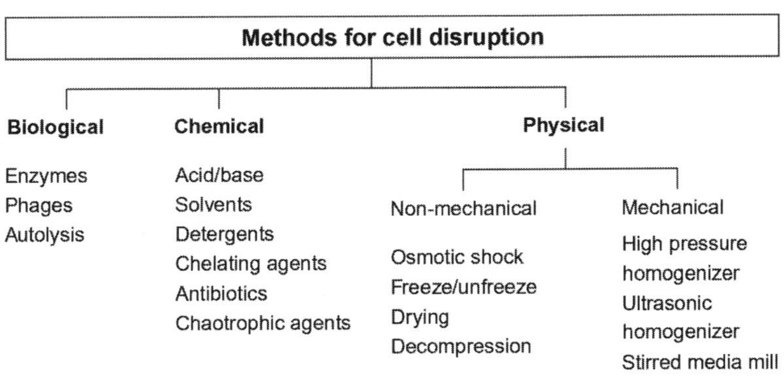

Figure 3.5. Overview of most common cell disruption methods (3).

certain enzyme. The measurement of conductivity can also serve as an indicator for the cell disruption result as long as the intracellular salt concentration is different from the one in the medium. The relative extent of concentration of liberated intracellular substances is specified as the degree of cell disruption A (Eq. 3.4).

$$A = \frac{c - c_0}{c_{\max} - c_0} \cdot 100 \qquad (3.4)$$

In this equation, c is the concentration of the measured cell compound, c_0 is the start concentration of this cell compound and $c_{\max}$ is the maximum concentration of cell compound after infinite time (without denaturation of this compound).

3.4.2 Cell Disruption Mechanisms in Mechanical Methods

When applying mechanical methods for cell disruption, the resulting stresses and strains disrupt the cell walls of the microorganisms.

3.4.2.1 Mechanical Stress. In mechanical disruption devices load is transmitted to the cells. According to Rumpf (44) three different types of mechanical stress can be defined (Fig. 3.6). In mechanical cell disruption, cells are mainly stressed by shear and compressive stress and stress transmitted from the fluid medium.

The effect of compressive stress on single cells is briefly discussed in the first sections of hervorheben this chapter. The effect of impact stress and stress on single cells due to the surrounding fluid has yet to be investigated. But from high pressure homogenizer experiments, conclusions can be made for the importance of these types of stress.

3.4.2.2 Stress by Cavitation. Cavitation appears when using an ultrasound homogenizer as well as a high pressure homogenizer. It is assumed that it is very important for cell disruption. The word cavitation simply means formation of bubbles (including expansion and collapse of the bubble) in a fluid (45). For cell disruption, hydrodynamic and acoustic cavitation are important.

Hydrodynamic cavitation in pipes and valves is caused by a reduction of the static pressure to below the vapor pressure of the contained fluid. For example, this occurs when fast-flowing fluid exits a valve. The sudden expansion of the cross section creates a reduction of pressure directly behind the valve. Fast flowing fluid can also create cavitation when flowing through a bend as the centrifugal force leads to a pressure drop on the inner side of the bend.

Acoustic cavitation is caused by ultrasound waves (20 kHz to 1 MHz) in aqueous media. The oscillating sonotrode pushes the fluid away, and then retracts immediately afterwards, causing a low pressure field at the front end of the sonotrode.

If the pressure drops below the vapor pressure of the fluid, the fluid rips open and the forming bubble is immediately filled with vapor. When the growing bubble reaches a critical size it collapses and the bubble implodes very fast (46). The vapor condenses and the water flows into the bubble from all sides. Upon collapse, conditions of very high temperature (up to 5000 K) and pressure (up to 500 bar) occur in the very last spot of the bubble. Furthermore, turbulences with high shear forces emerge from the collapsing bubble. If a bubble implodes in close proximity to a wall, fluid cannot flow in from all directions, and a fluid impulse is caused in the direction of the wall. This liquid jet (also called microjet), which can reach velocities up to 100 m/s, is responsible for the cavitation corrosion on marine propelling screws.

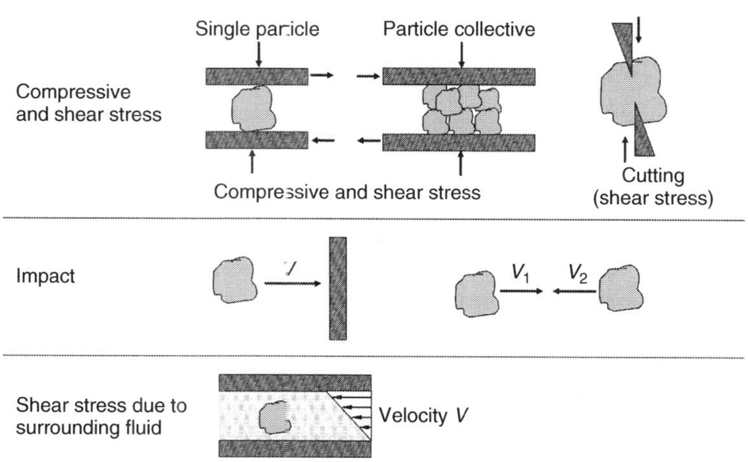

Figure 3.6. Types of mechanical stress [according to Rumpf (44)].

There are several hypotheses concerning how cavitation causes cell disruption. Some authors hold microjets responsible for the opening of cells. According to this theory the fluid jet hits the cell and leaves a hole in the cell wall. Others state that the shear forces of turbulences, resulting from imploding cavitation bubbles are responsible for cell disruption. Another theory assumes that the cells are ideal starting spots for cavitation bubbles, in which case a bubble forms in contact with the cell wall, causing the cell to burst due to the vapor pressure of the enclosed liquid.

3.4.2.3 ***Turbulent Stress.*** Apart from cavitation, turbulence fields of high energy density can also cause cell disruption. Depending on their size, vortices of the turbulent flow can have different effects on microorganisms. If the vortices are much bigger than the microorganisms, a cell that gets into the vortex is stressed by a laminar velocity gradient. Small, dissipative vortices with a length scale similar to that of the bacteria create the greatest shear forces.

Whether or not the bacteria are able to withstand the vortices can be estimated with the Weber number. The Weber number (Eq. 3.5) is derived from the spray and emulsification technique and describes the ratio of inertial force to surface force.

$$\mathrm{WE} = \frac{\rho \cdot w^2 \cdot 1}{\sigma} \tag{3.5}$$

In this case the cell is considered as a droplet with a surface tension σ that corresponds to the stability of the cell wall.

3.4.3 Cell Disruption Using a High Pressure Homogenizer

High pressure homogenizers are the most used devices for cell disruption. Manufacturers of cell disruption devices estimate these to account for around 80% of cell disruption devices, with an even greater percentage within the pharmaceutical industry. They belong to the group of flow dispersion devices that work based on the principle of pressure release. Two functional units are typical for such devices: a high pressure plunger pump and a homogenizing valve. The high pressure plunger pump produces a pressure independent and almost pulsation-free volume flow with the help of several plungers. Pressures up to 1500 bar (in ultra high pressure homogenizers up to 4000 bar) can be generated. The output of pilot plant stations and production lines ranges between 30–8000 L/h.

Most high pressure homogenizers work with valves similar to what is shown in Fig. 3.7. By forcing the valve seat onto the valve unit the homogenization gap is constricted until the desired pressure reduction is generated. The fluid passes through the valve seat and then flows radial through the homogenization gap, which may be only a few micrometers high. As a result, the velocity of the fluid is increased from about 6 m/s to 300 m/s. After leaving the homogenization gap the fluid hits the impact ring and leaves the valve through the outlet.

The construction of the high pressure homogenizer valve is essential for the cell disruption result. More than 150 valve designs have been patented, some with radically different shapes: these include Z-shaped microchannels; microchannels which divide the flow and afterwards let them hit each other under high pressure; and valves which induce a stream reversal, letting the forward stream shear with the reversed stream.

Normally, valves are made of carbide. When applied to abrasive suspensions like inclusion bodies or sand containing fluids (e.g. wastewater) they consist of ceramics.

3.4.3.1 ***The Principle of Cell Disruption in High Pressure Homogenizers.*** In high pressure homogenizers the cells are stressed by many different forces: Shear forces occur due to the redirection of the flow, while elongational flow due to increasing flow velocity causes tractive forces. Because of the sharp redirection and the high velocity, cavitation occurs in the homogenizing gap and finally, after leaving the homogenizer gap, the cells hit the impact ring. Which stress is applied is crucial for the disruption of cells, which has been discussed by many

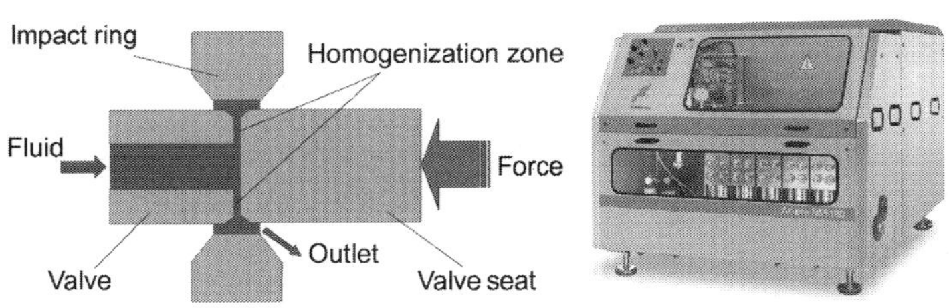

(a) (b)

Figure 3.7. (a) Schematic of a high pressure homogenizer valve assembly (3); (b) example of a process-scale homogenizer: GEA Niro Soavi, Ariete NS5180 (3000 L/h at 1500 bar). (This figure is available in full color at http://onlinelibrary.wiley.com/book/10.1002/9780470054581.)

authors, often using the example of yeast cells (47). The literature of emulsification can also be evaluated.

To determine the effect of shear flow, Tonnius (42) exposed yeast cells to a shear stress of 17000 Pa in a rotation viscosimeter. It was not possible to disrupt the yeast cells using this method. Since Tonnius calculated a shear stress of only 16000 Pa for the shear flow in the homogenization zone, he excluded the shear stress as a reason for the disruption of the yeast cells. The fact that the degree of disruption of yeast cells is independent of dry biomass content also supports this theory. If the shear flow was responsible for the disruption, the increasing viscosity should have an impact on the degree of disruption.

From emulsification techniques, it is known that in turbulent flow pressure differences on the surface of the fluid droplets appear, which lead to a strong deformation, and finally to a dispersion of the droplets. With the help of a nozzle without cavitation it was found that the result of an emulsification of a water/oil alloy is independent of whether there is laminar or turbulent flow in the gap. Only the energy input was proven to have an effect (48,49).

Büschelberger (50) constructed a homogenizer valve in which the nozzle was enclosed from a second, separately adjustable nozzle. With this arrangement pressure could be applied to the fluid directly after leaving the homogenization gap. Starting from a pressure of approximately 80 bar, the efficiency of disruption is reduced by about 30%. Büschelberger explained this with a reduction of cavitation due to the pressure.

Experiments in which impact rings with different internal diameters were used (51) showed that the distance between homogenization gap and impact ring is crucial in determining the cell disruption result. In experiments with *E. coli* the disruption degree decreased from 95–75% and with yeast it was even reduced to 50%. This clearly demonstrates that the impact stress at the impact ring has a big influence, and yet it is not the only factor influencing cell disruption. Furthermore these results indicate that cells that are relatively difficult to disrupt (in this case yeast cells) are primarily disrupted at the impact ring. In the case of these two microorganisms, different mechanisms may cause the disruption of the cells.

3.4.3.2 Disruption Kinetics in High Pressure Homogenizers. A popular model of disruption kinetics was developed in 1971 (52) by Hetherington.

$$\ln\left(\frac{100}{100 - A}\right) = k_1 \cdot N \cdot p^a \tag{3.6}$$

In Equation 3.6 the disruption degree (A) is dependent on the number of passes through the homogenizer (N) and the pressure (p). The two empirical parameters a and k_1 depend on several factors, like the type of the cell, its growth phase, composition of the growth medium in which the microorganism is cultivated, the design of the homogenization valve, the temperature, and the dry mass (53,54). Kleinig extended the right side of this equation by one term in order to consider the distance between the homogenization gap and the impact ring (51). A more physical prediction provides the so-called wall-strength model (55) (Eq. 3.7).

$$A = 1 - \int_0^{\infty} [1 - f_D(S)]^N f_s(S)\mathrm{d}S \tag{3.7}$$

In this equation the disruption degree (A) depends on the effective cell stability (S). $f_S(S)$ describes the distribution of cell stability and $f_D(S)$, the stress distribution in the homogenizer. As function for the distribution of the cell stability, a Gaussian distribution is used (56).

3.4.3.2.1 Specific Energy Input into High Pressure Homogenizers. The specific energy input based on the dry mass content is calculated as given in Equation 3.8.

$$E_{spec} = \frac{\Delta p}{c_{BDW}} \tag{3.8}$$

The specific energy input E_{spec} is directly proportional to the pressure drop Δp and the number of passes through the homogenizer N.

3.4.3.3 Parameters that Influence the Cell Disruption in High Pressure Homogenizers.

- *Biomass Dry Concentration.* The degree of disruption is independent of the dry biomass concentration within a wide range (3.5–24% dry biomass) (47). In order to improve the exploitation of the energy put into the system, dense bacterial suspensions should always be used.
- *Afflux Temperature.* The afflux temperature of the biomass suspension only has a small influence on the efficiency of cell disruption. However, it is important to consider that a pressure drop of 100 bar can raise the temperature of the suspension by 2.4°C, which can lead to denaturation of heat sensitive proteins.
- *Osmotic Effects.* As already mentioned above, osmotic effects have a considerable impact on the cell disruption result. The comparison between disruption of yeast cells in demineralized water and in a three percent sodium chloride solution shows that the disruption degree is lowered by 50% in the sodium chloride solution (50). A time dependency was also found for this effect, because the water has to diffuse out of the cells first. In order to improve the cell disruption result the biomass should be centrifuged after fermentation, and later resuspended in water with low salt content.

- *Viscosity*. Experiments with yeast cells whose viscosity was increased with the help of carboxymethylcellulose, and the same applied pressure drop, lead to the same disruption results (50).

3.4.4 Cell Disruption Using a Stirred Media Mill

The stirred media mill usually consists of a cylindrical, horizontally arranged, coolable grinding chamber with a stirring unit powered by an electric motor. The stirring unit can consist of pins or rotating disks with different shapes (Fig. 3.8). The grinding chamber is filled with grinding media, which is agitated by the stirring unit, and stresses the cells through compressive and shear stress. The cell suspension is pumped into the mill at the inlet. A sieve at the outlet prevents the escape of the grinding media.

The stirred media mill is even able to disrupt tough cells. It was built for the disruption of pigments, which are often very hard metal oxides. But it has the disadvantage of wear from the grinding media and the grinding chamber. Therefore its application in cell disruption is estimated by the manufacturers to be lower than 10%, and even smaller in the biopharmaceutical industry.

3.4.4.1 Specific Energy Input. The specific energy input (E_{spec}) into a stirred media mill is calculated as shown in Equation 9:

$$E_{spec} = \frac{(M_0 - M) \cdot 2\pi \cdot t}{V_{susp} \cdot c_{BDW}} \tag{3.9}$$

It is the quantity of the energy which is inserted into the grinding chamber but only a small fraction of the energy is transmitted to the product. The rest of the energy dissipates because of friction and grinding bead contacts that are irrelevant for disruption. The specific energy input can serve as a relevant parameter for the prediction of the disruption result.

Experiments showed that the specific energy contributes significantly to the result of the disruption. However, there is a difference in disrupting inorganic particles and cells. While an increase in stress energy (SE) leads to smaller average particle sizes with inorganic particles it is not important for the determination of the cell disruption result whether the cell envelope is disrupted or reduced to small pieces (57).

3.4.4.2 Model for the Characterization of Stress Intensity in Stirred Media Mills. Substantially the milling performance in stirred media mills can be described through the following parameters: specific energy (E_{spec}), number of stress events (SN), and stress energy (SE).

The SN is the absolute number of grinding media stressing cells. SE is the part of the specific energy that is transmitted to one particle of the grinding material during one stress event. The product of the SN and SE is proportional to the specific energy (58) (Eq. 3.10).

$$E_{spec} \propto \mathrm{SN} \cdot \mathrm{SE} \tag{3.10}$$

In order to reproduce a grinding result it suffices if two of the three parameters are constant. The SE that is transmitted to a particle differs within a mill. Grinding beads which are located near the outer area of the rotating disks are accelerated more than grinding beads close to the shaft. Hence in theoretical calculations, the SE is described with the help of distributions (59). For the calculation of the maximal SE acting on the grinding stock, the equation from Kwade (58) was extended by a term in order to consider the elastic modulus (E) of the grinding beads and the grinding stock by Becker (60) (Eq. 3.11).

$$\mathrm{SE} = d_{GM}^3 \cdot \rho_{GM} \cdot \left(\frac{E_{GM}}{E_{GM} + E_P}\right) \tag{3.11}$$

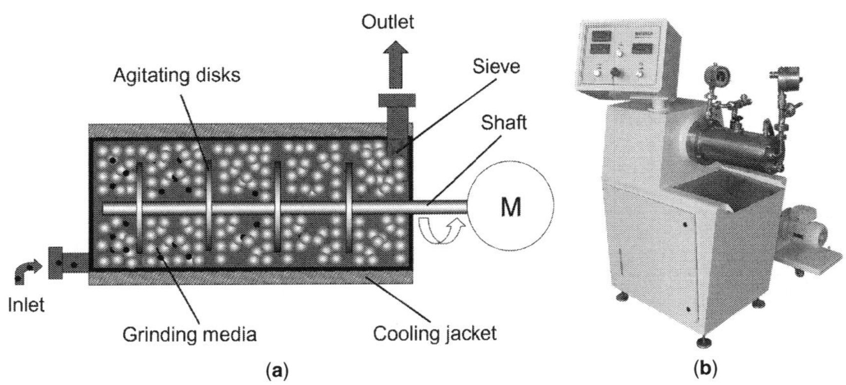

Figure 3.8. (a) Schematic of a stirred media mill (3); (b) example of a process- scale stirred media mill: NETZSCH- Feinmahltechnik GmbH, LME 4. (This figure is available in full color at http://onlinelibrary.wiley.com/book/10.1002/9780470054581.)

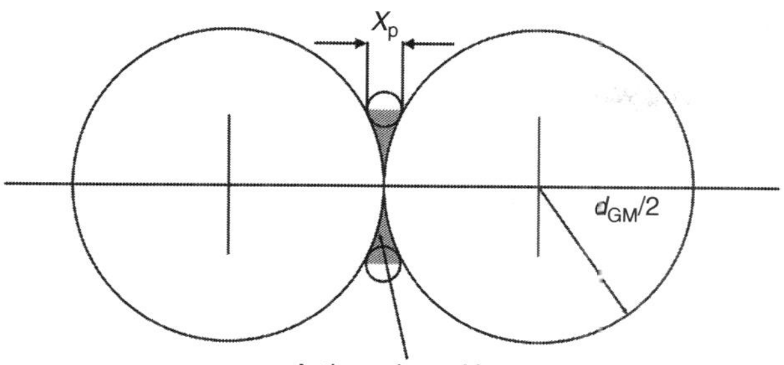

Figure 3.9. Example of an active volume between two grinding beads (3). (This figure is available in full color at http://onlinelibrary.wiley.com/book/10.1002/9780470054581.)

3.4.4.3 Active Volume. In order to stress a cell, the cell either has to be trapped between two grinding beads, or one grinding bead and the wall. But not all the volume between the grinding beads fulfils these requirements. The sum of the active volumes in the mill decreases with an increasing difference in size between the grinding beads and the particle (Fig. 3.9). When considering the elastic deformation of the cells, it becomes even smaller.

The active volume between two grinding beads is calculated as shown in Equation 3.12

$$V_{act,GM} = \pi \cdot \frac{x_p^2}{4} \cdot \left(d_{GM} + \frac{4}{3}x_p\right) \quad (3.12)$$

It can be seen that the active volume V_{act} between two grinding beads increases with the grinding bead and particle sizes. But adding the active volumes of all grinding media contacts in a mill, it decreases with increasing grinding media size as well as with decreasing sample particle diameter x_P.

3.4.4.4 Relation Between Grinding Media Size and Disruption Degree. While investigating the influence of grinding media sizes on yeast cell disruption, Bunge (57) found an optimum curve. Despite inputting the same amount of specific energy into the system, the disruption degree at a grinding media size of 750 μm was considerably better than at smaller or bigger grinding media sizes.

This observation can be explained by considering the SN and the SE. In Equation 3.11, the SE is cubically proportional to the diameter of the grinding media. Higher SE increases the probability of breakage and hence has a positive effect on the disruption result. But if more energy is applied to a cell than is necessary to disrupt it, the partial amount of energy is wasted and does not increase the cell disruption result. Moreover, because of the precondition of a constant energy input, an increase in SE leads to a decrease in the SN, which in turn has a negative effect on the disruption result. Another negative effect is the increase of wear from the grinding media, contaminating the product. If undersized grinding media is used, there are many stress events, which has a positive effect. But if the SE between two grinding beads is too low for cell disruption, the energy is wasted. These two contrary effects cause a curve with a wide optimum.

3.4.4.5 Influence of the Peripheral Stirrer Velocity. The peripheral stirrer velocity is referred to the outer radius of the agitating disks. A change in speed frequency affects the kinetic energy and therefore the SE as well (Eq. 3.11). However, the SE cannot be influenced in the same way by the stirrer velocity as by the grinding media size, because the SE is cubically connected to the diameter of the beads but only quadratic to the velocity of the stirrer. With the assumption of a constant specific energy input, the SN has to decrease if the speed frequency is raised. This can only be achieved with a reduction of the experiment duration since an increase in speed frequency also raises the SN per time. The effect of a change in the SN and SE has already been discussed in the section on the influence of the grinding media size.

The effect of a change in peripheral stirrer velocity depends on the working point of the system. If the SE already lies in the optimal range, the differences are very small because of the broad optimum. Outside the optimal range big differences in disruption degrees appear.

3.4.4.6 Influence of the Dry Biomass Concentration. An increase in biomass concentration has two effects. On one hand the probability increases that with a collision of two grinding beads one or more particles are included. On the other hand the viscosity of the suspension is increased so that the grinding beads are slowed down.

In general, however, a high biomass concentration is beneficial because the energy input into the system can be better exploited.

3.4.4.7 Influence of the Filling Rate. The filling rate is defined as the ratio of bulk volume of the grinding beads V_{GM} and net volume of the grinding chamber V_{GC} (Eq. 3.13):

$$\varphi_{GM} = \frac{V_{GM}}{V_{GC}} \quad (3.13)$$

Void space analysis with stationary grinding media consisting of tempered glass beads have resulted in an average of $\varepsilon = 0.41$. This is independent of the grinding bead diameter. In general the disruption degree rises with an increasing filling rate. In motion, however the bulk media expands, and a filling rate which is too high causes grouting of the mill. A value between $\varphi = 0.85$ and $\varphi = 0.9$ has proven to be optimal in various experiments. If the filling rate is too low, decomposition of the grinding media inside the grinding chamber can occur.

3.4.4.8 ***Disruption Kinetics in a Stirred Media Mill.*** In general cell disruption should be carried out no longer than necessary because the energy put into the system is almost completely dissipated as heat. This can eventually cause heat-labile proteins to denature, should the cooling be insufficient. Furthermore, proteins can lose their function due to mechanical stress. In general, the kinetics of disruption in stirred media mills are usually described with a first order reaction (61) (Eq. 3.14).

$$\ln\left(\frac{100}{100-A}\right) = k \cdot t \tag{3.14}$$

In this equation k is the rate constant of the disruption process. For continuous stirred media mills, the disruption kinetics can be described as a series of j completely mixed mills (54) with a certain residence time τ (Eq. 3.15).

$$\frac{100}{100-A} = \left(1 + \frac{k \cdot \tau}{j}\right)^{j} \tag{3.15}$$

3.4.5 Disruption of Cells by Ultrasonic Homogenization

Apart from cell disruption, the ultrasonic homogenizer is also used for deagglomeration, homogenization, emulsification, and for the cleaning of surfaces. Basically these devices are composed of three components: a generator that generates high-frequency voltage in the range of 15–40 kHz, a ceramic crystal made of piezoelectric material which transforms the electric impulses into mechanical vibrations, and the sonotrode that transmits these vibrations into the medium (Fig. 3.10).

Today's ultrasonic homogenizers have a wide range, starting from devices that can sonicate half a milliliter of suspension to homogenizers capable of stressing several hundred liters per hour with the help of series-connected sonotrodes. An important field of application of mechanical cell disruption through ultrasound lies in waste water treatment (62), in which ultrasound can help to improve the sedimentation characteristics of bulking sludges and floating sludges. It is also used for the treatment of the sludge to improve the anaerobic digestion for biogas production (63).

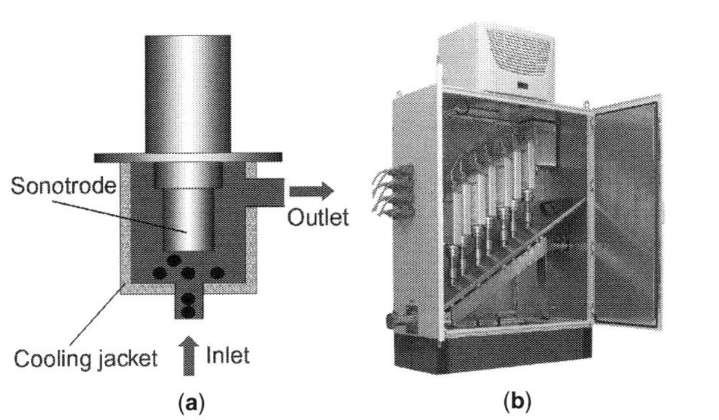

Figure 3.10. (a) Schematic of an ultrasonic homogenizer valve (3); (b) example for a process-scale homogenizer: Hielscher Ultrasonics GmbH, sonication system of 7 x UIP1000hd sonotrodes. (This figure is available in full color at http://onlinelibrary.wiley.com/book/10.1002/9780470054581.)

3.4.5.1 ***Specific Energy Input in Ultrasonic Homogenization.*** The specific energy input of the ultrasonic homogenizer is calculated as shown in Equation 3.16

$$E_{\text{spec}} = \frac{P \cdot t}{V_{\text{susp}} \cdot c_{\text{BDW}}} \tag{3.16}$$

3.4.5.2 ***Effect of Different Parameters on the Ultrasonic Cell Disruption Process.***

- *Pressure.* High pressure has a negative effect on the efficiency of disruption because cavitation is reduced under these conditions. Also large amounts of carbon dioxide under relatively low pressure can lead to a decrease of the disruption result because it can diffuse into the cavitation bubbles and raise the bubble's internal pressure. In that case the collapse of the cavitation bubble is clearly weaker. This phenomenon can be the reason for a lower disruption degree of microorganisms that oxidize carbon sources to a great extent.
- *Viscosity.* High viscosity (e.g. due to high biomass concentrations) causes a clear deterioration of the disruption efficiency. Usually the highest amount of bio dry weight is 40%.
- *Glass Powder.* In order to increase the number of cavitation events, cavitation germs like glass powder can be added to the suspension. The disruption result of yeast cells can be clearly improved with this method (64).
- *Frequency.* Neis and Tiehm investigated the connection between disruption result and frequency in the range of 41–3217 kHz (46). It was demonstrated that the best disruption result could be achieved with the lowest frequencies.
- *Amplitude.* An increase in amplitude leads to a noticeable acceleration in cell disruption. However, one has to take into account that with a higher energy input, the temperature of the suspension is raised faster. Furthermore more radicals are produced, and the introduction of scavengers (e.g. Glutathione) is recommended.
- *Constructive Details.* From dispersion investigations it is known that the distance between the front surface of a sonotrode and the inlet of the disruption chamber has

a great influence on the dispersion result (65), because the disruption process mainly happens in an area a few millimetres away from the front end. Narrowing this front end towards the inlet ensures that all cells will flow through this zone.

3.4.6 Other Devices for Mechanical Cell Disruption

Vibratory mills are suited for wet and dry grinding of brittle materials of all degrees of hardness and therefore are applied in many fields. In general, they are made up of a cylindrical grinding chamber which is put into vibration by an eccentric drive. The frequency of vibration ranges between 100 and 6000 waves per minute. The grinding chamber volumes of devices that are used for cell disruption, ranges between several millilitres and 1 L. The grinding beads that are located in the grinding chamber are set into motion by the vibrations. Through the motion of the grinding beads, the grinding media is stressed through pressure and shear forces, which leads to disruption of the cells. The cells are either put into the grinding chamber together with the grinding beads (discontinuous) or pumped through the mill. Grinding beads, that can range from 200 μm to 1 μm, are used for cell disruption. In general, the same laws for the size of grinding beads are applicable in vibratory mills as in stirred media mills. Due to their simple construction vibratory mills are also used for cryogenic grinding. Synthetic materials become brittle at low temperatures which make them easier to disrupt. In biology liquid nitrogen is used for DNA and RNA release in order to prevent the nucleic acids from breaking down.

Another device for cell disruption at the laboratory scale is the French Press. It was introduced as a discontinuously working cell disruption device by Millner, Lawrence, and French in 1950 (66). It is made up of a pressure resistant cylinder and a piston which tightly fits into the cylinder. There is a valve at the bottom of the cylinder. For cell disruption, the cell suspension is filled into the cooled cylinder while the valve is closed. With the help of the piston, a pressure of 2000 bar is applied to the suspension. Later the valve is opened a bit so that a small flow of the suspension can leave the cylinder. When flowing through the valve, the suspension is accelerated. The principle of disruption is very similar to the one of the high pressure homogenizer. Due to its small volume (about 50 ml), the French Press is only used in laboratories.

The Hughes Press which was first described by Hughes *et al*. in 1951, also consists of a pressure resistant cylinder and a piston. In contrast to the French Press, the Hughes Press works with a highly concentrated frozen suspension. With very high pressure (up to 5000 bar), the suspension is pressed through a narrow orifice in the bottom of the cylinder. The disruption is caused by high shear stress and the abrasive effect of the ice crystals. The Hughes Press was modified several times. Edebo, for example, modified it to the X-Press, in which the frozen suspension can be pressed back and forth between two chambers (67).

3.5 COMPARISON OF DIFFERENT CELL DISRUPTION DEVICES

All cell disruption devices have their advantages and disadvantages which makes them suitable for different applications. While high pressure homogenizers have a high throughput, medium energy consumption, and are available for work in pharmaceutical areas, they also have a damageable valve, which can be plugged and broken. Other problems may arise with the plunger pumps which also have valves. Working with filamentous fungi these valves may not close preventing the pressure from building up.

Stirred media mills can transfer the highest stress energies to the cells. But with this high energy, wear and tear occurs and will contaminate the product. Surprisingly the overall energy input in a mill is very low, so only a minor damage occurs due to heat. Furthermore, there are many parameters to modify until the optimal cell disruption parameters are found.

The ultrasonic homogenizer is very robust and easy to use. But it is not suitable for the disruption of very tough organisms. Moreover it transfers nearly all the energy into the suspension and generates excess heat. Furthermore, ear protection is required due to the intensity of the sound.

Taking these facts into account, a high pressure homogenizer is best used for Gram negative cells in a clean environment, such as in pharmaceutical production. The stirred media mill is suitable for tough Gram positive organisms and fungi which need more energy to disrupt or are growing in long hyphae. The ultrasonic homogenizer should be used in difficult environmental conditions such as in wastewater treatment plants.

3.6 CORRELATION OF MICROMECHANICAL RESULTS WITH CELL DISRUPTION RESULTS

Micromechanical measurements give an insight into the microprocesses which occur in cell disruption devices. Yet not all types of stress mechanisms are investigated. Also, only a few microorganisms have been tested. However, several qualitative correlations can be considered.

- *Stress Intensity.* Knowing the frequency distribution of the bursting forces and dividing each value by the cell volume, the required frequency distribution of stress intensities within a cell disruption device can be calculated. But it is very hard to meet these criteria in disruption devices because they have a broad range of stress mechanisms and therefore a broad range of stress intensities.

- *Maximum Stress Intensity.* Micromechanical tests have shown that only a small hole in the cell wall is enough to allow for protein release. It is not necessary to disrupt the cell envelope any further. Therefore a disruption device should not provide stress intensities, which are higher than necessary. This only creates excess heat and leads to denaturation of the biomolecule product.
- *Minimum Stress Intensity.* A device which does not meet the minimum stress intensities is not suitable for the disruption of the cells. All energy of these weak stress events is transferred into heat without an increase in the cell disruption result. It even decreases the cell disruption efficiency.
- *Multiple Stress Events.* As already mentioned above, multiple stress events which individually are not strong enough for the disruption of cells can increase the necessary bursting force. This makes weak stress intensities counterproductive. Also multiple stress events on already burst cells do not increase the cell disruption result. Therefore devices, providing single stress events of sufficient intensity, are best to disrupt cells.
- *Osmotic Pressure.* The effect of osmotic pressure on the disruption of yeast cells in a vibratory and in a stirred media mill was investigated by Arfsten (11). The cell disruption kinetics could be correlated with the bursting forces, which clearly indicates the convertibility of micromechanical results into disruption processes. These results concur with earlier findings, which revealed an easier cell disruption at low salt concentrations in a high pressure homogenizer (50).

The fusion of micromechanical results with macroscopic equipment design for biopharmaceutical application is still not fully understood. But both sides are approaching each other so the microprocesses will some day be understood and utilized for a more efficient and enhanced production of cell disruption devices for large-scale processes which manufacture biotechnological products.

NOMENCLATURE

c	protein concentration (kg/m[MK 3])
c_0	protein concentration at $t = 0$ (kg/m[MK 3])
c_{BDW}	concentration of bio dry weight (kg/m[MK 3])
c_{max}	maximum protein concentration (kg/m[MK 3])
d_{GM}	grinding media diameter (m)
$f_S(S)$	distribution of cell stability
$f_D(S)$	stress distribution in the homogenizer
i	indentation depth
j	number
k_1	constant
k_c	cantilever spring constant (N/m)
k_Z	cellular spring constant (N/m)
l	length (m)
p	pressure
t	time (s)
v_t	stirrer peripheral velocity (m/s)
w	fluid velocity (m/s)
x_P	particle size (m)
A	degree of disruption
E	Young's modulus (Pa)
E_P	Young's modulus of the particles (Pa)
E_{GM}	Young's modulus of the grinding media (Pa)
E_r	reduced Young's modulus (Pa)
E_{spec}	specific energy (J/kg)
E_Z	cellular Young's modulus (Pa)
F	force (N)
M	torque (Nm)
M_0	torque without grinding media and product (Nm)
N	number of cycles
P	power (J/s)
R	radius indenter (m)
S	slope
SE	stress energy (J)
SI	stress intensity (J/m^3)
SN	stress number
$V_{act,GM}$	active volume between two grinding media (m^3)
V_{GC}	net volume of the grinding chamber (m^3)
V_{GM}	bulk volume of the grinding beads (m^3)
V_{susp}	volume suspension (m^3)
WE	Weber number

3.7 GREEK LETTERS

ρ	density (kg/m^3)
ρ_{GM}	grinding media density (kg/m^3)
σ	surface tension (N/m)
π_{GM}	grinding media filling degree
υ	Poisson's ratio
Δp	pressure difference (Pa)
τ	mean residence time (s)
ν	rotation frequency (per s)

REFERENCES

1. Schlegel HG. Allgemeine Mikrobiologie: Georg Thieme Verlag Stuttgart; 1992.
2. Hull AS, Middelberg AP. Trans I Chem E 1993; 71(C): 264–266.

3. Kampen, I. Einfluss der Zellaufschlussmethode auf die Expanded Bed Chromatographie [dissertation]. Braunschweig: Technische Universität Braunschweig; 2005.
4. Höltje J-V. Microbiol Mol Biol 1998; 62: 181–203.
5. Butt H-J, Cappella B, Kappl M. Surf Sci Rep 2005; 59: 1–152.
6. Arnoldi M, Kacher CM, Bäuerlein E, Radmacher M, Fritz M. Appl Phys A 1998; 66: 5613–5617.
7. Hutter JL, Bechhoefer J. Rev Sci Instrum 1993; 64(7): 1868–1873.
8. Velegol SB, Logan BE Langmuir 2002; 18: 5256–5262.
9. Hertz H. J Reine Angew Math 1881; 92: 156–171.
10. Gaboriaud F, Bailet S, Dague E, Jorand F. J Bacteriol 2005; 187: 3864–3868.
11. Arfsten J. Mikromechanische Charakterisierung von Saccharomyces cerevisiae [dissertation], Braunschweig: Technische Universität Braunschweig; 2009.
12. Zhao L, Schaefer D, Xu H, Modi SJ, LaCourse WR, Marten MR. Biotechnol Prog 2005; 21: 292–299.
13. Zhao L, Schaefer D, Marten MR. Appl Environ Microbiol 2005; 71(2): 955–960.
14. Yao X, Walter J, Burke S, Stewart S, Jericho MH, Pink D, Hunter R, Beveridge TJ. Colloids Surf B: Biointerfaces 2002; 23: 213–230.
15. Sullivan CJ, Venkataraman S, Retterer ST, Allison DP, Doktycz MJ. Ultramicroscopy 2007; 107: 934–342.
16. Abu-Lail NI, Camesano TA. Colloids Surf B: Biointerfaces 2006; 51(1): 62–70.
17. Schär-Zammaretti P, Ubbink J. Biophys J 2003; 85: 4076–4092.
18. Vadillo-Rodriguez V, Beveridge TJ, Dutcher JR. J Bacteriol 2008; 190(12): 4225–4232.
19. Abu-Lail NI, Camesano TA. Biomacromolecules 2003; 4: 1000–1012.
20. Touhami A, Nysten B, Dufréne YF. Langmuir 2003; 19: 4539–4543.
21. Gaboriaud F, Parcha BS, Gee ML, Holden JA, Strugnell RA. Colloids Surf B: Biointerfaces 2008; 62: 206–213.
22. Shiu C, Zhang Z, Thomas CR. Biotechnol Tech 1999; 13: 707–713.
23. Arfsten J, Bradtmöller C, Kampen I, Kwade A. J Mater Res 2008; 23:(12): 3153–3160.
24. Arfsten J, Kampen I, Kwade A. Int J Mater Res 2009; 100(7): 978–983.
25. Mashmoushy H, Zhang Z, Thomas CR. Biotechnol Tech 1998; 12(12): 925–929.
26. Smith AE, Zhang Z, Thomas CR. Chem Eng Sci 2000; 55: 2031–2041.
27. Srinorakutara T. J Ferm Bioeng 1998; 86(3): 253–260.
28. Srinorakutara T. World J Microbiol Biotechnol 1998; 14: 719–725.
29. Smith AE, Moxham KE, Middelberg APJ. Chem Eng Sci 2000; 55: 2043–2053.
30. Smith AE, Zhang Z, Thomas CR, Moxham KE, Middelberg APJ. Biochem Eng J 2000; 29: 9871–9874.
31. Stenson JD, Thomas CR, Hartley P. Chem Eng Sci 2009; 64: 1892–1903.
32. Elias CB, Joshi JB. Adv Biochem Eng Biotechnol 1998; 59: 47–71.
33. Ghadge RS, Sawant SB, Joshi JB. Chem Eng Sci 2003; 58: 5125–5134.
34. Heydarian SM, Mirjalili N, Ison AP. Bioprocess Eng 1999; 21: 31–39.
35. Wang L, Ridgway D, Gu T, Moo-Young M. J Chem Technol Biotechnol 2003; 78: 1259–1266.
36. Kelly S, Grimm L, Hengstler J, Schultheis E, Krull R, Hempel DC. Bioprocess Biosyst Eng 2004; 26: 315–323.
37. Heydarian SM, Mirjalili N, Ison AP. Bioprocess Eng 1999; 21: 31–39.
38. Pilz RD. Partikelbeanspruchung in mehrphasig betriebenen Airlift-Reaktoren [dissertation]. Braunschweig: Technische Universität Braunschweig; 2006.
39. Hoffmann JF, Tralles S, Hempel DC. Chem Ing Tech 1992; 64: 953–956.
40. Mahnke EU, Büscher K, Hempel DC. Chem Eng Technol 2000; 23: 509–513.
41. Pilz RD, Mahnke EU, Hempel DC. J Chem Eng Japan 2004; 37: 955–961.
42. Tonnius FG. Verfahren zum Herstellen von Nucleinsäurearmen Proteinkonzentraten aus Bäckerhefe [dissertation]. Karlsruhe, 1982.
43. Müller J. Mechanischer Klärschlammaufschluß [dissertation]. Braunschweig: Shaker Verlag; 1996.
44. Rumpf H. Chem Ing Tech 1965; 37: 187–202.
45. Shah YT, Pandit AB, Moholkar VS. Cavitation reaction engineering. New York: Kluwer Academic/Plenum Publishers; 1999.
46. Neis U, Thiem A. In: Neis U, Thiem A, editors. Ultrasound in environmental engineering. Hamburg: GFEU; 1999. p 39–61.
47. Brookman JSG. Biotechnol Bioeng 1974; 16: 371–383.
48. Kiefer P. Der Einfluss von Scherkräften auf die Tröpfchenzerkleinerung beim Homogenisieren von Öl-in-Wasser-Emulsionen in Hochdruckhomogenisierdüsen [dissertation]. Karlsruhe; 1977.
49. Kurzhals H-A. Untersuchungen über die physikalisch-technischen Vorgänge beim Homogenisieren von Milch in Hochdruck-Homogenisiermaschinen [dissertation]. Hannover: Bönecke-Druck, Clausthal; 1977.
50. Büschelberger H-G. Untersuchung zum mechanischen Aufschluss von Mikroorganismen in Hochdruckhomogenisatoren [dissertation]. Karlsruhe: Technische Hochschule Fridericiana Karlsruhe; 1987.
51. Kleinig AR, Middelberg APJ. Chem Eng Sci 1996; 51: 5103–5110.
52. Hetherington PJ, Follows M, Dunnill P, Lilly MD. Trans I Chem E 1971; 49: 142–148.
53. Jahnke S. In: Müller J, Dichtl N, Schwedes J, editors. Klärschlammdesintegration. Braunschweig: Schmidt Buchbinderei & Druckerei; 1998. p 31–48.
54. Middelberg AP. In: Subramanian G, editor. Bioseparation and bioprocessing, band 2. Weinheim: Wiley-VCH; 1998. p 132–164.
55. Middelberg APJ. Trans I Chem E 1993; 71: 215–219.
56. Middelberg APJ, O'Neill BK. Biotechnol Prog 1993; 9: 109–112.
57. Bunge F. Mechanischer Zellaufschluss in Rührwerkskugelmühlen [dissertation]. Braunschweig: VDI Verlag; 1991.
58. Kwade A. Chem Eng Technol 2003; 26(2): 199–205.

59. Stender H-H. Einfluß von Größe und Bauart auf die Zerkleinerung in Rührwerkskugelmühlen [dissertation]. Braunschweig: TU Braunschweig; 2002.
60. Becker M. Zerkleinerung keramischer Rohstoffe und Simulation der Zerkleinerung in Rührwerkskugelmühlen [dissertation]. Braunschweig: TU Braunschweig; 1999.
61. Kula M, Schütte H. Biotechnol Prog 1987; 3(1): 31–42.
62. Nickel K, Thiem A, Neis U Müller J, Dichtl N, Schwedes J, editors. Klärschlammdesintegration. Braunschweig: Schmidt Buchbinderei & Druckerei; 1998. p 97–111.
63. Lehne G. Parameter der mechanischen Desintegration von Überschußschlämmen [dissertation]. Braunschweig: Cuvillier Verlag Göttingen; 2001.
64. Hughes DE, Wimpenny JWT, Lloyd D. Methods Microbiol 1971; 5: 2–53.
65. Pohl M, Hogekamp S, Hoffmannn NQ, Schuchmann HP. Chem Ing Tech 2004; 76(4): 392–396.
66. Milner HW, Lawrence NS, French CS. Science 1950; 6: 633–634.
67. Edebo L. Disintegration of cells. 1969.

4

CELL SEPARATION, YEAST FLOCCULATION

Eduardo V. Soares

Bioengineering Laboratory, Chemical Engineering Department, Superior Institute of Engineering from Porto Polytechnic Institute, Porto, Portugal

IBB-Institute for Biotechnology and Bioengineering, Centre for Biological Engineering, Universidade do Minho, Braga, Portugal

4.1 INTRODUCTION

Yeast flocculation is a process wherein yeast cells are gathered together into multicellular masses, which are composed of thousands or millions of cells (Fig. 4.1); these aggregates, called flocs, have usually macroscopic size and due to their mass sediment rapidly from the medium in which they are suspended. Since it was firstly described by Louis Pasteur, in 1876, yeast flocculation has had a renewed interest by the academy and industry, mainly by brewing industry, due to its importance in the production of a beer with bright appearance.

Saccharomyces genus, especially *S. cerevisiae* and its relatives, included in the *Saccharomyces sensu stricto* complex, are the most useful and industrially exploited microorganisms known by man (1). Besides the traditional applications of *S. cerevisiae* in the production of beer, wine, and bread, today, this yeast can be used in the production of fuels, food ingredients, various metabolic products, and in environmental applications. The advent of genetic engineering and recombinant DNA (rDNA) technology allowed the expansion of industrial uses of *S. cerevisiae*, as an efficient cell factory, in the production of heterologous products like hormones (insulin) and vaccines (hepatitis B and papillomavirus) (2). A large research has been carried out on biochemistry and physiology of *S. cerevisiae* and more recently on functional genomics using new tools like transcriptome, proteome, and metabolome analysis. These data originated a solid knowledge, both at global and at the molecular level, which can be used to a significant increase in the industrial exploitation of this yeast (2,3).

Yeast agglutination into cellular masses, as a natural method of cell separation from the medium, facilitates further downstream processing of many industrial applications. Flocculation is an off-cost and eco friendly process of cell separation and does not require energy input. In addition, it allows the use of high density cell reactors, which increase the efficiency of the processes and permit the use of different configurations of suspended biomass reactors without the risk of biomass washout. These advantages fit quite well in the concept of "White" (industrial) Biotechnology; this concept devises the optimization of the biotechnological processes with the reduction of wastes and energy consumption, which contributes to a more sustainable future (4).

This review focuses on the present knowledge of the molecular mechanism of flocculation in *S. cerevisiae* and in the genetics and environmental aspects that affect and control yeast flocculation. It further discusses the importance of yeast flocculation, as a separation process, in different biotechnological applications.

4.2 MICROBIAL AGGREGATION AND YEAST FLOCCULATION: SCOPE AND DEFINITIONS

Although microorganisms exist predominantly as a form of individual cells, the aggregation is a widespread phenomenon in the microbial world. Bacteria can aggregate under different forms such as biofilms (5), flocs in activated sludge (6) or dense granules in upflow anaerobic sludge blanket bioreactors (7). Other examples of microbial

Downstream Industrial Biotechnology: Recovery and Purification, First Edition. Edited by Michael C. Flickinger.

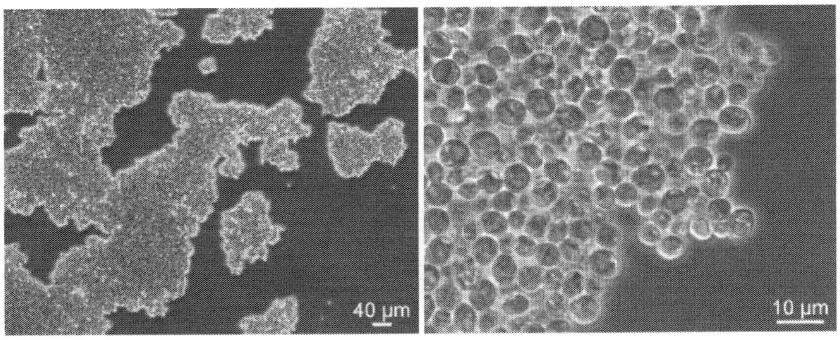

Figure 4.1. Photomicrographs of flocculent cells of *S. cerevisiae:* (a) phase-contrast observation of flocculated cells of the ale brewing strain of *S. cerevisiae* NCYC 1364. (b) observation of part of a floc with higher magnification. (This figure is available in full color at http://onlinelibrary.wiley.com/book/10.1002/9780470054581.)

aggregation are the formation of the "pellet" by *Penicillium chrysogenum* (8), the aggregation of the cellular slime mould *Dictyostelium* (9), the agglutination of the algae *Dunaliella salina* (8) or the aggregation due to the mating reaction in the protozoa *Paramecium caudatum* (10).

In the case of *S. cerevisiae*, three main different forms of aggregation can be found; these aggregation phenomena should be distinguished from flocculation. They are sexual aggregation, coflocculation, and chain formation. Sexual aggregation in haploid strains of *S. cerevisiae* results in the appearance of complementary molecules (proteins) on surface of complementary mating types, α and a, as a consequence of the exchange of pheromones, a and α factors, respectively (8,11). The binding of these two so-called "social" adhesins (12) to each other surfaces of cells with complementary mating type facilitates the fusion of the haploid cells (11). Another type of yeast aggregation in *S. cerevisiae* is usually known as *coflocculation* or mutual flocculation. Coflocculation is a heterotypic aggregation process among two strains, in which one of them is nonflocculent and the other weakly flocculent. Mixed together, in the presence of Ca^{2+}, the strains settle rapidly (13,14). Chain formation (a phenomenon found in same brewing yeasts) occurs as a consequence of the failure of the younger bud to separate from the mother cell, after cellular cycle; this results in an aggregate formation composed of 30–50 cells. These aggregates are covalently linked; thus, after mechanical dispersion, cells will not be able to stick together again. Although this phenomenon results in an aggregate formation, chains formation should not be considered a true cellular aggregation process or flocculation, since a reversible multicellular contacts among the cells are not established (8).

Yeast flocculation can be defined as a reversible, nonsexual, calcium-dependent and multivalent process of aggregation of yeast cells (11,15). Flocs can be dispersed by the action of chelating agents, such as ethylene-diamine-tetra-acetic acid (EDTA) or specific sugars (16).

Flocculation is usually associated with *S. cerevisiae* and more specifically with brewing yeast strains. However, this property is also found in other yeasts, namely: *Candida tropicalis* (17), *Hansenula anomala* (18), *Kluyveromyces bulgaricus* (19), *Kloeckera apiculata* (20), *Saccharomycodes ludwigii* (21), *Schizosaccharomyces pombe* (22) and *Zygosaccharomyces* sp. (23). Even if flocculation observable in these genera can differ from brewing yeasts flocculation in several aspects, such as in the cell–cell interaction mechanism.

4.3 GENETICS OF YEAST FLOCCULATION

Many genes have been reported to be involved in yeast flocculation and its regulation. The best known flocculation gene is *FLO1*, a dominant gene located on the right arm of chromosome I (24). *FLO1* is allelic with *FLO2* and *FLO4*, while *FLO5*, *FLO9*, and *FLO10* gene products are highly homologous to *FLO1*: 96%, 94%, and 58%, respectively (25). *FLO1* is a structural gene for flocculation and encodes for a lectin-like cell surface protein (26–28). The expression of *FLO1* provokes flocculation with Flo1 phenotype behavior (26,29), which is inhibited by mannose but not by glucose. It was proposed that *FLO1* gene is regulated at a transcriptional level (30). The gene *FLO1* seems to be inhibited by the suppressor genes *fsu1*, *fsu2*, and *fsu3* (flocculation suppressors) (25). In brewing lager yeasts, it was described a so-called *Lg-FLO1*, which is believed to encode to a flocculent lectin and causes flocculation with a NewFlo phenotype (29,31); for these strains, flocculation is inhibited by mannose and glucose. *Lg-FLO1*, in a lager-fermenting yeast, is located on chromosome VIII at the same position as the *FLO5* gene of the laboratory yeast strain *S. cerevisiae* S288c (32).

The genes *FLO3*, *FLO6*, and *FLO7* have been described as recessive/semidominant (25). *FLO8* gene encodes a transcriptional activator of *FLO1* and *FLO11* (33,34). Latter studies have shown that the protein kinase A pathway and the transcription factors FLO8p and Mss11p are the major regulators of the expression of *FLO1* gene (35,36). A recent study showed that the overexpression of *FLO1*, *FLO5*, *FLO9*, and *FLO10* genes induced flocculation; however, the different *FLO* genes displayed different characteristics in respect of floc size and carbohydrate inhibition (37).

The gene *FLO11* (also referred as *MUC1*) is involved in yeast flocculation, pseudohyphae formation, and invasion of substrates (38–40). Flo11 protein enables yeast cells to adapt to nutritional stresses, switching to a pseudohyphal growth, in diploid cells, under nitrogen starvation (39); haploid cells may invade substrate under glucose starvation (41). In both cases, this switching can be seen as a mechanism of cell adaptation to look for nutrients (12). FLO11 protein dependent flocculation is of FLO1 phenotype behavior (42). *FLO 11* gene expression is regulated by many signaling pathways, namely: cyclic adenosine monophosphate/protein kinase A (cAMP/PKA), mitogen-activated protein (MAP) kinase pathway, nutritional sensing pathways, a quorum sensing pathway and cyclins (12,43). In addition to transcriptional regulation, it was proposed that *FLO11* expression can be under control at posttranscriptional level (44,45).

The flocculation of *S. cerevisiae* is not a straightforward molecular mechanism. Many mutations, namely, those involving regulatory genes (*TUP1* and *SSN6*) or cell wall biosynthesis (*wal* and *abs*) have been described as bringing about yeast flocculation; these mutations and its pleiotropic effects were reviewed by Teunissen and Steensma (25).

4.4 MOLECULAR MECHANISM OF YEAST FLOCCULATION

4.4.1 Cell Wall, Surface Characteristics, and Flocculation

Cell wall surrounds periplasmic space and cytoplasmic membrane and determines the morphology of the cell Due to the combination of considerable mechanical strength and high elasticity, the wall confers mechanical stability and impairs the lysis in hypotonic environments; besides this skeletal function, the wall plays a role in the retention of periplasmic enzymes (46).

S. cerevisiae cell wall has a layered structure, consisting of an amorphous inner layer and a fibrilar outer layer. The inner layer is mainly composed by β-glucan and chitin, which are responsible for the mechanical strength of the wall. β-glucan is constituted by the fractions of $\beta(1 \rightarrow 3)$ and $\beta(1 \rightarrow 6)$-linked glucose residues and chitin is composed by linear chains of $\beta(1 \rightarrow 4)$-linked *N*-acetylglucosamine. Chitin is predominantly located in the bud scars, although some is dispersed throughout the wall. $\beta(1 \rightarrow 3)$ glucan is the major structural component of the wall (46,47). The fibrilar outer layer consists predominantly of α-mannan (highly glycosylated) associated with proteins (mannoproteins) (46). Several macromolecular components of the external layer allow the cells to recognize matting partners, flocculate, form biofilm, and grow pseudohyphally and invasively (48). For a detailed revision of the molecular structure and construction of yeast cell wall, reviews by Klis *et al*. (49) and Lesage and Bussey (50) can be consulted.

Yeast flocculation is an intrinsic surface characteristic. Flocculent cells retained their aggregation ability after being heat-killed (51,52); in the same way, walls prepared from flocculent cells keep flocculation capacity (53,54). The external mannoprotein layer is implicated in the definition of the flocculation characteristics of the strains. Proteases and chemical modifications of functional groups of amino acids lead to a irreversible dispersion of the flocs (53–56). On the other hand, the reversible inhibition of flocculation by mannose and mannosyl derivatives, as well as the prevention of flocculation by chemical modification of the cell surface sugars, clearly indicated that α-mannans are implicated in the flocculation process (56–59).

Among surface characteristics, the increase of cell wall hydrophobicity have been positively correlated with yeast flocculation (37,60–65). At physiological pH values, yeast cells walls have negative charge. Phosphate and carboxyl groups generate an electrical potential, which leads repulsion among negative charged cells and maintains the cells dispersed in suspension at a distance of the order of 10 nm from each other (60,66). The decrease of yeast surface charge should facilitate the cell–cell interaction and consequently yeast flocculation. However, no clear relationship between yeast surface charge and the onset of flocculation was found (60,66).

The mutual repulsion among negative charged cells prevents its approaching sufficiently close, acting as an effective barrier to floc formation. For flocculation to occur, mechanical agitation of yeast suspension is required to give enough energy to yeast cells to overcome this repulsion barrier (16). In the last years, more attention has been dedicated to the colloidal aspects of yeast flocculation (64,67); this attention has been particularly focused on the rate at which cells collide and associate and on the rheological properties of flocculent suspensions. For a review, see Jin and Speers (68).

In addition to mechanical agitation, a minimum threshold cell concentration is required for flocculation to occur. Below this minimum, the rate of aggregation is practically zero. This fact is due to the low number of cells per milliliter that limits the physical possibility of establishing a flocculent interaction among the yeasts (52,69–72).

4.4.2 Lectin-Like Theory

The lectin-like theory was formally proposed in the beginning of the 80 years (73). This is the model that prevailed until now, in almost all of its basic concepts. According to this model, flocculation of *S. cerevisiae* can be explained as the result of the interaction between specific proteins (lectins), only present on flocculent cells, with the sugar residues of the α-mannan (receptors) present on neighboring cells; calcium ions ensures the correct conformation of the lectins (73,74) (Fig. 4.2). Subsequent works proposed that, most likely, the receptors are the nonreducing termini of $\alpha(1 \rightarrow 3)$-linked mannan side branches, two or three mannopyranose residues in length (58,75).

The adhesion forces due to specific interactions (lectin-receptor) were estimated of 121±53 pN (76). These lectin-receptor interactions are most likely reinforced and stabilized by nonspecific interactions: hydrogen bounds and hydrophobic interactions (51,60,63,66,72).

4.4.3 Flo1 Protein

Flocculent lectins (also known as *flocculins*, *adhesins* or *zymolectins*) are coded by the so-called *FLO* genes. The open reading frame of the gene *FLO1* codes for a protein of 1537 amino acids (26). Flo1 protein has a hydrophobic N-terminus with a signal sequence and a C-terminal region, also hydrophobic, with a glycosylphosphatidylinositol (GPI) anchor consensus sequence. The middle part has repeated sequences rich in serine and threonine residues, which are potential sites for O-glicosylation and 14 potential N-glycosilation sites (12,26). Prediction of the secondary structure of Flo1 protein reveals that it is almost composed by β sheets and coils with N- and C-terminal regions with an α-helix conformation. The O-glycosilation of serine and threonine residues, in the middle part of the protein, increases the stiffness of the protein. This results in an extended conformation and allows the exposition of the N-terminus toward the cell surface (26,27). The region with repeated sequences seems to be important on the definition of the flocculation level: cells with greater number of repeats in *FLO1* gene show a higher flocculation (27,77). These repeats are highly instable and recombine at frequencies of about 10^{-5} per generation; this results in an increase or decrease of flocculation (78). It was proposed that the variation of repeating sequences enables yeast cell to adapt rapidly the adhesion properties to new environments (12,43,77). Recently, it has been shown that flocculating cells are more resistant to stress conditions induced by peroxide, ethanol, and amphotericin B than nonflocculent cells; stress resistance to amphotericin B increases with increasing number of tandem repeats in *FLO1* gene (79).

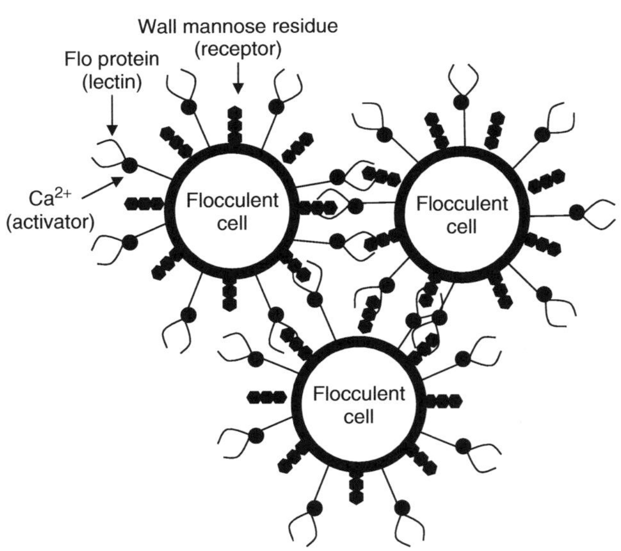

Figure 4.2. Lectin-like theory of flocculation in *S. cerevisiae*. Flocculent cells have specific surface proteins (lectins) that stick out of the yeast cell wall and bind to mannose residues (receptors) on neighboring cell walls. Calcium ions induce the correct conformation of the lectins.

Flo1 protein seems to be synthesized at the endoplasmic reticulum, glycosylated, and further transported through the secretory pathway; it is transiently anchored in the plasma membrane at the C-terminal region by a GPI anchor and stabilized by weak association with the cell wall by noncovalent bonding (28).

Immunofluorescent microscopy and immunoelectron microscopy showed that Flo1 protein is a structural protein located at yeast cell surface and straight involved in the flocculation process (27,28,80). Using avidin-fluorescein isothiocyanate (Avidin-FITC) probe, active cell surface lectins, can be visualized (63,81) (Fig. 4.3).

The N-terminal region of Flo1 protein contains the sugar recognition domain; more specifically, the region from 196 to 240 seems to be important for flocculation of Flo1 and NewFlo phenotypes (29).

4.4.4 Flocculation Phenotypes

Taking into account reversible inhibition studies induced by sugars, salt and low pH value, as well as proteases sensitiveness, two flocculation phenotypes were proposed (58): (i) Flo1 phenotype, for which flocculation is only inhibited by mannose and derivatives and (ii) NewFlo phenotype, for which flocculation is inhibited by mannose, maltose, glucose, and sucrose. Subsequent studies have shown that *FLO1* and Lg-*FLO1* genes encode to cell wall proteins. It is believed that these proteins govern flocculation ability of Flo1 and NewFlo phenotype, respectively (27–29,31,82).

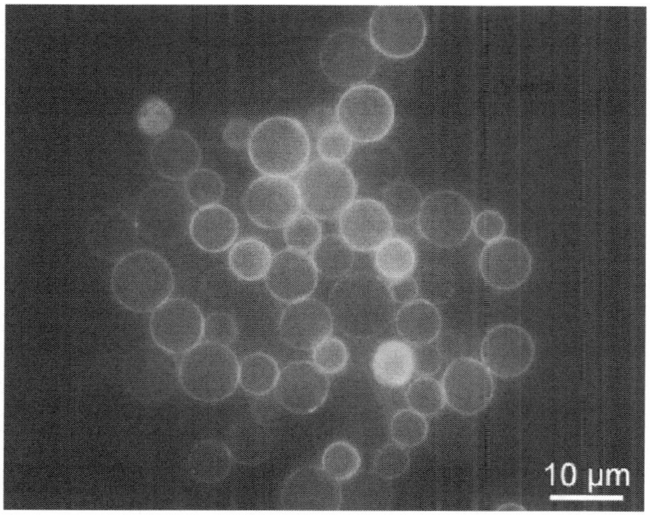

Figure 4.3. Photomicrograph of flocculent cells visualized by fluorescence microscopy. A green fluorescence (the brightest portions of the image are green fluorescence) is observed when a fluorescent probe of Avidin-fluorescein isothiocyanate (Avidin-FITC is a protein that contains a mannose side chain which functions as a binding site to cell wall flocculation lectins) is fixed on active lectins of flocculent cells of the ale brewing strain of *S. cerevisiae* NCYC 1195. (This figure is available in full color at http://onlinelibrary.wiley.com/book/10.1002/9780470054581.)

The analysis of the *N*-terminal region of Flo1 protein suggests that the domain formed by tryptophan 228 and its neighboring amino acids residues are of critical importance in the recognition of C-2 hydroxyl of the mannose, being not able to recognize the C-2 hydroxyl of the glucose. Similar analysis in Lg-Flo1 protein revealed that the domain formed by leucine 228 and its neighboring amino acid residues does not recognize the C-2 hydroxyl of mannose and glucose; in this protein, threonine 202 most likely interacts with C-2 hydroxyl of mannose and glucose, which allows its recognition (29).

Latter on, two other phenotypes have been described: MI phenotype, for which flocculation is not inhibited by sugars (83) and a flocculation phenotype, whose strains only flocculate in the presence of sufficiently high concentrations of ethanol (66,84). The strains, whose flocculation is induced by ethanol are top fermenting strains, do not require addition of calcium and can be mannose sensitive or insensitive (84). The exact mechanism of aggregation of these strains is far from being understood. Nevertheless, a lectin mechanism, as it has been reported for most brewing strains, should not be involved (66,84).

4.5 INDUCTIVELY VERSUS CONSTITUTIVELY FLOCCULENT STRAINS

With respect to flocculation properties, four classes of strains can be found: nonflocculent, constitutively flocculent, inductively flocculent, and strains constitutively flocculent after disruption of the *TUP1/SSN6* cascade. In the first case (nonflocculent), the strains lack an intact flocculation gene; the defect could be in the promoter, in the transcribed region or at both levels (30).

The inductively flocculent strains represent the ideal strains for brewing purposes and other fermentations operating in a batch or feed-batch mode. These strains belong to NewFlo phenotype and loss flocculation ability in the beginning of the growth recovering it toward the end of exponential phase (85,86) (Fig. 4.4). It was proposed that these strains correspond to a group of strains for which flocculation genes are transcriptionally regulated (30) by the protein complex Ssn6-Tup1 that acts as a general repressor in a regulatory cascade (87). Recent studies have shown that the loss of flocculation in starved cells or in the early phase of growth is an energy-dependent process. Flocculation loss is induced by several fermentable sources, including those found in wort: glucose, fructose, maltose, and sucrose (88–90). The loss of flocculation induced by glucose requires *de novo* protein synthesis and the presence of proteases; cycloheximide or phenylmethylsulphonylfluoride (PMSF) addition impairs the flocculation loss (89,90).

Since flocculation receptors seem to be present in all phases of growth (85,91,92), the onset of flocculation is most likely dependent of the presence of active lectins in yeast surface (93). In agreement with this supposition, Bony *et al*. (80), have shown that Flo proteins are not

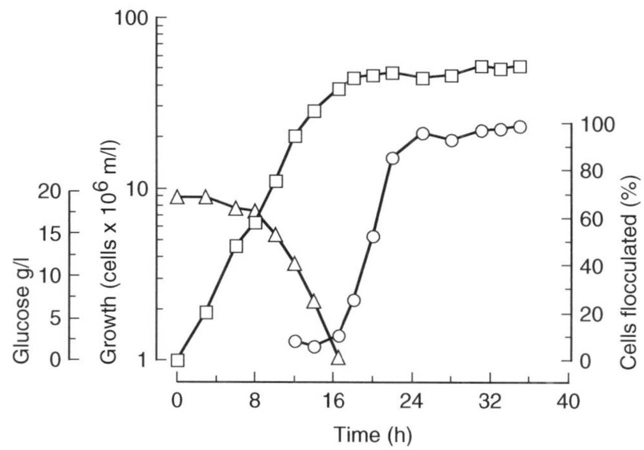

Figure 4.4. Onset of flocculation of a NewFlo phenotype strain of *S. cerevisiae* NCYC 1195. Cells were inoculated in yeast nitrogen base containing 2% (w/v) of glucose. Flocculation (circles) of washed cells in citrate buffer (pH 4.0, 50 mmol/l) containing Ca^{2+} (8 mmol/l) and glucose in culture medium (triangles) during the growth (squares) of the strain. The onset of flocculation occurs at the end of exponential phase of growth, which coincides with the attainment of the lowest amount of glucose in the culture medium. Reprinted with permission of the editor; from Sampermans *et al* (86).

permanently present in yeast surface; the amount of Flo proteins increases during the growth and can be correlated with the flocculation intensity. Flocculation properties seem to be defined primarily by the properties of specific flocculines rather than general cell wall properties (65).

The onset of flocculation occurs toward the end of fermentation, when low sugar and/or nitrogen concentration is reached in the culture medium (61,85,86) (Fig. 4.4). Other factors, namely, the shortage of sterols and unsaturated fatty acids seems to induce the flocculation in lager yeasts (62). The triggering of flocculation is an energetic dependent process (flocculation does not occur in the absence of a residual sugar concentration) and is influenced by the carbon source metabolism (flocculent phenotype was not observed in the presence of glycerol) (16,86). Nutrients depletion combined with the presence of ethanol can be the signal nutrition that induces the onset of flocculation (86,94).

Constitutively flocculent strains (Flo1 phenotype), flocculates during all phases of growth. This fact was verified when cells with different growth times were removed from the culture medium, washed and the flocculation was evaluated in standard conditions (85). Patelakis *et al*. (95) using a Avidin-FITC probe confirmed that a constant density of lectins was present during the growth of a strain belonging to Flo1 phenotype. Constitutively flocculent strains are insensitive to the presence of nutrients, namely, the surface inhibitory effect of sugars; in addition, they can flocculate over a wide range of pH (58,85,96). In these strains, the *FLO* genes are insensitive to the *TUP1/SSN6* cascade (30). Flo1 phenotype strains are the ideal ones to be used in bioreactors in a continuous mode. In these strains, cells are self-immobilized in pellets and there is no risk of washout of the biomass.

4.6 ENVIRONMENTAL FACTORS AFFECTING YEAST FLOCCULATION

Yeast flocculation is affected by a multiplicity of factors which can be clustered in genetic characteristics of the strain (*FLO* genes, suppressors, and activators already described) and in environmental factors. These last factors can act at different levels affecting flocculation genes expression, secretion of the lectins, cell–cell interactions or on yeast cell wall. Some of the complexity of flocculation control is due to the fact that some of these factors, such as the sugars, can act at more than one level: on *FLO* genes expression and on cell–cell interactions (Fig. 4.5).

4.6.1 Inorganic Salts

The effect of inorganic salts is one of the most studied and controversial aspects of yeast flocculation. For many strains, flocs can be dispersed with deionized water, after several washing steps, while for others strains flocs are only dispersed by the action of chelating agents such as EDTA. Addition of Ca^{2+} to EDTA dispersed cells promotes a rapid flocculation (15). Nowadays, there is a general consensus that among cations, Ca^{2+} is the most effective ion in the promotion of flocculation, by activation of the specific flocculation lectins (73,74,97,98). The amount of Ca^{2+} required to induce yeast flocculation seems to be strain dependent: for some strains, 10^{-8} mol/l of Ca^{2+} is sufficient to induce flocculation (97), while other strains require 5×10^{-4} mol/l of Ca^{2+} (99). It was shown that the available Ca^{2+} (i.e. the free and labile Ca^{2+}) is the only fraction able to induce the correct conformation of the lectins; the amount of available Ca^{2+} is influenced by the pH of the solution and the presence of complexing compounds (98,99).

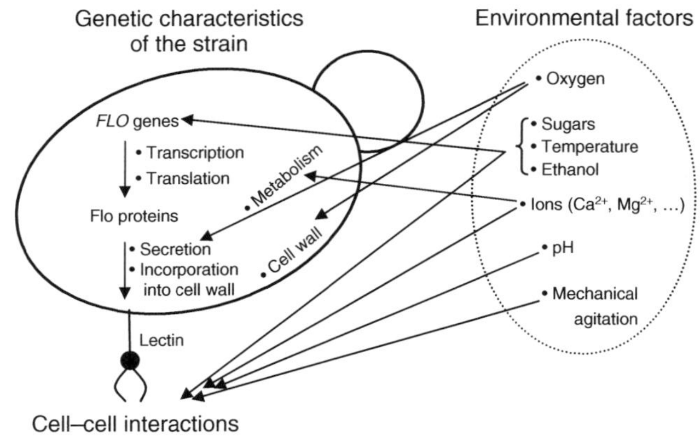

Figure 4.5. Diagram of the factors that influence flocculation of *S. cerevisiae*. Flocculation is mainly affected by genetic characteristics of the strains and by environmental factors that can act at different levels.

There is a strong evidence that the promotion of flocculation can be brought about by other ions, such as Rb^+, Cs^+, Fe^{2+},Co^{2+}, Cu^{2+}, Ni^{2+}, Zn^{2+}, Cd^{2+}, Al^{3+} and particularly by Mg^{2+} and Mn^{2+} (53,54,73,100). Conversely, the group 2 elements of periodic table, Ba^{2+} and Sr^{2+} and the heavy metal Pb^{2+} inhibited competitively flocculation promoted by calcium (53,57,81,101). Due to the similarity of their ionic ratio with Ca^{2+}, these ions compete most likely for the "calcium binding site" but they are not able to activate yeast flocculation lectin. Interestingly, Na^+ and K^+ induce flocculation at low concentrations, most likely by reducing the electrostatic repulsive forces between the cells and/or stimulating the leakage of intracellular Ca^{2+}; at high concentrations these ions provoke the antagonization of calcium-induced flocculation (53,74,101). It was proposed that the inhibition of yeast flocculation by high salt concentrations can be attributed to the distortion of the flocculation lectins caused by highly charged ions (16). Fe^{3+} inhibits irreversibly yeast flocculation probably due to a strong link to yeast flocculation lectins (102).

Besides the surface action, the presence of cations in culture medium (namely, Mg^{2+} or the Ca^{2+}/K^{+} ratio), as micronutrients, seems to be essential to the expression of flocculation (61,103).

4.6.2 pH Value

During wort fermentation, the pH value falls from 5.2–5.8 to 4.0–4.4, depending on whether it is a lager or an ale fermentation (104); for these pH values, yeast cells never lose their negative charge (16). Thus, it is generally accepted that, under brewing conditions, the pH value is not a determining factor of flocculation (16,105).

Nevertheless, the pH of the solution can have a deep effect on yeast flocculation and cause a reversible dispersion of the flocs, mainly at lower pH values; the modification of pH alters the ionization of amino acids which provokes a change in the conformation of the lectins (63,72,99). For several laboratory and industrial strains, flocculation occurs over a broad pH range, 2.5–9.0, while in many brewing yeasts, a subgroup of the NewFlo phenotype, only flocculate within a narrow pH range, 2.5–5.5. In both types of strains (wider or narrower pH range), the optimum pH value usually occurs between 3.0 and 5.0, depending on the strain (66,73,99,106).

In the case of strains that flocculate in a narrow range, the pH of culture medium is a very restrictive factor. Usually, these strains flocculates only when grown in wort, being nonflocculent in rich media yeast extract, peptone, dextrose (YEPD), for which pH at the end of the growth varies between 6.0 and 5.5. The same behavior occurs in unbuffered chemically defined media yeast nitrogen base (YNB), whose pH usually falls to 2.2 during the growth (106,107). In these culture media, flocculation can be restored by correcting the pH of the medium to a suitable value and for same strains by the simultaneous addition of calcium (Fig. 4.6) (99,106,107).

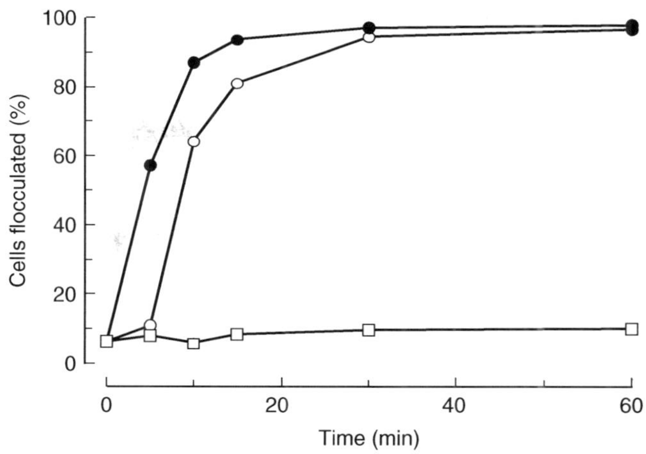

Figure 4.6. Induction of flocculation in cultures of a NewFlo phenotype strain of *S. cerevisiae* NCYC 1364, after 48 h of growth in yeast nitrogen base medium. Flocculation was induced (zero time of the figure) by: NaOH addition raising the pH of the culture medium to 4.0 (circles); Ca^{2+} (40 m*M*) addition and adjusting of pH of the culture medium to 4.0 (filled circles). Untreated control culture (squares). Reprinted with permission of the editor; from Soares and Seynaeve (99).

4.6.3 Sugars

Yeast flocculation can be reversible inhibited by specific sugars. This is the base of differentiation between Flo1 and NewFlo phenotypes, as it was described above. The sugars dispersive effect can be explained by its competition with the sugars of yeast cell wall (receptors) for the flocculation lectins, affecting directly the flocculent bond (57,58,73,83). It was suggested that during beer fermentations, the level of wort fermentable sugars and the occurrence of turbulent flow determine yeast flocculation (64).

On the other hand, the presence of fermentable sugars (including those found in wort), induce the loss of flocculation in the early phase of growth or in starved cells, via the metabolism of the yeast, most likely affecting the expression of *FLO* genes (88–90).

4.6.4 Oxygen

Moderate levels of oxygen have a positive effect on yeast flocculation (108,109). Conversely, a strong aeration (108,109) or anaerobiosis (62,69,109,110) leads to a reduction or a lack of flocculation in laboratory and industrial strains. In anaerobic conditions, *S. cerevisiae* is unable to synthesize unsaturated fatty acids and sterols, which affects the integrity, fluidity, and functionality of plasma membrane (111); the modification of plasma membrane functionality can affect the secretion of flocculation lectins and consequently yeast flocculation.

It was also suggested that reduction of flocculation of cells grown under anaerobiosis (110), with mitochondrial function affected by growing in the presence of antimycin (110) or acetylsalicylic acid (112) or in respiratory-deficient cells (110,113,114), may be due to modifications in the external layer of yeast cell wall. Cell wall mannoproteins seem to be differently expressed under aerobic or anaerobic conditions (115). Recently, it was proposed that the transition from aerobiosis to anaerobiosis, which occurs in brewing conditions, may be associated with modifications in the expression of genes that regulate or encode to flocculation lectins (116–118).

4.6.5 Temperature

Fermentation temperature can act at different levels and affect yeast flocculation, namely: yeast metabolism, cell–cell interaction (flocculent bond), and the expression of *FLO* genes.

In the first case, the lowering of fermentation temperature leads to a decrease of cellular metabolism resulting in a reduction of CO_2 production. As a consequence, turbulence of the medium declines and favors the settling of cells.

It is well described in the literature that the rise of temperature to 50–60°C promotes the reversible deflocculation of yeast cells (51,97), due to denaturation of flocculation lectins.

The growing of yeasts in supra-optimum temperatures (37°C) leads to a delay (119) or impair the flocculation expression (94,96). A brief heat shock (52°C, 5 min), in ale brewing strains, in exponential phase of growth, delayed the onset of flocculation (94), most likely by affecting, directly or indirectly *FLO* genes expression.

4.6.6 Ethanol

There is a considerable volume of work that evidences the positive effect of ethanol on yeast flocculation (60,63,66,72,79,84,86,90).

The mechanism through which ethanol influences flocculation, at a surface level, is still unclear. It was proposed that adsorption of ethanol to cell surface can decrease cell–cell electrostatic repulsions due to a reduction of the local dielectric constant (51,66). Another possibility is that the presence of ethanol allows the projection of polymer chains, which carry binding sites for nonspecific (hydrogen binding) or specific interactions (60,66). The positive effect of ethanol seems to be more pronounced in yeast strains with a strong surface hydrophobicity; thus, it was proposed that for these strains, organic compounds are better solvents than water (72). A slight increase of cell-surface hidrophobicity with the increase of ethanol concentration has also been reported (63).

Flocculent cells incubated in ethanol (4%, w/v), as sole carbon source, retain their flocculation properties, while the presence of fermentable sugars induce the loss of flocculation (89,90). In addition, the presence of ethanol at low concentration (2%, w/v) in the culture medium seems to contribute to the triggering of flocculation (86), while high ethanol concentration (10%, w/v) impairs the flocculation expression (94).

4.7 YEAST FLOCCULATION AND BIOTECHNOLOGICAL PROCESSES

4.7.1 Industrial and Environmental Applications of Yeasts

Yeasts have had a long and profitable association with food, biotechnology, and pharmaceutical industries. Historically, they are associated with the production of bread, alcoholic beverages (beer, wine and cider production), and fermented foods (120). Today, a wide range of yeast species have found industrial or environmental significance. Yeasts are used in the production of various metabolic compounds, such as: enzymes, vitamins (riboflavin), carotenoids, polysaccharides, and citric acid (121). Yeasts are also a source of ingredients and additives for food processing: antioxidants, aromas, flavors, and flavor enhancers (120). *S. cerevisiae* strains have the "status" of "generally recognized as safe" by the United States Food and Drug Administration (FDA); thus, the compounds produced by these yeasts are more acceptable in pharmaceutical and in food products for human consumption. Due to this fact, baker's and brewer's yeast preparations have been used in human dietary as a supplement nutrient (120). More than 2×10^{10} liters of ethanol per year are produced in the world by yeasts; this ethanol can be used for different industrial purposes like solvents for pharmaceutics or food ingredients as well as renewable energy source (122). Some yeast species have the ability to use hydrocarbons as sole sources of carbon and energy; due to this ability, they are used in biodegradation of oil spills on land and in aquatic environments (121). Another environmental application of yeast cells is their use in the removal of heavy metals from industrial wastewater as an alternative technology to classical physico-chemical methods (100,123). Nowadays, yeasts can be used as hosts for expression of recombinant DNA. Examples of heterologous (foreign) genes that have been cloned in *S. cerevisiae* and secreted by the recombinant yeast are human hormones (interleukin-α, calcitonin or insulin) and the hepatitis B virus surface antigen (HBsAg). The last resulted in the production of the first safe hepatitis B vaccine for use in humans (121).

4.7.2 The Importance of Solid/Liquid Separation Process

In most of the industrial and environmental applications mentioned above, after yeast growth and/or finishing its function, cells are in suspension and must be removed for further processing. Whatever the product may be (the biomass itself or an extracellular or an intracellular compound), the so-called downstream processing comprises the following steps: an initial separation of the bioreactor broth into a liquid and solid phase and a further extraction, concentration, and purification of the desired product. In all cases, the improvement and reducing of the costs associated with cell separation will benefit the overall efficiency and costs of the process.

The problem of biomass separation is due to the fact that yeast cells have a small size and density; the last one is only slightly greater than those of the reaction liquors in which they are suspended. Conventional solid–liquid separation methods are: sedimentation, filtration, centrifugation or flotation; however, these methods tend to be expensive and/or time consuming. In sedimentation process, large settling tanks are necessary due to extensive retention times required to obtain a cell-free supernatant (Fig. 4.7D–F). Cells can be removed by filtration. Continuous rotary vacuum filters are the most widely used in fermentation industries (124). The high costs associated with these equipments, such as those related with the capital investments of the filter systems, highly hinder their use in many industries or greatly increases the costs of the overall process; this point is particularly critical in the production of low cost, high volume products. Separation of yeast cells from fermented medium is also obtained by centrifugal forces; the two popular industrial centrifuges are: disc stack and tubular bowl centrifuges (124). Nevertheless, the supernatant obtained by centrifugation is not free of cells (10^3-10^5 cells/mL) and this separation process is considered a relatively expensive method taking into account the costs of maintenance and power demand per quantity of microorganisms removed (125). Flotation is another possibility of yeast cell separation. This process is based on the affinity of cells to air bubbles. The bubbling of air through the medium leads to the accumulation of cells on the surface where they can be collected (126). This cell separation process is relatively fast and efficient. Even so, the generation of bubbles to raise the particles to the surface needs energy input.

The precipitation of yeast cells have been attempted by the addition of natural (chitosan) or synthetic (cationic or anionic) polymers (127,128), inert powder (like nickel particles) (129), lectin Concanavalin A (130) or by the adhesion of fine magnetic particles followed by magnetic separation (131). Polyelectrolytes and polymeric particles provoke the yeast settling by interfering with cell surface charge; however, the addition of these compounds is not allowed in food industry because they are non-food grade constituents. Recently, another separation was described: nonflocculent yeast cells with flocculation receptors are settled by the addition of flocculent cells of *S. cerevisiae* in the fermented broth (92).

4.7.3 Immobilization of Yeasts

An alternative to traditional fermentation with free cells is the immobilization of yeasts. Cells can be immobilized in a porous polymeric matrix such as calcium and barium alginate, agar, polyacrylamide, polystyrene, polyvinyl alcohol, polyurethane or k-carrageenan beads (132,133). Adsorption of yeast cells to a support like (diethylamino)ethyl (DEAE) cellulose, pretreated glass, ceramics, wood chips or spent grains have also been described; cells are immobilized as a result of electrostatic attractions and hydrophobic interactions between the yeast cell wall and the support (132,134). In order to avoid desorption of the cells from the support, immobilization of yeast cells by covalent linkage, using a union agent such as glutaraldehyde, has been proposed (135).

Besides the costs associated with immobilization process, several works suggest differences, namely, related with the morphology and physiology, among immobilized and free cells; immobilization seems to affect yeast metabolic activity and flavor formation (136). Physiological changes can be due to modifications of physical and chemical environment around the immobilized cells or due to the immobilization process itself (137). The entrapment of yeast cells in polymer beads imposes mass transfer limitations (the diffusional limitations produce gradients of oxygen, substrates, and products), which are dependent of the type of polymer used as well as the porosity, texture, and the immobilized complex size (122). Alterations in cell wall and plasma membrane composition have also been reported in immobilized cells (137); these modifications seem to have a negative impact on the specific metabolic activity and stress resistance of the immobilized microorganisms (137). For a detailed review about yeast immobilizing processes, their uses in different fermentation systems and the impact of immobilization techniques on yeast cells and fermentation activity, the reader can consult the works published by Kourkoutas *et al*. (136), Strehaiano *et al*. (122) and Verbelen *et al*. (138).

4.7.4 Yeast Flocculation as a Self-Immobilization Process

Advantages of Yeast Flocculation. In the last years, there has been an increased and renewed interest in yeast flocculation as a spontaneous process of cell auto-immobilization. Yeast flocculation seems to have several advantages:

1. yeast growth is not significantly affected;
2. reuse of microorganisms for many cycles of production;
3. use of fermenters with high cell densities, which leads to high volumetric productivity and shorter fermentation times;
4. when operating in a continuous mode, there is a lower risk of external contamination due to the high cell densities and metabolic activity;
5. possibility of operation in a continuous mode with a minimized risk of biomass washout;
6. possibility of biomass control inside the bioreactor by simple purge of the flocs;
7. possibility of design of new bioreactors without sophisticated devices;
8. reduced bioreactor volume, which decreases capital costs;

9. easier separation of the biomass from the liquid phase at the end of the fermentation step, with the consequent reducing costs for equipment and energy demand.

Comparing with the most common immobilization techniques available, yeast flocculation is an easier and off-cost process of cell separation where no support (the main cost in immobilization procedures) or man-power is required. Yeast flocculation, is a natural and eco-friendly process of cell-broth separation, which does not require energy input (Fig. 4.7). These aspects, together with the advantages listed above, namely, the use of high cell density reactors, insert in the idea of "White" Biotechnology (4). This natural cell separation process contributes to sustainable industrial development since it increases the efficiency and reduces the energy consumption associated with the separation process. Yeast flocculation has a great potential of application

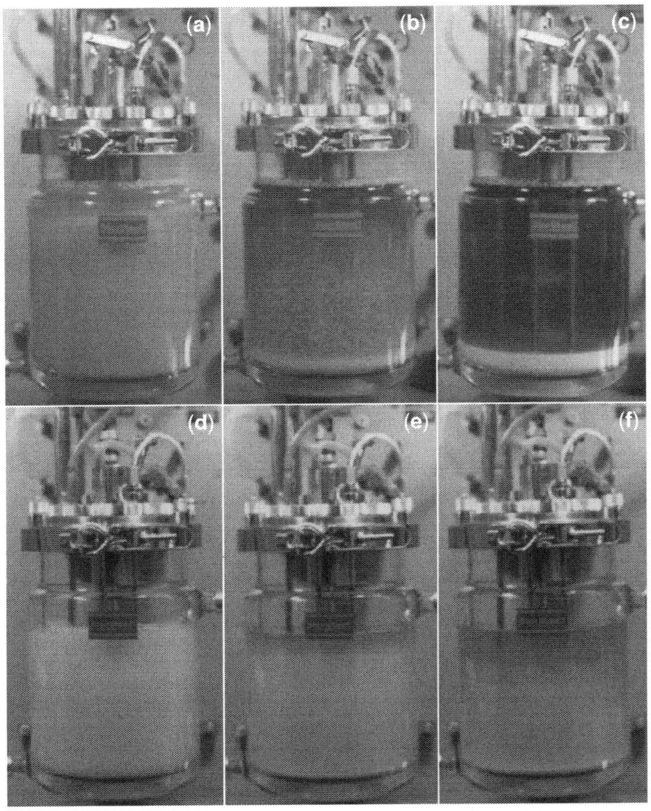

Figure 4.7. Laboratory fermentations carried out with a flocculent (upper panel) and a nonflocculent (lower panel) yeast strain. Upper panel: ale brewing yeast strain of *S. cerevisiae* NCYC 1195. (a) culture continuously aerated (0.1 vvm) and stirred (250 rpm); (b) and (c) 30 and 60 s, after aeration and stirring were stopped, respectively. Lower panel: strain of *S. cerevisiae* S646-8D. (d) culture continuously aerated (0.1 vvm) and stirred (250 rpm); (e) and (f) 1 and 2 h, after aeration and stirring were stopped, respectively. (This figure is available in full color at http://onlinelibrary.wiley.com/book/10.1002/9780470054581.)

in the production of high volume products, like the production of renewable fuels (ethanol), as it was recently described (139).

Taking into account all the benefits of yeast flocculation reported above, there is a significant financial advantage of extending the flocculation process behind to the brewing industry, to other industrial important processes involving yeasts and, in a wider way, involving microorganisms. This possibility has been attempted by different approaches: transference of the flocculation characteristic to other yeast strains or conversely, transference of important characteristics to flocculent strains.

4.7.4.1 Obtainment of Flocculent Strains with Desirable Properties. The transference of flocculation characteristics to other yeast strains by various classical genetic methods such as sexual mating or protoplast fusion of a nonflocculent industrial strain with a flocculent strain, was described (140). Nonflocculant industrial strains (such as alcohol production or brewing, and wine, whisky and sake making) and laboratory strains were modified by integrating the *FLO1* gene, using molecular biology techniques (141–144). Generally, *FLO1* gene is constitutively expressed in the transformed brewing strains (142). These transformed strains are not satisfactory for brewing proposes, because they slow down the fermentation rate (145); however, they can be useful in the production of ethanol in high cell density reactors operating in a continuous mode.

Another approach consists in the transference of important characteristic to flocculent strains. Javadekar *et al.* (146), described the construction of a flocculent yeast strain that produces a killer-toxin. In the same line, different recombinant strains of *S. cerevisiae* have been constructed with the ability to flocculate and secrete or display heterologous proteins in the surface (see below).

4.7.4.2 The Use of Inductively Flocculent Strains. The control of the triggering of flocculation can be particularly useful in many industrial processes. With modern gene technology, it is possible to insert *FLO1* gene in a nonflocculent strain under transcriptional control of appropriate promoters like the late-fermentation *HSP30* promoter (*HSP30p*) (147,148) or the glucose-repressible promoter (149); in these transformants, flocculation can be induced by heat shock treatment (15 min at 41°C), nitrogen depletion, pH fall, addition of ethanol (6%, v/v) (147,148) or glucose exhaustion (149), respectively.

Nowadays, the use and/or release of genetic modified (GM) yeasts by the industry requires a number of guarantees and are under various and complicated laws and guidelines. However, the main hindrance to industrial use in food industry lies with the negative public perception of GM organisms (no GM yeasts are used for food production) (147,150). In this context, the use of NewFlo phenotype strains can be an alternative solution. These strains show cyclic flocculation behavior (85,86) as it was described above (Inductively vs constitutively flocculent strains). In the metabolic active period, cells are in a nonflocculent state which avoids the problem of diffusional limitations on the transference of the substrate across the floc. The presence of flocs may affect mass and heat transfer characteristics and thus the productivity of fermentation (151). At the end of fermentation, cells spontaneously become flocculent, which facilitates further downstream process (Fig. 4.7A–C).

The NewFlo phenotypes strains, which flocculate only over a narrow pH range (99,106), can be particularly useful in industrial processes. For these strains, the pH switching allows the strain to be in a flocculate or a nonflocculate state. Thus, the strains can grow outside the favorable pH range as single cells; at the end of fermentation, the pH can be shifted by alkali or acid addition to an optimum pH range, that leads to a rapid induction of flocculation (Fig. 4.6). Thus, the control of onset of flocculation of these strains can be performed using this simple process of pH switching (99,152).

4.7.5 Examples of Applications of Yeast Flocculation

4.7.5.1 Brewing Industry. Flocculation characteristics are one of the most important properties to be considered when selecting a yeast strain for brewing purposes (153–155). For a brewer, two key aspects related with yeast flocculation should be considered: the degree of flocculation and the time during the fermentation process of the onset of flocculation.

Fermentation starts with the inoculation of the yeast cells (in beer terminology, known as *pitching*) into the sweet wort. The principal yeast function is the conversion of wort sugars into alcohol (*attenuation*), which is dependent of the concentration of suspended yeast, producing a "green" or immature beer. When the fermentation is judged to be completed, yeast cells are removed by self-flocculation or by centrifugation. Ideally, the onset of flocculation should occur when there is depletion of the sugars in the medium. The separation of yeast cells allows their future use, limits the wort nutrients to yeast cells and leads to a decrease of the number of cells to the green beer for further maturing process (this process is named *conditioning*, *aging*, or *lagering)*.

Ale brewing strains (*S. cerevisiae*) (1) classically rise to the top of fermenter most likely due to the affinity for the CO_2 bubbles (top fermentation). On the other hand, lager strains (*Saccharomyces pastorianus*, which due to historical reasons is referred as *Saccharomyces carlsbergensis*) (1) settle in the bottom of the fermenter (bottom fermentation). With the advent of the use of vertical cylindrico-conical fermentation vessels, yeasts with increased sedimentation ability (ale or lager) tend to sediment in the cone of the fermenter. After, they are removed for washing and subsequent *repitching* (154).

Among many brewing properties, yeast flocculation is generally seen as the most instable property (153). In the presence of high levels of sugars, early or premature yeast flocculation (PYF), leads to incomplete extract utilization (poor attenuation). PYF is a recurring problem in the brewing industry and one of the common causes of "hung" or "stuck" fermentations. This fact originates a sweeter and less fermented beers with low alcohol contents (153). PYF can be caused by: (i) a successive selection of more aged and flocculent yeasts (156) or the accumulation of respiratory-deficient mutants with an increased flocculation ability (157); (ii) bacterial infection, or (iii) by the action of wort components, more specifically high molecular weight polysaccharides rich in arabinose and xylose (158), which can act as a bridge between cells (16). It was suggested that these compounds are generated by an enzymatic degradation of barley husk due to the fungal contamination during malting (159).

Conversely, a delay or a lack of flocculation can cause problems in beer clarification. Difficulties during filtration can complicate the obtainment of a bright sparkling beer. Additionally, the presence of an excess of yeast during aging can cause off-flavors due to the yeast autolysis (153,154). A long-term study of a flocculent lager brewing strain, belonging to a NewFlo phenotype, showed that flocculation of this yeast tended to gradually decrease with successive generations. This is due to genetic alterations, such as chromosome deletion, *C*-terminal region deletion, partial deletion of the middle region of the Lg-*FLO1* gene or some defects in the posttranscriptional regulations (160). It was proposed that nonflocculent cells can grow dominantly through successive cultivation (160). Recently, Powell and Diacetis (161) showed that even though some brewing yeast strains are vulnerable to genetic variability, others are more stable. These authors did no observe any correlation between genomic variations and modifications of fermentation and flocculation characteristics among recent propagated ale or lager strains and after 98 or 135 serial repitchings, respectively.

In modern breweries, in the absence of flocculation, yeast cells can be removed from the green beer by filtration or centrifugation, apparently minimizing the importance of yeast flocculation. Nevertheless, these processes increase the operating costs and the passage of brewer's yeast cells through a disc stage centrifuge, at high *g*-force, have a negative impact on cell viability and physiology and beer physical stability (162). Hence, flocculation remains as an easy and off-cost method of cell separation that originates a clear and brilliant beer. Furthermore, for the brewer, yeast flocculation has an importance beyond the cell separation process because its absence can adversely affect the flavor characteristics, alcohol content, and thus, the final product.

4.7.5.2 Other Alcoholic Beverages Production. Yeast flocculation property has also been exploited in the production of wine and, particularly, in bottle-fermented sparkling wine.

In wine production, it is desirable that cells maintain a nonflocculent state in order to ensure a rapid fermentation rate. After consumption of sugars, the settling of cells in a flocculated state facilitates enormously their removal from the fermented product. This behavior allows a rapid and efficient clarification and reduces the handling of wine. Furthermore, flocculation seems to be associated with the enhancement of ester production (163). The fermenting of grape can be carried out by the native flora present in the winery or on the grape; nowadays, most wine fermentations are initiated by inoculating grape must with a pure commercial selected yeast strain (3). Flocculation is one of the main characteristics that the starter culture should have (163).

Another example of the use of flocculation properties is the production of sparkling wine. In the called "*Method Champenoise*," two fermentations occur: an initial fermentation, which is most often placed in stainless steel tanks, and a secondary fermentation in the bottle itself. During the secondary fermentation, yeast cells sediment. The removal of this sediment without removing the wine is a skilled process which requires a special know-how. The bottles are inverted and the settled cells accumulate on the top of the bottle neck. For the second fermentation, flocculent cells are most suitable since they facilitate the delicate phase of removing the yeasts from the bottles (164,165), a process called *dégorgement*.

4.7.5.3 Bio-Ethanol Production. In a general way, bio-ethanol industry uses a batch process, particularly in plants with small production capacities (166). Yeast cells are usually removed by centrifugation. In order to reduce the costs associated with the separation process, different processes of cell immobilization have been attempted. Immobilization of yeast cells, especially by gel entrapment, is not attractive for the production of bio-ethanol (139). Conversely, the use of flocculent yeast cells seems to be a very promissory technology, allowing to an economy of 16% of processing costs and 10% of installation costs associated with bio-ethanol production (167).

Among the techniques developed to achieve high cell density bioreactors, yeast flocculation presents several advantages regarding operational costs and simplicity. Flocculation bioreactors allow an increase of productivity keeping the substrate conversion and product concentration maximal (168). The use of flocculent cells in reactors of high cell density, reduce the number and size of required bioreactors, due to the increase of productivity. Many configurations of bioreactors with flocculent cells have been developed for continuous ethanol production: bubble columns, airlift reactors, packed and fluidized beds, being the first ones the most widely used (139,151). The different configurations of bioreactors and operation modes used with flocculent cells were discussed and reviewed by Domingues *et al*. (151).

Most of the works using high density cell reactors have been conducted at a laboratory scale (108,167,169–171). However, in 2005, a fuel ethanol plant, with an annual production of 200,000 tons, using flocculent yeast cells, has been started; the process consists in six bioreactors, arranged in a cascade mode, each one working with a volume of 1000 m^3 (139).

4.7.5.4 Production of Heterologous Proteins. Various heterologous proteins were displayed in *S. cerevisiae* cell surface, using recombinant DNA techniques (172). Yeasts with enzymes immobilized on their cell surface can also be used as whole-cell biocatalysts for several applications.

A flocculent strain of *S. cerevisiae* displaying a cell surface glucoamylase enables the production of ethanol from soluble starch without the previous stage of starch saccharification (173). Another possibility of using recombinant strains is the production of protein by secretion (174). The production of extracellular heterologous β-galactosidase using recombinant flocculent cells, in a high cell density bioreactor, operating in a continuous mode, was described (170). In both cases, cell separation is highly facilitated due to the flocculent properties of the recombinant strains.

4.7.5.5 Environmental Application. Heavy metal pollution mainly caused by the release of industrial effluents constitutes a worldwide environmental problem. The current technologies (precipitation, ion exchange, and reverse osmosis) have been shown inadequate or expensive. Therefore, different types of biomass (namely, yeast cells) have been proposed as the base of a new technology to remove heavy metals (175). A common difficulty related with the use of microorganisms is the rapid and efficient separation of the biomass, loaded with heavy metals, after the effluent had been treated.

The use of flocculent yeast biomass seems to be a promissory approach in the heavy metals biorremediation (52,100,176). Flocculent yeast cells have a higher ability of metal accumulation than the nonflocculent cells; most likely, heavy metals can also occupy lectin Ca^{2+} binding sites with the consequent increase in heavy metals uptake (100). This type of biomass combines an efficient removal of metals with an inexpensive and rapid separation of biomass after effluent treatment (52). As it was already mentioned, separation process using flocculent yeast does not require an energy input. This is a critical issue when we are looking for efficient and low cost technologies.

4.8 CONCLUSIONS

Microbial aggregation can be assumed to be widespread in nature. Yeast flocculation is a particular case of yeast aggregation. This property is well known in the brewing industry due to the fact that yeast flocculation characteristics facilitates the downstream process and, in a broader way, affects deeply beer properties. However, yeast flocculation is a more general phenomenon within yeasts, whose usefulness potential remains to be exploited.

The use of flocculent cells raises the possibility of exploiting different fermenter configurations and novel fermentation designs. Until now, the efforts of using yeast flocculent cells have been mainly concentrated in continuous-fermentations systems with constitutively flocculent strains. Another approach can be the use of appropriate strains with a NewFlo phenotype, in fermentations operating in batch or fed-batch mode. The onset of flocculation of these strains can be triggered by manipulating the culture conditions (adjusting the sugars concentration or the pH of the medium). Even though, modification of these fermentation parameters (or others like fermentation temperature or ethanol concentration) can only be slightly changed in beer production, as they affect beer characteristics, they can be useful in other industrial processes where the fermented broth is not the bio-product.

The use of genetic engineering raises the possibility of obtaining strains with the *FLO1* gene under the control of a suitable promoter. This manipulation allowed the expression of flocculation in controlled conditions. Until now, the use of genetically modified organisms obtained by recombinant DNA methods is not generally accepted by the consumer. Thus, they are being only devisable in other industries besides the food industry, like bio-ethanol industry.

The improvement of the knowledge of yeast flocculation control in its multiple aspects (namely, at a regulation of *FLO* genes level), creates the opportunity of manipulating this useful and fascinating property. The application of this tool enlarges the possibility of using flocculation, as a separation process, in other biotechnological industries, besides the well known alcoholic beverages production.

REFERENCES

1. Vaughan-Martini A, Martini A, Kurtzman CP, Fell JW, editors. The Yeasts: a taxonomic study. Amsterdam: Elsevier; 1998, pp 358–371.
2. Nielsen J, Jewett MC FEMS Yeast Res 2008; 8: 122–131.
3. Pizarro F, Vargas FA, Agosin E Yeast 2007; 24: 977–991.
4. Paula L, Birrer F J Agric Environ Ethics 2006; 19: 253–267.
5. Lasa I Intern Microbiol 2006; 9: 21–28.
6. Maximova N, Dahl O Curr Opin Colloid Interface Sci 2006; 11: 246–266.
7. Liu YQ, Liu Y, Tay JH Appl Microbiol Biotechnol 2004; 65: 143–148.
8. Calleja GB, Rose AH, Harrison JS, editors. The yeasts. London: Academic Press, Inc.; 1987. pp 165–238.
9. Kaushik S, Katoch B, Nanjundiah V Behav Ecol Sociobiol 2006; 59: 521–530.
10. Xu XH, Kumakura M, Kaki E, Takahashi M J Eukaryot Microbiol 2001; 48: 683–689.
11. Calleja GB Colloids Surf B: Biointerfaces 1994; 2: 133–149.
12. Dranginis AM, Rauceo JM, Coronado JE, Lipke PN Microbiol Mol Biol Rev 2007; 71: 282–294.
13. Stewart GG, Garrison IF Proc Am Soc Brew Chem 1972; 3: 118–131.
14. Nishihara H, Kio K, Imamura M J Inst Brew 2000; 106: 7–10.
15. Stewart GG Brewing Digest 1975; 50: 42–62.
16. Stratford M Adv Microb Physiol 1992; 33: 1–72.
17. Kang HY, Kim YS, Seo JH, Ryu YW J Microbiol Biotechnol 2006; 16: 1874–1881.
18. Saito K, Sato S, Shimoi H, Iefuji H, Tadenuma M Agric Biol Chem 1990; 54: 1425–1432.
19. Géhin G, Bonaly R, Coulon J FEMS Microbiol Lett 2001; 203: 229–233.
20. Farias ME, de Nadra MCM J Appl Microbiol 2003; 95: 457–462.
21. Stratford M, Pearson BM Lett Appl Microbiol 1992; 14: 214–216.
22. Johnson BF, Sowden LC, Walker T, Yoo BY, Calleja GB Can J Microbiol 1989; 35: 1081–1086.
23. Suzzi G, Romano P, Benevelli M Anton Leeuw Int J G 1992; 61: 317–322.
24. Teunissen AWRH, Van Den Berg JA, Steensma HY Yeast 1993; 9: 1–10.
25. Teunissen AWRH, Steensma HY Yeast 1995; 11: 1001–1013.
26. Watari J, Takata Y, Ogawa M, Sahara H, Koshino S, Onnela M, Airaksinen U, Jaatinen R, Penttilä M, Keränen S Yeast 1994; 10: 211–225.
27. Bidard F, Bony M, Blondin B, Dequin S, Barre P Yeast 1995; 11: 809–822.
28. Bony M, Thines-Sempoux D, Barre P, Blondin B J Bacteriol 1997; 179: 4929–4936.
29. Kobayashi O, Hayashi N, Kuroki R, Sone H J Bacteriol 1998; 180: 6503–6510.
30. Teunissen AWRH, Van Den Berg JA, Steensma HY Yeast 1995; 11: 435–446.
31. Sato M, Maeba H, Watari J, Takashio M J Biosci Bioeng 2002; 93: 395–398.
32. Ogata T, Izumikawa M, Kohno K, Shibata K J Appl Microbiol 2008; 105: 1186–1198.
33. Kobayashi O, Suda H, Ohtani T, Sone H Mol Gen Genet 1996; 251: 707–715.
34. Kobayashi O, Yoshimoto H, Sone H Curr Genet 1999; 36: 256–261.
35. Bester MC, Pretorius IS, Bauer FF Curr Genet 2006; 49: 375–383.

36. Fichtner L, Schulze F, Braus GH Mol Microbiol 2007; 66: 1276–1289.
37. Mulders SEV, Christianen E, Saerens SMG, Daenen L, Verbelen PJ, Willaert R, Verstrepen KJ, Delvaux FR FEMS Yeast Res 2009; 9: 178–190.
38. Lo W-S, Dranginis AM J Bacteriol 1996; 178: 7144–7151.
39. Lo W-S, Dranginis AM Mol Biol Cell 1998; 9: 161–171.
40. Guo B, Styles CA, Feng Q, Fink G Proc Natl Acad Sci USA 2000; 97: 12158–12163.
41. Cullen PJ, Sprague GF Proc Natl Acad Sci U S A 2000; 97: 13619–13624.
42. Bayly JC, Douglas LM, Pretorius IS, Bauer FF, Dranginis AM FEMS Microbiol Lett 2005; 5: 1151–1156.
43. Verstrepen KJ, Klis FM Mol Microbiol 2006; 60: 5–15.
44. Strittmatter AW, Fischer C, Kleinschmidt M, Braus GH Mol Genet Genomics 2006; 276: 113–125.
45. Fischer C, Valerius O, Rupprecht H, Dumkow M, Krappmann S, Braus GH FEMS Yeast Res 2008; 8: 225–236.
46. Klis FM, Mol P, Hellingwerf K, Brul S FEMS Microbiol Rev 2002; 26: 239–256.
47. Cabib E, Roh D, Schmidt M, Crotti LB, Varma A J Biol Chem 2001; 276: 19679–19682.
48. Smits GJ, Kapteyn JC, Van Den Ende H, Klis FM Curr Opin Microbiol 1999; 2: 348–352.
49. Klis FM, Boorsma A, De Groot PWJ Yeast 2006; 23: 185–202.
50. Lesage G, Bussey H Microbiol Mol Biol Rev 2006; 70: 317–343.
51. Mill PJ J Gen Microbiol 1964; 35: 61–68.
52. Machado MD, Santos MSF, Gouveia C, Soares HMM, Soares EV Bioresour Technol 2008; 99: 2107–2115.
53. Nishihara H Arch Microbiol 1982; 131: 112–115.
54. Sousa MJ, Teixeira JA, Mota M Biotechnol Lett 1992; 14: 213–218.
55. Nishihara H Arch Microbiol 1977; 115: 19–23.
56. Nishihara H, Miyake K, Kageyama Y J Inst Brew 2002; 108: 187–192.
57. Kihn JC, Masy CL, Mestdagh MM Can J Microbiol 1988; 34: 773–778.
58. Stratford M, Assinder S Yeast 1991; 7: 559–574.
59. Nishihara H, Toraya T Agric Biol Chem 1987; 51: 2721–2726.
60. Amory DE, Rouxhet PG, Dufour JP J Inst Brew 1988; 94: 79–84.
61. Smit G, Straver MH, Lugtenberg BJJ, Kijne JW Appl Environ Microbiol 1992; 58: 3709–3714.
62. Straver MH, Aar PCVD, Smit G, Kijne JW Yeast 1993; 9: 527–532.
63. Jin Y, Ritcey LL, Speers RA J Am Soc Brew Chem 2001; 59: 1–9.
64. Speers RA, Wan Y-Q, Jin Y, Stewart RJ J Inst Brew 2006; 112: 246–254.
65. Govender P, Domingo JL, Bester MC, Pretorius IS, Bauer FF Appl Environ Microbiol 2008; 74: 6041–6052.
66. Dengis PB, Nélissen LR, Rouxhet PG Appl Environ Microbiol 1995; 61: 718–728.
67. Hsu JWC, Speers RA, Paulson AT Biophys Chem 2001; 94: 47–58.
68. Jin Y, Speers RA Food Res Int 1998; 31: 421–440.
69. Miki BLA, Poon NH, Seligy VL J Bacteriol 1982; 150: 890–899.
70. Stratford M, Keenan HJ Yeast 1988; 4: 107–115.
71. Soares EV, Mota M J Inst Brew 1997; 103: 93–98.
72. Jin Y, Speers RA J Am Soc Brew Chem 2000; 58: 108–116.
73. Miki BLA, Poon NH, James AP, Seligy VL J Bacteriol 1982; 150: 878–889.
74. Stratford M Yeast 1989; 5: 487–496.
75. Stratford M Yeast 1992; 8: 635–645.
76. Touhami A, Hoffmann B, Vasella A, Denis FD, Dufrêne YF Microbiology 2003; 149: 2873–2878.
77. Verstrepen KJ, Jansen A, Lewitter F, Fink GR Nat Genet 2005; 37: 986–990.
78. Rando OJ, Verstrepen KJ Cell 2007; 128: 655–668.
79. Smukalla S, Caldara M, Pochet N, Beauvais A, Guadagnini S, Yan C, Vinces MD, Jansen A, Prevost MC, Latge JP, Fink GR, Foster KR, Verstrepen KJ Cell 2008; 135: 726–737.
80. Bony M, Barre P, Blondin B Yeast 1998; 14: 25–35.
81. Gouveia C, Soares EV J Inst Brew 2004; 110: 141–145.
82. Teunissen AWRH, Holub E, Van Der Hucht J, Van Den Berg JA, Steensma HY Yeast 1993; 9: 423–427.
83. Masy CL, Henquinet A, Mestdagh MM Can J Microbiol 1992; 38: 1298–1306.
84. Dengis PB, Rouxhet PG J Inst Brew 1997; 103: 257–261.
85. Soares EV, Mota M Can J Microbiol 1996; 42: 539–547.
86. Sampermans S, Mortier J, Soares EV J Appl Microbiol 2005; 98: 525–531.
87. Keleher CA, Redd MJ, Schultz J, Carlson M, Jonhson AD Cell 1992; 68: 709–719.
88. Soares EV, Duarte AA Biotechnol Lett 2002; 24: 1957–1960.
89. Soares EV, Vroman A J Appl Microbiol 2003; 95: 325–330.
90. Soares EV, Vroman A, Mortier J, Rijsbrack K, Mota M J Appl Microbiol 2004; 96: 1117–1123.
91. Stratford M Yeast 1993; 9: 85–94.
92. Mortier A, Soares EV World J Microbiol Biotechnol 2007; 23: 1401–1407.
93. Stratford M, Carter AT Yeast 1993; 9: 371–378.
94. Claro FB, Rijsbrack K, Soares EV J Appl Microbiol 2007; 102: 693–700.
95. Patelakis SJJ, Ritcey LL, Speers RA Lett Appl Microbiol 1998; 26: 279–282.
96. Soares EV, Teixeira JA, Mota M Can J Microbiol 1994; 40: 851–857.
97. Taylor NW, Orton WL J Inst Brew 1975; 81: 53–57.
98. Masy CL, Kockerols M, Mestdagh MM Can J Microbiol 1991; 37: 295–303.
99. Soares EV, Seynaeve J Biotechnol Lett 2000; 22: 1827–1832.
100. Soares EV, DeConick G, Duarte F, Soares HMVM Biotechnol Lett 2002; 24: 663–666.
101. Kuriyama H, Umeda I, Kobayashi O Can J Microbiol 1991; 37: 397–403.
102. Nishihara H, Fujita T, Yokoi N, Takao M J Inst Brew 1994; 100: 427–430.
103. Amri MA, Bonaly R, Duteurtre B, Moll M J Gen Microbiol 1982; 128: 2001–2009.

104. Lewis MJ, Young TW, Lewis MJ, Young TW, editors. Brewing. London: Chapman & Hall; 1995, pp 160–172.
105. Verstrepen KJ, Derdelinckx G, Verachtert H, Delvaux FR Appl Microbiol Biotechnol 2003; 61: 197–205.
106. Stratford M FEMS Microbiol Lett 1996; 136: 13–18.
107. Soares EV, Seynaeve J Biotechnol Lett 2000; 22: 859–863.
108. Kida K, Yamadaki M, Asno S, Nakata T, Sonoda Y J Ferment Bioeng 1989; 68: 107–111.
109. Soares EV, Teixeira JA, Mota M Biotechnol Lett 1991; 13: 207–212.
110. Iung AR, Coulon J, Kiss F, Ekome JN, Vallner J, Bonaly R Appl Environ Microbiol 1999; 65: 5398–5402.
111. Rose AH J Appl Bacteriol Symp Suppl 1993; 74: 110S–118S.
112. Strauss CJ, van Wyk PWJ, Lodolo EJ, Botes PJ, Pohl CH, Nigam S, Kock JLF J Inst Brew 2007; 113: 42–47.
113. Hinrichs J, Stahl U, Esser K Appl Microbiol Biotechnol 1988; 29: 48–54.
114. Ernandes JR, Williams JW, Russell I, Stewart GG J Am Soc Brew Chem 1993; 51: 16–20.
115. Abramova N, Sertil O, Mehta S, Lowry CV J Bacteriol 2001; 183: 2881–2887.
116. Smart KA, Lawrence SJ, Leclaire J, Davy S, Contribution 18, CD ROM 2006, Institute of Brewing and Distilling Asia Pacific, Hobart, 2006.
117. Gibson BR, Lawrence SJ, Leclaire JPR, Powell CD, Smart KA FEMS Microbiol Rev 2007; 31: 535–569.
118. Lawrence SJ, Smart KA J Am Soc Brew Chem 2007; 65: 208–213.
119. Williams JW, Ernandes JR, Stewart GG Biotechnol Tech 1992; 6: 105–110.
120. Fleet GH, Querol A, Fleet GH, editors. Yeasts in food and beverages. Heidelberg: Springer; 2006, pp 1–12.
121. Demain AL, Phaff HJ, Kurtzman CP, Kurtzman CP, Fell JW, editors. The yeasts, a taxonomic study. Amsterdam: Elsevier Science B.V.; 1998, pp 13–19.
122. Strehaiano P, Ramon-Portugal F, Taillandier P, Querol A, Fleet GH, editors. Yeasts in food and beverages. Heidelberg: Springer; 2006, pp 243–283.
123. Wang J, Chen C Biotechnol Adv 2006; 24: 427–451.
124. Shuler ML, Kargi F, Amundson NR, editors. Bioprocess engineering. New Jersey: Prentice Hall PTR; 2002, pp 329–384.
125. Hatti-Kaul R, Mattiasson B, Ratledge C, Kristiansen B, editors. Basic biotechnology. Cambridge: Cambridge University Press; 2001, pp 187–211.
126. Zouboulis AI, Matis KA Crit Rev Env Sci Technol 1997; 27: 195–235.
127. Weir S, Ramsden DK, Hughes J, Le Thomas F Biotechnol Tech 1993; 7: 199–204.
128. Kim JS, Akeprathumchai S, Wickramasinghe SR J Membr Sci 2001; 182: 161–172.
129. Weeks MG, Munro PA, Spedding PL Biotechnol Bioeng 1983; 25: 687–697.
130. Stratford M, Bond CJ Biotechnol Bioeng 1992; 40: 835–843.
131. Dauer RR, Dunlop EH Biotechnol Bioeng 1991; 37: 1021–1028.
132. Ramakrishna SV, Prakasham RS Curr Sci 1999; 77: 87–100.
133. Ting YP, Sun G J Chem Technol Biotechnol 2000; 75: 541–546.
134. Brányik T, Vicente AA, Machado-Cruz JM, Teixeira JA Biotechnol Lett 2001; 23: 1073–1078.
135. Jirku V J Ind Microbiol Biotechnol 1999; 22: 147–151.
136. Kourkoutas Y, Bekatorou A, Banat IM, Marchant R, Koutinas AA Food Microbiol 2004; 21: 377–397.
137. Shen HY, Moonjai N, Verstrepen KJ, Delvaux FR J Am Soc Brew Chem 2003; 61: 79–87.
138. Verbelen PJ, De Schutter DP, Delvaux FR, Verstrepen KJ, Delvaux FR Biotechnol Lett 2006; 28: 1515–1525.
139. Bai FW, Anderson WA, Moo-Young M Biotechnol Adv 2008; 26: 89–105.
140. Watari J, Kudo M, Nishikawa N, Kamimura M Agric Biol Chem 1990; 54: 1677–1681.
141. Watari J, Takata Y, Ogawa M, Murakami J, Koshino S Agric Biol Chem 1991; 55: 1547–1552.
142. Watari J, Nomura M, Sahara H, Koshino S, Keranen S J Inst Brew 1994; 100: 73–77.
143. Remize F, Schorr-Galindo S, Guiraud JP, Dequin S, Blondin B Biotechnol Lett 1998; 20: 313–318.
144. Wang FZ, Shen W, Rao ZM, Fang HY, Zhan XB, Zhuge J Biotechnol Lett 2008; 30: 97–102.
145. Hammond JRM Yeast 1995; 11: 1613–1627.
146. Javadekar VS, Sivaraman H, Gokhale DV J Ind Microbiol 1995; 15: 94–102.
147. Verstrepen KJ, Bauer FF, Winderickx J, Derdelinckx G, Dufour JP, Thevelein JM, Pretorius IS, Delvaux FR Cerevisia 2001; 26: 89–97.
148. Verstrepen KJ, Derdelinckx G, Delvaux FR, Winderickx J, Thevelein JM, Bauer FF, Pretorius IS J Am Soc Brew Chem 2001; 59: 69–76.
149. Cunha AF, Missawa SK, Gomes LH, Reis SF, Pereira GAG FEMS Yeast Res 2006; 6: 280–287.
150. Verstrepen KJ, Chambers PJ, Pretorius IS, Querol A, Fleet GH, editors. The yeast handbook. Heidelberg: Springer; 2006, pp 399–443.
151. Domingues L, Vicente AA, Lima N, Teixeira JA Biotechnol Bioprocess Eng 2000; 5: 288–305.
152. Stratford M Biotechnol Gen Eng Rev 1992; 10: 283–341.
153. Stewart GG, Russell I, Pollock JAR, editors. Brewing science, food science and technology. New York: Academic Press; 1981, pp 61–92.
154. Stewart GG, Russell I J Inst Brew 1986; 92: 537–558.
155. Donalies UEB, Nguyen HTT, Stahl U, Nevoigt E Food Biotechnol 2008; 111: 67–98.
156. Powell CD, Quain DE, Smart KA FEMS Yeast Res 2003; 3: 149–157.
157. Gyllang H, Martinson E Eur Brew Conv Proc Cong 1971; 13: 265–271.
158. Herrera VE, Axcell BC J Inst Brew 1991; 97: 359–366.
159. Van Nierop SNE, Cameron-Clarke A, Axcell BC J Am Soc Brew Chem 2004; 62: 108–116.
160. Sato M, Watari J, Shinotsuka K J Am Soc Brew Chem 2001; 59: 130–134.
161. Powell CD, Diacetis AN J Inst Brew 2007; 113: 67–74.

162. Chlup PH, Bernard D, Stewart GG J Inst Brew 2008; 114: 45–61.
163. Pretorius IS Yeast 2000; 16: 675–729.
164. Martinez-Rodriguez A, Carrascosa AV, Barcenilla JM, Pozo-Bayon MA, Polo MC Food Microbiol 2001; 18: 183–191.
165. Valles BS, Bedrinana RP, Queipo AL, Alonso JJM Food Microbiol 2008; 25: 690–697.
166. Wheals AE, Basso LC, Alves DMG, Amorim HV Trends Biotechnol 1999; 17: 482–487.
167. Andrietta SR, Steckelberg C, Andrietta MDS Bioresour Technol 2008; 99: 3002–3008.
168. Teixeira JA, Mota M, Goma G Bioprocess Eng 1990; 5: 123–127.
169. Sousa ML, Mota M, Teixeira JA Colloids Surf B: Biointerfaces 1994; 2: 181–188.
170. Domingues L, Lima N, Teixeira JA Process Biochem 2005; 40: 1151–1154.
171. Xu TJ, Zhao XQ, Bai FW Enzyme Microb Technol 2005; 37: 634–640.
172. Schreuder MP, Mooren ATA, Toschka HY, Verrips CT, Klis FM Trends Biotechnol 1996; 14: 115–120.
173. Kondo A, Shigechi H, Abe M, Uyama K, Matsumoto T, Takahashi S, Ueda M, Tanaka A, Kishimoto M, Fukuda H Appl Microbiol Biotechnol 2002; 58: 291–296.
174. Domingues L, Teixeira JA, Penttila M, Lima N Appl Microbiol Biotechnol 2002; 58: 645–650.
175. Volesky B Hydrometallurgy 2001; 59: 203–216.
176. Ferraz AI, Teixeira JA Bioprocess Eng 1999; 21: 431–437.

5

CELL WALL DISRUPTION AND LYSIS

F. A. P. GARCIA

Chemical Engineering Department, University of Coimbra, Coimbra, Portugal

5.1 INTRODUCTION

Microorganisms are important and efficient cell factories for the production of a variety of products. Many of these products are excreted to the medium but others remain intracellular or firmly bound to the membrane structures, and must be released into the suspending medium for further isolation and purification. Several strategies can be followed: the cells can be genetically engineered to excrete them, the wall can be made more permeable, the lysis of the cell may be induced, or the cell wall can be disrupted.

The cell wall is a rigid and complex structure protecting the cell from the aggressions of the environment, maintaining the cell shape and providing osmotic stability. However, the cell wall chemistry varies widely with the species in the microbial world and it is important to recognize the effects of these differences on the strength of the wall and the implications on the selection of the disruption techniques. Moreover, the cell age and the physiological state also determine important differences in the strength of a given species.

5.2 THE CELL WALL

5.2.1 Chemical Composition and Structure

The diversity of the cell wall composition and structure is well documented (1–5). Following the preparation of purified walls these can be hydrolyzed and different components can be isolated and identified.

5.2.1.1 ***Bacterial Cell Walls.*** The scaffold of the cell wall in both Gram-positive and Gram-negative bacteria is a supramolecular structure known as *peptidoglycan* or *murein*, a strong yet elastic structure that protects the underlying protoplasts from lysing under high internal osmotic stress. Peptidoglycan is a regular heteropolymer of alternating *N*-acetylglucosamine and *N*-acetylmuramic acid linked by β-(1-4) glycosidic bonds in linear chains of up to the order of 100 disaccharide units in *Bacillus subtilis*, with a mean in the range 25–35 in *Escherichia coli* (6). A peptide side chain, consisting of at least three amino acids (some of the unusual "D" series), is attached to the carboxyl group of the *N*-acetylmuramic acid. This peptide is most often L-Ala-D-Glu-m-A_2pm-(D-Ala)-(D-Ala) in the nascent peptidoglycan, losing the last one or two (D-Ala) in the mature state if the peptide chain is not involved in cross-linking (7). The peptide chain of one polysaccharide chain is linked to a peptide chain of an adjacent polysaccharide chain, either directly (the terminal D-alanine of one to the *meso*-diaminopimelic acid in position 3 of the other, as in *E. coli* or *B. subtilis*), or indirectly, through a pentaglycine peptide (the terminal D-alanine of one to the L-lysine in position 3 of the other, as in *Staphylococcus aureus*). Not all pairs of these peptide side chains are involved in the cross-linking of the glycan chains; the degree of cross-linking, as well as the nature of the amino acids in positions 2 and 3, being dependent on the strain and growth conditions.

The murein in Gram-positive bacteria, has a multilayered architecture, 20–80-nm thick. In addition, charged polymers like teichoic and teichuroic acids are associated with

Downstream Industrial Biotechnology: Recovery and Purification, First Edition. Edited by Michael C. Flickinger.

the murein on the external side of the wall and are unique to Gram-positive cells (not found in Gram-negative cells). Gram-positive cell walls are devoid of significant amounts of lipids and proteins.

In Gram-negative bacteria, the cell wall is thinner, more complex, and stratified. The mechanical resistance is provided by an inner layer of murein, which is much thinner (1–3 molecular layers thick) than in Gram-positive bacteria. This murein is covered externally by a bilayered lipid membrane, the internal leaflet being lipoprotein containing phospholipids, and the outer leaflet being lipopolysaccharide (LPS).

5.2.1.2 ***Fungi and Yeasts.*** The cell wall of fungi and yeasts, that was looked at in the past solely as an element of rigidity and protection from the environment, is now generally considered as an integral part of the cell physiology, dynamically varying in composition and thickness, depending on the developmental stage of the life cycle and the growth medium composition, temperature, pH, and oxygen tension (8). The walls of fungi and yeasts comprise glycoproteins and polysaccharides, and have a fibrillar structure due to the presence of mainly chitin, chitosan, and cellulose. Chitin is a structural polysaccharide, a polymer of β-(1-4)-linked *N*-acetylglucosamine residues, of great mechanical strength and is typically less abundant than either the glycoprotein or glucan portions of the wall. The individual chains run in an antiparallel way, held together by hydrogen bonds between adjacent sugars and forming microcrystalline microfibrils. Chitosan is a cationic polymer of glucosamine formed by deacetylation of chitin and probably exists with a large variation in the degree of acetylation. Cellulose is also a structural element of the wall, a β-1,4-linked glucan capable of combining into strong, straight, and crystalline microfibres.

These fibrillar and more crystalline compounds are embedded in a matrix of 1,3-β- and 1,6-β-glucans and glycoproteins, occupying the inner layer(s) of the wall. Glucans are the major structural polysaccharides of the fungal cell wall, representing approximately 50–60% of the wall's dry matter. The glucan component is predominately 1,3-β-glucan, long linear chains of β-(1-3)-linked D-glucose with side chains of D-glucose linked to the first ones through β-(1-6) bonds, covalently cross-linked in various degrees with other glucans and glycoproteins (9). The 1,3-β-glucan molecules have a flexible and helical shape, and due to the side chains the local association of 1,3-β-glucan chains is possible only through hydrogen bonds, therefore forming a continuous and elastic structural matrix that absorbs and distributes physical stresses and gives the cell increased resistance to mechanical disruption (10). Glycoproteins are mainly mannoproteins, peptidogalactomannans, and xylomannoproteins, tightly interwoven within the chitin and glucan-based structural matrix and extending outwards to the wall surface. In general, the fibrillar and crystalline material occupy the inner layer of the wall, being largely responsible for the mechanical strength of the wall and also providing the attachment sites for the proteins that form the outer layer of the wall. The chitin content in yeast cell walls (1–2% of the wall dry matter) is much less than in filamentous fungi cell walls and is localized mainly at septa or bud scars. A wide variety of β-1,3-glucans, β-1,6-glucans, chitin, and mannoproteins exists cross-linked in various and dynamic degrees of complexity.

5.2.1.3 ***Algae.*** The composition and structure of the algae cell walls vary significantly among the diverse lineage (5). Cellulose is again the main skeletal component of the algal cell wall but xylans and mannans (depending on the most abundant sugars) are also associated. The algae differ considerably from the fungi with reference to the cellulose content and crystallinity. The physical aspect is also fibrillar as in the other eucariots.

5.2.2 Effects of Upstream Conditions

The conditions that affect the synthesis of the skeletal wall constituents during the growth process will be reflected on the strength of the cell wall. Cells of *E. coli* grown in a complex medium are more resistant to disruption than the cells grown in a simple synthetic medium (11). *E. coli* cells are also weaker during the logarithmic phase and become stronger in the transition to the stationary phase (12). This corresponds to an increase in the murein thickness and cross-linking (13). A higher specific rate of growth, as attained in continuous culture of *Candida utilis*, led to a weaker cell wall, while a lower growth rate as in a batch operation, allowed the cells to build a stronger cell wall (14). Similar results were obtained for *E. coli* (15). The wall thickness and the resistance to disruption of baker's yeast increases also with the age, possibly due to the exhaustion of the carbon source (16). Bud scars, much more frequent in a younger yeast population, may introduce local areas of lower wall strength (14,16).

Despite this information, the process of cell disruption remains poorly understood in view of the scarce measurement data of the fundamental mechanical properties of the cells. The compressive force required for the disruption of brewer's yeast cells was measured using a micromanipulation technique, and ranged from 40 μN for cells in the exponential phase to 90 μN for cells in the stationary phase (17). The compressive force required to burst a yeast cell in the stationary phase using the same technique varied between 55 and 175 μN, with a mean value of $101 \pm 2\mu$N (18). Mammalian hybridoma TB/C3 cells, taken from a steady state continuous culture, are much weaker; single cells bursting under a compressive force in the range of only 1.5–4.5 μN, as they lack a true rigid wall (19). In the case of *Staphylococcus epidermis* from a batch culture, the bursting force at a compression speed of 6.2 μm/s varied from 3 to 34 μN, while for *E. coli* from a continuous culture, the bursting forces were in the range of 1–9 μN (20). Force deformation data allowed the determination of more fundamental mechanical properties: Young's modulus, $Y = 150 \pm 15$ MPa, maximum stress-at-failure, $\sigma = 70 \pm 4$ MPa, maximum strain-at-failure, $\varepsilon = 75\% \pm 8\%$ (21). The application of compressive forces to filamentous organisms is more difficult to perform and interpret, and tensile tests were performed to measure the mechanical properties of *Saccharopolyspora erythraea*, the following results were reported: the ultimate tensile strength was 24 ± 3 MPa and the elastic modulus was 140 ± 30 MPa for hyphae from both the midlogarithmic phase and stationary phase of a simple batch fermentation on defined medium (22) On the basis of a wall strength model for high pressure homogenization, the mean effective cell strength, $\overline{S}$, was correlated with average length of nonseptated cells, $\overline{Ln}$, and the degree of peptidoglycan cross-linkage, X_c, by the equation

$$\overline{S} = 34.2 \cdot X_c - 8.61 \cdot \overline{Ln} + 48.2 \tag{5.1}$$

Both $\overline{Ln}$ and X_c were measured experimentally (23,24).

5.3 ASSESSMENT OF THE DISRUPTION YIELD

Suitable analytical techniques are required to assess the extent of cell disruption to develop and validate models for process selection, scale-up, and optimization. Different methods have been used to quantify the degree of disruption at specified conditions, each one having its advantages and drawbacks. The number of disrupted cells have to be estimated and this can be done by direct counting of nondisrupted cells or, indirectly, by determination of any intracellular metabolite whose quantitative release into the medium is proportional to the number of disrupted cells.

Direct counting on suitably diluted samples can be achieved by plating out techniques (25,26), microscopic counting (27), electronic counting (28), or using an analytical disk centrifuge (29).

Nonviable intact cells can be present in a significant proportion and are uncountable using plating out techniques. The situation is even worsened if the cells have a tendency to agglomerate.

Microscopic counting is a faster and simpler alternative. Whole cells in suitably diluted samples of original and treated suspensions are counted in microscope hemocytometers (27). Proportional counts of stained smears may be advantageous, as for instance, in the case of yeasts (30), making it possible to distinguish among whole intact cells, broken cells contaminated with cytoplasm, and empty debris, at 1000× magnification. Image analysis was used to assess the viability of stained actinomycetes (31).

Electronic particle counters such as the Coulter counter, for instance, can be used to assess the concentration of yeast cells (28); but it is not a sensitive method for bacteria (32) and tends to be fouled by cellular debris (29).

The analytical disk centrifuge provides a size distribution in the homogenate and requires that the size distribution of the inclusion bodies is subtracted from homogenate sample distribution by numerical or graphical deconvolution of both distributions.

Indirect counting is based on the determination of certain compounds (e.g. soluble protein, enzyme activities) that are released into the solution when the cell wall is broken: the yield of disintegration is the percentage ratio of the released material, R, to the maximum amount, R_{max} that can theoretically be released into the solution. The most common material that has been used for this purpose is total soluble protein assayed in the sample supernatant (32–34).

In the case of enzyme activities (35) released into the medium, the assessed value of the disruption yield under fixed conditions depends largely on the enzyme selected for analysis, its location (12) within the cell, whether or not it is bound to membranous materials, and its heat or shearing lability (36), which can affect the values of both R and R_{max}. A "location factor" (LF), defined as the ratio of the rate constants for the release of a given enzyme and of the total protein, can be used as an indication of the intracellular location of the enzyme (37,38).

5.4 METHODS OF CELL WALL DISRUPTION

A ubiquity of methods exists for the microbial cell disruption that evolved over the years and are well documented in the literature (39–43). On the basis of the principles involved, there are mechanical and nonmechanical methods. Mechanical methods are based on shear forces that deform and rupture the cell wall. Nonmechanical methods induce the lysis of the cell by chemical, physical, or enzymatic means.

Many of these methods were developed for laboratory purposes with the objective of purifying biomolecules or organelles, while maintaining as much of their integrity as possible for fundamental studies. For preparative or commercial applications, the strategy is the maximization of the product release, keeping the biological activity and minimizing the formation of fine cell debris, which can interfere with further downstream operations. The mechanical methods, based on solid or liquid shear, are best suited for this purpose.

5.4.1 High Pressure Mechanical Homogenization

High pressure mechanical homogenizers, typified by APV Manton Gaulin, consist of a positive displacement pump capable of generating high pressures to force the cell suspension through an adjustable discharge needle valve under high shear stresses where the cells are disrupted. The body of the valve is positioned against a valve seat by a spring-loaded rod and the disrupting pressure is adjusted by the load that is applied, hydraulically or mechanically, to the spring. The treated suspension can either be collected or recycled for more passes through the valve.

The degree of disruption of baker's yeast in such a device (33) was proportional to the number N of passes through the valve according to the model ("first-order" kinetics relative to N)

$$\ln \frac{1}{1-X} = kNP^{\mathrm{a}} \qquad (5.2)$$

where $X = R/R_{\max}$ is the fraction of released protein, R is the amount of total protein released in the process, and $R_{\max}$ is the maximum amount of protein that would be released if all the cells were broken down, k is a dimensional, machine-dependent (it depends on the geometry of the needle valve) specific rate constant $[(k) = (P)^{-a} = MPa^{-a}]$; the constant a is a measure of the cell resistance to disruption (the value $a = 2.9$ was reported for baker's yeast and 2.2 for *E. coli*), and is dependent on the type of organism and the physiologic conditions and rate of growth (14).

The process has been found to be independent of cell concentration in the range 84–168 g dry matter/L and for pressures extending from 19.6 to 53.9 MPa, although lower protein recoveries for higher cell concentrations at higher pressures were obtained (33). Further studies extended the analysis of the influence of the cell concentration up to 240 g dry matter/L and no significant effect was detected (44). However, the disruption of *E. coli* in another type of homogenizer, Microfluidizer (Microfluidics Corporation, Newton, MA), was correlated by

$$\ln \frac{1}{1-\mathrm{X}} = \mathrm{k}\mathrm{N}^{\mathrm{b}}\mathrm{P}^{\mathrm{a}} \qquad (5.3)$$

where the exponent b, varied from 0.28 to 0.94, was found to be a function of both the cell concentration and the specific growth rate; a varied in the range 0.61–1.77, and k varied randomly between $0.27 \times 10^{-3} MPa^{-a}$ and $62.8 \times 10^{-3} MPa^{-a}$ depending on the strain and growth conditions, but independently of both the growth rate and the cell concentration (12).

The rate of cell wall disruption has been shown to be highly temperature-dependent: an increase of about 2–3°C (depending on cell concentration) can be expected for every 10 MPa increase in the backpressure of the needle valve (42,44). As a matter of fact, the suspension has to be precooled, to keep the temperature under control, and cooled immediately after leaving the homogenizer to avoid thermal denaturation of heat labile products. In most cases, no losses of enzyme activity or protein solubility were detected if the temperature was not allowed to rise above 30°C (14,33).

Increased productivity can be attained with higher pressures, allowing the recovery of more product with fewer number of passes, but at the expense of a higher energy consumption and, even worse, causing severe wearing problems in the valve seat. Erosion of the valve seat is, in fact, of great concern in using this type of equipment.

The breakage of the cell wall occurs as a result of normal stresses or shear forces developed during the passage of the suspension through the needle valve. One mechanism (shear-stress mechanism) considers the magnitude and the rate of pressure drop to be responsible for cell breakage (44). A hydrodynamic mechanism, in which turbulent eddies of smaller dimensions than the cells cause the intracellular liquid to oscillate with kinetic energy large enough to rupture the cell, was also proposed (45). Impingement of a high velocity jet against a flat surface, as it occurs in the needle valve when the direction of the jet is suddenly and abruptly changed, was shown to be the main cause of cell disruption (14,46). In the case of the Manton Gaulin type of homogenizers, this led to the redesign of the valve geometry (46); the so-called knife-edge configuration showed the best performance, while the flat valve configuration had the lowest yields in the disruption of baker's yeast. Cavitation was also considered important due to the high velocities in the valve and consequent regions of low pressure (47), and this was experimentally made evident. Cavitation occurs when the cavitation number, $C_{\mathrm{v}} = (P - P_{\mathrm{v}})/0.5V^2$, is less than 1.0, with free radical generation detected in the process (48,49). This could be avoided by applying a small backpressure up to 60 psi in the impingement zone (49).

All these mechanisms that cause cell disruption are based on models that describe experimental data and are system and culture specific; therefore they are nonpredictive. A more rational approach is the wall strength model that gives the disruption, D,

$$D = \int_0^\infty f_{\mathrm{s}}(S) \cdot f_{\mathrm{D}}(S) \cdot dS \qquad (5.4)$$

as a function of the distribution of the effective wall strength, $f_{\mathrm{S}}(S)$, adequately described by a simple Gaussian distribution, and the homogenizer stress distribution, $f_{\mathrm{D}}(S)$, given by

$$f_{\mathrm{S}}(S) = \frac{1}{\sigma\sqrt{2\pi}} \exp\left[\frac{-(S-\overline{S})^2}{2\sigma^2}\right] \qquad (5.5)$$

and

$$f_D(S) = \frac{(mP^n)^d}{S^d + (mP^n)^d} \quad (5.6)$$

In these equations, P is the pressure and S is the effective wall strength with mean $\overline{S}$ given earlier. The parameters m, n, and d are characteristic of the homogenizer used (but independent of the culture properties), and rely on experimental correlations without a physical meaning. Progress toward predictive models requires more fundamental data on the distribution of true strength of the cell walls and on the stress distributions in the disrupting valves. In the way of accomplishing this, computational fluid dynamics (CFD) code was used to model the flow in the inlet and valve gap regions, and in the impingement region of an APV Manton Gaulin homogenizing valve, the cell breakage being well correlated with the pressure gradient (50,51). The flow pattern in the whole valve geometry was also modeled using a CFD code (52). Recently, the flow pattern within a small new valve was modeled using also a CFD code, giving information on velocity profiles, cavitation and turbulence phenomena in this new valve (53).

The high pressure homogenizers have proved to be efficient for all types of microorganisms, with high degrees of disruption even with only one pass. Blockage has, nevertheless, remained a major concern in the use of high pressure homogenizers for the disruption of filamentous microorganisms (54). The use of a Manton Gaulin resulted in a severe degradation of plasmid recovered from *E. coli* (26).

Microfluidizers are a different type of homogenizers, where the suspensions are forced at constant high pressure (typically 20,000–30,000 psi) to flow along precisely defined fixed-geometry microchannels within a proprietary interaction chamber with an additional backpressure unit. As a result, the product stream accelerates to high velocities creating shear rates within the product stream that are orders of magnitude greater than in any other conventional means. The efficiency can be correlated by Equation (5.3), as aforementioned. When a *Streptococcus thermophilus* 143 culture was disrupted to release β-galactosidase activity, the efficiency of the Microfluidizer (83.7%) was somewhat lower as compared with the efficiency of a high pressure homogenizer and a bead mill (93.0–99.3%), but the specific enzymatic activity obtained was the highest, probably due to the ease of cooling (55). The conditions are highly reproducible and, as all the products experience identical processing conditions, the method is easily scalable.

5.4.2 Solid Shear Methods (Bead Mills)

Ball mills, originally designed for nonbiotechnological applications, have also been largely used to disintegrate microorganisms, particularly yeasts. They are the most suitable for scaling up and can be continuously or discontinuously operated. They consist of agitator disks mounted concentrically or eccentrically on a motor-driven central shaft inside a grinding chamber. During the operation the chamber is filled with small glass or stainless steel beads, which are necessary to promote an abrasive action to grind the cells. In continuous operation, a bead retainer (a sieve plate, a vibrating slot or a special rotating disk) prevents the beads from leaving the grinding chamber.

The mechanism of disintegration is very complex and results from both a liquid shearing and compaction when the cells are placed between colliding beads (56,57). The disruption process has been considered as following a first-order kinetics law (34,58) and, for a batch operation or continuous plug flow system, the integrated rate equation is

$$\ln \frac{R_{max}}{R_{max} - R} = kt \quad (5.7)$$

where R is the amount of protein released in time t, R_{max} is the maximum amount that could be released if 100% disruption had occurred, and k is the specific velocity of disintegration (protein released). For a continuous operation there is a certain degree of backmixing and, an n well agitated tanks in series model is a better representation of the hydrodynamics in the chamber. In this case a mass balance to the released protein gives (58):

$$\frac{R_{max}}{R_{max} - R} = \left[1 + (k\theta/n)^n\right] \quad (5.8)$$

where $\theta = V/Q$ is the mean residence time, V is the total free volume of the grinding chamber and Q is the flow rate of the microbial suspension.

Several process parameters affect the yield and the rate of the cell wall disruption. Besides the design characteristics of the mill (volume and geometry of the grinding chamber, the number of disks per unit volume of the mill, and geometry and construction materials of the disks), the disruption rate constant is influenced by a number of process parameters such as the agitator speed, cell concentration, bead size, bead load, bead specific weight, temperature, and flow rate of the suspension (34,56,58).

The *agitator disk speed* determines the frequency of bead collisions (57) and the intensity of shear and, within certain limits, the specific rate of disintegration is proportional to the disk tip speed, U (58):

$$k = KU \quad (5.9)$$

although, K tends to decline when U becomes larger (59). Practical limitations arising from high energy consumption, high heat generation, erosion of the beads, and inactivation of shear-labile products compel the disk tip speeds to lie between 5 and 15 m/s.

The influence of *cell concentration* requires a more detailed investigation as contradictory results have been reported. The disruption of baker's yeast was reported to be independent of the cell concentration in the range 4–20% w/v on a dry matter basis (35,60), in opposition to others (56,58), reporting a maximum of protein release for cell concentrations of 35–40% (wet weight), decreasing thereafter, depending on the disks materials (58), such effect being more pronounced when the agitator speed was low (56). The viscosity of the suspension increases with the cell concentration and is further increased with the progress of the disruption process, therefore increasing the rate at which the turbulence is damped out (58). However, insufficient data are generally generated or reported regarding the rheology of the suspension. The optimal cell concentration varies from case to case, and, considering also that less heat is generated but the power consumption per unit weight of treated cells is higher for lower cell concentrations, and that for low cell concentrations (<5% solids) increased wear is observed on the mill and beads, the optimum must be found by experiment and is generally in the range 30–60% or, more usually, in the range 40–50% (cell wet weight).

The *bead diameters* that have been reported range from 0.1 to 1.5 mm. Usually, the disruption is faster for smaller beads, but tendency of very small beads to float and difficulties in retaining them inside the grinding chamber impose practical limitations in the lower limits of sizes. In practice, the lower limit for batch operation of laboratory mills is 0.2 mm while for production equipment working continuously the lower sizes are ≥0.4 mm. For the same bead load larger particles correspond to less number of beads and reduced abrasive action or frequency of collisions of the beads. For the same bead load, there are also reports of an optimum bead size, which has also to be determined by experiment, and depends on the location (34,61) of the product (enzyme) inside the cells, with larger beads being preferred to recover enzymes located in the periplasmic region and smaller ones to recover cytoplasmic material.

The rate of protein release is also enhanced by increasing the *bead load*, α, (volume fraction of bead material relative to the total volume of the grinding chamber). A linear relationship, $k = c\alpha^2$, was found between the rate constant and the square of the bead load, α, where c is a constant that increases with the peripheral velocity (59). Bead loads ranging from 30 to 90% are reported in the literature but, depending on the bead diameter, the packing density inside the chamber is generally in the range 80–90%. The upper limit is generally imposed by the high power consumption and the difficulties in removing the heat release in the operation.

The rate constant is, obviously, dependent on the *temperature*, which results from the balance between the heat generated during the milling process and the heat removal by a cooling liquid, brine, or a cooling liquid circulating through a cooling jacket of the grinding chamber. The heat generated in the process, if accumulated, would rise the temperature to unacceptable levels and, as a matter of fact, the suspension has to be precooled and cooled immediately after leaving the chamber to avoid inactivation of heat labile products.

The effects of the *flow rate* should mirror the effects of the residence time of the cells inside the grinding chamber and considering a first-order process, the yield after one passage should decline as the feed rate increases. A careful criticism has, however, to be exercised as the degree of longitudinal dispersion or backmixing increases as the flow rate decreases and the yield of disruption is not a simple function of the flow rate. On the other hand, the power consumption per unit mass of processed cells strongly decreases as the feed rate increases and, for this reason, a high flow rate should be adopted, while the decline in the yield of disruption could be alleviated by recycling part of the suspension. It should be emphasized that by increasing the residence time the viscosity may be decreased as prolonged shear may cause degradation of the shear sensitive DNA polymer (62).

The disintegration rate was correlated with the "specific energy" in the case of a combination of parameters: small grinding beads, moderate to high cell concentrations, and/or low to moderate agitator speeds (61). In this case, the utilization of energy input reaches its maximum. In other words, high cell concentrations are energetically favorable. This result is considered as the basis for scale-up. For the cases of low cell concentrations in the suspension, high agitator speeds, and use of larger beads, the disintegration rate was no longer correlated with specific energy input, but was a function of the "stress frequency" (which is proportional to the product of the agitator speed and disintegration time), which was experimentally proved.

5.4.3 Ultrasonics

Ultrasounds have largely been used on the laboratory scale for analytical and preparative extraction and purification of cell constituents of bacteria or yeasts. The mechanism of cell disruption derives from the intensive shear induced by sonicating the suspension at sound frequencies above 20 kHz. A magnetorestrictive or a piezoelectric transducer converts the alternating current of an electric oscillator into mechanical waves that are transmitted to the suspension through a metallic probe (usually made from titanium) vibrating with the same frequency as the oscillator. The sound waves create many microbubbles at various nucleation sites in the suspension, which collapse implosively during the rarefaction period of the sound waves. This phenomenon of cavitation (formation, growth, and collapse of the vapour-filled bubbles) produces intense local shock waves, and intense local shear gradients are generated that cause the cells to deform beyond the limit of elasticity and rupture. The rate of microbial cell disintegration is usually low for preparative purposes and the inclusion of small stainless steel or glass (Ballotini) beads will help in the nucleation for cavitation, as well as with a "grinding" action (solid shear) that will increase the efficiency of disintegration.

The kinetics of disruption (protein release) has been reported to follow a first-order law (37,62) and for a well mixing pattern inside the disruption chamber the integrated equation is

$$1 - X = \exp(-kt) \qquad (5.10)$$

where X is the fraction of released protein and k is the rate constant. The *amplitude of vibrations* of the sound waves, directly related to the *acoustic power* supplied to the suspension, affects the rate constant according to Equation (5.11), where W is the acoustic power and W_0 is the cavitation threshold power (63).

$$k \propto (W - W_0)^{0.9} \qquad (5.11)$$

However, a linear relationship between the rate constant and the acoustic power was recently reported (37,64).

A number of variables can markedly affect either the sample or the efficiency of the sonication process. One of these is the *temperature rise*. To control the temperature within acceptable limits the suspension should be precooled to 0–5°C, and jacketed disruption vessels should be used with a continuous circulation of a cooling fluid. Alternatively or concomitantly, short periods of sonication can alternate with short periods of cooling. The ratio of the time of sonication to the time of cooling is known as *duty factor*.

The integrity of the bioproduct can also be adversely affected if *free radicals* are allowed to accumulate during the intense cavitation, which can be alleviated by the addition of radical scavengers (e.g. cysteine, histidine, or glutathione) or pregassing the suspension with hydrogen.

The *viscosity* of the suspension affects the hydrodynamics and the rate of energy dissipation and inhibits the phenomenon of cavitation. The initial *cell concentration* that affects the viscosity of the suspension was reported to have no effects on the rate constant (37).

Cavitation "unloading", that lowers the disruption efficiency may result from the presence of surface active agents that cause foaming problems, and from high working volumes that require high acoustic power. The *flow rate* in continuous operation determines the residence time of the cells in the vessel and affects the overall yield of the process. For tougher cells the inclusion of Ballotini beads is necessary. The *load and size of beads* have to be optimized in any case, but the value of the rate constant increases to a maximum and then decreases with increases in the bead diameter for the same bead load. The *shape of the probe* is of particular concern, and for a constant power level the amplitude of the sound waves at the probe tip is inversely proportional to the tip area. Manufacturing materials dictate the severity of the *wearing* problems that are also common with the probe tips. Titanium is a good material because of its good acoustic and mechanical properties and low toxicity to bioactive materials, but titanium probes are expensive and difficult to machine. Pitting erosion produces deep cavities on the tip surface, particularly when Ballotini beads are included, which lessen the efficiency of sonication. Reliability tests can be used to identify wearing problems (65). The tip can usually be detached for servicing to produce a new flat surface, which is difficult with titanium; stainless or hard steels are better alternatives despite their poorer acoustic and mechanical properties.

Sonicating equipment can be found in many laboratories but the application of this technique to large-scale purification of bioproducts is rather limited. In preparative applications the performance is generally not as good as with the high pressure homogenizers or bead mills. In the case of plasmid recovery from *E. coli*, even if the disruption yield was good, sonication failed in producing intact plasmids (26).

5.4.4 Other Methods

The mechanical methods are, generally, effective in disrupting tough microbial cell walls and much more efficient than the nonmechanical methods but they are energy intensive, generate high temperatures and shear forces that can damage labile products, are nonspecific with respect to the organisms and the product to be released, and cause intense cell fragmentation. Nonmechanical methods are nonefficient on their own and very often only increase the permeability of the cell envelope without breaking the wall. However, by weakening the cell wall, lytic methods can be combined with the more efficient mechanical methods and, therefore, less harsh conditions would be required.

Among the *physical methods* of lysing the cells, two have been reported more frequently although they are not production oriented: osmotic shock and freeze/thawing. Osmotic stresses are generated when a cell is placed in either a hypertonic or a hypotonic medium; in the first case the cell tends to shrink (plasmolyze) and in the second case to expand. In both cases the cells can suffer potentially serious injuries. The cells are placed in a concentrated solution of sucrose, supplemented with EDTA to release the LPS from the cell envelop in Gram-negative bacteria. If upon centrifugation, they are suddenly placed in a hypotonic medium (a medium of very low salt concentration), the cells suffer an *osmotic shock*, absorb water, and swell, eventually reaching a point of unsustainable internal pressure and bursting. This bursting process is effective and scaleable upwards with weak animal and blood cells. However, osmotic shocks are rather inefficient with tougher bacterial, fungal, or yeast cells, though some periplasmic enzymes and even some proteins/enzymes located near the inner side of the cytoplasmic membrane can be selectively released. Mild osmotic shock may be use to release recombinant proteins secreted to the periplasm of *E. coli*, and as a part of selective release procedures for disrupting yeast cells. Pretreatment of the *E. coli* cells with divalent cations (Ca^{2+} or Mg^{2+}) enhanced the release of LPS when the cells were subsequently treated with EDTA, thus increasing the permeability of the outer membrane. Osmotic shock is also frequently used to prepare membrane vesicles for studies of membrane proteins. The important operational parameter is the time to fully resuspend the cells in the hypotonic medium, which should be the shortest to avoid the osmo-adaptation of the organisms.

Heat-induced translocation of proteins and enzymes from the cytoplasm to the periplasmic space is now being exploited to improve the efficiency of the cell disruption, reducing the severity and the energy requirements of the process. The rate of translocation of β-galactosidase in heat stressed *Kluveromyces lactis* NCIM 3566, evaluated based on the rate of change of the "location factor", increased linearly with the temperature up to 45°C (66). The formation of insoluble protein aggregates that could be removed by centrifugation led to an increase in the specific enzyme activity.

Thermolysis is another method with potential for process-scale disruption (43), with results that depend on the organism and its growth phase. The cells are subjected to temperatures that cause the disruption of the outer membrane. Thermolysis was used to disrupt *E. coli* cells and recover a recombinant thermostable esterase at temperatures above 60°C (67). Several advantages were recognized in this technique (43): it may be operated to kill the host, preferential protease deactivation may occur if the product is not temperature sensitive, and large cellular debris result that may be easily separated from a soluble product. Disadvantages are significant increases in viscosity and the appearance of non-Newtonian viscoelastic behavior, while the large debris will be undesirable if the product is a solid inclusion body, making the separation by fractional centrifugation more difficult.

The release of intracellular components can also be induced by *freezing/thawing* the cells: the ice crystals that form and grow during the freezing step will mechanically disrupt the integrity of the inner cell membranes, which become permeable and allow several enzymes to be extracted into appropriate buffers during the thawing step. The efficiency of this method depends largely on the rates of thawing. With a large mass of frozen suspension, the rate of thawing is higher in the outside region while the interior of the block remains frozen, allowing chemical or enzymatic (eventually, proteolytic) reactions to proceed with different extensions in the different parts of the block. On the other hand, the lifetime of the transient pores (regions of discontinuity) formed in the membranes depends on the temperature of thawing that affects the rate of membrane resealing (43); this being longer for low temperatures. This may explain the lack of reproducibility of the results in the ball mill disintegration of brewer's yeast samples that were kept frozen and separately thawed (34). In view of that the freeze/thawing process is not easily scaled-up and finds application mainly in laboratory scale recovery of recombinant proteins.

Chemical methods of lysis are milder, rely on selective interaction of a chemical with components of the membrane, and often anticipate an autolytic enzymatic activity. Organic solvents (e.g. toluene, ethanol, *n*-butanol, isobutanol, acetone, methyl-ethyl-ketone, etc.) are believed to nonselectively extract lipids from the cell wall or cytoplasmic membrane, thereby increasing the permeability to intracellular components. However, organic solvents might cause protein denaturation and, by diffusing into the cell, they also destroy the membranes of intracellular organelles, thus allowing destructive enzymes, such as proteases, to be liberated. The use of concentrated organic solvents, however, imposes a fireproof environment.

Other chemicals, such as detergents (e.g. CTAB, SDS, Triton X-100 or Brij series), chaotropic agents (e.g. guanidine-HCl or urea), chelating agents (e.g. EDTA), phenethyl alcohol (PEA), or alkaline solutions have been used to increase the permeabilization of the cell walls. Addition of glycine in low concentrations to the medium at an appropriate pH could also increase the permeability of recombinant *E. coli* immobilized in *k*-carrageenan gel beads to α-amylase. All chemical methods suffer from long holding times and, thus, have long been considered not attractive for larger scale applications.

However, the recent progress in the recombinant DNA technologies renewed the interest in the development of methods to increase the permeability of the cell envelope to avoid a detrimental accumulation of plasmid-encoded proteins in the cytoplasm. Treatment of *E. coli* with guanidinium-HCl increases the permeability of the outer membrane, and periplasmically expressed proteins such as β-lactamase (68) can thus be leaked into the surrounding medium with high yield (better than 80% in 5 h), free of many intracellular contaminants. The release of β-lactamase was also enhanced by treating recombinant *E. coli* with EDTA and PEA at appropriate concentrations, while keeping the cell viability and the synthesis of the plasmid-encoded β-lactamase (69). Pretreatment with guanidine-HCl and Triton X-100 was also used in conjunction with a high pressure homogenization step to disrupt recombinant *E. coli* expressing human growth hormone inclusion bodies (70). Due to their synergistic effects, Triton X-100 was used to reduce the concentration of guanidine-HCl required to weaken the cell wall and to wash the inclusion bodies from contaminant proteins.

NaOH and NaOH/SDS solutions are commonly used for the extraction of proteins from purified cell walls. Pretreatment of *Candida lipolytica* with an alkaline solution resulted in a weakened cell wall and increased protein solubility, and further mechanical homogenization was thus facilitated at lower pressures than with nontreated cells (31). Pretreatment of *Pseudomonas putida* KT2442 with ammonia instead of sodium hydroxide, before high pressure homogenization, was used to increase the release of inclusion bodies, with the claimed advantage of being applied as a gas under pressure and the recovery of ammonia in a stripping column after the pressure release, therefore producing less wastewater (71).

Enzymic methods of cell wall hydrolysis are also becoming increasingly more attractive. The advantages associated with enzymic lysis include an increased selectivity of product release, an increased rate and yield of product release, the minimization of product damage, requirements for mild conditions of pH and temperature, and no debris being produced. More than any other methodology, enzymatic lysis requires a good understanding of the details of the chemical and physical structure of the cell walls.

No single enzyme is capable of fully degrading the cell walls. In Gram-positive bacteria, one single enzyme could lyse the cell, but a mixture of several enzymes (glycosidases (endoacetylmuramidase, endo-, and exo-acetylglucosaminidase), *N*-acetylmuramyl L-alanine amidase or peptidases), acting synergistically, would assist in the lysis of the bacterial peptidoglycan. Lysozyme from hen egg white is the best known and most used muramidase for bacterial cell disruption, but some Gram-positive bacteria are lysozyme-resistant. In Gram-negative species a pretreatment with a detergent, like Triton X-100, or a complexing agent, like EDTA, is usually required to remove the outer membrane to open access of lytic enzymes to attack the inner layer of peptidoglycan. Different strains of *Streptomyces* sp., *Pseudomonas* sp., *Staphylococcus* sp., *Cytophaga* sp., have a wide spectrum of bacteriolytic activities. Labiase, produced by *Streptomyces fulvissimus*, and achromopeptidase, produced by *Achromobacter lyticus*, are commercially available (72). Following extensive hydrolysis of the peptidoglycan the cells can then be osmotically disrupted in a hypotonic medium.

For yeast cells, a wall lytic protease and β-(1-3)-glucanase are essential, but the inclusion of β-(1-6)-glucanase, mannanase, and chitinase would assist the lytic process. Most lytic enzyme systems are active against pretreated nonliving cells (lyophilized or heat killed cells) or against cell walls.

The main restriction in using multienzyme systems is their availability and the high costs associated with their preparation. Different sources of yeast lytic enzymes have been investigated. Yeast lytic activity was identified in *Cytophaga* sp. and *Cellulosimicrobium cellulans*. The actinomycete *C. cellulans* (also known with the synonyms *Cellulomonas cellulans, Oerskovia xanthineolytica*, and *Arthrobacter luteus*) has been regarded as a major source of yeast lytic enzymes, particularly endo-β-1,3-glucanases, proteases, and mannanases, commercially available as Lyticase, Zymolyase, and Quantazyme (73).

The selection of the most appropriate lytic system depends on the nature of the wall to be lysed and the nature and the location of the product inside the cell. The composition of a multienzyme system determines the efficiency and the selectivity of the lytic process (74). If the product is proteinaceous, a high proteolytic system is undesirable; thus, if the organism to be lysed is a yeast, a high proteolytic activity should be provided in only the early part of the lytic process to attack the mannan–protein complex of the wall, and the protease should then be inactivated or a protease inhibitor should be added. However, the composition of lytic systems can in part be manipulated in the production phase, using different inducers and growth conditions. For instance, in the case of the lytic enzyme system from *Cytophaga* sp., high β-(1-3)-glucanase activities and high glucanase/protease ratios were obtained when produced in continuous culture at low dilution rates under carbon limited conditions. The quantitative data concerning the effects of pH, temperature, concentrations of lytic enzymes and cells on the lysis rate should also be obtained to optimize a lytic process (74).

As with all the nonmechanical methods, extensive enzymatic lysis is a very slow process but in conjunction with others the rate and the yield of product extraction can be much increased. Mutants that are defective in β-glucan can be produced by mutagenesis and are, therefore, more susceptible to a Zymolyase attack (62). Enhanced disruption of *C. utilis* was obtained by high pressure homogenization (using a Microfluidizer) when the cells were pretreated with Zymolyase (62). *Bacillus cereus* pretreated with cellosyl (a lytic enzyme produced by *Streptomyces coelicolor*) at pH 5.5 and 30°C, was disrupted with 98% yield in one single pass in a Manton Gaulin at 7×10^7 Pa, as compared to less than 40% when the cells were untreated. The same enzymatic treatment facilitated the disruption in a bead mill (75).

Autolysis is the process by which the cells use their own lytic capabilities and produce lytic enzymes. Autolysis, as a self-digestion process, is very slow and often requires several days, which is very inconvenient (because of that, it is not a common method of cell disruption, but yeasts are autolyzed on an industrial scale for the production of yeast autolysates). However, it can be speeded up by drying the cells or by freeze–thawing the cells. As mentioned above, autolysis can also be induced by plasmolysis with surfactants, thiol reagents, and organic solvents (e.g. toluene or butanol) or by ageing or temperature. A key factor is the value of pH, which generally has an optimum of around 8.0 when yeast is plasmolyzed by organic solvents or detergents (76).

β-Lactam antibiotics are known to cause the activation of autolytic enzymes (*autolysins*) and this has also been suggested to be the mechanism by which surfactants induce the autolytic process. Autolysins are important enzymes for bacterial cell growth and reproduction but once they are deregulated they attack the bacterial peptidoglycan. They can be classified as muramidases, glucosaminidases, *N*-acetylmuramoyl-L-alanine amidases, and endopeptidases according to the hydrolytic activity expressed. Autolysins are involved in many cellular processes including, among others, vegetative growth, cell wall turnover, peptidoglycan maturation, cell division, but the control of the autolysin activity remains unknown (77). The best characterized autolysins are those from *B. subtilis*. Gram-positive lysozyme-resistant bacteria can be lysed by cognate autolysins or by autolysins from related bacteria.

Microbial cells differ in their potential for autolysis and they often lack the genetic information to produce lytic enzymes. In this case, the genetic coding must be introduced from outside on the chromosomes or plasmids. Lytic enzyme systems that can be cloned include not only the autolysins but also colicin enzymes (a class of antibiotic proteins) and phages (78). *Lysins* (endolysins) are phage encoded lytic enzymes that function in a similar way as autolysins, acting by digesting the peptidoglycan at the terminal stage of the phage reproduction cycle, expressing endo *β*-*N*-acetyl glucosaminidases, *N*-acetylmuramidase, endopeptidase, or amidase. Lysins are cytoplasmic, not translocated through the cytoplasmic membrane, and gain access to the peptidoglycan under the control of *holin*, which is considered to degrade or alter the cytoplasmic membrane (72). Induction of *E. coli* cell lysis mediated by the coexpression of one of the T4 phage genes *e* (Gpe) that encodes a lysozyme, or *t* (Gpt) that encodes a holin, with the T7 lysozyme gene, was studied for the production of intracellular *β*-glucuronidase as a model target protein: the expression of gene *e* weakened the cell wall but did not lead to cell disruption, while the expression of gene *t* immediately induced cell lysis (79). Coexpression of constitutive T7 lysozyme gene and inductive DAAO (D-amino acid oxidase) in a recombinant *E. coli* BL21 (DE3) strain led to the cell lysis and secretion of coexpressed green fluorescent protein, used as the model target protein, to the medium (80).

5.5 EFFECTS OF CELL DISRUPTION ON DOWNSTREAM OPERATIONS

One of the most challenging problems in solid–liquid separation is the removal of the cell debris, which includes cell wall fragments, cell membranes, and protein precipitates. The disruption or lytic treatments cause a reduction in the sedimentation properties of the cells and fine debris (<0.1 μm), leading to inefficient centrifugation and/or filtration due to changes in the surface properties, high compressibility of the solids, high viscosity of the disrupted suspension, and small differences between the densities of the solids and the suspending medium (81). *Centrifugation* has been used, but it is unsatisfactory when the particles are smaller than 0.5 μm, particularly in viscous homogenates, therefore requiring very low flow rates (order of 10 times less than for whole cells). Examples of such separation problems are the recovery of inclusion bodies and viruslike particles (VLPs). A previous flocculation can improve the centrifugation efficiency. A highly selective *flocculation* of cell debris by borax (sodium tetraborate) was improved by further addition of poly(ethylene imine) to the clarified homogenate supernatant (82). *Rotary vacuum precoat filtration* was used for the clarification of disrupted yeasts but *cross-flow microfiltration* has the advantage of preventing cake formation and avoiding the need for expensive filter aids. The separation of cell debris is, however, the most complex problem that cross-flow microfiltration has been tested against. Membrane fouling and concentration polarization are some of the difficulties that contribute to a rapid flux decline and limit the concentration of the debris that can be attained. Apart from the influence of the operational parameters (pressure, feed velocity, fouling factors) on the flux and transmission, which are well documented in the literature, we have to consider that strong adsorption of proteins to cell debris may occur in the lysate, depending on the concentration, temperature, pH, or ionic strength. *Aqueous two-phase extraction* is attractive to recover whole cells and debris from the protein solutions, but involves high costs and recovery of reagents is not easy. Phase components include water-soluble polymers and electrolytes, the best known systems being PEG-Dextran and PEG-Phosphate.

The separation techniques depend on many specific aspects but the average particle size and the size distribution are crucial to all of them: the maximum amount of released protein increased with increasing micronization of the cell debris (54), but these become increasingly more difficult to separate.

The mechanical methods are generally effective in disrupting tough microbial cell walls. The major difference among them lies in the different sizes of the cell fragments that are generated; some methods disrupting the cell envelope with less overall destruction than the others. This is significant since the size of the fragments has an important impact on downstream processing when they are removed from the lysate or separated from other subcellular species (26); the submicron particles are easily resuspended in the supernatant after centrifugation, hinder the fractional precipitation, and tend to foul chromatographic columns or membranes that follow (26,82). Therefore, the influence of cell disruption conditions on product release, cell debris particle size and sizes distribution, and operating costs should be investigated.

No significant differences in the yield and rate of release of protein were reported when *E. coli* was disrupted by bead milling and high pressure homogenization (81); but the characteristics of the debris (particle size distribution, separability by centrifugation and dead-end-filtration) were quite dependent on the disruption method. The mean residence time in the bead mill did not significantly affect the particle size distribution, but the number of passes in the mechanical homogenizer broadens the debris sizes distribution, drastically reducing the centrifugal degree of separation. In the case of the Microfluidizer high pressure homogenizer, the cell concentration had only a minor influence on protein releases, while an increase in the biomass content reduced the separability of the cell disintegrate both in filtration and in centrifugation.

The effect of operating conditions on the cell debris size distribution was investigated for baker's yeast disrupted by high pressure homogenization. A two parameter model was developed, based on data fitting by a Boltzmann-type distribution and incorporating both the operating pressure and the number of passes, to simulate the particle size distribution in the homogenate. A threshold pressure of 11.5 MPa was determined below which no disruption of yeast cells will occur (83). Besides, the yeast growth conditions affect the cell debris particle size distribution (16): in batch cultivation the median diameter, d_{50}, was less in the exponential phase and larger in the stationary phase; in continuous culture the median diameter was less for higher dilution rates (younger cells).

High pressure homogenization of treated *E. coli* cells with a combination of guanidine-HCl and Triton X-100 produced cell debris of larger sizes than when the pretreatment was made with only guanidine-HCl, which is highly desirable for debris separation by cross-flow microfiltration.

Protease inhibitors are often required and added to the cell slurry to avoid unwanted degradation of labile products liberated from the disrupted cells, but are rarely used for biopharmaceutical processes due to toxicity concerns and cost (84). Lowering the temperature, shortening hold times to a minimum, or addition of complexing agents such as EDTA are methods of choice at large scale. Antioxidants will be useful in avoiding the oxidation of S–S bonds. An appropriate buffer should be included to maintain the pH in the desired value. All the chemicals that are added to the system in all phases of the process (starting from the medium preparation and fermentation steps) should take into account the effects they may have on the product and on the subsequent operations.

Evolving from the recent developments in the recombinant DNA technology, a large number of proteins of therapeutical importance, now in use or undergoing clinical evaluation, are produced as inclusion bodies (or "refractile" bodies) in the cytoplasm, mainly in *E. coli*, where they accumulate due to the absence of a secretory mechanism. These inclusion bodies are dense and amorphous aggregates of protein, consisting of misfolded (frequently denatured) polypeptides, often as large (0.8–1.5 μm) as a normal bacterium cell, with a relatively higher density (~1.2–1.35 kg/L) than the rest of the cell material, requiring the use of chaotropic agents or detergents to be solubilized in an aqueous environment (85). The standard procedure for the purification of a protein expressed as an inclusion body steps from the isolation of the inclusion bodies, which involves in the first instance, the disruption of the cell. The aggregates are then separated from the soluble components and from the debris of the lysed cells usually by centrifugation, typically using disc-stack centrifuges, at 5000–10,000 g for 5–10 min, the lighter cell debris being removed in the supernatant (85,86); cross-flow membrane filtration has also been investigated (87). The pellets are then washed with either dilute detergents or chaotropic reagents such as urea or guanidine-HCl to remove contaminants; guanidinium-HCl, though more expensive, is generally preferred in view of its better chaotropic properties and also because urea may contain or produce cyanate that can carbamylate the protein amino groups (88). Often, reducing agents (e.g. dithiothreitol or β-mercaptoethanol) are added to disrupt incorrect disulphide bonds. The solubilized and reduced protein is then ready for subsequent purification and refolding. Inclusion bodies processing is well documented and excellent reviews were made recently (84,88,89).

5.6 PROCESS INTENSIFICATION BY INTEGRATION OF DISRUPTION WITH PROTEIN CAPTURE

Manufacturers of biological products, particularly recombinant proteins for therapeutic use, under increasing pressure owing to market rules, seek competitive advantage through process intensification, yielding pure target proteins in fewer process steps, using less and more compact equipment. This has been done by coupling the disruption/permeabilization step with primary capture techniques such as fluidized bed/expanded bed adsorption (EBA) and aqueous two-phase extraction.

As aforementioned, the removal of cell debris continues to be one of the most challenging problems in solid/liquid separation and, concerning the degree of separation and the throughput, imposes severe demands on the operations of centrifugation and filtration. On the other hand, holdup times between operations are detrimental to the stability of the target product due to inactivation or degradation by hydrolytic enzymes or harsh physical conditions of the suspension, and product losses due to protein–debris interactions (90). These interactions are apparently enhanced by the degree of disruption when micronized debris are generated in bead mills at low flow rates (90). Direct coupling of cell disruption interfaced with fluidized bed adsorption shortens the process operating time with gains in product yield and quality at lower costs. The feasibility of combining the mechanical cell wall disruption in a bead mill with direct product adsorption in a fluidized bed contactor, operated immediately downstream of the disruptor, to recover G3PDH from waste brewer's yeast was demonstrated (91); but the inadequacy of the adsorbent used (Cibacron Blue 3GA derivative of a kieselguhr-agarose composite of low density, $p = 1.3$ g/cm^3) made the biomass throughput of the bead mill to feed the contactor rather limited to 15% w/v yeast disruptate. The application of a higher density adsorbent ($\varrho = 2.65$ g/cm^3) improved the process (90). In view of the elevated viscosities of the cell disruptates the adsorbent particles should have densities greater than 1.35 g/cm^3 (92).

Other problems require further developments in the design of adsorbents: cell, cell debris, and nucleic acids binding nonspecifically to the adsorbent matrix lead to aggregation of adsorbent particles resulting in higher degrees of backmixing and unstable beds (93). This interaction with the resin may compromise the resin performance and cleanability (84). In this respect, Cibacron Blue 3GA is very unspecific to target molecules, and the development of polymer (e.g. polyvinyl pyrrolidone) shielded dye-ligand to target elimination of nonspecific forces lead to improved efficiency (94). Capture and retention of cells and debris by the ion-exchange matrix result mainly from electrostatic forces, which can be expressed in terms of the ζ-potential (zeta potential) of both the biomass material and the adsorbent and the size of the biomass particles (93,94). These interactions depend on the particle size of the cell debris and its surface charge, which in turn depend on the extent of disruption used to release the intracellular products (95). The mean size of cell debris reduced dramatically during the cell disruption, and the absolute values of the ζ-potential also decreased significantly as the mean size reduced. More exhaustive cell disruption results in lower biomass adsorbent interaction during anion exchange EBA, and could potentially improve the stability of expanded bed and increase the adsorption capacity.

The potential of aqueous two-phase extraction to separate enzymes from cell debris has long been demonstrated. Integration of a bead mill disruption process with an aqueous two-phase system into a new unique process has been under current investigation: the bead mill reactor working both as a cell disrupter and as an extractor (96,97,98). The two-phase forming reagents (PEG and salts) were added into the cell suspension and then the mixture was fed to the bead mill. The product yield (alcohol dehydrogenase from baker's yeast) and selectivity of the integrated process were improved as compared with the traditional sequential process, while the disruption efficiency was not significantly altered by the aqueous two-phase forming reagents. A kinetic mechanism was established, and the kinetic parameters were calculated to characterize the recovery of ADH, LDH G6PDH, and total protein.

REFERENCES

1. Neidhardt FC, Ingraham JL, Schaechter M. Physiology of the bacterial cell. A molecular approach. Sunderland, Massachusets: Sinauer Associates Publishers; 1990.
2. Bartnicki-Garcia S. Annu Rev Microbiol 1968; 22: 87–108.
3. Bowman SM, Free SJ. BioEssays 2006; 28: 799–808.
4. Klis FM, Boorsma A, De Groot PWJ. Yeast 2006; 23: 185–202.
5. Domozych DS. Algal cell walls. Encyclopedia of Life Sciences. John Wiley & Sons, Ltd; 2006.
6. Scheffers D-J, Pinho MG. Microbiol Mol Biol Rev 2005; 69: 585–607.
7. van Heijenoort J. Glycobiology 2001; 11:25R–36R.
8. Aguilar-Uscanga B, Francois JM. Lett Appl Microbiol 2003; 37: 268–274.
9. Ross IK. Fungal cell walls. Encyclopedia of life sciences. John Wiley & Sons, Ltd.; 2001.
10. Klis FM, Mol P, Hellingwerf K, Brul S. FEMS Microbiol Rev 2002; 26: 239–256.
11. Wang DIC, Cooney CL, Demain AL, Dunnill P, Humphrey AE, Lilly MD. Fermentation technology. New York: John Wiley & Sons, Inc.; 1979. p. 241.
12. Sauer T, Robinson CW, Glick BR. Biotechnol Bioeng 1989; 33: 1330–1342.
13. Leduc M, Frehel C, Siegel E, van Heijenoort J. J Gen Microbiol 1989; 135: 1243–1254.
14. Middelberg APJ, O'Neil BK, Bogle IDL, Gully NJ, Rogers AH, Thomas CJ. Trans Inst Chem Eng 1992; 70(Part C): 213–218.
15. Middelberg APJ, O'Neil BK. Biotechnol Prog 1993; 9: 109–112.
16. Engler CR, Robinson CW. Biotechnol Bioeng 1981; 23: 765–780.
17. Balasundaram B, Harrison STL. Biotechnol Prog 2006; 22: 907–913.
18. Siddiqi SF, Bulmer M, Ayazi Shamlou P, Titchener-Hooker NJ. Bioprocess Eng 1996; 14: 1–8.
19. Roberts DA, Zhang Z, Young TW, Thomas CR. Proceedings of the Institution of Chemical Engineers Annual Research Event. London: University College; 1994. pp. 73–75.
20. Mashmoushy H, Zhang Z, Thomas CR. Biotechnol Lett 1998; 12: 925–929.

21. Zhang Z, Ferenczi MA, Lush AC, Thomas CR. Appl Microbiol Biotechnol 1991; 36: 208–210.
22. Shiu C, Zhang Z, Thomas CR. Biotechnol Tech 1999; 13: 707–713.
23. Smith AE, Moxham KE, Middelberg APJ. Chem Eng Sci 2000; 55: 2043–2053.
24. Stocks SM, Thomas CR. Biotechnol Bioeng 2001; 73: 370–378.
25. Woodrow JR, Quirk AV. Enzyme Microb Technol 1982; 4: 385–389.
26. Carlson A, Signs M, Liermann L, Boor R, Jem KJ. Biotechnol Bioeng 1995; 48: 303–315.
27. Heim A, Czerski R. Bioprocess Eng 1995; 13: 23–30.
28. Magnusson KE, Edebo L. Biotechnol Bioeng 1974; 16: 1273–1282.
29. Middelberg APJ, O'Neil BK, Bogle IDL, Snoswell MA. Biotechnol Bioeng 1991; 38: 363–370.
30. Stocks SM, Thomas CR. Biotechnol Bioeng 2001. 75: 702–709.
31. Lee CH, Tsang SK, Urakabe R, Rha RA. Biotechnol Bioeng 1979; 21: 1–17.
32. Engler CR, Robinson CW. Biotechnol Bioeng 1979; 21: 1861–1869.
33. Hetherington PJ, Follows M, Dunnill P, Lilly MD. Trans Instn Chem Eng 1971; 49: 142–148.
34. Marffy F, Kula M-R. Biotechnol Bioeng 1974; 16: 623–634.
35. Mogren H, Lindblom M, Hedenskog G. Biotechnol Bioeng 1974; 16: 261–274.
36. Follows M, Hetherington PJ, Dunnill P, Lilly MD. Biotechnol Bioeng 1971; 13: 549–560.
37. Kuboi R, Umakoshi H, Takagi N, Komasawa I. J Ferment Bioeng 1995; 79: 335–341.
38. Balasundaram B, Pandit AB. Biotechnol Bioeng 2001; 75: 607–614.
39. Hughes DE, Wimpeny JWT, Lloyd D. The disintegration of microorganisms. Volume 5B. In Norris JR, Ribbons DW, editors Methods in microbiology. London: Academic Press; 1971. pp. 1–54.
40. Coakley WT, Bater AJ, Lloyd D. Adv Microb Physiol 1977; 17: 279–341.
41. Chisti Y, Moo-Young M. Enzyme Microb Technol 1986; 8: 194–204.
42. White MD, Marcus D. Disintegration of microorganisms. In: Mizrahi A, editro. Downstream processes: equipment and techniques. New York: Alan R. Liss; 1988. pp. 51–96.
43. Middelberg APJ. Biotechnol Adv 1995; 13: 491–551.
44. Brookman JSG. Biotechnol Bioeng 1974; 16: 371–383.
45. Doulah MS, Hammond T, Brookman JSG. Biotechnol Bioeng 1975; 17: 845–858.
46. Keshavarz Moore E, Hoare M, Dunnill P. Enzyme Microb Technol 1990; 12: 764–770.
47. Save SS, Pandit AB, Joshi JB. Chem Eng J 1994; 55:B64–B72.
48. Shirgaonkar IZ, Lothe RR, Pandit AB. Biotechnol Prog 1998; 14: 657–660.
49. Lander R, Manger W, Scouloudis M, Ku A, Davis C, Lee A. Biotechnol Prog 2000; 16: 80–85.
50. Kleinig AR, Middelberg APJ. Chem Eng Sci 1996; 51: 5103–5110.
51. Kleinig AR, Middelberg APJ. AIChE J 1997; 43: 1100–1107.
52. Stevenson MJ, Chen XD. J Food Eng 1997; 33: 151–165.
53. Floury J, Belletre J, Legrand J, Ddesrumaux A. Chem Eng Sci 2004; 59: 843–853.
54. Keshavarz E, Bonnerjea J, Hoare M, Dunnill P. Enzyme Microb Technol 1990; 12: 494–498.
55. Geciova J, Giesova M, Jelen P, Plockova M. Milchwissenschaft Milk Sci Int 2002; 57: 509–511.
56. Limon-Lason J, Hoare M, Orsborn CB, Doyle DJ, Dunnill P. Biotechnol Bioeng 1979; 21: 745–774.
57. Schütte H, Kroner KH, Hunstedt H, Kula M-R. Enzyme Microb Technol 1983; 5: 143–148.
58. Melendres AV, Unno H, Shiragami N, Honda H. J Chem Eng Japan 1992; 25: 354–356.
59. Melendres AV, Honda H, Shiragami N, Unno H. Bioseparation 1991; 2: 231–236.
60. Canales M, Buxadó JA, Heynngnezz L, Enriquez A. Enzyme Microb Technol 1998; 23: 58–63.
61. Bunge F, Pietzsch M, Müller R, Syldatk C. Chem Eng Sci 1992; 47: 225–232.
62. Baldwin CV, Robinson CW. Biotechnol Bioeng 1994; 43: 46–56.
63. Doulah MS. Biotechnol Bioeng 1977; 19: 649–660.
64. Ho CW, Chew TK, Ling TC, Kamaruddin S, Tan WS, Tey BT. Process Biochem 2006; 41: 1829–1834.
65. Kapucu H, Gülsoy N, Mehmetodlu Ü. Biochem Eng J 2000; 5: 57–62.
66. Farkade VD, Harrison S, Pandit AB. Biochem Eng J 2005; 23: 247–257.
67. Ren X, Yu D, Yu L, Gao G, Han S, Feng Y. J Biotechnol 2007; 129: 668–673.
68. Naglak TJ, Wang HY. Enzyme Microb Technol 1990; 12: 603–611.
69. Ryan W, Parulekar SJ. Biotechnol Prog 1991; 7: 99–110.
70. Bailey SM, Blum PH, Meagher MM. Biotechnol Prog 1995; 11: 533–539.
71. van Hee P, Middelberg APJ, van der Lans RGJM, van der Wielen LAM. Biotechnol Bioeng 2004; 88: 100–110.
72. Salazar O, Asenjo JA. Biotechnol Lett 2007; 29: 985–994.
73. Ferrer P. Microb Cell Fact 2006; 5: 10–18.
74. Asenjo JA, Andrews BA, Hunter JB, LeCorre S. Process Biochem 1985; 20: 158–164.
75. Vogels G, Kula M-R. Chem Eng Sci 1992; 47: 123–131.
76. Breddam K, Beenfeldt T. Appl Microbiol Biotechnol 1991; 35: 323–329.
77. Smith TJ, Blackman SA, Foster SJ. Microbiology 2000; 146: 249–262.
78. Dabora RL, Cooney CL. Adv Biochem Eng Biotechnol 1990; 43: 11–30.
79. Morita M, Asami K, Tanji Y, Unno H. Biotechnol Prog 2001; 17: 573–576.
80. Chien L-J, Lee C-K. Biochem Eng J 2006; 28: 17–22.
81. Agerkvist I, Enfors S-O. Biotechnol Bioeng 1990; 36: 1083–1089.
82. Clarkson AI, Lefevre P, Titchener-Hooker NJ. Biotechnol Prog 1993; 9: 462–467.
83. Siddiqi SF, Titchener-Hooker NJ, Ayazi Shamlou P. Biotechnol Bioeng 1996; 50: 145–150.

84. Graumann K, Premstaller A. Biotechnol J 2006; 1: 164–186.
85. Mukhopadhyay A. Adv Biochem Eng Biotechnol 1997; 56: 61–109.
86. Wong HH, O'Neill BK, Middelberg APJ. Bioseparation 1997; 6: 361–372.
87. Bailey SM, Mragher MM. Biotechnol Bioeng 1997; 56: 304–310.
88. Vallejo LF, Rinas U. Microb Cell Fact 2004; 3: 11–22.
89. Lee GH, Cooney D, Middelberg APJ, Choe WS. The economics of inclusion body processing. Bioprocess Biosyst Eng 2006; 29: 73–90.
90. Bierau H, Zhang Z, Lyddiatt A. J Chem Technol Biotechnol 1999; 74: 208–212.
91. Bierau H, Hinton RJ, Lyddiatt A. Bioseparation 2001; 10: 73–85.
92. Jahanshahi M, Sun Y, Santos E, Pacek A, Franco TT, Nienow A, Lyddiatt A. Biotechnol Bioeng 2002; 80: 201–212.
93. Hubbuch JJ, Brixius PJ, Lin D-Q, Ollerup I, Kula M-R. Biotechnol Bioeng 2006; 94: 543–553.
94. Ling TC, Lyddiatt A. J Biotechnol 2005; 119: 436–448.
95. Lin D-Q, Dong J-N, Yao S-J. Biotechnol Prog 2007; 23: 162–167.
96. Balasundaram B, Harrison STL, Li J, Chase HA. Biotechnol Bioeng 2007, DOI: 10.1002/bit. 21586, 2007.
97. Chang B, Su Z. J Chem Technol Biotechnol 2006; 81: 454–460.
98. Su Z-G, Feng X-L. J Chem Technol Biotechnol 1999; 74: 284–288.

6

EXPANDED BED CHROMATOGRAPHY, SURFACE ENERGETICS OF BIOMASS DEPOSITION

MARCELO FERNÁNDEZ LAHORE
Downstream Bioprocessing Laboratory, School of Engineering and Science, Jacobs University, Bremen, Germany
OSCAR AGUILAR
Departamento de Biotecnología e Ingeniería de Alimentos, Centro de Biotecnología Tecnológico de Monterrey, Monterrey, México
RAMI REDDY VENNAPUSA AND MUHAMMAD AASIM
Downstream Bioprocessing Laboratory, School of Engineering and Science, Jacobs University, Bremen, Germany

6.1 INTRODUCTION

Industrial biotechnology depends on the availability of cost-effective downstreaming and bioprocessing strategies in order to consistently deliver products like proteins, enzymes, and nanoplexes. The impact is enormous: consider, for example, the impact of modern biopharmaceutical industry in the quality of human life or the impact of green chemistry in alleviating environmental burden. Among the mentioned bioproducts, proteins for example, "biotherapeutics" constitute an ever-growing important class of pharmaceutical products. Since the 80s genetic engineering has changed the nature of medicines that can be manufactured, through the development of the technology to produce virtually any protein by cloning and expression in various microbial or cellular hosts. Fermentation or cell cultivation procedures coupled to molecular engineering approaches has increased bioproduct titers enormously in the last decade (1–3).

As said, considerable efforts were invested on the genetic manipulation of cells and on the improvement of biosynthetic strategies. As a consequence, the productivity level of current biotechnological systems has risen to such levels that current downstream processes have to be reengineered to match such "upstream" performance. This has shifted the bottleneck in biopharmaceutical manufacturing and process development to the purification of such products (4,5). The main challenge today is the establishment of efficient recovery and purification steps that is, cost-effective processes with minimal losses of target protein, and capable of handling higher production levels of pharmaceutical proteins.

In the case of pharmaceutical proteins that have to be extensively purified before use, several strategies have been developed to reduce costs associated with downstream processing. Downstreaming is the most cost-intensive component of biopharmaceutical manufacture, which accounts for $\geq$50–80% of the total processing cost (3,6). The overall production cost is mainly determined by the efficiency of the initial capture and purification steps where (complex) feedstock volumes are large. Volume reduction is achieved once biomass solids and eventually oils are removed and protein is concentrated (7). An economic evaluation of the recovery of a transgenic protein from corn estimated 28% of the operative cost to be associated to the initial capture step on Diethylaminoethyl (DEAE)-Sepharose (8).

Among the many strategies on the development platform, *process integration* (substitution of two or more operation units for a single) and *process intensification* (processing of higher biological material using the same equipment capacity) have received a special interest. They have been an active research topic in biochemical engineering over the last decades and these efforts continue today. The reason

Downstream Industrial Biotechnology: Recovery and Purification, First Edition. Edited by Michael C. Flickinger.

for this is that integration and intensification could be one of the keys for the rational, cost-effective, and productive design of the recovery and purification processing steps.

More than a decade ago, Draeger and Chase (9) presented a novel integrated concept based on fluidized adsorbent beads for the direct sequestration of bioproducts. Fluidized beds have been widely used in the industry for several years for the recovery of antibiotics including batch-processing techniques for recovery of streptomycin (10) and semicontinuous systems for novobiocin (11). This technology allows direct capture of targeted species from an unclarified feedstock for example, a cell containing fermentation broth eliminating the need for particulate removal. This unit operation has the potential to *integrate* several operation units such as solids removal, product concentration, and partial purification in a single processing step. Early work on the capture of protein utilizing fluidized adsorbent beads was performed with standard Sepharose beads (12). The introduction of quartz-core-weighted particles followed; the population of such particles also promoted "classification" of the fluidized bed. The concept of expanded bed adsorption (EBA) was born (13–15).

The operating principles and real case examples demonstrating the advantages of the EBA approach to bioprocessing are well documented in the open literature (16–18). The reduction of unit operations seems to have a direct impact on the economics of EBA-based processes. A few authors have seriously addressed this issue by placing numbers to such claimed advantage. Walker and Feuser (19) have reported an economic advantage of 50%–70% cost benefit using EBA technology, excluding the cost of adsorbents. This is the direct result of having just one unit operation replacing the more traditional purification scheme, which consisted of three to four recovery and purification steps (19).

An important issue that has to be considered in relation to separation power, capacity, validations issues, and cost is the expected lifetime of the chromatographic adsorbents. In the particular case of EBA, (bio) process chromatography engineers faced, for the very first time, the challenge of operating a "chromatography" in the presence of biomass and other particulate components present in a crude feedstock. In other words, they shifted from the "fixed-packed-bed-with-clarified-inflow" situation to the more challenging one of feeding a (not always stable) fluidized-classified ("expanded") bed with an incoming complex and non-Newtonian cell (or cell debris) suspension. This may have detrimental consequences to the whole system in terms of increased adsorbent fouling, decreased adsorbent binding capacity, and detrimental hydrodynamic properties. Biomass and other suspended biological material can be very sticky, depending on the type of adsorbent employed and the solution chemistry involved and thus, can create severe limitations during processing. In summary, it can be stated that, considerably, variations can be expected during EBA depending on the characteristics of the feedstock and the regeneration conditions employed. This, in turn, could be detrimental to resin lifetime and overall process economics (19,20).

For EBA to be efficient during the primary sequestration of bioproducts, the perfect classification of the fluidized bed has to be protected against hydrodynamic alterations like mobile phase channeling, dead-water zones, or agglomerate formation. This can be done only when biomass adhesion to the fluidized beads can be avoided. Research in the last years has focused on mechanical and physicochemical tools to prevent such adhesion, with varying degrees of success. This chapter will emphasize the application of thermodynamic approaches to understand better the underlying biomass adhesion phenomena, which occurs during EBA processing. Additionally, surface energetics will be presented as a way to guide adsorptive matrix design for improved performance in EBA applications. A new generation of "biomass-repelling" adsorbents and improved process diagnostic tools may lead to the rebirth of EBA as a main downstream bioprocessing alternative in the industry.

6.2 TECHNICAL CHALLENGES IN EBA

The majority of unit operations applied in the downstream processing have a long history in the field of chemical and process engineering. All the platform technologies used in the traditional chemical industry are well established and optimized. However, in the more recent life science industry, the *bio*process engineering scenario is more challenging since many of the technologies currently in use are not exploited to their full potential or are not completely suitable to the complexity of the biological feedstock involved. Additionally, as previously mentioned, established manufacturers are considerably improving product titers from cell cultures, which have increased from a few milligrams a few years ago to (multi) grams per liter today (21,22). These improvements in mass productivity, however, do not translate directly into benefits when it comes to the downstream processing of such products. For example, fouling of synthetic microporous membranes that are commonly utilized during early recovery steps (e.g. for solids removal and dewatering) is a significant problem adding to the operational costs (21,23).

EBA has been proposed as an attractive alternative for the direct sequestration of bioproducts from a complex feedstock, therefore circumventing the need for a solid–liquid separation type of unit operation for example, centrifugation or tangential microfiltration. The latter which usually demands a long processing time is an absolute prerequisite for any packed-bed chromatography. EBA could then offer a significant reduction of time and cost compared to traditional processes, where chromatography follows purification (24–26).

Cells types commonly used in biotechnology as hosts for the production of recombinant proteins include bacteria, yeast, filamentous fungi, plant cells, insect cells, and mammalian cells. The different adsorbent types used in EBA (EBA modes) for product sequestration include ion exchange materials (cation- or anion- exchangers), hydrophobic interaction beads, and immobilized ion-metal affinity systems (IMAC) supports, among others like immuno- and pseudo-affinity. When a biological feedstock is loaded into the EBA system with the aim of selectively capturing the bioproduct of interest, it may be observed that bed hydrodynamics and stability is compromised due to biomass attachment to the chromatographic beads (27,28). Biomass attachment may derive from cell-to-support interactions and eventually from cell-to-cell aggregation (29). Biomass interactions and aggregations could lead to the impaired hydrodynamics in expanded bed systems, as illustrated in Fig. 6.1. In any case, there is a direct link between the quality of the hydrodynamic condition within the fluidized bed and the dynamic sorption performance of the system for example, dynamic capacity is reduced in the presence of a poor bed hydrodynamic condition. Further, reduction in bed capacity can occur, at the particle level, by masking of binding sites or pore clogging due to biomass and cell-derived components. The extent on biomass interference with sorption performance, both at the systemic (bed) and at the local (particle) level, depends on the particular combination of the biomass type to the adsorbent mode ("interaction") and on the tendency of biomass particles to aggregate among them ("aggregation"). Both phenomena are influenced by the solution chemistry and other operation parameters for example, superficial velocity and fluid phase mixing.

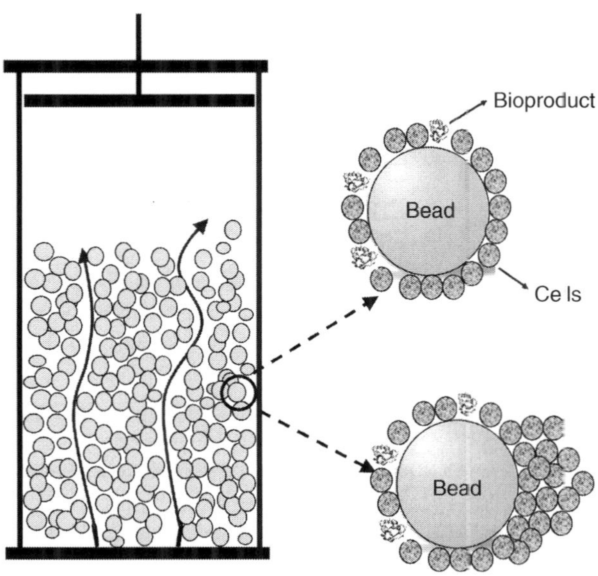

Figure 6.1. Biomass attachment, at the particle level, during processing or a complex feedstock can cause an impaired hydrodynamic condition of the fluidized bed.

After the commercial introduction of EBA in the early 90s, it became evident soon that technical limitations were present in day-to-day operations. In some extreme cases, a complete collapse of the fluidized bed was observed upon operating the system with an unfortunate combination of feedstock type and adsorbent mode. This situation triggered a more detailed investigation of such phenomena.

The influence of various biomass types like *Saccharomycess cervisiae*, yeast cell homogenate, *Escherichia coli* homogenate, and mammalian cells on several commercial adsorbents of the Streamline® family has been systematically studied. Combinations of the above mentioned biological materials and ion exchangers (DEAE, SP), hydrophobic interaction beads (Phenyl), immobilized metal ion supports (Chelating-Ni^{2+}), and affinity (Protein A) materials were tested (30–32) by residence time distributionanalysis (RTD) (with fluorescent tracers) and breakthrough curve determination (with model proteins). In general, biomass detrimental effects were observed to not be restricted to a single cell (biomass) type or adsorbent chemistry but to the combination between them under the processing conditions imposed by the adsorption mode. Adverse effects of different biomass types on various adsorbent effects are also very well described in the literature and technical notes (17,18,30,33). Despite the preceding facts, EBA is still considered as a feasible and efficient alternative to establish an initial capture process, substituting operations such as centrifugation, filtration, ultrafiltration, ammonium sulfate precipitation, and batch adsorption. There are numerous technical examples that can clearly demonstrate EBA potential (28,34–36).

This biomass deposition phenomenon is hampering the industrial utilization of EBA since its introduction in 1994 (20). Since then, many studies were performed to analyze the extent of biomass–adsorbent interactions, although utilizing descriptive and somewhat superficial approaches to the problem. The methods proposed included finite bath contactors, biomass-pulse response techniques, and RTD analysis (37). All these techniques provided an overall (systemic) indication of the extent of biomass "interaction" with the fluidized adsorbent; RTD provided an insight into the hydrodynamic condition of the bed possibly including also "aggregation" effects in selected cases.

One paradigmatic case of cell-to-support "interaction" is given by the deposition of yeast cells (negatively charged) onto DEAE Streamline beads (positively charged). Here, under the influence of Coulomb-type attraction a strong deposition can be noticed. In order to define these conditions better, some studies have relied on zeta potential measurements, which, not surprisingly, becomes the most significant factor affecting interaction (38). The concept of zeta potential has been used by others to describe the cell surface charge properties (39) and bioadhesion phenomena (40,41) and thus can be used partially, to explain electrostatic adhesion in expanded bed using ion-exchange adsorbents. Since the size of the interacting bodies is also known to affect interaction, a parameter containing size and charge information was defined and found to positively correlate with yeast cell interaction onto ion-exchange beads (38,42).

However, approaches to biomass adhesion during EBA which are solely based on zeta potential determinations would not be able to explain the biomass interactions observable with hydrophobic interaction and IMAC supports, as these processes are operated in buffers with high salt concentrations where charge effects are negligible (43,44). Early work by Brixius (45) that mainly dealt with charge-mediated attraction forces onto anion exchangers, addressed the potential role of some other forces like Van der Waals and hydrophobic forces in the adhesion of biomass to EBA supports, apart from the purely EL ones (45). Others also provided indications on physicochemical parameters for example, Lifshitz–Van der Waals (LW), acid–base (AB) and electrostatic components—with a potential to affect the adhesion of biomass (46). Minimization of cell (or cell fragments) interaction mediated by charge has been alleviated by adjusting operation parameters, and more specifically mobile phase conductivity by adding salt, within operational windows where biological particles show decreased deposition but where product binding is still possible (29). The introduction of heavier (denser) ceramic-composite particles, which require higher fluidization fluid velocities, has also contributed to alleviate biomass deposition probably due to an increased shear stress exerted by the fluid phase on attached and suspended cells. However, Arpanaei *et al.* (47) reported that added salt might increase DNA binding to ceramic Q HyperZ beads leading to a reduction in expanded bed adsorption performance (47).

Another approach to overcome biomass interference during direct sequestration by EBA has been the introduction of novel adsorbent materials, which are less prone to biomass fouling (48). Moreover, techniques like confocal laser microscopy have been adapted to observe fouling of chromatographic beads (49). Other methods that might contribute to explore biomass deposition phenomena are atomic force microscopy (50,51) and surface Plasmon resonance (52). A number of bead composites have been designed in the last years; however, complete information on their performance and fouling characteristics remains to be gathered (53,54). Equipment manufacturers have also modified parts like flow distribution devices to increase mixing, and therefore, increase cell or cell debris detachment in turbulent or transition regimes, however, an increase in turbulence may lead to broad RTD, affecting system performance (55–57). The latter possibility was already envisioned in batch contacting systems, which were run under gentle stirring conditions (29).

Despite the many technical improvements at the hardware and adsorbent levels, the limitations associated with potential adsorbent fouling and bed collapse cannot be completely ruled out under real processing conditions. In other words, the direct capture of the product from the complex feedstock could be hindered by the presence of cells, cell debris, protein aggregates, and nucleic acids (29,37,58). For this reason, a better understanding at the local level (Fig. 6.2) of overall biomass deposition, including the two separate cell-to-bead ("interaction") and cell-to-cell ("aggregation") phenomena has to be obtained to open the way for safer EBA operations. A better understanding of biomass interference in EBA can also help in defining suitable operational windows and in guiding adsorbent design. Such understanding can be gained by probing the EBA system at the local (particle) level and utilizing to this end, surface energetics (59).

Once the attachment of biological particles onto the fluidized support can be alleviated or controlled, EBA hydrodynamics comes close to ideal plug flow conditions, allowing the process to be modeled by utilizing numerical or analytical solutions already developed for packed-bed systems (60,61). Under such conditions, EBA design can also profit from approaches being developed for protein chromatography such as, for example, process proteomic routines (62), tailored ligand design for increased selectivity (63), and for protein tagging (7).

6.3 SURFACE THERMODYNAMICS OF BIOMASS DEPOSITION IN EBA

6.3.1 A General Approach to Understand Biomass Deposition

Biomass deposition is a relevant—but not fully understood—phenomenon in biochemical and environmental engineering, including EBA processing (64). Biomass adhesion on several types of surfaces has been studied by applying Classical Derjaguin, Landau, Verwey, and Overbeek (CDLVO) theory and Extended Derjaguin, Landau, Verwey, and Overbeek (XDLVO) theory (65,66).

For EBA, the XDLVO theory was found appropriate to determine "interaction" and "aggregation" onto the adsorbent surface in its various modes. XDLVO calculations were performed *via* experimental determination of contact angles (CAs) and zeta potential values for the interacting surfaces or particles. Therefore, our main goal has been to assess the interfacial forces acting between biomass

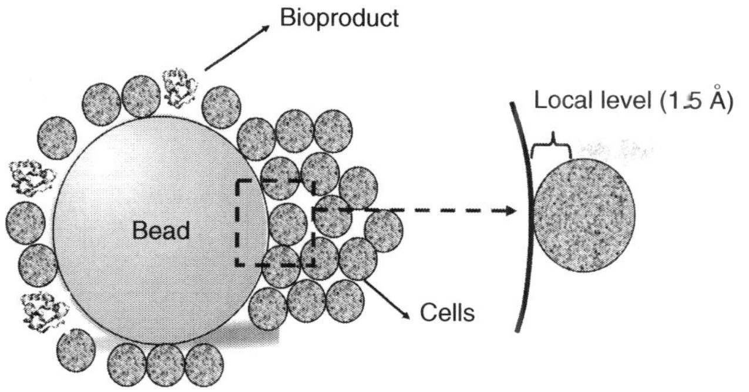

Figure 6.2. Biomass attachment onto the adsorbent beads may be due to cell-to-bead "interaction" or cell-to-cell "aggregation" phenomena, depending on the EBA system and the operational conditions.

(or biomass-derived particles) and chromatography adsorbents. Such forces can be of a different nature such as LW, hydrophobic attractive, hydrophilic repulsive and electrostatic. All these forces became operational at more- or- less short distances between the interaction surfaces; that is, at nanometer scale. By understanding these phenomena at the local level we would be capable of improving EBA process performance; it would become more robust and less complex to operate. Additionally, the understanding of these fundamental forces could help us to propose a series of heuristic rules as universal tools for bioprocess designers when a process/material for direct sequestration is in focus. The experimental XDLVO information generated has been correlated with actual EBA process performance and, additionally, by independent experimental methods like colloid deposition observation and laser diffraction measurements (67,68). A detailed description of the XDLVO-based procedure was described elsewhere (40,59).

6.3.2 Contact Angle

CA is the angle formed by a liquid at the three-phase boundary where a liquid, gas, and solid intersect. It is a quantitative measure of the noncovalent forces between the liquid and the first monolayer of the tested material. The three forces of interfacial tension shown in the Fig. 6.3 control the shape of the drop on the plain substratum.

To perform XDLVO calculations, CA measurements have to be determined with three diagnostic liquids that is, two high-energy polar liquids like water, formamide, and one high-energy nonpolar liquid like 1-bromonaphtalene. CA values are determined on lawns of hydrated biomass or crushed agarose beads. CA determination can also be performed on the immobilized layer of proteins on ultrafiltration membranes ("hydrated" condition) or on glass/plastic slides ("dehydrated" condition). For the estimation of the CA values, the sessile drop technique was utilized (69). A goniometric system can be used to measure the CAs (e.g. OCA 20 from Data Physics Instruments GmbH, Filderstadt, Germany). Data obtained can be acquired and analyzed utilizing commercial software (e.g. SCA20). Surface energy parameters (tensions) can be determined from CA measurements on the surfaces of interest utilizing the LW–AB approach (59,64,66).

The experimental information obtained from CA measurements is useful to calculate the LW and AB components of the free interfacial energy (ΔG), at closest distance of approximation. The interaction energy values (U), corrected by geometry, as a function of distance are shown in Fig. 6.4. The mathematical equations, which are utilized for such calculations are tabulated in Table 6.1.

6.3.3 Zeta Potential

The zeta potential is the electrostatic potential influenced by the accumulation of ions at the surface of a colloidal particle. A double layer structure can be recognized: the stern layer and the diffuse layer. As the particle moves through the solution, either due to gravity or an applied voltage, the ions (both counter- and co-) move with it. At some distance from the particle there exists a "boundary," beyond which

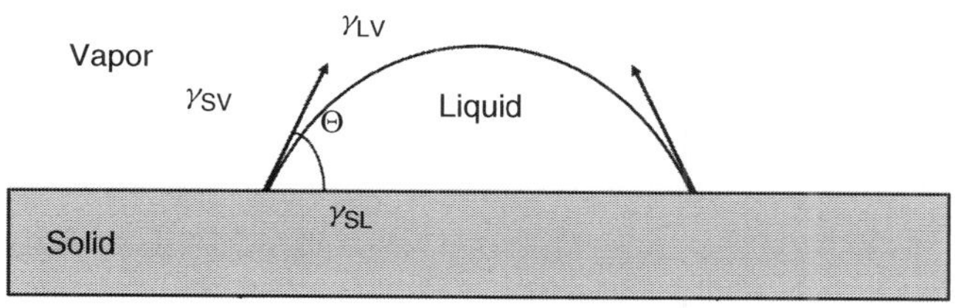

Figure 6.3. Illustration showing the contact angle of a liquid on a substratum. Contact angle measurements, with three diagnostic liquids, allow for the calculation of the LW and the AB components of the interaction energy between two bodies.

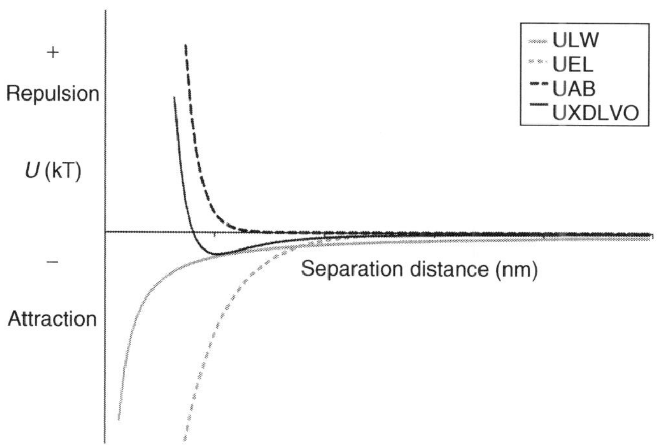

Figure 6.4. Interfacial free energy components and the total interaction energy, as a function of distance, according to the XDLVO theory.

TABLE 6.1. Mathematical Equations for XDLVO Simulations

Interaction Energy	Configuration: Sphere–Sphere	Configuration: Sphere–Plane
ΔG_{LW}	$U_{\mathrm{cwc}}^{\mathrm{LW}}(h) = \dfrac{-AR_{\mathrm{c}}R_{\mathrm{m}}}{6h(R_{\mathrm{c}}+R_{\mathrm{m}})}$	$U_{\mathrm{mwc}}^{\mathrm{LW}}(h) = -\dfrac{A}{6}\left[\dfrac{R_{\mathrm{c}}}{h} + \dfrac{R_{\mathrm{c}}}{h+2R_{\mathrm{c}}} + \ln\left(\dfrac{h}{h+2R_{\mathrm{c}}}\right)\right]$
$\Delta G_{\mathbf{AB}}$	$U_{\mathrm{cwc}}^{\mathrm{AB}}(h) = \pi R_{\mathrm{c}}\lambda\ \Delta G_{AB} \exp\left[\dfrac{h_0-h}{\lambda}\right]$	$U_{\mathrm{mwc}}^{\mathrm{AB}}(h) = 2\pi R_{\mathrm{c}}\lambda\ \Delta G_{AB} \exp\left[\dfrac{h_0-h}{\lambda}\right]$
$\Delta G_{\mathbf{EL}}$	$U_{cwc}^{EL}(h) = \dfrac{\pi\varepsilon_0\varepsilon_{\mathrm{r}}R_{\mathrm{c}}R_{\mathrm{m}}(\zeta_{\mathrm{m}}^2+\zeta_{\mathrm{c}}^2)}{(R_{\mathrm{c}}+R_{\mathrm{m}})} \left[\dfrac{2\zeta_{\mathrm{c}}\zeta_{\mathrm{m}}}{\zeta_{\mathrm{c}}^2+\zeta_{\mathrm{m}}^2}\ln\dfrac{1+\exp(-\kappa h)}{1-\exp(-\kappa h)} + \ln\{1-\exp(-2\kappa h)\}\right]$	$U_{\mathrm{mwc}}^{\mathrm{EL}}(h) = \pi\varepsilon_0\varepsilon_{\mathrm{r}}R_{\mathrm{c}}(\zeta_{\mathrm{m}}^2+\zeta_{\mathrm{c}}^2) \left[\dfrac{2\zeta_{\mathrm{m}}\zeta_{\mathrm{c}}}{\zeta_{\mathrm{m}}^2+\zeta_{\mathrm{c}}^2}\ln\dfrac{1+\exp(-\kappa h)}{1-\exp(-\kappa h)} + \ln\{1-\exp(-2\kappa h)\}\right]$

cwc cell water cell.
mwc matrix water cell.

the ions do not move with the particle. This is known as the surface of the hydrodynamic shear, or the *slipping plane*, and exists somewhere within the diffuse layer. The potential that exists at the slipping plane is defined as the *zeta potential*. A hypothetical representation of such layers is illustrated in Fig. 6.5.

Zeta potential values can be determined by measuring the dynamic electrophoretic mobility. From electrophoretic mobility data, zeta potentials are calculated according to the Smoluchowski's equation, in the case of colloidal particles, or as per the Henry's equation, in the case of globular proteins. The electrophoretic mobility can be determined utilizing a commercial available instrument for example, Zetasizer Nano ZS from Malvern instruments (Worcestershire, United Kingdom).

The experimental information obtained from zeta potential measurements is useful to calculate the electrostatic component of the free interfacial energy (U), which is shown as a function of distance in Fig. 6.4. The mathematical equations that are utilized for such calculations are tabulated (Table 6.1).

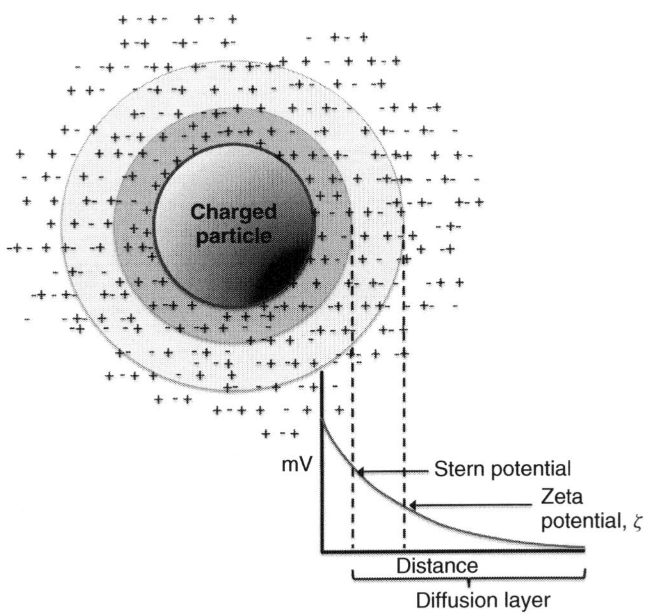

Figure 6.5. Illustration showing zeta potential of a charged particle in a suspension. This parameter is important to calculate the electrostatic component of the interaction energy between two entities.

6.3.4 Biomass Deposition and Surface Energetics

As previously mentioned, biomass interference during (direct) primary recovery may pose a serious technical challenge. Recently, our group has been systematically evaluating the physicochemical properties of several interaction pairs to better understand biomass deposition behavior in EBA (59,68,70). In doing so, the surface characteristics of biomass or biomass-derived material—taken as colloidal particles—versus the surface properties of the adsorbent beads (deposition surface) are determined indirectly via CA and zeta potential measurements. Subsequently, total (free) interfacial interaction energy values can be calculated as a function of surface distance in aqueous media for example, process buffer. This approach is universal and can be applied to any type of biomass or adsorbent surface. It is based on simple experimental determinations out of which relevant parameters are obtained in order to further calculate force or energy values, at the nanometer scale. Calculations cover a broad range of interaction forces, like LW, electrostatic, hydrophobic attractive, and hydrophilic repulsive. The latter are particularly relevant in the case of most protein chromatography supports in use today.

Several important findings were drawn from the above mentioned approach, as applied to a range of prototypical cell and adsorbent surfaces. The extent of interaction of freshly cultivated yeast cells (*Saccharomyces cerevisiae* FY 86, wild type, haploid) with anion- and cation- exchangers

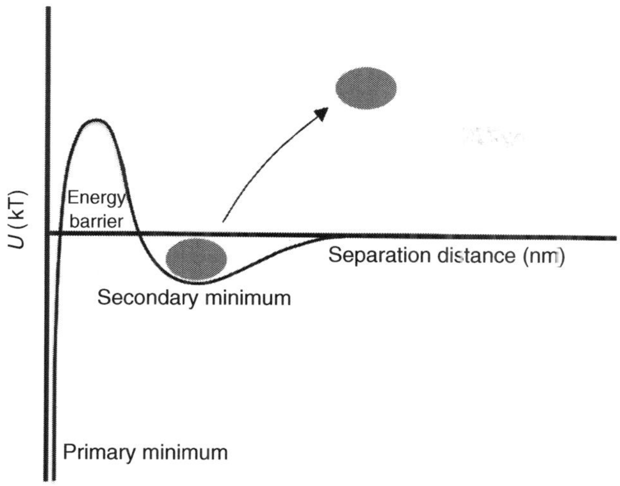

Figure 6.6. Secondary energy minimum where reversible cell adhesion can occur.

was explained by the calculated secondary energy minima values for example, reversible adhesion can be predicted as can be seen in Fig. 6.6. The degree of such interaction is also predicted to be dependent on the changes in buffer pH and conductivity, as expected from known process behavior (29). A positive correlation was found between energy minima values and cell deposition, as evaluated by the so-called cell transmission index (CTI) (59). Coulomb-type interactions, which are related to the measured zeta potential of the interacting bodies, were confirmed to be dominant (38). The XDLVO approach also gave us a clear idea about influence of biomass particle size on the extent of fouling. The total forces acting during biomass adhesion are greatly altered with the size of the cell or cell fragments, a fact reported previously (37,71). The particle size and charge have a clear influence on the probability of biomass attachment after collision with the adsorbent surface, a fact that is correlated with a change in the secondary energy minima values (59). Therefore, secondary interfacial energy minima values were experimentally validated via colloid deposition experiments (CDE). CDE experiments are useful to determine the attachment deposition coefficient (α), a lumped parameter reflecting biomass interaction and aggregation phenomena. CDE was proposed as a simple, straightforward and automatable diagnostic tool to evaluate biomass deposition onto EBA supports (66). CDE were sensitive enough to observe for the first time, changes in biomass deposition onto cation exchangers, which happened in buffers of low conductivity (1.0–10 mM), as a result of double layer compression effects within the "non-favorable" system under study (72). In the case of ion exchange EBA, due to the solution chemistry involved (typically, diluted buffers of neutral pH) no cell-to-cell aggregation was neither theoretically predicted nor experimentally observed.

Even though much attention has been paid in the literature to biomass interference with anion exchange EBA systems, the biomass challenge is also relevant in the context of hydrophobic interaction EBA (32,68). As hydrophobic interactions are expected to occur under high salt concentrations, a marginal contribution of charge-mediated effects is anticipated under such chromatographic mode of operation. At this point, the information provided by zeta potential measurement loses importance. However, a universal approach based on the XDLVO theory still remains useful and valid. In this case, the interaction of biomass with the chromatographic beads is mainly driven by LW and AB interaction forces. Indeed, biomass deposition onto Hydrophobic Interaction Chromatography (HIC) supports was correlated to the development of a (reversible) secondary energy minimum, which was observed to arise from LW and AB forces. Calculations showed the presence of only moderate interactions between yeast cells and phenyl adsorbent beads in the presence of high ammonium sulphate concentrations (0.4–1.6 M), which is customary in this type of separation process (73). On the other hand, XDLVO calculations were able to predict the existence of second phenomena: cell-to-cell aggregation taking place in the context of hydrophobic interaction chromatography. Due to the high salt concentrations involved, the repulsive electrostatic forces were negligible thus, allowing yeast cells to interact with each other; the balance between LW and AB forces was relevant to the observed aggregation phenomena. Buffer pH and conductivity have been found to influence cell-to-bead interaction and cell-to-cell aggregation. In both cases, predictions based on the XDLVO approach were validated by independent experimental measurements like biomass deposition experiments, laser diffraction spectroscopy, and confocal microscopy (68).

The XDLVO approach was extended to IMAC. Recent reports indicate that this pseudo-affinity mode, when applied to EBA, may promote adverse effects regarding biomass interference (31). EBA with "chelating beads" is usually run at moderate to high salt concentrations for example, mobile phases normally contain 0.25 M to 0.75 M sodium chloride. The information obtained from XDLVO principles was again relevant to explain the adverse effects on these supports at their operating conditions. The interaction of biomass with IMAC-Cu^{2+} was observed to be related to the development of a (reversible) secondary energy minimum. An influence of buffer pH and conductivity was observed within normal operational windows. From XDLVO calculations it was concluded that favorable interaction of yeast cells with chelating beads took place at pH $\geq$ 8. However, biomass deposition experiments failed to confirm such prediction that is, a decrease in deposition coefficient at pH 8 was observed. This anomalous behavior was explained

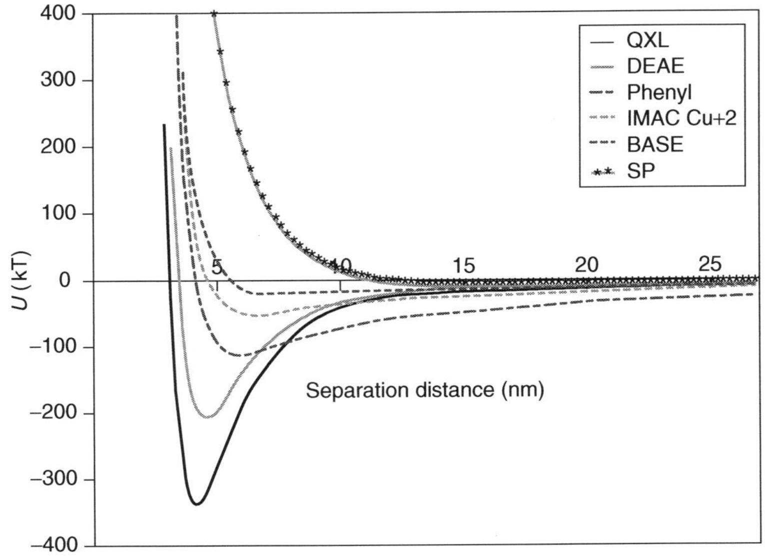

Figure 6.7. Total free energy of interaction as function of distance between intact yeast cells and several types of adsorbent beads. Physicochemical conditions for example, mobile phase composition, are assumed to be the ones commonly employed in each chromatography mode.

considering that Cu^{2+} can be actually sequestrated from the Chelating beads by yeast cells, a fact usually exploited for biosorption of metal ions from wastewaters (74). Other than this, XDLVO calculations were in full agreement with process behavior and experimentally observed biomass deposition. No cell-to-cell aggregation was neither theoretically predicted nor experimentally observed (70). The XDLVO profiles of the interactions of yeast on various adsorbent surfaces are depicted in Fig. 6.7.

6.3.5 Counteracting Biomass Deposition

Dainiak *et al*. (74) have modified a commercial anion exchanger intended for EBA by adding a layer of poly(acrylic acid) on the surface of such adsorbent beads. This procedure generated a shielding layer which marginally decreased the binding capacity of the ion exchanger for proteins but protected against interaction with yeast cells. The cell-resistant Poly (acrylic acid) (PAA)-coated anion exchanger was successfully employed and used for isolation of a model protein in the presence of biomass. Moreover, the layer of PAA was stable under mild elution conditions and the modified adsorbent could be used in the repeated purification cycles (75). The same concept was applied in another study (76) in which lactate dehydrogenase was recovered from a crude porcine muscle extract utilizing a Cibacron Blue 3GA adsorbent. A higher recovery of the target enzyme was obtained from the polymer shielded column as compared to the unshielded column. However, in the above mentioned approaches the polymer shielding layer is merely supported by electrostatic interactions and can be lost upon salt-promoted elution of the product which becomes contaminated.

Jahanshahi *et al*. (24) reported the use of particles characterized by a porous agarose pellicle cast upon commercial adsorbent beads. The presence of such (inert) polymer coat acted as a sieve reducing the nonspecific binding of cells and cell debris without diminution of selective capture of target protein from a complex feedstock. It was also reported that the presence of an inert polymer layer on native adsorbents not only prevents nonspecific binding of cells, but also guarantees better stability of the expanded beds by diminishing the formation of stagnant zones and flow channels (24).

Since the above mentioned studies have shown promising results, it was interesting to explore a range of shielding additives for commercial EBA supports and to utilize the XDLVO approach to explain (i) to predict strong binding shielding polymer on the bead surface and (ii) to evaluate the effect of the polymer layer on cell deposition. Testing of a number of chemicals resulted in the selection of polyvinylpirrolidone (PVP 360) as a useful shielding polymer for anion exchanger beads. The XDLVO clearly predicted the formation of a stable layer of the PVP onto an agarose bead (large energy minima values observed) and a decreased cell deposition of a native cell onto the shielded adsorbent (77). These assumptions were confirmed by actual EBA run where less hydrodynamic challenge was observed and protein dynamic capacities remained unaltered.

The XDLVO approach as applied to EBA has the potential, not only to provide a sound explanation of observed bioprocess behavior, but also to help in alleviating current EBA bottlenecks either by the selection of appropriate operational windows or by guiding novel adsorptive materials design, or by improving the available adsorbents.

6.4 SURFACE ENERGETICS AND PROTEIN ADSORPTION

Expanded Bed Chromatography, as any other type of chromatographic separation process relies on the capacity of selectively sequestrating the targeted product from a complex mixture. In EBA, this mixture contains not only soluble macromolecules but a range of suspended material with sizes ranging from the nano- to the micro- scale. Some nanosized entities for example, virus-like particles could be products themselves. Therefore, the question whether surface energetics would be useful to asses the adsorption of macromolecules and nanoparticles has a practical significance. As seen in the preceding section, XDLVO calculations were actually able to predict PVP covering of agarose beads.

Thermodynamic approaches to assess protein adsorption onto chromatography materials were already proposed. Some of the techniques utilized included isothermal titration microcalorimetry (78–80) and zeta potential measurements (81,82). Protein molecular interactions leading to separation from the solution phase for example, crystallization has been assessed with the help of virial coefficients (83–87). XDLVO calculations were shown to closely predict protein precipitation in the presence of ammonium sulphate (88). The latter approach may allow the study of proteins, macromolecules, and nanoplexes onto chromatography materials. This strategy could provide information on the adsorption of glycosylated proteins, which is difficult to gather by other experimental or computational methods. Indeed, glycan side chains can induce a combination of electrostatic, steric, and hydrophilic effects impacting on interfacial properties and deposition on adsorbent surfaces (89).

As opposed to chromatography systems where adsorption is mediated by charge (ion exchange and hydroxyapatite) hydrophobic interaction chromatography cannot be easily explained. Here, the XDLVO approach to understand protein adsorption behavior could facilitate the design of more effective downstream bioprocesses.

Recent studies focusing on the interaction of proteins with phenyl sepharose beads were performed in our group (unpublished observations). Following a similar experimental routine as the one described in the discussion on cell deposition, CA determinations were performed on layer of hydrated (on ultrafiltration membranes) and dehydrated (on glass and plastic) model proteins. This distinction was necessary since in the presence of high concentrations of ammonium sulphate, proteins are assumed to be in a dehydrated state. The "dehydrated" condition resembles the chemical environment found during protein adsorption in HIC; the opposite is true for protein elution (low salt; hydrated state).

After performing protein-support interaction calculations according to the XDLVO theory, it was found that model proteins in the dehydrated state (high salt concentration) generally presented a value for (free) interfacial energy minimum which was deeper than the values observed for proteins in the hydrated state. We can conclude from this observation that for the proteins in dehydrated state, the interaction energy (protein-support) becomes much larger at high salt concentration than the LW contribution to the total energy and consequently, the Hamaker's constant becomes higher. This phenomenon can explain why dehydration causes protein precipitation.

Calculated XDLVO energy versus distance profiles correlated with the observed chromatography behavior of a group of model proteins on phenyl sepharose. As illustrated in Fig. 6.8, the population of proteins under study was segregated into two subgroups according to surface energy calculations and elution behavior during chromatography. Moreover, calculations were able to discriminate early or late elution from a gradient chromatography experiment. It can be concluded that the higher the calculated interaction energy, the stronger the protein binding, and therefore a larger elution time will be observed.

In the case of proteins, interfacial energy of interaction with phenyl sepharose is very small in comparison to the values obtained with whole cells— ~ 0.16 kT and ~

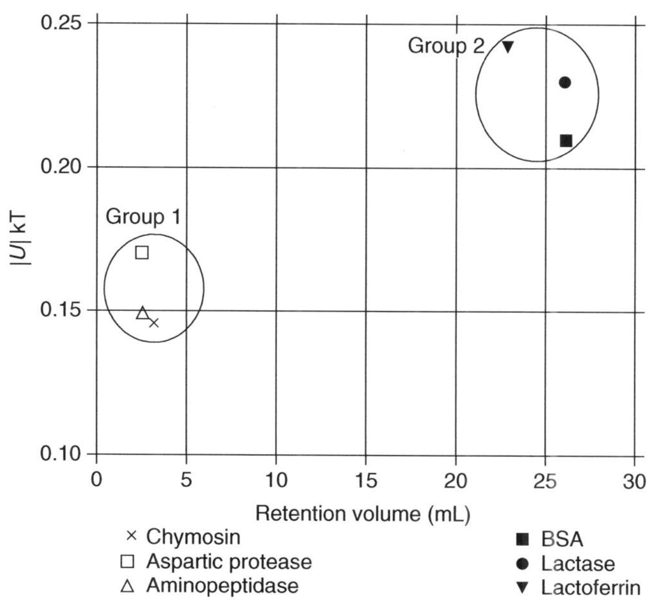

Figure 6.8. Correlation between retention volume and U for different class of proteins onto hydrophobic interaction chromatography surface.

80 kT, respectively. The reason is the influence of "particle" radius on the energy calculations—yeasts are typically 5–10 μm in diameter while globular proteins have a hydrodynamic radius within the range 2–20 nm. In summary, the XDLVO approach could also be useful in understanding protein chromatographic behavior particularly when there is a nondominant force leading to adsorption.

6.5 CONCLUSION

The biomass deposition/fouling phenomena that reduce EBA efficiency inevitably compromise its claimed advantages, as it could be the elimination of a clarification step before adsorption. However, the moderate cost and robustness of ion exchange and hydrophobic ligands makes them suitable for direct product recovery in EBA. Therefore, biomass deposition remains the main challenge that has to be solved in order to reach the highest productivity of the downstream operation.

As per the XDLVO theory, free interfacial energy versus distance profiles were calculated for a large combination of model biomass and adsorbent bead types. The energy minima values (absolute value/U) obtained for each of the analyzed cases were correlated with the corresponding deposition coefficient values (α). A positive correlation was obtained. It was concluded that total interaction energy $U \leq -25$ to -50 kT and biomass deposition parameter $\alpha \leq 0.15$ would be a safe region for EBA operation for example, biomass interference can be neglected (Fig. 6.9). Deviation in the correlation was found in such cases where moderate or high cell-to-cell aggregation occurs, for example, in hydrophobic interaction systems. In the later cases, laser diffraction studies were necessary to fully characterize the system.

The XDLVO approach was able to predict binding behavior of various model proteins on hydrophobic interactions beads. The routine presented here could be anticipated as a general approach to increase the understanding of biomass adhesion and protein adsorption onto process surfaces. The influence of factors like the composition of the mobile phase, the size of the interacting entities, and the nature of the functional groups present on the adsorbent surface can be adequately described. The approach is thus, useful during process and adsorptive material design and development.

The general principles of biomass interaction/aggregation and protein adsorption are beginning to be identified and understood for better bioprocess performance at industrial scale. With the help of colloid-chemistry principles, biomass interaction and protein sorption dynamics can now be observed from a different perspective.

6.6 NOMENCLATURE

AB	acid–base
CDE	colloidal deposition experiments
CTI	cell transmission index [—]
DEAE	diethylaminoethyl

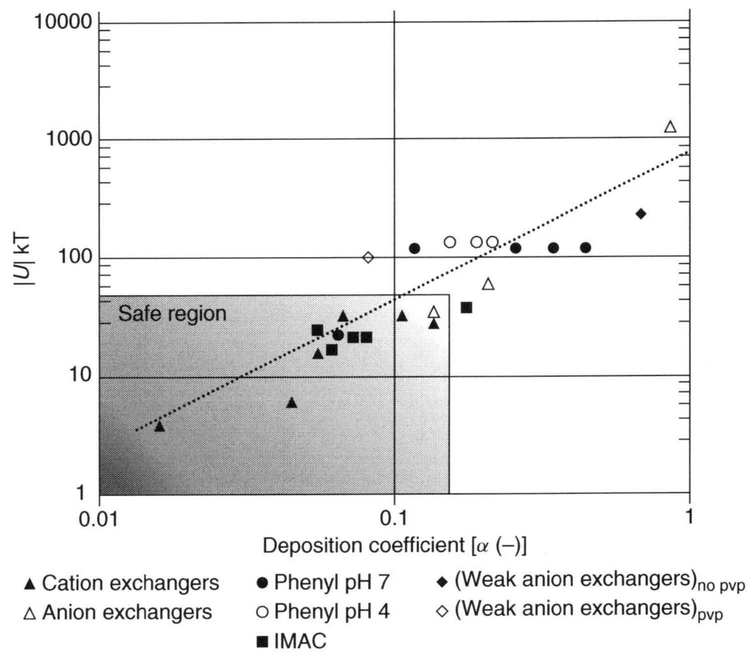

Figure 6.9. General correlation between the experimental deposition coefficient (α) and calculated free energy of interaction (U) for intact yeast cells onto various types of adsorbent beads.

DLVO	Derjaguin, Landau, Verwey, and Overbeek theory
EBA	expanded bed adsorption
EL	electrostatic
HIC	hydrophobic interaction chromatography
IY	intact yeast cells
IMAC	immobilized metal affinity chromatography
LW	Lifshitz–Van der Waals
PAA	poly(acrylic acid)
PVP	polyvinylpyrrolidone
RTD	residence time distribution
SP	sulphopropyl
U	total interaction energy [kT]
XDLVO	extended Derjaguin, Landau, Verwey, and Overbeek theory

6.6.1 Greek Letters

α	deposition coefficient
A	Hamaker constant [kT]
H	distance [m]
h_0	closest distance of approximation [1.57 Å]
ε_0	permittivity of vacuum [$\mathrm{Jm}^{-1}V^2$]
ε_r	relative permittivity or dielectric constant of water [—]
ξ_{m}	zeta potential of adsorbent
ξ_{c}	zeta potential of cell or protein
k^{-1}	double layer thickness (Debye length) [m]
R_{c}	radius of the cell or protein [m]
R_{m}	radius of the adsorbent [m]
λ	characteristic decay length [m]
ΔG	total interfacial free energy at closest distance [$\mathrm{mJ/m^2}$]
γ_{LW}	apolar or Lifshitz–Van der Waals component of surface tension [$\mathrm{mJ/m^2}$]
γ_{AB}	polar or acid–base component of surface tension (Lewis base) [$\mathrm{mJ/m^2}$]
γ^-	electron donor component of surface tension (Lewis base) [$\mathrm{mJ/m^2}$]
γ^+	electron acceptor component of surface tension (Lewis acid) [$\mathrm{mJ/m^2}$]

REFERENCES

1. Walsh G. Nat Biotechnol 2006; 24: 769–776.
2. Fischer R, Emans N, Schillberg S. In: Erickson L, Yu W-J, Brandle J, Rymerson R, editors. Molecular farming of plants and animals for human and veterinary medicine. The Netherlands: Kluwer Academic Publishers; 2002. pp. 259–285.
3. Twyman RM, Stoger E, Schillberg S, Christou P, Fischer R. Trends Biotechnol 2003; 21: 570–578.
4. Smith C. Nat Methods 2005; 2: 71–77.
5. Thiel KA. Nat Biotechnol 2004; 22: 1365–1372.
6. Gupta MN, Mathiasson B. Chem Ind 1994; 17: 673–675.
7. Menkhaus TJ, Bai Y, Zhang C, Nikolov ZL, Glatz CE. Biotechnol Prog 2004; 20: 1001–1014.
8. Evangelista RL, Kusnadi AR, Howard JA, Nikolov ZL. Biotechnol Prog 1998; 14: 607–614.
9. Chase HA. Trends Biotechnol 1994; 12: 296–303.
10. Barthels CR, Kleinman G, Korzon NJ, Irish DBA. Chem Eng Prog 1958; 54: 49–52.
11. Belter PA, Cunningham FL, Chen JW. Biotechnol Bioeng 1973; 15: 533–549.
12. Draeger MN, Chase HA. I Chem Eng Symp Ser 1990; 118:12.1–12.12.
13. Hjorth R, Kämpe S, Carlsson M. Bioseparation 1995; 5: 217–223.
14. Kämpe S, Hjorth R, Nyström L-E. 6th European Congress on Biotechnology; 1993; Florence.
15. McCormick DK. Biotechnology 1993; 11: 1059.
16. Mazumder J, Zhu J, Bassi AS, Ray AK. Biotechnol Bioeng 2009; 104: 111–126.
17. HealthCare GE. Application note STREAMLINE expanded bed adsorption. 2001–04. pp. 1150–1121 AA.
18. HealthCare GE. Application note STREAMLINE expanded bed adsorption. 2002–11. pp. 1144–1187 AB.
19. Walter JK, Feuser J. In: Extended Reports from the 4th International Conference on Expanded Bed Adsorption; 2002; Florida. Downstream EBA '02. Sweden: Amersham Biosciences; 2002. pp. 37–39.
20. Curbelo DR, Garke G, Guilarte RC, Anspach FB, Deckwer WD. Eng Life Sci 2003; 3: 406–415.
21. Low D, O'Leary R, Pujar NS. J Chromatogr B 2007; 848: 48–63.
22. Wurm FM. Nat Biotechnol 2004; 22: 1393–1398.
23. Expanded Bed Adsorption—Principles and Methods, Amersham Biosciences, Uppsala; 1998.
24. Jahanshahi M, Partida-Martinez L, Hajizadeh S. J Chromatogr A 2008; 1203: 13–20.
25. Dong W, Wang Y-D, Hu H, Su ZG. J Chem Technol Biotechnol 2007; 82: 135–142.
26. Ling T, Lyddiatt A. Biotechnol Bioprocess Eng 2006; 11: 84–87.
27. Anspach FB, Curbelo D, Hartmann R, Garke G, Deckwer WD. J Chromatogr A 1999; 865: 129–144.
28. Hubbuch J, Thommes J, Kula MR. Adv Biochem Eng Biotechnol 2005; 92: 101–123.
29. Fernandez-Lahore HM, Geilenkirchen S, Boldt K, Nagel A, Kula MR, Thommes J. J Chromatogr A 2000; 873: 195–208.
30. Feuser J, Walter J, Kula MR, Thommes J. Bioseparation 1999; 8: 99–109.
31. Poulin F, Jacquemart R, De Crescenzo G, Jolicoeur M, Legros R. Biotechnol Prog 2008; 24: 279–282.
32. Smith MP, Bulmer MA, Hjorth R, Titchener-Hooker NJ. J Chromatogr A 2002; 968: 121–128.
33. Brixius PJ, Lin D-Q, Mollerup I, Hubbuch JJ, Kula M-R. In: Extended Reports From the 4th International Conference on Expanded Bed Adsorption; 2002; Florida. Downstream EBA '02. Sweden: Amersham Biosciences: 2002. pp. 6–8.
34. Kaleas KA, Schmelzer CH, Pizarro SA. J Chromatogr A. DOI: 10.1016/j.chroma.2009.07.023.
35. Ng MYT, Tan WS, Abdullah N, Ling TC, Tey BT. J Biotechnol 2008; 138: 74–79.

36. Draeger MN, Chase HA. Bioseparation 1991; 2: 67–80.
37. Hubbuch JJ, Brixius PJ, Lin D-Q, Mollerup I, Kula M. Biotechnol Bioeng 2006; 94: 543–553.
38. Lin D-Q, Zhong LN, Yao SJ. Biotechnol Bioeng 2006; 95: 185–191.
39. William W, Wade MM, Holman SC, Champlin FR. J Microbiol Methods 2001; 43: 153–164.
40. Hermansson M. Colloids Surf B Biointerfaces 1999; 14: 105–119.
41. Tur KM, Ch'ng H-S. Int J Pharm 1998; 160: 61–74.
42. Lin D-Q, Brixius PJ, Hubbuch JJ, Thömmes J, Kula M-R. Biotechnol Bioeng 2003; 83: 149–157.
43. Gallardo-Moreno AM, Gonzalez-Martin ML, Perez-Giraldo C, Garduno E, Bruque JM, Gomez-Garcia AC. Appl Environ Microbiol 2002; 68: 2610–2613.
44. Klotz SA, Drutz DJ, Zajic JE. Infect Immun 1985; 50: 97–101.
45. Brixius PJ. PhD thesis. Dusseldorf, Germany: Institue of Enzyme Technology, Heinrich Heine University; 2003.
46. Vergnault H, Mercier-Bonin M, Willemot RM. Biotechnol Prog 2004; 20: 1534–1542.
47. Arpanaei A, Mathiasen N, Hobley TJ. J Chromatogr A 2008; 1203: 198–206.
48. Viloria-Cols ME, Hatti-Kaul R, Mattiasson B. J Chromatogr A 2004; 1043: 195–200.
49. Siu SC, Boushaba R, Topoyassakul V, Graham A, Choudhury S, Moss G, Titchener-Hooker NJ. Biotechnol Bioeng 2006; 95: 714–723.
50. Brant JA, Childress AE. Environ Eng Sci 2002; 19: 413–427.
51. Camesano TA, Logan BE. Environ Sci Technol 2000; 34: 3354–3362.
52. Råvika M, Cimanderb C, Elofssonc U, Veidea A. J Biochem Biophys Methods 2007; 70: 595–604.
53. Gao D, Lin D-Q, Yao S-J. Biochem Eng J 2008; 38: 355–361.
54. Gao D, Lin D-Q, Yao S-J. J Chromatogr B 2007; 859: 16–23.
55. Jahanshahi M, Najafpour G. In: Najafpour G, editor. Biochemical engineering and biotechnology. The Netherlands: Elsevier; 2006. pp. 390–415.
56. Hubbuch JJ, Heebøll-Nielsen A, Hobley HT, Thomas ORT. Biotechnol Bioeng 2002; 78: 35–43.
57. Van Der Meer AP, Blanchard CMRJP, Wesselingh JA. Chem Eng Res Des 1984; 62: 214–222.
58. Brixius P, Mollerup I, Jensen O, Halfar M, Thommes J, Kula M. Biotechnol Bioeng 2006; 93: 14–20.
59. Vennapusa RR, Hunegnaw SM, Cabrera RB, Fernandez-Lahore M. J Chromatogr A 2008; 1181: 9–20.
60. Fernandez-Lahore HM, Kleef R, Kula M-R, Thömmes J. J Biotechnol Bioeng 1999; 64: 484–496.
61. Chen W-D, Dong X-Y, Sun Y. J Chromatogr A 2003; 1012: 1–10.
62. Cabrera RB, Fernandez-Lahore HM. J Biochem Technol 2008; 1/1: 1–5.
63. Marani M, Camperi S, Cascone O, Iannucci N, Albanesi G, Fernández-Lahore HM. Sep Sci Technol 2005; 40: 3277–3287.
64. Absolom DR, Lamberti FV, Policova Z, Zingg W, Van Oss CJ, Neumann AW. Appl Environ Microbiol 1983; 46: 90–97.
65. Bos R, Van der Mei HC, Busscher HJ. FEMS Microbiol Rev 1999; 23: 179–230.
66. Tari C, Vennapusa RR, Cabrera RB, Fernandez-Lahore M. J Chem Technol Biotechnol 2008; 83: 183–191.
67. Vennapusa RR, Tari C, Cabrera RB, Fernandez-Lahore M. Biochem Eng J 2008; 43: 16–26.
68. Sharma PK, Rao KH. Adv Colloid Interface Sci 2002; 98: 341–463.
69. Vennapusa RR, Aasim M, Cabrera RB, Fernandez-Lahore M. Biotechnol Bioprocess Eng 2009; 14: 419–428.
70. Vergnault H, Willemot R-M, Muriel M-B. Process Biochem 2007; 42: 244–251.
71. Redman JA, Walker SL, Elimelech M. Environ Sci Technol 2004; 38: 1777–1785.
72. Queiroz JA, Garcia FAP, Cabral JMS. J Chromatogr A 1995; 707: 137–142.
73. Ting Y-P, Sun G. J Chem Technol Biotechnol 2000; 75: 541–546.
74. Dainiak MB, Galaev IY, Matiasson B. J Chromatogr A 2002; 942: 123–131.
75. Garg N, Galaev IY, Matiasson B. Bioseparation 1996; 6: 193–199.
76. Vennapusa RR, Aasim M, Shad N, Fernandez-Lahore M. *Abstracts from 15th International Conference on Biopartitioning and Purification*, Uxbridge, U.K., Brunel University, 0–27, (2009).
77. Lira RA, Minim LA, Bonomo RCF, Minim VPR, da Silva LHM, da Silva MCH. J Chromatogr A 2009; 1216: 4440–4444.
78. Chen W-Y, Liu Z-C, Lin P-H, Fang CI, Yamamoto S. Sep Purif Technol 2007; 54: 212–219.
79. Dias-Cabral AC, Ferreira AS, Phillips J, Queiroz JA, Pinto NG. Biomed Chromatogr 2005; 19: 606–616.
80. Schubert S, Freitag R. J Chromatogr A 2009; 1216: 3831–3840.
81. Faude A, Zacher D, Müller E, Böttinger H. J Chromatogr A 2007; 1161: 29–35.
82. Dumetz AC, Snellinger-O'Brien AM, Kaler EW, Lenhoff AM. Protein Sci 2007; 16: 1867–1877.
83. Payne RW, Nayar R, Tarantino R, Terzo SD, Moschera J, Di J, Heilman D, Bray B, Manning MC, Henry CS. Biopolymers 2006; 84: 527–533.
84. Jia Y, Narayanan J, Xiang-Yang L, Yu L. Biophys J 2005; 89: 4245–4251.
85. Valente JJ, Verma KS, Manning MC, Wilson WW, Henry CS. Biophys J 2005; 89: 4211–4218.
86. Warren PB. J Phys Condens Matter 2002; 14: 7617–7629.
87. Van Oss CJ, Good RJ, Chaudhury MK. J Protein Chem 1986; 5: 385–405.
88. Van Oss CJ, Moore LL, Good RJ, Chaudhury MK. J Protein Chem 1985; 4: 245–263.
89. Kreuß M, Strixner T, Kulozik U. Food Hydrocolloid 2009; 23: 1818–1826.

FURTHER READING

Mullick A, Griffith CM, Flickinger MC. Biotechnol Bioeng 1998; 60(3): 333–340.

Mullick A, Flickinger MC. Biotechnol Bioeng 1999; 65(3): 282–290.

7

FILTER AIDS

Tony Hunt

Advanced Minerals Corporation, Santa Barbara, California

7.1 INTRODUCTION

Among the most capable and attractive means of microparticulate (greater than 0.1 μ) separation used in biopharmaceutical processing today is the use of highly porous powdered media in dynamic systems. Powdered media, often called filter aids (Fig. 7.1), provide versatility, high solids-loading capacity, high product recovery, low cost, and ease of scale-up in any filtration process. Moreover, recent technical advances have stimulated a much greater breadth of applications that use this media filtration in biopharmaceutical processes, with a concomitant exponential increase in product innovation.

7.2 PROCESS CLARIFICATION

The first unit operation in downstream processing is clarification. It involves the removal of cells, cell debris, or precipitated components from a fermentation broth or process supernatant. Four types of solid–liquid separations are common to the industry: (*1*) The removal of whole cells in fermentation broths where the product is expressed into the supernatant; (*2*) The removal of cell debris from process broths after cell disruption and extraction to release the product of interest; (*3*) The removal of precipitated contaminants (usually proteins) from a process fluid, for example, in the removal of misfolded forms of recombinant proteins; and (*4*) The selective precipitation of the target protein and the capture of this precipitate from the process supernatant.

Most processes use either centrifugation or filtration for solid–liquid separations. Centrifugation, although widely accepted, has some disadvantages in biopharmaceutical applications. The challenge in many applications is to remove both cellular and subcellular debris without applying excessive shear forces on the solids. Centrifuges depend on differences in density and on the centrifugal force applied for solid–liquid separation. However, centrifugation becomes less practical when the product and waste possess similar densities. In addition, centrifuges can be difficult to maintain, clean, and sterilize. Finally, it is important to avoid the generation of aerosols when dealing with biological fluids, and this is often difficult to avoid when using centrifuges for this unit operation.

Depth filters containing diatomite, perlite, and cellulose are suited to low solids applications, because these products tend to blind quickly when subjected to moderate to high levels of solids. Their capacity can be increased by the addition of filter aid (as body feed) into the unfiltered broth, thus extending the lifetime of the pad.

The use of powdered media offers many advantages in biopharmaceutical applications. These systems are much more flexible and dynamic because powdered media addition can be adjusted based on changes in incoming solids level. In addition, they are extremely efficient at removing suspended solids and are particularly effective at clarifying supernatants heavily contaminated with colloids and small particles. The media can be used in all four applications described above, including the selective precipitation of the target protein and the capture of this precipitate from the process supernatant. Once captured on the filter media, the protein or product of interest can be redissolved by adjusting the pH or salt concentration in the wash buffer and recirculating this solution through the filter media bed.

Downstream Industrial Biotechnology: Recovery and Purification, First Edition. Edited by Michael C. Flickinger.

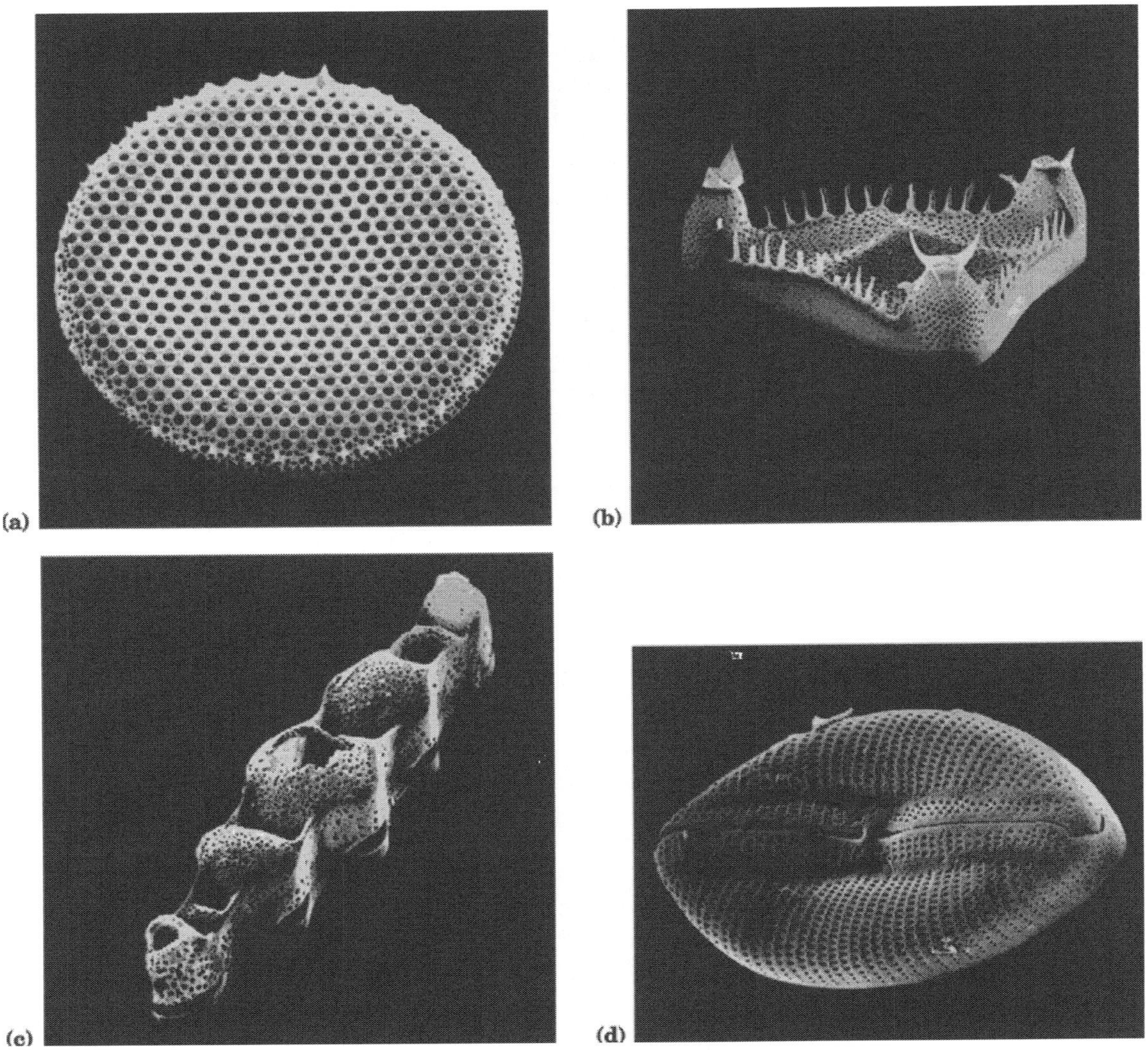

Figure 7.1. Size and shape of diatoms commonly found in diatomaceous earth deposits.

After resolubilization, the product can be eluted from the media with approximately two bed volumes of buffer.

Capital costs for powdered-media-based systems are relatively low when compared with centrifugation or other filtration systems. Because the media are disposable, media cleaning and lifetime studies do not need to be performed, hence the cost of validating the filtration process is substantially reduced. In addition, the ability to postwash the filter cake helps to maximize product recovery, a process that is extremely expensive or impractical with centrifuges. Finally, with improvements in both powder containment and filter equipment design, complete containment is now easily achieved both in the media preparation area and in GMP suites.

The combination of these improvements in powder containment, filter design, and powdered media technology position filter aids as a very cost-effective and economical approach to clarification.

7.3 POROUS MEDIA IN DYNAMIC PROCESS FILTRATIONS

Diatomite, perlite, and cellulose are the most widely used porous media in dynamic process filtrations, with a high percentage of applications using diatomite.

7.3.1 Diatomite

Diatomite is often used in combination with cellulose; however, diatomite is often the key ingredient of static or fixed-bed filters such as filter sheets and filter pads. Diatomite products are especially characterized by an inherently intricate and highly porous structure composed primarily of silica. These products are obtained from diatomaceous earth, a sediment greatly enriched in biogenic silica in the form of siliceous frustules of diatoms, a diverse array of microscopic, single-celled golden-brown algae of the class Bacillariophyceae. Surprisingly, these frustules are sufficiently durable to retain much of their ultrastructure nearly intact through long periods of geologic time when preserved in conditions that maintain chemical equilibrium.

7.3.2 Perlite

Perlite is a naturally occurring volcanic glass that thermally expands upon processing. After milling, porous, complicated structures result (Fig. 7.2) that are also useful in dynamic process filtrations of biopharmaceuticals. Because its structure is not as intricate as that of diatomite, perlite is better suited to the separation of coarse microparticulates from liquids having high solids loading. Like diatomite, perlite is also useful as a functional filtration component of filter sheets and filter pads.

7.3.3 Cellulose

Cellulose, like perlite, possesses a less intricate structure than diatomite. As a result, the use of cellulose is generally limited to coarse filtrations or in specialized applications where a fibrous precoat on the septum is required.

The rest of this chapter will focus attention on the use of diatomite in solid–liquid separations.

Figure 7.2. Scanning electron micrograph of perlite.

7.4 FUNDAMENTAL PRINCIPLES OF DIATOMITE FILTRATION

A discussion on solid–liquid separations can cover a wide range of techniques and applications, but each method is often defined by the relative size of solids being removed or collected from a given process stream. This chapter will focus on microparticulate filtration, where the solids are typically greater than 0.1 μ. The key objective is to remove unwanted solids from a process stream, so that the refined solution (filtrate) is adequately clarified and qualifies for further downstream processing.

7.4.1 Solids Composition

In general, solids can either be rigid or compressible in nature. Solids that form rigid cakes have some degree of inherent permeability, dictated by particle size distribution and packing arrangement, but most solids in biopharmaceutical applications are typically compressible. These often gelatinous, highly compressible solids retain a degree of permeability if low flow rates and low differential pressures are used for filtration. Unfortunately, these characteristics are contrary to typical processing requirements where this unit operation needs to be completed quickly and economically.

7.4.2 Dynamic Filtration

The problem in static fixed-bed filtration is that unwanted solids have very limited permeability when they collect or accumulate onto the filter septum (pad, paper, or fabric or metallic or plastic woven wire screen). If these solids build up on a septum, then eventually these solids will lack sufficient permeability for fluid drainage and filtration terminates. This behavior highlights a mistake frequently applied in filtration practice; often, higher differential pressure does not guarantee faster filtration rates precisely because these solids collapse to form an impermeable cake.

Introduction of a filter medium changes the composition of the accumulated cake, and therefore its filtration behavior. Powdered media (filter aids) are essential to this technology because they provide two functions: (*1*) as a precoat applied before the start of a filtration cycle and (*2*) as body feed added to the unfiltered feed throughout the filtration cycle. The precoat layer protects the filter septum, preventing penetration and blinding by the unwanted solids. This layer also facilitates septum cleaning and provides for a high-quality filtrate from the beginning of the filtration cycle.

The addition of powdered media (as body feed) to the process feed increases permeability in the accumulating filter cake, restricts solids movement, provides channels for filtrate recovery, and extends cycle length (Fig. 7.3). The

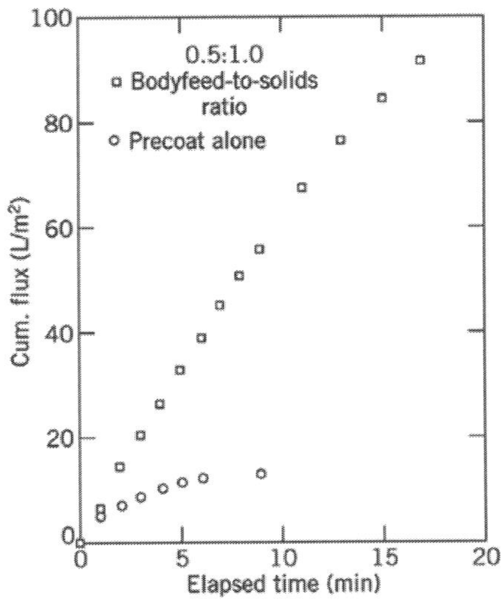

Figure 7.3. Addition of body feed to a filtration process extends the filtration cycle length.

extended permeability afforded by these media slows the rise in differential pressure for constant flow processes and retards the drop in flow for constant pressure operations. At the conclusion of a filtration cycle, product recovery can be maximized by rinsing or purging the accumulated cake in place. It is extremely important to maximize the body feed in a process. By adding too little, the solids quickly blind the filter cake and the filtration is terminated. Adding too much is not only wasteful but leads to a rapid increase in filter-cake thickness, with a corresponding increase in resistance to flow and a subsequent reduction in cycle length (1).

7.4.3 Filtration Theory Overview

The following section is not intended to be a comprehensive treatment of filtration theory, but rather an overview to describe how the permeability of an accumulated filter cake (solids and filter aid) ultimately affects filtration performance (rate). A thorough discussion of filtration mechanics can be found in numerous reviews (2–4).

Any discussion on porous media filtration can start with a simplified form of the Darcy equation (equation 1), which is used to describe laminar (nonturbulent) fluid flow through a homogeneous, porous medium.

$$k = \frac{Q \times \Delta x \times \mu}{A \times \Delta P} \tag{7.1}$$

The relative permeability (k) is measured in darcy units, and it is described in terms of the instantaneous volumetric flow rate (Q) and viscosity (μ) of the fluid, the thickness (Δx) and cross-sectional area (A) of the porous medium (accumulating filter cake), and the pressure drop or differential pressure (ΔP) associated with flow. The flow rate (Q) can also be represented by the differential change in volume over time (dV/dt). A value of 1 darcy corresponds to the permeability through a filter media 1-cm thick that allows 1 cm^3 of fluid with a viscosity of 1 cP to pass through an area of 1 cm^2 in 1 s under a pressure differential of 1 atm (i.e., 101.325 kPa). Although fluid viscosity varies with temperature and shear forces, values for it and the other variables can be obtained from tables and experimental observations. Consequently, the permeability (k) is an empirical value (constant) used to characterize the filter cake composition and not a property that can be readily altered by changing the process flow rate or differential pressure. In practice, permeability may appear to be influenced by flow rate or pressure, but those responses are strictly consequences of composition, and they are often inconsistent with the behavior described by equation 1. The type and relative usage rate of filter aid coupled with the nature and concentration of solids define composition, which in turn controls permeability and filtration performance.

The constancy of k explains why (*1*) the differential pressure increases as filter cake accumulates (Δx) for constant flow (Q_0) processes (equation 7.2)

$$k = \frac{Q_0 \times \mu}{A} (\text{constant}\ ) \times \frac{\Delta x}{\Delta P} \tag{7.2}$$

and (*2*) flow decreases as cake accumulates for constant pressure (ΔP_0) processes (equation 3)

$$k = \frac{\mu}{A \times \Delta P_0} (\text{constant}\ ) \times Q \times \Delta x \tag{7.3}$$

Permeability measurements are not essential in the application of dynamic filtration technology. In fact, once the unfiltered feed is introduced to the filter, the overall permeability (as measured in darcies) of the filter cake can be reduced by orders of magnitude. Ultimately, the manipulation of permeability (k), through filter aid selection and usage, describes the optimization of this solid–liquid separation process.

As an example, initial (laboratory scale) testing develops enough information to determine the permeability (k_{app}) in a constant pressure application. In fact, a key laboratory test is to determine the elapsed time (t_s) it takes to produce the maximum cake thickness (Δx_f; limited by the physical capacity of the filtration equipment, typically 3 cm) at optimum filter aid usage and terminal pressure (ΔP_f). Hence, the Darcy equation can be put to better use for process scale-up (equation 4):

$$[k_{app}\ ] = \frac{Q_{0\,lab}\ \times \Delta x_f\ \times \mu}{A_{lab}\ \times \Delta P_f} = \frac{Q_{0\ plant}\ \times \Delta x_f\ \times \mu}{A_{plant}\ \times \Delta P_f} \tag{7.4}$$

Notice that differential pressure (ΔP_f) and filter cake thickness (Δx_f) need not be affected by the scale of operations. Differences in allowable cake thickness (Δx_f)

must be accommodated by adjusting filtration area (A_{plant}) requirements. A physical capacity test also determines the total volume of process feed that can be filtered on the given area. In the simplest case of equal cake thickness, scale-up becomes a matter of sizing filter area to accommodate production rate (equation 5).

$$\frac{Q_{0\ \mathrm{lab}}}{A_{\mathrm{lab}}} = \frac{Q_{0\ \mathrm{plant}}}{A_{\mathrm{plant}}} \tag{7.5}$$

7.5 GRADE SELECTION AND OPTIMIZATION

An optimized filtration process maximizes throughput and product recovery and minimizes pressure drop while maintaining the desired filtrate clarity, all in a reasonable time. There are five main parameters one needs to consider when selecting an appropriate grade of powder media for a given process.

7.5.1 Product Stability

Careful attention should be paid to soluble metals when selecting a grade. High concentrations of soluble metals in reagents can lead to contamination of biopharmaceutical products, for example, aluminum contamination of albumin (5–7). Also, high concentrations of soluble metals are known to oxidize proteins and enzymes and can activate proteases in fermentation broths (8–10). Therefore, the use of high-purity reagents for filtration will help reduce or eliminate product contamination and degradation issues further downstream.

The industry demand for high purity and high performance reagents has led to the development of a new generation of powdered media. These products, sold under the trade name Celpure™ (Advanced Minerals Corporation, Calif.) offer greater filtration capacity with a corresponding reduction in powdered media consumption and reduction in disposal costs (Fig. 7.4).

Filtration processes with Celpure grades typically use less media because of a combination of the higher solids loading capacity and improved flow properties of the media. This results in a reduction in overall processing times compared with conventional grades of diatomaceous earth, such as the widely used Celite® grades (Fig. 7.5). Finally, the products are extremely pure and have very low levels of extractable metals such as aluminum and iron (Table 7.1) and correspondingly low electrical conductivities.

7.5.2 Filtrate Clarity

Most processes have a clarity specification that needs to be met or exceeded. Achieving stringent clarity specifications can extend the life of downstream filters and protect chromatography columns. The filtrate clarity achieved is dictated by the grade selected and the nature of the turbidity removed. Once the grade is chosen, its level of usage (body feed addition) combined with the available differential pressure to induce flow will control the volume of unfiltered feed that can be processed by a given filtration area.

7.5.3 Product Throughput

Product throughput and filtrate clarity are tightly linked when it comes to grade selection. The goal is to select a grade that achieves the desired clarity and maximizes throughput. By selecting a grade that is too fine, the clarity specification can be exceeded, but the throughput rate may be extremely low with correspondingly high differential pressures.

7.5.4 Operating Pressure

Actual operating pressure can be limited by shear sensitivity of the product or feedstream by equipment constraints or by overall plant design. Many processes have mechanical pressure limits of approximately 30 to 45 psi or 2 to 3 bar.

7.5.5 Product Recovery

This can be sometimes overlooked in initial filtration studies. At the conclusion of any filtration cycle, the accumulated solids should be washed with a product-compatible buffer to maximize product recovery. Upon completion of the filtration cycle (including rinse), it is important to confirm that the product of interest does not interact with the filtration media. Product recovery issues can often be improved by selecting a more permeable grade of filter aid. As you increase the porosity of a filter aid, the available surface area decreases and any nonspecific interaction decreases. All these factors are tightly linked; therefore, we recommend a systematic approach to grade selection and optimization.

7.6 SYSTEMATIC METHODS DEVELOPMENT APPROACH TO GRADE SELECTION

As stated above, the overall goal is to achieve optimum clarity and maximum throughput for a process while

TABLE 7.1. Typical Celpure™ Properties (Applicable to All Grades)

Property	Value
Soluble aluminum (mg Al/kg)	
Extraction in fermentation broth, pH 4.5.	<3
Soluble iron (mg Fe/kg)	
Extraction in fermentation broth, pH 4.5	<3
Conductivity in H_2O (μS-cm)	<12.5

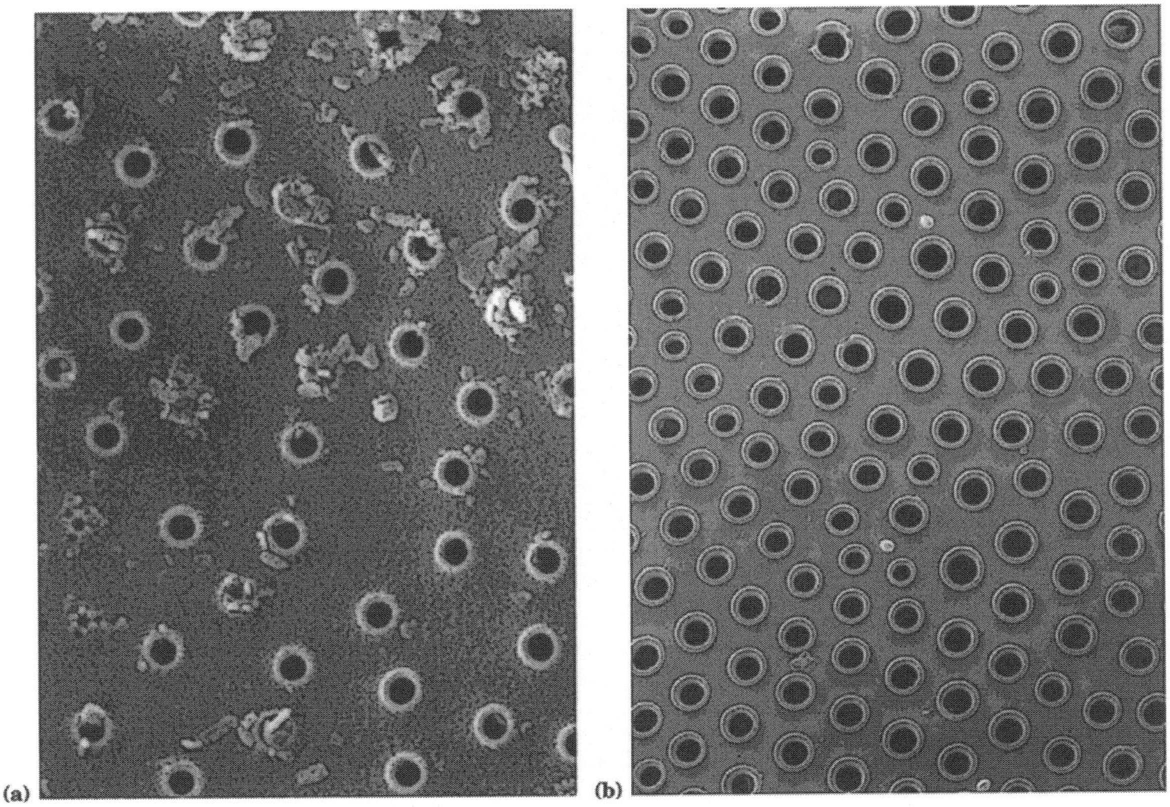

Figure 7.4. Scanning electron micrograph of the surface of an individual diatom from a conventional Celite grade (**a**) and a Celpure grade (**b**).

minimizing pressure and product losses. The following is a systematic approach to grade selection. Figure 7.6 shows a typical experimental setup for laboratory-scale constant rate filtration experiments. To perform constant pressure filtrations, a pressurized vessel is required to deliver unfiltered feed to the filtration housing.

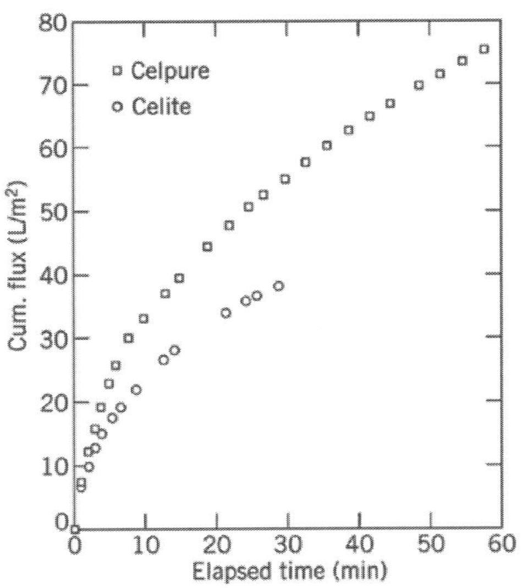

Figure 7.5. Equal weight comparison of Celpure and Celite grades in the clarification of a fermentation broth with 0.5:1.0 body-feed-to-solids ratio. Runs were terminated at minimum (equal) flow for reasonable production.

7.6.1 Determine the Percent of Solids in Your Feedstream

Any information on distribution and size of feed solids is useful in grade selection. If this information is not available, a recommended procedure is outlined below.

Determine percent solids by weight in your fermentation broth. Filter a small aliquot onto a preweighed filter disk, dry in an oven, and reweigh to determine percent solids in unfiltered feed.

1. Use a Whatman 934-AH filter paper as a septum with the filtration housing. This is an open-pore septum and enables you to precoat without having to custom make a wire screen.
2. Resuspend 1.5 g of media in 100 to 150 mL of buffer (similar in pH to your sample solution; 1.5 g of most grades gives a precoat of approximately 2 mm on 20 cm^2 of surface area). Start with the coarsest grade, Celpure™ 3000 or Celite® 545 AW. By starting with the coarsest grade, you will always achieve maximum throughput and optimal clarity for a given process.

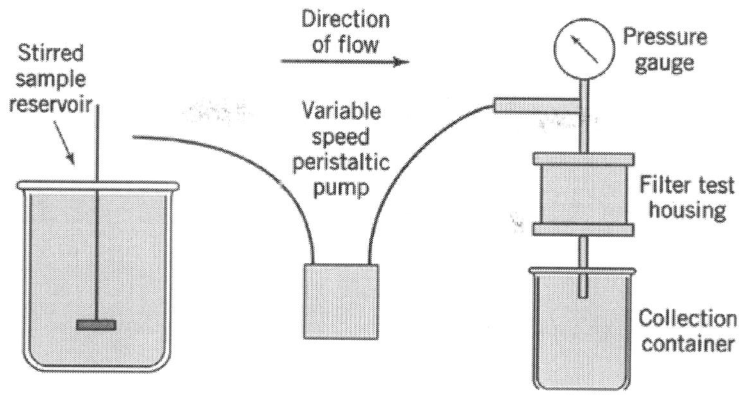

Figure 7.6. Experimental setup for constant rate experiments.

3. Recirculate the solution until filtrate is clear (typically 2 to 3 mins). The recommended flux rate is 40 L/m^2/min. On a 20-cm^2 surface area, this is equivalent to 80 mL/min.
4. Once the precoat has been formed, reduce the volume of buffer in the chamber until 2 to 3 mm above surface of precoat.
5. Add media to the unfiltered feed solution. As a starting point match the percent of solids in your feed with powdered media addition (2% solids by weight, add 2% by weight of powdered media). Add product/ media solution to filter chamber at a flux rate of 1 to 5 L/m^2/min. On a 20-cm^2 surface area, this is equivalent to 2 to 10 mL/min.
6. Measure initial flow rate and monitor both flow rate and pressure throughout the coarse of the filtration. Allow the pressure to rise to 30 psi or 2 bar (or maximum system pressure), and maintain this pressure for the duration of the filtration by reducing flow or until flow rate reaches 20% of initial flow.
7. At the end of the filtration cycle, wash the filter bed with two bed volumes of compatible wash buffer. This will maximize the product recovery for a process. Finally, measure the total volume processed, the time of filtration, filtrate clarity, and the bed height in the filter housing. Save a sample for product recovery analysis.

If clarity is not acceptable, then repeat steps 1 to 7 with a less permeable grade of filter aid. Powdered media are rated based on their Darcy permeability. Celpure 3000 and Celite 545 AW are considered the coarsest grades and Celpure 65 and Celite Filter-Cel® the finest.

For constant pressure filtrations, follow above procedure through step 4. Then switch to a pressurized vessel containing unfiltered broth mixed with appropriate grade of powdered media. Ramp the pressure up to the desired output and monitor decay in flow rate. Repeat experiments with different grades as outlined above.

7.6.2 Body Feed Optimization

Once grade selection is completed, the next step is to optimize the body feed addition. The level of body feed addition controls the permeability and volume of accumulating cake. Higher pressure does not mean greater throughput, especially when body feed usage coupled with the handling characteristics of the solids dictate filter cake compressibility. The dilatant nature of diatomite resists cake compression and thereby retains permeability. Feed properties (pH, viscosity, etc.) also affect filtration performance; therefore, these factors must be considered in process development and optimization studies.

Past work with biological solutions at high solids levels (greater than 5% by weight) suggests that constant flux (filtration) rates above 10 L/m^2/min exhibited steep rises in differential pressure because of compression and subsequent loss of cake permeability. Modest flux rates (5.0 L/m^2/min) extended filtration cycles, often producing better economy. Although dilution using a compatible solution can reduce shear and also make filtration easier, the goal to expedite processing implies that it will be used only after others options have failed.

As described earlier, an excellent starting point is to match the percent solids in the process stream on a weight basis with percent media also on a weight basis. If this ratio of 1:1 is not sufficient, i.e., the filtration terminates quickly because of a rapid rise in pressure and subsequent blinding of filter cake with solids, then consider increasing your body feed addition. Repeat the above steps using body feed additions of 1.5:1 or 2:1.

If a body feed addition of 1:1 allows you to process your batch with little or no rise in differential pressure, then you can optimize the body feed by reducing the

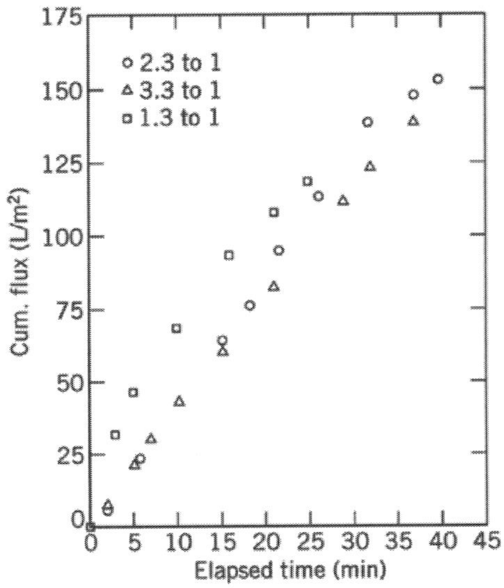

Figure 7.7. Body feed optimization study for a biological broth with body-feed-to-solids ratios.

ratio of powdered media to solids. Reduce the body feed until an acceptable throughput rate is achieved (Fig. 7.7). This approach will allow you to size an appropriate filter and minimize the size of filter needed to process a given batch.

7.6.3 Process Scale-Up and Filter Selection

Candle elements, pressure filters, and filter presses are all used in biopharmaceutical applications. Your choice of filter will depend on a number of factors, including plant design, equipment compatibility with existing systems, clean-in-place and steam-in-place considerations, scale of operation, and overall capital costs. All these factors need to be considered in selecting an appropriate filter. Once the filter design has been selected, the data generated in the methods development stage can be used to predict filter size. A smaller filter can be considered if the cycle time is short, and multiple cycles (including time for cleaning and setup) can be performed, in a typical 8 hour shift, to clarify the batch.

7.7 SUMMARY

The use of advanced filtration media in dynamic filtration systems is a flexible, efficient, and attractive means for the solid–liquid separation of microparticulates from various process streams. Powdered media provide versatility, high solids-loading capacity, high product recovery, low cost, and ease of scale-up in any filtration process. These systems are cost effective and extremely flexible and efficient at removing suspended solids and clarifying supernatants rich in colloids and small particles. Newer products such as Celpure™ demonstrate improvements in purity and performance and further address the needs of the biopharmaceutical industry.

In addition to the advances that have been made in powdered media, improved systems for media handling and containment are now available from vendors that readily meet the needs of the industry. Finally, a wide variety of economical filter housings of various designs, including candle elements, pressure filters, and filter presses, are manufactured to biopharmaceutical specifications.

REFERENCES

1. R.H. Rees and C.W. Cain, Chem. Eng. 72–74 (August 1990).
2. J. Kiefer, Brauwelt International IV, 1991, pp. 300–309.
3. J. Bear, Dynamics of Fluids in Porous Media, Dover Publications, New York, 1988, pp. 119–195.
4. J. Hermia and S. Brocheton, Filtration Separation 721–725 (November 1994).
5. W.P. Olson and R.S. Kent, Transfusion, 29: 86–87 (1989).
6. S.W. King, M.R. Wills, and J. Savory, Res. Commun. Chem. Pathol. Pharmacol. 26: 161–169 (1979).
7. M. Inoue, Y. Gion, H. Itoh, K. Ikariya, K. Takechi, S. Tomioka, K. Furuta, and S. Yabushita, Vox Sang. 66: 249–252 (1994).
8. W.A. Banks, L.M. Maness, M.F. Banks, and A.J. Kastin, Neurotoxicol. Teratol. 18: 671–677 (1996).
9. S.A. Hulea, E. Wasowicz, F.A. Kummerow, Biochim. Biophys. Acta 1259: 29–38 (1995).
10. N. Ahmed, J. Liggins, A.J. Furth, Biochem Soc. Trans. (England) 21: 93S (1993).

8

PROTEIN ADSORPTION, EXPANDED BED

SIDDARTHA GHOSE
Aston University, Birmingham, United Kingdom

8.1 INTRODUCTION

Traditional biochemical processes can be characterized into three major operational divisions: *(1)* the upstream process, which deals primarily with preparation and pretreatment of the raw material from which the specific target product is to be directly or indirectly obtained, *(2)* the actual production phase, in which the chemical transformation of the raw material takes place to produce the end product in which the desired target product is obtained along with associated by-products, and *(3)* the downstream processing, which is specifically involved with the recovery and purification of the target product. Progress in the area of recombinant DNA technology has made it commonplace to clone and express specific products for diagnostic and therapeutic applications within a variety of host microorganisms. This advancement has resulted in a pace of development of production phase technology that has far surpassed that of the recovery and purification areas. Recent years have seen significant advances in this latter area, and, with the incorporation of affinity-based interactions, it is now possible to include more powerful and specific operations for improving the overall efficiency of the recovery process (1).

Downstream processing refers to the collective set of operations involved in obtaining the target product in its required end form. The name derives from its relative position to the actual production stage in the process flow sheet. Factors such as the starting concentration of the target product, its final required purity, and the type of impurities to be removed dictate the exact choice of operations, and the sequence is necessarily adapted for each particular process. The use of microorganisms in biochemical processes makes cell harvesting, disruption (for the product residing within the cytoplasmic spaces as an intracellular product), and solid–liquid separation essential initial stages in the sequence of operations. Processes in which the product is expressed as an extracellular component require no cell disruption stage, although clarification of the feedstock from whole cells or cellular debris is always essential before subsequent treatment. The common solid–liquid separation methods traditionally used are centrifugation (2,3) and cross-flow microfiltration (4). The use of aqueous two-phase systems for primary treatment of feedstock (5) and cell debris removal has also been reported (6–9).

Biochemical product recovery nearly always involves one or more packed bed chromatographic processes as an often essential step of the separation protocol, usually in the secondary and final purification stages (10). It has now developed into a powerful tool for product separation at the analytical and preparative (11) scale of operation. The wide variety of modes of interaction between the solubilized product and ligands on the solid adsorbent (11–13), ease of scale-up and cleaning-in-place procedures, commercial availability of equipment and consumables, and the well-documented operation have given this technology wide acceptance in academic and commercial research and development. There is now a significant amount of information on the optimization of chromatographic separations (14). However, full utilization of its potential implies clarification of the biological feedstock before application onto a packed bed, failing which bed blockage is inevitable.

This chapter describes in some detail the use of expanded bed chromatography, which has been described as the only

Downstream Industrial Biotechnology: Recovery and Purification, First Edition. Edited by Michael C. Flickinger.

new unit operation that the bioprocess industry has seen for decades (15). This technique offers the possibility of directly applying the unclarified feedstock onto the bed, bypassing the conventional solid–liquid separation techniques while simultaneously capturing the desired product from the liquid phase. A typical sequence of separation operations is shown schematically in Figure 8.1, in which the relative position of expanded bed chromatography is also indicated. As will become clear from the description of its operation, the incorporation of this unit will enable facilities to replace cost-intensive and often problematic solid–liquid separation techniques such as centrifugation and cross-flow microfiltration. In addition, there is the added benefit of a chromatographic step early in the separation protocol. One must bear in mind that this technology is still in its development phase, and it is too early yet to conclude that it will become an integral part of a downstream processing protocol. However, as witnessed from the Second International Conference on Expanded Bed Adsorption recently held in the United States, there has been a growing awareness of the potential of the technology and it may indeed prove to have a promising future in bioseparations.

Before a detailed description of various aspects of expanded bed systems, a point of explanation should be made regarding the use of the terms *fluidized bed* and *expanded bed*. During the initial development of this technology, the terms were often used exclusively of one another. In classical chemical engineering operations, fluidized beds were associated with catalytic reactors in which a high degree of solid–fluid mixing was maintained to ensure good heat and mass transfer. Expanded beds, on the other hand, would benefit product capture if good plug-flow-type hydrodynamics could be obtained. Clearly mixing needed to be minimized and bed stability ensured. The two terms were clearly in contradiction with each other; to maintain the existence of the critical difference, the use of the term *fluidized beds* was avoided in the context of biological separations by *expanded beds*. Over the years, however, this distinction has eroded, and now both are interchangeably used to mean essentially the same thing. The same applies in this chapter.

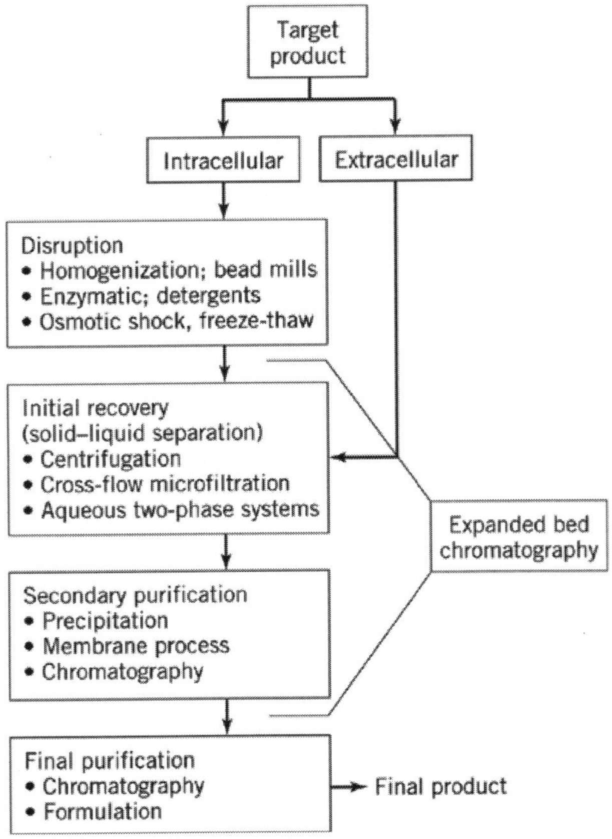

Figure 8.1. Typical sequence of operation in the recovery and purification of proteins from biological feedstock. The relative position of expanded bed as a viable alternative operation is also indicated.

8.2 THEORY

To understand the physical process by which an expanded bed is obtained and some of the basic expressions describing its various states, we begin with a settled bed of adsorbent packed in a column with no top flow adapter. When an upward flow of liquid is introduced at the lower end, there is a tendency for the adsorbent particles to move in the direction of flow. This movement occurs only when the liquid velocity exceeds the minimum fluidization velocity of the particles. The bed then undergoes a transient expansion phase, at the end of which the adsorbent particles attain a steady position when the upward force due to the liquid flow exactly counterbalances the downward gravitational force on the particles. The bed is then said to be fluidized and will remain in this state as long as the two forces counterbalance each other.

When the fluid velocity equals the minimum fluidization velocity of the particles, the bed is said to be in a state of incipient fluidization. At this point the bed can be considered packed for the purpose of estimating the pressure drop across it and fluidized for calculating the buoyant weight of the particles. By using the Ergun equation for pressure drop and equating the two forces, we obtain an expression for the minimum fluidization velocity:

$$150\,\frac{(1-\varepsilon_{\mathrm{mf}})^2}{\varepsilon_{\mathrm{mf}}^2}\cdot\frac{\mu\cdot u_{\mathrm{mf}}}{d_{\mathrm{p}}^2}+1.75\,\frac{1-\varepsilon_{\mathrm{mf}}}{\varepsilon_{\mathrm{mf}}^2}\cdot\frac{\rho u_{\mathrm{mf}}^2}{d_{\mathrm{p}}} = (1-\varepsilon_{\mathrm{mf}})(\rho_{\mathrm{p}}-\rho)g \qquad (8.1)$$

The pressure drop term on the left-hand side of equation 8.1 has two parts; the first term denotes the contribution due to laminar flow, and the second is due to turbulent flow. Depending on the flow regime, one could simplify the expression by neglecting the appropriate term. In the laminar flow region, the minimum fluidization velocity can be derived from equation 8.1 as

$$u_{\mathrm{mf}} = \frac{\varepsilon_{\mathrm{mf}}^2}{150(1 - e_{\mathrm{mf}})} \cdot \frac{1}{\mu} d_{\mathrm{p}}^2(\rho_{\mathrm{p}} - \rho)g \tag{8.2}$$

The physical parameters that affect the velocity at which a settled bed of particles begins to fluidize can be seen from equation 8.2 and are the viscosity of the fluid, the particle size of the adsorbent beads, and the density difference between the adsorbent and the liquid. An accurate estimate of the bed voidage at the point of minimum fluidization is also important for estimating u_{mf}. For spherical particles, 0.4 is always a good first approximate valid when the packing is not overly compact. Wen and Yu (16) have found a value of 0.42 to be satisfactory.

Once the bed is in a fluidized state, a simple relation exists between the bed voidage and the fluid velocity. This relation is valid over a given range of the latter, and the most commonly used relationship is that described by Richardson and Zaki (17), which relates the superficial fluid velocity, the terminal settling velocity of the particle, and the bed voidage. The relationship is expressed as

$$\frac{u}{u_{\mathrm{t}}} = \varepsilon^n \tag{8.3}$$

where n is the Richardson–Zaki coefficient. The determination of n and u_{t} is usually obtained from a logarithmic plot of experimentally determined values of the bed voidage and the corresponding flow velocity and forms a fundamental part of the study of these systems.

However, bed voidage is not an easily measured parameter and would in turn require further experimentation. A more convenient approach is to use the relationship between the bed voidage at a given bed expansion, ϵ, in terms of the expanded bed height, H_{exp}, the initial (or settled) bed height, H_{o}, and the voidage of the settled bed ϵ_{o}. Thus we have

$$\varepsilon = 1 - \frac{H_{\mathrm{o}}}{H_{exp}} \cdot (1 - \varepsilon_{\mathrm{o}}) \tag{8.4}$$

The ratio $H_{\mathrm{exp}}/H_{\mathrm{o}}$ is referred to as the degree of bed expansion. Experimentally, one would measure the stable expanded bed height, H_{exp}, at a given flow velocity through the column (18–20). The data can then be plotted to obtain n and u_{t} for the bed. Analysis of equations 8.3 and 8.4 shows us that a high u_{t} for a particle results in a lower bed expansion and vice versa.

The terminal velocity is the upper limit of validity of the Richardson–Zaki expression, indeed of fluidization in general. At a fluid velocity equal to the terminal velocity of the particles, the particles are no longer suspended in the fluid but travel along with it. In the context of expanded bed chromatography, this leads to what is known as elutriation, where the adsorbent particles are physically removed from the column. Clearly, the latter phenomenon is highly undesirable, and it is therefore essential that for effective operation of expanded beds, the velocity range lies well within the limits of the minimum fluidization velocity at the lower end and the terminal velocity at the higher. The value of u_{t} can also be determined theoretically from Stokes law; thus

$$u_{\mathrm{t}} = \frac{d_{\mathrm{p}}^2(\rho_{\mathrm{p}} - \rho)g}{18\mu} \tag{8.5}$$

The coefficient n has been estimated by Richardson and Zaki (17) to be a function of both the particle to column diameter and the Reynolds number. The correlations are valid for uniform spherical particles >100 μm in size. Garside and Al Dibouni showed that a different set of correlations are required when the fluidized bed voidage is greater than 0.85. Two expressions were proposed by them—one general type and a second similar to the Richardson–Zaki type of formulation. Both are valid over the entire range of Reynolds number. Some common expressions developed by different workers for estimating the values of n and u_{t} are shown in Table 8.1. Values determined experimentally or through the correlations are used for comparative evaluation of different adsorbents for their potential use as fluidized bed adsorbents. The equations are a simple and effective starting point for evaluating some of the fundamental characteristics of an expanded bed.

8.3 PRINCIPLES OF OPERATION

The operational procedure for protein recovery using an expanded bed is similar in principle to that of a conventional packed bed column and is schematically depicted in Figure 8.2. As mentioned in the previous section, the initial step involves formation of the stable expanded bed from its settled or packed state. This is done by first moving the top flow adapter to a sufficient height to permit free movement of the top of the bed in the upward direction of liquid flow. When flow of the equilibrating buffer is initiated, bed expansion occurs and the top of the bed rises. The bed height increase is usually characterized by a linear phase followed by a progressively reducing nonlinear rate of change until a final stable height is achieved. This expanded height depends on a number of physical factors of the system, which are described in the next section. Once a stable bed has been obtained, the top flow adapter is usually repositioned to lie anywhere between 1 and 15 cm

TABLE 8.1. Fluidization Parameters for Uniform-Sized Spherical Particles

Reference	Validity Range	Equations
Richardson and Zaki (17)	100–6,350 d_p (μm)	$\frac{u}{u_t} = \varepsilon^n$ where
	1.06–11.25 ρ_p (g/mL)	
	0.81–2.89 ρ_l (g/mL)	
	10^{-3}–113.10^{-3} μ(Ns/m^2)	$n = 4.65 + 20\frac{d_p}{d_c}(\text{Re} < 0.2)$
		$n = \left(4.4 + 18\frac{d_p}{d_c}\right)\text{Re}^{-0.03}(0.2 < \text{Re} < 1)$
		$n = \left(4.4 + 18\frac{d_p}{d_c}\right)\text{Re}^{-0.01}(1 < \text{Re} < 200)$
		$n = 4.4\,\text{Re}^{-0.1}(200 < \text{Re} < 500)$
		$n = 2.4(\text{Re} > 500)$
Garside and Al Dibouni (xx)	167–3,070 d_p (μm)	$\frac{(u/\varepsilon \cdot u_t) - A}{B - (u/\varepsilon u_t)} = 0.06\,\text{Re}^{(\varepsilon+0.02)}$ where
	2.8 ρ_p (g/mL)	
	1.0 ρ_l (g/mL)	$A = \varepsilon^{4.14}$
	10^{-3} μ(Ns/m^2)	$B = 0.8\varepsilon^{1.28}$ for $\varepsilon \leq 0.85$
		$B = \varepsilon^{2.65}$ for $\varepsilon > 0.85$
Wen and Yu (16)	200–350 d_p (μm)	$\varepsilon^{4.7}\left(\frac{d_p \cdot g \cdot (\rho_p - \rho_l)\rho_l}{\mu^2}\right) = 18\,\text{Re} + 2.7\,\text{Re}^{1.687}$
	2.37–7.84 ρ_p (g/mL)	
	1.0 ρ_l (g/mL)	$0.01 < \text{Re} < 10^4$
	10^{-3} μ(Ns/m^2)	

above the top of the bed. The actual position depends on the mode of operation employed (see later section). With the top adapter in position, the bed is essentially ready for initiation of feedstock load.

The flow is then switched from the buffer to the unclarified feedstock, usually by an appropriate valve mechanism. During the feedstock load, the solid matter, consisting of cell debris or sometimes whole cells, travels directly through the column while some of the solubilized products are adsorbed onto the column. Ideally there will be no retention of the particulate debris within the column, but in practice one could well observe a degree of nonspecific binding to the adsorbent particles. This issue becomes of particular relevance during cleaning-in-place and column regeneration. Feed load is usually terminated when the outlet concentration of the product being recovered is approximately 5–10% of its concentration in the feed. At this point the inlet flow is then switched back to the equilibrating buffer to wash out the residual particulate matter from the expanded bed.

When the outlet liquid is completely clear of insoluble material, the product can be eluted from the column, most commonly in the packed mode and after the bed is first allowed to settle by terminating the flow of liquid. The top flow adapter is then lowered onto the settled bed, bringing it to a packed mode. The bound products are then eluted by the appropriate elution buffers, usually in a single step because specific separation is not the primary objective at this stage of the recovery process. When elution is complete, the column is regenerated and where required cleaned. The bed is then expanded with the equilibration buffer and made ready for the next cycle of feedstock load.

8.3.1 Bed Stability

The sequence of steps in the operation of an expanded bed is relatively straightforward and simple to incorporate. As described earlier, the first step is obtaining a stable expanded bed; although not immediately obvious, this is a critical requirement with implications for overall process efficiency. A stable bed can be defined as one in which the liquid–liquid and solid–liquid mixing is low. Bed stability becomes a necessary requirement for good plug-flow-type hydrodynamics, which in turn affects the product binding characteristics. This is best quantified through residence time distribution (RTD) studies by analyzing the outlet profile of a tracer pulse in exactly the same manner in which it would be done for a packed bed system. Tracers such as acetone are effectively used in the absence of any particulate material. However, RTD experiments conducted in the absence of such particulates only indicate the state of the bed before feedstock load is initiated. Unlike packed bed chromatography, in which the hydrodynamics of the bed can be assumed to be more or less the same throughout

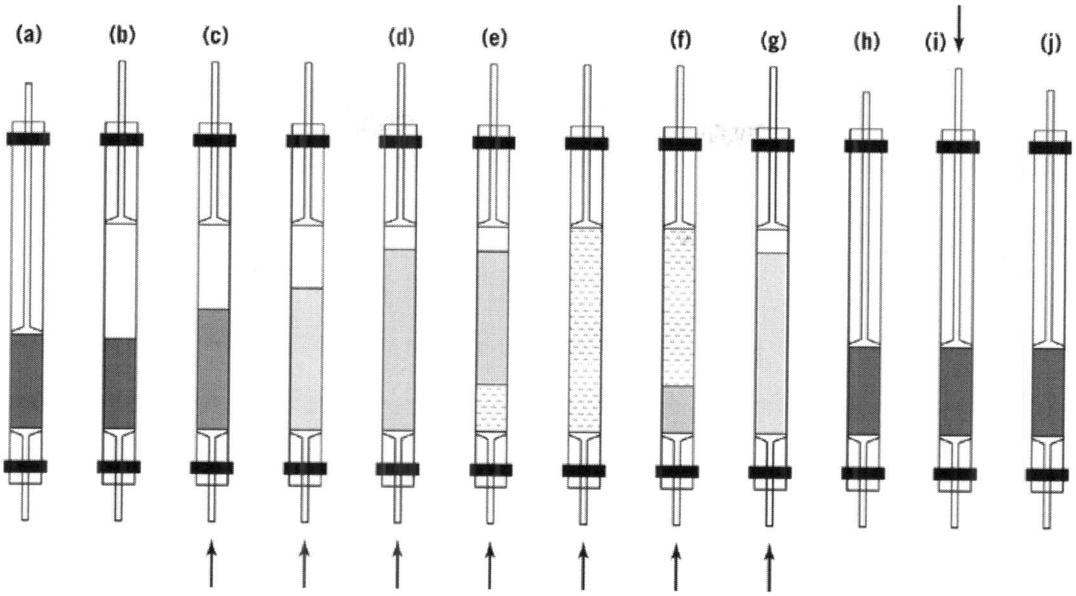

Figure 8.2. The various stages of operation of expanded bed chromatography from start-up to column regeneration. **(a)** Starting with a settled bed of adsorbent, **(b)** the adapter is moved up to remove any limitations on the movement of the bed. **(c)** Liquid flow through the column is then initiated from the bottom in the upward direction, during which time the bed undergoes a transient phase until **(d)** it finally reaches a stable steady-state height. The bed is equilibrated with buffer and **(e)** feedstock load initiated until the column is saturated, as seen from the breakthrough of bound product from the column. **(f)** The load is terminated and buffer wash started until **(g)** all the homogenate is washed out from the column. **(h)** The flow is then terminated, the bed allowed to settle, and the flow adapter lowered to attain a packed bed configuration, after which **(i)** the products are eluted, usually in the reverse direction. **(j)** The column is then cleaned and regenerated and made ready for the next run.

the load phase, expanded beds are more than likely affected to a certain degree by the presence of the particulate matter. Although the influence need not necessarily be significant, it is nevertheless an important point to remember when conducting stability tests in these systems.

The tracer outlet profile can be evaluated by the moments rule to obtain an estimate of the Peclet number or the vessel dispersion number (21), defined as

$$\text{Pe} = \frac{u}{D_{\text{ax}} \cdot L} \tag{8.6}$$

where D_{ax} is the liquid axial dispersion coefficient. Values of Pe greater than 40 are considered acceptable in terms of the flow hydrodynamics. The Peclet number is used interchangeably with the Bodenstein number, defined in the same manner and used extensively by Karau *et al.* (20) and Thommes *et al.* (22).

For the experienced user of expanded bed chromatography columns, the power of visual observation, although highly qualitative and impossible to validate, can often prove to be sufficient in ascertaining the extent of bed stability. Liquid–liquid (and thus solid–liquid) mixing is often a result of radial flow inequalities. Some of the common causes are poor distribution at the inlet to the column (which often transmits through the full length), inaccurate vertical alignment of the column (which is particularly important in scale-down systems), and aggregation of the adsorbent particles (which leads to significant variation in the bed voidage along the length of the column and thus affects the flow pattern). The consequential gross flow inadequacies can be more than sufficiently detected by the naked eye, and although an accurate determination of the Peclet number is not possible, visual observation often gives sufficient indication when a bed has yet to reach a stable configuration. Experience clearly plays a significant role in the use of visual observation as a tool for determining bed stability, and it has only limited applicability. Nevertheless, it may often prove sufficient when a rigorous residence time distribution experiment is not essential.

Stability of expanded beds is achieved through the inherent particle size distribution of the adsorbent. From the theory of fluidization outlined in the previous section, smaller particles fluidize more readily than larger ones in a given liquid stream, assuming that the particles are uniformly dense. For fluid velocity ranging between the minimum fluidization velocity of the smaller particle to the terminal velocity of the larger one, different situations arise, as depicted in Figure 8.3. As the flow velocity is increased such that $u_{mf,large}$, $u_{t,small}$, $u_{t,small} > u > u_{mf,small}$ the smaller particles fluidize and attain a steady-state position somewhere along the length of the column. Continual increase in the flow until $u_{t,large}$, $u_{t,small} > u > u_{mf,small}$, $u_{mf,large}$ results in both the small and large particles being fluidized. However, the increased flow velocity requires a longer column length for the inertial forces to drop to the level to be sufficiently counterbalanced by the buoyant force of the smaller particles. It therefore attains its steady state higher up along the column relative to both its initial position and that of the larger particle. A further increase such that $u > u_{mf,large}$, $u_{t,small}$ would wash the small particles out from the column and move the steady-state position of the larger particles higher up. Finally, when $u > u_{t,large}$, the larger particles are also elutriated.

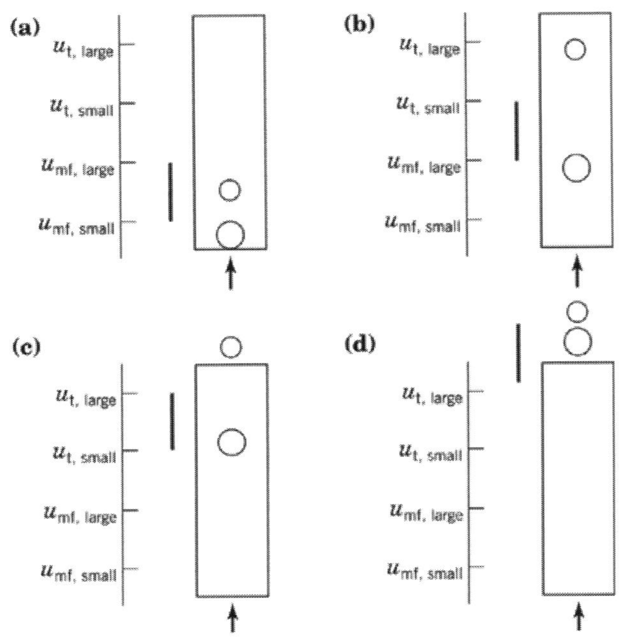

Figure 8.3. Schematic diagram showing the effect of flow velocity on the fluidization of two particles of different sizes. **(a)** As the flow velocity is increased such that $u_{t,large}$, $u_{t,small}$, $u_{mf,large} > u > u_{mf,small}$ the smaller particles fluidize and attain a steady-state position somewhere along the length of the column. **(b)** Continual increase in the flow until $u_{t,large}$, $u_{t,small} > u > u_{mf,large}$, $u_{mf,small}$ results in both the small and large particles being fluidized. The increased flow velocity, however, requires a longer column length for the inertial forces to drop to the level to be sufficiently counterbalanced by the buoyant force of the smaller particle. It therefore attains its steady state higher up along the column relative to both its initial position and that of the larger particle. **(c)** A further increase such that $u_{t,large} > u > u_{t,small}$, $> u_{mf,large}$, $u_{mf,small}$ would wash the small particles out from the column and move the steady-state position of the larger particles higher up. Finally, when $u > u_{t,large}$ the larger particles are also elutriated. The vertical line shows the flow velocity range on a relative scale between the limits.

Extending the preceding discussion to an expanded bed of adsorbent with a defined particle size distribution means that the particles will be stratified along the length of the column. This is schematically depicted in Figure 8.4 for a bed of adsorbent in the settled and stable expanded states. In the absence of gross flow inequalities, each particle remains in its steady-state position, with the larger ones at the bottom or inlet to the column and the smaller ones near the top (23). This configuration can be maintained indefinitely. In reality there is localized movement of the particles, but it contributes little to liquid–liquid mixing within the system.

The other most notable method for stabilizing fluidized beds is through the use of magnetic fields (24–27). A magnetic material is incorporated within the adsorbent beads, and the superimposition of the magnetic field acts to limit the movement of the adsorbent particles. The flow of particulate-laden feedstock can often be conducted at relatively higher flow velocities, enhancing the efficiency of its wash through the column.

8.3.2 Modes of Operation

Although stability has been discussed in terms of liquid–liquid and solid–liquid mixing, bed height movement must be considered. Because the top of the bed is not restricted in any way, changes in the flow or physical properties of the inlet liquid directly, although not always instantly, influence

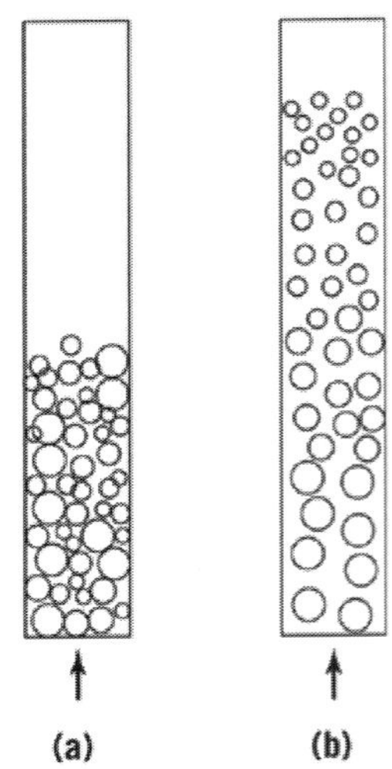

Figure 8.4. Schematic diagram showing the stratification of particles based on size along the length of the column going from **(a)** the randomly distributed settled bed state to **(b)** the stable expanded one.

the bed height. Thus, when switching between buffers of different viscosities, densities, or both, the bed exhibits a concomitant movement to achieve a stable height corresponding to the new physical conditions to which it has been subjected. This movement is also seen when feedstock load onto a stable expanded bed is initiated. The significantly different physical properties of the particulate containing feedstock from the equilibrating buffer implies that on switching to feedstock, the bed height will change to a new stable position, assuming that the flow rate through the entire operation is kept constant. Consequently, there are two distinct possibilities for modes of operation of the bed:

1. In a *constant flow rate* mode, the bed is operated with the flow rate maintained at its initial value, allowing the bed height to vary if necessary. Because biological feedstocks are in general more viscous and dense than the standard buffers commonly used in biochemical processes, switching from buffer to feedstock load is manifested in terms of the expanded bed height increasing through an initial transient phase to a new position. This increase is further influenced by the presence of particulate matter.
   Two operational implications arise as a result. Firstly, the position of the top flow adapter needs to be managed appropriately, either through continuous repositioning during the course of the run or by selecting an initial position significantly higher than the top of the bed before the feedstock load. In this mode the gradually increasing bed voidage during the course of feedstock load would be beneficial in the context of solids removal. Protein adsorption could be adversely affected, however, because of increasing mass transport distances, likely making it a critical rate limiting step in the process. The latter effect would be more pronounced for high molecular weight proteins.
2. In a *constant bed height* mode the bed is maintained at its initial expansion, and any tendency for the bed height to change is corrected by appropriate adjustment in the flow rate. An obvious advantage is that there is no need to adjust the position of the top flow adapter, simplifying equipment design. It is imperative, however, to be able to monitor the top of the expanded bed and any changes that it undergoes to effectively control flow rate.
   Protein adsorption in this case would perhaps be more efficient but would be offset by slower solids removal because of the progressive reduction in the flow rate to maintain the bed height at its initial level. The extent to which flow rate adjustment is required depends on the percentage of solids in the feedstock, with their removal being hindered at the higher solid content.

8.4 EQUIPMENT

Because of the nature of the feedstock typically used in expanded bed chromatography applications, the equipment varies from that used in conventional packed bed chromatography, although in principle the equipment undergoes only minor modifications. Certain design considerations with respect to the column and flow distributor would need to be satisfied, as discussed in the following sections.

8.4.1 Column

Expanded bed chromatography is a low-pressure operation that requires minimal modification of conventional packed bed columns. A principle design consideration is the inlet flow distributor, which must be able to handle particulate material in the feedstock being loaded onto the column. This aspect is discussed in detail later. Another point of practical relevance is that the column must be of sufficient length to handle the desired degree of bed expansion. Two- to threefold bed expansions are typical; for a settled bed height of 20 cm, this translates to a minimum of 60-cm column length. In practice, columns are usually 1 m long so that the top flow distributor can be suitably positioned to allow for an increase in bed expansion when unclarified feedstock is being processed in the column.

The commercially available columns now range from 2.0 to 1,200 cm in diameter. The first column to be marketed was the STREAMLINE 50™, which is extensively used by researchers today. Given the potential for commercial application, the subsequent generation of columns was scaled up to 20, 60, and 120 cm in diameter. Although useful for process scale operations, the size clearly limited their use for methods development and process validation work. Scale-down of the expanded bed columns followed, and column sizes of 2.0 and 2.5 cm are now available. I conducted the first scaled-down validation, and data from columns sizes of 0.5, 1.0, 5.0, and 25.0 cm in diameter showed excellent correlation in expansion properties and protein adsorption breakthrough.

Table 8.2 shows the major suppliers of the expanded bed chromatography columns and available specifications. Included in the table is the small glassware company that tailor-made the small-diameter columns I used in the scaled-down studies described earlier. These columns are fitted with a sintered glass disk as the inlet and outlet flow distributor. Columns up to 20 cm in diameter are constructed of glass, and larger ones are made from stainless steel. Glass columns have the advantage of allowing visualization of the process. This observation is particularly useful when column blockage results in excessive back pressure, reducing the flow through the column and causing the bed to rapidly settle.

TABLE 8.2. Commercial Suppliers of Columns for Expanded Bed Chromatography

Supplier	Column Diameters (cm)	Materials	Distributor Type
Amersham Pharmacia Biotech AB.	2.5, 5.0, and 20.0	Glass	Stainless steel disk and mesh arrangement
Amersham Pharmacia Biotech AB.	60 and 120	Stainless steel	Stainless steel disk and mesh arrangement
UpFront Chromatography A/S.			Distribution through stirred baffles in the column base

8.4.2 Flow Distributor

The importance of the flow distributor at the inlet has been a neglected issue in column design. There are several parallel requirements, beginning with the need for a uniform distribution of the liquid at the column inlet. A degree of variation from this ideal situation is always encountered, and this particular parameter is not considered to be critical in determining the overall performance of the bed. One of the reasons is the higher ratio of length to diameter, which offers a dampening effect for poor flow profiles that result from maldistribution at the inlet. A more relevant requirement of flow distributors in the context of expanded bed chromatography is that it should allow free passage of the particulate material present in the feedstock. The high voidage of the bed itself permits almost unhindered flow, and bed blockage is therefore more likely to be a problem at the flow distributor. Although larger pore sizes are clearly beneficial in overcoming this limitation, it is also important to remember that the adsorbent particles themselves must be retained within the column.

The most common distributor type is the perforated stainless steel plate fitted with a stainless steel wire mesh on top. This type is found in all of the Amersham Pharmacia Biotech (APB) columns available commercially. The number of holes in the perforated plate depends on the column diameter it fits. The wire mesh serves to prevent the liquid from jetting into the bed at the column inlet by offering a distribution surface. It also has a dual purpose of preventing the adsorbent particles from leaking out from the bottom of the column. De Luca *et al.* (19) have investigated bed expansion characteristics using distributors with various numbers of orifices and arrangements from the center. The settled bed height and bed expansion were found to be important parameters that influence mixing within the bed.

Although all expanded bed work using APB columns relies on these distributors, little has been published about the problems of blockage when using unclarified feedstock. This is not to say that the problem is nonexistent. Column back pressure buildup occurs because of gradual accumulation of the particulate material in either the pores of the perforated disk or the space between the disk and the mesh. In new columns and those seeing their first feedstock challenge subsequent to cleaning and regeneration, the buildup is relatively slow, and a much longer load time is possible before the column needs to be back flushed to clear the appropriate blockage. Back flushing is usually done by reversing flow direction for a short period, often at an enhanced rate. When the loading is resumed, the back pressure will be at its initial value; however, the buildup to the next back flush condition often occurs more quickly. The procedure can in principle be repeated until the end of the load cycle. The exact length of the initial and subsequent load periods depends on the parameters of the systems in question, including the solids concentration in the feedstock, the type of flow distributor, and the flow rate. Thus no reports on investigations of the influence of different parameters on the load efficiency of flow distributors have been published.

An interesting approach to flow distributor design has been adopted by UpFront Chromatography (Denmark). Its novel method essentially bypasses the conventional type of distributor and mesh from the flow path of the feedstock. In contrast to the commonly used method for load described earlier, the UpFront system introduces the feedstock from a side inlet and above the bottom support mesh. Distribution of the material is ensured through the use of a stirring device located immediately above the bottom support mesh. Although distributor blockage is clearly avoided in this system, the stirring speed needs to be carefully selected, balancing the requirement for uniform distribution over the inlet area with the need to avoid vortex formation and increased liquid–liquid mixing in the column itself. The scale-up potential of this distributor type needs to be addressed. The company manufactures 2-, 5-, and 20-cm-diameter columns using this technology.

Bioprocessing (England) has another approach to obtaining uniform flow distribution. Standard chromatography columns in which beds of glass ballotini are placed directly above the bottom support mesh are used. The expanded bed adsorbent is then added on top. When flow of buffer is initiated through the column, the bed of adsorbent fluidizes but the glass ballotini do not because of their significantly higher density. Uniform distribution is obtained, and the mixing of adsorbent and ballotini beads is avoided. Proper selection of ballotini size is important. The smaller ones have been shown to offer better distribution, equivalent to the use of a perforated disk with many orifices. However, one must remember

TABLE 8.3. Commercial Suppliers of Tailor-Made Adsorbents for Expanded Bed Chromatography

Supplier	Adsorbent Trade Name	Ligands	Density (g/mL)	Size	Matrix Type
Amersham Pharmacia Biotech AB.	STREAMLINE™	DEAE SP Phenyl Recombinant protein A	1.2	100–300 μm, mean size ~200 μm	Cross-linked agarose with crystalline quartz core
Bioprocessing Ltd.	PROSEP™	Protein A Thiosorb	1.3	100 μm; pore size ~70–1,000 Å	Porous, dense glass beads
UpFront Chromatography A/S.	Mimo™	Mixed mode	1.0–1.6		Cross-linked agarose with core of one glass particle
	FastMabs™	Antibody	1.0–1.5		
	UFC	Ion exchange	1.0–1.6	40–300 μm	
		Hydrophobic	1.0–1.6		
		Metal chelate	1.0–1.6		
Biosepra	HyperD				Mineral oxide based

that smaller particles have a lower minimum fluidization velocity, and the selection needs to balance the need for small-sized ballotini with maintaining a larger difference in the minimum fluidization velocity between it and the adsorbent beads.

8.4.3 Adsorbent

As with packed bed chromatography, the adsorbent must offer a basis for interaction with the solubilized components of the feedstock to enable product capture. In addition, expanded bed adsorbents must also possess good fluidization characteristics. During the flow of particulate bearing feedstock through an expanded bed, the adsorbent should, under the flow conditions employed, remain in a fluidized state and not tend to elutriate from the column. This can be achieved if the terminal settling velocity of the adsorbent is significantly higher than any of the other solid matter present in the system. The three physical properties of the system that most significantly influence the terminal velocity of the adsorbent—particle diameter, density difference between the particle and the liquid, and liquid viscosity—can be seen from the Stokes equation, described earlier.

Liquid viscosity is essentially dependent on the feedstock and is for the most part not an adjustable parameter, although some feed pretreatment is certainly possible. However, regarding the adsorbent, one can essentially experiment with the particle diameter and the density difference between the adsorbent and the liquid. The problem of suitably dense adsorbents can be offset to some extent by using size as a parameter for increasing the terminal velocity of an adsorbent particle. Thus a material of lower density and higher particle size can be used in expanded bed chromatography. However, a significant disadvantage lies in the resulting increase in mass transport resistance. The diffusion time is related to the square of the particle diameter; pore diffusion thus becomes the rate-limiting step in larger particles. Another disadvantage of expanded beds is increased interparticle distances, which may lead to slower transport of solute from the bulk to the surface of the adsorbent. The presence of particulate debris may cause further mass transport limitations and pose problems of adsorbent fouling and pore blockage.

The optimal adsorbent for use in expanded bed chromatography should therefore be small and very dense, and, as with any chromatography adsorbent, the dynamic bed capacity (DBC) should be acceptably high. There is an intrinsic link between these different parameters for effective expanded bed operation. DBC can be improved by using smaller particle sizes; however, efficient removal of the particulate debris present in the feedstock is an essential process requirement, and the two conditions can be simultaneously satisfied through the use of high-density adsorbents. Although good adsorbent design is critical in expanded bed operation, only limited work has been undertaken on its improvement, and most applications of this technology thus far have centered on the use of commercially available material. Only a few such materials exist currently. Table 8.3 summarizes some of the more common adsorbents currently used in expanded bed applications.

8.5 APPLICATIONS

The growing realization of the potential of expanded bed chromatography is borne out by the rapidly increasing number of reported applications of this technology in biological product recovery. The growth of this subject area over the last decade can be judged by following the literature for reported work during that period. The pioneering work in this field by Chase and coworkers relied on the use of conventional packed bed chromatographic adsorbents (28,29). Yeast cell homogenates were also commonly used in model systems. They are a suitable starting material given the many proteins present, any one of which can be targeted for recovery by expanded bed operation and offering the scope for use with ion exchange and affinity adsorbents (30,31).

Applications of expanded bed chromatography for the recovery of product directly from feedstock has moved on from the yeast homogenate systems that were studied a few years ago. There are now examples of applications for the recovery of both native and recombinant proteins from yeast (32) and bacterial cells (33–35), milk (36), and monoclonal antibodies from hybridoma cells.

The examples mentioned are but a few of the growing list of new applications that are being reported. The recently held Second International Conference on Expanded Bed Chromatography gives further evidence of the growing awareness of the potential of this technology in biological product recovery (see the book of abstracts for further information). In following the work in this area, it is obvious that the applications are for the most part based on conventional operational procedures and differ only in the system for which the recovery is being studied. Although the theory-oriented investigations have strived to tackle interesting issues of this technology, the applications mostly been bound by conventional operating protocols, similar to what has been described in the previous sections.

Two applications warrant specific mention because of their novel and interesting approach and are briefly highlighted. The work by Chase and coworkers on the use of a continuous countercurrent extractor (37–39) extends the principles of expanded bed chromatography to continuous protein separation. A perfluorocarbon-based adsorbent (40) was pumped between four different contactors, each representing in sequence one of the four stages (i.e., load, wash, elution, and regeneration). Crude unclarified feedstock was pumped into the first contactor and product obtained from the third. The use of a specific affinity ligand permitted highly purified product to be collected at the outlet of the elution stage.

Another interesting methodology has been demonstrated by Lyddiatt and coworkers in which a fluidized bed was coupled with batch fermentation for direct product removal through either continuous or intermittent recycle of the reactor contents through the fluidized bed. Inherent problems of holding the fermentation material outside the controlled reactor environment were addressed through the use of a multibed manifold comprising smaller beds. Each bed was operated until the adsorbent was saturated and then removed from the flow loop for product recovery, cleaning, and regeneration.

Expanded bed protein adsorption has now reached a stage in its development where a good understanding of the potential and drawbacks can be effectively assessed. Issues related to the bed design, performance, and application have been addressed in the literature. Although the potential of this technology is well accepted, it is dangerous to assume that it will offer the solution to many recovery and separation problems. Many aspects are still either not considered important or not understood. The worth of the technology will be seen only when they become an integral part of bioprocesses. Toward this goal, the first step may already have been taken: There is now at least one report of a biotechnology company having patented a process for plasmid DNA production in which expanded bed adsorption is incorporated in the purification protocol.

REFERENCES

1. J.E. Ramirez-Vick and A.A. Garcia, Separation Purif. Methods 25: (1996).
2. A.I. Clarkson, M. Bulmer and N.J. Titchener-Hooker, Bioprocess Eng. 14: 81–89 (1996).
3. J.P. Maybury, K. Mannweiler, N.J. Titchener-Hooker, M. Hoare, and P. Dunnill, Bioprocess Eng. 18: 191–199 (1998).
4. A. Pessoa and M. Vitolo, Appl. Biochem. Biotechnol. 70: 505–511 (1998).
5. D.P. Harris, A.T. Andrews, G. Wright, D.L. Pyle, and J.A. Asenjo, Bioseparations 7: 31–37 (1997).
6. S.G. Walker and A. Lyddiatt, J. Chromatogr., B 711: 185–194 (1998).
7. G.M. Zijlstra, M.J.F. Michielsen, C.D. de Gooijer, and L.A. van der Pol, Bioseparations 6: 201–210 (1996).
8. M. Ritopalomares and A. Lyddiatt, J. Chromatogr., B 680: 81–89 (1996).
9. G.M. Zijlstra, M.J.F. Michielsen, C.D. de Gooijer, and L.A. van der Pol, and J. Tramper, Biotechnol. Bioeng. 12: 363–370 (1996).
10. I.J. Isaacs, Australasion Biotechnol. 6: 88–92 (1996).
11. S.R. Narayanan, J. Chromatogr., A 658: 237–258 (1994).
12. M. Leonard, J. Chromatogr. 699: 3–27 (1997).
13. E. Boschetti, J. Chromatogr., A, 658: 207–236 (1994).
14. Q.M. Mao and M.T.W. Hearn, Biotechnol. Bioeng. 52: 204–222 (1996).
15. D.K. McCormick, Biotechnology 11: 1059 (1993).
16. Y.U. Wen and Y.H. Yu, Chem. Eng. Prog. 62: 100–111 (1966).
17. J.F. Richardson and W.N. Zaki, Trans. Inst. Chem. Eng. 32: 35–53 (1954).
18. G.S. Finette, Q.M. Mao, and M.T.W. Hearn, J. Chromatogr., A, 743: 57–73 (1996).
19. L. De Luca, D. Hellenbroich, N.J. Titchener-Hooker, and H.A. Chase, Bioseparations 4: 311–318 (1994).
20. A. Karau, C. Benken, J. Thömmes, and M.-R. Kula, Biotechnol. Bioeng. 55: 54–64 (1997).
21. O. Levenspiel, Chemical Reaction Engineering, 2nd ed., Wiley, New York, 1972.
22. J. Thommes, M. Weiher, A. Karau, and M.-R. Kula, Biotechnol. Bioeng. 48: 367–374 (1995).
23. H.A. Chase, TIBTECH 12: 296–303 (1994).
24. M. Goto, T. Imamura, and T. Hirose, J. Chromatogr. 690: 1–8 (1995).

25. L. Nixon, C.A. Koval, R.D. Noble, and G.S. Slaff, Chem. Mater. 4: 117–121 (1992).
26. B.E. Terranova and M.A. Burns, Biotechnol. Prog. 5: 98–104 (1989).
27. A.S. Chetty and M.A. Burns, Biotechnol. Bioeng. 38: 963–971 (1991).
28. H.A. Chase and N.M. Draeger, J. Chromatogr. 597: 129–145 (1992).
29. Y.K. Chang, G.M. McCreath, N.M. Draeger, and H.A. Chase, Trans. Inst. Chem. Eng. 71: 299–303 (1993).
30. Y.K. Chang, G.E. McCreath, and H.A. Chase, Biotechnol. Bioeng. 48: 355–366 (1995).
31. Y.K. Chang and H.A. Chase, Biotechnol. Bioeng. 49: 204–216 (1996).
32. F. Raymond, D. Rolland, M. Gauthier, and M. Jolivet, J. Chromatogr. 706: 113–121 (1998).
33. M. Hansson, S. Ståhl, R. Hjorth, M. Uhlén, and T. Moks, Biotechnology 12: 285–288 (1994).
34. G. Maurizi, V. DiCioccio, G. Macchai, P. Bossu, C. Bizzarri, U. Visconti, D. Boraschi, A. Tagliabue and P. Ruggiero. Prot. Expression Purif. 9: (1997).
35. H.J. Johansson, C. Jagersten, and J. Shiloach, J. Biotechnol. 48: 9–14 (1996).
36. A. Degener, M. Belew, and W.H. Velander, J. Chromatogr. 799: 125–137 (1998).
37. R.O. Owen, G.E. McCreath, and H.A. Chase, J. Mol. Recognit. 9: 575–584 (1996).
38. R.O. Owen, G.E. McCreath, and H.A. Chase, Biotechnol. Bioeng. 53: 427–441 (1997).
39. R.O. Owen and H.A. Chase, J. Chromatogr., A 757: 41–49 (1997).
40. G.E. McCreath and H.A. Chase, J. Mol. Recognit. 9: 607–616 (1996).

FURTHER READING

C.M. Griffith, J. Morris, M. Robichaud, M.J. Annen, A.V. McCormick, and M.C. Flickinger, J. Chrom. A 776: 179–195 (1997).

PART III

PROCESS DEVELOPMENT IN DOWNSTREAM PURIFICATION

9

SCALEDOWN OF BIOPHARMACEUTICAL PURIFICATION OPERATIONS

ANURAG S. RATHORE AND VARSHA S. JOSHI
Chemical Engineering Department, Indian Institute of Technology Delhi, Hauz Khas, New Delhi, India

9.1 INTRODUCTION

Biotech-based therapeutics have emerged as the growth engine for the pharmaceutical industry (1). Successful development and commercialization of biotech processes require us to understand the relationships between the various attributes of the product and the product's clinical safety and efficacy, as well as the effect of the different process parameters on the product's attributes (2). Considering the complexity of the biotech products, understanding the latter requires a lot of experimentation. Working at small scales offers a potential solution to this problem.

There are many drivers for working at small scales (3–5):

- Smaller the scale, lesser the requirement of feed material. This is often an issue during early stages of product development.
- Parallelization and automation offer an enormous increase in productivity when working at small scale when compared to manufacturing scale operations. This allows us to examine a lot more parameters than what we can otherwise.
- Higher productivity may also result in accelerated timelines, something that can be very critical in the pharmaceutical industry. In case of a blockbuster drug (sales greater than $1 billion/year), each extra day delay in development can result in excess of $1 million in lost sales revenue (3).
- Newer initiatives such as Quality by Design (QbD) (6–8) and Process Analytical Technology (PAT) (9–11) demand a deeper understanding of the process and the product. This in turn requires more extensive experimentation than what has been traditionally done.

When using small-scale models for process development, it is important that the small-scale model performs similar to the corresponding manufacturing scale unit operation. This ensures that the data from experimentation and the conclusions reached from the data are applicable to the manufacturing scale (12–14). Such small-scale models may be used for a variety of purposes:

- process development and optimization;
- process characterization to support process validation (15–18). These include range finding studies to set the acceptable ranges for the operating parameters, spiking studies to establish clearance of host cell impurities [host cell proteins (HCPs) and DNA] and other product-related impurities, and establishing viral clearance capability (19–23). The latter is a requirement for regulatory approval;
- establishing lifetime for reusable media for chromatography and membrane separation steps (19,24,25);
- manufacturing support for licensed commercial processes to address manufacturing investigations, postapproval process change for comparability, and evaluate raw material changes postlicensure.

Downstream Industrial Biotechnology: Recovery and Purification, First Edition. Edited by Michael C. Flickinger.

9.2 GENERAL CONSIDERATIONS

9.2.1 Use of Real Feed Streams and Raw Materials

When performing scale-down experiments, it is recommended that we use feed streams that are representative of manufacturing scale (24). It is best to use process intermediates produced at the manufacturing scale itself. If this is not possible then feed streams should be produced at pilot scale that is representative of the manufacturing scale. Furthermore, the raw materials used for the experiments (buffer, resin, etc.) should be taken from the cGMP-released inventory. This ensures that any additional variability in the experimentation originating from use of noncomparable raw materials or feed materials is avoided. Knowledge of the storage and stability of the feedstream material is acquired beforehand. Storage-related degradation of the inventory could affect the outcome of the small-scale runs (24).

9.2.2 Qualification of a Scale-Down Model

The qualification of the scale-down model must be carried out as per current good manufacturing practice (cGMP) validation guidelines that require a written qualification protocol describing qualification studies to be conducted and should include preapproved acceptance criteria for key output parameters. For qualification, the runs are performed in triplicate (or more) to assess the reproducibility of the scale-down model and for meaningful comparison to full-scale. In most cases, multiple performance parameters need to be compared across scales to establish comparability of performance and qualify the scale-down model. Compared to the univariate comparison that is usually done, the application of multivariate analysis (MVA) tools in performing this comparison has been recently proposed (26). This allows for a more comprehensive comparison of the data and an examination of all important inputs and output variables at the same time. This allows for correlations between variables to be more easily identified. In addition, MVA projection tools condense complex multivariate data sets into easily interpretable graphical projections (26,27).

In the following sections, we discuss major unit operations currently used in biopharmaceutical industry for downstream processing (Fig. 9.1) with respect to the scale-down methodology and the review literature.

9.3 CENTRIFUGATION

With the recent developments in mammalian and microbial cell culture, solids concentrations of up to 30% (w/v) are readily achievable (28). For such cases, centrifugation is often the method of choice for liquid–solid separation because of its cost effectiveness, robust operation, and applicability to continuous operation (29,30). Harvesting

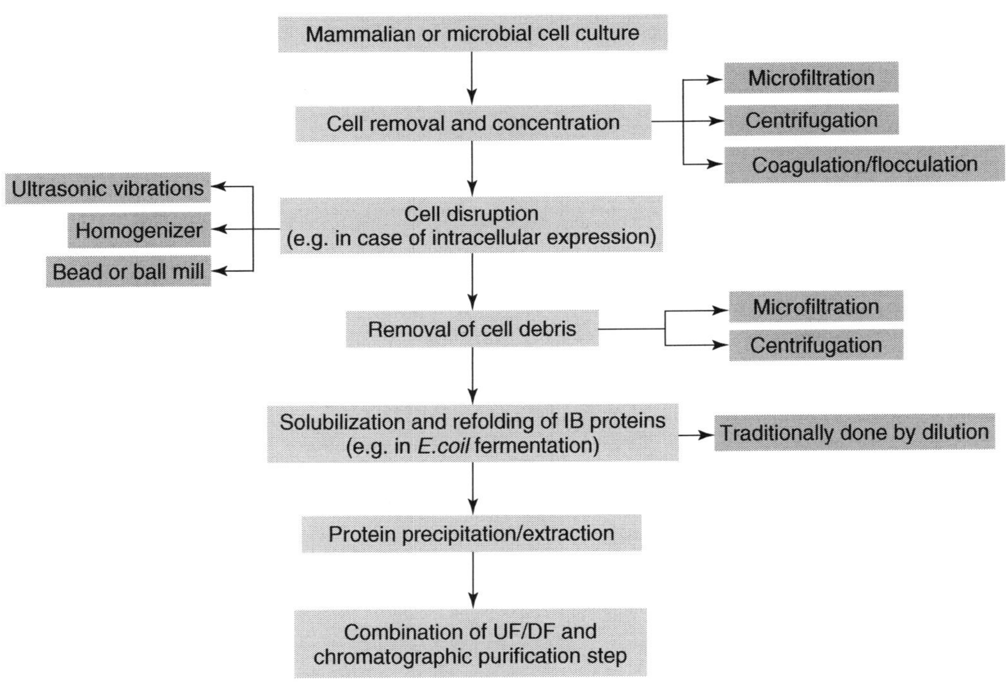

Figure 9.1. Process flow diagram for a typical biotech process used for manufacturing of a biotech-based therapeutics. The diagram is meant to be illustration and not extensive, as it only shows some of the options.

of bacterial and mammalian cells is commonly done using stacked disc centrifuges (31).

In this unit operation, the centrifugal force is used to separate mixtures of particles of varying masses or densities suspended in a liquid. When a vessel is rotated at high speeds, the angular momentum yields an outward force to each particle that is proportional to its mass. This force is countered by the so-called "drag" force that the liquid exerts on the particle. The net effect of "spinning" the sample in a centrifuge is that larger/denser particles move outward faster than smaller/lighter particles. The rate of centrifugation is determined by the angular acceleration applied to the sample, typically measured in comparison with the gravitational force (g). The sedimentation rate of a particle is proportional to the liquid flow rate through the centrifuge (Q) divided by the quantity sigma ($\sum$). The latter is the area of a simple gravity settling tank of equivalent sedimentation characteristics to that of the centrifuge. Theoretically, the performance of two centrifuges processing the same feed material will be equivalent if the $Q/\sum$ value is kept constant. The cell culture solution is continuously fed to the outside of the centrifuge and flows through the spaces between the stacked discs toward the center of the centrifuge. Cells and particles are thrown radially outward by the centrifugal force toward the underside of the upper disc and slide down the disc surface to the sediment holding space. The collected solids are either continuously or discontinuously discharged from the centrifuge. The liquid flows continuously to the top of the centrifuge and is discharged under pressure.

Unique expressions for $\sum$ exist for each centrifuge design. For a disc stack centrifuge (28),

$$\sum = \frac{2\pi n_d \omega^2 (r_2^3 - r_1^3)}{3g \tan\theta} f_1 \qquad (9.1)$$

where, n_d is the number of active discs, r_1 and r_2 are the inner and outer disc radius (in meters), g is acceleration due to gravity (in meters per square second), and θ is the half disc angle. Furthermore, $\omega = 2\pi N$, where ω is the angular velocity and N the rotational speed (in revolutions per second). Also, f_1 is the correction factor for spacer caulks determined using the following expression:

$$f_1 = 1 - \left(\frac{3Z_L B_L}{4\pi r_2}\right)\left(\frac{1-\left(\frac{r_1}{r_2}\right)^2}{1-\left(\frac{r_1}{r_2}\right)^3}\right) \qquad (9.2)$$

where Z_L is the number of caulks on a disc and B_L the caulk width (25).

For the laboratory centrifuge,

$$\Sigma = \frac{V\omega^2(3-2x-2y)}{6g \ln\left(\frac{2R_o}{R_o+R_i}\right)} \qquad (9.3)$$

where $\sum$ is the equivalent settling area of a laboratory centrifuge, V is the sample volume, ω is the angular velocity, x and y are fractional times required for acceleration and deceleration, respectively (measured for different centrifugation speeds using a stopwatch), R_i is the inner radius (the distance between the center of rotation and the top of the liquid), and R_o is the outer radius (the distance between the center of rotation and the bottom of the tube) (32).

The performance of a continuous full-scale centrifuge is often quantified by the amount of the solids remaining after centrifugation and can be described using the following relationship (4):

$$S = f\left(\frac{Q_c}{\sum_c}\right) \qquad (9.4)$$

where S is the solids remaining, Q_c the volumetric flow rate, and $\sum_c$ the equivalent settling area.

9.3.1 Scale-Down Modeling Approach

Scale-down of the centrifugation step can be achieved by keeping $Q/\sum$ constant (33,34). Sigma theory modeling can be generically applied to a wide number of centrifuge types using calibration factors to account for changes in flow patterns and centrifuge geometry (35). The feed flow rate, centrifuge angular velocity, and system back pressure can be suitably controlled. The temperature of the feed and centrate is kept identical to that used in the manufacturing scale. The smallest disc stack centrifuges available for scale-down have $\sum$ values of 1800 m^2 and can process feed volumes of 2–20 L at flow rates of 10–60 L/h with a 0.5 L solid holding volume. Production-scale disc stack centrifuges can have sigma values as high as 1.3×10^5 m^2. Thus, scale-down factors of 70-fold are achievable (12).

9.3.2 Scale-Down Qualification

The performance of the centrifugation step can be measured in terms of step yield, the clarity of the centrate, filterability of the centrate, particle size distribution, and extent of cell lysis. The clarity of the centrate can be measured using particle counters, ultraviolet (UV) spectroscopy, or turbidimeter. Yield can be determined by analyzing the feed material and the output for method by high pressure liquid chromatography (HPLC).

9.3.3 Literature Review

Researchers have proposed the use of a benchtop centrifuge as a scale-down model for pilot-scale equipment (3). The focus of the investigation was on properties of the protein precipitate, the fraction of solids recovered, and the extent of dewatering achieved. Good agreement was obtained at bench scale (a 1000-fold scale-down factor) for all these parameters when compared to pilot-scale, batch-feed operation. In addition, the methodology developed here allowed for identification of the extent of breakup that occurs in continuous-feed centrifuges when processing shear-sensitive materials such as protein precipitates. Other researchers have used the laboratory scale-down model to predict the performance of disc stack centrifuge for the recovery of yeast cell debris particles (34). The scale-down model was also used to investigate the extent of shear-sensitive protein aggregates during recovery in continuous high speed centrifuges. The authors concluded that such breakup can lead to over 10-fold reduction in separator capacity. Mannweiler and Hoare (36) have described the methodology for 10-fold scale-down of a disc stack centrifuge. Scale-down was achieved by the reduction of the number of discs available for separation purposes and by careful positioning of these discs in the overall disc stack. The authors used a combination of dye tracer and particle separation studies to optimize the disc stack configuration. The resulting grade efficiency curve was an accurate reflection of that for the full-scale centrifuge, especially in the critical design region where near complete particle recovery has been proposed. To date, several models that make good predictions at the 10–100 mL scale have been proposed (37–39). However, in recent years, there has been an increased interest in models that are based on microtiter plate technology (40–45). The use of a high speed rotating shear device to reproduce shear conditions in combination with benchtop centrifugation has successfully been used to demonstrate that an ultra-scale-down (USD) model using 15-mL centrifuge tubes can give an accurate prediction of the expected clarification of a mammalian cell culture in industrial disc stack centrifuges (46). Tustian *et al*. developed and verified a methodology in which USD curves for centrifugal clarification were corrected by diluting the feed material by 2%. After dilution, they found that the USD curves accurately predicted pilot-scale clarification performance of high cell density broths of *Saccharomyces cerevisiae* and *Escherichia coli* cells (28). Zaman *et al*. (47) have developed a USD-based approach accounts for both gravitational and shear effects on clarification performance.

In summary, despite advancements in scale-down of centrifugation, the models that exist at present are suitable for knowledge generation and preliminary examination of the effects of the various process parameters. Final determination of the acceptable ranges or the process validation acceptance criteria still requires experimentation with the manufacturing scale centrifuge. This is due to the complexity of the fluid dynamics in an industrial centrifuge.

9.4 HOMOGENIZATION

E. coli is among the most frequently used microbial hosts to produce recombinant therapeutic products. An intrinsic characteristic of *E. coli* and some other similar expression systems is that they retain the expressed proteins within the cell (48). Protein release in such cases requires cell lysis induced chemically (49,50), enzymatically (51), or mechanically. A robust and rapid way to achieve the latter is by mechanical cell disruption via homogenization by performing mechanical cell disruption followed by cell debris removal by centrifugation (52,53). Options for industrial-scale mechanical cell disruption include high speed bead mills (54–57), high pressure homogenization (HPH) (58–61), and ultrasonication. Most popular of these in manufacturing of biotech-based therapeutics is HPH.

In this unit operation, cell suspension is forced by pressures of several hundred bars to pass through an adjustable discharge valve, which has a restricted orifice. The combination of high shear, the high pressure drop, and the impact forces in the discharge valve results in cell disruption (62). Homogenizers may use discrete multiple passes through a recycle loop to effect the multiple passes needed for continuous processing. Disruption occurs via a first-order process with respect to the number of passes with the rate constant being a strong function of pressure (61) and the valve and valve seat design (63). In addition, the rate constant is affected by the conditioning of the cell suspension (e.g. chemical pretreatment) and the nature of the fermentation growth media. Cells grown in complex media are more difficult to disrupt compared with those grown in defined media (64).

9.4.1 Scale-Down Modeling Approach

Scale-down of an HPH-based process is complicated by the fact that varying the flow rate by orders of magnitude results in significant variation in the gap opening of the homogenization valve, with severe consequences on the fluid dynamics and mechanical interaction between the valve and cells. A procedure for calculating the pressure drop as a function of gap opening has been reported. This procedure has been validated for Stansted high pressure homogenizers, which exhibit a quadratic dependence on process medium flow rate (65). A typical industrial scale homogenizer needs at least 30 L of the feed volume, while a typical scale-down model may need up to 40 mL of the feed volume. Thus, scale-down factors of 750-fold are achievable. Inactivation depends not only on applied pressure of homogenization and number of passes but also on the microbial strain. For instance, in a single homogenization pass at 250 MPa, six log cycles of inactivation can be achieved for *E. coli*, five log cycles for *S. cerevisiae* but only one log cycle for

Lactobacillus delbrueckii. The latter is characterized by a thicker and more resistant cell membrane. Lethality can be increased not only by increasing the pressure but also through multiple-pass treatments (66). Since temperature rises during homogenization, one should aim at keeping the temperature rise similar so as to get similar step performance.

9.4.2 Scale-Down Qualification

The desired product recovery and purity after homogenization can be calculated using HPLC. Generation of aggregates can be monitored using size-exclusion high pressure liquid chromatography (SE-HPLC). The log cycle reduction in the feed should be determined at the end of the homogenization cycle along with particle size distribution of the cell debris and viscosity of the product. Temperature and pressure should be monitored throughout the process.

9.4.3 Literature Review

Literature is scarce on the topic of scale-down model of high pressure homogenizers. Donsi *et al*. (66) have compared the homogenization efficiency at two scales (laboratory scale and pilot scale) and demonstrated that the scale of the equipment does not affect the level of inactivation attained.

In summary, like centrifugation, homogenization too is a challenging unit operation to scale-down. While the existing models can be used to generate some fundamental knowledge about the behavior of cells under shear stress, the flow dynamics in most scale-down systems is quite different from that of their large scale counterparts. This necessitates that the final experimentation be done on the manufacturing equipment itself for setting process validation criteria and acceptable ranges.

9.5 REFOLDING

Protein refolding includes solubilization and reduction of inclusion bodies (IBs) followed by renaturation (slow removal of denaturant). In most cases, performance of the refolding step is critical to overall recovery of the process. There are many factors, both chemical (67–69) and physical (70–73), that can affect protein refolding. Refolding by dilution is still the preferred technology for large-scale refolding. The first step is solubilization of the IBs in a mixing vessel. After any remaining undissolved IBs are removed by centrifugation, the dissolved IBs are further reduced and renatured in an agitated tank. Aggregation is the major issue confronted during refolding and can be caused by improper mixing and/or high local protein concentration. Rapid and uniform dilution is necessary to avoid aggregation.

9.5.1 Scale-Down Modeling Approach

It has been observed that the refolding yield is inversely proportional to the protein concentration (67,69,74). Furthermore, slower addition of the denatured protein to refolding buffer has been known to result in higher yields as the influence of macroscopic concentration gradients is minimized even under poor mixing conditions (75,76). It is recommended that a temperature-controlled and agitator-driven closed vessel (such as a bioreactor) be used as scale-down model. This way, engineering considerations can be applied to mimic conditions at the manufacturing scale. Chemical conditions such as amounts and concentration of the various additives, sequence of addition, temperature, pH, conductivity, and protein concentration should be kept identical (Table 9.1). Physical aspects such as mixing and distribution of oxygen in the tank could be maintained similar by keeping the mass transfer coefficient ($k_{L}a$) same across scales. Similar quenching protocols followed by proper storage of the refolded protein should be followed at both the scales.

9.5.2 Scale-Down Qualification

Product quality can be monitored during and after refolding by a variety of generic (e.g. solubility and folding) or specific (e.g. activity) methods (77). Functional assays (bioassays, immunoassays, etc.) can be very informative. Analytical techniques such as intrinsic fluorescence, circular dichroism (CD, secondary structure), dynamic light scattering (DLS), turbidity measurements, or chromatographic analysis (HPLC) can be used as indicators to monitor performance of the refolding step (77–80). SE-HPLC can be used to determine the amount of dimers or aggregates generated in the refolding process. CD can be used to measure protein secondary structure and DLS to measure protein aggregation so as to verify protein structure (77,78). Crystallogenesis offers another approach for product quality measurement as only properly folded proteins with a uniform aggregation state (monodispersed sample) yield well-ordered crystals. While the method of choice will depend on the target protein or project, it is generally recommended that more than one method be employed (80).

9.5.3 Literature Review

Mannall *et al*. (81) have studied the USD approach to protein refolding in microwell plates and demonstrated the applicability and scalability of this microwell screening for the development of protein refolding processes. Qoronfleh *et al*. (82) have suggested the use of high throughput, automated refolding screens. Ordidge *et al*. (83) have used an automated robotic platform to develop a dilution refold process-screening platform on which a hierarchical set of assays rapidly determine optimal refolding conditions at the microscale.

In summary, the typical bioreactor offers a platform for performing refolding under conditions that mimic manufacturing conditions. Furthermore, high throughput process development (HTPD) platforms for performing refolding have also been successfully demonstrated. We expect these to be adopted by the biotech industry to expedite process development, optimization, and characterization of protein refolding step.

9.6 PRECIPITATION

Precipitation offers a low cost, high yield unit operation that can be used for separation of the various host-cell-related and process-related contaminants (84). Although the use of this step is somewhat limited in production of therapeutic proteins because of the concerns related to product stability, it is the step of choice for certain applications (85). A common application is as a primary isolation procedure in which the amounts of contaminating proteins (and other species) are decreased along with a significant reduction of process volume (86).

The basis of separation in this unit operation is reduction in the solubility of the product to induce precipitation. This can be achieved via several methods: ionic precipitation (e.g., ammonium sulfate and sodium chloride), temperature, pH, metal ions (e.g. Cu^{2+}, Zn^{2+}, and Fe^{2+}), nonionic polymers (e.g. polyethylene glycol), organic solvents (e.g. ethanol and acetone), tannic acid, heparin, dextran sulfate, cationic polyelectrolytes (e.g. protamines), short-chain fatty acids (e.g. caprylic acid), trichloroacetic acid (TCA), lectins (e.g. concanavalin A), group-specific dyes (e.g. Procion Blue), and ligand–antibody interaction (87). Of these methods, ionic precipitation utilizing inorganic salts is the most commonly used method.

The precipitation step is generally used at an early stage of the purification procedure, often on the clarified extract, to obtain initial purification and concentration. It entails manipulation of the feed solution to favor protein precipitation, mixing of the vessel contents as the precipitation takes place, and subsequent solid–liquid separation to separate the two phases. Mixing is a critical aspect of this unit operation and impacts apparent protein solubility, protein structure in precipitates, particle size, morphology, and recoverable activity (22). Aggregate levels and characteristics in the precipitation output impact the performance of the subsequent centrifugal separation. Protein solubility is affected primarily by the type and concentration of precipitant, pH, and ionic strength (67, 88–91). The physical properties of the aggregates depend on several factors: the type of precipitation reactor (90), choice, concentration, and rate of addition of precipitating agent (89); initial protein concentration; and nature and extent of mixing and residence time in the reactor (68,69,71).

Batch reactors are the simplest type of precipitation reactor and have been widely used in the industry. The precipitating agent is slowly added to the protein solution under mixing. Since the particles are exposed to a wide range of shear stresses for a long period of time, the aggregated protein particles tend to be compact, dense, and mechanically stable.

9.6.1 Scale-Down Modeling Approach

Parameters that are likely to impact the performance of the precipitation step need to be held constant (Table 9.1). These include pH, conductivity, final precipitant concentration, and temperature. Design parameters that need to be maintained same across scales include tank geometry (height-to-diameter ratio), impeller type, diameter of the impeller relative to that of the tank, placement of the impeller, baffle geometry, mean velocity gradient, aging parameter (Camp number), and residence time (4). In cases where the small-scale system does not mimic the manufacturing design, dimensionless parameters such as the mass transfer coefficient (k_La) can be used to target identical mixing conditions across scales (Table 9.1) (92).

9.6.2 Qualification of Scale-Down Model

Product purity (measured by specific activity or chromatographic methods), product and total protein yield, and impurity levels [assessed by sodium dodecyl sulfate polyacrylamide gel electrophoresis (SDS-PAGE)] can be used to compare step performance across scales. Additional measurements could include measurement of ion concentrations (sodium content), turbidity of the supernatant to gauge the extent of precipitate removal, and the phase ratio and/or fraction of solids in the precipitate. The kinetics of precipitation can be measured to establish equivalence across scales (12).

9.6.3 Literature Search

Chhatre *et al*. have investigated the integrated application of USD techniques and economic modeling as a means for identifying optimal bioprocess operating conditions. An economic model of this process was used to identify where greatest product loss occurs and hence where the largest opportunity cost was being incurred. Scale-down analysis was used to characterize the influence of the critical steps, identified as precipitation and centrifugation, on recovery of lactoperoxidase (LPO). The results demonstrate the power of combining USD data with models of economic and process performance in order

TABLE 9.1. Summary of Scale-Down Approach for Different Purification Unit Operations—Input Parameters That are Held Constant During Scale-Down, Input Parameters That are Altered During Scale-Down, and Performance Parameters That are Monitored to Ensure Qualification of the Scale-Down Model

Purification Step	Operational (Input) Parameters That are Kept Identical	Operational (Input) Parameters That are Altered	Performance Parameters for Model Qualification
Centrifugation	Linear flow rate, temperature, sigma factor, feed characteristics	Process volumes, volumetric flow rate, centrifugal speed	Product concentration and recovery, turbidity, cell lysis
Homogenization	Linear flow rate, temperature, solution pH, conductivity, protein concentration, feed characteristics	Process volumes, volumetric flow rate, homogenizer pressure, number of passes	Product concentration and recovery, turbidity, cell lysis, product-related impurities (aggregates)
Refolding	Mass transfer coefficient (k_La), temperature, chemical stock concentration, feed characteristics	Agitation rate	Product concentration and recovery, conductivity, pH, product-related impurities (aggregation, deamidation, methylation, etc.)
Microfiltration	Protein loading (g product/m^2 membrane area), number of diavolumes, volumetric concentration factor, feed characteristic, temperature, linear flow velocity, cross flow rate per channel, transmembrane pressure	Process volumes, volumetric flow rate, membrane area	Product concentration and recovery, turbidity, filter capacity
Precipitation	Mass transfer coefficient (k_La), precipitant concentration, precipitant-to-feed-volume ratio, feed characteristics	Mixing speed, precipitant feed rate, process volume	Product concentration and recovery, product-related impurities, process-related impurities, host-cell-related impurities, turbidity
Chromatography	Bed height, buffers (pH, conductivity, etc.), resin, linear flow rate, protein loading (g product/L resin), pooling criteria, temperature, feed characteristics	Column diameter, column volume, volumetric flow rate	Product concentration and recovery, chromatographic profiles, HETP, asymmetry, conductivity, pH, product-related impurities (aggregation, deamidation, methylation etc.), process-related impurities (protein A, antifoam, etc.), host-cell-related impurities (host cell impurities, DNA, etc.)
Ultrafiltration/Diafiltration	Protein loading (g product/m^2 membrane area), number of diavolumes, volumetric concentration factor, feed characteristics, temperature, linear flow velocity, cross flow rate per channel, transmembrane pressure	Process volumes, volumetric flow rate, membrane area	Product concentration and recovery, turbidity, filter capacity, pH, conductivity, excipient concentration, product-related impurities, host-cell-related impurities
Viral inactivation	Mass transfer coefficient (k_La), temperature, contact time, chemical stock concentration, feed characteristics	Agitation rate	Product concentration and recovery, product-related impurities
Virus removal filtration	Protein loading (g product/m^2 membrane area), feed characteristics, temperature, linear flow velocity	Process volumes, volumetric flow rate, membrane area	Product concentration and recovery, product-related impurities (aggregates)
Membrane adsorption	Bed height, buffers (pH, conductivity, etc.), resin, linear flow rate, protein loading (g product/L resin), temperature, feed characteristics	Column diameter, column volume, volumetric flow rate	Product concentration and recovery, conductivity, pH, product-related impurities (aggregation, deamidation, methylation, etc.), process-related impurities (protein A, antifoam, etc.), host-cell-related impurities (host cell impurities, DNA, etc.)

to establish the best overall operating strategies for biopharmaceutical manufacture (93). Boychyn *et al*. (4) have used a pilot-scale batch stirred-tank reactor and a multichamber-bowl centrifuge to predict the performance of larger reactors and centrifuges.

In summary, the lack of literature on the scale-down reflects the sparse usage of precipitation in production of therapeutic proteins. This may gain more interest in the future, as researchers develop new approaches toward mainstream application of precipitation for achieving cost effective way of protein purification.

9.7 CHROMATOGRAPHY

Chromatography is the primary workhorse of the biotech industry for product purification (94). In the commonly used bind and elute mode of chromatography, the first step often involves binding of the product to the adsorbent. Impurities that do not bind to the column will just flow through and go to waste. Then, a wash is conducted to wash off remaining unbound and/or weakly bound impurities. Finally, elution is performed by altering the conditions of the mobile phase in the manner that differential binding of the protein of interest and other impurities is exploited to achieve purification of the product. Alternatively, the column can be operated in flow through mode, whereby the impurities bind to the column and the product flows through. A variety of chromatographic techniques are employed in purification processes, mainly differing in the nature and structure of the stationary phase (resin particles). These techniques exploit differences in properties of proteins such as (i) size and shape (gel filtration/size-exclusion chromatography); (ii) net charge and distribution of charged groups (ion exchange chromatography); (iii) isoelectric point (chromatofocussing); (iv) hydrophobicity (hydrophobic interaction chromatography, reversed phase chromatography); (v) metal binding (immobilized metal ion affinity chromatography); (vi) content of exposed thiol groups (covalent chromatography); and (vii) biospecific affinities for ligands, inhibitors, receptors, and antibodies (affinity chromatography) (95). In recent years, the concept of multimodal, or mixed-mode adsorption chromatography, is receiving an increasing amount of attention. These stationary phases allow for more than one "mode" of chromatography to be conducted in a single step. In all cases, separation can be influenced by factors such as the nature of ligand and matrix, solvent pH, temperature, and size of the beads and pores (96). In the following sections, we discuss some of the key considerations when performing scale-down of chromatography steps.

9.7.1 Size of Scale-Down Model

Chromatography is one of the more straightforward unit operations to scale-down (13). The size of the scale-down system varies with the intended use of the system. For performing binding studies, a 96-well format HTPD platform for chromatography may suffice and is likely to be the most efficient (5). For performing resin screening studies, the above mentioned HTPD platform or small columns (~ 2.5 cm height, 1 mL) can be used to screen for various matrices and mobile phase compositions in order to perform preliminary optimization (97). Viral clearance studies are often done with shorter columns (5–10 cm bed height) because of the difficulties and expenses associated with creating virus spikes (98,99). Final process characterization studies that form the basis for setting the process validation acceptance criteria and the acceptable ranges for operating parameters are often performed using a column with diameter of at least 1.6 cm and same bed height as at the manufacturing scale (97). Column diameters of less than 1.0 cm are generally not used in these studies since the ratio of system dead volume to column volume can result in pool concentrations significantly lower than those seen at large scale (13). It should be noted that when compared to the higher end of column size in manufacturing scale (2 m), the scale-down factors may range from 1:100 to over 1:1000000 (12).

9.7.2 Chromatographic Equipment

Automated small-scale chromatography systems, such as the äKTA Explorer, excel at these scale-down studies. They offer precise, accurate, and reproducible control of flow rates and gradients. However, accuracy and precision of in-line detectors (pH, conductivity, temperature) should be calibrated at reasonable frequency by using off-line probes (13).

9.7.3 Scale-Down Modeling Approach

Column scale-down is normally performed by maintaining bed height while reducing column diameter, which maintains the residence time (Table 9.1). In this way, the linear flow rate is maintained, while the volumetric flow rate is reduced (100). The International Conference on Harmonization (ICH) guideline on viral safety states that the validity of the scaling down should be demonstrated and that the column bed height should be shown to be representative of commercial scale manufacturing (101). Thus, if the bed height is different at small scale, as in case of viral clearance studies, the linear flow velocity should be adjusted appropriately to keep the residence time constant (102).

The operating conditions of the experiments should be identical to those in manufacturing. This is so as subtle changes in temperature, pH, or conductivity could affect the retention time of proteins on chromatographic resins. Heat exchangers on the inlet tubing and column jacketing maybe necessary for accurate, consistent, and precise temperature control during scale-down. For example, hydrophobic interaction chromatography has been reportedly sensitive to temperature fluctuations and could give rise to product retention and selectivity with variation in temperature (12). Some of the significant operational parameters include the following:

- Protein loading (g protein/L resin) should be identical. In size-exclusion chromatography, loading should be based on volume percentage of the column bed volume.
- Gradient lengths (in number of column volumes) and slopes (e.g. increase in percent limit buffer per unit time) should be kept identical.
- Temperature, pH, and conductivity should be maintained identical. Differences in temperature across scales can result in significant changes in conductivity and thus impact step performance.
- Bed height of the chromatography column and the linear flow velocity are generally kept constant across scales.
- Pooling criteria should be the same as in manufacturing. They can be online (conductivity or absorbance) or off-line (gels or HPLC). If the pooling criteria are UV based, it is a good practice to ensure that the UV detectors of the small-scale systems correlate well with those of the large-scale system. UV detectors may be inaccurate or nonlinear over the optical density (OD) ranges required. If the absorbance range is nonlinear at 280 nm, it may be prudent to measure the online OD at 300 nm or some other wavelength that has a lower extinction coefficient (13).

9.7.4 Qualification of Scale-Down Model

Attributes that need to be compared would depend on the objective of the step and could include chromatogram, product concentration, product recovery, quality attributes (concentration of the various product-related, process-related, and host-cell-related impurities), and other feed stream characteristics (such as pH and conductivity).

Factors at the large scale such as column-header design and packing can be difficult to control and may result in reduced accuracy of the scale-down model (13). Differences in column packing at the two scales may have an impact on separation and result in differences in chromatograms. Height equivalent to theoretical plate (HETP) and asymmetry factor (A_s) are typically used to evaluate quality of column packing (103,104). While these values are likely to be different across scales, they should be monitored to ensure consistency of packing quality during small-scale experimentation. Furthermore, differences in equipment such as tubing length and diameter, monitors, and pumps should also be carefully examined to ensure comparability of performance across scales (12).

Table 9.2 summarizes the literature on scale-down or process chromatography. Researchers have used microliter resin wells (as in 96-well plates) for performing chromatography. A lot of effort has also undergone into modeling of chromatography so that the data generated from scale-down models can be applied toward prediction of performance at manufacturing scale. The optimal balance between the high throughput that the microdevices offer and the higher level of comparability that comes in the laboratory-scale columns is likely to be different for different applications. For most cases, however, the best would be to use the HTPD platforms for examining a large number of process parameters in wide ranges followed by use of the traditional laboratory-scale chromatography columns for performing the final process characterization experiments that support process validation acceptance criteria. In the future, the HTPD platforms are likely to continue to evolve and gradually with increased rigor they may be able to replace the laboratory-scale experimentation completely.

9.8 MICROFILTRATION AND ULTRAFILTRATION/DIAFILTRATION (UF/DF)

9.8.1 Microfiltration

Microfiltration (MF) membranes can be used to harvest cells from mammalian and bacterial cell culture to provide cell-free conditioned medium (containing the product) to the downstream purification steps. The pore sizes of typical MF membranes range from 0.1 to 0.8 μm. MF membranes retain cells and cellular debris by a sieving mechanism at the membrane surface. Pressure is applied to the feed stream to force permeate through the membrane. Proteins and small-molecular-weight species (media salts) pass through the membrane and are collected in the permeate stream (12).

Ultrafiltration (UF) and diafiltration (DF) are commonly used in downstream processing for product concentration and buffer exchange, respectively. In some cases, it is also used for fractionation or purification. Membranes are rated by molecular weight cutoffs, called *nominal molecular weight cutoffs* (NMWC). Membrane structure can influence the quality and robustness of separation. Minimizing macrovoids enables the use of higher pressures and higher temperatures and gives better separations because of sharp NMWC. Cassettes are the first choice for most protein processing. Hollow fibers are more appropriate when the product is sensitive to shear or a high viscosity solution

TABLE 9.2. Review of Literature for Development of Scale-Down Models for Process Chromatography

Topic	Summary	References
General	Different formats of microscale devices for scale-down of chromatographic separations. Advice on how best to operate these devices, examples of use, comparison with respect to metrics such as cost, usability, and role in QbD	(94)
	Scale-down experimentation integrated with general rate modeling, a 2.5 cm height, 1 mL column model calibrated against 20 cm height columns from 40 mL to 160 L volume. Simulations in agreement with experimental results. Overall objective to reduce the number of experiments, shorten development time, and reduce costs	(98)
	Novel ultra-scale-down approach that uses empirical correlations derived from conductivity changes during operation to correct chromatographic profiles for differences in dispersion and retention. Tested successfully with 1 mL column to predict elution profiles of a chimeric monoclonal antibody obtained from protein A chromatography columns of 3 mL and 18.3 L scale	(105)
Affinity chromatography	Development of software that uses data obtained from scale-down column studies to relate yield to protein load and the number of matrix reuses to enable decision making for chromatographic separations of antibodies	(106)
	Mass balance of protein and Fab fragments, measurement of capacity from batch uptake kinetics and use of an imaging technique to investigate the impact of varying column loading and resin reuse on matrix performance. Using a 20-fold scale-down, number of reuses before recovery starts decreasing was established, as in current large-scale manufacturing operations	(107)
Cation exchange chromatography	Influence of pH for optimized binding and removal of HCP. Initial screening was done by batch-binding capacities, followed by scale-down to 96-well plate format and proved that assay miniaturization still provided reliable data	(108)
	Use of scale-down and scouting studies to assist in scaling up a method for isolating potentially clinically relevant monoclonal antibody C595 from heterogeneous feedstocks	(109)
Expanded Bed chromatography	The hydrodynamic model was in agreement with experimental results obtained with 1- and 5-cm column diameters with buffer solutions of different viscosities. Supported by visual observations and scale-down model predictions, it was concluded that operating conditions (liquid viscosity and superficial velocity) resulting in a bed-void fraction between 0.7 and 0.75 would provide the optimal separation efficiency in terms of plate number	(110)
	Suitability of expanded bed chromatography as a scale-down technique to predict the behavior at large-scale was examined. The performance of three columns of diameters 5.0, 1.0, and 0.5 cm were compared in terms of the bed expansion, hydrodynamics, and breakthrough for lysozyme adsorption onto STREAMLINE-SPTM. Performance was similar across a scale-down factor of a 100-fold	(111)
Hydrophobic interaction chromatography (HIC)	Development and scale-up of a preparative HIC step for a therapeutic protein, removal of product- and process-related impurities, and direct transferability of the scale-down optimization experiments via the pilot scale to the final production scale	(97)
Annular chromatography	Mathematical model to estimate the minimum radius of an annular chromatograph. The minimal geometry of an annular chromatograph was calculated for a model system immunoglobulin and bovine serum albumin.	(112)

needs to be processed such as adenoviral vectors and influenza vaccines (113).

Size is the basis of separation in filtration in most cases. The solution is contacted with the membrane under an applied pressure (114). The applied pressure forces the solvents or buffer salts and the smaller molecules to migrate through the membrane. The membrane retains larger molecules such as proteins. MF and UF operations can be performed in a tangential flow filtration (TFF) mode or normal flow filtration (NFF) mode. An example of NFF would be dead-end or cartridge filters where cells are retained at the membrane surface or within the membrane structure. As cells build up at the membrane surface, the filtration rate, or flux, decreases as the resistance to permeate flow through the membrane increases. In TFF, the feed solution is passed tangentially over the membrane surface. Cells build up on the membrane surface as the permeate flows through the membrane and are swept off the membrane and recirculated back to the feed tank (51). UF is typically conducted in a TFF mode where the feed solution is recirculated over the membrane and then returned to the feed vessel. UF membranes are available in nominal molecular weight limits of 1000–1,00,000 Da (115–118). A typical system consists of a feed pump, strainer, rack equipped with filtration modules, filtrate/backwash tank, and connecting piping and valves. A separate pump is used to add DF buffer to the recirculation vessel. Control parameters may include feed and retentate pressure (the permeate pressure is usually atmospheric for UF applications), temperature, and the feed flow rate per feed channel.

9.8.2 Scale-Down Modeling Approach

TFF has traditionally been scaled up by maintaining the following parameters constant: filtrate volume to membrane surface area ratio, transmembrane pressure, cross flow velocity, membrane material and pore size, channel height, flow path geometry, and retentate and filtrate pressures (Table 9.1). The membrane area is linearly decreased as per the desired feed volume such that the filtrate volume to membrane surface area ratio stays constant (119,120). In the case of hollow fibers, linear scaling with hollow fibers is readily accomplished by producing scale-down modules in which fiber length is equal to that of industrial scale systems. Equal flow distribution between fibers and cartridges is easily accomplished through proper piping manifold and cartridge housing design. Flat sheet designs, however, provide a suitable geometric format for a wide range of linear-scale UF.

The challenge with using scale-down systems for performing MF or UF/DF is that the small-scale devices that are commercially available may differ from their large-scale counterparts with respect to their design. While cassette designs are based on parallel addition of membrane channels, the feed-port channels are added in series. This directly impacts the fluid dynamics of the manufacturing scale cassettes and the scale-down devices, resulting in differences in flow distribution, channel height compression, and entrance/exit effects. All these factors impact the performance of the step. Equal flow distribution throughout the system is important to maintain both the system average and system limit values of feed channel pressure drop (ΔP_L), transmembrane pressure (TMP; ΔP_{TM}), filtrate flux (J_f), and protein concentration at the membrane wall (C_w) constant at all scales of operation. Channel height compression is dependent on the physical deformation characteristics of membrane, spacer screen, encapsulant, and gaskets. Variations in channel height will have a significant impact on both pressure losses and mass transfer coefficients. Hence, process performance will vary at different scales of operation if consistent compression is not achieved (119).

9.8.3 Qualification of Scale-Down Model

Since MF and UF/DF steps do not alter the product purity (in most cases) and are generally used for clarification or concentration or buffer exchange, step recovery may be determined using UV absorbance at 280 nm. In cases where the sample matrix interferes with the UV signal, use of HPLC or product-specific assays may be required. Mass balance is required for determining if the product is being lost into the permeate stream, left behind in the holdup volume, or irreversibly bound to membrane. A comparison of the flux versus TMP curve across scales provides a good comparison of performance. If the UF system is being used for purification, analysis of the impurity concentration is a key measure of step effectiveness. It is important that appropriate cleaning is performed on the small-scale system so as to achieve consistent performance over long-term use (12).

9.8.4 Literature Review

A comprehensive methodology for implementation of linear scale-up and scale-down of TFF process has been developed by Reis and coworkers (119). The researchers modified the cassette manufacturing procedures and tolerances to achieve linear scale-up of UF processes for recovery for a ribosomal deoxyribonucleic acid (rDNA)-derived human therapeutic. A 400-fold linear scale-up was achieved without intermediate pilot-scale tests. In another publication, novel pleated scale-down devices ($A_m = 1.51$ to 15.1×10^{-3} m^2) have been designed and fabricated. It was shown that these can more accurately predict the performance of industrial scale single-use pleated membrane cartridges ($A_m = 1.06$ m^2) that are commonly used within biopharmaceutical manufacture (121). The authors found that single-use scale-down cartridges retain the same pleat characteristics of the larger cartridges but require a reduced feed volume by virtue of a substantially diminished number of active membrane pleats. A 1000-fold scale-down was achieved (121). A method to mimic flux and transmission performance in a laboratory scale cross flow operation by an USD rotating disc filter (RDF) has also been reported (122). The RDF has been modified by building in inserts to allow the flexibility of the chamber volume, so that only 1.5 mL of processing material is required for each DF experiment. Wall shear rate correlations have been established for at both the laboratory-scale cassette and the USD device, and identical performance was achieved by operating both scales under conditions with equivalent-averaged shear rates. Predicted flux and transmission data agreed well with the experimental results of a laboratory-scale DF where the cassette resistance was considered (122). In another publication, the authors have described operation of a continuous mode laboratory filter to prepare filter cakes and filtrate similar to what may be achieved at the industrial scale (123). Analysis of the filtration rate profile indicated the filter cake to have changing properties (compressibility) with time. Results from the USD filter were used to predict how the cake compressibility changes with time and thus allowed for accurate predictions about the flux profile in large-scale filtration. Furthermore, the USD filter allowed the production of materials (cake and filtrate) representative of a large-scale continuous filter.

In summary, there has been significant progress in creating scale-down devices for UF/DF unit operation in the past two decades. Two aspects remain that require some consideration from the users. First is to ensure that the design of the scale-down device is comparable to that of the large-scale device so that the flow pattern is similar if not identical. Second, the dead volume of the system (volume of the connected tubings, pump, and other parts of the setup besides the cassette) is always significant when compared to the cassette volume at small scale. Very high dead volume will result in underperformance of the UF/DF step and limit the usefulness of the scale-down model. The user should carefully select the equipment that has a low dead volume.

9.9 VIRAL CLEARANCE VIA CHROMATOGRAPHY AND FILTRATION

Assuring the viral safety of plasma-derived biologicals and biopharmaceuticals is critical for safety of health care consumers. Incidences of contamination of products derived from human plasma in the past have adversely impacted the health of hundreds of patients and tainted the image of certain segments of the health care industry (124). Owing to stringent measures taken by the industry and regulators to mitigate viral safety risks, similar incidents have not occurred in the history of biotech-based therapeutics in the recent decades (124). Viral clearance can occur either via process chromatography (the virus not binding to the resin while the product binds) or nanofiltration (viruses cannot pass through the nanofilters). The aspects related to process chromatography are already covered in the above section. In this section, we focus on viral clearance via nanofiltration.

Nanofiltration is a pressure-driven process where permeate is forced through a semipermeable membrane with typical pore size in the range of 15–50 nm (newer products have pores in the narrower range of 15–20 nm). This pore size distribution allows for passage of the product and buffer ingredients while retaining viral particles. Viral clearance devices are used once and then discarded. Viral clearance and design considerations have been reviewed in several publications (125–127).

Virus clearance studies are always performed with scale-down models of the manufacturing process steps. It is not feasible to perform viral clearance studies at the manufacturing scale because it would be inappropriate to introduce infectious virus into a cGMP manufacturing facility. Also, the volumes of virus needed to achieve a satisfactory spiking level at the manufacturing scale would be impractical and prohibitively expensive. Thus, in order for the virus clearance studies to be extrapolated to the manufacturing scale, it is imperative that the scale-down model is a true representation of the full-scale manufacturing process. Typically, the virus is spiked into the relevant intermediate, and the spiked material is processed across the scaled down unit operation (both for process chromatography and nanofiltration). The reduction in the virus load by the unit operation demonstrates the effectiveness of the process step for virus removal or inactivation. The virus spike used in viral clearance studies should be representative of a potential contaminant to the extent achievable. It is important not only to select appropriate relevant or model viruses but also to consider the properties of the virus spike (124). For example, the presence of serum in a virus spike may be problematic for a validation study of a serum-free manufacturing scheme. It is important that contaminants in the virus spike itself do not impact key or critical performance parameters in a way that makes the scale-down model unrepresentative of the large-scale process (116). The product fraction post filter is then assayed for residual virus infectivity. This enables a reduction factor to be calculated as follows:

$$\text{Viral reduction} = \log_{10} \frac{\text{Total virus measured in spiked load}}{\text{Total virus recovered in the product fraction}} \quad (9.5)$$

9.9.1 Scale-Down Modeling Approach

The loading, as measured in volume of load material filtered per unit surface area of the filter, should be kept constant across the two scales. A typical load material from an earlier processing step obtained from manufacturing should be used for the scale-down studies. Since this step is usually carried out under constant pressure, the inlet pressure should be matched for the two scales and similar flush volumes in terms of liters of buffer flushed to recover any protein held up in the pores per unit surface area should be used. An important output parameter is the average volumetric flux as a function of volumetric capacity.

The membranes used for viral clearance can be operated in direct flow or TFF mode. For viral clearance applications, it is required that viral clearance devices have an integrity test that is correlated to virus removal performance. The manufacturer generally provides this correlation. The operation of scale-down viral clearance systems can present complications. When using multilayer membrane disks for evaluation, specifically designed holders may have to be used to provide the required sealing between the membrane sheets to prevent feed material from bypassing layers. Scale-down viral clearance devices may have the same membrane in the production scale devices, but the feed-side flow geometries of the scale-down devices may also not be consistent with the production modules. Feed-side path lengths may not be consistent and feed flows may have to be adjusted to provide equivalent shear forces at the membrane surface. Similar to UF application, air–liquid interfaces should be minimized to prevent protein denaturation and aggregation (12).

9.9.2 Scale-Down Qualification

The step recovery and purity for viral clearance can be determined using HPLC or product-specific assays. Aggregate generation can be monitored using SE-HPLC. When pure proteins are used, UV absorbance at 280 nm can be used for yield measurements. Creating a mass balance is useful to estimate losses coming from dead volume in the system or irreversible binding to the membrane. A comparison of the process flux as a function of time and pressure over the whole process is also an indicator of comparability of performance.

9.9.3 Literature Review

Zhou *et al*. (128,129) have evaluated different scale-down models and proposed a new model to mimic the liquid flow path found in the large-scale capsule. They were able to achieve same capacity at small scale as at larger scale. Results from a four-model virus study with a redesigned Sartobind Q absorber scale-down model at the new process capacity were presented. The authors have discussed problems faced with the Q-membrane scale-down model as well as the proposal of the new model. The proposed Q-membrane scale-down model had a process capacity greater than 3000 g/m^2 or 10.7 kg/L with a log reduction in viability (LRV) over 5 for four-model viruses.

In summary, scale-down modeling for viral clearance through process chromatography and nanofiltration is widely practiced as every biotech company does these studies as per the regulatory requirements. The existing models work reasonably well, but as discussed in the references mentioned in this section, there is scope of further improvements with respect to the scale of operation. If the scale can be further reduced, the cost of the studies will be significantly lowered as well.

9.10 VIRAL INACTIVATION

Viral inactivation, along with other steps for viral clearance discussed above, plays a key role in biotech processes in demonstrating virus clearance capability. It is a part of the monoclonal antibody platform in most companies (130). Most commonly used approach for viral inactivation is acidic pH/low pH inactivation, and this is the focus of this section.

The inactivation of enveloped viruses by low pH is a very common step in a monoclonal antibody purification process. The working principle is that some viruses, when exposed to a low pH, denature spontaneously. Typically, low pH inactivation step is performed after the protein A affinity chromatography step. The product from the protein A affinity column is titrated to a pH of 3.8 or lower and incubated for a period of 15–60 min, depending on the stability of the monoclonal antibody under consideration. This brief exposure to low pH effectively inactivates most lipid-enveloped viruses. Following this incubation, the protein solution is then titrated upward to a pH of 5.0 or higher to prepare for the next step. While the inactivation of viruses is favored by low pH, there is an inherent risk of aggregating the target protein at these low pH conditions. Moreover, a strongly acidic solution that is used as a titrating solution for the low pH step can cause localized low pH conditions, if not adequately mixed. This could potentially cause aggregation of the protein solution.

9.10.1 Scale-Down Modeling Approach

Scale-down of a low pH inactivation step is relatively easy. Care must be taken to ensure adequate mixing at this step during the addition of the inactivating agent such that no undue aggregation of the target protein results across this step. As mentioned before, the mass transfer coefficient ($k_L a$) can be used to define mixing conditions at small scale that will mimic conditions used in the manufacturing scale. The scale-down factors for these types of virus inactivation steps may be large, as there are few limitations to designing a temperature-controlled contacting vessel, which simply needs to maintain well-mixed solution. Test tubes or small vessels may be used, if they are immersed in a temperature-controlled heating block, and either mixed before or during contact. The level of the heating solution should be above the liquid level in the test vessel to ensure all the test solution contents have reached the target temperature. A high temperature short-time treatment can be employed using microwave assistance, but this system needs careful scale-down using miniature model of the process equipment, which could limit the scale-down to more modest ratios. The mixing of the vessel contents plays important role in low pH assisted inactivation. Standard mixing studies must be employed and the kinetics of inactivation should be measured, as it is a regulatory requirement. The addition rate of inactivation chemicals should match the manufacturing process, and the final product concentration of acid should be same as that of full scale. Scale-down variables that should be controlled throughout the inactivation process are pH, temperature, contact time, and the final inactivation chemical concentrations.

9.10.2 Scale-Down Qualification

The aggregate content of the protein during scale-down studies is an important quality attribute to match across scales for qualification of the scale-down model. The chemicals used for inactivation should show concentration profile similar to that used in the manufacturing scale (12).

9.10.3 Literature Review

Not much has been written on the topic of scale-down of viral inactivation, perhaps because of the simple design of such a system.

In summary, scale-down of viral inactivation is quite a simple exercise. However, care should be taken that the mixing is equivalent so that the end results are comparable across scales.

9.11 MEMBRANE ADSORBERS

Membrane chromatography is increasingly becoming popular as an alternative to certain chromatographic steps, offering higher productivity (131–138). In the membranes, binding sites are located along the through pores rather than nestled with long diffusive pores. Even in the newer chromatography resins with "through" pores or convective pores (139), the liquid has the choice to flow through the particle or around it. Unlike this, in a stacked membrane, there is no choice but to flow through the pores. This reduces the mass transport resistance for the solute to the matrix by eliminating pore diffusion, leaving film diffusion from the core of the liquid to the membrane surface in the interior of a through pore as the only transport resistance. As film diffusion is usually orders of magnitude faster than pore diffusion, mass transport limitations are drastically reduced in membrane chromatography, shifting the limitation of the process more to the properties of the matrix–solute interactions.

9.11.1 Scale-Down Modeling Approach

Most of the scale-down systems that are available in the market do not have a liquid flow path similar to that found in the large-scale capsule and thus tend to generate extremely high operational backpressure when a high flow rate is applied ($\geq$ 450 cm/h). The axial velocity of a mobile phase is faster in the center of the membrane adsorber (MA) unit than near the edges of adsorptive bed (106). The decrease in apparent porosity from the center of the bed to the gasket area or the edges of the adsorptive bed may account for the decrease in velocity observed adjacent to the gasket or edges (106,140). In contrast, radial flow adsorbers, prepared by spirally winding a flat sheet membrane over a porous cylindrical core, are used in a capsule for large-scale process (114). Distorted or poor inlet flow distribution has less effect in a large-scale capsule, thus leading to a smaller operational backpressure drop for large-scale operation (17,18). Despite being very complex to model in a radially outward direction, radial flow adsorbers are claimed to be suitable for large-scale application (114). The scalability and operating backpressure, therefore, are main issues for the MA scale-down model. An extremely high operational backpressure observed with the scale-down model can lead to considerably less capacity and an oversized membrane at full scale. This oversizing may lead to an erroneous economic calculation. Larger scale membranes do not have backpressure issues. However, the use of a large-scale MA in a scale-down model is also not practical, especially for a viral clearance study because of the high cost of viruses and feedstock.

9.11.2 Scale-Down Model Qualification

These aspects are same as in process chromatography.

9.11.3 Literature Review

Zhou *et al*. have redesigned scale-down models for Q-membrane, Q125 and Q40 (128,129). These scale-down models mimic the liquid flow path found in the large-scale modules and achieve performance parameters comparable to those of the larger Q units. However, even with improved scale-down devices, the operational backpressure for Q-membrane chromatography still influences process capacity. Several researchers have focused on creating an improved scale-down model for membrane chromatography (128,129).

In summary, the newer scale-down models seem to have addressed the most critical of the problems that one faces. However, not all manufacturers offer such devices. Careful comparison of the performance of the device should be done in order to understand the differences between the two scales, and the resulting data should be suitably interpreted.

9.12 SUMMARY

In the era of QbD and PAT, use of scale-down models has become an integral part of characterizing and validating industrial processes. Characterization studies form the backbone of experimentation intended to provide process knowledge. These studies are used to identify critical parameters and to establish acceptable ranges, process validation acceptance criteria, and specifications. Process validation studies such as evaluation of viral clearance and impurity clearance require such scale-down models, as it is not feasible from a safety perspective to perform them at manufacturing scale. This chapter presents general guidelines and relevant parameters affecting scale-down of the various unit operations involved in purification of proteins. The key parameters, assessment methods, assessment techniques, and challenges involved in scale-down are discussed. The quality of the existing scale-down models varies from unit operation to unit operation. Factors that complicate the scale-down process include difficulties in mimicking the hardware design resulting in differences in hydrodynamics across scales (mixing, shear, etc.), higher holdup volumes at smaller scales, and differences in materials used for making the equipment and the separation media. A lot of progress has been made in each of these areas but with the growing importance of scale-down models, scope for more work remains. We hope the chapter would be valuable to those engaged in small-scale development, optimization, and characterization of biotech processes.

LIST OF ABBREVIATIONS

CD	circular dichroism
cGMP	current good manufacturing practice
DLS	dynamic light scattering
DNA	deoxyribonucleic acid
HCP	host cell proteins
HETP	height equivalent to theoretical plate
HPH	high pressure homogenization
HPLC	high pressure liquid chromatography
HTPD	high throughput process development
IBs	inclusion bodies
ICH	International Conference on Harmonization
LPO	lactoperoxidase
MA	membrane adsorber
MF	microfiltration
MVA	multivariate analysis
NFF	normal flow filtration
NMWC	nominal molecular weight cutoffs
OD	optical density
PAT	process analytical technology
RDF	rotating disc filter
rDNA	ribosomal deoxyribonucleic acid
SDS-PAGE	sodium dodecyl sulfate polyacrylamide gel electrophoresis
SE-HPLC	size-exclusion high pressure liquid chromatography
TCA	trichloroacetic acid
TFF	tangential flow filtration
UF	ultrafiltration
USD	ultra-scale-down
UV	ultraviolet

NOMENCLATURE

$\sum$	equivalent settling area of a laboratory centrifuge (m^2)
$\sum_c$	equivalent settling area (m^2)
A_s	asymmetry factor
B_L	caulk width (m)
Ca	camp number
C_w	protein concentration at the membrane wall (kg/m^3)
ΔP_L	feed channel pressure drop (kg/ms^2)
ΔP_{TM}	transmembrane pressure (Pa)
f_1	correction factor for spacer caulks
g	acceleration due to gravity (m/s^{-2})
G	average value of velocity gradient
J_f	filtrate flux (m^3 /s)
$k_L a$	mass transfer coefficient
LRV	log reduction in viability
N	rotational speed (rps)
n_d	number of active discs
ω	angular velocity (radians/s)
P	power dissipated (W)
P_o	power number
Q	liquid flow rate through the centrifuge (m^3 /s)
Q_c	volumetric flow rate (m^3 /s)
r_1	inner disc radius (m)
r_2	outer disc radius (m)
Re	Reynolds number
R_i	inner radius (the distance between the center of rotation and the top of the liquid) (m)
R_o	outer radius (the distance between the center of rotation and the bottom of the tube) (m)
S	solids remaining (kg)
θ	half disc angle
TMP	transmembrane pressure (Pa)
μ	dynamic viscosity (Pa·s)
V	sample volume (L)
x	fractional time required for acceleration (s)
y	fractional time required for deceleration (s)
Z_L	number of caulks on a disc

REFERENCES

1. Wheelwright SM. J Biotechnol 1989; 11: 89–102.
2. Rathore AS. Trends Biotechnol 2009; 27: 698–705.
3. Willoughby N, Martin P, Titchener-Hooker N. Biotechnol Bioeng 2004; 87: 641–647.
4. Boychyn M, Doyle W, Bulmer M, More J, Hoare M. Biotechnol Bioeng 2000; 69: 1–10.
5. Bhambure R, Kumar K, Rathore AS. Trends Biotechnol 2010; 29: 127–135.
6. Rathore AS, Winkle H. Nat Biotechnol 2009; 27: 26–34.
7. Rathore AS. Trends Biotechnol 2009; 27: 546–553.
8. Xu Z, Li J, Zhou JX. Prep Biochem Biotechnol 2012; 42: 183.
9. Read EK, Park JT, Shah RB, Riley BS, Brorson KA, Rathore AS. Biotechnol Bioeng 2010; 105: 276–284.
10. Read EK, Park JT, Shah RB, Riley BS, Brorson KA, Rathore AS. Biotechnol Bioeng 2010; 105: 285–295.
11. Rathore AS, Bhambure R, Ghare V. Anal Bioanal Chem 2010; 398: 137–154.
12. Godavarti R, Petrone J, Robinson J, Wright R, Kelley BD, Bolton GR. In: Rathore AS, Sofer G, editors. Process validation in manufacturing of biopharmaceuticals. Boca Raton (FL): Taylor & Francis Group LLC; 2005. pp. 69–142.
13. Rathore AS, Krishnan R, Tozer S, Smiley D, Rausch S, Seely J. BioPharm Int 2005; 18: 58–60.
14. Tait AS, Aucamp JP, Bugeon A, Hoare M. Biotechnol Bioeng 2009; 104: 321–331.
15. Cecchini DJ. In: Rathore AS, Mhatre R, editors. Quality by design for biopharmaceuticals. Hoboken, New Jersey John Wiley and Sons; 2008. pp. 127–142.
16. Kozlowski S, Swann P. In: Rathore AS, Mhatre R, editors. Quality by design for biopharmaceuticals. Hoboken, New Jersey John Wiley and Sons, 2008, pp. 9–30.
17. Points to Consider in the Manufacture and Testing of Monoclonal Antibody Products for Human Use. Center for Biologics Evaluation and Research, F.D.A.; 1997.
18. Banerjee A. Biopharm Int 2010; 23: 26.
19. Adner N, Sofer G. BioPharm 1994; 7: 44–48.
20. Leonard MW, Sefton L, Costigan R. In: Kelley BD, Ramelmier RA, editors. Validation of biopharmaceutical manufacturing processes. Washington (DC): American Chemical Society; 1998. pp. 55–68.
21. Breece T, Gilkerson E, Schmelzer C. BioPharm 2002; 15: 16–20.
22. Kelley BD. In: Sofer G, Zabriskie D, editors. Biopharmaceutical process validation. New York: Marcel Dekker; 2000. pp. 29–59.
23. Box GEP, Hunter WG, Hunter JS. Statistics of experiments. New York: John Wiley and Sons; 1978.
24. Seely R, Wright H, Fry H, Rudge S, Slaff G. BioPharm 1994; 7: 41–48.
25. Rathore AS, Sofer G. In: Rathore AS, Sofer G, editors. Process validation in manufacturing of biopharmaceuticals. Boca Raton (FL): Taylor & Francis Group LLC; 2005. pp. 169–204.
26. Pieracci J, Yusuf-Makagiansar H. In: Rathore AS, Sofer G, editors. Process validation in manufacturing of biopharmaceuticals. Boca Raton, FL, USA CRC press, Taylor and Francis Group; 2012.
27. Wold S, Geladi P, Esbensen K, Öhman J. J Chemom 1987; 1: 41–56.
28. Tustian AD, Salte H, Willoughby NA, Hassan I, Rose MH, Baganz F, Hoare M, Titchener-Hooker NJ. Biotechnol Prog 2007; 23: 1404–1410.
29. Axelsson H. In: Flickinger MC, Drew SW, editors. Encyclopedia of bioprocess technology. New York: John Wiley and Sons; 1999. pp. 513–531.
30. Hanle DD. In: Flickinger MC, Drew SW, editors. Encyclopedia of bioprocess technology. New York: John Wiley and Sons; 1999. pp. 553–559.
31. Kempken R, Preissmann A, Berthold W. J Ind Microbiol 1995; 14: 52–57.
32. Maybury JP, Mannweiler K, Titchener-Hooker NJ, Hoare M, Dunnill P. Bioprocess Biosyst Eng 1998; 18: 191–199.
33. Leung WWF. Industrial centrifugation technology. New York: McGraw Hill; 1998.
34. Maybury JP, Hoare M, Dunnill P. Biotechnol Bioeng 2000; 67: 265–273.
35. Mosqueira FG, Higgins JJ, Dunnill P, Lilly MD. Biotechnol Bioeng 1981; 23: 335–343.
36. Mannweiler K, Hoare M. Bioprocess Eng 1992; 8: 18–25.
37. Betts JI, Baganz F. Microb Cell Fact 2006; 5: 21.
38. Gill NK, Appleton M, Baganz F, Lye GJ. Biochem Eng J 2008; 39: 164–176.
39. Pampel L, Mcnerney T, Wypych J. Pap Am Chem Soc 2004; 227: 264–BIOT.
40. Duetz WA. Trends Microbiol 2007; 15: 469–475.
41. Fernandes P, Cabral JMS. Biocatal Biotransformation 2006; 24: 237–252.
42. Islam RS, Tisi D, Levy MS, Lye GJ. Biotechnol Prog 2007; 23: 785–793.
43. Islam RS, Tisi D, Levy MS, Lye GJ. Biotechnol Bioeng 2008; 99: 1128–1139.
44. Micheletti M, Lye GJ. Curr Opin Biotechnol 2006; 17: 611–618.
45. Micheletti M, Barrett T, Doig SD, Baganz F, Levy MS, Woodley JM, Lye GJ. Chem Eng Sci 2006; 61: 2939–2949.
46. Hutchinson N, Bingham N, Murrell N, Farid S, Hoare M. Biotechnol Bioeng 2006; 95: 483–491.
47. Zaman F, Allan CM, Ho SV. Biotechnol Prog 2009; 25: 1709–1716.
48. Sommer B, Friehs K, Flaschel E, Reck M, Stahl F, Scheper T. J. Biotechnol 2009; 140: 194–202.
49. Ariga O, Watari T, Andoh Y, Fujishita Y, Sano Y. J Ferm Bioeng 1989; 68: 243–246.
50. Zhao FS, Yu JY. Biotechnol Prog 2001; 17: 490–494.
51. Malamy MH, Horecker BL. Biochemistry 1964; 3: 1889–1893.
52. Clarkson AI, Lefevre P, Titchener-Hooker NJ. Biotechnol Prog 1993; 9: 462–467.
53. van Hee P, Middelberg APJ, van der Lans RGJM, van der Wielen LAM. Biotechnol Bioeng 2004; 88: 100–110.
54. Kloosterman J, Vanwassenaar PD, Slater NKH, Vanderpadt A. Chem Eng J 1988; 37: B47–B54.
55. Limon-Lason J, Hoare M, Orsborn CB, Doyle DJ, Dunnill P. Biotechnol Bioeng 1979; 21: 745–774.
56. Ricci-Silva ME, Vitolo M, Abrahao-Neto J. Process Biochem 2000; 35: 831–835.

57. Schutte H, Kula MR. Biotechnol Appl Biochem 1990; 12: 599–620.
58. Engler CR. Bioprocess Technol 1990; 9: 95–105.
59. Hetherington PJ, Follows M, Dunnill P, Lilly MD. Trans Inst Chem Eng 1971; 49: 142–148.
60. Middelberg APJ. Trans IChemE 1993; 71C: 215–219.
61. Sauer T, Robinson CW, Glick BR. Biotechnol Bioeng 1989; 33: 1330–1342.
62. Keshavarz-Moore E, Hoare M, Dunnill P. Enzyme Microb Technol 1990; 12: 764–770.
63. Kelly WJ, Muske KR. Bioprocess Biosyst Eng 2004; 27: 25–37.
64. Gray PP, Dunnill P, Lilly MD. In: Terui G, editors. Fermentation technology today. Japan: Society of Fermentation Technology; 1972. pp. 347–351.
65. Floury J, Bellettre J, Legrand J, Desrumaux A. Chem Eng Sci 2004; 59: 843–853.
66. Donsì F, Ferrari G, Lenza E, Maresca P. Chem Eng Sci 2009; 64: 520–532.
67. Clark EDB. Curr Opin Biotechnol 1998; 9: 157–163.
68. Hevehan DL, De Bernardez Clark E. Biotechnol Bioeng 1997; 54: 221–230.
69. Yasuda M, Murakami Y, Sowa A, Ogino H, Ishikawa H. Biotechnol Prog 1998; 14: 601–606.
70. Boyle DM, Buckley JJ, Johnson GV, Rathore AS, Gustafson MG. Biotechnol Appl Biochem 2009; 54: 85–92.
71. Buswell AM, Middelberg AP. Biotechnol Bioeng 2003; 83: 567–577.
72. Lee C, Buswell A, Middelberg AP. Chem Eng Sci 2002; 57: 1679–1684.
73. Mannall GJ, Titchener-Hooker NJ, Chase HA, Dalby PA. Biotechnol Bioeng 2006; 93: 955–963.
74. Goldberg ME, Rudolph R, Jaenicke R. Biochemistry 1991; 30: 2790–2797.
75. Katoh S, Katoh Y. Process Biochem 2000; 35: 1119–1124.
76. Katoh S, Sezai Y, Yamaguchi T, Katoh Y, Yagi H, Nohara D. Process Biochem 1999; 35: 297–300.
77. Cowieson NP, Wensley B, Listwan P, Hume DA, Kobe B, Martin JL. Proteomics 2006; 6: 1750–1757
78. Vincentelli R, Canaan S, Campanacci V, Valencia C, Maurin D, Frassinetti F, Scappucini-Calvo L, Bourne Y, Cambillau C, Bignon C. Protein Sci 2004; 13: 2782–2792.
79. Willis MS, Hogan JK, Prabhakar P, Liu X, Tsai K, Wei Y, Fox T. Protein Sci 2005; 14: 1818–1826.
80. Rathore AS, Sharma A, Chillin D. BioPharm 2006; 19: 48–57.
81. Mannall GJ, Myers JP, Liddell J, Titchener-Hooker NJ, Dalby PA. Biotechnol Bioeng 2009; 103: 329–340.
82. Qoronfleh MW, Hesterberg LK, Seefeldt MB. Protein Expr Purif 2007; 55: 209–224.
83. Ordidge GC, Mannall GJ, Liddell J, Dalby PA, Micheletti M. Biotechnol Prog 2012; 28: 435444
84. Fisher RR, Glatz CE. Biotechnol Bioeng 1988; 32: 777–785.
85. Glatz CE, Hoare M, Landa-Vertiz J. AIChE J 1986; 32: 1196–1204.
86. Rothstein F. Protein process engineering. New York: Marcel Dekker; 1994.
87. Bell DJ, Dunnill P. Biotechnol Bioeng 1982; 24: 1271–1285.
88. Smoluchowski M. Z Phys Chem 1917; 92: 129–168.
89. Rushton JH, Costich EW, Everett HJ. Chem Eng Prog 1950; 46: 467–476.
90. Bell DJ, Brunner K-H. Filt Sep 1983; 20: 274–286.
91. Mannweiler K, Titchener-Hooker NJ, Hoare M. IChemE Symp, Adv Biochem Eng/Biotechnol 1989; 1: 105–117.
92. Garcia-Ochoa F, Gomez E. Biotechnol Adv 2009; 27: 153–176.
93. Chhatre S, Pampel L, Titchener-Hooker NJ. Biotechnol Prog 2011; 27: 998–1008.
94. Chhatre S, Titchener-Hooker NJ. J Chem Technol Biotechnol 2009; 84: 927–940.
95. Watanabe E, Tsoka S, Asenjo JA. Ann N Y Acad Sci 1994; 721: 348–364.
96. Errson B, Ryden L, Janson JC. In: Janson JC, editor. Protein purification: principles, high resolution methods, and applications. New Jersey: John Wiley and Sons; 2011. pp. 8–9.
97. Graumann K, Ebenbichler A. Chem Eng Technol 2008; 28: 1398–1407.
98. Gerontas S, Asplund M, Hjorth R, Bracewell DG. J Chromatogr A 2010; 1217: 6917–6926.
99. Rubino M, Bailey M, Baker JC, Boose JA, Metzka L, Moore V, Quertinmont M, Wiler W. In: Rathore AS, Sofer G, editors. Process validation in manufacturing of biopharmaceuticals. Boca Raton (FL): CRC Press; 2005. pp. 545–564.
100. Smith TM, Wilson E, Scott RG, Misczak JW, Bodek JM, Zabriskie DW. In: Kelley BD, Ramelmier RA, editors. Validation of biopharmaceutical manufacturing process. Washington (DC): American Chemical Society; 1998. pp. 80–92.
101. Viral safety of biotechnology products derived from cell lines of human or animal origin, ICH Viral Safety Document. International Conference on Harmonization; 1998.
102. Yamamoto S, Nomura M, Sano Y. J Chromatogr 1987; 409: 101–110.
103. Barry A, Chojnacki R. BioPharm 1994; 7: 43–47.
104. Sofer G, Hagel L. Handbook of process chromatography: a guide to optimization, scale-up, and validation. New York: Academic Press; 1997.
105. Hutchinson N, Chhatre S, Baldascini H, Davies JL, Bracewell DG, Hoare M. Biotechnol Prog 2009; 25: 1103–1110.
106. Chhatre S, Thillaivinayagalingam P, Francis R, Titchener-Hooker NJ, Newcombe AR, Keshavarz-Moore E. Biotechnol Prog 2007; 23: 888–894.
107. Thillaivinayagalingam P, O'Donovan K, Newcombe AR, Keshavarz-Moore E. J Chromatogr B Analyt Technol Biomed Life Sci 2007; 848: 88–96.
108. Stein A, Kiesewetter A. J Chromatogr B Analyt Technol Biomed Life Sci 2007; 848: 151–158.
109. Denton G, Murray A, Price MR, Levison PR. J Chromatogr A 2001; 908: 223–234.
110. Fenneteau F, Aomari H, Chahal P, Legros R. Biotechnol Bioeng 2003; 81: 790–799.
111. Ghose S, Chase H. Bioseparation 2000; 9: 21–28.
112. Uretschlager A, Jungbauer A. J Chromatogr A 2000; 890: 53–59.

113. Hagel L, Jagschies G, Sofer G. Handbook of process chromatography: development, manufacturing, validation and economics. Netherland: Academic Press; 2008.
114. Mulder M. Basic principles of membrane technology. Dordrecht, The Netherlands: Kluwer Academic Publishers; 1996.
115. Ma G, Aucamp J, Gerontas S, Eardley-Patel R, Craig A, Hoare M, Zhou Y. Biotechnol Prog 2010; 26: 466–476.
116. Reynolds T, Boychyn M, Sanderson T, Bulmer M, More J, Hoare M. Biotechnol Bioeng 2003; 83: 454–464.
117. Englard S, Seifter S. Methods Enzymol 1990; 182: 285–3006.
118. Paul EL, Rosas CB. Chem Eng Prog 1990; 8: 17.
119. Stavrinides S, Ayazi Shamlou P, Hoare M. In: Ayazi Shamlou P, editor. Processing of solid–liquid suspensions. London: Butterworth Heinman; 1993. pp. 118–158.
120. Harrison RG. Protein purification process engineering. New York: Marcel Dekker; 1993.
121. Foster PR, Dunnill P, Lilly MD. Biotechnol Bioeng 1976; 18: 545–580.
122. Iyer HV, Przybycien TM. AIChE J 1994; 40: 349–359.
123. Iyer HV, Przybycien TM. Biotechnol Bioeng 1995; 48: 324–332.
124. Asher D, Brorson K, Hotta J, Hughes J, Martin J, Ruppach H, Sofer G, Wisher M, Willkommen H, Yang B Preparation of virus spikes used for virus clearance studies. Technical Report No. 47. University of Iowa: Parenteral Drug Association, Inc.; 2010.
125. Levy RV, Philips MW, Lutz H. In: Meltzer TH, Jornitz MW, editors. Filtration in biopharmaceutical industry. New York: Marcel Dekker; 1998. pp. 619–646.
126. Kelley BD, Petrone J. In: Wang WK, editor. Membrane separations in biotechnology. New York: Marcel Dekker; 2001. pp. 351–396.
127. Huang PY, Peterson J. In: Wang WK, editor. Membrane separations in biotechnology. New York: Marcel Dekker; 2001.
128. Zhou JX, Tressel T, Gottschalk U, Solamo F, Pastor A, Dermawan S, Hong T, Reif O, Mora J, Hutchison F, Murphy M. J Chromatogr A 2006; 1134: 66–73.
129. Zhou JX, Tressel T. Biotechnol Prog 2006; 22: 341–349.
130. Shukla AA, Hubbard B, Tressel T, Guhan S, Low D. J Chromatogr B Analyt Technol Biomed Life Sci 2007; 848: 28–39.
131. Rathore AS, Shirke A. Prep Biochem Biotechnol 2011; 41: 307–315.
132. Thommes J, Kula MR. Biotechnol Prog 1995; 11: 357–367.
133. Brandt S, Goffe RA, Kessler SB, O'Connor JL, Zale SE. Bio/Technology 1988; 6: 779–782.
134. Briefs KG, Kula MR. Chem Eng Sci 1992; 47: 141–149.
135. Charcosset C. J Chem Technol Biotechnol 1998; 71: 95–110.
136. Lutkemeyer D, Bretschneider M, Buntemeyer H, Lehmann J. J Chromatogr A 1993; 639: 57–66.
137. Reif O-W, Freitag R. J Chromatogr A 1993; 654:29–41.
138. Riese U, Lutkemeyer D, Heidemann R, Buntemeyer H, Lehmann J. J Biotechnol 1994; 34: 247–257.
139. Afeyan NB, Fulton SP, Gordon NF, Maszaroff I, Varady L, Regnier F. Bio/Technology 1990; 8: 203–206.
140. Fisher RR, Glatz CE, Murphy PA. Biotechnol Bioeng 1986; 28: 1053–1063.

10

ADSORPTION IN SIMULATED MOVING BEDS (SMB)

CESAR C. SANTANA
School of Chemical Engineering, State University of Campinas, Campinas-SP, Brazil
IVANILDO J. SILVA JR AND DIANA C. S. AZEVEDO
Department of Chemical Engineering, Federal University of Ceará, Fortaleza-CE, Brazil
AMARO G. BARRETO JR
Escola de Química, Universidade Federal do Rio de Janeiro, Rio de Janeiro-RJ, Brazil

10.1 INTRODUCTION

10.1.1 Adsorption and Chromatography as Separation Tools

A large impact in the life sciences has been addressed by advances in the broad area named *Biotechnology*. Separations needs have an important role in this area and are related to the pharmaceutical, biomedical, and other biotechnological industries and provide additional opportunities to translate progresses in separations technology, named generally as bioseparation or downstream processes (1).

The products of biotechnological origin possess great diversity and are generally present in fermentation broths and cultures of cells in low concentrations. Diluted systems coupled to considerable amounts of chemical species that interfere in the recovery processes, concentration, and final purification turn those difficult and onerous tasks answering, in most of the processes, for the largest percentage of the costs of biomolecules production. In the special case of proteins, the percentages of the purification costs reache values of the order of 60% in relation to the total costs of the production process and can even stay in the range of up to 80% to 90% for original fermentation products of recombinant DNA (2).

A general theme in the field of bioseparation processes is the improvement of the selectivity of solutes. The power of modern synthetic chemistry and the production of biomolecules in conjunction with separation science and technology can be used to develop new separating agents and equipments with an enhanced function for selective separation. Adsorption processes are concentration-controlled separations and are based on the differential affinity of various soluble molecules that are selectively transferred to a surface of a solid adsorbent. The solid phase is named the *resin* or *stationary phase* and the soluble molecules are in the fluid phase often called the *mobile phase*. Adsorption is a thermodynamically spontaneous process in which energy is released during the process. The reverse process by which the adsorbed molecules are removed from the surface to the bulk fluid phase is called *desorption*. Chromatographic separations are based on the phenomena of different degrees of interaction due to the differential affinity between the mixture components of the fluid phase and the stationary phase, which is generally packed in a column, resulting in different migration velocities inside the bed and consequently in different times for each component to reach the exit of the column. The separation methods using the adsorption phenomena and chromatographic arrangements present low energetic consume,

Downstream Industrial Biotechnology: Recovery and Purification, First Edition. Edited by Michael C. Flickinger.

allowing the use of appropriate adsorbents to promote components separation and purification on a competitive economic basis in the preparative and production scale (3).

Despite the dominance of adsorptive chromatographic systems in the production of biopharmaceuticals in recent times, this technology was not consolidated into standardized large-scale bioprocesses until the 1990s. Authoritative publications concerned with fundamentals and applications of adsorption processes and liquid chromatography (LC) in bioprocesses (4,5), put in perspective the enormous potential and the numerous uses of this technique in the separation and purification field. In conventional batch LC used for adsorptive separation, a small pulse of the feed mixture in then injected into the column, followed by the continuous inlet of a desorbent or solvent. Since different solutes migrate at different speeds, they are separated as they migrate through the column. The individual bands are then collected as products at the outlet ports. Any overlapping bands are either recycled or discarded as waste. To obtain a high purity (>99%) and high yield (>99%), complete separation is required, which involves the consumption of a large amount of solvent. Since a significant portion of the column is not used during the batch operation, the column utilization is also inefficient.

10.1.2 Classification of Chromatography

For the sake of processes nomenclature and to facilitate the communication among researchers, it is usual in the literature of fundamentals and applications of chromatographic processes to disclose a classification of these processes, as shown in Table 10.1, which resume a well-accepted classification for chromatographic processes (6). From Table 10.1 it is important to point out some characteristics of the operation modes in discontinuous chromatography. The usual chromatographic discontinuous system is depicted in Fig. 10.1, with the typical peaks originating from the interaction of solutes and the stationary phase.

TABLE 10.1. A Classification of Chromatographic Processes

Principle for Classification	Nomenclature
Physical state of both phases	Gas–liquid chromatography (GLC)
	Gas–solid chromatography (GSC)
	Liquid–liquid chromatography (LLC)
Physical state of mobile phase	Gaseous chromatography (GC)
	High performance liquid chromatography (HPLC)
	Supercritical fluid chromatography (SFC)
Scale	Analytical/Preparative/ Production scale
Operation modes	Discontinuous: frontal/displacement/elution continuous: multicolumn (SMB)/annular
Mechanism of separation	Partition/adsorption (normal and reversed phase)/ion exchange/size exclusion/affinity
Equilibrium relationship	Linear/nonlinear

In frontal chromatography, a sample is fed continuously into the chromatographic bed and no additional mobile phase is used, while in displacement chromatography the sample is fed into the system as a finite slug and the mobile phase contains a compound (the displacer) more strongly retained than the components of the sample. Elution chromatography is characterized with a sample fed into the system as a finite slug and the mobile phase is continuously passed through the chromatographic bed. The largest disadvantage of the usual chromatographic separations lasts in the discontinuity of the process and in the dilution of the product, consequently leading to low productivity values.

10.1.3 True Moving Bed Chromatography

It is a well-known fact in the operations of adsorption that continuous systems in which the solid phase is contacted in the direction opposed to the one of the flowing phase in such a way that the profile of mass-transfer stays stationary and the adsorbent is used in a more efficient way.

The process called *true moving bed* (TMB) allows a continuous operation, distinctly of the classic chromatography elution process. In a TMB (Fig. 10.2a), the liquid and solid phases flows are carried out in opposite directions. The inlet (feed and desorbent) and outlet (extract and raffinate) ports are fixed along the unit. According to the position of the inlet and outlet streams, four different operation sections can be distinguished: section I located between the eluent and extract streams, section II located between the extract and feed streams, section III located between the feed and raffinate streams, and section IV located between the raffinate and desorbent streams. The net flow rate has to be selected in each section in order to ensure the regeneration of adsorbent in section I, the desorption of the less strongly adsorbed component in section II, the adsorption of the more strongly adsorbed component in section III, and the regeneration of the desorbent in section IV. The need for circulation of not only the fluid but also the solid in the TMB process brings a series of disadvantages to the processes such as short life of the adsorbent due to attrition, fluid velocities limited by fluidization phenomena, and lack of efficiency. In this way, efforts have been driven to develop processes that maintain the advantages of the countercurrent operation but avoid the circulation of the solids. The first solution to solve the problems with TMB came from a patent from (7) with the proposal of using simple fixed bed columns and simulating the solid movement by a

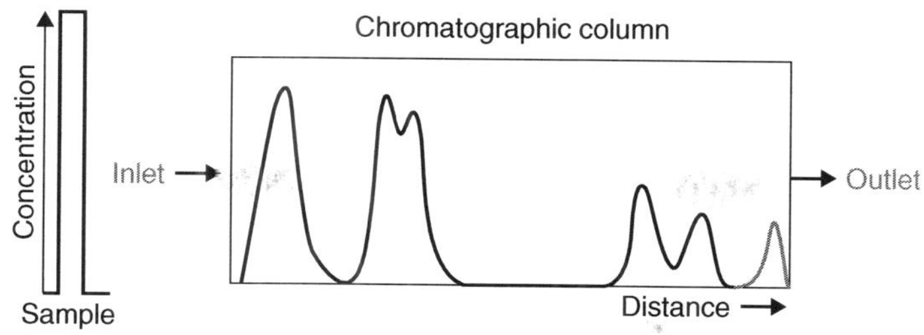

Figure 10.1. Sketch of a discontinuous chromatographic system. (This figure is available in full color at http://onlinelibrary.wiley.com/book/10.1002/9780470054581.)

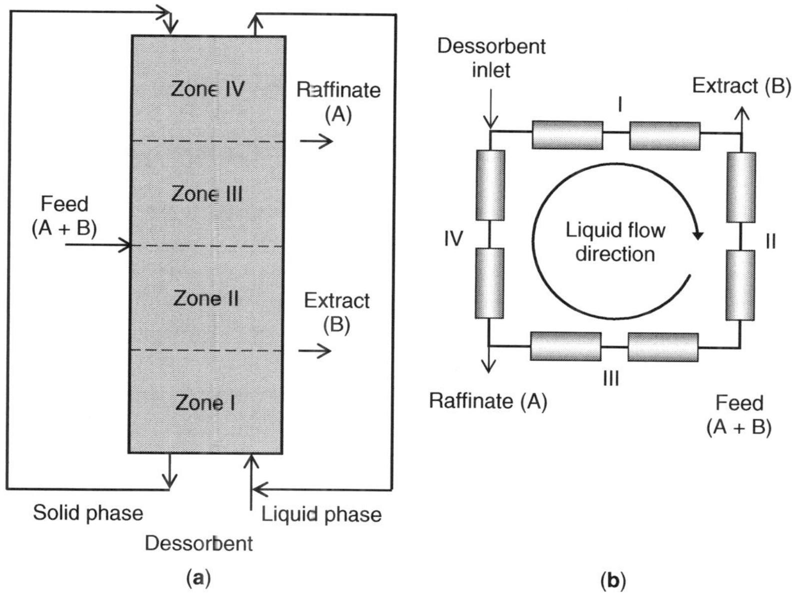

Figure 10.2. Schematic representations of (a) true moving bed and (b) simulated moving bed units.

synchronous shift of all inlet and outlet ports in the direction of the fluid flow (Fig. 10.2b).

It has been demonstrated experimentally (8) that these conditions will guarantee the success of the separation, as the more retained component moves to the extract port with the solid phase and the less retained component moves to the raffinate port with the liquid phase.

These ideas of simulated solid movement lead to the simulated moving bed (SMB) concept; an alternative to a countercurrent flow is to simulate adsorbent movement by periodically moving the input and output ports through a ring fixed bed while keeping the bed stationary.

10.1.4 Simulated Moving Bed Chromatography

In Fig. 10.2, the two different processes of adsorption are presented in a countercurrent mode of operation. In most of the innovations in that sense, the movement of the solids is obtained by periodic changes in the feeding and discharge in a system of multiple columns resulting in the outline of a well-known process such as the SMB.

The SMB chromatography has been applied since the 1960s by Universal Oil Products (UOP) for large-scale separations in the petrochemical industry. Nowadays, the application of SMB for preparative and production scale separation of sugars, fine chemistry, and pharmaceutical products, and in particular enantiomers, is gaining increasing importance. Successful examples are the separation of glucose and fructose (9) and the resolution of enantiomers on chiral stationary phases (10–14). New challenges for the SMB technology are reported by Li *et al.* (15) concerning its application to the separation and purification of biomolecules. Examples of products that are considered for SMB separation and purification are therapeutical proteins, antibodies, nucleosides, and plasmid DNA. The process SMB presents economic advantages over other chromatographic systems for several reasons: it is a continuous process and it allows the separation of similar compounds starting for example, from a racemic mixture, allowing high productions and low-solvent consumption. In general, in that system type the volume of requested adsorbent is approximately 25% of the requested quantity in batch chromatography (16,17).

A number of reviews on SMB systems is provided in the literature as a set of articles and chapters with references from (18–32), herein considered as very representative of the state of art of the subject.

As depicted in Fig. 10.2, SMB uses a series of columns of adsorption (e.g. 8 columns or 12 columns) with an appropriate adsorbent. The columns are connected to recipients that contain the feeding and the eluent and that receive the currents of exit of the product through lines controlled by a group of valves of multiple positions. That group of control valves allows that they are alternate, in regular intervals of time, the points of entrance of the feeding, of the eluent, and of the exit currents. The system therefore changes the positions between the entrance points and exit, simulating the countercurrent flow.

From the point of view of the operational variables, the project of SMB is relatively complex because it involves at least 10 specific parameters to know: diameter of the columns, four lengths of separation zones, four flowing currents, and an average velocity associated to the control of the opening of the valves of multiple positions. SMB is usually used for a mixture that contains two similar products, which attempts the separation. The use of SMB in the separation of multi-component mixtures is not still very well-known. The main claim of this separation method consists in its ability to separate mixtures of difficult resolution and for products of high added value.

SMB is a continuous apparatus, whose principle of operation can be best described with reference to equivalent TMB. In TMB (Fig. 10.2a), the liquid and solid phases flows are carried out in opposite directions. The inlet (feed and desorbent) and outlet (extract and raffinate) ports are fixed along the unit. According to the position of the inlet and outlet streams, four different operation sections can be distinguished: section I located between the eluent and extract streams, section II located between the extract and feed streams, section III located between the feed and raffinate streams, and section IV located between the raffinate and desorbent streams. The net flow rate has to be selected in each section in order to ensure the regeneration of adsorbent in section I, the desorption of the less strongly adsorbed component in section II, the adsorption of the more strongly adsorbed component in section III, and the regeneration of the desorbent in section IV. These conditions will guarantee the success of the separation, as the more retained component moves to the extract port with the solid phase and the less retained component moves to the raffinate port with the liquid phase.

The major problem in TMB operation associated with the movement of the solid phase was overcome by the introduction of SMB technology. An SMB opened loop unit (Fig. 10.2b) consists of a set of interconnected columns in series by valves and tubing to form a circuit. This circuit is divided into four zones with two inlet ports (feed and desorbent) and three outlet ports (raffinate, where a low-affinity solute A is removed, an extract, where a high-affinity solute B is removed, and pure desorbent). The inlet and outlet ports are periodically moved in the liquid flow direction by multiple position valves, causing an apparent countercurrent movement between the liquid and the solid phase. As in batch chromatography, solute A migrates faster than solute B in the liquid flow direction. In a four-zone SMB, solute A adsorption occurs in zone IV, while it's desorption occurs in zone II. Solute B adsorption occurs in zone III, and desorption occurs in zone I.

SMB units exhibit important advantages, in comparison to batchwise preparative chromatography. In particular, these are due to the continuous nature of the operation and to an efficient use of the stationary and mobile phases, which allows the decrease of the mobile phase (desorbent) requirement and the improvement of the productivity per unit time and unit mass of stationary phase. Moreover, high performances can be achieved even at rather low values of selectivity and with a relatively small number of theoretical plates. These features of SMB units are due to the fact that, contrary to preparative chromatography, the concentration profiles of the components to be separated are allowed to overlap along the adsorption beds, the requirement being that the components are pure only at the extract and raffinate outlet locations. Owing to these positive features, SMB is particularly attractive in the case of enantiomer separations, since it is difficult to separate enantiomers by conventional techniques considering the low selectivity factors normally found with these systems.

10.2 FUNDAMENTALS OF CHROMATOGRAPHIC SEPARATIONS

10.2.1 Influence of the Equilibrium Adsorption Isotherms

The unitary cell of a SMB—regardless of the chosen valve switching scheme and the number of inlet/outlet streams—is the chromatographic column. Therefore, all factors that influence the shape of a concentration front traveling inside a packed column will also have an impact on the performance of the continuous chromatographic unit. Such factors include fluid dynamics inside the packed bed, mass transfer phenomena and, most importantly, the equilibrium of the adsorption at the temperature of the system (31). The adsorption equilibrium is determined by the isotherm, which expresses the correlation between the loading of the solute on the adsorbent q at given fluid phase concentration c at a fixed temperature and other fixed variables, which are often crucial in biological systems, such as pH and ion strength (33,34).

At this point it is convenient to make a difference between two situations in batch chromatography: (i) that of a finite and relatively small amount of solute which is injected into the packed column under a constant flow of mobile phase and (ii) that of a finite and relatively large pulse of solute—very often highly concentrated—that is pumped into the column, usually preceded and followed by *clean* mobile phase at the same flowrate. The first situation is typical of analytical chromatography and the solute travels inside the column in such dilute concentrations (injection volume is typically less than 1% bed volume) that adsorption equilibrium essentially follows Henry's Law, that is, a linear adsorption isotherm. The principles of analytical chromatography are out of the scope of this chapter and may be found in several comprehensive works in the literature (35,36).

The second situation is commonly found both in batch and continuous chromatography and the basic principles that govern the displacement of concentration profiles inside the column may be analyzed in light of the equilibrium theory (37,38). If all nonideal (hydrodynamics and mass-transfer) effects are neglected, the mass balance on an infinitesimal volume element inside the column is described in Equation 10.1:

$$\int u_i \frac{\partial C_i}{\partial z} + \left[1 + \frac{1-\varepsilon}{\varepsilon} f'(C_i)\right] \frac{\partial C_i}{\partial t} = 0 \qquad (10.1)$$

In Equation 10.1 the variable u_i is the interstitial velocity, C_i is the fluid phase concentration of solute i, q_i is the adsorbed phase concentration of solute i, ε is the bed void fraction, f' is the derivative of the isotherm equation, z and t are axial and time coordinates, respectively.

From mathematical reasoning, one easily finds that each solute concentration c_i travels inside the bed with a characteristic velocity given by Equation 10.2 which states that, in the absence of nonideal effects, concentrations will move inside the column with a characteristic velocity that is inversely proportional to the first derivative of the adsorption isotherm.

$$\left.\frac{\partial z}{\partial t}\right|_c = u_c = \frac{u_i}{1 + \dfrac{1-\varepsilon}{\varepsilon} f'(C_i)} \qquad (10.2)$$

Although there are various adsorption models reported in the literature, they may be grouped into three general types: linear (f' is constant); favorable (f' increases with increasing C_i) and unfavorable (f' decreases with increasing C_i). Systems with favorable isotherms will result in a concentration front that travels inside the columns as a shock wave upon adsorption and as a spreading wave upon desorption. Likewise, for systems with unfavorable isotherms, concentration fronts will travel as a spreading wave upon adsorption and as a shock wave upon desorption. For linear isotherms, all concentrations will move at the same speed, which are only a function of the interstitial velocity, bed void fraction and the constant of adsorption. Figure 10.3 summarizes these effects.

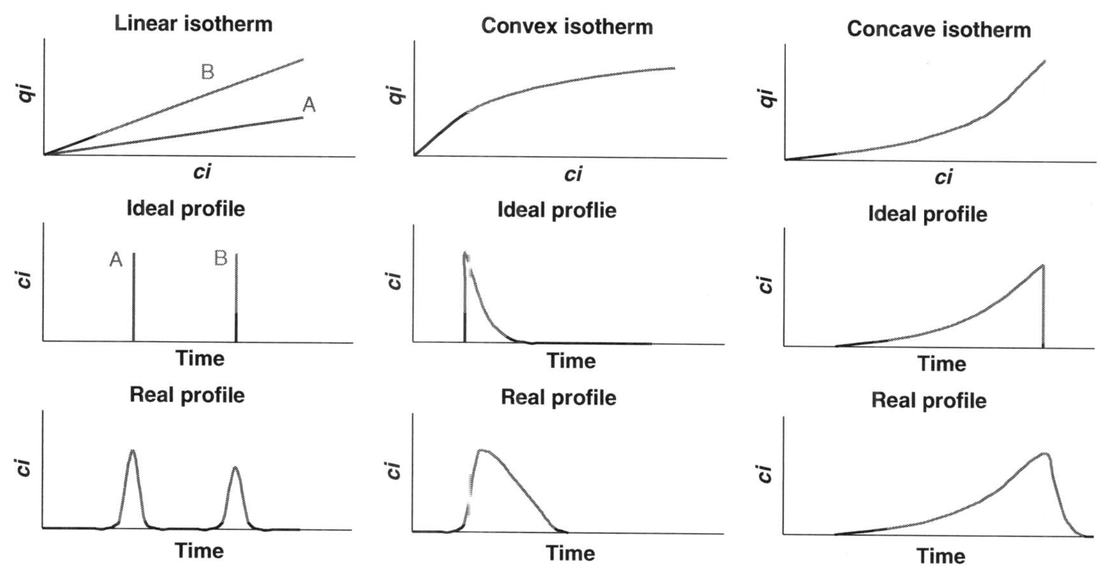

Figure 10.3. Influence of the type of isotherm on the chromatogram. [Reproduced from Ref. 30]. (This figure is available in full color at http://onlinelibrary.wiley.com/book/10.1002/9780470054581.)

The presence of nonideal effects—such as mass-transfer resistances and axial/radial dispersion—tend to smooth and broaden these band profiles inside the columns. They will be discussed in the forthcoming section. Nevertheless, the main features of how concentration fronts move inside chromatographic columns may be easily tackled by the equilibrium theory (39). This highlights the fact that the adsorption isotherm is the main parameter governing preparative-scale chromatographic separations and therefore, a precise experimental determination of adsorption equilibrium is essential for the the proper design of separation chromatographic processes.

10.2.1.1 Experimental Determination of Adsorption Isotherms. Adsorption isotherms are nonlinear under overloaded conditions, which are usual in preparative chromatography, most of them exhibiting favorable behavior in biological systems. The effects of nonlinear equilibrium on concentration fronts inside chromatographic columns have been summarized in Fig. 10.3 and further details may be found in the literature (37,38,40). In linear chromatography, mostly applied for analytical purposes, axial dispersion and mass transfer-effects are the main factors that cause peak broadening (39). In preparative chromatography, these effects are less important as compared to the effects of equilibrium nonlinearity (41). In Kaspereit *et al*. (42) we find a demonstration on how adsorption isotherm parameters may impact severely in purity and productivity of an SMB unit for various feed concentrations. Therefore, the knowledge of the adsorption isotherm over a wide concentration range may contribute to elucidating the retention mechanism and may help propose strategies of improving separation and, hence, productivity in preparative-scale separation devices. With the increasing cost of selective stationary phases (adsorbent), this should be of particular interest in enantiomeric separations (42), for instance.

Several experimental methods have been proposed in the literature (40,43,44) to determine adsorption isotherms for a single component and mixtures (competitive isotherms). These methods may be generally classified as static and dynamic. Static methods or immersion methods are not implemented in column manner, but in closed vessels, inside which a given amount of fluid phase of known concentration is placed in contact with a given amount of adsorbent. Adsorbed phase concentration is determined by mass balance considering the initial state and equilibrium state. Dynamic methods are based on the concentration curve (as a function of time) at the column outlet as a response to well-defined changes in concentration at the inlet of the column (40). The changes may be an infinite impulse (such as a Dirac delta function), a square pulse, a positive or a negative step change (single or stepwise).

Among the main disadvantages of static methods, it may be argued that they are time-consuming, there is always some degree of uncertainty as to whether equilibrium has actually been reached and they generally require relatively large amounts of solute and adsorbent for concentration changes to be precisely measured (45). Dynamic methods reproduce the mode of operation of chromatography and various experimental procedures and analytical methods have been developed along the last 50 years. These include frontal analysis (FA), frontal analysis from characteristic points (CPFA), characteristic point elution (CPE), perturbation methods (PM), and inverse method (IM) (46). The first three have widespread use. However, CPE and CPFA may only be used for single-component determinations. Only FA and PM may also be used for multicomponent adsorption measurements, FA being extensively reported in the literature for the determination of competitive adsorption isotherms (34,38,47–51).

10.2.1.1.1 Frontal Analysis. In FA, the experimental procedure for the determination of adsorption isotherms generally comprises the following steps (34,38). The column is initially equilibrated with the mobile phase and then the feed containing the adsorbate is pumped into the column, so that a step change in concentration at the inlet is implemented. The concentration of the adsorbate at the exit of the column is monitored up to complete saturation. After equilibrium whenthe feed concentration has been reached, another solution of (higher/lower) known concentration is pumped into the column up to saturation. This procedure is repeated several times for increasing or decreasing steps in feed concentration, generating stepwise breakthrough curves (concentration histories at the exit of the column). From mass balance, each breakthrough allows the calculation of the adsorbent loading in equilibrium with a given feed concentration, which is one isotherm point.

For a binary mixture, samples may be collected at the outlet of the column and concentrations may be determined by an adequate analytical method, so that the individual breakthrough of each of the adsorbates (under competitive conditions) may be plotted. Alternatively, if on-line detection (IR, UV–Vis, etc.) is placed at the column outlet, the thus obtained breakthrough curve (which is the sum of the detection signals of both adsorbates) has two waves separated by an intermediate plateau. If the column has been initially equilibrated with pure mobile phase, only the less retained adsorbate is eluted at this intermediate plateau, generally at a higher concentration than in the feed. If an elution (desorption) step is performed after saturation, the same characteristic plateau is observed, but this time corresponding to the strongest adsorbed species (43). An extent to ternary systems was provided by Lisec *et al*. (52).

Adsorption equilibrium data for the less and more strongly adsorbed species may be determined by mass balances both from the breakthrough (adsorption) and elution (desorption) curves, respectively, according to the following Equations 10.3 and 10.4 (43):

$$q_i^* = \frac{c_i\left(V_{F,1+2} - V_M\right) - c_{i,pi}\left(V_{F,1+2} - V_{F,1}\right)}{V_a} \tag{10.3}$$

$$q_i^* = \frac{c_i\left(V_{F,1+2} - V_M\right) + c_{i,pi}\left(V_{F,1+2} - V_{F,2}\right)}{V_a} \tag{10.4}$$

In Equations 10.3 and 10.4 c_i and $c_{i,pi}$ are the concentrations of component i in the feed and in the intermediate plateau, respectively; V_M is the dead volume in the chromatographic system; and V_a is the volume of adsorbent packed in the chromatographic column. For Equation 10.3, $V_{F,1}$ and $V_{F,1+2}$ are the retention volumes in the first and second inflection points of the breakthrough curve; for Equation 10.4, $V_{F,1+2}$ and $V_{F,2}$ are the retention volumes of the first and second inflection points in the elution curve.

10.2.1.1.2 Perturbation Methods. The PM method may be easily performed in regular High-performance liquid chromatography (HPLC) equipment. The experimental procedure consists in equilibrating the column at successive concentration plateaus of increasing concentration and performing analytical-size injections under each of these plateau conditions (53). These injections may be either a blank (pure mobile phase) or a solution of different concentration from that of the plateau. Such injection perturbs the previously established equilibrium between mobile and stationary phase and the peak elutes at a retention volume given by Broughton and Gerhold (7):

$$V_{r,i} = \frac{L}{u}\left(1 + \frac{1-\varepsilon}{\varepsilon}\frac{\mathrm{d}q_i}{\mathrm{d}C_i}\right) \tag{10.5}$$

In Equation 10.5 $\mathrm{d}q_i/\mathrm{d}C_i$ is the derivative of the adsorption isotherm at the plateau concentration C_i. The complete adsorption isotherm may be obtained by integration of the dependence of the retention volume of the perturbations as a function of the plateau concentration, solving Equation 10.5 for q. If a binary mixture (for instance a racemic mixture) is injected, there will be two perturbation peaks corresponding to each enantiomer. Equation 10.5 will be slightly modified for a mixture of n components so that the retention volume of peak i is a function of the total derivative, defined as in Equation 10.6:

$$\frac{\mathrm{d}q_i}{\mathrm{d}c_i} = \sum_{j=1}^{n} \frac{\partial q_i}{\partial c_j} \cdot \frac{\mathrm{d}c_j}{\mathrm{d}c_i} \tag{10.6}$$

Solving these equations requires a priori the assumption of an isotherm equation and by numerical techniques it should be possible to find the isotherm parameters that best fit the experimental data (54). The main advantage of the PM over FA is that it does not require an exact calibration of the UV detector and it gives data that are affected by sample impurities.

10.2.2 Models for Adsorption Equilibria

10.2.2.1 Competitive Langmuir Model. The Langmuir model considers that adsorption takes place on a surface comprising finite numbers of energetically equivalent sites; each molecule being adsorbed on one site up to the complete coverage of a monolayer (47). This is the most widely used nonlinear model for favorable isotherms, although most of its assumptions are not fulfilled in real stationary phases, for instance, those with chiral selectivity (46,50), especially with regards to the issue of an energetically homogeneous surface and monolayer formation. Nevertheless, the Langmuir equation has been an extremely useful equation to fit experimental equilibrium data and to be used in the analysis and modeling of adsorption process. Unlike other well-known empirical models (e.g. Freundlich), the Langmuir model has sound thermodynamic consistency (45). If molecules 1 and 2 are adsorbed on a homogeneous surface with q_m monolayer capacity, the extended form of the Langmuir model, also called competitive Langmuir model is given by Equation 10.7.

$$q_i^* = q_m \frac{b_i c_i}{1 + b_1 c_1 + b_2 c_2} = \frac{H_i c_i}{1 + b_1 c_1 + b_2 c_2} \tag{10.7}$$

In Equation 10.7 q_i^* is the adsorbed phase concentration of component i; b_i is an isotherm parameter that may be estimated from monocomponent experiments. Henry's constant (H_i) for component i is given as the product $q_s \cdot b_i$. Another remarkable advantage of this model is the small number of required parameters, all of which bear a qualitative physical significance. In fact, for adsorption of a binary mixture, only three parameters are required to describe competitive adsorption (55), as shown in Equation 10.3.

10.2.2.2 Competitive bi-Langmuir (dual site) Model. Most stationary phases used in analytical and preparative chromatography are expected to be heterogeneous surfaces. Chiral stationary phases, for instance, are thought to have a bimodal energy distribution, which means that their surface includes two types of sites with different energy/strength. One type of site would be nonspecific and adsorb both enantiomers indistinctively with the same adsorption energy. The other type of site would be enantioselective and adsorb each enantiomer with a different energy (with the same or different saturation capacity). Therefore, equilibrium constants would be the same for both enantiomers in the case of nonspecific sites. This model is expressed in Equation 10.8 and includes five parameters. It is an extension of the competitive Langmuir model, considering that the two kinds of sites coexist on the surface of the stationary phase, and have been used by a number of authors working on chromatography applied to the separation of chiral mixtures (46,49,55,56).

$$q_i = q_{\mathrm{ns}} \frac{b_{\mathrm{ns},i} c_i}{1 + b_{\mathrm{ns},i}(c_1 + c_2)} + q_{\mathrm{s}} \frac{b_{\mathrm{s},i} c_i}{1 + b_{\mathrm{s},1} c_1 + b_{\mathrm{s},2} c_2}; \quad i = 1, 2 \qquad (10.8)$$

In Equation 10.8 $b_{\mathrm{ns},i}$ is the equilibrium constant of component i (1 or 2) for the adsorption on nonselective sites, $b_{\mathrm{s},i}$ is the equilibrium constant of component i for the adsorption on enantioselective sites, q_{ns} is the maximum loading (saturation capacity) of nonselective sites and q_{s} is the saturation capacity of enantioselective sites.

*10.2.2.3 **Modified Competitive bi-Langmuir Model.*** This is a particular case of the dual site (bi-)Langmuir model, which has satisfactorily described adsorption equilibria in a number of chiral stationary phases (57–60). The amount of adsorbed species on nonspecific sites is linearly dependent on the fluid phase concentration, as may be seen in Equation 10.9. The nonlinear term on the right-hand side accounts for the enantioselective sites.

$$q_i^* = H_i c_i + q_{\mathrm{s}} \frac{b_i c_i}{1 + b_1 c_1 + b_2 c_2}; \quad i = 1, 2 \qquad (10.9)$$

In Equation 10.9 H_i is the linear adsorption constant for component i, q_{s} is the maximum adsorption loading of enantioselective sites and b_i is the nonlinear adsorption constant.

10.2.3 Nonideal Factors: Influence of Mass-Transfer Resistance

Although for most applications of preparative chromatography, nonlinear equilibria plays a much more important role than nonideal effects, these must be precisely taken into account since they tend to broaden concentration fronts and therefore induce product contamination and reduce stationary phase productivity. Generally speaking, nonidealities in chromatography include such diverse effects as imperfect packing, axial dispersion and mass transfer resistances (both between and inside adsorbent particles). One of the easiest and most straightforward ways to assess these effects quantitatively is the concept of a number of equivalent equilibrium stages (perfectly mixed tanks) that would reproduce the behavior of a chromatographic column. The number of theoretical "plates" or the number of equilibrium stages of a column is directly proportional to the column efficiency, which is a well-known concept in analytical chromatography. The height of a theoretical plate (HETP) is defined as the length of a bed fraction that would correspond to an equilibrium stage and thus, it is the ratio between the total length of the column and the number of theoretical plates. Analytical chromatographic columns may reach thousands of theoretical plates, whereas preparative columns will ordinarily have only hundreds.

Moment analysis combined with pulse experiments is a traditional tool to determine the HETP on a column. Axial dispersion coefficients and mass transfer parameters may be easily determined from such experiments using dilute solutions of both non-adsorbable (tracers) and adsorbable components (59,61). The first statistical moment of a chromatographic peak corresponds to the retention time of the injected substance and is related to its interaction/adsorption with the stationary phase.

The second moment is associated with peak broadening, which may either be caused by nonideal effects or by isotherm nonlinearity. The latter may be ruled out if experiments are performed under dilute conditions (Henry's Law region). Additionally, if the chromatographic peak has a gaussian shape, the second statistical moment is equal to the variance σ^2 (62,63). Detailed experimental procedures describing pulse experiments and the calculation of the first and second moments from the obtained peaks are described in a number of publications (28,38,64).

Under linear equilibria, the first moment of the chromatographic peak of an adsorbable species is related to the adsorption (Henry's Law) constant as follows (65):

$$\mu = \frac{L}{u}\left[1 + \left(\frac{1-\varepsilon}{\varepsilon}\right) K\right] + \frac{t_p}{2} \qquad (10.10)$$

and

$$K = \varepsilon_p + \left(1 - \varepsilon_p\right) H \qquad (10.11)$$

In Equations 10.10 and 10.11 L is the bed length; u is the superficial velocity; ε and ε_{p} are the void fractions between and inside adsorbent particles, respectively; t_p is the injection time and H_i is the Henry's Law adsorption constant. Note that if a nonadsorbable tracer is injected into the column, $H = 0$. If the substance is large enough not to penetrate the adsorbent pores, the first moment allows the determination of the bed void fraction ε. If the tracer is small enough to diffuse into the pores μ will provide the total porosity of the bed.

The second moment of the chromatographic peak is related to axial dispersion and mass-transfer effects shown in Equation 10.12 as follows (64):

$$\sigma^2 = \frac{2L}{u}\left\{\frac{D_L}{u^2}\left[1+\left(\frac{\varepsilon}{1-\varepsilon}\right)\frac{1}{K}\right]^2 + \left(\frac{\varepsilon}{1-\varepsilon}\right)\frac{1}{Kk_m}\right\} + \frac{t_p^2}{12} \quad (10.12)$$

where D_L is the axial dispersion coefficient and k_m is a global mass-transfer coefficient, which lumps most diffusive mechanisms present in adsorption and chromatography (external/film, pore/molecular, and surface diffusion). The statistical moments also provide a straightforward measurement of the number of theoretical plates N of a chromatographic column (4), as shown in Equation 10.13.

$$N = \frac{L}{\text{HETP}} = \left(\frac{\mu}{\sigma}\right)^2 \quad (10.13)$$

By combining Equations 10.10 and 10.11, a very useful result relates the HETP with the ongoing nonideal effects in a chromatographic column (Eq. 10.14). Subscripts i have been omitted.

$$\text{HETP} = \frac{L}{N} = \frac{\sigma^2}{\mu^2} = \frac{2D_L}{u} + 2u\left(\frac{\varepsilon}{1-\varepsilon}\right)\frac{1}{Kk_m} \times \left[1+\left(\frac{\varepsilon}{1-\varepsilon}\right)\frac{1}{K}\right]^{-2} \quad (10.14)$$

Equations 10.10 and 10.12 provide a simple way of estimating N, HETP, and parameters for axial dispersion and global mass transfer. It should be stressed, however, that the concept of HETP and the validity of Equations 10.10 through 10.14 is restricted to linear chromatography.

Note that if pulse experiments are performed with a tracer that does not penetrate into the stationary phase, Equation 10.14 may be solved for D_L using the first and second moments of the thus obtained peak. If experiments are performed with the target adsorbate under nonbinding conditions (which may be easily achieved in biological systems by tailoring such conditions as pH and ion strength), $K = \varepsilon_p$ and Equation 10.14 may be solved for k_m.

Nonideal effects may also have a significant impact on the performance of multicolumn chromatographic setups, although SMBs are recognized as being able to yield high recoveries and productivities despite the low efficiency of preparative columns (65). The classical design theory for binary separations in SMB requires only adsorption equilibrium data and feed concentration (66). However, band broadening produced by nonideal effects may be so severe so as to cause product contamination and reduce the window of operating conditions considerably, as compared to those defined by the equilibrium theory (67,68).

10.3 DESIGN OF OPERATING CONDITIONS

10.3.1 General Aspects

Operating conditions optimization is the use of specific methods to determine the most cost-effective and efficient solution. This task consists of choice decision variables to reach desirable performance variables. In complex systems and with many parameters, as SMB, pure empirical method is hardly possible. Therefore, adequate mathematical description is used to represent of dominants mechanism of the process.

Thus, the knowledge of mathematic model on different complexities of the process and physics parameters estimation are fundamental steps to be reached before the optimization task. The initial step for the optimization task is to gather information about the balance variables of the system to freedom degree evaluation. It consists of knowledge of specified variables, decision variables, performance variables, criteria, and constrains.

The decision variables are those selected as key-variables, which will be evaluated to the specified criteria for the performance variables, keeping specified variables constants and assisting constrains imposed by the physical characteristics, expressed for the mathematical model of the process.

On the SMB process, the performance variables are defined in accordance to Equations 10.15–10.18.

$$\text{PU}_i = \frac{\text{Mass rate of component } i \text{ in the stream}}{\text{Total mass rate in the stream}} \quad (10.15)$$

$$\text{RC}_i = \frac{\text{Mass rate of component } i \text{ in the stream}}{\text{Mass rate of component } i \text{ in the feed stream}} \quad (10.16)$$

$$\text{SC}_i = \frac{\text{Flowrate of solvent present in the feed and eluent streams}}{\text{Mass rate of component } i \text{ obtained}} \quad (10.17)$$

$$\text{PR}_i = \frac{\text{Mass rate of component } i \text{ in the stream}}{\text{Stationary phase volume}} \quad (10.18)$$

For evaluation of conditions operating SMB scope, the decision variables set consists frequently to the operating variables (Fig. 10.4). Geometric and thermodynamic variables are established usually as specified variables, but in many cases can be included on the set of decision variable. Obviously, the solution of optimization problem is dependent on your complexity degree and for SMB process the following aspects influence this:

- *Dynamic Behavior Description.* SMB units are essentially dynamics and a true steady state is not really approached. The performance variables

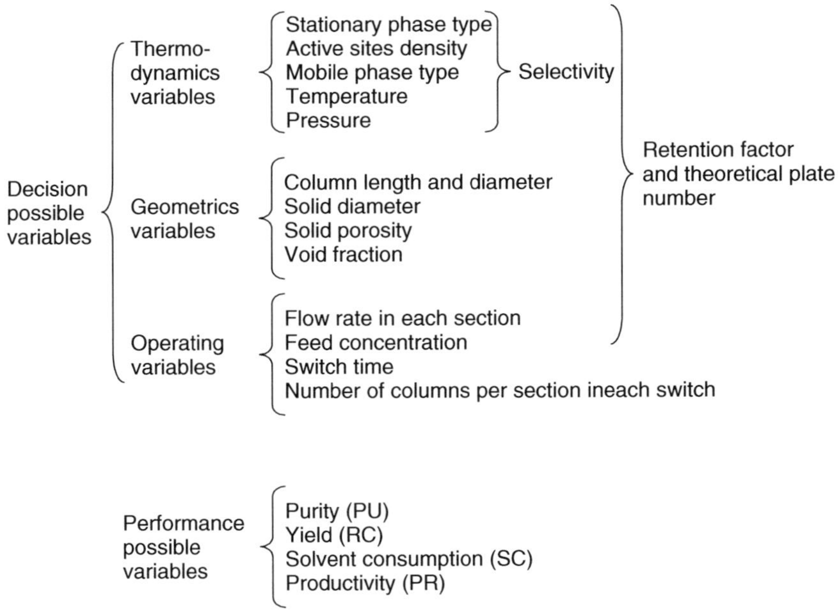

Figure 10.4. Set of possible decision and performance variables in SMB systems.

reach the called cyclic steady state (CSS), they are time-dependent, but for a long time, they present a systematic pattern. Thus, description of the performance variables through their mean value can reduce the complexity of the optimization procedure significantly. This can be made using the analogy of the SMB behavior with TMB behavior on steady state (see section titled "Fundamentals of Chromatographic Separations"). Such an approach is quite satisfactory when the number of columns for the section is large and constant along the operation.

- *Mathematical Column Model Complexity.* The accuracy of the mathematical models are dependent of the physical mechanisms described by them. Like this, the choice of the appropriate model is made in agreement with a comparison among results of models that describe different types of transfer mechanisms and adsorption. The main types of the mathematical models for the representation of chromatographic columns are classified in Table 10.2 and arein accordance with transport characteristics of the chromatographic column as ideal or nonideal, and as linear or nonlinear in accordance with thermodynamic equilibrium characteristics.

The adequate analytical methods to reach the solution for each set of the equation that quantitatively describe the phenomena showed in Table 10.2 (69–73) are different and

TABLE 10.2. Comparison Among Mathematic Model Forms

Mathematical Models	Mass-Transfer and Hydrodynamic Effects	Adsorption Equilibrium	Solution Method
Ideal and linear	Convection only	Henry isotherm	Analytical (69)
Ideal and nonlinear	Convection only	For example, various isotherm types such as Langmuir and competitive Langmuir	Analytical for Langmuir isotherms (70) and numerical for other isotherms (71)
Nonideal and linear	Convection combined with axial dispersion and/or external film diffusion and/or intraparticle diffusion	Henry isotherm	Different analytical solutions for each nonideal description (69), and (71)
Nonideal and nonlinear	Convection combined with axial dispersion and/or external film diffusion and/or intraparticle diffusion	For example, various isotherm types such as Langmuir and competitive Langmuir	Numerical (72) and (73)

dependent on Partial Differential Equations (PDE) type. Therefore, the analytical methods are available only for set equation with linear isotherms and some specific types of nonlinear isotherms. For most types of isotherms, only numerical methods are possible increasing the complexity degree of the optimization problem. Thus, the analysis of the advantages and disadvantages of each model type will orientate the choice of the appropriate mathematical model.

Clearly, there are many ways to describe quantitatively the phenomena before discussed (and showed in Table 10.2) moreover, there are other phenomena possibly not discussed in this text, such as kinetic adsorption effects, but those are not the objective of this text and will not be treated here. For more information, see Michel *et al*. (28).

- *Choice of Performance Variables.* Frequently, the selection of conditions of operation of chemical processes involves the adoption of conflicting performance variables, for example, it cannot be optimized in an independent and simultaneous way; for instance, the maximization of a variable takes obligatorily to the minimization of another performance variable—this behavior being observed typically between productivity and purity in SMB. This is a characteristic of procedures of multiobjective optimization; in this case, the values of the decision variables are reached as commitment solutions to assist the criteria for the performance variables. There is no single solution, but all of the reached solutions are the best ones possible for the group of performance variables simultaneously. The simplest method to solve this problem is the use of a function objective defined by the weighted sum of the performance variables, in which the weights describe the relative importance of the performance variables, but there are more sophisticated methods that treat the different performance variables independently as a multiobjective problem (57).
- *Dynamic Behavior of the Decision Variables.* The implementation of the problem optimization considering the dependence in relation to the time of variables of decision, as flow of each section (Powerfeed), number of columns for section (Varicol), and feed concentration (Modicon) transforms the parametric optimization in a dynamic optimization problem (74). In this form, as the decision variables are the time function, the responses of optimization procedure are policies of these variables that frequently take the performance variables to value better when compared to parametric optimization. It increases significantly the complexity of the optimization problem.

Thus, the solution of the problem optimization requires simplifier hypothesis about column mathematical description, as well as about the optimal search of the problem. In this context, the design of SMB operating conditions do not consist optimal process evaluation, but to reach a better possible solution for a set of constrains of the problem. Following it, we present increasing the complexity the approaches of selection of operation conditions in units available SMB in the literature.

10.3.2 Modeling of SMB Units

10.3.2.1 TMB Approach. The approaches based on the TMB analogy correspond to the description of the CSS of SMB from its similarity. In this case, the variables of decision are defined as: liquid flow-rates and of solid flow-rate in each section. All the other variables are specified. The main advantage is the reduction of the number of equations that describe the behavior of the unit. This analogy is used for the identification of necessary conditions for the movement of both phases to reach the complete separation of the substances of interest.

For a mixture A + B to be separated in a SMB or equivalent TMB, certain flow conditions for each of the species must be valid; such conditions presented in Fig. 10.5 establish that in the sections 2 and 3 the component more retained (A) will move in direction of the solid phase, while the component less retained (B) will move in the same direction of the liquid phase movement. Besides, in the section 1 the more retained will move in the same direction as that of the liquid and in the section 4 the less retained will move in the same direction as that of the solid.

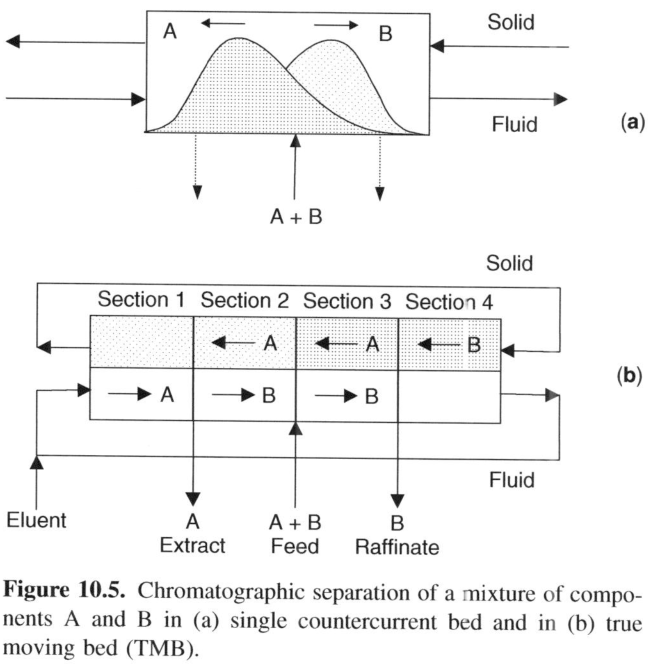

Figure 10.5. Chromatographic separation of a mixture of components A and B in (a) single countercurrent bed and in (b) true moving bed (TMB).

Mathematically, if A is the more retained species and B, the weakly adsorbed one, these flow restrictions may be stated as follows:

$$\frac{Q'_1 \times C_{A,1}}{Q'_S \times q_{A,1}} > 1 \quad (10.19)$$

$$\frac{Q\_2' \times C\_A,2}{Q\_S' \times q\_A,2} < 1 \text{ and } \frac{Q'_2 \times C_{B,2}}{Q'_S \times q_{B,2}} > 1\text{d} \quad (10.20)$$

$$\frac{Q\_4' \times C\_B,4}{Q\_S' \times q\_B,4} < 1 \quad (10.21)$$

In Equations 10.19–10.21 $Q'_2 < Q'_4$. C and q are the concentrations in bulk fluid phase and in the average adsorbed phase, respectively. Q'_j and Q'_S stand for the fluid flowrate in section j and solid flow-rate in a TMB, respectively.

The profiles of concentration of each species at end of the column are not evaluated directly. However, the internal profiles for each component are obtained in the steady state (Fig. 10.5). This analogy is the base for the "Triangle Theory" (75) and Separation Volumes Analysis (76).

10.3.2.1.1 Triangle Theory. The simplest case may be formulated for SMB systems with linear uncoupled adsorption isotherms. The analysis of the equivalent representation of a TMB under an equilibrium model leads to explicit inequal relations between solid and liquid flow rates in the four TMB sections (see Table 10.2). Therefore, for linear systems, the SMB system is exclusively a flow-controlled process. That is to say, its design does not depend on the feed concentration in the frame of equilibrium theory.

For isotherm described in Equation 10.22, the ratios of the liquid and solid flowrates in each section are described in Equations 10.23 to 10.25:

$$\text{Isotherm equation: } q_i^* = K'_i C_i \text{ where } i = \text{A, B} \quad (10.22)$$

$$K'_A < m_1 < \infty \quad (10.23)$$

$$K'_B < m_2 < m_3 < K'_A \quad (10.24)$$

$$0 < m_4 < K'_B \quad (10.25)$$

Figure 10.6 resumes the diagram for location of m_2 and m_3.

Nonlinear behavior of the isotherms adds to the dependency of the flow rates with the feed concentration as an influencing factor. Storti *et al*. (66) addressed the issue of SMB design for a system whose equilibria was described by the constant selectivity stoichiometric Langmuir isotherm, while mass-transfer resistances and axial mixing were neglected. The authors made use of the analytic solution proposed by Rhee *et al*. (77) for a single moving bed section and derived explicit expressions for

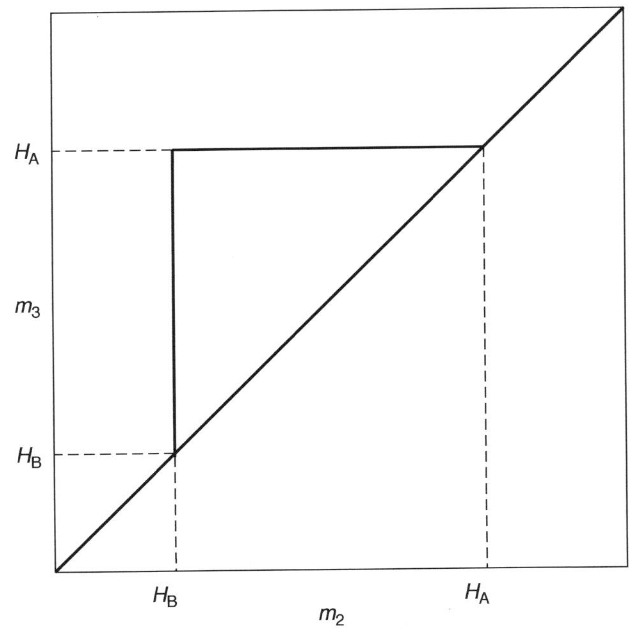

Figure 10.6. Operating conditions for the complete separation under the equilibrium theory. Linear adsorption isotherms [from Ref. 75].

the boundaries of the complete separation region in the $m_2 \times m_3$ plane. The variables m_2 and m_3 are defined as the ratio between the net fluid flowrate and the adsorbed phase flowrate in sections 2 and 3 of a TMB. The role of sections 1 and 4 was found to be less important and the constraints on them were much simpler. Following this analysis provided by the equilibrium models, numerous subsequent publications have reported the derivation of explicit relations to define the region of complete separation in the $m_2 \times m_3$ plane. In Reference (78) an analogous work is developed in order to define the separation region for adsorption equilibria described by a constant selectivity nonstoichiometric Langmuir model. These studies have been carried out in view of their applicability to the separation of hydrocarbons, as a key example of SMB use in the petrochemical industry. For such systems, the adsorbent regeneration is carried out by displacement of the species to be separated by a stronger adsorbing species, the desorbent. Hence, the equilibrium isotherm of the desorbent is an essential input parameter in the design of the SMB adsorber in the frame of equilibrium theory.

With the increasing number of applications of the SMB in chiral separations, more recent work has been published on the design of such systems (79–81). Most of the isotherms that describe the adsorption of optical isomers into adsorbents with chiral resolution are of the modified or extended Langmuir type. The isotherm parameters of the individual isomers constitute relevant information for the design in the frame of the equilibrium theory. In Reference (82) proposed explicit relationships to locate the boundaries of the separation region in a $m_2 \times m_3$ plane for variable selectivity modified Langmuir isotherms (Fig. 10.6) were found. The main finding provided by the equilibrium theory is certainly the explicit definition of the boundaries of separations regions in terms of the solid and fluid flow rates in SMB sections. Furthermore, it is always a useful check and source of comparison for design strategies that incorporate nonideal effects. The resulting equations reported in the literature (79,80,82) for the most common equilibrium isotherms found in SMB elution chromatography are Equations 10.26 to 10.40.

$$\text{Isotherm equation: } q_i^* = \frac{\lambda_i C_i}{1+b_A C_A + b_B C_B} \quad \text{where } i = \text{A, B} \tag{10.26}$$

$$\lambda_A = m_{1,\min} < m_1 < \infty \tag{10.27}$$

$$m_{2,\min}(m_2, m_3) < m_2 < m_3 < m_{3,\max}(m_2, m_3) \tag{10.28}$$

$$\frac{-\varepsilon_p}{1-\varepsilon_p} < m_4 < m_{4,\max}(m_2, m_3) \tag{10.29}$$

$$= \frac{1}{2}\left\{\lambda_B + m_3 + b_B C_B^F (m_3 - m_2) - \sqrt{\left[\lambda_B + m_3 + b_B C_B^F (m_3 - m_2)\right]^2 - 4\lambda_B m_3}\right\} \tag{10.30}$$

Boundaries of the complete separation region in the (m_2, m_3) plane:

In Figure 10.7 we can distinguish the straight lines wr and wb as well as the curve ra.

Straight line wr:

$$\left[\lambda_A - \omega_G\left(1 + b_A C_A^F\right)\right] m_2 + b_A C_A^F \omega_G m_3 = \omega_G\left(\lambda_A - \omega_G\right) \tag{10.31}$$

Straight line wb:

$$\left[\lambda_A - \lambda_B\left(1 + b_A C_A^F\right)\right] m_2 + b_A C_A^F \lambda_B m_3 = \lambda_B\left(\lambda_A - \lambda_B\right) \tag{10.32}$$

Curve ra:

$$m_3 = m_2 + \frac{\left(\sqrt{\lambda_A} - \sqrt{m_2}\right)^2}{b_A C_A^F} \tag{10.33}$$

Straight line ab:

$$m_3 = m_2 \tag{10.34}$$

The coordinates of the intersection points are given by:

$$\text{Point } a \quad (\lambda_A, \lambda_A) \tag{10.35}$$

$$\text{Point } b \quad (\lambda_B, \lambda_B) \tag{10.36}$$

$$\text{Point } r \quad \text{Point r} \frac{\omega_G^2}{\lambda_A} \quad \frac{\omega_G\left[\omega_F\left(\lambda_A - \omega_G\right)\left(\lambda_A - \lambda_B\right) + \lambda_B \omega_G\left(\lambda_A - \omega_F\right)\right]}{\lambda_A \lambda_B\left(\lambda_A - \omega_F\right)} \tag{10.37}$$

$$\text{Point } w \quad \left(\frac{\lambda_B \omega_G}{\lambda_A}, \frac{\omega_G\left[\omega_F\left(\lambda_A - \lambda_B\right) + \lambda_B\left(\lambda_B - \omega_F\right)\right]}{\lambda_B\left(\lambda_A - \omega_G\right)}\right) \tag{10.38}$$

with

$$\omega_G > \omega_F > 0 \tag{10.39}$$

which are given by the roots of the following quadratic equation

$$\left(1 + b_A C_A^F + b_B C_B^F\right)\omega^2 - \left[\lambda_A\left(1 + b_B C_B^F\right) + \lambda_B\left(1 + b_A C_A^F\right)\right]\omega + \lambda_A \lambda_B = 0 \tag{10.40}$$

In the above equations, C_A^F and C_B^F are the feed concentrations of species A and B, respectively.

Figure 10.7 refers to flowrate conditions under which a SMB adsorber may achieve 100% purity for both extract and raffinate products, independently on column number and dimensions. In practice, the results are only applicable for columns with a sufficiently high plate number.

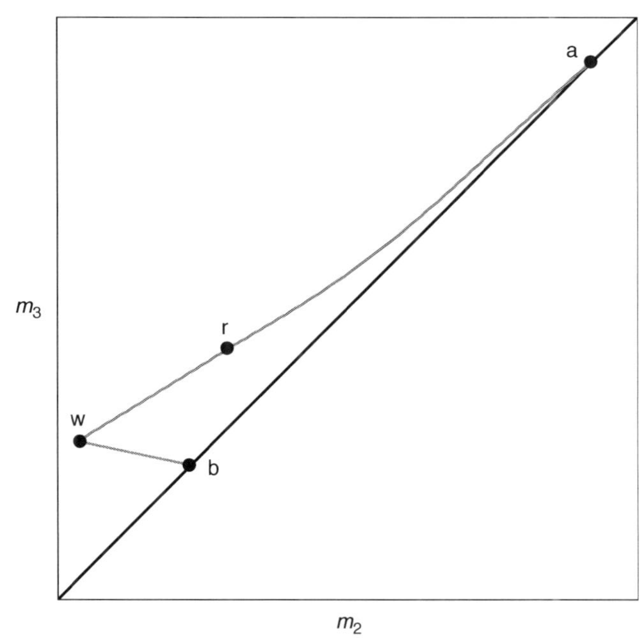

Figure 10.7. Operating conditions for the complete separation under the equilibriumtheory. Competitive langmuir adsorption isotherm [from Ref. 66].

10.3.2.1.2 Separation Volumes. Fortunately, for most applications, fairly inefficient columns may safely be used in a SMB operated under the conditions stated by the equilibrium theory (65). Alternatively, adsorbent particle size and column dimensions may also be tailored to overcome such effects as axial mixing and mass-transfer resistances, which are sources of deviation from the results of the equilibrium theory. Nevertheless, some authors (59) have attempted to refine the scope of the equilibrium theory, so as to include those nonideality effects into the design of operating conditions of SMB adsorbers. In Reference (59) the authors shaped the regions of separation under nonideal effects by using a detailed model with intraparticle mass transfer being described with a simple linear driving force (LDF) approximation. It was shown that the set of values of fluid/solid flowrate ratios in the $m_2 \times m_3$ plane is considerably reduced when mass transfer effects are present, even for a constraint of 99% on product purities. Similar results were obtained by Migliorini *et al*. (79). By using a complete detailed model, they defined the regions of separation in the $m_2 \times m_3$ plane for decreasing purity requirements. For SMB columns having the number of theoretical stages above a threshold value, the regions of separation would enlarge with decreasing purity requirements as compared to the region defined by the equilibrium theory. On the other hand, for less efficient columns, the regions of separation will virtually "shrink" in comparison with the ideal region and they may eventually not exist whether the constraint on purity is too strict or the columns have a plate number far below the limiting value. Both these works attributed values to the flow-rate ratios in sections 1 and 4 in accordance with the explicit relations defined in the equilibrium theory with a given safety margin. Azevedo (76) observed the impact of varying the values of the constraints for these sections on the resulting separation regions obtained for less-efficient columns. Instead of a two-dimensional (2D) parameter space, a 3D parameter space is used to present the obtained separation regions. This concept, which became known as the *separation volume analysis*, is based on two strategies of optimization for linear equilibrium binary mixtures which take mass-transfer effects into consideration.

In accordance with it, the flow-rate ratio constraints on sections 2 and 3 of an SMB are dependent the flow-rate ratios in neighboring sections, on the kinetic time constant, on zone length and on the solid velocity. The vertical axis γ_j is represented in terms of velocity ratio instead of flow-rate ratios, being $\gamma_j = \frac{1-\varepsilon}{\varepsilon} m_j$. As a result, the limits of the region of separation in a $\gamma_2 \times \gamma_3$ plane may not be evaluated explicitly but should be assessed through successive simulation. As expected, narrower ranges of $\gamma_2 \times \gamma_3$ values were found when compared to the values predicted from an ideal equilibrium model. The influence of mass-transfer effects on the constraints of zones 1 and 4 was also evident from numerical simulations.

10.3.2.2 SMB Approach. In the dynamic SMB approach, beds remain stationary and the periodic port movement is taken into account from dynamic inlet column changes. In this case, CSS is reached if all decision variables maintained constants during cycle time. This approach is more realist than TMB approach when the column number in each section is small according to Migliorini *et al*. (80).

The simplest method for this approach is Standing Waves Design (SWD). This was developed by Ma and Wang (81) for determining the optimal values of operating variables of nonideal SMB systems with linear isotherms. Later, the SWD was modified for nonideal and nonlinear SMB systems (8,11). Furthermore, unlike methods that rely on the equilibrium theory, SWD does not require time-consuming simulations of SMB processes to ensure that the purity and yield requirements are satisfied.

Moreover, methods based on the dynamic representation of column concentration behavior in SMB units are used to carry out the modulation of certain decision variables, instead of keeping them constant. This could largely improve performance variables. Particularly relevant in this context were the Varicol approach, the flow rates modulation (currently referred to as PowerFeed), and the ModiCon approach. The increase of the degree of freedom for the problem significantly improves the process performance with respect to the classical approach (constant decision variables).

10.3.2.2.1 Standing Waves Analysis. In accordance with standing waves analysis (11) to achieve binary separation in moving bed systems, the following constraints for concentration wave velocities in each zone should be satisfied.

$$u_c^j = \left(1 + P\delta_2^j\right)u + \frac{\beta_2^j}{L^j}\left(D_L + \frac{Pu^2\left(\delta_2^j\right)^2}{K_m^j}\right) \quad (10.41)$$

for j = I and III zone

$$u_c^j = \left(1 + P\delta_1^j\right)u + \frac{\beta_1^j}{L^j}\left(D_L + \frac{Pu^2\left(\delta_1^j\right)^2}{K_m^j}\right) \quad (10.42)$$

for j = 2 and 4 zone

Where β_i^j is the natural logarithm of the ratio of the highest concentration to the lowest concentration of standing-wave component i in zone j; L_j is the length of zone j; E_b is the axial dispersion coefficient; and K_f is the lumped mass-transfer coefficient, which can be estimated from the particle radius (R_p), the intraparticle diffusivity (D_p) represents the effective retention factor for ith component and for jth zone, defined by Xie *et al.* (11) as:

$$\delta_2^{\mathrm{I}} = \varepsilon_p + (1 - \varepsilon_p) H_2 + \frac{\mathrm{DV}}{\mathrm{PLS}\varepsilon} \quad (10.43)$$

$$\delta_1^{\mathrm{II}} = \varepsilon_p + (1 - \varepsilon_p) \frac{H_1}{(1 + b_2 C_{\mathrm{s},2})} + \frac{\mathrm{DV}}{\mathrm{PLS}\varepsilon} \quad (10.44)$$

$$\delta_2^{\mathrm{III}} = \varepsilon_p + (1 - \varepsilon_p) \frac{H_1}{(1 + b_1 C_{\mathrm{s},1} + b_2 C_{\mathrm{s},2})} + \frac{\mathrm{DV}}{\mathrm{PLS}\varepsilon} \quad (10.45)$$

$$\delta_1^{\mathrm{IV}} = \varepsilon_p + (1 - \varepsilon_p) \frac{H_1}{(1 + b_1 C_{\mathrm{s},1})} + \frac{\mathrm{DV}}{\mathrm{PLS}\varepsilon} \quad (10.46)$$

Equations 10.43–10.46 describe the differences between the port velocity and the key wave velocities, which can focus the waves toward the zone boundaries to counter wave spreading and maintain high product purity and high yield in a nonideal system. For a system without any mass transfer effects and linear isotherm the same equations 10.43–10.46 can be applied (81).

For a given feed flow rate, the five operating parameters (four-zone flow rates and port velocity) can be determined using Equation 47 in addition to Equations 43–46, where F_{feed} is the feed flow rate and S is the cross-sectional area of the column.

$$Q_{\mathrm{F}} = \varepsilon S \left(u_c^{\mathrm{III}} - u_c^{\mathrm{II}}\right) \quad (10.47)$$

A notable feature of the software design (SWD) is a system of algebraic equations that link product purity and yield to zone lengths, zone interstitial velocities, port-switching time, isotherms, and mass-transfer parameters (axial dispersion and lumped mass-transfer coefficients). Solving this system of equations provides a systematic procedure for achieving the desired purity and yield in the presence of mass-transfer effects and a pressure limit (11).

10.3.2.2.2 Dynamic Optimization. Optimal operation of SMB processes requires determination of the optimal operating parameters with respect to an economical objective function (Table 10.3). For this purpose, a single-objective optimization problem is formulated for a multistage SMB process as:

$$i = 1, \ldots, n_{\mathrm{stages}} : \min_{Q_{\mathrm{D}}(t), Q_{\mathrm{F}}(t), Q_{\mathrm{D}}(t), Q_{\mathrm{Raf}}(t), \delta t, t^*, c_{\mathrm{SMB}}}$$

Specificcost (*performance variables*)

Subject to the following set of equations:

$$\Gamma\left(c_{\mathrm{SMB}}\left(t^*\right) - c_{\mathrm{SMB}}(0)\right) = 0 \quad \mathrm{PU}_{\mathrm{Raf}} \geq \mathrm{PU}_{\mathrm{Raf,min}}$$

$$0 \leq Q_{\mathrm{col}}^{i,j} \leq Q_{,\max} \quad (10.48)$$

The aim is to find a cyclic steady state (CSS) with operating conditions that lead to minimal separation costs, while satisfying the purity requirements and plant restrictions. An additional constraint takes the maximal allowable pressure drop into account. The main difficulty of the optimization problem (set Eq. 10.48) comes from the large dimension of the CSS equations when a rigorous first-principles plant model is used (66) for example, for an eight-column SMB process 800 state variables have to be considered. Moreover, SMB processes exhibit a strongly nonlinear behavior due to competitive multicomponent adsorption and the interplay between continuous and hybrid dynamics. Only model-based approaches can exploit the full optimization potential and deal with the large number of optimization variables. Another advantage of model-based approaches is the simple inclusion of further design parameters such as the column length or the column diameter. Logical or integer-valued design parameters, for example, the number of columns, however, lead to complex mixed integer nonlinear problems (MINLP) which will not be discussed here. (set Eq. 48) comprises a complex dynamic optimization problem the solution of which essentially depends on an efficient and reliable computation of the CSS.

10.3.2.2.3 Automatic Control. In industry, the control of SMB processes is widely understood as the way in which pumps are appropriately adjusted or how the pressure of the SMB loop is stabilized. Automatic control of the flow rates or the switching time to meet purity specifications is difficult due to the extremely long delays and complex dynamics described by nonlinear distributed parameter models, and mixed discrete and continuous dynamics, leading to small operating windows and a strong nonlinear response to input variations. To reach the desired product purities, the flow rates are usually varied manually. Modification of the operating parameters is based either on heuristic rules or relies on operator expertise (74). Anita (83) proposed the following practical scheme:

- Start with low feed concentrations to achieve linear separation conditions.
- Increase V_1 to a large value and decrease V_4 to a low value so that the design criteria for the sections 1 and 4 are satisfied by a large margin. Attention is then focused on the appropriate choice of flow rates in the central sections 2 and 3.
- Increase the concentration of the feed in steps. Determine which outlet is polluted and correct the flow rates according to predefined rules. This can be repeated until the feed concentrations reach their upper-limits.
- Once the flow rates in the central sections are chosen appropriately, increase V_4 and decrease V_1. This ensures that a minimal flow of eluent is used and thus nearoptimal process performance is reached.

TABLE 10.3. Comparison Among Different Methodologies for Operating Conditions SMB Optimization

Methodologies	Hypothesis	Observations
Triangle theory	• TMB analogy • Ideal model • Linear or nonlinear isotherms: Constant selectivity isotherms or bi-Langmuir isotherm	• Closed analytical solution per parts for ratio flow rates • Iterative solution for zone flow rates, switch time, column volume and porosity
Volume separation	• TMB analogy • No ideal model • Linear or nonlinear isotherms: any isotherm type	• Numerical solution of the Ordinary Differential Equations (ODE) set for ratio flow rates • Iterative solution for zone flow rates, switch time, column volume and porosity
Standing waves design	• SMB evaluation on cyclic steady state • Ideal or no ideal model (LDF approach) • Linear or nonlinear isotherms: Henry or Langmuir isotherm type • Time-invariant decision variables	• Closed analytical solution for productivity using linear isotherm • Numerical solution for productivity of the algebraic equation set using nonlinear isotherm • Can be used to optimize units with few columns for zone • Very robust: very fast even with no ideal and nonlinear problems
Dynamic optimization	• SMB evaluation on cyclic steady state • Ideal or No ideal model (LDF approach) • Linear and nonlinear isotherms: any isotherm type • Decision variables as time functions	• Numerical solution of Partial Differential Equations (PDE) equation coupled to optimal control problem • Decision variables are time function and invariant time • Significant improvement process when compared to classical SMB

Anita (83) suggested that these heuristic rules are included in a fuzzy controller to achieve full automatic control of SMB processes, but no applications have been described so far (Fig. 10.7).

In recent work, a periodic Kalman filter that reconstructs the process state, was included in the robust model predictive control (RPMC) controller (84–86). The applicability of this scheme in the presence of strong nonlinearities as they occur in enantiomer separations and in the reactive case is an open question. Klatt *et al*. (87) proposed a two-layer control architecture where, on the upper level, the optimal operating regime is calculated at a low sampling rate by dynamic optimization based on a rigorous process model, and applied a similar approach to batch chromatography. Model parameters are adapted based on online measurements. The low-level control task is to keep the process on the optimal trajectory despite disturbances and plant/model mismatch. A disadvantage of this two-layer concept is that the stabilized front positions do not guarantee product purities if plant/model mismatch occurs. Thus, an additional purity controller is required (88). Mazotti *et al*. (78) recently proposed a control scheme where nonlinear model based on optimization is performed online and applied successfully to the control of a three-section reactive SMB process for glucose isomerization. The key feature of this approach is that the production cost is minimized online while product purities are considered as constraints. Fig. 10.8 shows the control structure used for it. Product purity is measured online in order to correct the actual operating point. The nonlinear model predictive control (NMPC) controller uses a rigorous general rate type process model, the parameters of which are re-estimated online during plant operation, which reduces plant–model mismatch and enables one to compensate for a drift or sudden changes in plant parameters.

10.4 APPLICATIONS

10.4.1 Introduction

SMB is a multicolumn, continuous adsorption separation process that increases throughput, purity, and yield relative

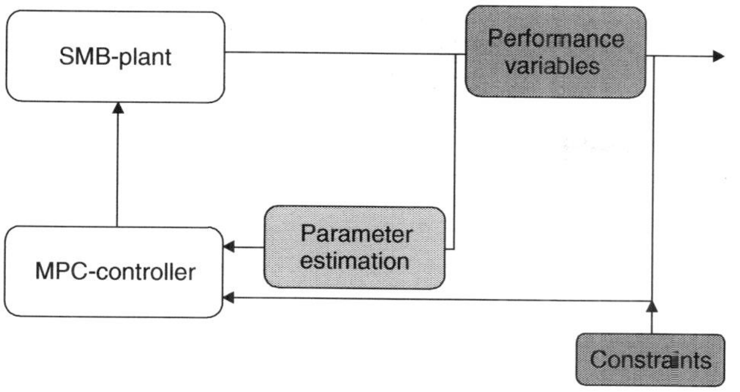

Figure 10.8. Online optimizing control structure [adaptec from Ref. 74].

to batch chromatography (89). The SMB technology was developed in the early 1960s by UOP for large-scale separations in the petrochemical and then the sugar industry (90). Nowadays, SMB technology had been applied not only to hydrocarbons and sugars, but also to various biotechnological and pharmaceutical mixtures. The application of SMB for preparative and production scale separation of fine chemical and pharmaceutical products, and in particular enantiomers, is gaining increasing importance. Successful examples of the SMB applications are the separation of *p*-xylene from a mixture of C8 isomers (91–93) the separation of glucose and fructose (6,8,94), and the resolution of enantiomers on chiral stationary phases (10–14). New challenge for the SMB technology is its application to the separation and purification of biomolecules. Some examples of bioseparations using the SMB technology are therapeutical proteins and aminoacids (15,95–104), organic acid (105) antibodies (106,107)), nucleosides (89,108), and plasmid DNA (109,110).

10.4.2 Separation of Proteins

Recent researches on the separation and purification of proteins by SMB technology have been performed in various chromatography modes, such as size-exclusion SMB (SE-SMB) chromatography, ion exchange SMB (IE-SMB), reversed phase SMB (RP-SMB) and affinity SMB (A-SMB).

Design of SE-SMB is very simple due to absence of ligand in these particles; the distribution coefficients of proteins in SE-SMB chromatography columns depend only on the accessible porosity in the particles (15). In this way, the SE-SMB unit can be designed based on the equilibrium theory and the complete separation region construction requirements of the distribution coefficient. On the other hand, the limitation imposed by mass-transfer effects can be taken into account mainly for large particles diameters.

The work done by Horneman *et al*. (106) employed the concept of SE-SMB chromatography to separate a binary mixture of bovine serum albumin (BSA) and myoglobin

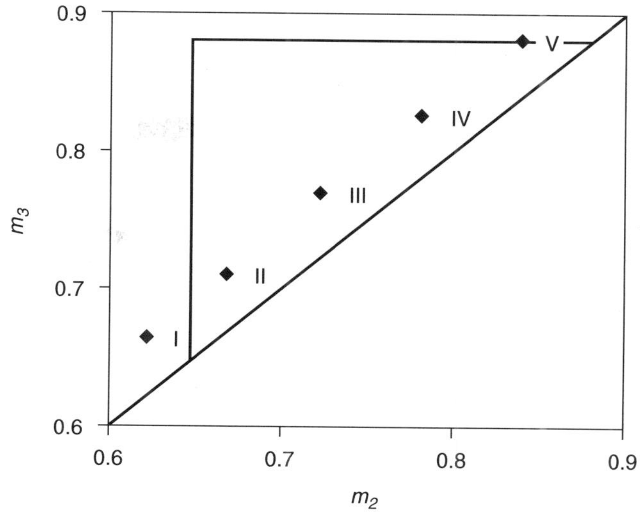

Figure 10.9. Position of experimental points (♦) relative to the region of complete separation for myoglobin and BSA.

on Sepharose big beads. The authors used the equilibrium theory to obtain the operating conditions in the region of complete separation (Fig. 10.9).

The flow-rate ratios m_j of each sections of a set of experiments as well the purity in the extract and raffinate streams are shown in Table 10.1. It was showed that a large protein has a smaller distribution coefficient, as weakly retained component, and will elute in the raffinate stream (BSA); in contrast, a small protein has a bigger distribution coefficient, as strongly retained component, will elute from the extract stream (myoglobin). It is easy to obtain a large protein molecule with high purity from the raffinate stream, but it is difficult to recover the smaller protein molecule with high purity protein in the extract stream because of the limitation of the mass-transfer resistance of the large protein molecule. These results can easily be observed in Fig. 10.10.

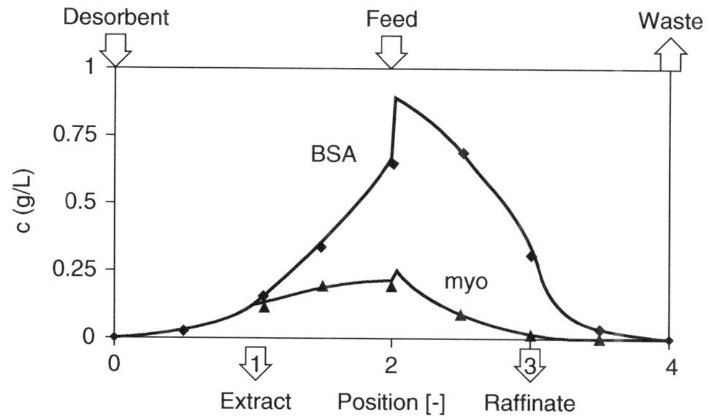

Figure 10.10. Experimental (symbols) and simulated (lines) concentration profiles along the laboratory-scale SMB in experiment III.

Houwing *et al*. (102) also investigated the effect of salt gradient in the separation of BSA and myoglobin in IE-SMB using Shepharose Q. Recently, Li *et al*. (15,111) reported a theoretical study in the linear and nonlinear regime respectively, based on previous results discussed by Houwing *et al*. (102,103). These authors investigated three configurations of the gradient SMB process: open loop, closed loop, and closed loop with a holding vessel. It was demonstrated by mathematical models that the separation and purification of proteins can be performed effectively by salt gradient IE-SMB chromatography.

From the time of the study reported by Gottschhlich and Kasche (112), some innovations in the field of the separation and purification of monoclonal antibodies by SMB have come into being. For example, Horneman *et al*. (106) presented the purification of monoclonal immunoglobulin G (IgG) from its heavy chain contaminant. The purification is performed using traditional size-exclusion chromatography (SEC) and surfactant-aided SEC (SASEC), testing two different surfactants ($C_{12}E_{23}$ and Tween20) and two different gels (Sephacryl S200HR and Sephacryl S300 HR) were performed by computational simulation. The simulation and performance parameters are shown in Tables 10.2 and 10.3, respectively. According to the results presented in Table 10.3, comparing the classical SEC-SMB and SASEC-SMB, the simulation shows a large increase in the productivity, considerable reduction in the solvent consumption, and more concentrated final product for SASEC-SMB than SEC-SMB. This study demonstrates an efficient and promising system for the purification of monoclonal antibodies.

Another innovation in the field of SMB separation of biomolecules was reported by Kebler *et al*. (107). These authors used a nonconventional SMB configuration (a three-zone open-loop gradient-SMB) to separate IgG from lysozyme and bone active dimeric morphogenetic protein-2 (BMP-2) from its monomeric form (Fig. 10.3). According to Molnár *et al*. (97,98), three-zone open loop SMB is preferred in systems with a high selectivity coefficient, when the less binding component has a low capacity factor almost running together with the mobile phase. The experimental results showed a satisfactory implementation of the SMB unity.

An important innovation in the separation of monoclonal antibody variants is presented in (113,114) through the use of a multicolumn solvent gradient purification (MCSGP) that is particularly suited for applications in the field of bioseparations. The separation of three monoclonal antibody variants was performed on a conventional cation exchange resin and the experimental process performance was compared to simulations based on a lumped kinetic model (Fig. 10.11).

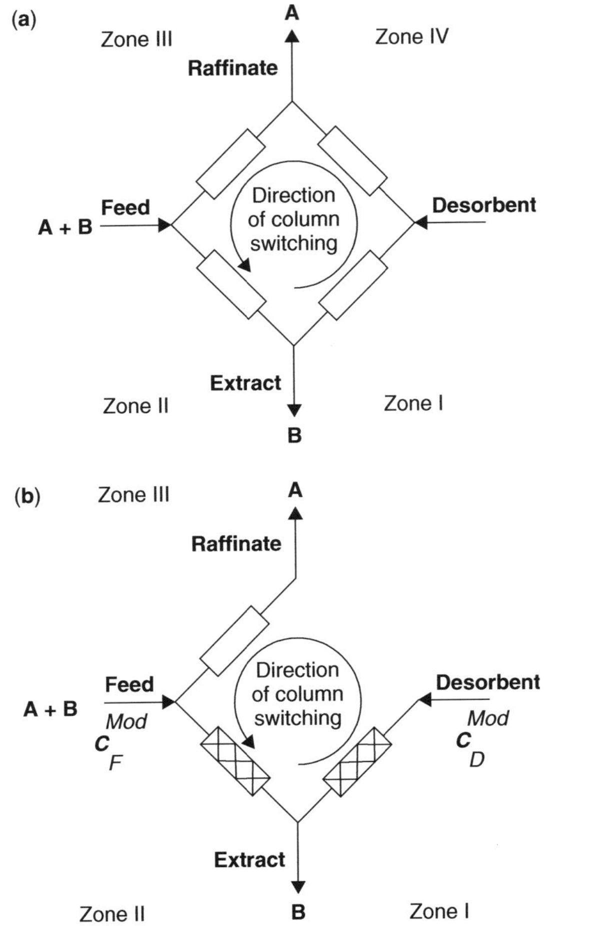

Figure 10.11. Schematic representations of a classical, closed-loop isocratic four-zone SMB process (a) and the open-loop gradient 3-zone SMB process (b) used in this study. In (b), the position of the first gradient plateau (high elution strength) is indicated by the crossed lines in the columns of zones I and II.

10.4.3 Chiral SMB Chromatography

Chirality is a prominent feature of most biological processes, and the bioactive molecules enantiomers often exhibit different biological effects. The phenomenon of enantioselectivity in biological action is not restricted to pharmaceuticals but is characteristic of all biologically active agents, including insecticides, herbicides, flavors and fragances, food additives, and so on. (115).

Chirality represents an intrinsic property of the "building block of life", such as amino acids and sugars, and therefore, of pepitides, proteins, and polysaccharides. As a consequence, metabolic and regulatory processes mediated by biological systems are sensitive to stereochemistry and different responses can be often observed when comparing the activities of a pair of enantiomers (116).

Owing to the different biological properties of each enantiomer that could be present in the biochemical process, enantiomer separation became a great challenge for researchers and the pharmaceutical industry. The accumulated experience acquired over the years lead to the international regulatory agencies having to introduce hard clinical drug control and to require the production and commercialization of chiral drugs in the form of pure enantiomer by the pharmaceutical industry. The US regulatory agency Food and Drug Administration (FDA), has pointed rigorous exigencies to release new patents of chiral drugs and requires a complete documentation of pharmacological and pharmacokinetic effects of individual enantiomers and their combinations. Table 10.4 presents some advantages of the commercialization of pure enantiomers, as mentioned by Ching-Joe (117).

A successful design of SMB chiral separation must be achieved for a high level of purity in both extract and raffinate streams as well as high productivity, high enrichment, high recovery, and low solvent consumption. These performance parameters of SMB unit (Table 10.5) were described in the section titled "Design of Operating Conditions".

The first chiral separation using the concept of SMB was reported by Negawa and Shoji (118). This publication compared the performance of SMB and batch chromatography showing the superiority of SMB technology that achieves high productivity (61:1 SMB/batch) and low solvent consumption (1:87 LMS/batch).

Over the years the use of SMB for chiral separations came into being and nowadays it is a consolidate technique. Most publications relating the application of SMB unit for chiral separations have confirmed the improvement in the performance parameters when compared with the conventional batch chromatography. For example, Yu and Ching (12) reported the chiral separation of fluoxetine in four-zone SMB with β-cyclodextrin columns. In this work, the authors evaluated the effect of feed flow-rate and feed concentration in the purity, enrichment, recovery, and productivity. It was demonstrated that the purity, enrichment and recovery decreases and the productivity increases with increasing feed concentration and feed flow-rate, for both extract and raffinate streams. These results are shown in Fig. 10.12. The transient and stationary concentration profiles are also presented in Fig. 10.13.

TABLE 10.5. Parameters Used in the Simulation of the Separation of IgG and BSA in Conventional SEC-SMB and in SASEC-SMB

	SEC-SMB	SASEC-SMB
m_{I}	0.3	2
m_{II}	0.1	0.95
m_{III}	0.15	3
m_{IV}	0.05	0.95
c_D [%, w/w]	0	9.5
c_{feed} [%, w/w]	0	13
T (s)	90	90
$c_{IgG\ feed}$ [g/L]	1	1
$c_{BSA\ feed}$ [g/L]	1	1

Recently, Grill *et al.* (119) compared the three preparative chromatographic techniques in the chiral resolution of a racemic pharmaceutical intermediate: batch HPLC, steady-state recycling (SSR), and SMB. These authors concluded that SMB thecnique was more powerful than the others, being able to process 247 kg of racemate with 4100 g of racemate/kg of Chiral Stationary Phases (CSP) per day, 98.4 of enantimeric excess, and 0.11 L of solvent/g of racemate. Miller *et al.* (120) also compared batch HPLC and SMB in enantiomeric separation. In this work, 1070 kg of racemate was used in six experiments (five experiments in SMB and one experiment in HPLC). The authors showed high productivity and low solvent consumption for SMB technology as reported in Tables 10.6 and 10.7.

As demonstrated in this comparison between chromatographic techniques, SMB technology is significantly more robust than preparative chromatography or SSR because it requires a smaller number of theoretical plates to achieve

TABLE 10.4. Experimental Conditions, Outlet Concentrations, and Calculated Purities in the Extract and Raffinate Streams

Experiment	I	II	III	IV	V
m_1	1.02	1.02	1.02	1.02	1.02
m_2	0.62	0.66	0.72	0.78	0.84
m_3	0.66	0.70	0.76	0.82	0.88
m_4	0.33	0.31	0.30	0.31	0.31
c_{myo} extract [g/L]	0.99	0.10	0.11	0.10	0.09
c_{BSA} extract [g/L]	0.22	0.17	0.12	0.10	0.07
c_{myo} raffinate [g/L]	0.00	0.01	0.01	0.02	0.04
c_{BSA} raffinate [g/L]	0.28	0.31	0.31	0.34	0.35
c_{myo} waste [g/L]	0.00	0.00	0.00	0.00	0.00
c_{BSA} waste [g/L]	0.01	0.01	0.00	0.01	0.01
Purity extract	0.30	0.38	0.47	0.50	0.54
Purity raffinate	0.99	0.98	0.96	0.94	0.90

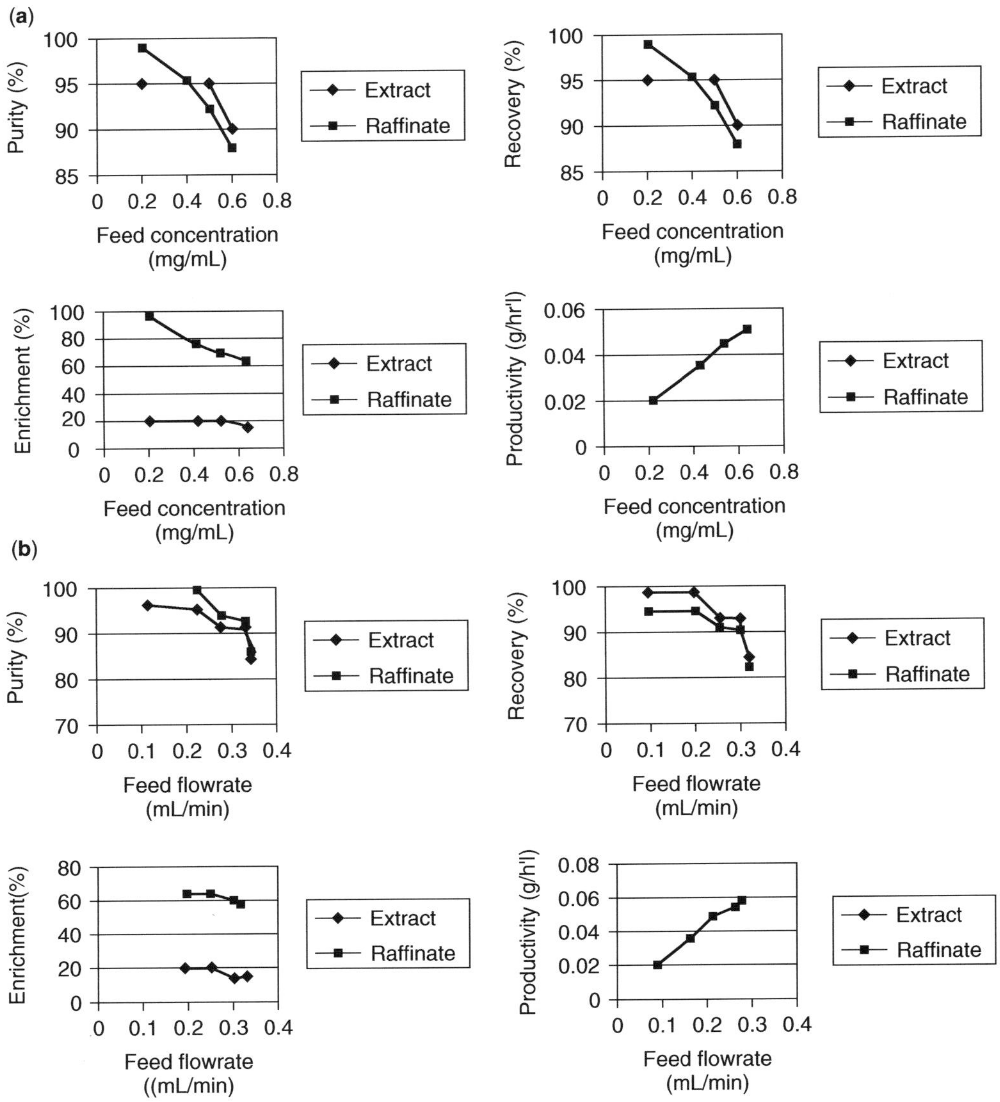

Figure 10.12. Effect of feed concentration (a) and feed flow-rate (b) on the performance parameters.

the same product purity. Moreover, the nature of countercurrent contact between the fluid and solid phases of an adsorption process maximizes the mass transfer driving forces maximized, resulting in minimization of the adsorbent and solvent requirements for a given separation.

Preparative chromatography, for economic reasons, normally is performed at overloaded concentration conditions and equilibrium isotherms are rarely linear. The SMB operation in nonlinear conditions is more favorable to preparative applications because of the possibilities of the more efficient use of CSPs. However, under these conditions the retention behavior of enantiomers depends on its concentration in the solid phase being described by competitive adsorption isotherm models. Most of these models used to represent the equilibrium adsorption in CSPs are empirical competitive models of Langmuir, modified Langmuir, and bi-Langmuir.

Since in preparative chromatography higher concentrations are of interest, additional parameters become essential and also decide whether a separation process can be performed economically or not. In particular, the course of the distribution equilibria at higher concentrations (including aspects of competition between the components of the feed) and constraints related to restricted solubility become decisive in the choice of optimal operating conditions (zone flow rates and switch time of SMB unity) to achieve the desired separation performance (40,82).

Normally, a high level of enantiomeric purity leads to a significant drop in the SMB productivity (Fig. 10.6). The extension of this productivity drops is strongly dependent

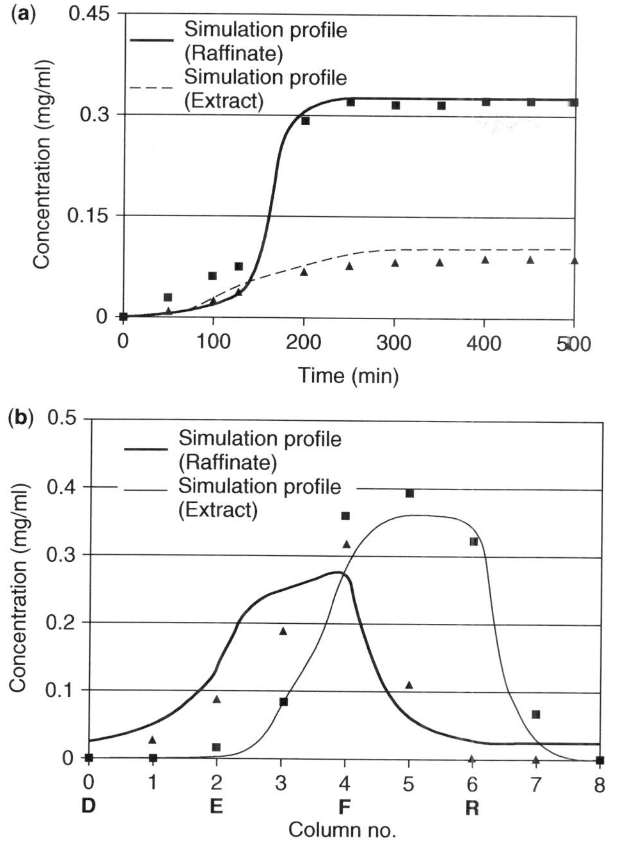

Figure 10.13. (a) Experimental (points) and theoretical (curves) transient changes in concentration of (S)-Fluoxetine in extrat and (R)-Fluoxetine in raffinate. (b) Experimental (points) and simulation (curves) steady state concentration at the exit of every column.

TABLE 10.6. Comparison of Yield (*Y*), Productivity (PR), Solvent Consumption (CS), Purity (Pu), and Concentration (c_{IgG}) Using SEC-SMB and SASEC-SMB

	SEC-SMB	SASEC-SMB
Y	0.99	0 99
PR [kg IgG/m^3/d]	3.1	129
CS [L/g BSA]	19.7	1 3
Pu [%]	99	99
$c_{IgG\ feed}$ [g/L]	0.50	1 95

on thermodynamic parameters. Alternatively, the crystallization has been used as an auxiliary technique to enhance the enantiomeric enrichment (40,121–123). The crystallization is based on the knowledge of two isomers of a raceme mixture having the same physical–chemistry properties and the crystallization could be practically the same for both enantiomers. However, one isomer crystallizes readily to a high purity (Fig. 10.14) when the other isomer is present at a low concentration (124).

10.4.4 Separation of Sugars

For the last 25 years, SMB processes have been applied in the carbohydrate industry. The term SAREX is the United Oil Products (UOP) name for separation of fructose from mixed sugars (125) and was recently used for the separation of betaine (126). The SAREX process belongs to a family of SMB processes named SORBEX and separates fructose-glucose solutions, resulting from enzymatic conversion of corn starch, in order to produce "high" fructose corn syrup. Since fructose index of sweetness is about twice as much as that of glucose, the separation of fructose from this mixture and recycling of glucose for enzymatic isomerization is of great commercial importance. Figures from the *Corn Refiners Association* (1993) showed that, in the early 1990s, around 14% of all corn produced in the United States was used in the production of corn sweeteners, which in turn accounted for nearly 53% of the market of nutritive sweeteners. The first patent issued by UOP for glucose–fructose separation (127) used zeolite Y, as the adsorbent, ion-exchanged with cations of metals K, Cs, Mg, Co, Sr, Ba+K, and Ba+Sr. The adsorbent selectivity for fructose ranged from 1.4 to 6.2. Subsequent patents introduced the use of ion-exchange resins (128) and zeolite X exchanged for potassium (127). In the latter, selectivity was greater than one for glucose, rather than for fructose.

Another important application of the SMB technology in the carbohydrate field is the separation of fine sugars and aminoacids from such feedstocks as molasses and biomass hydrolizates. Molasses is a by-product from the production process of sucrose, which may be obtained either from beet or sugarcane. It is the final liquor, from which additional sucrose cannot be economically crystallized, although it consists of 40–50% sucrose by weight. It accounts for nearly 15% of all sucrose present in the starting material and, since it is sold for animal feed or fermentation feedstock, it represents a loss of potential income. This material is rich not only in sucrose but in nonsugar salts, betaine (in the case of beet sugar molasses) and other saccharides of commercial interest. Companies such as Amalgamated Sugar Co. (USA), Nitten (Japan), and Organo (Japan) currently make use of the principles of SMB technology to separate sugars from nonsugars (129) and isolate fine chemicals such as raffinose (130) and betaine (131) respectively, summarize some relevant patents issued in the last 20 years, which illustrate the widespread use of SMB technology in the field of carbohydrate separations.

Barker and Critcher (132) and Ching and Ruthven (75,133–135) have devoted a great deal of attention to the separation of fructose and glucose using a SMB. They analyzed the performance of a three-section SMB having a pre-feed, post-feed, and purge sections. The equivalent moving bed representation was used to model the process under the approach of stages in equilibrium. The problem was addressed for the steady state (133) and

TABLE 10.7. The Advantages of Pure Enantiomers in the Pharmaceutical Applications

Chiral Drug Properties	Advantages of Pure Enantiomer Chiral Drugs
Only one of each enantiomer can be active	Reduction of dosage and load in the metabolism
One of the enantiomers can be toxic	Restriction less rigid in dosage and increase in drug usage
Enantiomers can exhibit different pharmokinetics properties.	Better control of kinetics dosage
Enantiomers can be metabolized at different rates in each person	Reduction on the variability of patient response
One of the enantiomers can be metabolized at different rates in the population	Better confidence on dosage protocols
One of the enantiomers can exhibit tendency of intromission in desintoxing routes	Reduction of interaction with other common drugs
One enantiomer can be agonist and the other antagonist	Increase of activity and dosage reduction
Enantiomers has variability in the spectra of pharmacological action and tissue specifity	Increase in specifity and reduction of colateral effects

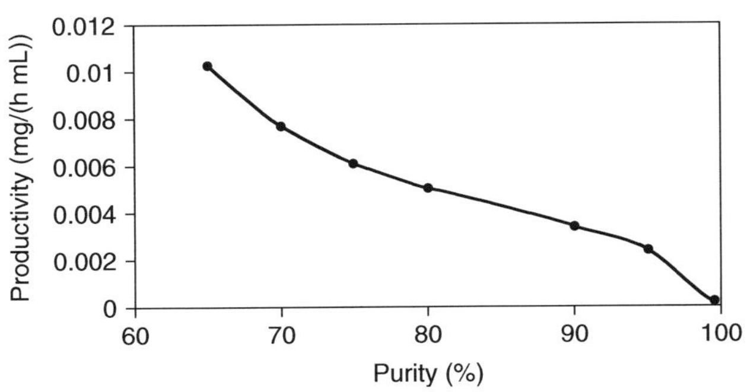

Figure 10.14. Productivity dependence to purity for optimal operating conditions in the enantiomer separation of mandelic acid (46).

transient response (75). Following this series of papers, the performance of a Sorbex-like system, having four distinct zones, was analyzed experimentally (75). McCabe-Thiele diagrams were used to make proper choice of operating conditions. The four-section SMB proved to yield a less diluted extract than the three-section equipment. The possibility of increasing extract concentration was also examined by applying a temperature profile to the system (134,135). By maintaining a temperature difference of 30–35°C across the bed, an extract product having a fructose concentration 10% higher than in the feed may be obtained.

Other authors have been working on the topic of glucose–fructose separation focusing on such aspects as modeling strategies, choice of best operating conditions, and equipment optimization. Hashimoto *et al.* (19) were one among the first authors to compare models for a real SMB and a TMB using a detailed model with external diffusion film-dependent mass transfer. Lameloiseand (136) used the analytical solution of an equilibrium-dispersive model for linear isotherms applied to a TMB section. Mallmann *et al.* (8) have also published a work in which a detailed model is proposed for the TMB equivalent and solved it by the theory of *Standing Wave Analysis*, as introduced by Erdem *et al.* (85). An optimization procedure for fructose–glucose separation in SMB was recently introduced by Beste *et al.* (137). Both TMB and SMB detailed models were included in the package, whose objective was to find, for a given SMB equipment, the best conditions to yield a desired performance in terms of criteria as purity, yield, productivity, and dilution.

Zhong and Guiochon (67) introduced the *separation volume methodology*, which accounts for the effect of the net flow in section I or IV on the separation regions, in the presence of mass-transfer resistance. These authors demonstrated that the constraints in each section do not depend only on the adsorption equilibrium, but also on the solid flow-rate, bed size, and the mass-transfer resistances. In this way, the constraints on zones I, II, and III are more restrictive than those derived from the equilibrium model, whereas the constraint on zone IV was less affected. The determination of the separation volumes requires the numerical solution of the model equations for each set of operating parameters. Azevedo and Rodrigues (138) reported the separation of fructose/glucose in a SMB pilot unit using 12 strongly acid cationic resin of gel type (Ca^{2+} form) Dowex Mono-sphere. In this study, these authors applied the design methodology of separation volumes to generate the operating conditions from the 3D plot (separation volume instead the 2D plot triangle separation). This region is showed in Fig. 10.15. The operating conditions obtained from this procedure were used to operate the SMB unit.

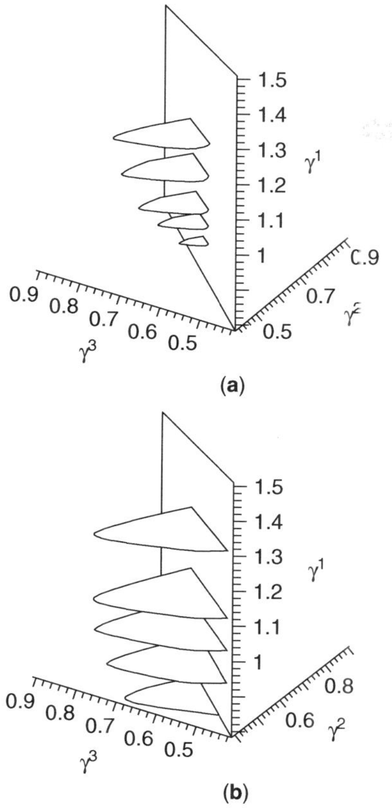

Figure 10.15. Separation volumes (both product purities above 90%) at (a) 30°C and (b) 50°C, for fructose–glucose separation on LICOSEP 12–26 SMB pilot unit using Dowex Monosphere cationic resin.

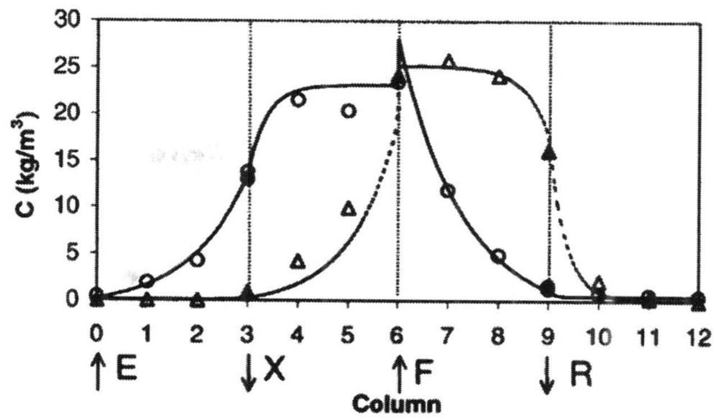

Figure 10.16. Internal concentration profile: experiment at the 15th cycle versus simulation using a TMB steady-state model. The curves are simulated data. △ and ○ are glucose and fructose concentrations, respectively, sampled at 50% of each period of the 15th cycle; ▲, ● are glucose and fructose concentrations, respectively, measured from the extract and raffinate products collected for a full cycle.

The experimental results are compared by the simulation strategies based on a true countercurrent and a real SMB with good agreement. Figure 10.16 shows the experimental and simulation results from SMB model for operating conditions presented in Table 10.8.

Nee (139) modified the conventional configuration of SMB unit and applied the two-section SMB to the separation of an aqueous mixture of glucose and fructose at high concentration up to 500 kg/m^3 using Dowex 50W-X12 resin of Ca^{2+} form as an adsorbent and water as an isocratic eluent. The authors concluded that the two-section SMB has the same performance like that of the three-section SMB unit. Also, the apparatus is more economical than the three-section SMB (Tables 10.9 and 10.10).

10.5 ADVANCES OF SMB TECHNOLOGY

10.5.1 Improving SMB Technology Efficiency

To improve SMB processes two general approaches are utilized aiming to reduce production costs and to increase efficiency. This effort results in several configurations of SMB equipment designs to provide more degreesof freedom than the classical SMB processes. The first general modification is the use of a small number of high-efficiency columns thus reducing the cost of stationary phase. In this way there is a tendency in built SMB units with a maximum number of six columns as an alternative to the conventional eight columns set-up. The second approach is based on efficiency enhancement which can be obtained both by modulating the solute capacity of adsorption in the different sections of the unity and also by the implementation of more complex operational conditions.

Modulation of solute adsorption can be acquired by the use of supercritical eluents (148–150), implementation of temperature gradient (151) and solvent gradient, both in the several sections of the SMB unit (152–155). Temperature gradient in SMB has been applied for sugar separations (135), while pressure gradients are being implemented in supercritical fluid SMB systems (156–158).

More complex dynamic conditions for the unit operation are exemplified by the processes named *VariCol* (159–161), *PowerFeed* (162–165), and *Modicon* (166). Details of progresses of SMB technology is presented below.

10.5.2 Supercritical Fluid SMB Chromatography

It has been demonstrated by Jusforgues *et al.* and Villeneuve *et al.* (167,168) that the use of supercritical fluids (SF) as eluents in chromatographic processes brings some advantages due to the solvation power of SF for organic compounds. The improvement of solubility in mild conditions of pressure and temperature convey to the management of more concentrated solutions contributing to the enhancement of productivity of the separation

TABLE 10.8. Comparative Results Among Batch and SMB Chromatography Techniques

Techniques	Processed Amount (kg)	Productivity (g of racemate/ kg CSP per day)	Solvent Consumption (L/g of racemate)
Batch	77	1410	0.550
SMB (run 1)	70	2050	0.174
SMB (run 2)	37	2800	0.127
SMB (run 3)	297	1650	0.240
SMB (run 4)	289	1670	0.260
SMB (run 5)	304	1050	0.286

TABLE 10.9. Main Applications of the SMB Technology in Carbohydrate Separation

Application	Company/Reference	Date Issued
Fructose–glucose separation (latest improvement in SAREX process)	UOP, Inc.,USA 140	11 Oct, 1983
Separation of psicose from a mixture of monosaccharides	UOP, Inc.,USA 141	14 Nov, 1989
Separation of glucose, maltose, and oligosaccharides from starch hydrolizate	Organo Corp., Japan 142	21 Feb, 1995
Separation of a water-soluble polydextrose from a glucose-based reacting mixture	Shin Dong Bang Corp., Korea 143	03 Nov, 1998
Separation of L-Arabinose from sugar beet pulp	Cultor Corp., Finland 144	04 Mar, 1999
Demineralization of a beet-sugar solution	Organo Corp., Japan 145	12 Aug, 1999
Recovery of betaine from sugar-beet molasses with a continuous SMB	Organo Corp., Japan 146	08 Aug, 2000
Recovery of betaine from sugar-beet molasses with a sequential SMB	Dänisco Finland Oy–Finland 147	25 Jul, 2000

TABLE 10.10. Operating Conditions and Process Parameters for Experimental and Simulation Results from Fig. 10.8

SMB Operation Flow Rate $\times 10^7$ m^3/s	Performance Parameter	Experimental	Predicted
$Q_1 = 6.21\gamma_1 = 1.00$	PUX (%)	94.9	98
$Q_2 = 4.65\gamma_2 = 0.50$	PUR (%)	92.7	92.1
$Q_3 = 5.21\gamma_3 = 0.68$	PRX (kg/m^3 h)	6.66	6.58
$Q_4 = 4\gamma_4 = 0.29$	PRR (kg/m^3 h)	6.54	6.68
$Q_E = 2.21$	CX_{FR} (kg/m^3)	13	12.99
$Q_R = 1.56$	CR_{GL} (kg/m^3)	16.19	16.93

process. Besides these advantages, the SF fluids have lower viscosity than liquids resulting in lower pressure drops in the column, which allows the use of small particles (5–10 μ) of stationary phase, which generates higher column efficiency. Most of the processes of this kind use CO_2 as SF fluid mainly due to convenient conditions of pressure and temperature and also because it is a nonexplosive and nontoxic fluid. For, polar substances, separation is common to add modifier solvents such as toluene, ethanol, or methanol.

The implementation of SF fluids in SMB technology has been patented (169,170) and demonstrated experimentally

(171) in the separation of a number of compounds as α-tocopherol from oleic oil, *cis*, and *trans*-phitol, *R* and *S* bi-naftol, *R* and *S* ibuprofen, respectively. A recent review on SMB-SF systems (172) points out the high potential of this technique, besides the facility for the integration of purification and collection units and makes a comparison between batch-SF and SMB-SF systems in the separation of pharmaceutical compounds.

10.5.3 VariCol, Powerfeed, and Modicon

10.5.3.1 VariCol. Conventional SMB is characterized by the syncrous movement of inlet and outlet stream fluxes, maintaining precise equivalence relationships with TMB. The number of columns in each section remains constant which means that the section length does not change with time, which makes it possible to characterize an SMB process specifying the manner of columns distribution among the sections. For example, a configuration 2-2-2-2 represents a system with a total of eight columns containing two columns per section (Fig. 10.17).

The VariCol (variable column length) process was patented by Novasep (155), and presents the innovation of continuous units operated by the advance of feed and withdraw streams on an asynchronous mode (173). With this innovation, the column length and configuration are not maintained constant alongside the period.

The operation of a VariCol unit is depicted in Fig. 10.18 (a) and (b). The process configuration at the instant t (Fig. 10.18a) is characterized by the absence of column in section I, the presence of two columns in section II, and one column in sections III and IV. No column separates the dessorbent inlet point from the extract outlet point. This column distribution remains constant during half period. In the time $t + 0.5\Delta t$, the outlet of the streams extract and raffinate changes simultaneously while the streams of dessorbent and feed inlet does not change. The new configuration (Fig. 10.18b) is characterized by the absence of any column in section IV and by the presence of two columns in section III and one column in sections I and II. This configuration remains constant up to the end of period Δt and in this moment the inlet of dessorbent and feed changes with the streams of extract and raffinate remaining in the same positions, leading to the original configuration (Fig. 10.17a).

As pointed out by O. Ludemann-Hombourger *et al.* (160), it is important to emphasize that between each column the two outlet streams must be connected to the recycle line before the dessorbent and feed streams, following the direction of recycle streams (Fig. 10.19). In this way, the contamination of the raffinate and extract streams by the feed stream when there is no column in sections II and III is avoided and the dilution of extract

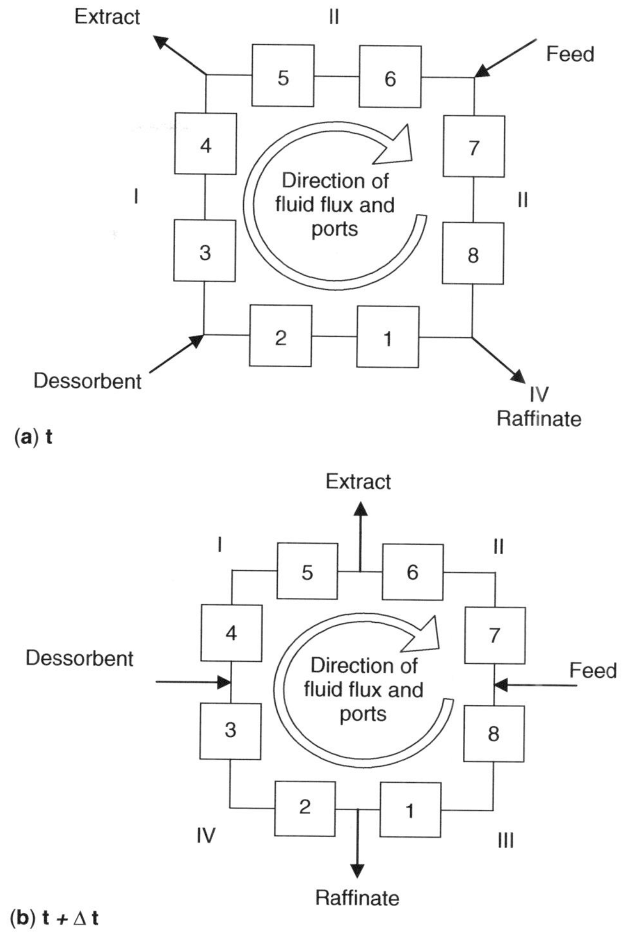

Figure 10.17. Streams motion in conventional SMB (configuration 2-2-2-2), (a) and (b). The number of columns remains constant for each switch period Δt.

and raffinate by the dessorbent when there are no columns in sections I and /or IV is also avoided.

As a result of the exposed mechanism, while the conventional SMB shows a limited number of configurations with a minimum of one column per section, the VariCol process does not exhibit this limitation and presents an infinite number of configurations. This fact makes the VariCol unit very flexible and powerful when compared to the conventional SMB especially in cases when a small number of columns are present.

10.5.3.2 Powerfeed and Modicon. The process *PowerFeed* exhibit variables flow rates for the inlet and outlet streams alongside the time. Different configurations of the PowerFeed operation mode has been investigated by simulation both in linear and nonlinear regime of operation and in some cases is considered more efficient than the conventional SMB (174).

The process ModiCon was recently proposed and is based on the cyclic modulation of feed concentration while

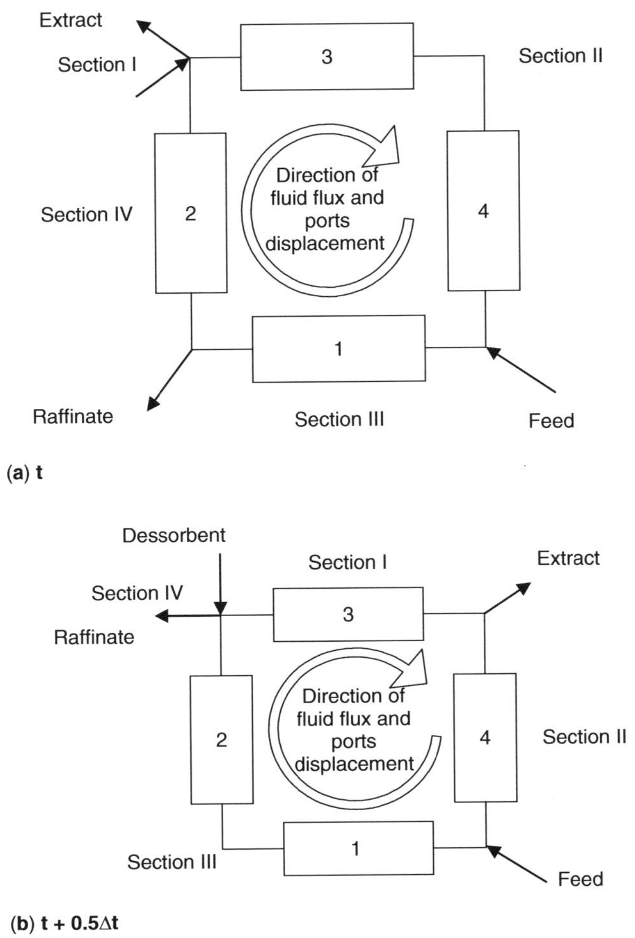

Figure 10.18. Scheme of a VariCol unit. (a) No column in section I, two columns in section II and one column in sections III and IV. (b) One column in sections I and II, two columns in section III, and no column in section IV.

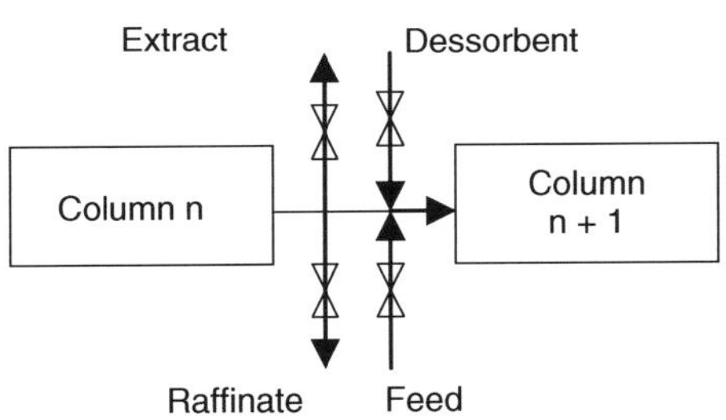

Figure 10.19. Connection of inlet and outlet streams in the VariCol process.

the flow rates of inlet and outlet streams and columns configuration are kept constant. It was experimentally demonstrated that a gain in productivity and a decrease of solvent consumption can be achieved in the nonlinear regime of operation (175).

Figures 10.20a, 10.20b, and 10.20c illustrate an example of comparison of Varicol, *PowerFeed*, and ModiCom, respectively, with the conventional SMB for columns configuration, variation of inlet and outlet streams, and modulation of feed concentration.

As an example of a large industrial SMB plant, a six-column unit for pharmaceuticals continuous chromatography separation including advances in the SMB technology is depicted in Figure 10.21. This unit was implemented by Aerojet Fine Chemicals in California, USA, for production of up to100 tons of enantiomer per year.

10.5.3.3 Outlet Streams Swing, Ternary Mixtures Separation, and Effect of Adsorbent Aging. Additional recent developments in SMB technology are the outlet streams swing (OSS), separation of ternary mixtures by pseudo-SMB chromatography, and operations strategies for SMB in presence of adsorbent aging. The first of these techniques (OSS) is based on the dynamic collection fronts from the equivalent TMB, which improves the product outlet purities (176). The separation of ternary mixtures is well exemplified by the division of the process cycle in two steps, according to the JO process of Japan Organo Co. (177). Step 1 is characterized by the introduction of feed and eluent streams in the system as an equivalent to a series of preparative columns, with the production of the intermediate component. In Step 2, similar to SMB, there is no feed and the less adsorbed species is collected in the raffinate while the more retained species is collected in the extract.

The operation strategies for SMB in the presence of adsorbent ageing is addressed by Sá Gomes and Rodrigues (178) through the use of a reduced contact time when high solids velocity are used and optimization of the Varicol concept.

10.6 CONCLUDING REMARKS

Separation is critical to every biochemical process and, typically, more than half of the invested capital in a plant is dedicated to separation and purification. Adsorption and chromatographic are concentration-controlled operations which have seen steady growth since the phenomenally successful scale-up of continuous processes using the concept of SMB. The rise of biotechnological products within the marketplace for pharmaceuticals is but one example of how recent developments in this area are making a major impact. Considering the pharmaceutical industry in particular, it is worth to note that process technologies such as SMB and SF are gaining acceptance and a considerable number of custom manufacturers are adding them to their tool chests. A recent report (17) points out the fast adoption of higher technology in separation of chiral molecules and drug-antibody conjugates by several companies. Research activities cover most of these separation methods, and researchers and companies have

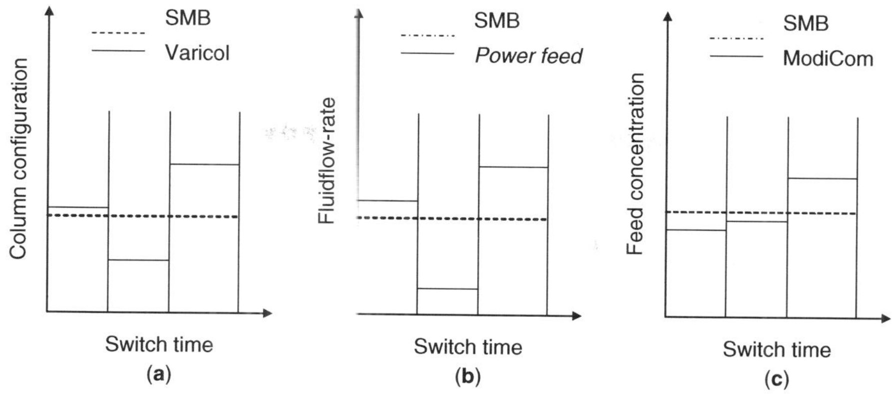

Figure 10.20. (a) Comparison of columns configuration in conventional SMB and Varicol. (b) Comparison of fluid flow rates in SMB conventional and *PowerFeed*. (c) Comparison of feed concentration of conventional SMB and ModiCom.

Figure 10.21. Large scale SMB plant: a six-column unit for pharmaceuticals continuous chromatography separation including advances in the SMB technology. This unit was implemented by Aerojet Fine Chemicals in California, USA, for production of up to 100 tons of enantiomer per year. (This figure is available in full color at http://onlinelibrary.wiley.com/book/10.1002/9780470054581.)

expanded the range of industrial problems introducing commercial-scale versions of what had only been offered at the laboratory scale. The advent of more specific adsorbents and process integration offers significant potential for future improvement of these processes.

10.7 NOMENCLATURE

b_i	Langmuir isotherm parameter
$b_{ns,I}$	equilibrium constant of component i for the adsorption on nonselective sites
$b_{s,i}$	equilibrium constant of component i for the adsorption on enantioselective sites
C_i	fluid-phase concentration of solute i
$c_{i,pi}$	concentration of component i in the intermediate plateau (FA)
D_L	axial dispersion coefficient
DV	extra-column dead volume
dq_i/dC_i	first derivative of the adsorption isotherm.
f'	derivative of the isotherm equation
HETP	height equivalent to a theoretical plate
H_i	Henry's Law constant
k_m	global mass-transfer coefficient
L	bed length
M	Ratio between fluid and solid volumetric flow rates in a TMB section
N	number of equilibrium stages in a chromatographic column
P	phase ratio, defined as $(1-\varepsilon)/\varepsilon$
Q_j	flow rate of section j
q_i^*	adsorbed phase concentration of component i
q_i	adsorbed phase concentration of solute i
q_m	Adsorption monolayer capacity
q_{ns}	maximum loading (saturation capacity) of non-selective sites
q_s	saturation capacity of enantioselective sites
S	cross-sectional area of a column
T	time coordinate
t^*	switching time
t_p	injection time
U	superficial velocity
u_c	velocity of concentration c in a chromatographic column
Ui	interstitial velocity
V_a	volume of adsorbent packed in a chromatographic column

$V_{F,1}$	retention volume of the first inflexion point in a breakthrough curve (FA).
$V_{F,1+2}$	retention volume of the second inflection points of the breakthrough curve or the first inflection point in an elution curve (FA).
$V_{F,2}$	retention volume of the second inflection point in an elution curve (FA)
V_M	dead volume in a chromatographic system
Z	axial coordinate
Greek Letters	
δ_i^j	effective retention factor for ith component and for jth section
δt	fraction of switch time
ε	bed void fraction
ε_p	adsorbent particle void fraction
γ	ratio between fluid and solid interstitial velocities in a TMB
Subscript	
Col	Column
D	Desorbent
Ext	Extracted
F	Feed
Raf	Raffinated

REFERENCES

1. Walsh G, Headon D. Protein biotechnology. New York: John Wiley & Sons; 1994.
2. Blanch HW, Clark DS. Biochemical engineering. New York: Marcel Dekker; 1997.
3. Ganetsos G, Barker PE, editors. Preparative and scale chromatography. New York: Marcel Dekker; 1993.
4. Ruthven D. Principles of adsorption and adsorption processes. New York: John Wiley & Sons; 1984.
5. Ahuja S, editor. Handbook of bioseparations. San Diego (CA): Academic Press; 2000.
6. Ahuja S. Chromatography and separation science. New York: Academic Press; 2003.
7. Broughton DB, Gerhold CG. US patent 2,985,589. 1961 May 23.
8. Mallmann T, Burris BD, Ma Z, Wang NHL. AIChE J 1998; 44: 2628–2646.
9. Azevedo DCS, Rodrigues AE. AIChE J 2001; 47: 2042–2051.
10. Zenoni G, Pedeferri M, Mazzotti M, Morbidelli M. J Chromatogr A 2000; 888: 73–83.
11. Xie Y, Hritzko B, Chin CY, Wang NHL. Ind Eng Chem Res 2003; 42: 4055–4067.
12. Yu HW, Ching CB. Adsorption 2003; 9: 213–223.
13. Santos MAG, Veredas V, Silva IJ Jr, Correia CDR, Furlan LT, Santana CC. Braz J Chem Eng 2004; 21: 127–136.
14. Veredas V, Carpes MJS, Correia CRD, Santana CC. J Chromatogr A 2006; 119: 156–162.
15. Li P, Xiu G, Rodrigues E. AIChE J 2007; 53: 2419–2431.
16. Gottschlich N, Weidgen S, Kasche V. J Chromatogr A 1996; 719: 267–274.
17. Mullin R. Chem Eng News 2007; 85: 49–53.
18. Rodrigues AE, Tondeur D, editors. Percolation processes: theory and applications, NATO science series. Alphen aan den Rijm: Sijthoff and Noordhoff; 1981.
19. Rodrigues AE, LeVan MD, Tondeur D, editors. Adsorption: science and technology, NATO science series E: Berlin, Germany: Springer; 1989.
20. Ruthven DM, Ching CB. Chem Eng Sci 1989; 44: 1011–1038.
21. Hashimoto K, Adachi S, Shirai Y, Morishita M. In: Ganetsos G, Barker PE, editors. Preparative and production scale chromatography. New York: Marcel Dekker; 1993. pp. 273–300.
22. Blehaut J, Nicoud RM. Anal Mag 1998; 26: 60–70.
23. Juza M, Mazzotti M, Morbidelli M. Tibtech 2000; 18: 108–118.
24. Nicoud RM. In: Ahuja S, editor. Handbook of bioseparations. San Diego (CA): Academic Press; 2000. pp. 475–509.
25. Imamoglu S. In: Scheper T, editor. Volume 76, Advances in biochemical engineering. Berlin: Springer Verlag; 2002. pp. 211–231.
26. Chin CY, Wang NHL. Sep Purif Rev 2004; 33: 77–155.
27. Schulte M, Wekenborg K, Wewers W. In: Schmidt-Traub H, editor. Preparative chromatography of fine chemicals and pharmaceutical agents. Weinheim: Wiley-VHC Verlag; 2005. pp. 173–214.
28. Michel M, Epping A, Jupke A. In: Schmidt-Traub H, editor. Preparative chromatography of fine chemicals and pharmaceutical agents. Weinheim: Wiley-VHC Verlag; 2005. pp. 215–312.
29. Pais LS, Mata VG, Rodrigues AE. In: Cox GB, editor. Preparative enantioselective chromatography. Oxford: Blackwell; 2005. pp. 176–198.
30. Santana CC, Azevedo DCS, Rodrigues AE. In: Pessoa A Jr, Kilikian BV, editors. Purification of biotechnological products (in Portuguese). São Paulo: Editora Manole; 2005. pp. 280–313.
31. Schulte M, Epping A. In: Schmidt-Traub H, editor. Preparative chromatography of fine chemicals and pharmaceutical agents. Weinheim: Wiley-VCH; 2005. pp. 9–49.
32. Sá Gomes P, Minceva M, Rodrigues AE. Adsorption 2006; 12: 375–392.
33. Lim BG, Ching C-B, Tan RBH. Sep Technol 1995; 5: 213–228.
34. Jacobson JM, Frenz JH, Horvath CG. Ind Eng Chem Res 1987; 26: 43–50.
35. Freitag R, Horvath C. In: Fietcher A, editor. Advances in biochemical engineering/biotechnology: Berlin, Germany: Springer-Verlag; 1995. pp. 17–59.
36. Heftmann E. Chromatography –fundamentals and applications. Amsterdam: Elsevier; 2004.
37. Helfferich FG, Carr PW. J Chromatogr A 1993; 629: 97–122.
38. Guiochon G, Shirazi SG, Katti AM. Fundamentals of preparative and nonlinear chromatography. Boston (MA): Academic Press; 1994.
39. Seidel-Morgenstern A, Guiochon G. Chem Eng Sci 1993; 48: 2787–2797.

40. Seidel-Morgenstern A. J Chromatogr A 2004; 1037: 255–272.
41. Miyabe K, Guiochon G. J Chromatogr A 1999; 849: 445–465.
42. Kaspereit M, Jandera P, Škavrada M, Seidel-Morgenstern A. J Chromatogr A 2002; 944: 249–262.
43. Mihlbachler K, Kaczmarski K, Seidel-Morgenstern A, Guiochon G. J Chromatogr A 2002; 955: 35–52.
44. Lenz K, Beste YA, Arlt W. Sep Sci Technol 2002; 37: 1611–1629.
45. Jacobson J, Frenz J, Horváth C. J Chromatogr A 1984; 316: 53–68.
46. Felinger A, Zhou D, Guiochon G. J Chromatogr A 2003; 1005: 35–49.
47. James F, Sepúlveda M, Charton F, Quiñónez I, Guiochon G. Chem Eng Sci 1999; 54: 1677–1696.
48. Piatkowski W, Antos D, Gritti F, Guiochon G. J Chromatogr A 2003; 1003: 73–89.
49. Zhou D, Cherrak DE, Kaczmarski K, Cavazzini A, Guiochon G. Chem Eng Sci 2003; 58: 3257–3272.
50. Gritti F, Guiochon G. J Colloid Interface Sci 2003; 264: 43–59.
51. Khattabi S, Cherrak DE, Fischer J, Jandera P, Guiochon G. J Chromatogr A 2000; 877: 95–107.
52. Lisec O, Hugo P, Seidel-Morgenstern A. J Chromatogr A 2001; 908: 19–34.
53. Kabir H, Grevillot G, Tondeur D. Chem Eng Sci 1998; 53: 1639–1654.
54. Heuer C, Kusters E, Plattner T, Seidel-Morgenstern A. J Chromatogr A 1998; 827: 175–191.
55. Cavazzini A, Felinger A, Kaczmarski K, Szabelski P, Guiochon G. J Chromatogr A 2002; 953: 55–66.
56. Cherrak D, Khattabi S, Guiochon G. J Chromatogr A 2000; 877: 109–122.
57. Zhang Z, Mazzotti M, Morbidelli M. J Chromatogr A 2003; 1006: 87–99.
58. Haag J, Wouwer V, Lehoucq S, Saucez P. Control Eng Pract 2001; 9: 921–928.
59. Pais LS, Loureiro JM, Rodrigues AE. Chem Eng Sci 1997; 52: 245–257.
60. Pais LS, Loureiro JM, Rodrigues AE. J Chromatogr A 1998; 827: 215–233.
61. Miyabe K, Suzuki M. AIChE J 1992; 38: 901–910.
62. Arnold FH, Blanch HW, Wilke CR. Chem Eng J 1985; 30: B25–B36.
63. Arnold FH, Blanch HW, Wilke CR. J Chromatogr 1985; 330: 159–166.
64. Ruthven DM. Principles of adsorption and adsorption process. New York: Wiley; 1984.
65. Charton F, Nicoud RM. J Chromatogr A 1995; 702: 97–112.
66. Storti G, Mazzotti M, Morbidelli M, Carrá S. AIChE J 1993; 39: 471–492.
67. Zhong G, Guiochon G. Chem Eng Sci 1997; 52: 3117–3132.
68. Biressi G, Ludemann-Hombourger O, Mazzotti M, Nicoud RM, Morbidelli M. J Chromatogr A 2000; 876: 3–15.
69. Lapidus L, Amundson NR. J Phys Chem 1952; 56: 984–988.
70. Helfferich FG, Klein G. Multicomponent chromatography–theory of interference. New York: Marcel Dekker Inc.; 1970.
71. Guiochon G, Lin B. Modeling for preparative chromatography. London: Academic Press; 2003.
72. Levenspiel O, Bischoff KB. Adv Chem Eng 1963; 4: 95–150.
73. Gu T. Mathematical modeling and scale-up of liquid chromatography. New York: Springer Verlag; 1995.
74. Toumi A, Engell S. Advanced control of simulated moving bed process. Preparative chromatography of fine chemicals and pharmaceutical agents. New York: Wiley-VHC; 2005.
75. Ching CB, Ruthven DM. Chem Eng Sci 1985; 40: 1411–1417.
76. Azevedo DCS, Rodrigues AE. AIChE J 1999; 45: 956–966.
77. Rhee H-K, Aris R, Amundson N. Philos Trans R Soc Lond 1971; A269: 187–205.
78. Mazzotti M, Storti G, Morbidelli M. AIChE J 1996; 42: 2784–2796.
79. Migliorini M, Gentilini A, Mazzotti M, Morbidelli M. Ind Eng Chem Res 1999; 38: 2400–2410.
80. Lehoucq S, Verhève D, Wouwer AV, Cavoy E. AIChE J 2000; 46: 247–256.
81. Ma Z, Wang N-H. AIChE J 1997; 43: 2488–2508.
82. Mazzotti M, Storti G, Morbidelli M. J Chromatogr A 1997; 769: 3–24.
83. Antia F. Chromatogr Sci Ser 2003; 88: 173–202.
84. Kloppenburg E, Gilles ED. J Process Control 1999; 9: 41–50.
85. Erdem G, Abel S, Morari M, Mazzotti M, Morbidelli M, Lee JH. Ind Eng Chem Res 2004; 43: 405–421.
86. Abel S, Erdem G, Mazzotti M, Morari M, Morbidelli M. J Chromatogr A 2004; 1033: 229–239.
87. Klatt K-U, Hanisch F, Dünnebier G. J Process Control 2002; 12: 203–219.
88. Hanish F. PhD Dissertation: Dortmund, Germany: University of Dortmund; 2003.
89. Rodrigues RCR, Canhoto TJSB, Araújo JMM, Mota JPB. J Chromatogr A 2008; 1180: 42–52.
90. Azevedo DSC, Rodrigues AE. Sep Sci Technol 2005; 40: 1761–1780.
91. Kurup AS, Hidajat K, Ray AK. Ind Eng Chem Res 2005; 44: 5703–5714.
92. Bae Y-S, Im S-H, Lee K-M. Sep Sci Technol 2005; 40: 2183–2204.
93. Minceva M, Rodrigues AE. Comput Chem Eng 2005; 29: 2215–2228.
94. Silva VMT, Minceva M, Rodrigues AE. Ind Eng Chem Res 2004; 43: 4494–4502.
95. Andersson J, Mattiasson B. J Chromatogr A 2006; 1107: 88–95.
96. Park B-J, Lee C-H, Mun S, Koo Y-M. Process Biochem 2006; 41: 1072–1082.
97. Molnár Z, Nagy M, Aranyi A, Hanák L, Argyelán J, Pencz I, Szánya T. J Chromatogr A 2005; 1075: 77–86.
98. Molnár Z, Nagy M, Aranyi A, Hanák L, Szánya T, Argyelán J. Chromatographia 2004; 60: 75–80.
99. Xie Y, Farrenburg CA, Chin CY, Mun S, Wang NHL. AIChE J 2003; 49(11): 2850–2863.
100. Mun S, Xie Y, Wang NHL. AIChE J 2003; 49: 2039–2058.

101. Mun S, Xie Y, Wang NHL. Ind Eng Chem Res 2003; 42: 3129–3143.
102. Houwing J, Jensen TB, van Hateren SH, Billiet HAH, van der Wielen LAM. AIChE J 2003; 49(3): 665–674.
103. Houwing J, van Hateren SH, Billiet HAH, van der Wielen LAM. AIChE J 2003; 49(5): 665–674.
104. Xie Y, Wu D, Ma Z, Wang NHL. Ind Eng Chem Res 2000; 42: 4055–4067.
105. Lee H-J, Xie Y, Koo Y-M, Wang NHL. Biotechnol Prog 2004; 20: 179–192.
106. Horneman DA, Ottens M, Keurentjes JTF, van der Wielen LAM. J Chromatogr A 2007; 1157: 237–245.
107. Kebler LC, Gueorguieva L, Rinas U, Seidel-Morgenstern A. J Chromatogr A 2007; 1176: 69–78.
108. Abel S, Bäbler MU, Arpagaus C, Mazzotti M, Stadler J. J Chromatogr A 2004; 1043: 201–210.
109. Paredes G, Mazzotti M, Stadler J, Makart S, Morbidelli M. Adsorption 2005; 11: 841–845.
110. Paredes G, Mazzotti M. J Chromatogr A 2007; 1142: 56–68.
111. Li P, Yu J, Xiu G, Rodrigues AE. Sep Sci Technol 2008; 43: 11–28.
112. Gottschlich N, Kasche V. J Chromatogr A 1997; 765: 2001–2206.
113. Aumann L, Morbidelli M. European Patent EP 05405421.8. 2005.
114. Muller-Spath T, Aumann L, Melter L, Strohlein G, Morbidelli M. Biotechnol Bioeng 2008; 100(6): 1166–1177.
115. Aboul-Enein HY. J Chromatogr A 2001; 906: 185–193.
116. Maier N, Franco P, Linder W. J Chromatogr A 2001; 906: 3–33.
117. Ching-Joe I. PhD Dissertaion: Delft, Netherlands: Delft University, Table II; 1997, 10.
118. Negawa M, Shoji F. J Chromatogr A 1992; 590: 113–117.
119. Grill CM, Miller L, Yan TQ. J Chromatogr A 2004; 1026: 101–108.
120. Miller L, Grill C, Yan T, Dapremont O, Huthmann E, Juza M. J Chromatogr A 2003; 1006: 267–280.
121. Kaspereit M, Gedicke K, Zahn V, Mahoney AW, Seidel-Morgenstern A. J Chromatogr A 2005; 1092: 43–54.
122. Ströhlein G, Schulte M, Strube J. Sep Sci Technol 2003; 38(14): 3353–3383.
123. Lorenz H, Sheehan P, Seidel-Morgenstern A. J Chromatogr A 2001; 908: 201–214, 2003:267–280.
124. Pynnonen B. J Chromatogr A 1998; 827: 143–160.
125. DeRosset AJ, Neuzil RW, Korous DJ. Ind Eng Chem Process Des Dev 1976; 15: 261–266.
126. Giacobello S, Storti G, Tola G. J Chromatogr A 2000; 872: 23–35.
127. Neuzil RW, Priegnitz JW. US patent 4,442,285. 1984.
128. Fickel RG. US patent 4,319,929. 1982.
129. Rearick DE, Kearney M, Costesso D. Chemtech 1997; 27: 36–40.
130. Sayama K, Kamada T, Oikawa S, Masuda T. Zuckerindustry 1992; 117: 893–898.
131. Hashimoto K, Adashi S, Noujima H, Marayuyama H. J Chem Eng Jpn 1983; 16: 400–415.
132. Barker PE, Critcher C. Chem Eng Sci 1960; 13: 82–92.
133. Ching CB, Ruthven DM. Chem Eng Sci 1985; 40: 887–885.
134. Ching CB, Ruthven DM. Chem Eng Sci 1986; 41: 3063–3071.
135. Ching CB, Ho C, Ruthven DM. AIChE J 1986; 32: 1876–1880.
136. Lameloiseand ML, Viard V. Trans IChemE 1993; 71C: 27–32.
137. Beste YA, Lisso M, Wozny G, Arlt W. J Chromatogr A 2000; 868: 169–188.
138. Azevedo DCS, Rodrigues AE. AIChE J 2001; 47(9): 2042–2051.
139. Lee KN. Korean J Chem Eng 2003; 20(3): 532–537.
140. LeRoy CF. US patent 4,409,033. 1983. (to UOP, Inc.).
141. Chin-Hsing C. US patent 4,880,920. 1989. (to UOP, Inc.).
142. Takayuki M, Kuniaki K, Isamu M. US patent 5,391,299. 1995. (to Organo Corp.).
143. Cheon AS, Hoe DM, Soon JH. US patent 5,831,082. 1998.
144. Antila J, Ravanko V, Walliander P. WO patent 99/10542. 1998. (to Cultor Corp.).
145. Kikuzo K, Takayuki M, Kouji T, Makoto T, Fumihiko M. WO patent 994022. 1999. (to Organo Corp.).
146. Kikuzo K, Takayuki M, Kohei S, Kouji T, Fumihiko M. US patent 6,099,654. 2000. (to Organo Corp.).
147. Heikkillla H, Hyoky G, Kuisma J. US patent 6,093,326. 2000. (to Dänisco Finland Oy).
148. Perrut M. Fr. patent 8,209,649. 1982.
149. Di Giovanni O, Mazzotti M, Morbidelli M, Denet F, Hauch W, Nicoud RM. J Chromatogr A 2001; 919: 1–12.
150. Denet F, Hauck W, Nicoud RM, Di Giovanni O, Mazzotti M, Jaubert JN, Morbidelli M. Ind Eng Chem Res 2001; 40: 4603–4609.
151. Migliorini C, Wendlinger M, Mazzotti M, Morbidelli M. Ind Eng Chem Res 2001; 40: 2606–2617.
152. Jensen TB, Reijins TGP, Biliet HAH, Van Der Wielen LA. J Chromatogr A 2000; 873: 149–162.
153. Antos D, Seidel-Morgenstern A. Chem Eng Sci 2001; 56: 6667–6682.
154. Abel S, Mazzotti M, Morbidelli M. J Chromatogr A 2002; 944: 23–39.
155. Adam P, Nicoud RM, Bailly M, Ludemann-Hombourger O. US patent 6,136,198. 2000.
156. Ikeda H, Negawa M, Soji F. US patent 5,770,088. 1988.
157. Depta A, Giese T, Johannsen M, Brunner G. J Chromatogr A 1999; 865: 175–186.
158. Johannsen M, Pepper D, Depta A. J Biochem Biophys Methods 2002; 54: 85–102.
159. Ludemann-Hombourger O, Nicoud RM, Bailly M. Sep Sci Technol 2000; 35: 1829–1862.
160. Ludemann-Hombourger O, Pigorini G, Nicoud RM, Ross DS, Terfloth G. J Chromatogr A 2002; 947: 59–68.
161. Zhang ZF, Hidajat K, Ray AK, Morbidelll M. AIChE J 2002; 48: 2800–2816.
162. Toumi A, Engell S, Ludemann-Hombourger O, Nicoud RM, Baylli M. J Chromatogr A 2003; 1006: 15–31.
163. Kearney MM, Hieb KL. US patent 5,102,553. 1992.
164. Zang YF, Wankat PC. Ind Eng Chem Res 2002; 41: 2504–2511.

165. Zhang ZF, Mazzotti M, Morbidelli M. AIChE J 2004; 50: 625–632.
166. Schramm H, Kaspereit M, Kienle A, Seidel-Morgenstern A. Chem Eng Technol 2002; 25: 1151–1155.
167. Jusforgues P, Shaimi M, Barth D. In: Anton K, Berger C, editors. Supercritical fluid chromatography with packed columns: techniques and applications. New York: Marcel Dekker; 1998.
168. Villeneuve MS, Miller LA. In: Cox GB, editor. Preparative enantiosective chromatography. Oxford: Blackwell Pub; 2005.
169. Perrut M. US patent 447,820. 1983.
170. Perut M. Eur patent 0,099,765. 1984.
171. Peper S, Lubbert M, Johannsen M, Brunner G. Sep Sci Technol 2002; 37: 2545–2566.
172. Peper S, Johannssen M, Brunner G. J Chromatogr A 2007; 1176: 246–253.
173. Pais LS, Rodrigues AE. J Chromatogr A 2003; 1006: 33–44.
174. Zhang Z, Mazzotti M, Morbidelli M. AIChE J 2004; 50: 625–632.
175. Schramm H, Kaspereit M, Kienle A, Seidel-Morgenstern A. J Chromatogr A 2003; 1006: 77–86.
176. Sá Gomes P, Rodrigues AE. Sep Sci Technol 2007; 42: 223–252.
177. Mata VG, Rodrigues AE. J Chromatogr A 2001; 939: 23–40.
178. Sá Gomes P, Rodrigues AE. Sep Sci Technol 2007; 42: 3555–3591.

FURTHER READING

Cox GB, editor. Preparative enantioselective chromatography. Oxford: Blackwell Publishing; 2005.

Rathore AS, Velayudhan A, editors. Scale-up and optimization in preparative chromatography. Basel: Marcel Dekker; 2003.

Schmidt-Traub H, editor. Preparative chromatography of fine chemicals and pharmaceutical agents. Weinheim: Wiley-VCH; 2005.

Subramanian G, editor. Chiral separations technique, a practical approach. Weinheim: Wiley-VCH; 2001.

Yamamoto S, Nakanishi K, Matsuno R. Ion-exchange chromatography of proteins. New York: Marcel Dekker; 1988.

11

ADSORPTION OF PROTEINS WITH SYNTHETIC MATERIALS

Joseph McGuire
Department of Chemical Engineering, Oregon State University, Corvallis, Oregon
Omkar Joshi
Bayer HealthCare LLC, Berkeley, California

11.1 INTERFACES

Interfaces, as well as the interactions that take place in interfacial regions, can be complex. In fact, the interface has been described as the fourth state of matter (1) The properties of atoms or atomic groups at a material surface are different from those of the bulk material. The first layer of atoms, in contact with the fluid phase, is particularly unique. Chemical composition, molecular orientation, and properties relevant to crystallinity differ at the surface. In addition, surfaces have different electrical and optical properties, and can be characterized by atomic- or molecular-level textures and roughnesses. Surfaces have wettabilities, or hydrophobic–hydrophilic balances, related to the factors named above. Further, surfaces are generally energetically heterogeneous. For example, while a surface may be assigned a particular wettability, it would most likely be the result of a distribution of surface regions of varying wettabilities.

In spite of this complexity, many researchers have met with success in describing some aspect of protein adsorption in terms of one or several surface properties. The effects of charge distribution, surface energy (i.e. whether it is "high" or "low"), and surface hydrophobicity, for example have received much attention (2–7). From a purely thermodynamic standpoint, the extent of protein adsorption, or biological adhesion in general, could be determined purely by surface energetics, that is the surface energies of the synthetic material, liquid medium, and adsorbates involved. Such an approach would imply that the free energy of adsorption is minimized at equilibrium. Adsorption would be favored if it caused the free energy function to decrease, and would not be favored if it caused the function to increase. In the absence of electrostatic and specific receptor–ligand interactions, the change in free energy upon adsorption could be written as

$$\Delta F_{\text{ads}} = \gamma_{\text{AS}} - \gamma_{\text{AL}} - \gamma_{\text{SL}} \qquad (11.1)$$

where $F_{\text{ads}}(J/m^2)$ is the free energy of adsorption per unit surface area, and $\gamma_{\text{AS}}, \gamma_{\text{AL}}$, and $\gamma_{\text{SL}}(J/m^2)$ are the adsorbate–solid, adsorbate–liquid, and solid–liquid interfacial energies, respectively.

If all the required interfacial energies of Eq. (11.1) could be estimated, one could predict the relative extent of adsorption among different surfaces. This would lead to a distinction between two situations (8,9), depending on whether adsorbate surface energy is greater than or less than the surface energy of the suspending liquid. Concerning protein adsorption from aqueous media, Eq. (11.1) would predict increasing adsorption with decreasing surface energy. In other words, a given protein would be expected to adsorb with greater affinity to hydrophobic as opposed to hydrophilic surfaces.

The importance of hydrophobic–hydrophilic balance in protein adsorption has prompted numerous investigators to

Downstream Industrial Biotechnology: Recovery and Purification, First Edition. Edited by Michael C. Flickinger.

develop techniques for measurement of this property at solid surfaces. Contact angle methods have been prominent in this regard (10). Contact angle analysis is inexpensive, rapid, and fairly sensitive. However, contact angle data can be difficult to interpret and the technique is subject to artifacts owing to macroscopic, energetic heterogeneities in the surface, hysteresis, and drop-volume effects among others. Still, useful conclusions regarding biological interactions with surfaces have been based on the results of contact angle analysis in areas of red blood cell adhesion, platelet adhesion, bacterial adhesion, and protein adsorption (10,11). Surface properties have been correlated to biological responses using other methods as well, including electron spectroscopy for chemical analysis (ESCA), secondary ion mass spectroscopy (SIMS), infrared and vibrational methods, and scanning probe microscopies. These methods and their relevance to biomedical technology have been reviewed by Ratner and Porter (10).

In summary, properties of a synthetic material's surface play a large role in dictating any biological response it may evoke. However, although much is known about selected surface property effects on protein adsorption, in a quantitative sense, we know very little about how the molecular properties of protein influence its adsorption. Interfacial behavior is a cumulative property of a protein, influenced by many factors: among these are its size, shape, charge, and thermodynamic (thermal, structural, or conformational) stability. Experimentally observed differences in interfacial behavior among different protein molecules have been very difficult to quantify in terms of these, because proteins usually vary substantially from one another in each category. The following discussion is an attempt to summarize the salient results from a wide range of experiments, focused on study of surface, solution, and protein effects on adsorption. Note that many observations have been explained in terms of a protein's charge, its tendency to unfold, and contact surface hydrophobicity.

11.2 PROTEINS AT INTERFACES

11.2.1 General Features of Adsorption from Single-Protein Solutions

11.2.1.1 Solution Chemistry Effects on Adsorption. Proteins are subject to a variety of dissimilar solution chemistries during manufacturing. Purification steps, in particular, involve buffer exchanges and salt gradients and these factors affect protein interfacial behavior. The net charge of a protein in solution depends on the difference between the pH of the solution and the isoelectric point (pI) of the protein. If the pH of the solution is greater than the pI, the net charge of the protein would be negative; and if the pH is less than the pI, the net charge of the protein would be positive. It is generally accepted that maximum adsorption occurs at the pI. As the out-of-balance charge of a protein increases, it will be in a more extended form than when the net charge is zero (12).

Norde and Lyklema (13) suggested that the degree to which pH affects the adsorption of a protein is determined by its conformational stability. They found that plateau values of adsorbed mass were independent of pH for structurally stable proteins, whereas those of less stable proteins varied considerably, apparently because less stable proteins were able to change structure with solution conditions. The effect of pH on protein adsorption and desorption can depend on solution "history" as well (14). Kondo and Higashitani (15) studied the adsorption of model proteins with wide variation in molecular properties. They explained the pH dependence of adsorbed mass in terms of lateral interactions. In particular, they suggested that lateral interactions between large protein molecules are stronger than those between small molecules. Large proteins would thus be expected to show maximum adsorption around their pIs, whereas the effect of pH on smaller proteins would be less pronounced.

The degree to which ionic strength affects protein adsorption is a function of the role electrostatics plays in the adsorption driving force. At low ionic strength, protein surface charge contributes fully to the total electrostatic interaction (12). At high ionic strength, the surface charges of proteins are shielded, reducing electrostatic interactions between proteins, whether attractive or repulsive (14). Luey *et al.* (16) showed that ionic strength effects on adsorbed mass are very much related to solid surface properties. They observed that increased ionic strength reduced the electrostatic repulsion between negatively charged β-lactoglobulin molecules and the hydrophilic, negatively charged surface they studied, increasing adsorbed mass. By contrast, increased ionic strength resulted in little change in adsorbsubsed mass at hydrophobic surfaces.

11.2.1.2 Surface-Induced Conformational Changes. The loss of biological activity through adsorption and surface-induced structural alteration is a significant problem in the production and administration of therapeutic proteins. Surface-induced structural alteration of plasma proteins adsorbed to biomedical implant surfaces evoke platelet adhesion and eventual activation of the coagulation cascade and in short, awareness of these kinds of events is essential to understanding and control of protein adsorption in a variety of circumstances. It is well accepted that a given protein can exist in multiple adsorbed conformational "states" on a surface (17–21). These states can be distinguished by differences in occupied area, binding strength, propensity to undergo exchange events with other proteins, and catalytic activity, or function. All of these features of adsorbed protein are interrelated, and can be time dependent. For example, decreases in surfactant-mediated elution of proteins from an adsorbed layer (an indirect measure of binding strength) are observed as protein-surface contact time increases (22). This time dependence is illustrated in Fig. 11.1. As conformational change proceeds, the likelihood of desorption decreases.

It has been observed that the extent of conformational change experienced by adsorbed fibrinogen increases with contact surface hydrophobicity (23). This is consistent with the findings of Elwing *et al*. (24) who used ellipsometry to make inferences regarding conformational changes experienced by complement factor 3, a plasma protein, on hydrophilic and hydrophobic silica surfaces. The results of Elwing *et al*. also showed that greater values of adsorbed mass were found on hydrophobic as opposed to hydrophilic surfaces. Protein molecules are assumed, in general, to change conformation to a greater extent on hydrophobic surfaces. This is due to the effect of hydrophobic interactions between the solid surface and hydrophobic regions in the protein molecule. In fact, surface-induced unfolding is often characterized as entropically driven, as the hydrophobic protein's interior associates with hydrophobic regions of the surface during unfolding. This unfolding and entropic gain is recognized as one of the driving forces of protein adsorption (25,26). Multiple contacts with a surface may form owing to the large size of protein molecules, leading to strong binding and irreversible loss. These interactions can give the molecule an extended structure, covering a relatively large area of the surface. If the repulsive force normally acting between native protein molecules is decreased for such structurally altered molecules, one would expect to measure a greater adsorbed mass on hydrophobic as opposed to hydrophilic surfaces. On the other hand, adsorption of positively charged protein to hydrophilic (negatively charged) silica can result in greater conformational change than adsorption of the same protein to hydrophobic silica, even with a greater extent of adsorption being observed at the hydrophobic silica surface (27). It is thus important to recognize that multiple factors affect the extents of protein adsorption and conformational change. One of these relates to the action of "bound water", or the hydration layer in the vicinity of the surface, with which the protein molecules interact (28). In this regard, protein adsorption is hypothesized to depend on the affinity of water for the surface: a hydrophilic surface would be expected to show less adsorption than a hydrophobic surface, as bound water is less readily removed from a hydrophilic surface.

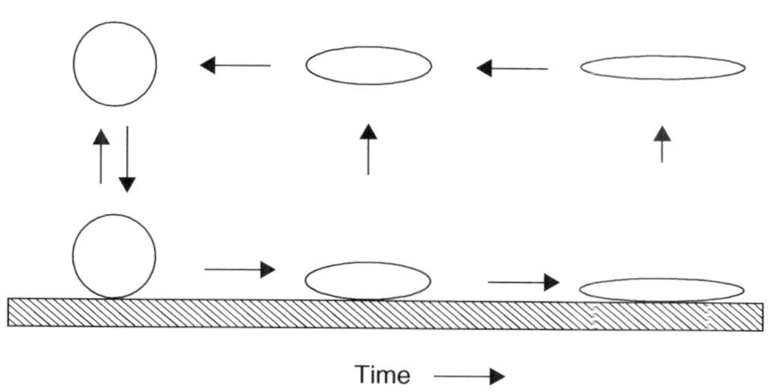

Figure 11.1. Surface-induced conformational changes undergone by adsorbed protein, resulting in multiple noncovalent bonds with the surface and coverage of greater interfacial area per molecule.

The concept that adsorbed proteins can exist in multiple states on a surface plays a role in interpretation of most, if not all, experiments in protein adsorption. Biophysicists rather gain information relevant to protein structure in solution routinely with circular dichroism (CD). It would be attractive to use CD in structural studies of adsorbed protein as well. One innovation that has made CD more applicable to study of structural changes during adsorption is the use of colloidal silica particles or nanoparticles (17–21). In these tests, particles have ranged from less than 10 to about 30 nm, and are small enough not to interfere with the CD spectra. Individual molecules are allowed to adsorb to nanoparticles, resulting in a stable suspension of adsorbed protein. In this way, structural changes upon adsorption have been unambiguously measured.

Work by Billsten *et al*. (20) and Tian *et al*. (21) has provided the most direct illustration of the effect of stability on structural rearrangements at a solid surface. Using site-directed mutants of bacteriophage T4 lysozyme, these investigators showed that both the rate and extent of secondary structure loss upon adsorption to colloidal silica were clearly related to protein thermal stability. With the same mutants, Fröberg *et al*. (29) used the interferometric surface force technique to study the structural characteristics of adsorbed layers of T4 lysozyme. The results demonstrated that less stable mutants lose their tertiary structure upon adsorption while more stable mutants retain their globular shape.

11.2.1.3 Steady-State Adsorption Behavior. A great deal is known about how various conditions affect the steady-state adsorbed mass of protein. Numerous protein adsorption isotherms have been constructed and compared on the basis of temperature, pH, ionic strength, conformational stability of the protein in solution, and solid surface charge and hydrophobicity. The effects of protein conformational stability and solid surface properties are perhaps best revealed with reference to effects of pH and ionic strength.

In general, the effect of pH and ionic strength on protein adsorption depends on the type of interactions that predominate (e.g. electrostatic, hydrophobic, or van der Waals interactions). At a negatively charged surface, if electrostatic interactions predominate, adsorbed mass should be greater at pH values below the pI relative to pH values above it. Below the pI, the protein and surface would be of opposite charge, whereas both the protein and surface would be negatively charged at pH values greater than the pI. As ionic strength increases, the electrostatic interaction would be reduced owing to the shielding of the protein by counter ions; consequently, increasing the ionic strength should decrease adsorbed mass at pH values less than the pI and increase the adsorbed mass at greater values of pH. The relationship between adsorbed mass and changes in pH and ionic strength becomes inextricably linked to protein conformational stability. In general, pH and ionic strength conditions that lead to a less stable conformation for the protein in solution will lead to an increased adsorbed mass, assuming that the protein molecule would be more stable on the solid surface (16).

Another observation of importance is that protein adsorption is often an apparently irreversible process, at least in the sense that is often irreversible to dilution, or buffer elution. The adsorbed mass remains constant or decreases very little when the solution in contact with the solid surface is depleted of protein. This irreversibility is more pronounced as protein-surface contact time increases. However, although spontaneous desorption is not generally observed, adsorbed protein can undergo exchange reactions with similar or dissimilar protein molecules adsorbing from solution (30). Such exchange reactions are shown schematically in Fig. 11.2. Adsorbed protein exchange rates are likely state-dependent, being slower for more conformationally altered protein.

Some researchers have reported that protein adsorbs onto a solid surface in more than one layer. Arnebrant *et al*. (31) studied adsorption of β-lactoglobulin and ovalbumin on hydrophilic and hydrophobic chromium surfaces using ellipsometry and electrical potential measurements. On hydrophilic surfaces, their results showed that a highly hydrated layer is obtained, which can be partially removed

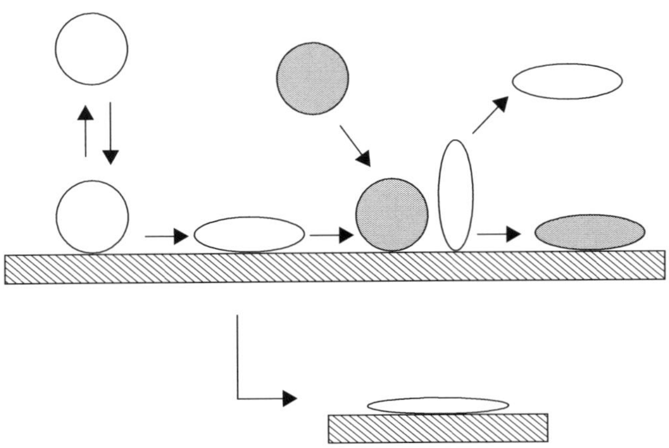

Figure 11.2. Exchange reaction between a conformationally altered, adsorbed protein and a dissimilar protein adsorbing from solution.

by rinsing. They suggested that the protein adopts a bilayer formation on the surface, with the layer in direct contact with the surface being unfolded and attached by strong "polar" bonds. Rinsing showed that the outer protein layer is loosely attached, which would imply that molecules in the outer layer have a structure closer to that of their native state. This adsorption behavior was described in terms of surface-induced conformational changes and charge interactions between the protein and surface. In particular, there are always polar amino acid side chains that can interact strongly with a surface, even if both the protein and surface are negatively charged. Such binding might be expected to result in the unfolding of the protein. A consequence of this might be exposure of hydrophobic regions into aqueous solution; therefore, adsorption of a second protein layer would reduce the interfacial free energy. In the case of protein in contact with a hydrophobic metal surface, values of adsorbed mass were found to be consistent with formation of a monolayer. Arnebrant and Nylander (32) reported possible bilayer formation upon adsorption of oligomeric units of insulin as well.

11.2.1.4 Adsorption Kinetics. In considering the kinetics of any interfacial process, the question of transport versus reaction control must be addressed. Protein adsorption at an interface depends not only on the intrinsic kinetic rate which is a function of protein, solution and surface properties, but also on the rate of protein transport from the bulk solution through the concentration boundary layer near the interface.

Proteins are macromolecules, and they can possess domains that differ chemically and physically. Diffusion coefficients may vary widely among proteins, depending on their concentration and the electrostatic condition of the solution (33). The initial adsorption rate of protein molecules at an interface can be transport limited either at low or high concentration. The diffusion limitation exists as long as there is a significant concentration gradient near the solid surface. With careful design of an experimental system to minimize the transport-limited period, however, an intrinsic adsorption kinetic rate can be estimated. Still, relatively little is known about the nature of the adsorbed layer, and predictive models to describe any aspect of adsorption as a function of protein and interfacial properties are lacking. Protein adsorption is characterized by the likely presence of a time dependence in the development of bonds with the surface, a time dependence in the lateral mobility and exchangeability of the protein molecules, and time-dependent conformational changes; it is thus very difficult to describe mathematically.

Many experimental observations have indicated that a major portion of the final adsorbed amount had been adsorbed within the first few minutes of contact. Soderquist and Walton (34) proposed the existence of three distinct processes contributing to protein adsorption kinetics on polymeric surfaces. First, rapid and reversible adsorption of the proteins occurs during a short period of time. Up to 50–60% surface coverage, there is a random arrangement of adsorbed molecules, but then some form of surface transition occurs that is probably in the direction of surface ordering, thereby allowing further protein adsorption. Second, molecules on the surface undergo structural transitions as a function of time that occur in the direction of optimizing the protein-surface interaction. Third, as time increases, the probability of desorption decreases and the adsorption becomes irreversible to dilution.

11.2.1.5 History-Dependent Adsorption. We have established that proteins can adsorb to surfaces, and undergo structural alteration as well as exchange reactions with proteins from solution. Another important, post-adsorptive behavior involves their lateral movement and "clustering." A number of macromolecular species, including proteins, exhibit history-dependent adsorption behavior, owing to the slow relaxation of nonequilibrium structures at the interface (35,36). That is, at a given surface loading of protein, the rate of adsorption depends on the formation history of that adsorbed layer. This is because the rate at which proteins adsorb at interfaces depends not only on solution concentration and the amount already adsorbed, but also on the structure and arrangement of those proteins already attached. A very simple illustration of this is presented in Fig. 11.3. In Fig. 11.3, the adsorbed mass is the same in each case (i.e. six "protein molecules" per unit area), but adsorbed layer structure is different owing to different formation histories. One would expect a greater adsorption rate for the case illustrated in Fig. 11.3b, if the intrinsic adsorption rate is high, as there would be a higher probability of an incoming protein landing in a "cavity" on the surface free from other proteins.

Adsorption rate data can thus provide important information relevant to adsorbed layer structure, For example, Tie *et al.* (36) studied the adsorption of fibronectin, cytochrome c, and lysozyme on $SiTiO_2$, using optical waveguide light-mode spectroscopy in multistep mode, where an adsorbing surface is alternately exposed to a protein solution and a solution free of protein. In general, the initial adsorption rate during the second step exceeded that observed at the same surface coverage during the first step. They postulated that, for a given mass density at an interface, if proteins were arranged in "clusters" or aggregates, more bare surface area would be available for further adsorption relative to proteins being randomly distributed. On the other hand, if the adsorbed protein films were at equilibrium, one would expect the same adsorption rates during each cycle, since the proteins would have identical structural characteristics. Joshi *et al.* (37) used this kind of approach to evaluate post-adsorptive molecular rearrangements (clustering) of adsorbed fibrinogen, and determined that higher solution concentration and longer adsorption time in the first adsorption step tended to lead to more rearrangement. They found that rearrangement was more pronounced on heparinized silica relative to bare silica, and could attribute this to an enhanced lateral mobility of fibrinogen at the heparinized surface.

The results of multistep experiments, in which the adsorbed amount is recorded continuously while the solution in contact with the solid surface is alternated between one containing protein and protein-free buffer, can be interpreted to yield quantitative information on structural parameters governing the process, as well as state-dependent protein adsorption and desorption rate constants (36). The "states" just referred to may represent different conformations, orientations, or extents of aggregation, and may result immediately following adsorption or involve post-adsorptive rearrangement of the adsorbed protein. The main structural parameter is the cavity function, defined as the fraction of the surface where the center of an incoming molecule could adsorb without overlapping a previously adsorbed molecule; it depends on the adsorbed mass, the adsorption mechanism, and the initial condition of the system (35,36). The cavity function is thus different from the "fractional surface coverage" (a quantity determined to be fixed for any given mass density at an interface) used in the more traditional, Langmuirian approach to kinetic modeling. This kind of information is potentially very important because it provides a quantitative understanding of the early,

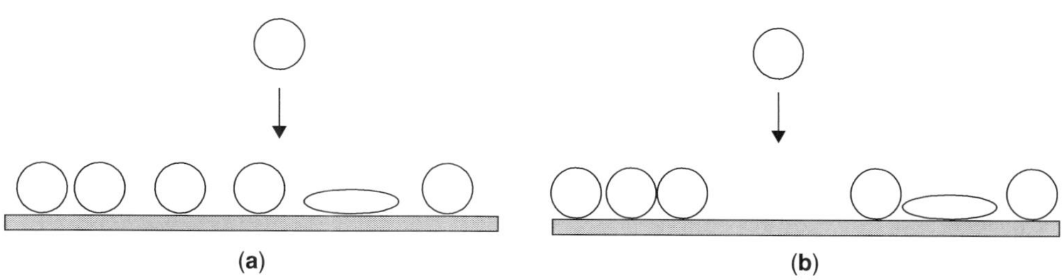

Figure 11.3. Protein adsorption to surfaces of equivalent mass density: (a) adsorption prior to surface rearrangement (or clustering); (b) adsorption after surface rearrangement.

time-dependent, post-adsorptive structural transitions that affect the performance of synthetic materials in contact with protein-containing solutions.

11.2.2 Competitive Adsorption

11.2.2.1 Molecular Structure and Interfacial Behavior. Study of molecular influences on protein adsorption has received much attention owing to its relevance to better understanding of adsorption competition in multiprotein mixtures. Important contributions to current understanding of molecular influences on protein adsorption have evolved from several comparative studies of protein interfacial behavior, in which similar or otherwise very well-characterized proteins (38–41), genetic variants (42,43), or site-directed mutants (20,21,27,29,44–46) of a single protein had been selected for study. A number of factors are known to affect protein adsorption, and these studies have stressed the importance of protein charge, hydrophobicity, and structural stability in interfacial behavior.

Shirahama *et al*. (40) studied hen lysozyme, ribonuclease A, and α-lactalbumin adsorption to hydrophilic and hydrophobic, polystyrene-coated silica (both negatively charged surfaces). At hydrophilic silica, they found that adsorbed mass increased with increasing charge contrast between the surface and protein. At the hydrophobic surface, they found electrostatic interaction to have a lesser effect, in that the adsorbed mass was not clearly related to charge contrast between the surface and protein. Arai and Norde (38) described adsorption from single-protein solutions of hen lysozyme, ribonuclease A, myoglobin, and α-lactalbumin to synthetic materials varying in surface charge density and hydrophobicity. They concluded that at a given surface, adsorption of a globular protein is related to its structural stability. That is, proteins of high stability behave like "hard" particles at a surface, with the interactions governed by surface hydrophobicity and electrostatics, whereas adsorption of proteins of low stability ("soft" proteins) may be influenced by structural rearrangement, allowing adsorption to occur even under conditions of electrostatic repulsion.

Horsley *et al*. (42) compared isotherms constructed for hen and human lysozymes at silica derivatized to exhibit negatively charged, positively charged, or hydrophobic surfaces. Differences in adsorptive behavior observed between the two lysozyme variants were largely explained with reference to the fact that human lysozyme contains one less disulfide bond and is less thermally stable than hen lysozyme. Xu and Damodaran (43) compared adsorption kinetic data measured for native and denatured hen, human, and bacteriophage T4 lysozymes at the air–water interface. Their results showed substantial differences in adsorption dynamics among the three variants, as influenced by their structural state, and the physical and chemical nature of the protein and surface.

Kato and Yutani (44) evaluated the interfacial behavior of six site-directed mutants of tryptophan synthase α-subunits, produced by amino acid substitution in the protein's interior, using surface tension, foaming and emulsifying property measurements. The stability of these subunits, as measured by their free energy of denaturation in water, varied from about 5 to 17 kcal/mol. They were able to attribute differences in interfacial behavior to protein stability with good success. In particular, they observed that less stable mutants were most surface active, that is they more rapidly adsorbed and/or more readily unfolded at the hydrophobic interfaces studied in that work.

11.2.2.2 Multicomponent Systems. Managing protein adsorption in the presence of other, dissimilar proteins is important in a number of processes, including immobilized biocatalyst operations, upstream processing of recombinant proteins, and issues surrounding the biocompatibility of blood-contacting materials. Research with multiprotein systems has yielded results enabling a better quantitative understanding of competitive adsorption. Shirahama *et al*. (40) used hen lysozyme, ribonuclease A, and α-lactalbumin to study sequential and competitive adsorption at hydrophilic and hydrophobic polystyrene-coated silica. At hydrophilic silica, they found that once an adsorbed layer of a given protein was formed, an almost total displacement of that protein would occur upon introduction of a second protein to solution if the second protein had a more favorable capacity for electrostatic attraction with the surface (otherwise sequential adsorption was not observed). In addition, when adsorbed from a mixture, the protein was capable of a more favorable electrostatic attraction with the surface preferentially adsorbed, essentially to the exclusion of the other protein. At the hydrophobic surface, they found that once an adsorbed layer of a given protein was formed, only a partial displacement of that protein would occur upon introduction of a second protein to solution even if the second protein had a more favorable capacity for electrostatic attraction with the surface. Moreover, the eventual makeup of a film adsorbed from a mixture was not related to charge contrast between the protein and surface. Other experimental observations (47,48) indicated that adsorbed protein molecules undergo exchange with protein from solution more readily on hydrophilic than on hydrophobic regions of a surface. Arai and Norde (39) studied the sequential and competitive adsorption behavior exhibited by hen lysozyme, ribonuclease A, myoglobin and α-lactalbumin, and concluded that whether introduced in sequence or in a mixture, adsorption of a globular protein is related to its structural stability. In particular, interfacial behavior of proteins of high stability is governed by surface hydrophobicity and electrostatics, while that of proteins of low stability are more influenced by structural rearrangement.

The elution of adsorbed protein by a surfactant has been used to provide a measure of protein binding strength (22,27,49–58). The essential steps of this type of experiment are illustrated in Fig. 11.4. Adsorption is allowed to occur, followed by rinsing with protein-free buffer. A surfactant solution is then introduced, after which adsorbed protein is displaced or "solubilized" (52). This is followed by rinsing and comparison of the amounts of protein present before surfactant addition, and after the final rinse. Elution by dissimilar protein has been used as a measure as binding strength as well. Slack and Horbett (59) evaluated the strength of attachment of fibrinogen to solid surfaces by measuring its time-dependent elution in plasma, and modeled fibrinogen adsorption with reference to its rate of conversion from a weakly bound (exchangeable) to a tightly bound (nonexchangeable) state. Wahlgren and Arnebrant (52,53) used *in situ* ellipsometry to continuously monitor the different effects of cationic and anionic surfactants on the elution of β-lactoglobulin and lysozyme from hydrophilic and hydrophobic surfaces, as well as adsorption from protein/surfactant mixtures. The elution studies allowed postulation of four mechanisms for surfactant-mediated elution of adsorbed protein. With the aim of relating elutability to protein molecular properties, Wahlgren *et al*. (54) studied removal of well-characterized proteins from silica surfaces using dodecyltrimethylammonium bromide. Some general trends regarding molecular property effects on elutability emerged from their work, but clear correlations between molecular properties and elutability remained difficult to quantify. By contrast, similar tests conducted with synthetic mutants of bacteriophage T4 lysozyme showed a clear correlation between protein stability and elutability (27). In particular, less stable proteins are more resistant to elution, presumably because they are more able to alter their conformation at the surfaces.

11.2.2.3 Modeling the Process. A number of mathematical models of protein adsorption at air–water and solid–water interfaces have been constructed (45,60–65). The problem is generally approached as an issue of molecular diffusion through a potential gradient, a reaction–diffusion problem involving interactions between diffusive–convective protein transport from the bulk solution and competitive adsorption and exchange kinetics on the surface, or as a kinetically controlled phenomenon involving adsorption, unfolding, and exchange. Such models and the kinetic simulations they allow provide a framework with which the complexity of protein adsorption can be better understood and quantified. In addition, they can be used to provide direction for further experiments, particularly involving surface modification.

Lundström (66) presented an equilibrium model of protein adsorption on solid surfaces. The model described the fractional surface coverage of adsorbed molecules as a function of equilibrium concentration, and allowed for reversible adsorption and conformation change. Later, Lundström and Elwing (30) described a model that allowed for bulk–surface exchange reactions among proteins in single-component and binary mixtures. The work featured manipulation of the equations describing the fractional surface coverage of protein in specific states, and simulations of total surface coverage as a function of equilibrium concentration and of time. Although no experimental data were presented, the shapes of the curves were in qualitative agreement with experimental observations. Currently, there is no adequate method to directly monitor changes in fractional surface coverages of protein in different adsorbed states. A less complex model that can be statistically compared with the available data would be useful, as it would enable individual rate constants to be related to surface, solution, and protein properties.

Past work with synthetic mutants of bacteriophage T4 lysozyme have involved *in situ* ellipsometry and surfactant-mediated elution (27,58,67,68), radioisotope labeling (68,69), air–water tensiometry (45,70), the interferometric surface force technique (29), CD (20,21), and spectrophotometric assays of bound enzyme activity (46,71). These studies have shown that structural alterations definitely occur upon adsorption, with the extent and rate of structural change being related to thermal

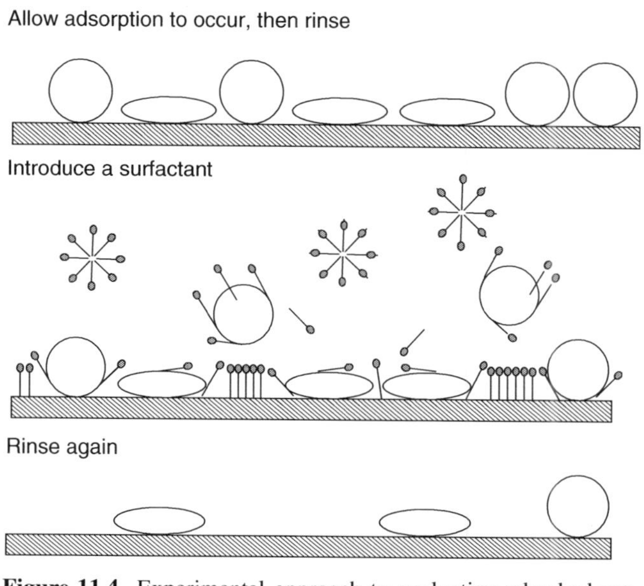

Figure 11.4. Experimental approach to evaluating adsorbed protein binding strength using surfactant-mediated elution. (This figure is available in full color at http://onlinelibrary.wiley.com/book/10.1002/9780470054581.)

stability. In addition, mutants exhibited resistances to surfactant-mediated elution that were proportional to thermal stability, and consequently, related to the extent of structural change. Finally, concerning a number of T4 lysozyme variants, results could be explained by modeling adsorption as occurring such that molecules adopt one of only two states, each varying in binding strength and occupied area, with differences in behavior among the molecules attributable to the relative populations in each state.

The simplest adsorption mechanism consistent with the fact that adsorbed proteins can exist in multiple states would include two adsorbed states. Figure 11.5 shows such a mechanism. Rate constants k_1 and k_2 govern adsorption into states 1 and 2, respectively. Although the mechanism is drawn to depict molecules adsorbing directly into states 1 and 2 from solution, a more accurate and detailed mechanism might include a multistep path to state 2. However, for modeling purposes, the actual path to state 2 is not consequential; we need only account for the different rates of generation of two functionally dissimilar forms of adsorbed protein. If adsorption of practically relevant proteins can be adequately described in this way, extension to the case of competitive protein adsorption would be straightforward.

Figure 11.6 shows a mechanism for competitive adsorption (between two proteins A and B) based on Fig. 11.5. In each case, all associated rate constants can be determined a priori. The protein-specific k1C and k2C of Fig. 11.5 can be obtained from single-component kinetic data and the various exchange constants can be determined through sequential adsorption experiments (72). Figure 11.6 can be easily redrawn to depict competitive adsorption of three proteins A, B, and C, or more. So long as sequential adsorption data are available for each pair permutation of A, B and C for example, an a priori estimate of all rate constants can be made and the adsorption competition can be simulated. Such comparisons would provide a basis for design of further experiments to better resolve molecular effects on competition in complex mixtures, subsequently enabling more quantitative prediction of adsorbed layer effects on practically relevant phenomena.

11.2.3 Surfactant Modulation of Protein Adsorption

11.2.3.1 General Observations. Surfactants can modulate both protein adsorption and surface-induced structural alteration by their own action at interfaces as well as by their participation in the formation of protein–surfactant associations. The use of surfactants in upstream and downstream processing as well as formulation is thus very common. For example, weak, nonionic surfactants such as Tween 80 are used commercially to minimize adsorption loss and aggregation, and preserve native structure and activity. But the specific mechanisms underlying Tween action in this context are not well understood. In particular, even for formulations that are considered "optimized" for chemical and physical stability of the protein, the effectiveness of the surfactant will depend very strongly on the chemistries of the interfaces present (whether gas–liquid, liquid–liquid, and solid–liquid) in a given circumstance.

An important goal in process development and formulation engineering is to minimize protein loss that occurs through colloidal and interfacial mechanisms, for example aggregation and adsorption (73,74). In order to achieve this, a fundamental understanding of the mechanisms underlying surfactant effectiveness is needed. In particular, better understanding of the specific roles of surfactant, protein,

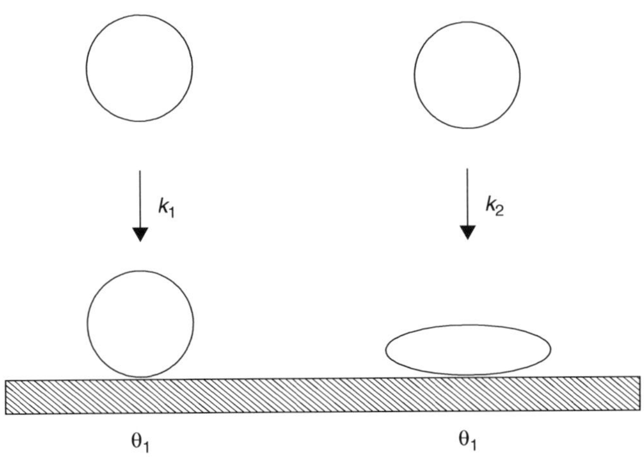

Figure 11.5. A simple mechanism for adsorption from a single-protein solution, allowing adsorption into one of two states exhibiting different resistances to elution and occupying different interfacial areas.

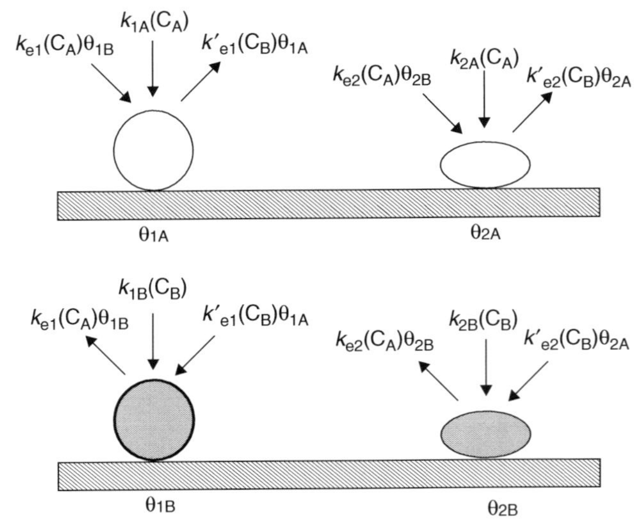

Figure 11.6. A mechanism for competitive adsorption between two proteins A and B, based on the single-component mechanism of Fig. 11.5.

and the surfactant–protein complex in modulating interfacial behavior will generate a basis to provide direction for much needed process improvements in the manufacture and finishing of therapeutic proteins.

A number of experimental investigations of the interfacial behavior of surfactant and protein mixtures have been conducted, and these have identified three possible adsorption outcomes: complete hindrance, reduced amounts, or increased amounts of protein adsorption. Complete hindrance is attributed to the faster diffusion of the (smaller) surfactant molecules as compared to protein molecules, with the adsorbed surfactant layer sterically preventing protein adsorption. Reduced and increased amounts are usually attributed to the formation of surfactant–protein complexes with reduced or increased surface affinity, respectively, and in any case different from that for pure protein or pure surfactant in solution.

The sequential introduction of a surfactant after protein adsorption may result in the removal of adsorbed protein owing to the formation of surfactant–protein complexes and subsequent solubilization of these complexes, and/or replacement of adsorbed protein by surfactant on account of a stronger surfactant–surface association. The extent of surfactant-mediated removal of adsorbed protein depends on protein, surfactant, and surface properties among other factors (27). In general, the difference in adsorbed protein elution by anionic, cationic, and nonionic surfactants corresponds with the strength of surfactant binding to protein in solution (75). Nonionic surfactants that are known to bind rather weakly to proteins are least effective in removing adsorbed protein molecules from the interface. Nonionic surfactants, when introduced to an adsorbed protein layer, do not generally affect the amount adsorbed on hydrophilic surfaces but do have an effect on the amount adsorbed on hydrophobic surfaces, presumably because of the difference in surfactant binding strength at the interface (76).

Joshi and McGuire (77) have described the interaction of the well-characterized, globular protein lysozyme, with the nonionic surfactant Tween 80 at solid–water interfaces. The concentration of surfactant as well as the method of surfactant and protein introduction to the surfaces (in sequence or combined) was varied in order to identify the separate roles of protein, surfactant, and the protein–surfactant complex in determining adsorption outcomes. They recorded a decrease in lysozyme adsorption on hydrophobic silica upon addition of Tween 80, and this reduction in adsorbed protein increased with Tween 80 concentration in solution. Sequential adsorption experiments showed that, at sufficiently high concentration, Tween 80 was able to remove adsorbed lysozyme from a hydrophobic surface. In addition, if Tween 80 is introduced to the hydrophobic surface prior to lysozyme addition, lysozyme adsorption can be reduced or even prevented. On the other hand, adsorption of lysozyme on hydrophilic silica showed no dependence on the presence of Tween 80 in solution. And, sequential adsorption experiments showed the presence of Tween 80, whether introduced to the interface before or after lysozyme, had no effect on lysozyme adsorption. These observations were attributed to surface-dependent differences in Tween-binding strength and emphasize the importance of the direct interaction between surfactant and solid surface, relative to surfactant–protein association in solution, in the modulation of protein adsorption by Tween 80. Accordingly, the rapid diffusion of surfactant (relative to protein) to the interface is likely to contribute to a reduction in protein adsorption only if surfactant–surface affinity is sufficiently high.

11.2.3.2 Practical Relevance. Lysozyme is a much used "model" protein for study of adsorption phenomena in a number of well-controlled circumstances, but results of such work have contributed to forming a foundation for understanding the behavior of more complex therapeutic proteins in surfactant-containing formulations. The adsorption, structural alteration, and biological activity of a recombinant factor VIII (rFVIII) was investigated in the presence of the surfactant Tween 80, at hydrophilic and hydrophobic solid–water interfaces (78). At the hydrophobic surface, the presence of Tween 80 in the protein solution resulted in a reduction in amount of protein adsorbed, whereas rFVIII adsorption at the hydrophilic surface was entirely unaffected by the presence of Tween 80. These observations were attributed to high binding strength between Tween and the hydrophobic surface, and low binding strength between Tween and the hydrophilic surface. Colloidal particles bearing hydrophilic and hydrophobic surfaces, each with net positive or negative surface charge densities, were used as substrates for rFVIII adsorption in the evaluation of tertiary structure change and biological activity retention at interfaces. Fluorescence emission spectroscopy showed that rFVIII tertiary structure was changed upon exposure to hydrophobic nanoparticle surfaces. Similarly, the biological activity of rFVIII (based on the activated partial thromboplastin time) was reduced at hydrophobic surfaces. At high surfactant concentration, these properties were better preserved, a likely result of Tween adsorption sterically inhibiting rFVIII adsorption. While relatively high rFVIII adsorption was recorded at hydrophilic surfaces, these surfaces did not induce large changes in structure or activity. This was attributed to the formation of a tightly packed, ordered adsorbed layer on these surfaces, governed by electrostatic attraction and not mediated by the rFVIII active site (78).

In the absence of a surfactant, proteins can be expected to adsorb with high affinity to hydrophobic surfaces as well as to negatively charged, positively charged, and electronically neutral surfaces. Substantial reduction in adsorption can be seen with surfactant addition under appropriate circumstances or, in general, with application of so-called "nonfouling" coatings (79–82), such as those exhibiting pendant poly(ethylene oxide) [also referred to as poly(ethylene glycol)] chains. It is probable that steric repulsion is a requirement for eliminating unwanted protein adsorption. Steric repulsion would explain the beneficial effect of nonfouling coatings, the beneficial effect of added surfactant in the presence of surfaces for which surfactant–surface binding is strong, and the absence of any beneficial effect of added surfactant in the presence of surfaces for which surfactant–surface binding is weak.

REFERENCES

1. Duke CB. J Vac Sci Technol A 1984; 2: 139–143.
2. Andrade JD. In: Andrade JD, editor. Surface and interfacial aspects of biomedical polymers. Volume 2, Protein adsorption. New York: Plenum Press; 1985. pp. 1–80.
3. Norde W. Adv Colloid Interface Sci 1986; 25: 267–340.
4. Horbett TA, Brash JL. In: Brash JL, Horbett TA, editors. Proteins at interfaces: physicochemical and biochemical studies. ACS Symp. Ser. 343, Washington, DC: American Chemical Society; 1987. pp. 1–35.
5. Brash JL, Horbett TA. In: Horbett TA, Brash JL, editors. Proteins at interfaces II: fundamentals and applications. ACS Symp. Ser. 602. Washington, DC; 1995. pp. 1–23.
6. Andrade JD, Hlady V, Feng L, Tingey K. In: Brash JL, Wojciechowski PW, editors. Interfacial phenomena and bioproducts. New York: Dekker; 1996. pp. 19–56.
7. McGuire J, Bower CK, Bothwell M. Encyclopedia of surface and colloid science. 2nd ed. New York: Taylor & Francis; 2006. pp. 5192–5205.
8. Absolom D, Lamberti F, Policova Z, Zingg W, Van Oss C, Neumann A. Appl Environ Microbiol 1983; 46: 90–97.
9. Absolom DR, Hawthorne LA, Chang G. J Biomed Mater Res 1988; 22: 271–285.
10. Ratner BD, Porter SC. In: Brash JL, Wojciechowski PW, editors. Interfacial phenomena and bioproducts. New York: Dekker; 1996. pp. 57–83.
11. Baier RE, Meyer AE. In: Brash JL, Wojciechowski PW, editors. Interfacial phenomena and bioproducts. New York: Dekker; 1996. pp. 85–121.
12. Lee S, Ruckenstein E. J Colloid Interface Sci 1988; 125: 365–379.
13. Norde W, Lyklema J. J Colloid Interface Sci 1978; 66: 257–265.
14. Bagchi P, Birmbaum S. J Colloid Interface Sci 1981; 83: 460–478.
15. Kondo A, Higashitani K. J Colloid Interface Sci 1992; 150: 344–351.
16. Luey J, McGuire J, Sproull RD. J Colloid Interface Sci 1991; 143: 489–500.
17. Kondo A, Oku S, Higashitani K. J Colloid Interface Sci 1991; 143: 214–221.
18. Kondo A, Murakami F, Higashitani K. Biotechnol Bioeng 1992; 40: 889–894.
19. Norde W, Favier JP. Colloids Surf 1992; 64: 87–93.
20. Billsten P, Wahlgren M, Arnebrant T, McGuire J, Elwing H. J Colloid Interface Sci 1995; 175: 77–82.
21. Tian M, Lee W-K, Bothwell MK, McGuire J. J Colloid Interface Sci 1998; 200: 146–154.
22. Bohnert JL, Horbett TA. J Colloid Interface Sci 1986; 111: 363–377.
23. Lu DR, Park K. J Colloid Interface Sci 1991; 144: 271–281.
24. Elwing H, Welin S, Askendal A, Lundström I. J Colloid Interface Sci 1988; 123: 306–308.
25. Norde W, Lyklema J. J Biomater Sci Polym Ed 1991; 2: 183–202.
26. Norde W, Zoungrana T. Biotechnol Appl Biochem 1998; 28: 133–143.
27. McGuire J, Wahlgren MC, Arnebrant T. J Colloid Interface Sci 1995; 170: 182–192.
28. Krishnan A, Liu YH, Cha P, Allara D, Vogler EA. J Biomed Mater Res A 2005; 75: 445–457.
29. Fröberg JC, Arnebrant T, McGuire J, Claesson PM. Langmuir 1998; 14: 456–462.
30. Lundström I, Elwing H. J Colloid Interface Sci 1990; 136: 68–84.
31. Arnebrant T, Ivarsson B, Larsson K, Lundström I, Nylander T. Prog Colloid Polym Sci 1985; 70: 62–66.
32. Arnebrant T, Nylander T. J Colloid Interface Sci 1988; 122: 557–566.
33. Cussler EL. Diffusion: mass transfer in fluid systems. Cambridge, MA: Cambridge University Press; 1989.
34. Soderquist ME, Walton AG. J Colloid Interface Sci 1980; 75: 386–397.
35. Calonder C, Tie Y, Van Tassel PR. PNAS USA 2001; 98: 10664–10669.
36. Tie Y, Calonder C, Van Tassel PR. J Colloid Interface Sci 2003; 268: 1–11.
37. Joshi O, Lee HJ, McGuire J, Finneran P, Bird KE. Colloids Surf B Biointerfaces 2006; 50: 26–35.
38. Arai T, Norde W. Colloids Surf 1990; 51: 1–16.
39. Arai T, Norde W. Colloids Surf 1990; 51: 17–28.
40. Shirahama H, Lyklema J, Norde W. J Colloid Interface Sci 1990; 139: 177–187.
41. Wei A-P, Herron JN, Andrade JD. In: Crommelin DJA, Schellekens H, editors. From clone to clinic. Amsterdam: Kluwer Academic Publishers; 1990. pp. 305–313.
42. Horsley D, Herron J, Hlady V, Andrade JD. In: Brash JL, Horbett TA, editors. Proteins at interfaces: physicochemical and biochemical studies. ACS Symp. Ser. 343. Washington, DC: American Chemical Society; 1987. pp. 290–305.
43. Xu S, Damodaran S. J Colloid Interface Sci 1993; 159: 124–133.
44. Kato A, Yutani K. Protein Eng 1988; 2: 153–156.
45. Wang J, McGuire J. J Colloid Interface Sci 1997; 185: 317–323.
46. Bower CK, Xu Q, McGuire J. Biotechnol Bioeng 1998; 58: 658–662.
47. Elwing H, Welin S, Askendal A, Nilsson U, Lundström I. J Colloid Interface Sci 1987; 119: 203–210.
48. Elwing H, Askendal A, Lundström I. Prog Colloid Polym Sci 1987; 74: 103–107.
49. Rapoza RJ, Horbett TA. J Colloid Interface Sci 1990; 136: 480–493.

50. Rapoza RJ, Horbett TA. J Biomed Mater Res 1990; 24: 1263–1287.
51. Ertel SI, Ratner BD, Horbett TA. J Colloid Interface Sci 1991; 147: 433–442.
52. Wahlgren MC, Arnebrant T. J Colloid Interface Sci 1991; 142: 503–511.
53. Wahlgren MC, Arnebrant T. J Colloid Interface Sci 1992; 148: 201–206.
54. Wahlgren MC, Paulsson MA, Arnebrant T. Colloids Surf A Physicochem Eng Aspects 1993; 70: 139–149.
55. Wahlgren MC, Arnebrant T, Askendal A, Welin-Klintström S. Colloids Surf A Physicochem Eng Aspects 1993; 70: 151–158.
56. Krisdhasima V, Vinaraphong P, McGuire J. J Colloid Interface Sci 1995; 161: 325–334.
57. Vinaraphong P, Krisdhasima V, McGuire J. J Colloid Interface Sci 1995; 174: 351–360.
58. McGuire J, Wahlgren MC, Arnebrant T. J Colloid Interface Sci 1995; 170: 193–202.
59. Slack SM, Horbett TA. J Colloid Interface Sci 1989; 133: 148–165.
60. Narsimhan G, Uraizee F. Biotechnol Prog 1992; 8: 187–196.
61. Krisdhasima V, McGuire J, Sproull R. J Colloid Interface Sci 1992; 154: 337–350.
62. Lu CF, Nadarajah A, Chitter KK. J Colloid Interface Sci 1994; 168: 152–161.
63. Wahlgren MC, Arnebrant T, Lundström I. J Colloid Interface Sci 1995; 175: 506–514.
64. Dejardin P, Cottin I. Colloids Surf B Biointerfaces 1995; 4: 111–120.
65. Lee WK, McGuire J, Bothwell MK. J Colloid Interface Sci 1999; 213: 265–267.
66. Lundström I. Prog Colloid Polym Sci 1985; 70: 76–82
67. Lee WK, McGuire J, Bothwell MK. J Colloid Interface Sci 2002; 252: 473–476.
68. Podhipleux N, McGuire J, Bothwell MK, Horbett TA. Colloids Surf B Biointerfaces 2003; 27: 277–285.
69. Lee WK, McGuire J, Bothwell MK. J Colloid Interface Sci 2004; 269: 251–254.
70. Podhipleux N, Damodaran S, McGuire J, Bothwell MK. Colloids Surf B Biointerfaces 1999; 13: 167–177.
71. Bower CK, Sananikone S, Bothwell MK, McGuire J. Biotechnol Bioeng 1999; 64: 373–376.
72. Daeschel MA, McGuire J. Biotechnol Genet Eng Rev 1998; 15: 413–438.
73. Chi EY, Weickmann J, Carpenter JF, Manning MC, Randolph TW. J Pharm Sci 2005; 94: 256–274.
74. Jones LS, Kaufmann A, Middaugh CR. J Pharm Sci 2005; 94: 918–927.
75. Arnebrant T, Wahlgren MC. In: Horbett TA, Brash JL, editors. Proteins at interfaces II: fundamentals and applications. ACS Symp. Ser. 602. Washington, DC: American Chemical Society; 1995. pp. 239–254.
76. Elwing H, Askendal A, Lundstrom I. J Colloid Interface Sci 1989; 128: 296–300.
77. Joshi O McGuire J. Appl Biochem Biotechnol. 2009; 152: 235–248.
78. Joshi O, McGuire J, Wang DQ. J Pharm Sci. 2008; 97: 4741–4755.
79. Desai NP, Hubbell JA. J Biomed Mater Res 1991; 25: 829–843.
80. Li JT, Caldwell KD. Langmuir 1991; 7: 2034–2039.
81. Li JT, Carlsson J, Huang SC, Caldwell KD. In: Glass JE editor. Hydrophilic polymers. Performance with environmental acceptability. Washington, DC: ACS; 1996. pp. 61–78.
82. Paulsson M, Kober M, Freij-Larsson C, Stollenwerk M, Wesslen B, Ljungh A. Biomaterials 1993; 14: 845–853.

12

AFFINITY FUSIONS FOR PROTEIN PURIFICATION

SUSANNE GRÄSLUND AND MARTIN HAMMARSTRÖM

Department of Medical Biophysics and Biochemistry, Structural Genomics Consortium, Karolinska Institutet, Stockholm, Sweden

12.1 INTRODUCTION

In the postgenomic era, the focus has shifted from studies of genomes to high-throughput analysis of entire proteomes. Since proteins are so much more chemically and structurally diverse than nucleic acids, they do not naturally lend themselves to high-throughput analysis. By employing common affinity tags, however, high-throughput protein production using general and robust protocols is enabled as opposed to highly customized procedures used in conventional chromatographic purification.

The use of biointeraction mechanisms to create affinity fusion systems is a quite young technology that has gained an enormously widespread use in the biotechnology field the last decades, both at academic labs and in industrial environments (1). Affinity fusions are primarily used to simplify recovery of a fused target gene product, but are also widely employed for detection, immobilization, stabilization, and to increase the solubility or to prolong the life and effect of antigen proteins. Most affinity tags can be used for several of these purposes while others have a more specific use. To date, a large number of different gene fusion systems that are capable of selective interaction with a ligand immobilized onto a chromatography matrix have been described (2–4). In such systems, different types of interactions have been utilized such as enzymes–substrates, transporter proteins–transported molecule, bacterial receptors–serum proteins, polyhistidines–metal ions and antibodies–antigens. Recently, gene technology in the format of combinatorial protein engineering has allowed *in vitro* selection of novel proteins that selectively bind to a desired target molecule. This new emerging technology enables the *de novo* creation of purpose-designed ligands suitable for affinity chromatography applications (5). The size of the fusion partners range from a few amino acids to whole proteins, and the immobilized ligand can vary in size as well, from small molecules to intact proteins. Often it is desirable to keep the fusion partner as small and inert as possible to allow functional studies or downstream applications without the necessity of fusion partner removal. In other cases, it can be beneficial with a larger fusion partner that helps in increasing the recovery of functional protein, by promoting production levels, solubility, or stability. However, large fusion partners must usually be removed before several applications, for example crystallization or antibody production. Large immobilized protein ligands are also often disadvantageous as binding capacity is reduced and they are more susceptible to impaired functionality upon repeated usage.

There is a great interest in developing methods for fast and convenient purification of large number of proteins in parallel, for example to facilitate functional and structural studies. The conditions for purification differ from system to system and the environment tolerable by the target protein is an important factor for deciding which affinity fusion partner to choose. Also, other factors including costs for the affinity matrix, buffers, the possibility to remove the fusion partner, and the possibility for automation are important to consider. Here, the most common systems (Table 12.1) for different applications is described and both advantages and drawbacks with the various available systems are summarized. For a better overview, we have also tried to categorize the different systems regarding

Downstream Industrial Biotechnology: Recovery and Purification, First Edition. Edited by Michael C. Flickinger.

TABLE 12.1. Commonly Used Affinity Fusion Systems

Fusion Partner	Size (kDa)	Ligand	Elution Conditions	Comment
His tag	~1	Chelated divalent metal ion	Imidazole/low pH/EDTA	Very small, both denaturing and native conditions
GST	25	Glutathione	Reduced glutathione	Slow binding kinetics
StrepII	1	Modified streptavidin	Desthiobiotin	Very small, low capacity, expensive matrix
Z	7	hIgG	Low pH	Low pH at elution
Protein A	31	hIgG	Low pH	Low pH at elution
ABP	5–25	HSA	Low pH	Low pH at elution
MBP	41	Amylose	Maltose	Quite large, low capacity
FLAG	1	MAbs M1/M2	EDTA/low pH	Very small, very low capacity, mainly used for detection

whether they are mostly being used for purification, stabilization, detection, or other purposes, but many of them belong to more than one category.

Historically, recombinant proteins have most commonly been produced in the gram-negative bacterium *Escherichia coli*, which provides a simple, fast, and cost-effective production system. However, approximately 50% of all proteins are insoluble when produced in *E. coli* (6,7). Proteins that fail to be expressed in a soluble form may be toxic to the host or could reflect lack of proper modifications or interaction partner or an inability to fold in the bacterial system. Therefore, other expression systems have become increasingly important even at a preparative scale, that is yeast, insect cells, and mammalian cell cultures as well as *in vitro* translation systems (8,9).

12.2 SYSTEMS FOR RAPID PROTEIN CAPTURE

There are many desired properties of an ideal affinity fusion system. It has to be highly specific so that the background is kept low and it must have a reasonable range of affinity, not too low and not too high, since the latter will cause problems with elution. The binding capacity must be high to reduce column bed volume and it is an advantage if the matrix is reusable. Furthermore, the buffer systems used (binding, wash, and elution) must suit a broad range of proteins and also be compatible with downstream applications. Cost is, of course, also an important parameter and both the materials (matrix and buffer components) and the column size, as well as the reusability, have a great impact on it. Since all affinity tags have their strengths and weaknesses, it is often hard to find a single tag with all desired properties for a certain application. Combinatorial tagging using the strengths of several tags might be the best solution. Also sandwich tagging, when two different tags are placed at the N and C termini, respectively, can be used to purify only full-length proteins. Figure 12.1 shows the basic principle of affinity purification using competitive elution.

12.2.1 Poly-Histidine Tags (His Tags)

The by far most commonly used affinity system is immobilized metal-ion affinity chromatography (IMAC) (10) for purification of proteins fused to a poly-His tag (11). Ni(II)-nitrilo-triacetic acid (Ni-NTA), which exhibits a high affinity for adjacent histidine residues, is the most commonly used chelating matrix for IMAC. Other common matrices are Ni(II)-imino-diacetic acid (Ni-IDA), Ni-sepharose, and Talon (Fig. 12.2). All these matrices can also be loaded with other divalent transition metal ions such as cobalt, copper, manganese, and iron to modulate

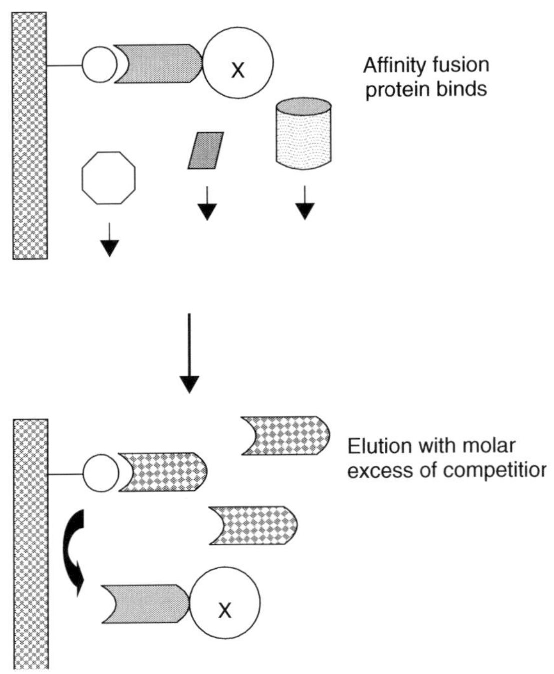

Figure 12.1. An illustration of the concept of affinity purification using competitive elution. Binding and washing is performed in the presence of no or low amounts of competitive binder, and elution is performed by adding an excess of the competitive binder.

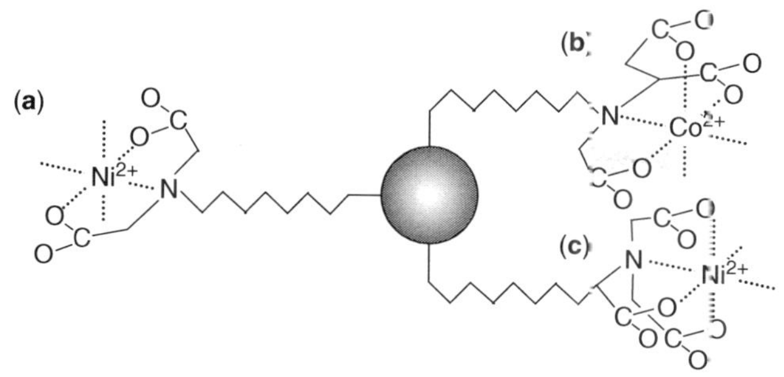

Figure 12.2. Structures of the chelating groups of different IMAC resins. (a) Ni-IDA, (b) Talon, and (c) Ni-NTA. The linkers between the resin and the chelating groups are only schematically drawn. Both Talon and Ni-NTA occupy four of the six coordination sites of the metal ion leaving two sites available for interaction with the histidines of the His tag. Ni-IDA occupy three coordination sites of the metal ion leaving three sites available for interaction with the His tag.

affinity and binding capacity. His tags have the advantages of a small size, a simple, robust and cheap chromatographic matrix, high binding capacity, mild elution conditions, possibility for stringent binding and washing using low amounts of imidazole in all buffers, and ability to withstand multiple regeneration cycles. Moreover, the His tag can be used for purification also under denaturing conditions adding yet another dimension to its versatility. Even if the protein is insoluble or aggregated in inclusion bodies, it can be purified under denaturing conditions and refolded. The only apparent drawback of His tags/IMAC is the high background of native histidine-rich *E. coli* proteins observed when purifying low-expressing proteins (12). Also, it does not enhance solubility. It is possible that the naturally derived histidine affinity tag (HAT), comprising 6 histidines dispersed among 12 other residues, is an alternative, which is less prone to affect solubility (13). However, it has not yet been shown that the His tag in itself has a general, negative effect on solubility—only that other tags or fusion proteins can have a more positive effect on solubility, and it is conceivable that other factors such as linker sequences and vector design have a larger effect than the His tag in itself (14).

All structural genomics centers use the His tag for capture purification in some way and only the details vary from site to site (15). IMAC is sometimes used as a single purification step but most commonly in combination with at least one further polishing step such as gel filtration or ion exchange chromatography. IMAC is also used for subtractive purification after proteolytic removal of the tag (see below).

Several different vendors provide commercial IMAC systems with slightly different binding capacities and binding strengths (Qiagen, GE Healthcare, SIGMA, Clontech, Invitrogen, among others). Also, several systems for purification screening in a 96-well plate format are available from the same vendors. Furthermore, IMAC procedures can be automated on the most advanced ÄKTA system, for example ÄKTAxpress(TM) (GE Healthcare) (16), and thus used for high-throughput parallel protein purification. As a consequence of IMAC being more and more widespread and its incompatibility with ethylenediaminetetraacetic acid (EDTA), EDTA-free protease inhibitors have become available from many companies. This is an example of how alternative products are developed to suit the requirements of a commonly used purification system. Another similar adaptation is the engineering of the thioredoxin fusion protein to introduce a cluster of histidines on the surface of the protein to allow IMAC purification (17). Yet another example is the insertion of a hexa-histidine tag into a surface loop of the green fluorescent protein (GFP) to develop a dual functional GFP, useful for both monitoring and purification of recombinant proteins (18).

12.2.2 Glutathione *S*-Transferase Tag

A historically very popular affinity tag is glutathione *S*-transferase (GST), which is a reasonably large protein tag. GST normally conjugates reduced glutathione (GSH) to a variety of ligands. Immobilized GSH is used for affinity purification of GST, and competitive elution is done by adding excess GSH. GST was once the most widespread affinity purification system (19) and has been reported to be functional in numerous expression hosts, but has now fallen a bit into shadow due to some drawbacks of the system. The main disadvantages are that GST shows quite slow binding kinetics, that it is a homodimer (20), which can impose dimerization of the fused protein as well, and that elution is performed under reducing conditions. Also, four exposed cysteines in each subunit of the GST dimer can lead to a significant degree of oxidative aggregation (20). Nevertheless, GST has been frequently used as a solubilizing/stabilizing fusion partner although the solubilizing effect has been challenged recently when comparative studies have been done (21–23). Since GST has been such a widespread system, several commercial kits for small-scale parallel purification are available, and like IMAC, GST purification using GSTrap columns (GE Healthcare) can also be automated on ÄKTA systems.

12.2.3 Streptag II

Another small and promising affinity system uses Streptag II. It consists of a streptavidin recognizing octa-peptide (WSHPQFEK) that was selected for its improved affinity. Proteins containing this tag can be affinity purified using a matrix with a modified streptavidin and eluted with a biotin analog (desthiobiotin) (24,25). It has not yet been systematically used, but it is promising for challenging proteins where the His tag fails since this system has much higher specificity, although the binding capacity is much lower (26). It is also well suited for purification from expression systems where histidine-rich proteins are more abundant (27). Another advantage is its independence from metal ions in the purification, an aspect often of interest when metalloproteins are to be studied or when paramagnetic impurities must be avoided for nuclear magnetic resonance (NMR) (24). Recently GE Healthcare released prepacked Streptag II (StrepTrap) columns, which are compatible with ÄKTA systems thus enabling purification automation. The main disadvantages are that Streptag II requires a quite expensive resin (in the order of 15 times as expensive per milligram of purified protein compared to a standard IMAC matrix) and that it does not enhance solubility.

12.2.4 Protein A

Staphylococcal protein A (SpA) is an immunoglobulin-binding surface receptor found in the gram-positive bacterium *Staphylococcus aureus*. SpA binds to the Fc part of certain immunoglobulins and has found extensive use in immunological and biotechnological research (28). The protein comprises five highly homologous domains all capable of binding to IgG (29). Protein A-sepharose (GE Healthcare) has been extensively used for purification of monoclonal antibodies (30) and coimmunoprecipitation to study protein–protein interactions (31).

Single IgG-binding domains are suitable fusion partners for the production of recombinant proteins and the most widely used has been Z, engineered from the B domain of SpA (32), which can be purified using IgG-sepharose. Z, or the more commonly used dimeric form ZZ, which shows a 10-fold increase in binding strength (33), has several advantages being quite small protein tags that are highly stable, can be secreted, and can be produced in soluble form at very high levels. The major drawback of the system has been the need for elution at low pH (~3), which can yield biologically inactive products. Low-pH elution can be circumvented by competitive elution with an engineered competitor protein (34), but this has not been widely used.

12.3 STABILIZATION OF EXPRESSED PROTEINS

Several different small-scale (96-well plate format) studies have shown the usefulness of solubility-enhancing fusions (21,23,35–37) but not as much have been systematically evaluated on larger sets of proteins in preparative scale. Several of the most frequently used solubility-enhancing fusions were originally developed for affinity purification purposes, for example maltose-binding protein (MBP) on amylose resin (38), SpA derivatives on IgG resin (39), GST on GSH resin (19), and also GB1 domain of streptococcal protein G on IgG resin (40). Now the number of fusion proteins/tags with claimed solubility-enhancing properties is ever increasing. Perhaps the best studied and most thoroughly validated are MBP (41) and N-utilization substance A (NusA) (42). Also Thioredoxin from *E. coli* (TrxA) has been quite widely used (43). Although several of these fusion proteins can be used for affinity purification on their own, they are usually combined with a small affinity tag, for example the His tag to enable IMAC purification.

12.3.1 Maltose-Binding Protein

The MBP of *E. coli* is normally located in the periplasm and involved in transportation of maltose across the cytoplasmic membrane. The affinity for maltose can be used for affinity purification of MBP fusions on resins with immobilized amylose and elution using free maltose. This purification method is, however, not used often as the specificity and binding capacity are low. Instead MBP has been more frequently used as a solubility enhancer. Although it has been firmly established that certain highly soluble proteins such as MBP and NusA can function as general solubility enhancers in the context of a fusion protein, very little is known about their ability to promote proper folding. A common problem is that proteins that can be purified in soluble form precipitate and become insoluble when the tag is removed thus indicating that the passenger protein was not properly folded in the fusion state (15). A system that circumvents this problem has been reported by Waugh and coworkers, where the protein is first produced as a fusion to a MBP and His tag combination and coexpressed with a protease that cleaves off the MBP part *in vivo* leaving only His tagged protein to be purified if it remains soluble after MBP removal (44,45).

12.3.2 SUMO Tag

Ubiquitin-based tags have been used to increase expression levels, but also to aid in protein folding. One of these tags, small ubiquitin modifying protein (SUMO), has emerged as an alternative for production of otherwise intractable proteins. As fusions to SUMO, insoluble proteins have been shown to fold properly and become soluble (46,47). SUMO cannot be used as an affinity tag so it has to be combined with another tag that enables affinity purification. The SUMO tag can be removed using a specific SUMO protease, but this also makes SUMO mostly constrained to *E. coli* since highly conserved SUMO proteases are present in eukaryotes and may cleave the fusion protein during production. Also, the high ratio of SUMO protease required to process the fusion protein makes the system not so well suited for large-scale production.

12.4 DETECTION OF PRODUCED PROTEINS

12.4.1 FLAG(TM) Tag

The FLAG(TM) tag was originally developed as an affinity purification tag, but has also gained widespread use for detection. The FLAG tag consists of a short hydrophilic octa-peptide (DYKDDDDK) (48). The FLAG peptide binds to a monoclonal antibody M1 in a calcium-dependent manner which allows gentle elution by the addition of a chelating agent (49). A limitation, however, is that the M1 antibody can bind to the FLAG peptide only when it is located at the extreme N terminus of the fusion protein. Alternatively, a different antibody (M2), capable of binding to the FLAG peptide also when it is fused to the C terminus, is available (50). However, the M2 antibody interacts with the FLAG peptide in a non-calcium-dependent manner and then low pH or competition with an excess of synthetically produced FLAG peptide is used for elution (50). The FLAG system has been used in a variety of cell types. The major drawback is that the antibody purification matrix is not as stable as others, for example Ni-NTA, and therefore the system has more been used for detection strategies than purification. In general, small tags can be readily detected with specific monoclonal antibodies with great sensitivity. Finally, the FLAG tag can be removed by treatment with enterokinase, specific for the five C-terminal amino acids (DDDDK) of the peptide sequence (48).

12.4.2 Lumio Tag

Another recently developed system for tag detection is the biarsenical ligands binding to tetracysteine motifs for protein labeling *in vitro* and *in vivo* (51). The system has been commercialized under the name of the Lumio(TM) tag by Invitrogen. The Lumio recognition sequence is a small, unobtrusive six–amino acid fusion (CCPGCC), and can be added to either the N- or C terminus for downstream monitoring. Lumio detection reagents bind this tag with high specificity and affinity, resulting in a bright fluorescent signal, which can easily be detected using a standard ultraviolet (UV) transilluminator, laser-based scanner, fluorometer, or a fluorescent filter on a microscope. The Lumio tag can be used both for in-gel detection of denatured proteins and for *in vivo* detection of proteins in a broad array of cell types including mammalian cells (52).

12.4.3 S-Tag

Another tag used mostly for detection is the S-tag. It is a fusion peptide that allows detection by a rapid sensitive-homogeneous assay or by colorimetric detection in Western blots. The system is based on the strong interaction between the 15-residue-long S-tag and the 103-residue-long S-protein, both of which are derived from RNaseA (53). The complex is dependent on pH, temperature, and ionic strength, and the elution conditions needed are very harsh (i.e. pH 2) (54). However, the discovery of a hypersensitive flourogenic substrate for RNaseA made the system very interesting for detection (55).

12.5 REMOVAL OF AFFINITY TAGS

Since all affinity tags have the potential to interfere with the biological activity of a protein, it is often important to be able to remove the fusion protein/tag. For therapeutic proteins, it is sometimes even required to leave only native residues on the produced target protein. Historically, Factor Xa, thrombin, and enterokinase have been used but were sometimes hampered by low efficiency or unspecific cleavage of the target protein. Now, the most frequently used proteases are viral proteases that normally function as to cleave the viral polypeptide to generate mature proteins. They are all highly processive under permissive condition and have high specificity, but efficiency of cleavage can still vary depending on the target protein. Most commonly used is probably the 49 kDa protease of tobacco etch virus (TEV), referred to as the *TEV protease* (56), which has the ENLYFQ*S recognition sequence. It can be produced routinely in any lab (57,58). The TEV protease can, to some extent, tolerate different amino acids at the terminal position in the cleavage site, which sometimes can be useful when it is critical to get the native form of the target protein. Also frequently used is a viral 3C protease; Rhinovirus 3C with the recognition sequence LEVLFQ*GP (59), commercially available as a GST fusion under the name PreScission (GE Healthcare).

Another system that can be used to remove the N-terminal tag or linker sequences is the TAGZyme(TM) from Qiagen (60). It is based on the dipeptidyl aminopeptidase (DAPase) activity of a recombinant peptidase. The enzyme will remove dipeptides from the N terminus until it encounters (i) a lysine or arginine at the N terminus (position 1) or (ii) a proline at position 2 or 3. With two auxiliary enzymes and careful planning of the sequence of the tag, complete and precise removal of all nonnative residues can be achieved. This is especially useful when it is critical to obtain the target protein in its native form, without any added residues.

Inteins are self-cleavable proteins and this feature is being used in the intein-based IMPACT system from New England Biolabs (61). The system uses a combination of a chitin-binding domain and an intein domain fused to the N- or C terminus of the target protein. Binding to a chitin matrix is followed by an on-column cleavage induced by addition of high levels of a reducing agent. This system can be used to remove either N- or C-terminal fusion partners. The disadvantage of the IMPACT system lies in the need for a strong reducing environment. However, other intein systems are available, which do not require the reducing agent. The advantage of the intein system lies in that it abolishes the need for adding protease to remove affinity fusion partners and that the intein cleavage can be activated in a controlled way. Some intein systems have been combined with special fusion partners that enable purification without affinity resins (62). One example is the polyhydroxybuterate (PHB) polymer forming system in which the *E. coli* cells, or other expression systems, are used to produce large intracellular PHB granules as well as a combined PHB binding and intein fusion protein (63). With this system, the granules can be recovered and used as a crude affinity resin in one single step followed by a change in pH to activate the intein which releases the unfused target protein. Another example is based on the aggregation-prone elastin-like peptide (ELP) consisting of numerous repeats of a peptide motif that undergo reversible transition from soluble to insoluble upon temperature upshift (64,65). Addition of salt and a temperature-shift to 30° induces aggregation of this fusion partner, allowing it to be collected by centrifugation after the cell lysis and clarification step. Activation of the intein releases the target protein, and the residual ELP and intein fusion are removed by centrifugation. Both the PHB and the ELP system could, of course, potentially be used with endopeptidases as well.

The availability of affinity tagged proteases also makes it possible to use the proteolytic removal of a tag as a polishing purification step. After the first capture step using the affinity tag, the tagged protease is added to the eluate to remove the affinity tag. Following a buffer change or dilution, to remove the competitive binder used for elution, the protein mixture is then run over another affinity matrix, which will bind the cleaved tag, any uncleaved target protein, as well as the tagged protease (Fig. 12.3). Additionally; any contaminating copurified proteins that interact unspecifically with the affinity resin will also be retained on the column leaving the cleaved protein with high purity in the flow-through fraction. This purification scheme has been successfully used at different structural genomics sites (66) to give protein of purity high enough for crystallographic studies.

Proteolytic release of the target protein can also be used as an alternative elution method when the normal elution conditions are unsuitable for the purified proteins or the downstream applications. Also, in this case, it is highly desirable to have an affinity tagged protease as a means to remove the protease after proteolysis; however, in this case it is tagged with another affinity tag to avoid trapping the protease on the column before proteolysis has occurred.

Fusions at the C terminus are often more problematic to remove as proteases cleave at the C-terminal end of the recognition sequence, thus leaving a number of residues still attached to the target protein. Carboxypeptidases could facilitate polishing of these remaining residues (67), but there is no system that is commonly used.

12.6 UTILIZATION OF FUSION PROTEINS AS ANTIGENS

Gene fusion technology might also be useful in prolonging the *in vivo* half-life of pharmaceutical proteins. Using fusion partners such as the Fc part of IgG (68) or bacterial protein receptors capable of binding serum albumins (69,70), the serum half-lives of different proteins have been shown to increase substantially. The latter system has been used to prolong the half-lives of antigens in order to get higher titers of antibodies upon immunization in combination with potent adjuvants (70,71). A large-scale project

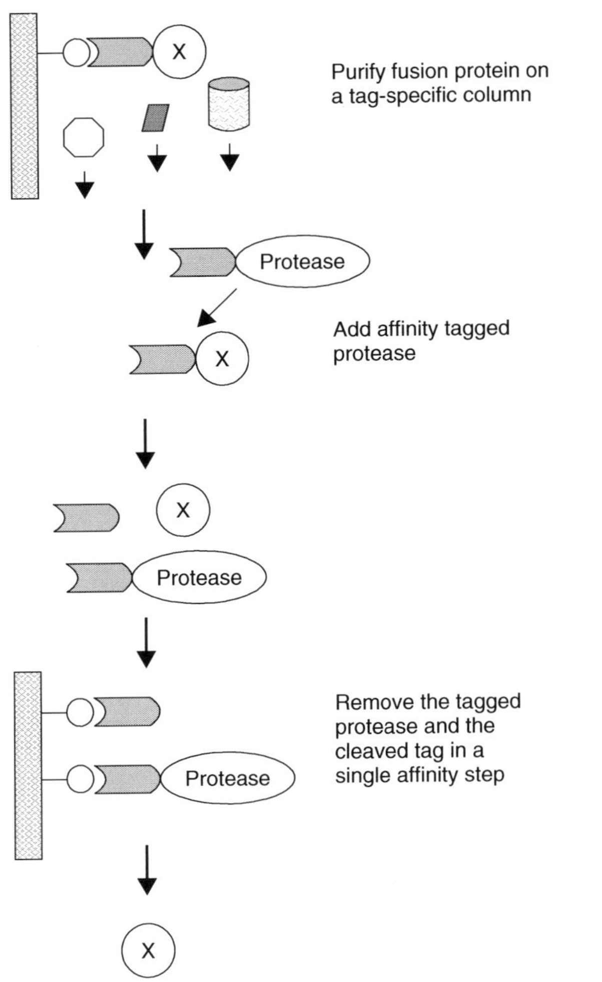

Figure 12.3. An illustration of the concept of affinity purification, tag cleavage, and a second round of subtractive affinity purification.

to produce polyclonal antibodies toward different human proteins, human protein atlas (HPA), has employed this approach to increase the response of the immune system (72,73). Protein epitope signature tags (PrESTs) of approximately 50–100 residues designed by low homology to other human proteins are produced as fusions to a His tag and the albumin-binding protein (ABP) derived from Streptococcal protein G. Purification is performed under denaturing conditions using the His tag, and the renatured PrESTs are then immunized with adjuvants to generate specific antibodies. Monospecific antibodies are then purified from sera using the immobilized PrEST as affinity ligand. ABP and other recombinant variants of protein G have also been extensively used for purification of human serum albumin. However, as low pH is required for elution and the tag is quite large, protein G derivatives have become less favored as compared to other available systems.

12.7 SUBUNIT IMMUNOGENS FOR VACCINE RESEARCH

Immunogenic proteins and peptides that are produced as fusion proteins have been frequently used for immunization purposes and represent a strategy to construct vaccines against infectious diseases. An important aspect regarding the production of antigens by recombinant DNA technology is the purification of the gene product from contaminating host components. The approach to express peptides or proteins as one part of a fusion protein where the other part constitutes an affinity handle allows rapid and efficient affinity purification of the gene product (74). The influence of the affinity tag on the immune responses to antigens expressed as fusion proteins has to be considered. It might, however, be possible to select affinity fusion partners with inherent positive immune stimulating effects, such as carrier function or adjuvant capacity (75).

Within the field of vaccine development, strategies for production of recombinant immunogens fused to hydrophobic peptides or lipid tags to improve their capacity to be incorporated into an adjuvant formulation, for example, immune stimulating complexes (iscoms), have also been reported (76). The strong interaction between biotin and streptavidin has also been evaluated for coupling of recombinant immunogens to the iscom matrix (77). Another group working with DNA vaccination has reported the use of a small viral DnaJ-like peptide as a fusion partner, which binds to the Hsp73 protein and thus triggers the immune response (78). Another example is a vector containing a dual fusion tag that combines the enzymatically active but nontoxic, CTA1-subunit of cholera toxin with a B cell targeting moiety, D, derived from SpA. Antigens could be incorporated into or just admixed with the new vector (79). These examples suggest that gene fusion techniques might be useful ways to improve the immunogenicity and *in vivo* stability of pharmaceutical protein drugs and antigens. It may also provide a novel way for the generation of potent and safe mucosal DNA vaccines. Finally, it may also be possible to combine these approaches with new methods for drug targeting and/or slow drug release.

12.8 CONCLUSIONS

Many research laboratories have now shifted their focus from single proteins or small families to larger and more diverse sets of proteins. Likewise, method development and protocol optimizations have shifted toward solutions that are applicable and economically viable for larger sets of proteins with a minimal need for protein-specific adaptations. In this chapter, we have described the most common affinity fusion systems for various applications such as purification, enhancing solubility and stability, as well as prolonging *in vivo* half-life of therapeutic proteins, with a special focus on systems that are generally and widely used, and also highlighted some new systems that show promise. Although several tags have been used extensively over the years, His tags are the most commonly used affinity tags for protein purification. However, no tag is ideal from every standpoint and several other tags are used for enhanced solubility or folding or other specific purposes. Therefore, combinatorial tagging appears to be the best way of deriving maximum possible benefit from affinity fusion systems.

12.8.1 Acknowledgments

The Structural Genomics Consortium is a registered charity (number 1097737) that receives funds from the Canadian Institutes for Health Research, the Canadian Foundation for Innovation, Genome Canada through the Ontario Genomics Institute, GlaxoSmithKline, Karolinska Institutet, the Knut and Alice Wallenberg Foundation, the Ontario Innovation Trust, the Ontario Ministry for Research and Innovation, Merck & Co., Inc., the Novartis Research Foundation, the Swedish Agency for Innovation Systems, the Swedish Foundation for Strategic Research, and the Wellcome Trust.

REFERENCES

1. Uhlén M, Nilsson B, Guss B, Lindberg M, Gatenbeck S, Philipson L. Gene 1983; 23: 369–378.
2. Arnau J, Lauritzen C, Petersen GE, Pedersen J. Protein Expr Purif 2006; 48: 1–13.
3. Esposito D, Chatterjee DK. Curr Opin Biotechnol 2006; 17: 353–358.
4. Waugh DS. Trends Biotechnol 2005; 23: 316–320.

5. Gräslund S, Eklund M, Falk R, Uhlén M, Nygren PÅ, Ståhl S. J Biotechnol 2002; 99: 41–50.
6. Braun P, LaBaer J. Trends Biotechnol 2003; 21: 383–388.
7. Büssow K, Scheich C, Sievert V, Harttig U, Schultz J, Simon B, Bork P, Lehrach H, Heinemann U. Microb Cell Fact 2005; 4: 21.
8. Shimizu Y, Inoue A, Tomari Y, Suzuki T, Yokogawa T, Nishikawa K, Ueda T. Nat Biotechnol 2001; 19: 751–755.
9. Endo Y, Sawasaki T. Curr Opin Biotechnol 2006; 17: 373–380.
10. Porath J. Protein Expr Purif 1992; 3: 263–281.
11. Hochuli E. Genet Eng (N Y) 1990; 12: 87–98.
12. Bolanos-Garcia VM, Davies OR. Biochim Biophys Acta 2006; 1760: 1304–1313.
13. Chaga G, Bochkariov DE, Jokhadze GG, Hopp J, Nelson P. J Chromatogr A 1999; 864: 247–256.
14. Woestenenk EA, Hammarström M, van den Berg S, Härd T, Berglund H. J Struct Funct Genomics 2004; 5: 217–229.
15. Gräslund S, Nordlund P, Weigelt J, Bray J, Gileadi O, Knapp S, Oppermann U, Arrowsmith C, Hui R, Ming J, dhe-Paganon S, Park HW, Savchenko A, Yee A, Edwards A, Vincentelli R, Cambillau C, Kim R, Kim SH, Rao Z, Shi Y, Terwilliger TC, Kim CY, Hung LW, Waldo GS, Peleg Y, Albeck S, Unger T, Dym O, Prilusky J, Sussman JL, Stevens RC, Lesley SA, Wilson IA, Joachimiak A, Collart F, Dementieva I, Donnelly MI, Eschenfeldt WH, Kim Y, Stols L, Wu R, Zhou M, Burley SK, Emtage JS, Sauder JM, Thompson D, Bain K, Luz J, Gheyi T, Zhang F, Atwell S, Almo SC, Bonanno JB, Fiser A, Swaminathan S, Studier FW, Chance MR, Sali A, Acton TB, Xiao R, Zhao L, Ma LC, Hunt JF, Tong L, Cunningham K, Inouye M, Anderson S, Janjua H, Shastry R, Ho CK, Wang D, Wang H, Jiang M, Montelione GT, Stuart DI, Owens RJ, Daenke S, Schutz A, Heinemann U, Yokoyama S, Büssow K, Gunsalus KC. Nat Methods 2008; 5: 135–146.
16. Bhikhabhai R, Sjöberg A, Hedkvist L, Galin M, Liljedahl P, Frigård T, Pettersson N, Nilsson M, Sigrell-Simon JA, Markeland-Johansson C. J Chromatogr A 2005; 1080: 83–92.
17. Lu Z, DiBlasio-Smith EA, Grant KL, Warne NW, LaVallie ER, Collins-Racie LA, Follettie MT, Williamson MJ, McCoy JM. J Biol Chem 1996; 271: 5059–5065.
18. Paramban RI, Bugos RC, Su WW. Biotechnol Bioeng 2004; 86: 687–697.
19. Smith DB, Johnson KS. Gene 1988; 67: 31–40.
20. Kaplan W, Husler P, Klump H, Erhardt J, Sluis-Cremer N, Dirr H. Protein Sci 1997; 6: 399–406.
21. Hammarström M, Hellgren N, van Den Berg S, Berglund H, Härd T. Protein Sci 2002; 11: 313–321.
22. Hammarström M, Woestenenk EA, Hellgren N, Härd T, Berglund H. J Struct Funct Genomics 2006; 7: 1–14.
23. Dyson MR, Shadbolt SP, Vincent KJ, Perera RL, McCafferty J. BMC Biotechnol 2004; 4: 32.
24. Skerra A, Schmidt TG. Methods Enzymol 2000; 326: 271–304.
25. Witte CP, Noel LD, Gielbert J, Parker JE, Romeis T. Plant Mol Biol 2004; 55: 135–147.
26. Lichty JJ, Malecki JL, Agnew HD, Michelson-Horowitz DJ, Tan S. Protein Expr Purif 2005; 41: 98–105.
27. Prinz B, Schultchen J, Rydzewski R, Holz C, Boettner M, Stahl U, Lang C. J Struct Funct Genomics 2004; 5: 29–44.
28. Uhlén M, Moks T. Methods Enzymol 1990; 185: 129–143.
29. Moks T, Abrahmsén L, Nilsson B, Hellman U, Sjöquist J, Uhlén M. Eur J Biochem 1986; 156: 637–643.
30. Fuller SA, Takahashi M, Hurrell JG. In: Ausubel FM, Brent R, Kingston RE, Moore DD, Seidman JG, Smith JA, Struhl K, editors. Current protocols in molecular biology. Hoboken (NJ): John Wiley & Sons; 2001. Chapter 11: Unit 11.11.
31. Yaciuk P. Methods Mol Med 2007; 131: 103–111.
32. Nilsson B, Moks T, Jansson B, Abrahmsén L, Elmblad A, Holmgren E, Henrichson C, Jones TA, Uhlén M. Protein Eng 1987; 1: 107–113.
33. Ljungquist C, Jansson B, Moks T, Uhlén M. Eur J Biochem 1989; 186: 557–561.
34. Nilsson J, Nilsson P, Williams Y, Pettersson L, Uhlén M, Nygren PÅ. Eur J Biochem 1994; 224: 103–108.
35. Braun P, Hu Y, Shen B, Halleck A, Koundinya M, Harlow E, LaBaer J. Proc Natl Acad Sci U S A 2002; 99: 2654–2659.
36. Shih YP, Kung WM, Chen JC, Yeh CH, Wang AH, Wang TF. Protein Sci 2002; 11: 1714–1719.
37. Korf U, Kohl T, van der Zandt H, Zahn R, Schleeger S, Ueberle B, Wandschneider S, Bechtel S, Schnolzer M, Ottleben H, Wiemann S, Poustka A. Proteomics 2005; 5: 3571–3580.
38. Kapust RB, Waugh DS. Protein Sci 1999; 8: 1668–1674.
39. Nilsson B, Abrahmsén L. Methods Enzymol 1990; 185: 144–161.
40. Huth JR, Bewley CA, Jackson BM, Hinnebusch AG, Clore GM, Gronenborn AM. Protein Sci 1997; 6: 2359–2364.
41. Fox JD, Waugh DS. Methods Mol Biol 2003; 205: 99–117.
42. Davis GD, Elisee C, Newham DM, Harrison RG. Biotechnol Bioeng 1999; 65: 382–388.
43. LaVallie ER, Lu Z, Diblasio-Smith EA, Collins-Racie LA, McCoy JM. Methods Enzymol 2000; 326: 322–340.
44. Nallamsetty S, Kapust RB, Tozser J, Cherry S, Tropea JE, Copeland TD, Waugh DS. Protein Expr Purif 2004; 38: 108–115.
45. Donnelly MI, Zhou M, Millard CS, Clancy S, Stols L, Eschenfeldt WH, Collart FR, Joachimiak A. Protein Expr Purif 2006; 47: 446–454.
46. Butt TR, Edavettal SC, Hall JP, Mattern MR. Protein Expr Purif 2005; 43: 1–9.
47. Chatterjee DK, Esposito D. Protein Expr Purif 2006; 46: 122–129.
48. Hopp TP, Prickett KS, Price VL, Libby RT, March CJ, Cerretti DP, Urdal DL, Conlon PJ. Biotechnology (N Y) 1988; 6: 1204–1210.
49. Prickett KS, Amberg DC, Hopp TP. Biotechniques 1989; 7: 580–589.
50. Brizzard BL, Chubet RG, Vizard DL. Biotechniques 1994; 16: 730–735.
51. Adams SR, Campbell RE, Gross LA, Martin BR, Walkup GK, Yao Y, Llopis J, Tsien RY. J Am Chem Soc 2002; 124: 6063–6076.
52. Martin BR, Giepmans BN, Adams SR, Tsien RY. Nat Biotechnol 2005; 23: 1308–1314.
53. Karpeisky M, Senchenko VN, Dianova MV, Kanevsky V. FEBS Lett 1994; 339: 209–212.
54. Connelly PR, Varadarajan R, Sturtevant JM, Richards FM. Biochemistry 1990; 29: 6108–6114.

55. Kelemen BR, Klink TA, Behlke MA, Eubanks SR, Leland PA, Raines RT. Nucleic Acids Res 1999; 27: 3696–3701.
56. Kapust RB, Tozser J, Fox JD, Anderson DE, Cherry S, Copeland TD, Waugh DS. Protein Eng 2001; 14: 993–1000.
57. Blommel PG, Fox BG. Protein Expr Purif 2007; 55: 53–68.
58. van den Berg S, Löfdahl PÅ, Härd T, Berglund H. J Biotechnol 2006; 121: 291–298.
59. Walker PA, Leong LE, Ng PW, Tan SH, Waller S, Murphy D, Porter AG. Biotechnology (N Y) 1994; 12: 601–605.
60. Pedersen J, Lauritzen C, Madsen MT, Weis Dahl S. Protein Expr Purif 1999; 15: 389–400.
61. Evans TC Jr, Xu MQ. Biopolymers 1999; 51: 333–342.
62. Banki MR, Wood DW. Microb Cell Fact 2005; 4: 32.
63. Banki MR, Gerngross TU, Wood DW. Protein Sci 2005; 14: 1387–1395.
64. Wu WY, Mee C, Califano F, Banki R, Wood DW. Nat Protoc 2006; 1: 2257–2262.
65. Meyer DE, Trabbic-Carlson K, Chilkoti A. Biotechnol Prog 2001; 17: 720–728.
66. Kim Y, Dementieva I, Zhou M, Wu R, Lezondra L, Quartey P, Joachimiak G, Korolev O, Li H, Joachimiak A. J Struct Funct Genomics 2004; 5: 111–118.
67. Hochuli E, Bannwarth W, Döbeli H, Gentz R, Stüber D. Biotechnology (N Y) 1988; 6: 1321–1325.
68. Capon DJ, Chamow SM, Mordenti J, Marsters SA, Gregory T, Mitsuya H, Byrn RA, Lucas C, Wurm FM, Groopman JE, Broder S, Smith DH. Nature 1989; 337: 525–531.
69. Nygren PÅ, Flodby P, Andersson R, Wigzell H, Uhlén M. In: Chanock RM, Ginsberg HS, Brown F, Lerner RA, editors. Vaccines. Cold Spring Harbor (NY): Cold Spring Harbor Laboratory; 1991. pp. 363–368.
70. Sjölander A, Nygren PÅ, Ståhl S, Berzins K, Uhlén M, Perlmann P, Andersson R. J Immunol Methods 1997; 201: 115–123.
71. Libon C, Corvaia N, Haeuw JF, Nguyen TN, Ståhl S, Bonnefoy JY, Andreoni C. Vaccine 1999; 17: 406–414.
72. Agaton C, Galli J, Höidén Guthenberg I, Janzon L, Hansson M, Asplund A, Brundell E, Lindberg S, Ruthberg I, Wester K, Wurtz D, Höög C, Lundeberg J, Ståhl S, Pontén F, Uhlén M. Mol Cell Proteomics 2003; 2: 405–414.
73. Uhlén M, Pontén F. Mol Cell Proteomics 2005; 4: 384–393.
74. Hansson M, Ståhl S, Hjorth R, Uhlén M, Moks T. Biotechnology (N Y) 1994; 12: 285–288.
75. Sjölander A, Ståhl S, Perlmann P. Immunomethods 1993; 2: 79–92.
76. Wikman M, Friedman M, Pinitkiatisakul S, Andersson C, Hemphill A, Lövgren-Bengtsson K, Lundén A, Ståhl S. Vaccine 2005; 23: 2331–2335.
77. Wikman M, Friedman M, Pinitkiatisakul S, Hemphill A, Lövgren-Bengtsson K, Lundén A, Ståhl S. Biotechnol Appl Biochem 2005; 41: 163–174.
78. Riedl P, Fissolo N, Reimann J, Schirmbeck R. Methods Mol Med 2006; 127: 41–53.
79. Lycke N. Cell Microbiol 2004; 6: 23–32.

13

BIOSEPARATION, MAGNETIC PARTICLE ADSORBENTS

URS ALEXANDER PEUKER
TU Bergakademie Freiberg, Institute for Mechanical Process Engineering and Mineral Processing, Freiberg, Germany
OWEN THOMAS
University of Birmingham, Biochemical Engineering, Birmingham, United Kingdom
TIMOTHY JOHN HOBLEY
Technical University of Denmark, Systems of Biology, Lyngby, Denmark
MATTHIAS FRANZREB
Karlsruhe Institute of Technology, Institute for Functional Interfaces, Eggenstein-Leopoldshafen, Germany
SONJA BERENSMEIER
Technische Universität München, Institute of Biochemical Engineering, Garching, Germany
MARIA SCHÄFER
TU Bergakademie Freiberg, Institute for Mechanical Process Engineering and Mineral Processing, Freiberg, Germany
BIRGIT HICKSTEIN
Clausthal University of Technology, Institute of Chemical Process Engineering, Clausthal-Zellerfeld, Germany

13.1 INTRODUCTION

The need for new approaches to improve the downstream processing (DSP) of biological products has been recognized for many years and has spawned a variety of new bioseparation unit operations. Many of these focus on reducing steps in the recovery part of the downstream process by combining product capture, intermediate purification, and concentration in one step, thus increasing process yield. Some of these, such as monolith adsorbents and expanded-bed adsorption (EBA) have been more or less successfully commercialized in recent times, while others such as aqueous two-phase extraction have never reached commercialization, despite showing considerable promise. Still others, such as magnetic particle-based processes fall into the category of upcoming new technologies for which the verdict of successful commercialization, or not, has yet to be given. Successful commercialization and wide-spread adoption of new technologies in the bioprocess industries demands that scientific proof of potential are available and that the equipment and consumables can be obtained on demand at the right price for beyond the foreseeable future.

Currently, the future of magnetic particle-based bioseparations looks promising. Scientific proofs of principle of bioprocessing at lab-scale using magnetic adsorbents have been reported in the literature for over 10 years (1–11), although only a very few examples of small pilot-scale studies have been reported (12). The most common examples involve a functionalized magnetic adsorbent which is mixed with the liquor to be processed thus, binding the product, then the adsorbent is captured by a magnetic separator, washed, and eluted in a process called *high gradient magnetic fishing* (HGMF) (10,11). However, other ways of using magnetic beads for bioseparations may also be possible. There exist many permutations and combinations of possible magnetic adsorbent types (e.g. nano sized,

Downstream Industrial Biotechnology: Recovery and Purification, First Edition. Edited by Michael C. Flickinger.

micron sized, porous, nonporous) and their use (e.g. with various high gradient magnetic separator types, magnetic centrifuges, magnetic extractive fluids) in order to deliver the productivity and purifying power needed.

Functional magnetic particles from the nano- to micron-sized range are currently applied in a range of diagnostic, analytical, and laboratory purposes, such as magnetic resonance imaging, cell sorting and others, and plasmid purification during molecular biology. In principle, the same basic concept of using a magnetic adsorbent functionalized with an active surface can be extended to solve a variety of large-scale DSP challenges.

13.1.1 Downstream Processing

DSP is a critical part of the manufacturing of biotech products encompassing all the steps after receipt of a feedstock to give the final formulated product with the purity, activity, and concentration needed. The downstream process is commonly referred to contribute between 50% and 80% of the product cost, and many share the belief that the bottleneck of biopharmaceutical production is located within the downstream process (13,14).

13.1.1.1 The Fundamental Problem. Even for the most simple bioproducts, such as industrial enzymes or single-cell protein, the downstream process is very complex and sophisticated and comprises at least four steps. Typically it is divided into four stages (14): product recovery, concentration, purification and polishing, and product formulation (Fig. 13.1a). The basics of these stages have been considered in detail elsewhere (15,16). The number of individual unit operations varies greatly depending on the feedstock and product. However, using monoclonal antibodies as an example, which are one of the hottest current biotech products, comparison of six different commercial downstream processes shows that between seven and nine single process steps are typically used (17). Product losses occur in each step, which can dramatically decrease the final yield. In a downstream process, losses of between 50% and more than 90% of the product can easily occur, even when the step yields are high (Fig. 13.1b). Furthermore, the demands being placed on DSP continue to increase. For example, monoclonal antibody concentrations of up to 5 g/L are expected in the fermentation broth in the future (18). The ability of the current platform technologies, which are based on chromatography with protein A, to cope is doubtful and the relationship between upstream and downstream costs will shift further to the disadvantage of the downstream side, not only for antibodies, but for a range of products. It is therefore clear that the demand for improved DSP of all types of biotech products has not decreased over the last years.

13.1.1.2 Optimization of Downstream Processing. Several approaches have been studied to optimize bioseparation processes. These range from improved

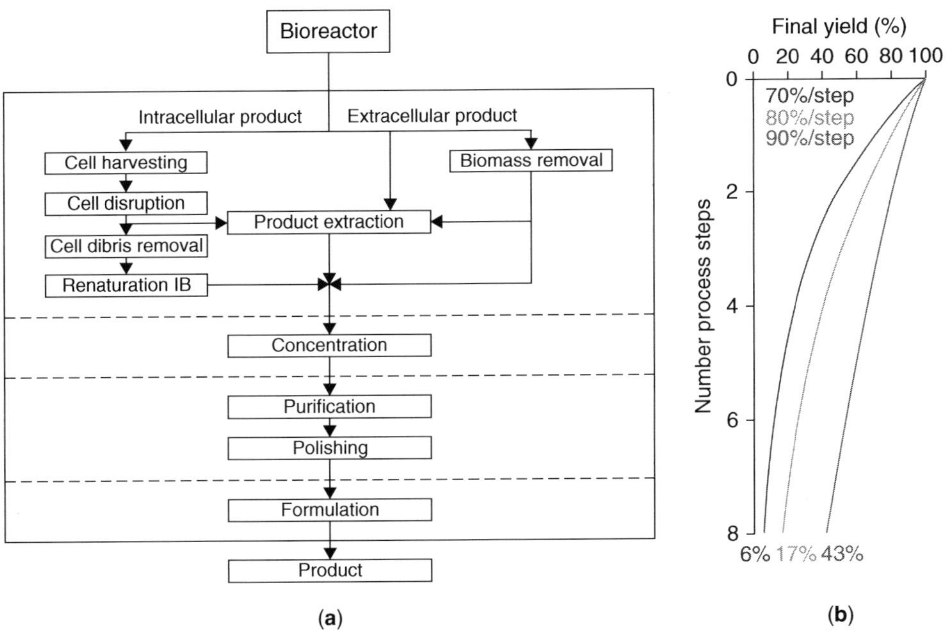

Figure 13.1. Generalized block diagram of DSP reduction of DSP process steps → increase in final DSP yield. (This figure is available in full color at http://onlinelibrary.wiley.com/book/10.1002/9780470054581.)

process development, to intensified use of computational resources for the modulation of separation steps, to innovative techniques on the downstream side. Still other options advocate integration of upstream and downstream for increased productivity, such as *in situ* product removal directly from the fermentation vessel (19). Nevertheless, most of these approaches are centered around downstream processes with chromatographic steps at their core and several additional other unit operations. Given that for the most part they focus on process optimization of existing techniques, they do not contribute to the leap in innovation that is needed to achieve a breakthrough in improved DSP.

From an engineering point of view, it is important to lower the number of process steps that are needed to come to the desired purity of the product without tremendous yield reductions. This can be achieved by integration of a number of single-unit operation steps in one step. Schügerl *et al.* (20) and Hubbuch *et al.* (13) have suggested focusing on the front end of the downstream process, such that clarification of the product from a crude medium, its concentration, and its purification are conducted in one single-unit operation. There are a number of potential ways to do this, such as by adsorptive or extractive methods as well as by crystallization. According to Schügerl *et al.* (20), extraction is preferred while purifying primary and secondary metabolites of low molecular weights. The most promising techniques developed in recent times for the recovery of higher molecular weight products such as proteins are adsorptive techniques, in particular EBA (21) and HGMF. Both of these competing techniques are able to combine solid–liquid separation from crude liquids with adsorptive purification and concentration. EBA has been commercialized, although with varying success. The technique has been used for purification of a very wide variety of proteins and has been reviewed in a number of recent articles (22,23).

13.1.1.3 High Gradient Magnetic Fishing for Integration of Unit Operation Tasks. HGMF is an integrated process for purifying proteins by using magnetic adsorbents combined with high gradient magnetic separation (10,11,24,25). The magnetic adsorbents contain ligands attached to the surface allowing capture of the target molecule when mixed with the feedstock to be processed. A magnetic force is then used which acts selectively on the magnetic beads allowing separation from a crude suspension, washing and elution of the bound target molecule in a semicontinuous way (Fig. 13.2). With the selection of appropriate magnetic beads with ligands chosen for the separation task, selective product adsorption, concentration, and purification is possible. In this process, it is necessary to be able to redisperse the adsorbents after magnetic capture, and thus they must be superparamagnetic, that is, they are highly magnetizable but retain no magnetic properties when removed from an external field. The current generation of magnetic separators impose the limitation of semicontinuous rather than continuous processing. Since the adsorbents are mixed with the solution to be processed, and the separators are compatible with particulates, such as whole cells, aggregates, and

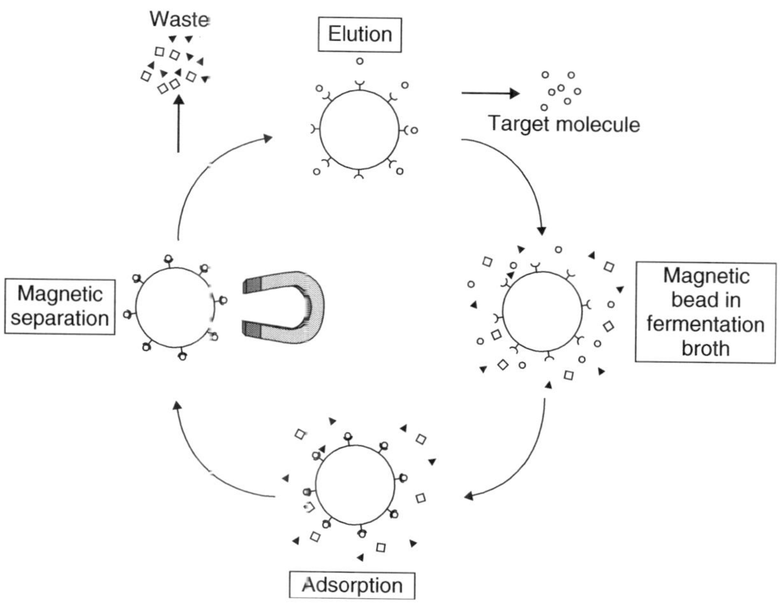

Figure 13.2. Principle of high gradient magnetic fishing with magnetic beads as magnetic adsorbents. (This figure is available in full color at http://onlinelibrary.wiley.com/book/10.1002/9780470054581.)

precipitates, clarification prior to use is not necessary. It is thus possible to integrate the unit operations typically used for clarification, concentration, and purification in one single-process step. Many proofs-of-principle examples have been published using affinity and ion-exchange magnetic adsorbents, such as the recovery of trypsin from crude porcine pancreatin (9), lectins from legume extracts (10), lactoferrin, lactoperoxidase, and superoxide dismustase from whey (10,11,26).

13.1.2 Magnetic Beads

13.1.2.1 Principles of Magnetic Bead Technology. Magnetic adsorbents (magnetic beads) use the same adsorption principles and ligands as conventional chromatographic resins. This is convenient since it allows transfer of ligand systems to magnetic adsorbents, which have been identified as promising during chromatography method development. Furthermore, the use of established chromatographic ligands on magnetic adsorbents may be expected to ease acceptance of such a radical new method. However, before HGMF can be accepted as a unit operation worthy of consideration during downstream process development a number of key technological consumable and equipment milestones must be met:

- Magnetic beads have to be commercially available at acceptable prices with various functionalities for appropriate separation tasks.
- Large quantities of magnetic beads have to be delivered.
- The process equipment for HGMF has to be available.

The majority of the research on use of magnetic beads for DSP has focused on micron-sized, nonporous adsorbents composed of encapsulated nano-particles, so called nano-composites. Hubbuch and Thomas (9) detailed cost-effective and easily scaled manufacturing methods for delivering such adsorbents for bioprocess scale applications. The most important features of these materials are: (i), the average particle size (>0.5 μm) and high magnetization ($M_s > 35$ Am^2 kg^{21}) encourage the use of HGMSs of only moderate field strength and, in turn, low cost, (ii), the particle size distribution is relatively narrow; as the magnetic velocity, u_{mf}, of a particle depends on the square of its diameter D_p [Eq. 13.1 (ERGUN-equation)]. The minimum fluid velocity required for fluidization, u_{mf}, can be estimated by assuming that the pressure drop along the bed is equal to the weight of the particles, reduced by the buoyancy, and divided by the bed cross-sectional area. This narrow size distribution will impact beneficially on the predictability of collection by HGMS, (iii), crucially, the adsorbents are superparamagnetic, (iv), the method of manufacture generates a highly irregular, nonporous, binding surface of exceptionally high area (>110 m^2/g), yielding maximum protein binding capacities greater than 100 mg/g of adsorbent, regardless of the *ligand-target* system, (v), their nonporous surfaces support easy cleaning, very fast adsorption/desorption kinetics and good utilization of immobilized ligands; and (vi), the ultra-thin surface polymer coating covering the magnetic core is easily derivatized to high ligand density, yielding efficient batch adsorbent materials with high affinity for their respective target molecule and very low levels of nonspecific binding.

$$u_{mf} = \frac{(\rho_s - \rho_f) \cdot g \cdot D_p^2 \cdot \epsilon_b^3}{150 \cdot pf \cdot \mu \cdot (1 - \epsilon_b)} \quad (13.1)$$

where ε_b is the bed voidage, μ is the viscosity of the fluid and ρ is the density of the fluid (f) and the solid phase (s).

However, such beads are not necessarily the most appropriate for DSP, particularly if other methods of handling them other than high gradient magnetic separation are to be used. It is thus of interest to consider how magnetic adsorbents have been prepared and used to date.

13.1.2.2 History of Magnetic Separation and Magnetic Beads. The first reports of magnetic immobilization and separation of biomolecules were described in the early 1970s in the group of Dunnill and Lilly (27,28). These workers examined the separation of L-asparaginase and β-galactosidase from crude *Escherichia coli* homogenate with magnetic bioaffinity materials. The potential of magnetic adsorbents quickly became apparent and a number of other research groups also examined the use of magnetic adsorbents for bioseparation purposes. For example, by the end of the 1970s Mosbach *et al*. reported the preparation of magnetic polymers for application in affinity chromatography (29). Even by the beginning of the 1980s, a diverse range of applications of magnetic separations in biotechnology had been reported such as cell fractionation, enzyme immobilization, magnetic affinity chromatography, immunoassays, and the extraction of impurities and were reviewed by Whitesides *et al*. (30). Within the following decades until now, numerous publications have followed covering a large variety of potential applications of magnetic adsorbents. Uhlen was one of the first, who reported the application of magnetic separation for DNA, especially by using the noncovalent biotin–streptavidin interaction (31). Further, diverse applications followed such as the specific separation of cells (32), isolation of plasma membranes for plasma membrane proteome research (33), applications in molecular biology for DNA hybridization (34), or the usage of appropriate magnetic particles for removal of heavy metals (35), lanthanides, and actinides (36) from waste water. Tartaj *et al*. (37) reviewed how superparamagnetic nanoparticles are used for *in vivo* biomedical applications such as for hyperthermia, as contrast agents for the enhancement of nuclear magnetic resonance imaging and as magnetic drug delivery systems. In combination with fluorescent dyes, quantum dots, or tagged bioaffinity biomolecules like streptavidin or antibodies, magnetic particles have been manufactured for application in life sciences and for immunoassays (38,39). Most of the reports on the application of magnetic adsorbents for the separation of proteins that have appeared in the last 10 years have used micron-sized beads and high gradient magnetic separation. An overview concerning automatic magnetic separator devices suitable for bioprocessing was given recently by Berensmeier (40). The paper shows the high research potential in the field of magnetic separation. There is an increasing number of publications on materials with magnetic- and specific-binding properties. A better selectivity and a higher degree of automation will allow the analysis of higher sample volumes at shorter time. But at the moment it is not possible to scale up these processes to large quantities.

The commercial potential of magnetic beads has not gone unnoticed and today more than 50 different magnetic beads are commercially available from a number of companies such as Bioclone Inc with BcMag® magnetic beads (41), Roche offering MagNa® Kits (42), Invitrogen with Dynabeads® (43), BD Biosciences with BD IMag® Sets and Reagents, Chemagen with chemagic® Kits and Micromod Partikeltechnologie with nanomags®. The commercially available magnetic beads are mainly offered for the isolation and purification of nucleic acids, and for protein and cell separation purposes in the laboratory. They are often supplied in kits for a specific purpose such as analytical, diagnostic, and mini preparative applications and are used in combination with proprietary magnetic separation tools or automated systems exclusively in the laboratory scale. It is therefore not possible to obtain these commercial beads in large quantities and their prices are also prohibitive for pilot- or production-scale use.

13.1.3 Economic Constraints on Magnetic Adsorbents

High prices can be demanded for commercial magnetic beads used in medical, diagnostic, and laboratory applications, since the amounts required are typically very small. However, in pilot- and large-scale bioseparation processes, the current prices of magnetic beads are prohibitive. To approximate what they may cost for industrial use some simple calculations can be made.

The costs of conventional downstream process lines and their active materials can be estimated as following: in general, one can assume that about 5–10% of the price of a drug that is commercially available account for the production costs (44,45). Half of these costs are caused by upstream, half by DSP. As a rule of the thumb, 33% of downstream costs are needed for the active material which is responsible for the product capturing and purification. Table 13.1 lists the calculation of the costs for commercial beads.

TABLE 13.1. Results of Calculations of Acceptable Costs for Magnetic Beads for an Application in Industrial Downstream Processes

Process Item	Calculation Basis	Cost
Commercial beads/kg	20 ml with 5 mg/ml	402,29 $
Commercial beads/kg	$4,05 \times 10^6$ \$/kg	
Production costs	5% of selling price	201 000 $/kg
Downstream costs	50% of production costs	100,500 $/kg
Material costs in DSP	33% of DSP costs	33 000 $/kg
Acceptable costs for DSP material	33 $/g	

That means that magnetic beads that recover 1 g product are not allowed to exceed costs above 32 $/g. Assuming an average capacity of 100 mg product per gram magnetic bead, we end up at a price of 3 $/g magnetic bead—provided that the magnetic beads are reusable like commercial adsorbent materials are. In comparison to magnetic beads that are nowadays available, the need in reducing magnetic bead syntheses and material costs for an application in industrial bioseparation becomes obvious. Commercially available beads are with a factor of 1000 or higher, too expensive for an application of this technique in industrial DSP. To overcome these challenges, engineering efforts have to focus on a process scheme that is able to manufacture inexpensive magnetic beads in large scale.

13.1.4 Synthesis Procedures of Magnetic Beads

There is a large variety of magnetic beads, the sizes of which vary with regard to their application and synthesis procedures from less than 200 nm (e.g. used for intravenous injection) up to 100 μm for crude separation applications (46). Despite the large number of manufacturing protocols, the synthesis procedures can be subdivided with regard to the general structure of the magnetic bead. The possible basic structures of magnetic beads are shown in Fig. 13.3. According to this, magnetic beads are available (a) as polymer matrices with attached magnetic particles, (b) as beads with a single core of magnetic material covered by a polymer matrix, (c) as several magnetic particles being embedded in a polymer matrix, or (d) as several magnetic particles deposited inside the pores of a polymer matrix.

To synthesize the structures described, several methods and materials have been used in the literature. Table 13.2 contains an overview of the procedures, matrices, and magnetic materials that have been used commercially as well as in scientific publications. The data is ordered

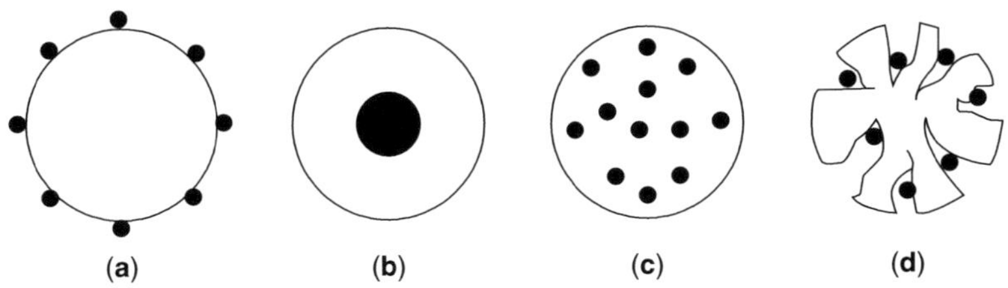

Figure 13.3. Structures of magnetic beads adopted from Yuan *et al.* (46). The black dots represent the magnetic material, the matrix material is colored in white. (a) Magnetic beads with tagged magnetic particles on the surface of a polymer matrix, (b) magnetic core within a polymer matrix, (c) magnetic particles embedded into a polymer matrix, (d) magnetic particles deposited within the pores of a polymer matrix.

according to the general structure of the particles due to the models a–d as described in Fig. 13.3. In almost every example commercial or self-made, Fe_3O_4 is used as the magnetic material. The only exception is the magnetic beads presented by Martin *et al.* (47) who used commercial stainless steel microspheres. In most cases, Fe_3O_4 is prepared by an alkaline precipitation from a solution of Fe(II) and Fe(III) salts. Tong *et al.* used an oxidation precipitation with H_2O_2 as the precipitating agent (48). The Fe_3O_4 is used as synthesized or it is stabilized with amphiphilic surfactants like oleic acid.

Table 13.3 shows that structure "a," where the magnetic material is attached on the outer surface of a matrix particle, is not represented. This suggests that such beads are unsuitable, probably due to the risk of dislodging the magnetic material but also due to the low amount of magnetic material in such beads (and thus a low magnetic velocity). Similarly, type "d" is essentially not represented, most likely for the same reasons. The only exception is the commercial Dynabead® which is a mixture between structure "c" and "d." During the precipitation process of Fe_3O_4, the material resembles structure "d" whereas in the final magnetic bead the magnetic material is distributed and fixed like in structure "c." The integration of Fe_3O_4 in magnetic beads is thus mainly done in two ways: either it is coated with polymers or it is applied in a polymerization procedure together with diverse monomers.

Type "c" beads mainly derive from a polymerization process, such as by suspension, dispersion, emulsion, miniemulsion, or a specific spraying suspension polymerization (as reported by Yang *et al.* (49)), which are started by radicals. When considering Table 13.4 it becomes evident that with polymerization procedures structure "c" is created whereas structure "b" is formed while coating Fe_3O_4 particles or aggregates. Furthermore, beads being synthesized with polymerization procedures often have sizes in the micrometer scale whereas beads originating from the coating of Fe_3O_4 have sizes in the nanometer scale. As matrices several polymers have been used like polystyrene (PS), polyvinyl alcohol (PVA), polyamide (PA), polyglycidyl methacrylate (PGMA), polymethyl methacrylate (PMMA) and polysaccharides. The functionalizations are as wide as the applications of magnetic beads: simple ion-exchange groups such as sulphuric acid residues, carboxylic acid, or amino groups are used as well as affinity ligands such as protein A and G, streptavidin and biotin, oligo dT, monoclonal antibodies, or cibacron blue. Some research groups use commercial magnetic beads and modify them for very complex applications, for example, Hoshino *et al.* for the separation of blood cells (50). A more detailed overview about the different polymerization methods for the synthesis of magnetic beads is given by Ma *et al.* (51).

The large majority of magnetic beads to date are either polymer coated magnetic cores or derived by polymerization in the presence of magnetic particles. Beads formed by suspension polymerization have sizes in the range of several hundred micrometers with broad size distributions. Emulsion and miniemulsion polymerization produce smaller particles with narrow-size distributions. As their particle sizes are typically within the nanometer scale, the resulting magnetic saturation is often small. Particles synthesized by dispersion polymerizations are characterized by large sizes, wide size distributions, and low magnetic contents. Several of the magnetic beads produced for lab-scale applications do not entirely fit to all of the specifications for industrial application such as: moderate size (e.g. 1–2 microns) of the beads to provide a sufficient capacity, high magnetite content to provide high magnetic forces during separation acceptable roughness/inner porosity to allow cleaning and reuse.

13.2 SELECTED SCALABLE SYNTHESIS PROCEDURES

13.2.1 Magnetic Beads Synthesis by Emulsion-Polymerization

Large-scale emulsion polymerization has been practiced for many years for production of a wide variety of porous

TABLE 13.2. Summary of Commercial Magnetic Beads in Literature

General Structure	Magnetic Material	Synthesis Magnetic Material	Matrix	Synthesis Magnetic Bead	Functionalization	Bead Size	Application	Reference
b		Commercial stainless steel microspheres	PS-DVB	Coating of MS with PS-DVB in suspension polymerisation	Sulfuric group (CEX)	50–70 μm	As chromatographic resin	Martin *et al.* (47)
b		Commercial thermoresponsive magnetic NP with streptavidin (Magnabead Inc. Chiba, Japan)		Coupling of biotinylated mAbs to NP	mAbs	NP: 170 nm aggregates: 1 470 nm	Cell separation: purification of neutrophils from a mixture of inflammatory cells	Hoshino *et al.* (50)
b	Fe_3O_4	Oxidation-PC from Fe(II), PVA + H_2O_2/NaOH	PVA-GA	Coating of Fe_3O_4 core by crosslinking PVA with GA	CB coupling	Fe_3O_4 core: 20 nm; 10 μm after crosslinking	Protein adsorption (LYZ from pure solution and ADH from yeast homogenate)	Tong *et al.* (48)
c		Commercial Fe_3O_4	PA	Suspension polymerization of PA + Fe_3O_4	Soybean Trypsin Inhibitor (STI)	60–600 μm	MSFB chromatography, separation of trypsin + chymotrypsin	Cocker *et al.* (78)
c	Fe_3O_4	PC from Fe(II) + Fe(III) in alkaline medium	PGMA	Dispersion polymerization of GMA and Fe_3O_4	–	100 nm—2 μm	Study focuses on variation of polymerization parameter	Horak *et al.* (79)
c		Commercial Fe_3O_4	PVB	Solvent evaporation method of emulsion	CB coupling	100–300 μm	Removal of heavy metals from aqueous solution [Cu(II), Cd(II), Pb(II)]	Denizli *et al.* (80)
c	Fe_3O_4	PC from Fe(II) + Fe(III); stabilized with OA	P (MMA-DVB-GMA)	Spraying suspension polymerization	$-NH_2$ groups	ca. 10 μm	Protein ad/desorption from pure solutions (BSA)	Yang *et al.* (49)
c	Fe_3O_4	PC from Fe(II) + Fe(III); stabilized with OA	PS	Miniemulsion polymerization	–	80 nm	–	Ramirez *et al.* (81)
c	Fe_3O_4	PC from Fe(II) + Fe(III); stabilized with OA	PVB, PVA, PMMA, PVAc	Spraying suspension polymerization	$-N(CH_3)_3^+$(AEX), $-SO_3^-$(CEX), IDA-Cu^{2+}(IMA)	ca. 5 μm	Protein adsorption (LYZ or BSA from cultivation suspensions)	Käppler *et al.* (24)

(*continued*)

TABLE 13.2. (*Continued*)

General Structure	Magnetic Material	Synthesis Magnetic Material	Matrix	Synthesis Magnetic Bead	Functionalization	Bead Size	Application	Reference
				Commercial Magnetic Beads				
b	Fe_3O_4	PC from Fe(II) + Fe(III)	PS: dextran, starch, chitosan	Coating of primary Fe_3O_4 aggregates with PS; crosslinking with silica structures	Carboxylic acid groups and PEG	50–200 nm (irregular shape)	DNA purification, protein detection, separation and purification; retrovirus detection	Nanomag® Micromod (83)
c		Commercial Fe_3O_4 is preferred (products like Bayferrox)	PVA-Silan	Emulsion polymerization of Fe_3O_4, PVA, crosslinker	Protein A, G, L, Ab, SAV, BT, etc.	0,5–10 μm	Isolation of nucleic acids	Chemagen beads (82)
c/d	Fe_3O_4	*In-situ* PC from Fe(II) + Fe(III) in polymer MS	PS	Emulsion/dispersion polymerization; followed by swelling of MS in OS for *in-situ* PC	Protein A, G, Ab, SAV, BT, -COOH, NH_2, Oligo dT etc.	1–5 μm	Isolation, purification and separation of cells, proteins, organelles and nucleic acids	Dynabeads® Dynal (57)

Note: Ab, antibodies; ADH, alcohol dehydrogenase; BSA, bovine serum albumin; BT, biotin; CB, Cibacron Blue 3GA; Cc, cytochrome c; CEX, cation exchanger; CT, chymotrypsinogen; DVB, divinylbenzene; GA, glutaraldehyde; GMA, glycidylmethacrylate; IEX, ion exchanger; LYZ, lysozyme; PA, polyacrylamide; PL, phospholipid layer; mAb, monoclonal antibody; MMA, methylmethacrylate; MS, microspheres; MSFB, magnetically stabilized fluidized bed; MYO, myoglobin; NP, nanoparticles; OA, oleic acid; OS, organic solvent; P, poly; PC, precipitation; PSa, polysaccharide; PVA, polyvinyl alcohol; RA, ribonuclease A; SAV, streptavidin; STI, soybean trypsin inhibitor.

TABLE 13.3. Measured Ion-Exchange Activities of PVB and PMMA-beads with Different CEX Contents

	Activity with Different Matrix Polymer (meq/g)		
Content CEX (%)	PVB	PMMA	Content Polymer (%)
10	1.01	0.77	70
20	1.45	1.56	60
30	2.03	1.488	50
40	2.34	2.3	40
50	2.59	2.82	30
100	4.58	0	

The Magnetic Beads Contain a Fixed Magnetite Content of 20 %

TABLE 13.4. Maximal Capacities q_m and Dissociation Constant K_d of Adsorption Isotherms of Fig. 13.12 (*related to the 40 wt% mass fraction of CEX within the PVB beads)

q_m(mg/g)	K_d(mg/L)	Desorption (%)	
Beads (LYZ/beads)	77.6	47.864	83.8
Beads (LYZ/CEX)*	193.9	47.864	–
CEX(LYZ/CEX)	183.3	31.06	68.5

From ref. (78–83). Adsorption either with PVB Beads (40/20/40 CEX) or CEX at pH 7, desorption pH 12.5

particles and adsorbents, especially for chromatography. It has also been used for small commercial production runs (e.g. Chemagic beads from Chemagen). Typically, a ferrous fluid (magnetic particle sizes of 10–200 nm) is dispersed together with an aqueous solution of PVA ($M_n \approx$ 23000–22400) in a common vegetable oil (viscosity of 130–190 mPas at 20°C) (see Fig. 13.4) (52). The emulsion is stabilized by several surfactants and the concentration of the emulsifiers in the oil phase is usually 2–6% (V/V). Because the magnetic colloid is suspended in the organic phase, the magnetic particles become encapsulated by the polymer. To guarantee a high porosity and good stability, the beads are cross-linked. In most cases glutaraldehyde is used together with an acid catalyst since the reaction is rapid. The formation of the magnetic beads is typically completed within 10–20 min (53). Afterwards, the beads are washed with n-hexan, 2-butanone, and deionized water. The ligands such as ion-exchange groups [carboxyl- and Diethylaminoethyl-(DEAE)-groups] are introduced by a catalyzed grafting technique (52,54). There is a variety of activation and modification possibilities for these beads, allowing application to capture of a wide variety of products. The hydrophilic PVA base is suitable for an adsorbent to be used for protein capture, but due to the limited reactivity makes derivatisation more difficult (52–54). The beads produced are typically large, (at least 2–3 microns Bergemann *et al*. (53) and often much larger Müller-Schulte *et al*. (52)) and rely on their porous nature for achieving acceptable binding capacities. However, this in turn limits the magnetic content and thus magnetic susceptibility of the adsorbent. These characteristics would be considered a limitation in a HGMF type process for DSP, demanding a modification of the manufacturing procedures to give suitable adsorbents. However, Chemagic adsorbents are typically used for diagnostic and analytical applications. For example, the magnetic particles provide a basis for immunoassays and they are used for cell separation. In these cases, antibodies against relevant antigens are chemically coupled to the Chemagic beads (53,54).

13.2.2 Magnetic Beads Synthesis by the Method of Activated Swelling

Particle seeds are first produced via an emulsion polymerization process (55) (Fig. 13.5). As an example a mixture of methyl methacrylate, glycidyl methacrylate, and ethyleneglycol dimethacrylate is dispersed in water. After rapid stirring the polymerization starter [e.g. $(NH_4)_2S_2O_8$] is added and at 65°C the polyreaction is carried out over several hours (56). The volume of these particles is approximately 1–2% of the final particles. Subsequently, the seeds are swollen in the presence of water or a mixture of water and organic solvent. A further polymerization is then started in the swollen particles by addition of an oil-soluble initiator, which delivers highly mono-dispersed stable macroporous polymer particles. The polymer used results in covalently coupled oxidative groups ($-NO_2$ or $-ONO_2$ from Nitrous Acid or Nitric Acid) that are distributed throughout the particles as iron-binding groups (55,56). Introduction of magnetic properties is accomplished by dispersing the particles in a solution of ferrous salts. The ferrous ions are transported into the particles and there the precipitation of magnetic iron oxy-hydroxy compounds takes place. After heating, grains of magnetite or maghemite are formed which are of such small size that they are superparamagnetic. An example of commercial adsorbents produced in this way are DynaBeads® from Dynal (57). These are typically marketed for use in diagnostic and analytical kits, for example, in biological separation, peptide, and protein binding.

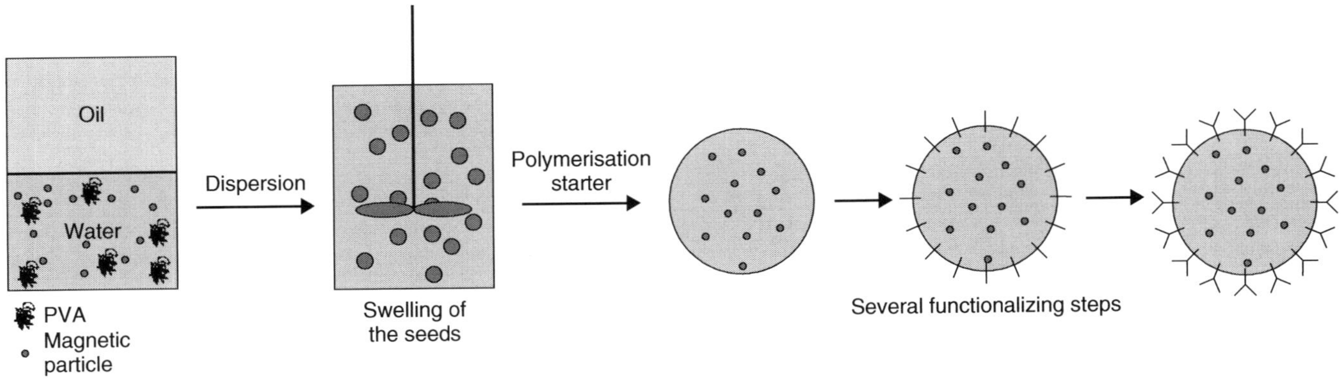

Figure 13.4. Process chain emulsion polymerization phase. (This figure is available in full color at http://onlinelibrary.wiley.com/book/10.1002/9780470054581.)

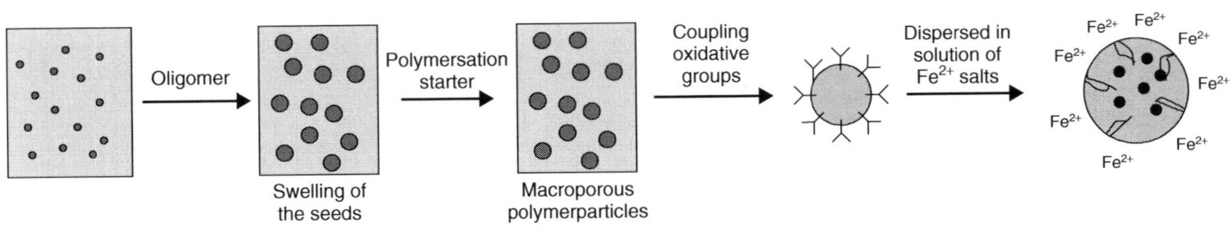

Figure 13.5. Method of activated swelling. (This figure is available in full color at http://onlinelibrary.wiley.com/book/10.1002/9780470054581.)

13.2.3 Magnetic Beads Synthesis by Solution Process (SOLPRO)

One of the manufacturing methods that may be most amenable to cheap large-scale production of finished magnetic adsorbents is the so-called solution process. This is because the three different single characteristics required, namely the superparamagnetism, the functionality, and the polymeric structure are delivered by individual components and then combined in a simple spray drying process. An example of the modular process scheme is presented in Fig. 13.6 (58). Each component is synthesized individually in a scalable process. The resulting composite magnetic beads (1–5 μm) consist of nanosized-superparamagnetic particles (10–20 nm), and nanoscale functional polymer particles (about 200 nm), which are embedded into the polymer matrix. The nanosized-functional polymer particles capture the target molecules which diffuse into the adsorbent.

The magnetite nano particles used are synthesized by a precipitation process and are stabilized in an organic solvent. The functional polymer particles are produced via emulsion polymerization with the following functionalization step (59–61). The matrix polymer is dissolved in an appropriate solvent, together with the magnetite and polymer nanoparticles. The resultant suspension is spray dried during which time the magnetite particles and the functional polymer particles are embedded into the polymeric matrix (24,58,59,61). The ratios of the three components can easily be varied, however, a typical composition is: 20% magnetite, 40–50% functionalized polymer-nano particles, and the remaining percentage consists of the matrix polymer. Furthermore, it is simple to vary the types of functionalized particles incorporated as well as the type of polymer used for making the final composite, permitting great flexibility in the type of adsorbents produced. For example, anion-exchanger (AEX) or cation-exchanger (CEX) particles with a diameter of about 200 nm can be incorporated in different matrix polymers, such as PMMA, polyvinyl butyral (PVB), PVA, and polyvinyl acetate (PVAc) (24,58,59).

13.2.4 Superparamagnetic Properties

One of the most important aspects of magnetic adsorbents to be used in bioprocessing is their superparamagnetic properties. For the adsorbent to be superparamagnetic, it needs to contain magnetic particles that are below a critical particle size. According to Lu *et al*., (62) the critical diameter for a spherical single domain magnetite particle (Fe_3O_4) is 128 nm, but if the single domain exceeds a certain particle size of about 15 nm time constants of the relaxation becomes much too long to allow demagnetization within sufficiently short process times in the seconds range. Therefore, the specification for

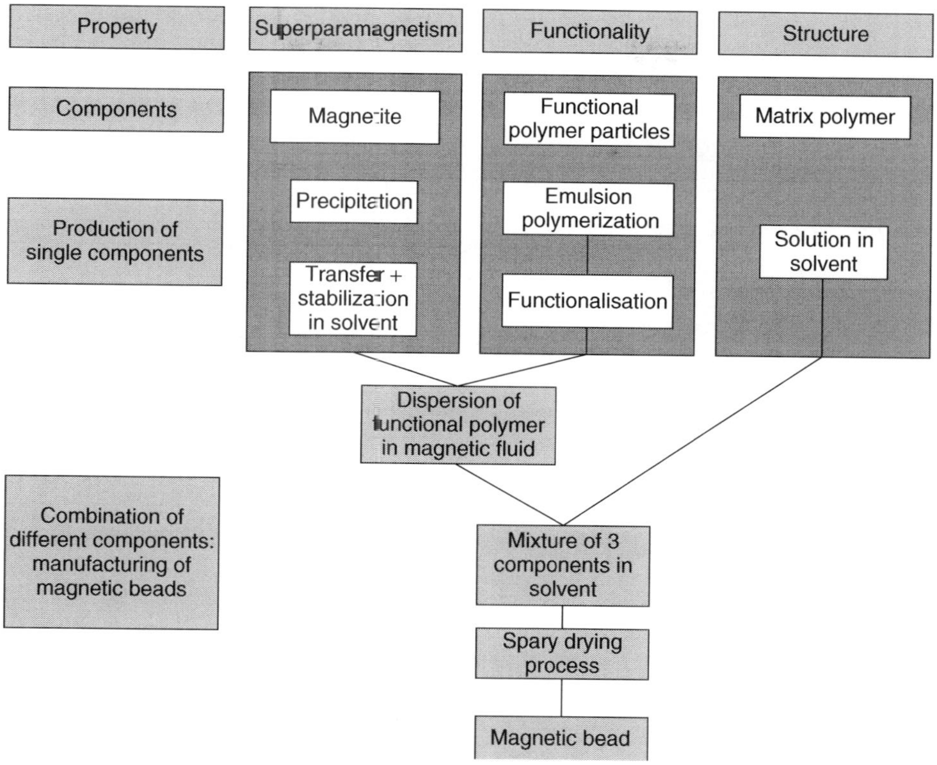

Figure 13.6. Process scheme of the synthesis of magnetic beads.

the incorporated magnetite nano-particles within most of the magnetic beads is a particle size of 10–15 nm.

These particles then consist of one single Weiss domain of superparamagnetic material. In a magnetic field the small single domains align, creating magnetic properties, however, when the external magnetic field is removed, the single domains lose their orientation immediately and the beads are no longer magnetic. If there are more than one magnetic domain in a single particle, they will remain aligned when removed from the magnetic field and have remnant magnetic properties. TEM pictures of the magnetite particles embedded in a polymer matrix from adsorbents produced by the SOLPRO process are shown in Fig. 13.7. The primary particle sizes that can be observed are between 10 and 15 nm, and thus it can be assumed that the adsorbent as a whole is superparamagnetic. The magnetization and the superparamagnetic properties of particles can be easily confirmed and examples of magnetization curves for different beads produced via solution process and one commercial bead type are shown in Fig. 13.8. For all of the particles curves are seen that have no hysteresis as the magnetic flux density is varied, indicating that there is no magnetic remanence and the composite adsorbent is superparamagnetic. In all of the particles shown less than 0.2 T is needed to elicit the maximum magnetization. The magnetic saturation depends, among other things, on the amount of superparamagnetic particles embedded in the polymer matrix.

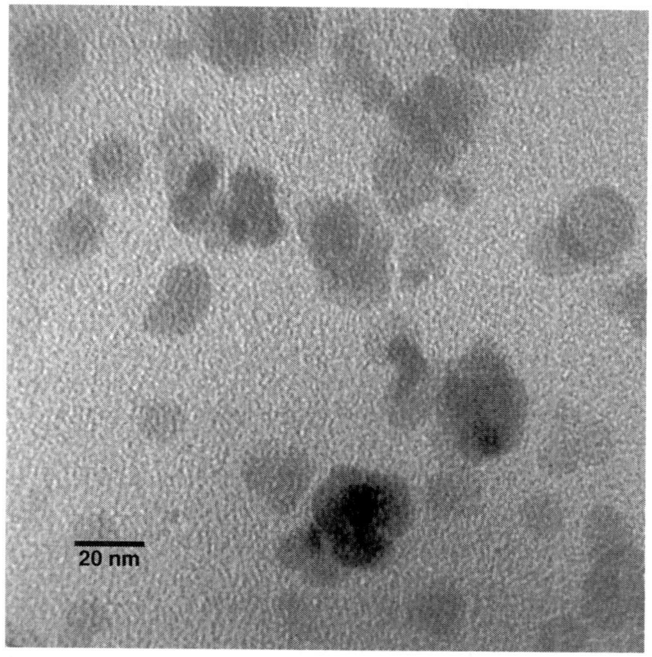

Figure 13.7. TEM of magnetite embedded in polymer matrix by the solution process (PMMA matrix with 30 wt% magnetite and 0.7 g fatty acid per gram magnetite).

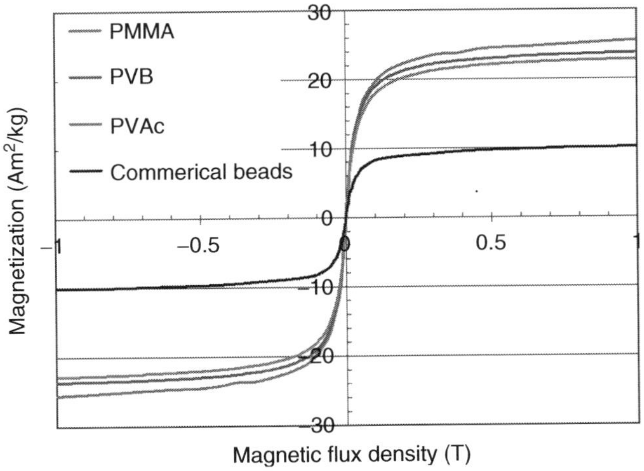

Figure 13.8. Magnetization curves of different bead types produced via solution process. (This figure is available in full color at http://onlinelibrary.wiley.com/book/10.1002/9780470054581.)

In Fig. 13.9 SEM pictures of several beads are shown. The particle size of magnetic beads differs by the different synthesis processes. The commercial beads are nearly monosized around 1 μm. The magnetic beads produced by spray drying process (SOLPRO) show a median particle size of about 2 μm. Furthermore, the ragged surface of these beads is clearly visible (Fig. 13.9). This structure increases the specific surface area of the bead compared to a smooth surface by a factor of 3 and with it the protein binding capacity per gram of beads.

13.3 MAGNETIC ADSORBENTS FOR LABORATORY SEPARATIONS

13.3.1 Screening and use of SOLPRO magnetic beads

The capacity of a magnetic adsorbent is the main property for any bioseparation. It depends on the ligand density as well as on the accessibility of the ligands by the target molecule. However, given that the ligands used are typically one of the most expensive parts of a magnetic adsorbent it is important to optimize the levels of functionalization needed. For SOLPRO beads, the basic ligand is located on the functionalized nano-particles, which are incorporated within the polymeric composite adsorbent. As an example of adsorbent optimization, the effects of incorporating different quantities of CEX nanoparticles on the ion-exchange capacity of the produced magnetic beads have been investigated.

For the overall charge and selective capacity, the contribution of the matrix polymer has to be taken into consideration (Table 13.4). Matrix polymers with a high density of functional groups have a certain impact on the overall ion-exchange capacity. This may lead to an unspecific binding or a binding with different binding strength and with it to product losses during washing steps Therefore, the selection of the matrix polymer has to be carefully chosen.

Adsorbents synthesized with 40 wt% PVB, 40 wt% CEX particles, and 20 wt% magnetite were tested for their ability to capture and recover either β-galactosidase or lysozyme. A single batch adsorption step with 1g/L of adsorbents was conducted in a 10 mM KH_2PO_4/K_2HPO_4 buffer with either 0.1–1g/L β-galactosidase at pH 4 (elution at 0.01 M KH_2PO_4/K_2HPO_4 buffer pH 8) or 0.1–1g/L lysozyme at pH 7 (elution at 0.01 M KH_2PO_4/K_2HPO_4 buffer pH 12.5). Fig. 13.10a shows a very steep binding isotherm for lysozyme, which is typical of an ion-exchange adsorbent. The maximum binding capacity amounts to about 160mg/g_{CEX} (Fig. 13.10a) and above 60mg/g_{beads} (Fig. 13.10c). The relation between the two values demonstrates that all CEX particles within the beads for the solution process are active in adsorption. Both adsorption isotherms can be approximated by a Langmuir fit. This approximation relies on the following fundamental assumptions (63).

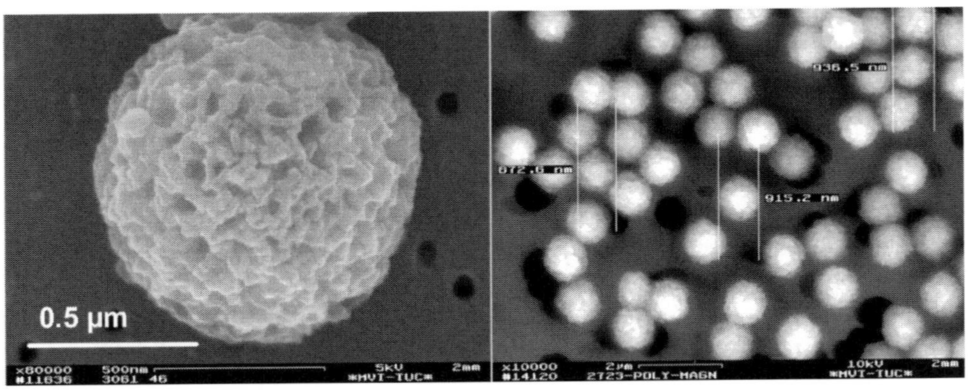

Figure 13.9. SEM from left to right: self-made PVB beads by solution process (40 wt% PVB, 20 wt% magnetite, 40 wt% functional particles), Dynabeads ©.

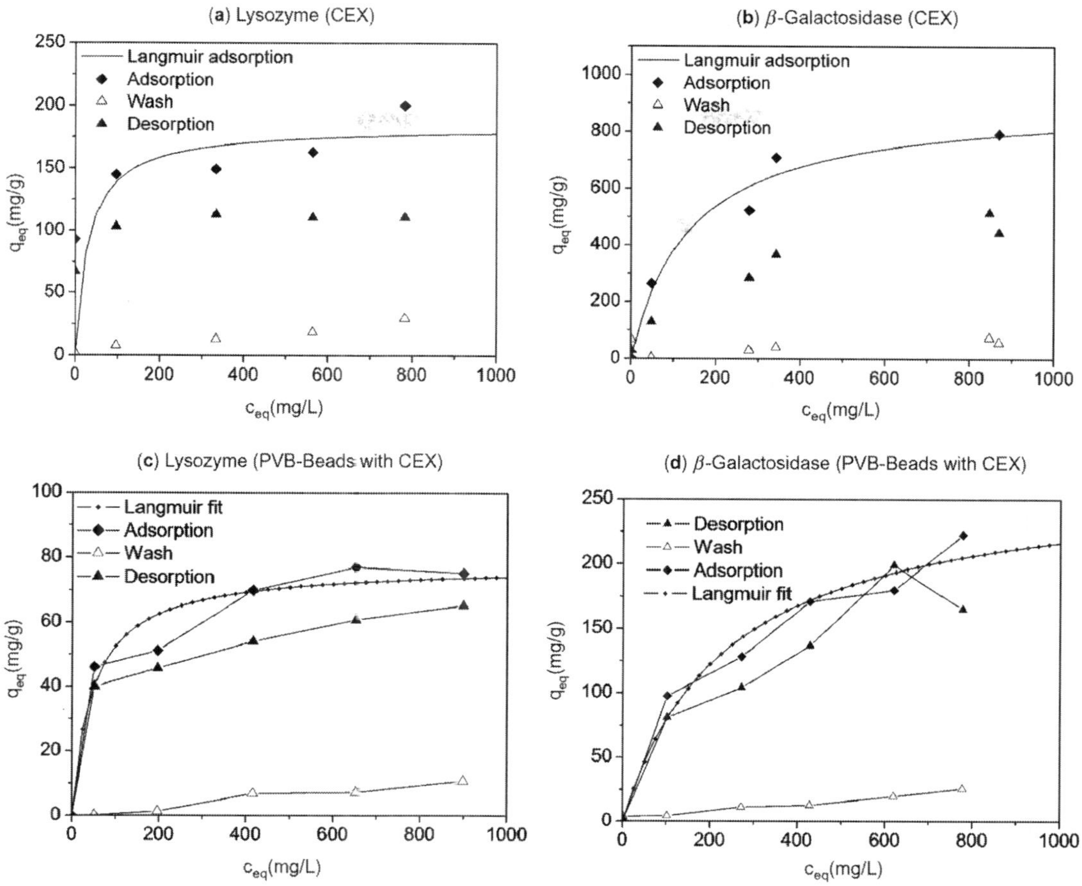

Figure 13.10. Langmuir isotherm and measured data of adsorption/desorption experiment a) of lysozyme with CEX. Adsorption: pH 7 (♦), washing pH 7 (△), desorption pH 12.5 (▲) b) of β-Galactosidase with CEX. Adsorption: pH 4 (♦), washing pH 4 (△), desorption pH 8 (▲) c) of lysozyme and PVB beads (40/20/40 with CEX). Adsorption (♦) pH 7; washing (△) pH 7, desorption (▲) pH 12.5. Langmuir parameter: q_m: 77.5 mg/g, K_d 44.864 mg/L, c(beads): 1mg/mL. and d) of β-galactosidase and PVB beads (40/20/40 with CEX). Adsorption (♦) pH 4; washing (△) pH 4, desorption (▲) pH 8. Langmuir parameter: q_m: 267.7mg/g, K_d 237.63 mg/L. c(beads): 1mg/mL.

1. Adsorption leads to the formation of a monolayer on the adsorbent surface.
2. Adsorption enthalpy is equal for every adsorption place and is independent from the adsorption ratio.
3. There is no interaction between adsorbed molecules.

The Langmuir equation correlates q_{eq}, which is the equilibrium capacity with c_{eq}, which is the equilibrium concentration using K_d, which is the dissociation constant and q_m the maximal capacity (Eq. 13.2). The latter are characteristic for the theoretical maximal adsorptive load.

$$q_{eq} = \frac{q_m \cdot c_{eq}}{K_d + c_{eq}} \tag{13.2}$$

The Langmuir equation is widely used to fit experimental data for adsorption of proteins. It is applied for instance by Xue *et al*. (64), Heeboll-Nielsen *et al*. (10), Hubbuch *et al*. (9), Tong *et al*. (48), Yang *et al*.(65), Peng *et al*. (66), Liao *et al*. (67) and Hoffmann (68).

Very little protein was released during washing with buffer and on average more than 75–80% of the bound protein could be recovered in one elution step. At optimal adsorption conditions, (pH 4) β-galactosidase showed similar adsorption properties as lysozyme. The initial gradient is less steep, which is due to the higher molecular weight of β-galactosidase, which occupies more surface area on the beads per molecule. With a corresponding procedure, it was possible to remove more than 90% of the adsorbed β-galactosidase from the beads (Fig. 13.10d), even though

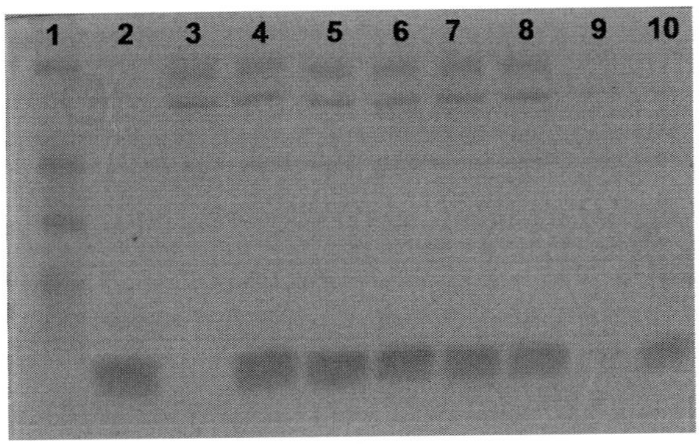

Lane	Sample
1	Marker
2	Lysozyme (LYZ)
3	β-galactosidase (β-Gal)
4 + 5	LYZ + β-Gal before adsorption (pH 8)
6	LYZ + β-Gal t = 0
7	LYZ + β-Gal t = 20 min
8	LYZ + β-Gal t = 60 min
9	Wash pH 8
10	Elution LYZ pH 12

Figure 13.11. Progress of β-galactosidase activity and concentration in the supernatant before adsorption ($t = 0$), after adsorption at pH 4 (adsorption), after washing with pH 4 (wash) and after desorption with pH 8 (desorption) with PVB beads (40/20/40 with CEX). c(β-galactosidase): 200 mg/L, c(beads): 1 mg/(mL), qeq(β-galactosidase): 35,6 mg/g.

the adsorption properties of the pure CEX particles showed a lower desorption rate (Fig. 13.10b).

Magnetic separations are believed to be a quick and gentle method and to confirm that enzymes are not inactivated by the procedures used, activity measurements using ONPG (o-Nitrophenyl-β-D-galactopyranoside) as substrate were made at each stage during β-galactosidase separation. The results in Fig. 13.11 show that the specific activity was increased in the elution fraction compared to the starting material used in the adsorption step indicating that purification had occurred. An activity balance suggests that more activity could be recovered during elution than expected, further confirming that purification from inhibiting compounds rather than enzyme damage had occurred.

The next step in screening the use of the adsorbents is to test them for the separation of lysozyme and β-galactosidase from each other. The adsorbents have cation-exchange functionality and can thus be expected to bind both proteins at pH values below their isoelectric points (pI for lysozyme is 9-10 and the pI for β-galactosidase is 3–4). However, the difference in isoelectric point can be exploited for selective washing/elution steps. The magnetic adsorbents were added (final concentration 4g/L) to a solution of 500mg/L lysozyme and 500mg/L β-galactosidase at pH 8 and adsorption allowed to progress for up to 1 h before being washed at pH 8 and eluted at pH 12. Analysis by SDS-PAGE showed that the adsorbents were overloaded with both proteins, which could be identified in the supernatant after adsorption. Activity measurements were used to determine that the capacity was 48.3 mg/g_{beads} of lysozyme was bound to the adsorbents, which is consistent with the isotherm in Fig. 13.10. Washing at pH 8 released neither protein and is also consistent with the results in Fig. 13.10. However, lysozyme could be selectively eluted at pH 12 without any contamination by β-galactosidase (Fig. 13.12, lane 10). The binding characteristics of the adsorbents used are given in Table 13.4.

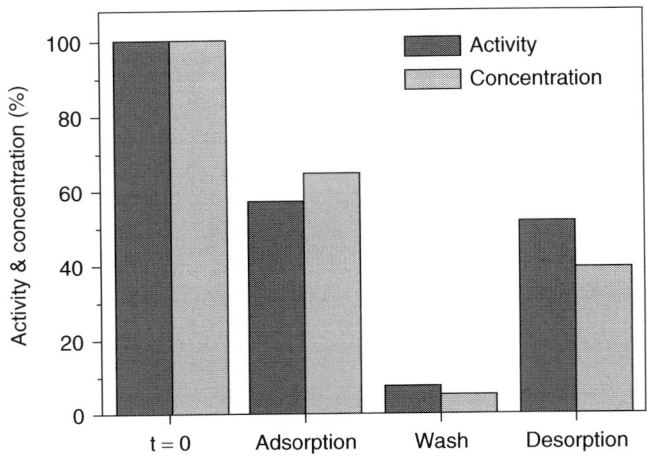

Figure 13.12. SDS-PAGE of selective separation of lysozyme from binary lysozyme/β-galactosidase mixture. PVB beads (40/20/40 CEX) were used as adsorbents. Concentrations: 500 mg/L lysozyme, 1000 mg/L β-galactosidase, 4g/L beads. Adsorption/desorption conditions: 60 min adsorption pH 8, 30 min washing pH 8, 45 min elution pH 12. SDS-PAGE gel was self-made (4% stacking gel and 12% resolving gel). Gel was loaded with 10 µL marker and 30 µL sample per slot and finally run with 20 mA/gel in stacking gel and 40 mA/gel in resolving gel. The binding characteristics of the adsorbents used are given in Table 13.4. (This figure is available in full color at http://onlinelibrary.wiley.com/book/10.1002/9780470054581.)

13.3.2 HGMF Product Recovery from Unclarified Feedstocks

In the above example, screening was conducted in micro test tubes using a bar magnet to accomplish magnetic adsorbent separation. That is an easy and straightforward way for screening of adsorbents and buffer types, but is not directly scalable and a more efficient method of handling the magnetic adsorbents is needed. Hubbuch and Thomas (9) successfully applied HGMF at pilot scale for trypsin recovery from an unclarified bioprocess liquor (crude porcine pancreatic feedstock). Potent trypsin-binding magnetic adsorbents ($q_{max} = 120$ mg/g; $K_d = 0.3$ μM), based on the synthetic serine protease inhibitor, benzamidine, were produced using activation and coupling methodologies commonly employed in the preparation of commercial chromatography media. Essentially nonporous micron-sized agglomerates of magnetic nanoparticles were used for construction of the adsorbents (9).

Benzamidine-linked magnetic particles were stirred with crude pancreatin for 0.5 h at room temperature. Thereafter, the particle/feedstock suspension was applied to a high gradient magnetic filter canister, vertically positioned in a 0.4 Tesla field. In the batch adsorption phase, the use of insufficient adsorbent resulted in the capture of only 79% of the available trypsin; thus, in addition to α-amylase, which does not bind to the adsorbents, low levels of trypsin were detected in the exit flow during system loading and washing (Figs 13.13 and 13.14). Nonspecifically adsorbed and physically entrapped materials were then removed by washing within a closed-system loop, with the field switched off. Three washing cycles were performed; collectively, these dislodged approximately 3% of the initially adsorbed trypsin. Specifically bound material was then eluted from the adsorbents in three cycles (Figs 13.13 and 13.14), using the recycle loop filled with 0.1 M Glycine–HCl, pH 2.6. In this way, over 80% of the initially bound trypsin was recovered with a purification factor of approximately 3.5. Although about 80% of the magnetic adsorbent could be recovered simply by flushing and recycling at high flow rates, complete adsorbent recovery could only be achieved by dismantling the filter canister. An increasingly important consideration in the design of bioprocess equipment is that efficient cleaning and sterilization procedures can be applied without dismantling the system. Clearly, in the context of bioprocessing, future HGMF apparatus will need to be modified to allow complete particle recovery in-line and implementation of efficient Cleaning in place (CIP) and Separating in place (SIP) procedures.

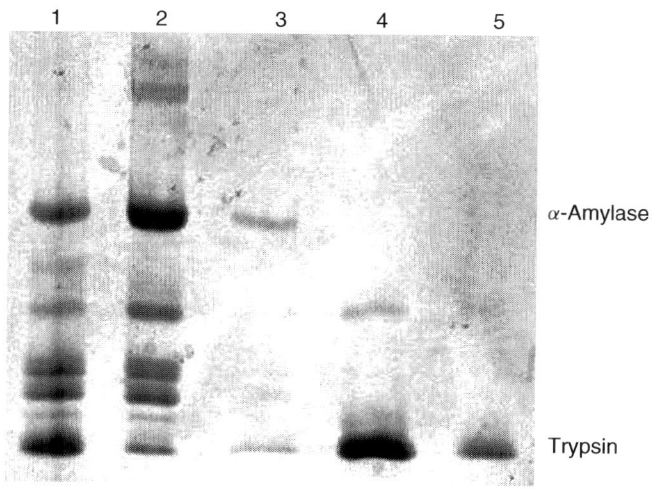

Figure 13.14. SDS-PAGE analysis of fractions collected at different stages of operation during high gradient magnetic affinity separation of trypsin from crude porcine pancreatin (Fig. 13.13). Lane 1, crude porcine pancreatin; Lane 2, during loading; Lane 3, during loading; Lane 4, 1st elution cycle; Lane 5, 2nd elution cycle [with permission, from J.J. Hubbuch and O.R.T. Thomas, *Biotechnol. Bioeng*. **79**, 301–313, Copyright © 2002 and John Wiley & Sons, Inc., New York].

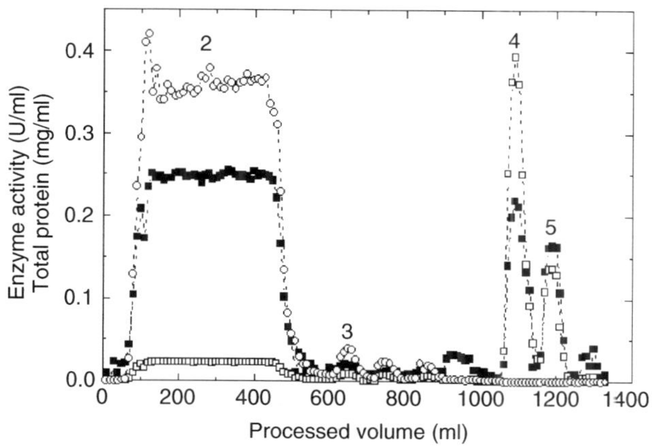

Figure 13.13. High gradient magnetic affinity separation of trypsin from crude porcine pancreatin. Key: trypsin activity (□), α-amylase activity (○), total protein content (■). The numbers (2–5) correspond to fractions analyzed in the SDS-polyacrylamide gel shown in Fig. 13.14 [with permission, from J.J. Hubbuch and O.R.T. Thomas, *Biotechnol. Bioeng*. **79**, 301–313, Copyright © 2002 and John Wiley & Sons, Inc., New York].

13.4 MAGNETIC SEPARATION TECHNIQUES

Magnetic separations can be split into two broad categories: (i) those in which materials are separated from one another on the basis of differences in *intrinsic* magnetic moment and (ii) those in which one or more components in a mixture have been *rendered* magnetically susceptible. While both types of magnetism are exploited in large-scale industrial magnetic separation, most biotechnological applications involve rendered magnetism, that is, bestowing magnetism upon a nonmagnetic (diamagnetic) species by, for example, attachment of, or association with a magnetically responsive particle. Rare exceptions include erythrocytes and magnetic bacteria, which are intrinsically magnetic, as they contain high concentrations of paramagnetic hemoglobin and small coated crystals of magnetic oxides or sulfides, respectively.

The efficiency with which magnetic species can be separated or manipulated using magnetic fields depends on the relationship between the magnetic, hydrodynamic, gravitational, and interparticle forces. The resulting force balance is determined by the nature of the material being separated, the type of magnetic device, and operating parameters employed. The critical parameters in magnetic separations are the magnetic susceptibilities of the materials being separated, the magnetic field strength in the separation region, the magnetic field gradient (dH/dx, i.e. the variation in the magnetic field with position), and the volume of the particles. The magnetic field gradient becomes increasingly important as the volume of the magnetic particles decreases, such that for very small particles high values of dH/dx are required to create forces which are sufficiently strong to elicit fast separations. For paramagnetic particles the absolute strength of the externally applied field, H, is also crucial as it influences the induced magnetization. That said the key factor in magnetic separations is identifying practical ways of generating high magnetic field gradients. In the following sections, the two main magnetic separation techniques used in DSP, magnetic extraction, and high gradient magnetic separation are briefly described.

13.4.1 Magnetic Enhanced Two Phase Extraction

Two-phase systems, based on the incompatibility of two polymers (or of a polymer and a salt) in aqueous solution, have been employed for the fractionation of cells, organelles, cell debris, viruses, nucleic acids, and proteins. These aqueous two-phase systems (ATPS) are typically characterized by rapid attainment of equilibrium and minimal mixing; the latter being due to the low interfacial tension between the two phases. In stark contrast, phase separation on standing is slow and centrifugation is often required to accelerate it. If many partitioning steps are required, centrifugally accelerated phase separation can impose serious economic limitations on the process. Wikström and coworkers (69) described how seeding one phase with a magnetically susceptible material (e.g. ferrofluids or magnetite) made it possible to enhance phase separation through application of a magnetic field. Phase droplets doped with magnetic material moved in the magnetic field to a position where they collapsed into the bulk phase. Magnetically assisted phase separation is easily scaled-up and should prove a useful addition to existing phase separation techniques in biotechnology, especially in problem applications (i.e. phases with similar densities, high viscosity, high volumetric phase ratios, and systems containing emulsion stabilizers), where gravitational or centrifugally assisted separation are either unfeasible or impracticably slow (70).

Recently, the concept of magnetic extraction was developed further by using functional magnetic micro- or nano adsorbents in combination with thermoresponsive ATPS (71). Thus, besides the benefit of accelerating the phase separation with the help of magnetic fields, protein partitioning is also enhanced by adsorbing the target selectively onto the magnetic adsorbents, which accumulate in the dispersed phase of the system. In addition, polymer recycling is facilitated by using thermoresponsive ATPS (generally aqueous solutions of nonionic surfactants), which separate above the so-called cloud point temperature (TCP) into two coexisting phases, one enriched (containing >99% of total surfactant), the other depleted in surfactant (containing <1% of total surfactant). The magnetic adsorbents adsorb the target protein and pull it out of the *surfactant-depleted* phase into the *surfactant-rich* phase. Hence, by reusing this phase, nearly all of the phase-forming polymer may be recycled (Fig. 13.15). The feasibility of the process was shown by applying it to model protein solutions of hen egg white lysozyme and ovalbumin (72).

13.4.2 High Gradient Magnetic Separation (HGMS)

When a ferromagnetic material (or collector) is positioned in a magnetic field, the net magnetic force it exerts for para- and ferromagnetic materials suspended in a medium is positive and attractive. The relative strength of attraction depends on both the magnetic susceptibility of the particles and their size. A very wide variety of magnetic separation devices have been employed in numerous, large-scale, industrial applications (Svoboda (73) and Hoffmann (74)). However, if complete removal of very small magnetic micro- or even nano-particles is required, then high gradient magnetic separators (HGMS) are the instruments of choice. HGMS can recover even weakly magnetic substances from fast flowing process liquors. The basic principle of HGMS, illustrated in Fig. 13.16, is simple and similar to that of deep-bed filtration. A canister, filled with a magnetizable separation matrix, is introduced into the area of an external magnetic field. The matrix, symbolized by a single wire cross-section in Fig. 13.16, may consist of wire meshes stacked on top of one another or of perforated discs or plates. The matrix bundles the external magnetic field in its surroundings and thus generate, on its surfaces, areas which strongly attract magnetic particles.

During the separation cycle, the magnetic micro particles within the feed are deposited on the separation matrix, such that fluid exiting the separator is either particle-free or loaded with nonmagnetic solids only. If the capacity of the separation matrix is exhausted, the supply of feed solution is suspended. The magnetic field is then switched off and collected particles are recovered from the separation

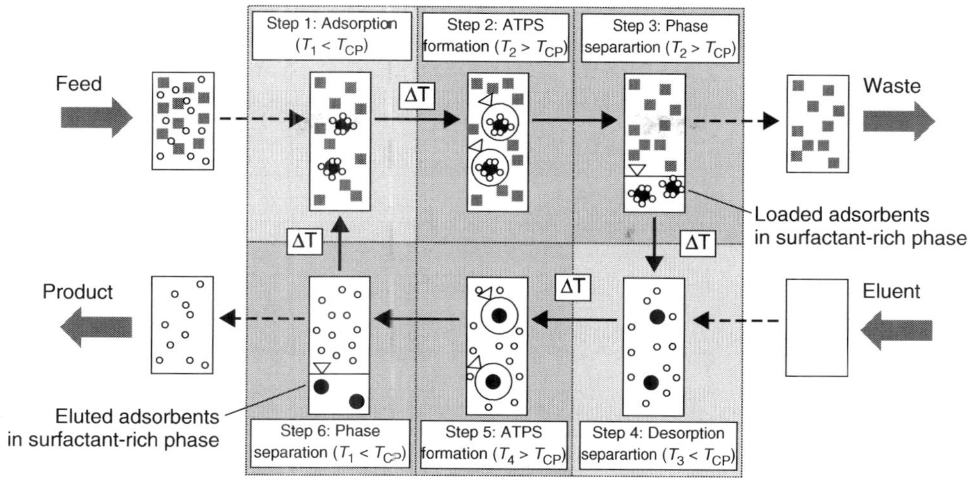

Figure 13.15. Process scheme for using SMEP (Smart Magnetic Extraction Phases) for protein separation. Flow diagram illustrating integration of adsorption, desorption, phase formation, and phase separation steps. (●) Magnetic adsorbents, (○) the target molecules, and (■) impurities.

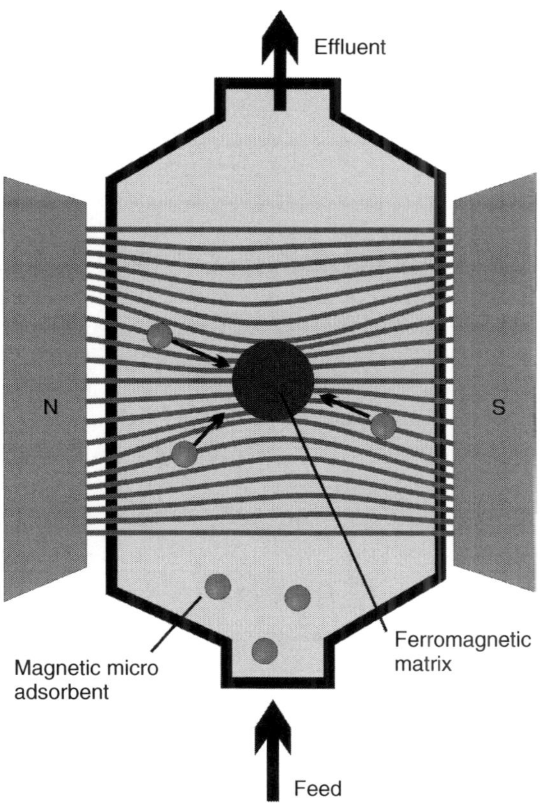

Figure 13.16. Separation principle of high gradient magnetic separation (HGMS).

matrix with a short, intensive wash, conducted in counterflow mode. The separated particles are collected as a concentrate, with solids content rarely higher than 5%.

Assuming the validity of the Stokes equation for hydrodynamic resistance, the theoretical particle velocity, u_m, caused by the magnetic force in the immediate vicinity of a wire, of radius a, may be approximated as:

$$u_{\mathrm{m}} = \frac{1}{18} \cdot \mu_0 \cdot \Delta\chi \cdot M_{\mathrm{w}} \cdot H_0 \cdot \frac{D_{\mathrm{p}}^2}{a \cdot \mu} \quad (13.3)$$

where $\Delta\chi$ is the difference between the susceptibilities of the particle and fluid, D_p the particle diameter, M_w the magnetization of the wire, H_0 the external magnetic field strength, μ_0 the permeability of free space and μ the permeability of the medium. In a number of theoretical and experimental investigations, the ratio of this magnetic velocity to the actual flow velocity, u_0, in the separator (i.e. u_m/u_0) proved important in predicting magnetic separation efficiency. At values of $u_m/u_0 >> 1$, the separation behavior of an HGMS is similar to that of a classical, deep-bed filter. This means that a relatively sharp loading front is formed within the separation matrix. With time, this front shifts along the filter toward the outlet. At values of $u_m/u_0 < 1$, rather extended loading fronts are obtained and magnetic separation is ineffective.

HGMS can operate efficiently at throughputs more than an order of magnitude greater than conventional separation processes (see Fig. 13.18). Additionally, practically 100% of the magnetic material, even those possessing only weak magnetic susceptibility can be recovered and the technology is also suitable for application in sterile environments. However, classical HGMS system adopted from, for example, the steel industry showed severe problems with a complete recovery of the magnetic particles from the matrix during flushing (Franzreb *et al.* (75), Meyer *et al.* (76) and (77)). While this can be tolerated in most applications of magnetic separation, the efficient use in DSP requires

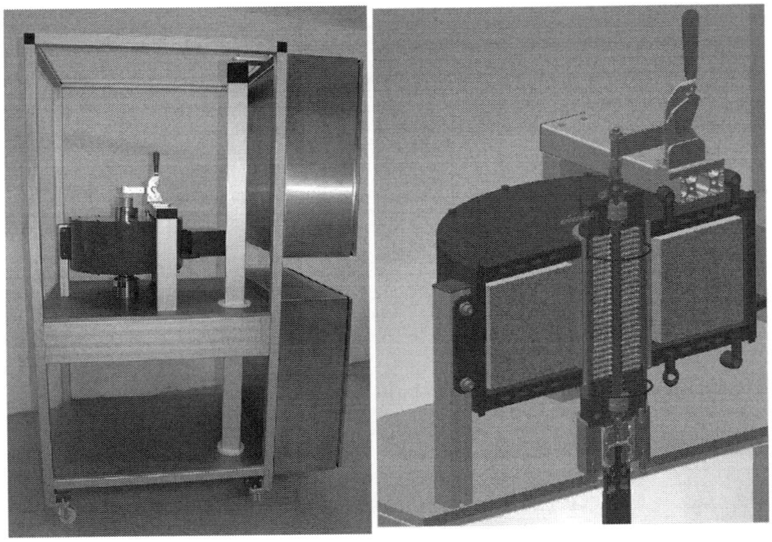

Figure 13.17. Rotor–stator magnetic separator (RSMS) a) whole separator b) filtration chamber with rotor–stator discs from chemagen Biopolymer Technologie AG (www.chemagen.de). (This figure is available in full color at http://onlinelibrary.wiley.com/book/10.1002/9780470054581.)

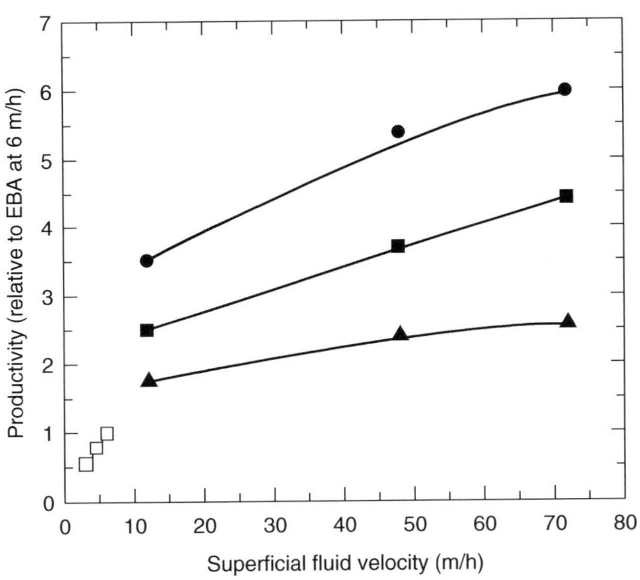

Figure 13.18. Effect of superficial fluid velocity on the productivity of EBA and HGMF for the recovery of Savinase® from a *Bacillus clausii* fermentation broth. Key: EBA (□); and HGMF after 1 (●), 2 (■), and 3(▲) batch adsorption steps [adapted from J.J. Hubbuch, D.B. Matthiesen, T.J. Hobley and O.R.T. Thomas, 2001, *Bioseparation*, **10**, 99–112].

a complete particle resuspension during washing and elution steps. Therefore a special type of HGMS, specially adapted for use in DSP, has been developed and recently commercialized (Fig. 13.17a).

In this so-called rotor–stator magnetic separator (RSMS), the adhesive forces of the particle agglomerates on the matrix are overcome by increasing the relative velocity between the separation matrix and the liquid by a rotation of the matrix. To prevent the liquid from rotating as well, the separation matrix of the RSMS is designed according to a rotor–stator principle. The separator housing shown in Fig. 13.17b does not only accommodate fixed matrix discs, but also matrix discs that are attached to a rotating shaft. In the separator, a large number of both types of discs are arranged alternately. While particle separation takes place with a stationary shaft, particle resuspension is initiated by rotating the shaft at high speed. The resulting shear forces between the closely adjacent rotating and stationary discs cause the particles to be sheared off and entrained by the liquid flow, thus allowing efficient washing and elution steps but also complete particle recovery at the end of the bioseparation cycle.

13.5 SUMMARY

The use of magnetically responsive supports permits selective manipulation and separation of an adsorbent in the presence of other suspended solids. This makes it possible to selectively recover the magnetic adsorbent, loaded with adsorbed target species, from crude biological process liquors containing suspended solids and fouling materials, thereby eliminating the multiple stages of pretreatment which normally precede application to packed bed adsorption chromatography columns. Magnetic separations are fast, gentle, freely scalable, easily automated, cost effective, can achieve separations that would be impossible or impractical using other techniques, and have demonstrated credibility in an exceptionally wide range of disciplines, including minerals engineering, waste water treatment, molecular biology, cell sorting, and clinical diagnostics.

The abilities of magnetic fishing technologies and the connected processes of magnetic separations and particle production have been proven in the lab and semipilot scale. The step into industrial-scale application, however, has a realistic chance of being realized in the upcoming decade. The union of low-cost magnetic adsorbents with specialized process equipment is expected to generate potent, economic, and highly selective adsorptive separation techniques, geared to rapid, large-scale processing of problematic bioprocess liquors to provide more and cheaper products from biotechnical syntheses.

REFERENCES

1. Malini KA, Anantharaman MR, Sindhu S, Chinnasamy CN, Ponpandian N, Narayanasamy A, Balachandran M, Pillai VNS. J Mater Sci 2001; 36: 821–824.
2. Marik J, Lau DH, Song AM, Wang XB, Liu RW, Lam KS. J Magn Magn Mater 2003; 264: 153–157. DOI: 10.1016/S0304-8853(03)00179-3.
3. Martin C, Cuellar J. Ind Eng Chem Res 2004; 43: 475–485. DOI: 10.1021/Ie0302239.
4. Lu SL, Cheng GX, Pang XS. J Appl Polym Sci 2003; 89: 3790–3796. DOI: 10.1002/App.12530.
5. Safarik I, Safarikova M. Monatsh Chem 2002; 133: 737–759.
6. Safarik I, Safarikova M. Chem Pap 2009; 63: 497–505. DOI: 10.2478/s11696-009-0054-2.
7. Safarik I, Safarikova M, Forsythe SJ. J Appl Bacteriol 1995; 78: 575–585.
8. Hubbuch JJ, Matthiesen DB, Hobley TJ, Thomas ORT. Bioseparation 2001; 10: 99–112.
9. Hubbuch JJ, Thomas ORT. Biotechnol Bioeng 2002; 79: 301–313.
10. Heeboll-Nielsen A, Dalkiaer M, Hubbuch JJ, Thomas ORT. Biotechnol Bioeng 2004; 87: 311–323.
11. Heeboll-Nielsen A, Justesen SFL, Thomas ORT. J Biotechnol 2004; 113: 247–262.
12. Safarik I, Ptackova L, Safarikova M. Biotechnol Lett 2001; 23: 1953–1956.
13. Hubbuch J, Kula MR. J Non-Equilib Thermodyn 2007; 32: 99–127.
14. Nfor BK, Ahamed T, van Dedem GWK, van der Wielen LAM, van der Sandt EJAX, Eppink MHM, Ottens M. J Chem Technol Biotechnol 2008; 83: 124–132.
15. Subramanian G. Bioseparation and bioprocessing. Weinheim: Wiley-VHC; 2007.
16. Wheelwright SM. Protein purification - design and scale up of downstream processing. New York: Wiley-Interscience; 1994.
17. Sommerfeld S, Strube J. Chem Eng Process: Process Intensificat. 2005; 44: 1123–1137.
18. Allgaier H. European downstream technology forum. Germany: Sartorius College, Göttingen; 2008.
19. Einicke WD, Mauersberger P, Handel J, Breuel A. Chem Tech 1995; 47: 261–268.
20. Schügerl K, Hubbuch J. Curr Opin Microbiol 2005; 8: 294–300.
21. Hubbuch J, Thommes J, Kula MR. Technol Trans Biotechnol: Lab Ind Prod 2005; 92: 101–123. DOI: 10.1007/B98917.
22. Arpanaei A, Mathiasen N, Hobley TJ. J Chromatogr A 2008; 1203: 198–206. DOI: 10.1016/j.chroma.2008.07.052.
23. Jungbauer A, Hahn R. Curr Opin Drug Discovery Dev 2004; 7: 248–256.
24. Käppler TE, Hickstein B, Peuker UA, Posten C. J Biosci Bioeng 2008; 105: 579–585.
25. Käppler T, Cerff M, Ottow K, Hobley T, Posten C. Biotechnol Bioeng 2009; 102: 535–545.
26. Heeboll-Nielsen A, Justesen SFL, Hobley TJ, Thomas ORT. Sep Sci Technol 2004; 39: 2891–2914.
27. Robinson PJ, Dunnill P, Lilly MD. Biotechnol Bioeng 1973; 15: 603–606.
28. Dunnill P, Lilly MD. Biotechnol Bioeng 1974; 16: 987–990.
29. Mosbach K, Andersson L. Nature 1977; 270: 259–261.
30. Whitesides GM, Kazlauskas RJ, Josephson L. Trends Biotechnol 1983; 1: 144–148.
31. Uhlen M. Nature 1989; 340: 733–734.
32. Safarik I, Safarikova M. J Chromatogr B: Biomed Sci Appl 1999; 722: 33–53.
33. Lawson EL, Clifton JG, Huang F, Li X, Hixson DC, Josic D. Electrophoresis 2006; 27: 2747–2758.
34. Chung TH, Pan HC, Lee WC. J Magn Magn Mater 2007; 311: 36–40.
35. Kaminski MD, Nunez L, Visser AE. Sep Sci Technol 1999; 34: 1103–1120.
36. Matthews SE, Parzuchowski P, Garcia-Carrera A, Grüttner C, Dozol JF, Böhmer V. Chem Commun 2001; 417–418.
37. Tartaj P, Del Puerto Morales M, Veintemillas-Verdaguer S, Gonzãlez-Carreno T, Serna CJ. J Phys D: Appl Phys 2003; 36: R182–R197.
38. Mulvaney SP, Mattoussi HM, Whitman LJ. BioTechniques 2004; 36: 602–609.
39. Rudershausen S, Grüttner C, Frank M, Teller J, Westphal F. Eur Cells Mater 2002; 3: 81–83.
40. Berensmeier S. Appl Microbiol Biotechnol 2006; 73: 495–504.
41. Flaschel E. Chem Eng Technol 2008; 31: 809.
42. Kokpinar O, Harkensee D, Kasper C, Scheper T, Zeidler R, Reif OW, Ulber R. Biotechnol Progress 2006; 22: 1215–1219.
43. Strube J, Grote F, Ditz R. Fachausschuss Biotechnology. Bremen ProcessNet; 2008.
44. Thierolf C. In: Schöffski O, Fricke F-U, Guminski W, editors. Pharmabetriebslehre. Berlin, Heidelberg: Springer; 2008. p. 117–128.
45. Grote F, Ditz R, Strube J, Achema 2009 Frankfurt am Main, 2009.
46. Yuan Q, Williams RA. China Particuol 2007; 5: 26–42.
47. Martin C, Ramirez L, Cuellar J. Surf Coat Technol 2003; 165: 58–64.

48. Tong XD, Xue B, Sun Y. Biotechnol Progress 2001; 17: 134–139.
49. Yang C, Liu H, Guan Y, Xing J, Liu J, Shan G. J Magn Magn Mater 2005; 293: 187–192.
50. Hoshino, A, N Ohnishi, M Yasuhara, K Yamamoto, Kondo A. Biotechnol Progress 2007; 23: 1513–1516.
51. Ma Z, Liu H. China Particuol 2007; 5: 1–10.
52. Müller-Schulte D, Brunner H. J Chromatogr A 1995; 711: 53–60.
53. Bergemann C, Müller-Schulte D, Oster J, à Brassard L, Lübbe AS. J Magn Magn Mater 1999; 194: 45–52.
54. Müller-Schulte D. U. Patent6204033. 2001.
55. Ugelstad J, Berge A, Ellingsen T, Schmid R, Nilsen TN, Mørk PC, Stenstad P, Hornes E, Olsvik Ø. Progress Polym Sci 1992; 17: 87–161.
56. Ugelstad J, Ellingsen T, Berge A, Helgee O. U. Patent 4774265. 1988.
57. Dynal Magnetic Beads. Available at http://www.invitrogen.com/site/us/en/home/brands/Dynal.html.
58. Hickstein B, Peuker UA. Biotechnol Progress 2008; 24: 409–416.
59. Banert T, Peuker UA. Chem Eng Commun 2007; 194: 707–719.
60. Hickstein B, Cecilia R, Kirschning A, Kunz U, Peuker UA. Chem Ing Tech 2007; 79: 2089–2097.
61. Hickstein B, Peuker UA. J Appl Polym Sci 2009; 112: 2366–2373.
62. Lu AH, Salabas EL, Schüth F. Angew Chem Int Ed 2007; 46: 1222–1244.
63. Atkins PW. Physikalische chemie. Weinheim: Wiley-VCH; 1987.
64. Xue B, Sun Y. J Chromatogr A 2002; 947: 185–193.
65. Yang CL, Liu HZ, Guan YP, Xing JM, Liu JG, Shan GB. J Magn Magn Mater 2005; 293: 187–192.
66. Peng ZG, Hidajat K, Uddin MS. J Colloid Interface Sci 2004; 271: 277–283.
67. Liao MH, Chen DH. Biotechnol Lett 2002; 24: 1913–1917.
68. Hoffmann C. Einsatz magnetischer Separationsverfahren zur biotechnologischen Produktaufbereitung TU Karlsruhe 2003.
69. Wikström P, Flygare S, Larsson P-O. In: Fisher, D, Sutherland IA, editors. Separations using aqueous phase systems. New York: Plenum Publishing Corporation; 1989. p. 445–461.
70. Flygare S, Wikstrom P, Johansson G, Larsson PO. Enzyme Microb Technol 1990; 12: 95–103.
71. Franzreb M, Becker J. 2007. Patent No. 10007020220.4.
72. Becker JS, Thomas ORT, Franzreb M. Sep Purif Technol 2009; 65: 46–53.
73. Svoboda J. Magnetic techniques for the treatment of materials. Netherlands: Springer; 2004.
74. Hoffmann C, Franzreb M, Höll WH. IEEE Trans Appl Supercond 2002; 12: 963–966.
75. Franzreb M, Ebner N, Siemann-Herzberg M, Hobley TJ, Thomas ORT. In: Shukla A, Etzel MR, Gadam S, editors. Process scale bioseparations for the biopharmaceutical industry. Boca Raton: CRC-Press, Taylor and Francis Group; 2007. p. 83–121.
76. Meyer A, Hansen DB, Gomes CSG, Hobley TJ, Thomas ORT, Franzreb M. Biotechnol Progress 2005; 21: 244–254.
77. Meyer A, Berensmeier S, Franzreb M. React Funct Polym 2007; 67: 1577–1588.
78. Cocker TM, Fee CJ, Evans RA. Biotechnol Bioeng 1997; 53: 79–87.
79. Horak D, Benedyk N. J Polym Sci Part A: Polym Chem 2004; 42: 5827–5837.
80. Denizli A, Tanyolac D, Salih B, Özdural A. J Chromatogr A 1998; 793: 47–56.
81. Ramrez LP, Landfester K. Macromol Chem Phys 2003; 204: 22–31.
82. chemagic Kits. Available at http://www.chemagen.com/fileadmin/downloads/chemagic_Kits.pdf. Accessed 2009 July 11.
83. nanomag. Available at http://www.micromod.de/scripts/t1.asp?sid=256261753&lng=e. Accessed 2009 July 11.

14

HIGH THROUGHPUT TECHNOLOGIES IN BIOPROCESS DEVELOPMENT

TRENT CARRIER
Invitrogen, part of Life Technologies, Grand Island, New York
EVA HELDIN, MATTIAS AHNFELT AND EGGERT BREKKAN
GE Healthcare Bio-Sciences AB, Uppsala, Sweden
RICHARD HASSETT AND STEVE PEPPERS
Invitrogen, part of Life Technologies, Grand Island, New York
GUSTAV RODRIGO
GE Healthcare Bio-Sciences AB, Uppsala, Sweden
GREG VAN SLYKE AND DAVID (XIAOJIAN) ZHAO
Invitrogen, part of Life Technologies, Grand Island, New York

14.1 INTRODUCTION

Bioprocess development is a relatively new field for the application of high throughput technologies (HTT). Similar to historical applications in drug discovery, these new tools are generally being applied to increase throughput and efficiency for either improved performance by testing a larger experimental design space or improved speed for development (1). Many of the HTT used for bioprocess development, such as flow cytometry, were originally developed for research and discovery work. The primary challenge for practitioners of these tools is "where do HTT bring value to bioprocess development?"

In most cases, companies will have established a platform workflow that is used to progress biopharmaceutical candidates as quickly as possible through early development. Figure 14.1 illustrates a workflow that begins with the identification of a lead therapeutic candidate, followed by generation of a recombinant cell line. The cell line and resulting product stream are then evaluated under a series of process conditions to identify parameters that provide the desired process and product performance, including yield, product quality, and impurities. The process is then scaled up to manufacturing, including the final form for use in the pharmacy. Materials confirmed as having acceptable product quality attributes are then made available for clinical use. The steps described are common to most bioprocess development workflows and, in fact, have not changed much in several decades of scientific study; however, the outcomes of today's bioprocess development programs are substantially improved and have become a point of competitive advantage for companies. Herein lies the opportunity for the bioprocess scientist—how can we accomplish more from the same basic process?

The authors of this chapter have endeavored to answer this question across the full workflow of bioprocess development and provide guidance on the basis of their acquired knowledge of how to deploy HTT in a practical sense. In surveying the general application of these technologies in the biotechnology industry, it becomes clear that HTT developers have targeted a few key steps within the bioprocess workflow. Although the definition

Downstream Industrial Biotechnology: Recovery and Purification, First Edition. Edited by Michael C. Flickinger.

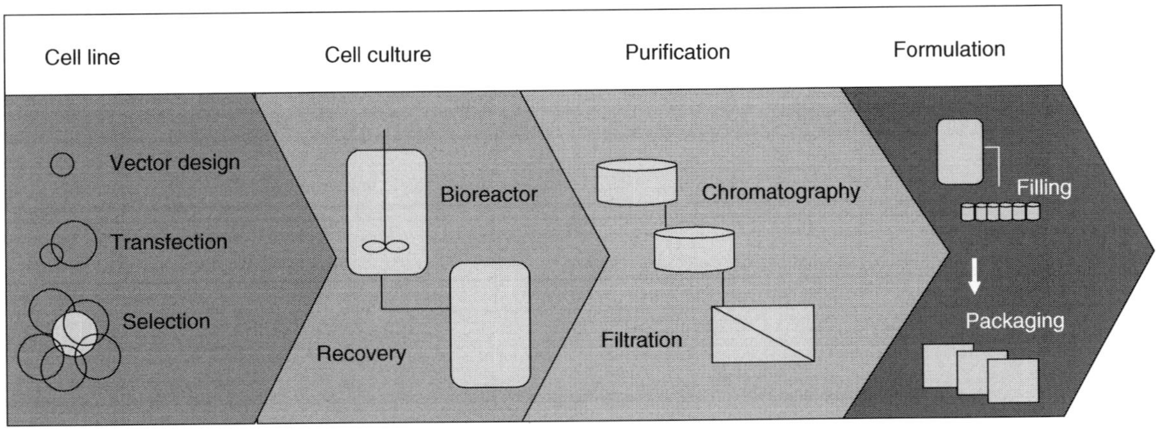

Figure 14.1. Biopharma development workflows. (This figure is available in full color at http://onlinelibrary.wiley.com/book/10.1002/9780470054581.)

of what constitutes "high throughput" is highly relative to the specifics of each application, the collective experience and the traction of these technologies in the biotechnology industry suggest that there are benefits from using HTT in bioprocess development.

14.2 HTT APPLIED TO UPSTREAM CELL CULTURE DEVELOPMENT

14.2.1 Overview

Upstream cell culture development is commonly defined by the steps encompassing cell line, media, and bioreactor process development. Each of these steps demands productivity and throughput to meet the increasing requirements of biopharmaceutical development. The implementation of HTT has become a technique used to shorten the time frame and improve outcomes for bioprocess development. HTT have also been used to reduce the risk of failure in bioprocess development, in particular, as related to Quality by Design initiatives (2).

14.2.1.1 Cell Line Development

Technology	Developer
ClonePix Fl	Genetix
LEAP™	Cyntellect
FACS	Multiple developers

The majority of new biopharmaceutical products are produced in cell culture systems and, for most of these products, stable cell line development is the first step in the bioprocess development workflow. Stable cell line development can be subdivided into different activities, including DNA sequence selection, codon optimization, vector design, transfection/amplification, clone selection, expression stability study, and cell banking. Within cell line development, the primary opportunity for HTT is in clone selection, where screening for high titer, product quality and cell growth has traditionally been a bottleneck of the cell line development process. Traditional manual screening for clone selection is labor intensive and time consuming. With this method, it is possible to screen only a fraction of clones generated within a stable pool. As a result, many new technologies and systems have been developed to accelerate and automate the clone screening process. These new HTT are designed to increase the number of clones that can be screened through automation to analyze cell characteristics in an HT manner.

The ClonePix Fl technology, developed by Genetix, capitalizes on a system in which mammalian cells are cultured in a semisolid media and then imaged using fluorescence assays. A fluorescent readout for the secreted protein can be multiplexed with up to five other fluorescent wavelengths and/or white light to allow other factors, such as cell growth and cell viability, to be taken into account when picking the desired colonies. The application of this technology in the isolation of clonal cell populations based on the expression of secreted proteins have been successfully adapted by many biotech companies, including Invitrogen.

Another approach was developed combining *in situ* capture and measurement of individual cell protein secretion, followed by laser-mediated elimination of poorly secreting cells. An automated, HT instrument (LEAP™) is used to image and locate every cell (3), quantify the cell-associated and secreted antibody surrounding each cell, eliminate all undesired cells from a well via targeted laser irradiation, and then track clone outgrowth and stability. This technology has been applied to cell line development for several years; however, the complexity of the technology has limited its broader adoption by many biopharmaceutical companies.

In comparison with these two proprietary HTT platforms, flow cytometry is a more traditional technology that has been adapted to screen recombinant cells from cell populations. Several specific applications have been developed for recombinant proteins using immobilization or encapsulation based on the gel microdrop method (4) or "affinity capture" methods (5,6). More recently, several novel approaches for the selection of high producing subclones have been conceived with flow cytometry (5,7–10). The use of flow cytometry instruments for stable cell line development has had mixed success, with adoption limited by both method IP and instrument validation.

14.2.1.2 Cell Culture Media Development

Technology	Developer
Well plates	Hamilton
TubeSpin	Sartorius, TPP
μ-24 Bioreactor	Applikon

As cell culture media moves toward chemically defined formulations to accommodate tighter regulatory requirements and demands for improved process robustness, the development of solutions that may contain over 60 components has become a bottleneck for media optimization. The traditional cell culture platforms, such as bench top bioreactors and shake flasks, are limited in their ability to evaluate very complex media formulations with multiple combinations of components. As such, companies have begun to adopt multifactorial experimental designs that necessitate HTT.

Developing culture media for suspension cell culture in multiwell plates has been difficult to accomplish due to its limitations in process control and sampling. Because of these limitations, many companies have adopted staged approaches to media development, where an initial media screen is performed in multiwell plates or microbioreactors followed by further process refinement in parallel bioreactor systems. Liquid handling systems, such as the STARplus system from Hamilton, are capable of formulating hundreds of media variants in less than a day using multiwell plates. Likewise, a variety of microbioreactor systems with limited parameter monitoring and processing capability have also been developed for media screening. TubeSpin technologies, which are based on orbital shaking, are used in both the CultiFlask 50 disposable bioreactor from Sartorius Stedim Biotech (11) and bioreactor tubes developed by Techno Plastic Products, TTP (12,13). Finally, the μ-24 Bioreactor system from Applikon can provide individual control and monitoring of reactor conditions. The system runs up to 24 simultaneous experiments with independent control of each reactor's gas supply, temperature, and pH (14).

14.2.1.3 Cell Culture Process Development

Technology	Developer
SimCell™	Seahorse Biosciences (previously supplied by BioProcessors)
DASGIP bioreactors	DASGIPAG

In addition to media screening, HTT are being used more frequently to increase the throughput of cell culture process development (15–18). Traditionally, small-scale systems such as shake flasks and spinner flasks, have been used to screen experimental conditions; however, due to limitations in mimicking process conditions, these models can miss critical factors influencing process performance.

The SimCell microbioreactor technology is a submilliliter cell culture platform supplied by Seahorse Biosciences that has the ability to mimic larger-scale bioreactor conditions, such as pH, temperature, and dissolved oxygen. This technology makes extensive use of automation and can perform hundreds of cell culture experiments simultaneously, thus potentially reducing the full process development time. The SimCell microbioreactor system has been shown in several laboratories to be advantageous in media and feed development with multiple cell lines and multiple media platforms (19–21).

The DASGIP technology is one of several minibioreactor platforms designed specifically to support parallel cultivation processes up to laboratory scale. Through automated control and monitoring, these bioreactors closely approximate scale-up conditions while also enabling rapid turnaround and low volume requirements.

14.2.1.4 Future Trends in Bioprocess Development. Much of the need for HTT in upstream developments arises from the variations in cell clones. With further optimization of mammalian host cell lines and vector expression systems, systematic optimization to reduce this variation is on the horizon. Direct cell engineering has been applied for many years to enhance cell lines through the manipulation of single genes that play important roles in key metabolic and regulatory pathways. Only recently has consideration been given to the use of indirect cell engineering approach, utilizing genomic and proteomic techniques and tools, to aid the discovery of novel targets for metabolic manipulation within the host cell line. It is likely that the use of these new systems combined with HTT will greatly advance the understanding of cellular mechanics to improve overall productivity in the manufacturing environment (22–26).

14.2.2 Selecting and Integrating High Throughput Technologies

HTT allow the study of many factors and parameters in parallel; however, this capability does not mean every application needs HTT. Before generating and then drowning in too much data, scientists need to make sure that all requirements are clearly defined and that the tools are appropriate. What is the objective of the study and how will the data support it? A summary of the process for selecting and integrating HTT is presented in Fig. 14.2. When selecting HTT, identifying the right technology to solve the problem is the first step to success. In cases where outcomes are dependent on the interactions of multiple parameters, evaluating many variables simultaneously and thereby increasing the experimental space with HTT can have great benefit. In contrast, there may be situations with HTT where the experimental space is definite and success is defined by the ability to shorten timelines through parallel processing. Both situations are good objectives for HTT, as long as it is clearly defined from the outset if increasing speed or experimental scope is the goal. Careful evaluation of needs and impact upfront can maximize the positive impact of HTT.

The following case study involving the choice of ClonePix FL as a clone screening technology illustrates the process of selecting HTT.

14.2.2.1 Define the Workflows for Clone Screening. The traditional process of dilution cloning has been well established for selecting recombinant cell lines and is summarized in Fig. 14.3. In this traditional approach, cells are diluted to a calculated density of between 0.5 and 1 cell per well, and each well is then observed for cell growth. Wells containing growth are assessed by ELISA to identify the clones with the highest production levels. The top performing clones are expanded and screened further to assess performance in larger-scale systems.

14.2.2.2 Identify the Bottleneck(s) in Clone Screening. When considering the pros and cons of implementing HTT, it is important that ample consideration be given to whether the process is a good fit for the technology. Two types of benefits are often the rationale for HTT investment: (i) the benefit of a larger experimental design space or (ii) the benefit of parallel processing. To increase the chance of identifying the highest producing clones, the cell screening process may require evaluating thousands of clones. To achieve this level of screening, laboratory staff may need to manually perform the same operations hundreds of times in a sequential fashion. Considering this workflow, an ideal HTT technology would both move the burden of the repetitive actions to a machine and enable parallel screening of clones. The following chart in Fig. 14.4 illustrates how HTT can impact workflow efficiency alone by comparing the manual process with an automated version.

In the example of clone screening, the goal is to screen the greatest number of clones within a set process to identify the rare, exceptionally high productivity cells. With HTT, labor input can be reduced for a larger number of activities, thereby increasing the number of clones screened. By contrast, in a workflow where the experimental design is well understood and a limited number of conditions need to be tested, increasing the replicates or broadening the range of conditions through HTT may not yield benefit. Each HTT must be considered only after the desired benefits (speed or increased scope) for the application are understood.

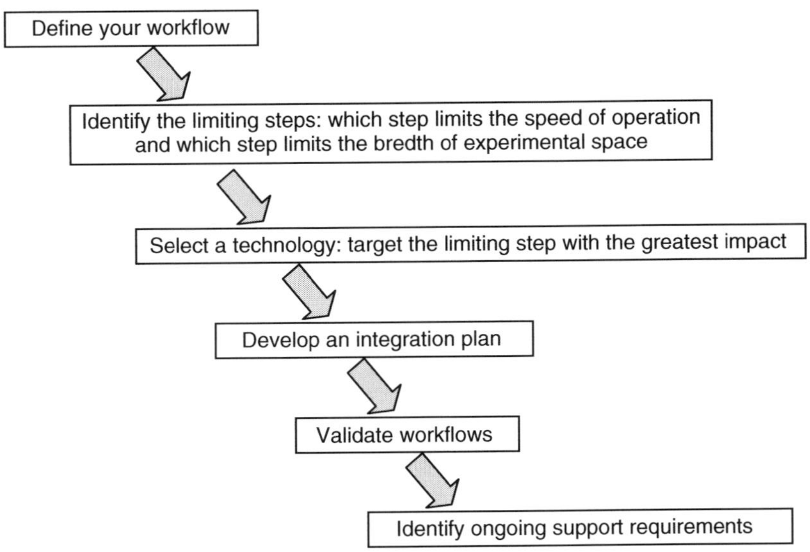

Figure 14.2. Process for selecting and integrating HTT. (This figure is available in full color at http://onlinelibrary.wiley.com/book/10.1002/9780470054581.)

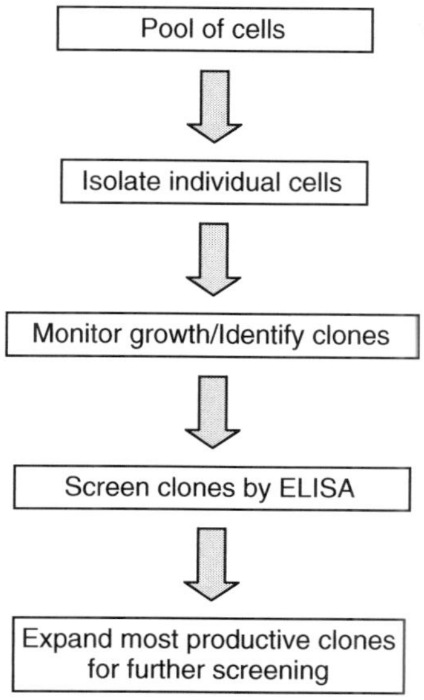

Figure 14.3. Process map for dilution cloning. (This figure is available in full color at http://onlinelibrary.wiley.com/book/10.1002/9780470054581.)

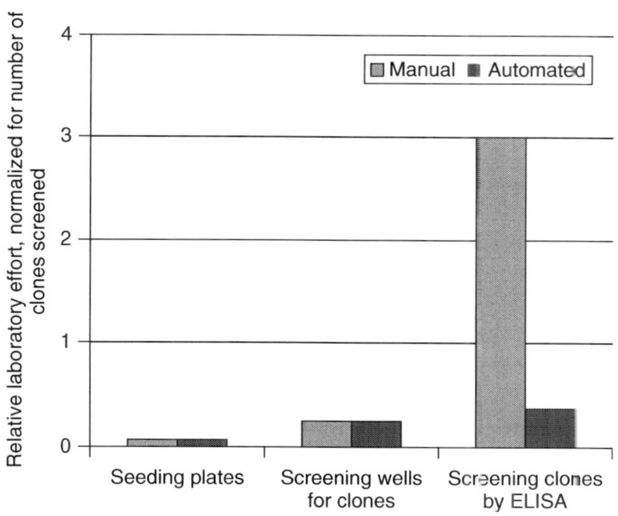

Figure 14.4. Chart of effort at each step. (This figure is available in full color at http://onlinelibrary.wiley.com/book/10.1002/9780470054581.)

14.2.2.3 Select a Technology for Clone Screening. Once the bottleneck of clone screening is identified, the next step in the process is to assess the technology options. A comparison of the technologies available is presented in Table 14.1. As each workflow will have specific needs, the technologies under consideration need to be gauged for the ability to meet particular goals. The workflow requirements were laid out so that any additional items could be identified, such as custom laminar flow cabinets. This assessment allowed for effective comparison between systems and accurate estimates of total ownership costs before technology selection. These costs totals included accounting for the sometimes substantial costs of specialized disposables. When comparing systems, it is also important to recognize that each HTT falls somewhere along the spectrum between "plug and play" to infinitely customizable. Plug-and-play systems may require little development on the HTT system, but the process flow often has to be modified to fit the technology. The time and effort devoted to the new workflow development may offset some of the savings realized by adopting the plug-and-play option.

An essential attribute of an effective HTT system is its comparability to the scaled-up systems. For example, the ability of ClonePix FL to yield highly productive and growing clones was successfully compared with that of plate-based methods and flask performance. In another example, the ability of SimCell system to yield reliable cell growth and bioproduction data was successfully evaluated through parallel cultures in shake flasks and bench-size stirred-tank bioreactors (e.g. DASGIP 1-L cultures). While in both examples, simple preliminary screening studies were used to initially qualify the equipment, a more rigorous validation was subsequently performed to ensure the output would consistently reflect the targeted process.

During HTT evaluation it may become clear that other parts of the workflow should also be changed. Removing an old bottleneck in the process usually means another one becomes capacity limiting. In considering the available technologies for clone screening, several of the options listed in the earlier cell line development section were evaluated. In each case, new workflows that would be required for each HTT were mapped out and evaluated for how each option matched the requirements.

Ultimately, the evaluation led to the selection of the ClonePix FL technology. The ClonePix FL technology best met the process needs by automating and consolidating clone isolation/productivity screening steps, thereby increasing the number of cells evaluated (as many as 10,000 per day) and shortening the time compared with traditional dilution cloning. With this technology, the workflow was slightly different from the traditional method of dilution cloning in well plates. Rather than isolating each cell in its own well, the cells were immobilized in a semisolid medium and allowed to grow into distinct colonies. Using image analysis, each colony was evaluated in place by fluorescent *in situ* detection of secreted product. It was determined that these process changes could be implemented without major disruption to the workflow. Once the HTT has been selected, it is important to consider how to integrate the system. In this discussion, a case study with the Hamilton STARplus is used to illustrate the elements to consider when undertaking HTT integration.

TABLE 14.1. Comparison of Technologies for Cell Line Screening

	Liquid Handling Robot	Automated Colony Picking Robot (ClonePix FL)
Process changes required	Minor—automation of existing process	Significant—adapt process to semisolid medium validation of *in situ* fluorescent detection
Number of clones screened at one time	1000–2000	5000–10,000
Process timeline	Same as manual method ~4 wk	Saving of ~1 wk due to screening and isolating clones in one step ~3 wk

14.2.2.4 Develop an Integration Plan for Hamilton STARplus. In considering an integration plan for a new technology platform, it is important to clearly match the operations on the instrument with external interfaces. Beyond simply generating a list of interfaces, identifying the critical parameters of each interface, such as time sensitivities and material handling, provides a good guideline for assessing the status of readiness of the integration. For a technology like the Hamilton STARplus platform, an interface map similar to that mentioned in Table 14.2 should be produced to serve as a useful guide when developing an integration plan.

In this particular application, the Hamilton STARplus was to be used to assemble panels of >100 cell culture media variants for optimizing cell performance using factorial designs, as shown in Fig. 14.5; however, a similar structure may be used for creating deletion media for QC assays or screening dose responses to new drug candidates. The core application of the technology is the assembly of the various media conditions and dispensing them into

TABLE 14.2. Checklist for Interface Map for Hamilton STARplus

Points to Consider	Questions to Ask
Compatibility	Is the system compatible with your cell-based system? Can you readily modify the system to switch to new media development platforms/designs? How fast can the system create the panels and is that level sufficient for required throughput?
Sterility	Is the system able to create a cell culture media aseptically? What is the rate of sterility failure? What are the key points of the system that could contribute to a sterility failure? Is the system cleanable?
Liquid handling	How does the system the aspirate and dispense liquids? Is the pipette technology pneumatic piston or are their liquid lines connected to the pipette channels? Is the system able to accurately dispense liquid at high and low volumes? Is the system capable of delivering volumes accurately and reproducibly? Does the system run over a long period of time without a fault?
Data handling	Does the system provide a log file to review if there is an error during the run? Can the data generated on the instrument be readily imported and exported into a database file? Is the system capable or reading and executing a program from an Excel spreadsheet?
Software	Does the system provide and easy to use interface? Is the Software upgradeable? Does the system provide a simulator mode that allows the programmer to test the program before an actual run? Does the require high level programming capabilities? Is it easy to create and change programs? Does the system come equipped with a vast library of reagent holder and assay plate configurations including 24-well, 48-well, 96-well, and 384-well plates for a variety of suppliers?
Hardware	Does the vendor provide adequate hardware for their instrument (reagent carriers, plate holders, pipette tip racks, etc.)? Is the system able to run for long periods of time without generating liquid handling errors during a run? Does the system need to have parts replaced often (pipette channels, Hepa Filters, etc.)?
System maintenance and support	Does the system require daily maintenance and calibrations? Does the vendor provide adequate training and technical support?
Equipment requirements	Are the pipette tips unique to the vendor? Is the cost of those tips an issue? If they are single sourced are there supply issues? How many backup supplies or manufactures do they have? Are the tips gamma irradiated? Are there multiple size tips to fit your application? 10, 20, 50, 100, 200, 500, 1000 μL, or higher?

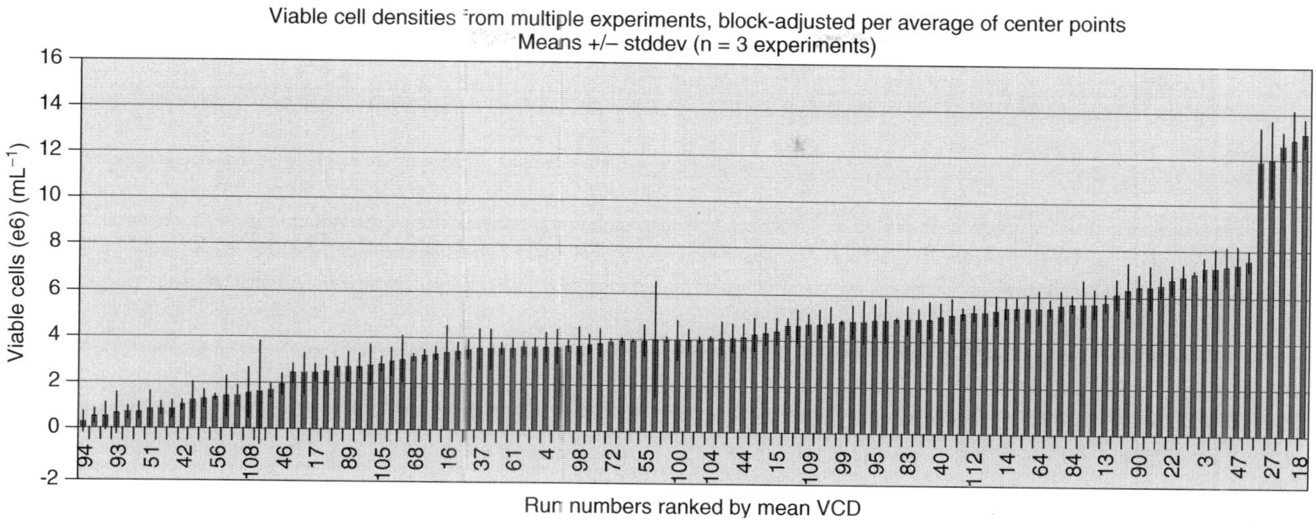

Figure 14.5. Growth of recombinant CHO cells in a panel of 111 different culture media assembled using the Hamilton STARplus technology. (This figure is available in full color at http://onlinelibrary.wiley.com/book/10.1002/9780470054581.)

either 24-well or 96-well cell-based assay plates in which cell performance is measured.

14.2.2.5 Validate Workflows for Assembly of Media Panels. In addition to validating the core technology as described previously, it is equally important to validate the adjacent workflows. The validation of workflows in this context is focused on confirming workflow integrity rather than defining limits of operation, as it may be difficult to foresee the full range of demands on a technology used in R&D programs. For the media panel workflows illustrated in Fig. 14.6, the process begins with preparing stock solutions, followed by media "building" on the instrument and measuring cell performance. Each of these includes several steps, which were prospectively identified on the basis of the list of interfaces.

The prospective validation plan for the media panel workflows followed a standard validation protocol requiring the assembly and testing of three sets of plates within three independent experiments. Upon completion of this validation, future programs involving a similar set of operations and inputs would be covered. As mentioned above, the workflow validation was not designed to predefine the operating limits of technology, but rather to confirm workflow integrity by testing run-to-run accuracy and consistency from within one panel. With a sound plan in hand, the validation studies were initiated and some imbedded dependencies that were not clearly elucidated were quickly highlighted.

As the first validation run was underway, it became apparent that there were several critical experimental

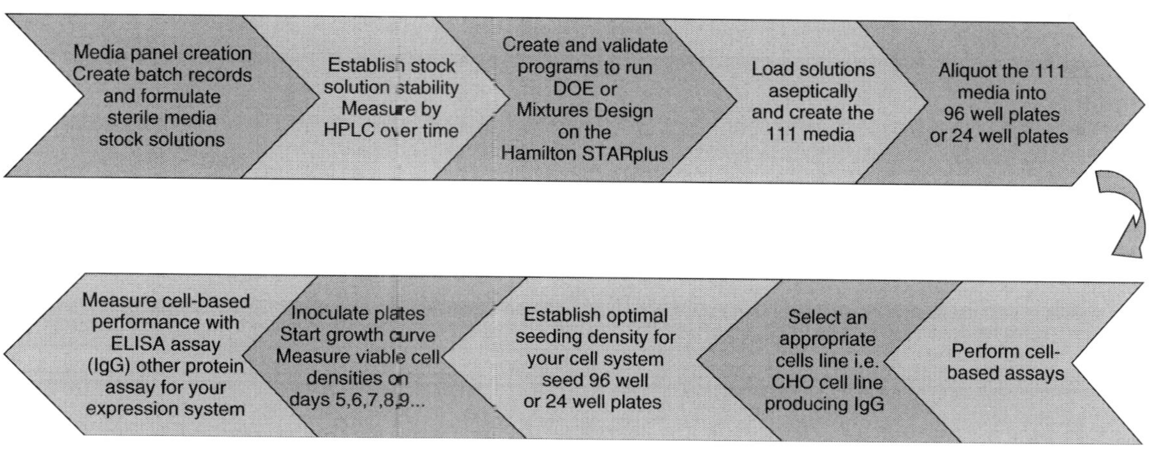

Figure 14.6. Media panel workflows. (This figure is available in full color at http://onlinelibrary.wiley.com/book/10.1002/9780470054581.)

parameters that had not been fully vetted. For example, pH (within ± 0.2 pH units) and osmolality (within −5 to +20 mOsm range) adjustments on each media condition had not been identified as a critical step in the workflows. In a traditional experiment, both of these parameters are typically measured at the end of the media building process and adjustments made from there, but translating that process into a media panel with >100 media with volumes too small for an accurate/sterile reading was a challenge. Additionally, where it is possible to take a time course of samples from a more traditional flask-scale study, the media panel volumes are so small that to measure cell performance requires either duplicate plates for measuring multiple days, or selecting a single endpoint.

Finally within workflow integration there is a need to carefully consider the mechanism and workflow of interfacing with other laboratory equipment instrument. How easily can the samples and experimental data translate between the systems? Just because each instrument works well standalone does not mean that they will work well together. For example, while growing cells in a 24-well plate format, there should be a clear plan of how to transfer and track samples in a 96-well plate sample analysis platform. Bioprocess scientists integrating new HTT should be aware of situations like these, where the technology format forces a different way of thinking about how to design and conduct experiments. As a large amount of data and many different kinds of data will be generated using a HTT system, a cell culture database with cell culture applications and very strong data mining capability integrated into the workflow can be helpful. The database should maintain data integrity and support analysis of information from different instruments.

14.2.2.6 ***Identify Ongoing Support Requirements.*** Fortunately, with many of today's HTT systems, implementation of the technology is substantially assisted by the hardware developers through imbedded support with service contracts. Users should take full advantage of this support, while some degree of self-management in troubleshooting the equipment and automation is constructive. Additionally, as many of the newer process technologies are being launched by early stage companies that may or may not be around in several years, learning the vendor's troubleshooting methods will protect the long-term investment in the platform.

With this in mind, a supply agreement should be secured that covers proprietary parts and consumables to ensure against times of limited supply and to provide a baseline for future part prices. This should also enable critical updates in key parts of equipment and prevent outdated hardware.

14.2.2.7 ***Key Takeaways.*** Performing a process analysis to identify the implications of HTT integration is best carried out before the technology is acquired. This analysis may include process mapping and careful consideration of current practices. Bottlenecks in the process should be identified and compared with those anticipated after integration. Adjustments to the workflow may be required to make an HT process more effective. An integration plan, similar to Fig. 14.7, can incorporate the elements of equipment selection and qualification, workflow validation, and future technical support with supply guarantees.

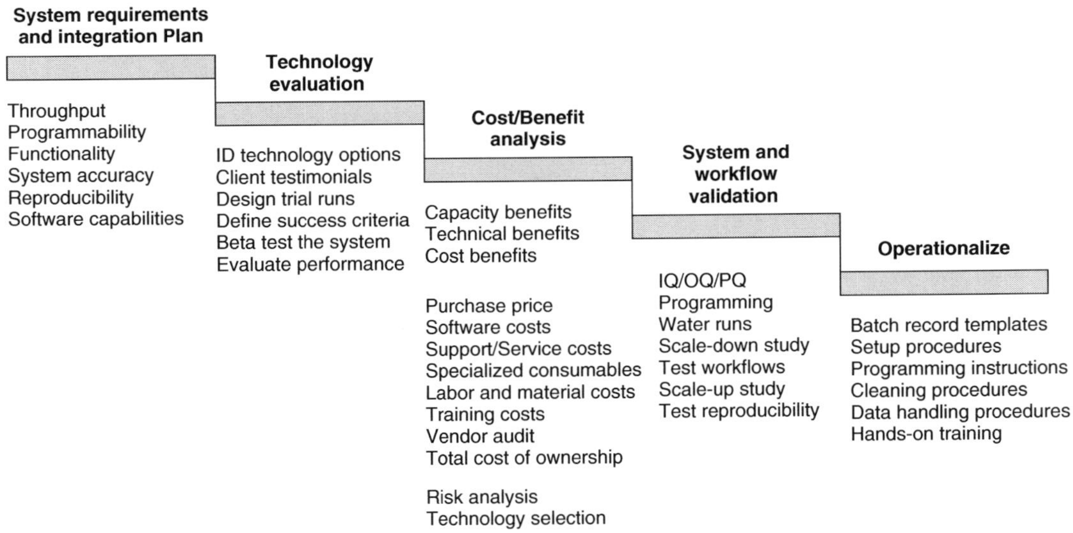

Figure 14.7. Gantt chart for HTT integration. (This figure is available in full color at http://onlinelibrary.wiley.com/book/10.1002/9780470054581.)

14.3 HTT APPLIED TO DOWNSTREAM PURIFICATION DEVELOPMENT

The majority of purification protocols require more than one step to achieve the desired product purity. The key to successful and efficient protein purification is to select the most appropriate techniques, optimize their performance, and combine them in a logical way. The use of complementary techniques to generate the highest purity while minimizing the number of steps required to maximize yield are key goals to strive for. Simply by organizing the techniques in a logical sequence may help eliminate some conditioning steps, resulting in a more efficient process.

Various techniques are included in the downstream purification, but the main challenges concern screening and optimization of the chromatographic steps included. High-throughput process development (HTPD) is a frequently used term for HTT applied to downstream purification development.

14.3.1 Overview

The purification process can be considered as a three-stage process for removal of product contaminants as shown in Fig. 14.8 (in addition to 2-steps for viral removal). The purpose of the first stage is to capture the target protein and concentrate it, the second is to remove most of the impurities, and the final step is to remove remaining trace amounts of impurities (polishing step). In practice, one stage does not always correspond to one chromatographic step. There are situations where two stages may be combined in only one operation, but on the other hand, one stage may require two separation steps to achieve the purity required.

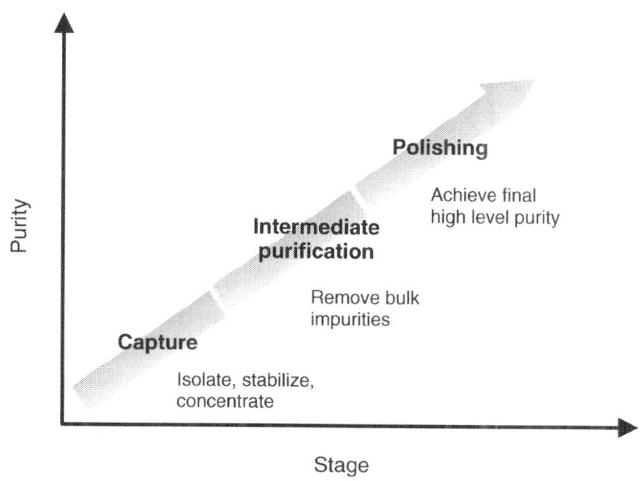

Figure 14.8. Schematic strategy for chromatographic purification of biopharmaceuticals. (This figure is available in full color at http://onlinelibrary.wiley.com/book/10.1002/9780470054581.)

If a generic purification process that fits a large number of target molecules can be found, it would only need to be slightly modified as each new target molecule is discovered. This may be a realistic approach in a facility working with monoclonal antibodies (mAbs) expressed in a standard cell host, as the mAbs will be closely related, although not identical (27). Although a completely generic process may not be attainable, a platform approach for downstream processing of mAbs has proven highly successful (28).

To take a biopharmaceutical drug candidate rapidly through the development phases, the purification system should be chosen at an early stage in the process. At this stage, typically sample material is available only in limited quantities and, thus, it is important that the screening and optimization methods involved require as small a sample amount as possible.

Lately parallel formats (both column and batch formats) have been introduced to speed up the development effort (29–33). Using the parallel formats for a first screening of chromatography resins and conditions enables exploration of a large experimental space within limited time and at a reasonable cost of materials. Once a more narrow experimental space has been identified, the optimization should be done in traditional column format and then further upscaled to the process columns (Fig. 14.9 and section titled "Downstream Design Case Study: Optimizing Conditions for Antibody Purity and Yield"). The increased knowledge gained in the screening phase will be of great value later on in troubleshooting situations and is in line with the recommendations put forward by the FDA initiative, Quality by Design (34).

In addition to the speed in development using the parallel format, a large amount of information can be obtained from each gram of protein. Typically, a binding study exploring 15 different conditions in 1 mL columns will require about 1 g of protein. The corresponding experiment performed

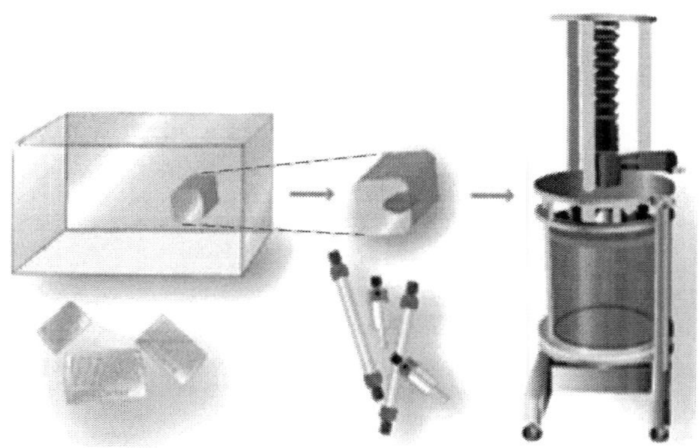

Figure 14.9. Conceptual visualization of a workflow for process development. (Reproduced with permission from GE Healthcare.) (This figure is available in full color at http://onlinelibrary.wiley.com/book/10.1002/9780470054581.)

with three replicates in a 96-well filter plate format containing 2 μL resin per well would need about 3–5% of that amount, approximately only 30–50 mg.

14.3.2 High Throughput Formats in Downstream Development

Frequently, a downstream purification process contains more than one chromatographic step. Optimizing two or more steps simultaneously or optimizing the steps one by one should be considered. So far, the most common approach has been to optimize the steps one by one and to designate each step to a specific task (Fig. 14.8).

Experiments in batch format such as a finite bath format have gained interest in the fundamental studies of chromatographic principles (35–38). By scaling down the experimental liquid volumes to a 96-well format compatible with liquid handling equipment including multipipettes and robotic systems, the batch format has evolved as an interesting tool for process development. Early publications (29,39) have demonstrated the possibility of using the batch format as the initial screening tool. The amount of chromatographic resin used in these experiments was 100 μL, but lately significantly lower resin volumes have proven useful.

Batch experiments may be performed in two different types of 96-well formats, which are wells with (29) and without (39) a filter at the bottom. With filters present, liquids may easily be removed from the wells and collected in 96-well collection plates. The experimental setup mimics the chromatographic steps; in other words, equilibration of resin, loading of sample, wash after loading, and finally elution, which may be followed by a strip step. Prefilled filter plates (*PreDictor*™, GE Healthcare) with high reproducibility within plate of resin volume are available. Plates with both equal and varied resin volumes from 2 μL to 50 μL have been developed. Plates may also be filled in-house, and depending on type of study and need for precision in data, the reproducibility in resin volume within the plate should be carefully considered.

Downscaled chromatographic columns filled with chromatographic resin are also available as tools for parallel screening, although they were initially developed for sample preparation purposes. Small pipette tips filled with resin may be bought from PhyNexus and Millipore. The latter has mainly been used for desalting before mass spectrometric analysis, while the PhyNexus tips have proven useful for purification of viruslike particles (40).

The packed bed format has also been miniaturized, both as in-house solutions and as marketed products. The commercially available columns from Atoll (Atoll Gmbh, Weingarten, Germany) are available in different resin volumes, 50–200 μL and can be run in parallel as sets of eight (33).

14.3.2.1 Column Format: Important Considerations. The possibility to perform experiments in a parallel format enables more experiments to be carried out in less time. The experimental outline of the studies is similar to conventional column format, but there are some considerations when downscaling.

When using conventional (nonparallel) columns, a detector is included in the system monitoring the effluent, enabling instant response to the changes in protein concentrations. In the parallel format, the liquid volumes involved are low and protein concentration in liquids is determined offline in collected fractions of liquid. The commercially available eight-in-a row columns have been used together with a robotic system from Tecan (Männedorf, Switzerland). The liquid handler included in this system is operated by liquid, not air, a prerequisite for controlled delivery of liquid, in a situation where back pressure is created by the columns. Care must be taken to control dispensing velocity and still, on the effluent side, the accuracy in collected volume can be less than perfect. One drop of liquid is approximately 25 μL (depending on solvent composition). Fraction volumes collected are typically in the 100–350 μL range and thus, if performing analysis on the total liquid volume, for example, UV absorbance, the liquid level in the well has to be monitored. For this purpose, the Tecan system is equipped with conductivity probes integrated in the dispensing unit.

When gradient elution is performed in the small parallel column format, an approach slightly different from the conventional one is required. In a robotic system, each column is connected to a single pump via a pipetting needle and, thus, a change in liquid composition has to be made in steps and cannot be performed continuously.

14.3.2.2 Batch Versus Column Format: How Does It Work? A column is packed with chromatographic resin and a liquid is pumped through this packed bed. Sample molecules introduced into the liquid stream are distributed between the resin and the liquid phase depending on the physiochemical properties of the components involved. During a chromatographic run, this partitioning takes place several times, although the result is monitored only in the effluent of the column. Thus the resulting chromatogram is a consequence of a large number of sequential distributions. If considering only a very small section of the column, this volume should be considered at a specific time as having a certain distribution of the protein between the beads and the liquid phase. This distribution can also be obtained in a small vessel, in other words, in a batch format. Thus, the column experiment can be looked upon as a system comprising a large number of consecutive batch experiments (Fig. 14.10).

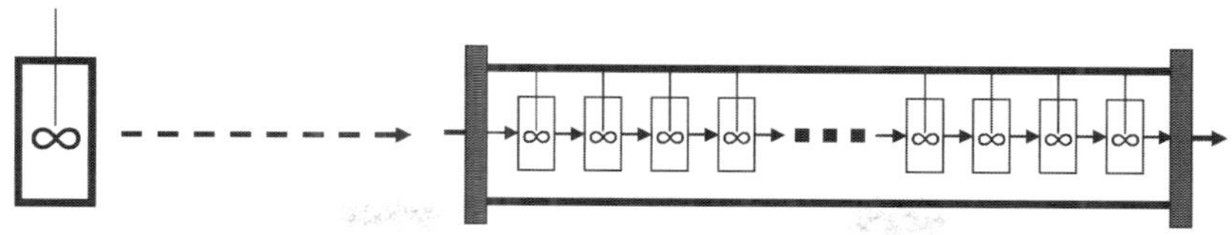

Figure 14.10. Schematic relation between a batch system and a chromatographic column. (This figure is available in full color at http://onlinelibrary.wiley.com/book/10.1002/9780470054581.)

The distribution of a given target protein between a chromatographic resin and the liquid phase is strongly dependent on a number of factors that are often in focus to optimize a purification process. The factors can, for example, be resin type, liquid composition, contact time (or flow rate), and temperature. It is important to have good control of all these factors independent of whether experiments are performed in a batch or in column format.

14.3.2.3 Batch Format: Principles and Important Considerations. In a finite bath (batch) system, the target protein is either found bound to the resin or free in solution. Initially, the protein is only present in the liquid phase, but with time, the distribution to the solid phase increases. In Fig. 14.11, the uptake is illustrated both as a change in concentration with time in the (a) liquid and (b) solid phases. Thus, if only one incubation time is to be used in a study, it is very important to set and control that time carefully, especially if short incubation times are used where a small time difference has a large effect on capacity.

Batch experiments may be performed in any vessel with low protein adsorption properties. The liquid phase should be particle free when analyzing for protein concentration, and if not using filter plates, care has to be taken only to remove liquid, not resin, in each step. Removing the liquid through the bottom of the well using vacuum or centrifugation is a straightforward approach. Each step in the chromatographic process can be mimicked in the filter plate, as the liquid phase is easily removed between each step, see Fig. 14.12. If all fractions are collected and analyzed, the mass balance can be calculated and thus the quality of data can be assessed and the total recovery in the bind–elute system used can be estimated.

The amount of protein added can be found either as unbound protein or as bound, in other words:

$$m_{\text{added}} = m_{\text{bound}} + m_{\text{unbound}} \Leftrightarrow m_{\text{added}} = m_{\text{FT}} + m_{\text{wash}} + m_{\text{eluate}} \qquad (14.1)$$

where, m corresponds to mass and FT to flow through.

The recovery can be calculated from the amount protein found and the amount added.

$$\text{Recovery (\%)} = 100 \times \frac{m_{\text{bound}} + m_{\text{unbound}}}{m_{\text{added}}} \qquad (14.2)$$

$$\text{Recovery (\%)} = 100 \times \frac{m_{\text{FT}} + m_{\text{wash}} + m_{\text{eluate}}}{m_{\text{added}}} \qquad (14.3)$$

For the batch experiments in general, it is important to have sufficient agitation of the sample, as diffusion in the liquid phase will be a rate limiting step in the distribution to the resin. For experiments performed in the small 96 wells, using an orbital shaker is recommended. The efficiency of the mixing is a result both of liquid volume and intensity in agitation, as illustrated in Fig. 14.13. For PreDictor

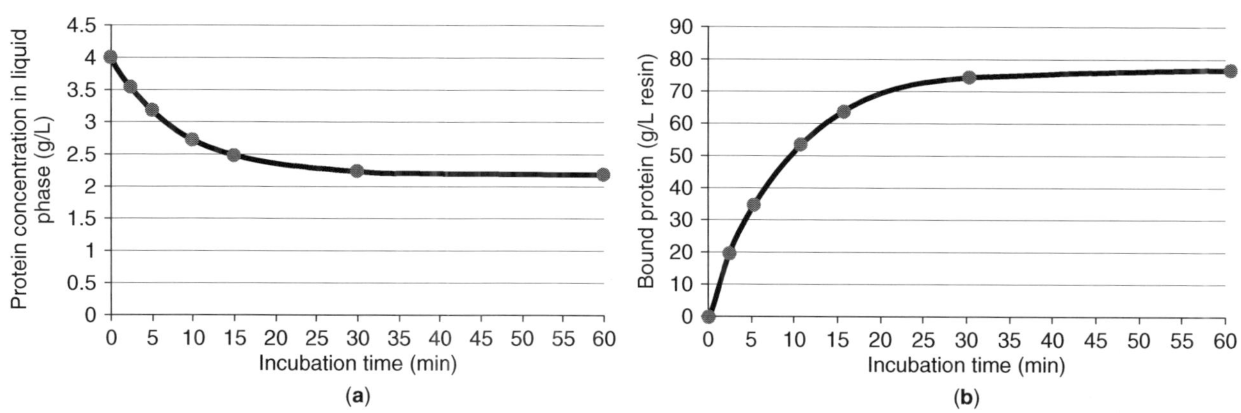

Figure 14.11. Illustration of change in protein concentration (a) in liquid phase, (b) bound to chromatographic resin. (This figure is available in full color at http://onlinelibrary.wiley.com/book/10.1002/9780470054581.)

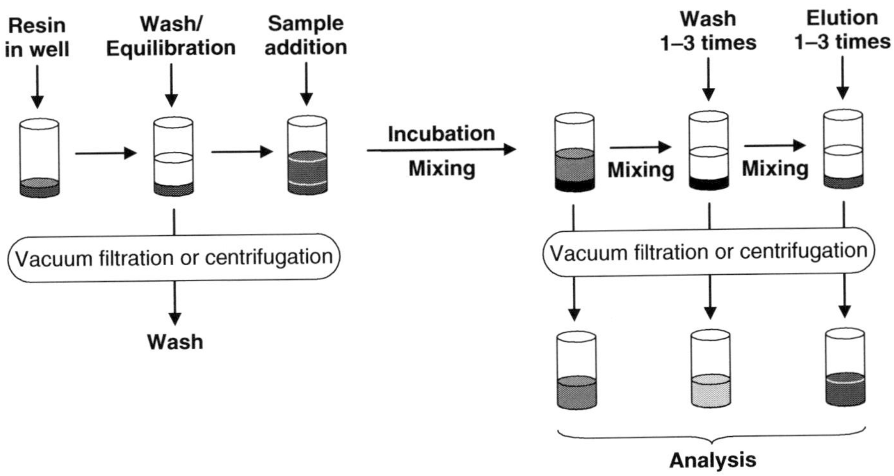

Figure 14.12. Schematic illustration of the workflow of a batch experiment occurring in the wells of a PreDictor™ plate showing the same steps as in a column experiment: equilibration, sample addition, wash, and elution. (Reproduced with permission from GE Healthcare.) (This figure is available in full color at http://onlinelibrary.wiley.com/book/10.1002/9780470054581.)

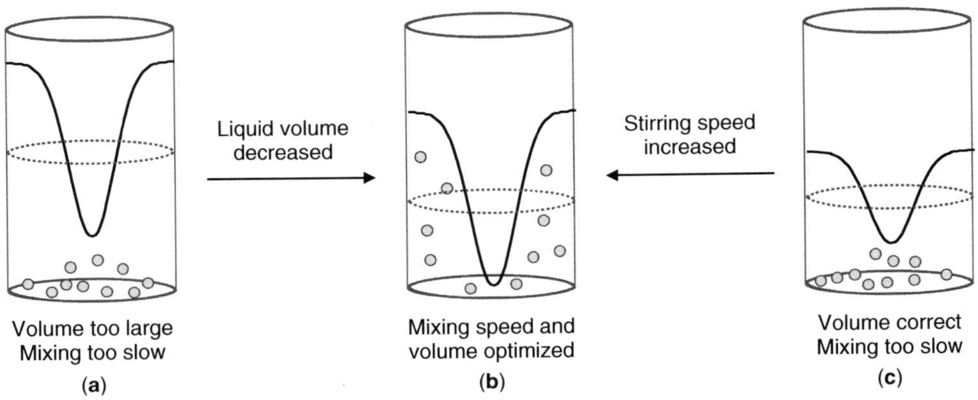

Figure 14.13. Schematic illustration of the effect of liquid volume and agitation speed on mixing efficiency. The situation before (red) and during (blue) mixing is shown. (a) Liquid volume too large, (b) liquid volume and agitation speed optimized, (c) agitation speed too low. (Reproduced with permission from GE Healthcare.) (This figure is available in full color at http://onlinelibrary.wiley.com/book/10.1002/9780470054581.)

plates having a well volume of 800 μL, a liquid volume of 100–300 μL, and an agitation speed of 1100 rpm with a 3 mm centripetal movement are recommended.

14.3.2.4 Binding Studies in Batch Format. Binding studies are performed for the following reasons:

1. to compare binding capacities for different resins;
2. to find the optimal conditions for binding (or non-binding in flow-through mode) of protein;
3. to study binding kinetics;
4. to determine the binding thermodynamics, which will give information on maximum capacity and binding strength.

The ability to perform a large number of experiments in parallel enables planning of studies that give answers to a number of different questions simultaneously. Thus, it is possible to combine in the same experiment, all the points listed above. However, the binding thermodynamics are often studied separately or with only a limited variation in other factors.

Although the determination of binding thermodynamics is not a prerequisite for performing binding and elution experiments, the quality in data, and fundamental knowledge of the separation system are significantly increased if these are performed. The thermodynamics of protein adsorption is described by the adsorption isotherm, in other words, a relation between concentration of protein

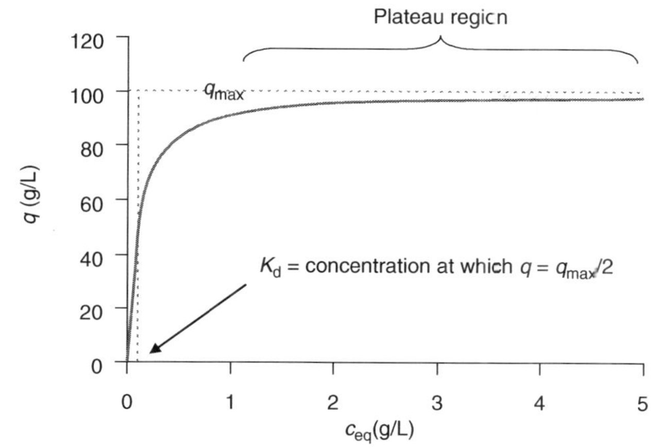

Figure 14.14. Langmuir isotherm, where $q_{max} = 100$ g/L and $K_d = 0.1$ g/L. Plateau region is indicated where the capacity is relatively independent of equilibrium concentration. (This figure is available in full color at http://onlinelibrary.wiley.com/book/10.1002/9780470054581.)

in the solid and liquid phases at equilibrium. For protein adsorption, the Langmuir isotherm is often used. A typical isotherm is shown in Fig. 14.14. It can be divided into the linear part where, binding capacity, q, is proportional to protein concentration in liquid phase at equilibrium, c_{eq} and a plateau part where q is relatively independent of c_{eq} and gives the maximum capacity q_m of the resin. The experiments can be performed either by varying the amount of protein added to the wells (by varying concentration or volume) or by varying the resin volume.

The binding strength, expressed as the equilibrium dissociation constant K_d, is equal to that liquid equilibrium concentration (c_{eq}), where capacity, q, is equal to half the maximum capacity ($q = q_m/2$), as can be derived from the Langmuir equation

$$q = \frac{q_m c_{eq}}{K_d + c_{eq}} \tag{14.4}$$

To reach equilibrium, long incubation times are required. Incubation times from 2 h (32) to overnight have been used (41). In many cases, an incubation of 60 min may suffice to obtain a satisfactory approximation of maximum capacity (Fig. 14.11b).

When performing binding studies, experiments should be performed in the plateau region of the adsorption isotherm if looking for maximum capacity q_m of the resin. Thus, the binding experiments are preferably performed with an excess of protein compared with resin volume in the well, in order to achieve saturation of the solid phase. To achieve this with a minimal use of protein, as low a resin volume as possible should be used. As a rule of thumb, the protein concentration should be no less than 50% of the initial concentration, when interrupting the equilibration after a defined time.

The result of a binding study for two different proteins on Capto™ S is shown in Fig. 14.15. The study was performed in a 96-well filter plate containing 2 μL of resin per well. In this study, 15 different loading conditions were investigated; three different pH; and five different salt concentrations. Each condition was repeated in triplicate, and the error bars show that the reproducibility of the results was good. As can be seen, the binding pattern obtained is protein dependent. When repeating the same study in the conventional column format (although only as single determinations), a similar result was achieved, proving the correlation between the two formats.

The kinetics in protein uptake to a chromatographic resin can be investigated in batch format by adding protein to the plate using a predefined time scheme, see Fig. 14.16. Adding operating conditions such as pH, buffer type to the plate experiment will provide valuable information of how these conditions affect the dynamic binding capacity (DBC). The prediction of DBC in column from batch data acquired in 96-well plates is enabled by using mathematic models (42).

14.3.2.5 Wash and Elution Studies in Batch Format.

After loading the sample to be purified, some of the impurities can be removed by including an intermediate wash step. In an analogous manner, all impurities do not have to be eluted in the elution pool. Studies on these two steps may thus improve the yield and purity of the chromatographic step.

14.3.2.5.1 A Wash Study with MabSelect SuRe™. To improve the mAb purity in the elution pool, an intermediate wash step with 17 different wash solutions was investigated. Analysis of both mAb and host cell proteins (HCP) in the elution pool was performed to maximize mAb yield and purity. As HCP levels were expected to be low, a resin volume of 20 μL was used to enable quantitative analysis by a generic ELISA method. And, it should be noted that for this type of study with a complex sample, it is important to mimic the column load of the resin. To reach a sample loading of 23 μg/μL resin as in the chromatography column format, it was necessary to perform multiple loadings of the mAb feed with a titer of 1.3 μg/μL. Three times loadings of 300-μL mAb feed were used, and for each load, the incubation time was 20 min. When comparing the results achieved in batch with corresponding experiments performed in 1-mL columns, the same ranking of conditions was obtained (Fig. 14.17).

14.3.2.5.2 Elution from MabSelect SuRe. In the wash study described above, the solution chosen to elute the

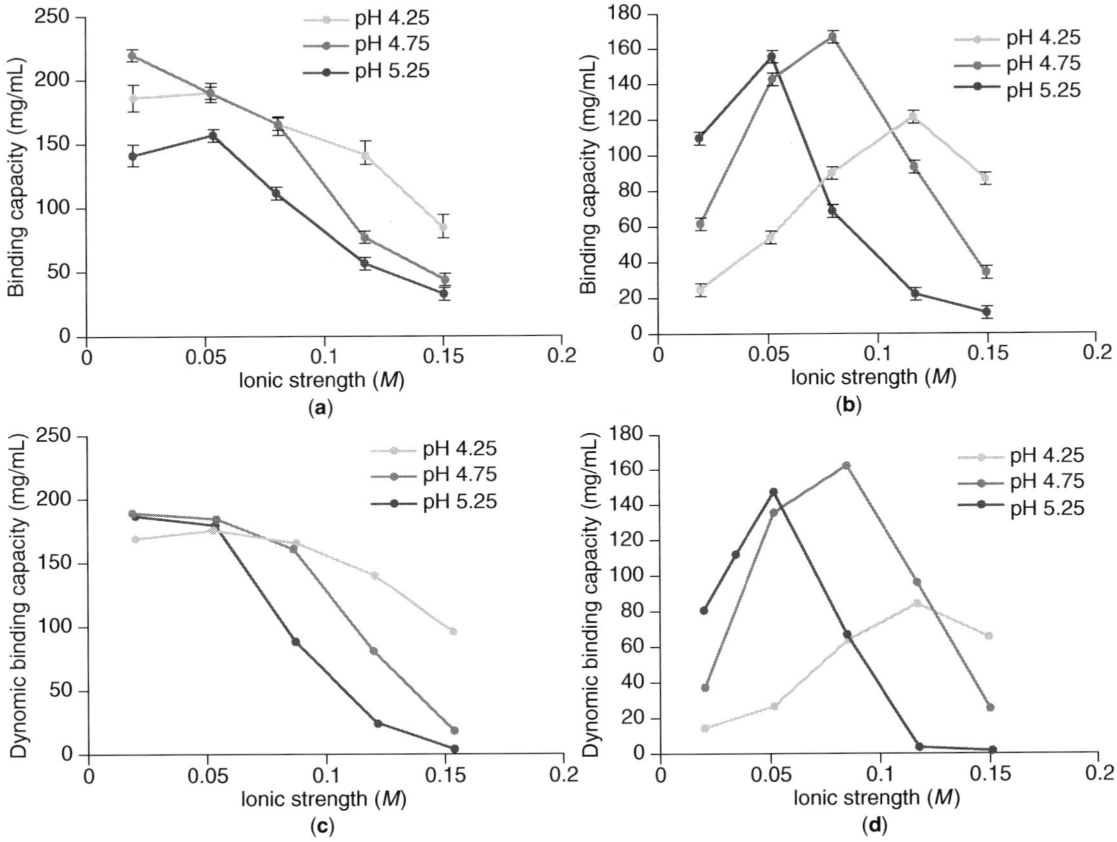

Figure 14.15. Determination of loading conditions for α-chymotrypsin and conalbumin on Capto S. Binding capacities of (a) α-chymotrypsin and (b) conalbumin, respectively, after 60 min incubation in PreDictor Capto S 2 μL plates. Dynamic binding capacities at 10% breakthrough for (c) α-chymotrypsin and (d) conalbumin, respectively. The residence time was 2 min and the column was a Tricorn™ 5/100 (CV 2 mL). (Reproduced with permission from GE Healthcare.) (This figure is available in full color at http://onlinelibrary.wiley.com/book/10.1002/9780470054581.)

mAb had a generic pH known to elute all mAbs. However, to improve stability of the mAb, it is advantageous to use a less acidic buffer if possible. In this study, five different mAbs and their elution at 15 different pHs (0.1 pH units successive difference) were studied. For four of five of the mAbs, the elution profile was as shown in Fig. 14.18. The elution started at about pH 4.4 and was complete at pH 3.6, the generic pH used in the wash study. The fifth mAb gave a slightly different result (not shown). It started to elute at a higher pH and was completely eluted at pH 4.4, and the pH interval for partial elution was broader compared with the other four mAbs. These results were confirmed in a column study, where a pH gradient was applied to the five mAbs. The chromatogram for four of the mAbs were identical (one of them is shown in Fig. 14.18), while the elution position and peak width of the fifth mAb differed in accordance with the plate results (elution at higher pH and broader elution peak).

This study also showed that the gradient elution obtained in a column may be mimicked by a set of wells with slightly varied pH.

14.3.2.6 Cleaning-in-Place(CIP) Study. Cleaning in place (CIP) and sanitization in place (SIP) are important parts of the operational cycle when preparing for reuse of the chromatographic column. During CIP remaining noneluted impurities are removed from the column and during SIP disease-causing microorganisms are reduced. Often, the conditions used for CIP and SIP are very similar, the major difference being a higher concentration of, for example, sodium hydroxide in the SIP compared with that in the CIP. It is therefore useful to monitor how the resin withstands these conditions, monitored as the protein capacity after treatment. The result from a study on two Protein A resins, MabSelect™ and MabSelect SuRe are shown in Fig. 14.19. NaOH and n-propanol concentrations were varied between 0.01 and 1.09 M

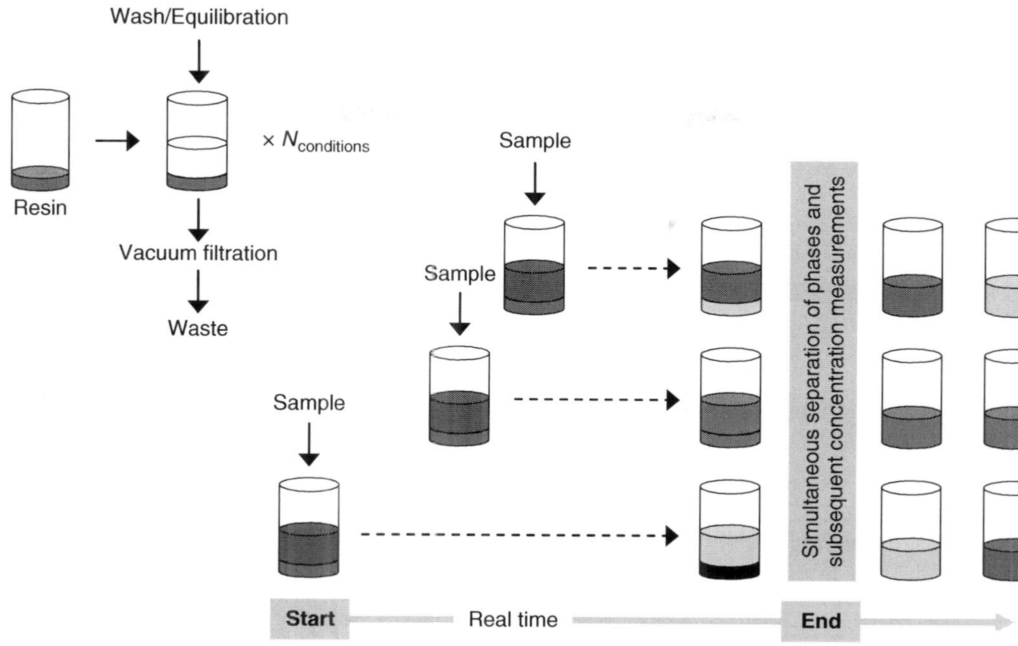

Figure 14.16. Outline of experimental protocol for the HT batch uptake method with variable incubation time. (Reproduced with permission from GE Healthcare.) (This figure is available in full color at http://onlinelibrary.wiley.com/book/10.1002/9780470054581.)

and 0–10%, respectively. To further stress the resin, the experiment was performed at 40°C in addition to the more normal 20°C. The media were stored in the CIP solution for 18 h, which corresponds to 180 cycles with 30 min CIP/SIP every fifth cycle. It could be concluded that the MabSelect SuRe resin, which contains alkali-stabilized Protein A, could be cleaned with 0.5 *M* NaOH and still retain good (>90%) protein-binding capacity.

14.3.3 Tips and Tricks when Planning and Performing a Study

Although the principles for performing batch studies are well established, there are several practical considerations to take into account. Below are some of the most common choices and issues that need to be addressed when planning a study.

14.3.3.1 Manual or Automated Systems. The first decision to make is whether to work manually or with robotics. For an inexperienced user of batch experiments, it is recommended to perform them manually. Once it is established that multiple experiments will be required, it is generally worth changing to robotics. Setting up a robotic system for any type of experiment takes time and experience but once up and running, the robotic system will free up personnel, although it should be stressed that it will not decrease the experimental time.

14.3.3.2 Replicates of Single Samples. In a 96-well plate, 96 different conditions can be used. It is however, wise to have at least duplicates, preferably triplicates, to track experimental errors. If only looking for trends, replication can sometimes be omitted (see the application example presented in the section titled "Downstream Design Case Study"). An upfront experiment can be performed to understand variance of replicates within an HTT system and will allow researchers to determine what levels of replicates are required for specific levels of precision in the results.

14.3.3.3 Resin Volume. The best resin volume to use depends on the type of study. For the evaluation of maximum binding capacity, the resin volume should match the expected binding capacity to reach saturation of the resin with reasonable low protein consumption. For wash and elution studies on the other hand, a higher resin volume may be needed primarily to obtain high enough concentrations of impurities for analysis. It can sometimes be necessary to use multiple additions of protein sample for these wash and elution studies, to obtain a representative sample load. For the PreDictor plates, recommendations are available for which resin volume to use depending on the resin and type of study. For binding studies, 2 μL and 6 μL resin volume per well is recommended, while for wash and elution studies 20 μL resin is recommended as a first alternative and if analysis demands are not fulfilled, 50 μL.

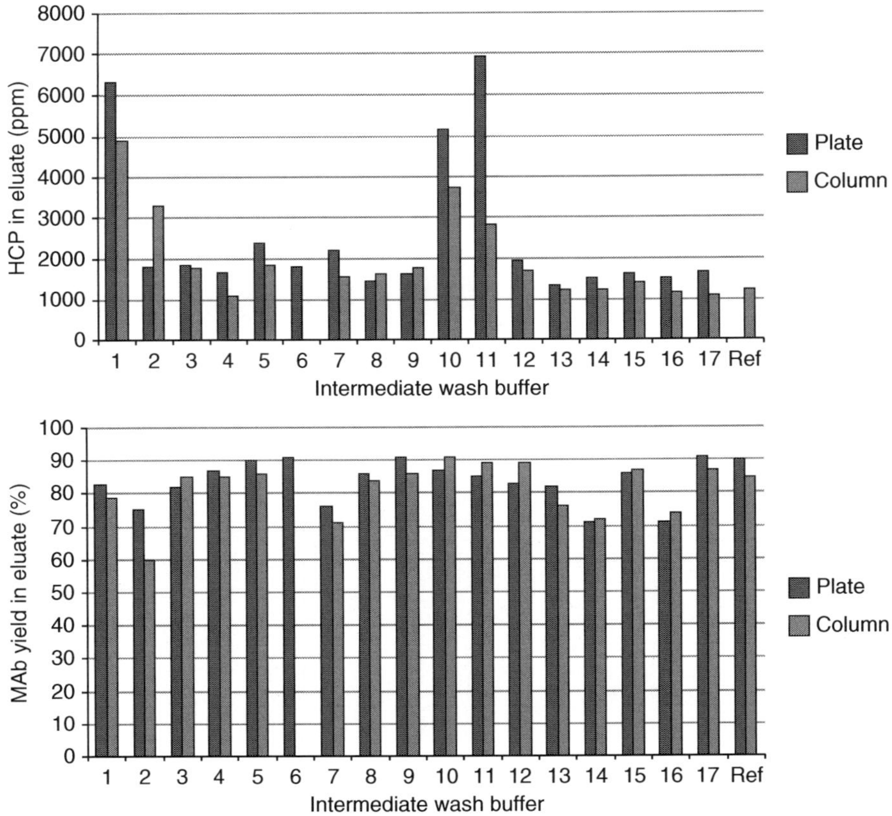

Figure 14.17. Effect of different wash buffers on the HCP levels and monoclonal antibody yield in the elution pool from MabSelect SuRe. Results from plate experiments are compared with column experiments. (Reproduced with permission from GE Healthcare.) (This figure is available in full color at http://onlinelibrary.wiley.com/book/10.1002/9780470054581.)

14.3.3.4 Experimental Outline for Dispensing of Liquid. Independent of choice of manual or robotic operation of the experiments, the experimental outline in the 96-well plate is very important for ease of operation. There are two options: (i) Make a mirror image of the plate when preparing buffers, so that 8, 12, or 96 solutions may be pipetted at a time. (ii) Plan the dispensing into the plate so that conditions are outlined in rows or columns.

14.3.3.5 Choice of Incubation Time. The incubation time in a batch experiment will influence the amount of protein bound to the resin, see Fig. 14.11. It is essential not to compare batch incubation time with residence time used to define column experiments. The residence time for a column study is often only a few minutes, although in practice this means that the contact time is considerably longer. As an approximation, it is better to compare batch incubation time with the sample loading time for a column.

14.3.3.6 Liquid Volume in Analysis Plate. For all analyses utilizing the total amount of liquid in the collection plate, the precision in the analysis will be improved if the exact amount of liquid is monitored. The plate readers from Molecular Devices have a patented well volume control called Pathcheck®, which normalizes the well absorbance to a path length of 1 cm (in other words, standard cuvette path length). The Pathcheck option may be used if well volume exceeds 100 μL.

Using a robotic system like the one by Tecan, well volumes may be monitored by the conductivity sensor present in the dispensing unit (33).

14.3.3.7 Liquid Composition: Limitations. Using the filter plate format may put some limitations on the types of liquids that can be used. Detergent type and concentration can cause leakage of liquid through the filter membrane, especially after repeated additions and incubation steps, and must be carefully monitored. To minimize risks of leakage, the drip tips of the plates should not be in contact with a surface. Capillary forces can otherwise increase the risk of leakage.

Also, high protein concentrations may cause foaming when applying vacuum. Centrifugation is recommended in this situation.

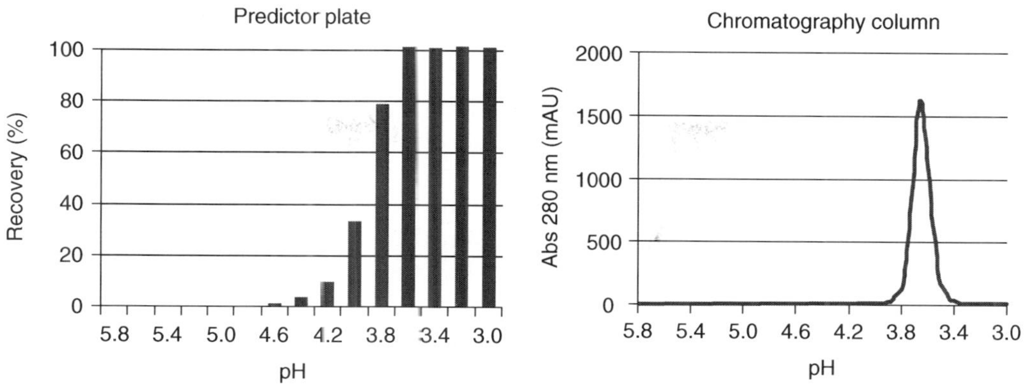

Figure 14.18. Comparison of elution patterns on MabSelect SuRe, PreDictor plates relative to column chromatography. The bars in the histograms correspond to the cumulative relative amount of mAb recovered after elution at different pH. The corresponding chromatogram is shown to the right. (Reproduced with permission from GE Healthcare.) (This figure is available in full color at http://onlinelibrary.wiley.com/book/10.1002/9780470054581.)

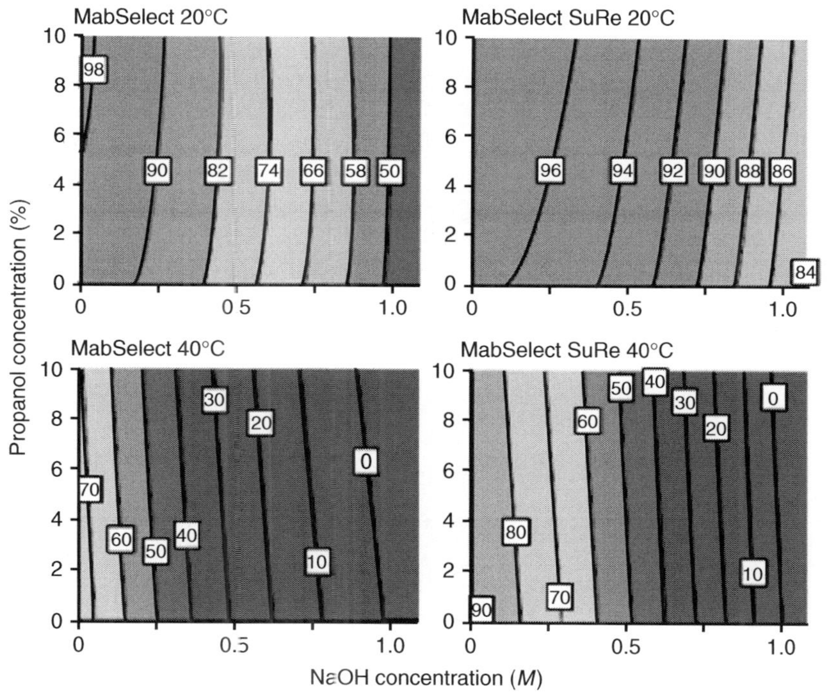

Figure 14.19. Contour plots showing the remaining IgG binding capacity (%) after exposure to CIP chemicals at two different temperatures. (Reproduced with permission from GE Healthcare.) (This figure is available in full color at http://onlinelibrary.wiley.com/book/10.1002/9780470054581.)

14.3.3.8 Number of Factors and Levels. The plate format enables screening at a large number of conditions in one single plate. This and the fact that often three or at most four factors are important for the result make planning of the experiments quite straightforward. The designs or plate layouts can vary from simple multilevel designs to the more advanced design of experiments (DOE) layout.

14.3.3.9 Planning of Experiments and Storage of Data and Results. Performing a 96-well study and analyzing a number of the steps will result in large numbers of data to collect, store, and evaluate. Calculations can be made in spreadsheets, for example, in Excel, and results are preferably presented as 2D and 3D plots. Simple mass-balance calculations and inspection of data can be performed on

Excel sheets, while statistical model evaluations may be performed using dedicated software. Still, the collection of data and computation of capacities, recoveries, yields, and so on require good spreadsheet design to be able to get a good overview of the results. To facilitate planning of experiments, as well as parts of the evaluation (excluding statistical modeling), the software Assist has quite recently (2009) become available from GE Healthcare. Experimental layout in the 96-well format, storage of data, raw data quality assessment, and mass-balance calculation are some of the features included in the program (Fig. 14.20).

14.3.3.10 Further Improvements in Quality of Data by Holdup Corrections. To obtain most accurate results in the computation of data, the retained liquid in the holdup volume should also be taken into account. After the removal of liquid from the filter plate by centrifugation or vacuum filtration, a small volume of liquid still remains within the pores of the resin and also in the filter itself. For example, when performing a binding study, the flow through liquid will not contain all unbound protein since a fraction of unbound protein remains in the holdup volume. If not accounted for, the consequence will be an overestimation of the binding capacity for the conditions used. This is most important when performing experiments for determination of the adsorption isotherms. The magnitude of the retained volume is about 50–70% of the resin volume, according to one publication (32), while the volume retained in the filter varies with filter plate used.

14.4 ANALYSIS NEEDS FOR HIGH THROUGHPUT FORMATS

With HT bioprocess development HTPD, a large number of samples are produced that must be tested and analyzed. If not addressed proactively, the analysis steps can become a surprising new bottleneck for workflows involving HTT. When planning the experimental work, one must decide the critical responses to be measured, how the samples are collected and analyzed, and how the data are collected and trended across studies. If done properly, the data collected from HT development can be leveraged, both to improve immediate outcomes, as well as make more intelligent de novo decisions about primary and secondary bioprocess factors in the future.

14.4.1 Identifying Critical Responses

Identifying the critical responses is a very important part for success in process development. The analysis strategy strives to optimize the number of analyses to perform on each sample and to reduce sample preparation before analysis. In HTT systems, researchers need to consider "nice to haves" versus "must-haves" as they design the experiment. The challenges met during analysis method development at the different stages in the process originate in the nature of the development objective. Upstream, the main challenge is to perform analysis on a rather crude yet consistent sample matrix, while downstream, the challenge may be to ensure that the varied sample matrix does not influence the analysis result from a purified protein sample.

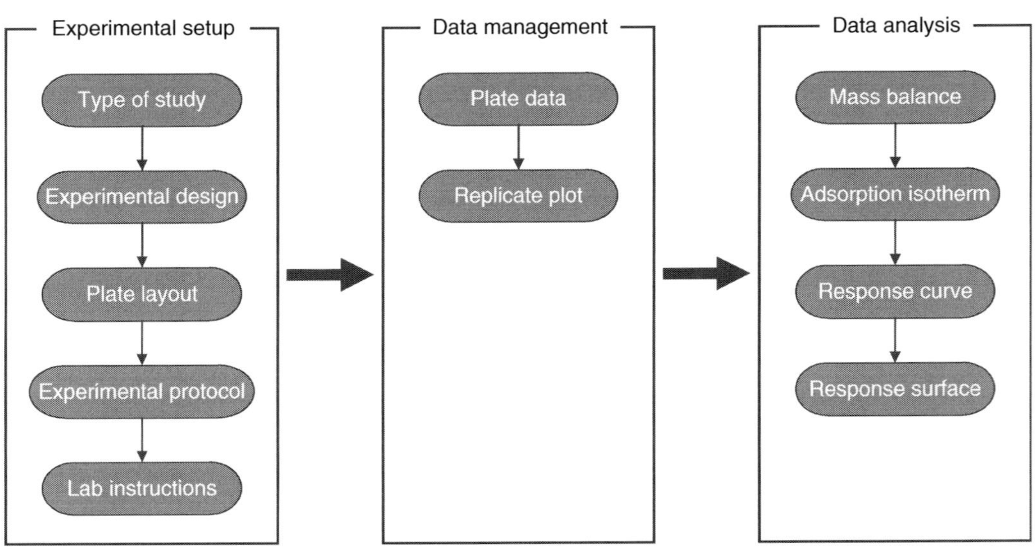

Figure 14.20. Workflow for planning and execution of HT experiments in 96-well plates as depicted by the Assist software. (Reproduced with permission from GE Healthcare.) (This figure is available in full color at http://onlinelibrary.wiley.com/book/10.1002/9780470054581.)

14.4.2 Sample Analysis as a Bottleneck

Given the order of magnitude increase in samples to be analyzed with HTT, the appropriate analytical techniques must be matched with the throughput of the preceding HTT, otherwise, analysis becomes a bottleneck. Similar to the HTT discussed previously, many of the analytical techniques applied to HT development fall into a category of either enabling parallel assays for higher throughput or performing sequential activities in a shorter time.

Parallel assays are typically based on microtiter plate formats that allow testing a large number of samples. One legacy of small molecule drug screening has been the development of analytical methods with HT, such as absorbance- and fluorescence-based assays (43). Thus, analysis of total protein using, for example, Bradford and Lowry, as well as target-specific enzymatic assays (ELISAs and equivalent techniques), are quite straightforward to perform in a parallel format. Similarly, cell culture technologies (such as the Guava ViaCount and Invitrogen Countess) have been developed that allow for multiwell testing of cell viability and culture conditions. While useful as screening tools, the parallel assays can be limited in quantitative accuracy due to the trade-off of sample purity for assay capacity. In situations where the sample matrix is highly variable or requires subsequent purification, sequential assays may be used.

For sequential assays used in tandem with HTT, the goal is to reduce the cycle time per analysis. Additionally, sequential assays such as HPLC can also require sample volumes that may exceed the material available from an HTT and therefore, reducing assay scale becomes a secondary criterion. Despite advances in the sequential techniques, these assays often restrict on the type of studies that can be performed in a HT platform. For example, the chromatographic analysis of aggregate levels is strongly dependent on the column length, flow rate (and bead pore structure) and can take 10–20 min per analysis. Similarly, throughput and cost have traditionally limited the use of HTT for analyzing protein glycosylation responses (44), although recent advances in lectin arrays (45) and mass spectrometry (46) have increased throughput by paralleling and reducing cycle times, respectively.

One route for minimizing the analysis work in the HT domain is to save more thorough investigations on samples for later steps in the development chain, see Fig. 14.21 (47), in other words, when the experimental space has been narrowed down. This approach is effective for transitioning between analysis that can be performed in parallel to identify general trends and analysis requiring sequential assays to obtain more in depth characterization of specific responses.

Another approach is to perform a more thorough investigation only of the samples (wells) in the high-throughput format that fulfill, the critical prerequisite. For example, conditions resulting in high-binding capacity or high-recovery in elution is an approach suggested by Coffman *et al*. (32).

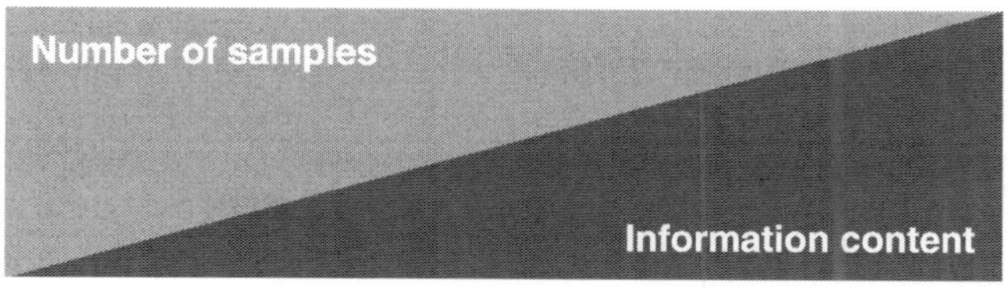

Figure 14.21. Tentative analysis strategy in HT screening. Correlation between number of samples to analyze and expected content of information for each sample. (This figure is available in full color at http://onlinelibrary.wiley.com/book/10.1002/9780470054581.)

Once sample analysis is complete, matching the experimental conditions from HTT with the analytical result can become a frustrating effort if the logistics of the sample identification through the workflow is not developed ahead of time. This coordination is not hard, but a plan for maintaining sample identity between a randomized sample set coming out of the HTT and a separate analytical instrument is needed, both of which will likely have differing nomenclature and database structures. This inefficiency can become very time consuming for scientists who just want to analyze their data.

14.4.3 Data Collection and Trending

The utilization of HTT can generate a great deal of data but how can researchers ensure the information is useful? It is generally challenging to handle such rich data to identify the best bioprocess condition. Furthermore, with data sets generated from instruments with different data formats, matching and interpreting information can be very challenging. For example, technologies such as SimCell can have hundreds of microbioreactors running concurrently under different conditions with different cell lines and media. Information such as cell culture conditions, cell line histories, cell culture media, feed and supplements, and processes must be recorded and analyzed over time together with cell performance data. One of the most common mistakes while using HTT is to become overwhelmed with data to a point where one ignores a critical response to process variations. Over time, a well-designed database with searching and trending capabilities becomes necessary to fully take advantage of these technologies.

To support scientists with process optimization, data systems should be able to simplify the information infrastructure by

- consolidating storage of data;
- streamlining data input from different analytical systems;

- integrating statistic experimental design and data analysis tools such as DOE software;
- providing an easy-to-maintain integration infrastructure;
- providing an extensible and adaptable architecture for unique workflows;
- ensuring data are stored in a secure manner;
- making sure the right data are available to make decisions and move forward.

14.5 EXPERIMENTAL DESIGN FOR HTT

A summary of HTT would only be partially complete without also discussing how to use these tools in concert with experimental design. Experimental design is the third leg of the stool, complementing the technologies and the analytical competencies required to capitalize on the possibilities of HTT. Only a systematic strategy can identify which of the variables are critical factors or interactions to be optimized.

14.5.1 Introduction to Design of Experiments

A successful development approach should determine optimal levels of critical variables and acceptable ranges of all variables. Methods frequently used for optimizing a development approach fall into the general discipline called design of experiments (DOE). Although there are more comprehensive and mathematically precise definitions within this field of study, a DOE is a statistically sound method of generating high quality information about responses to controlled changes of input variables. It often includes an element of efficiency following a strategy of testing the fewest number of individual run conditions needed for obtaining an acceptable quality of response data.

14.5.1.1 Two-Level Factorial Analysis. One of the most efficient tools for screening and optimizing is the two-level factorial, in which each input variable is tested at each of the two levels in a combinatorial pattern (48). Including the various types, the two-level factorial is probably the most often applied DOE method in biopharmaceutical development. Fractional two-level factorials can provide very high efficiency for testing many factors, but they sacrifice some information about interactions. For example, 15 factors can be screened at two levels each, using only 16 individual runs, which is called a *saturated factorial design*. Compared with 32,768 runs for a full factorial (2^{15}), this 2^{15-11} design seems much more efficient but only if none of the factors has any interactions with other factors. In the 2^{15-11} example, every single-factor effect is aliased (confounded) with seven different two-factor (2-F) interactions. The researcher does not have any way of determining if an effect is due to a single factor or to a 2-F interaction, or to some combination. Fractional two-level factorials are geometric, meaning that the number of runs (N) increase by the power of 2 (2, 4, 8, 16, 32, etc.). Each interaction in a geometric design is aliased to a single individual effect. Therefore, it is possible to build additional experiments upon a fractional factorial to resolve specific interactions of interest.

A similar type of two-level factorial, but where N is a multiple of 4 (4, 8, 12, 16, 20, etc.) is a family of nongeometric designs called Plackett–Burman designs. Unlike the geometric designs, every single-factor effect is partially aliased with all 2-F interactions not involving it. For example, to test 15 factors in 16 runs using this design, each main effect is aliased with 91 different interactions. There isn't any systematic and simple method to resolve important interactions from these designs. A researcher should be very careful in using Plackett–Burman designs and always follow up with other studies to confirm analyzed results. It seems surprising that many publications in biotechnology report results from this design, with little attention to biochemical complexity and interactions inherent to biological systems.

Living cells grown in complex environments naturally display many important interactions among input variables. For example, the effect of iron concentration upon cell growth may depend upon the concentration of an iron chelator compound. Likewise, an interaction may exist between the iron species compound and other molecules or physical parameters, such as temperature, pH, or oxygen tension. Optimal levels of each interacting factor cannot usually be predicted from prior knowledge because of the complexity of the interactions and association with other interactions. Rather, optimal levels of interacting factors within biological systems are determined empirically through properly designed experiments. Saturated factorial designs may appear adequate to screen input variables in the auto industry, but larger fractions of full factorials are needed in the biopharmaceutical industry.

Although true three-factor (3-F) interactions in biological systems may exist, these are generally believed to be rather rare and of low magnitude. Two individual 2-F factorials with one factor in common are found often, but this situation is different from a true 3-F interaction. It is important to identify all 2-F interactions during product development. The smallest two-level factorial design that can resolve each 2-F interaction from other 2-F interactions is called *a resolution V design*. Screening 15 factors by a resolution V fractional factorial (2^{15-7}_{V}) would require 256 individual runs. While it seems a large number compared to the 16 runs of the saturated design (2^{15-11}_{III}), the resulting data analysis is immediately clear regarding all interactions.

Alternatively, a smaller resolution IV designs (2-F interactions aliased only with other 2-F interactions) can be used in sequential combination to resolve selected sets of aliased interactions. These folding-over designs can isolate certain 2-F interactions, depending on the previous design and which group of interactions is to be targeted. The cost is primarily associated with the time for sequential experiments. Given that biological cell growth experiments typically require 1–3 weeks and that numerous interactions are likely to be detected, the folding-over approach may cost an additional month beyond an initial resolution V experiment. On the other hand, for product purification experiments, which require far less time, the sequential folding-over approach to identify important 2-F interactions may be preferred. The choice between these approaches also depends upon access to technology for experiment setup, execution, and analysis.

A relatively new partial factorial has been developed (49), called a "minimum run" factorial, which further reduces the number of runs needed for resolution V by minimizing redundant information about higher level interactions. The statistical software program Design-Expert (Stat-Ease, Minneapolis, MN) includes these minimum run resolution V designs for 6–50 factors. Testing 15 factors requires only 122 runs to resolve all 105 possible 2-F interactions.

Other experimental designs for resolving 2-F interactions include the D-optimal designs. These can be generated and analyzed using many statistical software packages. They are usually not orthogonal designs, meaning that effect estimates are not independent and may be correlated. Reasons for selecting a D-optimal design include the need to reduce the number of runs and accommodate a substantial constraint in the design space (e.g. levels of certain factors in combination are not feasible). D-optimal designs may be better used as a response surface method to evaluate curvatures. For more discussion on D-optimal designs and augmenting the computer generated design points, a good publication is available (50).

The 2-level factorial design identifies important effects and interactions when the levels of each factor are selected appropriately. There are three major cautions about selecting the factor levels: (i) If the two levels do not cover a sufficiently wide range, a researcher may miss an important effect that may be encountered during scale-up. (ii) If one of the factor levels is too high (such as toxicity) or too low (for example, essential nutrient), the measured responses at those levels may be drastically low, which hinders the ability to detect other effects. (iii) If there is substantial curvature in the measured response when moving from the low to high level for any factor, the model may often underestimate the impact of changing the factor level. Whereas the first two of these can be abrogated by prior knowledge or preliminary titration experiments, curvature with the response space is not easily predicted within a large factorial.

For this reason primarily, when setting up two-level factorials, it is common practice to include a replicated set of center points (CPs). A CP is an experimental condition with all factors at their midpoint levels; slight variations are sometimes needed for blocked designs or when using a categorical input variable (for example, incubator unit). If there isn't any curvature within the design space, the CP response for the linear model is predicted to be the average among all factor responses. But if the response space includes significant curvature, the linear model generated by ANOVA will differ from the measured average of the CP runs. In that situation, the researcher is alerted to the existence of significant curvature during analysis, but it is not possible with this data to determine which factor(s) are responsible for curvature.

14.5.1.2 Response Surface Analysis. A frequently employed method of measuring curvature within each individual factor response is to apply a central composite design (CCD). This is composed of a two-level factorial with replicated CPs, plus additional points called axial or star points. An axial point is a design point whereby the level of one factor is either high or low, but all other factors are at their CP levels. For each of the factors in the design, one high and one low axial point are generated, often further away from the CP than are the factorial levels. However, a face-centered CCD includes axial points assigned the same levels as the factorial (Fig. 14.22). Face-centered CCDs are preferred when it is impossible or impractical to extend the design beyond the boundaries of the two-level factorial, but they still enable measurement of individual factor curvature. It is often thought more important to replicate the axial points than the factorial points because their contribution to ANOVA is not derived from averaging with other design points. For each factor there are $2n$ axial points, where n = number of replicates selected. In the example mentioned earlier, for testing 15 factors in CCD, the number of duplicated axial point runs would be 60, ($2 \times 2 \times 15$). Along with 6–12 replicates of CPs, the minimum run resolution V type of CCD would contain a total of 188–194 runs. Unlike factorial design points, axial design points cannot be fractionated into smaller standalone experiments. The entire CCD can be performed together or only the additional axial and CPs can be run separately from the two-level factorial and analyzed together in two blocks. Replicated CPs must be included in each of the two blocks to link them during analysis. Alternatively, the entire CCD may be run at the same time if technical challenges are not too great for the number of runs required

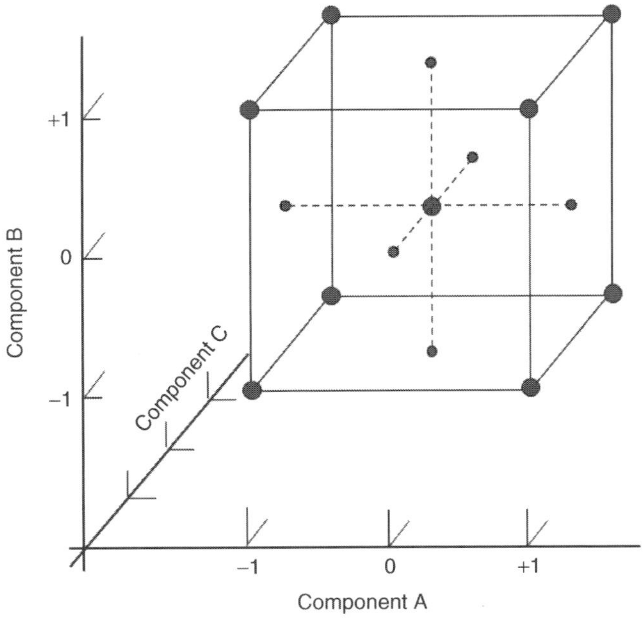

Figure 14.22. Schematic illustration of design points for a three-factor face-centered central composite design. (This figure is available in full color at http://onlinelibrary.wiley.com/book/10.1002/9780470054581.)

Analysis of CCD falls into a category called *response surface analysis* (RSA), also referred to as *response surface methods*, *response surface methodology*, or *response surface modeling* (RSM). In an RSA type of DOE, the selection of significant factors, interactions, and curvature terms are often guided by a statistical software package, resulting in an ANOVA model or a polynomial regression model (51,52). Compared with two-level factorials, RSA provides a better picture of how a measured response is likely to change because of variations in a factor level, including interactions among other factors. The responses predicted for two different factors at a time can be depicted on a three-axis plot with response surface in the vertical axis. For more factors, the statistical software can reveal which 2-F combinations appear to be most critical and/or most interesting. The model-predicted responses to these two input variables may be depicted while holding the other factors constant. The model-generated predictions can then be validated in subsequent experiments.

Another RSA method is a three-level factorial. Testing four factors at three levels each would require 3^4 or 81 conditions. Unlike two-level factorials, there is no clean way to fractionate a three-level design because 2-F interactions become aliased in complex ways. Therefore, true three-level factorials are only practical when the number of factors to be tested is low. The Box–Behnken design (BBD) is a useful three-level design for generating response surface information that can be obtained from multiple factors with a relatively few set of conditions. Rather than a true three-level factorial, a BBD combines a three-level factorial with a balanced incomplete block design (50). Fractional BBDs are efficient for testing many factors and generally require only slightly more runs than a minimum run resolution V type of CCD.

An important category of RSA methods is the mixture type of design. Whereas factorial designs have levels of factors that are independent of one another, mixture designs have factor levels that are not independent because different volumes or quantities are diluted to various proportions among other blended factors. For example, different mixtures of two culture media blended together in different proportions will have individual chemical component concentrations determined by these proportions and the two media formulations. A wide variety of mixture DOEs with various numbers of mixture factors and constraints are possible (53). The analysis of the mixture design response generally employs polynomial regression modeling. Mixture designs can quickly identify a suitable blend for further optimization, but it cannot give information about individual media component effects and interactions. Also, none of the component concentrations can exceed the maximum already present in one of the mixture factors, which limits the ability to explore optimal levels.

Researchers sometimes need to employ other statistically sound methods or general factorial approaches not restricted to 2-F levels. By following good statistical practices (GSP) a researcher can minimize bias and variability, regardless of the experimental design (54). The selection of a specific DOE should always be balanced with the key questions the researcher needs to answer, as illustrated in the case studies below.

Various simplex methods for optimizing multiple factors are nonstatistical tools that are self-directing (55). Simplex methods are iterative and find optimal solutions after several cycles. For example, after measuring results of the first set of test conditions, the algorithm derives the second set of conditions to test. Then the new data are used to derive the third set of conditions to test. An optimal set of conditions is discovered after several iterations. The greatest advantage of the simplex method is the relatively low number of conditions per run ($k+1$, where k = number of factors). However, there are at least three major disadvantages of iterative simplex methods: (i) To arrive at an optimum set of conditions may require a long time, especially if each of the iterations were 1–3 weeks long, as in cell growth assays. (ii) The iterative simplex method also does not directly measure effects and interactions among factors, important to understand and establish robust manufacturing processes. (iii) The method may indicate a local optimum, but miss finding the overall best combination within acceptable limits of the variables. However, the simplex method is a valuable tool for *in silica* derivation of the most desirable set of conditions when multiple response variables have already been analyzed. For example, using the Design-Expert software, each response can be given a relative importance level for calculating overall desirability. The maximum desirability is then derived by the simplex method and can be viewed graphically (52).

14.5.1.3 *Points to Consider.*

Using DOE to build a more complete understanding may lead to a more predictable and robust manufacturing process, but there are also some disadvantages or cautions to consider. For researchers unaccustomed to large DOEs, there can be a learning barrier associated with certain types of misapplications. For example, the concept of being able to test multiple factors simultaneously and with little, if any, true replication may seem strange. To compensate, researchers sometimes replicate every design point of a fractional factorial rather than choose single runs in a greater fraction design. As long as the design is able to resolve all 2-F interactions, this usually makes little difference. But if a low resolution design were selected because of space or resource limitations and the design was required to be replicated, the analysis could be difficult to interpret, possibly resulting in a follow-up experiment. Unnecessary replication may also restrict the design to fewer factors testable within a single experiment.

That being said, replication of design points is useful in at least three situations. First, when a process is believed to have high inherent variability, a lower resolution design could be replicated and then followed by another experiment to resolve interactions if needed. This replication lends higher certainty to detection of true differences (power of the test) and greater guidance in handling of statistical outlier data. Second, true replication enables the measurement of pure error during process or instrument validation. An example of this is presented in the upstream case study (below), part of the validation of the SimCell system. Third, process variability itself can be one of the response variables being optimized in a propagation of error (POE) experiment (52). Useful in Six Sigma initiatives, analysis of POE can reveal the response surface region of lowest inherent variability. Then by using the desirability function, POE results can be combined with other response data to predict the factor space of greatest robustness during scale-up and transfer to operations.

Another example of misapplying DOE methods involves generating a wide range of responses, but selecting only the best performing conditions without applying ANOVA or regression analysis. It is often tempting to display in a chart all the response values generated for the conditions, sometimes ranking them by response magnitude. There may be good reasons for doing that, such as during experiment process validation of repeated experiments. The problem occurs when a researcher or manager selects the highest performing condition and establishes this as the next best control. There is always more than 50% probability (usually much more) that this condition when retested will perform lower than its current level. After all, it was selected from the top of a random error distribution about an average effect. In special cases, as with the multilevel designs in the plate experiments in the downstream case study (mentioned later in the text), trends can be assessed without modeling because of the large number of factor levels within the specified range of the factor.

Another application problem relating to the learning curve involves the selection of appropriate and important factors and their levels. Selection of the best levels draws from prior knowledge of the process, as mentioned above. When many factors are desired, bundling them into factor groups may be needed. Determining which factors go together and the specific process of preparing reagents and cells for the experiment tend to be substantially more work than for the one-factor-at-a-time method of experimentation. As a result, the first complex DOEs nearly always take much more time from concept to start than initially thought. Subsequent experiments are usually faster to execute because many of the difficulties have been resolved and workflows established.

Sometimes there can also be a problem simply related to the generation of such a large amount of information. Although interesting interactions and curvatures may be discovered during DOEs, one should keep in mind the main objective, which is to make the most desirable outcome of the process predictable in scale-up and robustness. When many effects and interactions are measured, there is often an odd result merely because of random error in real data, but the focus should be directed toward larger high impact effects and interactions, allowing the small ones to vary. Too much information can also slow the decision making process; therefore, reports to managers not accustomed to DOE should emphasize the most important findings and give recommendations for next steps.

14.5.2 Upstream Case Study: Optimizing a Fed-Batch Cell Culture Process

Chinese hamster ovary (CHO) cells are currently the dominant choice within the biopharmaceutical industry for mammalian cell expressed human recombinant proteins. Whenever the internal biochemistry of a cell is manipulated, it is likely that the cellular requirements for nutrients and growing conditions change in response (56,57). Rates of growth and recombinant protein production would likely be influenced by such changes in recombinant CHO cells.

Researchers at Invitrogen planned an experiment to measure the impact of several parameters on growth and productivity of a recombinant CHO clone expressing a recombinant IgG. The goal of the experiment was to address four questions: (i) Will preconditioning of this CHO clone in glutamine-free medium prior to the experiment result in a different pattern of growth responses among the bioreactor conditions? (ii) What impact does pH have in fed-batch culture output? (iii) Which of four glutamine-free starting media work best in combination with which other conditions? (iv) How does the choice of feed solutions affect growth and protein production?

14.5.2.1 ***Experimental Design.*** Four factors were selected for testing in a general factorial type of DOE. Factor A-Cells was the preconditioning of the cells and held two-factor levels. CHO cells were split into two flasks and expanded for 2 weeks in either Medium A without glutamine ("adapted" cells) or in Medium A containing GlutaMAX® ("unadapted" cells). Factor B-pH was the pH during the SimCell culturing and it held 2-F levels (6.85 and 7.15). Factor C-Starting Medium was the culture medium into which the cells were inoculated into the microbioreactors and held four factor levels (coded as media A–D). Factor D-Feed, the feed solution, was administered on days 2, 4, 6, and 8 with 6%, 12%, 12%, and 10% feed per original volume, respectively. There were two feed supplements ("feed A" and "feed G").

Three technical constraints were considered in selecting the specific design for this experiment. First, there were 13–15 microbioreactor arrays (MBAs) budgeted for this experiment, with six 650 μL chambers per MBA. Second, two MBAs would be reserved for a non-fed set of controls to be harvested at an earlier date than the fed-batch cultures. Third, researchers wanted four replicates per fed-batch condition for measuring pure error and to obtain extra sampling volumes as a contingency regarding the analytical method for measuring IgG concentration. Another reason for replication was the fact that this experiment was one of the validation experiments for operating the SimCell system at Life Technologies.

As a result of the goals, factor levels, and technical constraints, a general factorial plan was developed (Table 14.3). Factors A-Cells, C-Starting Medium, and D-Feed were set into a full factorial with all 16 combinations. Factor B-pH was paired with Medium C and set into a fractional factorial with factors A-Cells and D-Feed. It was thought that the effect of pH was better understood than the other factors at that time, so pH interactions were disregarded to accommodate the constraints and still meet the goals. The selection of batch controls, as a partial factorial, included only three of the starting media and was intended only as a

TABLE 14.3. Description of Conditions Tested in SimCell Microbioreactors by Fed-Batch and Batch-Culture Methods

General Factorial Conditions

Factor A	Factor B	Factor C	Factor D	
Cell Type	pH	Starting Medium	Feed	Type of Bioreactor Run
Adapted	6.85	Medium A	Feed A	Fed-batch, $n = 4$
Adapted	6.85	Medium A	Feed G	Fed-batch, $n = 4$
Adapted	6.85	Medium B	Feed A	Fed-batch, $n = 4$
Adapted	6.85	Medium B	Feed G	Fed-batch, $n = 4$
Adapted	6.85	Medium C	Feed A	Fed-batch, $n = 4$
Adapted	6.85	Medium C	Feed G	Fed-batch, $n = 4$
Adapted	6.85	Medium D	Feed A	Fed-batch, $n = 4$
Adapted	6.85	Medium D	Feed G	Fed-batch, $n = 4$
Unadapted	6.85	Medium A	Feed A	Fed-batch, $n = 4$
Unadapted	6.85	Medium A	Feed G	Fed-batch, $n = 4$
Unadapted	6.85	Medium B	Feed A	Fed-batch, $n = 4$
Unadapted	6.85	Medium B	Feed G	Fed-batch, $n = 4$
Unadapted	6.85	Medium C	Feed A	Fed-batch, $n = 4$
Unadapted	6.85	Medium C	Feed G	Fed-batch, $n = 4$
Unadapted	6.85	Medium D	Feed A	Fed-batch, $n = 4$
Unadapted	6.85	Medium D	Feed G	Fed-batch, $n = 4$
Adapted	**7.15**	Medium C	Feed G	Fed-batch, $n = 4$
Unadapted	**7.15**	Medium C	Feed A	Fed-batch, $n = 4$
Adapted	6.85	Medium B		Batch, $n = 2$
Adapted	**7.15**	Medium C		Batch, $n = 2$
Adapted	6.85	Medium D		Batch, $n = 2$
Unadapted	6.85	Medium B		Batch, $n = 2$
Unadapted	**7.15**	Medium C		Batch, $n = 2$
Unadapted	6.85	Medium D		Batch, $n = 2$

verification of cell inoculation and as a general comparison to shake flask controls.

The locations of the 72 fed-batch cultures ($n = 4$) and 12 batch cultures ($n = 2$) were randomized among 14 MBAs. Six shake flasks were also prepared as batch controls for a limited set of conditions. The harvest of fed-batch cultures occurred at day 12 postinoculation for fed-batch cultures and at day 7 for all batch cultures. Total cell density of each SimCell culture was measured using the on-line optical density method. Integral cell growth was calculated using a trapezoidal summation method through day 12. Recombinant IgG concentrations in SimCell and shake flask cultures were measured in harvested cultures using the 96-well plate device Octet (Forte Bio).

All values of cell growth and IgG production were transformed into the percentages of the average among the analyzed raw data. For each of the measured responses, an ANOVA model was generated that included all possible main effects and interactions. Using Design-Expert software residual analysis was performed to screen for outliers and to verify acceptable distribution of random error. The first four SimCell run numbers were identified as statistical outliers in both growth and production responses. These extremely low values, thought to be caused by poor priming of cells before inoculation, were excluded from the raw data. The remaining 68 runs were then subjected to ANOVA and subsequent residual analysis. Data were then expressed as percentage of the overall ANOVA mean.

14.5.2.2 Data Analysis. The ANOVA output for integral cell growth is depicted in Table 14.4. The standard deviation due to residual error is tight at approximately 6%. A graph of predicted versus actual values, presented in Fig. 14.23) illustrates how well the model appears to describe the measured integral cell growth data. The data distribution appears acceptable.

The ANOVA output for IgG Concentration is depicted in Table 14.5. The standard deviation due to residual error is approximately 10%, within reasonable variability for the method of measurement. The graph of predicted versus actual values, presented in Fig. 14.24, depicts a rather puzzling spread of data near the top. This spread appears to be random with respect to sequence, position on MBAs, and design condition. Transformation of IgG data into a log scale would provide a slightly improved fit, but without perceptible difference in graphs of means or in final conclusions. Therefore, the data are not transformed for this analysis.

A four-part composite, shown in Fig. 14.25, illustrates changes in growth and IgG production in response to the three most significant and/or interacting factors: cell conditioning, starting medium, and feed solution. The largest influence in cell growth appears to be whether or not the cells are preconditioned to grow without supplied L-glutamine. Unadapted cells (not preconditioned) grew more quickly and reached a higher cumulative integral cell density than adapted cells. Among the unadapted cells, there are slight differences in growth versus starting medium, with Medium D being lowest. No such differences due to starting medium are apparent among the conditions of unadapted cells.

On average, cells preconditioned into glutamine-free medium (adapted) appear to produce a higher concentration of IgG than unadapted cells. The specific feed solution

TABLE 14.4. ANOVA of Integral Cell Growth for Fed-Batch Cultures

ANOVA Table and Descriptors: **Integral Cell Growth** (% of Average)

Source	Sum of Squares	df	Mean Square	*F* Value	*p*-value Prob > *F*
Model	14,759.33	17	868.78	24.92	<0.0001
A-Cells	12,210.04	1	12,210.04	350.24	<0.0001
B-pH	196.82	1	196.82	5.65	0.0214
C-Starting Medium	1,230.44	3	410.15	11.77	<0.0001
D-Feed	678.72	1	678.72	19.47	<0.0001
AB	118.69	1	118.69	3.40	0.0709
AC	610.08	3	203.36	5.83	0.0017
AD	0.22	1	0.22	0.01	0.9366
CD	78.79	3	26.26	0.75	0.5256
ACD	6.74	3	2.25	0.06	0.9784
Pure error	1,743.07	50	34.86		
Cor total	16,512.40	67			

SD	5.90	R-squared	0.8944
Mean	100	Adj R-squared	0.8585
CV %	5.90	Pred R-squared	0.8033
PRESS	3,248.22	Adeq precision	15.7204

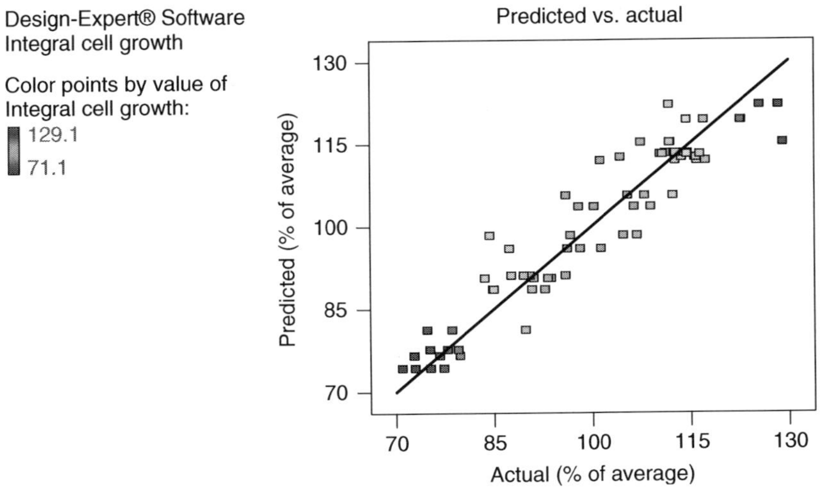

Figure 14.23. Integral cell growth predicted (ANOVA model) versus actual values. (This figure is available in full color at http://onlinelibrary.wiley.com/book/10.1002/9780470054581.)

TABLE 14.5. ANOVA of IgG Production for Fed-Batch Cultures

ANOVA Table and Descriptors: **IgG Production** (% of Average)					
Source	Sum of Squares	df	Mean Square	*F* Value	*p*-value Prob > *F*
Model	52,931.0	17	3,113.6	30.83	<0.0001
A-Cells	23,232.8	1	23,232.8	230.02	<0.0001
B-pH	4,841.6	1	4,841.6	47.93	<0.0001
C-Starting Medium	9,745.5	3	3,248.5	32.16	<0.0001
D-Feed	5,948.5	1	5,948.5	58.89	<0.0001
AB	323.5	1	323.5	3.20	0.0796
AC	204.4	3	68.1	0.67	0.5718
AD	5,930.4	1	5,930.4	58.71	<0.0001
CD	6,036.5	3	2,012.2	19.92	<0.0001
ACD	1,123.5	3	374.5	3.71	0.0174
Pure error	5,050.5	50	101.0		
Cor total	57,981.2	67			
SD	10.050		R-squared		0.9129
Mean	100		Adj R-squared		0.8833
CV %	10.05		Pred R-squared		0.8409
PRESS	9,223.05		Adeq precision		16.6138

appears to make a large difference in IgG production. The highest IgG concentrations across all media are observed when Feed G is used during the fed-batch culture with adapted cells. An equally high productivity is obtained when using Feed A, but only in combination with the starting medium C.

The effect of pH was not illustrated here, but for both growth and production there were no detectable interactions involving pH. The change in pH from 6.85 to 7.15 resulted in a 7% decrease in growth and a 37% decrease in IgG concentration. The graphs in Fig. 14.25 depict values at model-averaged pH levels.

Cells preconditioned to glutamine-free growth prior to the experiment appear to grow more slowly, but produce more IgG than do cells not preconditioned. As long as cells can be preconditioned and fed using Feed G, any of the four starting media perform equally well for generating the highest productivity. But if a researcher was not able to precondition the clone, the data suggest that the best combination is to use Feed A with Medium C to increase the IgG concentration.

Results of non-fed-batch control cultures (data not shown) revealed a similar trend, whereby cells preconditioned in glutamine-free medium grew more slowly and

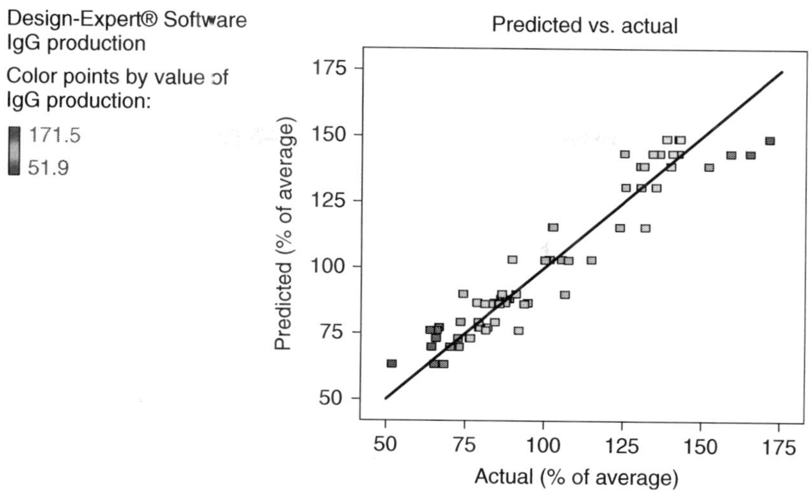

Figure 14.24. IgG concentration predicted (ANOVA model) versus actual values. (This figure is available in full color at http://onlinelibrary.wiley.com/book/10.1002/9780470054581.)

produced higher amounts of IgG than the unadapted cells. Similarly, pH 6.85 outperformed pH 7.15 in growth and IgG production. No interactions were detected, possibly because of a higher residual error term and a low number of samples. These controls were in the same range as other parallel controls in shaker flasks (data not shown).

14.5.2.3 Combining Multiple Responses into Overall Optimization. This case study exemplifies how growth and productivity do not always parallel each other in response to different factors. The question then arises, as to how to find the best combination of factor levels in the presence of multiple divergent responses. Researchers should first determine criteria describing an optimal solution. Using the numerical optimization function of the Design-Expert software, for each dependent response there is an option to ignore it or to include the response with a goal to maximize, minimize, assign acceptable range, or assign a target value. Each included response can then be assigned a relative importance ranking. The software uses a simplex method to solve for the highest desirability function according to the selected criteria. In addition to desirability function, several other compromise methods have been used (50).

Table 14.6 depicts the criteria selected and the conditions of highest desirability for this case study results. The response of productivity was set to be maximized with an importance level of 5 (highest). Growth response was also set to be maximized, but with an importance level of 2. It was thought that sufficient growth is essential for developing robust scalable processes, but it was not as important as IgG concentration. The outcomes are listed below the selected criteria and are ranked from the highest to the lowest desirability (only the top 12 from the 22 generated are depicted here). Added into the right columnis relative specific productivity values calculated by dividing the IgG concentration (percentage of average) by the Integral Cell Density (percentage of average). The six highlighted rows emphasize the highest desirabilities, whereby the IgG production levels are all rather high.

It is shown that pH 6.85 is better than pH 7.15 and that adapting the cells into glutamine-free medium before the bioreactor run are preferred for optimal results. Similar results were obtained in bench-scaled stirred-tank bioreactor runs (data not shown). Among the preadapted set of conditions, it also seems that Feed G may be used in combination with any of the four starting media with only minor differences in predicted IgG production. If specific productivity is considered more important than the original criteria, or if simply used as a tiebreaker, then medium D may be preferred because of a lower cell growth. An interesting twist is that for cells that were not preconditioned in glutamine-free medium, Feed A may be used but only with C-Starting Medium.

14.5.2.4 Key Takeaways

1. The specific DOE plan for this experiment was selected as a balance between the questions to be answered and the constraints of available technology and analytical methods. Four replicates of a partial factorial were selected rather than fewer replicates of a full factorial because of technical risks at that time. Sacrificing information about interactions involving pH was deemed acceptable in this case because of prior knowledge. Using ANOVA to derive mean effects and interactions within the factorial, accuracy

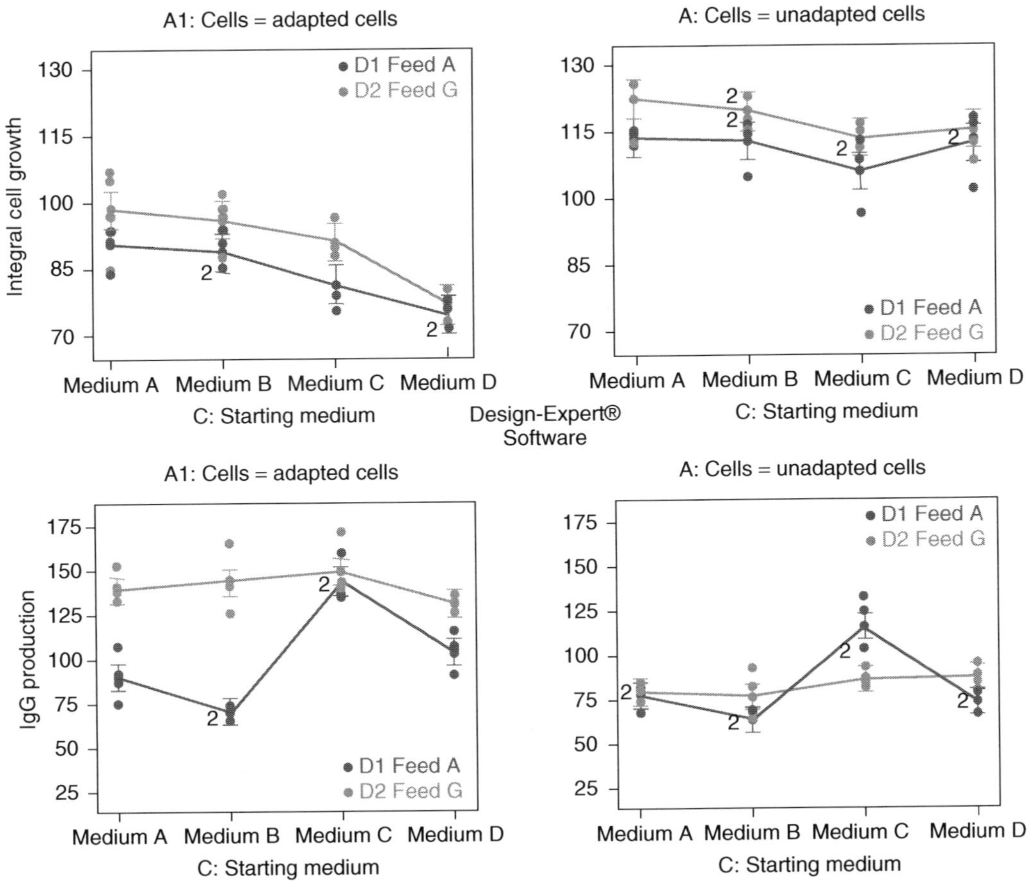

Figure 14.25. Comparison of cell growth and IgG production in response to cell conditioning, starting medium, and feed solution. Means (squares or triangles) and data points (circles) are depicted. Numbers to left of data points indicate overlayed points. Error bars are ±1/2 least significant difference (LSD). (This figure is available in full color at http://onlinelibrary.wiley.com/book/10.1002/9780470054581.)

is negligibly affected by this decision because of the large residual degrees of freedom.

2. Randomization of conditions among sequential locations is always important in DOE to minimize bias during analysis, bringing more accuracy to the predictive model by distributing experimental error throughout design. In this case study, the first four chambers, identified as statistical outliers, are different test conditions. If instead they were replicates, analysis would be uncertain at best and possibly disqualified altogether.
3. Factorials often reveal surprises among interaction effects, such as the ability for Feed A in combination with C-Starting Medium to enable high productivity with cells not previously adapted to glutamine-free culture. Although needing verification in larger scale, this could lead to lower variability among stock cultures during bioreactor scale-up trials.

14.5.3 Downstream Case Study: Optimizing Conditions for Antibody Purity and Yield

Capture purification of antibodies is usually achieved using protein A affinity resins that provide high purity and yield after one chromatography step. Subsequent chromatography polishing steps used to remove contaminants and/or antibody aggregates can, for example, contain a combination of anion and cation exchangers and hydrophobic interaction chromatography resins. One approach is to design a two-step process with a multimodal chromatography resin as the second step. In this study, a two-step antibody process is considered, in which the MabSelect SuRe (Protein A resin) is used as the capture step and Capto adhere (multimodal anion exchange resin) is used in the second step. The focus of the second step is the selective removal of antibody aggregates, as the feed contains high levels of aggregates. The study describes the optimization of conditions to be used with Capto adhere for maximum antibody monomer

TABLE 14.6. Numerical Optimization of IgG and Integral Cell Growth

Fed-Batch Conditions—Numerical Optimization

Criteria Selected

Name	Goal	Lower Limit	Upper Limit	Lower Weight	Upper Weight	Importance
Cells	Is in range	Adapted	Unadapted	1	1	3
pH	Is in range	6.85	7.15	1	1	3
Starting medium	Is in range	Medium A	Medium D	1	1	3
Feed	Is in range	Feed A	Feed G	1	1	3
IgG production	Maximize	51.914	171.503	1	1	5
Integral cell growth	Maximize	71.067	129.144	1	1	2

Number	Cells	pH	Starting Med	Feed	IgG Production	Integral Cell Growth	Desirability	Spec.Prodn. (% of Avg.)
1	Adapted	6.85	Medium A	Feed G	136.97	98.48	0.633	139.1
2	Adapted	6.85	Medium B	Feed G	138.14	95.69	0.620	144.4
3	Unadapted	6.85	Medium C	Feed A	120.75	107.59	0.590	112.2
4	Adapted	6.85	Medium C	Feed G	145.75	87.84	0.590	165.9
5	Adapted	6.85	Medium C	Feed A	141.93	81.82	0.504	173.5
6	Adapted	6.85	Medium D	Feed G	138.52	76.79	0.410	180.4
7	Adapted	7.15	Medium A	Feed G	99.78	91.73	0.387	108.8
8	Adapted	7.15	Medium B	Feed G	100.95	88.94	0.378	113.5
9	Unadapted	6.85	Medium C	Feed G	84.90	113.60	0.365	74.7
10	Adapted	7.15	Medium C	Feed G	108.56	81.09	0.355	133.9
11	Unadapted	6.85	Medium A	Feed G	81.32	121.65	0.353	66.8
12	Unadapted	6.85	Medium B	Feed G	81.26	119.39	0.348	68.1

purity and yield. The goal is to find conditions for the Capto adhere step that give <1% of aggregates in the final sample (>99% purity) with acceptable antibody monomer yield (>85%). A sequential experimental approach is described where the PreDictor plates are used in the initial screening phase, followed by exploration of a smaller experimental space in HiScreen™ columns with DOE, Monte Carlo simulation on DOE results, and finally verification of the results in a large-scale column.

14.5.3.1 Flow-Through Experiments in Plates. In the flow-through experiments the goal is to find conditions that provide minimal binding of monomeric antibody, while aggregate binding is maximized.

A PreDictor Capto adhere 6 μL plate was used. The buffers were prepared in an automated Tecan Freedom EVO-2 200 robotic system, but the experiments (e.g. sample addition) were performed manually. The experimental protocol was as recommended in the "Instructions for the PreDictor plates" (GE Healthcare). A conditioned 200 μL aliquot of sample (concentration 5 g/L) was applied to each well. A 60-min incubation time was used.

Total antibody concentration was quantified by use of absorbance and fraction of aggregates, and monomers in the samples were determined by use of size exclusion chromatography. Monomer and aggregate capacities (Q) were calculated with Equation 14.5:

$$Q = (C_{\text{ini}} - C_{\text{FT}})\,\frac{V_{\text{sample}}}{V_{\text{resin}}} \tag{14.5}$$

where V_{sample} and V_{resin} are the sample and resin volumes, respectively, and C_{ini} and C_{FT} are the initial concentration and final flow-through concentration of monomer or aggregate, respectively.

The factors studied were loading pH (4–7.5) and NaCl concentration (0–550 mM). Buffering substance was 50 mM citrate and 50 mM phosphate for the pH ranges 4–6 and 6.5–7.5, respectively (Fig. 14.26). The responses measured were monomer and aggregate capacities.

From the capacities, interpolated contour plots were constructed and are presented in Fig. 14.27). The highest aggregate capacity (best clearance of aggregates) is obtained at high pH and low NaCl concentration, but this is also true for monomer capacity. Therefore, for high monomer purity, monomer yield must be sacrificed.

From the plate data predictions of column performance, with respect to antibody yield and purity, were carried out. Assuming that plate capacities equal dynamic binding capacities (most likely valid for longer residence times), then purity and yield can be calculated using

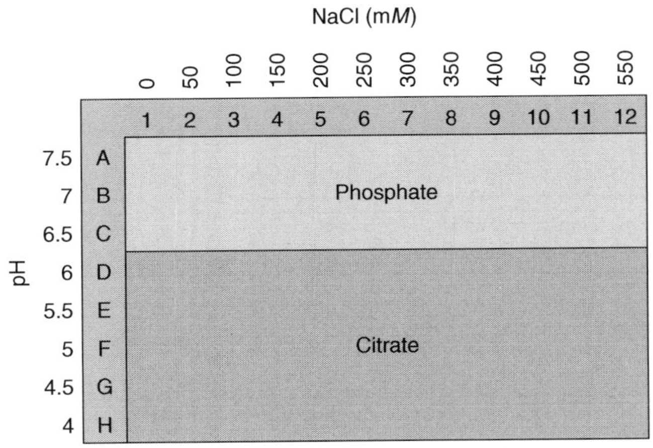

Figure 14.26. Plate layout. (This figure is available in full color at http://onlinelibrary.wiley.com/book/10.1002/9780470054581.)

Equations 5–7:

$$\text{Yield} = \frac{V_{\text{load}} \times C_{\text{ini,m}} - \text{CV} \times \text{DBC}_{\text{m}}}{V_{\text{load}} \times C_{\text{ini,m}}} \tag{14.6}$$

$$\text{Purity} = \frac{V_{\text{load}} \times C_{\text{ini,m}} - \text{CV} \times \text{DBC}_{\text{m}}}{V_{\text{load}} \times \left(C_{\text{ini,m}} - C_{\text{ini,a}}\right) - \text{CV} \times (\text{DBC}_{\text{m}} - \text{DBC}_{\text{a}})} \tag{14.7}$$

where V_{load} is volume loaded, C_{m} is monomer concentration, C_{a} is aggregate concentration, ini represents initial, FT represents flow through, CV is column volume, and DBC_{m} or DBC_{a} are the dynamic binding capacities for monomer and aggregates, respectively.

An optimal area for purity (>98% monomer, box) was found and coincided with the area where the yield (60–65%) was the lowest (Fig. 14.28).

The low yield was unacceptable for a large-scale process and therefore, an elution study was set up to explore the possibility of selectively eluting the bound monomer to increase the monomer yield.

14.5.3.2 Selective Elution Study in Plates. A PreDictor Capto adhere 6 μL plate was used. On the basis of the flow-through experiments, the loading condition selected was 50 m*M* sodium phosphate, 50 m*M* NaCl, pH 7. The elution conditions studied were 50 m*M* sodium phosphate, pH 6–7 and 0–550 m*M* NaCl. Sample concentration and volume were as in the flow-through study. Monomer and aggregates were analyzed in eluate fractions. The plate layout was essentially as shown in Fig. 14.26 except that the pH and salt concentrations were elution pH and salt concentrations, respectively.

Cumulative elution profiles were created. Selective elution of the monomer was successful, since the monomer was eluted while most of aggregates remained bound with the tested conditions (Fig. 14.29). Total monomer yield was increased from approximately 60% (flow through) to >80% (flow through + elution), by using the best conditions (pH 6, 250–300 m*M* NaCl).

14.5.3.3 Optimization in a Small Column. On the basis of the results of the plate experiments, an optimization study was set up using a HiScreen™ column (4.7 mL column). The goal of the study was to find the best conditions for monomer purity (>99% in final sample) and acceptable monomer yield (>85%).

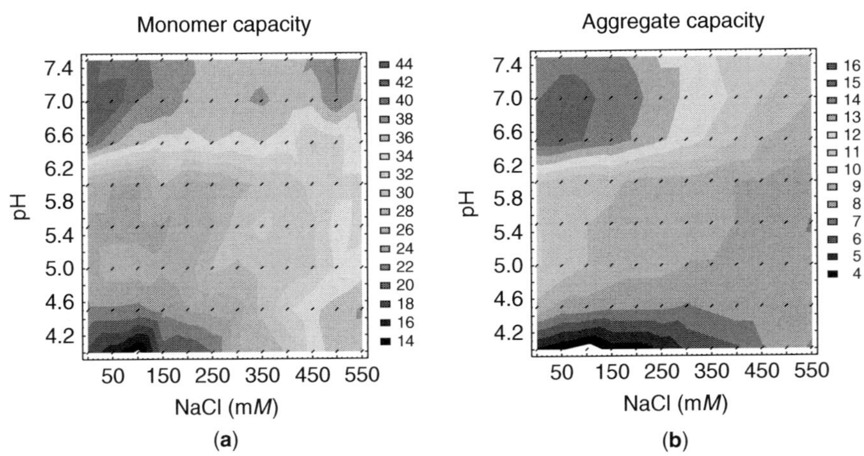

Figure 14.27. Contour plots showing (a) the antibody monomer capacity and (b) the aggregate capacity under different conditions. (Reproduced with permission from GE Healthcare.) (This figure is available in full color at http://onlinelibrary.wiley.com/book/10.1002/9780470054581.)

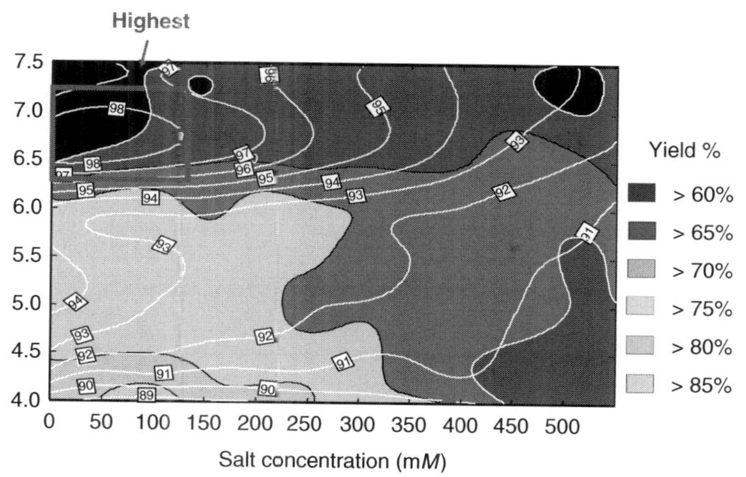

Figure 14.28. Column prediction of purity (iso-lines) and yield (color map). (Reproduced with permission from GE Healthcare.) (This figure is available in full color at http://onlinelibrary.wiley.com/book/10.1002/9780470054581.)

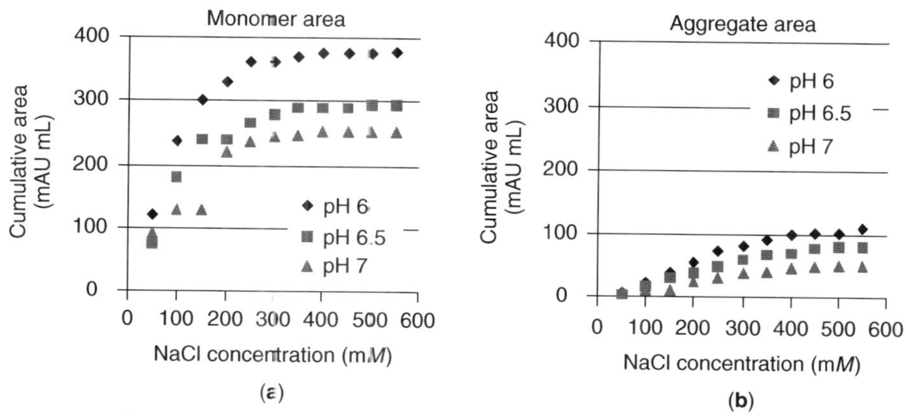

Figure 14.29. (a)Monomer and (b) aggregate elution profiles from PreDictor plate data. (a) Monomer elutes while most of the aggregates remain bound. (Reproduced with permission from GE Healthcare.) (This figure is available in full color at http://onlinelibrary.wiley.com/book/10.1002/9780470054581.)

TABLE 14.7. HTT in Bioprocess Development

Factors	Factor Ranges
Start aggregate level	9–14%
Start protein concentration	5–15 g/L
Load	60–100 g/L
Elution pH	6.1–6.5
NaCl in elution buffer	150–450 m*M*

The DoE software used for planning and evaluation was MODDE 8.0 by Umetrics, Sweden. The investigated factors are listed in Table 14.7. The chosen design was a face-centered CCD with the objective of response surface analysis. With 16 experiments in the 2_{V}^{5-1} factorial part, together with 10 experiments to quantify second-degree curvature, this design resulted in 26 design runs plus replicated center points.

The load was performed according to the optimal conditions found in the screening experiments with a residence time of 5 min. The responses were monomer yield and purity. The IgG sample used in the study was composed of two MabSelect SuRe elution pools containing 9% and 14% of aggregates, respectively. These were mixed (1:1) to obtain center points. The columns used were Capto adhere HiScreen, with a column volume of 4.7 mL.

Good prediction models were obtained for both responses, as is shown in the observed versus predicted plots presented in Fig. 14.30.

The coefficient plots presented in Fig. 14.31 show the significant factors affecting yield and purity. For yield,

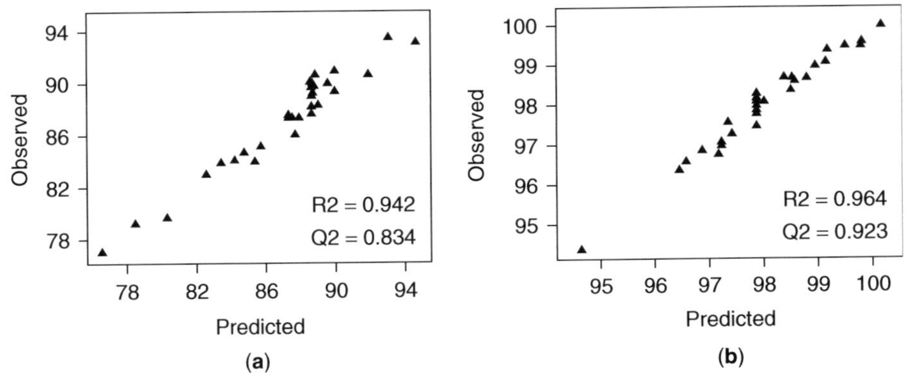

Figure 14.30. Observed versus predicted plot for (a) monomer yield and (b) purity, respectively. R2 is the fraction of variation of the response explained by the model. Q2 is the fraction of the variation of the response predicted by the model according to cross validation. (Reproduced with permission from GE Healthcare.)

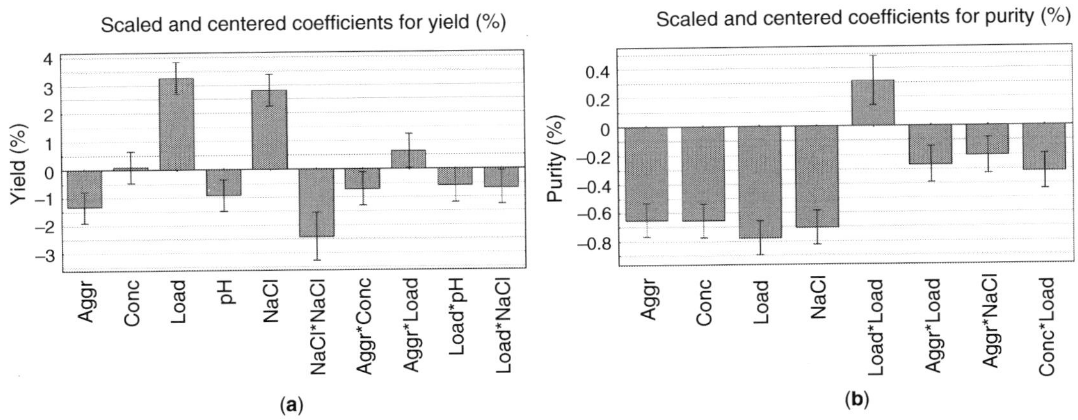

Figure 14.31. Coefficients plots for (a) monomer yield and (b) purity. The coefficients are scaled and centered. The height of the bars corresponds to the impact, while the sign of the bars indicate whether the factor is positively or negatively correlated with the response. (Reproduced with permission from GE Healthcare.) (This figure is available in full color at http://onlinelibrary.wiley.com/book/10.1002/9780470054581.)

four of the five main factors studied were significant. The nonsignificant main factor, start concentration, was kept in the model, since one of the significant interaction factors contained concentration. In addition to the main factors, interaction effects (Aggr*Conc, Aggr*Load, Load*pH, Load*NaCl) and a quadratic effect (NaCl*NaCl) were found to have significant effects on the yield. For purity, all the main factors except pH were significant and in addition, three interaction effects (Aggr*Load, Aggr*NaCl, Conc*Load) and one quadratic effect (Load*Load) were significant.

The contour plots for the yield and purity models are shown in Fig. 14.32a and b.

To estimate the process robustness with variation in all five factors simultaneously, Monte Carlo simulation (58) was performed using the DOE transfer functions for purity and yield. The following distributions for the five factors were assumed:

- Start aggregate levels: 9%–12% (uniform distribution)
- IgG start concentration: 4.5–5.5 g/L (triangular distribution)
- IgG load: 60–65 g/L (triangular distribution)
- Elution pH: 6–6.2 (triangular distribution)
- NaCl concentration for elution: 230–270 m*M* (triangular distribution).

The estimated variation in yield and purity was 83.5–87.4% and 98.9–99.9% respectively. The probability of obtaining a purity >99% was 99.98%.

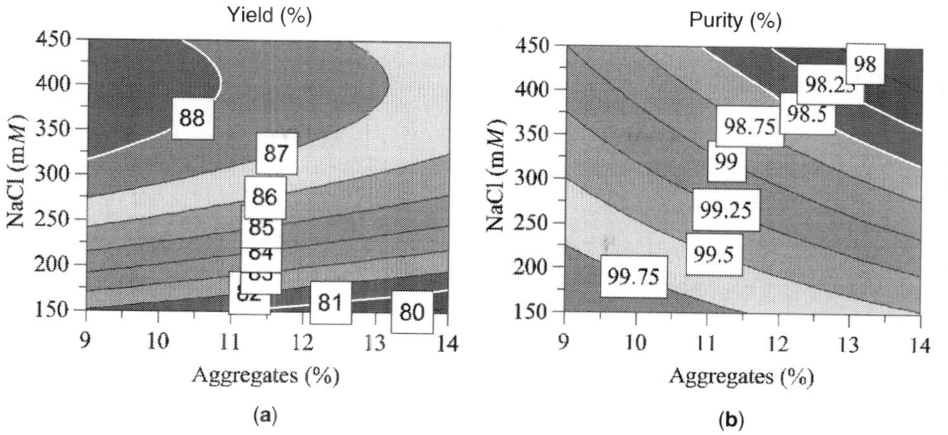

Figure 14.32. Response surface plots for (a) monomer yield (b) and purity. The factor levels for protein concentration, pH and load where locked at 5, 6.1, and 65 g/L, respectively. (Reproduced with permission from GE Healthcare.) (This figure is available in full color at http://onlinelibrary.wiley.com/book/10.1002/9780470054581.)

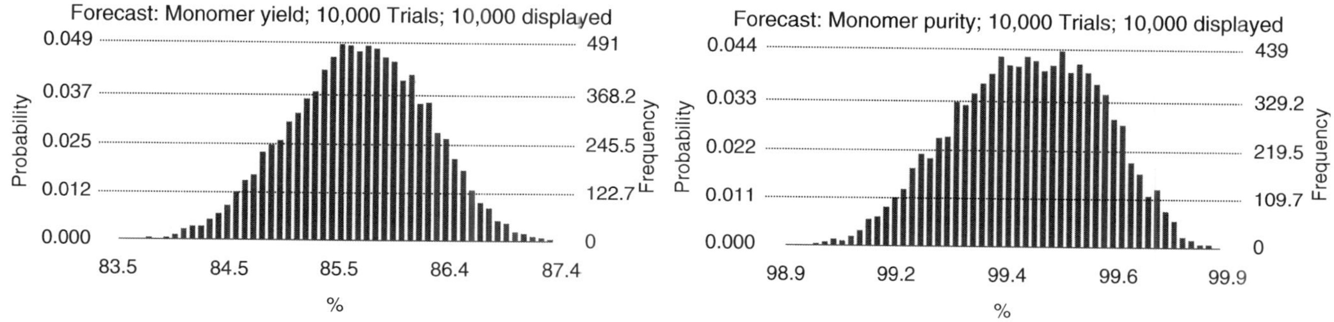

Figure 14.33. Monte Carlo simulations of the responses (a) monomer yield and (b) purity. (Reproduced with permission from GE Healthcare.) (This figure is available in full color at http://onlinelibrary.wiley.com/book/10.1002/9780470054581.)

14.5.3.4 Large-Scale Column Verification. A large-scale verification run was setup (Fig. 14.34). Capto adhere was packed in an AxiChrom™ 70/300 column (14.1 cm bed height, 543 mL). The conditions used were those identified as the best from the plate and HiScreen column experiments. Relatively low load was used (60 g/L), as the yield was expected to be somewhat higher at scale than in the HiScreen columns due to smaller dead volumes at scale.

The yield was 86% and the aggregates were reduced from 12% to 0.6% giving a purity of 99.4%, which was good considering the high starting aggregate content. The results correlated well with the predicted results from the DOE models and the Monte Carlo simulations.

14.5.3.5 Key Takeaways. A large experimental space was first explored by using a rapid and inexpensive HT technique (PreDictor 96-well filter plates). This allowed a significantly reduced experimental space to be explored during the DOE optimization in small columns. Finally, the DOEs and Monte Carlo simulations allowed for identification of large-scale process conditions and estimation of process robustness, and large scale results verified thesebreak conclusions.

The combination of HT techniques and statistical methods, DOE and Monte Carlo simulations, significantly reduced both time and resources needed for finding good process conditions and maximized the amount of information obtained about the process.

14.6 CONCLUSION

In this chapter, the current state-of-the-art HT bioprocess developments were reviewed, and suggestions for how to best deploy HTT for individual applications through a case study approach were provided. The application of HTT to bioprocess development has yielded tangible benefits for

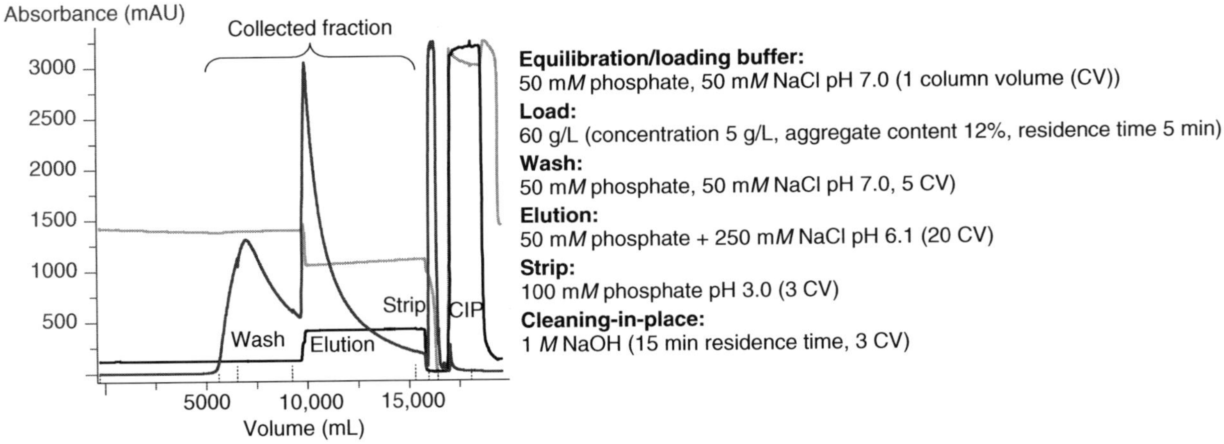

Figure 14.34. Chromatogram from the Capto adhere step. A_{280} trace is shown in blue, pH in green, and conductivity in brown. Chromatogram from the verification run in a 543-mL Capto adhere column (AxiChrom™ 70/300). (This figure is available in full color at http://onlinelibrary.wiley.com/book/10.1002/9780470054581.)

early adopters, particularly where capacity and timeline restrictions create workflow bottlenecks.

It is important to end this chapter by reinforcing that good science remains the most important factor in achieving good outcomes. The ability to generate large data sets through HT strategies only further amplifies the importance of having well-planned experiments on the basis of a solid knowledge of the underlying science. Not all projects will benefit from HTT, so the goals of the scientific investigation need to be understood. The scientist should always be cautious of the desire to use technologies just because they are available, rather than because they add value. Applied properly, HTT is an important tool in meeting the bioprocess challenges of today and tomorrow.

REFERENCES

1. Gryseeis T. Considering cell culture automation in upstream bioprocess development. Bioprocess Int 2008: 12–16.
2. Neway J. Process excellence: five critical elements of quality by design. Bioprocess Int 2008: 18–22.
3. Hanania E, Fieck A, Stevens J, Bodzin L, Palsson B, Koller M. Automated *in situ* measurement of cell-specific antibody secretion and laser-mediated purification for rapid cloning of highly secreting producers. Biotechnol Bioeng 2005; 91 (7)
4. Gray F, Kenney JS, Dunne JF. Secretion capture and report web: use of affinity derivatized agarose microdroplets for the selection of hybridoma cells. J Immunol Methods 1995; 182: 155–163.
5. Holmes P, Al-Rubeai M. Improved cell line development by a high throughput affinity capture surface display technique to select for high secretors. J Immunol Methods 1999; 230: 141–147.
6. Manz R, Assenmacher M, Pfluger E, Miltenyi S, Radbruch A. Analysis and sorting of live cells according to secreted molecules, relocated to a cell-surface affinity matrix. Proc Natl Acad Sci U S A 1995; 92: 1921–1925.
7. Böhm E, Voglauer R, Steinfellner W, Kunert R, Borth N, Katinger H. Screening for improved cell performance: selection of subclones with altered production kinetics or improved stability by cell sorting. Biotechnol Bioeng 2004; 88: 699–706.
8. Borth N, Zeyda M, Kunert R, Katinger H. Efficient selection of high-producing subclones during gene amplification of recombinant Chinese hamster ovary cells by flow cytometry and cell sorting. Biotechnol Bioeng 2000; 71: 266–273.
9. Carroll S, Al-Rubeai M. The selection of high-producing cell lines using flow cytometry and cell sorting. Expert Opin Biol Ther 2004; 4: 1821–1829.
10. Yoshikawa T, Nakanishi F, Ogura Y, Oi D, Omasa T, Katakura Y, Kishimoto M, Suga K. Flow cytometry: an improved method for the selection of highly productive gene-amplified CHO cells using flow cytometry. Biotechnol Bioeng 2001; 74(5): 435–442.
11. Disposable bioreactors gaining favor- new components and systems improve process reliability and reduce cost (2006). Genet Eng Biotechnol News 2006; 26(12)
12. Wurm FM. Production of recombinant protein therapeutics in cultivated mammalian cells. Nat Biotechnol 2004; 22: 1393–1398.
13. Puskeiler R, Kaufmann K, Weuster-Botz D. Development, parallelization, and automation of a gas-inducing milliliter-scale bioreactor for high-throughput bioprocess design (HTBD). Biotechnol Bioeng 2005; 89: 512–523.
14. Baldi L, Muller N, Picasso S, Jacquet R, Girard P, Thanh HP, Derow E, Wurm FM. Transient gene expression in suspension HEK-293 cells: application to large-scale protein production. Biotechnol Prog 2005; 21(1): 148–153.
15. Chen A, Chitta R, Chang D, Amanullah A. Twenty-four well plate miniature bioreactor system as a scale-down model for cell culture process development. Biotechnol Bioeng 2008. DOI 10.1002/bit.

16. Kensy F, John GT, Hofmann B, Büchs J. Characterization of operation conditions and online monitoring of physiological culture parameters in shaken 24-well microtiter plates. Bioprocess Biosyst Eng 2005; 75: 75–81.

17. Maharbiz MM, Holtz WJ, Howe RT, Keasling JD. Microbioreactor arrays with parametric control for high-throughput experimentation. Biotechnol Bioeng 2004; 86(4): 485–490.

18. Micheletti M, Barrett T, Doig SD, Baganz F, Levy MS, Woodley JM, Lye GJ. Fluid mixing in shaken bioreactors: implications for scale-up predictions from microliter-scale microbial and mammalian cell cultures. Chem Eng Sci 2006; 61: 2939–2949.

19. Anderlei T, Buüchs J. Device for sterile online measurement of the oxygen transfer rate in shaking flasks. Biochem Eng J 2001; 7(2): 157–162.

20. Knorr B, Schlieker H, Hohmann H, Weuster-Botz D. Scale-down and parallel operation of the riboflavin production process with Bacillus subtilis. Biochem Eng J 2007; 33: 263–274.

21. Carrier T. High throughput process development promises, myths, and truths. Bioprocess Int 2008: 54–58.

22. Schreyer H, Miller S, Rodgers S. High-throughput process development - microbioreactor system simulates large bioreactor process at submilliliter volumes. Genet Eng Biotechnol News 2007; 27(17)

23. Kuystermans D, Krampe B, Swiderek H, Al-Rubea M. Using cell engineering and omic tools for the improvement of cell culture processes. Cytotechnology 2007; 53(1–3): 3–22.

24. Al-Rubeai M. Apoptosis and cell culture technology Adv Biochem Eng Biotechnol 1998a; 59: 225–249.

25. Al-Rubeai M, Singh RP. Apoptosis in cell culture. Curr Opin Biotechnol 1998b; 9: 152–156. [PubMed].

26. Cotter TG, Al-Rubeai M. Cell death (apoptosis) in cell culture systems. Trends Biotechnol 1995; 13: 150–155.

27. Shukla AA, Hubbard B, Tressel T, Guhan S, Low D. Downstream processing of monclonal antibodies – application of platform approaches. J Chromatogr B 2007; 848: 28–39.

28. Shukla AA, Hinckley P. Host cell protein clearance during protein A chromatography: development of an improved column wash step. Biotechnol Prog 2008; 24: 1115–1121.

29. Kramarczyk J. MS thesis, Tufts University, Medford, MA, Department of Chemical and Biological Engineering; 2003.

30. Rege K, Pepsin M, Falcon B, Steele L, Heng M. High-throughput process development for recombinant protein purification. Biotechnol Bioeng 2006; 93(4): 618–630.

31. Staby A, Jensen RH, Bench M, Hubbuch J, Dünweber DL, Krarup J, Nielsen J, Lund M, Kidal S, Hansen TB, Jensen IH. Comparison of chromatographic ion-exchange resins: VI. Weak anion-exchange resins. J Chromatogr A 2007; 1164: 82–94.

32. Coffman JL, Kramarzcyk JF, Kelley BD. High-throughput screening of chromatographic separations: I. Method development and column modeling. Biotechnol Bioeng 2008; 100(4): 605–618.

33. Wiendahl M, Wierling PS, Nielsen J, Christensen DF, Krarup J, Staby A, Hubbuch J. High throughput screening for the design and optimization of chromatographic processes – miniaturization, automation and parallelization of breakthrough and elution studies. Chem Eng Technol 2008; 31(6): 893–903.

34. ICH – Quality: International Conference on Harmonisation – Quality Q8 (R1) Pharmaceutical Development, Revision 1.

35. Arve BH, Liapis AI. Modelling and analysis of biospecific adsorption in a finite bath. AIChE J 1987; 33(2): 179–193.

36. Chase HA. Prediction of the performance of preparative affinity chromatography. J Chromatogr 1984; 297: 170–202.

37. Hunter AK, Carta G. Protein adsorption on novell acylamido-based polymeric ion exchangers: II. Adsorption rates and column behavior. J Chromatogr A 1984; 897(1–2): 81–97.

38. Wesselingh JA, Bosma JC. Protein ion-exchange adsorption kinetics. AICheE J 2001; 47(7): 1571–1580.

39. Thiemann J, Jankowski J, Rykl J, Kurazawski S, Pohl T, Wittman-Leibold B, Schluter H. Principle and applications of the protein-purification-parameter screening system. J Chromatogr A 2004; 1043: 73–80.

40. Wenger MD, DePhillips P, Price CE, Bracewell DG. An automated microscale chromatographic purification of virus-like particles as strategy for process development. Biotechnol Appl Biochem 2007; 47: 131–139.

41. Tscheliessnig A, Hahn R, Jungbauer A. *In situ* determination of adsorption kinetics of proteins in a finite bath. J Chromatogr A 2005; 1069: 23–30.

42. Bergander T, Nilsson-Välimaa K, Öberg K, Lacki KM. High-throughput process development: determination of dynamic binding capacity using microtiter filter plates filled with chromatography resin. Biotechnol Prog 2008; 24: 632–639.

43. Hertzberg RP, Pope AJ. High-throughput screening: new technology for the 21st century. Curr Opin Chem Biol 2000; 4: 445–451.

44. Patel TP, Parekh RB, Moellering BJ, Prior CP. Different culture methods lead to differences in glycosylation of a murine IgG monoclonal antibody. Biochem J 1992; 285(Pt 3): 839–845.

45. Rosenfeld R, Bangio H, Gerwig G, Rosenberg, R, Aloni R, Cohen Y, Amor Y, Plaschkes I, Kamerling J, Ruth M. A lectin array-based methodology for the analysis of protein glycosylation. J Biochem Biophys Methods 2007; 70(3).

46. Gillmeister M, Tomiya N, Jacobia S, Yuan C, Gorfien S, Betenbaugh M. An HPLC-MALDI MS method for N-glycan analysis using small size samples: application to monitor glycan modulation by media conditions. Glycoconj J, 2009; May 2. [Epub ahead of print].

47. Bensch M, Wierling PS, von Lieres E, Hubbuch J. High throughput screening of chromatographic phases for rapid process development. Chem Eng Technol 2005; 28(11): 1274–1284.

48. Montgomery D. Design and analysis of experiments. 7th ed. New York: John Wiley and Sons; 2009.

49. Oehlert G, Whitcomb P. Small, efficient, equireplicated resolution V fractions of 2^k designs and their application to central composite designs. Proceedings of 46th Fall Technical Conference of the ASQ and ASA; 2002.

50. Myers R, Montgomery D. Response surface methodology—process and product optimization using designed experiments. 2nd ed. New York: John Wiley and Sons; 2002.

51. Myers W. Response surface methodology. In: Chow S-C, editor. Encyclopedia of biopharmaceutical statistics. 2nd ed. New York: Marcel Dekker; 2003. pp. 858–869.

52. Anderson M, Whitcomb P. RSM simplified— optimizing processes using response surface method for design of experiments. New York: Productivity Press; 2005.
53. Cornell John. Experiments with mixtures— designs, models, and the analysis of mixture data. 3rd ed. New York: John Wiley and Sons; 2002.
54. Chow S-C. Good statistics practice. In: Chow S-C, editor. Encyclopedia of biopharmaceutical statistics. 2nd ed. New York: Marcel Dekker; 2003. pp. 858–869.
55. Panda T, Naidu G. Rotating simplex method of optimization of parameters for higher product of extracellular pectinases in bioreactor. Bioprocess Eng 2000; 23: 47–49.
56. Butler M. Animal cell cultures: recent achievements and perspectives in the production of biopharmaceuticals. Appl Microbiol Biotechnol 2005; 68: 283–291.
57. Chee Furng WD, Tin Kam WK, Tang GL, Kiat HC, Gek Sim YM. Impact of dynamic online fed-batch strategies on metabolism, productivity and N-glycosylation quality in CHO cell cultures. Biotechnol Bioeng 2005; 89: 164–177.
58. Robert CP, Casella G. Monte carlo statistical methods. 2nd ed. New York: Springer; 2005.

15

LARGE-SCALE PROTEIN PURIFICATION, SELF-CLEAVING AGGREGATION TAGS

Iraj Ghazi and David W. Wood

Department of Chemical and Biomolecular Engineering, The Ohio State University, Columbus, Ohio

15.1 INTRODUCTION

The development of recombinant DNA technologies in the 1970s revolutionized many aspects of biological sciences, including the expression of arbitrary proteins in nonnative hosts. Recombinant protein expression now consists of a simple process of designing and constructing a DNA vector and transferring it into a suitable expression system (e.g. bacteria, yeast, insect, plant, or mammalian cells). Over the last few decades, the demand for active and highly purified peptides and proteins has grown rapidly in areas such as biomedical research, biotechnology, and the pharmaceutical industry. An important aspect of this scientific field is that once expressed, recombinant target proteins must generally be purified before they can be studied or applied. The purification of recombinant proteins, especially at large scale, is an expensive process and can account for 50–80% of their overall production costs (1,2). The development of more reliable, fast, and efficient methods for protein purification is essential for many new developments in biotechnology and has led to a focus on novel bioseparation technologies (3).

One of the most powerful protein isolation techniques is affinity purification, which is based on highly specific biorecognition by the target protein. This is possible in cases where a protein's structure leads to specific affinities toward a known ligand. In affinity purification techniques, the ligand can be a specific small molecule such as a substrate, cofactor, or inhibitor of an enzyme; or other proteins such as hormones or antibodies; or more general molecules such as carbohydrates, dyes, or metal ions. The specific and unique interaction of a protein with its particular ligand enables a novel separation method—affinity chromatography, which is capable of separating the target protein from a cell lysate containing thousands of contaminants in a single step (4,5).

Specific affinities between protein targets and their ligands provide a highly effective method for protein purification; however, these methods still suffer from some major limitations. Finding or making an appropriate ligand is often difficult for a protein with unknown function, and it could even be more complicated and more expensive than purifying the target protein through conventional chromatographic means. In addition, the ligand must resist the conditions of chemical immobilization, which can also be an expensive and complex process at the preparative scale. Further, even after a suitable affinity resin is made, affinity chromatography still has many of the same restrictions of common chromatographic methods. These include the necessity to optimize each new target protein, high cost and difficulties associated with scale-up, and relatively low productivity.

Discovery and improvement of simple and more economical methods in protein purification is one of the main objectives of research in this area, with the ultimate goal of completely eliminating chromatographic steps in bioseparations. The production of soluble recombinant proteins via exclusively nonchromatographic methods has great potential to impact biotechnology from laboratory to manufacturing scales. This goal may be realized through the use of self-cleaving aggregation tags, where the

Downstream Industrial Biotechnology: Recovery and Purification, First Edition. Edited by Michael C. Flickinger.

tagged precursor protein is composed of the target protein, self-cleaving intein (INTervening protEIN sequences), and an aggregation tag designed for purification using nonchromatographic techniques. Two of these tags are the elastin-like polypeptide (ELP) and the polyhydroxybutyrate (PHB) binding tag known as a *phasin*. Purification processes using these tags are very effective at reasonable yields, and typically require only cycles of centrifugation, resuspension, and a self-cleaving step.

15.2 CONVENTIONAL AFFINITY-TAG TECHNOLOGY

The strength of affinity separations can be generalized to arbitrary target proteins by engineering fusion proteins through recombinant DNA technology. This method is generally referred to as *affinity tag* technology, and involves the recombinant target protein being fused to a "tag" or "handle" to allow simple purification. The tag can be another protein, a protein domain, or a peptide, which has been previously optimized for simple purification through a readily available means. This technique is frequently used in different areas of protein sciences such as analytical methods, design of biosensors, immobilization, expression, refolding, purification, proteomics, and structural studies. The great strength of this method is that the construction of the tagged protein is relatively straightforward, and can be done using conventional recombinant DNA techniques without knowledge of the specific attributes of the target protein. Once constructed, the purification of the target is done using a single, known method, allowing effectively any target to be purified using an established protocol without the need for individual target-specific optimization.

A significant limitation of this technology is that, in general, the fusion protein must be cleaved to release the target protein from the associated purification tag. There are some chemical methods that utilize reagents such as cyanogen bromide or hydroxylamine to cleave the peptide bond within the fusion proteins (6,7). However, the use of chemical methods is restricted for several reasons. First, the specificity of the cleavage in a chemical method is determined mostly by a single amino acid (usually at methionine, tryptophan, aspartic acid, or cysteine residues), which are naturally present at several locations in almost all proteins. Moreover, chemical methods need harsh reaction conditions, for example, elevated temperature or extreme pH, which may lead to denaturation or side chain modification of amino acids in the target protein.

Thus, the most common and available techniques to cut a tag from a recombinant fusion protein are enzymatic methods, which are more specific and cleavage is achieved usually under mild conditions. Further, many endo- and exopeptidases are now available for the controlled cleavage of various fusion proteins (8). To make the process more controlled, the fusion protein is designed to include a unique amino acid sequence between the tag and target that is susceptible to efficient cleavage by a highly specific protease. Enterokinase (9), thrombin, and factor Xa (10) are among the most frequently used proteases for tag cleaving (11), and more efficient and specific proteases are in the process of development in many laboratories and companies.

Despite all the advantages of enzymatic cleaving in recombinant fusion proteins, it still suffers from some limitations. The first is that the pure proteolytic enzymes are expensive and for a complete cleaving at a reasonable time, a high ratio of enzyme to protein is required. This aspect makes proteolytic cleaving of the tag prohibitive at commercial scales. The second is the undesired, nonspecific cleaving that can occur at sites within the target other than the main cleaving site, resulting in degradation in the target protein. In addition, many protease enzymes require one or two exogenous amino acids to be added to the N-terminus of the target, resulting in a somewhat modified product protein. The presence of extra amino acid(s) can be a problematic issue for therapeutic proteins and its effect on the structure and properties of the protein must be studied completely. Finally, the cleaved tag and the protease must be separated from target protein, thus imposing an additional step to the purification process.

15.3 SELF-CLEAVING IN PROTEINS

Posttranslational protein self-splicing was first discovered during studies on the *Saccharomyces cerevisiae* VMA1 gene (12). In this process, a multidomain precursor protein is translated from its mRNA, which is followed by and an internal protein domain removing itself from the precursor. The removal is accompanied by ligation of the two flanking amino- and carboxy-terminal fragments to restore the active host protein (Fig. 15.1). In the case of the yeast VMA1 protein, a 120-kDa precursor encoded by the VMA1 gene is spliced into the 70-kDa catalytic subunit of the vacuolar H^+-ATPase and an internal 50-kDa DNA endonuclease (13). The internal spliced proteins are referred to as *inteins*, while the external, spliced proteins are referred to as *exteins* (14). In general, the splicing reaction is catalyzed entirely by the internal spliced protein.

Mechanistic studies of protein splicing have shown that different splicing precursors possess highly conserved amino acid sequence motifs at the intein–extein junctions (15–18). As shown in Fig. 15.1, the first step in the splicing reaction is the nucleophilic attack of a cysteine or serine residue located at the N-terminus of the intein, to the carbonyl group of the N-extein. The result of this acyl shift is generation of an ester or thioester bond between the intein and the N-extein. The second step is a transesterification

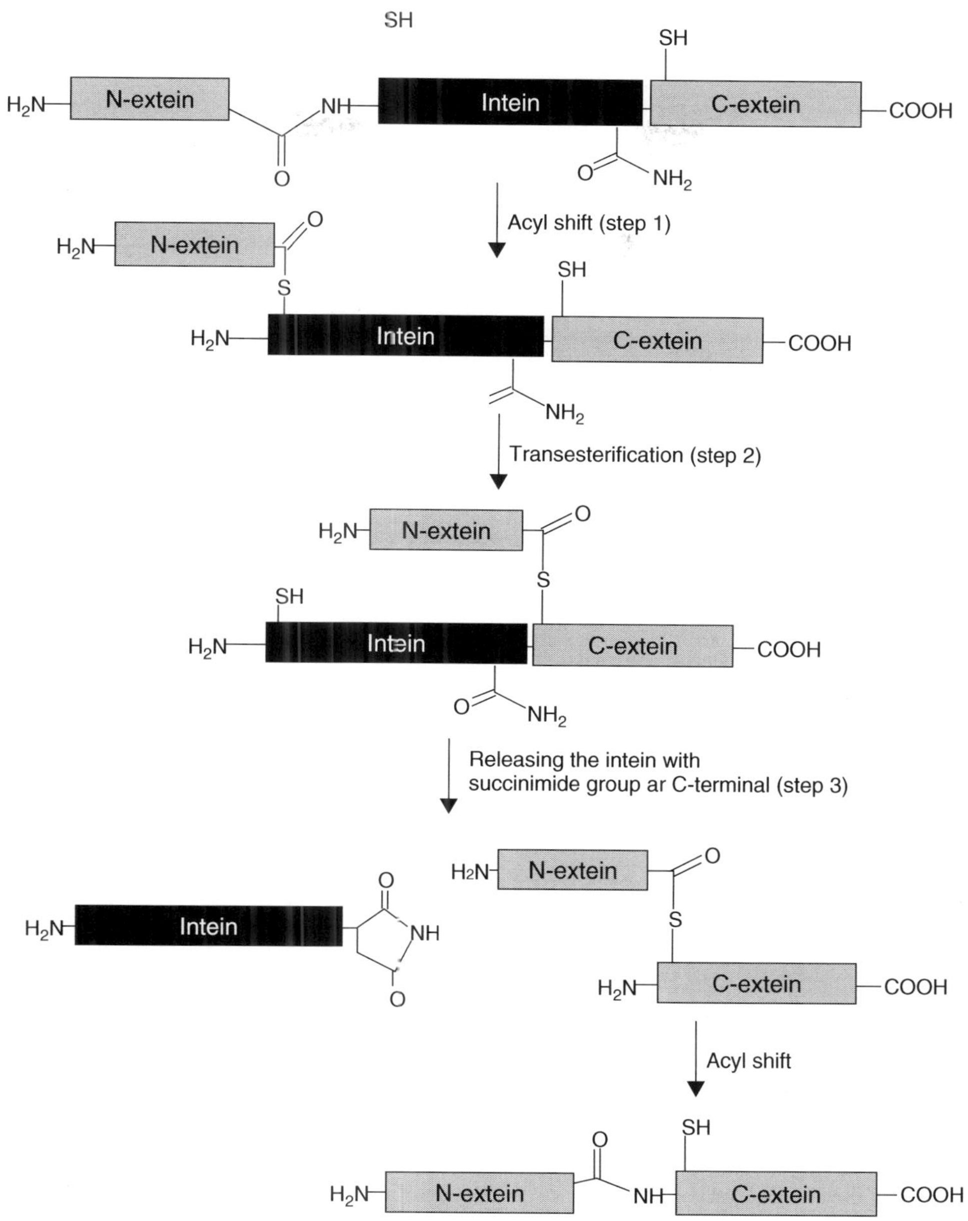

Figure 15.1. Mechanism of the canonical intein splicing reaction. Splicing takes place with steps described in text. Step1: splicing is initiated by initial acyl shift. Step 2: exteins are ligated by C- to N-terminal transesterification. Step 3: C-terminal cleavage by succinimide formation. Step 4: restoration of native host protein by acyl shift.

reaction between the nucleophilic part of the amino acid residue immediately after the C-terminal of the intein (cysteine, serine, or threonine) and the ester or thioester bond made in the first step. This transesterification then connects two N- and C-exteins through an ester or thioester bond, while the intein is still attached to the C-extein by a peptide bond. In the third step of the splicing process, a highly conserved asparagine residue at the C-terminus of the intein forms a succinimide ring, releasing the intein with a succinimide group at its C-terminus. Finally, the third step is followed by a spontaneous acyl shift in which the thioester linkage between the two exteins is transformed to a peptide bond, producing the native, spliced host protein.

Most inteins are bifunctional proteins, with two separate structural domains for the self-splicing property and the endonuclease activity (19,20). Endonuclease activity in large inteins is not necessary for the splicing process and functional mini-inteins have been constructed by deletion of the endonuclease domain from the splicing domain (21). Inteins as splicing elements have been widely identified in many organisms, including eucarya, bacteria, and archaea (22). Many inteins retain their splicing properties in foreign proteins or cells, and can often splice *in vitro*, and from nonnative host proteins (23–26). From approximately, 420 intein sequences that have been reported to date, only about one-fourth of them have been experimentally shown to splice outside their native environment (22). Most of the remaining inteins are putative, and many of them cannot splice outside the native host or need modifications for nonnative splicing (27,28).

Some inteins have been modified to allow control of their splicing characteristics (29), allowing them to be used in recombinant protein purification through combination with conventional affinity tags (30). Intein technology and *in vitro* self-splicing of proteins have made inteins important tools in protein engineering, allowing advances in many areas of biotechnology such as protein purification (31,32), proteomics (33), drug discovery (34), protein structure and function (35), and gene therapy (36). An important specific application of inteins is the generation of self-cleaving affinity tags through a modification of the splicing reaction to yield controllable N-terminal and C-terminal cleaving inteins.

15.4 CONVENTIONAL SELF-CLEAVABLE AFFINITY TAGS

Introduction of an affinity tag to a target protein may have positive or negative effects on the biochemical properties of the target protein. Increasing the protein solubility (37), assisting protein folding (38), and improvement in expression and protein yield (39) are among the positive effects of the presence of affinity tags. In contrast, the affinity tags can reduce protein yield (40) or change the protein conformation and alter its biological activity (41). In most cases, however, the tag must be removed to acquire a native target protein for study or use in various applications.

Inteins can be used to construct self-cleavable affinity tags from effectively any conventional affinity tag. The self-cleaving mechanism for an intein-based affinity tag is a little different from that shown in Fig. 15.1. In these cases, suitable mutation(s) at the intein–extein junction residues suppress different steps of the splicing reaction, resulting in isolated N- or C-terminal cleaving activities. Thus, in some cases, where a three-domain precursor is designed for affinity separations, the target protein can be fused to either the N- or C-terminus of an appropriately modified intein (target–intein–tag or tag–intein–target), while the affinity tag is generally attached to the opposite intein terminus.

For a target protein fused to the N-terminus of an intein (target–intein–tag), the thioester linkage generated by the first acyl shift (Fig. 15.1, Step 1) can be cleaved by an exogenous thiol compound (R-SH) such as dithiothreitol (DTT), releasing the target protein. In this case, the asparagine residue at the C-terminus of the intein can allow unwanted C-terminal cleavage to take place, leading to cleavage at both ends of the intein (42). For this reason, this residue is typically mutated to alanine. If the target protein is at the C-terminus of the intein, the splicing reaction and N-terminal cleavage can be suppressed by mutating the initial intein residue to alanine. This mutation prevents the initial acyl shift and thioester formation of the splicing reaction, but still allows succinimide formation and isolated C-terminal cleavage of the intein. In general, however, this mutation leads to very slowly cleaving inteins that are not fast enough to be reasonable for protein purification methods. This problem was solved through a directed evolution approach on the *Mycobacterium tuberculosis* RecA intein (43). This intein was modified by first deleting its endonuclease domain, and then an internal conserved aspartic acid residue was mutated to glycine. The result was a small, fast-cleaving intein, which has been used with several affinity and other tags for protein purification.

Combination of an appropriate affinity tag with a modified self-cleaving intein can thus provide a fast and convenient technique to purify a target protein using a single-step affinity method. As shown in Fig. 15.2, following the expression of the three-domain fusion protein in a suitable host, the precursor is separated from the contaminants in cell lysate using the specific binding activity of the affinity tag and the ligand on the solid matrix. Once the precursor is purified, the intein is induced to cleave off the product protein, allowing it to be collected in a highly purified native form.

Many affinity tags or carriers are available now to produce and purify fusion proteins in various host organisms (44,45), but only a few of them have been combined with inteins to make self-cleaving affinity-tagged fusion proteins (Table 15.1). Although conventional self-cleaving affinity-tag techniques have greatly simplified protein purification at laboratory scale, the high cost for equipment and consumable resins can be a limitation for large-scale purifications of recombinant proteins.

15.5 SELF-CLEAVING AGGREGATION TAGS

Despite the advantages and success of conventional and self-cleaving affinity-tag fusion technologies in the production of recombinant proteins, they still suffer from some significant disadvantages. Peptide-based affinity tags have a low binding capacity and some of them are toxic to the hosts (55). Moreover, these technologies use chromatographic separation methods that can be time consuming and difficult to scale-up, and depend on expensive affinity matrices (often more than US$1000 per liter of bed volume depending on the ligand and tag). These aspects make it

TABLE 15.1. Affinity Tags Used in Self-Cleaving Systems for Protein Purification

Tag	Size (Amino Acid)	Comments	References
Polyhistidine (PolyHis)	5–15	Protein purification under native or denaturing conditions The most common affinity tag Purification matrix: immobilized divalent metal (Ni^{2+}, Co^{2+}, Cu^{2+}, Zn^{2+}), capacity 5–15 mg/mL	46,47
Chitin-binding domain (CBD)	52	Used in the first commercially available intein-based purification system (IMPACT, from New England Biolabs) The ability to label the N- or C-terminus of the target protein Matrix: chitin functionalized sepharose, capacity 2 mg/mL	48–50
Maltose-binding domain (MBD)	396	Increases expression and solubility of eukaryotic proteins in bacteria Matrix: cross-linked sepharose, functionalized with amylose, capacity 3 mg/mL	31,51,52
Glutathione *S*-transferase (GST)	218	Increased expression and solubility of target protein Matrix: glutathione functionalized sepharose, capacity 10 mg/mL	53
Cellulose-binding module (CBM)	27–189	Irreversible binding of some CBDs to cellulose, useful for immobilization Matrix: cellulose	54

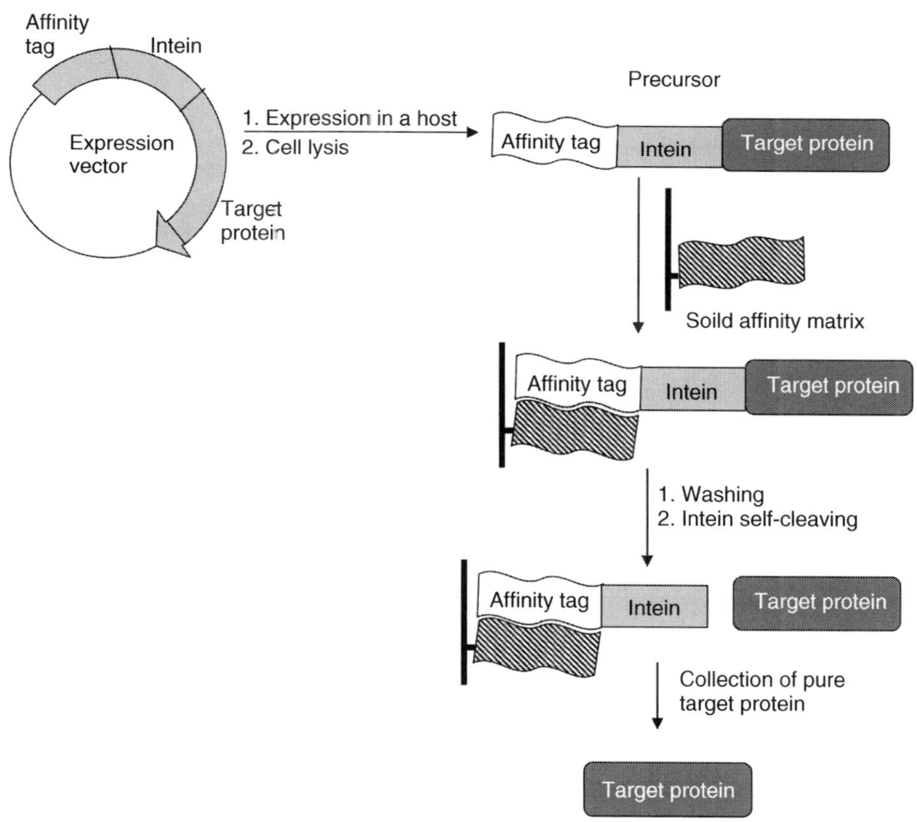

Figure 15.2. Protein purification by conventional affinity-tag self-cleaving method. An expression vector is designed to encode a three-domain, tagged precursor protein and transformed into a suitable host (e.g. E. coli or yeast). The precursor protein is then expressed and separated from the cellular contaminants by specific binding of the affinity tag to a solid matrix. Finally, the target protein is released from the precursor by the intein self-cleaving reaction, induced by changing the pH or addition of thiol compounds, and collected in a purified form in the column effluent.

difficult to use even self-cleaving affinity tags as a routine process for large-scale recombinant protein production. For this reason, the development of nonchromatographic alternatives has gained widespread interest. Two of these are the self-cleaving ELP tag, which uses reversible precipitation of the tagged target protein for separation, and the PHB system, where the tagged protein is bound to macroscopic granules of PHB coexpressed in the host cell.

15.5.1 The Self-Cleaving Elastin-Like Polypeptide (ELP) System

ELPs are artificial biopolymers composed of repeats of pentapeptide units Val-Pro-Gly-X-Gly, where the X (guest residue) can be any natural amino acid with the exception of proline (56). ELPs are the synthetic versions of a structural motif found in mammalian elastin protein. Because no posttranslational modifications are required for ELPs to be functional, they can be produced easily and efficiently by fermentation processes in various hosts such as *Escherichia coli* or yeast (57,58). Moreover, the resulting polypeptides have shown no immunogenic characteristics, making them very good nominees for biomedical applications including tissue engineering (59,60) and drug delivery (61).

The most important property of ELPs that make them perfect candidates for protein purification is their unique thermally induced phase transition behavior. Below the ELP protein's inverse transition temperature (T_t), it is highly solvated and soluble in aqueous solutions, but at temperatures higher than T_t, the interactions between the nonpolar regions of the polymeric chain increase and the ELP molecules aggregate and separate from solution (62). This behavior is reversible with cooling, that is, the insoluble ELP (i.e. aggregation tag) will resolubilize if the temperature of the suspension is lowered to below the T_t of the ELP. The transition temperature of ELP is dependent on several parameters including its structural properties, such as molecular weight (63), and type and concentration of the guest residue (64). Further, the buffer conditions, including the concentration of ELP, buffer ionic strength, the concentration and type of salt in the buffer (65), and pH of the buffer (66), can also influence T_t. For example, the incorporation of lysine residues into an ELP tag in a thioredoxin–ELP fusion protein was shown to increase salt sensitivity and reduce the amount of salt necessary for precipitation of the precursor (67).

Interestingly, the reversible phase transition behavior of free ELP is retained when ELP is a part of a fusion protein (68,69). Addition of an ELP tag to the C- or N-terminus of a target protein has been used to facilitate the purification of the target through rounds of heating, centrifugation, and resuspension, and this method is now known as *inverse transition cycling* (ITC). The ITC technique has been utilized by many researchers to purify ELP-tagged fusion proteins selectively and with high efficiencies without the necessity for any chromatographic step (70–72). In one example, the ELP fusion tag was used to purify ultralow levels (15–20 pmol/L culture) of ELP fusion proteins, corresponding to 2–3 protein molecules per cell (73). This was accomplished by adding excess free ELP to the cell lysate, which enhanced the precipitation of the tagged target. Because of its simplicity and lack of expense, ELP purification methods are an excellent alternative for other affinity methods, and can provide enough protein for mass spectrometry, radioactive assays, or other analytical techniques.

ELP-tagged fusion proteins can be purified and used directly for some applications such as ELP–enzyme immobilization onto solid supports for biocatalysis (74–76) and therapeutic or diagnostic purposes (77). In cases where a native target protein is required, however, the ELP tag should be cleaved from the fusion protein. As with conventional affinity tags, the target protein can be liberated from the fusion protein by enzymatic methods using a cleavage site placed between the ELP tag and target protein (78), but this method suffers from the same limitations as proteolytic removal with conventional affinity tags. Fortunately, in 2005, this method was enhanced by combining ELP-based purification methods with self-cleaving inteins (79,80). This self-cleaving aggregation-tag method eliminated the disadvantages of the enzymatic cleavage procedures, reducing the cost of the method and eliminating the need to remove the protease after tag removal. In the self-cleaving ELP-tag technique, the target protein gene's N- or C-terminus is fused to a gene encoding an engineered self-cleaving intein and ELP tag.

As inverse transition temperature (T_t) is a key parameter in the ELP protein purification, the temperature should be controlled in all steps of the protocol from expression to the self-cleaving stage. Figure 15.3 shows the production and purification steps for a target protein from an intein-mediated ELP-tagged precursor. The expression vector for the precursor is transformed into an appropriate host (*E. coli*, yeast, or plant cells) and overexpressed to produce a soluble precursor. Cell lysis and separation of cell debris from the soluble precursor is carried out at a temperature lower than the T_t of the fusion protein. A salt, usually sodium chloride or ammonium sulphate, is then added to the precursor to a previously optimized final concentration and the solution is heated to a temperature above its T_t. At this step, the ELP part of the precursor self-associates and forms a precipitate that can be separated using centrifugation (81) or microfiltration (82) at temperature above T_t. For some ELP-tagged fusion proteins, multiple rounds of ITC can increase the purity of the precursor. In this case, the aggregated precursor from an ITC step is dissolved in a buffer at temperature below the T_t, centrifuged, and the soluble fraction is reused for another round of ELP purification.

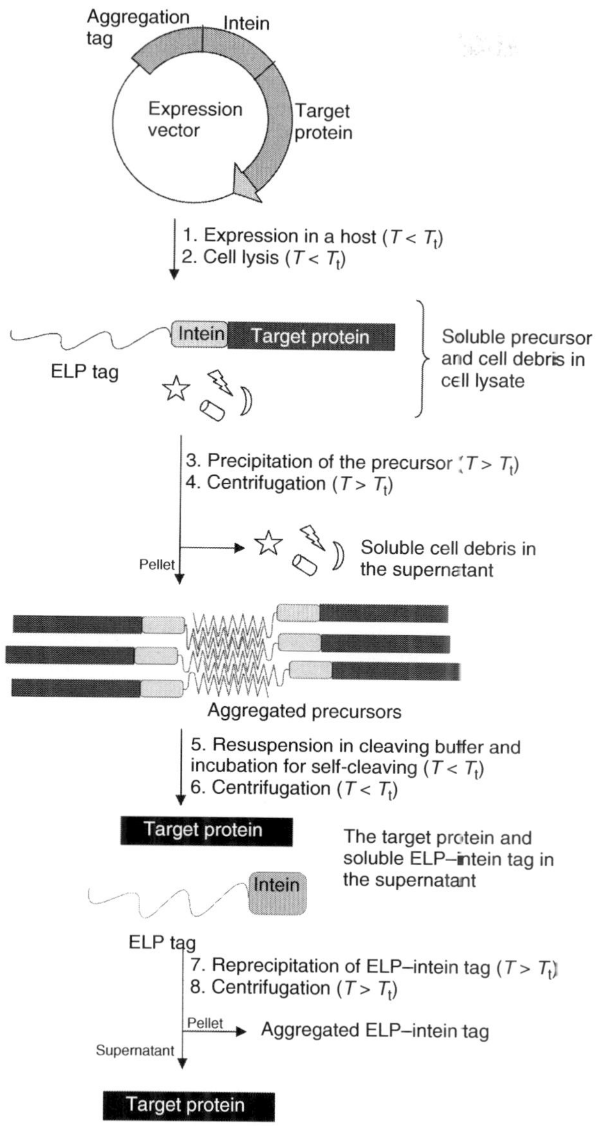

Figure 15.3. Production and purification of a native target protein using intein-mediated ELP-tagged protein method. The process consists of the following steps: (1) The ELP–intein–target protein precursor is overexpressed in a suitable host at low temperature to suppress premature cleavage of the intein. (2) Cell disruption at low temperature releases the soluble precursor and cell debris. (3 and 4) Soluble ELP-tagged precursor is separated from the rest of contaminants in the cell lysate by salt addition and/or heating and centrifugation. (5 and 6) The aggregated precursor is resuspended in a cleaving buffer at low salt and incubated to induce the self-cleaving reaction. (7 and 8) The cleaved ELP–intein tag is aggregated and removed by another round of precipitation and centrifugation. The purified target protein is recovered from the supernatant.

Self-cleaving of the intein to release the target protein starts after redissolving the aggregated precursor in a suitable low-salt buffer. Cleavage is induced by addition of thiol compounds in the case of N-terminal cleaving inteins (80), or by a pH change in the case of C-terminal cleaving inteins (79). Thiol-induced cleavage is the method of choice for some systems with high premature cleavage in their expression step. The purification process is completed by an additional cycle of salt addition, heating, and centrifugation to remove the ELP–intein tag. Ultimately, the pure and native target protein is recovered from the whole cell lysate without any chromatographic step.

15.5.2 The Polyhydroxybutyrate (PHB) System

15.5.2.1 PHB–Intein-Mediated Protein Purification. Polyhydroxyalkanoates (PHAs) are naturally occurring, biodegradable linear polyesters composed of 3-hydroxy fatty acid monomers (HO–C(R)–CH_2–COOH) that can be produced in various expression systems such as bacteria, yeast, and transgenic plant cells (83–86). The biosynthetic pathway of a PHA consists of three different enzymes: β-ketothiolase (phaA), acetoacetyl-CoA reductase (phaB), and PHA synthase (phaC). Biosynthesis of PHA is usually induced by limiting essential elements such as nitrogen, phosphorus, trace elements, or oxygen in the growth medium. Addition of different renewable carbon sources such as fatty acids and sugars is typically utilized to increase PHA production, which can account for as much as 80% of the cell's dry weight. PHB ($[O{-}CH(CH_3)CH_2CO]_n$) is the most common type of PHA, and has been produced in various expression systems such as *E. coli* (87,88), *S. cerevisiae* (89), and transgenic plant cells (90,91).

The most important characteristic of PHB granules, which makes them an outstanding tool in protein purification, is that they can act as an affinity carrier for *in vivo* immobilization of a tagged coexpressed target protein. Typically, the PHB is accumulated within the cells as granular inclusion bodies with diameter of 0.2–0.5 μm. These granules are macroscopic in size with a relatively high density, allowing them to be easily recovered and purified using a variety of mechanical means such as centrifugation or membrane filtration. The affinity tag that binds to PHB granules is derived from a class of PHB regulatory proteins known as *phasins*, which have been identified in several microorganisms (92–96). Phasins are low-molecular-weight proteins whose role in PHA synthesis is not very well understood, but are known to have specific binding affinity to the surface of PHB granules (97,98). During *in vivo* formation of PHAs, phasins accumulate and bind to the surface of granules and promote additional PHB synthesis (99–102). They are the major proteins on the surface of PHAs and can accumulate up to 5% of the total cell protein (103).

During fermentation, the phasin-tagged target protein and PHB granules are coproduced in the expression host. The target protein is immobilized *in vivo* onto the surface of PHB granules via the phasin tag. PHB granules associated with the phasin-tagged target protein are recovered from the cell lysate and cleaned using a series of centrifugation and resuspension steps. In one example, the N-terminal domain of the PhaF phasin from *Pseudomonas putida* was employed as a polypeptide tag (BioF) to anchor β-galactosidase to a PHA. The PHA bioplastic carrying the β-galactosidase fusion protein could be used directly for some applications, or alternately, the soluble BioF fusion protein could be obtained in relatively pure form by a mild detergent treatment of the granules (104). In another example, the substrate-binding domain (SBD) of PHA depolymerase was used instead of phasin for the immobilization of proteins onto the surface of PHA, and was subsequently applied for immunoassay experiments (105). In a third example, protein engineering techniques were used to fuse a β-galactosidase to the N-terminus of PHA synthase from *Pseudomonas aeruginosa*, demonstrating the production of PHA granules with immobilized enzymes covalently linked to the surface of the granules (106). Biocatalysts produced by this method can be used and easily recycled for biotransformations and fine-chemical production.

As with ELP tags, phasins can be combined with self-cleaving inteins to generate a self-contained nonchromatographic purification method (107–109). In effect, this technology combines the fermentation process to generate a tagged target protein with the binding step of affinity chromatography. All steps for the production and purification of the tagged target protein, including formation of affinity carrier and the binding of the tagged protein to the affinity media, take place inside the expression host. In practice, a plasmid carrying all the enzymes required for PHB biosynthesis is first transformed into a proper expression system such as *E. coli*. A second transformation introduces a plasmid encoding a fusion protein composed of the phasin protein (phaP), intein self-cleaving element, and the target protein. The double-transformed expression strain is then grown on a suitable growth medium supplemented with lactate, glucose, sucrose, fructose, xylose, or other cheap and renewable carbon sources to produce the PHB granules. Overexpression of the fusion protein is induced during early stationary phase, when the granules are also being produced. Cells are harvested and lysed under conditions that suppress the intein cleavage reaction (pH 8.5 for temperature-sensitive inteins and no DTT for thiol-activated inteins). The PHB granules, with a density of about 1.20 mg/mL, are then separated from the cell lysate and washed using either centrifugation or microfiltration (110). After washing the PHB granules with the fusion protein on the surface, intein self-cleavage is induced by shifting the pH from 8.5 to 6.5–7.0 or by addition of an appropriate amount of DTT (typically 30–50 m*M*). The cleaving reaction releases the target protein and it can be easily recovered in the supernatant, in a native and pure form, by a final centrifugation (Fig. 15.4).

Different proteins such as β-galactosidase, chloramphenicol acetyltransferase (CAT), NusA protein, maltose-binding protein (MBP), and green fluorescent protein (GFP) have been purified effectively, with 30–50 mg of cleaved, pure protein per liter of shake flask culture. In one example, GFP was used as a model protein to show the effective sequestration of the target proteins to the surface of PHB granules (109). In this case, fluorescence microscopy and sucrose density gradient fractionation techniques were utilized to show the localization of the target proteins (GFP, phasin–GFP, or phasin–intein–GFP) in the cell lysates. This work demonstrated that GFP fused to a phasin tag or phasin–intein is localized to the PHB granules in the pellet. In another example it was shown that PHA can act as a carrier in classic affinity chromatography, where the tagged target protein and PHB granules are produced separately and combined *in vitro* (111). The rest of the purification process is similar to PHB–intein protein purification method.

The PHB–intein-mediated strategy for protein purification has significant economical advantages in comparison with other conventional affinity-tag chromatographic systems. Concurrent *in vivo* production of the affinity carrier (PHB) and the target protein eliminates the costs related with the affinity matrix. To produce the PHB granules, the only additional cost is the addition of a cheap and renewable carbon source supplement to the fermentation process. The resulting PHB granules are biodegradable and can be easily disposed after recovery of the target protein. Commercial affinity resins available for intein systems such as chitin resins are very expensive with low protein loading capacities (about \$2000 L^{-1} resin and capacity 2 g/L of bed volume). In addition, the regeneration of the used resins is time consuming after each purification and costly for large-scale preparations. Because the cleaving reaction releases only the target protein from the granule surface, any contaminant proteins bound to the granules remain bound, and are removed along with the granules in the final centrifugation. This system, by combining high-level of expression of recombinant proteins with a simple purification method, has the potential to facilitate the large-scale production of various peptides or proteins such as enzymes, vaccines, and therapeutics.

15.6 ADVANTAGES, ECONOMY AND FUTURE PROSPECTS OF SELF-CLEAVING AGGREGATION-TAG TECHNOLOGIES

The ELP and PHB systems used in self-cleaving aggregation-tag technologies provide simple, efficient, and

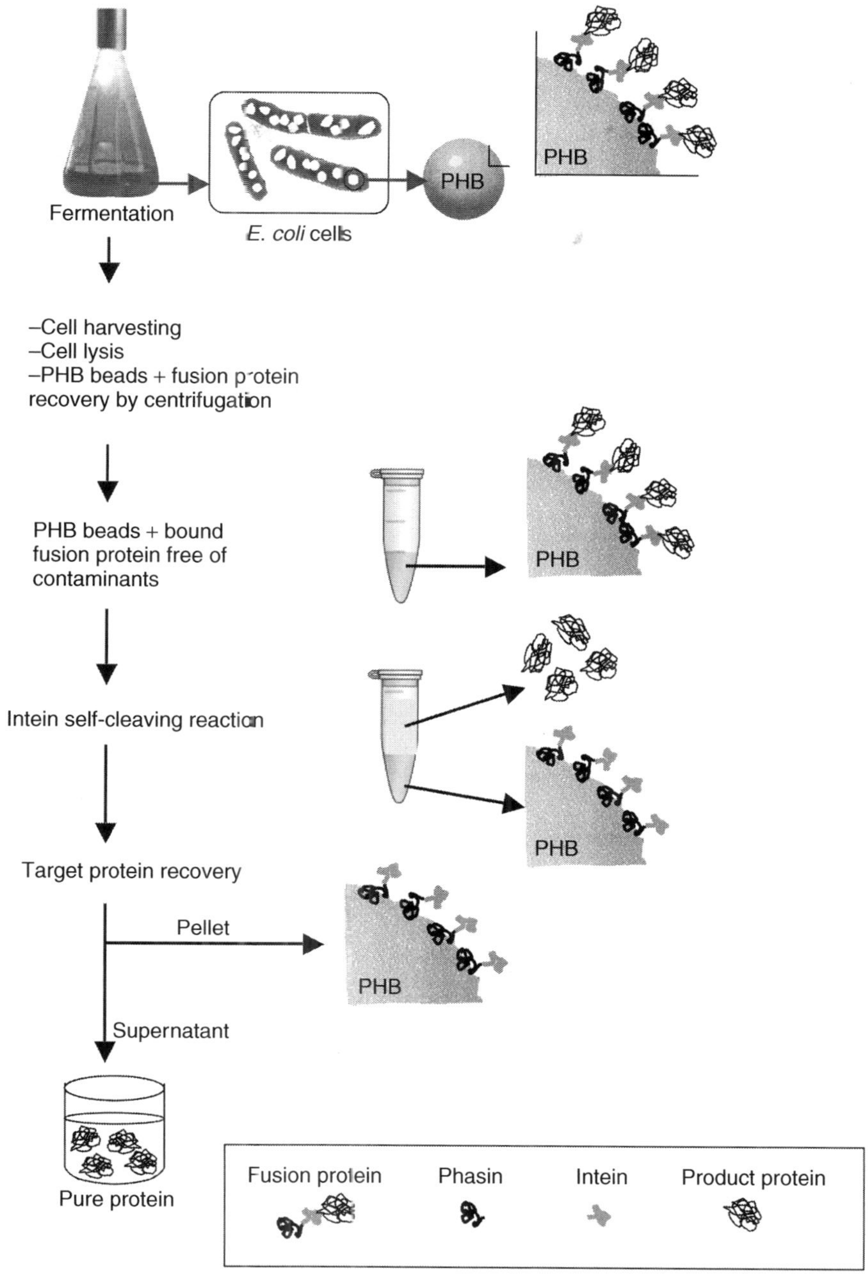

Figure 15.4. PHB–intein-mediated protein purification method. In the fermentation step, PHB beads and the phasin–intein tagged product protein are produced in a suitable expression host. In this step, the fusion protein is immobilized *in vivo* on the surfaces of the PHB granules. Harvested cells are lysed and centrifuged to separate soluble contaminants. The insoluble PHB granules with the fusion protein attached to the surface are resuspended and washed with a suitable buffer to obtain contaminant-free PHB granules. The target protein is released during incubation in a cleavage-inducing buffer. A final centrifugation step separates the PHB granules carrying the phasin–intein on the surface from the cleaved target protein in the supernatant.

economical means for the purification of recombinant proteins. Both ELP–and PHB–intein-mediated fusion proteins can be produced in various engineered expression systems such as bacteria, yeast, or transgenic plant cells. Developing mammalian cells capable of expressing ELP and PHB could open a new window for the production and purification of protein pharmaceuticals.

The ELP–and PHB–intein-mediated techniques require mild conditions for protein expression and recovery (e.g. expression variables and buffer compositions). Thus, the final product is recovered with the minimum risk of degradation or irreversible damage to the target protein. The intein self-cleaving step is nonenzymatic and leaves no extra amino acids on the target, making it ideal for large-scale production of pharmaceutical proteins. Development of new inteins with better characteristics (faster and more controllable cleaving rates) could further simplify the ELP–and PHB–intein-mediated systems of protein purification.

Recently, the material cost analysis for a large-scale protein purification process (1 kg pure protein) using nonchromatographic ELP–and PHB–intein-mediated technologies, compared with other conventional affinity chromatographic techniques, has shown a dramatic decrease in the production cost for these novel nonchromatographic methods (112). In this study, five different purification methods have been compared. The ELP-tagged intein-mediated method is the most economical process with a cost of \$75,000 kg^{-1} of pure product protein. The material cost for other methods are \$150,000 kg^{-1} for PHB–intein system, \$990,000 kg^{-1} for IMPACT™ Kit from NEB (New England Biolabs), \$840,000 kg^{-1} for His-tag system from Novagen, and \$9,700,000 kg^{-1} for pMAL™ fusion system from NEB. The major decrease in the cost for ELP and PHB systems is due to the higher yields of protein production and the elimination of the affinity resin and related costs. In the case of pMAL fusion system, which is a very expensive method (130-fold increase related to ELP system), the major portion of the cost (87%) is because of the price of proteolytic enzyme.

In spite of the high protein purities obtainable with His-tag fusion and pMAL fusion systems, they are usually used for small, lab-scale preparations and have not yet been commonly used for large-scale protein production. The product protein prepared from ELP– or PHB–intein systems can be directly used for large-scale industrial applications, where low levels of impurities are acceptable. If a highly pure product is required (e.g. for pharmaceutical applications), the product from ELP or PHB systems can be polished using common methods to higher levels of purity. In practice, ELP and PHB protein purification protocols are simple and sensitive, which make them ideal tools for high-throughput applications and can be applied to large libraries of targets in high-throughput methods.

REFERENCES

1. Freitag R, Horvath C. Adv Biochem Eng Biotechnol 1996; 53: 17–59.
2. Kalyanpur M. Mol Biotechnol 2002; 22: 87–98.
3. Hunt I Protein Expression Purif 2005; 40: 1–22.
4. Roque AC, Lowe CR. Methods Mol Biol 2008; 421: 1–21.
5. Yannis DC. J Chromatogr A 2006; 1101: 1–24.
6. Crimmins DL, Mische SM, Denslow ND. Curr Protoc Protein Sci 2005, Chapter 11: Unit 11 4.
7. Richardson JP, Macmillan D. Org Biomol Chem 2008; 6: 3977–3982.
8. LaVallie ER, McCoy JM, Smith DB, Riggs P. Curr Protoc Mol Biol 2001, Chapter 16: Unit16 4B.
9. Zheng XL, Kitamoto Y, Sadler JE. Front Biosci (Elite Ed) 2009; 1: 242–249.
10. Jenny RJ, Mann KG, Lundblad RL. Protein Expression Purif 2003; **1**: 1–11.
11. Charlton A. Methods Mol Biol 2008; 421: 211–228.
12. Hirata R, Ohsumk Y, Nakano A, Kawasaki H, Suzuki K, Anraku Y. J Biol Chem 1990; 265: 6726–6733.
13. Gimble FS, Thorner J. Nature 1992; 357: 301–306.
14. Perler FB, Davis EO, Dean GE, Gimble FS, Jack WE, Neff N, Noren CJ, Thorner J, Belfort M Nucl Acids Res 1994; 22: 1125–1127.
15. Mujika JI, Lopez X, Mulholland AJ. J Phys Chem B 2009; 113: 5607–5616.
16. Perler FB. IUBMB Life 2005; 57: 469–476.
17. Anraku Y, Mizutani R, Satow Y. IUBMB Life 2005; 57: 563–574.
18. Mills KV, Dorval DM, Lewandowski KT. J Biol Chem 2005; 280: 2714–2720.
19. Duan X, Gimble FS, Quiocho FA. Cell 1997; 89: 555–564.
20. Belfort M, Roberts RJ. Nucl Acids Res 1997; 25: 3379–3388.
21. Derbyshire V, Wood DW, Wu W, Dansereau JT, Dalgaard JZ, Belfort M. Proc Natl Acad Sci USA 1997; 94: 11466–11471.
22. New England Biolabs Inc., web information. Available at http://www.neb.com/neb/inteins.html. Accessed 2009 30 Aug.
23. Perler FB. Nucl Acids Res 2002; 30: 383–384.
24. Dori-Bachash M, Dassa B, Peleg O, Pineiro SA, Jurkevitch E, Pietrokovski S. Funct Integr Genomics 2009; 9: 153–166.
25. Liu XQ, Hu Z. FEBS Lett 1997; 408: 311–314.
26. Wang S, Liu XQ. J Biol Chem 1997; 272: 11869–11873.
27. Appleby JH, Zhou K, Volkmann G, Liu XQ. J Biol Chem 2009; 284: 6094–6199.
28. Culley AI, Asuncion BF, Steward GF. ISME J 2009; 3: 409–418.
29. Lockless SW, Muir TW. Proc Natl Acad Sci USA 2009; 106: 10999–10004.
30. Porte M, Chong S. Anal Biochem 2008; 381: 175–177.
31. Gillies AR, Hsii JF, Oak S, Wood DW. Biotechnol Bioeng 2008; 101: 229–240.
32. Mee C, Banki MR, Wood DW. Chem Eng J 2008; 135: 56–62.
33. Lue RY, Chen GY, Hu Y, Zhu Q, Yao SQ. J Am Chem Soc 2004; 126: 1055–1062.
34. Cheriyan M, Perler FB. Adv Drug Delivery Rev 2009; 61(11): 899–907.
35. Muralidharan V, Muir TW. Nat Methods 2006; **3**: 429–438.
36. Li J, Sun W, Wang B, Xiao X, Liu XQ. Hum Gene Ther 2008; 19: 958–964.
37. Nallamsetty S, Waugh DS. Protein Expression Purif 2006; 45: 175–182.

38. Planson AG, Guijarro JI, Goldberg ME, Chaffotte AF. Biochemistry 2003; 42: 13202–13211.
39. Ashraf SS, Benson RE, Payne ES, Halbleib CM, Gron H. Protein Expression Purif 2004; 33: 238–245.
40. Hammarström M, Woestenenk EA, Hellgren N, Hörd T, Berglund H. J Struct Funct Genomics 2006; 7: 1–14.
41. Goel A, Colcher D, Koo JS, Booth BJ, Pavlinkova G, Batra SK. Biochim Biophys Acta 2000; 1523: 13–20.
42. Chong S, Mersha FB, Comb DG, Scott ME, Landry D, Vence LM, Perler FB, Benner J, Kucera RB, Hirvonen CA, Pelletier JJ, Paulus H, Xu MQ. Gene 1997; 192: 271–281.
43. Wood DW, Wu W, Belfort G, Derbyshire V, Belfort M. Nat Biotechnol 1999; 9: 889–892.
44. Michelle, EK, Sondek J. Curr Protoc Protein Sci 2004, Chapter 9: Unit 9 9.
45. Terpe K. Appl Microbiol Biotechnol 2003; 60: 523–533.
46. Xu CG, Fan XJ, Fu YJ, Liang AH. Protein Expression Purif 2008; 59: 103–109.
47. Goodin JL, Raab RW, McKown RL, Coffman GL, Powell BS, Enama JT, Ligon JA, Andrews GP. Protein Expression Purif 2005; 40: 152–163.
48. Srinivasa BK, Antony A, Muthukumaran T, Meenakshisundaram S. Protein Expression Purif 2008; 57: 201–205.
49. Zhuo-Yu L, Jun-Hu F, Jing-Ming Y. Biotechnol Lett 2002; 24: 1723–1727.
50. Wu W, Wood DW, Belfort G, Derbyshire V, Belfort M. Nucl Acids Res 2002; 30: 4864–4871.
51. Wood DW, Derbyshire V, Wu W, Chartrain M, Belfort M, Belfort G. Biotechnol Prog 2000; 16: 1055–1063.
52. Wiese A, Wilms B, Syldatk C, Mattes R, Altenbuchner J. Appl Microbiol Biotechnol 2001; 55: 750–757.
53. Valiyaveetil FI, MacKinnon R, Muir TW. J Am Chem Soc 2002; 124: 9113–9120.
54. Hong J, Wang Y, Ye X, Zhang YH. J Chromatogr A 2008; 1194: 150–154.
55. Deo SK, Daunert S. Anal Chem 2001; 73: 1903–1908.
56. Chow D, Nunalee ML, Lim DW, Simnick AJ, Chilkoti A. Mater Sci Eng R Rep 2008; 62: 125–155.
57. McPherson DT, Morrow C, Minehan DS, Wu J, Hunter E, Urry DW. Biotechnol Prog 1992; 8: 347–352.
58. Schipperus R, Teeuwen RL, Werten MW, Eggink G, de Wolf FA. Appl Microbiol Biotechnol 2009; 85(2): 293–301.
59. Srokowski EM, Woodhouse KA. J Biomater Sci Polym Ed 2008; 19: 785–799.
60. Dreher MR, Simnick AJ, Fischer K, Smith RJ, Patel A, Schmidt M, Chilkoti A. J Am Chem Soc 2008; 130: 687–694.
61. Adams SB Jr, Shamji MF, Nettles DL, Hwang P, Setton LA. J Biomed Mater Res Part B 2009; 90, 67–74.
62. Li B, Daggett V. Biopolymers 2003; 68: 121–129.
63. Meyer DE, Trabbic-Carlson K, Chilkoti A. Biotechnol Prog 2001; 17: 720–728.
64. Ribeiro A, Arias FJ, Reguera J, Alonso M, Rodrrguez-Cabello JC. Biophys J 2009; 97: 312–320.
65. Fong BA, Wu WY, Wood DW. Protein Expression Purif 2009; 66: 198–202.
66. Li B, Daggett V Biopolymers 2003; 68: 121–129.
67. Lim DW, Trabbic-Carlson K, Mackay JA, Chilkoti A Biomacromolecules 2007; 8: 1417–1424.
68. Yamaoka T, Tamura T, Seto Y, Tada T, Kunugi S, Tirrell DA. Biomacromolecules 2003; 4: 1680–1685.
69. Meyer DE, Chilkoti A. Nat Biotechnol 1999; 17: 1112–1115.
70. Ge X, Yang DSC, Trabbic-Carlson K, Kim B, Chilkoti A, Filipe CDM. J Am Chem Soc 2005; 127: 11228–11229.
71. Lao UL, Mulchandani A, Chen W. J Am Chem Soc 2006; 128: 14756–14757.
72. Joensuu JJ, Brown KD, Conley AJ, Clavijo A, Menassa R, Brandle JE. Transgenic Res 2009; 18: 685–696.
73. Christensen T, Trabbic-Carlson K, Liu W, Chilkoti A. Anal Biochem 2007; 360: 166–168.
74. Kang HJ, Kim JH, Chang WJ, Kim ES, Koo YM. J Microbiol Biotechnol 2007; 17: 1751–1757.
75. Shimazu M, Mulchandani A, Chen W. Biotechnol Bioeng 2003; 81: 74–79.
76. Lee J, Kim O, Jung J, Na K, Heo P, Hyun J. Colloids Surf B 2009; 72: 173–180.
77. Mackay JA, Chilkoti A. Int J Hyperthermia 2008; 24: 483–495.
78. Trabbic-Carlson K, Liu L, Kim B, Chilkoti A. Protein Sci 2004; 13: 3274–3284.
79. Banki MR, Feng L, Wood DW. Nat Methods 2005; 2: 659–661.
80. Ge X, Yang DS, Trabbic-Carlson K, Kim B, Chilkoti A, Filipe CD. J Am Chem Soc 2005; 127: 11228–11229.
81. Wu WY, Mee C, Califano F, Banki R, Wood DW Nat Protoc 2006; 1: 2257–2262.
82. Ge X, Trabbic-Carlson K, Chilkoti A, Filipe CD. Biotechnol Bioeng 2006; 95: 424–432.
83. Singh M, Patel SK, Kalia VC. Microb Cell Fact 2009; 8: 38.
84. Li R, Zhang H, Qi Q. Bioresour Technol 2007; 98: 2313–2320.
85. Zhang B, Carlson R, Srienc F. Appl Environ Microbiol 2006; 72: 536–543.
86. Suriyamongkol P, Weselake R, Narine S, Moloney M, Shah S. Biotechnol Adv 2007; 25: 148–175.
87. Nikel Pi, de Almeida A, Melillo EC, Galvagno MA, Pettinari MJ. Appl Environ Microbiol 2006; 72: 3949–3954.
88. Yu H, Yin J, Li H, Yang S, Shen Z. J Biosci Bioeng 2000; 89: 307–311.
89. Leaf TA, Peterson MS, Stoup SK, Somers D, Srienc F. Microbiology 1996; 142: 1169–1180.
90. Petrasovits LA, Purnell MP, Nielsen LK, Brumbley SM. Plant Biotechnol J 2007; 5: 162–172.
91. Wrobel M, Zebrowski J, Szopa J. J Biotechnol 2004; 107: 41–54.
92. Hänisch J, Wältermann M, Robenek H, Steinbüchel A. Microbiology 2006; 152: 3271–3280.
93. Liebergesell M, Schmidt B, Steinbuchel A. FEMS Microbiol Lett 1992; 99: 227–232.
94. McCool GJ, Cannon MC. J Bacteriol 1999; 181: 585–592.
95. Pieper-Furst U, Madkour MH, Mayer F, Steinbuchel A. J Bacteriol 1995; 177: 2513–2523.
96. Prieto MA, Buehler B, Jung K, Witholt B, Kessler B. J Bacteriol 1999; 181: 858–868.

97. Jurasek L, Marchessault RH. Biomacromolecules 2002; 3: 256–261.
98. York GM, Junker BH, Stubbe J, Sinskey AJ. J Bacteriol 2001; 183: 4217–4226.
99. Neumann L, Spinozzi F, Sinibaldi R, Rustichelli F, Pötter M, Steinbüchel A. J Bacteriol 2008; 190: 2911–2919.
100. Wieczorek R, Pries A, Steinbüchel A, Mayer F. J Bacteriol 1995; 177: 2425–2435.
101. York GM, Stubbe J, Sinskey AJ. J Bacteriol 2001; 183: 2394–2397.
102. York GM, Stubbe J, Sinskey AJ. J Bacteriol 2002; 184: 59–66.
103. Potter M, Madkour MH, Mayer F, Steinbuchel A. Microbiology 2002; 148: 2413–2426.
104. Moldes C, Garcra P, Garcra jL, Prieto MA. Appl Environ Microbiol 2004; 70: 3205–3212.
105. Lee SJ, Park JP, Park TJ, Lee SY, Lee S, Park JK. Anal Chem 2005; 77: 5755–5759.
106. Peters V, Rehm BH. Appl Environ Microbiol 2006; 72: 1777–1783.
107. Gillies AR, Banki MR, Wood DW. Methods Mol Biol 2009; 498: 173–183.
108. Banki MR, Gerngross TU, Wood DW. Protein Sci 2005; 14: 1387–1395.
109. Barnard GC, McCool JD, Wood DW, Gerngross TU. Appl Environ Microbiol 2005; 71: 5735–5742.
110. Resch S, Gruber K, Wanner G, Slater S, Dennis D, Lubitz W. J Biotechnol 1998; 65: 173–182.
111. Wang Z, Wu H, Chen J, Zhang J, Yao Y, Chen GQ. Lab Chip 2008; 8: 1957–1962.
112. Banki MR, Wood DW. Microb Cell Fact 2005; 11(4): 32.

16

LIPOPOLYSACCHARIDE, LPS REMOVAL, DEPYROGENATION

PÉROLA O. MAGALHÃES
Department of Pharmacy, School of Health Sciences, University of Brasília, Brasília, DF, Brazil
ADALBERTO PESSOA JR.
Department of Biochemical and Pharmaceutical Technology, School of Pharmaceutical Sciences, University of São Paulo, Brazil

16.1 INTRODUCTION

Advances in biotechnology have opened up numerous possibilities for the large-scale production of many biomolecules that are important for research, pharmaceuticals, and industrial applications. The development of techniques and methods for the separation and purification of proteins and other biomolecules has been of paramount importance for many of these advances in the biotechnology industry (1,2). Such biomolecules may be expressed as soluble extracellular form, soluble intracellular form, or as membrane constitute or inclusion bodies. In all three cases, use of cellular rupture procedures for the recovery of the biomolecule of interest, which lead to the release of a large quantity of endotoxins called *lipopolysaccharide* (*LPS*), is necessary.

LPS, the cell wall component of gram-negative bacteria, is recognized by the immune system, and it elicits a wide variety of pathophysiological effects (3). In conditions where the body is excessively or systemically (as when small concentrations of LPS enters the blood stream) exposed to LPS, a systemic inflammatory reaction can occur, leading to multiple pathophysiological effects, such as endotoxin shock, tissue injury, and lethality (4,5). However, endotoxin does not act directly against cell or organs, but through activation of the immune system, especially through monocytes and macrophages, with the release of a range of proinflammatory mediators, such as tumor necrosis factor-α (TNF-α), Interleukin 6 (IL-6), and Interleukin 1β (IL1β), (4). Pyrogenic reactions and shock are induced in mammals upon intravenous injection of endotoxins at low concentrations (1 ng/mL) (6). The threshold level of endotoxins for intravenous applications of pharmaceutical and biologic products is set to five endotoxin units (EU) per kilogram of body weight and hour by all pharmacopoeias (7). The term *EU* describes the biological activity of an endotoxin. For example, 100 pg of the standard endotoxin EC-5 or 120 pg of endotoxin from *Escherichia coli* O111:B4 have activity of one EU (8). Meeting this threshold level has always been a challenge in biological research and pharmaceutical industry (9).

In the biotechnology industry, gram-negative bacteria are widely used to produce recombinant DNA products such as peptides and proteins. Many recombinant proteins are produced by using the gram-negative bacteria *E. coli*. These products are always contaminated with endotoxins, which may cause side effects when administered to animals or humans. For this reason, special attention must be paid when producing proteins from gram-negative bacteria, aiming a product as free as possible of endotoxin. Endotoxins can be considered temperature and pH stable, rendering their removal as one of the most difficult tasks in downstream processes during protein purification (10,11). Routinely, temperatures of 180–250°C and acids or alkalis of at least 0.1 *M* must be employed

Downstream Industrial Biotechnology: Recovery and Purification, First Edition. Edited by Michael C. Flickinger.

to destroy endotoxins in laboratory equipment. The removal of endotoxins becomes more challenging when associated with labile biomolecules, such as proteins (12). A number of approaches are typically utilized to reduce endotoxin contamination of protein preparations, including ion exchange chromatography (13,14), affinity adsorbents, such as immobilized L-histidine, poly-L-lysine, poly(γ-methyl L-glutamate), and polymyxin B (15–17), gel filtration chromatography, ultrafiltration, sucrose gradient centrifugation, and Triton X-114 phase separation (18,19). The success of these techniques in separating LPS from proteins is strongly dependent on the properties of the target protein (20).

The objective of this review is to discuss the relevant aspects with regard to the endotoxin removal techniques used in the removal of endotoxins from biotechnological preparations, considering its chemical and biological properties. This review does not concentrate on the extracorporeal removal of endotoxin *in vivo*, which is the subject of other reviews (21–23).

16.2 ENDOTOXINS: CHEMICAL AND PHYSICAL PROPERTIES

Endotoxins, also called *lipopolysaccharides*, are the major component of the outer membrane of gram-negative bacteria. They are composed of a hydrophilic polysaccharide moiety, which is covalently linked to a hydrophobic lipid moiety (Lipid A, LipA) (5,8,24). LPS of most species is composed of three distinct regions: the O-antigen region, a core oligosaccharide, and LipA (4). The LipA is the most conserved part of endotoxin (9,25), and is responsible for most of the biological activities of endotoxin, that is, the toxicity. Endotoxin is composed of β-1,6-linked D-glucosamine residues, covalently linked to 3-hidroxy-acyl substituents with 12–16 carbon atoms via amide and ester bonds. These can be further esterified with saturated fatty acids. This hydrophobic part of endotoxin adopts an ordered hexagonal arrangement, resulting in a more rigid structure compared to the rest of the molecule (9,20). The core oligosaccharide has a conserved structure with an inner—3-deoxy-D-manno-2-octulosonic acid (KDO)—heptose region and an outer hexose region. In *E. coli* species, five different core types are known, and *Salmonella* species share only one core structure. The core region close to LipA and LipA itself are partially phosphorylated ($pK_1 = 1.3$, $pK_2 = 8.2$ of phosphate groups at LipA), thus endotoxin molecules exhibit a net negative charge in common protein solutions (9,26). The O-antigen is generally composed of a sequence of identical oligosaccharides (with three to eight monosaccharides each), which are strain-specific and determinative for the serological identity of the respective bacterium (9).

The molar mass of an endotoxin monomer varies from 10 to 20 kDa due to the variability of the oligosaccharide chain; even extreme masses of 2.5 (O-antigen-deficient) and 70 (very long O-antigen) kDa can be found. It is well known that endotoxins form various supramolecular aggregates in aqueous solutions because of their amphipathic structures. These aggregates result from nonpolar interactions between lipid chains and bridges generated among phosphate groups by divalent cations (4). The aggregate structures have been studied by numerous techniques such as electron microscopy, X-ray diffraction, Fourier transform infrared (FTIR) spectroscopy, and nuclear magnetic resonance (NMR). Results from these studies have shown that, in aqueous solutions, endotoxins can self-assemble in a variety of shapes, such as lamellar, cubic, and hexagonal inverted arrangements, with diameter up to 0.1 μm, molar mass 1000 kDa, and high stability depending on the characteristics of the solution (pH, ions, surfactants, etc.) (27,28). It is proposed that proteins may also shift equilibrium by releasing endotoxin monomers from aggregates (8,9). According to molecular dynamics, the three-dimensional structure of endotoxin, especially the long surface antigen, is much more flexible than the globular structure of proteins (9).

Endotoxins are released in large amount upon cell death as well as during growth and division. They are highly heat stable and are not destroyed under regular sterilizing conditions. Endotoxins can be inactivated when exposed to a temperature of 250°C for more than 30 min or 180°C for more than 3 h (27). Acids or alkalis of at least 0.1 *M* can also be used to destroy endotoxins in laboratory scale (29).

16.3 MECHANISM OF ENDOTOXIN ACTION

Endotoxins elicit a wide variety of pathophysiological effects, such as endotoxin shock, tissue injury, and death (5). Endotoxins do not act directly against cells or organs but through activation of immune system, especially the monocytes and macrophages, thereby enhancing immune responses. These cells release mediators, such as TNF, several ILs, prostaglandins, colony stimulating factor, platelet activating factor, and free radicals (30,31). The mediators have potent biological activity and are responsible for the side effects upon exposure to endotoxin. These include alterations in the structure and function of organs and cells, changes in metabolic functions, increased body temperature, activation of the coagulation cascade, modification of hemodynamics, and induction of shock. Many approaches have been made to prevent or treat the deleterious effects of endotoxins on immune cells, such as the use of antiendotoxin antibodies and endotoxin partial structures for blocking endotoxin receptor antagonists.

Nevertheless, the interaction of endotoxins with immune cells is mediated not only by specific receptors but also by nonspecific intercalation of endotoxin molecules into the membranes of the target cells (32).

Studies about rise of body temperature after intravenous administration of certain solutions are dated before the nineteenth century. By the end of the nineteenth century, the designation *injection fever* was generally used to express the fever reactions observed after intravenous administration of several solutions. In the twentieth century, the administration of pharmaceuticals via intravenous route increased the number of such accidents, leading several researchers to develop a series of evaluating works about this subject. In 1912, Hort and Penfold created the name "pyrogenic" to designate the "waters" which, when injected, cause "hyperthermia". Such designation was retaken further, in 1923, by Florance Seibert, who called pyrogenic the *hyperthermizing* substance, which contained either dead bacteria–intact or disintegrated, pathogenic or not–or more often the bacterial metabolic products, such as the denatured protein, endotoxins, or exotoxins (29,33). The major impulse given to increase the knowledge about pyrogens occurred between 1925 and 1945. In particular, Co-Tui, helped by Schrift, deserves special credits for showing that gram-negative bacteria are the most dangerous producers of pyrogens (29).

According to Westphal (1945), the pyrogens, which will really be feared in the pharmaceutical preparations, correspond to the endotoxins of gram-negative bacteria, and such LPS complexes are found in the outer layer of the bacterial cell wall (34). Essentially, pyrogens are originated in the microorganisms from the Enterobacteriaceae family and are thought to be the main contaminant of an injectable solution prepared without the proper disinfecting and sterilizing processes. About two decades later, a collaborative study was developed by the US National Institutes of Health and 14 pharmaceutical industries to establish an animal system that would be adequate to evaluate the "pyrogenicity" of solutions. Such study culminated in the development of the first official pyrogen test in rabbits, which was incorporated into United States Pharmacopeia XII (USP XII), in 1942. In parallel, other efforts to purify and characterize the endotoxins have taken place, and isolated pyrogenics were obtained by several researchers (29,34).

Shear and Turner (1943) were the first to use the term *lipopolysaccharide* to name the endotoxin extract; a term that describes the nature of the endotoxin, and that has been adopted by the scientific community (35).

Finally, it should be mentioned that endotoxins may also have beneficial effects. They have been used in artificial fever therapy to destroy tumors and to improve, nonspecifically, the immune defense. The uncertainty about its role for human health was once described by Bennett (36) On the other hand, any superfluous endotoxin exposure must be strictly avoided to prevent complications. This is especially true for intravenously administered medicines.

16.4 TECHNIQUES APPLIED FOR ENDOTOXIN REMOVAL

Many recombinant proteins are produced using the gram-negative bacteria *E. coli*. Contamination of therapeutic products with endotoxins is therefore a primary concern for the manufacturers. The removal of LPS from these recombinant proteins can be a complicated but essential process especially if the proteins are destined for therapeutic uses.

A number of biomolecules such as lipopolysaccharide-binding protein (LBP), bactericidal/permeability-increasing (BPI) protein, amyloid P component, cationic protein (37,38), or the enzyme employed in the biological endotoxin assay (anti-LPS factor from *Limulus* amebocyte lysate (LAL)) (39) show interactions with endotoxins. These proteins are directly involved in the reaction of many different species upon administration of endotoxin (40,41). Molecular recognition can be viewed as interactions with antiendotoxin antibodies and proteinaceous endotoxin receptors (e.g. CD14, CD16, and CD18) (42). Other proteins such as lysozyme (43) and lactoferrin (44), which are basic proteins ($pI > 7$), interact with endotoxins even having no strong links to a biological mechanism; electrostatic interactions can be assumed as the main driving force. Irrespective of the mechanism that proves to be most significant, these interactions result in hiding endotoxin molecules, and consequently these molecules are not removed in the removal procedures. A typical example is described by Karplus *et al*. (45).

Therefore, owing to protein–endotoxin interactions, endotoxin removal from protein solutions requires techniques, such as affinity chromatography, which are able to generate strong interactions with endotoxins. Alternatively, a specific dissociation of protein–endotoxin complexes may improve the availability of endotoxin molecules. In view of the large variety of products, it is not possible to develop one general method for endotoxins removal. The question about how endotoxin removal can be carried out in an economical way has attracted the attention of many investigators and has been–although not published–the reason for process rearrangements in many cases. However, this item has not yet been solved satisfactorily. The discussion of relevant aspects of endotoxin removal from biological preparations and a critical review of the existing approaches are mandatory to make future studies easier.

In the pharmaceutical industry, several alternative routes are known to generate products with low endotoxin levels. However, their diversity indicates the dilemma in endotoxin removal. Taking advantage of the characteristics of the production process, several procedures were developed for pharmacoproteins, which are tailored to suit specific product requirements. Therefore, each procedure addresses the problem in a completely different way; none of them turns out to be broadly applicable.

Besides, endotoxins are considered to be temperature and pH stable, which renders their removal as one of the most difficult tasks in downstream processes during protein purification (10,11). The depyrogenation and removal of endotoxins from pharmaceutical products during parenteral manufacturing become more challenging when associated with labile biomolecules, such as proteins (12). A number of approaches, including heat, distillation, ionizing radiation, chemical inactivation (46), ion exchange chromatography (13,14), affinity adsorbents (such as immobilized L-histidine, poly-L-lysine, poly(γ-methyl L-glutamate), and polymyxin B) (15–17), gel filtration chromatography, ultrafiltration, sucrose gradient centrifugation, and Triton X-114 phase separation (18,19) are typically utilized to reduce endotoxin contamination of pharmaceutical preparations. The success of these techniques in separating LPS from proteins is strongly dependent on the properties of the target protein (20). Moreover, in many cases, it is necessary to involve the combination of several methods to remove endotoxins from the preparations.

Among the several methods described in the literature, some commonly used techniques for endotoxin contaminants removal are ion exchange chromatography, gel filtration chromatography, ultrafiltration, and affinity chromatography (47). However, we may also find some relevant results on the literature describing endotoxin removal using two-phase aqueous micellar systems (10,48).

As endotoxins are negatively charged, anion exchangers such as DEAE chromatographic matrices or DEAE membranes or matrices functionalized with quaternary amino groups (9) are employed for their adsorption from protein-free solutions. However, decontamination of negatively charged proteins would be accompanied by a substantial loss of the product due to adsorption (10,15,26). Also, net-positively charged proteins form complexes with endotoxins, causing the endotoxin to move along the column, consequently minimizing the endotoxin removal efficiency (15).

Hirayama and Sakata (8) assumed that endotoxin aggregates form supramolecular assemblies with phosphate groups as the head group, and exhibit a negative net charge because of its phosphate groups that originate from LipA (8). These characteristics suggest that ionic interaction plays an important role in the binding between the cationic adsorbent and phosphate groups of the endotoxins. When hydrophobic adsorbents are used in protein solutions, it is suggested that there is also hydrophobic binding between the adsorbent and the lipophilic groups of endotoxins. These binding processes depend on the properties of proteins (net charge and hydrophobicity) and the solution conditions (pH and ionic strength).

During the course of many of the projects developed with *E. coli* by Lin *et al*. (20), the LPS removal from LPS-binding proteins was carried out using denaturing hydrophobic interaction chromatography (HIC) and ethanol, isopropanol, or detergent washes of proteins immobilized on ion-exchange chromatographic resins (20). Alcohol and detergent washes, during ion-exchange chromatography were effective in reducing the protein-associated LPS levels while poor separation of the LPS from the proteins was observed by the denaturing HIC procedure, while detergents were usually more effective washing agents than the alcohols (20).

Alkanediols were shown to be effective agents for the separation of LPS from LPS–protein complexes during chromatography with ionic supports. Their effectiveness in reducing the protein complexation with LPS is dependent on (i) the size of the alkanediol, (ii) the isomeric form of the alkanediol, (iii) the length of the alkanediol wash, (iv) the concentration of alkanediol, and (v) the type of ionic support used, cationic or anionic. Alkanediols are nonflammable and as such are safer alternatives when compared to alcohols (ethanol or isopropanol) that have also been used to remove LPS from protein–LPS complexes (20). LPS removal is more efficient on cationic exchangers than on anionic exchangers.

In recent years, affinity chromatography has been developed for endotoxin removal and has proven to be unique and highly effective. This technique permits a purification process based on biological functions rather than on individual physical or chemical properties (47). To remove endotoxin from recombinant protein preparations, the protein solution can be passed through an affinity chromatography column containing polymyxin B immobilized on Sepharose 4B, in the hope that contaminating endotoxins will bind to the gel. Similarly, histidine immobilized on Sepharose 4B has also the capability to capture endotoxin from protein solutions (16). Polymyxin B affinity chromatography is effective in reducing endotoxin in solutions (49). Polymyxin B, a peptide antibiotic, has a very high binding affinity for the LipA moiety of most endotoxins (50). Karplus *et al*(45) reported an improved method of polymyxin B affinity chromatography in which endotoxin could be absorbed effectively after dissociation of the endotoxin from the proteins by a nonionic detergent, octyl-β-D-glucopyranoside.

The aforementioned methods are reasonably effective for the removal of endotoxins removal from protein solutions with relatively high protein recoveries. However, these affinity phases cannot be cleaned with standard depyrogenation conditions of strong sodium hydroxide in ethanol (51). Anspach and Hillbeck (15) revealed that such supports suffer from considerable efficiency decrease in the presence of proteins. Hence, they are not generally applicable for the above mentioned problem (15).

On the other hand, study by Zhang *et al.* (47) reasserted the applicability of affinity chromatography using silica gel as a matrix for preparing removal adsorbent. An affinity media for endotoxin removal based on silica gel was prepared by activation with silane coupling agent and subsequent conjugation with histidine as a ligand. The influence of pore size and particle dimension of silica gel were also studied and the results showed that the silica gel with a particle size of 200 μm and a pore size of about 12 nm was a good carrier material for the preparation of the affinity adsorbent (47).

Gel filtration chromatography is based on the molecular weight cutoff rather than on the size of the filtration pores. The basic subunit size of LPS is about 10 to 20 kDa. It can, therefore, be effectively removed from solutions by a 10 kDa molecular weight membrane. Such technique has been usually applied to ultrapure water laboratory gel systems and to remove endotoxin from solutions of products if the products are of low molecular weight (glucose, salts, etc.) (9). However, the unaggregated monomeric form of LPS is seldom, if ever, found in aqueous solutions generating problems to the process (52).

In the same way, ultrafiltration as a method for removing pyrogens has been successfully applied to a large number of low-to-medium-molecular-weight drugs and solutions. Endotoxin-contaminated antibiotics have been successfully depyrogenated without significant loss of antibiotics (46) and the process has also been utilized for large-scale production. High molecular weight solutions contaminated with aggregated endotoxin of a similar size may also be successfully ultrafiltrated if the endotoxin can be disaggregated through the use of agents such as chelators or surface active detergents. A more difficult situation arises when the endotoxin aggregates and the solute are of a similar molecular size. In such cases, the two may still be separated by ultrafiltration if a means of manipulating the endotoxin aggregated size can be found via the removal of factors such as cations, detergents, and chelators that decrease aggregate size. Aggregation between protein molecules can be suppressed, for instance, by arginine, which is as common additive in refolding (53). Arginine has also been demonstrated to dissociate proteins from protein–protein complexes by significantly facilitating the elution of monomeric antibodies from a Protein A matrix (54). L-Arginine is a normal metabolite in animals and man and has a low order of toxicity. Low concentrations in the final product are therefore not a problem, thus simplifying downstream applications (55).

Nevertheless, although effective in removing endotoxins from some products, ultrafiltration is an inefficient method in the presence of different proteins that can be damaged by physical forces. In recent years, the interest in the use of two-phase aqueous micellar systems for the purification or concentration of biological molecules, such as proteins and viruses, has been growing (10,48).

Fiske *et al.* (6) examined a number of approaches such as the use of the zwitterionic surfactants Zwittergent 3-12 (Z3-12) and Zwittergent 3-14 (Z3-14) for the dissociation of endotoxin from the purified UspA2 protein and the subsequent separation of endotoxin from UspA2 using either ion exchange or gel filtration chromatography. UspA2 protein is a potential vaccine candidate for preventing otitis media and other diseases caused by *Moraxella catarrhalis* (6). The approach that proved successful for the dissociation of endotoxin from UspA2 was the replacement of the Triton X-100 by a zwitterionic surfactant. The success of both Z3-12 and Z3-14 in dissociating the endotoxin–UspA2 complex may reside in the charge characteristics of the surfactants. Triton X-100 is a nonionic surfactant containing no charged moieties while the Zwittergents contain zwitterionic head groups with both negatively and positively charged moieties. Most zwitterionic surfactants are effectively neutral; however, in some cases, strong polarization exists (56). The charge characteristics of Z3-12 and Z3-14 and the interaction of the surfactant with either the endotoxin or the protein may aid in the dissociation of the endotoxin from protein (in this case UspA2). Structural differences between the surfactants may also play a role in effective dissociation of endotoxin and protein. Whatever the mechanism, the use of the Zwittergent surfactant was proved to be quite suitable for the removal of LPS from UspA2 without disrupting the immunogenic properties of the protein. Prior to endotoxin reduction, the UspA2 preparations contained as much as 1.58×10^{-4} EU/mg. However, following chromatography in the presence of Z3-12, Fiske *et al.* (6) achieved levels of approximately 7.2×10^{-9} EU/mg. The endotoxin removal process has been successfully implemented following the GMP to produce UspA2 subunit vaccine for clinical trials.

The levels of endotoxin appear to be much higher in recombinant proteins derived from soluble or cytoplasmic fractions than in proteins derived from insoluble fractions or inclusion bodies. This is consistent with the belief that LPS present in the cell wall are solubilized during the cell lysis procedure. Schnaitman (57) demonstrated that treatment of *E. coli* with the combination of Triton X-114, EDTA, and lysozyme resulted in solubilization of all LPS from the cell wall.

Reichelt *et al.* (10) tested whether the removal of endotoxin could be achieved during chromatography purification with the use of Triton X-114 in the washing steps. The application of 0.1% Triton X-114 in the washing steps was successful in decreasing the concentration of endotoxins during histidine and GST (resin GST Sepharose) fusion protein purification, whereas washing steps lacking surfactant were ineffective in eliminating endotoxins.

The recombinant human CXCL8(3–7)K11R/G31P (hG31P) expressed in *E. coli* was further purified and LPS was removed by a simple one-step SP-Sepharose column eluted with 1% Triton X-114 added in the washing buffer. The purity of the recombinant hG31P protein was more than 95% (58).

In contrast to purified materials (concentration: 2500–34,000 EU/mg) employing the standard protocol, purified recombinant proteins treated with Triton X-114 contained concentrations as low as 0.2–4 EU/mg (less than 99% of initial endotoxin content). Residual endotoxins in solubilized inclusion bodies can reach levels of 8×10^6 EU/mL despite the fact that endotoxin levels were found to be higher in recombinant proteins, which are isolated from soluble fractions (45).

Endotoxins have been shown to form complexes with proteins of different isoelectric points (15) where electrostatic interactions are thought to be the main driving forces. As a result, the removal of endotoxins from basic proteins is more difficult than that from acidic proteins (9). Reichelt *et al*. (10) studied whether the use of Triton X-114 in washing steps could eliminate endotoxins from proteins with a p*I* above 8.5. They found that washing with Triton X-114 coupled with affinity chromatography effectively removed endotoxins from negatively charged proteins (SyCRP and NdhR). The minimal endotoxin concentration achieved was less than 0.2 EU/mg; protein recovery and yield were close to 100% (10).

Using Triton X-114, Adam *et al*. (59) showed a 100-fold endotoxin reduction in two steps with a final endotoxin content of 30 EU/mg and 50% loss in bioactivity of the exopolysaccharide. In addition, about 100-fold endotoxin reduction was shown by Cotten *et al*. (60) from plasmid DNA preparation with a final endotoxin content of 0.1 EU in 6 μg DNA.

A comparison of affinity adsorption and Triton X-114 two-phase extraction for the decontamination of the recombinant proteins cardiac troponin I, myoglobin, and creatine kinase isoenzymes is described by Liu *et al*. (19). They concluded that phase separation to be the most effective method, reducing the endotoxin content by 98–99% with remaining amounts of 2.5–25 EU/mg, depending on the protein. However, Cotton *et al*. (60) observed slightly better removal efficiency with a polymyxin B sorbent.

Aida and Pabst (18) reported a method to reduce endotoxin in protein solutions using Triton X-114, in which the surfactant aids in the dissociation of endotoxin from the protein, while also providing a convenient phase separation capability for removing the dissociated endotoxin. According to these same authors, phase separation using Triton X-114 was effective in reducing endotoxin from solutions of three different proteins (cytochrome *c*, albumin, and catalase). The first cycle of phase separation reduced endotoxin contamination by 1000-fold. Further cycles of phase separation resulted in complete removal of endotoxin.

The detergents, even though they were also very effective in reducing the LPS levels, are relatively expensive, would add significant cost to a manufacturing process, and may affect the bioactivity of the protein of interest. Alternative chemicals are desired that could safely and cost effectively be used in place of the alcohols or detergents as washing agents for the separation of LPS from proteins during chromatographic unit operations. Ideally, these chemicals would be relatively inexpensive, chemically well defined, present minimal safety issues, and have minimal impact on the bioactivity of the protein in question when implemented into a process (20). However, the removal of the surfactant may be an additional problem in this purification process (15).

16.5 ENDOTOXIN REMOVAL IN BIOTECHNOLOGY MANUFACTURING PROCESSES

Biotechnology manufacturing processes are intended to produce therapeutics products that meet regulatory and company standards for safety, purity, and efficacy. Recombinant expression systems present a challenging array of biological impurities that must be removed during purification and prior to final fill. However, their diversity indicates a dilemma in endotoxin removal. Modern techniques used to remove endotoxins from therapeutics products in many cases involve the combination of several of methods discussed above.

REFERENCES

1. Rodrigues EMG, Milagres AMF, Pessoa A. Process Biochem 1999; 34: 121–125.
2. Rodrigues EMG, Pessoa A, Milagres AMF. Appl Biochem Biotechnol 1999; 79: 779–788.
3. Minutoli L, Altavilla D, Bitto A, Polito F, Bellocco E, Lagana G, Fiumara T, Magazu S, Migliardo F, Venuti FS, Squadrito F. Eur J Pharmacol 2008; 589: 272–280.
4. Anspach FB. J Biochem Biophys Methods 2001; 49: 665–681.
5. Ogikubo Y, Norimatsu M, Noda K, Takahashi J, Inotsume M, Tsuchiya M, Tamura Y. Biologicals 2004; 32: 88–93.
6. Fiske MJ, Fredenburg RA, VanDerMeid KR, McMichael JC, Arumugham R. J Chromatogr B Biomed Sci Appl 2001; 753: 269–278.
7. Daneshian M, Guenther A, Wendel A, Hartung T, von Aulock S. J Immunol Methods 2006; 313: 169–175.
8. Hirayama C, Sakata M. J Chromatogr B Analyt Technol Biomed Life Sci 2002; 781: 419–432.

9. Petsch D, Anspach FB. J Biotechnol 2000; 76: 97–119.
10. Reichelt P, Schwarz C, Donzeau M. Protein Expr Purif 2006; 46: 483–488.
11. Sharma SK. Biotechnol Appl Biochem 1986; 8: 5–22.
12. Kang Y, Luo RG. J Chromatogr Sci 1998; 809: 13–20.
13. Mitzner S, Schneidewind J, Falkenhagen D, Loth F, Klinkmann H. Artif Organs 1993; 17: 775–781.
14. Weber C, Henne B, Loth F, Schoenhofen M, Falkenhagen D. ASAIO J 1995; 41: 430–734.
15. Anspach FB, Hilbeck O. J Chromatogr A 1995; 711: 81–92.
16. Matsumae H, Minobe S, Kindan K, Watanabe T, Sato T, Tosa T. Biotechnol Appl Biochem 1990; 12: 129–140.
17. Sakata M, Kawai T, Ohkuma K, Ihara H, Hirayama C. Biol Pharm Bull 1993; 16: 1065–1068.
18. Aida Y, Pabst MJ. J Immunol Methods 1990; 132: 191–195.
19. Liu S, Tobias R, McClure S, Styba G, Shi Q, Jackowski G. Clin Biochem 1997; 30: 455–463.
20. Lin MF, Williams C, Murray MV, Ropp PA. J Chromatogr B Analyt Technol Biomed Life Sci 2005; 816: 167–174.
21. Tetta C, Bellomo R, Inguaggiato P, Wratten ML, Ronco C. Ther Apher 2002; 6: 109–115.
22. Shoji H. Therap Apher Dial 2003; 7: 108–114.
23. Shimizu T, Endo Y, Tsuchihashi H, Akabori H, Yamamoto H, Tani T. Transfus Apheresis Sci 2006; 35: 271–282.
24. Raetz CR, Ulevitch RJ, Wright SD, Sibley CH, Ding A, Nathan CF. FASEB J 1991; 5: 2652–2660.
25. Vaara M, Nurminen M. Antimicrob Agents Chemother 1999; 43: 1459–1462.
26. Hou KC, Zaniewski R. J Parenter Sci Technol 1990; 44: 204–209.
27. Gorbet MB, Sefton MV. Biomaterials 2005; 26: 6811–6817.
28. Darkow R, Groth T, Albrecht W, Lutzow K, Paul D. Biomaterials 1999; 20: 1277–1283.
29. Kaneko TM, Pinto TJA, Ohara MT. Controle Biológico de Qualidade de Produtos Farmacêuticos Correlatos e Cosméticos. São Paulo: Atheneu; 2000.
30. Rietschel ET, Kirikae T, Schade FU, Mamat U, Schmidt G, Loppnow H, Ulmer AJ, Zahringer U, Seydel U, Di Padova F, et al. FASEB J 1994; 8: 217–225.
31. Forehand JR, Pabst MJ, Phillips WA, Johnston RB Jr. J Clin Invest 1989; 83: 74–83.
32. Schromm AB, Brandenburg K, Loppnow H, Moran AP, Koch MH, Rietschel ET, Seydel U. Eur J Biochem 2000; 267: 2008–2013.
33. Probey TF, Pittman M. J Bacteriol 1945; 50: 397–411.
34. Westphal O. Int Arch Allergy Appl Immunol 1975; 49: 1–43.
35. Brunn GJ, Platt JL. Trends Mol Med 2006; 12: 10–16.
36. Bennett IL Jr, Beeson PB. J Exp Med 1953; 98: 477–492.
37. Beamer LJ, Carroll SF, Eisenberg D. Protein Sci 1998; 7: 906–914.
38. de Haas CJ, Haas PJ, van Kessel KP, van Strijp JA. Biochem Biophys Res Commun 1998; 252: 492–496.
39. FC Pearson. Pyrogens: endotoxins, LAL testing and depyrogenation. New York: Marcel Dekker; 1985.
40. Koizumi N, Morozumi A, Imamura M, Tanaka E, Iwahana H, Sato R. Eur J Biochem 1997; 248: 217–224.
41. Hoover GJ, el-Mowafi A, Simko E, Kocal TE, Ferguson HW, Hayes MA. Comp Biochem Physiol B Biochem Mol Biol 1998; 120: 559–569.
42. Morrison DC, Kirikae T, Kirikae F, Lei MG, Chen T, Vukajlovich SW. Prog Clin Biol Res 1994; 388: 3–15.
43. Ohno N, Morrison DC. Eur J Biochem 1989; 186: 629–636.
44. Elass-Rochard E, Roseanu A, Legrand D, Trif M, Salmon V, Motas C, Montreuil J, Spik G. Biochem J 1995; 312(3): 839–845.
45. Karplus TE, Ulevitch RJ, Wilson CB. J Immunol Methods 1987; 105: 211–220.
46. Williams KL. Endotoxins: pyrogens, LAL, testing and depyrogenation. New York: Marcel Dekker; 2001.
47. Zhang Y, Yang H, Zhou K, Ping Z. React Funct Polym 2007; 67: 728–736.
48. Magalhaes PO, Lopes AM, Mazzola PG, Rangel-Yagui C, Penna TC, Pessoa A Jr. J Pharm Pharm Sci 2007; 10: 388–404.
49. Issekutz AC. J Immunol Methods 1983; 61: 275–281.
50. Morrison DC, Jacobs DM. Immunochemistry 1976; 13: 813–818.
51. McNeff C, Zhao Q, Almlof E, Flickinger M, Carr PW. Anal Biochem 1999; 274: 181–187.
52. van Reis R, Zydney A. Curr Opin Biotechnol 2001; 12: 208–211.
53. Arakawa T, Tsumoto K. Biochem Biophys Res Commun 2003; 304: 148–152.
54. Arakawa T, Philo JS, Tsumoto K, Yumioka R, Ejima D. Protein Expr Purif 2004; 36: 244–248.
55. Ritzen U, Rotticci-Mulder J, Stromberg P, Schmidt SR. J Chromatogr B Analyt Technol Biomed Life Sci 2007; 856: 343–347.
56. Hjelmeland LM, Klee WA, Osborne JC Jr. Anal Biochem 1983; 130: 485–490.
57. Schnaitman CA. J Bacteriol 1971; 108: 553–563.
58. Cheng HT, Huang KC, Yu HY, Gao KJ, Zhao X, Li F, Town J, Gordon JR, Cheng JW. Protein Expr Purif 2008; 61: 65–72.
59. Adam O, Vercellone A, Paul F, Monsan PF, Puzo G. Anal Biochem 1995; 225: 321–327.
60. Cotten M, Baker A, Saltik M, Wagner E, Buschle M. Gene Ther 1994; 1: 239–246.

17

POROUS MEDIA IN BIOTECHNOLOGY

MANUEL MOTA
IBB, Centro de Eng. Biológica, University of Minho, Portugal
ALEXANDER YELSHIN AND INNA YELSHINA
Technological Dept, Polotsk State University, Novopolotsk, Belarus

17.1 INTRODUCTION

The word pore derives from the Greek word $\pi o \rho o \sigma$ (porous) meaning passage. According to (1), a pore is a "A hole or cavity in a body; if communicating with the surface it is an open pore; if 'shut' to prevent flow or transport of a fluid through it, it is a closed pore. However, for permeable media, the 'open' pores on the inlet and outlet surfaces must also be interconnected or communicate to provide a continuous, usually tortuous, passageway."

Biological systems need to interact with the environment and at the same time they need to set up barriers to protect themselves. In order to satisfy these two conditions—protection and controlled interaction with the environment—biological systems need to have: (i) membranes and walls and (ii) pores through which useful substances can be captured and harmful substances excreted.

In other words, no biological system can be devoid of pores, inasmuch as no biological system can be deprived of membranes and/or walls to establish its own border.

As a matter of fact, pores are ubiquitous in nature, being present not only in biological but also in nonbiological systems.

Porous media have to have at least two phases: a solid phase and a fluid phase occupying the pores. The solid phase may be continuous or discrete and simple or composite. Many biological porous systems are continuous and composite, such as the bone matrix, and many of the nonbiological porous systems are discontinuous and simple. In the latter case, discrete, independent particles—sand, gravel, rings, or saddles made out of several materials (ceramics, glass, metal)—are assembled in a confined space. This situation is rather common in separation processes, such as cake filtration, chromatography, settling, and gas cleaning, where a porous medium is packed inside a column through which a fluid is passing. Composite solid phases may be found both in natural and artificial porous systems, for instance, in multilayered dialysis membranes.

To understand the transport phenomena involved in porous media, three key concepts need to be understood: porosity, permeability, and tortuosity.

Porosity (ε) is defined as the ratio of the void volume to the total volume of the porous medium. Porosity must always be less than one, but in some cases it may be very close to one, such as in fibrous materials.

Permeability *(k)*, measures how easily a porous material allows the passage of a fluid. As smaller pores are more resistant to the fluid passage than larger ones, permeability may be reduced if the proportion of smaller pores is higher, even when the average porosity is the same. Furthermore, permeability depends upon the intrinsic properties of the passing fluid, such as viscosity and specific gravity, and on the solute under consideration. For example, a membrane can be permeable to water and impervious to ions, a usual case in biological membranes. In other words, porosity is an intrinsic property of the porous medium, whereas permeability depends on both, the porous medium and the fluid passing through it. In later sections, equations relating permeability and porosity will be presented.

Downstream Industrial Biotechnology: Recovery and Purification, First Edition. Edited by Michael C. Flickinger.

The third concept is tortuosity (τ). In nature, it is unusual to find straight, nontortuous pathways. The vast majority of natural pores and channels, be they biological or not, are tortuous. In general, if a particle moving in a fluid is pushed through a porous medium, its pathway is not rectilinear, thus leading to a concomitant pressure drop.

17.2 GENERAL DEFINITIONS

A porous medium is composed of a solid phase containing pores that are distributed in a matrix and filled with a fluid. This definition comprises two categories of porous media: impermeable and permeable. When a fluid is not able to cross a porous medium network under external forces, such as a hydraulic pressure drop, concentration gradient, or thermal gradient, the porous medium is impermeable to that fluid. The porous medium is permeable if a fluid is able to pass through the interconnected network of pores. Both the solid matrix and the pore network are assumed, for a matter of simplicity, to be continuous and forming two interpenetrating continua (2,3). Several permeable porous media are represented in Fig. 17.1 and will be the subject of further analysis.

Porous media can be mono- or poly-disperse, homogeneous or nonhomogeneous, composite, including different solids or combining different structural types (4–6). Many porous media in biotechnology are compressible and deformable. Compression corresponds to the reduction of the volume of porous media by applying pressure (restructuring, rearrangement), or to the state attained by treated porous media in this way (osmosis, swelling, etc.). Deformation is related to stress causing a change in geometry or shape that might lead to damage, disfigurement, or spoilage of porous media components. However, when deformation is elastic, porous media return to the original shape and geometry. Most biological porous media are elastic under mild conditions.

Porous media containing living components and living tissue are sensitive to external and internal factors, and react not only mechanically (as nonliving porous media) but also physiologically. This is the main difference between nonliving and living porous media (7).

17.3 CHARACTERISTICS OF POROUS MEDIA

A porous medium is characterized by its porosity (ε), tortuosity (τ), pore or particle size distribution, as well as by the physicochemical properties of solid and fluid phases. Transport properties are defined by permeability (k), diffusivity or by effective diffusion coefficient (D_e), and chemical kinetics—Thiele modulus (φ) and effectiveness (η).

17.3.1 Porosity

Porosity is defined as the ratio of the void volume to the total porous medium volume. Since in reality porous media can contain isolated pores (cavities), and a part of the pores may be represented by dead-end pores that form the nonconducting pore space of the medium, the definition of porosity needs to be refined. The ratio between the pores' volume participating in a certain transport phenomenon to the total porous medium volume is defined as the effective porosity. For instance, when the convective flow of a continuous fluid is considered, dead-end pores are stagnant sites that have no influence on transport. In contrast, when mass transfer of some fluid component (solute) takes place, the diffusion process must be taken into consideration, and dead-end pores become involved in the transport and affect the solute dispersion, as is observed in

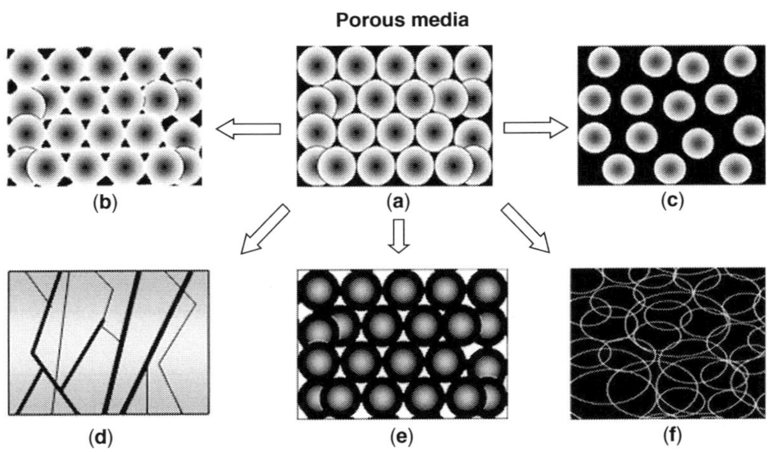

Figure 17.1. Several types of porous media, where voids are represented in black and the matrix (solid) in white to gray tones. (a) granular porous medium; (b) impermeable porous medium, isolated pores; (c) dispersed system, solid is not a continuum; (d) network pore system; (e) foam; and (f) fibers or gel-like porous media.

chromatography. Frequently, effective porosity is identified by the term *porosity*.

For low porosity materials exhibiting a pore network, a large part of the pore void may be nonconducting due to an increasing number of dead pores. In this case there is a percolation threshold, defined as the minimum porosity when the medium becomes impermeable (8–11). For a two-dimensional (2D) simulated diffusion of proteins in the plasma membrane, the percolation threshold occurs for $\varepsilon = 0.22$ (12).

Biotechnology often operates with multifractional, granular porous media (in terms of particle size): porous media with immobilized cells, biomaterials, cakes, and so on. Porosity of granular mixed beds has been discussed in numerous publications using experimental, geometrical, and statistical approaches (13–20).

In multifractional granular porous media, porosity can be calculated with the application of the fractional porosity approach (21). Fractional or particular porosity of i-th particular fraction $\varepsilon_i(x_i)$, is a ratio of residual free space, $v_i(x_i)$ after placing i particle fractions into a porous medium unit volume (v_0), to the residual free space $v_{i-1}(x_{i-1})$. Thus $\varepsilon_i = v_i(x_i)/v_{i-1}(x_{i-1})$, where x_i is a volume fraction of the i-th fraction in the composition, $i = 1, \ldots, m$. Hence, for m-fractional composition, v_m corresponds to the final free space in the porous medium unit volume. The generalized relation of the unit porous medium volume $v_0 = 1$ becomes:

$$\varepsilon = v_m(x_m)/v_0 = v_m(x_m)/1 = \prod_{i=1}^{m} \varepsilon_i(x_i), \sum_{i=1}^{m} x_i = 1 \qquad (17.1)$$

The dependence $\varepsilon_i(x_i)$ is a complex function of particle size ratios $\delta_i = d_i/d_{i-1}$ $(d_i < d_{i-1})$ and their shape (22–24). At δ less than 0.1 (25), the packing density of each fraction approaches the monosize packing density ε_i^0 and Equation 17.1 reaches a limit $\varepsilon = \prod_{i=1}^{m} \varepsilon_i^0$. Analysis (22,24,26) shows that binary porous media obtained by varying particle size ratio and fractional content, enable build-up of structures with a wide range of transport properties (27,28).

17.3.2 Tortuosity

Conventional geometrical tortuosity is defined as the ratio of the pore length L_e to the porous medium thickness (distance) as shown in Equation 17.2 (29–33).

$$\tau_G = L_e/L, \tau_G \geq 1 \qquad (17.2)$$

Sheffield (34) considers tortuosity as a ratio of the actual streamline trajectory length to the length of the bed (τ_P). In practice, this definition is preferable. In the porous medium model formed by tortuous nonintersecting capillaries, the pathway tortuosity is close to the geometrical tortuosity. Quite often, the tortuosity is an average value for the porous medium. Frequently, average tortuosity is calculated from the experimentally determined effective diffusivity, from electro-conductivity, or from permeability when porosity and pore size distribution are known. The tortuosity distribution results in different contributions from pores of various tortuosities to the total porous medium conductivity (32). Johnston (35) relates tortuosity and pore size distribution as $\tau = \varepsilon^{-1} + 1.196\,(\sigma/d)$, where σ is the standard deviation of the pore diameter, d.

Attention must be paid to the fact that in some models the tortuosity definition may have other forms (2,36), such as $(L_e/L)^2$, invert ratio L/L_e, or $(L/L_e)^2 \leq 1$.

Depending on conditions, tortuosity may change from a purely geometrical concept to a kinematic property (2,26). In general, tortuosity is the consolidation of a number of factors represented in Fig. 17.2. In the examples shown in Fig.17.2a and Fig. 17.2b, 1, geometrical tortuosity (τ_G) fully describes linear, curved, and folded capillary pores. The increase in the number of folds leads to the highly folded pores shown in Figure 17.2b, 17.2, and folding on different scale levels results in a fractal pore shown in Fig. 17.2b, 17.3. Highly folded and fractal pores may also be characterized by geometrical tortuosity but, in special cases, pore folding leads to additional effects even when τ_G has the same value (Fig. 17.2c). When shaped micro-objects diffuse or flow in narrow pore channels, they recognize a highly folded channel as more "tortuous" for their movement than small molecules (27,37). This effect may be taken into consideration by introducing a "folding" factor or by considering a pore channel fractal dimension. Additional effects can be generated by pore cross-section shape variation and constrictions, as shown in Figs. 17.2b, 17.4 through 17.2b, 17.6. Pathway trajectory may depend on a flow regime (38,39) or on external forces, such as the magnetic, electric, and so on (Fig. 17.2d). These are the reasons why, in similar porous media but under different conditions, the tortuosity estimated through an object's motion may vary widely (31).

Different types of network models are proposed (40,41). For all these reasons, it is better to consider tortuosity as a complex factor rather than as simply geometrical. The tortuosity factor, usually denoted as τ, comprises both the tortuosity of the average pathway and the variation of other conditions τ_C such as pore shape, constriction, diameter, and so on (42–48). Hence, tortuosity should rather be written as a function $\tau = \tau(\tau_G, \tau_P, \tau_C)$, where τ_C is the correction factor.

The tortuosity τ closely relates with the porosity and, depending on porous media, presents numerous correlation equations: $\tau = 1.5 - \varepsilon/2$ (49); $\tau = (2 - \varepsilon)^2/\varepsilon$ and $\tau = (3 - \varepsilon)^2/(4\varepsilon)$ (50); $\tau^2 = 0.2408 \log(r_{gr}/r_{eff}) + \log 1.2$

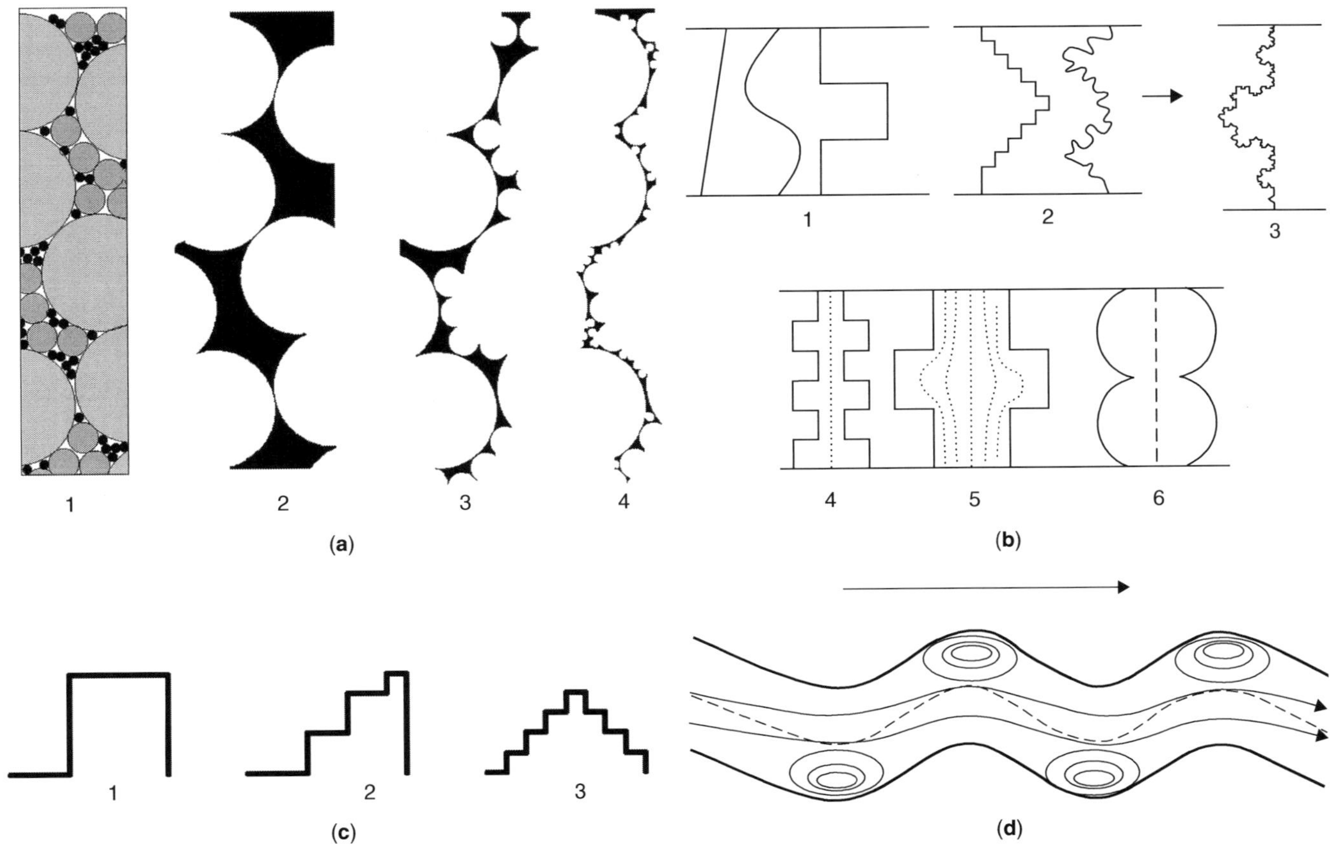

Figure 17.2. Simplified scheme of pore channels variety. (a) Pore channel structure depending on fractional composition of porous media: 1- ternary porous media; 2- large mono-size particles; 3- large+intermediate size; 4- ternary mixture fractality. (b) Pore configurations: 1—linear, curved, and folded channels; 2—highly folded pores; 3—fractal pore channel; 4, 5, and 6—channels with a shaped cross-section. (c) Curves with the same length and geometrical tortuosity ($\tau_G \sim 2.0$) but different number of folds: 1—3 folds; 2—7 folds; 3—15 folds. (d) Dependence of streamline tortuosity on fluid flow regime: dotted curve—laminar flow; solid curves—tortuosity decrease due to a fluid stagnant zone.

(51), where r_{eff} is the effective hydraulic pore radius, and r_{gr} is the radius of grains; $\tau = 1 + p\ln(1/\varepsilon)$ (52,53), where p is a parameter that depends on the shape of the particles and their mean orientation in the bed (for spheres $p = 0.49$–0.5, for fiber bed $p = 1.0$) (54).

The most frequently used tortuosity is a power–law relationship with a fitting constant n (29,55):

$$\tau = 1/\varepsilon^n \tag{17.3}$$

For granular porous media, $n = 0.4$ through 0.5 (25).

Equation 17.3 can be adopted for living tissues. In a previous study (7), it was shown that cells have the unique ability to control mass-transfer processes by adjusting the extracellular space (ECS) configuration, which is achieved by varying porosity and tortuosity in a way that can be described by a three-parametric model.

$$n = n_0 + a\varepsilon + \varepsilon^2 \tag{17.4}$$

This situation does not occur in inert materials, for which the dependence of porosity on tortuosity may be defined by Equation 17.4 at $n =$ constant.

The tortuosity of multicomponent porous media is calculated by means of Equations 17.1 and 17.3, hence the following series can be written (21):

$$\tau = 1/\varepsilon^n = \prod_{i=1}^{m} (1/\varepsilon_i^n(x_i)) = \prod_{i=1}^{m} \tau_i(x_i), \sum_{i=1}^{m} x_i = 1 \tag{17.5}$$

where $\tau_i = 1/\varepsilon_i^n$ is the fractional tortuosity contribution from the i-th particle size fraction to the overall mixture tortuosity τ. In the extreme case of $\delta \to 0$ the maximum tortuosity approaches $\tau_{\max} = \prod_{i=1}^{m} \tau_i^0$. For shaped particles, the value n can vary from one fraction to another, hence, in Equation 17.5 $\tau_i = 1/\varepsilon_i^{n_i}$.

17.3.3 Determination of Characteristics of Porous Media

There are several ways of determining the characteristics of porous media. Porosimetry has been applied for many decades, to the analysis of porous media properties. The technique involves the use of high pressure—up to 400 MPa—to force a nonwetting liquid, usually mercury, into the pores. Several assumptions have to be made, namely, that pores are cylindrical and that the pores' geometry is immune to pressure. These assumptions fail quite often with biological materials and with foams and aerogels, as was proven by Pirard *et al*. (56). Furthermore, the samples must be dry. Only those pores that reach the sample's surface are assessed. An alternative to porosimetry is the use of nitrogen adsorption. A comparison between these two methods is available in the work of Milburn and Davies (57). There are many problems when using these techniques with biological porous media, related not only with the determination of tortuosity but also with other porous media characteristics such as pore size distribution. Therefore, new techniques came up and are increasingly being used.

The first example is nuclear magnetic resonance (NMR). A pioneering work was presented by Mair *et al*. that reported the results of gas-diffusion nuclear magnetic resonance (GD-NMR) applied to study random packs of glass beads (58). The authors emphasised that simple NMR techniques should not be used in porous media imbibed with liquids. The use of GD-NMR allowed the authors to overcome the drawbacks presented by simple NMR and they were able to characterize a wide pore distribution starting at micron-sized pores and ending at millimeter-sized pores. The authors also noted that GD-NMR could successfully be applied to study wet foams and biological systems. The suggestion was picked by several researchers who applied this technique to study soil porosity and permeability (59) and to study porous characteristics of bone (60). As diffusion coefficients of gases are several orders of magnitude greater than those of liquids, which might reduce the spatial resolution, Acosta *et al*. proved that, instead of a single noble gas such as ^{129}Xe, binary mixtures of ^{3}He, ^{129}Xe, and other gases (e.g. SF_6) could be used to significantly improve GD-NMR resolution (61). The technique is since then being used to study lung porous characteristics and pulmonary function in general (62).

The combination of computational fluid dynamics (CFD) with magnetic resonance imaging (MRI) led to interesting studies on cardiovascular performance. Tortuosity of the superficial femoral artery was evaluated in this way by Wood *et al*. (63) and the geometry of carotid bifurcation was similarly studied by Lee *et al*. (64).

Besides NMR or NMR/CFD techniques, another approach that has gained popularity is microcomputed tomography (μ-CT). A work reporting high resolution results with this technique was presented by Beckmann *et al*. (65), who obtained 3D images from rat cerebrum at 8–15 μm spatial resolution using synchrotron radiation. The recent commercial offer of bench-top CT systems with improved resolution, propelled the utilization of this technique especially in the field of Tissue Engineering (66–68). A curious work reported by Chaunier *et al*. used μ-CT to study the permeability and the expanded structure of baked product crumbs (69).

An evenfurther improvement in CT resolution was recently reported by Weinekoetter (70); the nanotomography allows 3D images with submicron resolution. This new technique has already been applied by Brunke and Sieker to 3D analysis of textile structures (71).

All these new advances provide new tools for studies, for example, on microfluidics, and are also paving the way to make 3D analysis of cellular ultrastructures an exciting new field of research.

17.4 TRANSPORT PHENOMENA IN POROUS SYSTEMS

Models describing transport phenomena in porous media include complex characteristics of porous media such as porosity, tortuosity, particle or pore size, and correction parameters accounting from particle or pore shape and physicochemical properties on a solid or fluid interface. In a transport process, a specific flux $\mathbf{j}$ may be generally presented in the form (2,72) $\mathbf{j} = -K grad(X)$, where K and X are the generalized conductivity and driving force, respectively. In particular, for mass transfer we may write Flux $= K(c^* - c)$ where K will be the effective diffusion coefficient De, and c^* and c are the equilibrium and current concentrations of the transported solute. The equation may thus be written as Flux $= D_e(c^* - c)$. In the case of fluid flow, K represents the hydraulic conductivity and X a pressure drop.

17.4.1 Permeability

As discussed, the permeability characterizes the phenomenon of a fluid flowing through a porous medium (Eq. 17.6).

$$u = \frac{k}{\mu} \cdot \frac{\mathrm{d}p}{\mathrm{d}x} \tag{17.6}$$

where u is the apparent fluid velocity; $\mathrm{d}p/\mathrm{d}x$ is the pressure gradient in the direction x; k is the porous medium permeability, 1/m; μ is the fluid viscosity, and $Pa \times s$ is the fluid dynamic viscosity, measured in Pascal seconds.

Depending on the porous medium model, permeability can be represented in different manners. The simplest is the capillary model (2), where capillary tubes are oriented in one single direction: $k = \varepsilon d_p^2/(32\tau_G^2)$; for a network of capillaries oriented in orthogonal directions $k = \varepsilon d_p^2/(96\tau_p^2)$. Here d_p is the equivalent capillary diameter, averaged by the cross-section variance and size distribution. The granular bed capillary model may be described by the Kozeny-Carman equation:

$$k = \varepsilon^3/[K_0\tau^2(1-\varepsilon)^2 S^2] = d^2\varepsilon^3/[36K_0\tau^2(1-\varepsilon)^2] \tag{17.7}$$

where S is the specific surface area based on the solid volume; $d = 6/S$ is the equivalent spherical particle size; ε is the porosity; $K_0\tau^2$ is a factor that includes tortuosity τ (ratio of average pathway to the porous medium thickness); and K_0 is a coefficient dependent on the shape of the pore cross-section (for cylindrical pores $K_0 = 2$).

For nonspherical particles, Equation 17.7 may be partially corrected by introducing a sphericity factor Φ of a particle, which is defined as the ratio of the surface area of a sphere S (with the same volume as the given particle) to the surface area of the particle S_p: $\Phi = S/S_p$, hence, $d_p = 6/S_p = 6\Phi/S = \Phi d$. In this case, fractal behavior may affect all or part of the variables (73): pore fraction (ε), solid fraction $(1-\varepsilon)$ or ratio $\varepsilon/(1-\varepsilon)$, and tortuosity ($\tau$).

In general, the tortuosity depends on porosity and may be presented in the form $\tau = a/\varepsilon^n$ ($n \geq 0$; $a \geq 1$, conventionally $a = 1$, Eq. 17.3). Substituting τ and Φ in Equation 17.7 and normalizing the permeability (k/d^2) leads to the following relation (Eq. 17.8):

$$k/d^2 = (\Phi^2/K_0)[\varepsilon^3/(36\cdot(a/\varepsilon^n)^2(1-\varepsilon)^2)] \tag{17.8}$$

The value of K_0 depends on a pore cross-section configuration that, in turn, reflects the particles' shape. Since the exact relations between Φ, K_0 and ε are unknown, $\Phi^2/(K_0 a^2)$ can be considered as the fitting coefficient $1/A$.

Structural properties derived from fractal analysis must be introduced under the form of ΔD and $\Delta_1 D\cdot\varphi(\varepsilon)$ as fractal measures for void and solid phases, where $0 \leq \Delta D \leq 1$, $0 \leq \Delta_1 D \leq 1$, and $\varphi(\varepsilon)$ is a function accounting for deviation of porous media from granular packing, $\varphi(\varepsilon) = 1 + \varepsilon_0 - \varepsilon$; the function is $\varphi(\varepsilon) = 1$, when $\varepsilon = \varepsilon_0$ where ε_0 has a value in the range of 0.3 through 0.45 and is defined as a fitting parameter. Finally, dependence of normalized permeability on porosity becomes (74):

$$k/d^2 = \frac{\varepsilon^{3\Delta D}}{36A(1/\varepsilon^n)^2(1-\varepsilon)^{2\Delta_1 D\cdot\varphi(\varepsilon)}} = \frac{\varepsilon^{3\Delta D+2n}}{36A(1-\varepsilon)^{2\Delta_1 D\cdot\varphi(\varepsilon)}} \tag{17.9}$$

Equation 17.9 can describe a wide range of porous media: granular, fractal pore surface, and fractal tortuosity.

Other models of porous media permeability can be found in several scientific works: poly-disperse bed (75), ternary bed (24), binary packing (76–78), network model (79), and fibrous media (80) as in Equation 17.10 for which:

$$k = d^2/\{64(1-\varepsilon)^{1.5}[1 + 56(1-\varepsilon)^3]\},\ \varepsilon > 0.6 \tag{17.10}$$

17.4.2 Diffusivity

The term *diffusivity* is generally used as an alternative word for diffusion coefficient (81) and will be used here as meaning a "capacity to allow diffusion" (6). Diffusion coefficient of solute molecules in porous media is different from diffusion in bulk solution D_0, due to the presence of obstacles, solid phase and pore tortuosity, and is characterized as the effective diffusion coefficient in Equation 17.11:

$$D_e = D_0\varepsilon/\tau^2 \tag{17.11}$$

The ratio $\chi = D_e/D_0 = \varepsilon/\tau^2$ is defined as the relative diffusion conductivity of a porous medium, where it can be admitted that for diffusion in a single channel $\chi = 1/\tau^2$. Using the superposition principle described in Equation 17.1 for porosity and Equation 17.5 for tortuosity, the overall relative diffusion conductivity for multicomponent homogeneous porous media becomes:

$$\chi = D_e/D_0 = \prod_{i=1}^{m}\chi_i(x_i) = \prod_{i=1}^{m}\frac{\varepsilon_i(x_i)}{\tau_i^2(x_i)},\ \sum_{i=1}^{m}x_i = 1 \tag{17.12}$$

Equation 17.12 fits well to numerous experimental data of cells homogeneously immobilized in gel systems, leading to a correlation $\chi = (1-\phi_c)^\alpha$, where ϕ_c is the cell volume fraction and α has in general a value range of $\alpha = 1.8$ through 2.25 (4).

For a layered system and neglecting convection, the overall mass transfer coefficient K is related to layers by $1/K = \sum_{j=1}^{n} 1/K_j$, where K_j is the j-th layer system mass transfer coefficient and n is number of layers in the system. Consequently, $1/K_j = l_j\tau_j^2/(D_0\varepsilon_j)$ and $K = D_e/L = D_0\varepsilon/(\tau^2 L)$, where l_j is the thickness of the j-th layer, L is the total layered system thickness, and ε and τ represent the system equivalent porosity and tortuosity, respectively. With the ratio χ, the overall system mass transfer coefficient can be written as $K = D_0\chi/L$ and the relation becomes $L/\chi = \sum_{j=1}^{n} l_j/\chi_j$, which after normalization (6) becomes:

$$1/\chi = \sum_{j=1}^{n} y_j/\chi_j \text{ or } \chi = 1/\sum_{j=1}^{n} y_j/\chi_j \tag{17.13}$$

where $y_j = l_j/L$ is the linear fraction of the j-th layer thickness in the total system thickness, $\sum y_j = 1$.

The most complex case is encountered when each layer in the system is a composite porous medium and, as a result, models represented by Equations 17.12 and 17.13 must be considered together (5,6).

17.4.3 Hindered Diffusion

The so-called hindered or restricted diffusion is encountered when the size of solute molecules to be transferred, adsorbed, or separated is comparable to the pore size. Restricted diffusion gives rise to reduced effective diffusion coefficient due to steric and physicochemical effects. Usually, restricted diffusion models are built assuming smooth cylindrical pores although, in reality, porous media solute passages are narrow and tortuous.

When diffusion is considered in the whole porous medium Equation (17.14) is used (48,82,83).

$$D_e = D_0 \frac{\varepsilon}{\tau^2} F_1(\lambda) F_2(\lambda) \tag{17.14}$$

otherwise $D_e = D_0 F_1(\lambda) F_2(\lambda)/\tau^2$ for a single channel, where λ is the ratio of the Stokes-Einstein diameter of the diffusing micro-object d_m to the equivalent pore diameter d_{por}, $\lambda = d_m/d_{por}$. The parameter $F_1(\lambda)$ is the steric partition coefficient, which is defined as the pore cross-sectional area available to the solute molecule divided by the total pore cross-sectional area (Eq. 17 15).

$$F_1(\lambda) = (1 - \lambda)^2 \tag{17.15}$$

The correction factor $F_2(\lambda)$ accounts for the effect of the pore wall on the solvent properties (an increase in the local solvent viscosity near the pore wall) and is often represented by a polynomial series $F_2(\lambda) = 1 - a\lambda + b\lambda^3 - c\lambda^5$ (83,84), for instance, $F_2(\lambda) = 1 - 2.1044\lambda + 2.089\lambda^3 - 0.948\lambda^5$ (85), by an exponential function (82,86), or by a combination of both (87).

For shaped macromolecules, a shape factor is often introduced as a coefficient before the Stokes-Einstein diameter (84,85,87,88). Havsteen (88) analyzed nonclassical flow through narrow-pored membranes and concluded that when passages through the membrane have a size close to the size of the percolating molecules, fractal properties of the pore must be accounted for.

Giona *et al*. (89) underline that the topology complexity of a fractal pore network modifies the scaling properties of diffusion. The effective diffusion coefficient in fractal pore media may be represented as $D_e = l^{2-d_w}$, where l is the characteristic size of the structure, and d_w is the walk dimension, $d_w \geq 2$. A 2D simulation of porous media using a measure of pore fractal dimension leads to results that can be outlined as follows (90):

1. The restriction effect on molecular diffusion depends on pore topology, which, in the case of packed beds, is correlated with the packing type.
2. Starting from large pores with an aspect ratio of micro-particle size to pore size $\lambda > 0.001$, molecules (or any other test object) recognize the pore volume as a partially restricted space with a slightly reduced fractal dimension.
3. In turn, a dramatic reduction of pore fractal dimension is observed when λ overcomes 0.01 and approaches λ ~0.1 in a 2D approach, meaning that the test object recognizes the pore as 1D rather than a2D space; in turn, a 3D system will be recognized by the test object as a 2D system. Therefore, it is possible to foresee that the movement of asymmetric micro-objects inside tortuous channels might be affected.

These results suggest that diffusion of high molecular weight linear or branched biomolecules, such as DNA or polysaccharides, through narrow-pored media will be seriously affected.

17.5 POROUS MEDIA IN BIOPROCESSES

Bioreactors are a central piece of biotechnological equipment and have a variety of constructions (72,91,92). Many of them use porous media in different forms, immobilized cells or enzymes in or on porous support, adsorbents and catalysts, biofilms, layered porous systems and membranes, flocculated or aggregated systems, and so on. All the previously mentioned characteristics of porous media play a significant role in reaction, mass, and heat transfer. Figure 17.3 shows the main types of granular porous media applied in bioreactors as well as in chromatographic columns.

In case 6 (Fig. 17.3), composite particles packing with a significant difference in the size of particles, may be replaced
with a binary packing if there is no limit to hydraulic pressure. In case 4 (Fig. 17.3), giga-pores ensure convective flow through a particle. When small/large particle size ratio is in the range $0.4 > \delta > 0.1$ and reduction in the pore's specific area is not a problem, biporous particles may be replaced with a binary packing. With the same range of δ, the permeability of binary packing is larger than monosize packing because of the distortion effects in binary packing (77) (see Fig. 17.3, 17.9a–17.9c). At $\delta < 0.1$ through 0.15, the binary packing permeability becomes smaller than the permeability of monosize packing.

With developments in biomedicine and nanotechnology, bioreactor applications moved to the study of microreactors for tissue engineering and super-molecule synthesis. For instance (93,94), nanobiocatalysis, in which enzymes are incorporated into nanostructured materials, has emerged

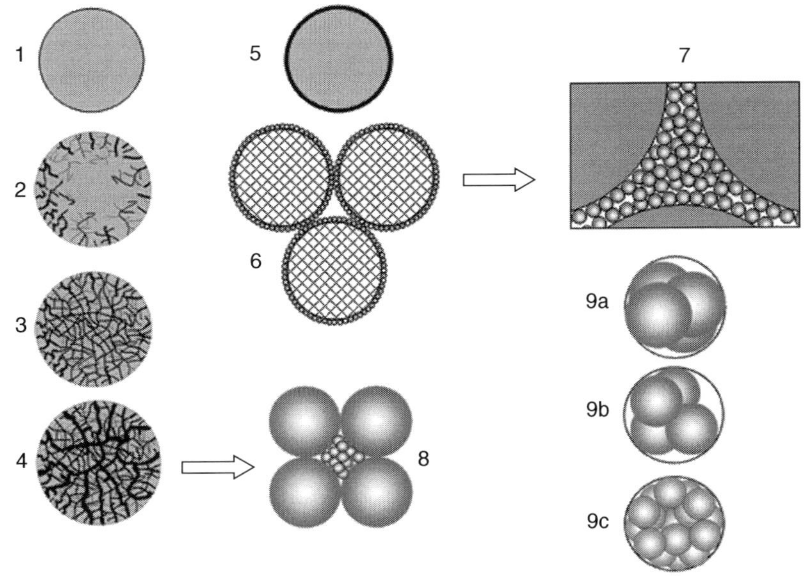

Figure 17.3. Particular stationary phase types: 1—nonporous particles, 2—with porous surface, 3—microporous, 4—biporous with micro- and giga-pores, 5—particle covered by a functional layer, 6—composite particle: nonporous beads covered by a micro/nano-spheres layer, 7—binary packing substitutes of 6, 8—substitute of 4 when $0.4 > \delta > 0.1$, and 9a through 9c respectively, $\varepsilon = 0.645$, 0.53, and 0.31.

as a rapidly growing area. Nanostructures, including nanoporous media, nanofibers, carbon nanotubes, and nanoparticles, have manifested great efficiency in the manipulation of the nano-scale environment of the enzyme and promise exciting advances in many areas of enzyme technology. In this situation, given the previously discussed increasing role of pore topology, the possibility of its control becomes crucial. Controlled pore topology is able to significantly affect mass transfer processes in microfilms, nanocomposites, and living tissues (7,95) and has a broad application in biotechnology, in particular, systems containing living cells (4–6): thick-films and immobilized cells, biofilms, biosensors (96–98), components of bioelectronic devices (99,100), and biocatalytic coatings (101,102).

17.5.1 Porous Media in Biological Systems

Porous biological systems may be separated into three types. The first one concerns biological objects (microorganisms, cells, tissues on a scaffold, biofilms, enzymes, etc.) immobilized onto an inert porous or nonporous support or into a porous matrix. Porous media can be in the form of grains, surface, and bed, permeable or semipermeable membranes. It must be emphasized that cell tissues may be treated as this kind of biological system. Attention must be paid to the way bio-objects are distributed. For instance, in Figure 17.4 the void occupied by a set of spheres in pores (a) and (b) is the same but the porosity, tortuosity, and transfer characteristics are completely different.

The second type concerns porous media formed by microorganisms or cell agglomeration, flocculation, growth, and differentiation.

The third type corresponds to porous systems composed of case 1 and case 2 types (Fig. 17.2); the most visible examples are activated sludge and biofilms. Components of porous systems display a variety of sizes and possible spatial distributions: homogeneous, nonhomogeneous, in the form of colonies, layers, and so on (4–6,91).

Mass transfer in biosystems is related with complex biochemical reactions (91) that may be characterized by the dimensionless Thiele modulus ϕ. The Thiele modulus is

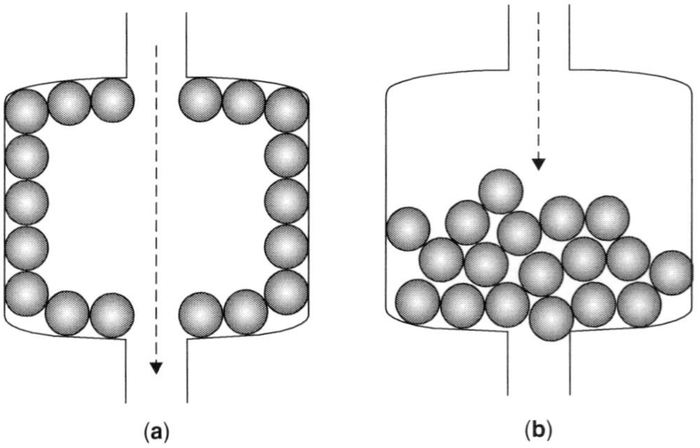

Figure 17.4. Example of two different types of 18-spheres arrangement inside a pore void that results in different mass transfer mechanisms; (a) diffusion and (b) convection.

equal to the ratio of the intrinsic chemical reaction rate in the absence of mass transfer limitations to the rate of diffusion through the porous medium. The effectiveness factor $\eta = \tanh(\phi)/\phi$ is the ratio of the actual diffusion affected reaction rate to the rate of the same reaction taking place without diffusion resistance. For a plate, the porous medium Thiele modulus is $\phi = \sqrt{k_s L^2/D_e}$, where k_s is a reaction rate constant and L is the porous medium thickness.

The presence of living cells in a biological porous system, raised the question of whether they are able to react to environmental changes by adjusting the porous media space in a manner similar to the one observed for a living tissue (7) (Eq. 17.3 and 17.4). To exemplify this case, a simulation is shown in Fig. 17.5, where glucose diffusion ($D_0 = 6/10^{-10}$ m^2/s) was simulated through a biofilm of $L = 2.3/10^{-5}$ and m and $k_s = 0.17$ s^{-1} was assumed. Four situations were considered: (i) constant tortuosity $\tau = 1.5$; (ii) inert granular packing, Equation 17.3, $n = 0.5$; (iii) and (iv) two "flexible" relations (7): $n = 0.26 + 0.3\varepsilon + \varepsilon^2$ for a "topologically" dense cell arrangement, and $n = 0.2 + \varepsilon^2$ for a loose configuration. For the same reaction rate, a living tissue is able to sustain the effectiveness factor at a higher level with increasing packing density than the level obtained with the conventional packing model.

17.5.2 Porous Media in Downstream Processing

Numerous processes are developed in the industrial and laboratory scales for purification, separation, or concentration of bioproducts. The choice of a particular downstream process depends, among other factors, on the nature and amount of the product, its physicochemical characteristics, concentration, and cost (103,104). Sedimentation and centrifugation are based on gravitational and centrifugal forces, respectively, and the liquid content in the sediment is an important economic parameter. For a fugate (liquid phase) volume V a fitting function can be used, adequately describing the dependence of the liquid decanted volume V at time t for the centrifugation as well as for sedimentation (105–107):

$$v = V/V_\infty = k_s t^a/(1 + k_s t^a) \qquad (17.16)$$

where V_∞ is the decanted liquid volume at equilibrium ($t = \infty$), and k_s, a are the model parameters (Eq. 17.16). Parameters a and k_s depend on a set of conditions, namely, the coagulation or flocculation pretreatment and the ratio of organic or nonorganic components in the solid. For process control and optimization, the definition of a liquid accumulation velocity may be useful (Eq. 17.17):

$$w = \mathrm{d}v/\mathrm{d}t = (k_s a t^{a-1})/(1 + k_s t^a)^2 \qquad (17.17)$$

A large group of downstream processes is represented by filtration and membrane processes where porous media play the role of a separator (103,108,109). Depending on the inflow organization, filtration is classified as dead-end, cross-flow, or centrifugal filtrations; filtration on rotating disks, and so on. By analyzing filter media pore size, we can distinguish deep bed filtration, conventional filtration, micro-, ultra-, and nanofiltration, and reverse osmosis separation (110–112). Regular patterned sieves and filters with comparable molecular dimensions, hold great promise

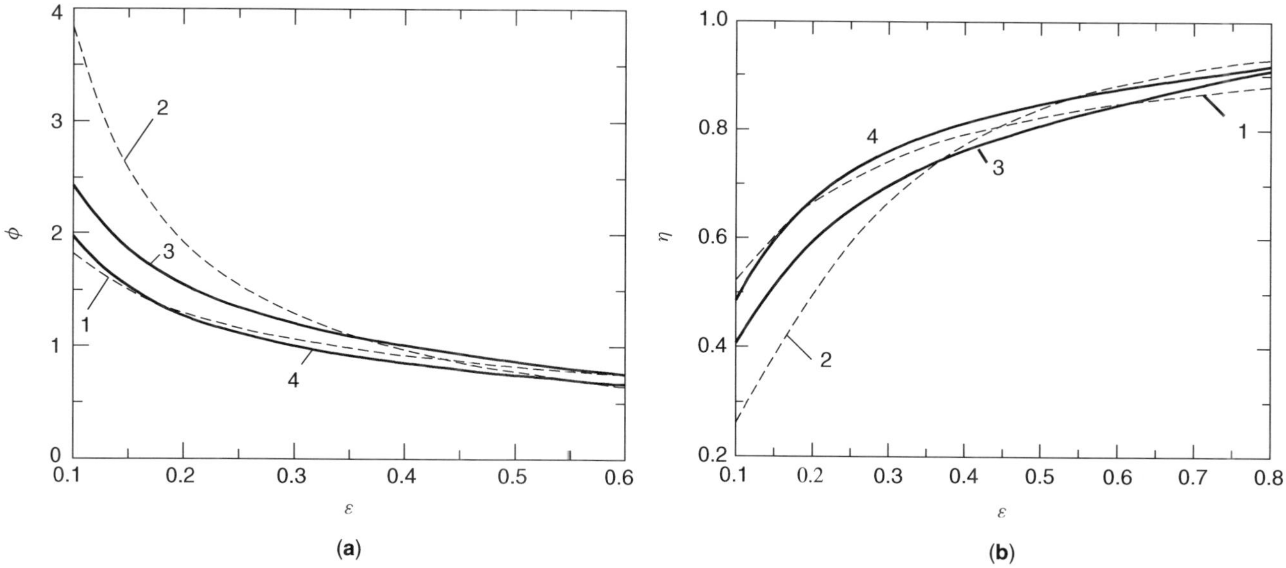

Figure 17.5. Simulated Thiele modulus (a) and η (b) versus biofilm porosity ε for fixed biofilm thickness. 1– $\tau = 1.5$; 2–Equation 17.3, $n = 0.5$; 3–$n = 0.26 + 0.3\varepsilon + \varepsilon^2$; 4–$n = 0.2 + \varepsilon^2$.

as an alternative to conventional polymeric gels and fibrous membranes, to improve biomolecule separation (113). Some examples are shown in Fig. 17.6.

In many cases, controlling the inlet slurry concentration is important because cake properties of colloidal systems are dependent on the slurry concentration. These phenomena are related to particle aggregations in suspension and the changes in cake structure arrangement. Furthermore, slurry concentration control can also control, at least partially, the cake compressibility (114–116).

Numerous chromatographic methods are available for the fractionation of macromolecules, biopolymers, and (nano) microparticles that are designed to sense differences in size, shape, flexibility, charge, polarity, or biological function of the compounds to be separated. In many cases, especially for biopolymers, the spatial structure of a macromolecule is essential for its biological activity. Therefore, chromatographic conditions similar to the physiological environment, such as aqueous buffer solutions at neutral pH, are the most appropriate for the preservation of biological activity. Consequently, the most successful separation techniques are those designed to have selective interaction with the biopolymers to be separated (108,117,118).

The research on "smart" polymers application is leading to new types of porous media for separation of biomolecules (119). Stimuli-responsive polymers undergo conformational rearrangement in response to small changes in their environment, such as temperature, pH, ultraviolet (UV) irradiation, ionic strength, or electric field. These environmental changes alter the structure of stimuli-responsive polymers and increase or decrease their overall hydrophobicity, resulting in reversible collapse, dehydration, or hydrophobic layer formation.

The development of macromolecule and biopolymer chromatographic separation is moving toward a growing use of small-sized particles (1–3 μm) in column packing (120–123). Optimization of particle dimensions for high efficiency in capillary electrochromatography, shows that a pore size of 30 nm may be optimal with respect to pore-to-interstitial flow ratio, required ionic strength, and separation efficiency (124) but, in some applications, columns packed with particles of size 20 nm and smaller may be suitable.

Miniaturization is leading to the fabrication of 2D nanocolumns for liquid chromatography with a regular structure of nanochannels (125) that may be considered a 2D porous medium. Applied fabrication methods may be based on microlithography and other processes already applied in microelectronics.

Recently, it was proposed that the fabrication of electrochemical sensors be done by utilizing micron-sized electrically conducting beads to form a disposable electrode, as well as nonconducting beads to form renewable layers of immobilized enzymes (126). An automated system for performing rapid immunoassays, kinetic measurements, and affinity ranking of biomolecular interactions using fluorescence-labeled ligands was developed (127). Its distinctive feature is the automated

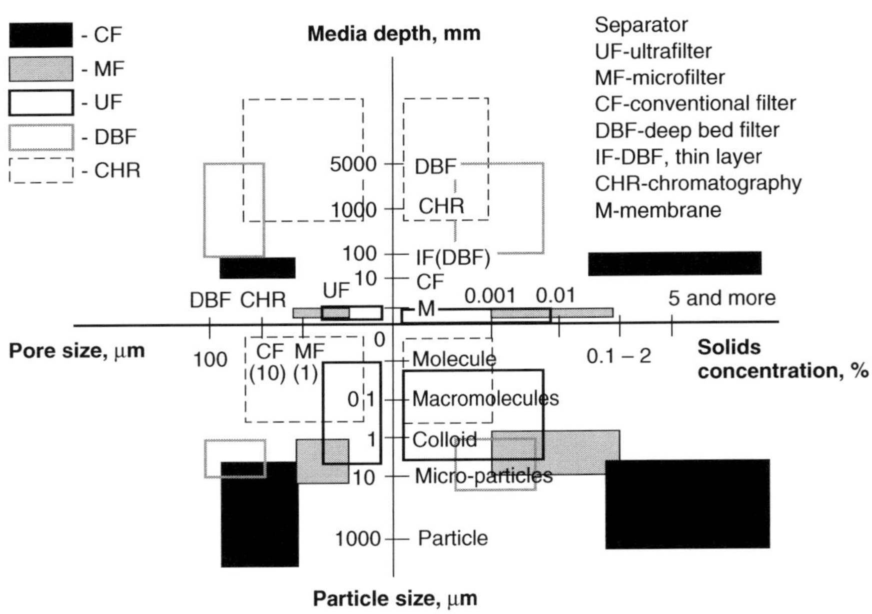

Figure 17.6. Scheme of separation processes using porous media. The following dimensions are displayed: pore size (μm), porous media thickness (mm), size of separating substrates (μm), and handling concentration of substrates (vol/vol wet concentration) to be separated. All size dimensions are displayed using a log scale.

renewal of solid phase for each measurement, which avoids the need for regeneration of the sensing surface.

The number of separation methods based on functionally modified surfaces of porous media is increasing with new findings in colloidal, polymer, and nanosciences. Transport properties involved in these novel methods will play an important role for improving process and cost efficiency.

Among chromatographic methods, size-exclusion chromatography (SEC), hydrodynamic chromatography (HDC), and slalom chromatography (SC) comprise a set of chromatographic modes in which the separation of components takes place without a specific interaction with the stationary phase (27,37,112). These methods may be characterized as follows (Fig. 17.7): SEC—this mode is appropriate to resolve differences in molecular size, the selectivity being provided by the pore size distribution in packing particles; the solute penetrates into the pore voids and a pore/solute size ratio controls the resolution; HDC—differences in solute size are set up by the steric exclusion region close to the inner capillary surface of pore channels and the excluded volume is higher if the object size is bigger; because of the parabolic flow velocity profile inside the capillary, a separation is developed: larger objects travel more rapidly through the capillary than smaller ones; SC—selection is based upon differences in length and flexibility of solute molecules, the selectivity being provided by the pore diameter of the packing combined with flow rate (117); this chromatographic mode is particularly useful for the separation of DNA fragments, the repetition of the DNA fragments across the porous medium being caused by the channel tortuosity following the flow direction (128).

It is interesting to note that the application of binary packing in HDC chromatographic column instead of monosize packing yields, improved resolution of microparticulate systems (27,37) as well as of macromolecules (28). The example shown in Fig. 17.8 was obtained from the work of Mota (129) for monosize (glass beads 111.5 μm) and binary packing (mixture of 111.5 and 337.5 μm glass beads, $\varepsilon = 0.347$, volume fraction of large beads 0.55).

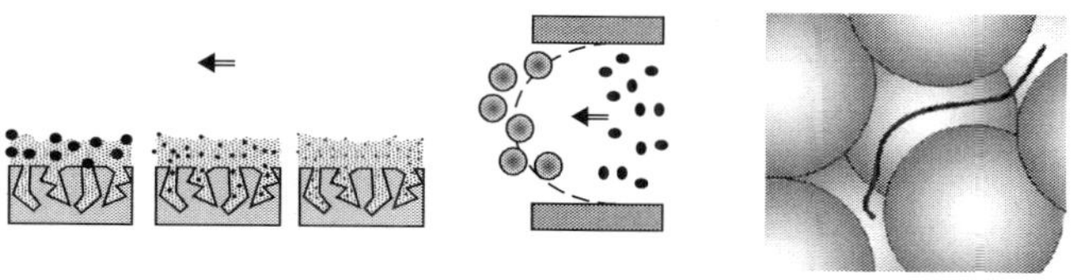

Figure 17.7. Sketch of SEC, HDC, and SC modes of separation. The solute objects flow through a porous medium which may withstand effects related to pore topology and particle shape, namely, (i) ratio of particle size to the pore size (λ), (ii) abrupt changes in pore cross-section area, and (iii) channel tortuosity.

17.5.3 Porous Media in Biomedical Applications

One of the conventional ways of drug delivery is by using the drug distributed or encapsulated in porous media. For this purpose, gels are often used as porous media. Modern gel systems provide controlled drug release by adjusting porous structure to near physiological conditions. Mechanisms of drug release involve drug diffusion, hydrogel matrix swelling and chemical reactivity of the drug or matrix (130). The release process can be controlled by time, salt concentration, pH or temperature (131). New multiresponsive core–shell microgels, with a thermoresponsive core and a glucose-responsive shell, have been applied for insulin encapsulation (132,133).

A promising approach is the production of drug nanoparticles. The major advantage of this technology is its general applicability to most drugs. Particulate drug delivery systems have become important in experimental pharmaceutics and clinical medicine (134). Microparticles, nanoparticles, pellets, hydrogels, and others, all have their own advantages and disadvantages. It is possible to combine the advantages of several classes of vehicles in the same drug delivery system (135). For instance, the application of temperature and pH-responsive polymeric composite membranes filled by nanoparticles, improves the controlled delivery of proteins and peptides (136).

Porous nanosystems represent new complex porous material that need further investigation (137,138). The control of parameters involved in the process of classic bridging flocculation allows the preparation and fine-tuning of this class of hybrid nanomaterials with respect to size, composition, and morphology (139). The

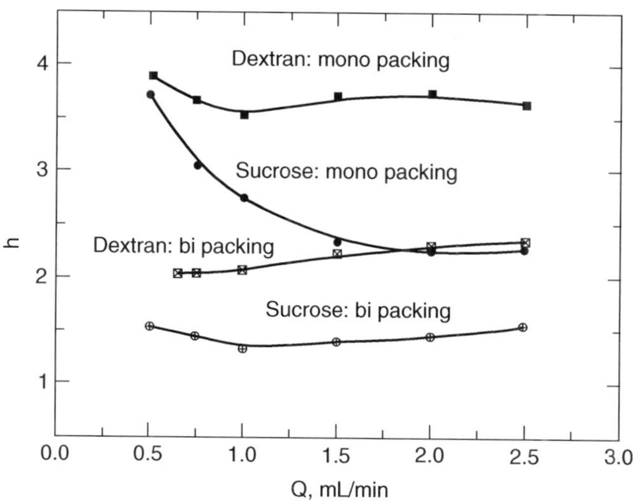

Figure 17.8. Reduced plate height h versus flow velocity Q for sucrose and dextran 2000 kDa. Column inner diameter 2.5 cm; packing length 48 cm; sample 1 mL; solute concentration of 0.6 g/L. [Presented experimental data obtained by Dr Ricardo Dias in the framework of the project (129)].

resulting nanoparticle-filled "nanobags" are obtained in aqueous suspension by mixing three basic components in different ratios: a polyelectrolyte, a multivalent ion, and nanoparticles. The size range in which nano- and micropouches may be prepared seems to start at about 25 nm; these are oligonanoparticle aggregates whose size is clearly related to the size of the nanoparticles themselves and seem to extend up to about 5 μm.

In many cases, to diffuse in living tissues drug macromolecules or nanoparticles need to overcome the blood-tissue barrier. One approach considers the barrier as a porous composite membrane, whereas in another approach porous media are represented by a complex system depending on the nature of the tissue.

Due to the variety of living tissue structure, only some examples are given below. Modeling diffusion in brain ECS is more complicated than the diffusion in porous media because ECS has a rather complicated physical 3D structure, and brain tissue responds dynamically to changes in environmental conditions (7). Analysis of numerous experimental data and models, has led to the conclusion that under external conditions, such as oxygen depletion, the ECS porosity decreases and cells (presumably through membrane rearrangements) adjust the void space to keep the diffusion within a defined range, which gives the living tissue the ability to support the diffusion level up to two or more times higher than what happens in conventional granular bed packing.

This finding resulted in a tortuosity correlation model expressed by Equations 17.3 and 4. Diffusion through ECS channels described through the ratio $D_e/D_0 = 1/\tau^2$ vs. ECS porosity is shown in Fig. 17.9, where $1\text{–}n = 0.5$ (inert porous media); $2\text{–}n = 0.2 + \varepsilon^2$ (loose cells configuration); and $3\text{–}n = 0.26 + 0.3\varepsilon + \varepsilon^2$ ("topologically" dense cell arrangement). Using this approach, the existence of three clusters was identified: a region of normal brain function, both for young and adult brains, for values between $\varepsilon = 0.15$ and 0.30; and two regions of abnormal brain behavior below and above the normal region, corresponding to different behaviors—aging, tumors, anoxia, brain death, and so on.

Skin is a multilayered system which is considered a porous medium (140). A comprehensive model of skin is important for topical drug delivery, wound repair, as well as for design of new patches. The development of tissue-engineered models that mimic human skin have provided novel experimental systems to study the behavior of normal and altered human stratified squamous cells (141).

The transport of fluid and solute molecules in the tumor interstitium is governed by the permeation and diffusion mechanisms described by porous media transport models (142,143). For instance, to reach cancer cells in a tumor, a blood-borne therapeutic agent must make its way into the blood vessels of the tumor, cross the vessel wall into the

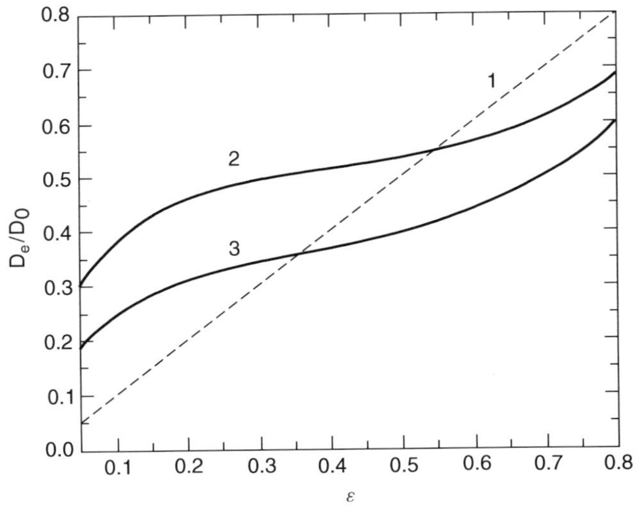

Figure 17.9. ECS (extracellular space) channel diffusivity versus porosity. $1\text{–}n = 0.5$; $2\text{–}n = 0.2 + \varepsilon^2$; $3\text{–}n = 0.26 + 0.3\varepsilon + \varepsilon^2$.

interstitium, and finally migrate through the interstitium. Unfortunately, tumors often develop ways that hinder each of these steps (144,145).

Tissue growing methods are, in many cases, based on porous scaffold with immobilized cells placed in bioreactors of appropriate construction (146). Ideally, engineered tissues should provide nutrient transport, mechanical stability, coordination of multicellular processes, and a cellular microenvironment that promotes phenotypic stability. To achieve this goal, many engineered tissues require both macro- (∼cm) and micro- (∼100 μm) scale architectural features (147).

Hydrogel networks are highly desirable as 3D tissue engineering scaffolds for cell encapsulation due to their high water content and ability to mimic the native extracellular matrix. Good results have been obtained with a hydrogel based on interpenetrating polymer networks of dextran and gelatin for vascular tissue engineering (148). Tissue engineering of dermal substitutes based on porous copolymer scaffolds has been successfully performed by dynamic seeding (149).

Biomaterials in cardiac tissue engineering and myocardial tissue in particular, combine isolated functional cardiomyocytes, and a biodegradable or nondegradable biomaterial to repair diseased heart muscles (150).

Biological tissues serve as templates to construct inorganic structures with different pore hierarchy that, in turn, have application in biotechnology as sorbents, catalysts, and immobilizing matrices. Examples of the biological templates used are shells, skeletons, living cells, diatoms (151,152), bacterial superstructures (153), and wood cellular structure (154,155). As an example, a variety of microstructures of biomorphic ZrO_2 ceramics obtained by a sol-gel process based on pine and rattan woods can be found (156). An additional example is given (155), where controlled synthesis of monolithic hierarchical porous materials use wood as a template with the assistance of supercritical carbon dioxide.

17.6 CONCLUSION

The complexity of porous media in biotechnology applications has been discussed. As was shown, it is important before describing transport phenomena, to analyze carefully porous media nature and structure. Any change in the model's postulates, as well as in the initial and boundary conditions, might lead to a dramatic change of the model solution. As a result, the investigation on the mechanisms acting in porous media at micro- and nanoscale needs further insight to provide sound knowledge for successful applications in this challenging field.

17.6.1 Acknowledgments

The authors wish to acknowledge the Post-Doctoral grant awarded by the Portuguese Fundação para a Ciência e Tecnologia to Dr Alexander Yelshin. This work is dedicated to Dr Alexander Yelshin, who collaborated in this study until March 2009, when he suddenly passed away. The study proceeded with the collaboration of his daughter, Dr Inna Yelshina, a researcher at the Polotsk State University, Belarus.

REFERENCES

1. Tarleton S, Wakeman R. Dictionary of filtration and separation, Filtration Solutions. Exeter; 2008.
2. Bear J. Dynamics of fluids in porous media. New York: Dover Publications; 1972.
3. Wikipedia. Porous medium. web information. Available at http://en.wikipedia.org/wiki/Porous_medium. Accessed 2008 Dec 18.
4. Mota M, Teixeira JA, Yelshin A. Biotechnol Prog 200_; 17: 860–865.
5. Mota M, Teixeira JA, Yelshin A. Biotechnol Prog 2002; 18: 807–814.
6. Mota M, Yelshin A, Fidaleo M, Flickinger MC. Biochem Eng J 2007; 37: 285–293.
7. Mota M, Teixeira JA, Keating JB, Yelshin A. Biotechnol Appl Biochem 2004; 39: 223–232.
8. Kirkpatrick S. Rev Mod Phys 1973; 45: 574–588.
9. Klemm A, Kimmich R, Weber M. Phys Rev E 2001; 63: 041514–041511–041514–8.
10. Hastedt JE, Wright JL. Pharm Res 2006; 23: 2427–2440.
11. Ofir A, Dor S, Grinis L, Zaban A, Dittrich T, Bisquert J. J Chem Phys 2008; 128: 1–9.
12. Sung BJ, Yethiraj A. J Phys Chem B 2008; 112: 143–149.
13. Ouchiyama N, Tanaka T. Ind Eng Chem Fundam 1981; 20: 66–71.
14. Ouchiyama N, Tanaka T. Ind Eng Chem Fundam 1984; 23: 490–493.
15. Hulewicz ZZ. Int Chem Eng 1987; 27: 566–573.
16. MacDonald MJ, Chu C-F, Guilloit PP, Ng KM. AIChE J 1991; 37: 1583–1588.
17. Yu AB, Standish N. Ind Eng Chem Res 1991; 30: 1372–1385.
18. Yu AB, Zou RP, Standish N. J Am Ceram Soc 1992; 75: 2765–2772.
19. Yu AB, Zou RP, Standish N. Ind Eng Chem Res 1996; 35: 3730–3741.
20. Zou RP, Xu JQ, Feng CL, Yu AB, Johnston S, Standish N. Powder Technol 2003; 130: 77–83.
21. Mota M, Teixeira JA, Bowen R, Yelshin A. Proceedings of 8-th World Filtration Congress, 2000 Apr 3–7; Filtration Society, Brighton, UK ; 2000. pp. 57–60.
22. Mota M, Teixeira JA, Yelshin A, Bowen WR. Miner Eng 2003; 16: 135–144.
23. Mota M, Teixeira JA, Dias R, Yelshin A. Proceedings of 9th World Filtration Congress, 2004 Apr 18–22; New Orleans, Louisiana, USA, AFS ; 2004; pp. 1–30.
24. Dias R, Mota M, Teixeira JA, Yelshin A. Trans Filtr Soc 2005; 5: 68–75.
25. Dias R, Teixeira JA, Mota M, Yelshin A. Sep Purif Technol 2006; 51: 180–184.
26. Mota M, Teixeira JA, Yelshin A. Sep Purif Technol 1999; 15: 59–68.
27. Mota M, Teixeira JA, Yelshin A, Cortez S. J Chromatogr B 2006; 843: 63–72.
28. Dias RP, Fernandes CS, Mota M, Teixeira JA, Yelshin A. Carbohydr Polym 2008; 74: 582–587.
29. Currie JA. Br J Appl Phys 1960; 11: 318–324.
30. Giddings JC. Dynamics of chromatography. Part I. principles and theory. New York: M. Dekker; 1965.
31. Satterfield CN. Mass transfer in heterogeneous catalysis. Cambridge: M.I.T. Press; 1970.
32. Tye FL. Chem Ind (London) 1982; 10: 322–326.
33. Cussler EL. Diffusion: mass transfer in fluid systems. Cambridge: Cambridge University Press; 1984.
34. Sheffield RE, Metzner AB. AIChE J 1976; 22: 736–744.
35. Johnston PR. Fluid sterilization by filtration. Boca Raton, FL: Interpharm Press; 1992.
36. Dullien FAL. Chem Eng J 1975; 10: 1–34.
37. Mota M, Teixeira JA, Yelshin A, Cortez S. J Chromatogr B 2008; 864: 178 .
38. Dybbs A, Edwards RV, Bear J, Corapcioglu MY. Fundamentals of transport phenomena in porous media, NATO ASI series E. The Netherlands: Nijhoff; 1984. pp. 201–257.
39. Mauret E, Renaud M. Chem Eng Sci 1997; 52: 1819–1834.
40. Burganos VN, Sotirchos SV. AIChE J 1987; 33: 1678–1689.
41. Rege SD, Fogler HS. AIChE J 1988; 34: 1761–1772.
42. Aris R. Volume 1, The mathematical theory of diffusion and reaction in permeable catalysts. Oxford: Clarendon Press; 1975.
43. Satterfield CN. Heterogeneous catalysis in practice. New York: McGraw-Hill; 1980.
44. Welty JR, Wicks CE, Wilson RE. Fundamentals of momentum, heat, and mass transfer. Singapore: John Wiley & Sons; 1984.

45. Rosner DE. Transport processes in chemically reacting flow systems. London: Butterworths; 1986.
46. Sharma RK, Cresswell DL, Newson EJ. Ind Eng Chem Res 1991; 30: 1428–1433.
47. Walas SM. Chemical reaction engineering handbook of solved problems. Amsterdam: Gordon & Breach Publishers; 1995.
48. Blanch HW, Clark DS. Biochemical engineering. New York: M. Dekker; 1996.
49. Riley MR, Muzzio FJ, Buettner HM, Reyes SC. Biotechnol Bioeng 1996; 49: 223–227.
50. Yang J, Volesky B. J Chem Technol Biotechnol 1996; 66: 355–364.
51. Pape H, Riepe L, Schopper JR. Colloids Surf 1987; 27: 97–122.
52. Ho F-G, Strieder W. Chem Eng Sci 1981; 36: 253–258.
53. Mauret E, Renaud M. Chem Eng Sci 1997; 52: 1807–1817.
54. Wolf JR, Strieder W. AICHE Symp Series: Diffusion and Convection in Porous Catalysts 1988; New York, 84: 23–27.
55. Zhang TC, Bishop PL. Water Res 1994; 28: 2279–2287.
56. Pirard R, Blacher S, Brouers F, Pirard JP. J Mater Res 1995; 10: 2114–2119.
57. Milburn DR, Davies BH. Ceram Eng Sci Proc 1993; 14: 130–134.
58. Mair RW, Wong GP, Hoffmann D, Hürlimann MD, Patz S, Schwartz LM, Walswoth RL. Phys Rev Lett 1999; 83 (16): 3324–3327.
59. Baumann T, Petsch R, Fesl G, Niessner R. J Environ Qual 2002; 31: 470–476.
60. Lan HY, Grine F, Ni Q, Rubin C, Qin YX. Determination of bone porosity by non-invasive nuclear magnetic resonance. Proceedings IEEE 29 th Annual Bioengineering Conference; 2003; pp. 213–214.
61. Acosta RH, Agulles-Pedrós L, Komin S, Sebastiani D, Spiess HW, Blümler P. Phys Chem Chem Phys 2006; 8: 4182–4188.
62. Swift AJ, Wild JM, Fichele S, Woodhouse N, Fleming S, Waterhouse J, Lawson RA, Paley MN, Van Beek EJ. Eur J Radiol 2005; 54 (3): 352–358.
63. Wood NB, Zhao SZ, Zambanini A, Jackson M, Gedroyc W, Thom SA, Hughes AD, Xu XY. J Appl Physiol 2006; 101: 1412–1418.
64. lee SW, Antiga L, David Spence J, Steinman DA. Stroke 2008; 39: 2341 .
65. Beckmann F, Bonse U, Busch F, Günnewig O. J Comput Assist Tomogr 1997; 21 (4): 539–553.
66. Hopper TAJ, Wehrli FW, Saha PK, Andre JB, Wrgiht AC, Sanchez CP, Leonard MB. J Comput Assist Tomogr 2007; 31 (2): 320–328.
67. Bolland BJRF, Kanzler JM, Dunlop DG, Oreffo ROC. Bone 2008; 43 (1): 195–202.
68. Malafaya PB, Reis RL. Acta Biomater 2009; 5 (2): 644–660.
69. Chaunier L, Chrusciel L, Delisee C, Della Valle G, Malvestio J. Food Biophys 2008; 3 (4): 344–351.
70. Weinekoetter C. X-Ray nanofocus CT: visualizing of internal 3D-structures with submicrometer resolution. International Conference on Applications of Computerized Tomography, AIP Conference Proceedings, Volume 1050; 2008; pp. 3–14.
71. Bruncke O, Sieker F. 3D analysis of textile structures with high-resolution computed tomography. Proceedings 9th International Conference on Textile Composites; 2008; pp. 291–297.
72. Bailey JE, Ollis DF. Biochemical engineering fundamentals. Singapore: McGraw-Hill; 1986.
73. Xu P, Yu B. Adv Water Res 2008; 31: 74–81.
74. Mota M, Yelshin A. In: Ferreira EC, Mota M, editors. Proceedings of the 10th international chemical and biological engineering conference - CHEMPOR 2008 Braga, Portugal, 2008 Sep 4–6. Universidade do Minho, Braga; 2008. pp. 61–66.
75. Li Y, Park C-W. Ind Eng Chem Res 1998; 37: 2005–2011.
76. Mota M, Teixeira JA, Yelshin A. Trans Filtr Soc 2001; 1: 101–106.
77. Dias R, Fernandes CS, Mota M, Teixeira JA, Yelshin A. Int J Heat Mass Transfer 2007; 50: 1295–1301.
78. Dias RP, Fernandes CS, Teixeira JA, Mota M, Yelshin A. J Hydrol 2008; 349: 470–474.
79. Chapuis RP, Marcotte D, Aubertin M. Can Geotech J 2006; 43: 110–114.
80. Wakeman RJ, Tarleton ES. Filtration: equipment selection, modelling and process simulation. Oxford: Elsevier; 1999.
81. Kizilyalli M, Corish J, Metselaar R. Pure Appl Chem 1999; 71: 1307–1325.
82. Suzuki M. Adsorption engineering. Tokyo: Kodansha-Elsevier; 1990.
83. Limbach KW, Wei J. AIChE J 1990; 36: 242–248.
84. Cannell DS, Rondelez F. Macromolecules 1980; 13: 1599–1602.
85. Deen WM. AIChE J 1987; 33: 1409–1425.
86. Baltus RE, Anderson JL. Chem Eng Sci 1983; 38: 1959–1969.
87. Anderson JL, Kathawalla IA, Lindsey JS. AIChE Symp Series: Diffusion and Convection in Porous Catalysts, New York; 1988; 84: 35–39.
88. Havsteen BH. Adv Colloid Interface Sci 1993; 45: 79–213.
89. Giona M, Schwalm WA, Schwalm MK, Adrover A. Chem Eng Sci 1996; 51: 4717–4729.
90. Mota M, Yelshin A. In: Ferreira EC, Mota M, editors. Proceedings of the 10th international chemical and biological engineering conference - CHEMPOR 2008 Braga, Portugal; 2008 Sept 4–61. Braga: Universidade do Minho; 2008. pp. 671–676.
91. Cabral JMS, Mota M, Tramper J, editors. Multiphase bioreactor design: Taylor and Francis, London: 2001.
92. Oksmah-Caldentey KM, Barz WH, editors. Plant biotechnology and transgenic plants. New York: Marcel Dekker; 2002.
93. Wang P. Curr Opin Biotechnol 2006; 17: 574–579.
94. Kim J, Grate JW, Wang P. Trends Biotechnol 2008; 26: 639–646.
95. Lee EJ, Holmes JW, Costa KD. Ann Biomed Eng 2008; 36: 1322–1334.
96. D'Souza SF. Biosens Bioelectron 2001; 16: 337–353.
97. Dubey RS, Upadhyay SN. Biosens Bioelectron 2001; 16: 995–1000.
98. Fang Y, Govind R. Chin J Chem Eng 2008; 16: 277–286.
99. Simpson ML, Saylor GS, Fleming JT, Applegate B. Trends Biotechnol 2001; 19: 317–323.
100. Willner I, Willner B. Trends Biotechnol 2002; 19: 222–230.

101. Gosse JL, Engel BJ, Rey FE, Harwood CS, Scriven LE, Flickinger MC. Biotechnol Prog 2007; 23: 124–130.
102. Flickinger MC, Schottel JL, Bond DR, Aksan A, Scriven LE. Biotechnol Prog 2007; 23: 2–17.
103. Holdich RG. Fundamentals of particle technology. Shepshed: Midlend Information Technology and Publishing; 2002.
104. Weuster-Botz D, Hekmat D, Puskeiler R, Franco-Lara E. Trends Biotechnol 2007; 105: 205–247.
105. Mota M, Teixeira JA, Abaev GN, Yelshyna I, Yelshin A. Proceedings of International Conference FILTECH 2005; 2005 Oct 11–13; Weisbaden, Germany ; 2008; pp. I-241–I-248.
106. Mota M, Teixeira JA, Abaev GN, Yelshyna I, Yelshin A. Filtration 2008; 8: 80–86.
107. Mota M, Yelshin A, Yelshyna I. 10th World Filtration Congress; 2008 Apr 14–18; Leipzig, Germany; 2008; pp. I-101–I-105.
108. Flickinger MC, Drew SW, editors. The encyclopedia of bioprocess technology: fermentation, biocatalysis, and bioseparation. Toronto: John Wiley & Sons; 1999.
109. Svarovsky L, editor. Solid-liquid separation: Elsevier Butterworth-Heinemann; Oxford, 2001.
110. Porter MC, editor. Handbook of industrial membrane technology. Westwood , NJ: Noyes Publications; 1990.
111. Baker RW. Membrane technology and applications. Chichester: John Wiley & Sons; 2004.
112. Saxena A, Tripathi BP, Kumar M, Shahi VK. Adv Colloid Interface Sci 2009; 145: 1–22.
113. Fu J, Mao P, Han J. Trends Biotechnol 2008; 26: 311–320.
114. Mota M, Teixeira JA, Yelshin A. Sep Purif Technol 2002; 27: 137–144.
115. Yelshin A, Teixeira JA, Bowen WR, Mota M. Trans Filtr Soc 2002; 2: 45–48.
116. Mota M, Teixeira JA, Yelshin A. Proceedings of 9th World Filtration Congress; 2004 Apr 18–22; New Orleans AFS, Louisiana, USA ; 2004; pp. 1–20.
117. Huber CG. In: Meyers RA, editor. Encyclopedia of analytical chemistry. Chichester: John Wiley & Sons; 2000. pp. 11250–11278.
118. Yao K, Shen S, Yun J, Wang L, Chen F, Yu X. Biochem Eng J 2007; 36: 139–146.
119. Maharjan P, Woonton BW, Bennett LE, Smithers GW, DeSilva K, Hearn MTW. Innov Food Sci Emerg Technol 2008; 9: 232–242.
120. Ceriotti L, de Rooij NF, Verpoorte E. Anal Chem 2002; 74: 639–647.
121. Broyles BS, Jacobson SC, Ramsey JM. Anal Chem 2003; 75: 2761–2757.
122. Wu N, Clausen AW. J Sep Sci 2007; 30: 1167–1182.
123. Anspach JA, Maloney TD, Colón LA. J Sep Sci 2007; 30: 1207–1213.
124. Stol R, Poppe H, Kok WT. Anal Chem 2003; 75: 5246–5253.
125. He B, Tait N, Regnier F. Anal Chem 1998; 70: 3790–3797.
126. Mayer M, Ruzicka J. Anal Chem 1996; 68: 3808–3814.
127. Willumsen B, Christian GD, Ruzicka J. Anal Chem 1997; 69: 3482–3489.
128. André C, Guillaume YC. Chromatographia 2004; 59: 487–492.
129. Mota M. Influência das Propriedades dos Meios Porosos no Desempenho de Bioseparações . Report POCI/ EQU/ 58337/ 2004, 2008.
130. Hamidi M, Azadi A, Rafiei P. Adv Drug Deliv Rev 2008; 60: 1638–1649.
131. Horkay F, Basser PJ. J Polym Sci Part B Polym Phys 2008; 46: 2803–2810.
132. Zhang Y, Guan Y, Zhou S. Biomacromolecules 2007; 8: 3842–3847.
133. Lapeyre V, Ancla C, Catargi B, Ravaine V. J Colloid Interface Sci 2008; 327: 316–323.
134. Müller RH, Jacobs C, Kayser O. Adv Drug Deliv Rev 2001; 47: 3–19.
135. Kohane DS. Biotechnol Bioeng 2007; 96: 203–209.
136. Zhang K, Wu XY. Biomaterials 2004; 25: 5281–5291.
137. Lou XW, Archer LA, Yang Z. Adv Mater 2008; 20: 3987–4019.
138. Amgoune A, Krumova M, Mecking S. Macromolecules 2008; 41: 8388–8396.
139. Schneider GF, Decher G. Nano Lett 2008; 8: 3598–3604.
140. Amsden BG, Goosen MFA. AIChE J 1995; 41: 1972–1994.
141. Garlick JA. Adv Biochem Eng Biotechnol 2006; 103: 207–239.
142. Jain RK. Cancer Res 1987; 47: 3039–3051.
143. Syková E, Mazel T, Vargová L, Vorisek I, Prokopova-Kubinová S. Prog Brain Res 2000; 125: 155–178.
144. Jain RK. Ann Biomed Eng 1996; 24: 457–473.
145. Drummond DC, Meyer O, Hong K, Kirpotin DB, Papahadjopoulos D. Pharm Rev 1999; 51: 691–744.
146. Kretlow JD, Mikos AG. AIChE J 2008; 54: 3048–3067.
147. Tsang VL, Bhatia SN. Adv Biochem Eng Biotechnol 2006; 103: 189–205.
148. Liu Y, Chan-Park MB. Biomaterials 2009; 30: 196–207.
149. Wang HJ, Bertrand-de Haas M, Riesle J, Lamme E, van Blitterswijk CA. J Mater Sci Mater Med 2003; 14: 235–240.
150. Chen Q-Z, Harding SE, Ali NN, Lyon AR, Boccaccini AR. Mater Sci Eng R 2008; 59: 1–37.
151. Anderson MW, Holmes SM, Hanif N, Cundy CS. Angew Chem Int Ed 2000; 39: 2707–2710.
152. Cai X, Zhu G, Zhang W, Zhao H, Wang C, Qiu S, Wei Y. Eur J Inorg Chem 2006; 18: 3641–3645.
153. Zhang B, Davis SA, Mendelson NH, Mann S. Chem Commun 2000; 9: 781–782
154. Dong A, Wang Y, Tang Y, Ren N, Zhang Y, Yue Y, Gao Z. Adv Mater 2002; 14: 926–929.
155. Li J, Xu Q, Wang J, Jiao J, Zhang Z. Ind Eng Chem Res 2008; 47: 7680–7685.
156. Rambo CR, Cao J, Sieber H. Mater Chem Phys 2004; 87: 345–362.

18

PROTEIN AGGREGATION AND PRECIPITATION, MEASUREMENT AND CONTROL

CATHERINE H. SCHEIN

Structural Biology, University of Texas Medical Branch, Galveston, Texas

18.1 INTRODUCTION

Proteins in solution exist in a precarious equilibrium. Some have even speculated that the unstable structures of important proteins may contribute to their cellular role (e.g. p53 (1)). However, most proteins form oligomers or aggregate at high density in solution, or when exposed to surfaces (2), a phenomenon that can to some extent be prevented by solvent additives (3,4) and mutations targeted to surface areas involved in aggregate formation (5). Solvent additives have proven invaluable in areas ranging from formulation of protein therapeutics to sample preparation for high resolution structure determination techniques. Despite many studies, choosing co-solvents is still a matter of trial and error, with little framework for predicting their effects or even optimizing concentration. This is partly due to the inherent problems in following the early stages of protein aggregation.

This chapter summarizes biophysical techniques useful for measuring solubility and following precipitation. Maintaining solubility depends on subtle interactions of side chains and backbone atoms with the solvent and any added co-solvents or salts. Determining surface properties of proteins can aid in determining basic solution conditions. Traditionally, proteins are presented as unique but ill-characterized globules, remarkable only with respect to size and net surface charge. Direct methods to determine the characteristics of the protein surface may allow a better accounting for protein–solvent interactions in the presence of co-solvents. Energy functions which account for the various forces on the protein have been derived; their application to different solvents is now being tested. Interpreting empirical data to derive atomic solvation parameters (ASP) or bulk solvent parameters may lead to a thermodynamic or kinetic accounting for the effect of co-solutes on protein stability and solubility.

Designing mutations that prevent aggregation can be greatly helped by a high resolution 3D structure of the protein–protein complex. There are now many good computer graphics programs that allow one to visualize 3D protein structures. Faster and more efficient computational methods have greatly expanded our abilities to model protein structures by homology to proteins of known structure or *ab initio* using protein threading and other algorithms from the primary sequence. Thus a section is also dedicated to methods for obtaining and using structural information to predict areas of the protein that may be involved in oligomerization and aggregation.

18.2 COMBINING METHODS TO FOLLOW AGGREGATION AND PRECIPITATION AND DETERMINE THE STRUCTURE OF COMPLEXES

The size of small complexes in solution can be accurately determined by equilibrium centrifugation (EC), native gel electrophoresis, or gel sizing chromatography (GSC) (e.g. using Sephadex beads or HPLC). However, these methods are not particularly useful for studying aggregated protein. *In vitro* aggregation can have many forms, extending from

---

Downstream Industrial Biotechnology: Recovery and Purification, First Edition. Edited by Michael C. Flickinger.

a drastic, sudden precipitation of large flocculent masses of protein, often visible to the naked eye, to the gradual formation of tiny invisible particles. The most valuable experiments look at the initial phases. The simplest direct quantitation of aggregation is thus to measure the absorption of visible light by the sample, if the particles are small enough to remain in solution, or to follow the precipitation by the drop in UV absorption of a protein solution where the particles are allowed to settle or are removed by centrifugation. If aggregation is related to unfolding, it may be possible to follow the increase in OD_{280} (as the tryptophan residues become exposed upon unfolding) and then its decrease as the protein aggregates.

While rapid onset suggests aggregation as a noncontrolled process, there are reports of sequence specificity based on the pattern of higher molecular weight aggregates seen on gels (6). An approximately similar pattern of bands was seen, whether a protein formed inclusion bodies alone or in the presence of another aggregating protein. Although this would indicate that each protein was only aggregating with itself, the authors could only study aggregates small enough to enter a nondenaturing gel, and a substantial portion of the protein in all the experiments remained at the top of the polyacrylamide gel (PAG). More recent data using computer simulations and light scattering suggest a cluster–cluster mechanism rather than a stepwise monomer addition polymerization (7). The aggregation of polypeptide chains during refolding occurred by multimeric polymerization, in which two multimers of any size could associate to form a larger aggregate and did not require a sequential addition of monomeric subunits.

Generally combinations of methods are applied to the study of aggregated protein.

Table 18.1 thus summarizes some common biophysical tools and methods that have been applied to study the interaction of proteins, Table 18.2 is a list of some of the protein complexes and aggregates that have been studied by combining biophysical methods

18.3 SPECTRAL METHODS FOR MEASURING SOLUBILITY AND PROTEIN ASSOCIATION

For details about the molecular interactions involved in complex formation, many methods measure the absorption or scattering of some form of electromagnetic radiation by the sample (UV/vis and IR/Raman spectroscopy, light scattering, laser light scattering, neutron diffraction and X-ray crystallography) (100). A beam of radiation sent through the sample can:

1. pass through the sample unchanged;
2. be absorbed by the sample and emitted at the same wavelength;
3. be absorbed by the sample and emitted at a different wavelength;
4. be scattered by the sample.

Various methods can be used to detect the effects of the sample on the beam (8). Figure 18.1 is a summary of methods aligned according to the electromagnetic radiation spectrum. At opposite ends of this spectrum lie the two highest resolution methods, X-ray crystallography and NMR spectroscopy, which yield complete structural information at the atomic level. However, combining results from lower resolution methods can be used to develop a working model of complexes and to obtain information about the aggregation process.

In the simplest and most commonly used method to measure protein concentration and aggregation, a coherent beam of visible or UV light is passed through the sample. The light emerging on the other side of a cuvette of known path–length is measured and related to the concentration of the sample. Some information about the nature of the sample can be obtained by observing how much light is absorbed when the wavelength is varied over the whole spectrum (often called UV/vis scan). In large proteins, the tryptophan side chains within the protein may be shielded, and as the protein unfolds this tryptophan absorption at 280 nm becomes more pronounced. Absorption in the visible range is usually due to the presence of a chromophore (e.g. Haem absorption) or particulates. Fluorescence spectroscopy measures the energy of the emitted light, and laser light scattering measures the angle of the reflected light to determine particulate size (101). Many other, more advanced methods, summarized in Table 18.1, have been used to measure the quaternary state of proteins in solution.

Aggregates and fibrils are notoriously difficult to characterize structurally. However, some methods, such as CD, Fourier Transform Infrared Spectroscopy (FTIR), and NMR can be used to follow the early stages of aggregation, and disaggregation in the presence of compounds. Electron microscopy (EM) and AFM can also be used to directly visualize and characterize the effects of the compounds on the fibril morphology. CD spectra indicate overall secondary structure, and can be used to detemine whether the inhibitors interfere with conformational changes that accompany aggregation. The data is particularly useful when analyzed with programs such as CDsstr2 (102) to quantitatively estimate percentage helix, strand and disordered structure (103,104). FTIR is somewhat more complicated to interpret, but can give a better estimation of the different secondary structures in the protein (105).

18.3.1 Protein Aggregation in Disease States

Many studies of the physicochemical properties of aggregated protein were stimulated by the recognition of the

TABLE 18.1. Physical Methods for Monitoring Changes in Protein Structure and Interprotein Interactions Involved in Aggregation with Respect to the Protein Requirements, Impedance by Common Buffers, and the Type of Information Obtainable about the Protein in Solution or Precipitate. For more Details, See Related References in Table 18.2; for General Information about the Methods, See Ref. 8, Cited Specific References or Other Books on Protein Characterization Techniques

Method (Abbreviation)	Brief Description	[Protein] Needed[a]	Sample preparation/buffer Interference[b]	Best used for:
Equilibrium (ultra)-Centrifugation (EC)	Protein complex is centrifuged at high speed in a density gradient and its equilibrium position used to determine the size of the molecule or complex	High	Gradient components may affect protein association	Determining the oligomeric or aggregation state of soluble complexes. Insoluble aggregates settle at the bottom of the tube or float on the surface of the gradient.
Gel sizing chromatography (GSC)	Soluble proteins are passed over a column of porous polymer. Molecules small enough to enter the pores are held back and elute according to size, larger ones pass through unimpeded with the void volume. Can be done with capillary HPLC columns on the microscale.	Moderate to high	sample should be concentrated for application, a step which may induce aggregation or oligomerization	Sizing small, soluble complexes of proteins; large aggregates either elute with the front or precipitated at the top of the column (latter may slowly redissolve, giving unreliable data).
Polyacrylamide gel electrophoresis (PAGE)	Proteins are separated according to size by electrophoresis through acrylamide gels of varying density. Proteins may be detected in a variety of ways: by staining, auto-fluorescence, or after transfer to a membrane, by reaction with a specific antibody or nucleic acid fragment.	Low	Concentrated sample required, but small amounts of protein may be visualized by silver or gold staining.	Sizing small, soluble complexes of proteins; large aggregates "stick in the slot". If the sample is not to be eluted for further analysis, apply less protein and use more sensitive dyes for accurate size determination.
UV/Vis-absorption (UV-abs)	A beam of coherent light is passed through the sample and the absorption by the protein constituents measured. Proteins absorb between 210–220 nm due to orbital transitions of the peptide backbone, which are overwhelmed by that of the side chains of D, N, E, Q, and R, and between 260–280 nm due to aromatic side chains of F, Y, and primarily W (9); chromophores in the protein may absorb throughout the visible range as well.	Low	Only for protein in solution; samples should be diluted to be in the linear range of the spectrophotometer.	Can be used to study the denaturing effect of co-solvents on proteins; to follow aggregation by the decrease in the fluorescence absorption as the protein aggregates.
Fluorescence transfer (10)	A fluorescence acceptor and a donor group are coupled to a protein. When excited by light, the quenching of donor fluorescence by the acceptor serves as a measure of the distance in the complex between them.	High	Only for protein in solution, some buffers can interfere with the fluorescence; requires coupling of donor and acceptor molecules usually to free cysteine groups in the proteins.	Distance between labeled side chains in a single molecule or oligomer; most useful with protein complexes with a well established 3D structure.

(*continued*)

TABLE 18.1. ***(Continued)***

Method (Abbreviation)	Brief Description	[Protein] Needed[a]	Sample preparation/buffer Interference[b]	Best used for:
Optical Rotatory Dispersion (ORD)/Circular dichroism (CD)	In ORD, the rotation of a beam of polarized light by the sample is measured by a coupled polarizer/photomultiplier on the exit side of the cuvette. The angle of rotation of the second polarizer to obtain the highest intensity of emitted light is plotted as a function of the wavelength. In CD, molar ellipticity is measured as the difference in absorbence by the sample of an incident beam rotated to the left or right (I_L and I_R) by a timed electric field change. A photomultiplier on the exit side measures absorbence as a function of time and a microprocessor interprets the data to determine the mean residue ellipticity (which is proportional to $I_L - I_R$) as a function of wavelength.	Moderate	Only for protein in solution, buffer interference is possible.	Measures secondary structure of proteins in various media; can be used to follow changes in the secondary structure as a function of denaturation or aggregation. Computer aided spectral interpretation can be used to obtain more complicated data about the protein configuration.
Electron microscopy (EM)	In vacuo, a beam of electrons is passed through a target with the (usually stained or coated) object affixed to it or imbedded in a matrix to obtain a 2D image; measures the electron density of the sample after coating to obtain a 2D image.	Low	Fixation/staining may be needed.	Determining the ultra-structure of large, electron dense, complexes; particularly useful to determine the shape of particles and their size distribution. Current resolution is in the range of 10 Å.
Cryo-Electron Microscopy (Cryo-EM)	Cryo-EM techniques use unstained samples embedded in vitreous ice. High speed calculations can be used to phase many images of particles (11,12).	Low	Consistent sample preparation, crucial to the Cryo-EM technique, can be quite difficult.	One of the few methods to visualize unstained samples of virus particles and large complexes; resoluton can be as low as 6–9 Å
Electron diffraction (ED)	By tilting the target during EM and measuring reflection of the electron beam at various angles from the incident, a 3D image can be reconstructed.	Low to high	2D crystals or membrane patches, viral particles.	Obtaining ultrastructural data about regularly structured monolayers such as the membranes of purple bacteria.
Scanning tunneling microscopy (STM)	The electron stream from the object is detected across a vacuum "barrier" by a sensitive probe. The name came from the Tunnelcurrent, which flows when there is a small electric current between the tip of the probe and the conducting surface to be scanned (which should be less than an nm away).	Low	Resolution is higher for conductive samples. Fixation/Staining may be needed for proteins and nucleic acids.	Determining the shape and size of fixed and coated large protein or protein–DNA complexes; resolution is theoretically at the atomic level; suited to the study of larger protein aggregates as it gives a three-dimensional image directly without the image reconstruction methods used for 3D transmission EM images.
Atomic force microscopy (AFM)	Technique (also called *scanning probe microscopy*) related to STM and using similar apparatus, more suited to biological samples as it is done under a solvent phase rather than in vacuo (13). Similar to "hydration STM", based on the electrical conductivity of molecularly thin water layers which adsorb to the sample surfaces in a humid atmosphere.	Low	Attachment to a surface necessary; can be done in a variety of solvents.	Direct viewing of the ultrastructure of large, not necessarily regular biomolecular complexes without staining. Methods are being developed for viewing protein–RNA complexes.

Fourier Transform- Infrared Absorption (FTIR)	Absorption of light in the infrared region (2.5–250 μm) by the sample is measured and the output beam resolved by Fourier transformation *IR difference spectra*, deconvoluted in the amide I band region (between 1620 and 1700 cm^{-1}) can show secondary structure in terms of %, α–helix, β–sheet, etc (14,15).	High	Can also be used for precipitates	Determining secondary structure elements and alterations thereof under different conditions.
Infrared attenuated total reflection spectroscopy (IR-ATR)	Infrared beam is reflected through an optically transparent germanium plate coated with sample; internal reflections along the plate increases sensitivity.	Low to moderate	Membrane patches attached to a special plate.	Membrane protein interactions with each other and external molecules.
Scanning fluctuation correlation spectroscopy (FCS)	Measures particle number concentrations by monitoring spontaneous equilibrium fluctuations in the local concentration of fluorescent species in a small (femtoliter) subvolume of a sample.	Low	Little problem with buffer interference.	Detect molecular aggregation for dilute, submicromolar samples by directly "counting particles."
Elastic (classical, static) laser light scattering: (SLLS)	For elastic light scattering, the scattering of a beam of polarized laser light as it passes through the sample is measured and the obtained "scatter factor" can be used to calculate both the weight-average molecular weight of the particles in solution as well as the radius of gyration.	Moderate	Scattering methods are usually not greatly affected by the buffer.	To determine the size of aggregates and precipitates. To obtain the "scattering weight" of a particle, one measures the scattering factor at several different scattering angles (q) and then plots the scattering factor as a function of q. Extrapolating back to $q = 0$ will yield the scattering weight.
Quasi-elastic (dynamic) laser light scattering: (DLLS)	As above, but changes in the frequency due to the translational ("Brownian") movement of the scattering particles are also measured. The broadness of the intensity distribution of the emitted light for frequencies around the primary monochromatic beam frequency is directly related to the diffusion coefficient of the particles, which can then be related to the hydrodynamic radius if a model for the particle shape is available.	Moderate	The beam strength should be controlled and exposure time limited so as not to damage the sensitive biological specimen.	Measure kinetics of particle coagulation by following the decrease in diffusion coefficient as the particle size increases (16); determine particle molecular weight as a function of time; to follow course of aggregation in any solvent; compare samples in different buffers if viscosity is accurately measured.
Raman spectroscopy, Raman resonance (RRS)	Light scattered at the incoming frequency, called Raleigh scattering, is measured in classical ("elastic") and quasi-elastic (dynamic) light scattering. A small fraction of the laser light is scattered at frequencies higher (anti-Stokes) and lower (Stokes) than the incident beam. These scattered beams are measured in Raman scattering, which has many variations.	High	Sample fluorescence may interfere with Raman resonance measurement of precipitates.	Follow quaternary interactions of soluble complexes and to compare the secondary structures of proteins under various conditions; to compare the effects on secondary structure of precipitating agents.

(*continued*)

TABLE 18.1. (*Continued*)

Method (Abbreviation)	Brief Description	[Protein] Needed[a]	Sample preparation/buffer Interference[b]	Best used for:
Neutron scattering (NS)	Scattering of a neutron beam by the sample is measured.	High	Beam time is limited	Determining the size of larger proteins and complexes.
Electron spin resonance (ESR)	Microwaves are absorbed by paramagnetic substances, and the change in the energy level of the electron spins measured.	Moderate	—	Structure and dynamics of lipid membranes; free radical detection in proteins; state of transition metals in proteins.
Nuclear magnetic resonance (NMR)	A concentrated solution of sample, possibly isotopically enriched for nuclei with a magnetic moment different from 0 (^{1}H, ^{13}C,^{15}N,^{31}P), is subjected to a high magnetic field. In the most common method, after the sample is pulsed, the decay in amplitude of the emitted radiowaves with time is measured and converted to a frequency spectrum by Fourier transformation. The magnitude and direction of the pulse can be altered to detect certain types of interactions between nuclei. The strength of the local magnetic field around an individual atom is decreased by the presence of covalently bound or spatially close neighboring groups, causing changes in the position of the resonance lines ("chemical shifts") in the output frequency spectrum. Peak intensities in NOESY spectra reflect the distances between individual spins. Distance geometry methods and restrained molecular dynamics calculations are used to calculate the three-dimensional structure of proteins or protein–protein complexes.	High	Most solvents interfere with measurements. Labeling of the sample with stable isotopes may be necessary. Oligomers may be artificially generated by high concentration required for measurement.	Determining the exact tertiary structure of small proteins, nucleic acids, and complexes. Combining results from different NMR experiments can be used to monitor changes in the resonance frequencies upon complex formation, and thus give specific information on binding sites. Various techniques (e.g. resonance transfer) can be used to measure interactions between two molecules even without establishing a structure. See (18) for a discussion of NMR techniques to study protein complexes.
Mass spectrometry (MS)	The magnetic field induced deflection of a vaporized, ionized molecule is used to determine its mass. Types are characterized by the method used to ionize and detect the sample. Advanced computational methods can be used to interpret structural data from the "cracking pattern" of larger molecules. For Hydrogen-deuterium exchange (HDXMS) methods, the protein complex or aggregate can be rapidly moved from one solution to the other to determine sites that are less solvent exposed.	Low (high for Deuterium exchange)	The protein can be in solution (for Ion spray) or bound in a matrix on a sample spot (Maldi).	Very accurate molecular weight of monomers; use in combination with chemical cleavage methods may aid in determining tertiary and quaternary structure. Electrospray ionization (ESI-MS) has been used to characterize complexes of proteins and protein/nucleic acids. (17). HDXMS can be used to estimate interaction sites.

Protease mapping (PM) (19)	Fragments left after cleaving with a site-specific protease are analyzed (by gel chromatography and sequencing or MS/MS sequencing) to determine surface accessible areas.	High	Need to have specific sites for proteases in the protein; buffers should not interfere with techniques used to identify protein fragments.	Tertiary structure of an isolated protein or protected areas in complexes.
Chemical based cleavage (CC)	Like PM, but an activated metal ion cleavage agent is incorporated at a free sulfhydryl group and the cleavage is triggered by chemical or physical stimuli.	High	Buffers used cannot stimulate cleavage or interfere with subsequent methods to determine the size and constitution of fragments; need free sulfhydryl groups in the protein.	Tertiary structure of an isolated protein or areas of contact in complexes.
Surface plasmon resonance biosensors (20)	A surface plasmon is an oscillation of free electrons that propagates across the surface of a conductor, typically a thin film of gold or silver. The sample to be measured is immobilized on one side of the gold film, for example in a hydrogel, and plasmon excitation is started with a beam of light. The intensity of plasmon generation (measured as the reflected light reaching the detector) is related to the refractive index of the sample. This can be altered by flowing a binding substance over the surface where it is immobilized.	Low to moderate	No labeling of the biomolecules is required, but the two media must have different refractive indices. To avoid rebinding artifacts, the bound protein should not be overloaded.	Real time analysis of (unlabeled) protein binding reactions (21) such as antigen-antibody interactions
Total internal reflection fluorescence (TIRF)	Fluorescently labeled proteins in aqueous solution interact with a solid phase (e.g. fused silica) coated with bilayers of synthetic phospholipids, unlabeled proteins or even whole cells.	Moderate	Two media must have different refractive indices.	Determining interactions between membrane bound and free proteins.
Small angle X-ray scattering (SAXS)	The scattering of X-rays at small angles can be used to obtain greater resolution of X-ray crystal structures and to measure the size of complexes.	Moderate	Crystals required and special detection/computation systems	Resolving the structure of large complexes to higher resolution than possible with diffraction data alone.
X-ray diffraction (XRD)	A protein crystal is bombarded with a high intensity X-ray beam of known wavelength and polarity. The diffraction pattern of the emitted energy, measured as spots on X-ray film or using an area detector, can be used to determine the position of individual atoms in the 3D structure of the protein or complex.	High	Crystals required and extensive experimentation must be done to solve the phase problem.	The only method now in use that can give the 3D structure of large, regular protein complexes at the atomic level. Adaptations are now being developed to use it to determine dynamics as well.

[a]Protein concentration required for measurement. Low is in the nM–μM range, high means 0.1 mM or higher. Note that highly concentrated protein solutions are usually required to obtain crystals and that a lot of protein is wasted during the process of crystallization. Thus one needs to start with about 100 mg of purified protein for X-ray analysis, although only a few μgs may find their way to the beam.

[b]Buffers used for sample preparation for all methods should be sterile and ultrafiltered (0.45μ or smaller pore diameter filter) to remove particulate matter. Buffer requirements for various methods are discussed in the references given.

TABLE 18.2. Model Systems for the Study of Protein Aggregation that are Amenable to Characterization by Several Methods. See Table 18.1 for Abbreviations Details and References on the Methods

Model System	Studied with	Examples and Key Results
A. Multiprotein complexes:		
Ribosomes	XRCrys, SANS, EM, neutron diffraction	Early crystals of Ribosomes from thermophilic organisms diffracted to > 20 Å were used for comparing the exterior structure with EM photos. A 37 Å resolution 3D structure of ribosomes from rabbit reticulocytes was calculated from electron micrographs of uranyl-acetate stained single-particle specimens. (24), while cryo-EM and NS data can be used to model the protein/nucleic acid interactions (25). Crystals diffracting to <5 Å were used, combined with image reconstruction, to determine structural details (22). Neutron diffraction of ribosome crystals was used to localize the Ribonucleic acid (RNA within the core of the particles using difference maps prepared at different D_2O/H_2O (altering the relative concentrations of D_2O and H_2O in the solvent changes the relative diffraction due to RNA or protein in the crystals) (23). Freezing crystals stabilized them in the X-ray beam.
Nucleosomes	EM, XRCrys	The crystal structure of the nucleosome core particle was developed to high resolution using a variety of isolation and protein techniques and modern image analysis (26).
Photosynthetic reaction center and other membranes	EM, XRCrys, NMR	Crystals diffracting to ~3 Å resolution required the development of small amphiphilic molecules (27). Other techniques include NMR of ^{19}F-labeled bacteriorhodopsin fragments in a membrane-like milieu (28) and *IR difference spectra* to monitor structural changes in ^{15}N-labeled or deuterated bacteriorhodopsin (29).
	"electron diffraction"	EM of purple membranes from bacteria are a natural 2-D crystalline array of lipids and bacteriorhodopsin. As the array is only one unit cell thick, it is not suitable for X-ray analysis, but ED patterns with spots to beyond 4 Å resolution can be obtained. By tilting the sample, a series of patterns or images can be converted to a 3D model of the structure of bacteriorhodopsin in the membranes (30,31). EM images of the hexagonally packed regular protein monolayer from the cell wall of the bacterium *Deinococcus radiodurans* (32) could also be used to obtain fine structure.
Virus assembly	CD, EM, XRCrys,SAXS, STM	CD showed the membrane associated coat protein of bacteriophage M13 (5.2 kD) changed from a α-helical conformation when it was in a membrane environment (i.e. in solution with cholate or in the capsid) to a ß-sheet form when aggregated (aggregation followed with PAGE) (20). NMR studies of detergent-solubilized M13 coat protein, in combination with the results of sedimentation equilibrium (SE), Raman and CD studies, indicated two conformers that represent the nonequivalent monomers of an asymmetric dimer (33) STM was used to view bacteriophage T4 polyheads (34). *Small-angle X-ray scattering* was used to follow the rapid assembly of tobacco mosaic virus protein induced by a temperature jump. The minimum counting time was 7.5 seconds (35). Assembly of the icosahedral shells of the dsDNA bacteriophages, herpesviruses, acid adenoviruses requires proteins not found in the mature virion, termed *scaffolding proteins*. The bacteriophage P22 precursor procapsid contains approximately 300 scaffolding molecules within a shell composed of 420 coat protein subunits (36). Cryo-EM has been used to characterize the scaffolding/assembly of herpes simplex virus particles (37).
Antibody-antigen interactions	XRCrys, NMR, SLLS, DLLS, STM	Crystal structures of antibodies or bacterially produced antibody fragments with peptide antigen (38) or proteins can diffract to <2.5 Å (39) and can be used to identify the atomic features of the binding site. NMR studies of binding of an antibody fragment (Fab) to a peptide (40). As the on/off rate of the peptide is short compared to the spin-lattice relaxation time of the Fab and peptide protons, a NOESY spectrum in the presence of a vast excess of peptide has extra cross peaks ("transfer NOEs") which are due to magnetic exchange between the bound and free peptide fractions and are not present when the peptide and Fab are present in a 1:1 ratio (41).

TABLE 18.2. ***(Continued)***

Model System	Studied with	Examples and Key Results
		Dynamic light scattering measurements of antibody–antibody complex aggregate formation indicated "fractal" dimensions rather than a defined geometric shape (42). STM ultrastructural images of unstained immunoglobulin G, air dried on the surface of highly oriented pyrolytic graphite (HOPG) required only 1 μl of a 0.5 mg/ml solution of protein in 50 m*M* Tris buffer (43).
Casein micelles	SANS, light scattering, EM	Neutron scattering measurements of suspensions of casein micelles were combined with EM data to characterize the structure as packed micelle monomer subunits (44).
Inclusion body formation	EM, SLLS, FTIR	The traditional method to determine the presence of aggregation after differential centrifugation is gel electrophoresis (also coupled with immunoblotting or isoelectric focusing (45)). Inclusion bodies contain protein-polymers (46,47) resistant to SDS and ß-mercaptoethanol solubilization. IBs (at least the polymeric forms that will enter a PAG) appear to consist of only one protein, not nonspecific co-aggregates with other proteins (6). More recent data using computer simulations and light scattering suggest a cluster-cluster mechanism rather than a stepwise monomer addition polymerization (7). Inclusion bodies (48) can be seen as electron dense particles in EM-thin sections of whole *E. coli* cells; the morphology of isolated IBs was studied by scanning and freeze etch-EM (49); content of properly folded protein decreases at higher growth temperature for *E. coli* (50,51).
ß-amyloid fibril formation	CD, static and dynamic light scattering, FTIR, NMR, solid state NMR, AFM, EM	ß-amyloid peptide (Aß) is a 42 residue peptide that is a major component of senile plaques in Alzheimer's disease patients. Assembly of synthetic Aß into amyloid fibrils in phosphate-buffered saline was followed by static and dynamic light scattering. CD was used to show Aß's structure varied when dissolved in organic solvents commonly used in neurotoxicity assays. In pure Dimethyl sulfoxide (DMSO), Aß had no detectable β-sheet content; in 0.1% trifluoroacetate, the peptide contained one-third β-sheet; and in 35% acetonitrile/0.1% trifluoroacetate, Aß was two-thirds β-sheet, equivalent to the fibrillar peptide in physiological buffer (52). Metal binding and its role in Aβ has been characterized with NMR (53,54). Solid state NMR structures of fibrils, formed from specifically labeled monomers, have also been determined (55–57), hydrogen deuterium exchange can also be used for structure definition (58). Nanolayers of an amyloid forming peptide have also been followed with IR and AFM (59). Prion protein aggregates and oligomers have been visualized with transmission EM and the secondary structure related to the conditions for fibril formation with Attenuated Total Reflectance Fourier Transform Infrared Spectroscopy (ATR-FTIR) (60). Strains of yeast prions were distinguished with hydrogen exchange/ NMR (61).
Bacterial flagella assembly, microtubules	EM and cryo-EM	Images of specimens stained with phosphotungstic acid can be used for *in vitro* reconstruction of organelles such as *Salmonella* flagellar filaments (62); similar for images of microtubules (frozen-hydrated or glutaraldehyde/tannic acid fixed and stained) (63).
Lipoxygenase substrate binding	EMR, X-ray absorption spectroscopy, infrared CD	Spectroscopic studies, including electron magnetic resonance, X-ray absorption spectroscopy, infrared circular dichroism, and magnetic circular dichroism, have been applied to compare lipoxygenases from varied sources and with different substrate positional specificity. Soybean lipoxygenase-1 contains a novel three-turn π-helix near the iron center (64).
Nucleotide binding	Magnesium NMR	The conformation of MgATP bound at the active site of Salmonella typhimurium 5-phospho-α-D-ribose 1-diphosphate synthetase (PRibPP synthetase) (65) and conformations of the adenosine moiety of MgADP and MgATP bound to rabbit muscle creatine kinase (66) have been investigated by two-dimensional transferred-NOE spectroscopy (TRNOESY).
Fibrinolase/Zn binding	NMR and CD	The Zn binding enzyme fibrolase (pI is 6.7) shows a minimum in its solubility profile (which can be followed with CD or NMR spectroscopy as a structural change) below pH 5. Zn loss, induced by rising temperature or lowered pH or adding chelators (EDTA, DTT), leads to a rapid loss of α-helical structure, exposure of hydrophobic residues, and aggregation (67).
Calmodulin/Ca	XRCrys, SANS	Small angle X-ray and neutron scattering were used to follow complex formation between of calmodulin with two peptides (68).

(continued)

TABLE 18.2. ***(Continued)***

Model System	Studied with	Examples and Key Results
B. Specific protein interactions		
Profilin/Actin	XRCrys	Crystals of profilin: actin diffracting to 1.8 Å resolution required ATP hydrolysis to form and a 3.2 M ammonium sulfate/ATP bathing solution for stability (69).
RecA protein/DNA	EM, SANS, STM	RecA protein complexes with itself and with DNA were characterized under various conditions by combining EM and small angle neutron scattering (70) STM images to a resolution of a few nM have been obtained for coated specimens of RecA bound to DNA (71).
Other protein DNA complexes	EM, STM	A combination of electron microscopy, image analysis, three-dimensional reconstruction (72) and STM were used to show that the active form of the DnaB protein, the primary replicative helicase in *Escherichia coli*, is a hexamer. Depending on the presence of different nucleotide cofactors (ATP, ATP γ S, AMP–PNP or ADP), two different hexameric forms of the DnaB ring were found, one with a threefold symmetry and one with sixfold symmetry. EM and image analysis was used to establish the relative orientation of *E. coli* RuvA and RuvB bound to model Holliday structures (73,74).
Clotting factors with inhibitors	XRCrys	Crystals structures of the complex of thermitase (279 amino acids) with the inhibitor Eglin C (70 amino acids), resolved to <2 Å resolution (75) and human α-thrombin with its inhibitor hirudin to 2.3 Å resolution (76).
Cyclosporin/Cyclophilin	XRCrys, NMR	NMR of the binding of cyclophilin to cyclosporin indicates that cyclosporin A and FK506 (but not rapamycin) undergo drastic conformational changes when bound to their respective binding proteins (77).
α-Crystallins	EM, SLLS, DLLS	Dynamic light scattering data (determines particle molecular weight as a function of time) were combined with EM to determine shape of aggregates of bovine α-crystallin. Static light scattering data indicated that the sample preparation methods used for (negative staining) EM had induced some particle breakdown (78).
1) Making lac repressor tetramer into a dimer or monomer	SAXS, SANS, XRCRYS, Site directed mutagenesis	Tetrameric (4×360 aa) wild type lac repressor was modeled based on scatter data (size and shape of tetramer) initially until a crystal structure was available. Targeted mutagenesis generated novel proteins that remained monomeric, formed preferentially dimers or did not associate, while retaining DNA binding function. A point mutation (Y282D) was sufficient to disrupt tetramer formation and generate monomeric protein (79). IPTG (inducer) binding reduced the fluorescence emission intensity of the repressor in solution. Low concentrations of IPTG reduced the fluorescence quenching effect of KI on the emission intensity (80).
2) Mechanism of isopropyl 1 thio-ß-D-galactoside (IPTG) binding	Equilibrium sedimentation; Fluorescence emission quenching	
Making a dimer from monomeric staphylococcal nuclease	XRcrys Sedimentation equilibrium	A serendipitous mutant of staphylococcal nuclease (6 aa in a surface loop, Δ114–119) forms a stable dimer (Kd < 10^{-8} M). The carboxy-terminal helices of each monomer swap places in the dimer (81).
C. *In vitro* aggregation	Light scattering, laser light transmission fluctuation	For monoclonal antibody aggregation in the presence of antigen, the estimates of the radius of gyration in solution, as measured by both classical and dynamic light scattering, agreed very well with those estimated from electron micrographs of the same aggregates (82). The aggregation of low density lipoprotein (LDL) induced by acetylation, carbamylation, maleylation or oxidation, was evaluated by laser light transmission fluctuation. Aggregated LDL, but not unmodified monomer, stimulates the uptake of cholesteryl ester in arteriosclerotic cells (83).
Protein denaturation	CD, FTIR, IR-ATR, RS, SLLS, NMR	*Raman Spectroscopy* has been used extensively to study the quaternary interactions of hemoglobin and to compare the secondary structures of proteins under various conditions (84,85). For example, 10%–20% solutions of α-chymotrypsin and chymotrypsinogen were used for a Raman spectroscopic analysis of secondary changes induced by pH and pressure (86).

TABLE 18.2. ***(Continued)***

Model System	Studied with	Examples and Key Results
		A relationship between the salt type and concentration used for precipitation of α-chymotrypsin and the disordering of the secondary structure was determined from Raman spectroscopy (87). When the method was extended to another 11 proteins, the general result was that the precipitated proteins contained, on average, a higher percentage of ß-sheet structure as determined by the amide I band intensity and location; the effect was more pronounced when the chaotrophic salt KSCN was used compared to the "structure stabilizing" salt, Na_2SO_4 (88).
Salt precipitation of protein	RS, SANS, Fluorescence "Scatter peak", SLLS, DLLS	SANS in combination with ultracentrifugation has been used to determine the particle weight of malate dehydrogenase from a halophilic bacteria in different salt solutions. In 1–2 M NaCl solution, the protein is a stable dimer and binds much more water and NaCl than in low salt solution, where the "salt-loving" protein unfolds (89). Stopped-flow turbidimetry has been used to derive a kinetic model for the precipitation of α-chymotrypsin by different salts (90).
Crystallization pathways	Fluorescent light scattering, EM, SLLS, DLLS	Quasi-elastic light scattering studies of the effect of salts on lysozyme and concanavalin A showed that precipitating salts decrease the translational diffusion coefficient while those furthering crystallization do not change it and solutions remain monodisperse with increasing salt (91). Light scattering studies of solutions of canavalin at a fixed salt concentration were used to follow the formation of rhombahedral crystals (92).
		Stereo-electron microscopy of frozen specimens used to characterize crystallization methods for lysozyme and tomato bushy stunt virus particles (330 Å). Crystallization intermediates had a more compact structure than aggregates that lead to amorphous precipitates (93).
		A symmetric dimer of a polypeptide in an organic solvent suggested as a model for the effects of crystal packing and solution dynamics on structure determination by X-ray crystallography and solution NMR methods (94).
Complexes with phospholipids and surfactants	Solid state NMR	The complete structure determination of a polypeptide in a lipid bilayer derived from solid-state NMR observations (95).
	NMR	IR has also been used to study the binding of CO_2 to carbonic anhydrase and mutants thereof (96) and combinations of 31P-NMR and ATR-IR have been used to study pulmonary surfactants (97,98).
	ATR/IR	The interaction of phospholipid with (R)-3-hydroxybutyrate dehydrogenase, a phosphatidylcholine-requiring membrane enzyme, has been studied using ESR spectroscopy of spin-labeled lipids, both as ordered multibilayers and in lipid vesicle suspensions (liposomes) (99).
	ESR	

role that protein aggregates play in many diseases (106–108). Alzheimer's disease (AD) and related dementias are degenerative disorders characterized by the accumulation in the brain of plaques of aggregated amyloid β protein (Aβ), that play a central role in pathogenesis (107,109–112). AD is part of the group of Protein Misfolding Disorders in which the key event is misfolding, aggregation, and tissue deposition of an otherwise normal, natively-folded protein (113). These include typical diseases of aging, such as Parkinson's disease, as well as transmissible spongiform encephalopathies, Huntington's disease, amyotrophic lateral sclerosis and cerebral ataxias (113–116) Insoluble aggregates of the Prion protein play a role in scrapies and acquitted Creutzfeld–Jakobs syndromes (117). Amyloid deposits can also occur in many organs of the body, such as those in the pancreas of type 2 diabetics, caused by islet amyloid polypeptide (IAPP) (118,119). Proline conformation was implicated in β2-microglobulin amyloid formation (120) and a single mutation to proline inhibits fibril formation in hIAPP (121). We still do not understand how to create, *in vitro*, infectious fibrils from prion proteins (122); polyanions may be needed to stimulate the aggregation process (123). Ron Wetzel and coworkers determined, experimentally, the effect of mutating each residue in Aβ to Pro or Ala on fibril formation and peptide denaturation curves (124,125) and found that many proline mutations destabilized fibril formation. On the other hand, an Asp23Ala mutant of Aβ was completely insoluble. This mutation would disrupt a salt bridge between Asp 23 and Lys 28 seen in the solid state NMR structure of an amyloid fibril (55). Their results also showed that two morphologically distinct quaternary

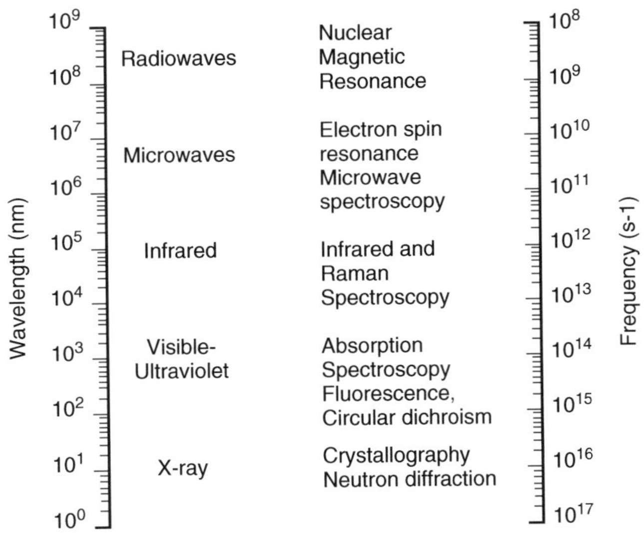

Figure 18.1. Methods used for analyzing the structure of biological molecules and complexes thereof as a function of position in the electromagnetic spectrum.

structures may exist in the fibrils, depending on conditions of fibril growth (55).

18.3.2 Peptide and Small Molecule Inhibitors of Amyloid Aggregation

Several strategies have been tested to minimize the amyloid pathway to AD (reviewed in (126,127)). A variety of low molecular weight compounds prevent or reverse Abeta misfolding and aggregation, including hexadecyl-N-methylpiperidinium bromide (128), congo red (129), small sulfonated anions (130), rifampicin (131), melatonin (132), nicotine (133), anthracycline 4′-iodo-4′-deoxydoxorubicin (134), estrogen (135), nitrophenol (136), tetracycline (137), glycosaminoglycans mimetics (138) and benzofuran-based compounds (139). Libraries of compounds, based on simple building blocks, have been explored and their ability to prevent toxic oligomers demonstrated (140,141). Many of these compounds are highly toxic and/or have obvious side effects, as they do not bind Aβ exclusively.

18.3.2.1 ***Chemical Aggregators.*** At this point, it is wise to mention this term for hydrophobic organic compounds as they reveal behavior in solution that may also pertain to basic mechanisms of protein aggregation. Some small molecule inhibitors of enzymes may work by forming aggregates that in turn bind proteins nonspecifically. The protein is deactivated simply by attachment to micelle-like aggregates of the compounds in solution (142,143). Originally, compounds that clearly had multiple targets and inhibited only at relatively high (30–100 μM) concentrations were termed *aggregators*. Now, any compound that has a greatly decreased inhibitory activity in the presence of 0.01% Triton-X 100 could be considered to work as an aggregator, even if the measured K_i is as little as 5 μM. Most (95%) of the compounds identified as β-lactamase inhibitors by a high throughput screen were, by this definition, aggregators (142). Troubling is that many commercially viable and useful drugs, such as clotrimazole, could be classified as aggregators, as well as many of the compounds that have been reported to disaggregate amyloid fibrils (144,145).

18.3.2.1.1 Peptide Inhibitors of Aggregation. This problem is probably not pertinent to one group of well characterized peptides, for the most part based on the primary sequence of the aggregating proteins that have been shown to disaggregate amyloid aggregates. The theory here is that the peptides bind to surfaces on the proteins that would otherwise interact with each other, and prevent/destabilize the intereactions that would lead to aggregates. Peptides related to the sequence LPFFD (114,115,146,147), based on the Aβ sequence, prevent aggregation of Abeta, as measured by the thioflavin T binding assay, a commonly used fluorometric method to quantify amyloid (Table 18.2). A peptidomimetic series revealed that essentially all parts of this peptide were required; the middle Pro residue could not be altered at all (so LHFFD was completely inactive). Molecular docking and modeling were used to determine compounds with similar activity to these peptides in a variety of assays. In a reverse approach, a peptide based on the insulin sequence, VEALYL, stimulated insulin aggregation into fibrils but reduced toxicity (148,286), suggesting that the soluble aggregates were the more problematic form of the molecule. Here, the effectiveness of the peptides could be related to their ability to increase the surface tension of the solvent (148–150).

18.3.3 Chemical Modifiers of Protein Folding that can be Used *in vivo*

Chemical treatments are now being developed for a variety of diseases involving mutant forms of proteins that render them less likely to fold properly, and hence more prone to aggregation (151). One exciting application of small molecule aggregation inhibitors outside the amyloid area has been the development of stabilizers of the tumor suppressor protein p53 as cancer therapies (1,152). Mutations of p53 have been detected in more than 50% of cancers. Many of the mutations characterized in aggressive cancers induce the protein to accumulate as aggregates in tumor cells. A small maleimide analogue, Mira1, restored the pro-apoptitic function of mutant p53 in tumor cells, and, *in vitro* at 25 μM concentration, stabilized the mutant p53 against denaturation at 37°C (153).

18.3.3.1 Formulating Pharmaceutical Proteins. Cytokines for human treatment can be physically modified for use in treating human diseases (154), for example by pegylation or even modification of their amino acid content (155). Many preparations of pharmaceutically active proteins contain human serum albumin; it may be possible to replace this expensive protein by modifying the formulation buffer and adding excipients, that will also be less likely to cause side effects (4).

18.3.4 Proteins for Structural Studies

Maintaining high concentrations of proteins in solution is particularly important for NMR studies (156,157), and to prepare crystals (158,159). Possibly relating crystal conditions to sequence (160), using special solvent conditions (optimized for T7 RNA polymerase and lysozyme crystals (161)) and a general first approach to crystallization (162) can help.

18.4 UNDERSTANDING PROTEIN-SOLVENT INTERACTIONS PROTEIN STABILITY IN A PRACTICAL SENSE

As Privolov has pointed out (163), if one assumes that all deviations from the X-ray crystal structure represent a new and separate state of a protein, the net thermodynamic stability of a protein at any temperature above a few degrees Kelvin is essentially zero. The thermodynamic framework works well for small molecules which can assume a limited number of stable conformations. However, proteins change conformation continuously; phenylalanine side chains are often free to rotate around the Cß-phenyl bond and as many as 40% of the amide protons of a protein exchange with the solvent in a few seconds (164). Solvent effects make these dynamic considerations even more complicated; the rate of hydrogen exchange of a protein changes when the isotopic nature of the solvent is changed (165).

To simplify the problem, thermodynamic stability has been defined as the energy needed to switch the protein structure from a folded, active "native" state (in reality a collection of substrates around a mean native structure) to a completely unfolded, "denatured" state. Formation of the denatured state in a dilute protein solution can be induced by slowly increasing temperature until the transition is indicated by extreme changes in the protein's activity or protease susceptibility or in the protein's characteristic UV-absorption, CD, or more recently, NMR spectrum. The temperature at which this change occurs is referred to as the *thermal denaturation temperature*, or T_d. Thermodynamic stability data can be collected in a few hours, but the measurements cannot be done at high protein density as aggregation usually precedes total denaturation. The T_d varies considerably from one protein to another and is a function of the pH and other properties of the solution in which the measurement is made. The free energy of unfolding (ΔG° unfolding), which, assuming other information is available, such as the ΔCp, can be calculated from the T_d, is approximately 12 ± 5 kcal/mol at 30°C for all proteins. This number is equated with the free energy of stabilization of proteins in aqueous solution according to the two state model (166).

As native, properly folded structures aggregate less than unfolded, denatured ones, there is an intimate relationship between solubility and thermodynamic stability. However, while extreme differences in the T_d can be used as a direct indication of the stability of the protein structure (peptides may loose their secondary structure at $T < 12$°C while thermophilic proteins retain it in a boiling water bath), the T_d of typical proteins is in too narrow a range to serve as an indicator of structure or purity. We know that proteins vary greatly in their stability in solution or the dry state. It has been shown that the $\Delta G_{\text{unfolding}}$ for RNase T1 can be almost doubled by adding 0.2 M Na_2HPO_4 or certain other salts, but the increase amounts to only 5 kcal/mol.10 Many additives that stabilize proteins against aggregation do not affect the T_d. Aggregation occurs at temperatures well below the T_d for proteins, indicating that a protein need not unfold completely to aggregate.

18.4.1 Empirical Definition of Protein Stability in Solution

Thus other methods must be used to determine protein stability in solution under nonextreme conditions. A two state model ignores the less abrupt (and therefore less easy to measure) changes in the protein structure that precede complete unfolding. Such changes, coupled with an increased rate of aggregate-producing encounters, probably lead to temperature dependent protein precipitation. Stability for practical purposes is the empirically determined resistance of a protein to oligomerization or to inactivation by other mechanisms. Stability can then be equated with the persistence of solubility or activity, and determined kinetically as the protein half-life (measured by protein concentration or activity after centrifugation) in a solution of specified pH, solute concentration and temperature (3).

The kinetic method of measuring stability requires several days and moderate amounts of protein. To shorten the time required for kinetic measurements of solution stability, an additional stress can be put on the system by agitating it or adding a protease. For example, resistance to mechanical shaking was found to be a useful indicator of solution half-life for 10 enzymes present in liver extracts (167). Use of more sensitive methods to detect protein aggregation should reduce the protein required for stability measurements. Osmolytes useful for formulating recombinant human growth hormone and interferon were detected by their ability to protect the native state of the protein against the stress of emulsification in organic solvent and re-extraction into buffer. Recovery of native protein was determined by native size exclusion HPLC followed by CD, both rapid techniques which require relatively little protein (168).

18.5 DETERMINING THE SURFACE CHARGE AND HYDROPHOBICITY OF A PROTEIN

Again, practicality dictates the use of chromatographic methods to determine the approximate surface charge and hydrophobicity of a protein in solution. Because aggregation of integral membrane proteins after transfer to an aqueous environment is a well described phenomenon (169), there is a tendency to regard precipitation in aqueous solvents as an indication that a protein is "hydrophobic". In reality, all proteins are to some extent hydrophobic, with tightly packed cores that exclude water (170,171) and even proteins with many polar residues at their surfaces, such as interferon-γ and RNase A, will aggregate depending on conditions. A protein is more than the sum of its parts, so analyzing the solubility of individual amino acid side chains, such as that of phenylalanine, which has negligible solubility in water, does not explain how a protein with many aromatic residues can ever stay in solution. On the other hand, the effect of certain osmolytes may be explained by their ability to stabilize side chains that have little favorable interaction with water, such as the aromatics, or by their forcing backbone atoms to avoid the solvent, driving the protein into a more compact fold (172,173).

As with stability, the surface charge of a protein is not easily defined. An isoelectric point (pI) measured by isoelectric focusing or chromatofocusing indicates the pH at which the protein has no net charge at very low salt concentrations (high ionic strength prevents proper electrophoresis and interferes with the establishment of the ampholyte-pH gradient). These methods cannot be used for proteins (e.g. T7RP) that precipitate in low salt buffers. Further, the effective pI of a protein is greatly affected by buffer conditions. Due to the binding of salt, one cannot assume that a protein in solution will be negatively charged at pH's above its pI in a solution containing salts or metal ions (e.g. acidic caseins bind Ca^{2+} and appear positively charged at pH 7[7]). Oligomerization can alter the apparent isoelectric point of a protein by more than a pH unit, as has been shown for insulin (174).

At pH 7.5 and 50 mM salt, most proteins will bind to DEAE-coupled (positively charged) resins if they are negatively charged and to phosphate and other negatively charged resins if they are positively charged. The degree of surface charge can be estimated from the salt concentration required for elution from DEAE or CM-conjugated resin columns; this can also give indirect evidence about the degree of surface exposure of altered residues. For example, when positively charged residues (Arg10 or Lys7) were changed to uncharged glutamine by site directed mutagenesis (SDM) from the N-terminus of RNase A, the mutants eluted at lower salt from CM-Sepharose at pH 8 then the wild type; the double mutant did not bind to the column at this pH (175). Similarly, C-terminal deletions of human interferon-γ eluted at progressively lower salt concentrations from CM-Sephadex as more positively charged residues (Lys (K) and Arg (R)) were deleted (Table 18.3).

Later in the purification, after the solubility of the protein has been better defined, chromatofocusing in an appropriately stabilizing buffer can be used to more accurately estimate the protein's pI. Once one has enough purified protein, the pI can be determined by measuring the amount of KOH necessary to increase the pH of a solution of protein in buffer from 2.5 to 11 compared to that needed to titrate the buffer alone. Gel methods for following the changes in surface charge during protein folding and aggregation have also been developed (45).

Similarly, binding to resins coupled with hydrophobic groups, for example, Phenyl- or Octyl-Sepharose® (Pharmacia) indicates the presence of hydrophobic residues at the protein surface. Proteins are applied in high salt (0.7–1 M ammonium sulfate), which increases hydrophobic interactions, and then eluted with a decreasing salt gradient. Most proteins elute between 0.5 and 0.1 M salt; very hydrophobic proteins will not elute into low salt buffer unless the polarity is decreased by adding ethylene glycol. If a protein does not bind to Phenylsepharose, it either has a very hydrophilic surface (e.g. RNase A) or is aggregated.

TABLE 18.3. Proteolytically Cleaved Human IFN-γ Elutes from CM-Sephadex Elute values were calculated at progressively lower salt concentrations, depending on the number of positively charged residues removed from the C-terminus, after overnight cleavage at room temperature with 1 unit of either the protease factor Xa (Biolabs) (for the δ 12 protein) or endopeptidase arg-C (σ) for the δ 14 protein. Samples were diluted 5-fold with 10 mM NaHPO4 buffer pH 5.7 and applied to small columns of CM-Sephadex C25 (Pharmacia) (176,177). The elutes were concentrated by centrifuging in Centricon-3 microconcentrators (Amicon) after automated Edman degradation, which indicated that the N-terminus of the proteins was unaltered; the number of amino acids deleted was determined by electrospray ionization mass spectroscopy

Human IFN-γ	C-Terminal	[NaCl] Eluted
Full length	–PAAKTGKRKRS QMLFRGRRASQ	300–500 mM
IFN-$\gamma\Delta$12	–PAAKTGKRKR	200–300 mM
IFN-$\gamma\Delta$14	–PAAKTGKR	80–100 mM

One can determine the hydrophobicity of a purified protein or follow changes in exposure of hydrophobic groups during folding by measuring interaction with a hydrophobic dye or radioactive tracer (e.g.1-anilino-8-napthalenesulfonate (178), or 125I-TID, 3-(trifluoromethyl)-3-(m-125iodophenyl)diazirine (179)).

18.6 EMPIRICAL MODELS FOR SALTING IN AND PRECIPITATION WITH VARIOUS AGENTS

Generally, charged proteins can be "salted in" or "out" by counter ions. This is due to ions binding to the protein and neutralizing the net charge at its surface (see Table 18.5 for dielectric constants of common ions). The solubility of lysozyme, a positively charged protein, was shown to vary more with the nature of the anion added than the cation; the anion dependence followed the Hofmeister series (180). On the other hand, the solubility of caseins with pI between pH 3 and 5 varies with the nature of the cation: sodium, potassium, and ammonium caseinates are soluble; calcium caseinates form micelles (181). Casein solubility varies inversely with the atomic radius of the divalent cations added; Zn^{2+} and Cu^{2+} are more effective precipitants than Co^{2+} (182). There are now many different theoretical accounts for the effects of salts and osmolytes on proteins, to which the reader can be recommended (183–186).

Early models of the effects of ions and osmolytes on protein solubility postulated that they caused specific changes in water structure. This was not an inreasonable view, considering that even small amounts of NaCl, for example, cause a drastic drop in the dielectric constant. Molecules could, according to this theory, be classified as either "kosmotrophs", which would enhance long range solvent ordering, or "chaotrophs", that should interfere with chains of hydrogen bonds within water. However, direct experimental measurements with X-ray absorption spectra have indicated that salts have few long range effects on water structure, as their effects do not extend beyond the first interaction shell (186). The experimentally supported view is that water structure is quite flexible, and can easily incorporate small ions and even quite hydrophobic compounds, such as benzene. Proteins and large hydrophobic molecules cannot be as easily accommodated.

A theory that could account for the effects of salts and osmolytes must thus allow for direct interactions with salts and co-solvents with both the protein surface and water molecules in the immediate vicinity (186,187). These molecules alter the interprotein interactions that would otherwise lead to aggregation, either by replacing them with protein-cosolvent ones, or by binding to the protein surface directly.

Interaction of salts as a function of pH with very soluble proteins can be accounted for mathematically, on a case by case basis. Detailed studies have been reported for a few proteins, including porcine pancreatic amylase (188), lysozyme (189) and ovalbumin (190). The latter study found that at 30°C, the solubility of ovalbumin (C_o) in ammonium sulfate solutions of varying concentration C_A could be described by the simple quadratic equation:

$$\log_{10}(C_o) = 5.06 - 0.006t - 0.205\ C_A + 0.5(pH - 4.58) + 1.1(pH - 4.58)^2$$

where concentrations of both ovalbumin and ammonium sulfate are expressed in g/100g H_2O and 4.58 is the isoelectric point of ovalbumin. The data agreed quite well with earlier measurements made at 18°C if the temperature correction was added.

Precipitation of proteins by larger metal ions can be quite specific. For example, Bovine serum albumin precipitated quickly with copper, but not with calcium, cobalt, nickel or aluminum (191). While small ions (Na, K) have mostly hydrostatic interactions, the larger divalent ions or transition elements have specific patterns of preferred binding sites in proteins. Mg and Ca, for example, prefer carboxyls (Asp and Glu residues), while "soft metals" such as Zn, Cu (and Ni), bind imidazole (His) and thiolates. Data for tight ionic complexes of various metals with protein groups are summarized in Table 18.5 bottom. This data should be considered when designing purification protocols for a specific sequence.

One major problem with using salts to precipitate proteins is the possibility of formation of tight complexes with the protein. Heavy metals can be quite a problem in the later stages of purifying a pharmaceutical grade protein. Nickel columns are routinely used to purify proteins that have been expressed with a poly-histidine tail. However, removal of the Ni ions after this process can be quite difficult, requiring extensive dialysis against imidazole buffer. Some proteins may have a natural affinity for Ni (192) (usually those with many His residues) that may prevent efficient elution from the column. One should also be conscious of the trace heavy metals that can be found in such common, nonmetallic salt protein precipitants such as ammonium sulfate. Precipitations with the latter should be avoided in industrial processes, both for this reason and the cost involved in using very high concentrations of any salt.

18.6.0.0.1 Modeling Precipitation with PEG and other High MW Polymers. As precipitation models are typically based on data for large, quite soluble proteins with the polymer used for precipitation of the same size and at about the same density as the protein being precipitated, it is difficult to apply the results directly to most lab scale processes. Here, typically, the polymer greatly outweighs the protein in solution and the intrinsic solubility of the protein is much less than the "standard" proteins.

Precipitation as a means of purifying or storing a protein in a dry state must be done in such a way that after removal of the precipitant, a soluble and active product is obtained. Organic solvents that perturb the aqueous shell around the protein (ethanol, acetone, and even trichloroacetic acid (193)) are used industrially for large-scale protein precipitation, as they can maintain a low temperature during the procedure, but the most useful precipitants for lab use are those that do not denature the protein, such as polyethylene glycol (PEG). Models for precipitation with such compounds are usually based on "excluded volume", meaning that the protein is forced into a progressively smaller area as the polymer occupies more of the solvent until the critical solubility limit of the protein is reached.

Models of precipitation usually involve common proteins (bovine serum albumin, human serum albumin, and ovalbumin) that have high solubility, and assume:

1. the protein is globular and has homogenous surface properties;
2. the precipitating polymer is linear, nonionic and homogenous (e.g. PEG of various average polymer length).

With regard to the first assumption, the model is only useful for charged proteins if the ionic strength of the solvent is adjusted to minimize electrostatic protein–protein interactions (194). To predict the phase diagram, the Gibbs free energies (evaluated using a second-order perturbation theory around a hard sphere as the reference) of protein in the solid and liquid phases were equated. Protein solubility was lowest (concentration of protein in the solid phase increased) for higher polymer concentrations, larger protein sizes, higher molecular weight of polymer, weaker protein-polymer interactions, poorer solvent, higher ionic strengths, and pH values close to the isoelectric point (195).

18.7 MODELS FOR DETERMINING THE EFFECT OF CO-SOLVENTS ON PROTEIN FOLDING

Accounting for the interaction between parts of the protein and the surrounding solvent is one of the most rapidly developing areas in protein folding (196). There are several theories to explain the effects of denaturants and osmolytes on protein solubility. The eventual outcome of this work is to understand how a protein folds, with an eye to predicting tertiary from primary structure, and secondly the effects of osmolytes and denaturants on protein stability in solution, hoping thereby to be able to predict *a priori* what co-solutes are best suited to a protein of a given composition.

Osmolytes have strikingly different effects on different proteins. One cannot simply add any osmolyte to a protein solution and assume it will increase solubility; indeed, addition of carbohydrates or other osmolyte to a concentrated protein solution will often induce precipitation. The osmolytic amino acids betaine and proline, which protect *E. coli* against urea and salt stress (197), can destabilize Deoxyribonucleic acid DNA helices (198). Betaine, depending on the chemically generated side chains, could be altered to be protective either against high salt or urea (197). Trehalose, which appears to be particularly effective in stabilizing proteins at high temperatures, also can interact with DNA. Thus its effects in enhancing PCR reactions can be accounted for by several different mechanisms (199). The same is true for carbon nanotubes, which also may enhance PCR, but their effects may not always be benificial (200).

Curiously, although osmolytes have the general property of increasing the net solubility of proteins in solution, that is, one can achieve a higher concentration of the protein before aggregation occurs, none of the individual amino acids are more soluble in the osmolytic solutions for which there is data (0.5 M sucrose (201); 1 M trimethylamine-N-oxide (TMAO) (202)). Specific protein groups, such as amides, are considereably less soluble in TMAO (172). There is also no really significant increase, regardless of side chain, in amino acid solubility in the presence of 2 M urea, although a cyclic glycine dipeptide designed to mimic the backbone unit is about 20% more soluble (202).

Solvents such as trifluoroethanol may induce helical secondary structures that may be nonnative in proteins (203–211). However, the osmolyte TMAO seems to specifically stabilize the native and active fold of proteins, without affecting the overall secondary structure. Simulations have indicated that TMAO can stabilize water structure (172), predictions that have to some extent been supported by NMR studies. These indicate that TMAO slows the rate of proton exchange with the solvent (212,213), which may aid in its ability to stabilize some proteins in solution. In addition, TMAO directly stabilizes the overall fold (214), an effect that is probably easiest to see with smaller proteins or peptides (215). As TMAO can aid in achieving the active configuration during refolding (216,217), it should certainly be included in solvent optimization schemes.

18.7.0.0.2 Thermodynamic Approaches to Predicting Stability in the Presence of Co-Solvents. Several models seek to account for the relative ability of molecules to stabilize the native state or the denatured state by using a thermodynamic framework, based on the absolute stability of the

3D structure in a given solvent. As one cannot explain the specificity of osmolytes by difference in solubilities of the amino acid side chains, the "preferential hydration" theory suggests that osmolytes are repelled from the surface of the protein, while denaturants bind. The basic theory resembles the excluded volume theory of protein precipitation, and deals with the protein as a hydrated ion. The binding of water molecules to the surface of the protein serves to shield charges and protect the protein from interaction with other proteins. According to the model, the concentration of the co-solvent in the immediate vicinity of the protein is either less than (for osmolytes) or greater than (for denaturants) that in the bulk solution. Thus, the solvent around a protein in osmolytic solution will be relatively richer in water molecules than the bulk solution. The theory assumes first that such a concentration gradient surrounding the protein exists and that this can serve as a driving force for stabilization. This leads to the question: what about the co-solvent molecules relative to the protein surface? After all, these are at a much lower concentration overall than the water molecules.

The concentration of co-solvent near the protein surface is not easily measured. Experimental results to demonstrate that this is different from that in the bulk solution are based on measurements of the partial molar volume of the protein in various solutions or more recently, infrared spectroscopy to measure O-H stretching rates in the vicinity of the protein (218). Barrier effects caused by the presence of high concentrations of co-solvents in the binding of CO to myoglobin have been analyzed by frequency resolved calorimetry (219). To develop more general ways to account for the effects of solvent at the protein surface, computer models have been developed. Explicit modeling of solvent, by inserting the protein into a shell of solvent of varying thickness, is computationally inefficient. Drastic perturbations have to be imposed in the unfolding simulation such as high temperature or pressure or by applying special algorithmic means to unfold a protein (220–222), which are likely to induce pathways not found in Nature. Alternatively, one can implicitly model solvent molecules, in an approach based on the continuum approximation and the calculation of the solvent accessible surface area (SASA).

This approach was experimentally validated by a recent statistical data analysis which showed that for 45 proteins, the measured free energy and heat capacity of unfolding correlates directly with the change in SASA during the process (223). The SASA can be explicitly calculated from any Protein database PDB format file of a 3D protein structure at a convenient webpage, GETAREA (224). By changing the protein-solvent interaction via the SASA, and extrapolating to zero co-solvent concentration, a folding pathway that should have more physiological meaning can be induced and monitored.

In this case, the protein–solvent interaction term E_{hyd} is computed in the continuum approximation by:

$$E_{hyd} = \sum_{i=1,\text{ atoms}} \sigma_i A_i$$

where A_i is the SASA of atom i and σ_i is a "solvation-parameter" (225–227) depending on the atom type and solvent composition. A variety of data is available for the determination of σ_i values for individual protein atoms. Several such "atomic solvation parameter (ASP; σ_i) sets have been derived (225,226,228–232) based on a measured "transfer free energy"(ΔG_{tr}) for moving organic molecules (usually amino acids or their analogues) from one medium to another. Determination of the ΔG_{tr} can be calculated from the relative solubility of the groups in the two media. In the latter case, solubilities in the two solutions (C_w and C_s for maximal solubility in water or alternative solvent, respectively) yield ΔG_{tr},by assuming that the ratio of the activity coefficients is ~1, according to the equation (233):

$$\Delta G_{tr} = RT \ln \frac{C_w}{C_s}$$

These ΔG_{tr} values can be directly related to the accessible surface area of each amino acid (A_i) by the equation:

$$\Delta G_{tr} = \Delta\sigma_i A_i$$

and it is assumed for parameter extraction that the ΔG_{tr} of a residue R is the sum of the free energies of each atom:

$$\Delta G_{tr}^{R} = \sum_{\text{atomsj}} \Delta\sigma_j A_j$$

Eisenberg and McLachlan (225), in one of the earliest systematic studies, used data from Fauchère and Pliska on the energy of transfer of amino acid side chains from n-octanol to water (234) to derive a set of 5 ASP values: $(\Delta\sigma(C), \Delta\sigma(N/O), \Delta\sigma(O^-), \Delta\sigma(N^+), \Delta\sigma(S))$. Their calculated ΔG_{tr} for the side chains using this set was linearly related to the observed values. Others have shown that the parameter sets, with slight modification, can drive disturbed structures back to their native state (224,235,236,236,237). While one can also expand the parameter list by distinguishing aliphatic, methylene or carbonyl-C, and backbone or guanidyl nitrogen atoms, too many parameters may be detrimental in determining accurately the effect of the interaction. Other data well suited to the derivation of ASPs are those for the transfer potential of side chain analogues to a vapor phase (238), or based on measurement of the hydration enthalpy of N-acetyl amino acid amides (239) or the isoentropic pressure coefficients of tripeptides (240). Data are also available to derive ASP sets for solvents containing denaturants (241) or osmolytes (201).

Other data are available for the solubility of protein components in ethanol, octane, and other solvents that could be used to derive ASPs for calculation purposes. Another source of information is that for purifying proteins with aqueous ethanol extraction or following the unfolding of proteins or peptides in various solvent conditions (242,243). It was also demonstrated that water miscible organic solvents solubilize naphthalene in concentration that can be related to those at which they denature well known proteins (trypsin, cytochrome c, chymotrypsinogen, chymotrypsin, laccase, and myoglobin) (244). This indicates that ASP sets derived for one solvent may be related to other solvents according to simple correlation constants.

In support of using such ASPs, the hydrations seen around atom types in protein crystals, (Table 18.4) show specific chemical interactions of the individual atoms with water which are unaffected by the crystal lattice. Neutral and negatively charged oxygen atoms are the most hydrophilic, followed by positively charged nitrogen atoms, neutral nitrogens, carbons, and sulfurs. ($O^{-} \sim O > N+ > N \gg C \sim S$, where $\sim$ means approximately the same).

Interactions between protein and bound water are fundamentally similar to those between protein and bulk fluid water, measured in partition experiments (245). The linear correlation coefficient of the data from Table 18.4 with side-chain free energy of transfer between cyclohexane and water (238,246) was $r = -0.95$ and with residue retention on reversed phase HPLC using a pH 7.5 average scaler$=$ -0.93 and $r = -0.86$ for the data of Fauchère (234).

18.7.0.0.2.1 A KINETIC APPROACH TO CO-SOLVENT EFFECTS. Aggregation, the major barrier to obtaining clear thermodynamic data, accelerates with increasing protein concentration. Co-solvents may also affect aggregation by altering bulk solution properties: viscosity, macroscopic dielectric constant, and the chemical potential, all of which affect also the kinetics of unfolding, refolding, and transitions between sub-states. This strongly suggests that protein precipitation is directly related to the rate of protein–protein interaction. Kinetically, the role of osmolytes in maintaining proteins in equilibrium solution can be described by equations initially based on Newton's law of viscosity and Archimedes' principle. A protein in solution is in buoyant equilibrium, with water structure/viscosity based forces acting to counteract the forces accelerating its precipitation, which are a function of the specific density of the protein itself and the gravitational constant. These interactions can be quite complex; the reader is referred to books on transport phenomena (247). One way to estimate the overall energy needed to maintain this equilibrium is to assume that to remain in aqueous solution the protein must inhabit a cavity in the solvent structure somewhat larger than its surface area. The energy to create such a cavity has been estimated, based on the surface tension of water being 72 dyn/cm, to be 104 cal/$Å^2$/mol (248). For a soluble protein dissolving in pure water, the sources of this energy are the binding of water molecules to the charged groups on the protein molecule, and desolvation energy, where aliphatic and aromatic residues turn away from the solvent toward the interior of the protein. This may explain why more highly charged proteins, such as RNase A, are quite stable in solution in the presence of moderate amounts of salt, in contrast to proteins with large hydrophobic surface areas. A compact globular shape has the lowest surface area. Unfolding the protein increases its surface area (and area for interaction with other molecules) and lowers its specific density.

The protein is not static in solution, and rotates or "tumbles" in solution. The tumbling time (τ_{tumble}), which is important for example in determining line broadening in NMR spectra of proteins, is calculated as the reciprocal of the rate constant for Brownian motion, or the rotational correlation time, ϕ:

$$\phi \approx 1/\tau_{\text{tumble}} = V\eta/kT$$

where V is the molecular volume, η is the viscosity, k is the Boltmann's constant, and T is the absolute T. Increasing the viscosity or the molecular volume (i.e. the protein size) decreases the tumbling time, which may affect aggregation. However, other effects are also introduced as the viscosity increases.

The second term in kinetic effects in protein unfolding states simply that the viscosity of the solution will affect the time required for folding, an effect summarized (249) as:

$$\bar{t}(n) = t_0 e^{\left(\Delta E(n)/k_bT\right)^2}$$

TABLE 18.4. Hydrations Seen in Crystal Structures

Protein Atom Group	Atom Types Included in Group	Water Ions/ Occurrence
Neutral O	Q,N,D,T,Y, and all carbonyl O atoms	0.53
Negative O	D and E side chain O	0.51
Positive N	H,K,R side chain N except Arg Nε, which is fairly neutral	0.44
Neutral N	W, N, and Q side chain N and Arg Nε, plus all main chain N	0.35
Carbon/sulfur	All carbon and sulfur atoms	0.08

where $t(n)$ is the lifetime of a typical microstate and t_0 is the timescale for a typical motion of a large segment of the chain, which is a factor of the local barriers and the solvent viscosity. For the case of viscosity (friction) effects on geminate binding reactions, please see (250) for a restatement of Kramer's law in order to apply it to proteins.

In addition, one must account for changes in the dielectric constant of the medium and for the polarizability (α) of the protein and its co-solvents on the interaction energy between protein particles as a function of distance (R). Classic thermodynamics says that this energy, for an oscillating system with no limits on the sphere of motion, should be:

$$\Delta E(R)_{\text{dis}} = -3h\nu/4(\alpha^2/R^6)$$

where the oscillator energy $h\nu$ can be equated with the ionization potential (IP). Thus the energy of interaction decreases as the distance between particles decreases to the sixth power and the polarizability increases. As the distance between particles decreases directly with the concentration of the protein, small increases in concentration will greatly accelerate the aggregation rate. On the other hand, a suitable co-solvent could substitute for interactions with another protein molecule and lower the rate of aggregation inducing encounters.

With respect to the IP, the constitutent atoms of proteins, with the exception of hydrogen, have very similar electronegativity values (Table 18.5). Hydrogen bonds of varying strengths maintain protein tertiary and quaternary structure. Note (Table 18.4) that hydration is seen in crystal structures at precisely the atoms in proteins where the ionic character of the bonds is most pronounced. Obviously, substituting any of the alkali metals for a hydrogen atom will increase the local dipole moment within the protein. The polarizing effect of substituting a larger ion is intermediate. These considerations may be used as the basis of analysis of salt interactions with proteins, if the surface charge is known. While such models will still require experimentation, they may be used as an optimization framework that may reduce the data needed.

18.8 COMPUTER DESIGN OF MORE SOLUBLE PROTEINS

Mutants of proteins can now be generated to alter oligomerization and solubility (Table 18.2 gives some examples). Residues that could be involved in aggregation can, for example, be identified with the help of a detailed 3D structure. These could be targeted with site directed mutagenesis to test which residues could be altered without affecting function or the global fold of the protein (254). Previously, the limitation on this methodology was having a reliable 3D structure to use for predicting surface exposed and thus potentially interactive surfaces. Two major forces have now intervened to provide the needed 3D data. First, computational advances now permit one to model a given protein, based on sequence homology to other proteins of known structure. This sort of modeling can be implemented to solve problems even during the initial isolation of a protein, if the sequence has high (>35% identity, depending on length) to a protein of known structure (255–259). The second major development is the structural genomics initiative, which seeks to determine the structure of novel proteins likely to have a fold different from those in the PDB repository (260–262).

As discussed elsewhere (5), there are a few empirical rules for identifying areas of proteins that cause aggregation. The few good examples of controlling aggregation of proteins through alteration of primary structure started from high resolution X-ray crystallographic or NMR structures of protein complexes or oligomers. It is now possible to view the 3D structure of proteins whose coordinates have been deposited in the PDB. Even when the fold of the protein is known, it is still not easy to identify aggregation sites; however, one can identify surface exposed amino acid stretches that might be involved in aggregation or inter-molecular contacts rapidly and test the effects of site directed mutagenesis at these positions on solubility. For example, the InterProSurf program allows rapid identification of surfaces that, according to statistical analysis, are most likely to interact with other proteins in known complexes (263,264). New methods for mimicking protein folding by computer indicate we may soon be able to identify "problem areas" in the primary sequence. Two basic assumptions may greatly simplify calculations of the protein aggregation problem:

1. Protein folding proceeds first by folding into a series of domains, which then interact with each other specifically to form the final stable global structure.
2. During this process certain domains may form inter- rather than intra-molecular contacts. Once two molecules are so intertwined, their residual domains may be forced to associate with other molecules, leading rapidly to a mass of partially folded, interlocked protein monomers. This idea, based on the domain swapping hypothesis of Eisenberg and coworkers is based on structural analysis of oligomeric proteins, including Diphtheria toxin, Staphylococcal nuclease, α-spectrin, and anti-sialidase, with swapped domains ranging from 14 to 252 amino acids (265).

The problem would then become identifying probable domains within the protein that are sufficiently separated during folding so as to allow another molecule to intercalate during the folding process. One can view surface groups on proteins, searching for areas of exposed hydrophobic surface, for example, that may be involved

TABLE 18.5. Electronegativity Values for Protein Atoms and Common Ions Associated with Proteins

Electronegativity of protein atoms		Ionic character of bond and (percent ionic)	
H	2.1	C–H, S–H	0.4 (4%)
C	2.5	C–0	1.0 (22%)
N	3.0	N–C	0.5 (7%)
O	3.5	0H	1.4 (39%)
S	2.5	NH	0.9 (251)
P	2.1	P–O	1.4 (39%)
Cl	3.18		
F	3.98		
Electronegativity of Ions:		Most common protein ligands[a] (252)	Average bond length in Å (253)
Li	1.0	No data	No data
Na	0.9	Water; carbonyl; carboxylate	2.42; 2.46; 2.3
K	0.8	Carbonyl; water	2.8; 2.82
Ca	1.0	Water; Carboxylates; Main chain carbonyl;	2.4; 2.33; 2.36
Mg	1.31	Water; Carboxylates; Main chain carbonyl	2.09;2.08; 2.26
Mn	1.55	Carboxylates; Water; imidazole	2.12; 2.22;2.16
Cu	1.9	Imidazole;thiolate	2.02;2.15
Zn	1.6	Imidazole;thiolate (water; Carboxylate)	2.04; 2.34 (2.06;2.01)
Fe	1.8	Thiolate;imidazole (carboxylate; water)	2.3; 2.03 (2.17; 2.10)

[a]In order of importance, based on number of observations.

in aggregation. There are also specific programs that allow one to view the dynamics of interaction between molecules, for example, the DOCK program, from the group of Irwin Kuntz at UCSF, which can be obtained over the Internet. DOCK is most suitable for viewing the interaction between small sections of the molecule and small ligands. For example, the 3D structure of neurotrophin (NTR also called p75NGFR or LANR) was modeled by its homology to regions in the receptor of tumor necrosis factor receptor-1(TNFR-I). Hypothetical complexes of this NTR model structure with nerve growth factor (266) were then modeled with DOCK3.5. Viewing of hundreds of predicted orientations with MIDAS revealed similarities between motifs in ligand-binding domains of the neurotrophin (NTR) and TNFR receptor, and pinpointed differences that could account for specificity (19) Autodock is another excellent and well maintained program for docking (287,288).

However, experimental structural data is not available for most proteins, and structures of complexes of proteins are even rarer. For a significant percentage of newly identified protein sequences, which have homology to a protein of known sequence homology, one can predict with some degree of confidence their overall topology and stretches of amino acids likely to be surface exposed. This does not mean that such models can replace experimental data. The protein folding problem is mathematically described as a "system with frustration", as the number of computing steps required to solve the problem increases faster than any power of the size of the system (267). This section covers some of the possibilities for using sequence data and molecular graphics tools to generate model 3D structures. Some programs from the Internet are presented in addition to the self-correcting distance geometry (SECODG) approach to the modeling being developed in the computational biology group of the UTMB. Of course, a model must be validated experimentally and theoretically. Tests for structure validity can be based on a "binary code" where all residues are considered polar except for seven nonpolars (CILMFWV). The arrangement of hydrophobic and polar residues alone as evaluated by a scoring scheme ("hydrophobic fitness score") (268), recognized the native fold out of 2000 near-native structures generated for each of five small monomeric proteins. Other tests for structural validity based on identifying structures with minimum thermodynamic energy are being developed.

18.9 AUTOMATIC HOMOLOGY MODELING

If a structure for a protein with significant (> 30–40%) sequence identity does exist, a 3D structure model can be easily obtained by a variety of methods, either using templates selected from fold recognition servers and sorted by hand (103,269) or using automatic methods, which have reached a very high level of accuracy (255,256). One of the oldest automatic servers, SWISSMODEL, uses the PDB to obtain structural information and automatically generates a model for sequences which share significant similarities with at least one protein of known 3D structure. Given an unknown sequence, SWISSMODEL uses a FastA and BLAST search to determine what protein in the PDB could be a possible structural homologue based on the degree of primary sequence similarity (270). However, the program only works for proteins where a high degree of sequence homology exists to protein of known structure.

Other methods may be used for proteins with lower identity to proteins of known structure, but where the sequence contains regular patterns of residue occurrence, or even physicochemical properties (258). For example, one can use the TOPITS program, which "threads" protein sequences onto model structures (271). Other fold recognition servers that also can generate models automatically include Phyre (http://www.sbg.bio.ic.ac.uk/phyre/) and I-TASSER (272) (http://zhang.bioinformatics.ku.edu/I-TASSER/)

One problem with using simple models to design mutagenesis experiments is that, while the program will be able to predict a global fold for the protein, local conformations that can account for specific properties are less easy to predict. For this reason, developing a working model (as of the date of this chapter) means using manually interactive programs that require more work on the investigator side.

18.10 MODELING USING SELF-CORRECTING DISTANCE GEOMETRY WITH THE PROGRAMS CLUSTAL, MASIA, NOAH, DIAMOD, AND FANTOM, TO DEVELOP A 3D MODEL OF A PROTEIN

Figure 18.2 summarizes methods used by our group for structure prediction based on a unique algorithm based on "Self-correcting distance geometry" (SECODG) from Dr Werner Braun and coworkers. The SECODG based program suite has the advantage that experimental results can be directly incorporated into the modeling process by manual insertion of "distance constraints" (DCs), that is, limits on the distance between two atoms in a 3D structure. Experimental DCs can be derived experimentally from NMR or crystal data or from other types of biophysical data, for example, from knowing where secondary structure elements lie or which two cysteine residues form a disulfide bond, or which residues lie within the active site (which means they must lie within a certain distance from each other). The method starts with BLAST searches, multiple sequence alignment, and fold recognition to determine whether a structure exists for a protein with reasonable sequence homology. If a homologous structure is available, the PDB coordinates can be regularized and DCs extracted automatically by related software programs.

Details of the method are included below for our search to identify the role of a 103 residue protein, temptin, involved in mating of the mollusk *Aplysia* (104). We previously characterized the NMR structure of a small protein, attractin, which had a novel, disulfide stabilized helical structure (273). Our coworkers, Sherry Painter, Gregg Nagel, and Scott Cummins identified temptin and another protein, enticin (103), as two factors needed to coordinate with attractin to exert signaling in this animal. While enticin appeared to have a 3D structure somewhat like attractin, temptin had no high degree of identity to any other protein. To characterize the probable 3D structure of temptin, the following protocol was used:

Step 1. Proteins with a significant identity pattern were searched for in the "nr" (nonredundant) protein sequence data base using Psi-Blast, and when no suitable template emerged, with fold recognition servers. The position-specific scoring matrix (PSSM) server, as well as others, indicated that the Temptin had some similarity to the EGF repeat region of human Fibrillin.

Step 2. CLUSTALW (downloaded from the internet) was used to align the best matched proteins in a block to identify areas of high homology with other proteins. An alignment of two Temptin sequences with human, chicken, and fish fibrillin and the sequence from the PDB of the crystal structure of two EGF repeats of fibrillin1 is shown in Fig. 18.2.

```
Hufibrillin1   ECPFG--YILAG---ECVDTDECSV-GNPCG--NGTCKNVIGGFECTCEEGFEPGPMMTC
Hufibrillin2   ECPMG--YNLDYTGVRCVDTDECSI-GNPCG--NGTCTNVIGSFECNCNEGFEPGPMMNC
Hufibrillin3   ECPFG--YSLDFTGINCVDTDECSV-GHPCG--QGTCTNVIGGFECACADGFEPGLMMTC
Chickfibrill   ECPYG--YILQET--QCVDTDECAV-GNPCG--NGTCRNVVGGFECTCVDGFEPGPMMTC
Pufferfish     ECPMG--YQLDIS--GVTYTNECST-GNPCG--NGTCTNAIGGFECACDDGFEPGSMMTC
TemptinCal     LNPFGEDFKAAGK—QWT-TDLCDMDSDGDGRSNG---VELGDPECVWSQGETPARTTDL
TemptinBra     LNPFGEDFKAAGR—QWT-TDLCEKDSDGDGLTNG---QELGDPECVWSQGETPARTTGL
```

Figure 18.2. Alignment of the central region of Temptin from *Aplysia califonica* and *A. brasiliana* with the central conserved EGF-like domains from fibrillins of human, chicken, and pufferfish. Conserved residues that are identical in all temptins and all fibrillins are red; similar residues are shaded gray. Bold residues include those involved in the common disulfide bond (connected by the top bracket), two tryptophan residues that lie near the two disulfides in the final model, and a probable metal binding loop (bottom bracket).

Step 3. The predicted structure was compared with biophysical data for temptin. In our case, a CD spectrum that indicated that protein was primarily β-strand, with some helical structure. We also knew the position of two disulfide bonds, determined experimentally using limited proteolysis and mass spectroscopy. One of these bonds was correctly predicted according to the alignment (the line under the alignment in Fig. 18.2 connects these two Cys residues); the other connected the N- and C-termini of the protein.

Step 4. A consensus alignment of temptin with the sequence of the PDB template, 1EMN, was then prepared, and the EXDIS program was used to generate a list of distance and dihedral angle constraints (this step is now done automatically when using the MPACK server). These constraints can be combined with results from other biophysical methods, for example, knowledge of where disulfide links and active site residues are in proteins. In our case, we specified the interior disulfide bond.

Step 5. The SECODG based package of NOAH/DIAMOD then used these constraints to generate a family of structures which best fit these constraints. The NOAH program is a "structure based filter", which means in simple terms that it uses a rough 3D structure to determine whether a list of DCs is internally consistent.

Step 6. The local structure was refined using the energy minimization program FANTOM, to determine structures with lowest energy. In our case we had to insert the disulfide bond that connected the C and N-termini by using another program, Sybyl-Biopolymer, as the structure in this area had to be deviate considerably from the template. We then ran another minimization of the protein in Biopolymer before analyzing the structure of the model with respect to conserved residues and known mutants.

For other examples of how this method can be applied to real situations, please see (236).

18.10.1 Modeling Complexes

Modeling the interaction of several molecules is important for our understanding of aggregation. Extensive modeling studies for protein complexes has been done, for example, to analyze the allosteric mechanism of the tetrameric lac repressor. Here, years of site-directed mutagenesis studies were done before a crystal structure of the repressor was available for study (274).

To go back to the temptin model, the implication of the model was that functionally, temptin might bind to attractin and perhaps enticin, to enable them to bind to mollusk scent receptors. A series of direct assays for this indicated that not only did mixing temptin with attractin cause a distinctive gel-shift indicative of a complex, but these complexes were similar to those found in Western blots of whole extract from the mollusk (104). We used two different programs to identify possible structures for this complex. First, the webserver InterProsurf (275) was used to predict residues on the surface of both proteins that would most likely interact (Fig. 18.3a). Secondly, the Zdock program was used to

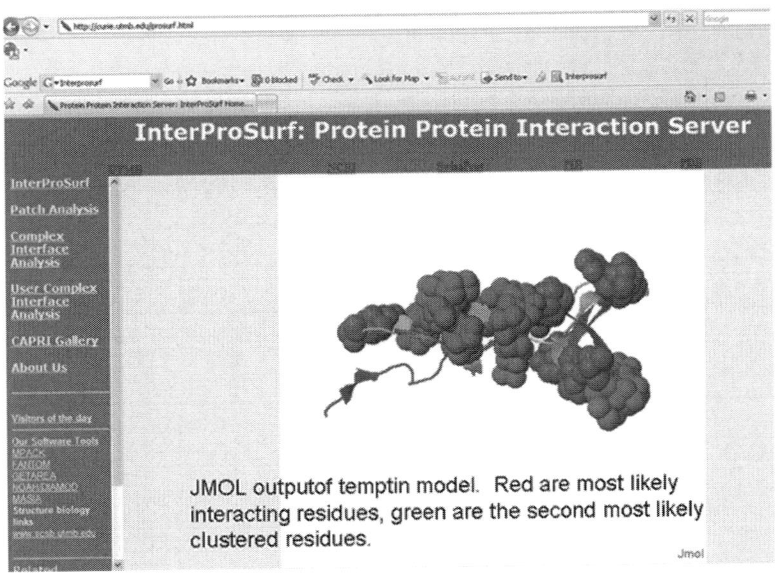

Figure 18.3. Interprosurf prediction for the temptin model, prepared based on the alignment of Figure 18.2 with constraints taken from hu Fibrillin1, indicating areas of the protein likely to be involved in complex formation with attractin (104).

dock the two proteins to one another, to determine the most likely complex structure (276), which also suggested these areas of temptin as having the highest potential to bind attractin.

This example shows how computational modeling can be used both to predict 3D structure and possible function, and how this information can be used to define protein complexes.

18.10.2 Protein Design

The field of protein design, that is, designing proteins that fold in a certain manner, is proving to have nearly as much impact on our understanding of protein folding as the direct measurement of natural proteins. Proteins can be designed to have a specific overall structure with defined secondary elements and selected for solubility (277,278). There have been several recent protein design projects that have illustrated how well we can plan a primary sequence that will form a desired structure. While the design of α-helical proteins is fairly advanced, β-sheet design is still hindered by solubility problems. The Paracelsus prize (small in cash value but honorable in the extreme) was claimed many years after it was offered to a group who were able to change protein G, which formed a β-sheet, to a four-helix bundle by substituting residues at discrete positions from the ROP protein. Helix forming residues were inserted and residues known to favor sheet formation removed (279). One biologically relevant example of such a transition is that aggregation of amyloid proteins is accompanied by a conversion from a largely helical conformation to a β-sheet based structure in the aggregate (280,281). Structures of these fibrils have now been visualized by solid state NMR (55) and other methods (Table 18.2).

Computational methods have also been introduced into the area of protein design. For example, a protein that contained no histidine, cysteine or chelated metal ion was designed that exactly fit the X-ray crystal structure of the Zinc finger protein. The structure of the synthesized model protein, which required no metal ion for structure stabilization, was determined by NMR to be identical to the original Zn finger region (282). Presumably, soon similar modeling algorithms will allow the design of proteins that match in terms of structure and functional abilities (e.g. DNA binding (283,284)).

Converting structures by using different solvents. The structure of a protein also depends on the solvent composition, and adding organic solvents can induce configurations of proteins that would not occur in an aqueous environment. For example, the structure of a "death domain" was greatly altered by the presence of 45% methyl pentanediol, a solvent often added (but generally at much lower concentrations) to the mother liquor during crystallization (285).

18.11 CONCLUSIONS

The study of protein oligomerization with various methods is leading to a better understanding of the atomic basis for aggregation. This is particularly important for understanding many diseases that involve protein aggregates. Different forms of biophysical data can be used to develop better models of the processes. Understanding how the interactions between solvents and the structure of proteins affect inter-protein contact will help design better methods for controlling aggregation and processes dependent on resolubilizing proteins after precipitation.

Acknowledgements

The author wishes to thank all my coworkers at the UTMB, especially Dr Scott Cummins (who prepared the alignment of Fig. 18.2 and ran the experiments needed to substantiate temptin's role in inter-protein interactions) and to acknowledge grants that enabled the preparation of this work, especially from the NIH (R01AI064913-01), US-Environmental Protection Agency EPA, and the Mitchell Center for Neurodegenerative Disorders at the UTMB. I also thank those who sent reprints, and apologize in advance to authors whose papers, due to my oversight, are not adequately cited here.

REFERENCES

1. Joerger AC, Fersht AR. Structure-function-rescue: the diverse nature of common p53 cancer mutants. Oncogene 2007; 26(15): 2226–2242.
2. Daly SM, Przybycien TM, Tilton RD. Aggregation of lysozyme and of poly(ethylene glycol)-modified lysozyme after adsorption to silica. Colloids Surf B Biointerfaces 2007; 57(1): 81–88.
3. Schein CH. Solubility as a function of protein structure and solvent components. Biotechnology 1990; 8: 308–317.
4. Hawe A, Friess W. Formulation development for hydrophobic therapeutic proteins. Pharm Dev Technol 2007; 12(3): 223–237.
5. Schein CH. Controlling oligomerization of pharmaceutical proteins. Pharm Acta Helv 1994; 69: 119–126.
6. Speed M, Wang D, King J. Specific aggregation of partially folded polypeptide chains - the molecular basis of inclusion body composition. Nat Biotechnol 1996; 14: 1283–1287.
7. Speed M, King J, Wang D. Polymerization mechanism of polypeptide chain aggregation. Biotechnol Bioeng 1997; 54: 333–343.
8. Galla H-J. Spektroskopische methoden in der biochemie. Stuttgart, New York: Georg Thieme Verlag; 1988.
9. Edelhoch H. Spectroscopic determination of tryptophan and tyrosine in proteins. Biochemistry 1967; 6: 1948–1954.
10. Matouschek A, Serrano L, Meiering EM, Bycroft M, Fersht AR. The folding of an enzyme. V:H^2H exchange-nuclear

magnetic resonance studies on the folding pathway of barnase : complementarity to and agreement with protein engineering studies. J Mol Biol 1992; 224: 837–845.

11. Mancini E, de Haas F, Fuller S. High-resolution icosahedral reconstruction: fulfilling the promise of cryo-electron microscopy. Structure 1997; 5: 741–750.
12. Zhou Z, Chiu W, Haskell K, Spears HJ, Jakana J, Rixon F, Scott L. Refinement of herpesvirus B-capsid structure on parallel supercomputers. Biophys J 1998; 74: 576–588.
13. Haggerty L, Lenhoff AM. STM and AFM in biotechology. Biotechnol Prog 1993; 9: 1–11.
14. Casal HL, Köhler U, Mantsch HH. Structural and Conformational changes of ß -lactoglobulin B: an infrared spectroscopic study of the effect of pH and temperature. Biochim Biophys Acta 1988; 957: 11–20.
15. Hester RC, Austin JC. Current highlights in spectroscopic studies of biological systems. In: Schmid ED, Schneider FW, Siebert F, editors. Spectroscopy of biological molecules: new advances. Chichester: John Wiley & Sons; 1988. pp. 3–10.
16. Versmold H, Härtl W. Kinetics of coagulation by dynamic light scattering. J Chem Phys 1983; 79: 4006–4009.
17. Veenstra TD, Benson L, Craig T, Tomlinson AJ, Kumar R, Naylor S. Metal mediated sterol receptor-DNA complex association and dissociation determined by electrospray ionization mass spectrometry. Nat Biotechnol 1998; 16: 262–266.
18. Wand A, Englander SW. Protein complexes studied by NMR spectroscopy. Curr Opin Biotechnol 1996; 7: 403–408.
19. Chapman BS, Kuntz ID. Modeled structure of the 75-kDa neurotrophin receptor. Protein Sci 1995; 4: 1696–1707.
20. Spruijt RB, Wolfs CJAM, Hemminga MA. Aggregation-related conformational change of the membrane associated coat protein of bacteriophage M13. Biochemistry 1989; 28: 9158–9165.
21. Myszka DG. Kinetic analysis of macromolecular interactions using surface plamon resonance biosensors. Curr Opin Biotechnol 1997; 8: 50–57.
22. Yonath A, Wittmann HG. Challenging the three dimensional structure of ribosomes. Tibs 1989; 14: 329–335.
23. Eisenstein M, Sharon R, Berkovitch-Yellin Z, Gewitz HS, Weinstein S, Pebay-Peyroula E, Roth M, Yonath A. The interplay between X-ray crystallography, neutron diffraction, image reconstruction, organo-metallic chemistry and biochemistry in structural studies of ribosomes. Biochemie 1991; 73: 879–886.
24. Vershoor A, Frank J. Three-dimensional structure of the mammalian cytoplasmic ribosome. J Mol Biol 1990; 214: 737–749.
25. Mueller F, Brimacombe R. A new model for the three-dimensional folding of Escherichia coli 16 S ribosomal RNA. II. The RNA-protein interaction data. J Mol Biol 1997; 271: 545–565.
26. Luger K, Mader A, Richmond R, Sargent D, Richmond T. Crystal structure of the nucleosome core particle at 2.8 A resolution. Nature 1997; 389: 251–260.
27. Deisenhofer J, Michel H. The photosynthetic reaction center from the purple bacterium Rhodopseudomonas viridis. EMBO J 1989; 8: 2149–2170.
28. Bystrov VF, Arseniev AS, Barsukov AL, Lomize AL, Abdulaeva GV, Sobol AG, Maslennikov IV, Golovanov AP. 2D-NMR for 3D-structure of membrane spanning polypeptides: gramacidin A and fragments of bacteriorhodopsin. In: Jardetsky O, editor. Protein structure and engineering. New York: Plenum Press; 1989. pp. 111–138.
29. Rothchild KJ, He YW, Gray D, Roepe PD, Pelletier S, Brown RS, Herzfeld J. Fourier transform infrared evidence for proline structural changes during the bacteriorhodopsin photocycle. Proc Natl Acad Sci U S A 1989; 86: 9832–9835.
30. Ceska TA, Henderson R. Analysis of high-resolution electron diffraction patterns from purple membrane labelled with heavy atoms. J Mol Biol 1990; 213: 539–560.
31. Henderson R, Baldwin JM, Ceska TA, Zemlin F, Beckmann E, Downing KH. Model for the structure of bacteriorhodopsin based on high-resolution electron cryo-microscopy. J Mol Biol 1990; 213: 899–929.
32. Amrein M, Wang Z, Guckenberger R. Comparative study of a regular protein layer by scanning tunneling microscopy and transmission electron microscopy. J Vac Sci Technol 1991; 9: 1276–1281.
33. Henry GD, Sykes BD. Detergent-solubilized M13 coat protein exists as an assymetric dimer. Observation of individual monomers by 15N,13C, and 1H nuclear magnetic resonance spectroscopy. J Mol Biol 1990; 212: 11–14.
34. Amrein M, Dürr R, Winkler H, Travaglini G, Wepf R, Gross H. STM of freeze-dried and Pt-Ir-C-coated bacteriophage T4 polyheads. J Ultrastruct Mol Struct Res 1989; 102: 170–177.
35. Hiragi Y, Inoue H, Sano Y, Kajiwara K, Ueki T, Nakatani H. Dynamic mechanism of the self-assembly process of tobacco mosaic virus protein studied by rapid temperature-jump small-angle X-ray scattering using synchrotron radiation. J Mol Biol 1990; 213: 495–502.
36. Greene B, King J. Scaffolding mutants identifying domains required for P22 procapsid assembly and maturation. Virology 1996; 225: 82–96.
37. Zhou Z, Macnab S, Jakana J, Scott L, Chiu W, Rixon F. Identification of the sites of interaction between the scaffold and outer shell in herpes simplex virus-1 capsids by difference electron imaging. Proc Natl Acad Sci U S A 1998; 95: 2778–2783.
38. Stanfield RL, Fieser TM, Lerner RA, Wilson IA. Crystal structures of an antibody to a peptide and its complex with peptide antigen at 2.8 Å. Science 1990; 248: 712–719.
39. Boulot G, Eisele J-L, Bentley GA, Bhat TN, Ward ES, Winter G, Poljak RJ. Crystallization and preliminary X-ray diffraction study of the bacterially expressed Fv from the monoclonal anti-lysozyme antibody D1.3 and of its complex with the antigen, lysozyme. J Mol Biol 1990; 213: 617–619.
40. Levy R, Assulin T, Scherf T, Levitt M, Anglister J. Probing antibody diversity by 2D-NMR: comparison of amino acid sequences, predicted structures and observed antibody-antigen interactions in complexes of two antipeptide antibodies. Biochemistry 1989; 28: 7168–7175.
41. Anglister J. Use of deuterium labelling in NMR studies of antibody combining site. Q Rev Biophys 1990; 23: 175–203.
42. Rarity JG, Seabrook RN, Carr RJG. Light-scattering studies of aggregation. Proc R Soc Lond A 1989; 423: 89–102.
43. Leatherbarrow RJ, Stedman M, Wells TNC. Structure of immunoglobulin G by scanning tunneling microscopy. J Mol Biol 1991; 221: 361–365.
44. Stothart P. Subunit structure of casein micelles from small-angle neutron scattering. J Mol Biol 1989; 208: 635–638.

45. van den Oetelaar PJM, de Man BM, Hoenders HJ. Protein folding and aggregation studied by isoelectric focusing across a urea gradient and isoelectric focusing in two dimensions. Biochim Biophys Acta 1989; 995: 82–90.

46. Schein CH, Noteborn MHM. Biotechnology 1988; 6: 291–294.

47. Albiges-Rizo C, Chroboczek J. Adenovirus serotype 3 fibre protein is expressed as a trimer in Escherichia coli. J Mol Biol 1990; 212: 247–252.

48. Perrot G, Cheng B, Gibson KD, Vila J, Palmer KA, Nayeem A, Maigret B, Scheraga HA. MSEED: A program for the rapid analytical determination of accessible surface areas and their derivatives. J Comput Chem 1992; 13: 1–11.

49. Bowden GA, Paredes AM, Georgiou G. Structure and morphology of protein inclusion bodies in Escherichia coli. Biotechnology 1991; 9: 725–730.

50. Ami D, Natalello A, Gatti-Lafranconi P, Lotti M, Doglia SM. Kinetics of inclusion body formation studied in intact cells by FT-IR spectroscopy. FEBS Lett 2005; 579(16): 3433–3436.

51. Doglia S, Ami D, Natalello A, Gatti-Lafranconi P, Lotti M. Fourier transform infrared spectroscopy analysis of the conformational quality of recombinant proteins within inclusion bodies. Biotechnol J 2008; 3(2): 193–201.

52. Shen C, Murphy R. Solvent effects on self-assembly of beta-amyloid peptide. Biophys J 1995; 69: 640–651.

53. Danielsson J, Pierattelli R, Banci L, Graslund A. High-resolution NMR studies of the zinc-binding site of the Alzheimer's amyloid beta-peptide. FEBS J 2007; 274(1): 46–59.

54. Hou L, Zagorski MG. NMR reveals anomalous Copper(II) binding to the amyloid beta; peptide of Alzheimer's disease. J Am Chem Soc 2006; 128(29): 9260–9261.

55. Petkova AT, Yau WM, Tycko R. Experimental constraints on quaternary structure in Alzheimer's beta-amyloid fibrils. Biochemistry 2006; 45(2): 498–512.

56. Petkova AT, Leapman RD, Guo Z, Yau WM, Mattson MP, Tycko R. Self-propagating, molecular-level polymorphism in Alzheimer's beta-amyloid fibrils. Science 2005; 307(5707): 262–265.

57. Tycko R. Solid-state NMR as a probe of amyloid structure. Protein Pept Lett 2006; 13(3): 229–234.

58. Luhrs T, Ritter C, Adrian M, Riek-Loher D, Bohrmann B, Dobeli H, Schubert D, Riek R. 3D structure of Alzheimer's amyloid-{beta}(1-42) fibrils. Proc Natl Acad Sci U S A 2005; 102(48): 17342–17347.

59. Lepere M, Chevallard C, Hernandez JF, Mitraki A, Guenoun P. Multiscale surface self-assembly of an amyloid-like peptide. Langmuir 2007; 23(15): 8150–8155.

60. Moore RA, Hayes SF, Fischer ER, Priola SA. Amyloid formation via supramolecular peptide assemblies. Biochemistry 2007; 46(24): 7079–7087.

61. Toyama BH, Kelly MJS, Gross JD, Weissman JS. The structural basis of yeast prion strain variants. Nature 2007; 449(7159): 233–237.

62. Ikeda T, Asakura S, Kamiya R. Total reconstruction of Salmonella flagellar filaments from hook and purified flagellin and hook-associated proteins in vitro. J Mol Biol 1989; 209: 109–114.

63. Wade RH, Chrétien D, Job D. Characterization of microtubule protofilament numbers. How does the surface lattice accomodate? J Mol Biol 1990; 212: 775–786.

64. Gaffney B. Lipoxygenases: structural principles and spectroscopy. Annu Rev Biophys Biomol Struct 1996; 25: 431–459.

65. Jarori G, Murali N, Switzer R, Rao B. Conformation of MgATP bound to 5-phospho-alpha-D-ribose 1-diphosphate synthetase by two-dimensional transferred nuclear Overhauser effect spectroscopy. Eur J Biochem 1995; 230: 517–524.

66. Murali N, Jarori G, Landy S, Rao B. Two-dimensional transferred nuclear Overhauser effect spectroscopy (TRNOESY) studies of nucleotide conformations in creatine kinase complexes: effects due to weak nonspecific binding. Biochemistry 1993; 32: 12941–12948.

67. Pretzer D, Schulteis B, Smith C, Vander Velde D, Mitchell J, Manning M. Fibrolase. A fibrinolytic protein from snake venom. Pharm Biotechnol 1993; 5: 287–314.

68. Trewhalla J, Blumenthal DK, Rokop SE, Seeger PA. Small angle scattering studies show distinct conformations of calmodulin in its complexes with two peptides based on the regulatory domain of the catalytic subunit of phosphorylase kinase. Biochemistry 1990; 29: 9316–9324.

69. Schutt CE, Lindberg U, Myslik J, Strauss N. Molecular packing in profilin:actin crystals and its implications. J Mol Biol 1989; 209: 735–746.

70. DiCapua E, Schnarr M, Ruigrok RWH, Lindner P, Timmins PA. Complexes of RecA protein in solution. A study by small angle neutron scattering. J Mol Biol 1990; 219: 557–570.

71. Amrein M, Stasiak A, Gross H, Stoll E, Travaglini G. Scanning tunneling microscopy of recA-DNA complexes coated with a conducting film. Science 1988; 240: 514–516.

72. Stasiak A, Tsaneva I, West S, Benson C, Yu X, Egelman E. The Escherichia coli RuvB branch migration protein forms double hexameric rings around DNA. Proc Natl Acad Sci U S A 1994; 91: 7618–7622.

73. Yu X, West S, Egelman E. Structure and subunit composition of the RuvAB-Holliday junction complex. J Mol Biol 1997; 266: 217–222.

74. Yu X, Jezewska MJ, Bujalowski W, Egelman EH. The hexameric E. coli DnaB helicase can exist in different quaternary states. J Mol Biol 1996; 259(1): 7–14.

75. Gros P, Betzel C, Dauter Z, Wilson KS, Hol WGJ. Molecular dynamics refinement of a thermitase-Eglin-c complex at 1.98 Å resolution and comparison of two crystal forms that differ in calcium content. J Mol Biol 1989; 210: 347–367.

76. Rydel TJ, Ravichandran KG, Tulinsky A, Bode W, Huber R, Roitsch C, Fenton JW. The structure of a complex of recombinant hirudin and human a-thrombin. Science 1990; 249: 277–280.

77. Braun W, Kallen W, Mikol V, Walkinshaw MD, Wüthrich K. Cyclophilin A review. FASEB J 1995; 9: 63–72.

78. Schurtenberger P, Augusteyn RC. Structural properties of polydisperse biopolymer solutions: a light scattering study of bovine α-crystallin. Biopolymers 1991; 31: 1229–1240.

79. Chen J, Matthews KS. Subunit dissociation affects DNA binding in a dimeric lac repressor produced by C-terminal deletion. Biochemistry 1994; 33: 8728–8735.

80. Chang WI, Matthews KS. Role of Asp274 in lac repressor: diminished sugar binding and altered conformational effects

in mutants. Biochemistry 1995; 34: 9227–9234.

81. Green SM, Gittis AG, Meeker AK, Lattman EE. One step evolution of a dimer from a monomeric protein. Nat Struct Biol 1995; 2: 746–751.
82. Murphy RM, Slayter H, Schurtenberger P, Chamberlin RA, Colton C, Yarmush ML. Biophys J 1988; 54: 45–56.
83. Tertov VV, Sobenin IA, Gabbasov ZA, Popov EG, Orekhov AN. Lipoprotein aggregation as an essential condition of intracellular lipid accumulation caused by modified low density lipoproteins. Biochem Biophys Res Commun 1989; 163: 489–494.
84. Rousseau DL, Ondrias MR. Raman scattering. In: Rousseau DL, editor. Optical techniques in biological research. Orlando (FL): Academic Press; 1984. pp. 65–132.
85. Williams RW. Methods Enzymol 1986; 130: 311–331.
86. Heremans L, Heremans K. Raman spectroscopic study of the changes in secondary structure of chymotrypsin: effect of pH and pressure on the salt bridge. Biochim Biophys Acta 1989; 999: 192–197.
87. Przybycien TM, Bailey JE. Structure-function relationships in the inorganic salt-induced precipitation of a-chymotrypsin. Biochim Biophys Acta 1989; 995: 231–245.
88. Przybycien TM, Bailey JE. Secondary structure perturbations in salt-induced protein precipitates. Biochim Biophys Acta 1991; 1076: 103–111.
89. Zaccai G, Eisenberg H. Halophilic proteins and the influence of solvent on protein stabilization. Trends Biol Sci 1990; 15: 333–337.
90. Przybycien TM, Bailey JE. Aggregation kinetics in salt-induced protein precipitation. AIChE J 1989; 35: 1779–1790.
91. Mikol V, Hirsch E, Giege R. Diagnostic of precipitant for biomacromolecule crystallization by quasi-elastic light-scattering. J Mol Biol 1990; 213: 187–195.
92. Kadima W, McPherson A, Dunn MF, Jurnak FA. Characterization of precrystallization aggregation of canavalin by dynamic light scattering. Biophys J 1990; 57: 125–132.
93. Durbin SD, Feher G. Studies of crystal growth mechanisms of proteins by electron microscopy. J Mol Biol 1990; 212: 763–774.
94. Pascal S, Cross T. Polypeptide conformational space. Dynamics by solution NMR disorder by X-ray crystallography. J Mol Biol 1994; 241: 431–439.
95. Ketchem R, Lee K, Huo S, Cross T. Macromolecular structural elucidation with solid-state NMR-derived orientational constraints. J Biomol NMR 1996; 8: 1–14.
96. Krebs J, Rana F, Dluhy R, Fierke C. Kinetic and spectroscopic studies of hydrophilic amino acid substitutions in the hydrophobic pocket of human carbonic anhydrase II. Biochemistry 1993; 32: 4496–4505.
97. Rana F, Mautone A, Dluhy R. Surface chemistry of binary mixtures of phospholipids in monolayers. Infrared studies of surface composition at varying surface pressures in a pulmonary surfactant model system. Biochemistry 1993; 32: 3169–3177.
98. Rana F, Harwood J, Mautone A, Dluhy R. Identification of phosphocholine plasmalogen as a lipid component in mammalian pulmonary surfactant using high-resolution 31P NMR spectroscopy. Biochemistry 1993; 32: 27–31.
99. Klein K, Rudy B, McIntyre J, Fleischer S, Trommer W. Specific interaction of (R)-3-hydroxybutyrate dehydrogenase with membrane phosphatidylcholine as studied by ESR spectroscopy in oriented phospholipid multibilayers: coenzyme binding enhances the interaction with phosphatidylcholine. Biochemistry 1996; 35: 3044–3049.
100. Schein CH. Physical methods and models for the study of protein aggregation. In: Georgiou G, de Bernardez-Clark E, editors. Protein refolding. Washington (DC): A.C.S. Books; 1991. pp. 21–34.
101. Berne BJ, Pecora R. Dynamic light scattering. New York: John Wiley & Sons; 1976.
102. Miles AJ, Whitmore L, Wallace BA. Spectral magnitude effects on the analyses of secondary structure from circular dichroism spectroscopic data. Protein Sci 2005; 14(2): 368–374.
103. Cummins S, Xie F, Misra M, Amare A, Jakubowski J, de Vries M, Sweedler J, Nagle G, Schein C. Recombinant production and structural studies of the Aplysia water-borne protein pheromone enticin indicates it has a novel disulfide stabilized fold. Peptides 2007; 28(1): 94–102.
104. Cummins SF, Xie F, de Vries MR, Annangudi SP, Misra M, Degnan BM, Sweedler JV, Nagle GT, Schein CH. Aplysia temptin - the 'glue' in the water-borne attractin pheromone complex. FEBS J 2007; 274(20): 5425–5437.
105. Oberg KA, Ruysschaert JM, Goormaghtigh E. The optimization of protein secondary structure determination with infrared and circular dichroism spectra. Eur J Biochem 2004; 271(14): 2937–2948.
106. Hoppener J, Lips CJ. Role of islet amyloid in type 2 diabetes mellitus. Int J Biochem Cell Biol 2006; 38: 726–736.
107. Irie K, Murakami K, Masuda Y, Morimoto A, Ohigashi H, Ohashi R, Takegoshi K, Nagao M, Shimizu T, Shirasawa T. Structure of beta-amyloid fibrils and its relevance to their neurotoxicity: implications for the pathogenesis of Alzheimer's disease. J Biosci Bioeng 2005; 99(5): 437–447.
108. Selkoe DJ. Toward a comprehensive theory for Alzheimer's disease. Hypothesis: Alzheimer's disease is caused by the cerebral accumulation and cytotoxicity of amyloid beta-protein. Ann N Y Acad Sci 2000; 924: 17–25.
109. Shaked GM, Kummer MP, Lu DC, Galvan V, Bredesen DE, Koo EH. A beta induces cell death by direct interaction with its cognate extracellular domain on APP (APP 597-624). FASEB J 2006; 20(8): 1254–1256.
110. Schroeder BE, Koo EH. To think or not to think: Synaptic activity and A beta release. Neuron 2005; 48(6): 873–875.
111. Mori C, Spooner ET, Wisniewski KE, Wisniewski TM, Yamaguchi H, Saido TC, Tolan DR, Selkoe DJ, Lemere CA. Intraneuronal A beta 42 accumulation in Down syndrome brain. Amyloid-J Protein Fold Disord 2002; 9(2): 88–102.
112. Chen X, Yan SD. Mitochondrial Abeta: a potential cause of metabolic dysfunction in Alzheimer's disease. Iubmb Life 2006; 58(12): 686–694.
113. Soto C. Unfolding the role of protein misfolding in neurodegenerative diseases. Nat Rev Neurosci 2003; 4(1): 49–60.
114. Soto C, Kindy MS, Prelli F, De Beer FC, Frangione B. In: Becker R, Giacobini E, editors. Peptide inhibitors of amyloidogenesis in Alzheimer's disease. Boston (MA): Birkhauser; 1996.
115. Soto C. Alzheimer's and prion disease as disorders of protein conformation: implications for the design of novel therapeutic approaches. J Mol Med 1999; 77(5): 412–418.

116. Soto C, Saborio GP. Prions: disease propagation and disease therapy by conformational transmission. Trends Mol Med 2001; 7(3): 109–114.

117. Soto C, Kascsak RJ, Saborio GP, Aucouturier P, Wisniewski T, Prelli F, Kascsak R, Mendez E, Harris DA, Ironside J, Tagliavini F, Carp RI, Frangione B. Reversion of prion protein conformational changes by synthetic beta-sheet breaker peptides. Lancet 2000; 355: 192–197.

118. Janson J, Laedtke T, Parisi JE, O'Brien P, Petersen RC, Butler PC. Increased risk of type 2 diabetes in Alzheimer disease. Diabetes 2004; 53: 474–481.

119. Lin CYGT, Haataja L, Hsueh WA, Butler PC. Activation of peroxisome proliferator-activated receptor-gamma by rosiglitazone protects human islet cells against human islet amyloid polypeptide toxicity by a phosphatidylinositol 3'-kinase-dependent pathway. J Clin Endocrinol Metab 2005; 90: 6678–6686.

120. Eakin CM, Berman AJ, Miranker AD. A native to amyloidogenic transition regulated by a backbone trigger. Nat Struct Mol Biol 2006; 13(3): 202–208.

121. Abedini A, Meng FL, Raleigh DP. A single-point mutation converts the highly amyloidogenic human islet amyloid polypeptide into a potent fibrillization inhibitor. J Am Chem Soc 2007; 129(37): 11300–11130.

122. Baskakov IV. The reconstitution of mammalian prion infectivity de novo. FEBS J 2007; 274(3): 576–587.

123. Deleault NR, Harris BT, Rees JR, Supattapone S. From the Cover: formation of native prions from minimal components in vitro. Proc Natl Acad Sci U S A 2007; 104(23): 9741–9746.

124. Williams ADSS, Wetzel R. Alanine scanning mutagenesis of Abeta(1-40) amyloid fibril stability. J Mol Biol 2006; 357: 1283–1294.

125. Williams ADPE, Kheterpal I, Guo JT, Cook KD, Xu Y, Wetzel R. Mapping abeta amyloid fibril secondary structure using scanning proline mutagenesis. J Mol Biol 2004; 335: 833–842.

126. LeVine H, Scholten JD. Screening for pharmacologic inhibitors of amyloid fibril formation. Methods Enzymol 1999; 309: 467–476.

127. Poduslo JF, Curran GL, Kumar A, Frangione B, Soto C. Beta-sheet breaker peptide inhibitor of Alzheimer's amyloidogenesis with increased blood-brain barrier permeability and resistance to proteolytic degradation in plasma. J Neurobiol 1999; 39(3): 371–382.

128. Wood SJ, MacKenzie L, Maleeff B, Hurle MR, Wetzel R. Selective inhibition of Abeta fibril formation. J Biol Chem 1996; 271(8): 4086–4092.

129. Lorenzo A, Yankner BA. Beta-amyloid neurotoxicity requires fibril formation and is inhibited by congo red. Proc Natl Acad Sci U S A 1994; 91(25): 12243–12247.

130. Kisilevsky R, Lemieux LJ, Fraser PE, Kong XQ, Hultin PG, Szarek WA. Arresting amyloidosis in-vivo using small molecule anionic sulfonates or sulfates - implications for alzheimers disease. Nat Med 1995; 1(2): 143–148.

131. Tomiyama T, Asano S, Suwa Y, Morita T, Kataoka K, Mori H, Endo N. Rifampicin prevents the aggregation and neurotoxicity of amyloid beta protein in vitro. Biochem Biophys Res Commun 1994; 204(1): 76–83.

132. Pappolla M, Bozner P, Soto C, Shao HY, Robakis NK, Zagorski M, Frangione B, Ghiso J. Inhibition of Alzheimer beta-fibrillogenesis by melatonin. J Biol Chem 1998; 273(13): 7185–8188.

133. Salomon AR, Marcinowski KJ, Friedland RP, Zagorski MG. Nicotine inhibits amyloid formation by the beta-peptide. Biochemistry 1996; 35(42): 13568–13578.

134. Merlini G, Ascari E, Amboldi N, Bellotti V, Arbustini E, Perfetti V, Ferrari M, Zorzoli I, Marinone MG, Garini P, Diegoli M, Trizio D, Ballinari D. Interaction of the anthracycline4′-iodo-4′-deoxydoxorubicin with amyloid fibrils: inhibition of amyloidogenesis. Proc Natl Acad Sci U S A 1995; 92(7): 2959–2963.

135. Hosoda T, Nakajima H, Honjo H. Estrogen protects neuronal cells from amyloid beta-induced apoptotic cell death. Neuroreport 2001; 12(9): 1965–1970.

136. De Felice FG, Houzel JC, Garcia-Abreu J, Louzada PRJ, Afonso RC, Meirelles MN, Lent R, Neto VM, Ferreira ST. Inhibition of Alzheimer's disease beta-amyloid aggregation, neurotoxicity, and in vivo deposition by nitrophenols: implications for Alzheimer's therapy. FASEB J 2001; 15(7): 1297–1299.

137. Forloni G, Colombo L, Girola L, Tagliavini F, Salmona M. Anti-amyloidogenic activity of tetracyclines: studies in vitro. FEBS Lett 2001; 487(3): 404–407.

138. Gervais F, Chalifour R, Garceau D, Kong X, Laurin J, Mclaughlin R, Morissette C, Paquette J. Glycosaminoglycan mimetics: a therapeutic approach to cerebral amyloid angiopathy. Amyloid 2001; 8(1): 28–35.

139. Allsop D, Gibson G, Martin IK, Moore S, Turnbull S, Twyman LJ. 3-p-Toluoyl-2-[4′-(3-diethylaminopropoxy)-phenyl]-benzofuran and 2-[4′-(3-diethylaminopropoxy)-phenyl]-benzofuran do not act as surfactants or micelles when inhibiting the aggregation of beta-amyloid peptide. Bioorg Med Chem Lett 2001; 11(2): 255–257.

140. Cohen T, Frydman-Marom A, Rechter M, Gazit E. Inhibition of amyloid fibril formation and cytotoxicity by hydroxyindole derivatives. Biochemistry 2006; 45(15): 4727–4735.

141. Maezawa I, Hong H-S, Wu H-C, Battina SK, Rana S, Iwamoto T, Radke GA, Pettersson E, Martin GM, Hua DH, Jin L-W. A novel tricyclic pyrone compound ameliorates cell death associated with intracellular amyloid-β oligomeric complexes. J Neurochem 2006; 98(1): 57–67.

142. Babaoglu K, Simeonov A, Irwin JJ, Nelson ME, Feng B, Thomas CJ, Cancian L, Costi MP, Maltby DA, Jadhav A, Inglese J, Austin CP, Shoichet BK. Comprehensive mechanistic analysis of hits from high-throughput and docking screens against beta-lactamase. J Med Chem 2008.

143. Giannetti AM, Koch BD, Browner MF. Surface plasmon resonance based assay for the detection and characterization of promiscuous inhibitors. J Med Chem 2008; 51(3): 574–580.

144. Feng BY, Toyama BH, Wille H, Colby DW, Collins SR, May BCH, Prusiner SB, Weissman J, Shoichet BK. Small-molecule aggregates inhibit amyloid polymerization. Nat Chem Biol 2008; 4(3): 197–199.

145. Rishton GM. Aggregator compounds confound amyloid fibrillization assay. Nat Chem Biol 2008; 4(3): 159–160.

146. Kim CWA, Berg JM. Thermodynamic beta-sheet propensities measured using a zinc-finger host peptide. Nature 1993; 362(6417): 267–270.

147. Wood SJ, Wetzel R, Martin JD, Hurle MR. Prolines and amyloidogenicity in fragments of the Alzheimer's peptide beta/A4. Biochemistry 1995; 34(3): 724–730.

148. Gibson TJ, Murphy RM. Inhibition of insulin fibrillogenesis with targeted peptides. Protein Sci 2006; 15(5): 1133–1141.
149. Kim JR, Gibson TJ, Murphy RM. Predicting solvent and aggregation effects of peptides using group contribution calculations. Biotechnol Prog 2006; 22(2): 605–608.
150. Kim JR, Gibson TJ, Murphy RM. Targeted control of kinetics of beta-amyloid self-association by surface tension-modifying peptides. J Biol Chem 2003; 278(42): 40730–40735.
151. Welch WJ. Role of quality control pathways in human diseases involving protein misfolding. Semin Cell Dev Biol 2004; 15(1): 31–38.
152. Bykov VJN, Selivanova G, Wiman KG. Small molecules that reactivate mutant p53. Eur J Cancer 2003; 39(13): 1828–1834.
153. Bykov VJN, Issaeva N, Zache N, Shilov A, Hultcrantz M, Bergman J, Selivanova G, Wiman KG. Reactivation of mutant p53 and induction of apoptosis in human tumor cells by maleimide analogs. J Biol Chem 2005; 280(34): 30384–30391.
154. Schein CH. The shape of the messenger: using protein structure information to design novel cytokine-based therapeutics. Curr Pharm Des 2002; 8(24): 2113–2129.
155. Ricci MS, Brems DN. Common structural stability properties of 4-helical bundle cytokines: possible physiological and pharmaceutical consequences. Curr Pharm Des 2004; 10(31): 3901–3911.
156. Bagby S, Tong KI, Ikura M. Optimization of protein solubility and stability for protein nuclear magnetic resonance. Nuclear magnetic resonance of biological macromolecules Pt B. Methods Enzymol 2001; 339: 20–41.
157. Bagby S, Tong KI, Liu DJ, Alattia JR, Ikura M. The button test: a small scale method using microdialysis cells for assessing protein solubility at concentrations suitable for NMR. J Biomol NMR 1997; 10(3): 279–282.
158. Chernov AA. Crystals built of biological macromolecules. Phys Rep 1997; 288(1–6): 61–75.
159. Blagova EV, Kuranova IP. Crystallization and preparation of protein crystals for X-ray diffraction analysis. Crystallogr Rep 1999; 44(3): 513–531.
160. Lorber B. Perspectives for pure and applied protein crystallogenesis studies. Cryst Growth Des 2005; 5(1): 17–19.
161. Jeruzalmi D, Steitz TA. Use of organic cosmotropic solutes to crystallize flexible proteins: Application to T7 RNA polymerase and its complex with the inhibitor T7 lysozyme. J Mol Biol 1997; 274(5): 748–756.
162. McPherson A, Cudney B. Searching for silver bullets: an alternative strategy for crystallizing macromolecules. J Struct Biol 2006; 156(3): 387–406.
163. Privalov PL. Stability of proteins, small globular proteins. Adv Protein Chem 1979; 33: 167–241.
164. Wagner G. Characterization of the distribution of internal motion in the basic pancreatic trypsin inhibitor using a large number of internal NMR probes. Q Rev Biophys 1983; 16: 1–57.
165. Parker MJ, Clarke AR. Amide backbone and water-related H/D isotope effects on the dynamics of a protein folding reaction. Biochemistry 1997; 36(19): 5786–5794.
166. Jaenicke R. Stability and self organization of proteins. Naturwissenschaften 1988; 75: 604–610.
167. Mann DF, Shah K, Stein D, Snead GA. Protein hydrophobicity and stability support the thermodynamic theory of protein degradation. Biochim Biophys Acta 1984; 788: 17–22.
168. Cleland JL, Jones AJ. Stable formulations of recombinant human growth hormone and interferon-gamma for microencapsulation in biodegradable microspheres. Pharm Res 1996; 13(10): 1464–1475.
169. McCloskey M, Poo MM. Protein diffusion in cell membranes: some biological implications. Int Rev Cytol 1984; 87: 19–81.
170. Rose GD, Geselowitz AR, Lesser GJ, Lee RH, Zehfus MH. Hydrophobicity of amino acid residues in globular proteins. Science 1985; 229: 834–838.
171. Wright PE, Dyson HJ, Lerner RA. Conformation of peptide fragments of proteins in aqueous solution:implications for initiation of protein folding. Biochemistry 1988; 27: 7167–7175.
172. Zou Q, Bennion BJ, Daggett V, Murphy KP. The molecular mechanism of stabilization of proteins by TMAO and its ability to counteract the effects of urea. J Am Chem Soc 2002; 124(7): 1192–1202.
173. Auton M, Bolen DW. Predicting the energetics of osmolyte-induced protein folding/unfolding. Proc Natl Acad Sci U S A 2005; 102(42): 15065–15068.
174. Kaarsholm NC, Havelund S, Hougaard P. Ionization behavior of native and mutant insulins: pk perturbation of B13-Glu in aggregated species. Arch Biochem Biophys 1990; 263: 496–502.
175. Boix E, Nogués MV, Schein CH, Benner SA, Cuchillo CM. Reverse transphosphorylation by ribonuclease A needs an intact p2 binding site. Point mutations at Lys-7 and Arg-10 alter the catalytic properties of the enzyme. J Biol Chem 1994; 269: 2529–2534.
176. Schein CH, Haugg M. The role of the C-terminus of human interferon-γ in RNA binding and activation of the cleavage of ds-RNA by bovine seminal ribonuclease. Biochem J 1995; 307: 123–127.
177. Schein CH. An enzymatic, cell free assay for selecting physiological inhibitors of interferon-γ in vitro. In Vitro Toxicol 1997; 10: 275–285.
178. Goto Y, Fink AL. Coformational states of β-lactamase: molten globule states at acidic and alkaline pH with high salt. Biochemistry 1989; 28: 945–952.
179. Mitchell RD, Simmerman HKB, Jones LR. Ca2+ binding effects on protein conformation and protein interactions of canine cardiac calsequestrin. J Biol Chem 1988; 263: 1376–1381.
180. Ries-Kautt MM, Ducruix AF. Relative effectiveness of various ions on the solubility and crystal growth of lysozyme. J Biol Chem 1989; 264: 745–748.
181. Kinsella JE. Milk proteins: physicochemical and functional properties. CRC Crit Rev Food Sci Nutr 1984; 21: 197–262.
182. Farrell HM, Kumosinski TF, Pulaski P, Thompson MP. Calcium-induced associations of the caseins: a thermodynamic linkage approach to precipitation and resolubilization. Arch Biochem Biophys 1988; 265: 146–158.
183. Shulgin IL, Ruckenstein E. Solubility and local structure around a dilute solute molecule in an aqueous solvent: From gases to biomolecules. Fluid Phase Equilibr 2007; 260(1): 126–134.

184. Pierce V, Kang M, Aburi M, Weerasinghe S, Smith PE. Recent applications of Kirkwood-Buff theory to biological systems. Cell Biochem Biophys 2008; 50(1): 1–22.

185. Mazo RM. A fluctuation theory analysis of the salting-out effect. J Phys Chem B 2006; 110(47): 24077–24082.

186. Ball P. Water as an active constituent in cell biology Chem Rev 2008; 108: 74–108.

187. Shulgin IL, Ruckenstein E. Local composition in the vicinity of a protein molecule in an aqueous mixed solvent. J Phys Chem B 2007; 111(15): 3990–3998.

188. Boistelle R, Astier JP, Marchis-Mouren G, Desseaux V, Haser R. Solubility phase transition kinetic ripening and growth rates of porcine pancreatic α-amylase isoenzymes. J Cryst Growth 1992; 123: 109–120.

189. Moon YU, Anderson CO, Blanch HW, Prausnitz JM. Osmotic pressures and second virial coefficients for aqueous saline solutions of lysozyme. Fluid Phase Equilibr 2000; 168(2): 229–239.

190. Judge R, Johns M, White E. Solubility of ovalbumin in ammonium sulfate solutions. J Chem Eng Data 1996; 41: 422–424.

191. Lee VE, Schulman JM, Stiefel EI, Lee CC. Reversible precipitation of bovine serum albumin by metal ions and synthesis, structure and reactivity of new tetrathiometallate chelating agents. J Inorg Biochem 2007; 101(11–12): 1707–1718.

192. Suck R, Weber B, Schaffer B, Diedrich E, Kamionka T, Fiebig H, Cromwell O. Purification strategy for recombinant Phl p 6 is applicable to the natural allergen and yields biochemically and immunologically comparable preparations. J Chromatogr B 2003; 787(2): 357–368.

193. Weissmann C, Schein CH, Biogen NV, assignee. Recovering products produced by host organisms. Eur patent Appl. 61250. 1982.

194. Guo M, Narsimhan G. A model for prediction of precipitation curves for globular proteins with nonionic polymers as the precipitating agent. Sep Sci Technol 1995; 3 (13): 1777–1804.

195. Guo R, Guo M, Narsimhan G. Thermodynamics of precipitation of globular proteins by nonionic polymers. Ind Eng Chem Res 1996; 35: 3015–3026.

196. Schellman JA. Fifty years of solvent denaturation. Biophys Biophys 2002; 96(2–3): 91–101.

197. Randall K, Lever M, Peddie BA, Chambers ST. Natural and synthetic betaines counter the effects of high Nacl and urea concentrations. Biochim Biophys Acta 1996; 1291(3): 189–194.

198. Rajendrakumar CSV, Suryanarayana T, Reddy AR. Dna helix destabilization by proline and betaine-possible role in the salinity tolerance process. FEBS Lett 1997; 410(2–3): 201–205.

199. Spiess AN, Mueller N, Ivell R. Trehalose is a potent PCR enhancer: lowering of DNA melting temperature and thermal stabilization of Taq polymerase by the disaccharide trehalose. Clin Chem 2004; 50(7): 1256–1259.

200. Cui DX, Tian FR, Kong Y, Titushikin I, Gao HJ. Effects of single-walled carbon nanotubes on the polymerase chain reaction. Nanotechnology 2004; 15(1): 154–157.

201. Liu Y, Bolen DW. The peptide backbone plays a dominant role in protein stabilization by naturally occurring osmolytes. Biochemistry 1995; 34: 12884–12891.

202. Wang AJ, Bolen DW. Effect of proline on lactate dehydrogenase activity-testing the generality and scope of the compatibility paradigm. Biophys J 1996; 71(4): 2117–2122.

203. Luidens M, Figge J, Breese K, Vajda S. Predicted and trifluoroethanol-induced alpha-helicity of polypeptides. Biopolymers 1996; 39(3): 367–376.

204. Miskolzie M, Kotovych G. The NMR-derived conformation of neuropeptide AF, an orphan G-protein coupled receptor peptide. Biopolymers 2003; 69: 201–215.

205. Crescenzi O, Tomaselli S, Guerrini R, Salvadori S, D'Ursi AM, Temussi PA, Picone D. Solution structure of the Alzheimer amyloid beta-peptide (1-42) in an apolar microenvironment. Similarity with a virus fusion domain. Eur J Biochem 2002; 269: 5642–5648.

206. Coadou G, Evrard-Todeschi N, Gharbi-Benarous J, Benarous R, Girault J. HIV-1 encoded virus protein U (Vpu) solution structure of the 41-62 hydrophilic region containing the phosphorylated sites Ser52 and Ser56. Int J Biol Macromol 2002; 30: 23–40.

207. Daly N, Hoffmann R, Otvos LJ, Craik D. Role of phosphorylation in the conformation of tau peptides implicated in Alzheimer's disease. Biochemistry 2000; 39: 9039–9046.

208. Chia B, Carver J, Mulhern T, Bowie J. Maculatin 1.1, an anti-microbial peptide from the Australian tree frog, Litoria genimaculata solution structure and biological activity. Eur J Biochem 2000; 267: 1894–1908.

209. Saviano G, Crescenzi O, Picone D, Temussi P, Tancredi T. Solution structure of human beta-endorphin in helicogenic solvents: an NMR study. J Pept Sci 1999; 5: 410–422.

210. Steinert P, Candi E, Tarcsa E, Marekov L, Sette M, Paci M, Ciani B, Guerrieri P, Melino G. Transglutaminase crosslinking and structural studies of the human small proline rich 3 protein. Cell Death Differ 1999; 6: 916–930.

211. Brinkworth C, Carver JA, Wegener KL, Doyle J, Llewellyn LE, Bowie JH. The solution structure of frenatin 3, a neuronal nitric oxide synthase inhibitor from the giant tree frog, Litoria infrafrenata. Biopolymers 2003; 70(3): 424–434.

212. Jaravine VA, Rathgeb-Szabo K, Alexandrescu AT. Microscopic stability of cold shock protein A examined by NMR native state hydrogen exchange as a function of urea and trimethylamine N-oxide. Protein Sci 2000; 9: 290–301.

213. Foord R, Leatherbarrow RJ. Effect of osmolytes on the exchange rates of backbone amide protons in proteins. Biochemistry 1998; 37: 2969–2978.

214. Tulla-Puche JGI, Woodward C, Barany G. Native-like conformations are sampled by partially folded and disordered variants of bovine pancreatic trypsin inhibitor. Biochemistry 2004; 43: 1591–1598.

215. Schein C, Oezguen N, Volk DE, Garimella R, Paul A, Braun W. NMR structure of the viral peptide linked to the genome (VPg) of poliovirus. Peptides 2006 July; 27(7): 1676–84.

216. Henkels C, Kurz JC, Fierke CA, Oas TG. Linked folding and anion binding of the Bacillus subtilis ribonuclease P protein. Biochemistry 2001; 40: 2777–2789.

217. Gursky O. Probing the conformation of a human apolipoprotein C-1 by amino acid substitutions and trimethylamine-N-oxide. Protein Sci 1999; 8: 2055–2064.

218. Demmel F, Doster W, Petry W, Schulte A. Eur Biophys J 1997; 26: 327–335.

219. Kleinert T, Doster W, Leyser H, Petry W, Schwarz V, Settles M. Solvent composition and vicosity effects on the kinetics

of CO binding to horse myoglobin. Biochemistry 1998; 37: 717–733.

220. Daggett V, Levitt M. A model of the molten globule state from molecular dynamics simulations. Proc Natl Acad Sci U S A 1992; 89: 5142–5146.

221. Hao M, Pincus M, Rackovsky S, Scheraga H. Unfolding and refolding of the native structure of bovine pancreatic trypsin inhibitor studied by computer simulations. Biochemistry 1993; 32: 9614–9631.

222. Hunenberger P, Mark A, van Gunsteren W. Computational approaches to study protein unfolding: hen egg white lysozyme as a case study. Proteins Structure Funct Genet 1995; 21: 169–213.

223. Myers JK, Pace CN, Scholtz JM. Denaturant m values and heat capacity changes: relation to changes in accessible surface areas of protein unfolding. Protein Sci 1995; 4: 2138–2148.

224. Fraczkiewicz R, Braun W. Exact and efficient analytical calculation of the accessible surface areas and their gradients for macromolecules. J Comput Chem 1998; 19: 319.

225. Eisenberg D, McLachlan AD. Solvation energy in protein folding and binding. Nature 1986; 316: 199–203.

226. Ooi T, Oobatake M, Némethy G, Scheraga HA. Accessible surface areas as a measure of the thermodynamic parameters of hydration of peptides. Proc Natl Acad Sci U S A 1987; 84: 3084–3090.

227. Vila J, Williams RL, Vasquez M, Scheraga HA. Empirical solvation models can be used to differentiate native from near-native conformations of bovine pancreatic trypsin inhibitor. Proteins 1991; 10: 199–218.

228. Wesson L, Eisenberg D. Atomic solvation parameters applied to molecular dynamics of proteins in solution. Protein Sci 1992; 1: 227–235.

229. Makhatadze GI, Privalov PL. Contribution of hydration to protein-folding thermodynamics. J Mol Biol 1993; 232: 639–659.

230. Delarue M, Koehl P. Atomic environment energies in proteins defined from statistics of accessible and contact surface areas. J Mol Biol 1995; 249: 675–690.

231. Evans JS, Chan SI, Goddard WA III. Prediction of polyelectrolyte polypeptide structures using Monte Carlo conformational search methods with implicit solvation modeling. Protein Sci 1995; 4: 2019–2031.

232. Cummings MD, Hart TN, Read RJ. Atomic solvation parameters in the analysis of protein-protein docking results. Protein Sci 1995; 4: 2087–2099.

233. Wang AJ, Bolen DW. A naturally occurring protective system in urea-rich cells—mechanism of osmolyte protection of proteins against urea denaturation. Biochemistry 1997; 36(30): 9101–9108.

234. Fauchère JL, Pliska V. Hydrophobic parameters of amino acid side chains from the partitioning of *N*-acetyl-amino acid amides. Eur J Med Chem Chim Ther 1983; 18: 369–375.

235. von Freyberg B, Richmond TJ, Braun W. Surface area included in energy refinement of proteins: a comparative study on atomic solvation parameters. J Mol Biol 1993; 233: 275–292.

236. Mumenthaler C, Braun W. Folding of globular proteins by energy minimization and Monte Carlo simulations with hydrophobic surface area potentials. J Mol Model 1995; 1: 1–10.

237. von Freyberg B, Braun W. Efficient search for all low energy conformations of Met-enkephalin by Monte Carlo methods. J Comput Chem 1991; 12: 1065–1076.

238. Radzicka A, Wolfenden R. Comparing the polarities of the amino acids: side-chain distribution coefficients between the vapor phase, cyclohexane, 1-octanol, and neutral aqueous solution. Biochemistry 1988; 27: 1664–1670.

239. Barone G, Della GG, Del VP, Giancola C, Graziano G. Hydration enthalpy of model peptides: N-acetyl amino acid amides. Biophys Chem 1994; 51(2–3): 193–199.

240. Hedwig GR, Hoiland H. Thermodynamic properties of peptide solutions. Part 11. Partial molar isentropic pressure coefficients in aqueous solutions of some tripeptides that model protein side-chains. Biophys Chem 1994; 49(2): 175–181.

241. Myers JK, Pace CN, Scholtz JM. Denaturant m values and heat capacity changes: relation to changes in accessible surface areas of protein unfolding. Protein Sci 1995; 4: 2138–2148.

242. Karle I, Flippen-Anderson J, Uma K, Balaram P. Apolar peptide models for conformational heterogeneity, hydration, and packing of polypeptide helices: crystal structure of hepta- and octapeptides containing α-aminoisobutyric acid. Proteins Struct Funct Genet 1990; 7: 62–73.

243. Karle I, Flippen-Anderson J, Uma K, Balaram P. Unfolding of an a-Helix in peptide crystals by solvation: conformational fragility in a heptapeptide. Biopolymers 1993; 33: 827–837.

244. Rosell CM, Vaidya AM, Halling PJ. Prediction of denaturing tendency of organic solvents in mixtures with water by measurement of naphthalene solubility. Biochim Biophys Acta 1995; 1252(1): 158–164.

245. Thanki N, Thornton JM, Goodfellow JM. 1988. Distributions of water around amno acid residues in proteins. J Mol Biol 1988; 202: 637–657.

246. Gibbs PR, Radzicka A, Wolfenden R. The anomalous hydrophilic character of proline. J Am Chem Soc 1991; 113: 4714–4715.

247. Bird RB, Stewart WE, Lightfoot EN. Transport phenomenon. New York: John Wiley & Son; 1960.

248. Ferscht A. Enzyme structure and mechanism. New York: Freeman; 1985.

249. Bryngelson J, Onuchic J, Socci N, Wolynes P. Funnels, pathways, and the energy landscape of protein folding: a synthesis. Proteins Struct Funct Genet 1995; 21: 167–195.

250. Ansari A, Jones CM, Henry ER, Hofrichter J, Eaton WA. Conformational relaxation and ligand binding in Myoglobin. Biochemistry 1994; 33: 5128–5145.

251. Meadows R, Post CB, Luxon BA, Gorenstein DG. MORASS program. Galveston (TX): University of Texas Medical Branch; 1996.

252. Harding MM. The geometry of metal-ligand interactions relevant to proteins. Acta Crystallogr 1999; D55: 1432–1443.

253. Harding M. Small revisions to predicted distances around metal sites in proteins. Acta Crystallogr D 2006; 62(6): 678–682.

254. Schein CH. Solubility and secretability. Curr Opin Biotechnol 1993; 4: 456–461.

255. Das R, Qian B, Raman S, Vernon R, Thompson J, Bradley P, Khare S, Tyka MD, Bhat D, Chivian D, Kim DE, Sheffler WH, Malmström L, Wollacott AM, Wang C, Andre I,

Baker D. Structure prediction for CASP7 targets using extensive all-atom refinement with Rosetta@home. Proteins Struct Funct Bioinform 2007; 69(S8): 118–128.

256. Kryshtafovych A, Fidelis K, Moult J. Progress from CASP6 to CASP7. Proteins Struct Funct Bioinform 2007; 69(S8): 194–207.

257. Arnold K, Bordoli L, Kopp J, Schwede T. The SWISS-MODEL workspace: a web-based environment for protein structure homology modelling. Bioinformatics 2006; 22: 195–201.

258. Ivanciuc O, Oezguen N, Mathura V, Schein CH, Xu Y, Braun W. Using property based sequence motifs and 3D modeling to determine structure and functional regions in CASP5 targets. Curr Med Chem 2004; 11(5): 583–593.

259. Simons K, Bonneau R, Ruczinski I, Baker D. Ab Initio protein structure prediction of CASPIII targets using ROSETTA. Proteins Struct Funct Genet 1999; 37 Suppl 3: 171–176.

260. Wunderlich Z, Acton TB, Liu JF, Kornhaber G, Everett J, Carter P, Lan N, Echols N, Gerstein M, Rost B, Montelione GT. The protein target list of the Northeast Structural Genomics Consortium. Proteins Struct Funct Bioinform 2004; 56(2): 181–187.

261. Bhattacharya A, Tejero R, Montelione GT. Evaluating protein structures determined by structural genomics consortia. Proteins Struct Funct Bioinform 2007; 66(4): 778–795.

262. Hughes RC, Ng JD. Can small laboratories do structural genomics? Cryst Growth Des 2007; 7(11): 2226–2238.

263. Negi S, Schein CH, Oezguen N, Power TD, Braun W. InterProSurf: a web server for predicting interacting sites on protein surfaces. Bioinformatics 2007; 23(24): 3397–3399.

264. Negi SS, Kolokoltsov AA, Schein CH, Davey RA, Braun W. Determining functionally important amino acid residues of the E1 protein of Venezuelan equine encephalitis virus. J Mol Model 2006; 12(6): 921–929.

265. Bennett M, Schlunegger M, Eisenberg D. 3D domain swapping: a mechanism for oligomer assembly. Protein Sci 1995; 4: 2455–2468.

266. Watts NR, Misra M, Wingfield PT, Stahl SJ, Cheng N, Trus BL, Steven AC, Williams RW. Three-dimensional structure of HIV-1 Rev protein filaments. J Struct Biol 1998; 121: 41–52.

267. Hansmann UHE, Okamoto Y. Comparative study of multicanonical and simulated annealing algorithms in the protein folding problem. Preprint SC-94-20, Konrad Zuse Zentrum Berlin (available via anonymous-ftp: ftpserv01zib-berlindeSC94-20cd/pub get README) 1994.

268. Huang E, Subbiah S, Tsai J, Levitt M. Using a hydrophobic contact potential to evaluate native and near-native folds generated by molecular dynamics simulations. J Mol Biol 1996; 257: 716–725.

269. Ivanciuc O, Oezguen N, Mathura VS, Schein CH, Xu Y, Braun W. Using property based sequence motifs and 3D modeling to determine structure and functional regions of proteins. Curr Med Chem 2004; 11(5): 583–593.

270. Peitsch MC. ProMod and Swiss-Model: internet-based tools for automated comparative protein modelling. Biochem Soc Trans 1996; 24: 274–279.

271. Rost B. Predicting one-dimensional protein structure by profile based neural networks. Methods Enzymol 1996; 266: 525–539.

272. Zhang Y. I-TASSER server for protein 3D-structure prediction. BMC Bioinformatics 2008; 9: 40.

273. Garimella R, Xu Y, Schein CH, Rajarathnam K, Nagle GT, Painter SD, Braun W. NMR solution structure of attractin, a water–borne protein pheromone from the mollusk Aplysia californica. Biochemistry 2003; 42(33): 9970–9979.

274. Barker SA, Caldwell KK, Hall A, Martinez AM, Pfeiffer JR, Oliver JM, Wilson BS. Wortmannin blocks lipid and protein kinase activities associated with PI 3-kinase and inhibits a subset of responses induced by Fc epsilon R1 cross-linking. Mol Biol Cell 1995; 6(9): 1145–1158.

275. Negi SS, Schein CH, Oezguen N, Power TD, Braun W. InterProSurf: a web server for predicting interacting sites on protein surfaces. Bioinformatics 2007; 23(24): 3397–3399.

276. Wiehe K, Pierce B, Mintseris J, Tong W, Anderson R, Chen R, Weng Z. ZDOCK and RDOCK performance in CAPRI rounds 3, 4, and 5. Proteins 2005; 60: 207–213.

277. West MW, Hecht MH. Binary patterning of polar and nonpolar amino acids in the sequences and structures of native proteins. Protein Sci 1995; 4: 2032–2039.

278. Roy S, Helmer K, Hecht MH. Detecting native-like properties in combinatorial libraries of de novo proteins. Fold Des 1997; 2: 89–92.

279. Dalal S, Balasubramanian S, Regan L. Protein alchemy: changing b-sheet into a-helix. Nat Struct Biol 1997; 4: 548.

280. Prusiner SB, Telling G, Cohen FE, Dearmond SJ. Prion diseases of humans and animals [Review]. Semin Virol 1996; 7(3): 159–173.

281. Safar J, Roller PP, Gajdusek DC, Gibbs CJ Jr. Thermal stability and conformational transitions of scrapie amyloid (prion) protein correlate with infectivity. Protein Sci 1993; 2(12): 2206–2216.

282. Dahiyat B, Sarisky C, Mayo S. De novo protein design: towards fully automated sequence selection. J Mol Biol 1997; 273: 789–796.

283. Pomerantz JL, Wolfe SA, Pabo CO. Structure-based design of a dimeric zinc finger protein. Biochemistry 1998; 37: 965–970.

284. Kim JS, Pabo CO. Getting a handhold on DNA: design of poly-zinc finger proteins with femtomolar dissociation constants. Proc Natl Acad Sci U S A 1998; 95: 2812–2817.

285. Xiao T, Gardner KH, Sprang SR. Cosolvent-induced transformation of a death domain tertiary structure. Proc Natl Acad Sci U S A 2002; 99(17): 11151–11156.

286. Chen D, Martin Z, Soto C & Schein C.H Computational selection of inhibitors of Abeta aggregation and neuronal toxicity. Biorg. Med. Chem. 2009; 17: 5189–5197.

287. Morris G.M, Goodsell D.S., Huey R & Olson A.J Distributed automated docking of flexible ligands to proteins: parallel applications of AutoDock 2.4. J. Comput. Aided Mol. Des. 1996; 10: 293–304.

288. Chen D.L, Menche G., Power T.D., Sower L., Peterson J.W. & Schein C.H. Accounting for ligand-metal ions in docking small molecules on adenylyl cyclase toxins. Proteins-Structure Function and Bioinformatics 2007; 67: 593–605.

PART IV

EQUIPMENT DESIGN FOR DOWNSTREAM RECOVERY AND PROTEIN PURIFICATION

19

CLEANING AND SANITATION IN DOWNSTREAM PROCESSES

GAIL SOFER
GE Healthcare, Piscataway, New Jersey
CRAIG ROBINSON
GE Healthcare, Westborough, Massachusetts
JONATHAN YOURKIN
GE Instruments, Boulder, Colorado
TINA PITARRESI AND DARCY BIRSE
Fast Trak Biopharma Services, GE Healthcare, Piscataway, New Jersey

19.1 INTRODUCTION

Downstream processing of biopharmaceuticals typically includes chromatography and filtration unit operations. Figure 19.1 depicts a typical downstream process for production of a monoclonal antibody (mAb). Downstream processing enables the production of highly purified biological therapeutic products for preclinical testing, clinical, and commercial use. The design and implementation of suitable cleaning protocols are part of regulatory commitments ensuring the consistent production and control of manufacturing conditions for biotherapeutics.

In current biomanufacturing processes, in certain scale production scenarios, the use of disposables has allowed for increased facility and unit operation flexibility resulting in reducing the effort, time, and cost that is spent on cleaning and cleaning validation. The decision as to where to best introduce disposable technology in a process is based on clinical stage and process development stage of the candidate biopharmaceutical product. Process development improvements, development, cost, process volumes, time, staff training, and existing facility design are key drivers toward implementation of disposables. Not all bioprocess unit operations, facility, and layout and design permit for the effective implementation of disposable technologies due to engineering, cost, or scale constraints. In current modern biomanufacturing, cleaning and validation remain critical concerns for industrial biotechnology products and processes.

In this chapter, the review is focused on cleaning and sanitization of chromatography media, membranes, and equipment for the production of biotherapeutics and *in vivo* diagnostics. The cleaning and cleaning validation principles described in this review apply to the production of a range of biotechnology products including biotherapeutics such as vaccines [nucleic acid (DNA, RNA), viruses, antigens, and polysaccharides], mAbs, recombinant proteins (rProteins), and peptides.

19.2 DESIGNING AN EFFECTIVE CLEANING PROTOCOL FOR DOWNSTREAM BIOPROCESSES

Cleaning is defined as the physical removal of soil, organic debris, and particulates from surfaces; whereas, sanitization is defined as removal or elimination of vegetative bacterial cells (1). It is common to combine

Downstream Industrial Biotechnology: Recovery and Purification, First Edition. Edited by Michael C. Flickinger.

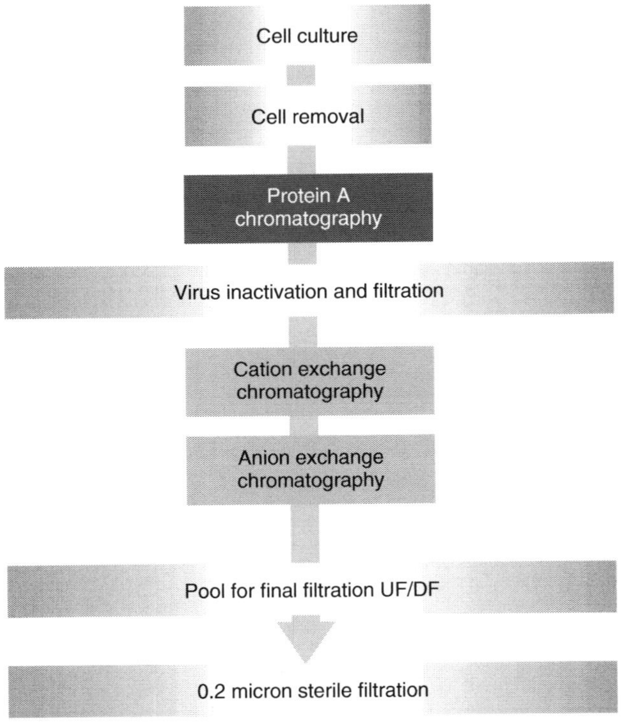

Figure 19.1. Typical monoclonal antibody process platform. (This figure is available in full color at http://onlinelibrary.wiley.com/book/10.1002/9780470054581.)

cleaning and sanitization applications in downstream processing procedures. Key points to consider during the protocol design phase, development of experimental conditions, and validation studies are to include appropriate targets, measurements, and acceptable parameter limits to quantify the quality and robustness of the intended study. Factors that are important to define early during design of the cleaning protocol include upstream product source and feed stream, intended use of product (biotherapeutics, diagnostic, research material, etc.), position of equipment, materials, and unit operation in the processing train, phase of product development, and compatibility with suitable cleaning reagents. The ability to control bioburden should be demonstrated and supported with experimental and process data defining the robustness and reproducibility of the process. A logical stepwise approach will minimize protocol design time and result in improved effectiveness (Table 19.1).

TABLE 19.1. Questions to Answer in a Stepwise Approach to Cleaning Protocol Development

What are you trying to remove?
What surface are you trying to clean?
Where is the unit operation located in the downstream process train?
What is the phase of product or clinical development?
How is cleaning effectiveness measured?
What level of detail is included in the protocol?
How do you establish acceptance limits?

19.2.1 What are You Trying to Remove?

The primary step toward developing an effective cleaning protocol is to assess and evaluate the containment profile identifying what is critical to remove from the process and its risk impact to the product integrity or safety profile in patients. When working with a new biologic source or complex raw materials, it can be a significant challenge to define the contaminants that need to be removed and assessing a risk profile for the remaining contaminants. Complex materials include cell culture growth media, plasma sources, and supplement reagents (antifoams, cell growth milieu, leachables, etc.). For recombinant biotechnological products, the most common source materials are mammalian cells [such as Chinese hamster ovary (CHO), NS0, PER.C6, and HEK293 cells], microbial cells (such as *Escherichia coli* and *Lactococcus*), and yeast (*Bacillus subtilis*, *Saccharomyces cerveisiae*). All of these feed streams may contain lipids or hydrophobic proteins that adhere to chromatography media, membranes, and equipment surfaces and in general, any wetted contact point in a process stream. Biological feed streams contain high levels of negatively charged impurities (e.g. host cell DNA, lipids, membrane components, and endotoxins). Process additives (such as stabilizing agents or inorganic compounds from upstream unit operations) may be difficult to remove without compromising the productivity or efficiency of the overall process. Impurities from cell culture bioreactors or fermentors should be assessed for contamination of endogenous retroviruses or gram-negative microbial systems for endotoxins.

19.2.2 What Surface are You Trying to Clean?

Whether designing cleaning protocols for bioreactors, packed chromatography columns, empty columns, membranes, holding tanks or filtration and chromatography skids, the chemistry of the surfaces being cleaned should be carefully assessed. Applying high salt concentration containing solvents with strong ionic strength may be highly effective in removing residual protein or DNA, but stainless steel may not tolerate these physical—chemical conditions over long repeated exposure times, and may become compromised resulting in reduced equipment lifetime. Sodium hydroxide (NaOH) has been shown to be an effective combination of cleaning and sanitizing agent and is considered the industrial biopharmaceutical standard for these applications. It should be noted that, at high concentrations, NaOH may damage membranes or chromatography media, such as those with labile protein ligands or complex coupling chemistries. When searching for ideal cleaning or sanitizing agents, it is possible to find a selection of excellent single purpose reagents, but it is critical to understand the effect the reagent has on the equipment, membranes, or media in terms of compatibility and longevity. A specific example, peracetic acid, illustrates the importance of generating data when selecting the most suitable cleaning reagent. Peracetic acid was investigated as a sanitizing reagent based on its cost and effectiveness; following further experimental challenges, it was revealed that peracetic acid was responsible for the oxidization of 0-rings and several other wetted-parts and components in filtration skids, chromatography columns, and systems.

Chemical compatibility of equipment on wetted-part surfaces and media is valuable information that should be provided by suppliers on all parts or products involved in a bioprocess prior to the development of a cleaning protocol. Collection and experimental investigation of compatibility data with defined procedures can aid to expedite cleaning protocol development and suitable cleaning and sanitization regimes for a given bioprocess.

For many bioprocesses, not all chemical or physical interactions of feed stream components with the range of materials are understood or defined. An example of this unknown process challenge may involve a hydrophobic protein adhering to hydrophobic media and equipment wetted-part surfaces. If a high salt concentration containing solution is used as the preliminary step in a cleaning protocol, a hydrophobic protein contaminant may precipitate and generate a larger residue that must be cleaned. These type of nonspecific interactions, along with chemical compatibility, are key criteria to be taken into consideration when developing cleaning protocols, including defining the cleaning reagents, contact time, temperature, sequence of wash and rinse steps, number of cycles for a given bioprocess and lifetime expectations of materials and equipment.

19.2.3 Where is the Unit Operation Located in the Downstream Process Train?

Upstream unit operations such as bioreactors, hold tanks, and harvest stages early in the bioprocess, are likely to be exposed to the most complex materials often requiring more stringent cleaning and sanitization process rigor. This challenge is best addressed early in the process development stage to allow for informed decisions on media selection and representative equipment configurations suitable for the successful transfer to manufacturing scale. Final product purification stages, known as *polishing steps*, target the removal of trace impurities and may provide secondary buffer exchange for final product formulation. The high level of product purity at these later stages in the bioprocess involve highly purified product where contaminant concerns are focused around bioburden risk and final aseptic processing demands for sterility. Cleaning is critical at the later stages of the process.

19.2.4 What is the Phase of Product or Clinical Development?

It is valuable to recognize and consider that fouled equipment, membranes, or media are not suitable for development applications including early process design. Fouled components can create artifact impurity profiles, poor performance metrics such as dynamic binding capacity (DBC), break-through, flow rates, unwanted back pressure, and unnecessary or not representative bioburden risk. Constructing a robust and well-defined cleaning protocol can build in improved risk reduction and enable the more effective transfer of a given bioprocess to manufacture scale saving, time, cost, and risk of process failure.

For the manufacture of products intended for human use, compliance with current good manufacturing practices (cGMPs) is a regulatory requirement, in adherence to this guidance, cleanliness, and sanitization are key criteria. For commercially marketed products, cleaning and cleaning validation regimes are formal regulatory requirements. The legal requirements for US licensed products are found in the code of federal regulations (CFR), specifically 21 CFR 211.67 and 21 CFR 600.11.

19.2.5 How is Cleaning Effectiveness Measured?

In conducting studies to select the most effective assays to judge cleaning methods, data-driven assays are instrumental in providing quantifiable metrics to monitor robustness and reproducibility of cleaning and sanitization procedures. The selection of assay type, frequency of cleaning procedure, and testing routine should be developed and defined early in the process and validated. Historically, cleaning protocol development, routine monitoring, and cleaning validation have been problematic in downstream processing, specifically for membranes and media used in multiproduct facilities using common equipment. In legacy processes, cleaning strategies were modeled on the expectation that product-specific assays could be used to detect carryover in conjunction with risk-based assessments of any detectable carryover to determine how carryover, if any, may affect the safety and potency of the final product. Cleaning experts in the biopharmaceutical sector debated that product-specific assays would not detect product carryover downstream of a membrane or column that had been cleaned with harsh cleaning or sanitizing agents; following comprehensive review of multiple sources of testing data, it was observed that there is a risk of cross-reacting material as detected in

ELISAs (2,3). To allow for improved testing rigor and the inclusion of orthogonal methods in cleaning regimes, in May 2005, the US FDA acknowledged that total organic carbon (TOC) is an acceptable method for evaluating cleaning effectiveness. TOC methods applied to cleaning studies have demonstrated data useful in detecting residue contaminants used to challenge cleaning effectiveness protocols. The adequacy of TOC in measuring contaminant residues was published in the inspection guide on cleaning validation (4).

The FDA commented "TOC or TC can be an acceptable method for monitoring residues routinely and for cleaning validation. In order for TOC to be functionally suitable, it should first be established that a substantial amount of the contaminating material(s) is organic and contains carbon that can be oxidized under TOC test conditions. This is an important exercise because some organic compounds cannot be reliably detected using TOC."

Further orthogonal assays used for determining cleaning effectiveness include monitoring high performance liquid chromatography (HPLC), sodium dodecyl sulfate polyacrylamide gel electrophoresis (SDS-PAGE) and total protein (mass balance). Blank, or null, process runs measuring UV detection can be useful to monitor for trace impurities where wash buffers can be benchmarked to standard data on water for injection (WFI), pH, and conductivity measurements.

19.2.6 What Level of Detail is Included in the Protocol?

Protocols are designed and implemented to ensure control of a given process through cleaning regime rigor. An operator must be able to clearly understand and execute all the instructions in the cleaning protocol to effectively demonstrate control of the process. The chronological sequence of adding cleaning reagents and rinsing agents may be critical to the overall cleaning and sanitizing performance. Process parameters such as flow rates, contact time, temperature, and pressure will be specified if important to cleaning effectiveness. Cleaning reagent quality, preparation, concentration, and expiration dates must be clearly defined and detailed in standard operating procedures (SOPs). Reagent temperature and contact time are critical elements in cleaning processes and must be defined in operational ranges that are appropriate for manufacturing at the intended commercial scale with the respective equipment and facility design. The frequency of executing a cleaning protocol should be defined and validated including hold times prior to, and after, cleaning protocols. To adequately demonstrate control of cleaning processes, robust protocols are required to include the respective assays for monitoring that can detect the defined limits of contamination such as bioburden, product carryover, and residual host cell contaminants. To allow for this level of cleaning scrutiny to ensure control of the cleaning process, in-line or at-line measurements of cleaning effectiveness must be designed into the protocols to permit operators to assess if the cleaning procedure met specifications.

19.2.7 How do You Establish Acceptance Limits?

Defining acceptable limits for a given cleaning procedure can be challenging due to the impact that cleaning regimes have on productivity, equipment wear and tear, cost, reagent effectiveness, and time. For any given process, this requires a definitive balance to permit adequate cleaning, design-in of cleaning robustness and ensuring overall control of the process. In selecting the appropriate contaminant detection assay, it is critical to design the procedures to be within specification of risk-based detection limits. A question to pose in developing cleaning procedures, is that if a contaminant is not detected, or is outside the limits of detection with a given assay or instrument, can the cleaning process be demonstrated as being effective? Factors that influence specification limits for cleaning procedures include assay detection sensitivity and correlation of relevant data with safety studies used in clinical production. As a model for cleaning validation in biopharmaceutical processes, multiproduct drug manufacturing practices (5) provide a reference point to develop the appropriate cleaning strategies that may be effective for biologics production. Prior to developing and validating a cleaning strategy, a risk-based assessment is essential to define the necessary acceptance criteria toward establishing a robust testing regime. Based on the ICH document on risk management Q9, the risk to product and process quality should be based on scientific knowledge and linked to patient safety (6). The data-driven experiments and clinical observations including patient population, product indication, dosing range, and dose frequency should be factored into the cleaning assessment. When TOC approaches are utilized to set carryover specification limits, the total level of carbon measured will be assigned as the benchmark level associated with the highest risk factors. This analytical method measures total protein content and organic components including buffers and impurities.

In establishing an effective cleaning strategy, experimental data is necessary to develop the most efficient and cost-effective cleaning methods, reagents, and protocols suitable for each unit operation. A logical evaluation, aided by a design of experiments (DoE) approach may expedite cleaning protocol development and identifying, early in the process, important factors for effective cleaning. To ensure robustness of a defined cleaning protocol on a given process, the cleaning effectiveness should demonstrate through testing specification limits of the worst-case contaminant conditions.

19.3 CHROMATOGRAPHIC MEDIA

Current commercially available chromatography media are engineered to tolerate a broad range of relatively harsh cleaning conditions. There is a clear trend in the biopharma sector to improve cleaning rigor and effectiveness, ultimately improving control of safety and reducing risk in a given process, by chromatography media innovators to develop products with protein ligands capable of cleaning-in-place (CiP) under alkaline conditions with NaOH. Figure 19.2 compares the cleaning performance and alkaline tolerance of two commercially available Protein A affinity ligand products (GE Healthcare's, MabSelect™, and MabSelect SuRe). The Protein A affinity ligand in the latter product is genetically engineered replacing alkaline-sensitive amino acids with those that are base-stable at higher pH ranges. In the study, the experimental data generated provides an evaluation of DBC to challenge with cleaning performance and robustness; the results of the study demonstrate that alkaline-tolerant media performs consistently under routine process conditions cycling with 15 min contact times to 0.1 M NaOH beyond120 cycles.

For covalent or chemical linked ligands on chromatographic media, factors that influence media stability include chemical cross-linking of the sugar moiety backbone (agarose-based), ligand stability, and ligand conjugation stability. Suppliers will provide relevant data and product-specific details on commercially available products. In addition, a supplier can provide, often upon request (contractual), a confidential regulatory support file (RSF) which may be used for a regulatory filing. As the end user of the chromatographic media, it is important to collect data and supplier information to support a product regulatory filing. This documentation can be additionally supported if the media supplier has a drug master file (DMF) on record with the regulatory authorities [e.g. the FDA or European Medicines Agency (EMEA)]. The company making the product filing to the respective regulatory body is required to provide supporting data on cleaning strategy and validation, typically in accordance with the supplier's recommendations, and is responsible for the overall control of the process, including cleaning and performance of the media, independent of the suppliers documentation. These studies should include experimental data demonstrating process control by monitoring and detection methods, specifically in the case of Protein A media, data addressing ligand leakage important to media lifetime, and cleaning procedures.

As previously described, the chemical nature of cleaning and sanitization agents has an impact on media and equipment lifetime. Media lifetime studies are scale-down processes designed to model overall unit operations to derive number of useable cycles of media defined by performance specifications (DBC, flow rate, pressure, regeneration, etc.), and data be used to predict the number of overall cycles in a given process adhering to cleaning regimes. Typically, these cleaning processes are validated prospectively as part of a master validation plan (MVP), with some processes being validated concurrently, as the development of the process moves toward commercial manufacture. When compiling and implementing cleaning validation protocols, media and

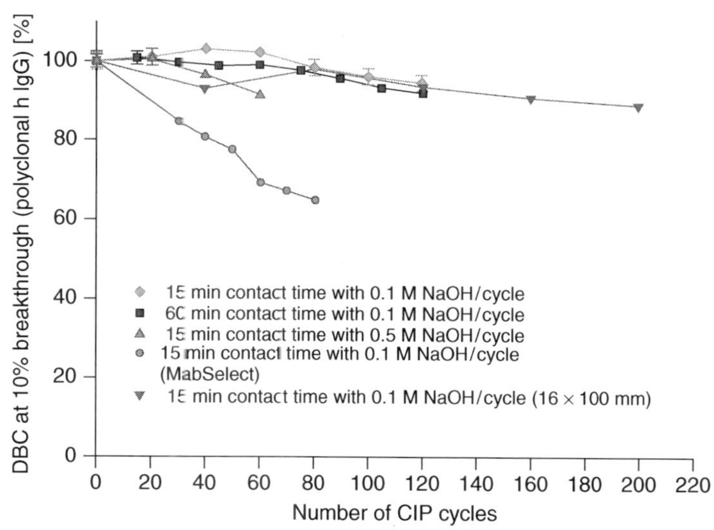

Figure 19.2. Comparison of cleaning performance of two protein A affinity media, MabSelect™ and alkaline-tolerant MabSelect SuRe. The alkaline-tolerant MabSelect SuRe maintains dynamic binding capacity under routine process conditions cycling with 15 min contact times to 0.1 M NaOH beyond 120 cycles. (This figure is available in full color at http://onlinelibrary.wiley.com/book/10.1002/9780470054581.)

buffer storage is a key part of the lifetime studies. An anomalous cleaning effect is often observed following column storage cycles of packed media; the elution of tightly bound contaminants, for example, residual DNA, denatured proteins, cell debris, or lipids bound to anion exchangers (AEX) is released from the column. Following a cleaning cycle, the column effluent should be tested to monitor for carbon-containing or UV-absorbing materials prior to storage to establish a reference baseline then retested using the same assays when the column returns to production cycles. The premise for the release of contaminants off-cycle is that the cleaning regime is a function of contact time and that chromatographic media operates as a dynamic process, with variable binding rates and efficiencies under diverse conditions, such as under cleaning and sanitization conditions. To avoid these issues, it is recommended to ensure that columns are extensively rinsed following storage conditions when returning to manufacturing campaigns and compared to postcleaning baseline contaminant levels prior to bringing on-line in production setting. Once these conditions have been identified, they should be challenged, developed, and validated as part of the cleaning validation strategy.

After determining which cleaning reagents are effective for a given process, it is important to evaluate compatibility of all equipment wetted-parts in contact with the cleaning reagents. Evaluate cleaning protocols based on effectiveness of cleaning, process robustness, and impact on product and equipment. It is also important to understand the stage of the bioprocess in terms of variability of the upstream unit operations. If changes in the upstream bioreactor feed stream changes such as growth media, growth time, or supplements of chemical additives, there can be a profound effect of contaminant profile and levels resulting in a measurable compromise in cleaning effectiveness. As with any element of a bioprocess, it is important to monitor for the clearance of the cleaning, sanitization, and storage reagents themselves. The testing criteria should be selected to allow for adequate cleaning and rinsing defining column volumes (CVs), contact time, and volumetric/linear flow rate. If feasible, in-line analysis using validated methods may provide valuable cleaning process throughput based on experimentally established acceptance criteria specifications.

When conducting cleaning validation protocols, studies on cleaning efficacy are required with the correlative sufficiently sensitive assays to monitor and detect contaminant profile changes and levels. To support this application, TOC has evolved as a powerful tool in the biopharm industry for helping to identify cleaning effectiveness. The tool is useful for a large number of bioprocesses and qualifying processes to determine if TOC is an adequate analytical tool based on the evaluation of noncarbon containing materials, and materials, such as buffers and resins, that may have a high carbon-content background. In some cases, the use of TOC for cleaning evaluation of a packed column may not be appropriate. Ideally, contaminant profiles and levels should be challenged through orthogonal nonspecific test methods such as: SDS-PAGE, HPLC, pH, conductivity, and UV analysis. There may be applications where total protein (mass balance) endotoxin detection assays may be applied. In bioprocesses, an important indicator of bioburden fouling is observed as an increase in back pressure often indicating a buildup of material on a chromatography column or membrane, resulting in decaying rates of stabilization during regeneration and reequilibration steps. It is highly recommended to develop and implement an orthogonal assay approach to best challenge your process and rigor of your cleaning strategy generating specific and nonspecific data.

19.3.1 Residual DNA

Residual DNA, introduced by the host cell source (cell culture), is a problematic contaminant to remove due to its highly negatively charged nucleic acid backbone. These highly charged biomolecules have strong negative electropotential and, if unfolded, can become quite large relative to the biotherapeutic that is being purified. The combination of high negative charge and large mass can result in nonspecific binding to chromatographic media, membranes, and even the target biotherapeutic, such as a MAb. As aguidance to the challenges associated with removing residual DNA, early experimental work generated data with the objective of evaluating cleaning reagents. Studies were performed on Diethylaminoethyl (DEAE) Sepharose™ Fast Flow using radio-labeled calf thymus DNA to examine the cleaning effectiveness of NaCl and NaOH targeting DNA removal. The study demonstrated NaOH to be a more effective cleaning reagent compared to NaCl alone, emphasizing the importance of screening several cleaning reagents under various conditions. Screening of several cleaning reagents to monitor for the effective removal of nucleic acids from a quaternary amine anion exchanger (Q Sepharose™ Fast Flow) challenged using different control samples (7); summary of the data is shown in Tables 19.2 and 19.3. Mass balance analysis of residual DNA was used to monitor the effectiveness of cleaning. Again, NaOH was shown to be an effective cleaning reagent. In order to do a complete mass balance, DNase may be used to test for residual DNA remaining on the chromatography column as shown in Table 19.3.

19.3.2 Endotoxins

Cleaning and sanitization protocols on chromatography columns and membrane must demonstrate adequate cleaning and robustness over multiple cycles to generate a consistent intermediate product that meets all defined critical quality attributes and specifications. To accomplish

TABLE 19.2. Mass Balance of Calf Thymus DNA Clearance from Q Sepharose® Fast Flow

Fraction Tested	DNA (μg)	% of Total DNA
Starting material	699.75	100
Buffer A wash	0.79	0.11
100% B eluate	253.80	36.27
Acid wash	0.64	0.09
1 M NaOH/1 M NaCl wash	449.60	64.25
Water wash	3.92	0.56
Total DNA eluted from column	708.75	101.2

TABLE 19.3. Mass Balance of Total DNA Clearance from a Monoclonal Antibody

Fraction Tested	DNA (μg)	% of Total DNA
Total DNA in mAb	481.50	100
Flow through	0.31	0.06
Wash	0.12	0.02
Peak 1	10.00	2.08
Peak 2	270.68	56.21
100% B wash	6.08	1.26
Acid wash	4.00	0.83
1 M NaOH/1 M NaCl	64.00	13.29
2 M NaOH	3.21	0.67
3 M NaCl	0.65	0.13
Water wash	0.29	0.05
Subtotal	359.34	74.61
DNase 1 treatment	125.00	25.96
Total DNA eluted	485.9	100.6

this process control, characterization of the product, the impurity profile, and their respective levels is key to assessing purity and challenging the rigor of designed cleaning protocols. To illustrate the depletion and removal of endotoxins in a bioprocess from a microbial source, an example study is provided for a capture step from an *E. coli* homogenate. The chromatography media, a strong anion exchanger (Capto™ Q), was cleaned for 4 h with 1.0 M NaOH after each purification cycle. Column packing consistency (plates per meter), DBC (Q_B 10%), carryover of a specific impurity (endotoxin), back pressure, ionic capacity, and visible appearance of media were evaluated for 79 cycles (Table 19.4 and Fig. 19.3). Qualitatively, the results of the study show that no discoloration of the media was observed after repeated use. Quantitatively, the ionic capacity decayed from 0.19 mmol Cl^-/mL in the unused media to 0.17 mmol Cl^-/mL following the 79th cycle and the DBC decreased by 2.6% after 39 cycles and by 10% after the 79th cycle with a total cleaning time of 316 h in contact with 1.0 M NaOH.

19.3.3 Viral Inactivation and Clearance

Regulatory authorities have provided guidance on the retention of viruses on chromatography columns and membranes stated in the International conference on harmonization (ICH) Guideline on Viral Safety, "*Assurance should be provided that any virus potentially retained by the production system would be adequately destroyed or removed prior to reuse of the system. For example, such evidence may be provided by demonstrating that the cleaning and regeneration procedures do inactivate or remove virus*" (8). To challenge viral inactivation and clearance, orthogonal methods assessing infectivity (plaque assays) and clearance [Quantitative-polymerase chain reaction (Q-PCR) assays] are used to evaluate the effectiveness, robustness, and reproducibility of cleaning procedures on virus inactivation and removal steps in a bioprocess. Infectivity assays generate data on detection of viable infectious viral particles and Q-PCR analytical assays provide measurements of viral depletion in terms of both viable and inactive viruses. In conventional bioprocesses, performing viral inactivation and clearance

TABLE 19.4. Performance Measurements and Carryover Measurements after Cycling. Each Cycle Included Cleaning for 4 h with 1 M NaOH

Cycle No.	QB10% (mg/mL Capto™ Q)	Endotoxin in Blank Runs (EU/mL[a])	Column Packing Efficiency (plates per meter)
0	116	21	2,986
1	120	9.6	3,134
11	116	90	3,146
20	115	36	3,186
29	116	20	3,155
39	113	6.5	3,139
50	108	72	3,126
59	109	20	3,328
69	105	307[b]	3,180
79	104	11	2,986

[a]Clarified *E. coli* homogenate: ≈ 2.5 × 10^6 EU/mL.
[b]Thought to be due to contamination.

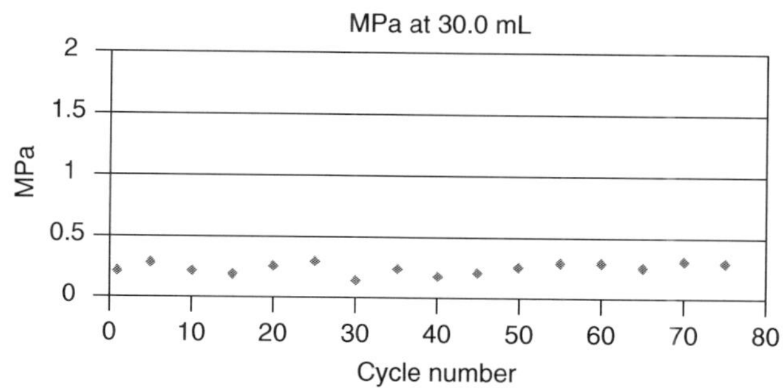

Figure 19.3. Effect of cleaning and sanitization protocol on Capto™ Q column performance. Capto™ Q was cleaned in place for 4 h with 1.0 M NaOH after each purification cycle. There was no impact on the column's back pressure profile during the 79 cycle study (total cleaning time: 316 hours in 1.0 M NaOH.) (This figure is available in full color at http://onlinelibrary.wiley.com/book/10.1002/9780470054581.)

assays on chromatographic media packed in a column or on contained membranes is a formidable challenge as there is no feasible way to measure if all viral particles have been inactivated or depleted; specifically, direct assays on the chromatographic media and contained membranes is not practical, so process models must be designed to challenge the cleaning regimes driven by relevant statistical analysis to ensure the required process control. Orthogonal solution studies are applied to demonstrate viral inactivation, and viral clearance is demonstrated analyzing column effluent by measuring both infectivity and Q-PCR following cleaning and sanitization procedures.

Viral challenge studies, including viral inactivation and clearance involve exposing a process to spiked virus to demonstrate cleaning effectiveness. The introduction of viruses into a manufacturing facility could cause contamination of product(s) and endanger the safety of operators. To develop and experimentally challenge cleaning methods, viral spiking studies are performed in specialized laboratories using scale-down models.

19.3.4 Evaluation of Cleaning Procedures at Scale-Down and Production-Scale

Cleaning procedures are typically designed using scale-down process models and validated at production-scale. Scale-down models allow for the flexibility and agility to develop and challenge cleaning procedures with a range of reagent conditions, process parameters, and different contaminants. The screening of cleaning procedures allows for the evaluation of robust protocols ensuring control is designed in to the process.

To complement the cleaning regime validated at production-scale, routine monitoring is required during commercial manufacture of the product to maintain diligence and ensure data-driven control of the process.

To demonstrate understanding and control of bioprocesses, blank (null) runs, are performed at production-scale to challenge process parameters in the absence of product. Blank runs are performed with the objective to simulate production parameters and preview the impact of key process steps, challenge procedures, troubleshoot bottlenecks, and execute sampling strategy. The detection of impurities that remain on a chromatography column or membrane can be quantified and set as reference levels.

When a production-scale chromatography column is packed with media and qualified, the maintenance and cycling frequency of the column should be assessed to ensure consistent performance and most effective management of cost. The design and implementation of robust cleaning and sanitization protocols in conjunction with routine monitoring can have an impact on extending column and media lifetime resulting in reduced production costs. In some cases, depending on process scale,

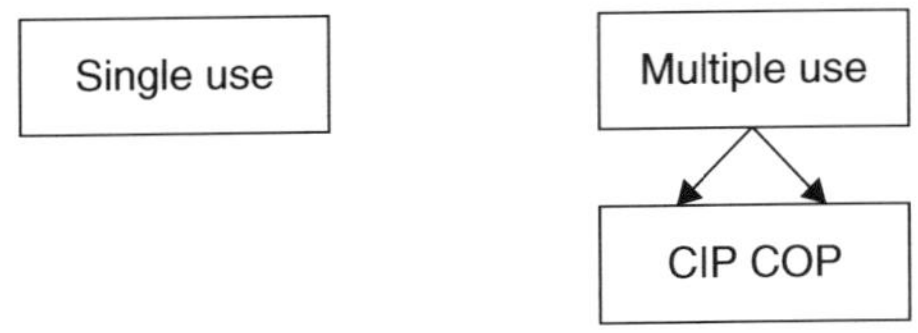

Figure 19.4. Cleaning strategies: disposable single use strategies versus multiple use strategies. Multiple-use strategies can include both Clean-in-Place (CIP) and Clean-out-of Place (COP) techniques.

development stage and process economics, single-use chromatographic media or membranes may be preferred, where following process run, media or membrane are discarded eliminating the need for cleaning procedures. Cleaning regimes on packed chromatography columns or contained membranes utilize a process termed CiP where the procedures are executed on the intact unit operations (Fig. 19.4).

19.4 CROSS-FLOW FILTRATION (CFF)

In current bioprocesses, the emergence of disposable devices to service many aspects of the downstream processing purification train is steadily becoming more prevalent as development and production data is made available. Common products implemented in disposable bioprocessing at production-scale are cross-flow filtration (CFF) devices. The use of CFF disposable devices have been demonstrated in unit operations applying both single-use and cleaning strategies, where in the reuse strategy, the device is validated for either a defined period of time or a specific number of process batches. The decision to implement disposable CFF technologies is based on cost models and risk assessments. A disposable CFF device differs from conventional nominal-flow filtration (NFF) (also described as dead-end filtration) devices in that NFF devices are typically single-use, disposable products, whereas the CFF membranes have been demonstrated to perform consistently following chemical cleaning between each batch of product for a given number of cycles. It should be noted, that independent of the nature of the unit operation, either disposable technologies or products which require defined cleaning regimes, the principle of demonstrating control over the process is the same; specifically, unit operations are required to be validated at scale for commercial manufacture.

Current trends in the Biopharma sector have moved toward the consolidated use of chemical reagents. Cleaning, sanitization, and storage reagents can be common for multiple unit operations and development screening on varied parameter and conditions such as reagent concentration, temperature, flow rate, and contact time, should be evaluated based on the required effectiveness needed for the respective step. The cleaning and sanitization of CFF membranes should be developed to effectively eliminate any bioburden that could result in lot-to-lot contamination and to restore membrane performance by removing contaminating material coating the membrane surface or trapped within the device.

CFF membrane devices fall into two design categories, open channel and screen spacer format. Open channel filtration devices can be round in shape (hollow fiber designs) or rectangular in shape (plate, sheet, and frame designs). The screen spacer configuration separates two layers of membranes to enhance flow distribution at the membrane surface, resulting in a reduction in recirculation flow and increase in flux. Filtration screens in both CFF formats enhance product separation performance enabling high flux applications and conservation of recirculation flow; screens can also trap suspended biomass; in general, open channel CFF designs are more responsive to cleaning regimes removing biomass and restoring membrane performance characteristics.

CFF membrane devices also fall into two application categories, microfiltration (MF) and ultrafiltration (UF). MF membranes are frequently used in feed stream clarification processes for the removal of cells and/or cellular debris in an open channel design. UF membranes are typically used in diafiltration and concentration steps. The principles for the maintenance and cleaning regimes for both MF and UF membranes are similar. When developing effective cleaning strategies, consideration should be given to the intended application and nature of the contaminant. Due to the asymmetric structure and design, UF membranes are typically more easily cleaned in direct comparison with MF membranes.

The primary cleaning strategy for CFF membrane devices is to chemically decompose biomass that could lead to contamination or cause a loss of filtration performance. The chemical cleaning reactions most commonly used on membranes follow first order kinetics, where, for every 10°C increase in temperature, the reaction rate is doubled. In developing chemical cleaning regimes, temperature plays an important role in reaction kinetics where cleaning rigor is less effective at ambient temperature compared to the same reaction at 50°C. Contact time is a key parameter in cleaning processes, where the duration the cleaning reagent is in contact with the membrane has a direct impact on cleaning effectiveness. As a comparison of parameters, recirculation flow and pressure have little to no effect on cleaning effectiveness.

Independent of CFF membrane design or application, it is recommended to execute cleaning procedures immediately following the process unit operation step. Efforts to rapidly reduce biomass and contaminant levels in the system have a direct impact on cleaning effectiveness. As a cleaning process operational step, it is recommended that initial CFF membrane rinse material should be directed to waste avoiding recirculating contaminants back to the system.

The operational practice of backwashing membrane filters is common in the industrial sector (beyond Biopharma). In general, hollow fiber filtration designs can withstand backwash procedures without significant risk of compromising membrane integrity, whereas spiral cartridges and some plate and frame design filtration devices have low tolerance levels to back pressure flux resulting in significant loss of integrity. In some processes where high levels of particulate biomass are observed, a short backwash cycle will improve the effectiveness of the cleaning process. To develop the most effective and robust cleaning regime, adequate development of methods on the various products and applications is a critical step to ensure cleaning rigor and demonstrate process control.

19.4.1 Chemical Cleaning of CFF Membrane Devices

Effective chemical cleaning of CFF membrane devices requires an understanding of the contaminants to be removed and of the membrane materials, design, and construction of the device. It is recommended to refer to the user manual or validation guide from the filtration supplier to evaluate the recommendations for the specific feed stream being processed. The supplier documents will typically provide experimental data on temperature, pH, and pressure operating parameters, as well as chemical compatibilities.

Similar to chromatographic media, the range of CFF membrane applications in bioprocesses are in contact with common contaminants such as host cell proteins, DNA, RNA, lipids, cellular debris, viruses, synthetic antifoams, and chemical supplements. Typically, chemical cleaning of CFF unit operations have characteristics previously described with chromatographic cleaning requirements, such as utilizing, cleaning, sanitization, and storage reagents that are commonly used in multiple unit operations; analytical assays to monitor and detect changes in contaminant profile or level; and cleaning procedures that are practical in terms of impact on CFF membranes and equipment and taking a risk-based approached to developing, implementing, and validating cleaning protocols.

The list of candidate cleaning, sanitization, and storage reagents to be screened for a given cleaning regime can be categorized as alkaline pH, oxidizers, acidic pH, enzymes, surfactants, and solvents (e.g. low molecular weight alcohols).

19.4.1.1 Alkaline pH. The most commonly used reagent for cleaning CFF membranes is NaOH, followed by potassium hydroxide (KOH). Alkaline solutions have been demonstrated as being highly effective as sanitizing agents. NaOH as a cleaning reagent is considered inexpensive at production scale (relative to other cleaning reagents), and can be effectively rinsed, analyzed, and neutralized. As previously mentioned, contact time and temperature are important factors for cleaning effectiveness. Alkaline pH solutions act to break down residual contaminants enabling predictable restoration of CFF membrane performance. A concern to cleaning regime development that can impact cleaning effectiveness and membrane performance is the supplement of lipid or antifoam agents to bioreactors during cell culture. Feed streams with lipids and silicone-based antifoams can be problematic to clean even at high basic pH ranges.

Synthetic membranes made of polyethersulfone (PES) and polysulfone (PS) are routinely cleaned using 0.5 N NaOH or 0.5 N KOH at temperatures ranging from 25–50°C. Typically, cellulose-based membranes have limited tolerance to strong basic cleaning reagents and conditions; typically performance can be restored by cleaning with 0.1 N NaOH at ambient temperature allowing for 1–2 h contact times.

19.4.1.2 Oxidizers. When the use of NaOH as a cleaning reagent does not restore CFF membrane performance in terms of flux, a common subsequent reagent to consider is NaOCl (sodium hypochlorite), which has been demonstrated to remove both DNA and lipids. There are several issues to consider when using NaOCl as part of the cleaning scheme including assessing the compatibility of the reagent with the filtration devices (including stainless steel hardware), detection after rinsing, and dissipation with time. To minimize pitting damage of stainless steel as a result of harsh chemical cleaning conditions, the cleaning reagents can be buffered to pH of 9–10. For applications where depyrogenation is critical, it has been demonstrated that NaOCl is more effective than alkaline pH alone. A typical NaOCl concentration for effective cleaning is in the range of 100–300 ppm active chlorine, which should be monitored and redosed to maintain the concentration over a 1–2 h cleaning period.

19.4.1.3 Acidic pH. Phosphoric acid (H_3PO_4) as a cleaning reagent is an alternative that has been shown to be effective in cleaning UF membranes in the food industry, specifically in dairy processing. The H_3PO_4 may be a suitable cleaning reagent for a given process. H_3PO_4 cleaning processes work in the pH range of 3–4 at a temperature of 50°C for 1–2 h cleaning period.

19.4.1.4 Enzymes. Commercial manufacturing bioprocesses typically avoid cleaning regimes that use supplemented enzymes. Though enzymes are highly effective in dissolving biomass and contaminants, there is reasonable risk associated with clearance issues of the enzyme by rinsing, as well as the risk of introducing trace contaminant with formulated enzyme supplement. There is, however, a precedent in veterinary vaccine products, where 2% Terg-A-Zyme®[1] has been demonstrated to be an important component in some cleaning protocols. Temperature ranges for enzymatic cleaning should be maintained at 50°C with a contact time of 1–2 h cleaning period. Extensive rinsing and enzyme inactivation are recommended following enzymatic cleaning procedures.

19.4.1.5 Surfactants. Cleaning reagents including 0.1% solution of Tween-80 at a temperature of 50°C may be effective for removing some types of lipid contamination. The surface active chemistries of this detergent may promote surfactant to bind to the surface of the membrane or other materials (wetted-parts) within the system. As a result, a protracted rinse may be needed. As with enzyme cleaning, a sensitive assay is critical to support the validation of complete rinsing.

19.4.1.6 Alcohol. A dilute solution of either isopropyl alcohol (IPA) or ethanol may have a similar cleaning effect as a surfactant on lipid-based contamination by significantly decreasing the surface tension of the solution. It is highly recommended that a risk-based assessment is performed prior to developing cleaning methods with alcohols. Concerns about flammability explosions, facility design, and layout are key parameters in the feasibility of using alcohols as cleaning reagents. In general, alcohol solvents are rarely implemented in cleaning CFF systems in commercial biomanufacturing processes.

19.4.2 Cleaning Standard Operating Procedures (SOPs)

Once an effective cleaning reagent has been identified, and the cleaning regime been developed, a cleaning protocol SOP can be crafted. The SOP for cleaning will include instructions on cleaning describing the various parameters important to effective cleaning such as reagent preparation, concentrations, contact time, flow rate, temperature, rinsing steps, assays for monitoring, and detection.

Cleaning SOPs may contain information such as recommendations from cassette suppliers to use 125–150% of the process cross-flow rate to enhance removal of any particulate contamination while running at low pressure to promote separation of the polarization layer from the membrane surface. Operationally, the back pressure valve should be fully open and closing or restricting the filtrate flow may accelerate the cleaning process. If the cleaning solution has high turbidity, it may be necessary to discard the cleaning solution and replace it with a fresh solution to avoid redepositing the polarizing material back on the membrane surface.

The CFF process is designed to use recirculation flow through a number of parallel pathways. In a dynamic flow path, fluids will flow through the path of least resistance, this fluid behavior is important to consider when developing cleaning procedures so that all areas receive the necessary flow to ensure effective cleaning. Flow meters and pressure gauges are required to monitor that the appropriate flow rate has been reached with each CFF membrane device.

When rinsing the system in preparation for a new production campaign, it is recommended to fully drain all areas of the filtration system and device to minimize the amount of hold-up solution needed to restore the conductivity to its required value. Table 19.5 summarizes common cleaning schemes and the contaminants they have demonstrated removing.

19.4.3 Demonstration of Cleaning Effectiveness

Robust cleaning regimes provide for contaminant removal and restoration of the original membrane performance. The most common measurement of cleaning efficacy is a simple water flux test that is performed with a new device and repeated after each cleaning. The water flux test measures the frictional resistance of water passing through the membrane under a set of controlled conditions. The viscosity of water varies with temperature, which may impact cleaning conditions and effectiveness. To adequately control this variable, equipment must be controlled and monitored to provide consistent comparisons. The following calculation is used to compare water flux on new and used devices to be used in developing cleaning procedures:

$$\text{Water Flux (WF) Recovery (\%)} = \text{WF after cleaning/Initial WF} \times 100\%$$

If the water flux test is performed under consistent conditions and the pressure readings indicate an unobstructed flow path. If the water flux is lower than expected, it is unlikely that the device is completed blocked, alternatively, the explanation is that the membrane may have residual material coating the surface adding to the frictional resistance of the flow. This results in a lower water flux reading.

It has been shown that there is limited precision in water flux testing in a process setting, and that designating the specification empirically at 100% is not recommended. From development stages of cleaning procedures through to validation testing of cleaning regime, data should be generated to determine the expected flux range with a given piece of equipment and specify that postcleaning water flux values fall within that range. To aid in monitoring use of a CFF membrane over a number of cycles, it is recommended to follow trend data that will provide information of performance decay and bioburden carryover. A decrease in water flux is often the first indication that the cleaning is not fully effective.

Cleaning effectiveness can also monitor for process time consistency. In general, there is a return to close to the initial flux values, indicating a restoration of the water flux. As processes increase in number of cycles or process time, the process flux may start to decay resulting in prolonged process times. If this trend is observed, investigate to see if lipid-based polarization is the root case, a resolution may utilize an additional cleaning agent in the cleaning process to remove lipids or antifoams.

A supplemental test to consider, that is useful in cleaning procedure development and cleaning validation, is to fill the cleaned device with a buffer solution for a prolonged period of time, sample the buffer, and then analyze the solution for any evidence of residual contaminant. Typically, biomass contains proteins that exhibit different surface interactions as a function of the ionic strength and pH of the surrounding solution. When there is only a marginal level of contaminant, a high purity water rinse may indicate the device is cleaned, whereas the process buffer may enable detection of additional contaminants.

As a final challenge to cleaning procedure development, a destructive test for confirmation of cleaning effectiveness in cassettes and spiral cartridges may be investigated. By recirculating a suitable stain throughout a CFF device,

TABLE 19.5. Commonly Used Cleaning Agents and Conditions for Removing Typical Contaminants in Downstream Bioprocessing

Type	Agents	Contaminants	Conditions
Alkaline solutions	NaOH	Proteins, vaccines, bacterial cells, pyrogen, etc.	0.1–0.5 *M* 35–50°C
	NaOH-NaOCl	Nucleic acids	0.3–0.5 *M* NaOH 200–400 ppm NaOCl
Acids	HNO_3 H_3PO_4	Nucleic acids, inorganic, etc.	0.1 *M* 35–50°C
Surfactants	SDS, Triton X-100, Tween 80	Precipitated, proteins, lipids, antifoams	0.1%, pH 4–9

[a] A combination of ALCONOX® and protease enzyme; A registered trademark of Alconox, Inc.

[b] Sterile chromatography processes are rare. Currently, superheated water is being investigated for those processes which must maintain sterility.

areas that are not fully cleaned can be detected. With a brightly colored stain and a careful disassembly process, it can be demonstrated that a given cleaning procedure is either effective or that there are areas that require more stringent cleaning methods to improve cleaning effectiveness.

19.5 EQUIPMENT

In general, there are significantly more choices for cleaning equipment than for cleaning chromatographic media or membranes (Fig. 19.5). Equipment cleaning may involve not only chemical cleaning, but may also include physical methods. Chemical cleaning methods that are harsher than those used for chromatography media, packed columns, and membranes may also be appropriate. Steam sterilization is common for some downstream processing equipment, although the need for such treatment is typically not warranted.

There are several elements to consider in cleaning downstream equipment. These include the intended used of the equipment (dedicated or multi-products), equipment design, facility design, compatibility with cleaning reagents, and analytical methods for determining cleaning effectiveness.

19.5.1 Dedicated Versus Multi-Use Equipment

As described with chromatographic media and membranes, these are materials that are dedicated to a single biopharmaceutical product; whereas, equipment hardware is commonly used for multiple products. Such variable use of equipment requires stringent cleaning, monitoring, and demonstration of the absence of any potentially harmful carryover from one product to the next emphasizing control over the process. Where feasible, replacement of components for each new product can help to minimize carryover risk. Readily replaceable wetted-parts, such as tubing, column gaskets, and screens are often dedicated to a single product. These components can be stored between production campaigns, or introduce new parts each time product is manufactured. Replacement of wetted-parts may be the more economical choice when the costs of cleaning and storage validation are considered along with available storage space.

If columns are to be unpacked and repacked for either dedicated or multiproduct use, a visual inspection should be made as a qualitative assessment that there is no obvious carryover of previously used chromatographic media.

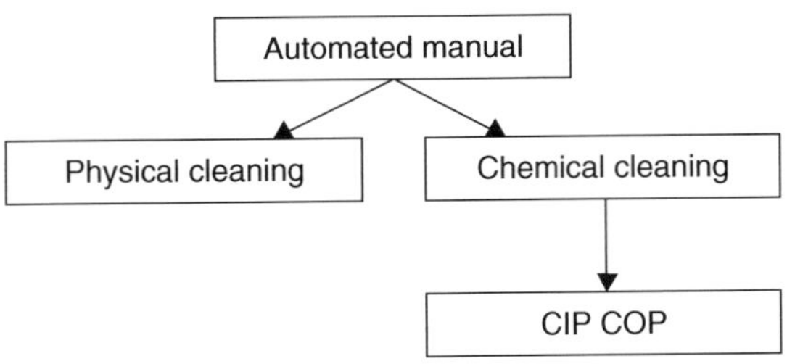

Figure 19.5. Equipment cleaning strategies: equipment cleaning may involve chemical and physical cleaning methods and can include both Clean-in-Place (CIP) and Clean-out-of Place (COP) techniques.

19.5.2 Equipment Design

Equipment design has a major impact on cleaning efficacy. Sanitary connections are used wherever feasible. Sufficient cleaning of chromatography media and columns is dependent on the chromatography skid and column design (type and size). For columns, that design should minimize crevices and dead spaces and enable good flow distribution. Equipment design has been discussed by the PDA Biotechnology Cleaning Validation Committee (9).

19.5.3 Compatibility with Cleaning Protocol

As noted in section titled "Designing an Effective Cleaning Protocol for Downstream Bioprocesses", it is essential to know what wetted surfaces are being cleaned and their compatibility with cleaning reagents. For cleaning equipment, one must sometimes address not only chemical compatibility but evaluate pressure constraints as well. The most effective chemical cleaning reagents should be sought and they should be cost-effective. If reagents such as detergents are needed, consider how they will be removed and that removal demonstrated. A detergent might provide an excellent cleaning effect for some contaminants but, as noted earlier, might also be difficult to remove from some surfaces. Ethanol in large quantities may require an explosion-proof environment. As with media and membranes, the quality of the cleaning reagents should be defined and disposal issues addressed.

19.5.4 Methods for Determining Cleaning Effectiveness

Evaluating cleaning effectiveness for chromatography media and membranes is limited practically to rinse fluid analysis; with equipment, there are more options available for physical sampling and testing. Swabbing is one option that is commonly used. Inherent in swabbing, there are potential inaccuracies that can lead to erroneous assumptions of cleaning effectiveness including operator techniques, ability of the swab to remove a given contaminant, ability to extract from the swab, physical access to swab site, etc. A sampling strategy is key to the development, and scalability of a cleaning regime.

Rinse water analysis might not be sufficient to analyze cleaning efficacy in the hard-to-reach areas in any given piece of equipment. And swabbing those hard-to-reach spots may be difficult or impossible. One solution is the use of coupons (10). Coupons are cut out pieces of equipment, usually available from an equipment supplier or subsupplier. Coupons can be treated with a worst-case soil and allowed to sit for an extended time. They may even be placed in the location of concern (e.g. an elbow in a pipe or a pipe thread). The cleaning protocol is then applied to determine its effectiveness. Swabbing and rinse analysis can then be applied to determine the efficacy of the cleaning protocol on the coupon soil.

For hazardous soils, such as those containing potentially harmful virus or Transmissible Spongiform Encephalopathy (TSE) agents, coupons offer a reasonable solution. Spiking with hazardous agents is inappropriate unless it is done in a controlled and safe environment. To minimize risk and reduce cost, such studies are performed on scale-down models using coupons.

Whether dedicated or intended for production of multiple products, downstream processing equipment must be routinely monitored for cleaning effectiveness. On-line TOC units are used to monitor and validate cleaning of chromatography skids and column hardware. Figure 19.6 illustrates sampling for TOC analysis of a column for cleaning validation for a manual direct (swab) sampling process. The sampling is performed at the sites that are most difficult to clean or classified as "worst-case." Sampling the sites can be difficult at times and exposes the equipment to potential contaminants. Therefore, more and more bioprocess manufacturing facilities are implementing on-line TOC analysis with supporting in-line monitoring methods including UV, pH, and conductivity. Some TOC instruments are specifically designed to handle fluctuations in operating parameters such as pH, conductivity, flow, and temperature. Such instruments have a selective membrane-conductometric detection system used in the determination of true TOC content within the sample. The patented technology is unique and is not subject to interfering compounds that contain nitrogen, sulfur, or halogens such as chlorine that may be present during the cleaning process (11).

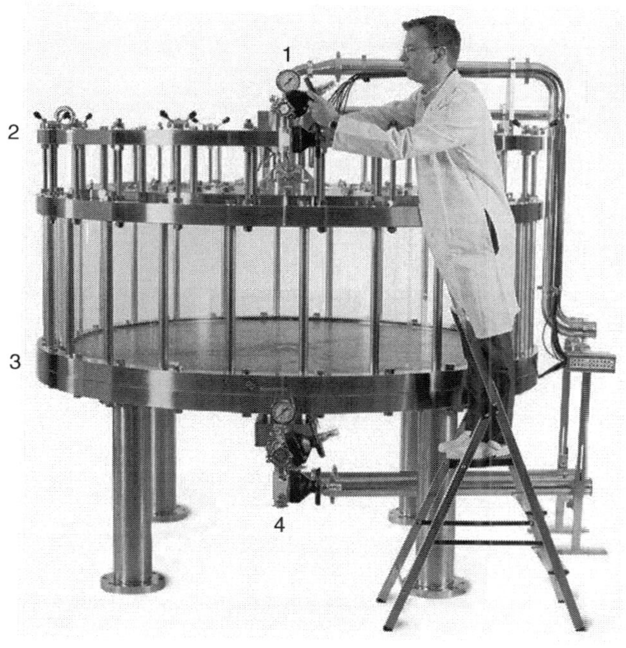

Figure 19.6. Cleaning validation sampling of chromaflow column for total organic carbon (TOC) analysis. Sample sites 1 through 4 were classified as "worst-case" sites or the most difficult to clean. (This figure is available in full color at http://onlinelibrary.wiley.com/book/10.1002/9780470054581.)

19.6 SANITIZATION AND STERILIZATION

Sanitization removes or otherwise eliminates vegetative microorganisms that may proliferate during storage and result in contamination. Typically, the sanitization process is also intended to depyrogenate (e.g. endotoxin removal).

As noted in section titled "Designing an Effective Cleaning Protocol for Downstream Bioprocesses" above, it is common to combine cleaning and sanitization applications in a cleaning strategy. But even when microorganisms are removed, they may leave behind toxins or other harmful substances. There is a concern that a low level of bioburden (e.g. bacteria and fungi) might increase during processing when growth conditions are favorable. Protein deposits and neutral pH conditions can provide an ideal environment for microbial growth. With few exceptions, downstream processes are aseptic, not sterile. Routine cleaning helps maintain low levels of bioburden.

Challenges with microorganisms are often performed by equipment suppliers. These studies are intended to identify those areas to which special attention should be paid for routine cleaning and sanitization, and they typically demonstrate conditions that are suitable for sanitization. In these studies, the microorganisms used as a challenge are the same as those used to test WFI (12). As previously discussed with viral inactivation and clearance studies, introducing microorganisms into a biopharmaceutical manufacturing facility is unacceptable due to contamination risk, so any microorganism spiking studies have to be carried out away from that production environment. Since the real risk to a process is in the environment in which it is run, routine monitoring is the most feasible approach.

Equipment packed chromatography columns, and membranes should be monitored for bioburden prior to use and should have established limits. The use of rapid micro methods (RMMs) may enable less processing at risk for downstream processing unit operations (13).

It is common practice to store both packed chromatography columns and membranes, and even entire systems, in aqueous solutions that minimize microbial growth. NaOH has become the standard storage solution for packed columns and filtration systems. However, there may be situations where NaOH is inappropriate due to labile ligands or equipment components that are not compatible with alkaline solutions. Ethanol is also sometimes used to inhibit microbial growth, but explosion-proof issues may arise depending on ethanol concentration and volume. For filtration, 0.05–0.1 M NaOH can typically be used for up to 6 months' storage at room temperature and 0.1–0.5 M NaOH is, in most cases, appropriate for up to 1 year of storage at 4°C.

With systems in daily use or in the case of a system which has been in storage for an extended period of time, it is best to sanitize prior to reuse to eliminate any potential source of microbial contamination. Fortunately, there has been very good success with the use of the same solutions for cleaning and sanitization.

Specifically, for filtration, to maximize the effectiveness of the sanitization process, the recirculation flow should be kept high and all air must be voided from the system. Open all appropriate valves to ensure fluid contact with all product contact surfaces. As in the case of chromatography equipment and all other process components, swab testing and microbial monitoring of the process stream will confirm that the sanitization scheme has been successful.

19.6.1 Sterilization of CFF Equipment

There are certain process applications requiring that the entire CFF system be sterilized.[2] These are most often for upstream operations or for the systems used in the final formulation of large molecules (e.g. viruses or vaccines). For sterile CFF operations, the equipment must be either supplied as sterile or be capable of being rendered sterile by the user in a validated process.

As disposable equipment becomes more popular, it may become common practice to employ presterilized components (a closed system which includes membrane devices, tubing, and bags) which have been preassembled in a hood or other controlled environment to produce a suitable sterile system. With the use of specialized connectors, these systems can be safely connected to bioreactors and other process equipment.

For production-scale processes with more than a square meter of membrane area, it may be more practical to sterilize the entire system. The two most common sterilization methods are autoclaving or steam-in-place (SIP). In both cases, it is critical to designate a system design with suitable materials of construction that can withstand the high temperatures used in these operations. In some cases, SIP is performed on systems without the filtration modules in place when the devices themselves are not compatible with high temperatures. This would not be a sterile process but rather employed as a precaution to eliminate any possible sources of microbial contamination.

19.7 CLEANING VALIDATION

If a scale-down study is used for cleaning validation, then it must represent manufacturing scale. This requires a validated small-scale model. Wetted materials, system configurations, flow cell size proportions, and flow distribution systems are likely to be slightly different at different scales. The small-scale downstream processing systems, therefore, are unlikely to be identical to those used in production but it should provide the same performance. Performance is determined by evaluating purity, impurity profiles, and yield. The feed stream should be obtained from manufacturing scale and buffers and cleaning solutions prepared the same as in manufacturing. In order to carry out cleaning validation, all the assays that will be used must be validated along with the relevant sampling protocols.

Cleaning validation is typically performed prior to Phase III clinical trials, once the process is finalized. However, there are certain aspects of cleaning validation that are carried out prior to beginning any human studies. Where there is a patient safety risk, that risk must be mitigated. It must, therefore, be demonstrated that any harmful agents will be inactivated or removed. The use of disposables during early clinical trials can minimize cleaning validation and ensure patient safety.

19.8 CONCLUSIONS

There is a great deal of available information on cleaning, sanitization, storage, and cleaning validation. Basic principles are described in several publications, but much of the information is dated (14,15). The publications provide general information but do not address the specifics of cleaning chromatographic media and membranes. In downstream processing, the first decision to make is the intended scale of the process and the economics at production-scale to assess the needed equipment and facility design and process unit operation configuration. The decision to move forward with disposable components or to recycle has a significant impact on cleaning validation studies and risk-based approaches to efficient biomanufacturing. Once a decision is made to recycle, it is essential to design a robust cleaning protocol that is tested with production feed stream and validated, typically by a combination of scale-down and production runs. A critical element is selection of practical analytical tools to monitor cleaning performance. Setting specifications for cleaning effectiveness is a challenge and requires an in-depth risk assessment. Implementation of scientific and technological advances will enable more in-process measurements and feedback controls for cleaning of packed columns and multiproduct systems.

REFERENCES

1. Lowry SA. Designing a contamination control program. In: Prince R, editor. Microbiology in pharmaceutical manufacturing. Bethesda (MD): PDA/DHI; 2001. pp. 203–265.
2. Hale G, Drumm A, Harrison P, Phillips J Repeated cleaning of Protein A affinity column with sodium hydroxide. J Immunol Methods 1994; 171: 15–21.
3. Seely RJ, Wight HD, Fry HH, Rudge SR, Slaff GF. Validation of chromatography resin useful life. Biopharm 1994; 7: 41–48.
4. http:// www.fda.gov/cder/guidance/cGMPs/equipment.htm# TOC.
5. Fourman GL, Mullen MV. Pharm Technol 1993; 17: 54–60.
6. ICH Q9 Quality Risk Management. www.ich.org.
7. Dasarathy Y. Biopharm 1996; 9: 41–44.
8. International Conference on Harmonization. Guidance on viral safety evaluation of biotechnology products derived from cell lines of human or animal origin. Fed Regist 1998; 63: 51074–51084.
9. PDA Biotechnology Cleaning Validation Committee. Cleaning and cleaning validation: a biotechnology perspective. Bethesda (MD): PDA; 1996.
10. Campbell J. Validation of a filtration step. In: Rathore AS, Sofer G, editors. Process validation in manufacturing of biopharmaceuticals: CRC Press; 2005. pp. 205–275.
11. GE Water and Process Technologies, Analytical Instruments Application Notes 1 and 2.
12. Available from ATCC (American Type Culture Collection), www.ATCC.org.
13. Moldenhauer J. Rapid microbiological methods and the PAT initiative. Biopharm 2005; 18: 31–46.
14. 21 CFR 211.67: Equipment cleaning and maintenance.
15. U.S. FDA. Guide to Inspections of Cleaning Validation; 1993.

20

CLEAN-IN-PLACE

PHIL J. BREMER AND RICHARD BRENT SEALE

Department of Food Science, University of Otago, Dunedin, New Zealand

20.1 INTRODUCTION

Clean-in-place (CIP) is the cleaning of the assembled plant (vats, fermenters, tanks, processing equipment, probes) and pipeline circuits (values, flowmeters, gaskets) by the jetting or spraying of surfaces or the circulation of cleaning solutions under conditions of high temperature, turbulence, and flow velocity (2). CIP involves little or no manual involvement from an operator. Cleaning efficiency is due to mechanical, thermal, and chemical activity.

20.2 THE REQUIREMENT FOR CIP SYSTEMS

In most biotechnology and pharmaceutical plants, production generally involves a series of bioreactors and equipment for downstream processing. Raw materials, intermediate and final products move through the plant in transfer piping under the control of a wide range of flow devices (pumps, valves, and probes). In the majority of plants bioreactors are operated in batch mode under tightly controlled conditions to produce products of tightly controlled specifications. To maintain the quality, consistency, and safety of the final product all product contact surfaces need to be maintained to exacting hygienic requirements. An effectively designed, validated, operated, and monitored CIP system enhances cleaning and helps to ensure the production of consistent high quality products.

CIP systems originated in the dairy industry where prior to the 1950s, manual cleaning methods were used to clean increasingly complex and production driven plants (3). Manual cleaning methods typically involved the complete disassembly of process piping systems and machinery, followed by the scrubbing of product contact surfaces and the application of detergent solutions. This was followed by rinsing with water, the application of a sanitizing agent, another rinse and finally the reassembly of the piping system and equipment. This process was very labor intensive, not very reliable, and required long periods of plant downtime between production runs.

A CIP system consists of piping for distribution and return of cleaning agents, reservoirs for the storage and standardizing of cleaning solutions, fluid distribution devices such as heat exchangers, spray heads, pumps, valves, sensors, gauges and recording devices, and a programmable control unit which enables automation. CIP systems while representing a significant capital investment have a number of advantages over manual cleaning systems.

Compared to manual cleaning methods, CIP systems increase cleaning effectiveness and reproducibility, reduce cleaning costs, and enhance operator safety. The effectiveness of cleaning is enhanced by the elimination of human error and the assurance of uniformity and reproducibility possible with automation and well-designed monitoring regimes. Accidental product contamination due to operator error is prevented by foolproof system design.

Savings occur due to reductions in water and chemical usage and discharge, since a well-operated CIP regime generally requires a lower volume of water, chemicals, and steam than manual cleaning. In addition, as manual disassembly of the plant is not required and as the cleaning processes are automated, costs associated with labor and plant downtime due to cleaning are dramatically reduced. The safety of production and cleaning personnel is enhanced, as

Downstream Industrial Biotechnology: Recovery and Purification, First Edition. Edited by Michael C. Flickinger.

there is little operator exposure to cleaning chemicals and there is no requirement for operators to physically dismantle or clean the equipment.

20.3 GENERAL OUTLINE OF A CIP REGIME

The operating parameters and chemicals used in a CIP regime vary tremendously depending on the design of the facility, the materials being cleaned, the nature of the product residues (soil) that need to be removed, the level of hygiene required, the microorganisms of concern, and constraints that exist around safety or costs. A typical CIP cycle is a sequential series of steps that are arranged to achieve different cleaning or process objectives. The cycle may be designed to rinse and clean, or to rinse clean and steam (sanitize) and are built on the following steps.

Program Initiation. This step is the start of the CIP regime. It establishes that the product cycle is complete and the CIP circuit boundary and flow devices (pumps, valves, pipelines) are available and on line.

Product Recovery. This is the removal of bulk product from the production system immediately after production ceases. It is carried out in order to reduce product losses.

Prerinse. The purpose of this step is to remove as much residual media, product, or waste material as possible prior to cleaning. Removal of most of the organic fat, carbohydrate, or proteinaceous soil is generally accomplished using ambient or warm water. Viscous solutions may require scrapping or, in pipelines, the use of a pig (solid material plug propelled along the pipe to push out product).

Cleaning (often Alkaline or Acid Washes). The purpose of this step is to remove as much product soil and associated microorganisms as possible from the product contact surfaces. Since relatively long contact time is required, recirculation of cleaning solution is essential for economical operation.

Intermediate Drain. This step is used between CIP phases to actively transfer spent wash and rinse solutions from the circuit boundary to enhance a clear transition between dissimilar phases.

Postrinse with Water. This step is designed to rinse away most of the cleaning solution. The postrinse water is sometimes recovered for the prerinse in the next CIP cycle.

Sanitizing Step. The purpose of this step is to kill any bacteria that have not been removed or inactivated by the cleaning steps. This can involve either a chemical treatment or heating by the application of steam, normally used in biopharmaceutical and biotechnology plants.

Final Rinse Phase. The final rinse is performed with water of suitable quality to meet product specification. This step ensures that product contact surfaces do not contain unacceptable levels of chemical residues. In many pharmaceutical applications, the final rinse water is tested to confirm that it is not contaminated with residual cleaning chemicals or soil. Testing usually involves monitoring the conductivity of the rinse water to provide additional assurance that the process equipment has been adequately cleaned. The postrinse is sometimes heated to speed up drying.

Final Drain Phase. (*Gravity*.) This phase begins by opening all drain valves, providing for gravity drainage to and from all CIP circuit low points.

Program Completion. This step confirms that the circuit boundary is clean and isolated from the product processes and switches all devices (valves, transfer systems, and pumps) to their safe state. A generalized CIP for a single-use system is shown in Table 20.1.

TABLE 20.1. Generalized CIP Program

Step	Action
Program initiation	Confirm CIP circuit boundary and flow devices are available and on line
Prerinse	Flush to remove all free material
Intermediate drain	Drain return side, CIP supply side remain charged
Alkaline wash	Establish circuit recirculation, add in alkaline solution (1.0% NaOH at 65°C, 10 min)
Gas blow and drain	Clear CIP line of chemicals, drain lines for effective minimum volume rinse
Postrinse with water	Flush circuit of spent alkali
Intermediate drain	Drain return side, CIP side remain charged
Acid wash	Establish circuit recirculation, add in nitric acid solution (1.0% $H_2\,NO_3$ at 75°C, 10 min)
Gas blow and drain	Clear CIP line of chemicals, drain lines for effective minimum volume rinse
Postrinse with water	Flush circuit of spent acid
Intermediate drain	Clear CIP line of chemicals; CIP side remains charged
Final rinse phase	Flush with high quality water to defined endpoint
Gas blow and drain	Clear CIP line of water, drains lines
Final drain phase (gravity)	Gravity drain of CIP low points
Program completion	Release clean CIP boundary

20.4 CIP CHEMICALS

Cleaning is designed to reduce surface chemical fouling and the number of microorganisms on the surface. Cleaning is carried out via the addition of detergent formulations and the application of mechanical (flow) and thermal energy. Sanitizing is designed to kill any microorganism remaining after the completion of the cleaning step. Sanitizing occurs through exposure to heat (usually steam) and/or disinfectants (sanitizers). There are a number of factors to be considered when selecting particular cleaning agents for use in a CIP system, including the nature of the soil to be removed, the materials used for construction, cleaning temperature, the time available for cleaning, the effectiveness of the chemicals and their cost and environmental impact (2,4,5).

The nature of the material that needs to be removed by the CIP regime is dependent on the product being processed, the treatment applied to it, and the processing environment. Like the media streams they are formed from, deposits can consist of carbohydrates, lipids, proteins and minerals, and microorganisms, either from the raw material or more usually added from stock cultures. The type of contaminating material plays a large role in dictating the most appropriate cleaning compound (Table 20.2).

20.4.1 Cleaning Chemicals

The effectiveness of a CIP regime is greatly dependent on chemical action. The most effective step in a CIP system is the alkaline step because, as a general "rule of thumb" most residues are easily removed with a dilute (1% or less), formulated alkaline cleaner. The three steps required for effective cleaning are as follows (6):

1. separation of the soil from the substrate;
2. dispersion of the soil in the detergent medium;
3. prevention of soil redepositing on the substrate.

The most common and aggressive alkali cleaner is sodium hydroxide (NaOH). It is typically used in concentrations between 0.15 and 1.0% at temperatures in the range of 70–80°C for 10 to 30 min. However, for heavily soiled surfaces, especially those containing burnt on protein, such as found in plate-type and tubular heat exchangers, concentrations up to 5% are used. The active part in the caustic is the hydroxyl ions (OH^-), which form the bulk of the cleaning fluid. These ions rapidly transfer to the solid–liquid interface on the surface to be cleaned where the high pH of the solution causes deprotonation of functional groups present within proteins, leaving a net negative charge. These negative charges repel one another causing protein to swell and become translucent (7). The subsequent removal of this swelled surface is by the mass transfer of material into the bulk of the solution. In order to increase the effectiveness of the caustic wash step, other compounds can be added to the basic caustic (8). The addition of oxidizers and complexing agents can increase efficiency by a factor of 10 compared with a pure 0.25% sodium hydroxide. Sodium hypochlorite is added to alkaline solutions to enhance the removal of both fat and protein soiling, and chelating agents can be used in systems where hard water is a problem. Silicates and wetting agents can also be added to alkaline cleaners to improve cleaning and to reduce their potential to cause corrosion.

TABLE 20.2. Types of Cleaning Compounds for Soil Deposits

Type of Residue	Required Cleaning Compound
Inorganic material	*Acid-type cleaner*
Oxidized iron films, hard water scale, milk stone	Nitric acid, phosphoric acid, acetic acid
Water-soluble material	*Water*
Inorganic salts, sugars, starches	
Organic material—nonpetroleum	*Alkaline-type cleaner*
Fatty acids, blood, proteins, Fats	NaOH, sodium metasilicate, sodium carbonate
	Chlorine-based cleaner
	Sodium hypochlorite, calcium hypochlorite
	Synthetic detergents
Organic material	*Solvent-type cleaner*
Petroleum	Ether or alcohol based

20.4.2 Acid Detergents

An acid detergent wash generally follows an alkaline wash in order to aid in the removal of any traces of alkaline product from equipment surfaces, enhance draining and drying, provide bacteriostatic conditions that delay the growth of organisms that can be found in the water supply and very importantly, for some systems, to remove mineral deposits such as hard water stone, beer stone, calcium oxalate, or milk stone. The acid step requires critical temperature-concentration combinations to be completely effective. The most common acid detergent is nitric acid, which is generally used at a concentration of 0.5–1.0% under either ambient or heated conditions (55–80°C) for 5 to 20 min (9). For material where corrosion is a concern organic acids such as lactic or acetic acid may be preferred. As with caustic chemicals, acid detergents can be formulated to contain compounds such as surfactants which improve their surface wetting, soil penetration, and cleaning properties.

20.4.3 Sanitizers

Sanitizers are used primarily to kill vegetative cells that remain on the surface after "cleaning" (10). In CIP systems, the main sanitizers used are either chlorine (calcium or sodium hypochlorite) or organic acid based (acetic, peroxyacetic, lactic, propionic, and formic acid).

Sodium hypochlorite based sanitizers are widely used as they have many features which make them desirable for CIP applications. They are active against a wide range of microorganisms, including gram-positive and gram-negative bacteria, bacterial spores, and viruses. Additionally, they are nonfoaming, do not leave an active residue and are widely available and relatively inexpensive. However, they have a number of disadvantages. They can be corrosive to a wide range of materials, including stainless steel, are unstable at the concentration they are usually used at (200–400 ppm), are inactivated by inorganic material, may give rise to taint problems in industries such as brewing, and need to be handled with care by operators as they are irritating to skin and eyes.

Organic acid sanitizers neutralize excess alkalinity that remains after alkaline cleaning, prevents the formation of alkaline deposits as well as sanitizing. Their efficacy is dose and time dependent and they can be added into the final rinse of a CIP regime, they are effective against a broad range of microorganisms, in particular, yeasts, viruses, and psychotropic bacteria, they are relatively unaffected by organic matter and have combined acid cleaning, free rinsing, and sanitization properties. They can also be held overnight with little danger of corrosion being enhanced. While some acid sanitizers have the tendency to foam, nonfoaming acid synthetic detergent sanitizers have been developed for CIP systems. Disadvantages of acid based sanitizers are that their effectiveness decreases as pH increases above pH 3 and they are more expensive than chlorine-based sanitizers.

Peroxyacetic acid or peracetic acid (PAA) contains as its major ingredients PAA and hydrogen peroxide. In recent years the use of PAA as a sanitizer has increased, as it is highly effective against a broad spectrum of bacteria and spores. It is also nonfoaming, has low corrosiveness, functions extremely well under cold conditions (4°C), which means that sanitization can be carried out on processing equipment at ambient temperatures and it has tolerance to hard water. PAA also has favorable biodegradability, breaking down to acetic acid (vinegar), water, and oxygen. Disadvantage associated with the use of PAA in CIP regimes is that it is more expensive to apply than hypochlorite, it has a pungent odor, is highly toxic, a potent irritant, and a powerful oxidizer. Thus, great care must be taken in its use.

The other common sanitizers such as quaternary ammonium compounds, iodine, and acid anionics are not suitable for use in CIP systems as they form excess foams and in some cases can form films on surfaces, making them difficult to rinse off and leading to possible contamination problems.

20.5 CIP DESIGN AND CONSTRUCTION

To achieve the most effective CIP results, it is necessary to design the production process and the CIP components and circuits simultaneously, giving equal consideration to production and cleaning requirements. CIP is seldom efficient as an afterthought. All bioreactor, processing, and pipe circuit surfaces need to be designed to be accessible to the CIP solutions and made of corrosion resistant material and elastomers. Stainless steel surfaces are commonly used in pharmaceutical processing lines. Typically stainless steel type 304 is used for non–product contact surfaces with the more corrosion resistant 316L being used for surfaces that come into contact with the product and will therefore be exposed to high temperatures and cleaning solutions. Product contact surfaces should be free of crevices or pits, which could protect soil or bacteria from the shear forces required for removal. Typically, product contact surfaces are polished to achieve an arithmetic roughness (R_a) value between 0.4 and 0.5 μm. Electropolished surfaces, which have a R_a value of 0.3 μm, should be used for surfaces used in sterile operations.

All bioreactors, processing equipment, and interconnecting piping must be drainable to prevent the formation of standing pools of water which can support microbial growth and dilute and reduce the temperature of chemical solutions. Corners in tanks, vessels, or bioreactors should be rounded with a minimum radius of 25.4 mm. Flat bottomed tanks should have a minimum slope of 1 in 45 from the back to the outlet nozzle. The side to center slope should be 1 in 24. All parts of piping and ductwork must be continually sloped to drain points at a ratio of 5–10 mm (vertical drop) per 1 m (horizontal length).

In stainless steel transfer systems, welded joints should be used where possible for permanent connections, and welds should be continuous, smooth, and crevice free. Butt welds should be ground flush with internal surfaces and lap welds should be contoured to ensure satisfactory drainage. If clamp-type joints are unavoidable, they should be designed so that the gasket is positioned with a flush interior surface to prevent product buildup.

To ensure removal of soil from pipeline internal surface during CIP operation high flow rates are required. Typically, a minimal flow velocity within transfer piping should be 1.5 m/s, which provides a Reynolds number (Re) of around 10,000, which is well above that required to achieve turbulent flow ($Re > 2100$). In most systems flow rates equating to Re values of around 30,000 are preferred (11). Turbulent flow ensures a high surface shear rate which helps lift and remove soil from a surface and promotes good radial mixing, heat transfer, and mass transfer.

Piping should be free of dead spaces and pockets, such as dead ends and branches, which prevent the removal of soil. If unavoidable, branches or T's must be either located in a horizontal position and limited to less than two pipe diameters (length/diameter ratio of 2) to ensure fluid mixing, or be cleaned-through during the CIP. Vertical dead ends are undesirable in fluid processes because entrapped air prevents cleaning solution from reaching the upper portion of the fitting. The minimum radius for pipe bends should be equal to or exceed pipe diameter. The support system provided for the piping and ductwork should be of rigid construction to maintain pitch and alignment under all operating and cleaning conditions.

A well-designed CIP system will automatically clean the processing equipment and interconnecting piping with the minimum number of cleaning circuits (Fig. 20.1). This is achieved by utilizing process lines as CIP supply/return lines, combining lines in series in a circuit, or by combining lines in parallel in a circuit. These so called integral systems typically yield superior cleaning results compared to other piping configurations, as the number of connection points between the dedicated cleaning and processing piping is minimized. In all cases it is important to isolate dedicated CIP piping from process piping and protect against product contamination from cleaning fluids via the use of fail safe valves or removable transfer lines. If valves are used for isolation they should be arranged in a double block and bleed configuration, so that there are two valves in series between the process piping and the dedicated CIP piping (the double block). The space between the valves is vented to the atmosphere during isolation (the bleed). When linking separate process lines in series during cleaning, piping should ideally be of similar diameter to ensure hydraulic balancing and effective cleaning, or pumps should be sized to ensure minimum flow requirements in the largest diameter pipes despite flow restrictions imposed by the presence of smaller diameter pipes in the circuit.

All pumps, valves, and fitting (flow, temperature and conductivity sensors) used in CIP and process systems should be self-cleaning and of sanitary design. For hygienic applications in CIP and process systems, either rising stem or diaphragm flow control valves are generally used due to their long history of reliable use and their ability to be cleaned.

The CIP of vessels, such as bioreactors and storage tanks, is carried out by spraying solutions onto the product contact surfaces via pressure spray devices located inside them (Fig. 20.2). These spray devices can either be static or dynamic (rotating or oscillating) and are typically used to clean all areas of a particular unit. Static spray heads are typically shaped as balls (spray balls), but may take the shape of tubes or bubbles. They have no moving parts and therefore usually provide trouble free operation with minimal maintenance.

Dynamic spray balls have the advantage of being able to clean a much larger area (vessel) than a single static spray ball. Further, as they can provide 360° coverage, the quantity of solution required to coat all surfaces can be considerably less than required for a fixed spray ball where virtually all the solution is applied to the top head and

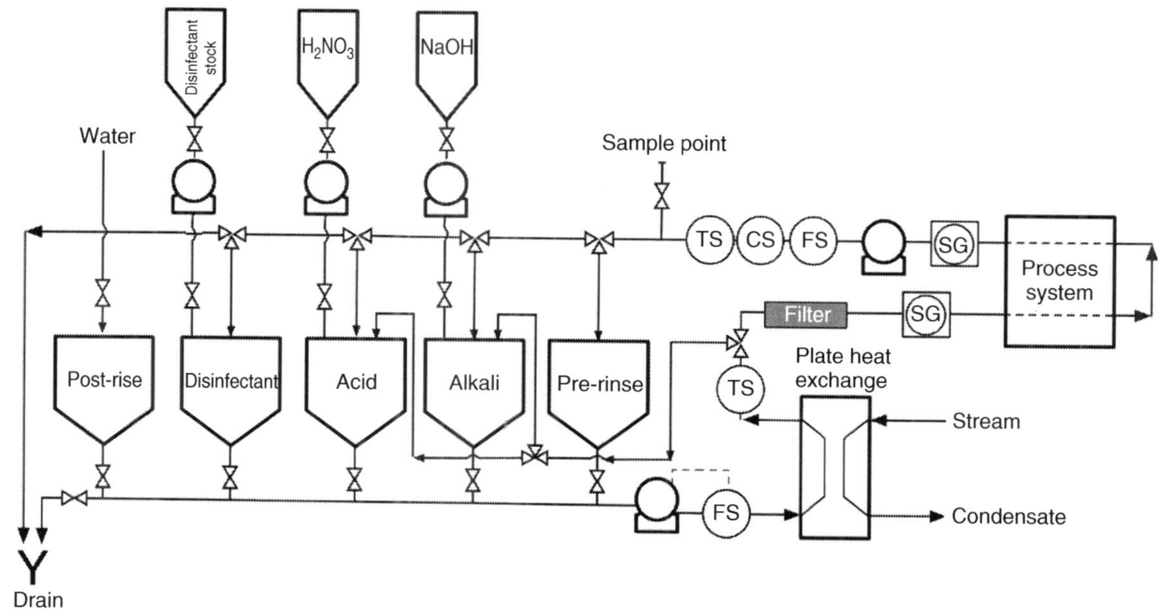

Figure 20.1. A typical multiple tank CIP system with a pumped return. FS, flow sensor; CS, conductivity sensor; TS, temperature sensor; SG, sight glass.

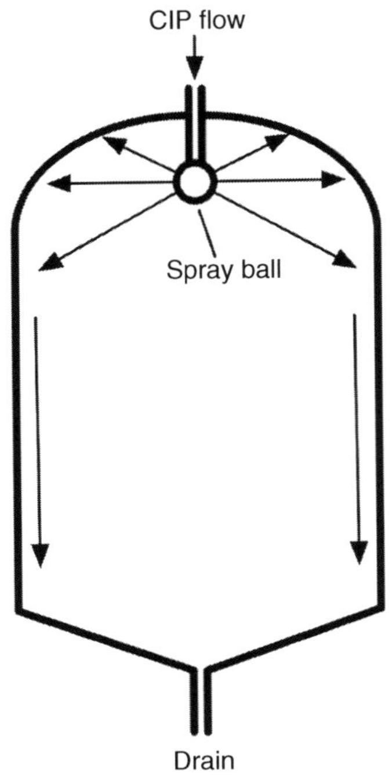

Figure 20.2. Side view of a vessel with spray device positioned at the top to create a spray and cascade action along with a sloped bottom to ensure draining.

allowed to flow down the side walls. Dynamic spray balls need to be self-cleaning, particularly when recycled wash solutions, which may contain foreign particles, are in use. If spray balls stick or wear they may give a distorted spray pattern, causing incomplete cleaning.

Spray balls are designed to suit specific vessel volume and configuration (Fig. 20.3). They are usually self-draining and contain drilled holes in a particular pattern in the head to allow a spray pattern to cover all areas of a vessel including probes, agitators, nozzles, manways, feed tubes, coils, baffles, and areas shadowed by them. In many cases more than one spray ball may be required (Fig. 20.4). Spray heads should be positioned at the top of the tank as close as possible to the vertical central axis. In vertical cylindrical tanks without internal projections the recommended spray ball flow rate is 25 Lpmm (liters per minute per meter) of the tank circumference (12). Generally, spray heads are designed to spray onto the upper third of the tank with the remainder of the tank being cleaned by the fluid film running down the sidewalls and across the bottom. In vessels that contain a large number of head nozzles and other interruptions to a smooth surface flow, directionally drilled spray heads should be used to specifically target nozzles, agitators collars, probes, and so on. It is recommended to use 3.8–5.7 Lpm for each head nozzle and 25 Lpm/m of circumference based on diameter for manway and agitators collars. These flows are in addition to the 25 Lpm/m of vessel circumference, meaning that the total flow rate through spray heads for vessels typically average around 38 Lpm/m of circumference.

Spray balls typically operate at pressure of 20–25 psig, higher pressures may create a fine spray (atomization), which is ineffective at scouring the surface and therefore reduces cleaning effectiveness. Too low a pressure will reduce the force of jet impingement decreasing the effectiveness of cleaning. High-pressure dynamic spray heads rely on the scouring action of the jet for cleaning. Because the jet continuously moves in three dimensions, the entire vessel wall is not irrigated, which results in a lower flow rate. Movement of the jet is essential for effective cleaning and therefore motion detectors may need to be installed to ensure effective functioning of the dynamic cleaning heads. Jet spray cleaning requires higher pressures, typically 30–40 psig.

Filters should be incorporated into the CIP system as it is important to reduce the possibility of spray ball becoming blocked, especially those in relatively inaccessible places. The CIP inlet pipe connecting to the spray ball and the region directly above the spray ball also needs to be cleaned. Typically, those areas are cleaned by an upward discharge of CIP fluids from engineered clearances between the spray ball sleeve and the inlet pipe (Fig. 20.3). It should be noted that adequate venting of vessels such as bioreactors is important in order to avoid collapse of the vessel, due to the vacuum created during the CIP when a cold water rinse immediately follows a hot wash step.

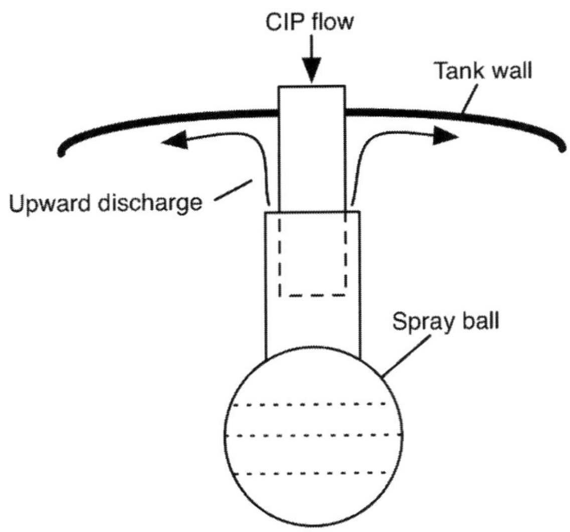

Figure 20.3. A stationary spray ball. The outer surface of the supply pipe and tank wall directly above the spray ball are cleaned via the upward discharge of the cleaning fluid through specially designed clearances.

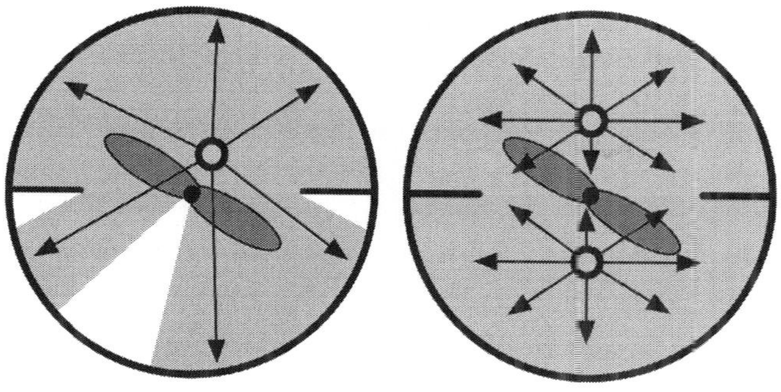

Figure 20.4. (a) Top view of a vessel displaying the shadowing of spray to areas of a tank due to offset spray ball placement and obstructions. (b) Double spray ball placement to ensure that all areas of a tank are receiving adequate cleaning. (This figure is available in full color at http://onlinelibrary.wiley.com/book/10.1002/9780470054581.)

The exterior of the CIP system should also be designed with hygienic design principles in mind. It must be constructed and maintainable to ensure that the equipment can be effectively and efficiently cleaned and sanitized over the life of the equipment. Construction materials must be completely compatible with the product, environment, cleaning and sanitizing chemicals, and the methods of cleaning and sanitation. Electrical, hydraulics, steam, and air connections must be sealed, accessible, and cleanable. All parts of the equipment should be free of niches such as pits, cracks, corrosion, recesses, open seams, gaps, lap seams, protruding ledges, inside threads and bolt rivets, and readily accessible for inspection, maintenance, cleaning and/or sanitation. Maintenance enclosures (e.g. electrical control panels, chain guards, belt guards, gear enclosures, junction boxes, pneumatic/hydraulic enclosures) and human machine interfaces (e.g. push buttons, valve handles, switches, touch screens) must be designed, constructed and be maintainable to ensure that water, or product liquid does not penetrate into, or accumulate in or on the enclosure and interface.

20.6 CIP CONFIGURATION

CIP systems can be categorized as being centralized, local, or satellite. In centralized systems, CIP circuits run from the CIP storage tanks used to hold and monitor the cleaning fluids and reuse rinse, to the processing equipment. While this system is generally effective, it becomes problematic in larger plants where the long pipe runs required result in increased energy costs for pumping and maintaining temperature, and buildup of residual rinse water can cool or dilute the cleaning fluids. A further disadvantage of a centralized system is that the whole plant is dependent on the central station, and if it fails, the whole plant is affected.

Local systems employ a number of CIP tanks, pumps, and pipeline circuits to clean individual or related parts of the processing equipment. While this system protects against a complete failure, it requires a greater number of controllers, pumps, heating, units and tanks. The satellite system is a combination of central and local systems. In the satellite system there are central solution tanks and the local units draw the solution from them, generally heating it locally.

CIP systems are further classified as single-use, reuse, or multiuse systems, depending on whether the same cleaning solution is used for one, many, or a few cycles of cleaning. In single-use units, a minimum amount of CIP chemicals are used once and discarded. This system is best suited for cleaning small plants, heavily fouled pieces of plant such as heat exchangers, or where cross contamination is a concern. While these systems are simple to install and operate, they can have high costs in terms of heat input, cleaning solution and effluent discharge.

The reuse system provides for the recovery and reuse of cleaning and rinse solutions. Rinse solutions are often subsequently used as a prerinse or a flush during the next cleaning event. The recirculated cleaning solution is stored for the next cleaning event. More detergent is added to counteract the loss of detergency due to use, with the solution being discarded once it becomes too dirty for use. If only the cleaning solution is used, then the original supply tank may be sufficient for storage. However, if the rinse water is also collected, larger collection tanks may need to be added, resulting in the need for larger floor space for the entire CIP system.

Recycling CIP systems are preferable in plants with larger diameter pipe circuits, where soil loads are low and cleaning events are frequent (i.e. once a day), as there is a saving in the cost of cleaning chemicals and effluent discharge. The multiuse system is a compromise between the single-use and reuse system. The final rinse water and solution are used for a few cleaning cycles before being discharged.

20.7 AUTOMATION

CIP systems require strict control over run times, chemical concentrations, and temperatures and flow rates in order to optimize cleaning and reduce costs. While manual operation of these parameters by an operator may be sufficient in a small operating plant, this may not be adequate for larger facilities. For this reason, automation is typically used to control these parameters in precise sequence.

Control systems for CIP regimes involve a central microprocessor, memory, and an input/output interface for operator access. This allows the operator to control the process-oriented equipment such as pump motors, valve solenoids, pressure sensors, level probes, flowmeters, and temperature switches. Typically the microprocessor will be programmed with several different cleaning programmes. The automation enables the operator to monitor the status of the many valves needed to direct the flow of chemicals through a positive feedback loop. In the event of a failure, the system can be programmed to shut down the cleaning process in order to contain cleaning chemicals to a particular section of the plant. The cleaning process can be stopped and recommenced at any time by the operator through the user interface. While the internal CIP programme might be complex, the cleaning process is kept simple through an operator interface.

20.8 VALIDATION AND VERIFICATION

A validated CIP system is a certified system, which ensures that all aspects of the system have been subjected to a full successful testing regime with each separate test certified once it has been completed. Validation involves a process to ensure that cleaning effectiveness and reliability are challenged and compared to a set of predetermined criteria that cover all the requirements for a CIP. These steps include

1. functional design specifications to confirm the systems ability to meet the design requirement;
2. factory acceptance test to show the user that the equipment meets the design specifications;
3. site acceptance test to demonstrate that the CIP system has been correctly installed on site to function as intended;
4. process qualification is performed with the vendor on site during the first operating run to ensure that the equipment can operate to required performance criteria;
5. continued validation and verification needs is carried out to ensure that the required performance criteria is maintained overtime.

Prior to testing, all critical instruments such as those that measure temperature, flow rate, and concentration should be calibrated. If these instruments have not been calibrated, then the information gathered by them during subsequent testing run may be invalid.

Verification is the process that occurs to ensure that the CIP system is operating correctly. It may involve direct observation of equipment in the plant during operation to ensure that systems such as valves, pumps, and instruments occur in the correct order and that correct temperatures, runtimes, and cleaning solution concentrations have been obtained. A visual inspection of the contact surfaces is also an important initial step. More complex tests may include verification that surfaces have been adequately cleaned. This may involve include testing surfaces by swabbing, or testing the final rinse water for the presence of microbial contamination or product residues. Riboflavin is also commonly used to aid in the visual inspection of spray coverage. Riboflavin or vitamin B_2 is an aqueous soluble compound that fluoresces under a UV lamp (black light). Generally a riboflavin solution (100–200 ppm) is applied to the surfaces to be spray cleaned. An alternative approach is to apply a riboflavin/dextrose solution (50 mg/L/10 wt%) to a surface and leave it to dry overnight to simulate a more challenging condition. After the CIP regime has been carried out, the equipment is inspected using a UV lamp to verify that none of the fluorescent residue remains on any of the internal surfaces including probes, baffles, and agitators. If fluorescence is observed that cannot be removed by modification of the spray ball arrangement or the cleaning cycle, the system will either need to be redesigned or a manual cleaning step will need to be incorporated into the cleaning cycle to remove the contamination.

REFERENCES

1. Sieberling DA. Clean-in-place for biopharmaceutical processes. New York: Taylor and Francis Ltd.; 2007.
2. Changani SD, Belmar-Beiny MT, Fryer PJ. Engineering and chemical factors associated with fouling and cleaning in milk processing. Expe Therm Fluid Sci 1997; 14: 392–406.
3. Stewart JC, Seiberling DA. The secrets out: clean in place. Chem Eng 1996; 103: 72–79.
4. Boulangé-Petermann L, Jullien C, Dubois PE, Benezech T, Faille C. Influence of surface chemistry on the hygienic status of industrial stainless steel. Biofouling 2004; 20(1): 25–33.
5. Lelievre C, Antonini G, Faille C, Benezech T. Cleaning in place: modelling of cleaning kinetics of pipes soiled by *Bacillus* spores assuming a process combining removal and deposition. Inst Chem Eng 2002; 80(Part C): 305–311.
6. Russell MJ. Live long and prosper. Food Eng 1992; Dec: 77–80.
7. Mercadé-Prieto R, Sahoo PK, Falconer RJ, Paterson WR, Wilson DI. Polyectrolyte screening effects on the dissolution of whey protein gels at high pH conditions. Food Hydrocolloids 2007; 21: 1275–1284.
8. Bremer P, Fillery S, McQuillan AJ. Laboratory scale Clean-In-Place (CIP) studies on the effectiveness of different caustic and acid wash steps on the removal of dairy biofilms. Int J Food Microbiol 2006; 106: 254–262.
9. White JC, Rabe GO. Evaluating the use of nitric acid as a detergent in dairy cleaned-in-place systems. J Milk Food Technol 1970; 33(1): 25–28.
10. Dunsmore DG, Thomson MA. Bacteriological control of food equipment surfaces by cleaning systems. 2. Sanitizer effects. J Food Prot 1981; 44(1): 21–27.

11. Chisti Y. Modern systems of plant cleaning. Encyclopedia of food microbiology: process hygiene. Academic Press, London; 1999; 3: 1806–1815.
12. Franks JW, Seiberling DA. CIP spray device design and application. In Seiberling DA, editor. Clean-in-place for biopharmaceutical processes. New York: Taylor and Francis, Ltd.; 2007.

FURTHER READING

Chisti Y, Moo-Young M. Clean-in-place for industrial bioreactors: design, validation and operation. J Ind Microbiol 1994; 13:201–207.

Voss J. Cleaning and cleaning validation: A biotechnology perspective. CRC Press, USA; 1996.

21

LARGE SCALE CHROMATOGRAPHY COLUMNS, MODELING FLOW DISTRIBUTION

Zhiwu Fang

Amgen Inc., Systems Informatics, Thousand Oaks, California

21.1 INTRODUCTION

Column packing involves one or more of the following operations: mechanical compression, flow packing and dynamic axial compression packing. Resin particles are first mixed with solution to make slurries. The suspension is allowed to settle under gravity overnight in some practice. The top layer of the fluid is removed before the packing process. During mechanical compression, a compression headplate is placed on the top of columns and a hydraulic force applied to lower the headplate. The fluid is discharged through the tubes at the top or the bottom of the column. The compressive force transmitted from the headplate to the column bottom compresses the column bed. In the case of flow packing, a flow is introduced and distributed through the frit head, flowing through the column from top to bottom, and is discharged through the tubes from below. The viscous drag force serves as the compressive force on the column bed. During dynamic axial compression packing, flow packing is applied first and then followed by mechanical compression.

Knowing the distribution of the porosity of the packed bed and the internal porosity of the resins is the key to understanding the flow in the columns. It is well known that the flow in chromatographic columns is not uniform (1–5). A number of factors are known to cause nonuniformity in the columns. First, the column walls cause a concentration gradient of resin particle distribution in a boundary layer with a thickness of a few particle diameters (3,6). Radial nonuniformity due to "wall effects" is more severe in smaller columns than in large ones. Secondly, imperfect design of the header frit for mechanical compression and the flow distributor for flow packing can produce nonuniform compressing force on the column, giving rise to nonuniform packing throughout the column (7). Thirdly, resins commercially available for packing chromatographic columns are composed of polymer particles of different strength, from rigid to soft. These resins may exhibit very different mechanical responses to the process conditions, ranging from being elastic to viscoelastic to plastic (8,9). Even the same resin may go through all these responses at different stages of the packing process. Agarose-based resins are often selected for their highly porous, selective, chemically inert, and hydrophilic nature. However, these mechanically soft particles are highly compressible. The resulting excessive compression under the prevailing flow conditions undesirably reduces the porosity of the column bed and limits throughput and separation efficiency. The complex rheological responses of resin particles under compression may give rise to nonuniform displacement along the column (10–12), the longer the column, the larger the gradient of porosity distribution (13,14).

Chromatography engineers are interested in predicting the behavior of the resin and packing quality under all various process conditions. Several empirical models have been proposed (15–17). The model by Mohammad *et al*. provided quantitative predictions on pressure-flow curves only for laboratory-scale columns. The model by Stickel and Fotopoulos wasevaluated at several scales and can provide quantitative predictions of critical velocity (the maximum

Downstream Industrial Biotechnology: Recovery and Purification, First Edition. Edited by Michael C. Flickinger.

velocity on the pressure-flow curve), a key variable used to develop a flow packing method. These two empirical models are restricted to flow packing and cannot be extended to describe packing behaviors by other packing methods. Furthermore, these models can only serve as a quick reference tool because they provide no information on porosity or flow distribution within the beds. Physics-based models have also been proposed, mostly by adapting the Biot formulation for the three-dimensional soil consolidation problem (15,18,19) and elasticity theories (20–22). The strain is related to solid stress and water pressure through the elasticity modules for the solid skeleton. The flow in the skeleton is then described by Darcy's law. These models are applicable for column packing with rigid particles. More recently, Keener *et al*. (20,21) have proposed a one-dimensional model on the basis of the large deformation elasticity theory. This model is able to provide information on the axial distribution of porosity and stress.

Meanwhile, chromatography practitioners are more interested in knowing the impact of imperfect process conditions on separation efficiency and possible remedies. Guiochon and his coworkers have done extensive studies on band profiles in chromatographic columns under the influence of hardware design, packing quality, and wall effects (7,23–29). They found that the radial diffusion coefficient decreases progressively with increasing time as the edges of the sample band get close to the wall. In contrast, the axial diffusion coefficient remains constant. They ascribed the increase in local height equivalent to a theoretic plate (HETP) in the region close to the wall to increasing mass transfer resistance and the degree of heterogeneity of the bed. They confirmed the occurrence of viscous fingering flow instability when the viscosity difference of solute and mobile phase is sufficiently large. They also found that the size and porosity of the frit has an important role on the homogenization of the distribution of the sample concentration at the column inlet. Yuan *et al*. (30) employed a linear elasticity model to predict the flow distribution in an axis-symmetrical geometry, predicting that the wall region would likely be more densely packed than the center of the column and that radial variations in stress decrease as the column gets wider. Based on the model predictions, they proposed a new header design that consists of two regions, collimator and manifold, both porous solids, to achieve the desired uniform velocity profile as well as uniform holdup time for all fluid elements.

Current computer capacity still does not allow one to perform pore (particle) level simulations on realistic systems (31). Most computer models are designed to derive macroscopic effective properties, such as viscosity, permeability, and particle pressure, from the observed microscopic structures in a periodic system on the scale 10,000 particles (32–34). These derived macroscopic models are useful for modeling the realistic porous systems, for example, packed chromatography beds, on the continuum level. Furthermore, advances in computer hardware make it possible for one to study the impact of process conditions on chromatography separation efficiency using macroscopic models. Such models are based either on equilibrium theory or on more realistic mathematical models that take into account the relevant physics of, for example, axial dispersion, interfacial mass transfer between the mobile and stationary phases, intraparticle diffusion, and multicomponent isotherms.

Computational fluid dynamics (CFD) technique is one of the applications of mathematical modeling of fluid flow on the continuum level. It is capable of predicting the flow dynamics and the associated species transport. A few researchers have utilized CFD to investigate the separation efficiency of large-scale high performance liquid chromatography (HPLC). Wu (35,36) investigated the effect of column heterogeneity and frit quality (FQ) for a conventional column configuration that has a sudden expansion and a sudden contraction section and a frit placed at the inlet and the outlet of the column. He found that a real column with *radial* variation in porosity produces a higher HETP compared with an ideal homogeneous column, but performs better than a fritless homogeneous column. This discovery highlighted the critical importance of the frit in promoting good separation. Tan and Koo (37) further considered the effect of the inlet and outlet geometry by replacing a sudden expansion at the inlet and sudden contraction at the outlet with a more gradual expansion and contraction. They discovered that having higher permeability expanding and contracting regions results in better performance and lowers pressure drop, thus making it possible, by adjusting the inlet and outlet geometrical configuration, to achieve the same separation efficiency as a conventional column with a frit of high FQ by using a frit of low FQ. Ching *et al*. (38) investigated the thermal effect on the performance of HPLC. They found that suitable temperature differences between the wall and the inlet could improve the separation efficiency by accelerating the elute velocity near the wall due to decreased local viscosity.

The purpose of this chapter is to provide mathematical modeling tools and to apply these models to each chromatography process. This chapter first reviews the challenges of the scale-up of liquid chromatography packing and then provides a one-dimensional elasticity theory–based model that the author believes can facilitate the column scale-up process. Next, it provides a dimensional analysis to illustrate the scale-dependent wall effect. The second part of the chapter presents case studies on modeling flows in large chromatography columns and separation efficiency analysis.

21.2 CHALLENGES OF SCALING UP CHROMATOGRAPHY

Several parallel phenomena take place during bed consolidation (9,29,39,40): the compression of the individual particles, the compression of the particle network, and the migration of the mobile phase escaping from the interstitial voids between the particles. When the excess mobile phase initially contained in these voids flows out under the influence of a transitory hydraulic pressure gradient, the volume is reduced by the compression of the particle network. The kinetics of the expulsion of the mobile phase from the consolidated bed takes place rapidly in chromatography, faster than is typical in soil mechanics, because of the high porosity of the column bed and its small dimensions. The compressibility of particles is a very important characteristic of soft materials, although often negligible for rigid particles. For resins composed of rigid particles, the deformation of the particle network has the strongest impact on consolidation. There are distinctive differences for soft and rigid particles in terms of rheology. For rigid particles, elasticity before yield and then plasticity afterwards are the dominant responses (9,40,41) while soft particles exhibit viscoelastic responses to the applied force (18). At small deformation, the packed bed recovers to varying degrees upon removal of the applied force. With increasing force, the deformation for both soft and rigid particles becomes irreversibly plastic.

Modeling the packing of compressible biochromatographic resins and predicting the coupled flow distribution present many challenges because the performance of columns depends on the column packing method, the resin properties, and the column geometry. Very often, the performance of columns exhibits strong scale dependence; therefore, it is difficult to set up practical packing procedures based solely on bench-scale experimental data. Figure 21.1 shows the flow-pressure drop curve, flow-porosity profile, and flow-deformation curve. The columns used in the experiments are bench scale $D_c = 3.2$ cm, pilot-scale $D_c = 10$ cm and product scale $D_c = 80$ cm. The initial porosity after gravitational settlement and before compression is $\varepsilon = 0.38$. The columns are packed at different initial lengths, ranging from short (9.1 cm) to long (38.5 cm). Raw experimental data for a large-scale column with $D_c = 80$ cm and initial column length $L_c = 24.5$ cm is listed in Table 21.1.

Sepharose fast flow (SPFF) resins are relatively rigid so that the volume loss during compression can be neglected. Columns packed with SPFF resin can experience much higher flow rate than columns packed with softer resins, such as butyl 4FF. In any case, all experimental data exhibit the following general trends for flow packing:

1. The pressure drop diverges at a critical velocity U_{crit}.
2. The mean porosity decreases with increasing flow rate.
3. The column height decreases linearly with increasing flow rate until critical velocity.
4. All measurable quantities—pressure drop, column height change, and mean porosity—depend on column geometry. The critical velocity is lower for larger columns. Shorter columns have higher critical velocity.

The last plot in Fig. 21.1 is the Stickel curve (17), which correlates the product of the critical velocity and initial column length with column aspect ratio. The intercept of the curve provides a numerical indication of the compressibility of a medium at given operating conditions, while the slope is an indicator of the wall effect. The degree of scattering in the plot indicates a strong dependence of the packing performance on the scale and column geometry, which is undesirable for the purpose of scale-up and makes scale-up based on this kind of empirical correlation unreliable.

To understand the underlying physics for the observed packing behaviors, the following analysis on the force balance for the stationary and mobile phase in the packed beds and the effect of the presence of physical bounding walls is performed.

21.3 ANALYSIS ON THE WALL EFFECT

Let us start with the conservation equations for a two-phase flow. The general form of the force balance equation for the solid phase reads

$$\rho_s \mathbf{a}_s = \nabla \cdot \boldsymbol{\sigma} + \rho_s \mathbf{g} + \mathbf{F}^{\text{SF}} + \mathbf{F}^{\text{ext}} \tag{21.1}$$

in which ρ_s is the density of the solid, $\mathbf{a}_s$ is the acceleration, $\boldsymbol{\sigma}$ the stress of the solid phase, $\mathbf{g}$ the gravity, $\mathbf{F}^{\text{ext}}$ the external forces, and $\mathbf{F}^{\text{SF}}$ the interaction force between the solid and the fluid phase, typically, it is the viscous drag force. The momentum balance for fluid phase states

$$\rho_f \frac{du_f}{dt} = -\nabla P + \nabla \cdot \tau + \rho_f \mathbf{g} - \mathbf{F}^{\text{SF}} \tag{21.2}$$

At the steady state, we have

$$\nabla \cdot \boldsymbol{\sigma} + \nabla \cdot \boldsymbol{\tau} + \rho \mathbf{g} + \mathbf{F}^{\text{ext}} - \rho_f \mathbf{u}_f \cdot \nabla \mathbf{u}_f = 0 \tag{21.3}$$

in which ρ is the density of the mixture. Performing a nondimensional analysis on the above equation, using the following parameters

$$x \sim R_0, \mathbf{u}_f \sim u_c = Q/A, P \sim \eta_f u_c R_0/d_p^2, \tau \sim \eta_f u_c/R_0$$

$$\boldsymbol{\sigma} \sim E, Re = \rho_f u_c R_0/\eta_f, \hat{E} = E R_0/u_c \eta_f,$$

$$\xi = R_0/d_p \boldsymbol{\Theta} = \rho_f \mathbf{g} R_0^2/u_c \eta_f, \boldsymbol{\Phi} = F^{\text{ext}} R_0^2/u_c \eta_f$$

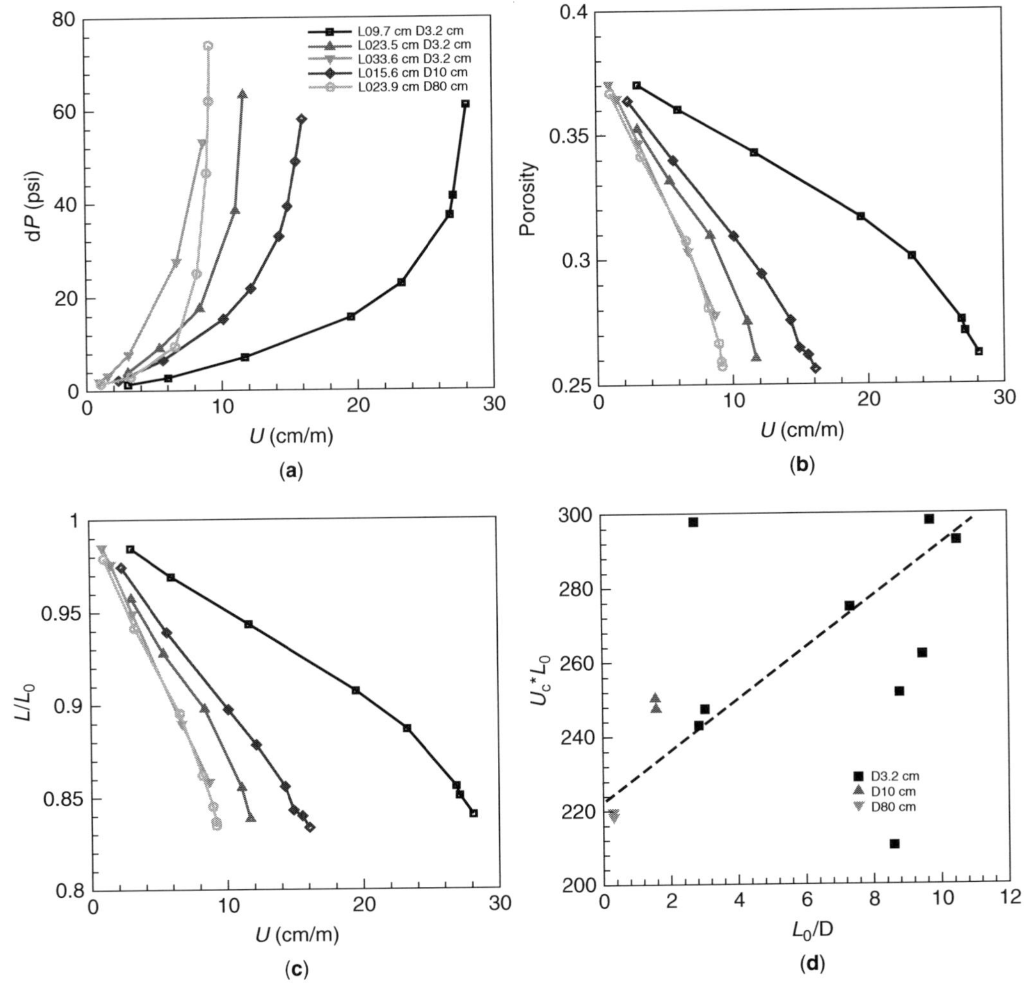

Figure 21.1. SPFF column packing behavior. (a) Flow-pressure and (b) flow-porosity, (c) flow-compression, and (d) Stickel curve for SPFF. Initial porosity is 0.4. Three columns are used, $D = 3.2$ cm, 10 cm and 80 cm. (This figure is available in full color at http://onlinelibrary.wiley.com/book/10.1002/9780470054581.)

yields, after dropping the lower order terms,

$$\nabla \cdot \boldsymbol{\sigma} + \rho \mathbf{g} + \mathbf{F}^{\text{ext}} = \nabla P \tag{21.4}$$

The force balance equation for fluid phase at steady state reads

$$-\nabla P + \nabla \cdot \boldsymbol{\tau} + \rho_f \mathbf{g} - \mathbf{F}^{\text{SF}} = 0 \tag{21.5}$$

At dilute limit, solid concentration $\phi \ll 1$, the interaction between the fluid and solid phase is simply the Stokes drag, $\mathbf{F}^{\text{SF}} = 3\pi \eta_f d_p \mathbf{U}$, where η_f is the fluid viscosity and $\mathbf{U}$ is the relative velocity of the particles, but for the concentrated suspensions or the porous media flows, multiple particle interactions should be considered and usually the interaction force is modeled by the means of permeability k, $\mathbf{F}^{\text{SF}} = \eta \mathbf{U}/k$.

There are many correlations proposed for the permeability of the porous media. A few correlations on the permeability are plotted as a function of porosity (42) in Fig 21.2. Different correlations may give permeability several orders of magnitude difference. The most often used one is the Blake–Kozeny correlation that is valid for rigid spherical particles. Alazmi gives a comprehensive comparison on the variants within the porous media transport models in a review (43).

If neglecting the viscous stress and incorporating the gravity into the pressure term, we have Darcy's law,

$$\nabla P = \frac{\eta}{k} \mathbf{U} \tag{21.6}$$

where k is the permeability, a function of particle volume fraction ϕ or porous porosity ε, and $\mathbf{U}$ is the superficial velocity of the fluid.

TABLE 21.1. Raw Experimental Data for a Large-Scale Column (80 cm in Diameter and Initial Length 24.5 cm)

q (mL/min)	h (cm)	delta P Total (psi)	Prosity
5,300	23.9	1	0.364435
13,700	23.3	5	0.348069
31,700	21.9	18.5	0.306393
40,000	21.1	35.5	0.280095
43,000	20.85	47.5	0.271463
43,900	20.55	60	0.260827
44,600	20.45	69.5	0.257213
44,800	20.35	74.5	0.253563

The first column is the flow rate; the second, length after compression; the third, the pressure; and the last, calculated porosity.

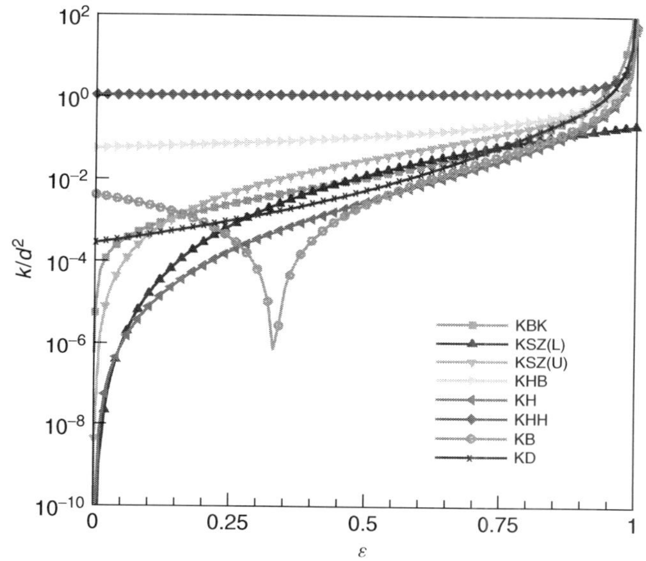

Figure 21.2. Permeability as a function of porosity. (This figure is available in full color at http://onlinelibrary.wiley.com/book/10.1002/9780470054581.)

The viscous stress term is shown to be much smaller than other terms in the dimensional analysis, but this is not true in the boundary layer near the column wall. Let us check Equation 21.5 for the case of fully developed steady state in a cylindrical coordinate, which is written as

$$\frac{\mathrm{d}p}{\mathrm{d}z} = -f(\varepsilon)\frac{\eta_f}{d_p^2}u_z + \frac{\eta}{r}\frac{\mathrm{d}}{\mathrm{d}r}\left[r\frac{\mathrm{d}u_z(r)}{\mathrm{d}r}\right] - \rho_f g \qquad (21.7)$$

After nondimensionalizing it using the following parameters,

$$p \sim \rho U_0^2, x \sim D, u \sim U_0 = Q/A$$

where D is the diameter of the column, we have

$$\frac{\mathrm{d}p}{\mathrm{d}z} = -\frac{1}{Re}\left[\frac{1}{\xi^2}f(\varepsilon)u_z - \frac{1}{r}\frac{\mathrm{d}}{\mathrm{d}r}\left(r\frac{\mathrm{d}u_z}{\mathrm{d}z}\right)\right] - \hat{g} \qquad (21.8)$$

where $Re = \rho_f U_0 D/\eta_f$ is the Reynolds number, $\xi = d_p/D$ is the relative particle diameter and $\hat{g} = gD/U_0^2$ is the nondimensional gravity force.

The first term is the Darcy term, the second is the viscous shearing stress term, and the last, the gravity force. Let us proceed with dimensional analysis as follows.

1. Since $\xi \ll 1$, the highest order term is the first term, $Re^{-1}\xi^{-2}\phi f(\varepsilon)u_z$. The above equation is reduced to the Darcy's law. The pressure drop across the column bed is balanced by the viscous drag force.
2. The first and second term have the same order of magnitude in a boundary layer of the depth of particle diameter.
3. The gravity force term has the same order of magnitude as the leading order term only when $U_0 \sim \rho g d^2/\eta f(\varepsilon)$.
4. If $Re \ll 1$, then

$$\frac{\mathrm{d}p}{\mathrm{d}z} = -\frac{1}{Re}\left[\frac{1}{\xi^2}f(\varepsilon)u_z + \frac{1}{r}\frac{\mathrm{d}}{\mathrm{d}z}\left(r\frac{\mathrm{d}u_z}{\mathrm{d}r}\right)\right]$$

5. Reynolds number Re increases linearly with D, but relative particle size ξ decreases linearly with D, therefore, the first term in Equation 21.8 increases as D/d_p^2 while the second term decreases as D^{-1}. This is consistent with the observation that the wall support decreases with increasing column diameter.

The Darcy's law, Equation 21.6, describes the flow through a rigid unbounded porous media. However, the Brinkman equation should be used to take into account boundary effects, and Brinkman–Forchhemer equation for inertial corrections.

In a cylindrical coordinate system, the Brinkman equation reads,

$$\frac{\mathrm{d}P}{\mathrm{d}z} = \eta\left[\frac{1}{r}\frac{\mathrm{d}}{\mathrm{d}r}\left(r\frac{\mathrm{d}u(r)}{\mathrm{d}r}\right)\right] - \eta\frac{u(r)}{k(\varepsilon)} \qquad (21.9)$$

and the Forchheimer equation

$$\frac{dP}{dz} = \eta\left[\frac{1}{r}\frac{d}{dr}\left(r\frac{du(r)}{dr}\right)\right] - \eta\frac{u(r)}{k(\varepsilon)} - \frac{\rho F(\varepsilon)}{\sqrt{k(\varepsilon)}}u^2(r) \tag{21.10}$$

where P is the fluid pore pressure, η is the fluid viscosity, ρ the fluid density, $u(r)$ the axial velocity of the fluid, and $k(\varepsilon)$ is the permeability of the porous medium, which is a function of the porosity ε. Please note both viscosity η and porosity ε are assumed constant in the following derivation.

The following parameters are chosen to nondimensionlize the above equations:

$$r \sim R_0, u \sim u_c = -\frac{R_0^2}{\eta}\frac{dP}{dz}, k \sim R_0^2, \tau = \eta\frac{u_c}{R_0}$$

The Brinkman equation is then recast into a dimensionless form

$$\frac{1}{r}\frac{d}{dr}\left[r\frac{du(r)}{dr}\right] - \frac{u(r)}{k(\varepsilon)} + 1 = 0 \tag{21.11}$$

subject to the boundary conditions

$$u|_{r=1} = 0, \left.\frac{du}{dr}\right|_{r=0} = 0$$

An analytical solution to the above equation is obtained:

$$u(r) = k\left[1 - I_0\left(\frac{r}{\sqrt{k}}\right)/I_0\left(\frac{1}{\sqrt{k}}\right)\right] \tag{21.12}$$

The shear stress on the wall is given by

$$\tau_w = \frac{dP}{dz}\sqrt{k}I_1\left(\frac{1}{\sqrt{\tilde{k}}}\right)/I_0\left(\frac{1}{\sqrt{\tilde{k}}}\right) \tag{21.13}$$

in which $\tilde{k} = k/R_0^2$ is nondimensional permeability.

It is clear that $\tau_w/(dP/dz)$ is in the order of $\sqrt{k}$ since $I_1(1/\sqrt{\tilde{k}})/I_0(1/\sqrt{\tilde{k}}) \simeq 1$, therefore, unless $k \sim O(1)$, the boundary effect is negligible.

21.4 MODEL THE COUPLING BETWEEN BED COMPRESSION AND FLOW

21.4.1 A One-Dimensional Model

For a chromatographic engineer, a one-dimensional formula would be useful for the task of scaling up. One such model can be obtained when neglecting the variation in the radial direction (20,21).

In a cylindrical coordinate system, Equation 21.4 is written as

$$\frac{1}{r}\frac{\partial}{\partial}(r\sigma_{rr}) - \frac{\sigma_{\theta\theta}}{r} + \frac{\partial(\sigma_{rz})}{\partial z} = \frac{\partial P}{\partial r} \tag{21.14}$$

$$\frac{1}{r}\frac{\partial}{\partial r}(r\sigma_{rz})\frac{\partial(\sigma_{zz})}{\partial z} - \rho g + F^{ext} = \frac{\partial P}{\partial z} \tag{21.15}$$

Integrating Equation 21.15 across the cross section of column, such as

$$\overline{f} = \frac{2}{R^2}\int_0^R f(r)r dr \tag{21.16}$$

and relating the shear stress on the wall $\overline{\sigma}_{rz}|_{r=R}$ to the average axial stress $\overline{\sigma}_{zz}$ through Mohr–Coulomb failure criterion for elastic materials,

$$\sigma_{rz} = C + \tan\phi\sigma_{rr} \tag{21.17}$$

we arrive at

$$\frac{d\overline{\sigma}_{zz}}{dz} + \frac{2}{R}\tan\phi\sigma_{rr}|_{r=R} = 0 \tag{21.18}$$

Here ϕ is the material angle of friction, and C is the cohesion of the material (C is assumed zero in the derivation). If assuming radial normal stress gradient is small and correlating the average radial normal stress to the averaged axial normal stress for a radially confined sample undergoing uniaxial compression, as

$$\overline{\sigma}_{rr} = \frac{\nu}{1-\nu}\overline{\sigma}_{zz} \tag{21.19}$$

where ν is the Poisson ratio for the material, we then have the following model

$$\frac{d\overline{\sigma}_{zz}}{dz} = -\frac{4}{D}\mu_f\frac{\nu}{1-\nu}\overline{\sigma}_{zz} \tag{21.20}$$

where $\overline{\sigma}_{zz}$ is the axial effective stress on the material, μ_f is the wall friction coefficient, and D the column diameter.

For flow packing, we have

$$\frac{d\overline{\sigma}_{zz}}{dz} = -\frac{4}{D}\mu_f\frac{\nu}{1-\nu}\overline{\sigma}_{zz} - \frac{dP}{dz} \tag{21.21}$$

This approach follows the soil mechanics theory and a number of similar models (4,5) have been proposed. Given appropriate boundary condition, Equations 21.20 and 21.21 can be solved to provide the distribution of the stress along column axis. Supplemental rheological constitutive equations, which correlate the stress to strain, are needed to know the distribution of porosity. The simplest one is based on the linear elasticity

$$\begin{aligned}
\sigma_{rr} &= 2G\gamma_{rr} + \lambda(\gamma_{rr} + \gamma_{\theta\theta} + \gamma_{zz}) \\
\sigma_{\theta\theta} &= 2G\gamma_{\theta\theta} + \lambda(\gamma_{rr} + \gamma_{\theta\theta} + \gamma_{zz}) \\
\sigma_{zz} &= 2G\gamma_{zz} + \lambda(\gamma_{rr} + \gamma_{\theta\theta} + \gamma_{zz})
\end{aligned} \tag{21.22}$$

where G and λ are the Lame constants for the material and are related to the Yong's modulus, E and Poisson ratio, ν.

For a radially confined column, it can be proven that

$$\sigma_{zz} = (2G + \lambda)\gamma = \frac{(1-\nu)E}{(1+\nu)(1-2\nu)}\gamma \tag{21.23}$$

where γ is the area-averaged strain in the axial direction, which is related to the column deformation $\mathrm{d}u/\mathrm{d}z$.

Keener *et al.* (20,21) considered the nonlinearity of the column large deformation and provided the following model:

$$\sigma_{zz} = \frac{(1-\nu)E}{(1+\nu)(1-2\nu)} f(s)s \tag{21.24}$$

where $s = \frac{\mathrm{d}u}{\mathrm{d}z} - \frac{1}{2}\left(\frac{\mathrm{d}u}{\mathrm{d}z}\right)^2$ is the axial Eulerian strain and function $f(s)$ assume an empirical form

$$f(s) = f_0 e^{ks} \tag{21.25}$$

The model works very well for both flow packing and mechanical compression. Through a regression based on both column length change and pressure drop, the author was able to find a universal set of fitting parameters of f_0 and k for the datasets listed in the previous section. Figure 21.3 shows the comparison between model predictions and the experimental data on the pressure drop and column length change as a function of the flow rate for all scales, d = 3.2 cm, d = 10 cm, and d = 80 cm. It is clear that the model predictions agree with experimental data reasonably well.

However, the author noticed that the fitting parameters f_0 and k somehow depend on the type of applied stress or compression approach. This dependence is likely due to the wall effect. As indicated in the above derivation, the wall effect is not present in the one-dimensional model because any variation in the radial direction is ignored due to the area-averaged process. Also demonstrated in the previous section, the wall effect is more important for small-scale columns; therefore, two- or three-dimensional models should be used to capture the variations in the radial direction.

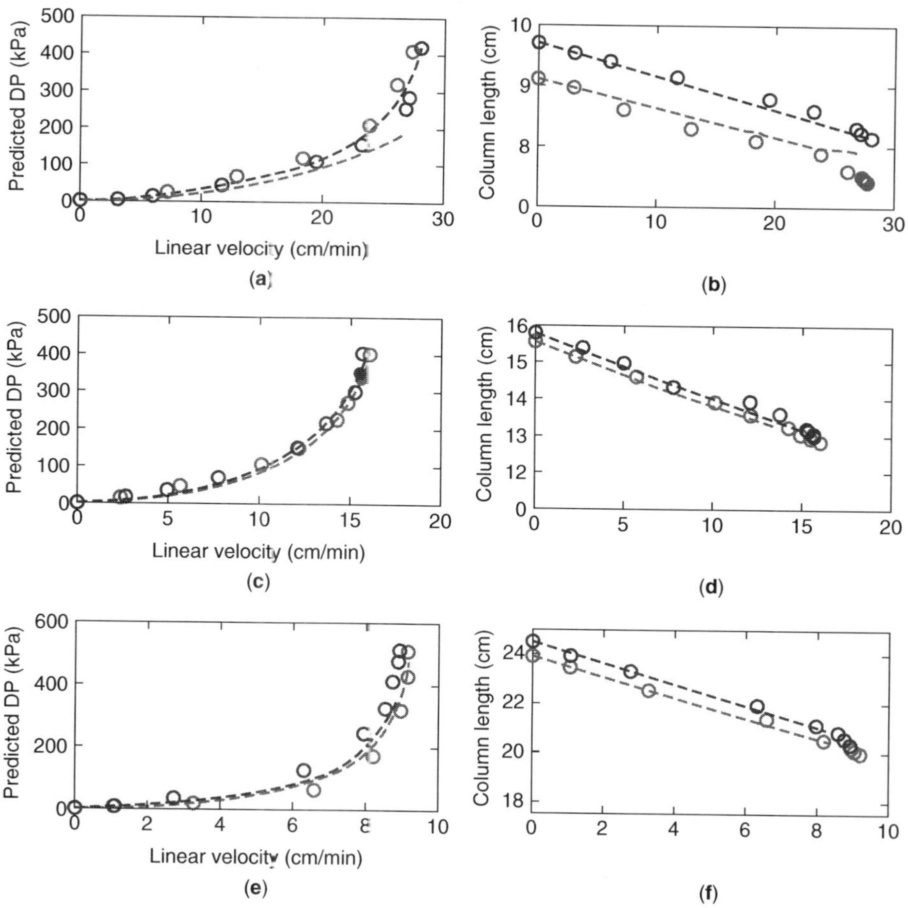

Figure 21.3. Comparison between model predictions and experimental data on the pressure drop (a, c, e) and the column length change (b, d, f). (a, b) is for column D = 3.2 cm, L_0 = 9.7 cm, (c, d) the middle row for D = 10 cm, L_0 = 15.6 cm, and (e, f) the bottom D = 80 cm, L_0 = 24.5 cm. (This figure is available in full color at http://onlinelibrary.wiley.com/book/10.1002/9780470054581.)

McCue *et al*. (44) set up a two-dimensional model based on linear elasticity theory and adopted Keener correlation, taking into account the changes to the particle porosity, displacement, and stress in the both the axial and radial directions. They applied the model to two-dimensional column geometry over a range of column diameters and bed heights and temperatures. In all cases, the model predictions for the pressure-flow behavior agreed well with experimental data. The model was able to accurately predict the pressure-flow behavior of large columns (column diameters >20 cm ID), and to accurately describe particle compression during the flow packing process for the different stationary phases, using a single expression for the empirical modulus.

21.4.2 Considerations on Constitutive Modeling

Elasticity theories are proper for small deformation of the columns (22,45,46). However, as pointed out in the literature, chromatographic column consolidation always exhibits plastic, viscoelastic, or viscoplastic behaviors (9,47). Structured fluid or soft solid materials such as paste and particulate suspensions usually possess a shear yield stress (48–50).

The compression of column bed generally involves the following processes: (i) rearrangement of particles in the packed bed under the applied stress, (ii) deformation of particles in the packed column bed, and (iii) both effects taking place simultaneously. Low strength column bed deforms immediately upon the exertion of compressive force without particle rearrangement, in contrast, when compressing a high strength column bed, particle rearrangement occurs before particle deformation. Once the interstitial void space has been lost, the resulted solid matrix is much more resistant to further deformation.

The relationship between the state of stress and the state of strain or deformation is described by a constitutive equation. In the case of a linear elastic material, in which the strain is proportional to the stress, this relationship is known as *Hooke's law*. A material that behaves elastically is one in which the deformation will be completely recovered upon the removal of the stress, much like an elastic band. Many materials behave elastically up to yielding or failure, at which plastic deformation occurs. For columns packed with silica-based materials, the particles can be assumed to behave in a nonlinear elastic manner up to the point where particle breakage begins to take place, after which plastic response is assumed. With polymer-based particles, large levels of local strain is probable, causing significant plastic deformation of the particles that is irreversible upon the removal of applied stress, accompanied by a decrease in the bed permeability.

Column beds exhibit creeping and relaxation during the compression process, a sign of viscoelasticity. The viscoelastic response is probably the result of the microstructure change of the whole bed after deformation of each individual particle due to their high internal porosity. Consequently, the consolidation may never be achieved instantaneously, but spends considerable time at intermediate metastable states, lasting a few minutes to a few hours (25). A quasi-exponential decay may follow with a pseudo time constant on the order of hours for a small preparative column. It is common to observe a slow, small shrinkage of a column bed overnight after the headplate is lowered and locked at the desired column height in a mechanical compression.

Among other factors for any modeling effort are the choice of permeability, consideration of the loss of the volume of resin particles during compression, and boundary conditions. The permeability of the porous media is characterized by the specific properties of the medium, such as the bed porosity, the pore-size distribution, the size and shape of the particles, and the surface area exposed to the fluid. One of the most popular correlations is the well-known Kozeny–Blake equation,

$$k = \frac{\varepsilon^3}{k_0 S_0^2 (1-\varepsilon)^2} \tag{21.26}$$

in which S_0 is the specific surface area of the particles, $S_0 = 6/d_p$ for rigid spherical particles with a diameter d_p, and k_0 is a constant. However, for a porous media consisting of soft particles, the correlation needs to be modified to take into consideration the effects of particle deformation (51),

$$k' = \frac{\varepsilon^3}{k_0' S_0'^2 (1-\varepsilon)^2} \tag{21.27}$$

Here, $S_0' = \Omega S_0$ is the effective specific surface area of the soft particles after compression, and k_0' is the modified Kozeny constant for extremely low bed porosity conditions.

When the change of particle volume is not negligible, we can relate column deform to the porosity such as

$$\frac{du}{dz} = \frac{\varepsilon - \varepsilon_0}{1-\varepsilon} + \frac{1-\varepsilon_0}{1-\varepsilon}\frac{\delta V_s}{V_s^0} \tag{21.28}$$

in which ε_0 and V_s^0 are the porosity of bed and the volume of particles before deformation, respectively, and δV_s is the change of particle volume.

If solid particles conserve their volume during compression, $\delta V_s = 0$, the above Equation 21.28 is reduced to

$$\frac{du}{dz} = \frac{\varepsilon - \varepsilon_0}{1-\varepsilon} \tag{21.29}$$

Setting appropriate boundary condition in the models for column packing is very important but unfortunately most often does not get enough attention. There exist two types of slippage—physical slip and apparent slip of suspensions on the wall. In the former case, velocity is discontinuous on the wall, the solids/suspensions slide relatively to the wall (52,53); while in the latter case, a thin layer of fluid exists next to the wall with the particles either not interacting with the wall or interacting weakly, so-called apparent slip (54–56). What occurs is that at sufficiently high shear rates, the velocity of the fluid next to the wall is higher than that of the bulk material (3,57). In the extreme case, the particle-lean fluid near the wall flows, while the bulk material does not deform at all. The shear rate seems discontinuous near the wall (54). The shear amplitude is much larger in the thin fluid layer than in the rest of the material. This phenomenon has been observed for particulate systems such as foams, emulsions, suspensions, polymers and gels (2,58–62). However, the magnitude of slip velocity and the thickness of the boundary layer are extremely difficult to measure, and they are generally believed to depend on the stress on the wall (both the shear stress and the normal stress therefore pressure (52)), and particle concentration (63,64). Including slip boundary conditions will surely complicate the computation of stress on the wall.

21.5 IMPACT OF HARDWARE DESIGN ON THE FLOW IN LARGE COLUMNS

As indicated in the introduction, the hardware design has a great impact on the flow distribution and hence the separation efficiency. This section presents a case study on the flow distribution in a commercial scale chromatographic column that has a unique header design and its impact on separation efficiency. Flow is introduced into the header zone from four 1-inch pipes, aiming four 2.5-inch diameter antijet disks that sit on a layered screen mesh. The author used a commercial CFD package FLUENT 6.2.2 to model the flow distribution in the whole column, including the header zones. Dye study was also performed both numerically and experimentally. Model prediction on dye distribution in packed bed shows reasonable agreement with experimental data. The goal of the study was to

1. monitor flow dynamics in the headplate and evaluate the effects of the snap ring, slurry port, and the antijet plates on the distribution of flow in the column;
2. seek process optimization by changing hardware design; and
3. perform dye studies to validate the CFD model with experimental results and to have a better understanding of the dye propagation that can be used to design better dye studies in the future.

21.5.1 Hardware Configuration

There are two major components of the experimental system, an Eastern Rivers 1.6-m stainless steel chromatography column and a chromatography skid used to deliver flow to the column. The skid has the following components relevant to the study: (i) feed pump, (ii) mass flow meter, (iii) buffer filter path piping, (iv) conductivity meter upstream of the column (used for HETP related measurements), (v) conductivity meter downstream of the column (used for HETP related measurements), (vi) pressure indicator upstream of the column and (vii) pressure indicator downstream of the column. Two 2-inch hoses connect the skid to the column assembly.

The column assembly consists of an automatic dye/HETP spike solution injection device that delivers a set volume of dye to the column. This device is connected to the top column distributed port. The column assembly also consists of a two-way valve assembly connected to the column bottom plate distribution port. This two-way valve assembly uses a bleed butterfly valve and a diaphragm valve to seal off the bottom plate distribution port. The column top and bottom headplates have screens to contain the resin within the column and a product distribution port. The product distribution port has four distribution piping segments. Each of these segments is piped to each of the four column quadrants. Both the top and bottom headplates have slurry valves located in their centers.

The headplate is provided with four inlets and each inlet has an antijet plate juxtaposed in a manner to "splash" the incoming fluid in the radial direction in the headplate. This helps to distribute flow in the headplate and the packed bed. A polypropylene mesh is provided to prevent contact between the headplate and the antijet plates. A layered mesh screen is placed just below the antijet plates and serves as the barrier for the resin in the packed bed. The layered mesh screen is fitted to the headplate by a "snap ring" that contacts the packed bed. The column is symmetric across the mid plane and can be run in up-flow or down-flow conditions. A schematic of the column and the headplate design is shown in Fig 21.4.

21.5.2 Dye Study

Dye testing studies have been the standard approach used for such evaluations and utilize the quality of mass distribution of an injected dye tracer to assess the flow distribution. In the period of dye testing, study execution equating mass distribution to flow distribution is a reasonable assumption. In brief, dye is injected in an equilibrated column and allowed to flow for a predetermined time. Subsequently the column is dismantled to expose the resin bed. The bed is excavated in a predetermined fashion to expose the dye band and get a picture of the mass and flow distribution in the column. The studies are resource intensive and can only

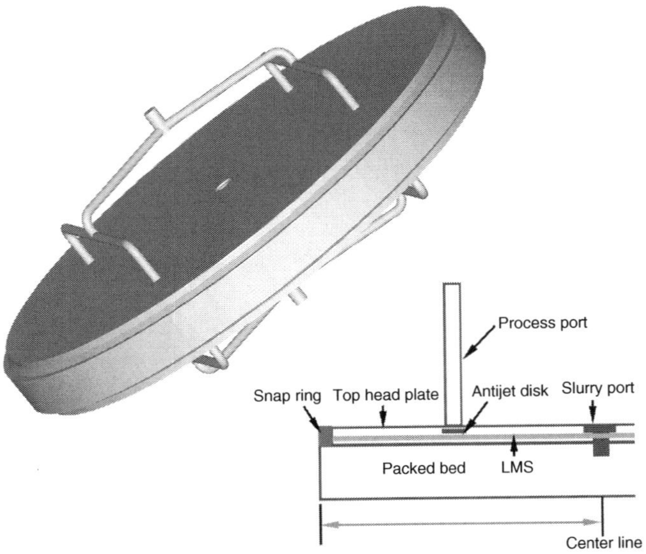

Figure 21.4. Geometry of a commercial scale chromatography column. (This figure is available in full color at http://onlinelibrary.wiley.com/book/10.1002/9780470054581.)

be performed after column fabrication, thus making it difficult to make hardware changes, if any, that are proposed by dye testing results.

In these studies, a volume of dye equivalent to 1–2% of a CV was injected into a packed column previously equilibrated with 25 m*M* Tris, 45 m*M* NaCl, pH 9.0 using the dye injection device. The flow rates of the running buffer during the dye injection were typically 3–20 standard liters per minute (slpm) for a 1-m column and 7–51 slpm for a 1.6-m column. Either single or multiple injections were made followed by excavation. A compression factor of 18% was targeted for all the packs used for spike injection. During most repack operations, water for injection (WFI) was pumped into the column and the resin was slurred followed by the compression packing operation. Following the compression pack, the column was conditioned with 2CV up-flow and 2CV down-flow of WFI. The resin was then equilibrated with buffer (25 m*M* Tris, 45 m*M* NaCl, pH 9.0) in preparation for the spike injection. The equilibration of the packed bed with the buffer was critical since the dye color is dependent on pH and the color was the primary means of deriving conclusions regarding mass and flow distribution.

21.5.3 Modeling the Flow in the Packed Bed

Since the focus is on the design of header and antijet disks and we are dealing with columns with large aspect ratio (D/H), we will neglect the heterogeneity of the packed bed and assume the porosity of packed bed is uniform. However, it should be noted that due to the size of snap ring and penetration of slurry ports into the packed bed in some configurations, the bed directly beneath the snap ring and slurry port will be more compressed compared with the bulk of the packed bed. In addition, the local porosity should be lower, resulting in even slower flow in these regions.

The governing equations to describe the flow in the whole system are given by

$$\frac{\partial \rho}{\partial t} + \nabla \cdot \rho \mathbf{u} = 0 \tag{21.30}$$

$$\frac{\partial (\rho \mathbf{u})}{\partial t} + \nabla \cdot \rho \mathbf{u}\mathbf{u} = -\nabla P + \nabla \cdot [\eta(\nabla \mathbf{u} + \nabla^T \mathbf{u})] + \rho \mathbf{g} + \mathbf{S} \tag{21.31}$$

$$\frac{\partial (\rho \mathbf{u}\phi)}{\partial t} + \nabla \cdot \rho \mathbf{u}\phi = \nabla \cdot (D\nabla\phi) \tag{21.32}$$

In the momentum balance equation, the source term is the extra pressure drop due to the flow in a porous media and it is modeled as

$$\mathbf{S} = -\frac{\eta}{k}\mathbf{u} + \frac{1}{2}C\rho|\mathbf{u}|^2 \tag{21.33}$$

where η is the viscosity of the liquid, $\mathbf{u}$ the velocity vector, C is a constant for inertial effects and k the permeability of the porous media characterized by the specific properties of the medium, such as the bed porosity, the pore-size distribution, the size and shape of the particles, and the surface area exposed to the fluid. The inverse of permeability is called *flow resistance*.

The flow field is obtained by solve Equations 21.30 and 21.31, while the transport of the dye is obtained by performing transient simulation and solving the species transport Equation 21.32.

Figure 21.5 shows the axial and radial velocity in the column and in the headplate respectively. The flow in the headplate shows a large velocity gradient in both radial and axial direction, especially when coming close to the antijet disks. On the one hand, the flow in the bulk of the packed bed, away from the slurry port and antijet disks, is uniform to a certain degree and is parallel to the column axis. On the other hand, the flow near the wall and process ports has significant radial flow component.

The impact of the hardware design on the flow distribution is clearly seen in Fig 21.6. Snap ring holds the layered mesh screen and is essentially embedded in the packed bed. As shown in the plot, a bulky snap ring leaves a significant footprint, leading to a low flow region of significant size and giving rise to a smiley face near the column wall for dye testing. The modified snap ring, which was much sleeker in design, resulted in significant improvement in flow distribution. Upon implementation of this hardware modification, dye testing confirmed the improved flow distribution.

Model simulations also clearly indicate that antijet disks exert a great impact on the flow distribution. Changing the

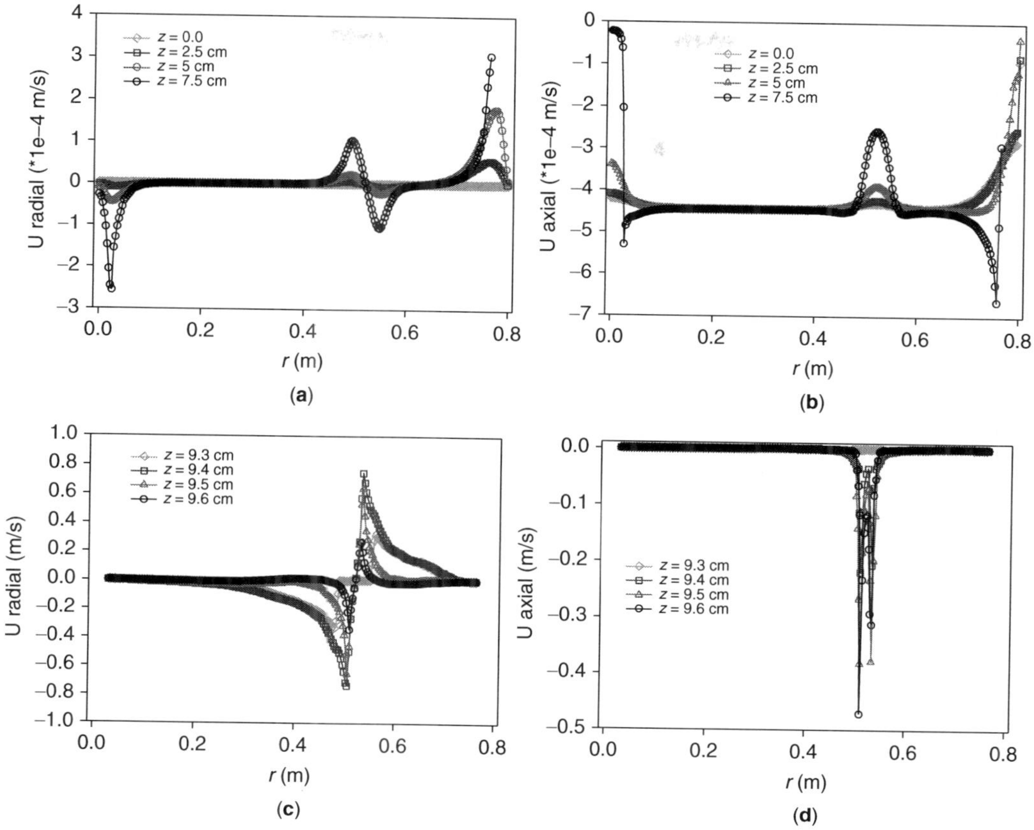

Figure 21.5. Axial and radial velocity in the column (a and b) and in the headplate (c and d). The inlet velocity is 51 LPM, column diameter 1.6 m, and packed bed 18 cm. (This figure is available in full color at http://onlinelibrary.wiley.com/book/10.1002/9780470054581.)

size and the porosity of the antijet plate results in improved flow distribution, which is clearly shown by the dye distribution profiles (65). The role of antijet disks on the flow distribution and consequentially on the HETP is further demonstrated in the following section.

21.6 MODELING THE TRANSPORT OF ELUTION AND HETP ANALYSIS

In this section, the fundamental physics of the elution process is first briefly discussed and then the mathematical models and implementation are presented.

Chromatography is a separation process based on the distribution equilibrium of the feed components between two different phases—the stationary phase and the mobile phase—which involves a combination of selective, convective, and diffusive transports.

One of these two phases must be a fluid to permit the differential migration needed in chromatography. The second phase is most often a solid and is usually stationary. Among the many possible solid–liquid equilibria, ion exchange and adsorption equilibria are the two types most often used in practical applications. To model a separation at high concentration, it suffices to know the equilibrium isotherms of the feed components between the two phases. The isotherms of the feed components are most often nonlinear in the concentration range used in preparative chromatography while behave linearly in analytical chromatography merely because the nonlinear effects are too small to measure. There are many models of isotherms for pure compounds (66,67), among which the Langmuir isotherm model has proven to be a reasonable accurate empirical equation. This model assumes that both the solution and the adsorbed layer are ideal, that there are no adsorbate–adsorbate interactions, and that adsorption takes place in a monolayer. It takes the following form:

$$q = \frac{K_{eq} Q_{max} c}{1 + K_{eq} c} \qquad (21.34)$$

where q is the bound protein concentration, c liquid-phase concentration, that is in equilibrium with q, Q_{max} the maximum bound protein capacity, and K_{eq} the equilibrium constant.

The transport of elute along the column, hence their separation, is controlled by the flow velocity of the mobile

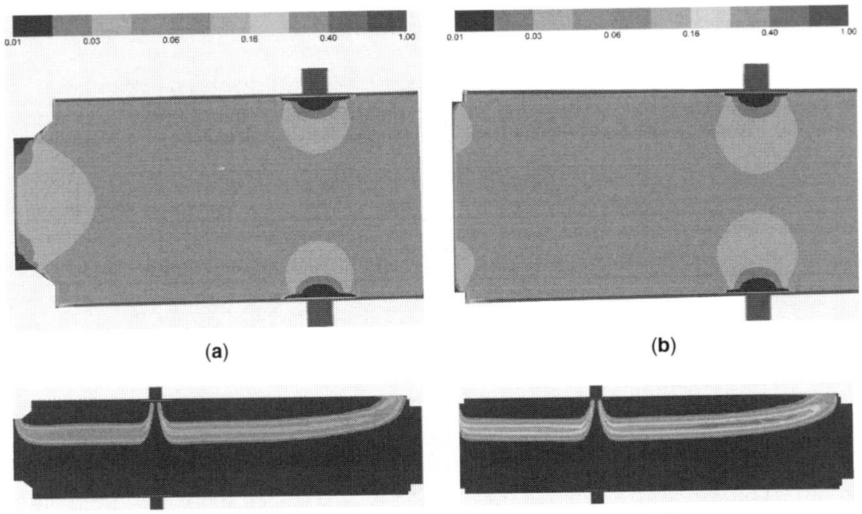

Figure 21.6. Reduced snap ring improves the flow distribution as indicated by the velocity magnitude and dye distribution in a vertical plane cut through the slurry ports. The velocity magnitude distribution, old bulky ring (a) and modified ring (b). The dye distribution, old bulky ring (c) and modified snap ring (d). (This figure is available in full color at http://onlinelibrary.wiley.com/book/10.1002/9780470054581.)

phase percolating through the column bed. The percolating flow of the mobile phase through a liquid chromatography column bed is always creeping flow with very low Reynolds number based on the particle size. This velocity is described by Darcy's equation, Equation 21.6, or Brinkman equation, Equation 21.9, when viscous effects are included.

The separation process also involves diffusion and dispersion. In a stream flowing through a porous medium, diffusion includes two contributions—one related to molecular diffusion in the axial direction and the other, called *eddy diffusion*, to the heterogeneity of the distribution of the streamlines in a porous medium. When particles are porous, there is an additional contribution to band spreading from the finite character of the kinetics of mass transfer between the two phases of the system that is often called *apparent dispersion* in the literature. Dispersion under linear conditions is characterized by the HETP in the equilibrium-dispersive model. The HETP concept is strictly valid only when the rate of the mass transfer kinetics is relatively fast and the profile of the elution peak is Gaussian. In a first-order approximation, the first two contributions are additive and

$$\mathbf{D}_a = \tau_L \mathbf{D}_m + \alpha d_p \mathbf{u} \tag{21.35}$$

where $\mathbf{D}_m$ is the molecular diffusivity, d_p the average particle diameter and $\mathbf{u}$ the mobile phase velocity, τ_L and α are two geometrical parameters. In particular, τ_L is related to the tortuosity of the packed bed. In a packed column, molecular diffusion is slowed down by the presence of the solid particles that prevent straight trajectories of molecules along any significant distance compared with the particle diameter and force the molecules to diffuse around them. However the channels open to the molecules are extended and vary in cross section, and these restrictions result in reduced apparent dispersion.

Through lattice-Boltzmann simulations, Koponen *et al.* (31,68,69) have found correlations between the average tortuosity of the flow paths and the porosity of the substance. A simple correlation is given by

$$1/\tau_L = 1 + 0.8(1 - \varepsilon) \tag{21.36}$$

where ε is the porosity of the porous media. The tortuosity is valued approximately at 0.7 for chromatography column.

It should be pointed out that the radial heterogeneity of the packed bed causes extra contribution to the apparent dispersion. The heterogeneity induces mobile phase velocity fluctuation locally. Maier *et al.* (70,71) performed pore-scale simulations of monodisperse sphere packing and fluid flow in cylinders and demonstrated that the heterogeneities in packing density induced by the presence of physical bounding walls would enhance hydrodynamic dispersion and that the degree of enhancement is related to the cylinder radius. The effective axial dispersion constant D_a^{eff} is larger than in the bulk. For dense random packed beds, the D_a^{eff} can be much larger than the bulk value, especially for axial component.

The second term in Equation 21.35 accounts for the contribution of eddy dispersion to the apparent dispersion. The eddy dispersion, which originated from the so-called

Saffman dispersion (72), is a result of the stochastic distribution of particles in the beds of packed columns, causing variations in the permeability and hence local velocity fluctuations, and contributes to the acceleration of radial mass transfers in the mobile phase (73).

When a sudden concentration change occurs in a region of a biphasic system at equilibrium, the equilibrium cannot be restored instantaneously everywhere, but takes time to relax to a new equilibrium state, depending on the rate at which the mass transfer takes place in the system. The new equilibrium state may never be achieved in the column because the band keeps migrating and the concentration keeps changing in the mobile phase. It has been demonstrated that the contributions of the mass transfer resistance is equivalent to band broadening and an apparent dispersion (74), and illustrated by the presence of mass transfer resistance terms in the Van Deemter and in the other plate height equations (66).

A layer of stagnant fluid surrounding each particle in the packed bed prevent mass transfer directly from the mobile phase stream to the stagnant mobile phase inside the pores of the particles, but through molecular diffusion. Its thickness, which controls the rate of this contribution to the mass transfer kinetics, depends on the velocity of the fluid. The film mass transfer kinetics is usually described by the concentration gradient:

$$\frac{\partial Q}{\partial t} = k_f A(c - c^*) = \frac{6k_f}{d_p}(c - c^*) \quad (21.37)$$

where k_f is the effective mass transfer coefficient, A the external surface area of the particles per unit volume ($a = 6/d_p$ for spherical particles), Q the concentration in the adsorbed phase, c the solute concentration in the mobile phase, and c^* the equilibrium value of c.

The mass transfer kinetics through a particle depends on pore diffusion and surface diffusion. The effective intraparticle diffusivity is given

$$D_e = D_p + \rho_P K D_s \quad (21.38)$$

where D_P is the intraparticle pore diffusivity and is often correlated in such as

$$D_P = [\varepsilon_p/(2 - \varepsilon)]^2 D_m \quad (21.39)$$

in which ε_p is the intraparticle porosity. D_s is the surface diffusivity, ρ_p the particle density, and K the Henry constant of adsorption.

The last phenomenon is the adsorption of the molecules from the mobile phase and then the release back to the mobile phase. In fact, the net retention time is the integral of all these incremental times spent by molecules in the adsorbed state. Among the several kinetic models available to account for this kinetics is the Langmuir kinetic model (66,75,76) which is written as

$$\frac{\partial Q}{\partial t} = k_a(Q_{\max} - Q)c_p - k_d Q \quad (21.40)$$

where $Q_{\max}$ is the column saturation capacity, c_p the concentration in the pore, k_a and k_d are the rate constants of adsorption and desorption of the molecule.

Although the actual influence of the mass transfer resistance should never be neglected, often it can be taken into account by including it with axial diffusion into an apparent axial dispersion term, thus the transport of elute can be described by an equilibrium-dispersive model. This model assumes that mass transfer across the column is instantaneous and that the mobile phase and stationary phase, at any point in the column, are constantly in equilibrium (28,77,78). It reads

$$\frac{\partial c}{\partial t} + F\frac{\partial q}{\partial t} + \mathbf{u} \cdot \nabla c = \nabla \cdot D_a \nabla c \quad (21.41)$$

where c and q are the solute concentration in the mobile and stationary phase, respectively, $F = (1 - \varepsilon)/\varepsilon$ is the phase ratio with ε being the total column porosity, $\mathbf{u}$ is the mobile phase flow velocity, and D_a is the apparent dispersion coefficient.

21.6.1 Model Setup and Implementation

The column used for case study on the elute transport and HETP analysis has all the features as for the one studied in the previous section, including the inlet and outlet pipes, the headplate zones, antijet disks, and the packed bed (Fig 21.7, but is simplified to allow us to perform two-dimensional axis-symmetrical simulations. Fine grid resolutions require a computer memory beyond our current capacity for three-dimensional simulations. The model, Equations 21.41 and Equation 21.45, was implemented and executed in Comsol (version 3.5a).

The following simplifications and assumptions are made in this case study:

1. The heterogeneity of the packed bed is neglected, the porosity and the permeability of the packed bed is assumed constant.
2. All processes are taken to be isothermal.
3. Diffusion between different species is neglected.
4. The flow is incompressible. The various physical properties are independent of pressure.
5. The solute, solvent, and mixtures of the two in any proportions have the same physical properties; therefore, the velocity field is independent of the composition of the fluid.

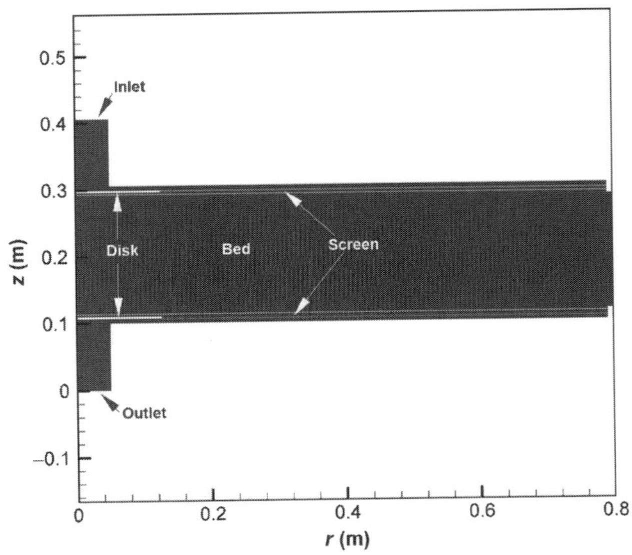

Figure 21.7. Geometry of a modeled column, a simplified version of a 1.6-m chromatography column. (This figure is available in full color at http://onlinelibrary.wiley.com/book/10.1002/9780470054581.)

The governing equation for saturated porous media and entirely fluid and solid system is given by Equation 21.41, in which the apparent diffusivity is given by tensor

$$D_a^{rr} = \frac{1}{\varepsilon}\left[\alpha_1 \frac{u_r^2}{|\mathbf{u}|^2} + \alpha_2 \frac{u_z^2}{|\mathbf{u}|^2} + \tau_L\right] D_m \quad (21.42)$$

$$D_a^{zz} = \frac{1}{\varepsilon}\left[\alpha_1 \frac{u_z^2}{|\mathbf{u}|^2} + \alpha_2 \frac{u_r^2}{|\mathbf{u}|^2} + \tau_L\right] D_m \quad (21.43)$$

$$D_a^{rz} = D_a^{zr} = \frac{1}{\varepsilon}(\alpha_1 - \alpha_2)\frac{u_r u_z}{|\mathbf{u}|^2} D_m \quad (21.44)$$

where α_1 is the dispersivity parallel to the flow direction and α_2 transverse direction. Here D_m is the molecular diffusion coefficient, and τ_L is the tortuosity factor.

The velocity field is given by the Brinkman equation,

$$\frac{\rho}{\varepsilon}\frac{\partial u}{\partial t} + \frac{\eta}{k}\mathbf{u} = -\nabla \cdot \left[p\mathbf{I} + \frac{1}{\varepsilon}\{\eta(\nabla \mathbf{u} + \nabla^T \mathbf{u})\}\right] \quad (21.45)$$

in addition to the continuity equation, Equation 21.30.

The Langmuir model, Equation 21.34, was adopted to represent the equilibrium isotherm. In this study, the author simulated three systems, namely, no adsorption, affinity, and ion exchange, with the constants $Q_{\max}$ and K_{eq} varied accordingly.

The following boundary conditions and initial conditions are assumed:

1. No-slip, no species flux boundary condition at column wall and antijet disks.
2. $U_n = U_0$, $c = c_0{}^*(t \le t_0)$ at the inlet.
3. $P = 0$, and $\mathbf{n} \cdot (D_a \nabla c) = 0$ at the outlet.
4. Axial symmetrical boundary condition on $r = 0$.

The steady-state momentum balance equation, Equation 21.45, was solved first to obtain the velocity field. Then the solute transport equation, Equation 21.41, was activated to provide transient solute concentration profiles. At the inlet, the elute concentration is set to be $c = 0$ at $t = 0$, then after a period of $t_0 = V_0/U_0$, it is set to be zero. V_0 is the specified volume of injected elute. Model parameters and material properties are given in Table 21.2.

Four hardware configurations are set up to study the influence of hardware design and process conditions on the flow distribution and the separation efficiency in terms of HETP. Table 21.3 provides the details on the geometry. No adsorption, ion exchange, and affinity system are modeled to see the difference in performance under these hardware configurations. The Langmuir model is used for all three systems with different constants. Values are given in Table 21.2.

21.6.2 Data Analysis Method

The method of moments (66) is used to calculate the retention time and the HETP.

TABLE 21.2. Physical Parameters and Process Conditions Used in the Model

Parameter	Symbol	Value
Fluid density	ρ	1000 kg/m^3
Fluid viscosity	η	0.001 Pa s
Mixture diffusivity	D_m	1×10^{-10}m^2/s
Particle density	ρ_p	1 kg/m^3
Bed porosity	ε_b	0.263
Bed permeability	k_b	1.308e^{-10}m^2
Bed tortuosity	τ_b	0.629
Screen porosity	ε_s	0.35
Antijet disk porosity	ε_d	0.5
Screen tortuosity	τ_d	1
Screen permeability	k_s	1.211e^{-10}m^2
Antijet permeability	k_d	(0.01–0.1) k_s
Inlet velocity	U_0	4.678–37.42 cm/s
Injected elute volume	V_0	1.51 L
Inlet elute concentration	c_0	1 mol/m^3
Axial dispersivity in bed	α_1	220
Transverse dispersivity in bed	α_2	15
Langmuir constant	K_{eq}	(No retentation) 0 m^3/mol (Ion exchange) 1 m^3/mol (Affinity) 100 m^3/mol
	$Q_{\max}$	(No retentation) 0 mol/m^3 (Ion exchange) 0.7 mol/m^3 (Affinity) 0.5 mol/m^3

TABLE 21.3. Four Simulated Hardware Configurations

	Config1 (full)	Config2 (half)	Config3 (hp001)	Config4 (hp01)
Column diameter (cm)	160	80	80	80
Antijet disk diameter (cm)	25.4	25.4	25.4	25.4
Antijet disk thickness (cm)	0.64	0.64	0.64	0.64
Antijet permeability compared to screen permeability	0	0	0.01X	0.1X
Inlet/outlet pipe diameter (cm)	10.16	10.16	10.16	10.16
Inlet/outlet pipe length (cm)	10	10	10	10
Snapring width (cm)	0.93	0.93	0.93	0.93
Snapring height (cm)	1.66	1.66	1.66	1.66

The first moment is the retention,

$$t_r = \int_0^\infty ct\mathrm{d}t \Big/ \int_0^\infty c\mathrm{d}t \tag{21.46}$$

the second is half width of the peak,

$$\sigma^2 = \int_0^\infty c(t - t_r)^2\mathrm{d}t \Big/ \int_0^\infty c\mathrm{d}t \tag{21.47}$$

and the third measures the skewness, that is, the asymmetry of the peak, given by

$$\gamma = \int_0^\infty c(t - t_r)^3\mathrm{d}t \Big/ \int_0^\infty c\mathrm{d}t \tag{21.48}$$

The number of theoretical units (NTU) is given by

$$\mathrm{NTU} = t_r^2/\sigma^2 \tag{21.49}$$

and HETP is given by

$$\mathrm{HETP} = \frac{L}{\mathrm{NTU}} \tag{21.50}$$

where L is the length of the fixed bed.

A Gaussian distribution leads to $\gamma = 0$. When $\gamma > 0$, the peak exhibits tailing. Conversely, when $\gamma < 0$, fronting is observed. Usually, a peak with a lower value of HETP will also have a smaller value of γ. A large value of γ indicates severe tailing of the peak and retention of traces of solute in the column, which is undesirable as minute amounts of solute remains in the column while the majority has been eluted, making it difficult to separate mixtures free of a contaminating species.

The surface-averaged transient elute concentration at the bottom of the column is used to compute these quantities. Time step is chosen to be 0.1 s in the beginning the elution and 5 s toward the end.

21.6.3 Results

The typical flow distribution in the column is illustrated in Fig 21.8, in which the streamlines are plotted. The antijet disks change the flow direction from axial to radial direction, serving the purpose of distributing the flow. However, it is clearly seen that the existence of the antijet disks distort the streamlines in the packed bed. The antijet disks affect the flow between them to a more severe degree than those near the column wall where there is a small sudden expansion and contraction. This poses the serious challenge of creating uniform flow in the column bed. Intuitively increasing the ratio of the disk to the column could help to distribute the flow in the headspace and in the packed bed better. However, the model prediction on the HETP indicates this is not the case, unless the disks are made porous. The plug flow assumption used in some practice would grossly underestimate this effect.

Figure 21.9 shows the snapshots of the elute concentration distributions at various time instances. At the flow rate of 91 L/min, the radial flow created by the antijet disks is not strong enough to cover the full range of the column diameter. On the other hand, the short bed height (therefore short elution time) does not allow elute to diffuse all the way to the column wall, thus giving rise to a highly nonuniform distribution of the elute concentration (Fig 21.10. In

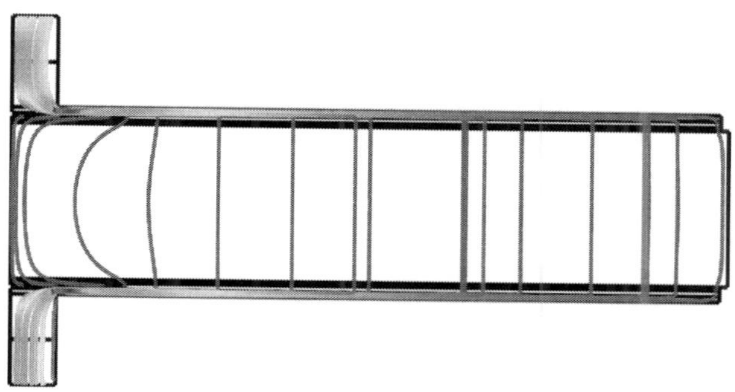

Figure 21.8. Streamlines distribution colored by velocity magnitude. Purple is the faster and red the slowest. Inlet linear velocity $U = 0.18712$ m/s. No adsorption is included. (This figure is available in full color at http://onlinelibrary.wiley.com/book/10.1002/9780470054581.)

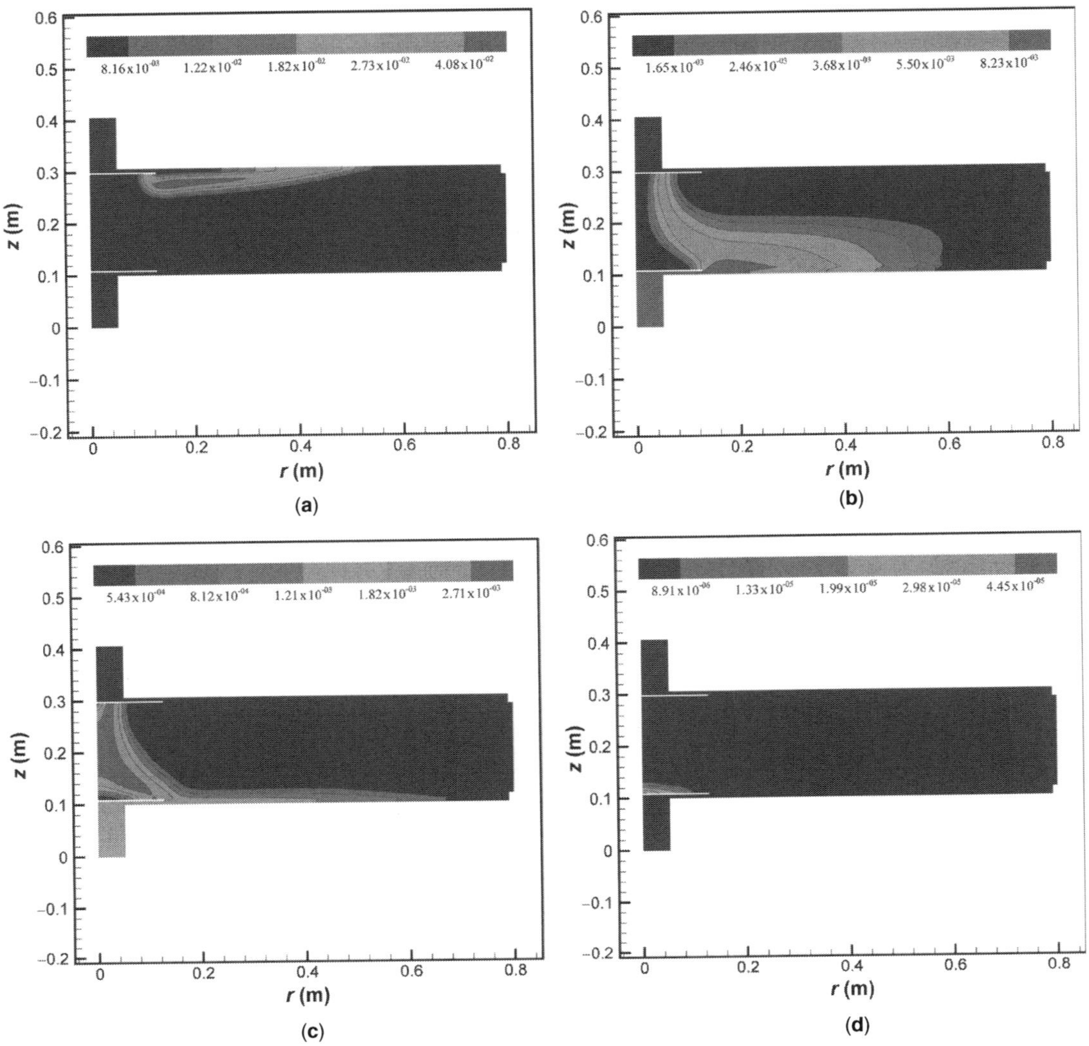

Figure 21.9. Snapshots of the elute concentration profile. Elute concentration distribution at $t = 5$, 60, 120, and 600 s, assuming no adsorption. (This figure is available in full color at http://onlinelibrary.wiley.com/book/10.1002/9780470054581.)

the column, the elute concentration distribution is primarily determined by the convection because the diffusion is a much slower process. The distance that elute can diffuse during the whole elution period is $2\sqrt{D_a\tau}$. The best case is the affinity system that has the longest residence time. The estimated diffusion distance is about 6 cm, given the total apparent diffusivity is 2×10^{-8} m^2/s. Indicated in the last snapshot of the series is that a substantial residual of the elute lingering in the area between the antijet disks and right above the bottom antijet disk for a long time. The observed nonparabolic elution front is the direct consequence of nonuniform flow in the column. This lingering caused by the presence of antijet disks can be alleviated by turning the disks porous.

Inlet velocity is varied from 20 to 360 L/min to see the effect of flow rate on the separation efficiency. Figure 21.11 presents the calculated HETP as a function of the linear velocity. It can be seen that HETP initially decreases with increasing flow rate and eventually increases with increasing flow rate. An optimal flow rate is around 0.3 m/s.

Among the three systems, the nonadsorption system has the sharpest peak and the least retention time (Fig. 21.12). A small second peak is observed when the column diameter is reduced, making the profile non-Gaussian. The second peak goes away when the disk is changed to be porous. This indicates that flow distribution in configuration 2 is the worst. The highest value of HETP is also a clear indication of extreme nonuniform elution.

The retention time, peak width, NTU, HETP, and peak skewness are presented in Table 21.4, from which a few observations can be made. For full-sized column, HETP slightly decreases from nonadsorption to affinity, but the

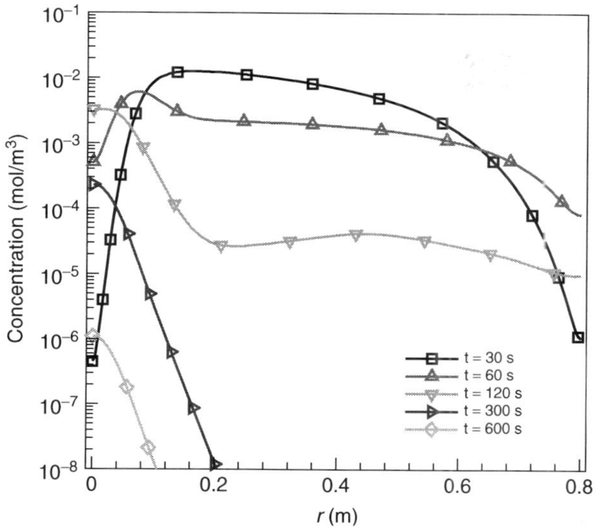

Figure 21.10. Elute concentration at the bed centerline at time t = 5, 60, 120, and 600 s, assuming no adsorption. (This figure is available in full color at http://onlinelibrary.wiley.com/book/10.1002/9780470054581.)

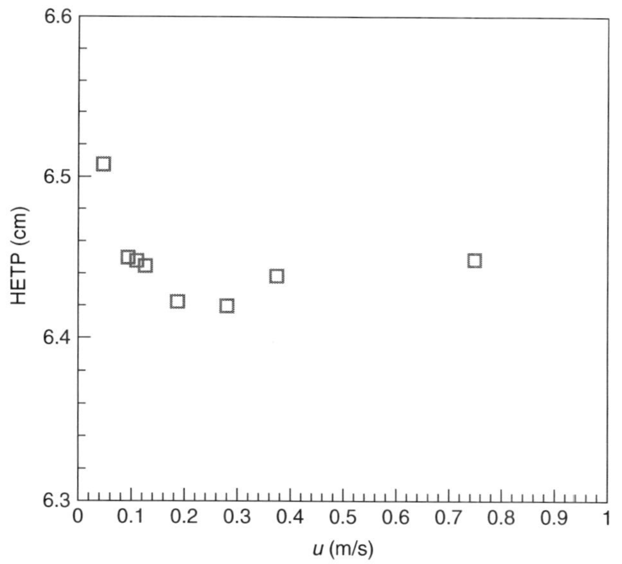

Figure 21.11. Calculated HETP as a function of the linear velocity. No adsorption is included. (This figure is available in full color at http://onlinelibrary.wiley.com/book/10.1002/9780470054581.)

trend is reversed for three other configurations. The retention time drops when column size is reduced, and more so if the disks are porous, for each modeled system. Similar effects can be seen on peak spread and skewness. The difference in the magnitude of retention time emphases the importance of using realistic physical models that incorporate appropriate binding kinetics, which is the author's current research.

21.7 SUMMARY

The chapter modeled the flow in large chromatography columns in each stage using physical mathematical models. It first reviewed a one-dimensional, elasticity-based theoretical model and then applied this model to scale up column packing from laboratory scale to commercial scale. Dye study in a 1.60-m column was performed using CFD technique. Finally, the transport of elute in large columns was modeled using an equilibrium-dispersive model, assuming that mass transfer between the mobile phase and the stationary phase at any point in the column is constantly in equilibrium. It should be pointed out that columns of such scale with this type of flow distributor have never been studied before.

It is challenging to scale up chromatographic column packing. Empirical phenomenological models are inadequate to capture the underlying complex rheological behaviors. Analysis of the wall effects demonstrates that the presence of the bounding column wall exerts an influence on the flow distribution close to the wall in the range of a few particle diameters. The wall effects should be taken into consideration for small to medium scale columns. The case studies on large commercial size columns indicate the flow distribution and thus separation efficiency depend on the hardware design to a great degree. Flow distributor in the large-scale columns is extremely important to distribute even flow through packed bed. Uniform elute front cannot be achieved in the studied configurations. The case studies also demonstrate that process optimization is achievable through mathematical modeling and simulation approach. CFD technique shows its promise as an important quality by design tool.

Acknowledgments

The author wishes to thank his former colleagues in Amgen Process Development and Research. Special thanks are given to Dr. Suresh Vunnum for the discussions on modeling column elution process, and to Dr. Arleen Paulino and Dr. Mark Durst for their help with the manuscript.

LIST OF ABBREVIATIONS

CFD	computational fluid dynamics
HETP	height equivalent to a theoretic plate
FQ	frit quality
HPLC	high performance liquid chromatography
NTU	the number of theoretical units

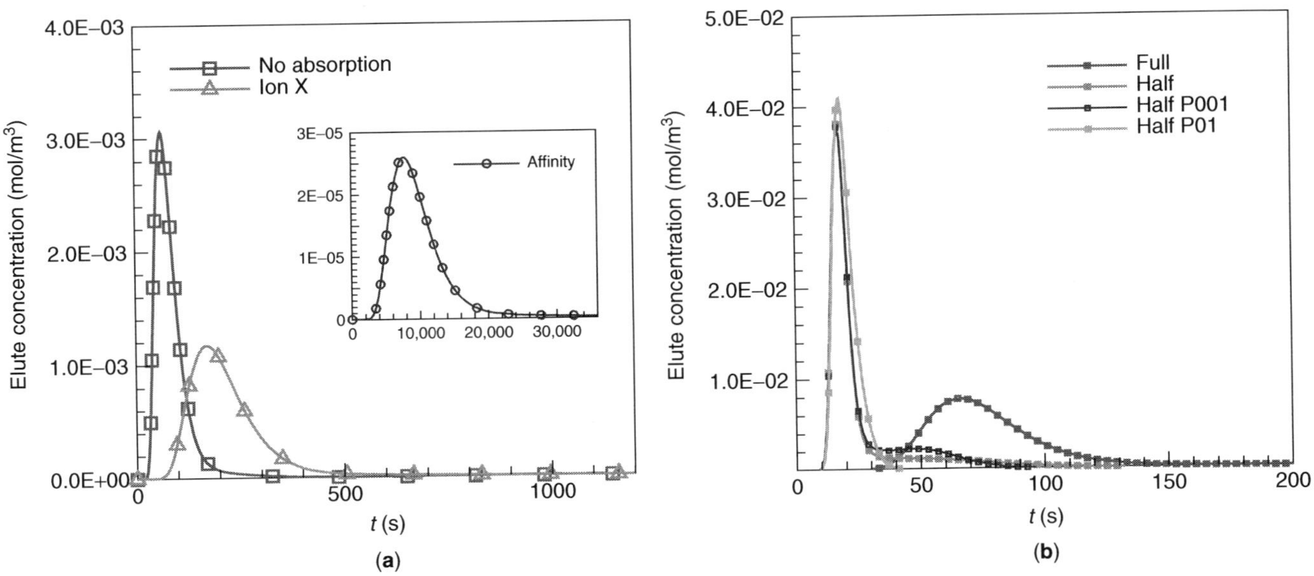

Figure 21.12. (a) Elution profiles for three modeled systems: nonabsorbent, affinity and ion exchange. (b) Elution profiles for nonabsorbent system in the configuration of FULL, HALF, HP001, and HP01. (This figure is available in full color at http://onlinelibrary.wiley.com/book/10.1002/9780470054581.)

TABLE 21.4. Retention Time, Sigma, NTU, and HETP for All Modeled Systems

System	τ	NTU	γ	σ	HETP(cm)
No retention	81.92	2.80	5.07	48.93	6.42
NR half	26.31	1.72	2.75	20.08	10.48
NR hp001	25.56	2.80	1.93	15.27	6.42
NR hp01	20.06	19.16	0.94	4.58	0.94
Ion exchange	222.31	2.81	4.81	132.70	6.41
IonX half	70.53	1.70	2.53	54.13	10.60
IonX hp001	68.45	2.65	1.86	42.04	6.79
IonX hp01	53.46	16.82	1.01	13.04	1.07
Affinity	10,103.15	2.82	4.61	6,017.34	6.39
AFF half	3,189.62	1.66	2.43	2,473.88	10.83
AFF hp001	3,089.44	2.54	1.84	1,939.04	7.09
AFF hp01	2,405.76	15.02	1.03	620.70	1.20

NOMENCLATURE

a	acceleration
A	external surface area of particles per unit volume
c	concentration in liquid phase, in the pore (p)
C	constant
c^*	equilibrium value of c
D	diffusion, effective (e), intraparticle pore (p), molecular (m), surface (s)
D_c	column diameter
d_p	particle diameter
E	Yong's modulus
F	force
F	phase ratio, $(1-\varepsilon)/\varepsilon$.
g	gravity
G	the Lame constants
K	the Henry constant
k	permeability of porous media, or constant parameter
k_a	rate constant of adsorption
k_d	rate constant of desorption
K_{eq}	equilibrium constant
k_f	effective mass transfer coefficient
L_c	column length
P	fluid pressure
q	bounding protein concentration
Q	concentration in the absorbed phase
Q_{max}	maximum bound protein capacity
R	column radius

r, θ, z	cylindrical coordinates
Re	Reynolds number
s	axial Eulerian strain
S	specific surface area of particles
t_r	retention, related to HETP
u	displacement of solid
$\mathbf{U}$	superficial velocity, relative velocity
$\mathbf{u}$	velocity of fluid
U_{crit}	critical velocity of flow packing
V	volume of particles
x, y, z	spatial coordinates

Greek Symbols

Ω	effective specific surface area parameter
α	geometrical parameter for apparent diffusion, related to convection
ε	porosity
ϕ	particle volume fraction
γ	area-averaged strain in the axial direction, or skewness of elution, related to HETP
η	fluid viscosity
λ	Lame constants
μ_f	wall friction coefficient
ν	the Poisson ratio
ρ	density of solid (s) or fluid (f)
σ	half width of the peak, related to HETP
ρ	stress of solid
τ	fluid stress
τ_L	geometrical parameter for apparent diffusion, related to tortuosity of packed bed
τ_w	shear stress on the wall
ξ	relative particle diameter

REFERENCES

1. Bey O, Eigenberger G. Chem Eng Sci 1997; 52: 1365–1376.
2. Cohen Y, Metzner AB. J Rheol 1985; 29: 67–102.
3. Park JC, Raghavan K, Gibbs SJ. J Chromatogr A 2002; 945 (1–2): 65–81.
4. Zou RP, Yu AB. Chem Eng Sci 1995; 50: 1504–1507.
5. Zou RP, Yu AB. Chem Eng Sci 1996; 51: 1177–1180.
6. Hardin RA, Burmeister LC. J Appl Polym Sci 1993; 48: 625–637.
7. Broyles BS, Shalliker RA, Guiochon G. J Chromatogr A 1999; 855 (2): 367–382.
8. Stanley BJ, Sarker M, Guiochon G. J Chromatogr A 1996; 741: 175–184.
9. Yew BG, Ureta J, Shalliker RA, Drumm EC, Guiochon G. AIChE J 2003; 49 (3): 642–664.
10. Cherrak DE, Al-Bokari M, Drumm EC, Guiochon G. J Chromatogr A 2002; 943 (1): 15–31.
11. Cherrak DE, Guiochon G. J Chromatogr A 2001; 911 (2): 147–166.
12. Inci MN, Erman B, Okay O, Durmaz S. Polymer 2001; 42: 3771–3777.
13. Koh JH, Broyles BS, Guan-Sajonz H, Hu MZC, Guiochon G. J Chromatogr A 1998; 813 (2): 223–238.
14. Koh JH, Guiochon G. J Chromatogr A 1998; 796 (1): 41–57.
15. Colby CB, O'Neil BK, Middelberg APJ. Biotechnol Prog 1996; 12: 92–99.
16. Mohammad AW, Stevenson DG, Wankat PC. Ind Eng Chem Res 1992; 31: 549–561.
17. Stickel JJ, Alexandros F. Biotechnol Prog 2001; 17: 744–751.
18. Lu WM, Tung KL, Hung SM, Shiau JS, Hwang KJ. Powder Technol 2001; 116: 1–12.
19. Soriano GA, Titchener-Hooker NJ, Shamlou PA. Bioprocess Eng 1997; 17: 115–119.
20. Keener RN, Maneval JE, Fernandez EJ. Biotechnol Prog 2004; 20: 1146–1158.
21. Keener RN, Maneval JE, Ostergren KCE, Fernandez EJ. Biotechnol Prog 2002; 18: 587–596.
22. Ostergren KCE, Tragardh AC, Enstad GG, Mosby J. AIChE J 1998; 44 (1): 2–12.
23. Broyles BS, Shalliker RA, Guiochon G. J Chromatogr A 2001; 917 (1–2): 1–22.
24. Farkas T, Guiochon G. Anal Chem 1997; 69: 4592–4600.
25. Guiochon G, Drumm E, Cherrak D. J Chromatogr A 1999; 835 (1–2): 41–58.
26. Shalliker RA, Broyles BS, Guiochon G. Anal Chem 2000; 72 (2): 323–332.
27. Shalliker RA, Broyles BS, Guiochon G. J Chromatogr A 2000; 888 (1–2): 1–12.
28. Shalliker RA, Broyles BS, Guiochon G. J Chromatogr A 2003; 994 (1–2): 1–12.
29. Shalliker RA, Wong V, Broyles BS, Guiochon G. J Chromatogr A 2002; 977 (2): 213–223.
30. Yuan QS, Rosenfeld A, Root TW, Klingenberg DJ, Lightfoot EN. J Chromatogr A 1999; 831 (2): 149–165.
31. Koponen A, Kataja M, Timonen J. Int J Mod Phys C 1998; 9 (8): 1505–1521.
32. Brady JF, Bossis BG. Annu Rev Fluid Mech 1988; 20: 111–157.
33. Kim AS, Stolzenbach KD. J Colloid Interface Sci 2002; 253: 315–328.
34. Sierou A, Brady JF. J Fluid Mech 2001; 448: 115–146.
35. Wu YX, PhD thesis: "Computational fluid dynamics study of high performance liquid chromatography", University of Singapore; 2002.
36. Wu YX, Ching CB. Chromatographia 2003; 57 (5–6): 329–337.
37. Tan SN, Khoo BC. In the MIT Report High Performance Computation for Engineered Systems (HPCES) ; 2003.
38. Ching CB, Wu YX, Lisso M, Wozny G, Laiblin T, Arlt W. J Chromatogr A 2002; 945: 117–131.
39. Guiochon G, Farkas T, GuanSajonz H, Koh JH, Sarker M, Stanley BJ, Yun T. J Chromatogr A 1997; 762 (1–2): 83–88.
40. Guiochon G, Sarker M. J Chromatogr A 1995; 704: 247–268.
41. Melekaslan D, Gundogan N, Okay O. Polym Bull 2003; 50: 287–294.
42. Happel J, Brenner H. Low Reynolds Number Hydrodynamics: with special application to particulate media. The Hague: Martinus Nijhoff Publishers; 1983.

43. Alazmi B, Vafai K. J Heat Transf 2000; 122: 303–326.
44. McCue JT, Cecchini D, Chu C, Liu W-H, Spann A. J Chromatogr A 2007; 1145 (1–2): 89–101.
45. Ostergren KCE, Tragardh C. Chem Eng J 1999; 72 (2): 153–161.
46. Zhao J, Wang C-H, Lee D-J, Tien C. J Colloid Interface Sci 2003; 262: 60–72.
47. Yew BG, Drumm EC, Guiochon G. AIChE J 2003; 49 (3): 626–641.
48. Meeten GH. Rheol Acta 2000; 39: 399–408.
49. Meeten GH. Rheol Acta 2001; 40: 279–288.
50. Meeten GH. Rheol Acta 2002; 41: 557–566.
51. Tung KL, Lin YL, Shih TC, Lu WM. J Chin Inst Chem Eng 2004; 35 (1): 101–110.
52. Hatzikiriakos SG, Dealy JM. J Rheol 1991; 35: 497–523.
53. Hatzikiriakos SG, Dealy JM. J Rheol 1992; 36: 703–741.
54. Barnes HA. J Non-Newtonian Fluid Mech 1995; 56: 221–251.
55. Barnes HA. J Non-Newtonian Fluid Mech 2000; 94: 213–217.
56. Barnes HA, Nguyen QD. J Non-Newtonian Fluid Mech 2001; 98 (1): 1–14.
57. Liu S, Masliyah JH. J Non-Newtonian Fluid Mech 1999; 86: 229–252.
58. Cohen J, Metzner AB. J Rheol 1985; 29, 67–102.
59. Kalyon DM, Gevgilili H. J Rheol 2003; 47 (3): 683–669.
60. Kalyon DM, Yaras P, Aral B, Yilmazer U. J Rheol 1993; 37 (1): 35–53.
61. Lawal A, Kalyon DM, Yilmazer U. Chem Eng Commun 1993; 122: 127–150.
62. Walls HJ, Caines SB, Sanchez AM, Khan SA. J Rheol 2003; 47 (4): 847–868.
63. Jana SC, Kapoor B, Acrivos A. J Rheol 1995; 39 (6): 1123–1132.
64. Yilmazer U, Kalyon DM. J Rheol 1989; 33 (8): 1197–1212.
65. Pathak N, Norman C, Kundu S, Nulu S, Fang Z. Bioprocess Int J 2008; 6 (9): 72–81.
66. Guiochon G, Golshan-Shirazi S, Katti AM. Fundamentals of preparative and nonlinear chromatography. Boston (MA): Academic Press; 1994.
67. Ruthven DM. Principles of adsorption and adsorption processes. New York (NY): Wiley; 1984.
68. Koponen A, Kataja M, Timonen J. Phys Rev E 1996; 54 (1): 406–410.
69. Koponen A, Kataja M, Timonen J. Phys Rev E 1997; 56 (3): 3319–3325.
70. Maier RS, Kroll DM, Bernard RS, Howington SE, Peters JF, Davis HT. Philos Trans R Soc Lond A Math Phys Eng Sci 2002; 360: 497–506.
71. Maier RS, Kroll DM, Bernard RS, Howington SE, Peters JF, Davis HT. Phys Fluids 2003; 15 (12): 3795–3815.
72. Tallarek U, Bayer E, Guiochon G. J Am Chem Soc 1998; 120: 1494–1505.
73. Weber SG, Carr PW. High performance liquid chromatography. New York: Wiley; 1989.
74. Giddings JC. Dynamics of chromatography. New York (NY): M. Dekker; 1965.
75. Guiochon G, Lin B. Modeling for preparative chromatography. San Diego (CA): Academic Press; 2003.
76. Gu T. Mathematical modeling and scale-up of liquid chromatography. Berlin: Spinger-Verlag; 1995.
77. Smith MS, Guiochon G. J Chromatogr A 1998; 827 (2): 241–257.
78. Yun T, Smith MS, Guiochon G. J Chromatogr A 1998; 828 (1–2): 19–35.

22

PUMPS, INDUSTRIAL

Bob Stover and Ed Domanico
Tri-Clover, Valencia, California

22.1 INTRODUCTION

Liquid transfer is a vital component of most biotechnology processes. There is possibly more information available on pumps than any other single piece of equipment (pumps being a very common piece of rotating equipment, second only to the electric motor) existing in a typical plant, so this section must be a rough introduction to hydraulic theory. This chapter provides a basic understanding of pump mechanics, nomenclature, and principles and sufficient information to begin the selection process of a new pump.

22.2 THEORY

Before a specific selection can be initiated, the operating conditions the pump will be expected to meet must be understood. Breaking down the system into clearly defined and agreed-upon components will not only simplify calculations, and help guarantee nothing is overlooked, but also make it easier to evaluate the consequences of changes to any part of the system. There are universally understood and applied ways of breaking down a system, recognized by all pump manufacturers and users.

22.2.1 Flow

Capacity is expressed in flow/unit of time, commonly gallons per minute (gpm) or per hour (gph), or as cubic meters per hour (m^3/h).

To convert from volume to gallons:

$$1\ ft^3 = 7.48\ gallons$$

$$1\ m^3 = 264.2\ gallons$$

It is important to be realistic with estimates, because accurate pressure loss calculations depend on an accurate flow rate. Overestimating will result in equipment that will need to be valved back or slowed down to produce the demands actually required or in oversized/overpriced pumps.

22.2.2 Head

Head is the accepted hydraulic term for pressure. It is most commonly expressed in pounds/inch^2 (psi), or feet of head. The concept of a linear unit being used to express pressure can be most easily visualized as taking a resistance to flow and mentally converting it to a vertical column of water, such that the weight of this imaginary column is equivalent to the resistance of the considered section. Any pump that can push fluid against a column of water of the height thus calculated will be able to deliver flow against the system used to make the calculation.

To convert from psi to feet:

$$1\ psi = 2.31\ ft\ of\ water$$

$$1\ ft\ of\ water = .433\ psi$$

Most pipe friction loss charts express their losses as feet of head per 100 feet run of pipe, and all losses given as psi

Downstream Industrial Biotechnology: Recovery and Purification, First Edition. Edited by Michael C. Flickinger.

can be converted to feet by the above conversion factors (see Specific Gravity for adjustments given liquids heavier or lighter than water).

This linear unit has become the most common choice for hydraulic pressure (in the sense of resistance to flow) because:

- It is easily visualized.
- A linear unit fits into most hydraulic formulas (velocity, horsepower, specific speed, etc.).
- It compensates for various specific gravities before inclusion.
- It provides a common unit for all components of the total head that can then be added together to arrive at total resistance.

22.2.2.1 Types of Head. The total head referred to above is made up of four components. Any resistance to flow will fall into one of the following categories.

22.2.2.1.1 Static Head (ΔZ). Static head refers to the elevation differential between the source level of the liquid and the elevation at the destination. In other words, the elevation differential the pump must overcome. Caveats to this are:

- If a line enters at or near the bottom of a destination tank, the liquid level *in* the tank is considered the appropriate one, not where the pipe makes its entry (the pump must overcome the weight of the column *inside* the tank as part of its pressure capability).
- The vertical ups and downs are irrelevant (they will be accounted for later under Friction Head); the elevations at the beginning and end of the line are the ones used for ΔZ.
- The pump's position in relationship vertically to either the source or destination point is also irrelevant.

Static head is easily expressed in feet.

22.2.2.1.2 Pressure Head (ΔP). Pressure head is the differential in pressure at the ends of the system, the inlet and outlet, expressed in feet. An example would be a vacuum tank at the source; it provides resistance to flow (the vacuum holds back the fluid) and must be overcome by the pump. Resistance to flow from vacuum can be converted to feet of head by the following:

$$1 \text{ inch mercury (Hg)} = 1.133 \text{ ft of water}$$

Pressure at the discharge end (a pressurized tank or injection into a pressurized line) also needs to be added into the total and is included in ΔP.

22.2.2.1.3 Velocity Head (H_V). Velocity head is that energy necessary to overcome inertia. It is calculated by

$$H_V = V^2/2g$$

where V = velocity through the piping lines (in feet/second), and $g = 32.2$ ft/s^2, the acceleration caused by gravity at sea level.

Velocity head is reasonably estimated by taking the average velocity in a system and applying the formula. However, given a maximum average velocity of less than 8 ft/s in most systems, H_V is usually a relatively small number (less than 1 ft in the average 8 ft/s example above) in a typical hydraulic calculation and almost always ignored.

22.2.2.1.4 Friction Head (H_F). Friction loss is usually the largest individual component of total head for most calculations. There are formulas available for its calculation given various types of pipe, fittings, valves, and filters in texts, but fortunately there are charts in existence giving the calculated heads. Pipe losses themselves are usually expressed in feet of head/unit of length (usually 100 ft). Fitting losses are expressed either in equivalent feet of straight pipe to be added to the run of actual straight pipe for what is called an *equivalent feet of pipe* to be used in the calculations. Or, charts may give the actual loss per each fitting. Losses through valves, filters, and such are often expressed in psi or feet of loss per valve. Each component in a line must be considered, and a grand total for the system is arrived at as a total of the individual pieces.

The friction loss numbers will always vary considering the flow through the line; higher flows mean more drag and therefore more resistance. Filters will vary with flow and also condition of the filter element. It is common practice to use worst-case conditions.

22.2.2.1.5 Total Head (H_M). The four individual elements listed previously are added together for the *total mechanical head* (H_M, also known as mechanical head, total dynamic head [TDH], or simply total head). They are the components for Bernoulli's formula, modified for the pump industry, given as:

$$H_M = \Delta Z + \Delta P + H_V + H_F$$

22.2.2.2 Adjustments to Head. There are qualities a liquid can possess that will affect the H_M.

22.2.2.2.1 Specific Gravity. Specific gravity (SG) is the direct numeric ratio of the weight of a particular liquid to the weight of water. Given that at certain standard conditions water weighs approximately 8.34 lb/gallon, anything heavier than that would have a specific gravity greater than 1.0 (for example, a liquid weighing 16.68 lb/gallon would have an SG of 2.0), anything lighter less than 1.0. Specific gravity will affect a conversion from psi to feet as follows:

$$\text{psi} = (\text{feet of water} \times \text{SG})/2.31$$

or

$$\text{feet of water} = (\text{psi} \times 2.31)/\text{SG}$$

Specific gravity will also directly affect horsepower because of the extra torque needed to overcome the additional weight of the liquid (as will be shown under Horsepower).

The second quality a liquid possesses that directly affects total head calculations is its *viscosity*. Most pump selection charts assume waterlike viscosity (1 centipoise [cps] or 31 Saybolt second universal [SSU]). The viscosity of liquids thicker than water affects pumps in several ways:

- It affects friction loss calculations; losses are higher with more viscous liquids. There are separate charts used for liquids more viscous than water.
- It affects horsepower. Manufacturers often provide specific charts to give the added torque needed to overcome viscosity.
- It affects the very capability of a pump to even *deal* with a product. Centrifugal pumps especially have limited abilities to handle viscous products.

Viscosity is usually expressed in the referenced units (cps or SSU), but there are many others. Viscosity can vary under dynamic conditions. Specifically:

- If a liquid becomes thinner under shear, it is *thixotropic* (example: paint).
- If it becomes thicker under shear, it is *dilatant* (example: taffy).
- If it becomes thinner only after exceeding a certain minimum threshold of shear, it is *plastic* (example: ketchup).
- If a liquid is unaffected by shear, it is *Newtonian* (example: water).

As mentioned above, viscosity affects centrifugal pumps in their very ability to pump a liquid; with viscous fluids their pressure capability falls off, the flow capability falls off even more, and their efficiency may fall off to a point where their use is totally unacceptable. Positive displacement pumps are affected in the speed they can turn and still move product without starving at the inlet. Manufacturers provide charts giving maximum speed for varying viscosities, for adjustments needed to horsepower to drive the unit with a thicker product, and perhaps for minimum port sizes necessary to allow product to get in quickly enough.

If the viscosity of a product is unknown, many pump manufacturers provide services wherein a liquid's viscosity will be tested, often at no charge, in exchange for the opportunity to quote their pumps.

22.2.3 Horsepower

Given the results of the total head calculations above, it is possible to make accurate calculations of the horsepower necessary to drive a particular pump.

The basic horsepower formula is

$$\text{HP} = (\text{gpm} \times H_{\text{M}} \times \text{SG}) \div (3,960 \times \text{eff.})$$

where gpm = flow (in gallons/min); H_{M} = total mechanical head (in feet); SG = specific gravity; eff. = efficiency (expressed as a decimal; e.g., 50% would be .50). *Note*: 3,960 is a constant, derived from the horsepower units and the weight of water.

A variation of this formula, commonly used for positive displacement pumps wherein specific gravity has already been calculated into the total head before being expressed as psi, is

$$\text{HP} = (\text{gpm} \times \text{psi}) \div (1,715 \times \text{eff.})$$

The 3,960 constant has been changed to 1,715 to provide for psi as a pressure unit by dividing by 2.31.

Often, viscosity will affect the horsepower as calculated above; in those cases manufacturers will provide correction factors for differing viscosity. The Hydraulic Institute has generic charts for centrifugal pumps that seem to work fairly well for almost all manufacturers, although some especially high-efficiency pumps claim to outperform the charts and therefore provide their own correction factors.

22.2.4 Net Positive Suction Head

Net positive suction head (NPSH) is the hydraulic term referring to the pressure under which liquid enters the pump, not only in terms of what the existing system is able to provide but also in reference to what any particular pump needs in order to function properly. It is an absolute unit, always expressed relative to absolute zero (as opposed to gauge units, whose values are always in reference to normal atmospheric pressure, called zero gauge). If a pump does not receive the liquid under sufficient pressure, it will cavitate.

22.2.4.1 ***Cavitation.*** Liquids will stay in liquid form only if sufficient atmospheric or system pressure is available to prevent it from going off to vapor. Even cold water, for example, will flash off into vapor in a strong enough vacuum, outer space as an obvious example. All liquids have a predictable vapor pressure necessary to keep them in their liquid state, varying up or down respectively as the liquid is heated or cooled.

This is of particular interest in hydraulics because of the often high vacuum present at the inlet of a pump. If the pressure provided by the system is less than the liquid's vapor pressure, the pump will undergo a physical process called cavitation. This action varies in specifics from centrifugal to positive displacement type pumps, but the general process is always approximately the same:

- The liquid finds itself in an area of low pressure, the pump's inlet port.
- If the pressure is *too* low (less than the liquid's vapor pressure at the operating temperature), it undergoes a state change from liquid to vapor. Because the vapor is not free to escape, it stays entrained in the liquid and assumes a bubble (spherical) shape.
- The vapor bubbles are carried further into the pump with the flow of surrounding liquid.
- Eventually the low-pressure bubbles and surrounding liquid find themselves in a higher pressure area as they near the pump discharge.
- The higher pressure causes the vapor bubbles to implode (collapse). This implosion is called cavitation.

The implosion of the low pressure vapor bubbles makes a very identifiable sound; in centrifugal pumps it produces a rattle, sounding like gravel is being carried through the housing; in positive displacement pumps the sound is more percussive, often *very* loud. In both cases the implosion of the vapor bubbles is damaging to the pump itself and often to the product. Centrifugal impellers and positive displacement pump rotors will actually have material removed by the very powerful implosions, shafts can be deflected to the point of fatigue, higher horsepower is required, and performance will be negatively affected.

22.2.4.2 ***Net Positive Suction Head Required.*** Manufacturers can test their pumps to see exactly how much positive suction pressure is necessary at various flows to keep their pumps from cavitating. This is called the net positive suction head required (NPSH_R). Typically, the greater the flow, the more pressure necessary because of the higher velocities going through the housing. A possible exception: at extremely low flows, the slippage caused by recirculation within the housing in some centrifugal pumps may create low pressure areas, sometimes low enough to fall under a liquid's vapor pressure and allow cavitation. This is called either suction or discharge recirculation cavitation, depending on where it physically occurs in the pump housing. Manufacturers determine the NPSH_R from empirical data. They set up a test skid wherein they can gradually restrict flow into a pump, constantly monitoring the performance with pressure gauges and flow meters. Eventually they find a point at which the restrictions affect performance an agreed-upon minimum amount (3% below normal curve is the current standard). The pump is considered to be cavitating, and this point is registered on a chart. The same process is performed for a number of flows, and an NPSH_R curve is generated. NPSH_R curves are assumed to be drawn using 60°F water; adjustments can be made later for other liquids.

22.2.4.3 ***Net Positive Suction Head Available.*** Net positive suction head available (NPSH_A) is the pressure any particular system makes available at a pump's inlet under operating conditions. It is a total of several factors. After calculation (or determination by an actual physical test), it will be compared to the pump's NPSH_R. Also by previous agreed-upon standard, NPSH_A must exceed NPSH_R by a minimum of 2 ft for a pump to be considered acceptable. This is a controversial issue; some people believe that this differential is insufficient, that the NPSH_R chart represents the point at which the pump is *already* cavitating, a 2-ft differential is often too low; and that a sliding scale of minimal percentages by which NPSH_A should exceed NPSH_R would serve the user better. However, at this time the 2-ft differential is still accepted by most hydraulic engineers.

NPSH_A is calculated by the formula:

$$\text{NPSH}_\text{A} = H_\text{ATM} \pm \Delta Z - H_\text{F} - H_\text{VAP}$$

where H_ATM = atmospheric pressure in absolute units, expressed in feet (atmospheric pressure at sea level = 14.7 psi = 33.9 ft); ΔZ = elevation differential from liquid level to pump's inlet (in feet). If the liquid level is above the pump inlet, the number will be positive (+), if below it will be considered a negative (−); H_F = friction loss in suction line only (in feet); and H_VAP = vapor pressure of liquid at operating temperature (in feet).

H_ATM and H_VAP are available from various reference books; H_F and ΔZ are calculated as in the section on Friction Head.

NPSH_A can be determined on an operating pump by installing either a compound or vacuum gauge at the pump's inlet and by taking this reading and subtracting the liquid's vapor pressure (which will not be indicated on the gauge: H_ATM, H_F, and ΔZ will all be indicated by the gauge reading).

As stated earlier, the results of the calculation (or gauge reading $-H_{VAP}$) should be compared to the manufacturer's $NPSH_R$, a minimum 2-ft differential necessary to guarantee a pump will not cavitate. The $NPSH_A$ calculation is valuable for several reasons:

- It can show us beforehand whether a pump will cavitate after installation.
- It indicates the area most easily altered to eliminate existing cavitation and how much it needs to be altered. For example, if H_F is high, it may be lowered with larger pipe or fewer elbows. If ΔZ is low (or even negative), we can pick up positive head by raising the source tank.

If a system is still cavitating, and the $NPSH_A$ calculation does not indicate an easy correction, it may be that a pump with a lower $NPSH_R$ is needed. There are whole classes of pumps considered low NPSH that typically have requirements below 1 ft or 2 ft.

22.3 CENTRIFUGAL PUMPS

Pumps generally are divided into two main classifications, centrifugal and positive displacement (PD). There are many subclassifications. Several will be listed here later according to their suitability for use in bioprocessing, including a few that seem to cross the boundaries of centrifugal and PD. The vast majority of pumps in use are centrifugal because of the many advantages of centrifugal design (to be listed later), but bioprocessing has unusual demands, and the percentages of positive displacement pumps in use is significantly higher than in other industries.

22.3.1 Principals of Operation of Centrifugal Pumps

Centrifugal pumps (Fig. 22.1) consist of

- A rotating impeller
- An impeller housing (volute) to channel the flow from the suction port to the discharge port
- A power source, usually an electric motor
- A connection from the motor to the impeller; either the motor shaft itself, or sometimes a separate shaft extension
- A mechanical seal assembly to seal liquid inside the liquid-end assembly at the point where the shaft goes through the backplate.

There are many designs of impeller, having to do with whether the vanes are enclosed by shrouds or not (open or enclosed; open is cleaner, but less efficient; enclosed has less slip and is therefore more efficient, but is much harder to clean and to inspect); number and angle of the vanes themselves (influencing the shape of the pump curve); and diameter and thickness of the vanes themselves (deeper vanes give more flow, larger diameters mean more pressure capability). Manufacturers can create impeller designs for very specific sets of conditions. A centrifugal pump's simple design means that

Figure 22.1. Centrifugal pump. *Source*: Courtesy of Tri-Clover Inc., Kenosha, Wis.

- Centrifugal pumps are usually less expensive (often *considerably* less expensive) than PD pumps.
- Centrifugal pumps can be close-coupled, meaning they can be mounted directly on the face of specially designed close-coupled motors, a rare feature on PD pumps, helping keep them inexpensive (and also compact).
- Their flow is relatively pulsation free.
- Because of the inherent slip caused by the generally open nature of the impeller/volute package, there is a maximum pressure (shut-off head) a centrifugal pump can develop. This also means they can be easily controlled by regulating valves and often do not need pressure relief valves or controls in the system because HP loads actually decrease when flow is valved back.
- They are easy to maintain and repair.
- They are easily adapted to sanitary standards.

There are also disadvantages possessed by centrifugal pumps that are often especially significant in bioprocessing.

- Their inherent slip means they do not handle viscous materials well. The rule of thumb is anything over 2,000 SSU (about 500 cps) is usually considered too viscous for a centrifugal pump.

- Because pressure is generated by vane tip speeds, centrifugal pumps are not by nature easily adapted to high-pressure applications. This can be overcome to a degree by running them especially fast (designs exist that run at over 10,000 rpm) or by incorporating multistage designs wherein a chain of impellers is mounted on a common shaft, each impeller boosting the pressure of the impeller preceding it.
- Centrifugal pumps must be run at relatively fast speeds to develop sufficient impeller vane tip velocity to generate reasonable head (usually a minimum of 1,200 rpm). PD pumps, on the other hand, are often run efficiently at well under 100 rpm. This minimum speed means that centrifugal pumps are not usually good in low shear applications.
- The minimum speed also means that centrifugal pumps are not efficient with low-flow applications; turning slowly to limit flow means all pressure capability is gone. Even very small centrifugal pumps are often rated at 10 gpm or more; they will run at lesser flows by regulating them back in any of various fashions (regulating valves, inverters, bypass lines), but they are not efficient at very low flows.
- The presence of a mechanical seal means that abrasive products are a potential problem, not only because of the limited life of the components, but also for the danger of materials wearing off the faces and entering the product stream. There are designs that remove seals from the product flow (double mechanical seals) or eliminate seals altogether (magnetically driven pumps, or pumps with gas injection seals), but they are often expensive and sometimes not available in sanitary designs.

22.3.2 Types of Centrifugal Pumps

22.3.2.1 ***End Suction.*** By far the most popular general classification of centrifugal pump is the end-suction design. Liquid enters through the suction port in the center of the housing, is picked up by the leading edge of the impeller, has energy imparted by the impeller's rotation, and exits tangentially at the discharge port on the volute's periphery. End-suction centrifugal pumps are available as close coupled (mounted directly onto a motor's face) or pedestal mounted (the pump is a separate entity with its own shaft/bearing housing, connected to the motor by a flexible coupling, the assembled package bolted to a common base). Close-coupled design is generally preferred because of lower cost, its more compact design, and its inherent alignment between components, but pedestal-mounted pumps often have more space for special seal options and often will adapt more easily to any special motor requirements.

22.3.2.2 ***Magnetically Coupled.*** One way to arrive at a seal-less design and remove the difficulties of seal compatibility (see Mechanical Seals) is to adapt the standard end-suction design. Rather than the motor shaft extending directly through the mechanical seal and into the liquid end of the pump, it drives a large, hollow rotating magnet. The liquid end of the pump is in a sealed housing and fits up against the motor with a magnetized impeller extended into the hollowed-out area of the drive magnet. As the motor shaft rotates the hollow drive magnet, the impeller is pulled along. Because of the lack of a direct connection between the motor and impeller, magnetically driven pumps can handle only limited viscosity or specific gravity before the impeller decouples (uncouples). They can be very inexpensive in smaller designs, but the price escalates quickly as they get larger. Often, soft-start motors are necessary to prevent decoupling. At present, there are no sanitary-design magnetically driven pumps, but they will undoubtedly be available soon.

22.3.2.3 ***Multistage.*** Centrifugal design is normally best suited for medium-to-high flow, low-head applications such as liquid transfer. When the many advantages of centrifugal pumps makes them the design of preference, but the total head necessary exceeds that available from standard models, multistage pumps may provide an alternate to PD pumps. Impellers are staged, one after another in a line, and are driven by a long common shaft. Internal channels see that the first impeller's discharge is fed to the inlet side of the second, which discharge feeds the third, and so forth. Pressure is cumulative, but the flow is limited to that which can be carried by any individual impeller. The only restriction to the number of stages, and therefore the total pressure possible, is the housing-pressure limitations. This design carries all the advantages of standard centrifugal pumps, except they are not as easy to repair (impellers must be removed in series, and clearances can be difficult to adjust), and they are not necessarily inexpensive. There are sanitary designs available.

22.4 POSITIVE DISPLACEMENT PUMPS

22.4.1 Design and Principles of Operation

Positive displacement pumps make-up the second major classification of pumps used in the bioprocess industry today. In contrast to centrifugal pumps, discussed earlier, PD pumps function on a very different principle: that compressing a fluid volume will increase the ability of the fluid to do work. This change in volume takes place within the pump cavity as the pump drives or pushes that volume to the outlet, thereby increasing the pressure on that volume of fluid to be equal to the pressure at the pump outlet or discharge. An example of this principle is a simple syringe. As the syringe plunger is pushed in, the fluid within the syringe will begin to be exposed to a higher pressure as the volume holding the fluid decreases. The higher pressure then drives the fluid out the end of the syringe. In both the syringe and a pump, the pressure that the pump will generate is controlled by the resistance to flow that the fluid is exposed to as it leaves the pump discharge. Unlike dynamic or centrifugal pumps, this creates a situation whereby the PD pump can become overpressurized. The PD pump will continue to drive fluid forward regardless of the level of resistance. This overpressurization will only be relieved by either a pressure relief device on the discharge side of the pump or through the slip that is experienced as the pump operates.

Slip, in a positive displacement pump, is defined as an efficiency calculated by taking the amount of fluid that escapes the encapsulated volume as the pressure of the fluid volume increases and dividing by the amount of fluid delivered. The escaping fluid is driven back to the inlet or low-pressure side of the pump through clearances within the pump. By knowing the amount of slip that the pump is experiencing, one can calculate the pump efficiency. Typically, PD pump efficiencies are relatively high, 80 to 90%, and will increase as the fluid viscosity, discussed earlier, increases.

22.4.1.1 ***Advantages.*** Advantages that can be found in using PD pumps in the bioprocess industry include:

- The ability to deliver a constant flow rate; the discharge rate for a PD pump is relatively constant, against a varying discharge pressure.
- They function well at a low-flow rate; the PD pump can deliver a low-flow rate at a high discharge pressure and still maintain a high efficiency as compared to centrifugal pumps.
- They function well at high viscosity; the PD pump can easily be adjusted to handle high viscosity liquids and, in fact, will run at a higher efficiency as the higher viscosity acts to reduce the amount of slip experienced by the pump.
- They have the ability to self-prime; most PD pumps will act to create a vacuum at the pump inlet. This vacuum will then pull liquid into the pump cavity.
- PD pumps can easily be made to sanitary design specifications.

22.4.1.2 ***Disadvantages.*** Disadvantages that can be found in using PD pumps in the bioprocess industry include

- Overpressurization can occur; a flow restriction on a PD pump will build pressure until that pressure is relieved. An internal or external pressure relief valve may be required for safe operation of the pump.
- Pulsing flow will occur; the design of PD pumps is such that usually they will cause pressure pulsation as they maintain flow rate. This pressure pulsation can act to damage discharge piping caused by excessive vibration. A pulsation dampener may be required.
- High flow rates are hard to maintain; high flow rates may be hard to achieve even with large pumps because the maximum pump speed may be limited by the viscosity and suction port size.
- Power limitation; because the discharge pressure is directly proportional to the horsepower required, very high discharge pressures may require a very high horsepower, which may exceed the physical strength of the pump.

22.4.2 Types of Positive Displacement Pumps

22.4.2.1 ***Gear Pumps.*** Gear pumps (Fig. 22.2) comprise, perhaps, the most common PD pump in industry. They have wide acceptance because of low cost and simple construction. In general, gear pumps are made in two very distinct designs. The first is the external gear pump. This design uses two meshed gears that rotate in opposite directions. The first gear is driven by the pump motor, referred to as the driven gear, and the second gear is driven by the first and is called the pump gear. Fluid enters and exits the gear housing perpendicular to the gear mesh, with the incoming fluid being trapped within the gear teeth as the opening mesh creates a vacuum pulling liquid in. This trapped liquid rotates on the outside of the gears and is expelled at the pump discharge port as the gear teeth again start to mesh. The fluid is forced out of the gear housing at the higher pressure of the discharge side of the pump.

The other type of gear pump used in industry is the internal gear pump. The internal gear pump works under the same principle as the external gear pump; however, the

Figure 22.2. Gear pump. *Source*: Photo used by permission of Roper Pump Company.

gears are configured such that a smaller, off-center, internal gear is placed inside a planetary gear. The internal gear meshes with the inner teeth of the planetary gear to create the pumping action. In this design, the larger planetary gear is the motor driven gear, with the inner gear acting as an idler. The gap created by mounting the internal gear off center is filled with a crescent-shaped scraper. This scraper acts to seal the inlet port from the discharge port and reduces the amount of slip in the pump. This pump is slightly more expensive, but is quieter and often longer lived.

22.4.2.2 Air-Driven Double-Diaphragm Pumps. Double-diaphragm pumps (Fig. 22.3) are a unique type of PD pump. The design of a double-diaphragm pump is made up through the porting of four check valves and two diaphragms mounted in parallel. The two diaphragms are connected to an air-driven piston. The operation of the pump has liquid entering a common inlet port and the liquid separating and alternately flowing through one of two check valves, the check valve that is exposed to the opening diaphragm. At the same time, on the opposite side of the closed check valve, the closing diaphragm is expelling fluid at a discharge pressure (limited by the air supply pressure that powers the pump). The fluid being expelled pushes open the check valve exiting the closing diaphragm section and pushes closed the check valve on the opening diaphragm. This action cycles back and forth as the air piston that drives the two diaphragms moves in and out. Because both diaphragms are attached to the same shaft, when one diaphragm is opening, the other is closing.

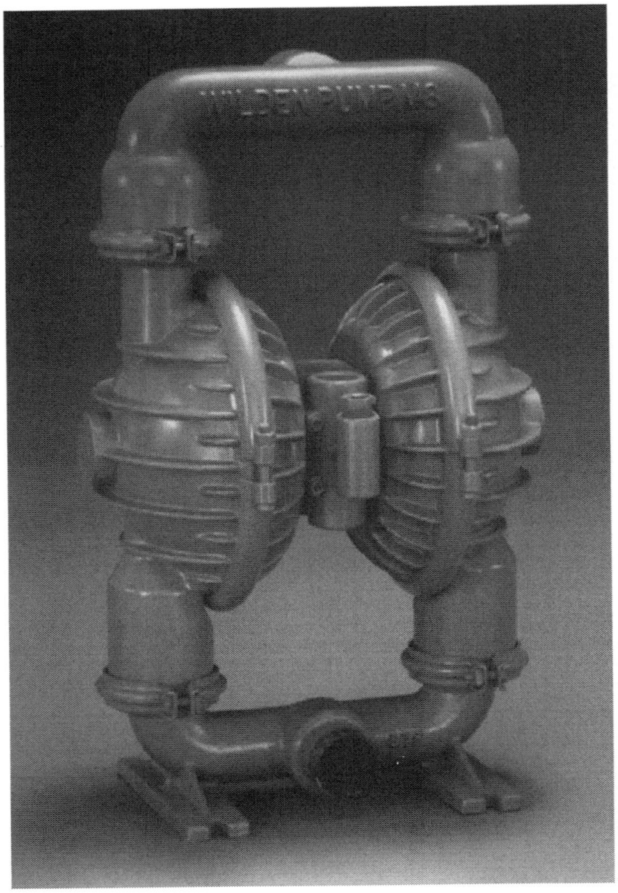

Figure 22.3. Double-diaphragm pump. *Source*: Courtesy of Wilden Pump Company.

22.4.2.3 Progressing Cavity Pumps. Progressing cavity pumps (Fig. 22.4) have a special design that allows for very gentle pumping of highly viscous liquids. Most progressing cavity pumps are made up from a single helical (metal) rotor rotating inside a (elastomeric) double helical stator. Cavities formed by this combination of rotor and stator progress down the length of the rotating shaft as the shaft turns. The function of the rotor is very similar to that of a screw pump slowly pushing the pumped liquid forward toward the discharge of the pump. Pressure is developed because close clearances between the rotor and stator minimize the amount of slip that is found in the pump. Further, because the pump is well suited to pumping high-viscosity liquids, this action is even better utilized.

Figure 22.4. Progressing cavity pumps. *Source*: Photo used by permission of Roper Pump Company.

22.4.2.4 Flexible Impeller Pumps. Flexible impeller pumps (Fig. 22.5) make up a class of very inexpensive PD pumps that can be easily adapted to many applications. Flexible impeller pumps work on the principle of a multivaned rubber impeller rotating in a housing that is cammed such that the vanes are pushed close as the impeller rotates. This closed impeller expels fluid out the discharge of the pump and at the same time pulling liquid into the suction port. This action creates a vacuum at the suction port. Vacuum values as high as 22 in Hg are easily achievable. The major drawback of this style of PD pump is the inability of the pump to run dry. Dry running of this type of pump will quickly erode the rotor.

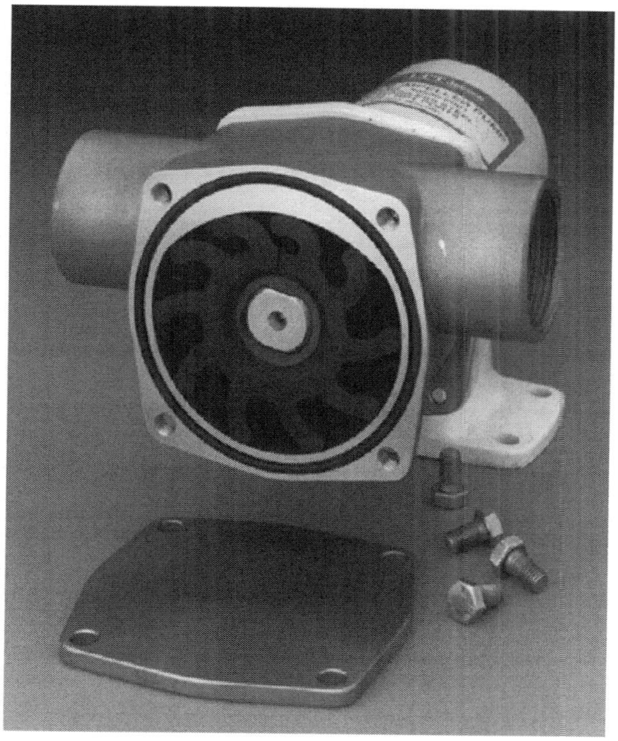

Figure 22.5. Flexible impeller pump. *Source*: Courtesy of ITT Jabsco.

22.4.2.5 ***Circumferential Piston Pumps.*** Circumferential piston pumps (Fig. 22.6a) are used in the bioprocess industry to transfer a wide range of both raw material and finished products. The circumferential piston pump works with two opposing hemispherical rotors spaced 180° apart. These two rotors turn in opposite directions much like that of the external gear pump; however, the action of the circular piston moving past the suction port creates a vacuum that draws fluid into the pump cavity. This fluid, which is trapped between the sides of the rotating piston and the pump housing, travels around the pump casing and is discharged at the pump outlet as the piston of the opposing rotor pushes the trapped liquid into the outlet of the pump. This action alternates from pump rotor to driven rotor as trapped fluid is pushed into the pump outlet. This pump design requires that the rotors of the pump not come in contact with one another. To accomplish this, the rotors are driven with a gear housing that contains supporting shafts, bearings, and timing gears. It is therefore very clean and ideally adds no metal to the product.

22.4.2.6 ***Lobe Pumps.*** Very similar in design to circumferential piston pumps, lobe pumps (Fig. 22.6b) also use a gear housing containing supporting shafts, bearings, and timing gears. The rotors of the lobe pump rotate in opposite directions, creating a low pressure or vacuum at the pump inlet as the lobes separate. The fluid drawn into the pump cavity is trapped in the valley of the lobe and rotated to the outlet of the pump where the trapped fluid is forced into the pump outlet as the opposing rotor lobe rotates into the lobe valley containing the trapped liquid. This action alternates from rotor to rotor similar to that found on the circumferential piston pump. The lobe pump is somewhat less shear intensive as compared to the circumferential piston pump and is more easily cleaned when cleaning a pump in place.

22.4.2.7 ***Peristaltic Pumps.*** Peristaltic pumps (Fig. 22.7) are very common throughout the bioprocess industry. They can be found in all size ranges, handling flows on the laboratory bench of 3 to 4 mL/min to flow rates on the production floor of 30 to 40 gpm. The basic design of the pump functions through the progressive squeezing of tubing by rollers mounted to a rotating plate. The rollers trap fluid within the tubing and rotate it to the outlet. They are self-priming and have the major advantage that the fluid only contacts the pump tubing so sealing the pump is not

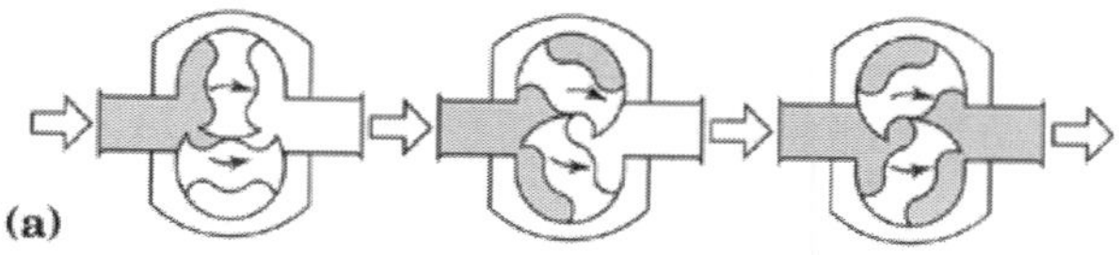

(a)

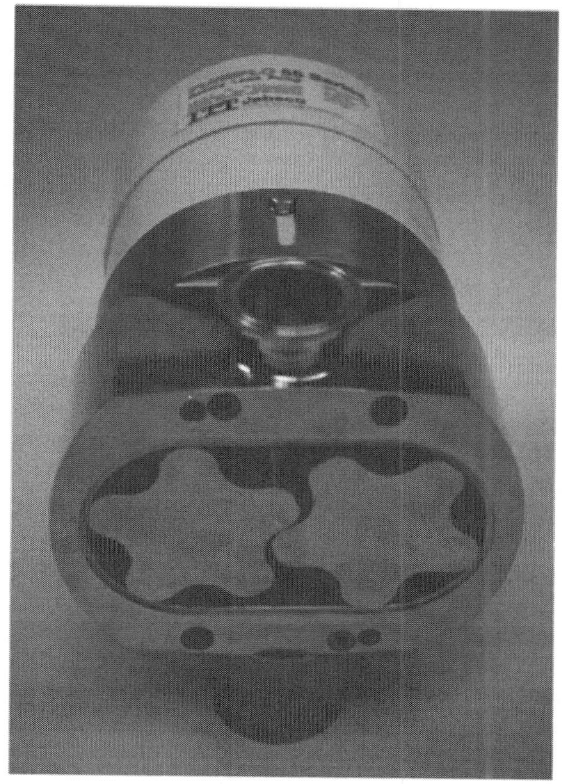

(b)

Figure 22.6. (a) Circumferential piston pump; (b) Lobe pump. *Source*: (a) courtesy of Waukesha Fluid Handling; (b) courtesy of ITT Sherotec.

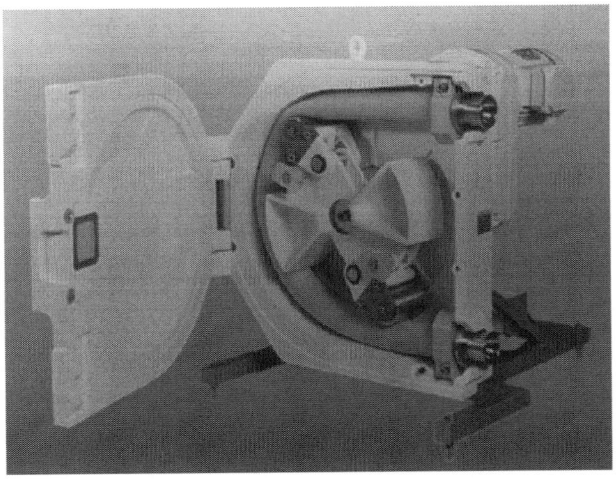

Figure 22.7. Peristaltic pump. *Source*: Courtesy of Watson-Marlow Inc.

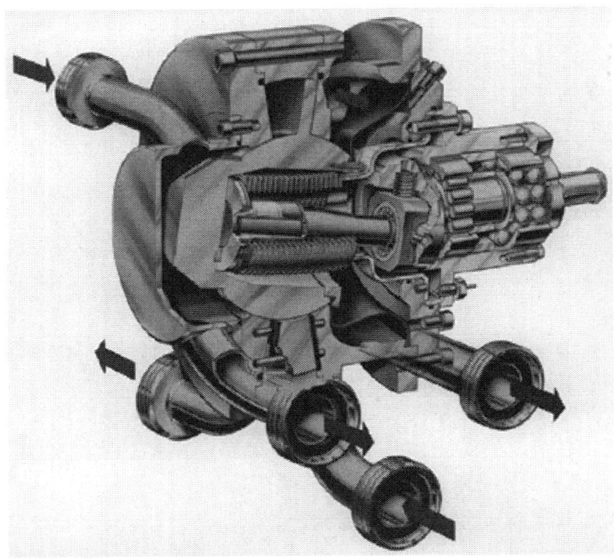

Figure 22.8. Nutating disk-type pump. *Source*: Courtesy of KSI.

required. The action also minimizes the amount of shear that the pump fluid is exposed to.

22.4.2.8 ***Nutating Disk-Type Pumps.*** Nutating disk-type pumps (Fig. 22.8) are new to the bioprocess pump market and provide for advantages typically found in several different types of older pump designs. The principle that drives this class of PD pumps is the action of a wobbling disk within the pump cavity. This disk can be used to trap a volume of fluid between it and the stationary pump housing. The trapped fluid is pushed toward the pump outlet through this wobbling action and the use of a scraper gate that rides on the width of the rotating disk.

22.5 DRIVERS

The most popular choice of drivers for pumps, by far, is the AC motor. They have become the standard because they are inexpensive and long lived, have relatively few moving parts, and are available from a large variety of sources, in a large variety of styles and enclosures. They are relatively quiet (typically between 70 and 85 decibels), efficient (often over 90%), and institutions exist (National Electrical Manufacturer Association [NEMA]) that guarantee a certain degree of interchangeability from manufacturer to manufacturer. Typically, only if an AC motor is unacceptable are other choices considered. Considerations include style, electrical specifications, and enclosure.

22.5.1 Style

Close coupled, foot-mounted motors are common for centrifugal pumps that will accept this. Standard horizontal foot-mounted motors are needed when pedestal-mounted centrifugal pumps have been chosen. PD pumps usually need motors with gear boxes integrally mounted to achieve the low speed necessary. Because most pump suppliers sell their pumps as motor and pump packages, the user is not usually concerned with physically matching the components, but should understand the principles.

22.5.2 Electrical Specifications

Unlike the general motor style, it is the user's responsibility to know the electrical systems the drive will be expected to operate under. Concerns the supplier must address will be:

- *Phase*: Single phase is available for small horsepower drivers (below 5 hp), and common for fractional horsepower, but three phase is available for fractional and large integral motors.
- *Voltage*: in the United States, standards are 115 or 230 V for single phase and 230 or 460 V for three phase, but others voltages exist. Because motors are good for ±10%, there is some flexibility.
- *Frequency*: 60 Hz is the standard in the United States, but 50 Hz exists all over the world. Motors tolerate only ±5% frequency variation.

22.5.3 Enclosure

The desired enclosure is an important consideration. Standard types available are open drip-proof for environments where the motor will be protected from atmospheric dangers; totally enclosed, fan cooled (TEFC) for areas where the motor will be exposed to light moisture; explosion proof where a UL rating is necessary and the motor's ability to keep sparks from escaping is crucial for safety considerations; and common in bioprocessing, the wash-down enclosure where the area the unit is to be mounted in will be periodically hosed down.

NEMA is an organization that has written a standardized list of motor dimensions such that a user may substitute one brand for another in confidence that the new motor will physically interchange. This is done by a series of frame designations, such that a brand A motor in, for example, frame 145TC will exactly fit in place of a brand B motor of the same frame.

An accessory often considered today is the frequency inverter, which varies frequency and voltage in such a fashion as to allow an AC motor to act as a variable speed device. This has many advantages in addition to allowing a user to dial in more precisely the desired flow, such as the ability to control speed remotely, and to be able to start up a pump slowly to minimize start-up damage to the pump and the motor.

Other options to an AC motor are gasoline or diesel engines for areas where electrical power is unavailable, and more and more commonly, compressed air drives. Air motors have the advantages of being inherently explosion proof and being able to stall without incurring damage, but have the disadvantages of needing very dry compressed air and of being relatively inefficient. DC power is used in some very small pump packages, but is uncommon.

22.6 SPECIAL CONSIDERATIONS FOR BIOPROCESSING PUMPS

22.6.1 Sanitary Standards

The main standard for the sanitary design and construction of pumps to be used in the bioprocess industry is the 3A Standards (Code of Federal Register Title 21, part 177). These standards govern certain design modifications that are key elements in maintaining sanitary conditions within a pump. These guidelines cover surface finish, materials of construction, fabrication, and other areas. The sanitary design of a pump must also comply with all other applicable standards, including FDA, cGMP, and the ANSI/ASME standards for food, drug, and beverage equipment.

22.6.2 Materials of Construction and Surface Finishes

The most commonly found material of construction for pumps in the bioprocess industry is austenitic stainless steel alloys. Austenitic stainless steel alloys used in the construction of sanitary stainless steel pumps include AISI types 304, 304L, 316, and 316L. Analysis of the makeup of these alloys reveals that a major element in all stainless steel alloys is iron. Iron alone is quickly attacked by most fluids, but in the presence of chromium in correct combinations of Ni, C, Mn, P, Mo, Si, and S, the ability of stainless steel to withstand corrosion is excellent. This, plus its low carbon content and its ability to be easily welded, machined, and polished, makes 316L the most popular material of construction for pumps in bioprocessing. Product-contact surface finish is critical in most bioprocess systems because of the need to preserve sterile or sanitary conditions. All product contact surfaces must be free of cracks, crevices, and occlusions that can harbor and promote bacteria growth. Surface finish in most pumps is obtained through the use of mechanical polishing and electropolishing. This surface treatment will result in a surface finish that would measure at approximately 15 Ra surface roughness.

22.6.3 Mechanical Seals

*22.6.3.1 **Theory.*** Mechanical seals (Fig. 22.9) are a standard product used in the manufacture of most pumps to seal between the rotating pump shaft and the pump cavity. The type of mechanical seal used to seal the pump cavity is a key component of most pumps used in the bioprocess industry. The mechanical seal that is placed into pumps must serve two functions:

- Provide a barrier to the atmosphere
- Contain the fluid in the pump cavity without affecting fluid quality

Bioprocess pump seals are designed to be easily cleansed and not harbor bacteria. However, for them to be used in pumps for sterile fluids, particular modifications are required. These modifications include changes in the seal design, placement of the seal within the pump housing and selection of seal materials.

Figure 22.9. Mechanical seals. *Source*: Courtesy of the Durametallic Company.

A discussion on mechanical seals is best started with a brief background on how mechanical seals work. A single mechanical seal works with two very flat surfaces, one a rotating face element and the other a stationary face element. These faces ride on one another with an ultrathin layer of fluid between the seal faces. When the fluid enters the zone between the two faces, the fluid temperature rises owing to the frictional resistance of one face on the other. If the seal is functioning correctly, the fluid will vaporize as it exits this zone and enters the atmosphere. This principle applies to most single mechanical seals in bioprocess pumps. The key to this functioning correctly is to have proper force pushing the rotating element on to the stationary seat. The force can be obtained either mechanically or hydraulically. The mechanical force is provided by a compressed spring. The hydraulic force, found in a hydraulically balanced seal, uses the pressure inside the pump head to push the rotating element to the stationary seat. Both of these methods are used in bioprocess pumps seals to produce a force that will generate enough friction between the faces to vaporize the fluid just before it leaves the seal faces without causing the faces to wear too quickly or generate too much heat.

If the mechanical seal is to function as described and provide a sterile barrier between the atmosphere and the fluid within the pump cavity, certain modifications are required to the seal design. First, a secondary mechanical seal is needed behind the primary inboard seal. This combination is referred to as a double mechanical seal. Second, both of these seals are placed into a stuffing box or gland that is attached to the pump where the rotating shaft enters the pump head. This is done so that a flushing fluid can be run between the two seals. The fluid barrier will guard against any contaminants from the atmosphere penetrating into the fluid within the pump cavity. The flushing fluid is pushed through the stuffing box under pressure so that a positive pressure on the outboard seal stops contaminants from entering the stuffing box. A secondary function of the seal flush liquid is to provide a certain amount of cooling to the seal faces.

The requirement that the seal contain the fluid within the pump cavity without affecting the fluid quality necessitates changes in the way the inboard seal functions. Fluid in the pump cavity is kept pure by ensuring that a positive pressure is maintained on the pump side of the inboard seal. This pressure functions in the same manner as that on the outboard seal; however, it should be greater than the stuffing box pressure to ensure that any leakage through the seal passes from inside the pump into the stuffing box. This will prevent the passage of contaminants from the stuffing box into the pump cavity. The seal designed for this configuration creates some problems.

First, if the inboard seal fails, product can be lost though the flush outlet of the stuffing box. Second, if the fluid in the pump contains active organisms that must be contained, these organisms could pass into the seal flush. The common flushing procedure for double mechanical seals in this situation is to maintain the stuffing box at a higher pressure than the pump side inboard seal pressure. In the case of seal failure, the flush fluid would move into the pump head, and no product would be lost. This arrangement can present problems if particulate, worn from the seal faces, is forced into the pump cavity by the high stuffing box pressure, thereby contaminating the fluid in the pump. Applications should be reviewed to determine which flushing configuration is most applicable.

The mechanical seal chosen for an application should be constructed of the highest quality FDA-approved materials of construction to provide a reliable, long-lasting seal. The face and rotating element materials should be selected for fluid compatibility, hardness, and flatness. Commonly used materials include tungsten carbide, silicone carbide, ceramics, carbon graphite, and stainless steel. The supporting items of the seal, including the springs, cups, O-rings, drive collar and set screws, should also be selected for fluid compatibility and performance characteristics. These pieces will function as a system to provide a reliable seal.

22.6.4 Static Seals

Static seals within pumps used in the bioprocess industry fall into two categories. These categories are elastomer O-rings and flat gaskets. O-rings are more commonly used in modern pump designs because they can be sized to properly seal flat or round pump components within the pump design. Further, O-rings and their corresponding O-ring grooves can be easily configured to minimize any fluid entrapment and provide a cleanable crevice-free seal. O-rings are commonly used to seal pump shafts to mechanical seals and pump coverplates to pump housings. The O-ring material is selected based on the required durometer (hardness) and material compatibility with the sealed fluid. Commonly found materials include Viton®, EPDM, Buna-N, Silicone, and Kalrez®.

Flat gaskets are less common in modern pump designs; however, they are used when maintaining clearances within the pump design is critical. As with O-rings, they can also be constructed of an elastomeric compound, but it is also common to see them made from such materials as paper or PTFE (Teflon®).

22.6.5 Aseptic Modifications

Sterile fluids require systems that are capable of being steamed-in-place (SIP). To minimize downtime and preserve the integrity of the sterile atmosphere within the pumping system, pumps can be modified so they can be fully steamed-in-place. Most pumps can be steamed-in-place if they are designed in accordance with 3A standards. These standards, as mentioned earlier, will specify proper surface finish, impeller, and seal design. Further, the pump must utilize a seal that will allow for full steam penetration under pressure. Because proper steam flow is required to maintain saturated steam within the pump being sterilized, steam flow is enhanced in pumps by providing a means of condensate removal. As steam sterilizing the pump cools, it will form condensate. For proper sterilization to occur, this condensate must be removed from the pump quickly, preventing cold spots from forming. Hence, there is often the need for the pump to be self-draining by adding a tangential bottom drain.

22.6.6 Water-for-Injection Modifications (Centrifugal Pumps)

The design of centrifugal pumps make them very well suited to the application of pumping sterile water for injection (WFI). Because WFI must be kept moving at high temperature and high velocity within a piping system, a standard sanitary centrifugal pump can be modified to handle this application. The first modification is to the surface finish and materials of construction. The pump should be constructed of AISI Type 316L stainless steel that has been finished so that the surface roughness measures approximately 15 Ra. Second, all elastomeric components of the pump should be selected to be compatible with the WFI. This requires matching compatibility for temperature. Typical materials include steam resistant Viton® or EPDM. The third measure is the correct configuration of the mechanical seal to be used. This should be selected to reduce the chances of contamination. The last typical modification includes the addition of a casing drain to promote the removal of condensate during the steam sterilization of the WFI pump.

22.7 TROUBLESHOOTING

There is a general theme in successful troubleshooting. Specifically:

1. Make sure the system is fully understood (to eliminate oversights).
2. Narrow down or isolate potential sources of the problem.
3. Determine all possible causes within these areas.
4. In order of probability, go through them one by one.

The difficult step is often number 3, but all four hold potential hazards and solutions.

1. System analysis can be tedious. Fear of asking what appear to be obvious questions can tempt one to give this step short shrift. However, if problems began after a change in the system, system analysis will usually hold the answer. Walking through a typical flow pattern and questioning each step is often helpful. Keep in mind that flow at any division seeks equilibrium; flow will divide such that resistance to flow on one side given its gallonage will exactly equal the resistance to flow on the other side with the remainder of flow. Double checking that the pump in question will, in theory, meet the system conditions may be also be appropriate.
2. In many cases, narrowing down whether a problem is in the system or the pump can be affected by switching pumps in an installation containing one that is working well. If the problem follows the pump, it is in the pump; if it stays with the line, it is in the system. In other cases, noises can be isolated by separating various components of a system to determine if they continue or cease. The motor sometimes can be separated from the pump, for example, or suspect components temporarily removed. Valving can sometimes isolate pumps or piping sections to check the results.
3. There are usually more possible explanations for a problem than a typical manufacturers' lists provides. Following is a sampling of potential causes for a few specific complaints. Creativity is essential in this step, with nothing *ever* taken for granted. Often, a marginal explanation tentatively or casually offered up will be immediately recognized by an engineer and the problem solved.
4. There will always be possibilities (listed in the following section) that are more likely to occur than the others, depending on the symptoms. Of these, the most easily checked should be the first inspected, and then the more difficult. At times there are two simultaneous flaws in a system, making checking of individual items fruitless. In these cases, logic will be of more value than experiment.

Suggestions for a few common problems follow.

22.7.1 Low Flow

22.7.1.1 Centrifugal Pumps Only

1. Reverse rotation (*Note*: a centrifugal impeller flings the liquid as opposed to scoops it, if there is ever a question as to rotational arrows. A centrifugal pump will pump if turning in reverse, but with low flow and pressure and higher horsepower because of lowered efficiency).

2. More head than originally calculated.
3. Clogged impeller.
4. Wrong impeller diameter for conditions.
5. Cavitation (see earlier section).
6. Impeller clearances worn too large.
7. Motor speed slower than nameplate indicates.
8. Impeller spins loosely on shaft.
9. Broken coupling.
10. Viscosity has increased.
11. Vortexing in source tank.
12. Air in tank introduced from splashing or churning close enough to tank outlet to allow air to enter with stream.
13. Wrong pump for conditions.
14. If pump is above source tank, air leak in suction pipe.

22.7.1.2 Positive Displacement Pumps Only

1. Excess wear on components.
2. Stuck internal check valves, if any.
3. Torn diaphragms, if present.
4. Stuck internal (or external) pressure relief valve.
5. Speed too slow.
6. Cavitation (see preceding section).
7. Wrong pump for conditions.
8. If belt driven, a slipping belt.
9. Broken coupling.
10. Speed not slow enough for viscosity, leading to cavitation.
11. If pump is above source, air leak in inlet pipe.

22.7.2 Cavitation

Note: Cavitation is indicated by crackling noise for centrifugal pumps, by banging for most PD pumps, or by finding that TDH has fallen at least 3% below the standard curve for the particular centrifugal pump.

1. Blocked inlet.
2. Too much friction loss in suction line.
3. Elbow too close to pump suction port (*Note*: The rule-of-thumb is that the elbow should be located no closer than 6 × the suction port diameter).
4. For centrifugal pumps only, running at extreme left end of the curve causing low pressure areas to develop.
5. Suction line with high spot (as opposed to continually rising), therefore, air trap to disturb or restrict flow.
6. Liquid too hot and therefore vapor pressure too high.
7. Elevation differential too great (too much lift).
8. If drawing from closed tank, vacuum level too high, or vacuum level that builds to an unacceptable point as liquid is pumped out.
9. Excess prerotation at inlet of centrifugal, causing flow to enter in wrong pattern for design (*Note*: Rotation against impeller rotation is a positive factor and will move the curve upward, but the more common prerotation in the same direction as the impeller is harmful and may drop the curve).

22.7.3 Motor Overheats

22.7.3.1 Centrifugal Pumps Only

1. Too much flow; therefore, further out on curve than calculated.
2. Wrong rotation.
3. Oversized impeller.
4. Unbalanced impeller.

22.7.3.2 Positive Displacement Pumps Only

1. Too much backpressure.
2. Binding or rubbing components.
3. In case of progressive cavity, flexible impeller, or other pump with rubbing components, insufficient motor starting torque to get up to full speed.

22.7.3.3 Any Pumps

1. Voltage above ±10% allowed by manufacturers.
2. Frequency beyond ±5% allowed by manufacturers.
3. Bad motor bearing(s).
4. Three-phase motor not getting power to all three lines (single phasing).
5. Cavitation.
6. Misaligned motor/pump shaft, or misaligned pulleys.
7. Undersized cable to motor.
8. Loose connection at motor outlet box or control box.
9. Insulation has broken down through time and use.
10. Cable too long to handle amperage.
11. Ambient temperature too high.
12. Overgreased bearings, causing binding when bearings heat up.
13. On TEFC design, broken or slipping fan.
14. Speed too high for pump.
15. Not enough air movement to keep motor cool because of confined or poorly vented area.
16. If single phase motor, starting windings do not disengage.
17. Viscosity higher than calculated.
18. Pipe torque, causing misalignment and binding.

19. Motor stops and starts too often (consult manufacturer for maximum stops and starts for applicable horsepower, it is probably lower than you think!).
20. Dirty surfaces not allowing efficient cooling.
21. On drip-proof design, internal diffuser or fans not channeling air flow properly.

22.7.4 Noise

Note: Various designs have various degrees of noise considered standard and acceptable. The manufacturer will be able to supply the numbers if given operating conditions.

1. Motor or pump bearings worn.
2. Cavitation.
3. If a PD pump, components hitting or rubbing.
4. If rigidly piped, amplification of what might be considered normal noise through lines into abnormal amounts.
5. If mounted on channel steel base, amplification of noise from pocket of air beneath motor/pump package.
6. Coupling hitting coupling guard.
7. Motor fan hitting fan cover guard.
8. Check valves chattering as they stick.
9. Water hammer (valves must close slowly enough for momentum in line to be absorbed). A rough rule of thumb is $t = 2L/a$, where t = time in seconds for valve to close (minimum); L = length of pipe via longest route, in feet; a = speed of pressure wave in feet/second (an average for water is 4,000 ft/s). An example would be a 200-ft length of pipe handling water, which would demand a valve closing in no less than $2 \times 200/4{,}000 = 0.1$ s to avoid water hammer.

22.7.5 Abnormal Seal Wear

Note: A mechanical seal in a continuous-duty, clean application can be expected to last several years, as a minimum, to as many as 10 years or longer if under absolutely ideal conditions.

22.7.5.1 Seat Fracture

1. Thermal shock from dry running followed by sudden introduction of cooler liquid, or any sudden drastic change in the temperature or the liquid. Usually appears as a hairline fracture.
2. Seal head installed incorrectly, causing the head to strike the seat.
3. Physical shock from a blow such as being dropped.
4. Water hammer.
5. Pipe misalignment, twisting the bracket holding the seat.

22.7.5.2 Carbon Washers Wear Prematurely

1. Abrasives (clearly visible by scored faces).
2. Chemical incompatibility between the chemical and either the washer itself, or the binder holding the carbon together.
3. Improper installation, usually with the seal head cocked.

22.7.5.3 Elastomer Problems

1. Chemical incompatibility. Elastomer may become brittle, softened, swollen, or even totally eaten away.
2. Dry running. Elastomer may look burned or cracked or become brittle and hard.

ADDITIONAL READING

Hydraulic Institute Standards, Hydraulic Institute.

I. Karassik, W. Krutzsch, W. Fraser, and J. Messina eds. The Pump Handbook, 2nd ed., McGraw Hill, New York, 1985.

R. Stover, in B. Lydersen, N. D'Elia, and K. Nelson eds. Bioprocess Engineering: Systems, Equipment and Facilities, Wiley, New York, 1994, pp. 253–315.

V. Streeter, Handbook of Fluid Dynamics, 1961.

C.R. Westaway and A.W. Loomis, Cameron Hydraulic Data, 16th ed., Ingersol Rand, 1984.

PART V

DOWNSTREAM cGMP OPERATIONS

23

AFFINITY CHROMATOGRAPHY OF PLASMA PROTEINS

Mirjana Radosevich and Thierry Burnouf
Human Protein Process Sciences, Research and Development Department, Lille, France

23.1 INTRODUCTION

Since long ago, protein biochemists working on protein isolation and purification have strived to develop improved methodologies to separate and polish proteins extracted from complex raw biological materials, such as human and animal plasma. The most usual systems of protein separation were initially based upon chemical or physical precipitation and partitioning tools, using ammonium sulphate, caprylic acid, ethanol, polyethylene glycol, or pH/temperature variations. Later on, chromatography was introduced as a more selective and refined fractionation and purification procedure, initially mostly relying upon the use of ion-exchange, hydrophobic interaction, and size-exclusion chromatography (1,2). Although, such methods were far more selective than precipitation steps and generated great interest for the purification of trace proteins from complex feedstock, still, achieving higher degree of specificity in protein capture was required for some applications. Affinity chromatography is such a purification procedure which exploits the ability of molecules to bind to others; it makes use of specific biological interaction which may exist between protein and antibody, enzyme and substrate or co-factor, or receptor and ligand. Affinity chromatography has emerged as an efficient tool to purify human plasma biopharmaceuticals (3).

Until the end of the 1980s, the human plasma industry depended almost exclusively on the application of the cold ethanol precipitation process for the preparation of major protein therapeutics, such as albumin and immunoglobulins G (IgG) (4). Coagulation factor VIII (FVIII) preparations were obtained mostly using a cold precipitation of whole plasma called cryoprecipitation (5), followed by polyethylene glycol or glycine precipitation steps to partially remove protein contaminants such as fibrinogen (6–8). Such production processes implemented in plasma fractionation plants in the 1970s yielded the so-called "low- or intermediate- purity" FVIII products that gradually replaced, in industrialized countries, the single-donor cryoprecipitate fractions produced by blood banks. These early FVIII concentrates soon became obsolete as contaminating proteins were feared to induce protein overload and possible immunological dysfunctions in hemophilia A patients receiving frequent substitution therapy. These preparations could also lead to severe haemolytic events due to the presence of residual blood group A and B isoagglutinins. Similarly, crude factor IX (FIX) complex fractions, also called prothrombin complex concentrates (PCC or PPSB), were isolated in the late 1970s from the supernatant of cryoprecipitate by unspecific anion-exchange batch chromatographic adsorption using soft gels made of agarose or cellulose matrices (9). Contamination of PCC by unwanted proteins (e.g. activated factors) and coagulant-active phospholipids (10,11) was recognized very early to lead to significant risks of thrombotic complications in particular in hemophilia B patients (FIX-deficient individuals) subjected to intense substitutive treatment during surgical procedures (12). Efficient purification procedures for FIX, in part based on affinity methods combined with ion-exchange chromatographic procedures, were therefore developed to alleviate these complications. Similarly, the gradual emergence of recombinant plasma proteins expressed by genetically modified mammalian cell cultures has

Downstream Industrial Biotechnology: Recovery and Purification, First Edition. Edited by Michael C. Flickinger.

prompted the development of sophisticated purification methodologies relying upon highly specific affinity-based separation tools. By contrast to therapeutic products obtained from feedstock of human origin, DNA-engineered protein products need to have almost absolute purity levels to avoid possible risks of immunological reactions in patients exposed to non-human proteins. One approach to achieve high purity of recombinant proteins in a minimal number of purification steps is to tag proteins by genetic modification, creating a fusion protein that possesses affinity for a specific ligand. The tag could be another protein or a short amino acid sequence, which is recognized by an antibody. The antibody would then bind to the protein whereas it would not have done so before. Examples of tags are His-tags and glutathione-S-transferase (GST) tags which have an affinity for nickel or cobalt ions or glutathione, respectively (13). One other very successful application of affinity chromatography is the use of protein A for the large-scale purification of recombinant monoclonal antibodies (14).

Over the last few years, impressive developments in ligand chemistry and peptide synthesis as well as in bioinformatics (15) have led to the availability of a wide range of affinity gels or membranes for extracting and purifying plasma-derived (PD) and recombinant protein therapeutics. This chapter provides an overview of the main achievements accomplished in the use of affinity matrices in bioprocessing of PD and recombinant therapeutic plasma proteins and addresses main issues related to the quality control of plasma biopharmaceuticals purified by affinity method. Recent promising developments on affinity technologies applicable to human plasma proteins will also be presented.

23.2 LIGANDS AND SUPPORTS FOR AFFINITY PURIFICATION

23.2.1 Naturally Occurring Ligands

For several decades, industrial-scale affinity chromatography of plasma proteins has been somewhat restricted to the availability of naturally occurring inhibitors or substrates such as heparin, dextran sulfate, and gelatin. These bio-ligands were immobilized most often on agarose-based matrices. Their use has allowed major progress in protein purification; currently, they still represent a major tool in modern industrial purification of clinically important coagulation factors, anti-thrombotics, and protease inhibitors (16). However, in some cases, the selectivity of these ligands, which may, like heparin or dextran sulphate, also act as ion-exchangers, is insufficient as proteins with similar affinity/binding profiles may co-purify with the target protein. Use of such non-specific ligands therefore may require the implementation of additional purification steps either upstream or downstream, or the development of optimized elution conditions to isolate the target proteins while also eliminating protein contaminants. Use of venom components as ligands has also been described for the production of antivenoms immunoglobulin fragments (17), but the technology remains very confidential in this field.

23.2.2 Antibody Ligands

Advances in hybridoma cell cultures (fused B-cells containing the target antibody to myeloma tumor cells) and in large-scale purification of monoclonal antibodies opened the way for pharmaceutical immunoaffinity systems. Indeed, improvement in cell culture and antibody purification strategies favored, on one side, the availability of large quantities of antibodies and, on the other side, their efficient isolation with higher yield and safety margin for pharmaceutical applications. Such systems rapidly gained in popularity thanks to their much higher selectivity in capturing target proteins from relatively crude and complex extracts, or for final polishing. They were initially implemented in plasma fractionation for the production of immunopurified FVIII and FIX concentrates, prior to their use for the purification of recombinant analogue products. Although antibodies used as ligands, have larger tolerance to low pH range than other plasma proteins, they are naturally sensitive to aggregation when subjected to shear forces as during ultrafiltration. This procedure may induce as high as 20% IgG aggregation which implies the use of an additional chromatographic step for removing aggregates, thus increasing the cost of their production as ligands (18). They are also susceptible to degradation by proteases or bacteria present in cell cultures or in the biological feedstocks. For the same reason, immunoaffinity systems do not withstand harsh cleaning procedures needed for sanitization and complete removal of strongly bound proteins. Consequently, they are limited to a much lower number of chromatographic cycles (an average of 10 to 20) than synthetic adsorbents. Besides, due to their manufacturing and immobilization processes which are more demanding than that of non-biologic ligands, they can be more costly than other affinity systems, and are considered to be ten times more expensive than ion-exchange matrices (19).

23.2.3 Synthetic Ligands

Other affinity technologies have been developed or optimized, giving rise to novel synthetic and biosynthetic ligands and matrices (20). Affinity ligands have evolved from enzymatic substrates, co-enzymes, hormones, lectins, co-factors, antibodies, nucleic acids, effectors, and inhibitors to a variety of peptides, polypeptides, peptide fragments, and other synthetic structures. At present, the two most selective ligands for affinity systems appear to be monoclonal antibody and peptide-affinity adsorbents.

Synthetic ligands include dye molecules, which have been known for a long time and considered as one of the important alternatives to biologic ligands for specific affinity chromatography. Dye-ligands are able to bind most types of proteins, in some cases in a remarkably specific manner (21). The technology using biomimetic dyes which mimic biological ligands and differ from immunoaffinity or metal chelate affinity chromatographies is often referred to as "pseudo-affinity." If standard dye-ligands do not exhibit high specificity for a given protein, newly designed ones (e.g. mimetic ligands) are chemically modified to ensure high specificity for a variety of plasma proteins, in particular for IgG. They have the advantage of being resistant to chemical and biological degradation. They can bind a wide range of enzymes and proteins in a selective and reversible manner, mimicking the structure of substrates, co-factors, or binding agents (22,23). Most of these reactive dyes consist of a chromophore, like azo dyes, anthraquinone, or phathalocyanine, which is linked to a reactive group, often a mono- or a di-chlorotriazine ring structure. The risk of dye leaching has been a drawback to their use in the purification of biologicals, especially when high volumes of end-products (e.g. albumin) are likely to be infused to patients.

The discovery of new ligands with interesting properties has been accelerated with the availability of phage display (24,25) and synthetic combinatorial libraries which may contain multiple randomized structural variants for a given target protein (26,27). These methods could identify highly selective ligands for the isolation and purification of human kallikrein and thrombin from complex mixtures of plasma proteins. However, phage display technologies provide biosynthetic structures (polypeptides) which are restrained, like antibodies, to mild processing conditions.

On the contrary, combinatorial chemistry strategies have been shown to provide novel fully synthetic ligands in an efficient manner if combined with rapid and sensitive screening methods. Future improvements will probably arise from automated chemical sensor technologies. Nevertheless, affinity peptide ligands can be generated by classical methods as the solid-phase peptide synthesis. In this way, an affinity peptide (GPRP) for binding human fibrinogen and a GAQGHTVEK peptide for capture of human albumin were successfully produced for analytical purposes (28). The dynamic binding capacities were about 10 and 19 mg/ml gel for fibrinogen and albumin, respectively. These studies showed that peptide-based affinity columns derived from standard solid-phase peptide synthesis could also represent a viable alternative to the use of monoclonal antibodies in the manufacture of high-preformance affinity supports. Thanks to moderate affinity for the target protein, this approach is claimed to allow higher recoveries under mild elution conditions and 10-fold higher binding capacity in the case of albumin. Later on, another peptide ligand, Phe-Leu-Leu-Val-Pro-Leu (FLLVPL), was also used for possible production purposes to isolate fibrinogen with high purity and biological activity (29).

There are several advantages to the purification of pharmaceutical products, most specifically immunoglobulins, using synthetic affinity ligands. A major advantage is that these are most likely not derived from biological sources and, thus, carry less risks of contamination by infectious agents. Hence, regulatory and medical concerns related to the presence of unknown contamination and infectious agents in the final product are circumvented. However, given the complex supply chains for chemical raw materials, appropriate certification of the absence of biological sources in the production of these ligands should be obtained from the suppliers. Other major advantages of these ligands include (i) lower manufacturing cost, (ii) possibility of immobilization under a wide variety of coupling chemistries to an extensive range of commercially available supports, (iii) amenability to ligand modifications to enhance specificity or stability, and (iv) higher resistance to denaturation (30).

Stability and binding capacity in affinity chromatography are usually related to the size and molecular complexity of the ligand. It was found that ligands derived from Camelid antibodies show an inherent chemical stability that may be due to its small size and to their peculiar characteristics of lacking the light chains. These new ligands are IgG fragments that are complexes of two heavy chains with a large range of specificities and which can display tunable affinity and high selectivity for binding antigens like monoclonal antibodies from cell cultures and important proteins from plasma. As they appear to have multiple binding sites for multiple epitope recognition, their affinity for human Fc could be useful in the purification of polyvalent human IgG. Such affinity matrices are thus considered to represent improved options for bioprocessing IgG and other proteins (31).

23.2.4 Affinity Supports

Generally, ligands are immobilized to the matrix by covalent linkage. Most often, the affinity support is loaded into a chromatographic column for better control of chromatographic performance. Common supports for ligand immobilization are agarose, cellulose, dextran, silica, and more recently, synthetic materials like methacrylate-based sorbents have been made available (32). Ligands can also be immobilized onto membranes or magnetic beads, although their industrial implementation has so far been limited. Choice of the support for ligand immobilization is critical as are the coupling methods. Density and orientation of ligands may influence the dynamic binding capacity of the affinity matrix (33). Thus, an optimal ligand density has to be defined not only for maximum target binding but also for mild protein desorption. The ligand-target binding efficiency may be increased by a physical extension of the ligand from the matrix through a hydrocarbon spacer arm, thus preventing steric hindrance (34).

23.3 APPLICATION OF AFFINITY CHROMATOGRAPHY IN PLASMA PROTEIN PROCESSES

Nowadays, the manufacturing process of almost all therapeutic plasma proteins depends to some extent upon chromatographic methods, including affinity. Table 23.1 indicates ligands used for the industrial purification of PD- and recombinant therapeutic plasma protein products described at laboratory scale. Figure 23.1 shows a typical plasma fractionation process, pointing out more specifically the manufacturing steps typically based on affinity chromatography. Industrial processing experience reveals the benefits provided by affinity chromatography in improving protein purity, in increasing yields, and/or in enhancing overall performance (reduced processing time and purification steps). The following sections describe affinity methods reported for various plasma protein classes.

23.3.1 Coagulation Factors

The usefulness of affinity chromatography is well illustrated in the manufacturing of coagulation factors fractionated from plasma or, even more, when they are purified from recombinant systems. The development of purification methods for recombinant coagulation factors has been and still is a challenge as the processing of these complex glycoproteins generally implies the use of (i) cell cultures of mammalian or human origin, (ii) murine antibodies, and (iii) viral vectors, which are potential sources of unwanted proteins, non-human nucleic acids, immunogenic components, and infectious agents.

23.3.1.1 Factor VIII. Haemophilia A patients are deficient in, or have a defective, FVIII molecule, and are at risk of bleeding, unless receiving regular substitution therapy over their life span. FVIII products should be free from deleterious components such as activated proteins, hemolysins, viral pathogens, and chemical contaminants that may be used during processing. Affinity chromatography has contributed to alleviate those risks.

Since FVIII is bound in plasma to another procoagulant protein, the Von Willebrand factor (VWF), a natural stabilizer of FVIII activity, affinity ligands can be directed toward either one of these two molecules. Common ligands (e.g. heparin and dextran sulfate) which have anticoagulant properties, also have affinity for FVIII/VWF complex and thus can be linked to a matrix to produce chromatographic resins (37). Another ligand, di-methyl-aminopropyl-carbamyl-pentyl, was also tested at pilot-scale for direct capture of the coagulant complex from plasma (74) but further use was prevented by the inability to recover other major therapeutic proteins from the chromatographed plasma. However, the most evaluated ligands have been antibodies that recognize specific FVIII or VWF epitopes and can be immobilized on a matrix (35,36). Such immunoaffinity procedures are still largely used at industrial scale to manufacture highly purified FVIII concentrates. The purified FVIII fraction needs to be stabilized by the addition of human albumin. FVIII immunoaffinity procedures contribute to viral safety but are not considered as robust as dedicated viral inactivation or removal treatments (75) since viral removal by chromatography is difficult to predict and to monitor (76). Drawbacks of this method are the use of murine antibodies (rodent proteins), which may leach and thus must be eliminated by subsequent adsorption chromatographic step, and the relative high cost.

The manufacture of several current therapeutic recombinant FVIII (r-FVIII) preparations often depends on immunoaffinity, as a tool for both the capture of proteins and removal of nucleic acids derived from cell cultures. However, recent advances in peptide ligand research illustrate the current trends to replace immunoaffinity systems, which require stringent quality control measures to keep chromatographic performance and product safety. A purification method of commercial B-domain deleted r-FVIII uses a peptide ligand named TN8.2 (disulfide-bonded peptide with 27 amino acid sequence of ~2800 Da), chosen from a phage display library, and coupled to a Sepharose resin. Validation has shown three to four log reduction of hamster cell proteins and DNA by this peptide affinity step; however, three additional chromatographic steps are needed to further reduce the level of contaminants. Advantages over the immunopurification step are improved dynamic capacity, better stability during column regeneration, and higher step-yield (38). A dye-ligand affinity system used in cascade for direct IgG capture from plasma is able, at experimental pilot-scale (40), to bind also the complex FVIII/VWF with high VWF recovery and specific activity but low FVIII purity. The designed ligand (MAbsorbent A2P) has a triazine structure with amine substituted groups that closely resembles the Phe-Tyr- dipeptide binding site in Protein A. More recently, the properties of a smaller peptide

TABLE 23.1. Examples of Immobilized Ligands Used in Industrial and Laboratory-Based Affinity Purification Processes of Plasma-Derived (PD) and Recombinant Plasma Proteins

Ligand Protein	Heparin	gelatin	MAb	Thiol	IMAC (Copper)	Dextran sulfate	Protein A	Synthetic Ligand	Dye-Ligands	Bioligand
FVIII	(*c*)[a]		(35,36)[a,b]			(37)[a]		(38,39)[c]	(40)[c]	
FIX	(41,42)[a]		(43)[a]		(44)[a]	(45)[a]				
VWF		(46,47)[a]				(37)[c]			(40)[c]	
FXI	(48)[a]									
FVIIa			(49,50)[a]							
Fibrinogen								+(29)[c]		
Fibronectin	(51)[c]	(52–54)[a]								
MBL						(55)[a,d]				
Kallikrein								(32)[c]		
ATIII	(3,56)[a]									
AAT				(57)[c]					(58)[c]	(59)[c]
ITI	(60)[a]									
C1-Inh					(61)[c]					(62)[c]
Protein C	(63)[a]		(64)[a]		(65)[c]					
APC			(66)[a]							
TFPI	(67,68)[c]		(67,68)[c]							(69)[c]
Albumin								(70)[c]	(3,71)[c]	
IgG							(14)[a]	(72)[c]	(40)[c]	
IgG fragments (antivenoms)						(73)[a,c]				

[a] applied at industrial scale to prepare therapeutic products
[b] used both for plasma cryoprecipitate derived- and recombinant-FVIII
[c] laboratory-based research
[d] Mixture of dextran/ agarose (Superdex 200)

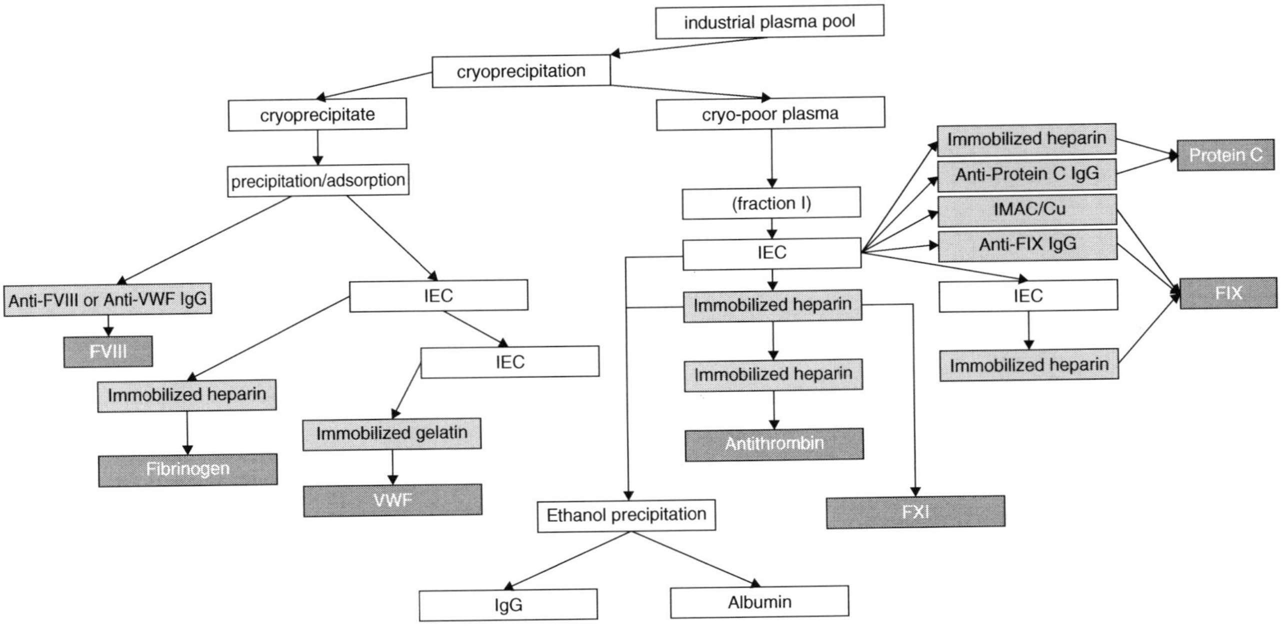

Figure 23.1. A typical plasma fractionation process indicating the steps (light grey boxes) where affinity chromatography has been implemented, and the therapeutic protein products obtained (deep grey boxes). (This figure is available in full color at http://onlinelibrary.wiley.com/book/10.1002/9780470054581.)

ligand grafted on Toyopearl AF-Epoxy-650M, a synthetic polymeric resin, has been described for research-scale FVIII purification. This one-step chromatography eluted FVIII from a commercial preparation which was diluted in culture medium containing 10% bovine serum. The eluted FVIII had 78% overall yield and high purity, as evidenced by sodium dodecyl sulfate-polyacrylamide gel electrophoresis (SDS PAGE) and immunoblotting. The small ligand, designated as a peptidomimetic ligand L4 and obtained by standard solid-phase synthesis, was protease-resistant and, thus, FVIII adsorption/desorption could be achieved under mild conditions, like when using a standard ion-exchanger for plasma FVIII purification (39).

23.3.1.2 Von Willebrand Factor (VWF). VWF, the physiological carrier of FVIII, is a high-molecular mass coagulation factor comprised of monomers of a molecular mass of 260,000 Da, which form multimers of a molecular mass above 10,000,000 Da. Substitutive therapy is clinically needed for some patients with congenital deficiency or a defective molecule. Industrial extraction of therapeutic PD-VWF is carried out from plasma that undergoes cryoprecipitation and ion-exchange chromatography, for example, using DEAE-Toyopearl 650M, for the separation of fibrinogen, VWF, and FVIII. The VWF eluate is further concentrated and purified on a second DEAE-Toyopearl column. The single major contaminant, fibronectin, is removed by binding on an affinity gel, gelatin Sepharose, yielding a highly purified VWF concentrate (46,47). An updated characterization of this therapeutic concentrate, now subjected to nanofiltration and dry-heat, was reported (77).

23.3.1.3 Factor IX. This coagulation factor is missing or dysfunctional in haemophilia B patients, who also depend on life-long substitution therapy to avoid or control bleeding episodes occurring spontaneously or as a result of accidents, pathological situations, or surgery. Various types of ligands can be used at industrial scale for isolating PD-FIX: heparin (41,42), copper immobilized metal affinity chromatography (IMAC), (44) and dextran sulphate (45) as well as murine monoclonal antibodies directed against human FIX (43). In all cases, prior to affinity chromatography, cryoprecipitate-poor plasma is subjected to DEAE ion-exchange capture of the coagulation factors from the prothrombin complex and of some anti-thrombotic proteins which have similar biochemical properties (16). A typical production sequence for a PD-FIX comprises cryoprecipitation, capture of FIX complex (FII, VII, IX, X, and other proteins) by a soft gel (DEAE Sephadex-A50), initial fractionation of the complex on a DEAE Sepharose FF column, followed by separation of FX and FIX by immobilized heparin chromatography (41). A typical processing of an immunopurified FIX product includes, after cryoprecipitation and DEAE Sephadex, specific binding on an immunoaffinity column and further polishing on Aminohexyl Sepharose to remove leached murine antibodies. Use of DEAE Spherodex prior to immunopurification has also been described (43) but, to our knowledge, this process has not been implemented at a large scale. Immunoaffinity systems are also used for the capture of r-FIX from crude cell cultures. The operational procedure is basically the same as for immunopurified PD-FIX products.

23.3.1.4 Factor XI. Deficiency of this coagulation factor is less frequent in human populations than that of FVIII or FIX and its clinical symptoms (bleeding problems and sequels) are usually less critical. FXI deficient patients used to depend exclusively on plasma infusions as purified concentrates were not available prior to the mid-1980s. The first preparation, from England, was obtained using an immobilized heparin adsorption step similar to that used for antithrombin (AT) production (48). However, heparin-bound affinity gels may not be ideal for the purification of this coagulation factor as it is now known that the activated form (FXIa) also binds to heparin and even more strongly than FXI through its A3 domain which contains the binding site for heparin (78). More recently, another therapeutic FXI was purified by adsorptive filtration and cation-exchange chromatography (79). It was shown to be free of contaminating plasma proteases and activated factors, illustrating that conventional ion-exchange chromatography can be used to prepare therapeutic products of higher purity than those obtained using unspecific affinity-based processes.

23.3.1.5 Activated Factor VII (FVIIa). This coagulation enzyme is very effective to counter balance the presence of anti-FVIII inhibitors in haemophilia A patients (80). A large-scale process to extract PD-FVIIa from cryo-poor plasma includes anion-exchange for preliminary capture, and monoclonal anti-FVII antibody affinity columns for achieving high purification factor. The purified FVII is auto-activated on an anion-exchange resin, and nanofiltered for virus removal (49). A commercial r-FVIIa is purified from mammalian cell cultures through four chromatographic steps including immunopurification with murine anti-FVII antibodies (50). The r-FVIIa has evidenced good clinical efficacy in massive bleeding unresponsive to conventional therapy as well as in haemophilia patients with IgG inhibitors (81). Safety of this r-FVIIa concentrate has been shown to be generally good but thromboembolic events have been reported recently (82).

23.3.1.6 Fibrinogen. Therapeutic preparations of fibrinogen, a protein involved in haemostasis and healing processes, can be administered to patients having post-partum bleeding. It can also be manufactured as a component of fibrin sealant which, when activated by thrombin, allows topical use for hemostatic, sealing, or wound healing applications (83). Fibrinogen purification has traditionally been obtained from side-fractions of the purification of FVIII from cryoprecipitate using heparin affinity adsorption (84), or directly from plasma through a series of ethanol precipitation steps (16,85,86). Recently, a tetrapeptide ligand (GPRP), identified from a peptide library, was claimed to allow, at experimental scale, isolation of fibrinogen under a highly pure and functional state. Such affinity-purified fibrinogen retained a FXIII crosslinking activity claimed to be required to prepare elastic and high tensile-strength fibrin sealants. The peptide ligand was more efficient when immobilized at high density on a matrix (29).

23.3.2 Protease Inhibitors

23.3.2.1 Alpha-1-antitrypsin (AAT). Also called alpha 1-proteinase inhibitor, this human plasma protein has been purified for the treatment of pulmonary emphysema and cystic fibrosis to neutralize elastase, which is implicated in the occurrence of these pathologies. Integrated ethanol precipitation and ion-exchange chromatographic processes are commonly used for the industrial purification of this protein but albumin has tendency to co-elute under these conditions due to comparable molecular mass and isolectric point. However, experimental work has shown that immobilized chymotrypin (59) and Affi-Gel Blue can be of value for binding AAT, if a highly purified and homogeneous preparation is required and when the degree of albumin contamination interferes with its elastase inhibitory activity (58). Thiol–disulfide exchange chromatography has also been described for the capture of AAT at the research scale (57).

23.3.2.2 C1-Inhibitor (C1-Inh). This serine protease inhibitor is used to treat hereditary angioedema, a pathology where, due to C1-Inh deficiency, the Component C1 of the complement system can get activated, causing a life-threatening condition. Other potential applications of this inhibitor have been highlighted in pathologies involving systemic inflammation and multiple organ failure, such as septicaemia and disseminated intravascular coagulation (DIC) (87). Purification from plasma is achieved after prothrombin complex adsorption using ion-exchange chromatography (88) but earlier purification strategies used at laboratory scale included the use of affinity matrices such as zinc ion chelate bound to agarose (61) or immobilized jacalin, a lectin from jack fruit, followed by an ion-exchange step to improve stability of the purified C1 inhibitor (62). A r-C1 inhibitor concentrate is produced in transgenic rabbit mammary glands (89) but, to our knowledge, the purification process from milk has not been published. Phase II and III clinical trials have shown efficacy of the recombinant product but higher therapeutic doses may be needed as half-life is shorter than that of the PD-protein (90).

23.3.2.3 ***Inter-alpha-trysin inhibitor (ITI).*** A member of serine protease inhibitors, ITI expresses its biological activity by releasing bikunin which has evidenced its anti-inflammatory effects in various experimental models. This plasma inhibitor was purified at large scale as a by-product of the PCC process and included ion-exchange and immobilized heparin chromatographies (60). Use of Cibachrom Blue affinity ligand has been described at experimental scale (91). Epoxy-activated monolithic convective interaction media (CIM) disks used for immobilization of protein ligands including a monoclonal antibody against human inter-alpha inhibitor proteins has also been published: this antibody, immobilized on monolithic disk, was used for very rapid isolation of inter-alpha trypsin inhibitor (92). The heavy chains of this inhibitor were recently shown to be beneficial in down-regulation of various types of cancer processes (93).

23.3.3 Anticoagulants/Anti-Thrombotics

23.3.3.1 ***Antithrombin (AT).*** This glycoprotein regulates blood clotting and is used for substitution therapy in deficient patients or for acquired deficiency leading to recurrent thromboembolic episodes (94). It is a physiological co-factor of heparin; it is therefore not surprising that heparin affinity chromatography has proven crucial for several decades to isolate PD-AT and to segregate between active and denatured AT molecules generated during heat viral inactivation treatments (3,56). The purification process of a r-AT preparation isolated from milk of transgenic goats, also includes heparin affinity chromatography (95).

23.3.3.2 ***Protein C (PC).*** Protein C (PC) is a plasma protein anticoagulant. It also exhibits anti-thrombotic and profibrinolytic properties and, thus, PC therapeutic preparations are used in PC-deficient patients susceptible to thrombotic events and in patients with deep-venous thrombosis (DVT) and lung embolia. Highly purified therapeutic concentrates are prepared from human plasma as a by-product of PCC and FIX productions. One of them is purified by a combination of classic purification methods and immunoaffinity using murine monoclonal antibodies and vapor-heated for viral inactivation (64). This PC concentrate was shown to be efficient to treat purpura fulminans in patients with sepsis-induced coagulopathies (96). Later on, another highly purified PC preparation, obtained using ion-exchange and immobilized heparin matrices (63), had also been made available in Europe. Multiple clinical applications of this concentrate have demonstrated its efficacy and safety in the treatment of various thrombosis-associated pathologies (97). The therapeutic safety of both products reveals the essentially complete elimination of potentially thrombogenic PCC factors that can be achieved by affinity chromatography. Other experimental methodologies for purifying PC were applied to an ethanol-rich plasma fraction (Cohn fraction IV-1) and comprised either single-chain variable IgG fragments, produced in E. Coli cells and coupled to a Sepharose gel, or an IMAC column using imidazol-containing buffers (65). However, to our knowledge, it has not been implemented for production. Also, r-PC expressed in adenovirus-transformed cell lines was purified by a chromatographic method, termed "pseudo-affinity", capable of resolving closely related molecules (98).

23.3.3.3 ***Activated Protein C (APC).*** Activated Protein (APC) is an anti-thrombotic serine protease having anti-coagulant, profibrinolytic, and anti-inflammatory activities. A production process for a highly-purified, functionally active and stable PD-APC concentrate has been described; it is prepared by enzymatic activation and purified by three chromatographic steps, including immunoaffinity (66). A commercial r-APC is available for therapeutic use in the treatment of sepsis. The efficacy and safety of this r-APC preparation compared to therapeutic heparin was evident in a randomized prospective double-blind trial with patients presenting DIC. Mortality rate was significantly lower in the APC group versus the heparin group and there was no increase in bleeding (99). Another medical study demonstrated that therapy combining APC and tissue-plasminogen activator (t-PA) was more efficient than heparin after thrombolysis in patients with acute myocardial infarction (100).

23.3.3.4 ***Tissue Factor Pathway Inhibitor (TFPI).*** Tissue Factor Pathway Inhibitor (TFPI) is a powerful physiological anticoagulant, present in vascular endothelium, platelets, and plasma, with the ability to inhibit coagulation at both the intrinsic and extrinsic pathways. Initially, it was purified from plasma for characterization using hydrophobic, ion-exchange, and affinity chromatographies and identified as being a glycoprotein associated to low density lipoprotein (LDL) and to apolipoprotein A-II (HDL) (69). A related TFPI form releasable by heparin and found in human plasma was also shown to have binding affinity on immobilized heparin (67). There is no licensed PD-TFPI but a recombinant product expressed in prokariotic cells as a non glycosylated molecule and comprising an added N-terminal alanine (101). This r-TFPI was clinically evaluated in a randomized trial but was found to increase bleeding risk in severe sepsis (102), possibly due to interaction with heparin as observed with other therapeutic anticoagulants (103). More recently, another r-TFPI having preserved molecular structure was produced in CHO cells and purified using monoclonal antibodies and immobilized heparin (68).

23.3.4 Albumin

Ethanol precipitation steps, sometimes combined with ion-exchange and/or size-exclusion chromatography, remain the main method used industrially for the fractionation of plasma albumin, as reviewed before (3). Until now, no industrial process applies affinity chromatography to extract albumin for infusion purposes, although research and pilot-scale works have demonstrated capacity of albumin to bind to various dye-ligands (3,71). Recently, specific, high-affinity camelid antibody fragments derived from immune libraries have been developed and could have benefits for the large-scale purification of albumin as an alternative fractionation method (70), although, one or two additional purification steps may be needed to eliminate leached ligands or fragments.

23.3.5 Immunoglobulin G

Human plasma continues to be regarded as a unique source of potential therapeutics and, among them, polyvalent and hyperimmune IgG are of increasingly recognized medical value. The market potential for IgG preparations are currently expanding worldwide as their multiple clinical indications are still steadily increasing for treating infectious as well as autoimmune, inflammatory, and neurological diseases (104–106). Development of synthetic ligands, such as the Protein A mimetic peptides, may be useful for more convenient and less expensive affinity systems than the bacteria-derived proteins (Protein A, Protein G, and Protein L) currently employed as ligands for the large-scale purification of therapeutic monoclonal antibodies (72). In the plasma fractionation area, immobilized protein A affinity chromatography (14) is not used for the production of immunoglobulin preparations, possibly due to regulatory and/or cost issues, as well as due to the need to fractionate other products simultaneously (16). However, a dye-ligand affinity system, used in a cascade sequence of columns, has shown a potential at pilot-scale for direct IgG capture from plasma. Other important plasma proteins such as IgA, IgM, albumin, protease inhibitor, and the complex FVIII/VWF could also be isolated as by-products of this process and purified with high yields compared to standard fractionation methods (40). This method using mimetic ligands appears particularly interesting for VWF purification which is reported to be obtained at over 40% yield (versus ~20% for current industrial yield) and comparable specific activity (~80 IU/mg) to that achieved by ion-exchange chromatography from cryoprecipitate (47). FVIII activity is also obtained at relatively high yield of ~40% (versus ~15% for standard plasma fractionation methods) but purity is similar to that of a crude cryoprecipitate. The target molecules in this technology are the immunoglobulins for which the recovery is 98%; however, additional chromatographic steps are still needed to improve protein purity, therefore lowering the yield.

More recent developments in affinity ligands are illustrated by the discovery of antibodies from the Camelidae family as an interesting source of IgG composed of a single variable domain and lacking the common IgG light chains in classical antibodies. Low-molecular mass fragments could be isolated from such camelid antibodies and generate ligands of high affinity and selectivity, showing improved stability to chromatographic conditions and storage. These new generation of ligands are intended for the large-scale purification of human IgG, monoclonal antibodies, Fab fragments, and even viruses (107).

Low-molecular weight acidic affinity ligands have also been shown at experimental scale to bind all IgG sub-classes as well as Fab fragments by a combination of hydrophobic/aromatic, isoelectric point and hydrogen bonding interactions at near neutral pH values, elution being performed by increasing the pH to a range between 8 and 9. This approach does not require dilution of the feedstocks and is amenable to implementation in both packed bed and high-density expanded bed adsorbents (EBA) (108). High-density mixed-mode ligands EBA materials that can be eluted in the expanded bed mode (109) could process large volumes of unclarified raw materials under high flow-rates and, with enhanced specificity (110). Also, it may potentially offer an advantageous alternative to the core ethanol plasma fractionation process, in particular, for the extraction of IgG.

23.3.6 Other Plasma Proteins

23.3.6.1 Fibronectin. This glycoprotein is present in plasma but also in connective tissues and extracellular matrix. It is involved in cell adhesion, tissue organization, and wound healing. Although there is essentially no licensed PD-product, several clinical reports have demonstrated, since the early 1980s, clinical benefits for healing corneal ulcers in situations where other treatments failed (111). The most common experimental way to purify fibronectin uses binding on immobilized gelatin followed by elution with a highly concentrated urea or amino compounds (52). High quality column desorption can be achieved using arginin buffers under less denaturing conditions (53). Combination with immobilized heparin chromatography has also been described (51). Fibronectin can be adsorbed and polished on Arginin-Sepharose matrix, a step that allows removing some immunogenic components, yielding a highly purified preparation. Covalently immobilized gelatin on poly(2-hydroxyethyl methacrylate) (PHEMA) microspheres was shown to have an increase in fibronectin binding capacity to ca. 22 mg/gram of gel (54). A biotin-affinity gel was useful for the purification of r-fibronectin when the expressed protein was biotinylated *in-vivo*. The method provided a bioactive fibronectin as measured by cell adhesion assay and proved to be effective in preventing liver failure in an endotoxemia experimental model (112).

*23.3.6.2 **Mannan-binding lectin (MBL).*** Mannan-binding lectin (MBL) is a lectin present in human plasma that exhibits antimicrobial properties and is involved in the innate immunity. Deficiency of MBL predisposes children and adults to infectious diseases, particularly in patients with immature or compromised immune systems (113). A European experimental concentrate was purified from plasma fraction III using an affinity resin, the Superdex 200, which is a mixture of dextran and highly cross-linked agarose. Presumably, the D-glucose moiety of the dextran structure behaves as a ligand for the calcium-dependent binding of MBL. The protein is desorbed from the resin with a mannose-containing buffer, nanofiltered, and solvent-detergent treated for viral safety. The downstream steps include ion-exchange and gel filtration chromatographies (55). The product underwent phase I clinical studies with good tolerance and safety records (114). More recently, another MBL preparation has been obtained by gene recombinant technologies. The respective phase I clinical studies showed satisfactory results with only mild side-effects and no antibody formation (115).

*23.3.6.3 **Kallikrein.*** A synthetic ligand (an inhibitor of plasma kallikrein) coupled to a Toyopearl resin, a large-pore methacrylate support, allowed the purification of kallikrein with high purity and recovery. Subsequent to the affinity step, a DEAE-Toyopearl 650 M purification step allowed kallikrein isolation in 2 steps from plasma, with a 1720-fold purification factor (32).

23.4 QUALITY CONTROL OF AFFINITY-PURIFIED PROTEINS

Bioprocessing of PD- and recombinant plasma proteins should follow the stringent current regulatory requirements applying to medicinal drugs of biological origin. Guidelines and recommendations on Good Manufacturing Practices (GMP) and relevant quality control and quality assurance measures to guarantee product quality are prepared by international organizations and regulatory bodies such as the World Health Organization (116) the European Medicine Agency (117,118), the US Food and Drug Agency, and the Pharmaceutical Inspection Convention/Pharmaceutical Inspection Co-operation Scheme (119). One key aspect is the fulfilment of GMP which provides a legally binding insurance, verified by national regulatory authorities, ensuring that production processes are done in a consistent manner and controlled according to previously defined quality specifications of the product.

Examples of common quality and safety testing for crucial manufacturing steps of fractionated plasma proteins were described in a former review (120) and a thorough description of those applicable to recombinant proteins was also reported earlier (38). Besides the regular quality control methods, the most critical aspects to be checked when dealing with affinity systems for industrial protein purification are mentioned below:

23.4.1 Changes in Structural Conformation

Changes in structural conformation of recombinant proteins may originate at the stage of cell line production and maintenance as subtle changes in production parameters may occur over multiple culture cycles and lead to variations in the synthesis or translation of the target component. Other changes include protein cleavages by proteases present in cell cultures, starting plasma or in-process purification fractions, either causing activation and, thus, loss of biological activity, or generation of degradation fragments which could be immunogenic. Some processing operations, like concentration steps, may also induce protein alterations. Most antibodies appear to be particularly sensitive to shear forces of long-lasting ultrafiltration processes that generate aggregated proteins (18). Aggregated antibodies if present in therapeutic preparations may cause acute allergic reactions in patients through activation of the complement system. In principle, affinity chromatography can also generate structural changes of target proteins when strong ligand affinity requires harsh elution conditions. Similarly, the protein-ligand interaction may potentially lead to changes in conformation affecting functionality or pharmacokinetic features (e.g. *in-vivo* half-life of a therapeutic protein).

23.4.2 Leakage

This major concern when using affinity chromatography has to be validated using extreme operational conditions (known as "worst-case scenario") in such a way that if drifts happen from the regular procedure, the affinity sorbent will still tolerate them and keep the eluted target protein free from ligands, fragments of it, chemical components from the affinity support, or grafted spacer arms that may potentially lead to *in-vivo* toxic effects. Leached monoclonal antibodies and peptide ligands may co-elute with the target protein but can be subsequently removed by ion-exchange chromatographic steps.

23.4.3 Viral Safety

All protein processes that include biological (human or animal) components as source material or as purification tools should integrate validated viral reduction methods. Viruses may originate from the starting raw materials, such as human plasma or animal cell cultures used to express recombinant proteins. In addition, purification of many genetically engineered proteins depends on immunoaffinity procedures using animal-derived antibodies, implying a risk of adventitious viral transmission to patients. Two methods of viral inactivation or viral removal are usually applied during the manufacture of PD-therapeutic proteins, and increasingly, for recombinant preparations, as well.

23.4.4 Residual Impurities

Toxic chemicals used for viral reduction procedures, or deriving from raw materials used for cell culture, chromatographic supports, and buffers should be removed below acceptable levels. Common residual impurities include murine antibodies, nucleic acids, detergents (such as triton X-100, tween-80, or sodium cholate), solvents (such as TnBP or ethanol), and heavy metals (such as aluminium or baryum).

23.4.5 Neoantigen Formation

The chemical interaction with affinity ligands may cause molecular modification of the target proteins (e.g. change in conformation), in particular, if the ligand has a synthetic or a biosynthetic (polypeptides) structure. Such protein modification may imply a change in its conformation or in its glycosidic moiety which may provoke the appearance of new protein antigenic determinants, called neoantigens, and the risk of synthesis of neoantibodies by the immune system of the patient. Such neoantibodies in the case of FVIII replacement therapy are inhibitors neutralizing the infused FVIII molecules and rendering the treatment quite inefficient. In other situations, molecular modifications affect protein behaviour (e.g. reduced binding ability of FVIII for its natural VWF carrier and stabilizer). For instance, r-FVIII preparations, which appear to be more immunogenic than fractionated plasma FVIII products (121), were also found to contain ~20% of its antigen molecule unable to associate with VWF (122).

23.4.6 Standards for Quality Control

Potency assessment of both PD- and recombinant-protein products requires a precise standardized assessment of its functionality, relative to the antigen content and overall protein content. This helps to guarantee a correct product dosage to avoid overdose or insufficient product administration in prophylactic and therapeutic procedures. Thus, quality control assays for product potency definition are done by comparison with known standard preparations calibrated for specific products and run in parallel during the respective determinations. Such standards are available either as plasma preparations or as purified concentrates for coagulation factors (120), anticoagulants, proteases inhibitors, and other therapeutics which are prepared as WHO International Standards available from the National Institute for Biological Standards and Controls (NIBSC, London). An updated list of international reference preparations for both fractionated and recombinant plasma preparations can be found on the NIBSC (http://www.nibsc.ac.uk/products/cataloguefull) and WHO (http://www.who.int/biologicals/reference_preparations/en/) web sites.

23.5 CONCLUSIONS

Affinity chromatography using naturally occurring ligands and murine monoclonal antibodies has already played a major contribution in the purification of therapeutic PD and DNA-engineered plasma proteins. As such, it has become an established and robust industrial purification tool which, when properly monitored by manufacturers and implemented following GMP, is generally well accepted by regulatory authorities. Still, novel biosynthetic and synthetic ligands for protein affinity purification appear to be an attractive alternative to the immunoaffinity systems. They are shown to be more reliable than monoclonal antibodies, offering better stability, longer shelf-life, improved robustness, and increased cost-effectiveness. Such new systems avoid the risks of leakage of potential immunogenic antibodies and eliminate the needs for process validations for pathogenic (viruses and prions) and nucleic acid contaminations. Moreover, there is now a potential for applying peptide affinity technologies to process human plasma for producing current and new therapeutic components at higher yields and reduced costs. However, industrial implementation of such new ligands would require substantial validation and pre-clinical studies to establish safety by both suppliers and users to meet the stringent regulatory requirements applying to the manufacturing processes of biotherapeutics. Such requirements may substantially increase the cost and slow down the introduction of such promising downstream technologies.

REFERENCES

1. Burnouf T. J Chromatogr B Biomed Appl 1995; 664: 3–15.
2. Johnston A, Adcock W. Biotechnol Genet Eng Rev 2000; 17: 37–70.
3. Burnouf T, Radosevich M. J Biochem Biophys Methods 2001; 49: 575–586.
4. Cohn E, Strong L, Hughes W, Mulford D, Ashworth J, Melin M, Taylor H. J Am Chem Soc 1946; 68: 459–475.

5. Pool JG, Gershgold EJ, Pappenhagen AR. Nature 1964; 203: 312.
6. Thorell L, Blomback B. Thromb Res 1984; 35: 431–450.
7. Brodniewicz-Proba T, Beauregard D. Vox Sang 1987; 52: 10–14.
8. Wagner BH, McLester WD, Smith M, Brinkhous KM. Thromb Diath Haemorrh 1964; 11: 64–74.
9. Josso F, Menache D, Steinbuch M, Blatrix C, Soulier JP. Bibl Haematol 1970; 34: 18–22.
10. Pejaudier L, Kichenin-Martin V, Boffa MC, Steinbuch M. Vox Sang 1987; 52: 1–9.
11. Giles AR, Nesheim ME, Hoogendoorn H, Tracy PB, Mann KG. Blood 1982; 59: 401–407.
12. Sakuragawa N, Takahashi K, Hoshiyama M, Niiya K, Itoh M, Matsuoka M, Ohnishi Y. Thromb Res 1977; 10: 315–318.
13. Nilsson J, Stahl S, Lundeberg J, Uhlen M, Nygren PA. Protein Expr Purif 1997; 11: 1–16.
14. Goding JW. J Immunol Methods 1978; 20: 241–253.
15. Clonis YD. J Chromatogr A 2006; 1101: 1–24.
16. Burnouf T. Transfus Med Rev 2007; 21: 101–117.
17. Sullivan JB Jr. Ann Emerg Med 1987; 16: 938–944.
18. Aldington S, Bonnerjea J. J Chromatogr B Analyt Technol Biomed Life Sci 2007; 848: 64–78.
19. Charlton H. In: Antibody development & production conference (IBC, Ed.). Carlsbad, California; 2006.
20. Labrou NE. J Chromatogr B Analyt Technol Biomed Life Sci 2003; 790: 67–78.
21. Denizli A, Piskin E. J Biochem Biophys Methods 2001; 49: 391–416.
22. Lowe CR, Burton SJ, Burton NP, Alderton WK, Pitts JM, Thomas JA. Trends Biotechnol 1992; 10: 442–448.
23. Lowe CR, Burton SJ, Pearson JC, Clonis YD, Stead V. J Chromatogr 1986; 376: 121–130.
24. Dunn IS. Curr Opin Biotechnol 1996; 7: 547–553.
25. Cortese R, Monaci P, Luzzago A, Santini C, Bartoli F, Cortese I, Fortugno P, Galfre G, Nicosia A, Felici F. Curr Opin Biotechnol 1996; 7: 616–621.
26. Ellman J, Stoddard B, Wells J. Proc Natl Acad Sci U S A 1997; 94: 2779–2782.
27. Hogan JC Jr. Nature 1996; 384: 17–19.
28. Pingali A, McGuinness B, Keshishian H, Jing FW, Varady L, Regnier F. J Mol Recognit 1996; 9: 426–432.
29. Kaufman DB, Hentsch ME, Baumbach GA, Buettner JA, Dadd CA, Huang PY, Hammond DJ, Carbonell RG. Biotechnol Bioeng 2002; 77: 278–289.
30. Baumbach GA, Hammond DJ. Biopharm 1992; 5: 24–29.
31. Klooster R, Maassen BT, Stam JC, Hermans PW, Ten Haaft MR, Detmers FJ, de Haard HJ, Post JA, Theo Verrips C. J Immunol Methods 2007; 324: 1–12.
32. Tada M, Wanaka K, Okamoto S, Okamoto U, Nakaya Y, Horie N, Hijikata-Okunomiya A, Tsuda Y, Okada Y. Biol Pharm Bull 1998; 21: 105–108.
33. Doyle ML, Myszka DG, Chaiken IM. J Mol Recognit 1996; 9: 65–74.
34. Cuatrecasas P. J Biol Chem 1970; 245: 3059–3065.
35. Rotblat F, O'Brien DP, O'Brien FJ, Goodall AH, Tuddenham EG. Biochemistry 1985; 24: 4294–4300.
36. Liu S, Addiego J, Gomperts E, Kessler C, Garanchon L, Neslund G, Foster V, Berkebile R, Courter S, Lee M, Kingdon H, Griffith M. Colloq INSERM 1989; 175: 263–270.
37. Harrison P, Saundry RH, Savidge GF. Thromb Res 1988; 50: 295–304.
38. Kelley BD, Tannatt M, Magnusson R, Hagelberg S, Booth J. Biotechnol Bioeng 2004; 87: 400–412.
39. Knor S, Khrenov A, Laufer B, Benhida A, Grailly SC, Schwaab R, Oldenburg J, Beaufort N, Magdolen V, Saint-Remy JMR, Saenko EL, Hauser CAE, Kessler H. J Thromb Haemost 2008; 6: 470–477.
40. Chen T, Allen S, Baines D, Betley J, Blackman D, Hayes T, Schmidt D, Busby T, Harris G, Lobezoo B, Watson K. In: 4th Plasma Product Biotechnology Meeting. Porto Elounda, Crete, Greece; 2005.
41. Burnouf T, Michalski C, Goudemand M, Huart JJ. Vox Sang 1989; 57: 225–232.
42. Hoffer L, Schwinn H, Josic D. J Chromatogr A 1999; 844: 119–128.
43. Lutsch C, Gattel P, Fanget B, Véron JL, Smith K, Armand J, Grandgeorge M. Volume 227, Biotechnology of plasma proteins. Nancy, France: INSERM; 1993. pp 75–80.
44. Feldman PA, Bradbury PI, Williams JD, Sims GE, McPhee JW, Pinnell MA, Harris L, Crombie GI, Evans DR. Blood Coagul Fibrinolysis 1994; 5: 939–948.
45. Menache D, Behre HE, Orthner CL, Nunez H, Anderson HD, Triantaphyllopoulos DC, Kosow DP. Blood 1984; 64: 1220–1227.
46. Burnouf-Radosevich M, Burnouf T. Vox Sang 1992; 62: 1–11.
47. Burnouf-Radosevich M. J Tissue Cult Methods 1994; 16: 223–226.
48. Smith JK, Winkelman L, Evans DR, Haddon ME, Sims G. Vox Sang 1985; 48: 325–332.
49. Tomokiyo K, Yano H, Imamura M, Nakano Y, Nakagaki T, Ogata Y, Terano T, Miyamoto S, Funatsu A. Vox Sang 2003; 84: 54–64.
50. Jurlander B, Thim L, Klausen NK, Persson E, Kjalke M, Rexen P, Jorgensen TB, Ostergaard PB, Erhardtsen E, Bjorn SE. Semin Thromb Hemost 2001; 27: 373–384.
51. Allary M, Amoignon T, Lafargue O, Sene C, Boschetti E, Saint-Blancard J. Rev Fr Transfus Hemobiol 1989; 32: 107–113.
52. Vuento M, Vaheri A. Biochem J 1978; 175: 333–336.
53. Vuento M, Vaheri A. Biochem J 1979; 183: 331–337.
54. Kayirhan-Denizli F, Arica MY, Denizli A. J Biomater Sci Polym Ed 2001; 12: 479–489.
55. Laursen I, Houen G, Hojrup P, Brouwer N, Krogsoe LB, Blou L, Hansen PR. Vox Sang 2007; 92: 338–350.
56. Heger A, Grunert T, Schulz P, Josic D, Buchacher A. Thromb Res 2002; 106: 157–164.
57. Myerowitz RL, Handzel ZT, Robbins JB. Clin Chim Acta 1972; 39: 307–317.
58. Finotti P, Pagetta A. Clin Chim Acta 1997; 264: 133–148.
59. Twining SS, Brecher AS. Proc Soc Exp Biol Med 1975; 150: 98–103.
60. Michalski C, Piva F, Balduyck M, Mizon C, Burnouf T, Huart JJ, Mizon J. Vox Sang 1994; 67: 329–336.
61. Prograis LJ Jr, Hammer CH, Katusha K, Frank MM. J Immunol Methods 1987; 99: 113–122.

62. Donaldson VH, Falconieri MW. J Immunol Methods 1993; 157: 101–104.
63. Radosevich M, Zhou FL, Huart JJ, Burnouf T. J Chromatogr B Analyt Technol Biomed Life Sci 2003; 790: 199–207.
64. Johann E, Ludwig P, Hans-Peter S. Vol. Eur. patent EP 0533210 B1. 1991.
65. Rezania S, Ahn DG, Kang KA. Adv Exp Med Biol 2007; 599: 125–131.
66. Orthner CL, Ralston AH, Gee D, Kent R, Kolen B, McGriff JD, Drohan WN. Vox Sang 1995; 69: 309–318.
67. Novotny WF, Palmier M, Wun TC, Broze GJ Jr, Miletich JP. Blood 1991; 78: 394–400.
68. Kamei S, Kamikubo Y, Hamuro T. Vol. US Patent N° 6300100. 2001.
69. Novotny WF, Girard TJ, Miletich JP, Broze GJ Jr. J Biol Chem 1989; 264: 18832–18837.
70. Klooster R, Maassen BTH, Stam JC, Hermans PW, ten Haaft MR, Detmers FJM, de Haard HJ, Post JA, Theo Verrips C. J Immunol Methods 2007; 324: 1–12.
71. Allary M, Saint-Blancard J, Boschetti E, Girot P. Bioseparation 1991; 2: 167–175.
72. Fassina G, Verdoliva A, Odierna MR, Ruvo M, Cassini G. J Mol Recognit 1996; 9: 564–569.
73. Smith DC, Reddi KR, Laing G, Theakston RG, Landon J. Toxicon 1992; 30: 865–871.
74. te Booy MP, Riethorst W, Faber A, Over J, Konig BW. Thromb Haemost 1989; 61: 234–237.
75. Horowitz B, Minor P, Morgenthaler JJ, Burnouf T, McIntosh R, Padilla A, Thorpe R, van Aken WG. World Health Organ Tech Rep Ser 2004; 924: 1–232, backcover
76. Burnouf T. Dev Biol Stand 1993; 81: 199–209.
77. Mazurier C, Poulle M, Samor B, Hilbert L, Chtourou S. Vox Sang 2004; 86: 100–104.
78. Ho DH, Badellino K, Baglia FA, Walsh PN. J Biol Chem 1998; 273: 16382–16390.
79. Burnouf-Radosevich M, Burnouf T. Transfusion 1992; 32: 861–867.
80. Goudemand J. Transfus Clin Biol 1998; 5: 260–265.
81. Selin S, Tejani A. In: (www.ccohta.ca), C. C. O. f. H. T. A. Issues in emerging health technologies. Ottawa: Canadian Coordinating Office for Health Technology Assessment; 2006.
82. Diringer MN, Skolnick BE, Mayer SA, Steiner T, Davis SM, Brun NC, Broderick JP. Stroke 2008; 39: 850–856.
83. Radosevich M, Goubran HA, Burnouf T. Vox Sang 1997; 72: 133–143.
84. Burnouf T, Burnouf M. Centre regional de transfusion sanguine de lille. European Patent Office; 1989; EP0359593B2.
85. Burnouf T, Burnouf-Radosevich M, Huart JJ, Goudemand M. Vox Sang 1991; 60: 8–15.
86. Burnouf-Radosevich M, Burnouf T, Huart JJ. Vox Sang 1990; 58: 77–84.
87. Caliezi C, Wuillemin WA, Zeerleder S, Redondo M, Eisele B, Hack CE. Pharmacol Rev 2000; 52: 91–112.
88. Poulle M, Burnouf-Radosevich M, Burnouf T. Blood Coagul Fibrinolysis 1994; 5: 543–549.
89. Longhurst H. Curr Opin Investig Drugs 2008; 9: 310–323.
90. Choi G, Soeters MR, Farkas H, Varga L, Obtulowicz K, Bilo B, Porebski G, Hack CE, Verdonk R, Nuijens J, Levi M. Transfusion 2007; 47: 1028–1032.
91. Dubin A, Potempa J, Travis J. Prep Biochem 1990; 20: 63–74.
92. Lim YP, Josic D, Callanan H, Brown J, Hixson DC. J Chromatogr A 2005; 1065: 39–43.
93. Hamm A, Veeck J, Bektas N, Wild PJ, Hartmann A, Heindrichs U, Kristiansen G, Werbowetski-Ogilvie T, Del Maestro R, Knuechel R, Dahl E. BMC Cancer 2008; 8: 25.
94. Vinazzer H. Semin Thromb Hemost 1999; 25: 257–263.
95. Edmunds T, Van Patten SM, Pollock J, Hanson E, Bernasconi R, Higgins E, Manavalan P, Ziomek C, Meade H, McPherson JM, Cole ES. Blood 1998; 91: 4561–4571.
96. Schellongowski P, Bauer E, Holzinger U, Staudinger T, Frass M, Laczika K, Locker GJ, Quehenberger P, Rabitsch W, Schenk P, Knobl P. Vox Sang 2006; 90: 294–301.
97. Dreyfus M, Ladouzi A, Chambost H, Gruel Y, Tardy B, Ffrench P, Bridey F, Tellier Z. Vox Sang 2007; 93: 233–240.
98. Yan SC, Razzano P, Chao YB, Walls JD, Berg DT, McClure DB, Grinnell BW. Biotechnology (N Y) 1990; 8: 655–661.
99. Aoki N, Matsuda T, Saito H, Takatsuki K, Okajima K, Takahashi H, Takamatsu J, Asakura H, Ogawa N. Int J Hematol 2002; 75: 540–547.
100. Sakamoto T, Ogawa H, Takazoe K, Yoshimura M, Shimomura H, Moriyama Y, Arai H, Okajima K. J Am Coll Cardiol 2003; 42: 1389–1394.
101. Gustafson ME, Junger KD, Wun TC, Foy BA, Diazcollier JA, Welsch DJ, Obukowicz MG, Bishop BF, Bild GS, Leimgruber RM, Palmier MO, Matthews BK, Joy WD, Frazier RB, Galluppi GR, Grabner RW, Protein Expr Purif 1994; 5: 233–241.
102. Abraham E, Reinhart K, Opal S, Demeyer I, Doig C, Rodriguez AL, Beale R, Svoboda P, Laterre PF, Simon S, Light B, Spapen H, Stone J, Seibert A, Peckelsen C, De Deyne C, Postier R, Pettila V, Artigas A, Percell SR, Shu V, Zwingelstein C, Tobias J, Poole L, Stolzenbach JC, Creasey AA. JAMA 2003; 290: 238–247.
103. Fourrier F. Med Mal Infect 2006; 36: 524–533.
104. Siberil S, Elluru S, Graff-Dubois S, Negi VS, Delignat S, Mouthon L, Lacroix-Desmazes S, Kazatchkine MD, Bayary J, Kaveri SV. Ann N Y Acad Sci 2007; 1110: 497–506.
105. Misra N, Bayry J, Ephrem A, Dasgupta S, Delignat S, Van Huyen JP, Prost F, Lacroix-Desmazes S, Nicoletti A, Kazatchkine MD, Kaveri SV. J Neurol 2005; 252 (Suppl 1):I1–I6.
106. Over J. Blood Bank Transfus Med 2003; 1: 60–66.
107. Detmers F, Hermans P, ten Haaft M. Genet Eng Biotechnol News 2007; 27: 5–6.
108. Noel R, Bendix Hansen M, Vaarst Andersen I, Pontoppidan M, Lihme M. In: Antibody Production and Downstream Processing. Dublin, Ireland; 2006.
109. Lihme A, Zafirakos E, Hansen M, Olander M. Bioseparation 1999; 8: 93–97.
110. Noel R, Bendix Hansen M, Lihme A. Bioprocess Int 2007; 5: 58–61.

111. McCulley JP, Horowitz B, Husseini ZM, Horowitz M. Trans Am Ophthalmol Soc 1993; 91: 367–386, discussion 386-390.
112. Wu Y, Chen YZ, Huang HF, Chen P. Acta Pharmacol Sin 2004; 25: 783–788.
113. Bouwman LH, Roep BO, Roos A. Hum Immunol 2006; 67: 247–256.
114. Valdimarsson H, Vikingsdottir T, Bang P, Saevarsdottir S, Gudjonsson JE, Oskarsson O, Christiansen M, Blou L, Laursen I, Koch C. Scand J Immunol 2004; 59: 97–102.
115. Petersen KA, Matthiesen F, Agger T, Kongerslev L, Thiel S, Cornelissen K, Axelsen M. J Clin Immunol 2006; 26: 465–475.
116. World Health Organization (WHO). Volume 2, Quality assurance of pharmaceuticals. A compedium of guidelines and related materials. 2nd updated ed. Good manufacturing practices and inspection; Geneva, Switzerland 2007.
117. CPMP. ICH Q5A. Note for guidance on quality of biotechnological products: viral safety evaluation of biotechnology products derived from cell lines of human or animal origin. London: European Medicine Agency; 1997. *CPMP/ICH/295/95, April*. http://www.emea.eu.int.
118. CPMP. Note for guidance on plasma -derived medicinal products: The European Agency for the Evaluation of Medicinal Products. London: 2001. *CPMP/BWP/269/95 rev.3*. http://www.emea.eu.int.
119. Guide to good manufacturing practice for medicinal products. Part I. PE009-8. 15 January 2009. Ed: Pharmaceutical Inspection Convention/Pharmaceutical Inspection Co-operation Scheme, 14 rue du Roveray CH-1207 Geneva (Note: this reference has been updated in 2009) 2003.
120. Radosevich M, Burnouf T. Curr Pharm Anal 2007; 3: 83–94.
121. Goudemand J, Rothschild C, Demiguel V, Vinciguerrat C, Lambert T, Chambost H, Borel-Derlon A, Claeyssens S, Laurian Y, Calvez T. Blood 2006; 107: 46–51.
122. Lin Y, Yang X, Chevrier MC, Craven S, Barrowcliffe TW, Lemieux R, Ofosu FA. Haemophilia 2004; 10: 459–469.

24

ANTIBODY PURIFICATION, MONOCLONAL AND POLYCLONAL

JAMES J. REILLY AND MICHIEL E. ULTEE
Laureate Pharma, Inc., Princeton, New Jersey

24.1 INTRODUCTION

The healthcare industry uses monoclonal antibodies (MAbs) extensively in the diagnosis and therapy of disease. Antibody and antibody derivatives are thought to constitute nearly 1/3 of the biopharmaceutical products in development currently (1). In addition, there has been renewed interest in the use of polyclonal derived therapeutics, and the expression of fully human polyclonal antibodies using transgenic animals has recently been reported (2). The efficient and cost-effective manufacturing of antibodies will be critical to the future success of the biotech industry (3–6).

Antibodies or immunoglobulins consist of pairs of disulfide-linked heavy and light chains containing variable (V) and constant (C) domains. There are five different heavy chains (γ, α, μ, δ, and ε) and two different light chains (κ and λ) determining the immunoglobulin classes IgG, IgA, IgM, IgD, and IgE. For more detailed information on the structure and function of immunoglobulins, the reader is directed to the numerous textbooks and reviews that have been written on the subject. This chapter will focus instead on the numerous techniques and approaches to monoclonal antibody purification, with an emphasis on monoclonal IgG purification for biopharmaceutical applications.

Given the dominance of MAbs in the industry, this review will address their purification. This being said, however, there are selected applications for polyclonal antibodies. In particular, they are often used for *in vitro* diagnostics in immunoassays, where their broad selectivity can be advantageous for heterogeneous antigens. The techniques suitable for monoclonal antibody purification are generally applicable to the purification of polyclonal antibodies as well. As the resulting purified antibody will actually be isolated as an immunoglobulin fraction, only some of which will bind the target antigen, antigen-affinity based purification is sometimes employed. The reader is referred to appropriate references on this subtopic, especially those from companies specializing in this technique (7–9).

24.2 APPROACH TO DOWNSTREAM PROCESSING

At small-, or bench-scale, precipitative techniques such as ammonium sulfate, capyrilic acid, and polyethylene glycol (PEG) are still commonly used. These techniques are easy to perform, require little in the way of instrumentation, generally have a concentrating effect and have modest purification power (10–12). The requirement for centrifugation and the need to perform aseptic operations make these approaches less suited for the preparation of biopharmaceuticals, especially in large-scale manufacturing operations (12,15). Very recently, however, some companies are reexamining precipitative techniques as a cost-effective method of primary purification (16). The requirement that therapeutic antibodies reduce contaminants such as DNA, host cell protein (HCP), product-related contaminants, endotoxin,

Downstream Industrial Biotechnology: Recovery and Purification, First Edition. Edited by Michael C. Flickinger.

and virus-like particles to the ppm range (17,18), make precipitative techniques alone insufficient for the production of an industry-accepted product. In contrast, conventional adsorption chromatography of various modes such as affinity, ion exchange chromotography (IEX), hydrophobic interaction chromatography (HIC) has been successfully employed for the production of highly pure MAbs (19,20). The methodology employed for the purification of antibodies, like any protein, depends on the physical characteristics of the antibody (isoelectric point, hydrophobicity, charge density, etc.), the desired yield and purity and the throughput and costs associated with the process (21). The culture media formulation used to grow the antibody-producing cells may also be taken into account when selecting both the modes that will be utilized and also their relative sequencing. Serum-free media (SFM) are preferred for MAb production, as the presence of serum adds complexity to the purification (22), increased risk of viral contamination, and potential variance to the host cell growth and productivity rates (23). Even more desirable from the standpoint of purification are protein-free media, as these only contribute easily removed components of low molecular weight to the feedstock.

The typical model used when constructing a downstream purification process is CIPP or Capture, Intermediate Purification, Polishing. This model is not meant to imply that a purification process contains three chromatographic steps or separations. Instead, it is meant to highlight the common approach to purification. In practice, Intermediate Purification and Polishing are closely related. They are meant to remove the trace impurities that were not removed from the product stream during the product's initial capture from the crude feedstock. Depending on the relative purity of the product after capture, only a single polishing step may be needed. In fact, many process development scientists are actively pursuing two-step process platforms, usually involving an affinity capture and an ion-exchange polishing step. Regardless of how many separations are performed in a process, most of the separations employed today can be grouped into one of several categories, defined by the manner in which they interact with the product or its contaminants.

24.3 AFFINITY CHROMATOGRAPHY

Protein A, G, or L affinity chromatography is a standard technique for the purification of MAbs. Numerous cells and viruses have proteins on their surfaces that bind to the Fc part of immunoglobulins. Designated as Fc receptors, these surface proteins were identified as virulence factors in bacteria and studied extensively because of their implications to pathogenesis. In the time since their initial discovery, they have become invaluable tools in the purification of immunoglobulins. The best known of these Fc receptors are Protein A from *Staphylococcus aureus* and Protein G from *Streptococcus* spp. Although Protein A and Protein G share no sequence homologies, they bind to the same portion of IgG Fc regions. While more versatile than Protein A due to its wider specificity (IgG from different sources), the tighter binding of Protein G can make elution under mild conditions difficult to achieve (24–27).

Protein A affinity chromatography continues to remain the dominant affinity technique, and most frequently employed capture step in the purification of MAbs. The dominance of this particular resin is a consequence of several factors. First and foremost, its selectivity (affinity constant of $10^8/M$) results in highly purified product with exceptional yield. Protein A has been well characterized (28–30). Several groups have published operational characteristics for many of the commercially available protein A resins (dynamic capacities, adsorption isotherms, etc.) (31–34). Protein A is obtainable in large amounts from recombinant bacteria. It is stable over a wide range of pH (2–11) and is able to refold after treatment with urea and guanidinium salts (21).

The use of Protein A is not without its drawbacks. Ligand leakage from the affinity support can occur, necessitating analytical demonstration that the resulting product is free from any leached Protein A (LPA), prior to its therapeutic use in humans. The protein A ligand itself is a bacterial product that must be produced and conjugated to an affinity support matrix, a fairly expensive and time-consuming process that is reflected by its high cost. There is also concern regarding the stability of Protein A resins to cleaning agents. Utilizing an optimized cleaning regime, however, it is possible to reuse Protein A resins up to 300 cycles while maintaining good yield, thereby decreasing the overall long-term cost (35–38). In addition to optimized cleaning and regeneration regimes, an alkaline resistant Protein A resin is available (39). In this Protein A resin, selected asparagine residues have been engineered out of the amino acid sequence to reduce the protein's sensitivity to deamidation under alkaline conditions.

The low pH typically required for antibody elution from a Protein A column, and particularly a Protein G column (pH <3) (12,15,40), has the potential to cause antibody denaturation, aggregation, and loss of biological activity (41). In cases where an antibody is acid sensitive, it is still desirable to use Protein A or G; a number of potential alternative elution conditions have been reported, including the use of basic pH conditions (42), chaotropic ions or basic pH followed by ethylene glycol (43), and binding site mimic peptides (44–46).

Currently, the most common Protein A adsorbents are agarose-based. While these adsorbents generally work well, the relatively large agarose beads are compressible and exhibit slow intraparticle diffusion for macromolecules such as antibodies. There are beaded alternatives, such as those based on porous divinylbenzene polymer, and controlled-pore glass, but agarose-based absorbents remain popular. This is appropriate both to their general lack of matrix-induced interaction with most biomolecules, and to technical improvements to address their compressibility and slower flow properties. The continued dominance of Protein A affinity chromatography will likely result from its well established track record of use in antibody purification, its excellent selectivity, and its ability to meet the demands of high dynamic binding capacity and high throughput to keep pace with increasing bioreactor volumes (20,000 L) and cell-culture expression levels (5 g/L). The newest forms of Protein A supports also allow it to be more vigorously cleaned with sodium hydroxide than earlier versions. The pressure to reduce the cost of biologics such as antibodies, however, will continue to push alternatives to Protein A.

In recent work done by Jungbauer (47–49), many of the commercially currently available Protein A resins were evaluated for their mass transfer characteristics. In terms of dynamic capacity, MabSelect Xtra™, MabSelect Sure™, and ProSep®—vA Ultra had capacities above 30 g/L with 3-min residence times (200 cm/h linear velocity). The current expectation is that cell-culture expression levels will reach the 8–10 g/L within the next 5 years (50). With these capacities, the processing times would be extremely long. Theoretically, the maximum dynamic binding capacity of a Protein A agarose resin is 70 g/L (50). In an effort to close the gap between the 30 g/L and the theoretical 70 g/L being seen nowadays, several approaches are being investigated. These approaches include higher ligand density, particle size, pore size and distribution, and the use of different support matrices. Also being developed are chromatographic methodologies that allow a higher percentage of column load capacity to be utilized—sequential multicolumn chromatography (SMCC) (51) and simulated moving bed chromatography (SMB) (52).

The limitations discussed above and the emerging importance of other classes of immunoglobulins, besides IgG, has prompted research focused on the identification of novel synthetic ligands that could be utilized in place of the traditional Protein A and Protein G ligands. Compared to Protein A and Protein G, these synthetic ligands should display not only reduced production costs at large scale, but also increased resistance to chemical and biological degradation, no leaching of Protein A or G, high capacity for large-scale purification and extended selectivity for other classes of immunoglobulins not recognized by Proteins A or G (53). Several different Protein A mimetics are now on the market with varying selectivity, capacity, and performance. While the use of Protein A and Protein A mimetics are by far the most popular method of affinity chromatography employed currently, several other methods have been reported, including Histidine ligand chromatography (53), metal-affinity chromatography (53–56), thiophilic interaction chromatography (53,57–61), and lectin affinity chromatography (53,62–66). More recently mixed-mode chromatographic media (discussed later in this review), employing, for example, both HIC and ion exchange, have allowed very high selectivity for antibodies rivaling that of Protein A (67–70).

24.4 ION-EXCHANGE CHROMATOGRAPHY

Apart from affinity chromatography, the most common mode of chromatography employed in the purification of MAbs is IEX. The high cost and low throughput associated with Protein A resins have led some development scientists to forgo the use of an affinity capture in place of a cation-mediated capture. Furthermore, most processes still employ one or more IEX steps for intermediate and/or polishing purification. While less specific than the affinity resin Proteins A/G, IEX resins can often provide purities of >80% in one step for feedstocks based on protein-free media, a purity level that is adequate for many purposes (71–74). Given the right combinations of nonaffinity modes (IEX, HIC, etc.) and the relative sequence in which they are employed, it is entirely possible to achieve a high level of purity, without the use of an affinity resin such as Protein A for the initial capture of a monoclonal antibody. An example would be the commercial process for Humira® (75), which utilizes an ion exchange resin for the initial capture and does not use Protein A. There are several advantages to the use of an IEX resin for the initial capture of a monoclonal antibody. The capacities are generally much higher then affinity resins and cleaning is typically much simpler. However, unlike an affinity resin, the clearance of HCP and viral contaminants is much more dependent upon the operational parameters (pH, conductivity, wash and elution buffer composition, residence time, etc.).

In respect to polishing chromatography, both cationic and anionic IEX can be used, in either binding or flow-through modes. The choice of exchanger and mode of operation depends on the relative ionic characteristics of the antibody and the desired impurities to be removed. In a flow-through mode, an anion exchanger efficiently removes nucleic acids, virus and many HCPs. This is done by performing the anionic IEX under mobile-phase parameters such that the antibody product does not bind whereas the contaminants do bind and are therefore removed from the product stream. Alternately, a cation exchanger can be used to capture the product allowing for the removal of contaminants such as residual Protein A, from the column used earlier, or solvents and detergents introduced earlier in the product stream for the purpose of viral inactivation.

Generally, for most human or humanized IgG, the impurities described above could be removed by the sequential execution of anion exchange and cation exchange at neutral pH and at low conductivity (pH range of 5.5–6.5). Nonetheless, care needs to be taken when selecting the parameters that an ion exchanger will be operated under, not only from the standpoint of resolution, but also of capacity. In particular, IEX separations are much more sensitive to the effects of pH, salt concentration, and ionic strength of the chromatographic buffers. Furthermore, if conditions are chosen where both the antibody and impurities bind, the capacity of the resin for antibody could be dramatically reduced. Typically, the first point considered in selecting parameters such as operational pH and conductivity is the isoelectric point (pI) of the antibody. The pI of an antibody, or any protein for that matter, is a guideline to a protein's behavior on IEX. The charge distribution on the surface of the protein and a number of other minor effects also play a role. Nonetheless, pIs are a good first approximation and are usually easily determined using standard techniques or already known.

While the industry trend is to use serum-free and protein-free media formulations during cell culture, some antibody products are purified directly from serum or plasma. Two common plasma contaminants are albumin and transferrin, the two proteins that are stable and abundant in plasma. Albumin has a pI of about 4.5–5.0 (depending on the species) (76,77). The pI of transferrin ranges from 5.5 to 6.0, depending on the species and degree of iron saturation (78). Thus, it is possible to have an antibody that is acidic enough to be difficult to separate from transferrin and even albumin by IEX.

While conventional ion-exchange column chromatography steps are effective and reliable, they suffer the same drawback common to all conventional adsorption chromatography methods and low product throughput (kg processed/hour). The capacity of IEX has come a long way in the last 10 years, but the diffusive path lengths characteristic of the pores of preparative beads or gels are difficult to overcome. Work is being done to improve the rigidity, porosity, and charge density of many IEX resins. Nonetheless, one alternative that slowly gaining industry acceptance is membrane ion-exchange chromatography.

For conventional ion exchange beads, the protein must diffuse into the porous beads due to a concentration gradient. This requirement for diffusion results in a diffusion limited separation, and as a consequence, long processing times. On an ion-exchange membrane, the binding sites are immobilized in the membrane pores. As a result, the mass transfer of the product to the binding sites relies on the bulk convective flow rather than the pore diffusion. In practical terms, this allows the binding capacity of membrane ion exchangers to remain independent of flow rate (79–82) whereas the capacity of conventional adsorption-based chromatography beads is inversely related to flow rate. The reliance on convective flow reduces the resistance to mass transfer, allowing the binding kinetics to dominate the adsorption process.

There are now a number of commercially available large-scale membrane adsorbents of both the anion and cation exchange variety. At this time, the capacities of the ion-exchange membranes, while independent of flow rate, are still much less than the capacities of conventional ion-exchange resins. Nevertheless, the use of an ion-exchange membrane in place of a conventional column has several advantages. First and foremost, as discussed previously, the throughput of the membrane is superior to that of the column. Perhaps, more importantly, this increase in throughput does not come at the cost of decreased capacity. The ion-exchange membranes are disposable, eliminating the need for the cleaning or reuse validation studies that are often required with a conventional column. Unlike a conventional column, an ion-exchange membrane does not need to be packed and qualified prior to use. This translates into savings when it comes to buffer production (labor) and usage (materials). Proper flow distribution in a production scale column requires a minimum bed height (83). As a result, many production columns are oversized, that is, they are sized for speed of processing not capacity. The trade-offs to this approach are a greatly increased cost and complexity for the IEX step.

The current generations of membranes are multilayered, with bed volumes much smaller than a conventional column. The multiple-layer configurations of the modern membrane adsorbers help minimize some of the initial drawbacks evident when they first appeared, some 10–15 years ago. Of the initial concerns, poor inlet flow distribution (84–86), nonidentical membrane pore-size distribution (87–89), uneven membrane thickness, (89) and capacity (84–89), only capacity remains a significant issue. The issue of capacity, however, subsides when the membrane adsorber is used in a flow-through mode, especially when used as a polishing step in the purification process.

Given the high capacity that would be desired in an ion exchanger that is being used in a traditional capture mode, there is no clear advantage in using an ion-exchange membrane over a conventional column. In fact, given the gap between the capacities of the two, a conventional column would likely be the most appropriate choice. The low binding capacity of the membrane adsorbers is attributed to a low surface-to-bed ratio as well as flow distribution problems, both of which are difficult to overcome (90). As mentioned previously, many processes utilize an anion-exchange step, operated in a flow-through mode, principally for the removal of nucleic acids, virus, and HCP. An anion-exchange membrane is well suited to this purpose since the removal of these trace contaminants requires only a small amount of adsorbent. As a result of flow rate limitations and throughput requirements (91), commercial purification schemes often use large anion exchange columns. As an alternative, a small membrane with a breakthrough that is independent of flow rate may be desirable.

24.5 HYDROPHOBIC INTERACTION CHROMATOGRAPHY

Since antibodies tend to be among the most hydrophobic of the proteins in the crude feedstock (92,93), HIC is another technique that can be used for the capture and purification of immunoglobulins. HIC is based on the surface hydrophobicity of biomolecules and the solute–adsorbent interactions that occur in the presence of high concentrations of salts. The product often elutes at salt concentrations that require buffer exchange prior to the next purification step. As a result, HIC purification has not gained the prevalence, in monoclonal antibody purification, that the various IEX modes have. Nonetheless, given the proper optimization, it can be a powerful tool to remove MAb process impurities, particularly aggregates and low molecular weight species related to the product.

Since there is variation in manufacturing, it is difficult to reliably comment on the binding strengths of the various HIC resins. Within an individual manufacture's line, however, usually the longer the alkyl chain length, the stronger the binding. Nevertheless, this relationship needs to be experimentally confirmed with the specific product of interest. Selection of an appropriate HIC media can be somewhat more difficult than the selection of IEX media. The choice of potential binding strengths is very broad. Not only is there supplier-to-supplier variation in ligand density and surface chemistry (also seen in IEX), there are a large number of binding groups to consider (propyl, butyl, phenyl, ether, octyl, phenylether, etc.).

Two parameters of a particular HIC resin that can be modulated are selectivity and retention. In general, the stronger the lyotropic effect of the buffer salt, the stronger the induction of hydrophobic binding (93,94). Ammonium sulfate is the most common salt used for this purpose, but has the disadvantages of outgassing ammonia (above pH 7.5), being corrosive and being difficult to dispose of properly in large quantities. Sodium sulfate and potassium phosphate are also good choices, but are somewhat limited in solubility (93–95). Generally, a pH above 9 and below 4, retention increases probably due to the denaturation of the protein. Solution modifiers such as sugars, alcohols, PEG, and urea can also affect both selectivity and retention (93–95).

The binding of proteins to HIC adsorbents can vary with temperature. This binding is entropy driven, with the van der Waals attraction forces at work in hydrophobic interactions increasing with temperature (96). While this understanding would seem to imply that protein binding to HIC adsorbents is better achieved at higher temperatures, that is not always what is observed. The changes in confirmation and solubility that a protein may undergo in response to increased temperature can work to counter the predicted beneficial binding. For this reason, the effect of temperature on HIC can be complex, and it should not be assumed that what works well at 20°C will work comparably at 2–8°C.

24.6 CERAMIC HYDROXYAPATITE CHROMATOGRAPHY

As discussed earlier, Protein A affinity chromatography is able to clear the majority of HCPs, nucleic acids, endotoxin, and viruses from the product stream. Subsequent to Protein A chromatography, the major impurities and contaminants at issue, in most processes, are product aggregates, HCPs, and LPA. The LPA contamination of the product is introduced by the Protein A resin itself. The aggregate impurities may already be present in the feedstock, prior to chromatography, or caused by the low pH elution conditions often employed during separation. Even if the aggregates are present in the feedstock, the Protein A chromatography can concentrate them with the product. While subsequent anion exchange chromatography can further reduce DNA, endotoxin, and virus contamination, it is not particularly effective at reducing HCP or LPA. Optimization can enhance the ability of an anion exchanger to reduce HCP, but LPA, because of its acidic properties, is much more difficult to clear from the product stream by anion exchange. Cation exchange chromatography is particularly effective at reducing LPA contamination. Nevertheless, if a single ion exchange resin was capable of addressing all of these contaminants, it would be a good candidate for inclusion in a two-stage process platform, or even a third step added for robustness. Work done in the last 10 years has provided a better understanding of the complex interaction between proteins and ceramic hydroxyapatite (CHT), making it an attractive candidate for these roles.

CHT, a mineral form of calcium phosphate, is a mixed mode chromatography support. CHT can theoretically bind protein through anion exchange (with positively charged calcium), by metal affinity (with calcium), by cation exchange (with phosphate), and by hydrogen bonding (to crystal hydroxyl groups). Generally, antibodies (and most proteins) bind to CHT through a combination of metal affinity and phosphoryl cation exchange (97). The metal affinity of proteins for CHT is thought to result from closely neighboring carboxyl groups approximating the carboxyl configuration of chelating agents, such as ethylenediaminetetraacetic acid (EDTA) (98–100). Conceptually, there is another difference between CHT and the commonly used agarose-based ion-exchange resins. Whereas other resins are composed reactive ligands or functional groups attached to a matrix such as agarose, CHT is both the ligand and the matrix.

CHT has been around for more then 20 years, but its acceptance in the industry has been hindered by considerable practical disadvantages. Historically, CHT has been fragile, with column lifetimes considerably below that of conventional ion exchangers. It has suffered from high backpressures, resulting in longer processing times due to linear velocity limitations. The strong, sometime irreversible binding of antibodies has often produced broad, diluted, and elution peaks. The capacities of CHT resins have been poor, in comparison to silica or organic-based resin matrices. Advances in synthesis methods have resulted in a much better quality CHT than was originally available. Modern resins are composed of particles sintered at high temperatures to form a stable, porous, ceramic mass. This results in a macroporous structure that provides a large surface area, limited mass transfer resistance, high mechanical resistance, and resistance to alkaline conditions (101).

CHT's ability to remove product aggregates and LPA from the product stream and its orthogonal reduction in DNA, HCP, endotoxin, and virus, make it a very attractive companion to Protein A chromatography (101). While the capacity of CHT can closely rival that of Protein A when used for polishing, the conductivity of crude feedstocks often reduces this capacity. For lab scale purification, this can easily be compensated for by a dilution of the feedstock. However, at large scale, such dilutions pose problems when considering bioreactor volumes in the thousands, even tens of thousands of liters, that may need to be diluted four or fivefold prior to loading onto a CHT capture column. As a result, CHT may not be a practical alternative to Protein A. However, a new class of resins, mixed mode sorbents, is showing some promise in this regard.

24.7 MIXED MODE CHROMATOGRAPHY

Most single mode chromatography resins contain an element of a separate, second mode; however, mixed mode chromatography (MMC) resins are intentionally designed in this way. These most common combinations seen nowadays are resin-based on both a hydrophobic and ion-exchange mode. This mixed mode capacity allows the resins to use a combination or aromatic, hydrophobic, ionic, and hydrogen-bonding groups. The combinations of these groups may recognize specific targets on the surface of proteins. Both anion- and cation-based MMC resins are now available. The combinatorial modes of these resins provide the potential for clearance of contaminants, including product aggregates, in both the traditional capture and flow-through modes. Like conventional ion-exchange resins, some optimization is required; perhaps, more so for MMC resins. However, the ability to load these resins in moderate to high conductance and elute from them based upon *either* pH or conductance are attractive benefits.

One of the primary disadvantages to HIC has been the requirement for the use of lyotropic salts. Besides the cost associated with production, disposal, and corrosion associated with these salts, the product stream often needs to have these salts removed prior to subsequent ion-exchange steps or formulation. This later concern, however, can be alleviated with good process design (optimal sequencing of chromatographic steps, compatible load/elution conditions for separations adjacent to one another in the process, etc.) and/or the incorporation of diafiltration operations. MMC resins allow a HIC mode to be utilized, without necessarily having to deal with the issues inherent to these salts. In addition, the wider operation range for both pH and conductivity allow for a more efficient and streamlined processing, potentially eliminating the needs for conditioning steps such as dilutions or tangential flow filtration (TFF) operations.

Closely related to MMC resins, but in a class of their own, are the resins used for hydrophobic charge induction chromatography (HCIC). Similar to other MMC resins, these resins utilize both ion-exchange and HIC modes. There are currently only two of them commercially available, MEP HyperCel® and MBI HyperCel® (67–70). These resins are based on a charge inducible ligand that becomes repulsive at low pH, thereby converting *in situ* from an HIC to an AEX mode. Because of the two orthogonal separation methodologies being applied sequentially on the same support matrix, HCIC can achieve similar selectivity to Protein A chromatography, with two important differences. The IgG binding capacity is independent of subclass and elution is performed under less acidic and therefore much milder conditions (pH 4.0), potentially decreasing the likelihood of aggregation or inactivation.

Unlike the MMC resins, where adsorption of the product to the resin may be based on IEX, HIC or both, the adsorption of protein to HCIC resins is primarily based on mild hydrophobic interactions, at near physiological pH, independent of lyotropic salts. Elution is achieved by reducing the pH below that of the pKa (4.8 for MEP HyperCel) of the ionizable ligand. Once ionized, the positively charged ligand repels the bound proteins, which under these conditions are now also positively charged.

Since both the capacity and contaminant clearance that can be achieved using MMC and HCIC resins is dependent on pH and conductivity, the optimization of these resins can be critical. To this end, it is worthwhile that a formal design of experiment (DoE) be performed.

24.8 PURIFICATION OF IgM

Historically, IgM MAbs have been underrepresented in human therapeutic development due to production difficulties and other manufacturing constraints largely resulting from their molecular size (discussed later). IgMs are much less stable than IgG. They are prone to aggregation at low temperature, pH extremes and low conductivity (below physiologic conductivity). Native IgM exists as a pentamer or hexamer of IgG monomers and is heavily glycosylated (7.5–12%) relative to IgG (~2%). In addition to the five IgG subunits, pentameric IgM contains a J chain. Found only in IgM and IgA, the J chain was once thought to be a requirement for the polymerization of the IgG subunits (IgA is a dimer). This has subsequently been shown to be untrue and it is now widely believed that the J chain is required for secretion of the antibodies into the mucosal tissues.

Pentameric (the more common) and hexameric IgM are large proteins (0.96 Mda and 1.15 Mda respectively). As a result, they have much slower diffusion constants than IgG (2.6×10^{-7} for IgM compared to 4.9×10^{-7} for IgG) (102). The slower diffusion constant has plagued purification efforts, resulting in reduced capacities and diminished resolution, particularly with older resins that rely on diffusion for mass transport. IgM also has no affinity for Proteins A/G, leaving process development scientists without the affinity mode that has come to dominate monoclonal IgG purification. Over the years, several affinity approaches have been developed but none have risen to a level where they are accepted as a true analog to Protein A. Most early approaches at purification have been based on the reversible cryoprecipitation (103,104), preparative scale SEC (or Gel Filtration), and IEX. HIC chromatography can substitute for IEX, but carries the risk of inducing precipitation due to the changes in conductance required for this mode of separation. SEC, while gentle, suffers from low throughput and manufacturing limitations at scale due to load volume limitations. These limitations can be compensated for but can result in larger capital costs (e.g. the use of multiple small columns in series). The polymeric nature of IgM results in a molecule that is highly charged, when compared to IgG. However, the size, charge, and heavy glycosylation of IgM make the selection and optimization of an IEX resin more difficult. Complimenting the chromatographic separation difficulties, the large size of IgM makes it more susceptible to shear forces and aggregation during the filtration operations.

Advances in recent years have the potential to alleviate some of the difficulties in IgM purification. As discussed earlier, CHT resins have been vastly improved over the years. CHT chromatography allows for the capture and elution of IgM under relatively mild conditions with minimal sample modification. Optimization of both NaCl and phosphate in the elution buffer has been shown to enhance the clearance of product aggregate, DNA, HCP, endotoxin, and virus (105). Similarly, the inclusion of macromolecular crowding agents such as PEG can also enhance clearance of contaminants (106). The development of membrane-based ion exchanges also hold promise for IgM purification. As mentioned earlier, the low diffusion constant of IgM creates problems for diffusion-based separation modes, such as those performed on particle-based IEX resins. Membrane ion exchangers are not diffusion based, but convection based. Similar to the approach taken with IgG, these membrane ion exchangers can be used in either a capture or flow-through mode. However, just as with the purification of IgG, use in a capture mode is not particularly effective in regards to capacity and resolution. This is in part because the capacities of membrane ion exchangers are much lower in comparison to particle-based ion exchangers (discussed previously). Poor resolution subsequent to capture is a result of a single chromatographic plate, often within a large housing, that results in uneven flow distribution and mixing. When used in a flow-through mode for contaminant removal during IgM purification, membrane ion exchangers show great promise.

Lastly, monolithic ion exchange resins have great potential to replace particle-based IEX resins in regards to IgM purification. These resins, sometimes referred to as *continuous bed* or *convection interaction media* (107), can be made of synthetic organic or inorganic materials. The monolithic resin matrix is typically a sponge-like, interconnected, open pore (50–300 μ) structure. This monolithic matrix has readily accessible surfaces on to which the ion-exchange functional groups diethylaminoethyl (DEAE), quaternary ammonium (Q), carboxylmethyl (CM) and sulfopropyl (SP) are easily immobilized. Similar to membrane ion exchangers, mass transport in monolithic resins is primary based on convection rather than diffusion. There is a very small amount of pore diffusion (108–110) as well as some film diffusion (111,112) that offers resistance to mass transfer. Nonetheless, they offer a good compromise between the throughput of a membrane ion exchanger and the capacity of particle-based beads. In regard to large macromolecules with low diffusion constants (such as IgM), they have clear advantages over particle-based resins.

24.9 PLATFORM PROCESSES

As mentioned in the introduction to this chapter, the market for MAbs is increasing rapidly, and expected to continue to grow in the future. Most monoclonal antibody therapeutics have doses in the gram range, creating an enormous demand to increase production. The increase in cell titers from milligram/Liter to multigram/Liter ranges has been the most significant contribution toward meeting this demand. There is also great pressure to reduce the high cost of biopharmaceuticals and MAbs, with their high dosage, which are the natural focus of efforts to drive down the costs from the current $1000/g to $100/g (113). In order to meet these demands, the productivity and efficiency of downstream processes will need to increase and keep pace with the increasing titers. One approach to meeting these demands is the adoption of process platform technologies.

A process platform technology is a process that is designed to take advantage of the similar characteristics of a class of potential therapeutics. MAbs, while all unique in their antigen-binding sites, are still members of the immunoglobulin family of homologous proteins. With respect to IgG, the already established dominance of Protein A in the initial capture of MAbs, lends itself well to a process platform. By adopting a process platform, scientists can decrease the costs and time associated with process development, decrease the costs and time needed to enter clinic, and as a result, more efficiently screen potential therapeutics. Validation, materials control, column life cycle, and reuse studies and quality assurance activities can potentially be simplified as well. Platform processes *per se* do not address high titer feedstocks, but in designing them, one of criteria can be selection of component steps that offer high capacity and throughput.

There is a danger in becoming comfortable with the idea of "one shoe fits all" and platform process design should be undertaken and applied carefully. One of the benefits to adopting process platforms is the amount of data that can be generated across multiple products in regards to the performance of a process. These data could then be used for improvements to the platform.

24.10 CONCLUSION

Antibody purification has evolved enormously from the early blood-fractionation procedures to the modern chromatographic and filtration-based processes achieving high yields and purity at ppm levels of impurities. The growing dominance of antibodies and antibody-based molecules in biopharmaceuticals, along with antibodies' homologous structure and properties, have encouraged process development scientists to continue to improve their processes. Manufacturers of separation materials have correspondingly improved their products to provide increased capacity and selectivity, higher linear velocities, and enhanced stability. Continuing challenges of high titer feedstocks, pressures to reduce costs, and more rapid process development are being actively addressed by process development scientists throughout the biopharmaceutical industry, academia, and government institutions.

REFERENCES

1. Glennie MJ. Immunol Today 2000; 21: 403–410.
2. Robl JM, Kasinathan P, Sullivan E, Kuroiwa Y, Tomizuka K, Ishida I. Theriogenology 2003; 59: 107–113.
3. Reichert J, Pavlou A. Nat Rev Drug Discov 2004; 3: 383–384.
4. Dutton G. Genet Eng News 2006; 26(14): 1.
5. Monoclonal Antibody Therapies. Entering a new competitive era. Minnetonka (MN): Arrowhead Publishers; 2004.
6. Molowa DT. The state of biologics manufacturing: part 2. New York, NY: JP Morgan Securities Equity Research; 2002.
7. Newcombe C, Newcombe AR. J Chromatogr B Analyt Technol Biomed Life Sci 2006; 168(93): 686.
8. Pierce Chemical Company. Web information. Available at http://www.piercenet.com/Objects/View.cfm?Type=Page&ID=84DF5176-0BE6-40CC-B0D8-C0768D6BFDF1. Accessed 2008 Dec 15.
9. KPL Protein Research Products. Web information. Available at http://www.kpl.com/docs/techdocs/How%20KPL%20Purifies%20Its%20Antibodies.pdf. Accessed 2008 Dec 15.
10. Harlow E, Lane D. Antibodies, a laboratory manual. Cold Spring Harbor (NY): Cold Spring Harbor Laboratory; 1988. pp. 283–318.
11. Polson A, von Wechmar MB, van Regenmortel MHV. Immunol Commun 1980; 9(5): 475–493.
12. Neoh SH, Gordon C, Potter A, Zola H. J Immunol Methods 1986; 91: 231–235.
13. Phillips AP, Martin KL, Horton WH. J Immunol Methods 1984; 74: 385–393.
14. Thompson PW, Kenney AC, Moulding P, Wormald D. Ann N Y Acad Sci 1990; 529: 529–539.
15. Fraser CM, Lindstrom J. The use of monoclonal antibodies in receptor characterization and purification. In: Venter JC, Harrison LC, editors. Receptor biochemistry and methodology. New York: Alan R. Liss; 1984. pp. 1–30.
16. Glynn J. Biopharm Int 2008; Suppl: 12–16.
17. Jiskoot W, Van Hertrooij JJ, Hoven AM, Klein-Gebbinick JW, Van der Velden-de Groot T, Crommelin DJ, Beuvery EC. J Immunol Methods 1991; 138: 273–283.
18. Wolter T, Richter A. Bioprocess Int 2005; 3: 40–46.
19. Necinam R, Amatschek K, Jungbauer A. Biotechnol Bioeng 1997; 60: 689–698.
20. Graf H, Rabaud JN, Egly JM. Bioseparation 1994; 4: 7–20.
21. Huse K, Bohme H-J, Gerhard H, Scholtz J. J Biochem Biophys Methods 2002; 51: 217–231.
22. Broedel SE, Papciak SM. Bioprocess Int 2003; 1: 56–58.
23. Fletcher T. Bioprocess Int 2005; 3: 30–36.
24. Björck L, Kronvall G. J Immunol 1984; 133(2): 969–974.
25. Surolia A, Pain D, Khan MI. Trends Biochem Sci 1982; 7: 74–76.

26. Walker BW. Use of immobilized protein G to isolate IgG. In: Boyle MDP, editor. Volume 2, Bacterial immunoglobulin binding proteins. San Diego (CA): Academic Press, Inc.; 1990. pp. 255–368.
27. Äkerstrom B, Bjorck L. J Biol Chem 1966; 261(22): 10240–10247.
28. Langone JJ. Adv Immunol 1982; 32: 157–252.
29. Langone JJ. J Immunol Methods 1982; 55: 277–296.
30. Deisenhofer J. Biochemistry 1981; 20: 2361–2370.
31. McCue JT, Kemp G, Low D, Quinones-Garcia I. J Chromatogr A 2003; 989: 139.
32. Fahrner RL, Iyer HV, Blank GS. Bioprocess Eng 1999; 21: 289.
33. Fahrner RL, Whitney DH, Vanderlaan M, Blank GS. Biotechnol Appl Biochem 2003; 30: 121.
34. Hahn R, Schlegel R, Jungbauer A. J Chromatogr B 2003; 790: 35.
35. Healthcare GE. Downstream 39, 2005.
36. Brorson K, Brown J, Hamilton E, Stein KE. J Chromatogr A 2003; 989: 155.
37. Hale G, Drumm A, Harrison P, Phillips J. J Immunol Methods 1994; 171: 15–21.
38. Francis R, Bonnerjea J, Hill CR. In: Pyle DL, editor. Separations for biotechnology 2. London: Elsevier; 1990. pp. 491–498.
39. Linhult M, Gulich S, Graslund T, Simon A, Karlsson M, Sjoberg A, Nord K, Hober S. Proteins 2004; 55: 407.
40. Villemez CL, Russell MA, Carlo PL. Mol Immunol 1984; 21(10): 993–998.
41. Gulich S, Uhlen M, Hober S. J Biotechnol 2000; 76: 233–244.
42. Croze EM. Eur patent Appl 453,767. 1991 Oct 30. (to E.R. Squibb and Sons, Inc., now Bristol-Myers Squibb, Inc.).
43. MacKenzie AP. Therapeutic peptides and proteins: formulation, delivery and targeting. Cold Spring Harbor (NY): Cold Spring Harbor Laboratory; 1989. pp. 17–21.
44. Bywater R, Eriksson G, Ottosson T. J Immunol Methods 1983; 64: 1–6.
45. Bywater R. Chromatogr Synth Biol Polym 1978; 2: 337–340.
46. Yarnell M, Boyle MDP. Biochem Biophys Res Commun 1986; 135(3): 1105–1111.
47. Hahn R, Schlegel R, Jungbauer A. J Chromatogr B Analyt Technol Biomed Life Sci 2003; 790: 35–51.
48. Hahn R, Bauerhansl P, Shimahara K, Wizniewski C, Tscheliessnig A, Jungbauer A. J Chromatogr A 2005; 1093: 98–110.
49. Hahn R, Shimahara K, Steindl F, Jungbauer A. J Chromatogr A 2006; 1102: 224.
50. Low D, O'Leary R, Pujar NS. J Chromatogr B 2007; 848: 48–63.
51. MCC information. Available at www.novasep.com. Accessed 2008 Dec 15.
52. SMB information. Available at www.tarponbiosystems.com. Accessed 2008 Dec 15.
53. Fassina G, Ruvo M, Palombo G, Verdoliva A, Marino M. J Biochem Biophys Methods 2001; 49: 481–490.
54. Porath J. Protein Expr Purif 1992; 3: 263–281.
55. Hale JE, Beidler DE. Anal Biochem 1994; 222: 29–33.
56. Boden V, Winzerling JJ, Vijayalakshmi M, Porath J. J Immunol Methods 1995; 181: 225–232.
57. Zachariou M, Hearn MTW. J Chromatogr A 2000; 890: 95–116.
58. Porath J, Maisano F, Belew M. FEBS Lett 1985; 185: 306–310.
59. Maisano F, Belew M, Porath J. J Chromatogr 1985; 321: 305–317.
60. Belew M, Juntti N, Larsson A, Porath J. J Immunol Methods 1985; 102: 305–317.
61. Oscarsson S, Porath J. Anal Biochem 1989; 176: 330–337.
62. Robertson ER, Kennedy JF. Bioseparation 1996; 6: 1–15.
63. Peng Z, Arthur G, Simons E, Becker AB. Vet Immunol Immunopathol 1993; 36: 83–88.
64. Roque-Barreira MC, Campos-Neto A. J Immunol 1985; 134: 1740–1743.
65. Roque-Barreira MC, Praz F, Halbwachs-mecarelli L, Greene LJ, Campos-Neto A. Braz J Med Res 1986; 19: 149–157.
66. Kabir S. J Immunol Methods 1998; 212: 193–211.
67. Schwartz W, *et al*. J Chromatogr A 2001; 908: 251.
68. Ferreira GM, Dembecki J, Patel A, Arunakumari A. Biopharm Int 2007: 32–43.
69. Guerrier L, *et al*. J Chromatogr B 2000; 755: 37.
70. Boschetti E. J Biochem Biophys Methods 2001; 49: 361.
71. Gemski MJ, Doctor BP, Gentry MK, Pluskal MG, Strickler MP. Biotechniques 1985; 3(5): 378–384.
72. Deschamps JR, Hildreth JEK, Derr D, August JT. Anal Biochem 1985; 147: 451–454.
73. Clezardin P, McGregor JL, Manach M, Boukerche H, Dechanvanne M. J Chromatogr 1985; 319: 67–77.
74. Pavlu B, Johansson U, Nyhlén C, Wichman A. J Chromatogr 1986; 359: 449–460.
75. Turner B. BioLOGIC USA 2003, Boston, MA, U.S.A., 2003.
76. Hanna LS, Pine P, Reuzinsky G, Nigam S, Omstead DR. Biopharm 1991: 33–37.
77. Righetti PG, Caravaggio T. J Chromatogr 1976; 127: 1–28.
78. Baker EN. Perspect Bioinorg Chem 1993; 2: 161–205.
79. Kubota N, Konno Y, Miura S, Saito K, Sugita K, Wantanabe K, Sugo T. Biotechnol Prog 1996; 12: 869.
80. Dancette OP, Taboureau J, Tournier E, Charcosset C, Blond PJ. Chromatogr B 1999; 723: 16.
81. Gerstner JA, Hamilton R, Cramer SM. J Chromatogr 1992; 596: 173.
82. Camperi SA, Navarro del Canizo AA, Wolman FJ, Smolko EE, Cascone O, Grasselli M. Biotechnol Prog 1999; 15: 500.
83. Yuan QS, Rosenfeld A, Root TW, Klingenberg DJ, Lightfoot EN. J Chromatogr A 1999; 831(2): 149–165.
84. Roper DK, Lightfoot EN. J Chromatogr A 1995; 702(1–2): 3–26.
85. Reif OW, Freitag R. J Chromatogr A 1993; 654(1): 29–41.
86. Lightfoot EN, Coffman JL, Lode F, Yuan QS, Perkins TW, Root TW. J Chromatogr A 1997; 760: 139–149.

87. Frey DD, Walter RV, Zhang B. J Chromatogr A 1992; 603(1–2): 43–47.

88. Sauer PW, Burky JE, Wesson MC, Sternhard HD, Qu L. Biotechnol Bioeng 2000; 67(5): 585–597.

89. Ghosh R. J Chromatogr A 2002; 952(1–2): 13–27.

90. Zhou JC, Tressel T. Biotechnol Prog 2006; 22: 341–349.

91. Knudsen HL, Fahrner RL, Xu Y, Norling LA, Blank GS. J Chromatogr A 2001; 907: 145–154.

92. Gagnon P, Cartier PG, Maikner JJ, Eksteen R, Kraus M. LC GC 1993; 11(1): 26–34.

93. Gagnon P, Grund E. Biopharm 1996: 54–64.

94. Fausnaugh JL, Kennedy LA, Regnier FE. J Chromatogr 1984; 317: 141–155.

95. Arakawa T, Narhi LO. Biotechnol Appl Biochem 1991; 13: 151–172.

96. Troy DB, Beringer P. The science and practice of pharmacy. Philadelphia: Lippincot Williams and Wilkins; 2005. p. 620.

97. Gagnon P, Ng P, Zhen J, Aberinm C, Hie J, Mekosh H, Cummings L, Zaidi S, Richieri R. Bioprocess Int 2006: 50–60.

98. Gorbunoff M. Anal Biochem 1984; 136(2): 425–432.

99. Gorbunoff M. Anal Biochem 1984; 136(2): 423–439.

100. Gorbunoff M. Anal Biochem 1984; 136(2): 440–445.

101. Gagnon P, Ng P, Aberrin C, Zhen J, He J, Mekosh H, Cummings L, Richieri R, Zaidi S. Bioprocess Int 2006; 4(2): 50–60.

102. Gagnon P, *et al*. Recent advances in the purification of IgM monoclonal antibodies. 3rd Wilbio Conference on Purification of Biological Products; 2007; Waltham, MA. See www.validated.com for a copy of this and related presentations.

103. Middaugh CR, *et al*. Proc Natl Acad Sci U S A 1978; 75: 3440.

104. Middaugh CR, *et al*. J Biol Chem 1978; 255: 6532.

105. Gagnon P, *et al*. Practical issues in the industrial use of hydroxyapatite for purification of monoclonal antibodies. 232nd Meeting of the ACS; 2006; San Francisco, CA.

106. Gagnon P, *et al*. Nonionic polymer enhancement if aggregate removal by ion exchange and hydroxyapatite chromatography. 12th Waterside Conference; 2007; San Juan, Puerto Rico.

107. Cabrera K, *et al*. J High Resolut Chromatogr 2000; 23: 106.

108. Unarska M, *et al*. J Chromatogr A 1990; 519: 53.

109. Thommes J, Kula M. Biotechnol Prog 1995; 11: 357.

110. Belenkii BG, Maltsev VG. Biotechniques 1995; 18: 288.

111. Josic D, Starncar A. Ind Eng Chem Res 1999; 38: 333.

112. Iberer G, Hahn R, Jungbauer A. LC GC 1999; 17: 998.

113. Farid S. Adv Biochem Biotechnol 2006; 101: 1–4.

25

CHROMATOGRAPHIC PURIFICATION OF VIRUS PARTICLES

PETE GAGNON
Validated Biosystems, San Clemente, California

25.1 INTRODUCTION

Virus particles of a given species, like other complex biomolecules, exhibit unique characteristic ranges of physicochemical characteristics, some of the most fundamental of which include specific numbers of positive and negative charges, distribution of charges, and their individual titration characteristics; also numbers and distribution of surface aliphatic and aromatic hydrophobic residues; and size. As such, each virus species represents a defined entity that can be discriminated from other such entities by tools with the ability to fractionate molecules on the basis of these features. Chromatography media were conceived from the beginning for this type of application, and have evolved collectively into a sophisticated toolbox with versatile and powerful capabilities for biomolecule purification. Traditional chromatography media have mostly been optimized for purification of molecules much smaller than viral particles, chiefly proteins, and it has been a fortunate accident that they have proven to have utility for virus purification. They have, nevertheless, served to demonstrate that virus particles can be fractionated by the same chemical mechanism as proteins, and have thus provided a bridge to a wealth of pertinent process development, manufacturing, and regulatory experience already in place with therapeutic proteins. With a new generation of virus-optimized media now available, researchers and commercial process developers are better equipped than ever before to exploit the full capabilities of chromatography for virus purification. This chapter focuses on the practical issues of selecting the chromatography media most appropriate for purification of virus particles, and development of multistep purification procedures capable of meeting the research, economic, and regulatory requirements of this rapidly expanding field.

25.2 CHROMATOGRAPHIC SEPARATION METHODS

25.2.1 Size Exclusion Chromatography (SEC)

Size exclusion chromatography (SEC) is used widely in the field of virus purification (1–14) and deserves special recognition for several reasons, first because of its selectivity. With the exception of biospecific affinity chromatography, SEC offers more effective removal of proteins, DNA fragments, and other "small" molecules from virus particles than any other single method. But in contrast to affinity, for which a given ligand is applicable to only a single species, SEC is universal: it can be used for purification of all virus species. Second, the separation mechanism is independent of buffer composition. This makes SEC unique in the field of separation methods. It endows it with the flexibility to accommodate samples without respect to buffer composition, and to elute fractionated virus particles in a wide range of buffer formulations. An associated benefit is that it can support fractionation under near-physiological conditions that favor conservation of virus viability (1–8). Finally, SEC deserves recognition

Downstream Industrial Biotechnology: Recovery and Purification, First Edition. Edited by Michael C. Flickinger.

because it is the one chromatography technique for which protein-optimized media are ideally suited to virus purification. Protein-optimized SEC media are arguably even better suited to purification of virus particles than to proteins.

SEC is performed on columns packed with porous particles. The particles of a given SEC medium have a defined range of pore sizes. The largest pores define the *exclusion limit*. Solutes too large to enter the largest pores, such as virus particles, are said to be *excluded*. They are able to flow only through the space between particles, the *void volume*. Excluded solutes elute earliest from the column, in the so-called void peak. Solutes that are able to enter some of the pores in the media, proteins for example, are said to be *included*. The time it takes for included solutes to enter and exit the pores causes them to elute later than the void fraction. Smaller solutes are able to enter a larger proportion of the pores and elute later. Very small solutes such as salts and sugars are able to enter all the pores and elute last. Figure 25.1 is a generalized SEC elution profile illustrating the distribution of virus particles and smaller solutes. Figure. 25.2 illustrates SEC fractionation of influenza H1N1 virus following capture by anion exchange chromatography.

The mechanisms by which solutes are transported through the column are critical to the most effective use of the technique. Included solutes enter and exit the particle pores by a process called *diffusion* (15). Diffusion can be defined as random thermal movement from an area of high concentration to an area of low concentration. Its hallmark is that it is slow, and increasingly so for large solutes (Table 25.1). As a consequence, when SEC is used to fractionate among included solutes, for example, among proteins of varying size, the flow rate must be slow enough to permit them to diffuse into and out of the pores, in equilibrium with the buffer as it passes down the column. If the flow rate is too fast, a subpopulation of a given solute will not have time to enter all of the available pores and will elute in advance of the main population. Another subpopulation will not have time to exit the pores in equilibrium with the main population and will elute later. This is why increasing flow rate in SEC creates wider, more dilute peaks that are more poorly resolved from one another (15). They reflect the inefficiency of diffusive mass transport.

Solutes restricted to the void volume move primarily by a process called *convection*. Convection can be defined as

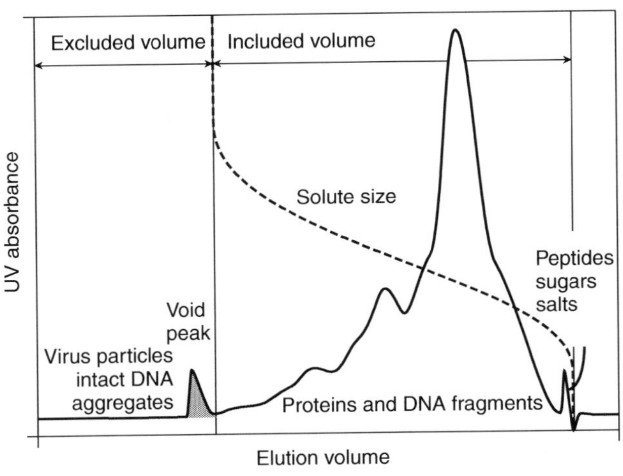

Figure 25.1. Generalized hypothetical SEC elution profile of virus-containing cell culture supernatant. The aggregates in the void peak may include viral aggregates, protein aggregates, or heteroaggregates. The dashed curve shows relative solute size distribution throughout the profile. Partial solute size distribution curves are usually provided by vendors, but they typically illustrate only the linear portion within the included volume. Exclusion limits are often not defined but can be inferred from the fractionation range suggested by the manufacturer. Refer to the text, Refs 7, 15, and 16 for additional discussion.

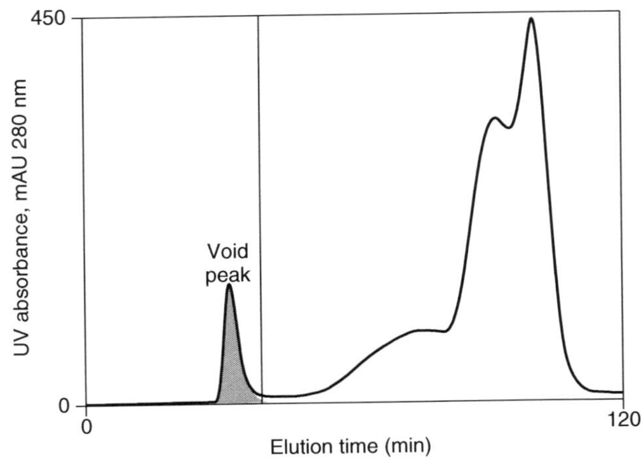

Figure 25.2. SEC of influenza virus H1N1 on Sepharose™ CL4B following initial purification by anion exchange chromatography (Fig. 25.9). The virus-containing fraction is highlighted in gray. Refer to Ref. 17 for additional information, [Redrawn from Ref. 17, with permission].

TABLE 25.1. Diffusion Constants for Selected Solutes

Solute	Size	K_{diff}
IgG, light chain	23 kDa	9.1×10^{-7}
BSA	66 kDa	6.7×10^{-7}
IgG	150 kDa	4.9×10^{-7}
Urease	480 kDa	3.5×10^{-7}
IgM	960 kDa	2.6×10^{-7}
ETX	2 MDa	2.1×10^{-7}
CMV	5 MDa	1.2×10^{-7}
TMV	40 MDa	5.0×10^{-8}
DNA_1	4.4 kbp	1.9×10^{-8}
DNA_2	33.0 kbp	4.0×10^{-9}

ETX, endotoxin; CMV, cucumber mosaic virus; TMV, tobacco mosaic virus.
Diffusion constants in square centimeters per second.

movement induced by an external force, in this case the flow of buffer through the column. Diffusion also occurs in the void volume but contributes mainly to random mixing and not to the passage of solutes down the column. The efficiency of convective mass transport is unaffected by either flow rate or molecular size of solutes, which means that the composition and concentration of the void peak is likewise unaffected. To the extent that virus particles are restricted to the void volume, flow rates can be significantly faster than employed in SEC of proteins. Theoretically, the flow rate could be increased to the pressure tolerance limits of the column but, in practice, flow rate remains partially restricted by the diffusional limitations of included solutes. As noted above, when flow exceeds a velocity that permits an included solute to achieve full diffusional equilibrium, a subpopulation of that solute is unable to enter the pores and elutes earlier than the main population. The faster the flow rate, the larger the subpopulation that elutes prematurely. For included solutes that elute close to the void volume at ideal flow rates, this manifests as spill-over into the void volume, reducing the purity of the virus fraction. This leads to the practical recommendation that the most effective SEC media for purification of a given virus species will be those with the highest exclusion limits that do not include the target virus. Such media reduce the population of solutes that elute close to the void volume. Accordingly, the potential for spill-over of normally included solutes into the void volume is reduced, and the flow rate can be increased by a larger increment. The extent to which the flow rate can be increased can only be determined experimentally. A linear flow rate of 50 cm/h is a reasonable place to start, which is at least double the flow rate commonly used for preparative SEC of proteins. Increase it if possible; reduce it if necessary; according to the ability of other steps in the process to remove spill-over contaminants.

It is also important to appreciate how resolution relates to column bed height. For included solutes, resolution is inversely related to the square root of column height (15). This explains why SEC of proteins is most commonly conducted in columns 60–100 cm high. High resolution fractionation of included solutes is not the objective of virus applications, however. The objective is separation of the excluded volume (void peak) from the included volume, and this parameter is much less sensitive to bed height. Thus, the height of SEC columns for virus purification can be reduced to 10–30 cm, depending on the needs of a particular application (7,18). A shorter column of the same volume will have a larger diameter and inlet surface area, which means that even if linear flow rate is held constant, the volumetric flow rate will increase linearly with the increase in surface area. This decreases process time and increases overall productivity.

The third important factor in performance of SEC separations is column loading. When conducting separations among included proteins, column loads are seldom higher than 5% of the column volume (CV), and often much lower. This is Strike 3 for most developers of protein purification processes. The combined burden of slow linear flow rate, tall columns, and low column loads becomes economically prohibitive (19). SEC columns for virus fractionation can theoretically be loaded up to 40% of the bed volume, which corresponds to the void volume in packed particle columns. If the plan is also to increase linear flow rate and reduce bed height, it will be prudent to reduce actual loading to be within the range of 10–30% (7,18). Even with this reduction, the combined productivity improvement of faster flow rates, shorter beds, and higher loading may exceed 10 times the efficiency of protein applications, making SEC far more attractive for virus purification than it is for proteins.

The relative inertness of SEC to buffer conditions enables two valuable enhancements. The most well known is its ability to conduct buffer exchange coincident with fractionation (15). The virus fraction elutes in the buffer to which the column is equilibrated. The small molecules of buffers and salts in which the sample was loaded elute much later, at the end of the included volume. While compelling, this application is not as broadly applicable as could be hoped. The perfect SEC medium would be absolutely inert to chemical interactions with solutes. In fact, most SEC media are weakly hydrophobic, most include negatively charged carboxyl or sulfo groups, some include positively charged groups, and all are hydroxyl rich (15,20–23). These groups pose all manner of opportunities for nonspecific interactions between solutes and SEC media. At low salt concentrations, charged groups can act as ion exchangers, which may bind or retard the virus, and partially or wholly prevent its elution in the void volume (15,24). Nonspecific charge interactions may also force smaller solutes into the void volume. For example, negatively charged DNA fragments will be repelled from negatively charged media surfaces at low conductivity, despite being small enough to enter the pores and elute in the included volume. This phenomenon is known as *ion exclusion*, and probably accounts for many instances of small DNA fragments unexpectedly appearing in void peaks. Charge-based nonspecific retention or exclusion can usually be overcome by increasing the salt concentration (15,25,26). This leaves the sample less prepared for direct application to a subsequent ion exchange step, but modest dilution will usually be sufficient to restore forward compatibility. Arginine has recently been shown to substantially improve resolution and recovery in SEC of proteins, and may be expected to behave similarly with virus applications (27,28). It is a strong hydrogen donor/acceptor capable of suspending hydrogen bonds, and its guanido side group weakens hydrophobic interactions (HIs).

The other important enhancement relates to contaminants that may be bound to virus particle surfaces. Recent publications on protein A affinity chromatography have revealed that most of the contaminants that co-elute with IgG do so because they form stable complexes with the product during cell culture production (29,30). DNA liberated from lysed cells acts as a liquid phase cation exchanger (CX). More than 1 *M* sodium chloride may be required to dissociate DNA:product complexes. Histone proteins, also liberated from lysed cells, can act as liquid phase anion exchangers (AXs), and while they may have a binding preference for DNA, they bind other proteins as well and may require as much as 1.5 *M* sodium chloride for dissociation. Product complexation with other contaminants may also occur via hydrophobic interactions, hydrogen bonding, or metal affinity interactions. These complexes can persist throughout the course of an entire purification process, imposing an artificial ceiling on the level of final product purity that can be obtained (29,30). There is no reason to presume that virus particles should be immune from this phenomenon. Given their relative complexity and size, they should be expected to be even more prone to transporting hitchhikers through a process than proteins, and indeed some studies show that virus complexation with DNA can be strong enough to create large aggregate masses that severely complicate purification (31–33). As with proteins, these complexes can often be dissociated with sodium chloride. Hydrogen bonding can be suspended by inclusion of strong hydrogen donors and/or acceptors, such as urea or arginine. Both also weaken hydrophobic interactions, as do protein-stabilizing organic additives such as propylene glycol or low molecular weight polyethylene glycol, such as PEG-400, or those smaller. Use of decomplexant washes with bioaffinity has been observed to improve host cell protein clearance by a factor of 10 or more (29,30), but this is less important than the fact that the contaminants they displace may otherwise persist throughout an entire purification process. SEC does not support the application of washes per se, but there are two variations that support the same benefits without sacrificing buffer exchange into whatever formulation is desired. One is to add decomplexants directly to the sample. The other is to preload the column with a decomplexant buffer. The combined volume of the decomplexant and sample must be small enough not to encroach on the void volume. Otherwise, buffer exchange capability will be compromised. The former strategy allows a larger volume of sample to be applied to the column but direct addition of decomplexants to the sample will dilute it somewhat. The latter strategy avoids dilution but sample application volume will be reduced by the volume of previously applied decomplexant. With either strategy, the decomplexants are included in all of the media pores and pass slowly down the column. Virus particles pass through the decomplexant zone as they migrate more rapidly through the void volume, and elute in whatever buffer the column was equilibrated to.

The ultimate irony of using SEC for virus purification is that it generally does not support size-based fractionation of intact monodisperse virus particles from partially assembled capsids, degradation fragments, or viral aggregates. Recent work with proteins has suggested that for diffusive mass transport to be unrestrictive, the pore size should be at least 10 times larger than the solutes to be purified (34). This becomes problematical with viruses because extremely large pore lumens come at the expense of thinner pore walls. Thinner walls reduce particle stability. Even if structural stability could be assured with an appropriately robust polymer, the dramatic flow rate reductions required to support effective diffusive mass transport of virus particles would make separation times prohibitively long. SEC separation times typical of protein applications would be expected to compromise resolution and virus recovery. This probably explains the poor recovery of Lily symptomless virus from Sephacryl™ S-100 SF in comparison with Superdex™ 200 HR (35). In addition, maximum resolution would require sacrificing the benefits of shorter columns and higher sample loads that make SEC attractive for virus purification in the first place.

25.3 ADSORPTION CHROMATOGRAPHY

Adsorption chromatography includes techniques such as ion exchange, hydrophobic interaction (HIC), hydroxyapatite (HA), and affinity chromatography. Chromatography media for conducting adsorptive methods are available on a wide diversity of solid phase supports, including porous particles, membranes, and monoliths. Porous particle-based media dominate, mostly because the field has evolved to support the needs of protein purification, but in contrast to SEC, where protein-optimized particles support highly efficient virus purification, porous particles impose serious limitations on the effectiveness of adsorptive methods. The pore size distribution of most particle-based adsorption media is in the range of 60–100 nm (18). This supports high protein binding capacities at reasonably rapid flow rates, but many viral particles are too large to diffuse into such pores and are limited to binding on the external particle surfaces (Table 25.2, Fig. 25.3). This severely truncates virus binding capacity. Smaller virus particles may have access to a small proportion of media pores, but suffer from highly limited pore access due to diffusional limitations. Published protein binding capacities for porous particles therefore, overestimate virus binding capacities. The relationship of solute size to capacity is represented in Table 25.3. Diffusion of eluted virus particles out of the pores suffers the same limitations but with other negative consequences, including peak broadening, loss of resolution, and depressed

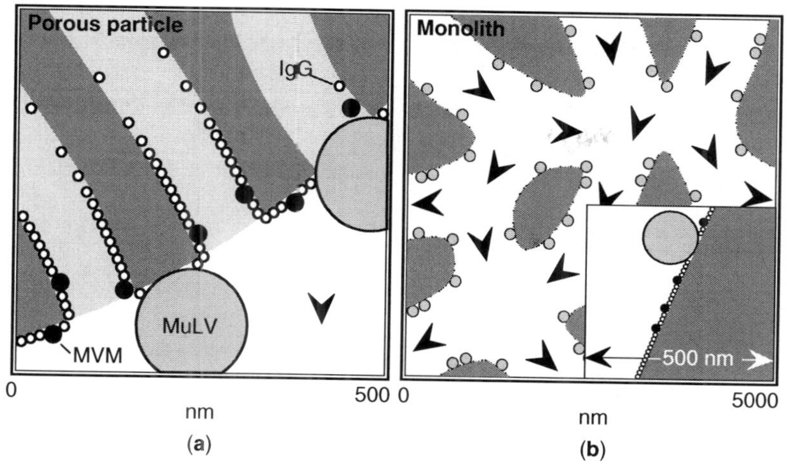

Figure 25.3. Virus and protein size relative to particle pores and monolith channels. Size comparison of particle pores versus monolith channels (drawn to scale). Illustrated pore size (a) is 100 nm. Channel size (b) is 1 μm (1000 nm). MuLV, diameter 150 nm, gray circles. MVM, diameter 25 nm, black circles. IgG, hydrodynamic diameter 12 nm, white circles. White areas indicate areas of convective mass transport. Gray areas indicate areas of diffusive mass transport. Arrows indicate direction of flow. As indicated, MuLV cannot enter the pores and has access only to the surface of the particle. Surface accessibility is unrestricted in the monolith, as it is in membranes. 100 nm is the upper end of the pore size range for most porous particles. One μm is about half the size of the smallest channels in industrial monoliths. Binding capacity for both virus species is restricted on the particle; by low efficiency of diffusional mass transport for MVM, and by inaccessibility of internal pore surface area for MuLV. Both panels illustrate the important point that proteins, other contaminants such as DNA, can reduce virus binding capacity by competing for available surface area. [Redrawn from Ref. 36, with permission].

TABLE 25.2. Approximate Diameters of Selected Virus Particles

Virus	Diameter (nm)
AAV	20–26
MVM	25
Rhinovirus	30
HBV	42
Adenovirus	59–67
EBV	80–100
IVA	80–120
HIV	100–120
HSV	110–200
MuLV	120–150

AAV, adeno-associated virus; MVM, minute virus of mice; HBV, hepatitis B virus; EBV, Epstein–Barr virus; IVA, influenza virus A; HIV, human immunodeficiency virus; HSV, herpes simplex virus; MuLV, murine leukemia virus.

TABLE 25.3. Dynamic Binding Capacities of Selected Solutes on Porous Particle and Monolithic Chromatography Media

Solute	Method	Monoliths	Particles
BSA	AX	20–25 (81[a])	75–300
IgG	protein A	10–12	25–60
IgG	AX/CX	20–25	50–150
IgM	AX/CX	20–50	10–50
DNA	AX	12–15[b]	0.5–3[b]
ETX	AX	115–150[b]	9–15[b]
IVA	AX	10–100x[c]	1x[c]

ETX, endotoxin; IVA, influenza virus A. All values in milligrams per milliliter except influenza virus.
[a]Ref. 37.
[b]Ref. 38.
[c]Ref. 17. Other data reproduced from Ref. 39.
Note the antiparallel relationship between solute size and capacity for monoliths versus particles. Refer to the text, Figs 25.3, 25.6, and 25.7 for the underlying technical rationale.

product recovery, all of which become more severe with increasing flow rate.

The void volume of particle-based columns contributes two additional effects, both of which have negative consequences for virus purification. Buffer flows preferentially through the void, but friction at particle surfaces leads to formation of interstitial vortices called *eddies*. The resultant turbulent mixing is known as *eddy dispersion*, and it contributes directly to peak broadening (15,40–42). This reduces resolution between peaks and dilutes their contents. Eddy dispersion is independent of flow rate, which means that even though an eddy may spin faster at higher

Figure 25.4. Eddy dispersion and sheer in packed particle columns. Gray areas indicate particles. White area indicates the void space between particles. Arrowheads indicate direction of flow. Black arrowheads indicate primary flow. White arrowheads mark eddy flow. Black crescents mark zones of adjacent countercurrent flow where sheer occurs. Refer to the text for discussion. [Redrawn from Ref. 36, with permission].

flow rates, its net effect on mixing remains the same. The greater concern for virus applications is that eddies create zones of adjacent countercurrent flow that produce strong sheer forces (Fig. 25.4). Unlike dispersion, sheer increases linearly with flow rate, which means that risk of physical damage to complex biological structures increases with flow rate (41–43).

Some porous particle media are designed to support a degree of convective mass transport within the particles. These media are referred to as perfusive particles, and in contrast to traditional particles which are characterized exclusively by shallow dead-end pores, perfusive media include so-called through-pores: larger channels that transect the particle (44). Protein-based comparisons of traditional and perfusive particles confirm that perfusive particles support a higher proportion of convective mass transport than traditional particles but only modestly so (Fig. 25.5). Access to the interior of a particle should also increase the virus-accessible surface area. It is therefore reasonable to expect that perfusive particles should offer modestly higher virus binding capacity than traditional porous particles. Perfusive media retain the burdens of the void volume however—peak broadening and sheer from eddy formation. Perfusive particles for virus applications are presently limited to ion exchangers.

Monoliths are chromatography media that are cast as a single unit. They are characterized by a highly interconnected network of channels, sometimes likened to a sponge. There are no particles. There is no void volume. Since there is no void volume there is no eddy dispersion. Flow is laminar and there is no molecular sheer (40). The lack of eddy dispersion supports elution of

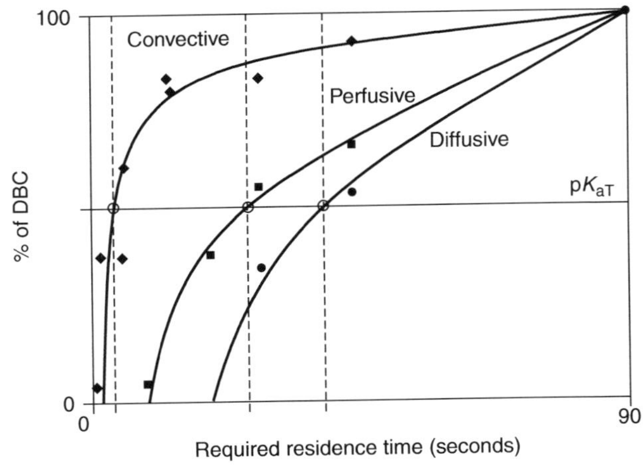

Figure 25.5. Residence time required to achieve 50% binding: diffusive particles, perfusive particles, and monoliths. This plot reveals the relative efficiency of convective and diffusive mass transport. Data are derived from experiments performed with purified IgG on 1-mL protein A affinity columns at various flow rates. The time required for the monolith to achieve 50% binding is slightly less than 4 s. The time required for perfusive media is more than six times greater, 26 s. Time required for diffusive media (38 s) is nearly 10 times greater than the monolith. Since viruses have lower diffusion constants, the differentials would be even more dramatic. Refer to text for additional discussion. [Data replotted from Ref. 36].

sharper more concentrated peaks, which in turn increases resolution (34,40,45–48). Channel size in the present generation of industrial monoliths is in the range of 2–5 μm, which is about 20–50 times greater than the largest pores in porous particle media, and substantially greater than most virus particles (40). These channels are too large to support efficient mass transport by diffusion but they support highly efficient mass transport by convection. This has several important ramifications for virus purification. Efficiency of convective mass transport is unaffected by either molecular size or flow rate. This means that capacity and resolution are conserved at high flow rates, and without elevated risk of damage to viral particles (34,45–48). Not surprisingly, one of the most compelling applications of monoliths is purification of live recombinantly attenuated vaccines (17,49–51).

Protein binding capacities are often published for monoliths, and as with porous particles, they are grossly unrepresentative of virus binding capacity, but with an important distinction: protein binding capacities on monoliths *underestimate* virus binding capacity. Figure 25.6 illustrates the basis for this phenomenon. Fewer large particles can fit on a fixed two-dimensional surface, but the mass of those particles increases in proportion to the three-dimensional space they occupy (34,52,53). This explains the data shown in Table 25.3, with monoliths offering 10 to 100 times higher virus binding capacity than

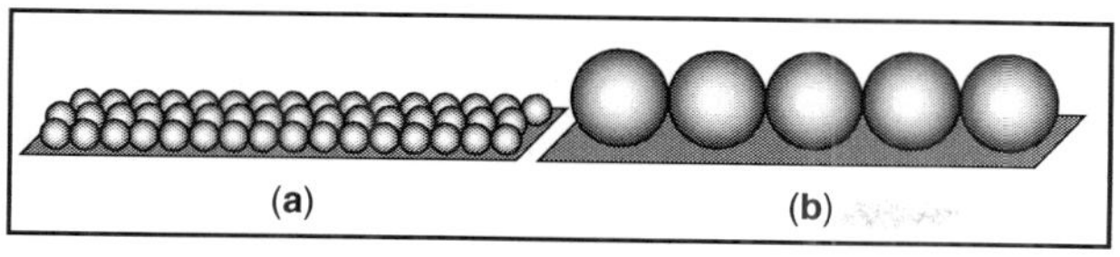

Figure 25.6. Binding capacity as a function of surface area and solute size. Illustration (a) represents binding of globular proteins to a fixed surface area. Illustration (b) represents viral particles bound to a surface of identical area. Molar capacity is greater for the protein, but mass capacity is greater for the virus. Refer to the text and Refs 52–54 for additional discussion.

porous particles. The importance of this differential cannot be overstated. It translates directly into monolith volume requirements of 1–10% the volume of porous particle columns, corresponding reductions in buffer volume, with corresponding reductions in eluted fraction volumes, and proportional increases in eluted product concentration. The smaller footprint imposed by reduced column and buffer volumes also increases productivity per unit of expensive GMP manufacturing space. On packed particle columns, a 10-fold reduction in process volume would not reduce process time, since linear flow rates are conserved across process scales. The independence of convective mass transport from flow rate in monoliths however supports much higher flow rates than are possible with porous particles, which means the economic benefits of lower monolith volume are compounded by shorter process times. Monoliths are fabricated from the same polymers used for porous particles, and are available with numerous surface chemistries, including a variety of ion exchangers, HIC, and preactivated supports for immobilizing affinity ligands.

Strictly speaking, adsorptive membranes are monoliths, at least to the extent that they are cast as a single unit and are characterized by large channels instead of pores. However, their extremely shallow bed height and the physical formats in which they are applied endow them with a different suite of operating characteristics. Channel diameters are generally smaller in membranes than monoliths: ~0.35–1 μm, while bed heights are fractions of a mm (18). Membranes are often stacked to provide more capacity. This is a less effective format than the continuous structure of monoliths, and its limitations are compounded by poor flow distribution in housings that are frequently burdened by large dead volumes. Uncontrolled mixing that occurs in these dead volumes is comparable to eddy dispersion in porous particle systems (15,54). This modestly reduces capture efficiency, but strongly reduces elution efficiency. Eluted peaks are broad, dilute, and poorly resolved in comparison to both porous particles and monoliths. As a result, the application of membranes is most advantageous in situations where neither high purity nor high eluted product concentration is required; or in situations where they selectively bind contaminants but permit the virus to flow through. Adsorptive membranes are often fabricated from regenerated cellulose, which is sometimes used in porous particle media, or from more exotic synthetic polymers that may exert secondary hydrophobic effects. Available surface chemistries are limited for the most part to ion exchangers, but a phenyl membrane for hydrophobic interactions chromatography (HIC) has been introduced recently (Sartorius), and some membrane manufacturers will immobilize affinity ligands on a custom basis.

25.4 ION EXCHANGE CHROMATOGRAPHY

Ion exchange chromatography is the workhorse of bioseparations, including virus purification, with more applications than any other method (1–14,55–88). Ion exchangers are also available on a wider diversity of solid phases than any other method: porous particles synthesized from many kinds of polymers, perfusive particles, monoliths, and membranes. AXs are positively charged, repel positively charged solutes, and bind negatively charged solutes. CXs are negatively charged, repel negatively charged solutes, such as DNA, and bind positively charged solutes. Both AXs and CXs include classifications of exchange ligands that are referred to as weak or strong. The term *strong exchangers* refers to ion exchange ligands that maintain their charge over a wide range of pH (89). For example, quaternary amine-based AXs such as Q, QA, QAE, and TMAE maintain their positive charge even at pH 13 (1 *M* sodium hydroxide). Weak anion exchange groups such as DEAE lose charge as pH increases above neutrality, and become essentially uncharged by about pH 9. Among CXs, sulfonic acid ligands such as SO_3, SE, and SP are strong exchangers and maintain their charge down to pH values as low as 3. Carboxymethyl (CM) is an example of a weak cation exchange ligand, and progressively loses charge as pH descends below about 5. For any given separation, strong and weak exchangers exhibit different degrees of resolution between adjacent peaks, and sometimes different elution orders among peaks. Sometimes a given solute will bind more strongly to a strong exchanger than to a weak exchanger. This has given rise to the misconception that the term *strong exchanger* pertains to how strongly it binds a particular solute. It does not. When a weak ion exchanger is fully charged, a given solute may bind as strongly, or more strongly, than it does to a strong exchanger. The terms strong and weak pertain exclusively to the pH titration characteristics of the ligand.

There are several practical reasons why it makes sense to favor strong exchangers, and go no further unless necessary. The first is that because of their charge variability with respect to pH, weak ion exchangers add another variable to separation processes. Routine minor pH variations from lot-to-lot of buffer that would not appreciably affect capacity or selectivity of a strong ion exchanger may significantly affect reproducibility of capacity and purification performance on a weak exchanger. Second, weak exchangers act as solid phase buffers, and bind/release significant concentrations of counterions from solution within the pH ranges typically applied for bioseparations. For example, DEAE will bind high concentrations of hydroxide ions, even at neutral pH. When a sodium chloride gradient is applied, the higher affinity of chloride for DEAE will displace previously bound hydroxide ions into solution and may elevate pH by 2 or more units. The magnitude and duration of the increase depends on the concentration of sodium chloride that was introduced. The pH change is reversed on CXs, where pH may descend by 2 or more units with the introduction of salt on CM (90,91). At best, such pH transients represent a loss of process control, but the greater concern is the potential for damage to the virus. On a CM CX equilibrated to pH 4, for example, pH might drop below 2. pH excursions are also observed with strong exchangers but they are of lesser magnitude and duration. The magnitude and duration pH of excursions can be reduced with higher buffer concentrations, but they cannot be eliminated. The final reason to prefer strong exchangers is that weak exchangers typically require substantially greater volumes of equilibration buffer to achieve their initial operating pH.

Ion exchange behavior is strongly affected by buffer composition. This applies equally to capacity, selectivity, and reproducibility. Buffers fall into three different categories according to their charge configuration: anionic (net negative charge), cationic (net positive charged), and zwitterionic (net neutral, bearing both negative and positive charges). Anionic buffers include acetate, citrate, phosphate, and borate. Cationic buffers include Tris. Zwitterionic buffers include MES and HEPES. Among anionic and cationic buffers, the buffer chosen for a particular application should have the same charge as the exchanger. This causes the buffering ion to be repelled from the exchanger surface and ensures that it will provide the mobile phase with the intended buffer capacity. Using a buffer that is charged opposite to an exchanger results in a significant proportion of the buffering ions being bound to the exchanger, which proportionately reduces pH control in the mobile phase. These guidelines are frequently violated. For example, phosphate is often used as a buffer with AXs. It may work, but there is a price in process control, and its important to be aware of it so that experiments can be designed to validate that process control is adequate despite compromising the system. A minor liability with both anionic and cationic buffers is that they elevate conductivity, which weakens interactions between solutes and exchangers. This may be manifested as weaker retention or reduced binding capacity, and it is the reason buffer concentrations are commonly very low in ion exchange applications. Zwitterionic buffers stand out for two important reasons, first that they have no inherent conductivity. The only conductivity in a zwitterionic buffer comes from the sodium hydroxide used to titrate it to operating pH. This means that they can be used at higher concentrations than anionic or cationic buffers, but without compromising virus binding capacity. Zwitterionic buffers are also immune from binding to charged groups, either positive or negative, which means that they can be used interchangeably with either CXs or AXs.

Buffer preparation is also important, as illustrated by an example. When making Tris, with an intended pH of 8, start with Tris base, and titrate pH down with hydrochloric acid. If you begin with Tris-HCl and titrate pH up with NaOH, the conductivity of the buffer will be significantly higher, which will weaken binding of solutes to the exchanger. It is also important not to back-titrate. If you accidentally titrate pH below the target, discard it and start over. Adding NaOH to compensate will inadvertently raise conductivity and weaken solute retention. Since back-titrations are seldom documented, it may be next to impossible to trace a process deviation caused by the higher conductivity, or to reproduce the performance obtained with that particular buffer lot. The same guidelines apply to other ion exchange buffers. All buffers must be filtered to 0.22 μm before being applied to chromatography columns. This applies to all chromatography methods.

Solute isoelectric point (pI) is often suggested as a predictor of ion exchange behavior (92,93), but detailed retention studies have revealed that it is most often misleading, sometimes grossly so (92,94). This reflects a deficiency in the conceptual model, and it has special significance for virus particles. The basic model states that a given solute should bind to AXs at pH values more than a unit above its pI, but it should not bind at pH values close to or below its pI. The pattern should be reversed on CXs. In practice, proteins often bind to AXs a full pH unit below where they should, or fail to bind a full pH unit above where they should, and the opposite occurs on CXs. The discrepancy is explained by charge distribution. The traditional model assumes that the properties of all charged residues are expressed in a single point. In the real world, charges are distributed unevenly on biomolecule surfaces. Negative charges are sometimes mixed with positive charges, tending to neutralize binding to both AXs and CXs. In other cases, most of the positive charges or negative charges may be localized in a charge-dense domain that dominates the ion exchange binding characteristics of the entire molecule. This gives rise to phenomenon known as *preferential orientation*: molecules orient themselves so as to present their most complementary surface to an ion exchanger (95,96). Since proteins are amphoteric, the location and composition of the preferred binding site can change with pH, depending on the titration characteristics of the specific charge residues at the binding interface (96,97). Thus elution order and resolution may change in unexpected ways at different pH values. Preferential orientation has not been studied specifically in viruses but given that viral surfaces may carry a variety of proteins, charged carbohydrates, calcium ions, or lipid phosphoryl residues, and that their arrangement on the surface may favor or disfavor the interaction of specific subsets of charged residues, it seems prudent to anticipate that virus retention behavior may deviate substantially from expectations based on the simplistic model of pI. If you want to find out the ion exchange binding characteristics of a particular viral species, and how it relates to the retention of contaminants, the most reliable way to do so is to conduct retention mapping over an appropriate range of conditions.

Ion exchangers are most commonly eluted with conductivity gradients produced by increasing the concentration of a salt at a fixed pH. Sodium chloride is used most widely but sodium acetate is frequently substituted because it is less corrosive to stainless steel buffer vessels and chromatography skids. Ion exchangers may also be eluted with pH gradients. Selectivities may be superficially similar to salt gradients but more commonly exhibit significant differences. The two formats also have different process ramifications. Salt gradients separate on the basis of absolute charge. pH gradients separate on the basis of solute titration characteristics. For example, in a salt gradient at pH 4, a protein whose retention is defined by a cluster of six histidine residues (pKa about 6) will likely co-elute from the CX with a protein whose behavior is defined by a cluster of six lysine residues (pKa about 10), while a protein that is bound by an intermixed cluster of six lysine and six histidine residues will elute later than either. At pH 6, the histidines will have lost about half of their charge, so the 6-his protein will elute first, followed by the 6-lys protein, and then the 6/6 protein. At pH 8, where histidine residues will have lost most of their charge, the 6-his protein will probably not bind, and the 6-lysine protein will probably co-elute with the 6/6 protein. In an increasing pH gradient, 6-his will elute first, while the other two will probably co-elute later, but with a critical difference: all the proteins will elute at a very low conductivity value, also at a higher pH than they would have eluted in a fixed-pH salt gradient. This will facilitate application of the CX eluate to an AX column. pH gradients have been used successfully to purify phages (86). Among proteins, pH gradients usually offer higher resolution among product-related variants and impurities (16,93,98). This suggests that pH gradients may be better suited to the same task with viral particles. but remains to be evaluated experimentally.

Ion exchange and other adsorptive methods are commonly conducted in either of two modes: bind-elute or flow-through. *Bind-elute* mode is characterized by binding the product of interest and selectively fractionating it from contaminants by changing the buffer conditions. *Flow-through* mode, also called *negative chromatography*, is characterized by the ideal of selectively binding contaminants while the product flows through the column. Bind-elute applications usually offer the best overall process performance. They have the ability to concentrate product from dilute feed streams, and they offer high resolution fractionation of contaminants that elute either earlier or later than the product. Flow-through applications are attractive because they require fewer process steps, use fewer buffers, and lower buffer volumes, but any contaminants with weaker retention characteristics than the product will co-elute with it. It is also difficult or impossible to develop flow-through buffer conditions that reproducibly support effective removal of contaminants that elute immediately after the product, without compromising product recovery. Flow-through applications also fail to concentrate product. For all of these reasons, flow-through applications are at their best when the feed stream is fairly concentrated and the key contaminants have much stronger retention characteristics than the product. Such applications represent the best use of membrane-based media. If higher resolution separation of contaminants with retention properties similar to the product is required, bind-elute applications on monoliths or porous particle media will usually provide better results.

25.5 HYDROPHOBIC INTERACTIONS CHROMATOGRAPHY

HIC has important applications in virus purification (63,64,83), though far fewer than ion exchange. The two most commonly used HIC ligands are phenyl and butyl. Phenyl ligands have a higher affinity for proteins with a preponderance of aromatic amino acid residues; butyl ligands have a higher affinity for proteins with a preponderance of aliphatic residues. Differences in selectivity between the two may be considerable and can be expected to carry over to viral particles. Strong HIC ligands, including both phenyl and butyl also have a documented reputation for denaturing labile proteins (99–104). In many cases the proteins revert to their native structure upon elution so this may not result in viral inactivation, but it will be prudent to check. Stronger HIC ligands like hexyl and octyl are destructive to most proteins and likely to be destructive to viral structures. Monoliths are presently limited to butyl, and membranes to phenyl, so head-to-head comparison of selectivity among different hydrophobic ligands will require the use of porous particle-based media.

Most of the HIC literature describes the use of ammonium sulfate for promoting HIs, usually at concentrations ranging from 1 to 2 *M*. Ammonium sulfate is simple and effective but bears several burdens that make it unattractive for manufacturing applications. Ammonium ions convert to ammonia gas at alkaline pH. Dissolved ammonia is sufficiently alkaline that it may cause hydrolysis of the product. Gradual release of ammonia gas will likely affect pH control. Released ammonia gas may be a safety hazard at large scale. Municipal authorities may place limits or penalties on disposal of ammonium salts. Some of these problems can be overcome by substituting other salts that strongly promote hydrophobic interactions. Potassium phosphate, sodium sulfate, and sodium citrate are all candidates, but the biggest liability with ammonium sulfate, and one that it shares with all of these alternatives, is that it may precipitate the product when the sample is equilibrated before being introduced to the column. Loading a precipitated product may be tolerated in small scale experiments, but it will cause serious problems with scale-up and manufacturing. This problem and most of those above are suspended by the counter-intuitive approach of using a salt that is a weak promoter of hydrophobic interactions, such as sodium chloride. This approach generally requires two to three times the molar concentration of sodium chloride compared to ammonium sulfate, but the molecular weight of ammonium sulfate is about 2.6 times higher than sodium chloride, so the difference in grams per liter is more modest. Whatever salt is employed, some virus species may be partially inactivated by exposure to high salt solutions (105).

Some viruses may bind so strongly to phenyl or butyl ligands, even in sodium chloride, that they elute only partially or not at all. In such cases, an additive that weakens hydrophobic interactions may cause the product to elute earlier and with better recovery. Candidates include up to 25% propylene glycol and low molecular weight polyethylene glycol polymers, such as PEG-400, up to 10%. Both compounds are protein-stabilizing and unlikely to adversely affect virus structures. In addition, both are nonionic, which means that they will not interfere with a subsequent ion exchange step, and both will flow through an ion exchanger and thus be removed when the product binds. PEG-400 is also an approved inactive ingredient for intravenous administration (106). 1–2 *M* urea or 100–200 m*M* arginine may also cause earlier elution, but care must be taken to ensure that they do not inactivate virus. Arginine is an approved injectable, but it has also been shown to inactivate some viruses (105–109). Alcohols are occasionally suggested for weakening hydrophobic interactions in HIC but bear a significant risk of product denaturation. Another approach to strongly hydrophobic virus species is to evaluate more weakly hydrophobic media, such as ToyoPearl PPG (immobilized polypropylene glycol, Tosoh Biosciences) or the even more weakly hydrophobic ToyoPearl Ether (immobilized ethylene glycol, Tosoh).

25.6 MULTIMODAL METHODS

Multimodal methods, or mixed modes, refers to chromatography supports that employ combinations of discrete well-defined mechanisms. The multimodal supports most frequently used for virus purification include HA and immobilized heparin. The benefit of multimodal systems is that they produce unique selectivities that can confer valuable process features. The liability is that method development is more complex.

25.6.1 Hydroxyapatite

HA is a mineral of calcium and phosphate. The two dominant binding mechanisms are cation exchange on negatively charged HA phosphate groups, and metal coordination with HA calcium groups. The latter is often referred to as calcium affinity. Two different development pathways have been developed for HA. The strategy most represented in the technical literature, especially in virus purification, is elution with simple phosphate gradients (67,110–114). The ionic nature of phosphate elevates conductivity and elutes the cation exchange component of binding. Its strong affinity for calcium elutes the calcium affinity component. A strategy that has emerged more recently is the application of sodium chloride elution gradients while phosphate concentration is held constant. Calcium affinity is largely unresponsive to sodium chloride, so it primarily affects the cation exchange component, producing completely different selectivities than phosphate gradients. Lipid enveloped viruses in particular tend to bind strongly to HA. Their phosphate residues bind strongly to HA calcium and their calcium residues bind strongly to HA phosphate. This suggests that phosphate gradients may be more suitable for lipid enveloped viruses and chloride gradients more suitable for nonenveloped viruses, but this remains to be shown and recommends that both strategies be evaluated for any given separation challenge. It is likely that chloride gradients will be more effective than phosphate gradients for discriminating among size-differentiated product-related impurities. This has proven to be the case with proteins such as IgG, where chloride gradients are dramatically more effective than phosphate gradients for removing fragments and aggregates (115). On the other hand, phosphate gradients have proven capable of discriminating compositional variants of IgG, such as light chain variants, including bispecific antibodies (116–119), which may portend well for separating product-related variants such as full and empty capsids.

HA is available commercially in several forms, the differences among which may affect the efficiency of a given application. The original form of HA consists of thin flat crystals that fracture easily and are marginally suitable for chromatography. A more suitable form employs fragments of such crystals embedded in porous particles of agarose (HA Ultrogel, Pall). Most applications are developed on so-called ceramic HA. This material is fabricated from needle-shaped HA nanocrystals that are aggregated into microspheres, then bonded by sintering at high temperatures. Type I has an average pore size of 60–90 nm. Type II has an average pore size of 80–120 nm. Both types are available in particle sizes of 20, 40, and 80 μm (120). Forty micrometers is the usually the best compromise. Twenty micrometers is too small for the frits on many industrial columns. Eighty micrometers offers the lowest back pressure but has only about half the capacity of 40 μm. Fluorapatite (FA) is also available commercially (Type II, 40 μm). It is chemically similar to HA, differing mainly in the replacement of HA hydroxyl groups with fluoride residues. FA is about four times mechanically stronger than HA, and also more stable chemically. It can be operated at pH values down to about 5, whereas HA is limited to pH values greater than pH 6.5. Both are stable for thousands of hours in 1 *M* NaOH. Preliminary results indicate that the same buffer strategies can be applied to both FA and HA, but the selectivities are distinct, with most solutes binding more weakly to FA.

25.6.2 Heparin

Heparin is primarily a mixed weak and strong CX ligand, containing both sulfo and carboxyl groups, as well as numerous hydroxyl groups. Like conventional CXs, immobilized heparin is usually eluted with salt, but the level of salt required to elute virus particles is often more than 1 *M* (87,121–125), which is roughly double than what would be needed to elute a strong-binding virus from either sulfo or carboxy CXs. This may reflect the contribution of hydrogen bonding, which has been shown in well-characterized cases to substantially increase the strength of binding (126). The branched linear structure of heparin also creates the possibility of it extending out from the solid phase like a tentacle, potentially interacting with solutes at a larger number of points.

Method development is typically confined to simple salt gradients at near-neutral pH, but the potential exists to evaluate a broader range of possibilities. As with traditional CXs, binding should become stronger with decreasing pH, weaker with increasing pH. The contribution of hydrogen bonding can be manipulated experimentally by the inclusion of strong hydrogen donors/acceptors in the buffers. Urea is ideal because it is nonionic, making it possible to control hydrogen bonding independently from the cation exchange component. Concentrations below 2 *M* are unlikely to be denaturing. Various sugars can be evaluated if a particular virus proves not to tolerate urea.

25.7 OTHER MULTIMODAL METHODS

Another family of multimodal chromatography methods is represented by ligands that combine electrostatic and hydrophobic interactions with potential for hydrogen bonding. Examples include Capto adhere™ a hydrophobic AX (GE Healthcare), and Capto MMC, a hydrophobic CX. These products have been introduced only recently and their value for virus purification remains to be determined, but they are able to bind proteins at higher salt concentrations and over a wider range of pH than conventional ion exchangers. This makes them potentially well suited for virus capture from crude feed streams. Their chief liability is that well-defined method development pathways are yet to emerge.

25.8 BIOSPECIFIC AFFINITY CHROMATOGRAPHY

Biospecific affinity, as exemplified by immunoaffinity, is the most seductive of chromatography methods, and it delivers on many of its promises, but it also imposes some serious liabilities. It generally receives the most recognition for its ability to achieve a high degree of purity in a single step, but deserves equal recognition for its ability to concentrate a dilute product and accommodate feed streams with little or no requirement for adjusting pH or conductivity. Method development requirements are also far less than for physicochemical methods, but far less does not mean nonexistent; a quarter century of industrial experience with protein A affinity purification of IgG has shown that immense benefits are realized by looking beyond its superficial simplicity. The two principal benefits have been development of secondary wash formulations that substantially improve eluted product purity, and development of elution formulations that substantially improve recovery of product mass, activity, and stability. Both should serve virus applications. As noted in the discussion of SEC, stable complexes can form between biological products and various contaminants during cell culture production. Secondary washes are able to dissociate such complexes and more effectively remove the contaminants from a product bound to an affinity support (29,30). Immunoaffinity and similarly strong interactions are sufficiently robust that they can tolerate potent combinations of decomplexants, such as 1.5 *M* sodium chloride, 2 *M* urea, and 10 m*M* EDTA, which collectively suspend nonspecific electrostatic complexes, weak hydrophobic interactions, hydrogen bonding, and metal affinity interactions. Care must be taken to ensure that the treatment does not inactivate virus, but contact duration is short (2–5 CV) and usually at neutral pH, both of which should help to ameliorate potentially destructive effects.

Improvements in elution conditions have mostly been in the form of additives that moderate the pH required for elution. Low pH (~2.3–3.3) although widely used historically, is known to cause permanent conformational changes in some proteins, causing loss of activity, formation of aggregates, and reducing long-term product stability (16). Elution of protein A with 0.1–0.2 *M* arginine at pH 3.8 has proven to achieve higher mass recovery, full recovery of immunoreactivity, no apparent aggregate formation, and better long-term stability. Arginine apparently accomplishes this by a combination of characteristics. It is a strong hydrogen donor and acceptor, capable of suppressing hydrogen bonds, and its guanido side group is capable of weakening hydrophobic interactions. Arginine may not be suitable for all viral applications because of its potential inactivation effects (127,128), and where it is applied, it may be prudent to limit its concentration and duration of contact. But where arginine will not serve, other additives may, for example, propylene glycol or polyethylene glycol (PEG). Inclusion of sodium chloride concentrations up to 1 *M* may also support elution at higher pH. This is undesirable with protein products because it restricts the eluted sample from being loaded directly onto methods that require low conductivity, such as ion exchange. With virus products, the ability of SEC to accommodate any feed stream composition, and simultaneously buffer exchange the product into a more compatible formulation bypasses this issue. Whatever elution-enhancing additives may be employed, it is worthwhile to evaluate linear pH gradient elution to identify the most moderate pH at which elution occurs. Another reason to evaluate linear pH gradients is that bioaffinity media sometimes have higher affinity for aggregates than monodisperse product, and higher affinity for monodisperse product than fragments. If a pH gradient supports effective fractionation of these populations, it can simplify subsequent purifications steps. Otherwise, convert the linear gradient to a step at the highest effective pH value.

The most conspicuous liability associated with bioaffinity is cost. In the field of antibody purification, protein A affinity media, although effective and widely used, has emerged as the most costly single component in the manufacture of IgG, accounting for about 3% of the total manufacturing costs. Availability can also be a limitation. The only commercial bioaffinity ligand currently available for virus purification is directed to Adeno-associated virus (AAV) (129,130). Developing affinity media from scratch is even more expensive. Patient safety and associated regulatory requirements further expand development requirements and expense. Ligand leakage must be measured and toxicology studies conducted to ensure that leached ligand will not pose a safety risk to patients (131,132). Downstream methods must be developed to remove leached ligand from the product. This can be deceptively difficult since leached affinity ligands remain bound to the product under the conditions used for most separation methods. Media maintenance is also an issue. Protein A will tolerate repeated exposure to 0.05 NaOH, as will commercial ligands for purification of AAV, but few other affinity ligands are so robust, and even 0.1 *M* NaOH is insufficient to fully sanitize these media.

25.9 PROCESS DEVELOPMENT

Process development is laborious, expensive, and time consuming. And it's worth it. A good process will reward you every time you run it. A bad process will punish you every time you run it. Still, there is no reason to put more work into it than it reasonably requires. The system summarized in the following pages has been refined over decades of successful commercial purification process development and represents a rapid, efficient, and resource-kindly way to conduct process development.

25.10 SAMPLE DEFINITION

Variations in cell culture media formulation, product concentration, and so on, can have profound effects on both sample preparation and chromatographic performance. This might be inferred to suggest that cell culture production processes should be finalized before purification process development begins, but there are practical limitations to this notion. Few purification process developers can afford to wait for finalization of the cell culture process and are therefore compelled to begin development with "evolutionary" product feed streams. This inevitably leads to surprises, for example, when a new media additive doubles the concentration of virus particles but inadvertently reduces purification process yield by half due to product aggregation during preparation for the first chromatography step. This is arguably a healthy situation, however, because it makes purification issues part of the selection criteria for the eventual media formulation. The earlier that cell culture and purification development groups integrate their operations, the faster process development will proceed for both, and likely with results far superior to isolated development efforts.

25.11 SAMPLE PREPARATION

The use of chromatography generally requires that the sample be filtered through a 0.45 μm membrane; a 0.22 μm membrane is better but can be excessively limiting for larger virus species. Anything that will not go through such filters will foul any chromatography media to which it is applied. At best, fouling will increase backpressure, but increased backpressure is only a secondary indicator of pore or channel occlusion, which translates into reduced chromatographic performance. It contributes to reduced capacity, reduced separation between peaks, and reduced product recovery, not to mention reduced column life. It may also impair the cleanability and sanitizability of the media, leading to accumulation of debris that can ultimately provide substrate for bacteria, which may release proteases, glycosidases, and other components that could affect the product.

In the event that filtration results in unexpected loss of virus particles, it may be worthwhile to filter a series of samples with increasing amounts of sodium chloride added to dissociate charge complexes that may have formed between virus particle surfaces and DNA, proteins, cell membrane fragments, or other cell culture constituents (31–33). If postfiltration virus recovery is still lower than expected, other additives may be helpful, including those discussed above as nonspecific dissociating agents. Alternatively, or in addition, nonionic detergents (Tween, Triton) or zwitterionic detergents (CHAPS, CHAPSO, octaglucoside) may be helpful, but with two potential pitfalls. Once present, detergents bind to protein and virus surfaces, and may be impossible to remove without damaging the product. They also bind to chromatography media surfaces, where their removal may require washing with organic solvents that represent a fire hazard. It is prudent to avoid detergents unless they will be included in the final product formulation. Even then, use only the detergent that will be used in formulation. The second potential pitfall pertains to detergent purity: detergent solutions should be water-clear. If they have a yellowish color they are probably contaminated with peroxides that can cause uncontrolled oxidative damage to the product. Ultrapure detergents are available from Pierce Chemical Company (SurfactAmps™ Rockford, Illinois, USA). As noted above, any dissociating agent has the potential to adversely affect virus stability, so their investigation should be undertaken with care. They may also affect chromatography methods in various ways, so it is best to develop sample treatments before beginning chromatographic process development, and essential that sample treatment protocols be finalized before final specifications are set for the purification process. Despite the extra complication of developing prefiltration dissociating formulations, and accommodating their effects on downstream purification methods, they offer an important side benefit: they can alleviate or eliminate the effects of changes to cell culture media formulations during upstream process development. In the event that filtration fails utterly to yield a suitable sample, centrifugation may suffice, or at least buy time to further investigate decomplexation treatments. Note that virus samples treated with formaldehyde or glutaraldehyde may not be suitable for chromatographic purification, since the treatment may introduce covalent cross-links among virus particles and other solutes that create sufficiently large masses to clog most chromatography media.

Nuclease treatment (before membrane filtration) is optional but usually beneficial. Intact nucleic acids bind more strongly to AXs than fragments but they can be difficult to remove (133,134). Strong AXs maintain a positive charge even in 1 *M* sodium hydroxide. As a consequence, a small amount of DNA remains bound from each run and gradually accretes on the surface of the exchanger, even if 1M sodium hydroxide is combined with 2M sodium chloride. This alters the ability of the exchanger to remove DNA, among other things, which makes it a process control and reproducibility issue. Nuclease treatment of cell supernatants tends to produce fragments over a wide range of sizes. This creates a higher probability that some will co-elute from AXs with the virus product, but it also makes it easier to quantitatively remove DNA from the AX. The presence of polydisperse DNA fragments will also broaden the DNA elution zone on HA, and increase the probability of co-elution with virus. Nuclease digestion probably has little effect on DNA elimination by HIC, CXs, or heparin since it fails to bind to HIC and is actively repelled by CXs and heparin. Digestion should be expected to increase the efficacy of DNA removal by SEC. Finally, there is the consideration that nuclease digestion in virus purification procedures is very familiar to regulatory agencies, and may be viewed favorably during the product approval process (131,132). If DNA digestion will be performed in the final process, it makes sense to develop the process that way from the beginning.

Processes can be developed exclusively from an unpurified sample but it unnecessarily and massively inflates the time and amount of work to develop a process. The many protein components of cell culture supernatants create complex chromatograms from which it is usually impossible to discern the virus product. Extensive secondary testing of dozens of chromatography fractions, in the form of electrophoresis, blots, ELISA, PCR, or other assays is required just to find the product. Secondary testing can be limited to a much more manageable subset of samples and tests if sample is available that is virus-dominant and substantially free of proteins and other "small" contaminants. This can make it possible to detect the virus peak visually from chromatograms, which dramatically reduces the number of fractions that need to be tested, and limits initial testing to the most expedient method. Once product behavior has been defined under a particular set of conditions, it may be possible to dispense with secondary testing in subsequent chromatography experiments under closely related conditions, for example, in conjunction with gradient optimization.

SEC is well suited to producing such a sample, especially following membrane concentration. The system need not be optimized at this point. Use a column with a bed height of about 20 cm, load concentrated filtered virus to 20% of CV, run at 50 cm/h. Equilibrate the column with a buffer with pH near neutrality, such as 20 m*M* Hepes or phosphate, 50 m*M* sodium chloride, pH $\sim$ 7. This salt concentration is usually sufficient to maintain product solubility and suppress interference with virus binding to ion exchange groups that may reside on the surface of the particles. Increase the salt concentration if necessary. The low buffer concentration minimizes pH interference with subsequent ion exchange experiments.

25.12 INITIAL SCREENING

25.12.1 Ion Exchange

Monoliths greatly accelerate screening and early method optimization. Their low dispersion characteristics also produce sharper, more concentrated elution peaks than other media, which enhances visual identification of virus peaks from chromatograms. The small size of screening monoliths (0.34 mL) also helps to conserve sample, which is often a limited resource during early process development. As explained above, initial screening can be limited to a strong AX and a strong CX. Conditions for screening conductivity gradients are given in Table 25.4. Conditions for screening pH gradients are given in Table 25.5. Most nonenveloped viruses will likely tolerate the full range of suggested conditions. Enveloped viruses can be expected to be more labile. Screening can be restricted to near–physiological conditions, although ultimately, the only way to establish the acceptable range is to see what actually works. A monolithic anion exchange elution profile of adenovirus, previously fractionated by SEC, is illustrated in Fig. 25.7.

25.12.2 Hydrophobic Interaction

Phenyl and butyl HIC ligands on the same solid phase are available only on porous particle media, and there are many of them. At this stage it is probably not productive to screen the large diversity of products available on the market. Instead, select a butyl and a phenyl product that have the same polymer backbone, particle size, and porosity. Pack the media in 1 mL columns. Screening more weakly hydrophobic media can be deferred, pending initial results on phenyl and butyl. Screening conditions are given in Table 25.6.

25.12.3 Hydroxyapatite

Ceramic HA CHT™ type II will likely offer higher virus binding capacity than type I. Forty-micrometer particles offer the best compromise between capacity and compatibility with industrial column frits. Pack the media in 1–2 mL columns. Screening other types, particle sizes, or FA (CFT™ type II, 40 μm) can be deferred, pending initial results on CHT type II, 40 μm. If the virus binds very strongly on HA, it will probably elute under more moderate conditions from FA. Screening conditions for phosphate gradients, and chloride gradients at constant phosphate, are given in Table 25.7.

25.13 BIOSPECIFIC AFFINITY

Other than SEC, biospecific affinity is the one method where loading a raw sample will not be a liability. Given that secondary wash conditions carry a risk of virus inactivation, it is prudent to begin with simple equilibration and wash buffers, like phosphate buffered saline at neutral pH. As noted above, linear pH gradient elution may reduce risk of product denaturation as well as support higher recovery of higher quality product than a step to an excessively severe low pH step. A linear gradient can be formed easily with two buffers: (A) 20 m*M* sodium phosphate, 20 m*M* sodium citrate, pH 7.0, and (B) the same formulation at pH 2.5. Phosphate and citrate are both triprotic. Their respective p*K*s are spread fairly evenly over this pH range and form reproducible linear gradients when mixed appropriately. Equilibrate the column in A, apply sample, wash the column in A, then elute in a 10 CV linear gradient to B. Hold at B until the pH comes to equilibrium.

25.14 INTERPRETATION OF INITIAL RESULTS

The first priority of screening is to identify a capture method that (i) involves minimal modification of the feed stream, (ii) supports high binding capacity so that it can concentrate the product and reduce product volume, (iii) provides a high degree of product purity, and (iv) elutes the product in a formulation that is compatible with potential subsequent purification steps. Minimal requirement for feed stream modification and high binding capacity can both be inferred from where the virus elutes in a gradient. The later it elutes, the more tolerant it will be of loading conditions, and the higher its relative capacity. Gross purification potential, which is all that is needed at this point, can be estimated visually by comparing chromatograms of unpurified sample and post-SEC enriched sample. This will be sufficient to determine if most of the contaminants elute earlier or later than the product, and in what relative proximity. Compatibility with potential downstream steps is relative. Products that elute at fairly low conductivity will be fairly easy to prepare for subsequent ion exchange steps. Products that elute at low to moderate conductivity will be easily

TABLE 25.4. Conditions for Initial Screening of Ion Exchangers with Conductivity Gradients Buffers

A1: 20 m*M* acetate, pH 4.0
A2: 20 m*M* MES, pH 5.5
A3: 20 m*M* Hepes, pH 7.0
A4: 20 m*M* Tris or Bicine, pH 8.5
B1: 20 m*M* acetate, 1 *M* sodium chloride, pH 4.0
B2: 20 m*M* MES, 1 *M* sodium chloride, pH 5.5
B3: 20 m*M* Hepes, 1 *M* sodium chloride, pH 7.0
B4: 20 m*M* Tris or Bicine, 1 *M* sodium chloride, pH 8.5

Sample preparation. For anion exchange, dilute 1 part filtered sample with 4 parts buffer A4. For cation exchange, dilute 1 part filtered sample with 4 parts buffer A1.

Chromatography media: monolithic strong anion or strong cation exchanger, 0.34 mL.
Flow rate: 4 mL min
Equilibrate column: until effluent pH equals buffer pH
Inject sample: As much as necessary to produce a visible product peak
Wash: 12 CV (1 min) buffer A
Elute: 60 CV (5 min) linear gradient to corresponding buffer B
Hold/Clean: buffer B until conductivity is stable.

Note: Use the same conditions for membrane ion exchangers. The same buffer conditions can be applied to weak exchangers, although it may be prudent to increase buffer concentration to 50 m*M* to counteract pH excursions. For porous particle exchangers, use a 1-mL column with a flow rate of 100–300 cm/h, reduce wash volume to 10 CV, and gradient volume to 20 CV. The same buffer conditions can be applied to weak exchangers, although it may be prudent to increase buffer concentration to 50 m*M* to counteract pH excursions.

TABLE 25.5. Conditions for Initial Screening of Ion Exchangers with pH Gradients Buffers

A1: 10 m*M* MES, 10 m*M* Hepes, 10 m*M* Tris (or Bicine), pH 5
A2: 10 m*M* sodium citrate, 10 m*M* sodium phosphate, pH 4
B1: 10 m*M* MES, 10 m*M* Hepes, 10 m*M* Tris (or Bicine) pH 8.5
B2: 10 m*M* sodium citrate, 10 m*M* sodium phosphate, pH 7
C: 1 *M* sodium chloride, buffered or not, pH per convenience.

Sample preparation. See Table 25.4.

Recommended media: monolithic strong anion or cation exchanger, 0.34 mL.
Flow rate: 4 mL min
Cation exchange
Equilibrate column: buffer A1 or A2 until effluent pH equals buffer pH
Inject sample: As much as necessary to produce a visible product peak
Wash: 12 CV (1 min) buffer A
Elute: 60 CV (5 min) linear gradient to buffer B1 or B2
Hold: buffer B until pH is stable
Clean: buffer C

Note: Do not mix zwitterionic and anionic buffers. Use A1 with B1, or A2 with B2. If desired, 10 m*M* boric acid can be added to buffer B2 and the pH raised to 9 to extend the range of the gradient. Borate is not human-injectable and may interfere with some methods of carbohydrate analysis.

Anion exchange
Equilibrate column: 10 CV buffer B1 until effluent pH equals buffer pH
Inject sample: As much as necessary to produce a visible product peak
Wash: 12 CV (1 min) buffer B1
Elute: 60 CV (5 min) linear gradient to buffer A1
Hold: buffer A1 until pH is stable
Clean: buffer C

Note: The A1/B1/C series may be applied with either anion or cation exchange. The A2/B2/C series is suitable for cation exchangers only. Borate can be deleted from buffers A2/B2 if the pH is not greater than 7.5. These formulations will produce linear gradients with strong ion exchangers such Q and S, but cannot be relied upon to do so with weak exchangers such as DEAE and CM.

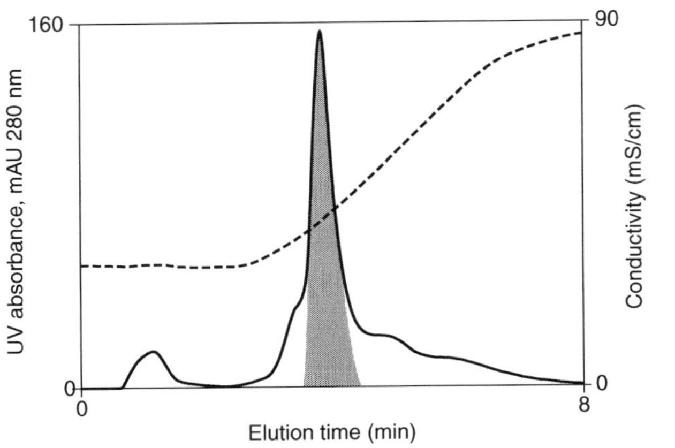

Figure 25.7. Salt gradient elution of SEC-fractionated adenovirus Ad5-GFP from a monolithic AX (CIM QA). The dashed curve illustrates conductivity throughout the separation. The gray zone marks the main product fraction. The high resolution and short separation time are typical of monoliths. A shallower gradient would likely have improved the purity of the main component. This figure also illustrates two important points that (i) although SEC provides very substantial purification of viral particles from cell lysates, the eluted virus fraction is far from homogeneous and, (ii) estimates of recovery based on UV absorbance of the SEC virus fraction are likely to be highly inflated. Refer to Ref. 135 for conditions and additional discussion. [Redrawn from Ref. 135, with permission].

TABLE 25.6. Conditions for Initial Screening of Hydrophobic Interaction Chromatography

A: 20 m*M* sodium phosphate, 4 *M* sodium chloride, pH 7
B: 20 m*M* sodium phosphate, pH 7
C: 2 *M* guanidine-HCl, pH 5

Sample preparation. Dilute one part sample with three parts buffer A.

Recommended media: Phenyl and butyl porous particles
Linear flow rate: 200 cm/h (1 mL/min on a 5 mm × 50 mm column)
Equilibrate column: 10 CV buffer A, or until conductivity is stable
Inject sample: As much as necessary to produce a visible virus peak
Wash: 5 CV buffer A
Elute: 15 CV linear gradient to buffer B
Hold: buffer B until conductivity is stable
Clean: 5 CV buffer C

compatible with HA or heparin. Products that elute at high conductivity will be easily compatible with HIC.

If strong product binding is observed on a CX, it is often the best of capture options, since most contaminants fail to bind, or bind only weakly. A possible liability with strong product binding is that the product will tend to elute at high salt concentrations. This can interfere with binding in the next purification step, for example, when going from a CX to an AX step. In many cases however, the elution conductivity can be moderated by changing the elution pH. For example, binding can be conducted at pH 6 or lower to obtain high capacity, than the column washed and reequilibrated at a higher pH where less salt will be required for elution. Higher elution pH on a CX usually corresponds to more effective contaminant removal as well. This highlights an important practical point: the pH that supports the most effective product capture may be different from the pH that supports the most effective elution. In this case, bind at the pH that supports the best capacity, and elute at the pH that supports the best purity. Taking advantage of such opportunities may make it possible to reduce the overall number of process steps, and even if not, it may make the process more robust or more economical.

Initial capture by anion exchange dominates the virus purification literature but it usually imposes the disadvantage of binding the majority of protein contaminants and DNA. This diminishes binding capacity for the virus species, and makes the media susceptible to fouling, and in turn, difficult to clean. If strong virus binding can be obtained at fairly low pH and/or fairly high salt concentrations, anion exchange may nevertheless serve adequately as a capture step. This marks an exception where the weak AX DEAE may be more suitable than a strong AX. At pH values below 7, DEAE will be fully charged and should exhibit binding capacity roughly equivalent to strong AXs at the same pH. In sodium hydroxide, DEAE completely loses its charge, which makes it easier to clean than strong AXs.

HIC is seldom used for capture, even in the world of protein applications. The expense of large volumes of concentrated salts is generally prohibitive, not to mention the costs associated with disposing them.

HA is generally unattractive as a capture method mostly because of the presence of metals and chelating agents in cell culture supernatants. Metals bind to and discolor HA. This has been shown not to significantly affect separation performance (136), but is a red flag to regulatory agencies and will likely require extra validation. Chelators are a greater concern because they remove calcium from the structure of HA and lead to its gradual dissolution. This will reduce the number of cycles for which a column can be used. DNA, even if digested by nuclease enzymes, binds strongly to HA and will likely limit virus binding capacity. In contrast to strong AXs however, mass balance of DNA can be achieved by washing the column with 500–600 m*M* phosphate at pH 7.

The high salt tolerance of heparin binding makes it a natural candidate for capture, but also a likely victim of competition for binding substrate by contaminating proteins. Placement as a capture step provides more opportunities to remove leached heparin that may co-elute with the product. Although this is unlikely to be a safety risk, heparin's potent anticlotting abilities make leachate removal a point of interest for regulatory authorities. Anion exchange and

TABLE 25.7. Conditions for Initial Screening of Hydroxyapatite

A: 10 m*M* sodium phosphate, pH 7
B1: 10 m*M* sodium phosphate, 1 *M* sodium chloride, pH 7
B2: 500 m*M* sodium phosphate, pH 7

Sample preparation. Add phosphate to sample so that the phosphate concentration is 10 m*M*. This assumes that the sample was prepared by SEC as discussed in the text. If the phosphate concentration is already higher than 10 m*M*, dilute as necessary to achieve 10 m*M*.

Recommended media: Ceramic hydroxyapatite CHT™, type II, 40 μm
Linear flow rate: 200 cm/h (1 mL/min on a 5 mm × 50 mm column)
Equilibrate column: 10 CV buffer A
Inject sample: As much as necessary to produce a visible virus peak
Wash: 5 CV buffer A
Elute (chloride gradient): 20 CV linear gradient to buffer B1
Hold buffer B1 until conductivity is stable
Clean: B2
Or, Elute: phosphate gradient: 20 CV linear gradient to B2
Hold/Clean: B2

Note: Sample must not contain chelating agents. Use phosphate monohydrate or hexahydrate. Avoid anhydrous phosphate (120). If the virus does not elute in a chloride gradient at 10 m*M*, increase the phosphate concentration in A and B1 to 20 m*M*. If it still does not elute, increase it to 40, etc. until it does. The best separation performance will usually be observed at the minimum phosphate concentration that supports virus elution in less than 1 *M* sodium chloride.

HIC should both be effective for leached heparin removal, the former because its strong attraction to heparin, the latter because the high salt operating conditions will likely suspend heparin's charge attraction to the product, allowing it to be washed away.

SEC is superficially unattractive as a capture step because of its low capacity and slowness relative to adsorptive methods, but it can be practical with preconcentration of the feed stream, short columns, large sample loads, and high flow rates. As noted above, it also facilitates process continuity by offering buffer exchange in tandem with product fractionation. For example, it can be used to remove most of the contaminants that might compete with virus for binding capacity on AXs, and simultaneously equilibrate the sample for direct application to AXs following elution from the SEC column. It may serve equally well as a precursor to any other method. The additional capability to remove product-complexed contaminants makes it even more attractive, wherever it may be placed in a process.

Biospecific affinity chromatography, if available for a given virus species, usually best fulfills the various needs of a capture step. It is sometimes suggested that it may be better placed later in a process to protect the expensive media from fouling, but commercial experience with direct IgG capture on protein A has shown that it can consistently meet process specifications for hundreds of cycles. Placement as a capture step also provides more opportunities downstream for removal of leached affinity ligand.

The second priority of screening is to identify one or more intermediate purification candidates that bind a product over a wide range of loading conditions and remove a subpopulation of contaminants different from the capture method. This step is likely to be the focal point in the process for removing product-related impurities such as aggregates, fragments, or empty capsids. If bioaffinity was used for initial capture, an intermediate step will also be necessary to remove leached affinity ligand. Ion exchange, HIC, and HA, are all good candidates for this application and will offer their best capabilities in bind-elute mode.

The third priority of screening, as it applies to adsorption chromatography methods, is to identify one or more candidates that fail to bind the product under a given set of conditions, while selectively removing a subset of stronger binding contaminants. Such methods are candidates for flow-through applications. As noted, flow-through applications may be more attractive than bind-elute applications for their relative convenience, but they are less able to reproducibly support high resolution separations, and are completely unable to fractionate virus from more weakly binding contaminants. Still, if they are able to remove a key contaminant or subset of contaminants, they may be useful.

Preliminary screening will usually reveal candidates in all three priority categories, and probably suggest process sequences that support good continuity without requiring extensive dilution or diafiltration. For example, a virus population may elute at high salt from a CX capture step. CX would also be effective for DNA removal since it is repelled from the negatively charged surface of the exchanger. SEC would remove remaining protein contaminants, eluting the sample in a neutral or slightly alkaline buffer at a low salt

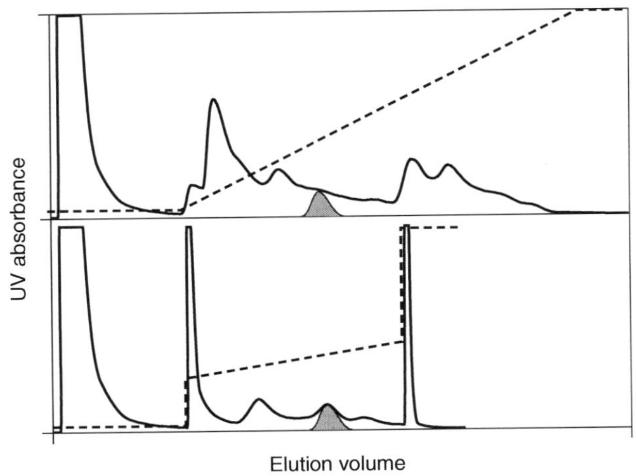

Figure 25.8. Initial screening profile versus roughly optimized profile, ion exchange. Upper trace illustrates results from the screening run. Virus elution position, determined from injection of SEC-fractionated sample, is indicated in gray. Note the impossibility of visually detecting the virus peak without the aid of the SEC reference. The lower profile indicates the results of rough optimization. A step is added after loading to eliminate the majority of contaminants. Another step is added after the virus elutes to eliminate remaining contaminants. As indicated, both steps can be optimized further, but final optimization and conversion to steps is better left until after the final media have been selected and loading capacity determined. Refer to text for further discussion.

concentration that supports eluted sample application with little or no dilution to an AX column. Screening results may also suggest an alternative sequence with the same methods. For example, the virus population, bound at pH 6 to a CX, may elute at pH 7.5, facilitating direct application to the AX column, which is eluted with a salt gradient, and the sample applied to an SEC column and eluted in the intended final product formulation. Before modeling and comparing these sequences however, it will usually be worthwhile to do some rough optimization. Figure 25.8 provides an example. The upper profile shows initial screening results of crude sample on a CX at pH 6. The lower profile illustrates a roughly optimized profile in which early contaminants are eluted in a step, the linear gradient interval is shorter and shallower, providing better resolution, and the remaining contaminants are removed in a cleaning step. Developing these conditions will require conducting a series of experiments, but with monoliths, such a series can be competed in a morning or an afternoon. The AX step can be similarly refined. The SEC step may be modified to include a contaminant decomplexation treatment as discussed above. At this point, the entire process can be modeled. And, at this point, perhaps for the first time, it will make sense to do comprehensive testing of key fractions.

If one or more model sequences provide promising results, this will be the time to initiate more comprehensive process optimization. A key part of that optimization is to explore other media, if desired, since the steps from this point forward need to be developed on the media that will be used in the final process. Evaluate alternative media first for selectivity. Initially apply the same binding and elution conditions used from the rough optimization. It is best to continue with linear gradients at this point because they will support more meaningful comparison between media. Virus peaks may elute slightly earlier or later from one chromatography product to another but should still elute within the gradient. If not, extend the gradient as necessary. Be prepared also to encounter significant differences in the ability of various media to achieve fine separations, for example, among product-related variants and impurities.

With media selected, the next step is to determine virus binding capacities. Whatever media are ultimately chosen, binding capacity depends not only on accessible surface area and mass transport efficiency, but also on the surface properties of the virus, the buffer conditions, and how much resolution is required. The practical result is that capacity can only be determined empirically. One way to approach capacity determination is to run each step with an arbitrary load, then double the load in each of a series of experiments until the method fails to deliver the required purification performance. Run the process at the highest load that fulfills specifications. Data from this series of runs can also be used to estimate column volume requirements at each step of the process.

The dramatic discrepancy between virus binding capacity on porous particles and convective media (monoliths and membranes) strongly favors the latter, but the low virus capacity of porous particle media could be advantageous in at least one special case. A porous particle column of a particular type may be used as a precolumn for high capacity removal of proteins and other "small" contaminants, allowing most of the virus population to flow through to be captured on the convective support. This treatment can be applied with any method but particularly suggests itself for enhancing the effectiveness of anion exchange as a capture step. Removing most of the proteins and nuclease-digested DNA fragments would conserve the bulk of the binding capacity for virus on the succeeding convective AX. The particle column would be removed from the flow stream prior to eluting the monolith. Figure 25.2, an SEC profile of a sample initially purified by anion exchange, provides a suggestion of the potential of this approach. A particle-based Q precolumn could potentially have removed most of the contaminants eluting after the void volume. Judging from the profile, this appears to be at least 90% of the total UV signal, suggesting that virus binding capacity of the monolith might have been increased by a factor of 10, and its size reduced by the same factor. The main liability of this approach is the inevitable loss of virus on the particle column. This can be minimized by reducing its size to only the volume required to bind the "small" contaminants. A highly contaminant feed stream will also be helpful since many of the particle surface charged sites that could bind virus would instead be occupied by proteins and DNA. A secondary liability is that the relatively slow flow requirements of the particle column would impose the same restriction on the otherwise more rapid convective media. A wide-diameter shallow bed column could be used to increase volumetric flow rate through the particle media. If the particle column is still limiting, it can be run independently from the monolith. Used particle media may optionally be discarded after use, potentially saving development, validation, and manufacturing costs associated with its cleaning, sanitization, storage, and reuse.

With capacity determined, linear gradients can finally be converted to step gradients (Fig. 25.9). The reason it makes sense to wait until this juncture is that column load partly defines elution conditions. The higher the load, the lower the concentration of eluent required to displace product from the column. Thus, a wash step determined on an underloaded column may inadvertently elute some product on a fully loaded column. This is a key cautionary point as well, because it highlights the importance of column load in maintaining lot-to-lot reproducibility of a purification process. Conditions for wash and elution steps should be evaluated over a sufficient range of loads to generate confidence that lot-to-lot variations in feed stream content, buffer conditions, and chromatography media will not affect the ability of the process to yield consistent product quality.

It is not possible to predict the number of steps a purification process will require. It will depend on the target specifications, the composition of the feed stream,

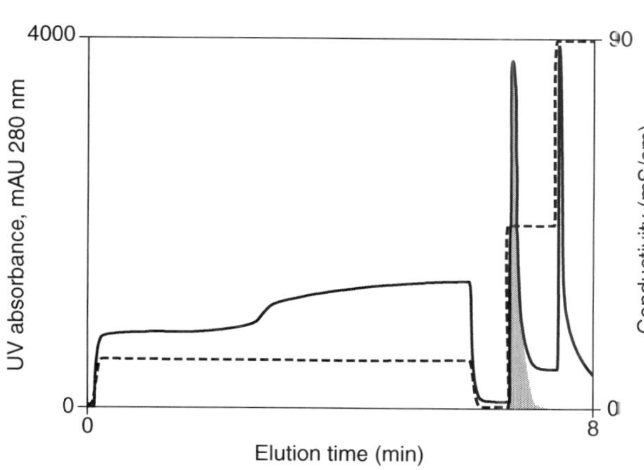

Figure 25.9. Capture of influenza virus H1N1 with an anion exchange monolith (CIM QA). The dash line indicates conductivity. Purity of the main virus fraction (shown in gray) can be inferred from the corresponding SEC profile in Fig. 25.2; probably in the range of 5–10%. See Ref. 50 for conditions. [Redrawn from Ref. 50, with permission].

the retention characteristics of the particular virus, and the ability of the various methods to remove contaminants and product-related impurities. For nontherapeutic applications, one-step purifications may be feasible. Two-step procedures may be sufficient if there is no need to remove contaminating virus or product-related impurities. More stringent specifications will probably require at least three steps.

25.15 CONCLUDING REMARKS

The above discussion makes process development sound like a linear process. This sometimes proves to be the case but there is usually an iterative component. For example, running the first generation process models may reveal deficiencies that require evaluation of additional models from the initially screened methods. If these methods fail to produce the required performance, it may be necessary to screen a wider diversity of media, and once again run complete process models. It may occur that a change in the cell culture production process imposes unanticipated changes on a purification process. It may also occur that unexpected product behaviors are encountered, such as precipitation at low salt concentrations or spontaneous inactivation under certain conditions, either of which may restrict the range of tools and conditions that can be used for purification. In all cases, the above suggestions should be considered as guidelines. Use them where they serve; ignore them where they don't.

Several examples in the text have been drawn from the field of protein purification, which raises the question, how much guidance do proteins really provide for virus purification. Some aspects of the technology are transparent to the particular application, for example, the chemical mechanisms by which solutes bind to various chromatographic supports, and the strategies by which they can be eluted. Also like proteins, intact monodisperse virus particles of a given species will usually elute in a single peak. If there is heterogeneity, there is a reason for it, and it may have clinical significance. Likewise, every virus species will exhibit unique retention characteristics on each chromatographic method, while retention behaviors among species may range over the entire spectrum. Closely related virus species may be expected to exhibit similar chromatographic behavior, but maybe not: the field of IgG purification has taught that industry the hard lesson that even antibodies of the same class and subclass can exhibit remarkable levels of variation from one clone to another. Probably, the greatest practical difference between virus and protein purification lies in the suitability of different chromatography supports. The diffusive limitations and sheer effects of porous particle media are tolerable for proteins, but they impose immense disadvantages on virus purification. Monoliths and membranes, which are disadvantaged by relatively low capacity for protein purification, support high capacity, high resolution, and low sheer performance that are ideal for virus purification. The successful history of porous particle-based virus purification to date is a testament to the improvements that can be expected with the growing application of convective supports.

This chapter overall may create the impression that exclusively chromatographic purification is the goal towards which the industry should be evolving. The goal is fictional to begin with. At the very least, filtration methods are fully integrated with chromatography methods for virtually all commercial purification processes; for removal of particulates (microfiltration), for product concentration (ultrafiltration), and for buffer exchange (diafiltration). Chromatography methods can also be integrated effectively with precipitation or density gradient separation methods. Chromatography undeniably supports levels of purification that cannot be achieved by other methods, better process control, better reproducibility, easier scalability, and faster throughput. All of these features serve the goals of regulatory compliance and/or economical manufacture, and are the main drivers for chromatography's dominance in the field of protein purification. But even in protein purification, there is a small but vocal contingent that espouses a philosophy of ABC: Anything But Chromatography. At the end of the day, select the combination of methods that most effectively allows you to achieve the specifications you require for your particular product.

25.16 RECOMMENDED READING

The reviews by Rodriguez *et al*. (1) Segura *et al*. (87), and Morenweiser (7) all address the trend toward chromatographic processing in virus purification, and are outstanding points of entry into the virus purification literature. The articles by Gavin and Gagnon (131,132) provide an introduction to regulatory issues pertinent to chromatographic purification of viruses for human injectable applications.

REFERENCES

1. Rodriguez T, Carondo M, Alves P, Cruz P. J Biotechnol 2007; 127: 520–541.
2. McGrath M, Witt O, Pincus T, Weissaman I. J Virol 1998; 25: 923–927.
3. Slepushkin V, Chang N, Cohen R, Gan Y, Jiang B, Deausen E, Berlinger D, Binder G, Andre K, Humeau L, Dropulic B. Bioprocess J 2003; 2: 89–95.
4. Transfiguracion J, Jaalouk D, Ghani K, Galpeau J, Kaman A. Hum Gene Ther 2003; 14: 1139–1153.
5. Transfiguracion J, Jorio H, Meghrous J, Jacob D, Kaman A. J Virol Methods 2007; 142: 21–28.
6. Peixoto C, Sousa M, Silva A, Carondo M, Alves P. J Biotechnol 2007; 127: 452–461.
7. Morenweiser R. Gene Ther 2005; 1(12 Suppl): S103–S110.
8. Volkin D, Burke C, Marfia K, Oswald C, Wolanski B, Middaugh C. J Pharm Sci 2007; 86: 666–677.
9. Kalbufuss B, Wolff M, Morenweiser R, Reichl U. Biotechnol Bioeng 2007; 96: 932–944.
10. Kim S, Jeong H, Park S, Kim H. J Virol Methods 2007; 139: 24–30.
11. Smith R, Ding C, Kotin R. J Virol Methods 2003; 114: 115–124.
12. Kamen A, Henry O. J Gene Med 2004; 6(Suppl 1): S184–S192.
13. Burova E, Ioffe E. Gene Ther 2005; 12(Suppl 1), S5–S17.
14. Smith R, Yang L, Kotin R. Methods Mol Biol 2008; 434: 37–54.
15. Hagel L, Janson J-C, Ryden L,editors. Protein Purification: Principles, High resolution Methods, and Applications, New York: VCH Publishers; 1989. pp. 63–106.
16. Gagnon P. Purification Tools for Monoclonal Antibodies, Tucson: Validated Biosystems; 1996.
17. Maurer E, Brazzale T, Muster T, Gassner M, Seper H, Gelhart F, Banjac M, Lah B, Kramberger P, Meznar N, Urbas L, Barut M, Peterka M. An industrial platform for purifying influenza virus particles, Poster, IBC conference on Next Generation Vaccines, Baltimore, 2008
18. Gagnon P. BioProcess Int 2008; 6(Suppl 6): 24–30.
19. Aldington S, Bonnerjea J. J Chromatogr B 2006; 848: 67–78.
20. Bruck C, Derbin J, Glineur C, Portatelle D. Methods Enzymol 1986; 121: 587–595.
21. Scopes R. Anal Biochem 1987; 165: 235–246.
22. Jones K. LC-GC 1991; 4(9): 32–37.
23. Fujita T, Suzuki Y, Mauti J, Takagahara I, Fujii K, Yamashita J, Horio T. J Biochem 1980; 87: 89–100.
24. Rubenstein M, Familletti P, Miller RS, Waldman A, Pestka S. Proc Nat Acad Sci USA 1979; 76: 640–644.
25. Pfankoch E, Lu K, Regnier F, Barth H. J Chrom Sci 1980; 18: 430–441.
26. Janado M, Shimada K, Nishida T. J Biochem 1976; 79: 513–520.
27. Ejima D, Yumioka R, Arakawa T, Tsumoto K. J Chromatogr A 2005; 1094: 49–55.
28. Tsumoto K, Ejima D, Senczuk A, Kita Y, Arakawa T. J Pharm Sci 2007; 96: 1677–1690.
29. Shukla A, Jiang C, Rubaca M, Flansburg L, Lee S. Biotechnol Prog 2008; 24(3): 615–622.
30. Luhrs K, Harris D, Summers S, Parsehgian M. J Chromatogr B 2009; 877: 1543–1552.
31. Wright J, Le T, Prado J, Bahr-Davidson J, Smith P, Zhen Z, Sommer J, Pierce G, Qu G. Mol Ther 2005; 12: 131–178.
32. Konz J, Lee A, Lewis J, Sagar S. Biotechnol Prog 2005; 21: 466–472.
33. Floyd R, Sharp D. Appl Environ Microbiol 1978; 35: 1084–1094.
34. Jungbauer A. J Chrom A 2005; 1065: 3–12.
35. Wang R, Wang J, Li J, Wang Y, Xie Z, An L. J Virol Methods 2007; 139: 125–131.

36. Gagnon P, Richieri R. Productivity improvements in the capture and initial purification of monoclonal antibodies, Oral presentation, Second Annual Conference on Purification of Biological Products, Thousand Oaks, CA September 18-20, 2006. http://validated.com/revalbio/pdffiles/PUR_MassTrans.pdf
37. Frankovic V. Characterization of a grafted weak anion methacrylate monolith, Oral presentation, 3rd International Monolith Symposium, Portoroz, May 30-June 4 2008.
38. Gagnon P, Richieri R, Aolin F, A comparison of microparticulate, membrane, and monolithic anion exchangers for polishing applications in purification of monoclonal antibodies, Poster, BioProcess International Conference and Exhibition, Boston, October 1-4, 2007. http://validated.com/revalbio/pdffiles/IBCBOS07a.pdf.
39. Gagnon P. Eliminating the downstream processing bottleneck with monoliths and simulated moving bed chromatography, Oral presentation, BioProcess International Conference and Exhibition, Anaheim, Sept. 23-26, 2008. http://validated.com/revalbio/pdffiles/LithSMB.pdf
40. Strancar A, Podgornik A, Barut M, Necina R. Adv Biochem Eng Biotechnol 2002; 76: 49–85.
41. Giddings J. Dynamics of Chromatography: Part I., New York: Marcel Dekker; 1965.
42. Hearn MT. Adv Chromatogr 1982; 20: 1–64.
43. Kirkland J, Truszkowski F, Dilks C Jr, Engel G. J Chromatogr A 2000; 890: 3–13.
44. Afeyan N, Gordon N, Mazaroff J, Varaday C, Fulton S, Yang Y, Regnier F. J Chromtogr 1990; 519: 1–29.
45. Iberer G, Hahn R, Junbauer A. LC-GC Int 1999; 11: 998–1005.
46. Svec F, Tennikova T, Deyl Z. Monolithic Materials: Preparation Properties and Applications, Amsterdam: Elsevier; 2003.
47. Podgornik A, Strancar A. Biotechnol Ann Rev (Suppl) 2005; 11: 281–333.
48. Hahn R, Panzer M, Hansen E, Mollerup J, Jungbauer A. Sep Sci Technol 2002; 37:(7): 1545–1565.
49. Maurer E. Influenza vaccine purification platform, Oral Presentation, 3rd International Monolith Symposium, Portoroz, May 30-June 4 2008.
50. Maurer E, Peterka M, Gassner M, Seper H, Gelhart F Jarc M, Lah B, Kramberger P, Strancar A, Muster T. Influenza virus purification platform, Poster, Wilbio Conference onViral Vectors and Vaccines, Austin, October 29-31, 2007.
51. Banjac M, Kramberger P, Lah B, Strancar A, Maurer E, Gelhart F, Seper H, Muster T, Peterka M. Comparison of ion exchange ligands for purification of different influenza viruses, Poster, 3rd International Monolith Symposium, Portoroz, May 30-June 4, 2008.
52. Etzel M, Svec F, Tennikova T,editors. Monolithic Materials, Amsterdam: Elsevier; 2003. p. 213.
53. Yamamoto S, Kita A. Trans ICHemE, Part C 2005 84: 72–77.
54. Etzel M. Charged membranes and monoliths in chromatography, Oral Presentation, 3rd International Monolith Symposium, Portoroz, May 30-June 4, 2008.
55. Qu G, Barr-Davidson J, Proado J, Tai A, Craniag F, McDonnell J, Zhou J, Hauck B, Luna J, Sommer J, Smit P, Zhou S, Colosi P, High K, Pierce G, Wright J. J Virol Methods 2007; 140: 183–392.
56. Vicente T, Sousa M, Peixoto C, Mota J, Alves P, Carondo M. J Memb Sci 2008; 311: 270–283.
57. Specht R, Hahn B, Wickramasinghe S, Carlson J, Czermak P, Wolf A, Reif O. Biotechnol Bioeng 2004; 88: 465–473.
58. Kalbfuss B, Wolff M, Geisler L, Tappe A, Wickramasinghe R, Thom V, Reichl U. J Memb Sci 2007; 299: 252–260.
59. Wu C, Soh K, Wang S. Hum Gene Ther 2007; 18(7): 665–672.
60. Urthaler J, Schegl R, Podgornik A, Strancar A, Jungbauer A, Necina R. J Chromatogr A 2005; 1065: 93–106.
61. Smrekar F, Ciringer M, Peterka M, Podgornik A, Strancar A. J Chromatogr B 2005; 861: 177–180.
62. Boratynski J, Syper D, Weber-Dabrowska B, Lusiak-Szelchowska M, Pozniak G, Gorski A. Cell Mol Biol Lett 2004; 9: 253–259.
63. Vellekamp G, Porter F, Sujipto S, Cutler C, Bondoc L, Liu Y-H, Wylie D, Cannon-Carlson S, Tang J, Frei A, Voloch M, Zhuang S. Hum Gene Ther 2001; 12(15): 1923–1936.
64. Huyghe B, Liu X, Sujipto S, Sugarman B, Horn M, Shepard H, Scandella C, Shabram P. Hum Gene Ther 1995; 6(11): 1403–1416.
65. Kaludov N, Handelman B, Chiorni J. Hum Gene Ther 2002; 13(10): 1235–1243.
66. Kramberger P, Petrovic M, Strancar A, Ravnikar M. J Virol Methods 2004; 120: 51–57.
67. O'Riordan C, LaChapelle A, Vincent K, Wadsworth S. J Gene Med 2000; 2: 444–454.
68. Kramberger P, Petrovic N, Strancar A, Ravnikar M. J Virol Methods 2004; 120: 51–57.
69. Kramberger P, Peterka M, Boben J, Ravnikar M, Strancar A. J Chromatogr A 2007; 1144: 143–149.
70. Brument N, Morenweiser R, Bloin V, Toublanc E, Rimbaud I, Chere Y, Folliot S, Gaden F, Boulanger P, Kroner-Lux G, Moullier P, Rolling F, Salvetti A. Mol Ther 2002; 6: 678–686.
71. Wu C, Ker Y, Wang S. Hum Gene Ther 2007; 18: 665–672.
72. Rodrigues T, Carvalho A, Carmo M, Carrondo M, Alves P, Cruz P. J Gene Med 2007; 9(4): 233–243.
73. Rodrigues T, Carvalho A, Roaldo A, Carrondo M, Alves P, Cruz P. J CHromatogr B 2006; 837: 59–68.
74. Strauss D, Gorell J, Plancarte M, Blank G, Chen Q, Yang B. Biotechnol Bioeng 2009; 102: 168–175.
75. Konz J, Livingood L, Bett A, Goerke A, Laska M, Sagar S. Hum Gene Ther 2005; 16: 1346–1353.
76. Konz J, Pitts L, Sagar S. Methods Mol Biol 2008; 434: 13–23.
77. Yamada K, McCarthy D, Madden V, Walsh C. Biotechniques 2003; 34(5): 1074–1080.
78. Peixoto C, Ferreira T, Sousa M, Carrondo M, Alves P. Biotechnol Prog 2008; 24(6): 1290–1296.
79. Trilisky E, Lenhoff A. J Chromatogr A 2007; 1142: 2–12.
80. Zolotukhin Z, Potter M, Zolotukhin I, Sakai Y, Loiler S, Fraites T Jr, Chiodo V, Phillpsberg T, Muzyczka N, Hauswirth W, Flotte T, Byrne B, Snyder R. Methods 2002; 2: 158–167.
81. Davidoff A, Ng C, Sleep S, Gray J, Azam S, McIntosh J, Karimpoor M, Nathwani A. J Virol Methods 2004; 121: 209–215.

82. Debelak D, Fisher J, Iulano S, Sesholtz D, Sloane D, Atkinson E. J Chromatogr B 2000; 740: 195–202.
83. Chahal P, Aucoin M, Kamen A. J Virol Methods 2007; 139: 61–70.
84. Lee D, Kim B, Seol D. Biochem Biophys Res Commun 2009; 378: 640–644.
85. Coleen A, Huang J, Scott M, Kierstend D, Beaupre I, Gao G, Wilson J. Hum Gene Ther 2002; 13: 1921–1934.
86. Brorson K, Shen H, Lute S, Perez J, Frey D. J Chromatogr, A 2008; 1207: 110–121.
87. Segura M, Kamen A, Garnier A. Biotechnol Adv 2006; 24: 321–337.
88. Kaludov N, Handelman B, Chiorini J. Hum Gene Ther 2002; 13: 1225–1243.
89. Karlsson E, Ryden L, Brewer J, Janson J-C, Ryden L,editors. Protein Purification: Principles, High Resolution Methods, and Applications, New York: VCH Publishers; 1989. pp. 107–148.
90. Ghose S, McNerney T, Hubbard B. Biotechnol Prog 2002; 18: 530–537.
91. Perez J, Frey D. Biotechnol Prog 2005; 21: 902–910.
92. Perez J, Frey D. Ion Exchange Chromatography: Principles and Methods, Uppsala: Pharmacia; 1980.
93. Perez J, Frey D. FPLC Ion Exchange and Chromatofocusing: Principles and Methods, Uppsala: Pharmacia; 1991.
94. Brautigan D, Ferguson-Miller S, Margoliash E. J Biol Chem 1978; 253: 130–139.
95. Regnier F. Science 1987; 238: 319–323.
96. Dismer F, Hubbuch J. J Chromatogr A 2007; 1149: 312–320.
97. Dismer F, Petzold M, Hubbuch J. J Chromatogr A 2008; 1194: 11–21.
98. Sluyterman L, Elgersma O. J Chromatogr 1978; 150: 31–44.
99. Zizkovsky V, Strop P, Lukesova S, Korkacova J, Dvorak P. Oncodev Biol Med 1981; 2: 323–330.
100. Gagnon P, Grund E, Lindbäck T. BioPharm 1995; 8:(3): 21–27.
101. Hofstee B. Biochem Biophys Res Commun 1975; 63(3): 618–624.
102. Rosengren J, Pahlmann S, Glad M, Hjerten S. Biochim Biophys Acta 1975; 412: 51–61.
103. Kunitani M, Cunico R, Staats S. J Chromatogr 1988; 443: 205–220.
104. Wu H, Figueroa A, Karger B. J Chromatogr 1986; 371: 3–27.
105. Utsunomiya H, Ichinose M, Tsujimoto K, Katsuyama Y, Yamasaki H, Koyama A, Ejima D, Arakawa T. Int J Pharm 2009; 366: 99–102.
106. FDA FDA CDER Inactive Ingredient Search for Approved Drug Products, 2009, http://www.accessdata.fda.gov/scripts/cder/iig/index.cfm
107. Arakawa T, Kita Y, Koyama A. Biotechnol J 2009; 4: 174–178.
108. Yamasaki H, Tsujimoto K, Koyama A, Ejima D, Arakawa T. J Pharm Sci 2008; 97: 3067–3073.
109. Katsuyama Y, Yamasaki H, Tsujimoto K, Koyama A, Ejima D, Arakawa T. Int J Pharm 2008; 361: 92–98.
110. Kuiper M. Biotechnol Bioeng 2002; 80(4): 445–453.
111. Tusuru S. Bio-Med Mat Eng 1991; 1: 143–147.
112. Srivastava A. Trop Med 1987; 29(4): 187–194.
113. Cynthia S. J Chromatogr 1985; 326: 191–197.
114. Raptis L. Biochim Biophys Acta 1981; 653(3): 331–343.
115. Gagnon P, Ng P, Aberin C, Zhen J, He J, Mekosh H, Cummings L, Richieri R, Zaidi S. BioProcess Int 2006; 4(2): 50–60.
116. Juarez-Salinas H, Ott G, Chen J. Methods Enzymol 1986; 1221: 615–622.
117. Vola R, Lombardi A, Mariani M. BioTechniques 1993; 14: 650–655.
118. Tarditi L, Camagna M, Parisi A, Vassarotto C, Delmonti L, Letarte M, Malvasi F, Mariani M. J Chromatogr 1992; 599: 13–20.
119. Ford C, Osborne P, Mathew A, Rego B. J Chromatogr B 2001; 754: 427–435.
120. Bio-Rad Laboratories CHT Ceramic Hydroxyapatite Instruction Manual, LIT611 rev E. 2007.
121. Segura M, Kamen A, Trudel R, Garnier A. Biotechnol Bioeng 2005; 90(4): 391–404.
122. Segura M, Kamen A, Garnier A. Methods Mol Biol 2008; 434: 1–11.
123. Segura M, Kamen A, Garnier A. Biotechnol Adv 2006; 24(3): 321–337.
124. Segura M, Kamen A, Lavoie M, Garnier A. J Chromatogr B 2007; 846: 124–131.
125. O'Keefe R, Johnston M, Slater N. Biotechnol Bioeng 1999; 62(5): 537–545.
126. Thompson L, Pantoliano M, Springer B. Biochemistry 1994; 33: 3831–3840.
127. Ejima D, Yumioka R, Tsumoto K, Arakawa T. Anal Biochem 2005; 345: 250–257.
128. Arakawa T, Kita Y, Tsumoto K, Ejima D, Fukada H. Prot Pept Lett 2006; 13: 921–927.
129. Detmers F, Hermans P, ten Haaft M. LC-GC 2007; 7(9): 13–17.
130. GE Healthcare. AVB Sepaharose High Performance Data file 28-9207-54AA, 2007.
131. Gavin D, Gagnon P. BioProcess Int 2006; 4(10): 22–30.
132. Gavin D, Gagnon P. BioProcess Int 2006; 4(11): 28–34.
133. Dasarathy Y. BioPharm 1996; 9(8): 41–44.
134. Ng P, McLaughlin V. Bioprocess Int 2007; 5(5): 52–56.
135. Eglon M, Banjac M, Strappe P, O'Brien T, Lah B, Strancar A, Peterka M, Development of a fast and reliable chromatography method for adenoviral vector purification using methacrylate monoliths, Poster, 3rd International Monolith Symposium, Portoroz, May 30-June 4, 2008.
136. Shepard S, Brinckman-Stone C, Scrimsher L, Koch G. J Chromatogr A 2000; 891: 93–98.
137. Fausnaugh-Politt J, Thevenon G, Janis L, Regnier F. J Chromatogr 1988; 443: 221–228.
138. Kopaciewicz W, Rounds M, Fausnaugh J, Regnier F. J Chromatogr 1983; 266: 3–21.
139. Scopes R, Algar E. FEBS Lett 1979; 106: 239–242.
140. Fägerstram L, Söderberg L, Whalström L, Fredriksson U, Plith K, Walden E Peeters H,editors. Protides of the Biological Fluids. Volume 30, Oxford: Pergamon Press; 1982. pp. 621–628.

26

CHROMATOGRAPHY, HYDROPHOBIC INTERACTION

Per Kårsnäs

Institute of Biology and Chemical Engineering, Mälardalens högskola, Eskilstuna, Sweden

26.1 INTRODUCTION

Hydrophobic interaction chromatography (HIC) is an important chromatography technique used for the separation of biomolecules. The interaction is based on physicochemical characteristics of the molecules other than the charge, as in ion-exchange chromatography (IEC), or the size, as in gel filtration. The importance of the technique is high also in industrial applications. The technique is certainly very powerful when optimal conditions for a certain separation are determined, but to achieve this is still often far from easy. One reason is that the mechanism of the interaction is complex, depending on several properties of the separation medium as well as of the separated substances, and not yet fully understood. There are a number of hypotheses that approach the phenomenon of hydrophobic interaction from different points of view, but the practitioner will still find that there is no simple theory that will always facilitate the optimization of a separation, as is the case when using other chromatography techniques. Hydrophobic interaction is favored by high salt concentration and is thus a logical purification step after salt precipitation or IEC, where the environment has a high ionic strength.

Chromatography–hydrophobic interaction (HIC) has been used as a group separation method, rapidly separating what retards on the column from what is not interacting, and as a powerful high-resolution technique with gradients of descending salt concentration.

In addition to what is usually designated as HIC, there are several related chromatography techniques based entirely or partly on hydrophobic interaction. Reversed-phase chromatography (RPC) is the most important, in which strong hydrophobic interaction is often achieved by high substitution with long aliphatic chains. The principal difference between HIC and RPC is methodological in nature. In HIC, the interaction is increased by salt in order to retard the sample on the column and, basically, a return to native conditions will elute the compounds, whereas in RPC the sample is retarded on the column and eluted by addition of organic solvents. The methods based on what are described as salt-mediated and thiophilic interactions gain in importance, whether they should be considered as HIC techniques is perhaps a matter of semantics. Moreover, the binding in affinity chromatography and to dye–ligand matrixes often have significant elements of hydrophobic interaction.

26.2 HYDROPHOBIC INTERACTION

26.2.1 Basic Theory

There are several approaches to theoretically elucidate the nature of hydrophobic interaction, ranging from thermodynamics over surface tension to van der Waals forces (1–3). From different angles, they all try to explain the observed influence on hydrophobic interaction by ions with salting-out effect, chaotropic ions, type of salt, temperature, pH, and additives. No single theory, however, manages to fully cover the mechanisms of interaction. In HIC, the added complexity, contributed by matrixes, spacers, ligands, and the effects of buffer environment on different species of biomolecules, implicates that, apart from very

Downstream Industrial Biotechnology: Recovery and Purification, First Edition. Edited by Michael C. Flickinger.

general recommendations, the design of a HIC procedure still has to be founded on careful practical and empirical studies.

26.2.2 Salting Out and Hydrophobic Interaction

There is definitely a relationship between the salting-out process used for the precipitation of proteins and hydrophobic interaction as used in chromatography. The Hofmeister series lists anions and cations according to decreasing ability to precipitate and increasing ability to dissolve proteins, and the combinations (salts) from the beginning of the list are well known to promote hydrophobic interaction (4–7) (Fig. 26.1). Probably precipitation by salting-out and HIC are based on the same phenomenon, the principal difference being that in precipitation the hydrophobic interaction has to take place between molecules of the same kind or others that happen to be present in the solution, whereas in chromatography a well-defined counterpart is hydrophobic enough to bind some of the passing compounds, preferably at lower salt concentration than is needed for precipitation.

26.2.3 Thermodynamic Theory

The main and very elegant theory of hydrophobic interaction (8) refers directly to the Gibbs free energy equation ($\Delta G = \Delta H - T\Delta S$), which describes the change in energy when a chemical system is transferred from one state to another:

Anions: Increasing "salting-out" effect ←

PO_4^{3-}; SO_4^{2-}; CH_3COOH^-; Cl^-; Br^-; NO_3^-; ClO_4^-; I^-; SCN^-

(a)

Cations: Increasing chaotropic effect →

NH_4; Rb^+; K^+; Na^+; Cs^+; Li^+; Mg^{2+}; Ca^{2+}; Ba^{2+}

(b)

Salts promoting the molal surface tension of water:

$Na_2SO_4 > K_2SO_4 > (NH_4)_2SO_4 > Na_2HPO_4 > LiCl > KSCN$

(c)

Figure 26.1. The Hofmeister series of anions (a) and cations (b). The *arrows* indicate increasing salting-out effect and chaotropic effect, respectively. Salts with high ability to promote hydrophobic interaction are combinations of ions found to the left in the series. The same salts are found early in the series of salts listed in order of decreasing ability to promote the surface tension of water (c). This indicates the relationship between hydrophobic interaction and surface tension. Which salt to chose depends on the solubility and the production cost of the salt as well as the cost of preventing pollution of the environment by waste buffers.

Where [$\Delta G = G_2 - G_1$] is the change in Gibbs free energy, [$\Delta H = H_2 - H_1$] is the change in enthalpy or heat content, [T] is the temperature in Kelvin, and [$\Delta S = S_2 - S_1$] is the change in entropy or disorder of the system. If [ΔG] is negative, the second state has a lower energy and the system will adapt to it, provided that there are no energy barriers. In hydrophobic interaction, the entropy change is of greatest importance (8). If hydrophobic molecules such as aliphatic carbon chains are immersed in water, the monolayer of water molecules in contact with a carbon chain will be in higher order than those in the bulk of the water. If two or more hydrophobic structures come together, the surface that has to be covered by ordered water molecules decreases, some water molecules join the less-ordered bulk water, and the entire system gains in entropy and thus, according to equation 26.1, decreases its free energy (Fig. 26.2). This state will be favorable for energetic reasons and thus promoted. If some of the hydrophobic molecules are attached to a chromatographic matrix as ligands, the setup for HIC is obvious (Fig. 26.3).

The phenomenon of increased interaction in the presence of salting-out ions is explained by a higher gain in entropy when water is transferred from the surface of a nonpolar molecule to the bulk of water. Whether this is achieved because the surface water in the presence of salt is initially more ordered or the bulk water less ordered or both is not quite clear.

Equation 26.1 reveals that at a higher temperature an increase in entropy will give a higher energy gain because the entropy factor of the equation will be multiplied by a higher number. Consequently, if the binding to a medium has been shown to increase with the temperature, this has been taken as an indication that the interaction is really hydrophobic.

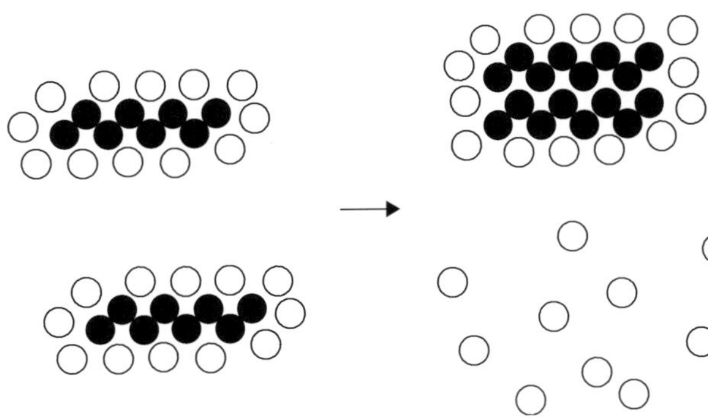

Figure 26.2. Hydrophobic interaction. Two aliphatic carbon chains (*black circles*) are immersed in water. Only the water molecules (*white circles*) with high order in the monolayer adjacent to the nonpolar molecules are indicated. When two chains are in contact, the surface area that is covered with ordered water molecules is decreased and some water is in a state of less order in the bulk of the water. The free energy of the system decreases, and thus the binding together of the two molecules is favored.

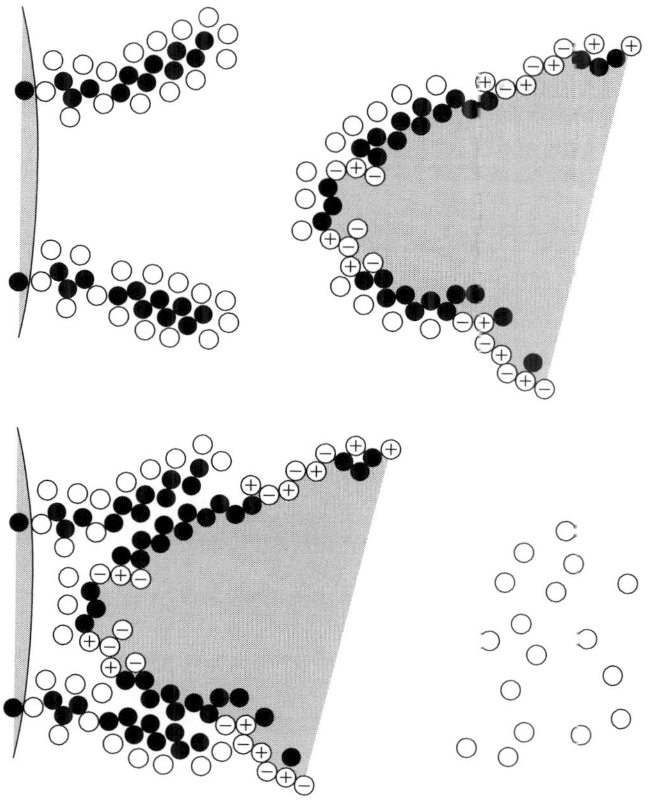

Figure 26.3. Hydrophobic interaction used for chromatography. Two octyl chains are coupled to a chromatographic medium. An energy gain, equivalent to that described in Fig. 26.1, is achieved by the interaction of the octyl chains with nonpolar moieties on a protein. The carbon atoms of the hydrophobic parts of the protein are indicated with *black circles*, the oxygen atoms of the water with *white circles*, and charged atoms or groups of the protein with *white circles* containing plus or minus signs. The *hatched area* indicates the interior part of the protein.

26.2.4 Surface Tension, van der Waals Forces

Another theory (4) suggests that the surface tension of the water surrounding a nonpolar structure and the tendency to minimize it are responsible for the interaction. If salt is added to the water, it has been shown that the surface tension increases. As a consequence, the energy gain in minimizing the surface exposed to the hydrophobic molecules is larger and thus the interaction is stronger. Some salts that influence the surface tension are listed in order of decreasing effect in Fig. 26.1. Although the salts that increase the surface tension most are among those that should promote hydrophobic interaction according to the Hofmeister series, the order is not quite the same.

A third theory suggests that van der Waals forces (5) are responsible for the hydrophobic interaction in HIC. This is supported by the fact that these forces should be increased as the order of water increases in the presence of salt. Probably the theories mentioned here all describe one part of the complicated interaction, and a combination of them is more relevant to describe the phenomenon of hydrophobic interaction.

26.2.5 Interaction based on mixed mechanisms

Several investigators confirm that the interaction in chromatography is really a mixed mechanism depending on all the different parameters including the very individual properties of a certain protein (9–11).

26.3 HYDROPHOBIC INTERACTION CHROMATOGRAPHY

26.3.1 Development of the Technique

As early as 1948, Tiselius presented what he called "adsorption separation by salting-out" (12), but not until the beginning of the 1970s were reports on the synthesis of media published. Some researchers were working with a mixed mode because of their medias' content of charges (13–16), but charge-free hydrophobic media were also synthesized (6,17,18), and the first commercially available products were presented some years later (19). At present, a wide range of products, including HIC-media for high performance liquid chromatography (HPLC), are available but phenyl, octyl, butyl, and alkyl ligands are still by far the most common (20,21). Expanded bed HIC-media for the purification of proteins from crude extracts are also available (59,60).

26.3.2 Factors Influencing HIC

In addition to the environmental parameters, including ionic strength, temperature, and pH that influence the hydrophobic interaction as such, there are several factors that have to be controlled to use the hydrophobic interaction for chromatographic purposes. The chromatographic medium is constructed by coupling hydrophobic ligands to a chromatographic matrix, often by use of a short spacer to enhance the sterical possibilities of interaction with large molecules. The type of ligand and the ligand density affect the characteristics of the medium. The choice of buffer ions has been found to change the interaction pattern more profoundly than generally anticipated. The optional use of organic solvents and detergents has to be taken into consideration. The molecules themselves have certain characteristics and will often change chromatographic interaction behavior with the buffer salts in an individual manner (9). Finally, the choice of dimensions of the chromatographic column and the mode of elution will depend on the size range of biomolecules that are subject to processing and the desired throughput of the separation step. One approach, to describe the complex interaction of HIC from a practical point of view, has been to characterize a HIC medium by the interaction pattern of a significant set of protein molecules and to evaluate the result by the use of principal component analysis (9). As described in later sections, the media can be classified by this approach, and the importance of different characteristics of the proteins can be correlated to the interaction of different media.

26.3.3 The Matrix

Agarose-based matrixes with varying degrees of cross-linking are most commonly used for HIC, but organic resins and silica-based media are also used, especially in HPLC applications. All common chromatographic matrixes, such as cross-linked agarose (22,23), organic polymers (24), and silica (25), can be derivatized with hydrophobic groups to give HIC media. Normally, the influence of the by-design hydrophilic chromatographic matrix will just add slightly, if at all, to the performance of the medium, but it is a well-known fact that unsubstituted agaroses intended for gel filtration will expose what can only be characterized as hydrophobic interaction at extremely high salt concentrations (26–28). This allows for using HIC even for rather hydrophobic proteins that would stick irreversibly to conventional hydrophobic gels. It has to be noted, though, that under normal salt concentrations (e.g. those used in gel filtration), no hydrophobic interaction will be taking place because the gels were successfully designed to be as hydrophilic as possible.

26.3.4 The Ligands

Alkyl, butyl, phenyl, and octyl groups are the most common ligands in commercially available HIC media (Fig. 26.3). They are listed in what is commonly considered as in order of increasing hydrophobicity, but as mentioned earlier there is no linear or single dimension relationship between their interaction with different species or sets of molecules. It is obvious that the aromatic part of the phenyl group can bind with π-to-π binding as well thus adding to the complexity of the interaction. Other ligands of interest include mercapto derivates, which have yet another spectrum of interaction. They have been claimed to work partly with a special thiophilic interaction, but there is some doubt whether their content of sulfur is really responsible for their interaction pattern or whether they are just another type of hydrophobic medium among the others and working by a combination of the same mechanisms as other hydrophobic media.

26.3.5 The Coupling Reactions

Glycidyl ether is the coupling agent used most often to immobilize hydrophobic ligands (20), but in principle all the common coupling procedures can be used to attach the ligands to the chromatographic matrix. An extensive review of coupling methods is presented in the work by Carlsson *et al*. (29). The coupling procedure will almost inevitably result in the ligands being attached via an often very short spacer, which is an advantage because conformational restrictions when large protein molecules bind are decreased. On the other hand, the choice of coupling chemistry will always influence the performance of the chromatographic medium by introducing hydrophobic or charged moieties, and the selectivity and the capacity may be very different even if the coupled ligand happens to be the same.

26.3.6 Binding Capacity

The binding capacity of an HIC medium depends partly on the ligand density (29–32), but it is important to remember that any figures given will always be related to the specific protein that was used to determine the capacity and to the conditions under which the determination was performed. The capacity can be determined as total capacity, with the protein exposed to the medium without flow, or as dynamic capacity, where the linear flow and the residence time of the molecules in the column will influence the result. There may well be a difference of magnitude of 10 to 20 times albumin between the capacities for different proteins, a fact that has to be considered when choosing a medium from a catalog where the dynamic capacity is often determined using serum proteins like albumin.

The capacity that can be used is dependent on the contamination picture as well and whether the main part of the impurities are more or less hydrophobic than the desired substance, so in practice, it is necessary to determine the dynamic capacity in every specific application; this should be seen as a normal part of an optimization process.

Much work is being done to optimize the dynamic capacity of media, both by suppliers and by users at research sites, but the question is whether the most convenient and cost-efficient way of increasing the capacity of a column is to increase the amount of gel.

26.3.7 HIC and RPC and Column Design

HIC and RPC are both based on hydrophobic interaction, and the borderline between the techniques is more defined by difference in methodology and the type of sample than by the interaction as such.

In RPC, the hydrophobicity of the media is higher because the ligands are more hydrophobic and the ligand density is higher. When the sample is run through an RPC column at conditions in the range of physiological pH and salt concentrations, most compounds are retarded on the column. Elution has to be achieved by using increasing gradients of organic solvents. Commonly used solvents are, in order of increasing elution ability, methanol, isopropyl alcohol, and acetonitrile.

In contrast, in HIC very little of the sample is retarded at physiological conditions. The ionic strength of the sample has to be increased to promote attraction, and the elution is normally done by decreasing salt gradients. If some components are still attached to the matrix in weak buffer, a mild organic solvent such as ethylene glycol will be used to wash them out.

As a consequence, HIC is more often used for protein separation where the task often is to retain the tertiary conformation and the biological activity of the separated molecules, whereas RPC is often used for smaller molecules, such as peptides, which lack tertiary structure and subsequently will withstand the elution conditions without denaturation.

Historically, RPC was developed as an HPLC technique to separate small molecules using mainly silica and organic polymer-based matrixes, and HIC was developed as a technique for the separation of proteins, as an alternative to IEC and gel filtration and using the common agarose matrixes as carriers of the hydrophobic ligands. The difference in molecular size range of the molecules to be separated leads to several important differences in the methodologies of the two techniques (33). The differences in behavior between small and large molecules are important in all attraction techniques, such as IEC, HIC, and RPC. But because of the restrictions to expose large molecules to organic solvents, the borderline between separation of small and large molecules happens to coincide more or less with the borderline between the two hydrophobic techniques. But there are other technical reasons behind the difference of methodology based on molecular size (33).

Small molecules interact with the matrix by single point attachment. A molecule is bound to the gel for a certain fraction of the time and is free to move with the buffer the rest of the time. Another way to express the same fact is that at a certain time one fraction of the molecules attach to the column while the rest are moving with the flow. Even if the equilibrium is shifted far toward binding, the macroscopic effect will be that small substances move down the column at some rate in a broad range of salt concentrations, although the speed may be rather low. In a mixture of substances, the rates of movement down the column will be different, as described by the individual equilibrium constants, and a separation will take place. To allow for the molecules to separate, a long path length is needed, and thus long columns and keeping the separating zones sharp are essential for high efficiency. This fact has led to a hunt for high plate numbers, a feature that can be achieved by using small matrix particles. This in turn will cause a high backpressure and the need to work under HPLC conditions. The full utilization of the selectivity of the column will be achieved by elution under isocratic conditions or by shallow gradients. The resulting extremely resolving techniques have been very successful and impressive and have been used for peptide fingerprinting and other types of qualified analysis of small organic molecules.

Proteins, on the other hand, interact with the matrix by multipoint attachment. Even if at some salt concentration, some of the interacting parts of the large molecules may be unbound, there are often several points of interaction left that keep the protein attached to the column. Not until a certain elution condition is reached, such as higher ionic strength in IEC or lower salt concentration in HIC, where all binding sites of the protein are simultaneously unbound, will the protein move down the column. The range between when the protein is not moving at all and moving at full speed is very narrow compared to when small molecules are separated. Isocratic elution is not possible because at constant buffer concentration, some proteins will move without retardation and some will hardly move at all. A gradient is needed to elute the proteins in order. Because the resolution is not very dependent on the effectiveness of the column, an HIC step where proteins are separated should be performed in short columns, and the linear flow should be kept low by increasing the diameter of the column. In large scale, the use of short and wide columns in protein separation based on IEC and HIC opens up the possibility to run with a high total flow and throughput and still keep the linear flow, low enough to allow for the adsorption and desorption to take place without being too far from equilibrium.

26.4 CLASSIFYING OF MEDIA AND MODELLING OF CHROMATOGRAPHIC RESULTS

26.4.1 Classifying HIC Media by Multivariate Analysis

The performance of HIC media can be monitored by multivariate analysis, using a set or probe of protein molecules with different hydrophobic properties. The interactions of the molecule set with the HIC media could be correlated to molecular weight, isoelectric point, fraction of hydrophobic amino acids, surface charge, and several hydrophobicity indexes (31) without the need for detailed knowledge about the parameters or the mechanisms involved. Instead, the different combinations of media and chromatography conditions can be fingerprinted by their performance in practice and easily compared. As well as artificial mixtures of proteins, serums with 15 to 20 of the proteins identified have been used as probes to determine adsorption patterns (34).

Inversely, biomolecules can be classified by their hydrophobic characteristics as measured by their interactions with a standard set of different hydrophobic media. Molecules with similar adsorption patterns over the media should have similarities on a molecular level.

The results hint that in the tested buffer system, there are at least two main mechanisms that have an impact on HIC (9,35). One mode of interaction was typical on highly hydrophobic media, for example, with an octyl chain as ligand, where the binding seems to be entirely correlated to the fraction of hydrophobic amino acids and to the hydrophobicity index of the protein molecule. An explanation may be that a higher content of hydrophobic amino acids automatically leads to more hydrophobic amino acids being exposed on the surface of the protein. But because the fraction of exposed hydrophobic areas of proteins is always in the vicinity of 40% to 50% (36,37), another possible explanation takes into consideration that proteins are not rigid bodies with rigid surfaces but that the conformation should be considered as a statistical mean around which the molecule must have the property to breath or change the microconformation, thus allowing a long hydrophobic chain to come in contact with hydrophobic structures that statistically are not always exposed at the surface.

An interesting finding is that the binding to the octyl medium is highly correlated with the molecular weight as well. This phenomenon is logical, though, taking into consideration that larger protein molecules generally have to have a higher fraction of hydrophobic amino acids to keep the tertiary structure.

The other mode of action does not correlate with the content of hydrophobic amino acids of the proteins at all, but instead correlates inversely with the surface charge of the molecules as found in biochemical handbooks. This interaction was displayed on weakly hydrophobic matrixes with short alkyl chains but also with some thiophilic media. An explanation is that charged groups will mask the type of hydrophobic interaction taking place on these gels under the tested conditions. Probably this interaction will be influenced significantly by altering pH.

Very commonly used HIC media, based on phenyl groups and butyl groups as ligands, were shown to work with a mixed mode, that is, combination of the two pure mechanisms. This fact adds to the explanation of the difficulty, especially using these HIC media, to find simple rules after which the separation should be optimized.

It should also be noted that the performance of a HIC medium will always be the result of the combination of ligand, spacer, and matrix. Phenyl gels from different manufacturers have quite different patterns of interaction if tested by a set of proteins as described earlier. This has to be remembered when, after the optimization in laboratory scale, alternative media are tested before scaling-up of a HIC procedure. This optimization is only valid for the medium used. There is thus a risk that competing media, which might be advantageous, will be tested under nonoptimal conditions and rejected under false premises.

26.4.2 Prediction of Chromatographic Results

Several attempts have been made to predict, for example, retention times based on protein properties. Examples of studied parameters are the amino acid content (38), the protein surface area (39) applying of flexible molecular docking B, as well as more general approaches including protein and buffer properties as mentioned above (40,41). Methods for finding the mathematical correlations between protein parameters and retention times are presented (42,43). However, the majority of predicting methods demand the full knowledge of protein properties such as the amino acid content or even the full structure. Although these findings definitely cast light on the mechanisms involved in HIC still the practical approach for optimization of a HIC separation will be the highway to good results

26.5 THE CHROMATOGRAPHY CONDITIONS

26.5.1 Ionic Strength

A high ionic strength is needed by definition to attach biomolecules to the column. Very often, 1.5 to 1.7 *M* ammonium sulfate is used as initial buffer concentration without further trials to optimize. Another common approach is to equilibrate the column in weak Tris-buffer to preserve the pH and increase the ionic strength by the addition of sodium chloride to a concentration of up to 3 *M*. It has to be noted, though, that ammonium sulfate, at a certain concentration, is between three and four times as effective in promoting the hydrophobic interaction compared to sodium chloride, so if the substances in question need a stronger interaction to be retarded on the column than can be achieved by almost saturated sodium chloride, ammonium sulfate is a natural choice. Using ammonium sulfate extends the range of HIC to be used for the separation of less hydrophobic molecules.

In production-scale operations, the use of ammonium sulfate has been questioned because of the effect the nitrogen of used buffers may have on the environment by increasing the overnutrition. The choice between overnutrition and possible toxic effects of substitute buffers such as sodium sulfate may not be easy, and all the factors, from the ease of the separation procedure to the cost of purifying the sewage, will have to be considered.

26.5.2 Buffer Ions

Recently, it has been shown that the choice of buffer ions greatly influences the function of an HIC medium (34,44). The overall behavior of a buffer-ligand-spacer-matrix system could be switched in a wide range, such as from typical phenyl-like to typical butyl-like or vice versa, by just changing the buffer ions. Optimization work by testing different buffers might then be more beneficial than usually anticipated but, as with pH, the task still includes much work based on trial and error.

26.5.3 pH

It has been shown that hydrophobic interaction between proteins and an octyl gel generally increases with lowered pH and most often decreases with high pH, but the pH dependence in this case is not very profound from pH 6 to pH 8.5. For some proteins, though, the interaction increases with an increase in pH (22). One reason for the different results might be that sodium sulfate was used above pH 8.5 and ammonium sulfate buffer was used below pH 8.5. According to Oscarsson and Kårsnäs (34), the adsorption behavior of serum proteins on octyl gel is quite different in the two environments. The reason for this complexity again has certainly to do with the multimechanism nature of HIC, with irregularities implied by different kinds of interactions between buffer ions and protein surfaces (21) and thus, secondarily, interaction with the HIC matrix.

In practice, changes in pH to get a desired separation can be very powerful because it may alter the entire adsorption spectrum, but, because of the trial-and-error nature of the task, pH is not the primary parameter to use for an optimization.

26.5.4 Temperature

High temperature promotes the binding, and HIC is preferably run at room temperature if the compounds to be separated can withstand this condition without losing activity (5,8,10,45). In room temperature, slightly lower buffer concentrations can be used than if the separation is made in a cold room. Elution of hardly bound substances may also be facilitated by performing the elution step at cold room temperature. In practice, however, the increase of temperature at the reequilibration of the column in room temperature inevitably produces air bubbles that will destroy the packing of the column, so if a two-temperature scheme is chosen, repacking of the column will most certainly have to be included in the separation cycle.

26.5.5 Flow, Residence Time

Hydrophobic interaction shows signs of being a slower process compared to other interactive chromatography techniques. The desorption of compounds from the column is often run under nonequilibrium conditions, resulting in long tailing peaks that seriously damage the resolution.

It is also generally found that the adsorption of molecules sometimes seems to be a two-step reaction, where the initial binding is weaker and more readily broken under the elution conditions. After some time, the bound molecules seem to adapt themselves to the column, involving conformational changes (46,47). More harsh conditions may have to be introduced for the elution, and there is a risk that the recovery will decrease. On the other hand, there are reports showing that an elution pattern can be quite reproducible even when the bound substances have been immobilized on the column for a long time (22), and HIC is generally considered a high-recovery technique (48).

26.5.6 Elution

Generally, elution of substances bound to an HIC column can be achieved by altering one of the parameters that promote binding. It is possible to elute by increasing pH, if the adsorption was performed at low pH, or by lowering the temperature, although the effect of this may be too small. But most commonly, elution is performed by a descending salt gradient or, in large scale, by stepwise lowering of the salt concentration (6,17). A powerful technique for the elution is the use of displacement agents such as nonionic surfactants (61) or amino acids (62,63).

If substances remain on the column at very low buffer concentration, an organic additive such as ethylene glycol or a detergent can be used for the elution. The ethylene glycol will alter the water structure and weaken the hydrophobic interaction, and the detergent will act by competing with and displacing the protein on the chromatographic medium.

26.6 REGENERATION AND CLEANING-IN-PLACE

For process economy, the regeneration of the HIC column is essential. Fouling of the column and irreversible binding of contaminants to the column are common obstacles in HIC. There are three degrees of regeneration: removal of strongly bound or precipitated proteins, removal of hydrophobic substances like lipids and lipoproteins, and sanitation, which means removal or deactivation of microorganisms. Preferably, the regeneration is performed as a cleaning-in-place (CIP) procedure without costly and cumbersome repacking of the column. All washing steps should be performed with reversed flow to give the contaminants the shortest distance out of the column and avoid secondary interaction with the rest of the column. First, all conditions that will decrease possible hydrophobic interaction can be used for the regeneration. Detergents that will compete with the attached substances and ethanol and isopropyl alcohol, buffers normally used in RPC, are obvious choices NaOH, 0.5 to 1 M, has been shown to almost universally sanitize the column and has a good cleaning ability as well. It is important to note that washing with narrow zones of the washing solution is often more effective in removing the column contaminants than exposing the medium for a constant concentration of the washing agent for an extended time. It is not unusual when cleaning HIC media that material will be washed out both in the beginning and at the end of a washing zone where the concentration is increasing and decreasing, respectively. Probably, a lot of equilibria are present, including hydrophobic interaction and interactions involving charges, and if one type of binding is minimized another may be enhanced and keep the contaminants on the column. By wobbling the concentration from low-to-high ionic strength and vice versa, the differences of reaction velocities are probably utilized, with elution as the result.

26.7 OPTIMIZATION PROCEDURES

Although all the parameters mentioned above can be used to optimize a HIC step, for many of them the effect of a change is not easy to predict because it will also be dependent on the individual properties of the proteins subject to separation. Thus, a pH change may alter the entire elution pattern or just influence the binding of a single component. Other parameters that may change the entire interaction picture are the buffer ions and the chosen chromatography medium.

26.7.1 Choice of Medium

Ideally, it is preferable that the desired components attach to the medium under moderate salt concentrations (e.g. up to 1 *M* ammonium sulfate), and a screening of the most common commercially available media is always advisable. This screening is easily performed in small scale and should include a test of whether elution with a high recovery is possible with water or weak buffer. If a lower salt concentration is sufficient on a certain gel, this is a preferable choice for economical and environmental reasons provided that the resolution obtained is still acceptable.

26.7.2 Determination of the Starting Buffer Concentration

An HIC run is normally performed by equilibrating the column in a starting buffer, applying the sample, and eluting with a stepwise descending ionic strength or a descending continuous gradient.

Optimizing the starting concentration, or the concentration at which the proteins are adsorbed to the column, is important to really utilize the HIC step (Fig. 26.4). In laboratory scale and analytical experiments, where the resolution can be achieved by gradient elution, the ionic strength

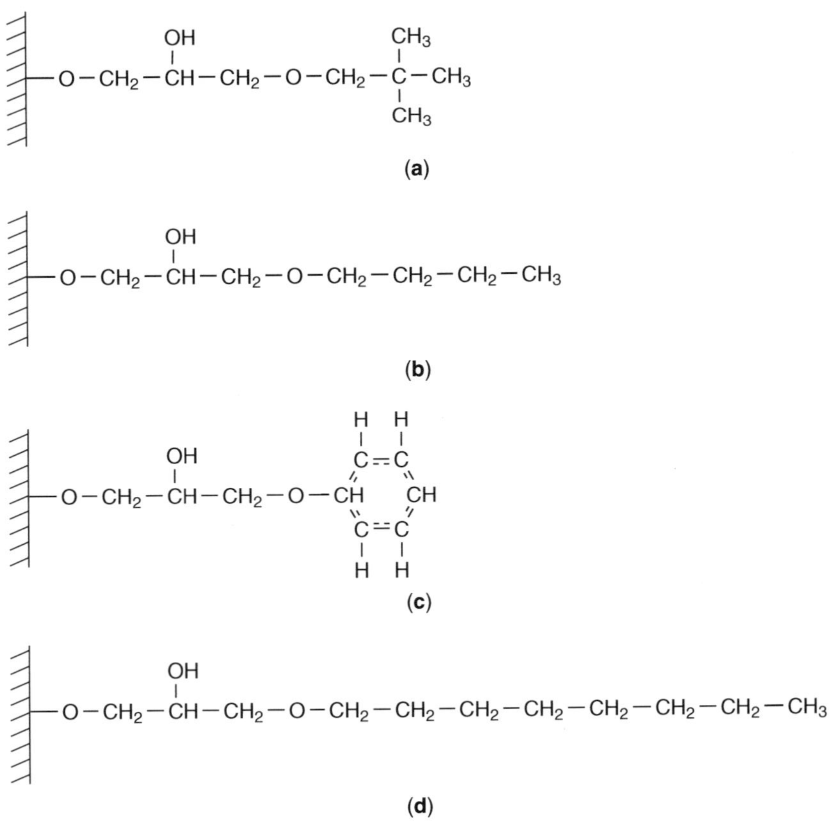

Figure 26.4. Common hydrophobic ligands. Alkyl (a), butyl (b), phenyl (c), and octyl (d) ligands coupled to a chromatographic matrix by glycidyl ether.

should be high enough to retard the desired component and to let it elute somewhere in the middle of the descending gradient. In this way, the combination of the capacity for the component of interest and the resolution will be favorable. In large-scale runs, well-defined and reproducible gradients are more difficult to achieve, so the elution is preferably done by stepwise decrease of the ionic strength. In some cases, it will work beautifully to chose an ionic strength that is just enough to retard the substance of interest, but in other cases, the peak containing the desired substance will broaden unacceptably at this condition, and the salt concentration will have to be increased (Fig. 26.5).

26.7.3 Gradient Volume

A gradient volume between 5- and 10-column volumes will give enough resolution in most cases. By increasing the gradient volume, the resolution will increase, but the components will be eluted in a larger volume.

26.7.4 Selectivity of the On and Off Reactions

It is often found that the reaction of binding to the column is rather rapid and selective, whereas the elution of a compound from the column is slower. As a result of this, tailing of zones at elution is very often experienced and the resolution of substances is jeopardized. A way of optimizing a method fully, utilizing the selectivity of the adsorption, is to use the following scheme:

1. Equilibrate the column with a starting buffer that has a lower concentration than is needed to retard the desired substance, adjust the sample accordingly, and run it through the column.
2. Elute everything that was retarded on the column with water or low salt. Check for activity.
3. Add some salt to the buffer and the sample and repeat steps 1 to 3 the desired substance is eluted from the column. To speed up the procedure, a prerun with a gradient starting at high salt concentration will hint at a proper initial concentration when step 1 is run the first time.

An optimized separation scheme includes as its first step running the column in a salt concentration just below the one where the desired substance was found to be retarded. All impurities that bind harder than the desired substance attach to the column and should be eluted. The next step is to bind the desired substance using a slightly higher salt concentration. If this is done carefully, very few contaminants will bind, and subsequently the desired substance can be eluted in high yield by a steep descending gradient, if any contaminating substances were retarded as well, or by a

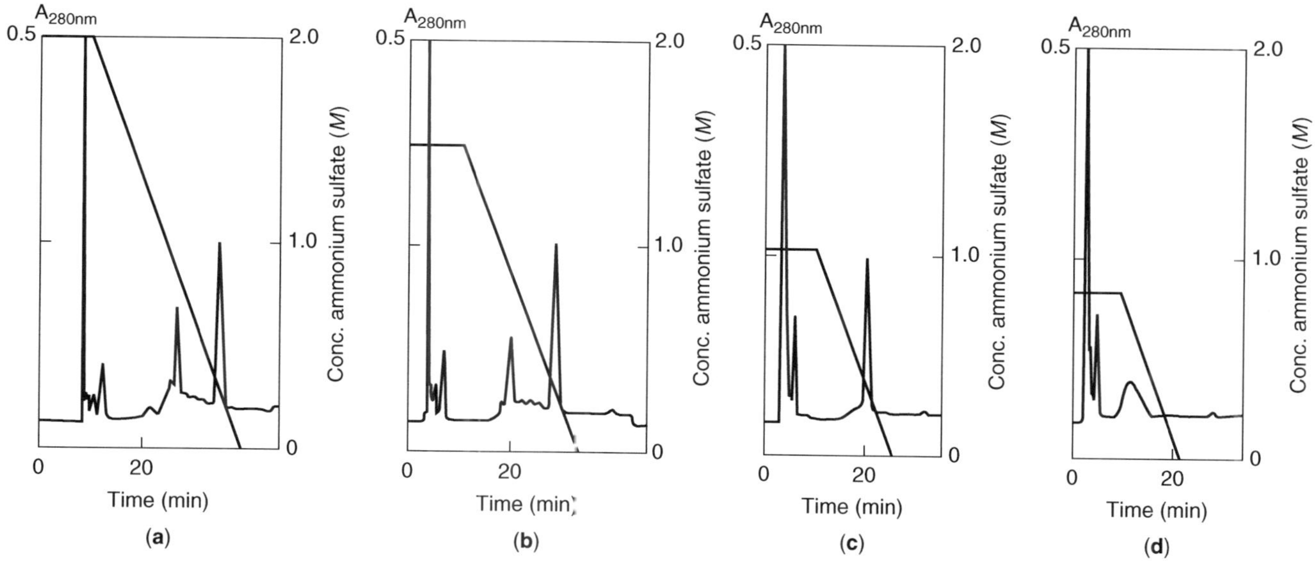

Figure 26.5. The effect of starting concentration. 100 μL anti-CEA MAB (−IgG1) from mouse ascites fluid, in 0.8 *M* $(NH4)_2SO_4$ (AS), applied to an Alkyl Superose HR 5/5 column with a flow rate of 2.5 cm/min. The different concentrations of AS in the initial buffer were mixed in 0.1 *M* sodium phosphate, pH 7. (a) The sample is applied in 2 *M* AS. Both albumin and IgG are accumulated on the column. (b) The sample is applied in 1.5 *M* AS. Less albumin is bound, which indicates a weaker interaction and less binding capacity. (c) In 1 *M* AS only IgG is binding, and (d) in 0.8 *M* AS the IgG is still retarded but eluted in a broad peak. *Source*: Figure reprinted from Ref. 20 by kind permission of Pharmacia Biotech, Uppsala, Sweden.

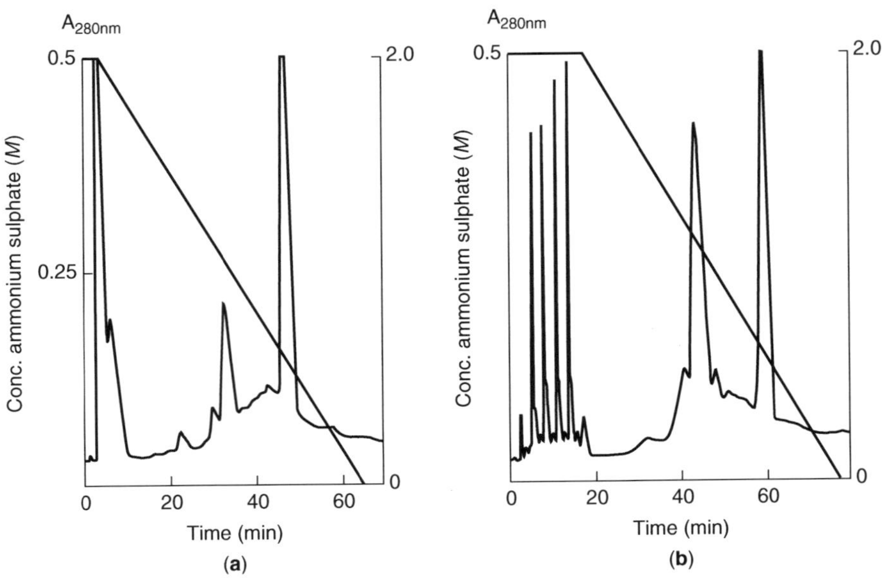

Figure 26.6. The effect of the suggested method of injecting the sample in diluted aliquots intermissioned by zones of concentrated buffer. In (a) 500 μL anti-CEA MAB (−IgG1) from mouse ascites fluid, in 0.9 M $(NH_4)_2SO_4$ (AS), was injected on a column equilibrated in 2 *M* AS in 0.1 *M* sodium phosphate, pH 7. In (b) the same sample was injected in 100 μL aliquots. 1.3 mL of 2 *M* AS were injected between every injection of sample. The capacity of the column was fully utilized without the risk of precipitating the sample in tubings and adapters. *Source*: Figure reprinted from Ref. 20 by kind permission of Pharmacia Biotech, Uppsala, Sweden.

step of very diluted buffer. The described method is advantageous not only because the selectivity of the interaction is fully utilized but also because of the simplicity to recondition sample and buffers by just adding salt. Moreover, the capacity of the column is utilized to bind mostly the compound of interest and very little contaminants. A drawback may be that the capacity of the column for the substance of interest might be lower because the binding is taking place near the elution concentration.

26.7.5 Equilibration of the Column and Sample Application

There is always a risk that the high salt concentration needed to attach the sample to the column will cause precipitation in tubings, pumps, columns, and detectors. This may be prevented by the following scheme: The column should be equilibrated with the proper buffer concentration but the sample with slightly lower salt concentration. This is easily accomplished by dilution of the sample. Instead of injecting the whole sample volume at one time, the injection should be done in aliquots divided by zones of the equilibration buffer. If the volume of the aliquots is kept less than 10% of the column volume and the volume of the intermediate zones with full buffer concentration exceeds 10% of the column volume, the sample condition will adapt to that of the equilibrated column. In this way, the precipitation in wrong places is avoided and the interaction with the column is still taking place in the correct environment (Fig. 26.6).

26.7.6 Scaling-up and scaling-down

To get a similar result in large scale as in laboratory scale, it is important to keep some parameters constant. The column length, the linear flow, and the ratio between gradient and column volumes belong to this category. A constant column length leads to the fact that the desired capacity will be proportional to the cross-sectional area of the column and actually decide it, because the same proportion of the column length should be occupied by the sample in both scales.

Because of the relationship between the volumetric flow, the linear flow, and the cross-sectional area of the column on one side and the gradient-to-column-volume ratio on the other side, the gradient volume will be decided when the diameter of the column is set. In principle, a gradient elution in process scale will then take the same time as in laboratory scale. In practice, a decrease in resolution caused by a more ineffective solvent distribution system at the inlet of a process scale column will sometimes have to be compensated for by a lower flow.

A very powerful way of optimizing an existing separation step is to scale down the experiment to a small laboratory scale (49,50). This opens the opportunity to perform a wide range of optimization experiments without spending large amounts of material and chemicals. Afterwards the scaling-up can be performed as mentioned above.

26.8 APPLICATIONS

HIC has been used for the purification of serum proteins (19,51), nuclear proteins (52), membrane-bound proteins (53), and viruses and cells (54), and recombinant proteins (55) and receptors (56,57) have been successfully prepared. Interesting applications include the use of HIC media as biochemical reactors by coupling enzymes by hydrophobic interaction (58) and a method for the exchange of detergents bound to membrane proteins (21).

For further applications and more discussion about large-scale processing using HIC, the works by Pharmacia Biotech and Eriksson (20,21) are recommended.

REFERENCES

1. Queiroz JA, Tomaz CT, Cabral JMS. J Biotechnol 2001; 87(2): 143–159.
2. Tanford C. The hydrophobic effect: formation of micelles and biological membranes. New York: Wiley; 1973.
3. Crighton TE. Proteins, structures and molecular properties. New York: W.H. Freeman; 1984.
4. Melander W, Horwath C. Arch Biochem Biophys 1977; 183: 200–215.
5. Srinivasan R, Ruckenstein E. Sep Purif Methods 1980; 9: 267–370.
6. Porath J, Sundberg L, Fornstedt N, Olsson I. Nature 1973; 245: 465–466.
7. Porath J. J Chromatogr 1986; 376: 331–341.
8. Hjertén S. J Chromatogr 1973; 87: 325–331.
9. Kårsnäs P, Lindblom T. J Chromatogr 1992; 599: 131–136.
10. Jennisen HP. J Chromatogr 1978; 159: 71–83.
11. Salgado JC, Rapaport I, Asenjo JA. J Chromatogr A 2005; 1098(1–2): 44–54.
12. Zhou P, Tian F, Li ZL. Sci China B 2007; 50(5): 675–682.
13. Hofstee BHJ. Anal Biochem 1973; 52: 430–448.
14. Shaltiel S, Er-el Z. Proc Natl Acad Sci U S A 1973; 70: 778–781.
15. Halperin G, Breitenbach M, Tauber-Finkelstein M, Shaltiel S. J Chromatogr 1981; 215: 211–228.
16. Wilchek M, Miron T. Biochem Biophys Res Commun 1976; 72: 108–113.
17. Hjertén S, Rosengren J, Pahlman S. J Chromatogr 1974; 101: 281–288.
18. Påhlman S, Rosengren J, Hjertén S. J Chromatogr 1977; 131: 99–108.
19. Jansson J-C, Lååš T. In: Roger E, editor. Chromatography of synthetic and biological macromolecules. Chichester: Ellis Horwood; 1978.
20. Pharmacia Biotech. Hydrophobic interaction chromatography-Principles and Methods. Uppsala: GE Health Care; 1992.
21. Eriksson K-O. In: Jansson J-C, Rydén L, editors. Protein purification: principles, high resolution methods, and applications. Cambridge: VCH Publishers; 1989. pp. 207–226.
22. Hjertén S, Yao K, Eriksson K-O, Johansson B. J Chromatogr 1974; 359: 99–109.
23. Maisano F, Belew M, Porath J. J Chromatogr 1985; 321: 305–317.
24. Kato Y, Kitamura T, Hashimoto T. J Chromatogr 1986; 360: 260.
25. Fausnaugh JL, Pfannkoch E, Gupta S, Reigner FE. Anal Biochem 1984; 137: 464.
26. von der Haar F. Biochem Biophys Res Commun 1976; 70: 1009–1013.
27. Adachi K. Biochim Biophys Acta 1987; 912: 139.
28. Porath J. Nature 1962; 196: 47.
29. Carlsson J, Jansson J-C, Sparrman M. In: Jansson J-C, Rydén L, editors. Protein purification: principles, high resolution methods, and applications. Cambridge: VCH Publishers; 1989. pp. 275–329.
30. Tiselius A. Ark Kemi 1948; 26B: 1–5.
31. Fausnaugh JL, Reigner FE. J Chromatogr 1986; 359: 131.
32. Jennissen HP, Heilmeyer IMG. Biochemistry 1975; 14: 754–760.
33. Ekström B, Jacobson G. Anal Biochem 1984; 142: 134–139.
34. Oscarsson S, Kårsnäs P. J Chromatogr 1988; 803: 83–93.
35. Melander WR, El Rassi Z, Horvath C. J Chromatogr 1989; 469: 3–27.
36. Miller S, Janin J, Lesk AM, Chotia C. J Mol Biol 1987; 196: 641.
37. Lee B, Richards FM. J Mol Biol 1971; 55: 379–400.
38. Salgado JC, Rapaport I, Asenjo JAJ. J Chromatogr A 2005; 1098(1–2): 44–54.
39. Katti A, Maa Y-F, Horváth CS. Acta Chromatographica 1987; 24(1): 646–650.
40. To BC, Lenhoff AM. J Chromatogr A 2007; 1141(2): 191–205.
41. To BC, Lenhoff AM. J Chromatogr A 2007; 1141(2): 235–243.
42. Lienqueo ME, Malm A, Vásques L, Asenjo JA. J Chromatogr A 2002; 978(1–2): 71–79.
43. Lienqueo ME, Malm A, Vásques L, Asenjo JA. J Chromatogr A 2003; 1009(1–2): 189–196.
44. Arakawa T, Narhi LO. Biotechnol Appl Biochem 1991; 13: 151–172.
45. Parsegian VA, Ninham BW. Biophys J 1970; 10: 664–674.
46. Jennissen HP. J Colloid Interface Sci 1986; 111: 570.
47. Haimer E, Tscheliessnig A, Halm R, Jungbauer AJ. J Chromatogr 2007; 1139(1–2): 184–194.
48. Scopes RK. Protein purification, principles and practice. New York: Springer; 1994.
49. Ghose S, Chase H. Bioseparation 2000; 91: 21–28.
50. Ghose S, Chase H. Bioseparation 2000; 91: 29–36.
51. Hrkal Z, Rejnkova J. J Chromatogr 1982; 242: 385–388.
52. Comings DE, Miguel AG, Lesser HH. Biochim Biophys Acta 1979; 563: 253–260.

53. McNair RD, Kenny AJ. Biochem J 1979; 179: 379–395.
54. Hjertén S. In: Glick D, editor. Methods of biochemical analysis. New York: Wiley; 1981. pp. 89–108.
55. Belew M, Yafang M, Bin L, Berglof J, Jansson J-C. Bioseparation 1991; 1: 397–408.
56. Kuehn L, Meyer H, Reinauer H. Proceedings 2nd International Insulin Symposium. Aachen; 1980. pp. 243–250.
57. Levy A, Boyle DM, van der Walt LA. Biomed Chromatogr 2005; 5(2): 62–67.
58. Caldwell KD, Axen R, Porath J. Biotechnol Bioeng 1975; 17: 613–616.
59. Klimchak RJ, Wang S. Biotechnol Tech 1997; 11(7): 497–501.
60. Lewin S. Displacement of water and its control of biochemical reactions. New York: Academic Press; 1974.
61. Rosengren J, Påhlman S, Glad M, Hjertén S. Biochim Biophys Acta 1975; 412: 51–61.
62. Rukhadze MD, Sebiskveradze MV, Makharadze TG, Sidamonidze NS. Biomed Chromatogr 2003; 17(8): 538–542.
63. Tsumoto K, Ejuna D, Nogase K, Arakava TJ. J Chromatogr A 2007; 1154(1–2): 81–86.

27

CHROMATOGRAPHY, RADIAL FLOW

TINGYUE GU

Department of Chemical and Biomolecular Engineering, Ohio University, Athens, Ohio

27.1 INTRODUCTION

The production of a modern biotechnology product, frequently a recombinant protein, requires a multistage downstream process because the feedstock is usually a liquid that is dilute in product concentration and contains many impurities some of which are unknown chemical compounds. Such a process typically centers on two or more liquid chromatography steps in order to achieve the desired purity. As production scales escalate, the chromatography columns used become larger and more expensive. It is not uncommon to have column bed volumes of hundreds of liters at industrial scales.

Radial flow columns were first used for gas–solid catalytic reactions in large packed beds. They were designed to increase gas flow rate and reduce pressure by increasing the cross-sectional flow area. Radial flow chromatography (RFC) columns first entered the commercial biotechnology market in the mid-1980s (1). They were marketed as an alternative to the conventional axial flow chromatography (AFC) for preparative- and large-scale applications. These columns were not configured for analytical applications because RFC columns do not offer any advantages for such purposes. In an RFC column (Fig. 27.1), the mobile phase flows in the radial direction, not in the axial direction. The mobile enters from the outside tube and merges into the center tube (Fig. 27.2). In comparison to a slim AFC column, an RFC column provides a larger flow area and a shorter flow path (i.e. radial bed length). It allows a higher volumetric flow rate with a lower bed pressure. The effect is equivalent to using a short pancake-like AFC column (Fig. 27.3). Pancake-like AFC columns are quite common in industrial applications. They are available from most major commercial column vendors. In scale-up to accommodate large feed loads, it is impractical to increase the column height of AFC columns by too much because excessive bed pressure drop will compress the gel too much and make fluid flow more difficult. This is especially troublesome when soft gels are used. Thus, pancake-like AFC columns are used despite their short flow paths.

This chapter is devoted to the discussion of the applications of RFC columns for bioseparations and RFC modeling and scale-up issues. Applications examples will be provided based on the existing open literature. A general rate model will be presented for the modeling of RFC. Experimental and theoretical comparisons between AFC and RFC will be discussed.

27.2 RADIAL FLOW COLUMN CONFIGURATIONS

In 1947, Hopf (2) described a radial chromatography device. The device had a feed pipe in the center. The outward liquid flow in the radial direction was driven by the centrifugal force when the device was rotated. Because of this, he called this device a chromatofuge. Such a device was obviously too complex and expensive for large-scale industrial application and thus it was not adopted in biotechnology applications.

To date, only a few commercial companies have marketed RFC columns. The first one was the CUNO, Inc. in Connecticut. They marketed the Zetaffinity series

Downstream Industrial Biotechnology: Recovery and Purification, First Edition. Edited by Michael C. Flickinger.

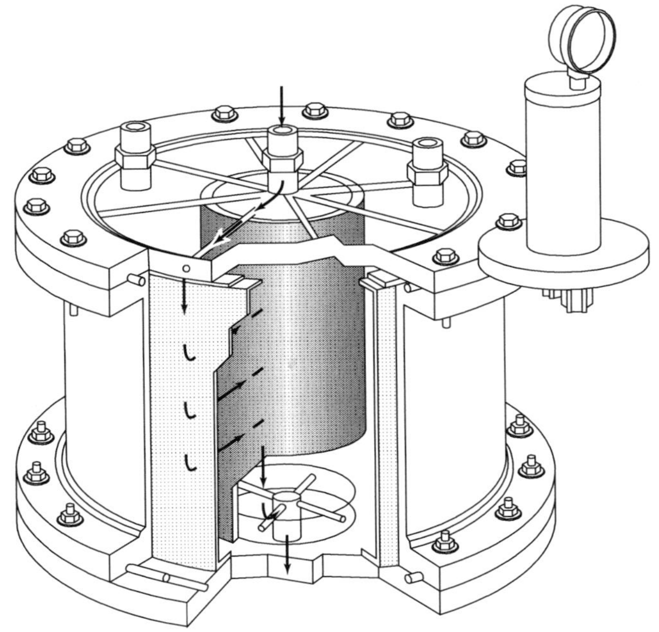

Figure 27.1. Diagram of an acrylic Superflo column. (Courtesy of Sepragen Corp.)

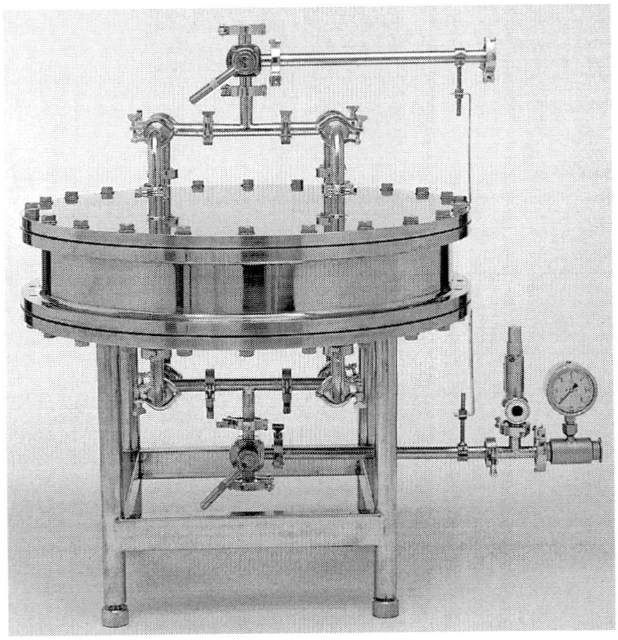

Figure 27.3. A large-scale pancake-like axial flow column. (Courtesy of General Electric. ©General Electric.)

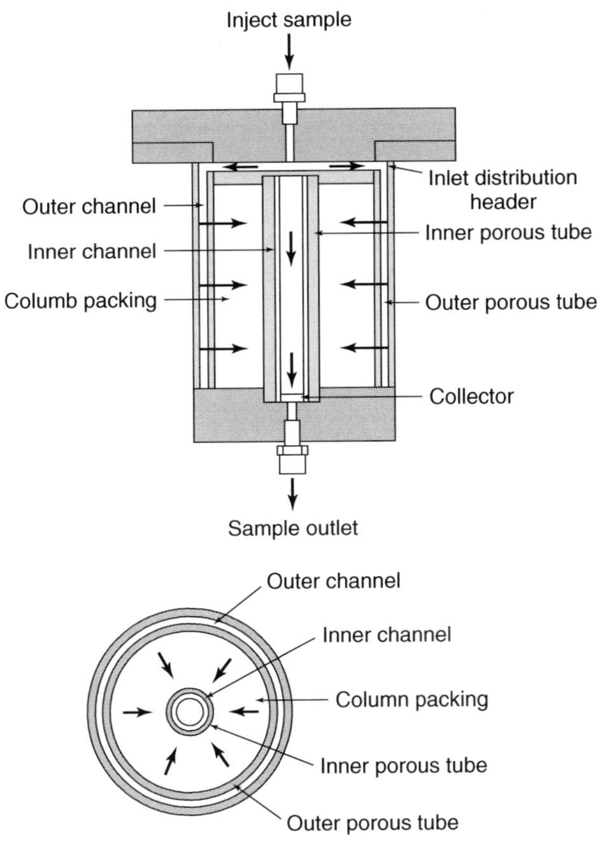

Figure 27.2. Anatomy of a Superflo column (Courtesy of Sepragen Corp.)

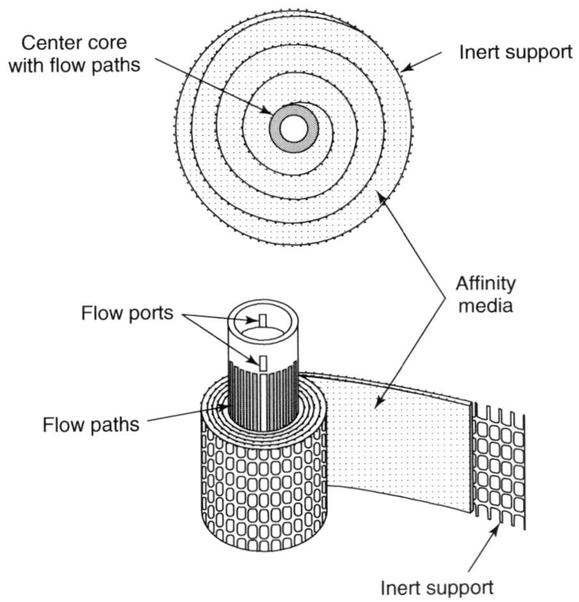

Figure 27.4. Structure of a Zetaffinity cartridge.

of preparative-scale radial flow cartridges that looked like a spiral-wound microfiltration cartridge (Fig. 27.4). Such a design obviously was rooted in the fact that the CUNO is a major manufacturer of industrial filtration systems. The packing for CUNO's RFC cartridges was fabric-like modified cellulose (3) instead of adsorbent particles. These cartridges were used for affinity chromatography

Figure 27.5. Superflo columns. (Courtesy of Sepragen Corp.)

with a trade name of Zetaffinity. CUNO discontinued its Zetaffinity product line in 1991.

The second vendor and manufacturer of RFC columns is the Sepragen Corporation (http://www.sepragen.com) in Hayward, California. Their RFC product line carries the trade name Superflo. Sepragen markets unpacked RFC columns ranging from 50 mL to 200 L. Figure 27.5 shows several Superflo columns with a similar diameter but different column heights. Superflo columns come in stainless steel, acrylic, polycarbonate and polyethylene. They can sustain a pH range of 2–12 and a maximum pressure of 50 psi. They are used by some biopharmaceutical companies at production scales. Vinit Saxena of Sepragen owns a US patent (4) on the design of unpacked RFC columns. Figure 27.1 shows an acrylic RFC column with inward radial flow. The feed stream enters from a center input port at the top. It is then distributed through several flow channels to the outer shell and subsequently enters the packing media in the radial direction toward a center collection tube. The effluent exits the tube through an outlet port at the bottom. The two ports next to the center input port at the top in Fig. 27.1 are packing ports used during media packing. A bubble trap with a pressure gauge sits on the edge of the column's top lid. Figure 27.2 is an anatomical view of a Superflo column. Superflo columns usually use inward flow instead of outward flow because it is more difficult to distribute outward flow without increasing flow distortion caused by gravity. Another reason for using inward flow is that it provides slightly sharper peaks than outward flow based on computer simulation (5). Outward flow is usually used for the packing or regeneration of an RFC column.

Founded in 2002, PROXCYS Downstream Biosystems (http://www.proxcys.nl) in Emmen, The Netherlands, is a new supplier of preparative- and large-scale radial flow chromatographic columns. PROXCYS offers radial flow columns ranging from 50 mL to 1200 L. Figure 27.6 shows a CRIO Series CA-601 S 5-L radial flow column. PROXCYS also markets AXCIS columns that are hybrid radial flow columns that allow particle-containing crude feeds much like the so-called expanded bed chromatography columns that were first commercialized by Pharmacia in Piscataway, New Jersey (now part of GE Healthcare). The difference is in flow direction with the latter being axial flow.

Figure 27.6. PROXCYS CRIO Series CA-601 S 5-L radial flow column. (Courtesy of PROXCYS Downstream Biosystems.)

Both Sepragen and PROXCYS market small "slice" radial flow columns. They look like a carved out portion of an RFC column resembling a slice of a cake with its center beveled. Figure 27.7 shows a 100-mL CRIO-MD 121 column from PROXCYS. The CRIO-MD series columns look much like Sepragen's Wedge columns. Such columns can be used for small-scale applications and in scale-up and scale-down investigations. They are somewhat like a conical column with axial flow that was studied by Pfeiffer (6) because both types of columns have a narrower bed downstream.

In addition to the three companies above, Ngo and Khatter (7) mentioned the use of 100- and 500-mL Avid AL radial flow columns made by BioProbe International in California, USA. The columns are no longer being marketed. BIA Separations (http://www.biaseparations.com) in Austria markets 8, 80, and 800 mL tube-shaped monolithic columns for ion exchange, reversed phase,

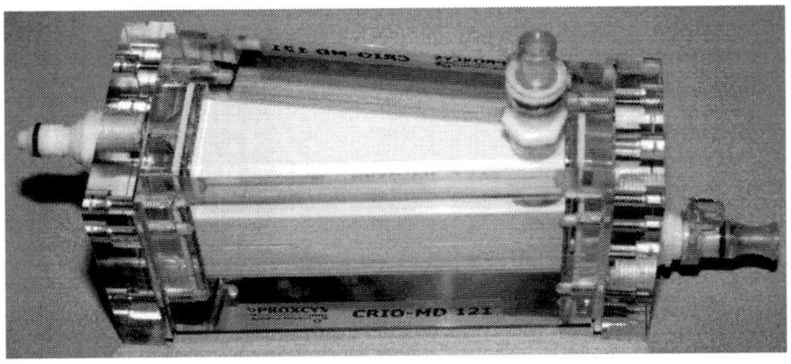

Figure 27.7. PROXCYS CRIO-MD 121 (100 mL) radial flow column. (Courtesy of PROXCYS Downstream Biosystems.) (This figure is available in full color at http://onlinelibrary.wiley.com/book/10.1002/9780470054581.)

affinity, and other applications. Harrison Research, Inc. (http://www.harrisonresearch.com) in California, USA, sells preparative centrifugally accelerated thin-layer chromatography devices with radial flow.

27.3 PACKING PROCEDURES FOR RFC COLUMNS

Because of their structure, CUNO's Zetaffinity cartridges do not require packing by the end user. Sepragen's Superflo columns are packed by first displacing air using outward flow with a buffer solution. Packing slurry is then pumped into the column through the two packing ports on the column top cover (Fig. 27.1). Excess buffer is squeezed out of the column through the input port on the column's top cover. A detailed packing procedure for Superflo columns was provided with illustrations by Wallworth (8). PROXCYS' columns are slurry packed using an automated process using a pump. Toward the end of the packing process, the bed pressure starts to take off sharply, this is when packing should stop. Unlike Superflo columns, PROXCYS' columns have packing ports very close to the center frits. PROXCYS claims that this packing port arrangement achieves better packing results. Packing of large radial flow columns remains a technical challenge. Users may shy away from RFC columns that do not resolve the undesirable channeling problem during packing. Munson-McGee (9) investigated packing problems in a 1.5-L Superflo 1500 radial flow column packed with ion exchange resins, and also mathematically modeled pressure and velocity distribution.

27.4 PRESSURE DROPS OF RFC COLUMNS

Because of a short flow path and a low pressure drop from the column inlet to outlet, both CUNO's and Sepragen's RFC devices exhibit a highly linear relationship between

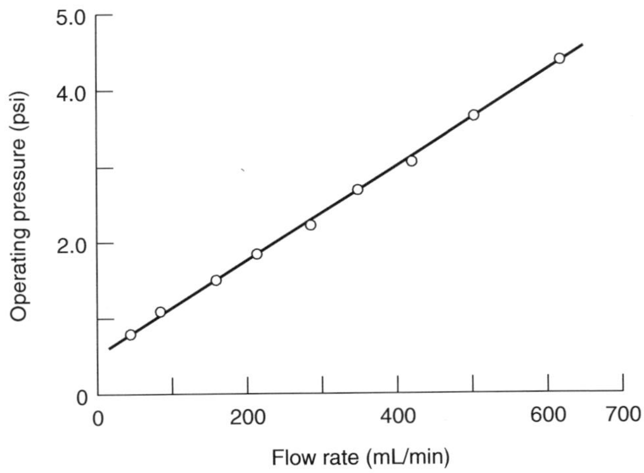

Figure 27.8. Pressure drop versus flow rate for an 800 mL Zetaffinity cartridge.

bed pressure drop and flow rate. Figure 27.8 shows a linear relationship between bed pressure and flow rate for an 800 mL CUNO's Zetaffinity cartridge studied by Huang *et al*. (3).

Figure 27.9 shows the pressure drops for a Superflo 20 L column and a Superflo 50 L column packed with several different soft-gel chromatographic media (10). The figure indicates that the pressure drops are quite low at high flow rates. At lower flow rates, the pressure drop curve is linear. This behavior is similar to that of short pancake-like AFC columns.

It would be interesting to measure the pressure distribution inside an RFC column. Unfortunately, such data are not available. With a constant volumetric flow rate, the interstitial linear fluid velocity increases in inward flow

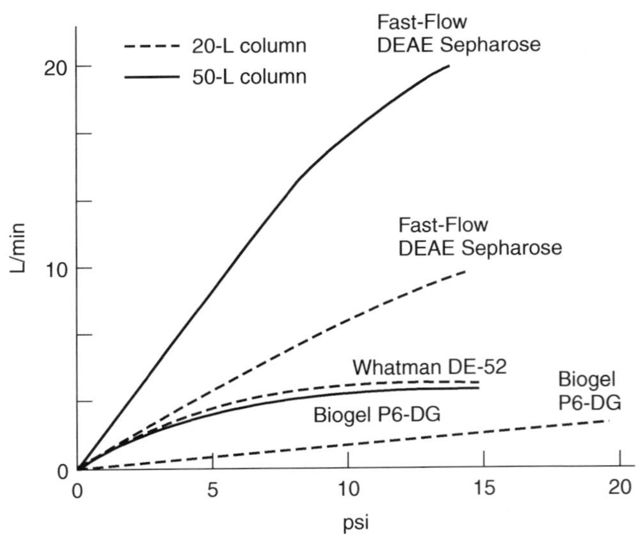

Figure 27.9. Pressure drop versus flow rate for 20 and 50 L Superflo columns. (Courtesy of Sepragen Corp.)

toward the center because the cross-sectional area for flow becomes narrower. The increased velocity results in higher kinetic energy for the fluid. The kinetic energy gain by the fluid is offset by a reduced local pressure as dictated by Bernoulli's principle in fluid mechanics. On the other hand, the increased velocity considerably increases friction loss that requires a larger pressure drop to overcome. The consequences from the increased velocity compromise each other, and thus the net effect on local pressure is unknown. For soft-gel columns, bed compression is unavoidable even in low-pressure RFC. If the pressure drop per unit radial length progressively increases toward the center in inward flow, gel density increases. This helps balance the gradual reduction of gel volume toward the center. Pfeiffer (6) demonstrated this for a conical column with its narrower end downstream. The same phenomenon in RFC columns needs to be proven experimentally.

27.5 COMPARISON OF RADIAL- AND AXIAL FLOW COLUMNS

Saxena and Weil (11) did an experimental case study to compare RFC with AFC. They used a 100-mL axial flow glass column (Econocolumn) with 2.5 cm i.d. from Bio-Rad Laboratories in Hercules, California (http://www.bio-rad.com), and a Superflo-100 RFC column from Sepragen with a bed volume of 100 mL. Both columns were packed with quaternary aminoethyl (QAE) cellulose with the following procedures. The Superflo column was packed using 25% slurry (with 0.5 *M* NaCl) that was pumped in through the two packing ports at a flow rate of 30 mL/min. The final QAE cellulose density after packing was 6 mL per dry gram. The Superflo column was packed in about 20 min. The AFC column was packed using 50% slurry added from the top of the column. After the liquid was drained and the bed was settled, additional slurry was added. This process was repeated until the same amount of QAE cellulose was packed into the AFC column. The two columns were used to separate two 10 mL ascites fluid samples. The samples were pretreated by dialyzing the ascites fluid with 10 m*M* phosphate buffer of pH 8.0 for 2 days with three changes. Precipitates and debris were then removed by centrifugation. Stepwise salt gradients were used. Figure 27.10 shows the comparison between the 100-mL AFC column and the 100-mL Superflo-100 RFC column. From Fig. 27.10, it can be seen that the RFC column achieved similar separation results with much less time. In this case, the two columns were equivalent only in bed volume. A stricter comparison should use a pancake-like short AFC column with its bed height about the same as the packing thickness in radial direction for the RFC column, and both columns should have equal bed volumes. The case study described next is close to satisfying these conditions.

Tharakan and Belizaire (12,13) used a 50-mL RFC column with a bed height of 0.95 cm and packing thickness

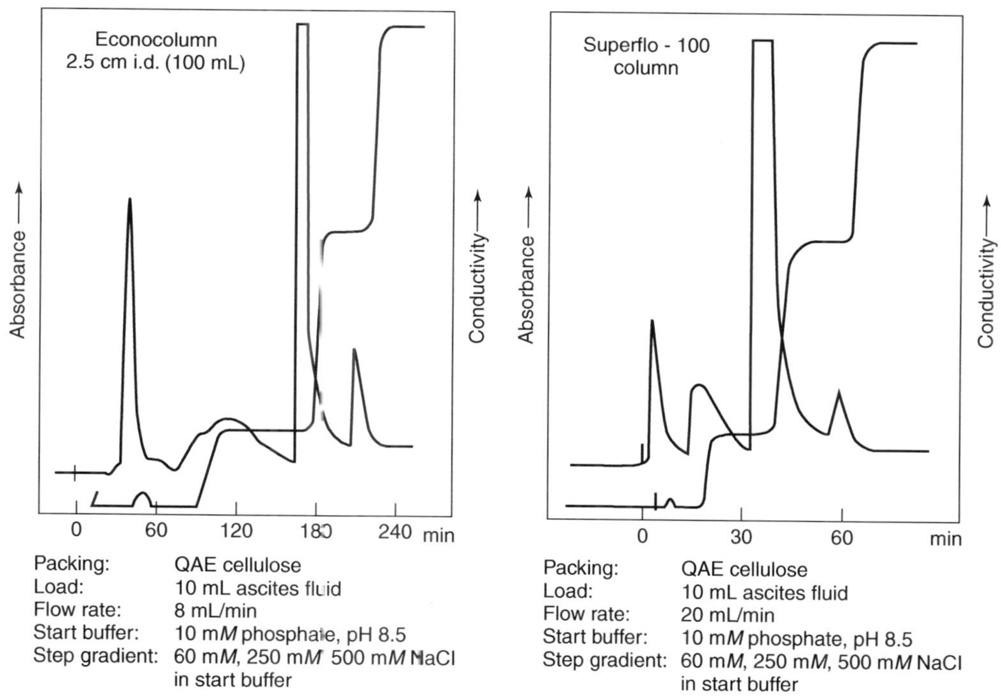

Figure 27.10. Comparison of AFC and RFC columns. (Courtesy of Sepragen Corp.)

of 3.0 cm in the radial direction and a 50-mL AFC column with a bed height of 2.8 cm. Both columns were packed with Sepharose CL2B resin containing monoclonal antibody from Pharmacia. They studied the purification of a protein called *factor IX*. Their experimental results indicated that the two columns gave similar purification results. This was expected because the AFC column was like a pancake. Tharakan and Belizaire (12) also packed the same column with S-200 Sephacryl size exclusion gel from Sigma Chemical Co. (St. Louis, Missouri). The protein band was more diffused using RFC than AFC.

Cabanne *et al*. (14) compared a 120-mL AFC column with a 120-mL RFC column (CRIO-MD 62 from PROXCYS). Both columns had an identical flow path length of 6 cm. They showed slightly better performances for the RFC column.

Lane and coworkers (15) compared the performance of a Superflo-100 RFC column and a 6.6 × 4.4 cm i.d. AFC column for the separation of egg-white proteins. Both columns had a nominal bed volume of 100 mL. They were packed with the same media for comparison. Two anion exchange cellulose media were tested. One was Whatman DE52 and the other was QA52 from Whatman Specialty Products Division (Maidstone, UK). Egg whites were first separated from fresh hen eggs and then treated with a buffer. After being treated with a cell debris remover, egg-white suspensions were filtered using a filter paper. The samples for chromatography had a protein concentration of 14 mg/mL. The sample loading volume was 40 mL. After sample loading, the column was washed with a buffer. Elution was carried out using a linear gradient of 0–0.5 *M* NaCl in 0.025 *M*-Tris/HCl buffer at pH 7.5. Various flow rates were tested ranging from 5 to 50 mL/min for the AFC column and 5 to 150 mL/min for the RFC column. Figures 27.11 and 27.12 represent typical results obtained by Lane *et al*. (15). They indicate that the AFC column gave slightly sharper peaks and faster elution times for both DE52 and QA52 media at a flow rate of 25 mL/min.

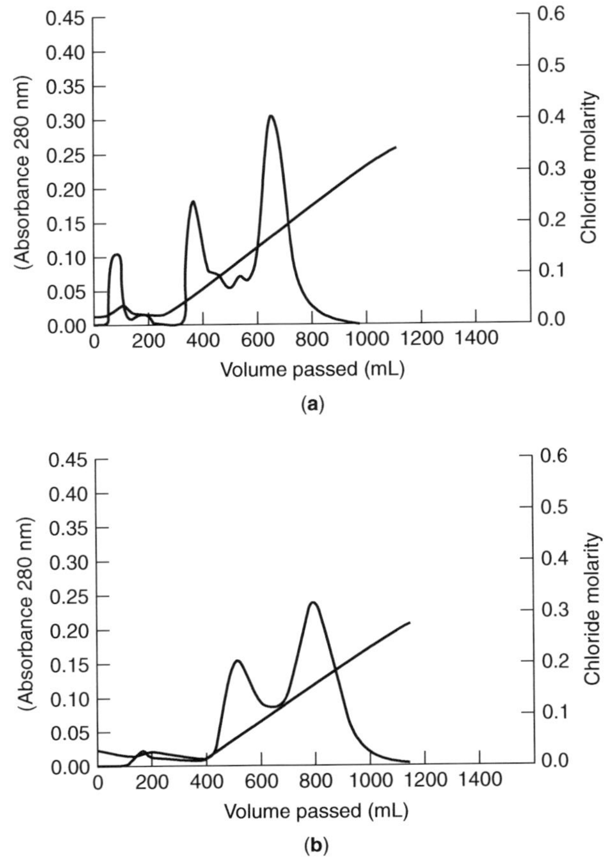

Figure 27.11. Comparison between a 100-mL AFC column (a) and a 100-mL RFC column (b) packed with DE52 in the separation of egg-white proteins.

27.6 PROS AND CONS OF RFC COLUMNS

RFC columns provide a short flow path and a large cross-sectional area. This has the same effect as short pancake-like AFC columns. However, RFC columns occupy considerably less floor space. Both RFC and pancake-like AFC columns face flow distribution problems. According to Sepragen, its Superflo columns have better flow distribution than typically large pancake-like columns.

Compared to long AFC columns, RFC columns produce smaller pressure drops, and thus enabling larger volumetric flow rates. If soft gels are used as separation media, the low pressure drop of RFC columns helps relieve bed compression (16,17). RFC is especially suitable for affinity chromatography using soft gels. In affinity chromatographic operations, dilute feeds are typically used. Owing to the extremely high affinity binding between the product and the gel matrix, very high flow rates are permitted without sacrificing column resolution. RFC is also a good choice for strong reversed phase, hydrophobic interaction and ion exchange media.

Scale-up of RFC columns is more straightforward since it is usually done by increasing the column height that is comparable to increasing the column diameter of an AFC column. To a certain extent, this does not seem to increase flow distortion problem in practice. If an AFC column is scaled up by increasing diameter, the flow distribution behaves quite differently from a small AFC column for which plug flow is easy to achieve. Thus, predicting actual performance of a large pancake-like AFC column from a much thinner one is problematic.

The disadvantage of an RFC column is primarily its limited resolution due to a short flow path. If the flow path is increased to a large extent, there will be flow distribution problems in the radial direction due to gravity. High resolution for more demanding separations can only come from using AFC columns with a sufficiently large column length.

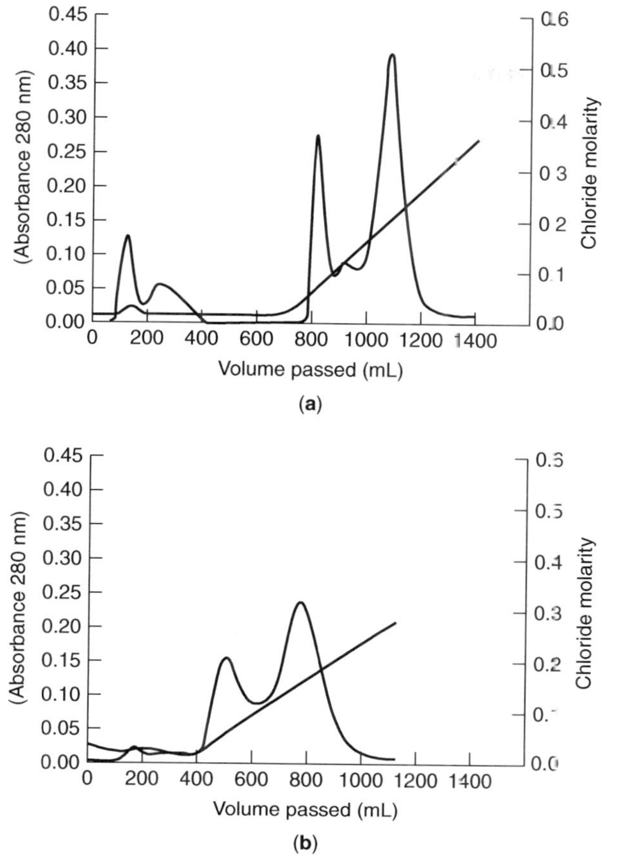

Figure 27.12. Comparison between a 100-mL AFC column (a) and a 100-mL RFC column (b) packed with QA52 in the separation of egg-white proteins.

This is precisely the reason why RFC has no use in analytical high performance liquid chromatography (HPLC). RFC is not suitable for separations in which solute–stationary interactions are weak or there is no specific binding. For example, RFC is not a suitable choice for size exclusion chromatography (SEC), because SEC depends strongly on the length of flow path for its resolution with no specific binding. RFC's short flow length cannot meet the demand.

If mechanically strong packing materials, such as silica-based particles, are used, a longer column flow path can be used because the bed can sustain a much higher pressure. If a relatively high resolution is desired, RFC would be at a disadvantage compared to AFC since RFC is limited by its short flow path.

27.7 APPLICATION EXAMPLES

27.7.1 Application Examples Using RFC Columns Packed with Membrane Sheets

Huang *et al*. (3) studied several Zetaffinity cartridges from CUNO with modified cellulose-based affinity media.

TABLE 27.1. Dimensions of CUNO Radial Flow Cartridges Tested by Huang *et al*. (3)

Parameter	Small	Medium	Large
Nominal size (ml)	250	800	3200
Bed volume (ml)	210	810	3020
Outer diameter (cm)	7.0	12.7	12.7
Inner diameter (cm)	0.6	0.9	0.9
Height (cm)	6.4	6.4	23.8

Table 27.1 lists the dimensions of three Zetaffinity cartridges tested by Huang *et al*. Figure 27.13 shows a chromatogram for the removal of proteases from human plasma using an 800 mL Zetaffinity cartridge with a flow rate of 100 mL/min (3). The cartridge contained modified cellulose with *p*-aminobenzamidine (PAB) as ligand. Protease removal was achieved with 70% efficiency in a single pass. The treated plasma was expected to have a threefold increase in stability. The same type of 800 mL Zetaffinity cartridge loaded with 1260 mg PAB ligand was also used to purify crude trypsin purchased from Sigma Chemical. Figure 27.14 shows the results obtained by Huang *et al*. (3) using a flow rate of 295 mL/min.

Planques *et al*. (18) used a 250-mL Zetaffinity cartridge from CUNO packed with modified cellulose as chromatography media. They first chemically treated the cartridge to couple the media with L-lysine. This affinity chromatography media was able to bind with a protein called *human plasminogen*. After centrifugation and microfiltration, human plasma was diluted with a buffer and then fed to the cartridge at a flow rate of 20 mL/min. After washing and elution, recovery yield of 85% and an increase of 110-fold in specific activity were achieved.

Wang and coworkers (19) fabricated a radial flow column with an i.d. of 7 cm and height of 2 cm to purify

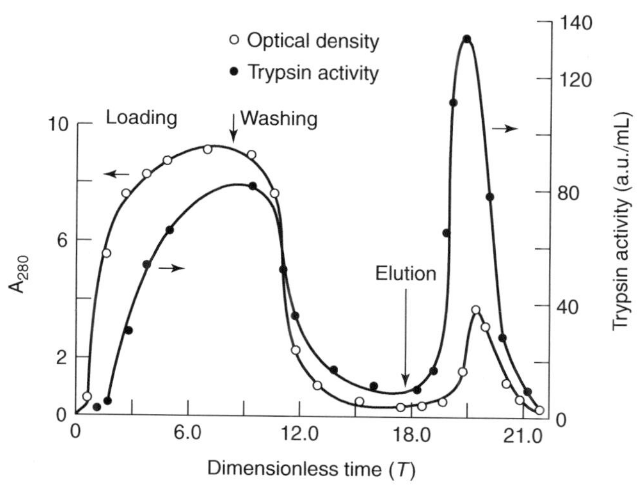

Figure 27.13. Protease removal from human plasma using a Zetaffinity cartridge.

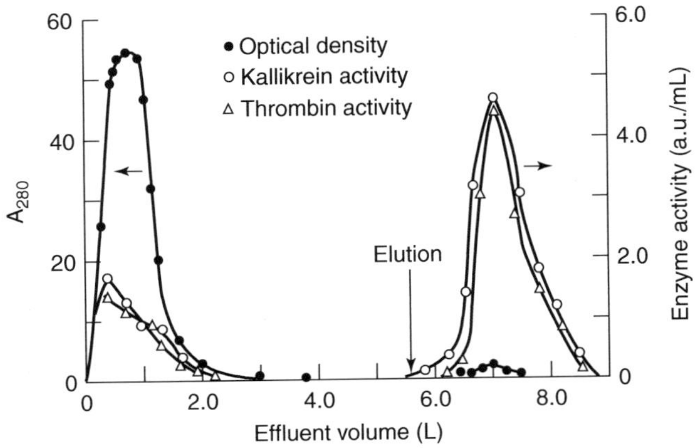

Figure 27.14. Trypsin purification using a Zetaffinity cartridge.

human urinary kallikrein from pretreated human urine. The perforated center cylinder was wrapped with a QAE ion exchange membrane sheet. Thus, the column was similar to a CUNO's Zetaffinity cartridge. Sun *et al*. (20) used a small RFC column packed with cellulose membrane covalently linked with diethylaminoethyl (DEAE) for the purification of human prothrombin from Nitschmann fraction III. The column was purchased from the Institute of Chemical Physics, Chinese Academy of Sciences.

27.7.2 Applications Examples Using Sepragen's Columns

Akoum and coworkers (21) used a Sepragen's Superflo-400 RFC column packed with histidyl-Sepharose gel for the purification of myxalin, a glycopeptide with anticoagulant property. The feed for the column was obtained from *Myxococcus xanthus* fermentation broth. Before the feed was applied to the column, it was clarified and concentrated by using centrifugation, microfiltration, and reverse osmosis. A relatively short processing time was achieved.

Strætkvern and coworkers (22) used a 60 mL, 2.2 cm i.d. AFC column, a Superflo-250 column (250 mL bed volume), and a 2500-mL AFC column for the separation of deoxyribonuclease (DNase) from the extracts of cod pyloric caeca. The packing medium was Q-Sepharose Fast Flow anion exchange gel from Pharmacia. Column dimensions and operating conditions are listed in Table 27.2. Table 27.3 is a summary of their experimental results. Their results indicate that the RFC column required much less time and achieved a higher productivity. The superior performance in this case should not be interpreted as a fixed rule here because the AFC columns they used were not equivalent pancake-like columns.

Weaver and coworker (23) used a 10-L Superflo RFC column packed with Q-Sepharose Fast Flow medium from Pharmacia for the separation of uridine phosphorylase from total crude extracts of *Escherichia coli*. After fermentation,

TABLE 27.2. Columns Used by Strætkvern *et al*. (22)

Parameter	Small Column	Medium Column	Large Column
Column type	Axial flow	Radial flow	Axial flow
Column volume (ml)	60	250	2500
Cross-sectional flow area (cm^2)	3.8	120 (outer)	154
Volumetric flow rate (h^{-1})	0.6	17.4	20.9
Sample volume (L)	0.033	0.135	1.32
Protein concentration (mg/mL)	2.6	2.6	2.4
Scale-up factor	1	4	40

TABLE 27.3. Purification Results Obtained by Strætkvern *et al*. (22)

Parameter	Small Column	Medium Column	Large Column
Elution volume (L)	0.20	0.87	3.57
Protein (mg)	10	44	350
Total Activity (units $\times 10^{-6}$)	20	78	783
Specific Activity (units $\times 10^{-6}$ mg^{-1})	2.00	1.77	1.35
Yield (%)	100	107	76
Purification (fold)	20	17	13
Cycle time (h)	1.3	0.25	1.4
Productivity (units $\times 10^{-6}$ $h^{-1} \cdot mL^{-1}$ gel)	0.256	1.25	0.135
Productivity (mg$\cdot h^{-1} \cdot mL^{-1}$ gel)	0.13	0.70	0.1

375-L broth was concentrated to 10-L. It was then washed with 50-L of 20 m*M* K_3PO_4 buffer containing 1 m*M* $MgCl_2$. After adding 20 μg of lysozyme per milliliter of broth, 100 mg DNase, and 100 mg RNase, the cells were homogenized using a bead mill. The supernatant was diluted to 30 L and then applied to the 10-L RFC column. The feed was recirculated back to the column at a flow rate of 1.3 L/min for 3 h and then discharged. The column was washed with three different buffer solutions to remove bound lipid and hydrophobic proteins. Elution was then carried out using a buffer containing 0.225 mM NaCl. The collected effluent was 50 L. The final product after dialysis had a purity of 85% and a recovery yield of 82%.

McCartney (24) used two 100-mL Superflo-100 RFC columns packed with S-Sepharose FF (Pharmacia) ion exchange media in tandem for the purification of an undisclosed recombinant protein from *E. coli*. The system was able to process 4 L feed in less than 2 h with a concentration factor of 64 times.

Saxena and coworkers (17) used a Superflo-1500 column (1500 mL in bed volume) packed with immobilized Protein-A Sepharose in the purification of an anti-melanoma IgG2A antibody from ascites fluids. The sample was loaded at a flow rate of 104 mL/min. After loading, the column was washed with a buffer at a flow rate of 170 mL/min. Elution was carried out at a flow rate of 92 mL/min. An actual recovery of 3.1 g of the antibody with a product purity greater than 97% was achieved in 3.5 h. A Superflo-1500 packed with riboflavin immobilized on a Sepharose 4B matrix via an epoxy linkage was used by the same researchers to obtain a crude preparation of a riboflavin-binding protein (17). The entire run was carried out using a flow rate of 350 mL/min. They also used a Superflo-200 column packed with anti-ricin B chain antibody immobilized on cross-link agarose (17). The flow rate used was 45 mL/min. Yield was achieved with a product amount of 2.1 g and a purity of 100%.

Sun *et al*. (25) used a 15-mL AFC column a 50-mL Superflo radial column, both packed with DEAE Sepharose Fast Flow media, to separate human prothrombin from Nitschmann fraction III derived from fractionation of human plasma for albumin and IgG with the Nitschmann method. The AFC column was loaded with 50 mL sample and RFC column with 200 mL sample. A slightly better peak resolution was achieved using the RFC column. The same researchers also used the same columns to purify human fibrinogen from Nitschmann fraction I (26). Tseng *et al*. (27) successfully utilized a 100-mL Superflo radial flow column from Sepragen packed with DEAE-52 cellulose for the purification of human salivary cystatin SN (CsnSN), a member of the cystatin superfamily of cysteine proteinase inhibitors. Levison (28) reported a comparison of the performances of a 100-mL Superflo-100 column, a 10-L Superflo-10L column, and a 100-mL AFC column, all packed with DE52 for the separation of egg-white proteins. The chromatograms resemble each other indicating a linear scale-up from 100 mL to 10 L.

27.7.3 Applications Examples Using PROXCYS' Columns

So far, the aforementioned paper by Cabanne *et al*. (14), which compared a 120-mL AFC column with a 120-mL RFC column (CRIO-MD 62 from PROXCYS), appears to be the only published example. Because PROXCYS is a newcomer in the RFC market, more published case studies using their columns have not appeared in the open literature yet.

27.7.4 Applications Examples Using RFC Columns with Monolithic Packing

In recent years, several research groups have attempted to pack RFC columns with monolithic media that have proven to reduce band spreading in AFC. Gustavsson and Larsson (29) fabricated a 65 mL RFC column packed with a single piece of superporous agarose gel (considered a type of monolithic chromatography media). After the gel was derivatized with Cibacron Blue 3GA affinity ligands, the column was used to separate dehydrogenase from a 200 mL crude bovine lactate dehydrogenase extract with excellent results. Yang *et al*. (30) obtained a monolith from the polymerization of glycidyl methacrylate and ethylene dimethacrylate in the presence of porogens. The polymer was modified with DEAE weak anion exchange groups before being packed into a 38 mL radial flow column. The pressure drop of the column showed a linear relationship with flow rate and its value was 1.7 MPa at the highest tested flow rate of 50 mL/min.

Hahn *et al*. (31) studied dispersion effects in a preparative polymethacrylate monolith with radial flow having dimensions of 15 mm outer in diameter, 1.5 mm in inner diameter, and 45 mm in bed height. It should be noted that for rigid monoliths, RFC unlikely would provide any major advantages over AFC.

27.7.5 An Example of Continuous Radial Flow Chromatography

Lay *et al*. (32) reported the use of RFC in continuous mode. They constructed an RFC column with rotating annulus and eight input ports equally spaced on the outer cylinder as shown in Fig. 27.15. The bed packed with 500 mL of DEAE Sepharose Fast Flow anion exchanger was divided into four neighboring zones in a full circle: feed zone, wash zone, elution zone, and a second wash zone. The column was used to separate bovine serum albumin (BSA) from lactoferrin achieving a separation factor of 4.78 for BSA. Continuous RFC is the equivalent of continuous annular chromatography that was reviewed by Hilbrig and Freitag (33). They differ in flow direction with the latter using axial flow.

27.8 MATHEMATICAL MODELING OF RADIAL FLOW CHROMATOGRAPHY

In 1950, Lapidus and Amundson (34) proposed a simplified theoretical model for RFC. Their model ignores radial

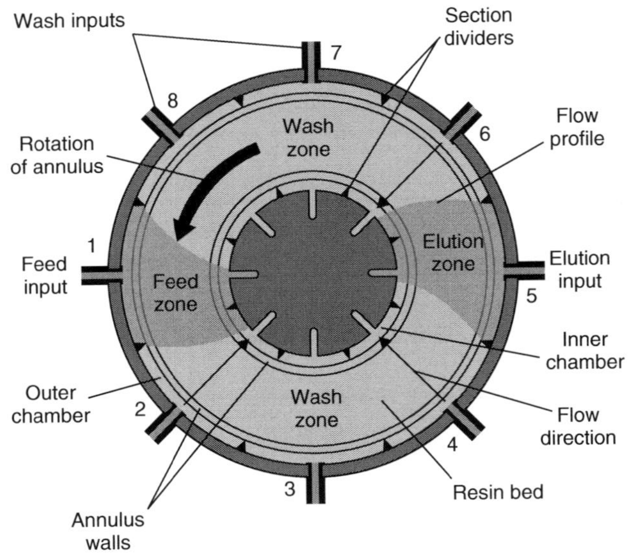

Figure 27.15. Schematic of a continuous RFC column. (Source: Ref. 32.)

diffusion in the bulk-fluid phase and intraparticle diffusion. It is similar to a model used by Rachinskii (35). Inchin and Rachinskii (36) subsequently included molecular diffusion in the bulk-fluid phase. Lee *et al*. (37) proposed several single component rate models for the comparison of statistical moments for RFC and AFC. They included radial dispersion, intraparticle diffusion, and external mass transfer effects. Kalinichev and Zolotarev (38) performed an analytical study on moments for single component RFC in which they treated the radial dispersion coefficient as a variable. Duong and Shallcross (39) presented a model for ion exchange RFC, which considered radial dispersion, external film mass transfer mechanisms, and binary cation exchanges. The model was used to predict experimental breakthrough curves.

A rate model for nonlinear single component RFC was solved numerically by Lee (40) by using the finite difference and orthogonal collocation methods. His model considered radial dispersion, intraparticle diffusion, external film mass transfer, and nonlinear isotherms. It used averaged radial dispersion and mass transfer coefficients instead of treating them as variables. A nonlinear model of this kind of complexity has no analytical solution and must be solved numerically. Recently, Lay *et al*. (32) modeled a continuous RFC column shown in Fig. 27.15. Their model that considered dispersion in radial and angular directions, intraparticle diffusion, and interfacial film mass transfer and a second-order kinetics involving BSA and NaCl was solved using the explicit finite difference method implemented in MATLAB (http://www.matlab.com). The second-order kinetics leads to the Langmuir isotherm when the forward and backward reactions are at equilibrium. The radial and angular dispersion terms were found to be negligible for their column with a short radial flow path of 3 cm, but intraparticle diffusion was important. Model predictions matched experimental data satisfactorily for the separation of BSA from lactoferrin.

Realistic modeling of RFC should treat the radial dispersion and external film mass transfer coefficients as variables rather than as constants because the linear flow velocity (v) in the RFC column changes continuously along the radial coordinate of the column. Without this distinctive feature the curvature in the flow path is lost, and thus the column can be imaginatively cut and spread out to become exactly like a pancake-like AFC column.

27.8.1 General Rate Model for Multicomponent RFC

Gu and coworkers (41,42) presented a general rate model for RFC in which radial dispersion and external mass transfer coefficients were treated as variables rather than as constants. The model was solved numerically. Figure 27.16 shows the anatomy of an RFC column for the purpose of modeling. The following basic assumptions are made in order to formulate a general rate model for RFC:

1. The column is isothermal.
2. The porous particles in the bed can be treated as spherical and uniform in diameter.
3. The concentration gradients in the axial direction are negligible. This means that the maldistribution of radial flow is ignored.
4. The fluid inside particle macropores is stagnant; that is, there is no convective flow inside macropores.
5. An instantaneous local equilibrium exists between the macropore surfaces and the stagnant fluid in the macropores.

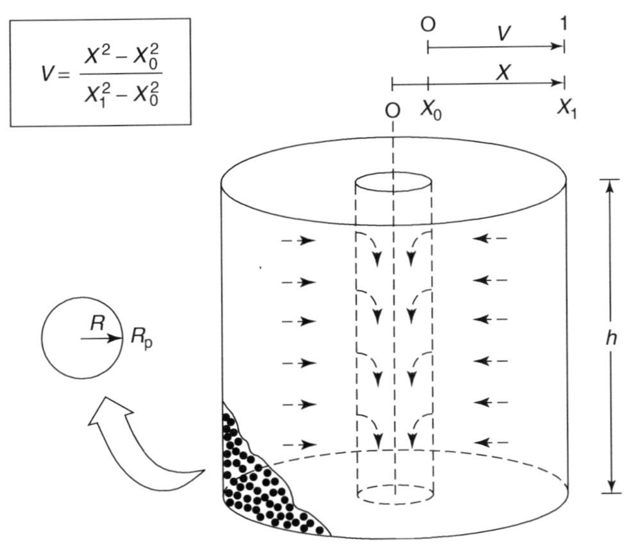

Figure 27.16. Anatomy of an inward flow RFC column.

6. The film mass transfer theory can be used to describe the interfacial mass transfer between the bulk-fluid and particle phases.
7. The diffusional and mass transfer coefficients are constant and independent of the mixing effects of the components involved.

On the basis of these basic assumptions, Equations (27.1) and (27.2) are formulated from the differential mass balance for each component in the bulk-fluid and particle phases, respectively. In Equation (27.1), "+v" represents outward flow and "−v" inward flow.

$$-\frac{1}{X}\frac{\partial}{\partial X}\left(D_{bi}X\frac{\partial C_{bi}}{\partial X}\right) \pm \mathrm{v}\frac{\partial C_{bi}}{\partial X} + \frac{\partial C_{bi}}{\partial t} + \frac{3k_i(1-\varepsilon_b)}{\varepsilon_b R_p}(C_{bi} - C_{pi,R=R_p}) = 0 \quad (27.1)$$

$$(1-\varepsilon_p)\frac{\partial C^*_{pi}}{\partial t} + \varepsilon_p\frac{\partial C_{pi}}{\partial t} - \varepsilon_p D_{pi}\left[\frac{1}{R^2}\frac{\partial}{\partial R}\left(R^2\frac{\partial C_{pi}}{\partial R}\right)\right] = 0 \quad (27.2)$$

In Equation (27.2), C^*_{pi} is related to C_{pi} via isotherm (41).

The initial conditions for the partial differential equation (PDE) system are as follows:
At $t = 0$,

$$C_{bi} = C_{bi}(0, X) \quad (27.3)$$

and

$$C_{pi} = C_{pi}(0, R, X) \quad (27.4)$$

The boundary conditions are as follows:

At the inlet X position : $\partial C_{bi}/\partial X = (v/D_{bi})[C_{bi} - C_{fi}(t)]$ (27.5)

At the outlet X position : $\partial C_{bi}/\partial X = 0$ (27.6)

Equations (27.1) and (27.2) can be written in dimensionless forms as follows:

$$-\frac{\partial}{\partial V}\left(\frac{\alpha}{Pe_i}\frac{\partial c_{bi}}{\partial V}\right) \pm \frac{\partial c_{bi}}{\partial V} + \frac{\partial c_{bi}}{\partial \tau} + \xi_i(c_{bi}, c_{pi,r=1}) = 0 \quad (27.7)$$

$$\frac{\partial}{\partial \tau}\left[(1-\varepsilon_p)c^*_{pi} + \varepsilon_p c_{pi}\right] - \eta_i\left[\frac{1}{r^2}\frac{\partial}{\partial r}\left(r^2\frac{\partial c_{pi}}{\partial r}\right)\right] = 0 \quad (27.8)$$

In Equation (27.7), the dimensionless variable $V = (X^2 - X_0^2)/(X_1^2 - X_0^2) \in [0,1]$ is based on the local volume averaging method (41). Parameter $\alpha = 2\sqrt{V+V_0}(\sqrt{1+V_0} - \sqrt{V_0})$ is a function of V, in which $V_0 = X_0^2/(X_1^2 - X_0^2)$.

The dimensionless initial conditions are as follows:

$$\text{at } \tau = 0, \quad c_{bi} = c_{bi}(0, V) \quad (27.9)$$

and

$$c_{pi} = c_{pi}(0, r, V) \quad (27.10)$$

The dimensionless boundary conditions are as follows:

$$\partial c_{bi}/V = Pe_i[c_{bi} - C_{fi}(\tau)/C_{0i}] \quad (27.11)$$

At the inlet V position, for frontal adsorption, $C_{fi}(\tau)/C_{0i} = 1$; for elution,

$$C_{fi}(\tau)/C_{0i} = \begin{cases} 1 & 0 \le \tau \le \tau_{imp} \\ 0 & \text{otherwise} \end{cases}.$$

After the introduction of a sample in the form of a rectangular pulse,

if component i displaced, $C_{fi}(\tau)/C_{0i} = 0$;

if component i is a displacer, $C_{fi}(\tau)/C_{0i} = 1$.

At the outlet V position, $\partial c_{bi}/\partial V = 0$. For the particle phase governing equation, the boundary conditions are as follows:

$$\text{At } r = 0, \quad \partial c_{pi}/\partial r = 0 \quad (27.12)$$

$$\text{At } r = 1, \quad \partial c_{pi}/\partial r = Bi_i(c_{bi} - c_{pi,r=1}) \quad (27.13)$$

Note that all the dimensionless concentrations are based on C_{0i}, the maximum of the feed profile $C_{fi}(\tau)$ for each component.

The radial dispersion coefficient D_{bi} depends on the linear velocity v. In liquid chromatography, it can be assumed (5,38,42) that $D_{bi} \propto v$. Thus, $Pe_i = v(X_1 - X_0)/D_{bi}$ can be considered constant in liquid RFC. The variation of Bi_i values observes the following relationship:

$$Bi_i \propto k_i \propto v^{1/3} \propto (1/X)^{1/3} \propto (V + V_0)^{-1/6} \quad (27.14)$$

If $Bi_{i,V=1}$ values are known, Bi_i values anywhere else can be obtained from Equation (27.15).

$$Bi_{i,V} = [(1 + V_0)/(V + V_0)]^{1/6} Bi_{i,V} = 1 \quad (27.15)$$

Parameter ξ_i can be calculated from Bi_i using its definition $\xi_i = 3Bi_i\eta_i(1-\varepsilon_b)/\varepsilon_b$.

27.8.2 Numerical Solution

The PDE system of the governing equations is first discretized to become an ordinary differential equation (ODE) system. The finite element and orthogonal collocation methods are used to discretize the bulk-fluid phase and the particle phase governing equations, respectively. The resulting ODE system is then solved using a public domain ODE solver called *double-precision variable-coefficient ordinary differential equation solver* (DVODE) developed by Brown *et al*. (43). The Microsoft Windows-based executable software program is free to any academic researchers for teaching and research. Information on the software is available at http://www.ent.ohiou.edu/~guting/CHROM/.

A study of the effects of treating D_{bi} and k_i as variables compared to treating them as constants was carried out by Gu *et al*. (5,42). The comparison between RFC and AFC was also studied through computer simulation. Figure 27.17 shows that inward flow in RFC gives slightly sharper concentration profiles (5,42) than outward flow. The small difference is theoretically predicted, but it may not show up experimentally, especially in columns with a large X_0 or V_0 value. The figure also shows that RFC gives similar concentration profiles as AFC with equivalent physical parameters. These theoretical results support the experimental results obtained by Tharakan and Belizaire (12).

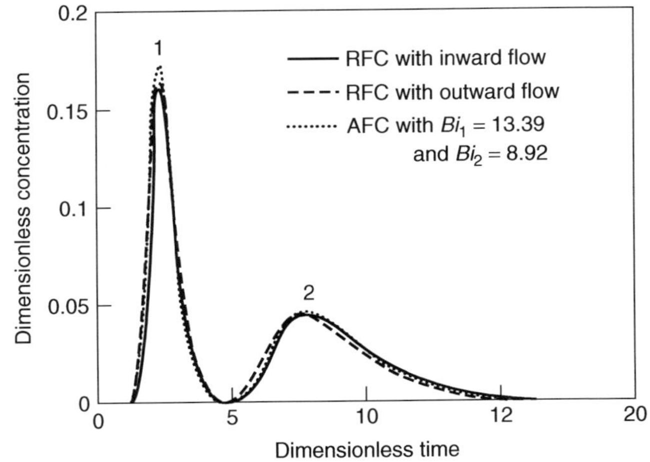

Figure 27.17. Simulated comparison for inward and outward flow RFC and AFC.

27.9 SCALE-UP OF RFC COLUMNS

One of the claimed advantages of RFC columns is the relative ease for scale-up. In Sepragen's Superflo column series, increasing the column height is a rather safe way to accommodate an increase in sample size to a certain degree. However, one must keep in mind that flow distribution in radial flow may deteriorate. If the bed thickness in the radial direction is increased, bed pressure usually increases proportionally. Figure 27.18 shows that a 15-fold

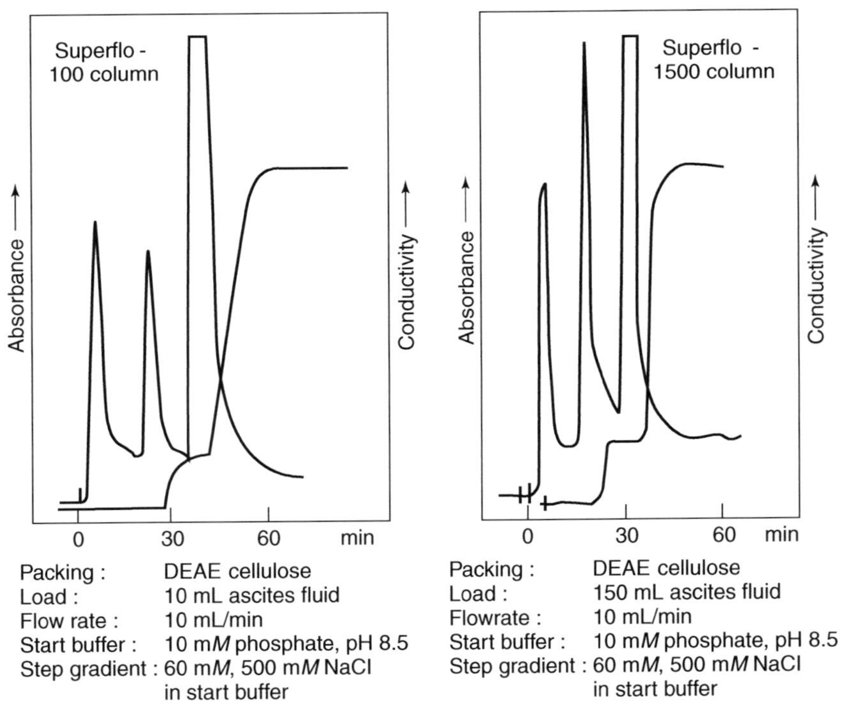

Figure 27.18. A 15-fold scale-up example using Superflo columns.

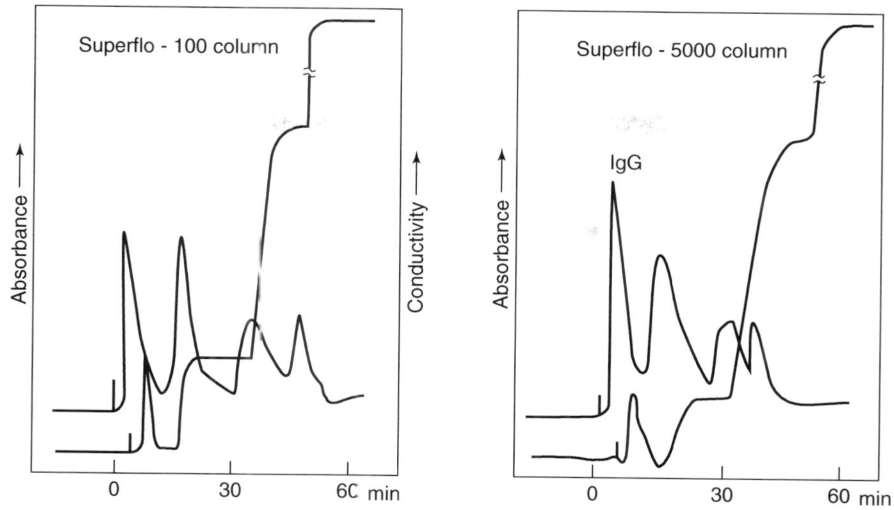

Packing : DEAE cellulose
Load : 10 mL cell culture fluid (murine IgG)
Flow rate : 10 mL/min
Start buffer : 10 m*M* phosphate, pH 8.5
Step gradient : 60 m*M*, 250 m*M*, 700 m*M* NaCl in start buffer

Packing : DEAE cellulose
Load : 10 mL cell culture fluid (murine IgG)
Flow rate : 10 mL/min
Start buffer : 10 m*M* phosphate, pH 8.5
Step gradient : 60 m*M*, 250 m*M*, 700 m*M* NaCl in start buffer

Figure 27.19. A 50-fold scale-up example using Superflo columns.

increase in sample load and bed column produced similar performances when DEAE cellulose was used to separate an ascites fluid (11,44). Figure 27.19 is an example with a 50-fold increase in sample load and bed volume (11). These two examples are very successful examples.

The performance prediction will be difficult if the RFC column diameter is increased, which is equivalent to increasing column length in AFC. This kind of scale-up will change elution times and peak shapes. The mathematical model described above can be used to help the scale-up process. Before a column is bought off the shelf or custom built, its performance can be predicted using computer simulation-based different column dimensions and operating conditions. Isotherm data must be obtained experimentally or be supplied by the vendor of the packing material. To simplify the problem, only one or two key components should be used in computer simulation. Various mass transfer parameters used in the model can be estimated by using existing correlations (41).

27.10 CONCLUSIONS

RFC columns have a short flow path and a large flow area, resulting in small bed pressure. They are specially suited for pressure sensitive soft gels, although they are also used for rigid particles. Owing to its limited resolution, RFC should only be used in preparative- and large-scale chromatographic separations based on strong solute–stationary phase interactions, such as affinity chromatography, strong anion or cation exchange, strong reversed phase, and strong hydrophobic interaction chromatography. RFC is not suitable for SEC and other forms of chromatography with weak solute–stationary phase interactions. RFC is not a replacement for AFC, but rather an alternative to AFC in preparative- and large-scale separations. Both experimental results and theoretical modeling indicate that an RFC column behaves much like a pancake-style AFC column with the same packing volume and with its bed height about the same as the packing's radial thickness in RFC column. However, RFC has a much smaller footprint, and it seems to handle flow distribution better than very wide pancake-like columns.

Acknowledgments

The author wishes to thank Mr Marcel Raedts of PROXCYS Downstream Biosystems and Mr Sanjeev Saxena of Sepragen Corp. for providing product information on their radial flow columns.

NOMENCLATURE

Bi_i	Biot number of mass transfer for component i, $k_i R_p/(\varepsilon_p D_{pi})$
C_{0i}	concentration used for nondimensionalization, $\max\{C_{fi}(t)\}$
C_{bi}	bulk-fluid phase concentration of component
C_{fi}	feed concentration profile of component i, a time dependent variable

C_{pi} concentration of component i in the stagnant fluid phase inside particle macropores

C_{pi}^* concentration of component i in the solid phase of particle (based on unit volume of particle skeleton)

c_{bi} C_{bi}/C_{0i}

c_{pi} C_{pi}/C_{0i}

c_{pi}^* C_{pi}^*/C_{0i}

D_{bi} axial or radial dispersion coefficient of component i

D_{pi} effective diffusivity of component i, porosity not included

k_i film mass transfer coefficient of component i

Pe_i Peclet number of radial dispersion for component i, $v(X_1 - X_0)/D_{bi}$

R radial coordinate for particle

R_p particle radius

r R/R_p

t dimensional time ($t = 0$ is the moment a sample enters a column)

v interstitial velocity

V dimensionless volumetric coordinate, $(X^2 - X_0^2)/(X_1^2 - X_0^2)$

V_0 $X_0^2/(X_1^2 - X_0^2)$

X coordinate in the radial direction for an RFC column

GREEK LETTERS

α $2\sqrt{V + V_0}(\sqrt{1 + V_0} - \sqrt{V_0})$ for RFC

ε_b bed void volume fraction

ε_p particle porosity

η_i dimensionless constant, $\varepsilon_p D_{pi} L/(R_p^2 v)$

ξ_i dimensionless constant for component i, $3Bi_i\eta_i(1 - \varepsilon_b)/\varepsilon_b$

τ dimensionless time, vt/L

τ_{imp} dimensionless time duration for a rectangular pulse of the sample

REFERENCES

1. McCormick D. Biotechnology (NY) 1988; 6: 158–160.
2. Hopf P. Ind Eng Chem 1947; 39: 938–940.
3. Huang SH, Roy S, Hou KC, Tsao GT. Biotechnol Prog 1988; 4: 159–165.
4. Saxena V, inventor; Sepragen Corporation. US patent 4,627,918. 1986 Dec 9.
5. Gu T, Tsai G-J, Tsao GT. In: Fiechter A editor. Advances in biochemical engineering/biotechnology. Berlin-New York: Springer; 1993. pp. 73–95.
6. Pfeiffer W. J Chromatogr A 2003; 1006: 149–170.
7. Ngo T, Khatter N. Appl Biochem Biotechnol 1991; 30: 111–119.
8. Wallworth DM. Downstream processing of proteins: methods and protocols. In: Desai MA, editor. Volume 9, Methods in biotechnology. Berlin-New York: Springer; 2000. pp. 173–184.
9. Munson-McGee SH. Sep Sci Technol 2000; 35: 2415–2429.
10. Saxena V, Dunn M. Biotechnology (NY) 1982; 7: 250–255.
11. Saxena V, Weil AE. BioChromatography 1987; 2: 90–97.
12. Tharakan JP, Belizaire M. J Liq Chromatogr 1995; 18: 39–49.
13. Tharakan JP, Belizaire M. J Chromatogr 1995; 702: 191–196.
14. Cabanne C, Raedts M, Zavadzky E, Santarelli X. J Chromatogr B Biomed Appl 2007; 845: 191–199.
15. Lane L, Koscielny ML, Levison PR, Toome DW, Butts ET. Bioseparation 1990; 1: 141–147.
16. Ernst P. Aust J Biotechnol 1987; 1: 22–26.
17. Saxena V, Weil AE, Kawahata RT, McGregor WC, Chandler M. Am Lab 1987; 19: 112–120.
18. Planques Y, Pora H, Menozzi FD. J Chromatogr 1991; 539: 531–533.
19. Wang H, Li T, Zou H, Zhang Y, Chao J, Chao L. Biomed Chromatogr 1996; 10: 139–143.
20. Sun T, Chen G, Liu Y, Bu F, Wen M. J Chromatogr B Biomed Appl 2000; 742: 109–114.
21. Akoum A, Devichi F, Kalyanpur M, Neff JP, Vijayalakshmi MA, Sigot M. Process Biochem 1989; 24: 55–59.
22. Strætkvern KO, Raae AJ, Folkvord K, Næss BA, Aasen IM. Bioseparation 1991; 2: 81–93.
23. Weaver K, Chen D, Walton L, Elwell L, Ray P, BioPharm 1990; July/August: 25–29.
24. McCartney JE. BioTechniques 1991; 11: 648–649.
25. Sun T, Chen G, Liu Y, Bu F, Wen M. Biomed Chromatogr 2000; 14: 478–482.
26. Sun T, Chen G, Liu Y, Bu F, Wen M. Biotechnol Tech 1999; 13: 831–835.
27. Tseng C-C, Tseng C-P, Levine MJ, Bobek LA. Arch Biochem Biophys 2000; 380: 133–140.
28. Levison PR. J Chromatogr B Biomed Appl 2003; 790: 17–13.
29. Gustavsson P-E, Larsson P-O. J Chromatogr A 2001; 925: 69–68.
30. Yang C, Wei Y, Zhang Q, Zhang W, Li T, Hu H, Zhang Y. Talanta 2005; 66: 472–478.
31. Hahn R, Tscheliessnig A, Bauerhansl P, Jungbauer A. J Biochem Biophys Methods 2007; 70: 87–84.
32. Lay MC, Fe CJ, Swan JE. Food Bioprod Process 2006; 84(C1): 78–73.
33. Hilbrig F, Freitag R. J Chromatogr B Biomed Appl 2003; 790: 1–15.
34. Lapidus L, Amundson NR. J Phys Colloid Chem 1950; 54: 821–825.
35. Rachinskii VV. J Chromatogr 1968; 33: 234–242.
36. Inchin PA, Rachinskii VV. Russ J Phys Chem 1977; 47: 1331–1333.
37. Lee W-C, Huang SH, Tsao GT. AIChE J 1988; 34: 2083–2087.
38. Kalinichev AI, Zolotarev PP. Russ J Phys Chem 1977; 51: 871–873.

39. Duong H, Shallcross DC. Ind Eng Chem Res 2005; 44: 3681–3691.

40. Lee W-C, PhD thesis, West Lafayette, IN: Purdue University; 1989.

41. Gu T. Mathematical modeling and scale-up of liquid chromatography. Berlin-New York: Springer; 1995, pp. 102–106, 116.

42. Gu T, Tsai G-J, Tsao GT. Chem Eng Sci 1991; 46: 1279–1288.

43. Brown PN, Byrne GD, Hindmarsh AC. SIAM J Sci Stat Comput 1989; 10: 1038–1051.

44. Saxena V, Subramanian K, Saxena S, Dunn Michael. BioPharm 1989; March: 46–49.

28

DRYING, BIOLOGICAL MATERIALS

Chung Lim Law
Department of Chemical and Environmental Engineering, The University of Nottingham, Malaysia Campus, Selangor, Malaysia
Arun S. Mujumdar
Department of Mechanical Engineering, National University of Singapore, Singapore

28.1 INTRODUCTION

Most biotechnological products appear in the form of liquids or cultures that require refrigeration for storage and distribution, thereby adding handling costs and inconvenience. However, these biotechnological products may be dehydrated to eliminate the disadvantages associated with refrigeration and liquid storage. Biotechnological products in the form of dry powders are superior to liquid or frozen state with reference to sterility and stability. Furthermore, transport and storage costs of dry products are relatively lower than liquid or frozen products.

Drying, by definition, involves removal of a liquid (generally water, but in many bioprocessing applications it could be an organic solvent or an aqueous mixture) from a solid, semisolid, or liquid material to produce a solid product by supplying thermal energy to cause a phase change, which converts the liquid to vapor. In the exceptional case of freeze-drying, the liquid is first solidified and then sublimed. Biotechnological products are produced by microbial action, and are related to living organisms. Biotechnological products are a subset of a broader generic definition of biomaterials, which includes wood, coal, biomass, foods (biopolymers), vegetables, fruits and so on. This chapter is limited to such biotechnological products as whole cells (e.g. baker's yeast, bacteria, blood, plasma, vaccines, fungi), fermented foods (e.g. yogurt, cheese), synthetic products of both low molecular weight (e.g. amino acids, citric acid), high molecular weight (e.g. antibiotics, xanthene), carbohydrates, and enzymes.

All of these products are characterized by their high thermal sensitivity; they are damaged or denatured and inactivated by exposure to certain temperatures specific to the products. Some are inactivated by mechanical stress (e.g. shear stress, etc.), surface tension, or damage caused to the cell walls during the drying operation. These products are often produced in smaller qualities in batch mode. Further to this, they are typically high-value products, such that the cost of drying is often secondary to quality constraints. It is therefore, not unusual to use more expensive drying techniques (e.g. freeze-drying, vacuum drying etc.) even when less-expensive techniques such as heat pump drying could be applied successfully. Of course, some biotechnological products are produced in bulk in continuous operation using conventional drying technologies, such as spray drying or fluidized bed drying.

The activity of water in a biotechnological product is determined by the state of water in it. Free water represents the intracellular water in which nutrients needed by the living cells are in solution. Bound water is built into cells or the biopolymer structures. It is held more strongly to the solid matrix, and is also resistant to freezing. The ratio of the vapor pressure expected by the water in the product to the equilibrium vapor pressure of pure water at the same temperature is referred to as the *water activity*. For safe storage, the objective of a drying process is to reduce the product moisture content so as to lower its activity below a threshold value safe for storage. During thermal drying, biotechnological materials may undergo some changes such

Downstream Industrial Biotechnology: Recovery and Purification, First Edition. Edited by Michael C. Flickinger.

as destruction of cell membranes, denaturation of proteins or enzymes, or even death of cells.

Selection of a dryer for the processing of biotechnological products is of paramount important as it involves heat and mass transfers that give rise to the problems mentioned above. A suitable dryer produces desirable final product qualities, which includes higher cell viability, high cell biomass, preservation of active ingredients, desirable final moisture content, particle/granule size and so on.

28.2 DRYING OF BIOTECHNOLOGICAL PRODUCTS

There are many types of yeast applied in various industries, such as fermentation of sugar, bread production, beer fermentation, wine fermentation, xylitol production, production of ethanol, and bioremediation (e.g. degradation of palm oil mill effluent, fatty acids, fats, oils, etc.). Yeast is also a source of health and probiotic supplements. It is an excellent source of protein and vitamins, especially B-complex vitamins. Yeast extracts are used as food additives or flavors. Table 28.1 lists some of the industrial applications of yeast.

Bacteria have many properties that are useful to industry, for instance, biotransformation. It has been applied in many industrial applications such as fermentation of foods, waste processing, bioremediation, biological pest control, degradation of pesticides and herbicides, production of chemicals, pharmaceuticals, and agrichemicals, and microbial mining. Bacterial survival during drying process and storage is affected by various factors such as initial concentration, protective agent, rehydration, storage conditions, species kinetics, and operating parameters (1–4). Thus, the selection of a dryer is vital to maximize the storage stability, viability, and activity of the bacterial cells. The protective agent that is used during freeze-drying is another factor that affects the bacteria survival and cell viability, for instance, adonitol, betaine, glycerol, lactose, sucrose, skim milk and dimethyl sulfoxide, trehalose, sorbitol, and mannitol. There are many strains of bacteria that have been used in many industrial applications, including lactic acid bacteria (LAB) (*Lactococcus* and *Lactobacillus*), acetic acid bacteria, recombinant bioluminescent bacteria, avirulent bacteria and so on. Table 28.1 also lists some applications of these bacteria.

Proteins and amino acids are primary products in food industry. Appropriate handling and processing of proteins and amino acids are important to preserve or improve their nutritive and functional properties such as digestibility and solubility. Improper drying and excessive heating may cause change of structure, thus reducing their digestibility or absorbability (5–7). During drying, chemical and physical reactions occur and they are detrimental to the digestibility of the products due to protein denaturation (8–11). A change in operating or environmental conditions, such as temperature may cause cross-linking interactions among protein molecules. This in turn results in aggregation, coagulation, and finally precipitation (12,13).

In recent years, there have a considerable rise in applications of enzymes as industrial catalysts, pharmaceutical products, clinical diagnostic chemicals, and applications in molecular biology. Enzymes are protein catalysts produced by plants or microorganisms. Industrial enzymes are produced in bulk such as proteases, amylases, and pectinases (14), whereas analytical enzymes are produced in small quantities for chemical analysis purposes. Because most enzymes are not stable in water, dehydration is used to stabilize them. Table 28.2 lists the application of some common enzyme groups in industry. Proper selection of a dryer

TABLE 28.1. Some Industrial Applications of Different Types of Yeasts and Bacteria

Biotechnological Products	Type	Industrial Applications
Yeast	Various yeast strains	Food production, beer and wine fermentation, health and probiotic supplement, biological control in agriculture and horticulture, and bioremediation
	Yeast extract (β-D glucan)	Pharmaceutical yeast product with immunostimulatory activity
Bacteria	Lactic acid bacteria: *Lactococcus*, *Lactobacillus*	Production of flavor ingredients, exopolysaccharides, fermented milks products, dairy starter cultures, probiotics, and silage preservatives
	Acetic acid bacteria	Oxidation of alcohols and sugars into commercial foods and chemical products such as vinegar, cellulose, sorbose, gluconic acid, etc.
	Recombinant bioluminescent bacteria	Toxicity monitoring system, toxicity biosensors, genetically engineered bacteria, *E. coli*
	Avirulent bacteria (*Bordetella pertussis*)	Skim milk production

TABLE 28.2. Industrial Applications of Various Types of Enzyme Groups

Enzyme	Application
Protease	Detergents, dairy, bakery, and leather industries
Amylase	Detergents, starch, distilling, bakery, and textile industries
Lipase	Detergents, and dairy industries
Cellulose	Detergents, wine and juice, textiles, and animal feed industries
Lactase	Dairy industry
Pectinase,cellobiase, polyphenol oxidase	Wine and juice industry
Glucose oxidase	Wine and juice, and bakery industries
Glucose isomerase	Starch industry
Glucoamylase	Starch, and distilling industries
Catalase	Textiles industry
Phytase	Animal feed industry
Xylanase	Pulp and paper industry
Tannese	Tea industry
Acetoacetate decarboxylase, β-glucanase	Brewery industry

for the production of enzymes, and optimization of operating conditions is vital in producing dried enzymes that pose desirable qualities such as retention of activity, solubility and dispersibility, stability, purity, color, and odor. In addition, properties of the dried enzymes such as flowability, powder size, size distribution, homogeneity, and density are equally important.

Storage and transport of blood platelet concentrates is a major problem as the platelets are activated by refrigeration at low temperature. Hence, it can only be stored at temperatures higher than 22°C. In a blood transfusion center, platelet-rich plasma concentrates are stored in blood bags at 22°C, with a shelf life of not more than 5 days. On the other hand, platelets stored in the liquid state tend to rapidly lose their functionality and vitality (15). Moreover, storage of platelets at ambient temperature may result in growth of bacteria. One way to preserve blood platelets is to freeze dry them using protective agents.

28.3 EFFECT OF DRYING ON BIOTECHNOLOGICAL PRODUCT QUALITY

Quality attributes of biotechnological products cover a wide range of aspects including cell vitality, survival rate, active ingredients content, color, texture, organoleptic properties, nutritional values, taste, flavor, final moisture content, and so on. Bacterial survival during the freeze-drying process is also affected by species (16), initial concentration (17,18),

TABLE 28.3. Possible Quality Changes During Biomaterial Drying

Material	Change Type	Effect
Yeast	Biochemical	Atrophy of cells
Bacterial	Biochemical	Atrophy of cells
Molds	Biochemical	Atrophy of cells
Enzymes	Enzymatic	Loss of activity
Vitamins	Enzymatic	Loss of activity
Proteins, fats, carbohydrates antibiotics	Chemical	Loss of activity, nutritive contents
Other	Physical/chemical/ biochemical	Solubility, rehydration, loss aroma, shrinkage

growth and drying medium (19), drying parameters (20), rehydration (19), and storage conditions. Low initial cell concentration is reported to be detrimental to the survival of freeze-dried biotechnological products (17,19). Palmfeldt *et al*. (21) found that the optimal initial cell concentration for freeze-drying of *Pseudomonas chlororaphis* was between 1×10^9 and 1×10^{10} CFU/mL when sucrose was used as protective solute.

According to Beker and Rapoport (22), it is necessary to reduce the moisture content of Baker's yeast from 65% to 70% to a final moisture content of 4–6%. Bayrock and Ingledew (23,24) found that the viability of pressed yeast was not affected by the drying temperature, when moisture contents were higher than 15%. However, cell viability was poor when the moisture level was lower than 5–8%. This is due to the fact that irreversible damage occurs to metabolic functions when the bound water is removed (25).

Numerous and varied undesirable changes can occur in the product during drying. The changes may be physical, chemical, biochemical, or enzymatic. In the worst-case scenario, one may obtain a dry but totally inactivated product. Table 28.3 summarizes such changes for various biomaterials and their effects on product quality. Various indices are used to quantify changes in quality, and their choice clearly must depend on the product, but it is beyond the scope of this chapter to discuss this important issue. Briefly, typical quality criteria may be as follows:

- For food biopolymers, criteria include color, texture, organoleptic properties, nutritional value (vitamin content), taste, and flavor.
- For "live" products (e.g. bacteria, yeast) or products such as enzymes or proteins that are thermally destabilized or inactivated, quality indices may be used.

As an example of the diverse quality criteria used in practice for a biotechnological product, Table 28.4 lists quality indices that are often used to define suitability of

TABLE 28.4. Quality Changes: Drying of Protein-Containing Compounds

Quality Indexes
Nitrogen solubility index (NSI)
Protein dispersibility index (PDI)
Water dispersed protein (WDP)
Water-soluble protein (WSP)
Nitrogen solubility curve (NSC)
Protein precipitate curve (PPC)

For fruits, vegetables, and other foods, other criteria apply, including color, texture, taste, flavor, nutrition, organoleptic properties, etc.

dried proteins or protein-containing compounds. Not all of these criteria are used for a given product, however.

For biomaterials such as foods, fruits, and vegetables, numerous other quality attributes such as structural properties (density, porosity, pore size, specific volume), optical properties (color and appearance), textural properties (hardness, stickiness, chewiness, etc.), thermal properties, sensory properties (aroma, taste, flavor), nutritional properties (vitamins, proteins contents), and rehydration properties (rehydration and rate and capacity) are used. These quality attributes are also applicable to biotechnological products (26). The attributes are classified into four categories, viz physical, chemical, biological, and nutritional, as shown in Table 28.5.

TABLE 28.5. Classification of Quality Attributes

Quality	Quality Attributes	Remarks
Physical	Color	Caused by Browning reaction, Maillard reaction, caramelization, oxidation, etc.
	Visual appearance	Caused by changes in color, shape (shrinkage)
	Porosity	Depends on drying methods, affects rehydration properties
	Texture	For example, hardness, stickiness, chewiness, etc.
	Rehydration properties	Depends on drying methods
Chemical	Flavor—Odor	Good storage and packaging practice to preserve flavor and avoid off-flavor
	Water activity	Below 0.65 to prevent growth of microbes and yeasts
	Chemical stability	
Biological	Microbial	Avoid infection of moulds, fungus, etc.
	Free from pests/contaminants	Stored below 5°C
Nutritional	Retention of nutrients	For example, proteins, lipids, carbohydrates, vitamins, minerals, etc.

28.4 BASIC PRINCIPLE OF DRYING

Figure 28.1 shows a typical drying curve. As the moisture content decreases, the drying rate varies. Some materials exhibit a short period of initial transient where the drying rate increases as the moisture content decreases. This is due to the fact that part of the heat is transferred to the drying materials to raise their temperature. After the initial transient, it is followed by constant rate period if the surface of the materials is covered by a thin layer of moisture. Evaporation is the dominating transport process during the constant rate period. The constant rate period ends when the materials surface is partially dry. The drying rate then starts to drops. This is due to the rate of internal diffusion of moisture being slower than the rate of evaporation on the surface. The moisture content that marks the onset of falling rate is known as *first critical moisture content*, X_{cr1}. During falling rate period, diffusion dominates the mass transport. Some materials only exhibit falling rate period; some have two distinct falling rate periods. The second critical moisture content, X_{cr2} distinguishes the first and second falling rate periods. Detailed discussion on drying periods can be obtained in Mujumdar and Davahastin (27), Law and Mujumdar (28), and Monlar (29).

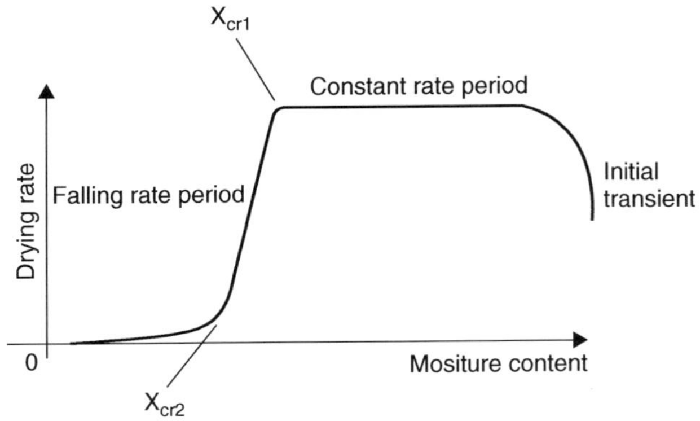

Figure 28.1. Drying curve.

TABLE 28.6. Commonly Used Dryers and Emerging Drying Technologies Suitable for Biotech Products

Conventional	Dryers Emerging Dryers
Spray dryer	Heat pump dryers (below/above freezing point)
Spray fluid-bed (two-stage)	Intermittent batch dryer
Freeze dryer	Vacuum fluid-bed dryer
Vacuum tray	Low-pressure spray dryer (plate or turbo dryer) with ultrasonic atomizer
Continuous tray dryer	Sorption dryer
Drum dryer/vacuum	Pulse combustion dryer
Indirect vacuum	Cyclic pressure/vacuum dryer High electric field (HEF) dryer Superheated steam dryer at low pressures

TABLE 28.7. Choice of Dryers and Drying Conditions for Biotech Products Depending on Specific Constraints

Restrictive Criterion when Drying Biotech Products	Possible Dryer/ Drying Conditions Conditions
Highly heat sensitive; thermally inactivated, or damaged	Dehumidified air drying (heat pump or adsorption dehumidifier) at low temperatures Vacuum drying with indirect heat supply Intermittent batch drying Cyclic vacuum/pressure drying Freeze-drying
Damaged by oxidation	Convective drying in N_2 or CO_2 Vacuum drying Freeze-drying
Product subject to destabilization (e.g. enzymes)	Addition of sugars, maltodextrin, salts, etc. to stabilize some enzymes Control of pH change during drying
Product affected by physical processing	Use of gentle drying (e.g. packed bed or continuous tray as opposed to fluid bed) Better drying of some products in one type of dryer than others (e.g. yeast in spouted bed vs fluid bed)

28.5 COMMONLY USED DRYERS

Often, the wet biotechnological product to be dried is in the form of wet solid, sludge, filter cake, suspension, or solution. Mujumdar and Menon (30), Mujumdar (31), and Mujumdar (32) presented a classification scheme for the numerous dryer types and their selection criteria in a general way. Suffice it to say that the choice of dryers for biotechnological products is constrained mainly by the ability of the dryer to handle the material physically, while the choice of the operating conditions is determined by the thermal sensitivity of the material. Table 28.6 lists some of the conventional dryers, as well as some emerging drying techniques for heat-sensitive biotechnological products, many of which are already commercialized, but not commonly offered by vendors yet. Table 28.7 summarizes the key restrictive criteria that determine suitability of a given drying technology for biotechnological products. Note that aside from heat, such products may be damaged by the presence of oxygen. Some products may have to be stabilized by additives such as sugars or salts, as in the case of drying of some enzymes. Certain cryoprotective chemicals are used in freeze-drying of live cells to avoid rupture of the cell walls. The rate of drying may have a direct or indirect effect on the quality as well as on the physical handling of the product. Spray drying and freeze-drying are some of the most common drying technologies used for drying of biotechnological products, although fluidized bed, batch and continuous tray dryer, spin-flash, and vacuum dryers are also common. Pilosof and Terebiznik (33) reviewed the literature on the drying of enzymes using spray- and freeze-drying.

Multistage drying systems (e.g. spray dryer to remove surface moisture followed by a fluidized or vibrated bed to remove internal moisture at milder drying conditions over an extended period) are often used to speed up the overall drying process while maintaining product quality. Low-pressure fluidized bed drying can be used to achieve drying of particulate solids at lower temperatures, although it is not a commonly used process. Freeze-drying (lyophilization) is used extensively in the industry to dry ultra-heat-sensitive biomaterials (e.g. some pharmaceuticals). Some $200 billion worth of pharmaceutical products are freeze dried worldwide each year. It is a very expensive dehydration process, justified by the high value of the product.

28.5.1 Spray Dryer

Spray dryers are used to convert suspension/slurry to powders. Mujumdar (31) described and discussed various methods of powders formation from suspensions and pastes, which include spray drying. Figure 28.2 shows the schematic diagram of a typical spray drying system. The drying system consists of a drying chamber and a dust/powders separation unit. Nozzle is normally mounted on top of the chamber although it can be placed at the side of the drying chamber which is in the case of horizontal spray dryer (34). Liquid atomization by nozzles produces droplets which are then dried by drying medium to form

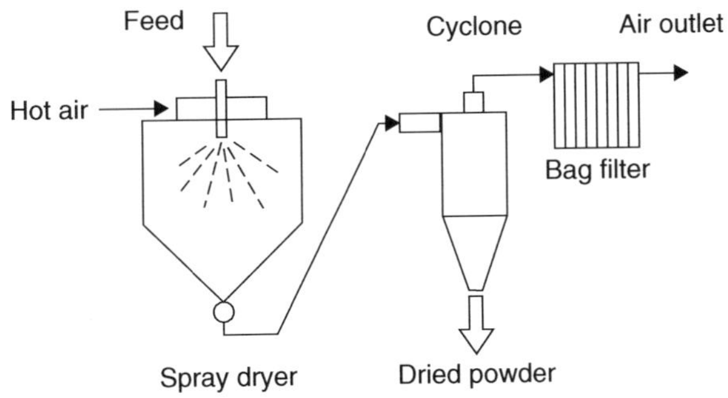

Figure 28.2. Spray dryer.

powders. Hot air is normally used as the drying medium. After the water in the droplets evaporates, powder is formed and drops on the bottom of the drying chamber. The powders are then entrained with the exhaust air and discharged from the drying chamber. The gas-powders mixture is then charged into a dust/powders separation system. Cyclone is normally used for the first stage of powders gas separation. Coarse powders are separated in the cyclone, but some fine powders may entrain with the cyclone exhaust gas. Thus, secondary dust separation such as bag filter or water scrubber may be installed to remove the fine powders from the gas stream. If water scrubber is used, the fine powders are dissolved in the water. The solution is then recycled and mixed with feed stream for powder formation in the spray dryer.

Spray drying is becoming common for the production of biotechnological products in powder form from liquid/suspension. Many research works have been carried out to compare their drying performance with reference to the conventional freeze-drying, and their product quality, which includes cell viability. Products that have been successfully tested with spray drying are brewer's yeast (35,36), where its viability was improved; glucan particles extracted from Baker's yeast (37), where the native state of the extract was preserved; *Enterococcus faecium* (38), where the dry particles were well encapsulated; α-lactalbumin and β-lactoglobulin (39), where the solubility of both proteins were not affected by medium outlet temperatures but decreased when the temperature is high; and carboxymethyl chitosan/β-cyclodextrin microspheres (40), where high product yield was obtained.

Spray drying is four to seven times cheaper than freeze-drying (41), and it is more energy efficient. Master (42) gave comprehensive accounts on various topics related to spray dryers including design and description of various industrial spray dryers. Huang and Mujumdar (34) discussed the features of spray dryers and presented simulation of spray dryers, as well as their classification. Filkova *et al.* (43) presented detailed accounts on atomization, various arrangements of spray drying systems, and their classification.

28.5.2 Spray Fluidized Bed Dryer

When powders are formed in a spray dryer, they contains internal moisture. If the spray dryer is used to remove the internal moisture content, the operating cost would be relatively high, as the thermal efficiency of a spray dryer is normally low. This is due to the fact that the removal of internal moisture is dependant on diffusion of internal moisture. Thus, enhancing external operating conditions does not enhance the rate of internal moisture removal. Hence, the removal of internal moisture tends to take a longer time.

An alternative way to remove the internal moisture would be using a cost-effective and lower operating cost dryer. A fluidized bed dryer is a suitable candidate for this purpose, as its operating cost is relatively lower than a spray dryer and it allows a longer operating time without incurring huge operating costs. As such, a spray fluidized bed dryer can be deployed to dry solutions/slurries that form powders with high-internal moisture content. Figure 28.3 shows the schematic diagram of a spray fluidized bed dryer. The powders formed in a sprays, which contain internal moisture are transported into a fluidized bed dryer attached beneath the spray dryer. The internal moisture content is then removed in the fluidized bed dryer where fluidization enhances the contacting efficiency between the powders and the fluidizing gas. Longer residence time can be set by prolonging the length to width ratio of the fluidized bed dryer. Cool air can be used to cool down the powders to avoid condensation that might occur during packaging of powders. A sieve separator can be used to screen the undesirable product sizes. The coarse product is ground to form a smaller product size and recycled, whereas the fine product is dissolved in the solvent and recycled to the spray dryer for formation of powders.

Wang *et al.* (44) reported that bioproperties of product powders from bovine serum albumin (BSA) and skim milk (with avirulent bacteria *Bordetella pertussis*) formulations were well maintained after being spray-freeze-dried. They found that the percentage of α-helix of the BSA was unaffected and the survival of *B. pertussis* was more than 90% for atmospheric spray-freeze-dried powder. The drying time was appreciably less than freeze-drying. Jinapong *et al.* (45) found that the flowability and wettability of instant soy milk powders formed in a spray dryer were very poor due to dominating cohesive forces occurring between fine powders (particles size <25 μm). It was also found that fluidized bed agglomeration can be used to improve the quality by using maltodextrin solution as binder. Mounir and Allaf (46) reported that application of instant controlled pressure drop between the spray drying and final drying can further enhance the product functional quality.

A brief account on spray fluidized bed dryer and the principles of how this dryer works can be found in

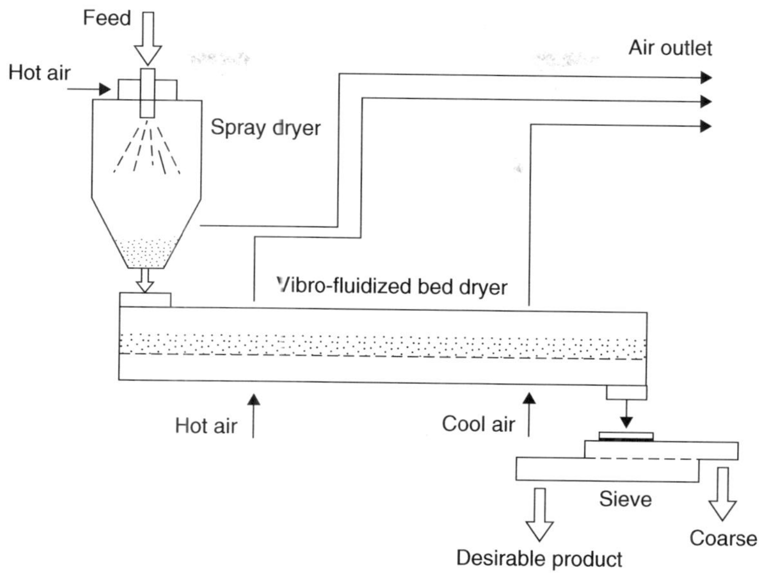

Figure 28.3. Spray fluidized bed dryer.

Huang and Mujumdar (34), Law and Mujumdar (47), and Filkova *et al.* (43).

28.5.3 Freeze Dryer

The freeze dryer is traditionally used to dehydrate blood cell and plasma, as it can preserve the viability of the cell. It has now been applied for the dehydration of bio-origin products, pharmaceutics, as well as nanomaterials (48). Industries that are dealing with bio-origin products, especially those that have a high margin find that the freeze dryer is the best dryer to produce products with high quality such as better vitamin C retention, textural properties, and color preservation (49). However, the operating cost of a freeze dryer is very much higher than other dryers as this drying technology involves freezing and drying at low pressure. Moreover, the drying time is long and thus, requires high energy input. As a result, the freeze dryer is feasible in the pharmaceutical industry and the biotechnology industry where the product margin is high.

The freeze dryer is one of the dryers that operates dehydration at low temperature and thus, suitable for heat-sensitive products. It involves three stages of operation. The first stage is freezing. At this stage, the operating temperature is reduced below the melting point of the solvent (normally free water) in the drying materials. The solvent (free water) is frozen and becomes solid (ice if the solvent is free water). At this stage, bound water remains in the liquid phase. The second stage is primary drying, where the pressure is reduced until vacuum is achieved. The solid form of the solvent is sublimed. Theoretically at the end of primary drying, all free water is removed. Second drying involves the removal of bound water. The removal of bound water is accomplished by heating the product under vacuum.

For the freeze-drying of LAB, which include *Lactococcus* and *Lactobacillus*, the survival of the LAB strains can be protected or increased by using protective agents and cryoprotectants. It was reported that the stability of probiotic microorganisms during freeze-drying and storage may be enhanced by the addition of protective agents such as adonitol, betaine, glycerol, lactose, skim milk, and dimethyl sulfoxide (50). Carvalho *et al.* (51) suggested that skim milk powder should be selected as the drying medium for freeze-drying of LAB. Lactose and sucrose are two of the sugars that have been tested for their protective effect during drying and subsequent storage of LAB (52,53). Carvalho *et al.* (51) reported that skim milk increased the survival rate and addition of carbohydrates improved resistance to freeze-drying stress. Trehalose and combinations of trehalose with other protective media were found to greatly enhance the survival rate of bacteria (54), blood platelets (55,56), and water–oil–water multiple emulsions (57). Gu *et al.* (58) also found that trehalose is the best cryoprotectant for freeze-drying of genetically engineered *E. coli*.

Liapis and Bruttini (59,60) gave detailed account on freeze-drying technology, which includes processes, possible improvements to the conventional freeze-drying, and its classification.

28.5.4 Vacuum Dryer

Vacuum drying is another method in which the removal of water is carried out at low temperature. As the environment pressure is reduced, sensible heat that is required to vaporize the water is reduced due to lower boiling point. It should be noted that convective heat transfer becomes insignificant as the drying medium is absent in vacuum drying. The heat transfer mode in vacuum drying is typically by radiation. Hence, conductive heat transfer is normally required to increase the drying rate. Heating plates/trays and heating chamber walls are usually installed in a vacuum dryer to allow conductive heat transfer for enhancing the total heat transfer efficiency.

28.5.5 Tray Dryer

The tray dryer is a conventional dryer where the product is placed on plates/trays and subject to convective and conductive heat transfers. It can be operated at atmospheric pressure or vacuum. Oven is a typical tray dryer. A tray dryer may be operated under vacuum such as vacuum dryer; or use a different drying medium such as low-temperature dehumidified air in heat pump dryer.

28.5.6 Drum Dryer

The drum dryer is also one of the conventional dryers. A rotating drum is used where a layer of liquid/suspension is coated on the drum. The adhering thin layer is dried conductively by the heating medium inside the drum. The drying rate can be further enhanced by blowing hot air on the surface of the thin layer. The drum dryer can be operated in vacuum to reduce the boiling point of water, which is suitable for drying of heat-sensitive products. Daud (61) gave a detailed account on the operations of a drum dryer and its classification.

28.5.7 Fixed Bed Dryer

The fixed bed dryer is another type of conventional dryer. It is used to dry powders or granules. Powders or granules are packed in a column to form a fixed bed/packed bed. Hot air is then charged into the packed bed column typically from the bottom. The drying air passes through the bed of powders/granules and carries away moisture.

28.5.8 Fluidized Bed Dryer

The fluidized bed dryer is similar to the packed bed dryer, but operated at higher air velocity. At higher air velocity, powders/granules in the fluidized bed column are suspended and this in turn increases the powders' surface that is exposed to the drying medium. As a result, heat and mass transfers and thus, their drying rates are enhanced. Figure 28.4 shows a typical well-mixed fluidized bed drying system. As a fluidizing gas stream may carryover some fine powders, a dust separation system is therefore required.

Strumillo *et al*. (62) simulated the fluidized bed drying of biosynthesized products, and found that plug flow fluidized bed dryer is better than well-mix fluidized bed dryer in term of final product quality. Strasser *et al*. (63) found that freeze-drying of *E. faecium* resulted in 80% in survival rate compared to fluidized bed drying that resulted in a 70% survival rate. However, fluidized bed drying consumes less time and energy than freeze-drying.

Various drying strategies such as vacuum drying, heat pump drying, and intermittent drying can be incorporated with fluidized bed drying. Law and Mujumdar (28,47) gave a comprehensive overview of fluidized bed dryers and a detailed account on various modified fluidized bed dryers that apply various drying strategies to overcome different operational difficulties.

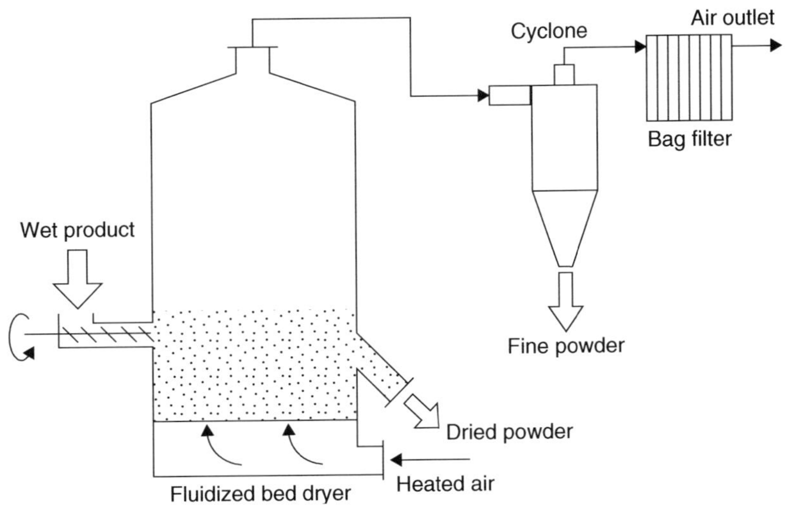

Figure 28.4. Well-mixed fluidized bed dryer.

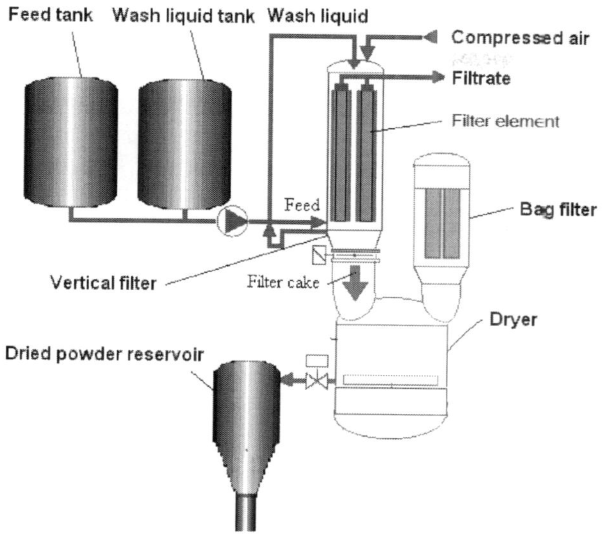

Figure 28.5. Vertical filter dryer. (This figure is available in full color at http://onlinelibrary.wiley.com/book/10.1002/9780470054581.)

28.5.9 Filter Dryer

The filter dryer combines filtration and drying. It is used to dry suspension. Because filtration is a dewatering process without involving phase change, the energy demand for filtration is relatively low compared to drying. Thus, filtration dewatering followed by drying tends to give better energy performance. Figure 28.5 shows the schematic diagram of a vertical filter dryer. Feed is charged into the filter dryer and fills up the filter vessel. Compressed air is charged into the vessel to give positive pressure, whereas the filtrate is withdrawn from the vessel to create negative pressure. Both positive and negative pressures exert a driving force for the filtrate to permeate the filter element. A layer of filter cake is formed and encircles the filter element. Water is then sprayed on the filter cake to remove impurities. After the cake washing, the cake is subjected to thermal drying to remove the captive liquid in the cake. Thermal drying, however, only removes the surface moisture; a second stage of drying is required to remove the internal moisture. Finally, the filter cake is removed from the filter element and introduced into a secondary dryer, for example, a fluidized bed dryer or a vacuum dryer. Mujumdar and Law (64) discussed various types of filter dryers that are suitable for liquid extraction as well as drying of the filter cake.

28.6 SOME EMERGING DRYING TECHNOLOGIES

Numerous new drying techniques proposed and tested over the past decades have potential for various applications to biotechnological products. Extensive discussion of the basic principles, advantages, and limitations of each of these is beyond the scope of this chapter. Table 28.6 lists some of the emerging technologies. Table 28.7 suggests some of the possible dryers or drying conditions for some restrictive criteria where drying of biotechnological products is concerned.

Drying is an energy extensive operation. The operating cost of drying is forever increasing due to increasing fuel price and high demand on product quality. This has driven drying experts and researchers to conceive ideas, innovations, and design new dryers to tackle these problems. These ideas and innovations are classified into the following categories:

- Drying strategy
- Drying medium
- Handling of drying materials
- Mode of heat input.

28.6.1 Drying Strategy

This classification includes innovations, ideas, and strategies on how the heat input is set and manipulated, how the feed stream is charged and exposed to the drying medium, and how the external operating conditions are manipulated to enhance the drying performance and cost-effectiveness.

28.6.1.1 Intermittent Batch Drying. Conventional drying applies the drying medium continuously throughout the drying process. It is well known that drying toward the end of the process is controlled by internal diffusion. Thus, continuous drying at constant operating profile is not an optimized drying strategy. Intermittent drying involves the application of thermal energy to the drying material intermittently rather than continuously. The period where the thermal energy is not supplied is referred to as *tempering period*. Moisture within the solids is allowed to redistribute during the tempering period. Intermittent drying is a drying strategy that can be applied to any direct dryer such as a conveyor dryer, fluidized bed dryer, and spouted bed dryer. Various intermittency (fraction of cycle when drying is carried out) can be applied depending on the characteristics of the drying materials (65). Law *et al.* (66) gave a comprehensive overview of intermittent drying and discussed some of the latest developments in this drying strategy for bio-origin materials.

For batch drying, intermittent supply of energy is an especially interesting concept if the bulk of the drying takes place in the falling rate period. Jumah *et al.* (67) explained the principle with application to a novel intermittently spouted and intermittently heated spouted bed dryer for grains; it was shown that appreciable

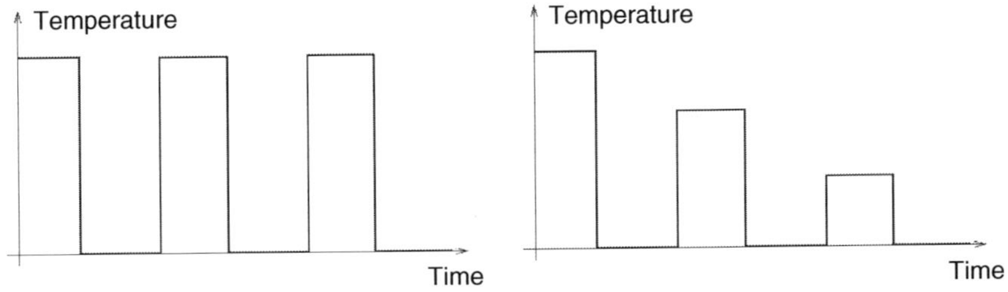

Figure 28.6. Temperature profile in intermittent drying.

reductions in energy and air consumption could be made while enhancing product quality due to lower product temperature attained, as well as reduced mechanical handling of the grain due to intermittent spouting. This idea has been extended to fluidized beds as well. Again, no direct biotechnological applications have been reported, but the concept is fundamentally sound and is expected to find new applications.

Apart from intermittent drying, variable intermittency profiles such as stepwise change of drying temperature during drying and tempering can be applied to further enhance the drying performance. This strategy combines intermittent drying and variable operating conditions. Figure 28.6a shows the temperature profile of intermittent drying, and Fig. 28.6b shows the profile of variable intermittent drying. Further to this, multiple heat inputs can be used to remove both surface and internal moistures simultaneously. Here, intermittency can be applied to one or all of the heat inputs.

Besides the on/off intermittency and time varying temperature profile given in Fig. 28.6, intermittent drying can also apply time varying heat input, multivariable heat input or cyclic variation of temperature pressure or gas velocity. Cyclic pressure is further discussed in the section titled "Cyclic Pressure Vacuum Dryer". Chua *et al*. (68) demonstrated experimentally and by mathematical modeling, the superior performance of intermittent drying of heat-sensitive fruits in terms of quality parameters such as color and ascorbic acid content. The drying time may be increased marginally.

Intermittency can also be applied to a spouted bed dryer (69). Figure 28.7 shows the schematic diagram of a rotating jet spouted bed dryer. The rotating jet, which has two gas spouting outlets is driven by a motor and installed at the bottom of the spouted bed dryer. The two gas outlets supply spouting gas through the bed of powders, thus creating two spouting regions in the bed. As the jet rotates around the central axis, the outer spouting region rotates as well. The spouting regions enable rapid heat and mass transfers, whereas the static region allows tempering where internal moisture is redistributed inside the drying materials.

28.6.1.2 Pulse Combustion Dryer. Pulse combustion is a process where a combustible mixture of fuel and air is ignited and discharged periodically in a pulse combustor. The periodic combustion of the fuel mixture is accompanied by periodic pressure oscillations. Typical pulse combustion consists of the following processes in sequence:

- ignition and combustion of fuel mixture
- expansion of combustion of gaseous product
- discharging and purging of combustion gaseous product
- recharging of fuel mixture and compression of the mixture.

The idea of pulse combustion drying has been proposed and revisited several times over the past two decades with limited success. In principle, even highly heat-sensitive products such as vitamins, enzymes, and yeasts can be dried by direct injection into the highly turbulent pulse combustor exhaust tailpipe, despite the ultrahigh temperatures of the exhaust. Rapid heat and mass transfer rates and fine atomization of the feed (slurry or dilute paste) by highly turbulent flow allow drying in a fraction of a second and without thermal degradation.

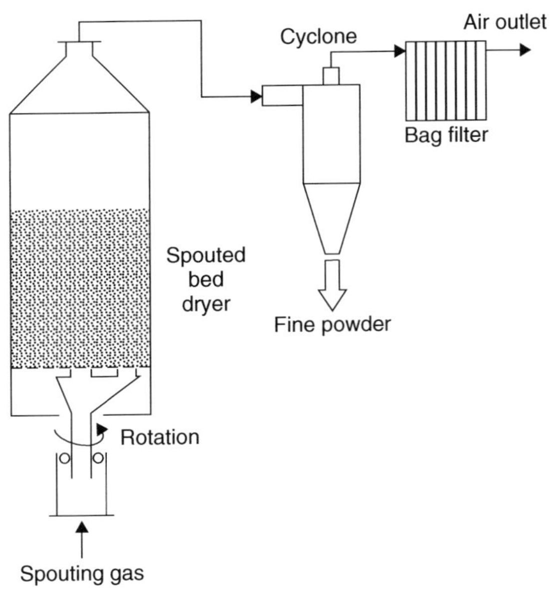

Figure 28.7. Rotating jet spouted bed dryer.

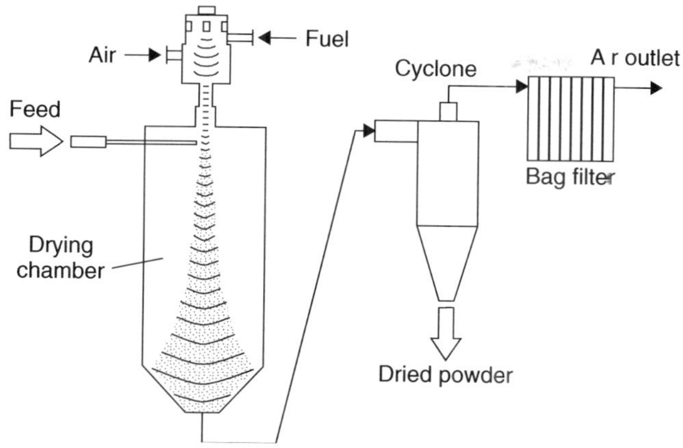

Figure 28.8. Pulsed combustion dryer.

Figure 28.8 shows the schematic diagram of a typical pulse combustion dryer. The pulse combustor is mounted on the drying chamber. The combustion gaseous product discharged from the combustor is mixed with the product liquid, and the mixture is then charged into the drying chamber. Rapid drying is carried out in the drying chamber where powders are formed and drop on the bottom of the chamber. The products together with the rapidly moving gas stream exit the chamber and enter dust/powder separation system.

Pulse combustors can be installed in a spray dryer to replace nozzle for the atomization of liquids (70). This in turn eliminates the traditional operational problems with nozzles that become jammed or worn. It, therefore, requires regular repair and replacement. The unstable rapid oscillating hot air flow from the combustor tail pipe atomizes the liquid that is introduced near the tail pipe. The atomization is affected by air flow oscillation frequency, velocity, density, and properties of liquid.

The process has not been a commercial success yet, possibly due to problems of noise, scale-up, and capital cost. Much R&D is needed before the process can be successfully commercialized. Mujumdar and Wu (71) gave a comprehensive overview of the pulse combustion dryer and a detailed account on various types of combustors. A detailed account on this drying technology can also be found in Kudra and Mujumdar (72).

28.6.1.3 Impinging Stream Dryer. The impinging stream dryer is an emerging dryer that is suitable for drying of suspensions/pastes/sludges. Two high-velocity streams, one containing feed materials, are collided to allow rapid heat and mass transfers (73). The powders formed in the impingement zone are then entrained by the gas stream. The high intensity of turbulence, and the rapid, unsteady particle motion yields very high heat and mass transfers, thus enhancing the drying rate as well as reducing its drying time (74). This drying technique is useful in removing free water on the powder surface. If the drying kinetics are controlled by the internal diffusion, the high heat transfer in an impingement dryer may denature the products, if they are heat sensitive, such as biotechnological products. This dryer may be combined with other types of drying operations such as granulation, agglomeration, and chemical reaction. Various configurations of impinging streams are possible. Kudra and Mujumdar (75) have given a detailed account on this type of dryer as well as various configurations of impingement stream.

28.6.1.4 Cyclic Pressure Vacuum Dryer. The cyclic pressure vacuum dryer allows the pressure in the vacuum chamber to be fluctuated within a range. This in turn changes the intensity of convective heat transfer. Fluctuation of pressure in the vacuum dryer accelerates the discharging of water vapor from the dryer. This enhances the driving force of mass transfer as the vapor pressure is reduced. Sadykov *et al.* (76) have proposed an interesting technique to dry bioactive materials. It involves cycling the operating pressure in a batch mode. Heat is supplied convectively at atmospheric pressure for a certain length of time, and then the moisture is flashed off in a subsequent cycle when vacuum is applied to the chamber for a given, but different, length of time. This process may be repeated several times. Heat can be supplied indirectly by conduction through the chamber walls. For heat-sensitive products, intermittent application of high-pressure and low-pressure environments reduces the operating temperature. The process must be operated in batch mode, however. One way to achieve intermittency of high and low pressure is to place the drying material in a cylindrical chamber for which volume (and pressure) can be altered cyclically at a desired frequency (or cycle time) by a tight-fitting reciprocating piston. The cyclic pressure vacuum dryer is still new to the industry and more research works are indeed required.

28.6.1.5 Spray-Freeze-Dryer. The spray-freeze-dryer is suitable for drying of solutions/pastes that are heat sensitive, which require a long drying time in a freeze dryer. In spray-freeze-drying, the solution is atomized in a cryogenic medium such as liquid nitrogen to form freeze sprayed droplets. The droplets are then subjected to freeze-drying under vacuum. As the droplets' surface area is relatively larger than the solution itself, its freeze-drying time is comparatively shorter. However, because it involves freeze-drying, its drying time will be longer than spray drying. Filkova *et al.* (43) reported some latest findings on this technology as well as their classification.

28.6.1.6 Atmospheric Freeze Dryer. The atmospheric freeze dryer operates the drying stages at atmospheric pressure instead of vacuum. With the absence of vacuum, the freeze-drying is simplified and vacuum operating cost is saved. Liquids and foods have been tested using this technique (77–79). Atmospheric freeze-drying has a relatively long residence time due to internal diffusion. It is recommended that the product size should not exceed 2 mm (80). This technique can be combined with the fluidized bed dryer (81). Rahman and Mujumdar (82) presented a novel atmospheric freeze dryer, where a fluidized bed and an absorbent are used. Drying with the assistance of an absorption agent is further discussed in the section titled “Absorption Agent—Sorption Dryer”.

28.6.1.7 ***Vacuum Fluidized Bed Dryer.*** As stated earlier, vacuum drying lowers the boiling point of water, thus allowing the drying to be conducted at a lower temperature. Likewise, vacuum can be applied to fluidized bed drying to combine the advantages of both drying strategies. Fluidized beds can increase the powder's surface exposed to the drying medium, and thereby, its drying rate, whereas vacuum can lower the operating temperature. As such, this drying technology is suitable for the drying of heat-sensitive materials such as biotechnological products.

28.6.1.8 ***Low-Pressure Spray Dryer.*** Similarly, low pressure may be applied in a spray dryer to reduce its operating temperature, and thus make it suitable for the drying of heat-sensitive products. It has been reported that vacuum spray drying of probiotic bacteria at temperatures as low as 80°C gave comparable bacterial survival rate and better storage stability (83).

28.6.2 Drying Medium

Heated atmospheric air is normally used in conventional dryers. As the atmospheric air contains oxygen, it is detrimental to drying materials that contain active ingredients, which are oxidative. Oxidation denatures the drying materials and damages the product quality. In addition, some drying materials may generate combustible vapor, thus posing the risk of combustion. As such, other drying mediums, which do not contain oxygen, can be applied to avoid oxidation and combustion.

28.6.2.1 ***Superheated Steam—Superheated Steam Dryer.*** Superheated steam can be used to replace heated air in direct dryers. As it does not contain oxygen, oxidative or combustion reactions can be avoided. Furthermore, it eliminates the risk of fire and explosion hazards. The quality of superheated steam-dried products tends to be better, as oxidation is eliminated. Mujumdar (84) discussed the principles, advantages, and limitations, as well as diverse applications of this technology. As oil prices have been skyrocketing over the years, this technology provides strong incentives in saving operating costs as well as limiting carbon emissions.

Superheated steam drying is a concept that has been around for over a century, although commercial products appeared on the market only two decades ago for such products as pulp, waste sludge, hog fuel in the paper industry, beet pulp, and so on. For heat-sensitive materials that are damaged in an atmosphere containing oxygen, superheated steam drying is possible only at low operating pressures. This technique has been shown by Chen *et al*. (85) to be successful for drying of silkworm cocoons. The resulting silk is also found to be stronger and brighter. More recent laboratory studies have focused on drying of vegetables, but the results are tentative. No work has been reported on biotech product drying to date. Owing to the fact that most biotech products are made in small quantities and in batch mode, it is unlikely that superheated steam drying will be a major contender in this application area.

28.6.2.2 ***Low-Temperature Dehumidified Air—Heat Pump Dryer.*** Low-temperature drying can avoid product denaturation (8) and vitamins loss (86). Krokida *et al*. (87) found that the aroma retention is mainly due to lower product temperature in drying of apples.

One of the methods in conducting drying at low temperature is to use low-temperature dehumidified air as the drying medium. This type of drying system is known as *heat pump drying*. A heat pump is used to dehumidify the recycled moist air at its dew point. The dehumidified air is then heated to a higher level, but relatively lower than normal hot air. Heat pump drying is normally used in convective drying of heat-sensitive materials. Figure 28.9 shows the schematic diagram of a typical heat pump drying system where the evaporator of the heat pump system serves as condenser to dehumidify the recycled moist air and condenser of the heat pump system is used to elevate the temperature of the dehumidified recycled air.

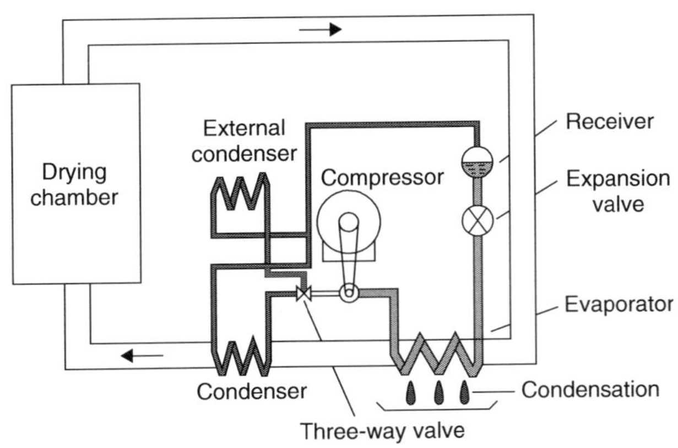

Figure 28.9. Heat pump dryer. (This figure is available in full color at http://onlinelibrary.wiley.com/book/10.1002/9780470054581.)

Tsaousi *et al*. (88) recommended 32°C for the thermal of drying of the yeast, *Saccharomyces cerevisiae*. They found that the dry yeast's quality at low-temperature thermal drying is comparable with the freeze-dried yeast in terms of cell viability. Thermal drying is relatively cost-effective than freeze-drying as its operating cost is lower and the drying time is shorter.

Chua *et al*. (68) provided a comprehensive overview of the numerous variants possible, including those involving multistage heat pumps; multistage dryers; drying below the freezing point; heat pump drying with supplemental heat input by conduction, radiation, or dielectric (microwave or radio frequency fields); and use of cyclical variation of the drying air temperature for batch drying of heat-sensitive products. The use of cyclical variation of drying air temperature is a type of intermittent drying, which is elaborated in the section titled "Intermittent Batch Drying".

Heat pump dryers are further subdivided into the following classifications (89):

- batch or continuous operation
- operating temperature and pressure
- continuous, cyclic, or interrupted heat input
- coupled with conventional dryer such as tray dryer, fluidized bed dryer, and so on.
- with/without auxiliary heat input; continuous or intermittent conduction/radiation; microwave, infrared, or radio frequency
- single or multistage dryer (any convective type)
- single-stage or multistage heat pump system.

It should be noted that not all of heat pump dryer variants mentioned above have been tested at laboratory or pilot scales. Several of these are of interest when drying highly heat-sensitive biotechnological products, as they are more cost-effective than freeze dryers.

The two-stage heat pump dryer designed by Alves-Filho and Strommen (90), in which the first stage is a fluid-bed freezer/freeze dryer at atmospheric pressure, and the second stage is a fluid-bed dryer operated with dehumidified air but above the freezing point, can successfully compete with the freeze dryer for certain products. It yields dried product properties that resemble those obtained by the much more expensive freeze-drying process. Proteins have been tested using this technology, and it was found that the product quality is better than single-stage drying (91). When the biomaterial to be dried is very sticky due to the presence of proteins, fats, or sugars, special drying techniques may be necessary. Such problems, however, must be solved on an individual product basis.

Besides the heat pump system, zeolite is utilized to produce dehumidified air (92). The temperature of the dehumidified air is then elevated to lower its relative humidity. It was reported that multistage zeolite absorption can be used to improve the efficiency of zeolite drying.

28.6.2.3 Inert Gas—Inert Medium Drying. An inert gas such as nitrogen can be used to replace hot air as the drying medium. As oxygen is not present in the inert gas, oxidative reactions are prevented in inert gas drying.

28.6.2.4 Supercritical Fluid Drying. Supercritical drying is suitable for drying of aerogels and biotechnological products that are porous and detrimental to surface tension. Surface tensions occur when liquids dry in pores/voids and vaporize, thus causing distortion and shrinkage to the pores/voids. Under supercritical conditions, surface tension is eliminated. As this drying technology operates at high temperature, it is suitable for those materials that are thermally resistant. Jovanović *et al*. (93) reported that supercritical fluid drying can be used to produce stable microparticulate protein powders from human serum immunoglobulin by adjusting the process conditions. Jovanović *et al*. (94) also found that stable, sugar-based protein formulations can be obtained from lysozyme solutions with and without the addition of sucrose or trehalose by using critical fluid drying. Supercritical fluid spray drying has been reported to dry ethyl cellulose microparticles (95).

28.6.2.5 Absorption Agent—Sorption Dryer. Most dryers apply gaseous heating medium, for instances hot air, superheated steam, low-temperature dehumidified air, and pure nitrogen, to transfer the heat to the material and remove the moisture from the materials. In sorption drying, inert solids are used as the heating medium as well as the absorption agent. Inert solids of high thermal conductivity enhance the conductive heat transfer, whereas inert solids, which are hygroscopic, enhance the contact moisture transfer. The most promising absorption agents are heated hygroscopic materials that allow simultaneous contact heat and mass transfer. These include bentonite, zeolite, chabazite, and synthetic materials of high sorption capacity resistance to thermal shock (75).

When a product to be dried is used in a mixture, one of the components of the mixture can be used as a carrier. This drying method is known as *contact-sorption drying*. The carrier can have different roles:

- If the product is a liquid suspension, the particulate-form carrier disperses it, thus, providing a large interfacial area for evaporation of the moisture while producing a granulated product.
- The presence of the carrier effectively reduces the hygroscopicity of the material.

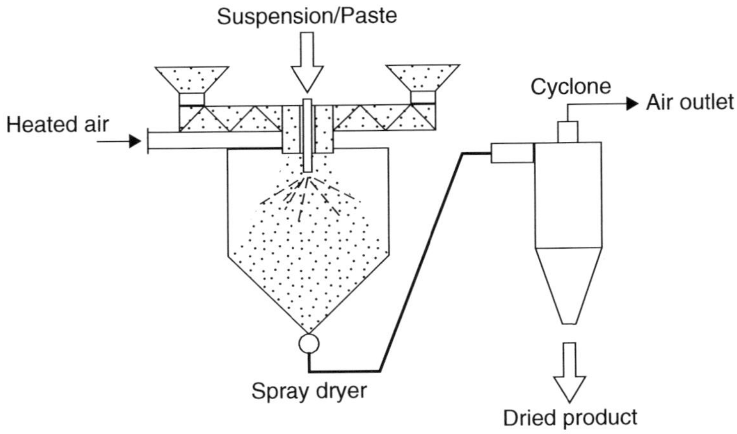

Figure 28.10. Spray dryer with dispersion of carrier.

- Dispersion of the liquid on a "dry" substrate makes the mixture easier to handle (e.g. fluidize, convey, feed), thus permitting the use of a number of conventional dryers.

Sorption dryers of various designs have been reported in the literature, ranging from single or multistage fluidized bed dryers to cocurrent spray dryers, in which the carrier is dispersed in the zone with the drying air in the atomizer zone. Figure 28.10 shows the schematic diagram of a concurrent spray dryer with dispersion of the carrier. The carrier is charged into the spray drying chamber by a screw conveyor, while the biotechnological product in the form of a suspension or a paste is sprayed in the drying chamber. Droplets of the product mix with the dispersed carrier in the chamber and form a mixture. The mixture is exposed to the drying medium (can be heated air or low-temperature dehumidified air). The mixture dries and forms powders. Likewise, the powders are separated with the air stream in a cyclone. Figure 28.11 shows the schematic diagram of spray fluidized bed dryer with the dispersion of carrier. This drying system uses the fluidized bed dryer as a secondary dryer to remove internal moisture, which is difficult and expensive to remove, in the spray chamber. Another spray nozzle can be installed to spray the suspension/paste on the mixture, which is sprayed from the primary nozzle. Dry fluidizing gas stream is introduced from the bottom of the drying chamber and passes through a perforated distributor plate. The plate is slanted at a few degrees to ensure smooth powders flow.

28.6.3 Handling of Drying Materials

Conventional drying methods normally place the drying materials on a tray or pack the powders in a packed bed. The contacting efficiency between the drying materials and the drying medium is generally unsatisfactory, results

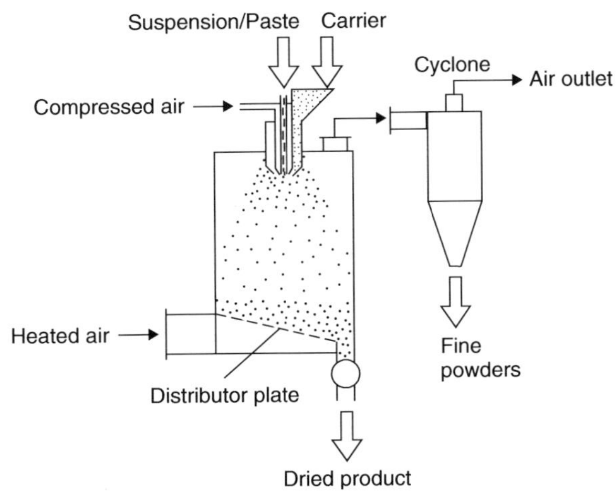

Figure 28.11. Spray fluidized bed dryer with dispersion of carrier.

in poor heat and mass transfers and low drying rate. Therefore, conventional dryers normally take longer drying times to accomplish the operation. Various advancements in handling drying materials have been reported that greatly enhance the contacting efficiency between the drying medium and materials.

28.6.3.1 ***Spouted Bed Dryer.*** The spouted bed dryer is normally used to dry coarse particles, which are relatively difficult-to-fluidize in a fluidized bed dryer. Fluidization quality of coarse powders is generally poor, as the fluidizing gas tends to form big bubbles and bypasses the bed of particles without having efficient contact with the particles. This in turn reduces the efficiency of heat and mass transfers. Figure 28.12 shows the schematic diagram of a spouted bed. A powerful gas stream is charged into the central region of the bed of particles where the particles together with the gas stream are mixed and move upwards vigorously. The particles are thrust at the bed surface and

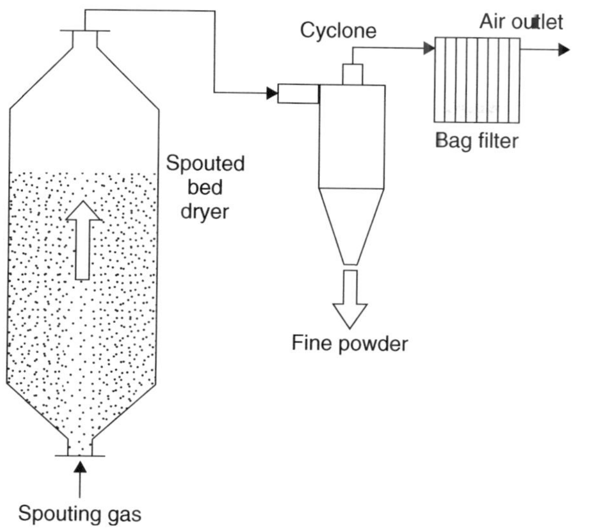

Figure 28.12. Spouted bed dryer.

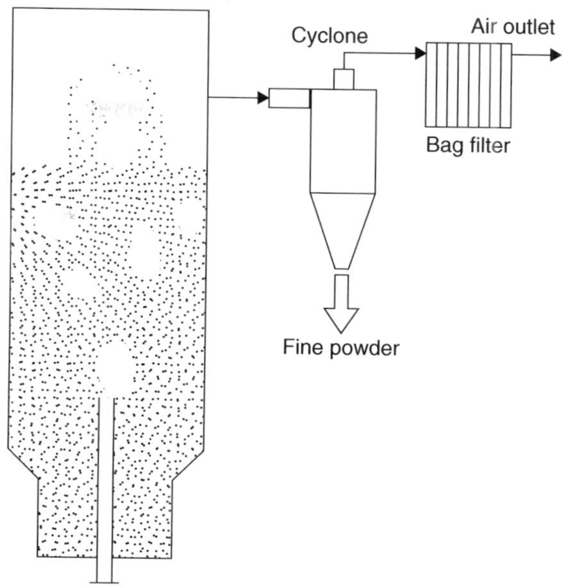

Figure 28.13. Jet spouted bed dryer.

fall on the region near to the wall. The particles at the wall region slowly move toward the central region of the bed for another round of spouting. Rapid spouting at the center and slow movement of particles at the wall region, can be regarded as a type of intermittent drying where rapid heat and mass transfers occur during spouting and tempering during the slow movement of particles at the wall region. During the tempering period, internal moisture redistributes within the particles.

Inert particles can be placed into a spouted bed to assist drying. The inert particles may serve as a conductive heat transfer medium or absorption agent. This technique has been used for the drying of chemical products, animal blood, herbal extracts (96), antibiotics, yeast (97), bacteria (98), pastes (99), and gelatin capsules (100). A spouted bed dryer can be combined with other types of dryers such as fluidized bed dryers to improve the overall drying performance. Grabowski *et al*. (101) found that two-stage drying viz. spouted bed dryer (reduced the moisture content from 70% to 35%) followed by fluidized bed dryer (reduced the moisture content to 6–8%) gave the best final product quality.

28.6.3.2 Jet Spouted Bed Dryer. Figure 28.13 shows a typical jet spouted bed. This type of dryer is modified from spouted bed, where a large tube is inserted into the bed of particle to replace the spouting gas stream at the center of the bed. One distinctive feature between this dryer and spouted bed is that bubbles are formed rather than dilute phase at the center region of the bed.

28.6.3.3 Vibrating Fluidized Bed Dryer. Vibration can be applied to fluidized bed dryers to dry difficult-to-fluidize powders such as fine and coarse powders. Vibration can

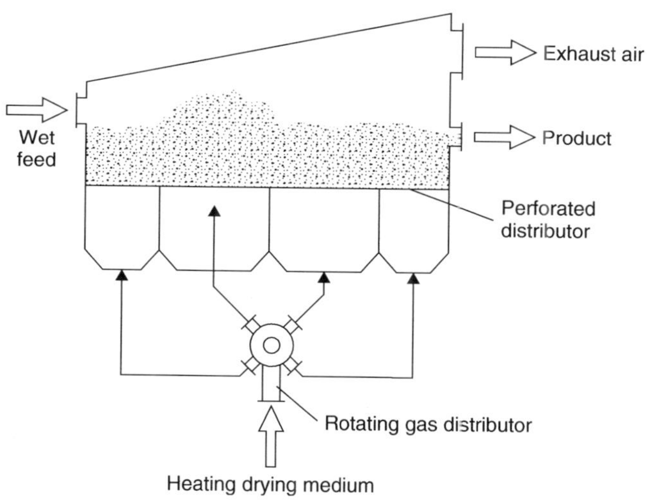

Figure 28.14. Pulsating fluidized bed dryer.

disintegrate agglomerates of fine powders and pseudofluidize coarse powders. It, therefore, improves the fluidization quality of these powders. This in turn enhances their heat and mass transfers. Normally vibration is applied to horizontal fluidized bed dryer where its length to width ratio is typically higher than conventional fluidized bed dryers. Longer ratio gives longer residence time. For a brief account on vibrated fluidized bed, refer Law and Mujumdar (47).

28.6.3.4 Pulsed Fluid-Bed Dryer. Pulsating fluidized bed drying can be regarded as a type of intermittent drying. Figure 28.14 shows the schematic diagram of a typical pulsating fluidized bed dryer. The fluidizing gas stream is relocated to fluidize part of the bed at different times (102).

A rotating gas distributor is used to charge the fluidizing gas into a portion of the bed of powder at one time. When the portion of the bed of powders is fluidized, rapid heat and mass transfers occur, whereas the rest of the bed is in tempering where internal moisture is allowed to migrate from the interior to the surface of the powders. This dryer is typically used to dry powdery materials (103), but it is also applicable to suspensions (104). Microwaves can be applied to enhance volumetric heating (105). Detailed account of this technology can be found in Kudra and Mujumdar (72).

28.6.4 Mode of Heat Input

Heat input of conventional dryers includes convective and conductive heat transfers. Both heat transfer modes only allow heat transfer to the surface of the drying materials, thus making them inefficient in removing internal moisture. New heat input methods such as microwave and high electric field (HEF) enable heat to penetrate to the interior of the drying materials. This mode of heat transfer is also known as *volumetric heating*. Both convective and conductive heat transfer modes can be combined with volumetric heating methods to remove the surface and the internal moistures simultaneously.

28.6.4.1 High Electric Field. HEF drying is a relatively new application for a well-known technique. Kulacki (106) discussed the fundamental principles of electrohydrodynamics and the effect of electrical field on heat and mass transfer. In the HEF technique, wet materials can be dried at ambient temperature and pressure (or at lower temperatures and pressures) using an alternate current HEF (107). Unlike microwave or radio frequency, heat is not generated in the material, so no loss of color, nutrients, or texture occurs during drying. The apparatus is very simple, consisting of point and plate electrodes. The main cost is that of electrical power consumption. Bajgai and Hashinaga (108) reported the high quality attained in HEF drying in a field of 430 KV/m of chopped spinach. The drying rates were very low, but the dried product quality was very high.

It was reported that pulse electric field can be used as one of the pretreatment methods for convective drying (109,110), where trains of pulses at the duration of few microseconds and intervals of hundreds of seconds are applied to drying materials. Each train interval is in the order of milliseconds to tens of seconds. The trains of pulses are applied to the materials intermittently at isothermal condition.

Although not tested for biotech products, this technique could have the potential for drying smaller batches of materials. Further research is needed to evaluate and compare the technoeconomics of this technique with competing drying methods.

28.6.4.2 Microwave Field—Microwave Drying. Microwave fields can be applied to drying materials apart from the popular conventional methods based on conductive, convective, and radiative heat transfers. Dielectric heating heats a material containing a polar compound volumetrically. Heat is generated due to friction of the excited molecules with asymmetric charges, for example, water when an alternating electromagnetic field is applied to a dielectric material. The microwave frequency permitted for industrial applications ranges from 915 to 2450 MHz.

Microwave drying offers advantages, which include uniform energy and high thermal conductivity to the interior of the material, space utilization, shorter drying time, and prevention of enzymatic reaction. In addition, microwave

TABLE 28.8. Dryer Selection with Reference to the Physical Form of Biotechnological Products

Liquids		Biotechnological Products				
Solution and slurry	Pastes	Cakes	Free Flowing Solids/Fiber	Fragile Solids	Formed Solids	Nanomaterials
Fluidized bed (FB) dryer	Spray dryer	Flash dryer	Belt conveyor dryer	Belt dryer	Belt dryer	Tray dryer
Spray dryer	Drum dryer	FB dryer	Flash dryer	Tray dryer	Batch tray dryer	Spray dryer
Drum dryer	Freeze dryer	Rotary dryer	Fluidized bed dryer			Freeze dryer
Freeze dryer	Tray dryer	Tray dryer	Rotary dryer			Microwave dryer
Jet spouted bed (SB) dryer	FB dryer		Tray dryer			Supercritical dryer
Impingement dryer			Drum dryer			
			Superheated steam dryer			
			Impingement dryer			
			Spouted dryer			
			Vacuum dryer			
			Paddle dryer			

TABLE 28.9. Conventional Versus Innovative Drying Techniques in Food Processing

Feed Type	Conventional Dryer	New Techniques
Liquid suspension	Drum dryers	Inert solids fluidized bed dryers
	Spray dryers	Inert solids spouted bed dryers
	Freeze dryers	Spray fluidized bed dryers
		Low-pressure spray dryers
		Vacuum belt dryers
		Pulse combustion dryers
		Atmospheric freeze dryers
		Spray-freeze-dryers
Paste/sludge	Drum dryers	Inert solids spouted bed dryers
	Spray dryers	Fluidized bed dryers with solids back-mixing
	Paddle dryers	Low-pressure spray dryers
		Superheated steam dryers
Particulate	Cabinet dryers	Vibrated bed dryers
	Conveyor dryers	Heat pump dryers
	Rotary dryers	Vacuum fluidized bed dryers
	Fluidized bed dryers	
Chip/piece form	Tray dryers	Impinging jet dryers
		Vibrated bed dryers
		Heat pump dryers
		Microwave dryers
Continuous sheets	Multicylinder contact dryers	Combined impingement/radiation dryers
	Impinging stream dryers	Combined impingement and through dryers
		Combined impingement/microwave dryers
		Combined impingement/radio frequency dryers

drying can be operated in vacuum to enhance its drying performance (111). It was found that vacuum microwave drying of sea cucumber (112) and potato chip (113) gave good product quality and relatively shorter drying time as compared to hot air drying and freeze-drying. Wu and Mao (9) found that microwave-dried samples showed lower fat loss, higher protein solubility, and lower anisidine values than hot air-dried samples, thus preventing lipid oxidation.

Table 28.8 gives the selection of dryers with reference to the physical form of the biotechnological products in question. Table 28.9 lists some commonly used dryers with reference to the physical form of feed and some new techniques that can be applied to replace the commonly used dryers. It should be noted that new techniques are not necessarily better than conventional ones. Feasibility studies should be carried out and carefully studied.

28.7 CLOSING REMARKS

Energy consumption for thermal dehydration depends to a great extent on the dryer or drying system chosen and on the wet feed and properties of the dried product to some extent. Sometimes, a lower thermal efficiency dryer is chosen for a given application, as the alternative higher efficiency dryers yield a lower quality product. Clearly, it is impossible to overview all of the emerging drying technologies that are relevant to drying of the diverse and ever increasing numbers of biotechnological products. This chapter covers a wide assortment of dryers that are suitable for the dehydration of heat-sensitive biotechnological products. It should be noted that selection of a dryer is also geography specific. A dryer that works at one location might not work well at another location, for the same drying materials.

The reader is to refer to recent books authored by Kudra and Mujumdar (72) for advanced drying technology, by Mujumdar (114) for new drying technologies and information on biotechnological product drying, and a handbook edited by Mujumdar (115) for detailed information on a wide range of industrial dryers. The proceedings of the biennial International Drying Symposium (IDS) now represent a wealth of technical literature providing the latest information on emerging drying techniques and research and development. Researchers interested in drying will find these proceedings invaluable for their work. For archival information on the subject, Drying Technology—An International Journal (Marcel Dekker) remains the premier periodical for both academic and industrial practitioners.

REFERENCES

1. Wang Y, Yu R, Chou C. Int J Food Microbiol 2004; 93: 209–217.
2. Schwab C, Vogel R, Ganzle MG. Cryobiology 2007; 55: 108–114.
3. Schoug A, Olsson J, Carlfors J, Schnurer J, Hakansson S. Cryobiology 2006; 53: 119–127.
4. Ziadi M, Touhami Y, Achour M, Thonart P, Hamdi M. Biochem Eng J 2005; 24: 141–145.
5. Malumba P, Vanderghem C, Deroanne C, Béra F. Food Chem 2008, Food Chemistry 2008; 111: 564–572.
6. Sarwar G. J Nutr 1997; 127: 758–764.
7. Savoie L, Charbonneau R, Parent G. Plant Foods Hum Nutr 1989; 39: 93–107.

8. Anandharamakrishnan C, Rielly CD, Stapley AGF. Drying Technol 2007; 25: 799–807.
9. Wu T, Mao L. Food Chem 2008; 110: 647–653.
10. Eckhoff SR. In: Wrigley C, Walker CE, editors. Volume 2, Encyclopedia of Grain Science. Amsterdam: Elsevier; 2004. pp. 225–241.
11. Finot PA. In: Damodaran S, Paraf A, editors. Food protein and their application. New York: Marcel Dekker; 1997. pp. 551–577.
12. Pelegrine DHG, Gasparetto CA. LWT 2005; 38: 77–80.
13. Terebiznik MR, Buera MP, Pilosof AMR. LWT 1997; 30: 513–518.
14. Pilosof AMR, Sanchez VE. In: Mujumdar AS, editor. Handbook of Industrial Drying. 3rd ed. New York: CRC Press; 2008. pp. 981–991.
15. Moroff G, Holme S. Transfus Med Rev 1991; 5: 48–59.
16. Miyamoto-Shinohara Y, Imaizumi T, Sukenobe J, Murakami Y, Kawamura S, Komatsu Y. Cryobiology 2000; 41: 251–255.
17. Bozoglu TF, Ozilgen M, Bakir U. Enzyme Microb Technol 1987; 9: 531–537.
18. Heckly RJ. Dev Ind Microbiol 1985; 26: 379–395.
19. Costa E, Usall J, Teixido N, Garcia N, Vinas I. J Appl Microbiol 2000; 89: 793–800.
20. Souzo H. Volume 2, Encyclopedia Microbiol. New York: Academic Press; 1992.
21. Palmfeldt J, Radstrom P, Hahn-Hagerdal B. Cryobiology 2003; 47: 21–29.
22. Beker MJ, Rapoport AI. Adv Biochem Eng Biotechnol 1987; 35: 128–171.
23. Bayrock D, Ingledew WM. Food Res Int 1997; 30: 407–415.
24. Bayrock D, Ingledew WM. Food Res Int 1997; 30: 417–425.
25. Koga S, Echigo A, Nunomura K. Biophys J 1966; 6: 665–674.
26. Krokida M, Maroulis Z. In: Mujumdar AS, Suvachittanont S, editors. Volume II, Developments in Drying. Bangkok: Kasetsart University Press; 2000. pp. 149–195.
27. Mujumdar AS, Devahastin S. In: Mujumdar AS, editor. Guide to Industrial Drying. Mumbai: Colour Publications; 2004. pp. 1–22.
28. Law CL, Mujumdar AS. In: Mujumdar AS, editor. Guide to Industrial Drying. Mumbai: Colour Publication; 2004. pp. 75–141.
29. Molnar K. In: Mujumdar AS, editor. Handbook of Industrial Drying. 3rd ed. New York: CRC Press; 2008. pp. 33–52.
30. Mujumdar AS, Menon AS. In: Mujumdar AS, editor. Handbook of Industrial Drying. 2nd ed. New York: Marcel Deker; 1995. pp. 1–40.
31. Mujumdar AS. In: Devahastin S, editor. Mujumdar's Practical Guide to Industrial Drying. Montreal: Exergex Corp.; 2000. pp. 37–71.
32. Mujumdar AS. In: Mujumdar AS, editor. Handbook of Industrial Drying. 3rd ed. New York: CRC Press; 2008. pp. 3–32.
33. Pilosof AMR, Terebiznik VR. In: Mujumdar AS, Suvachittanont S, editors. Volume II, Developments in Drying. Bangkok: Kasetsart University Press; 2000. pp. 71–94.
34. Huang L, Mujumdar AS. In: Mujumdar AS, editor. Guide to Industrial Drying. Mumbai: Colour Publications; 2004. pp. 143–174.
35. Luna G, Salgado MA, Garcra MA, Rodrrguez GC. Volume C, Proceedings of International Drying Symposium, Thessaloniki: Edition Ziti; 1998. pp. 1815–1821.
36. Luna G, Salgado MA, Garcra MA, Rodrrguez GC. J Food Proc Eng 2000; 23: 453–462.
37. Hromádková Z, Ebringerová A, Sasinková V, Sndula J, Hrbalová V. J Omelková Carbohydr Polym 2003; 51: 9–15.
38. Millqvist-Fureby A, Malmsten M, Bergenståhl B. J Colloid Interface Sci 2000; 225: 54–61.
39. Anandharamakrishnan C, Rielly CD, Stapley AGF. LWT 2008; 41: 270–277.
40. Zhang WF, Chen XG, Li PW, Liu CS, He QZ. Drying Technol 2008; 26: 108–115.
41. Horaczek A, Viernstein H. Biol Control 2004; 31: 65–71.
42. Master K. Spray Drying Handbook. New York: Longman; 1991.
43. Filkova I, Huang LX, Mujumdar AS. In: Mujumdar AS, editor. Handbook of Industrial Drying. 3rd ed. New York: CRC Press; 2008. pp. 215–256.
44. Wang ZL, Finlay WH, Peppler MS, Sweeney LG. Powder Technol 2006; 170: 45–52.
45. Jinapong N, Suphantharika M, Jamnong P. J Food Eng 2008; 84: 194–205.
46. Mounir S, Allaf K. Drying Technol 2008; 26: 452–463.
47. Law CL, Mujumdar AS. In: Mujumdar AS, editor. Handbook of Industrial Drying. 3rd ed. New York: CRC Press; 2008. pp. 173–201.
48. Chen GH, Wang W. Drying Technol 2007; 25: 29–35.
49. Shitanda D, Wanjala NV. Drying Technol 2006; 24: 95–98.
50. Kearney L, Upton M, Loughlin A. Appl Environ Microbiol 1990; 56: 3112–3116.
51. Carvalho AS, Silva J, Ho P, Teixeira P, Malcata FX, Gibbs P. Int Dairy J 2004; 14: 835–847.
52. Leslie SB, Israeli E, Lighthart B, Crowe JH, Crowe LM. Appl Environ Microbiol 1995; 61: 3592–3597.
53. Carvalho AS, Silva J, Ho P, Teixeira P, Malcata FX, Gibbs P. J Appl Microbiol 2003; 94: 947–952.
54. Zayed G, Roos YH. Proc Biochem 2004; 39: 1081–1086.
55. Wolkers WF, Walker NJ, Tablin F, Crowe JH. Cryobiology 2001; 42: 79–87.
56. Wolkers WF, Tablin F, Crowe JH. Comp Biochem Physiol A 2002; 131: 535–543.
57. Choi MJ, Briancon S, Bazile D, Royere A, Min SG, Fessi H. Drying Technol 2007; 25: 809–819.
58. Gu MB, Choi SH, Kim SW. J Biotechnol 2001; 88: 95–105.
59. Liapis AI, Bruttini R. In: Mujumdar AS, editor. Handbook of Industrial Drying. 2nd ed. New York: Marcel Deker; 1995. pp. 309–344.
60. Liapis AI, Bruttini R. In: Mujumdar AS, editor. Handbook of Industrial Drying. 3rd ed. New York: CRC Press; 2008. pp. 257–283.
61. Daud WRW. In: Mujumdar AS, editor. Handbook of Industrial Drying. 3rd ed. New York: CRC Press; 2008. pp. 203–213.

62. Strumillo C, Grabowski S, Kaminski W, Zbicinski I. Chem Eng Proc 1989; 26: 139–145.
63. Strasser S, Neureiter M, Geppl M, Braun R, Danner H. J Biotechnol 2007; 131:S194.
64. Mujumdar AS, Law CL. Chem Ind Dig 2006: 54–61.
65. Jumah R, Al-Kteimat E, Al-Hamad A, Telfah E. Drying Technol 2007; 25: 1421–1426.
66. Law CL, Waje SS, Thorat BN, Mujumdar AS. Stewart Postharvest Rev 2008; 4: 1–23.
67. Jumah R, Mujumdar AS, Raghavan GSV. Can J Chem Eng 1996; 74: 479–486.
68. Chua KJ, Mujumdar AS, Chou SK, Hawlader MNA, Ho JC. Drying Technol 2000; 18: 907–936.
69. Bon J, Kudra T. Drying Technol 2007; 25: 523–532.
70. Xiao Z, Xie X, Yuan Y, Liu X. Drying Technol 2008; 26: 427–432.
71. Mujumdar AS, Wu Z. In: Mujumdar AS, editor. Guide to Industrial Drying. Mumbai: Colour Publication; 2004. pp. 253269.
72. Kudra T, Mujumdar AS. Advanced Drying Technologies. New York: Marcel Dekker; 2002.
73. Sathapornprasath K, Devahastin S, Soponronnarit S. Drying Technol 2007; 25: 1121–1128.
74. Kudra T, Mujumdar AS. Drying Technol 1989; 7: 219–266.
75. Kudra T, Mujumdar AS. In: Mujumdar AS, editor. Handbook of Industrial Drying. 3rd ed. New York: CRC Press; 2008. pp. 453–517.
76. Sadykov RA, Pobcdimsky DG, Bakhtiyarov ER. Drying Technol 1997; 15: 2401–2420.
77. Claussen IC, Strømmen I, Hemmingsen AKT, Rustad T. Drying Technol 2007; 25: 853–865.
78. Di Matteo P, Donsi G, Ferrari G. J Food Eng 2002; 59: 267–275.
79. Boeh-Ocansey O. Drying Technol 1983–84; 2: 389–405.
80. Claussen IC, Ustad TS, Strømmen I, Walde PM. Drying Technol 2007; 25: 957–967.
81. Wolff E, Gibert H. Drying Technol 1990; 8: 385–404.
82. Rahman SMA, Mujumdar AS. Drying Technol 2008; 26: 393–403.
83. Chavez BE, Ledeboer AM. Drying Technol 2007; 25: 1193–1201.
84. Mujumdar AS. In: Mujumdar AS, editor. Handbook of Industrial Drying. 3rd ed. New York: CRC Press; 2008. pp. 439–452.
85. Chen SR, Chen JY, Mujumdar AS. Drying Technol 1992; 10: 251–260.
86. Nicoleti JF, Silveira V Jr, Telis-Romero J, Telis VRN. Drying Technol 2007; 25: 891–899.
87. Krokida MK, Philippopoulos C. Drying Technol 2005; 23: 799–830.
88. Tsaousi K, Dimitrellou D, Koutinas AA. Food Chem 2008; 110: 547–553.
89. Chou SK, Chua KJ. In: Mujumdar AS, editor. Handbook of Industrial Drying. 3rd ed. New York: CRC Press; 2008. pp. 1103–1131.
90. Alves-Filho O, Strommen I. In: Strumillo C, Pakowski Z, editors. Proceedings of International Drying Symposium, Drying '96. Poland: Lodz Technical University; 1996. pp. 405–415.
91. Alves-Filho O, Eikevik TM, Goncharova-Alves SV. Drying Technol 2008; 26: 470–475.
92. Djaeni M, Bartels P, Sanders J, van Straten G, van Boxtel AJB. Drying Technol 2007; 25: 1063–1077.
93. Jovanović N, Bouchard A, Hofland GW, Witkamp GJ, Crommelin DJA, Jiskoot W. Eur J Pharm Biopharm 2008; 68: 183–190.
94. Jovanović N, Bouchard A, Sutter M, Speybroeck MV, Hofland GW, Witkamp GJ, Crommelin DJA, Jiskoot W. Int J Pharm 2008; 346: 102–108.
95. Li B, Zhang Y, Zhang W, Hua Z. Drying Technol 2008; 26: 464–469.
96. Marreto RN, Freire JT, Freitas LAP. Drying Technol 2006; 24: 327–338.
97. Markowski AS. Drying Technol 1993; 11: 369–387.
98. Oliveira AC, Moretti TS, Boschini C, Baliero JCC, Freitas LAP, Freitas O, Favaro-Trindade CS. Drying Technol 2007; 25: 1687–1693.
99. Passos ML, Massarini G, Freire JT, Mujumdar AS. Drying Technol 1997; 15: 605–624.
100. Oliveira HVA, Peixoto MPG, Freitas LAP. Drying Technol 2005; 23: 2039–2053.
101. Grabowski S, Mujumdar AS, Ramaswamy HS, Strumillo C. Drying Technol 1997; 15: 625–634.
102. Gawrzynski Z, Glaser R. Drying Technol 1996; 14: 1121–1172.
103. Gawrzynski Z, Glaser R, Kudra T. Drying Technol 1999; 17: 1523–1532.
104. Reyes A, Herrera N, Vega R. Drying Technol 2008; 26: 122–131.
105. Reyes A, Campos C, Vega R. Drying Technol 2006; 24: 1469–1480.
106. Kulacki FA. In: Mujumdar AS, Mashelkar RA, editors. Volume 2, Advanced Transport Protocols. New York: Wiley; 1982. pp. 105–147.
107. Hashinaga F, Bajgai TR, Isobe S, Barthakur NN. Drying Technol 1999; 17: 479–495.
108. Bajgai TR, Hashinaga F. In: Proceedings of 12th International Drying Symp. Amsterdam; 2000. p 155.
109. Shynkaryk MV, Lebovka NI, Vorobiev E. Drying Technol 2008; 26: 695–704.
110. Amami E, Khezami L, Vorobiev E, Kechaou N. Drying Technol 2008; 26: 231–238.
111. Jaya S, Durance TD. Drying Technol 2007; 25: 2005–2009.
112. Duan X, Zhang M, Mujumdar AS. Drying Technol 2007; 25: 2011–2019.
113. Song X, Zhang M, Mujumdar AS. Drying Technol 2007; 25: 2021–2026.
114. Mujumdar AS. Drying of Products of Biological Origin, New Hampshire: Science Publishers; 2004.
115. Mujumdar AS, editor. Handbook of Industrial Drying. 3rd ed. New York: CRC Press; 2008.

29

FREEZE-DRYING, PHARMACEUTICALS

JINSONG LIU
Product Development, Abraxis BioScience, Melrose Park, Illinois

29.1 INTRODUCTION

Freeze-drying or lyophilization is a drying technology that has been widely utilized in the pharmaceutical industry. The freeze-drying process takes place at low temperatures, which in general minimizes the in-process deterioration, resulting in dried products with improved stability and ease of handling, such as shipping and storage, as compared to products in the liquid state. Therefore, it is especially applicable to labile drugs. Since the freeze-dried product is often a cake with high specific surface area, the freeze-drying process is also used to enhance the dissolution property of the product. Applications of freeze-drying in the pharmaceutical industry include active pharmaceutical ingredient (API), and mostly, finished products in final dosage forms. Although in some cases, freeze-drying is used to generate oral solid products with fast-dissolution property (1), the most common application of freeze-drying involves the manufacture of injectable products.

The freeze-drying process for pharmaceutical injectable products starts with filling the liquid or semiliquid product aseptically into vials, which are then partially stoppered with elastomeric closures (stoppers). The vials are then transferred to a freeze-dryer and loaded onto the shelves in the product chamber of the freeze-dryer. By circulating heat transfer fluid through the internal channels in the shelves, the product is cooled to a very low temperature, converting the majority of water in the solution into ice and the solutes into solids. This stage is known as *freezing*. In the freezing stage, sometimes, a thermal treatment or annealing step is added. In order to sublimate the formed ice crystals, the chamber pressure is reduced to at least below the vapor pressure of ice at the product temperature, while the shelf temperature is increased to supply the energy for the sublimation process. Because the majority of water in the product is removed by sublimation, this stage is called *primary drying*. The following stage, that is, *secondary drying*, is intended to further reduce the unfrozen water in solid matrix to a target level by desorption at a relatively high temperature. The three stages, *freezing, primary drying, and secondary drying*, constitute the freeze-drying process, which is also known as the *freeze-drying cycle*. A representative freeze-drying cycle is shown in Fig. 29.1.

At the end of secondary drying, the chamber pressure is either kept under vacuum, or often increased to partial or full atmosphere pressure, by backfilling with an inert gas, normally sterile dry nitrogen. Following backfilling, the vials are fully stoppered within the product chamber by hydraulic compression of the shelf stack. The product vials are then unloaded and transferred for inspection and labeling.

A freeze-dried product is expected to have a nearly full recovery of the original chemical or biological potency after reconstitution, rapid and complete dissolution, appropriate residual moisture level, and pharmaceutically acceptable cake appearance. These quality attributes, amongst others, should be maintained through the shelf life.

The success of developing and manufacturing a freeze-dried injectable product is determined by a number of factors, including the freeze-drying process, the formulation, the container-closure system, and the freeze-dryer. This chapter provides an overview of these fundamental aspects for pharmaceutical freeze-drying, including the rationale for freeze-drying process design and control, the

Downstream Industrial Biotechnology: Recovery and Purification, First Edition. Edited by Michael C. Flickinger.

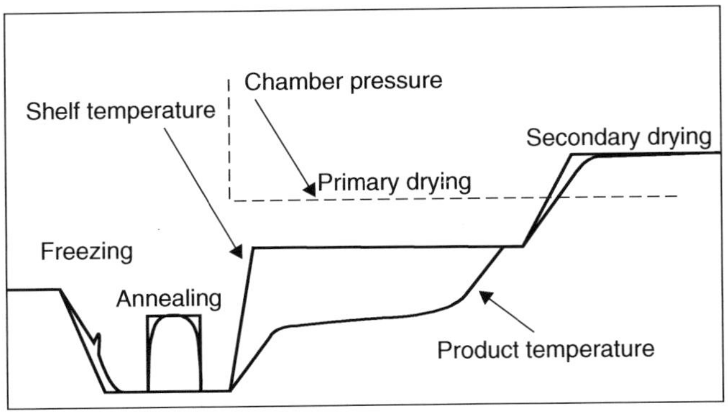

Figure 29.1. Schematic of a typical lyophilization cycle with annealing step.

basics of formulation development for freeze-drying, the importance of the container-closure system, and a general description of pharmaceutical freeze-dryers.

In recent decades, with the advent of biotechnology, development and manufacture of biopharmaceutical products such as peptides, proteins, and other biologics has been dramatically increasing. Since most macromolecules suffer from marginal stability, freeze-drying is the first choice for the development of these biopharmaceutical products. Both the fragile nature and the high cost of the biopharmaceutical products warrant optimal design and control of the freeze-drying process to ensure the quality of the resulting product. Therefore, in recent years, some particular challenges in freeze-drying are being addressed, and some advancements in freeze-drying-related science and technology have been made. Description and discussion of these challenges and new advancements comprise the second part of this chapter.

29.2 PHARMACEUTICAL FREEZE-DRYING: FUNDAMENTALS

29.2.1 Freeze-Drying Process Design

29.2.1.1 Freezing.

29.2.1.1.1 Phase Transitions during Freezing. Freezing is generally the first step in a freeze-drying process, in which nearly 90% of the water is converted to ice crystals while all solutes in the formulation are solidified into a matrix either in amorphous or crystalline state, or in a mixture (Fig. 29.2). The conversion from water to ice crystals starts with ice nucleation, which is followed by ice crystal growth. The ice nucleation is dependent upon a number of process and formulation variables as well as surface characteristics

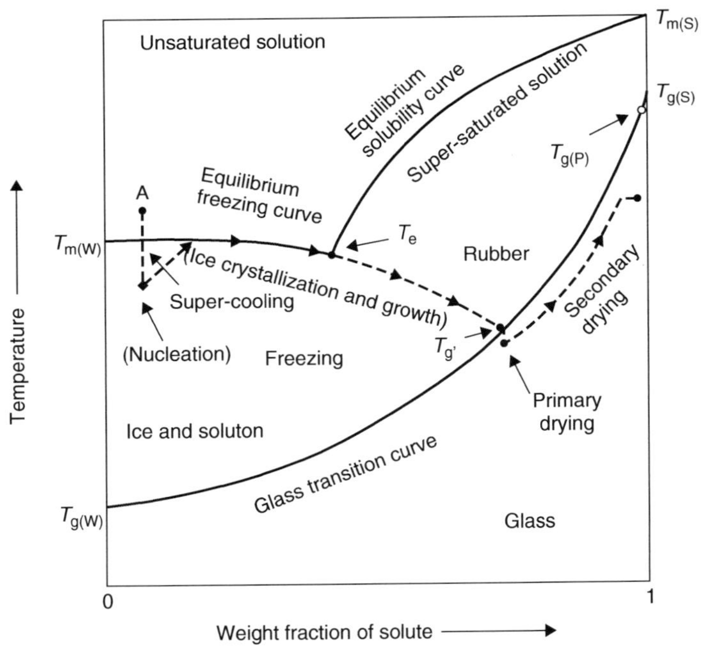

Figure 29.2. Supplemented phase diagram for a binary formulation system where the solute (S), such as sugar, does not crystallize during freeze-drying. Arrows indicate the freeze-drying process. *A* represents formulation aqueous solution in the fill for freeze-drying; $T_{m(W)}$ and $T_{m(S)}$ denote melting temperatures of water and solute, respectively; $T_{g(W)}$ and $T_{g(S)}$ denote glass transition temperatures of water and solute, respectively; $T_{g'}$ and $T_{g(P)}$ denote glass transition temperatures of freeze concentrate of the formulation and the freeze-dried product, respectively; T_e denotes the eutectic temperature [Source: Reprinted from Ref. 2, with permission from Taylor & Francis].

of the container. For a typical pharmaceutical formulation, the ice nucleation temperature is often in the range of 10–15°C or more below the equilibrium freezing point; a phenomenon is referred to as *super-cooling*. Due to the stochastic nature of ice nucleation and ice crystal growth, ice formation during freezing in a vial and in a batch is essentially a heterogeneous process.

The solidification of solutes in the freezing stage is completed through a crystallization or/and vitrification process. In the latter, the solutes that do not crystallize, for example, proteins are converted to amorphous solids when the temperature drops below the glass transition temperature ($T_{g'}$) of the maximally concentrated solute (or freeze concentrate). Figure 29.2 illustrates a supplemented phase diagram of a model binary system of sugar and water for freeze-drying (2). The morphologies of both the ice crystals and the solutes formed during freezing greatly impact the ice-sublimation during primary drying and the water desorption during secondary drying, and consequently, the quality of the finished product. Therefore, freezing is a very critical stage in a freeze-drying process.

29.2.1.1.2 Process Design and Control. During freezing, the chamber pressure is slightly lower than the atmospheric pressure due to low temperature, or/and prereduction of chamber pressure to enhance the sealing of the chamber door. The stage of the process is generally controlled by the shelf-cooling/heating rate, shelf-holding temperatures, and holding times. Note that the cooling rate of the formulation solution is not necessarily correlated to the shelf-cooling rate.

29.2.1.1.2.1 PREFREEZING HOLD. In order to facilitate relatively uniform ice nucleation and ice crystal growth, the product vials on the shelf are held at a temperature lower than room temperature before cooling down. This temperature is generally the loading temperature, for example, 5°C. For formulations with high super-cooling temperature, holding at an even lower temperature (a few degrees higher than the ice nucleation temperate) is more appropriate.

29.2.1.1.2.2 COOLING DOWN TO THE FINAL FREEZING TEMPERATURE. Cooling the product to a terminal (final) freezing temperature facilitates the ice nucleation/growth and solute solidification. If a super-cooling hold is applied, a relatively faster cooling is generally helpful for intra-vial uniformity of ice formation. Note that even though a much higher cooling rate can be achieved in a laboratory-scale freeze-dryer, only a moderate cooling rate, not higher than 2°C/min, is feasible in production freeze-dryers, unless equipped with a liquid nitrogen-driven refrigeration system. The final freezing temperature should be low enough and hold-time should be long enough to ensure the completion of solute solidification.

29.2.1.1.2.3 ANNEALING. In general, annealing, or thermal treatment refers to a step in the freezing stage intended to facilitate crystallization of a bulking agent, such as glycine or mannitol. Crystalline solids of a bulking agent provide not only elegant cake appearance, but also accelerate the subsequent primary drying process. Since the melting temperature of the crystalline solid (T_e) is in general much higher than the glass transition temperature of the freeze concentrate ($T_{g'}$) in the amorphous phase of the formulation, primary drying can be executed at a product temperature between T_e and $T_{g'}$. In this case, even though microscopic form of collapse is developed in the dried material, the crystalline matrix retains a good cake structure. Therefore, the ice-sublimation rate can be substantially accelerated due to both the higher product temperature and the low resistance to vapor flow resulting from micro-collapse (3).

The annealing temperature and duration should be designed based upon characteristics of the formulation. The annealing temperature is a temperature higher than the onset temperature of crystallization, characterized by, for instance, differential scanning calorimetry (DSC). The duration at such an annealing temperature should ensure the completion of crystallization of the bulking agent. Incomplete crystallization of the bulking agent results in continued crystallization during the subsequent primary drying stage, secondary drying stage, and/or even during the shelf life. Continued crystallization of, typically mannitol, at accelerated rate due to the increased temperature during ramping to, or at the primary-drying stage can result in rapid volume expansion, leading to “vial breakage,” which is very troublesome in freeze-drying production (4).

In the freezing stage, a so-called “ice-annealing” step is sometimes incorporated as well. That is, after freezing for solidification, the temperature of the frozen formulation is raised up to a higher temperature between $T_{g'}$ and the ice melting point, and held for a certain period of time before refreezing below the $T_{g'}$. During the annealing, because of “ripening effect” the unfrozen water diffuses through the frozen matrix and the ice crystals grow in size. Also, more uniform distribution of ice size and better ice crystal interconnections are obtained. Therefore, ice-annealing helps to accelerate ice sublimation and to overcome the vial-to-vial heterogeneity of the drying rate and product quality attributes in a batch, which results from the stochastic ice nucleation and ice crystal growth (5,6).

After annealing, the product temperature is generally lowered to a final temperature and held long enough to complete solidification.

29.2.1.2 Primary Drying. The ice crystals formed in the freezing stage are removed by sublimation in the primary drying stage. That is, the product is “dried” at relatively low temperature and low pressure, and the solute structure formed during freezing will be retained to ensure acceptable physical attributes, for example, elegant cake appearance.

29.2.1.2.1 Process Design Principle: Mass and Heat Transfer for Ice Sublimation. The driving force for the ice sublimation process is the difference between the equilibrium vapor pressure of ice (P_0) at the temperature of the frozen product (T_P) and the chamber pressure (P_C) (7). The ice sublimation rate, *dm/dt*, which is the mass of ice sublimated per unit time per vial, can be described as follows:

$$\frac{dm}{dt} = \dot{m} = \frac{P_0 - P_C}{R_P + R_S} \qquad (29.1)$$

where R_P and R_S are the resistance of the dried layer of product and the stopper to the vapor transport during the sublimation, respectively. The vapor pressure of ice (P_0, in mTorr) depends upon the temperature of the frozen product (T_P, in °C), which can be described as follows (8):

$$P_0 = 2.7 \cdot 10^{13} \cdot \exp\left[-6145/(T_P + 273.15)\right] \qquad (29.2)$$

In order to compensate the heat loss due to ice sublimation and maintain the product temperature, heat must be continuously supplied from the shelf to the product vial. The heat transfer rate from the shelf to the product vial may be described by Equation (29.3)

$$\frac{dQ}{dt} = A_V \cdot K_V \cdot (T_s - T_P) \qquad (29.3)$$

where A_V is the cross-sectional area of the vial, and K_V is the heat transfer coefficient from the shelf surface to the vial, therefore, strongly depends on the chamber pressure P_C. T_S is the shelf temperature, or to be more precise, the surface temperature of the shelf on which the product vial rests, and T_P is the product temperature at the vial bottom.

For a steady-state sublimation process, the heat required for sublimation is ($\frac{dm}{dt} \cdot \Delta H_S$), where ΔH_S is the heat of ice sublimation, and the mass transfer and heat transfer is coupled as follows:

$$\Delta H_S \cdot \left(\frac{P_0 - P_C}{R_P + R_S}\right) = A_V \cdot K_V \cdot (T_s - T_P) \qquad (29.4)$$

From the above equations, it can be seen that chamber pressure, shelf temperature, and product temperature are interrelated, and integrally determine the ice sublimation rate. Figure 29.2, redrawn in a simplified sketch based upon the experimental results from Chang and Fischer (9), provides an excellent illustration of the functional relationships amongst the sublimation rate, the product temperature, and the two independently controlled process variables, shelf temperature, and chamber pressure. Evidently, while keeping the chamber pressure constant, an increase in shelf

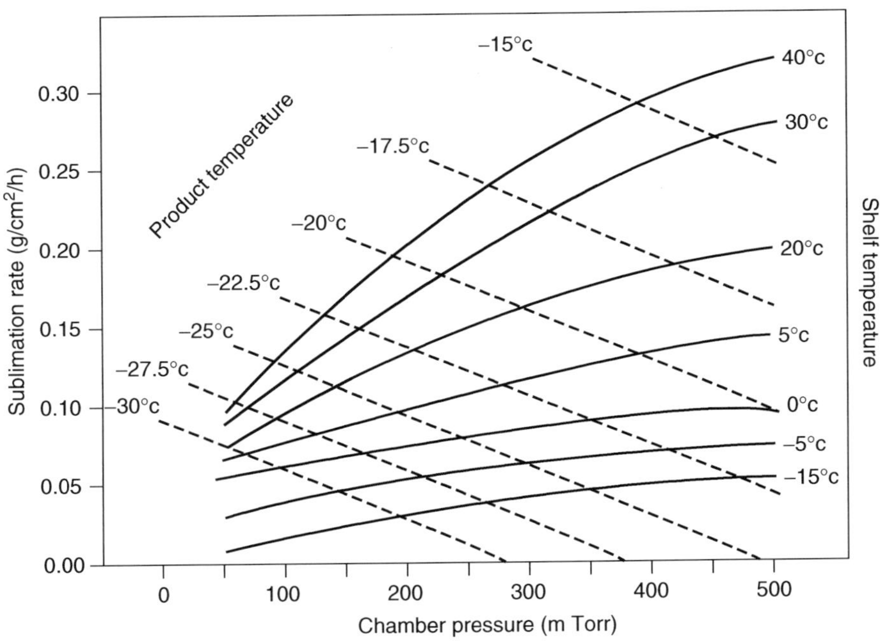

Figure 29.3. Effect of chamber pressure, shelf temperature on the sublimation rates in freeze-drying a protein formulation. Dotted lines show the isothermal curves of product temperature [Source: Redrawn from Ref. 9, with permission from Springer Science and Business Media].

temperature leads to an increased ice sublimation rate. Similarly, at a constant shelf temperature, an increase in chamber pressure also results in an accelerated ice-sublimation, since the heat transfer coefficient from the shelf to the vial greatly increases as the chamber pressure increases.

29.2.1.2.2 Process Design and Control.

29.2.1.2.2.1 TARGET PRODUCT TEMPERATURE AND STRUCTURAL COLLAPSE. During primary drying, if the product temperature exceeds the eutectic melting temperature (T_e) or the glass transition temperature ($T_{g'}$), the eutectic crystalline solids melt down or the viscosity of the amorphous solute phase decreases. While eutectic melting normally causes immediate loss of pore structure in the dried matrix, a decrease in the viscosity of the amorphous solid phase does not necessarily cause structural loss, or collapse at the time-scale of primary drying. Collapse temperature for an amorphous system refers to the temperature, above which the dried region adjacent to ice loses its structure (10), that is, collapse temperature (T_c) can be higher than the glass transition temperature ($T_{g'}$). The difference is about a few degrees, as shown in Table 29.1 (2,11–13). However, for a formulation with high concentration of proteins, T_c can be much higher than T_g, up to 15°C (14).

Primary drying is normally the most time-consuming stage. Since an increase in T_P by 1°C leads to a reduction in primary drying time by more than 10% (15), an optimized process must operate near the maximum allowable product temperature, that is, T_C for an amorphous system or T_e for a crystalline system, while the product remains a porous cake that has good rehydration property and acceptable appearance. In general, a safety margin should be kept during primary drying, that is, the product temperature (T_P) should be 2–5°C below T_C or T_e. This safety margin accounts for (i) the product temperature increases, in general, between 1 and 3°C, due to the increase of resistance from the dried-layer as drying progresses; (ii) the heterogeneity of product temperature across the shelf and from shelf to shelf; and (iii) the differences of product temperature due to scaling-up and/or transfer (see details in section "Freeze-Drying Scale-Up and Transfer").

For protein formulations, one might cautiously conduct primary drying by maintaining T_P below $T_{g'}$, instead of T_c which is generally higher, based on the assumption that protein drugs are unstable above $T_{g'}$ (15). However, recent work demonstrated that at the practical time-scale for freeze-drying, the unfolding of a protein at a temperature between $T_{g'}$ and T_c should not be a concern (16).

However, even though the physical properties allow the primary drying to be conducted under very aggressive conditions, the heat and mass transfer capabilities of the freeze dryers, particularly at production scale, may constrain these aggressive conditions. The massive vapor flow resulting from the ice sublimation under aggressive conditions may overload the condenser or be "choked" in the vapor tube, which will be discussed further in the section titled "Freeze-Drying Scale-Up and Transfer". It was suggested by Tang and Pikal (16) that the product temperature, in general, should not be higher than −15°C.

TABLE 29.1. Glass Transition Temperatures of Freeze Concentrate ($T_{g'}$) and Collapse Temperatures (T_C) of Selected Proteins and Excipients Commonly Used in Freeze-Dried Pharmaceutical Products[a]

Components		$T_{g'}$ (°C)	T_C (°C)
Proteins	Galactosidase	−29	−15
	Ovalbumin	−11	−10
	rhuMAB HER	—	−20
Sugars	Sucrose	−32	−31
	Trehalose	−29	−28.5
	Lactose	−28	−30.5
	Maltose	−30	—
Polyols	Glycerol	−65	—
	Sorbitol	−46	—
	Mannitol	−35	−1.4[b]
Polymers	Dextran (70 K)	−10	−10
	Polyvinylpyrrolidone (PVP) (40 KD)	−21	−24
	Ficoll	−19	−20
	Gelatin	−9	−8
	Hydroxyethyl starch	−12	>−5
Amino acids	Glycine	−62	−3.5
	Alanine	−65	—
	Histidine	−33	—
Salts	Sodium acetate	−64	—
	Sodium citrate	−41	—
	KH_2PO_4	−55	—
	K_2HPO_4	−65	—
	NaCl	−60	—
	$CaCl_2$	−95	—
	$ZnCl_2$	−88	—

[a]From (2,11–13).
[b]In crystalline state.

29.2.1.2.2.2 CHAMBER PRESSURE. Based upon the mass and heat transfer equations, also as illustrated in Fig. 29.3, a combination of a high shelf temperature and a low chamber pressure leads to a higher sublimation rate than the combination of a low shelf temperature and a high chamber pressure, provided that both combinations result in the same product temperature, therefore, a most efficient primary drying condition for the product in a given vial should be a combination of a "very high" shelf temperature with a "very low" chamber pressure. However, in practice, reducing the chamber pressure to a very low level increases the heat transfer heterogeneity amongst the vials in a batch, raises the risk of volatile component desorption from stoppers and chamber walls, and causes problems like back-streaming of pump oil. Therefore, the chamber pressure is generally not lower than 50 mTorr. As a general rule, a chamber pressure, approximately between 10% and 50% of ice vapor pressure at the target product temperature (7,17), is chosen, which generally falls into a range between 50 and 200 mTorr. In order to minimize the vial-to-vial heterogeneity in heat transfer, wherever possible, a chamber pressure between 100 and 200 mTorr is advised (7,18).

29.2.1.2.2.3 Shelf Temperature. Once the chamber pressure has been chosen, the shelf temperature can be determined experimentally, which in combination with the chosen chamber pressure, leads to the target product temperature. Some methods for mass and heat transfer modeling, such as manometric temperature measurement (MTM), can be employed to facilitate the process of shelf temperature determination. According to the above mass-heat transfer equations, the shelf temperature is higher than the product temperature and sometimes can be much higher, up to 40°C, depending upon the ice sublimation rate under the designed primary drying conditions.

Even for a given formulation, if there are changes in product freezing behavior, container-closure system, or freeze-dryer, the same combination of shelf temperature and chamber pressure may not lead to the same target product temperature. Consequently, the freeze-drying profile of the product can be different. In this case, the cycle parameters, either shelf temperature, or the chamber pressure, or both, will have to be adjusted. This is often seen in cycle scale-up or transfer, which is further discussed in section titled "Freeze-Drying Scale-Up and Transfer".

29.2.1.2.3 Process Monitoring and Control: Determination of Endpoint of Primary Drying. Once the shelf temperature and chamber pressure for primary drying are determined, the primary drying process, essentially including two steps, can be designed:

29.2.1.2.3.1 Ramp from Freezing to Primary Drying. After evacuation to reduce the chamber pressure to the target level, the shelf temperature is ramped up to the target value. The ramp rate should not be too high, normally less than 1°C/min. For a highly loaded batch, the ramp rate from the temperature at which the vapor pressure is higher than the chamber pressure, to the target shelf temperature should be set at a low value, less than 0.5°C/min. During this initial period of sublimation, ice sublimation rate can be quite high, since the resistance in the product is nearly zero. If the ramp (heating) rate is too high, the massive vapor flow can overwhelm the condenser during this period.

29.2.1.2.3.2 Duration of Primary Drying. Once the shelf temperature reaches the set point, the primary drying continues under the controlled temperature and chamber pressure conditions until the end of the ice sublimation. The duration of primary drying is determined by the ice sublimation rate, the characteristics of formulation solution and the fill volume, and thus, can be roughly estimated theoretically by calculations based upon the mass and heat transfer equations. However, in practice, the duration of primary drying or the end-point of ice sublimation is normally determined by monitoring the drying progression.

In a regular pharmaceutical freeze-dryer, a number of process variables are measured to monitor the progression of the freeze-drying process, particularly to detect the end-point of primary drying. Routinely, product temperatures are measured by thermocouples, the chamber pressure is measured either by a capacitance manometer gauge or/and a thermal conductivity (Pirani) gauge, and condenser temperature is measured by a resistive temperature detector (RTD) sensor. Optionally, condenser pressure is measured by another capacitance manometer, the rate of nitrogen flow for chamber pressure control is monitored by flow meter, the dew point or vapor composition is measured by a humidity sensor, or vapor flow rate is assessed by pressure rise test, which measures the chamber pressure rise by closing the isolation valve for a short period of time.

At the end of primary drying, that is, the completion of the ice sublimation in a given vial, the product temperature as measured by thermocouple shows a steep increase as it approaches the shelf temperature. Even though the product temperature measurement by thermocouple is often regarded as the "gold standard" for monitoring the process progression, the freeze-drying profile indicated by thermocouple often does not represent the batch as a whole, unless an ice annealing step is implemented, because the insertion of a thermocouple in the product vial changes the ice nucleation, which consequently alters the ice-sublimation rate in the product vial containing the thermocouple. Differences in the placement of the thermocouple in a product vial and in the location of the vial containing a thermocouple may also prevent the temperature profile from being representative of the drying progression in a whole batch. In this case, a soak time, which is about 10–20% of the total time for ice sublimation measured by thermocouples, is suggested to be added to the duration of primary drying.

Unlike the pressure reading of the capacitance pressure gauge, the pressure reading of the thermal conductivity pressure or Pirani gauge depends upon the composition of the gas-phase in product chamber. During primary drying, the vapor in the product chamber is essentially water vapor, so the reading in the Pirani gauge is generally 1.5 times that of the actual pressure, as read by the capacitance pressure gauge. When approaching the end of ice sublimation, the nitrogen percentage increases as the water vapor from sublimation decreases dramatically, therefore, the reading of Pirani gauge declines and approaches the reading of the capacitance gauge, which remains constant. Meanwhile, the flow rate of dry nitrogen increases, and the dew point or humidity in the product chamber sharply decreases. If a pressure rise test is performed, the pressure rise rate decreases as the primary drying nears the endpoint, and approaches a constant, indicating the completion of ice sublimation. If the batch size is large enough, for example, in a pilot-scale or production-scale, the condenser temperature often increases during the primary drying stage, due to the massive ice load on the condensation coils or plates, then decreases, indicating the near end of ice sublimation.

Figure 29.4 shows an example of the correlations between pressure rise and other process variables, including product temperature measured by thermocouples, comparative pressure (capacitance manometer versus Pirani gauge), pressure drop between the chamber and condenser, N_2 flow rate, and condenser coil temperature. Pressure rise tests demonstrate an accurate method for the determination of the end point of ice-sublimation (primary drying) and for the automatic control of the cycle advancement from primary drying to secondary drying. However, when designing the pressure rise testing parameters, great care should be taken such that the pressure perturbation from valve isolation does not lead to product collapse.

29.2.1.3 Secondary Drying. When all ice crystals are removed from the product by sublimation, the dried product contains a fairly high amount of "unfrozen water" (5–20% in the solid content), which is either absorbed on the surface of crystalline solids, existing as hydrate water in a crystalline hydrate, or dissolved in amorphous solids. In the secondary drying stage, the unfrozen water is further reduced to a desired, much lower level at a higher temperature.

The glass transition temperature (T_g) of the dried formulation is a function of the moisture content, which is governed by the Gorden-Taylor equation. Therefore, the T_g changes sharply with the decrease of moisture during the ramp from primary drying to secondary drying, and during secondary drying (Fig. 29.2).

29.2.1.3.1 Process Design and Control.

29.2.1.3.1.1 CHAMBER PRESSURE FOR SECONDARY DRYING. It has been demonstrated that the water desorption rate is not affected by the chamber pressure change between 50 and 200 mTorr, but very sensitive to product temperature (19). As discussed above, in most cases, the optimized chamber pressure in primary drying is in this range (50–200 mTorr), therefore, there is no need to change chamber pressure for secondary drying.

29.2.1.3.1.2 RAMP FROM PRIMARY DRYING TO SECONDARY DRYING. During the ramp from primary drying to secondary drying, in order to avoid structural shrinkage or deformation of the cake of the final product, the product temperature should always be lower than the T_g during the ramp. In general, a very slow ramp rate of 0.1–0.2°C/min

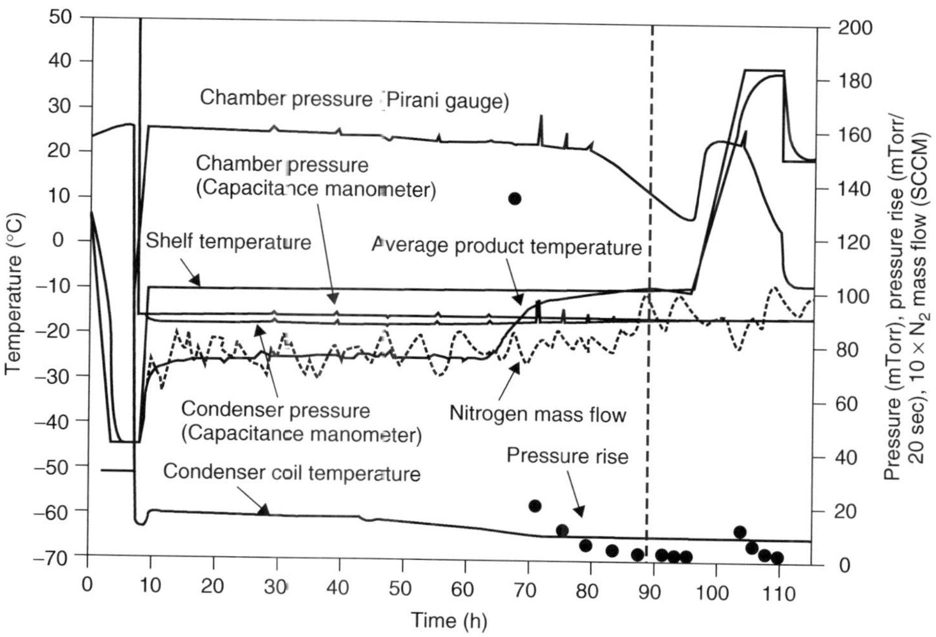

Figure 29.4. Correspondences of process monitoring profiles: freeze-drying of 15% (w/w) sulfobutylether 7-beta-cyclodextrin solution [Source: Reprinted from Ref. 6, with permission from Taylor & Francis].

is recommended for a product in amorphous solids, while a relatively high ramp rate, up to 0.5°C/min should be safe for the product in crystalline solids.

29.2.1.3.1.3 Shelf Temperature and Hold-time at Secondary Drying. In the secondary drying stage, the product temperature is slightly lower than the shelf surface temperature. At a given temperature, water content decreases dramatically during the first few hours, and then, approaches a plateau level. As the temperature increases, the time for water content to reach a plateau level sharply decreases (19). For the sake of efficiency, if the stability of product allows, the shelf temperature for the secondary drying should be high, normally between 30°C and 50°C. Therefore, the product can be dried at an efficient desorption rate, generally 4–10 h, in order to reach the target moisture level.

It was not an uncommon opinion that for protein products, in order to avoid denaturation, the temperature in secondary drying should not be too high (20). However, it was argued that after primary drying, the denaturation of protein, being in the relatively dry solid state, is not an issue below 100°C (16), which justifies the practice of conducting the secondary drying for protein product at a relatively high temperature, for example, 40°C.

The secondary drying time, which is needed for the product to reach the target moisture level under a given condition, can be determined by removing samples from the batch periodically and measuring the moisture content. Some advanced process analytical technologies (PATs), such as MTM (21), have made on-line measurement of moisture content possible.

In order to establish the target moisture content for a given freeze-dried product, empirical study on the dependency of the stability on the moisture content has to be performed. For freeze-dried products with small molecules, in general, the stability is inversely correlated with the moisture level. Therefore, the product is usually freeze dried to a very low level of residual moisture (less than 1%). However, this has been found to be not always true for proteins or other biologics, most likely due to their high structural complexity (22). When the target moisture content is an intermediate level, a lower shelf temperature, in combination with a longer secondary drying time, will be better than a higher shelf temperature in combination with a shorter time, because the former will be more robust for control of moisture uniformity in a batch and for scaling up.

29.2.2 Formulation Development for Freeze-drying

A successful formulation for a freeze-dried product, among others, should maintain the chemical or biological activity and stability of the drug molecules through routine manufacturing, including freeze-drying, distribution, storage and delivery processes. Also, the final product should have desired physical attributes of quality, such as pharmaceutically acceptable cake appearance and reconstitution properties.

A number of functional excipients may be included in the formulation for freeze-drying. Preferably, each of the selected excipients should have good freezing/drying characteristics and be incorporated in appropriate concentrations for an efficient freeze-drying cycle. Detailed discussion on formulation development is certainly beyond the scope of this chapter, and the reader is referred to some excellent review chapters (12,23,24) and book chapters (25,26). Here, the impact of formulation properties on the freeze-drying process is discussed, and the destabilization stresses during the freeze-drying process as well as the formulation approaches to mitigate these stresses are briefly outlined, along with the formulation strategy for achieving high storage stability and manufacturing viability.

29.2.2.1 Impact of Formulation Properties on the Freeze-Drying Process. In freeze-drying of pharmaceuticals, the formulation and the process are interrelated. Therefore, the formulation properties may have considerable impact on the freeze-drying process.

As discussed before, crystallization of one or more components in the formulation, the bulk agent for instance, may enable the primary drying to be conducted under aggressive conditions. For a formulation in a purely amorphous state, collapse temperature or glass transition temperature ($T_{g'}$) can have significant impact on the efficiency of freeze-drying. For example, a pharmaceutical formulation with a $T_{g'}$ lower than −40°C not only requires a very high cooling capacity to solidify the solute in the freezing stage, but also requires a very long time for ice sublimation. Practically speaking, such a formulation is unsuitable for freeze-drying, even though it may have good stability. In this case, it is necessary to increase the $T_{g'}$ of the formulation to a level which is suitable for freeze-drying. The improvement can be realized by, for example, simply lowering the buffer concentration without compromising the buffering function, or adding a so-called *collapse modifier*, which refers to a high molecular weight excipient with a high $T_{g'}$, dextran for instance. The modification is based on the fact that the glass transition of the formulation depends upon the chemical composition of the amorphous phase. It was demonstrated that the glass transition temperature $T_{g'}$ of the freeze concentrate of a formulation can be estimated by the Fox equation, with sufficient accuracy for practical use (27):

$$1/T_{g'} = w_1/T_{g'1} + w_2/T_{g'2} \tag{29.5}$$

where w_i is the weight fraction of component i, and $T_{g'i}$ is the $T_{g'}$ of pure component i, and Equation 29.5 can be generalized to systems of more than two components.

Table 29.1 lists the glass transition temperature ($T_{g'}$) and collapse temperate (T_C) of excipients commonly used for pharmaceutical formulations. Freeze-drying a formulation with a high $T_{g'}$ usually results in a final product with a high glass transition temperature T_g, which is favorable to product storage stability as well (Section titled "Formulation and Process Impact on Storage Stability").

29.2.2.2 Stresses from Freeze-Drying Process and Stabilization.

29.2.2.2.1 Stresses during Freezing and Stabilization. Drug molecules, particularly proteins, in a formulation experience a number of stresses at the freezing and drying stages, due to the dramatic temperature and phase changes. Extensive studies on the stresses during freezing and corresponding stabilization strategies have been conducted, particularly for protein formulations (28).

29.2.2.2.1.1 FREEZE CONCENTRATION, SOLUTE REDISTRIBUTION, CRYSTALLIZATION, AND PHASE SEPARATION. These destabilization stresses may occur to both small molecules and proteins in the formulations during freezing. While the majority of water is converted to ice crystals, the solute concentration and viscosity of the unfrozen fraction increase dramatically. Therefore, one may expect some degradation reactions to be accelerated, although the acceleration may be mitigated by the effect of decreased temperature.

Freeze concentration can induce the redistribution of solutes, particularly when the formulation concentration is high and the cooling rate is low (6). The extent of solute redistribution within the freeze-concentrate depends on its solubility within the solution phase and its ability to diffuse away from the interface. Therefore, addition of different salts has shown various effects on the solution redistribution during freezing. For protein formulation, it is generally found that the protein tends to concentrate at the surface of the filled solution in the vial, dependent on freezing rate, concentration, and formulation composition (29). Addition of surfactant can help to minimize such phenomenon as surface concentration.

Freeze concentration during freezing may lead to super-saturation of amorphous solutes, and consequently, crystallization. Crystallization of amorphous solutes can occur to buffer salts, stabilizers, and bulking agents. Intentional crystallization of bulking agents by annealing in the freezing stage has already been discussed. Crystallization of one of the buffer components during freezing, which often results in pH shifts up or down depending on the buffer selection and concentrations, other excipients in the formulation, and the freezing process (28). The stability of most freeze-dried products, particularly proteins, can be strongly influenced by the pH changes; crystallization of buffer salts may lead to significant in-process degradation, as well as reduced storage stability. To minimize the pH shift and its impact on the stability, (i) an appropriate buffer system should be selected. For example, for the phosphate buffer system, potassium phosphate is more favorable than sodium phosphate, because the pH shift of sodium phosphate during freezing is generally much larger than potassium phosphate; (ii) a high ratio of solutes, such as stabilizers, if possible, should be formulated to inhibit the crystallization of buffer components; (iii) the concentration of the buffer components should not be higher than necessary.

Freezing may also induce crystallization of some stabilizers, such as polymers and sugars. For example, polyethylene glycol crystallized in frozen solutions, consequently losing its stabilization effect (30). In cases as such, the freezing process, such as cooling rate, temperature and annealing conditions, should be carefully designed to avoid the crystallization, or the formulation should be modified, because crystallization of a stabilizer is also strongly dependent on the presence of other solutes.

A more common type of phase separation phenomenon occurring with stabilizers during freezing is the separation of active drug molecules, frequently proteins, from the stabilizers into different amorphous phases, which can directly or indirectly influence the stability by changing the component interaction. Such phase separation is often observed in formulations with polymers or other high molecular weight compounds (31). An appropriate choice of formulation compositions coupled with process controls (cooling rate) may eliminate the occurrence of phase separation.

29.2.2.2.1.2 COLD DENATURATION AND CRYSTAL–LIQUID INTERFACIAL DENATURATION. Cold denaturation and ice–aqueous interfacial denaturation are the two stresses that only occur to proteins in the formulation during freezing. When freezing to a temperature below −40°C, or even lower, spontaneous unfolding of the protein may occur, which is often known as *cold denaturation*. Cold denaturation may lead to in-process degradation, and moreover, also impact the storage stability. Unlike thermal denaturation, cold denaturation in relation to freeze-drying has been much less studied. Recently, it was revealed (32) that the protein cold denaturation temperature is dependent on the pH of the formulation, the protein concentration, and the presence of excipients in the formulation. Stabilizers, such as saccharides and polyols can greatly minimize protein cold denaturation.

During freezing an ice crystal–liquid interface is generated. Proteins can be adsorbed to the interface, resulting in surface-induced denaturation, which is common in freeze-drying of protein formulations (33,34). Not surprisingly, the crystal surface-induced perturbation is not limited to ice crystals. It was reported (35) that crystals of the bulking agent, glycine, in the formulation can cause stress to the protein during the freezing as well. Surfactants can be very effective at inhibiting protein aggregation caused by crystal–liquid interfacial denaturation during freezing, since they preferentially interact with the crystal–liquid interface, minimizing protein structural perturbations.

29.2.2.2.2 Stresses during Drying Process and Stabilization. During primary drying, the ice crystals in the formulation matrix in the vial are being removed by sublimation and the formulation matrix is in partially frozen and partially dried solid state. In the frozen portion, the stresses formed during freezing continue, while in the dried portion, significant desorption of unfrozen water can also occur along with the sublimation. That is, the secondary drying process actually already starts in the primary drying stage, albeit at a low rate. Therefore stresses induced by water desorption from the formulation, or dehydration, should already be experienced in the primary drying stage. Recent work by Luthra *et al*. (36) demonstrated that as long as protein is protected from destabilization induced by freezing stresses, significant loss only occurs during secondary drying or water desorption process. That is, the stress during drying process essentially results from water removal or dehydration.

Since removal of unfrozen water from the formulation solid matrix can further increase the concentration, some stresses resulting from freezing concentration may also be induced or enhanced, because the product temperatures at the primary drying stage, and particularly the secondary drying stage are much higher than the freezing stage. However, such stresses can usually be mitigated by the cryoprotectants used in the freezing stage, and by additional stabilizers, such as saccharides, which form a good glass matrix to minimize the molecular mobility.

For protein formulations, removal of unfrozen water may lead to protein unfolding, since a certain amount of water is essential for retaining the native structure of the protein (34). Protein unfolding may not only lead to aggregation, but also may reduce storage stability in the finished product. To lessen such a stress, stabilizers, such as sucrose and trehalose, are often used. That is, the stabilizer molecules are able to substitute the water molecules, which are being removed from the binding sites of the protein, to minimize the perturbation of protein structure. Therefore, saccharides, particularly sucrose and trehalose, are good choices of lyoprotectants, which provide both kinetic and thermodynamic stabilization to proteins during drying.

29.2.2.3 Formulation and Process Impact on Storage Stability (Shelf Life). A formulation, which results in a high activity recovery after freeze-drying, does not necessarily lead to a freeze-dried final product with a good storage stability or shelf life. The destabilization stresses encountered in the drying stage, in principle, remain in the storage, and at a much larger time-scale, though the product temperature during storage can be much lower than that during drying.

In addition to the impact on in-process stability, the freeze-drying process or the thermal history may have significant impact on the storage stability as well, through altering the glass dynamics of the freeze-dried formulation solid matrix (37,38). Interesting results in this aspect were reported recently by Luthra *et al*.(39), showing that annealing the final product after freeze-drying at relatively high temperatures might help improve the storage stability.

The general guidance for formulation development to achieve acceptable storage stability can be summarized below; however, some measures may have already been taken to mitigate the in-process stresses as discussed above. (i) select an appropriate buffer system and concentration that results in a pH optimal for stability, moreover, with no or minimum shift due to the freeze-drying process; (ii) use excipients to prevent some major, specific degradations from occurring, if any. For example, an antioxidant and/or metal chelator should be included in the formulation if the major degradation is oxidation; (iii) choose stabilizers, such as trehalose or sucrose, to form a glass matrix with low molecular mobility, usually a high glass transition temperature (T_g), to not only provide kinetic stabilization to the drug molecules, but also, in the case of proteins, retain the protein conformation as much as possible throughout the shelf life (24).

However, protein formulation development is not as easy as it appears. A formulation developed under such a general guidance does not necessarily provide good stability, because each protein is often, to some extent, unique both chemically and physically, and the current understanding of protein stabilization mechanisms in the solid state is still incomplete. For example, sucrose in some cases stabilizes proteins better than trehalose, which cannot be explained either by the molecular mobility of the glass former or the retention of the protein structure (40).

Some of the most recent developments may help us toward further understanding the protein stabilization in the freeze-dried formulation solids. For example, studies of fast dynamics in freeze-dried formulation solids showed, in many cases, that at a storage temperature well below T_g, the "β-relaxation" process might dominate the diffusion of small molecules or local motion of molecules leading to denaturation and degradation of proteins (41). Also, it was reported that in amorphous matrix with comparable water content and structural relaxation times, the "Hammett acidity" of the solid matrix significantly impacts the differences in the chemical reactivity (42). As for the protein structure retention, further probing the differences in protein structures at levels higher than secondary structure, such as tertiary structure, may also help elucidate better the stabilization mechanisms of protein in the solid state, since retention of protein secondary structure may not be sufficient to the stability (24). Only with a full understanding of the protein stabilization mechanisms in the solid state will the real rational design in protein formulation development toward acceptable storage stability be expected in the pharmaceutical industry.

29.2.2.4 Other Formulation Considerations. Development of a formulation for freeze-drying should also take into account the commercial manufacturing viability. For example, the formulation should have enough hold-time stability. Particularly for protein products, agitating, filtration, and filling can also cause deterioration to the protein molecules. Nonionic surfactants are often included in protein formulations to minimize the interfacial denaturation and aggregation during agitation, filtration, and filling.

Freeze-dried products, particularly those for subcutaneous (SC) and intramuscular (IM) doses, should be formulated with an isotonicity modifier, such as sodium chloride and glycerol, to ensure the isotonicity of the reconstituted product. However, since such agents greatly reduce the collapse temperature of the formulation for freeze-drying, it is recommended, wherever possible, the isotonicity modifier should be provided in the diluent, rather than included in the freeze-dried product. Table 29.2 lists the excipients frequently used in formulations for freeze-drying, including stabilizers for freezing, drying, or/and storage, antioxidants, nonionic surfactants, metal ion chelators, bulking agent, collapse temperature modifiers, and isotonicity modifiers (25,26).

29.2.3 Container-Closure System for Freeze-Dried Pharmaceutical Products

In freeze-dried injectable pharmaceutical products, the container-closure system should be regarded as an important part of the product, because it not only protects the dried cake from moisture/outgas ingressing, but also influences the heat and mass transfer during the freeze-drying process and the throughput of the process.

The heat transfer from the shelf to the vial, and subsequently, the ice sublimation rate, is influenced by glass type, vial diameter, fill volume, as well as configuration, such as bottom radius and concavity (43,44). Typically, the containers for freeze-dried products are vials made of type I glass. The vial size (diameter) should be selected based on a number of factors, such as the targeted fill volume, reconstitution volume, and so on. While the fill volume can be larger or smaller than the reconstituted volume at the end-use, it is generally less than half of the total internal volume of the vial, because a high fill-depth leads to a prolonged drying time, and increases the possibility of vial breakage.

While both tubing glass and molded glass are used for manufacturing vials for freeze-drying, the tubing glass vial is usually preferred, because of its elegancy, high uniformity in configuration, and heat transfer properties. In addition to glass vials, coated vials or polymer vials are also used for freeze-dried products. Freezing behavior of water in these vials can be significantly different than the glass vials, which impacts the subsequent drying rates as well.

The elastomeric closures used for freeze-dried products are usually butyl or halogenated butyl rubber stoppers, which have characteristics, such as low moisture permeation, low moisture adsorption, low extractables, low absorption/adsorption, and so on. The volatile contents in the stoppers should be extremely low, since the wide ranges of temperature and high vacuum and possibly long cycle time can facilitate the transfer of volatiles from the stoppers to the final product. Coatings are sometimes applied to the stoppers in order to resolve the problem of the stopper sticking on the shelves during the stoppering after completion of the freeze-drying cycle. It is often claimed that stoppers do not prolong the drying time, which is actually not accurate. The resistance to mass transfer also includes the finite openings in the partially stoppered vials. Even for the same size of stopper, the

TABLE 29.2. Functional Excipients Commonly Used in Formulations for Freeze-Drying

Function	Excipients
Buffers (pH range of use)	Citrate (pH 2.5–6.0), histidine (pH 6.2–7.8), phosphate (6.0–8.2), tris(hydroxymethyl)aminomethane (6.8–7.7)
Stabilizers/cryo-, lyo-protectants	Sugars (Sucrose, trehalose, lactose, maltose), amino acids (glycine, arginine), polyols (mannitol, sorbitol, glycerol), polymers (PEG, dextran, PVP)
Antioxidant	Ascorbic acid, glutamate, sulfite, bisulfite
Metal chelators	EDTA (disodium salt), citric acid/sodium citrate
Nonionic surfactants	Polysorbate 20, polysorbate 80
Bulking agents	Glycine, mannitol, sucrose, lactose, arginine, histidine
Collapse temperature modifiers	Dextran, gelatin, ficoll, hydroxyethyl starch
Isotonicity modifiers	Sodium chloride, glycerol

change in opening may change the ice sublimation rate up to 10% (45).

Even though most stoppers for freeze-dried product have low moisture transmission rate and internal stoppering under dry nitrogen enables minimization of moisture uptake in product until sealed, it is not uncommon to observe moisture increase, sometimes to a substantial degree, in the dried product cake during storage. This is due to the fact that moisture is occluded in the rubber stopper during the sterilization and desorbed from the stoppers to the product cake during the storage. The amount of moisture that may be occluded in the stopper is determined by the type of rubber formulation, the size of the stopper, and the sterilization condition. In this regard, caution should be taken, especially to the freeze-dried product with low cake-mass, which can be particularly sensitive to absorption of occluded moisture, because the amount of desorbed moisture can cause a significant increase of moisture percentage in the dried cake. Selection of appropriate stoppers and/or further drying the stoppers under appropriate conditions after sterilization are the practical ways to minimize the moisture increase due to the desorption of the occluded moisture (46).

29.2.4 Pharmaceutical Freeze-Dryers

In general, a freeze-dryer for pharmaceutical products consists of (i) a product chamber containing shelves, through which the heat transfer fluid is circulated for temperature control, and onto which the filled product vials are loaded for freeze-drying; (ii) a system for pumping, heating, and cooling the fluid for shelf temperature control; (iii) a vacuum pumping system, which, with aid of bleeding an inert dry gas (N_2) in a controlled manner, enables maintenance of the chamber pressure at a set point; (iv) a condenser for capturing the vapor sublimated from the product; and (v) a refrigeration system for cooling the condenser (47). The condenser can be internally installed inside the product chamber, but in most cases, they are separately located, but connected by a vapor tube or duct. Most production freeze-dryers are also equipped with systems for defrosting, clean-in-place (CIP), and sterilization-in-place (SIP). A typical pharmaceutical freeze-dryer is shown schematically in Fig. 29.5.

From development to production, laboratory-scale, pilot-scale, and production-scale freeze dryers are utilized. Even though there is no firm definition of size (shelf area) for each of the three scales, a laboratory dryer typically has one–three shelves, each of only about 1–2 square feet, a pilot dryer is about a factor of 10 larger than the laboratory-scale and typically is of the size used to prepare Phase I clinical trial materials, and a production dryer is about a factor of 10 larger than the pilot dryer (48). From laboratory-scale to production freeze-dryer, the capability, configuration, and building materials are different, thus the characteristics of freeze-dryers for heat and mass transfer can be significantly different, which has to be considered with great caution when scaling up. Table 29.3 presents a comparison of characteristics between typical laboratory-scale, pilot-scale, and production-scale freeze-dryers (48). Even for the similar scale freeze-dryers, the configuration can vary widely from vendor-to-vendor, which sometimes also presents challenges when transferring a process from one freeze-dryer to another. Evaluation of performance capabilities and capacities of the freeze-dryer to be used becomes critical to the scale-up and/or transfer of a freeze-drying process.

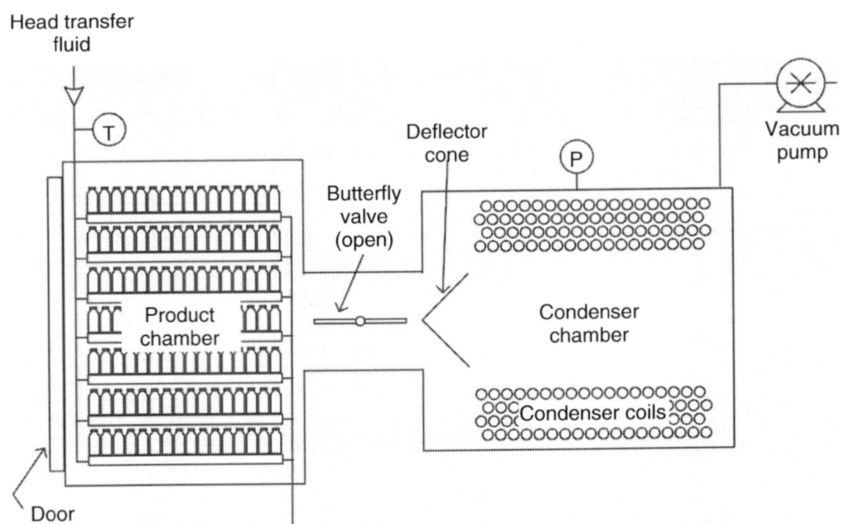

Figure 29.5. Schematic of a typical pharmaceutical freeze-dryer.

TABLE 29.3. Characteristics of Typical Laboratory-, Pilot- and Manufacturing-Scale Freeze-Dryers

Characteristics		Laboratory Durastop (FTS)	Pilot Lyofast (Edwards)	Manufacturing Lyomax (Edwards)	Manufacturing Stokes
Total shelf area (m^2)		0.38	2	39	24.2
Condenser surface (m^2)		0.64	2	40	24.6
Chamber to condenser pathway	Diameter (m)	0.05	0.25	0.91	0.9
	Length (m)	0.27	0.75	1.5	0.9
Shelf heat transfer coefficient $K_S \cdot 10^3$ (cal/s/cm^2/K)		8.0 ± 2.3	18.1 ± 4.3	—	13.9 ± 8.5
Standard resistance constants[a]	K_P	0.020	0.011	0.003	—
	K_C	0.450	0.360	0.262	—
	K_R	0.064	0.066	0.097	—
Emissivity	Wall	0.75	0.66	0.65	—
	Door	0.90	0.65	0.35	—

[a] K_P, heat transfer resistance of the chamber to the condenser; K_C, the resistance to transport of water vapor in the condenser and conversion to ice; K_R, the resistance of refrigeration system to removal of heat, resulting from the conversion of vapor to ice.

29.3 CHALLENGES AND NEW ADVANCEMENTS IN FREEZE-DRYING

29.3.1 Freeze-Drying Scale-Up and Transfer

It is not uncommon that a freeze-drying cycle, which works fine in the laboratory-scale freeze-dryer, does not yield reproducible product in the pilot-scale or production-scale freeze-dryer. Freeze-drying scale-up is to generate a robust freeze-drying cycle, which proves to yield reproducibly and uniformly product at a larger scale with nearly the same quality as at a smaller scale (49). Transfer of a freeze-drying process or a freeze-dried product can be from a research and development site to a manufacturing site, which involves the process scale-up, or from a production site to another production site or even from one freeze-dryer to another freeze-dryer of the same scale at the same production site. Freeze-drying scale-up and transfer can be very challenging, particularly for protein formulations, which are more sensitive to changes in process variables. Extensive discussion on this topic can be found in recent literature (49–55). Here, only major challenges and general strategies to address them are described.

29.3.1.1 Challenges.

29.3.1.1.1 Difference in Ice Nucleation Temperature between the Laboratory and the Sterile Production. In the research and development stage, freeze-drying is generally carried out in a pharmaceutical laboratory environment, from which significant particles can be introduced into the formulation solution in the vials, even though the solution is filtered through a sterile-grade filter and the vials have been washed as in production. In contrast, in the production site, the filling and loading of product vials are operated in a Class 100 clean room. Therefore, the ice nucleation temperature for the formulation solution in the sterile production freeze-dryer can be significantly lower than the one in the laboratory or pilot freeze-dryer for development work.

As already mentioned, the ice nucleation temperature can greatly impact the subsequent drying process and product quality. A lower ice nucleation temperature leads to smaller size of ice crystals, which means smaller pore sizes and thus a longer primary drying time and a shorter secondary drying time. Searles *et al*. (56) reported that a decrease of 1°C in ice nucleation temperature caused about a 3% increase in the primary drying time. Also, the heterogeneity of ice formation in a batch causes large heterogeneity in both primary drying and secondary drying rates.

The ice nucleation difference presents one of the major challenges in freeze-drying scale-up, which cannot be fully addressed, until a controlled ice-nucleation technology comes to practice in the pharmaceutical production freeze-dryers. This will be discussed in Section titled "Freezing Control". However, the impact of the difference in ice-nucleation temperature can be minimized. During development, particularly in the scale-up and transfer stage of a pharmaceutical freeze-drying process, container and closure components should be washed and processed as in the production. Furthermore, the formulation solution preparation and freeze-drying development batch should be carried out in as clean an environment as possible. On the other hand, an ice-annealing step can be incorporated in the freezing stage in order to reduce the heterogeneity of ice nucleation.

29.3.1.1.2 Challenge from Freeze-Dryer Equipment Differences.

29.3.1.1.2.1 DIFFERENCES IN DESIGN.

Radiation Effect. In freeze-drying of pharmaceutical products in vials, the heat to the product is supplied via direct conduction from the shelf to the vial at the contact points, thermal radiation, and conduction through the gas between the shelf surface and the vial bottom. Since the freeze-drying is operated at low temperature, thermal radiation is generally not the dominant mechanism for heat transfer. However, for the product vials on the edges of shelves, radiation from the door and the walls of the chamber can be significantly higher than the center vials. Therefore, freezing behaviors in the edge vials can be different, and particularly, much higher sublimation rates of product in these vials are expected as compared to the center vials on the same shelf. If the thermocouples are placed in these edge vials for process monitoring and control, which is generally the case in production, an atypical freeze-drying profile will be obtained.

The radiation effect or "edge effect" varies depending upon the difference in design and configuration of freeze-dryers, including shelf spacing, the wall/door temperature and the distance between the wall/door to the edge vials, the emissivity of surfaces, and so on. More evident radiation effect is expected for laboratory freeze-dryers, because the chamber doors for these freeze-dryers are normally made of Plexiglas or similar material with high emissivity, while the doors in production freeze-dryers are made of stainless steel with low emissivity (Table 29.3). Radiation effect is generally one of the major issues in freeze-drying scale up and transfers (57).

Resistance of the Pathway from the Chamber to Condenser. With some exceptions where internal condensers are used, most pharmaceutical freeze-dryers are constructed with external condensers. In the latter, the product chamber and the condenser are connected by a duct, which provides a pathway for the vapor flow. The resistance to the water vapor flow sublimated from the product to the condenser during the primary drying comprises mainly resistance of the dried layer of product, resistance of the vial stopper, and the resistance of the chamber to the condenser pathway. Since the resistance of such a pathway depends upon the size (radius and length) of the duct (Table 29.3), the difference in duct sizes can lead to varied freeze-drying profiles for the product when scaling up or transferring a freeze-drying process.

Condenser Capacity and Choked Flow. Formulations with high collapse temperatures, for example, those containing bulk agents in crystalline, or high concentration protein, can be freeze-dried under very aggressive primary drying conditions without compromising product quality. However, if the heat and mass transfer capacities of the condenser are not able to meet the need to convert the generated water vapor to ice crystals at a temperature low enough to maintain the chamber pressure control, the condenser capacity will be overloaded and the freeze-drying process will be out of control. During scaling-up and transfer of a freeze-drying process, dynamic condenser capacity of the objective freeze-dryer can be potentially limiting. Appropriate design and execution of sublimation tests in the operational qualification (OQ) to evaluate the performance of the condenser are critical in scaling up and transfer freeze-drying process, particularly for those with aggressive primary drying rates (48).

The condenser capacity can be greatly improved by a liquid nitrogen-driven refrigeration system (58). However, even if a freeze-dryer with a high capacity condenser, which can handle the massive vapor condensation properly is utilized, an aggressive ice sublimation process can still lead to loss of pressure control in the product chamber, because the vapor can be "*choked*" in the vapor pathway or the duct. This extreme scenario is the so-called *choked flow* (59).

The basic cause of a choked flow is related to the nature of gas flow dynamics. The velocity of vapor flow through the vapor tube is proportional to the pressure difference between the chamber and the condenser, and also depends upon the sublimation rate, loading capacity, and the size of the vapor tube. If the velocity reaches a maximum value (Mach1) at some point during the primary drying, the vapor flow becomes choked (27). This choked point becomes a limitation for the maximum attainable drying rate for a given freeze-dryer. Awareness of this potential risk should be appreciated when scaling up and transferring an aggressive freeze-drying process, particularly when the batch size increases and/or the vapor tube size decreases.

29.3.1.1.2.2 DIFFERENCES IN TEMPERATURE AND PRESSURE CONTROL. The cooling and heating rates of a freeze-dryer may vary with the manufacturer and unit size. Ideally, performance capabilities of the production freeze-dryer should be available or evaluated before designing the freeze-drying cycle to be used for a large-scale production. Very often, the reality is that the freeze-drying process development work is completed before a production site is identified. It is important for a development scientist in the laboratory or pilot plant to understand what capacities are typical for production-scale freeze-dryers. A range of typical capacities for heating/cooling of shelf temperature, control of shelf temperature and pressure, that might be expected in large-scale production freeze-dryers, are listed in Table 29.4, as a general guideline (49). Note that these ranges can vary depending upon the manufacturer, model, size, age, condition, and so on.

From one freeze-dryer to another, particularly from a laboratory freeze-dryer to a production freeze dryer, the temperature sensors, and the locations of the sensors for shelf temperature monitoring and/or the method for controlling can be different. Such differences may result in different shelf surface temperatures, the impact of which

TABLE 29.4. Typical Ranges of Capacities for Shelf Heating/Cooling Rates, Shelf Temperature Control, and Chamber Pressure Control in Pharmaceutical Manufacturing Freeze-Dryers

Function	Minimum	Maximum
Cooling rates; fastest (°C/min)	0.5	1.0
Heating rates; fastest (°C/min)	0.5	1.0
Shelf temperature control (°C)	−50	70
Shelf temperature variability; widest (± °C)	1.0	3.5
Shelf temperature uniformity; widest (± °C)	1.0	2.0
Pressure control range; highest (± mTorr)	500	10,000
Pressure control variability; widest (± mTorr)	2	20

will be discussed below. For pressure monitoring and control, the same type of gauges should be used, preferably the capacitance manometer gauges. Like the temperature sensors, the location of a pressure sensor can also influence the pressure reading. Therefore, the location difference of the pressure gauges can also impact the freeze-drying process when scaling up or transferring from one freeze-dryer to another.

29.3.1.1.3 Shelf Surface Temperature. Shelf temperature is generally reported as one of process parameters describing a freeze-drying cycle. Depending upon the freeze-dryer, the shelf temperature is controlled based on the temperature of the heat transfer fluid, normally, silicone oil at the inlet of the shelf. In fact, it is the temperature of the shelf surface that determines the heat transferred to the product vials. Thus, theoretically speaking, the process should be controlled and monitored by the shelf surface temperature (50).

Depending on the freeze-dryer design and the shelf temperature control, the shelf surface temperature may be different in different freeze-dryers, even though the set point of the shelf temperature (the temperature of heat transfer fluid) is kept the same. Moreover, even for the same dryer, particularly for a large scale one, the surface temperature across a shelf can be different, and such a variation can be different from shelf to shelf. The differences and variations depend upon the freeze-dryer design, heat transfer characteristics, and the sublimation rate. In the production freeze-dryer, larger temperature variations, for example, 2–5°C, across shelves are not uncommon, especially during heavy heat exchange periods, such as freezing ramp and initial period of primary drying, or primary drying at a lower temperature.

To address this challenge, a shelf temperature mapping study should be included in the OQ sublimation test, by which the difference between the set point of shelf temperature and the shelf surface temperature, the surface temperature variation across a shelf, and the variation from shelf to shelf can be determined. The gained information is critical for process adjustment, if needed, during scale-up and transfer of a freeze-drying cycle. For example, shelf temperature may be lowered and the duration may be extended in the primary drying stage to accommodate the "hot" and "cold" spots in the shelves, which are identified by shelf temperature mapping.

29.3.1.2 Freeze-Drying Development at Laboratory-Scale and Pilot-Scale. Successful scale-up and/or transfer of a freeze-drying process starts with development of the formulation, since the formulation characteristics greatly impact the freeze-drying process. During initial freeze-drying development, physical properties and chemical stability have to be well characterized. For example, for a robust formulation a higher collapse temperature is preferred and physical properties and chemical stability in the frozen state and solid state should be less sensitive to small changes in temperature and in hold-time during freeze-drying. Information gained from formulation characterization also provides scientific basis for process development and optimization at the scale-up and transfer stages (2). With a well-characterized formulation, freeze-drying cycle development starts in a laboratory freeze-dryer. In this stage, the scale-up issues as discussed above should be taken into account or kept in mind.

For scale-up and transfer, pilot-scale studies are always useful. As compared to the laboratory-scale dryer, the design and control of a pilot-scale dryer is much closer to the production dryer. In this stage, the process parameters need to be carefully selected to account for the limitations in the production freeze-dryers. Vials and stoppers should be identical to the ones to be used in production, and processed in the same way as for the production batches. With well-designed pilot-scale freeze-drying studies and effective use of the results, the number of engineering runs at a production freeze-dryer can be minimized.

In the laboratory or pilot development stage, particularly for expensive materials such as proteins, the freeze-drying is sometimes performed with partially loaded batches. It was demonstrated that the loading condition or vial packing density strongly affects the freeze-drying profiles, and thus, care has to be taken when developing a cycle with a partially loaded batch, especially with partially empty shelves in a small unit (60).

29.3.1.3 Process Modeling and Engineering Runs. Modeling of freeze-drying processes, particularly on mass and heat transfer of ice sublimation during vial lyophilization has been extensively reported and several modeling techniques have been developed (61–64). Use of appropriate modeling may help optimize the cycle and evaluate the robustness during the scaling up and transfer of a freeze-drying process, and thus save a significant amount of time and effort, as well as provide a scientific basis for understanding the process changes.

However, in most cases, predictions based upon appropriate modeling should not completely replace the direct execution of development batches, the so-called engineering runs, in the production freeze-dryers. Engineering runs at the production-scale under full load should be performed to determine the final dynamics of the process under the real production conditions, evaluate the scale-related issue between different dryers, and mostly to characterize the freeze-drying cycle in the intended production dryer, which yields the product with targeted quality and acceptable uniformity (51,55).

29.3.1.4 Surrogate Formulation Development for Freeze-Drying Scale-Up. For saving cost, or due to the unavailability of raw material (API), a placebo or surrogate formulation is often used for development batches in the pilot or production scale freeze-dryer. If the content of the API is low, the placebo can be simply the same formulation without API content. However, if the content of API in a formulation is significantly high, which is not uncommon in protein formulation, a placebo that simply removes API can cause issues in terms of freeze-drying characteristics. For example, the freeze-drying profile in the placebo vials can be significantly different from the vials containing active formulation. In a batch with placebo vials spiked with active vials, the product temperature, freezing rate, and drying rate in the active vials can be significantly influenced by the surrounding placebo vials.

In this case, a suitable surrogate formulation needs to be developed (52). This surrogate formulation should have physical characteristics similar to the active formulation. For example, the microstructure in the frozen and dried state, the collapse temperature and the resistance to the ice sublimation, consequently, the freeze-drying behavior, and the final product quality, such as cake appearance and moisture content of the surrogate formulation should be comparable to the active formulation. Surrogate formulation can be developed in laboratory-scale freeze-dryer, and preferably demonstrated in a pilot-scale freeze-dryer.

29.3.2 Freezing Control

Freezing is not only a critical step in freeze-drying processes, but also regarded as the most difficult step to control. In pharmaceutical production freeze-dryers with compressor-based refrigeration systems, the attainable shelf-cooling rate is, in general, low (not more than 1°C/min) and dictated by the temperature operation range, that is, the lower the operating range is, the slower the cooling rate will be. In recent years, liquid-nitrogen refrigeration system has been increasingly used in pharmaceutical freeze-dryers, which not only enhances the condenser capacity, but also makes it possible to cool the shelf at a faster, more constant rate, toward a lower freezing temperature, as compared to the compressor-based freeze-dryer. This partially meets some challenges faced in freeze-drying formulations with biologics, nonaqueous solvents or in high fill depth/volume (58).

And yet, the ice nucleation remains the major challenge in freezing control, due to its spontaneous and stochastic nature. The control of the ice nucleation appears to be one of the key factors in the cycle optimization for pharmaceutical freeze-drying production. In recent years, a number of technologies for controlled ice nucleation have been reported, for example, vacuum-induced freezing (6), electro-freezing (65), and ultrasound-induced nucleation (66,67). However, it seems difficult to implement either of them in a production freeze-dryer in a sterile environment. The so-called "ice-fog" technique (68) seems to be a more promising approach for pharmaceutical freeze-dryer. This technique involves lowering the shelf temperature and cooling the samples to the desired temperature for nucleation, and then, introducing a flow of very cold nitrogen gas into the product chamber, in which ice crystals form and enter into the vials, resulting in the nucleation of the solution at the desired temperature. Most recently, another possibly practical technology is reported, in which the ice nucleation in the formulation at a controlled temperature is introduced by a pressurization–depressurization procedure (69). However, even being successful in laboratory freeze-dryers, it may take some time for these technologies to be demonstrated and eventually be implemented in production freeze-dryers in the pharmaceutical industry.

29.3.3 Use of Cosolvents for Freeze-Drying

Nonaqueous cosolvents have been increasingly used in freeze-drying development to modify the physical properties of formulations, for example, to increase the wettability and/or the solubility, or to alter the physical state of the molecule of interest, or simply to facilitate the sublimation rate.

Amongst others, *tert*-butyl alcohol (TBA), with a high freezing temperature and high vapor pressure, is one of the top choices, especially to facilitate the drying process. However, in some cases, it is difficult to reduce the residual TBA to a satisfactory low level (70), and TBA may have strong disruptive effect on the protein structure and bioactivity during freeze-drying (71). Another cosolvent that has been receiving increased attention is ethanol, which is a typical organic solvent and easily mixed with water. Ethanol has been used to change the physical state of active molecules or excipients in a freeze-dried formulation. However, ethanol, along with most nonaqueous solvents present in the formulation may freeze or remain as unfrozen liquid residues distributed throughout the ice matrix, depending upon the percentage and the other formulation components. This results in substantially different thermal and structural properties of formulations in the frozen state, and thus affects the rate of drying (72). It was reported that the freeze-drying rate of formulations with ethanol/water cosolvent systems was slower than the corresponding aqueous system, because of incomplete frozen systems (73). Indeed, depending upon the nature and percentage of a cosolvent, the vapor composition, the heat transfer in freeze-drying a formulation with a cosolvent can be very different from the one without cosolvent (74). Therefore, while one may gain some benefits in formulation and processing, use of nonaqueous solvents for freeze-drying presents additional challenges in the choice of plant design, the production capacity, and the product quality. The reader is referred to an excellent review paper for more details (75).

29.3.4 Freeze-Drying Products in Primary Packaging Systems Other than Vials

While the most common primary containers for freeze-dried pharmaceutical products are glass vials, syringes and cartridges are also used as the primary containers for lyophilized product (76,77). Particularly, freeze-dried products in dual chamber prefilled syringes are of great convenience to the end user, such as patients and/or clinicians, who have to perform reconstitution and administration. Therefore, the dual chamber syringe has been increasingly used in pharmaceutical freeze-drying.

Most of the principles governing the freeze-drying process design for the vial are applicable to syringe freeze-drying as well. However, the different configuration does provide unique challenges from freeze-drying process development and scale-up of pharmaceutical products in such different containers. For the dual chamber syringe system, since the formulation solution is filled in the middle of a syringe, there is a lack of direct thermal contact with the shelf, and mass and heat transfer conditions are significantly different as compared to traditional glass vials. The temperature difference between the product and the shelf can be higher than usual and relatively high chamber pressure is needed as well. Studies on syringe freeze-drying are limited in literature. Hottot *et al*. (78,79) reported that though significant inter-syringe heterogeneity was observed, the intra-syringe ice formation uniformity during freezing was surprisingly high, probably due to the small size of syringe, and the convection effects around the syringe walls were probably more important for ice sublimation in syringe freeze-drying.

29.3.5 Process Analytical Technologies and Quality-by-Design in Freeze-Drying

The initiative of Process Analytical Technology (PAT) was launched by the FDA a few years ago, which, in general, aims to promote a better understanding and control of process by in-line or on-line monitoring, as well as to facilitate the quality decision process during the manufacturing process.

In fact, since the formulation and process are strongly interrelated in freeze-drying, even well before the PAT initiative, on-line monitoring technologies have been continuously explored to analyze, or characterize the formulation and/or the process. In formulation development for freeze-drying, a number of analytical or characterization techniques have been used for *in situ* characterization of samples under pseudo-freeze-drying conditions. For example, freeze-drying microscopy (80), freeze-drying X-ray (81), and freeze-drying SEM (scanning electron microscopy) (82) were utilized to characterize the physical property changes during the freeze-drying process, while infrared spectroscopy (83) and infrared microscopy (84) were also employed to characterize the protein structure changes *in situ*. These technologies, in general, employed a special freeze-drying stage, instead of a conventional freeze-dryer, in direct contact with the instrument. If possible, PAT under real freeze-drying conditions is preferred to understand physical changes of formulations (85).

For process monitoring, a number of technologies have been proposed. In general, monitoring methods can be categorized into two groups, as summarized in Table 29.5 (86–102). One group comprises the methods, which measure the changes of physical properties of the product in a vial during freeze-drying, so they are essentially "single-vial" monitoring technologies. In this group, some of the methods are invasive to the product vial, while others are not. The other group consists of the so-called "batch monitoring" technologies, which monitor the process variables of the entire batch during freeze-drying and are noninvasive to the product vial. Detailed discussion on these technologies, particularly the advantages and limitations of each technology, can be found in the respective references, as well as some excellent reviews (103,104).

In light of the FDA initiative, PAT in freeze-drying would not only monitor the freeze-drying process, but also create feedback loops for control. Because of heterogeneities of heat and mass transfer in a freeze-drying

TABLE 29.5. Process Analytical Technologies (PAT) for Freeze-Drying

Characteristics		Technologies	Process Variables to be Monitored	References
Single-vial monitoring	Invasive	Temperature probes: thermocouple, resistance thermal detector (RTD), etc.	Product temperature	Tradition
		Wireless temperature measurement	Product temperature	86
	Noninvasive	Microbalances	Vial weight change	87,88
		NIR	Residual moisture	89
		Raman	Residual moisture	90
Batch monitoring	Noninvasive	Gas Flow Controller	N_2 flow	91
		Windmill	Vapor flow	92
		Comparative pressure measurement	Pirani/capacitive differential pressure	93
		Moisture probe (Electronic hygrometer)	Change of moisture level in the product chamber	94–96
		Calorimetry	Heat transfer rate and total energy transferred	97
		Pressure rise test	Sublimation dynamics	6,93
		Manometric temperature measurement (MTM), pressure rise analysis	Product temperature determined by analyzing data by using a mathematical model (indirect)	98,99
		Mass spectroscopy	Partial vapor pressure	100
		Plasma emission spectroscopy	Humidity change in the product chamber	101
		Tunable diode laser absorption spectroscopy (TDLAS)	Vapor mass flux from chamber to condenser	102

batch, it may be misleading to rely on individual vial measurements to control the whole batch. In this aspect, the "batch monitoring" technologies are preferred. Amongst others, MTM and tunable diode laser absorption spectroscopy (TDLAS) are probably the two technologies that have been studied most extensively. The MTM method has been demonstrated to be a useful PAT tool for process monitoring, cycle development, robustness evaluation, and scale-up. The idea of automatic development of an "optimized" freeze-drying process based on the MTM led to the commercialization of SMART freeze-dryer (21), a laboratory-scale freeze-dryer for development. While the use of MTM as a PAT tool in pilot or production has yet to be demonstrated, the "transfer" of TDLAS from laboratory-scale to production scale freeze-dryers seems straightforward, since it only involves a modification of the duct and should not be difficult to establish in a sterile production environment.

In general, implementation of a feasible PAT in a sterile operation of freeze-drying requires the technology to be noninvasive, compatible with sterile practices, for example, steam sterilization, CIP, leak control, and so on. These challenges, along with the high cost of a batch, particularly in freeze-drying of biopharmaceuticals may call for more extensive demonstration than usual before practically implementing a PAT in freeze-drying of a commercial batch.

Quality-by-design (QbD), a new regulatory philosophy, was proposed by the FDA a few years ago based upon the principles of the ICH Q8, Q9, and Q10, that is, quality cannot be "*tested into*" product, but should be *built-in* or *designed into* product. PAT, design space, and risk management constitutes the QbD. Through *PAT* monitoring and control of a process within the established *design space*, a product of predetermined quality is consistently generated, and the manufacturing flexibility is justified by utilizing quality *risk management*.

The concept of design space is a key element of QbD, which is a multidimensional combination and interaction of input variables and process parameters that have been demonstrated to provide assurance of quality. The quality of a freeze-dried product is determined by freeze-drying cycle parameters, product design (formulation, container-closure system, and fill volume), and equipment characteristics of the freeze-dryer employed for manufacturing the product, including the freeze-dryer design, capacity, operating, and control. Therefore, while the QbD is an excellent opportunity to apply science and new technologies in freeze-drying to achieve consistent product quality, it also presents a significant challenge to pharmaceutical scientists and engineers. For example, appropriate development and characterization of design space for freeze-drying not only requires in-depth knowledge of the product, freeze-drying process and equipment, but also warrants tremendous work in development and scale-up (105). A recent chapter by Nail and Searles (106) presents an excellent illustration on how the design space might be applied to the development, scale-up and transfer of freeze-drying processes for injectable pharmaceuticals.

Acknowledgment

The author would like to thank Drs. Enrico Bellomo, Freeman Stanfield (Abraxis BioScience), and Beena Uchil (APP Pharmaceuticals) for their help in reading the manuscript and their comments, as well as Mr Terry Hendricks and Mr William Lu (Abraxis BioScience) for their help in preparing some of the figures.

REFERENCES

1. Seager H. J Pharm Pharmacol 1998; 50: 375–382.
2. Liu J. Pharm Dev Technol 2006; 11: 3–28.
3. Overcashier DE, Patapoff TW, Hsu CC. J Pharm Sci 1999; 88: 688–695.
4. Jiang G, Akers M, Jain M, Guo J, Distler A, Swift R, Wadhwa MV, Jameel F, Patro S, Freund E. PDA J Pharm Sci Technol 2007; 61(6): 441–451.
5. Randolph TW, Searles JA. Am Pharm Rev 2002; 5(4): 40–45.
6. Liu JS, Viverette T, Virgin M, Anderson M, Dalal P. Pharm Dev Technol 2005; 10: 261–272.
7. Pikal MJ, Roy ML, Shah S. J Pharm Sci 1984; 73: 1224–1237.
8. Janso G, Pupezin J, van Hook WA. J Phys Chem 1970; 74: 2984–2989.
9. Chang BS, Fischer NL. Pharm Res 1995; 12:, 831–837.
10. Pikal MJ, Shah S. Int J Pharm 1990; 62: 165–186.
11. Her LM, Nail SL. Pharm Res 1994; 11: 54–59.
12. Wang W. Int J Pharm 2000; 203: 1–60.
13. Chang BS, Randall CS. Cryobiology 1992; 29: 632–656.
14. Blue J, Yoder H. Am Pharm Rev 2009; 12(1): 90–96.
15. Pikal MJ. Biopharm 1990; 3: 18–28.
16. Tang X, Pikal MJ. Pharm Res 2004; 21: 191–200.
17. Franks F. Eur J Pharm Biopharm 1998; 45: 221–229.
18. Liveseyand RG, Rowe TW. J Parenter Sci Technol 1997; 41: 169–171.
19. Pikal MJ, Shah S, Roy MJ, Putman R. Int J Pharm 1990; 60: 203–217.
20. Chang BS, Patro SY. In: Costantino HR, Pikal MJ, editors. Lyophilization of biopharmaceticals. Arlington (VA): AAPS press; 2004. pp. 113–138.
21. Tang XC, Nail SL, Pikal MJ. Pharm Res 2005; 22: 685–700.
22. Hsu CC, Ward CA, Pearlman R, Nguyen HM, Yeung DA, Curley JG. Dev Biol Stand 1992; 74: 255–270.
23. Carpenter JF, Pikal MJ, Chang BS, Randolph TW. Pharm Res 1997; 14: 969–975.
24. Chang LL, Pikal MJ. J Pharm Sci 2009; 98: 2886–2908.
25. Akers MJ, Vasudevan V, Stickelmeyer M. In: Nail SL, Akers MJ, editors. Development and manufacture of protein pharmaceuticals. New York (NY): Kluwer Academic/Plenum Publishers; 2002. pp. 47–127.
26. Costantino HR. In: Costantino HR, Pikal MJ, editors. Lyophilization of biopharmaceticals. Arlington (VA): AAPS press; 2004. pp. 139–228.
27. Rambhatla S, Pikal MJ. In: Costantino HR, Pikal MJ, editors. Lyophilization of biopharmaceticals. Arlington (VA): AAPS press; 2004. pp. 75–109.
28. Bhatnagar BS, Bogner RH, Pikal MJ. Pharm Dev Technol 2007; 12: 505–523.
29. Millqvist-Fureby A, Malmsten M, Bergenstahl B. Int J Pharm 1999; 191: 103–114.
30. Izutsu K, Yoshioka S, Kojima S, Randolph TW, Carpenter JF. Pharm Res 1996; 13: 1393–1400.
31. Hancock BC, Shalaev EY, Shamblin SL. J Pharm Pharmacol 2002; 54: 1151–1152.
32. Tang XC, Pikal MJ. Pharm Res 2005; 22: 1167–1175.
33. Chang BS, Kendrick BS, Carpenter JF. J Pharm Sci 1996; 85: 1325–1330.
34. Jiang S, Nail SL. Eur J Pharm Biopharm 1998; 45: 249–257.
35. Liu W, Wang DQ, Nail SL. AAPS PharmSciTech 2005; 6: 150–157.
36. Luthra S, Obert JP, Kalonia DS, Pikal MJ. J Pharm Sci 2007; 96: 61–70.
37. Liu J, Rigsbee DR, Stotz C, Pikal MJ. J Pharm Sci 2002; 91: 1853–1862.
38. Abdul-Fattah AM, Truong-Le V, Yee L, Nguyen L, Kalonia DS, Cicerone MT, Pikal MJ. J Pharm Sci 2007; 96: 1983–2008.
39. Luthra SA, Hodge IM, Utz M, Pikal MJ. J Pharm Sci 2008; 97: 5240–5251.
40. Shamblin SL, Hancock BC, Pikal MJ. Pharm Res 2006; 23: 2254–2268.
41. Cicerone MT, Soles CL, Chowdhuri Z, Pikal MJ, Chang L. Am Pharm Rev 2005; 8(6): 22–24.
42. Chatterjee K, Shalaev EY, Suryanarayanan R, Govindarajan R. J Pharm Sci 2008; 97: 274–286.
43. Pikal MJ. PDA J Pharm Sci Technol 1985; 39: 115–139.
44. Cannon A, Shemeley K. Pharm Res 2004; 21: 536–542.
45. Willemer H. In: Rey L, May JC, editors. Freeze-drying/lyophilization of pharmaceutical and biological products. New York (NY): Marcel Dekker, Inc.; 1999. pp. 79–122.
46. Donovan PD, Corvari V, Burton MD, Rajagopalan N. PDA J Pharm Sci Technol 2007; 61: 51–58.
47. Gatlin LA, Nail SL. Bioprocess Technol 1994; 18: 317–367.
48. Rambhatla S, Tchssalov S, Pikal MJ. AAPS PharmSciTech 2006; 7(2): 61–70.
49. Trappler EH. Am. Pharm Rev 2001; 4(3): 55–60.
50. Jennings TA. Am Pharm Rev 2002; 5(1): 34–40.
51. Speaker SM, Teagarden DL. Am Pharm Rev 2008; 11(5): 54–61.
52. Sane SV, Hsu CC. Am Pharm Rev 2007; 10(1): 132–136.
53. Tsinontides SC, Rajniak P, Pham D, Hunke WA, Placek J, Reynolds SD. Int J Pharm 2004; 280: 1–16.
54. Jameel F, Paranandi M. Am Pharm Rev 2006; 9(2): 53–55.

55. Tchessalov S, Dixon D, Warne N. Am Pharm Rev 2007; 10(2): 88–92.
56. Searles JA, Carpenter JF, Randolph TW. J Pharm Sci 2001; 90: 860–871.
57. Rambhatla S, Pikal MJ. AAPS PharmSciTech 2003; 4(2): 22–31.
58. Liu JS, Rouse D. Bioprocess Int 2005; 3(2): 28–31.
59. Searles J. Am Pharm Rev 2004; 7(2): 58–69.
60. Gieseler H, Lee G. Pharm Res 2008; 25: 302–312.
61. Pikal MJ, Cardon S, Bhugra C, Jameel F, Rambhatla S, Mascarenhas WJ, Akay HU. Pharm Dev Technol 2005; 10: 17–32.
62. Kuu WY, Hardwick LM, Akers MJ. Int J Pharm 2005; 302: 56–67.
63. Hottot A, Peczalski R, Vessot S, Andreiwu J. Dry Technol 2006; 24: 561–570.
64. Kramer T, Pikal MJ, Petre WJ, Shalaev EY, Gatlin LA. J Pharm Sci 2009; 98: 307–318.
65. Petersen A, Schneider H, Rau G, Glasmacher B. Cryobiology 2006; 53: 248–257.
66. Nakagawa K, Hottot A, Vessot S, Andrieu J. Chem Eng Prog 2006; 45: 783–791.
67. Passot S, Tréléa IC, Marin M, Galan M, Morris GJ, Fonseca F. J Biomech Eng 2009; 131: 074511.
68. Rambhatla S, Ramot R, Bhugra C, Pikal MJ. AAPS PharmSciTech 2004; 5(4): 54–62.
69. Sever R. In: First Annual Meeting of the Midwest Chapter of the International Society of Lyophilization - Freeze Drying 2009; April 16; Oak Brook (IL).
70. Wittaya-Areekul SS, Nail SL. J Pharm Sci 1998; 87: 491–495.
71. Zhang Y, Deng Y, Wang X, Xu J, Li Z. Int J Pharm 2009; 371: 71–81.
72. Seager H, Taskis CB, Syrop M, Lee TJ. J Parenter Sci Technol 1985; 39: 161–179.
73. Takada A, Nail SL, Yonese M. Pharm Res 2009; 26: 1112–1120.
74. Daoussia R, Séverine V, Julien A. Chem Eng Res Des 2009; 87: 899–907.
75. Teagarden DL, Baker DS. Eur J Pharm Sci 2002; 15: 115–133.
76. Mosharraf M, Malmberg M, Fransson J. Int J Pharm 2007; 336: 215–232.
77. Roth C. In: AAPS National Biotechnology Conference; 2007 Ju 24–27; San Diego (CA).
78. Hottot A, Andrieu J, Vessot S, Shalaev E, Gatlin LA, Ricketts S. Dry Technol 2009; 27: 40–48.
79. Hottot A, Andrieu J, Vessot S, Shalaev E, Gatlin LA, Ricketts S. Dry Technol 2009; 27: 49–58.
80. Nail SL, Her LM, Proffitt CP, Nail LL. Pharm Res 1994; 11: 1098–1100.
81. Cavatur RK, Suryanarayanan R. Pharm Dev Technol 1998; 3: 579–586.
82. Meredith P, Donald AM, Payne RS. J Pharm Sci 1996; 85: 631–637.
83. Remmele RL, Stushnoff C, Carpenter JF. Pharm Res 1997; 14: 1548–1555.
84. Schwegman JJ, Carpenter JF, Nail SL. J Pharm Sci 2007; 96: 179–195.
85. Romero-Torres S, Wikström H, Grant ER, Taylor LS. PDA J Pharm Sci Technol 2007; 61: 131–145.
86. Schneid S, Gieseler H. AAPS PharmSciTech 2008; 9: 929–739.
87. Gieseler H, Lee G. Pharm Dev Technol 2008; 13: 463–472.
88. Roth C, Winter G, Lee G. J Pharm Sci 2001; 90: 1345–1355.
89. Brülls M, Folestad S, Sparén A, Rasmuson A. Pharm Res 2003; 20: 494–499.
90. De Beer TRM, Alleso M, Goethals F, Coppens A, Heyden YV, Lopez De Diego H, Rantanen J, verpoort F, Vervaet C, Remon JP, Baeyens WRG. Anal Chem 2007; 79: 7992–8003.
91. Chase DR. Pharm Eng 1998; 18(1): 92–98.
92. Couriel B. Bull Parenter Drug Assoc 1977; 31: 227–236.
93. Nail SL, Johnson W. Dev Biol Stand 1991; 74: 137–151.
94. Roy ML, Pikal MJ. J Parenter Sci Technol 1989; 43: 60–66.
95. Bardat A, Biguet J, Chatenet E, Courteille F. J Parenter Sci Technol 1993; 47: 293–299.
96. Genin N, Rene F, Corrieu G. Chem Eng Prog 1996; 35: 255–263.
97. Jennings TA, Duan H. PDA J Pharm Sci Technol 1995; 49: 272–282.
98. Milton N, Pikal MJ, Roy ML, Nail SL. PDA J Pharm Sci Technol 1997; 51: 7–16.
99. Chouvenc P, Vessot S, Andrieu J, Vacus P. PDA J Pharm Sci Technol 2005; 59: 298–309.
100. Connelly JP, Welch JV. J Parenter Sci Technol 1993; 47: 70–75.
101. Yeresse Y, Veillon R, Sibille PH, Nomine C. PDA J Pharm Sci Technol 2007; 61: 160–174.
102. Gieseler H, Kessler WJ, Finson M, Davis SJ, Mulhall PA, Bons V, Debo DJ, Pikal MJ. J Pharm Sci 2007; 96: 1776–1793.
103. Wiggenhorn M, Presser I, Winter G. Am Pharm Rev 2005; 8(1): 38–44.
104. Gieseler H. In: The Freeze drying of Pharmaceuticals and Biologicals Conference; 2008 Aug 6–9; Breckenrideg (CO).
105. Sane SV. In: Peptalk: Lyophilization 2009 Jan 12–14; San Diego (CA).
106. Nail SL, Searles JA. Biopharm Int 2008; 21(1): 44–52.

30

FREEZING, BIOPHARMACEUTICAL

PHILIPPE LAM AND JAMIE MOORE
Pharmaceutical Development Genentech, Inc., South San Francisco, California

30.1 INTRODUCTION

The production of commercial therapeutic proteins is typically conducted in campaign mode where multiple bioreactors are harvested and the protein of interest is purified, ultimately yielding several hundred liters of formulated drug substance. Since the drug is not immediately filled into the final vial for market, frozen storage is often employed to extend protein shelf-life and to provide flexibility prior to fill finish operations. While freezing is a routine laboratory practice for preserving small samples, the freezing of commercial size batches of protein solutions present additional challenges.

The physics of freezing for protein solutions is very similar to the freezing of regular aqueous sugar or salt solutions such as sucrose or sodium chloride. However, the presence of significant amounts of protein in the solution can impact the freezing dynamics due to the very different properties of macromolecules as compared to small size solutes. The following sections present the salient aspects of freezing and thawing as applicable to solutions in general and discuss the consequences of freezing and thawing on protein and protein stability relevant to large-scale commercial therapeutic bulk processing.

30.2 FREEZING OF SOLUTIONS

30.2.1 Freezing Point Temperature

For a pure substance such as water, the freezing temperature is well defined and is the same as the melting temperature. When dealing with aqueous solutions the *freezing point* actually refers to the *freezing point depression* of the normal freezing temperature of the pure solvent by the solutes. Freezing point depression is a colligative property of the solution, and when a value is given, it applies only to the initial solution. Table 30.1 gives freezing point values for several solutions, including some formulated pharmaceutical IgG1 monoclonal antibody (mAb) products. Note that for solutions that are near physiological conditions, the freezing point is not very much lower than that of pure water. The freezing point depression is contributed primarily by salts and buffer components, proteins have minimal effect unless they are present at very high concentrations.

During the freezing process, solutes are concentrated in the liquid phase as they are excluded from the growing ice crystals. Hence, for solutions, the freezing point actually varies throughout the solidification process as more and more solute accumulate in the remaining liquid.

30.2.2 Nucleation Temperature and Supercooling

Freezing of solutions, which is mostly the crystallization of water into ice, is a phase transition and requires nuclei to proceed. The *nucleation temperature* refers to the temperature at which the first crystal nucleus forms. For most practical industrial biotechnology processes, ice-nucleation occurs at or very near the cooled surfaces contacting the liquid and at temperatures significantly lower than the bulk freezing point. Nucleation can also be triggered by particles or surface defects of the containers. The nucleation temperature for pharmaceutical products, which are required to have very low particulate content, can be as low as −20°C. Furthermore, these solutions, when filled into small

Downstream Industrial Biotechnology: Recovery and Purification, First Edition. Edited by Michael C. Flickinger.

TABLE 30.1. Freezing Point of Selected Solutions. Freezing Point Values are Measured Using a Freezing Point Osmometer. The Examples of mAb Solutions are Commercial Biopharmaceutical Products

Solution	Freezing point (°C)
Water	0
0.150 M NaCl	−0.5
mAb A (~25 mg/mL protein, 60 m*M* trehalose and 5 m*M* histidine–histidine.HCl buffer)	−0.1
mAb B (~25 mg/mL protein, 150 m*M* NaCl and 25 m*M* citrate buffer)	−0.7
mAb C (~150 mg/mL protein, 200 m*M* arginine chloride and 20 m*M* histidine–histidine.HCl buffer)	−0.8

smooth-wall containers such as glass vials, can exist as a liquid at these *supercooled* temperatures for extended periods. The term *supercooling* (sometimes *subcooling* is also used) refers to cooling a liquid below its freezing point. A solution that has been supercooled will remain in a metastable liquid state until the formation of an ice nucleus, then the temperature will rapidly rise back to the solution freezing point as ice crystals grow.

In cases involving very clean, low particulate solutions that have been cooled gradually, all of the liquid in the container may exist in a supercooled state. Upon ice-nucleation, the entirety of the volume appears to freeze nearly instantaneously. This phenomenon is exemplified in the sequence of photographs shown in Fig. 30.1 where the vial on the left turned from liquid to solid in less than half a second.

In actuality, only a fraction of the water was frozen in that instant. The amount of liquid solidified during the nucleation/rapid ice growth event is proportional to the magnitude of supercooling. For example, an energy balance calculation reveals that for a 15 degree supercooling, at most 20% of the water would be turned to ice. The supercooled liquid acts as a heat sink, warming up as it absorbs the latent heat of solidification released by the rapidly growing dendritic ice crystals. Since the entire volume of liquid is supercooled, ice forms throughout. This rapid ice growth event ends when the temperature reaches the freezing point of the liquid fraction. Thereafter, the remaining liquid freezes by the same process as described in subsequent sections. In Fig. 30.1, where the samples are cooled from below, the normal freezing front is clearly visible in the right most vial.

In the case of very rapid freezing, where the process is completed in seconds or minutes, *nonequilibrium* freezing dynamics may be encountered (1). In freezing *far-from-equilibrium*, solution constituents may be kinetically trapped in a transient state and subsequently undergo a relaxation that cause significant changes in the local environment experienced by the solute molecules, possibly impacting protein stability. This freezing regime, dominated by kinetics, is mostly applicable to small samples (up to a few milliliters) that have undergone a high degree of supercooling such as conditions that are routinely encountered in lyophilization processes (2). In the case of typical industrial freezing-for-storage, much larger sample volumes are involved (liters to hundreds of liters) so that freezing times may be in the orders of hours to days. Hence, for this application freezing generally occurs at or very close to equilibrium. The discussion of freezing process in subsequent sections will be mostly restricted to the equilibrium freezing regime.

30.2.3 Impact of Supercooling on Protein Stability

During cooling and supercooling protein solutions can be exposed to temperatures below 0°C for extended periods of time. While lower temperatures may slow down some chemical and physical reactions, there are other issues that

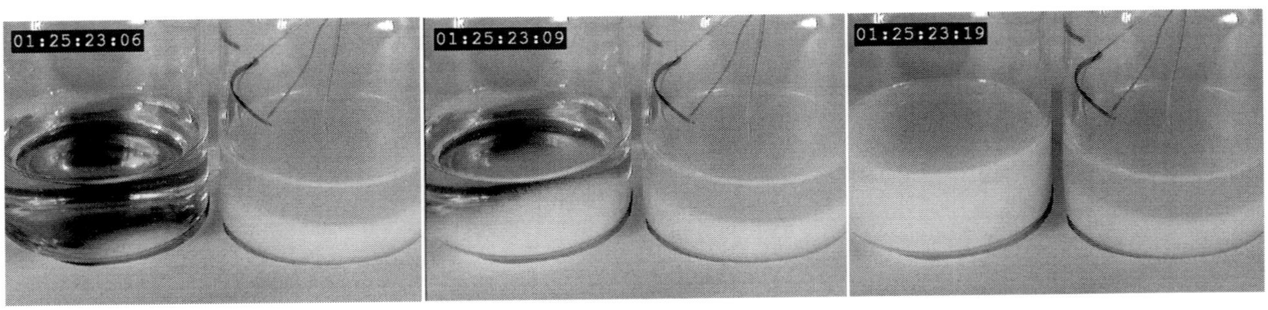

Figure 30.1. Supercooling and nucleation of a protein solution. Vials containing a pharmaceutical protein are cooled from below. The vial on the left supercooled to about −17°C before the ice-nucleation event took place. Rapid dendritic ice growth fills the entire liquid volume in less than 0.5 s. The sequence of images was recorded at 30 frames/s, the last two digits of the time stamp (in top left corner of each frame) indicate frame numbers. (This figure is available in full color at http://onlinelibrary.wiley.com/book/10.1002/9780470054581.)

may arise including cold denaturation and liquid–liquid phase separation.

Cold denaturation is the spontaneous unfolding of proteins. This phenomenon has been extensively reviewed in the literature; it is essentially a thermodynamic event that occurs at low temperatures where the unfolded nonnative state is thermodynamically favored over the native folded state (3–6). Studies with staphylococcal nuclease have demonstrated that maximal stability (maximum $\Delta G^{\circ}{}_{D}$) is observed at room temperature and that protein stability decreases with both higher and lower temperatures (7). A recent NMR study monitored the stability of titin I28 in the liquid state from −16 to 65°C and a similar parabolic stability profile was observed with maximal stability around room temperature (8). This appears to be a general phenomena exhibited by many proteins including β-lactoglobulin, phosphoglycerate kinase, ribonuclease, staphylococcal nuclease, lactate dehydrogenase, myoglobin, and chrymotrypsinogen (6,7,9–12).

Cold denaturation of proteins is often a reversible event. However, if unfolding does occur during the freeze process, long-term storage at lower temperatures may result in further protein denaturation. In addition, there is a concern of incomplete protein refolding after thawing. As such, recent studies have investigated the impact of stabilizers on cold denaturation (13,14). The results of these studies demonstrated that stabilizers such as sucrose and trehalose can significantly shift the cold denaturation temperature to lower values, by as much as 35°C. Additionally, stabilizers can increase the unfolding half-life to time-scales (years) (13) that are not relevant to the freeze/thaw process. Other factors such as pH and protein concentration can also influence the cold denaturation temperature.

Under certain solution compositions, protein solutions have been observed to undergo phase separation upon cooling. Recently an IgG mAb was observed to separate into two liquid phases at temperatures below 0°C. The liquid–liquid phase separation was dependent on protein concentration, ionic strength, and pH. The two phases were shown to be composed of different protein concentrations and slightly different solution conditions. While the two phases reverted to a single phase upon warming, protein aggregation, and denaturation could occur at the higher protein concentrations (15). In addition, protein partitioning could result in unfavorable levels of stabilizers and cryoprotectants.

While both cold denaturation and liquid–liquid phase separation can result from supercooling or prolonged liquid storage at temperatures below 0°C, it has been shown that stabilizers and modifying formulation conditions can prevent these phenomena from occurring and should be evaluated during formulation development.

30.2.4 Physics of Freezing of Solutions

Freezing of a solution is a solidification as well as an unmixing process. The presence of solutes has a profound effect on the structure of the ice crystals and the overall freezing dynamics but heat transfer also has a major impact since the solidification of water is exothermic. As a result, freezing is a complex process, involving simultaneous geometry dependent mass and heat transfer.

For the purpose of discussion, it is beneficial to look at each principal direction of heat transfer separately, assuming a cylindrical shaped container. Insights may be gained from this exercise with understanding that in practice, it is likely that all events are coupled to some extent.

30.2.5 Freezing from Bottom

Ice formed from degassed pure water is essentially transparent with a smooth freezing front (16). The situation is much more complex as soon as another component is introduced. Figure 30.2 illustrates the sequence of events for freezing of a static solution cooled from the bottom:

1. A layer of liquid adjacent to the bottom cooled surface, which is maintained at a constant temperature, is supercooled below the solution freezing point. Depending on the magnitude of supercooling, there may be some thermal convective flow due to the density difference between the liquid near the cooled surface and the solution bulk.
2. Nucleation occurs on the cooled surface and ice crystals propagate rapidly into the bulk of the solution as the latent heat of solidification is absorbed by the adjacent supercooled liquid warming it back to the solution freezing point. The thermal supercooling condition is said to have been "relaxed" (17).
3. The ice crystals grow rapidly at the base where there is contact with the cooling surface but some melt back at the tips occurs due to heat conduction from the warmer solution bulk. If the ice formed faster than solute diffusion, then there are likely solute pockets trapped in the initial ice layer. The temperature at the ice front is at or near the bulk solution freezing point. Freezing slows down as the ice now covers all of the cooled surface and the ice liquid interface, at this point, is relatively flat. As freezing continues, the solute that is rejected from the ice lattice accumulates at the growing ice front.
4. After some time, small disturbances can lead to ice crystals growing slightly faster at certain locations on the ice front. These instabilities then propagate, leading to a condition that is observed in the freezing of solutions but not pure substances (16). This is referred to as *unstable solidification*, and is characterized by

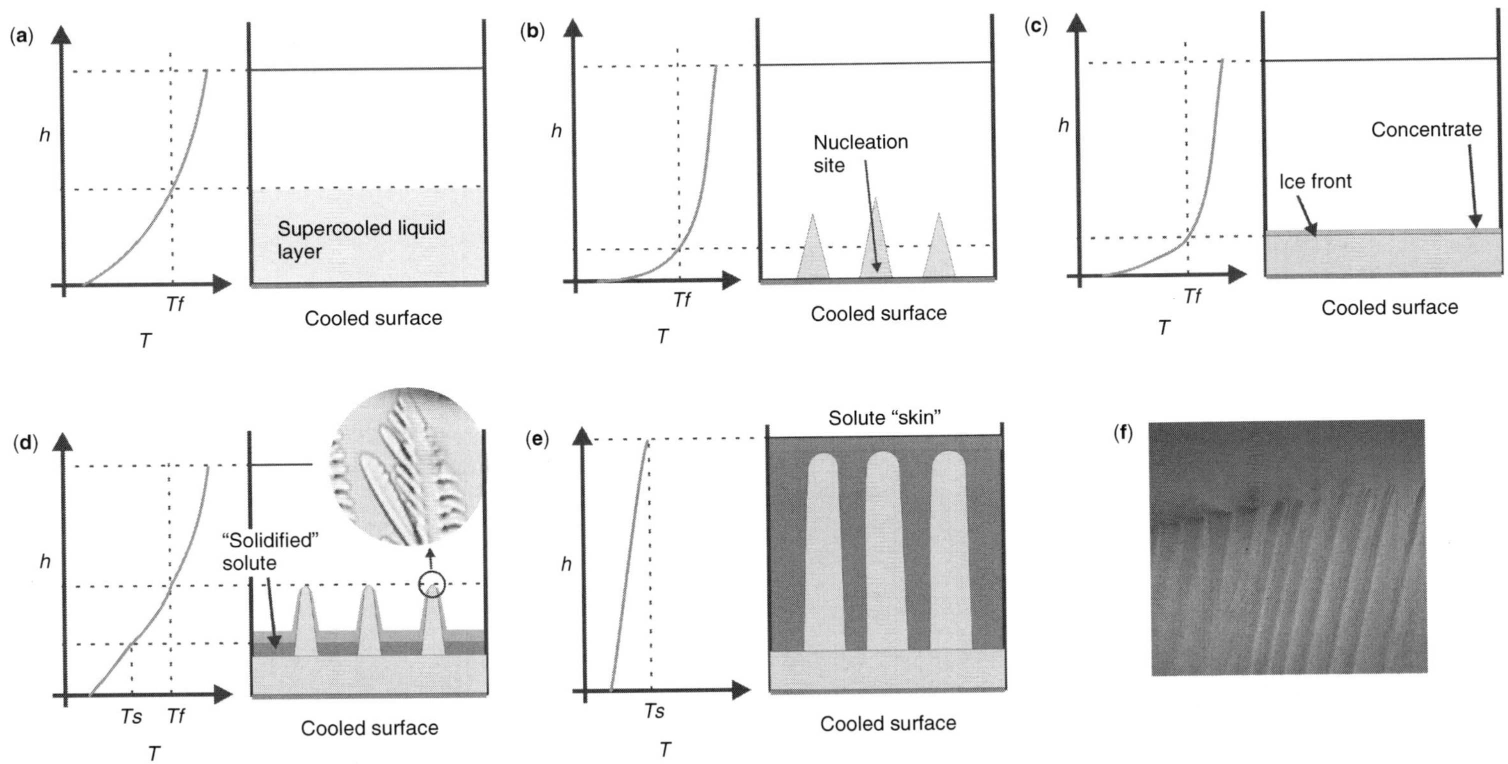

Figure 30.2. Illustration of freezing of a solution cooled from the bottom. (a) Layer of supercooled liquid; (b) ice nucleates on surface; (c) a planar ice front is formed and grows, solute begins to concentrate at the ice front; (d) instabilities cause the ice to form dendrites, this is characteristic of freezing of solutions; (e) freezing is complete; (f) close up photograph of dendritic ice front, the solute collects in between the ice crystals. Food dye was added to improve image contrast. (This figure is available in full color at http://onlinelibrary.wiley.com/book/10.1002/9780470054581.)

a rough freezing front where the ice crystals have a dendritic morphology (17–19). This dendritic ice growth is due to the solute being more concentrated at the base and sides of the ice crystals, causing the local freezing point to be lower there as compared to the tip where the solute concentration is closer to the initial bulk concentration. This phenomenon is known as *constitutional supercooling*, though this term is somewhat misleading since ice is already present. The local freezing point is determined by the local solute concentration and is actually the true freezing temperature for that location. Therefore, the rough freezing front is due to the tendency of the ice crystals to grow faster at the tip, where the constitutional supercooling is lowest, than at the edge or base. Their length does not extend infinitely in the solution because the latent heat of solidification must still be conducted away from the freezing front back down to the cooled surface.

As more water is solidified, the ice crystals grow both in length and width, rejecting solute which accumulate in the intercrystalline space. Depending on the temperature, part of the concentrated solute (also referred to as *cryoconcentrate* or *freeze-concentrate*) "solidifies", either because it forms an eutectic (as is the case for sodium chloride, for example), or because it is cooled below the glass temperature (as is the case for sugar solutions, for example).

The reason ice formed from aqueous solutions appears opaque is due to light scattering caused by the difference in refractive indices between the crystalline water ice phase and the cryoconcentrate phase.

5. Often times, solutions that have been frozen from bottom up exhibit a top "skin" which can appear glossy, crystalline, or powdery. This thin layer is formed from solute concentrate that has been squeezed out of the water ice intercrystalline space due to the expansion of water upon solidification.

30.2.6 Freezing from Side

When freezing occurs at a vertical cooling surface, the sequence of events is similar to that depicted in Fig. 30.2. However, in this case, thermal and constitutional convective flows (20) become more prevalent and can affect the dynamics of freezing. Thermal convection loops bring warmer liquid to the freezing front, slowing ice growth at that location while constitutional convective flow, illustrated in Fig. 30.3, influences the way the solute is distributed in the final frozen mass.

The situation depicted in Fig. 30.3 corresponds roughly to frame (d) of Fig. 30.2 in the freezing sequence. At some point after the establishment of the rough freezing front, solute that has been rejected into the ice intercrystalline space (*microscopic cryoconcentrate*) becomes sufficiently concentrated and dense (but not to the point of solidification), as to drain out of the porous frozen mass, to be replaced by fluid of lower density drawn in from the bulk unfrozen solution. The denser cryoconcentrate pools at the bottom of the freezing front (18,21,22). As a result, on a macroscopic level, the final frozen mass will not have uniformly distributed solute throughout. The regions where *macroscopic cryoconcentrate* accumulate, will have, on a volume basis, greater solute content than elsewhere. This macroscopic solute distribution is observable by taking ice core samples in the final frozen mass (23).

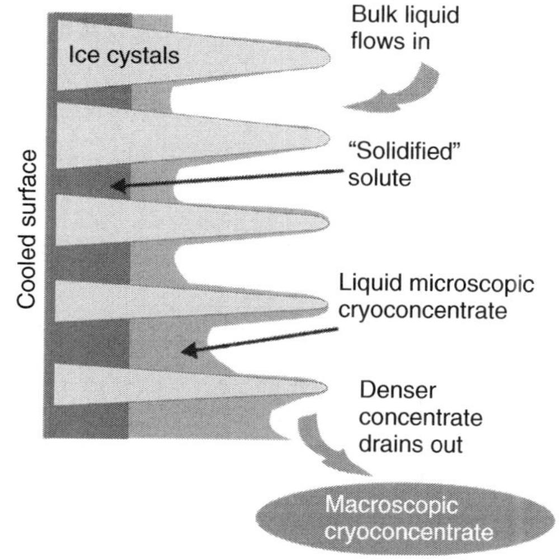

Figure 30.3. Illustration of freezing from the side. Mechanism of constitutional convection. (This figure is available in full color at http://onlinelibrary.wiley.com/book/10.1002/9780470054581.)

Note that macroscopic cryoconcentration does not occur readily in the previously discussed case of freezing from below since the dense cryoconcentrate cannot easily flow to areas that are already frozen. Furthermore, the absolute extent of solute concentration (after water has been converted to ice) is not necessarily greater in the macroscopic cryoconcentrate pool. However, for multicomponent systems, the ratio of solutes may be different in different parts of the final frozen mass. These issues will be discussed in greater details in subsequent sections.

Figure 30.4 shows the combined effects of convective flow in a sequence of photographs taken during an experiment using an apparatus designed to freeze from the left side only. As it can be seen, the freezing front, which is moving from left to right, is not vertical but rather pear shaped. The ice growth at the top is hindered by warmer fluid brought from the right by a thermal convection cell. At the bottom, the cryoconcentrate draining from the frozen mass lowers the local freezing point resulting in the curvature in the freezing front there. Food dye was added

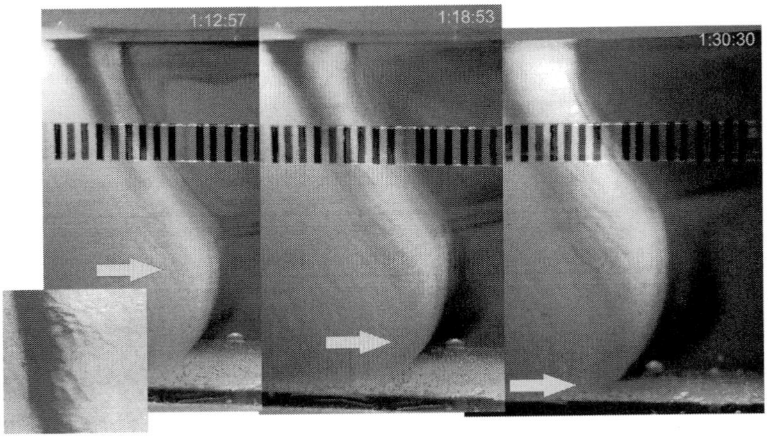

Figure 30.4. Freezing from the side of a 20% wt. of arginine chloride. Sequence of images showing the effects of thermal and constitutional convective flow on the ice front. Food dye was added to improve image contrast. Inset shows details of the "drainage channels". Arrows indicate the progression of the red dye in the ice mass. (This figure is available in full color at http://onlinelibrary.wiley.com/book/10.1002/9780470054581.)

to enhance contrast and accentuates the cryoconcentrate "drainage" channels visible in the frozen mass.

Solute motion through ice is a well known phenomenon for sea ice (24) and has even been exploited for desalination purposes (25).

30.2.7 Freezing from Above

Freezing from above, such as in the case of oceanic ice sheets, involves much of the same physics as described previously. When heat removal medium is air, the dendritic ice crystals form in many orientations and the frozen layer is not cohesive until a substantial thickness is attained. Similar to the previous case, cryoconcentrated solutes remain mobile in the ice matrix and percolate out of the frozen mass through channels (26,27). For most practical application where the material is contained in vessels of standard shapes, heat transfer from above represents only a small fraction of the total freezing heat load.

30.2.8 Impact of Freezing on Protein Stability

There are several implications to protein stability resulting from the formation of the freeze-concentrate including redistribution of solutes, altered reaction rates and increased viscosity, and the induction of solute crystallization.

30.2.9 Freeze-Concentration and Solute Redistribution

The distribution of solutes during the freezing process is related to several factors including proximity to ice crystals, convective flow, and formulation composition. The gradient that forms at the ice-interface will depend on solubility, diffusion, and surface tension of the solutes present.

A study was conducted to monitor the concentration changes during freezing in a manufacturing scale 300 L freeze-thaw vessel (discussed in a later section on Manufacturing scale equipment). Liquid samples were taken at the lowest possible location inside the vessel using a custom positive displacement aspirating apparatus. The samples were then analyzed for concentration by measuring the osmolality and protein concentration. The intent of these experiments was to probe the extent of macroscopic cryoconcentration during freezing. Figure 30.5a and b shows the result for two different mAbs and their respective placebos. The effect of macroscopic cyroconcentration over time is displayed for both a sugar–based (trehalose) formulation, mAb A, and salt-based (NaCl) formulation, mAb B. Note that the ordinate (y-axis) scale for Fig. 30.5b is 10 $\times$ that of Fig. 30.5a.

The data of Fig. 30.5 demonstrates that in the salt-based formulation there is a larger degree of cryoconcentration observed. This is due to the higher mobility of the salt-based formulation even at higher protein concentrations. The trehalose formulation likely became more viscous with increasing concentrations and lower temperatures, restricting its mobility. Interestingly, the presence of protein significantly lowers the observed macroscopic cryoconcentration (Fig. 30.5b) suggesting that protein concentration has mobility reducing effects as does the sugar excipient.

Furthermore, the data also reveals that the relative ratio of excipients to protein is not constant over the course of the freezing. This may be due to the very different mobilities between the large protein molecules and the small excipients, but the individual diffusion rates for the molecules

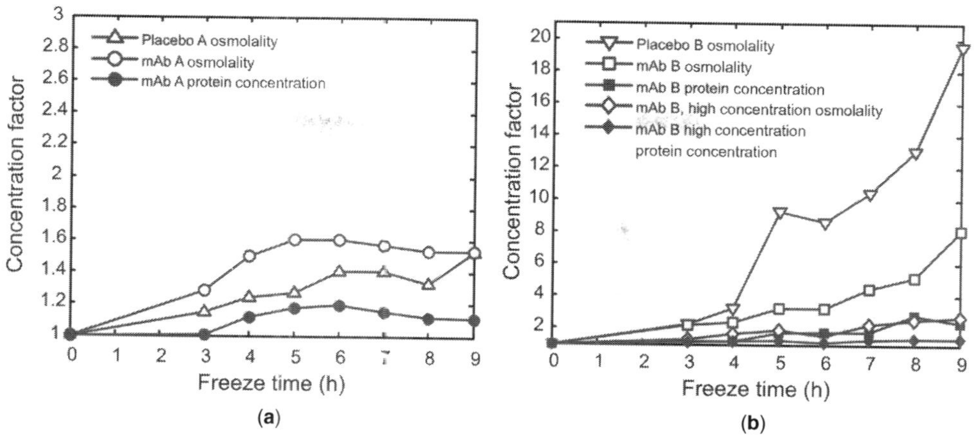

Figure 30.5. Estimating the extent of macroscopic cryoconcentration. Results from at-scale experiments conducted in a 300 L freeze-thaw vessel with two monoclonal antibodies. (a) mAb A at 25 mg/mL, in 60 m*M* trehalose, and 5 m*M* histidine–histidine. HCl buffer and (b) mAb B at 25 mg/mL, in 150 m*M* NaCl, and 25 m*M* citrate buffer. The higher concentration mAb B is at ~ 55 mg/mL in the same buffer/salt formulation as the lower concentration. Note the *y*-scale on pane (b) is 10 × that of pane (a). (This figure is available in full color at http://onlinelibrary.wiley.com/book/10.1002/9780470054581.)

are difficult to ascertain because the exact local conditions (composition, concentration, temperature) experienced by the solutes are not known. This behavior has also been observed at smaller scales (23). Depending on the magnitudes of the changes, proteins could be deprived of stabilizing agents needed for adequate stabilization or conversely, protein damage may be induced by an excess of excipients such as salts and buffer salts.

30.2.10 Freeze-Concentration and Altered Reaction Rates

Lower temperatures, cyroconcentration and solute redistribution during the freezing process leads to increased viscosity and therefore reduced molecular mobility. Lowered temperatures also reduce reaction rates for both physical and chemical reactions. But there are competing elements: while higher concentrations can enhance physical degradation rates for bimolecular reactions, the reduced molecular mobility and low temperatures both decreases reaction rates. It is assumed that, in general, chemical reaction rates are independent of concentration and overall rates will be reduced based on lower temperatures and diminished molecular mobility. For physical reactions, it is believed that there is some partial coupling of reaction rates that can potentially amplify the overall physical degradation rate (28,29).

30.2.11 Freeze-Concentration and Solute Crystallization

Another consequence of cyroconcentration is the supersaturation of additives that may induce crystallization. There are several stresses resulting from excipient crystallization that may impact protein stability. It has been well documented that the crystallization of buffer components such as sodium phosphate and potassium phosphate triggers large pH shifts in the frozen system (30–33). In sodium phosphate buffers, the dibasic form is less soluble than the monobasic salt and therefore more susceptible to solubility issues leading to a drop in pH of the overall frozen system. In potassium phosphate buffers, an increase in pH has been observed due to the precipitation of the monobasic salt. Given that proteins stability is strongly dependent on pH, crystallization of buffer salts can have deleterious affects on protein stability.

Other excipients prone to supersaturation and subsequent crystallization during freezing include glycine, mannitol, sorbitol, raffinose, trehalose, and polyethylene glycol. Crystallization of these additives can lead to protein instability such as aggregation. Sorbitol crystallization led to protein aggregation of an Fc-Fusion protein (34) during long-term storage at −30°C. Compromised protein stability was also linked to crystallization of raffinose, mannitol, and trehalose. There are several explanations for this loss of protein stability. Piedmonte *et al* speculated that the reduction in stabilizer concentration, sorbitol, led to protein aggregation. Others have suggested that the additional surface area provided by the crystals could be responsible for surface-induced denaturation (35). The heat of crystallization could also lead to local denaturation. Increases in protein concentration due to solute crystallization may also lead to further protein denaturation.

Regardless of the molecular mechanism for denaturation, preventing excipient crystallization can be achieved by modifying formulation components, process parameters, and storage conditions. Decreasing the concentration of crystallizable excipients and increasing the concentration of other formulation components that are more soluble, including bulking agents and protein can obviate the crystallization event. In addition, modifying the freezing temperature and storage temperature has also been shown to prevent excipient crystallization and maintain protein stability long-term (34).

30.2.12 Ice-Induced Protein Denaturation

The ice-water interface presents another stress to proteins during the freezing process. Using fluorescence, Strambini *et al* monitored conformational changes of azurin, alkaline phosphatase, and liver alcohol dehydrogenase as a function of ice-surface area (36,37). They observed increased perturbations in tertiary structure as the ice-surface area became larger. Studies with phosphofructokinase, lactate dehydrogenase, and glutatmate dehydrogenase also correlated protein denaturation with increasing ice-water interface. Recent studies using infrared microscopy have added to this growing body of literature, providing additional evidence of partial protein unfolding at the ice-water interface (38–40). The mechanism of ice-induced protein denaturation is not fully understood.

However, surfactants, such as polysorbate 80 and polysorbate 20, can mitigate this problem. Several studies have demonstrated that in the absence of surfactants, proteins tend to concentrate at the ice-interface inducing aggregation. In addition, it has also been shown that a higher degree of protein integrity and tertiary structure were retained during the freezing process for solutions that contained surfactants versus simple buffer solutions (40–42).

30.2.13 Protein Stabilization/Destabilization During Freezing

There are many stresses that impact protein stability during the freeze process. This includes exposure to cold temperatures, freeze concentration, and ice-formation. Since it is difficult to study these stresses individually and attribute protein damage to any one phenomenon, it is recommended that one empirically assesses protein stability during formulation development to identify buffer and solution conditions that will provide adequate protein stabilization during freezing. This may include several stabilizers and known cyroprotectants.

Sugars such as sucrose and trehalose are one class of stabilizers that provide cryoprotection. Although there is much debate in the literature about the true mechanism of stabilization, it has been suggested that these cryoprotectants provide both thermodynamic and kinetic stabilization. In liquid formulations, it is postulated that sugars shift the equilibrium from the unfolded nonnative state toward the native state through preferential exclusion also referred to as *preferential hydration*. This essentially means that in the unfolded state with more surface area and contact with the solvent, there is greater preferential solute exclusion and therefore a greater increase in the free energy of the unfolded state compared to the native state. This shifts the thermodynamic equilibrium in favor of the native folded protein. This same theory has been extended to the frozen state. However, kinetic contributions are also likely. As solutions are frozen, the freeze-concentrate and lower temperatures increases viscosity and reduces mobility which subsequently slows all physical and chemical reactions. Recent studies investigating the impact of stabilizers on cold denaturation demonstrated that both the cold denaturation temperature and the kinetics of unfolding were significantly altered, supporting the idea that both thermodynamic and kinetic stabilization are important (13,14).

A second class of cryoprotectants is surfactants. Surfactants such as polysorbate 80 and polysorbate 20 are known for their ability to protect proteins against surface-induced denaturation (43). This same property of surfactants appears to also prevent ice-induced denaturation as demonstrated with phosphofructokinase, lactate dehydrogenase, and glutatmate dehydrogenase (44).

Contrary to sugars and surfactants, there are compounds that are known to destabilize proteins. These include commonly used antimicrobial preservatives such as phenol, methylparaben, and m-cresol. While in the liquid state, these substances are in sufficiently low concentrations so as to not significantly impact protein stability over the shelf-life of the product. However, during freezing, the cryoconcentration effect increases solute concentration to levels that cause protein aggregation (45). Hence, generally, protein pharmaceutical liquid products that contain these types of preservatives cannot be frozen.

30.2.14 Rate of Freezing

Freezing rates are commonly reported as an advancing distance over time of the observed freezing front (cm/min, mm/min etc...). However, as discussed in previous sections, for aqueous solutions the freezing front and the frozen mass are not uniform, so care must be exercised when interpreting results and drawing conclusions based on freezing rate data.

Since freezing is essentially a heat transfer process, freezing rate is proportional to the temperature driving force ($\Delta T/\Delta x$) between the freezing interface and the periphery of the vessel containing the liquid where heat is removed from the system. As freezing proceeds the latent heat of solidification must be conducted from the freezing front to the cooled surface through an increasing thickness of ice/solute layer which presents resistance

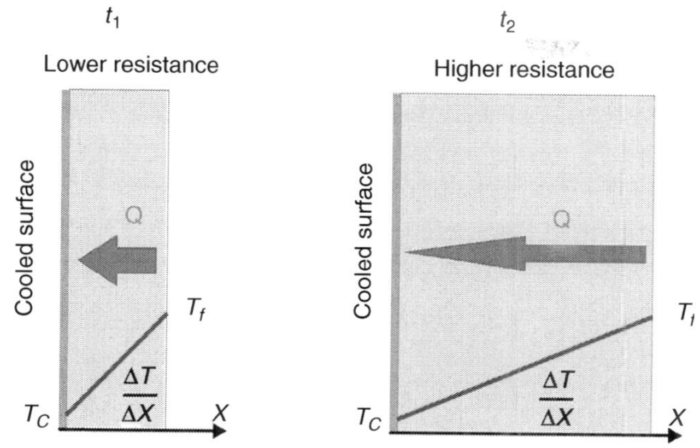

Figure 30.6. Illustration of heat transfer driving force and resistance. Resistance to heat transfer increases as more ice is formed. (This figure is available in full color at http://onlinelibrary.wiley.com/book/10.1002/9780470054581.)

to heat transfer (as illustrated in Fig. 30.6). Hence, for a fixed temperature at the cooling surface, the rate of freezing changes throughout the freezing process, faster at the onset and slower toward the end. The difference in the initial and final freezing rate is strongly dependent on the geometry of the vessel and the location of the cooling surfaces.

It is well known that ice crystal size is a function of crystal growth rate (46). The faster the rate of freezing, the smaller the crystals and vice versa. However, once formed, fine ice crystals may coalesce into larger ones over time, depending on the temperature. This "annealing" phenomenon has been exploited in lyophilization processes as a means to decrease the water vapor mass transfer resistance during the primary drying sublimation step (47). Smaller ice crystals also imply smaller intercrystalline space which is more resistive to bulk flow. Hence, fast frozen solutions tend to be less affected by constitutional convective flows as compared with solutions frozen slowly.

It is important to note that generally the absolute extent of cryoconcentration is a function of temperature and composition but not of freeze rate (for equilibrium freezing conditions) or initial concentration (provided that the relative proportion of solutes does not vary). That is, the cryoconcentrate is in thermodynamic equilibrium with the frozen water; the solute concentration in the liquid fraction is not dependent on the size of the ice crystals. For a given final temperature, a rapidly frozen solution has many small ice crystals in equilibrium with many small cryoconcentrate regions while a solution frozen slowly has fewer but large ice crystals in equilibrium with fewer but larger cryoconcentrate regions (see Fig. 30.7). The concentration of the solute in the cryoconcentrate is the same in both cases.

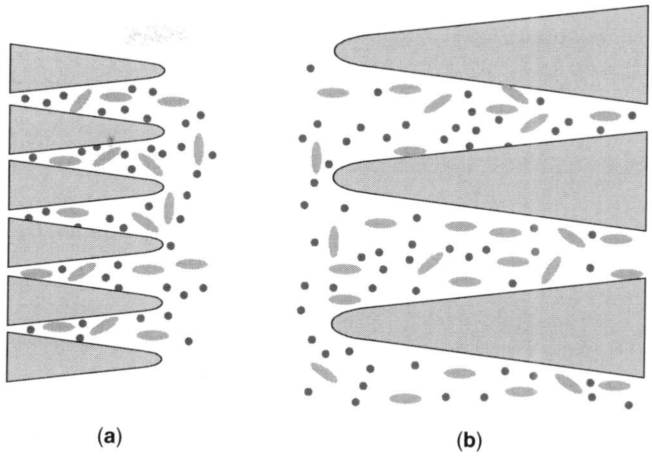

Figure 30.7. Illustration of ice crystal size. Freezing rate affect the size of the ice crystals. (a) faster rates produce smaller ice crystals with smaller cryoconcentrate regions; (b) slower freezing rate result in large ice crystals and cryoconcentrate regions but for a given temperature, the amount of cryoconcentration is the same in both cases. (This figure is available in full color at http://onlinelibrary.wiley.com/book/10.1002/9780470054581.)

This is illustrated in the example using the familiar sodium chloride-water solution. Figure 30.8, shows the phase diagram for this binary mixture. Starting at point A, a 5 wt% solution initially at 20°C is to be cooled to −30°C.

In segment [AB], the liquid is simply cooled by removal of the sensible heat until point B, at −3°C, is reached. Point B is the normal freezing point for this solution where, after ice is nucleated, a pure water ice phase is in equilibrium with liquid phase of 5 wt% sodium chloride. This is roughly equivalent to the conditions shown in Fig. 30.2b. If the temperature of the cooling surface is simply held at −3°C, no more ice forms. If the temperature of the cooling surface is lowered either gradually or rapidly to −30°C then freezing proceeds along paths [BD] (roughly equivalent to the illustration of Fig. 30.2c), as more water is sequestered in the ice phase, the concentration of sodium chloride in the liquid phase increases. The NaCl-water mixture forms an eutectic at −21°C (point D); above this temperature the ice phase coexist with liquid NaCl-water (brine), below −21°C the ice phase coexist with another solid phase formed by sodium chloride dihydrate (shown in Fig. 30.2d as "solidified solute"). Beyond point D, no more composition change occurs, further lowering of the temperature simply cools the two existing solid phases to points E and F.

The differences in the freezing rates are reflected in the size of the ice crystals and the time duration spent in the states of Fig. 30.2c to Fig. 30.2d. With the gradual cooling scheme, more time will be spent in the "slushy" state (Fig. 30.2c), whereas in rapid cooling the system will transition to the state of Fig. 30.2d quickly. Note that the above

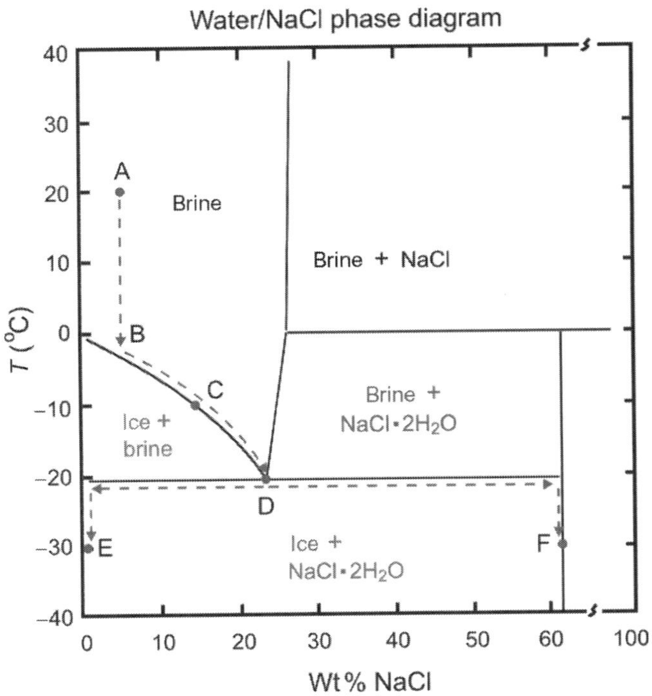

Figure 30.8. Phase diagram of sodium chloride–water system. Cooling of a 5% wt. NaCl solution from 20 to−30°C follow the outlined path. Figure is for illustration purpose, the values may be approximate. (This figure is available in full color at http://onlinelibrary.wiley.com/book/10.1002/9780470054581.)

discussion is valid for local equilibrium. The contents of an actual freezing vessel experience all these conditions simultaneously, depending on the location.

For simple salts systems, the rate of freezing is inconsequential. However, for more complex mixtures involving salts, buffer components, excipients and proteins, the actual freezing process may impact protein stability.

30.2.15 Impact of Freeze-Rates on Protein Stability

The stresses induced by variable freeze-rates include ice-formation, differential ice-surface area, and cold temperatures. Most of these stresses were discussed in earlier sections.

Rapid freezing rates results in numerous, small ice crystals, creating a larger ice-aqueous interfacial surface area. Increased ice-aqueous surface area has been correlated with protein conformational changes and increased protein aggregation. Studies by Cao *et al* demonstrated that at faster freezing rates (>5°C/min) and lower thawing rates (<10°C/min), protein denaturation increased and recovery of active proteins decreased; conversely, slower freezing rates and faster thawing rates, resulted in increased active protein recovery (48). Similarly, Hsu *et al*. describe the aggregation of proteins following rapid freezing and suggest that protein molecules that come into contact with the interfacial ice surface may unfold and aggregate (41).

Increased ice-aqueous surface area may also provide additional nucleation sites that can stimulate the crystallization of metastable additives. This crystallization event can then in turn lead to protein aggregation.

With an optimal protein formulation, the impact of freeze-rates on protein stability can be greatly reduced. A study conducted with mAb1, formulated in 5 m*M* Histidine-HCl, pH 6.0, 60 m*M* trehalose, and 0.01% polysorbate 20, investigated both short-term and long-term affects to protein stability as a function of varying freeze-rates including slow (~100 hours), medium (few hours), and fast (few seconds) freezing. To enable a slow freeze, samples were frozen by placing protein sample vials in a lyophilizer on precooled shelves and then holding at −1°C for 24 hours. Ice was nucleated by applying Dry Ice® to the side of each vial until ice-formation was observed and then the temperature was linearly ramped to −40°C over 144 hours. A medium freeze rate was established by directly placing protein sample vials in a −20°C freezer. Ice-nucleation was controlled to prevent super cooling and to ensure that all samples experienced the same freezing rates. Finally, fast freezing rates were induced by quench cooling 50 μl sample aliquots drop-wise into liquid nitrogen. After protein samples were frozen, protein quality was monitored over time (Fig. 30.9). There was no appreciable increase in soluble or insoluble aggregates (data not shown) as a function of freeze-rates. In addition, regardless of freeze-method, protein stability was maintained after storage at −20°C for up to 12 months. The results of this study demonstrate that stabilizers such as trehalose used in proper ratios, and additives like polysorbate 20 can protect proteins from stresses experienced during the freezing process and negate the affects that may arise from varying freeze-rates.

Table 30.2 represents some additives/excipients that are commonly used in protein formulations. This list is not comprehensive but provides a good starting point for development studies. Proteins behave unpredictably, even related molecules, differing in only a few residues, can exhibit very different characteristics (physical or chemical stability, viscosity) under the same conditions. Therefore, the results generated from experiments that used a surrogate "model" protein in lieu of the active interest must be carefully scrutinized.

30.3 THAWING OF SOLUTIONS

30.3.1 Thawing Point (fusion) Temperature

A pure substance has a well defined fusion temperature which coincides with the solidification temperature. For frozen solutions, the thawing point is neither coincident

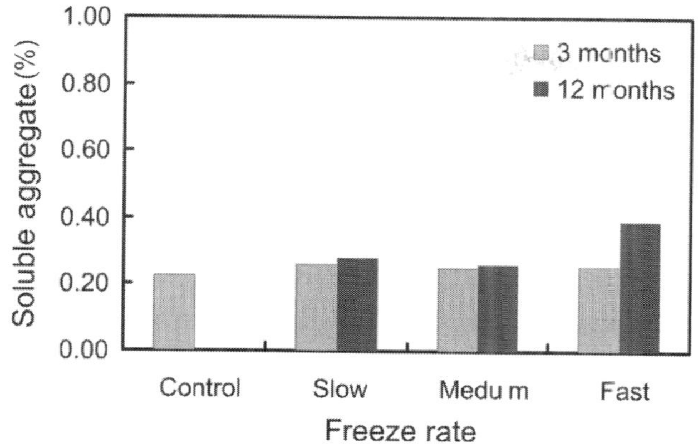

Figure 30.9. Impact of variable freeze-rates on long-term protein stability. mAb A, formulated in 5 m*M* Histidine-HCl, pH 6.0, 60 m*M* trehalose, 0.01% polysorbate, was frozen at three different rates then placed at −20°C for long-term storage. Product quality was monitored by measuring percent soluble aggregate using size-exclusion chromatography. (This figure is available in full color at http://onlinelibrary.wiley.com/book/10.1002/9780470054581.)

TABLE 30.2. Common Formulation Components

Components	Function	Notes
Sodium chloride	Tonicifier, stabilizer	Forms eutectic at −21°C
Potassium chloride	Tonicifier, stabilizer	
Sucrose	Tonicifier, cryo protectant, stabiliizer	
Trehalose	Tonicifier, cryo protectant, stabiliizer	
Mannitol	Tonicifier, bulking agent	
Glycine	Stabilizer, bulking agent	
Polysorbate 20, 80	Surfactant	
Mono sodium phosphate	Buffer	May crystallize leading to pH shifts
Di sodium phosphate		
Phosphoric acid		
Sodium citrate	Buffer	
Citric acid		
Histidine-HCl	Buffer	
Histidine Acetate		
Sodium acetate	Buffer	

with the freezing point nor is it well defined. Furthermore, while a liquid solution is uniform in composition at the onset of freeze, a frozen solution is inherently inhomogeneous, composed of at least two distinct phases (17).

The practical significance of the thaw temperature of a solution is to determine proper storage conditions as it is normally desired to maintain the frozen material below this value if possible.

The extent and temperature at which the solute phase begins to "liquefy" depends on the concentration and composition of that phase. For systems involving more than two components, there is a strong possibility that more than two phases are present, each with distinct characteristics.

Theoretically, it should be possible to estimate the onset of thaw using calorimetry (49) since phase changes are normally accompanied by thermal energy exchange between the samples and the surroundings. Figure 30.10 shows differential scanning calorimetry (DSC) thermograms for several substances including water, a mAb solution, and a 5 wt% sodium chloride solution. For pure water ice (curve a), the thaw onset temperature, is easily determined by drawing tangent lines in front of the sharp thermal transition peak. In the case of a protein solution (curve b), there is a gradual rise in the baseline leading to the transition peak. Consequently, the location of the thaw onset point is more subjective since it is unclear as to where the tangent line should be drawn. This is a refection of the inhomogeneity of the frozen mass where the different phases or regions begin to thaw at different temperatures. In the sodium chloride-water binary system (curve c), an eutectic melt peak is also present before the main ice thaw event. For such cases, there are some debates about where the thaw onset should be set, whether it is in front of the eutectic or the ice melt peak. Note that the location of the peak maximum is different for each of the examples shown. However, peak maximum location not only depends on composition but on the DSC instrument heating rate setting as well, complicating data analysis.

For solutions that form amorphous phases, the use of modulated DSC help can identify the glass transition temperature (*Tg'*), if one is present, in the temperature range of interest. Glass transition events are relatively weak as compared to the water melting peak and are difficult to observe on a thermogram. It is believed that a substance that is below its *Tg'* has significantly diminished molecular mobility but the magnitude of molecular mobility change from pre-*Tg'* to post-*Tg'* is not well characterized.

30.3.2 Physics of Thawing of Solutions

The thawing of a solution is not the inverse of freezing because the starting conditions are very different (17). A frozen solution mass is composed of different regions on microscopic and macroscopic scales. These regions have different properties and therefore, behave differently during the thaw process. At the beginning of the thaw, heat is transferred into the frozen mass by conduction. Depending on the rate of heat transfer (temperature driving force), thermal

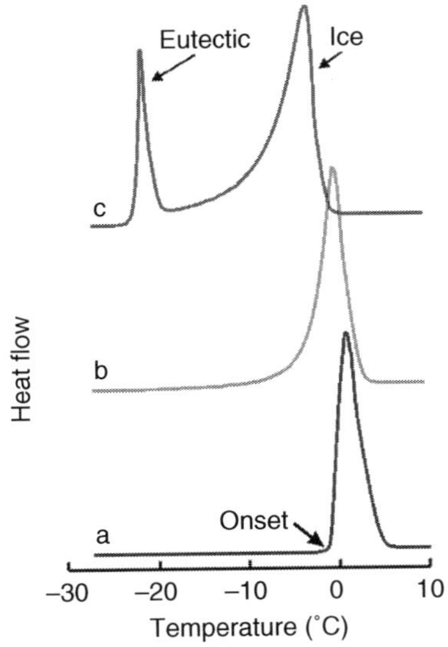

Figure 30.10. DSC thermograms for several solutions and onset of thaw temperature. (a) Water; (b) A therapeutic antibody solution with 25 mg/mL protein; (c) 5% wt. NaCl in water. (This figure is available in full color at http://onlinelibrary.wiley.com/book/10.1002/9780470054581.)

convection heat transfer may become dominant when sufficient amount of material has been liquefied (50).

30.3.3 Thawing from Sides and Bottom

In practice, the majority of the heat transfer during thaw occurs from the sides and bottom of a container. Since the resulting thaw dynamics are similar for these two directions of heat input, this section will discuss both together. Figure 30.11 shows the main sequence of event during thaw in the absence of external agitation:

1. Heat is conducted into the frozen mass from the sides and bottom of the container, warming the contents. The solute cryoconcentrate phase begins to be mobile.
2. The regions adjacent to the container walls are liquefied, the more concentrated and denser solute collects at the bottom of the container. Thermal convective flow becomes more prominent as the layer of liquid near the walls is heated.
3. Thaw is well underway, thermal convection now carries the majority of the heat from the container walls to the thawing frozen mass which is now "slushy". Since the ice block is porous, the denser cryoconcentrate drains out and pools at the bottom, creating additional constitutional driven convective flow.
4. Thaw is complete, but the liquid is not uniform in composition. A concentration gradient exists as the solute at the bottom of the container is more concentrated than at the top. This stratification can persist for an extended period of time unless mechanical agitation is applied to mix the contents.

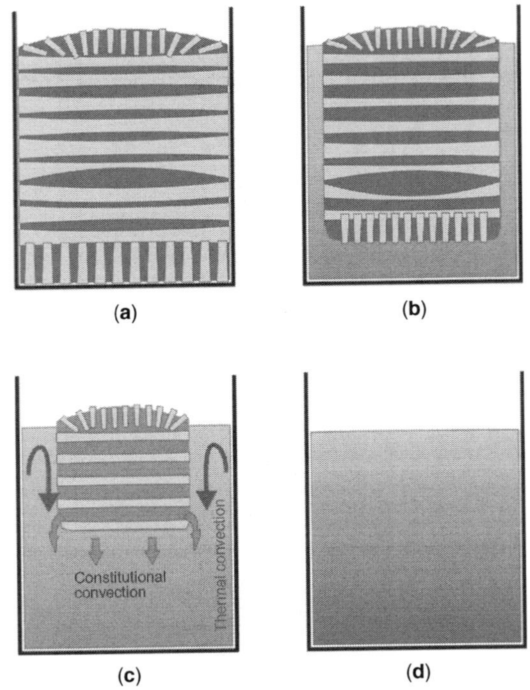

Figure 30.11. Illustration of thawing from bottom and sides. (a) conduction heat transfer dominates at the beginning of thaw; (b) sufficient liquid is present for thermal convection to occur; (c) both thermal and constitutional convection are present; (d) thaw is completed but there is a significant solute concentration gradient present in the liquid. (This figure is available in full color at http://onlinelibrary.wiley.com/book/10.1002/9780470054581.)

30.3.4 Thawing from the Above

This direction of heat transfer is not important for practical industrial processes since the efficiency is low. Warmer material stratifies toward the top of the container and does not create any significant thermal convective flows.

30.3.5 Rate of Thawing

Similar to freezing, temperature differential provides the driving force and controls the rate of thaw. However, because thermal convection greatly enhances the heat transfer rate (51), only moderate temperatures (10 to 20°C) at the vessel walls are needed to achieve efficient thawing. Lower temperatures (5°C) can be used but the thaw time will be greatly lengthened due to thermal convection being much weaker under these conditions (52).

Thermal convective flows can also provide some additional mixing. Containers that have been thawed faster at moderate temperatures typically have been found to be more homogenous than those thawed slower under low temperatures.

30.3.6 Effects of Thawing on Protein Stability

During the thaw process, protein solutions can experience exposure to cold temperatures (<0°C) and concentration gradients. At slower thaw rates, both the duration at lower temperatures and the nonuniformity of solution would be exacerbated. There are several examples demonstrating that slower thaw rates led to reduced enzyme activity (39,48). As discussed earlier, prolonged low temperature exposure can lead to cold denaturation or liquid–liquid phase separation. However, it has been demonstrated that stabilizers and other formulation additives can greatly reduce the risk of both of these phenomena (13,14). The affect of heterogeneous solutions post-thaw could expose proteins to different formulation conditions that may impact protein stability such as varying ionic strength and pH conditions. The exposure to these varying formulation conditions and increased protein concentration could promote protein–protein interactions such as aggregation (See more detailed explanation in Physics of Freezing section).

Another consideration during thawing is exposure to heat. Depending on the container and volume, faster thawing rates could unnecessarily expose protein solutions to higher temperature which could in turn lead to protein aggregation. For example, a case study evaluating different thaw parameters observed increased levels of aggregation in IgG2 samples that were thawed rapidly compared to IgG2 samples that were thawed slowly (53). On the basis of control samples it was determined that the increase of aggregation was due to the higher thaw temperature (55°C) and not the result of faster thaw rate.

30.3.7 Mixing

After thaw, the solution generally shows stratification where the bottom layer is more concentrated than the surface layer of the liquid. For a container size of a few liters or less, gently inverting the bottle end to end for a couple of cycles is usually sufficient to achieve mixing. Larger containers may require mechanical agitation such as the use of an orbital shaker. It is also possible to agitate the vessel during the thaw operation which provides mixing as well as an increase in convective heat transfer efficiency, hence, decreasing thaw time. For large vessels, it may be more practical to use a pump, withdrawing the denser liquid from the bottom and reintroducing it near the top via a special port that allows the fluid to be run along a dip tube or the vessel wall to minimize potential foaming. Mixing is achieved by the turn over of fluid volume and gentle impact of the pumped liquid returned to the surface of the liquid in the vessel. The benefit of this technique is that mixing can be started as soon as there is sufficient liquid to operate the pump and carried out concurrent to thaw, thereby achieving a mostly homogeneous vessel content by the end of the thaw operation and reducing the risk of protein instability.

30.4 FREEZE-THAW SCALE UP

30.4.1 Container Scale Up

For a rectangular slab with cooled surface(s) at the large face(s) such as the case of a tray, it is straight forward to scale up to larger volumes by increasing the width and the length, keeping the thickness constant. Since the heat transfer distance is fixed, the freeze rate is independent of volume. Unfortunately, for containers of other shapes such as cylinders, the freeze and thaw processes are scale and geometry dependent. Given the same temperature driving force, a larger vessel will always require a longer time to complete a freeze or thaw operation as compared with one that is smaller (of similar shape and aspect ratio). However, operation time is not linearly related to volume. For example, if a 1 L vessel can be fully frozen in 2 h, a 10 L vessel does not necessarily require 20 h to complete the same operation under the same conditions. This is a vexing problem for biopharmaceutical manufacturing operations, not only in terms of equipment design but more importantly, in terms of process characterization.

The high value of biologicals prohibit using actual product for conducting extensive at-scale process testing and so small-scale "models" are used. For freeze and thaw operations, at-scale dynamics can not be exactly duplicated using small-scale devices since substance properties such as heat capacity and thermal conductivity are intrinsic and scale independent (the bulk properties are identical at 1 mL or at 1000 L). For example, for freezing, it is not possible to mimic both the freezing rate and the freezing temperature of at-scale vessels with a simple small container in the same experiment. Given a temperature set point, the small-scale vessel will freeze at a faster rate; conversely, for a given freeze rate, the small-scale vessel will require a higher temperature set point to maintain the same freeze rate as the at-scale case. One possible solution to this dilemma is to design small-scale studies that bracket conditions that are encountered at-scale by conducting separate experiments that deliberately freeze very slowly and very rapidly.

30.4.2 Manufacturing Scale Equipment

Samples of up to 20 L can be frozen in standard off the shelf polymeric containers. Bottles made from polypropylene, polycarbonate, or PETG (glycol-modified polyethylene terephthalate) are routinely used as these materials remain sufficiently ductile under typical storage conditions. In addition, the freezing operation can be carried out in normal stand alone, walk-in, or blast freezers. Containers can be frozen at the lower temperatures (e.g. −50°C) to reduce the freezing time and subsequently moved to the higher temperature storage freezer (e.g. −25°C) if desired. The freezing of samples in conventional or walk-in freezers is said to be *nondirectional* or *uncontrolled*. In actuality, freezing is always directional since heat must travel from warm to cold, but the principal direction of heat flow may change unpredictably throughout the process because the convective heat transfer field is not well defined in these freezers. By contrast blast freezers, which are widely used in the frozen food industry, can typically achieve much better freezing rates and maintain directionality due to their strong convective flow field.

While plastic containers have the advantage of being inexpensive and disposable, they may not be the best solution when large production batches (hundreds or thousands of liters) are involved. For example, antibody therapeutics typically require large doses to be efficacious; hence, these compounds are routinely produced in batches of 300 –1000 L. A minimum of 150 containers would be needed if a 300 L batch is to be stored in 2 L bottles, creating a challenging batch tracking problem. Using larger volume carboys would reduce the number of containers but handling would be more difficult and the freezing time would also increase significantly. To meet these manufacturing challenges, specialized equipment have been developed (54). These consist of a dedicated skid that handles the heat transfer to which reusable portable metal vessels of various sizes can be attached for the freeze/thaw operations.

The vessels are essentially jacketed pressure-rated tanks with outer insulation. Cooling coils inserted through the top provide additional heat transfer surfaces to the center of the volume. Some modest increase in heat transfer efficiency may be achieved by attaching fins to the cooling coils. Figure 30.12 shows a photograph of such vessels with capacities of 120 L (left) and 300 L (right). These freeze-thaw tanks have stainless steel or higher alloy product contacting surfaces, which are designed to be clean-in-place and steam sterilized-in-place. Once filled and frozen, the vessels and their contents can be moved to a large freezers and stored until needed. They have been designed to accommodate standard commercial shipping methods such as refrigerated trucks or insulated air cargo containers.

The freeze-thaw skid circulates either cooled or heated heat transfer fluid (HTF) through the vessel's jacket and coils to conduct freezing or thawing. The temperature can be controlled via the skid's automation to provide reproducible freezing or thawing conditions. A typical industrial freeze-thaw unit is capable of freezing a 300 L vessel in less than 20 h. Note that the freezing temperature (e.g. −50°C) is typically significantly lower than the storage temperature (e.g. −25°C). This allows better thermal driving force, reducing the operational freezing time. Thawing can be achieved in about 12 h (with the HTF set at 25°C). Smaller vessels can be processed even faster. Figure 30.13 shows a schematic of a freeze-thaw vessel attached to a skid.

These specialized equipment, freeze-thaw skids and vessels, require significant capital investment. However, these systems provide a robust validated solution for performing freezing, thawing, and long-term storage. The cost of such equipment is justified when very high value products are involved.

There are several vendors that specialize in turn-key freeze-thaw equipment targeted at the biopharmaceutical industry. The European based Zeta Christ Water Technology Group offers manufacturing scale freeze-thaw systems,

(a) (b)

Figure 30.12. Photograph of commercial freeze-thaw vessels. Capacities are 120 L (a) and 300 L (b). The outlines of personnel give a sense of scale. (This figure is available in full color at http://onlinelibrary.wiley.com/book/10.1002/9780470054581.)

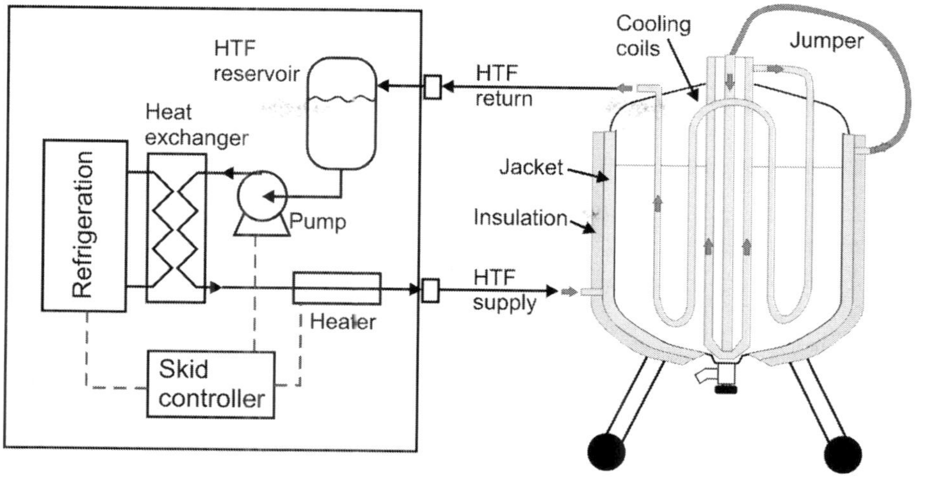

Figure 30.13. Schematic of a freeze-thaw skid with vessel attached. The HTF flow path is indicated by the arrows. (This figure is available in full color at http://onlinelibrary.wiley.com/book/10.1002/9780470054581.)

including skids and vessels of various sizes. An extensive line of freeze-thaw equipment ranging from laboratory (milliliter scale) up to manufacturing process scale (several hundred liters) is also available from Sartorius Stedim Biotech. In addition to vessels and skids, Sartorius Stedim also sells freezable bag systems of for volumes up to 16.5 L. The main advantage of bags as compared with bottles is better freezing times due to a more favorable aspect ratio of a bag. At this time, bag systems capable of handling larger volumes are not available though the technology is being developed (55).

30.5 CONCLUSION

The freezing of solution is complex, involving simultaneous mass, and heat transfer. During the freezing solutes molecules are excluded from the forming ice lattice and concentrate in the liquid phase between ice crystals. This microscopic cryoconcentration phenomenon is unavoidable for all practical processes. Depending on the geometry and freezing rate, macroscopic cryoconcentration may also occur. The thaw process is equally complex and begins with a system that comprises multiple phases. Therefore, mixing is required at the completion of thaw to obtain a homogeneous solution.

Lower temperatures and freeze-concentration increases the viscosity of the frozen solution which limits molecular mobility. In addition, frozen storage temperatures typically decreases the rates of most chemical reactions (e.g. oxidation, deamidation, hydrolysis, etc.) and some physical reactions (e.g. aggregation) that affect proteins. For these reasons, liquid formulations are generally frozen and stored at low temperatures ($\leq -20^{\circ}$C) to preserve protein stability. However, there are physical conditions inherent of the freezing and thawing process that can lead to protein degradation. While it is possible to mitigate potential freeze/thaw stresses by selecting proper freezing, storage, and thawing strategies (e.g. freezing/thawing rates, storage temperature), the impact of these stresses can be greatly diminished with robust formulation development. Furthermore, an optimally formulated protein will afford much greater latitude in manufacturing freeze/thaw operations.

REFERENCES

1. MacKenzie AP. Philos Trans R Soc Lond B Biol Sci 1977; 278: 167–189.
2. Gomez G, Pikal MJ, Rodriguez-Hornedo N. Pharm Res 2001; 18: 90–97.
3. Franks F, Hatley RH, Friedman HL. Biophys Chem 1988; 31: 307–315.
4. Jaenicke R. Philos Trans R Soc Lond B Biol Sci 1990; 326: 535–551. discussion 551–3.
5. Franks F, Hatley RHM. Pure Appl Chem 1991; 63: 1367–1380.
6. Privalov PL. Biochem Mol Biol 1990; 25: 281–306.
7. Griko YV, Privalov PL, Sturtevant JM, Venyaminov SY. Proc Natl Acad Sci U S A 1988; 85: 3343–3347.
8. Sanfelice D, *et al*. J Am Chem Soc 2009; 131(33): 11662–11663.
9. Hatley RH, Franks F. FEBS Lett 1989; 257: 171–173.
10. Franks F. Biophysics and biochemistry at low temperatures. Cambridge: Cambridge University Press; 1985.
11. Griko YV, Venyaminov SY, Privalov PL. FEBS Lett 1989; 244: 276–278.
12. Zhang J, Peng X, Jonas A, Jonas J. Biochemistry 1995; 34: 8631–8641.

13. Tang XL, Pikal MJ. Pharm Res 2005; 22(7): 1167–1175.
14. Tang XL, Pikal MJ. Pharm Res 2005; 22(7): 1176–1185.
15. Cromwell MEM, *et al*. Opalescence in antibody formulations is solution critical phenomenon. In: 236th ACS National Meeting. A.o. Papers, Editor. Philadelphia (PA); United States: 2008.
16. Harrison JD, Tiller WA. Ice and snow: properties, processes and applications. Cambridge (MA): Massachusetts Institute of Technology: The M.I.T. Press; 1963.
17. Hayashi Y. In: Cheng KC, Seki N, editors. Freezing and melting heat transfer in engineering: selected topics on ice-water system and welding and casting processes. New York: Hemisphere Publishing Co.; 1991. pp. 491–519.
18. Nagashima K, Furukawa Y. Physica D 2000; 147: 177–186.
19. Nagashima K, Furukawa Y. J Phys Chem B 1997; 101: 6174–6176.
20. Gross GW, Wong PM, Humes K. J Chem Phys 1977; 67: 5264–5274.
21. Nagashima K, Furukawa Y. J Cryst Growth 2002; 237-239: 81–85.
22. Nagashima K, Furukawa Y. Can J Phys 2003; 81: 99–105.
23. Webb SD, *et al*. Biopharm 2002; 15: 2–8.
24. Kingery WD, Goodnow WH. Ice and snow: properties, processes and applications. Cambridge (MA): Massachusetts Institute of Technology: The M.I.T. Press; 1963.
25. Adams CM, French DN, Kingery WD. Ice and snow: properties, processes and applications. Cambridge (MA): Massachusetts Institute of Technology: The M.I.T. Press; 1963.
26. Worster MG, Wettlaufer JS. J Phys Chem B 1997; 101: 6132–6136.
27. Wettlaufer JS, Worster MG, Huppert HE. J Fluid Mech 1997; 344: 291–316.
28. Parker R, Ring SG. Cryo Letters 1995; 16: 197–208.
29. Pikal MJ. Volume 137, Freeze-Drying/Lyophilizationof Pharmaceutical and Biological ProductsMechanisms of protein stabilization during freeze-drying and storage: the relative importance of thermodynamic stabilization and glassy state relaxation dynamics. New York: Marcel Dekker, Inc; 2004.
30. Larsen SS. Arch Pharm Chem Sci 1973; 1: 41–53.
31. Van Den Berg L, Rose D. Arch Biochem Biophys 1959; 81: 319–329.
32. Murase N, Franks F. Biophys Chem 1989; 34: 293–300.
33. Shalaev EY, *et al*. Pharm Sci 2002; 19(2): 195–201.
34. Piedmonte DM, *et al*. Pharm Res 2007; 24: 136–146.
35. Randolph TW. J Pharm Sci 1997; 86: 1198–1203.
36. Strambini GB, Gabellieri E. Biophys J 1996; 70: 971–976.
37. Gabellieri E, Strambini GB. Biophys J 2006; 90: 3239–3245.
38. Chang BS, Kendrick BS, Carpenter JF. J Pharm Sci 1996; 85: 1325–1330.
39. Bhatnagar BS, Bogner RH, Pikal MJ. Pharm Dev Technol 2007; 12: 505–523.
40. Schwegman JJ, Carpenter JF, Nail SL. J Pharm Sci 2009; 98: 3239–3246.
41. Hsu CC, *et al*. Pharm Res 1995; 12: 69–77.
42. Jiang S, Nail SL. Eur J Pharm Biopharm 1998; 45: 249–257.
43. Levine HL, Ransohoff TC, Kawahata RT, McGregor WC. J Parenter Sci Technol 1991; 45: 160–165.
44. Carpenter JF, Crowe JH. Cryobiology 1988; 25: 244–255.
45. Maa YF, Hsu CC. Int J Pharm 1996; 140: 155–168.
46. Woinet B, *et al*. J Food Eng 1998; 35: 395–407.
47. Searles JA, Carpenter JF, Randolph TW. J Pharm Sci 2001; 90: 872–887.
48. Cao E, *et al*. Biotechnol Bioeng 2003; 82: 684–690.
49. Ablett S, Izzard MJ, Lillford PJ. J Chem Soc Faraday Trans 1992; 88: 789–794.
50. Sparrow EM, Patankar SV, Ramadhyani S. J Heat Transf Trans ASME 1977; 99: 520–526.
51. Kemink RG, Sparrow EM. Int J Heat Mass Transf 1981; 24: 1699–1710.
52. Yen Y-C. In: Cheng KC, Seki N, editors. Freezing and meltin heat transfer in engineering: selected topics on ice-water systems and welding and casting processes. New York: Hemisphere Publishing Co.; 1991. pp. 261–314.
53. Kueltzo LA, *et al*. J Pharm Sci 2008; 97(5): 1801–1812.
54. Wisniewsky R, Wu V. in: Avis KE, Wu V, editors. Biotechnology and biopharmaceutical manufacturing, processing, and preservation. Englewood (CO): IHS Health Group; 1996. pp. 7–60.
55. Lam P, Sane S. Biopharm Int 2007; Suppl. Nov: 6–16.

31

MEMBRANE CHROMATOGRAPHY

JOHN PIERACCI AND JÖRG THÖMMES

Department of Process Biochemistry, Biogen Idec, San Diego, California

31.1 INTRODUCTION

The key to the successful development of a chromatographic procedure for the industrial scale purification of proteins is to establish a high productivity method which delivers a product of sufficient purity. Laboratory-scale protocols are usually set up with special regard to resolution, which is maximized by choosing high performance methods based on small adsorbent particles and high pressure equipment. At the process scale, high throughput and process operability are additional considerations. Therefore, small adsorbent particles usually have to be replaced by larger ones upon process scale-up due to pressure-drop limitations at high fluid velocity within large columns. Furthermore, capacity per unit volume of adsorbent becomes an issue, thus requiring the use of porous adsorbents which offer a high internal surface area for product adsorption. When using comparatively large porous adsorbents (d_p 50–200 μm) the process performance is mainly determined by the rate of protein transport to the binding sites situated on the internal surface. As large (protein) molecules are characterized by slow diffusion coefficients in free solution ($D = 1-10/10^{12}$ m^2/s), and since "hindered" transport within a porous structure leads to even lower "effective" diffusion coefficients, protein chromatography in these media has been described as being limited by slow intraparticle transport (1). Large-scale operations therefore seem to be characterized by the compromise between low pressure drop at high fluid velocity and slow mass transport employing larger adsorbent particles.

The driving force for new developments in stationary phase design has been the need to reduce the limitation of intraparticle mass transport. This may be achieved by increasing the influx velocity into the particle by enhancing diffusive transport (2) or by perfusing the stationary phase in order to support (or even replace) diffusive movement of the solutes to the ligands by convection. Perfused particles with very short diffusive pores are one option to improve transport (3). An even more effective way is the use of a continuous stationary phase, where the internal surface is completely accessible by convection. Adsorbents manufactured in this manner may be short in length (L) as the concept of convective solute transport essentially speeds up the sorption process and permits short residence times $t_C(= L/v)$. Sufficient capacity (adsorbent volume) is supplied by units of increased diameter, which has up to now resulted in a membrane design characterized by a low *L/d* ratio. This concept was first realized by the employment of modified microfiltration membranes of variable design (hollow fiber, flat sheet or membrane stack) (4). Therefore, the description *membrane chromatography* is justified despite the fact that a microfiltration membrane is not the only possible stationary phase fulfilling the criteria discussed above. Macroporous monoliths are stationary phases that are cast as a single block into diverse geometries, using standard polymer chemistry techniques. Monoliths cast with a low *L/d* ratio have mass transport properties that should be very similar to those of membranes. Consequently, their application can be considered a subset of membrane chromatography. As such, they will not be discussed separately from membranes in this review. For more detailed review of monoliths and their application in protein purification, the reader is directed to a recent review chapter (5). The increasing importance of membrane adsorbers is evident

Downstream Industrial Biotechnology: Recovery and Purification, First Edition. Edited by Michael C. Flickinger.

by the number of reviews that have been written (6–10). In the following sections, the basic ideas behind membrane chromatography will be elucidated from a conceptual point of view supported by a short theoretical analysis. Existing membrane formats, membrane materials, and ligand chemistries that have been utilized and investigated will be discussed. Possible limitations of the technology will be addressed followed by an attempt to position the membrane adsorber within a train of protein purification methods. Finally, applications of adsorptive membranes at small and pilot scales will be used to demonstrate the large potential of this technology as a highly productive method of purifying proteins on a process scale.

31.2 THE BASIC CONCEPT

31.2.1 Improved Mass Transport

As discussed above, membrane adsorbents try to overcome the diffusion limitation present in porous particles by allowing convective transport of solutes to ligands. This is achieved by attaching ligands to the internal surface of stationary phases which contain mainly through-pores and only few or no dead-end pores, such that solutes are convectively transported to their binding site. Slow diffusion in dead-end pores, as is the case with porous adsorbents, is thus avoided. The situation is depicted schematically in Fig. 31.1. Since a continuous stationary phase is used, packing problems common to particle-based media that necessitate a minimum bed length (in order to neutralize packing irregularities) are circumvented. As a result of these characteristics, adsorbents with low *L*/*d* ratio may be constructed that can achieve high productivity at short residence times due to fast mass transport in a continuous matrix. It is critical that the adsorbate solution is homogeneously distributed over the whole frontal area of the membrane, otherwise flow irregularities will neutralize the advantages mentioned above. This will be discussed below in more detail.

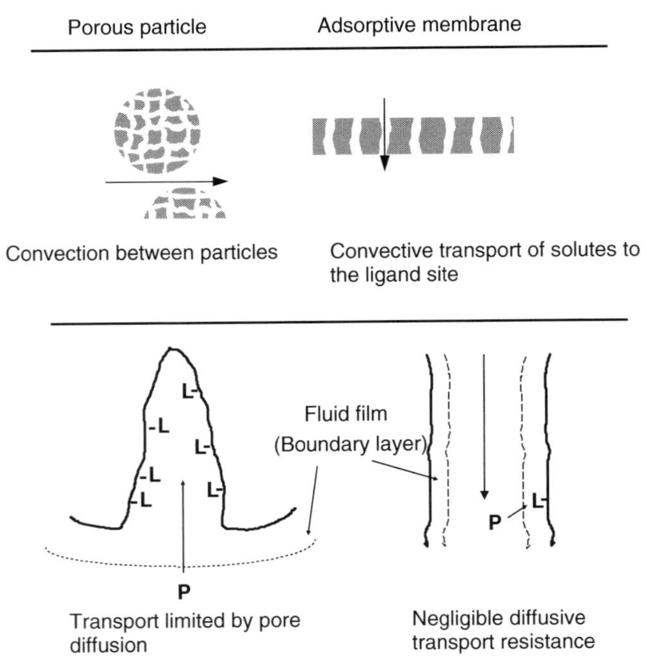

Figure 31.1. Schematic representation of the major transport characteristics of adsorptive membranes and porous particles. P represents the adsorbing particle and L represents the ligand.

31.2.2 Membrane Formats

Membrane adsorbents fulfilling the criteria developed above may be arranged in different formats. As shown in Fig. 31.2, it is possible to differentiate between four basic concepts: single or stacked flat sheet membranes (a), cartridges packed with continuous stationary phases in the form of membranes or rods (b), radial flow cartridges containing spiral-wound membranes (c), and hollow-fiber modules (d).

Single-sheet membranes may be employed in suitable filter holders as long as sufficient care is taken to equally distribute the process liquid across the frontal membrane area,

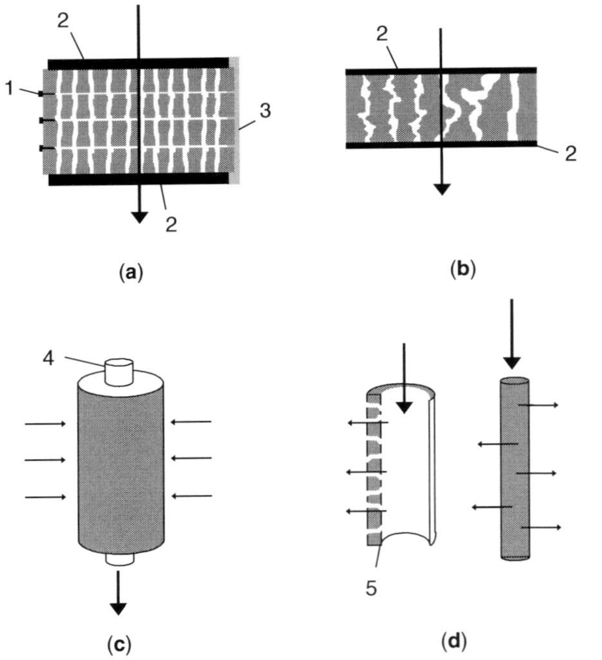

Figure 31.2. Schematic representation of the four major classes of membrane adsorbents: (a) stack of flat sheet membranes; (b) monolith or continuous cartridge with through-pores; (c) radial flow unit with spiral-wound membranes; (d) hollow-fiber membrane; 1 individual gasket; 2 distributor and collector plate; 3 complete sealing around a membrane stack (e.g. silicone); 4 central tube for liquid collection; 5 cross-cut through an individual hollow fiber.

and to efficiently collect the liquid after passage through the membrane. The arrangement of membranes in stacks allows the increase of the capacity of an individual unit as well as an adjustment of the length of the adsorbent bed. The sheets must be separated by proper sealing in order to avoid fluid flow around the membranes instead of through them. This may be achieved by single gaskets or by embedding the stack in a suitable sealing substance (e.g. silicone). Alternatively, a continuous adsorbent packing with thoroughgoing pores (Fig. 31.2b) such as a monolith may be used, thus making the stacking of individual sheets obsolete. Radial flow cartridges contain internal tubing for fluid collection, which is surrounded by a spiral-wound membrane sheet. The liquid is transported in a radial direction from outside to inside; this keeps the column aspect ratio L/d small, as is the case in a flat sheet membrane. The same is true for hollow-fiber modules where the fluid permeates through the membrane, perpendicular to the influx direction. L is defined here as the wall thickness of a single fiber. These modules may be operated in a dead-end mode with the column outlet sealed such that all the liquid must permeate through the membrane. Alternatively, a cross-flow mode may be chosen with the possibility of recycling the process solution. The total capacity of a hollow-fiber adsorption module can be adjusted by bundling several fibers into a single module.

A literature review of membrane chromatography used for protein purification, by Ghosh (9), reported that flat sheet membranes were the most utilized membrane format (>70%), with hollow-fiber and radial flow formats far behind. This is driven by a few considerations: flat sheets are easier to cast, allow for easier and more consistent surface functionalization, and are more easily miniaturized to reduce feedstock volumes as compared to hollow fibers. The dominance of the flat sheet format is reflected in the choice of commercially available membrane adsorbers (Table 31.1). Small-scale formats based on flat sheets are currently available for laboratory-scale separations while spiral-wound capsules or cartridges are the format for larger-scale applications (pilot-scale or manufacturing scale). In general, there are only a few membrane adsorbers available for commercial use. Most of the membranes available are based on ion-exchange chemistries and are marketed for impurity removal.

31.2.3 Membrane Matrix

There are a number of ideal physical and chemical properties that membranes should possess when used in membrane chromatography. Membranes should have high mechanical strength to withstand pressure drops during operation. High chemical resistance is needed to

TABLE 31.1. Commercial Membrane Adsorbers

Product Name	Membrane Material/ Ligand Chemistry	Format	Manufacturer
Sartobind MA5, MA15, MA75, MA100	Stabilized reinforced cellulose, strong and weak cation exchange, strong and weak anion exchange, Protein A, Protein G, iminodiacetic acid, epoxy-activated, aldehyde-activated, cibachron blue, p-aminobenzamidine	Flat sheets, encapsulated in a ready-to-use device	Sartorious
Sartobind SingleSep	Stabilized reinforced cellulose strong cation and anion exchange	Radial flow capsule	Sartorious
Sartobind System	Stabilized reinforced cellulose, strong and weak cation exchange, strong and weak anion exchange	Radial flow cartridge	Sartorious
Mustang	Modified hydrophilic polyethersulfone, strong and weak cation exchange, anion exchange (polyethylenimine)	Flat sheets, encapsulated in a ready-to-use device	Pall
Mustang	Modified hydrophilic polyethersulfone, strong and weak cation exchange, anion exchange (polyethylenimine)	Radial flow capsule and cartridge	Pall
Immunodyne	Modified nylon	Flat sheets	Pall
Biodyne	Modified nylon, anion exchange, cation exchange, amphoteric	Flat sheets	Pall
Ultrabind	Modified polyethersulfone	Flat sheet	Pall
Vivapure	Strong and weak cation exchange, strong and weak anion exchange	Flat sheets, encapsulated in a centrifugal device	Vivascience

tolerate acidic (pH 1–3) and basic (pH 10–14) cleaning agents that are standard in chromatography processes. The membrane surface should have high hydrophilicity and low nonspecific protein adsorption. High specific surface area is desirable to maximize ligand density and adsorption capacity. An open microporous structure is required to allow passage of large biomolecules, and a narrow pore-size distribution is important to minimize flow irregularities. Another important consideration is keeping membrane cost comparable to that of conventional chromatography adsorbents in order for them to be competitive. This requires that an appropriate measuring stick be used to compare chromatography beads and membrane adsorbers because chromatography beads are priced based on volume, whereas membranes have traditionally been priced based on surface area.

Most of these stated requirements are generic to membranes used in any application in the biotechnology field. The one required property that is somewhat unique to membranes used in membrane chromatography is the need for functionalizable groups on the membrane surface, to allow the attachment of the ligands. The relative ease and the extent to which the membrane surface can be functionalized dictate the adsorption capacity, and hence, how competitive membranes can be with conventional chromatography media.

Relatively few polymers have been successfully cast into flat sheet membranes or spun into hollow fibers that are able to provide many of the desirable properties listed above. Most of the commercially available membrane adsorbers (Table 31.1) and membranes created or used by researchers (Tables 31.2–31.7) are based upon regenerated cellulose, aliphatic polyamides

TABLE 31.2. Membrane Chromatography Based on Protein A and G Ligands

Ligand	Membrane Material	Target Protein	Membrane Format	References
Protein A/G	Sartobind epoxy	Rabbit IgG	Flat sheet	20
Protein A	Polysulfone	Human IgG	Hollow fiber	21
Protein A/G	Polyamide	Human IgG	Flat sheet/hollow fiber	22
Protein A	Polyethersulfone/ polyethyleneoxide	Human IgG	Hollow fiber	23
Protein A/G	Poly(methylmethacrylate); poly(acrylonitrile), poly(sodium methallyl sulfonate)	Mouse IgG; human IgG	Flat sheet	15
Protein A	Nylon	Human IgG	Flat sheet	24
Protein A	Sartobind epoxy	Human IgG	Flat sheet	25
Protein A	Sartobind epoxy, Sartobind aldehyde, nylon	Human IgG	Flat sheet	26
Protein A	Sartobind Protein A	Human IgG	Flat sheet	27

TABLE 31.3. Membrane Chromatography Based on Immunoaffinity Ligands

Ligand	Membrane Material	Target Protein	Membrane Format	References
Rabbit IgG	Sartobind epoxy	Protein A	Flat sheet	20
Monoclonal antibody to Interleukin-2	Hydrazide	Humanized antibody, rombinant interleukin-2 (IL-2), IL-2 immunotoxin	Hollow fiber	28
Interleukin-2 receptor	Hydrazide	Human IL-2	Hollow fiber	29
Rabbit anti-BSA	Nylon-6	BSA	Hollow fiber	30
Anti-IgE (rabbit IgG)	Regenerated cellulose	Rat IgE	Flat sheet	11
Human IgG	Poly(gycidyl methacrylate–co-ethylene dimethacrylate)	Protein G	Flat sheet	12
Peptides	Poly(gycidyl methacrylate–co-ethylene dimethacrylate)	IgG	Flat sheet	31
Anti-low-density lipoprotein antibody	Poly(hydroxyethyl-methacrylate–co-methacryloylamido-phenylalanine)	Cholesterol	Flat sheet	32

TABLE 31.4. Immobilized Metal Affinity Membrane Chromatography

Ligand	Membrane Material	Target Protein	Membrane Format	References
Iminodiacetic acid (IDA) Zn^{2+}, Co^{2+}, Ni^{2+}, Cu^{2+}	Sartobind IDA	HSA Lysozyme, cytochrome C, myoglobin, chymotrypsinogen, recombinant fusion protein	Flat sheet	33
IDA/Cu^{2+}	Nylon-6	Ovalbumin, concanavalin A, lysozyme	Flat sheet	34
IDA/Cu^{2+}	Polyvinylidiene fluoride	Lysozyme, hepatocyte growth factor	Flat sheet	35
IDA/Cu^{2+}, N,N,N-tris(carboxymethyl) ethylenediamine/Cu^{2+}	Regenerated cellulose	Lysozyme, BSA, γ-globulin	Flat sheet	36
IDA/Cu^{2+}	Regenerated cellulose	Penicillin G acylase	Flat sheet	37
Iminodiacetic acid Zn^{2+}, Co^{2+}, Ni^{2+}, Cu^{2+}	Poly(ethylenevinyl alcohol) (PEVA)	Mouse IgG	Hollow fiber	38
IDA Zn^{2+}, Co^{2+}, Ni^{2+}, Cu^{2+}	PEVA	Human proinsulin	Hollow fiber	39
IDA/Ni^{2+}, tris(2-aminoethyl) amine/Ni^{2+}	PEVA	Human IgG	Hollow fiber	40
IDA/Ni^{2+}	Regenerated cellulose	Group-specific protein (VP3) of infectious bursal disease virus	Flat sheet	41

TABLE 31.5. Pseudo-Affinity Membrane Chromatography Based on Amino Acids and Dye Ligands

Ligand	Membrane Material	Target Protein	Membrane Format	References
Cibacron blue F3-GA Procion red HE-3R Procion yellow HE-4R	Nylon	BSA Pyruvate decarboxylase Formate dehydrogenase	Flat sheet	42
Cibacron blue F3-GA	Sartobind Blue 2	Glucose-6-phosphate dehydrognease	Flat sheet	43
Cibacron blue F3-GA	Polyethersulfone–Chitosan	Human serum albumin (HSA)	Flat sheet	44
Cibacron blue F3-GA	Chitosan	HSA	Flat sheet	45
Procion red HE-3B Procion blue MX-R	Chitin	BSA		
Procion red HE-3B	Polyethylene	Lysozyme	Hollow fiber	46
Cibacron blue F3-GA	Polyamide	HSA	Hollow fiber	47
Procion brown MX-5BR Procion green H-4G	Poly(2-hydroethylmethacrylate)	Lysozyme	Flat sheet	16
Procion red HE-3B Procion blue MX-R	Poly(2-hydroethyl-methacrylate)	Lysozyme	Flat sheet	48
L-Phenylalanine, L-tryptophan	Polyethylene	Bovine γ-globulin	Hollow fiber	13
L-Histidine	Poly(ethylenevinyl alcohol)	Human IgG	Hollow fiber	17
L-Histidine	Poly(hydroethylmethacrylate-co-methacrylolyamido)	Human IgG	Flat sheet	14
L-Histidine	Poly(ethylenevinyl alcohol)	Oligoglucuronans	Hollow fiber	15,22,23,49–51
Phenylalanine	Nylon-6	Human γ-globulin	Flat sheet	52
L-Phenylalanine, L-tryptophan, L-histidine	Poly(vinylidene fluoride)	Human γ-globulin	Hollow fiber/flat sheet	53
L-Lysine, polymyxin B, L-histidine, hystamine, α-amylase	Nylon-6	Endotoxin	Flat sheet	54
L-Lysine	Nylon-6	Tissue plasminogen activator	Flat sheet	55

TABLE 31.6. Hydrophobic Membrane Chromatography

Ligand	Membrane Material	Target Protein	Membrane Format	References
Octyl groups, dodecyl groups	Poly(glycidyl methacrylate-co-ethylene dimethacrylate)	Lysozyme, ribonuclease A, ovalbumin	Flat sheet	18
Octyl groups; dodecyl groups	Poly(glycidyl methacrylate-co-ethylene dimethacrylate)	Chymotrypsinogen A; myoglobin, lysozyme, ribonuclease A, chicken egg albumin	Flat sheet	56
Phenyl groups	Polyethylene	BSA	Hollow fiber	57
Phenyl groups	Polyethylene	BSA	Hollow fiber	58
Vinylidene fluoride groups	Poly(vinylidene fluoride)	Humanized IgG, BSA	Flat sheet	59
Vinylidene fluoride groups	Poly(vinylidene fluoride)	Humanized IgG, HSA	Flat sheet	60
Vinylidene fluoride groups	Poly(vinylidene fluoride)	Humanized IgG, HSA	Flat sheet	61
Vinylidene fluoride groups	Poly(vinylidene fluoride)	Humanized IgG, BSA	Flat sheet	62
Vinylidene fluoride groups	Poly(vinylidene fluoride)	Humanized IgG	Flat sheet	63

(nylon), polyvinylidenefluoride (PVDF), polyethylene (PE), polysulfone or polyethersulfone. Membranes have also been created out of more specialized polymers such as polyethylenevinylalcohol (PEVA) (11–14) and methacrylates or methacrylate copolymers (11,12,15–19). Membranes made from cellulose are very hydrophilic and have low nonspecific protein adsorption that makes them good candidates for protein applications. However, early cellulosic membranes had poor flow properties because they had to be constructed with small pores and a low porosity in order to compensate for the poor mechanical strength of the polymers. The use of regenerated cellulose has improved the mechanical strength and flow properties of cellulosic membranes such that they have become the base membrane for a major commercial product (Sartobind membranes, Table 31.1) and are widely cited in the literature (Tables 31.2–31.5,31.7). The surface hydroxyl and carboxyl groups can be functionalized by many well-established chemical means to attach epoxide or amine groups, onto which a variety of ligands can be bound.

The hydrophobic polymers (polyamides, PVDF, PE, and polysulfone or polyethersulfone) are very well suited to forming membranes; they provide high mechanical strength and high thermal and chemical resistance, but suffer from nonspecific protein adsorption. In addition, the surfaces do not have an abundance of functionalizable groups onto which ligands can be attached. Consequently, these membranes have typically been coated with hydrophilic polymers such as hydroxylethylcellulose (21–23,34,46,52,54), polyvinylalcohol (24,26,54) and polyethylenimine (42,52) to increase hydrophilicity and provide groups onto which ligands can be attached directly or through epoxide groups that are attached first. PE, due to its lack of active sidechains, requires more complex modification methods. The membrane is irradiated by electron beams and glycidyl methacrylate is grafted onto the surface (13,57,58,64,66,77). The ligands of interest are then covalently attached through the epoxide groups of the glycidyl methacrylate chains. A number of reviews have been written that can provide more details on commonly used membrane functionalization and ligand attachment techniques (8,93).

31.2.4 Ligand Chemistries

A wide variety of ligand chemistries have been attached to membrane adsorbers and investigated for the separation of proteins. The ligands can be classified into three main groups: affinity, ion-exchange (cationic and anionic), and hydrophobic interaction. Applications of these ligand chemistries are listed in this section.

The use of affinity membranes represents by far the largest area of investigation. This is a consequence of the very large variety of ligands that are known to have specific interactions with different proteins. Affinity ligands can be further subclassified into four main groups: Protein A/G, immunoaffinity, dyes and amino acids, and immobilized metals. Protein A/G ligands have been investigated because of their specificity to immunoglobulins (Table 31.2). Protein-A functionalized membranes have been created from a variety of base membrane matrices in both the flat sheet and hollow fiber format. Most investigations appear to have been proof-of-concept studies; this is suggested by the low level of functionalization imparted to the membrane, which resulted in low human immunoglobulin (hIgG) capacities (<5 mg/mL). However, Klein *et al*. (22) created Protein-A functionalized hollow-fiber membranes that were capable of reasonably high dynamic capacities for hIgG (~20–25 mg/mL membrane at 10% concentration

TABLE 31.7. Ion-Exchange Membrane Chromatography

Ligand	Membrane Material	Target Protein	Membrane Format	References
Anion exchange, diethylaminoethyl (DEAE)	Polyethylene	BSA	Hollow fiber	64
Anion/cation exchange Quaternary amine	Sartobind Q Sartobind S	Model proteins Antithrombin III	Flat sheet	65
Anion exchange, ethanolamino (EA)	Polyethylene	BSA	Hollow fiber	66
Anion/cation exchange; quaternary amine/sulfopropyl	Sartobind Q; Sartobind S	HSA	Flat sheet	67
Anion/cation exchange	Sartobind Q, Sartobind S	IgM	Flat sheet	68
Anion/cation exchange	Sartobind Q; Sartobind S	Model proteins; subtilisin Carlsberg, β-galactosidase	Flat sheet	69
Anion/cation exchange	Sartobind Q; Sartobind S	Mouse IgG; antithrombin III	Flat sheet; pleated module	70
Anion exchange	Quick disc (Knauer)	Human tumor necrosis factor	Flat sheet	71
Cation exchange	Sartobind S	Lysozyme, hemoglobin	Radial flow cartridge	72
Anion exchange, diethylaminoethyl groups	Polysulfone	β-galactosidase	Hollow fiber	73
Cation exchange	Sartobind S	BSA, lysozyme	Flat sheet	74
Anion/cation exchange	Ion-exchange beads encased in ethylene vinyl alcohol membrane	BSA	Flat sheet	75
Anion/cation exchange	Ion-exchange beads encased in ethylene vinyl alcohol membrane	BSA, hemoglobin	Flat sheet	76
Anion exchange	Polyethylene membranes with diethylamino and 2-hydroxyethylamino groups	Gelsolin	Hollow fiber	77
Anion exchange	Sartobind Q, Sartobind D	Whey proteins	Flat sheet	78
Cation exchange	Sartobind S	Whey proteins	Flat sheet	79
Anion/cation exchange	Vivapure ion-exchange spin columns	IgG, HSA, Human growth hormone, penicillin acylase	Flat sheet in centrifugal device	80
Anion exchange	Mustang Q (polyether sulfone)	Plasmid DNA	Flat sheet	19
Anion exchange	Mustang Q (polyether sulfone)	Supercoiled DNA	Flat sheet	81
Anion exchange	Polyethersulfone	Plasmid DNA	Flat sheet	82
Anion exchange	Mustang Q	Plasmid DNA	Flat sheet	83
Anion exchange	Mustang Q	Plasmid DNA	Radial flow	84
Anion/cation exchange	Sartobind Q, Sartobind S	Humanized monoclonal antibody, Chinese hamster ovary cell (CHO) host cell proteins, DNA	Flat sheet	85
Anion exchange	Polyvinylidene fluoride	Humanized monoclonal antibody, CHO DNA, CHO host cell proteins, endotoxin, mammalian viruses	Flat sheet	86

(*continued*)

TABLE 31.7. ***(Continued)***

Ligand	Membrane Material	Target Protein	Membrane Format	References
Anion exchange	Sartobind Q	Humanized monoclonal antibody, CHO DNA, CHO host cell proteins, mammalian viruses	Flat sheet, radial flow capsule	87
Anion exchange	Sartobind Q, Sartobind D	Human and equine influenza A virus	Flat sheet	88
Anion/cation exchange	Sartobind Q, Sartobind S, Sartobind C, Sartobind D	*Aedes aegypti* densonucleosis virus	Flat sheet	89
Anion exchange	Sartobind Q, Sartobind S, Sartobind C, Sartobind D	*Aedes aegypti* densonucleosis virus	Flat sheet	90
Anion/cation exchange	Sartobind Q; Sartobind S	Model proteins; *Aedes aegypti* densonucleosis virus	Flat sheet	91
Anion/cation exchange	Sartobind D	Rotavirus-like particles	Flat sheet	92

breakthrough, $C_o \sim 1$ mg/mL). Likewise, the Protein-A functionalized hollow fibers tested by Charcosset *et al.* (15) demonstrated hIgG capacities as high as ~25 mg/mL (C_o of 0.1 mg/mL). Despite these promising results, large-scale Protein-A functionalized membrane modules have not yet been commercialized. This might be a consequence of a combination of the membrane capacity still lagging behind the higher capacities achieved by typical Protein-A adsorbents (40–70 mg/mL at 10% concentration breakthrough for hIgG) (94), the limited data available demonstrating the ability to cycle these membranes to a high cycle number, and the cost of membranes. Immunoaffinity membranes have been created by the attachment of antibodies, peptides and receptors onto flat sheets and hollow fibers, for the purification of a wide range of biomolecules (Table 31.3). As with Protein-A functionalized membranes, much of the published work demonstrates proof of the concept (11,12,29–32). There are a number of membranes that are commercially available as base matrices for the immobilization of immunoaffinity ligands by the consumer (e.g. Immunodyne and Biodyne, Pall Filtron, Inc.). However, most immunoaffinity functionalized membranes are marketed for laboratory-scale protein purifications or as part of assay kits.

Immobilized metal affinity membrane chromatography (IMAMC) is based on the coordination of lone electron pairs of metal ions (electron-pair donors) with amino acid groups (electron-pair acceptors) of proteins. A number of factors affect the capacity of immobilized metal affinity membranes (IMAM), but the most important are the choice of chelating agent and metal ion. The most popular chelators are multidentates, which provide multiple attachment points to metal ions. Four types of dentates are commonly used: didentates, tridentates, tetradentates, and pentadentates. Based on the papers reviewed in this section and a more comprehensive review written by Suen (95), the tridentate chelator, imino diacetic acid (IDA), appears to be the most widely used chelator for IMAMC. This reflects its relative low cost and commercial availability as compared to other chelators. In addition, tridentate chelating agents should also have stronger protein adsorption because of the potentially greater number of coordination sites available on the metal (less taken up by chelator) when compared to the higher dentates (96,97). However, a disadvantage is that there is a weaker coordination of the metal ions, which may result in more metal leakage. The example publications in Table 31.4 and in a review by Suen (95) show that the most commonly used metal ions are Cu^{2+} and Ni^{2+}. However, the choice of metal ion can affect protein separation and selectivity (38,39). IMAM have been created using all of the major membrane polymer matrices and have been evaluated using a wide variety of biomolecules including model proteins (lysozyme, bovine serum albumin or BSA, myoglobin, and Y-globulin), enzymes, human and mouse IgG, and human proinsulin (Table 31.4). Researchers have shown that IMAMC can speed up separations compared to immobilized metal affinity chromatography (IMAC) resins and in some cases, offer the same selectivity (39). However, IMAMC does not appear to be widely used in large-scale biotechnology applications, as is evident by the lack of large-scale commercial IMAMC modules. This could be the consequence of the nature of the IMAC ligand and load capacity limitations when compared to other modes of chromatography. The group specificity of IMAC ligands might not be able compete with Protein A sorbents in delivering high product purity in most instances. While the use of histidine tags can increase specificity, challenges associated with histidine tag removal and the potential need for harsher elution conditions complicate their use. In impurity clearance applications, the higher binding capacity of ion-exchange sorbents makes IMAC as a whole less competitive. However, for laboratory applications, IMAC (and hence IMAMC) may be more competitive and find some niche applications.

Table 31.5 shows the pseudo-affinity membranes that have been created by the attachment of dyes and amino acids. Chlorinated triazine dyes taken from the textile industry have long been used as chromatographic ligands because of the wide range of dyes available, their low cost, and the group specificity they have to different classes of biomolecules (98,99). These dyes interact with proteins through their polar (sulfate, carboxyl, hydroxyl, amine, etc.) and nonpolar (e.g. aromatic ring) groups. Triazine dye functionalized membranes have been shown to be effective in the purification of lysozyme from egg whites (16,46,48), enzymes from Baker's Yeast (42,43), and HSA from plasma (45,47). Membranes functionalized with histidine have been used to purify immunoglobulins from human serum (17,48,52,53) and polysaccharides from crude bacterial fractions (49–51). Tryptophan (Trp) and phenylalanine (Phe) membrane adsorbers have been applied to the purification of human (53) and bovine immunoglobulins (13). In a less conventional application, Petsch *et al.* (54) demonstrated that L-lysine and L-histidine membrane adsorbers could depyrogenate buffers and protein solutions.

HIC has not received as much attention as the other modes of chromatography despite the well-documented use of hydrophobic interaction adsorbents in the purification of proteins (Table 31.6). Early work focused on the creation of HIC or reverse-phase membranes and their use in the separation of model proteins on a more analytical scale. Tennikova (18,56) separated model proteins by membrane adsorbers made of methacrylate copolymers containing epoxide groups, onto which octyl and dodecyl hydrocarbon chains were attached. The studies demonstrated that the resolution of the membrane adsorbers were comparable to reverse-phase column chromatography. Kubota *et al.* attached phenyl groups onto PE membranes through radiation-induced grafting of methacrylates containing epoxide groups (57,58). BSA was adsorbed onto the membranes in high ammonium sulfate concentrations (2 M) and recoveries were reported to stay consistent (~80%) over 10 cycles of use. Recent work by Ghosh *et al.* has focused on characterizing the performance of hydrophilized PVDF membranes by separating hIgG from BSA and HSA, and by the capture and purification of hIgG from mammalian cell culture media (59–63). Membrane capacities were determined to be high, ~60 mg/mL, at high ammonium sulfate concentrations (~1.7 M) for solutions of pure hIgG (60). hIgG was separated from HSA (63) or BSA (59) at high recoveries (>95%) and high purities (>97–98%).

Ion-exchange membrane adsorbers are available for both anion and cation exchange chromatography. Anion exchange membranes have either strong (quaternary amine) or weak (ethanolamino or EA and diethylaminoethyl or DEAE) groups attached to their surface, while cation exchange membranes are based on either strong cationic exchange groups like sulfonic acid or weak cationic groups like carboxylic acids. While membrane adsorbers with the other ligand chemistries mentioned above are scarcely available commercially or must be created by the researcher through various chemical modification techniques, ion-exchange membranes are readily available in multiple formats from a large number of manufacturers (Table 31.1). These commercial membranes are primarily strong anion and cation exchangers attached to regenerated cellulose or polyethersulfone membrane matrices. As a consequence of their availability, much of the research performed in this area has used these commercial membranes (Table 31.7). Furthermore, large-scale radial flow capsules and cartridges are available almost exclusively for ion-exchange applications. This speaks of the greater success ion-exchange membranes have found in competing with conventional column chromatography in large-scale manufacturing processes as compared to affinity and hydrophobic membranes. Researchers have also created weak anion exchange membranes by attaching EA groups onto PE hollow fibers (66) or diethylamino groups onto PE (64) and polysulfone (73) hollow fibers. Avramescu *et al.* cast membranes from a copolymer of ethylene and vinylalcohol with anion and cation exchange beads trapped within the porous matrix (75,76).

Ion-exchange membranes have been used to separate model proteins solutions (64–66,68,69), enzymes from bacterial lysates (65,73,80), human and animal immunoglobulines from mammalian cell culture supernatants (68,70), whey proteins (78,79), and plasma proteins (72,74,77). The membranes have also been found useful in the purification of large biomolecules such as DNA (19,81–83) and viruses (88–92).

31.2.5 Static Capacity

As discussed in the introductory section, the capacity per unit volume of adsorbent available for purification becomes an important issue on a preparative scale since adsorbent cost as well as plant-space requirements of the whole purification unit have to be taken into economic considerations. As the ligands are attached to the internal surface of stationary phases, their specific surface area a (m^2/m^3) determines the equilibrium capacity to a first approximation. Champluvier and Kula measured a for nylon membranes (d_p 1.2 μm) as 250 m^2 per m^2 of frontal area (42). Assuming an average membrane thickness L of 200 μm, an a of 1.2 $\times 10^6 m^2/m^3$ is obtained. Conventional porous particles for protein chromatography have a specific area that is significantly larger than this value (e.g. 1.5 $\times 10^7$ m^2/m^3, own measurements with the silica-based matrix Bioran CPG, Schott, Germany, 200 μm average particle diameter, 100 nm average pore diameter). Therefore, specific surface area appears to be a major drawback of membrane-based stationary phases. In fact, membrane adsorbents with a monolayer coverage of ligands on their internal pore surface have suffered from reduced static binding capacity, as has been summarized by Roper and Lightfoot in their review (6). An approach to circumvent reduced capacity is the introduction of a three-dimensional binding layer on the pore surface. This is achieved by graft polymerization of monomers containing functional groups, which may be derivatized to protein-binding ligands (100). The resulting functionalized polymer layer works as a three-dimensional binding space with a gel-like structure that is able to take up adsorbate molecules in multiple layers. This is depicted schematically in Fig. 31.3. This technique has been applied to hollow-fiber membranes with great success (13,52,54,57,64,66,77,101–108) and is also used in commercially available flat sheet membrane adsorbents (Sartobind, Sartorius, Göttingen, Germany) yielding a capacity comparable to porous adsorbent particles (109). The increase in static capacity for standard proteins compared to (theoretical) monolayer coverage, has been reported to be between 4- and 44-fold for hollow-fiber modules (64,100) and up to 100-fold for flat sheet membranes (110). As will be discussed below, the gel-like structure of the layer, however, may change the mass-transfer characteristics of the membrane adsorbents, a fact which has to be included in the analysis of the sorption performance.

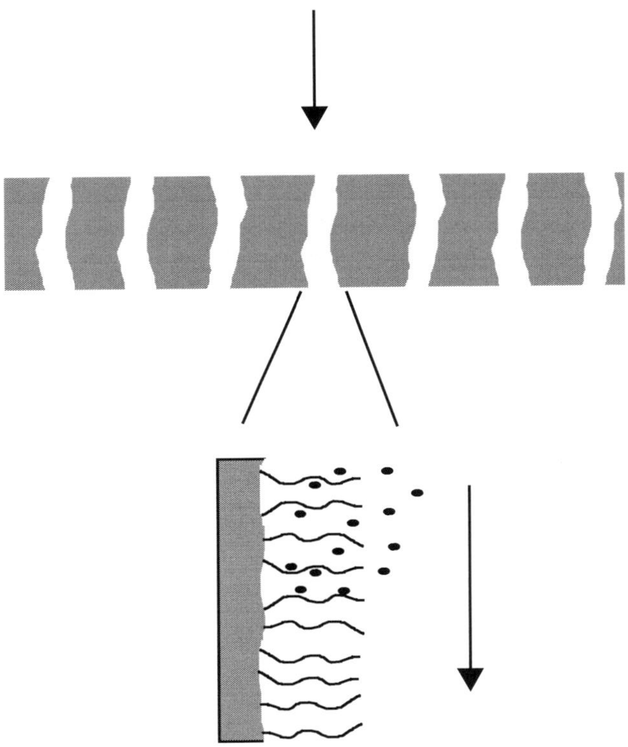

Figure 31.3. Schematic representation of the modification of the internal pore surface with grafted polymer ligands.

Despite this improvement, in most cases the capacity of membranes for small biomolecules will be lower than that for porous particles. Kubota *et al*. compared the BSA equilibrium capacity of hollow-fiber membranes modified with a layer of EA groups, to an agarose-based adsorbent with DEAE groups and found that the membrane had ~3 times lower capacity than the weak anion exchange porous beds (25 g/L compared to 80 g/L) (66). Knudsen *et al*. determined that the MAb breakthrough capacity of cation exchange beads was 2–4 times higher than membranes at typical column chromatography velocities (5–15 CV/hr), but the difference decreased significantly at higher velocities (85).

In the case of very large biomolecules, the inherent advantage of the large internal surface area of beads may become neutralized because of the hindered transport into the small pores. This was demonstrated by Yang who examined the dynamic and static capacity of a small protein (α-lactabumin, 14.4 kDa, 3.5 nm diameter) and a large protein (thyroglobulin, 650 kDa, 20 nm diameter) on anion exchange membranes and beads (111). The α-lactalbumin static and dynamic capacities for the beads were the same, but the thyroglobulin dynamic capacity was ~40 times lower than the static capacity. This suggested that the larger protein encountered hindered transport into the pores, such that much of the internal surface area was inaccessible. The static membrane capacity for either protein was the same as its dynamic capacity, which demonstrated the advantage of convective transport of the membranes. The α-lactalbumin dynamic capacity of the membranes was ~25 times lower than the beads, which is not surprising given the inherent lower surface area of the membranes. However for thyroglobulin, the dynamic capacity of the membranes was ~2.5 times greater than that of the beads. This suggested that membranes might be more advantageous for even larger molecules, such as plasmid DNA and viruses, which are more likely to be completely excluded from the internal bead surface area. Ljunglof *et al*. used confocal microscopy to confirm that plasmid DNA adsorption occurred primarily on the surface of typical anion exchange beads (112). Teeters *et al.* (19) observed a 6.1-kb plasmid DNA capacity of ~10 mg/mL on anion exchange membranes; this was ~2 times larger than the 6.9-kb plasmid DNA capacity of commercial small particle beads and ~5 times larger than commercial large-pore beads (113). It should be noted that pore exclusion is not restricted to only porous beads, but also occurs in membranes. Confocal microscopic studies following the adsorption of BSA, lysozyme, and thyroglobulin onto commercial ion-exchange membrane adsorbers demonstrated that the larger thyroglobulin did not have the same binding pattern as the smaller proteins. Coupled with the significantly lower static thyroglobulin capacity when compared to the smaller proteins, this strongly suggested that thyroglobulin had less accessibility to the internal membrane surface area. Overall, a large biomolecule size can reduce adsorption in both beads and membranes but the impact is likely to be greater for bead adsorbents. By restricting adsorption to the outside surface of beads, adsorption capacity of beads and membranes become more comparable.

There are purification applications in which static capacity is not necessarily a controlling factor. Polishing chromatography steps are tasked with reducing low levels of impurities, including host cell protein (HCP), DNA, and viruses. Anion exchange column chromatography performed in the flowthrough mode has been demonstrated to be particularly robust at impurity clearance. However columns are often oversized, not because of concerns over capacity but rather, to achieve a high throughput (85). The low mechanical strength of adsorbents restricts fluid velocities because of limitations imposed by column pressure. Therefore, if productivity is measured in processed volume per unit time, a membrane adsorber might prove to be well adapted to this application due to its inherently superior throughput.

31.2.6 Process Performance Considerations

In order to judge the suitability of any stationary phase design for the large-scale purification of proteins, performance benchmarks should be established. These benchmarks should be generic to any adsorbent and capable of comparing them on as equal a basis as possible. This section is meant to be a general discussion on factors suggested as potential benchmarks for process performance but is not specific to any one adsorbent.

Two factors stand out that can be used to assess process performance: (i) process productivity or the amount of product obtained within a certain time with a given amount of adsorbent and (ii) product purity. While the impact of each factor on performance can be assessed separately, they are not independent of each other. There is usually a trade-off between productivity and resolution; often higher resolution requires that productivity be sacrificed. The productivity is limited by the permeability of the stationary phase (mechanical limitation) and the efficiency of the sorption process (dynamic limitation), that is, the extent to which the static capacity of the adsorbent is used under the process conditions chosen. The permeability, B, of any adsorbent is dictated by the porosity of the matrix and the pressure drop across the device, whether the device is a column packed with adsorbent particles, a monolith, a single membrane, or a stacked configuration of multiple membranes. The porosity of a monolith or packed bed is a function of the pore diameter and the interstitial void fraction, while the porosity of a membrane is a function of pore diameter and the number of pores. The pressure drop in the adsorption unit may then be calculated for a given set of operating conditions (L, v, η) after Equation 31.1.

$$\Delta p = \frac{\mathrm{v} \cdot L \cdot \eta}{B} \qquad (31.1)$$

The dynamics of the sorption process may be determined from frontal adsorption experiments. The deviation of measured breakthrough curves from the ideal step function shows how efficiently the static capacity of the adsorbent is utilized. Measuring the breakthrough capacity Q_D (at 1, 5, or 10% breakthrough) under different experimental conditions (e.g. by varying L and v) allows the definition of the process conditions that make the most use of the static capacity of the adsorbent. In addition, a theoretical analysis will yield characteristic quantities, such as numbers of transfer units, plate heights, or transport and dispersion coefficients, which allow a prediction of the process performance according to standard models (110,114–116).

The overall productivity, however, must take the wash, elution, and regeneration steps into consideration; this can be expressed as the total number of bed volumes required for all of these operations (α). Productivity may be defined as amount of product per unit time and unit adsorbent volume, as shown in Equation 31.2 (110). The product yield is considered in the recovery ratio R_E.

$$P = R_\mathrm{E} \cdot \frac{1}{\left(\dfrac{Q_\mathrm{D}}{C_0} + \alpha\right)} \cdot \left(Q_\mathrm{D} \cdot \frac{\mathrm{v}}{L}\right) \qquad (31.2)$$

Finally, the product purity can be addressed by introducing the resolution R_s of the membrane-based separation under process conditions. An attempt to optimize R_s will often show that a compromise has to be found between maximizing the throughput at low L/d and maintaining high resolution because the short bed lengths make adsorptive membranes low plate number systems (117). However, Coffman *et al.* have shown that most protein separations can be achieved using membrane adsorbents because separation will occur according to an on–off separation mechanism rather than by differential migration (118).

31.3 LIMITING FACTORS IN MEMBRANE ADSORPTION PROCESSES

As with all processes based on fixed beds, sorption dynamics of adsorptive membranes may be limited by mixing, mass transport, and the kinetics of the adsorption and desorption. In order to discuss these effects precisely, an attempt to define them according to the specific details of membrane adsorption units shall be made. Kinetic limitations arise from the rate of formation of the adsorbate–ligand complex (k_a) and the rate of dissociation of the same complex (k_d). Taken together, these constants make up the dissociation constant of the binding equilibrium $K_D(k_d/k_a)$. Slow adsorption or desorption can have significant effects on the sorption dynamics. Further limitations are caused by slow transport of adsorbate molecules from the bulk solution to the binding site; for example, diffusive motion through the fluid boundary layer above the binding surface. Although the existence of thoroughgoing pores in membrane adsorbents can overcome slow diffusion in dead-end pores, pore diffusion still needs to be considered in the description of sorption dynamics because the membrane production process may result in adsorbents with some pores of this type. Additionally, as mentioned above, the attachment of a three-dimensional adsorbing layer to the internal pore surface must be regarded as a potential mass transport barrier. Mixing in adsorptive processes should be split into three major effects: Fickian dispersion, nonideal fluid flow within the adsorbent, and extra-column mixing (119). The first effect describes the dispersive motion “against” the direction of convective flow, the second effect accounts for irregularities within the adsorbent leading to uneven fluid flow through the packing, while the third summarizes contributions of tubing, extra-column dead volumes, as well as liquid distributors, collectors and so on to zone broadening. Very often, these different contributions are lumped together in a single dispersion coefficient D_{ax}, which is used to characterize the entire experimental set-up rather than the adsorbent itself. In the following sections, the respective importance of these limitations will be discussed in detail in order to provide a basis for the prediction of the performance of adsorptive membranes in the purification of biomolecules.

31.3.1 Mixing

The influence of overall mixing on performance, during frontal adsorption, was investigated in a theoretical study by Suen and Etzel (115). These authors used the dimensionless column Peclet number Pe to quantify dispersion according to Equation 31.3 (v_e = interstitial flow velocity).

$$Pe = \frac{v_e \cdot L}{D_{ax}} \tag{31.3}$$

Peclet numbers above 40 were described to be sufficient to exclude a deteriorating influence of mixing on the breakthrough behavior during frontal adsorption to membrane stacks. As Pe is proportional to the bed length, augmented mixing represented by increased D_{ax} may be compensated for by increasing L. Similarly Liu and Fried (120) reported Pe above 25 to be sufficient for neglecting dispersion as a possible limitation of the sorption process. It is important to note that these considerations were set up for frontal adsorption of a single component system; stronger boundary conditions may be required for the separation of a multicomponent system.

Roper and Lightfoot (119) analyzed peak broadening in stacked flat sheet membranes under nonbinding conditions with a specific flow reversal technique in order to differentiate between Fickian dispersion, nonideal fluid flow within the adsorbent, and extra-column mixing. They used the HETP concept to determine the individual contributions to the overall peak broadening and found the plate heights h (resulting from Fickian dispersion) to be 1–5 μm for a range of flow velocity v between 0.5 and 4 cm/min. Peak broadening by a factor of 1.7 occurred due to extra-column dispersion. The plate heights were in accordance with theoretical considerations which predicted h as a function of both, the dispersion coefficient D_{ax} and v, according to Equation 31.4.

$$h = \frac{2 \cdot D_{ax}}{v} \tag{31.4}$$

However, values for h reported in literature for adsorptive membranes were between 20 and 200 μm and showed a slight increase with fluid velocity. This difference could be ascribed to the inclusion of extra-column effects and uneven flow distribution in the stack in the later results. A similar analysis was performed by Teeters *et al*. on a different commercial flat sheet system using a number of model proteins (121). The membrane plate height was determined to be 2 orders of magnitude smaller than 15 mm porous beads and less affected by changes in velocity. The plate height changed by order of magnitude when the direction of elution was changed; that was deemed a consequence of nonuniform flow distribution. Gebauer *et al*. (110) found that the mixing in stacks of commercially available flat sheet membranes, represented by an overall D_{ax}, was 10-fold higher compared to an estimation based on literature correlations. This difference was attributed to extra-column effects. Overall, these studies suggested that nonuniform fluid distribution and extracolumn effects dominated mixing in membrane systems while Fickian dispersion made only a small contribution.

The effect of large mixing on the performance of membrane adsorbents was demonstrated in a scale-up study on the isolation of HSA from plasma using stacks of ion-exchange membranes (109). The increased mixing found with large-scale modules as compared to laboratory-scale units, led to a more than 3-fold reduction of the productivity of the process. This was attributed to the large dead volume of the distributor, which was twice as large as the membrane volume. Reif *et al.* (69) also demonstrated the importance of reduced dead volume behind the membrane, as a large mixing zone led to deteriorated resolution in the analytical separation of standard proteins. Josic *et al.* (122) investigated the influence of liquid distribution at the column inlet of cartridges with a continuous stationary phase of type b (compare with Fig. 31.1). In the case of nonoptimized inlet design, a significant part of the fluid passed through the central region of the unit, thus leaving large parts of the membrane unreachable by convective transport. The uneven flow distribution led to a significant loss of performance of the chromatographic separations employing these units. With optimized inlet design, resolution was comparable to established HPLC methods.

Nonuniform flow distribution may also be caused by variations in the pore size of adsorptive membranes. Kim *et al.* described the modification of hollow fibers with grafted polymer chains bearing Phe and Trp as pseudo-affinity ligands for immunoglobulin adsorption (13). The deviation of protein breakthrough from ideal behavior with these adsorbents was also attributed to a wide distribution in pore length. The distribution in pore diameter in Protein-A-modified hollow-fiber modules was described by Klein *et al.* as the reason for incomplete use of the static capacity of these adsorbents in immunoglobulin capture (123). Suen and Etzel (115), as well as Liu and Fried (120), demonstrated by computer simulations that a wide pore-size distribution may lead to significantly flattened breakthrough curves during frontal protein adsorption and thus, to a strongly reduced useful adsorption capacity. Similar effects were shown for variations in membrane thickness. As a consequence, stacking of membranes was recommended to reduce these effects.

To summarize, as long as sufficient care is taken during process design to minimize extra-column effects as well as nonuniform flow, mixing phenomena are unlikely to be the limiting factors in membrane chromatography. However, it has to be kept in mind that adsorptive membranes are used in a chromatographic process and therefore, standard criteria of filtration module design will not meet the requirements of a sorption process and may possibly lead to a loss of the advantages of the membrane-based process.

31.3.2 Mass Transport

When dealing with porous particles for protein purification, there are two potential sources of transport resistance to adsorption: transport of adsorbate from the bulk of the solution to the external particle surface, as well as transport within the particle to the binding site. The general perception is that fluid-side transport is an order of magnitude faster than transport within the adsorbent. Therefore, film transport resistance is often neglected in a simple analysis. Band-spreading in frontal analysis as well as peak-broadening during chromatography, is attributed to slow, often hindered, intraparticle diffusion. The use of membranes with thoroughgoing pores for protein adsorption was supposed to remove the transport resistance by reducing the characteristic length of protein diffusion to the ligands. Brandt *et al.* (4) provided a general estimation of the importance of mass transport for the efficiency of an adsorptive membrane by relating the characteristic diffusion time in a membrane pore $t_D (t_D = d_p^2/D)$ to the contact time $t_C (t_C = L/v)$ during passage through the membrane. Efficient usage of the static capacity was supposed to take place if $t_C >> t_D$. Inserting standard values for adsorptive membranes into this equation ($d_p = 3/10^6$ m, $L_{\text{single sheet}} = 2/10^4$ m, $v = 8.3/10^4$ m/s) and assuming an average diffusion coefficient D of $5/10^{11}$ m^2/s, a t_D of 0.18 s is obtained while the t_C for a single membrane is 0.24 s. Therefore using a stack of 10 membranes, which is often found in commercially available modules, easily fulfills the necessary condition and allows the diffusive transport resistance to be ignored. This is in contrast to standard porous particles, where t_D is based on the particle radius and therefore can be estimated as 40.5 s for the same protein in a particle of $9/10^5$ m diameter.

The absence of dead-end pores and the concomitant disappearance of intraparticle transport resistance should make diffusion through the fluid boundary layer above the ligands the only transport step in membrane chromatography (Fig. 31.1). The rate of transport through this layer is dependent on the free diffusion coefficient of the adsorbate, as well as on the thickness of the layer. As interstitial flow velocities in narrow pores of microfiltration membranes are comparably high and boundary layer thickness may therefore be low, this transport resistance is commonly regarded as not likely to be limiting the sorption process. Suen and Etzel (115) demonstrated the validity of this approach for membranes with small pore diameter by comparing the characteristic time needed to diffuse through the fluid boundary layer to the average residence time in a module. They were able to show that even for very small diffusion coefficients (i.e. large protein molecules), the residence time in a correctly designed membrane stack is much higher than the diffusion time through the fluid film. As such, the importance of this transport resistance is low when compared to other possible limitations. In cases of much larger pore diameter these considerations may no longer be valid. Adisaputro *et al.* (124) measured the dynamics of whey protein uptake in membranes of very large pore diameters (50–300 μm) and found reduced capture efficiency with decreasing residence time. Calculation of the fluid film diffusion time for an average protein showed that film transport was likely to be the dominant limitation.

As mentioned above, the introduction of a three-dimensional ligand layer onto the pore surface may also introduce mass transport resistance. Gebauer *et al*. (110) investigated commercially available flat sheet ion-exchange membranes with respect to their sorption dynamics and found that the sorption efficiency was limited by diffusion. Depending on the length and density of the ligand layer, a pore or solid diffusion mechanism of protein transport was identified. The breakthrough capacity, however, could be maintained on a very high level for a wide range of flow velocity. This was attributed to convection-aided mass transport into the flexible chains of the ligand layer. Attaching polymer-grafted ligands to the internal pore walls of hollow fibers also yielded stationary phases with three-dimensional adsorbing layers. No dependency of breakthrough behavior from residence time in the module was detected, thus supporting the idea of fast protein influx into the ligand sphere (13,106–108,125).

In summary, mass transport limitations are not as dominant in membrane adsorption as they are in standard protein chromatography using porous particles. The complete neglect of transport resistance, however, is only possible under certain conditions. Therefore, transport in adsorptive membranes should be analyzed carefully during process design based on these stationary phases.

31.3.3 Binding Kinetics

If axial mixing and mass transport may be excluded as potential limiting factors of sorption efficiency under certain circumstances, the kinetics of the protein–ligand interaction comes into focus. Brandt *et al*. (4) used a second-order rate expression to estimate the reaction time t_R for interaction of BSA with an MAb affinity ligand. They obtained a t_R of 5 s, which is significantly higher than the average diffusion time calculated above for membrane pores (0.18 s). From this estimation it can be deduced that slow association kinetics may limit the performance of affinity membrane adsorbents. A similar situation was predicted by Suen and Etzel (115) from a thorough theoretical analysis, which was experimentally verified later with the Pepstatin A or Pepsin affinity system (126). Kinetic limitations were also found to be responsible for deviations of protein breakthrough from an ideal step function in dye ligand or dehydrogenase pseudo-affinity systems employing stacks of flat sheet membranes (43). Briefs and Kula (127) reported the onset of peak-broadening due to kinetic limitation with a similar pseudo-affinity interaction from t_C below 20.5 s. For ion-exchange interactions, however, no kinetic limitation could be verified within the range of experimental conditions ($t_C = 20.5$–5.1 s). An investigation on kinetically limited membrane chromatography was performed by Nachman *et al*. (28,29,128) employing hollow-fiber modules with affinity ligands attached to the internal pore surface. In this study, it was shown that reducing the residence time in the module reduced the capture efficiency of the adsorbents. Under all experimental conditions, t_C was well above t_D, which had been calculated based on the membrane geometry and diffusion coefficient of the adsorbate. The capture efficiency was also independent of initial concentration, as can be predicted for kinetic limitation under conditions of a rectangular isotherm (110). Kinetically limited adsorption may be exploited in membrane chromatography to separate adsorbate molecules which have identical binding isotherms but different association rate constants. In standard porous adsorbents, these pairs cannot be separated due to the fact that slow diffusion in the pores leads to the formation of a local equilibrium between adsorbate in the pore liquid and bound adsorbate. Membrane adsorbents are sensitive to differences in adsorption kinetics, provided neither mass transport nor mixing "hide" the kinetic differences. This was theoretically predicted (116) and proven experimentally (126) for an affinity interaction based on stacks of flat sheet membranes.

In summary, adsorptive membranes offer a rare opportunity to investigate effects of binding kinetics in adsorptive processes. This fact underlines the excellent mass transport characteristics of these matrices, which allow very fast protein purification under conditions in which protein–ligand interactions are the rate-limiting step in the sorption process. However, this seems to be limited to affinity interactions, which show comparatively slow association rates with ligands attached to planar surfaces due to the higher steric requirements. With fast ion-exchange interactions, even adsorptive membranes have not been able to demonstrate kinetic effects during protein sorption.

31.4 OPTIMIZING THE PERFORMANCE OF ADSORPTIVE MEMBRANES

If the performance of a membrane adsorption unit is defined as desired product purity at maximized productivity, then optimization has two main goals: the achievement of a certain resolution and the simultaneous increase of throughput. The problem of maximizing throughput requires information on mechanical stability of the stationary phase, the quality of fluid flow through the unit, and knowledge of the overall kinetics of the sorption process. The latter point may in a first approximation, be reduced to the question of residence time within the adsorbent phase. As discussed in the section titled "Mass Transport", adsorptive membranes are supposed to have superior transport properties even if the static capacity is increased by introducing a three-dimensional ligand layer on the internal membrane surface. The enhanced transport reduces the residence time ($t_C = L/v$) needed for efficient usage of the available capacity. A measure of this efficiency is the breakthrough capacity which increases with residence time and is dependent on the performance of the stationary phase with regard to the overall dynamics of the sorption process, be it limited by transport processes, dispersion or binding kinetics. The better the dynamic properties of the adsorbent, the shorter the residence time can be without compromising the breakthrough capacity, and hence, the greater the throughput. In the case of adsorptive membranes, enhanced transport allows the design of adsorbent beds of short length (L) that can be operated at high fluid velocity (v). The work of Yamamoto *et al*. (129) and Equation 31.2 predict that the productivity increases with the breakthrough capacity and decreases with t_C. However, as breakthrough capacity increases with t_C, there is an optimum t_C with respect to productivity. The advantage of adsorptive membranes is that efficient usage of the static capacity is already obtained at short t_C. Therefore, membrane-based adsorption methods can be anticipated to have a high productivity.

The improved mass transport of membrane adsorbers can lead to limitations in the fluid distribution within membrane modules that may affect performance. The short residence times result in membrane module configurations with a small column–aspect ratio (*L/d*). This geometry makes the design of membrane modules capable of evenly distributing fluid flow across the entire membrane surface more challenging, especially as the diameter is increased for large-scale applications. In the case of particle-shaped adsorbents, uneven distribution of fluid at the column inlet will be neutralized during passage through the comparatively long bed by the lower part of the column that acts as an additional distribution unit. In the case of membrane adsorbents, this is not possible because of the very short bed length. Thus, a very careful design of the distribution unit of large-scale equipment is essential. Furthermore, the fabrication of the membrane adsorbent itself has to be taken into account. Wide distribution in pore size will lead to a preferential passage of fluid through certain regions of the membrane resulting in a broad distribution of residence times that will affect process performance. This was analyzed theoretically by Suen and Etzel (115), who demonstrated that the stacking of flat sheet membranes helped homogenized flow through membranes with a large variance in porosity. Despite the challenges in membrane module design, several large-scale radial flow cartridges have been developed and are currently marketed. Their volumes range from several hundred mL to as large as 5 L. A number of FDA-approved drugs, such as Aldurazyme (130) and CMAPATH-1H (131) use purification processes that employ large-scale anion exchange radial flow cartridges to clear DNA and viruses. This indicates that some of the concerns raised above and in the section titled "Mixing" have been addressed in the current generation of membrane adsorber modules.

The lower resolution of adsorptive membranes compared to porous beads might be expected to limit their use in purification processes. However, as shown by Coffman *et al*. (118), most protein separations are operated in an on–off mode during elution chromatography. This means that the selectivity of the separation relies on the differences in the interaction between the proteins to be separated and the adsorbent, at various concentrations of the eluting agent. Separation by differential migration, which is caused by the different retardation of the individual molecules during their movement through the adsorbent, plays only a minor role. Therefore only a few theoretical plates are needed for the satisfactory separation of proteins by ion exchange, hydrophobic interaction, or reversed phase methods as here the intrinsic separation factor of the different proteins is high enough to justify separation according to an on–off mechanism. Consequently, short membrane adsorbents allow very good separations and the resolutions obtained are mostly comparable to HPLC separations, where long columns yield very high plate numbers under high pressure conditions. Coffman *et al*. reported about 50 stages to be sufficient for most separation tasks. Therefore, HETP values between 20 and 200 μm, which have been found for adsorptive membranes, necessitate bed lengths of only 10 mm for good resolution in protein separation. Similar findings on the bed length necessary for satisfactory results in protein separation using adsorptive membranes were presented by Dubinina *et al*. (132). In their analysis, the authors demonstrated that only a short bed is needed if proteins are separated according to an on–off mechanism. The influence of gradient slope, fluid velocity, and bed length were investigated and it was shown that short adsorptive membranes may be used for high throughput separation with good resolution.

As increasing throughput (either by increasing the load or by reducing t_C) is likely to reduce resolution, the same criteria for optimizing *L*/v apply as in the case of the breakthrough capacity of membrane adsorbents. The good mass transport capabilities then allow the use of high flow rates at retained resolution when using well-designed membrane units. This is supported by the findings of many workers, who reported excellent resolution over a wide range of fluid velocity during the separation of proteins with these devices (18,56,65,68–70,122,133–135).

In summary, the optimization of membrane adsorber process productivity is a compromise between maintaining required resolution while maximizing throughput as defined by finding the optimum in capacity and residence time.

While decreasing residence in the name of increasing productivity can have a detrimental effect on resolution, for most protein separations, membrane adsorbers should have suffient plate numbers in order to be used successfully.

31.5 POSITIONING OF MEMBRANE ADSORPTION IN A PURIFICATION TRAIN

After characterizing adsorptive membranes according to their basic properties, their potential location in a standard protein purification scheme can be discussed. Wheelwright differentiated between initial recovery, low resolution, and high resolution purification in a downstream purification process (136). Depending on the design of the adsorption unit, adsorptive membranes can accommodate all of the tasks mentioned. The versatility of adsorptive membranes in serving the needs of the different stages of a downstream process is summarized in Fig. 31.4.

Initial recovery is product capture out of clarified or unclarified cell culture broths. When capturing out of the latter feedstock, two separate steps, that is centrifugation or depth filtration followed by column chromatography, are combined into a single step. Whether captured out of clarified or unclarified solutions, the product protein is concentrated and delivered in a reduced process volume. Given the potentially large cell culture volumes that need to be processed, throughput becomes important. The good mass transport characteristics and fast binding kinetics of ion-exchange membranes make them well-suited for achieving high throughput (17,57,73,103,107,137). Membranes with hydrophobic interaction (HIC) and pseudo-affinity ligands (IMAC, dyes, and so on) could also be options (46,138,139). Affinity (20,21,140), immunoaffinity (28), and receptor affinity membranes (29) also belong to this category despite the more traditional use of these ligands in high resolution applications. They may be used early in a purification train in order to introduce a high selectivity step in the very beginning of the downstream process. Fast binding kinetics favor the use of very dilute feed solutions as high affinity adsorption may be operated with large throughput in membrane devices even if the feed concentration is in the order of magnitude of the dissociation constant K_D of the adsorbate–ligand interaction (128).

Low resolution purification should lead to a significant enrichment of the target molecule and the removal of major impurities. Anion exchange membranes have been demonstrated to remove process-related impurities (HCP and DNA) and contaminants (viruses and endotoxins) (54,85–87). Membranes with HIC or pseudo-affinity ligands should also be capable of low resolution High resolution purification refers to the removal of minor impurities with properties similar to the desired product, as well as to the final polishing (e.g. separation of correctly processed molecules from false variants, multimers and so on). The use of membranes for high resolution purification was realized in HPMC (High Performance Membrane Chromatography) (18,56,122). Here the properties of continuous stationary phases of type b (Fig. 31.2) are used with ion-exchange hydrophobic interactions as well as with reversed-phase ligands. Affinity (20,21,140), immunoaffinity (28), and receptor affinity membranes (29) also belong to this category.

31.6 APPLICATIONS

The scope of this section is to discuss applications from each of the positions within a purification scheme, but it

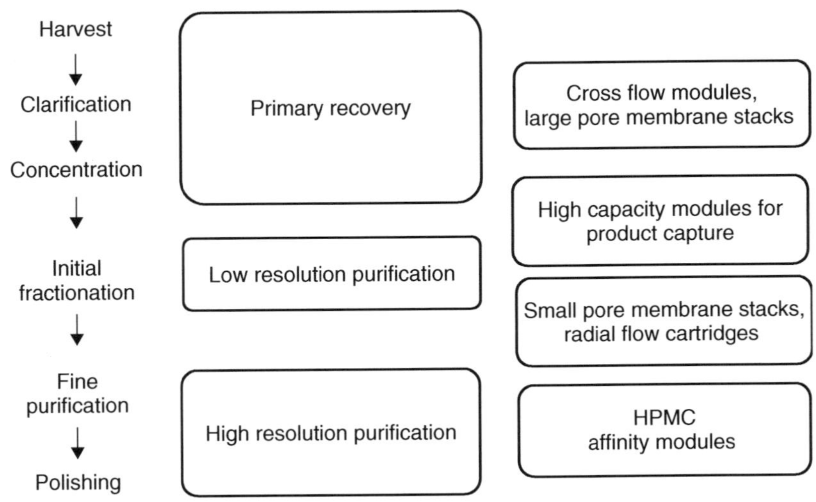

Figure 31.4. Positioning of membrane adsorbents in a purification train.

TABLE 31.8. Selection of Some Typical Productivity Values Obtained with Membrane Adsorbents

Ligand	Geometric Format	Target Protein	Productivity $(g/L \cdot h)^a$	References
Q	Stack	Formate dehydrogenase	36	142
Q	Pleated module	Antithrombin III	21.5	70
Q	Stack	HSA	220	109
Q	Radial flow capsule	pDNA	5.7	84
Q	Stack	MAb	5680	85
Q	Stack	MAb	5000	86
DEAE	Hollow fiber	Urease	200	107
DEAE	Hollow fiber	Gelsolin	26	77
S	Pleated module	IgG	75.2	70
S	Stack	IgG	29	143
MAb	Hollow fiber	recombinant Interferon-α2a	2.7	28
Histidine	Hollow fiber	IgG	6.7	17
Procion yellow	Stack	Pyruvate decarboxylase	40.3	127
Procion red	Stack	Formate dehydrogenase	24	142
Procion red	Hollow fiber	Lysozyme (from homogenized egg yolks)	12	46
Cibacron blue	Cross-flow module	Malate dehydrogenase (from unclarified broth)	1.1	138
IDA-Ni^{2+}	Stack	Structural group-specific protein (VP3) of the infectious bursal disease virus	50	41
Protein A	Hollow fiber	IgG	35	21
Protein A	Stack	IgG	10.8	20

[a]Productivity was calculated from the cited references by calculating the total time of the respective purification cycles (including washing and elution when available) and relating the amount of product purified to the cycle time and the adsorbent volume used.

is not meant to be an all-encompassing review. Therefore the reader is referred to recent review articles for additional applications (6–10,141). The examples discussed were selected to demonstrate the wide potential applications of membrane adsorbers and provide evidence that productivity of processes employing membrane adsorbents are superior at retained resolution. Table 31.8 summarizes the productivity obtained for selected applications.

31.6.1 Primary Recovery

As microfiltration membranes are suitable for the clarification of culture broth or homogenates, the attachment of ligands to the internal pore surface would allow a combination of protein adsorption and solid–liquid separation in a single step. However, very few examples in the literature exist concerning the direct capture of proteins out of unclarified cell culture feedstocks. Krause *et al.* (138) used Cibacron-blue-modified microfiltration membranes (Sartobind Blue, Sartorius, Göttingen) for the isolation of malate dehydrogenase directly from unclarified *E. coli* homogenate. The membrane adsorbent had a static capacity similar to commercially available particles with the same ligand and was used in a cross-flow mode with recycling of the filtrate, that is, a batch adsorption mode. The overall productivity was 1,200,000 units/L·h, while clarification, capture, and initial purification into a single step. Data on the long-term stability of this method were lacking; therefore, the influence of traditional shortcomings of microfiltration membranes due to fouling, layer formation and so on, cannot be judged. A different field of application is the capture of whey proteins from unfiltered whey by employing large-pore membrane stacks (124) (Productiv S and QM, BPS Separations, Spennymoor, UK). Adisaputro *et al.* replaced the traditional method of batch adsorption of whey proteins, by membrane stacks. Based on their results, the authors predicted an increased productivity of the membrane method at the short residence times which become possible due to fast mass transport characteristics. The size of the pores, however, was a point of concern because a large pore size was required to process the unfiltered whey, but very large pore diameters could cause fluid-side boundary layer problems. Vogel *et al.* (55) used flat sheet membranes functionalized with lysine seated in a rotating shear-cell device for the capture of tissue plasminogen activator (t-PA) from unclarified Chinese hamster ovary (CHO) cell culture. The device format was designed to create a hydrodynamic force away from the membrane surface and a drag force parallel to the surface to inhibit membrane fouling and cake-layer formation during the clarification of cell culture broths. The membrane device was demonstrated to reproducibly clarify CHO broth and capture t-PA to a high recovery. When cycled 12 times, the transmembrane pressure drop increased only modestly, indicating the absence of a secondary resistance layer (cake layer) on the membrane surface despite the cell density increasing by a factor of 25. The capacity and recovery remained consistent from cycle to cycle. The t-PA purification factor was 16.7 with an overall recovery of 86% and a purity of 65%. The cell viability remained almost constant, which indicated that the device could potentially be used in continuous perfusion processes.

31.6.2 Capture

31.6.2.1 Small Biomolecules. Ion-exchange membrane adsorption has been described as an efficient method to capture directly from clarified broth, possibly preceded by dilution in order to reduce ionic strength. Jungbauer *et al*. (144) used Zetaprep QAE radial flow cartridges (LKB, Bromma, Sweden) for this purpose and performed a scale-up study from 50 to 930 mL total adsorbent volume. The productivity reported was 0.2 g/L·h and a certain enrichment of the MAb was demonstrated by SDS (sodium dodecyl sulfate) gel electrophoresis. Cation exchange membrane stacks as well as a pleated module were used by Lütkemeyer *et al*. for the isolation of IgG from diluted hybridoma supernatant on a laboratory scale (28 mL of adsorbent, feedstock volume 5.5 L) (70). The high static capacity of the adsorbent (50 mg/mL) together with short residence times in the short adsorbent bed yielded a productivity of 75.2 g/L·h, which was 13 times higher compared to the same process performed using a packed bed of porous cation exchange particles. In a later publication, a similar process was shown at the pilot scale with undiluted hybridoma supernatant. Even at the high ionic strength the overall productivity was 29 g/L·h. During the experiments, 100 L of supernatant were processed in 2 h in a unit of 4-mm bed length (Sartobind S, Sartorius, Göttingen, Germany; $V = 220$ mL); the process volume was reduced 44-fold at 100% recovery (55). A similar approach was chosen by Wang *et al*. for the purification of a recombinant immunofusion protein secreted by *E. coli* (139). The same stack of cation exchange membranes was used for capture of the target protein from 200 L of clarified and 10-fold diluted culture broth. The reported productivity was up to 1.5 g/L·h at 10-fold reduction of process volume. The capture of urease to a hollow fiber which had been modified by the attachment of tentacle polymer chains with anion exchange ligands was described by Matoba *et al*. (107). The productivity (expressed as urease collection rate) found in these investigations was up to 200 g/L·h. Grasselli *et al*. purified lysozyme from homogenized egg yolks using a hollow-fiber membrane functionalized with the dye Procion red HE-3B (46). The membrane provided a productivity of 12.1 g/L·h, which was 6 times higher than a commercial cation exchange bead. The purities and yields were comparable (~92–94% yield and ~90% purity) for both systems. A hollow-fiber membrane with attached weak anion exchange groups (DEAE groups) was created by Hagiwara *et al*. and used to purify Gelsolin, an actin-binding protein in plasma, from bovine plasma with a productivity of 26 g/L·h (77). The membrane performance was consistent over three cycles of membrane use. Hu *et al*. compared the ability of (Ni^{2+}) membranes and IMAC (Ni^{2+}) agarose gels to purify a structural group-specific protein (VP3) of the infectious bursal disease virus (41). The productivity of the membrane was high (50 g/L·h) and similar to the gel (47 g/L·h). The recoveries (~87%) and purities (98–99%) were similar in both cases. All these examples demonstrate how the efficiency of membrane adsorbents may be exploited for high-productivity capture of target proteins from culture supernatants.

31.6.2.2 Large Biomolecules. Membranes have been investigated for the purification of large biomolecules, such as viruses and plasmid DNA (pDNA), out of cell culture supernatants because they can provide inherent advantages over adsorbent beads. The large cell culture lysate volumes that are typical for these molecules can be processed at a high throughput using membranes due to their superior transport and lack of pressure drop (84). Due to the compressible nature of the adsorbent beads, a high pressure drop that can limit throughput can be seen in the adsorbent columns. The size of the biomolecule can also neutralize the larger surface area, and hence capacity, of adsorbent beads. Large biomolecules can experience hindered diffusion into the small pores of most bead adsorbents or be excluded completely (section titled "Improved Mass Transport"). This can severely decrease the capacity of adsorbent beads such that membrane capacity becomes comparable.

Anion exchange membranes have been shown to be well-suited to virus purification by virtue of their positive charge. Specht *et al*. reported that the dynamic and static capacity of Sartobind anion exchange Q membranes (Sartorius, Göttingen) for *Aedes aegypti* densonucleosis virus (AeDNV) captured out of distilled water were very similar (1.35×10^{10} particles/mL compared to 3.1×10^{10} particles/mL, respectively) (89). This suggested that most probably, the membranes did not suffer from hindered transport into the pores. Using the same anion membranes, Kalbuss *et al*. captured the human influenza virus from cell culture supernatants (88). The virus activity recovery was 72% and a volume reduction of 5-fold was achieved at a productivity of 67 L/m^2·h, but the purity was low due to the complete recovery of DNA in the product fraction. Vicente *et al*. developed a process to purify rotavirus-like particles (RLP) from a culture of *Spodoptera frugiperda* Sf-9 cells that was clear of HCP and DNA (92). The clarified concentrate was purified over a weak anion exchange membrane (Sartobind DEAE, Sartorious, Göttingen) to capture and concentrate the virus and remove HCP and DNA. This was followed by gel filtration to clear impurities of low molecular weight.

The RLP recovery on the membrane was 55%; the overall process delivered a 98% pure product at 46% recovery and a volume reduction of 134-fold.

Plasmid DNA (pDNA) have been investigated as gene-therapy drugs or vaccines in a variety of diseases. While the large size of pDNA and the large bacterial lysate volumes present obvious problems for adsorbent beads (as discussed above), another potential limitation is the delivery of high purity products. While pDNA adsorption will be relatively limited to the outside of the beads, major impurities like endotoxins and HCP may have access to the large internal surface area due to their smaller size. The high internal surface area of beads, which can be advantageous with small biomolecules, can lead to higher impurity adsorption. In this case membranes, by virtue of their larger pores and lower surface area, might be better suited to deliver high purity. Endres *et al*. (83) purified 6.3 kbp pDNA from *E. coli* lysate using a laboratory-scale strong anion exchange membrane (Mustang Q, Pall Corp., East Hills, NY). Anion exchange membranes could be loaded to a dynamic capacity of 15–18 mg pDNA/mL at high velocities (5.5–55 MV/min); this was 5–25 times higher than the static equilibrium capacity of a 4.3 kbp pDNA captured out of a pure solution by four commercial anion exchange beads as reported by Ferreira *et al*. (145). A larger scale purification of a 4.5 kbp pDNA from 71 L of *E. coli* lysate was performed by Zhang *et al*. using a commercial strong anion exchange radial flow membrane capsule (84) loaded to 5.7 mg/mL. The pDNA was concentrated by a factor of 10 out of the lysate, at a recovery of 95%. The product pool purity was estimated between 85 to 90%; RNA, endotoxin, and DNA fragments were the main remaining impurities. The operation took less than 1 h; the overall productivity was ~5.7 g/L·h. This example illustrates the productivity potential of membrane adsorbers for pDNA applications. However, it also demonstrates the difficulty in achieving high purity at the same time.

31.6.3 Low Resolution Purification

There is no clear demarcation between capture and low resolution purification membrane processes because both separate the target molecules from impurities. One way of differentiating them is with regard to their main task, which is reduction of process volume in the first case and isolation of the main component from major impurities in the latter. Given this classification, optimizing t_C with regard to resolution under high throughput conditions, as outlined in section titled "Optimizing the Performance of Adsorptive Membranes", becomes important.

31.6.3.1 General Protein Purification. A typical example for low resolution purification is the isolation of HSA in a plasma fractionation process based on ion-exchange adsorbents. This was first described by Lacoste-Bourgeac for the valorization of fraction IV paste which originated from the traditional ethanol precipitation during plasma fractionation (146). A series of anion exchange and cation exchange spiral-wound membrane cartridges (Cuno, Cergy-Pointoise, France) was used to recover albumin at 66% yield from the paste. Thus the high content of protein of this fraction, which otherwise was reported to be only of little value, could be exploited. A process for the isolation of HSA employing stacks of flat sheet membranes was presented by Gebauer *et al*. and was compared to the well-established ion-exchange process based on porous beads (109). A series of strong anion and cation exchange steps was used with a productivity of up to 220 g/L·h on the laboratory scale. This value was 6- to 8-fold higher than that of the ion-exchange column process (Sepharose FF, Pharmacia, Uppsala, Sweden). However when the process was increased in scale based on commercially available equipment, to 110 mL of adsorbing volume, the productivity decreased by a factor of 3. Additionally, a significant dilution of the product was observed. The deteriorated performance of the large-scale module was attributed to overall mixing effects in the module which originally was a filtration unit and had not been optimized with regard to chromatographic needs. This supports the conclusions from the section on Mixing: that poor membrane module design could lead to a performance loss due to the sensitivity of short-bed adsorbents to extra-column effects. A series of anion exchange and hydrophobic interaction membranes was used by Luksa *et al*. for the purification of recombinant TNF-α from clarified *E. coli* extract (71). Comparing the results obtained with commercially available cartridges of type b (Quick Disk, Knauer, Berlin, Germany) to a conventional method based on porous particles showed similar product purity at half the process time. This demonstrated that increased throughput at retained resolution may be obtained with adsorptive membranes. The performance of another commercially available stack of anion exchange membranes (Memsep DEAE, Millipore, Bedford, USA) was evaluated by Gerstner *et al*. (133). The authors were able to separate three component mixtures of standard proteins at a productivity of 58 g/L·h. The separation of whey proteins in strong ion-exchange membranes was reported by Splitt *et al*. (147). The influence of bed length was shown to be important in the separation of the two genetic variants of β-lactoglobulin from whey, which became possible when a minimum bed length of 2 mm was chosen by coupling of commercially available flat sheet units. The capture of whey proteins from diluted solutions was achieved at a productivity of 45.5 g/L·h with more than 60-fold reduction of the feed volume. Dye ligand membranes were used by Champluvier and

Kula (42) to isolate dehydrogenases from clarified yeast extracts. A productivity of 150,000 units/L·h was found for glucose-6-phosphate dehydrogenase with Cibacron blue immobilized to nylon flat sheet membranes (43). The throughput was described to be limited by the kinetics of the association reaction. In a sequential arrangement of ion-exchange and dye-ligand-modified membranes, formate dehydrogenase was purified with a productivity between 24 and 37 g/L·h. The protein was purified 4.6-fold with 27% overall yield by this sequence, where the eluate from one step was directly applied to the next membrane without intermediate buffer exchange (142).

31.6.3.2 Monoclonal Antibody Purification. The use of anion exchange membranes to clear low level impurities such as HCP, endotoxin, DNA, and virus in the production of MAb has shown great promise in being widely implemented at the manufacturing scale. The step is operated in flowthrough mode in which, impurities are adsorbed while the product molecule flows through and is collected. The low level of impurities creates a production scenario in which process throughput, not capacity, is the limitation (148). This scenario favors membranes over adsorbent particles because of their faster transport and low pressure drop (section titled "Mass Transport"). Columns of adsorbent particles must typically be significantly oversized in order to achieve high flowrates because of pressure drop concerns. A large column diameter is used and long bed heights are needed to overcome any fluid distribution issues that can arise because of nonuniform packing or suboptimal fluid distribution in the column header. Not only is the column underutilized, but large volumes of buffers are required. The small device format of membranes, capable of delivering high volumetric flowrates at minimal pressure drops, is ideally suited for these applications.

Knudsen *et al*. demonstrated the throughput advantage of anion exchange membranes as compared to adsorbent particles, during the polishing of CHO MAb process streams (85). Both, a Sartobind Q membrane (Sartorius, Göttingen) and a column of Q-Sepharose FF adsorbent were capable of clearing low levels of HCP (10.6 ppm) to <2 ppm, but the membrane processed 300 times more protein mass (15,000 g/L vs. 50 g/L) at 8 times the velocity (~38 mL/min vs. 0.4 mL/min). The productivity of the membrane, 5680 g/L·h, was more than 3 orders of magnitude higher than that of the column (4.7 g/L·h). At higher load HCP levels, the membrane had the same HCP clearance performance as the adsorbent column at the same antibody loading, but with 8 times higher throughput. The membrane capacity for solutions of purified sperm-whale DNA was 8–9 g/L, which was consistent for a wide pH (4–9) and conductivity range (<40 mS/cm). Phillips *et al.* demonstrated the robustness of strong anion exchange membranes to changes in load pH and flowrate in the clearance of DNA, HCP, and endotoxin (86). Quaternary amine functionalized flat sheet membranes were capable of consistently clearing 3–4 logs of DNA and 4–5 logs of endotoxin from pure solutions over a wide range of pH (4–8) and up to a high flowrate of 1000 cm/h (~167 MV/min.). The clearance of both impurities was sensitive to load NaCl concentration; endotoxin and DNA clearance decreased significantly above 100 and 300 m*M* NaCl concentration in the pool, respectively. The membrane was able to reduce HCP by ~1 log from a pure MAb feed spiked with HCP and loaded out to 6000 mg/mL at 600 cm/h (~100 MV/min). The productivity of this run was ~5000 g/L·h. The Q membrane also showed high clearance of three mammalian cell viruses: xenotropic murine leukemia virus (MuLV) (>5.5 logs), mouse minute virus (MMV) (>5.2 logs), and simian virus 40 (>5.5 logs), at low solution NaCl concentration (50 m*M*). Zhou *et al*. also reported excellent viral clearance reduction factors using strong anion exchange membranes (87). Sartobind Q membranes were loaded to ~10,700 g/L (3000 g/m^2) of membranes at 7.5 MV/min and were able to clear >5.3 logs of MuLV, >6 logs of MMV, >5.5 logs of pseudorabies virus, and >6.9 logs of respiratory enteric orphan virus-III.

These studies demonstrate that membranes can outperform column chromatography in antibody production applications by delivering the necessary impurity clearance at significantly higher productivity. However, ultimately, the acceptance and widespread use of membranes in this area will be predicated on the membrane cost relative to bead adsorbents. Zhou *et al*. performed a cost comparison between an anion exchange membrane adsorber and an anion exchange adsorbent column used as the flowthrough purification step in standard antibody production process (148). Their example was extrapolated from 2-kL scale data and standard process and cost assumptions. They assumed a 15-kL scale process with a titre of 1 g/L and a 90% recovery prior to the examined process step (2700 L of feed). A 220 L, Q-Sepharose FF column (GE Healthcare, NY) loaded to 70 g/L at 76 cm/h was compared to a ~4.5 m^2 Sartobind Q membrane cartridge configuration (Sartorius, Göttingen) loaded to 3000 g/m^2 at 450 cm/h. The resin was cycled 100 times, while the membrane was only single use. The column process required 30 times more buffer than the membrane process and the processing time was 3 times longer. On a per cycle basis, the membrane process cost (~\$11,700) was higher than the column (~\$9200). Over 80% of the membrane process cost was the membrane cartridges themselves, while the resin cost made up on ~30% of the overall cost. The cost was very sensitive to membrane loading; decreasing the membrane loading to 2000 g/m^2 (33%

lower) increased the cost to $16,200 a cycle. This illustrates the importance of properly pricing the membranes. However, the initial cost comparison did not account for the advantages of the single-use nature of membranes. Since the adsorbent is cycled, additional adsorbent re-use studies, column packing studies, and adsorbent cleaning and lifetime validation studies would need to be included. When these additional column costs were added in and the processes compared over a 10-year production time frame, the membrane cost was significantly lower ($472,000 vs $611,000). Again, the membrane cartridge cost made up a high percentage of the process cost (>70%) and was 8 times higher than the resin cost. Overall, membranes appear to be well-positioned to compete with columns on a cost basis in polishing chromatography applications so long as the membrane manufacturers price the membranes modules.

31.6.4 High Resolution Purification

Membrane-shaped stationary phases were introduced into high resolution purification with the concept of HPMC (18,56,122,135,149). By choosing continuous stationary phases of a membrane type of geometry, separations could be performed at resolutions comparable to HPLC methods. The improved transport characteristics of the continuous phase allowed higher throughput in spite of the low *L/d* ratio which was used reduced pressure drop along the columns. A thorough analysis of the parameters influencing the performance of HPMC was provided by Tennikova and Svec (56). An interesting result of their investigations was an increased resolution of anion exchange HPMC during the separation of standard proteins with increasing flow velocity, whereas the separation in HIC-mode was not affected. As in the case of low resolution purification employing adsorptive membranes, step gradients yielded better separations than a linear gradient.

Attaching affinity ligands to membranes offers another opportunity to use them in high resolution purification. As shown by Nachman *et al.* (28,29,128), affinity membrane adsorbents are well suited for the fast purification of complex molecules from diluted feedstock. By coupling receptors or antibodies to commercial hollow-fiber modules (Sepracor, Marlboro, USA), automatic separations were performed at high productivity (2–6 g/L·h). Similar results were obtained with protein-A-modified hollow-fiber modules from the same manufacturer (21). A dimensionless analysis of IgG adsorption to this matrix revealed a kinetic limitation of the sorption efficiency, where breakthrough capacity increased with t_C but was independent from C_0 (23). As an alternative to hollow fibers, flat sheet membranes were modified with Protein A by Langlotz *et al.* (20) and were used to purify IgG from hybridoma supernatant. High resolution and high productivity were obtained at the same time (10.8 g/L·h).

31.7 CONCLUDING REMARKS

Over the past 20 years, membrane chromatography has been shown to have potential in the purification of a large number of biomolecules. An ample number of laboratory- or pilot-scale membrane formats functionalized with variety of chemistries are currently marketed. In addition, concern around the poor fluid distribution in large-scale modules that may have plagued early prototypes has appeared to be addressed by a number of manufacturers. Despite these advancements, relatively few large-scale applications of the technology exist. This may partly be due to the traditional reluctance to use new matrices in the industrial-scale purification of proteins, where the proof of long-term stability and reproducibility is essential for the success of the downstream process. The cost of membrane processes compared to column chromatography is another likely cause. From a cost standpoint, in order for membranes to compete with adsorbent beads in bind and elute applications despite their lower intrinsic capacity, they will have to be cycled, their performance must be proven to be maintained over the multiple cycles, and they must be priced competitively. The need for cycling, however, will make them less attractive as an alternative to adsorbent columns. The two areas in which membrane adsorbers might have the best opportunity to gain widespread use at large scales are the clearance of trace impurities and the capture and purification of large biomolecules such as DNA and viruses. Researchers have demonstrated the advantage that the more open pore structure of membranes has in binding large biomolecules when compared to bead adsorbents. As the development of gene-based therapies using DNA or viruses increase, membrane adsorbers may be properly poised to become the platform technology in this area. A number of publications have also illustrated the advantage of membrane adsorbers for impurity removal in terms of delivery of the required clearance factors at significantly higher productivity. In this application area, the key for manufacturers will be appropriately pricing modules such that their single-use nature, which can provide a clear practical processing advantage compared to column adsorbents, remains cost-effective. Finally, another consideration, but one that is less critical, is the availability of large-scale membrane modules to minimize both cycling in bind and elute applications and the footprint of the membrane process in general. However to date, only a single manufacturer makes very large modules (>5 L). More work is needed to provide manufacturers with greater process flexibility.

NOMENCLATURE

a	Specific surface area (m^2/m^3)
α	Number of bed volumes required for wash, elution, and regeneration during chromatographic cycle
B	Permeability of a packing (m^2)
C	Initial concentration (kg/m^3)
D	Diffusion coefficient (m^2/s)
D_{ax}	Axial dispersion coefficient (m^2/s)
D	Adsorbent bed diameter (m)
d_p	Pore diameter (m)
ε	Void fraction (-)
η	Dynamic viscosity [kg/(m · s)]
HETP	Height equivalent of a theoretical plate (m)
HIC	Hydrophobic interaction chromatography
HPLC	High performance liquid chromatography
HPMC	High performance membrane chromatography
HSA	Human serum albumin
k_a	Association rate constant (second order) [l/(mol · s)]
k_d	Dissociation rate constant (per s)
L	Adsorbent bed length (m)
L/d	Column aspect ratio (-)
MAb	Monoclonal antibody
N	Plate number in HETP concept (-)
P	Productivity (g/L · h)
Pe	Peclet number (-)
Phe	Phenylalanine
Q_D	Breakthrough capacity (kg/m^3)
R_E	Recovery ratio
R_s	Resolution
t_C	Residence/contact time (s)
t_D	Diffusion time (s)
t_R	Reaction time (s)
Trp	Tryptophan
V	Adsorbent volume (m^3)
v	Flow velocity (m/s)
v_e	Interstitial flow velocity (m/s)

REFERENCES

1. Janson J, Peterson T. Large-scale chromatography of proteins. In: Ganetsos G, Barker PE, editors. Preparative and production scale chromatography. New York: Marcel Dekker; 1993. pp. 559–590.
2. Boschetti E, Guerrier L, Girot P, Horvath J. J Chromatogr B 1995; 664: 225.
3. Afeyan NB, Fulton SF, Mazsaroff I, Regnier FE. Biotechnology 1990; 8: 203.
4. Brandt S, Goffe RA, Kessler SB, O'Connor JL, Zale SE. Biotechnology 1988; 6: 779.
5. Jungbauer A, Hahn R. J Chromatogr A 2008; 1184: 62.
6. Roper DK, Lightfoot EN. J Chromatogr A 1995; 702: 3.
7. Thömmes J, Kula MR. Biotechnol Prog 1995; 11: 357.
8. Charcosset C. J Chem Technol Biotechnol 1998; 71: 95.
9. Ghosh R. J Chromatogr A 2002; 953: 13.
10. Boi C. J Chromatogr B 2007; 848: 19.
11. Adachi T, Mogi M, Harada M, Kojima K. J Chromatogr B 1995; 668: 327.
12. Kasper C, Meringova L, Freitag R, Tennikova TB. J Chromatogr A 1998; 798: 6.
13. Kim M, Saito K, Furusaka S, Sato T, Sugo T, Ishigaki I. J Chromatogr 1991; 585: 45.
14. Arica MY, Yalcin E, Bayramoglu G. J Chromatogr B 2004; 807: 315.
15. Dancette OP, Taboureau J, Tournier E, Charcosset C. J Chromatogr B 1999; 723: 61.
16. Arica MY, Yilmaz M, Yalcin E, Bayramoglu G. J Chromatogr B 2004; 805: 315.
17. Bueno SMA, Haupt K, Vijayalakshmi MA. J Chromatogr B Biomed Sci Appl 1995; 667: 57.
18. Tennikova TB, Bleha M, Svec F, Almazova TV, Belenkii BG. J Chromatogr 1991; 555: 97.
19. Teeters MA, Conrardy SE, Thomas BL, Root TW, Lightfoot EN. J Chromatogr A 2003; 989: 165.
20. Langlotz P, Kroner KH. J Chromatogr 1992; 591: 107.
21. Klein E, Eichholz E, Yeager D. J Memb Sci 1994; 90: 69.
22. Klein E, Yeager D, Seshadri R, Baurmiester U. J Memb Sci 1997; 129: 31.
23. Charcosset C, Su Z, Karoor S, Duan G, Colton CK. Biotechnol Bioeng 1995; 48: 415.
24. Castilho LR, Deckwer WD, Anspach FB. J Memb Sci 2000; 172: 269.
25. Castilho LR, Anspach FB, Deckwer W-D. Biotechnol Prog 2002; 18: 776.
26. Castilho LR, Anspach FB, Deckwer WD. J Memb Sci 2002; 207: 253.
27. Yu D, McLean MD, Hall JC, Ghosh R. J Chromatogr A 2008; 1187: 128.
28. Nachman M, Azad ARM, Bailon P. Biotechnol Bioeng 1992; 40: 564.
29. Nachman M, Azad ARM, Bailon P. J Chromatogr 1992; 597: 155.
30. Kugel K, Moseley A, Harding GB, Klein E. J Memb Sci 1992; 74: 115.
31. Platonova GA, Pankova GA, Il'ina IY, Vlasov GP, Tennikova TB. J Chromatogr A 1999; 852: 129.
32. Denizli A. J Chromatogr B 2002; 772: 357.
33. Reif O, Nier V, Bahr U, Freitag R. J Chromatogr A 1994; 664: 13.
34. Beeskow TC, Kusharyoto W, Anspach FB, Kroner KH, Deckwer WD. J Chromatogr A 1995; 715: 49.
35. Tsai Y-H, Wang M-Y, Suen S-Y. J Chromatogr B 2001; 766: 133.
36. Wu CY, Suen S-Y, Chen S-C, Tzeng J-H. J Chromatogr A 2003; 996: 53.
37. Liu Y-C, ChangChien C-C, Suen S-Y. J Chromatogr B 2003; 794: 67.
38. Serpa G, Augusto E, Tamashiro W, Ribeiro M, Miranda E, Bueno S. J Chromatogr B 2005; 816: 259.

39. Lins de Aquino L, Tunes de Sousa H, Miranda E, Vilela L, Bueno S. J Chromatogr B 2006; 834: 68.
40. Ribeiro M, Vijayalakshmi M, Todorova-Balvay D, Bueno S. J Chromatogr B 2008; 861: 64.
41. Hu H-L, Wang M-Y, Chung C-H, Suen S-Y. J Chromatogr B 2006; 840: 76.
42. Champluvier B, Kula M-R. J Chromatogr 1991; 539: 315.
43. Champluvier B, Kula MR. Biotechnol Bioeng 1992; 40: 33.
44. Zeng X, Ruckenstein E. J Memb Sci 1996; 117: 271.
45. Ruckenstein E, Zeng X. J Memb Sci 1998; 142: 13
46. Grasselli M, Camperi SA, Navarro del Canizo A, Cascone A. J Sci Food Agric 1999; 79: 333.
47. Kassab A, Yavuz H, Odabasi M, Denizli A. J Chromatogr B 2000; 746: 123.
48. Arica MY, Bayramoglu G. Process Biochem 2005; 40: 1433.
49. Delattre C, Michaud P, Hamze K, Courtois B, Courtois J, Vijayalakshmi MA. J Chromatogr A 2005; 1099: 121.
50. Pirlet AS, Pitiot O, Guentas L, Heyraud A, Courtois B, Courtois J, Vijayalakshmi MA. J Chromatogr A 1998; 826: 157.
51. Pirlet A-S, Guentas L, Pitiot O, Heyraud A, Vijayalakshmi MA, Courtois B, Courtois J. J Chromatogr A 1999; 841: 1.
52. Gan HY, Sheng Z, Wang J-D. J Chromatogr A 2000; 867: 161.
53. Sun H, Zhang L, Chai H, Yu J, Qian H, Chen H. Sep Sci Technol 2006; 48: 215.
54. Petsch D, Beeskow TC, Anspach FB, Deckwer WD. J Chromatogr B 1997; 693: 79.
55. Vogel JH, Anspach B, Kroner KH, Piret JM, Haynes CA. Biotechnol Bioeng 2002; 78: 806.
56. Tennikova TB, Svec F. J Chromatogr 1993; 646: 279.
57. Kubota N, Kounosu M, Saito K, Sugita K, Wantanabe K, Sugo T. J Chromatogr A 1995; 718: 27.
58. Kubota N, Kounosu M, Saito K, Sugita K, Wantanabe K, Sugo T. J Memb Sci 1997; 134: 67.
59. Ghosh R, Wang L. J Chromatogr A 2006; 1107: 104.
60. Wang L, Kanani DM, Ghosh R. J Immunol Methods 2006; 314: 1.
61. Ghosh R. J Chromatogr A 2001; 923: 59.
62. Ghosh R. J Memb Sci 2004; 237: 109.
63. Ghosh R. J Memb Sci 2005; 260: 112.
64. Tsuneda S, Saito K, Furusaki S, Sugo T. J Chromatogr A 1995; 689: 211.
65. Freitag R, Splitt H, Reif OW. J Chromatogr A 1996; 728: 129.
66. Kubota N, Miura S, Saito K, Sugita K, Watanabe K, Sugo T. J Memb Sci 1996; 117: 135.
67. Gebauer KH, Thommes J, Kula MR. Biotechnol Bioeng 1997; 54: 181.
68. Santarelli X, Domergue F, Clofent-Sanchez G, Dabadie M, Grissely R, Cassagne C. J Chromatogr 1998; 706: 13.
69. Reif OW, Freitag R. J Chromatogr A 1993; 654: 29.
70. Lutkemeyer D, Bretschneider M, Buntmeyer H. J Chromatogr A 1993; 639: 57.
71. Luksa J, Menart V, Milicic S, Kus B, Gaberc-Porekar V, Josic D. J Chromatogr A 1994; 661: 161.
72. Demmer W, Nussbaumer D. J Chromatogr A 1999; 852: 73.
73. Heng MH, Glatz CE. Biotechnol Bioeng 1993; 42: 333.
74. Gebauer KH, Thommes J, Kula MR. Chem Eng Sci 1997; 52: 405.
75. Avramescu M-E, Borneman Z, Wessling M. Biotechnol Bioeng 2003; 84: 564.
76. Avramescu ME, Borneman Z, Wessling M. J Chromatogr A 2003; 1006: 171.
77. Hagiwara K, Yoneda S, Saito K, Shiraishi T, Sugo T, Toyjo T, Katayma E. J Chromatogr B 2005; 821: 153.
78. Splitt H, Mackenstedt I, Freitag R. J Chromatogr A 1996; 664: 87.
79. Plate K, Beutel S, Buchholz H, Demmer W, Fischer-Fruhholz S, Reif O, Ulber R, Scheper T. J Chromatogr A 2006; 1117: 81.
80. Suck K, Walter J, Menzel F, Tappe A, Kasper C, Naumann C, Zeidler R, Scheper T. J Biotechnol 2006; 121: 361.
81. Haber C, Skupsky J, Lee A, Lander R. Biotechnol Bioeng 2004; 88: 26.
82. Tseng WC, Ho F-L, Fang T-Y, Suen S-Y. J Memb Sci 2004; 233: 161.
83. Endres HN, Johnson JAC, Ross CA, Welp JK, Etzel MR. Biotechnol Appl Biochem 2003; 37: 259.
84. Zhang S, Krivosheyeva A, Nochumson S. Biotechnol Appl Biochem 2003; 37: 245.
85. Knudsen HL, Fahrner RL, Xu Y, Norling LA, Blank GS. J Chromatogr A 2001; 907: 145.
86. Phillips M, Courmier J, Ferrence J, Dowd C, Kiss R, Lutz H, Carter J. J Chromatogr A 2005; 1078: 74.
87. Zhou JX, Tressel T, Gottschalk U, Solamo F, Pastor A, Dermawan S, Hong T, Reif O, Mora J, Hutchison F, Murphy M. J Chromatogr A 2006; 1134: 66.
88. Kalbfuss B, Wolff M, Geisler L, Tappe A, Wickramasinghe SR, Thom V, Reichl U. J Memb Sci 2007; 299: 251.
89. Specht R, Han B, Wickramasinghe SR, Carlson JO, Czermak P, Wolf A, Reif OW. Biotechnol Bioeng 2004; 88: 465.
90. Han B, Specht R, Wickramasinghe SR, Carlson JO. J Chromatogr A 2005; 1092: 114.
91. Wickramasinghe SR, Carlson JO, Teske C, Hubbock J, Ulbricht M. J Memb Sci 2006; 281: 609.
92. Vicente T, Sousa M, Peixoto C, Mota J, Alves P, Carrondo M. J Memb Sci 2008; 311: 270.
93. Klein E. J Memb Sci 2000; 179: 1.
94. Swinnen K, Krul A, Van Goidsenhoven I, Van Tichelt N, Roosen A, Van Houdt K. J Chromatogr B 2007 ; 848: 97.
95. Suen SY, Liu Y-C, Chang C-S. J Chromatogr B 2003; 7979: 305.
96. Porath J, Olin B. Biochemistry 1983; 22: 1621.
97. Gaberc-Porekar V, Menart V. J Biochem Biophys Methods 2001; 49: 335.
98. Clonis YD, Labrou NE, Kotsira VPh, Mazitsos C, Melissis S, Gogolas G. J Chromatogr A 2000; 891: 33.
99. Koch C, Borg L, Skjodt K, Houen G. J Chromatogr B 1998; 718: 41.
100. Kawai T, Saito K, Lee W. J Chromatogr B 2001; 790: 131.
101. Saito K, Ito M. Ind Eng Chem Res 1989; 28: 1808–1812.

102. Iwata H, Saito K, Furusaki S. Biotechnol Prog 1991; 7: 412.
103. Kim M, Saito K, Furusaki S, Sugo T, Ishigaki I. J Chromatogr 1991; 586: 27.
104. Shinano H, Tsuneda S, Saito K, Furusaki S. Biotechnol Prog 1993; 9: 193.
105. Kobayashi K, Tsuneda S, Saito K, Yamagishi H, Furusaki S, Sugo T. J Memb Sci 1993; 76: 209.
106. Tsuneda S, Shinano H, Saito K, Furusaki S, Sugo T. Biotechnol Prog 1994; 10: 76.
107. Matoba S, Tsuneda K, Saito K, Sugo T. Biotechnology 1995; 13: 795.
108. Tsuneda S, Saito K, Sugo T, Makuuchi K. Radiat Phys Chem 1995; 2: 239.
109. Gebauer KH, Thömmes J, Kula MR. Biotechnol Bioeng 1997; 54: 181.
110. Gebauer KH, Thömmes J, Kula M-R. Chem Eng Sci 1996; 52: 405.
111. Yang H, Viera C, Fischer J, Etzel MR. Ind Eng Chem Res 2002; 41: 1597.
112. Ljunglof A, Bergvall P, Bhikhabhai R, Hjorth R. J Chromatogr A 1999; 844: 129.
113. Levy MS, O'Kennedy RD, Ayazi-Shamlou P, Dunnill P. Trends Biotechnol 2000; 18: 296.
114. Frey DD, Water VD, Zhang BR. J Chromatogr 1992; 603: 43.
115. Suen SY, Etzel MR. Chem Eng Sci 1992; 47: 1355.
116. Suen S-Y, Caracotsios M, Etzel MR. Chem Eng Sci 1993; 48: 1801.
117. Yamamoto S. Presented at Recovery of Biological Products VIII, Tucson, 1996.
118. Coffman JL, Roper DK, Lightfoot EN. Bioseparation 1994; 4: 183.
119. Roper DK, Lightfoot EN. J Chromatogr A 1995; 702: 69.
120. Liu HC, Fried JR. AIChE J 1994; 40: 40.
121. Teeters MA, Root TW, Lightfoot EN. J Chromatogr A 2002; 944: 129.
122. Josic D, Reusch J. J Chromatogr 1992; 590: 59.
123. Klein E, Eichholz E, Yeager DH. J Memb Sci 1994; 90: 69.
124. Adisaputro IA, Wu YJ, Etzel MR. J Liq Chromatogr 1996; 19: 1437.
125. Kubota N, Konno Y, Miura S, Saito K, Sugita K, Watanabe K, Sugo T. Biotechnol Prog 1996; 12: 869.
126. Suen S-Y, Etzel MR. J Chromatogr A 1994; 686: 179–192.
127. Briefs KG, Kula M-R. Chem Eng Sci 1992; 47: 141–149.
128. Nachman M. J Chromatogr 1992; 597: 167–172.
129. Yamamoto S, Sano Y. J Chromatogr 1992; 597: 173–179.
130. In Atkinson S, editor. Membrane Technology. New York: Elsevier; 2004. pp. 2–3.
131. Presentation at Biomanufacturing Process IBC's Biopharmaceutical Week, San Diego (CA), 2001. pp. 127.
132. Dubinina NI, Kurenbin OI, Tennikova TB. J Chromatogr A 1996; 753: 217.
133. Gerstner JA, Hamilton R, Cramer S. J Chromatogr 1992; 596: 173.
134. Briefs KG, Kula M-R. Membrane chromatography. In: Ladisch M, Bose R, editors. Harnessing biotechnology for the 21st century. Washington (DC): Journal of the American Chemical Society; 1992. pp. 258–261.
135. Svec F, Tennikova TB. J Bioact Compat Polym 1991; 6: 393.
136. Wheelwright SM. J Biotechnol 1989; 11: 89.
137. Le Borgne S, Graber M, Condoret JS. Bioseparation 1995; 5: 53.
138. Krause S, Kroner KH, Deckwer WD. Biotechnol Tech 1991; 5: 199.
139. Wang WK, Lei S-P, Monbouquett HG, McGregor WC. Biopharm International 1995; June: 52.
140. Kucerova Z, Turkova J. Int J Bifurcat Chaos 1997; 2: 145.
141. Klein E. Affinity membranes. New York: John Wiley & Sons; 1991.
142. Champluvier B, Kula MR. Bioseparation 1992; 2: 343.
143. Luetkemeyer D, Ameskamp N, Tebbe H, Bracht K, Lehmann J. Direct capture of monoclonal antibodies using high capacity membrane ion exchangers in pilot scale. In: Carrondo MJT, editor. Animal cell technology-from vaccines to genetic medicine. MA, Norwell: Kluwer Academic Publishers; 1997. pp. 325–329.
144. Jungbauer A, Unterluggauer F, Uhl K, Buchacher A, Steindl F, Pettauer D, Wenisch E. Biotechnol Bioeng 1988; 32: 326.
145. Ferreira GNM, Cabral JMS, Prazeres DMF. Biotechnol Prog 2000; 16: 416.
146. Lacoste-Bourgeacq JF, Desneux C, Allary M. Chromatographia 1991; 532: 27.
147. Splitt H, Mackenstedt I, Freitag R. J Chromatogr A 1996; 729: 87.
148. Zhou JX, Tressel T. Biotechnol Prog 2006; 22: 341.
149. Josic D, Lim YP, Strancar A, Reutter W. J Chromatogr A 1994; 662: 217.

32

MEMBRANE SEPARATIONS

MANOHAR KALYANPUR

Consultant, Bioseparations & Pharmaceutical Validation, Plaisir, France

32.1 MEMBRANE SEPARATIONS

This chapter aims to cover the developments in the field of membranes used for separations and their particular application in the domain of biotechnology. Different kinds of membranes have been developed to date. These are described along with the criteria one needs to take into consideration before deciding to employ a specific type of membrane for a given application. The membranes can be used in different ways to suit the application and the goals such as product yield and quality set by the user. These methods are described in detail so as to provide the reader with the knowledge of the advantages of using one method as opposed to the other one.

32.2 INTRODUCTION

The term filtration refers to the separation of two or more components from a liquid or gas. Generally, it refers to the separation of solids from either liquids or gases. In membrane separations, one goes beyond the classical separation of solids from the other phases. Membranes can be and are used to separate dissolved solutes in a solution from one another and also to remove the solutes from the liquid in which they are dissolved. In this manner, membranes are used to either purify products or to concentrate them. This is particularly advantageous in biotechnology where the products of biological origin are very often proteinaceous in nature and are heat labile. The membranes offer a means of separation that occurs at or below ambient temperatures. This gives the biotechnologist a separation or purification technique that provides maximum assurance of product stability. In membrane based separations, the membrane acts as a barrier to the passage of certain components while others move freely across the membrane. Therefore, the end result of a simple membrane separation is the concentration in the upstream fraction of the components that do not cross the membrane barrier. This fraction is referred to as the *retentate*. The fraction which passes across the membrane is called the *permeate* or *filtrate* and contains the components of the solution in essentially the same concentration as the starting fluid. This is the simple basic *membrane separation*.

32.3 THREE MAJOR APPLICATIONS OF MEMBRANE SEPARATIONS

32.3.1 Product Purification

This is the most important application for membrane separations in biotechnology. Membranes are made of different materials and the biotechnologist needs to make his choice based on the goal of the application and the suitability of the membrane for the application. This chapter aims at providing the biotechnologist with information that will help him or her to select the best membrane and the manner in which to use the membrane in order to accomplish the separation and purification task easily. Membrane separations are an important part of both the upstream and downstream processing in biotechnology. A typical process can include a number of steps such as centrifugation followed by membrane based clarification of the supernatant, one or more

Downstream Industrial Biotechnology: Recovery and Purification, First Edition. Edited by Michael C. Flickinger.

chromatographic procedures and final membrane processing steps to accomplish further purification, and finally concentration.

32.3.2 Sterile Filtration

One of the most critical steps in the manufacture of injectable therapeutic agents is the sterilization of the product. The simplest method of sterilizing solutions is the one that employs heat and autoclaving is a typical example of such methods. However, products of the biotechnology industry are most often thermolabile and therefore, cannot be heat sterilized. Membranes provide an excellent alternative here. The membrane filters are today used for filtering out bacteria and other microorganisms from the final formulations rendering the products very safe for injection. This step is followed by lyophilization of products that are marketed as solid powders. Those products which are marketed as liquid formulations are filled straight into ampoules, vials or other appropriate containers after sterile filtration through membrane filters.

32.3.3 Virus Removal by Membrane Filtration

This is the most recent application of membrane separations in the biopharmaceutical industry. A new generation of membranes is today used to remove viruses from biological preparations. Therapeutic agents manufactured from blood, by cell culture processes or by the most modern method employing transgenic animals, can contain dangerous viral contaminants from the source materials. These products can be filtered through membrane devices to remove most of the viral contaminants including the human immunodeficiency virus (HIV) and the viruses known to cause different forms of hepatitis.

There is a point of differentiation here from the previously mentioned sterile filtration in which there is complete or total removal of microbiological contaminants from the solution. In virus filtration, the removal is not complete, especially when the viruses are very small. Therefore, for complete viral safety, the practice is to employ membrane filtration as a method complimentary to the longer known generic methods of viral inactivation such as treatment with a combination solvents and detergents, other chemical inactivating agents and pasteurization by mild heating for prolonged periods of time.

32.4 CLASSIFICATION OF MEMBRANES AND MEMBRANE PROCESSES

The more commonly used membrane processes in the biotechnology industry are microfiltration (MF), ultrafiltration (UF) and reverse osmosis (RO). All three are pressure driven processes and differ from each other in the size of the components they retain. The membrane itself, under the influence of applied pressure controls the retention and passage of individual components of the solution being processed.

32.4.0.1 Microfiltration. These processes retain particles in the submicron range that are suspended in the process fluid. The membranes are also used in removing particles and microorganisms from air and gases as in the fermentation processes in biotechnology and other industries. The particles removed by membrane filtration range in size from 0.1 μm to 10 μm. Conventional filtration processes such as those employing pad filters and filter presses in the more conventional pharmaceutical industry are generally capable of retaining particles only above the 10 μm size.

32.4.0.2 Ultrafiltration. These membranes generally retains macromolecules or particles from about 0.001 to 0.2 μm in size. By choosing one or more of the available membranes, one can separate larger macromolecules from smaller ones within the above size ranges. For example, separation of large proteins or polysaccharides from smaller molecules is a typical application for ultrafiltration membranes in the biotechnology industry.

32.4.0.3 Reverse Osmosis. This turns out to be one of the oldest known applications for membrane separations. This is the process where membranes employed are capable of retaining even smaller molecules than ultrafiltration membranes. Salts and other molecules below 1000 Dalton molecular weight which are not normally retained by ultrafiltration membranes, are retained by the reverse osmosis membranes. The first use of these membranes was in desalting sea water. The membranes are today used for concentrating small molecules while ultrafiltration is employed for fractionation and purification of macromolecules between a thousand and a million Dalton molecular weights. Figure 32.1 summarizes the point of use of membrane filters in biotechnology. Table 32.1 gives examples of products that can be retained by the three types of membranes. In a typical biotechnology process, microfiltration is the first membrane based step and is employed to separate suspended material such as cells from a bioreactor and other particulate matter including cell debris from the multitude of dissolved substances, which very often includes the specific product of interest expressed by the cell line. In many instances, clarifying filters made of a variety of materials are used for preclarification of the process stream prior to microfiltration. This usually helps to improve the performance of the microfiltration step and extends the life of the membrane filter. Honig and Schwartz (1) have given an excellent review of this subject. The filtrate or permeate from the microfiltration step contains the desired product elaborated

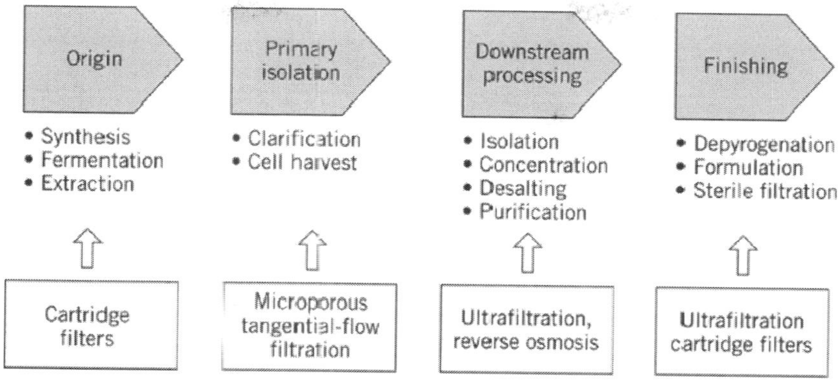

Figure 32.1. Different steps in downstream processing where membrane separations are employed.

TABLE 32.1. Products Retained by Microfiltration, Ultrafiltration and Reverse Osmosis Membranes

Size	Product	Membrane Process
100 μm	Pollen	
	Starch granules	
10 μm	Blood cells	Microfiltration
	Bacteria	
1 μm	Smaller organisms	
50,000–100,000 Dalton	Albumin, Immunoglobulins	
10,000–100,000 Dalton	Enzymes, other proteins,	Ultrafiltration
1000–10,000 Dalton	Other proteins, large oligopeptides.	
	Glucose,	
<1000 Dalton	Small peptides,	Reverse Osmosis.
	Salts	

by the process along with other proteins, carbohydrates and smaller molecules such as salts from the culture medium as well as those produced during the process. To separate the product of interest from these contaminants, the ideal membrane separation step is ultrafiltration. Membranes with the appropriate nominal molecular weight cut-off, a term used to identify and differentiate ultrafiltration membranes from one another, are next used in successive purification steps to facilitate recovery of the desired product of high purity.

Quite often, the manufacturing process also includes chromatographic or other suitable process steps in between membrane separations to achieve an overall process advantage in terms of product quality and yield. Figure 32.2 illustrates how membrane and chromatography procedures are integrated together step by step for upstream and downstream processing in the biotechnology industry.

32.5 MEMBRANE CHEMISTRY, STRUCTURE AND FUNCTION

The high volume microfiltration in the biopharmaceutical industry requires removal of particulate and colloidal substances in an economical and efficient manner. The nature of these particles and the interplay of filtration dynamics contribute to the difficulty of the task. It is important to balance filter performance against process costs in order to arrive at a system which provides the ideal separation solution at a low operating cost.

In general, the filtration process starts out with passing the process stream through a porous medium that entraps the solids in its matrix or retains them on its surface. The materials used in the manufacture of these filters are porous or fibrous such as paper, felt, woven yarns and fabrics, etc. Depending upon the basic filtration mechanisms, particle retention mode and construction characteristics, the different filters used in the biologicals industry are classified as *depth filters*, *surface filters* and *screen filters*. Figures 32.3, 32.4 and 32.5 show the mechanism of retention of particulates by these three types of filters.

Depth filters consist of fibrous, granular or sintered materials that produce a random porous structure. When a fluid passes through a depth filter, particles become entrapped in its tortuous network of flow channels. Particle retention mechanisms are random adsorption and mechanical entrappment throughout the depth of the filter matrix. The filtration media of these filters can be wound cotton, rayon or polypropylene as well as resin bonded laminates such as cellulose and fiber glass.

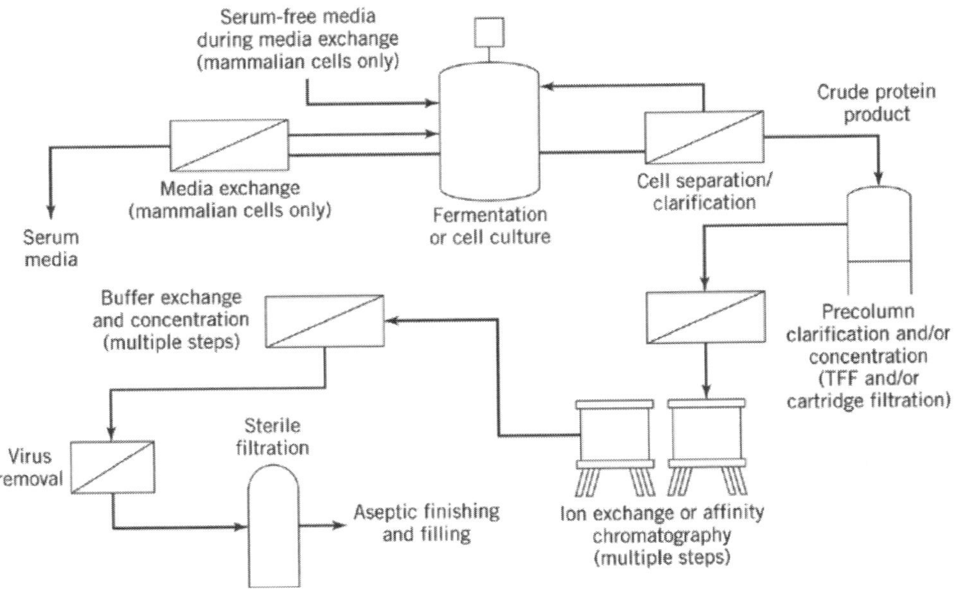

Figure 32.2. Integreation of membrane separations with other techniques in downstream product recovery operations in biotechnology.

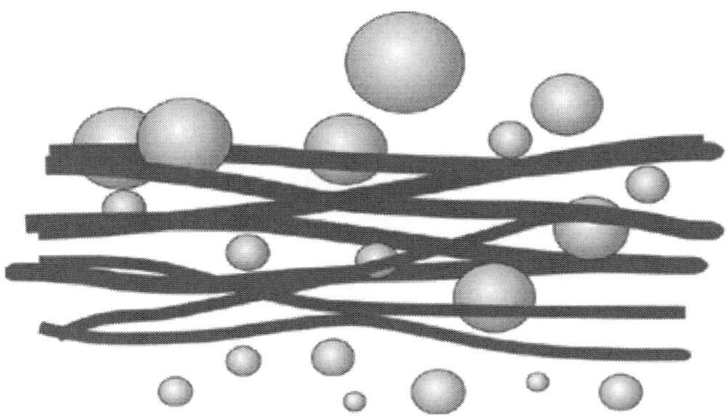

Figure 32.3. Mechanism of particle retention by depth filters.

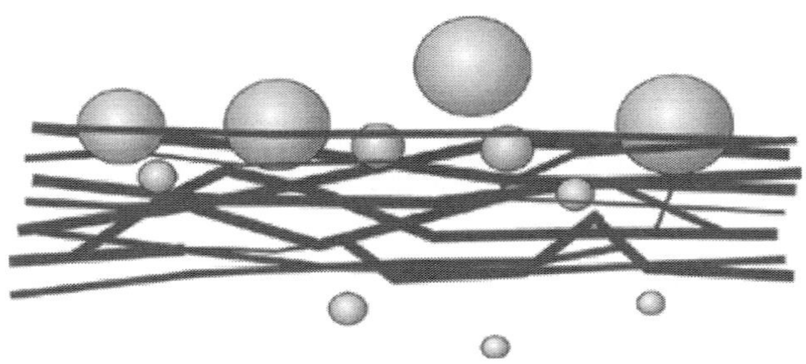

Figure 32.4. Mechanism of particle retention by surface filters.

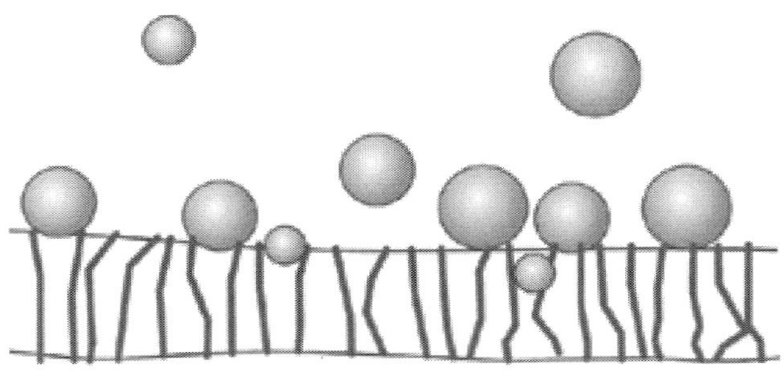

Figure 32.5. Mechanism of particle retention by screen filters.

Surface filters are made up of several layers of media like glass or polymeric microfibers. During filtration, particles larger than the spaces within the filter matrix are retained primarily on the filter surface. Other smaller particles can get entrapped within the matrix. Thus these filters have the property of both depth and screen filters. The surface filters are generally made of polymeric materials, resin bonded paper or fiberglass and paper.

Screen filters have a geometrically regular porous matrix and particles are retained on its surface primarily by a sieving or size exclusion mechanism. Small particles and microorganisms larger than the pore size are retained on the surface. The filtration process here is "absolute" because everything larger than the rated pore size cannot get through the filter. Examples of such filters include cast polymeric membranes used for all critical submicronic and macromolecular separations in the biotechnology industry. The structure of the filter material is more rigid and the manufacture is much better controlled than for the other types of filters.

Screen filters made of *microfiltration membranes* have a well defined pore size in the range of 0.1 to 1 μm. Since particles and microorganisms are retained at or near the surface of the membrane, there is very little possibility of the so called "*grow through*" phenomenon taking place when these filters are employed for sterile filtration applications in the manufacture of parenteral drugs. This is a truly major application for the microfiltration membranes in the general pharmaceutical and biotechnology industries. They also find use in another biotechnology application which is the harvesting of cells from bioreactors.

Ultrafiltration membranes are the second major group of screen filters and are often called *asymmetric* or *skinned* membranes. The skin of the membrane does the job of retaining the dissolved solutes larger than the membrane rating. the skin is supported by a more rigid support layer or backing. Unlike microfiltration membranes, ultrafiltration membranes are not designated by their pore size but according to the molecular weights of the dissolved solutes which the membrane will *retain* or *reject*. This as mentioned before is called the nominal molecular weight cut-off (NWCO) of the membrane.

32.6 METHODS OF MEMBRANE MANUFACTURE

Most membranes are made of organic polymers such as the natural polymer, cellulose and its derivatives like cellulose acetate and certain other synthetic polymers like poyethylene, polycarbonate, polysulfone and their derivatives. Some membranes are also made of inorganic materials like zirconium and aluminium oxide. These are often referred to as *ceramic membranes*.

In the manufacture of polymeric membranes, the organic polymers are dissolved in an appropriate solvent and combined with the non solvent or swelling agent. This makes up the membrane casting solution. The membranes are cast on casting machines which are almost always made by the membrane manufacturers themselves. Most membranes are made in the form of flat sheets but some are cast on the module support as in the case of tubular module membranes. Some membranes are also made directly in the form of hollow fibers. The membranes are then annealed.

Flat sheet membranes are subsequently incorporated into flat membrane plate and frame modules or made into spirally wound or pleated cartridges. Critical details of membrane manufacture are well guarded secrets and are proprietery processes developed and owned by membrane manufacturers.

Microfiltration membranes are made to a thickness of about 100 to 120 μm and the retention of particulate material takes place at the membrane surface or within the first few micrometers of the tortuous length of the membrane pores. *Ultrafiltration membranes* consist of a very thin skin which performs the task of separation of the solutes in solution. This skin is supported by a much thicker support material which contains voids, often referred to as "finger voids" due to their finger like shape. These voids facilitate the rapid flow of the fluid once it gets past the fine skin of the ultrafiltraton membrane. *Reverse osmosis membranes*, as per their classical definition, are meant to retain all solutes other than the solvent itself. These were originally developed for sea water desalination to make potable water. This is still by far the largest industrial application for reverse osmosis membranes. These membranes have since been made into somewhat modified versions which can retain smaller molecular weight solutes that are generally not retained by the ultrafiltration membranes with a low nominal molecular weight cut-off. The molecules retained are larger than the salts which can pass more easily through the membrane. These new generation reverse osmosis membranes find a niche application in the concentration of low molecular weight solutes such as peptides, antibiotics and other products in the biopharmaceutical industry. Table 32.2 summarizes the characteristics of the three membrane processes described.

32.6.1 Track-Etch Membranes

As described earlier, most polymeric membranes contain tortuous pores in a spongy material. The pores of these membranes can have a fairly wide pore size distribution as a result of their method of manufacture. A different type of membranes is made by a two step process which includes the manufacture of the polymer sheet membrane of a certain thickness followed by bombardment of the membrane sheet with massive energetic nuclei in a beam of U^{235} fission fragments. These nuclei go through the membrane leaving tracks which are then the pores formed in the membrane. After the bombardment step the membrane undergoes further process steps to make it antistatic and to modify its surface characteristics such as either hydrophobicity or hydrophilicity.

TABLE 32.2. Characteristics of Membrane Processes

Process	Permeate	Retentate
Microfiltration[a]	Solvent with all dissolved solutes	Particulates in suspension, colloidal material
Ultrafiltration[b]	Water with dissolved solutes of lower molecular weight	Dissolved solutes of higher molecular weight and colloidal material
Reverse Osmosis	Solvent, typically water	solutes

[a]In microfiltration, particulate retention generally depends upon the pore size of the membrane.
[b]In ultrafiltration, the passage or retention of solutes depends on the nominal molecular weight out-off of the membrane.

It is natural and quite logical to assume that such a process of making pores in a membrane would produce a membrane with highly uniform pore diameters and with a sharp nominal molecular weight cut-off. However, in actual practice, the random bombardment can as well result in two or more pores being formed close together thus giving rise to much larger pores than the individual pores. This problem of overlapping pores is supposedly avoided if the nuclear bombardment takes place at an angle of 29° between the the particle beam and the membrane sheet (2). There are also some other disadvantages to these membranes. Their extreme thinness of only about 15 μm which makes them difficult to handle. Also if the bombarding particles fail to penetrate completely through the membrane sheet, the apparent pores in the membrane are not true pores.

32.6.2 The New Ultrafiltration and Virus Removal Membranes

The ultrafiltration membranes were introduced in downstream processing in the biopharmaceutical industry many years ago. Typical membranes exhibited large voided structures and frequently the voids tended to break and create an open channel through the membrane. The product that was normally to be retained by the membrane would be lost through these openings. Membrane manufacturers tightened the retention profiles of the membranes that reduced the hydraulic capacity of the membranes. But the reduced permeability had a negative impact on process economics due to either increased processing times or a need to increase the membrane surface area.

With the emergence of new regulatory guidelines to reduce the level of viral contamination in both biological and biotechnology products, Millipore Corporation increased their effort to develop a new generation of membranes for virus filtration. The voids in the classical ultrafiltration membranes were eliminated by reducing the thickness of the active layer that also increased the hydraulic permeability of the membranes. Since thinner membranes process fluids faster, they improve process economics. The new membranes have sharper retention profiles than their predecessors and permit a greater specificity of the membranes for a given application.

Membranes with the reduced thickness called for better substrates upon which they were to be formed. Millipore Corporation developed a new process to cast these void free membranes on microporous substrates that protected the thin membrane. These new membranes are known as "composite" membranes that possess the distinct advantage of robustness and a high degree of tolerance for the rigors of device manufacture as well as everyday use in the biotechnology industry. The membranes are available from Millipore Corporation in their new range of "*Biomax*" and "*Ultracel*" ultrafiltration membranes and the "*Viresolve*" range of virus removal membranes. Now the industry has a wider choice of membrane devices to choose from for a specific application.

32.7 HOW MEMBRANE PROCESSES ARE OPERATED

Membrane processes are driven by pressure and they bring a great advantage to the biotechnology industry where a very large majority of the products are heat labile. The processes also consume less energy since they do not require either heat transfer or heat generation. They call for a pump to run the process fluid through the membrane under pressure as illustrated in Figure 32.6. Some of the processes can, however, be run with either compressed air or nitrogen and do not need pumps. Most membrane processes are run at ambient temperature with few exceptions such as those processes where extremely heat labile products call for the processes to be run at lower than ambient temperatures. Low temperature membrane processing is also common when the process time is long and there is the accompanying risk of product degradation or microbial contamination of the process fluid. At the other extreme, higher temperatures are employed when a fluid is very viscous and the product of interest is not quite heat sensitive. The higher temperature reduces the viscosity of the process stream and increases the rate of filtration with the overall advantage of significant reduction in process time.

In a well developed process in the biotechnology industry, it is quite common to find membrane filtrations operating in tandem with one or more chromatographic purification steps or other methods of separation and purification. Further, one may also use more than one membrane separation step i.e. microfiltration, ultrafiltration and reverse osmosis at appropriate points in the downstream process just as different chromatography procedures like ion exchange, size exclusion or affinity methods are used in a complimentary fashion. One single membrane separation step is not always enough to accomplish the end result. Centrifugation is sometimes used prior to the membrane processes for clarification of fluids containing a high biomass as in fermentation

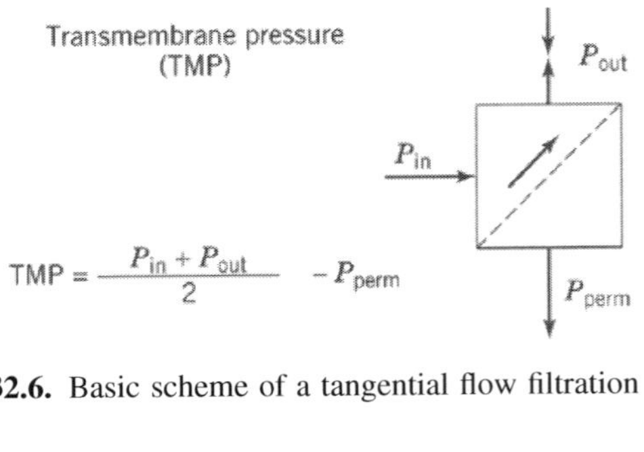

Figure 32.6. Basic scheme of a tangential flow filtration process.

processes. The supernatant of the centrifugation is then better handled by other methods of purification including membrane based processes.

32.7.1 Two Methods of Operation of Membrane Processes

The processes are run according to two distinctly different methods. Figure 32.7 illustrates these two methods of filtration. The first is the classical method of filtering a solution containing suspended particulates either to recover the particulates themselves, as in the recovery of a product crystallized from a solution or to obtain a clear filtrate when the product is in the filtrate. This method is known as *dead end filtration* or *normal flow filtration*.

When the solution being filtered contains a high concentration of particulates or bioburden, the filter can quickly get clogged due to the accumulation of solids at the surface of the membrane. This usually ends up with the deposition of a boundary layer at the membrane surface which then acts as a secondary membrane. This prevents free passage of the liquid and eventually the membrane is completely clogged. When this happens filtration cannot be continued and the only solution is to stop the process and replace the membrane filter with a new one. Such stoppages are time consuming and often result in loss of valuable product affecting the economics of the overall process.

A very practical and often successful alternative to the classical dead end filtration is the use of the membrane filters by a method known as *tangential flow filtration* or *cross flow filtration*. Figure 32.6 illustrates the circulation of the complete process fluid in a direction parallel to the membrane surface.

Different types of pumps are used in the process. The advantage of the method is that the particulates which would accumulate at the membrane surface in dead end filtration, are now kept in the body of the fluid. They stay away from the membrane surface permitting the liquid to flow relatively unhindered through the membrane pores. If the rate of filtration or *flux* drops significantly, the process fluid can be diluted by adding an appropriate pure solvent such as water or a buffer. This procedure helps to recover much of the original flux and the filtration can be continued as before. The addition of the diluting fluid

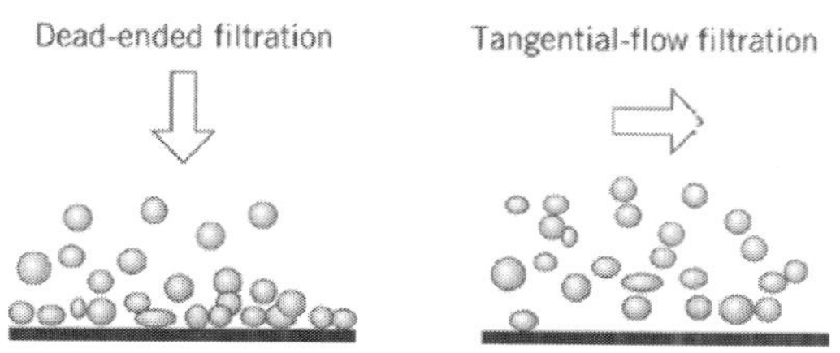

Figure 32.7. The two methods of using membrane filters.

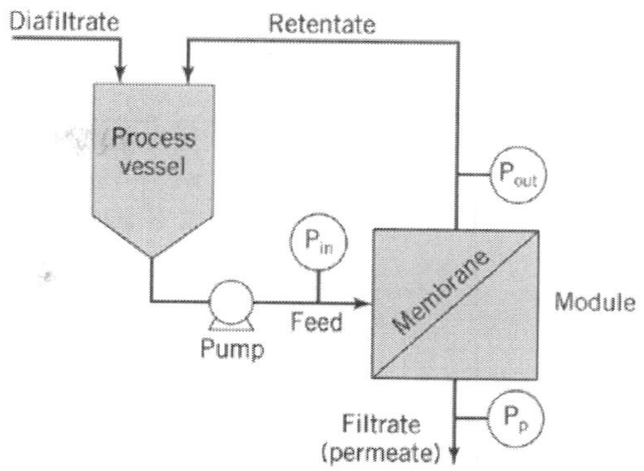

Figure 32.8. Schematic of a tangential flow filtration operation.

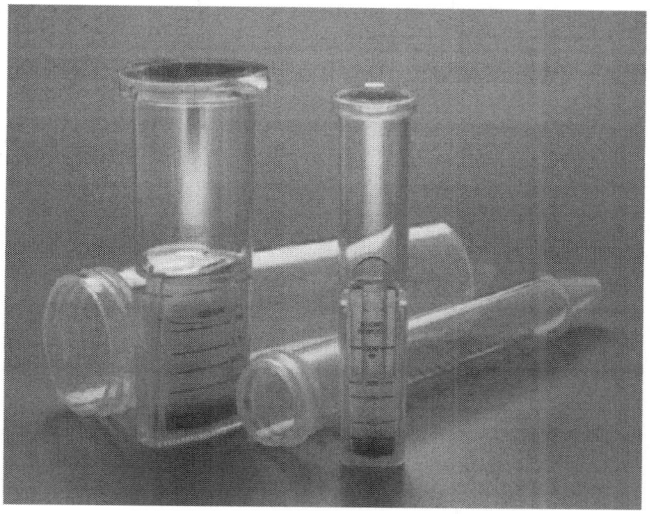

Figure 32.9. Small ultrafiltration devices used in centrifuges for small scale ultrafiltration. Photo courtesy Millipore S.A., Molsheim, France.

can take place without stopping the process as shown in figure. This step of *in process dilution* is referred to as *diafiltration*. A typical tangential flow filtration process scheme is illustrated in Figure 32.8. In case the desired product is in the filtrate which is now more dilute, the filtrate can be reconcentrated to the desired concentration by either ultrafiltration or reverse osmosis, choosing the appropriate membrane according to the molecular weight of the product of interest. The concentration and further purification can also be performed by a chromatographic method.

32.7.2 Equipment Employed in Membrane Filtration

Membrane filters in different configuration are required for different processes depending on the method of filtration used. The choice of a particular type of filter depends on several considerations such as process fluid composition,

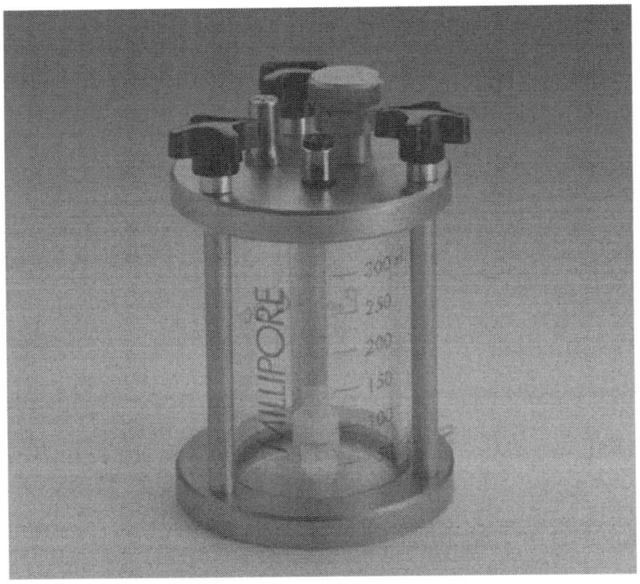

Figure 32.10. A bench top device for small volume ultrafiltration applications. Photo courtesy Millipore S.A., Molsheim, France.

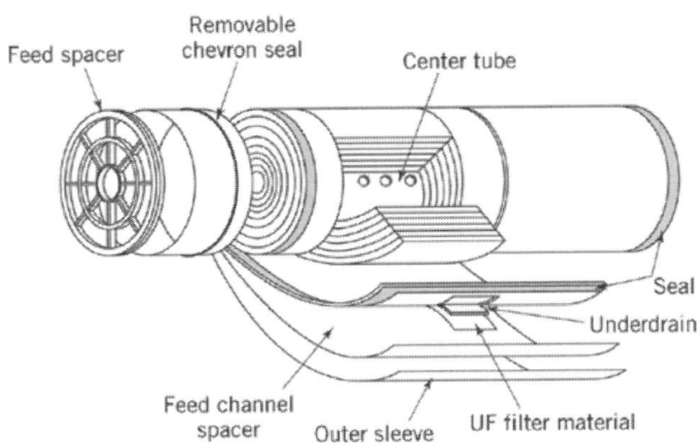

Figure 32.11. Components of a spiral wound ultrafiltration membrane module. Photo courtesy Millipore S.A., Molsheim, France.

goal of the process which would either be to recover the particulates or the filtrate and of course, the economics of the process.

32.7.3 Equipment for Microfltration in the Dead End or Normal Flow Filtration Mode

For clarifying small volumes of solutions as in a research and development environment, small filter discs of different diameters and pore sizes are employed. The filter discs are placed in appropriate holders of plastic materials or stainless steel. The stainless steel holders can be autoclaved for sterility before the filtration and can be used over and over again.

These less expensive filters are also quite useful in determining what type of filters are appropriate for a certain

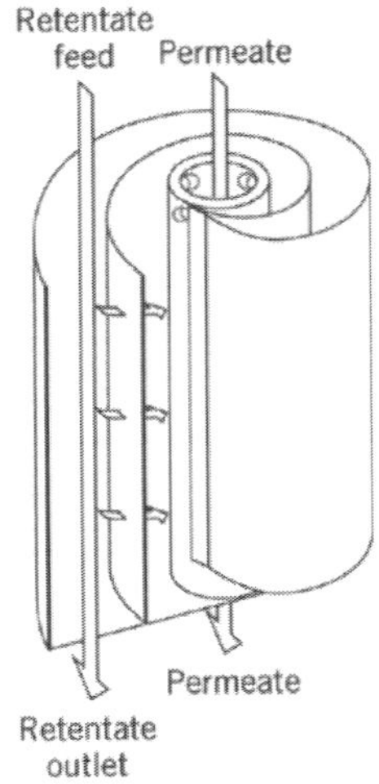

Figure 32.12. Flow schematic through a spiral wound ultrafiltration module. Photo courtesy Millipore S.A., Molsheim, France.

application. Filters are also available in the form of fabricated devices which can be placed at the end of a syringe. Such filters are ideal for filtering volumes from 1 ml to about 100 ml. For applications where large volumes of process fluids need to be filtered as in an industrial environment, cartridge filters placed in suitable stainless steel holders are ideal. The filters placed in such holders are easy to handle since and can be autoclaved or steam sterilised and sometimes cleaned for reuse.

Quite often, the particulate level in filtrations is high and the membrane filters can quickly get clogged. The common solution to this problem is to place in the filtration step prefilters and clarifying filters before the membrane filters. The economic advantage of extending the life of the usually more expensive final filter is far greater than the cost of the prefilters. In well designed processes, the filters are used in a sequence as follows:

Clarifying filter $\Rightarrow$ Prefilter $\Rightarrow$ Membrane filter

Each filter removes the suspended or colloidal material for the process stream to improve the performance of the filter next in line (1).

32.7.4 Equipment for Small Scale Ultrafiltration Including Tangential flow Filtration

For concentrating small volumes of 1 to 20 ml protein solutions, centrifugal filtration units (Figure 32.9) fitted with ultrafiltration membranes are available. Other devices such as the stirred cell (Figure 32.10) which can be used with a wide range of ultrafiltration membranes are also marketed by filter suppliers. These devices are ideal for processing volumes of a few milliliters to about half a liter. A stirring bar helps to stir the fluid above the membrane and keeps

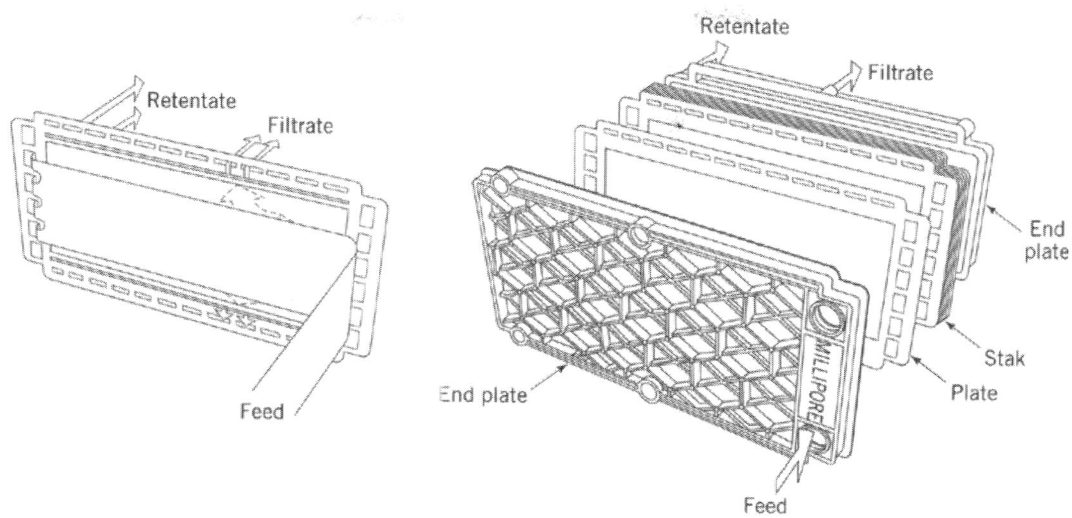

Figure 32.13. Flow schematic through an individual membrane plate (left) and through the entire Prostak plate and frame membrane module (right). Photo courtesy Millipore S.A., Molsheim, France.

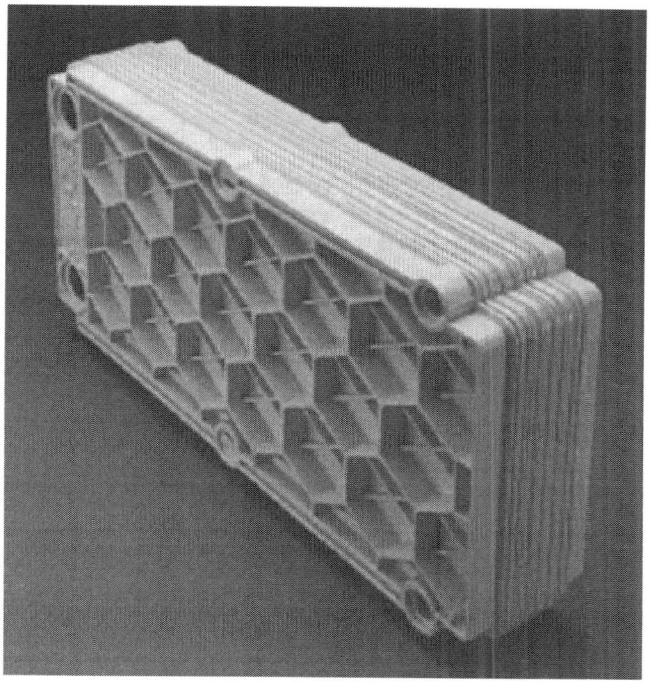

Figure 32.14. Photograph of a Prostak membrane module. Photo courtesy Millipore S.A. Molsheim, France.

the foulants moving in the body of the solution, reducing filter clogging and increasing filtration rates. Although this device will help to achieve small laboratory scale separations or protein concentrations, it will not give indications of the membrane area or process conditions to use when the process is eventually scaled up to large volume operations.

Since large scale ultrafiltration processes are usually conducted in tangential flow filtration systems, it is better to start with a device that operates in this mode at the research and development laboartory. Small scale devices for these applications are available from the filter suppliers. Devices that hold single sheets of membrane or small filter modules with more membrane area are also available. Here it must be borne in mind that it is important to proportionately increse membrane area of the filtration device as the volume of the process fluid increases. When the process is scaled up to the pilot plant level or to full scale manufacture, large surface area modules and systems are available and offer the convenience of proper scale up. The membrane filters in the following configurations are available from suppliers: *hollow fiber modules*, *spiral wound modules*, *plate and frame devices using flat sheet membranes*, and *tubular filters*. All these configurations have been developed by individual manufacturers based on sound hydrodynamic principles and rather complex mathematical models for module and system performance. The idea behind the development of each type of device is to improve its performance and ultimately reduce processing costs for the end user. The decision to use a particular device for a given application is usually made by taking into consideration the properties, more specifically, the composition of the process stream and the goal of the application and the results of the initial small scale trials. The impact of the filtration step on the overall process economics is undoubtedly an important factor.

Small devices operated on bench tops are ideal for early studies in the research and development laboratories. Properly performed experiments help to determine which membrane and membrane device are best suited for a particular application and also for correct optimization of the process. Most filter manufacturers help in proper planning of

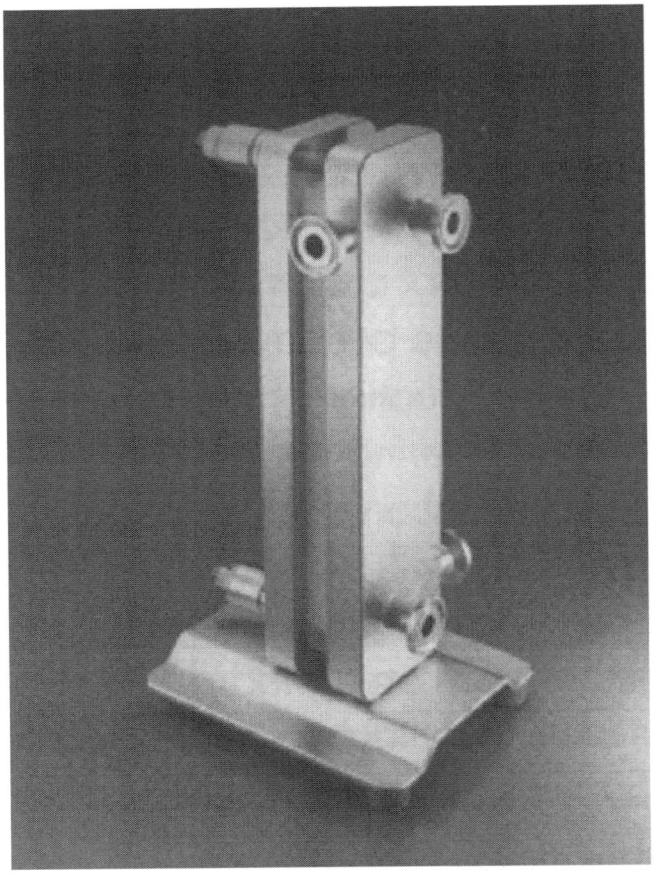

Figure 32.15. A simple bench top device for evaluation of membranes in tangential flow filtration in downstream processes.

these stuidies which are critical when one needs a quick answer to a bioseparation problem. Some companies even offer hands-on training for the operators. The result of the studies is the knowledge of which membrane pore size (in case of microfiltration membranes) or what nominal molecular weight cut-off (in case of ultrafiltration membranes) is best for the application.

Hollow fiber modules are sometimes employed in ultrafiltration processes. The active membrane which participates in the separation process is on the inside of the hollow fiber of a uniform bore diameter. The process fluid is pumped through the hollow fiber and the solvent and molecules which cross the membrane appear on the outside of the fiber. Bundles of many fibers from 50 to 3000 based on the internal diameter are sealed inside a transperent cylindrical cartridge with the two ends sealed with epoxy sheets. Each cartridge has permeate ports at either end and also feed and retentate ports. The ports can be connected to the piping of the system in which the module is installed by standard triclover clamps. Hollow fiber modules carrying ultrafiltration membranes of different cut-offs and surface area are available from several suppliers.

Figure 32.16. An assembly of Prostak modules giving several square feet of membrane area for use in a large system for bioseparations.

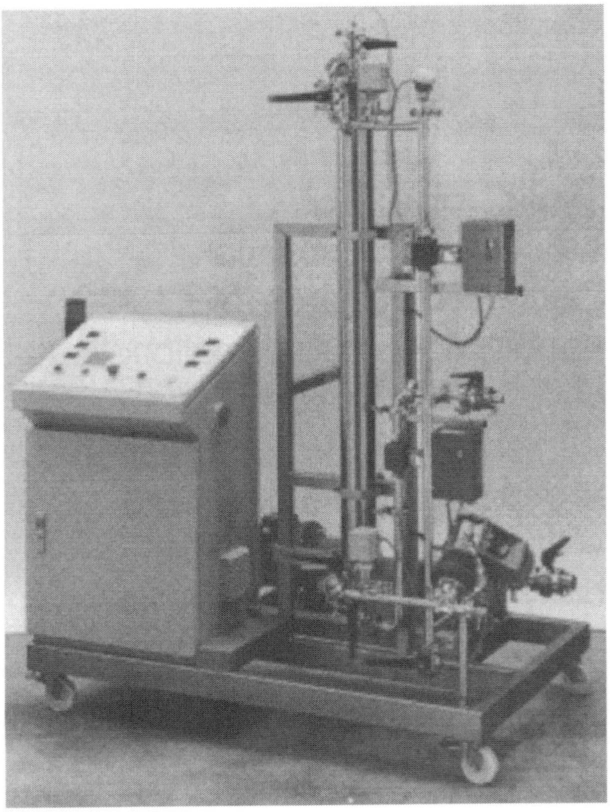

Figure 32.17. A small system with spiral wound membrane modules. Photo courtesy Millipore S.A., Molsheim, France.

Figure 32.18. A fully automatic system with several membrane cassettes for protein concentration. Photo courtesy Millipore S.A , Molsheim, France.

Figure 32.19. A fully automatic system used in downstream processing.

32.7.5 Spiral Wound Modules

These offer the biotechnology industry the most compact and least expensive of the ultrafiltration devices. The device is constructed of the same flat sheet membranes used in plate and frame devices. Each cartridge is made by wrapping alternate layers of membrane and separator screens concentrically around a hollow core as shown in the Figure 32.11. The flow schematic in a spiral wound ultrafiltration membrane cartridge manufactured by Millipore Corporation is shown in Figure 32.12. The product to be processed enters one end of the cartridge under pressure and flows tangentially down the central axis. The ultrafiltrate containing salts, water and molecules not retained by the membrane flows through the membrane into the permeate channel and spirals all the way to the central core of the cartridge. The permeate is continuously removed from the central core while the retentate containing the rejected Spiral cartridges made with different ultrafiltration membranes with varying cut-offs are available from filter suppliers.

32.7.6 Flat Sheet Cassettes and Plate and Frame Modules

This is a frequently employed configuration in tangential flow filtration and devices offered by suppliers contain microfiltration membranes of different pore sizes and ultrafiltration membranes with several cut-offs. The microfiltration membranes are the same ones offered in cartridges used in the dead end mode. Small devices which can be operated in the TFF mode with single sheets of membranes or modules are ideal. The work at this scale will help the user to understand problems of filter clogging or membrane polarization and fouling phenomena and how to address the problems rapidly by either choosing another type of membrane or changing the process conditions.

The *cassette filter* devices are made by stacking and bonding together identical membrane sheets and separator screens together in parallel. Devices with identical membranes contain the same materials of construction. Therefore development of a process can start with a small mini-cassette for the small volume experiment and is eventually scaled up to a several hundred liter process by simply stacking together many larger surface area cassettes in stainless steel or acrylic holders.

32.7.6.1 Plate and Frame Modules. As a process moves from the small bench top experiment in the research laboratory to the floor of the manufacturing plant, the industry needs equipment to process large volumes of process streams. The filter suppliers have responded to this need of the biotechnology and other industries by developing modules which not only carry large membrane areas of several square meters but are extremely robust. These can be used repeatedly after proper cleaning and sanitization for several years. Typical examples of such devices are the Prostak membrane modules available from Millipore Corporation. These are tangential flow stacked plate devices. The modules are solvent bonded and contain neither adhesives nor compression-dependent membrane seals. The modules containing Millipore's Durapore range of microporous membranes can be steam sterilized dozens of times. Figure 32.13 shows the flow path of the process fluid in the Prostak modules which are shown in Figure 32.14. Since these are available in both the microfiltration and ultrafiltration versions it is possible to use the same hardware for both steps in a downstream process giving a significant reduction in capital investment. Most filter suppliers offer either manually operated or fully automatic systems which contain the membrane devices. A small bench top filter evaluation module in a stainless steel holder is illustrated in Figure 32.15 while Figure 32.16 shows how a number of plate and frame membrane modules can be put together in several stainless steel holders and plumbed together to form an assembly of membrane modules which can be the part of a large process system. Figures 32.17, 32.18 and 32.19 illustrate a variety of typical small and large membrane systems used in the biotechnology industry. The systems carry the different membrane configurations described above.

32.8 CONCLUSION

The information provided above will give present and potential users of membranes for bioseparations a clear picture of how best to approach the task to achieve the desired end results. The manufacturers of membranes and membrane devices are always ready to give the users the necessary help in choosing the right product for each application. For detailed information on membrane separations, the readers can also refer to numerous review articles and text books on the subject such as the one on ultrafiltration by M. Cheryan (3). Another useful book edited by McGregor (4) includes chapters on a variety of specialised topics written by persons practising the art of membrane separations in biotechnology. The use of microfiltration in the tangential mode in the recovery of intracellular products has been reviewed by Le and Atkinson (5) while Short and Webster (6) have authored an excellent chapter on the use of ultrafiltration in the pharmaceutical industry.

REFERENCES

1. E.S. Honig and P.D. Schwartz, Filtration and Separation, Jan/Feb 1997.
2. M.C. Porter, Handbook of Separation Techniques for Chemical Engineers, published by McGraw-Hill, N.Y. (1979).
3. M. Cheryan, Ultrafiltration Handbook, published by Technomic Publishing Company, Inc. (1986).
4. W.C. McGregor, Membrane Separations in Biotechnology, published by Marcel Dekker, Inc. (1986).
5. M.S. Le and T. Atkinson, Process Biotechnology, February 1995.
6. J.L. Short and D.W. Webster, Process Biochemistry, March/April 1982.

33

PLASMID PURIFICATION

H.S.C. Barbosa and J.C. Marcos
Center of Chemistry, University of Minho, Campus de Gualtar, Braga, Portugal

33.1 INTRODUCTION

Gene therapy is one of the fields that in recent years has attracted great interest, and this is expected to grow more promptly in the future with the discovery of novel potential therapeutic targets in the Human Genome (1,2). The high potential of this therapy strategy is due to its potential to cure diseases by the introduction of nucleic acids into human cells, in order to restore, enhance or cancel a specific element of the genetic repertoire. Unlike other traditional treatments, gene therapy aims to remove the cause of a disease and not to correct the biochemical imbalance, or get rid of/attenuate the disease symptoms as is done in the traditional treatments (2).

Commonly, the nucleic acid used is the double-stranded deoxyribonucleic acid (dsDNA) encoding a specific protein but other nucleic acids such as the single-stranded deoxyribonucleic acid (ssDNA) or antisense RNA have also been described (3). Antisense RNA binds to a target sequence in the host cell and inhibits the expression of a specific gene by blocking either gene expression or its promoters. Alternatively, these transgenes can encode a target antigen that will elicit both, the humoral and cellular immune systems' response to the specific antigen. This process is designated by DNA vaccination (4).

Two of the critical challenges in gene therapy and DNA vaccines are the efficient and safe delivery of the nucleic acids into the target cells. Viral and nonviral delivery vectors have been developed for this purpose (5). The use of viral vectors (e.g. retrovirus, adenovirus) have been questioned due to the safety and regulatory concerns regarding their toxicity and the strong immune response after repeated administration (3). In addition, the possibility of activating oncogenes or suppressing tumor-suppressor genes has caused apprehension in the application of these vectors (5). In fact, viral vectors are responsible for some of biggest gene therapy setbacks as deaths have been reported during clinical trials (6).

The development of nonviral vectors has thus emerged as a safer alternative to viral vectors and led to the study of naked plasmid deoxyribonucleic acid (pDNA) or encapsulated/complexed pDNA as vectors for transferring prophylactic and/or therapeutic genes. It is generally simpler to develop when compared to the viral counterparts, and has the advantage of being free from most of the safety concerns associated with the virus (7).

The increasing usage of pDNA vectors in gene therapy/DNA vaccination trials has resulted in an urgent need for development of large-scale and cost-effective manufacturing processes for pDNA. Since pDNA usually has low transfection efficiencies, it is also expected that larger doses of the DNA-based product are needed or repeated administrations of the pDNA will be necessary in order to achieve successful gene transfer (3,8). The average dose size will differ between therapies and each disease is likely to have specific needs. It is expected that typical dose sizes will vary between 0.1 and 5 mg of pDNA, but full treatments may require larger quantities of therapeutic DNA (9). Thus, great amounts of pDNA have to be produced in order to supply the world's requirements (3,10).

As of March 2009, there are currently 246 ongoing gene therapy trials worldwide that utilize pDNA, of which two are already in clinical trial phase III (11,12). Thus, it is

Downstream Industrial Biotechnology: Recovery and Purification, First Edition. Edited by Michael C. Flickinger.

expected that these products, if they get approval for commercialization, will reach the market in 2–5 years.

In addition, when the DNA vaccines prove to be an efficient and safe alternative to the traditional vaccination methods, the high demand for these pDNA vectors will increase dramatically. Compared to conventional therapeutic methods, DNA vaccines can be cheaper to produce, more stable at ambient temperature, and safer to administrate (13). Moreover, DNA vaccines can be produced within a time period as short as 2 weeks, in contrast to the 8–9 months required by the current methods (14). This clearly represents a huge advantage since response to an epidemic outbreak can be made more rapidly.

33.2 THERAPEUTIC PLASMIDS

Plasmids are circular DNA molecules found in bacteria that replicate independently from chromosomal DNA. The pDNA can exist in three tertiary structures: supercoiled, open circular, and linear. As seen in Fig. 33.1 the plasmid structure denoted as supercoiled is a circular dsDNA structure where the DNA helix has been wound up. The US Food and Drug Administration (FDA) states that pDNA may be less effective in the open circular or linear form and thus requires almost all pDNA content to be in the supercoiled form, for its clinical approval. However, there is not yet a consensus on whether the therapeutic effectiveness in gene therapy is affected by the detailed topological structures, as efficient transfections have already been described using these tertiary forms of pDNA (15).

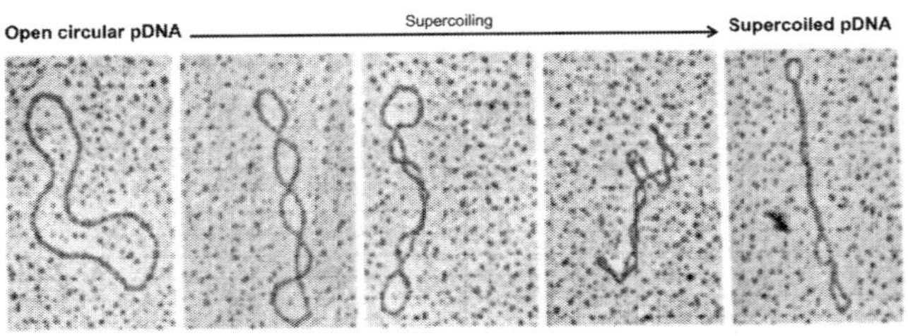

Figure 33.1. Plasmid DNA supercoiling [Adapted from Ref. 16].

The open circular and linear forms result from enzymatic or shear-induced breakage of the sugar phosphate backbone and also represent a critical and difficult purification challenge since they are treated as an impurity but share a nature identical to that of the supercoiled DNA (17).

The utilization of pDNA for therapeutic applications requires a stringent clearance of impurities. The high standard removal of all these impurities is essential for product approval as a therapeutic drug, and this leads to one of the major concerns on scaling-up the production of plasmids. The regulatory agencies limit the impurities content in each DNA formulation as summarized in Table 33.1.

Production of the pDNA usually encompasses three steps: vector construction and transformation of the host cells, cultivation of the transformed cells to amplify the plasmid, and finally, purification of the plasmid from the harvested cells. *Escherichia coli* (*E. coli*) strains are usually employed as host cells and the last step of the pDNA production may account for up to 80% of the total costs.

The high pDNA product specifications impose several constraints on its downstream processing, starting with the relatively low amounts of the target molecule in the *E. coli* bacterial cell lysates. pDNA accounts for less than 3% of the total content of a typical lysate, which increases the difficulty of its extraction. In addition, the pDNA is structurally related to other impurities such as RNA and the genomic deoxyribonucleic acid (gDNA). Both these impurities are in higher concentrations in the lysate and share some physicochemical properties, such as size and charge, with the pDNA.

Nonfragmented gDNA is a large molecule, which has a molecular weight of 2700 MDa, whereas pDNA may vary between 2 and 20 MDa, which is significantly smaller (19). However, inefficient lysis and shear effects during preparation of the lysate may fragment the gDNA into smaller portions with similar sizes and charge densities to pDNA, making the removal of gDNA particularly challenging. It has been observed that the size of pDNA and gDNA is a

TABLE 33.1. Impurities: Specifications and Content in Bacterial Cell Lysates[a]

Content of Bacterial Cell Lysates	Final Product: Issue in Focus	Final Product: Range of Acceptance	Final Product: Determined by
55% Proteins	Protein	<3 μg/mg pDNA	BCA (Bicinchoninic Acid) test
3% Host gDNA	gDNA	<2 μg/mg pDNA	TaqMan®-PCR
3% LPS	Endotoxins	<10 Endotoxin Unit/mg pDNA	LAL (Limulusamoebocyte lysate) test
<3% Plasmid	scDNA form	>97%	CGE (Capillary Gel Electrophoresis)
21% RNA[b]	RNA	<0.2 μg/mg pDNA	Analytical HPLC
15% Others			

[a] Source: Adapted from Ref. 18.
[b] 53% 23 S rRNA; 27% 16 S rRNA; 2% 5 S rRNA; 14% tRNA; 4% mRNA.

TABLE 33.2. Comparison of the Physicochemical Properties of Proteins and Plasmids

Characteristics	Biomolecule	
	Protein	Plasmid DNA
Building blocks	Amino acids	Nucleotides
Molecular mass (kDa)	10^3-10^5	10^6-10^{7a}
Stokes' radius	<5 nm	100 > 300 nm
Charge	Depending on pI/pH	Negative
Diffusion coefficient	Medium–high	Low
Viscosity	Low	High
Shear force sensitivity	Low	High

[a]The larger size of the plasmids strongly influences all other properties.

major constraint in the physical operations because of their susceptibility to fluid mechanical forces. This is probably one of the most critical bottlenecks in large-scale preparation of plasmid DNA in the spectrum of available pDNA downstream processing techniques (19).

Nonnucleic acid impurities consist mainly of endotoxins and proteins. Bacterial endotoxins are negatively charged (net charge) lipopolysaccharide elements of the outer membrane of gram-negative bacteria. When cells are broken down during cell lysis, endotoxin globules of widely varying sizes can appear. In humans, endotoxins can cause responses ranging from slight effects such as septicemia to severe effects such as intravascular coagulation (20). Thus, the removal of endotoxins from a DNA formulation is crucial and since they also exhibit sizes (and charges) similar to plasmids, they are considered difficult to remove.

Endotoxins have affinity to almost every material and/or solvent used in the purification process, and it has been observed that concentration of endotoxins decreases rapidly when a chain of purification systems is employed. Thus, at the end of a multistage purification process, they are usually sufficiently reduced to meet the regulatory demands without the need for a specific endotoxin removal operation (21,22). Nonetheless, affinity matrices such as polymyxin B (23), thiophilic aromatic chromatography (18), and hydrophobic interaction chromatography (HIC) (to some extent) (24) have been optimized for this purpose.

On the other hand, protein impurities as low as 1 ng per dose have been known to give rise to adverse reactions in humans, although the immunogenic response is highly dependent on specific proteins and recipients (25). Therefore, the efficient protein removal during a pDNA extraction process is also required by regulatory agencies which impose a maximum limit of 3 μg/mg pDNA. As seen in Table 33.2, proteins are physiochemically less similar to pDNA, and generally, efficient protein removal does not impose a lot of hurdles on plasmid downstream processing since selective conditions for pDNA can be found more easily.

33.3 CELL LYSIS

Cell recovery process followed by cell lysis marks the initial steps of the downstream processing of pDNA. After fermentation, the pDNA-containing cells are recovered usually by centrifugation or by filtration in order to separate them from the growth media. Bacterial cell walls are then disrupted for the pDNA release to the supernatant in order to be captured in the following purification steps. Cell disruption falls into two main categories: mechanical and chemical disruption (26).

The mechanical disruption (e.g. sonication, bead milling, heat etc.) is normally an excluded option since the mechanical stress can potentially damage the gDNA resulting in fragments with the same size as the pDNA, and complicating plasmid isolation from lysate impurities (19).

Alternatively, chemical lysis has been widely adopted as a variation of the alkaline lysis originally described by Birnboim *et al.* (27). Briefly, cells are resuspended in a neutral buffer, to which is added an alkaline 0.2 *M* NaOH solution containing 1% (w/v) sodium dodecyl sulfate (SDS). This step solubilizes the cell walls, releasing its contents to the supernatant. Within the pH range of 12.0–12.5 the pDNA does not fully denature and remain in solution, whereas within this pH range both gDNA and proteins get irreversibly denatured. This is a critical step since higher pH buffers will also irreversibly denature pDNA and may also affect the supercoiled form content of pDNA. The lysate shows marked non-Newtonian properties, exhibiting a rheological behavior that makes flow and handling of the material very difficult (19). Such pH extremes may occur in a mechanical stirred lysis reactor, given the high concentration of NaOH and SDS. At large scales, mixing can be very inefficient and consequently the lysate homogeneity in the vessel may be compromised. This may affect pDNA quality and make more difficult its further purification. The third stage is used to lower the pH of the lysate mixture to 5.5 by addition of ice-cold sodium acetate solution. The change in pH causes the flocculation of the gDNA and the precipitation of protein–SDS complexes and some RNA (15,19).

The separation of the precipitated and flocculated material from the pDNA-containing supernatant is usually achieved by low shear filtration at large scales. Whereas, at the laboratory and preparative scale, centrifugation on fixed-angle rotors are the most common operations (28). At large scales, the continuous feed flow centrifugation has also been used due to its high capacity, but the liquid entering the centrifuge can cause shearing effects that compromise the efficient removal of debris (29,30). After the removal of the debris, most gDNA and proteins are also removed, but the pDNA still represents less than 3% of the total lysate contents (Table 33.1).

Figure 33.2. System for industrial-scale lysis, neutralization, and clarification, for routine production of pharmaceutical-grade pDNA [Reprinted from Ref. 33].

Sometimes a clarification and concentration step is performed after this step. The clarification is mainly designed to remove remaining proteins in the supernatant and usually is performed by using chaotropic salts, such as, lithium chloride and ammonium acetate (31,32). pDNA can be concentrated using poly(ethylene glycol) (PEG) in preparation for the next purification steps and to remove some host nucleic acids. This step also has the advantage of enabling the buffer exchange, preparing the pDNA lysate for the subsequent steps (15). Alternate precipitation methods are discussed in more detail in the section titled "Precipitation."

Recently a system for the fully automated alkaline lysis, neutralization, and clarification of *E. coli* cells for the industrial production of pharmaceutical grade pDNA (33) (Fig. 33.2) was reported. The system comprises three different units. First, the homogeneous cell suspension is gently mixed with the lysis solutions in a reactor. This avoids local pH extremes and varying contact times. The formation of flocs during the neutralization and precipitation step takes place in a neutralization coil, which is designed to support mixing while avoiding mechanical stress. Gentle clarification is carried out by a combination of flotation and filtration, in a third special device. With this integrated device, yields up to 100% of pDNA were obtained as compared to a standard procedure where centrifugation was used for clarification. In addition, the lysate homogeneities obtained with this integrated process are similar to those obtained with gentle manual lysis. This device is routinely used for cGMP production of pharmaceutical grade pDNA from 200-L fermentations.

The patent landscape in this initial lysate preparation steps includes various methods and devices to perform alkaline lysis at large scale. A notable patent is one issued to Boehringer Ingelheim, which discloses a method and device for continuous flow lysis utilizing glass bead columns that act as lysis and clarification reactors (34). A high salt precipitation of impurities is described in patent WO2004060277 (35) to remove some of the RNA and much of the endotoxins.

33.4 CHROMATOGRAPHIC METHODS

Chromatography is a well-characterized and well-established method for protein purification in the pharmaceutical industry and it is considered the method with the highest resolution and reproducibility (36). Chromatography is therefore one of the techniques that has been widely explored for pDNA purification. Several physicochemical features of either the target pDNA, or its impurities, can be exploited via selective interactions with the solid chromatographic support. Properties such as size, charge, hydrophobicity, conformation, and accessibility of molecular groups have been studied. An extensive review of the current chromatographic matrices for the separation of pDNA has been given by Diogo *et al*. (37).

As a technological alternative to porous particles, monoliths are currently gaining increased significance in the pDNA purification process. Monoliths are a single piece of macroporous (10–4000 nm or larger) solid material with high mass transfer rates which make them more suitable for larger molecules such as pDNA, rather than traditional bead chromatography (38). This new type of chromatographic stationary phase has thus improved capacity (can reach more then 10 mg/mL) over the traditional chromatography supports (tenths of milligrams plasmid per milliliter), resulting in higher pDNA recovery yields (39).

Further, the optimal separation of large molecules can be achieved with short and fast runs without sacrificing resolution. Polymethacrylate-based short monolithic columns distributed under the trade namebreak "Convective Interaction Media (CIM)" [BIA Separations patent, Ref. 40] are the most common type of monolithic column. Several studies reporting the use of this material for pDNA purification have been published (39,41,42).

The difficulty in preparing high volume monolithic columns with uniform pore structure, owing to the exothermic nature of the polymerization, is probably the major drawback of this technique. This could be circumvented by preparing individual units and merging them to form one large monolith with uniform properties, which will behave as a single radial flow tube (36).

Figure 33.3 shows the current process flow for the production of a pDNA vaccine and the simplification that would result by incorporating a monolith.

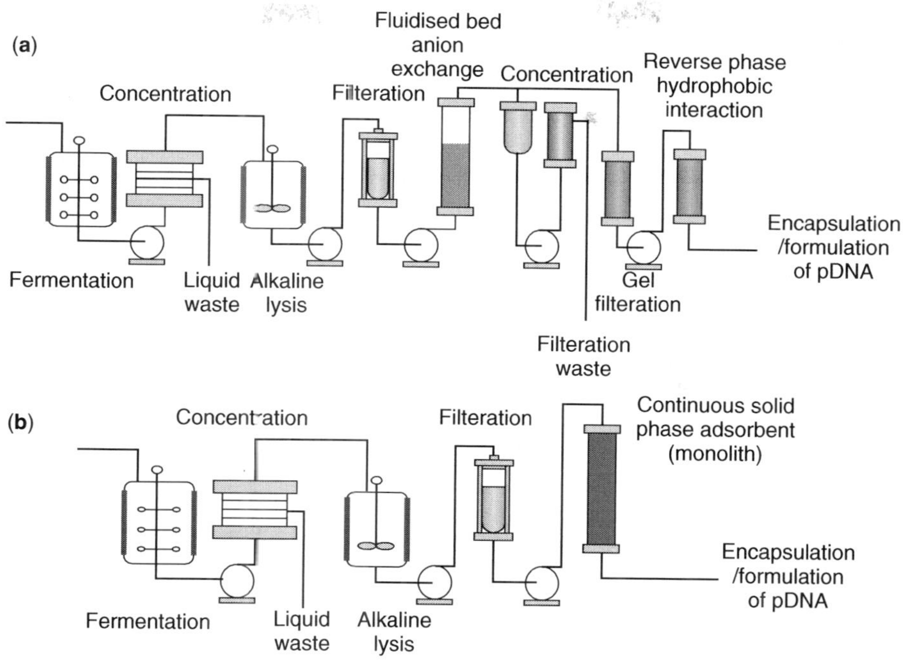

Figure 33.3. (a) A schematic flow diagram showing the current state of pDNA vaccine production. Several unit operations are incorporated into downstream processes, and (b) a process flow diagram showing improvements on current pDNA vaccine production technique by incorporation of a monolith [Reprinted from Ref. 43].

33.4.1 Anion Exchange

Anion-exchange chromatography (AEXC) is the most used method for the primary capture of pDNA with more than 81% of the patents pertaining to DNA purification methods (44). AEXC explores the electrostatics interactions between the negatively charged plasmids and the stationary phase bearing positively charged groups such as quaternary amines (45).

The main limitation of this technique is the copurification of anionic polymers of similar structure and charge with pDNA, such as gDNA, endotoxins, and some RNA (45,46). Therefore, additional purification steps are necessary, increasing the operational costs. Other disadvantages of AEXC chromatography are the relative low capacity of the pDNA and its elution in high salt buffers.

Nonetheless, some AEXC supports have been shown to be selective toward the supercoiled form of pDNA (which have higher charge density), eluting them later than the open circular forms (which have lower charge density) (30). This is one of the major advantages of this technique. The efficient removal of RNA (although usually with the pretreatment of the lysate with RNase), oligonucleotides, and some proteins are the other main advantages of this chromatographic method (37). Anion-exchange stationary phase is used in most of the lab kits used for pDNA purification, such as the commercial available ones from Qiagen or Promega.

Recently a comparison between four different AEXC, supports, including conventional media (Source 30 Q, Q Sepharose HP), a polymer grafted medium (Fractogel EMD DEAE (M)), media with large pores (Celbeads DEAE, PL SAX 4000 Å 30 mm), and a monolithic medium (CIM-DEAE or convective interaction media–2-chloro-*N*,*N*-diethylethylamine), showed that the last one has the fastest pDNA adsorption rate and the highest binding capacity with 13 mg pDNA/mL (47). Other monoliths derivatized with ligands bearing positive charges also show both, high capacities and pDNA adsorption rates (41,42,48–50). Polyacrylamide gel monolithic columns (pAAm-columns) grafted with different polycations showed that poly(2-(methacryloyloxy) ethyl)-trimethyl ammonium chloride (polyMETA) was the one with the highest selectivity for pDNA at intermediate NaCl concentration (0.5 *M*) (42). Given the large pores of the monolith, it was possible to apply a nonclarified bacterial lysate to the column and obtain purified plasmid with 99% yield, with practically undetectable levels of RNA.

A custom-made monolith obtained via free radical liquid porogenic polymerization of polyethylene glycol dimethacrylate (EDMA) and glycidyl methacrylate (GMA) was also derivatized with different amino groups (51). It was found that 2-chloro-*N*,*N*-diethylethylamine hydrochloride (DEAE-Cl) has the highest capacity (18.73 mg pDNA/mL) with a dissociation constant (K_D) of

0.11 mg/mL. The ligand also proved to be very stable as 90.5% of ligand density remained after 7 weeks of storage in 1.0 M NaOH, at room temperature. These results open up good prospects for the future implementation of cleaning-in-place (CIP) methodologies. With a 5-mL column of this monolith pDNA was purified from clarified bacterial lysates without RNA or protein contamination, in just 5 min (49).

33.4.2 Exclusion

Size exclusion chromatography (SEC) explores the different hydrodynamic sizes of the plasmids and their impurities, rather than the adsorption process, in order to purify the target molecule. To achieve good separation between two species, it is required that the two molecules differ in size twofold, otherwise coelution of the two species is obtained, and the resolution is limited (52). Although good removal of impurities such as RNA and proteins has been observed, the separation between the pDNA and the gDNA proves to be more difficult since both get eluted near the void volume (52) or the separation was dependent on the concentration of the two species (53).

Thus, the main disadvantages of this approach are the coelution of impurities with size similar to the pDNA; the limited scale-up at a manufacturing scale; and the high dilution factors that may result (37). Therefore, SEC is applied more often at later stages of the pDNA purification process for polishing and for accommodating the pDNA on the formulation buffers.

33.4.3 Hydrophobic Interaction

HIC has also been used to purify pDNA (24,54,55). HIC makes use of the more hydrophobic nature of single-stranded nucleic acid impurities (RNA, denatured gDNA, and denatured pDNA) and endotoxins to retard the flow of these impurities through a chromatography column with a hydrophobic support (e.g. 1,4-butanediol diglycidyl ether, butyl, phenyl, octyl) so that supercoiled DNA is eluted in the flowthrough (37). HIC needs high salt concentration to promote binding, thus it can be more conveniently used after AEXC, or when the plasmid pool has a high salt content. However, the loading of the clarified lysate directly into an HIC column had also been reported (56). In this case, two HIC columns, butyl-6PW (Tosoh) and octyl-6PW (Tosoh) were used in tandem. With high salt concentration (ammonium sulfate 2.0 M), the first column retains RNA and proteins while pDNA and gDNA are adsorbed to the second one. pDNA was then eluted by decreasing the salt concentration in the running buffers. Although the capacity of the second column was only 1.1 mg/mL, it achieved a pDNA yield of 90%. However, the pDNA eluate has to be further purified by AEXC since some contaminants remained in the eluate.

Despite the low plasmid binding capacities of HIC resins, yields as high as 95% have been reported with minimal solvent requirement (57). The main limitation of the technique is that it requires high salt concentrations (e.g. 1.5–2.5 M ammonium sulfate) and pDNA is eluted in high salt buffer. Additionally, in most cases, pDNA has to be pre-precipitated from the lysate using isopropanol (limited utilization at large-scale) and then solubilized in high salt concentration buffer prior to loading into the column, thereby increasing the operation time and costs (37,55). In a recent study, it was demonstrated that the commonly used ammonium sulfate salt could be substituted by the citrate salt (58). Although lower yields and purification factors were obtained, at large scales this salt replacement would result in a more environment-friendly process. It is estimated that the environmental impact of citrate is at least three times lower than ammonium sulfate (58,59).

An experimental vaccine against rabies had been produced with a purification procedure based on HIC (56). From 4.5 L of bacterial cell culture, 143 mg of plasmid with HPLC (high performance liquid chromatography) purity greater than 99% were obtained. The final pDNA preparation was in conformation with the required pDNA product specifications in terms of impurities and it was able to induce rabies virus-neutralizing antibodies in mice, protecting them against a rabies virus challenge. Several patents have been issued using HIC as the main pDNA purification process (60–62).

33.4.4 Affinity

The affinity capture of pDNA based on its biological function or individual chemical structure has been developed recently and is gaining more interest. Nonetheless, there is not yet an extensive literature available and three relevant approaches for the high selective recovery of pDNA have been developed: Triple helix, DNA-binding proteins, and amino acid interactions (63).

Triple helix affinity chromatography (THAC) is based on the sequence-specific interaction of a pyrimidine oligonucleotide with pDNA (64,65). The binding occurs via the major DNA groove and through the formation of Hoogsteen hydrogen bonds. The binding kinetics is generally slow (may need >1 h) and requires an acidic pH in order to stabilize the binding complex. Despite the discrimination of different plasmids in one step, based on their sequence, the recovery yields are typically low (best yield ~50% (65)) and it is necessary that the chromatography runs are long, due to the slow kinetics of triple helix formation. However, good elimination of the gDNA and RNA has been achieved.

The first demonstration of the usage of DNA-binding protein as affinity ligands on pDNA chromatographic purification was performed by Woodgate *et al.* (66). In this study, the bifunctional Zinc Finger (ZnF) DNA-binding protein fused to Glutathione-S-transferase (GST-ZnF) was used for direct isolation from crude lysate of pDNA bearing the ZnF recognition site. Using this approach, it was possible to distinguish plasmids that contained the ZnF recognition sequence (5′-GGGGCGGCT-3′). Nonetheless, very low yield was obtained (10%) and the authors did not demonstrate pDNA elution from the binding complex.

A similar strategy was developed within the same group exploiting a different heterofunctional protein, LacI-His_6-GFP (Green Fluorescent Protein), that binds to a specific sequence of the *lac* operon (67,68). This technique allowed the isolation of a pDNA bearing the *lac* operon, directly from crude cell lysates free of gDNA, RNA, proteins, and open circular pDNA. The major limitation of this chromatography process is the lack of capacity and difficulty in scaling up.

Amino acid–DNA affinity chromatography has been explored by using histidine (69) or arginine (70) amino acid ligands, usually referred as pseudobio affinity ligands. Both amino acids have been immobilized into a chromatography support and discrimination between supercoiled and open circular pDNA forms was observed in one operation step. The elimination of gDNA and endotoxins was within acceptable levels and no proteins or RNA were detected in the elution sample. However, these ligands have been shown to have relatively low yields and the interaction between pDNA and amino acid is dependent on the base composition, which may compromise future purification processes with different plasmids (63).

A 16mer peptide representing helix II of the DNA-binding domain of the lac repressor has also been used as a ligand for the purification of pDNA from bacterial lysates (71). The stationary phase employed was a novel monolith synthesized by the polymerization of EDMA and GMA. The utilization of the monolith, besides the increased capacity and the speed of the separation, allows an easier access of the pDNA to the ligand thereby improving mass transfer. Using this column, the recovery of 81% pDNA with 92% purity was achieved from clarified lysates.

33.5 NONCHROMATOGRAPHIC METHODS

Alternative nonchromatography processes for the purification of pDNA have been developed in order to circumvent some of the pDNA chromatography disadvantages. Although high resolution can generally be achieved using a chromatographic technique, the capacity for pDNA is usually low, resulting in relatively low pDNA yields. This is usually attributed to the fact that chromatographic supports were originally designed with small pore sizes for protein purifications and not for much larger molecules such as plasmids (37,66). It has been observed that in these chromatographic supports, the plasmids adsorb only at the bead's outer surface and thus, not all potential binding sites are used to bind to pDNA (72). The capacity is usually in the order of 200 mg of protein per mL of matrix, whereas when pDNA is the target biomolecule, only few hundreds of micrograms are usually obtained (72).

Although new supports are being developed with high binding capacities, such as the monoliths, chromatography processes are also difficult to scale-up and at larger scales, the resolution can be lost due to diffusion problems with large volume feeds (36). Chromatography-based processes are also acknowledged to be more expensive techniques in terms of global operation costs (columns, pumps, process controller, high volume of buffers, absorbents etc.) when compared to nonchromatographic approaches (73).

Chromatography-free processes are thus very attractive since they could circumvent the disadvantages of the chromatographic approaches and have a high potential to improve the overall pDNA purification processes. The cost of downstream processing of a biotech product is a major contributor to the overall process economics (estimated to be as high as 80% (74)) and thus attention is paid to this stage. There is great potential for substantial savings to be made in downstream processing, by improvements of existing methods. Eliminating the need for chromatographic steps will dramatically reduce these costs.

33.5.1 Precipitation

Precipitation techniques are a convenient and common way to separate DNA from soluble contaminants such as RNA and some proteins at the lab scale. Plasmids can be precipitated selectively by the addition of isopropanol (31), PEG (26,35), spermidine (75), and cetyltrimethylammonium bromide (CTAB) (76,77). For instance, CTAB was used in one of the first industrial applications of a nonchromatographic process for the purification of pDNA (78). Using CTAB at 20 g/L, an overall recovery of 83% of pDNA was obtained.

Although a good rate of removal of the main impurities in pDNA can be achieved, the precipitation approach is not easily scaled up since plasmids may need to be centrifuged out and the resulting pellet resuspended for further processing. In addition, some solvents are not recommended for large-scale pDNA manufacturing since some of them are not safe for human health or may need special facilities such as spark-proof ones.

Alternatively, precipitation could be achieved by the formation of polyectrolyte complexes with cationic polymers. This is a powerful technique for capturing pDNA as the precipitation does not depend on the property changes of the medium as in the case of addition of PEG or salts. This means that less polymers have to be added to the solution to precipitate pDNA. Poly (*N*,*N*-dimethyldiallylammonium) chloride (PDMDAAC) was used to selectively precipitate pDNA from clarified alkaline lysates with a recovery of 75–80% (79). A significant removal of RNA (95%) and protein (90%) was also achieved.

Other precipitation methods have been developed for the selective precipitation of pDNA impurities while pDNA remains in solution. For example, RNA can be precipitated by antichaotropic salts, such as ammonium acetate (80), ammonium sulfate (81), and lithium chloride (82) but the use of high concentrations of salts is also environmentally and economically problematic at large scales.

Of notable interest are the recently developed plasmid affinity precipitation reagents. Thermoresponsive elastin-like proteins (ELPs) (83,84) were conjugated with a DNA-binding protein resulting in a thermally reversible affinity precipitation system, specific for plasmids that incorporate the corresponding DNA-binding sequence. Also, a short single-stranded oligonucleotide sequence that recognized a specific sequence in the target plasmid was linked to an *N*-isopropylacrylamide oligomer (85). This conjugate binds to the plasmid in solutions, at low temperature (4°C) by triple helix formation. On increasing the temperature to 40°C the complex precipitates quantitatively, capturing the target plasmid. Yields of 70–90% pDNA were obtained after redissolution of the complex and pH shift to release the pDNA.

33.5.2 Membrane Processes

Adsorptive membranes offer yet another nonchromatographic technique for reducing the concentration of plasmid impurities and for the desalting and condition/formulation step. The main advantages of this technique are the increased loading capabilities of some of these membranes (10 mg/mL of membrane) and the potential to maintain high efficiencies with large molecules at high flow rates (86,87). Nonetheless, the relatively high cost of the membranes prevents their use at a large scale (88).

Different approaches have been developed for pDNA purification. Tangential flow filtration (TFF) and nitrocellulose-integrated processes have been optimized for pDNA purification with pDNA recoveries of virtually 100% and with significant removal of RNA, proteins, endotoxins, and gDNA (89). Charge-based membranes have also been developed (just like in AEXC) for rapid and high capacity (~10 mg/mL) purification of pDNA (87,88). However, contaminant RNA could not be removed completely without the use of ribonuclease (RNase). This is probably the main drawback of this technique that requires the pre-treatment of the feed stream to remove excess of RNA, either by using RNAse or extended alkaline lysis (1day) (90).

Representative examples of patents claiming the use of TFF in a pDNA purification system are the one issued to Butler *et al*. (91) and the one issued for utilization of an anion exchange membrane, Pall Mustang™ (92).

33.5.3 Liquid Extraction

Liquid extraction of pDNA has mainly been achieved by using aqueous two-phase systems (ATPS). These systems form when two polymers or a polymer and a salt are mixed in water above some critical concentrations. The different physicochemical characteristics of the two phases formed allow the separation of components from a complex mixture and it is specially advantageous for biomolecules given their high water content (80–95%) (73,93).

The recovery of pDNA by ATPS had been based on the utilization of the polymer PEG, and a second polymer or a salt-enriched bottom phase. Potassium phosphate, ammonium sulfate, and sodium citrate have been used as phase-forming salts, with comparable results (55,94,95). In most of the studies published using the PEG–salt systems, high recovery of pDNA is obtained in the salt phase with the elimination of most RNA, proteins, and endotoxins in the polymer phase. Better results are usually obtained using a PEG–salt system with low molecular weight PEG. However, the system composition must be carefully chosen as significant differences between the top and bottom phase compositions corresponding to high tie-line values, result in plasmid precipitation at the interface with the concomitant yield reduction (55).

Most of the host gDNA is also removed to the polymer phase in these PEG–salt systems as evaluated qualitatively by agarose gels (96). However, the precise quantification of the gDNA is difficult given the interference of phase components with the available quantification methods. Polymers have been shown to interfere with the chromatographic supports when gDNA is quantified by AEXC (97). On the other hand, high PEG concentration interferes with gDNA analysis by colorimetric southern slot blot and salts have been shown to interfere with quantitative PCR (polymerase chain reaction) (94). Agarose gel eletrophoresis is the only method that allows a semiquantitative analysis of gDNA in both phases. However, it should be noted that most of the gDNA precipitates after the addition of the neutralizing solution during the preparation of an alkaline bacterial cell lysate. The remaining gDNA is mostly single-stranded given the alkaline denaturation and usually partitions to the polymer phase due to the exposure of the hydrophobic bases. Consequently, a significant decrease of gDNA content is observed in the salt phase of most polymer–salt systems (94,96).

The PEG molecular weight in PEG–$(NH_4)_2SO_4$ systems seems to have a great effect on the pDNA partition. Results showed that pDNA could be directed toward the top phase (PEG-rich phase) (MW<400) or to the bottom phase (salt-rich phase) (MW>400) (55,94). For PEG 600–salt systems, the pDNA partitions totally to the bottom phase (yield 88.4%) whereas the protein from the bacterial cell lysate remains in the top phase (55).

Rahimpour *et al.* (98) reported 99% pDNA recovery in systems composed of PEG400–citrate at pH 6.9. Under these conditions, RNA removal was only about 68%. A recent study reported that while ammonium sulfate systems give a higher purity (~100%) when compared to citrate systems, a lower recovery yield is obtained (99). A compromise between these two parameters could be found using a mixture of the two salts. A mixture of 25% (w/w) ammonium sulfate and 75% (w/w) sodium citrate resulted in plasmid recovery of 91.1% and purity (17.2%). In this study, it was possible to show that ammonium sulfate could be partially replaced with citrate buffers without compromising performance (99).

As stated before, from an environmental point of view, the utilization of citrate is preferable to sulfate or phosphate systems, because it is biodegradable and nontoxic (98,100). On the other hand, the extraction of pDNA to the ammonium sulfate phase is advantageous since the salt phase could be applied directly to HIC for removal of remaining impurities (55).

Polymer–polymer systems are usually regarded as more expensive than polymer–salt systems (73). However, these systems could be advantageous given their low salt content.

Thermoseparating phase components such as the 50% ethylene oxide–50% propylene oxide copolymer (EOPO) have been used with dextran to form two-phase polymer systems (97). In EOPO–dextran systems, almost 100% of pDNA accumulated in the EOPO rich phase, which can be removed and heated to 55°C to form a second aqueous phase to elute the pDNA. With this approach, the removal of RNA is not so efficient although 80% reduction of this contaminant has been achieved (97). Another drawback of this approach is that the lysate has to be previously desalted in order to precisely control the salt composition of the systems, which in turn determines the partitioning of nucleic acids. In contrast in salt–polymer systems, the lysate could be directly added to the phase components (95).

Some of the aforementioned studies applied another purification step to recover the pDNA from the salt phase of ATPS and remove remaining impurities. Integration with HIC (55), membranes (96) or with more than two unit operations have been reported for this purpose (101).

The low selectivity that is usually found in these systems could be increased by the addition of affinity ligands. This approach has been widely explored for the purification of proteins (102–104) and recently was applied for pDNA purification in ATPS. The cationic polymer polyethyleimine (PEI) derivatized with PEG (PEGylated) was used as a nonspecific ligand for the purification of pDNA from crude cell lysates in a two-step ATPS process (105). The authors report up to 100% pDNA recoveries in the form of DNA–PEI polyplexes with no RNA contamination and only minor protein impurities. The purification of pDNA as polyplexes would be an advantage since after some polishing steps, the polyplexes would be ready for human administration.

Alternatively a ZnF transcription factor fused to a GST protein (GST-ZnF), was able to isolate pDNA bearing a ZnF recognition site from the bottom to the top phase of a PEG–dextran ATPS (106). The protein could be used in either its native or PEGylated form depending on its partition on ATPS. The experiments were performed with pure plasmids just to establish the proof-of-concept, and given the difficulty in eluting the pDNA from the protein–pDNA complex formed, no attempts were made to extract pDNA from a crude cell lysate.

33.6 INDUSTRIAL PROCESSES

Since some of the techniques described above cannot achieve high purity pDNA in one single operation, usually several combinations of different techniques are explored in order to obtain high standard product specifications. Prazeres *et al.* (45) designed several overall process flowsheets representing the current panorama in pDNA downstream processing and in all of them, multiple stage operations were assumed for efficient purification processes.

Several examples of combined operations have been also reported elsewhere, such as the combination of AEXC, triple helix chromatography and HIC that was developed by Centelion (Gencell), yielding pDNA within the product standards (9).

Urthaler *et al.* (72) have demonstrated an automated continuous system of HIC, AEXC, and SEC chromatography delivering high purity pDNA homogeneities (98%) of the supercoiled form in any desirable final buffer.

Cobra Biomanufacturing Plc (UK) uses three chromatographic stages, starting with an anion exchange, followed by size exclusion, and mixed mode chromatography, which use the commercial adsorbent, Poly-Flo™ that binds to endotoxins during purification (44).

Such combinations of multichromatography techniques are very common; however, they are not cost-effective for commercial-scale purification (9). However, it is unusual for an overall process to eliminate the use of packed bead chromatography totally and a polishing step using chromatographic processes is usually necessary (45). Alternatively, the combinations of chromatographic and nonchromatographic techniques in the same overall process have been also developed.

For example, the integration of nonchromatographic processes such as the utilization of PEG precipitating agents to enhance AEXC (107) have been performed. The authors claim that with the use of PEG-8000 (at a concentration of 1%) prior to AEXC chromatography, recoveries increased from about 20% to 80%.

Kepka *et al*. (101) integrated ATPS with membrane filtration and lid bead chromatography, also described in patent WO2004020629 (108). The process started with a volume reduction and lysate buffer exchanged using a hollow fiber membrane system. The concentrated samples were then applied to an EOPO–dextran system to remove most of the proteins and RNA. Finally, the eluted pDNA from the top phase was polished using lid bead chromatography. This new type of chromatography uses positively charged inner cores (for the binding of remaining RNA) and an inert surface layer. An efficient removal of all impurities was accomplished but the overall process yield was about 69%.

Frerix *et al*. (96) have used membranes as a polishing step after a PEG–Salt ATPS extraction (and totally avoiding a chromatographic step) to select the supercoiled form of pDNA (35 mg/95%) and remove it from the remaining nucleic acid impurities, such as open circular pDNA (<3%) and gDNA (<1%). This process has been also described in a patent issued in 2006, US2006286080 (109).

Frequently, companies have issued patents for only the individual stages of the downstream processing of pDNA which have to be integrated with third-party systems for complete downstream processing operations or for the upstream processing stage. GE Healthcare, for example, has developed the PlasmidSelect Xtra downstream purification package (110), which combines different proprietary technologies: the high fermentation process from Nature Technology Corporation and a proprietary thiophilic aromatic chromatographic media (9,111). *S*-Aryl ligands represent a selective tool to remove unwanted forms of plasmid DNA, thus they are highly selective for the supercoiled pDNA. The supercoiled concentration of the affinity column reached 96% from 45% in the original feed (15). Some companies, such as Boehringer Ingelheim and Merck have combination process platforms that couple high yield fermentation, cell lysis, and purification patent portfolios and can perform operations almost completely with their own licensed technologies (9).

Four cGMP processes described in the literature (43,112–114) are compared in Table 33.3. In all of them, alkaline lysis, as described in the section titled "Cell Lysis",is used for cell disruption. After lysis, the flock-like particles of gDNA, proteins, and cell debris may be removed by centrifugation or filtration. As some of them float, filtration is usually a more efficient method for its removal. In two of the processes described (113,114), a low pressure of approximately 500 mbar is applied to the lysate to make the precipitated particles float. With the escape of air from the solution, a compact bed of flocculent material is formed, allowing its filtration using a standard, commercially available capsule filter (113). The conditioning step depends on the subsequent chromatography type. Samples to be used in HIC must be concentrated by ultrafiltration and filtered after addition of ammonium sulfate (112). When using AEXC, an

TABLE 33.3. cGMP Purification of Plasmids for Therapeutic Applications

Plasmid	pRZ-hMCP1 4.9 kbp	pPING-MT 6.5 kbp	pDERMATT 3.8 kbp	pVR1020-PyMSP4/5
Therapeutic indication	None (model)	Melanoma	Melanoma	Malaria
Fermentation	Fed-batch fermenter	Shake flask	Batch fermenter	—
Cell recovery	Centrifugation	Centrifugation	Tangential flow filtration	—
Cell disruption	Alkaline lysis	Alkaline lysis	Alkaline lysis	Alkaline lysis
Clarification	—	Filtration	Filtration	Centrifugation
Conditioning	Ultrafiltration $(NH_4)_2SO_4$ addition Filtration	Qiagen buffer endotoxin removal (ER) addition	Endotoxin removal buffer addition	—
Chromatography	1. HIC 2. AEXC 3. SEC	Ultrapure 100 column	Ultrapure resin	Methacrylate monolithic column
Concentration	Ultrafiltration	Precipitation with isopropanol Washing with ethanol Vacuum drying	Tangential flow filtration Lyophilization	
Yield (%)	65	95	—	—
Supercoiled DNA (%)	96	98	98	92.5
gDNA	6 mg/mg	0.01%	<5%	<0.01%
RNA	<20 mg/mg	<1.3%	<4%	<0.01%
Endotoxins	0.02–0.04 mg/mg	<0.005 Endotoxin Unit/mL	<2 Endotoxin Unit/mg	0.28 Endotoxin Unit/mg DNA
Proteins	0.8 mg/mg	<10 mg/mg	<5 mg/mL	0.26%
References	112	113	114	43

endotoxin removal buffer is added to prevent endotoxins from binding to the column (113,114). One advantage of using monoliths is that no conditioning step is needed (43).

Three of the cGMP processes use packed bead chromatography as a main purification step (112–114). One of them is a multistep process and uses HIC, AEXC, and SEC sequentially, whereas the other two use only AEXC. In the multistep procedure, a comparison between using packed bead (Fractogel) and monolith (CIM) stationary phase was performed. It was concluded that while the productivity of the monoliths is 5–10 times higher, only small amounts of pDNA could be obtained given the limitation of the CIM sizes currently available. The cGMP protocol based on monoliths uses a methacrylate polymer derivatized with amino groups as stationary phase. The polymer was synthesized using a novel heat expulsion technique that assures a more uniform pore size distribution along the length of large-volume monoliths obtained via bulk polymerization (43).

A final concentration step is usually needed when manufacturing plasmids. This may be done by precipitation with isopropanol, washing with ethanol, and resuspension in an appropriate buffer after convenient vacuum dying (113). However, the use of potential toxic organic solvents in plasmid manufacturing intended for human use is not recommended. Membrane processes such as ultrafiltration and TFF are also used to concentrate the eluate. If higher concentration of the plasmid is needed, the final solution may be lyophilized. Concentration of the plasmid may then be adjusted by adding the required volume of buffer or formulation solution (114).

The final purity of the preparations are similar for the four cGMP processes. However, the contents of supercoiled pDNA are compliant with the strict specifications presented in Table 33.1 only for the ones that use the AEXC Ultrapure support from Qiagen.

33.7 CONCLUSIONS AND PERSPECTIVES

The crescent interest in nonviral gene therapy applications based on pDNA and DNA vaccines is increasing the demand for the production of large quantities of pharmaceutical grade pDNA. This demand should increase in the next few years when the first products finally reach the market.

The traditional purification methods based on packed-bed chromatography, although capable of supplying pDNA of high purity, have several limitations. The main ones are their high cost, difficulty in scaling up, and low capacity. The utilization of monoliths could circumvent some of these limitations, although they are currently limited by their size. Exciting developments are expected in this area, making possible the utilization of these supports at large scales and in industrial settings. The role of nonchromatographic methods such as membrane processes, precipitation, and liquid extraction in the purification of pDNA is also increasing. Traditionally, they had been used as an aid to chromatography. However, their crescent selectivity, achieved by using new materials and specific pDNA ligands, makes it possible to think of new, entirely nonchromatographic processes to purify pDNA. Some examples of this have already been demonstrated at the laboratory scale. The challenge is now to prove their potential to contribute to the production of pharmaceutical grade pDNA at large scales.

It is expected that the following years will witness the introduction of new cGMP processes for the production of pDNA. The most promising ones will be based on monoliths and nonchromatographic methods, hence making the production of these new pharmaceuticals more affordable. This is an essential point to make the advances in molecular medicines available to the world's population.

REFERENCES

1. Internatinal Human Genome Sequencing Consortium. Nature 2004; 431: 931–945.
2. Gillet JP, Macadangdang B, Fathke RL, Gottesman MM, Kimchi-Sarfaty C. Methods Mol Biol 2009; 542: 5–54.
3. Ferreira GN, Monteiro GA, Prazeres DM, Cabral JM. Trends Biotechnol 2000; 18: 380–388.
4. Kutzler MA, Weiner DB. Nat Rev Genet 2008; 9: 776–788.
5. El-Aneed A. J Control Release 2004; 94: 1–14.
6. Yi Y, Hahm SH, Lee KH. Curr Gene Ther 2005; 5: 25–35.
7. Parker SE, Vahlsing HL, Serfilippi LM, Franklin CL, Doh SG, Gromkowski SH, Lew D, Manthorpe M, Norman J. Hum Gene Ther 1995; 6: 575–590.
8. Luo D, Saltzman WM. Nat Biotechnol 2000; 18: 33–37.
9. Carnes AE, Williams JA. Recent Pat Biotechnol 2007; 1: 151–166.
10. Crystal RG. Science 1995; 270: 404–410.
11. Williams JA, Carnes AE, Hodgson CP. Biotechnol Adv 2009; 27: 353–370.
12. Wiley. Gene therapy clinical trials worldwide. 2009. Available at http://www.wiley.co.uk/genetherapy/clinical/.
13. Manoj S, Babiuk LA, Littel-van den Hurk SvanDrunen. Crit Rev Clin Lab Sci 2004; 41: 1–39.
14. Forde GM. Nat Biotechnol 2005; 23: 1059–1062.
15. Shamlou PA. Biotechnol Appl Biochem 2003; 37: 207–218.
16. Ferreira GNM. Chem Eng Technol 2005; 28: 1285–1294.
17. Kelly WJ. Biotechnol Appl Biochem 2003; 37: 219–223.
18. Lemmens R, Olsson U, Nyhammar T, Stadler J. J Chromatogr B Analyt Technol Biomed Life Sci 2003; 784: 291–300.
19. Levy MS, O'Kennedy RD, Ayazi-Shamlou P, Dunnill P. Trends Biotechnol 2000; 18: 296–305.
20. Karima R, Matsumoto S, Higashi H, Matsushima K. Mol Med Today 1999; 5: 123–132.

21. Hirayama C, Sakata M. J Chromatogr B Analyt Technol Biomed Life Sci 2002; 781: 419–432.
22. Liu S, Tobias R, McClure S, Styba G, Shi Q, Jackowski G. Clin Biochem 1997; 30: 455–463.
23. Anspach FB. J Biochem Biophys Methods 2001; 49: 665–681.
24. Diogo MM, Queiroz JA, Monteiro GA, Martins SA, Ferreira GN, Prazeres DM. Biotechnol Bioeng 2000; 68: 576–583.
25. Briggs J, Panfili PR. Anal Chem 1991; 63: 850–859.
26. Prazeres DM, Ferreira GN, Monteiro GA, Cooney CL, Cabral JM. Trends Biotechnol 1999; 17: 169–174.
27. Birnboim HC, Doly J. Nucleic Acids Res 1979; 7: 1513–1523.
28. Theodossiou I, Collins IJ, Ward JM, Thomas ORT, Dunnill P. Bioprocess Eng 1997; 16: 175–183.
29. Ferreira GN, Cabral JM, Prazeres DM. Biotechnol Prog 1999; 15: 725–731.
30. Prazeres DM, Schluep T, Cooney C. J Chromatogr A 1998; 806: 31–45.
31. Chakrabarti A, Sitaric S, Ohi S. Biotechnol Appl Biochem 1992; 16: 211–215.
32. Horn NA, Meek JA, Budahazi G, Marquet M. Hum Gene Ther 1995; 6: 565–573.
33. Urthaler J, Ascher C, Wohrer H, Necina R. J Biotechnol 2007; 128: 132–149.
34. Urthaler J, Necina R, Ascher C, Woehrer H. Patent ID: WO04085643. 2004.
35. Budahazi G, Goff B. Patent ID: WO04060277. 2004.
36. Przybycien TM, Pujar NS, Steele LM. Curr Opin Biotechnol 2004; 15: 469–478.
37. Diogo MM, Queiroz JA, Prazeres DM. J Chromatogr A 2005; 1069: 3–22.
38. Strancar A, Podgornik A, Barut M, Necina R, Scheper T, Freitag R. Advances in biochemical engineering/biotechnology, modern advances in chromatography. Heidelberg: Springer–Verlag; 2002.
39. Branovic K, Forcic D, Ivancic J, Strancar A, Barut M, Kosutic Gulija T, Zgorelec R, Mazuran R. J Chromatogr B Analyt Technol Biomed Life Sci 2004; 801: 331–337.
40. Necina R, Urthaler J, Strancar A, Jancar J, Merhar M, Barut M, Podgornik A. Patent ID: WO03051483. 2003.
41. Urthaler J, Schlegl R, Podgornik A, Strancar A, Jungbauer A, Necina R. J Chromatogr A 2005; 1065: 93–106.
42. Hanora A, Savina I, Plieva FM, Izumrudov VA, Mattiasson B, Galaev IY. J Biotechnol 2006; 123: 343–355.
43. Danquah MK, Liu S, Ho J, Forde GM, Wang L, Coppel RL. AIChE J 2008; 54: 2290–2998.
44. Forde GM. Doctor of philosophy [dissertation]: University of Cambridge; 2004.
45. Prazeres DMF, Ferreira GN. Chem Eng Process 2004; 43: 609–624.
46. Wicks IP. Hum Gene Ther 1995; 6: 317–323.
47. Tarmann C, Jungbauer A. J Sep Sci 2008; 31: 2605–2618.
48. Zochling A, Hahn R, Ahrer K, Urthaler J, Jungbauer A. J Sep Sci 2004; 27: 819–827.
49. Danquah MK, Forde GM. J Chromatogr B Analyt Technol Biomed Life Sci 2007; 853: 38–46.
50. Bencina M, Podgornik A, Strancar A. J Sep Sci 2004; 27: 801–810.
51. Danquah MK, Ho J, Forde GM. J Sep Sci 2007; 30: 2843–2850.
52. Li LZ, Liu Y, Sun MS, Shao YM. J Chromatogr A 2007; 1139: 228–235.
53. Bywater M, Bywater R, Hellman L. Anal Biochem 1983; 132: 219–224.
54. Diogo MM, Queiroz JA, Monteiro GA, Prazeres DM. Anal Biochem 1999; 275: 122–124.
55. Trindade IP, Diogo MM, Prazeres DM, Marcos JC. J Chromatogr A 2005; 1082: 176–184.
56. Diogo MM, Ribeiro SC, Queiroz JA, Monteiro GA, Tordo N, Perrin P, Prazeres DM. J Gene Med 2001; 3: 577–584.
57. Diogo MM, Ribeiro SC, Queiroz JA, Monteiro GA, Perrin P, Tordo N, Prazeres DM. Biotechnol Lett 2000; 22: 1397–1400.
58. Freitas SS, Santosa JAL, Prazeres DM. Sep Purif Tecnhol 2009; 65: 95–104.
59. Heinzle E, Biwer A, Cooney C. Development of sustainable bioprocesses: modeling and assessment. New York: John Wiley & Sons, Inc.; 2006.
60. Ramasubramanyan N. Patent ID: US20056953686. 2005.
61. Ramasubramanyan N. Patent ID: WO0073318. 2000.
62. Prazeres DM, Rodrigues F, Diogo MM, Queiroz JA. Patent ID: US20077169917. 2007.
63. Sousa F, Prazeres DM, Queiroz JA. Trends Biotechnol 2008; 26: 518–525.
64. Wils P, Escriou V, Warnery A, Lacroix F, Lagneaux D, Ollivier M, Crouzet J, Mayaux JF, Scherman D. Gene Ther 1997; 4: 323–330.
65. Schluep T, Cooney CL. Nucleic Acids Res 1998; 26: 4524–4528.
66. Woodgate J, Palfrey D, Nagel DA, Hine AV, Slater NK. Biotechnol Bioeng 2002; 79: 450–456.
67. Darby RA, Hine AV. FASEB J 2005; 19: 801–803.
68. Darby RA, Forde GM, Slater NK, Hine AV. Biotechnol Bioeng 2007; 98: 1103–1108.
69. Sousa F, Freitas S, Azzoni AR, Prazeres DM, Queiroz J. Biotechnol Appl Biochem 2006; 45: 131–140.
70. Sousa F, Prazeres DM, Queiroz JA. Biomed Chromatogr 2009; 23: 160–165.
71. Han Y, Forde GM. J Chromatogr B Analyt Technol Biomed Life Sci 2008; 874: 21–26.
72. Urthaler J, Buchinger W, Necina R. Acta Biochim Pol 2005; 52: 703–711.
73. Azevedo AM, Rosa PA, Ferreira IF, Aires-Barros MR. Trends Biotechnol 2009; 27: 240–247.
74. Roque AC, Lowe CR, Taipa MA. Biotechnol Prog 2004; 20: 639–654.
75. Mourich DV, Munks MW, Murphy JC, Willson RC, Hill AB. J Immunol Methods 2003; 274: 257–264.
76. Ishaq M, Wolf B, Ritter C. Biotechniques 1990; 9:19–20–22–24.
77. Lander RJ, Winters MA, Meacle FJ, Buckland BC, Lee AL. Biotechnol Bioeng 2002; 79: 776–784.
78. Lander RJ, Winters MA, Meacle FJ. Patent ID: US20046797476. 2004.

79. Wahlund PO, Gustavsson PE, Izumrudov VA, Larsson PO, Galaev IY. J Chromatogr B Analyt Technol Biomed Life Sci 2004; 807: 121–127.
80. Evans JK, Troilo P, Ledwith BJ. Biotechniques 1998; 24: 416–418.
81. Sparks RB, Elder JH. Anal Biochem 1983; 135: 345–348.
82. Cathala G, Savouret JF, Mendez B, West BL, Karin M, Martial JA, Baxter JD. DNA 1983; 2: 329–335.
83. Lao UL, Kostal J, Mulchandani A, Chen W. Nat Protoc 2007; 2: 1263–1268.
84. Kostal J, Mulchandani A, Chen W. Biotechnol Bioeng 2004; 85: 293–297.
85. Costioli MD, Fisch I, Garret-Flaudy F, Hilbrig F, Freitag R. Biotechnol Bioeng 2003; 81: 535–545.
86. Levy MS, Collins IJ, Tsai JT, Shamlou PA, Ward JM, Dunnill P. J Biotechnol 2000; 76: 197–205.
87. Teeters MA, Conrardy SE, Thomas BL, Root TW, Lightfoot EN. J Chromatogr A 2003; 989: 165–173.
88. Grunwald AG, Shields MS. Anal Biochem 2001; 296: 138–141.
89. Kendall D, Lye GJ, Levy MS. Biotechnol Bioeng 2002; 79: 816–822.
90. Kahn DW, Butler MD, Cohen DL, Gordon M, Kahn JW, Winkler ME. Biotechnol Bioeng 2000; 69: 101–106
91. Butler MD, Cohen DL, Kahn D, Winkler ME. Patent ID: WO0107599. 2001.
92. Nochumson S, Yang Y, Kinsey JL. Patent ID: WO0194573. 2001.
93. Albertsson PÅ. Partition of cell particles and macromolecules: separation and purification of biomolecules, cell organelles, membranes, and cells in aqueous polymer two-phase systems and their use in biochemical analysis and biotechnology. New York: Wiley; 1986.
94. Ribeiro SC, Monteiro GA, Cabral JM, Prazeres DM. Biotechnol Bioeng 2002; 78: 376–384.
95. Frerix A, Muller M, Kula MR, Hubbuch J. Biotechnol Appl Biochem 2005; 42: 57–66.
96. Frerix A, Geilenkirchen P, Muller M, Kula MR, Hubbuch J. Biotechnol Bioeng 2007; 96: 57–66.
97. Kepka C, Rhodin J, Lemmens R, Tjerneld F, Gustavsson PE. J Chromatogr A 2004; 1024: 95–104.
98. Rahimpour F, Feyzi F, Maghsoudi S, Hatti-Kaul R. Biotechnol Bioeng 2006; 95: 627–637.
99. Gomes GA, Azevedo AM, Aires-Barros MR, Prazeres DM. Sep Purif Technol 2008; 65: 22–30.
100. Freitas S, Canario S, Santos JA, Prazeres DM. Biotechnol J 2009; 4: 265–278.
101. Kepka C, Lemmens R, Vasi J, Nyhammar T, Gustavsson PE. J Chromatogr A 2004; 1057: 115–124.
102. Ku CA, Henry JD Jr, Blair JB. Biotechnol Bioeng 1989; 33: 1089–1097.
103. Sivars U, Abramson J, Iwata S, Tjerneld F. J Chromatogr B Biomed Sci Appl 2000; 743: 307–316.
104. Birkenmeier G, Vijayalakshmi MA, Stigbrand T, Kopperschlager G. J Chromatogr 1991; 539: 267–277.
105. Duarte SP, Fortes AG, Prazeres DM, Marcos JC. J Chromatogr A 2007; 1164: 105–112.
106. Barbosa H, Hine AV, Brocchini S, Slater NK, Marcos JC. J Chromatogr A 2008; O 1206: 105–112.
107. Murphy JC, Fox GE, Willson RC. J Chromatogr A 2003; 984: 215–221.
108. Kepka C, Rhodin J, Tjerneld F. Patent ID: WO04020629. 2000.
109. Muller M, Hubbuch J, Frerix A, Kula MR. Patent ID: US2006286080. 2006.
110. Ge H. Application Note 28; 2008. pp. 4094–4085.
111. Lemmens R, Nyhammar T, Berglof J, Stadler J. Patent ID: US20056916919. 2005.
112. Urthaler J, Buchinger W, Necina R. Chem Eng Technol 2005; 28: 1408–1420.
113. Przybylowski M, Bartido S, Borquez-Ojeda O, Sadelain M, Riviere I. Vaccine 2007; 25: 5013–5024.
114. Quaak SG, van den Berg JH, Toebes M, Schumacher TN, Haanen JB, Beijnen JH, Nuijen B. Eur J Pharm Biopharm 2008; 70: 429–438.

34

PROTEIN CHROMATOGRAPHY, MANUFACTURING SCALE

Joseph Bertolini
CSL Bioplasma, Broadmeadows, Victoria, Australia

34.1 INTRODUCTION

The scale-up of a chromatographic process from the laboratory involving the processing of milliliters of material, to pilot and manufacturing scales involving the processing of liters or hundred liters of material, often requires that the choice of chromatographic techniques and operational parameters be made so as to comply with specific economic, regulatory, and engineering constraints. In fact, the actual chromatographic process used at large scales is dictated equally by these factors as by the effectiveness of the process for purifying the product of interest (1,2,3).

34.2 SCALE-UP OF CHROMATOGRAPHY

34.2.1 Process Design Considerations

Chromatographic procedures can be complex, expensive, and time consuming. Considerable attention should be placed on devising an efficient process, which minimizes the number of chromatographic steps. Given the various modalities available, that is ion exchange, affinity, hydrophobic, and gel filtration, it is usually possible to devise an efficient process to address any purification requirement (4).

One important consideration when dealing with interactive chromatographic procedures is whether the chromatography will be performed in capture or flow-through mode. If the aim of a chromatographic step is to remove low levels of contaminating material, then the column is run in flow-through mode where chromatographic modality and conditions are chosen such that impurities are bound and the desired purified protein flows through. In this way, required column size can be reduced when compared to capturing a high abundance component and allowing the impurities to flow through.

Operating chromatography in flow-through mode can have a number of advantages. The recovered product is usually in a low conductivity buffer which may allow direct loading onto a subsequent column without requiring an intervening ultrafiltration step. Direct capture mode would require subsequent elution of the bound protein, often with high conductivity buffer, and thus subsequent ultrafiltration may be required. Avoiding capture of the protein of interest can often improve recoveries as there is no loss of material in the drop-through and there is no loss during elution of captured material. Often, in order to avoid extreme conditions of salt concentration or pH, this may not be complete.

Utilizing capture mode has an advantage in that it can allow the concentration of a protein from a dilute feedstock. Following passage of the dilute feedstock through the column, the bound protein can usually be recovered in a smaller volume of elution buffer, thereby achieving concentration.

Downstream Industrial Biotechnology: Recovery and Purification, First Edition. Edited by Michael C. Flickinger.

34.2.2 Technical Considerations

Chromatography is readily linearly scalable by ensuring that the following conditions are met:

1. column length is kept constant, increased capacity is achieved by increasing column diameter;
2. flow rate expressed in terms of linear velocity (centimeters per hour) is kept constant;
3. volume of sample and buffers (elution, equilibration) is kept in proportion to the bed volume.

Resin beads are available in various diameters. At bioprocessing scales, larger beads are preferable since they minimize the back pressure generated by the packed column. The resolution of a column reflects the number of beads, and is a function of the accessible bead surface area and the rate of diffusion of proteins into the pores of the beads. A 1-L column containing 100 μm beads would have the same number of beads as a 1-mL column with 10 μm beads. If the beads are porous, accessible surface area would also increase in parallel with the column volume and ensure comparable resolution following scale-up of the chromatographic process.

If column dimensions that do not follow the scale-up principles need to be changed, appropriate changes should be made in flow rate to achieve equivalent residence time of proteins within the column. Protein–resin interactions and hence binding and resolution are kinetic events dependent on diffusive and convective forces that are influenced by the allowed degree of interaction of proteins in the mobile phase, with the stationary (resin) phase.

One problem associated with manufacturing-scale columns is that the weight of the resin may distort the packing. At the laboratory scale, the column walls have a considerable supportive effect that is not present in columns with larger diameters. Thus, these columns have the tendency to become distorted and compacted in the middle, forming a concave surface. This can result in differential flow rates across the resin, leading to a reduction in column resolution. For this reason, the bed heights of large columns are maintained at levels that are generally less than 50 cm. In fact, it is more common for bed heights to not exceed 20 cm when columns have large surface area. If a longer length is required, individual columns can be connected in series.

34.3 IMPACT OF CHROMATOGRAPHIC PARAMETERS AT MANUFACTURING SCALE

When developing a chromatographic process, there should be an awareness from the beginning that the process is intended for scale-up and that the choice of operating parameters will impact the implementation of the process. This can manifest through equipment and reagent requirements, batch and cycle time, and staff required in operating the process. Software packages are available for modeling of chromatographic processes and allow the identification of possible bottlenecks in the process flow, process cycle time, and potential throughput.

The product yield of the developed process and the number of steps in the process have to be critically assessed. While at laboratory scale a low yield process may be tolerated because extra batches can be quickly manufactured, this is usually not the case once a large-scale process is in place. It would represent a drain on time and resources.

The number of chromatographic steps required in a process to effect purification needs to be minimized. Each additional step adds to expense in terms of column hardware and resin, operator time, and reagents. Furthermore, it can lead to lower recoveries of the desired product: three steps of 90% efficiency would result in an overall yield of 73%.

It is advantageous to minimize column size as it can have significant cost and operational implications. This includes column housing and associated infrastructure, resin requirements, processing times, buffer requirements, and pump requirements. The functional binding capacity of a resin affects the amount of resin required, and this in turn affects the size of the column housing. Therefore, optimization of resin performance is important. This involves establishing the extent of resin capacity utilization and operating conditions such as flow rate, loading protein concentration, pH, and conductivity. In some instances, increased capacity may be gained by ensuring that the protein of interest passes unimpeded through the column while binding the contaminant proteins present in lower concentrations.

Increased throughput is desirable and can be achieved by utilizing increased flow rate through the process, provided that residence time is maintained so that there is no adverse impact on binding and resolution. However there are other issues that need to be considered, including the fact that pressure capabilities of large-scale columns could be exceeded and that suitable pumps are required. The choice of pumps can be critical as the shear that is imparted during pumping can affect the integrity of proteins. A number of different pump types are available including peristaltic, membrane, and lobe pumps; these should be evaluated.

The type of resin chosen for a process becomes more critical with scale-up. For example if the resin is to be used in the manufacture of a pharmaceutical product, the resin should have suitable specification with respect to leachables. It should exhibit acceptable stability, especially in face of repeated treatment with sanitizing and regeneration solutions such as high concentrations of NaOH.

34.4 PRACTICAL ISSUES REGARDING CHROMATOGRAPHY AT MANUFACTURING SCALE

34.4.1 Resin

The choice of resin is a key determinant of the performance of a chromatographic process. High throughput screening systems using robotic liquid handling systems, 96 well plate formats and design of experiment (DoE) analysis approaches have been described which facilitate resin evaluation and optimization of operating conditions. The technical capabilities of a resin can have considerable economic and regulatory implications during the scale-up of a chromatographic process (5,6).

The cost of resins can also vary considerably. However, the properties of the resins offered should be considered: information on allowable flow rate, dynamic binding capacity, cleanability, performance with repeated use, functionable group stability, existence of leachables, and regulatory support needs to be assessed.

One of the major differences between resins is the composition of the base matrices. The rigidity of the matrix determines the flow-rate capabilities of the resin. In addition, its stability in the presence of a number of chemical agents can determine the useful life span of the resin. For example, NaOH is typically used with chromatographic processes to sanitize and clean chromatographic media because it inactivates bacteria and viruses, destroys endotoxins, and solubilizes lipids and hydrophobic proteins. However, not all matrices are equally stable to repeated exposure to NaOH. Thus, if NaOH is to be used, an appropriate choice of resin must be made. Alternatively, a different sanitization regime must be used. Resins are, however, generally resistant to a range of acids, detergents, organic solvents, and chaotropic agents. In fact the limiting factor can often be resistance of the column material itself to various chemical substances.

One of the key specifications of a resin is its dynamic binding capacity. This is taken as the loading to the point where absorbance increases by between 5 and 10%. This indicates a degree of saturation of binding sites resulting in the drop-through of protein. Dynamic binding capacity is often affected by flow rate, with increasing flow rates resulting in reduced dynamic binding capacity. This can be explained by the fact that within the resin, diffusion is involved in the transport of protein molecules to binding sites. During the shorter loading period at the higher flow rates, diffusion does not achieve from the feedstock the mass transfer that would be associated with the longer loading period of the slower flow rates.

It is important to examine the performance of the resin across the projected lifetime of the resin. This information is generally not available from resin manufacturers and must be defined for the target process. This involves scaling-down the process and subjecting the resins to repeated cycles of loading and regeneration. Total binding capacity is a useful parameter to monitor. This is more convenient to perform than dynamic binding capacity and provides a measure of the total number of functional groups present on the resin beads. Any decrease in total binding capacity would be reflected in a decrease in dynamic binding capacity. Total binding involves loading a model protein (e.g. albumin for anion exchange resins) to saturate the functional groups on the resin. The bound molecules are then eluted and quantified to derive a measure of the total binding capacity of the resin. Other parameters that can be monitored include HETP (height equivalent to a theoretical plate; discussed below), leaching of functional group and matrix components, packability of the resin, and evidence of fouling of the resin.

Another important factor in resin selection is availability of the resin over the projected lifetime of the product. Therefore, consideration should be given to the commercial viability of the manufacturer and its commitment to the product.

It is critical that the manufacturer has regulatory support documentation for the resin. This will greatly facilitate the acceptance by the regulatory authorities, of a process utilizing the chosen resin.

34.4.2 Columns

Modern process-scale columns are designed to handle the requirements of large-scale chromatographic purification. Today's chromatographic columns are convenient to use, scalable, and reproducible. Packing and unpacking of columns has been significantly automated. Columns can vary in dimensions from 280 up to 2000 mm and can handle pressures up to 5 bar, thereby allowing the utilization of high flow rates during processing.

The products are usually available with full validation support files which are essential if they are to be used in the manufacture of a registered biopharmaceutical product. Manufacturers provide a guide to solvent resistance of the materials used in the construction of their columns. Clearly, a column may be exposed to various chemicals in different processes, and this may arise during the course of regeneration and sanitization. Construction materials include 316L stainless steel or polyethylene and polypropylene components which are resistant to a wide range of chemicals and temperatures. This allows cleaning-in-place sanitization procedures to be used. The most susceptible components to degradation are the seals and O rings incorporated into the column. Although these are usually compatible with many aqueous solutions, they can be susceptible to a number of organic solvents. However, specialized silicone seals can be sourced.

Process-scale columns are generally compatible with most cleaning-in-place solvents such as NaCl (up to 1 *M*) and 2 *M* NaOH for up to 4 h. Detergents (such as Triton X-100) and chaotrophic agents (up to 8 *M* urea) can be safely used with most bioprocess columns, as can mild concentrations of acids (acetic acid up to 1.7 *M*, trifluoroacetic acid up to 0.1%, and nitric acid up to 0.1 *M*). Alcohol solutions such as ethanol (up to 20%) and isopropyl alcohol (up to 30%) can also be used. If conditions involve contact with high salt concentrations or strong acids (such as hydrochloric acid), polyethylene would be more suitable than stainless steel columns.

34.4.3 Sanitization and Regeneration of Columns

Preventing batch-to-batch contamination is a major concern in pharmaceutical manufacture and requires the regeneration of resins between batches and the sanitization of resins with respect to bacteria, viruses, and prions. Consequently, the regulatory authorities place heavy emphasis on regeneration and sanitization of manufacturing-scale chromatographic columns to ensure the quality and safety of products (7).

Regeneration of columns between batches is important not only to prevent batch-to-batch contamination, but also to prevent fouling and maintain column performance. This would include maintaining HETP and hence resolution, preventing back pressure buildup. The functional life of the resin can also be extended and this has clear financial benefits. Regeneration regimens need to be optimized for individual situations. An initial elution with high concentration of salt (1.0 *M* NaCl) followed by NaOH (0.5–1.0 *M*) is commonly used for ion exchange columns. However, with particular feedstocks, the use of detergents (e.g. Tween 80) or organic solvents (acetone, ethanol, isopropanol) may be required.

The effectiveness of the cleaning of a resin can be established by monitoring the presence of total organic carbon (TOC) in the equilibration buffer of the resin, following regeneration. Techniques such as X-ray photo electron spectroscopy (XPS) and time-of-flight secondary ion mass spectroscopy (Tof-SIMS) can be used to establish directly the degree of residual contaminants remaining on resin beads, following regeneration. These can be of great use in developing appropriate regeneration regimens.

The reduction of microbial load is typically achieved in industrial-scale chromatographic systems using chemical treatment before, during, and after operation. Aqueous mixtures of ethanol in neutral or acidic conditions are often used to inactivate microbes. Ethanol is most effective when used at concentrations of 60–70% v/v but in the manufacturing environment, 20% solutions are typical. However, at these lower concentrations, ethanol has very little effect on the viruses and little or no effect on pyrogens. For these reasons, ethanol is considered to be only bacteriostatic.

Sodium hypochlorite is the most well-known oxidizing agent that is used to sanitize columns. The chemical possesses excellent microbicidal properties and is harmless to human skin and mucosa. Its use as a chromatographic sanitizing agent has, however, been limited by its potential to react with the resins, resulting in formation of toxic, chlorinated by-products. Peracetic acid used in combination with ethanol (pH 5.5) has been demonstrated to be a highly effective antimicrobial agent. Additionally, peracetic acid is an effective antiviral agent. Another benefit of using peracetic acid is that it decomposes to the relatively innocuous acetic acid.

Sodium hydroxide is the most widely used agent in the pharmaceutical industry where it is used to both sanitize and regenerate chromatographic resins. It is effective against pyrogens, viruses, and prions. The latter are of particular significance if plasma products are being produced. Concentrations of use (0.1–1.0 *M*), incubation times (minutes to days), temperatures, and the frequency of column cleaning may vary depending on the sanitization or cleaning challenges that may be encountered. Validation studies need to be undertaken to confirm that the chosen conditions are effective in sanitizing and cleaning the column, and thus preventing batch-to-batch transmission of pathogens or impurities. Sanitization can be confirmed by using appropriate assays to monitor any residual pyrogen, virus, or prion after exposure of a resin that in turn, has been exposed to a spiked feedstock and various NaOH cleaning regimens.

34.4.4 Chromatographic Systems

Developers of chromatographic processes at laboratory scales would be well acquainted with chromatographic systems such as AKTA (GE Healthcare), Bio-Cad (Applied Biosystems), and Biologic (Bio-Rad).

Systems applicable to pilot-scale work and smaller manufacturing requirements are also available. For example, Bioprocess systems (GE Healthcare) and K-Prime systems (Millipore).

These systems allow computerized and automated operation of chromatographic processes. The software provided in these systems captures how runs were performed with respect to sequence details. This can include flow rates, which valves were activated, and time that the buffers were pumped.

Vendors can provide extensive validation support with respect to the control system (hardware and software), and with respect to the equipment encompassing electronic components, and items such as valves, pumps, and tubing. Compliance with Good Manufacturing Practice (GMP) is an important requirement if the unit is required for pharmaceutical manufacture.

The qualification of chromatographic systems is necessary to provide documented evidence that they are fit for their intended use and are in a state of maintenance and calibration consistent with the use. The qualification process should consist of four parts: (i) design qualification (DQ), (ii) installation qualification (IQ), (iii) operational qualification (OQ), and (iv) performance qualification (PQ) (8).

The DQ describes the requirements and defines the functional and operational specifications of the chromatographic system such as pump capabilities, detector features, and data analysis requirements. Other areas that may be included as designing elements are user instructions, maintenance requirements, validation procedures, and training. The DQ is used as the basis for the OQ evaluation.

The IQ ensures that the chromatographic system is received as designed and is properly installed in the selected environment. Things to be considered include ensuring that the equipment supplied is consistent to the purchase order, that all documentation has been included (operating manuals, maintenance instructions, Standard Operating Procedure (SOP), and validation certificates) and that the system has been installed correctly with all systems able to be powered up. IQ includes running test samples and generating chromatograms comparable to the manufacturer's reference chromatograms.

The OQ is used to demonstrate the instrument's ability to perform a designated task in a standard environment and verifies that the chromatographic system is able to comply with the operational and functional requirements set out in the DQ. When performing OQ on chromatographic systems, it is often better to perform the testing on the system in total rather than isolating individual systems within the instrument. Tests may include leak testing, baseline drift, injection volume precision, flow-rate precision, detector linearity, temperature accuracy, and wavelength accuracy.

The PQ demonstrates that the instrument is able to perform a function in a consistent manner, according to a series of preset specifications. Thus the PQ testing requires a number of cycles to be performed in order to demonstrate instrument consistency. The PQ is application specific and should therefore be performed under conditions that most closely resemble those to be routinely performed. That would include using the same buffers, feedstock, resin, bed height, and elution conditions. Measures would include peak widths, baseline noise, wavelength accuracy, and retention time data.

Once each of these qualification tasks has been completed, there is the requirement for documentation of the DQ, IQ, OQ, and PQ. This includes the procedures and test results used during OQ and PQ testing. Reference should be made to acceptance criteria for each parameter analyzed.

For larger chromatographic processes where column diameters of up to 2.0 m may be required, detailed scoping of specific requirements needs to be undertaken with respect to columns, pumps, detectors, and other necessary ancillary equipment. Such activities are best performed with multifunctional teams, which can include departments responsible for manufacturing, engineering services, research and development, and quality. Following this activity, the appropriate equipment can be sourced. Installation of the equipment should follow the same validation procedures detailed above.

34.4.5 General Operational Issues

A well-packed column with an appropriate regeneration regimen should provide consistent performance for an extended period of time. As has been mentioned, regeneration and sanitization of the column is important to prevent fouling and microbial growth. In addition, the integrity of the column can be safeguarded by prefiltering protein solutions prior to loading onto the column. This can serve to remove any insoluble particles which could become irreversibly embedded in the resin and thus affect the flow of solutions through the column. Often high protein concentration solutions need to be processed by chromatography. It should be recognized that these solutions will have a higher viscosity and that the rate at which the solution is pumped through the column needs to be reduced to prevent buildup of back pressure. Otherwise this could cause resin compression which would create a void in the column and hence, an increase in HETP and a decrease in resolution. Furthermore, as the limit of the diffusional capacity of the resin approached, the degree of interaction between the resin functional groups and the protein is decreased, which would again result in decreased resolution.

34.5 BASIC CHROMATOGRAPHY SETUP

34.5.1 Column and Associated Equipment

A typical arrangement for chromatographic procedures, in addition to the column containing the resin, would include a bubble trap to prevent air entering the column, pressure gauges to monitor differential pressure across the column, and a bypass line to facilitate loading of product (Fig. 34.1). Absorbance detectors are required to allow monitoring of protein binding and elution. Conductivity and pH meters should also be in place to confirm buffer changes as they occur during equilibration, sample loading, elution, and column regeneration. Heat exchanges may also be required if the temperature of the feedstock differs from that at which the chromatography will be performed. With integrated chromatographic control systems, monitoring and display of data is achieved through computer-based systems.

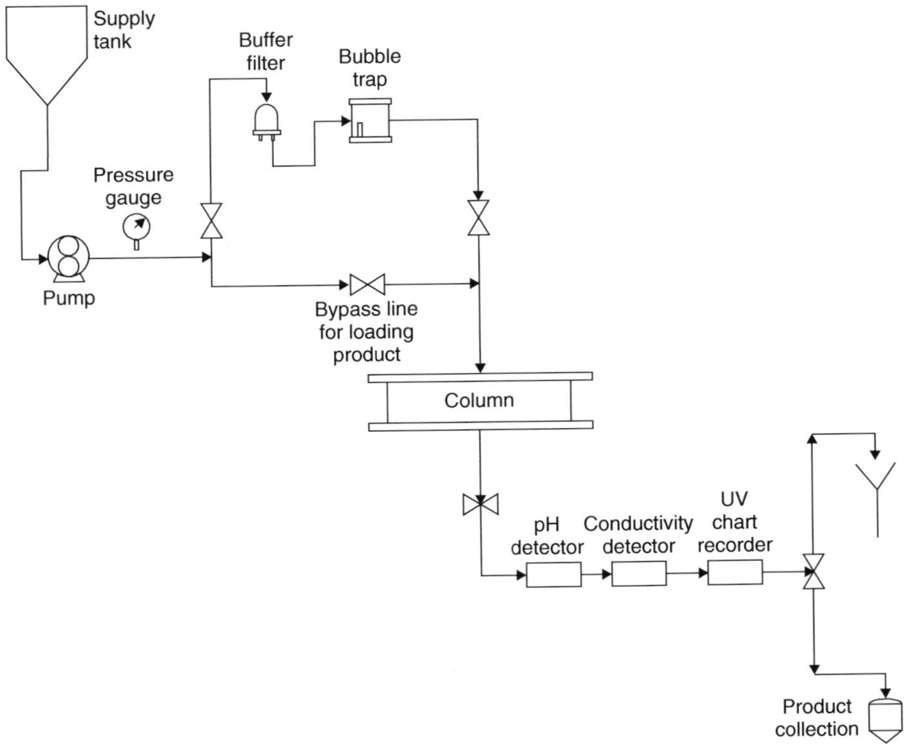

Figure 34.1. Setup of a chromatographic system.

34.5.2 Column Packing

Optimum column packing is essential for a good chromatographic separation. Poorly packed columns can lead to channeling or uneven flows, resulting in zone broadening and reduced resolution.

The flow rate used for packing is based on the equipment, the column type, and quantity and type of media to be used. Flow rate can be established empirically by generating a pressure–flow rate curve as described below. Usually, there is extensive relevant information available from the resin and column manufacture.

A typical procedure for manually packing small- to medium-sized process columns is described below. Typical of these are the BPG series from GE Healthcare, which range from 100 to 450 mm in diameter. Packing initially involves making a slurry of the resin (approximately 75%) before pouring into the column. For short bed heights in wider columns, the stirring is continued when the slurry is in the column so as to achieve even distribution of resin. Once the top plate is in place and any air is removed from the pump lines, pumping of solvent (e.g. 0.1 *M* NaCl) through the column is commenced at a low flow rate. The flow is progressively increased and the operating pressure recorded. As the media settles, a clear space between the bed and top plate develops. The height of the top plate should not be adjusted. The rate of pumping is increased until a plateau is reached where increased operating pressure is not associated with an increase in flow rate. A plot of pressure verses flow rate allows the determination of the optimal packing flow rate, which is 70–100% of the maximum flow rate. Modern rigid resins may not plateau, therefore a flow rate up to the allowable pressure limit for the column can be used.

Larger process columns of greater than 1 m diameter are usually packed to a height of 15–17 cm. These columns are designed without a plunger system. Therefore, packing involves attaching a flange extension to the column and then filling with an appropriate amount of slurried resin. The liquid within the column is drawn out by suction via a pump attached to the column outlet. This promotes settling and packing of the resin. When packing has been achieved, the flange is removed, any excess resin above the height of the column is scraped off, and the lid of the column is replaced. As packing of these columns is performed manually, the quality of the packing (as assessed by HETP and peak symmetry) can vary and may require repacking of the column.

Recently automated packing systems for columns have been developed for large applications which automatically slurry and transfer resin into a column, via a specially designed nozzle incorporated in the column top plate or plunger. The adjustment of the top plate is usually mechanized and compresses the resin to a defined pressure limit. These systems greatly facilitate the column packing process, reducing the labor required and ensuring consistent results.

34.5.3 Evaluation of Packing

The performance of a column is determined by how well it is packed. The method of choice for expressing column efficiency is HETP and the peak asymmetry factor (A_s). Such test methods are simple, quantitative, and do not involve the introduction of colored or contaminating substances into the column. HETP and asymmetry are typically determined by the application of a sample of NaCl (not suitable for ion exchange resins), acetone, or NaOH. Acetone is monitored by UV absorption at 280 nm, while NaOH and NaCl are monitored by conductivity. If the latter are to be used, the column needs to be equilibrated with solutions at lower concentration to decrease the background conductivity so that the passage of the test solution can be monitored. Typically, the volume of the HETP test solution should not be more than be 2.5% of the total bed volume. The sample should be applied as close to the column inlet as possible to avoid dilution. In particular any air-trap and buffer filter should be bypassed.

The HETP values should only be compared between columns packed with the same media under identical conditions. Preferably three test runs should be undertaken before a comparison is made. Once a certain performance standard has been set, future HETP values will serve as a guide to judge the acceptability of column performance. Acceptable HETP values are usually <0.06 cm.

The eluted peak should be symmetrical. Leading or trailing peaks can be an indication of poor packing or channeling within the resin bed and are the first indication of column deterioration. Peak asymmetry factor should be as close to one as possible.

An increased asymmetry factor (increased tailing and decreased leading peak) indicates a column that is too loosely packed. Alternatively, a decreased asymmetry factor (decreased tailing and increased leading peak) indicates a bed that is too tightly packed, or the presence of cracks in the resin bed. Double peaks (i.e. extra peaks before or after the main peak) are an indication of cracked bed due to poor packing or may be due to pump pulsations.

34.5.4 Calculation of Height Equivalent to a Theoretical Plate (HETP)

Equation 34.1 is used to calculate the HETP.

$$\text{HETP} = \frac{L}{N}(\text{cm}) \tag{34.1}$$

where L is column length in centimetres and N is the number of theoretical plates.

$$N = 5.54\left(\frac{V_e}{W_{1/2}}\right)^2 \tag{34.2}$$

V_e corresponds to elution volume (litres) measured as the volume immediately after the start of post-HETP equilibration buffer loading; $W_{\frac{1}{2}}$ is the width of the elution peak at half peak height. Peak height is measured as the vertical distance between the maximum point of the peak to the line tangential to the points of inflection of the trace. V_e and $W_{\frac{1}{2}}$ must be in the same units for the calculation of N. These values in centimetres can readily be measured from the chromatogram (Fig. 34.2).

Example of an HETP Calculation

Given that

$$V_e = 7.330 \text{ cm}$$
$$W_{1/2} = 0.845 \text{ cm}$$

Therefore,

$$\begin{aligned} N &= 5.54\left(\frac{V_e}{W_{1/2}}\right)^2 \\ &= 5.54\left(\frac{7.330}{0.845}\right)^2 \\ &= 5.54(8.674)^2 \\ &= 5.54(75.25) \\ N &= 417 \end{aligned}$$

Hence,

$$\begin{aligned} \text{HETP} &= \frac{L}{N}(\text{cm}) \\ &= \frac{15}{417} \\ &= 0.036 \text{ cm} \end{aligned}$$

34.5.5 Asymmetry Calculation

Figure 34.3 illustrates the measurement of a and b used in the calculation of the asymmetry factor a_f as follows:

$$a_f = b/a \tag{34.3}$$

where a_f is the asymmetry factor, a the leading side of the peak at 10% height, and b the tailing side of peak at 10% height.

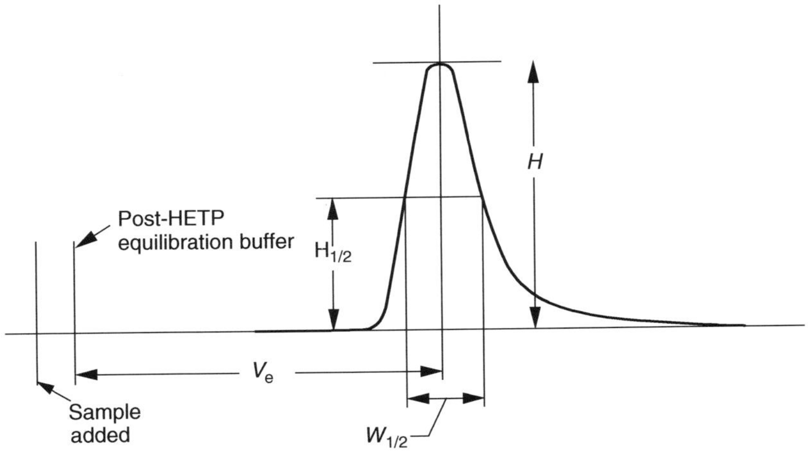

Figure 34.2. Analysis of a chromatogram for measurement of V_e (the elution volume) and $W_{\frac{1}{2}}$ (the width of the elution peak at half peak height) for the calculation of N (the number of theoretical plates).

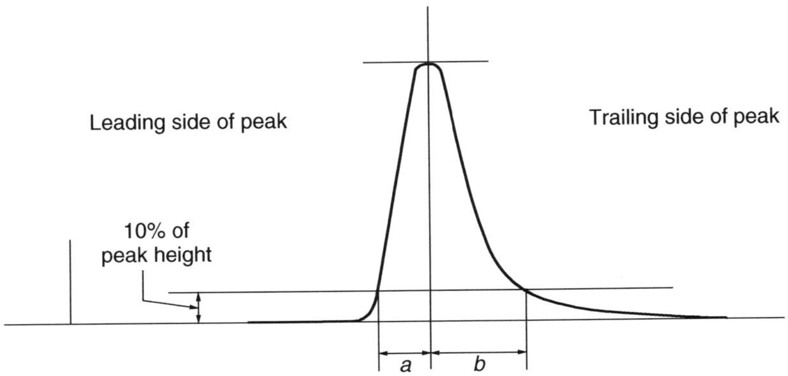

Figure 34.3. Analysis of a chromatogram for measurement of a (the leading side of the peak at 10% height) and b (the tailing side of the peak at 10% height) for the calculation of a_f (the asymmetry factor).

The asymmetry factor should be as close to 1 as possible. If $a_f > 1$ it is considered *tailing*; if $a_f < 1$, it considered *leading*.

REFERENCES

1. Janson J-C, Hedman P. Large-scale chromatography of proteins. Adv Biochem Eng 1982; 25: 43–99.
2. Wheelwright SM. Protein purification: design and scale-up of downstream processing. New York: Hanser Publishers; 1991.
3. Sofer G, Hagel L. Handbook of process chromatography. London: Academic Press; 1989.
4. Janson J-C, Ryden L. (editors) Protein purification: principles, high resolution methods, and applications. New York: VCH Publishers; 1989.
5. Dasarathy Y, Ramberg MM, Andersen M. A systematic approach to screening ion-exchange chromatography media for process development. Biopharm 1996; 9: 42–45.
6. Kelley BD, Switzer M, Bastek P, Kramarczyk JF, Molnar K, Yu T, Coffman J. Biotechnol Bioeng 2008; 100: 950–963.
7. Jungbauer A, Lettner HP. Chemical disinfection of chromatographic resins, part I: preliminary studies and microbial kinetics. Biopharm 1994; 7: 46–56.
8. Huber L. Systems qualification for high-performance liquid chromatography systems. Biopharm 1998; 11: 65–66.

35

PROTEIN CRYSTALLIZATION, KINETICS

GIANLUCA DI PROFIO
Institute on Membrane Technology (ITM-CNR), c/o University of Calabria, Arcavacata di Rende (CS), Italy
Department of Chemical and Materials Engineering, University of Calabria, Arcavacata di Rende (CS), Italy
EFREM CURCIO
Department of Chemical and Materials Engineering, University of Calabria, Arcavacata di Rende (CS), Italy
ENRICO DRIOLI
Institute on Membrane Technology (ITM-CNR), c/o University of Calabria, Arcavacata di Rende (CS), Italy
Department of Chemical and Materials Engineering, University of Calabria, Arcavacata di Rende (CS), Italy

35.1 PROTEIN MOLECULES IN SOLUTION

35.1.1 Protein Solubility, Precipitants, and Phase Diagram

Protein solubility, s (mole per liter), is governed by many factors including pH, surface charge distribution, molecular size, salt type, and salt concentration, whose effect is not always well understood. In water, protein molecules are zwitterions, that is, some of the polar groups in the side chains of the molecule tend to ionize. As a consequence, different parts of the molecular surface may be either positively or negatively charged, and thus the molecule as a whole, depending on solution pH. Under these conditions, Coulomb repulsion dominates the interaction between the protein ions. At the isoelectric point, pI, the net surface charge vanishes and the dominating force between the particles is attractive (van der Waals, hydrophobic, and multipole) so that solubility drops drastically, due to the lacks of electrostatic repulsions. Accordingly, protein solubility is a U-shaped function of pH with a minimum at the pI.

The dependence of solubility on pH may, in principle, be used to address crystallization. In practical attempts, however, crystallizing proteins by simply changing the pH is not well suitable. This is because at the pI attractive forces are usually too strong to induce orderly aggregation of molecules to form regular crystals and an amorphous precipitate generally forms. To reduce the net attractions, one attempts to keep some repulsive contribution to the interaction potential. This is typically achieved by setting the pH $\neq$ pI and fine-tuning the repulsion by adding precipitants that might for example, partly screen the protein surface charges.

In this respect, protein crystallization is generally carried out by adding inorganic salts or polymers, most often polyethylene glycol (PEG) with molecular weight ranging from 200 to 20,000 Da. For electrically neutral PEG, the mechanism that induces effective attraction between large biomolecules is due to depletion forces. These forces originate when access of small molecules (PEG) into the gap between the large macromolecules is limited, because the gap width is comparable to or smaller than the size of the small molecules. Then, the osmotic pressure resulting from the Brownian motion of these small species and their collisions with macromolecules on the outer sides of the latter will exceed the repelling pressure from the gap side. Here, the depletion forces push together the large proteins.

If the precipitant is an ionic salt, numerous interactions are involved, which makes the system much more complicated. The decrease of protein solubility with increasing ionic strength of the solution is termed *salting*

Downstream Industrial Biotechnology: Recovery and Purification, First Edition. Edited by Michael C. Flickinger.

out. This effect comes from "competition for water" between proteins, on one side, and salts on the other. Any ion in solution binds water to some extent thereby reducing the amount of water otherwise available for the solvation of the protein. In other words, water molecules are monopolized to form bonds with the small ions, thus the salt "dehydrates" the protein. When the concentration of ions becomes sufficiently high, the proteins are driven to neutralize their surface charges by interacting with one another, thus giving rise, eventually, to crystallization. Ions can be categorized into *kosmotropes* and *chaotropes* on the basis of the strength of ion–water interactions. Kosmotropes are small ions with high charge density that bind water molecules strongly, decreasing the mobility of nearby water molecules. Chaotropes are large monovalent ions with low charge density that bind water weakly relative to the strength of water–water interaction.

But ionic precipitants can promote attractive interactions among proteins to induce crystallization by *salting in*. At high ionic strength of precipitant, near the p*I*, counterion adsorption by proteins produces the screening of the electrostatic repulsion between the like-charged protein molecules (Debye screening). This effect is described by the Derjaguin, Landau, Verwey, and Overbeek (DLVO) theory of colloid stability (1–3), which is based on the balance between van der Waals attractions and electrostatic repulsions mediated by the electrolytes in solution. In the classical DLVO theory, the only ionic properties are the size and charge of the salt ions and protein precipitation does not depend on salt type. However, in the 1800s, Hofmeister showed that the concentration of a salt required to precipitate a protein solution depends strongly on the choice of the salt (4) and different salt types induce different effects on protein precipitation and phases (Hofmeister effect). Hofmeister established the lyotropic series that ranked various salts according to their ability to precipitate hen egg white proteins:

$$\text{Cations: } NH_4^+ > K^+ > Na^+ > Li^+ > Mg^{2+} > Ca^{2+}$$

$$\text{Anions: } SO_4^{2-} > HPO_4^{2-} > CH_3COO^- > C_6H_5O_7^{3-} > C_3H_2O_5^{2-} > HCO_3^- > CrO_4^{2-} > Cl^- > NO_3^- \gg ClO_3^- > SCN^-$$

The Hofmeister effect has been poorly understood and cannot be explained in the framework of the DLVO theory as this model lacks salt-specific input. Possible explanations have been proposed including salt-specific ion-dispersion forces between the salt ions and protein molecules (5), hydration forces, water structure, and dissolved gas (6). However, no detailed theory exists, which accounts for the effects of salt type on the phase diagrams of globular proteins.

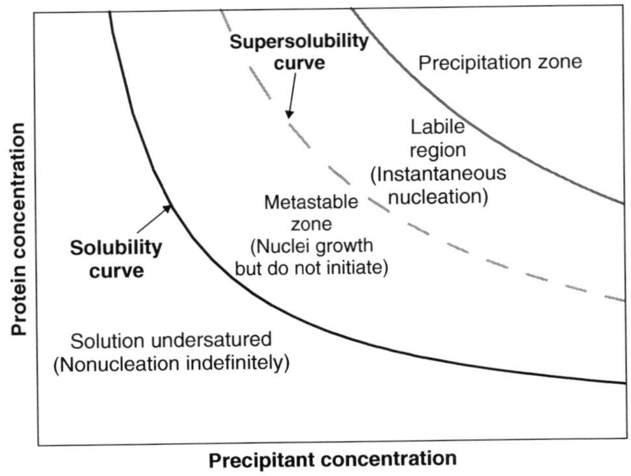

Figure 35.1. Phase diagram applying to crystallization of globular proteins. The solubility curve (solid, black) divides two regions: supersaturated solutions, in which crystallization proceeds and unsaturated solutions, where crystals will dissolve. The supersolubility curve (dashed, green) further divides the supersaturated region into higher supersaturation conditions, where nucleation and growth compete (labile phase) and lower levels, where only crystal growth will occur (metastable phase). An unstable zone, where proteins precipitate as an amorphous phase, exists (solid, red curve). (This figure is available in full color at http://onlinelibrary.wiley.com/book/10.1002/9780470054581.)

A typical phase diagram for globular proteins is shown in Fig. 35.1. The diagram can be subdivided into a number of regions. The region below the solubility curve is undersaturated and no crystallization can take place in it; above the solubility curve, crystallization can occur. In the metastable region (see section titled "Homogeneous Nucleation") nucleation rates are very small and can generally be neglected so only growth occurs; in the labile region, both nucleation and growth will occur. In extreme parts of the labile region, protein aggregation, probably as an amorphous precipitate, can be observed.

35.1.2 Optimum Conditions to Crystallize Proteins: The Crystallization Slot

Because protein solubility is not well understood, it is sometimes difficult to select optimum conditions to precipitate a target protein. The equilibrium solubility of globular proteins correlates well with the thermodynamic parameter B_{22} (7), the second viral coefficient. B_{22} is an integral measure of a quantity that relates to the strength of protein–protein interactions, the mean-force potential (8). This quantity provides a measure of the dilute protein–protein interactions that are Boltzmann-averaged over all distances and orientations in the liquid phase (8). Since B_{22} is related to the mean-force potential through equilibrium solution theory, its evaluation can give a qualitative measure of the protein–protein interactions in a dilute solution. B_{22} is defined as:

$$B_{22} = 2\pi \int_0^\infty \left(1 - e^{-U(\bar{r})/k_B T}\right) \cdot \bar{r}^2 d\bar{r} \qquad (35.1)$$

In Equation 35.1 $\bar{r}$ is the center-to-center distance of the particles, k_B is the Boltzmann constant, T is the absolute temperature and $U(r)$ is the pair interaction potential. In this equation, what is added to the protein solution to achieve a given value of B_{22} does not matter, once the strength of attraction has been achieved, the protein will have the same solubility.

Positive values of B_{22} generally indicate larger protein–solvent interactions compared to protein–protein interactions, that is, the net protein intermolecular forces are repulsive; on the other hand, negative values of B_{22} imply overall attractive forces between the protein molecules in solution. Therefore, B_{22} has a predictive character regarding protein crystallization and/or precipitation from solution. There is a region of B_{22} values, called the *crystallization slot*, where crystallization is favored. It corresponds approximately to the range: -1.0×10^{-4} to -8.0×10^{-4} mol mL/g^2 (9). Values outside this region do not appear to favor crystallization or lead to precipitation of amorphous phase (10). It is also known that these values of B_{22} correspond to protein–protein interactions that are short-ranged with respect to the size of the protein molecule. This condition typically corresponds to salt concentrations sufficiently large to screen the Coulomb repulsive interaction, so that the Debye screening length is small (11). The osmotic second virial coefficient correlates with nucleation kinetics (12). This means that kinetics of nucleation correlates with the protein solubility via the coefficient B_{22}. The crystallization slot, in terms of the osmotic second virial coefficient, could point to solution conditions that lead to "good" nucleation. As B_{22} is determined in a nondestructive manner from dilute solutions, B_{22} forms a useful guideline while attempting the crystallization of proteins by manipulating solution conditions to a place where proteins will readily crystallize.

The second virial coefficient is an experimentally measurable quantity determined from the osmotic compressibility:

$$\lim_{\rho_C \to 0} \left[\mathrm{d}\Pi/\mathrm{d}\rho_C\right] = k_B T\,(1 + 2B_{22}\rho_C + \cdots) \qquad (35.2)$$

where Π is the suspension osmotic pressure and ρ_c is the number density of particles. The osmotic compressibility, and hence B_{22}, can be determined through static light scattering (13) and by self-interaction chromatographic technique (14).

35.1.3 Static Light Scattering

Static light scattering (SLS) has been used to obtain information leading to estimation of the second osmotic virial coefficient for some proteins and for mapping the crystallization slot (15).

A polarized laser light beam is scattered by a molecule in solution, and the deflected light beam is detected by either a single photomultiplier that can be moved radially in a plane around the scattering volume or by multiple detectors that are arranged radially in a plane around the scattering volume enabling simultaneous measurements at multiple angles. The quantity that is determined through light scattering is the Rayleigh ratio, R_θ, which corresponds to the scattering intensity of protein molecules in excess to the solvent scattering:

$$R_\theta \equiv \frac{\bar{r}^2}{\sin^2\theta} \frac{I_S}{I_0} \qquad (35.3)$$

where θ is the scattering angle and I_S/I_0 is the ratio of the intensity of incident light to that of scattered light. For dilute solutions, the equation can be written as (16):

$$R_\theta = \frac{Kc}{1/MW + 2B_{22}c + 3B_{33}c^2 + \ldots} \qquad (35.4)$$

where MW is the molecular weight, c is the molar protein concentration, B_{22}, B_{33}, and successive terms are the second, third, and so forth virial coefficients and K is an optical constant of the system:

$$K = \frac{4\pi^2 n_0^2 (\mathrm{d}n/\mathrm{d}c)^2}{N_A \lambda^4} \qquad (35.5)$$

where n_0 is the refractive index of the pure solvent not including the scattering molecules, dn/dc is the refractive index increment, λ is the wavelength of the laser beam, and N_A is Avogadro's number. If the scattering molecules are small compared to the wavelength of the incident laser light beam, then the light scattering has no angular dependence and measurements are done at a single angle, usually with the detector placed at $\theta = 90^\circ$ to the incident light beam. To determine molecular parameters Equation 35.4 can be rewritten and truncated after the first-order term:

$$\frac{Kc}{R_\theta} = \frac{1}{MW} + 2B_{22}c \qquad (35.6)$$

Thus, a graph of Kc/R_θ versus c provides information on the molecular weight from the intercept and the second virial coefficient, which is the slope divided by 2.

35.1.4 Dynamic Light Scattering

Dynamic light scattering (DLS), also referred to as *quasielastic light scattering* or *photon correlation spectroscopy*, is a versatile optical method used for measuring mutual diffusion coefficients and mode of aggregation of proteins or other biomolecules in solution. In fact, this technique provides the protein's mutual diffusion coefficient, D_0, and the size of protein monomers or aggregates that can be obtained from D_0; in addition, the dependence of D_0 on protein concentration provides information pertaining to both protein interparticle and hydrodynamic interactions. Compared to other methods, DLS has the great advantage of requiring very small samples and short experimental times.

In typical DLS geometry, light coming from a laser is scattered by a sample and is collected at a given angle θ (usually 90°) by a photodetector. Rather than the absolute average intensity measured by SLS, in DLS the scattered intensity fluctuations, primarily due to the random diffusive motion of the protein molecules into and out of a focused laser beam, are measured. The scattering angle defines the direction of the scattering vector q:

$$q = \frac{4\pi n}{\lambda} \sin\frac{\theta}{2} \qquad (35.7)$$

where λ is the wavelength of light in vacuum and n is the refractive index of the sample. One way to mathematically describe the stochastic temporal fluctuations of the scattered electric field is via a correlation function $g(\tau)$:

$$g(\tau) = B\left[1 + a(g'(\tau))^2\right] \qquad (35.8)$$

where τ is the sample time, B is the baseline of the correlation function, a is a constant which depends on the spatial coherence of the detected scattered light, and $g'(\tau)$ is the normalized scattered electric field correlation function. In the limit in which particles are small compared to the length of the inverse scattering vector q^{-1}, thermodynamic fluctuation theory and hydrodynamics provide a theoretical framework that relates $g(\tau)$ to diffusion coefficients. The main point is that the fluctuations of the scattered electric field depend on concentration fluctuations, which dissipate by diffusion. For a binary solution, the normalized field correlation function is, to a good approximation, given by

$$g'(\tau) = \mathrm{e}^{-\Gamma\tau} \qquad (35.9)$$

where Γ is the decay constant, which contains the protein molecular information via $\Gamma = D(c)q^2$, $D(c)$ being the concentration-dependent translational diffusion coefficient of the protein. Larger values for $D(c)$ indicate smaller particle size and cause the correlation function to decay rapidly, whereas smaller values for $D(c)$ correspond to larger particles and result in a slowly decaying function. Since globular proteins are often assumed to be spherical in shape, the Stokes–Einstein relation can be used to estimate the apparent hydrodynamic radius, R_h(app), of the protein by

$$R_h\,(\mathrm{app}) = \frac{k_B T}{6\pi\eta D(c)} \qquad (35.10)$$

where η is the solution viscosity. Thus, analysis of $g(\tau)$ from DLS provides an appealing way of estimating the apparent size of the protein. The diffusion coefficient has a protein concentration dependence usually described by

$$D(c) = D_0(1 + k_D c + \cdots) \qquad (35.11)$$

where D_0 is the infinite dilution value for $D(c)$ and k_D is the diffusion virial coefficient. The true equivalent spherical hydrodynamic radius of the protein molecules can be estimated using D_0 in:

$$R_h = \frac{k_B T}{6\pi\eta D_0} \qquad (35.12)$$

35.2 HOMOGENEOUS NUCLEATION

According to classical concepts, nucleation is the initial step for first-order phase transitions during the birth of a new phase starting from an old phase which has a higher free energy. The main concepts in the area of nucleation were introduced in the nineteenth century by J.W. Gibbs (17,18), who considered the formation of nuclei of a fluid phase as a consequence of density fluctuations in another fluid phase. In further developments, some of the Gibbs's assumptions were transferred to the nucleation of ordered solids, such as crystals, from dilute or condensed fluids and the so-called *classical nucleation theory* (CNT) emerged. CNT assumes that the arrangement in a crystal's embryo is identical to that in a large crystal, and hence, the surface free energy of the nuclei will be equal to the crystal interface. It was also assumed that the size of the nuclei decreases smoothly as supersaturation increases. While CNT has been used rather successfully to describe the nucleation of small molecules, a specialized theory for the nucleation of proteins has not yet been developed. Therefore, the theoretical concepts of the classical nucleation theory are generally extended to proteins although, in some cases, they look poorly adequate to explain phenomena observed during protein crystallization processes (see section titled "Nonclassical Approach to Nucleation").

35.2.1 Classical Nucleation Theory

In crystallization, homogeneous nucleation takes place in a crystal-free solution when solute species come together to adopt an ordered arrangement corresponding to the minimization of their free energy. This process can be visualized by considering a certain number of molecules moving randomly in solution, among the molecules of the solvent, and colliding inelastically with each other. Some of them converge to form clusters of growth units, whose location is subjected to fluctuations. These aggregates, being monomers or oligomers, are the growth units of the crystals. The lifetime of these clusters is governed by the balance between the cohesion forces, that let the cluster to survive, and those that tend to disaggregate the growth units. The force F_V that maintains a cluster depends from the number of bonds saturated between the molecules and it is proportional to the volume of the cluster, while the force F_S, that tends to separate them, is proportional to the number of unshared bonds pointing toward the solution on the surface and it is dependent on the clusters surface (19,20). The free energy (work) change ΔG_{hom} (joule) necessary to create a protein cluster is given by

$$\Delta G_{hom} = \Delta G_{S,\,hom} + \Delta G_V \tag{35.13}$$

where $\Delta G_{S,\,hom}$ is the surface free energy change associated with the formation of the aggregate (a positive quantity) and ΔG_V is the volume free energy change associated with the phase transition (a negative quantity). The term $\Delta G_{S,hom}$, also known as *excess free energy of the cluster*, can be expressed as the product of the total surface area A_n (square meter) of the aggregate and the interfacial energy γ (joule per square meter) between the nucleating cluster and the crystallization medium:

$$\Delta G_{S,\,hom} = A_n\gamma = \beta l^2\gamma \tag{35.14}$$

where β (unitless) is a factor depending on the shape of the cluster and l (meters) is the characteristic length of the stationary cluster size distribution. The volume energy term ΔG_V, both for homogeneous and heterogeneous (see section titled "Heterogeneous Nucleation") nucleation, is given by the difference in chemical potential Δu of molecules in solution, μ_a, and in the solid phase, μ_b:

$$\Delta G_V = -\alpha l^3 \frac{1}{\upsilon}\Delta\mu \tag{35.15}$$

where α (unitless) is the volume shape factor and υ (cubic meter) is the macromolecular volume in the crystal. The chemical potentials of the molecules in the solution and in the solid phase can be expressed as a function of their respective activities a_a and a_b:

$$\mu_a = \mu_a^0 + k_B T \ln a_a \tag{35.16a}$$

$$\mu_b = \mu_b^0 + k_B T \ln a_b \tag{35.16b}$$

with k_B (joule per kelvin) the Boltzmann constant and T (kelvin) being the absolute temperature. Then the driving force of the aggregation process can be rewritten as:

$$\Delta\mu = k_B T \ln\left(\frac{a_a}{a_b}\right) = k_B T \ln\left(\frac{\gamma_a C_a}{\gamma_b C_b}\right) \tag{35.17}$$

where $\gamma_{a/b}$ are the activity coefficients and $C_{a/b}$ (mole per cubic meter) are the concentrations in solution and at the equilibrium, respectively. From the above equations, by defining supersaturation S (unitless) as the ratio of the actual activity of the solution divided by the activity at equilibrium, one obtains:

$$S = \Delta\mu/k_B T = \ln\left(\frac{a_a}{a_b}\right) = \ln\left(\frac{\gamma_a C_a}{\gamma_b C_b}\right) \tag{35.18}$$

If the activity coefficient is assumed to be independent from the concentration in the given concentration range, then $\gamma_a \approx \gamma_b$ and the supersaturation becomes:

$$S \approx \ln\left(\frac{C_a}{C_b}\right) = \ln\left(\frac{C}{s}\right) \tag{35.19}$$

where s (mole per cubic meter) is the protein solubility and C replaces C_a. A frequently used definition for the supersaturation is

$$S_r = \frac{C - s}{s} \tag{35.20}$$

A further simplification is to use:

$$S_t = \frac{C}{s} \tag{35.21}$$

or as a concentration difference:

$$S_d = C - s \tag{35.22}$$

Supersaturation may be created by various methods that regulate the solute activity. These include solvent removal (evaporation or freezing), addition of different salts with ions that participate in precipitation (salting out). Supersaturation can also be created by methods that regulate the solute solubility, such as temperature change, pH change, and addition of a solvent that lowers the solubility of the solute (drowning out).

Combining Equations 14–19 and substituting into Equation 35.13, one obtains:

$$\Delta G_{\text{hom}} = \beta l^2 \gamma - \alpha l^3 \frac{1}{\upsilon} k_B T \ln S \qquad (35.23)$$

If assuming that the nuclei are compact spherical objects $\alpha = 4\pi/3$ and $\beta = 4\pi$ and that they behave like small droplets of bulk phase, which means that the surface energy γ does not change with direction, the free energy of a spherical liquid droplet of radius r in a vapor becomes

$$\Delta G_{\text{hom}} = 4\pi r^2 \gamma - \frac{4}{3}\pi r^3 \frac{1}{\upsilon} k_B T \ln S \qquad (35.24)$$

This relation is known as the *Gibbs–Thompson expression*. The first term on the right-hand side of Equation 35.24 is the surface term while the second is the volume term. This formalism, although derived for the nucleation of liquid droplets from a high energy fluid, is also useful for the case of crystals growth from solution, if assuming that the interfacial energy and the shape factor do not depend on the cluster size and equals the values for macroscopically large crystals, that is, they are not a function of orientation. However, in case of a crystalline compound, the interfacial energy is anisotropic with respect to the surface orientation. Therefore, the excess energy of a crystalline cluster is a weighed average of the surface energy of all interfacial orientations present on the cluster. To consider the crystalline anisotropy of the different faces i, of total surface area A_i, of a crystal, γ should be replaced by an expression of the type: $\sum_i A_i \gamma_i$ (21).

The plot of $\Delta G_{\text{S, hom}}$, ΔG_{V} and ΔG_{hom} as function of the cluster size is given in Fig. 35.2. Note that ΔG_{hom} increases with r until a critical value, ΔG^*_{hom}, is reached and then decreases. This means that there is an energetic barrier that must be crossed to induce the formation of stable nuclei. Below this critical value, the surface energy prevails and clusters, due to thermal instabilities, form and redissolve continuously. Above the critical point, the volumetric term stabilize the clusters, which can grow to become macroscopic crystals. Hence, a critical size r^* for which F_{V} equals F_{S} exists. Any cluster with a size below r^* will tend to dissolve, while cluster larger than this critical size experiences a lowering of free energy when it grows, that is, the cluster becomes more stable as it grows. To reach clusters of that critical size the solution needs to overcome the activation barrier—the so said nucleation barrier—observed in the ΔG_{hom} versus the r curve.

To calculate the value of the critical radius r^* is needed only to differentiate Equation 35.24 with respect to r and set it to zero ($\partial \Delta G_{\text{hom}}/\partial r = 0$). Thus, the expression for the critical nucleus is obtained:

$$r^* = \frac{2\upsilon\gamma}{k_B T \ln S} \qquad (35.25)$$

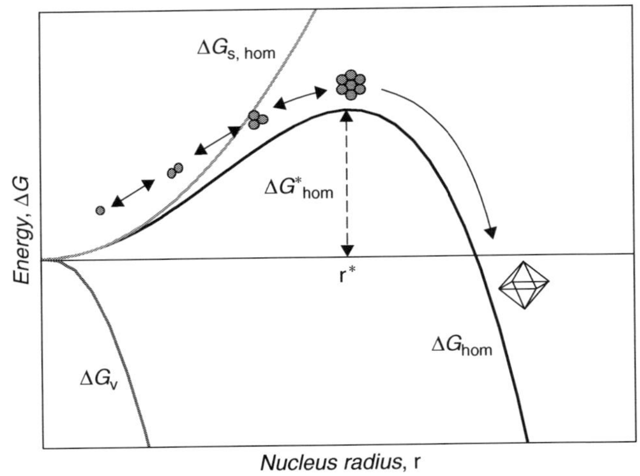

Figure 35.2. Dependence of the free energy ΔG_{hom}, the volume term $\Delta G_{\text{V}} \propto r^3$ and the surface term $\Delta G_{\text{S}} \propto r^2$ from the size r of the cluster. Note that for $r < r^*$ clusters form and redissolve continuously, while for $r > r^*$ the aggregates are stable and evolve toward the formation of macroscopic crystals. (This figure is available in full color at http://onlinelibrary.wiley.com/book/10.1002/9780470054581.)

The height of the nucleation barrier ΔG^* can be obtained by substituting the expression of r^* into Equation 35.24, yielding:

$$\Delta G^*_{\text{hom}} = \frac{16\pi \upsilon^2 \gamma^3}{3\,[k_B T \ln S]^2} \qquad (35.26)$$

From Equation 35.26 it is clear that the nucleation work ΔG^*_{hom} depends on two main parameters: the externally controlled supersaturation and the material surface/solution composition-dependent interfacial energy. The interfacial energy γ can be estimated for a spherical nucleus from the bulk solubility s and from the molecular volume υ according to (22):

$$\gamma = 0.514 k_B T \frac{1}{\upsilon^{2/3}} \ln \frac{1}{\upsilon s N_A} \qquad (35.27)$$

Figure 35.3 shows graphically the Gibbs–Thompson expression for different values of supersaturation. Note that the height of the nucleation barrier and the critical size for the formation of stable nuclei, vary inversely with supersaturation. This shows the strong dependence of nucleation on supersaturation. The existence of the energy barrier and its dependence on supersaturation explains the existence of the metastable zone in the phase diagram (see Figure 35.1). In fact, even if a solution has bring under that thermodynamic conditions appropriate for precipitation–above the solubility curve–it really does so only if a certain value of supersaturation is exceeded. This value being related to the nucleation energy barrier ΔG^*_{hom}

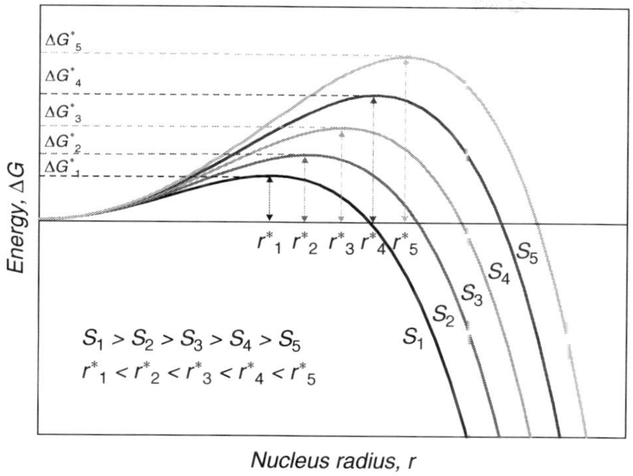

Figure 35.3. Dependence of the nucleation barrier from supersaturation: r^* and $\Delta G^* \rightarrow 0$ as $S \rightarrow \infty$. (This figure is available in full color at http://onlinelibrary.wiley.com/book/10.1002/9780470054581.)

by Equation (35.25). Moreover, according to the above descriptions, nucleation can be seen as a phase transition that results from fluctuations leading to local variations in the solution density whose lifetime is related to interaction between solute molecules. However, density fluctuations are phenomena whose spatial and temporal distributions are unpredictable. This intuitive description therefore explains why homogeneous nucleation is considered a probabilistic phenomenon.

35.2.2 Nucleation Rate

According to the classical nucleation theory, the stationary rate of nucleation, J (number per cubic meter per second), that is, the number of nuclei that exceed the critical size per unit volume and per unit time, can be described according to the following equation (35.22):

$$J = A \exp\left(\frac{-\Delta G^*_{\mathrm{hom}}}{k_{\mathrm{B}}T}\right) \tag{35.28}$$

The preexponential collisional factor $A(m^{-3}s^{-1})$ is a function of the molecular-level attachment-kinetics parameters. It is the product of the three terms: the number density of molecules ρ, the rate at which molecules attach to the nucleus causing it to grow, j, and the Zeldovich factor Z. The number density of molecules ρ is essentially the number of possible nucleation sites per unit volume, as for homogeneous nucleation, the nucleus can form around any one of the molecules present. For a dilute protein solution, $\rho = 10^{24}/m^3$. An upper bound on the rate at which molecules attach to the nucleus causing it to grow, j, is provided by the diffusion-limited flux onto the nucleus. The term j is then related to the number of molecules of the crystallizing phase in a unit volume and to the frequency ν of molecular transport at the nucleus–liquid interface. It depends on the viscosity of the solution, the molecular charge, the molecular volume, the density of the solution, and the fluid dynamic regime inside the crystallization system. The frequency of atomic or molecular transport at the nucleus–liquid interface can be related to the bulk viscosity, η (kilogram per meter per second), by the Stokes–Einstein relation:

$$\nu \approx \frac{k_{\mathrm{B}}T}{3\pi a_0^3 \eta\,(T)} \tag{35.29}$$

where a_0 (meters) is the mean effective diameter of the diffusing species. If the critical nucleus is approximated by a sphere of radius r^* then this upper bound is $j \approx \rho D r^*$, where D is the diffusion coefficient for the molecules. The Zeldovich factor Z (unitless) is the probability that a nucleus at the top of the barrier will go on to form a crystal; it is less than one. Therefore the rate at which a nucleus actually crosses the barrier and grows into a new phase is Zj. The Zeldovich factor Z is approximately equal to $[\partial^2(\Delta G/k_{\mathrm{B}}T)/\partial n^2]^{1/2}$ evaluated at the top of the barrier. Here n is the number of molecules in the nucleus. Within CNT, the second derivative of the free energy at the top of the barrier scales as $k_{\mathrm{B}}T/(n^*)^{4/3}$, where n^* is the excess number of molecules in the critical nucleus. Thus within CNT $Z \approx 1/(n^*)^{2/3}$, and so for nuclei with $n^* = 10 - 100$ molecules Z is of order 0.1 or 0.01.

There have been attempts to analytically derive an expression for the coefficient A in Equation 35.28. An expression for the preexponential factor A in the case of homogeneous nucleation and when the controlling mechanism for solute transport is the diffusion from the bulk (see section titled "Crystallization In Forced Solution Flow Regime") is:

$$A_{\mathrm{HON},D} = \left(\frac{k_{\mathrm{B}}T}{\upsilon_0^2 \gamma}\right)^{1/2} Ds \ln S \tag{35.30}$$

where D (square meter per second) is the diffusion coefficient and s (mole per cubic meter) the equilibrium solubility.

Substituting Equation 35.26 in Equation 35.28, one obtains for spherical nuclei:

$$J = A \exp\left(-\frac{16\pi \upsilon^2 \gamma^3}{3\,(k_{\mathrm{B}}T)^3\,(\ln S)^2}\right) \tag{35.31}$$

Note that the exponential factor typically varies more rapidly with supersaturation than the preexponential factor, and so A is often taken to be a constant.

Inspection of the equation above clearly suggests that the nucleation rate can be experimentally controlled by the following parameters: molecular or ionic transport, viscosity, supersaturation, solubility, solid–liquid interfacial tension, and temperature. The nucleation rate will increase by increasing the supersaturation while all other variables are constant. However, at constant supersaturation, the nucleation rate will increase with increasing solubility. Solubility affects the preexponential factor and the probability of intermolecular collisions. Furthermore, when changes in solvent or solution composition lead to increases in solubility, the interfacial energy decreases since the affinity between crystallizing medium and crystal increases. Consequently, the supersaturation required for spontaneous nucleation decreases with increasing solubility.

Several techniques have been used for detecting nucleation in protein solutions. These include optical microscopy for direct counting of protein crystals grown to optically detectable size in supersaturated solutions, atomic force microscopy, static and dynamic light scattering, small-angle X-ray or neutron scattering, turbidimetry, neutron magnetic resonance, Raman spectroscopy, electron microscopy, and differential scanning calorimetry.

35.2.3 Nucleation Time

In precipitation processes, the nucleation time $\overline{\tau}$ (seconds) is the time interval elapsed between the setting of a protein solution at a given supersaturation value and the formation of crystalline nuclei. It is composed by two terms: $\overline{\tau} = t_d + t_n$. The first term, t_d, is the time needed to achieve a stationary size distribution of precritical clusters while the second term, t_n, is the time required to form nuclei of critical size. The nucleation time is difficult to be measured experimentally as it is not possible to directly observe the critical nuclei as they form in solution. However, as they can only be observed after they have grown to a certain size, the waiting time for nucleation can be measured. This waiting time is the sum of the nucleation time and the term t_g which accounts for the time required for those nuclei to grow to a size large enough to be experimentally detected (24). Accordingly, the waiting time, which is referred to as *induction time*, t_{ind}, is defined as the time elapsed between the moment when supersaturation is created and the moment that crystals are observed, and it is often used as a macroscopic measure of the nucleation time. The induction time is one of the few parameters, related to nucleation, which can be assessed experimentally, at the condition to detach nucleation from growth. The induction time is affected by several parameters such as the initial supersaturation, temperature, pH, agitation speed, and the presence of additives/impurities (25–27). As t_{ind} depends on the technique used for the detection of nucleation, reliable methods for the determination of induction time periods are important. Several techniques such as measurement of solution conductivity (28), intensity of transmitted light (29), electronic microscopy (30), fluorescence (31), and turbidity (32,33) can be used for the experimental determination of the induction time. Most of the aforementioned techniques are more sensitive than visual detection of crystals by using simply an optical microscope.

When the formation of a stable nucleus is the rate-limiting step in crystallization, then the induction time is inversely related to the nucleation rate, J. Therefore, according to the classical nucleation theory, t_{ind} is related to the supersaturation, S, the temperature, T, and the interfacial tension, γ, according to (20):

$$J \propto \frac{1}{t_{ind}} \tag{35.32}$$

from which, taking into account Equation 35.31:

$$\ln\, t_{ind} = \overline{B} + \vartheta \left(\frac{\gamma^3 \upsilon^2}{(k_B T)^3}\right)\left(\frac{1}{\ln\, S}\right)^2 \tag{35.33}$$

where $\overline{B}$ is a constant and $\vartheta = 4\beta^3/27\alpha^2$ ($= 16\pi/3$ for spherical clusters). From Equation 35.33 it is inferred that an increase in the initial supersaturation will result in a decrease in the observed induction time.

According to Equation 35.33 a plot of $\ln(t_{ind})$ against $1/(ln\ S)^2$ should yield a straight line of slope $16\pi\gamma^3\upsilon^2/\ 3k_B^3 T^3$, from which γ can be estimated. A typical value of γ is 0.6 mJ/m^2 for a lysozyme cluster of about 5–10 mol (34). A representative plot of such graph is shown in Fig. 35.4. It shows two distinct regions with different slopes. The occurrence of these two regions can be attributed, for both organic (35) and inorganic (36) compounds, to the change in nucleation mechanism from heterogeneous to homogeneous as the supersaturation

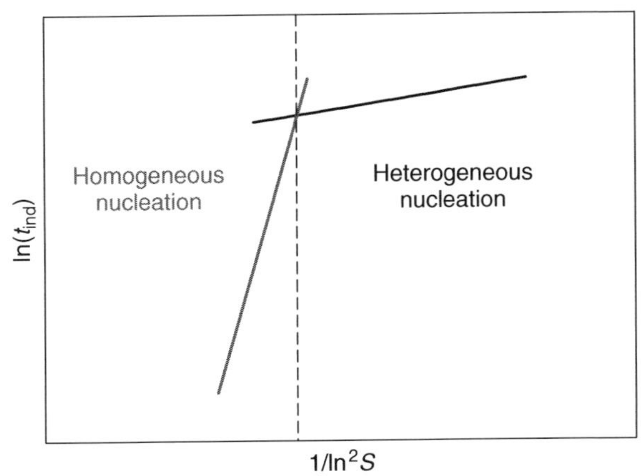

Figure 35.4. Homogeneous (high supersaturations) and heterogeneous (low supersaturations) nucleation regions. From the slopes of the curves the values for interfacial energy γ can be calculated. (This figure is available in full color at http://onlinelibrary.wiley.com/book/10.1002/9780470054581.)

increases. Such a type of graph is very common during crystallization processes because under laboratory conditions, in some circumstances, homogeneous nucleation might be unlikely to occur because of the existence in the solution of solid foreign surfaces, such as the container wells or impellers' faces, dust particles, large impurities, preassembled clusters existing in commercial preparation, already formed crystal of the molecule to be crystallized (seed crystals), or a different type of solid substance that has nucleation-inducing properties and which might therefore affect nucleation kinetics.

35.2.4 The Metastable Zone

The nucleation rate is very low at low supersaturation values and increases rapidly after a certain critical value S^* is achieved. This explains the existence in the solubility diagram (Fig. 35.1) of a region between $S = 1$ (saturation) and S^* (corresponding to the supersolubility curve in the phase diagram), termed the *metastable zone*, where there is a very low probability of nucleation. The metastable zone is therefore bounded by two curves of different natures. The lower boundary is the solubility curve whose location is fixed by the thermodynamic nature of the system. A solution located below this curve will be stable forever, the probability for a nucleation event to occur is 0 and the induction time for nucleation is infinite. The upper limit of the metastable zone is termed the *metastable limit* or *supersolubility curve* and, unlike the solubility curve, it has a kinetic nature. It is defined by the loci of the solubility diagram where the probability for a nucleation event to occur is 1, so that the nucleation can be considered instantaneous as the system approach this curve. Any solution inside the metastable zone will nucleate spontaneously given enough time, which is a function of the nucleation barrier, and therefore of the supersaturation: the higher the supersaturation the shorter the nucleation time. The supersaturation boundary depends on several parameters like for example, temperature, solution composition, rate of supersaturation generation, the presence of impurities, mechanical effects, fluid dynamics, and so on. Due to the kinetics nature of the supersolubility curve, the metastable zone width (MZW) depends on the experimental conditions. MZW is very important as its extent might have profound effect on the yield of a crystallization process (37). As the MZW increases with the rate of achievement of supersaturation (38–40), this affects the rate of both heat and mass transfers during crystallization and, finally, the properties of the crystalline product such as size, size distribution, shape (41), and the overall crystal quality (42).

35.2.5 Necessity to Control Nucleation

Unlike growing crystals of inorganic materials, protein crystallization is typically based on spontaneous nucleation from solutions rather than on growth from seeds. This is due to the small amount of biomolecular material often available, to the difficulty in handling fragile small seeds, and to the poisoning effect of the crystal surface by small amount of impurities, which can cause early growth cessation. In spite of its apparent simplicity, poor control of the spontaneously nucleated crystals, in terms of their number and perfection, predetermines poor control of the output of the whole crystallization experiment. For instance, the high supersaturation needed to achieve crystal nucleation leads to excess of nucleation and to the fast growth of poor-quality crystals (43). Large numbers of crystals means small sizes of each at the end of the process. While this material is inadequate for structural determinations, it is appropriate for the formulation of enzyme crystals to be used for biocatalytic applications, as for example, in the case of cross-linked enzyme crystals (CLECs) (44). If protein crystal nucleation could be enhanced, this would allow crystal nucleation to occur at lower supersaturations, at which growth of higher quality crystals may be expected. In other cases, new crystals are continuously nucleated throughout a crystallization run and may be incorporated into previously nucleated, larger crystals. Such incorporation may remain undetected and lead to mosaicity and lattice strain. Hence, for systems of this latter type, suppression of the secondary nucleation events should contribute to higher perfection. Control of protein crystal nucleation is needed in other health-related areas: production of protein-based crystalline pharmaceuticals (45), protein separation (46), and treatment of protein condensation diseases (47). This explains why the necessity to control protein nucleation is pretty essential in both fundamental and applied research in order to grow crystals with opportune properties with respect to their specific intended uses. Therefore, the possibility to enhance or suppress the rate of nucleation of protein crystals opens broad avenues in the area of protein crystallization for X-ray structure studies or for biocatalytic applications of proteins.

35.3 HETEROGENEOUS NUCLEATION

In the previous section it has been considered that the protein solution is homogeneous and therefore the probability of a given density fluctuation is identical over the whole volume of the solution. This is the ideal situation for what has been defined as homogeneous nucleation, which takes place fundamentally inside extremely pure solutions. Moreover, unlike small molecules, nucleation of protein crystals requires amazingly high supersaturation. This is because protein crystal nucleation is a process of highly precise self-assembly of macromolecules that require highly selective and exact directional interactions.

For this reason, a successful collision between the protein molecules, resulting in the formation of the crystalline bond, requires not only a sufficiently close approach of the two species but also their proper spatial orientation. As a consequence, in many crystallization experiments, because the free energy barrier to homogeneous nucleation is relatively large (of the order of 100 $k_B T$ or more), the required saturation levels are not reached so that nucleation does not occur. In practice, even in filtered or distilled solvents, a large number of particles that can act as template for nucleation are present. Therefore, homogenous nucleation is really unlikely in crystallization. These substances induce a higher local concentration of solute and thereby facilitate crystal formation. This mechanisms is said to proceed by heterogeneous nucleation, the process by which the surface of a foreign material lowers the nucleation barrier and facilitates aggregation in those conditions which would not be adequate for spontaneous (homogeneous) nucleation (48).

A first reason for the attractiveness of heterogeneous nucleation for protein crystal growers is that nucleation can occur inside the metastable zone. Because growth in the metastable zone affords kinetic advantages that often lead to the production of larger and better-ordered crystals than those grown at higher supersaturation, an aim of protein crystallizers is the possibility to induce heterogeneous nucleation in a controlled manner. Therefore, to create an environment that favors nucleation, the so-called nucleation-inducing agents (or nucleants) are introduced into the crystallization container and attempts to effectively use nucleants have become a common practice. Such a nucleant could help to enhance the chances of any single trial producing crystalline material, thus reducing the amount of starting materials to be used for screening, and/or to increase the nucleation rate, with consequent effect on crystal size and size distribution. Generally, both the induction time and protein concentration necessary for the nucleation decrease whereas the nucleation density increases on going from a prevalently homogeneous to a prevalently heterogeneous nucleation contribution.

The mechanisms of heterogeneous nucleation of proteins might arise from both physical and chemical interactions between the solute molecules and the nucleant. Although the exact mechanisms of interactions are still unclear, it is known that different surfaces may affect heterogeneous nucleation through different ways, for example: (i) introduction of spatial characteristics related to the crystalline lattice (49,50); (ii) modification of the supersaturation profile near the surface due to the concentration polarization and/or adsorption of the solute onto the surface by specific interactions; and (iii) the presence of a surface microstructure, for example roughness or porosity, conducive to facilitate nucleation.

The different interaction mechanism is dependent on the patches, with different chemical properties, which are available on a protein surface. Hydrophobic and hydrophilic spots, positively and negatively charged functional groups, and hydrogen-bonding moieties are known to provide affinity for almost any kind of nonbiological surface. Structural rearrangements of adsorbed macromolecules develop with protein–surface residence time and are considered as one of the driving forces of adsorption, which contributes significantly to nucleation free energy changes (51). This has induced difficulties and irreproducibility in the development of heterogeneous nucleation strategies in protein crystallization.

Control of heterogeneous nucleation has been attempted by using substrates as epitaxial nucleants for protein crystallization. Minerals (49,52) or lipid layers (53) with a close lattice match to the protein crystal might be effective to reduce induction time and to obtain single crystals suitable for X-ray diffraction studies. However, efficient epitaxial nucleation requires reversible interactions to concentrate and orient the protein molecules on the surface without loss of mobility (54). These interactions should be neither strong nor specific enough to restrict the degree of rotational freedom of the protein and the mode of protein–protein interaction. Therefore, in spite of the preliminary positive results, which have been pursued for several years to employ a variety of substrates, none have proved to be generally applicable as a "universal nucleant" for controlled heterogeneous nucleation of proteins.

Experimental evidence demonstrated that the protein nucleation is rather controlled by electric charges than from epitaxial constrains. In this respect, charged layers induce nucleation concentrating protein molecules on the surface by means of nonspecific electrostatic interactions. Additionally, the conformation of many proteins has been found to be adaptable, leading to the possibility of structural rearrangements near the surface (55). Some examples of heterogeneous nucleants whose effectiveness is due to the nonspecific attractive and local interactions between protein and the surfaces are poly-L-lysine-coated surfaces (56), chemically modified mica surfaces (57), silanized polystyrene flat-bottomed wells (58), polymeric film surfaces (59), and chemically treated glass surfaces (60). Crystallization of proteins from solutions at very low supersaturation thus using small initial amounts of macromolecular material, and the enhancement of the crystallization kinetics for proteins showing very long crystallization time, can be obtained by using these substrates.

Surface microstructure is another factor which might be effective in inducing heterogeneous nucleation of proteins. Porous media, such as porous silicon (61), porous glass surfaces (62), polymeric microporous membranes (50), and nucleation in gels (63) or in silicon devices displaying

Figure 35.5. Porcine pancreas trypsin crystals grown on the surface of a microporous hydrophobic polypropylene membrane. (Reprinted with permission from Ref. 65. Copyright 2005 American Chemical Society).

fractal structure (64), demonstrated interesting properties as heterogeneous nucleants for proteins. Surfaces containing irregular structures (Fig. 35.5) may confine and concentrate molecules and thereby encourage them to form crystalline nuclei.

From the classical nucleation approach, the energetics of nucleation concerns mainly the work to create a surface. If there is already a hydrophobic/hydrophilic substrate in the system, this will decrease the work required to create critical nuclei and will increase locally the probability of nucleation with respect to other locations in the system. Quantitatively the free energy (ΔG_{het}) required for the formation of two-dimensional nuclei is lowered by the presence of an appropriate substrate. This is described by the following equation (66):

$$\Delta G_{het} = \left(\frac{-\Delta\mu}{\Omega}\right) + \gamma_{12}A_{12} + (\gamma_{23} - \gamma_{13})A_{23} \quad (35.34)$$

where Ω is the molar volume ($\Omega = 4\pi r^3/3\upsilon$), υ is the molecular volume, $\Delta\mu$ is the driving force ($\Delta\mu = k_B T \; ln \; S$), γ is the interfacial energy per area, A is the surface area of the interfaces, and the subscript 1, 2, 3 represent the solute, the solvent, and the substrate, respectively. The total change in surface free energy will be lowered by favorable interactions between the aggregate and the substrate, and unfavorable interactions between the crystallization medium and the substrate, due to the negative value of the second term in the equation above. Consequently, nucleation will be enhanced by increasing the surface area of the substrate. The nucleation work depends on two main parameters: the externally controlled supersaturation ratio and the material surface/solution composition-dependent interfacial energy. According to Young's equation for an ideal smooth surface, interfacial energy γ_{12} can be estimated by

$$\frac{\gamma_{23} - \gamma_{13}}{\cos\alpha} = \gamma_{12} \quad (35.35)$$

where γ_{12}, γ_{13}, and γ_{23} are the nucleus–liquid, nucleus–substrate, and liquid–substrate interfacial energies and α is the angle that the interface between the nucleus and the bulk phase makes with the surface. The contact angle is determined by the interactions between the surface and the molecules in the nucleus. Attractions between the surface and the molecules that are stronger than those between the molecules in the nucleus will lead to a small angle α as the nucleus spreads into a thin droplet to maximize its contact area with the surface. However, if the surface tends to repel the molecules, then the nucleus is pushed away from the surface, resulting in a contact angle $\alpha > 90^\circ$. From the Young equations, the effective interfacial energy, γ_{eff}, for heterogeneous nucleation will be reduced by a factor $0 < \phi < 1$ compared to the interfacial energy γ_{12} for a pure homogeneous process. Because $\gamma_{eff} < \gamma_{12}$, the work of formation for heterogeneous nucleation is substantially reduced compared to that for a homogeneous process. Furthermore, the pre-exponential kinetic parameter, A_{het} in the Equation 35.31, is inversely proportional to the concentration of heterogeneous particles which is much smaller than the molecular volume, υ. Typically $A_{het} \approx 10^{15} - 10^{25} << A_{hom} \approx 10^{35}$. Therefore, heterogeneous nucleation on a substrate is generally energetically less demanding than homogeneous nucleation because of the lowering of the surface energy of the nucleus on the substrate upon interfacial contact (67):

$$\gamma_{eff} = \phi^{1/3}\gamma_{12} \quad (35.36)$$

$$\Delta G^*_{het} = \phi \cdot \Delta G^*_{hom} \quad (35.37)$$

where ϕ is the ratio between the heterogeneous and the homogeneous nucleation mechanism contribution. This ratio, considering the interaction between protein molecules with perfectly planar and chemically homogeneous substrate, can be expressed by

$$\frac{\Delta G^*_{het}}{\Delta G^*_{hom}} = \phi = \frac{1}{2} - \frac{3}{4}\cos\alpha + \frac{1}{4}\cos^3\alpha \quad (35.38)$$

Note that if the nucleus wets the substrate completely ($\alpha = 180^\circ$), $G_{het} = G_{hom}$; when the contact angle $\alpha = 90^\circ$, $G_{het} = 1/2G_{hom}$, and the smaller the contact angle α, the smaller the value of the activation energy for nucleation, which turns to be zero for $\alpha = 0$.

Equation 35.37 is no longer applicable when nucleation takes place on a porous substrate. In this case, a modified version of the equation, which takes into account the porous structure of the surfaces, has to be considered (50):

$$\frac{\Delta G^*_{\mathrm{het}}}{\Delta G^*_{\mathrm{hom}}} = \frac{1}{4}\,(2+\cos\alpha)\,(1-\cos\alpha)^2\left[1-\varepsilon\frac{(1+\cos\alpha)^2}{(1-\cos\alpha)^2}\right]^3 \tag{35.39}$$

where ε is the surface porosity, defined as the ratio of the total pore areas on the whole geometrical surface. If $\varepsilon = 0$, Equation 35.38 reduces to the form reported in the literature (Equation 35.37) for heterogeneous nucleation on nonporous surfaces. A similar equation applies for rough substrates, assuming the deposition of a nucleating solution with a spherical cap shape with the main radius R on a surface with a rough structure simulated by a series of n cones of height h and radius r (68):

$$\frac{\Delta G^*_{\mathrm{het}}}{\Delta G^*_{\mathrm{hom}}} = \frac{1}{4}\frac{\left[2\,(1-\cos\alpha)-\cos\alpha\sin^2\alpha\right]^3}{\left[(1-\cos\alpha)^2\,(2+\cos\alpha)+3\beta\sin^2\alpha-n\delta^2\beta\right]^2} \tag{35.40}$$

where $\beta = h/R$ and $\delta = r/R$. If $n = 0$, $\delta = 0$, and $\beta = 0$, Equation 35.39 can be converted again into Equation 35.37 for heterogeneous nucleation on an ideal surface.

Figure 35.6 shows the graphical representation of Equations 35.38 and 35.39 for various values of surface roughness (β) and porosity (ε). It can be seen from the figure that, if other solute/substrate interactions are not considered, the heterogeneous contribution to the free energy barrier for nucleation increases as both roughness and porosity of the surface increase.

Figure 35.7 shows a possible heterogeneous nucleation mechanism generated by an irregular surface topography. In this scheme, molecules dispersed in solution are first adsorbed on the surface. The irregular structure may physically block the lateral migration of the adsorbed protein molecules into the concaves, so that they are forced to be packed into compact aggregates. The trapping of molecules on the surface may result in a relatively higher local supersaturation, which would increase the possibility of nucleation compared with the nucleation on an ideally flat surface. Here, nucleation will follow with the formation of critical clusters comprising molecules forming suitable bond angles with their neighbors, while the molecules in a randomly packed compact structure may form a fractal cluster, which cannot work as a nucleus for crystal growth. Critical clusters then grow into crystals while the fractal clusters grow into larger clusters. This simple mechanism can be generalized to describe heterogeneous nucleation on irregular substrates such as porous or rough surfaces. From a technological point of view, while surface roughness is a characteristic that is quite difficult to control in the production of solid surfaces, the control of surface porosity is much easier, so that specific nucleants with the right value of porosity, which might be used to achieve the desired $\Delta G^*_{\mathrm{het}}/\Delta G^*_{\mathrm{hom}}$ ratio, can be produced. This is the case of protein crystallization by using membrane crystallization technique (50). With this approach, microporous hydrophobic membranes are used both as a physical support for solvent extraction in vapor phase and as a substrate for heterogeneous nucleation.

Although the mechanism described above is simple, in many cases the irregular structure of the substrate may not be sufficient enough to account for the effect on heterogeneous nucleation. More often than not, electrostatic/hydrophobic interactions between protein molecules moieties and the substrate play an additional role so that the resulting overall heterogeneous contribution to

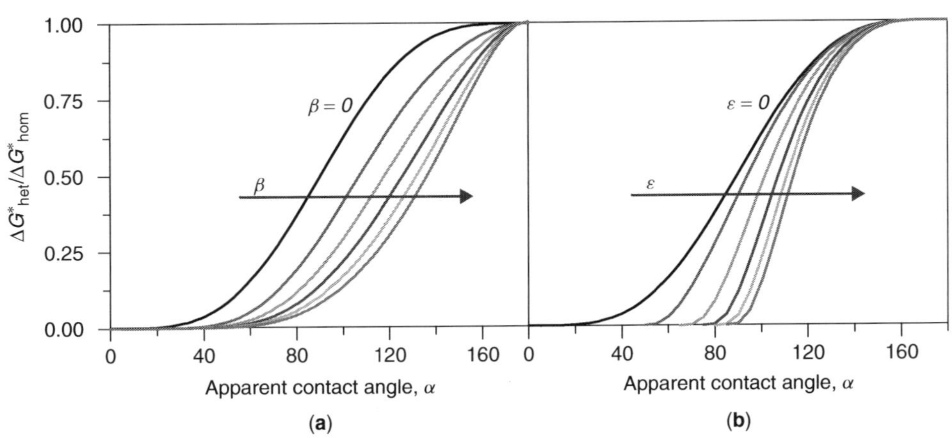

Figure 35.6. Effect of the surface roughness (β) and porosity (ε) on the ratio $\Delta G^*_{\mathrm{het}}/\Delta G^*_{\mathrm{hom}}$ in the heterogeneous nucleation of protein on solid surfaces. The energetic barrier to nucleation decreases as both roughness and porosity increase. (This figure is available in full color at http://onlinelibrary.wiley.com/book/10.1002/9780470054581.)

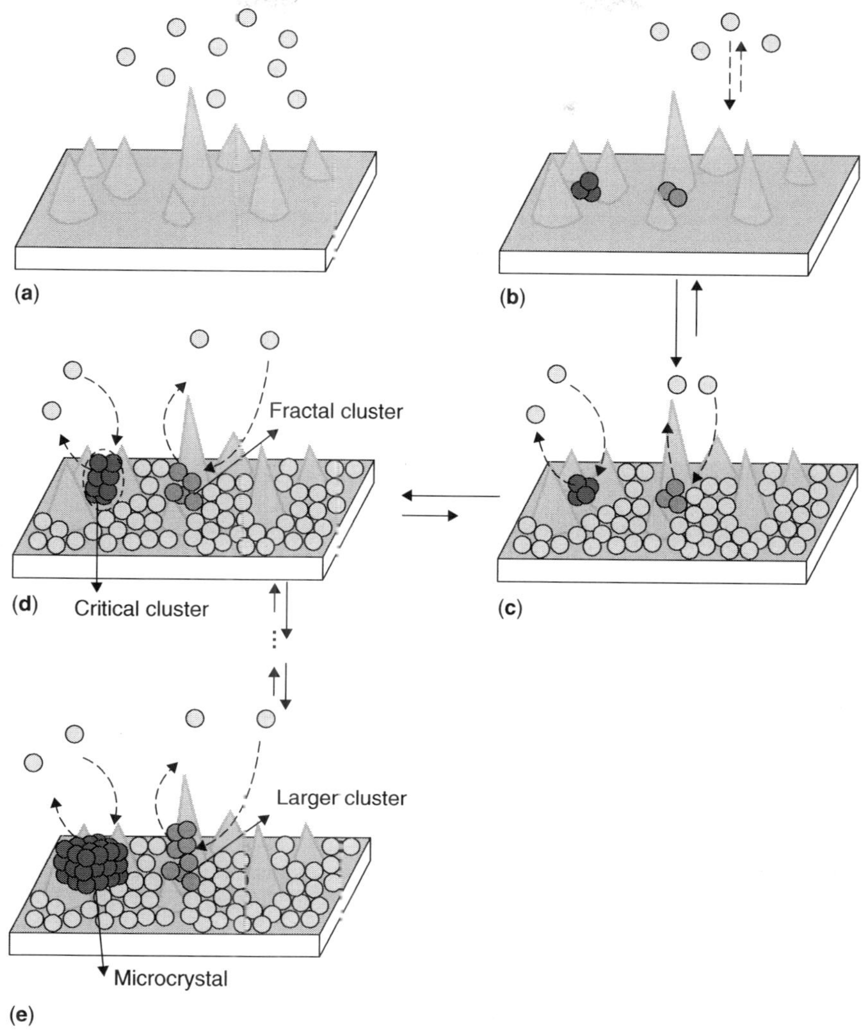

Figure 35.7. Schematic illustration of protein aggregation stimulated by irregular surfaces. (Reprinted with permission from Ref. 68. Copyright 2007 American Chemical Society). (This figure is available in full color at http://onlinelibrary.wiley.com/book/10.1002/9780470054581.)

nucleation is a combination of physical (entrapment) and chemical (electrostatic/hydrophobic interaction) effects.

35.4 NONCLASSICAL APPROACH TO NUCLEATION

35.4.1 Weaknesses of the Classical Nucleation Theory

Two of the main concepts of the classical nucleation theory are that: (i) the structure of the near-critical clusters is identical to that of large crystals and (ii) the clusters shape is determined by the minimization of the surface free energy.

The first assumption is a good approximation to reality only for large nuclei, which according to CNT, exist for relatively low supersaturated solutions. However, at high supersaturation, the size of the stable cluster is so small that there is no way to differentiate between the surface and the bulk. In the case of, for example, lysozyme crystallized with 4% NaCl as precipitant, the stable clusters can be as small as one to two protein molecules (69). This number of molecules is smaller than the number required to build the unit cell. Clusters so small cannot certainly have the structure of a large crystal, thus upon aggregation a crystalline structure does not necessarily follow; the most probable configuration being an amorphous material lacking translational order. This is the mechanism that originates the region of amorphous precipitation in the phase diagram of proteins.

Regarding the second assumption, the critical cluster sizes can range from few tent to 1–2 molecules as in the case of lysozyme (69). With such small clusters, rearrangements of the molecules within the aggregate will lead to high variations of its total free energy to take the system out from its critical region. As a consequence, in contradiction with the assumption (ii), it cannot be viewed as fluctuations of the shape of the critical cluster that will result in a shape minimizing its free surface energy.

In addition to these conceptual contradictions, a frequently observed phenomenon during protein crystallization and that cannot be explained by CNT is a phase separation, which occurs in some particular conditions. When present, this event has significant effects on the crystallization kinetics so that protein nucleation resembles much more the nucleation of colloidal system rather than the nucleation of atomic substances.

35.4.2 Liquid–Liquid Phase Separation

A complementary approach to describe protein nucleation, the so-called *nonclassical approach*, can be formulated by considering protein solutions as colloidal systems in which particles experience a highly anisotropic and short-range interaction potential (70). When the range of attraction is longer than the particle diameter, the phase diagram of the colloidal suspension resembles that of a typical atomic substance (23), with three stable phases coexisting in pairs and a triple point (71). However, when the range of interactions between particles is comparable to their size, the typical phase diagram of the colloidal system shows two-phase transitions upon a change in temperature (or other external condition): the formation of a condensed phase, which may have a regular structure (crystals) or be amorphous (aggregates) and a liquid–liquid phase separation (LLPS) (coacervation) formed by a colloid-rich and a colloid-poor liquid phases. The analogous counterparts of these phase transitions in simple molecular fluids is the fluid–solid transition for the solidification of colloids while gas–liquid coexistence is analogues to colloidal liquid–liquid phase separation. A striking difference, however, between the two cases is that when simple fluids are cooled, the order of phases observed is gas → liquid → solid. In colloidal solutions, instead, the coacervate usually transforms into a solid without passing through a liquid phase. This "anomalous" order of phases, while observed for colloid systems, directly relate to solutions of proteins. In fact, at high supersaturation, the range of interactions between protein molecules in solution is comparable to their size and, in addition, it is highly anisotropic, thus generating a phase diagram as that showed in Fig. 35.8, which is typical for globular proteins. In this representative graph the formation of a dense liquid phases have two-phase lines: the coexistence curve, or *binodal*, which is the boundary between the region where the protein solution remains homogeneous and the region in which LLPS occurs, and a *spinodal*, that separates the metastable region from the unstable region, below which the metastable dense liquid forms instantaneously (23). The

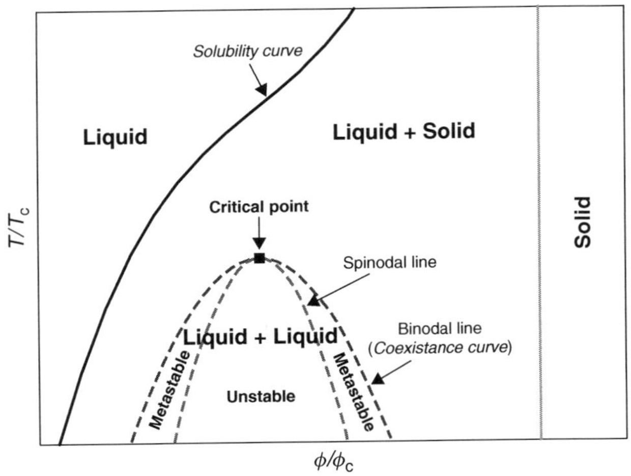

Figure 35.8. Theoretical phase diagram of a simple model protein, in the temperature–concentration plane, assuming hard spheres with short-range attractive interactions. The reduced temperature T/T_c(T_c is the critical temperature) is shown as a function of the volume fraction ϕ/ϕ_c. The liquid and solid lines mark the boundaries for the stable fluid and solid phases. The coexistence curve (dashed blue) lies below the liquid line. The solid black curve is the volume fraction of the dilute protein solution that coexists with the crystalline phase. The solid green curve is the volume fraction of the crystal. (This figure is available in full color at http://onlinelibrary.wiley.com/book/10.1002/9780470054581.)

binodal and the spinodal meet at the *critical point*, at which the two phases become identical. Beyond this point, LLPS is not possible. In the case of protein however, at higher protein concentrations, even above the critical temperature for LLPS, protein solutions often form a gel (24), due to the action of additional long-range attractive forces.

In protein solutions, the location of the metastable fluid/fluid boundary can be observed in many systems by rapidly quenching a protein solution below the fluid/solid phase boundary by rapidly lowering temperature or rapidly increasing the strength of interaction at a fixed temperature. In the case of a solution located near the liquid–liquid boundary, density fluctuations may induce the formation of the intermediate phase, consisting of high density protein drops surrounded by low concentrated solution, so that, under the influence of gravity, the drops may separate and the two liquid phases will be visible. The many small drops, which form scatter light, and the solution will therefore no longer be transparent, but becomes turbid. The coexistence curve can be therefore obtained for a macromolecular system by measuring the drop in the light intensity transmitted through the solution by the so-called "cloud-point method." The method includes tree steps: (i) shining of a light upon a solution of known protein concentration as it is cooled until the temperature at which the transmitted intensity falls to half of its initial value (*clouding temperature*);

(ii) once the transmitted intensity reaches zero, the solution has to be warmed until the transmitted intensity returns to half of its initial value (*clearing temperature*); (iii) the phase separation temperature can be approximated by the average between the clouding and the clearing temperatures.

35.4.3 Nucleation Mechanism in LLPS

When LLPS occurs, rather than one step characteristic of the classical approach, nucleation is said to proceed by two-step crystallization (TSC) mechanism. This mechanism has been observed not only for proteins but also during the crystallization of small molecules (72), for atomic systems (73,74), during the growth of nanocrystals (48), and in biomineralization (75). As the volume of the rich liquid phase is very small compared to the total volume of the solution, droplets with concentration as high as 400–500 mg/mL of protein (24) can form. At these tremendously high supersaturations, the resulting protein-rich phase may produce crystals with particular morphologies. Nucleation of lysozyme occurring while LLPS lead to the growth of tiny "sea urchin" or "spherulitic" crystals (Fig. 35.9) inside the high concentration drops. These protein crystals with needle morphology have been suggested to be presumably unstable or metastable polymorphs (76).

Owing to its significant implications in science and technology, TSC mechanism has been studied both theoretically (77) and experimentally (31). However, the understanding of TSC has, so far, remained poor. What is known is that the presence of a metastable fluid–fluid critical point drastically changes the pathway for the formation of a critical nucleus. The large density fluctuations produced by the proximity of the metastable spinodal near the critical point, reduces the nucleation barrier ΔG^*, which is in the order of 100 $k_B T$ or more, to about 50 $k_B T$ (23), so that the rate of crystal nucleation is enhanced. However, the enhanced rate of nucleation is moderated by the loss of the collective solute diffusivity at the critical point. As a result of the nonlinear coupling between these two opposing effects, the nucleation rate varies non-monotonically with supersaturation (78). As the enhancement in nucleation is maximum when the critical point is closest to the solubility curve, working under conditions such that the metastable critical point is closest to the solubility curve, results in the simultaneous occurrence of the maxima in both the nucleation rates under the conditions in which the average supersaturation is small. As a result, the thermodynamic driving force for crystallization is reduced so that crystal growth will be slow, thus enhancing the crystal order, which leads to better crystals.

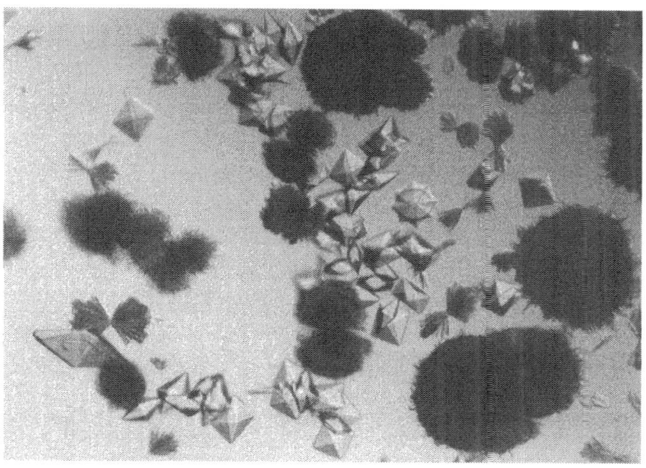

Figure 35.9. "Sea urchin aggregates" and unconventional lysozyme crystals obtained with NaCl 4.5% of precipitating agent in acetate buffer by membrane crystallization system (by G. Di Profio, unpublished work). (This figure is available in full color at http://onlinelibrary.wiley.com/book/10.1002/9780470054581.)

The difficulty in describing protein nucleation, and the limitations of the described TSC mechanism, is however apparent when considering some experimental observations such as the decrease in nucleation rate at temperatures deeper into the liquid-liquid (L-L) coexistence region and that, in some cases, numerous dense liquid droplets do not cause the nucleation of new crystals. This accounts for the fact that the presence of long lifetime droplets of dense liquid is not a prerequisite for enhanced nucleation (79). These considerations call for further efforts to define a specialized theory to describe protein nucleation.

35.5 CRYSTAL GROWTH

The elementary act for crystal growth from solution is generally divided into two separate steps that can be assumed to take place in series: (i) diffusion of growth units from the bulk solution toward the diffusion boundary layer near the crystal surface, and (ii) surface integration into the crystal lattice. This last step involves surface diffusion, reorientation, and partial or total desolvation of the growth unit before its integration into the crystalline lattice. For surface integration, there are two modes for the development of the faces of a crystal: *tangential layer growth* and *normal growth*. Different faces of a single crystal, being nonidentical, might simultaneously employ different mechanisms for development. Furthermore, a single face may exhibit more than one mechanism at the same time, and the type of mechanism may change as some experimental variable, such as temperature, is altered. Most crystals utilize all mechanisms at one time though some one mechanism may be favored.

35.5.1 Tangential Layer Growth Mode

The tangential layer growth mode occurs when the crystal/solution interface is smooth on a molecular scale, that

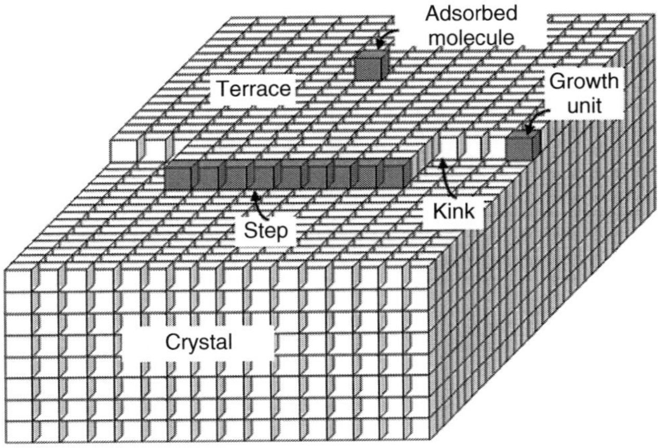

Figure 35.10. Schematic illustration of the structure of the surface of a growing faceted crystal.

is, it is essentially similar to a crystallographic plane that can be viewed by cutting the crystal lattice by a plane with simple Miller indices. In this mode, a new layer, typically one lattice spacing high, is deposited on the smooth surface of the previous lattice layer. The edges of the incomplete layers are called *steps*. The flat terraces between the steps are singular crystal planes. The steps themselves are also incomplete, containing kinks. The attachment of new growth units occurs at kinks, where the number of neighbors is half of that in the crystal bulk. The kinks are retained after the attachment, and the attachment it is assumed does not alter the surface free energy of the crystal (80). Thus, faceted crystals grow by the generation and spreading of layers (Fig. 35.10).

Molecules that attach at kinks make more bonds to neighboring molecules than the ones that attach to the terraces or to flat step edges. Consequently they are more likely to stick. Conversely, when molecules leave the crystal, they can do so more easily by detaching from kinks than from either complete step edges or from embedded sites in the terraces. As a result, the rate R at which molecules can be added to a faceted crystal, measured in a direction perpendicular to the growing face and for a given solute concentration, scales with the mean step density, p, defined as the step height, h, divided by the average distance, λ, between the steps:

$$R = p \cdot v = \frac{h}{\lambda} \cdot v \qquad (35.41)$$

where v is the step velocity. The interstep distance is determined by the mode of step generation and thus depends on free surface energy and supersaturation. In this section, however, the source of the steps will be ignored and just the step kinetics will be considered. The rate of step propagation during growth (or dissolution) is determined by the kink density, a/λ_k, and the velocity, v_k, of kink propagation along the step:

$$v = \frac{a}{\lambda_k} v_k = \frac{a}{\lambda_k} a\,(j_+ - j_-) \qquad (35.42)$$

where a is the lattice parameter (approximated to the typical molecular size), j_+ and j_- are frequencies of molecular attachment to and detachment from a kink, and $a(j_+ - j_-)$ is the rate of kink propagation along the step. Even at equilibrium, the steps have kinks due to thermally activated detachment of molecules (81). Consequently, the step edges are not static as molecules are constantly attaching and detaching. Growth from a supersaturated solution therefore occurs because the flux of molecules attaching to the crystal surface exceeds the flux of molecules detaching from the surface. The energy barrier to detaching a molecule from a step is exclusively determined by the strength of its bonds to adjacent molecules in the crystal. Since the strength of bonding is a function of temperature rather than the flux to the surface, the total flux from the surface is almost independent of concentration and it cannot be influenced just by altering solution composition. In contrast, the flux to the surface can depend on several experimental conditions and thus can be varied by, for example, promoting solution convection in the system (see section titled "Crystallization in Forced Solution Flow Regime"). The frequencies j_+ and j_- may be expressed considering the incorporation of molecules into kinks as a chemical reaction so that an activation barrier ΔG^* for their incorporation exists. This barrier is related to the energy involved in the water structure rearrangement around the incoming solute molecules (82):

$$v = \frac{a}{\lambda_k} a v_+ \exp\left(\frac{-\Delta G^*}{k_B T}\right) \Omega\,(C - s)$$
$$\equiv \beta\Omega\,(C - s) = \beta\Omega s\left(\frac{C}{s} - 1\right) \qquad (35.43)$$

Here, v_+ is the typical frequency of thermal attempts to join a kink by overcoming the barrier ΔG^*, Ω is the molecular volume, C is the solute concentration, and s is the equilibrium concentration (solubility). The parameter β (meter per second) is commonly referred to as the *kinetic coefficient* and is related to the first-order kinetic constant k for incorporation into a kink as $\beta = ak$. It can be considered as the rate at which molecules are incorporated into the lattice at supersaturation $\Omega\,(C - s) = 1$. β is also related to the free energy barrier ΔG^* as

$$\beta = \frac{a}{\lambda_k} a v_+ \exp\left(\frac{-\Delta G^*}{k_B T}\right) = \frac{a}{\lambda_k} a v_+ \exp\left(\frac{\Delta S^*}{k_B}\right) \exp\left(\frac{-\Delta H^*}{k_B T}\right) \qquad (35.44)$$

Derivation of Equation 35.43 have been presented elsewhere (82). Equation 35.43 means that the rate constant of growth of a crystal is determined by the density of kinks on the interface with the growth medium, and the barriers, both entropic and enthalpic, for incorporation of a molecule into a kink. Therefore, growth rates of crystals can be altered by either blocking kink sites or by roughening steps. Kinetic coefficients have been measured by atomic force microscopy and interferometry (83). According to Equations 35.40 and 35.42, the rate of face propagation is:

$$R = pv = p\beta\Omega(C - s) \tag{35.45}$$

Equations 43–45 show that the kinetic coefficient β, the step velocity v, and to a large extent, the crystal growth rate R, are determined by the kink density a/λ_k and the incorporation barrier ΔG^*. The kink density is determined by the mechanism of generation of kinks, while ΔG^* is determined by the mechanism of incorporation of molecules into kinks and the chemical interactions between solute/solvent and crystal/solvent. One of the main consequences of Equation 35.44 is that the solubility is the concentration at which the two fluxes are equal. This means that if two types of crystals are placed in solutions of the same supersaturation, the crystal that is more soluble will grow faster than the other, simply because there is a larger flux of molecules to its surface. In other words, step speed scales with solubility and one cannot assume that faster growth rates imply faster kinetics at the kink sites. Shifting the solubility automatically changes the growth kinetics, providing yet another means of altering crystal growth rates.

35.5.2 Modes of Steps Generation

New steps are generated by two mechanisms: (i) two-dimensional nucleation or (ii) a screw dislocation crossing the crystal face. The mechanism which prevails depends on supersaturation.

In two-dimensional (2D) nucleation, the nucleation of islands of new layers is followed by layer growth (84). This mechanism is showed in Fig. 35.11.

The 2D nucleation mechanism can be mononuclear or polynuclear. This mechanism operates at high supersaturations. Guided by the underlying lattice, molecules from solution adhere to the surface and organize themselves into ordered arrays consistent with the preexisting layer. These molecules can also leave the surface, but when the organized array exceeds a certain critical size, the addition is favored and the growth of the island expands by recruitment of molecules from the solution into its step edges. The first event, two-dimensional nucleation, provides growth normal to the surface, while the latter process is tangential growth. Because creation of new layers requires, in the absence of dislocations, the appearance of a new ordered arrangement atop the preexisting layer where none previously existed, it is a kind of nucleation event. This nucleation, which represents an activated process, generally occurs with difficulty

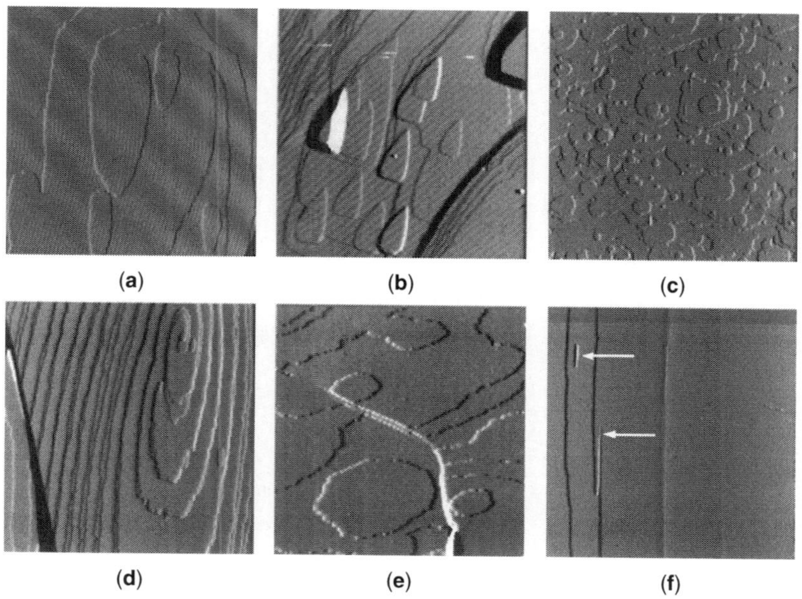

Figure 35.11. AFM images of: (a) two-dimensional nucleation on terraces formed by a dislocation spiral on the surface of a thaumatin crystal, (b) two- and three-dimensional nucleation, (c) two-dimensional nucleation on the surface of a thaumatin crystal at low and high supersaturation, respectively, (d–e) dislocation sources and two-dimensional nucleation on the surface of lysozyme crystal, (f) dislocation source and two-dimensional nuclei on the surface of a xylanase crystal. (Reproduced from Ref. 85 with permission of the Publisher).

and requires surmounting an energy or probability barrier. It is the slower and more difficult process in crystal growth. On the contrary, the recruitment of molecules into step edges and the extension of new layers over the surface is a much easier process because the incorporation of a new individual is essentially a cooperative process favored by both the molecules composing the existing step edge and the new recruit. Thus, once a nascent layer appears, its two-dimensional expansion may proceed unimpeded. Indeed, the surfaces of a crystal which has stopped growing, is flat as there are no islands or step edges remaining on the surface. Two-dimensional islands generally do not advance at equal rates in all directions. This is so because the step edges present a different structure, that is, display different bonding possibilities in different directions according to the energetics of association at those points and, therefore, at different rates. Impurities, which affect the rate of step advancement as well, are also incorporated with different affinities at the periphery of the growth islands, and these too might alter recruitment rates, leading to asymmetric shapes for the islands.

If steps are generated by 2D nucleation, for the mononuclear mechanism, the growth rate G (meter per second) perpendicular to the surface can be described as a function of supersaturation S according to

$$G_{MN} = d_0 A_{cr} J_{2D} = d_0 A_{cr} A_0 S \exp\left(-\frac{\pi a_0 \kappa^2}{(k_B T)^2 \ln S}\right) \quad (35.46)$$

In this equation, d_0 (meters) is the molecular diameter, A_{cr} (square meter) is the area of the crystal face, J_{2D} (square meter per second) is the 2D nucleation rate, $A_0(m^{-3}s^{-1})$ is the preexponential kinetic parameter, a_0 (square meter) is the area of a molecule in the monolayer and κ (joule per square meter) is the step free energy, which is related to the interfacial energy γ by $\kappa = k_s d_0 \gamma$ where k_s (unitless) is a shape factor.

For the polynuclear mechanism, the growth rate can be described as

$$G_{PN} = d_0 \left(\frac{1}{3}\pi v_s^2 J_{2D}\right)^{1/3} = d_0 \left(\frac{1}{3}\pi a_0 D c_e\right)^{2/3} A_0^{1/3} (S-1)^{2/3} S^{1/3} \exp\left(\frac{\pi a_0 \kappa^2}{3 (k_B T)^2 \ln S}\right) \quad (35.47)$$

where $v_s = a_0 D(c_0 - c_e)$ (meter per second) is the velocity of step propagation.

A common layer generation mechanism is by screw dislocations (86) piercing the growing facet (Fig. 35.12). These arise when, for some reasons like incorporation of a contaminant or misincorporation of one or several molecules, a discontinuity occurs along the direction normal to the surface. At such points, steps are continuously propagated in a supersaturated solution in a spiral about the dislocation, hence their name, screw dislocations. Therefore, in this growth mode, the screw dislocations provide steps for the attachment of new molecules that would be very difficult for

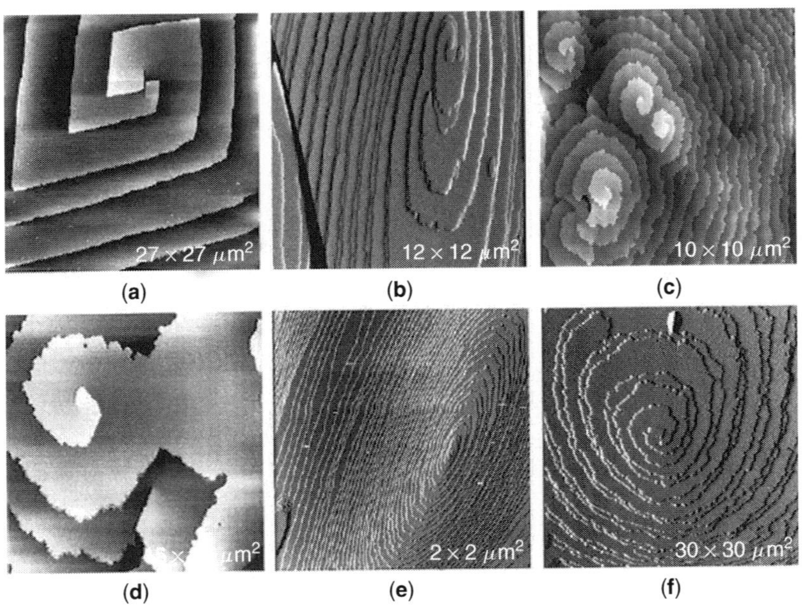

Figure 35.12. AFM images of screw dislocations on the surfaces of a variety of protein crystals: (a) glucose isomerase, (b) lysozyme, (c, d) canavalin, (e) lysozyme, and (f) beef trypsin. (Reproduced from Ref. 88 with permission of the Publisher).

a molecularly very smooth crystal surface (87). The growth rate perpendicular to the surface can be expressed as:

$$G_{BCF} = \frac{d_0 \upsilon_S}{\beta R^*} = \left(\frac{a_0 D c_e k_B T}{\beta d_0 \kappa}\right)(S-1)\ln S \quad (35.48)$$

where $\beta = 19$ (unitless) is a numerical factor and $R^* = a_0\kappa/\Delta\mu$ (meters) is the radius of a 2D nucleus. For low supersaturation, where $\ln S \approx (S-1)$, the growth rate equation becomes quadratic $G_{BCF} \approx (S-1)^2$.

The salient difference between growth by this mechanism and 2D nucleation is that face growth arising from spiral dislocation step generation is more readily facilitated since no two-dimensional nucleation event is necessary to initiate a new layer. The spirally produced step edges continually provide nascent layers and the crystal grows almost exclusively by tangential addition of molecules to step edges. Spirals may be left-handed or right-handed depending on the nature of the dislocation and a single crystal surface will often exhibit both. Spirals may be single or double at the dislocation and these more complex spirals may also have either hand.

35.5.3 Normal Growth Mode

Another mechanism for crystal development, called *normal growth*, does not proceed by layer addition, but by random recruitment of molecules at arbitrary sites on the surface. Molecules join the lattice everywhere on the surface at the crystal–solution interface. This leads to atomically "rough" surfaces as opposed to the atomically "smooth" surfaces yielded by layer growth. Roughness of an interface is a sign that the free interface energy is so low that it cannot prevent disorder induced by thermal fluctuations and the potential barrier for 2D nucleation vanishes. The major difference between the smooth and the rough interfaces and their growth modes lies in the density of the step and kink sites on the crystal surface. On the smooth faces, the kink and step density is less than $\sim 10^{-2} - 10^{-3}$ of total density of molecular sites on the crystal face. On the rough, disordered surfaces, steps and kinks are present at high density so that the average distance between the steps and kinks is comparable to the lattice spacing along the surface. Since new molecules join the crystal freely only at the kink sites, the disordered surfaces on which steps and kinks are easily generated by fluctuations grow much faster than the smooth ones for which step generation is a problem. Growth rate in this case weakly depends on surface orientation, because when a solute molecule arrives at the surface it is immediately integrated, so that the growing crystal has a rounded shape—spherical, ellipsoidal, and so on. Growth by this mechanism, in the regime of high supersaturation, is disorganized and produces, as one might expect, crystals of poor quality. Kinetic roughening is generally observed when the activation energy required for the formation of a 2D critical nucleus becomes of the order $k_B T$. For rough growth of a surface, the growth rate depends on the difference between the flux of growth units from the bulk that impinge on a growth position in the crystal surface, $f_s(s^{-1})$, and the flux of growth units that leave the surface, $g_s(s^{-1})$, according to

$$G_{RG} = d_0(f_s - g_s) = a_0 D c_e (S-1) \quad (35.49)$$

Rough growth is hence a linear function of the supersaturation ratio.

35.5.4 Three-Dimensional Nucleation

A mechanism that may be unique to macromolecular crystal growth, as it has not been described previously for conventional crystals, is the growth by three-dimensional nucleation (Fig. 35.13).

This mechanism is associated with the phenomena of liquid–liquid phase separation (see section titled "Nonclassical Approach To Nucleation") observed for macromolecular solutions in certain supersaturation range. In these conditions, below the threshold needed for 2D nucleation, droplets of dense liquid phase of the protein solution, several hundred nanometers in size, and composed of many thousands of molecules, and exhibiting short-range order exist. These droplets land on the crystal face and transform into a crystalline matter in perfect registry with the underlying lattice. When the droplets contact existing crystal surfaces, the lattice serves as an epitaxial substrate to guide and promote crystallization in the molecules above. These form a crystal layer, inspire crystallinity in the molecules above them, and so forth, propagating a continuous series of growth layers, a multilayer stack. The island of layers, several lattice parameters thick, spreads sideways generating several new steps. Each layer of the stack provides step edges and, therefore, sources for tangential growth and the formation of new layers. When tangential growth of layers proceeds simultaneously from several multilayer stacks on the surface of a crystal, the corresponding layers from the various stacks ultimately encounter one another, merge, and knit in a completely contiguous manner. This mechanism can in some cases be the dominant growth mechanism.

35.5.5 Effect of Impurities

Macromolecules, because they are derived from biological sources, are subject to an exotic variety of impurities. These include not only foreign proteins and denatured or degraded macromolecules, but also their aggregates and even their microcrystals. All of these, in addition to the dust particles

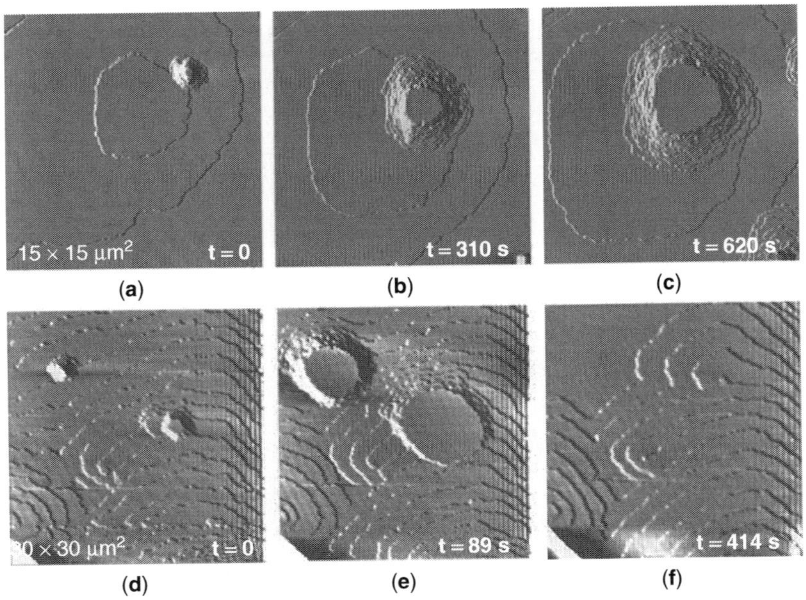

Figure 35.13. Sequences of images showing the development with time of three-dimensional nuclei on the surfaces of growing crystals. (Reproduced from Ref. 88 with permission of the Publisher).

and other foreign inorganic materials that may contaminate conventional mother liquors, can be incorporated in their entirety into growing crystals. Therefore, macromolecular crystals incorporate an extensive variety of impurities into their interiors.

Impurities may produce a diverse array of lattice defects, or they may create serious imperfections in the crystal. The extent of this imperfection is largely dependent on the nature of the macromolecule, the type of impurity, and the physical forces responsible for maintaining the lattice. An effect of impurity molecules incorporating into crystals is the roughness of step edges, though ideally they should be smooth. This is a consequence of the locally "pining" step edge progression, thereby resulting in gaps and teeth in advancing layers. Vacancies, empty unit cells, partially filled unit cells, and even lines of missing particles or unit cells are also common and may represent up to 1% of all of the unit cells in a protein crystal.

Generally, the effects of impurity may significantly effect the diffraction properties of the crystal. While this detrimental consequence is rather limited in the case of individual absences, it may be more influent when they are numerous. More serious imperfections that produce long-range disorder, affecting many molecules in a crystal, are planar defects. These are dislocation planes that may extend through hundreds or thousands of molecular layers in crystals and which displace great numbers of molecules in their immediate neighborhoods. Planar dislocations in large numbers can form networks and create a system of domains within crystals. The relative number and displacements of these domains determine the mosaic character of crystals and the consequent mosaicity of the diffraction patterns. In conventional crystals these planar defects, and the long-range discontinuities they produce, are commonly referred to as *grain boundaries*.

35.6 CRYSTALLIZATION IN FORCED SOLUTION FLOW REGIME

The process of crystallization from solutions occurs according with two sequential stages: (i) transport of growth units from the bulk of the mother liquor toward the crystal surface and (ii) incorporation into the crystal face. Depending on which of the two processes is prevalent, the growth is said to be governed by either a diffusive or an interfacial kinetics contribution, respectively. In the first case, the crystal growth rate depends on the rate at which the solute molecules diffuse toward the growing surface; since convection increases the rate of supply of solute, it is expected that increasing solution convection will increase crystal growth rate. When the controlling mechanism is interfacial kinetics, the growth rate is governed by the rate at which units are included in the growing face from the solid/solution interface; in these conditions fluid convection does not affect significantly the solute incorporation into the crystals (89) and, hence, the overall growth rate. Usually, either one of these two contributions dominates the overall transport mechanism during crystallization, although a mixed regime might also be possible. The general trend that emerges

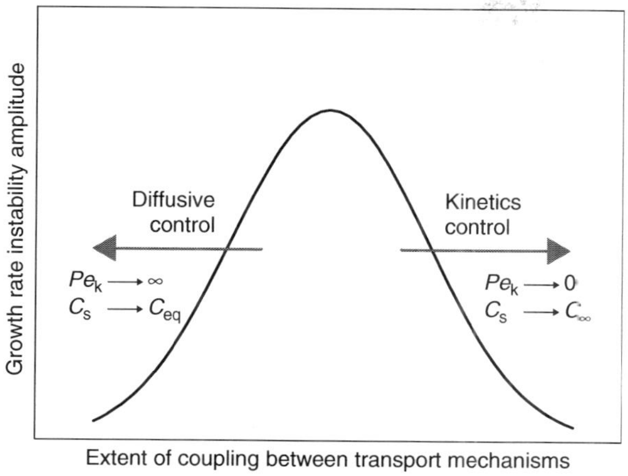

Figure 35.14. Fluctuations of the crystal growth rate with the coupling between the transport mechanism contribution to the overall crystallization process. In correspondence of the maximum of the curve, the maximum fluctuation amplitude is due to the coupling between both transport and interfacial kinetics processes. Under both pure interfacial kinetics or diffusive control, when the interfacial solute concentration C_S approaches respectively the bulk (C_∞) or equilibrium (C_{eq}) concentration, the perturbation amplitude to the growth decays. (This figure is available in full color at http://onlinelibrary.wiley.com/book/10.1002/9780470054581.)

from the description above is schematically illustrated in Fig. 35.14.

If growth proceeds under pure kinetics or transport control, where the interfacial concentration C_S approaches the bulk concentration C_∞ and equilibrium concentration (solubility) C_{eq} respectively, the system is stable, that is, perturbations are dampened. For mixed control, however, the system is unstable. Perturbations lead to fluctuations about a mean rate. The strongest instability (highest fluctuation amplitude) is expected for processes that operate under equal influence of the transport and kinetics components on the overall rate control (90). Under such solution conditions, the growth mechanism can be described as a combination of both interfacial kinetics and diffusive contribution. In this case, intrinsic instability of the growth process leads to a fluctuation in the growth rate, step density, and step velocity in the order of up to 80% of their respective average values (91). These fluctuations are associated with a decrease in the homogeneity of the crystals in the form of striate and to reduced crystal quality (90).

Based on the dependence of the amplitude of these fluctuations on the "working point" (relative weight of diffusive and interfacial kinetics contribution in the overall rate control), one can expect that shifts toward either pure interfacial kinetics or diffusive control should result in more steady growth. In the absence of fluid convection, molecules are transported toward the crystal only by diffusion. Here, the growing crystal is surrounded by its mother liquor, so that a depletion zone near the solid surface originates from the withdrawal of growth units out of the liquid into the solid state. The resulting gradient of concentration induces density changes that give rise to solute buoyancy-driven convection in the presence of gravity. Pure kinetic control can be therefore achieved by suppressing buoyancy-driven convection. The slow supply of molecules causes a reduction of protein concentration in the vicinity of a growing face, thus preventing the formation of new nuclei and giving to the protein molecules a better chance to reach a specific position in the lattice. This translates in the improvement of the quality of protein crystals. Moreover, the depletion zone around a growing crystal can help to improve its quality working as a filtration zone for impurities (92). Suppressing buoyancy-driven convection can be obtained for example in microgravity by using highly viscous gels and oils as dispersing media, and by applying magnetic/electric fields affecting molecular motion in solution. On the other side, pure diffusive control can be obtained by enhancing buoyancy-driven convection by shifting the working conditions from still to forced fluid convection. Such shift of the working point more steady growth results in improved crystal quality (93–96).

The control parameter for the individuation of the dominant transport mechanism in crystal growth is the kinetic Peclèt number (90):

$$Pe_k = \beta_f \frac{\delta}{D} \tag{35.50}$$

where β_f is the face kinetic coefficient (e.g. $\sim 10^{-6}$ cm/s for lysozyme), D is the solute diffusivity, and δ is a characteristic diffusion length (commensurate to the boundary layer thickness). The boundary layer thickness δ is evaluated in the case of laminar flow with a slip velocity v (97):

$$\begin{aligned} \delta &= \left(\frac{x}{0.282}\right) \mathrm{Re}^{-1/2}\, Sc^{-1/3} \\ &= \left(\frac{x}{0.282}\right) \left(\frac{vx}{\eta}\right)^{-1/2} \left(\frac{\eta}{D}\right)^{-1/3} \end{aligned} \tag{35.51}$$

x can be approximated by the crystal linear dimension, $Re = \rho u_s L/\eta$ is the Reynolds number, $Sc = \eta/(\rho D)$ is the Schmidt number, u_s (meter per second) is the particle slip velocity, ρ (kilogram per cubic meter) is the solution density, L (meter) is the crystal diameter and η (kilogram per meter per second) is the solution viscosity.

Values of Pe_k <<0.1, or >>1 indicate, respectively, purely kinetic or diffusion controlled growth, in which perturbations are dampened and growth is steady. Intermediate values characterize mixed growth control in which kinetics fluctuations and step bunching might occur. The correspondence between this criterion and the growth instability described above in terms of the surface concentration C_S becomes clear when considering that (81):

$$C_S \approx C_\infty - \left(C_\infty - C_{eq}\right)\frac{Pe_k}{1 + Pe_k} \qquad (35.52)$$

Thus, if $Pe_k \to 0$, $C_S \to C_\infty$, and if $Pe_k \to \infty$, $C_S \to C_{eq}$ (see Fig. 35.14). For a given crystallizing solution, in principle, according to Equation 35.49 all three components of Pe_k can be varied. However, inducing variations in β_f (e.g. by changing the concentration of other solution components), or D (by the addition of viscosity-increasing substances) may also lead to changes in the other two parameters, as indicated by Equation 35.50, thus introducing unaccounted variations in Pe_k. Therefore, the solution flow velocity is generally varied in order to control the dominant transport contribution.

Other than the crystal quality, forced solution convection affects the crystal morphology by acting on the overall crystal growth rate. When diffusion in solution is the rate-determining step for crystal growth, a concentration gradient between the bulk and the crystal surface is established; in this case, first and second Fick's laws can be used to describe the diffusion of a molecule, through the boundary layer, to a stationary particle. In condition of forced solution flow regime, diffusion is enhanced by convective mass transport, and the particle flux J toward a crystal growing in supersaturated environment can be expressed as:

$$J = k_x\left(C_\infty - C_{eq}\right) = \frac{D}{\delta}\left(C_\infty - C_{eq}\right) \qquad (35.53)$$

where k_x (meter per second) is the mass transfer coefficient. This yields to a convective growth rate G given by

$$G = \xi^{-1} k_x C_{eq}(S - 1) \qquad (35.54)$$

where ξ is the number of molecules precipitated per unit volume of solution and S the supersaturation. From this theory δ is predicted to be inversely proportional to the squared root of the velocity and that G should increase proportionally to $v^{1/2}$.

Figure 35.15 shows the effect of increasing solution convection (solution velocity v) on the crystal growth rate of bovine pancreas trypsin (BPT) and porcine pancreas trypsin (PPT) crystallized with ammonium sulphate and PEG 6000 as precipitants. The initial increase in the growth rate with v, until a maximum, is due to the enhancement in the convective supply of the growing-units to the solid/liquid interface in a regime of diffusive transport control. The maximum of the curve corresponds to a shifting of the operative working point to a regime with significantly faster transport and higher relative weight of interfacial kinetics effects in the overall control of the crystallization process

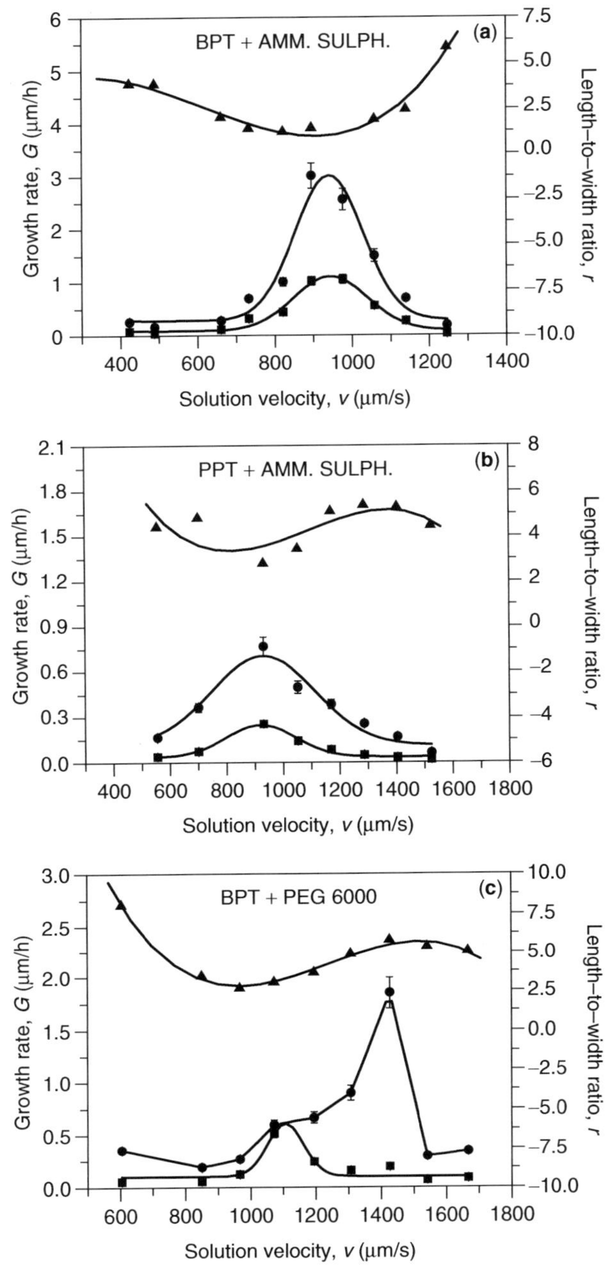

Figure 35.15. Crystal width (□) and length (•) growth rate and length to width ratio r (▲) as function of the solution velocity v, for three protein solutions: (a) bovine pancreas trypsin (BPT) + ammonium sulfate; (b) porcine pancreas trypsin (PPT) + ammonium sulfate; (c) bovine pancreas trypsin + PEG 6000. (Reprinted with permission from Ref. 65. Copyright 2005 American Chemical Society).

and a further increase in solute supply does not induce increase of growth rate as the depletion zone is saturated. Here the rate-limiting step is the integration of growth units from the depletion zone into the crystal surface and not the rate of their transport into the depletion zone. For values of solution velocities beyond the maximum, growth rate decreases with v up to complete cessation. The deceleration can be explained with the overpowering of the convection-enhanced supply of impurities that are incorporated into the crystal in an impeding manner for the growth kinetics (98) and/or to the high protein concentrations in the boundary layer, close to the solid/liquid interface, which might lead to secondary nucleation, which in turn gives rise to excess nucleation that suppresses growth (65).

REFERENCES

1. Derjaguin BV. Theory of stability of colloids and thin films. New York: Plenum; 1989.
2. Verwey EJW, Overbeek JTG. Theory of stability of lyophobic colloids. Amsterdam: Elsevier; 1948.
3. Derjaguin BV, Landau L. Acta Physicochim URSS 1941; 14: 633–662.
4. Hofmeister F. Arch Exp Pathol Pharmakol 1888; 24: 247–260.
5. Kunz W, Henle J, Ninham BW. Curr Opin Colloid Interface Sci 2004; 9: 1–18.
6. Boström M, Williams DRM, Ninham BW. Biophys J 2003; 85: 686–694.
7. Guo B, Kao S, McDonald H, Asanov A, Combs LL, Wilson WW. J Cryst Growth 1999; 196: 424–433.
8. Haas C, Drenth J, Wilson WW. J Phys Chem B 1999; 103: 2808–2811.
9. George A, Wilson WW. Acta Crystallogr D Biol Crystallogr 1994; 50: 361–365.
10. Wanka J, Peukert W. Chem Ing Tech 2006; 78: 273–278.
11. Gunton JD, Shiryayev A, Pagan DL. Protein condensation: kinetic pathways to crystallization and disease. Cambridge: Cambridge University Press; 2007.
12. Bhamidi V, Varanasi S, Schall CA. Cryst Growth Des 2002; 2: 395–400.
13. Bonnetè F, Finet S, Tardieu A. J Cryst Growth 1999; 196: 403–414.
14. Tessier PM, Lenhoff AM, Sandler SI. Biophys J 2002; 82: 1620–1631.
15. Velev OD, Kaler EW, Lenhoff AM. Biophys J 1998; 75: 2682–2697.
16. Tanford C. Physical chemistry of macromolecules. New York: John Wiley & Sons; 1961.
17. Gibbs JW. Trans Connect Acad Sci 1876; 3: 108.
18. Gibbs JW. Trans Connect Acad Sci 1878; 16: 343.
19. Gibbs JW. Collected works. Volume I, Thermodynamics New Haven (CT): Yale University Press; 1948.
20. Volmer M. Kinetic der phasenbildung. Leipzig: Steinkopff; 1939.
21. ter Horst JH, Kramer HJM, Jansens PJ. Cryst Growth Des 2002; 2: 351–356.
22. Mersmann A. J Cryst Growth 1990; 102: 841–847.
23. Kashchiev D. Nucleation, basic theory with applications. Oxford: Butterworth; 2001.
24. Söhnel O, Müllin JW. J Colloid Interface Sci 1988; 123: 43.
25. Gomez-Morales J, Torrent-Burgues J, Rodriguez-Clemente R. J Cryst Growth 1996; 169: 331.
26. Müllin JW, Zacek S. J Cryst Growth 1981; 53: 515.
27. Datta S, Grant DJW. Cryst Res Tech 2005; 40: 233.
28. Söhnel O, Müllin JW. J Cryst Growth 1978; 44: 377.
29. Kozlovskii MI, Wakita H, Masuda I. J Cryst Growth 1983; 61: 37.
30. Michinomae M, Mochizuki M, Ataka M. J Cryst Growth 1999; 197: 257.
31. Crosio MP, Jullien M. J Cryst Growth 1992; 122: 66.
32. Hu H, Hale T, Yang X, Wilson LJ. J Cryst Growth 2001; 232: 86.
33. Di Profio G, Curcio E, Cassetta A, Lamba D, Drioli E. J Cryst Growth 2003; 257: 359–369.
34. Galkin O, Vekilov PG. J Cryst Growth 2001; 232: 63–76.
35. Beckmann W. J Cryst Growth 1999; 198–199: 1307.
36. Söhnel O, Müllin JW. J Cryst Growth 1982; 60: 239.
37. Ulrich J, Strege C. J Cryst Growth 2002; 237–239: 2130.
38. He G, Bhamidi V, Tan RBH., Kenis PJA., Zukoski CF. Cryst Growth Des 2006; 6: 1175–1180.
39. Mersmann A, Bartosch K. J Cryst Growth 1998; 183: 240–250.
40. Garcìa-Ruiz MM. J Struct Biol 2003; 142: 22–31.
41. Kim K-J, Mersmann A. Chem Eng Sci 2001; 56: 2315–2324.
42. Talreja S, Kim DY, Mirarefi AY, Zukoski CF, Kenis PJA. J Appl Crystallogr 2005; 38: 988–995.
43. McPherson A. Crystallization of biological macromolecules. Cold Spring Harbor (NY): Cold Spring Harbor Laboratory Press; 1999.
44. Margolin AL, Navia MA. Angew Chem Int Ed 2001; 40: 2204.
45. Brange J. Galenics of insulin. Berlin: Springer; 1987.
46. Mahadevan H, Hall CK. AIChE J 1992; 38: 573.
47. Broide ML, Berland CR, Pande J, Ogun OO, Benedek GB. Proc Natl Acad Sci U S A 1991; 88: 5660–5664.
48. Mullin JW. Crystallization. 4th ed. Oxford: Butterworth-Heinemann; 2001.
49. Kimble WL, Paxton TE, Rousseau RW, Sambanis A. J Cryst Growth 1998; 187: 268–276.
50. Curcio E, Fontananova E, Di Profio G, Drioli E. J Phys Chem B 2006; 110: 12438.
51. Horbett TA. In: Ahern TJ, Manning MC, editors. Stability of protein pharmaceuticals, Part 1. New York; Plenum Press: 1992.
52. McPherson A, Shlichta P. Science 1988; 239: 385–387.
53. Edwards AM, Darst SA, Hemming SA, Li Y, Dkornberg R. Nat Struct Biol 1994; 1: 195–197.
54. Kornberg RD, Darst SA. Curr Opin Struct Biol 1991; 1: 642.
55. Roth CM, Lenho AM. Langmuir 1993; 9: 962.
56. Hemming SA, Bochkarev A, Darst SA, Kornberg RD, Ala P, Yang DSC, Edwards AM. J Mol Biol 1995; 246: 308–316.
57. Tang L, Huang YB, Liu DQ, Li JL, Mao K, Liu L, Cheng ZL, Gong WM, Hu J, He JH. Acta Crystallogr D 2005; 61: 53–59.
58. Paxton TE, Sambanis A, Rousseau RW. Langmuir 2001; 17: 3076.
59. Fermani S, Falini G, Minnucci M, Ripamonti A. J Cryst Growth 2001; 224: 327–334.
60. Tsekova D, Dimitrova S, Nanev CN. J Cryst Growth 1999; 196: 226–233.

61. Chayen NE, Saridakis E, El-Bahar R, Nemirovsky Y. J Mol Biol 2001; 312: 591–595.
62. Rong L, Komatsu H, Yoshizaki I, Kadowaki A, Yoda S. J Synchrotron Radiat 2004; 11: 27.
63. Vidal O, Robert MC, Boùe F. J Cryst Growth 1998; 192: 257–270.
64. Sanjoh A, Tsukihara T, Gorti S. J Cryst Growth 2001; 232: 618–628.
65. Di Profio G, Perrone G, Curcio E, Cassetta A, Lamba D, Drioli E. Ind Eng Chem Res 2005; 44: 10005–10010.
66. Fletcher NH. J Chem Phys 1963; 38: 237.
67. Bonafede SJ, Ward MD. J Am Chem Soc 1995; 117: 7853.
68. Liu Y-X, Wang X-J, Lu J, Ching C-B. J Phys Chem B 2007; 111: 13971–13978.
69. Vekilov PG. Cryst Growth Des 2004; 4: 671–685.
70. ten Wolde PR, Frenkel D. Science 1997; 277: 1975–1979.
71. Hagen MHJ, Frenkel D. J Chem Phys 1994; 101: 4093.
72. Veesler S, Revalor E, Bottini O, Hoff C. Org Proc Res Dev 2006; 10: 841.
73. ten Wolde PR, Frenkel D. Phys Chem Chem Phys 1999; 1: 2191.
74. Chen X, Samia ACS, Lou Y, Burda C. J Am Chem Soc 2005; 127: 4372–4375.
75. Addadi L, Raz S, Weiner S. Adv Mater 2003; 15: 959.
76. (a) Bhamidi V, Skrzypczak-Jankun E, Schalla CA. J Cryst Growth 2001; 232: 77–85 (b) Heijna MCR, Theelen MJ, van Enckevort WJP, Vlieg E. J Phys Chem B 2007; 111: 1567–1573.
77. Neilsen AE. In: Peiser S, editor. Crystal growth. Oxford: Pergamon; 1967. p. 419.
78. Dixit NM, Zukoski CF. J Colloid Interface Sci 2000; 228: 359–371.
79. Anderson VJ, Lekkerkerker HNW. Nature 2002; 416: 811.
80. Stranski IN. Z Phys Chem 1928; 136: 259–278.
81. Chernov AA. Volume 36, Modern crystallography. III. crystal growth, Springer series on solid state science. Berlin: Springer; 1984.
82. Chernov AA, Komatsu H. In: van der Eerden JP, Bruinsma OS, editors. Science and technology of crystal growth. Dordrecht, Netherlands: Kluwer Academic; 1995. pp 67–80.
83. Vekilov PG, Chernov AA. Solid State Phys 2002; 57: 1–147.
84. Stranski IN, Kaischew R. Z Phys Chem B 1934; 26: 100–113.
85. Malkin AJ, Kuznetsov YG, McPherson A.. J Cryst Growth 1999; 196: 471–488.
86. Burton WK, Cabrera N, Frank FC. Philos Trans R Soc Lond A 1951; 243: 299–310.
87. Vekilov PG. Prog Cryst Growth Charact Mater 1993; 26: 25–49.
88. McPherson A, Kuznetsov YG, Malkin A, Plomp M. J Struct Biol 2003; 142: 32–46.
89. Lin H, Rosenberger F, Alexander JID, Nadarajah A. J Cryst Growth 1995; 151: 153.
90. Vekilov PG, Alexander JID, Rosenberger F. Phys Rev E 1996; 54: 6650.
91. Vekilov PG, Monaco LA, Rosenberger F. J Cryst Growth 1995; 146: 289.
92. Tanaka H, Inaka K, Sugiyama S, Takahashi S, Sano S, Sato M, Yoshitomi S. Ann N Y Acad Sci 2004; 1027: 10.
93. Kadowaki A, Yoshizaki I., Adachi H, Komatsu H, Odawara O, Yoda S. Cryst Growth Des 2006; 6: 2398–2403.
94. Kadowaki A, Yoshizaki I, Rong L, Komatsu H, Odawara O, Yoda S. J Synchrotron Radiat 2004; 11: 38–40.
95. Adachi H, Takano K, Matsumura H, Inoue T, Moria Y, Sasaki T. J Synchrotron Radiat 2004; 11: 121–124.
96. Di Profio G, Curcio E, Drioli E. J Struct Biol 2005; 150: 41–49.
97. Schlichting H. Boundary layer theory. 6th ed. New York: McGraw-Hill: 1968.
98. Vekilov PG, Rosenberger F. J Cryst Growth 1998; 186: 251.

36

PROTEIN PURIFICATION, AQUEOUS LIQUID EXTRACTION

Maria-Regina Kula and Klaus Selber
Heinrich Heine University Düsseldorf, Jülich, Germany

36.1 INTRODUCTION

For the large-scale production of proteins, complex separation procedures are necessary after protein synthesis by living cells is completed. In comparison to downstream processing in the chemical industry, the separation of proteins must be very gentle to conserve the native structure, which means that in general near-neutral pH values and moderate temperatures (≤30°C) must be maintained. Important objectives in separation are high selectivity and yield because multistep procedures are usually required. Because of the polar character and high molecular mass of proteins, several unit operations common in the chemical industries cannot be applied, for example, distillation or extraction into organic solvents. Aqueous two-phase systems, however, can be used with advantage for protein isolation by liquid–liquid extraction. The technique is based on the incompatibility of two different hydrophilic polymers in the common solvent water or the reversible hydration of oxygen in polyethers.

In an aqueous liquid–liquid extraction process, cells and cell debris and other unwanted substances should remain in one phase while the desired protein is partitioned into the other phase, preferentially the top phase. The phases are separated and further processed until the desired level of product purity is reached. The extraction of proteins is performed preferentially at early stages of an isolation process using cell homogenates or whole fermentation broth because this allows the concomitant removal of solids, a product enrichment and concentration in a single step.

Partitioning of proteins by aqueous liquid–liquid extraction has been well described (1–6) and finds increasing application for large-scale processes in industry. For further reading the monographs by Albertsson (4) and especially by Walter *et al.* (5) are recommended. The latter gives detailed information about handling principle and applications. The Aqueous Two-Phase (a.k.a. Biphasic) Systems Community home page has a bibliography compiled by Harry Walter that contains more than 1,000 files of journal publications on aqueous two-phase systems published between 1956 and 1996: http://radar.ch.ua.edu/~aq2phase/

36.2 BIOCHEMICAL FUNDAMENTALS

Miscibility gaps exist in different aqueous systems; these are based on polymer–polymer interactions, polymer–salt interactions, or thermoseparation. A special case is affinity extraction, which can be viewed as a reactive extraction process. For further descriptions see "Alternative Two-Phase Systems".

36.2.1 Principles of Aqueous Two-Phase Extraction

Most of the hydrophilic, natural, and synthetic polymers are miscible with water and show phase separation in a mixture with another polymer or salt. The phase diagram of polyethylene glycol (PEG) and potassium phosphate in Figure 36.1 demonstrates the miscibility of polymer and salt phases. Above the curve (binodal) the system

Downstream Industrial Biotechnology: Recovery and Purification, First Edition. Edited by Michael C. Flickinger.

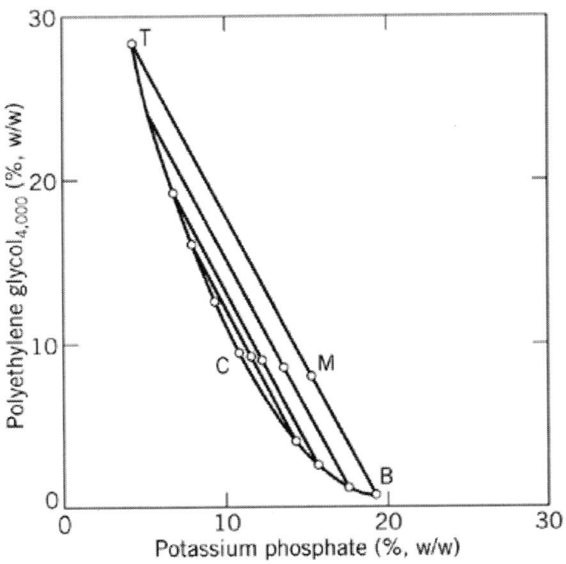

Figure 36.1. Phase diagram of PEG 4,000–potassium phosphate at 20°C. *Source*: Ref. 4.

separates into two liquid phases. If PEG and salt are mixed in a percentage represented by a point M above the binodal, the mixture separates along the tie-line. The composition of the two resulting phases are represented by the cross-points of the tie-line and the binodal. The volume ratio of the PEG-rich phase to the salt-rich phase is given approximately by the ratio of the tie-line sections $\overline{MB}$ and $\overline{TM}$. The critical point is reached when the tie-line is represented by only one point on the binodal. At the critical point, the addition of an infinitesimally small amount of water results in two phases of equal volume and composition; therefore, any added component should have a partition coefficient K of 1, where K is defined as the ratio of concentration in the top ($c_{i,\mathrm{T}}$) and bottom phase ($c_{i,\mathrm{B}}$)

$$K = \frac{c_{i,T}}{c_{i,B}} \tag{36.1}$$

With increasing length of the tie-line the composition of the phases at equilibrium diverge and consequently K changes.

It must be noted that phase diagrams are altered by addition of other components in sufficient concentration such as cell homogenates and different salts (7,8). Nevertheless, the phase diagram is useful for the selection of phase systems. Many phase diagrams of polymer–polymer and polymer–salt systems can be found in Refs 4 and 6. An extraction system is usually designed such that the wanted protein partitions predominantly to the top phase (PEG, in this case) and the undesired material including cells and cell debris partitions to the bottom phase (salt). The separation power of a system is characterized by the yield Y, the separation factor G, depending on the volume ratio R, and K. The yield in the top phase is defined as

$$Y_\mathrm{T} = \frac{n_\mathrm{T}}{n_\mathrm{T} + n_\mathrm{B}} \tag{36.2}$$

or consequently as

$$Y_\mathrm{T} = \frac{1}{1 + \left[\frac{V_\mathrm{B}}{V_\mathrm{T}} \cdot \frac{1}{K}\right]} \tag{36.3}$$

where n is the molar quantity of the desired protein in the volume of the top (V_T) and bottom phase (V_B), respectively. G is defined as the product of the partition coefficient K and the volume ratio R. The definition of the volume ratio is given in equation 36.5.

$$G = K \cdot R \tag{36.4}$$

$$R = \frac{V_\mathrm{T}}{V_\mathrm{B}} \tag{36.5}$$

Y, K, and R affect each other. Usually Y_T increases if K is increased at a constant R or if R is increased at a constant K (Figure 36.2).

K does not depend on the concentration of the protein i, so long as the concentration c_i is much smaller than the concentration of the phase-forming polymers, and no complex formation takes place. According to Bronstedt (9), there exists an exponential relation between K and the surface area (molecular mass) of i:

$$\ln K_\mathrm{i} = \frac{\alpha \cdot A_\mathrm{i}}{k \cdot T} \tag{36.6}$$

α is a factor related to the chemical potential, A_i is the surface for interaction, k is the Boltzmann constant, and T is the absolute temperature. Thus, cells and cell debris should

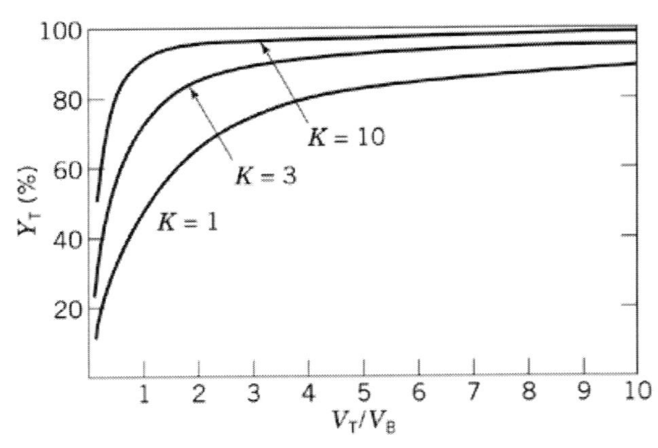

Figure 36.2. Interdependence of yield (Y), partition coefficient (K), and volume ratio (R). *Source*: Ref. 8.

exhibit one-sided partition approaching zero or infinity even at low values of α because of their large value of A, while metabolites should have K values approximately 1 and proteins usually between 0.1 and 10. The relation between $\ln K$ and A has been capitalized in a process to isolate hepatitis B surface antigen particles (molecular weight, $\simeq$1 to 2 million Da) as a hepatitis B vaccine from recombinant yeast cultures (10). Equation 36.1 is valid for the equilibrium between two liquid phases only; however, sometimes a solid phase seems to be involved and rather large values of K may result as a consequence of preferential solubility of i under extraction conditions in the top phase of PEG–salt systems (11). In such a case, K will increase with c_i.

36.2.2 Factors Determining Partitioning

The forces that cause partitioning can be explained in a first approximation by splitting the partition coefficient into increments:

$$\ln K = \ln K_{\text{ion}} + \ln K_{\text{hphob}} + \ln K_{\text{hphil}} + \ln K_{\text{conf}} + \ln K_{\text{lig}} \quad (36.7)$$

where K_{ij} denotes the partition coefficient from charges or hydrophobic or hydrophilic forces or its dependence on conformation or ligand interaction, respectively (12). Only $\ln K_{\text{lig}}$ is to some extent predictable. For this reason, affinity extraction is much studied, because the affinity interaction can be made to dominate $\ln K$. Even in the absence of affinity interactions many parameters will influence the partitioning of proteins in aqueous two-phase systems, affecting the partition coefficient and the selectivity.

These factors are as follows:

- Selection of component(s) to form a two-phase system
- Concentration of polymers used
- Molecular weight of the polymers
- Type and concentration of ions
- pH
- Temperature
- Addition of other agents (e.g., chaotropic agents) (13,14)
- Concentration of cells and cell debris added

The strong influence on the partition of only small changes in the system composition is shown in Figure 36.3, where the partition coefficient for L-2-hydroxyisocaproate dehydrogenase is changed significantly by addition of potassium phosphate and the resulting change of the tie-line.

The molecular mass of a given polymer can strongly influence the partition of a protein. The partition coefficient can be increased by increasing the molecular mass of the bottom phase forming polymer (for example, dextran 1)

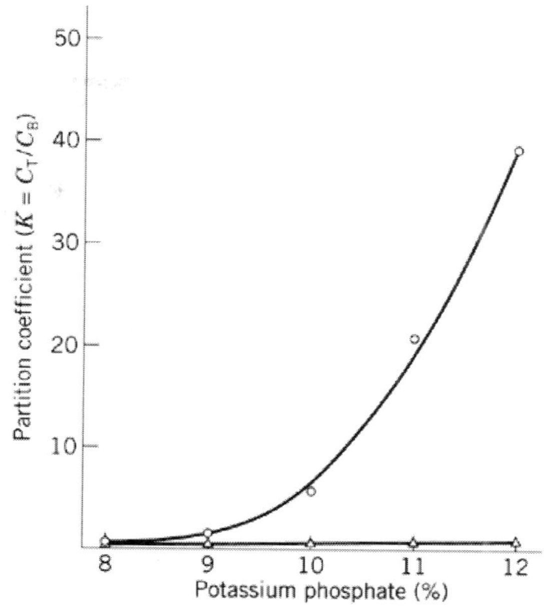

Figure 36.3. Partition of L-2-hydroxyisocaproate dehydrogenase (○) and D-lactate dehydrogenase (△) as a function of potassium phosphate concentration. (*Source*: Ref. 15.)

or by lowering the average molecular mass of the top-phase-forming polymer, most often PEG (1,4). The latter represents a fast and convenient way to increase protein yield in the top phase and is aided by the fact that the polymerization of ethyleneoxid can be easily adjusted to yield different molecular weight fractions of PEG. Several fractions are commercially available and are included in the pharmacopoeia of many countries. Changes in pH affect the ionization of side chains of the protein and therefore the charge and polar/hydrophobic balance of the surface; it will also alter the ionization of some weak and multivalent acids present in the extraction mixture leading to changes in K. pH values are easily manipulated in the range of protein stability. The addition of ions has a remarkable effect on K (1,4) by altering the length of the tie-line and the solvating power in PEG–salt systems as well as changes in the electrochemical potential across the interface in polymer–polymer systems.

The search for a separating system of a new protein is usually carried out by trial and error guided by the heuristic rules discussed previously. A series of phase systems differing in one parameter are produced in the scale of 5.0–10.0 g using a statistical design. A nice example of a statistical approach is shown by Hart *et al.* for the isolation of nonnative IGF-I (16). Y, K, and R are determined as well as protein concentration. From this information, a useful extraction system can be derived that can be further optimized with regard to specific goals, such as product yield and quality. Other criteria for optimization are biodegradability, ease of recycling, toxicity of phase-forming chemicals, physical properties of extraction systems, and last although not least the total costs (see "Economic Aspects").

For research purposes, the most investigated system is the PEG–dextran (1) combination. Both polymers are commercially available in different molecular weight fractions. For technical applications, fractionated dextran is too expensive; crude dextran may be used in low concentrations (17,18), but results in high-viscosity bottom phases. A less expensive possibility is PEG–salt systems, which are commonly used for large-scale applications (8). Other alternatives are reported (19,20). When substituting other polymers for dextran, for example, the limiting concentration for phase separation needs to be considered as well as product yield and quality. Combinations of the different systems are possible, like polymers and surfactants (21,22) or affinity partitioning combined with temperature-induced phase separation (23).

For large-scale operations, questions of waste disposal and the possibility of recycling auxiliary chemicals are of increasing interest for environmental as well as economic reasons. These topics are discussed separately under "Applications".

36.2.3 Thermodynamics

Phase separation in aqueous systems and partitioning of proteins are governed by thermodynamic forces. Because thermodynamics including mathematical models of aqueous two-phase systems cannot be explained in a few sentences, the reader is referred to the literature. For a broader introduction, the review by Cabezas (24) should be consulted.

Phase separation is based on the Gibbs free energy of mixing and will occur if ΔG^{mix} of two phases becomes energetically more favorable than ΔG^{mix} of a single phase:

$$\Delta G^{\mathrm{mix}} = \Delta H^{\mathrm{mix}} - T\Delta S^{\mathrm{mix}} \qquad (36.8)$$

where ΔH^{mix} and ΔS^{mix} are the enthalpy and entropy of mixing and T is the temperature. ΔG^{mix} is related to the excess enthalpy G^{e} by

$$\Delta G^{\mathrm{mix}} = G^{\mathrm{e}} + RT\sum_{i} x_i \ln x_i \qquad (36.9)$$

where x_i is the mole fraction of the different phase components i.

There is a large diversity of polymer–polymer models available, such as Edmond and Ogston-based models using McMillan–Mayer or Hill virial expansions (25,26), Flory–Huggins–based lattice models using the classical polymer solution theory (27,28), the UNIQUAC lattice model (29), integral equation theory-based models (30), the VERS model based on group contributions of salt and polymer (31), and the excluded volume theory based on statistical geometric arguments (32). A comprehensive discussion can be found in a recent paper by Johansson *et al*. (33). The problem for most of these models is that they include theories derived for gases or crystalline solids. They have advantages and disadvantages, such as too many parameters, time-consuming experiments to obtain parameters, and derived parameters that are valid only for a certain range of conditions.

An interesting approach to solve binodales for PEG–dextran systems is shown by Baughman and Liu (34). They used about 100 different binodales and protein partitioning data from other investigators as raw data for an expert system based on a neural network, resulting in quantitative predictions for unknown phase diagrams. To predict protein partition away from the isoelectric point the results were not satisfactory, however. In polymer–salt systems most of the electrostatic interactions are screened by the high salt concentration of the phases, which makes modeling more complicated. Only a few models have been developed based on virial expansions (31), integral equation theory (30), and a potential of mean force model including fluctuation solution theory (35).

Until now a reliable prediction of protein partitioning in aqueous two-phase systems for a random protein has not been possible. The systems are too complex, depending on pH, salt, polymer, protein structure, temperature, specific interactions, and charge distribution on the protein surface. The aim to minimize experiments for certain systems and predict trends is advancing. The problem with the model experiments is that they rely upon commercially available cheap proteins, which have strong specific interactions, however, such as lysozyme or α-chymotrypsin in which dimerization or self-digestion occurs. These effects in aqueous two-phase systems have not been studied extensively. Work has been done to predict peptide partitioning (27,(36–38)). Predictions have improved over recent years for certain peptides and amino acids, but depending on the sequence, predictions do not work well yet in general.

The diverse models describe phase separation reasonably well but a state-of-the-art model is not available (24). The interaction between proteins and solutions needs more attention to understand protein partition and describe it in quantitative terms based on fundamental principles. This goal will be elusive for quite some time to come. Fortunately it is very easy to determine K values experimentally so long as an appropriate specific assay is available.

36.3 ALTERNATIVE TWO-PHASE SYSTEMS

In addition to the commonly used polymer–polymer and polymer–salt systems, there are some alternative, also well-investigated, aqueous two-phase systems.

36.3.1 Affinity Extraction

Ligands covalently bound to one of the phase-forming components are useful to direct the extraction of certain proteins because the ligand will predominantly partition with the polymer in one phase. There exists a linked equilibrium of partition and complex formation, as shown in Figure 36.4. The extraction efficiency depends on the number of binding sites of the protein and the dissociation constant of the complex. The latter may be different in the top and bottom phase (39,40).

Affinity ligands can be cofactors or inhibitors of the target protein. General ligands based on textile dyes are more often used than natural ligands. Besides textile dyes metal chelators are attractive as general ligands. These are robust, cheap chemicals applicable in PEG–salt systems and offering useful selectivity. The syntheses of PEG derivatives has been described by Harris (41). Johansson *et al.* (42) investigated the partition of phosphofructokinase from yeast. The partition coefficient increased by a factor of 10^4 compared to physical extractions using the pseudo-affinity ligand cibacron-blue in an affinity process (Fig. 36.5).

The effectivity of affinity extraction was also demonstrated by Kroner *et al.* (43). With increase of the amount of PEG blue, the partitioning coefficient K increases. A yield of 98% could be obtained thereby with a single extraction step. Other parameters than the ligand concentration (43) are the polymer concentration (44), the kind of salt and its concentration (42), the pH value (45), the temperature (45), and the addition of free ligands (44,45). Useful models describing affinity extraction have been derived by Cordes *et al.* (40) and Suh and Arnold (46).

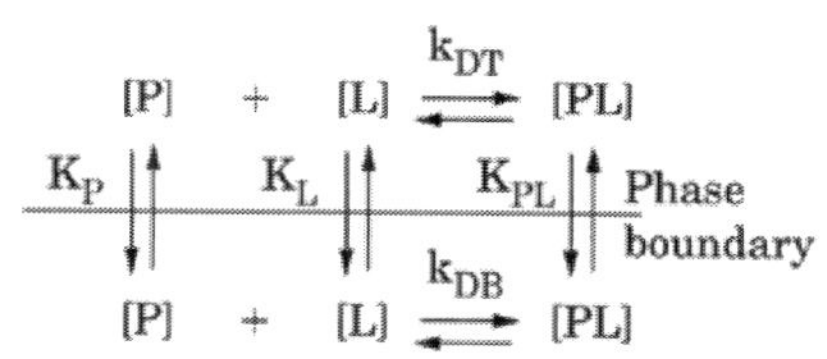

[P] Concentration of protein
[L] Concentration of ligand
[PL] Concentration of protein–ligand complex
K_P Partition coefficient of the protein
K_{PL} Partition coefficient of the ligand
K_L Partition coefficient of the complex
k_{DT} Equilibrium constant of the complex in the top phase
k_{DB} Equilibrium constant of the complex of the bottom phase

Figure 36.4. Equilibrium of partition and complex formation in affinity extraction processes. [P], concentration of protein; [L], concentration of ligand; [PL], concentration of protein–ligand complex; K_P, partition coefficient of the protein; K_{PL}, partition coefficient of the ligand; K_L, partition coefficient of the complex; k_{DT}, equilibrium constant of the complex in the top phase; and k_{DB}, equilibrium constant of the complex in the bottom phase.

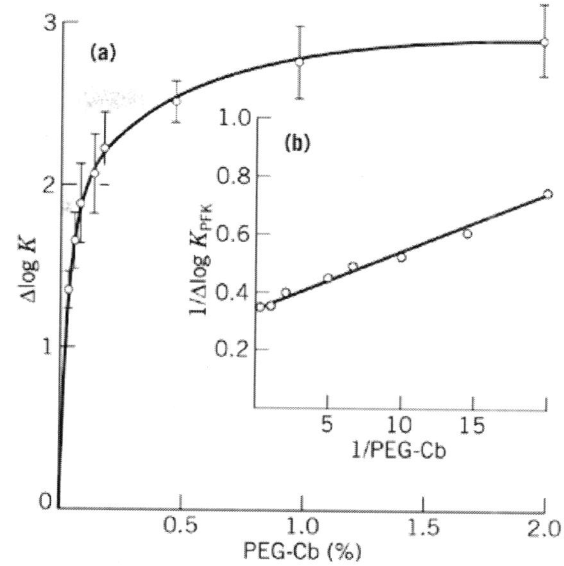

Figure 36.5. Changes of the partition coefficient of phosphofructokinase depending on the amount of added ligand. *Source*: Ref. 42.

An example of screening for a suitable dye can be found in a patent of Guliano and Szlag (47) describing experiments with the triazin dyes Cibacron-Blue FGF, Procion Turquoise H-A, Procion Green HE-4BDA, and Procion Red HE-3B to separate alcohol dehydrogenase. Large-scale applications of affinity extraction are listed later. Reviews of polymer–ligand systems have been published by Johansson (48) and Johansson and Tjerneld (49). It can be expected that specific ligands originally selected from libraries of suitable synthetic chemicals or peptides for affinity chromatography will find applications also in affinity extraction.

36.3.2 Detergent-Based System

A recent development is the two-phase extraction in aqueous two-phase systems based on nonionic detergents carrying ethylenoxide groups as the hydrophilic moiety. The protein is partitioned using only one auxiliary chemical. The principle is shown in Figure 36.6. At low temperature, a homogeneous phase containing micelles of the detergent and fermentation broth exists. The detergent interacts preferably with hydrophobic proteins. As soon as the temperature is increased above a certain system-dependent temperature, turbidity develops as an indicator of phase separation. The so-called cloud point depends mainly on the average number of ethylene oxide groups; a narrow range is desirable. A C12EO5 detergent, for example, has a cloud point between 23 and 25°C. The phase separation depends on the reversible, temperature-induced dehydration of the ether oxygens in the detergent and leads to a detergent-rich and a detergent-depleted phase in equilibrium. The micelles

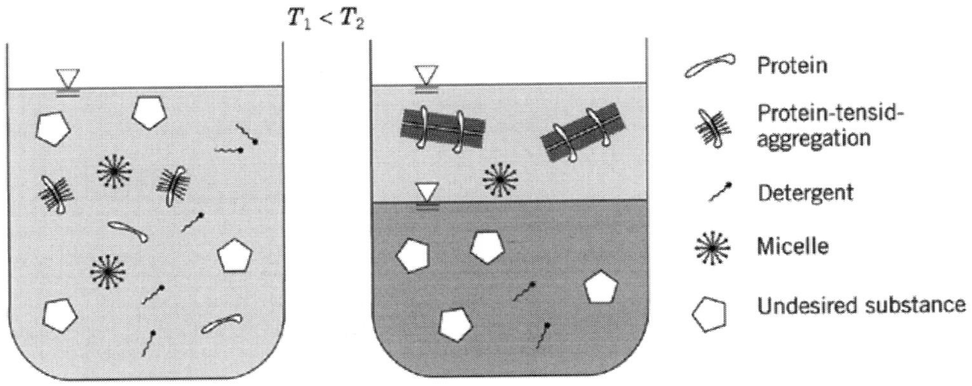

Figure 36.6. Schematic illustration of aqueous detergent solutions below and above the cloud point.

are believed to aggregate into laminar structures carrying hydrophobic proteins into the so-called coazervate phase, which may still contain 70–80% water (50).

Bordier (51), Minuth *et al*. (52), and others investigated cloud-point extraction of proteins. Nonionic or zwitterionic detergents are most useful for the separation of membrane proteins (53,54). A complete process design for the purification of cholesterol oxidase is presented in Figure 36.7. A scale-up study indicated that the rather low density difference in detergent-based systems limits phase separation by centrifugation in the initial step (56).

It has been shown that the detergent for extraction of hydrophilic proteins and membrane-bound proteins should be chosen with respect to its hydrophile–lipophile balance (HLB) (57). Table 36.1 shows partitioning results using Tritons with different HLBs for the separation of an endopeptidase in the presence of sodium chloride. Yield, concentration factor, and partitioning coefficient increased significantly with an increasing HLB.

36.3.3 Reversed Micelles

Because it is similar to aqueous liquid–liquid extraction, reversed micellar protein extraction is mentioned here briefly. Reversed micelles extract proteins into the water pool contained in the micelles present in an organic solvent depending on the isoelectric point of the protein and therefore pH and ionic strength in the process solution. The principle is shown in Figure 36.8. The organic solvent allows the organization of the reversed micelles and the separation of the bulk water from the water pool in the micelles.

Extraction into reversed micelles has been investigated for several industrially relevant proteins, but these are apparently not yet used in industry (58,59). The major remaining problem is the difficult reextraction of the bioactive protein into an aqueous phase and the reuse of reversed micelles.

TABLE 36.1. Influence of the Surfactant Hydrophile–Lipophile Balance (HLB) on the Partition Coefficient, Concentration Factor, and Yield of the Purification Process for the Example of Endopeptidase

Surfactant	HLB	Partition Coefficient	Concentration Factor	Yield (%)
Triton X-100	13.5	0.81	0.7	15
Triton X-165	15.8	4.17	3.65	49
Triton X-305	17.3	8.93	3.1	69
Triton X-405	17.9	53	6.63	100

[a] *Source:* Ref. 57.

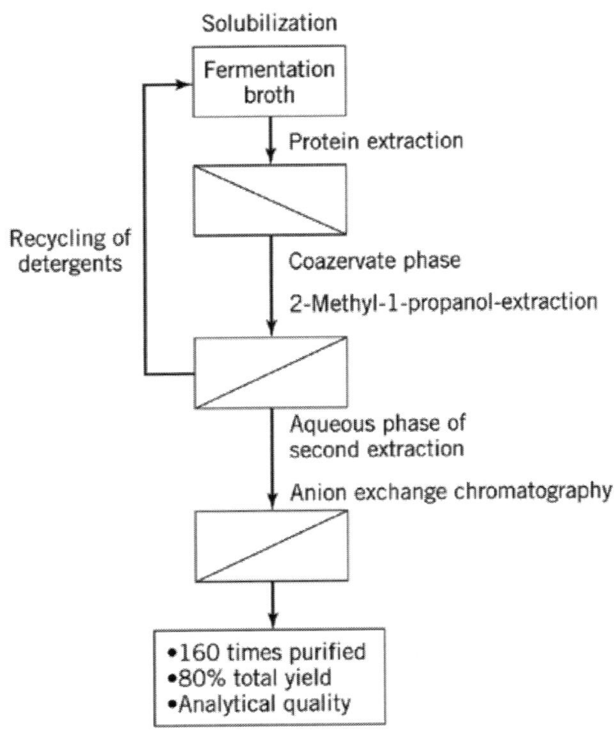

Figure 36.7. Optimal process design for the extraction of cholesterol oxidase from *Nocardia rhodochrous* using detergent-based systems in the initial step. (*Source*: Ref. 55.)

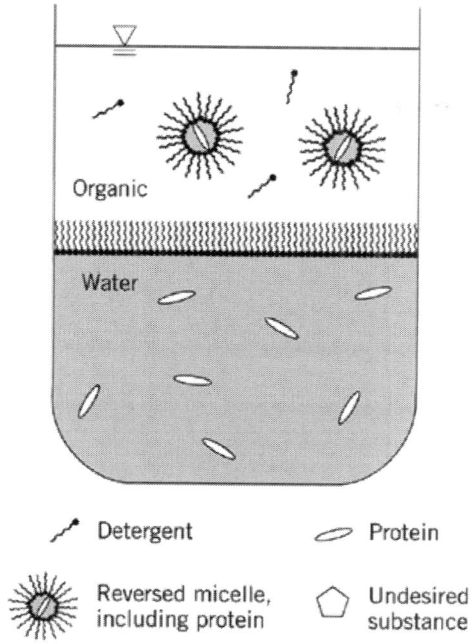

Figure 36.8. Operating principle of extraction into reversed micelles.

36.3.4 Thermoseparating Polymers

Certain polyethylene–polypropylenoxide copolymers also show reversible dehydration of the oxygen groups in an interesting temperature range and have been investigated by Alred *et al.* (60), Johansson *et al.* (61), and recently by Persson *et al.* (62) for application in aqueous two-phase systems. Usually a polymer–polymer system is used in the first step selected to partition the protein in the copolymer-rich phase. After primary phase separation, the solution is heated above the cloud point, which results in phase separation under conditions in which the protein is almost exclusively found in the polymer-depleted phase.

36.4 APPLICATIONS

36.4.1 Technical Aspects

Aqueous liquid–liquid extraction uses the same principles as the well-known chemical process of liquid–liquid extraction. Therefore similar, commercially available equipment may be used for mixing and phase separation. The physical properties of aqueous two-phase systems, however, are very different from those of solvent–water systems. The most notable difference is the dramatically lower interfacial tension (Table 36.2).

In general, for the purification of an intracellular protein the cells are disrupted, the phase-forming components added, and the pH adjusted; the suspension is then mixed to reach equilibrium. In large-scale processes this is done in a stirred vessel with baffles or using static mixers (63). An extremely low energy input is needed because of the low interfacial tension of the whole system. The approach to equilibrium is surprisingly fast ($\leq$30 s), as demonstrated by Fauqueux *et al.* (64), provided turbulent mixing is ensured. After equilibration is reached the phases should separate into a top and bottom phase. This separation can be done under unit gravity or can be speeded up by centrifugation. The latter is recommended if the viscosity is high or the density difference between the phases is low (8). Centrifugation gives higher yields as entrainment is avoided at higher *g* forces; optimal feed rates of the suspension up to 3 L/min corresponding to residence times of about 16 s have been reported using small disk stack separators (8).

After the initial separation step, the top phase is often mixed with salts to generate a secondary phase system. Conditions are selected such that the protein of interest is partitioned to the bottom phase and can be further processed after phase separation (Fig. 36.9). Other possible procedures for the isolation of proteins from polymer phases are centrifugation, adsorption, ultrafiltration, or electrophoreses (66). In large-scale processes, ultrafiltration or diafiltration is used to concentrate and condition the product from the secondary bottom phase. The top phase may be recycled (52,67). A secondary bottom phase may be applied under certain conditions to a chromatographic support, such as hydrophobic interaction matrices.

Extraction usually is performed in the batch mode, but often it may be more cost-effective to use a continuous process. In general there are three possibilities of continuous processing in production lines: the co-, cross-, and countercurrent flow. The three modes are illustrated in Figure 36.10, and a comparison of specific properties is shown in Table 36.3. In large-scale processes the counter-current method should be the most efficient because of a higher yield and lower consumption of

TABLE 36.2. Physical Properties of Aqueous Two-Phase Systems

System Composition	$\Delta\rho$ (g/cm^3)	η_T (mPa s)	η_B (mPa s)	$\eta_{dispersion}$ (mPa s)	δ (mN/m)	References
9% PEG 4000, 2% dextran T-500	1,010	3	94	3.4	—	16
7% PEG 4000, 1.25% crude dextran	1,046	2.7	2,000	4	—	16
16% PEG 4000, 13% phosphate	1,085	12	2.1	—	1.25	58

[a] *Source:* PEG, polyethylene glycol. Ref. 8.

TABLE 36.3. Advantages and Disadvantages of Different Flow Regimes in Continuous Processing

	Suitable For	Consumption Requirements	Yield	Purification Factor
Cocurrent	Low K	As cross-current	Medium	Medium
Cross-current	High K	As co-current	Lowest	Highest
Countercurrent	Low K	Lowest	Highest	Lowest

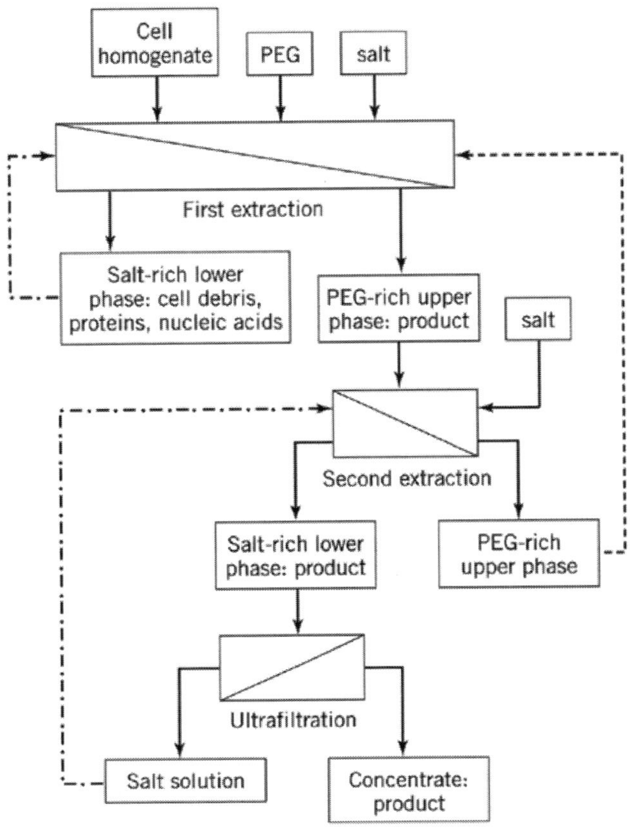

Figure 36.9. Flow sheet for product, chemicals, and waste flows in an extraction process using a PEG–salt aqueous two-phase system. ·–·–·–·, recycling of salt; – – – –, recycling of PEG. *Source*: Ref. 65.

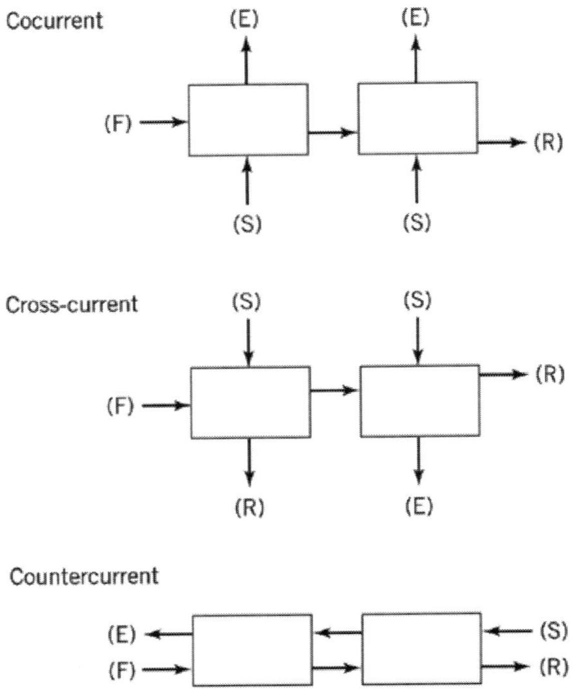

Figure 36.10. Principles of co-, cross-, and counter-current flow. F, feed; S, solvent; E, extract; R, raffinade.

chemicals (63). Countercurrent extraction has been studied for the multistep procedures discussed next. The continuous cross-current extraction has been investigated experimentally in pilot scale. Enzyme extraction has been successfully carried out using more than 500 kg yeast per day in a computer-controlled process (68). The procedure could also be applied to extract enzymes continuously from bacterial cell homogenates (69).

To achieve a higher purification, multiple partition steps would be required. Not every multistage apparatus used in the chemical industry for liquid–liquid extraction is suitable for operation with aqueous phases as the physical properties of the two phases are too similar and the interfacial tension too low. The performance of the Graesser contactor has been best investigated. For the separation of α-lactalbumin from whey, a space time yield of 29 g/L reactor volume and day was achieved (70). The speed of rotation in the "raining bucket contactor" had to be reduced from the usual 20 U/min to 2–5 U/min, working with PEG–salt systems (71). In addition, the rotational direction was reversed to reduce the intensity of mixing and to avoid flooding. Other apparatus investigated for multistage extractions are mixer-settler, Kühni columns (72), the Podbielniak extractor (1), and spray columns (73). All can be operated successfully with specific precautions against flooding. Changing operating conditions such as pH or the length of the tie-line, however, may be more effective for protein purification than multiple partition steps.

At Miles Inc., a conventional downstream process was compared to an operation using aqueous two-phase systems and *Bacillus* alpha-amylase as an example (74). The latter was carried out in a 50,000-L large-scale process

(Fig. 36.11). In this example the extraction process in aqueous two-phase-systems required fewer steps, resulting in higher yield and lower costs.

36.4.2 Recycling of Process Chemicals

To lower high chemical and waste disposal costs, recycling or recirculation of the auxiliary chemicals is recommended. Recycling of phosphate by crystallization at 6°C has been described from solid-free process streams (8). The successful recirculation of most of the PEG has been demonstrated by Papamichael *et al*. (75) and of detergents by Minuth *et al*. (55) (see Fig. 36.7). The recovery of salt from the primary bottom phase was accomplished by Greve and Kula (76) making use of miscibility gaps in the ternary-phase systems—water–salt–lower aliphatic alcohols (Fig. 36.9).

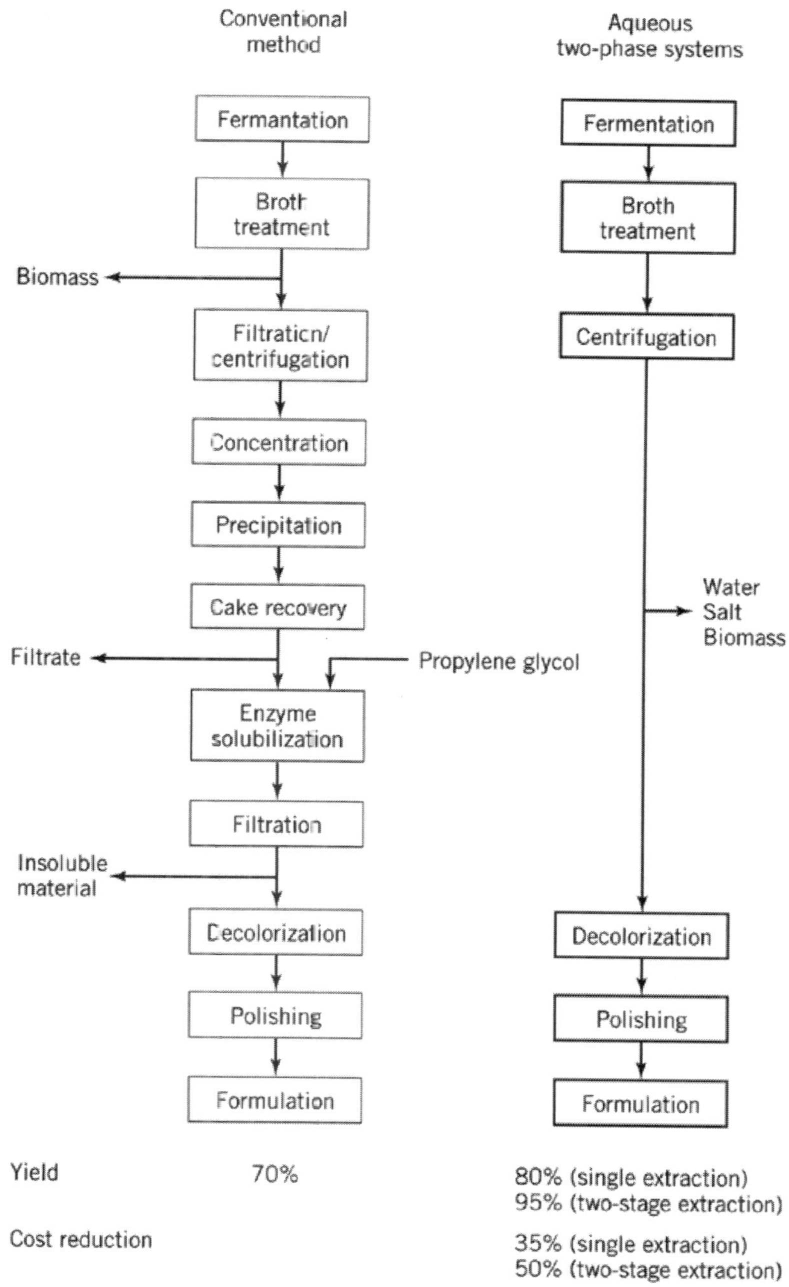

Figure 36.11. Comparison of the downstream process of alpha-amylase, using a conventional method (left column) and using aqueous two-phase systems (right column). *Source*: Ref. 74.

The authors also presented an economic analysis (76). In principle, the material balance of the auxiliary chemicals introduced into the process stream to accomplish phase separation can be closed more than 90% by recycling, yielding cost-effective and environmentally acceptable procedures.

36.4.3 Large-Scale Applications

36.4.3.1 Examples of Large-Scale Applications. Direct information about industrial applications of aqueous two-phase extraction is scarce. In Table 36.4, a list of patents and large-scale applications (beginning in 1982; minimum, 24-kg cells) described in the literature is compiled. The size of the operation reached 50,000 L or more (74). Usually centrifugation is preferred in the initial phase separation. (An example is illustrated in Fig. 36.11.)

36.4.3.2 Scale-Up. One big advantage of aqueous liquid–liquid extraction is the ease of scaling up. Hart *et al*. (89), Schütte *et al*. (93), and Kroner *et al*. (94) showed that scale-up is feasible by proportional linear increase of the amount of ingredients. The necessary centrifugation equipment is commercially available and can be selected using the Sigma concept. In Figure 36.12, scale-up data of Kroner *et al*. (94) are presented for the purification of formate dehydrogenase from *Candida boidinii* in PEG–potassium phosphate. A scale-up of nearly 40,000 was carried out for different separation steps without significant changes in yield.

36.4.3.3 Economic Aspects. The costs of the protein recovery and enrichment by aqueous liquid–liquid extraction is of course highly dependent on the actual system, scale, etc. As an example Kroner *et al*. (95) analyzed the cost structure for extraction and compared the economics with other downstream separation methods employed as alternatives for the purification of intracellular fumarase from *Brevibacterium ammoniagenes* in a 100-kg batch process. With a system of PEG and potassium phosphate, an overall yield of 70% was achieved by a two-step extraction. The cost structure of the extraction is shown in Table 36.5. Even if prices have changed meanwhile, the table still gives some idea about the relative cost distribution of material, labor, energy, and investment for the different process steps.

In Figure 36.13, a comparison of the cost factors and the total costs of the initial recovery operations is shown (95). Data in Table 36.5 and in Figure 36.13 are calculated from equation 36.10, which is derived from coefficients used for the cost calculation in the chemical industry and normalized for the yield:

$$T = \frac{1.13M + 2.6L + 1.13E + 0.13I}{Y} \quad (36.10)$$

where the letters represent costs for T, total production; M, material; L, labor; E, energy; I, investment (capital); and Y, yield.

Aqueous two-phase extraction is a low-energy, labor-requiring technique with comparatively low investment costs and high material costs. These high costs for auxiliary chemicals can be lowered by a good choice of volume ratio and by new phase systems and especially by the development of recycling processes (65). The process time can be decreased dramatically by centrifugal separation, resulting not only in high space time yields but also in improved product quality because of the absence of significant proteolytic degradation (96,97). Economic aspects of affinity extraction have been discussed by Johansson *et al*. (45) and Cordes *et al*. (18). In such cases, economics rely heavily on the initial costs and the reusability of the affinity ligand.

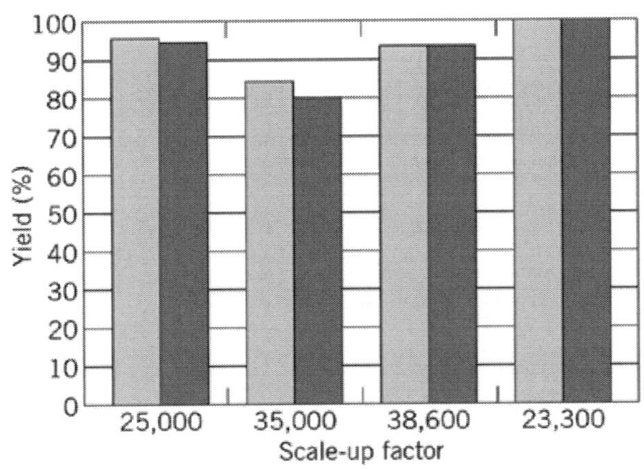

Figure 36.12. Scale-up of various extraction steps during the isolation and purification of formate dehydrogenase from *Candida boidinii*. *Source*: Ref. 94.

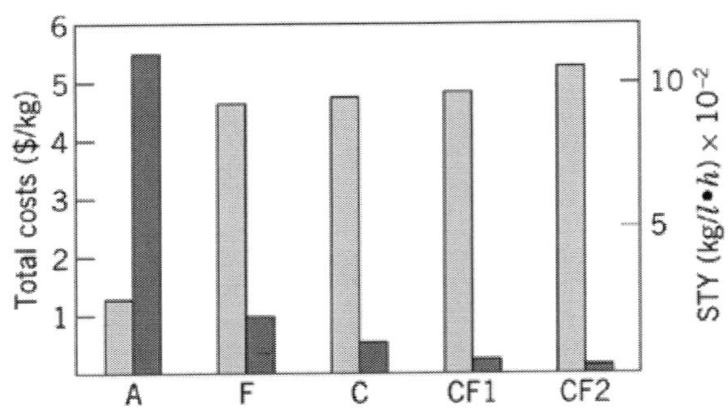

Figure 36.13. Cost factors and total costs comparing different initial isolation procedures. A, aqueous-phase system operation; F, filtration; C, centrifugation; CF, cross-flow filtration (1, retention coefficient $R=0$; 2, $R=0.7$). *Source*: Ref. 95.

TABLE 36.4. Patents and Large-Scale Applications of Protein Extraction Using Aqueous Two-Phase Systems

Literature		Author	Date	Enzyme or Protein	System Used	Ligand	Origin
77	Patent	Kula	1982	Interferon	PEG–dextran		Raw material
15	Journal article	Hummel *et al.*	1983	D-Lactate dehydrogenase	PEG–phosphate		*Lactobacillus cellubiosus*
78	Patent	Kim *et al.*	1985	Protease, amylase	PEG–dextran		*Aspergillus orizae*
79	Journal article	Schütte *et al.*	1985	L-Leucin dehydrogenase	PEG–phosphate		*Bacillus cereus*
80	Patent	Gustafsson *et al.*	1986	ADH, hexokinase	PEG–K_2HPO_4, KH_2PO_4	Yes[d]	Yeast
80	Patent	Gustafsson *et al.*	1986	Transferrin	PEG–K_2HPO_4, KH_2PO_4	Yes	Blood plasma
81	Patent	Paul *et al.*	1986	Dextran-sucrase	PEG–dextran		*Leuconostoc mesenteroides*
82	Patent	van Wijnendaele *et al.*	1986	Hepatitis B antigen (HBsAg)	PEG–$(NHY_4)_2SO_4$		Yeast
82	Patent	van Wijnendaele *et al.*	1986	Alpha-1-antitrypsin	PEG–$(NH_4)_2SO_4$		Yeast
83	Patent	Dove *et al.*	1987	Therapeutically active proteins[b]	PEG–KH_2PO_4		Blood plasma
44	Journal article	Tjerneld *et al.*	1987	Lactate dehydrogenase	PEG–aquaphase PPT	Yes	Pig muscle
84	Patent	Ananthapadmanabhar	1988	Alcaline protease	PEG–Na_2SO_4	Yes	
85	Patent	Brewer *et al.*	1988	Protease	PEG–Na_2SO_4		Whole fermentation beer
14	Patent	Sieron *et al.*	1991	Recombinant proteins	PEG–polyvinylalcohol	[c]	
86	Patent	Enfors *et al.*	1992	Human immunoglobulin G	PEG–phosphate		Staphylococcus
47	Patent	Giuliano *et al.*	1992	Alcohol dehydrogenase (ADH)	PVP–maltodextrin	Yes	Baker's yeast
87	Patent	Heinsohn *et al.*	1992	Chymosin	PEG–Na_2SO_4		*Aspergillus niger* var. *awamori*
88	Patent	Kirchberger *et al.*	1992	Alkalische phosphatase	PEG–dextran		Calf intestine
71	Journal article	Coimbra *et al.*	1994	Beta-lactoglobulin	PEG–phosphate		Cheese whey
71	Journal article	Coimbra *et al.*	1995	Alpha-lactalbumin	PEG–phosphate		Cheese whey
18	Journal article	Cordes *et al.*	1994	Formate dehydrogenase	PEG–phosphate	Yes	*Candida boidinii*
89	Journal article	Hart *et al.*	1994	IGF[b]	PEG–Na_2SO_4		*E. coli*
90	Patent	Lorch *et al.*	1994	EG[d]	PEG–$(NH_4)_2SO_4$		Cellulase mixture
13	Patent	Builder *et al.*	1995	IGF-I or mammalian polypeptide	PEG–citrate	[c]	*E. coli*
91	Patent	Heinssohn *et al.*	1995	Chymosin	PEG–Na_2SO_4		Bovine stomach
92	Patent	Lee and Kahn	1995	Hemoglobin	PEG–phosphate buffer		Bovine blood
57	Patent	Braunstein *et al.*	1995	Different lipases and proteases	Detergents		Different organisms

[a]Patents often cover wide ranges; therefore, in some cases only examples from the patent are given.
[b]Albumin, IgM, IgG, alpha-1-antitrypsin.
[c]Use of chaotropic substances.
[d]Adsorption at Sepharose particles.

TABLE 36.5. Cost Structure of an Extraction Process

Step	Feedstock ($/kg)	Material ($/kg)	Labor ($/kg)	Energy ($/kg)	Instrumentation ($/kg)	Total ($/kg)	Percentage of total
Fermentation	5.80					6.55	40.1
Disruption		0.03	0.92	0.03	2.76	2.82	17.2
1. Extraction		1.89	0.48	<0.01	0.94	3.52	21.5
2. Extraction		0.17	0.29	<0.01	0.94	1.08	6.6
Ultrafiltration		0.20	0.48	0.01	0.90	1.60	9.8
Waste water						0.78	4.8
Sum ($/kg)	5.80	2.29	2.17	<0.06	5.54	16.35	
Percentage of total	35.5	14.0	13.3	0.4	33.9		

For a valid evaluation, it is important to realize that extraction constitutes an integrative technology; it achieves removal of solids and an increase in the specific activity of the product in a single step. In addition, in most cases the product can be separated from high molecular weight nucleic acids and its concentration can be increased, improving the performance of subsequent separation steps. An industrial total cost comparison between a downstream process using aqueous two-phase systems and a conventional one is shown in Figure 36.11. A cost reduction of 35% to 50% was achieved for the separation of *Bacillus* alpha-amylase.

36.5 CONCLUSION

Liquid–liquid extraction of proteins using aqueous two-phase systems is a comparatively new but nevertheless well-investigated technique for the isolation and separation of proteins. It is apparently not yet widely used in industry, but the number of industrial applications seems to have increased in recent years due to the effectiveness and ease of operation on a large scale. Intensive research is going on worldwide with respect to practical applications and theoretical understanding. Together with increases in production size for pharmaceutical proteins, this will make liquid–liquid extraction in aqueous two-phase systems a more common unit operation in the biotechnological industry.

REFERENCES

1. M.-R. Kula, K.H. Kroner, and H. Hustedt, Adv. Biochem. Eng. 24: 73–118 (1982).
2. M.-R. Kula, Bioseparation 1: 181–189 (1990).
3. H. Walter and G. Johansson, Methods in Enzymology, vol. 228, Academic Press, San Diego, Calif., 1994.
4. P.A. Albertsson, Partition of Cell Particles and Macromolecules, Wiley, New York, 1986.
5. H. Walter, D.E. Brooks, and D. Fisher, Partitioning in Aqueous Two-Phase Systems, Academic Press, Orlando, Fla., 1985.
6. B.Y. Zaslavsky, Aqueous Two-Phase Partitioning, Dekker, New York, 1994.
7. M.-R. Kula, in A. Humphrey, C.L. Cooney eds., Comprehensive Biotechnology, vol. 2, Pergamon Press, New York, 1985, pp. 451–471.
8. H. Hustedt, K.H. Kroner, and M.-R. Kula, in H. Walter, D.E. Brooks, and D. Fisher eds., Partitioning in Aqueous Two-Phase Systems, Academic Press, Orlando, Fla., 1985, pp. 529–587.
9. J.N. Bronsted, Z. Phys. Chem. Abt. A 157: 257 (1931).
10. D.J. Kubek, in R.G. Harrison ed., Protein Purification Process Engineering, Dekker, New York, 1994, pp. 87–114.
11. U. Menge, M. Morr, U. Mayr, and M.-R. Kula, J. Appl. Biochem. 5: 75–90 (1983).
12. P.A. Albertsson, Endeavour 1: 69 (1977).
13. U.S. Pat. 5,407,810 (April 18, 1995), S. Builder, R. Hart, P. Lester, J. Ogez, and D. Reifsnyder (to Genentech, Inc.).
14. GDR Pat. DD 288 837 (April 11, 1994), R. Sieron, R. Wondraczek, K. Binder, H. Dittmar, K. Lambrecht, and H. Thessa (to Zentralinstitut für Mikrobiologie und experimentelle Therapie).
15. H. Schütte, W. Hummel, and M.-R. Kula, Appl. Microbiol. Biotechnol. 19: 167–176 (1984).
16. R.A. Hart, J.R. Ogez, and S.E. Builder, Bioseparation 5: 113–121 (1995).
17. K.H. Kroner, H. Hustedt, and M.-R. Kula, Biotechnol. Bioeng. 24: 1015–1045 (1982).
18. A. Cordes and M.-R. Kula, in H. Walter and G. Johansson eds., Methods of Enzymology, vol. 228, Academic Press, San Diego, Calif., 1994, pp. 600–617.
19. D.C. Szlag, K.A. Giuliano, and S.M. Snyder, J.F.P. Hamel, J. Hunter, and S.K. Sikdar eds., Downstream Processing and Bioseparation, ACS Symposium Series, vol. 419, American Chemical Soc., Washington, D.C., 1990, pp. 71–86.
20. M.-R. Kula, in H. Brauer ed., Biotechnology, vol. 2, VCH-Verlagsgesellschaft, Weinheim, 1985, pp. 725–760.
21. U. Sivars, K. Bergfeld, L. Piculell, and F. Tjerneld, J. Chromatogr. B 680: 43–53 (1996).
22. H.-O. Johansson, G. Karlström, and F. Tjerneld, Macromolecules 26: 4478 (1993).
23. P.A. Alred, A. Koslowski, and J.M. Harris, Bioseparation 2: 363–373 (1992).
24. H. Cabezas, Jr., J. Chromatogr. B 680: 3–30 (1996).
25. D. Forciniti and C.K. Hall, ACS Symp. Ser. 419: 53–70 (1990).
26. F. Döbert, A. Pfennig, and M. Stumpf, Macromolecules 28: 7860–7868 (1995).

27. A.D. Diamond, K. Yu, and J.T. Hsu, in Protein Purification, ACS Symposium Series, American Chemical Soc., Washington, D.C., 1990, pp. 52–65.
28. A. Gustafsson, H. Wernerström, and F. Tjerneld, Polymer 27: 1768–1770 (1986).
29. H. Hartounian and S.I. Sandler, Biotechnol. Prog. 7: 279 (1991).
30. C.A. Haynes, F.J. Benitez, H.W. Blanch, and J.M. Prausnitz, AIChE J. 39: 1539 (1993).
31. C. Großmann and G. Maurer, Fluid Phase Equilib. 106: 17–25 (1995).
32. Y. Guan, T.H. Lilley, and T.E. Treffry, Macromolecules 26: 3971 (1993).
33. H.-O. Johansson, G. Karlsström, F. Tjerneld, and C.A. Haynes, J. Chromatogr. B 711: 3–17 (1998).
34. D.R. Baughman and Y.A. Liu, Ind. Eng. Chem. Res. 33: 2668–2687 (1994).
35. M. Kabiri-Badr and H. Cabezas, Jr., Fluid Phase Equilib. 115: 39–58 (1996).
36. J. Bringmann, B. Keil, and A. Pfennig, Fluid Phase Equilib. 101: 211–225 (1994).
37. W.-Y. Chen, C.-G. Shu, J.Y. Chen, and J.-F. Lee, J. Chem. Eng. Jpn. 27: 688–690 (1994).
38. M.A. Eiteman, C. Hassinen, and A. Veide, Biotechnol. Prog. 10: 513–519 (1994).
39. M.-R. Kula, A. Walsdorf, and A. Cordes, Ber. Bunsen-Ges. Phys. Chem. 93: 968–970 (1989).
40. A. Cordes, J. Flossdorf, and M.-R. Kula, Biotechnol. Bioeng. 30: 514–520 (1987).
41. J.M. Harris, J. Macromol. Sci. C-25: 325 (1985).
42. G. Johansson, G. Kopperschläger, and P.A. Albertsson, Eur. J. Biochem. 131: 589–594 (1983).
43. K.H. Kroner, A. Cordes, A. Schelper, M. Morr, A.F. Bückmann, and M.-R. Kula, in T.C.J. Gribnau, J. Visser, and R.J.F. Nivard eds., Affinity Chromatography and Related Techniques, Elsevier, Amsterdam, 1982, pp. 451–501.
44. F. Tjerneld, G. Johansson, and M. Joelsson, Biotechnol. Bioeng. 30: 809–816 (1987).
45. G. Johansson and M. Andersson, J. Chromatogr. 303: 39 (1984).
46. S.-S. Suh and F.H. Arnold, Biotechnol. Bioeng. 35: 682–690 (1990).
47. U.S. Pat. 5,093,254 (March 3, 1992), K.A. Giuliano and D.C. Szlag (to the United States).
48. G. Johansson, in S.K. Sikdar and M. Bier eds., Frontiers in Bioprocessing, CRC Press, Boca Raton, Fla., 1989, pp. 271–284.
49. G. Johansson and H. Tjerneld, J. Biotechnol. 11: 135–142 (1989).
50. G.C. Terstappen, R.A. Ramelmeier, and M.-R. Kula, J. Biotechnol. 28: 263–275 (1993).
51. C. Bordier, J. Biol. Chem. 250: 1604–1606 (1981).
52. T. Minuth, J. Thömmes, and M.-R. Kula, J. Biotechnol. 38: 151–164 (1995).
53. R.P. Frankewich and W.L. Hinze, Anal. Chem. 66: 944–954 (1994).
54. W.L. Hinze and E. Pramauro, Crit. Rev. Anal. Chem. 24: 133–177 (1993).
55. T. Minuth, J. Thömmes, and M.-R. Kula, Biotechnol. Appl. Biochem. 23: 107–116 (1996).
56. T. Minuth, H. Gieren, H. Pape, C. Raths, J. Thömmes, and M.-R. Kula, Biotechnol. Bioeng. 55: 339–347 (1997).
57. Int. Pat. WO 96/23061 (January 27, 1995) E.L. Braunstein, N.T. Becker, G. Ganshaw, and T.P. Graycar (to Genencor International, Inc.).
58. B.D. Kelley, D.I.C. Wang, and T.A. Hatton, Biotechnol. Bioeng. 42: 1199–1208 (1993).
59. M.J. Pires and J.M.S. Cabral, J. Chem. Technol. Biotechnol. 61: 219–224 (1994).
60. P.A. Alred, A. Kozlowski, J.M. Harris, and F. Tjerneld, J. Chromatogr. A 659: 289–298 (1994).
61. H.-O. Johansson, G. Karlsström, B. Mattiasson, and F. Tjerneld, Bioseparation 5: 269–279 (1995).
62. J. Persson, L. Nyström, H. Ageland, and F. Tjerneld, J. Chromatogr. B 711: 97–109 (1998).
63. N. Papamichael and H. Hustedt, Methods Enzymol. 228: 573–584 (1994).
64. P.-F. Fauqueux, H. Hustedt, and M.-R. Kula, J. Chem. Technol. Biotechnol. 35B: 51–59 (1985).
65. A. Greve and M.-R. Kula, J. Chem. Technol. Biotechnol. 50: 27–42 (1991).
66. G. Johansson, in H. Walter and G. Johansson eds., Methods of Enzymology, vol. 228, Academic Press, San Diego, Calif., 1994, pp. 569–573.
67. H. Hustedt, K.-H. Kroner, U. Menge, and M.-R. Kula, Trends Biotechnol. 3: 139–144 (1985).
68. H. Hustedt, B. Börner, K.H. Kroner, and N. Papamichael, Biotechnol. Tech. 1: 49–54 (1987).
69. H. Hustedt and N. Papamichael, Enzyme Eng. 9: 135–139 (1988).
70. J. Dos Reis Coimbra, J. Thömmes, A. Meirelles, and M.-R. Kula, Bioseparation 5: 259–268, (1995).
71. J. Dos Reis Coimbra, J. Thömmes, and M.-R. Kula, J. Chromatogr. A 668: 85–94 (1994).
72. H. Hustedt, K.-H. Kroner, U. Menge, and M.-R. Kula, Enzyme Eng. 5: 45–47 (1988).
73. K.R. Jafarabad, S.B. Sawant, J.B. Joshi, and S.K. Sikdar, Chem. Eng. Science 47: 57–68 (1992).
74. C.Y. Kim, J.W. Brewer, C.E. Brothers, T.F. Farver, and E.K. Lee, 3rd Chemical Cong. of North America, Toronto, Canada, June 5–10, 1988.
75. N. Papamichael, B. Börner, and H. Hustedt, J. Chem. Technol. Biotechnol. 54: 47–55 (1992).
76. A. Greve and M.-R. Kula, Bioprocess Eng. 6: 173–177 (1990).
77. FRG Pat. DE 2943016 C2 (September 6, 1984), U. Menge, M. Morr, and M.-R. Kula (to the Gesellschaft für Biotechnologische Forschung mbH).
78. U.S. Pat. 4,508,825 (April 2, 1985), C.Y. Kim, T.F. Farver, and J.W. Brewer (to Miles Laboratories, Inc.).
79. H. Schütte, W. Hummel, H. Tsai, and M.-R. Kula, Appl. Microbiol. Biotechnol. 22: 306–317 (1985).
80. U.S. Pat. 4,579,661 (April 1, 1986), S.J. Gustafsson, P.O. Hedman, T.G.I. Ling, and B.G. Mattiasson (to Pharmacia AB, Sweden).
81. U.S. Pat. 4,591563 (May 27, 1986), F. Paul, P. Monsan and D. Auriol (to Societe Nationale Elf Aquitaine, France).

82. E.U. Pat. 0 199 698 B1 (August 28, 1991), F. van Wijnendaele, D. Gilles, and G. Simonet (to Smithkline Biologicals S.A.).
83. U.S. Pat. 4,684,723 (August 4, 1987), G.B. Dove and G. Mitra (to Miles Laboratories, Inc.).
84. U.S. Pat. 4,743,550 (May 10, 1988), K.P. Ananthapadmanabhan and E.D. Goddard (to Union Carbide Corporation).
85. U.S. Pat. 4,728,613 (March 1, 1988), J.W. Brewer, C.E. Brothers, T.F. Farver, C.Y. Kim, and E. Lee (to Miles Laboratories, Inc.).
86. Int. Pat. WO 92/07868 (May 14, 1992), S.-O. Enfors, K. Köhler, Ch. Ljundquist, B. Nilsson, and A. Veide (to Pharmacia AB, Sweden).
87. U.S. Pat. 5,139,943 (August 18, 1992), H.G. Heinsohn, J.D. Lorch, and R.E. Arnold (to Genencor International, Inc.).
88. FRG Pat. DD 298 424 (February 20, 1992), J. Kirchberger, G. Kopperschläger, M. Rockstroh, and K. Eisenbrandt (to Universität Leipzig, Germany).
89. R.A. Hart, P.M. Lester, H. Riefsnyder, J.R. Ogez, and S.E. Builder, BioTechnology 12: 1113–1117 (1994).
90. U.S. Pat. 5,328,841 (July 12, 1994), J.D. Lorch, K.A. Clarkson, E. Larenas, B.S. Bower, and G.L. Weiss (to Genencor International, Inc.).
91. E.U. Pat. 0 477 284 B1 (August 16, 1995), H.G. Heinssohn and K.J. Hayenga (to Smithkline Biologicals S.A.).
92. U.S. Pat. 5,407,579 (April 18, 1995) C.-J. Lee and P. Khan (to National Science Council, Taipei).
93. H. Schütte, K.H. Kroner, W. Hummel, and M.-R. Kula, Ann. N.Y. Acad. Sci. 413: 270–282 (1983).
94. K.H. Kroner, H. Schütte, W. Stach, and M.-R. Kula, J. Chem. Technol. Biotechnol. 32: 130–137 (1982).
95. K.H. Kroner, H. Hustedt, and M.-R. Kula, Process Biochem. 19: 170–179 (1984).
96. T.R. Pulliam, S. Winston, and W.E. Bentley, Enzyme Microb. Technol. 20: 46–51 (1997).
97. H. Hellebust, A. Veide, and S. Enfors, J. Biotechnol. 7: 185–198 (1988).

37

PROTEIN ULTRAFILTRATION

ROBERT VAN REIS
Genentech, Inc., South San Francisco, California
ANDREW L. ZYDNEY
Department of Chemical Engineering, The Pennsylvania State University, University Park, Pennsylvania

37.1 INTRODUCTION

Although protein ultrafiltration (UF) has been used in the laboratory since the early 1900s, large-scale industrial applications of UF only date back to about 1970. The key breakthrough was the development of the asymmetric membrane by Loeb and Sourirajan and the subsequent extension of this technique to the production of UF membranes by Alan Michaels and colleagues at Amicon. These asymmetric membranes have a very thin skin (approximately 0.5 μm thick), which provides the membrane its selectivity, and a more macroporous substructure, which provides the required mechanical and structural integrity (Fig. 37.1). The thin skin results in much higher permeation rates than obtainable with homogeneous membranes, significantly reducing the required membrane area and/or process time. Typical process fluxes in UF range from 20 to 150 L/(m^2h). Operating pressures are generally in the range of 0.2–4 bar, which are much lower than 25–80 bar required in reverse osmosis (RO) to overcome the large osmotic pressure of the retained salts. UF membranes are available with a variety of pore sizes and can be cast from a range of base polymers, each having unique physical and chemical characteristics. Electrically charged membranes, typically generated by posttreatment modification of the base polymer, provide opportunities for enhanced separation performance by exploiting electrostatic interactions between the charged protein and the membrane pores.

Most large-scale UF devices use tangential flow filtration (TFF), also referred to as a cross-flow configuration, in which the feed flow is parallel to the membrane and thus perpendicular to the filtrate flow (1,2). This allows retained species to be swept along the membrane surface and out the device exit, significantly increasing the process flux compared to that obtained with dead-end operation. A concentration boundary layer consisting of a high concentration of retained solutes typically develops at the upstream surface of the membrane (Fig. 37.2). This highly concentrated region reduces the effective pressure driving force and can cause membrane fouling due to protein adsorption, denaturation, precipitation, and/or aggregation. The performance of many TFF systems is determined almost entirely by the rate at which these retained solutes are transported away from the membrane and back into the bulk solution, a phenomenon referred to as concentration polarization (1). A variety of membrane modules have been developed to achieve the desired mass transport rates while maintaining high membrane packing densities, and these are discussed in some detail in "Modules and Devices." Most units are produced in modular form to allow easy replacement and scale-up. Membrane devices are available with surface areas as small as 50 cm^2 for bench-top operation to more than 300 m^2 for large-scale processing. Dead-end (or normal flow) filtration is used primarily for laboratory-scale separations and for systems in which the retained species are present at very low concentration. For example, dead-end microfiltration (MF) cartridges have

Downstream Industrial Biotechnology: Recovery and Purification, First Edition. Edited by Michael C. Flickinger.

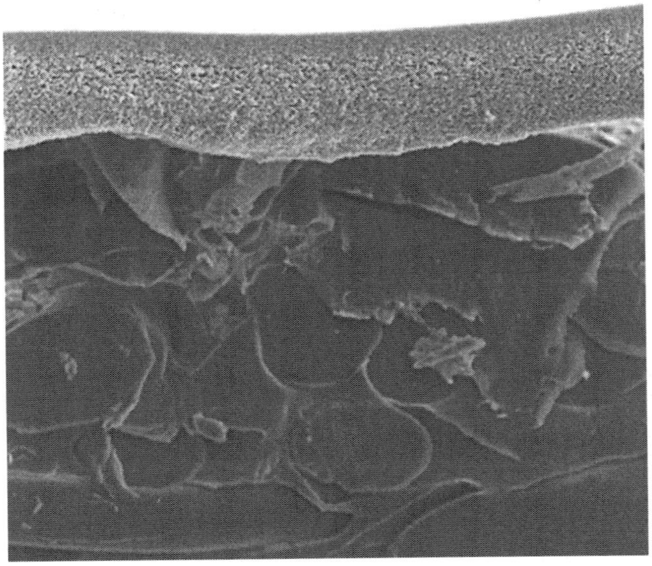

Figure 37.1. Scanning electron micrograph of the cross-section of an asymmetric regenerated cellulose ultrafiltration membrane cast on a nonwoven polypropylene support. Membrane skin is located at very top of image.

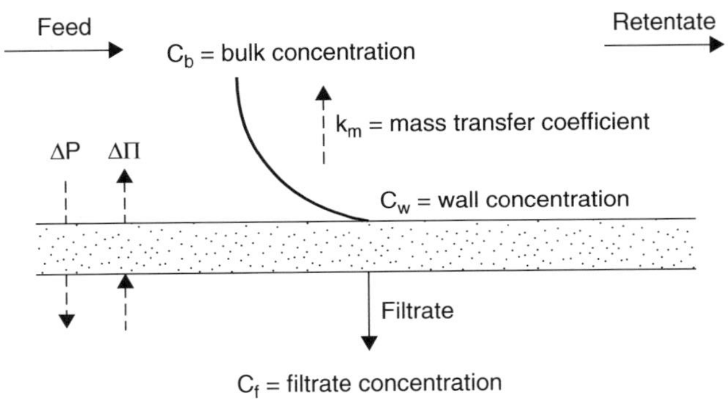

Figure 37.2. Schematic drawing of concentration polarization. The concentration of a retained solute at the membrane wall (C_w) is considerably larger than the bulk concentration (C_b). Adapted from van Reis *et al*. (3).

been used extensively for sterile filtration, in which the retained species are present at very low concentration. Similar modules can be employed for virus removal.

Conventional UF has traditionally been used to separate solutes that differ by more than 10-fold in size, making it ideal for protein concentration and buffer exchange (4). UF processes are thus used throughout the downstream purification, often to condition the product prior to, or immediately after, a chromatographic column. In addition, UF is the primary method for final product formulation to achieve the desired final concentration and buffer composition. The fairly limited selectivity of UF has generally been attributed to the wide pore size distributions in commercial membranes, the presence of significant bulk mass transfer limitations in existing modules, and membrane fouling phenomena. Over the past decade, it has become clear that membrane systems are capable of effecting actual protein–protein separations, even for proteins with very similar molecular weight, using a process known as *high performance tangential flow filtration* (HPTFF) (5). High selectivity values are obtained in HPTFF by choice of both membrane and buffer conditions, with the latter chosen to maximize the difference in hydrodynamic volume for the two proteins. Concentration polarization is exploited to enhance, rather than limit, solute separation. Membrane fouling is controlled with membrane chemistry and by using appropriate fluid dynamic and mass transfer conditions. High yields and purification factors are obtained using a diafiltration mode. The net result is that HPTFF provides a competitive separation process that complements chromatographic techniques for protein purification.

37.2 THEORETICAL PRINCIPLES

37.2.1 Membrane Transport

Theoretical analyses of solute and solvent transport through UF membranes are generally developed using either the Kedem–Katchalsky or Stefan–Maxwell descriptions of irreversible thermodynamics (6). The filtrate flux J (equal to the volumetric filtration rate normalized by the membrane surface area) is proportional to the effective pressure driving force (7):

$$J = L_p \left[\Delta P - \sum_i \sigma_i \Delta \Pi_i \right] \quad (37.1)$$

where L_p is the overall permeability (equal to the reciprocal of the overall resistance) and ΔP is the transmembrane pressure drop. L_p is only equal to the clean membrane permeability (evaluated from water flux data obtained with the clean membrane) in the absence of irreversible fouling. The osmotic reflection coefficient (σ_i) provides a measure of the membrane selectivity: $\sigma_i = 1$ for a solute that is completely retained and $\sigma_i = 0$ for a solute which is completely permeable. The osmotic pressure difference across the membrane ($\Delta\Pi_i = \Pi_w - \Pi_f$) is a function of the solute concentrations at the upstream surface (C_w) and in the filtrate solution (C_f) (Fig. 37.2). Although protein osmotic pressures are typically small for $C < 10$ g/L, $\Delta\Pi_i$ can be comparable to ΔP during UF due to the build-up of a concentrated region of retained protein at the membrane surface (see section titled "Concentration Polarization").

The solute flux N_s (in units of grams protein collected in the filtrate per membrane surface area per time) is conveniently expressed as (8):

$$N_s = J S_\infty C_w \quad (37.2)$$

where S_∞ is the intrinsic (or asymptotic) sieving coefficient, which is approximately equal to $1-\sigma_i$ based on Onsager reciprocity. Equation 37.2 is valid only under conditions where the solute flux is dominated by convection; a more detailed analysis accounting for the coupling between convection and diffusion is given by Opong and Zydney (8). The asymptotic sieving coefficient is determined by thermodynamic partitioning of the solute into the membrane and the relative rate of solute transport through the pore (9):

$$S_\infty = \frac{C_f}{C_w} = \phi K_c \quad (37.3)$$

where K_c is the hydrodynamic hindrance factor, C_f is the filtrate concentration, and C_w is the concentration in the retentate solution immediately adjacent to the membrane. The hydrodynamic hindrance factor accounts for the additional drag on the solute due to the presence of the pore wall. Theoretical analyses for a hard sphere in a cylindrical pore indicate that K_c only varies between 1.0 and 1.47 (10), thus S_∞ is determined primarily by the equilibrium partition coefficient (ϕ), which is a function of both steric (size) and long-range (e.g. electrostatic) interactions. The electrostatic energy of interaction between a protein and a pore has three distinct contributions (11). In UF systems, the dominant factor is typically the charge on the solute (σ_s). The partition coefficient decreases with increasing σ_s due to the increase in free energy associated with the distortion of the electrical double layer surrounding the protein caused by the presence of the pore wall. This can be interpreted as an increase in effective protein size (or hydrodynamic volume) associated with the diffuse ion cloud (12). This effect is the basis for the 100-fold reduction in S_∞ with decreasing salt concentration seen for bovine serum albumin (13). There are also contributions from the double layer surrounding the pore wall (which reduces the effective pore size) and from direct charge–charge interactions, with the latter causing an increase in ϕ when the solute and pore have opposite charge. These direct charge–charge interactions are particularly important when using charge-modified membranes for enhanced UF or HPTFF.

In the absence of long-range interactions, the sieving coefficient for a spherical solute in a cylindrical pore is (9):

$$S_\infty = (1 - \lambda)^2 K_c \quad (37.4)$$

where λ is the ratio of the solute (r_s) to pore (r_p) radius. Thus, even a membrane with a single pore size has a gradual decrease in S_∞ due to the reduction in accessible pore volume with increasing r_s. The retention characteristics of commercial UF membranes are more complex due to the broad distribution of irregularly shaped pores. A simple approximate analysis yields (14):

$$S_\infty = \exp\left(-r_s/s\right) \quad (37.5)$$

where s provides a measure of the mean pore size. In this case, S_∞ approaches zero asymptotically due to the presence of a small number of very large pores. The effects of different pore size distributions on S_∞ are discussed elsewhere (14).

37.2.2 Concentration Polarization

One of the critical factors determining the overall performance of any UF device is the rate of protein transport in the bulk solution adjacent to the membrane. The filtrate flow causes an accumulation of partially (or completely) retained solute at the upstream (feed-side) surface of the membrane, a phenomenon referred to as concentration polarization (Fig. 37.2). The solute concentration thus varies from its maximum value at the membrane surface (C_w) to its bulk value (C_b) over the thickness of the concentration boundary layer (δ). Most analyses of concentration polarization have employed the simple stagnant film model originally presented by Michaels (15):

$$J = k_m \ln\left[\frac{C_w - C_f}{C_b - C_f}\right] \quad (37.6)$$

where the solute mass transfer coefficient (k_m) is equal to the ratio of the solute diffusivity (D) to the boundary layer thickness (δ). More rigorous analyses of solute transport indicate that Equation 37.6 is valid at low degrees of concentration polarization, with the actual concentration dependence becoming more complex at high C_w/C_b (16). The wall concentration increases with increasing filtrate flux, causing an increase in $\Delta\Pi$ and a reduction in the effective pressure driving force (Eq. 37.1). The osmotic pressure increases very rapidly at high wall concentrations (approaching the solubility limit of the protein), with the net result that the flux approaches a nearly constant, pressure-independent value at high ΔP (Fig. 37.3). High wall concentrations can also reduce L_p through irreversible fouling or the formation of a protein "gel" on the membrane surface. In the latter case, the pressure-independent flux is determined by the maximum solubility or gel concentration for the protein. Although this "gel polarization" model has been used quite extensively, the calculated values of C_w are device-dependent and are often in poor agreement with independent estimates of protein solubility (10), suggesting that an actual "gel"

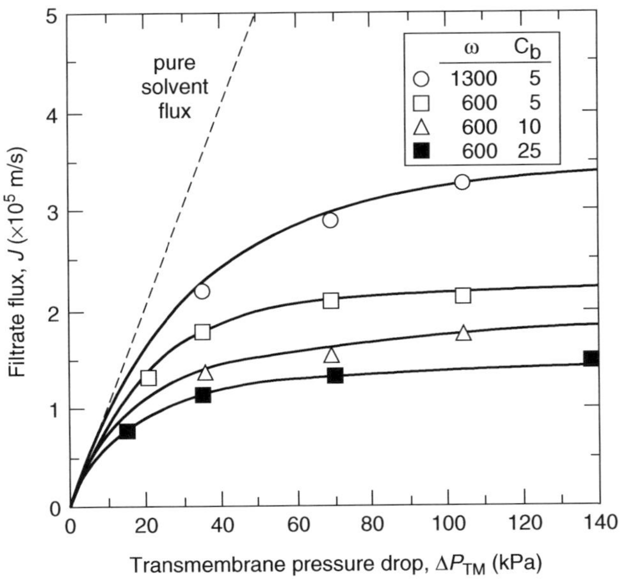

Figure 37.3. Filtrate flux as a function of transmembrane pressure for stirred cell UF of bovine serum albumin solutions of different bulk concentrations (C_b) using a constant stirrer speed (ω). Solid curves are model calculations based on Equations 37.1 and 37.6. Figure adapted from Opong and Zydney (8).

does not form on the membrane surface during UF. More detailed models of UF that account for electrostatic, van der Waals, and entropic interactions in the concentration polarization boundary layer have also been developed (17).

Theoretical analyses of solute mass transfer coefficients in laminar flow are generally based on the Leveque solution (1):

$$k_m = 0.807\left(\frac{D^2\gamma}{L}\right)^{1/3} \tag{37.7}$$

which predicts a 2/3 power dependence on the protein diffusivity (D) and a 1/3 power dependence on shear rate (γ). The protein diffusion coefficient in turn depends on the protein charge (pI and pH), buffer conductivity, and protein concentration. These effects can be quite pronounced, with the mass transfer coefficient varying by as much as a factor of 2 in different buffers due to the differences in protein charge and intermolecular electrostatic interactions. The diffusion coefficient is also proportional to the absolute temperature and inversely proportional to the viscosity, which itself is a strong function of temperature. This can result in large increases in mass transfer coefficient and hence filtrate flux at higher operating temperatures.

Empirical correlations for k_m have been developed from pressure drop data using the Chilton–Colburn analogy with a power law dependence on feed flow rate around 0.5 for spacer-filled channels and near 1.0 for turbulent flow (10).

The most accurate method for evaluating the mass transfer coefficient, particularly for modules with complex fluid flow characteristics, is to use actual filtrate flux data in the module of interest. The simplest approach is to evaluate k_m from the slope of a plot of the "pressure-independent" flux as a function of $\ln(C_b)$ using Equation 37.6 (assuming a fully retentive membrane so that $C_f = 0$). This requires minimal experimental data, and this approach has been used successfully in the recent implementation of C_w control strategies for monoclonal antibodies. The mass transfer coefficient can also be evaluated by fitting filtrate flux versus ΔP data to Eqs 37.1 and 37.6 simultaneously, which was the approach used to generate the solid curves in Fig. 37.3. This requires an estimate of the process membrane permeability, which can be evaluated from buffer flux data obtained with the fouled membrane, and an expression for the protein osmotic pressure as a function of C_w. It is also possible to evaluate both k_m and $\Delta\Pi$ simultaneously by analyzing J versus ΔP data obtained at different C_b (3).

The accumulation of retained solute at the membrane also effects protein retention. Equation 37.6 can be rearranged to evaluate the observed protein sieving coefficient (S) using $N_s = JC_f$, yielding (8):

$$S = \frac{C_f}{C_b} = \frac{S_\infty}{(1 - S_\infty)\exp(-J/k_m) + S_\infty} \tag{37.8}$$

At low filtrate flux, $S = S_\infty$ since C_w is approximately equal to C_b. The observed sieving coefficient increases with increasing flux, approaching a value of one (i.e. $C_f = C_b$) at very high flux as long as the membrane is at least partially permeable to the solute and the required wall concentration ($C_w = C_b/S_\infty$) is physically attainable. Typical experimental results for bovine serum albumin are shown in Fig. 37.4. The solid curve is the model calculation using Equation 37.8. The slight increase in S at very low filtrate flux is due to the diffusive contribution to protein transport through the membrane (8).

37.3 MEMBRANE MATERIALS, CHARACTERIZATION, AND FOULING

37.3.1 Membrane Chemistry and Morphology

Almost all UF membranes have an anisotropic or asymmetric structure with the membrane selectivity determined by a thin skin layer and the mechanical integrity provided by a much thicker macroporous support (Fig. 37.1). Most asymmetric UF membranes are prepared by immersion casting (10). A thin film of a polymer solution is spread on a suitable support (e.g. a nonwoven fabric). The cast film is then immersed in a nonsolvent (often water). Diffusional exchange occurs between the solvent and

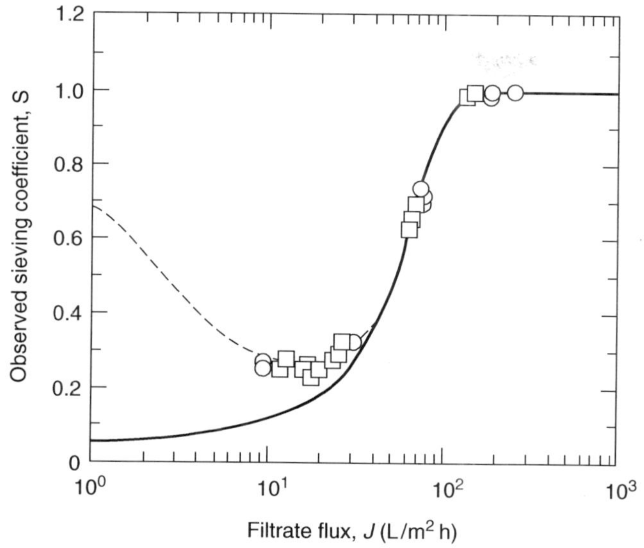

Figure 37.4. Observed sieving coefficient as a function of filtrate flux for UF of BSA through a 100 kDa polyethersulfone membrane. Solid curve is model calculation based on Equation 37.8. Dashed curve accounts for diffusive contribution. Figure adapted from Opong and Zydney (8).

nonsolvent, causing phase separation of the polymer solution and subsequent gelation of the polymer-rich phase. This gives the membrane a nodular morphology with the pore size determined by the size and spacing of the polymer-rich globules. The asymmetric structure develops because of mass transfer (kinetic) effects, e.g. growth and coalescence of polymer-rich globules prior to gelation (10). Large tear-shaped macrovoids often form beneath the membrane skin due to either interfacial instabilities or the growth of polymer-lean regions in the film (18). Membranes with macrovoids have lower maximum pressure limits. More importantly, defects can form if the macrovoids extend up to (or through) the skin. Hollow fiber membranes are also made by phase inversion with the polymer solution extruded through an orifice into a gelation bath while a second gelation medium is pumped into the fiber lumen. Skin layer(s) develop on the inner and/or outer surface depending on the properties of the gelation media. Composite membranes are produced using multiple casting steps. For example, very uniform UF membranes that are free of macrovoids can be produced by casting a thin (5–10 μm) polymer film on an MF substrate (19). These membranes are particularly attractive for virus filtration (VF) where the presence of even trace defects leads to poor virus removal. In addition, composite regenerated cellulose membranes have been developed that have excellent mechanical strength and cause very little fouling, providing higher flux and better retention characteristics for protein UF than other membrane chemistries (20).

UF membranes can be cast from a wide range of polymers including polysulfone, polyethersulfone, polyamides, polyimides, polyvinylchloride, polyacrylonitrile, polyvinylidene fluoride, regenerated cellulose, and cellulose acetate. Regenerated cellulose membranes have the lowest protein-binding but can be damaged by harsh cleaning protocols. The synthetic polymers have much greater chemical and thermal stability but are more susceptible to protein fouling. Ceramic membranes, produced by slip-casting of Al_2O_3 or acid leaching of SiO_2 (21), have very high thermal and chemical stability but tend to be considerably more expensive than polymeric membranes. Many polymeric membranes are modified during or after casting, for example, by inclusion of polyvinylpyrrolidone in the initial polymer solution, sulfonation of cast polysulfone, radiation-grafting of methacrylate monomers to polyamides, or coating with hydroxypropyl cellulose (22). These modifications increase membrane hydrophilicity and alter membrane surface charge, but they can reduce long-term chemical and thermal stability. Electrically charged membranes have very high retention of like-charged proteins, with significantly higher permeability at an equivalent protein retention (23). These membranes can be used for enhanced UF and HPTFF processes when the conductivity of the buffer solution is below 10 mS/cm.

37.3.2 Membrane Characterization

Information on UF membrane characterization most commonly provided to end users consists of water flux measurements, solute retention tests, selected protein binding data, chemical compatibility, and thermal limits. Water flux data provide a measure of the membrane permeability as determined by evaluating the filtrate flux (filtration rate normalized for surface area) as a function of the transmembrane pressure (ΔP):

$$\Delta P = \frac{P_{\text{F}} + P_{\text{R}}}{2} - P_{\text{fil}} \qquad (37.9)$$

where P_{F} is the feed pressure, P_R is the retentate pressure, and P_{fil} is the filtrate pressure. Normal flow filtration devices yield a linear relationship between flux (J) and transmembrane pressure as long as membrane compaction is avoided. Compaction of commercial UF membranes is usually negligible at pressures below 1 bar. Membranes can also be preconditioned at high pressure for a short period of time, which then minimizes any subsequent changes in permeability. The situation is more complex in tangential flow filtration (TFF) since the conversion of feed flow to filtrate flow creates a nonlinear pressure drop through the feed channel. A correct value for the permeability of a membrane in a TFF device can be obtained by measuring flux versus transmembrane pressure and extrapolating

$J/\Delta P$ versus ΔP to $\Delta P = 0$. Membrane permeability is primarily determined by the pore size distribution, porosity (pore density), membrane thickness, and solvent viscosity.

Several issues should be considered when judging membrane permeability. The type and quality of water used for the tests can have a significant influence on the results. The pH and ionic strength of the water can influence the flux due to charge interactions, and contaminants can foul the membrane during the test. It is also important to control and report the temperature of the test since permeability is linear with viscosity. As previously mentioned, membrane compaction and device configuration can also play a significant role in these tests. It should be noted that membrane water permeability data are of minimal value unless used in conjunction with other membrane characterization information. Water permeability data are not predictive of process fluxes due to significant differences in fouling behavior of various membrane chemistries. Polysulfone membranes, for example, typically have high water permeability values, but regenerated cellulose membranes typically have higher process permeability values due to their inherent low fouling characteristics with most biotechnology feed streams (20).

It is important to consider the intended use of membranes when selecting a retention test. The process objectives, and hence the criteria for membrane selection, are very different for protein concentration, buffer exchange, virus–protein separations, clarification, or protein purification. Retention testing of UF membranes involves measurements of solute transmission. Sieving coefficients (S) or retention coefficients (R) are determined by analyzing the solute in the feed and filtrate streams under specific process conditions.

$$S = C_{\mathrm{f}}/C_{\mathrm{b}} \tag{37.10}$$

$$R = 1 - S \tag{37.11}$$

Tests are carried out using either single solutes or multiple solutes in separate experiments. The most commonly used single solutes are narrow molecular weight distribution polymers (such as polyethylene oxides and dextran polysaccharides), proteins, and viruses. The purpose of the retention test is to provide nominal molecular weight cutoff values for each membrane. Single-solute retention tests are the easiest to perform and provide either the least or most useful data depending on the chosen solute and test conditions. Single-solute tests performed by membrane manufacturers typically involve the use of one or several experiments with solutes that are partially retained by the membrane. This information is of limited value for protein concentration and buffer exchange applications since the retention coefficients of partially retained solutes are not predictive of the retention values one would obtain for a highly retained solute. The retention of solutes with $R = 0.3$ and $R = 0.9$ for example do not provide guidance on the molecular weight of a solute with a retention coefficient of $R = 0.999$, which is what one would strive for in a protein concentration/diafiltration operation. In all applications, retention coefficients will also be highly dependent on fouling, buffer chemistry, and fluid dynamics. Furthermore, there is a lack of standardization of retention testing protocols between manufacturers. Different solutes, buffers, and fluid dynamic conditions are used. Zydney and Xenopoulos (24) have recently shown that device format and flux conditions can have a large effect on the measured dextran retention values. In addition, the protocols for assigning nominal molecular weight designations (e.g. the molecular weight of a solute that has a retention coefficient of $R = 0.9$) are not even standardized within a given company. It is therefore imperative that the end users perform their own retention testing on the solutes of interest under conditions that are a linear scale down of the intended process.

The inherent trade-off between the hydraulic permeability and protein retention, using bovine serum albumin as a model protein, is shown in Fig. 37.5 (23). The y-axis is simply the reciprocal of the protein sieving coefficient, which provides a measure of the selectivity of the membrane between a very small solute (sieving coefficient equal to one) and the protein of interest. The solid curve is a theoretical calculation accounting for purely size-based exclusion assuming a log-normal pore size distribution. Most of the experimental data cluster along this theoretical curve, defining the "upper bound" for currently available commercial UF membranes. The data points that lie well below the curve are primarily for track-etch membranes that have a lower porosity and greater thickness than the asymmetric UF membranes. The best combination of performance characteristics for a given application is determined by the specific process requirements and economics. The symbols

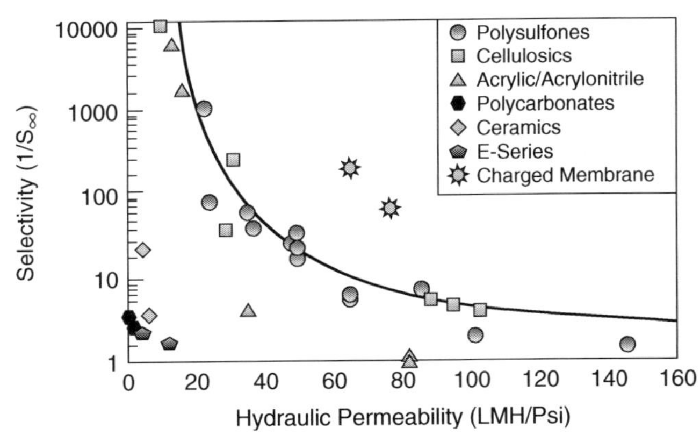

Figure 37.5. Selectivity-permeability trade-off for UF membranes using BSA as a model protein. The selectivity is defined as the reciprocal of the protein sieving coefficient. The solid curve is a model calculation accounting for steric interactions assuming a log-normal pore size distribution. Adapted from (23). (This figure is available in full color at http://onlinelibrary.wiley.com/book/10.1002/9780470054581.)

lying up and to the right of the upper bound represent data obtained with prototype negatively charged composite regenerated cellulose membranes. These membranes have much better performance characteristics (high permeability and high selectivity) than any of the commercially available membranes due to the strong electrostatic exclusion of the negatively charged protein.

The mixed dextran test (25) provides an alternative methodology for membrane characterization. A mixture of several polydisperse dextrans are used with the feed and filtrate streams analyzed using size exclusion chromatography (SEC) with an on-line refractive index (RI) detector. The SEC chromatograms represent the solute concentrations (RI absorbance) as a function of dextran molecular weight (elution time). A continuous retention curve (retention versus molecular weight) can hence be generated by constructing a curve using Eqs 37.10 and 11 with the feed and filtrate concentrations represented by the respective chromatograms. Because the SEC chromatograms provide elution times for every molecular weight fraction, it is not necessary to use dextrans with tight molecular weight distributions. Mixed dextran tests performed under conditions of constant mass transfer coefficient and flux are a useful tool in comparing the relative retention characteristics of various UF membranes. The method is a good quality control test for both membrane manufacturers and end users. The method can also be used to derive information on pore size distributions (26). Some predictive power can also be derived from this methodology by converting both dextran and protein molecular weights to equivalent hydrodynamic volumes (10).

Several enhancements to the mixed dextran test are incorporated into the so-called Flex Test (27). The sensitivity of the test is increased by about 10-fold by using fluorescent-tagged dextrans. The conventional dextran test is typically used to specify a retention at $R = 0.9$ ("R90") or $S = 0.1$, and sensitivities much above $S = 0.01$ require very careful attention to experimental detail ($S = 0.01$ corresponds to only 90% yield at $N = 10$ during a diafiltration process as discussed later in this chapter). In contrast, the Flex Test can easily provide sieving measurements down to $S = 0.001$ (corresponding to 99% yield at $N = 10$) due to the greater sensitivity of the fluorescent detection. In addition, the Flex Test has the option of using fluorescent-tagged charged dextrans to capture the effect of membrane charge on sieving. Chemical characteristics (charge and hydrophobicity) should be taken into account when selecting the fluorescent tag. For charged membranes, the sieving of a charged solute is usually measured with fluorescent-tagged dextrans (high sensitivity for low sieving values) whereas the neutral dextrans may not require a fluorescent tag since they likely represent passing markers. The resolving power of different membranes can be compared by evaluating the selectivity (ψ) and throughput ($J\Delta S$) parameters from data for the fluorescent charged and neutral dextrans (see section titled "High Performance Tangential Flow Filtration").

A more sensitive method of characterizing UF membranes involves the principle of liquid–liquid pore intrusion. Two highly immiscible liquids such as solutions of a sulfate salt and a polyethyleneglycol are contacted through mixing to reach equilibrium partitioning. The membrane to be tested is primed with one of the liquids so that all the pores are filled. After draining the feed channels, the second fluid is introduced into the system. The first fluid is then displaced out of the pores by the second fluid and the flow rate is measured as a function of transmembrane pressure. The resulting data provides information on pore size distribution and can be correlated with the nominal molecular weight cutoff. It has also been demonstrated that this methodology can be used as an excellent quality control test for membranes used in HPTFF applications since performance can be correlated with protein sieving data (28).

Protein binding studies are often performed to contrast the relative behavior of various membrane chemistries. The data generated in these studies can be useful as a general guide. It is important, however, to recognize that protein adsorption will be highly dependent on pH, ionic strength, and the specific isoelectric point and hydrophobicity of a particular protein. In addition, simply soaking the membrane in a protein solution can give misleading results since considerable adsorption can occur in the membrane support structure, a region that would be largely inaccessible to a retained protein during UF. It is also useful to note that most membranes will not cause significant protein losses due to adsorption because of the low binding capacities (1–10 mg/m^2), the low surface areas (0.01 m^2/g), and the high protein concentrations (1–150 g/L) obtained in modern biotechnology. This is even true for processes such as MF and VF, in which the product may be exposed to the internal surface area that is 10 to 500 times the frontal membrane area.

Chemical and thermal limits for membranes are determined by characterizing the membranes (using the techniques described herein) both before and after controlled exposure. Membrane manufacturers often perform additional tests to determine pore size distribution, porosity, morphology, thickness, charge, process flux, and mechanical strength (compaction, tensile strength, and delamination). Pore size distribution and porosity can be determined by mixed dextran, liquid–liquid intrusion, mercury intrusion, air–liquid intrusion (for large pore sizes), thermoporometry, scanning electron microscopy (SEM), or atomic force microscopy (AFM). Membrane morphology and thickness are determined by either SEM or transmission electron microscopy (TEM). Membrane charge can be determined by streaming potential measurements and is of particular use in HPTFF (20). Process fluxes provide some complimentary information to the water permeability values provided by manufacturers but are best carried out by the end user on the specific application of interest. Determination of mechanical strength should include testing of membrane compaction, tensile strength, and ability to withstand reverse pressure. Soft membranes are susceptible to membrane compaction under process conditions. Membrane compaction can easily be determined by measuring water permeability as a function of transmembrane pressure throughout the desirable ΔP range. Tensile strength is mostly relevant to the membrane manufacturers and is analyzed using specialized equipment that measures the elongation of the membrane as a function of applied force. The reverse pressure limits of membranes are important to the end user as unintended pressure spikes may occur during processing with resultant delamination of the membrane and subsequent product loss. Testing is done by subjecting the membrane to reverse pressure cycles at various pressures followed by retention testing as outlined previously.

Integrity testing of UF membranes can be performed by bubble point measurement, air diffusion flow measurement or using the liquid–liquid intrusion method just described. Bubble point determinations involve measuring a gas flow rate through a wetted membrane at increasing levels of pressure. The bubble point is determined by an inflection in the flow rate versus pressure curve indicative of liquid displacement out of the pores. Bubble point measurements are primarily used for MF membranes that have bubble point values in the 1–6 bar range (for 0.65 to 0.1 μm pore sizes) using water as solvent. Bubble point measurements of UF membranes are only practical when using much lower surface tension solvents since the bubble point values using water as a solvent often exceed the pressure ratings of either the membrane or device. UF membrane integrity is therefore commonly performed using air (or nitrogen) diffusion methods. Air diffusion measurements are done at a single pressure and strive to detect abnormally high flow rates due to convective flow through membrane defects that are large enough to permit the bubble point to be exceeded. The measured air diffusion flow rate is compared to the range specified by the manufacturer, which is determined from a population of membranes that have passed the appropriate quality control tests.

37.3.3 Membrane Fouling and Regeneration

Fouling refers to the irreversible alteration in membrane properties caused by specific physical and/or chemical interactions between the membrane and the various components present in the feed. Fouling can occur because of protein adsorption, deposition, precipitation, denaturation, and aggregation, all of which affect membrane performance. The rate and extent of fouling are determined by the morphological and chemical characteristics of the membrane, the fluid mechanics in the module, and the buffer chemistry (29). In addition, protein aggregation or denaturation can occur during repeated pump passes, with these species then deposited on the membrane during filtration. Static protein adsorption (in the absence of filtration) generally attains monolayer values on the accessible membrane surface (30), with significantly less than monolayer adsorption on hydrophilic membranes. As discussed previously, adsorptive losses are typically very small (<0.01% for products in the retentate and can usually be limited to <5% even for products in the filtrate that have access to the internal pore surface).

The convective filtrate flow during UF can cause additional fouling depending upon the membrane pore size and surface chemistry. Fouling increases with increasing filtrate flux (or transmembrane pressure) and can be almost entirely eliminated by operating below the critical flux (31). The initial fouling during MF is often due to physical deposition of large protein aggregates on the membrane surface (32), but this is much less important in UF due to the much smaller pore size and the fact that the feed is typically prefiltered at some point prior to the UF step. Protein fouling is generally maximum near the protein isoelectric point (33) and is enhanced under conditions where the protein and membrane have opposite charge. The extent of fouling is also a function of start-up conditions, with less fouling seen when the system is slowly ramped to the desired filtrate flux or transmembrane pressure (5). Composite regenerated cellulose membranes typically show little fouling during protein UF.

Extensive experience has demonstrated that protein aggregation in UF systems is primarily due to protein–oxygen interfaces caused by microcavitation in pumps and valves (34). Nitrogen-sparged solutions and nitrogen overlay virtually eliminate protein aggregation. Nitrogen overlay can also be used to control retentate pressure instead of a retentate control valve but operator safety must be considered (oxygen deprivation when opening large tanks with gas overlay). In practice, the amount of insoluble protein aggregates generated during UF is insignificant with respect to process yield. Turbidity of ultrafiltered protein solutions may still be high due to the dependence of scattered light on particle size. Light scattering intensity is proportional to r^3 while the dynamic light scattering signal is proportional to the r^6 where r is the particle radius. Fortunately, sterile filters provide an excellent means for removal of insoluble protein aggregates. While the amount of insoluble aggregates represents an insignificant amount of the total protein mass, these feed streams do require a fair amount of sterile filter surface area for processing. Experience with a large number of monoclonal antibodies demonstrates that generation of soluble multimers in UF is generally not an issue, with the measured multimer content before and after UF often within the accuracy of the assay (fractions of a percent).

Membrane cleaning or regeneration is performed whenever the flux and/or selectivity drop below some minimally acceptable level. In addition, membrane disinfection/sterilization must be performed between process runs even in the absence of significant fouling. Cleaning can be accomplished by physically removing the foulants from the membrane, for example, by fluid recirculation at zero transmembrane pressure, by backflushing with negative ΔP, or by mechanical scrubbing (generally limited to large diameter tubular modules). Chemical cleaning is done using appropriate detergents, acids, alkalis, enzymes, and/or chelating compounds. These cleaners must effectively displace, solubilize, and/or chemically modify the foulants while not exceeding the mechanical, thermal, or chemical limits of the membrane polymer or module. For example, cellulosic membranes are highly susceptible to damage by extremes of pH and temperature (35). They are, however, easy to clean with 0.1 *N* NaOH at moderate temperature, often providing 100% recovery of water permeability over multiple cycles. Polysulfone (PS) and polyethersulfone (PES) are much more stable, with some manufacturers reporting acceptable performance even after limited exposure to temperatures as high as 125°C (36). These membranes can also be operated over a wide range of pH (1–13), and can tolerate low levels of chlorine (up to 200 ppm). Alkali cleaners (e.g. NaOH) are particularly effective for removing biological foulants (37). Many alkali cleaners are used in combination with chelating agents (e.g. citrate or EDTA) to remove free Ca^{2+} and Mg^{2+} to prevent precipitation of saponified fats and oils (10). Detergents like sodium dodecyl sulfate (SDS) are also effective in many systems. Enzymatic cleaners can be used for membranes that are unable to withstand elevated temperatures, strong chemicals, or pH extremes, but these solutions tend to be very expensive and care must be taken to insure complete removal of residual proteases. Membrane cleaning remains very much an art, with the development of an optimal cleaning cycle for a given application determined largely by trial and error.

Sodium hypochlorite (NaOCl) is used extensively for chemical disinfection of membrane systems, and it is also a highly effective cleaning agent (typically used in combination with NaOH). Hypochlorite is most effective at low pH, but it is very corrosive under these conditions (38) and care must be taken to avoid generation of toxic chlorine gas. Disinfection is typically performed at pH≈10 using fairly concentrated solutions (0.25–4% NaOCl). Other oxidizing agents used for membrane disinfection include hydrogen peroxide (H_2O_2) and peracetic acid ($CH_3COCOOH$) (36). Peracetic acid is attractive since it is fast-acting, and it decomposes into the nontoxic acetic acid. Sodium bisulfite ($NaHSO_3$), a strong reducing agent, can be used for disinfection of membrane systems that would be degraded by strong oxidizing agents like hypochlorite (39).

Steam sterilization can be performed on disassembled membrane systems in an autoclave, but the membrane unit must then be aseptically attached to the remainder of the presterilized system (40). Steam-in-place (SIP) is the preferred method of sterilization, with the entire unit exposed to flowing steam as part of the completely assembled filtration system. Minimum requirements for effective steam sterilization are 15 min exposure to steam at 121°C and 1 bar pressure. Effective validation of SIP procedures is essential (41). Membrane systems can also be sterilized using ethylene oxide (EtO) or chlorofluorocarbons (CFC), but these gas sterilization procedures require fairly long incubation periods to insure complete deactivation of all microorganisms and spores (41). Gamma irradiation can also be used, although some membrane materials (e.g. PTFE and cellulosics) are damaged upon exposure to high radiation doses (41).

37.4 MODULES AND DEVICES

The membrane module is the physical unit that houses the UF membrane in an appropriately designed configuration. The module must provide physical separation of the retentate and filtrate streams, mechanical support for the membrane, high membrane packing densities, easy access for cleaning and/or replacement, and scalability. In addition, the feed channel must provide adequate bulk mass transfer rates with minimum operating (typically pumping) costs. Many of these criteria are to a large extent contradictory, for example, devices with large membrane packing densities are generally more susceptible to particulate fouling and more difficult to clean. A number of commercial UF modules have been developed, differing primarily in the size and shape of the feed (and filtrate) flow channels. These are conveniently classified into five distinct types: flat sheet cassettes, spiral wound modules, hollow fiber cartridges, tubular, and enhanced mass transfer modules. The latter category includes a wide range of system geometries and operating strategies. Some of the important characteristics of these modules are summarized in Table 37.1; more detailed descriptions of the economics, fluid mechanics, and mass transfer characteristics of these modules are available in Zeman and Zydney (10).

37.4.1 Flat Sheet Cassettes

Flat sheet (also known as plate-and-frame) cassettes were among the earliest configurations developed for large-scale commercial applications. The module uses a support plate

TABLE 37.1. Comparison of Different Module Configurations

Module Configuration	Channel Spacing (cm)	Packing Density (m^2/m^3)	Energy Costs (pumping)	Particulate Plugging	Ease of Cleaning
Hollow fiber	0.02–0.25	1200	Low	High	Fair
Tubular	1.0–2.5	60	High	Low	Excellent
Flat sheet	0.03–0.25	400	Moderate	Moderate	Good
Spiral wound	0.03–0.1	600	Low	Very high	Poor–fair

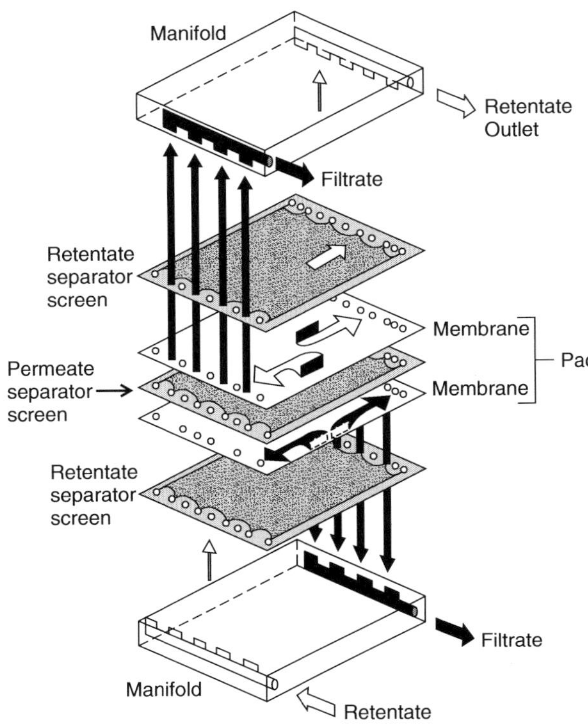

Figure 37.6. Schematic drawing of flat sheet cassette. Adapted from (10).

(which also defines the filtrate flow path), membrane, and channel spacer (which defines the retentate flow path) in a sandwich configuration (Fig. 37.6). The membranes can be sealed to the plates using gaskets, in which case the module (stack) is hydraulically clamped to form a tight fit. Alternatively, the membranes can be directly bonded or glued to the plates using heat sealing or an appropriate adhesive to form an integral membrane element. Several of these elements/plates are then stacked together and clamped to form a complete module. The feed flow is distributed among the different channels at one end of the device, and the retentate is collected at the opposite end. The filtrate is manifolded separately from the feed stream and can be combined within or outside the module depending upon the particular device design. The feed channel is typically 0.03 to 0.1 cm in height. The channel can either be fully open or it can employ an appropriate screen (typically a polypropylene mesh) to promote local mixing and improve mass transfer. The small channel height leads to relatively high membrane packing densities (generally 300–500 m^{-1}) and low hold-up volumes, which is particularly attractive for recovery of high value products. Open channel cassettes are less susceptible to plugging and operate with smaller pressure drops (feed minus retentate pressure). However, backflushing may not be possible since the membranes are effectively supported only on one side.

37.4.2 Spiral Cartridges

Spiral wound cartridges are constructed using flat sheet membranes in the form of a pocket consisting of two membrane sheets separated by a highly porous permeable mesh, which provides physical support for the membrane and defines the filtrate flow path (Fig. 37.7). The membranes are sealed along three edges to form a pocket using an appropriate epoxy or polyurethane adhesive. The open side of the pocket is attached (glued) to a central perforated tube, which collects the filtrate flow. Several pockets are wound in a spiral configuration around a single collecting tube using a feed-side (polypropylene) mesh as a spacer between the pockets to establish the desired feed channel thickness. The use of multiple pockets reduces the total path length and pressure drop for the filtrate flow. The entire spiral is generally wrapped with fiberglass tape or shrink wrap for added mechanical strength and to prevent unwrapping of the spiral under pressure. The spiral wound element is fitted into a tubular steel or plastic pressure vessel with an antitelescoping device placed at both ends to prevent distortion of the spiral during operation. This also allows several spiral elements to be inserted into a single tube. Pressurized feed solution flows through the plastic screens along the surface of the membranes with the retentate collected out the other end of the device. Filtrate flows into the closed membrane pockets and flows in a spiral path inward to the central tube. The region between the outer edge of the spiral element and the inside of the cylindrical housing is sealed to avoid fluid bypass. Spiral cartridges that direct the feed flow in the spiral direction and the filtrate flow in the axial direction have also been produced. Spiral wound cartridges have relatively high membrane packing densities and very effective mass transfer characteristics due to the feed-side spacers. The cartridges also have fairly low energy costs

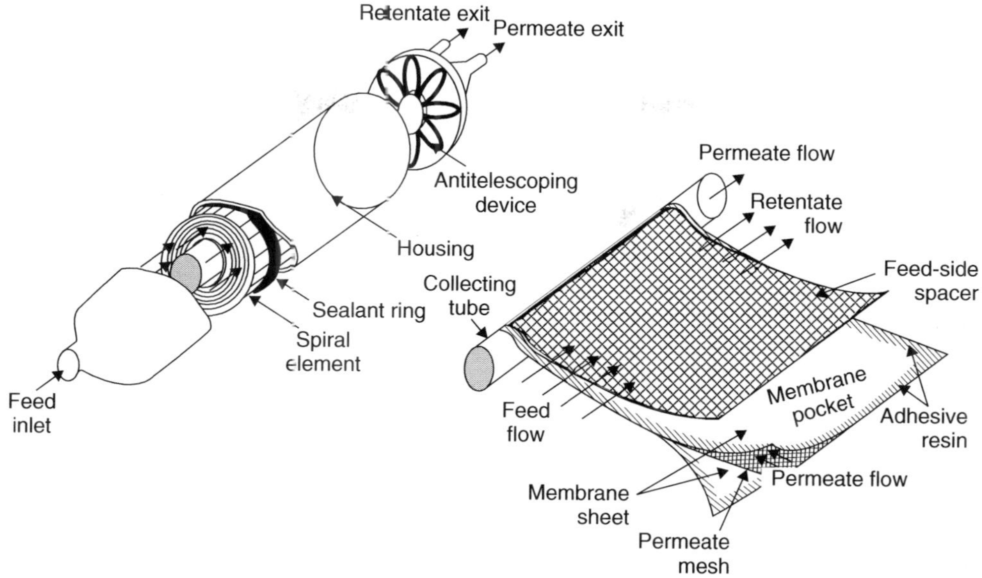

Figure 37.7. Schematic drawing of spiral wound module showing membrane pocket. Adapted from (10).

since they provide effective mass transfer at relatively low flow rates. The primary disadvantage of the spiral wound modules is that they are highly susceptible to particulate fouling due to the narrow and irregular flow path through the spacers. In addition, the dead space between the outer edge of the element and the cylindrical housing is difficult to clean and highly susceptible to bacterial contamination. This is sometimes overcome by allowing a very limited feed bypass to continuously flush the annular space. There can also be problems with the integrity and chemical resistance of the seals used to form the pockets, particularly for repeated applications using strong cleaning solutions.

37.4.3 Hollow Fiber Cartridges

Hollow fiber cartridges use an array of narrow bore (self-supporting) fibers typically with an anisotropic structure. Fiber diameters range from about 200 to 2500 μm, with the fiber wall around 200 μm in thickness The dense skin layer is typically at the lumen-side of the fiber, although it can also be placed on the outside. A parallel array of 50–10,000 fibers is potted at the ends in an epoxy or polyurethane resin within a cylindrical cartridge of plastic or steel to form a tube which is machined to expose the open bores (lumens) of the fibers (Fig. 37.8). These modules are generally operated with the feed flow into the fiber lumens, with the filtrate moving in a radial direction outward through the fiber walls. The cartridge is typically equipped with a single feed inlet, a retentate outlet, and two filtrate ports (one near each end of the unit). The feed and filtrate flows can be co-current or counter-current. Hollow fiber devices have high packing density due to the narrow diameter of the individual fibers and the ability to pack the fiber bundles tightly in the cylindrical cartridge (shell). The use of small diameter fibers can provide effective mass transfer at relatively low pumping costs. The absence of any spacers or multiple sealing procedures reduces labor costs involved in assembly. The hollow fiber membranes are self-supporting, which often allows them to be cleaned by backflushing of the device either by reversing the direction of the filtrate flow or by reversing the feed flow with the filtrate ports closed. One disadvantage of the hollow fiber modules is that they are susceptible to particulate plugging. Fiber breakage can also be a problem, thus it is critical to follow the manufacturer's recommendations with respect to operating pressures. Fiber breakage can often be difficult to detect, and several manufacturers recommend using a modified bubble-point test to detect ruptured fibers.

37.4.4 Tubular Modules

Tubular modules are very similar in design to the hollow fiber cartridges described in the last section but employ much larger diameter tubes (typically 0.3 to 2.5 cm). The tubular membranes are not self-supporting (with the exception of some inorganic membranes). The membranes are usually cast in place within a porous support tube made of fiberglass, ceramic, plastic, or stainless steel. The individual tubes can be placed inside a plastic or stainless steel sleeve to form a single tube cartridge or they can be packed together in small bundles, which are kept in place using an appropriate end plate (Fig. 37.9). The feed flows through the bore of the tubes, while the filtrate flows in a radial

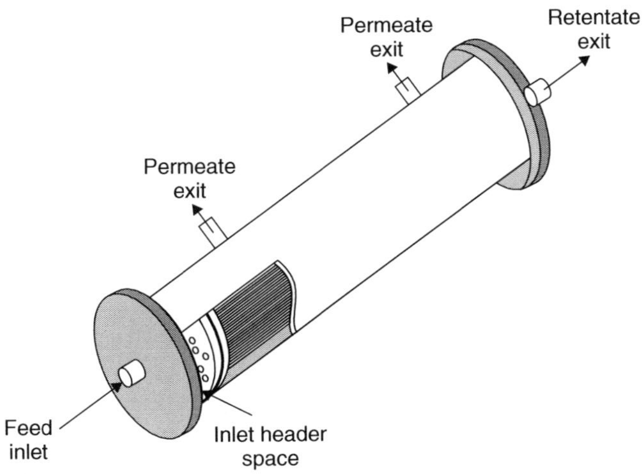

Figure 37.8. Schematic drawing of hollow fiber module. Adapted from (10).

direction outward across the membrane and support tube where it is collected from the filtrate outlet ports. The two streams may be co-current or counter-current. Inorganic membranes are typically made in a honeycomb monolith (21) in which the membranes are arranged in a parallel array. The unit is housed in an appropriate plastic or stainless steel cartridge. Tubular modules are typically operated in the turbulent flow regime to obtain effective mass transfer in the large diameter tubes. The primary advantages of the tubular modules are their resistance to particulate plugging and ease of chemical/mechanical cleaning. Physical cleaning of the membranes can be effected by passage of a rubber sponge, a scouring ball, or a narrow rod through the bore of the tubes. It is also possible to replace individual tubes in the multielement units in the field, resulting in a considerable reduction in membrane replacement costs and an increase in the overall unit lifetime. The primary disadvantages of the tubular modules are the low membrane packing densities ($<100\ m^{-1}$), high hold-up volumes, large feed flow rates, and high capital costs (including high floor space requirements).

37.4.5 Enhanced Mass Transfer Devices

Enhanced mass transfer modules exploit flow instabilities or turbulence to increase the transport of retained solutes. A variety of devices have been designed using pulsatile flow or membranes with pronounced surface corrugations (42). Alternatively, systems have been developed in which the feed flow generates secondary flow patterns or vortices. The most successful of these to date has been the rotating cylinder device originally proposed by Hallstrom and Lopez-Leiva (43). The feed flows into a thin (0.05–0.10 cm) annular region between two concentric cylinders, one (or both) of which is porous and has a membrane bound to the surface. The filtrate is collected in the central chamber (for a device with membrane on the inner cylinder) or in a separate annular filtrate region adjacent to the outer porous cylinder. The inner cylinder in these devices is rotated at high speed (typically >3000 rpm) to induce Taylor vortices, which are caused by the centrifugal forces acting on the rotating fluid as it moves around the cylindrical flow path. The vortices effectively mix the fluid near the surface of the membrane, increasing the feed-side mass transfer coefficients. Mass transfer in the rotating system is determined almost entirely by the rotation rate of the inner cylinder, which effectively de-couples the mass transfer characteristics from the feed flow rate. Consequently, rotating devices can be operated at very low flow rates. The use of low flow rates can be particularly attractive for many high viscosity applications. However, the rotating systems have very large capital costs, are difficult to scale-up, and there is concern about long-time performance of a device with so many moving parts. In addition, membrane packing densities are low, and it is difficult to develop and maintain effective seals in such a rotating system.

Belfort and co-workers (44) and Winzeler (45) independently proposed using a vortex-flow system based on Dean, rather than Taylor, vortices. Dean vortices are induced by

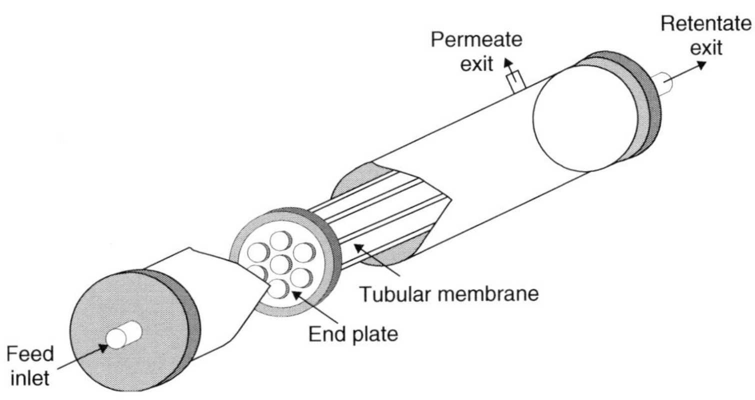

Figure 37.9. Schematic drawing of tubular module. Adapted from (10).

centrifugal forces exerted on a fluid as it flows inside a highly curved channel. This can be accomplished using spiral wound flat sheet membranes, with the channel curvature adjusted so as to maintain the Dean vortices throughout the system, or with hollow fiber membranes wound around a central core at an appropriate pitch in a screw-like geometry (46). Both systems generate counter-rotating vortices that effectively depolarize the concentration boundary layer, but without the need for any moving parts or seals. Available data on the Dean-vortex systems are very encouraging (44). These units have not yet been commercialized, so it is impossible to evaluate their economics and overall performance characteristics, and the total membrane area per unit volume is below that of conventional hollow fiber modules.

Another approach to achieving high mass transfer is to move the membrane, with the bulk of the fluid remaining largely stationary. For example, Vigo *et al*. (47) developed a module in which a cylindrical membrane is vibrated (longitudinally) inside a concentric cylindrical housing. Culkin and Armando (48) developed vibratory shear enhanced processing (VSEP) in which a stack of parallel circular membrane disks is moved back-and-forth at high speed inside a concentric cylindrical housing using a torsional spring. Shear rates at the membrane-feed interface can be as high as 150,000 s^{-1} for a 60 Hz oscillation. The high shear rates obtained in these systems can result in dramatic improvements in filtrate flux; however, important questions remain about the scalability and robustness of these modules as well as the overall economics.

High frequency transmembrane pressure pulsing can also be used to enhance mass transfer. In contrast to classical backflushing, in which the reverse filtration occurs after several minutes or hours of normal filtration, transmembrane pressure pulsing is performed with a frequency around 0.1–1 Hz (49). The backpressure pulse is applied using nitrogen or air pressurization, with the frequency and duration of the pulse set using computer-controlled solenoid valves (49). Very short forward filtration times (<1 min) are used to prevent significant concentration polarization or fouling, thereby maintaining high average filtrate flux. Data for filtration of albumin–immunoglobulin mixtures showed more than an order of magnitude increase in flux (50), but a corresponding reduction in protein retention. Although transmembrane pressure pulsing systems appear to have excellent performance characteristics, they have not yet been employed in large scale commercial applications. Gas sparging has also been used to increase mass transfer and filtrate flux in UF (51,52), although there are concerns that the extensive gas–liquid interface in these systems can increase protein aggregation and denaturation.

37.5 EQUIPMENT

Large-scale UF systems used in biotechnology were originally based on equipment used in the food and dairy industries. This equipment posed several significant limitations. Sanitary design was not suitable for processing of human pharmaceuticals, cleaning procedures were cumbersome, and the systems lacked the ability for linear scaling (see next section). Polysulfone membranes are mechanically robust but easily fouled and difficult to clean. Cellulose membranes are low fouling and easy to clean but prone to delamination with costly yield losses and plant down-time. All of these issues were resolved by (i) developing mechanically robust composite regenerated cellulose membranes, (ii) introducing simple and effective sanitizing (peroxyacetic acid) and storage solutions (0.1 N NaOH), and (iii) developing linear scale-down and scale-up versions of the modules (20). Housing design was converted from a horizontal to a vertical design, alignment rods were introduced to avoid offsets between cassettes, internal bores were increased to match processing flow rates, and ports were redesigned to adapt to industrial piping designs (20).

The advent of high dose products (>100 mg/mL protein concentrations) with annual metric ton production requirements has led to the development of larger housings (up to 4 inch internal bore with a 6 high holder design and 10 cassettes per side), larger systems (6-inch pipe and 2800 ft^2 surface area) and low hold-up volume designs (total minimum operating volumes as low as 66 ml/ft^2). A low hold-up volume design was achieved using several new design concepts (20). The recycle tank was re-designed so that it consists of a cylinder and cone with the height of the cylinder and the height of the cone equal to the radius of the cylinder. The outlet of the tank is at the bottom of the cone as in conventional tank design but the retentate tube enters through the bottom of the cone, instead of the traditional dip-tube design which normally enters through either the top or side of the cylinder, thereby minimizing hold-up volume. The retentate tube has a tee outlet instead of straight pipe. This aids in good mixing within the tank without short-circuiting back into the tank outlet and without air entrainment at low operating volumes. The mixer location, angle, and rotation (up-pumping instead of down-pumping) were optimized using pilot and full-scale mixing studies to ensure good mixing during concentration and diafiltration. The mixer is not used during final concentration. Feed and retentate piping were minimized by close-coupling all components in a vertical design. The tank outlet valve was connected directly to the feed pump via a 90° elbow (to facilitate pump maintenance). The pump was connected directly to a flow meter, cross (pressure switch and sample valve) and isolation valve mounted directly on the top of the cassette holder. Conventional designs have the feed entering through the bottom of the holder, which adds a significant amount of pipe. Priming of the vertical pipe design was verified, and air entrainment was eliminated, using buffer prior to introduction of protein into the system. Pressure transducers for feed, retentate, and filtrate were located directly on the holder. The need for long pipe lengths before and after the flow meter were eliminated by recognizing that the application only required accuracy at 50–100% of the flow rate range. The cassette holder was used as part of the recycle loop and the retentate line included only the holder isolation valve, product transfer/buffer addition valve, tank isolation valve with integrity test nitrogen valve, and a short pipe section to match the height of the feed line components. Upper and lower filtrate valves were used for process and drain phases, respectively. A product recovery valve was located on the bottom of the holder further eliminating pipe (normally inserted into recycle loop) and improving product recovery. Product recovery is accomplished by buffer displacement in a completely vertical design with lower density buffer chasing higher density protein solution. This new system has one-third less hold-up volume, with the most significant reductions in the tank and piping.

37.6 PROCESS CONFIGURATIONS

37.6.1 Batch

UF and HPTFF processes consist of various combinations of concentration and diafiltration steps. A concentration step is either the main objective of the process or may be used as a means to optimize a subsequent diafiltration step (see section titled "Design Equations"). Concentration steps are carried out in either batch or fed batch mode. In a batch mode, the entire volume of feed is contained within the recycle tank as illustrated in Fig. 37.10. Batch mode operations use a minimum of hardware, provide simple manual or automatic control, and provide the highest flux rates. While the initial and final retentate concentrations are the same regardless of process configuration, the batch operation gives higher flux rates than a fed batch operation since the bulk concentration follows a more dilute path from initial to final concentration (10). The disadvantages of a batch configuration can include less flexibility in using the same system for multiple processes, increased difficulty in designing a well-mixed system, and difficulties in obtaining high concentration factors. Mixing is important both to keep the bulk concentration at a minimum and for efficient buffer exchange during diafiltration. High concentration factors can in some cases be obtained in a batch mode using tanks with two or more sections of progressively decreasing diameters.

37.6.2 Fed Batch

A fed batch configuration utilizes an additional tank to feed into the recycle tank as shown in Fig. 37.11. Fed batch configurations are commonly used in industry to obtain high concentration factors, provide flexibility for use in multiple processes, and to provide well-mixed low hold-up retentate reservoirs for efficient diafiltration. The volume of the recycle tank is usually chosen based on the need to obtain good mixing in the final product solution. In choosing a fed batch configuration several trade-offs must be considered. For a given membrane area, process time will increase with the use of a fed batch configuration due to the increase in bulk concentration in the feed to the module compared to a batch operation. In addition, the number of times the retained solute passes through pumps and valves in the recycle line will increase both due to the longer processing time and due to the smaller retentate volume. An increased number of pump/valve passes can lead to shear induced cell lysis and protein denaturation and aggregation. Protein denaturation

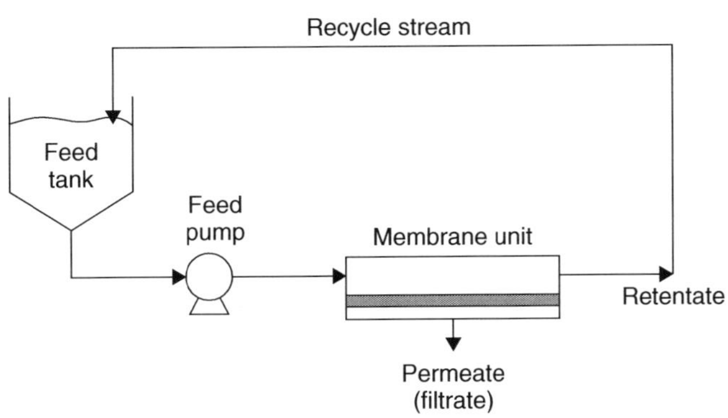

Figure 37.10. Batch-mode UF for a retained product.

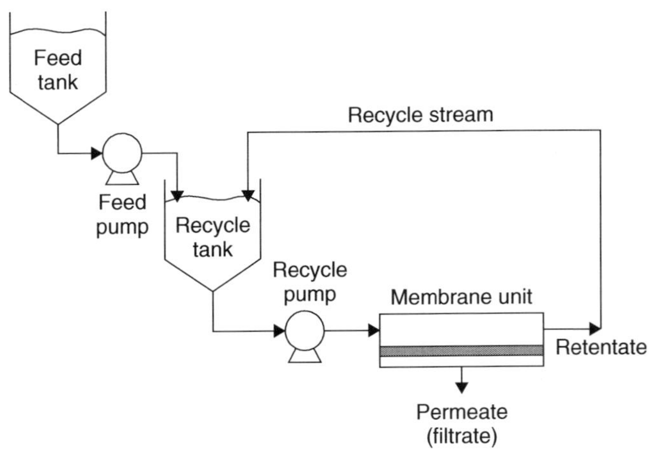

Figure 37.11. Fed batch mode UF for a retained product.

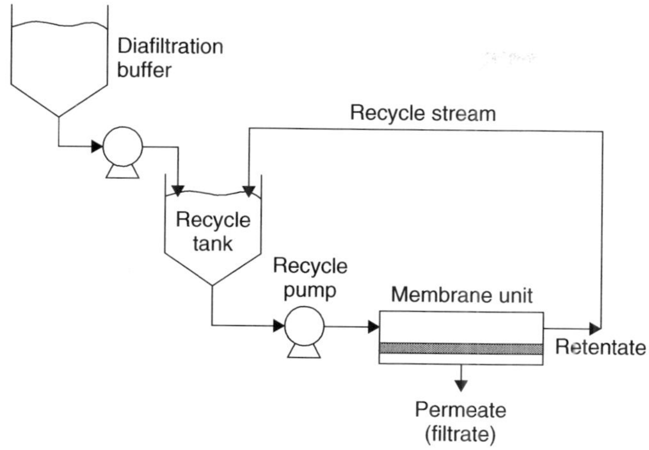

Figure 37.12. Diafiltration mode UF for a retained product.

in pumps and valves is believed to be due to protein–gas interfaces formed as a result of microcavitation (34).

37.6.3 Diafiltration

Diafiltration is an operation in which a solvent or solution (typically a buffer) is added to the recycle tank simultaneously with filtration (Fig. 37.12). The most common form is constant retentate volume diafiltration in which the buffer is added at the same rate as the filtrate rate. In UF processes, diafiltration is used to perform buffer exchange if the product is in the retentate, or to enhance yields if the product is in the filtrate. In HPTFF, diafiltration is used to achieve purification. In certain processes, it can be advantageous to perform simultaneous dilution or concentration with diafiltration (53). Optimization of diafiltration processes will be described in "Fluid Dynamics and Process Control" for UF and in "Process Optimization" for HPTFF.

37.6.4 Cascade Systems

In several plant designs, it may be advantageous to employ cascaded process configurations. These configurations include both open loop and closed loop designs. In both configurations, the filtrate from the first stage is used as a feed to the second stage. In the open loop design, the filtrate from the second stage is directed to drain while a replacement buffer is added to the first stage. In the closed loop design (Fig. 37.13), the filtrate from the second stage is used as the diafiltration buffer for the first stage, thereby significantly reducing the amount of buffer required for the process. Cascade systems can be used to couple multiple process steps into a single operation. For example, a microfiltration (MF) step used to harvest protein from a culture can be cascaded with a second stage UF step to achieve simultaneous harvest and protein concentration. In HPTFF, two-stage cascade systems can

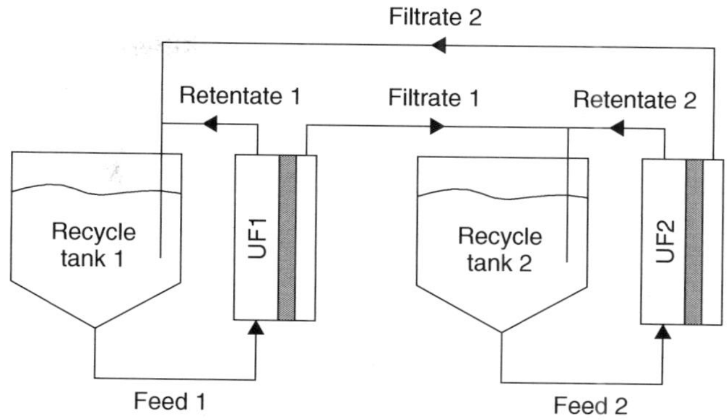

Figure 37.13. Two-stage closed-loop cascade configuration.

be used in a similar manner when the product of interest is in the first stage filtrate. Once a normal diafiltration has been performed on the first stage to achieve optimum buffer conditions, a closed loop configuration can be used to obtain a very large number of diavolumes by reusing a volume of buffer that is equal to the volume of retentate. It should be noted that the retentate volumes in the first and second stage tanks do not have to be the same. In cases where the buffer cost is very high one might also consider using such a configuration even when the product is in the first stage retentate if economically justified. Cascade configurations with more than two stages can also be employed in processes where it is desirable to separate a feed stream into more than two fractions. For example, Cheang and Zydney (54) used a two-stage process to recover both α-lactalbumin and β-lactoglobulin from whey.

37.7 PROCESS DESIGN—ULTRAFILTRATION

37.7.1 Buffer Chemistry

Buffer exchange is often used to condition a product for subsequent loading onto a chromatographic column or for final formulation. The buffer chemistries employed in UF are hence often dictated by the preceding and subsequent steps in the process. Due to the very large amounts of protein processed per surface area of membrane, UF can be used with a wide range of ionic strengths and pH values without significant adsorptive product loss. As discussed previously, adsorptive yield losses are usually insignificant (<0.01%) for products in the retentate and can be limited to less than 5% for products in the filtrate. While UF is routinely used to provide buffer exchange in preparation for subsequent processing, it can be advantageous to consider selecting buffer conditions for the sake of optimizing the UF step. Changes in pH and ionic strength can be made to effect mass transfer, protein solubility, stability, adsorption, and hydrodynamic volume. These parameters are directly linked to yield losses incurred at the membrane surface (see section titled "Fluid Dynamics and Process Control"). In addition, the mass transfer coefficient can also be significantly enhanced by changes in buffer composition resulting in enhanced fluxes, reduced membrane area, and minimal volumetric hold-up losses. In general, higher fluxes are obtained at pH values either above or below the protein isoelectric point and at low to moderate ionic strength.

The theoretical removal of low molecular weight components during diafiltration is discussed in "Design Equations." It is important to note that one cannot assume that the sieving coefficients for these components are unity despite the fact that the membrane may have a molecular weight cutoff an order of magnitude larger than the molecular weights of the components. Deviations in theoretical clearance of low molecular weight components can be due to poor mixing in the recycle tank but can also be due to buffer conditions. The use of detergents in either the feed material or in the diafiltration buffer can cause incomplete diafiltration if the critical micelle concentration is exceeded. Buffer components are entrained in the micelles, which may retained by the membrane depending on micelle size and membrane molecular weight cutoff. It is generally advisable to avoid the use of detergents in UF for these reasons. Charged membranes may cause Donnan exclusion effects that result in increased retention of ionic solutes of all sizes. It has also been observed that low molecular weight ionic solutes can have increased clearance, possibly as a result of Donnan exclusion by the retained (charged) solutes.

37.7.2 Membrane Selection

Membrane selection should start with the choice of a high quality vendor since robustness, reliability, and reproducibility of biotechnology manufacturing operations are of paramount importance. Consistent membrane and device characteristics can be as important to product quality, yield, and economics as the inherent differences between various membranes and devices. The most important engineering parameters to consider during membrane selection are product retention, process permeability, mechanical strength, and chemical compatibility. Adsorptive properties play an important role in determining process permeability but are often less important with regard to product loss. Membrane chemistry should therefore be chosen in conjunction with the choice of buffer chemistry. Membrane charge should also be considered with regard to potential Donnan exclusion effects as discussed previously. Water permeability values should not be used as a selection criterion since they are not predictive of process performance. Pore size distributions, porosity, morphology, and membrane thickness all affect the ultimate retention, permeability, and mechanical characteristics of the membrane. In addition, water permeability in cassettes is typically below that measured for membrane coupons due to loss of membrane area associated with the glues, seals, and spacers used in the cassettes. As previously discussed (in section titled "Membrane Characterization"), retention properties should be determined on the actual solute under process conditions since marker studies can be completely misleading. Rules of thumb for selection of a membrane molecular weight cutoff relative to a protein molecular weight should also be avoided since protein hydrodynamic volumes vary drastically with pH and ionic strength (12) and since membrane manufacturers have not standardized their membrane characterization and nomenclature for UF membranes. The mechanical strength of a membrane can be important both with regard to membrane compaction and reverse pressure tolerance. Membrane compaction can influence process permeability if retentate pressure limits are exceeded. The reverse pressure limits of a membrane are also important since reverse pressure spikes are very difficult to avoid in industrial UF systems. The resulting membrane delamination can have a catastrophic impact on process yields. Chemical compatibility needs to be considered for both the intended feed stream as well as for the regeneration chemicals. Polysulfone and modified polysulfone membranes tend to have much broader chemical compatibility than regenerated cellulose membranes but also require much harsher chemical conditions for regeneration due to their inherent fouling characteristics. Long-term stability of membranes in bacteriostatic storage solution must also be considered for economic reasons. The long-term stability of UF membranes should be evaluated by studying product retention as a function of membrane storage time in the chosen storage solution.

37.7.3 Fluid Dynamics and Process Control

UF processes can been performed using constant transmembrane pressure (ΔP), constant flux (J), or constant retentate wall concentration (C_w). The primary benefit of constant transmembrane pressure UF processes is the inherent simplicity of control, especially in manual systems. The feed rate is ramped up to the set point value and a retentate valve is partially closed to achieve a specified retentate or transmembrane pressure. If the feed stock viscosity changes during the process, one can adjust the retentate valve closure to maintain either constant retentate or transmembrane pressure. This is easily accomplished in both manual and automatic systems and often requires minimal adjustments after the initial set point has been achieved. The retentate pressure can also be controlled by using a gas overlay on the recycle tank. The feed rate and transmembrane pressure are empirically optimized by measuring the flux as a function of bulk solute concentration at various feed rates and transmembrane pressures. It is also important to measure solute sieving as a function of bulk concentration and flux. Even very small sieving coefficients may result in substantial product loss when a large number of diavolumes are required in diafiltration processes. For example, a sieving coefficient $S = 0.01$ will result in a 10% product loss during a 10 diavolume buffer exchange process (Eq. 37.17).

In microfiltration applications, it is not always possible to maintain constant transmembrane pressure without severe reductions in filtrate rate. It has been demonstrated that one can obtain higher overall throughput in these applications (such as clarification) by operating at constant flux (55). This can be done by regulating the retentate pressure control valve or by using a pump on the filtrate line. The latter alternative provides more precise control and is easier to use in manual systems. Optimization is typically done through empirical studies varying feed rate and flux and determining process throughput, filtrate clarity, and product sieving.

In a third method of process control, the flux is varied with both retentate concentration and mass transfer coefficient to maintain a constant wall concentration of retained species (evaluated from Equation 37.6). Control is performed using a control loop that measures flux and controls the transmembrane pressure with a retentate control valve to maintain a constant wall concentration (Fig. 37.2) throughout the concentration (3):

$$J = k_m \ln\left(\frac{C_w V}{C_o V_o}\right) \tag{37.12}$$

where J is the flux, k_m is the mass transfer coefficient, C_w is the wall concentration of the retained solute, V is the retentate volume, V_o is the initial feed volume, and C_o is the initial retentate concentration. The benefits of constant C_w control are that product yield is maximized, product quality is ensured, membrane area is minimized, and process time is consistent and independent of membrane permeability. Process control for constant C_w during a concentration phase requires an automated system. Constant retentate volume diafiltration using constant C_w, however, can easily be performed using either manual or automatic control since the C_w set point is maintained at a constant flux. Adjustments for changing mass transfer coefficients can be made during the diafiltration process. The optimum C_w is determined by minimizing yield losses attributable to sieving, adsorption, solubility, and volumetric hold-up (Fig. 37.14). The feed rate is optimized by evaluating a trade-off between increased mass transfer coefficient (which increases flux and reduces membrane area for a given C_w) and an increased number of pump/valve passes (which may increase protein denaturation). The mass

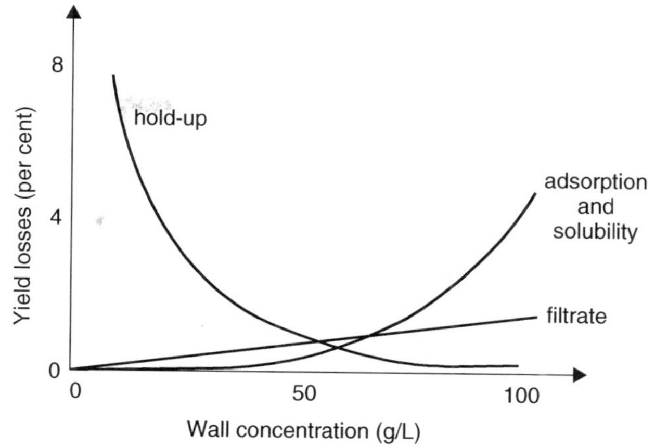

Figure 37.14. Schematic representation of yield losses as a function of the wall concentration (C_w). Hold-up losses decrease with increasing C_w due to the reduction in system size. Adapted from van Reis *et al*. (3).

transfer coefficient can be calculated as:

$$k_m = aQ^b D^{2/3} \tag{37.13}$$

where a and b are coefficients determined for the specific UF module, Q is the feed flow rate normalized for membrane area, and D is the solute diffusivity.

37.7.4 Design Equations

The yield and purification obtained in UF can be evaluated from a mass balance, typically assuming a constant sieving coefficient. The fractional retentate product yield in a batch process is:

$$Y_r = \left(\frac{V_f}{V_o}\right)^{-S} \tag{37.14}$$

where V_f is the final retentate volume, V_o is the initial retentate volume, and S is the sieving coefficient. The analogous yield for a product in the filtrate can simply be calculated as

$$Y_f = 1 - Y_r \tag{37.15}$$

For a fed batch process the retentate yield is given by

$$Y_r = \left[1 - (1 - S)\exp\left(-S\left(\frac{V_o}{V_f} - 1\right)\right)\right] - S\left(\frac{V_o}{V_f}\right) \tag{37.16}$$

and the corresponding filtrate yield can be calculated by combining Equations 37.15 and 37.16. The retentate yield during constant retentate volume diafiltration is derived as

$$Y_r = \exp(-S \cdot N) \tag{37.17}$$

where the number of diavolumes is given by:

$$N = V_D/V \tag{37.18}$$

where V_D is the diafiltration buffer volume and V is the constant retentate volume. It is often useful to combine and rearrange Equations 37.14 and 37.17 to determine the retention coefficient required to obtain a desired retentate yield for a batch UF and constant volume DF process as

$$R = 1 + \ln\left[\frac{Y_r}{\ln(V_f/V_o) + N}\right] \tag{37.19}$$

For a filtrate product the required retention coefficient is

$$R = 1 + \frac{\ln(1 - Y_r)}{\ln(V_f/V_o) + N} \tag{37.20}$$

Diafiltration processes are often used to remove undesirable low molecular weight components. The residual concentration can be calculated as

$$C = C_o \exp(-S \cdot N) \tag{37.21}$$

Buffer exchange may also involve the addition of new component:

$$C = C_i\left[1 - \exp(-N)\right] \tag{37.22}$$

where C_i is the bulk concentration of the component that is being added during diafiltration (assumed to have $S = 1$). Simultaneous addition and removal of a component may also occur in cases where the initial buffer and the diafiltration buffer have a common component. The concentration of such a component can be calculated by adding Equations 37.21 and 37.22. Process time and membrane area can be minimized by operating DF processes at an optimum bulk concentration (56):

$$C_b^* = \frac{C_w}{e} \tag{37.23}$$

By combining Equations 37.23 and 37.6, with $C_f = 0$ for a fully retentive membrane, it has been shown (3) that the optimum diafiltration flux at C_b^* can be derived as:

$$J^* = k_m \tag{37.24}$$

37.8 HIGH PERFORMANCE TANGENTIAL FLOW FILTRATION

UF has been limited to separating solutes that are at least 10-fold different in size. In contrast, HPTFF enables the separation of solutes without limit to their relative size. HPTFF can even be used to perform "reverse" separations in which the higher molecular weight solute passes through the membrane while the lower molecular weight solute is highly retained. High resolution separations are accomplished by exploiting concentration polarization, optimizing buffer chemistry and using highly selective UF membranes. The performance of HPTFF is quantified by two dimensionless numbers, Ψ and $N\Delta S$:

$$\Psi = S_2/S_1 \tag{37.25}$$

$$N\Delta S = \left(\frac{JAt}{V}\right)(S_2 - S_1) \tag{37.26}$$

where Ψ is the selectivity, S_1 and S_2 are the respective sieving coefficients for the less and more highly retained species, N is the number of diavolumes, A is the membrane area, t is the process time, and V is the retentate volume. Membrane area and process time can be chosen to satisfy economic and manufacturing criteria independent of other variables. The retentate volume can be optimized separately. It is therefore useful to initially evaluate the selectivity (Eq. 37.25) and $J\Delta S$ values which only depend on three simple experimental parameters:

$$J\Delta S = J(S_2 - S_1) \tag{37.27}$$

The following sections will describe how buffer chemistry, membrane selection, and fluid dynamics are optimized. The final section will describe the mathematical basis and use of optimization equations and diagrams.

37.8.1 Buffer Chemistry

Selecting buffer pH and ionic strength are of paramount importance to achieving optimum separation performance in HPTFF. By operating close to the pI of the lower molecular weight solute and far away from the pI of the larger molecular weight solute, one can take advantage of differences in hydrodynamic volume and charge (5). The electric double layer which surrounds charged solutes increases the effective hydrodynamic volume as determined by SEC (12). Direct charge effects can also be exploited by using a membrane that has opposite charge to the retained solute. In addition, the conformation of a molecule, and hence the hydrodynamic volume, may be changed under highly charged conditions. Utilization of

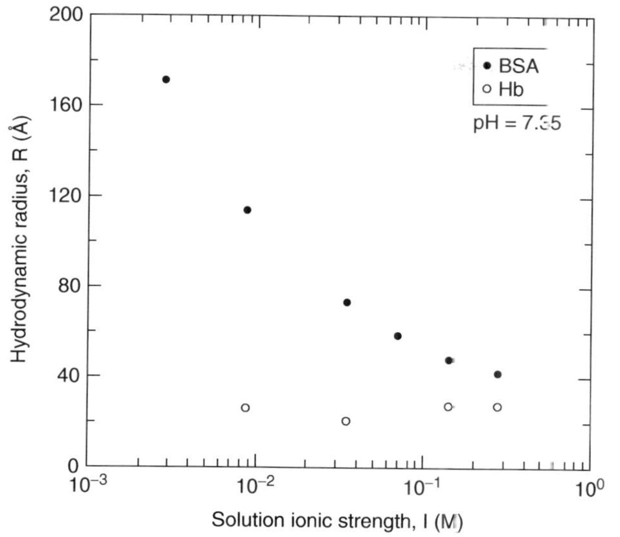

Figure 37.15. Effective hydrodynamic radius of BSA and hemoglobin (Hb) as a function of solution ionic strength. Calculated values of R were determined from retention time measurements in SEC using narrow molecular weight distribution dextran standards for calibration (12).

charge effects are optimized by operating at low ionic strength (Fig. 37.15). The optimum ionic strength is generally achieved by using the lowest possible value while still maintaining an adequate pH buffering capacity. The buffering capacity of proteins at high concentrations near the membrane wall must be taken into account. Even in cases where there is no difference in pI between the solutes, it is advantageous to optimize both pH and ionic strength since the hydrodynamic volumes can be adjusted to increase selectivity and $J\Delta S$ for a given membrane molecular weight cutoff (5). This also reduces the need to have a large selection of different molecular weight cutoff membranes for HPTFF. Hydrodynamic volume and charge effects can also be altered by using buffering ions that have different protein binding characteristics (57). Rapid process development can, however, often be accomplished by testing just a few pH values around the pI of the lower molecular weight solute with a single buffering species at a concentration of 10 mM. As in all purification process development, it is important to ensure that the buffer composition does not cause undesirable biochemical alterations to the product and that product solubility is maintained at the concentrations incurred at the membrane wall (see section titled "Fluid Dynamics and Process Control").

37.8.2 Membrane Selection

The general considerations for membrane selection in UF also apply to HPTFF. The selectivity and $J\Delta S$ that can be achieved with a membrane is determined by the pore size distribution and membrane chemistry. Pore size distribution affects both selectivity and $J\Delta S$ since the sieving coefficients depend on pore size and the filtrate flow distribution varies with the fourth power of the pore radius (10). Membrane chemistry has an indirect impact on the pore size distribution since membrane formation conditions are linked to the casting solution chemistry. Membrane chemistry can also play a significant role in fouling, which in turn influences the selectivity and $J\Delta S$ values. Direct charge effects can be exploited by using a membrane that has an electrical charge to increase the retention of all species with like polarity. Thus, a positively charged membrane will provide much greater retention of a positively charged protein than will a negatively charged or neutral membrane of the same pore size (58). Note that it may be possible to exploit electrostatic interactions even for solutes with identical pI due to the different charge pH profiles for the different species and the combined effects of protein charge and size on protein transmission through the membrane. It should also be noted that selective fouling of a membrane can actually be used to enhance separations since the pore size distribution can be reduced and the surface charge can be altered. Using these guidelines, it is often possible to narrow down the choice to one or two membranes before proceeding with fluid dynamic optimization.

37.8.3 Fluid Dynamics and Process Control

The intrinsic sieving coefficient for a solute is defined by Equation 37.3. The intrinsic sieving coefficient depends on the size and charge characteristics of both the solute and the membrane pores. Buffer optimization and membrane selection have hence focused on establishing conditions that provide the best differential intrinsic sieving coefficients for the solutes to be separated. The observed sieving coefficient of a solute, however, depends not only on the intrinsic sieving coefficient but also on the concentration of solute at the membrane wall. The wall concentration is determined by the mass transfer coefficient and the filtrate flux. The observed sieving coefficient can be related to the intrinsic sieving coefficient (S_∞), mass transfer coefficient (k_m), and flux (J) by Equation 37.8. The optimization of HPTFF is thus strongly influenced by concentration polarization effects. Concentration polarization has often been cited in the literature as one of the reasons why UF has not been successfully used for high resolution separations. It has been demonstrated (5), however, that concentration polarization can be used to enhance rather than limit the resolving power of UF membranes. Key differences used in HPTFF methodology (59,60) include operating below the transition point (Fig. 37.16) and minimizing the transmembrane pressure gradient along the length of the filtration module. While

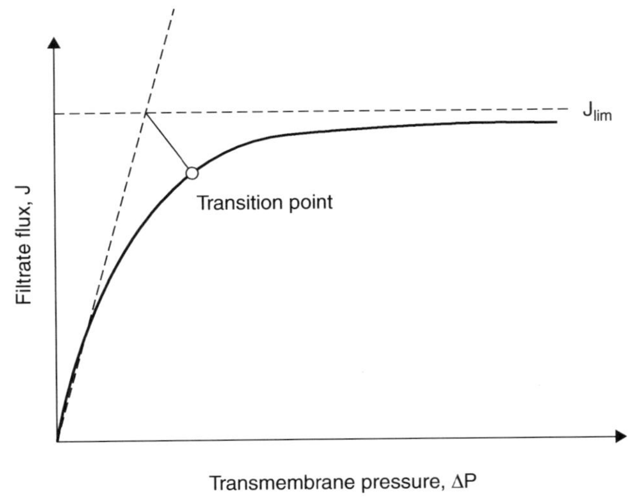

Figure 37.16. Typical flux versus transmembrane pressure curve showing the transition point flux at the inflection point (59). The optimum selectivity (Ψ) and differential mass flux ($J\Delta S$) occur below the transition point flux.

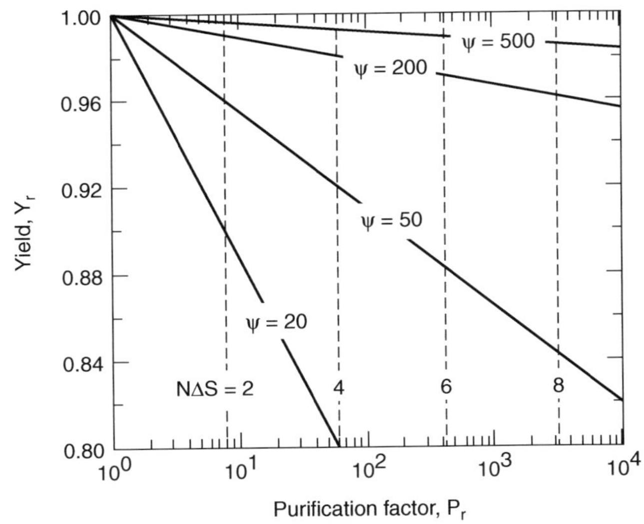

Figure 37.17. Optimization diagram for product in the retentate. The process follows a constant selectivity (Ψ) curve starting at a fractional yield $Y = 1$ and can be operated up to a maximum $N\Delta S$, which depends on the experimental $J\Delta S$ value and the ratio of membrane area (A), process time (t), and retentate volume (V). Adapted from van Reis and Saksena (61).

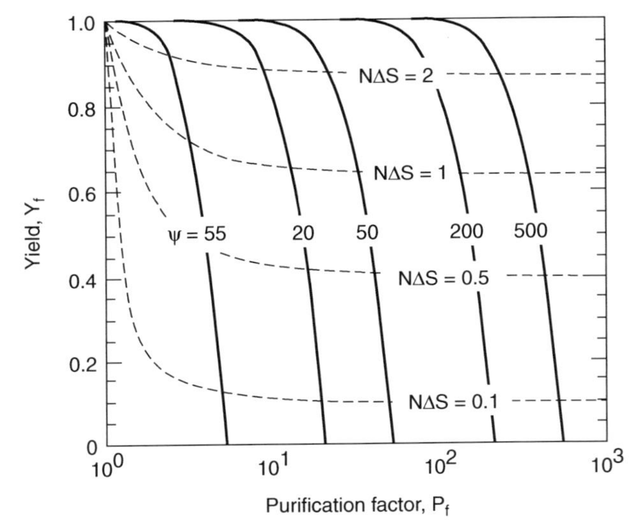

Figure 37.18. Optimization diagram for product in the filtrate. The process follows a constant selectivity (Ψ) curve starting at a fractional yield $Y = 0$ and can be operated up to a maximum $N\Delta S$, which depends on the experimental $J\Delta S$ value and the ratio of membrane area (A), process time (t), and retentate volume (V). Adapted from van Reis and Saksena (61).

conventional UF has often sought to maximize flux by operating above the transition point it has been shown that the optimum selectivity and $J\Delta S$ (a differential mass flux) occur below the transition point (5). It has also been demonstrated that membrane fouling is more consistent and can be controlled below the transition point (3). Optimization of HPTFF fluid dynamics can therefore be reduced to a set of simple experiments in which module geometry, feed flow rate and filtrate flux are investigated. The results are evaluated by measuring selectivity (Ψ) and differential mass flux ($J\Delta S$) as a function of these variables.

37.8.4 Process Optimization

The goal of process optimization is to achieve the best combination of yield and purification factor utilizing an economic amount of membrane area while providing an acceptable process time. All of these criteria can be evaluated by using an optimization diagram (61) for either retentate product (Fig. 37.17) or filtrate product (Fig. 37.18). The optimization diagrams consist of a family of curves with the dimensionless parameters selectivity and $N\Delta S$. The family of curves in the optimization diagrams are mathematical expressions which link the yield (fractional protein recovery) and purification factor (ratio of yield of desired product to yield of impurity) to the selectivity and $N\Delta S$ values. The equations are derived (61) as

$$P_r = Y_r^{1-\psi} = \exp(N\Delta S) \tag{37.28}$$

$$P_f = \frac{Y_f}{1-(1-Y_f)^{1/\psi}} = \frac{Y_f}{1+(Y_f-1)\exp(N\Delta S)} \tag{37.29}$$

where P is the purification factor and Y is the fractional yield in the retentate (r) and filtrate (f). Both retentate and filtrate product processes operate along lines of constant selectivity. In the case of a retentate product the process starts at a purification factor of 1 and a fractional yield of 1 and proceeds along a constant selectivity line up to the limit of $N\Delta S$ with an increasing purification factor and a decreasing yield. Conversely, the filtrate process starts at zero yield and a purification factor that numerically equals the selectivity (61) and proceeds along a constant selectivity curve with increasing yield and decreasing purification factor up to a limiting $N\Delta S$ value. Different experimental conditions (buffer chemistry, membrane chemistry and pore size, and fluid dynamics) can now be compared by evaluating the impact of selectivity and $J\Delta S$ on yield and purification factor. Selectivity and $J\Delta S$ often reach maxima under different experimental conditions. The best combination of selectivity and $J\Delta S$ depends on the overall process objectives and can be evaluated by comparing yield and purification factors for different combinations of these parameters using the optimization diagrams. As part of this analysis, $J\Delta S$ is converted to $N\Delta S$ by including membrane area, process time and retentate volume according to Equation 37.26. The membrane area and process times are chosen to provide acceptable economics, system hold-up volume, and process time. The retentate volume must be optimized. Decreasing the retentate volume (while maintaining a constant mass of solute) decreases the amount of buffer required for the process and can also decrease the amount of membrane area required for a given process time. This must be balanced against the increase in the number of pump and valve passes that occurs at increased concentrations, which can adversely impact the required downstream sterile filtration capacity (see section titled "Fluid Dynamics and Process Control").

HPTFF has been successfully demonstrated for the purification of both antigen binding fragments (Fab) and monoclonal antibodies (20). Data for monoclonal antibody purification from a harvested cell culture fluid show identical performance, as determined by nonreduced SDS-PAGE and ELISA for a nonaffinity monoclonal antibody purification process using cation exchange, anion exchange, and HPTFF compared to that of a process using Protein A affinity chromatography for the bulk of the purification (20).

37.9 SCALE-UP

Effective commercial implementation of UF and HPTFF requires accurate and dependable scale-up of the membrane process. It is also important to be able to effectively scale down existing manufacturing processes for further optimization, validation, and troubleshooting. Scale-up in UF and HPTFF obviously requires use of the same membrane material and pore size as well as the same module configuration and channel height at all scales of operation, with the membrane area scaled proportional to the filtrate volume. The most effective approach to scale-up of the membrane surface area is linear scaling, in which the pressure, fluid flow, and concentration profiles along the length of the filtration module are all kept constant when changing scale of operation (56). Since the local retentate velocity, solute concentration, and fluid pressure vary with position in the channel due to fluid removal and frictional pressure losses, linear scaling can only be achieved by keeping the channel length constant. Thus, the increase in membrane surface area required for large filtrate volume operation must be accomplished by increasing the number of parallel channels or fibers. It is also possible to increase the channel width in flat sheet cassettes, although some care must be taken to insure that this does not alter the fluid flow profiles in different size modules. Linear scaling of hollow fiber systems is relatively straightforward using modules with equal length but different numbers of fibers. Equal flow distribution between fibers and between fiber cartridges is easily accomplished with appropriate cartridge housings and piping manifolds. Linear scaling of flat sheet designs has also been achieved using appropriate system design to insure equivalent flow distribution, channel height compression, and entrance/exit effects at both small and large scale (62). Linear scale-up of hollow fiber and flat sheet devices over at least a 400-fold range of membrane area has been done successfully without intermediate pilot scale tests (62). Recent developments include automated manufacturing of robust third generation thermoplastic cassettes covering a range of linear scale modules from 100 cm^2 to 1 m^2 with built-in gaskets, tight tolerances for pressure drop, reduced parasitic pressure losses, and high pressure capability (up to 7 bar) (20).

Spiral wound modules have a more limited range of linear scalability since pressure losses on the filtrate side will vary as the membrane area is changed, particularly in the design of very low area (<0.1 m^2) modules. Linear scaling can also be developed for spiral devices in which the feed flow is in the spiral direction (with the filtrate flow in the axial direction), and these modules should provide enhanced mass transfer through the formation of Dean vortices. Linear scaling of the other enhanced mass transfer modules discussed in "Modules and Devices" can often be difficult to achieve over a wide range of membrane surface area due to practical limitations on system geometry and design.

NOMENCLATURE

a	Parameter in mass transfer coefficient correlation
C	Solute concentration (kg/m^3)
C_b	Solute (protein) concentration in bulk (feed) solution (kg/m^3)
C_b^*	Optimum bulk concentration for performing diafiltration (kg/m^3)

C_f	Solute (protein) concentration in filtrate solution (kg/m^3)
C_i	Concentration of solute added during diafiltration (kg/m^3)
C_o	Initial feed (retentate) concentration (kg/m^3)
C_w	Solute (protein) concentration at upstream surface of membrane (kg/m^3)
D	Solute diffusion coefficient (m^2/s)
J	Filtrate flux (m/s)
J^*	Optimum diafiltration flux (m/s)
K_c	Hydrodynamic hindrance factor for convective solute transport
k_m	Solute mass transfer coefficient (m/s)
L	Channel length (m)
L_p	Membrane permeability (m/s/Pa)
N	Number of diavolumes
N_s	Solute flux through membrane (kg/m^2/s)
ΔP	Transmembrane pressure (Pa)
P_F	Feed (inlet) pressure (Pa)
P_f	Purification factor for filtrate product
P_{fil}	Filtrate pressure (Pa)
P_r	Purification factor for retentate product
P_R	Retentate (exit) pressure (Pa)
Q	Feed flow rate normalized by membrane area (m/s)
R	Retention coefficient
r_p	Pore radius (m)
r_s	Solute radius (m)
s	Characteristic pore size for pore size distribution (m)
S	Observed sieving coefficient
S_∞	Asymptotic value of intrinsic sieving coefficient at high flux
V	Feed volume (m^3)
V_D	Total volume of diafiltration buffer (m^3)
V_f	Cumulative filtrate volume (m^3)
V_o	Initial feed (retentate) volume (m^3)
Y_f	Product yield in filtrate
Y_r	Product yield in retentate

Greek Symbols

γ	Wall shear rate in module (s^{-1})
δ	Boundary layer thickness (m)
λ	Ratio of solute to pore radii
$\Delta\Pi_i$	Osmotic pressure difference for solute i (Pa)
σ_i	Osmotic reflection coefficient for solute i
ϕ	Solute partition coefficient between membrane and bulk solution

REFERENCES

1. Blatt WF, Dravid A, Michaels AS, Nelsen L. In: Flinn JE, editor. Membrane science and technology. New York: Plenum Press; 1970. pp. 47–97.
2. Michaels AS. US patent 3,615,024. 1971.
3. van Reis R, Goodrich EM, Yson CL, Frautschy LN, Whiteley R, Zydney AL. J Memb Sci 1997; 130: 123–140.
4. Kurnik RT, Yu AW, Blank GS, Burton AR, Smith D, Athalye AM, van Reis R. Biotechnol Bioeng 1995; 45: 149–157.
5. van Reis R, Gadam S, Frautschy LN, Orlando S, Goodrich EM, Saksena S, Kuriyel R, Simpson CM, Pearl S, Zydney AL. Biotechnol Bioeng 1997; 56; 71–82.
6. Lightfoot EN. Transport phenomena in living systems. New York: Wiley; 1974.
7. Kedem O, Katchalsky A. Biochim Biophys Acta 1958; 27: 229–246.
8. Opong WS, Zydney AL. AIChE J 1991; 37: 1497–1510.
9. Deen WM. AIChE J 1987; 33: 1409–1425.
10. Zeman LJ. Zydney AL. Microfiltration and ultrafiltration: principles and applications. New York: Marcel Dekker; 1996.
11. Pujar NS, Zydney AL. J Colloid Interface Sci 1997; 192: 338–349.
12. Pujar NS, Zydney AL. J Chromatogr A 1998; 796: 229–238.
13. Pujar NS, Zydney AL. Ind Eng Chem Res 1994; 33: 2473–2482.
14. Mochizuki S, Zydney AL. J Memb Sci 1993; 82: 211–228.
15. Michaels AS. Chem Eng Prog 1968; 64: 31–44.
16. Zydney AL. J Memb Sci 1997; 23: 275–282.
17. Bown WR, Williams PM. Adv Colloid Interface Sci 2007; 134: 3–14.
18. Paulsen FG, Shojaie SS, Krantz WB. J Memb Sci 1994; 91: 265–282.
19. Tucelli R, McGrath PV. US patent 5,522,991. 1994.
20. van Reis R, Zydney AL. J Memb Sci 2007; 297: 16–50.
21. Hsieh HP. AIChE Symp Ser 1988; 84: 1–18.
22. Belfort G, Zydney AL. In: Malmsten M, editor. Interfacial behavior of biopolymers. New York: Marcel Dekker; 1998.
23. Mehta A, Zydney AL. J Memb Sci 2005; 249: 245–249.
24. Zydney AL, Xenopoulos A. J Memb Sci 2007; 291; 180–190.
25. Tkacik G, Michaels S. Biotechnology 1991; 9: 941–946.
26. Meireles M, Aimar P, Sanchez V. J Memb Sci 1991; 56: 13–28.
27. Mulherkar P, van Reis R. J Memb Sci 2004; 236: 171–182.
28. Gadam S, Phillips M, Orlando S, Kuriyel R, Pearl S, Zydney A. J Memb Sci 1997; 133: 111–125.
29. Belfort G, Davis RH, Zydney AL. J Memb Sci 1994; 96: 1–58.
30. Robertson BC, Zydney AL. J Colloid Interface Sci 1990; 134: 563–575.

31. Bacchin P, Aimar P, Field RW. J Memb Sci 2006; 281: 42–69.
32. Kelly ST, Opong WS, Zydney AL. J Memb Sci 1993; 80: 175–187.
33. Hanemaajer JH, Robbertsen T, van den Boomgard T, Gunnik JW. J Memb Sci 1989; 40: 199–212.
34. Narendranathan TJ, Dunnill P. Biotechnol Bioeng 1982; 24: 2103–2107.
35. McCray SB, Glater J. In: Sourirajan S, Matsuura S, editors. Reverse osmosis and ultrafiltration. Washington (DC): ACS Symposium Series; 1985.
36. Cheryan M. Ultrafiltration handbook. Lancaster (PA): Technomic; 1986.
37. Tragardh G. Desalination 1989; 71: 325–335.
38. Lawrence CA, Block SS. Disinfection, sterilization, and preservation. Philadelphia (PA); Lea and Febiger; 1968.
39. Smith KE, Bradley RL. In: Lund D, Plett E, Sandu C, editors. Fouling and cleaning in food processing. Madison (WI): Extension Duplicating, University of Wisconsin; 1985. p. 410.
40. Berman D, Meyers T, Chrai S. J Parenter Sci Tech 1986; 40: 119–121.
41. Goel V, Accomazzo MA, DiLeo AJ, Meier P, Pitt A, Pluskal M, Kaiser R. In: Ho WSW, Sirkar KK, editors. Membrane handbook. New York: Chapman and Hall; 1992. pp. 506–570.
42. Belfort G. J Memb Sci 1988; 35; 245–270.
43. Hallstrom B, Lopez-Leiva M. Desalination 1978; 24: 273–279.
44. Chung K-Y, Bates R, Belfort G. J Memb Sci 1993; 81: 139.
45. Winzeler HB, Belfort G. J Memb Sci 1993; 80: 35–47.
46. Gooding CH. Hollow fibers in a spiral wound configuration. Presentation at Membrane Technology and Separations Planning Conference; 1997 Oct; Newton (MA).
47. Vigo F, Uliana C, Ravina E. Sep Sci Technol 1990; 25: 63.
48. Culkin B, Armando AD. Filtr Sep 1992; 29: 376–378.
49. Redkar SG, Davis RH. AIChE J 1995; 41: 501–508.
50. Rodgers VGJ, Sparks RE. J Memb Sci 1992; 68: 149–168.
51. Cui ZF, Wright KIT. J Memb Sci 1997; 117: 109–116.
52. Smith SR, Cui ZF, Field RW. Ind Eng Chem Res 2005; 44: 7684–7695.
53. Lutz H. US patent 5,597,486. 1997 Jan 28. (to Millipore Investment Holdings Limited).
54. Cheang BL, Zydney AL. J Memb Sci 2004; 231: 159–167.
55. Sheehan JJ, Hamilton BK, Levy PF. ACS Symp Ser 1990; 419: 130–155.
56. Ng P, Lundblad J, Mitra G. J Sep Sci 1976; 2: 499–502.
57. Menon M, Zydney AL. Biotechnol Bioeng 1999; 63: 298–307.
58. Mehta A, Zydney AL. Biotechnol Prog 2006; 22: 484–492.
59. van Reis R. US patent 5,256,294. 1993 Oct 26. (to Genentech, Inc.).
60. van Reis R. US patent 5,490,937. 1996 Feb 13. (to Genentech, Inc.).
61. van Reis R, Saksena S. J Memb Sci 1997; 129: 19–29.
62. van Reis R, Goodrich E, Yson CL, Frautschy LN, Dzengeleski S, Lutz H. Biotechnol Bioeng 1997; 55: 737–746.

38

VIRUS RETENTIVE FILTERS

George Miesegaes and Scott Lute
Office of Biotech Products, Center for Drug Evaluation and Research, Food and Drug Administration, Silver Spring, Maryland
Hazel Aranha
GAEA Resources Inc., Northport, New York
Kurt Brorson
Office of Biotech Products, Center for Drug Evaluation and Research, Food and Drug Administration, Silver Spring, Maryland

38.1 PROCESS VIROLOGY OVERVIEW

38.1.1 Imperative

Viral safety of plasma-derived biologicals and recombinant DNA–derived biopharmaceuticals is critical for regulatory approval, marketing by industry, and ultimately safe use by consumers of these products (2). Viral contamination of products derived from human plasma (e.g. HIV in clotting factors) has adversely impacted the health of tens of thousands of patients in the past (3). Mammalian cell cultures used in the production of biotechnology products carry a potential threat of viral contamination, as well. Although bioreactor contamination of biologicals with microbes (i.e. bacteria) occurs at a higher frequency, the consequences of a viral contamination event are more severe (4,5).

In the 1990s, a series of minute virus of mice (MVM) contamination occurrences were reported by a large biotech plant in California (4). The source of the contamination was never pinpointed, but it was surmised that cell culture raw materials were the likely point of introduction. These events precipitated in the undesirable but necessary shutdown and extensive decontamination of the facility. Indeed, contamination events in manufacturing environments have drastic consequences; in addition to jeopardizing the availability of product supplies for patients, regulatory and economic consequences are significant in terms of product loss, product liability, and missed opportunity costs. Viral safety and robustness of viral clearance are therefore not only important aspects of biopharmaceutical product quality, but are major process economics considerations as well.

There are additional considerations associated with virus (vs microbial) contamination of biopharmaceuticals. Virus detection methods are not sensitive enough to detect low levels of viruses; consequently, it is impossible to establish the absolute absence of viral presence. Some viruses pose a real threat (e.g. plasma product contamination with pathogenic bloodborne viruses such as HIV and the hepatitis viruses). In other cases the threat is theoretical. For example, in the biotechnology industry, cell lines such as Chinese hamster ovary (CHO) cells that contain endogenous retrovirus at levels of $10^6 - 10^9$ particles/mL (as visualized by electron microscopy) are deemed acceptable, as the particles are noninfectious and pose primarily, a theoretical safety concern. Regulatory agencies emphasize a holistic approach directed at risk minimization, which along with process monitoring constitutes an appropriate risk management program.

38.1.2 Implementation Strategy

The current strategy for ensuring viral safety of biologicals involves multiple levels of control over the product and the

Downstream Industrial Biotechnology: Recovery and Purification, First Edition. Edited by Michael C. Flickinger.

process. This includes establishing a tiered cell banking system, screening of source material, and incorporating specific virus removal or inactivation steps into the manufacturing process. Validating the ability of the process to remove or inactivate viruses is key toward estimating the ability of the manufacturing scheme to clear viruses (in the unlikely event that contamination occurs), and in assessing whether the clearance capacity afforded by the various process steps in manufacturing is adequate to assure viral safety. It is important to understand, however, that the safety factor afforded by this strategy is nonabsolute. For known contaminants of continuous cell lines used in biopharmaceutical production, for example, endogenous retroviruses (which pose primarily, a theoretical safety concern) the acceptable safety factor is calculated based on particle load in a dose equivalent of cell culture harvest and the clearance capacity of the purification process.

38.1.3 Viral Clearance Methods

Viral clearance may be achieved as a consequence of routine processing and purification operations (serendipitous or fortuitous methods) or through strategies specifically designed for removing/inactivating viruses (deliberate methods), which are incorporated directly into the manufacturing process. Several methods used in the downstream purification of biopharmaceuticals serendipitously offer the added bonus of viral clearance. For example, virus inactivation can occur merely by the buffers used in manufacturing (e.g. low-pH incubation). Virus removal can also be demonstrated with downstream processing steps, designed to concentrate or purify proteins, such as precipitation and chromatography. These methods, if demonstrated to provide virus clearance through appropriate process clearance evaluation (validation) studies, are often used to calculate the overall virus clearance ability of the manufacturing process. Specific virus clearance steps include virus retentive filtration and solvent detergent inactivation. Overall, these steps, when well designed and implemented sequentially can cumulatively provide a $\log_{10}$ reduction value (LRV) of as much as 15–20 $\log_{10}$ for certain viruses (e.g. retroviruses). Refer to section titled "Validation (Virus Clearance Evaluation) Studies" for the calculation of LRV.

38.1.4 Removal Versus Inactivation

Viral clearance unit operations are broken down into two major categories: (i) removal steps, where the virus is partitioned/removed from the product; and (ii) Inactivation steps, where physicochemical treatment disrupts, denatures, or otherwise renders the virus noninfectious. Examples of physical treatments include low pH, solvent/detergent, and heat incubation. Unlike removal steps such as chromatography, partitioning, and virus filtration, for inactivation steps, dead/noninfectious viral particles and/or any residual components still remain in solution with the product. Chromatography, partitioning, and virus filtration on the other hand physically remove viruses from the product.

Robustness (where the viral inactivation/removal effectiveness is widely independent of variability in production) for certain operations has been verified experimentally and described in the scientific literature. For any given operation, clearance is typically dependent on a few critical unit operation parameters and is relatively less sensitive to others, so long as these parameters remain within the defined manufacturing ranges. For example, low-pH inactivation of murine retroviruses is highly dependent on time, temperature, and pH, but is relatively independent of the type of model protein or salt content in the matrix (6). Virus filtration is believed to be largely dependent on the pore size distribution of the filter and can be impacted, for instance, by overloading with submicron particulates (7). Establishing the mechanism of clearance by these unit operations is a regulatory requirement (see below) and enhances the confidence that changing parameters unrelated to their mechanism of clearance will not impact clearance efficacy.

38.1.5 Clearance Studies

Product-specific validation studies include (i) demonstration that an adequate level of virus clearance has been incorporated into the manufacturing process; and (ii) a description and understanding of the mechanism(s) by which virus clearance occurs, that is, whether it is by inactivation, removal, or a combination of both.

Viral clearance studies begin by designing scale-down models of the actual manufacturing unit operations. The objective of the scale-down model is to determine the performance and extent of viral clearance that can be expected of a unit operation at full scale. First, the scale-down model must match each key and critical operating parameter of the large-scale unit operation. Second, key and critical performance parameters, such as step yield, separation power, and chromatogram profile must be representative of the large-scale unit operation. Nonkey/noncritical operating parameters (e.g. column bed diameter and filtration volumetric throughput) are adjusted to a scale practical for lab studies. Model viruses are spiked into product intermediates. The total viral load of both the input and the output material are measured using either infectivity assays or quantitative polymerase chain reaction (Q-PCR). The $\log_{10}$ reduction factor of the individual step is defined as the ratio of the titer or total virus load in the input material divided by the titer or total virus in the output material. The result is expressed as a $\log_{10}$ titer reduction (LTR) or reduction value (LRV). When calculating the clearance factor of a complete manufacturing scheme, only those reduction factors that are greater than one are included.

38.1.6 Guidance Documents

The viral safety approach for biopharmaceuticals is described in three key regulatory documents; viral safety evaluation of biotechnology products derived from cell lines of human or animal origin Q5A (8); the Food and Drug Administration (FDA) Center for Biologics Evaluation and Research (CBER) Points to Consider in the Manufacture and Testing of Monoclonal Antibody Products for Human Use (9); and the European Medicines Evaluation Agency (EMEA) Guideline on virus safety evaluation of biotechnological investigational medicinal products (10).

38.2 PRINCIPLES OF OPERATION

38.2.1 Sieving Versus Other Mechanisms of Retention

Filters designed for the retention of particles in liquids are often assumed to act solely by a retention-based mechanism (particle sieving). In reality, there are a number of retention mechanisms involved, as described in an original comprehensive review of filtration (11). The mechanisms identified by Grant *et al.* (11) include gravitational settling, electrostatic/electrokinetic deposition, interception, impaction, diffusive transport followed by surface capture, as well as sieving. The extent of influence each has can vary across filter types and filtration modes (tangential vs normal flow vs depth filtration). Also, some mechanisms like gravitational settling and impaction generally do not occur with colloid-size particles <100 nm, such as viruses, which tend to stay in suspension; rather they occur with larger particles.

Grant *et al.* (11) covered the theoretical influence each mechanism has on particle retention from air and liquid feed stocks, the latter having been acknowledged as being much more complicated to model. For instance, many of the air filtration concepts are also applicable to liquid filters under ideal circumstances (12,13); retention, reviewed in Grant *et al.* (11), becomes increasingly dependent on sieving when performed under challenge conditions. This is primarily because the physical chemistry of particle retention in liquids is process dependent. Chemical interactions, for instance, between the target virus particle or the filter itself and the process fluid employed can significantly impact one or more retention mechanisms.

38.2.2 Factors Influencing Virus Clearance by Virus Retentive Filters

Factors influencing virus filtration may be categorized as being related to the virus (size), the filter, the feedstream, and the filtration conditions (mode). Filter properties with potential to impact virus retention include pore size and pore size distribution, membrane structure, surface chemistry, and thickness (including number of filter layers). Because filters are of divergent design, primary factors influential to performance may be different for different filters. For example, whereas thickness may be important to a filter that relies on retention through the depth of the filter, it is of secondary importance to a filter that relies primarily on surface capture. Feedstream-related issues include physiochemical attributes such as pH, osmolarity, ionic strength, temperature, formulation components and protein concentration, molecular weight and quality (e.g. aggregate content). The feedstream should be of like quality to that seen in the manufacturing process. Changes to this quality could compromise test validity. For example, a heavy protein aggregate load could lead to premature filter fouling. Filtration mode (dead ended vs tangential flow), flow rate or pressure differential (ΔP), cross flow/ΔP (for tangential flow), volume-to-area ratio, and postuse filter flushing are the key filtration conditions that should be modeled in a virus clearance study.

In cases of heavily fouling feedstreams, use of prefilters can improve virus filter performance and total achievable processing capacity. Development work is warranted to find the appropriate prefilter match for a particular process application.

38.2.3 Influence of Process Parameters and Filter Type

In virus retentive filters, that is, filters specifically designed for viral removal, sieving is considered to be the predominant mechanism. However, it is necessary to recognize that both product- and process-related factors will impact the virus retention capability of the filter. For example, a typical bioprocess fluid is a rather complex mixture containing salts, buffering chemicals, the protein of interest, and process-related impurities (e.g. host cell proteins, DNA). As expected, the relevant process parameters are numerous. For example, a recent study concluded that patterns in LRV relationship with cumulative load varied between filter types (7), arguing that the extent of impact each retention mechanism may have on viral clearance is also filter type–dependent.

38.3 PHYSICAL ATTRIBUTES

38.3.1 Chemistry and Configurations

Virus filters are cast polymeric membranes with a complex internal porous structure. Solutions pass through this pore network while viruses and other particles carried along by the fluid may be retained on the membrane surface or penetrate some distance into the membrane structure prior to entrapment. Virus filters, discussed below, target two broad classes of viruses: large viruses (e.g. retroviruses) and small viruses (e.g. parvoviruses). Currently available virus filter membranes are constructed of polymers (e.g. polyethersulfone, surface modified polyvinylidene fluoride, cuprammonium regenerated cellulose, etc.); surface modification of these polymers is often undertaken during filter manufacture to render them hydrophilic (if the polymer is hydrophobic). The housings for the membranes are, typically, cartridges also constructed of polymers (polycarbonate, polypropylene, etc.) The filter membrane, the filter housing, and/or the use of additives all contribute to the leachable load associated with the filter.

In general, virus filters possess an asymmetric membrane structure. Membrane orientation can have a significant impact on the performance characteristics of the virus filter (14). Most virus filters used in direct flow filtration (DFF) mode are oriented with the tight skin side facing away from the feed. It has been shown that skin side down membranes have higher total capacities, reflecting the different fouling mechanisms between different membrane orientations (14). tangential flow filtration (TFF) mode filters, in contrast, possess skin side up membranes. However, fouling in these membranes is less problematic as it is minimized during processing by the sweeping tangential liquid flow of the circulating retentate.

Depending on the manufacturer, membranes exist in the form of pleated sheets, flat sheets bonded to plates, or as bundled hollow fibers. It should be noted that because of their distinct chemistries, filter brands and models are not interchangeable. Additionally, consistent viral clearance performance is assumed only if the filter is used as per manufacturer instructions. A summary of marketed or soon-to-be marketed virus filters is provided in Table 38.1.

38.4 PROCESS CONSIDERATIONS

38.4.1 Minimizing Filter Leachables and Particulates

Leachables are soluble chemical components of a filter or other bioprocessing apparatus that are released under normal operating conditions. Filters can potentially release low levels of particulates or leachables into bioprocess streams. They can accumulate as a consequence of spontaneous chemical interactions between process buffers/fluids and the filter membrane, housing unit, or unit operation machinery components. Particulates can also stochastically arise in process fluids to a limited extent.

There are several levels of control for particulates and leachables. First, virus filters are constructed specifically of materials that contribute small, if any, levels of chemicals or fibers into aqueous solutions. Virus filters, unlike final sterilizing filters, are generally never the last step in bioprocessing, but are incorporated somewhere near the end of the biopharmaceutical product manufacture scheme. Even though there may be downstream manufacturing steps that can remove leachables and particulates, it is warranted as a best practice to minimize introduction of filter leachables and particulates into bioprocess intermediates. Biopharmaceuticals are essentially aqueous products; they do not solubilize plastics from housing or other filter materials. Buffer attributes such as pH or detergent content should be considered during a process-specific assessment of potential leachables. Prior to use, manufacturers typically flush the filter in order to ensure removal of storage solutions and to minimize particulates and leachables associated with the filter. Also, wetting the filter prior to use typically decreases the extent of protein adsorption to the filter and minimizes product loss on the filter.

38.4.2 Extractable Data Analysis

Extractables differ from leachables in that they result from manipulations that occur artificially under stress conditions (i.e. solvents, heat treatment, etc.). It is generally believed that analyses of extractables may serve to identify putative leachables (15). In this regard, leachables can be considered a subset of extractables, although this is not always the case (e.g. chemical degradation over time). Virus filters are disposable/single-use. Thus, extractables are less of a consideration for virus filters (vs UF/DF membranes) in that virus retentive filters are not subject to harsh chemical treatments during cleaning or storage. Nevertheless, extractable-related information is helpful from a regulatory standpoint and is generally supplied by filter vendors.

38.5 FILTRATION MODES

There are two major types of filtration modes in use for virus filters: direct flow and tangential flow. Each type of filtration is briefly discussed below.

38.5.1 Direct Flow Filtration (DFF)

In DFF, also referred to a *dead-end* or *normal* flow filtration, the fluid flow path is perpendicular to the filter membrane surface. The primary advantage of DFF is its ease of use. DFF operations also recover a very high fraction of the feed as permeate, especially when the product is followed by a buffer flush. DFF generally requires clean process fluids as particles can penetrate into and be retained throughout the pore structure; this could potentially result in clogging the filter if the particulate load is high.

TABLE 38.1. Currently Marketed Virus Retentive Filters

Filter Vendor	Filter Brand	Virus Target Class[a]	Mode[b]	Membrane Pore Symmetry	Layers	Membrane Chemistry
Millipore	Viresolve NFR	Retro	DFF	Asymmetric	Triple layer	Hydrophilic polyethersulfone (PES)
	Viresolve NFP	Parvo	DFF	Asymmetric	Triple layer	Hydrophilic polyvinylidene fluoride (PVDF)
	Viresolve 70	Parvo	TFF	Asymmetric	Single layer	Hydrophilic PVDF
	Viresolve 180	Retro	TFF	Asymmetric	Single layer	Hydrophilic PVDF
	Viresolve Pro	Parvo	DFF	Asymmetric	Double layer	Hydrophilic PES
Sartorius Stedim Biotech	Virosart CPV	Parvo	DFF	Symmetric	Double layer	Hydrophilic PES
Pall	Ultipor VF grade DV50	Retro	DFF	Symmetric	Triple layer	Hydrophilic acrylate-modified PVDF
	Ultipor VF grade DV20	Parvo	DFF	Symmetric	Double layer	Hydrophilic acrylate-modified PVDF
	Pegasus grade LV6	Retro	DFF	Symmetric	Double layer	Hydrophilic acrylate-modified PVDF
	Pegasus grade SV4	Parvo	DFF	Symmetric	Double layer	Hydrophilic acrylate-modified PVDF
Asahi Kasei	Planova 35N	Retro	DFF/TFF[c]	Asymmetric	Hollow fiber	Hydrophilic cuprammonium regenerated cellulose
	Planova 20N	Parvo	DFF/TFF	Asymmetric	Hollow fiber	Hydrophilic cuprammonium regenerated cellulose
	Planova 15N	Parvo	DFF/TFF	Asymmetric	Hollow fiber	Hydrophilic cuprammonium regenerated cellulose
	Planova BioEX	Parvo	DFF/TFF	Asymmetric	Hollow fiber	Modified PVDF

[a]"Target virus class" can be a subjective concept based on the application and product type (i.e. biotech vs plasma product). Some large virus retentive filters can retain viruses smaller than retroviruses. Consultation with filter vendors is warranted if clearance of a particular virus is a process requirement.

[b]DFF, direct flow filtration; TFF, tangential flow filtration.

[c]Can be configured by end users to run in either mode.

For this and other reasons, DFF systems are designed for single-use applications. Most DFF filters are now available as fully disposable units within encapsulated plastic holders or housings.

38.5.2 Tangential Flow Filtration (TFF)

TFF unit operations are more complex than direct flow. In TFF, the bulk of the fluid to be filtered is recirculated on the upstream side of the filter, so that the primary flow occurs tangential (or parallel) to the membrane surface. This "crossflow" constantly sweeps the surface of the membrane, keeping it clear of debris (such as particulates and aggregates/agglomerates) thereby reducing the rate and extent of clogging. Because of this, TFF filters are in general capable of handling higher particulate loads than DFF filters. The portion of the fluid that passes through the membrane and is collected is referred to as the *permeate*. As the filtration process progresses, an increasing volume of the initial fluid is drawn off into the permeate, and the volume of the recirculating retentate continually drops.

To maximize step yield, the product can be flushed into the permeate by continuous buffer addition to the retentate side during diafiltration. As a result, viable and nonviable contaminants that are retained by the filter are concentrated in the retentate. TFF filters, in theory, can be reused, but process economics and validation issues associated with reuse have discouraged firms from taking this approach.

38.6 FILTER CLASSES

38.6.1 Classification of Targeted Viruses

Viruses of relevance to the biopharmaceutical and plasma products industries generally fall into three size categories: large viruses (>80 nm, e.g. retroviruses), small viruses (<30 nm, e.g. parvovirus), and intermediate size viruses (30–80 nm, e.g. hepatitis B and C viruses, reoviruses). Virus filters target two broad classes of viruses: large viruses (e.g. retroviruses, 80–110 nm in size) and small viruses (e.g. parvoviruses, 18–26 nm in size). For removal of intermediate size viruses the small pore size virus retentive filters are used.

Viruses have been arbitrarily broken into these classes because the viruses of concern for biotechnology have been either retroviruses or parvoviruses. For plasma-derived products, clearance of intermediate sized viruses is also desirable. Some large virus filters have been claimed to be effective at clearing medium size viruses like SV40. On the basis of the targeted class and filter capabilities, filter vendors have developed ratings to allow for selecting the appropriate virus filter for an application.

38.6.2 Ratings Vary among Filter Manufacturers

In some cases, manufacturers have assigned a rating associated with a particular type of virus, for example, parvovirus or retrovirus (Table 38.2). In other cases, a virus size rating has been assigned based on either the retention of a particular model such as a bacteriophage at a given LRV or an average pore size rating is established from a mathematical model for the permeability. In yet other cases, a molecular weight retention or passage rating has been assigned. This system has recently been unified by a bacteriophage retention–based rating promulgated by the PDA (1).

38.6.3 Filter/Process Compatibility

The product and/or process may impact the chemical and physical attributes of the filter, including its removal performance. Potential product effects on the filter include the formation of a gel layer on the surface of the membrane, pore plugging, and adsorptive effects on pore surface properties.

A protein gel layer that forms on the membrane surface of TFF systems can act as a secondary, tighter separation layer and potentially can lead to an increased clearance. Thus, these filters are the opposite of the DFF filters, in that low protein loads are a more stringent challenge condition. Note, however, that the increased viral clearance, which could potentially result, is dependent on process-associated variables and is not considered robust.

For certain small virus retentive filter types, DFF systems (described below) with extended throughputs or high flow decay may show a decline of virus titer reduction (16). For these filter types, this is believed to be due to partial plugging of pores by particulates on the finer end of the membrane pore size distribution. According to this theory, the product flow is diverted through larger pores leading to lower viral titer reduction. Process fluid/spike preparation overloading and partial plugging of the filter during spike studies may cause flux restriction, resulting in lower LRVs. Consultation with filter vendors is warranted to avoid situations where unwanted passage (e.g. viruses) can occur.

TABLE 38.2. Ratings of Currently Marketed Virus Retentive Filters

Filter Vendor	Filter Brand	Targeted Classes (Viruses)	Vendor Ratings	PDA Ratings[a]
Millipore	Viresolve NFR	Large	≥6 LRV for Φ6	PR772-LRF6
	Viresolve NFP	Small	>4 LRV for Φ X174	PP7-LRF4
	Viresolve 70	Small	Based on 70 kDa protein passage	Not reported
	Viresolve 180	Large	Based on 180 kDa protein passage	Not reported
	Viresolve Pro	Small	>4 LRV for Φ X174	PP7-LRF4
Sartorius Stedim Biotech	Virosart CPV	Small	≥4 LRV for PP7	PP7-LRF4
Pall	Ultipor VF-DV50	Large	≥6 LRV for PR772	PR772-LRF6
	Ultipor VF-DV20	Small	≥3 LRV for PP7	Not reported[b]
	Pegasus LV6	Large	≥6 LRV for PR772	PR772-LRF6
	Pegasus SV4	Small	>4 LRV for PP7	PP7-LRF4
Asahi Kasei	Planova 35N	Large	JEV clearance	PR772-LRF6
	Planova 20N	Small	Parvovirus clearance	PP7-LRF4
	Planova 15N	Small	Parvovirus clearance	PP7-LRF4
	Planova BioEX	Small	Parvovirus clearance	PP7-LRF4

[a]Verified by third-party lab (23–25).
[b]Pall has generated data showing Ultipor VF DV20 also meets PP7-LRF4 per PDA methodology. Report available on request from vendor.

38.7 INTEGRITY TESTING

38.7.1 Methodology

From a biopharmaceutical process development standpoint, it is critical to demonstrate filter performance in terms of (i) consistent viral clearance, and (ii) high product recovery. In addition, a filter user must also be able to document that individual filters used are performing as expected. In principle the assessment of individual filter performance should correlate with virus retention capability. While a viral challenge is the only true test of a filter's performance, it is a destructive test and therefore precludes any subsequent use of the filter in manufacturing. Moreover, introducing a virus into the production facility is contrary to cGMP requirements. Thus, physical integrity tests are designed, developed, and specified by the filter manufacturer.

While some integrity tests for virus filters have been adapted from those used for integrity testing of sterilizing grade filters (e.g. forward flow test), others are simply not practical for use with filters designed for virus retention (i.e. bubble point tests would rupture the filter and/or the housing). The specifics of a given integrity test vary across filter manufacturer brand and/or model but fall into one of three categories: particle challenge tests, gas–liquid porosimetry tests, and liquid–liquid porosimetry tests (Table 38.3). Leak testing is considered to be an installation test; in general, these are performed at considerably lower pressure and detect only gross defects.

38.7.2 Noninterchangeability

Integrity testing is filter type–specific and must be conducted as per manufacturer recommendations. As part of the validation of the virus filter, to ensure that the filter performs as per the virus clearance claim, the filter manufacturer conducts testing and specifies the test type and conditions of the integrity test (which correlates filter performance in terms of virus removal capability with a physical integrity test).

38.8 PERFORMANCE

38.8.1 Published Performance Information

Performance data from multiple types of virus filters have been released in the scientific literature. Filters have been challenged with mammalian viruses and bacteriophages under normal challenge conditions and failure mode conditions. In general, filters performed robustly under normal challenge conditions. On the other hand, failure mode tests identified important limit parameters for performance and revealed unique (filter specific) performance limit qualities. Table 38.4 summarizes published virus filter studies.

38.8.2 Filtration Mode

TFF and DFF can be run under either a constant flux/flow rate or constant pressure (1). Note that both parameters exert a reciprocal influence on each other. Filters run in constant flow mode demonstrate a measurable increase in pressure over time to maintain flow as foulants accumulate on the incident surface. Filters run under constant pressure slowly develop a drop in flow rate (i.e. flow or flux decay). The patterns of flux decay over extended challenges vary across filter types and manufacturers (7). Other factors such as process stream composition (particle load, protein concentration, etc.) play a role as well.

38.8.3 Fouling and Flux Decay

Like all filters, virus retentive filters eventually foul. This occurs over time as protein and other particulates build up inside the filter and/or on its surface. One identified consequence of membrane fouling is a decrease in flux/flow rate, which can be monitored during processing. Filters run in DFF mode have been claimed to be more susceptible to fouling due to the perpendicular direction of flow through the membrane. Tangential flow tends to sweep away and reduce potential buildup during the run.

TABLE 38.3. Methods Currently Recommended by Filter Manufacturers for Integrity or Installation Testing

Integrity Test	Category	Methodology	Notes
Gold particle	Particle challenge	Sieving based on size of gold particles	Destructive test
Leak	Gas–liquid	Detects gross defects at a low pressure	Qualitative assay (pass/fail)
Forward/ diffusive flow test	Gas–liquid	Measures the migration of gas through a wetted filter as a function of pressure	Temperature-sensitive; solvent-dependent
Pressure hold/decay	Gas–liquid	Assesses rate of pressure decay as a function of gas diffusion through filter	Temperature-sensitive, upstream volume, and solvent-dependent
Liquid flow rate	Liquid–liquid	Measures the flow rate of one liquid through a membrane that is initially wetted with a second immiscible membrane	Temperature-sensitive, solvent-dependent

TABLE 38.4. Summary of Existing Literature Concerning Virus Retentive Filter Performance[a]

Filter Types	Model System[b]	Viruses Tested	Reported LRVs	References
Normal Challenge Conditions[c]				
Planova 35 N	A, B, C	BVDV, Calicivirus, HCV, HIV, PRV, Reo, Sindbis, SV40, X-MuLV	>5.3 to >7.4	31–35
Ultipor DV50	B, C	Φ6, HBV, HCV, HIV, HSV-1, Influenza A, PR772, Semliki Forest, Sindbis, SV40, T1 phage, X-MuLV	3.2 to >7.2	36–41
Viresolve NFR	B, C	Φ6, Ad2, MuLV, PR772, Reo	>6.1 to >6.8	42
Asahi Kasei prototypes "R","U"	C	HIV, JEV	>4.9 to >5.3	43
Planova 15N	B, C	BPV, Polio, PPV, PRV, SV40, X-MuLV	>4.4 to >7.0	32,34,44
Planova 20N	A, B	ΦX-174, B19, BVDV, CPV, EMC, HAV, HIV, MMV, PP7, PPV, PRV	>3.0 to 7.0	7,33,44,45
Ultipor DV20	B	ΦX-174, PP7	7.0	7
Viresolve NFP	A, B	ΦX-174, B19, BHV, BVDV, EMC, HAV, HIV, MMV, PP7, PPV, PRV, Reo, X-MuLV	3.9 to >6.1	7,46-48
Virosart CPV	B	ΦX-174, PP7	6.0–7.0	7
Pall Omega 300k RV Maximate[d]	C	Φ6	4.9	36
Pall prototype Polyacrylonitrile (50k)	C	Polio, PP7, T1 phage	4.6–6.5	49
Pall prototype Polyacrylonitrile (13k)	C	Polio, PP7, T1 phage	>6.5 to 7.6	49
Pall prototype Polyacrylonitrile (6k)	C	Polio, PP7, T1 phage	>6.6 to >7.7	49
Pall prototype Polysulfone (6k)	C	Polio, PP7, T1 phage	>6.4 to 7.4	49
Viresolve 70	B, C	Φ6, ΦX-174, EMC, Maedi-Visna, MuLV, Polio, Reo, Sindbis, SV40	3.3–7.5	50–54
Viresolve 180	B	Φ6, Sindbis	6.6–7.9	53,54
Planova 20N, Planova 15N, Virosart CPV, Pegasus SV4, Viresolve NFP	C	PP7, PR772	3.1–8.4	21
Virosart CPV	A, B, C	PPV	5.0–7.0	55
Failure testing[e]				
Planova 35 N	A, B, C	B19, BPV, EMC, HAV, MMV, Polio, PPV	0.8 to >6.2[f]	33–35,44,56
Viresolve 180	B, C	ΦX-174, EMC	2.8–3.1	53,54
Viresolve NFR	C	ΦX-174,	<0.2	42
Ultipor DV50	C	Polio, PP7	1.8–2.2	39–41
Asahi Kasei prototype "R"	C	JEV	>1.7	43
Planova 20 N, Virosart CPV, Ultipor DV20, Viresolve NFP	B	ΦX-174, PP7	0.1–4.5	7

[a]This table should not be viewed as a comparison of filter types or brands, but rather as a summary of studies reported in the literature. Clearance values reported in this table were compiled from the published literature from multiple, diverse sources. The values were reported for a wide range of products, under different process conditions and using different process fluid compositions. Further, the reported clearance values were generated using different viruses, assay formats (e.g. plaque forming vs $TCID_{50}$, etc.), and from different laboratories. No inferences should be made concerning comparative performance between individual filter brands, variable performance by individual filter types, or predicted ranges of clearance factors for individual virus filter products. This table is not a substitute for process validation; process-specific filter performance should be evaluated for each process fluid and application by individual filter end users.

[b]Model challenge solution contains: A, plasma-derived model protein; B, biotechnology-derived model protein; or C, artificial protein, cell culture media, or buffer.

[c]Under normal loading conditions.

[d]Marketed as an ultrafilter, but tested for viral clearance.

[e]Defined as running filters under virus overloading conditions or with extreme flux decay ($> V_{50}$), or by challenging a large virus filter with a small virus that it is not designed or claimed to remove.

[f]Unexpectedly high clearance was noted for bovine parvovirus.

For certain filter types, the instantaneous clearance capacity is tightly associated with instantaneous flow rate, relative to the initial flow rate (16). Thus, the extent of fouling is considered a critical process parameter to monitor in order to identify the point at which the retentive capabilities might be compromised for this filter type.

38.8.4 Vmax

Vmax is a mathematical extrapolation of the maximum amount of liquid that can be sent through a filter based on flux data. Filter manufacturers can provide estimates of Vmax for common model process fluids, and techniques to measure this. While this is useful information, Vmax should not serve as a sole basis for filter choice as other important attributes (LRV, filter area per cartridge, cartridge flow properties, scalability, ease of use, ease/extent of sterilization, etc.) are relevant for efficient and cost-effective design of a filtration operation.

38.8.5 Bacteriophage

When choosing a model virus for process development studies, bacteriophages may warrant consideration over mammalian viruses (17,18). Phages are benign in the event of human contact. Thus, the required laboratory safety level is typically less than that for mammalian viruses (e.g. BSL-1 as compared to BSL-2 or above). In addition, phages can be cultured to very high titers (in excess of 10^9-10^{10} pfu/mL) and can be purified on cesium chloride gradients to even higher titers ($>10^{12}$ pfu/mL) (7,19). The resulting preparations are cleaner compared to those typical of some mammalian viruses. The plaque assays used to detect their presence in the filtrate take 1 day and tend to be more sensitive than the infectivity assays employed for mammalian viruses.

ICH Q8 has established the concept of design space. The use of a phage as a surrogate model virus when establishing a product's design space is likely a good starting point for reasons mentioned above. Some phage species have been fairly well characterized and possess physical properties (i.e. size, isoelectric point, etc.) close to their mammalian counterparts (19–22). Design space conditions established with a phage will be predictive of filter clearance of the corresponding mammalian virus. It is important to note that current regulatory expectations for filter validation studies for submission to regulatory authorities use mammalian viruses.

38.8.6 PDA Virus Filter Rating

In 2002, the PDA organized the PDA Virus Filter Task Force to develop a common nomenclature and a consensus test method for classifying and identifying virus retentive filters. At that time, each filter manufacturer rated their filters differently based on pore size or functional criteria as described in Table 38.2. There is great interest within the filter and the biopharmaceutical industry to develop a single rating system. A single rating system would promote industry uniformity by developing a single reliable test to ensure a given level of viral clearance performance under consensus conditions for filters produced by all manufacturers.

In 2004, the PDA entered into a cooperative research and development agreement (CRADA) with the Center for Drug Evaluation and Research (CDER)/FDA, for the FDA to evaluate test methods and perform background laboratory work necessary to develop test methods. By 2005, the task force had successfully developed a rating system for large virus retentive filters based on 6 $\log_{10}$ retention of a 64–82-nm bacteriophage, PR772, and 95% transmission of plasma-derived immunoglobulin G (IVIG). Filters from three manufacturers (Pall Life Sciences Ultipor VF DV50, Millipore Viresolve NFR, Asahi Kasei Planova 35N) were tested according to the method and found to be within the method acceptance criteria (23,24).

Subsequently, the task force developed a general protocol for small virus retentive filters based on 4 $\log_{10}$ retention of a ~30-nm pseudomonas bacteriophage PP7 and 90% IVIG passage. To an even greater extent than large virus retentive filters, significant technical challenges are associated with small virus filters (e.g. the potential for passage) (7). No currently marketed small pore virus filters claim "absolute" retention of 20–25-nm viruses. Rather, these "nonabsolute" filters are understood to show a range of particle or microbial LRVs depending on fluid and process conditions. To support the method development, a series of feasibility studies were performed to decide important parameters such as test fluid composition, choice of test phage, filtration end point definition, and target LRV (21). The final method was set based on these studies and consensus within the task force. Filters from four manufacturers (Pall Life Sciences Pegasus SV, Millipore Viresolve NFP, Asahi Kasei Planova 15N and 20N, and Sartorius Stedim Biotech Virosart CPV) were tested according to the final method and found to be within acceptance criteria (25). After the initial testing to establish the method, additional filter types were tested by a third party according to the small virus filter retentive method and passed the acceptance criteria (e.g. Asahi Kasei Planova BioEX, Millipore VPro; Lute and Brorson; personal Communication).

The filter performance in the above-mentioned studies does not predict virus filter performance in a biopharmaceutical manufacturing setting. Manufacturers of biopharmaceuticals are required to conduct "representative" or product-specific testing in order to establish the viral clearance capacity of the filter under process conditions.

38.8.7 Breakthrough in Small Virus Filters

Small virus filtration is considered to be a significant technical challenge. This is primarily because the viruses targeted for removal are roughly twice the size (20–30 nm) of the protein pharmaceuticals (4–12 nm) that must be retained in the process fluid. Designing filters with narrow pore size distributions and a mean pore size tight enough to fractionate these two components is challenging, and under certain conditions the separation capacity can breakdown.

Virus filtration performance was evaluated in a phage passage study with small virus retentive filters from each commercial vendor tested (Millipore Viresolve NFP, Pall Ultipor VF DV20, Sartorius Stedim Biotech Virosart CPV, and Asahi Kasei Planova 20N (7). Each filter brand had a unique pattern of performance impact caused by overloading. In three filter brands, phage overloading accelerated flux decline to a variable degree. However, brand-specific differences were apparent even in the concavity pattern of flux decline relative to volumetric throughput. In some filter brands, monomer protein alone caused eventual flux decay, while in others a cumulative bacteriophage challenge load of $2-4 \times 10^{15}$ pfu/m^2 was required for near complete restriction of fluid flow. Finally, flux decay with another brand related less to phage challenge levels than to filter-to-filter variation.

For all filters, PP7 bacteriophage LRV_{init} was largely above 6–7 $\log_{10}$, but PP7 phage breakthrough occurred later in filtration, especially in the event of phage overloading. In one brand, a correlation of phage passage with flux decay was seen previously (16). In others, overloading appeared to be central for PP7 phage passage. On the basis of phage challenge levels, "protein-dominated" and "phage-dominated" zones could be defined where the pattern of ΔLRV varied between filter type (e.g. relationship with flux decay). One filter brand differed from the other brands in that PP7 LRV_{init} varied from filter to filter, and only modestly increased or decreased over the course of filtration.

Overall, these studies argued that small virus retentive filters should not be viewed as absolute in their capacity to clear a virus, and should not be viewed as interchangeable in that their performance characteristics differ considerably between brands.

38.9 VALIDATION (VIRUS CLEARANCE EVALUATION) STUDIES

38.9.1 Scale-Down Procedure

Because it is impossible to demonstrate effective removal/inactivation of viruses at a large scale, and one cannot introduce a virus into a cGMP environment, regulatory agencies mandate use of scale-down models. When performing scale-down clearance assays, a firm must be able to demonstrate a direct relationship between their model and the corresponding full-scale unit operation. Scale-down filtration models are typically downsized up to 1/4000 (26).

In order to properly model large-scale filtration, certain parameters must be held constant. These include volumetric throughput (generally), transmembrane pressure, filter type, and fluid composition. As noted above, for certain filter types, flux decay is a better end point definition than volumetric throughput. In addition it is worth noting that scale-down studies are not an exact replicate of the full-scale operation. Some virus filters are subject to excessive fouling under conditions (due to the nature of the study) beyond what would normally be observed during normal processing; this is likely due to low capacity for virus spike impurities. As a result, studies with some filters may underestimate the true retentive capabilities that would normally be achieved during production.

38.9.2 Calculation of Logarithmic Retentive (or Reduction) Value (LRV)

As described in Q5A and above (8), the removal and/or inactivation of a given virus is typically expressed as the $\log_{10}$ reduction in viral titer. It is referred to as the *log reduction value* (LRV) (see above) and is calculated as follows:

$$LRV = \log_{10}((Vol_{in} \times Titer_{in})/(Vol_{out} \times Titer_{out}))$$

where

$$Vol_{in} = \text{volume of input feedstream}$$

$$Vol_{out} = \text{volume of filter permeate}$$

$$Titer_{in} = \text{titer of virus in input feedstream}$$

$$Titer_{out} = \text{titer of virus in output permeate}$$

38.9.3 Model Viruses

There is no single indicator species to be employed for virus validation studies, as is the case with demonstration of filtration sterilization for bacterial removal. The product-specific validation must be conducted with a panel of viruses, the appropriateness of which is determined by considerations including the type of source material (plasma-derived biologicals vs cell-line derived) and the product phase for which viral clearance testing is conducted (2,4). In general, the panel of test viruses used should include relevant viruses (i.e. known/suspected viral contaminants), and model viruses. Relevant viruses are, for example, HIV and hepatitis B and C viruses, which can be found in human plasma, and are known contaminants of blood products. Some relevant viruses, for example, hepatitis B and C viruses are difficult to propagate *in vitro*; in these cases specific model viruses may be used. Specific model viruses are viruses that resemble known viral contaminants; for example, the bovine viral diarrhea virus (BVDV) and the Sindbis virus have been used as models for hepatitis C virus. Similarly, murine leukemia virus (MuLV) is often used as a model for noninfectious endogenous retroviruses associated with rodent cell lines. Additionally, nonspecific model viruses are also included in the test panel to characterize the theoretical clearance capability of the manufacturing process, that is, to assess the "robustness" of the process. This category includes viruses of different size and varied physicochemical and biophysical characteristics. They are not expected to be associated with the product, but are included to address theoretical safety concerns and add confidence that the process can handle unknown or undetected viruses.

Examples of the types of viruses employed in viral clearance studies are shown in Table 38.5.

38.10 FUTURE TRENDS

38.10.1 Quality by Design (Q bD), Phage, and Virus Filters

In filter evaluation and validation studies, virus spike concentrations should be sufficiently high to meet the LRV needs of the process but not too high to alter the filter behavior. In other words, it is necessary to stay within the zone where process fluid effects dominate the filter fouling characteristics ("protein-dominated" zone) (7). This is normally seen when the cumulative virus challenge is $<1 - 5 \times 10^{13}$ particles/m^2. Also of note, the similar size of phage relative to small mammalian viruses [25–30 nm for PP7 and Φ X-174 bacteriophage vs 20–26 nm for MMV (or PPV)] (21) and spike purity considerations lend advantages to bacteriophage spikes in filter validation studies. Remaining in the "protein-dominated" zone and at the same time measuring the full extent of the filter clearance capacity is relatively difficult to achieve with mammalian viruses for some filters. This is because commercially available mammalian virus preparations are often <0.1% infectious (28) and contain virus cultivation–related impurities. Thus, a typical 0.1–0.5% (v/v) spike with infectious mammalian viruses may actually constitute a spike with two or more orders of magnitude more total particles. This is already inside the zone where the virus particles themselves dominates the filter fouling behavior (i.e. overloading or "virus-dominated zone"). Other extraneous impurities from the spike may introduce additional unpredictable and problematic variables to filter performance (29).

Other key advantages of bacteriophages are the ease with which they are produced and are quantified. Bacteriophage titers can typically exceed that of mammalian viruses under comparable growth conditions, and the preparations themselves tend to have fewer contaminant particles. In addition, the overnight plaque assay facilitates quantitative analysis of multiple grab samples, as the nature of the assay allows for rapid turnaround and high throughput. Overall, the use of bacteriophages may be more suitable for enabling the extensive testing required to define a study's design space (30), and do so for example, on an extremely cost-efficient basis.

38.10.2 Generic and Modular Validation (i.e. Bracketing)

A generic clearance study is one in which virus removal and inactivation is demonstrated for several steps in the purification process of a model antibody that can then be extrapolated to other antibodies. To claim this however, each protein product must undergo the same purification and virus

TABLE 38.5. Viruses Typically Used in Filter Validation Studies

Virus	Family	Size (nm)[a]	Envelope	Application
MuLV	Retroviridae	80–110	Yes	Biotech industry
MMV	Parvoviridae	18–26	No	Biotech industry
PRV	Herpesviridae	120–200	Yes	Biotech industry
Poliovirus	Picornaviridae	22–30	No	Biotech industry
Reovirus 3	Reoviridae	60–80	No	Biotech industry
HIV	Retroviridae	80–110	Yes	Plasma product industry
Duck HBV	Hepadnaviridae	40–48	Yes	Plasma product industry
Sindbis	Togaviridae	70	Yes	Plasma product industry
PPV	Parvoviridae	18–26	No	Biotech industry; plasma product industry
SV40	Papovaviridae	40–50	No	Biotech industry
HAV	Picornaviridae	22–30	No	Plasma product industry
B19	Parvoviridae	18–26	No	Plasma product industry
BVDV	Flaviviridae	50–70	Yes	Biotech and plasma product industry

[a]Size information from International Committee for the Taxonomy of Viruses and ICH Q5A (8,27).

removal/inactivation scheme as the model antibody. A modular clearance study on the other hand, is one that demonstrates virus removal or inactivation of individual steps, such that a given step may be extrapolated to identical steps employed for other antibodies. Therefore, each "module" in the purification scheme may be treated (studied) independently of the other modules. In principle, under a modular approach, different model monoclonal antibodies may be used to demonstrate viral clearance in different modules, if necessary and as long as the operating conditions are the same. The FDA's monoclonal antibody Points To Consider (mAb PTC) document (9) specifies acceptable conditions for both generic and modular validation. A similar validation streamlining concept described in the mAb PTC is bracketing. In this case, sponsors may demonstrate virus removal/inactivation for a particular module at two different values of a given parameter (e.g. volumetric throughput or for some filters, flow decay). In the European Union, these approaches are termed *in house* data, although the concept is essentially very similar. While sterilizing grade filters are validated at both the upper and lower limit, due to the considerable cost associated with conducting viral clearance evaluation studies, filter users generally conduct these studies at the "worst-case" condition for the specific unit operation. This demonstrates the minimum clearance a unit operation could provide. Worst-case conditions will vary depending on the method used and are determined by those factors that influence the clearance mechanism.

The application of modular, generic and bracketing approaches requires a strong scientific justification based on small-scale studies, manufacturing experience, and consultation of the peer-reviewed scientific literature. Phage-based quality by design (QbD) studies as described above could provide a useful path forward. In summary, flexibility during product development is considered critical for the health of patients, for the biopharmaceutical industry, and for the overall product pipeline

38.10.3 Potential New Market Entrants

Product and process refinement by biotech and enabling technology firms is an on-going process. Implementation of new technology as it evolves is desirable if consistent and robust performance resulting in high quality active ingredient can be demonstrated. It is likely that standardization efforts such as those championed by PDA will shape technical and performance attributes of new filter products in the future.

38.11 CONCLUDING COMMENTS

The clinical acceptability of biologicals and recombinant deoxyribonucleic acid (rDNA)-based products is concomitant with risk assessment and must be guided by risk-benefit analysis (30). Biopharmaceutical safety is the result of multiple orthogonal barriers operating in concert. While each approach, individually, may have limitations, their use in an integrated manner provides overlapping and complementary levels of protection from putative viruses to recipients of plasma-derived products as well as recombinant and monoclonal products. Concerted efforts of and collaboration between regulatory and industry groups, industry experts, and filter manufacturers will go a long way toward ensuring the virological safety of biopharmaceuticals.

ACKNOWLEDGMENTS

The authors thank Carla Lankford, Ewa Marszal, Elena Gubina (FDA), Jerry Martin, Mohammed Haque, Oliver Triebsch (Pall Corp.), Jennifer Campbell (Millipore Corp.), Sherri Dolan, Anika Meyer (Sartorius Stedim Biotech), and Mike Morgan (Asahi Kasei) for reviewing this document. This work was supported by a CRADA between CDER/FDA and Eli Lilly and Company. Views expressed in this chapter are those of the authors and not necessarily views or official policy of the US Food and Drug Administration.

LIST OF ABBREVIATIONS

UF/DF	Ultrafiltration/Diafiltration
cGMP	Current Good Manufacturing Practices
HCV	Hepatitis C Virus
HIV	Human Immunodeficiency Virus
PRV	Pseudorabies Virus
JEV	Japanese Encephalitis Virus
SV40	Simian Vacuolating Virus 40
PR772	Coliphage
BPV	Bovine Papilloma Virus
PPV	Porcine Parvovirus
CPV	Canine Parvovirus
EMC	Encephalomyocarditis
HAV	Hepatitis A Virus
MMV	Mouse Minute Virus
TCID	Tissue Culture Infectious Dose
BSL	Biosafety Level
ICH	International Conference on Harmonization

REFERENCES

1. PDA. Technical report 41: virus retentive filtration. Parenteral Drug Association; Bethesda (MD): 2008.
2. Farshid M, *et al*. Curr Opin Biotechnol 2005; 16: 561–567.
3. Horowitz B, Ben-Hur E. Ann Med 2000; 32: 475–484.
4. Garnick RL. Dev Biol Stand 1998; 93: 21–29.

5. Brorson K, Swann P, Brown J, Wilcox B, Shapiro M. In: Knäblein J, editor. Volume 4, Modern biopharmaceuticals. Weinheim: Wiley-VCH; 2005. pp. 1637–1668.
6. Brorson K, *et al*. Biotechnol Bioeng 2003; 82: 321–329.
7. Lute S, *et al*. Biotechnol Appl Biochem 2007; 47: 141–151.
8. ICH. Q5A: viral safety evaluation of biotechnology products derived from cell lines of human or animal origin. International Conference on Harmonisation of Technical Requirements for Registration of Pharmaceuticals for Human Use; Geneva, Switzerland; 1999.
9. FDA. Points to consider in the manufacture and testing of monoclonal antibody products for human use. Department of Health and Human Services, Food and Drug Administration; Silver Spring (MD): 1996.
10. EMEA. Guideline on virus safety evaluation of biotechnological investigational medicinal products. Committee for Proprietary Medical Products, European Medicines Evaluation Agency; London; 2006.
11. Grant D, *et al*. J Environ Sci 1989; 32: 43–51.
12. Burns DB, Zydney AL. Biotechnol Bioeng 1999; 64: 27–37.
13. Menon MK, Zydney AL. Biotechnol Bioeng 1999; 63: 298–307.
14. Syedain ZH, Bohonak DM, Zydney AL. Biotechnol Prog 2006; 22: 1163–1169.
15. Markovic I. Expert Opin Drug Saf 2007; 6: 487–491.
16. Bolton G, *et al*. Biotechnol Appl Biochem 2005; 42: 133–142.
17. Aranha H. Biopharm Int 2001; 14: 28–35.
18. Aranha H. Biopharm Int 2001; 14: 32–43.
19. Lute S, *et al*. Appl Environ Microbiol 2004; 70: 4864–4871.
20. Brorson K, *et al*. J Chromatogr A 2008; 1207: 110–121.
21. Lute S, *et al*. PDA J Pharm Sci Technol 2008; 62: 318–333.
22. Strauss DM, Lute S, Tebaykina Z, Frey DD, Ho C, Blank GS, Brorson K, Chen Q, Yang B. Biotech Bioeng 2009; 104: 371–380.
23. Brorson K, Sofer G, Aranha H. PDA J Pharm Sci Technol 2005; 59: 341–345.
24. Brorson K, *et al*. PDA J Pharm Sci Technol 2005; 59: 177–186.
25. Brorson K, *et al*. PDA J Pharm Sci Technol 2008; 62: 334–343.
26. Phillips M, Bolton G, Krishnan M, Lewnard J, Raghunath B.. In: Shukala A, Etzel M, Gadam S, editors. Process scale bioseparations for the biopharmaceutical industry. Boca Raton (FL): Taylor and Francis; 2007. pp. 333–366.
27. Van Regenmortel M, Fauquet C, Bishop D, Carstens E, Estes M, Lemon S. Virus taxonomy: seventh report of the international committee on taxonomy of viruses. San Diego (CA): Academic Press; 2000.
28. Brorson K, *et al*. Biologicals 2002; 30: 15–26.
29. PDA. Preparation of virus spikes used for virus clearance studies. Parenteral Drug Association virus preparation standardization task force. Bethesda (MD); 2010.
30. ICH. Pharmaceutical development. International Conference on Harmonisation of Technical Requirements for Registration of Pharmaceuticals for Human Use; Geneva, Switzerland; 2008.
31. BurnoufT, Radosevich M, El-Ekiaby M, Satoh S, Sato T, Amin SN, *et al*. Vox Sang 2003; 84: 111–119.
32. Burnouf-Radosevich M, Appourchaux P, Huart JJ, Burnouf T. Vox Sang 1994; 67: 132–138.
33. Furuya K, Murai K, Yokoyama T, Maeno H, Takeda Y, Murozuka T, *et al*. Vox Sang 2006; 91: 119–125.
34. O'Grady J, Losikoff A, Poiley J, Fickett D, Oliver C. Dev Biol Stand 1996; 88: 319–326.
35. Troccoli NM, McIver J, Losikoff A, Poiley J. Biologicals 1998; 26: 321–329.
36. Aranha-Creado H, Fennington GJ Jr., PDA J Pharm Sci Technol 1997; 51: 208–212.
37. Aranha-Creado H, Peterson J, Huang PY. Biologicals 1998; 26: 167–172.
38. Oshima KH, Comans TW, Highsmith AK, Ades EW. J Acquir Immune Defic Syndr Hum Retrovirol 1995; 8: 64–65.
39. Oshima KH, Evans-Strickfaden T, Highsmith A. Vox Sang 1998; 75: 181–188.
40. Oshima KH, Evans-Strickfaden TT, Highsmith AK, Ades EW. Biologicals 1996; 24: 137–145.
41. Roberts P. J Virol Methods 1997; 65: 27–31.
42. Brough H, Antoniou C, Carter J, Jakubik J, Xu Y, Lutz H. Biotechnol Prog 2002; 18: 782–795.
43. Yamaguchi K, Hamamoto Y, Manabe S, Yamamoto N. J Electron Microsc 1991; 40: 337–345.
44. Asahi Kasei Corporation, Planova Division, Tokyo, Japan (Product Brochure).
45. Laursen I, Houen G, Hojrup P, Brouwer N, Krogsoe LB, Blou L, *et al*. Vox Sang 2007; 92: 338–350.
46. Kim IS, Choi YW, Kang Y, Sung HM, Sohn KW, Kim YS. J Microbiol Biotechnol 2008; 18: 1317–1325.
47. Parkkinen J, Rahola A, von Bonsdorff L, Tolo H, Torma E. Vox Sang 2006; 90: 97–104.
48. Zhou JX, Solamo F, Hong T, Shearer M, Tressel T. Biotechnol Bioeng 2008; 100: 488–496.
49. Oshima KH, Evans-Strickfaden TT, Highsmith AK, Ades EW. Can J Microbiol 1995; 41: 316–322.
50. DiLeo AJ, Allegrezza AE, Jr. Builder SE. Biotechnology (N Y) 1992; 10: 182–188.
51. DiLeo AJ, Vacante DA, Deane EF. Biologicals 1993; 21: 287–296.
52. DiLeo AJ, Vacante DA, Deane EF. Biologicals 1993; 21: 275–286.
53. Hughes B, Bradburne A, Sheppard A, Young D. Dev Biol Stand 1996; 88: 91–98.
54. Phillips MW, DiLeo AJ. Biologicals 1996; 24: 243–253.
55. Tarrach K, Meyer, A., Dathe, J.E., Sun, H. BioPharm Intl 2007; 20: 58–63.
56. Kreil TR, Wieser A, Berting A, Spruth M, Medek C, Polsler G, *et al*. Transfusion 2006; 46: 1143–1151.

PART VI

BIOPHARMACEUTICAL FACILITY DESIGN

39

BIOPHARMACEUTICAL FACILITY DESIGN AND VALIDATION

JEFFERY N. ODUM

CPIP Biotech Sector Lead & Director of Operations Integrated Project Services,

39.1 INTRODUCTION

The manufacture of biotherapeutic drug substances is based on a synergy between science, product, process, and facility. This relationship is often difficult for many outside the industry to understand; manufacturing drug products from living organisms is a complex effort that must focus priority on patient safety. The guidelines that provide the foundation of compliance and regulatory oversight are supported by a number of guidance documents that provide the basis for validation, the precursor to product licensure. Facility validation is recognized as the documented process that supports the verification that the facility and its systems, equipments, and features meet the established specifications and quality attributes defined by the manufacturing organization.

Biomanufacturing facilities are based on a simple, yet complex principle: protect the product. As the manufacturing process requires the introduction of living organisms into a mechanical-based process manufacturing system, the focus on product protection and environmental control is one of the key attributes of facility design.

There are a number of basic considerations of biomanufacturing facility design. There is no "universal" standard for good manufacturing practice (GMP). Each product, process, and facility has unique attributes that influence regulatory compliance as defined by the GMPs. As we will see, defining these attributes is a critical aspect of the validation process and one that regulators have placed a greater focus on today than in the past.

The product and manufacturing process have a great influence on GMP interpretation. The guidelines that are defined in 21 Code of Federal Regulations (CFR), parts 210 and 211 (1), are open to interpretation and influenced by a number of global guidance documents. For products marketed in countries outside the United States, compliance to global guidelines from Europe, Asia, and other regions, having a familiarity with the validation requirements, will be critical in the validation process.

While the majority of biologic drug substances are not sterile but controlled to a low, measurable level of bioburden, the principles of aseptic processing as defined by the US Food and Drug Administration (FDA)'s 2004 Aseptic Guidance (2) are generally followed. This higher level of manufacturing control is also one of the reasons that biologic products are more complex and costly to manufacture. The convergence of unique product attributes, complex process unit operations, fragile cell lines, and interpretive regulatory requirements becomes the backdrop of the validation challenge.

39.2 DESIGNING FOR COMPLIANCE

From the beginning of the facility design process, the focus on regulatory compliance leading to successful validation and product/facility licensure must be maintained. Every design activity will eventually have some form of impact on the validation process. Commissioning, qualification/verification, and process validation (PV) can all be influenced by what may seem the most insignificant decision at the time.

The FDA utilizes a systems-based approach for regulatory review (3). The six systems are quality, production,

Downstream Industrial Biotechnology: Recovery and Purification, First Edition. Edited by Michael C. Flickinger.

facilities, materials, packaging and labeling, and laboratory controls.

The routine inspection process will either be a full inspection of the quality system and three of the other defined systems or an abbreviated inspection of the quality system plus one other system. The fact that the facility is considered a critical system defines the importance of the relationship between facility design and validation.

Regulatory guidance for how drug manufacturing facilities will be reviewed and inspected comes from a number of different documents. In addition to the previously referenced current good manufacturing practices (cGMPs), the US FDA has issued Q7A that defines many of the requirements for facility and equipment compliance to the cGMPs (4).

The European Union facility requirements can be found in Annex 1 (5).

The current global guidance that is a reference foundation document has been issued by the International Committee on Harmonization (ICH) under its guidance Q7 (6). This document provides the base compliance requirements that have been accepted by all members of the ICH working groups, including the United States, Europe, and Japan.

39.2.1 Facility Considerations

Key to a successful validation effort is a well-defined manufacturing process that is seamlessly integrated with the facility. In defining the manufacturing process, there are a number of drivers that will have a significant impact on the facility design. The synergy between product–process–facility will become apparent.

The large majority of biomanufacturing operations are not designed to produce sterile drug substance. Even if aseptic manufacturing standards are implemented, the focus is on bioburden control. In bioburden-controlled processing, the distinction between upstream and downstream operations is important. The upstream operations (seed inoculation and cell culture fermentation) are commonly designed as axenic operations that imply the need for a greater level of product protection. Downstream operations (recovery, purification, and bulk filling) are operated as bioburden-control processes, again necessitating the need for contamination control, but to predetermined levels based on the product/process attributes.

Biomanufacturing process systems will be either defined as open or closed (7). Process definition in this case relates to how the various manufacturing unit operations are executed based on their ability to protect the product. Open systems require special design attributes that will allow for the reduced risk of product contamination due to exposure to the manufacturing environment. Open systems are allowed under the GMP regulations but must be integrated into the facility in a manner that minimizes risk in accordance with current regulatory expectations.

Closed systems are designed in a manner to ensure product protection by removing the environmental risk of exposed product. While the concept of a closed manufacturing system may seem like a simple undertaking, it is often very difficult to deliver. A simple example is the process of taking in-process samples. Many current sampling systems require the system to be placed in an "open" state to remove the sample from the system. If overlooked, GMP compliance is not only unmet but product safety may be compromised.

The GMPs make reference for the need to have an orderly flow of manufacturing operations to minimize product contamination risks and allow for an orderly flow of material, personnel, and equipment through the facility during manufacturing operations (8). There is also a need for "logic" in terms of room and area adjacencies to ensure manufacturing efficiency and provide for room/area segregations that also meet GMP requirements. The importance of these facility attributes is seen in the documentation required for licensure. FDA and other regulatory agencies require companies to submit specific documents as part of the application, including layouts and flow diagrams (9).

As facility design concepts are developed, the focus on functional adjacency becomes a critical aspect of GMP compliance. This concept includes the definition of layouts so that flows are unidirectional, segregation of clean and dirty operations is maintained, area environmental classifications support open/closed manufacturing operations, and air flows and room pressures support both product and environmental protection. Examples of these documents are given in Figs. 39.1–39.5.

There may also be an impact of facility validation that is related to the biosafety classification of the product/process. While the current GMPs do not specifically mention biosafety-related issues around containment, the attributes of the facility design and thus, the validation of support systems will be affected. For example, common biosafety level 3 design requires the building ventilation systems utilize segregated, single-pass air flow with negative room pressures to "contain" any particles that may escape to the environment that can prove detrimental to the workers in the facility. For this type of systems, the qualification protocols must be written to specifically address the issues around air flow volume, direction, and quality.

Another facility consideration that affects validation is the gowning philosophy/approach that is implemented. The definition of personnel flows will have a major impact on how gowning philosophy is translated into physical room layouts and air lock configurations. Figure 39.6 provides some different examples of how the questions around air lock configuration to meet defined gowning requirements can be addressed.

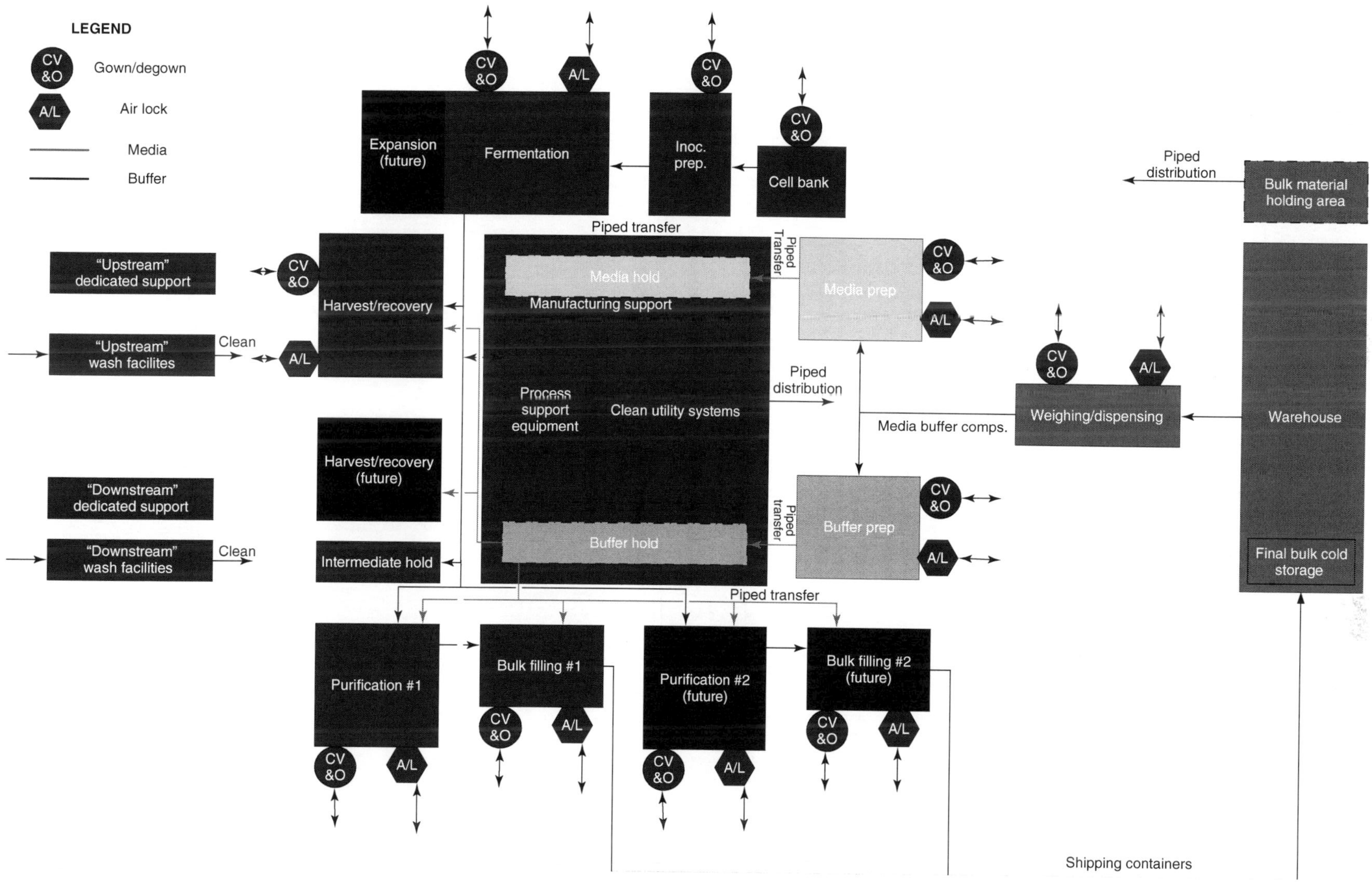

Figure 39.1. Block flow adjacency diagram.

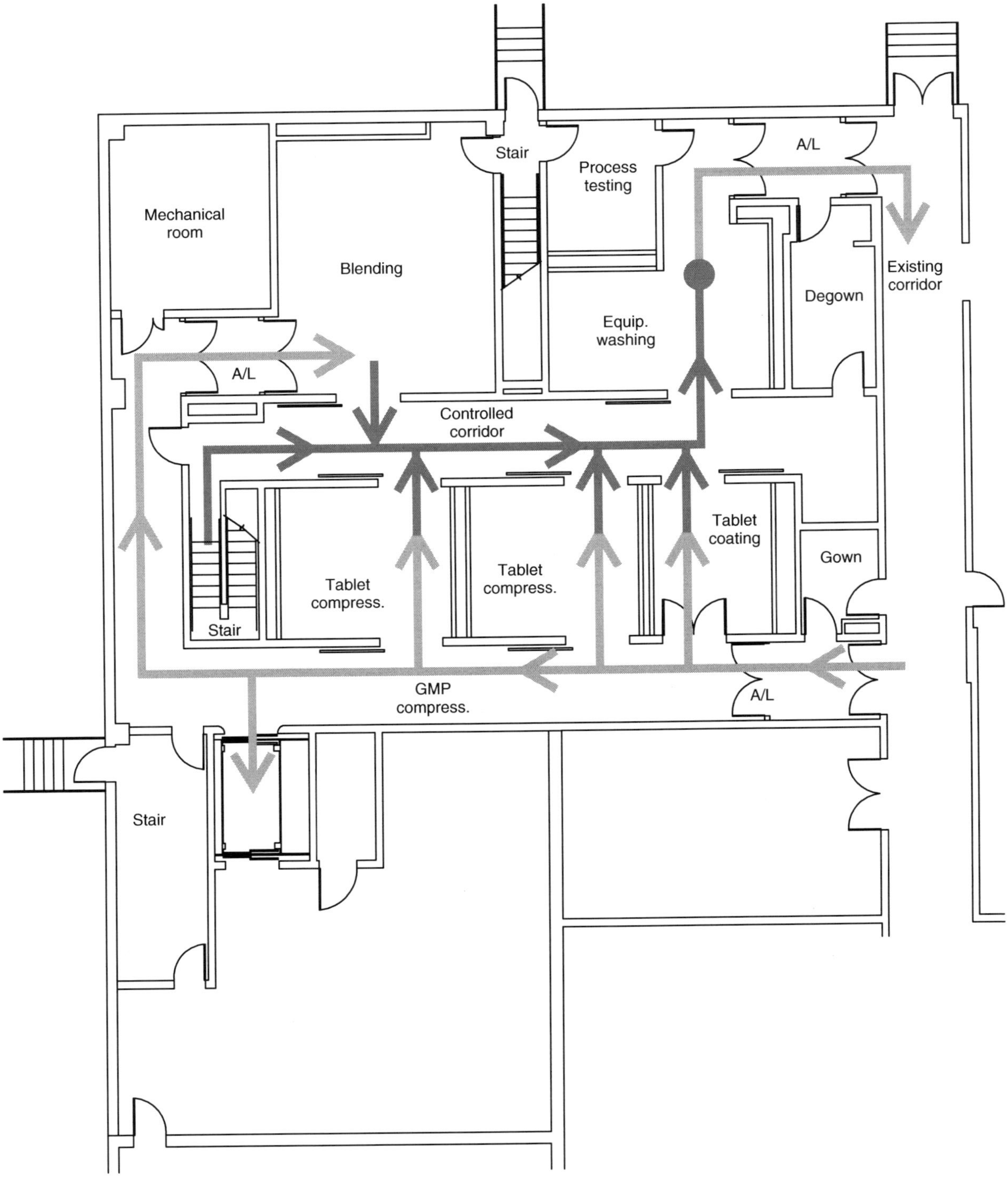

Figure 39.2. Equipment flow diagram.

Facility considerations also include design aspects such as materials-of-construction selection, finishes that support cleaning in accordance with the GMPs (10), and safety aspects based on personnel and/or environmental aspects. These include containment requirements for biosafety level compliance (11), hazardous material handling and storage, and unique cleaning requirements for equipment. Gowning approach and requirements will also affect facility design and be a key element in the overall compliance strategy.

39.2.2 Product–Process–Facility Integration

Designing for compliance in the manufacture of biological drug products is based on an industry continuum that

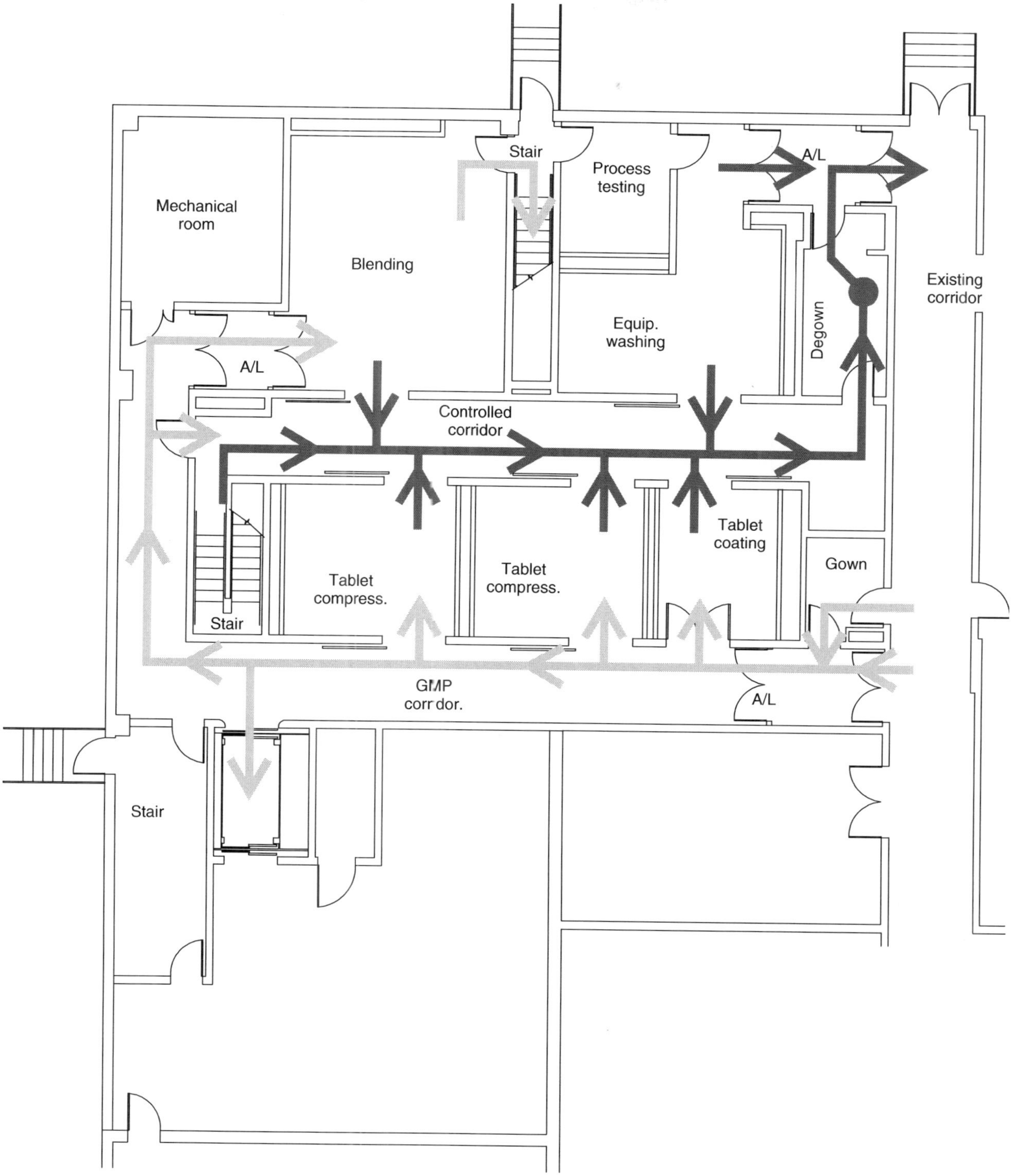

Figure 39.3. Personnel flow diagram.

recognizes the synergy between product and process. This synergy is aptly discussed in Chapter 1 of the Biopharmaceutical Facilities Baseline Guide (12). For a facility to be in compliance with the cGMP guidelines, there must be a well-defined link between the product attributes, the process unit operations, and the facility attributes. This link is discussed in more detail later.

39.2.3 The Role of Quality by Design

Quality by design (QbD) is defined in ICH Q8 as a systematic approach to product development that begins with

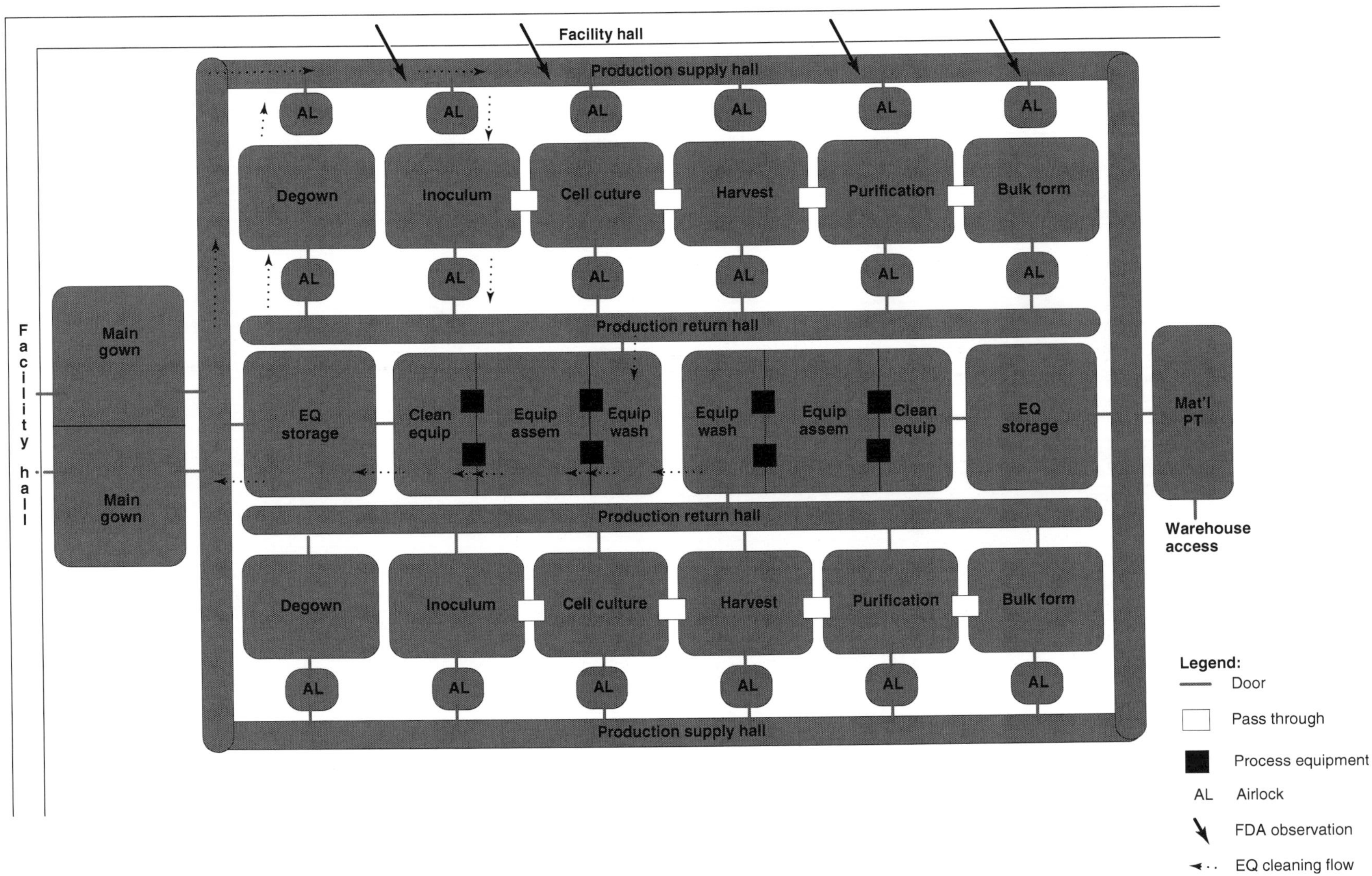

Figure 39.4. Segregation diagram (courtesy of International Society of Pharmaceutical Engineering).

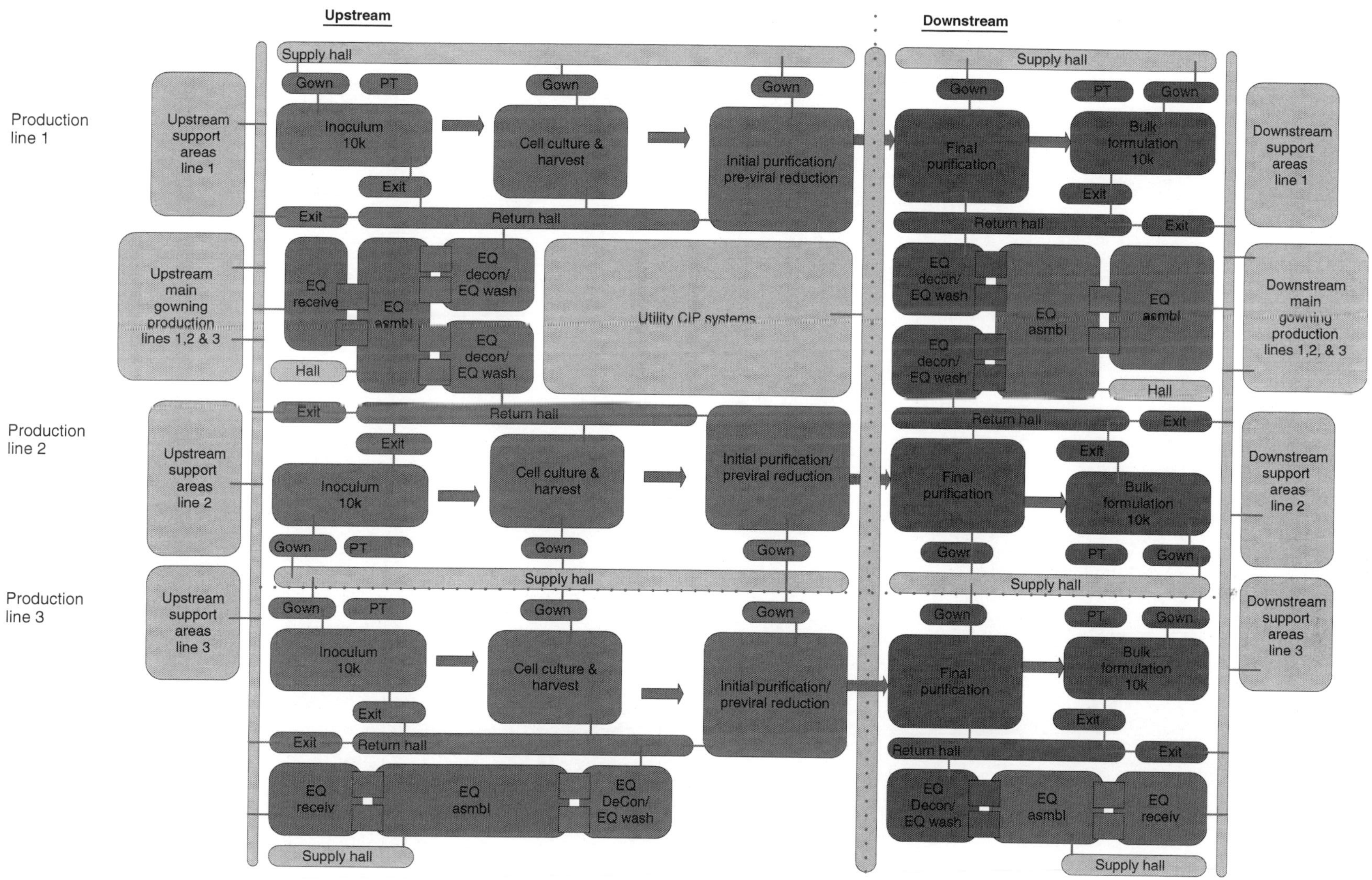

Figure 39.5. Facility layout adjacency diagram with flows.

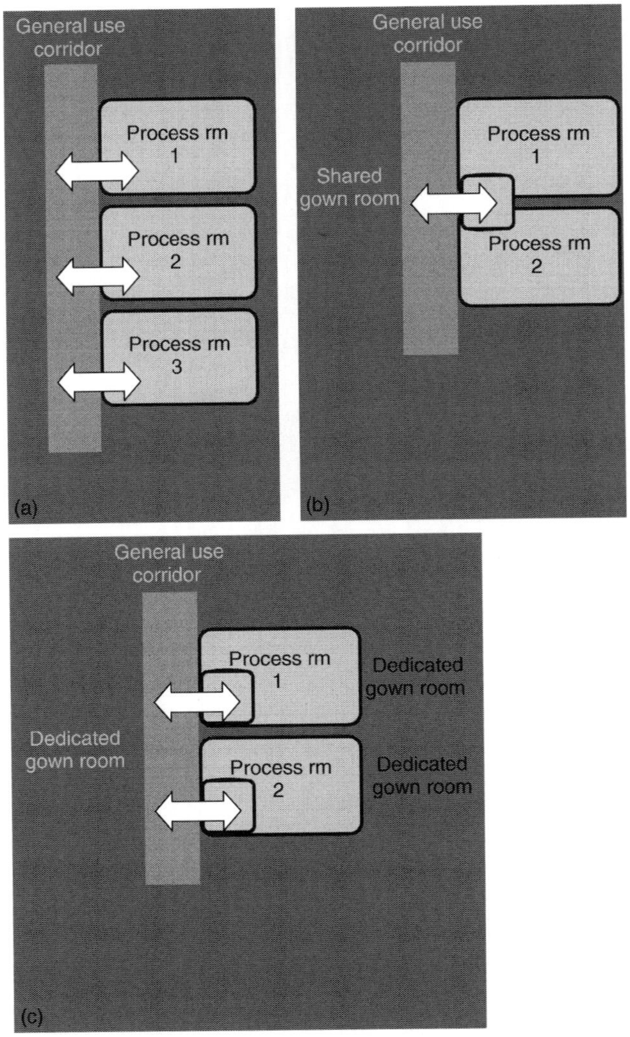

Figure 39.6. Air lock configurations.

predefined objectives and emphasizes product and process understanding and process control based on sound science and quality risk management (13). Current regulatory philosophy focuses on the implementation of QbD to support PV.

There are six steps in the QbD approach and are shown in Fig. 39.7.

The quality target product profile (QTPP) is a prospective summary of the quality characteristics of the drug product that should ideally be achieved to ensure the desired quality, taking into account product safety and efficacy. The QTPP includes the intended use, quality characteristics, delivery method, and attributes affecting pharmacokinetic attributes. A simple example of a QTPP is shown in Table 39.1.

Critical quality attributes (CQAs) are any physical, chemical, biological, or microbiological property or characteristic of a material that should be within an appropriate limit, range, or distribution to ensure the desired product quality. With respect to QbD, the need is to focus only on those attributes and the impact of those attributes across their ranges on safety and efficacy. For biological products, some common CQAs would include chemical purity, qualitative and quantitative impurities, and microbial quality.

TABLE 39.1. QTPP Example

Variable	QTPP Design Specs (Must-Haves)
Indications	Prostate cancer
Presentation(s)	100-mg/mL vial
Form(s)	Lyophilized product
Dose(s)	5 mg/kg/dose
Dosing regimen(s)	Q3
Storage conditions	24 months @ 2–8°C
Delivery system(s) and devices	IV infusion
Placebo required?	Yes

Critical process parameters (CPPs) are the independent parameters of the manufacturing process that are most likely to affect the CQAs of the product. Under current regulatory guidance as defined in ICH Q8, CPPs must be determined by sound scientific judgment and based on research, scale-up, and/or manufacturing experience.

CPPs must be controlled and monitored to confirm that the product profile is comparable to or better than the historical data from both development and manufacturing operations. In terms of the manufacturing process, these parameters should be controlled within a meaningful, narrow operating range to ensure that specifications are met. Common parameters would include temperature, pH level, oxygen content, or agitation speed.

The CQAs and CPPs are the key elements that are used to determine a product's design space, as defined in ICH Q8. Design space is the multidimensional combination of interaction of input variables and process parameters that have been demonstrated to provide assurance of product quality. The design space is proposed by the applicant (product manufacturer) and is subject to regulatory assessment and review.

Design space is a region where quality product can be produced. It includes material attributes and process parameters and includes an evaluation of scale and the equipment that is used in the manufacture of the drug product. It is arrived at by an iterative application of risk assessment and experimental design to the knowledge space. The steps in the development of a design space include (i) defining the CQAs, (ii) using prior knowledge and risk assessment to determine CPPs, (iii) understanding all the process variables that affect CQAs, (iv) relating product attributes to support process needs, and (vi) representing the understanding in the design space.

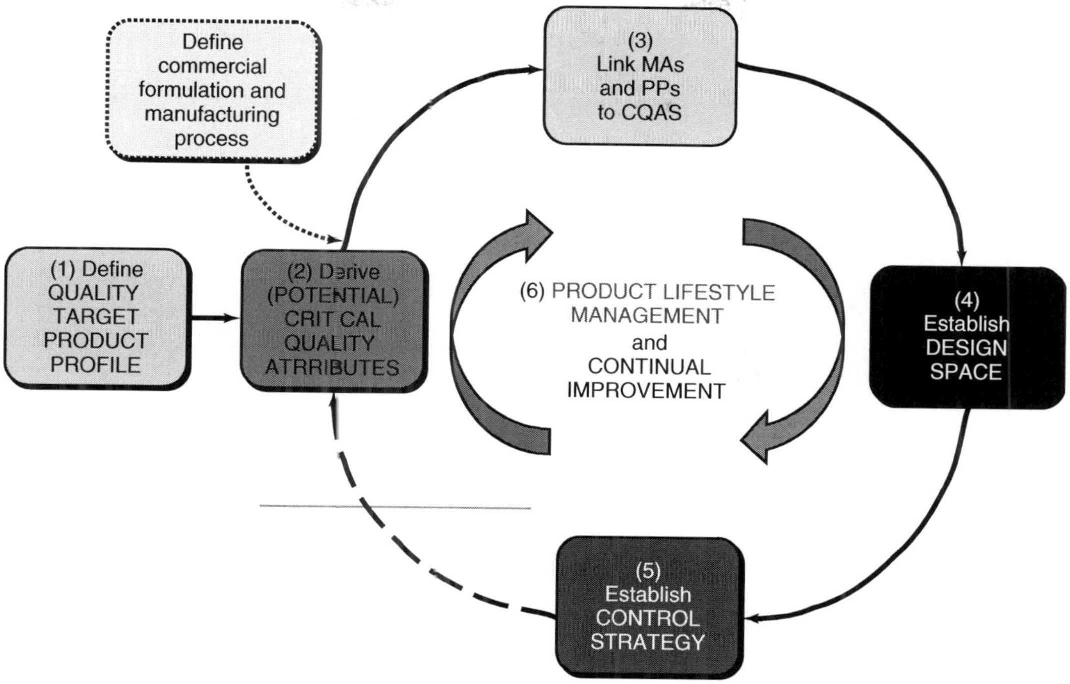

Figure 39.7. QbD approach steps.

Design space is visually represented as a three-dimensional model, as shown in Fig. 39.8.

Control strategy is defined as a planned set of controls, derived from current product and process understanding that assures process performance and product quality (14). The controls can include parameters and attributes related to drug substance and drug product materials and components, facility and equipment operating conditions, in-process controls, finished product specifications, and the associated methods and frequency of monitoring and control.

A control strategy can include (14) control of input material attributes, product specification(s), controls for unit operations, in-process or real-time release testing, and a monitoring program for verification of prediction models.

A holistic view of manufacturing control strategy is shown in Fig. 39.9.

A control strategy can be a complex and extensive document that contains elements that can include both conventional specifications as well as GMPs that are applied as a "baseline" approach. Some of these elements will contain both in-line and at-line testing during the manufacturing operations.

By using QbD principles, process requirements become well defined by the multidimensional design space and can be easily communicated to the project team to ensure that equipment and facility design will support PV requirements based on product and process requirements.

39.3 RISK MANAGEMENT

The FDA Twenty-First Century Initiative (15) placed a tremendous amount of emphasis on risk assessment and understanding. This emphasis is detailed in ICH Q9. This guidance document details the entire risk process that is represented in Fig. 39.10. Under ICH Q9, risk is defined as the combination of the probability of occurrence of harm and the severity of that harm as they relate to drug product quality. The focus is on understanding the risks to a product during manufacturing, how to identify those risks, and what steps can be taken to mitigate or eliminate the risk to ensure product quality, safety, and efficacy (16).

The risk process has three steps: risk identification, risk analysis, and risk evaluation. Identification focuses on what might go wrong during the manufacturing process and requires there be a clear identification of the risk in question. Analysis assesses the likelihood of the risk occurrence and evaluates the impact to the product. The evaluation then assigns levels of severity and defines actions that can be implemented. Figure 39.11 provides a sample risk matrix.

ICH Q9 recognizes that some risks are unavoidable. However, to reduce or control risks to the patient, consideration must be given to the following:

- Is the risk above an acceptable level?
- What can be done to reduce or eliminate the risk?
- What is the appropriate balance between benefits, risk, and resources?

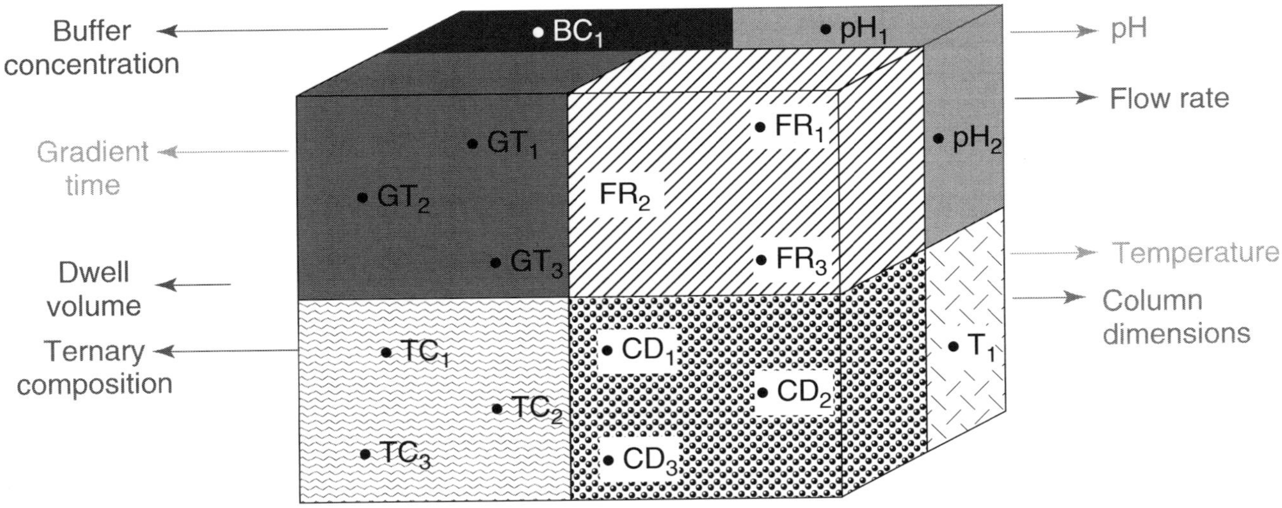

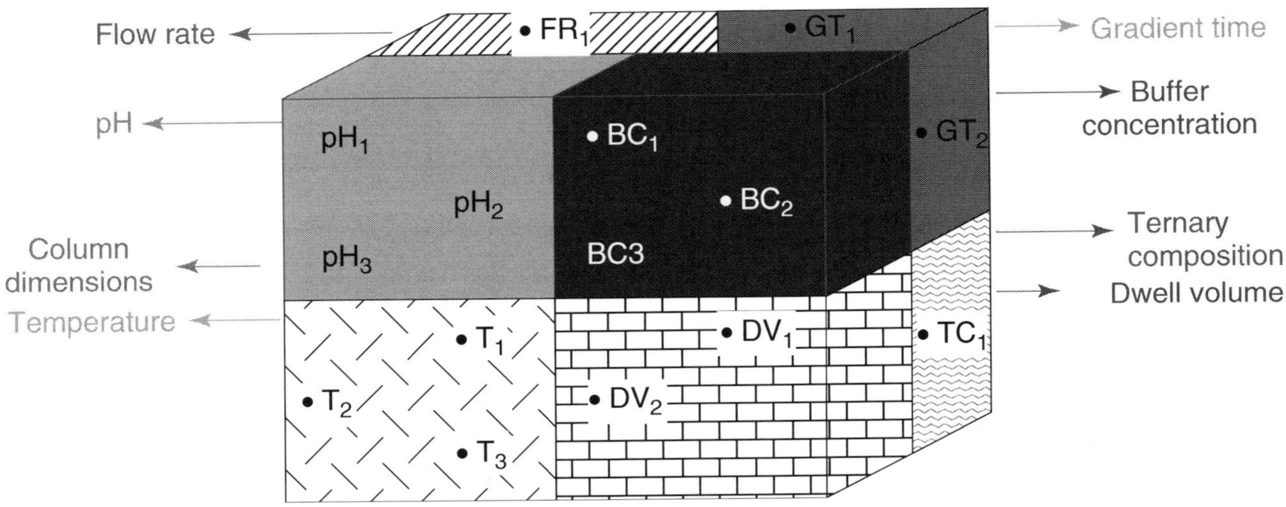

Figure 39.8. Design Space Model

- Are new risks introduced as a result of the identified risks being controlled?

The regulatory expectation is that any risk management program must provide clear, documented evidence that these questions have been answered. The risk management process must be systematic, well organized, and build clear decision making pathways that could withstand the scrutiny of a regulatory audit.

The following are two examples of risk assessment tools that are often used in industry and that are also recognized in ICH Q9. Both examples are for a production bioreactor (Figs. 39.12 and 39.13).

Evaluating risk involves placing a weight/value on each risk using severity and probability scores for each. This evaluation will identify the hazard/risk, the CQA that is affected, a score leading to a ranking of low (dark grey), medium (light grey), or high (black). Figure 39.13 provides an example of this type of risk evaluation tool.

To conduct effective risk analysis and provide the supporting data for validation, there are a number of recognized tools that can be used. These include control charts, design of experiments (DOEs), histograms, Pareto charts, and process capability analysis.

The impact of risk management on facility validation can be directly linked to the decisions made during facility design. To address process risks, it may become necessary to implement a particular technology or specific vendor's equipment in the process design. This decision would then

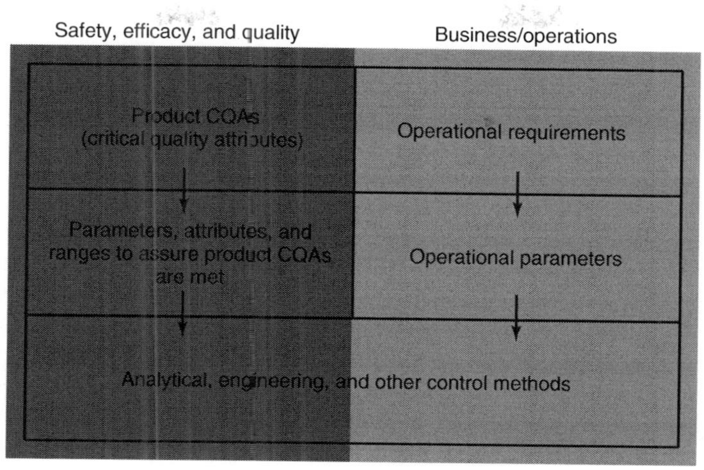

Figure 39.9. Manufacturing control strategy.

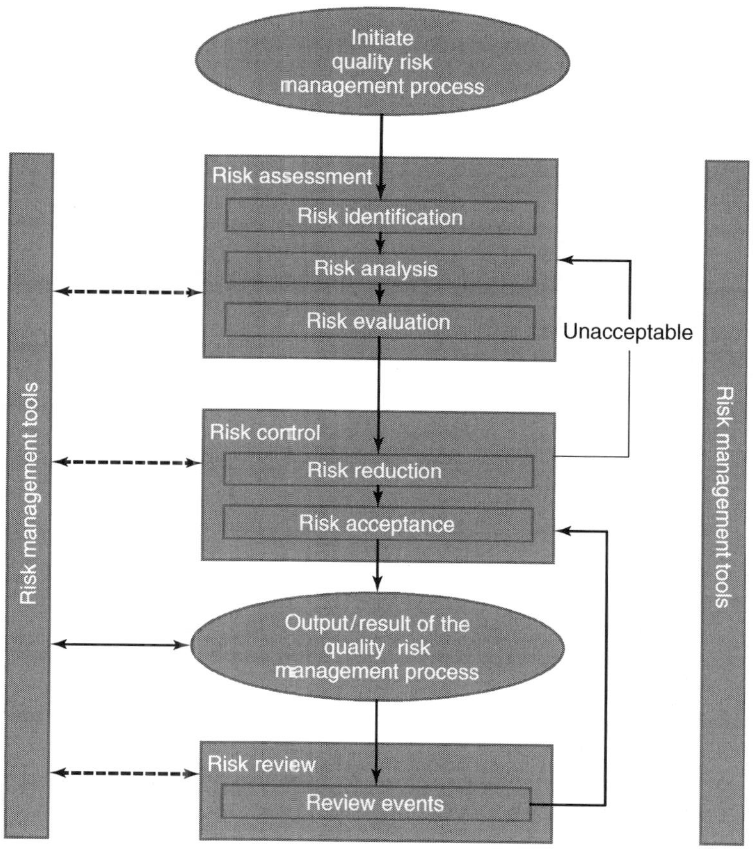

Figure 39.10. Chart from ICH Q9.

influence the development of protocols, the testing methods that are implemented, the training of operators, and even the overall configuration of the facility. That is why it is so important to address risk early; the impacts are often "trickle down" and can have a negative impact on the final completion efforts of qualification and validation activities.

39.4 QUALIFICATION/VERIFICATION

To understand the role that qualification plays in facility validation, it is necessary to look at a historical timeline showing the evolution of commissioning and qualification activities as recognized by global regulators (Fig. 39.14) (17). Qualification was defined as the series of activities

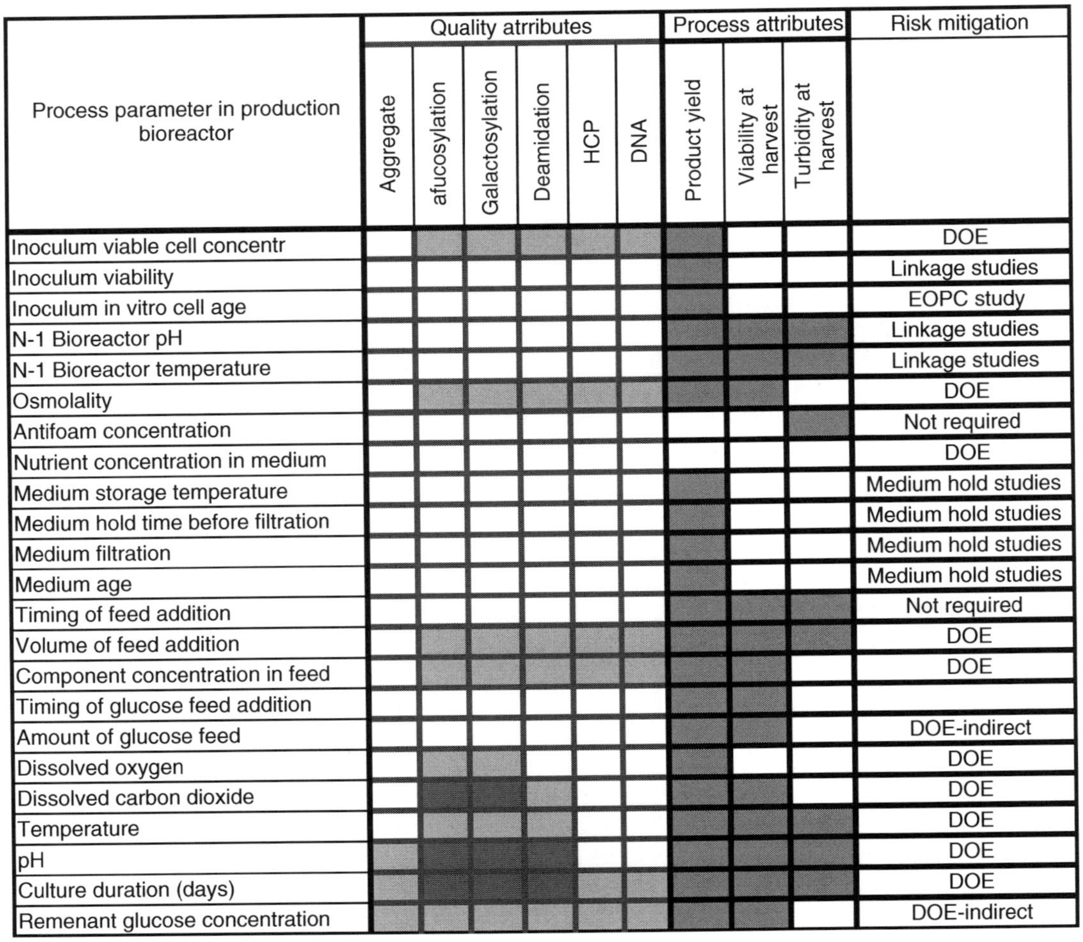

Figure 39.11. Sample risk matrix.

that were undertaken to demonstrate that facility, utility, and equipment systems were suitable for their intended use and would perform properly based on a predetermined set of operational (OQ) criteria. The primary focus was on quality control. Installation (IQ) and OQ protocols were engineering-based documents that defined specific design, IQ, and OQ requirements. These protocols were (and still today are) reviewed and approved by the quality unit and are focused on good documentation practices based on the development of a "check list" approach with approvals of execution activities.

Commissioning and qualification activities formed the foundation for PV as defined in the 1987 FDA Guide to Process Validation. System boundaries were defined as part of the Validation Master Plan. Each system was assigned an impact assessment level based on their potential impact to product quality, safety, and/or efficacy. The three levels of impact are (18) direct impact—system expected to have a direct impact on product quality; indirect impact—system not expected to have a direct impact on product quality but typically will support a direct impact system; and nonimpact—system that will not have any impact, direct or indirect, on the product quality.

ASTM (American Society for Testing & Materials) E 2500 (19) now provides the new standard for the qualification practices to support a risk-based validation system. In this new approach, the qualification activities are now based on a process of verification-based science and risk assessment to assure that cGMP equipment and systems. Quantification includes whether the activities are fit for use, can perform satisfactorily, and will they operate in a drug manufacturing, processing, packaging, or holding system in accordance with cGMP.

Verification is viewed in the ASTM E 2500 standard as a documented, life-cycle approach based on defined acceptance criteria, not a single event.

The foundation activities and life-cycle approach of an E 2500-based system are shown in Fig. 39.15 (19).

There are 10 recognized principles for executing a risk-based qualification process using the ASTM E 2500 guidance (20). They are

1. Focus on attributes that affect product quality,
2. User requirements specifications (URSs) are key to acceptability; IQ and OQ support the PV effort,

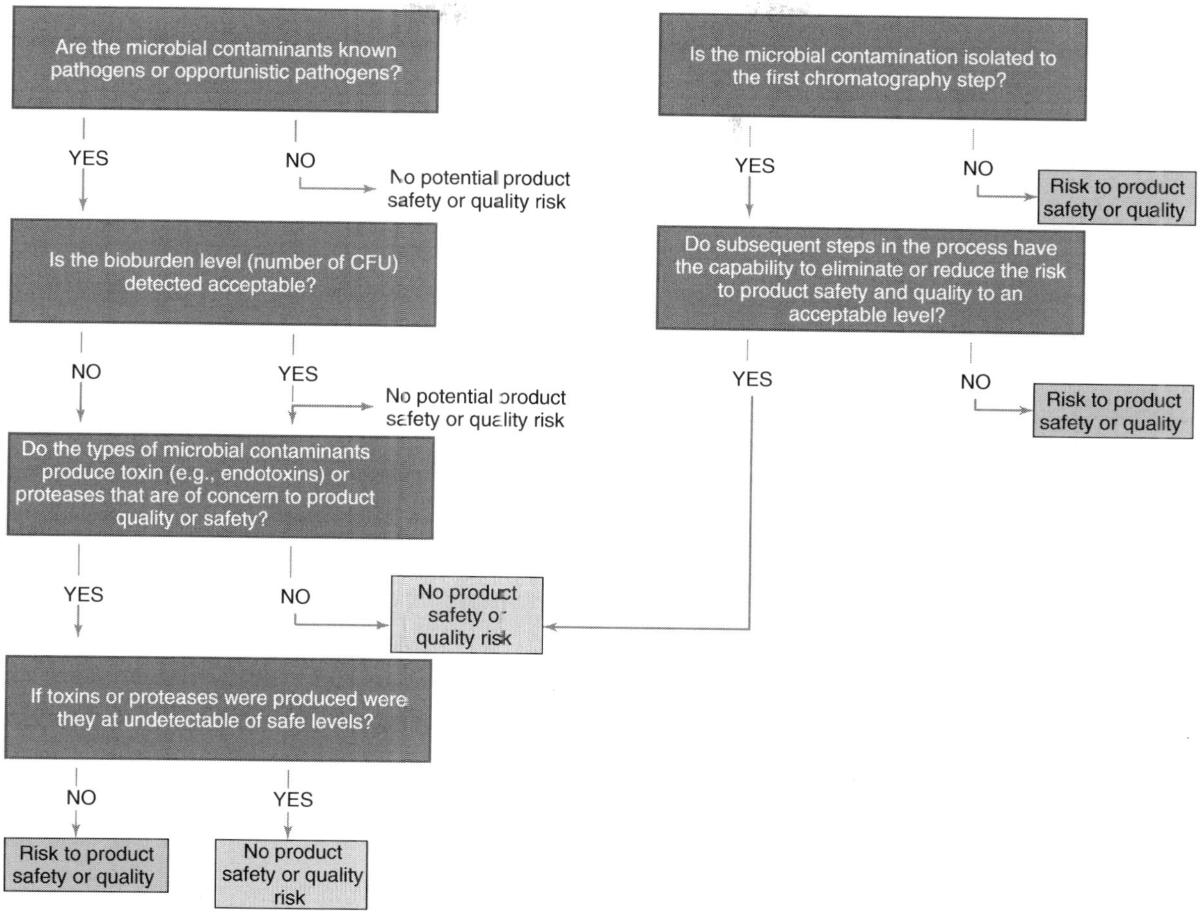

Figure 39.12. Risk assessment decision tree.

3. Risk assessments and process knowledge must be used to identify the critical elements,
4. Only critical features/functions need to be qualified,
5. All activities must contribute value,
6. Risk-based asset delivery; not "cookbook" requirements,
7. Value-added documents based on technical merit,
8. Use of supplier documentation is encouraged,
9. Test planning; test once,
10. Foster innovation.

There is value to looking at the differences in the traditional versus risk-based approach to qualification (Table 39.2). Today, both these approaches are still used and recognized by regulators. Migration toward the ASTM E 2500-based system will support the current PV guidance much better.

The verification process as defined in ASTM E 2500 provides some very distinct advantages. These can be summed up as follows. Verification is a life-cycle approach; not a "check" at the end of the qualification process. It is based on product and process knowledge, regulatory guidance

TABLE 39.2. Differences in Traditional Approach Versus Risk-Based Approach

For the traditional approach

- (Product) user requirements not formally documented
- Protocols developed from "templates"
- IQ/OQ protocols "preapproved"
- Commissioning not leveraged
- Engineering and "validation" personnel often distinct
- Emphasis on documents and not on system performance

For the risk-based approach

- Process requirements documented, approved
- Risk assessments determine critical aspects of design
- Engineering testing ("commissioning") verification
- All documents with technical merit used as evidence of fitness for use
- Emphasis on meeting process requirements

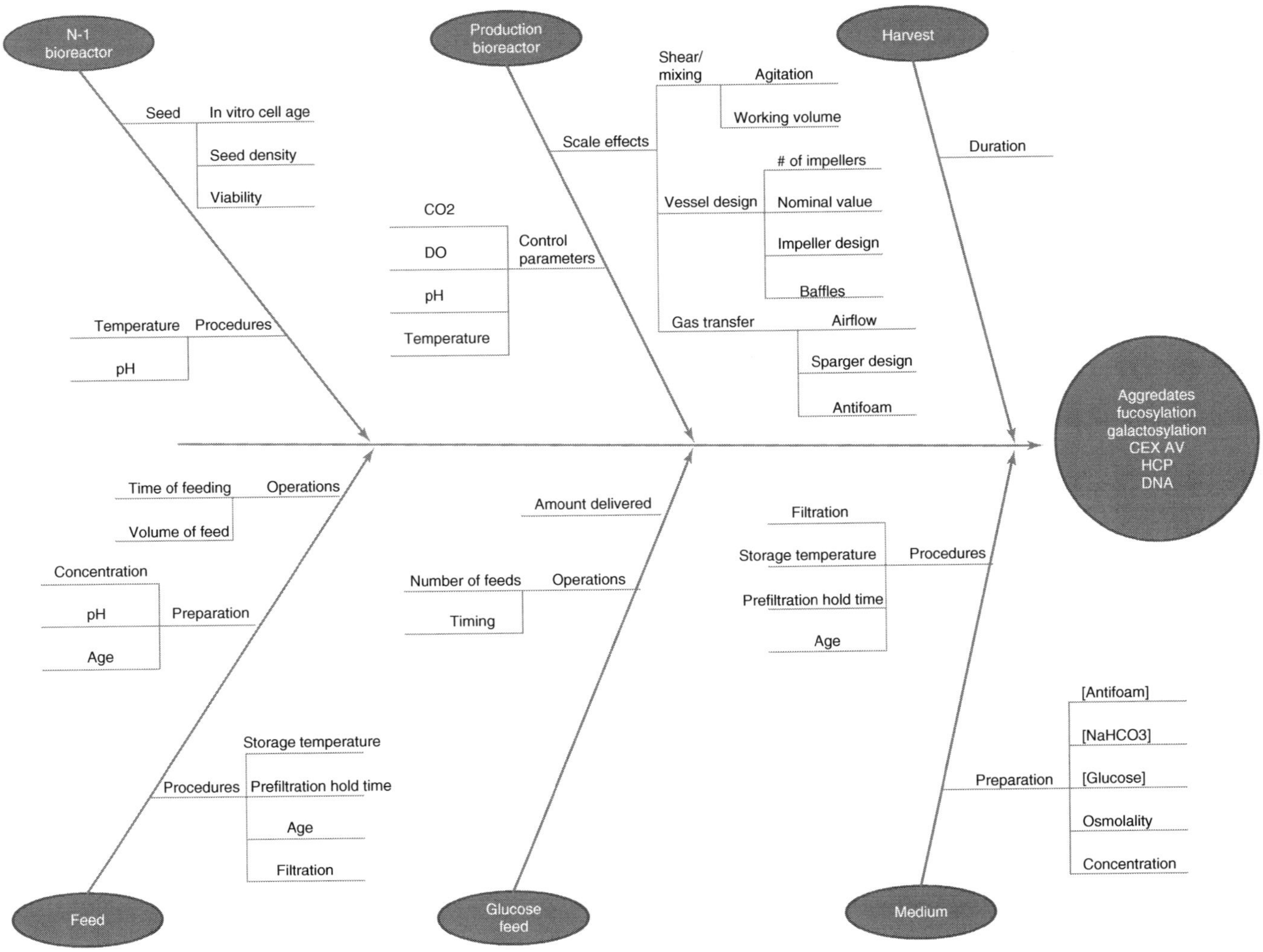

Figure 39.13. Ishikawa diagram of CPPs for a production bioreactor.

defined in the ICH documents, and internal quality standards. The approach must be documented through a verification plan that is approved by the quality unit. Acceptance criteria must be predetermined. There must be documentation that the equipment/system is fit for intended use [factory acceptance test (FAT) and site acceptance test (SAT)]. Acceptance and release must be documented and approved by subject matter experts (SMEs). SMEs are designated based on their combined educational and professional work experience and area of expertise.

The document types that support a risk-based verification approach include requirements document that defines the product and process quality requirements, project verification plan that defines verification strategy and roles/responsibilities and risk assessments of the product quality/patient safety, design review summary report that confirms design review(s), verification test plan that documents the testing plan and all the corresponding check sheets, and system acceptance and release report that confirms the system to be fit for intended use.

This verification process flow is shown in Fig. 39.16 (21).

39.5 PROCESS VALIDATION

In January of 2011, the FDA issued the first revision to the PV guidance since 1987. This document made significant changes to the regulatory expectations for validation and is based on the current risk-based approach that has been implemented in other guidance documents.

The most familiar definition of validation comes from the 1987 document.

> Establishing documented evidence which provides a high degree of assurance that a specific process will consistently produce a product meeting its pre-determined specifications and quality attributes. (22)

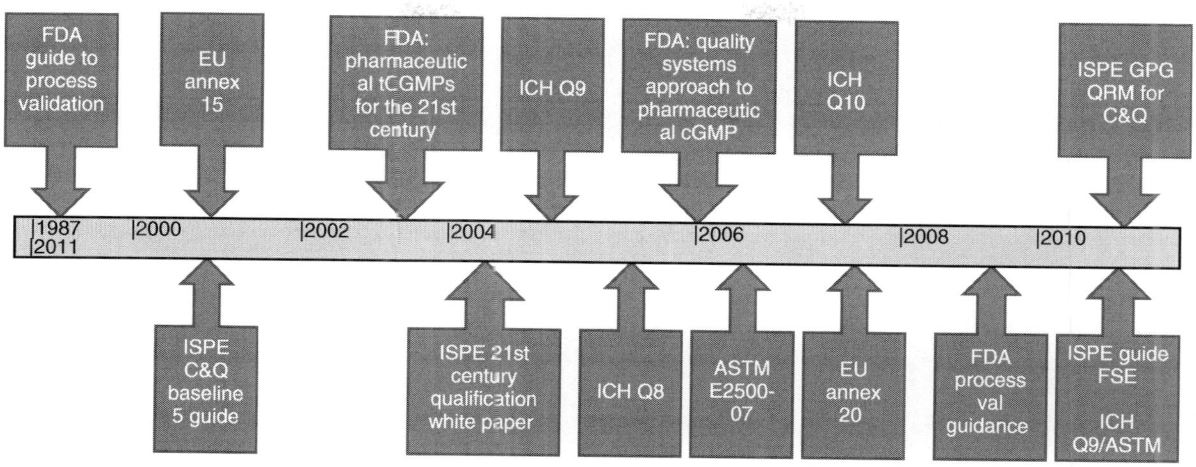

Figure 39.14. Evolution of commissioning and qualification.

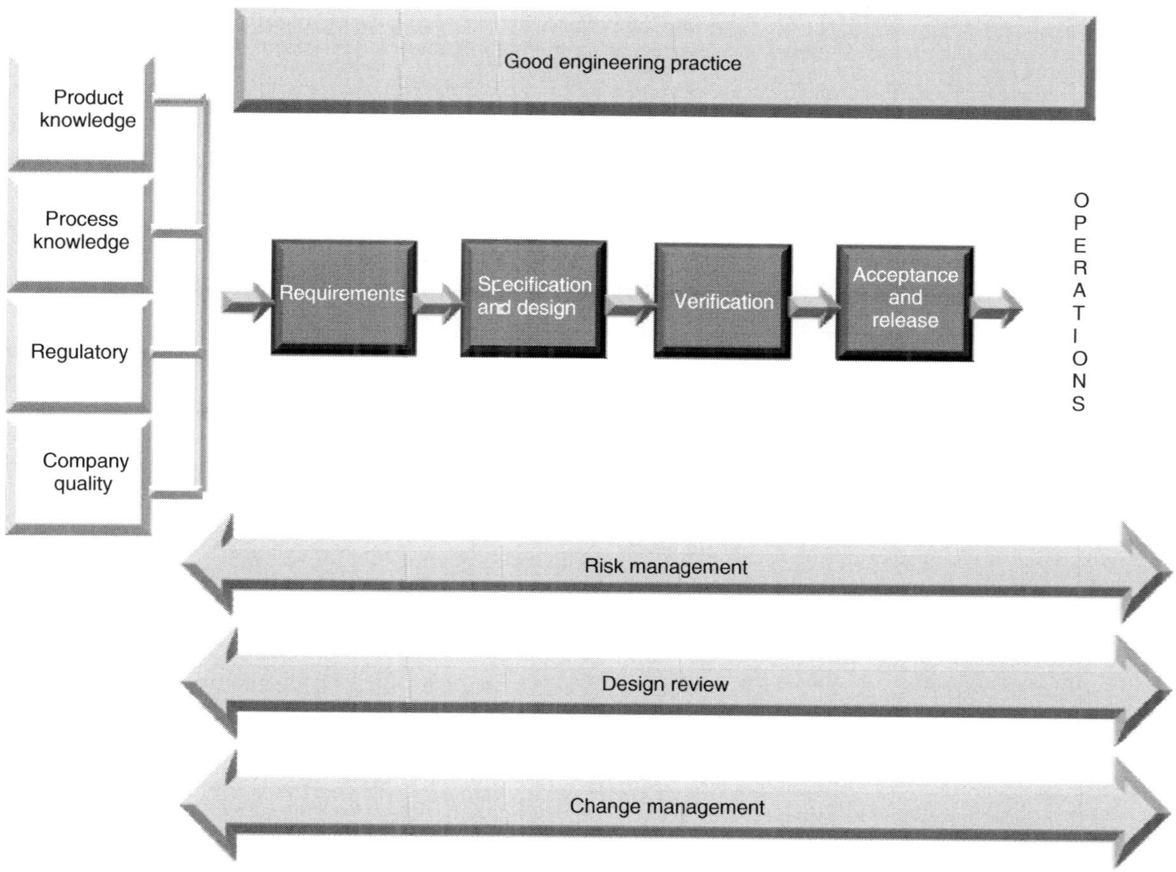

Figure 39.15. Good engineering practice.

Under the new PV guidance, this definition takes on a different approach:

> The collection and evaluation of data, from the design stage through production, which establishes scientific evidence that a process is capable of consistently delivering quality products. (23)

The established prerequisites for PV to occur can be divided into two general categories: process and facility elements. For the process, the requirements are as follows: equipment SOP (Standard Operating Procedure), approved process development reports and master batch records, and trained manufacturing personnel.

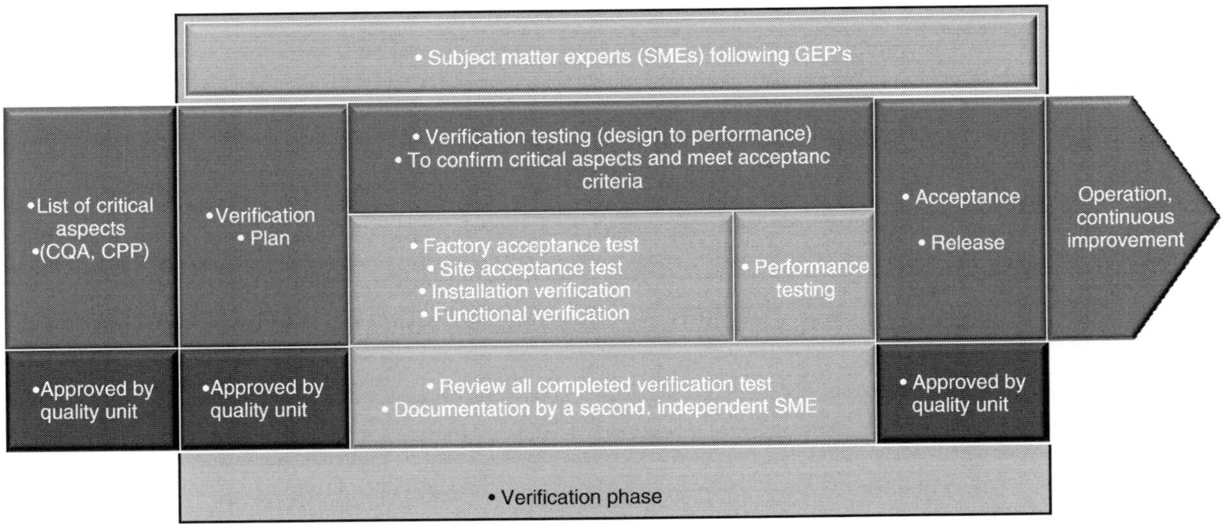

Figure 39.16. Verification process flow.

For the facility, these requirements are as follows: cGMP production and support systems in place; facility, utility, and production equipment qualified; preventative maintenance and calibration programs in place; cleaning, sanitization, and sterilization methods defined; and quality system in place.

The new PV paradigm has a three-stage process for validation. This process implements a risk-based approach to validation utilizing the concepts that are defined in ICH Q8 and Q9 and focuses on compliance around the Q7A guidance supported by the ASTM E 2500 methodology. The three stages are stage 1: process design, stage 2: process qualification, and stage 3: continued process verification.

Stage 1: process design. This stage defines the commercial process and focuses on development and scale-up experience. Also, the design space is created. In this stage, the process is defined based on the knowledge gained through the process development and scale-up activities. Studies undertaken at laboratory, pilot, and commercial scale will be the source of significant amounts of data. These data will be used to develop an understanding of the sources of variability in the process unit operations and their potential impact on the quality of the product.

During this stage, the regulatory expectation is that variability in the process can and will be managed commensurate with the risk(s) that have been defined. The use of DOE and process modeling are some of the methods that will define these criteria. The CPPs will be evaluated at the baseline.

Stage 2: process qualification. In this stage, the focus is on the confirmation that the process can reproduce product at the commercial scale. The facility, utility systems, and equipment must be qualified. This will be a qualification executed in accordance with the ASTM E 2500 requirements. The qualification of the process will have been completed and confirmation of the design space verified. This will require a series of tests that will support process confirmation.

To qualify the process there must be a qualification of the facility. This is accomplished through the IQ and OQ activities previously discussed. It also requires that all raw materials and components used in the manufacture of the drug substance/product have been approved. All written control procedures such as SOPs and batch records must be complete and approved. Finally, all personnel must have completed any required training.

Stage 3: continued process verification. In this stage, the focus is on process verification; does the process produce a consistent product that meets the design space requirements? The stage-3 activities are heavily focused on process monitoring, data collection, and the implementation of any process changes or improvements. The process will be in a constant state of monitoring during operation. To that end, the implementation of tools such as process analytical technology (PAT) will become even more prominent. Out-of-specification and out-of-tolerance conditions will be identified in real time and this feedback will be used to confirm the design space of the product.

Using this approach and methodology results in a significant amount of regulatory flexibility. On the basis of the science-based process knowledge, analysis of process risk, and defined design space boundaries, organizations may now be able to move away from the concept of "revalidation" and also no longer need to manufacture and hole for evaluation, three conformance batches of product before commercial release.

LIST OF ABBREVIATIONS

cGMP	Current good manufacturing practice
CPPs	Critical process parameters
CQAs	Critical quality attributes
FAT	Factory acceptance test
FDA	Food and Drug Administration
GMP	Good manufacturing practice
ICH	International Committee on Harmonization
IQ	Installation
ISPE	International Society of Pharmaceutical Engineering
OQ	Operational
PAT	Process analytical technology
PV	Process validation
QbD	Quality by design
QTPP	Quality target product profile
RTRT	Teal-time release testing
SAT	Site acceptance test
SMEs	Subject matter experts
URSs	User requirements specifications

Regulatory Documents Sited	Web Site
1. Title 21, *Code of Federal Regulations*, Good Manufacturing Practice Regulations	http://ecfr.gpoaccess.gov/cgi/t/text/text-idx?c=ecfr&rgn=div5&view=text&node=21:2.0.1.1.10&idno=21
2. "Guidance for Industry, Sterile Drug Products Produced by Aseptic Processing—Current Good Manufacturing Practice," US FDA, September 2004	http://www.fda.gov/ohrms/dockets/ac/05/briefing/2005-4136b1_04_Sterile%20Drug%20Products.pdf
3. "Guidance for Industry, Quality Systems Approach to Pharmaceutical cGMP Regulations," FDA, September 2006	http://www.geinstruments.com/library/industry-links-and-regulations/pharmaceutical-biopharmaceutical.html
4. "Guidance for Industry, Q7A Good Manufacturing Practice for Active Pharmaceutical Ingredients," FDA, August 2001	http://www.fda.gov/downloads/regulatoryinformation/guidances/ucm129098.pdf
5. EudraLex, "The Rules Governing Medicinal Products in the European Union," Annex 1, *Manufacture of Sterile Products*	http://freedownload.is/pdf/eudralex-the-rules-governing-medicinal-products-in-the-european-union-2071969.html
6. ICH Q7, "Good Manufacturing Practice Guide for Active Pharmaceutical Ingredients," November 2000	http://www.ich.org/products/guidelines/quality/quality-single/article/good-manufacturing-practice-guide-for-active-pharmaceutical-ingredients.html
7. *ISPE Baseline Pharmaceutical Engineering Guide*, Volume 6 "Biopharmaceutical Manufacturing Facilities," June 2004	http://www.ispe.org/
8. "Guidance for Industry, Q7A Good Manufacturing Practice," Section 4.13	http://www.fda.gov/downloads/regulatoryinformation/guidances/ucm129098.pdf
9. FDA form 356h, "Application to Market a New Drug, Biologic, or an Antibiotic Drug for Human Use," Item 15	http://www.fda.gov/downloads/AboutFDA/ReportsManualsForms/Forms/UCM082348.pdf
10. "Guidance for Industry, Q7A Good Manufacturing Practice," Section 4.1	http://www.fda.gov/downloads/regulatoryinformation/guidances/ucm129098.pdf
11. *NIH Guidelines for Research Involving Recombinant DNA Molecules*, Appendix K, October 2011	http://oba.od.nih.gov/oba/rac/guidelines/APPENDIX_K.htm

Regulatory Documents Sited	Web Site
12. ISPE Baseline Pharmaceutical Engineering Guide, Volume 6 "Biopharmaceutical Manufacturing Facilities," Chapter 1, June 2004	http://www.ispe.org/
13. "Guidance for Industry, Q8 (R2) Pharmaceutical Development, ICH," November 2009	http://www.fda.gov/downloads/Drugs/GuidanceComplianceRegulatoryInformation/Guidances/ucm073507.pdf
14. PQLI Guide, Part 1, "Product Realization using Quality by Design (QbD): concepts and Principles," ISPE, Chapter 18	http://www.ispe.org/
15. "Pharmaceutical cGMPs for the 21st Century—A Risk-Based Approach," US FDA, February 2003	http://www.fda.gov/drugs/developmentapprovalprocess/manufacturing/questionsandanswersoncurrentgoodmanufacturingpracticescgmpfordrugs/ucm071836
16. "Guidance for Industry, Q9 Quality Risk Management," US FDA, ICH, June 2006	http://www.fda.gov/downloads/Drugs/GuidanceComplianceRegulatoryInformation/Guidances/ucm073511.pdf
17. International Society of Pharmaceutical Engineering (ISPE) Training, March, 2012	http://www.ispe.org/
18. *Pharmaceutical Engineering Baseline Guides*, Volume 5, "Commissioning and Qualification," ISPE, March, 2001	http://www.ispe.org/
19. ASTM Standard for Specification, Design and Verification of Pharmaceutical and Biopharmaceutical Manufacturing Systems and Equipment, ASTM, July 2007	http://www.astm.org/
20. "Risk Based Qualifications for the 21st Century," ISPE White Paper, March 2005	http://www.ispe.org/
21. *Good Practice Guide, Applied Risk Management for Commissioning and Qualification*, ISPE, October 2011	http://www.ispe.org/
22. " Guidance for Industry, General Principles of Process Validation," US FDA, May 1987	http://www.fda.gov/

REFERENCES

1. Title 21, *Code of Federal Regulations*, Good Manufacturing Practice Regulations.
2. Guidance for industry, sterile drug products produced by aseptic processing—current good manufacturing practice. US FDA; September 2004.
3. Guidance for industry, quality systems approach to pharmaceutical cGMP regulations. FDA; September 2006.
4. Guidance for industry, Q7A good manufacturing practice for active pharmaceutical ingredients. FDA; August 2001.
5. EudraLex. The rules governing medicinal products in the European Union. Annex 1, *Manufacture of sterile products*.
6. ICH Q7. Good manufacturing practice guide for active pharmaceutical ingredients; November 2000.
7. *ISPE baseline pharmaceutical engineering guide*. Volume 6, Biopharmaceutical manufacturing facilities; June 2004.
8. Guidance for industry, Q7A good manufacturing practice, Section 4.13.

9. FDA form 356h. Application to market a new drug, biologic, or an antibiotic drug for human use, Item 15.
10. Guidance for industry, Q7A good manufacturing practice, Section 4.1.
11. *NIH guidelines for research involving recombinant DNA molecules*, Appendix K; October 2011.
12. ISPE baseline pharmaceutical engineering guide. Volume 6, Biopharmaceutical manufacturing facilities Chapter 1; June 2004.
13. Guidance for industry, Q8 (R2) pharmaceutical development, ICH; November 2009.
14. PQLI Guide, Part 1: Product realization using quality by design (QbD): concepts and principles, ISPE, Chapter18.
15. Pharmaceutical cGMPs for the 21st century—a risk-based approach'. US FDA; February 2003.
16. Guidance for industry, Q9 quality risk management'. US FDA, ICH; June 2006.
17. International Society of Pharmaceutical Engineering (ISPE) Training; March 2012.
18. *Pharmaceutical engineering baseline guides*. Volume 5, Commissioning and qualification'. ISPE; March 2001.
19. ASTM standard for specification. Design and verification of pharmaceutical and biopharmaceutical manufacturing systems and equipment. ASTM; July 2007.
20. Risk based qualifications for the 21st century. ISPE White Paper; March 2005.
21. *Good practice guide, applied risk management for commissioning and qualification*. ISPE; October 2011.
22. Guidance for industry, general principles of process validation. US FDA; May 1987.
23. Guidance for industry, process validation: general principles and practices. US FDA; January 2011.

40

CLOSED SYSTEMS IN BIOPROCESSING

Jeffery Odum

IPS, Morrisville (RTP), North Carolina

40.1 INTRODUCTION

The concept of a closed processing system makes common sense. What may be surprising is how closed systems are defined and referenced within the body of regulatory guidance documents and how companies differ in their implementation of a "closed manufacturing" concept. What is not surprising is the impact that closed systems have on facility design.

40.2 DEFINITION OF CLOSED SYSTEMS

A search of the current literature within the industry will result in very little direct reference to defining a closed system. For example, there is no reference to closed process systems and their definition in 21 Code of Federal Regulation (CFR)—Part 211. However, there are some reference materials that begin to give definition to the concept of closed systems.

Commission Directive 2003/94/EC of the EMEA (1) introduces the term *closed system* in reference to equipment and system design for production. *ICH Q7* (2) also references the use of "closed or contained systems" in section IV (Buildings and Facilities) and section V (Process Equipment). However, neither of these guidance documents provides a definition for a closed system.

The *ISPE Baseline Pharmaceutical Engineering Guide® for Biopharmaceutical Manufacturing Facilities* (3) does provide a definition of a closed process or system:

A process step (or system) that utilizes processing equipment in which the product is not exposed to the immediate room environment.

The Guide goes on to state that:

It is the Manufacturer's responsibility to define and prove closure for a process step.

Closed systems (Figure 40.1) are systems that prevent escape of the product and prevent entry of contaminants from the external environment into the product. To achieve this closed systems typically

- utilize processing equipment and components that do not expose product to the external environment;
- allow materials to enter or leave the system via predetermined control points;
- can be validated to prevent unwarranted material exchange between the process and the environment.

40.3 CLOSED SYSTEM DESIGN

The basic elements of containment design in the manufacture of a biological product provide a simple graphic representation of the foundation of closed systems and how they impact facility design. The implementation of primary containment elements (equipment and systems) are the primary means of both product and personnel protection.

Downstream Industrial Biotechnology: Recovery and Purification, First Edition. Edited by Michael C. Flickinger.

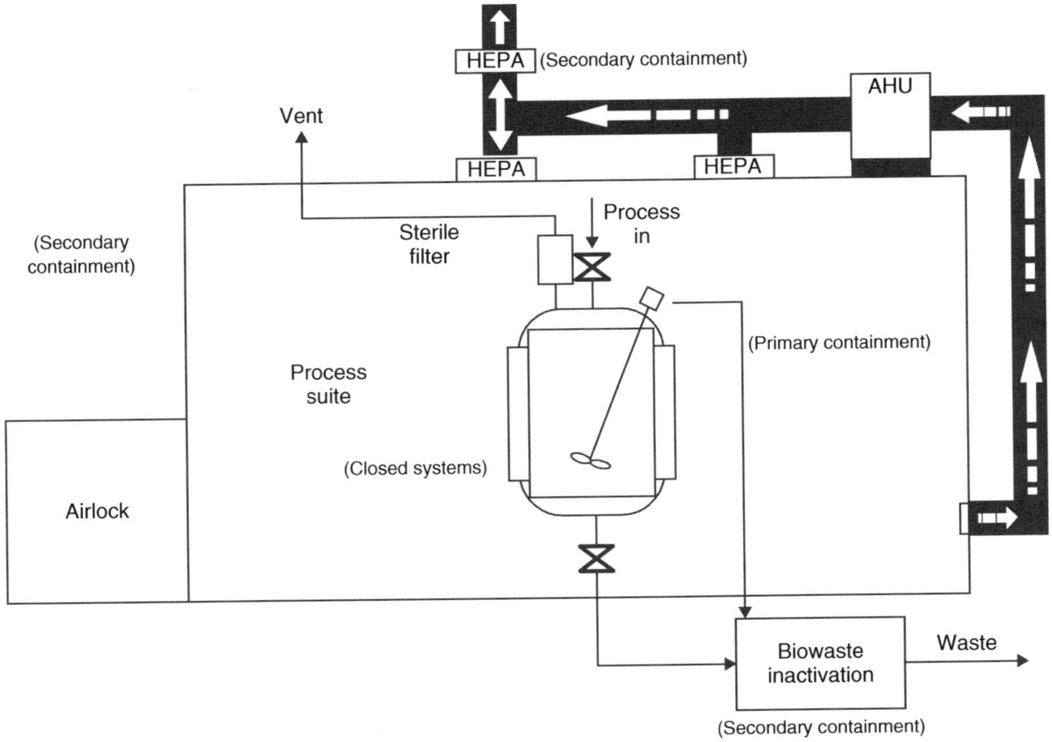

Figure 40.1. Primary and secondary containment boundaries are defined by elements of the facility design, and include the use of closed systems to provide both product protection and environmental containment. (This figure is available in full color at http://onlinelibrary.wiley.com/book/10.1002/9780470054581.)

It is critical to recognize that "locally protected" systems are not closed. Locally protected systems are those that do not meet the definition of closed but which provide greater protection than an open process or system. This would include those systems where product protection is achieved via some form of localized environmental protection such as a laminar flow curtain or biological safety cabinet/isolator system. It is also important to recognize that biotech processes are different; some lend themselves better to a closed system than others. Less refined processes requiring frequent in-process intervention are more difficult to close. Earlier stage, smaller scale operations typically more "open" in their design approach due to an often limited availability of small closed equipment and because it is easier to control the external environment at smaller scale.

The American Society of Mechanical Engineers (ASME) provides a comprehensive set of guidelines applicable to closed bioprocessing systems and equipment (4). The 2007 edition of the BioProcess Equipment Standard is the recognized industry guide for equipment design. The critical elements of the design standards covered in the guideline centers around sterility and cleanability. Each component is required to be cleaned and sterilized and designed to be operated in a clean and sterile manner. The accepted practices defined in the guidelines meet the intent of the GMPs for equipment design and construction (5). These include accepted practices for cleanability, materials of construction, connections and fittings, design for system drainability, seals and gaskets, and containment.

There are many different design approaches that are employed within the industry to meet the closed system design criteria. One example that provides a clear view of the complexities and options available to engineers involves the design of systems for collecting in-process samples while maintaining aseptic conditions.

Figure 40.2 illustrates one approach for a closed system design for in-process sampling, where the product transfer occurs via a series of automated and manual valves and the sample port is cleaned and sterilized using the Sterilize-in-Place (SIP) system. In this scenario, the sample assembly is cleaned prior to the product transfer and again upon completion of the transfer. There is a specific sequence of operation that must be followed for the cleaning and sampling process. Additionally, there are a number of parts and components that are required to complete the entire design.

Another approach to obtaining in-process samples in a closed system design is the implementation of closed disposable sampling. The NOVAseptic NovaSeptum system is

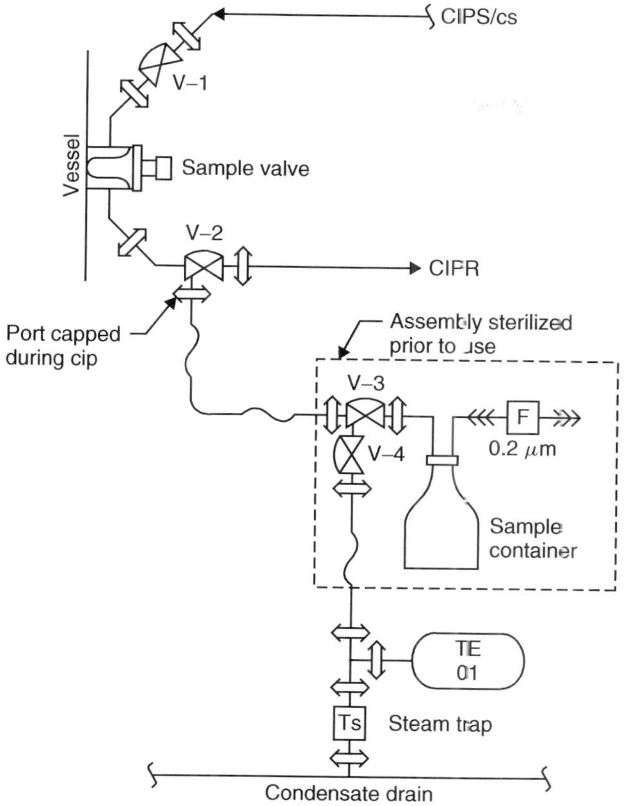

Figure 40.2. Closed system design example for in-process sampling. Drawing courtesy of CRB Consulting Engineers, Kansas City, Missouri.

a presterilized component that has been used to accomplish the same sampling as the previously described process. In this system (Fig. 40.3), components are connected to sample bags via a closed cannula and septum. The number of samples that can be taken is limited by the configuration of the bag holder, but the system has proven successful for many sampling functions. The surface of the septum will also be sterilized during the SIP cycle.

In selecting a sampling approach, there are also a number of variables that will impact the overall cost including the number of samples required, disposal cost, cleaning frequency, and the level of automation incorporated into the entire process.

40.4 IMPACT ON FACILITY DESIGN

If a system design is truly closed, it opens up a wide array of facility options and provides a very flexible approach to compliance from a regulatory perspective. This is clearly shown in the following quote from ICH Q7A:

> *Where the equipment itself (e.g., closed or contained systems) provides adequate protection of the material, such equipment can be located outdoors.*

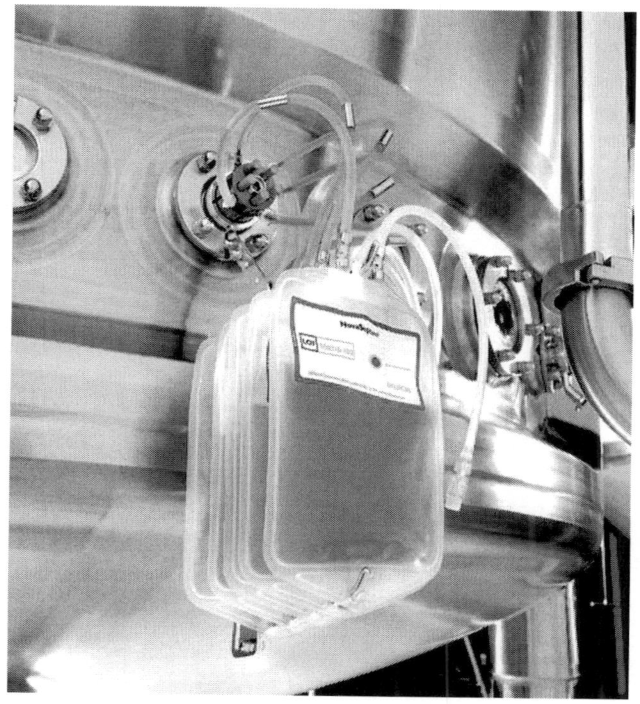

Figure 40.3. The NOVA septic NovaSeptum system with a presterilized component. Photograph courtesy of NOVA Septic-Millipore, Gothenburg, Sweden. (This figure is available in full color at http://onlinelibrary.wiley.com/book/10.1002/9780470054581.)

While no one expects to see this dramatic implementation in any facility design, it does provide architects and engineers with many options from which to approach facility and process system integration, especially as firms seek to find ways to reduce not only capital costs, but annual operating and maintenance costs in order to improve cost-of-goods figures and reduce the impact of validation.

Costs for biomanufacturing space are very expensive. Some facilities can average between $700 and $2000 per square foot and greater (6,7). It is always one of the major challenges that design teams face to find creative ways to reduce facility cost by reducing the amount of classified space that is defined in the design documents.

The traditional approach to clean room design is shown in Fig. 40.4. In this approach, facility cost factors include equipment and room finishes, Heating, Ventilation, and Air Conditioning (HVAC) equipment, routine maintenance of clean room space, cleaning and gowning materials, and increased access control requirements due to open system operations.

By choosing to implement a predominately closed system approach, the same layout would look dramatically different in Fig. 40.5. In this layout, the actual amount of classified space is dramatically reduced. Critical issues such

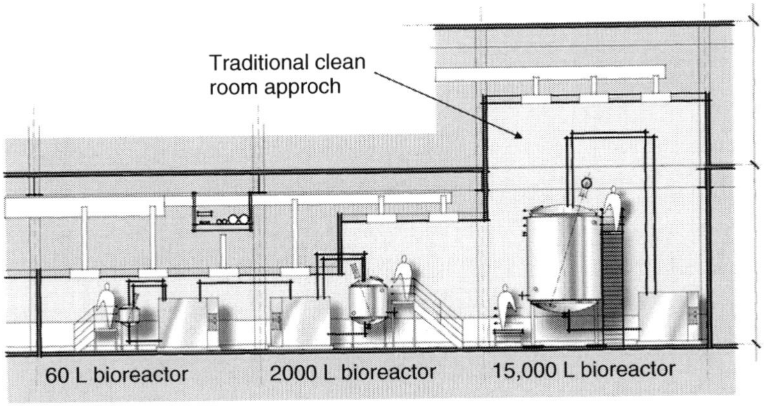

Figure 40.4. Traditional approach to clean room design; large areas of classified space where closed systems were often not implemented. Graphic courtesy of Flour Corporation, Greenville, South Carolina. (This figure is available in full color at http://onlinelibrary.wiley.com/book/10.1002/9780470054581.)

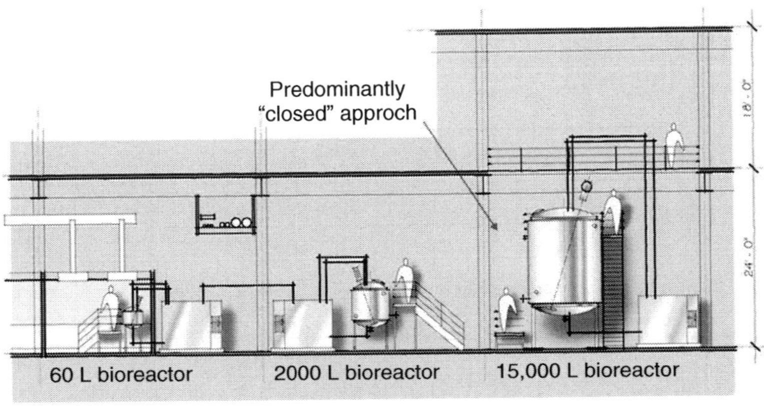

Figure 40.5. Implementing a predominately closed system approach: the layout of Fig. 40.4 looks different, as the actual amount of classified space is dramatically reduced. Graphic courtesy of Flour Corporation, Greenville, South Carolina. (This figure is available in full color at http://onlinelibrary.wiley.com/book/10.1002/9780470054581.)

as sampling and filter change-out must still be addressed, but overall construction, operation, and maintenance costs can be significantly less.

The *ISPE Baseline Guide®* addresses the options for implementing a closed process design as a means of developing manufacturing areas designated as *controlled, nonclassified* (CNC) space. It is recognized in the industry that a large majority of process operations occur in areas that could be considered CNC space, regardless of whether they are claimed as such by the manufacturer. As discussed in the *ISPE Baseline Guide*, CNC space can resemble a wide variety of configurations and appearances (8). But, the key attribute of CNC space is the fact that it begins with closed system design.

Figure 40.6 provides four different layout considerations for closed systems in CNC space (9). In the first case, the layout approach and classification philosophy that are described represent what is currently the primary approach taken by the industry; all process equipment is located in a classified space that is validated and under protocol control.

The second case resembles the approach often referred to as *gray space* implementation, where some of the mechanical and "maintenance intense" components of process equipment are located in a nonclassified space. Moving to the third case, a less-often used approach of placing only sample port locations inside a clean vestibule space is illustrated. Figure 40.7a and b provides a graphic and photo of one such installation.

The final case is the true implementation of complete closed system production. As shown, the amount of actual classified clean room space is dramatically reduced, and logically, so will the costs of operation and maintenance, gowning, and validation for this space be reduced as well.

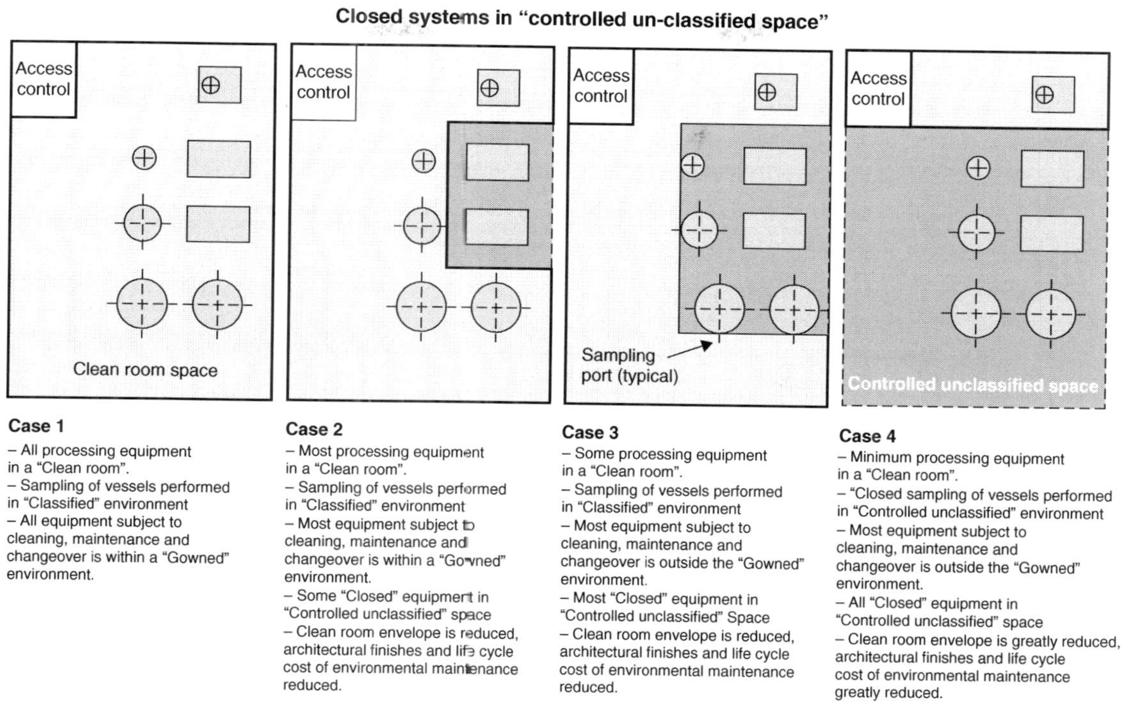

Figure 40.6. The evolution of classified space design, from large open areas (Case 1) to current implementation of closed system design philosophy as defined by ICH Q7. Graphic courtesy of ISPE, from the "Biopharmaceutical Facilities Baseline Guide, Volume 6." (This figure is available in full color at http://onlinelibrary.wiley.com/book/10.1002/9780470054581.)

40.5 IMPACT ON OPERATIONS

It is considered good engineering practice to design the main process stream as closed as possible during operation. This will substantially reduce the risk of product contamination from personnel or the facility environment. It will also provide protection of personnel from potential exposure to viruses or other pathogens that may be present.

There are products and processes that are better suited to closed system design. It is much easier to design closed systems for well-characterized products and well-defined processes that are robust. However, in many cases, such as clinical or process development or smaller scale operations, it may be much easier and cost effective to place the entire process into a classified environment (Case 1) and allow for more open process operations.

Upstream process operations such as seed inoculum and cell culture unit operations are typically designed as axenic operations, where the culture is free from living organisms other than the host cell. This, by its nature, implies a closed system design approach. Bulk and final filling operations are also designed to control contamination and bioburden, both in terms of product contamination and risk. Product contamination at this final stage of the process can have a disastrous impact economically to the manufacturer.

There are some process unit operations that are difficult to keep fully closed. Some forms of in-process sampling, column fractionation, and solids additions from bulk containers are examples. It is also important to understand that closed systems will be open at certain times. This will occur for maintenance and setup. For example, filter housings are open system components during setup, but are closed during operation. The loss of the closed state due to these routine activities does not negate closure as a key component of the facility design. It is necessary to have validated procedures for cleaning, assembly, and sterility testing to ensure that the closed state has been reinstituted. A pressure hold test to prove system integrity will also be used to validate closure of the system prior to operation.

As many companies continue to implement single-use or disposable technology into their manufacturing operations, the question of system closure must also be addressed. Many process unit operations have successfully been designed and validated for system closure. These include

- media, buffer, transfer, and storage
- product transfer, holding, and shipping
- bioreactors
- distribution and filling manifolds
- sampling systems.

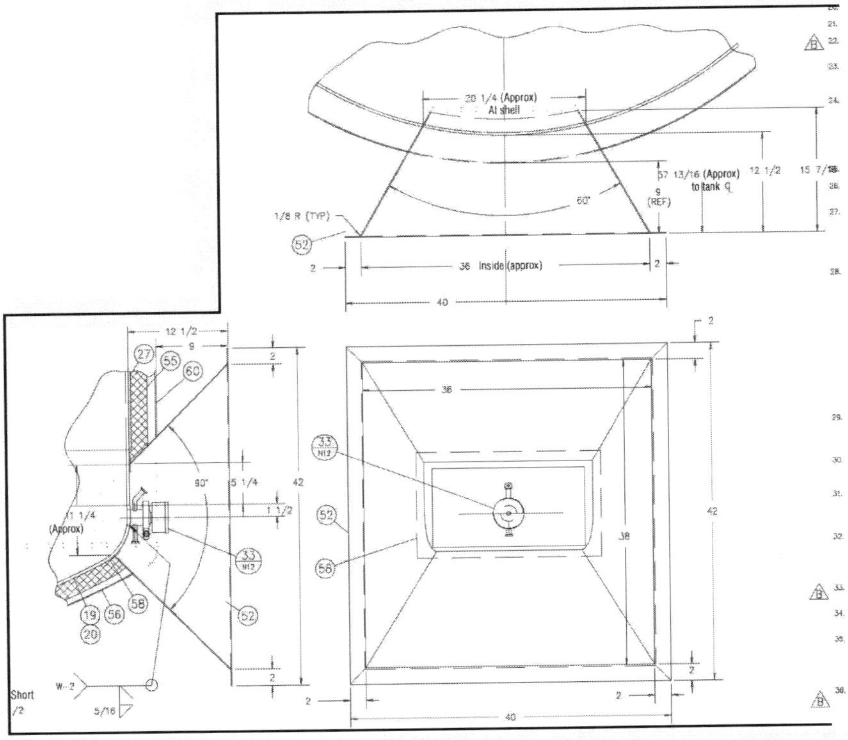

(a)

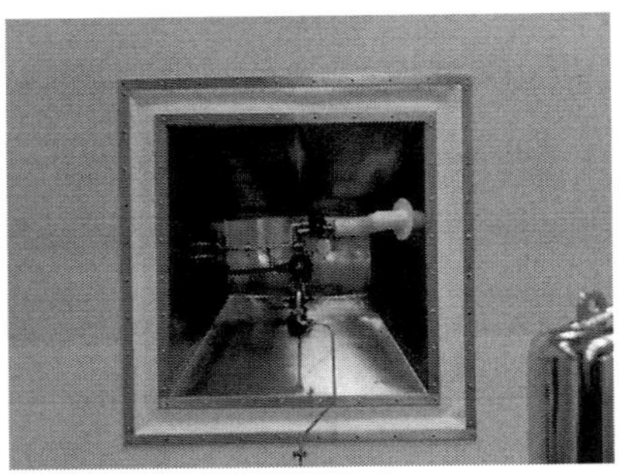

(b)

Figure 40.7. Sample port locations inside a clean vestibule space sketch (a) and photograph (b). The creation of the vestibule allowed for sample collection inside a classified clean room space while reducing overall area of clean manufacturing space. Sketch and photograph courtesy of Biogen IDEC, Research Triangle Park, North Carolina. (This figure is available in full color at http://onlinelibrary.wiley.com/book/10.1002/9780470054581.)

One example of where an enabling technology based on single-use design is being implemented is with the transfer of fluids under aseptic conditions. The rapid aseptic fluid transfer (RAFT) system, based on Biosafe Technology by Stedim allows aseptic zone-to-zone transfer of fluids via closed transfer system and is used by many companies (Fig. 40.8).

This type of component–system integration is becoming more common within the biopharmaceutical manufacturing industry.

40.6 SUMMARY

Closed systems are critical for meeting GMP requirements. International regulatory bodies and industry standard institutions recognize the importance of this fundamental practice. While it is not necessary that all production systems be completely designed around the closed system concept, good engineering design practice encourages closed systems to protect the product, the workers, and the environment. Closed system design is a key to risk minimization,

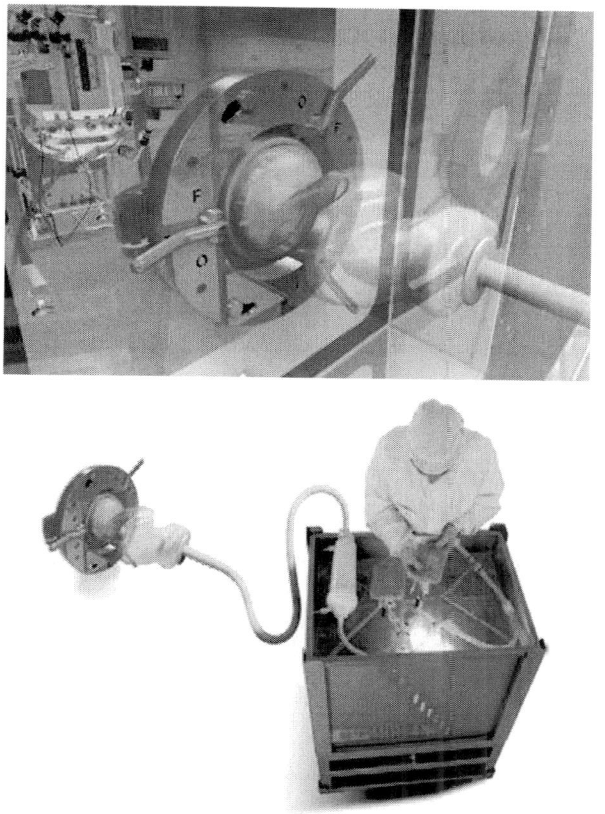

Figure 40.8. The RAFT (rapid aseptic fluid transfer) system, based on Biosafe Technology by Stedim, allows aseptic zone-to-zone transfer of fluids via closed transfer system and is used by many companies. Photograph courtesy of Sartorius Stedim Biotech, Copenhagen. (This figure is available in full color at http://onlinelibrary.wiley.com/book/10.1002/9780470054581.)

both in terms of facility capital cost and in operational expenditures.

REFERENCES

1. European Medicines Agency. Medicinal products and veterinary use: good manufacturing practice. Annex 2, Volume 4, Chapter 5; 2003 and 2008. p. 3.
2. International Conference on Harmonization. Guidance for industry: Q7 good manufacturing practice guidance for active pharmaceutical ingredients; 1998.
3. International Society for Pharmaceutical Engineering Baseline Pharmaceutical Engineering Guide for New and Renovated Facilities. Volume 6, Biopharmaceutical manufacturing facilities. 1st ed; 2004. p. 171.
4. American Society of Mechanical Engineers. ASME bioprocessing equipment standard, BPE-2007, 2007. pp. 9–57.
5. 21 CFR, part 211, subpart D–Equipment, 211.63 and 211.65.
6. Odum JN. Sterile product facility design and project management. 2nd ed. Florida: Boca Raton; CRC Press; 2004. p. 14.
7. Pavlotsky R. Approximating facility costs, Cleanrooms, Volume 18, Section 6.6.5 2004.
8. ISPE Baseline Pharmaceutical Engineering Guide for New and Renovated Facilities. Volume 6, Biopharmaceutical manufacturing facilities. 1st ed. Section 6.6.5. p. 101.
9. ISPE Baseline Pharmaceutical Engineering Guide for New and Renovated Facilities. Volume 6, Biopharmaceutical manufacturing facilities. 1st ed. Section 13, Figure 13.9. p. 156.

41

FACILITY DESIGN FOR SINGLE USE (SU) DOWNSTREAM MATERIALS

ROBERT Z. MAIGETTER

Centocor R&D, Spring House, Pennsylvania

TOM PIOMBINO

Integrated Project Services, Inc., Lafayette Hill, Pennsylvania

CHRISTIAN WOOD, TOM GERVAIS, CLAUDIO THOMASIN, BRYAN SHINGLE, DAVE A. WAREHEIM, AND DAVID CLARK

Centocor R&D, Spring House, Pennsylvania

41.1 INTRODUCTION

Single-Use (SU) technology has been available for decades, replacing lab-scale equipment such as glass shake flasks, pipettes, and containers. This eliminated the potential of cross-contamination, cleaning and handling requirements, and facilitated rapid experiment setup and turnaround The last 8–10 years have seen an acceleration of SU technology, especially into larger scales of operation. Because of this, SU technology has a significant impact on biotech facilities. This chapter will highlight the type of SU technologies currently available in the major process areas of cell culture and purification, for example, and illustrate how SU technology has affected facility design and process costs.

In the early 1970s, some companies began using SU technology at a medium scale up to 200-L SU bags for the storage of buffers, and eventually replacing 50-L glass vessels. Through the 1980s and 1990s, the demand for SU buffers continued and more companies were recognizing the potential that SU technology provided. The trend for more rapid SU expansion started around 10 years ago, when the 50-L SU bioreactors were introduced. In the last 5 years numerous articles have been published, especially in *Bioprocess International* and *BioPharm International*. In the latter, a supplement titled "The Evolution and Application of Single-Use Technology" featured numerous articles which focused on a wide variety of SU technologies. Eric S. Langer (1) authored the lead chapter, "Trends in Single-Use Bioproduction: What Users are Saying," stating that "...the use of disposables in biomanufacturing is moving forward. It's become less of a question of *if* disposables will be implemented than of *where* and *how*."

With this momentum, a series of outlooks by PricewaterhouseCoopers (PwC) sets the stage for technologies such as SU. The PwC highlights that changes are coming and suggests that many current medicines will expire in the next few years (2). To increase efficiency the pharmaceutical industry is expanding technology, especially in multiproduct facilities. There will be pressure on the pharmaceutical industry to utilize new technologies to survive and to include SU. The main drivers are speed to market, flexibility, capital and ongoing costs, and comparable product quality.

41.2 FACILITY DESIGN

41.2.1 Past Design Strategies and Challenges

The last 20 years of biotech facility design have been subject to either the customization of the building structure

Downstream Industrial Biotechnology: Recovery and Purification, First Edition. Edited by Michael C. Flickinger.

to meet the equipment design, or the customization of the equipment design to fit within the building structure. Depending on a company's available capital, design challenges have taken on many forms, but in the end, resulted, for most, in a series of compromises between quality, cost, speed, and maintenance. For every 1000 sq. ft. of manufacturing space, designers would typically allocate 300 to 500 sq. ft. of mechanical space to accommodate large mechanical and process utility support systems. For most sites where space is a premium, this meant going vertical with costly multistory facilities, large central utility plants, and heavily reinforced structures.

Only in the last couple of years have operating companies and engineers begun to challenge the cost and schedule limitations of the older design philosophies and look to technology for answers. This practical curiosity has led to innovation, changes in regulatory perspective, and cost reduction, effectively reducing the barrier of entry for smaller biotech companies looking to grow, nurture their own intellectual property, and be high impact contributors in drug development. The following ideas and their subsets have profound impacts on the way we think about and design facilities and processes:

- *Risk Assessments and Risk-Based Design.* Process of evaluating risk factors associated with various manufacturing methods.
- *Closed Systems and Processes.* Manufacturing product within closed piping systems that do not communicate with the surrounding room environment.
- *Gray Space.* Support space where skids and other nonaseptic support equipment are kept outside the good manufacturing practice (GMP) space.
- *Resurgence of Mono-Corridor Design.* Term for a single corridor GMP facility design where people and material cross paths outside their respective containment areas.
- *Temporal Separation.* Using time schedules to eliminate cross-contamination of the process or facility.
- *Just-In-Time Deliveries.* Technique of using suppliers storage space & manufacturing to order capacity to reduce onsite storage quantities and warehousing.
- *Electronic Batch Records.* Compliant storage of validated records in an electronic format versus hard copies.
- *Isolation Technology.* Performing aseptic manipulations within a sterile enclosure.
- *SU Manufacturing Technologies and Aseptic Connectors.* Utilizing specially manufactured equipment and components that are meant to be discarded after a limited quantity of uses (usually 1–3 times maximum).

As more operators challenge their processes and associated production risks, doors to reducing cost and increasing speed to market have opened. The time and money spent to keep closed systems and products from communicating with the surrounding environment can be used to reduce the need for clean manufacturing environments. For example, acceptance by the regulatory agencies has shifted stainless steel (SS) vessels from classified areas into gray space, and only critical sampling and addition ports remain within the classified areas. By relocating most of the stainless steel vessels to gray space, a nonclassified area, there is a modest reduction in heating, ventilation, and air conditioning (HVAC) and utility costs as well in the convenience factor of not having to don a gown in the gray space to get access to equipment for certain process manipulations and maintenance. Although a significant shift in the way we design, utilization of gray space has done little to reduce overall facility square footage and capital cost. Large stainless steel vessels, process utility generation systems, and cleaning distribution skids require a fixed quantity of space to allow for installation, maintenance, and testing. Additionally, many in the industry have not quite embraced the full potential of reducing the environmental requirements and therefore fail to take full advantage of the opportunity to save on energy and operating cost reductions.

Figure 41.1 provides an estimate of how innovations have impacted our business. For simplification, a scale of −2 to 0 and 0 to 2 has been used based on whether the impact increases or decreases the criteria listed on the legend compared to a zero baseline or traditional design approach. For the purposes of this analysis, gray space was the only risk-based design subset evaluated since the others listed above are related and/or complementary in use. The reader is requested to refer to the online version of this chapter for color indication

As indicated in Fig. 41.1, SU Technologies have and will continue to make the biggest impact in our process and facility design decisions because they increase convenience, and save time and money. Many in the industry have noted a reduction in overall operating cost in their manufacturing facility by using single-use (SU) systems, but most published data indicates that the savings in energy and utilities are still offset by the cost of consumables (3,4). As the use of these materials become more widespread and advancements in environmentally friendly disposal/recycling techniques are developed, these factors and market competition will inevitably begin to reduce their cost matrix and tip the scale in favor of reduced operating cost.

41.2.2 Impact of SU Technologies on New Facility Design

The next 20 years of facility design will allow for cost reduction, decreased energy use, reduced environmental

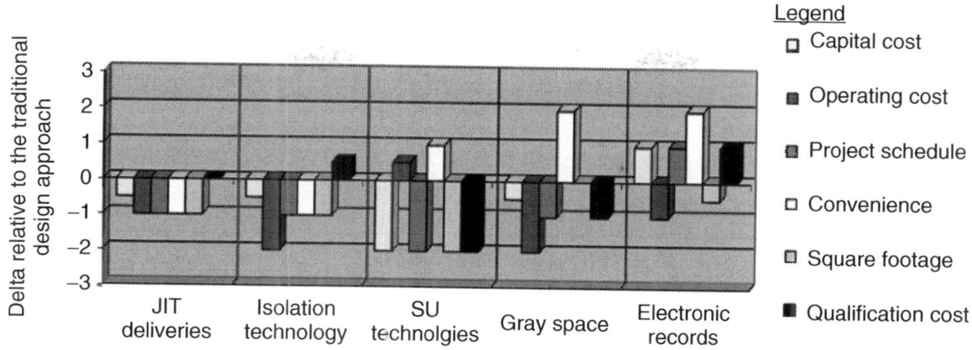

Figure 41.1. Major innovations in design and their impact on facility design. (This figure is available in full color at http://onlinelibrary.wiley.com/book/10.1002/9780470054581.)

TABLE 41.1. Integration Opportunities for SU Systems into Biotech Facilities

Single use equipment	Inoculum prep	Media/Buffer prep	Cell culture	Harvest	Purification	Bulk storage	Formulation/ compounding	Aseptic filling
Disposable flasks	■							
Disposable tubing	■	■	■	■	■	■	■	■
Aseptic connectors	■	■	■	■	■	■	■	■
Mixing systems		■	■	■			■	■
Bioreactors	■		■					
Bags / manifolds / bins		■	■	■		■	■	
Point of use filters	■	■	■	■			■	
Harvest filtration				■				
Disposable sensors			■	■	■			
Diafiltration systems					■			
Capture chrome columns					■			
Chrome capsules					■			
Sampling devices	■	■	■		■	■		
Syringes (small additions)	■		■					■
Filling nozzles/needles								■
SU integrity testers	■	■	■	■	■	■	■	■
Tangential filters			■	■				
Lyophilization trays								■
Dosing systems								■
Peristaltic tubing pumps	■	■	■	■	■		■	■
Rapid transfer systems	■	■	■	■	■	■	■	■
Flow meter liners		■	■	■	■		■	■

impact, and increased speed to market. Continued development of new SU technologies, ongoing downward cost pressure along with some innovative movements in how companies choose to develop and characterize their products may provide the impetus to achieve some or all of these outcomes. Table 41.1 provides an overview of the opportunities to incorporate SU systems as well as the stage in the manufacturing process that companies have been integrating them.

For years, engineers have been separating manufacturing facilities into two areas: upstream and downstream processing. Upstream processing has historically been characterized as inoculum prep to cell harvest/separation while downstream included chromatography, viral inactivation, ultrafiltration, and bulk fill. The approach to designing facilities to house upstream processes has always been relatively fixed due to larger immobile equipment that needs to be cleaned in place [cleaning-in-place (CIP)]. Downstream processing has been designed with more flexibility since operations can be cleaned out-of-place and mobile chromatography systems have been available for many years. Innovative SU technologies for upstream unit operations are now easily available and provide similar flexibility upstream. SU bioreactors, disposable filtration/separation systems, mixing systems for media preparation, and other equipment provide mobility and reduce cleaning requirements, both of which reduce piping networks and overall footprint.

Consider the flexibility of using the same equipment in a campaign mode for media preparation, buffer prep and cell culture, and being able to move this equipment around in a suite or in a facility as each unit operation is completed. As operating companies begin to examine these possibilities, the requirement to design floor space for every piece of equipment in use at the same time may greatly diminish, resulting in a reduction of allocated space, duplicate equipment, and utility hook-up locations. This leveraging of common space is similar to how downstream purification spaces have been designed, but takes space utilization to the next level with SU components. Many of these items do not need to be cleaned. Thus, decreasing utilities to fewer locations will begin to create significant savings in all parts of the facility. Additionally, companies may be able to decouple the facility design from the equipment design to gain an edge on getting their products to market. For example, architects and engineers require very specific process and equipment information to design a traditional mammalian cell culture facility. Tank sizes and cleaning requirements have a large impact on space, and without this information, the design of a facility would not be possible. With SU systems, many of the large manufacturing space considerations are reduced or eliminated. Operating companies, architects and engineers who are versed in SU technology and understand the scale of the process may have enough insight to design, construct, and validate a facility prior to finalizing the actual process development parameters. Having a facility that is essentially ready to receive a process or product would provide a competitive edge.

The industry continues to digest these concepts as new processes are designed around SU technology. With few exceptions, the majority of the SU installations are in existing manufacturing facilities where the initial first cost investment in traditional systems has already been made and conversion to SU yields savings, but they cannot be maximized without an invasive renovation. For example, an existing Clinical Stage Pilot Plant Facility designed to perform mammalian cell culture likely has an active water for injection (WFI) generation system, storage tank, and loop sized for CIP. As the facility migrates from SS bioreactors to SU bioreactors, it is unlikely that they would obtain a good return on investment (ROI) to renovate the WFI system to save space or operating cost. Having said that, companies evaluating their existing investments have many factors to weigh, have different ideas of what justifies a good ROI; or may have other motivations such as corporate energy reduction programs or lean manufacturing incentives. Without question, new facilities built based on SU systems would have the greatest opportunity to maximize both capital costs and operating expense.

We have found that many operating companies are and have been using SU bioreactors in their commercial facilities up to 50 L, while their clinical and research facilities have new systems up to 1000 L. It is clear that the next formulation or generation of products will be made commercially in facilities that not only use SU bioreactors, but other SU equipment and components that compliment them. Larger companies like Lonza, Centocor, Shire, Genentech (6), Wyeth and others are using the technology now within their existing facilities and will soon seek to build or renovate commercial facilities on a SU platform. Smaller companies are taking advantage of SU because of their cost effectiveness and scale. Novavax, a vaccine company in Maryland, is one of the first to respond, having already constructed a pilot plant and commercial launch facility for their Virus Like Particle (VLP)-based flu vaccine. Though they have not built a full commercial facility, they have published estimated savings of over 66% in building square footage and 90% in overall project cost (7). Some of these savings are credited to their VLP process over the typical egg-based production technique, but we have experience with and believe that target savings above 75% in equipment footprint in some specific process areas (Fig. 41.2) is achievable with an overall facility reduction in a space of 30% or more over traditional design techniques. The future of facility design will require an adaptable skill set and knowledge base in SU technologies that accepts lean design and flexibility, rather than the historical tendency toward conservatism.

In today's economic environment, operating companies are aggressively examining ways to use lean processes and

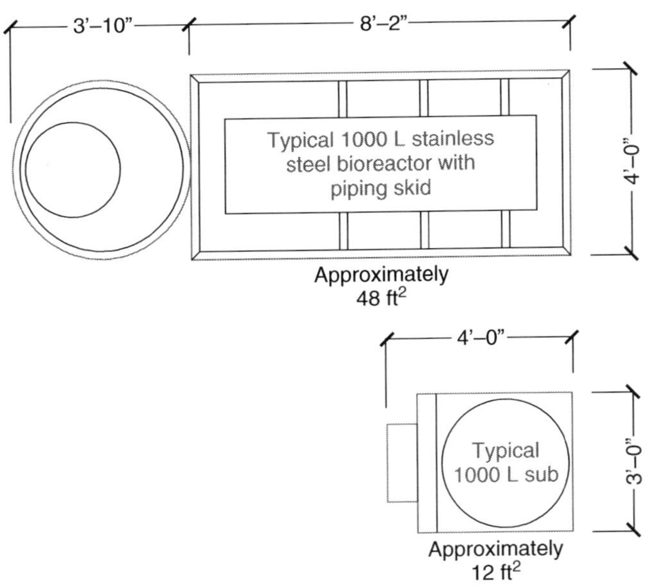

Figure 41.2. Bioreactor comparison for stainless steel and single-use. (This figure is available in full color at http://onlinelibrary.wiley.com/book/10.1002/9780470054581.)

costs, but it is difficult to break tradition and standards. SU technologies are not traditional and may force this transition, but companies looking to design or renovate a facility need to take a top-down approach to ensure that their managers and consulting engineering firms scrutinize their facility programming efforts. Opportunities to think differently and reduce cost will be brought to the table, but they must not be dismissed without a proper risk assessment. Even facility square footage can be inflated due to unknown factors that are deemed high risk, and so getting the facts about SU technologies is critical. In designing your next facility, based on our experience, we recommend considering the following if SU Technologies are used:

- Know your manufacturing culture. To implement change, know the impact on the "norm" before wasting time developing an unsellable concept.
- Get buy-in and continuous support from management. The team should leverage SU systems to include facility scope change and reduction.
- The introduction of SU equipment should reduce plant footprint over traditional fixed tanks, but in order to truly maximize the facility cost saving possibilities associated with SU technologies, we need to look at the temporal positioning of the equipment. The key to this is mapping the location of your single use equipment over the course of each production run (daily or hourly) and taking advantage of equipment portability and utilization of space.
- Analyze physical material transfers and understand their impact on storage, staging, and corridor circulation space.
- Leverage closed processing and final sterile filtration to reduce environmental conditions.
- Provide for warehouse storage of SU components, incoming spares, and outgoing waste.
- Understand what special material handling devices and unique operator ergonomics will be encountered. Although manageable, moving 500-L bags of material can have its challenges and will need to be planned.
- Make sure SU equipment is consistent with process control strategy. SU technologies are typically made to be lean and without robust control functionality: this may lead to some customization that impacts architectural layout and system cost
- Leverage the SU equipment manufacturers knowledge base and challenge them to provide life cycle resources that cover your needs from receiving Just-in-Time (JIT) bags at the dock to shipping bins of used plastics to a local recycling plant
- Challenge old metrics used by many in the industry for central utility requirements for "the process". Design rules of thumb are very helpful but can be a detriment to addressing the differences and maximizing value in SU technology.

TABLE 41.2. 1000-L SU Facility Reduction Potential versus Traditional 1000-L Stainless Steel (SS)

	Target Delta from SS(%)
Capital first cost	−30
Process/process utility equipment	−50
Facility design and construction	−38
Commissioning and validation	−25
Utility consumption	
Steam (Hp)	−35
Chilled water (tons)	−28
WFI (gallons/day)	−20
Consumables cost	+10

If your company is looking to invest in a new facility, or renovate an existing one, to accommodate a SU manufacturing platform, Table 41.2 below provides some targets to consider if you are developing budget figures. The proforma is based on a new interior fit-out of a 1000-L monoclonal antibody (Mab) facility using primarily manual product transfers and SU equipment for cell culture, harvest, hold steps, and media and buffer prep versus the equivalent 1000-L facility using stainless steel vessels, piping, and automated product transfers.

41.2.3 Future Challenges

Adaptation to SU technologies in the biopharmaceutical industry is a significant shift in itself, but operating companies who use SU equipment and the manufacturers who make them have an ongoing challenge to see how and if they can improve their drug development life cycle and reduce overall Cost of Goods (COGs) using SU systems. In terms of facility design, the biggest challenge is gaining the expertise to know their benefits and how they are used enough to maximize design and energy efficiency without compromising manufacturing reliability and product integrity. To date, operating companies investing in SU technology have proven their effectiveness in many areas of drug development. It will be up to the first generation SU facilities and their designers to maximize their potential.

41.3 CELL CULTURE

The implementation of SU technologies in cell culture processes has gained significant momentum in the biopharmaceutical industry over the last decade. It is now possible to create an upstream process from vial through a 2000-L production bioreactor using only SU components.

Certain applications, particularly small-scale development (<10 L) and seed train expansion have been using SU products for several decades. This includes items such as shaker flasks, T-flasks, tubes, filters, pipettes, syringes, and vials. Larger-scale cell culture operations are quickly following suit, with a range of options now commercially available, and several innovations providing unique advantages. Cell culture is often one of the first process areas to make the transition from reusable systems, such as SS tanks, to SU. SU mixing systems have been in the market for a number of years. As bioreactors are similar in basic design to mixing systems (e.g. an agitated vessel) they have naturally evolved after a few key modifications, such as the addition of gas exchange ports, exhaust filters, and methods for sensing pH and dissolved oxygen (DO).

There are several advantages that make implementation of SU cell culture systems particularly attractive. These typically include: flexibility, reduced pilot plant facility startup time, reduced turnaround time between operations, lower risk of cross-contamination, a lower capital investment, and in some cases, reduced operating costs. Along with the benefits, however, are some associated risks and limitations. These include: the risk of leachables, process comparability (to SS), product comparability (to SS), safety and operational concerns, and dependence on single source suppliers. These and other concerns and limitations are gradually being addressed as more development and testing of SU components is completed. Additionally, several organizations provide guidance and education on SU components, including how to implement them and on how to conduct testing (8).

Many of the bioreactors and cell culture systems described in this chapter are designed for both fermentation and cell culture. For the purpose of this chapter, "cell culture" will be the focus and the terminology used throughout.

41.3.1 Process Flow

41.3.1.1 Small-Scale Cell Culture Processing. SU materials have been used in small-scale cell culture processing for many decades. Beginning in the early stages of development, SU products are commonly used in clone generation and clone selection. Once a clone is selected, it can be grown in a well plate system and then expanded using SU T-Flasks, shaker flasks, centrifuge tubes, spinners, rocking type bioreactors, and many other options. When cells are ready for cell banking, they may be dispensed to SU cryovials or cryobags. Freezing the cells in disposable cryobags provides the advantage of being able to start a seed train at much higher cell density, reducing the overall duration of expansion to the production bioreactor. Additionally, since the cryobag can be tube welded to a seed vessel or bioreactor, it is possible to maintain a closed system from the beginning of the process.

Glass spinner flasks, commonly used for small-scale development and cell expansion, can now be replaced with SU plastic versions from multiple suppliers in sizes from approximately 100 mL to 3 L. The side arms can be fitted with diptubes to provide a closed process early on in a cell expansion. Various types of bench-scale, stirred SU bioreactors have also been recently released in sizes from approximately 1–3 L (Fig. 41.3). They often have similar dimensions and functionality as their glass counterparts. Whereas SU spinners and shake flasks can only provide surface aeration and pH control by addition of CO_2 in an incubator, these new SU bioreactors provide more efficient oxygenation through spargers and more accurate control of DO and pH with sensors. The additional cost of materials incurred by using SU spinners and bioreactors can be significant. However, the time saving and lack of cleaning and sterilization often more than make up the difference. In the case of a GMP process, validation of cleaning and sterilization would essentially be eliminated, reducing cost and time to start up. No change of the control environment is needed, so a switch to SU spinners and bench top bioreactors can be achieved immediately, with minimal to no pilot plant facility modification. The same incubators and bench top bioreactor control systems can be used.

Rocking motion SU bioreactors were first developed by Wave Biotech. These bioreactors, owing to their flexibility and ease of use, have been integrated into many existing facilities, and selected over glass or SS equipment in newly constructed facilities. They are currently available in sizes from approximately 1–500 L (working volume), and often used in small-scale development or cell expansion to provide inoculum for a larger bioreactor.

Rocking motion SU bioreactors offer several advantages over small-scale SS or glass bioreactors. They occupy a small footprint (at 1–50 L scale) and can be stacked vertically in cabinets (Fig. 41.4). The bags, as with any SU bags for mixing or bioreactor applications, can be customized with a wide array of tubing and material configurations to meet the needs of multiple processes and types of operation (batch, fed-batch, and perfusion). For a multiproduct pilot plant facility, the same equipment can be used for many different products, can reduce turnaround time, can reduce the risk of cross-contamination, and maximize the utilization of the equipment.

Operation is relatively simple, with very quick setup and turnaround. No cleaning or sterilization is required, since the bags come presterilized (gamma irradiation) from the suppliers, which eliminates the need for cleaning and sterilization validation for GMP facilities. These bioreactors are often operated with very simple controls (temperature, rocking speed, and rocking angle), and use a steady flow of air and CO_2 to maintain DO and pH. Conventional probes

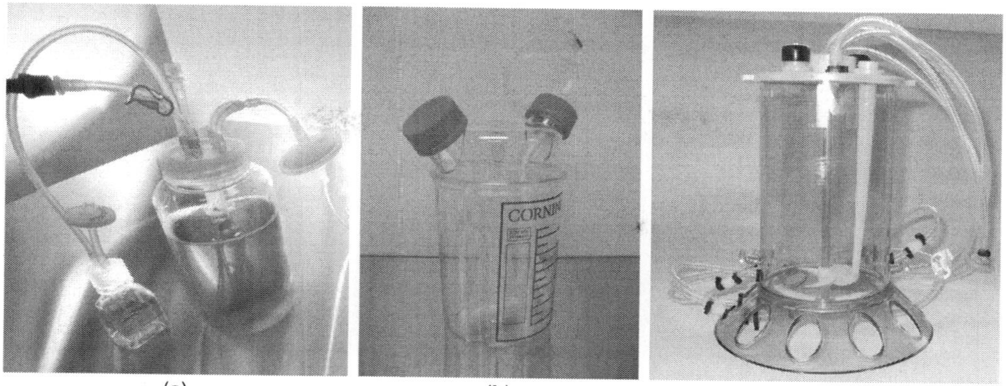

(a) (b) (c)

Figure 41.3. SU spinners and SU small-scale bioreactor. **(a)** Sartorius Stedim SuperSpinner D 1000 1-L Bioreactor [photo courtesy of Sartorius Stedim], (b) Corning Disposable 1-L Spinner Flask, (c) Millipore Mobius CellReady 3-L Bioreactor [photo courtesy of Millipore]. (This figure is available in full color at http://onlinelibrary.wiley.com/book/10.1002/9780470054581.)

Figure 41.4. Cabinet with Wave Biotech bioreactors. (This figure is available in full color at http://onlinelibrary.wiley.com/book/10.1002/9780470054581.)

are not typically used in rocking motion bioreactors. However, for more accurate control of pH and DO there are a number of SU sensors commercially available that can be integrated.

41.3.1.2 Large-Scale Cell Culture Processing. Larger scale SU bioreactors (for suspension or microcarrier cultures) were introduced to the market with Wave Biotech's 200-L and 500-L systems, which are scaled up versions of the 20-L rocking motion bioreactor system. The development of stirred tank SU bioreactors followed shortly after and they became commercially available in 2006 with Hyclone's 250-L SU Bioreactor (Single-Use Bioreactor (SUB)). Through joint collaboration between Centocor and Hyclone, the first 1000-L SU bioreactor was developed and implemented in Centocor's Development and GMP pilot plant facilities (9). Other SU bioreactors from multiple manufacturers (e.g. Xcellerex, Sartorius Stedim, ATMI/LevTech, and Cellexus) have since been developed, offering a wide array of innovations in mixing, gassing strategies, bag materials, support vessels, and use of sensors (Fig. 41.5). They are currently available up to 2000 L scale.

A typical SU bioreactor is composed of a cultivation chamber (i.e. a plastic bag) and a rigid support structure, such as a SS shell or holder, in which the cultivation chamber is secured. Gamma irradiation of the cultivation chambers is typically managed by the supplier. The SS shells are either jacketed to provide heating and cooling with water and steam, or supplied with an electric heating blanket if rapid cooling is not required. These bioreactors provide many of the same features and options found on SS bioreactors, including gas addition ports, liquid transfer ports, exhaust filters, and methods of monitoring of DO, pH, pressure, and temperature. SU bioreactors can also be easily linked to existing control systems.

In the past few years, the rapid expansion of SU components in cell culture processing has led to the emergence of several variations of SU bioreactors. While SS cell culture systems from different manufacturers are quite similar in terms of the major components, SU systems have recently led to many new innovations in mixing, gas addition, and

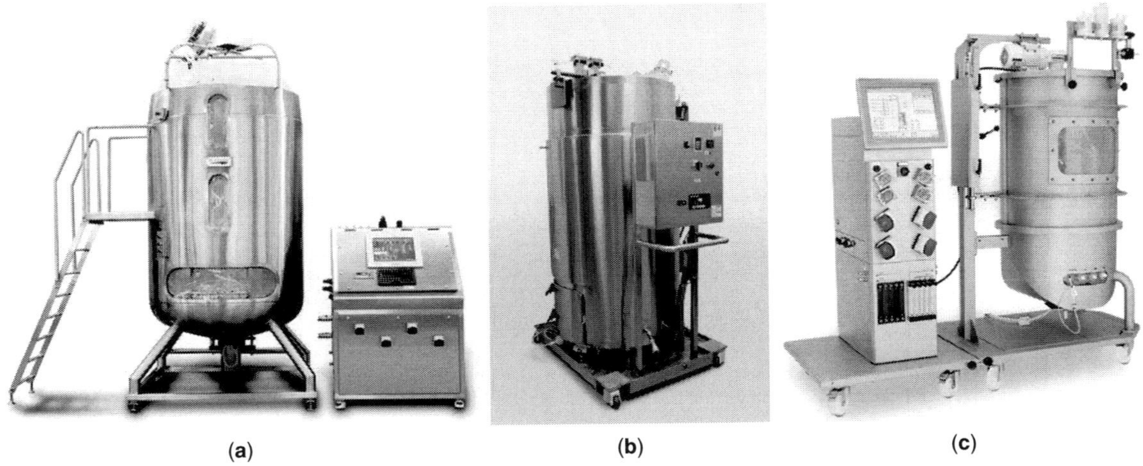

Figure 41.5. SU bioreactors. (a) Xcellerex 2000-L XDR2000 [photo courtesy of Xcellerex], (b) Hyclone 1000-L SUB [photo courtesy of Hyclone], (c) Sartorius Stedim 250-L BIOSTAT Cultibag STR [photo courtesy of Sartorius Stedim] (This figure is available in full color at http://onlinelibrary.wiley.com/book/10.1002/9780470054581.).

monitoring of key parameters. Rocking motion and stirred motion bioreactors have set the standard for SU cell culture systems currently in the market. Rocking motion was the first to arrive in the market and to be implemented in manufacturing, and stirred motion (with impellers) is the most closely adapted change from conventional SS bioreactors, and often a more easily adopted change in equipment from the traditional. Other innovative variations of mixing in SU bioreactors such as paddles, wheels, wands, oscillating plates, and orbital motion each with their own advantages have also been developed.

Mixing characteristics of SU systems, just as with SS systems, must be evaluated to insure homogeneous suspension of cells, minimize vortexing, and to prevent physical or thermal gradients. Calculation of mixing time, Reynolds number, and power input per volume are a few parameters commonly used for bioreactor scale-up, as well as comparative studies, and these influence the design and configuration of the impellers or agitation device. In addition to providing adequate mixing, it is also important to determine whether the mode of agitation is gentle enough on the cells to prevent cell death by mechanical stress. Typically, shear stress is determined from a calculation of an impeller tip speed. Several characterization studies have been performed on various SU bioreactors based on the parameters mentioned above, and have been shown to be comparable to SS systems (10).

The method of gas addition is another widely varying design aspect of SU bioreactors. Some SU systems employ separate gas addition ports or spargers, ranging from conventional designs such as open tubes and porous microspargers to other innovative solutions such as perforated membranes sealed onto a cultivation chamber. Other SU systems incorporate gas addition into mixing paddles or impellers, or simply on a large liquid-gas interface to provide surface area for gas exchange.

Providing sufficient oxygenation and removal of carbon dioxide are critical to the operation of a bioreactor, and are usually predicted based on mass transfer coefficients (k_La), which can be experimentally determined. These coefficients can be calculated for both the headspace and for sparged gases, over a range of operating conditions (e.g. agitation speeds, and gas sparge rates). Mass transfer studies have been performed on various SU bioreactor systems showing comparability to SS bioreactors (9,10).

Measurement of key parameters, such as DO, pH, and CO_2, is another aspect that has been adapted from SS bioreactors. Innovative methods of inserting conventional probes into SU bioreactors, such as the use of aseptic connection devices, made it possible to monitor key parameters using the same probes and control systems as SS bioreactors. This ability to monitor and control in a SU system significantly added to the acceptability and robustness of SU systems. SU sensors, which have been in development for a number of years (e.g. Polestar, Presens, and Finesse), have also added to the ease of use and flexibility of SU systems, eliminating the need to clean and sterilize probes. These sensors can be incorporated into almost any SU bag design and can often be linked to existing control systems.

SU systems specific to anchorage-dependent cells such as Corning CellCubes, Nunc Cell Factories, and roller bottles, which have been in use for many years, are also available. These systems, each with varying advantages and modes of operation, provide large surface area for cells to adhere and grow to high cell density. For cultures requiring greater cell densities, the cells can be grown on microcarriers in stirred tank or rocking motion bioreactors described above.

41.3.1.3 Liquid and Material Transfers. Liquid and material transfers from one process step or component to another represent potential risk for contamination, and require reliable techniques for connection and disconnection. Several methods have been developed to facilitate sterile (closed system) connectivity between SU components, including tubing welders, tubing sealers and crimpers, and SU connection devices. Terumo SCD's have been very widely used in hospitals and in biopharmaceutical processing for sterile connection of small diameter C-Flex and polyvinyl chloride (PVC) tubing (<1/8 ID). Other suppliers, such as Wave Biotech, Sartorius Stedim, Wescor Tubing, and Flex Concepts, have developed similar tube welders for connections of various types of thermoplastic tubing (e.g. C-Flex and PharMed) up to 1.25 ID. Some provide the option of welding dry and liquid filled tubing. Similar machines are available to seal tubing, which can be cut with scissors after the seal to provide a means of sterile disconnection of SU components. The tubing welders and sealers are generally restricted to certain types of tubing, depending on the suppliers.

Alternative means of connectivity are available in the form of aseptic connection devices (Fig. 41.6), such Pall's Kleenpak ACD, Millipore's Lynx STS connector, and GE Healthcare's ReadyMate DAC (formerly known as Bio-Quate DAC). These can be applied to many types of SU components, including bag and tubing assemblies. They are available from suppliers gamma irradiated (some are also autoclavable) in sizes from 3/8 in. ID to 3/4 in. ID. They can add flexibility to a process when a sterile tube welder is not available, or is unable to reach the point of connection. Alternatives to sterile sealing machines are also available, such as Millipore's NovaSeal crimping solution, in which a tube and pinch-pipe are simultaneously crimped closed, creating a sterile separation.

41.3.2 Advantages

41.3.2.1 Flexibility. SU bioreactors add significant flexibility to a process and pilot plant facility, particularly in a multiproduct environment. They are typically available on mobile skids (including load cells), making it possible to easily move them within a pilot plant facility. A single skid can be used for many different uses and types of processes. A SS bioreactor often has ports in fixed locations for specific purposes, and making changes to the bioreactor configuration and piping can be time-consuming and costly. On the other hand, a SU bioreactor bag is very easily customizable, and many features (ports, spargers, liquid transfer lines, etc.) can be added or removed to suit the needs of multiple processes. Many suppliers provide pre-assembled tubing and filter assemblies, which can be used to add more flexibility and connectivity to a SU bioreactor bag design. In a multiproduct pilot plant facility, this flexibility can help maximize the utilization of a bioreactor.

41.3.2.2 Time, Space, and Cost Savings. SU bioreactors typically occupy a much smaller footprint than SS bioreactors, due to the reduced piping, valves, and support structure (Fig. 41.7). As a result, more SU bioreactors can fit into a given space and more space is available for other operations, or a pilot plant facility can initially be designed to be much smaller. The savings in space, particularly for a classified lab space, can translate to significant savings in capital investment.

SU bioreactor bags are typically gamma sterilized by the supplier, and can be discarded after use. As a

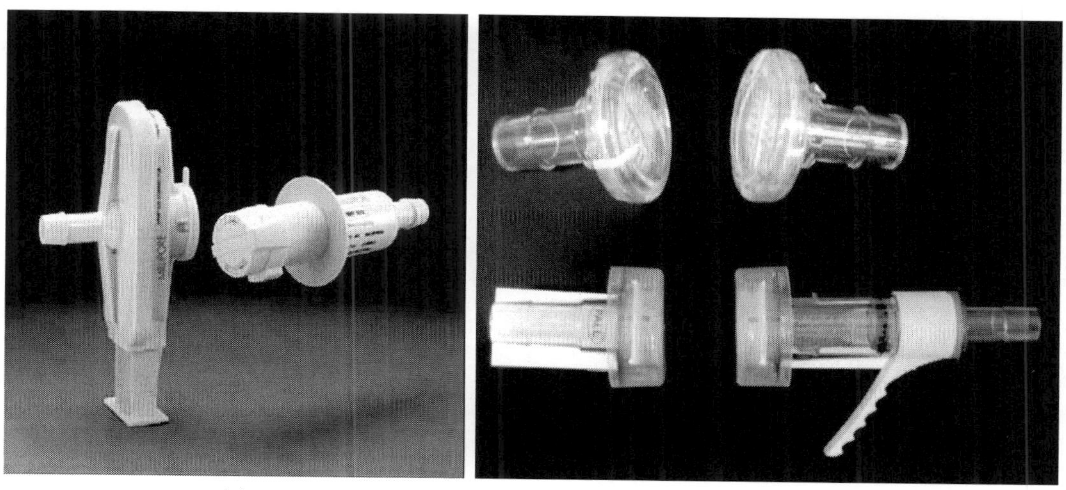

(a) (b)

Figure 41.6. Aseptic connection devices. (a): Millipore S2S Lynx [photo courtesy of Millipore], (b): GE ReadyMate DAC (top) and Pall Kleenpak ACD (bottom). (This figure is available in full color at http://onlinelibrary.wiley.com/book/10.1002/9780470054581.)

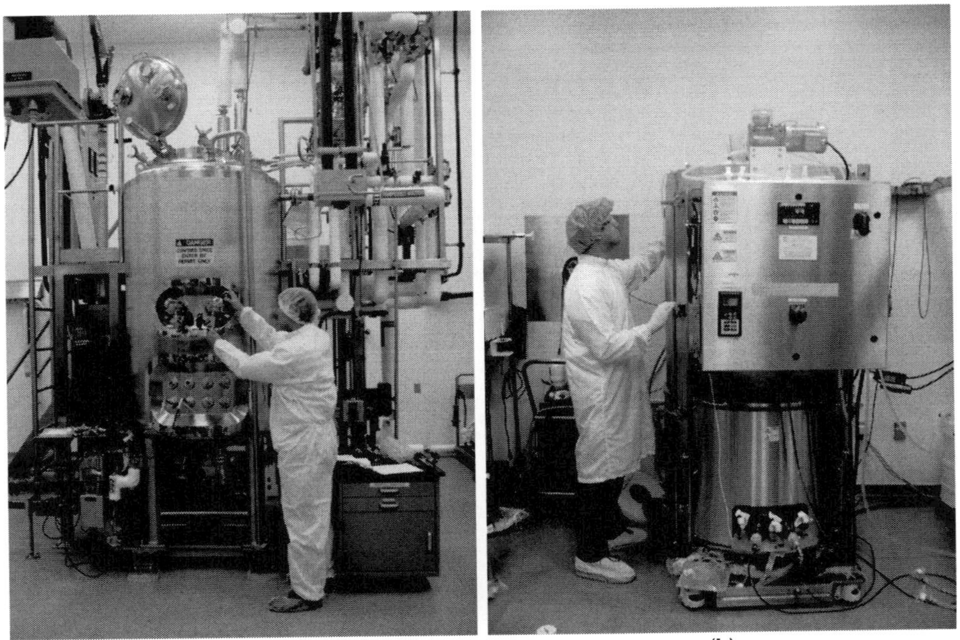

(a) (b)

Figure 41.7. Comparison of space usage: (a) 1000-L stainless steel bioreactor versus (b) 1000-L SU bioreactor. (This figure is available in full color at http://onlinelibrary.wiley.com/book/10.1002/9780470054581.)

result, no cleaning (CIP) and sterilization [autoclave or steaming-in-place (SIP)] are needed, the overall automation for the system is simplified, the utilities are reduced (e.g. water, steam, and compressed air), and the amount of piping and valves is greatly reduced. This effectively eliminates the cleaning and sterilization validation, simplifies the automation and mechanical qualification, and reduces the time to startup the process or pilot plant facility.

Another benefit of not having to clean and sterilize a SU bioreactor is the rapid turnaround between bioreactor runs, especially between different products, where a team may need to perform lengthier cleaning, verification of cleaning, and other changeover procedures associated with a SS bioreactor. There is less downtime between runs, only one or two days versus up to a week or more with SS systems, which allows for completion of more runs per year. In some cases, the turnaround costs may be much lower for a SU bioreactor, since the cleaning and sterilization of a SS bioreactor are often much more expensive than the material costs associated with SU bags.

The simple design of most SU systems also provides shorter lead times from suppliers, faster factory acceptance testing, installation, and validation, and overall results in faster pilot plant facility start-up.

41.3.2.3 Reduced Risk of Cross-Contamination. Another major benefit of not having to clean or sterilize SU bioreactors is the reduced risk of cross-contamination from product to product, a significant concern for a multiproduct pilot plant facility. Once a bioreactor run is complete, the cultivation chamber or bag is simply discarded into a biohazard container.

41.3.2.4 Advantages Vary Depending on Pilot Plant Facility. The many potential benefits of implementing SU bioreactors described above may vary in significance depending on the circumstances and purpose of a particular pilot plant facility. A single product pilot plant facility may have a very different outlook on implementing SU systems than a multiproduct pilot plant facility, which may operate with many additional concerns and complexities. Likewise, an older pilot plant facility with substantial investment in validated SS equipment may weigh the benefits of SU implementation differently than a new pilot plant facility. A hybrid pilot plant facility or process, which can continue to make use of existing reusable equipment and add on SU equipment as capacity expansion is needed, or as older reusable equipment becomes obsolete and is phased out, is often the most practical. In some cases, hybrid operations are necessary due to the fact that SU equipment for a specific operation may not yet be developed, or is not scaleable, or lacks sufficient data for immediate implementation. Other aspects of a pilot plant facility, such as waste disposal, utilities, and process space must also be considered, as they may be affected by switching from reusable to SU systems.

41.3.3 Disadvantages

41.3.3.1 Component Compatibility: Extractables and Leachables. Despite the many potential advantages of implementing SU cell culture systems, the technology is still in the relatively early stages of development, and there are risks and limitations to consider. One of the most prevalent concerns is the release of leachables and extractables into a process or drug product. Extractables are defined as "chemical entities that can be extracted from components of a material by exertion of an exaggerated force (e.g. organic solvents, extreme temperatures, ionic strength, pH, contact time, etc.)" (11). Leachables, a subset of extractables, are chemical entities that can migrate into a process or product under normal process and operating conditions (11). Detection, characterization, and quantitation for these chemical compounds are often critical to validating a process using SU systems.

Extractables and leachables may come from multiple sources, including any components in direct contact with the process stream or product, such as SU bioreactors, transfer tubing, filters, vials, and bags, as well as secondary packing (e.g. overwraps, labels). The manufacturing of many SU products includes the use of adhesives, coatings, additives, and other chemical components, which can potentially migrate into the product. Furthermore, exposing SU materials to steam or gamma sterilization may result in degradation of polymeric components, making them more readily leachable. These leachable compounds "raise a number of safety and efficacy concerns due to their potential to cause acute toxicity, as well as other adverse events (e.g carcinogenicity, chronic toxicity, immune reaction, endocrine dysfunction, etc.)" (12).

"Detection, identification, characterization, and quantification of leachables can be challenging, since these substances represent diverse chemical classes of organic and/or inorganic compounds that co-exist in complex mixtures at trace amounts" (11). The FDA and other global regulatory agencies, as well as organizations supporting the use of SU products (e.g. bio-process systems alliance (BPSA), recommend performing a risk assessment as part of leachables and extractables testing. Generally, a risk assessment is completed, which takes the following into account: all materials that have product contact, duration of contact, temperature, proximity of the material to the final product, surface area of exposed materials, and any pretreatment steps that could result in greater concentrations of leachables and extractables (13,14).

The cost and complexity associated with leachables and extractables testing for a multicomponent polymeric system can be significant. In the past, data was often nonexistant, or leachable and extratable components had only poorly characterized physico-chemical and toxicological profiles. However, with the rapidly increasing demand for SU systems in biological pilot plant facilities over the past decade, suppliers of these products are now often able to provide a starting point with preliminary extractables and leachables profiles for their products, and provide guidance to facilitate acceptance of their products by regulatory agencies.

Several organizations (e.g. BPSA, FDA, and others) provide guidance and education in the use and testing of SU components, as well as comprehensive lists of tests that may be performed for SU systems to demonstrate compatibility with a drug product (8,13).

41.3.3.2 Process and Product Comparability. Another common concern is the comparability of process and product between SU and reusable systems. A SU bioreactor must be able to produce the same product as a SS bioreactor, and in most cases must also maintain similar process conditions in order to do so. Currently, there is data available for SU bioreactor systems from multiple suppliers demonstrating that process conditions can be maintained comparably to a SS system, and that the resulting product (monoclonal antibody) is comparable (9,10).

In order to show process comparability, the growth, metabolic activity, and productivity are commonly compared. Parameters such as maximum cell density, cell viability, nutrient consumption, metabolic waste production, and product titer are often plotted side by side to determine any significant differences between SU and SS. Figures 41.8 and 41.9 show examples of comparable cell density and titer for a Hyclone 250-L SUB and a 250-L SS bioreactor. Several studies have applied design criteria such as vessel geometry, mixing, mass transfer, and thermal characteristics to show that a process environment is comparable.

It is also critical to determine product comparability between SU and SS bioreactors. Depending on the type of product, this is commonly determined by methods such as electrophoresis, high pressure liquid chromatography (HPLC), peptide mapping, oligosaccharide mapping, and mass spectroscopy.

41.3.3.3 Process Safety. The common safety concern associated with SU bioreactors revolves around maintaining the integrity of the bag. A SU bioreactor is a plastic bag, and is susceptible to (i) punctures or tears if handled aggressively or manufactured improperly, (ii) melting if excessive temperature is applied, and (iii) overpressurization if too much liquid or gas is added, or exhaust filters clog. SU bioreactor bags are available from multiple suppliers with a Certificate of Analysis (CoA), indicating that actions have been initiated to control

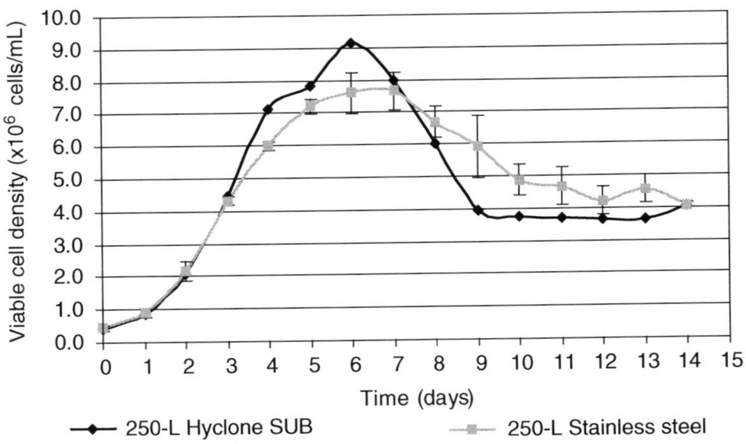

Figure 41.8. Comparison of cell density between a 250-L Hyclone SUB and a 250-L stainless steel bioreactor. (This figure is available in full color at http://onlinelibrary.wiley.com/book/10.1002/9780470054581.)

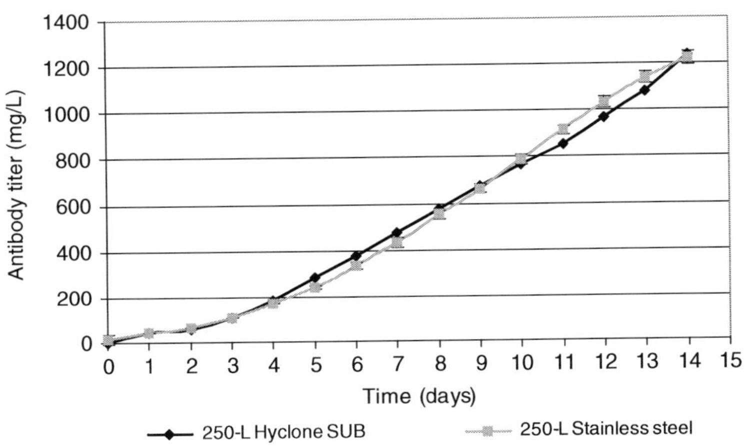

Figure 41.9. Comparison of titer between a 250-L Hyclone SUB and a 250-L stainless steel bioreactor. (This figure is available in full color at http://onlinelibrary.wiley.com/book/10.1002/9780470054581.)

and verify the steps in the construction of the bags. A CoA, along with in-house quality control and creation of standard operating procedure's (SOP) that include controls to insure the careful handling of the bags, are usually sufficient to nearly eliminate any puncture or tearing issues. Melting is another concern, since low-density plastics used to make bags can melt at relatively low temperatures. This potential safety issue can be avoided by implementing simple interlocks to shut off heat supply in the event of excessive temperatures being reached. This is a feature that can often be provided by a SU bioreactor supplier. Overpressurization of a bioreactor bag can occur if exhaust filters are clogged, or if too much liquid or gas is added. Filter clogging due to condensation on an exhaust filter can be prevented by adding an electrical heating jacket to the filter, which can often be provided by the bioreactor supplier. Overpressurization can be prevented by (i) adding a SU pressure sensor (e.g. PendoTech, Scilog) to monitor pressure, and (ii) adding a pressure relief pinch-valve that releases and vents at a specified pressure (e.g. PendoTech).

41.3.3.4 Scalability. Scalability of a development or small-scale process is critical to its success on a large scale. Each of the many different types of SU bioreactors have varying potential for maximum scale-up volume depending on modes of agitation and other design factors. Currently, SU bioreactors are commercially available in working volumes up to 2000 L. Larger volumes may be possible, but will be challenged by bag integrity constraints and the ability to provide adequate mixing and oxygenation. For many processes, SU products can be used from vial thaw through production bioreactor. For others, SU bioreactors must be limited to seed and cell expansion, and must be used to inoculate a larger SS production bioreactor.

41.3.3.5 ***Storage Space.*** Facilities that adopt SU processing often require additional storage space to maintain a continuous stock of components. For a process made up mainly of SU components, a significantly larger storage area or warehouse may be needed compared to a traditional process made up of reusable components. In some cases, a supplier can hold an inventory of a SU item, which can be shipped at regular intervals depending on usage to alleviate the end-user's space constraints.

41.3.3.6 ***Solid Waste Disposal.*** SU processing generates more solid waste than traditional processing with reusable components. A typical cell culture process made up of SU components may require disposal of shaker flasks, media bags, bioreactor bags, filters, several meters of tubing, and potentially many more items. This amounts to a large volume of materials to dispose, and adds to the overall operating costs. Depending on the volume and level of biohazard, various options are available for disposal, including: landfill, incineration, and recycling. Recycling is an environmentally friendly option, however, it is most often not possible since many SU components used in bioprocessing are made up of multiple layers of different materials and may be contaminated with biohazardous material. Cogeneration of heat, steam, or electricity during incineration of plastic materials is another potential option to reduce the environmental impact of disposal (15).

41.4 PURIFICATION

The adoption of SU technologies for the production of biologics has been driven by the increasing economic pressures on the pharmaceutical industry (16). SU systems are viewed as economically advantageous since they require less capital investment than traditional technologies and eliminate the need for costly CIP and SIP procedures. Additional benefits of SU systems include the reduction of cross-contamination risks in multiproduct facilities, process scale flexibility, and the reduction of project lead time (17).

Currently, downstream processing exists in hybrid facilities where SU systems are used in conjunction with SS hardware. Although SU technologies have been used for some time for purification applications, they have generally been limited to liquid storage, liquid transfer, and filtration operations (18). This section will provide an overview of systems in current use and of the recent developments in SU technologies for downstream processing.

41.4.1 Process Flow

41.4.1.1 ***Liquid Handling.*** SU containers, bags, and tubing manufactured from medical-grade polymers are widely used for the storage and transfer of process buffers and process intermediates. The containers and bags are generally fitted with ports, tubing, and connection fittings and arrive from the manufacturer presterilized by gamma irradiation. They are available in a wide range of volume capacities from 50 mL up to 3000 L from suppliers such as Thermo Fisher Scientific (HyClone) and Sartorius Stedim Biotech.

SU mixing systems have also been developed. These systems employ similar materials as SU bags, but they include internal agitation that is driven either magnetically across the surface of the bag or through the bag by an externally mounted system that is physically connected to the bag, but separated from the bag's interior by either rotating seals and/or a shaft sheath. In addition to being used for buffer preparation, they are used for making adjustments to process intermediates, such as pH titrations. There are several systems available with similar capabilities with volume capacities up to 3500 L offered by manufacturers such as Thermo Fisher Scientific (HyClone), Sartorius Stedim Biotech, Millipore, Xcellerex, and ATMI Life Sciences (19).

41.4.1.2 ***Filtration.*** A number of suppliers provide SU filter capsules for normal flow or "dead-end" filtration for processes such as clarification, sterilization, and the removal of adventitious viruses. SU cassettes and hollow fiber cartridges are also available for tangential flow filtration (TFF) operations, such as ultrafiltration/diafiltration and microfiltration. Suppliers include Millipore, Pall, Sartorius Stedim, CUNO, Asahi Kasei, and GE Healthcare.

The clarification of bioreactor harvest to remove cells and/or cell debris from product has been traditionally performed with the use of depth filtration in which pressurized SS housings contain lenticular filter support media with filter aids such as diatomaceous earth (20). SU polymeric housings that have made the entire housing and filter media SU, as demonstrated in the CUNO Zeta Plus™ filter series have been developed. Millipore has recently introduced the Pod SU depth filtration system that includes SU cassettes and a holder frame (Fig. 41.10). The Pod system provides process flexibility since the holder can accommodate a number of cassettes for increased surface area and since it offers the added advantage of not requiring cleaning validation as the process streams do not contact the holder frame. Historically, depth filtration has been limited by scale and clarification operations have utilized centrifugation in tandem with depth filtration to alleviate this constraint. With expected improvements in depth filter capacity, fully SU clarification operations will be possible at larger production scales.

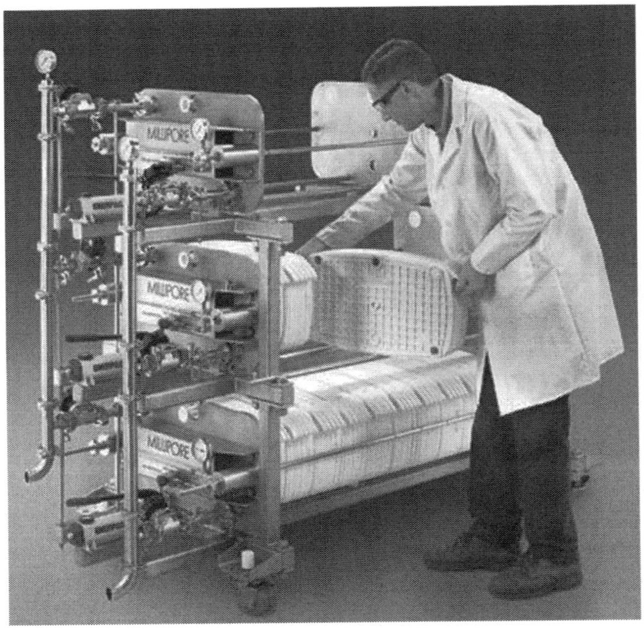

Figure 41.10. Millipore Pod system [Printed with permission from Millipore]. (This figure is available in full color at http://onlinelibrary.wiley.com/book/10.1002/9780470054581.)

Although single-use filter capsules, cassettes, and cartridges have been available, the supporting systems, such as pumps, piping, and process skids have not been designed for SU. Recently, integrated systems, such as the Millipore's Mobius® FlexReady system has been developed for both normal and TFF operations in which the entire flow path to the filter module is SU.

41.4.1.3 ***Chromatography.*** Chromatography is the core technology used for downstream processing because of its selectivity and robustness for the separation of active pharmaceutical ingredients (API) from chemical and biological impurities and contaminants (21). Packed bed chromatographic columns consist of SS or glass tubes packed with resins containing chemical functionalities that provide for separations through unique molecular interactions. The columns are operated in either bind-and-elute mode, where product is bound to the resin and selectively eluted to remove impurities, or flow-through mode, where the product flows through the column while impurities are retained. Typically, the packed bed of a chromatography column has a lifetime of hundreds of cycles before diminished performance is observed. After this decrease, the column is unpacked, the resin is discarded, and the tube is cleaned and reused. Although this particular technology is not readily adaptable for SU applications, efforts have been made to develop approaches to packed bed chromatography that reduce capital costs and cleaning procedures. SU technologies, such as membrane absorbers, have also been developed as a substitute to packed bed chromatography and have proven useful in certain applications (22).

The availability of prepacked, prequalified, and presterilized process scale chromatography columns is a recent development that applies some of the advantages of SU systems, such as process scale and flexibility, the reduction of cleaning validation costs, and cross-contamination risks. A pre-packed column consists of a tube packed with a resin. However, the tube is constructed from a polymer and it is intended that the entire column be disposed once diminished bed performance is observed. Although these systems are not SU, they can be considered single product. Prepacked columns are available from GE Healthcare in their ReadyToProcess™ format. These columns are packed with GE Healthcare's MabSelect™ SuRe, Capto™ Q, Capto™ S, Capto™ Adhere or Phenyl Sepharose™ 6 Fast Flow (low sub) resins in 2.5, 10 or 20 L volumes with fixed bed heights of 20 cm. Atoll GmbH produces a disposable column called the MediaScout® MaxiChrom 100-X which has a 10 cm inner diameter and can be packed to bed heights of 5 to 30 cm or volumes of 0.4 to 2.4 L, respectively. For this product, the intention is that the purchaser provides the vendor with the resin of interest and specifies the bed height. Atoll will pack, qualify, and sterilize the column prior to delivery. Similarly, BioFlash Partners, LLC offers BioFlash DFC™ disposable columns in 1.2 cm, 8 cm or 20 cm inner diameters which are then packed with the purchaser's resin of interest.

Similar to the implementation of SU flow paths for integrated filtration systems, chromatographic systems employing SU flow paths have also been developed. The AKTAReady™ system from GE Healthcare is such a system and is designed to operate with ready-to-use SU flow paths supplied in low and high flow-rate formats. These flow paths are constructed from disposable medical-grade polymeric tubing and connectors and include inlet/outlet manifolds, air trap/column bypass flow paths, and disposable process monitoring sensors. The AKTAReady™ system incorporates a peristaltic pump and valves external to the tubing so that product remains contained within the SU flow path. The AKTAReady™ system with a ReadyToProcess™ column is shown in Fig. 41.11. For simulated moving bed (SMB) applications, Tarpon BioSystems, Inc.'s BioSMB™ system incorporates a SU flow path containing a proprietary disposable valve cassette to direct fluid flow amongst the multiple columns required for SMB technology (23).

Membrane absorbers, referred to as membrane chromatography, are an alternative to traditional chromatography that is gaining acceptance for use in flow-through mode applications. They are especially attractive for use in polishing operations for the removal of trace impurities, such as host cell proteins (HCP), DNA, endotoxin (24), and virus removal (25). Membrane

Figure 41.11. AKTAReady™ System with ReadyToProcess™ column [printed with permission from GE Healthcare]. (This figure is available in full color at http://onlinelibrary.wiley.com/book/10.1002/9780470054581.)

absorbers employ thin, synthetic, porous membranes containing functional groups analogous to those used for chromatography resins affording similar separations (26). They do not require packing and are manufactured like the filtration capsules. The physical configuration of membrane absorbers allows for higher flow rates and lower back pressures when compared to packed bed chromatography leading to an additional advantage of decreased processing time. Although membrane chromatography can also be used in bind-and-elute applications, their utilization for this purpose has been limited because of reduced performance compared to packed bed chromatography.

41.4.2 Advantages and Disadvantages

Economic considerations have driven the increasing adoption of SU technologies for the production of biologics. Recent developments have been made to expand the application of SU systems to additional purification operations. These include disposable mixing devices, integrated systems to conduct unit operations with SU flow paths, membrane absorbers, and prepacked chromatography columns. Developments such as these may lead to the future possibility of a complete SU production facility in place of the current hybrid situation. However, limitations and uncertainties, such as increased costs at larger scales, regulatory requirements and leachable/extractable evaluation, will have to be addressed before this becomes a reality for commercial facilities. At present, pilot scale and clinical production facilities may be better situated to fully implement SU technologies.

41.5 APPLICATIONS IN A FILL AND FINISH PLANT

This section provides an overview of how SU technology is applied in the aseptic fill and finish process for final drug products. Traditionally, the aseptic filling process for medications has utilized SS equipment almost exclusively, allowing multiple use of the infrastructure. Over time, SU materials entered the manufacturing process gradually. Typically, steps involving filter cartridges and flexible tubing for drug product transfer were the most prominent areas for using disposable plastic materials. Currently, the fill and finish facility may best be described as a hybrid combining SU and reusable technology. However, it can be assumed that the level of SU technology applications will increase in a fill and finish plant especially as vendors develop more innovative SU equipment.

SU technology is strongly linked to the use of plastic material, which is not only applied as part of drug product manufacturing but already has a significant history as material for final drug product container in the fill and finish process. Facilities processing liquid drug formulations by blow-fill-seal-technology or by filling them into plastic IV-bags involve a larger degree of SU plastic materials because the drug product container is no longer made of glass. This same observation applies for aseptic manufacturing of prefilled syringes and vials made of cyclo-olefine polymers.

Moreover, we will see different levels of SU implementation if we compare a multiproduct, multi-format research and development fill and finish plant to a commercial facility dedicated to few products but larger volumes. This aspect has also been addressed by Vogt and Paust (27), who conclude that SU technology is best applied in the screening, development, and pilot stages with volumes ranging typically from milliliters up to a few liters, followed by hybrid technology for manufacturing of clinical material (up to 100 L) and steels-based infrastructure dominating the commercial production of large volumes (1000 L and more). Although such a classification is certainly meaningful for facilities manufacturing API, where bioreactors are of much larger volume than the typical batch size volumes encountered in drug product fill and finish facilities, it can also be applied as a guidance for the fill and finish area.

Finally, the drivers for applying SU technologies in fill and finish processes are similar to those for bulk facilities, such as fermentation and purification of API. These can be summarized as follows (28):

- reduce risk of (cross-) contamination
- reduce time for setup and production turnaround
- reduce space requirement
- reduce need for cleaning and and sterilization
- reduce plant validation efforts
- increase flexibility and simplicity.

Since the fill and finish process is the last step in the manufacturing process, one of the biggest concerns is the potential for leachables and extractables from materials that are in contact with the drug product. A comprehensive data package built from leachable and extractable studies, as well as from mixing and filtration trials is of crucial importance to avoid extra burden, time, and costs by duplicating efforts at the fill and finish site. In this context, a wise choice of materials and suppliers should become a strategic imperative for a drug product supply chain to optimize the benefit of SU technology both at the bulk API as well as at the fill and finish site. The qualification burden associated with leachable/extractable studies is currently considered one of the limitations of SU implementation and it to some extent outweighs the benefits. This cost/benefit aspect will certainly be a key parameter for companies to decide for or against SU implementation. Many vendors and manufacturers of SU equipment and components offer reports on their extensive leachables and extractables studies.

41.5.1 Role of Plastics in Fill and Finish Processes

As indicated above, the term "single-use" or "disposable" points to material that is discarded after use. In focusing on the fill and finish manufacturing process, we may differentiate between:

- A SU component in the fill and finish manufacturing train.
- A SU drug product container.

These two aspects will be discussed in further detail in the following sections.

41.5.2 Unit Operations in a Typical Biotech Fill and Finish Process: Steps Involving SU Material

When examining a typical fill and finish process, SU materials are encountered at the following steps (Fig. 41.12):

- *Incoming Materials.* Biotech-based drug formulations or preformulated bulk (PFB) are typically supplied by the bulk API facility in screw-cap bottles made of polycarbonate or polyethylene terephthalate (PET-G). In addition, more sophisticated technologies are available for controlled freezing, transport, and thawing of plastic bags are in use as well.
- *Pooling and Mixing Step.* This is an area where SU technology is growing. The use of bags with integrated mixing functionality either using magnetic stirring bars or integrated impellers (>50 L) have been available in the market for several years and offer flexibility to replace costly SS tanks. In addition, SU transfer systems to connect the pooled concentrated drug formulations with buffer solutions stored in bags and to dilute the formulation to its final concentration are well established and more or less identical with the features used in the API bulk facility. Material of choice for SU tubing or transfer systems is platinum cured silicon tubing.
- *Sampling for In-Process Testing (IPT).* presterilized, SU bags attached with sampling features (e.g. NovaSeptum as pictured in Fig. 41.13, AsepticProcessEquipment) allow to pull samples for IPT from the production process without the risk of contamination. This sampling technology is often used to access SS tanks via a dedicated sampling port. A typical field use is microbiological and endotoxin testing.
- *Filtration.* this is a core domain of SU equipment. Presterilized filter cartridges both for liquid filtration and venting or gas overlay have been available for years and are considered state-of-the-art in aseptic fill and finish processes.
- *Filling.* This is certainly the area where SU will require some further development. As of now, most of the fill and finish facilities, in the commercial production level, in particular, still use volumetric filling systems such as rotary piston pumps made of SS or ceramics. These pumps have excellent dosing accuracy, have a certified finishing, and are robust. Replacing SS with plastics might add difficulties to fabricate within the required tight tolerances, and excessive wear, and the risk for leaking would be another issue. These constraints currently exclude piston pumps to be operated as SU equipment. If any component of the dosing equipment could be developed as SU, the product contact pathway would be ideal. This aspect has become increasingly important since the industry has seen more and more biotech-based drug formulations entering the manufacturing process and scientists, engineers, and manufacturing experts have been confronted with

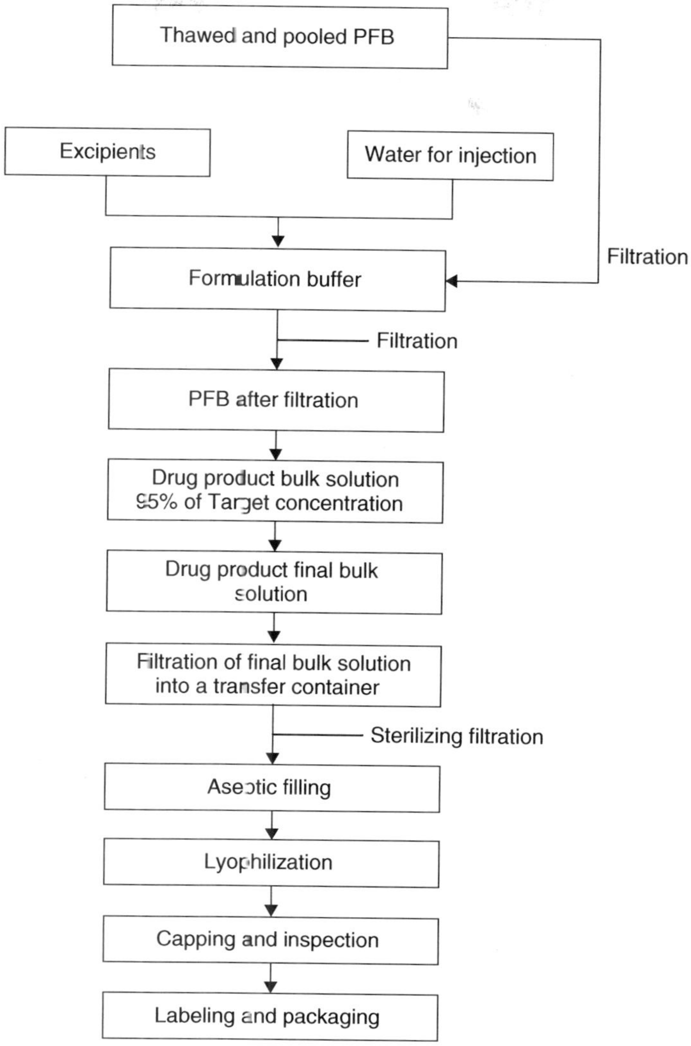

Figure 41.12. Overview of a typical fill and finish process for a biotech drug product. Unit operations possibly involving single-use materials and equipment are in italic. (This figure is available in full color at http://onlinelibrary.wiley.com/book/10.1002/9780470054581.)

new physico-chemical challenges posed by these large molecules. Shear sensitivity of biotech drugs has recently raised some concern in filling (29). The sensitivity of some protein-based formulations to metal microabrasion and the formation of subvisible particles or visible aggregates have put the focus on filling machine suppliers to improve their dosing systems and associated materials with direct product contact. Currently, we see increased interest to replace rotary piston pumps by less shear intense dosing systems particularly in the biotech area. Amongst those are filling technologies that combine low shear properties with the benefits of SU. Besides the well-established *peristaltic pump technology* or *time-pressure* based filling features, which both make use of disposable flexible tubing (silicon and teflon), new SU filling systems (e.g. Prevas™ by Bosch) have recently entered the market and will be discussed in more detail below. These systems also use SU filling needles made of polycarbonate or polytetrafluoroethylene (PTFE).

Recently, Fuller and Pora (30) reported from a case study, wherein a fill and finish contract manufacturing plant that replaced its SS systems with a ~90% disposable manufacturing train is discussed. The initiative was triggered by an incompatibility issue of one of their formulations with SS and oxygen. Besides resolving the stability issue of this particular formulation, the change to SU also resulted in a significant reduction of batch processing time. Not surprisingly, time savings were most prominent in cleaning, sterilization, and equipment setup.

Figure 41.13. Example of a NovaSeptum sample bag to access a stainless steel tank for IPC sampling [reprinted from Novaseptic]. (This figure is available in full color at http://onlinelibrary.wiley.com/book/10.1002/9780470054581.)

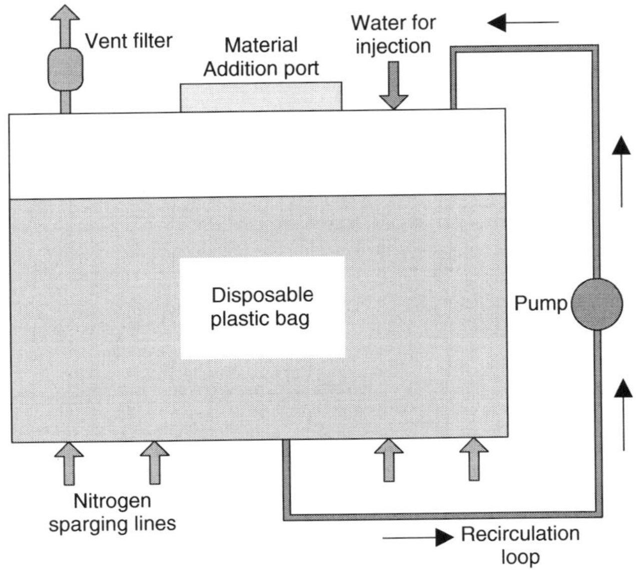

Figure 41.14. Compounding system based on SU components: disposable plastic bags fitted with plastic vent filters and flexible liners for nitrogen overlay replace standard stainless steel tanks [adapted from Fuller and Pora (30)]. (This figure is available in full color at http://onlinelibrary.wiley.com/book/10.1002/9780470054581.)

Figures 41.14 and 41.15 provide examples of how unit operations can be mapped into a SU manufacturing train philosophy.

Although attempts toward a complete SU fill and finish infrastructure will increase, it is unlikely for a fill and finish biotech facility to see the concept of a "fully disposable" factory to be put in place. The reason for this is that many of these biotech formulations require lyophilization to stabilize the drug formulation. CIP/SIP features on a state-of-the-art freeze dryer will require the infrastructure and utilities to supply the equipment with WFI and clean steam. Given the broad span of vacuum and pressurization required for the freeze-drying process and sterilization of the equipment, reusable SS infrastructure will remain a key asset in a biotech fill and finish facility. In addition, the washing and sterilization of primary packaging materials will require the same basic infrastructure features for water and clean steam as long as drug containers are based on glass and rubber stopper components. As a consequence, the fill and finish facility of the near future will continue to be a hybrid combining disposable and reusable technology.

41.5.3 Examples for SU Filling Technology

For years the peristaltic pumping principle (Fig. 41.16) has been the most versatile and probably the easiest filling system to operate with SU components. These peristaltic systems are typically comprised of a product hold bag, supply tubing, and a filling needle that are bagged together and gamma irradiated or steam sterilized.

Figure 41.17 provides an example of a filling needle setup with connectors and flexible tubing prior to sterilization. A large variety of tube connectors are available and the connections are typically fixed by cable tie. The presterilized assembly is removed from the pouch and connected to the peristaltic filling system, which is either located outside the filler or integrated into the filler platform. However, such technology is limited by speed and dosing accuracy due to wear of the flexible tubing. Additionally, scaleup to high speed lines may be limited as compared to piston pumps or time-pressure based dosing technology.

Time-pressure based filling systems partly operate with SU but the main limitation is that the technology requires pressurizing the surge tanks. Fabricating a surge tank as a SU bag would be possible but the material would not withstand the required pressurization for accurate dosing to take place (31).

A recent innovation was the commercial release of the Prevas™ presterilized, SU rolling diaphragm system (Figs 41.18–41.20). This setup includes plastic rolling diaphragm pumps, silicone tubing, and plastic fittings. The pumps and needles are made of injection molded medical-grade polycarbonate, partly reinforced by SS. The system is designed for Bosch filling lines, which may limit applications on equipment from other vendors.

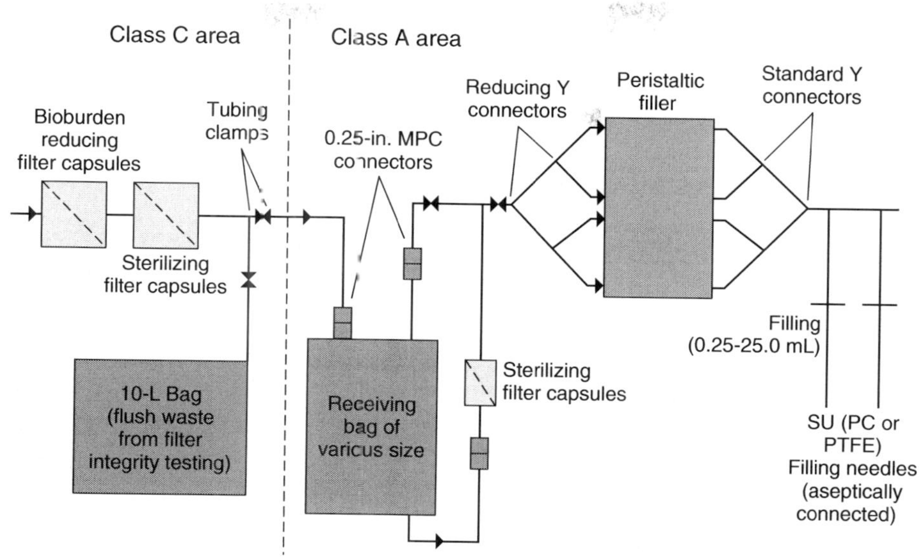

Figure 41.15. Overview of a SU manufacturing train in a fill and finish clean room environment [adapted from Fuller and Pora (30)]. (This figure is available in full color at http://onlinelibrary.wiley.com/book/10.1002/9780470054581.)

A comparable filling SU module was recently developed by Brandt. The Seripettor™ system (Fig. 41.21) builds on the well-established dispensing platform from the same company. The filling system is presterilized for aseptic use and will find applications with robotic fillers. While these technologies appear to be viable applications of SU for new filling facilities, they may be of limited value for retrofitting into existing filling lines.

41.5.4 Examples of Plastics Used as Drug Product Containers

Blow-fill-seal technology is widely applied in aseptic filing of unit dose ophthalmics and large and small volume parenterals. It has gained much market focus in recent years due to the increased emphasis on biologics, proteins, suspensions, and other complex solutions. These important products can often not withstand exposure to high temperatures for extended period of times without degradation. Aseptic blow-fill-seal allows processing of such formulations into a final plastic container at room temperature while excluding operator interference (32). This results in a fully automated process with a sterility assurance level superior to classical aseptic manufacturing. Lu and Frank (33) have reported their successful qualification work performed for filling a biotech drug by blow-fill-seal technology. An overview of the manufacturing steps are provided in Fig. 41.22.

The technology is mostly based on the use of polyethylene (PE) and polypropylene (PP) (examples shown in Fig. 41.23), which are considered inert by the FDA (32). Recently, more sophisticated co-extruded multilayer materials have been developed to limit oxygen permeability through PE and PP.

Apart from this well-established process platform, a new filling concept for closed plastic vials made of cyclo-olefine polymers was recently developed. This technique allows to fill solution into a presterilized, closed vial and to reseal the pierced stopper by laser. Similar, presterilized ready-to-fill plastic syringes supplied in plastic tubs are processed nowadays in fill and finish facilities. Finally, SU plastics will start taking a prominent role in the device and combination product area, where SU auto-injectors are used for application of parenteral medications.

41.5.5 Points to Consider in Introducing Disposables for Fill And Finish

The general benefits and drawbacks for introducing SU technology have been widely discussed in the literature and will not be further addressed. According to Philipps (34) no longer is there a need to convince end users of the value added through integration of disposables both in cost/scope control and simplicity/flexibility of process. It is no longer a question of "Should we use disposables?" as it is of "Which should we use?" and "How do we implement them?" Some specific aspects to be examined when it comes to the implementation of SU for aseptic processing in a fill and finish environment are discussed hereafter.

Typically, fill and finish of biotech drug formulations must rely on aseptic processing including sterilization filtration as the API's in most cases are very sensitive to irradiation in the final packaged drug product. As a consequence, the SU components in the manufacturing train

(a)

(b)

Figure 41.16. (a) Mobile two-head peristaltic pump system for use in clean rooms. (b) Opened pumping drive with inserted SU flexible. (This figure is available in full color at http://onlinelibrary.wiley.com/book/10.1002/9780470054581.)

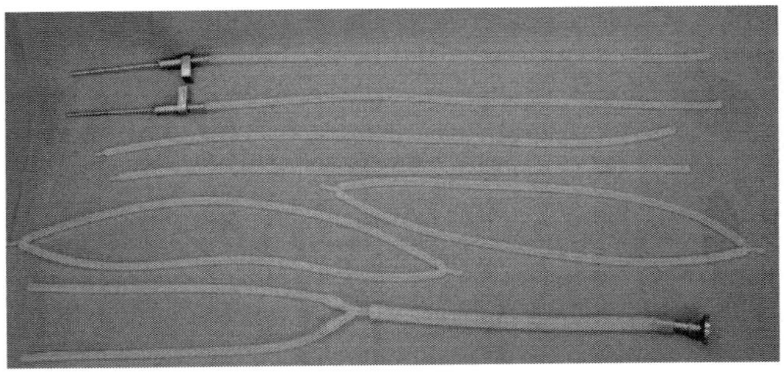

Figure 41.17. Filling needle setup with SU tubing and connectors for use with peristaltic pumps. Needles are also available as SU plastic. (This figure is available in full color at http://onlinelibrary.wiley.com/book/10.1002/9780470054581.)

(a) (b)

Figure 41.18. Rolling diaphragm pump Prevas™. (a) stainless steel and (b) single-use. (This figure is available in full color at http://onlinelibrary.wiley.com/book/10.1002/9780470054581.)

need to be delivered by the supplier in sterile conditions and packed aseptically to transfer material into the sterile suites or filling isolators. This shifts the sterilization validation to the supplier of the components and makes the end-user more dependent on the quality and supply chain stability of the vendor. Some companies are reluctant to outsource sterilization processes as they become more dependent from specific vendors and have less control on the aseptic quality.

41.5.6 Summary

Given the overall cost situation in health care, the SU topic has gained a lot of attention and momentum by the pharmaceutical industry since there is an ongoing pressure to optimize processes to reduce manufacturing costs for new medications. This need has also heavily promoted the marketing efforts of vendors offering SU materials and

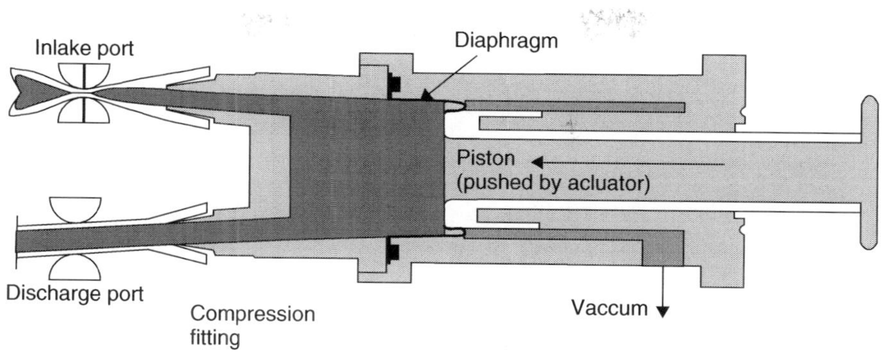

Figure 41.19. Cross-section of a rolling diaphragm pump highlighting the working principle. (This figure is available in full color at http://onlinelibrary.wiley.com/book/10.1002/9780470054581.)

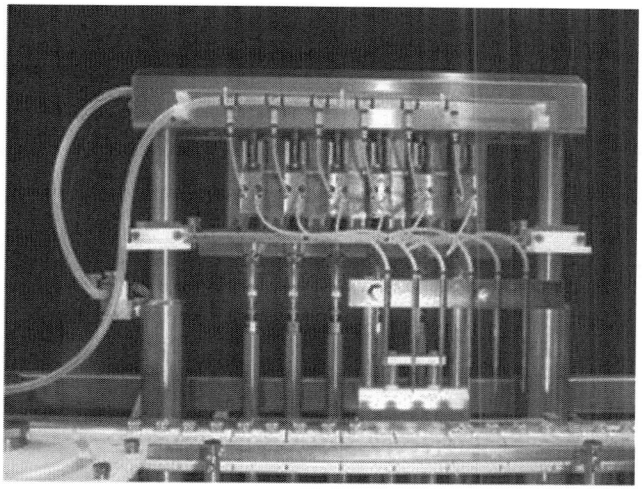

Figure 41.20. Overview of a single-use Prevas™ filling module. (This figure is available in full color at http://onlinelibrary.wiley.com/book/10.1002/9780470054581.)

technology to the pharmaceutical companies. The number of biotech based drug products will significantly increase over the next decade and whenever the manufacturing prerequisites for new product introductions need to be defined, emphasis will be put on a detailed analysis of required capital, resources and expenses. This analysis will strongly drive the decision for or against SU-based production in comparison to the currently still dominating SS pharma infrastructure. Besides, the success of SU implementation will heavily depend on the comfort level that the end-user will gain on SU product quality, process control, and the associated validation package provided by the vendor.

The applications and the experience with SU in the fill and finish area will steadily grow and largely benefit from the know-how generated over the past at the bulk API manufacturing sites. Not surprisingly, also the literature currently available on SU mostly still refers to SU applications in fermentation and purification of the API and only few point to aspects unique for the fill and finish area. This will change to some extent as more networking groups will focus on SU implementation for final drug product manufacturing as well. Amongst others, the International Society for Pharmaceutical Engineering (ISPE) and the Parenteral Drug Association (PDA) have installed communities of practice and interest groups to share information and to map the direction for SU in the pharmaceutical industry. However, it is to be assumed that the future in fill and finish will see a hybrid of reusable and SU, the latter dominating more in a multiproduct R&D environment requiring a higher level flexibility than in large-scale operations.

41.6 MEDIA BUFFER

Clinical manufacturing pilot plants are faced with changing production demands that result in changing time lines, and shifting priorities, especially for media and buffer preparation. A multiple product environment, fast turnaround requirements, and potential for cross-contamination drive the need for flexibility to provide a wide variety of media and buffer solutions. SU systems can meet many concerns for media/buffer preparation in multiproduct clinical manufacturing pilot plants.

41.6.1 SS Systems

Glass formulation containers, multiuse assemblies and SS vessels have traditionally been used for the production of media and buffer solutions. These containers, assemblies, and vessels typically require lengthy cleaning processes to ensure minimal carryover of product or contamination between formulations. There is a variety of equipment such as tubing, mixing shafts, valve assemblies, and connectors associated with each individual formulation process that are assembled, cleaned, segregated, and tracked. These manipulations require substantial amounts of time and materials to prepare and complete for each formulation.

Once these formulations are prepared using a conventional approach, the solutions are stored in specific holding

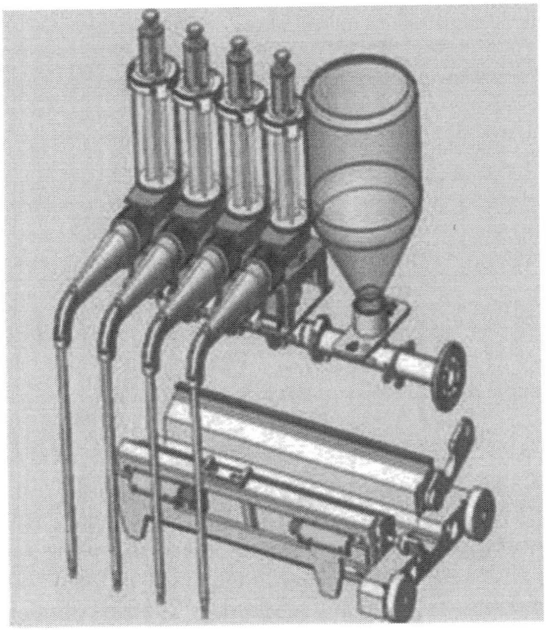

(a)

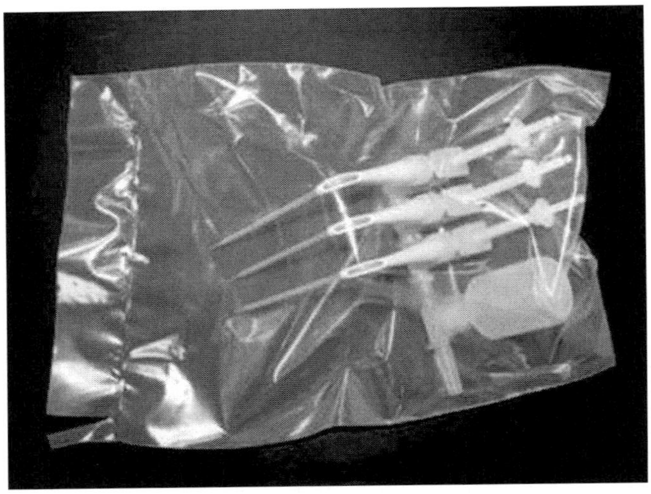

(b)

Figure 41.21. (a) Modular single-use Seripettor™ system with pump holder, (b) preassembled single-use Seripettor™ unit. (This figure is available in full color at http://onlinelibrary.wiley.com/book/10.1002/9780470054581.)

vessels or containers. Similar to the multiuse formulation systems, these holding vessels or containers require cleaning, sterilization, and testing for residuals. Often, the holding vessels or containers are dedicated for specific media or buffer solutions that can lead to higher capital costs and reduced flexibility.

41.6.2 SU Systems

If the facility is multiproduct and the process development environment does not follow a standard platform (i.e. each product may have different process steps utilizing unique media and buffer solutions), end-users need the flexibility of SU systems. Specifically in the media and buffer preparation area, a nonplatform process development environment leads to multiplicity of formulations and many different batch sizes. Economies of scale production are difficult to implement since the media and buffer preparation output is typically of small batch sizes and a variety of formulations (35).

Facility design is another factor in driving the implementation and use of SU systems. If the media and buffer formulation operation has been designed to prepare both animal component and animal component free media, plus a variety of buffer formulations and dedicated, SS vessels have not been provided, then, the SU mixers, transfer lines, filters, and containers offer the flexibility and adaptability to prepare a larger variety of powders and solutions required for multiple processes. A single SU vessel can be used for media and buffer preparation and that reduces facility footprint.

41.6.2.1 Advantages of Single-Use Systems. A significant benefit of utilizing SU systems in media and buffer preparation areas is the potential to eliminate or reduce different types of waste. Specifically, overproduction waste, producing in excess of what is actually required, can be greatly reduced by the use of SU systems. SU systems offer a broad working volume range, compared to fixed SS systems; therefore, overproduction of media and buffer solutions is less likely because batches are produced more closely to the volumes required.

Ready for use, SU assemblies can dramatically reduce motion waste. An example of motion waste occurs when preparing and sterilizing traditional formulation and filtration assemblies such as SS siphons, filter transfer lines, and glass formulation equipment. This excessive effort can result in: ergonomic issues, a substantial direct labor effort, and the utilization of costly utilities such as clean steam for sterilization. We have found that most assemblies can be prepared and sterilized by a supplier.

SU systems can reduce or eliminate inventory waste, such as spare parts and assembly components associated with SS formulation systems and assemblies. SU systems provide the flexibility to produce media and buffer solutions for specific operations with minimal inventory. A vendor-controlled inventory system can be implemented to reduce the amount of SU bag and assembly inventory held internally. In this process, the vendor manages the customer's inventory to ensure the correct amounts of SU systems are ready for use.

In order to maximize the flexibility of SU systems and minimize SU component inventory, universal bag systems (Fig. 41.24) have recently been developed for multiple configurations and "universally" applied to: filter different

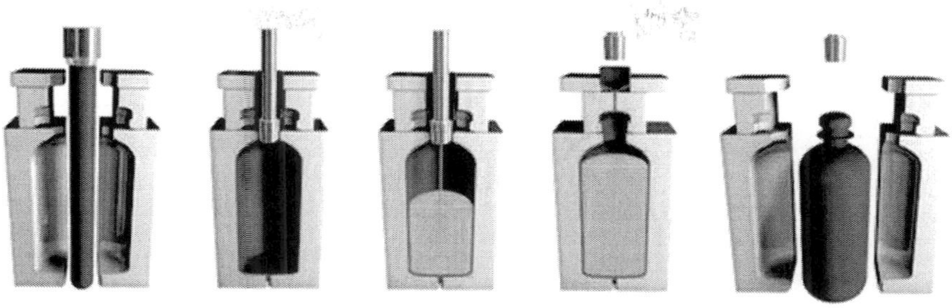

Figure 41.22. Processing steps of blow–fill–seal technology (left to right): Extrusion of the parison; closing the mold and cutting the parison; venting and filling; closing the mold to hermetically seal the container; opening the mold to remove the completed, filled container. (This figure is available in full color at http://onlinelibrary.wiley.com/book/10.1002/9780470054581.)

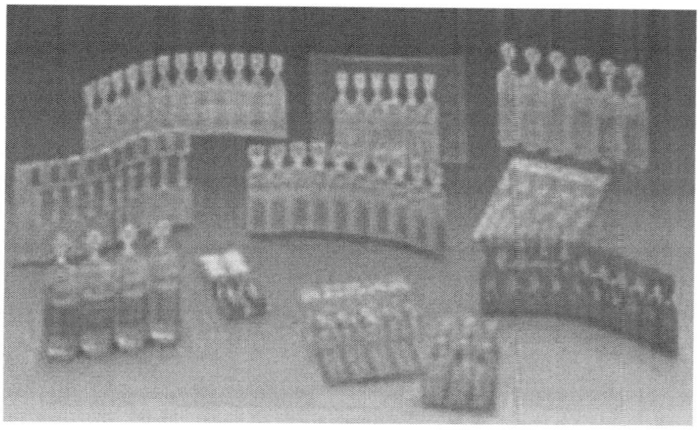

Figure 41.23. Examples of drug product-filled plastic containers made by blow–fill–seal technology. (This figure is available in full color at http://onlinelibrary.wiley.com/book/10.1002/9780470054581.)

media and buffer solution batch sizes, to store and transfer product, and to remove process waste streams. There are several advantages to this type of SU system. Some of them include selection of multiple bag sizes, incorporation of sterile connecting devices and filtration of liquids into multiple bags simultaneously. The result is a "plug and play" concept that greatly increases flexibility without adding unnecessary inventory. When implemented and utilized throughout the manufacturing process, this bag system design enables a more flexible and universally adapted approach in contrast to customizing each filter train or manifold system per process application. This universal concept results in a reduction in inventory from a large number of custom bag assemblies to a few different universal assemblies.

In addition to a smaller footprint, SU systems are more readily mobile. An example is a 750 L Hynetics system (Fig. 41.25) that consists of disposable formulation and powder feed bags, a nonproduct contact mixing mechanism, and disposable transfer lines. Since process piping and cleaning equipment are absent from SU mixing systems, clean room space requirements are decreased (36). These SU mixing system working volume ranges are expandable, whereas the traditional SS or glass mixing systems are limited to the working volume of the fixed shell. SU mixing systems can be fully integrated into a

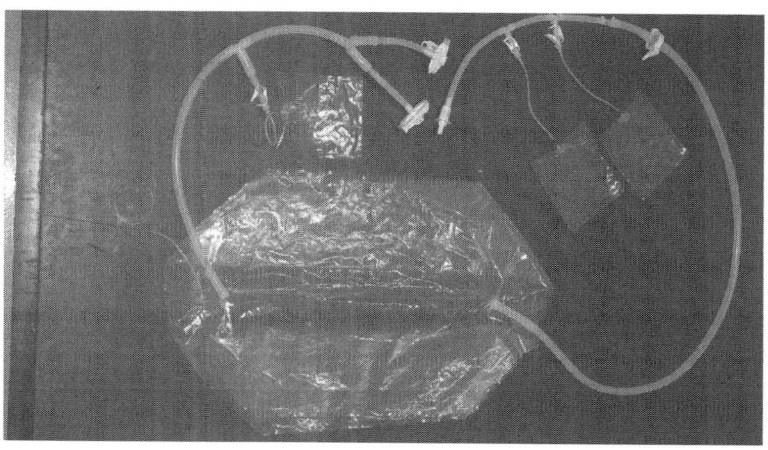

Figure 41.24. Universal Bag System [Millipore Corporation]. (This figure is available in full color at http://onlinelibrary.wiley.com/book/10.1002/9780470054581.)

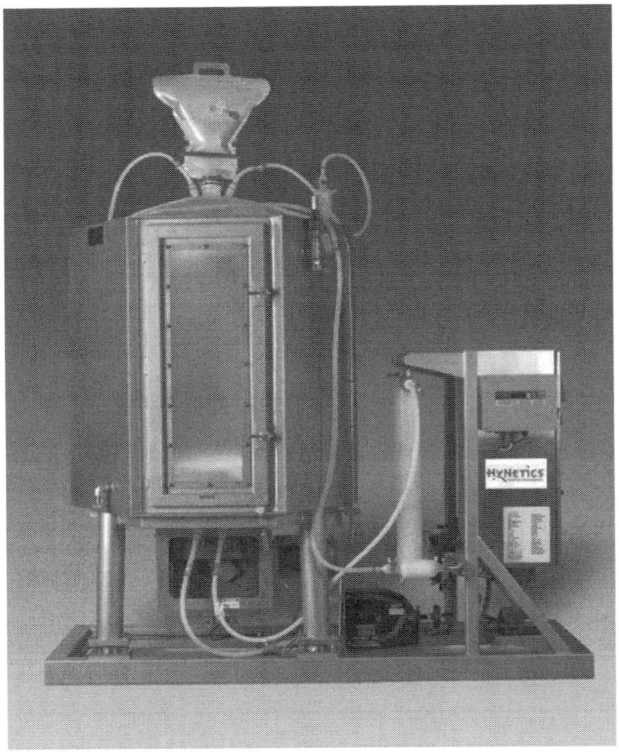

Figure 41.25. 750-L Single-Use mixing system [Hynetics Corporation]. (This figure is available in full color at http://onlinelibrary.wiley.com/book/10.1002/9780470054581.)

document control system, similar to traditional SS mixing systems.

Another advantage that SU materials and systems provide is an overall reduced labor effort during assembly. Figure 41.26 displays fully prepared, presterilized tubing and filter assemblies that are ready for use. Historically, individual components, such as tubing sets and filter assemblies were manually prepared and sterilized using conventional methods as part of the manufacturing process, and could be very complex in attachment to the process. Currently, there is an increasing trend to procure these assemblies fully prepared and sterilized directly from a supplier. The result is a reduction in clean utility usage and labor effort (37).

41.6.2.2 Disadvantages of SU Systems. Once the decision has been made to utilize SU systems, one is dependent on a supplier's ability to meet production times. If a supplier cannot meet a delivery date that was agreed upon, operations are delayed until the supply is available. Inventory and supply forecasting become critical to ensure that selected suppliers maintain inventory to meet the forecasted requirements.

The bags and assemblies can be damaged during shipping, punctured during setup, or have problematic manufacturing processes that lead to faulty materials. Product failures and defects may include: faulty port seals, pin-hole size defects, misaligned molded connections, and nonintegral plastic seams. SU system failures and defects can lead to costly reprocessing, product or facility contamination, and delays in manufacturing timelines. Suppliers adapt to these challenges by implementing corrective actions such as increasing integrity testing frequencies, collaborating with the film supplier to ensure defect free production, and partnering with end-users to address any product failures. It is important to understand that end-users receive a brand new SU system every time, compared to a SS system that is built once and used many times. Also, there are some bag assembly steps that are performed manually, and some that are semi-automated, that bring in a need for well-trained SU system assemblers, and a solid quality system.

Figure 41.26. Universal Bag System assemblies [Millipore Corporation]. (This figure is available in full color at http://onlinelibrary.wiley.com/book/10.1002/9780470054581.)

41.7 COSTS

Whether building a new pharmaceutical manufacturing plant or modifying an existing plant, the decision on how much SU equipment to incorporate needs to be made early in the design phase and has significant implications on the layout and facility specifications. Over the past 4 years the development of SU has included all of the process steps in biological process trains. Furthermore, SU equipment has been developed at scales of operation that enable a commercial pharmaceutical plant to strategically incorporate SU equipment.

The main drivers for SU are flexibility, lower capital costs, ease of use, and minimizing post-use cleaning. SU equipment generally requires less space and fewer utilities that translate into reduced capital costs as compared to a plant using SS. Additionally, a pilot plant facility using SU equipment can produce a monoclonal antibody at a lower cost than one employing fixed SS equipment. The following tables demonstrate the process economics of a SU plant as compared to a SS plant.

41.7.1 Total Capital Cost and Space Needs

Table 41.3 illustrates a SU solution for every SS step. Certain assumptions are: equipment costs are capital only; space cost is calculated at an estimated average of $800/sq. ft., SS vessels are $200/L, and bioreactors $300/L; Millipore SU Pod filters D0HC replace a SS centrifuge, and AIHC filters follow a centrifuge or D0HC filters; SU chromatography skids, an acrylic column, with SS heads/screens, 40 cm (20 L) SS, or 36 cm (20 L) SU columns are included (from one supplier); the SU chromatography skid flowpaths are $7500 each; for Media/Buffer prep, only one 1000 L Single-Use Mixer (SUM) is needed since SU turnaround is so rapid, and can be used for media and buffer; in-process testing room, office spaces, and corridors, are not included; any SU component is already clean and sterile before entering the facility; no CIP or SIP or autoclaving is performed in the SU facility, and WFI supply and use is process-only.

41.7.2 Economics for Selecting SS or SU for Bioreactors

The replacement of SS bioreactors with SU in an existing facility, or the selection of SU for a new installation is illustrated in Table 41.4 (38). The capital, qualification, Preventative Maintenance (PM), and ongoing costs, are included. For a 44 week production year, 50% more runs can be completed using SU, at a modest additional cost of 15% per year, with a reduction in COG. The costs of removing SS bioreactors from an existing facility are not included, as they may vary widely between organizations and geographical areas and would be incurred during the first year.

41.7.3 Affect on Cost of Goods from a Single Step Change

Table 41.5 provides a breakdown on major process steps and the COGs associated with each step (39). A single change is illustrated at the harvest step using Millipore Pods only, in place of a centrifuge. Our estimation focuses on direct costs, to include consumables, media, water, resins, and direct labor. Since a Millipore Pod AIHC was required postcentrifugation replacing the centrifuge with Millipore Pod D0HC membranes simply expanded the same technology. This change can be scaled to 20,000 L production bioreactor. Our case for Mabs assumes a $100/g COG, from thaw of working cell bank (WCB) vial through bulk substance.

Before executing such a change one would typically perform an Net Present Value (NPV), and Internal Rate of Return (IRR) analysis. These are standard algorithms in spreadsheet software. A positive NPV, and an IRR $>=$ 15%, would be acceptable to advance such a proposed change. Hurdles to advancement can include alignment with commercial operations (internal and Contract Manufacturing Organizations), consistency with platform technology, product quality, and differences in contaminants in the product stream. The example in Table 41.5 shows about a 4% change in COG, but avoiding the cost, complexity of operation, CIP, and SIP of a centrifuge could provide a significant incentive to use Pods.

Finally, we are confronted with the need to be environmentally friendly (40). When one harvests a 2000 L bioreactor, the plastic bag, tubing, filter capsules, connectors, disposable probes, the bag it came in, and the box are discarded. There is the cost to process the ashes, since the market to reuse does not currently exist. Using biohazard boxes, up to 7 cu ft each, the cost to transport and burn them is $1.87/lb (Table 41.4 Biohazard waste). If one chooses to use a drum, the cost increases by about 3×. Other publications have described the cost differences between reducing to ashes only, a 99% volume reduction, cogen, and pyrolysis. The impact on a community could be significant, if several producers of this waste stream could consolidate through a steam or electricity cogeneration plant.

41.7.4 Summary

Most SU systems are less than or equal to an equivalent SS system, for the original capital outlay. Advantages of SU that translate into cost savings are (i) greatly reduced turnaround times, (ii) significantly reduced time to first use (fully installed and qualified), (iii) reduced capital and operating costs, (iv) flexibility, (v) reduced process space, (vi) simplified automation, (vii) reduced validation costs, and (viii) reduced footprint. There are hurdles in achieving 100% SU. One is scale of operation and applies particularly to bioreactors, chromatography, and UF/DF.

TABLE 41.3. Space and Equipment Costs for an Entire Equipment Train in a New Facility are listed below

Equipment	Step	Space, GSF SS	Space, GSF SU	Equipment $ SS	Equipment $ SU
Shakeflask (glass or SU)– 20 L SS Bx or 20 L Bags	Preculture	300	300	200,000	100,000
50 L	Seed/prod Bx	40	12	150,000	40,000
50 L	Seed/prod Bx	40	12	150,000	40,000
250 L	Seed/prod Bx	45	15	200,000	60,000
250 L	Seed/prod Bx	45	15	200,000	60,000
2000 L	Seed/prod Bx	50	15	950,000	550,000
2000 L	Seed/prod Bx	50	15	950,000	550,000
Feed vessels × 4	Prod Bx feed	80	60	200,000	100,000
10 LPM, A1HC	Centrifuge and filtration	150	0	600,000	0
D0HC/AIHC	filtration	0	50	0	250,000
1000 L	Holding vessel	64	10	200,000	10,000
½" skid, 40 cm, resin	Affinity, $13 K/L	30	30	530,000	487,500
100 L × 1	Viral inactivation	8	4	50,000	500
100 L × 1	Neutralization	8	4	50,000	500
½" skid, 40 cm, resin	CEX, $2 K/L	30	30	350,000	237,500
100 L × 1	Holding vessel	8	4	50,000	500
½" skid, 40 cm, resin	AEX, $2 K/L	30	30	350,000	237,500
100 L × 1	Holding vessel	8	4	50,000	500
½" skid, 4 M^2	Virus filtration	30	30	50,000	10,000
100 L × 1	Holding vessel	8	4	50,000	500
½" skid, 5 M^2	Ultrafiltration	30	30	250,000	195,000
50 L × 1	Holding vessel	8	4	50,000	500
Cart and biosafety cabinet	Formulation/fill	30	30	50,000	50,000
50 L × 1	Holding vessel	8	4	50,000	500
Freezer(s)	Freeze	400	400	100,000	100,000
Weigh dispense	Med/buffer preparation	400	400	50,000	50,000
1000–200 L	Media preparation	64	10	200,000	120,000
1000–200 L	Buffer preparation	64	0	200,000	0
500–100 L	Buffer preparation	50	8	100,000	80,000
200–50 L	Med/buffer preparation	20	8	80,000	50,000
Washer		100	0	250,000	0
Autoclave		100	0	350,000	0
CIP skid		100	0	250,000	0
SIP		100	0	100,000	0
WFI		200	200	100,000	80,000
Mechanical space		1000	800	100,000	80,000
Storage	Warehouse	500	750	0	0
Total GSF	Sum(above)	4198	3288	0	0
Total Cost	Total GSF x $800	$3,358,400	$2,630,400	$7,610,000	$3,541,000
For SU, % savings			20–40%		50–60%

Note: Gross Square Feet (GSF), Clean In Place (CIP), Steam In Place (SIP), Water For Injection (WFI)

41.8 CONCLUSIONS

Flexibility of SU equipment in smaller facilities that is easy to use, minimizes cleaning, and has the potential to achieve competitive market entry is a goal that many drug suppliers are trying to achieve. With the explosion of SU equipment for all stages of biological drug development, the opportunity to modify or build SU pilot plants has become a reality. As SU technology further develops and process productivity increases, one can easily predict that we will see commercial plants being completely outfitted with SU equipment from cell culture through fill and finish.

The barriers to SU technology continue to fall as technical issues are overcome. There is still valid concern over SU because of the potential for leachables and extractables: this is an important concern that must be carefully addressed for

TABLE 41.4. Annual Costs for 2000-L SS versus 2000-L SU Bioreactors, including Capital, Installation, Qualification, and Operating Costs

Cost description or activity	SS Parameter	SS Costs	SU Costs	SU Parameter
Bx Capital Cost, installation, scales, pumps, controls		$950,000	**$500,000[a]**	
Qualify Bx	0.10 of cost	$950,000	$500,000	0.10 of cost
CIP skid	Fixed	$150,000		**None Required[a]**
Qualify CIP skid	0.10 of cost	$15,000		**None Required[a]**
Ongoing PM (annual), pump cal, gaskets, gauges	0.05 of cost	$55,000	$25,000	0.05 of cost
Annual capital 10 yr depreciation	0.10 of cost	$110,000	$50,000	0.10 of cost
System operation—assemble, setup, SIP for SS	12 hr	$1,800	$450	**3 hr[a]**
Run (8 hr/weekday, +12 hr w/e)	92 hr	$13,800	$13,800	92 hr
Inactivate, CIP for SS	8 hr	$1,200	$300	**2 hr[a]**
Changeover CIP for SS	8 hr	$1,200		
Biohazard waste, at $1.87/lb	5 lbs	$9	$56	30 lbs
Num people per activity	2 people			2 people
QC FTE hrs for Cleaning verification only	16 hr	$1,200	$0	0
Cost per hour per person	$75/hr			$75/hr
Quantity of *Media, Buffers* for all steps	4			4
Total vol, to incl all step	2000 L	$16,000	$16,000	2000 L
Avg Cost Per Liter (incl mat, water, bags, FTE)	$8.00/L			$8.00/L
Filters membranes, etc, per run	$1100ea	$1,100	0	$0ea
Change-out (o-ring, valve diaphragms, sparger, gaskets), per SS Bx run, or bags, bellows for SU	$4000ea	$4,000	$3,900	$3900ea
Water for CIP	4000 L	$40	$0	0 L
Avg Cost Per Liter (incl equip. depreciation)	$0.02/L			$0.02/L
CIP verification every run		$2,000	0	
Total One Time Equipment Costs, including Quality		**$1,210,000**	**$550,000**	**−120%[a]**
Without CIP Skid		**$1,045,000**	**$550,000**	**−90%[a]**
Annual Recurring—Assemble, setup, SIP for SS	**12 hr**	$26,400	$9,900	3 hr
Run (8 hr/weekday, +12 hr w/e)	92 hr	$202,400	$303,600	92 hr
Inactivate, remove bag, CIP for SS	8 hr	$17,600	$6,600	2 hr
Changeover CIP	8 hr	$0		
CIP verification for all SS runs		$46,933		
Biohazard waste, in boxes		$137	**$1,234[b]**	
Pallet spaces for CIP stock, $900/pallet/yr	1 pallet	$900	$2,700	3 pallet
Change-out parts for SS, or bags/bellows for SU	**NO[c]**	$16,133	$85,800	
Frequency of use	3.0 wks			**2.0 wks[a]**
Number of runs in 44 weeks	15 runs			**22 runs[a]**
All Media/buffers for 44 weeks		$235,253	$352,000	**50% More runs[a]**
Annual PM		$55,000	$25,000	
Annual Depreciation (over 10 years)		$110,000	$50,000	
Annual Cost, including PM and depreciation		**$710,757**	**$836,834[b]**	**15%[b]**
Total product per 44 weeks		29,333 g	**44,000 g[a]**	
Cost per gram (including PM and depreciation)		**24 $/g**	**19 $/g[a]**	

Note: Assuming a 1.0g/L product yield, or 2.0K g/batch, 250L seed/run, 900L production media, and 850L additions.
[a] Represents noteworthy advantages for SU vs SS.
[b] Represents noteworthy additional cost for SU vs SS media, and 850L additions.
[c] The "No" represents a conservative case of no change-out of elastomers between products. By entering "Yes" an additional CIP would be required and an increase in change-out costs would occur.

biological products. Additionally, its environmental aspect, mainly waste disposal, has not been adequately addressed. Continued development of SU technology will depend on close collaboration between the manufacturers of SU equipment and the biotech community.

Facility footprints can be significantly reduced since equipment and interstitial piping for cleaning is eliminated or minimized. As the productivity of monoclonal antibody-producing cell lines increase from 1–2g/L to 3–5g/L and beyond, 2000 L bioreactors could become

TABLE 41.5. An Example of a COG Point Change (a Single Step), from SS to SU

Process Step	Description	COG $100/g	Example Step Change	COG $96/g
Preculture	SF, wave bags, Bx	6	—	6
Prod Bx	Includes feeds	18	—	18
Harvest	Centrifuge and filtration tanks	6	Pod's only	2
Affinity	Column, skid, resin, tank(s)	22	—	22
Chemical VIN	Tank(s) and addition pumps	5	—	5
Freeze and store	Blast, or direct freeze	4	—	4
Transport	Geography change	2	—	2
Thaw	Tank	1	—	1
CEX	Column, skid, resin, tank(s), filter	11	—	11
AEX	Column, skid, resin, tank(s), filter	11	—	11
VF	Skid, capsule, tank	5	—	5
UF/DF	Feed ves, skid, membranes, tank	7	—	7
Form/fill/freeze/store	Tank, filter	2	—	2
		100		96

VIN, viral inactivation and neutralization; CEX, cation exchange chromatography; AEX, anion exchange chromatography; VF, virus filtration; UF, ultrafiltration; DF, diafiltration.

adequate for commercial scale operations. One can anticipate that engineers will solve the current mixing and aeration challenges that will enable 5000 L SU bioreactors. Since SU systems are more portable (i.e. not fixed) installing this enables ease of future SU replacements or staging within existing spaces.

Because of the high cost of chromatography resins coupled with their powerful separation capabilities, SU technologies may not replace our current separation operations for some time. However, most of the equipment, tubing, valves, bags, and sensors supporting column chromatography will be fairly rapidly replaced by SU technology.

We do not foresee the need to increase the scale of media/buffer preparation. There is a wide range of tubing/bag and tubing/tubing types and sizes to accommodate the typical Mab process steps. Where we do see a need for scale increase is SU UF/DF. The current Millipore and Stedim systems accommodate 5 m^2. With an increase to 3–5 g/L in the bioreactor, multiple purification cycles would be performed or a large square meter system (i.e. 10–20 square meters) would be required.

Many of these systems also help to reduce costs. As more SU projects are completed, the expectation will be that engineers and design firms align with this paradigm shift, and discover better ways to implement this technology. One aspect of SU implementation is that most connections are closed and one should be able to use less classified spaces, further reducing the facility costs. For potent compounds whose patient doses are 1–3 mg, a 1000–2000 L bioreactor or a set of up to six of them and with associated purification train could be a commercial system.

REFERENCES

1. Langer ES. Bioprocess Int 2009; 7: 6–8.
2. Pharma 2020. The Vision: Which path will you take? Pricewaterhouse Cooper, June 2008. Google: Pharma2020: The Vision: Which path will you take. Pricewaterhouse Coopers.
3. Article in an electronic publication: andrew Sinclair and Miriam Monge, Biopharm Services Issue: V06. Available at http://www.sartorius.or.kr/nh/fileadmin/sartorius_pdf/stedim/stedim/Quantitative_Analysis-of_Single_Use_Disposables.pdf
4. Article in an electronic publication: victor Papavasileiou, Charles Siletti, PhD, Demetri Petrides, BioPharm International.com PhD. Available at http://biopharminternational.findpharma.com/biopharm/Disposables+Articles/Systematic-Evaluation-of-Single-Use-Systems-Using/ArticleStandard/-Article/detail/566015
5. Article in an electronic publication: Andrew Sinclair and Mirian Monge. Available at http://biopharm-international.findpharma.com/biopharm/Disposables+Advisor/Disposables-Open-Up-Possibilities-in-Facility-Desi/ArticleStandard/Article/detail/53205.
6. Presentation at ISPE: Chris Lorenz from Genentech, "Technologies of the Future: Disposables at Genentech". San Francisco; 2008 Feb 28. Available at http://www.ispe.org/galleries/sanfrancisco-files/Lorenz_ISPE_Presentation_0-22808.pdf
7. Article in an electronic publication: James M. Robinson. Available at http://biopharminternational.-findpharma.com/biopharm/FeatureRotatingArticle/article/detail/575563
8. Martin J, Hartzel W, Vogel J, Mongue M, Repetto R. Bioprocess Int 2009; 7 Suppl 4:9–10, 12, 14, 15–16. (Organizing the Organizations of Single-Use Manufacturing).
9. Wood C, Richardson N, Bhatia R, Ozturk S. The evolution of a disposable stirred tank bioreactor at centocor –from concept to GMP implementation. Abstracts of Papers, 234th

ACS National Meeting; 2007 Aug 19–23; Boston, MA; BIOT-248.

10. Bhatia R, Wood C, Richardson N, Ozturk S. Development of a large-scale stirred-tank disposable bioreactor for cell culture processes. Abstracts of Papers, 232nd ACS National Meeting; 2006 Sep 10–14; San Francisco, CA, BIOT-155.
11. Markovic I. Am Pharm Rev 2009; 96: 98–101.
12. Markovic I. Expert Opin Drug Saf 2007; 6(5): 487–491.
13. Raymond R, Sette A, Martin J, St. Laurent J, Hartzel B, Hockstad M, Jahn D, Lehman T, Murphy T, Pembleton B, Potheir N, Stover J, Swisher L. Bioprocess Int 2008; 5 Suppl 3:28, 30, 32, 34, 36–40. (Recommendations for Extractables and Leachables Testing).
14. Bestwick D, Raymond R. Bioprocess Int 2009; 7 Suppl 1:88–90, 92, 94. (Extractables and Leachables from Single-Use Disposables).
15. Wells B, Boehm J, Cardona M, Gettings R, Gupta V, Hockstad M, Kossay B. Bioprocess Int 2008; 6 Suppl 3:24, 26–27. (Guide to disposal of single - use bioprocess systems: disposals subcommittee of the bio-process systems alliance).
16. Rao G, Moreira A, Brorson K. Biotechnol Bioeng 2009; 102(2): 348–356. (Disposable bioprocessing: the future has arrived).
17. Galka N. Filtr Sep 2007; 44: 18–21.
18. Low D, O'Leary R, Pujar NS. J Chromatogr B 2007; 848: 48–63.
19. Sinclair A, Monge M. Biopharm Int 2008; 22(2): 24–29.
20. Garbarino T, Odegaard D. Contract Pharm 2009; 11(4): 62–67.
21. Gottschalk U. Biotechnol Prog 2008; 24: 496–503.
22. Glynn J, Hagerty T, Pabst T, Annathur G, Thomas K, Johnson P, Ramasubramanyan N, Mensah P. Biopharm Int 2009 Mar Suppl: 16–20.
23. Bisschops M, Frick L, Fulton S, Ransohoff T. Bioprocess Int 2009; 7(S6): 18–23.
24. Clutterbuck A, Kenworthy J, Liddell J. Biopharm Int 2007; 20(5): 44–54.
25. Zhou JX, Solamo F, Hong T, Shearer M, Tressel T. Biotechnol Bioeng 2008; 100: 488–496.
26. Gottschalk U. Advances in biochemical engineering/biotechnology (Online). New York: Springer; 2009. Available at http://springerlink.com/content/5471kk5nx500302k/?p=f31f70c829234504aba8d438e5bfb0ce&pi=2.
27. Vogt R, Paust T. Bioprocess Int 2009; 7(1): 72–77.
28. PDA workshop on Disposables/Single-Use-Technology, Munich, 3-4 March 2009.
29. Bausch UJ. In edoc University Library of Basel (online). Available at http://edoc.unibas.ch/diss/DissB_8427, 2008.
30. Fuller M, Pora H. Bioprocess Int 2008; 6(10): 30–36.
31. Isberg E. In: Lysfjord J, editor. Volume 1, Practical aseptic processing-fill and finish. Bethesda, MD: PDA; 2009. pp. 167–180.
32. Reed CH. In: Lysfjord J, editor. Volume 2, Practical aseptic processing-fill and finish. Bethesda, MD: PDA; 2009. pp. 223–246.
33. Wu LV, Leo F. In: Avis EK, Wu LV, editors. Biotechnology and biopharmaceutical manufacturing, processing, and preservation. Boca Raton, Florida: CRC Press; 1996. pp. 265–293.
34. Philips C. Bioprocess Int 2008; Mar Suppl., 6–9.
35. Shingle BM. Case study: an innovative approach to the use of single-use products based on principles of lean manufacturing and process improvement. ISPE Conference on Implementing Development and Manufacturing Technology; 2009 Mar 2–5; Tampa Bay, FL.
36. Utlee ME, Mullen LJ. Pharm Process 2009; 24(3): 12–14.
37. Gonzalez MM. Am Pharm Rev 2007; 10: 78–82.
38. Sandstrom C. Society for biological engineering, Disposables vs. Traditional Equipment –a facility-wide view; 2009. pp. 30–35.
39. Monk J. Society for biological engineering, exploring new paradigms for bioengineering; 2009. pp. 19–22.
40. Rawlings B, Pora H. Bioprocess Int 2009; 7(3): 40–46.

42

cGMPs FOR PRODUCTION ROOMS

CLAUDE ARTOIS, JEAN DIDELEZ, PATRICK FLORENT, AND GUY GODEAU
University of Surrey, Guildford, Surrey, United Kingdom
SmithKline Beecham Biologicals, Rixensart, Belgium

Viral vaccine development began approximately fifty years ago with virus propagation in stationary cell culture. It was the ability to grow human virus outside a living host that led to the development of viral vaccines. This chapter describes design considerations for an inactivated viral vaccine bulk manufacturing facility. Many of the design features described in the text may also be applicable to manufacturing other biological products. The drawings are published by the courtesy of Dr. C. Vandecasserie, SmithKline Beecham Biologicals, Rixensart, Belgium. The authors thank Mrs. Doris L. Conrad and Dr. Michael Korczynski for their constructive comments. They were beneficial in preparing of this chapter.

The design of the manufacturing areas and the HVAC system is based on the concepts and technical data described in the section "General Points for Facility Design" and the section "Design of HVAC System parameters." The facility design could fit the requirements of various production processes involving cell culture, viral replication and viral inactivation, for instance, hepatitis A or noninfectious polio vaccines. The facility can accept a multistrain manufacturing strategy, as described later. The facility contains all of the equipment and support services necessary for cell culture, viral culture, purification, and inactivation of the vaccine. In the design of an industrial-scale facility for cell culture, viral replication, purification, and inactivation operations, engineering, manufacturing, maintenance, validation, quality control, and quality assurance disciplines are joined to establish a satisfactory production and control system. Due to the different perspectives and priorities of vendors, contractors, design engineers and users, some of the operational, safety, and quality requirements may be overlooked.

In addition, both facility and equipment designs should meet operator expectations and GMP requirements. The system features must aid the operator in achieving training, operational reproducibility, safety, product quality, and maintenance requirements.

This chapter attempts to provide the reader with a global view of design considerations. Emphasis is put on HVAC parameters and operational flows which satisfy cGMPs (1,2), regulatory (3,4), and biosafety requirements (5). The design considerations are illustrated by using an example of an inactivated viral bulk manufacturing facility that combines three major biotechnological challenges: cell culture, aseptic operations, and biosafety level 3 associated with handling pathogenic viruses.

42.1 GENERAL POINTS FOR FACILITY DESIGN

42.1.1 Points to Consider

Primary manufacturing of biotechnological products generally involves operations starting from cell culture, including biological expression of the product, extraction, and purification. A recombinant cell culture or viral replication system usually includes one or more viral clearance steps to ensure that product safety requirements are met.

In addition, a final sterile filtration step is usually included in the manufacture of the biological product to

Downstream Industrial Biotechnology: Recovery and Purification, First Edition. Edited by Michael C. Flickinger.

obtain a sterile product. The sterile filtration step may or may not be part of the primary manufacturing operations depending on the type of product, its final use, and the organization of the whole manufacturing process.

To comply with GMPs, each manufacturing area is designed keeping in mind the following biopharmaceutical facility features:

- separate access for personnel and raw materials
- separate exit for bulk product
- dedicated HVAC units
- dedicated water systems
- need for containment, decontamination, and waste handling
- dedicated multiuse or multistrain manufacturing strategy

Cell cultures are lengthy and highly susceptible to contamination. Viral replication offers an additional challenge, especially when a live pathogenic virus is handled. Segregation of areas, dedicated HVAC units, material decontamination, and effluent treatment systems represent additional engineering requirements that must comply with the various applicable codes.

The codes for Good Manufacturing Practices applicable to the manufacture of biological products are specific:

- CFR 21 Parts 600 "Biological Products" (United States).
- Annexe 2 of European Union, code of Good Manufacturing Practice for Medicinal Products: Manufacture of Biological Medicinal Products for Human Use.

If a multiuse or multiproduct manufacturing strategy is developed, then the design features will incorporate the principles described in the codes mentioned. The section "Dedicated, Multiuse or Multistrain Manufacturing Strategy?" describes how the manufacturing strategy impacts the design of the facility.

Sterile filtration involves all requirements associated with drug aseptic operations. Guidelines and regulatory requirements for Good Manufacturing Practices applicable to the manufacture of sterile drug products are also specific (see references at end of chapter). Facility design philosophy should attain primary containment in the production vessels and transfer piping by using closed systems. This maintains containment and protects the environment, personnel, and product. The piping slopes, welds, joints and seal materials used throughout the system are designed to eliminate leakage and maintain overpressure to assure sterility of the system. Sanitary valves need to be welded.

Secondary containment is achieved by maintaining the rooms, where live pathogenic viruses are handled, under negative pressure. Refer to the section "Design Features and Safety Precautions for Personnel Working in BL-2 and BL-3 Areas."

Sterility of cell culture or aseptic operations performed in open systems should be accomplished by using laminar air flow (Class 100) by personnel appropriately dressed and trained in aseptic operations and techniques.

The authors of this chapter propose a room class description (Table 42.1) that complies with EU GMP Annex 1 "Manufacture of Sterile Medicinal Products (January 1997)" and U.S. FDA Guidelines for Drug Aseptic Processing and U.S. Federal Standard 209E (6).

Figure 42.1 illustrates the personal, materials, and product flow concepts used to design the inactivated viral vaccine bulk manufacturing facility. Figure 42.2 provides a view of the layout of this facility.

42.1.1.1 Room Class Descriptions. In a pharmaceutical environment, room air pressures and directional air flows are designed and balanced to create a positive pressure cascade from more controlled or critical areas to less controlled areas (7,8). Rooms are classified according to their level of cleanliness (6–8).

In a biopharmaceutical environment where live pathogenic organisms are manipulated, HVAC systems should be segregated for production areas according to biosafety (5) and cGMP (3,4) requirements. HVAC systems should be segregated to prevent recirculation of air between these designated production areas (9). Airlocks are widely used to meet biosafety requirements and to maintain an aseptic environment. Rooms might be kept at negative pressure with respect to adjacent rooms. Exhaust air is HEPA filtered.

42.1.2 Personnel Flow

Personnel should access the facility through one general entrance. The facility should be designed so that rest rooms, ladies' and gentlemen's changing rooms, and administrative offices are located outside of the Production core. Entrance to the facility should be secured and limited to authorized personnel. This is usually achieved by an electronic lock requiring a coded badge entry. Personnel enter the production core through access to changing rooms. Three changing rooms will meet requirements for female personnel, male personnel, and visitors. Entrance to changing rooms should also include electronic locks requiring coded badge entry.

Before entering the production core, production personnel should remove their city garments and put on production garments. Personnel can then access various process rooms including aseptic production rooms. Personnel should move

TABLE 42.1. A Guide to Classify a Series of Processing Rooms Commonly Found in a Biopharmaceutical, Cell-Culture, Production Facility[a]

Production Area Classification						
					Microbial Requirements	
EU Class	US Class	Static	Dynamic	Type of Operation	per (m^3)	per (10 ft^3)
A	100	100	100	Aseptic operation during which sterile product is exposed to the environment	<1	<0.3
B	10 000	100	10 000	Aseptic area surrounding a class 100 A. Room pressure is positive versus adjacent rooms of lower cleanliness level	<10	<3
B	10 000	1000[b]	10 000	Aseptic area surrounding a class 100 A. Room absolute pressure is negative due to the handling of pathogenic microorganisms. The room is used for aseptic bulk cell culture and for viral replication.	<18	<5
C	100 000	10 000	100 000	Fermentation/Extraction, nonsterile media preparation, purification of antigens, capping of filled containers, storage of clean and sterilized Materials	<85	<25
D	100 000	100 000	100 000	Materials wrapping room equipped with a laminar flow hood Any room used for sterilization of equipment and materials	<85	<25
Unclassified	Unclassified	100 000	NA[c]	Materials Washing Room, nonsterile corridor, office, in-process control laboratory, visual inspection room, storage of materials, Product, and equipment	NA[c]	NA[c]

Notes:
[a] Areas are classified under static and dynamic conditions. Static condition: the room classification is determined at rest with all equipment in operation but no personnel inside the area. Dynamic condition: the room classification is determined while equipment is in operation and personnel perform production operations.
[b] The authors propose a class 1000 for a room kept under negative pressure. Acceptance of this concept will depend on the risk analysis that the designer of a new biotechnology facility performs during the conceptual study phase.
[c] Not applicable.

between areas of a biofacility according to detailed procedure established in compliance with biological regulatory requirements (10,11). Before entering aseptic rooms, production operators enter the gowning room (Personnel Air-Lock) where they remove their production garments and put on sterile gown, sterile boots, sterile hood, sterile gloves, and a mask or helmet (10).

Figure 42.3 illustrates the personnel flow in the viral vaccine bulk manufacturing facility. Personnel enter the building on level 1 (ground floor) where three changing rooms are available. Personnel access level 2 using a staircase or an elevator.

42.1.3 Design Features and Safety Precautions for Personnel Working in BL2 and BL3 Areas

Four biosafety levels (BL) are recommended for activities involving infectious microorganisms. The main features of the levels BL2 and BL3 are summarized in Table 42.2. This information is directly extracted directly from the publications of the U.S. Centers for Disease Control and Prevention (CDC) and the U.S. National Institutes of Health (NIH).

The features that were employed to meet the BL-3 requirements for the live virus manipulation areas of the inactivated virus vaccine facility are described as follows:

1. Primary barriers
   - The access to live virus manipulation areas is limited to personnel involved in production operations and supervision. Each entry into BL-3 area is recorded on a checklist. The use of an electronic controlled access system for each live virus manipulation area was not selected during the detail design phase because it was decided to equip the entrances to the facility and the changing rooms with electronic locks requiring coded badge entry. The facility manager controls access to the areas and restricts access to persons whose presence is required for production purposes.

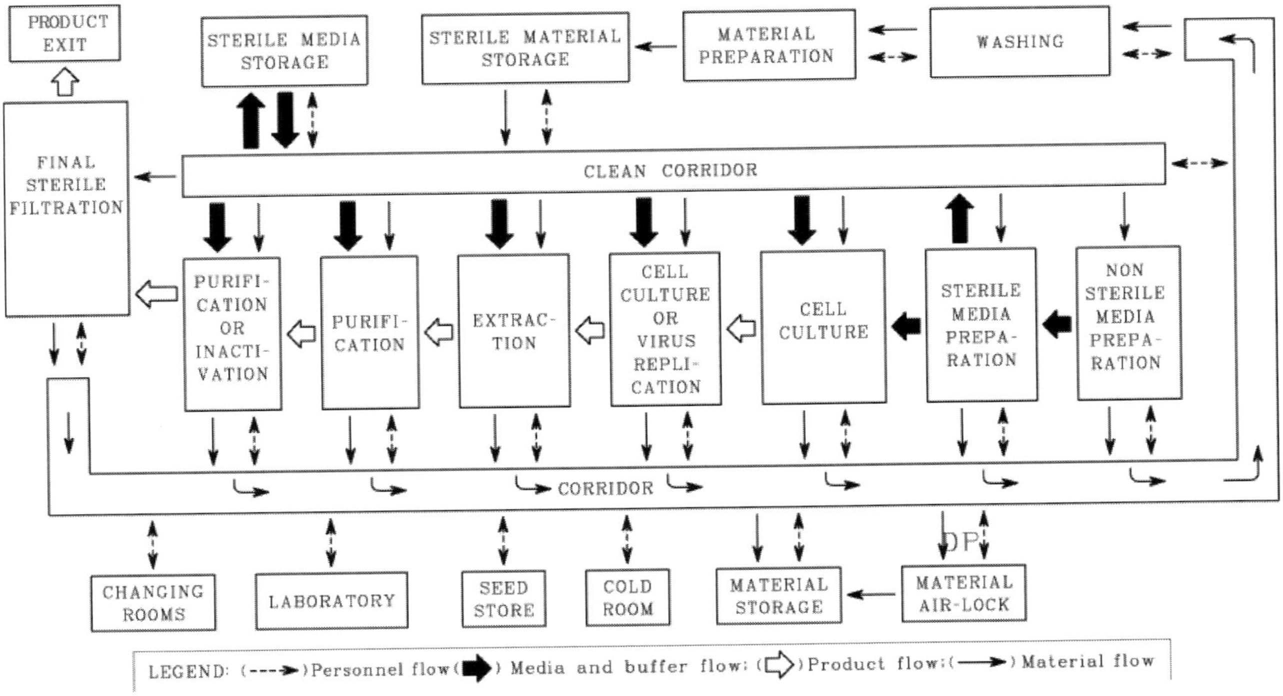

Figure 42.1. Facility design illustrating personnel, materials, and product flow Concepts.

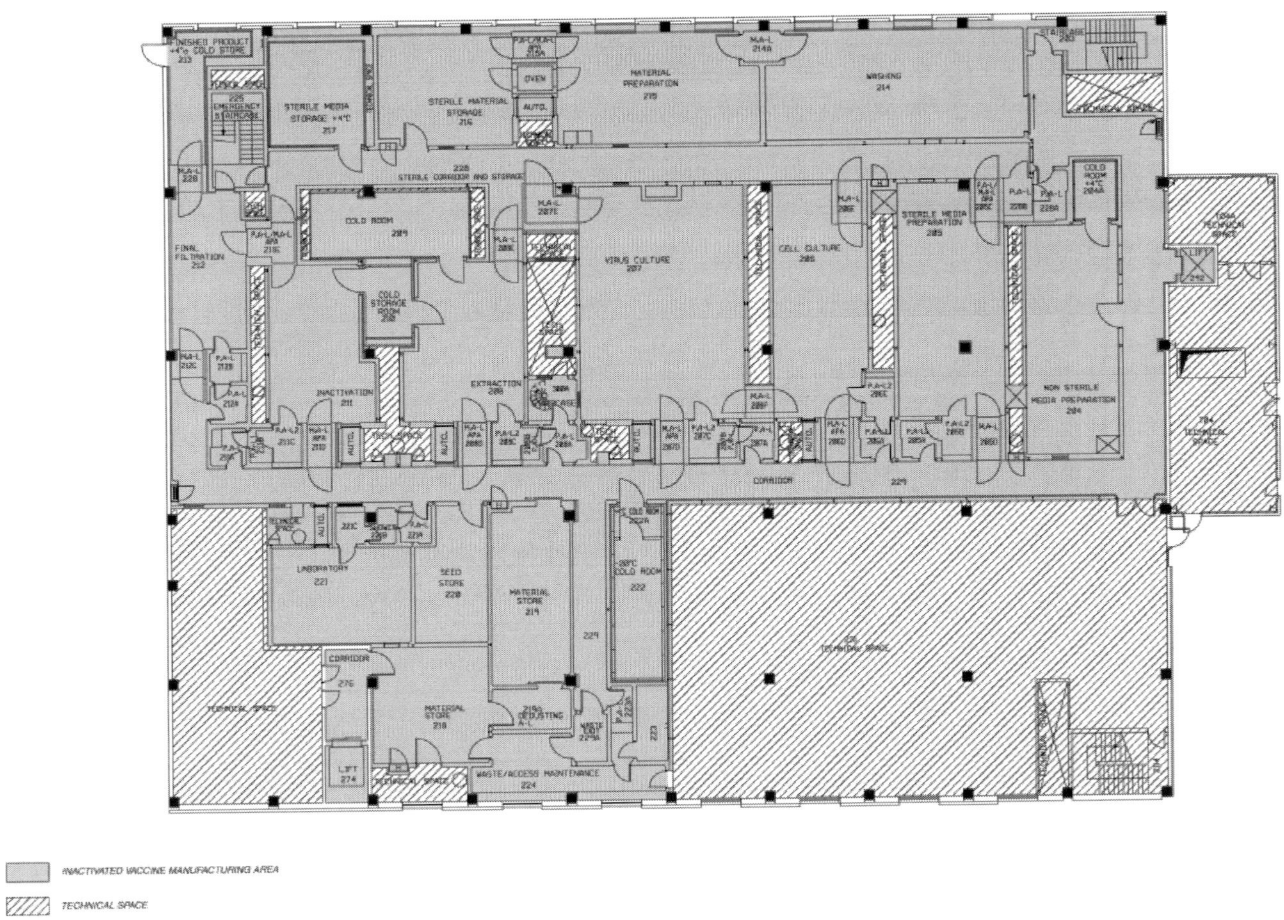

Figure 42.2. General layout.

TABLE 42.2. Summary of Recommended Criteria for Biosafety Levels 2 and 3

Biosafety Level	Agents	Practices	Safety Equipment (Primary Barriers)	Facilities (Secondary Barriers)
2	Associated with human disease hazard comes from autoinoculation, ingestion, mucous membrane exposure	Standard microbiological practices plus • biohazard warning signs. • "sharps" precautions. • biosafety manual defining any needed waste decontamination or medical surveillance policies	Primary barriers involving Class 1 physical containment devices used for all manipulations of agents that cause splashes or aerosols of infectious materials; personnel protection equipment involving protective lab clothing, gloves, respiratory protection as needed	Open bench top sink required plus autoclave available
3	Indigenous or exotic agents with potential for aerosol transmission; disease may have serious or lethal consequences.	BL-2 practice plus • controlled access. • physical • decontamination of all waste • decontamination of lab clothing before laundering • baseline serum	Similar to BL-2	BL-2 plus • Physical separation from access corridors • self-closing double door access • exhausted air not recirculated • negative airflow into laboratory

- Liquid wastes are decontaminated through dedicated effluent decontamination stations.
- Solid materials are decontaminated through double-door autoclaves installed in each live virus manipulation area. The autoclaves are designed to maintain the steam condensates within the autoclave chamber during the decontamination cycle.
- Personnel are gowned with overalls, boots, masks, and hoods which are decontaminated through autoclaves.
- Personnel exposure to live virus is reduced through the use of BL-3 Laminar Air Flow Bench (Cabinet), Closed system (Mobile or fixed tanks), and specific training in the procedures related to the operations.
- Personnel are vaccinated against the live virus manipulated in the facility.
- Floor and works surfaces are sanitized each day using a sodium hypochlorite solution.

2. Secondary barriers (Facility)
   - Segregation of air handling units: each live virus manipulation area is equipped with a dedicated air-handling unit.
   - Exhausted air is not recirculated. Air supply and return are HEPA filtered.
   - Air pressure differentials: the room pressures are negative at 30 Pa or −0.12 inch of water versus atmospheric pressure. Air locks have positive air pressure or are designed as decontamination air locks (see section on "Material Flow").
   - Each air locks is equipped with a self-closing double-door access.
   - Personnel air locks consist of three physically separate compartments (See section on "Design of HVAC System Parameters")
     Personnel air lock, first part: personnel remove production garment. Air pressure is positive (+30 Pa, or +0.12 inch of water).
     Personnel air lock, second part: this is equipped with a safety shower. Air pressure is negative (+30 Pa, or +0.12 inch).
     Personnel air lock, third part: personnel put overalls under laminar air flow. Air pressure is negative (−15 Pa or −0.6 inch).
   - Sealed windows and sealed penetrations.
   - HEPA filters are tested every twelve months.

42.1.4 Materials Flow

All raw materials and components should be sampled, tested, and released by the Quality Control Unit prior to entering the production facility. QA/QC released materials

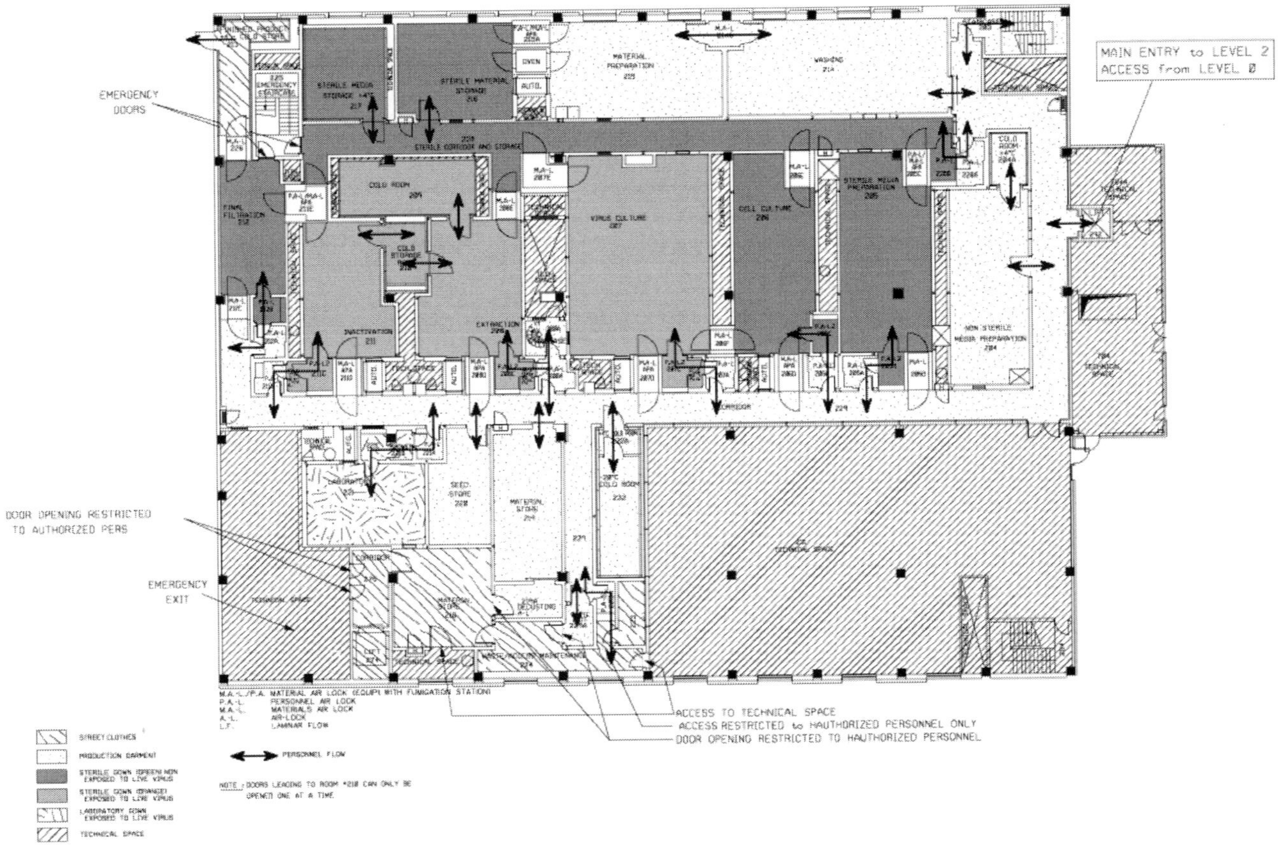

Figure 42.3. Flow of personnel. Personnel enter the building on the level 1 (ground floor) where three changing rooms are available (Female, male and staff/visitors). The entrances to changing rooms are equipped with electronic locks requiring coded badge entry. Personnel access level 2 using a staircase or an elevator. Before entering aseptic rooms (Sterile Media Preparation, Cell Culture, Virus Culture, Purification, Inactivation, Final Filtration, Sterile Material Storage, and Sterile Media Storage), production operators enter the gowning room (Personnel Air Lock 1) where they remove their blue production garments and put on a color-coded sterile gown, sterile boots, sterile gloves, and a mask.

should be transferred to Production using a dedicated Material Cardboard Lock. External packaging protection like cardboard and plastic films should be removed in the Air Lock. Cardboard and wood should not be introduced in the Production Core to minimize the risk of microbial, rodent, and insect ingress into the facility. The Production Core should include provisions for raw materials storage at controlled room temperature of 2 to 8°C or at adequate freezing temperatures. Exit of used raw materials from the Production Core might be achieved using the same Air Lock providing that a time-based segregation procedure is strictly followed by production operators.

Materials that are used in sterile conditions should be prepared and sterilized using validated equipment and procedures. The Code of Federal Regulations 21 CFR 600.11 (12) describes specific requirements for material sterilization procedures used in the manufacture of biological product: "The effectiveness of the sterilization procedure shall be no less than that achieved by an attained temperature of 121.5°C maintained for 20 minutes by saturated steam." That requirement is higher than the usual rule of 15 minutes at 121.5°C expected for sterile drug products.

To maintain the sterile status of wrapped materials, the production unit should be designed with a dedicated sterile material storage room. Sterilized equipment and materials should then be taken from the sterile material storage room to the appropriate manufacturing area via a clean corridor and the corresponding Material Air Lock, so-called "Air Lock In."

After use, soiled equipment and material should be appropriately removed from the processing rooms via corresponding Air Locks, so-called "Air Lock Out." Decontamination, if required, should take place in decontamination autoclaves or decontamination material

Air Locks. Decontamination Air Locks are used to decontaminate the external surfaces of mobile vessels when they are removed from a live virus manipulation area. After use, the mobile vessel is decontaminated-in-place (DIP) using clean steam. The mobile vessel is then transferred into the decontamination Air Locks in which a fumigation cycle is applied. Fumigation can be carried out using a peracetic acid or hydrogen peroxide generator.

Within the Production Core, materials should follow a unidirectional flow. Figure 42.4 illustrates the following flow of equipment, materials, and waste in the viral vaccine bulk manufacturing facility.

- Sterilized equipment and materials are taken from sterile material storage room 216 to the appropriate manufacturing area via corridor 228, and the corresponding Material Air Lock.
- The fixed vessels used for non sterile media preparation are cleaned-in-place per the appropriate procedure.
- After use, soiled equipment and material are appropriately removed and decontaminated from the processing rooms via corresponding fumigation Material Air Locks or decontamination autoclave, corridor 229, and are transferred to Washing Room 214.
- Materials are washed following appropriate cleaning procedures. Cleaned equipment and materials are then transferred into Materials Preparation Room 215 via Material Air Lock 214A.
- As appropriate, materials and equipment are wrapped, assembled, and sterilized by autoclave, dry heat oven, or in a fumigation air lock (215A) after which they are stored until use in room 216.
- Waste and single-use materials that are decontaminated are discarded via corridor 229, waste exit 229A, Waste/Access Maintenance 224, Material Store 218, corridor 276 and removed from the Production Core using lift 274.

42.1.5 Product Flow

When considering the design of a facility, care should be taken to address the following concerns about product flow:

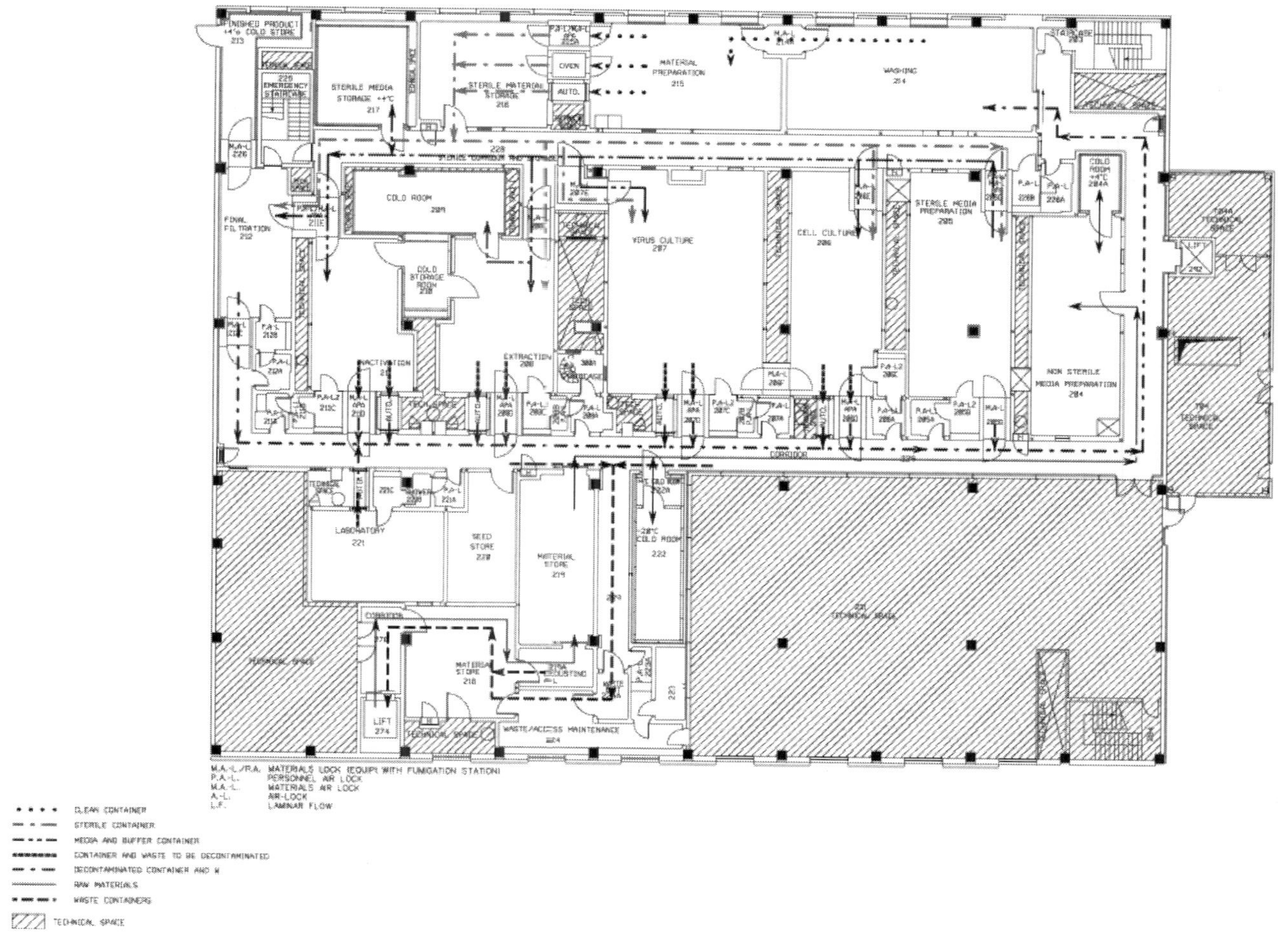

Figure 42.4. Flow of equipment, materials, and waste.

- dedicated storage equipment for Working Cell Bank and Working Seed.
- Unidirectional product flow from Media and Buffer preparation up to product exit.
- Segregation between areas where live organisms are handled and areas where no live organism is manipulated.
- Dedicated product storage room and product exit.

42.1.5.1 Dedicated Storage Equipment for Working Cell Bank and Working Seed. Master and Working Cell Banks and/or Seed should be stored in two separate buildings to minimize the risk of loss due to potential disaster. Within each building, care should be taken to store the biological material to prevent any cross-contamination or damage (13). The 21 CFR 600.11 standard states that cold or freezer rooms shall be maintained at temperatures within applicable ranges and shall be free of extraneous material that might affect the safety of the product. An accurate inventory log of the Master and Working Cell Banks and Seed must be maintained at each storage site.

The following practices are frequently observed in industry to prevent cross-contamination and mix-ups:

- Dedicated freezer or nitrogen tank for each type of micro organism strain.
- Dedicated freezer or nitrogen tank for bank and seed lot awaiting QA/QC release (Quarantine).
- Spare freezer ready to use in case of mechanical problems on the main storage units.
- Chemical sanitization of freezer internal chamber after each shutdown and maintenance intervention.
- Limited access to storage units.
- Log book to record entries in and removal from each storage unit.

42.1.5.2 Media and Buffer Flow. Media and buffers used for cell-culture operations are critical components. Right from the start of their preparation in sampling, weighing, and powder dissolution, strict procedures should be followed to minimize the risk of contamination. Mycoplasma, adventitious viruses, and some bacteria strains may be difficult to detect in cell cultures. Mycoplasma and viruses may not be retained on 0.2-μm sterile filter membranes. Therefore, the design of media and buffers preparation rooms should address the following concerns:

- Limited access to authorized personnel.
- Two-stage media preparation suite: nonsterile media and sterile media preparation areas.
- Specific gowning practices for nonsterile media preparation similar to those followed for aseptic operations.
- After sterile filtration, validated procedures and container/closure integrity systems should be used to protect media and buffers during storage, transfer to, and distribution into the processing rooms.
- Equipment design: media residues accumulate in badly drained or inaccessible locations. Residues dry out and become baked on in decontamination, cleaning, and sterilization procedures. Therefore, design of equipment should include provision for good draining of all equipment parts. Vessels should be made of stainless steel (316L) and should be vented using 0.2-μm air filters. Other materials, single-use irradiated plastic bags, for instance, can be successfully used providing that compatibility between materials and the cell or virus culture has been validated. Single-use sterile plastic bags can be used to store media and buffer. That reduces the need for stainless steel mobile vessels and the need for cleaning capacity associated with CIP stations (clean-in-place). All valves installed on equipment must meet sanitary design requirements. Stationary tanks are cleaned and sterilized using CIP/SIP stations.
- Process design: media and buffers are usually sterilized through 0.2-μm liquid sterile filters. The manufacturer will assess the needs for stationary and mobile tanks. The decision to use installed tanks instead of mobile vessels will be based on the volume of media, buffer, or product transferred between the various processing units. The higher the volume of one specific medium, the higher will be the need for a stationary tank. However, the complexity of CIP/SIP stations for big vessels and multiple transfer piping need to be evaluated against the benefit of using a fixed vessel. Containers of media and buffers that are partially used in areas where live organisms are manipulated should never be transferred back to the central storage room (14).

42.1.5.3 Product Flow. Within the Production Core, product flow should be unidirectional. Areas where cells are cultured should be segregated from other processing areas. Areas where live viruses are manipulated should also be segregated from other processing rooms. Figure 42.5 provides a view of the live virus manipulation areas.

42.1.6 Water Systems

Many of the industry questions about producing water for pharmaceutical purposes were clarified by the publication of the USP Workshop on Microbiology and Pharmaceutical Water in April 1997 (15). The USP publication, Monograph <1231> "Water for Pharmaceutical Purposes," provides guidance for compliance with the current Good Manufacturing Practice (cGMP) regulations for drug products codified

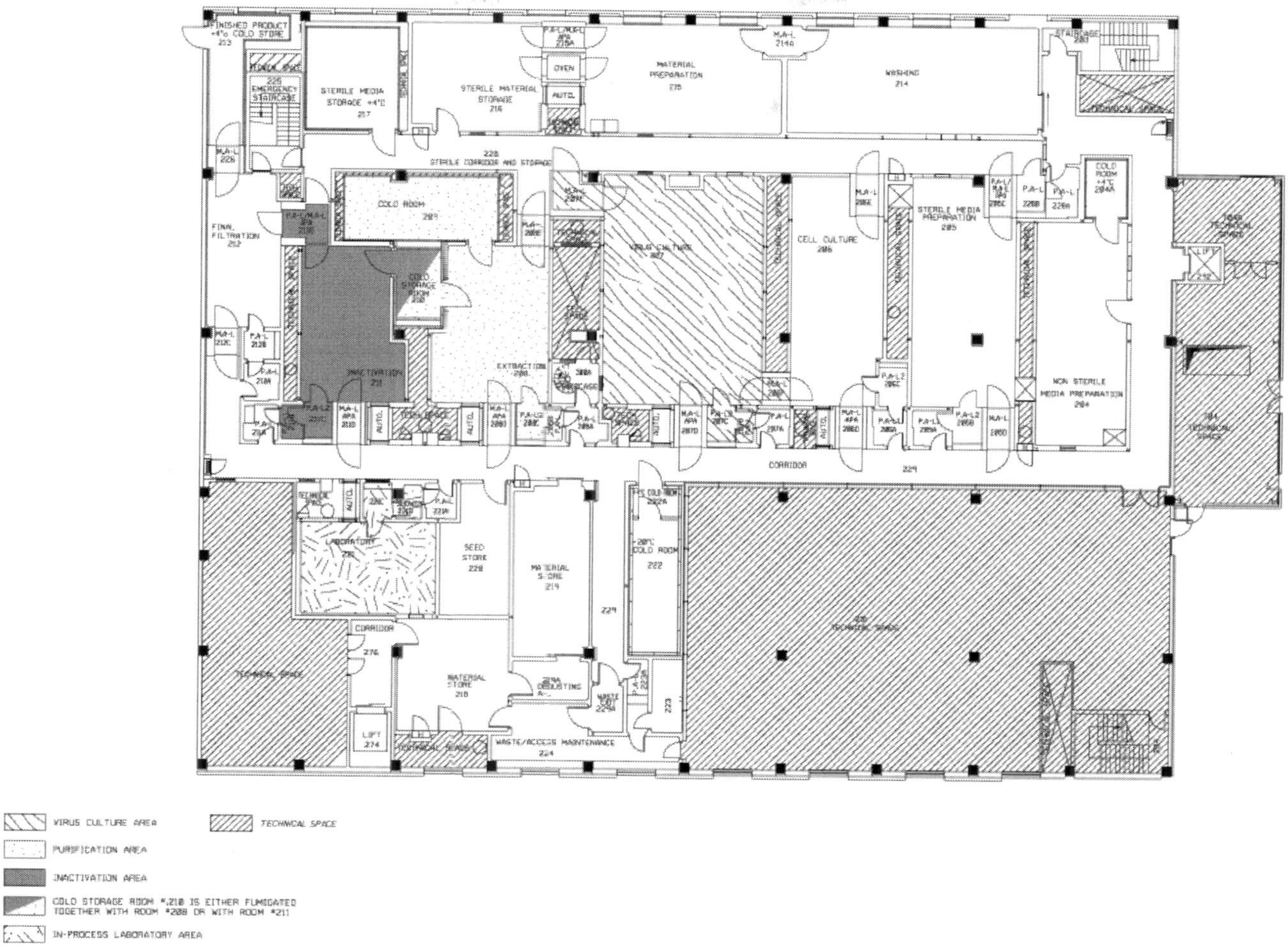

Figure 42.5. Live virus manipulation areas.

TABLE 42.3. Water Systems: Sources and uses of Water

Water Quality	Use	Source	Microbial Limits
Iron-free water	Production of chilling water	Municipal drinking water or treated well water	NA[a]
Purified water	In-feed water for clean steam and water for injection production Rinsing of materials and equipment before washing	Municipal drinking water or treated well water that complies with regulations 40 CFR 141 (National Primary Drinking Water Regulatory)	Less than 10000 CFU per 100 mL
Water for injection	Cleaning; last rinse water; water for media/buffer preparation.	Purified water	Less than 10 CFU per 100 mL

Note:
[a]NA = Not Applicable.

under Title 21 of the Code of Federal Regulations (CFR), Food and Drugs, Chapter 1, Parts 210 and 211 (21 CFR 211). Bearing in mind that cell cultures are highly susceptible to contamination, especially viral contamination, Water for Injection should be used at all process stages, including media/buffer preparation. Table 42.3 provides a summary of water systems commonly found in biotechnology facilities dedicated to cell culture.

42.1.7 Dedicated Multiuse or Multistrain Manufacturing Strategy?

42.1.7.1 Multiuse Facility. If a multiuse manufacturing strategy is developed, the facility will be designed for successive production of different strains of live organisms. Detailed and validated changeover procedures should then be followed to prepare the next product campaign. Depending on the type of organism handled during the previous campaign, the changeover procedures should address the following concerns:

- Removal of Working Seeds.
- Decontamination of all materials and equipment.
- Room cleaning and sanitization.
- Room fumigation (If environmental safety codes permit) using paraformaldehyde, peracetic acid or other validated fumigation method. Room fumigation using paraformaldehyde involves the sublimation of paraformaldehyde over a period of ±6 hours. HVAC unit is shut down during the fumigation. This phase is followed by a neutralization step involving the sublimation of ammonium carbonate. Cleaning the surfaces before room fumigation is recommended. The efficiency of this method is increased when the temperature and the relative humidity of the air are increased up to 25 ± °C 80% RH, for instance. The reader should note that fogging with formaldehyde is no longer permitted in the United State America. Other alternatives are available that involve peracetic acid, hydrogen Peroxide, or glutaraldehyde. Whatever the method, the efficiency of fogging against the microorganism should be validated both in laboratory experiments and production condition.
- Air and surface sampling to test for the absence of previously handled organisms.
- Room cleaning and sanitization.
- Equipment cleaning and sterilization or sanitization.
- Introduction of new organism strains.

42.1.7.2 Multistrain Facility. If a multistrain manufacturing strategy is developed, then the design of the facility should allow producing different strains concurrently in the different manufacturing areas within the limits defined by regulatory requirements (11,14). In addition to detailed and validated changeover procedures, the principles allowing concomitant presence of different strains in different manufacturing areas are

- Separate air handling units (HVAC): each room, together with its respective air locks, should be serviced by at least one HVAC which prevents air recirculation between rooms where different strains are handled (14).
- Dedicated air locks for personnel and material.
- Dedicated system for material decontamination and effluent treatment.
- Dedicated personnel: movement of personnel within the Production Core should be restricted to ensure segregation between the different areas (6,7).
- Validated cleaning procedures, especially for multiuse equipment.

42.1.8 Facility Finishes

Areas in a facility for bio production should be designed with smooth, readily cleanable surfaces (floors, walls, ceilings). Lighting fixtures, windows, doors and control panels should be flush, minimizing the potential for dust buildup.

42.2 DESIGN OF HVAC SYSTEM PARAMETERS

42.2.1 Introduction

In any biotechnological production facility, the HVAC system is crucial to good operations. The system must be capable of providing a uniform flow of conditioned air at sufficient velocity and pressure to meet both biosafety and GMP requirements. Requirements for maintaining room cleanliness have become more strict. Therefore, a series of operating parameter must be carefully fixed when designing modern facilities. The authors do not claim that the proposed description is the only way to comply with cGMPs and Biosafety Level requirements. The document is a translation of their experience in design, validation, and registration of biotechnology facilities.

Four major factors determine the operating parameters of the process rooms. The airflow is the first aspect to consider. Sufficient airflow to provide at least 20 to 40 air changes per hour is required to meet Class 100000 or 10000 requirements. The specification of 90-feet-per-minute air velocity for class 100 can be achieved only by using a vertical unidirectional flow hood or cleanroom. A vertical unidirectional flow Class 100 clean room requires 20 times more airflow than a Class 10000 clean room of the same size. Energy costs just to operate the blowers to provide the airflow make up an important part of clean room cost. The authors recommend the installation of laminar air flow hoods for cost and technical reasons associated with BL-2/BL-3 requirements. Secondly, heating and cooling capacities are required to maintain air temperature and relative humidity within a specified range: 18 to 22°C and 30 to 50 RH% for Class 100 and 10000, for instance. These values have been found satisfactory for controling microorganism growth with reasonable worker comfort. To maintain the values of temperature and relative humidity within the specified range, the engineers should take into account the variations of external ambient conditions, the number of persons, and the heat and moisture that the process may generate. The internal design of the room is the third major factor that determines the capacity to maintain the cleanliness of a production area. An adequate positioning of the air delivery inlets and the air exhaust outlets versus the positioning of the process equipment and personnel within the area will directly impact the capacity to meet the classification requirements. This capacity is measured by the decontamination time which represents the time required to recover the room class after contamination with a chemical aerosol. The test involves the use of nonviable particle counting devices which are placed at specific locations selected as representative of the clean room area (6). And last, differential pressures and air locks adjacent to manufacturing areas play a major role in protecting the cleanliness of the room and ensuring the containment of the live microorganisms within the area. Air locks are helpful in separating adjacent areas. They should be designed with sufficient airflow velocity and self-closing doors to make a real barrier; the interlocking system will maintain a minimum time between the closing and the opening of the two doors of the air lock. This system will ensure the efficacy of the so-called, "air shower." Differential pressures between adjacent manufacturing areas will be chosen to compensate for the variability of the air pressure and to ensure a unidirectional airflow.

42.2.2 Basic Description of Production Areas and their Operating Functions

1. Class 100 and 10000 (EU class A and B) kept at positive air pressure
   - Scope: Filtration of inactivated product, filling, formulation, sterile storage if included in a neutral aseptic production room, cellular culture, and nonpathogenic virus replication if aseptic processing, purification if required by the process, aseptic media preparation. The room is equipped with a Class 100 (EU class A) laminar airflow hood.
   - Access via:
     - Personnel Air Lock three doors: two parts
     - Materials Air Lock In
     - Materials Air Lock Out with fumigation for live organism handling areas.
2. Class 10000 (EU class B) kept at negative air pressure
   - Scope: Culture of pathogenic virus and inactivation of pathogenic virus if aseptic operations, purification if requested by the process. The room is equipped with a class 100 (EU class A) laminar airflow hood.
   - Access via:
     - Personnel Air Lock four doors: two parts
     - Materials Air Lock In
     - Materials Air Lock Out with fumigation.
3. Class 100000 (EU class C) kept at positive air pressure
   - Bioreactor culture and purification of nonpathogenic products, capping, clean materials storage, media preparation.
   - Access via Air Lock two doors: one part if access from a nonclassified production room.
   - Direct access if access from a 100000 (EU class D) production room.
   - The access to media preparation production rooms proceeds via an Air Lock if the facility is designed for cell culture and viral replication.
4. Class 100000 (EU class C) kept at negative air pressure
   - Scope: Bioreactor culture and purification of pathogenic products. If the production room is classified BL3,
     Access via:
     - Personnel Air Lock four doors: three parts
     - Materials Air Lock In
     - Materials Air Lock Out with fumigation.

     If the production room is classified BL2,
     Access via:
     - Air Lock two doors: one part
5. Class 100000 (EU class D)
   - Scope: Materials preparation with laminar flow, washing room, products storage, materials storage, corridor close to the production rooms.
6. Specific consideration
   Whatever the classification of a room, it can often be maintained at a low temperature but not exclusively between 2°C to 8°C to meet process requirements. In this case, the relative humidity needs to be kept lower than 90% to avoid condensation.

42.2.3 HVAC Characteristics

The characteristics of HVAC systems are described on the basis of dynamic room class and Biosecurity level (U.S. Standards). The following HVAC characteristics are described for each type of production area:

- Pressure.
- Air changes.
- Decontamination time (= time required to recover the room class after contamination by a chemical aerosol).
- Partial recycling of the air.
- Temperature.
- Relative humidity.
- Efficacy of air delivery filter (final filters).
- Efficacy of air exhaust filter (final filters).
- Location of air delivery inlets outlets.
- Location of air exhaust outlets.
- Fumigation.
- HVAC group in standby for the critical cold rooms.

42.2.4 Pressure Changes

The principle is a pressure decrease by steps of 15 Pa ±5 Pa (= 0.06 inch of water ±0.02).

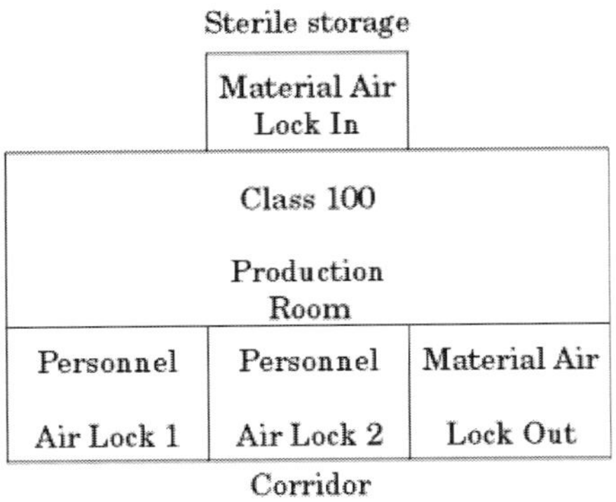

42.2.5 HVAC Technical Data

42.2.5.1 Class 100 and 10000 (EU Class A and B) Kept at Positive Air Pressure.

- Layout

42.2.5.2 Class 100 (EU Class B) Kept at Negative Air Pressure.

- Layout

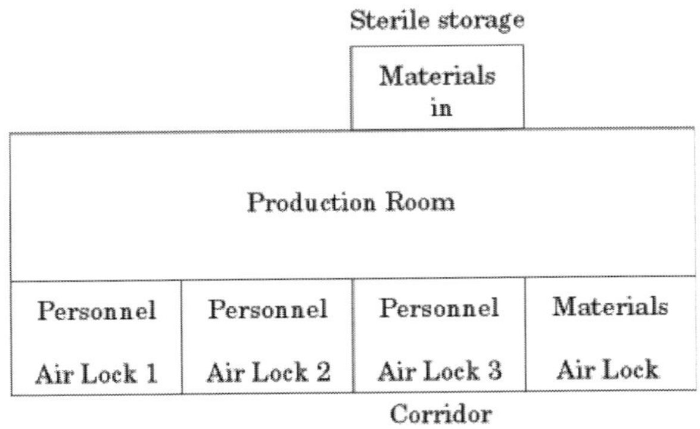

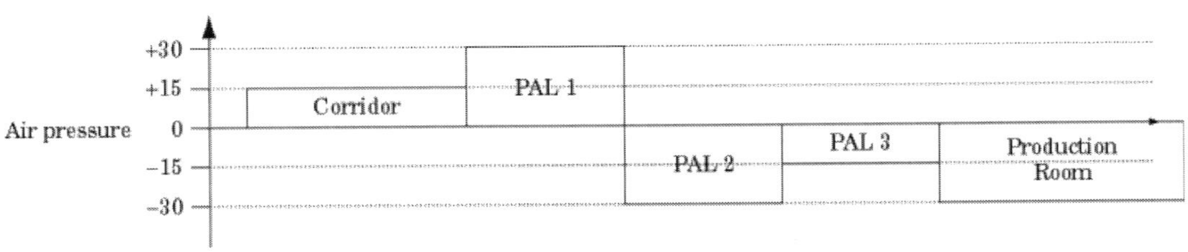

Operating parameters for Class 10000 (EU Class B) area kept at positive air pressure:

	Production Room	Materials Air Lock In	Materials Air Lock Out	Personnel Air Lock 1	Personnel Air Lock 2
Pressure (Pa)	+45	[a]	[b]	+30	[e]
Min. changes (AC/h)	40	200	200	20	600
Decontamination time (min)	15	3	3	12	1
Partial air recycling	Yes	100%	[c]	yes	100%
Temperature (°C)	18–22	NA	[d]	18–22	18–22
Relative humidity (%)	30–50	NA	[d]	40–60	40–60
Air delivery filter (% DOP retention)	99,995	99,995	99,995	99,995	99,995
Air exhaust filter	99,995	G 85	G 85	G 85	G 85
Location of delivery inlets	Ceiling	Ceiling	Ceiling	Ceiling	Ceiling
Location of exhaust outlets	Bottom of wall	Bottom of wall	Bottom of wall	Ceiling	Bottom of wall
Fumigation	Yes	Yes	Yes	No	Yes

Notes:
[a]Between sterile storage and production room by balancing.
[b]Between production room and corridor by balancing.
[c]Delivery and haust separated for the fumigation cycle.
[d]The parameter must be controlled during the fumigation cycle.
[e]Between PAL 1 and production room by balancing.

Operating parameters for Class 10000 (EU Class B) area kept at negative air pressure:

	Production Room	Materials Air Lock In	Materials Air Lock Out	Personnel Air Lock 1[e]	Personnel Air Lock 2[e]	Personnel Air Lock 3[e]
Pressure (Pa)	−30	[a]	[b]	+30	−30	−15
Min. changes (r/h)	40	200	200	20	20	600
Decontamination time (min)	15	3	3	12	12	1
Partial air recycling	No	100%	[c]	Yes	No	No
Temperature (°C)	18–22	NA	[d]	18–22	18–22	18–22
Relative humidity (%)	30–50	NA	[d]	40–60	40–60	40–60
Air delivery filter (% DOP retention)	99,995	99,995	99,995	99,995	99,995	99,995
Air exhaust filter	99,995	G 85	G 85	G 85	99,995	99,995
Location of delivery inlets outlets	Ceiling	Ceiling	Ceiling	Ceiling	Ceiling	Ceiling
Location of exhaust outlets	Bottom of wall	Bottom of wall	Bottom of wall	Ceiling	Ceiling	Bottom of wall
Fumigation	Yes	Yes	Yes	No	Yes	Yes

Notes:
[a]Between sterile storage and production room by balancing.
[b]Between production room and corridor by balancing.
[c]Delivery and exhaust separated for the fumigation cycle.
[d]The parameter must be controlled during the fumigation cycle.
[e]The authors have proposed and implemented an original approach with respect to the differential pressures of the Personnel Air Locks. The set points are fixed so that the differential pressures ensure two opposite directional air flows at two different locations: PAL 1 and PAL 3. Such design provides a double-barrier protection system working on two separate air handling units. The following schema illustrates the concept:

42.2.5.3 *Class 100000 (EU Class C) Area Kept at Positive Air Pressure.*

- Layout:

Production Room
Air lock

Operating parameters for Class 100000 (EU Class C) area kept at positive air pressure:

42.2.5.4 *Class 100000 (EU Class C) Area Kept at Negative Air Pressure (Biosafety Level 3).*

- Layout:

Production Room			
Personnel Air Lock 1	Personnel Air Lock 2	Personnel Air Lock 3	Materials Air Lock Out

42.2.5.5 *Class 100000—Pathogenic BL2 (EU Class C) Kept at Negative Air Pressure.*

- Layout:

Production Room
Air lock

Operating parameters for Class 100000 (EU Class C) area kept at negative air pressure:

	Production Room	Air Lock[a]
Pressure (Pa)	+30	[b]
Min. changes (r/h)	20	20
Decontamination time (min)	12	3
Partial air recycling	yes	yes
Temperature (°C)	18–22	18–22
Relative humidity (%)	40–60	NA
Air delivery (% DOP retention)	99,995	95
Air exhaust filter	G 85	G 85
Location of delivery inlets	Ceiling	Ceiling
Location of exhaust	Bottom of wall	Bottom of wall
Fumigation	Occasional	Occasional

Notes:
[a]An Air Lock is justified if the corridor is not classified (100000) or if the room is a production room for media preparation for cell culture and viral production.
[b]Between production room and corridor.

	Production Room	Air Lock[a]
Pressure (Pa)	−15	−30
Min. changes (r/h)	20	20
Decontamination time (min)	12	3
Partial air recycling	No	No
Temperature (°C)	18–22	18–22
Relative humidity (%)	40–60	NA
Air delivery filter (% DOP)	99,995	99,995
Air exhaust filter	99,995	99,995
Location of delivery inlets	Ceiling	Ceiling
Location of exhaust outlets	Bottom of wall	Ceiling
Fumigation	Yes	Yes

Note:
[a]Air lock on independent HVAC group.

Operating parameters for Class 100000 (EU Class C) area kept at negative air pressure:

	Production Room	MAL Out	PAL 1	PAL 2	PAL 3
Pressure (Pa)	−30	[a]	+30	−30	−15
Min. changes (r/h)	20	200	20	20	600
Decontamination time (min)	12	3	12	12	1
Partial air recycling	No	[b]	Yes	No	No
Temperature (°C)	18–22	[c]	18–22	18–22	18–22
Relative humidity (%)	40–60	[c]	40–60	40–60	40–60
Air delivery filter (% DOP)	99,995	99,995	99,995	99,995	99,995
Air exhaust filter	99,995	G 85	G 85	99,995	99,995
Location of delivery inlets	Ceiling	Ceiling	Ceiling	Ceiling	Ceiling
Location of exhaust outlets	Bottom of wall	Bottom of wall	Bottom of wall	Bottom of wall	Bottom of wall
Fumigation	Yes	Yes	No	Yes	Yes

Notes:
[a]Between production room and corridor by balancing.
[b]Delivery and haust separated for fumigation cycle.
[c]Control requested during the fumigation cycle.

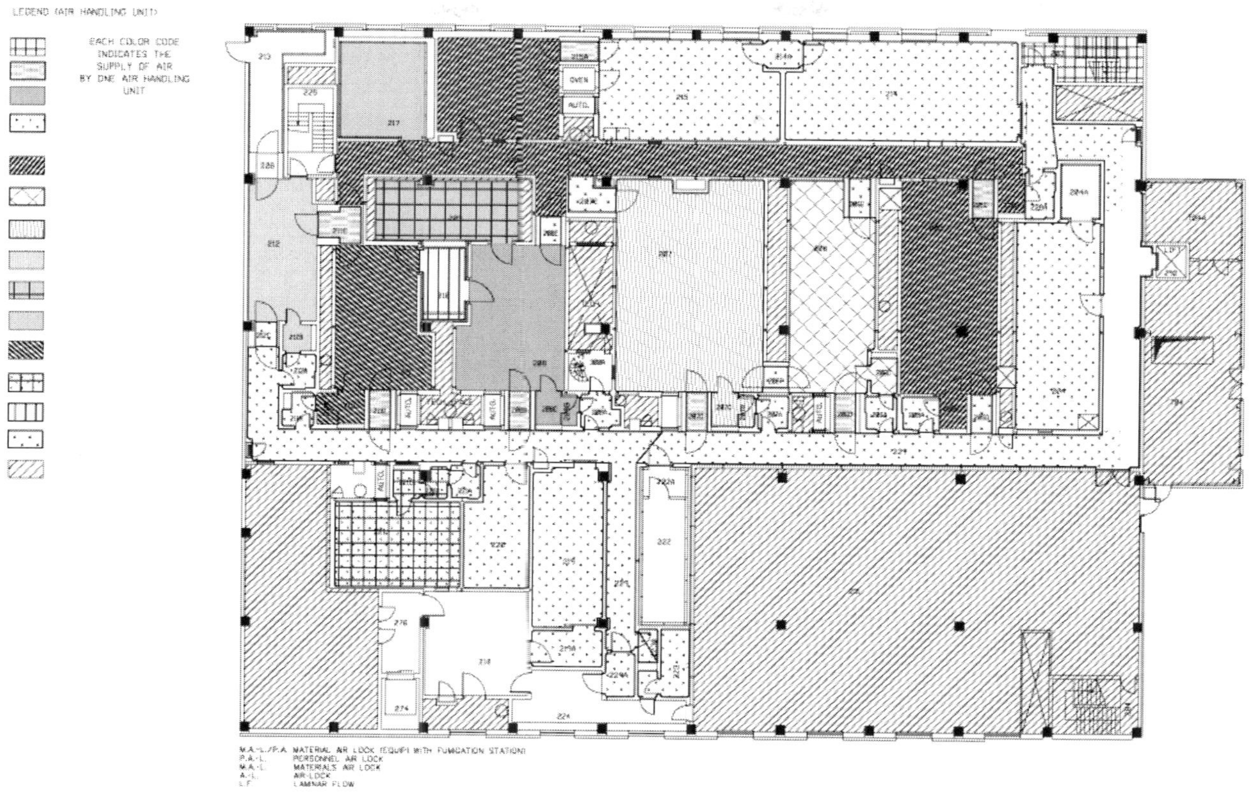

Figure 42.6. HVAC systems.

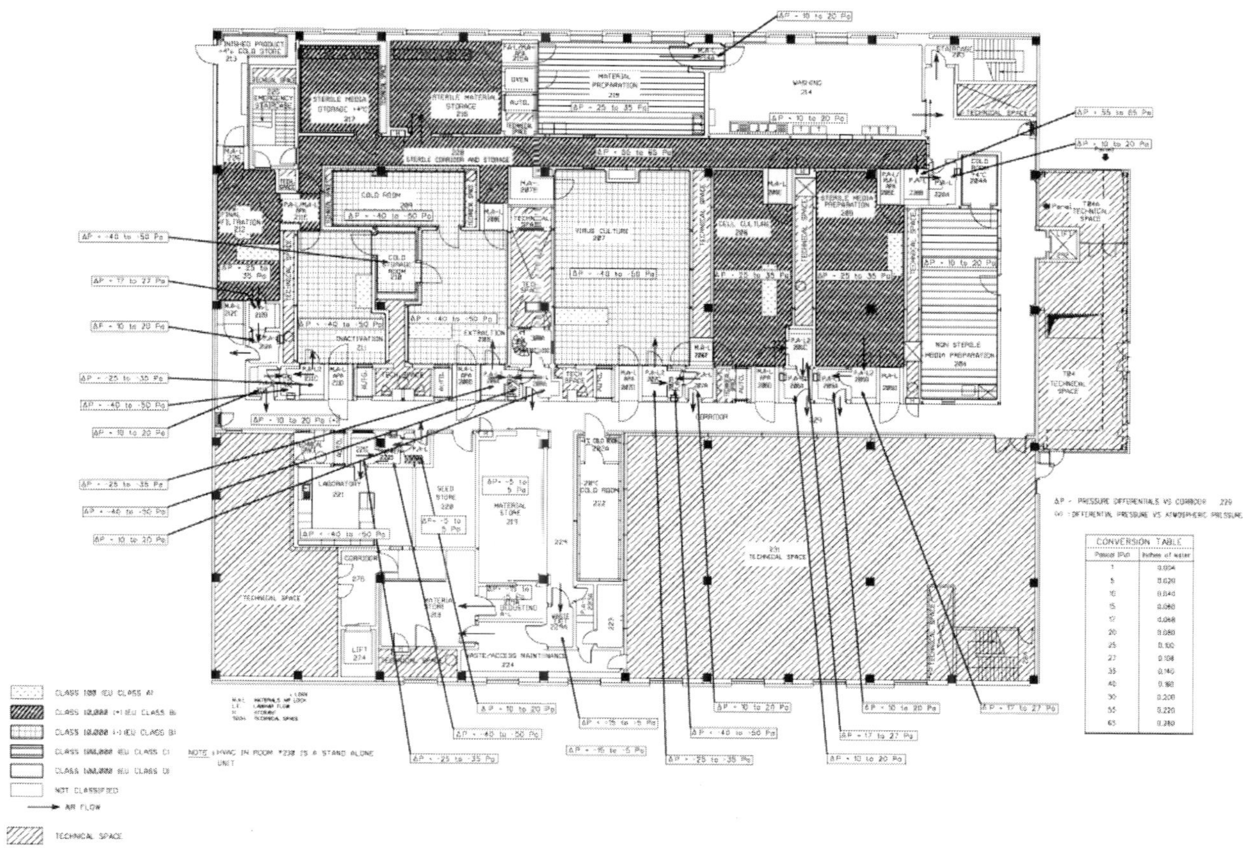

Figure 42.7. Room classification (dynamic conditions), pressure differentials, and Airflows.

42.2.5.6 Class 100000 (EU Class D).

	Production Room
Pressure (Pa)	Steps of 15 Pa
Min. changes (r/h)	20
Decontamination time (min)	3
Partial air recycling	Yes
Temperature (°C)	18–22
Relative humidity (%)	40–60
Air delivery filter (% DOP)	95
Air exhaust filter	G 85
Location of delivery inlets	Ceiling
Location of exhaust outlets	Ceiling
Fumigation	Occasional

42.2.6 HVAC Systems of the Viral Vaccine Bulk Manufacturing Facility

The production unit is serviced by twelve HVAC units as depicted in Figure 42.6. Each system has fresh air intake and all rooms classified as 10,000 or better are equipped with low wall air returns. Exhaust air is HEPA filtered.

- In production areas exposed to virus (Virus Culture area, Purification area, Inactivation area and Laboratory 221 previously described:
  - HVAC systems are segregated, and there is no recirculation of air between these designated production areas.
  - These areas are kept at negative pressure with respect to adjacent areas.
- In production areas with no live virus,
  - Room pressure and directional air flows are designed and balanced to create a positive pressure cascade from more controlled, or critical areas, to less controlled areas; see Figure 42.7 for pressure differentials and airflows.

REFERENCES

1. Code of Federal Regulations 21 Parts 210 and 211 (United States of America): Good Manufacturing Practices for Finished Pharmaceuticals, U.S. Government Printing Office, Washington, D.C.
2. Good Manufacturing Practices for Medicinal Products, European Union.
3. CFR 21 Parts 600.
4. Annexe 2 of European Union Code of Good Manufacturing Practice for Medicinal Products: Manufacture of Biological Medicinal Products for Human Use.
5. Biosafety in Microbiological and Biomedical Laboratories, Center for Disease Control and Prevention (CDC), U.S. Department of Health and Human Services (NIH), Washington, D.C.
6. Airborne Particulate Cleanliness Classes for Clean Room Production Rooms, U.S. Federal Standard 209E.
7. FDA Guidelines on Sterile Drug Products Produced by Aseptic Processing, Food and Drug Administration, Washington, D.C., 1987.
8. Annex 1 of the EU Guide to Good Manufacturing Practice: Manufacture of Sterile Medicinal Products, 1997.
9. Annexe 2 of EU Guide to Good Manufacturing Practice: Manufacture of Biological Medicinal Products for Human Use, Section "Premises and Equipment".
10. 21 Parts CFR 600.10, Section C.
11. Annexe 2 of EU Guide to Good Manufacturing Practice: Manufacture of Biological Medicinal Products for Human Use, Section 5 (Personnel).
12. 21 Parts CFR 600.11, Section (b).
13. 21 CFR 600.11, Section (a).
14. 21 CFR 600.11, Section (c).
15. USP XXIII Monograph <1231>.

ADDITIONAL READING

FDA Biotechnology Inspection Guide, November 1991, Food and Drug Administration, Washington, D.C., 1991.

ISO/FDIS 13408-1, Aseptic processing of health care products. Part 1. General requirements.

ISO/FDIS 14644-1, Cleanrooms and associated controlled environments. Part 1. Classification of airborne particles.

ISO/FDIS 14644-2, Cleanrooms and associated controlled environments. Part 1. Specifications for testing and monitoring to prove continued compliance with ISO 14644-1.

ISO/FDIS 14644-4, Cleanrooms and associated controlled environments. Part 4. Design, construction, and start-up.

ISO/FDIS 14698-1, Cleanrooms and associated controlled environments. Biocontamination control. Part 1. General principles.

ISO/FDIS 14698-2, Cleanrooms and associated controlled environments. Biocontamination control. Part 2. Evaluation and interpretation of biocontamination data.

ISO/FDIS 14698-3, Cleanrooms and associated controlled environments. Biocontamination control. Part 3. Measurement of the efficiency of processes of cleaning and/or disenfection of inert surfaces bearing biocontaminated wet soiling or biofilms.

USP XXVIII, Supplement 8, Section <1116>, Microbial Evaluation of Clean Rooms and other Controlled Environment.

43

HEATING, VENTILATION, AND AIR CONDITIONING

DENNIS DOBIE

Fluor Daniel, Marlton, New Jersey

43.1 INTRODUCTION

Biological products are derived from living sources, which presents special challenges for a heating, ventilation, and air-conditioning (HVAC) system serving a facility that processes and handles these products. The most important challenge is to prevent microbial contamination, which to a large measure is controlled by the HVAC system by utilizing classified spaces, providing containment, and maintaining pressurization. Biological facilities are subject to federal regulation. The Food and Drug Administration's (FDA's) Center for Biologic Evaluation and Research (CEER) is the primary regulating agency in the United States and its principal concern is the prevention of cross-contamination between different products produced in a multiproduct facility (1). Bioprocess procedures are generally conducted in environments that are cleaner than standard buildings. The HVAC system assumes a large part of the responsibility in maintaining these clean environments. Air used to condition a space must be highly filtered, properly directed, and able to maintain pressure relationships within and between adjacent spaces.

43.2 HVAC DESIGN PROCESS

43.2.1 Initial Effort

The HVAC design process begins with meetings with process engineers, architects, and representatives from the owner or facility user. The process and instrument diagrams (P&IDs) are reviewed, and a general understanding of the process is conveyed to all interested parties. Operation of the facility is reviewed, and any plans for future additions or modifications are discussed.

43.2.2 Basis of Design

After the initial meeting, a written basis of design is produced that describes the regulations and codes that will govern the design. Spaces are defined by function, and temperature and humidity requirements are determined. Room classifications are listed and adjacency of spaces and pressure relationships are documented. Any unusual or unique facility requirements must also be designed into the HVAC system at this time, such as emergency backup or redundancy for HVAC systems. This is also the stage of the design process during which alternate studies are conducted to compare options for the HVAC system. The cost of a backup or redundant HVAC supply system may be compared with the cost of product loss or experiment interruption should temperatures or airflow go out of control or specification. Heat recovery from exhaust systems and thermal storage are examples of other potential areas for study (2).

Airflow diagrams are produced that show areas served by a particular air handling system including supply, return, exhaust, and transfer air between spaces. The basis of design also describes major equipment to be used and the level of quality of components and construction material.

43.2.3 Cost Estimate

From the basis of design and a plan of the area, a factored estimate can be developed for budgeting purposes.

Downstream Industrial Biotechnology: Recovery and Purification, First Edition. Edited by Michael C. Flickinger.

Preparation of the estimate usually includes a dollar per cubic feet per minute (CFM) quote or a budget quote from vendors, as well as previous history to determine air handling unit costs. Estimates are also made for ductwork, as derived from pounds per CFM history and historic cost data. Costs for fans, high-efficiency particulate air (HEPA) filters, terminal boxes, and building management systems are also included. The level of effort expended to produce the basis of design will affect the accuracy of the estimate. Estimates from a basis of design can range from 15 to 25% accuracy. The estimate at this stage of the project is usually developed for capital appropriation.

43.2.4 Document Preparation

After owner or user approval of the basis of design, final plans and specifications are prepared using manual or computer-aided design tools for preparation of plans and sections. The utilization of three-dimensional design systems for document preparation is becoming increasingly popular. In a three-dimensional system, an electronic model is built that includes all disciplines, including HVAC, architectural, structural, electrical, piping, equipment, and so on. From the model interference, checks can be made that indicate physical conflict. These conflicts are resolved in the design process, avoiding potentially expensive field modifications and possible project delays.

43.2.5 Bidding and Award

When plans and specifications are completed they are sent to HVAC contractors for bid and eventual award. Design and build is also becoming a very popular option, in which design and construction are performed by a single entity with potential cost and schedule savings (3). When awarding a contract the engineer usually meets with the contractor to review the bid for completeness and technical accuracy, and to review the contractor's history on previous projects performing similar work. Since this is specialized HVAC work, it is necessary to have a contractor with sufficient resources and similar experience to avoid delays and potential rework that could increase costs.

43.3 REGULATION AND CODE CONSIDERATION

43.3.1 Code Compliance

Codes affect separation of air systems, gowning of people, airflow rates, exhaust requirements, construction materials, and many other aspects of an HVAC system for bioprocess facilities. If codes are not met, the result can be a danger to people working in the facility, products that do not meet specifications, and failure to license the facility.

43.3.2 Classified Spaces

Clean rooms provide clean air in aseptic spaces and are defined or classified in the United States by Federal Standard 209E of September 1992. The European Economic Community (EEC) published guidelines called "Guide to Good Manufacturing Practice for Medical Products in the European Community," which are more stringent than U.S. FDA regulations (4). Room classifications are established by measurement of the number of particles 0.5 μm and larger that are contained in 1 ft^3 of sampled air. Generally class 100 to 100,000 rooms are used in the bioprocess industry. Rooms may be classified as clean at class 1 or 10 for other applications, particularly in the microchip industry. Table 43.1 (derived from Federal Standard 209E) shows the air cleanliness classes.

EEC guidelines classify spaces by alpha character, as indicated in Table 43.2, and Table 43.3 shows the U.S. and EEC comparison.

43.3.3 Laboratory Spaces

Laboratory spaces are associated with bioprocess facilities. These classifications range from BL-1 through BL-4, with BL-1 being the least hazardous. BL-1 compares to a school

TABLE 43.1. Airbone Particulate Cleanliness Classes

Class		Measured Particle Size (μm)									
		0.1		0.2		0.3		0.5		5	
SI	English	m^3	ft^3	m^3	ft^3	m^3	ft^3	m^3	ft^3	m^3	ft^3
M 1.5	1	1,240	35	265	7.5	106	3	35.5	1	—	—
M 2.5	10	12,400	350	2,650	75	1,060	30	353	10	—	—
M 3.5	100	—	—	26,500	750	10,600	300	3,530	100	—	—
M 4.5	1,000	—	—	—	—	—	—	35,300	1,000	247	7
M 5.5	10,000	—	—	—	—	—	—	353,000	10,000	2,470	70
M 6.5	100,000	—	—	—	—	—	—	3,530,000	100,000	24,700	700

Note: Limits listed are maximum concentrations of particle sizes equal to and larger than the measured particle sizes shown.

TABLE 43.2. Classification of Clean Areas Defined by the European Economic Community

Grade	Maximum Permitted Number of Particles Per m^3 Equal to or Above 0.5 μm	Maximum Permitted Number of Viable Microorganisms Per m^3
A	3,500	<1
	Laminar airflow work station	
B	35,000	5
C	350,000	100
D	3,500,000	500

- Laminar airflow speed: 0.3 m/s for vertical flow 0.45 m/s for horizontal flow.
- Air changes per hour for grades B, C, and D should be higher than 20.
- U.S. Federal Standard 209E corresponding particle counts are

Class 100	(Grades A and B)
Class 10,000	(Grade C)
Class 100,000	(Grade D)

- It is accepted that sometimes conformity with particle counts may not be met at point of fill due to generation of particles or droplets from the product itself.

TABLE 43.3. Air Cleanliness Classification for Aseptically Produced Products

Operation	Parameters	United States	European Economic Community
Nonsterilized product or container (U.S.: controlled area)	Clean room type	Class 100,000	Grade C (Class 10,000)
	Maximum particle size 0.5 μm or above	100,000/ft^3 (3,531,000/m^3)	350,000/m^3 (9,900/ft^3)
	Maximum viable organisms	2.5/ft^3 (88.3/m^3)	100/m^3 (2.8/ft^3)
	Airflow rate	Minimum 20 air changes/h	Higher than 20 air changes/h
	Space pressurization	0.05 in water	Positive
Sterilized product or container (U.S.: critical area)	Clean room type	Class 100 environment in class 10,000 background	Grade A (class 100) environment in grade B (class 100) background
	Maximum particle size 0.5 μm or above	100/ft^3 (3,531/m^3)	3,500/m^3 (99.1/ft^3)
	Maximum viable organisms	0.1/ft^3 (3.53/m^3)	Class 100 room with grade A area: 1/m^3 (0.03/ft^3) Grade B background 5/m^3 (0.14/ft^3)
	Airflow rate	90 ft/min ± 20% (45.7 m/s)	Grade A: laminar flow work station Vertical: 0.3 m/s (59 ft/min) Horizontal: 0.45 m/s (88.6 ft/min) Grade B: higher than 20 air changes/h
	Space pressurization	0.05 in water	Positive

science laboratory, while BL-4 labs must provide ultimate containment and require glove boxes and double HEPA filtration. These laboratory classifications were formulated in the United States by the National Institutes of Health (NIH) and are published as guidelines in the *Federal Register*. The guidelines contain requirements for filtration of air, recirculation of air, containment, and biosafety cabinet requirements. There are various classifications of biosafety cabinets, depending on the level of containment required. The American Society of Heating, Refrigerating, and Air-Conditioning Engineers, Inc. (ASHRAE) guide should be consulted and owner requirements considered when selecting the required class and type of cabinet (5).

43.3.4 Other Regulations

Other regulations and guidelines may include those published by insuring agencies and corporate or plant guidelines formulated from specific industry experience and current good manufacturing practices (cGMPs) as published in the *U.S. Code of Federal Regulations* parts 210 and 211. The cGMPs apply to manufacture, holding, or processing of drugs to ensure safety, purity, and quality of the product. cGMPs regulate space temperature, and humidity and require a pressure differential between adjacent rooms of different cleanliness levels and a minimum air change rate of 20 air changes per hour. The regulation also addresses class 100 requirements at filling lines and microbiological testing sites, in addition to laminar flow control equipment parameters and testing at fill lines. Good practice usually places class 100 areas in class 10,000 backgrounds, which have between 35 and 60 air changes per hour in the space, depending on the activity and particulate generation within the space. Walls, floors, and ceilings for cGMP areas are to be constructed of smooth, cleanable surfaces, impervious to sanitizing solutions and resistant to chipping, flaking, and oxidizing. Horizontal ducts, pipes, or crevices that foster dust accumulation and cannot be easily cleaned are not permitted.

Figure 43.1 is a sample graphic depiction of an air classification drawing that is prepared to communicate to the design team and reviewing agencies the classification of each area. This is usually a full-size drawing (scale: 1/8 ft= 1 ft) and uses various shading symbols to identify each classified area.

43.4 TEMPERATURE, HUMIDITY, AND AIRFLOW

43.4.1 Space Temperature

Generally areas are designed to provide room temperatures from 67 and 77°F (19 and 25°C) with a control point of 72°F (22°C). Control of humidity from 40 to 55% is necessary for personal comfort, to prevent corrosion, to control microbial growth, and to reduce the possibility of static electricity. Calculations are made to determine internal heat gains from lights, people, and equipment in addition to transmission gains from adjacent spaces to determine if supply airflow quantities are adequate to provide design space temperature conditions. These calculations are compared with airflow quantities required by classification to establish the minimum air required to satisfy both the space cooling load requirements and air cleanliness classification. Fan heat from recirculating fans can also be a large heat contributor in clean spaces.

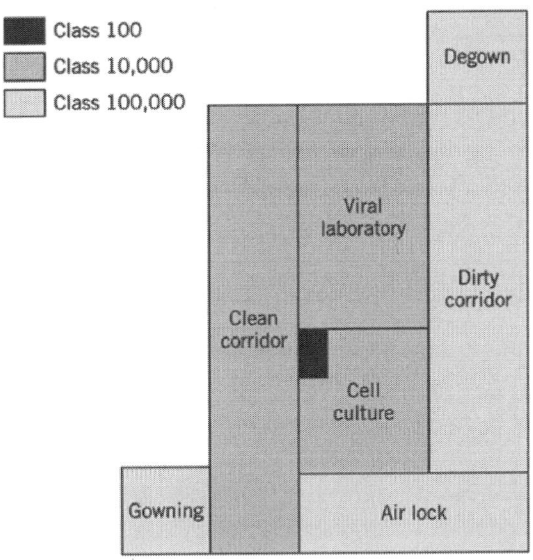

Figure 43.1. Air classification drawing.

Heat-loss calculations must also be made to determine heat loss through walls, roof, and floor. It may be necessary to add additional heat to a reheat coil to overcome these losses. No credit should be taken for process heat gain in this calculation, since the process could be dormant and the space would still need to be maintained at proper temperature. Lower space temperatures may be required where people are very heavily gowned and would be uncomfortable at "normal" room conditions. Humans are generally the largest contributors to airborne particulate in clean spaces; personal discomfort may cause perspiration, which significantly adds to the generation of airborne particulate. To achieve colder room conditions, the temperature of the supply air to the rooms may have to be colder than the normal 50–55°F (10–13°C), resulting in a greater dehumidification demand on the air-conditioning system. The amount of dehumidification required can be determined by use of a standard psychrometric chart by plotting room, mixed air, outdoor air, and coil conditions. Formulas to determine cooling coil loads are available from most coil manufacturers and air handling unit vendors (6). Cooling coil duty in British thermal units per hour (Btuh; W) is determined from the enthalpy difference between entering and exiting cooling coil air dry and wet bulb conditions.

43.4.2 Space Humidity

In some cases where products sensitive to moisture are handled, the room relative humidity requirement may be as low as 15 to 20% and may require the use of chemical dehumidifiers. It is necessary to establish this criteria early in the design process to allow selection of dehumidification equipment and add allowances to the construction-cost budget estimate. The lower the required temperatures and humidity, the more sophisticated the HVAC equipment becomes, resulting in higher start-up and operating costs. Range of temperature control also affects the cost of the control and HVAC delivery system. The closer the space temperature tolerances are, the more sensitive the controls must be to react. To achieve close temperature tolerances, the air distribution system must be more extensive and elaborate to prevent hot or cold spots. An HVAC room balance, shown in Table 43.4, is a sample of a preliminary design document

TABLE 43.4. HVAC Room Balance Sheet

Room Name and Number	Room Temperature	Relative Humidity (%)	Area (ft^2)	Room Height (ft)	Air Changes Per Hour	Hood, General Exhaust	Required Return CFM	Total Supply CFM	Transfer Air Out	Transfer Air In
Viral lab #1—100 (BL2, class 10K)	68	50	544	9	40	1,230	1,984	3,264	170	120
C/C lab #1—101 (BL2, class 10K)	68	50	453	9	40	345	2,323	2,718	170	120
Gowning 102 (BL2, class 10K)	68	50	125	9	40	0	385	750	365	0
Viral lab #2—103 (BL2, class 10K)	68	50	544	9	40	1,230	1,984	3,264	170	120
Clean corridor 104 (BL2, Class 10K)	68	50	1,300	9	40	0	6,905	7,800	1,320	425
Dirty corridor 105 (BL2, class 100K)	68	50	700	9	20	0	3,300	2,100	0	1,250
Degown airlock 106 (BL2, class 100K)	68	50	80	9	20	0	480	240	0	240

that indicates room temperature and humidity design conditions, in addition to dimensions. Table 43.4 also lists room classification air changes per hour, supply, exhaust, return, and transfer air requirements. This information is useful in communicating conditions to the design team and reviewing agencies. The table should be kept up to date, because it is useful to installing contractors and the construction supervision team as a check on ultimate system performance. When the project is near completion this information is usually transferred to the construction documents for ease of validation and agency approval.

43.4.3 Calculation of Air Changes

Table 43.4 indicates a typical range of air change rates generally used to achieve the desired room cleanliness classifications and to meet federal and local regulations. These air change rates vary widely in actual practice due to the level of activity, number and type of particulate generators in a room (such as people and equipment), and room size and quality of air distribution. It is generally best to use historic data to establish airflows, which is usually done with significant input from the owner based on past experience or preference. There is nothing sacred about an air change rate as long as minimum airflow rates required by code are maintained. The goal is to achieve desired particulate cleanliness levels and stay at or above a 20 air changes/h minimum.

To calculate airflow requirements from room air change rates, the following formula should be used:

$$\text{air flow (CFM)} = \frac{\text{room volume (ft}^3) \cdot \text{Number of air changes /h}}{60}$$

or (metrically)

$$\text{air flow (m}^3\text{/h)} = \frac{\text{room volume (m}^3) \cdot \text{Number of air changes /h}}{60}$$

43.4.4 Adjacency

In the early stages of design, the HVAC engineer must work with the architect, process engineer, user, and layout planner, if one is involved, to establish adjacency of spaces. It is suggested that a computerized drafting system with block layouts be used to allow the frequent manipulation that is required to arrive at an optimum arrangement. During this exercise the layout is primarily determined by process function and spaces are classified as clean, dirty, or contained. A contained space is used to contain a product or material that could be spilled into an occupied space or find its way into the building air system. From an HVAC standpoint it is desirable to keep similarly classified areas as physically close to each other as possible so they can be connected to the same air handling system, thereby minimizing duct runs, cost, and air system complexity. It is also imperative that spaces be arranged to allow people to move around without disrupting the cleanliness or containment of the spaces.

Ultimately, as the layout develops, the suites or groups of spaces will be established and groupings of rooms can be made for assignment to separate air handling systems. It is not desirable to mix dirty and clean systems or suites that may allow the possibility of cross-contamination from one suite to another. Leaks can develop in a filter, or some source of contamination could find its way through the air supply or return systems, providing a source for cross-contamination.

43.5 PRESSURIZATION

43.5.1 Requirement and Theory

Pressurization of spaces is required to keep products from being contaminated by particulate or to protect people from contact with harmful substances by physical means or inhalation. Where very severe potential exposure exists, air packs and breathing apparatus are employed.

Pressurization can best be explained by thinking of air leaking from a balloon. Pressure is built up inside the balloon and air escapes from a small leak in the balloon. If the hole stays the same size, a greater pressure differential between the inside and outside of the balloon will allow the air to escape at a higher velocity, which can be termed a higher level of pressurization. As the air in the balloon is depleted, the pressure differential drops, the rate of air escaping diminishes, and the velocity decreases. The same principle applies in bioprocess facilities where rooms are made as tight as possible to contain air, allowing the air handling system to build up or reduce pressure in rooms. These conditions are maintained by airflow, or leakage, between adjacent spaces caused by pressure differentials created by the air supply, return, and exhaust systems. Where major demarcations of pressure are required, air locks are used. These are small rooms with controlled airflows acting as barriers between spaces. Standard 209E recommends maintaining pressures of 0.05 in water (12 Pa) between adjacent spaces with doors closed. During facility operation as doors are opened the design differential is greatly reduced, but air must continue to flow from the higher to lower pressure space, even though at a reduced flow rate. To maintain a differential of 0.05 in water (12 Pa), a velocity of approximately 900 ft/min (4.7 m/s) should be maintained through all openings or leaks in the room, such as cracks around door openings. In theory the actual required velocity is less, but in actual practice it is prudent to use 900 ft/min. Table 43.5 gives the velocity pressures airflows from 0.01 to 0.20 in water (2.4–48 Pa), which can be used in approximating airflow between rooms. Limits on pressure drops must be set during the design (usually 0.05 in water) and observed. The pressure differential exerts a force on the door that can be calculated. If the force is too great (0.15 in water/36 Pa), the door may not close fully or may be difficult to open. This is particularly important in large complex facilities where many levels of pressurization may be required. Many facilities now use sliding doors, and it is essential that the seals be carefully designed to allow minimum leakage and proper containment or pressurization.

TABLE 43.5. Velocity Pressure to Velocity Conversion: Standard Air

VP	V	VP	V	VP	V	VP	V
0.01	401	0.06	981	0.11	1328	0.16	1602
0.02	566	0.07	1060	0.12	1387	0.17	1651
0.03	694	0.08	1133	0.13	1444	0.18	1699
0.04	801	0.09	1201	0.14	1499	0.19	1746
0.05	896	0.10	1266	0.15	1551	0.20	1791

Note: VP, velocity pressure; V, velocity.

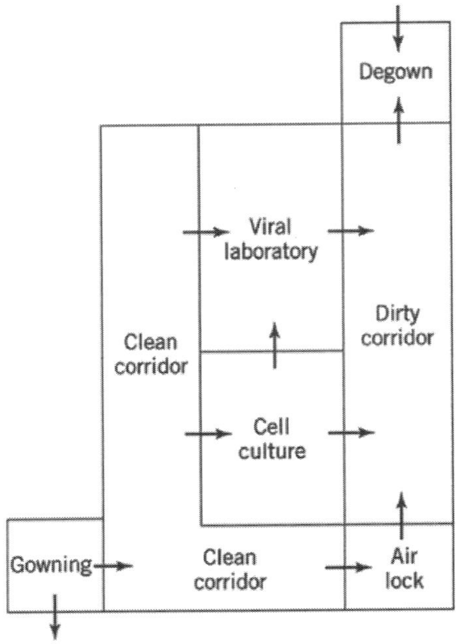

Figure 43.2. Pressurization diagram.

43.5.2 Pressurization Diagram

Figure 43.2 indicates levels of higher to lower pressurization with arrows showing the direction of airflow, (leakage) between the spaces. In actual practice the calculated airflow quantities are approximate since in constructing a facility it is very difficult to build everything absolutely airtight. A very small unexpected leak or a seal that is not as tight as expected in design can affect the pressurization and airflow required to maintain the design value of differential. Some flexibility must be built into the air handling system capacity and controls to compensate for these variations.

One technique to establish pressurization in separate rooms is to have a common reference for the control sensors. This common reference is usually a mechanical room or an interstitial space where the pressure is not affected by changes in the controlled spaces. It may be necessary to build an elaborate zero pressurization chamber to house the common sensor to negate the effects of wind or crosscurrents. Referencing spaces to each other can lead to problems if doors are frequently opened and upsets in pressure occur. The controls may never settle down and tend to constantly modulate, a problem termed *hunting*.

43.5.3 Room Seals and Doors

In most facilities the openings around the doors between rooms are where leaks occur due to pressure differentials between rooms. In making rooms tight any room openings must be sealed with a proper sealant that will not promote growth of organisms and can be easily cleaned. Areas to be sealed include ceiling tiles, lighting fixtures, pipe penetrations, telephone outlet penetrations, and any cracks or openings that appear in the structure. A typical door would have the following dimensions and crack area at the perimeter: door size, 3 ft (0.915 m) wide by 7 ft (2.41 m) high; cracks at top and sides, 1/8 in (3.2 mm) with an undercut of 1/4 in (6 mm). The calculated area around the door is equal to 0.24 ft^2 (0.024 m^2). To achieve 0.05 in water (12 Pa) pressure differential across the door, approximately 215 CFM (375 m^2/h) of airflow through the cracks is required. Door seals around the top and sides are usually made of closed cell neoprene and should generally be used to reduce the crack area. To reduce the undercut, a drop-type seal, which is commercially available, should be used. The drop type is preferred to a wipe type, since it will not mar or leave residue on the floor. Air used for pressurization must be accounted for in system calculations. Air through cracks or openings is accounted for as transfer air and shown in the HVAC room balance table.

Of particular importance is exhaust air from equipment and hoods that may be on or off at different times during occupied periods. These variations must be dynamically compensated for to maintain room pressurization. To maintain the required balance, numerous systems are employed using manual and automatic dampers, constant and variable volume air control boxes, and elaborate airflow sensing devices. These components are combined with control systems and sensing devices to ensure that the room pressurization is maintained.

43.6 AIR HANDLING SYSTEMS

43.6.1 Constant Volume Systems

After block layouts, adjacency layouts, room design criteria, and pressurization levels have been established, the number and type of air handling systems can be selected. Air handling systems most commonly used are constant and variable volume with and without reheat coils for control. For controlled spaces the most reliable system is a constant volume system with terminal reheat (CVRH). This system is not used in less critical or commercial applications. In a terminal reheat system the air leaving the cooling coil is set at a fixed temperature, and the terminal reheat responds to a space thermostat, turning on heat to satisfy the load. This can waste energy, since air is cooled and then reheated. Reheat systems require energy to both simultaneously heat and cool the air, as can easily be demonstrated on a psychrometric chart. Many energy codes prohibit this practice for comfort applications, however, where close control of temperature and humidity is required for process areas the energy conservation requirement is waived. The advantages of reheat systems are that humidity is always controlled (since dehumidification always takes place at the cooling coil) and each space or zone that needs temperature control can easily be accommodated by adding a reheat coil and thermostat. Another advantage of the CVRH system is that airflow is constant, which makes balancing and pressurization easier to maintain. A reheat system is probably the simplest and easiest of all systems to understand and maintain. Figure 43.3 is a typical depiction of an airflow diagram for a CVRH system.

43.6.2 Variable Air Volume Systems

A variable air volume (VAV) system is generally used in administrative areas and some storage spaces where

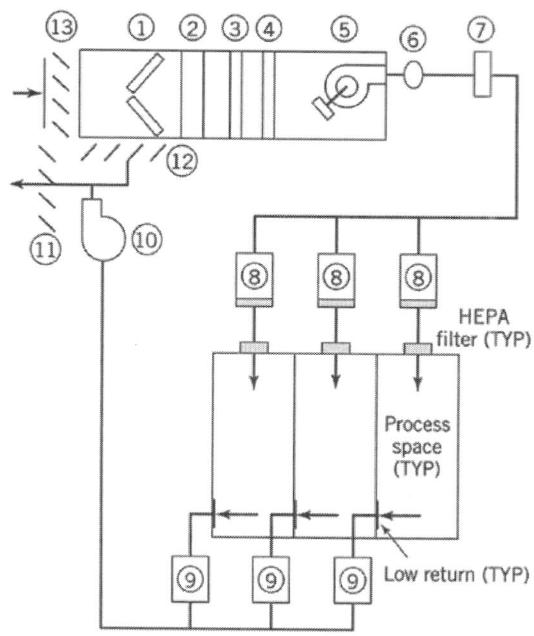

Figure 43.3. Airflow diagram: 1, 35% filters; 2, 85% filters; 3, preheat coil; 4, cooling coil; 5, fan section; 6, humidifier; 7, air monitoring device; 8, constant volume reheat box; 9, constant volume box; 10, return fan; 11, relief air damper; 12, return air damper; 13, outdoor air damper.

pressure control is not critical, humidity control is not essential, and some variations in space temperature can be tolerated. The VAV system works by delivering a constant-temperature air supply to spaces with reductions in airflow as cooling loads diminish. This reduction in air results in a reduced cooling effect, allowing the airflow or cooling to match the actual room load, which provides the design space temperature condition. This eliminates the energy used for reheat and saves fan energy, because the total amount of air moved is reduced. Some form of perimeter heating must be supplied for spaces with exterior walls or large roof heat losses. The perimeter heating can be baseboard radiation or some form of air heating using heating coils. Finned radiation or convection heating devices should not be used in clean spaces, since they are not easily cleaned and allow places for unwanted particulate buildup. Combinations of systems can be used, especially if variable quantities of supply and exhaust air are required for fume hoods or intermittent exhausts.

43.6.3 Recirculation Systems

In bioprocess facilities large quantities of air may be required to promote unidirectional flow and air cleanliness. This is particularly true in a class 100 space. In many cases the large quantities of air exceed the requirements for cooling, so it is desirable and possible to recirculate air within the space and only pass enough air through the air handling unit to perform the heating or cooling. In many cases the recirculating systems are supplied as a separate package installation. Numerous manufacturers provide these packages, which include supports, fans, and filters. Some vendors provide fans that are self-contained with each 2-;ft by 4-ft filter module.

43.6.4 Airflow Sheets

The airflow sheets should be developed on full-size drawings using computer drafting. They should list each space and show air quantities supplied, returned, and exhausted from each space. They also must show air transferred into and out of spaces, and, while quantities should be shown, they will probably require field modification to attain pressurization. The airflow sheet becomes a useful tool for transfer of information to the owner or user, for agency reviews, for transmission of information to HVAC designers, and for other engineering disciplines. These documents are also invaluable to construction contractors and for system checking by construction managers and balancing contractors. Airflow sheets provide a pictorial description of each air system and show how the elements comprising the system are related.

43.7 DEHUMIDIFIERS

43.7.1 Need for Dehumidification

In many cases low relative humidity is required, such as in drying or powder areas where hygroscopic materials are handled and would absorb unwanted moisture from the air. A chemical dehumidifier may be required to obtain the desired low humidity conditions. The need for a dehumidifier can be determined by utilizing a psychrometric chart, which will establish a leaving-coil dew point. Normal chilled water is supplied at 42–44°F (5.6–6.7°C) for use in cooling coils. Depending on the number of coil rows and fin spacing, a minimum dew point of about 50–52°F (10–11.1°C) can be obtained. This results in a minimum room relative humidity of approximately 50% at 70°F (21.1°C). To obtain lower relative humidity conditions chemical dehumidification is employed.

43.7.2 Chemical Dehumidification

Chemical dehumidifiers are commercially available air handling units that contain a sorbent material (desiccant) that can be a solid or liquid. Wet dehumidifiers use absorbents that change physically during the process. Lithium salt solutions are generally used to remove moisture from conditioned air and are then regenerated by heat, usually using a steam heat exchanger. Dry dehumidifiers use adsorbents that do not experience phase changes during the process. Silica gel and activated alumna are generally used. A rotating wheel is commonly used to remove moisture from the conditioned air. The wheel is regenerated by passing heated outdoor air over the wheel to dry it out. Steam or electric coils are usually employed for regeneration. Depending on the amount of dehumidification required and the amount of outdoor air (usually with a high moisture content), it may be best to combine the dehumidifier with a conventional air handling unit and only dehumidify a small portion of the air or just the outdoor air. The dehumidifier has a high initial cost compared with a conventional air handling unit. The size should be optimized to do only the required duty with an appropriate safety factor. Knowledgeable vendors in this specialized area should be consulted to find the best combination of dehumidification equipment, system arrangement, and control for the application. These systems also require considerable physical space, energy consumption, and service—important criteria to be considered in system selection.

43.8 HUMIDIFIERS

Air handling systems that bring in quantities of outdoor air for makeup require the addition of moisture to the airstream to provide desired space humidity levels. There are many commercially available devices for providing humidity to air, but the most commonly used are steam grid humidifiers. It is important to use clean steam (7), not plant steam, which may contain boiler chemicals and impurities from deteriorating piping and equipment. Humidifiers are controlled by modulation of a steam valve at the humidifier, and include a chamber to prevent condensation and water droplets in the duct. The valve is controlled by a signal located in the return or exhaust airstream or in a room humidstat. A high-limit stat is placed in the duct downstream of the humidifier to override the controlling stat and prevent condensation in the duct. Placement of the humidifier in the duct is critical and must follow the manufacturer's recommendations to prevent condensation and provide proper dispersion space. Duct downstream of the humidifier is usually constructed of stainless steel to prevent rust or corrosion should water be present.

43.9 AIR HANDLING UNIT APPLICATION

43.9.1 Air Handling Unit Construction

Conventional air handling units consist of filters, coils, and fans in a metal casing, with an insulation liner applied to the inside of the casing. For bioprocess applications the unit casing must be a sandwich of metal with insulation between the metal sheets to provide a smooth, cleanable interior surface that does not foster the growth of organisms. Chilled water or propylene glycol solutions are generally used for cooling and dehumidification. Direct expansion refrigerant, in which the refrigerant is in the air unit coil, may be used, but these systems are less reliable than chilled water or glycol and are more difficult to control in the narrow air temperature ranges required. Units should contain good access doors, view ports, electrical convenience outlets, and interior lighting for maintenance. The casings should be tightly sealed and designed for pressures that are higher than commercial applications due to the generally high system air pressures required for bioprocess applications. All sealants and lubricants exposed to the airstream should be food grade to minimize the chance of air contamination. Units designated as draw-through have the coils on the suction side of the fan. Blow-through units have the coils on the discharge side of the fan and have the advantage of a filter downstream of the coils, reducing potential contamination of the supply duct system. On blow-through units an air distribution plate must be installed to properly distribute air evenly over the filter and coils. It should be noted that if the unit contains a HEPA filter and the air is ducted directly to the clean space without a terminal HEPA filter, the duct should be constructed of a clean, smooth, nonflaking material, usually stainless steel.

43.9.2 Air Handling Unit Location

It is best to locate units indoors for ease of service and to extend the life of the units. They should also be located as close as possible to the main rooms they are serving to minimize larger and longer duct runs. Airflow measuring devices and humidifiers must also be installed, and unit location must allow enough straight run of duct for installation and proper operation of these devices. Location of outdoor air-inlet louvers must be carefully considered. Intakes are generally located on the building sidewall high off the ground to minimize dust intake. Intakes should also be away from truck docks or parking lots, where undesirable fumes and particulate are generated. In locating inlets the prevailing winds should also be considered, and any nearby exhausts or fume concentrations should be avoided to prevent recirculation of exhaust air back into the supply system.

43.10 RETURN AND EXHAUST FAN SELECTION AND LOCATIONS

43.10.1 Return Fans

Return and exhaust fans are integral parts of the air handling systems and also must be shown on airflow diagrams. Return fans are used on systems with long duct returns or that have return system pressure drops greater than 0.5 in water (120 Pa). This allows proper total system balance and minimizes suction pressure required from the supply fan. If a return fan is not used, the capacity of the supply fan can be overextended and it may be difficult to limit and properly control the amount of outside air being admitted to the unit. Outside air fluctuations are also more susceptible to exterior wind conditions. Return fans are also needed when required to provide a negative pressure in rooms that require containment. Return fans can be of standard centrifugal type or an in-line type, which works nicely for installation directly into return ducts in crowded equipment rooms. Return fans may also be required to handle varying quantities of air or provide a constant flow of air at varying pressure conditions. To achieve these conditions some form of damper control, inlet vane, or variable frequency drive motor control is generally used.

43.10.2 Exhaust Fans

Building exhausts are generally collected and ducted to exhaust fans in groups or clusters. Exhaust fans should be located as near to the building discharge as possible since this keeps the duct under a negative pressure and any leaks will be into the duct, and not contaminated air from the duct into an occupied space or mechanical room. For this reason roof locations of fans are preferred, even though this may make service difficult in severe weather conditions. When fans are located in mechanical rooms or interstitial spaces, it is essential to tightly seal the discharge duct before it exits the building in a roof vent or wall louver. Roof penetrations should be kept to a minimum to prevent leaks. Fumes and toxic exhausts should be extended through the roof and terminated well above the roof line in a suitable stackhead. Designs for stackheads are found in most duct manuals and are used to direct exhaust vertically at high velocities to enhance dispersion of fumes and prevent water from finding its way into the exhaust system (8). Extremely toxic or dangerous active biological agents may require HEPA filtration or other treatment, such as incineration, before exhaust to the atmosphere.

43.10.3 Redundancy

Another important consideration occurs when return or exhaust fans are used as part of the containment system. Should the fans be essential to maintaining containment, it may be desirable to have a backup fan or redundant system. This is essential if loss of containment can be harmful to humans or would result in an expensive loss of product. Airflow switches, which give a warning in case of fan system failures, are also desirable options for critical systems. The airflow sensing method to prove flow is preferred to an electrical motor indication since the motor could be running with a broken fan belt and the operator would not know that the fan is not moving air.

43.11 HEPA FILTERS

43.11.1 HEPA Filter Description

HEPA filters are used in the final cleaning stages of air to remove very fine particles. Rough and intermediate filters that are not as efficient as HEPAs are used to remove larger particulate matter. This prolongs the life of the more expensive HEPAs, significantly reducing the cost of filter element replacement. A HEPA filter, by definition, removes 99.97% of particles of 0.3-μm size. HEPA-type filters called ultraparticulate arrestences (ULPAs) can be even more efficient and are commercially available at a greater cost. Pressure drop across filters is rated for nominal flow at new clean conditions and is usually about 1 in (25.4 mm), and the systems are designed to change out filters at approximately 2 in (50.8 mm) or double the clean ratings, although filters can be operated at higher pressure drops, which some owners prefer. The higher operating pressures result in a greater use of electrical energy to power the fan to overcome the increased resistance that it must operate against.

43.11.2 Location

HEPA filters are usually located downstream of heating and cooling coils in air handling systems since coils are potential sources of contamination. The most popular HEPA filter location is in the room ceiling using standard laminar flow outlets nominally 24 in by 48 in (0.6 m by 1.2 m). These outlets contain manual control dampers, testing sample ports, a diffusion panel, and a HEPA filter element. They are commercially available in permanent or throwaway units where the entire assembly is disconnected from the flexible duct supply terminal and replaced when the filter is dirty. Sealing of filters to frames is an installation problem and is best solved by using a filter frame with a gel-like seal into which the filter fits. This material does not support biological growth, and these filters are offered by several manufacturers.

43.11.3 Bag-In Bag-Out Housings

Where danger of exposure to filter changing maintenance personnel exists, filters are placed in special enclosures called bag-in bag-out housings. These units are quite expensive and employ elaborate sealing methods and a double bag arrangement that allows changing of the filter element(s) without ever exposing the changer to the dirty filter or its collected contaminant.

43.11.4 Laminar Flow Workstations

HEPAs are also used in class 100 workstations in single or double filtration of air to provide sufficient cleaning before unidirectional (laminar) air distribution to the work space. These workstations are commercially available in horizontal or vertical flow patterns generally recirculating within the clean space. Filters are commercially available in many standard sizes and come with many frame arrangements with different methods used to form a seal between the filter element and the frame.

43.12 TERMINAL AIR CONTROL DEVICES

In most bioprocess applications the volume of airflow must be controlled to ensure proper air changes and provide air for pressurization. The choices to control airflow are (*1*) a variable flow responding to some pressure or temperature actuated signal or (*2*) constant volume to provide a uniform quantity of air under varying upstream pressure conditions. Many commercial devices, usually called boxes, or terminal control units, are available that utilize a variety of air dampening designs—from a simple blade type to pneumatic bladder type and conical type. Each type has its own application and associated cost implications.

The published damper design and test data for flow and repeatability should be studied against the suspected operation and matched to the proper application. Dampers are controlled from pneumatically controlled operators or electric motors responding to electric or electronic signals. Pneumatic dampers generally react more quickly to changing airflow requirements. Dampers and controls should be selected for linearity of response, repeatability, and percentage of error when compared with desired flow. The damper is generally housed in a device called a volume control box, which contains deflecting baffles and flow-measuring or pressure-sensing devices that are used for flow control. Accuracy of flow control boxes, combined with the controlling system, must be aggregated to ensure that the actual flow will be within acceptable limits to control pressurization. The accuracy of control devices can vary between 5 and 10% of maximum flow. This built-in inaccuracy must be calculated when establishing flow differential rates. For commercial applications, control boxes are lined with insulating acoustical material, which should be avoided on boxes used for critical or controlled areas to avoid dirt catchers or host locations for unwanted growth.

Some vendors offer complete control box systems, including controls, supply and return boxes, and a system to allow tracking. *Tracking* is a term used to describe when the control system monitors the supply flow and automatically adjusts the return or exhaust flow to maintain a fixed CFM differential between supply and return or exhaust. This maintains the space pressure differential.

43.13 AIR TERMINAL OUTLETS

43.13.1 Supply Terminals

In clean spaces the desired distribution of air is unidirectional or acts as a piston of air that reduces turbulence and eddy currents in the airflow. This piston of air carries particulate from the ceiling to the floor return and helps to prevent airborne particulate matter from recirculating and contaminating the work space. In most cases it is desirable to recirculate air within a space through a filter since the return air has less particulate than typical outdoor air and does not require extensive heating and cooling. Air terminals should be selected of materials that are nonflaking, nonoxidizing, and are easily wiped clean.

43.13.2 Location

Placement of outlets should be above work areas since the cleanest air should be introduced at these spaces. If the product at the workstation is hazardous to humans the air should be supplied at the operator's back and pulled into the work area to ensure that flow is away from the operator and any particulate will be captured in the airstream. Containment velocity at the inlet of a biosafe hood or enclosure should be about 100 ft/min (0.508 m/s). Airflow near hoods or enclosures should avoid any air disturbance that could cause eddy currents or air turbulence. These currents will upset capture patterns of the hood and make verification of hood face velocity difficult. Many of the cabinet manufacturers have extensively tested capture velocities and provide excellent data on the actual operating conditions. When locating terminals and enclosures it is important to consider movement of personnel within the space and identify frequently opened doors, which also tend to upset airflow patterns. Should the operator pose a threat to the product, the airflow pattern should be reversed, with the air supplied behind the product and blowing past the product to avoid contamination from the operator.

43.13.3 Return Terminals

Return terminals are also an important consideration and are generally located low in the walls for clean rooms. In class 10,000 to 100,000 rooms low cleanable wall registers are generally used. In cleaner areas low return wall systems, termed *air walls*, are used. The air wall is an almost continuous opening at the base of the wall with the air ducted up in the wall system and collected for return to the air handling system. Air wall inlets are generally located not more than 15 ft (4.5 m) in plain view from a supply terminal to reduce the likelihood of turbulence. Figure 43.4 shows a typical return air wall detail.

43.14 DUCT MATERIALS, PRESSURE, AND CLEANABILITY

43.14.1 Materials of Construction

Unlined galvanized steel ductwork is used in rectangular, round, and elliptical (or flat oval) configurations for the majority of the systems. Because galvanized duct can flake off or rust, it should not be used downstream of the HEPA filters to avoid contamination from the duct system itself. When the HEPA filter is located upstream of the room terminal and a long run of duct is present, the material of choice for the duct is stainless steel, but this is expensive and its use should be minimized. Many systems may also be fumigated or cleaned in place, and the duct material chosen should not be affected by the cleaning agent.

43.14.2 Duct Pressures

Ductwork systems in bioprocess facilities tend to have higher system pressure requirements than commercial

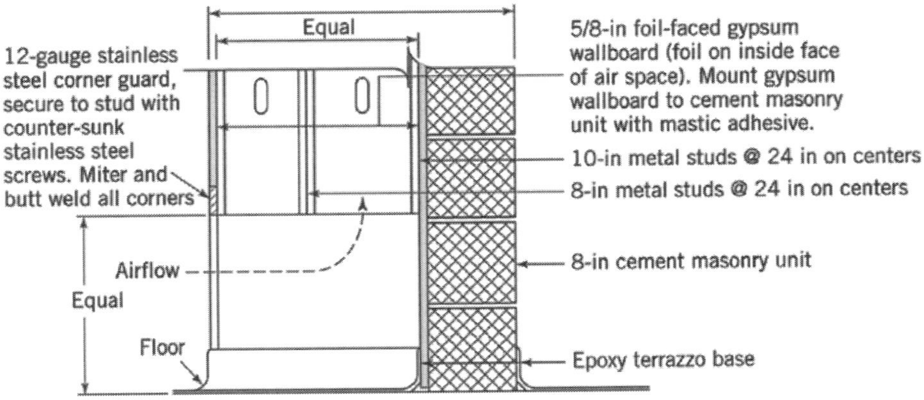

Figure 43.4. Return air wall detail.

systems due to extensive use of filters, volume control devices, and physically complex arrangements. The duct system pressures must be calculated and clearly stated on the contract documents to allow the fabricator to provide the proper metal thickness and construction methods for the required system pressures (9). System pressures will also change as the system is operated with filters that get dirty or space pressure conditions that vary. Duct systems must allow for these pressure fluctuations, and the fans may require speed controls, inlet vanes, or variable pitch blades to match the varying flow and pressure conditions.

43.14.3 Cleanability

Cleanability of duct systems is important to ensure that if an installed system gets dirty or contaminated it can be cleaned. In the design stage care must be taken to locate access doors in the duct, where they can be easily reached without compromising a process or violating a controlled space. It is best to minimize entry of facility system maintenance personnel into the clean spaces. All sealed duct shipped to the site should have only end seals broken, and then quickly resealed, during final installation. In very critical applications the duct is factory cleaned and sealed before shipment to the site. This step removes the oil and other contaminants present during duct construction but is expensive. It may be difficult to find sheet-metal fabricators willing to do this work, since they are not always set up for such procedures and it makes shipping the completed product difficult.

43.15 SYSTEM OPERATING PROCEDURES

43.15.1 Operating Procedure

When the airflow sheets are completed it is appropriate to finalize the system operating procedure. This is usually a written document stating how systems are turned on and off, and it gives a description of their frequency of operation. For clean spaces general practice is to operate the air system 15–30 min for purging before beginning production or use of the space. In some cases it is necessary to operate a system continuously to ensure that pressurization or containment is always maintained. When analyzing system operation it is important to assess how an unexpected system shutdown or flow reduction will affect an adjacent system or areas. Exhaust fans and fume hoods that may have intermittent or varying flows should be clearly noted in the operating procedure. It is essential that laboratory and production personnel clearly understand and follow the established procedures; otherwise, the air system may go out of balance if an exhaust or return system is inadvertently shut down.

43.15.2 System Alarms and Emergency Measures

Alarms that sound to indicate loss of pressurization are valuable features and essential in the HVAC design of critical areas. At this stage of design, procedures should be developed for the air system to accommodate a product spill or accident in a contained space. The ramifications of a spill on the air system, controlled space, and adjacent operations must be evaluated. Cleanup procedures could include fumigation of the air system, which would require operation of a relief connection to the ductwork for venting the fumigant.

43.15.3 Operations Review

The entire step of establishing operating procedures should ask as many what if questions as possible. System operation should ensure that the HVAC systems can maintain cleanliness and containment requirements. These conditions must be maintained while operating as close to normal as possible, as well as during periods of unexpected problems.

Generally the reviewing agencies ask many of these questions; therefore, it is essential that the design team be fully prepared with thoughtful, responsive answers.

43.16 EMERGENCY ELECTRICAL POWER

An essential step in the HVAC design process is coordination with the electrical design team. Motor lists for HVAC equipment must be prepared and reviewed with the electrical design team. The need for motors designated for emergency power, variable speed, reduced voltage starting, or other special characteristics must be communicated to the electrical designers early in the design process. The sizing of the emergency generator can be greatly affected by motors required on emergency power from the HVAC system. Fans, equipment, or sensing devices that require interlocks must also be picked up by the electrical designers. The motor list must be kept up to date from project inception through commissioning. The motor list is useful for a reviewing agency, a valuable tool in training plant operators, and a great aid in understanding the HVAC system.

43.17 BUILDING CONTROL AND AUTOMATION SYSTEMS

43.17.1 Sequence of Operation

The automatic control system that controls and monitors the HVAC system is called by many names: the automatic temperature control system (ATC), the energy management and control system (EMCS), the building automation system (BAS), or the building management system (BMS). The first element in the design of the system is the development of a sequence of operation, which is a written description of the HVAC and related systems operation. A separate sequence is usually written for each air handling system, describing the complete operation of the system from control of coils and humidifiers to control of the room temperature and humidity. Starting and stopping of the air handling unit fans is outlined, along with interlocking of exhaust or return fans in relation to the main air system fan operation. Generally all fans operate at the same time, which is necessary to maintain pressurization. The sequence also addresses abnormal occurrences such as a smoke detection alarm or failure of an exhaust fan. The sequence describes what happens to system components during an abnormal occurrence. It may be necessary to shut a supply fan down if a major exhaust fan should fail to prevent or minimize the loss of pressurization. The sequence also describes any energy management strategies to be included in the system, such as a night temperature setback or reduced ventilation and exhaust rates during unoccupied periods.

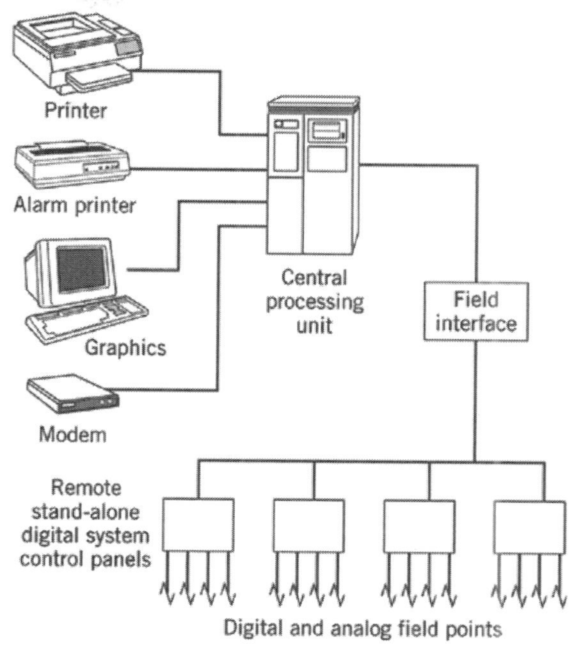

Figure 43.5. Direct digital control system architecture.

43.17.2 System Architecture

The control system of choice for major facilities, and even for some small systems, is a direct digital control (DDC) system. Most major control system vendors and many of the smaller vendors offer DDC systems that are similar but contain internal differences. The systems are computer based and have the ability to communicate within and outside the system by coded digital signals. System architecture refers to the major components of the DDC system and their interrelationship. The architecture is developed by determining what components are initially required, what may be required in the future, and how the system may expand as additional requirements are added. Figure 43.5 shows a typical DDC system architecture.

43.17.3 Points List

After the sequence of operation is completed and the airflow diagrams are defined, the next step is to develop the alarm, control, and monitoring points list. This is an all-inclusive list of points that are to be connected to the DDC system. There are two major types of points: digital and analog. A digital point is simpler, generally less expensive, and works on a simple on–off or contact principle. Digital points are used to start and stop fans, indicate an on–off condition, or anything that requires only a single contact. An analog point is used to measure variables such as temperature, pressure, and flow rate. These points generally use 4- to 20-mA signals that provide varying signals in response to the parameter measured. The electronic signals used by the BAS may be transduced from variable pneumatic or pressure signals.

The points list should include analog control points such as cooling coil valves and room temperatures. Monitoring points can be digital or analog, and can include fan run, room temperature indication, damper position, and room pressure indication. Alarm points can be either digital or analog and can include smoke detection in an air handling unit system, high or low environmental chamber temperature, high room humidity, or loss of room pressurization. Sensors located in clean spaces should be unobtrusive and installed in a cGMP acceptable manner.

43.17.4 Estimate of System Cost

The automatic control and monitoring system is a major cost element in the overall HVAC system for a bioprocess facility. After the points list is developed a good estimate can be made for the system cost. Several estimating numbers can be used in providing an educated guess of the cost, with a general range from a low of $500/point to as high as $1,200/point. Larger systems have lower dollar-per-point costs because the cost of the computers, monitors, and printers are diluted over many points. Smaller systems have higher costs because this dilution effect does not come into play. Systems with a higher percentage of analog points over digital points also tend to raise dollar-per-point costs. Control wiring is also a significant component of the BAS cost, and long runs of wiring tend to raise the dollar-per-point cost. When a floor plan is completed, the system architecture is established, and the points list is completed, this information should be given to one or more DDC vendors for a budget estimate. Vendors will make estimates based on the information provided, take off wiring runs, and suggest panel locations in arriving at the budget cost. At the completion of this exercise the cost of this major element can be locked in. Should the cost be greater than desired some points can be eliminated or the scope of the system reduced to bring the cost within the budget. When a system is bid it is recommended that component unit pricing be obtained from the vendor to lock in future costs should additions or modifications be desired. It is essential that vendors give classes and instructions on their systems before final turnover to the owner's operators.

43.17.5 Future System Flexibility

The trend is that after a DDC system is installed the number of points grows as the user becomes familiar with the system. Additional points may be added for maintenance, such as filter change out pressure devices. Run time logs set up in the system software are useful for lubrication schedules, belt checks, and so on. This requires software to log in and integrate data and to print out notices at preselected intervals of run time. Another useful feature is the phone modem, which would allow an operator to call in from a remote location and check on the system operation. This feature can save a trip to the facility if an alarm situation develops. Someone at the plant, such as the security guard who received the alarm, could place a call to the operator at home. The operator could dial into the BMS by modem and possibly analyze the problem by reading some key parameters. He or she could then determine the severity of the problem and either ignore it until the next shift or instruct someone on-site to take corrective action. Even though the original budget may not allow all features that are ultimately desired, it is good to look toward the future in anticipation of system expansion and ensure that the purchased system is capable of desired expansion.

43.18 TESTING, BALANCING, AND CLEANING

Testing and balancing for bioprocess systems is more critical than for commercial systems due to the very complex systems and higher pressures encountered. Associations of testing and balancing agencies publish excellent standard forms and procedures that are used in balancing air and water systems. Air balancing consists of setting terminal devices and air-moving equipment to deliver the proper quantities of air. For bioprocess facilities the additional requirement of establishing pressure differentials between adjacent spaces must also be set. These differentials are obtained by adjusting airflows, smoke tests, taking pressure readings, and setting controls. This effort can take some time as each facility is different and each room has different leakage characteristics that affect pressurization. As part of the balancing it may be discovered that duct systems or rooms are not as tight as designed and may require additional sealing to obtain the required pressure differentials. Recall that airflows shown on drawings are design values and generally require minor adjustment to achieve the required pressure differentials. A simple solution to many pressurization problems is to keep increasing outdoor air to the system. This can lead to problems if design values are exceeded, with heating or cooling coils not meeting this need, resulting in off-design room temperatures and humidity levels. The air handling unit coils will use available cooling capacity to condition excessive quantities of outdoor air, resulting in room supply temperature higher than as designed.

In these cases the rooms or systems should be tightened to get closer to design values. The optimum time to balance is when few construction workers or facility personnel are in the spaces. The balance should be done with all doors closed, since opening and closing result in system pressure upsets and make balancing difficult. Before the system is turned over to the owner, the rooms and ductwork system must be thoroughly cleaned to allow certification of room cleanliness conditions. During construction the work areas should be broom-cleaned after each day's work and open duct systems kept closed. If followed during construction, these procedures will result in an easier final cleanup before turnover.

43.19 VALIDATION

When a bioprocess facility is to be validated, the validating agency will peruse the HVAC documentation and should communicate with the design engineers to establish the validation protocol as it relates to the HVAC system. If the design is proper, the contractor has properly installed the system, and the components perform as specified, the systems should be easily validatable. The validator will follow a master plan and protocols to verify the actual system installation and operation against design values and intent (10). The physical parameters reported by the BAS system will be verified by measurements using calibrated instruments to verify accuracy. Airflow measurements will be made at supply and exhaust outlets, as well as traverses across the face of hoods, to verify proper flows and capture patterns. Filters will be physically tested or challenged to uncover leaks in the filter media or at seals between the holding frame and elements. Any leaks will be field corrected. Samples of air will also be taken by the validator to ensure that the cleanliness requirement for the controlled space is actually maintained by the installed operating air system. If problems occur they are usually in the form of pressurization deficiencies or filter leaks. Even after the mechanical contractor has completed his or her work, a hole for a pipe could be cut in a room and not properly sealed, resulting in a change in room pressurization. Filter elements may not be properly sealed with the frame, resulting in leaks, or small holes may be discovered in the filter element as a result of handling or damage during installation. Should any of these problems occur, and are not easily resolvable by the validator, it may be necessary for the HVAC contractor and engineer to lend assistance in making the system valid. Another important criterion at this stage is the calibration of airflow monitors, which should be installed in supply and return systems to critical areas to ensure that airflow rates are as designed. Temperature and humidity sensors at critical areas should also be checked for accuracy at this time by actually reading space conditions and checking against values reported by the BMS.

43.20 SUMMARY

The HVAC system is an essential component in the successful operation of a bioprocess facility and must be given a proper amount of attention during the planning and design stage. The space requirements for air walls, control boxes, extensive duct runs, and large air handling units must be established early in the design process to prevent unsatisfactory compromises later on. The owner and operators must work closely with the architect and engineer design team to agree on the HVAC system philosophy and the proposed operation of the system. The amount of energy management and control, as well as system redundancy, are among important economic decisions that should be agreed on during this design process. Compromises made during the design development should make economic sense, be mutually beneficial, and be understood by affected members of the project team. It is best from a cost perspective to resolve HVAC issues as early in the design process as possible to avoid design delays and more costly changes later in the design process. This avoids delays in the design process and minimizes the effect of changes on other engineering disciplines. All parties involved must believe in the HVAC system philosophy to avoid costly and time-consuming retrofits during construction or at commissioning, validation, or after the facility begins operating.

REFERENCES

1. A.J. Shahidi, R. Torregrossa, and Y. Zelmanovich, Pharm. Eng. 5: 72–82 (1995).
2. B.K. Lyderson, N.A. D'Elia, and K.L. Nelson, Bioprocess Engineering Systems: Equipment and Facilities, Wiley, New York, 1994, pp. 643–668.
3. J. Olsztynski, PM Eng. 2: 20–21 (1996).
4. M. delValle, Pharm. Eng. 15: 14–22 (1995).
5. Technical Committee 9.10, HVAC Applications, ASHRAE Handbook, American Society of Heating, Refrigerating, and Air-Conditioning Engineers, Inc., Atlanta, Ga. 1995, pp. 13.6–13.8.
6. Carrier Corporation, System Design Manual, vol. 1, Load Estimating, Carrier Air Conditioning Company, Syracuse, N.Y., 1972, pp. I-115–I-151.
7. B.K. Lyderson, N.A. D'Elia, and K.L. Nelson, Bioprocess Engineering Systems: Equipment and Facilities, Wiley, New York, 1994, pp. 588–593.
8. Committee on Industrial Ventilation, Industrial Ventilation, American Conference of Governmental Hygienists, Inc., Lansing, Mich., 1992, pp. 5–53.
9. Technical Committee 5.2, Fundamentals, ASHRAE Handbook, American Society of Heating, Refrigerating, and Air-Conditioning Engineers, Inc., Atlanta, Ga. 1993, pp. 32.1–32.40.
10. B.K. Lyderson, N.A. D'Elia and K.L. Nelson, Bioprocess Engineering Systems: Equipment and Facilities, Wiley, New York, 1994, pp. 2, 747–781.

44

STERILIZATION-IN-PLACE (SIP)

P.T. Noble
Fluor Daniel GmbH, Wiesbaden, Germany

44.1 INTRODUCTION

Sterilization-in-place (SIP) refers to the practice of sterilizing process equipment or materials in process at their place of use or installed location. Sterilization can be simply defined as a procedure for rendering a system free of living organisms. SIP is a very important technology, needed for microbiological control during production. Although the process of microbial sterilization is a well-studied area of science, SIP varies widely in its interpretation and industrial application. Steam is usually employed as the sterilizing agent, although many sterilizing methods are conceivable. The term SIP in some places is used loosely for a sanitization process whereby the goal is not necessarily achievement of a high assurance level of sterility (SAL).

44.2 APPLICATIONS

SIP supplements or replaces the use of process equipment whose purpose is to sterilize materials such as autoclaves, sterilization tunnels, and ovens. Introduction of the equipment into a sterilization chamber often is not practical. SIP is also preferred when several components can be sterilized together via SIP rather than separately treated and subsequently connected under aseptic conditions.

Clean piping systems, such as a water for injection (WFI) supply ring, will usually be sterilized in place before start-up and after service interruptions. Here, the goal is to begin or resume microbiological control but not achieve a specified sterility assurance level. Fermentors, especially those used to culture genetically altered cells, need to be sterilized before innoculation, which can be achieved to a high level of assurance by SIP. Fermentation media are sometimes first introduced into the fermentor and then sterilized with the fermentor. Aseptic manufacture of parenteral drugs and devices requires sterile process equipment, which is usually achieved by SIP. For example, modern filling equipment often undergoes an SIP of the filling lines and nozzles. It is also a current Good Manufacturing Practice (cGMP) standard that lyophilizers for parenteral products be outfitted with an SIP cycle to sterilize the drying chamber and process connections that are open to the product. Finally, as part of biocontainment practice (especially when using recombinant or hazardous organisms), processing equipment may receive an SIP after use but before opening to inactivate any residual microorganisms and prevent their release into the environment.

44.3 STEAM-IN-PLACE TECHNOLOGY

The use of saturated steam for sterilization has been relied on for many years. It is natural that SIP would evolve from this experience. One of the first publications on the use of steam for SIP was provided by Myers and Chrai (1) and later reviewed by Agalloco (2). The preference for steam as the sterilizing agent can also be attributed to its advantages in heat transfer, which are elaborated in the following.

Sterilization in general is a kinetic process and is a subject of sufficient breadth for textbooks (3). It follows first-order kinetics for individual microbial populations, whereby the rate constant is species specific as well as being dependent on the sterilizing conditions. When only

Downstream Industrial Biotechnology: Recovery and Purification, First Edition. Edited by Michael C. Flickinger.

water is present at the saturated steam condition, the temperature alone is sufficient for completely defining these conditions and hence the rate constant for a particular organism. The rate constants are exponentially dependent on temperature, and the dependencies are significant for temperature variations as small as 1°C.

For scientific studies, sterilization processes are usually quantified in terms of the *D* and *F* values. The *D* value provides the kinetic information and is simply related to the rate constant. It is species and temperature dependent, and its value is the time in minutes for a 10-fold reduction in the population. The *F* value gives the time in minutes for a desired reduction in the population, as calculated by $F = (\log N_0 - \log N)\, D$, where N_0 is the starting count and N is the desired count.

The F_0 value has become a standard for defining and comparing sterilization processes. It is based on the sterilization kinetics that have been found for heat-resistant organisms, such as *Bacillus stearothermophilus*. The F_0 value can be calculated from thermal data using the following integral equation, $F_0 = \int 10^{(T-121.1)/10}\, dt$, where T is the sterilization temperature in degrees Celsius and t is the sterilization time in minutes. SIP operations typically should demonstrate an F_0 value in the range of 8 to 20 min, depending on the application.

Steaming conditions other than with pure saturated steam have been found by experience to be less effective (3). Apparently, the presence of both phases, liquid and gas, is important because superheated steam at the same temperature is not as lethal. Also, the presence of air reduces sterilization effectiveness. If air is present, and the sterilizing pressure is kept constant, then the contribution of air to the total pressure lowers the partial pressure of the steam and hence its saturation temperature. The lower temperature has a direct effect on the kinetics. If an outside heating source raises the temperature, the system departs from saturated steam conditions; that is, water is present then only as superheated steam. If the total pressure is raised, so that the steam saturation temperature is maintained, such as with an overpressure autoclave cycle, then the kinetics predicted with saturated steam are followed.

44.3.1 Validation Issues

Important validation issues arise with SIP. The general problem is the estimation of sterility, but the additional process complexities associated with an SIP operation often add problems of control for the process parameters. Validation is needed to provide assurance that the probability of surviving organisms is small enough to be of insignificant risk to the operation. Because the initial bioburden (number and type of microorganisms) can only be partially known from past monitoring experience, kinetic models cannot predict with accuracy the outcome of SIP. Measurements of sterility are in general destructive tests; that is, they violate the integrity of the enclosed sterile system.

Accurate temperature and pressure measurement and control is of special concern for steam-in-place processes. It is especially important to know the location of the coldest points in the system to estimate conservatively the sterilization effect. An accurate measurement of pressure is needed to verify saturated steam conditions. Validation testing include extensive thermal measurements, the use of biological indicators (e.g., spore strips) and often sampling methods via swabbing or media contact. Such testing should yield consistent, favorable, and reproducible results to give such assurance.

Reflecting the uncertainties and risks associated with a sterilization process, regulatory bodies, such as the U.S. Food and Drug Administration (FDA) (4), have specified relatively severe process parameters to meet for general use. These conditions are often termed overkill. In the cited reference, 121.5°C is to be maintained with saturated steam for 20 min, or equivalency to these conditions must be demonstrated as by achieving the same F_0 value. Using the formula provided, this value would be F_0=21.9 min.

44.3.2 Process Description

A technical description and a theoretical modeling of the steam-in-place process can be found in Ref. 5 and is relied on here. A steam-in-place process follows the general operations of an autoclave, which have the following cycle steps:

1. Air purge
2. Heat-up
3. Sterilization holding period
4. Displacement and collapse of the steam with sterile air

SIP of a vessel or piping system usually accomplishes the air purge step by displacing the air with steam. The air exits the system through drain and vent openings, which are usually equipped with steam traps.

A residual presence of air is to be avoided because it introduces a poorly controllable heat transfer resistance, which is described next. Its contribution to the total gas pressure also lowers the effective steam pressure and hence saturation temperature, which affects directly the sterilization kinetics. Therefore, this first step is critical to the reproducibility and success of the operation. Unfortunately, displacement of air with steam is often not very effective. Some autoclaves utilize a series of vacuum pump-outs followed by steam repressurization to remove the air from the chamber. Vacuum pump-outs are seldom part of an SIP operation unless an efficient vacuum system is already present for process needs, as with freeze dryers.

Convective displacement of air with steam always results in some mixing of air with the steam. In branched piping systems with poor flow distribution or in large vessels, this mixing can be extensive. In theory (5), turbulent steam flow is desirable because the additional kinetic energy helps to break up any air pockets and transport the air (with steam) out. The air purge step can be successful when the piping design meets appropriate criteria (to be described later) and restrictions to flow are minimized such that maximal steam flow can be reached.

Using steam to heat an enclosure combines both heat and mass transport phenomena. In the gas phase, a comparable situation (condensation of vapor in the presence of a noncondensable gas) has been analyzed and modeled for psychrometric calculations (6). Geometric complications prevent an exact analysis of a typical SIP application. Additionally, steam-in-place must consider transport in the condensed phase as well as heat transport in the walls of the system.

The combined heat and mass transport created by steaming yields important advantages for heat transfer. When saturated steam contacts a colder surface, it condenses, and the resulting void created by the collapse of water vapor molecules into a condensed phase drives more steam to the region of the colder surface. The pumping effect of the condensation helps bring steam heat to the colder regions of the enclosure and thereby improves the heat transfer. Reliance on latent heat also reduces temperature drops, which are otherwise needed for sensible heat transfer.

During the heat-up step, the system pressure is allowed to build up to the steam supply pressure by restricting any vents or drains or via the functioning of the installed steam traps. Any trapped air left will be largely swept to the cold boundaries by the dragging effect of the moving steam. Thus, the residual air will not be uniformly distributed but will be concentrated near the colder boundaries. As a stagnant layer, through which the steam must pass to condense on the walls, residual air can bring a significant resistance to heat transfer. Enclosures with a large volume-to-surface ratio, such as tanks, are particularly prone to interference from trapped air. In Ref. 5, temperature drops of the order of 1°C are predicted from residual amounts of air equivalent in some cases to less than 0.1% of the original volume. Such exquisite sensitivity is possible in piping dead legs (to be described later) and is greatly amplified when the equipment is not thermally insulated. Insulation reduces heat transport requirements and thereby mitigates the sweeping of air into stagnant boundary layers.

Condensate removal during steam-in-place is a serious concern. Condensate forms on all cold surfaces and will subcool if allowed to accumulate. Because of its low thermal diffusivity, liquid water can maintain a substantial temperature difference between the steam phase and the equipment walls. Condensate layers will collect at all low points. Flow in these layers to the next steam trap is usually laminar. A one-dimensional analysis (5) is useful to predict order-of-magnitude effects. In an insulated pipe, a condensate layer 1 mm in depth will have a temperature drop of the order of 0.5°C. Without insulation, only a 0.2-mm depth would show such a temperature drop.

The boundary layer phenomena discussed here are not usually observable without extensive temperature monitoring, which may only be performed during validation as a thermal mapping activity. From this study the coldest locations are identified. Thermal data from these locations are then used to define a sterilization holding period to achieve equivalence to the overkill process, or a predefined F_0. In normal operations only one or a few sensors, which for practical reasons sometimes cannot be positioned at the coldest locations, are used to monitor temperature. Temperature nonuniformities bring an uncertainty into the estimation of the sterilization effect; their existence demonstrates the presence of heat transfer resistances, usually resulting from boundary layer phenomena, which are not under direct control. Statistical analysis of temperature data during process validation will be used to add additional time to the holding period to provide the desired degree of sterility assurance.

On completion of the sterilization process, the steam supply is shut, and the equipment can be cooled either actively or passively before use. The collapse of residual steam will ultimately create a vacuum in the enclosure, which is to be avoided because a vacuum increases the risk of recontamination of the system. Therefore, the steam is usually displaced with sterile-filtered process air while there is still overpressure present.

44.3.3 Equipment Design

Process requirements for steam-in-place operations have determined to a large degree the special equipment design specifications associated with bioprocesses. The sterilization conditions usually chosen are 121°C or greater, resulting in a saturated steam pressures above 1 bar gauge. Materials of construction and construction methods must be suitable for these conditions.

44.3.3.1 Materials of Construction. 316L stainless steel is usually chosen because of its strength and chemical resistance to the high-purity steam used. Vessels need to be pressure rated for these conditions and protected with safety relief valves or rupture disks.

Specifications for surface finishes and construction methods have the purpose of reducing the potential for the buildup of soil and deposits on the equipment surfaces. Surface deposits provide an additional barrier to the transport of heat and steam. They may also introduce an uncontrolled bioburden level. A high degree of polish will be required; electropolished is preferred. Corners and joints should meet with a generous radius of curvature. Ideally, sealing methods would be seamless. For this reason, welding is preferred, but the welds must be polished and passivated. Seiberling (7) provides general specifications of equipment to receive SIP.

All low points inside the enclosures must have drains; these are needed to drain the condensate and are also required to effectively clean the equipment so as to prevent soil buildup. Thermal insulation of the equipment is needed to reduce the temperature variations in the system. If the equipment is located in a clean room, the insulation must be clad with a housekeeping skin, so that particles from the insulation do not enter the clean rooms and the surfaces can be maintained clean. Stainless steel sheathing is usually used on vessels, and plastic is easier to install on piping. The enclosed insulation must not chemically release agents that would attack the equipment, such as chlorides, which are known to attack stainless steel.

44.3.3.2 Piping Design. Piping particularly should be thermally insulated because of its large surface-to-volume ratios. All piping must be sloped to drain condensate. Because the mass transport of steam and condensate are driven by heat transport and gravity, respectively, and not by large pressure gradients, it is important in the piping design to remove any uncertain flow paths. A clear example of this concern can be made with parallel flow arrangements. If two or more identical items such as heat exchangers are steamed in parallel, it is likely that one item will not receive the proper treatment because of poor steam distribution. In an extreme case, one item may stay filled with air or become blocked with accumulated condensate, effectively stopping the flow of steam.

Dead legs are to be avoided, but this is not usually possible. Dead legs are commonly understood to be sections of piping with no outlet. Connections that are closed with a valve during SIP are typical dead legs. Instrument ports can be another. Dead legs also need to be sloped to allow for draining. Depending on their length, flow of steam in and air out can be largely a diffusive process. Dependence on diffusion means some air is always trapped in dead legs during SIP, contributing to the difficulty of their sterilization (8). General design guidelines (7,9) limit dead legs in length by a scale factor expressed in terms of pipe diameters, but this rule cannot be defended in theory (5). The best guideline is to minimize their length because they will usually be the most difficult locations to sterilize. Practically, dead legs can easily be limited to four pipe diameters with commonly available pipe fittings.

Dead leg–free valves are commercially available (Fig. 44.1); they function by using the pipe wall on one side as the valve seating surface. The seating design employs a flexible diaphragm or membrane, which is pushed into the flow path to press against the seating surface. Such valves belong to a type, commonly referred to as diaphragm or membrane valves, which are preferred for equipment that receives SIP or clean-in-place (CIP). Such valves have no dynamic seals for moving parts such as valve stems that may enter or leave the sterile side of the valve. Dead leg–free valves are only such on one side of the valve. On the other side, the valve body itself is a short dead leg.

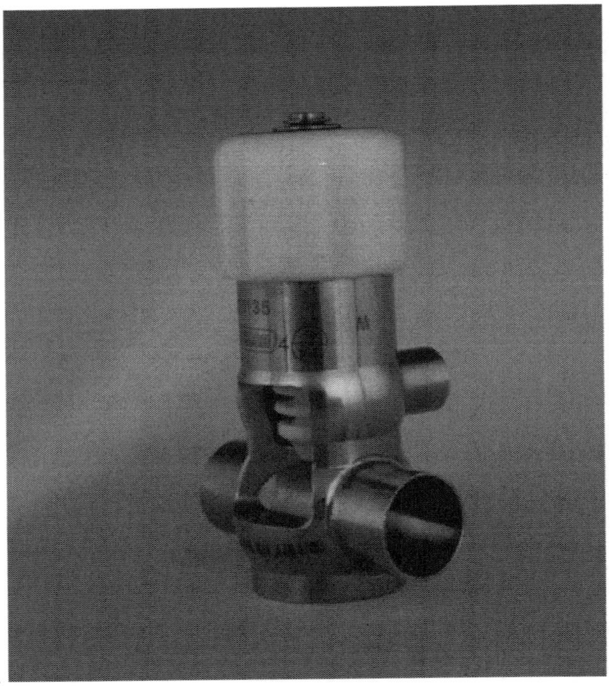

Figure 44.1. Dead leg–free valve. *Source*: Reproduced with permission of SÜDMO Schleicher AG, Riesbürg, Germany.

44.3.3.3 Instrumentation and Monitoring. The critical parameters to monitor are temperature and pressure. By measuring both, one can confirm that saturated steam conditions are present. Instrumentation must be of a sanitary design, which follows the general considerations given earlier. Temperature is usually measured with resistance temperature detectors (RTD) because their signal can be recorded, they are robust, and they can be accurately calibrated. A calibration accuracy of no more than $\pm 0.5°C$ is recommended because such variations significantly affect the sterilization effectiveness.

44.3.3.4 Clean Steam Supplies. If the SIP application is for parenteral pharmaceutical products, high-purity steam, commonly known as clean steam, must be specified. Product contact surfaces in such applications must meet only strictly controlled media. Clean steam is usually defined by the purity of its condensate, and this must usually have

WFI quality. Normal plant steam will contain more dissolved solids and potentially also some suspended debris. From these discussions, the steam supply should in any case be free of air, which is normally the case even with plant steam.

The steam supply system (boiler and distribution headers) normally is at higher pressure than the desired steam sterilization pressure. The steam connection to the equipment will then include a pressure-regulating valve to reduce the pressure. This design is preferred because the supply system then has reserve to supply large amounts of steam, without a serious loss of steam pressure. Too large a pressure reduction should not be expected from the pressure regulation valve, however. When too large a reduction is made just before entering the equipment, the steam acquires significant superheat. Superheated steam is known to be less effective as a sterilizing agent and as a heat transfer medium. Its presence affects the sterilization kinetics but is not easily observable. An industrially based guideline is to design the pressure reduction for 2:1 (10).

Sizing the steam supply for a steam-in-place is not straightforward, and there is no consensus in the industry. The greatest demand for steam occurs during the air purge and subsequent heat-up phases. The demand during the air purge is really determined by the restrictions to flow, often created largely by the vents and drains, as is discussed in the next section. During the heat-up step, the capacity of the steam supply could limit the heat-up rate, but the actual rate is not normally considered a critical variable.

44.3.3.5 ***Steam Traps and Vents.*** From the previous process description, it is clear that these components are critical to a successful steam-in-place operation. Steam traps are devices that are designed to not pass steam but rather air and condensate. There are many types available (11), and some models have been created especially for the biotech industry (Fig. 44.2). These devices operate as valves that are self-actuating, can sense by physical property differences whether they are passing steam, air, or water, and thus open or close appropriately. For steam-in-place applications, it is critical that they remove air and condensate. However, the steam trap designs that have evolved from other industries have the major goal of conserving steam. The types commonly selected for biotechnological applications allow some accumulation of condensate to occur upstream and may bring a significant restriction to the flow of air out of the system.

As a consequence, pharmaceutical companies have in the past dispensed with the use of steam traps and employed their own simpler solutions. The simplest solution is to install manual valves instead of the steam traps, which must then be manually adjusted during the course of the procedure. During the air purge step, the valves may be fully opened. They are subsequently throttled, but only so far

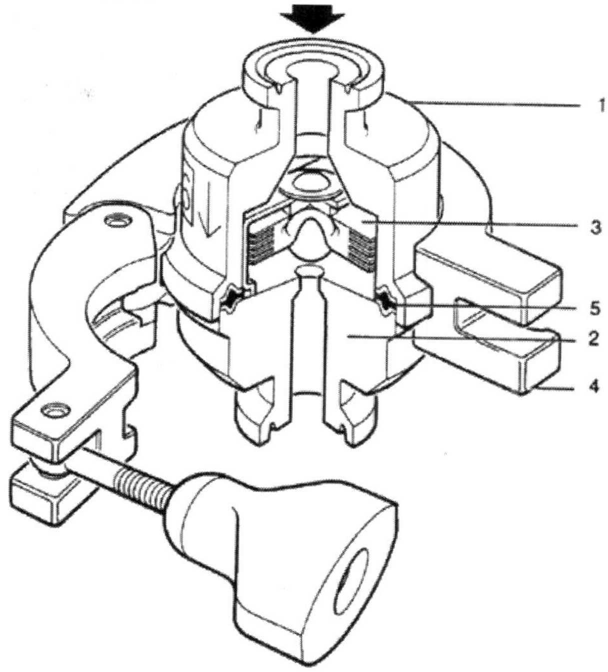

Figure 44.2. Sanitary blanced pressure steam trap. *Source*: Reproduced with permission of Spirax Sarco GmbH, Konstanz, Germany.

that steam continues to emerge, that is, condensate does not back up. A more automated approach is to employ restriction orifices instead (Fig. 44.3). During the air purge step, the orifice is bypassed with an open isolation valve to allow a relatively unrestricted flow of air and condensate. During the subsequent steps in the process, the isolation valve is closed, forcing remaining flow through the orifice. The orifice must be custom sized to ensure is large enough to allow all condensate to freely drain. Correctly sized, the orifice will be continuously passing (and wasting) steam and will prevent any buildup of condensate.

It is usual now to encounter sanitary steam traps of the type illustrated in Figure 44.2. Provided one realizes their limitations, successful steam-in-place operations can be performed with them. Typically, the steam traps are placed a significant piping distance from the limits of the system to be sterilized. The limit is usually an isolation valve, which will close immediately on completion of the operation. The piping downstream of this valve should then be noninsulated to allow for a sufficient condensation of steam and subcooling of condensate to keep the steam trap open. Steam traps must also be selected for the ability to pass air, and a sufficient number must be planned to allow for a successful air purging operation and condensate removal.

44.3.3.6 ***Filters.*** Sterilizable enclosures are almost always equipped with sterile filters, acting as barriers for sterile exchange of air or other media with the outside.

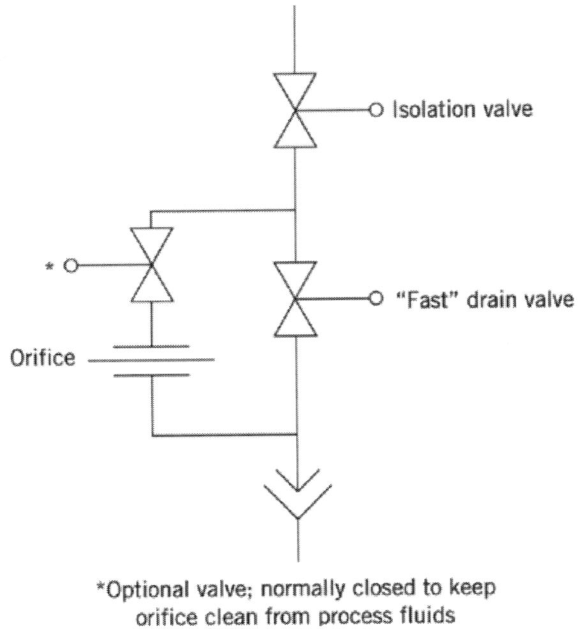

Figure 44.3. Use of orifices in place of steam traps.

Sterile vent filters are needed for pressure equalization, and process liquids are introduced though sterilizing liquid filters. Sterile filter installations are often sterilized together with the enclosure, although there is also no consensus on this practice. The filter element and enclosure may also be sterilized separately, but this raises concerns about ensuring that the separately sterilized components can subsequently be connected together aseptically.

The steaming-in-place of filters is not straightforward (12). Of special concern is the passage of steam, air, and condensate through the filter element. Also, the pressure drop across the filter element is limited by design. These concerns result in special piping designs and controls to ensure that both sides of the filter receive sterilizing conditions. It is not practical to attempt only a sterilization of the future sterile side of the filter, because without flow through of steam the filter will not be air purged successfully.

When steaming-in-place an enclosure, the vent filter will not be employed as the major vent for the air purge operation nor will it be used as the connection to the steam supply because it is typically too limited in flow capacity. It usually receives a secondary source of steam from the vessel during the sterilization process. Vent filter elements are typically nonwetting, which allows them to be quickly functional after steaming. A sterilizing liquid filter is sometimes steamed through to the enclosure (i.e., becomes the source of steam for the enclosure). Because of its limited ability to pass steam, the enclosure downstream is limited in size, often only piping. Liquid filters do not easily pass air after steaming because their hydrophilic pores become blocked with condensate.

44.4 ALTERNATIVE SIP TECHNOLOGIES

Equipment design requirements for steaming-in-place constrain opportunities for applying SIP, which raises interest in alternative technologies that are less demanding on mechanical designs. By employing chemical agents, sterilizing conditions can be reached without elevated temperatures or pressures. Sterilizing methods using radiation are not usually considered for SIP applications because the distribution of radiation throughout an extended enclosure is not practical.

When a new alternative SIP technology is chosen, a large amount of development work follows to make it a reliable industrial process. Much of this work involves scientific studies of the sterilization effect on the diverse microflora potentially present. A large amount of work is needed to develop monitoring and testing procedures for the validation and subsequent monitoring studies. Chalumeau has outlined the concerns raised with alternative sterilization methods in general (13). Some representative SIP methods receiving industrial attention are now used as examples for the technical issues encountered with alternative methods.

44.4.1 SIP with Superheated Water

The application of superheated water is an obvious extension to the use of saturated steam. This method can rely largely on the scientific knowledge base of the thermal sterilization effect with saturated steam. Häggström has reported its effectiveness (14), but it has not been widely practiced. The advantage of using superheated water over steam is that it can be pumped and reheated to distribute the thermal conditions needed. Häggström claims that no steam traps or vents are needed and that a more uniform temperature distribution is attained.

A theoretical analysis (15) points to a fundamental weakness of the technology when compared with steaming-in-place. Heat transport by conduction through water is less effective than the combined heat and mass transport occurring in steam. The analysis considered heat transport in dead legs of hot WFI systems, where significant temperature drops of the order of 10°C were predicted for typical dead legs. Although acceptable for WFI operation, such temperature variations are very unattractive for SIP. The advantages of superheated water have no benefit for the dead legs in the system, which are typically the most difficult to SIP. Also, to properly heat an enclosure with superheated water, it must be completely filled, or the headspace must be purged of air as in a classical steam-in-place process; this means that steam traps and vents are still required, although fewer would be needed. When the technical features are considered that would be needed to prepare and then discard large amounts of superheated water, it is not surprising that this method has not gained popularity.

44.4.2 SIP with Chemical Agents in a Liquid Phase

Chemical agents used for desinfection or sanitization are sometimes considered for SIP. Sanitizing solutions are not generally useful as sterilizing agents for SIP. To be useful they would have to have biocidal activity against the full spectrum of microorganisms yet be relatively noncorrosive and safe to handle. Commercially available agents that approach this ideal are often termed cold sterilants but still require many hours of exposure to achieve sterilizations.

In general, one must design a system to uniformly distribute the agent, monitor the operation for the proper exposure, and finally remove the agent, so that the equipment can be put back in operation. Monitoring of the operation must consider all parameters that can significantly affect the chemical reactions that determine the sterilization kinetics. Temperature may be important, but the concentration of the agent certainly is. Other material components or chemicals present can potentiate or interfere with the desired reactions. Unless a sensor is available, routine sampling and subsequent chemical analysis are required to verify sterilizing conditions.

High-purity water systems, such as those claiming to be sterile, have been reported using ozone as a circulating agent (16). Ozone has many advantages as a sterilant. It is extremely effective even at low concentrations because of its high reactivity, and it is easily generated electrically from ambient oxygen. In water systems, its decomposition product is also oxygen, which is usually not considered a contaminant. Before use of the media, residual ozone is catalytically decomposed with the aid of ultraviolet radiation. Widespread acceptance of this agent, even for simple microbiological control, has not been reached because of some underlying problems. First, it is an extremely aggressive agent that will degrade the piping system, especially gaskets. Second, it is difficult to remove it to levels sometimes desirable for end use. (See analyses in Ref. 15. Finally, penetration of chemical agents such as ozone into dead legs can be a difficult problem. Where diffusion dominates the transport, the relatively low liquid-phase diffusivities coupled with the low bulk concentrations result in very slow penetration of agent into dead legs (15).

At another extreme, caustic (sodium hydroxide) at relatively high concentrations can be considered as a sterilizing agent. Caustic is often used as a cleaning agent, such as in a CIP operation. If the concentration is high enough, it can be used as a sterilizing agent. Validation of the sterilization effect can be difficult because of the lack of scientific knowledge regarding such use of this agent with microorganisms. Also, the agent must be washed from the installation with large amounts of sterile water. However, it can be appropriate if the goal is to inactivate a biological agent left in equipment or in a waste stream before opening to the environment. Here, the removal of caustic after the operation is not critical; it will be subsequently removed in a washing step.

44.4.3 SIP with Chemical Agents in a Gaseous Phase

The use of gaseous chemical agents is probably the most active area of interest for alternative SIP technologies. Unlike liquid-phase agents, gaseous agents are easier to distribute and remove. There is also a long history of use, and consequently a significant knowledge base, with some of the biocidal gases. Formaldehyde and ethylene oxide have long been used for room and equipment disinfections in hospitals. As with all chemical agents, their use requires measurements of all important reaction conditions for the sterilization. One must also determine their reactivity with any system components and the ultimate fate of their decomposition products.

Both formaldehyde and ethylene oxide have become restricted in their use because of their toxicity and their activity as carcinogens. Ethylene oxide also has physical hazards; not only is it flammable, its chemical instability can lead to explosive decompositions. Recently, vaporized hydrogen peroxide as a sterilizing agent has been developed (17) to the point of a commercially available generator for SIP applications (Fig. 44.4). Its commercial success has been spurred by the active interest in the use of isolator technology for aseptic processing. Haas (18) has

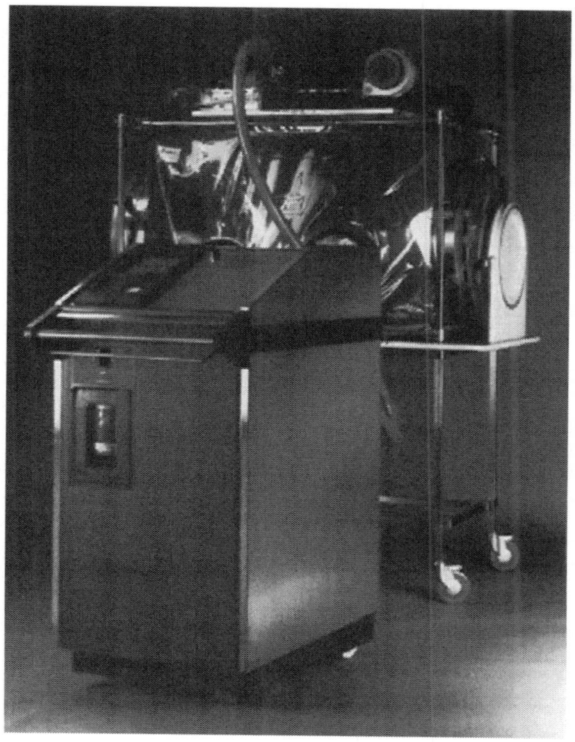

Figure 44.4. Vaporized hydrogen peroxide generator. *Source*: Reproduced with permission of AMSCO Scientific, Apex, North Carolina.

reviewed isolator design issues and listed gaseous agents that have been attempted for their SIP. At present, there does not appear to be a more promising chemical sterilant than hydrogen peroxide, which is likely to become the method of choice for pharmaceutical applications.

The AMSCO VHP™ systems have a basic three-step cycle: dehumidification, sterilization, and sterilant removal. Their commercial systems perform these functions on an attached isolator or enclosure. Often the isolator has operating components that assist the process. The reader is directed to the vendor for specific details on the operation of the commercial AMSCO VHP™ systems; the basic physical and chemical phenomena are considered here.

Hydrogen peroxide solutions have long been recognized for their disinfecting value. The greater lethality of vaporized hydrogen peroxide is a more recent discovery (17). H_2O_2 is an unstable material, available commercially in aqueous solutions up to 70%. High concentrations of H_2O_2 require proper handling because it can explosively decompose or cause fires (19). Its decomposition products are water and oxygen, and it will react with any oxidizable substance, particularly organic materials. Aqueous solutions of H_2O_2 are corrosive, but they have been found to be relatively benign to materials when only vapors are present (17).

A relatively safe solution concentration of 30% is usually the raw material consumed by a H_2O_2 generator. H_2O_2 has a lower vapor pressure than water, and considerable water vapor will also be generated when a 30% solution is vaporized. The water vapor released with the H_2O_2 limits the gas concentration of H_2O_2 achievable because water can easily reach saturation conditions in air. When water condenses, H_2O_2 is also lost from the gas phase because it will dissolve readily in the condensate. SIP with vaporized hydrogen peroxide must therefore control not only the concentration of H_2O_2 but also water in the air when the solution is vaporized. The first step in such an SIP is usually to dehumidify the enclosure so that reproducible initial conditions are established. The temperature of the isolator and the air within is also important because this has a strong effect on the relative humidity. The kinetics of the sterilization reaction do not appear to be strongly a function of temperature.

During the sterilization step, the H_2O_2 reagent is vaporized, and the vapors are introduced into the chamber. The chamber must be leak-tight to prevent emissions and exposures to personnel. Forced convection is needed to distribute the vapors throughout the enclosure and ensure a uniform exposure to, and concentration of, H_2O_2. To maintain the gas concentration, additional H_2O_2 must usually be generated for the duration of the sterilization to replace H_2O_2 that has decomposed. The sterilization step is timed to achieve the desired level of assurance of sterility. This activity on a closed system results in a steady rise in both water vapor content and air temperature. The commercial generators are successful in part because they continuously remove water vapor from the returning gas stream, as they are generating more H_2O_2 in the supply stream. They are thus able to achieve and maintain the necessary H_2O_2 concentrations needed for a rapid sterilization (0.5–2 mg/L) while avoiding reaching water saturation conditions.

As the air temperature rises during the sterilization, a temperature difference develops between the air and the enclosure and its fixtures. Saturation conditions are likely to occur at these surfaces, and process development efforts must focus on avoiding this condition by parameter adjustment. SIP with gas requires good gas circulation to achieve a uniform exposure. Modern isolator designs (18) usually have the necessary vigorous air circulation needed because they employ it to also remove airborne particles via HEPA filtration. Fixtures inside an isolator may have restricted spaces where air circulation is ineffective. Transport of biocidal gas into these spaces then is dependent on diffusion, and dead leg effects for transport can be expected.

Graham *et al.* (17) reported sterilization conditions could be reached up to 60 pipe diameters into a dead leg with a pulsed cycle, which included a vacuum pulldown between cycles. The vacuum pulldown ensures that transport into the dead leg is not entirely dependent on diffusion because breaking the vacuum with sterilant creates a pressure difference for convection up the tube. Most isolators are not designed for vacuum conditions, and reliance on diffusion reduces this impressive penetration result.

The analysis of diffusional transport of H_2O_2 into dead legs can be modeled analogously to the transport of ozone into dead legs (15). In the gas phase, diffusional transport is much more rapid, but the decomposition kinetics of H_2O_2 are also larger than for ozone. The diffusivity of H_2O_2 in air is reported to be 0.188 cm^2/s at 60°C. For such gas systems, the temperature dependence of the diffusivity should be of the order of $T^{1.5}$. This corresponds to a value of 0.167 cm^2/s at 35°C. In comparison to ozone in water, such gas-phase diffusivities are four orders of magnitude larger. The defining parameter for penetration is the diffusivity divided by the rate constant for decomposition (15). Because H_2O_2 decomposition is observable in terms of minutes and ozone decomposition in hours, one would expect that the penetration of H_2O_2 into dead legs would be dramatically better than predicted for ozone in water (15). An industrial guideline to use for limiting dead leglengths with H_2O_2 has not yet evolved but would be strongly affected by the specific conditions created by the enclosure, the sterilization cycle, and the material of the dead leg, and hence not generally useful.

Finally, it must be realized that the SIP of an enclosure or isolator has the main goal of sterilizing the exposed surfaces. The air in the enclosure is usually recirculated through a sterilizing HEPA filter to keep microbes out of the air, and it therefore does not have the major bioburden. Surfaces must be clean. Microorganisms encased in dirt or a biofilm will not be exposed to the gas-phase H_2O_2 concentration because of transport resistance through the film and decomposition reactions occurring with the film. Reactive surface materials will in general reduce the local concentration of agent at the surfaces.

After the sterilization step, a H_2O_2 removal step is needed before normal operations can resume. The AMSCO VHP™ systems employ a catalytic reactor in the recirculating air stream to the enclosure to accelerate the decomposition and removal of H_2O_2. Plastic components are generally porous to H_2O_2, and thus a significant length of the cycle time often is devoted to removing the outgassing amounts of H_2O_2. This phase is again a diffusional process; but now in a solid phase and hence relatively slow.

The permeability of plastics to biocidal gases, notably ethylene oxide, has been relied on to sterilize materials sealed in plastic bags via exposure from the outside. This method has also been tested successfully with H_2O_2, but a vacuum pulsed cycle is needed (17); this step is not normally considered an SIP operation. A sterilization chamber, similar in concept to an autoclave, would be needed to alternatively evacuate the air and introduce H_2O_2.

REFERENCES

1. T. Myers and S. Chrai, J. Parenter. Sci. Technol. 35: 8–12 (1981).
2. J. Agalloco, J. Parenter. Sci. Technol. 44: 253–256 (1990).
3. K.H. Wallhäuser, Praxis der Sterilisation Desinfektion-Konservierung, 4th ed., G. Thieme Verlag, Stuttgart, Germany, 1988.
4. FDA, *Code of Federal Regulations Title 21*, Ch. 1, part 606.60 (Nov. 18, 1975).
5. P.T. Noble, Biotechnol. Prog. 8: 275–284 (1992).
6. R.B. Bird, W.E. Stewart, and E.N. Lightfoot, Transport Phenomena, Wiley, New York, 1960.
7. D. Seiberling, in W.P. Olson and M.J. Groves eds., Aseptic Pharmaceutical Manufacturing, Interpharm Press, Prairie View, Ill., 1987, pp. 247–314.
8. J.H. Young and B.L. Ferko, J. Parenter. Sci. Technol. 46: 117–123 (1992).
9. C.E. Meyrick, Pharm. Eng. 9: 20–27 (1989).
10. B.D. Reeks, British Parenteral Society Tutorial Booklet No. 2: The Validation of Steam Sterilisers, P.S. Swindon, Wiltshire, 1990, p. 11.
11. D. Coleman and P. Smith, Pharm. Eng. 12: 8–14 (1992).
12. T. Myers and S. Chrai, J. Parenter. Sci. Technol. 36: 108–112 (1982).
13. H. Chalumeau, J. Parenter. Sci. Technol. 47: 9–15 (1993).
14. M. Häggström, Biotechnol. Forum Eur. 3: 164–167 (1992).
15. P.T. Noble, J. Parenter. Sci. Technol. 48: 108–112 (1994).
16. C. Nebel and W.W. Nezgod, Solid State Technol. 10: 185–193 (1984).
17. G. Graham, J. Rickloff, and J. Dalmasso, in PDA, Inc., Proceedings of the International Congress, Basel, Switzerland 1992, pp. 32–51.
18. P. Haas, Pharm. Technol. 2: 26–38 (1995).
19. J. MacKenzie, Chem. Eng. 6: 85–90 (1990).

PART VII

FDA cGMP REGULATORY COMPLIANCE

45

PHARMACEUTICAL BIOBURDEN TESTING

NATHANIEL G. HENTZ, PhD

North Carolina State University, Golden LEAF Biomanufacturing Training and Education Center, Raleigh, North Carolina

45.1 INTRODUCTION

Pharmaceutical manufacturing demands rigorous testing to meet various safety, quality, and regulatory requirements (1). This testing is employed throughout the entire pharmaceutical manufacturing process including raw materials, in-process, fill/finish, and final product. Bioburden or microbial limit testing, specifically, has taken on an increasing quality check role in key areas throughout the production process, including initial acceptance of incoming raw materials and components, evaluation of processes (assembly, cleaning, packaging, etc.) and manufacturing steps (including final fill), evaluation of changes to a product or process, validation of cleaning or sterilization processes, development and monitoring of aseptic operations, and monitoring and validation of water systems.

While control of microbial contamination is required for nonsterile products (2), bioburden testing is also especially important for biopharmaceutical products considering that the processes involve nutrient-rich materials that can sustain microbial growth. Since 1982, the number of biopharmaceutical (recombinant proteins such as monoclonal antibodies and nonrecombinant products such as blood and vaccines) has increased steadily from 1 approved product to 19 in 2009. Although there was a decline to 12 new approvals in 2011, the overall trend indicates that biopharmaceuticals will continue to play an increasing role in approved drugs (3). Furthermore, it is predicted that the number of biosimilars will increase over the next several years as several innovator biopharmaceuticals reach patent expiration.

For the purposes of this chapter, bioburden is defined as the degree of microbial contamination, microbial load, or the number of viable microorganisms contaminating an object. The object can be a solid (e.g. surface, device, and vessel), liquid (e.g. water growth medium), starting material (e.g. growth medium components and mammalian cells), or final product (powder or liquid). Bioburden results are typically reported as colony forming units (CFUs) per volume, weight, or area sampled. For example, Water for injection has US Pharmacopeia limits of <10 CFU/100 mL, while an aseptic processing suite may have an Alert Level of 1 CFU/25 mm^2. Bioburden is classified as a total viable count (TVC) test, which is the sum of total aerobic microbial count (TAMC) and total combined yeasts and molds count (TYMC), each of which requires different media to support specific microbial growth. It is important to point out that there are a variety of methods for detecting and counting specific organisms (4).

Bioburden and microbial testing extends into such diverse industries as space exploration (5), medical devices (6), clinical (7), animal and veterinary (8), food (9), pharmaceuticals/biopharmaceuticals (10), and cosmetics (11). This chapter focuses on bioburden and microbial testing in the context of pharmaceutical and biopharmaceutical manufacturing.

Like other raw material's tests, bioburden testing of constituents demonstrates to regulatory agencies the usage of proper control measures during preparation and handling and whether they follow the necessary prerequisites of British Pharmacopeia, European Pharmacopeia, Japanese Pharmacopeia, or US Pharmacopeia. Validation for raw materials or intermediates of finished products may also

Downstream Industrial Biotechnology: Recovery and Purification, First Edition. Edited by Michael C. Flickinger.

be carried out, as it is a pharmacopeial prerequisite for the bioburden testing to be considered valid. Individual aliquots of sample in diluents are spiked with known levels of specified organisms, which is then proceed with the normal testing process alongside a control. Recuperation of organism, as compared with the control, has to be within specification or the test is considered invalid.

45.2 BIOBURDEN CONSIDERATIONS

Since microorganisms are usually present at very low concentrations, it is imperative that the bioburden testing protocols truly reflect the amount of contamination. Factors such as sample collection (e.g. swabs, plate, and diluents), extractants (e.g. saline, low-nutrients buffer, and water), type of neutralizer (e.g. polysorbate 80 and lecithin), time between collection and retrieval, removal methods (e.g. diffusion, ultrasonics, or vortex mixing), whether the microorganism is plated or filtered, or media type can all impact accuracy. All these factors need to be considered when designing and validating a test plan. It is also imperative that the appropriate model organisms be tested under these conditions, as each manufacturing facility should be considered unique. Whatever methods are chosen, they need to be carefully studied, validated, and justified. Regulatory agencies expect current standard industry practices.

45.2.1 Sample Collection

There are many methods of collecting samples: swabbing, contact plates, strips, rinse, and air. Each method has its advantages and disadvantages. For example, swabbing is incredibly useful for uneven or curved surfaces, cracks, crevices, and so on. However, the swab design and material can heavily impact sample collection. Also, the swabbing pattern and sampling area need to be evaluated and consistent. Contact plates are convenient in that the sampling area is always defined and there is no transfer step. An added benefit is that the same plate design can often be used in air samplers (12,13). Rinse solutions are necessary for tanks, vessels, tubing, and so on, but the volume of rinse solution tested can impact the results. When using rinse solutions, the amount tested is always a subset of the actual sample and may not represent the actual microbial counts. Tape-lift systems are a direct method and can be convenient, but they can be difficult to handle and analyze. Once the sample is collected, the samples are affixed to a microscope slide and manually counted by a microbiologist.

45.2.2 Extractants

Of primary importance to swab sampling is that once swabbed, the microorganisms need to be removed and transferred to a growth medium. The type of extractant can negatively impact microorganisms that may be in a sensitive state of shock, especially if sampling occurred immediately following a cleaning or disinfection procedure. Furthermore, the microorganisms may come from an oligotrophic environment. Typical extractants are benign solutions such as saline, saline–peptone, Cary-Blair medium (14), Amies medium (15,16), Stuart medium (17), or universal transport medium (18). There is a fine balance in keeping the microorganisms alive, but not replicating.

45.2.3 Neutralizer

In clean room environments, surfaces are regularly cleaned and disinfected. During sampling, it is entirely possible to transfer the biocidal agent as well as any microorganism that can inhibit growth of viable microorganisms on the growth medium. Some classes of biocidal agents (e.g. phenolics, alcohols, and gluteraldehydes) can be neutralized simply by sample dilution. Other classes such as quaternary ammonium compounds can be neutralized by lecithin and polysorbate 80 and mercurial biocides by thioglycollate. These and other neutralizers have been demonstrated to be effective while minimally effecting microbial growth on nutrient agar (19,20). There are other types of neutralizers, but whatever is chosen, needs to be demonstrated as effective.

45.2.4 Storage Time

Once bacterial samples are collected, it is suggested that the samples be tested within 2 h; if that is not possible, the samples shall be refrigerated at 2–8°C for up to 12 h to maintain proper microbial attributes. It is recommended that samples be tested no longer than 48 h after collection (21) because of drop in viability.

45.2.5 Sample Removal

Sample removal concerns apply primarily to swab samples. Many swab designs and materials exist; however, they may not all behave the same. For instance, the newer nylon-flocked swabs have been shown to release 90% of the microorganisms picked up by the swab. However, different matrices such as food or bacterial species may alter the outcome. The typical method of sample removal from swabs is to vortex the swab in a solution of benign transport medium such as isotonic saline solution or saline–peptone. Depending on the sample matrix, sonication may be necessary. However, care must be taken to not damage the microorganism viability during sonication. Another method of sample removal is to make direct contact between the swab and the plate containing growth medium and agar. In this case, it is standard practice to swab, while rotating the swab tip, down the face of the agar plate. Finally, another consideration is how much of the transport liquid to sample once the swab has been mixed in the transport medium.

45.2.6 Agar Plate versus Membrane Filtration

Membrane filtration is inherently more sensitive because liquid samples are not limited by volume, only by clogging the pores by matrix components. Since 0.45-μm membrane filters are typically used, it is possible to lose smaller bacterial species into the filtrate. Agar plates, on the other hand, capture everything that is presented to them. For the pour plate method, the temperature of the molten agar cannot be more than 45°C. Even at that temperature, there is risk in altering microbial viability. The spread plate method, while gentle, can be limited in sensitivity because of the sample size restriction of about 100μL. Since the pour plate method can handle as much as 1 mL of sample, it is inherently more sensitive.

45.2.7 Growth Media Type

Since media affects the growth of microorganisms to such a large degree, the correct choice must be made depending on the situation. The most common general growth medium is tryptic soy or soybean casein digest (SCD). Most heterotrophic and mesotrophic bacteria grow well in this nutrient-rich medium. However, bacteria isolated from water may grow better in a low-nutrient medium where R2A medium (22) is commonly used in industry. Various selective media may be appropriate, especially if trying to broadly categorize microorganisms or when testing for *objectionable* organisms. One common selective medium is Sabouraud dextrose agar (SDA or malt extract agar), which allows growth of yeasts and molds, while minimizing bacterial propagation. Some other examples of such media include mannitol salt agar, which is selective for gram-positive staphylococci, and MacConkey agar and eosin methylene blue (EMB) agar, which are selective for gram-negative bacteria. Differential media, on the other hand, allows multiple microorganisms to be distinguished based on colony appearance. Typically, the media have an indicator dye sensitive to an active metabolic pathway specific to different microorganisms. An example of such is the lactose or mannitol sugar metabolism and subsequent pH change in EMB/MacConkey agar and mannitol salt agar, respectively. M-Endo is another common example (23–25) of a selective and differential medium, where coliforms are detected by the aldehyde formation from lactose metabolism. In this case, colonies with a metallic green sheen are visible and are thus classified as coliforms.

Blood agar contains mammalian blood (usually sheep) and is used to isolate fastidious organisms and detect hemolytic activity. The categories of hemolytic activity are β-hemolysis (complete hemolysis and digestion of blood cells), α-hemolysis (partial blood cell lysis and digestion), and γ-hemolysis (complete lack of hemolysis) (26).

45.3 STANDARD INDUSTRY METHODOLOGIES

45.3.1 Membrane Filtration Method

Liquid samples are filtered through a membrane with a nominal pore size of not greater than 0.45μm. In this method, liquid samples are typically diluted 1:10 by diluting 10 mL (or 10 g) of sample in a benign diluent such as phosphate buffer, pH 7.2, or other isotonic solution such as saline–peptone buffered to pH 7.0. ISO 6887-1 (27) recommends this medium as a diluent for the preparation of initial suspension for microbiological samples. It should be noted that the low concentration of peptone does not cause a multiplication of the organisms within 1–2 h of dilution of the sample. Alternatively, a SCD broth, otherwise known as tryptic soy broth (TSB), can be used as the diluent. Since SCD/TSB broths are rich in nutrients, microbial growth can be promoted, so dilution with either of these should be done just before filtration.

45.3.1.1 Procedure. There are two types of membrane filtration configurations. The first type is a conventional vacuum manifold capable of using standard 47-mm membrane filters, which is commercially available through many scientific suppliers. In this configuration, the vacuum manifold, membrane support, and funnel are manually assembled (Fig. 45.1) and need to be autoclaved before use. This type of arrangement can be in multiples of three to six funnel ports. Careful retrieval of membrane filter and placement on the agar with growth medium of choice is also needed. The other configuration uses preassembled, sterile components, and after the filtration step is finished, the membrane/funnel unit is snapped together with the presterilized cartridge containing medium of choice (Fig. 45.2). Although the cost per sample can be higher, this preassembled configuration minimizes potential contamination and reduces incidences of accidentally corrupting the membrane filter. Once the membrane filter and funnel are properly fixed, samples need to be diluted 1:10 in an isotonic solution, and the contents filtered through the membrane filter. The funnel and filter are then rinsed three times with sterile isotonic solution. After filtration and rinse, the filter is removed with sterile forceps and placed on agar medium (if using the in-house method) or the membrane/funnel cartridge is snapped into place with a cartridge preassembled with growth medium of choice, if using the MilliFlex® system or similar (Fig. 45.3). Since bioburden is a TVC test, it is important to remember to prepare enough sample for both TAMC [using tryptic soy agar (TSA)] and TYMC (using SDA). The TSA plates are

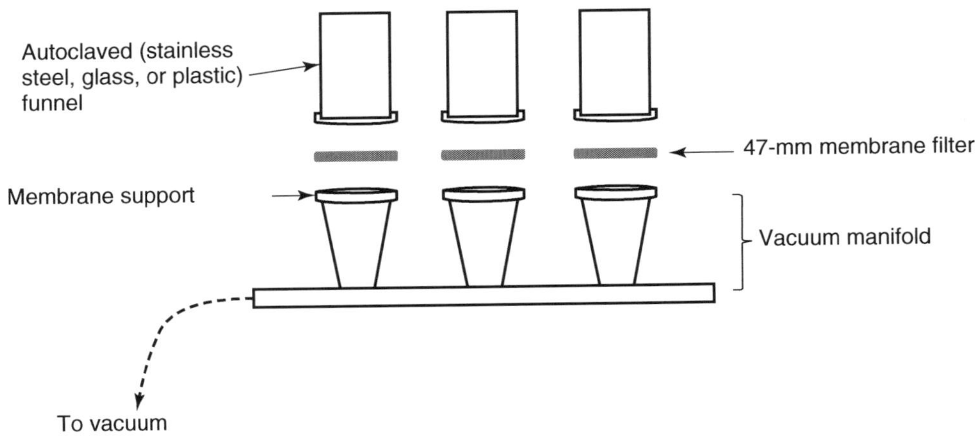

Figure 45.1. Example of manual membrane filtration device configuration.

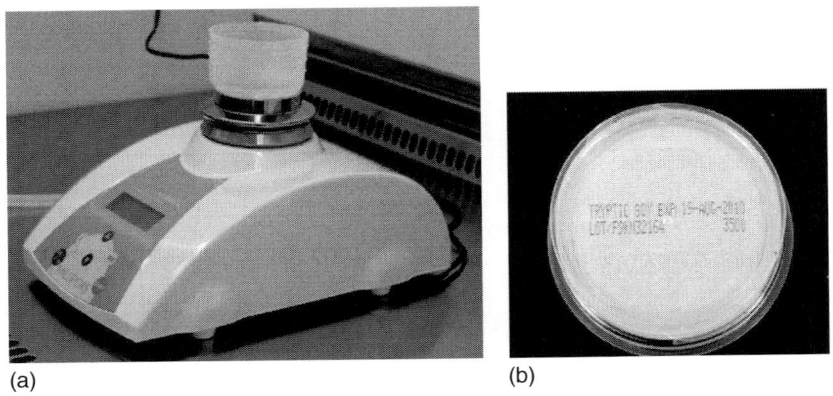

Figure 45.2. (a) Example of cartridge-based membrane filtration device configuration, including semiautomated vacuum housing, and (b) growth medium cartridge.

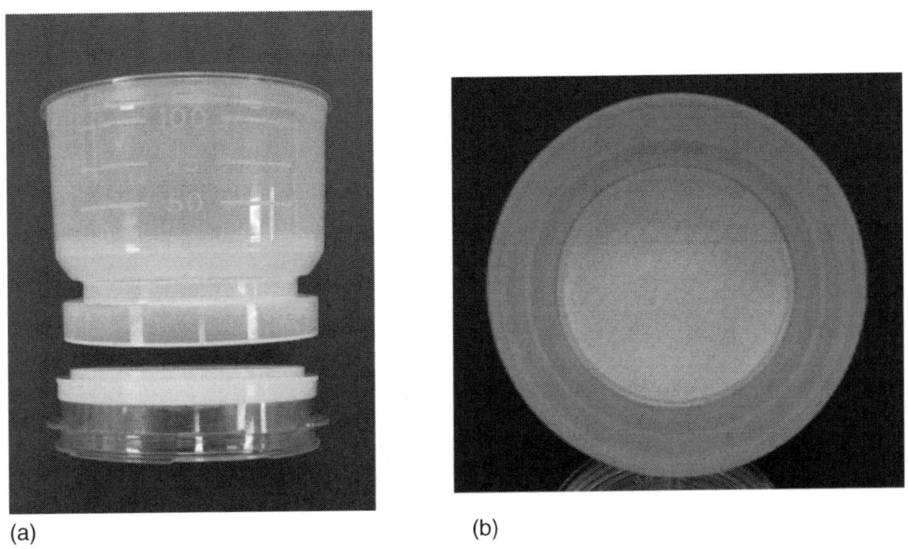

Figure 45.3. (a) Example of a presterilized membrane filtration cartridge assembly and (b) top view of the membrane filter with grid

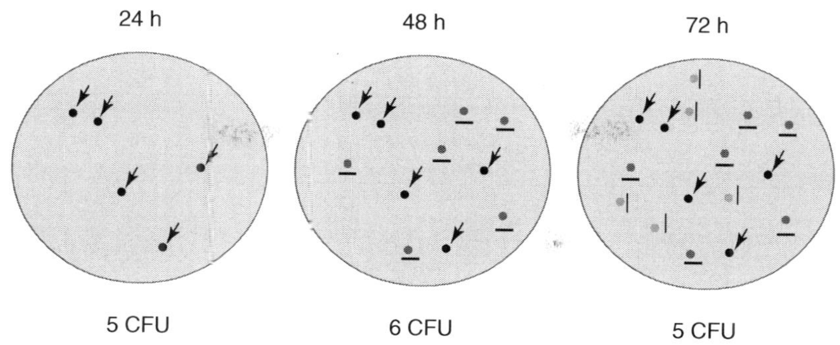

Figure 45.4. Example of colony counting over 3 days for a 10 mL sample mixed with 90 mL of diluent. The entire 100 mL of diluted sample was filtered through a membrane filtration device, rinsed and incubated at 30–35°C for 3 days. Step 1: Count colonies each day and mark counted colonies on plate; use a different designation or ink color for each day. Step 2: Add the colonies for each day to calculate the total microbial count: 5 + 6 + 5 = 16 CFU. Step 3: Divide by total volume of sample filtered and account for dilution factor: 16 CFU/10 mL = 1.6 CFU/mL

incubated at 30–35°C for 3–5 days, and the SDA plates are incubated at 20–25°C for 5–7 days. Finally, the colonies are counted daily over the incubation period and summed at the end of the designated incubation period (Fig. 45.4). Furthermore, there are instances requiring incubating one type of growth medium at different temperatures. In cases of low bioburden for organisms isolated from an oligotrophic environment, it is typical to incubate plates at 30–35°C for 2–3 days and then switch to 20–25°C for 5–7 days. If psychrophiles are of interest, plates may be incubated first at 20–25°C for 5–7 days, followed by incubation at 30–35°C for 2–3 days.

45.3.2 Plate Count Method

The plate count method consists of the spread plate and pour plate techniques and should be performed in duplicate (28) for each test medium. It is standard practice to use SCD/TSA for TAMC and SDA for TYMC, where the SCD plates used for TAMC are incubated at 30–35°C for 3–5 days and the SDA plates used for TYMC are incubated at 20–25°C for 5–7 days. Since it is recommended that the number of colonies counted be more than 30 and less than 300 per plate (29), samples may have to be diluted to accommodate.

45.3.2.1 ***Spread Plate Procedure.*** About 15–20 mL of molten growth medium containing SCD/TSA or SDA is added to each 9-cm diameter Petri dish and allowed to cool until it becomes a solid. After the plates are allowed to dry, no less than 0.1 mL of appropriately prepared sample is spread with a T- or L-shaped spreader over the surface of the agar and incubated at the desired temperature for 3–7 days (Fig. 45.5). It is standard practice to count colonies once a day over the 3–7 days, where colonies are marked by placing a dot of permanent ink over each colony as it is counted. Different color ink can be used to differentiate each day the plate is counted. At the end of the incubation period, the colonies from each day are summed and results are reported as the total CFUs per milliliter or gram of starting material. Counting each day over the incubation period can serve as an aid toward identification of the organism.

45.3.2.2 ***Pour Plate Procedure.*** This technique requires that 1 mL of appropriately prepared sample is added to each 100 × 15-mm (9-cm diameter as per USP) Petri dish followed by an overlay of 15–20 mL of molten SCD agar or SDA (of temperature not more than 45°C). When the agar solidifies, the plate is then incubated at the desired temperature for 3–7 days (Fig. 45.6). It is standard practice to count colonies each day for 3–7 days, where colonies are marked by placing a dot of permanent ink over each colony as it is counted. Different color ink can be used to differentiate each day the plate is counted. At the end of the incubation period, the colonies from each day are summed and results are reported as the total CFUs per mililiter or gram of starting material.

45.3.3 Most Probable Number Method

It is generally recognized that the most probable number (MPN) method is the least precise and accurate of various methods, but certain situations may require its use. This method relies on observed turbidity as the end point, where samples are diluted 1:10, 1:100, and 1:1000 in triplicate and incubated at 30–35°C for not more than 3 days.

45.3.3.1 ***Procedure.*** First, prefill each of 14 test tubes with 9 mL of SCD/TSA growth medium. Label two tubes as negative controls, four tubes as "0.1 mL" (one of which is additionally labeled "A"), four tubes as "0.01 mL" (one of which is additionally labeled as "B"), and three tubes as

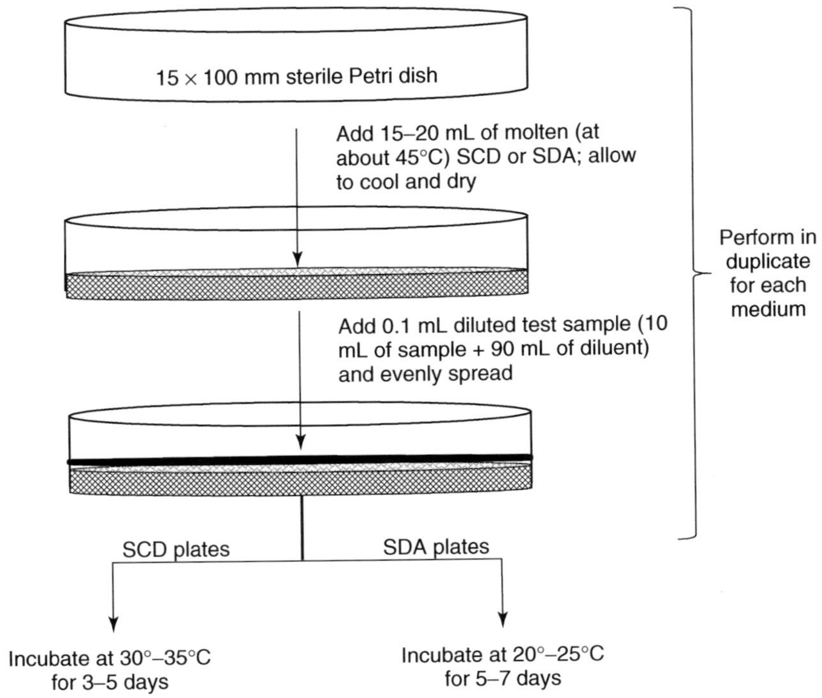

Figure 45.5. Schematic of the spread plate method

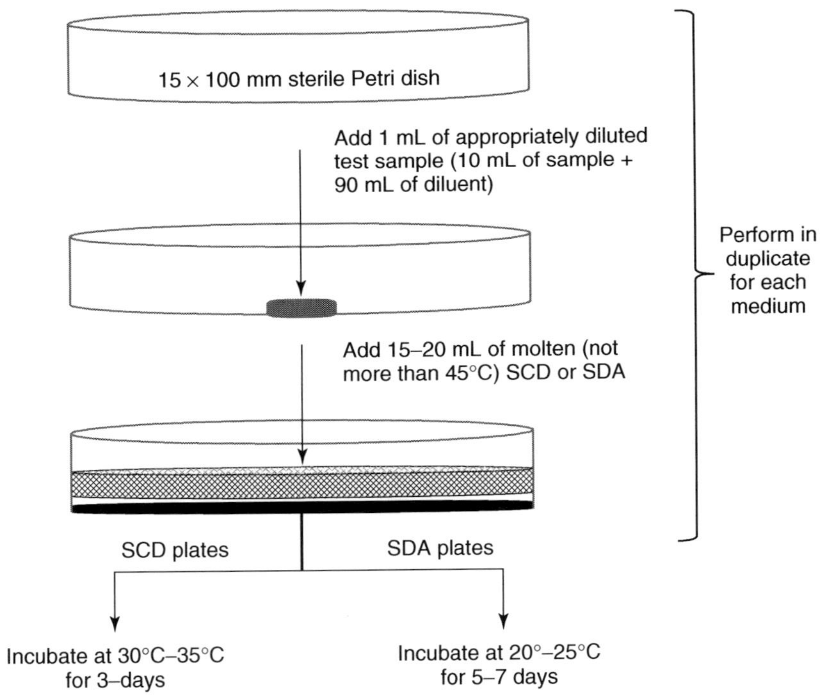

Figure 45.6. Schematic of the pour plate method

"0.001 mL". Next, prepare the sample by diluting 10 mL of neat sample in 90 mL of SCD in a separate sterile flask. To each of the "0.1 mL" test tubes, add 1 mL of the diluted sample and mix. From the "A" tube, dispense 1 mL of "0.1 mL" diluted sample into the "0.01 mL" tubes and mix. From the "B" tube, dispense 1 mL of the "0.01 mL" diluted sample into the "0.001 mL" tubes and mix. Discard Tubes "A" and "B" and incubate the remaining tubes, including the negative controls, at 30–35°C for not more than 3 days (Fig. 45.7). The results are interpreted as

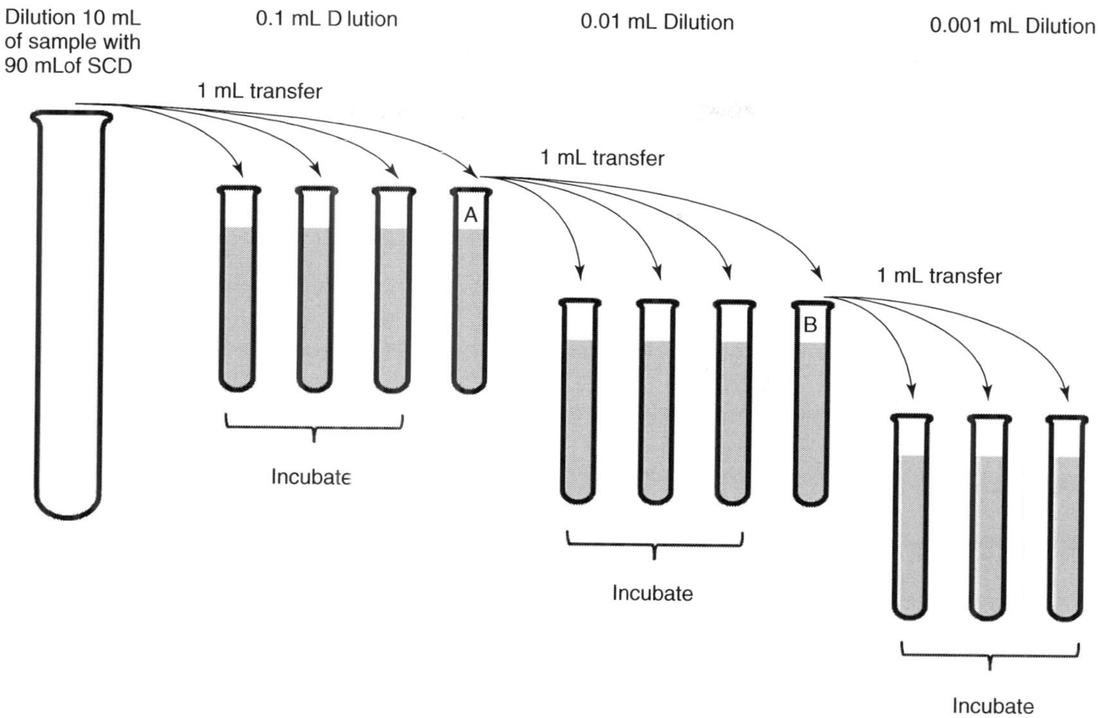

Figure 45.7. Schematic of the most probable number method. Step 1: Dilute 10 mL of sample with 90 mL of SCD medium. Step 2: Prefill all sample tubes with 9 mL of SCD medium and label tubes. Step 3: Perform sample transfer as indicated in schematic. Step 4: Incubate at 30–35°C for not more than 3 days. Step 5: Inspect tubes for turbidity and compare to Table 45.3

positive if the solution is turbid and microbial growth is visible. The MPN is estimated by comparing to Table 45.3 (most-probable-number values of microorganisms) (28) (refer to Fig. 45.8 for example of interpretation).

45.3.4 Surface Sampling Methods

Since it is not practical to sample 100% of the surfaces in the room to be monitored, the sampling that does take place must be enough to accurately represent the entire sampling area. To determine how many sampling sites a room shall have, ISO14644-1 (30) states that the number of sites (N) is equal to the square root of the area ($\sqrt{A}$) of the clean room in square meters. For example, consider a clean room with dimensions of 20 ft ×35 ft. First, convert the units to meters: 1 ft = 0.305 m, therefore 20 ft = 6.10 m and 35 ft = 10.7 m. Next, calculate the area: 6.10 m ×10.7m = 65.0 m^2. Then, take the square root of the calculated area: $\sqrt{65.0} = 8.06$. Finally, round to the nearest whole number: 8 total sampling locations within the example clean room. The exact location for each site should be determined based on several factors including work flow, air flow, and overall risk of contamination. Examples of relevant sampling sites include key boards, control panels, floor drains, sinks, handles, air vents, spaces around doors, and high traffic areas.

Once the location and number of sampling sites have been determined, the target microorganisms must be collected and recovered with high efficiency. Between collection and recovery, samples will need to be transported and requires an inert medium such that microorganisms are maintained alive, yet prevented from multiplying. Another consideration is that the microorganisms will likely be present in low numbers and stressed because the sampling environment has probably been subjected to a cleaning/decontamination regimen.

There are a number of designs for sample collection: contact (or RODAC™) plates, swabs, tapes, and rinses.

45.3.4.1 ***Contact Plates.*** The most common method for surface sampling is the RODAC™ (Replicate Organism Detection and Counting) or contact plate. The contact plate is a plastic dish shaped such that when filled with agar gives a convex surface with an area of 25 cm^2 (Fig. 45.9). This design allows for the total surface contact of the area being sampled. The plates are prefilled with culture media and may contain neutralizers (such as soy lecithin and/or polysorbate 80) reported to inactivate residual disinfectants where the sample is being collected (31–34). Contact plates are the preferred sampling method for flat surfaces such as tabletops, walls, benches, floors, garments, and gowned personnel. Once the plate comes into contact

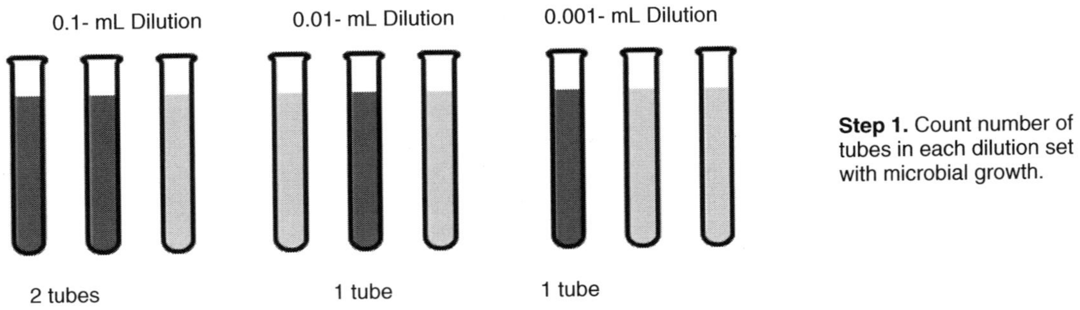

Step 2. Compare to Table 3 in USP 34, <61> or other harmonized compendial method.

Table3. Most Probable Number of Microorganisms (Partial tableadapted from USP 34, (61).

Number of Tubes with Observed Microbial Growth			MPN per mL of Sample	95% Conf. Limit
0.1 mL Set	0.01 mL Set	0.001 mL Set		
2	0	2	20	5-38
2	1	0	15	4-38
2	**1**	**1**	**20**	**5-38**
2	1	2	27	9-94

Step 3. Record MPN determine for each sample; in this example, the MPN is 20

Figure 45.8. Interpreting the most probable number method

Figure 45.9. Schematic and photograph of a RODAC (contact) plate. Note the agar protrudes from housing in a convex shape allowing a complete coverage on sample surface

with a surface and picks up resident microorganisms, the plates are then covered and incubated to promote growth. After 3–5 days at 30–35°C for TSA or 5–7 days at 20–25°C for SDA plates, the microorganisms are counted and results are reported as the number of CFU per area sampled.

45.3.4.2 Swabs (Wet and Dry). For areas that are not flat, smooth surfaces, swabs are commonly used (Fig. 45.10). These types of areas include sinks, vents, drains, corners, bioreactor interior, door handles, and exterior bioreactor components such as ports and probe inlets. For samples that are collected by swabbing, the challenges are threefold: sample collection, sample transport, and sample release.

Depending on the material the swab tab is made of, the tab moisture content, the nature of the surface being sampled, and the operator's technique, the efficiency of microbial contamination pickup and deposit onto a testing device or solution has been measured to be anywhere from 20% to 80%.

A fundamental weakness in the design of swabs is the ability to release collected sample. The microbiological community has generally recognized that woven swabs retain 50–70% of the sample (35,36). Newer designs include nylon-flocked swabs, where nylon flocking is a methodology in which strands of nylon are oriented perpendicular to the shaft creating microcapillaries that have been reported to release up to 90% of collected sample (37,38).

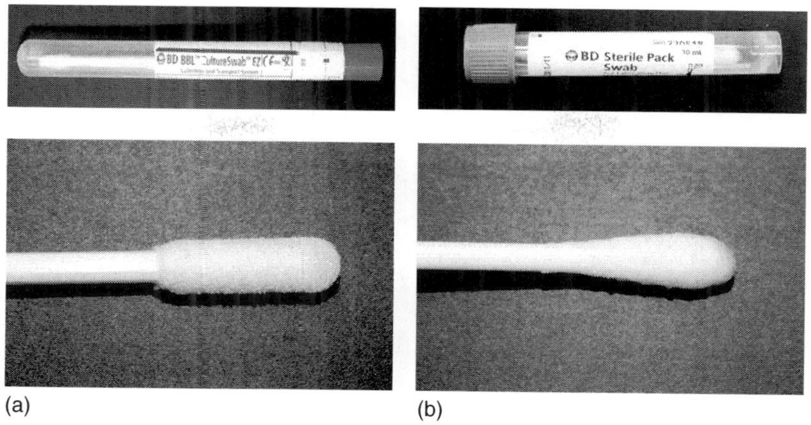

Figure 45.10. Examples of (a) dry and (b) wet swabs and their transport systems

A variety of swabs are commercially available, from wet and dry to different types of swab material (cotton, Dacron®/rayon, polyurethane, microfiber, macrofoam, polyester, flocked nylon, and calcium alginate) to different shapes (contoured, short, elongated, rounded, flattened, etc.) of the swab tip (39). Because of the enormous variety of swabs, a standard for the evaluation of microbial transport systems was published by the Clinical Laboratory Standards Institute (CLSI) (40). This allows swab products to be compared and evaluated using standard criteria. Swabbing pattern and sampling size also need to be considered in order to maintain sampling consistency. It is generally recognized that using a template reduces sample size variability, where an area of 5 cm ×5 cm (25 cm^2) or 10 cm ×10 cm (100 cm^2) is common (41). Templates are available commercially or prepared in house by cutting the desired geometry out of thin, autoclavable plastic (42). Then, an efficient means for actual sample collection must be considered. Industry practice is to roll the swab slowly over the sample surface in overlapping serpentine pattern throughout the template (Fig. 45.11) (3). In this case, the analyst would lay a sterile template of designated size over the sample surface and swab back and forth, up and down, and diagonally according to protocol.

Another consideration is a wide range of transport media available (e.g. Amies Stuart, Cary-Blair, buffered peptone water, Letheen broth, DE neutralizing broth, Butterfields, Copan SRK™ neutralizing solution, and UTM), which are used to maintain the viability of microbial cells during transport of swab to the microbiology laboratory. The transport media may also contain neutralizers such as polysorbate 80 and/or lecithin to counteract any antimicrobial agents used in a clean room environment.

It is common for dry swabs to be used on wet surfaces and wet swabs to be used on dry surfaces. For the dry swabs, once the sample has been collected, the swab tip is soaked and agitated in an isotonic solution. Following

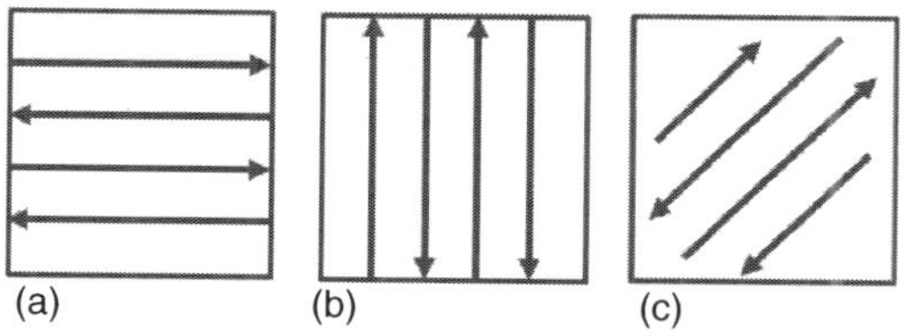

Figure 45.11. Bacterial swabbing pattern for pharmaceutical environmental monitoring. Typical swabbing procedure: (i) wet the swab with appropriate solution; (ii) set the 5 cm ×5 cm (25 cm^2) template on the area to be swabbed; (iii) swab the entire area in a (a) back-and-forth, (b) up-and-down, and (c) diagonal direction while rotating the swab head between the thumb and index finger—all three directions should be conducted on the same sampling area; (iv) after swabbing, return the swab to the originating swab vial; (v) vortex the swab and solution in the vial; and (vi) sample is now ready to be plated or filtered

this, the liquid can then either be filtered through a membrane filtration device or be prepared through the pour plate or spread plate method. For less quantitative analysis, the swab can simply be rolled over an agar-filled plate. In this case, the data indicates the presence or absence of microorganisms.

Wet swabs are convenient in that they are already packaged in an isotonic solution. In these cases, the swabs are removed from the transport tube and the excess liquid is squeezed out of the tip by rolling the tip on the inside of the transport tube. After the sample has been collected, the tip is simply returned to the transport tub and threaded to seal in place. The entire transport tube with tip is then agitated by vortex mixing. All or a portion of the liquid is then filtered through membrane filtration or processed by the pour plate or spread plate method.

For either the wet or dry swabs, the results are reported as number of CFUs per area sampled. For example, if using a 25-mm^2 contact plate or swab template, the results would be reported as CFU/25 mm^2.

45.3.4.3 ***Surface Rinse Method.*** For surfaces inside large tanks, bioreactors, kettles, pipes, or other fillable vessels, it can be advantageous to collect the rinse solution and test by membrane filtration. In these cases, the vessels are filled with rinse solution and, depending on the volume, all or a portion of the rinse solution is filtered through a membrane filter, which is then incubated as described previously.

45.3.4.4 ***Tape or Adhesive Strip.*** Tape-lift samples are collected using a 2–3″ strip of clear adhesive tape. The tape is placed over the sample and the surface of the tape is lightly pressed. After collection, the tape-lift samples are placed into a sterile, sealable plastic bag or adhered directly onto a microscope slide for transport and analysis. The tape can either be stained or be directly observed under a microscope (43,44). Bio-Tape™ (Zefon International) (45) is a slightly more standardized tape-lift collection system, whereby an adhesive area is already applied to a flexible plastic microscope slide. In this case, a protective film is removed from the adhesive area and the slide is pressed down onto the sample surface with the sticky side down.

45.3.5 Air Sampling

45.3.5.1 ***Nonviable Monitoring.*** Under regulated environments, it is necessary to demonstrate control over potential environmental contaminants including dust, microorganisms, and other particles. Of particular concern are particles between 0.5 and 5.0μm, which indicate environmental contamination. Nonviable particle counting is required for designated clean rooms, which have certain thresholds for different classification levels. For example, in the United States, particles greater than 0.5μm cannot exceed 100 particles per cubic foot for a Class 100 (ISO 5) clean room, 10,000 particles per cubic foot for a Class 10,000 (ISO 7) clean room, and 100,000 particles per cubic foot for a Class 100,000 (ISO 8) clean room (46).

The simplest method of counting particles is a microscope, where the user manually counts particles on a surface. Since the sample must adhere to a surface, requires manual counting, and does not measure the amount of air, it is tedious and not practical for large numbers of samples. Most laboratories use automated optical-based methods to monitor such environments. In these cases, air is pulled in through a device and channeled in front of a high intensity beam of light (typically a diode laser) and scattered, reflected, and refracted light are measured. In some instances, the shadow cast by a particle is measured and correlated to particle size. Particle counters are useful because they count and determine size of the particle (47).

45.3.5.2 ***Viable Monitoring.*** Clean rooms must also meet requirements for viable organisms. For example, the viable airborne count must not exceed 0.1, 0.5, and 2.5 CFU/ft^3 for Class 100 (ISO 5), 10,000 (ISO 7), and 100,000 (ISO8) clean rooms, respectively. There are multiple acceptable methods for monitoring levels of viable microorganisms.

- *Passive Sampling*. Perhaps, the most convenient and least expensive method for monitoring viable microorganisms in air is the air-settling plate. Air-settling plates are typically 9-mm or 14-mm in diameter and are prefilled with 20–30 mL of agar containing growth medium. The plates are placed in predetermined locations throughout a laboratory or clean room and allowed to remain in place uncovered for up to 4 h. After the collection period, the plates are covered with their lid and incubated for 3–5 days at 30–35°C for TSA plates and 5–7 days at 20–25°C for SDA plates. The disadvantage of air-settling plates is that is takes 4 h to collect samples and there is no measure of how much air has come into contact with the plate.
- *Active Sampling*. To mitigate these shortcomings a variety of manufacturers have designed active air monitoring devices. These air sampling devices force air to come into contact with plates or strips containing agar and growth medium or directly into liquid medium. There are several types of active air samplers: slit impactor, sieve impactor, surface vacuum sampler, centrifugal impactor, filtration, and liquid impingement. A detailed description of the various air sampler technologies including advantages and disadvantages has already been published (47). In these cases, results from conventional active air sampling devices are reported as CFUs per volume of air sampled. Rotary centrifugal air samplers are commonly used in the pharmaceutical industry. Some examples include the M Air T® air tester by EMD Millipore (48) and RCS PLUS/High Flow Air Samplers by Biotest Diagnostics (49). It is important to note that new technologies are continuing to improve quality and speed of analysis. Of particular note is the commercially available Instantaneous Microbial Detection (IMD-A®) system by Azbil BioVigilant (50). This particular technology relies on laser excitation and Mie light scattering to determine particle size and fluorescence emission to determine whether the particle is likely a microorganism because of its intrinsic fluoresce from NADH, NADPH, and riboflavin. Although the IMD reports number of microorganisms detected per volume of air sampled and categorizes the particle sizes, correlation to viable microorganisms still needs to be performed. Another option for real-time assessment of air samples is the BIOTRAK™ Real-Time Viable Particle Counter by TSI (51). In this case, the BIOTRAK™ particle counter

detects total and viable particle counts in real time by laser-induced fluorescence. In addition to the particle counting, the BIOTRAK™ has integrated particle collection onto a membrane filter for off-line culturable versus nonculturable studies and microbial identification.

Although no regulations exist that dictate what sampling methodology to use, it is imperative that the method is performed consistently from the sampling technique to the sample tool to the incubation. By using the same techniques and tools between analysts and over time, consistency and accuracy can be achieved. Sampling is consistent if the technique picks up and releases the same number of microorganisms found on a uniformly contaminated surface every time, while accurate sampling reflects the true value of the numbers of microorganisms on the surface.

Some key elements to consistent sampling include different swabbing protocols for each type of surface, same swab pattern and sampling area for each type of surface, same sampling pressure for swabs or contact plates, consistent swabs rotation/rolling when sampling or releasing, and rigorous hands-on training.

45.4 NEW AND RAPID TECHNOLOGIES

The conventional methodologies for bioburden testing include sampling, growing on a nutrient-rich agar, incubating for several days, and then manually counting. In between, there may be dilutions or rinses as necessary. The rate-limiting steps are microbial growth, followed by manual counting. To address these rate-limiting steps, researchers have developed a variety of rapid techniques for bioburden analysis. For the purposes of this chapter, rapid technologies are limited to techniques that determine the number of viable organisms in a shorter period of time compared to conventional methods. This definition does not include rapid microbial identification techniques. Also, it is important to note that there have been many technologies described in the literature (52–54), but only a few have been commercially viable. The availability of new, rapid microbial detection methods will increase as these technologies are approved by regulatory agencies such as the US Food and Drug Administration.

45.4.1 Flow Cytometry

Flow cytometry is a technique in which particles (cells in this case) are focused in a sheath of fluid before intercepting a beam of focused light. The resulting forward-scattered light correlates with the cell surface area or size, while side-scattered light provides information on the granular content (intracellular) or cellular complexity. This technique is useful for particles and cells in the range of 0.5–40μm in diameter. The BD FACSMicroCount™ (55) is based on sensitive and rapid flow cytometry technology, with reported claims of direct enumeration of microorganisms and quantitative results within 5 min. Samples are prepared for analysis by labeling microbes with a fluorescent nucleic acid dye in conjunction with a patented BRAG3 labeling compound as a viability stain. Fluorescence intensity decreases in the presence of membrane-compromised cells.

45.4.2 ATP Luminescence

Adenosine triphosphate (ATP) is a product of all living organisms, and its presence indicates active metabolism. ATP also happens to be a key energy source in the enzymatic luciferase reaction, where the substrate luciferin is converted to oxyluciferin plus emission of a photon (56). In this methodology, microorganisms are allowed to metabolize for a small period of time, followed by lysis in the presence of the luciferase/luciferin reagent (57–59). An elevated concentration of ATP correlates with active metabolism and is thus proportional to concentration of viable microorganisms. Since the reaction involves emission of photons and does require a light source, the technique is very sensitive. Compared to the traditional method of growing the organisms for several days, ATP luminescence can detect the presence of bacteria within minutes (60). A number of companies have taken advantage of this very sensitive reaction and commercialized it into microbial detection systems. The Pallchek™ Rapid Microbiology System (61) consists of a highly sensitive luminometer and is based on the measurement of ATP that is released from cells and converted to light their proprietary cocktail of luciferase, luciferin, and stabilizers.

Direct measurement of microbial contamination is also possible using membrane filters for liquid samples and swabs for surface contamination. The Milliflex® Rapid (62) is an automated platform for rapid detection of filterable samples. Detection is accomplished by monitoring ATP production using a proprietary bioluminescence reagent and enumeration via charge-coupled device (CCD) camera and image analysis (63).

A third alternative by Celsis, the RapiScreen/AkuScreen (64), also measures cell metabolism via ATP regulation.

In a related and expanded market, LuminUltra Technologies has developed stabilized formulations to allow microbial testing in the food and beverage, industrial process water, and waste water industries (65).

45.4.3 Electrical Impedance

Electrical impedance is the measure of the opposition that a circuit exhibits toward the passage of a current when a voltage is applied through a medium. In the case of bacterial detection, a reference chamber containing medium only is compared to a sample chamber in which component fluctuations occur because of microbial growth (66). The impedance change caused by bacterial metabolism in the medium is monitored over time and is produced by the release of ionic metabolites from viable microorganisms. Ionic flux in microorganisms is caused by two main methods: energy metabolism (catabolism) and ion exchange through the cell membrane. During catabolism, oxygen and a carbon source such as glucose produce lactate, which is then metabolized into carbonate. In this case, both lactate and carbonate increase conductivity. Although both conduct, the carbonate is more effective. During ion exchange, ions such as H^+, K^+, and Na^+ are actively transported across the membrane through ion channels in an effort to regulate membrane potential and osmotic flux. While both sources of ions contribute to the conductance, catabolic ions contribute most (66–70).

The RABIT (rapid automated bacterial impedance technique) system (71) operates in a direct mode, whereby metabolizing microorganisms increase the electrical conductance of the culture medium in the system and an indirect mode, where the amount of carbon dioxide produced by growing organisms in monitored. The basic premise of the technique is that a change in electrical impedance is correlated with the presence of microorganisms.

45.4.4 Intrinsic Fluorescence

Fluorescence from a biological cell mainly originates from its chemical components such as aromatic amino acid residues (tryptophan, tyrosine, and phenylalanine), nucleic acids, pyroxidine, riboflavin, and coenzymes such as NADH, NADPH, and FAD. Fluorescence due to reduced nicotinamide adenine dinucleotide species is especially important as they indicate active metabolism (72). The Growth Direct™ system by Rapid Micro Biosystems (73) captures the intrinsic fluorescence in all living cells. The IMD-A™ by BioVigilant takes advantage of the same intrinsic fluorescence and combines fluorescence signal with Mie light scattering to monitor viable microorganisms and quantitate particles in air samples.

45.4.5 Carbon Dioxide Evolution

Carbon dioxide is a component in the Krebs cycle in which glucose is metabolized in part to CO_2 as part of respiration. There are two primary modes of CO_2 detection: colorimetric and fluorometric. The BacT/ALERT® platform (74,75) is commonly used in environmental monitoring and relies on colorimetric readout. In this methodology, CO_2 generated by actively metabolizing microorganisms diffuses into a sensor and generates H^+ through dissolution into water. This pH change causes a color change in the sensor housing (76). The BACTEC™ FX system (77) takes advantage of a similar principle, although the pH change is measured by fluorescence. Another detection method relies on measuring gas-phase CO_2 in the headspace where the microorganisms are contained within a closed environment. In this technique, CO_2 is given off as a by-product of normal metabolism and infrared absorbance is used to determine concentration of CO_2, which is then correlated to microorganism concentration (78).

45.4.6 Fluorescence Staining

There are a number of ways to fluorescently stain microorganisms including immunostaining with fluorescently labeled antibody, nucleic acid dyes (such as the classic Hoechst 3342 or acridine orange dyes), or membrane-permeable fluorogenic substrates. Perhaps, the most useful tool is that of nonfluorescent ester derivatives that, on permeation across cell membrane, become cleaved by functional cytoplasmic enzymes (79). Fluorescence staining works well when combined with membrane filtration (80). The ChemScan® RDI system by bioMerieux (81,82) relies on capturing all viable microorganisms on a membrane, which are then labeled with a nonfluorescent substrate. Once the substrate crosses the cell membrane, it is enzymatically cleaved in the cytosol, whereby fluorescence increase. Only metabolically active microorganisms can cleave the substrate.

Similarly, liquid samples are filtered through a presterilized, filter unit in the Milliflex® Quantum platform (83). The membrane filter is then placed onto a Milliflex® Quantum R2A media cassette, which contains fluorescent dye in addition to growth medium. As soon as the dye is metabolized into a fluorescent product, the membrane is visualized using a fluorescent scanning device and colonies are counted.

45.4.7 Nucleic Acid Technologies

Polymerase chain reaction (PCR) is a sensitive and specific method that relies on amplification of DNA using short, synthetic primers that bind to complimentary sequences on the target microorganism's DNA. The amplified PCR product is then separated by gel electrophoresis that is stained with a fluorescent intercalating dye (such as ethidium bromide). If the target microorganism is present then a PCR band appears. To detect a panel of organisms, sets of primers, each unique to a species of microorganism, may be used. For more quantitative results, real-time polymerase chain reaction (RT-PCR) can be used. In this technique, a quenched fluorescent probe is added to the PCR mix. When the Taq polymerase extends the DNA and encounters the fluorescent probe, the probe is cleaved and released allowing fluorescence to occur. The fluorescence signal increases with each PCR cycle, which can then be correlated with the amount of DNA present and thus microorganism concentration in the original sample. Although the DNA indicates the presence of specific microorganisms, this technique cannot quantitate the level of viability. RT-PCR is not only quantitative but also very specific (84–86). There are plenty of commercial examples where nucleic acid tests are used to detect microorganisms, whether in a clinical (BD Affirm™ VPIII Microbial Identification Test (87), PANTHER® System/APTIMA Combo 2® Assay (88), NucliSENS EasyQ® platform (89), COBAS® TaqMan® MTB Test (90)) or pharmaceutical setting (MycoTOOL® Mycoplasma Detection (91) and MilliPROBE® Real-Time Detection System for Mycoplasma (92)).

45.5 CONCLUSION

Concurrent with the FDA's current position on improving quality of drugs, pharmaceutical manufacturers continue to develop systematic approaches to product development. Within this quality-by-design (QbD) approach, science and quality risk management are built into the process, rather than simply testing for compliance (93,94). Since the United States and international regulatory agencies are providing guidances for QbD (95), pharmaceutical and biopharmaceutical manufacturers have a clear need for rapid, online measurements. Conventional bioburden technologies rely on colony growth before detection, where it may take anywhere from 3 days for traditional bioburden to 28 days for mycoplasma testing. Since techniques requiring microbial growth are neither rapid nor online, alternate technologies are needed to fulfill the current stances on quality by regulatory agencies.

Two promising technology classes have been described whereby rapid and online detection may be feasible: nucleic acid technologies (NAT) and cytometry (flow and solid state). NATs are well documented and characterized with respect to performance, selectivity, and sensitivities and are beginning to be used for rapid in-process testing (96–98). It has been demonstrated that in-process decisions can be made within hours instead of days and even offer a significant advantage when considering current online monitoring using glucose, pH, or oxygen sensing (96). Although sensitive, these methods require off-line sample processing such as sample purification, strand denaturation, primer annealing, and strand extension.

Cytometric techniques offer a potential advantage in that they rely on flow and can conceivably be adapted to online measurements. For example, flow cytometry has been demonstrated to provide equivalent results to traditional growth-based methods (99). Although results are obtained in only a fraction of the time compared to conventional bioburden methods, cytometry has limited sensitivity, which ultimately affects usefulness for detecting low levels of bioburden. Solid-phase cytometry has improved sensitivity, whereby microorganisms are captured on a solid support (usually a membrane filter), followed by staining and fluorescence detection (100). While this methodology improves sensitivity, the sample processing steps are not conducted online. Both methodologies show promise but currently require several hours for immunofluorescence staining or capturing on a filter followed by metabolic staining.

Although no commercialized rapid microbial detection product to date is truly online, there are rapid microbial detection technologies that offer clear advantages in time over conventional growth-based techniques. By retrieving results quicker, thus allowing decision to be made sooner, process efficiencies can be realized.

REFERENCES

1. Current good manufacturing practice for finished pharmaceuticals. Code of Federal Regulations Title 21. Part 211; 2011. Available at http://www.accessdata.fda.gov/scripts/cdrh/cfdocs/cfcfr/CFRSearch.cfm?CFRPart=211. Accessed 2012 May 3.
2. Volume 1, United States Pharmacopeia and National Formulary (USP 34-NF 29). Rockville (MD): United States Pharmacopeia Convention; 2011. p 630–631.
3. Radar RA. BIOPHARMA: biopharmaceutical products in the U.S. and European markets (Online). Available at http://www.biopharma.com/approvals_2011.html. Accessed 2012 May 3.
4. Clontz L. Microbial limit and bioburden tests: validation approaches and global requirements. Boca Raton (FL): Taylor and Francis Group; 2009.
5. National Aeronautics and Space Administration web information. Available at http://www.nasa.gov/mission_pages/station/research/experiments/Environmental_Monitoring.html#overview. Accessed 2012 May 03.
6. FDA web information. Available at http://www.fda.gov/MedicalDevices/default.htm. Accessed 2012 May 03.
7. Centers for Medicare and Medicaid Services web information. Available at http://www.cms.gov/Regulations-and-Guidance/Legislation/CLIA/index.html. Accessed 2012 May 03.
8. FDA web information. Available at http://www.fda.gov/Food/default.htm. Accessed 2012 May 03.
9. FDA web information. Available at http://www.fda.gov/Drugs/default.htm. Accessed 2012 May 03.
10. FDA web information. Available at http://www.fda.gov/Cosmetics/default.htm. Accessed 2012 May 03.
11. IUL Instruments web information. Available at http://www.iul-inst.com/spin-air-air-sampler.html. Accessed 2012 May 3.
12. Rapid Microbiology web information. Available at http://www.rapidmicrobiology.com/test-methods/Air-Samplers.php. Accessed 2012 May 03.

13. Cary SG, Blair EB. J Bacteriol 1964; 88: 96–98.
14. Amies CR. Can J Public Health 1967; 58(7): 296–300.
15. Barry AL, Fay GD, Sauer RL, Appl Microbiol 1972; 24(1): 31–33.
16. Stuart RD. Public Health Rep 1959; 74(5): 431–438.
17. Chernesky M, Castriciano S, Jang D., Smieja M. J Clin Microbiol (2006; 44(3): 1084–1086. DOI:10.1128/JCM.44.3.1084–1086.2006.
18. Sutton SV, Proud DW, Rachul S, Brannan DK. PDA J Pharm Sci Technol 2002; 56(5): 255–266.
19. Johnston MD, Lambert RJW, Hanlon GW, Denyer SP. J Appl Microbiol 2002; 92: 784–789.
20. Volume 1, United States Pharmacopeia and National Formulary (USP 34-NF 29). Rockville (MD): United States Pharmacopeia Convention; 2011. p 787–808.
21. Eaton, AD, Clesceri LS, Greenberg AE, editors. Standard methods for the examination of water and wastewater, 19th ed. Washington, DC: American Public Health Association; 1995.
22. Rompré A, Servais P, Baudart J, de-Roubin M-R, Laurent P. J Microbiol Methods 2002; 49: 31–54.
23. TCI America web information. Available at http://www.tciamerica.com/product/bio-chem/B016.shtml. Accessed 2012 May 03.
24. Neogen web information. Available at http://www.neogen.com/Acumedia/pdf/ProdInfo/7724_PI.pdf. Accessed 2012 May 3.
25. Wisc-Online web information. Available at http://www.wisc-online.com/objects/ViewObject.aspx?ID=mby4307. Accessed 2012 May 3.
26. ISO 6887-1:1999: Microbiology of food and animal feeding stuffs—preparation of test samples, initial suspension and decimal dilutions for microbiological examination—Part 1: General rules for the preparation of the initial suspension and decimal dilutions. 1999.
27. United States Pharmacopeia and National Formulary (USP 34-NF 29), Microbial examination of nonsterile products: microbial enumeration tests. Rockville (MD): United States Pharmacopeia Convention; 2011; 1: p 56–56.
28. Madigan MT, Martinko JM, Dunlap PV, Clark DP. Brock biology of microorganisms, 12th ed. San Francisco (CA): Pearson Education, Inc.; 2009. p 154–155.
29. The Microbiology Network (online) web information. Available at http://www.microbiol.org/resources/monographswhite-papers/qualification-of-an-environmental-monitoring-program{-}1-selectionjustification-of-sample-sites/. Accessed 2012 May 7.
30. Quisno R, Gibby IW, Foter MJ. Am J Pharm 1946; 118: 320–323.
31. Erlandson AL, Lawrence CA. Science 1953; 118(3062): 274–276.
32. Brummer B. Appl Environ Microbiol 1976; 32(1): 80–84.
33. Murray PR, et al., editors. Manual of clinical microbiology, 6th ed. Washington, DC: American Society for Microbiology; 1995.
34. Rapid Test Methods web information. Available at http://www.rapidmicrobiology.com/test-methods/Swabs.php. Accessed 2012 May 3.
35. Dalmaso G, Bini M, Paroni R, Ferrari M. PDA J Pharm Sci Technol 2008; 62(3): 191–199.
36. Copan web information. Available at http://www.copanswabs.com/products/microrheologics/work.php. Accessed 2012 May 7.
37. Dalmaso G, Ferrari M, Paris A. Validation of the new Irradiated Nylon™ flocked QUANTISWAB™ for the quantitative recovery of micro-organisms in critical clean room environments. Available at http://www.biomerieux-usa.com/upload/QuantiSwab-Microorganism-Recovery-1.pdf.
38. Faoagali J. Microbiology Australia 2010: 133–136.
39. Body BA, Arbique JC, Bourbeau P, cavagnolo R, Miller JM, Poole FM, Sharples N. M40-A-Quality control of microbiological transport systems; approved standard, Wayne (PA): Clinical and Laboratory Standards Institute; 2003; 23:34
40. ISO 18593:2004(E): Microbiology of food and animal feeding stuffs—horizontal methods for samplings techniques from surfaces using contact plates and swabs. International Organization for Standardization, Geneva, Switzerland; 2004.
41. Copan USA web information. Available at http://copanusa.com/. Accessed 2012 May 7.
42. Khan NN, Wilson BL. J Environ Sci Health 2003; A38(12): 2759–2772.
43. Horner WE, Barnes C, Codina R, Levetin E. J Allergy Clin Immunol 2008; 121(3): 592–597.
44. Zefon web information. Available at http://www.zefon.com/store/bio-tape-surface-sampler.html. Accessed 2012 May 03.
45. ISO14644-1:1999 (E): Cleanrooms and Associated controlled environments. International Organization for Standardization, Geneva, Switzerland; 1999.
46. Kochevar SD. Basic guide to particle counters and particle counting, Boulder (CO): Particle Measuring Systems, Inc.; 2006.
47. EMD Millipore web information. Available at http://www.millipore.com/publications.nsf/a73664f9f981af8c852569b9005b4eee/f7a9d6cefa74f583852568f600496a60/$FILE/TB1000EN00.pdf. Accessed 2012 May 07.
48. Biotest Diagnostics web information. Available at http://cleanroom.net/brochures/RCS_HighFlowTouch_EN.pdf). Accessed 2012 May 07.
49. Biovigilant web information. Available at http://www.biovigilant.com/. Accessed 2012 May 03.
50. TSI web information. Available at http://www.tsi.com/. Accessed 2012 June 21.
51. Moldehauer J. In: Zourob M, Elway S, Turner APF, editors. Principles of bacterial detection: biosensors, recognition receptors and microsystems, Springer Science and Business Media, 2008. p 49–79.
52. Noble RT, Weisberg SB. J Water Health 2005; 3(4): 381–392.
53. Duguid J, Balkovic E, du Moulin GC. Am Pharmaceut Rev 2011; 14(7).
54. Becton Dickinson web information. Available at http://www.bdbiosciences.com/eu/instruments/facsmicrocount/index.jsp. Accessed 2012 May 03.
55. Chappelle EW, Levin GE. Biochem Med 1968; 2: 41–52.
56. Zeng AP, Deckwer WD. Biotechnol Prog 1995; 11(1): 71–79. DOI: 10.1021/bp00031a010.
57. Molin O, Nilsson L, Anséhn S. J Clin Microbiol 1983; 18(3): 521–525.

58. Satoh T, Kato J, Takiguchi N, Ohtake H, Kuroda A. Biosci Biotechnol Biochem 2004; 68(6): 1216–1220. Available at http://dx.doi.org/10.1271/bbb.68.1216.
59. Stanley PE. J Biolumin Chemilumin 1989; 4(10): 375–380.
60. Pall Life Sciences web information. Available at http://www.pall.com/pdfs/Biopharmaceuticals/USD2357a_Pallchek_ENG.pdf. Accessed 2012 May 03.
61. EMD Millipore web information. Available at http://www.millipore.com/catalogue/module/c10711 Accessed 2012 May 03.
62. Chollet R, Ribault S. In: Lapota D, editors, Bioluminescence—recent advances in oceanic measurements and laboratory applications. InTech; 2012. p 99–118, DOI:10.5772/2088. http://www.intechopen.com/.
63. Celsis Inc. web information. Available at http://celsis.com/products/pharmaceuticals. Accessed 2012 May 03.
64. LumenUltra Technologies web information. Available at http://www.luminultra.com/technology.html. Accessed 2012 May 03.
65. Cady P, Dufour SW, Shaw J, Kraeger SJ. J Clin Microbiol 1978; 7(3): 265–272.
66. Ur A, Brown DF. J Med Microbiol 1975; 8(1): 19–28. DOI:10.1099/00222615-8-1-19.
67. L. Yang and R. Bashir, Biotechnology Adv 2008; 26: 135–150.
68. Specter S, Throm R, Strauss R, Friedman H. J Clin Microbiol 1977; 6(5): 489–493.
69. Buckland A, Kessock-Philip S, Bascomb S. J Clin Pathol 1983; 36(7): 823–828.
70. Don Whitley Scientific Limited web information. Available at http://www.dwscientific.co.uk/rabit.php. Accessed 2012 May 03.
71. Sohn M, Himmelsbach DS, Barton FE, Fedorka-Cray PJ. Appl Spectrosc 2009; 63(11): 1251–1255.
72. Rapid Micro Biosystems web information. Available at http://www.rapidmicrobio.com/growth-direct/microbial-contamination. Accessed 2012 May 3.
73. Riedel S, Siwek G, Beekmann SE, Richter SS, Raife T, Doern GV. J Clin Microbiol 2006; 44(6): 2262–2264; DOI: 10.1128/JCM.00635-06.
74. bioMerieux web information. Available at http://www.biomerieux-industry.com/servlet/srt/bio/industry-microbio logy/dynPage?doc=NDY_BPA_PRD_G_PRD_NDY_1. Accessed 2012 May 03.
75. Thorpe TC, Wilson ML, Turner JE, DiGuiseppi JL, Willert M, Mirrett S, Reller LB. J Clin Microbiol 1990; 28(7): 1608–1612.
76. Becton Dickinson web information. Available at http://www.bd.com/ds/productCenter/BC-Bactec_allWhite_Papers.asp. Accessed 2012 May 03.
77. Threlkeld CH. J Food Sci 1982; 47(4): 1222–1225.
78. Veal DA, Deere D, Ferrari B, Piper J, Attfield PV. J Immunol Methods 2000; 243: 191–210.
79. Durtschi JD, Erali M, Bromley LK, Herrmann MG, Petti CA, Smith RE, Voelkerding KV. J Med Microbiol 2005; 54: 843–850. DOI:10.1099/jmm.0.46092-0.
80. AEI Chemunex a bioMerieux Company web information. Available at http://www.aeschemunex.com/microbial-testing-solutions-for-all-labs,6/33,scan-rdi-analyzer.html. Accessed 2012 May 03.
81. Smith R, von Tress M, Tubb C, Vanhaeckepda E. J Pharm Sci Tech 2010; 64: 356–363.
82. EMD Millipore web information. Available at http://www.millipore.com/publications.nsf/a73664f9f981af8c852569b9005b4eee/242d91c82f187f28852576d50059d816/$FILE/DS1072EN00.pdf. Accessed 2012 May 03.
83. Glynn B. In: Zourob M, Elway S, Turner APF, editors. Principles of bacterial detection: biosensors. Recognition receptors and microsystems. Springer Science + Business Media; 2008. p 603–628.
84. Mothershed EA, Whitney AM. Clin Chim Acta 2006; 363: 206–220.
85. Whiting I, Cook N, Hernández M, Rodrrguez-Lázaro D, D'Agostino M. In: Nollet LML, Toldra F, editors. Safety analysis of foods of animal origin, CRC Press, 2010, 811–821. DOI:10.1201/EBK1439848173-35.
86. Becton Dickinson and Company web information. Available at http://www.bd.com/ds/productCenter/MD-Affirm VPIII.asp. Accessed 2012 May 8.
87. Gen-Probe web information. Available at http://www.gen-probe.com/news/PressReleaseText.asp?releaseID=1692954. Accessed 2012 May 8.
88. bioMerieux web information. Available at http://www.biomerieux-diagnostics.com/servlet/srt/bio/clinical-diagnostics/dynPage?node=Nucleic_Acid_Testing. Accessed 2012 May 8.
89. Roche Molecular Diagnostics web information. Available at http://molecular.roche.com/assays/Pages/COBASTaqMan MTBTest.aspx. Accessed 2012 May 8.
90. Lonza web information. Available at http://www.lonza.com/products-services/pharma-biotech/rapid-microbial-detection/mycotool-mycoplasma-detection.aspx. Accessed 2012 May 8.
91. EMD Millipore web information. Available at http://www.millipore.com/catalogue/module/c66523. Accessed 2012 May 8.
92. Riley BS, Li X. AAPS PharmSciTech 2011; 12(1): 114–118.
93. Guidance for industry quality systems approach to pharmaceutical CGMP regulations. Rockville (MD)Food and Drug Administration; 2006. Available at http://www.fda.gov/downloads/Drugs/GuidanceCompliance RegulatoryInformation/Guidances/ucm070337.pdf. Accessed 2012 June 21.
94. ICH Q8(R2) pharmaceutical development. International Conference on Harmonisation of Technical Requirements for Registration of Pharmaceuticals for Human Use; 2009. Available at http://www.ich.org/fileadmin/Public_Web_Site/ICH_Products/Guidelines/Quality/Q8_R1/Step4/Q8_R2_Guideline.pdf. Accessed 2012 June 21.
95. Hentz NG. Evaluation of rapid off-line microbial contamination detection of fermentation processes. 19th Annual ISPE-CASA Life Sciences Technology Show; 2012 Apr 10; Raleigh, NC.
96. Shintani H, Sakudo A, McDonnel GE. Biocontrol Sci 2011; 16(1): 13–21.
97. Adams G, Berg H, Galbraith D, McCarthy P. BioProcess Int 2011; 9(8): 28–35.

98. McHugh IOL, Tucker AL. Cytometry A 2007; 71A: 1019–1026.
99. Baumstummler A, Chollet R, Meder H, Rofel C, Venchiarutti A, Ribault S. Lett Appl Microbiol 2010; 51(6): 671–677.

FURTHER READING

Bacteriological analytical manual. Rockwell (MD): Food and Drug Administration. Available at http://www.fda.gov/Food/ScienceResearch/LaboratoryMethods/BacteriologicalAnalyticalManualBAM/ucm071363.htm. Accessed 2012 May 8.

Easter MC. Methods in the pharmaceutical industry. Boca Raton (FL): CRC Press; 2003.

Guidance for industry: sterile drug products produced by aseptic processing—current good manufacturing practice. Rockville, (MD): Food and Drug Administration; 2004.

Kirsch L, editor. PDA J Pharm Sci Technol 2001; 55(5 Suppl TR13): 1–35.

FDA web information. Available at http://www.fda.gov/AnimalVeterinary/default.htm. Accessed 2012 May 03.

46

CHROMATOGRAPHY, INDUSTRIAL SCALE VALIDATION

SANDY WEINBERG
Clayton State University, Atlanta, Georgia
CARL A. ROCKBURNE
The Rockburne Group, Atlanta, Georgia

46.1 INTRODUCTION

The process of chromatography, used both in laboratory settings to definitively identify compounds and in the manufacturing operation to separate complex proteins, is well understood. The United States Food and Drug Administration (FDA) has developed a long familiarity with and general acceptance of chromatography in both settings. Routine inspections lead to inquiries about initial process validation; cleaning and coping with residuals; and issues arising from the reuse of media.

But chromatography systems are, by design and definition, also software-driven systems, and as such are subject to the more focused requirements of 21 CFR (Code of Federal Requirements) Part 11. Part 11 provides guidelines for system validation; archiving and retrieval of data; control of the software environment; the establishment and review of data audit trails; and more. These system validation requirements have increasingly become the target of FDA investigations, and should be addressed in all chromatography validations.

This chapter provides a checklist for applying 21 CFR Part 11 to industrial scale validation of chromatography systems, as well as an inspection checklist for good manufacturing practices (GMP) inspections of chromatography facilities.

46.2 SYSTEMS REQUIREMENTS

The FDA requirement 21 CFR Part 11 applies to all nonfinancial computer systems, including both chromatography systems (the internal code controlling and interpreting chromatographic processes and results) and the manufacturing execution systems (MESs) that manage the industrial chromatographic system. This checklist is designed to provide guidance for a self review of compliance with 21 CFR Part 11.

46.3 CFR PART 11 SOFTWARE EVALUATION CHECKLIST FOR CLOSED SYSTEMS WHICH DO NOT USE BIOMETRICS

Only those sections of 21 CFR Part 11 that describe technical controls required for 21 CFR Part 11 compliance of closed systems are included in this checklist. Sections that describe only procedural controls (11.10(i), (j), (k); 11.100(b), (c); 11.300(c) that cannot be implemented by a software product or additional controls for open system (11.30) are not included. Procedural controls can only be exercised during the implementation of a 21 CFR Part 11 compliant system, of which the software is a component.

Downstream Industrial Biotechnology: Recovery and Purification, First Edition. Edited by Michael C. Flickinger.

Section	Regulatory Requirements	Functionality to be Demonstrated	Observations
11.10(b)	Procedures and controls shall include the ability to generate accurate and complete copies of records in both human readable and electronic form suitable for inspection, review, and copying by the agency.	Demonstrate the functionality to generate accurate and complete copies of records in both human readable and electronic form, suitable for inspection, review, and copying by the agency. Include: methods sequences raw data results, both data and graphs reports other (?). In "review," can the agency regenerate results from raw data? How? Can the agency query the data (not simply visually inspect)? Demonstrate retention of metadata. What is the relationship of the "single file" to the database?	
11.10(c)	Procedures and controls shall include protection of records to enable their accurate and ready retrieval throughout the records retention period.	Demonstrate the functionality to accurately and readily retrieve archival records throughout the record retention period (e.g. backup and restore or archive/retrieve or other.) Include: methods sequences raw data results, both data and graphs reports calibrations standards event logs other (?). Can the agency regenerate results from raw data? How? Do all metadata remain? Links between files? Audit trails? Are records protected during record retention period? Accessible by database commands, SQL, and so on? Is original hardware and software required for access/query?	

Section	Regulatory Requirements	Functionality to be Demonstrated	Observations
11.10(d)	Procedures and controls shall include limiting system access to authorized individuals.	Demonstrate that functionality exists to limit user access to authorized individuals:	
		from the operating system (Windows NT/2000/XP, etc.)	
		from within the software	
		for application start-up	
		for direct access to files for edit, rename, delete.	
		Demonstrate setup of users and privileges.	
		Demonstrate that Admin changes to Users and Privileges are subject to audit trail.	
		Demonstrate that the logged in user ID is displayed on all screens.	
		Demonstrate that stored passwords are encrypted, and that encryption uses at least suggested standards.	
		Demonstrate that the Admin password can be changed.	
11.10(e)	Procedures and controls shall include use of secure, computer-generated time-stamped audit trails to independently record the date and time of operator entries and actions that create, modify, or delete electronic records. Record changes shall not obscure previously recorded information. Such audit trail information shall be retained for a period at least as long as that required for the subject electronic records and shall be available for agency review and copying.	Demonstrate that data cannot be overwritten.	
		Demonstrate that audit trails are secure.	
		Demonstrate that audit trails are created and maintained for:	
		date and time of operator entries and actions that create, modify, or delete electronic records (methods, sequences, raw data, results, reports, calibrations, standards, event logs)	
		Admin changes to privileges	
		Admin changes to passwords.	
		Demonstrate that audit trail is linked to data files during retention period.	
		Demonstrate that audit trail is available for agency review and copying.	
		Demonstrate that audit trail is available for query.	
11.10(f)	Procedures and controls shall include the use of operational system checks to enforce permitted sequencing of steps and events, as appropriate.	Demonstrate that the system uses operational system checks to enforce permitted sequencing of steps and events, as appropriate. Does the system enforce running blanks or standards prior to a sample?	

Section	Regulatory Requirements	Functionality to be Demonstrated	Observations
		Does the system employ "required" fields?	
		Does the system require all method and sequence data be defined before a run? (e.g. can the sample name, concentration, volume, etc. be changed after data is acquired?)	
11.10(g)	Procedures and controls shall include use of authority checks to ensure that only authorized individuals can use the system, electronically sign a record, access the operation or computer system input or output device, alter a record, or perform the operation at hand.	Demonstrate functionality for authority checks for:	
		system use (access)	
		electronic signature	
		access to computer system input or output device (Can input or output devices be altered without authority checks in a manner that will predictably affect results?).	
		record alteration	
		individual operation	
		Does the system require the use of a stored system user ID and password to access shared storage devices and perform system operations?	
11.10(h)	Procedures and controls shall include use of device (e.g. terminal) checks to determine, as appropriate, the validity of the source of data input or operational instruction.	Demonstrate that the system uses device checks to identify (and record) the source of input data.	
		Demonstrate that the system does not allow data acquisition from unidentifiable or incorrect sources.	
		Demonstrate that the system uses checks for the validity of operational instructions. (e.g. must instructions come from the application, or can they be overridden by a keypad?)	
11.50	Signed electronic records shall contain information associated with the signing that clearly indicates the following:	Demonstrate that the signed electronic records contain information associated with the signing that clearly indicate:	
	The printed name of the signer;	the printed name of the signer (not just the user ID)	
	The date and time when the signature was executed; and	the date and time when the signature was executed (traceable to the time zone)	
	The meaning (such as review, approval, responsibility, or authorship) associated with the signature.	the meaning of the signature.	
	These items are subject to the same controls as for electronic records and shall be included as part of any human readable form of the electronic record (such as electronic display or printout).	Demonstrate that the electronic signature information has access controls, data integrity, audit trails, and record retention.	

Section	Regulatory Requirements	Functionality to be Demonstrated	Observations
		Demonstrate that the name, date/time, and meaning are included as part of any human readable form of the electronic record, including display and printed report.	
11.70	Electronic signatures and handwritten signatures executed to electronic records shall be linked to their respective electronic records to ensure that the signatures cannot be excised, copied, or otherwise transferred to falsify an electronic record by ordinary means.	Demonstrate that electronic signatures are linked to their respective electronic records in a manner that prevents excision, copying, modifying, or otherwise transferring to falsify an electronic record by ordinary means (e.g. by opening in WordPad to edit, or by simple file operations).	
		Demonstrate that handwritten signatures executed to electronic records ("hybrid systems") are linked to their respective electronic records.	
		Demonstrate that the printed, hand signed copy has sufficient information to link the report to a unique electronic record (date, time printed, name of person printing report, file name, date/time file creation, unique file identification, location, etc.).	
11.100(a)	Each electronic signature shall be unique to one individual and shall not be reused by, or reassigned to, anyone else.	Demonstrate that user ID (an essential element of user ID/PW combination comprising the electronic signature) is not reusable by deletion/recreation, overwriting, or other means.	
		Demonstrate that the system does not allow redundant user IDs.	
11.200(a)(1)	Electronic signatures shall employ at least two distinct components such as an identification code and password.	Demonstrate that electronic signatures employ at least two distinct components (user ID and password).	
	When an individual executes a series of signings during a single, continuous period of controlled system access, the first signing shall be executed using all electronic signature components; subsequent signings shall be executed using at least one electronic signature component that is only executable by, and designed to be used only by, the individual.	Demonstrate that the user ID of the person logged on to the system is displayed across all screens that allow user inputs.	
	When an individual executes one or more signings not performed during a single, continuous period of controlled system access, each signing shall be executed using all of the electronic signature components.	Demonstrate that the first signing of a continuous session uses all electronic signature components.	

Section	Regulatory Requirements	Functionality to be Demonstrated	Observations
		Demonstrate that user ID is displayed at the time of application of the password to execute an electronic signature (i.e. at least one electronic signature component that is only executable by, and designed to be used only by, the individual).	
		Demonstrate that each signing not performed in a continuous session uses all electronic signature components.	
		Demonstrate that the system performs logout after a configurable interval to end an unattended session	
11.200(a)(2)	Electronic signatures shall be used only by their genuine owners.	Demonstrate that passwords (one of the two components of electronic signatures) can only be known to the genuine owners, and cannot be viewed by anyone, including administrators of the account (at operating system and application levels).	
		Demonstrate that Administrator password management privileges extend only to the ability to reset a password.	
		Demonstrate that the user must change the reset password at initial subsequent login.	
11.200(a)(3)	Electronic signatures shall be administered and executed to ensure that attempted use of an individual's electronic signature by anyone other than its genuine owner requires collaboration of two or more individuals.	See 11.200(a)(2). Refer also to demo that use of invalid password does not allow access to system or permit electronic signature.	
11.200(b)	Electronic signatures based on biometrics shall be designed to ensure that they cannot be used by anyone other than their genuine owners.	N/A—system does not employ biometrics.	
11.300(a)	Identification codes/passwords controls shall include maintaining the uniqueness of each combined identification code and password, such that no two individuals have the same combination of identification code and password.	Demonstrate (refer to earlier demo) that user IDs are unique (cannot be deleted or redundant).	

Section	Regulatory Requirements	Functionality to be Demonstrated	Observations
		If a user ID has been inactivated, can it be reactivated? Would these actions be audit trailed? If reactivation is not possible, how would a new user ID for a returning employee be linked to the past ID so all records created or signed by an individual could be queried? (Does the system provide a technical solution, or would this be handled by a procedure?)	
11.300(b)	Identification codes/passwords controls shall include ensuring that identification code and password issuances are periodically checked, recalled, or revised (e.g., to cover such events as password aging).	Demonstrate that controls include such configurable parameters as: the password expiration period whether reuse of a previous password is allowed the password minimum length whether numeric or special characters must be included in the password the number of failed login attempts allowed before system lockout occurs. Does the system allow configuration to exclude common (dictionary) terms from use as passwords?	
11.300(d)	Identification codes/passwords controls shall include use of transaction safeguards to prevent unauthorized use of passwords and/or identification codes, and to detect and report in an immediate and urgent manner any attempts at their unauthorized use to the system security unit, and, as appropriate, to organizational management.	Demonstrate that the system includes controls to detect multiple attempts at unauthorized use (e.g. repeated login attempts/failed password entry on login and electronic signature). Demonstrate that such attempts at unauthorized use can be reported in an immediate and urgent manner to the system security unit, and, as appropriate, to organizational management.	
11.300(e)	Identification codes/passwords controls shall include initial and periodic testing of devices that bear or generate identification code or password information to ensure that they function properly and have not been altered in an unauthorized manner.	Does the system employ devices that bear or generate ID codes? Does the system employ such codes/passwords for instruments? For the system, servers, other?	

Section I: Receiving Control	Response (Yes)	(No)	(N/A)	Reference (21 CFR)
1. Does receiving inspection check incoming shipments to requirements of the purchase order, specifications, and applicable drawings?				211.84
2. Do receiving inspection records indicate acceptance or rejection of incoming material?				211.84, 211.184
3. Are material weight checks made upon receipt?				211.84, 211.184 211.101
4. Do receiving inspection records reflect the reasons for rejection?				211.84
5. Are records of receiving materials properly stored?				211.184
6. Are inspected items properly segregated from the materials awaiting inspection?				211.42
7. Is inspected material adequately identified for acceptance or rejection?				211.80
8. Is rejected material adequately controlled?				211.89
9. Are the sampling plans employed by Quality Control sufficient?				211.84

Section II: Materials Storage and Handling	Response (Yes)	(No)	(N/A)	Reference (21 CFR)
1. Is access to storerooms and materials storage areas restricted to authorized personnel?				211.122(d)
2. Are materials handled, identified and stored in a manner that will prevent damage, contamination, mix-up, and/or loss?				211.80
3. Are accountability records kept to permit forward and backward traceability?				211.42 211.188
4. Are stocks reinspected and tested at intervals?				211.184 211.87

Section II: Materials Storage and Handling	Response (Yes)	(No)	(N/A)	Reference (21 CFR)
5. Are stocks rotated?				211.86, 211.150
6. Are raw materials and preweighed materials stored off the floor on pallets/shelving to permit cleaning and inspection?				211.80c
7. Is there a sufficient accountability system for labels?				211.125c
8. Are preweighed and original containers of raw materials properly identified; such as in-house code number batch/lot number, product disposition/status and other information required by procedure?				211.80d 211.101b
9. Are raw materials properly identified for status: approved, rejected, quarantined?				211.80d
10. Are all raw materials approved (released by Quality department) before issued to a batch for use?				211.84e 211.101c
11. Are reject materials identified and segregated from approved materials?				211.42, 211.80 211.89
12. Are raw materials stored under quarantine until examined for released or rejection:				211.82, 211.84, 211.89
13. Do raw material records list the identity and quantity of each shipment of each lot, the name of the supplier, the receiving in-house lot number, and the date of receipt?				211.84 211.184

Section III A: Production/In-Process Control. General	Response (Yes)	(No)	(N/A)	Reference (21 CFR)
1. Are all employees wearing appropriate clothing including proper head, face, and hand coverings?				211.28a
2. Are employees performing their assigned function in an orderly manner, according to instructions and procedures?				211.25a
3. Are there written procedures for production and process control?				211.100a
4. Are procedures being followed?				211.100b
5. Are training records current?				211.25a
6. Does the production equipment appear to be appropriately designed, constructed, and maintained?				211.63 211.65 211.67a

Section III A: Production/In-Process Control, General	Response (Yes)	(No)	(N/A)	Reference (21 CFR)
7. Are equipment cleaning/use logs being maintained?				211.67c 211.182
8. Are procedures for the cleaning, sanitizing, and maintenance of equipment and utensils being followed?				211.67b
9. Are the departmental instructions for general cleaning being followed?				211.56b
10. Are proper cleaning materials and solutions being prepared and used according to departmental instructions?				211.56b 211.67
11. Are utensils cleaned to prevent contamination between different product usages?				211.67
12. Where appropriate, are containers and utensils that require cleaning, and that are clean, properly identified, segregated, and stored in appropriate areas?				211.67b 211.105a

Section III B: Production/In-Process Control, General	Response (Yes)	(No)	(N/A)	Reference (21 CFR)
1. Are master and batch production records properly assembled and sufficient in the following content:				211.186, 211.188
A. Is each prepared, dated, and signed in full signature by one person and independently checked, dated and signed by a second?				211.186 211.188a
B. Are batch records kept at the work stations during operation?				211.100b
C. Have batch records been, or are being completed properly?				211.186a 211.188b
D. Is a complete list of components provided?				211.186(b)(3)
E. Are the weights and measures of each raw material recorded when measured?				211.188(b)(4)
F. Are information and signatures recorded on records immediately after performing an operation?				211.100(b)
G. Are there inventory records for the use of each raw material?				211.184(c)
H. Can a raw material be traced to its use in a particular lot of product?				211.184(c) 211.188(b)(3)

Section III B: Production/In-Process Control, General	Response (Yes)	(No)	(N/A)	References (21 CFR)
I. Are there records fully documenting any planned or unplanned variances from procedures or operations				211.100(b)
				211.192
J. Are raw materials reconciled against each in-house lot number?				211.184(c)
K. Are reconciliation records within established limits?				211.184(c)
L. Have any discrepancies in reconciliation records been investigated and documented?				211.192
M. Are calibrated stickers/tags present and dated on scales and balances?				211.68(a)
				211.160(b)
				211.194(d)
N. Are scales zeroed before weighing operations?				211.160(b)(4)
				211.194(d)
O. Are scales that are not in proper working order identified to prevent use?				211.160(b)(4)
2. Have time limitations on the holding of processing and in-process items been established and are they being adhered to?				211.111
3. Are changes to the issued batch records approved prior to the start of work and documented on the batch records?				211.100
4. Are proper gowning techniques and coverings employed?				211.28
5. Are nonfiber releasing filters used for liquid filtration in the manufacture of injectable drug products?				211.72
6. Are bagged or boxed components stored off the floor?				211.80
7. Are lubricants and coolants controlled properly such that they cannot come in contact with product containers, closures, in-process materials, or finished products?				211.65(b)
8. Have the sterilization processes been validated?				211.100

Section IV: Facilities and Metrology	Response (Yes)	(No)	(N/A)	References (21 CFR)
1. Are operations performed in defined areas to prevent contamination and mix-ups?				211.42(b,c)
2. Are production areas sufficient in size, construction, and location?				211.42(a)
3. Does the entire area meet an acceptable level of cleanliness, and are all areas in an orderly condition?				211.42, 211.46
				211.56, 211.58
				211.67

Section IV: Facilities and Metrology	Response (Yes)	(No)	(N/A)	References (21 CFR)
4. Are appropriate exhaust and vacuum systems employed in operations to minimize air contamination?				211.46
5. Are the air handling systems operating properly?				211.46(a,b,c)
6. Are all light fixtures operating properly?				211.44
7. Are waste and trash collected and removed when appropriate and are there written Standard Operating Procedures (SOPs)?				211.50
				211.56(a,b)
8. Is there any evidence of infestation by rodents, birds, insects, or vermin?				211.56(a)
9. Is there a formal pest control program and are records up to date?				211.56(c)
10. Are maintenance and cleaning logs or records accessible and completed properly?				211.63, 211.67(c)
				211.182
11. Are there adequate written calibration procedures?				211.68
Is there a record of the calibrations being performed?				211.194(d)
				211.160(b)(4)
12. Are drains to sewers designed with an air break to prevent back siphonage?				211.48(b)
13. Are the environmental systems adequate to control air pressure, humidity, temperature, microorganisms, and particulate matter?				211.42
				211.46
14. Are the adjacent exterior grounds manicured?				211.56
15. Are adequate washing facilities with hot/cold water soap, clean toilets, air driers, or single service towels provided?				211.52
16. Have support systems such as water, vacuum, and compressed air been validated to assure causal variability in the characteristics of in-process material and final product?				211.00
17. Is there a designated area for eating, drinking, smoking?				211.42
18. Is there an acceptable separate gowning area?				211.42(c)
19. Are the environmental control systems periodically inspected, and are such inspections documented?				211.42c(10)IV
20. Are tools, gauges, and test equipment identified to reflect the date calibrated and the next calibration date?				211.68
21. Does the calibration data reflect the person responsible for calibration?				211.68
22. Are adequate facilities used for storage of tools, gauges, and test equipment?				311.68
23. Are the positive atmospheric controls calibrated and monitored?				211.68
24. Are there positive pressure differentials with all doors leading into less clean areas?				211.42c(10)iii
25. Are drawings and blueprints adequately controlled?				211.42

Section V: Quality Control, Final Inspection	Response (Yes)	(No)	(N/A)	References (21 CFR)
1. Is there a final inspection performed by Quality Control on a batch by batch basis?				211.22
2. Does Quality Control have written procedures tc assure that production records are reviewed?				211.192 211.22(d)
3. Is there a thorough investigation of any discrepancies found during the final review of the production documents?				211.192
4. Is the product checked for correct expiration dating?				211.188(b)(12) 211.137
5. Are records of inspection and test data maintained?				211.160
6. Are adequate sampling plans used?				211.134 211.165
7. Do the sampling procedures specify the manner in which the samples are to be pulled and by whom?				211.160
8. Is there a classification of defects?				211.165 211.16
9. Are reserved samples of finished products maintained?				211.170
10. Has the accuracy, sensitivity, specificity, and reproducibility of the test methods been demonstrated via validation?				211.134 211.165(e)
11. Are the quality control facilities adequate?				211.22
12. Do the facilities have adequate equipment for the performance of Quality Control functions?				211.22
13. Are all components, manufacturing materials, in-process materials, packaging materials, and labeling tested prior to release to production?				211.42(c)(9) 211.184(a)
14. Are test procedures documented?				211.194
15. Are classifications of defects utilized during visual and dimensional inspections?				211.165
16. Are raw material file samples maintained for future reference?				211.84(b)
17. Are laboratory reagents and other chemical supplies identified, tested, and expiration dated?				211.194(c)
18. Are laboratory instruments requiring calibration maintained according to the calibration procedures?				211.194(d)
19. Is there a written microbial monitoring program for nonsterile products?				211.113

Internal Facility Audit Checklist	Response (Yes)	(No)	(N/A)	References (21 CFR)
Section VI: Quality Assurance				
1. Are the responsibilities and procedures applicable to Quality Control documented and approved?				211.22
2. Do written procedures exist governing a document change control system?				211.100
3. Are adequate controls in effect to assure drawings, change notices, and specifications in use at the time and place of the operation?				211.100
4. Are records maintained, which reflect a history of change incorporation?				211.100
5. Are the documents up to date?				211.100
6. Are the documents comprehensive in text?				211.100
7. Are the procedures employed by Quality Assurance documented?				211.22(d)
8. Does Quality Assurance have the authority to approve/reject plant equipment, process, and procedural changes?				211.22
9. Is an adequate product complaint/product failure handling system in effect and are investigations facilitated properly?				211.198
10. Is there an adequate stability program?				211.192 211.166
11. Is batch record data reviewed on an annual basis to determine continual acceptability of quality standards specifications, manufacturing, and control procedures?				211.180(e)
12. Does Quality Assurance approve/reject product manufactured, processed, packaged, or held under contract by another company?				211.22
13. Have in-process specifications been derived from process average and process variability estimates (validation)?				211.11
14. Is the individual responsible for Quality Assurance program not directly responsible for the performance of a manufacturing operation?				211.22
15. Are formal audits of the Quality Assurance program performed by individuals not having direct responsibility for the program?				211.180
16. Are documented corrective actions taken on noted audit deficiencies?				211.180
17. Are the procedures for performing audits formalized?				211.180
18. Are the procedures available in an acceptable format for review by an FDA inspector?				211.180
19. Are production records retained at least 1 yr beyond expiration date and 2 yr beyond the date of release?				211.180

Internal Facility Audit Checklist	Response (Yes)	(No)	(N/A)	References (21 CFR)
Section VII: Packaging, Shipping and Distribution				
1. Is a check list used to verify shipping requirements?				211.20
2. Is product configuration verified prior to shipment?				211.150 211.196
3. Do packing and shipping records identify the individual performing and inspecting the shipping operation?				211.25
4. Are adequate storage facilities available and in use to safeguard the quality of the product between final acceptance and shipping?				211.42
5. Is there a distinct and adequate size label room?				211.142 211.42
6. Is the label room designed to avoid mix-ups?				211.122 211.42
7. Is there a sufficient accountability system for labels?				211.122 211.125
8. Are there specified restrictions on authority for access to labels and other labeling?				211.122(d)
9. Are labels proofread for accuracy before being released to inventory?				211.122

Internal Facility Audit Checklist	Response (Yes)	(No)	(N/A)	References (21 CFR)
A: Biopharmaceutics				
1. Are the analytical lab facilities adequate to support its workload? Describe: ____________________ ____________________				CP-Bio
2. Does the lab have written Standard Operating Procedures (SOPs)?				CP-Bio
3. Are the SOPs adequate, followed and available to personnel?				CP-Bio
4. Are analytical lab personnel capable of performing the required testing?				CP-Bio
5. Were CVs of analytical lab personnel available and reviewed?				CP-Bio
6. Does the analytical lab receive blood or urine specimens for drug analysis? Describe: time and condition in prior clinic/lab or in transit, etc. ____________________ ____________________				CP-Bio
7. Were specimen storage conditions adequate before receipt at the analytical lab? Describe: special arrangements, shipping restriction, etc.____________________ ____________________				CP-Bio

Internal Facility Audit Checklist	Response (Yes)	(No)	(N/A)	References (21 CFR)
8. Was dry ice used to ship specimens to the analytical lab?				CP-Bio
9. Will specimens be accepted at the analytical lab 24 h/d, 7 d/wk? Describe special arrangements, shipping restrictions, etc.: ______ ______				CP-Bio
10. Is a receipt log used to record the receiving of specimens, including conditions of samples, quantities, etc.				CP-Bio
11. Was adequate equipment used to store specimens awaiting analysis? Describe: ______ ______				CP-Bio
12. Does the above storage equipment have a warning device when not operating properly?				CP-Bio
13. Does the above storage equipment have a temperature recording device? Describe: ______ ______				CP-Bio
14. Were the stored specimens in adequate condition? Describe: separation of samples, labels intact, broken, etc. ______ ______				CP-Bio
15. Was the analytical equipment used adequate? Describe: models, conditions, etc. ______ ______				CP-Bio
16. Were there written equipment operating procedures? Describe: adequate, available, etc. ______ ______				CP-Bio
17. Were there written calibration/standardization procedures for analytical equipment? Describe: adequate, used, frequency, etc. ______ ______				CP-Bio
18. Are specific instrumental conditions recorded anywhere?				CP-Bio
19. For *Antibiotic Studies*, are incubators available? Describe: type, dimensions, etc. ______ ______				CP-Bio
20. Are other conditions adequate? Describe: room temp. controlled, petri dishes, zone readers, autoclaves, etc. ______ ______				CP-Bio
21. Is the analytical lab involved in the analysis of drug standards or products employed in a biopharmaceutic study? Describe: adequate, etc. ______ ______				CP-Bio

Internal Facility Audit Checklist	Response (Yes)	(No)	(N/A)	References (21 CFR)
22. For an *in vivo* analytical method, are there adequate data to document/validate claims on specificity, sensitivity, precision, and stability?				CP-Bio
23. Did the analyst use coding techniques to blind samples?				CP-Bio
24. Was the order in which samples were analyzed adequate? Describe: randomized, in order, etc.____________				CP-Bio
25. Were test and reference samples run at the same time, under identical conditions?				CP-Bio
26. Were standard curves prepared for each batch of unknown samples? Describe: how often run, all reported, etc.____________				CP-Bio
27. Were blinded, spiked, control samples used in each CP-Bio run? Describe: who prepared, etc. ____________				CP-Bio
28. Was the source of blank biological fluid adequate? Describe: use subject's zero hour sample, pooled plasma, etc. ____________				CP-Bio
29. If samples were rerun, were control samples run simultaneously?				CP-Bio
30. Is there an adequate procedure defining whether an original or rerun value is reported?				CP-Bio
31. Are prepared reagents properly labeled? Describe: include chemist, storage condition, prep date, expiration date, etc. ____________				CP-Bio
32. For *Antibiotic Analyses*, were samples properly run through the incubator?				CP-Bio
33. Were control samples incubated at the same time, in the same incubator?				CP-Bio
34. Were other conditions adequate? Describe: temperature of incubator burner used to heat wire, zone readers used, etc. ____________				CP-Bio
35. For *Radiometric Analyses* was the equipment used adequate? Describe: scintillation counter, etc. ____________				CP-Bio
36. Is the background level determined? How? ____________				CP-Bio
37. Are bound notebooks used in the analytical lab?				CP-Bio

Internal Facility Audit Checklist	Response (Yes)	(No)	(N/A)	References (21 CFR)
38. Are the procedures for filling in the notebooks adequate? Describe sequential entries, ink, chemist sign/date, how often, supervisor review/sign, etc.: ________ ________				CP-Bio
39. Are all raw data (chromatograms, standard curves, etc.) retained? Describe: where chemist sign/date, supervisor review/sign, etc. ________ ________				CP-Bio
40. Are the procedures for maintenance of lab data adequate? Describe: "corrections," erasure, correction fluid, ink, record retention, etc. ________ ________				CP-Bio

Internal Facility Audit Checklist	Response (Yes)	(No)	(N/A)	References (21 CFR)
B: Computer Systems Validation				
1. Does the user feel that the computer system comprises hardware, software, and a regulated application that warrants formal validation?				CS-Val
2. Does the interpretation of a "regulated application" include computerized data used in the assessment of safety or efficacy of an investigational drug?				CS-Val
3. Has the user identified/described the computer system to a Validation Board or management, which agrees with the need for validation?				CS-Val
4. Is this computer system being purchased, in total or in part? Describe: ________ ________				CS-Val
5. For an existing computer system, has it ever been validated before, in total or in part?				CS-Val
6. Is this a new computer system just being developed?				CS-Val
7. Has an appropriate task force been identified to develop and validate this new system?				CS-Val
8. Was the system development life cycle approach used to develop the computer system?				CS-Val
9. Was a validation protocol developed up front to describe how the system would be validated?				CS-Val
10. Did the validation protocol identify all the necessary ingredients? Describe: ________ ________				CS-Val

Internal Facility Audit Checklist	Response (Yes)	(No)	(N/A)	References (21 CFR)
11. Does an adequate change control system exist? Describe: cover H/W, S/W, SOPs documented, etc.				CS-Val
12. Is the change control system based on change only? or also on time interval, whichever comes first? Describe. ______ ______				CS-Val
13. Were test data sets used to test the modules/total system?				CS-Val
14. Did the test data sufficiently "stress" the system? Describe: extreme values, erroneous values, replications, etc. ______ ______				CS-Val
15. If software is purchased, will the vendor give the user the program source code?				CS-Val
16. Is the vendor willing to "escrow" the source code with a third party?				CS-Val
17. Was training adequately addressed for the users of the system?				CS-Val
18. Was backup for the system adequately addressed?				CS-Val
19. Are there adequate SOPs covering the necessary aspects of the system?				CS-Val
20. Is there formal approval of the system by the user and management?				CS-Val
21. Was a final validation package (documentation) prepared for the computer system?				CS-Val
22. Was the validation package approved by the Validation Board or management?				CS-Val
23. Will the Validation Board continue to monitor the system/changes in the future?				CS-Val

46.4 GMP REQUIREMENTS

Assuring regulatory compliance with large-scale industrial chromatographic systems requires examination of the utilization environment as well as the internal and controller system software. This second checklist provides GMP guidance related to the general facility.

46.5 CONCLUSION

Taken together, these two checklists, 21 CFR Part 11 and Inspection of a GMP Facility, provide highly applicable guidelines for the design, management, validation, and utilization of an industrial scale chromatography system. They can be used to spotlight and address problems, and to ultimately audit and prove compliance.* When coupled with a traditional process validation of the chromatography system, including focus on cleaning and residuals, the end result is highly credible evidence in support of the reliance upon and use of the system.

*The FDA will generally accept (and review) the results of an independent audit; to date, over 300 of the author's audit reports and certifications have been accepted by the FDA as *de facto* evidence of compliance. The authors would like to thank Ms. Lisa Gonzales and Ms. Janine Cahill for their assistance in compiling the checklists.

REFERENCES

1. Chapter 21, CFR, Part 210, Part 211.
2. Weinberg S. Cost-effective compliance. Sci Comput Instrum 2003.
3. Weinberg S. The FDA vector: part 11 risk-based changes. Am Biotechnol Lab 2003.
4. Weinberg S, editor. Good laboratory practice regulations. Marcel-Dekker; 2007.
5. Weinberg S. Emergent FDA biodefense issues for microarray technology: process analytical technology. Expert Rev Mol Diagn 2004; 4(6): 779–81.
6. Weinberg S. Taking control of regulatory inspections. J Bioprocessing 2005.
7. www.fda.gov. The official website of the US food and drug administration, 2009.

47

GMPs AND GLSPs

Beth H. Junker
Bioprocess R&D, Merck Research Laboratories, Rahway, New Jersey

47.1 INTRODUCTION

Validation is the process of "establishing documented evidence that a system will do what it purports to do," a simple definition given by the US Food and Drug Administration (FDA) in 1977 (1). The progression of an organization with respect to its validation status has been described by a Validation Capability/Maturity Model (2). It starts with the "validation unaware," then shifts to the "validation aware" once knowledge of the theory is gained. The organization is then "validation active" when it is forced to participate by regulations and customer demand. It moves to the highest level of "validation enthusiast" when it possesses practical experience, recognizes benefits/limitations, and encourages others.

Validation has been presented as a complement to distinct limitations in the range of product testing techniques; in fact, validation was developed due to product failures that were not detected by product testing (3). One specific event was a sepsis outbreak caused by improperly autoclaved vials not detected despite sterility testing of a statistically significant number of vials (4). Further background has been compiled in the form of a history of early Good Manufacturing Practice (GMP) regulations for bulk pharmaceutical drugs (5), initial interpretations of validation guidelines (6), and a retrospective of validation in United States (1). The approaches to issues in biological and pharmaceutical manufacturing change as better analytical techniques evolve, improvements in equipment design and operation are implemented, and pilot approaches are tested and proved worthwhile over time.

Bulk pharmaceutical chemicals (BPCs) are made by chemical synthesis, fermentation (recombinant or secondary metabolite), and/or enzymatic reactions. For BPCs, FDA guidelines clearly state that the level of GMP rigor can be modulated by the position of the step in the overall process (early steps requiring less depth) and the specific product (for some products, early steps might require more depth) (7). A specific evaluation of the FDA's approach to BPCs has been documented with emphasis placed on determining at what step focused GMP attention should be given (8). Although the discussion involves a specific processing step, there are implications on the validation of the equipment, portion of the facility, and operating/testing personnel being used for that processing step.

Biologics are products made from living organisms (or isolated from blood products) that are complex proteins, viruses, antigens, or antibodies. Production may use recombinant or nonrecombinant bacterial, yeast, fungal, animal, insect or plant cells, viruses, or live animals. The guidelines for biologics GMPs include a history of GMPs for biologics (9), a series of early papers by former FDA inspectors (10–12), and a description of differences in validation approaches between conventional pharmaceuticals and biologics (13). Commentary concerning how the guidelines for drug inspections in effect at the time (14) applied to inspections of biopharmaceutical manufacturing facilities by the FDA was also published with clear emphasis on purification, formulation, filling, and lyophilization (15). These papers provided insight as to what the FDA was seeking regarding biologics facility design, validation and operation. Although they now are almost two decades old, the areas cited should still be addressed, but in a manner con-

Downstream Industrial Biotechnology: Recovery and Purification, First Edition. Edited by Michael C. Flickinger.

sistent with current GMP guidelines where changes have evolved. As a result of these writings, specific guidelines for biotech facilities were developed (16–19).

There are clear differences in approaches to GMPs between biologics and drugs in terms of both bulk finishing (tableting, granulation, sterile filling) and bulk synthesis (fermentation, chromatography) (20). The focus areas are: (i) raw material variability, specifically natural sources from corn steep liquor to fetal calf serum, and the potential for uncontrolled contaminants such as bovine spongiform encephalitis (BSE), (ii) product tracking and yield, specifically use of activity assays rather than mass assays, and the resulting inability to conduct material balances, (iii) incomplete step yield information and diagnostics making it difficult to account for step activity losses, and (iv) process complexity encompassing several interrelated processing steps (20). Recent progress defining well-characterized biopharmaceuticals utilizing advanced analytical techniques (21) has assisted in providing regulatory consistency based on product characteristics rather than whether the product was approved by the Center for Drug Evaluation and Research (CDER) or Center for Biologics Evaluation and Research (CBER) (22). It also has permitted processing equipment, procedure, and manufacturing site changes to occur later in the clinical development of a product (23).

Direct and indirect information is available to companies about the scope and nature of validation for GMP. Direct statements are present in the Code of Federal Regulations (CFR), specifically Sections 21 CFR 210 and 211, which focus on drugs/pharmaceuticals, and Section 21 CFR 600, which adds additional regulations relevant for biologics. These chapters discuss outcomes rather than methods to achieve them and consequently are subject to interpretation by each company for each type of process and product. This thematic approach might be analogous to varied interpretations of literature passages by individuals based on their individual lifetime experiences and goals with no single interpretation universally applicable. Interpretations of the regulations also can be found in various published "Points to consider" and "Guidelines" documents authored by FDA officials, often with substantial input from industry (24–31). Guidelines are available on demand at the FDA's website. Public conference presentations, panel discussions, and publications by FDA inspectors also offer interpretations but, as each situation can be different, it can be difficult for inspectors to speak specifically (or answer a "quick question") and for audience members to directly and correctly apply the approaches presented. The FDA publication entitled *Human Drug cGMP Notes* provides some up-to-date guidance regarding specific policy questions. Finally, the major indirect method for companies to obtain information is by reviewing published 483 violations and other public reports from inspections of other companies. However, these summaries (obtainable from the *The Pink Sheet* and *The Gold Sheet* published by F-D-C Reports) usually contain two- to three- sentence quotes out of situational context, thus subject to possible misinterpretation and misapplication. The most valuable information concerning "reduction to practice" arises from individual companies who present and publish their specific approaches to GMP. Implementation has been a challenge for companies as they evolve from research to a development focus; specifically of note is an early paper published by Biogen outlining these issues for interferon (32).

Risk-based approaches are being adopted by the FDA to make regulatory decisions and streamline inspections. Inspections are streamlined by maintaining/increasing the inspection level for drug manufacturing sites needing frequent oversight and reducing inspections for facilities with lower risk; consequently, inspections are targeted for high risk operations that have the greatest impact on public health (33). The FDA is promoting a transition to a more scientific approach to GMP regulation (34). Manufacturers have to rely on identification, analysis, and quantification of risk to arrive at cost-effective procedures and operating practices (35–37). Quality risk is being defined as the probability combined with the severity of impact for an occurrence (31). Risk-based approaches have been published for cleaning (38–41), suite changeover (42), laboratory computer systems (43), and process validation (44).

The FDA's initiative, pharmaceutical quality in the twenty-first century, encourages industry-wide adoption of quality control and risk management to better assure quality with less direct FDA oversight (33). Its basis asserts that a quality emphasis in process design, control, and testing directly leads to acceptable GMP operation (34).

Process analytical technology (PAT) (45), a second related FDA initiative, ensures final product quality via in-process measurements of critical quality and performance attributes. It helps in identifying the sources of variability and risk thereby improving process understanding, which leads to continuous improvement. PAT has been applied to several aspects of bioprocessing already, specifically (i) cleaning validation by on-line monitoring of rinse water for total organic carbon (TOC) and conductivity to minimize cleaning time and rinsing volumes (38), (ii) bioreactor control by measuring DO and pH in the broth, (iii) harvest operations by measuring centrate turbidity, and (iv) isolation steps by measuring elution stream UV absorption and buffer conductivity. In some cases, the PAT criterion is directly implemented as part of the validation protocol (40).

A third complementary initiative is quality by design (46) which emphasizes quality by design (desired state) versus quality by inspection (present state), and (ii) understanding of the process (including raw materials and controls), which leads to understanding of the impact of process changes on quality. In this framework, scalability and validatability is directly incorporated into process development over the product's lifecycle emphasizing sound experimental rationales and analytical measurements (47).

47.2 FACILITY REGULATORY GUIDELINES

47.2.1 Good Manufacturing Practice (GMP)

The following of GMPs regulations, as well as validation execution, is not a complete set of rules but rather a philosophy (48) that is largely dependent on the process, product, and the company. The goal of validation is to satisfy Section 211.42 of 21CFR, which states that the facility and equipment "must be appropriate and suitable for the intended function" (49). Fundamentally, validation is aimed at defining procedures and processes that minimize adverse effects on product quality, and then providing documented evidence that the procedures and processes are appropriate and repeatable. Although some may insist that all equipment used in a process must be validated, many others have debated about just how much validation is sufficient. It is most efficient to judge the need for additional validation through scientific justification for the knowledge demonstrated (50,51). Nevertheless, there can be a "keeping up with the Joneses' philosophy, both within and among companies, which can lead to wasteful testing.

Validation, along with verification, qualification, and commissioning activities, forms a significant percentage of the time and money expended for a new pharmaceutical facility. Initially, validation tasks essentially repeated installation, construction, and start up testing and became redundant tasks whose goals were merely to ensure contractors performed their jobs correctly (49). Later on, validation tasks were designed to complement rather than repeat prior testing. Validation scope, benefits, limitations, and intensity may not be appreciated initially. This lack of appreciation is most detrimental when full start up and commissioning is delayed, then causing compressed validation timelines (48). Appropriate validation testing will expose nonconformances and deficiencies in design, construction, or operation (48) before they impact process or product quality. Such revelations can be troublesome, particularly if they erupt in waves. If the end user is permitted to evaluate the potential impact of the problem and choose to accept operation with the limitation present, it may not be necessary to resolve every variation or discrepancy with a modification.

The validation master plan is essentially the facility's quality specification and can be as important as the facility design specification itself. It is a structured, detailed work plan that, although formally approved by the various subgroups involved, should be a working document linking all aspects of validation from start to finish (48). At a minimum, changes to these master plans should be tracked consistently (e.g. via redlining if permitted), and then formally updated periodically, typically in every 3–5 years. The plan contains a description of the facility especially its purpose, function, and limitations. A preliminary equipment and utility list, briefly describing each equipment item, stating what type of testing is required, and listing preliminary acceptance criteria, is useful for identifying validation test equipment as well as constructing preliminary manpower estimates (52). It also serves as a useful handbook describing the facility and its capabilities. Applicable policies and standard operating procedures (SOPs), both approved and those not yet developed, can be referenced, along with floor plans marked with clean/used equipment, raw material/product/waste, and personnel flow patterns. Finally, a sample protocol might be drafted at this point to document team agreement on a common template. Master plans also document information about ongoing validation programs for the facility, including policies associated with change control (52).

There are several segments and types of validation testing, and each of which encompasses various activities depending on the practice of the facility. Grouping of individual activities within these sections is less important than completing the required testing to insure proper and consistent equipment operation for the process (50). Redundancy of testing among various validation phases also should be avoided (50). Traceability matrices link testing deliverables with equipment specifications for tracking as well as minimize testing duplications and omissions. Electronic document management has the potential to further reduce nonvalue added effort (53). The major categories of validation testing are briefly described below. The early validation phases lead to later validation phases aimed at providing subsequent proof of product or process specific reliability.

The design qualification (DQ) focuses on a structured GMP review and verification to identify potential design weaknesses that may impact qualification/validation prior to ordering equipment and initiating construction (52,54,55). It affirms that the design conforms to expectations by relating each design element (often specified in a functional specification (FS) to specific regulatory or user requirements (55). To examine each aspect of the design (54), one approach is to conduct this GMP review using guidewords, such as *contamination, containment, validation, instrumentation, cleaning, and sterilization*. This exercise is best done by representatives of the unit (either internal or external) who conduct the validation. Some project leaders recommend a different unit conduct the validation than executed the design and/or construction to maintain appropriate checks and balances. Others believe that "design/build" strategies that encompass validation give a single-source responsibility to a single unit and thus avoid inefficient finger-pointing when errors are uncovered (35). Potential GMP omissions that are uncovered during this exercise then can be evaluated for their impact by the validation/design team. Appropriate solutions may involve additional procedures, extended validation testing, or design modification. The system requirements (SR) may be developed at this time to provide acceptance criteria for operational and performance qualification protocols in the form of limits, ranges, and/or tolerances (55).

Installation qualification (IQ) provides documented evidence that the equipment was both built and installed correctly according to design criteria and manufacturers' recommendations. Its primary focus is on documentation collection and review including manuals, purchase orders, pressure test data, passivation logs, spare parts lists, SOPs, hazard operability (HAZOP) results, calibrations/loop checks, and as-built piping and instrument diagrams (P&IDs)/isometric drawings. IQ insures that the ordered equipment was received and is safe to operate. Recommendations for the scope and collection of these have been documented (56). There is a balance between requesting sufficient documentation and cataloging of excess, redundant, or unnecessary documents (48). Results of factory acceptance testing (FAT) as well as site acceptance testing (SAT) also can be included or referenced in the IQ. Typically, the IQ is essentially a documentation review (49), but including physical inspection of the equipment and its installation, with some not willing even to power up the equipment until the IQ has been executed, reviewed, or even approved. However, such constraints might hamper timely identification and resolution of equipment problems.

Operational qualification (OQ) provides documented evidence that the equipment and facility operate correctly and have the sufficient range of capabilities over the expected span of operational parameters (48). OQ testing centers on functional tests of equipment features including alarm and interlock verification. It includes draft SOP suitability testing, some aspects of start-up testing if not included or referenced in the IQ, and also focuses on overall system capacity checks for utilities (49). Often, IQ, and in some cases OQ, can be performed by mechanical contractors and portions of these studies commonly are performed as FAT and SAT tasks at the vendor site prior to shipment of the equipment. The OQ testing needs to be completed satisfactorily to test the limits of the system, at a minimum over the intended ranges for equipment operation and over which, any subsequent performance qualification (PQ) testing will be done.

PQ provides documented evidence that the facility can repeatedly produce product within established specifications and quality attributes (48). It involves extensive testing of specific process conditions as well as the process equipment. This testing encompasses many types of qualification/validation including steam-in-place (SIP), clean-in-place (CIP), sterilizing and/or decontamination autoclave load patterns, glasswasher load patterns, environmental testing under processing conditions, and analysis of utility quality. Generally, triplicate runs are utilized to show reproducibility but can get extensive when combined with upper and lower processing boundary condition values (1). The concept of a "worst case" scenario is used to minimize testing, but the worst case should be within the boundaries of normal operating conditions that were previously validated for critical parameters and not necessarily an "edge of failure" value (1). Before or at least concurrent with PQ, key analytical methods utilized should be validated (57).

Performance qualification leads directly into process validation for which testing strategies are available (58,59). Process validation (PV) provides documented assurance that process routinely and consistently meets its goal of producing the same product with the same safety and efficacy (6,59,60). Major sources of process variability are identified and quantified to assess and then reduce potential risks to achieving established quality attributes. By improving process understanding, PV decreases processing risks that can lead to lost productivity and efficiency.

During facility operation, substantial additional testing effort is required. This includes ongoing monitoring of utility and environmental quality, periodic (typically annual) ongoing validation challenges, and revalidation or equivalency testing prompted by equipment and/or process modifications (i.e. change control). Appropriate pages and previous acceptance criteria from original protocols might be used *in lieu* of authoring new protocols. Ongoing validation protocols might be reused over several years provided they are written in a generic fashion referencing umbrella policy documents or SOPs for specific details that are subject to alteration. Once the scope of the ongoing validation effort is documented and agreed upon, the impact of the planned workload on both equipment downtime and manpower can be evaluated, and a system to manage and organize activities can be implemented. Initial and ongoing validation activities have been minimized (e.g. no cleaning validation) and/or transferred to the vendor (e.g. leachable/extractable testing, irradiation sterilization effectiveness) through the increased use of disposables in processing (61).

Commissioning, the proper documentation of facility construction and installation, can be the first step in the validation of critical equipment (62) or an attractive alternative to the validation of noncritical utilities that are not in product contact, such as cooling fluids or instrument air (52). Key aspects and strategies for commissioning have been documented (63). Every activity performed by construction, installation, or start-up personnel is documented, although a formal qualification protocol may not be followed. For full effectiveness, commissioning should include challenge and capacity testing (49). Interestingly, one author notes that when noncritical "commissioned" utilities unexpectedly fail or are problematic, their critical importance becomes obvious (49) suggesting that more rigorous testing might be appropriate.

The extent of equipment, facility, and process validation is notably dependent on the degree to which in-process controls and end-product testing can assure quality and safety. Validation thus is most extensive for biologics products (complex therapeutic proteins and vaccines) and less extensive for well-characterized proteins and secondary metabolites (22,64). References are available concerning validation best practices (65–67), general validation issues (64), potential validation problems and pitfalls (48,68), and validation streamlining (69–72). The details of validation stages, including example regulatory internal and external references, sample systems to be examined, and typical schedules also are available for model cases (52,73). A benchmarking survey on current practice in aseptic process validation, as well as investigation and additional testing strategies when results are unsatisfactory, is available; although it is oriented toward aseptic filling, its approaches might be applicable earlier to aseptic processing steps (74). In addition, descriptions of how GMP regulations have been applied to bioprocessing applications, such as the production of adenoviral vectors and stem cells, is available, indicating that GMPs continue to evolve to address current productions needs (75,76).

Validation protocols should cross-reference but not include items that are subject to change, such as equipment data sheets and equipment specifications, so that as data sheets are updated by engineering, the protocol is automatically updated (48,77). Protocols may be written and even approved prior to installation of the equipment especially when the facility operations group already is familiar with the item. For computer equipment, a computer system validation and system life cycle approach outlined in published standards has been adopted and sometimes integrated with equipment validation (78–80). The extent of validation depends on the complexity and customization of the computer system.

47.2.2 Good Large Scale Practice (GLSP)

GLSP (or GILSP) started as the title to a section added to the *Guidelines for Research Involving Recombinant DNA Molecules* issued by National Institutes of Health (NIH) in 1991 (81–83). The GLSP level of containment is recommended for large-scale research or production of biological materials that are well characterized and have an extended history of large scale safe use (84,85).

The major difference between GLSP and BL1-LS is the reduction of the requirement for closed system operation with inactivation of organisms, elimination of the need for exhaust gas treatment, and control of aerosols via procedure methods not physical devices (83). It refers to a level of physical containment, less rigorous than BL1-LS, which is recommended for recombinant organisms, composed of a nonpathogenic host cell, free of adventitious agents and possessing a history of safe use and a well-characterized vector/insert, free of harmful sequences. The resulting recombinant or genetically modified organism (GMO)should be safe to humans and the environment; specifically the inserted rDNA should not increase the drug resistance of disease-causing substances. The new GLSP section in the above guidelines requires the following (83): (i) institutional procedures for control of production of recombinant organisms (i.e. an evaluation that health and safety issues are addressed), (ii) SOPs and training to insure proper culture handling and maintenance of clean workplaces, (iii) suitable protective clothing and facility/equipment containment, (iv) proper handling of environmental discharges, (v) aerosol control when necessary, and (vi) spill containment procedures for large losses. Additional guidelines focus on good laboratory practice, omitting smoking, mouth pipetting, and eating/drinking, as well as the establishment of internal accident reporting procedures. Procedures for commercializing new or modified strains are set forth in the Toxic Substances Control Act (86).

An outline of BL1 through BL3 criteria is available (83,87) in which the LS designation is added for large scale, typically greater than 10 L, to the corresponding laboratory designation. For BL1-LS, a closed system to reduce the escape of organisms is required with a validated inactivation procedure prior to removing material from the closed system. Aerosol production is minimized through design (e.g. sealed sampling devices with filters). Systems are sterilized prior to maintenance and an emergency spill plan is in effect. BL2-LS designation includes all BL1-LS requirements as well as prevention of aerosol generation, enclosure of the closed system within a secondary containment device, and use of rotating mechanical seals to prevent leakage. Devices are installed to monitor containment integrity during operation. The closed system is tested for organism release with a nonrecombinant organism prior to use and the system is permanently identified with a biohazard symbol. For BL3-LS, all BL1-LS and BL2-LS requirements are applicable, as well as operating the production equipment at the lowest possible internal pressure to insure containment integrity. The area surrounding the closed system must contain air lock access, readily decontaminated surfaces, sealed penetrations to permit decontamination by gas and reduce cross-contamination into other areas, and hand washing facilities with valves that are not hand controlled.

Finally, there should be controlled air flow into the facility and negative pressure with respect to surrounding areas.

Generally speaking, GLSP guidelines cover NIH requirements for working with recombinant DNA cultures; however, GMP guidelines may require product protection for some aspects of a process that has been classified at a higher BL2 or BL3 level. Specifically, negative pressure can be required for BL2 to protect personnel, while positive pressure can be required for GMP to protect product. This conflict has been resolved in one instance using inlet HEPA filtration and higher pressure surrounding areas to provide a barrier as well as physical segregation of live virus areas (88). In a second instance, risk assessments were used to evaluate several areas of contradiction and then determine how best to design for both GMP and containment of live virus production (87)

A general outline of safety and containment issues and practices at various containment levels is available, which includes examples of monitoring of organism release and suggestions for what equipment and where on the equipment to monitor (89). A second outline describes biosafety practices and challenges for specific types of viruses, bacteria, and fungi utilized for the research and development of proteins and viruses (90). A goal distinction can be made spanning the range from product to personnel protection (89): "low" avoids contamination of product cultures, "minimal" targets typical laboratory levels attainable when good laboratory practice is utilized, and "prevent" aims for no release in the air, on surfaces, or on personnel. Specific goals vary considerably according to the facility location and organism containment level (91). In one facility in Ireland, inactivation of recombinant CHO cultures was required to be demonstrated within the facility prior to discharge (92). In a second facility in the Netherlands, simulated discharges using nonGMO organisms were used to demonstrate the extent of inactivation after release to the wastewater treatment system (93).

47.3 MAJOR FOCUSES OF FACILITY PERFORMANCE VALIDATION

47.3.1 Steam-In-Place (SIP)

47.3.1.1 Strategies. Typically, SIP refers to vessel sterilization, most commonly that of fermenters and associated tankage and also in some cases to bioburden reduction. It can also refer to sterilization of other items such as dead-end filters in their housings, transfer lines, and cross-flow microfiltration skids. The basic sequence for an SIP procedure involves air removal, steam injection, condensate removal, and the breaking of vacuum with sterile air (94). Typically, the vessel contents are heated with venting until 95°C, at which time back pressure is applied with continued jacket heating until the vessel temperature is slightly above 100°C. Steam then can be introduced through various internal ports, such as the sparger line, by turning on associated steam supplies to these ports, creating a forward flow of steam into the vessel. Alternatively, internal ports, such as addition ports without subsurface dip tubes, can be opened to external traps or cracked valves creating a backward flow of steam out from the vessel. Vessel back pressure may be applied by passing the exhaust through a trap, orifice plate, or through a back-pressure control valve, which permits fine adjustments of the sterilization hold temperature. To assure proper equipment operation, back-pressure transmitter values should be compared against vessel pressure gauge readings and should be in agreement with expected values for the sterilization hold temperature according to thermodynamic steam tables. An outline of example sterilization procedures for bioreactor, media sterilization filters, and inoculum transfer lines is available (95). In addition, a useful reference describing sterilization theory and practice has been published (96).

The extent of sterilization is calculated using the parameter F_0, the magnitude of which is directly influenced by the area under the sterilization hold temperature versus time curve. The definition and examples of the use of F_0 in the design of sterilization cycles are available (97,98). The F_0 value is a measure of the equivalent kill of a test microorganism, typically *Bacillus (Geobacillus) stearothermophilus* spores for saturated steam applications, at the vessel sterilization temperature relative to that at 121°C. Although commonly calculated including only those temperatures at or above 121°C, calculation beginning at temperatures below 121°C can be appropriate for larger vessels, heat-sensitive media, or difficult to heat loads. Accounting for this, heat up time can substantially augment the value of F_0, although the precise amount of this augmentation is complicated by the need to measure D-values for the biological indicator's spores in the medium under study at various temperatures. A target F_0 thus may be calculated based on temperature mapping results from the system under study and the D-value of spore suspension in the medium to be sterilized.

For most SIP systems, the expected temperature and time required for sterilization are known or can be estimated based on prior experience of the operations group. Thus, production cycle safety factors need not be excessive and might be reduced to as little as a five minute extension of the hold time or a 1°C rise in sterilization hold temperature. Conversely, for transfer line sterilizations where F_0 accumulation rises to above 100 min within 5 min poststeam introduction, it can be beneficial to extend the production cycle sterilization time to 20–30 min because there are few, if any, detrimental effects of this slight oversterilization.

47.3.1.2 Effectiveness Studies. Effectiveness of SIP cycles is evaluated overall by the rate of foreign contamination occurring in the specific vessel for the specific process over time. Because inoculated cultures might overgrow low level contaminants, a media challenge, preferably with the same medium to be used during processing, often is conducted. Often, a sterility medium is substituted for the process medium, but care should be taken that the medium ingredients are of similar quality as typical raw materials, current CIP procedures can remove them adequately, and the vapor pressure and solids content of the medium reflect the problematic aspects of the simulated process. A media hold test typically lasts 7–10 days and uses a growth temperature of 37°C for favoring bacterial contaminants but a lower temperature of 25–30°C to favor fungal contaminants. A compromise temperature of 33–35°C might be appropriate if only one test is desired.

For definitive proof of sterilization performance, a spore-spiked media challenge can be undertaken. Spores challenges (biological indicators, BIs), containing 10^6 spores, are used in the form of strips for nonsubmerged locations and ampoules for submerged locations. Ampoules contain spores suspended in the liquid being sterilized. When large particles are involved, spores can be mixed into gels typical of particle sizes, sterilized, and then minced to expose spores during sterility testing of the challenges. Successful kill of these challenges provides a sterility assurance of at least a 10^{-6} log reduction of spores. Challenges are placed, along with thermocouples, at the hardest-to-heat locations (i.e. those with the lowest accumulated F_o) based on the temperature distribution study results. Examples of challenged locations include the following: (i) condensate return lines for the agitator seal, sample valve, addition assemblies, and bottom valve, (ii) sparger and vent air filters, and (iii) liquid and headspace of the vessel. Typically, the side of the equipment not exposed to direct steam is the side challenged. A summary of spore challenge methodologies is available (99,100); many of these methodologies were established by the food as well as the pharmaceutical industry.

During processing, sterile samples are obtained from individual vessels on average once per day depending on the facility practices. Sampling might be more frequent if the vessel contents are to be transferred to other sterilized fermenters in the near future (e.g. inoculation, fed-batch addition). Sampling might be less frequent if containment (product, personnel, or environment) is an issue. Samples are evaluated for culture purity for an inoculated cultivation and for sterility for an uninoculated cultivation. Aliquots of samples can be subcultured in nutrient rich broth, incubated at 37°C to promote growth of bacterial contaminants, then streaked on a nutrient rich agar and reincubated at 37°C. Subcultures can be examined microscopically using Gram stain techniques. In some cases, aliquots can be placed into phenol red containing medium and any color change noted (typically yellow if contamination present). Loopfuls of samples can be directly plated onto a nutrient agar conducive to bacterial or fungal growth and incubated at 37°C or a lower temperature of about 25°C respectively. False positives on sterility testing have been noted at 0.25% (101). Automated sterility analysis may reduce the number of false positives as well as speed up testing by incorporating fluorescence or other color change endpoints based on metabolic end products (101).

SIP sterilization of hydrophobic sparger or vent filters can be complicated by insufficient air removal. Specifically, dual filters positioned in series on an SIP system may not attain sufficient air removal or condensate drainage due to suboptimal placement of steam/condensate piping (102). Filter integrity testing after the production SIP cycle (which may have the hold time extended by adding a safety factor relative to the fractional test cycle) is required to insure that the SIP process did not damage the filter. Although in some cases, the filters might be heat damaged if the sterilization cycle is unusually long or hot, typically damage is caused by a large pressure differential across the filter, often due to suboptimal cool down procedures poststerilization, which might not be revealed during SIP cycle development (103,104).

Sterilizing filters also may need to be integrity tested after batch use depending on the GMP requirements associated with the type of product and the processing step. Recording of filter lot numbers, as well as integrity testing of filters postuse, is an important aid in investigating contamination. Postuse integrity testing of liquid filters is complicated by the difficulty of removing adsorbed materials that can alter the bubble point and/or diffusion rate. These issues are best reviewed on case-by-case basis with the filter manufacturer who may have measured bubble points and/or diffusion points for different solutions in addition to those for water. Preuse integrity testing is conducted if the vendor testing certification is not sufficient for process step use (e.g. final filtration for sterile bulk), although passing results do not insure subsequent filter integrity throughout the batch. Dual series filters are used to minimize the need for preuse integrity testing, but sterilization strategies need to be adapted to accommodate this arrangement. Postuse integrity testing can be done by removing the filter from its housing, removing the entire housing from the equipment with the filter intact, or testing the filter while located on the equipment with appropriate tie-ins. The latter two methods provide proof of proper installation of the filter if that is required.

47.3.2 Investigations of Failed Results

Most SIP cycle deficiencies can be corrected via SOP changes (most desired) or equipment modifications (less desired). Equipment modifications are more difficult to accomplish for an automatically sterilized vessel. It should be confirmed that the act of testing is not inducing failures prior to implementing corrections; however, selected investigation strategies for SIP failures can also be applied to autoclave load pattern testing failures.

Failures during SIP validation testing might be due to test procedure faults rather than proposed SIP cycle deficiencies (72). The biological indicator and/or thermocouple may be blocking steam/condensate flow, particularly if large spore strips are used in small process lines. Because spore suspensions and strips are incubated in nutrient medium and inspected visually for cloudiness daily, the contaminant cultured should be identified to genus and species if possible to confirm that growth is due to the indicator organism. Contamination from handling can result in growth of *Staphylococcus* instead of *B. stearothermophilus* (105). Contamination after testing by contact with unkilled spores from positive controls or from bags used to enclose spore vials might also occur.

The most frequent causes of contamination during fermenter operation (106) include construction materials, mechanical seals, valves, poor or overly complex design, operator error, instrumentation (e.g. calibration, valve failure), process air, transfer/feed lines, contaminated inoculum, incorrect batch or continuous media sterilization, and inadequate procedures. A mechanical checklist for returning a fermenter to service after an initial or repeated contamination can be useful. Factors that are important in fermenter contamination (107–109) as well as in design issues and sterility risks (102,110) have been documented.

Incorrect procedures often can be responsible for contaminations. Specifically, direct charges of powders through the vessel manway or handhole need to be carefully rinsed into the tank to avoid accumulation of dry powders on the tank upper sidewall. Homogenization of clumpy hydrophobic media powders should be considered and sieving of large particles from protein meals (e.g. corn meal) should be conducted. Media which foams upon heat up during sterilization (e.g. medium which contains tomato paste and soy flour) can be avoided as these excessively rich laboratory media, developed for screening and initial growth of a wide variety of organisms, may not be necessary for growth of the desired culture (111). Use of modified production tank medium for the seed tank stage(s) has been successfully explored instead of using laboratory shake flask seed medium (112).

A contamination tracking system can be used to present trends in the most probable causes of contamination (109). These might be classified as follows: (i) a mechanical problem involving an equipment failure such as a leaky valve, leaky mechanical seal, gasket, probe breakage, or incorrect temperature calibration, (ii) an operational problem in which SOP procedures were not followed, (iii) a cross-contamination in which contamination detection was not timely and material transferred subsequently was later found to be contaminated, and (iv) a process problem in which the procedures specified were not optimal such as incorrect batch sheet instructions concerning sterilization hold time or temperature. Reports should be circulated regularly to the operations group at all management levels and mechanical staff, as well as reviewed by quality personnel.

Contaminating organisms should be identified both in a preliminary fashion by microscopic observation and Gram stain, as well as more definitively using commercial biochemical testing kits fatty acid analyses or genetically based methods. A record of the types of contaminants as well as the ages and processes contaminated should be maintained and evaluated at least yearly for trends (109). Yearly contaminants can be classified into categories by process, age of contamination (e.g. after sterilization (AS), after inoculation (AI), <48 h, 48–120 h, >120 h) and type of organism (cocci, gram- positive/negative rod, fungal, and miscellaneous). The sterility investigation experience of the facility can be then utilized to determine the likely areas where the contamination may have entered. Example guidelines suggest that (i) gram-positive rods might result from incomplete media sterilization, improper inlet air filter installation, or air filter integrity failure, (ii) gram-negative rods might result from incomplete water sterilization (perhaps of filter sterilized nutrient additions), jacket leakage into the vessel, or water present in the inlet air to the vessel, (iii) cocci might result from human contact either during laboratory seed preparation or seed fermenter inoculation, and (iv) fungal contaminants, though rare, might result from improper inlet air sterilization as well as inadequate cleaning of cracks/crevices from fermenter ports (109).

47.3.3 Clean-In-Place (CIP)

47.3.3.1 Strategies. Cleaning validation establishes that written equipment cleaning procedures are developed and validated using actual product and production equipment, then periodically revalidated during the equipment's lifetime of processing that particular product. It should be demonstrated that cleaning procedures consistently reduce product and cleaning agent residuals to predetermined levels that are acceptable for subsequent processing steps based on scientific rationale. Calculations for the permitted amount of carryover residues from the prior batch are used to set acceptance criteria (38,113). CIP was first developed by dairy industry (114) and this industry has conducted substantial research into effective strategies and appropriate CIP skid design. One major decision important for CIP skid design (115,116) is the issue of reuse of acid/base cleaning solutions which must be weighed against concerns of cross-contamination and resoiling of equipment. A second major decision is the degree of dismantling and manual cleaning operationally acceptable for each piece of equipment. Cleaning strategies must be developed for equipment skids that are clean-in-place, as well as equipment parts that are clean-out-of-place (e.g. using a glass/parts washer) or manually cleaned (113). The amount of time expected to elapse between the end of processing and before initiation of cleaning, along with the length of time equipment can sit idle after completion of cleaning and the next process step prior to recleaning, needs to be established based on expected operating and process requirements, then supported with appropriate test data (40). Verification of cleaning effectiveness involves testing the equipment before use until it is clean. It usually is done as an alternative to validation for short production campaigns.

Key factors influencing cleaning effectiveness are the discharge of cleaning solutions at their operating pH (if acceptable environmentally) to avoid precipitation of soil, adequate rinsing of cleaning solutions through all internals, adequate contact time and velocity of cleaning agent/rinses, maintenance of optimal temperature during cleaning solution hold time or recirculation time, and proper concentration of cleaning solution. The drainability of equipment is another key factor in assuring cleaning effectiveness. Common problem areas include minor dead legs of a few milliliters that are not rinsed through during the cleaning process, circulation pumps that are not completely drainable, shadowed areas in which impingement of a spray is prevented by vessel internals, and crevices caused by bolts and other fittings. In cases where failed batches are pasteurized or sterilized in the vessel prior to discard, dilution of the batch with water can assist subsequent cleaning efforts.

Often there are multiple points of entry of CIP liquid into a vessel (e.g. spray balls, air inlet, vent, and transfer lines) and liquid is returned by the reactor drain (115). Spray balls can be fixed or rotating, removable during processing, or remaining in batch contact. High pressures and flow rates can be required for proper impingement for best cleaning effectiveness leading to heavy utility loads, particularly for the water rinses. Although in many applications, two spray balls are located at the vessel top, additional spray balls can be aimed at hard to reach locations such as impeller undersides or behind baffles. The acceptable level of flooding of the vessel needs to be determined as the filling of the vessel to at least the lower impeller may enhance cleaning of lower baffles and bottom mechanical portion of the agitator (115).

A typical cleaning scenario (117) may include several steps depending on the product requirements. These steps are an initial thorough rinse with lower quality warm water to remove loose soil (particularly at the batch level), a caustic-based detergent solution to hydrolyze biological materials, a rinse with deionized water, an acid-based detergent solution to remove traces of mineral salts, a second rinse with deionized water, and final rinses with higher purity water. In some cases, a water boilout (heating of water to 80–90°C) is conducted prior to the final water rinses for vessels. Typically, the acid-based detergent is not necessary for fermentation applications, especially if deionized water is used for cleaning, and can be used on an as needed basis (115). Dilute phosphoric acid and sodium hydroxide can be substituted for acid/caustic-based detergents for some applications, although the presence of a surfactant can greatly enhance soil removal. Finally, an empty vessel sterilization, with condensate flowing out from the bottom valve, can aid in soil removal, provided it does not "bake on" these remaining soils.

For cleaning of nonvessels, the ability to document effectiveness becomes increasingly more difficult. Membrane filters often are restricted to single product use primarily because it is difficult to assure total removal of the prior product without destructive testing. Filters, as well as chromatography resins, can be reused for the same product, but sanitization, sterilization, and/or storage procedures must be evaluated to minimize unwanted bioburden. For homogenizers and cell disrupters, cleaning and visual inspection of the head is likely to require substantial manual disassembly and rebuild.

In the case of chromatography columns, due to significant resin expense, reuse is especially desirable for multiple products, but product- specific testing for remosval is exceedingly difficult. More commonly, chromatographic media is dedicated to one product (118), but equipment (e.g. columns, piping, pumps) is used for multiple products (119). Chromatographic media has been reused for multiple cycles (more than 100) of the same product, but validation of the number of multiple uses is required (59,120). Specifically, the media and equipment must be compatible with validated cleaning processes and consideration must be given to which substances bind reversibly and irreversibly to the resin (119). Cleaning and sanitization can occupy about 20% of each cycle period which encourages the use of larger versus smaller columns (119). Blank runs are conducted followed by a reduced cycle cleaning run and evaluation of the contaminant breakthrough potential (119). Finally, quantification of the functional stability of the resin after exposure to specific cleaning agents can be accomplished by measuring retention times of various test proteins (119).

47.3.3.2 Effectiveness Studies. Hardest-to-clean locations are identified based on prior experience with the existing equipment, initial cleaning studies of new equipment, or the best judgment of the processing group based on similar equipment designs. The criteria for sample site location should be documented. Some examples of common locations for fermenters include the air sparger, underside of impellers, batch level, sample valve, baffles, lifting holes for shaft, fermenter dome, ingold ports/plugs, threaded fittings and bolted internals, and piping dead legs on addition ports. Similar equipment can be grouped together with one item's cleaning undergoing a complete set of three validation tests and equivalent items undergoing only one validation test. The use of worst case scenarios, although a popular approach to reduce the amount of testing required for a multiuse facility, may lead to overly rigorous and operationally burdensome cleaning regimens for many nonworse case processes.

Visual inspection, although subjective, is a powerful asset available for the prompt analysis of cleaning effectiveness. Training of inspectors properly concerning where and how to inspect greatly improves the likelihood of identification of problem areas. Chrome mirrors with extended arms, powerful flashlights, and clean white cloths aid inspection. Tank imperfections should be documented ahead of time to avoid mistaking tank welds, external coiling coil welds, and/or the tank's intrinsic stains as cleaning failures. Unfortunately, visual inspection may not be possible for some pieces of equipment without excessive mechanical dismantling (e.g. disk-stack centrifuges, packed chromatography columns). The limit of visual detection has been determined using spiked coupons of typical process soils (113) or by comparing actual swab and visual inspection results (41). For example, the visual detection limit was 100 μg per 25 cm^2 swab area for an active pharmaceutical ingredient isolated powder (121).

The sampling of rinse water indirectly evaluates equipment cleanliness, but it assumes that the residual soil is neither insoluble nor occluded and introduces a large dilution factor (113). Sometimes, rinse water samples are taken throughout the cleaning process so that the progression of cleaning can be evaluated; this testing can be detrimental to sensitive assay equipment such as TOC analyzers, which are neither designed nor calibrated for excessively soiled samples. Solution pH can be especially useful to indirectly detect cleaning agent residuals, as most cleaning agents are concentrated acids/bases in addition to carbon-based surfactants (Table 47.1). In addition, a 100 ppm solution of CIP100 corresponds to a TOC of 4.8 ppm (114); other results are available for other concentrations (Table 47.1).

Direct sampling of the vessel interior is accomplished by a consistent surface swabbing procedure, typically a 25 cm^2 (5 cm × 5 cm square area or equivalent) (122). Drawbacks of swabbing include inaccessibility of some surfaces, possible recleaning because equipment is disassembled, and variability owing to its manual nature (113). Recoveries depend on the type of contaminant and type of surface being swabbed (80); high levels of soil adsorption to the swab can create difficulty in quantifying recoveries (113). Specifically, recovery data for a purified recombinant DNA protein (containing phosphate and lactic acid at pH 4.5) varied from 89% for stainless steel, 62% for polypropylene, and 55% for glass (123). Recovery data for stainless steel varied from 67% for whole cell homogenate, 82% for formulated protein (containing mannitol, phosphate and lactic acid at pH 4.5), and 105% for purified protein (123). Other recovery data for three types of media using different types of materials (stainless steel, Nalgene, Teflon, polypropylene, Pyrex) (Table 47.2a), (70) and for ten broths using two materials (stainless steel, Pyrex) (Table 47.2b) are available.

TABLE 47.1. Ability of Various Dilutions of Cleaning Agents CIP100 and CIP200 to Pass USP Chemical Tests

Solution	PH	Conductivity (μmho/cm)	TOC (ppm)
CIP100			
1% (10,000 ppm)	12.6	1927	25.7
0.05% (5000 ppm)	11.4	581	13.4
0.01% (1000 ppm)	10.7	123	3.5
0.001% (100 ppm)	9.4	8.6	0.57
0.0001% (10 ppm)	6.5	2.9	0.16
CIP200			
1% (10,000 ppm)	2.0	7530	N/A
0.05% (5000 ppm)	2.7	872	9.7
0.01% (1000 ppm)	3.2	179	2.6
0.001% (100 ppm)	4.1	21.4	0.41
0.0001% (10 ppm)	5.1	2.1	0.15

Rinse water and swab samples can be tested for TOC, protein content, specific product, residual cleaning agent, and overall rinse water quality roughly equivalent to that of water source used for cleaning (e.g. pH, conductivity, TOC, bioburden, LAL). Product-specific testing can include ELISA, SDS-PAGE, and other protein-specific assays. Attribute comparisons of various analytical methods are available (113). TOC analysis, which detects all carbon-based residuals, is becoming a low cost, rapid turn around, low level detection method for applications when a product-specific test is not required (124). Descriptions of the method, analyzer and its limitations are available (114,125). The sample's TOC content must be within the range of instrument calibration for accurate results. For an equipment that is to be used wet and/or without subsequent sterilization or sanitization, a bioburden study might be beneficial using both rinse water and swabbing tests. Such a study should be done after the equipment has been idle for a period of time greater than or equal to permitted idle periods during processing and after rinsing to remove bioburden accumulating in areas remaining wet after CIP. Procedures also have been documented for bioburden (126) and LAL (127) testing.

A survey of sampling methods (specifically visual, rinse water, and swab) and assay methods used for early and later phase processing steps was conducted for multiple companies as well as details of methods in place at Chiron (128). Case studies of cleaning validation are published from Regeneron (129), Immunex (130), Lonza (131),

TABLE 47.2a. Recovery Study Results (Fractional Recovery) for Media from Three Processes Tested in Absence of Cells

Material of Coupon	Recombinant Yeast/ Chemically Defined Medium A	Recombinant *E. coli*/ Chemically Defined Medium B	Mammalian Cell Culture/Williams E Medium
Stainless steel	0.86	0.80	0.92
Nalgene	0.28	0.58	1.07
Teflon	0.35	0.65	1.11
Polypropylene	0.36	0.86	1.05
Pyrex	0.26	0.61	0.76

TABLE 47.2b. Recovery Study Results (Fractional Recovery) for Various Broths

Material of Coupon	Culture/ Medium	Actinomyces	Culture/ Medium	Nonyeast fungal	Culture/ Medium	Yeast
Stainless steel	1	0.76	4	0.74	8	0.93
	2	0.83	5	0.84	9	0.30
	3	0.48	6	0.56	10	0.29
			7	0.53		
Pyrex	1	0.51	4	0.41	8	0.22
	2	0.41	5	0.60	9	0.68
	3	0.15	6	0.30	10	0.23
			7	0.087		

and other companies (132). As processing progresses into downstream isolation steps, methods for analysis must have greater product specificity (114), but for upstream steps nonspecific assays such as TOC are likely to be sufficient. In some cases, this product-specific testing requirement adds substantial cost and sampling burdens. Acceptance criteria are affected by the sensitivity of analytical equipment, as well as the amount of residue tolerable in the equipment for the specific processing step. Guidelines for carryover levels of product are established for specific products by some companies such as Eli Lilly (121).

Cleanability studies using different types of materials and soils can be conducted, typically to evaluate whether a new soil is easier or harder to clean than previously tested soils. The general procedure is to submerge a coupon in the soil matrix, evaporate to dryness, and then process it through cleaning solutions/rinses either in stagnant trays or using impingement (Table 47.3). Each surface is then visually evaluated and swab samples are submitted for TOC testing. Dye removal studies can be conducted, but these may not be representative due to different rinsing characteristics (115). Challenge studies also can be developed to document log reduction of TOC, bioburden (based on indicator organisms), and/or endotoxin by spreading a known amount of material on the equipment. Usually, more material is spiked than typically is present as a contaminant to have a quantifiable amount remaining at the conclusion of the simulated cleaning process. However, it is not clear that soil removal is proportional between the simulated and typical soil concentration ranges.

Validation of a backup cleaning agent is recommended in case material from the primary supplier becomes unavailable or unusable. This assignment might be accomplished by demonstration of equivalency during the initial validation or as part of a later changeover program. Evaluation might include cleanability studies on test coupons and

TABLE 47.3. Cleanability Study Results, Residual TOC after Cleaning of Dried Medium to Point of Visual Cleanliness up to a 10-min Rinse via Impingement

Material of Coupon	Recombinant Yeast/ Chemically Defined Medium 1	Recombinant *E. coli*/ Chemically Defined Medium 2	Mammalian Cell Culture/Williams E Medium
Stainless steel	0.10	0.13	0.08
Nalgene	0.16	0.08	0.07
Teflon	0.09	0.09	0.09
Polypropylene	0.10	0.10	0.08
Pyrex	0.01	0.14	0.10

equipment using worst case scenarios and/or conducting equivalency runs with the new agent in actual processing equipment.

47.3.3.3 Investigations of Failed Results. Investigations of failed cleanouts generally center upon procedural issues rather than testing interferences as in the case of SIP testing. Common deficiencies include insufficient hold time or hold temperature for the cleaning agent, inadequate contact of the cleaning agent and/or rinse water with all of the vessel's inner surfaces and internals, improper amount of cleaning agent used, and inappropriate cleaning pretreatment steps. Sufficient time must be devoted to training and clarification of procedures with operating staff. For example, the equipment must be properly vented to allow complete rinsing. As the rinse water quality depends both on the cleaning effectiveness and on the quality of the source water, proper water system operation also may impact the investigation. In rare cases, cleaning failures are attributable to swab or rinse water mishandling.

47.3.4 Laboratory Autoclave Load Patterns

Load pattern testing should be conducted for typical loads that are used for laboratory-sterilized materials. Items to be sterilized are placed in the autoclave such that they are not touching, and there is an adequate room for steam to surround each object. Temperature distribution testing of the load typically is conducted first. Even if the accumulated F_0 appears satisfactory, spore challenges still should be pursued because spores will be killed efficiently only if moist heat is present (133). Assuming both the temperature distribution and spore challenge test results are adequate, a production run is performed with the actual materials to be sterilized. As reasonable, each element in the load should be tested for sterility, either through a test usage for equipment or incubation then sterility testing of liquid-containing items. This production load sterility testing is particularly important to verify noninterference of equipment wrapping closures, ability of personnel to execute validated load patterns, and successful sterilization of solids-containing media.

For dry materials, the use of the prevacuum cycle is most desirable to remove air from the equipment promptly, and its effectiveness can be evaluated using a Bowie–Dick or similar test for noncondensable gases. If a prevacuum cycle is not available or not considered reliable due to equipment failure or fear of disrupting product closures, then a small amount of water (10 ml in a 4-L bottle) may be added to increase the amount of steam generated within the bottle. Sterilization of filter manifolds for nutrient addition is complicated by the fact that hydrophobic filters, especially more hydrophobic ones, may not be wet enough to generate sufficient steam for spore kill. Autoclaving of capsule filters in series is particularly problematic due to poor heat transfer through the plastic filter body as well as poor steam penetration through the tubing ends. Dual filters in series need to be tested carefully, especially when more than one set is used in a manifold arrangement that limits steam penetration. After each production cycle run of a validated autoclave load pattern, any filters in the load need to be integrity tested to insure that no damage has occurred during autoclaving. A complete reference on integrity testing of membrane filters is available (134).

For liquid cycles, liquid heat up time is greatly affected by the container size, fill volume, type of liquids, and material of construction of the container. There is a conflict between safety directives to remove glass from processing areas by replacing glass with plastic containers and the substantially longer heat up times observed with plastic (e.g. Nalgene) carboys due to their low heat transfer coefficient. Consequently, temperature mapping of liquid load patterns is useful prior to performing actual validation studies.

Decontamination of liquids or equipment should be conducted in a different autoclave unit than the sterilizing autoclave to avoid cross-contamination of raw materials/clean equipment with waste/used equipment. Decontamination load patterns also may undergo load pattern validation, although the heat sensitivity of the individual production organism should be considered during validation. Although "worst case" studies using *B. stearothermophilus* may be appropriate, typical BL1 or BL2 organisms and/or viruses may exhibit a substantially lower heat sensitivity, thus reducing the necessary decontamination load cycle time. This potential for reduction may become particularly important when processing larger liquid volume loads.

Generally, the vacuum break filter is steam-sterilized separately prior to the autoclave load. Periodically, after a specific number of autoclave loads have been sterilized, the filter sterilization cycle is run again. Because most vacuum break filters are not maintained under positive air pressure between the conclusion of the filter sterilization cycle and the beginning of the autoclave chamber cycle, then this filter does not necessarily remain sterile once the autoclave door is opened

47.3.5 Utilities

Regular monitoring of utility quality is a critical factor to assure product quality. Monitoring focuses on water, steam, and air systems in product contact. Frequencies can vary from daily in the case of WFI water systems to monthly in the case of clean steam systems. In a few applications, sampling just before, during, and/or after system use is warranted. Action limits are established based on product quality requirements, while alert limits are established based on historical data (if available) and/or a comfortable safety margin. Because the alert limit level does not impact product quality but rather serves to warn of the possibility for a quality problem to occur, it should be set considering prior or expected system behavior.

By far, the major testing focus is on water systems because impurity levels can become concentrated due to the large volumes used in processing and water is used in most processing steps. Testing may include Unites States Pharmacopeia (USP) chemical analysis, bioburden, and endotoxin depending upon the application. Although sampling of water systems is complicated by multiple use points, in many cases sampling is rotated among use points with each use point sampled on a regular basis. Typically, at least one bioburden and one endotoxin sample is taken daily from at least one point of use. Changes to the USP simplified purified water and WFI testing regimens (135), reducing required USP chemical testing to conductivity, pH, and TOC (114,136) analysis. Before these modifications, numerous chemical tests were required, some of which were qualitative in nature with difficult to interpret end points. The impact of these updated USP chemical testing procedures was evaluated and applied for existing water systems (137). Currently available on-line TOC, pH, and conductivity meters can assure proper system operation according to USP chemical testing guidelines, reducing costs associated with manual samples (138).

Bioburden evaluation is conducted to determine both the number and type of organism present. Water systems are subject to the formation of biofilms in which bacteria attach themselves to piping crevices/cracks. They create a polysaccharide extracellular matrix, which promotes long-term attachment of both the original and subsequently adhering bacteria. References on biofilm formation mechanisms and control are available (139–141). Like weeds on a lawn, once present biofilms are difficult to remove, thus operational, maintenance, and sanitization procedures are designed so as not to permit their formation initially.

LAL analysis for endotoxin, although primarily a test conducted on clean steam and WFI systems, also can be helpful for monitoring the operation of USP purified water systems. In fact, LAL, TOC, and bioburden results may be examined together to investigate excursions. All material present in water, which is quantified as TOC, is not necessarily endotoxin due to the specificity of the endotoxin assay for lipopolysaccharide. However, all endotoxin should be detected in the less specific TOC assay. Using the *Escherichia. coli* endotoxin standard, TOC levels were measured for 0.25, 0.5, and 1.0 EU/mL solutions, and a linear relationship was obtained as follows:

$$\text{TOC (mg/L)} = 72.2\ \text{LAL (EU/mL)} - 5.6\ (r^2 = 0.99). \quad (47.1)$$

Consequently, for an endotoxin limit of 0.25 EU/mL (a typical action limit for WFI systems), the TOC measured in the same sample might be as high as 14 mg/L (14 ppm). Thus, TOC analysis can be a good indicator of rising endotoxin levels and may be able to serve as confirmation of endotoxin assay results. Similarly, a rise in bioburden from 1 to 100 cfu/mL may not coincide with a noticeable rise in TOC because 1–100 cfu/mL corresponds to a substantially lower than usually detectable range of 10^{-4}–10^{-6} ppm TOC (assuming 10^{-12} g dcw/cell and 50% carbon content).

47.3.6 Controlled Environments

Controlled environments ensure that there is documented control of air quality that is appropriate for each processing step. Application of stricter downstream environmental classifications further upstream can lead to inefficiencies in investigating excursions that do not relate to actual product quality risks (142). Establishment of classified areas should be examined carefully to insure that classification at the proposed level is reasonable, as well as achievable, while the area is in operation. Specifically, areas that produce powders, mist, or steam may not be able to achieve class 100,000 conditions if samples were taken during truly worst case dynamic operation. Such areas might include rooms containing glasswashers, autoclaves, fermenters, or raw material subdivision operations where classified operation is being reevaluated in some organizations.

A comprehensive monitoring program should be developed to include items such as viable airborne particles, nonviable airborne particles, pressure differentials, airflow direction, temperature, relative humidity, and surface microbial contamination present on staff, equipment, work surfaces, walls, doors, floors, and curtains (143,144). Monitoring is conducted using viable and nonviable air samplers, contact plates, surface swabs, and settling plates (145). A regular monitoring schedule should be instituted with some tests taken during simulated or actual dynamic operating conditions (while work occurring in the area). When bringing an area in and out of GMP service, in addition to GMP restart efforts for equipment and personnel, initially there might be an intensive amount of monitoring, perhaps 3–5 consecutive days, to show consistency of air quality prior to initiation of GMP processing. A summary of environmental monitoring program characteristics, including additional references, is available (146).

Evaluation criteria for a laminar flow hood or biosafety cabinet is dependent upon the critical nature of the hood, specifically whether there exists a subsequent sterilizing step downstream for a sterile product. Sometimes, monitoring during actual operation is considered using either exposure plates or an air sampler. If monitoring is conducted directly within the work area, then it may interfere with the technician's ability to properly manipulate materials. Conversely, if monitoring is moved to occur in a corner of the work area, then testing is not actually sampling air directly exposed to product. Motorized air samplers can interfere with desired airflow patterns in biosafety cabinet/laminar flow hood by creating unwanted air currents as well as create noise, which can be a distraction to the technician. Settling plates give a qualitative outcome because a defined amount of air is not sampled (143). These plates may selectively attract larger particles and may need to be exposed for a far longer time than the length of the actual manipulation (143).

Action and alert limits should be set based on the nature of step, the ability of subsequent steps to remove/propagate contaminants, and the product exposure risk (open versus closed system) (143). Although a typical alert limit is 80% of the action limit, the alert limit may be set lower to insure a comfortable quality safety margin. Data should be trended over a year to detect seasonal variations. Often data does not monotonically increase toward action/alert limit but rather spikes to higher levels due to a system disturbance (e.g. unusual personnel activity). Thus, environmental monitoring is one key tool to assess the suitability of established procedures and controls (147).

For action and/or alert limit excursions, identification of the microorganism class can be useful (e.g. gram positive/negative rod, cocci, fungal), but identification down to the genus/species can be expensive and burdensome. A breakdown of typical cultures found from air and surface monitoring of a class 10,000/100,000 manufacturing area suggests filamentous fungi are found 48% of the time (particularly in class 100,000 environments), gram-positive *Bacillus* (spore formers) are found 25% of the time, gram-negative rods found 10% of the time, and cocci found 16% of the time (148). For any action limit excursion, there is a need to evaluate what processes may have been affected and what was the potential quality impact. Typically, such an event is considered as a processing abnormality and the resulting impact is evaluated. Culture identification may be useful to support a negligible quality impact assessment.

Corrective actions implemented in response to an excursion might include improved cleaning regimens, rotation of disinfectants (149), altered sanitization frequencies, monitoring of ultraviolet light intensities, and revised gowning procedures for the area. Testing of disinfectant effectiveness (e.g. disinfectant type and contact time) against a challenge organism also might be useful in troubleshooting reasons for elevated bioburden levels (Table 47.4) (148). Documented retraining or simple observation of operating personnel using the area, sampling personnel, and/or sample analysis personnel also may be useful. Additional testing to determine if the action limit value is a one-time only or reproducible occurrence is key to the investigation.

TABLE 47.4. Disinfectant Effectiveness Against Microbial Challenge

Disinfectant (dilution atio, v/v)	1 min	5 min	10 min	45 min
***Staphylococcus epidermidis* (3270 cfu/ml)**				
Klenzyme (1:64)	520	480	425	455
CIP100 (1:1280)	420	490	505	375
CIP100 (1:11)	345	345	295	245
Vesphene (1:128)	<10	<10	0	0
LpH (1:256)	25	<10	0	0
3M Compublend (1:2)	<10	<10	0	0
***Aspergillus niger* (200,000 cfu/ml)**				
Klenzyme (1:64)	48	43	38	46
CIP100 (1:1280)	57	55	46	47
CIP100 (1:11)	34	26	30	27
Vesphene (1:128)	0	0	0	0
LpH (1:256)	0	0	0	0
3*M* Compublend (1:2)	0	0	0	0

Such testing may consist of repeated sampling of the same locations or increased sampling locations vertically and horizontally within the area to localize the problem. Often, quality groups may not support additional sampling due to retesting concerns, but such sampling may be the only way to effectively and scientifically troubleshoot a problem. If the additional sampling plan is defined prior to execution, then randomness can be minimized. The initial action/alert limit value should not be discarded; rather all the data should be considered together in evaluating the most probable reason for the excursion.

47.3.7 Constant Temperature Units

Validation testing of constant temperature units centers on documentation of uniform temperature distribution and required humidity level throughout the chamber. Testing is relatively straightforward and inexpensive. Although the requirements for uniformity depend on the type of unit and its particular application, typical examples of set point deviations might include ±1.0°C for an incubator, ±2.0°C for a refrigerator, and ±10°C for a low temperature freezer. Testing with the unit loaded can identify problem areas such as reduced air circulation in a low temperature freezer and excessive heating from a rotary shaker motor in an incubator cabinet or room. Portions of the unit not satisfying the testing requirements can be excluded from use of the process unit if repairs are not chosen to be made.

47.3.8 Product Changeover

The purpose of a documented product changeover procedure is to assure that equipment, personnel, and facilities are ready for the next product (150). The major expense of time, money, and effort is expended in preparing the equipment. Equipment cleaning is evaluated using equivalency testing to assure that existing procedures can clean the next product with revalidation or cleaning verification recommended if major changes in procedures are required. Disposables that may be difficult to clean, such as diaphragms, probes, or gaskets, may be replaced, although the mechanical support and expense necessary to replace every diaphragm in product contact may be excessive. An evaluation of existing equipment SIP/CIP cycles, autoclave, and glass washer load patterns, as well as manual equipment washing procedures, to determine sufficiency for the new process should be conducted. This review should be followed by a cataloging of any changes in equipment, particularly those changes necessitating revalidation, that might be needed for the next process. The facility preparation spans from the removal of raw materials and equipment used in the previous process from the area to updating/revision of batch records and SOPs. Personnel need to be trained for the new process using batch sheets, newly developed load patterns, SIP, or CIP cycles, and altered SOPs. Product changeover requirements are minimized when platform processes are in place and adhered to.

Surface testing of the equipment and the facility might be conducted to document removal of the prior product. Verification testing after the first run of the new process (typically a practice or engineering run) might be conducted to document the ability of existing procedures to clean the new product from the area acceptably. This minimized testing only may be appropriate if (i) similarity of processing steps and their components (e.g. media, host cell organism, contaminating organisms) can be scientifically justified, with the ability to do this successfully varying according to the processing step, (ii) similarity with respect to TOC recoveries either by comparison testing or by scientific justification, and (iii) similarity with respect to cleanability by comparison testing or scientific justification.

47.3.9 Computer and Automation Systems

A history of computer validation shows that it has evolved and reached a maturity level similar to that of other types of validation (151). Although the emergence of 21 CFR Part 11 guidelines has generated intense discussion and interpretation, the fundamentals of computer validation lie in demonstrating basic quality attributes (152). Requirements for several aspects of computer system validation were compared among different published standards and were found to be essentially similar, differing mainly in the level of detail (153). Specifically, PQ of computer and automation systems is mainly driven by how the end user is using the software rather than the software's capabilities. Thus, it includes stress testing under worst case conditions (154).

Risk assessments (such as 21 CFR Part 11 assessments) are used to determine the extent of computer system validation required, starting first with a determination of whether the system handles regulated data or controls a regulated process (43,155). Other risk assessments can be then performed to determine whether testing can be reduced (i.e. test less than 100% of the application) or documentation can be streamlined (i.e. record results only when they differ from expected values) (72).

47.4 INVESTIGATIONS OF NONCONFORMANCES AND DEVIATIONS

There is an ongoing tension between facility productivity and the possibility of nonconformance and deviations. The results of nonconformance affect productivity, schedule, manpower, maintenance, and capital/retrofit costs. Table 47.5 lists some common deficiencies observed during internal or external reviews of facility and product testing. As written in the BPC guide for inspections

TABLE 47.5. Common Deficiencies Observed During Internal and External Reviews of Facility and Product Testing

Area	Examples	Method	Common Deficiency	Reference (Where Available)
Utility generation equipment and supply piping	Water Steam HVAC	Sampling/testing	Failure to test the utility in the manner in which it is used for the process	7
Constant temperature unit	Refrigerator Freezer Incubator	Temperature mapping	Empty unit profile substantially different from full unit profile	
Sterility	SIP systems Autoclaves Dry heat ovens Gamma irradiation	Temperature or radiation mapping/biological (spore) challenges	Failure to use validated load patterns when processing	
Decontamination	Autoclaves Vaporized hydrogen peroxide	Temperature mapping/biological (spore) challenges	Cross-contamination with clean materials	
Cleaning/ Sanitization	CIP systems Glass/parts washer Manual washing	Swab and rinse water testing	Lack of documented procedures Insufficient validation Insufficient training	7,113
Integrity	Container closure Filter	Media/spore challenge Bubble point/ diffusion rate	Incorrect filter test parameters entered into integrity testing unit	
Stability	Product Raw materials	Sampling/testing	Lack of sufficient or applicable testing	7
Laboratory instrumentation and analytical assays/methods	Raw material In process controls Product quality/ impurity testing	Sampling Testing	Insufficient validation Discard of out-of-specification (OOS) results/retesting	7,161–163
Abnormalities	Excursions Out-of-specification (OOS) results	Sampling/testing	Inadequate or untimely investigation	163

(156), the recommended starting point for inspectors is to review product failures (and by extension failed batches), excursions in monitoring results, and failed validation testing. It is thus of utmost importance to determine the most probable cause of failure, complete any necessary investigation reports, and document any resulting corrective actions. Guidelines for the investigation of testing and validation failures are available (105,157–159), which include recommendations about evaluating their significance using a risk-based framework (160). Pressure to obtain a most probable or assignable cause should not be imposed, and the cause should be permitted to be listed as "unknown" if truly no one cause emerges as most likely during the investigation. A higher than normal percentage of "unknown" assignments might then prompt reevaluation of investigative and/or testing procedures and eventually aid in obtaining the real trends in assignable causes for the facility. Failures believed to be due to poor equipment design, inadequate procedures, or suboptimal process development should precipitate the appropriate alterations, which then are evaluated according to established change control procedures. Investigations are always ongoing and, although hypotheses can be formed quickly within the short window permitted for an investigation, the accuracy of any hypothesis is tested and revised in the weeks/months that follow. Recognition of the scientific nature of any investigation of a nonconformance should be featured prominently in any investigation SOPs.

REFERENCES

1. Chapman K. Pharm Technol 1991; 15: 82–96.
2. Tayler C. Pharm Eng 1996; 16: 50–58.
3. Gold D. PDA J Pharm Sci Technol 1996; 50(1): 55–60.
4. Bruch C. Bull Parenter Drug Assoc 1974; 28(3): 105–121.
5. Larkin D. Pharm Eng 1989; 9(4): 27–30.
6. Chiu Y. Pharm Technol 1988; 12(6): 132–138.
7. Martrnez ER. In: Berry IR, Harpaz D, editors. Validation of active pharmaceutical ingredients. Denver (CO): HIS Health Group; 2001. pp. 97–132.
8. Martinez C. Pharm Eng 1994; 13: 8–14.
9. Beatrice M. Pharm Eng 1991; 10(3): 29–35.
10. Hill D, Beatrice M. Pharm Eng 1989; 9: 35–41.
11. Hill D, Beatrice M. Biopharm 1989; 2: 20–26.

12. Hill D, Beatrice M. Biopharm 1989; 2: 28–32.
13. Schoemaker T. Pharm Eng 1991; 11(3): 23–27.
14. Guide to Inspection of Bulk Pharmaceutical Chemical Manufacturing. Washington (DC): CDER, FDA; 1987.
15. Avallone H. Pharm Eng 1989; 9(5): 40–48.
16. Interferon Test Procedures: Points to Consider in the Production and Testing of Interferon Intended for Investigational Use in Humans. Washington (DC): Office of Biologics, FDA; 1983.
17. Supplement to the Points to Consider in the production and testing of new drugs and biologicals produced by recombinant DNA technology: Nucleic acid characterization and genetic stability. CDER, April 1992.
18. Points to Consider in the Characterization of Cell Lines used to Produce Biologicals. Washington (DC): CBER, FDA; 1993.
19. Points to Consider in the Manufacture and Testing of Monoclonal Antibody Products for Human Use. Washington (DC): CBER, FDA; 1994.
20. Fitzpatrick S, Ma'ayan A, Wagget J. Chem Eng Prog 1990; 86(12): 26–31.
21. Stern M. Genet Eng News 1998; 18(6): 10–32.
22. Little L. Biopharm 1997; 10: 8–9.
23. Orr T. Genet Eng News 1998; 18(5): 12–14.
24. Guideline on sterile drug products produced by aseptic processing. Washington (DC): Center for Drugs and Biologics and Office of Regulatory Affairs, FDA; 1987.
25. Guideline on the preparation of investigational new drug products. Washington (DC): CDER, FDA; 1991.
26. Guideline on the general principles of process validation. Washington (DC): Center for Drugs and Biologics, FDA; 1987.
27. Guideline for Inspections of Biotechnology Manufacturing Facilities. Washington (DC): FDA; 1991.
28. Guide to Inspections of Validation of Cleaning Processes. Washington (DC): FDA; 1993.
29. U.S. Department of Health and Human Services, Food and Drug Administration. Guidance for Industry. Quality systems approach to pharmaceutical CGMP regulations. Pharmaceutical cGMPs. September 2006.
30. U.S. Department of Health and Human Services, Food and Drug Administration. Guidance for Industry. Q7A Good manufacturing practice guidance for active pharmaceutical ingredients. ICH, August 2001.
31. Guidance for Industry. Q9 quality risk management. Washington (DC): ICH, FDA; 2006.
32. Kelley W. J Pharm Technol 1986; 10(3): 48–54.
33. Wechsler J. Biopharm Int 2007; 20(6): 22–24.
34. Clinton P. Biopharm Int 2003; 16: 26–32.
35. Signore A. Pharm Eng 1998; 18(1): 72–82.
36. Coburn J, Weddle G. Bioprocess Int 2006; 4: 22–24, 26–28.
37. Johanning H. Pharm Technol Eur 2006; 18(7): 14–16.
38. Mollah AH, White EK. Biopharm Int 2005; 18(11): 54–67.
39. Tidswell EC. PDA J Pharm Sci Technol 2005; 59(1): 10–32.
40. Tidswell EC. Am Pharm Rev 2005; 8(6): 28–33.
41. Forsyth RJ, Hartman JL, Van Nostrand V. Pharm Technol 2006; 30: 104–114.
42. Murray KS, Anderson DKS, Reichert HM. Bioprocess Int 2005; 3(3): 42–52.
43. McDowall RD. Pharm Regul Guid Book 2006; 1: 24, 26, 28, 30.
44. Hamid M. Bioprocess Int 2004; 2(9): 28–35.
45. Junker BH, Wang HY. Biotechnol Bioeng 2006; 95(2): 226–261.
46. Malhotra G. Pharm Process 2007; 23(2): 10, 12, 14, 16.
47. Kara B. Eur Biopharm Rev Autumn, 2006; 9: 114–120.
48. James P. Pharm Eng 1998; 18(1): 72–82.
49. Angelucci L. Pharm Eng 1998; 18(1): 40–44.
50. Sofer G, McEntire J, Akers J. Biopharm 1995; 8(5): 60–63.
51. Welch KA. J Validat Technol 2003; 9(3): 246–252.
52. Garvey W. Pharm Technol 2005; 29(12): 68–76.
53. Lord SV, McCarthy TJ, Aleem H, Zhao Y, Sharratt PN. Proc Inst Mech Eng 2003; 217(4): 287–293.
54. Fletcher N. Pharm Eng 1989; 9(6): 23–27.
55. Perry SD, Greiner-Powell D. Bioprocess Int 2007; 5(6): 14–20.
56. Tisak D, Koster R. J Validat Technol 1997; 3(4): 394–397.
57. DeSain C. Biopharm 1992; 5(5): 30+.
58. DeSain C. Biopharm 1992; 5(6): 22–24.
59. Junker B. In: Gad SC, editor. Handbook of Pharmaceuticals Biotechnology. New York: John Wiley and Sons; 2007. pp. 319–370.
60. PDA. PDA J Pharm Sci Technol 2005; 59(S-4): 1–28.
61. Uettwiller I. Genet Eng News 2006; 26(3): 54, 56.
62. DePalma A. Pharm Eng 2003; 23: 33–36, 38.
63. Wheeler W. Pharm Eng 1994; 4: 48–56.
64. Werner RG, Langlouis-Gau H, Walz F, Allgaier H, Hoffman H. Arzneimittelforschung 1988; 38(6): 855–862.
65. Collins B, Sides K. Pharm Eng 2006; 26(6): 80–92.
66. Aleem H, Zhao Y, Lord S, McCarthy T, Sharratt P. Proc Inst Mech Eng 2003; 217(2): 141–151.
67. Greenwald RF, Smit T. Pharm Eng 2007; 27(3): 8–24.
68. Akers J, McEntire J, Sofer G. Biopharm 1994; 7(1): 8–11.
69. ISPE San Francisco/Bay Area Chapter. Pharm Eng 1998; 18(1): 8+.
70. Junker B. Biotechnol Bioeng 2001; 74(1): 49–61.
71. Richman GB. Pharm Regul Guid Book 2006; 1: 54, 56, 57–60.
72. Lohrey K. Pharm Eng 2007; 27(2): 54–64.
73. Adamson J. Pharm Eng 1992; 12: 16–22.
74. Agalloco J, Akers J. J Pharm Sci Technol 1997; 51 Suppl: S1–S23.
75. Lusky M. Hum Gene Ther 2005; 16: 281–291.
76. Bosse R, Kulmburg P, von Kalle C, Engelhardt M, Dwenger A, Rosenthal F, Schulz G. Ann Hematol 2000; 79: 469–476.
77. Bauers J, Hargroves J. Pharm Eng 1996; 16(1): 36–42.
78. Alford J, Cline F. Biopharm 1990; 3(8): 32–36.
79. Junker B, Kardos P, Smizaski W, Brix T. In: Langer ES, editor. Advances in large-scale biopharmaceutical manufacturing and scale-up production. Washington (DC): ASM Press; 2004. pp. 510–554.
80. Junker B, Christensen J. In: Langer ES, editor. Advances in large-scale biopharmaceutical manufacturing and scale-up production. 2nd ed. Washington (DC): ASM Press; 2007.
81. Fed Regist 1986; 51(88): 16958–16985.
82. Fed Regist 1991; 56(138): 33174–33183.

83. Van Houten J, Fleming DO. J Ind Microbiol 1993; 11: 209–215.
84. J Van Houten. In: Ladisch M, Bose A, editors. Harnessing biotechnology for the 21st century (proceedings of the Ninth Internat'l Biotechnology Symposium). Washington (DC): American Chemical Society; 1992. pp. 415–418.
85. Cipriano ML. In: Fleming DO, Hunt DL, editors. Biological safety: Principles and practices. 4th ed. Washington (DC): ASM Press; 2007. pp. 561–577.
86. Shaw JL, Rogers SA. Chem Process 2006; 69(9): 25–34.
87. Halkjaer-Knudsen V. Eur J Parenter Pharm Sci 2005; 10(2): 35–42.
88. Hamers M. Biotechnology (N Y) 1993; 11: 561–570.
89. G Muijs. In: Collins C, Beale A, editors. Safety in industrial microbiology and biotechnology. Chapter 11. Oxford: Butterworth-Heinemann; 1992. pp. 214–238.
90. Meecham PJ, Gyuris J, Petuch BR, Chartrain MM, Herber WK. In: Fleming DO, Hunt DL, editors. Biological safety: Principles and practices. Chapter 30. 4th ed. Washington (DC): ASM Press; 2007. pp. 551–660.
91. Moran E, Fink R. Biopharm Int 2005; 18: 32–34, 36, 38, 40.
92. Heenan M, Looby M, McGowan S, Cullen S, Moran E. Biopharm Int 2005; 18: 70, 71–76.
93. Noordover JAC, Hofmeester JJM, van der Burg JP, deLeeuw A, van Dijck PWM, Luiten RGM, Groot GSP. J Ind Microbiol Biotechnol 2002; 28: 65–69.
94. Shahidi A, Torregrossa R, Zelmanovich Y. Pharm Eng 1995; 15: 72–83.
95. Christi Y. Chem Eng Prog 1992; Washington DC, 88(9): 80–85.
96. Bader F. In: Demain A, Solomon N, editors. Manual of industrial microbiology. Chapter 24: American Society for Microbiology; 1986. pp. 345–362.
97. Akers M, Attia I, Avis K. Pharm Technol Int 1978; 1: 45–49.
98. Enzinger M, DeVisser A, Townsend M. Pharm Technol Int 1978; 1(3): 41–43.
99. Enzinger M, Goodsir M, Korozynski M, Parham F, Schneier M. Validation of steam sterilization cycles, Technical Monograph No. 1. Philadephia (PA): Parental Drug Association, Inc.; 1978.
100. Guidance for industry for submission documentation for sterilization process validation. Washington (DC): CDER, FDA; 1994.
101. Reisman M. Pharm Cosmet Qual 1998: 22+.
102. Junker B, Brix T, Lester M, Kardos P, Adamca J, Lynch J, Schmitt J, Salmon P. Biotechnol Prog 2003; 19: 693–706.
103. Berman D, Myers T, Suggy C. J Parent Sci Technol 1986; 40: 119–121.
104. Hunter JN. Eur J Parent Pharm Sci 2004; 9(3): 79–81.
105. Borghese R. Biopharm 1993; 6(2): 28–32.
106. Manfredini R, Golzi Saporiti L, Cavallera V. Proceedings Conferences Chem. Biotechol. Biol. Active Nat. Prod., 1 Meet., 2, 1981. pp. 320–391.
107. Perkowski C. Biopharm 1987; 1: 62–65.
108. Perkowski C. J Parent Sci Technol 1990; 44(3): 113–117.
109. Junker B, Lester M, Leporati J, Schmitt J, Kovatch M, Borysewicz S, Maciejak W, Seeley A, Hesse M, Connors N, Brix T, Creveling E, Salmon P. J Biosci Bioeng 2006; 102(4): 251–268.
110. Sharma MC, Gurtu AK. In: Neidleman S, Laskin A, editors. Volume 39, Advances in Applied Microbiology. New York: Academic Press, Inc.; 1993. pp. 1–27.
111. Junker B. Biotechnol Prog 2007; 23: 767–784.
112. Junker B, Mann Z, Gailliot P, Byrne K, Wilson J. Biotechnol Bioeng 1998; 60: 580–588.
113. Cardot J-M, Beyssac E. Bioprocess Int 2007; 5(6): 66–76.
114. Jenkins K, Vanderwielen A, Armstrong J, Leonard L, Murphy G, Piros N. PDA J Pharm Sci Technol 1996; 50(1): 6–15.
115. Christi Y, Moo-Young M. J Ind Microbiol 1994; 13(4): 201–207.
116. Greene D. Pharm Eng 2003; 22: 1–7.
117. Wisiniewski R, Burman C. In Lubiniecki A, Vargo S, editors. Regulatory practice for biopharmaceutical production. Chapter 20. New York: Wiley-Liss; 1994. pp. 407–445.
118. Turton J, Moola Z. Biopharm 2002; 15: 24–30.
119. Adner N, Sofer G. Biopharm 1994; 7(3): 16–21.
120. Seely R, Wight H, Fry H, Rudge S, Slaff G. Biopharm 1994; 7(7): 41–48.
121. Fourman GL, Mullen MV. Pharm Technol 1993; 17: 54–60.
122. Glover C. PDA J Pharm Sci Technol 2006; 60(5): 284–290.
123. Lombardo S, Inampudi P, Scotton A, Ruezinsky G, Rupp R, Nigam S. Biotechnol Bioeng 1995; 48: 513–519.
124. Bristol P. Manuf Chem 2004; 75(4): 37–38.
125. Clark KA. Pharm Lab 2000: 16, 17, 20.
126. Amador C, Deschenes P, Dorminy M, Dudley B, Gordon B, Lampe C, McGarrah R, Prulello J, Wild R. J Parent Sci Technol 1990; 44(6): 324–331.
127. Held D, Mehigh R, Wooge C, Crump S, Kappel W. Biopharm 1997; 10(3): 32+J.
128. McArthur P, Vasilevsky M. Pharm Eng 1995; 15: 24–31.
129. Inampudi P, Lombardo S, Ruezinsky G, Baltrus T, Dugger J, Remsen P, Rupp R, Nigam S. Ann N Y Acad Sci 1996; 782: 363–374.
130. Geigert J, Klinke R, Carter K, Vahratian A. PDA J Pharm Sci Technol 1994; 48(5): 236–240.
131. Hill L. Biopharm 2000; 13: 24–27, 29.
132. Bousquet-Bedu M, Laban F, Albadine J, Barbu M, Bonwarlet MN, Collard M, Goncalves C, Leclerc A, Lemeland G, Morvan F, Oursel MH, Perreau M, Viguier-Freeman M. STP Pharm Prat 2005; 15(1): 28–52.
133. Macek T. In: Phillips G, Miller W, editors. Industrial sterilization. Durham (NC): Duke University Press; 1972. pp. 19–34.
134. Brantley J, Martin J. Genet Eng News 1997; 17(10): 24.
135. USP-NF Fifth Supplement 3442–3444, 3464–3467, 3482–3483, 3547–3555. 1996.
136. Bevilacqua AC. Eur J Parent Sci 2001; 6(1): 3–11.
137. Collentro W. Ultrapure Water 1994; 11(3): 22–29.
138. Martrnez JE. Bioprocess Int 2004; 2: 30, 32, 34, 36, 38.
139. Riedewald F. Pharm Eng 1997; 17(6): 8–20.
140. Mittelman M. Microcontamination 1985; 3(10): 51–55.
141. Lukanich J. Chem Process 1998; 61: 78+.
142. DePalma A. Genet Eng News 2006; 11(26): 40–42.

143. Roscioli N, Renshaw C, Gilbert A, Kerry C, Probst P. Biopharm 1996; 9(8): 32–40.
144. Barta J. Biopharm 1998; 11(5): 84+.
145. Whyte W. Eur J Parent Pharm Sci 2005; 10(2): 43–50.
146. Immel B. Biopharm 1998; 11(3): 46–49.
147. Akers J, Agallaco J. PDA J Pharm Sci Technol 2001; 55(3): 176–184.
148. Prince HN, Prince DL. Contract Pharm 2005; 5: 84, 86, 88, 90, 82, 94.
149. Conner D, Eckman M. Pharm Technol 1992; 16: 148–160.
150. Odum J. Pharm Eng 1995; 15: 8–20.
151. Budihandojo R, Coates S, Huber L, Matos JE, Schmitt S, Stokes D, Tinsley G, Rios M. Pharm Technol 2007; 31: 86, 88, 90–93.
152. Bissonnette D. Biopharm Int 2005; 18(12): 26–30.
153. Segalstad SH. Pharm Eng 2007; 27(2): 94–101.
154. Erickson J. Biopharm Int 2003; 16: 1–2.
155. Siconolfi RM, Bishop S. Drug Inf J 2007; 41(1): 69–79.
156. Guide to Inspections of Bulk Pharmaceutical Chemicals. Washington (DC): CDER, FDA; 1994.
157. Kuwahara S. Biopharm 1996; 9(1): 24–29.
158. Kuwahara S. BioPharm 1996; 9(3): 40+.
159. Sandberg J. Med Device Diagn Ind 1995; 17(5): 102–106.
160. Slatter A. GMP Rev 2004; 3(3): 12–16.
161. Porter DA. Pharm Technol 2007; 33: 146–154.
162. Winter W. Bioprocess Int 2006; 4(9): 46–48, 50.
163. Zaret EH. Pharm Formul Qual 2000: 18–20.

48

QUALITY BY DESIGN (QBD)

Rakhi B. Shah, Jun T. Park, Erik K. Read, Mansoor A. Khan, and Kurt Brorson
Office of Pharmaceutical Science, Center for Drug Evaluation and Research, United States Food and Drug Administration

48.1 BIOPHARMACEUTICAL DEVELOPMENT

In the biopharmaceutical industry, consistently manufacturing a safe and high quality protein drug substance is an equivalent or sometimes greater challenge than dosage from manufacturing of traditional medicines. Recently, a regulatory structure supporting Quality by Design (QbD) for pharmaceuticals and biopharmaceuticals has been described in International Conference on Harmonization (ICH) Quality documents (Q8 through Q10; available at www.ich.org). The ICH Q8(R1) guidance (1) can be used as the basis for biopharmaceutical development despite the fact that the ICH Q8(R1) document was written for pharmaceutical development in other words what is termed as *drug product* or *dosage forms*. Many of the concepts espoused in ICH Q8(R1) (e.g. quality target product profile (QTPP), product and process understanding, design spaces, control strategies with process analytical technologies (PAT) product life cycle, continuous process verification are universal and can readily be applied to biotech protein drug substance manufacture. Recently, a concept paper for a new ICH guideline document, called "Q11 for development and manufacture of drug substance for both chemical and biotech molecules" was released. This document is likely to provide additional guidance for implementation of QbD in protein drug substance manufacturing.

Quality is defined as the suitability of either a drug substance or drug product for its intended use, including attributes such as the identity, strength, and purity (1) All of these attributes are just as important for biotech protein products as for small-molecule medicines, despite often major differences in Active Pharmaceutical Ingredient (API) production or final dosage forms.

The important aspects of drug product development (drug substance, excipients, formulation development, overages, physicochemical and biological properties, manufacturing process development, container closure system, microbiological attributes, and compatibility) as well as the concept of QbD are presented in the original ICH Q8 guidance. In June of 2009, the ICH Q8(R1) document has also been published in the Federal Register for implementation. QbD is defined as "A systematic approach to development that begins with predefined objectives and emphasizes product and process understanding and process control, based on sound science and quality risk management."

Properties of the molecule can influence the drug product quality attributes and manufacturability (1). In the case of proteins, long-term stability is a critical issue for manufacturability and subsequent use by patients. Formulation development presents a significant challenge for long-term protein stability as well as delivery in the patients. Proteins are susceptible to several potential degradation pathways including clipping, aggregation, oxidation, deamidation, and denaturation. Formulation development (product understanding), including careful balancing of stabilizing excipients, buffering agents, tonifiers, bulking agents, detergents, and pH is critical for long-term protein stability. Processing issues such as mixing, filtration, sterilization, freeze-thawing, agitation, filling, and lyophilization also require a thorough understanding to prevent product variability. Drug substance

Downstream Industrial Biotechnology: Recovery and Purification, First Edition. Edited by Michael C. Flickinger.

issues, such as effective removal of host-cell enzymes during protein purification can also impact drug product stability. It is clear that principles stated in ICH Q8(R1) regarding formulation development and drug substance quality are key for the development of stable protein-based medicinal products.

A trend toward more complicated dosage forms is growing. For example, products initially marketed as solutions or lyophilized form in vials when first licensed are now available in liquid form prefilled syringes or autoinjectors (e.g. Enbrel). Lyophilization itself is a complicated process that can bring about changes in the protein structure due to several compositional and process variables. While more convenient for patients, the newer dosage forms are more complicated to manufacture and can be more prone to stability issues. In these cases, biopharmaceutical development studies should be consistent with the intended scientific purpose of the medicine.

This is illustrated by recent stability issues observed with interferons. Interferons are marketed for chronic administration to multiple sclerosis patients. The protein product can be stored in liquid form inside the prefilled syringe for several months. Any degradation of the protein, such as clipping or aggregation, could cause a loss of potency or enhanced immunogenicity, potentially causing adverse reactions. Tungsten alloys from syringe needles was recently found to cause stability issues for certain Interferon prefilled syringe dosage forms by catalyzing protein clipping/aggregation over time (2). The unpredictability of events such as these strengthen the case for a sound science background supporting application of QbD for biopharmaceutical development as described in ICH Q8.

The next sections describe steps to implementing QbD for biopharmaceutical product manufacturing and processes. We also include a hypothetical case study to illustrate the decision-making and experimental flow for implementing QbD in a specific unit operation, anion exchange chromatography (AEX).

48.2 KEY STEPS FOR IMPLEMENTING QbD

Implementation of QbD is a multistage, science-based, iterative process. Individual steps in this process are described in ICH guidance: Q8(R1) (1), Q9 (3), and Q10 (4). Implementation strategy will vary with product class, complexity, and risk. A potential flow chart for implementing QbD during biopharmaceutical development is below:

1. *Identifying the QTPP.* QTPP is defined as "a prospective summary of the quality characteristics of a drug product that ideally will be achieved to ensure the desired quality, taking into account safety and efficacy of the drug product" ICH Q8(R1) further stipulates that the QTPP should consider route of administration, dosage form and strength, delivery system, container closure system, attributes affecting pharmacokinetic characteristics, and drug product quality criteria.
2. *Identifying Critical Quality Attributes (CQAs).* Once the QTPP has been identified, the next step is identifying the relevant product CQAs. CQA is defined as "a physical, chemical, biological, or microbiological property or characteristic that should be within an appropriate limit, range, or distribution to ensure the desired product quality" (1). CQAs for drug substance, excipients, intermediates (in-process materials), and drug products can be determined based on either prior experience or from experimental designs. The key consideration is an assessment of the extent to which individual CQA variation can have an impact on the overall quality of the drug product.
3. *Assessing Quality Risks and the structure of Quality Risk Management (QRM).* Risk assessment is a valuable science-based process used in QRM (section titled "Quality Risk Management for Biopharmaceuticals" of this chapter). QRM can be used to identify which material attributes and process parameters can potentially have an effect on product CQAs.
4. *Defining Process Design Spaces.* Design Space is defined as "the multidimensional combination and interaction of input variables (e.g. material attributes) and process parameters that have been demonstrated to provide assurance of quality" (1). Design space is proposed by an applicant and is subject to approval by regulatory agency. As discussed in section titled "Design Space(s) for Biopharmaceuticals" of this chapter, working within a design space is not considered as a reportable change from a regulatory standpoint.
5. *Defining the Process Control Strategy.* Control Strategy is defined as "A planned set of controls, derived from current product and process understanding that assures process performance and product quality" (3). These include both drug substance and drug product controls. They can include parameters and attributes related to materials and components, facility and equipment operating conditions, in-process controls, finished product specifications, and the associated methods and frequency of monitoring and control.

48.3 QUALITY RISK MANAGEMENT FOR BIOPHARMACEUTICALS

The purpose of this section is to illustrate a systematic application of QRM concepts to facilitate QbD for biopharmaceuticals.

Risk has been classically defined as the combination of the probability of occurrence of harm and the severity of that harm (4). The QbD paradigm stipulates that biopharmaceutical quality should be maintained throughout its life cycle. QRM helps to identify and control potential quality issues from early stage development to marketing and large-scale manufacturing for biopharmaceuticals. This strategy is beneficial to the industry as well as regulatory agencies since QRM can facilitate assurance of a sponsor's ability to identify and mitigate the risks.

A schematic of the major components of QRM is presented in Fig. 48.1. The major stages for initiation and planning include (i) risk assessment, (ii) risk control, and (iii) risk review.

48.3.1 Risk Assessment

Risk assessment consists of the three steps.

1. Risk identification involves systematic use of information such as historical data, theoretical analysis, and informed opinions among others. Flow charts, Check sheets, process mapping, fault tree analysis, cause and effect (Ishikawa or fish bone) diagrams are useful tools to visualize and facilitate identification of risk factors in a process. Figure 48.2 shows an example of a fish bone diagram for a QbD case study for an anion exchange unit operation for a hypothetical biotech product (presented below). Some potential unit operation parameters that can potentially affect the viral clearance are listed. However, due to resource constraints, it is impractical to study them all in a systematic way such as using Design of Experiments (DoE) as described in the sectiontitled "Design Space(s) for Biopharmaceuticals". Thus, once the most likely risks are identified, risk prioritization and analysis follows the initial risk identification. A key part of risk identification is careful review of manufacturing experience and the scientific literature. This process can identify for inclusion process parameters already identified by others as critical or key for bioprocessing unit operation performance (5).
2. Risk analysis is a semiquantitative process of linking the likelihood of occurrence and severity of harm. This may be applied to process deviations, quality attribute changes, or other factors related to the drug substance or product. There are many different approaches to analyze and prioritize the risk and the method of choice would depend upon the internal procedures within a biotechnology firm. Failure Mode Effect Analysis (FMEA) or Failure Mode Effect and Criticality Analysis (FMECA) are commonly applied methods to prioritize risks. FMEA is an evaluation of potential failure modes for the manufacturing process or product quality

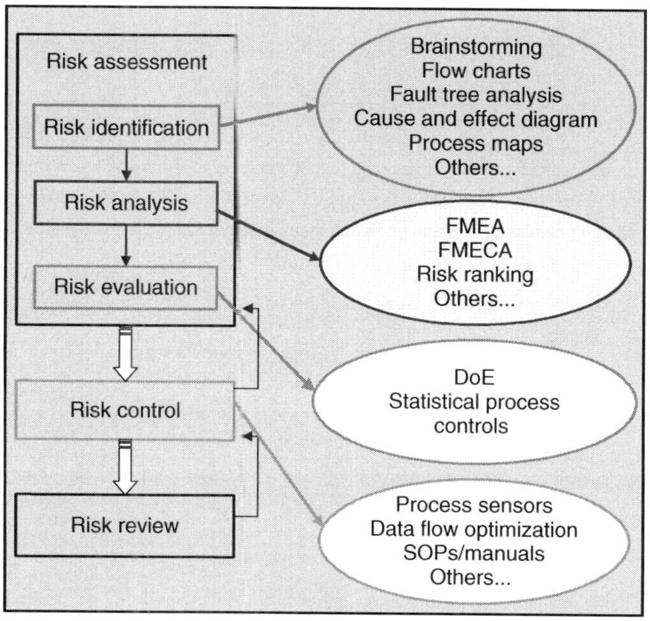

Figure 48.1. A quality risk management (QRM) model [modified from (4)]. (This figure is available in full color at http://onlinelibrary.wiley.com/book/10.1002/9780470054581.)

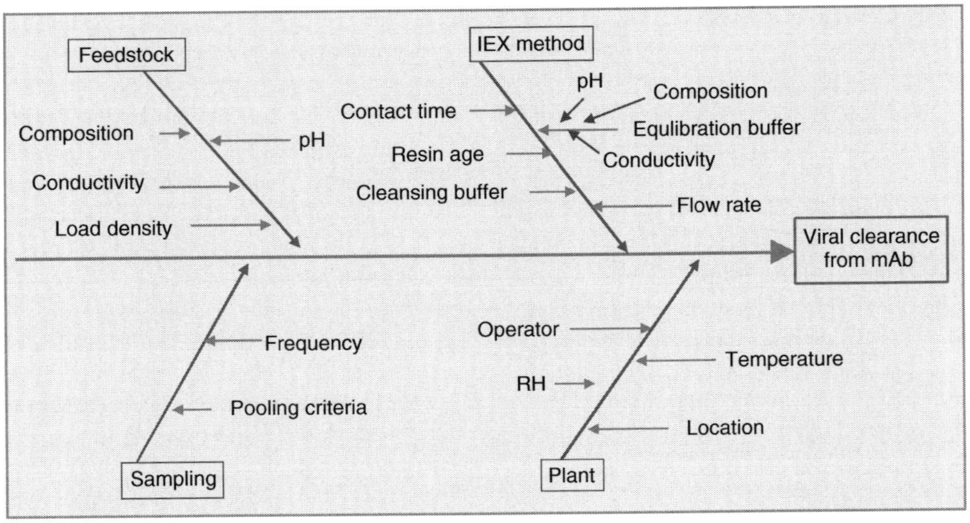

Figure 48.2. Fishbone (Ishikawa or cause and effect) diagram showing nonexhaustive list of factors affecting viral clearance from monoclonal antibody (mAb) by anion exchange chromatography (AEX). (This figure is available in full color at http://onlinelibrary.wiley.com/book/10.1002/9780470054581.)

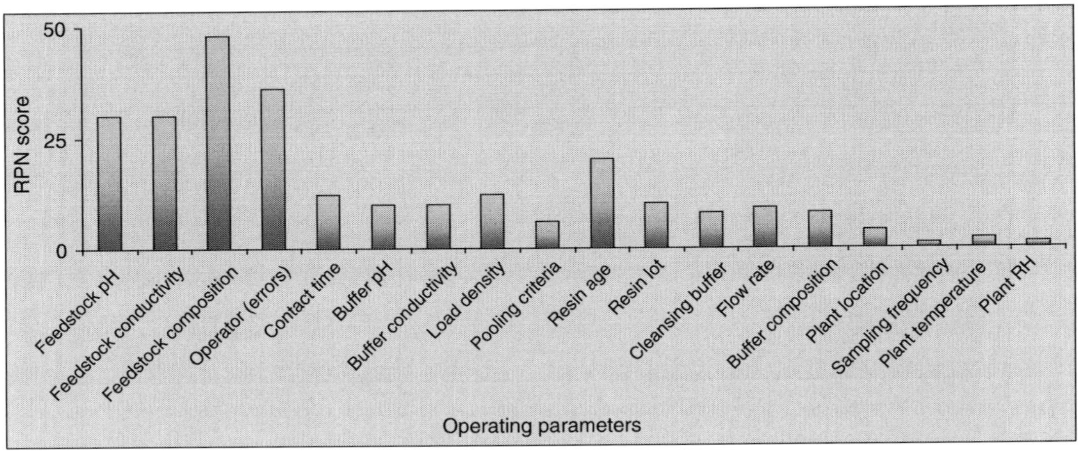

Figure 48.3. RPN scores for the nonexhaustive list of operating parameters for a viral clearance step in a mAb purification by AEX process. Parameters that had RPN scores higher than the cutoff can be further examined in process characterization. (This figure is available in full color at http://onlinelibrary.wiley.com/book/10.1002/9780470054581.)

attributes and their likely effect on outcomes and/or biotech product performance in patients or other settings. It relies on biotech product and process understanding, that is, if process step X fails, then A, B, and C are potential quality or patient safety outcomes. FMECA is an extension of FMEA which incorporates an investigation of the (i) degree of severity of the consequences, (ii) their probabilities of occurrence, and (iii) their detectability. The product of these three is termed as risk priority number (RPN). Figure 48.3 shows RPN values for the parameters affecting CQA of the QbD case study presented in section titled "QbD Case Study for a Biopharmaceutical: Anion Exchange Chromatography." FMEA application in the biotech industry is commonly utilized to assess potential failures and risks associated with manufacturing processes. The risk prioritization is explained in QbD case study in the section titled "QbD Case Study for a Biopharmaceutical: Anion Exchange Chromatography."

3. Risk evaluation compares the identified and analyzed risk against given risk criteria. Severity and probability of occurrence act as aids to the risk assessment.

Control charts, DoE, histograms, Pareto charts (i.e. a chart with bars displaying values in descending order, and a line graph showing cumulative totals of each category), and process capability analysis are various statistical tools to evaluate the risk.

There are other risk assessment tools such as Hazard Analysis and Critical Control Points (HACCP), Hazard Operability Analysis (HAZOP), Preliminary Hazard Analysis (PHA), Risk ranking, and filtering. There is no regulatory preference for risk assessment tools; the choice of the tool should be based on suitability for the intended application. Interested readers are encouraged to read ICH Q9 guidance for these various methods.

48.3.2 Risk Control

Risk control includes reducing or mitigating the risk to an acceptable level (4). The amount of effort used for risk control should consider the significance of the risk; this can be determined from a risk-benefit ratio. Care must be taken to avoid introduction of new risks as a result of controlling the identified risks. In the case study presented in section titled "QbD Case Study for a Biopharmaceutical: Anion Exchange Chromatography", careful scale-down modeling during virus removal validation studies and assurance of operation of the AEX step within approved ranges for buffer and load pH and conductivity would be key components to the risk control strategy.

48.3.3 Risk Review

A mechanism for periodic review or monitoring events is helpful to identify new risks and gauge the potential impact of residual risks. If a residual risk becomes unacceptable, the risk mitigation strategies might need careful evaluation (4).

48.3.4 Setting Specifications for a Biopharmaceutical

A "specification" is defined as a list of tests, reference to analytical procedures, and appropriate acceptance criteria that are numerical limits, ranges or other criteria (6). QbD principles including application of QRM can help set the specifications for a biotech drug substance or product based on science and risk. For certain quality attributes that are not directly measured, like microbial and viral safety, QRM can help set validation strategy. For others, like operator errors or process deviations in other stages of manufacture, QRM can determine an overall risk mitigation strategy considering process robustness and resulting product quality. In an ideal situation, clinically meaningful specifications are set based on CQAs.

Conformance to specifications means that the biotech drug substance and/or drug product, when tested according to the listed analytical procedure, will meet the listed acceptance criteria. Specifications are critical, but minimal, quality standards that are proposed by the manufacturer and approved by regulatory authorities as conditions of approval. Specifications should focus on those characteristics found to be useful in ensuring the safety and efficacy of the biotech drug substance and the drug product. Specifications are only part of assurance of biopharmaceutical quality. For example, process controls and verification strategies based on viral clearance studies and process validation as well as cell bank and unprocessed bulk harvest in-process tests are also key for ensuring quality standards. The experience and data obtained during development of a new biotech drug substance or product in part forms the basis for setting the specifications. It may also be possible to propose excluding or replacing certain tests on this basis. Some tests may be performed as an in-process test, or may be performed as a release test for either drug product and/or drug substance. DNA or host-cell proteins are an example where it is possible to perform them either at lot release or as an in-process basis. Others, like mycoplasma testing is routinely performed on bulk harvest in-process intermediate. Justification of specification and testing point should refer to relevant developmental data (linkage to performance, stability, safety, efficacy, etc. ICH or FDA guidance, pharmacopoeia standards, and test data). Additionally, a reasonable range of expected analytical and manufacturing variability should be considered. Justification is also needed if an alternative approach is proposed.

It should be noted that appropriate use of QRM does not obviate a biotech firm's obligation to comply with specific regulatory requirements as outlined in regulations, ICH or FDA guidance. However, effective QRM can facilitate more informed decisions as well as better use of resources, where it may be possible to reduce certain end-product tests when quality can be ensured to remain within agreed upon quality limits because of timely measurements (6).

48.4 DESIGN SPACE(S) FOR BIOPHARMACEUTICALS

48.4.1 Design Space Concept

The design space is defined as "The multidimensional combination and interaction of input variables (e.g. material attributes) and process parameters that have been demonstrated to provide assurance of quality" (1). The concept of "Design Space" as defined in ICH Q8, has been widely discussed as a tool for implementing QbD for biotech products. ICH Q8(R1) further states that "Working within the design space is not considered as a change. Movement out of the design space is considered to be a change and would normally initiate a regulatory post-approval-change process. Design space is proposed by the applicant and is subject to regulatory assessment and approval." Thus, ICH has provided a clear incentive to establish design spaces and reduce regulatory reporting requirements by remaining within them.

Thus, operating within a defined process design space provides the "assurance of the product quality." Generally, the concept of process design space is gaining increasing understanding and importance in the biotechnology and pharmaceutical industry.

48.4.2 Development of Design Space

Design Space development is a dynamic and iterative process; it begins at product conceptualization and continues to evolve throughout the entire lifecycle of the product. Design Space creation often initiates with a complete risk assessment, this identifies those product attributes and process parameters that are critical to product quality (with an underlying assumption of patient safety and efficacy). The risk assessment thus determines CQAs and then Critical Process Parameters (CPP) for evaluation (5), based on prior scientific knowledge from similar products, formulations, and processes.

A possible flow for defining the design space is summarized from ICH Q8(R1) and other recent publications (7,8). The major elements are listed in order below:

1. *Defining the QTPP.* Identify the desired product performance characteristic for intended clinical uses.
2. *Prior Knowledge and Risk Assessment.* This has been discussed in the section titled "Quality Risk Management for Biopharmaceuticals."
3. *Experimental Studies.* Experimental studies for process characterization are designed and executed (i.e. DoE) to evaluate process robustness. These studies identify critical formulation and process variables, and provide the basis for the design space of each unit operation with models for interaction studies. For practical reasons, these process characterization studies are performed at laboratory scale. Therefore, developing and justifying representative scale-down or scale-up models is crucial to the success of the DoE characterization studies. We will present a case study for a biotech process in section titled "QbD Case Study for a Biopharmaceutical: Anion Exchange Chromatography" of this chapter (i.e. a viral clearance study by a flow-through mode AEX), where a chart obtained from a hypothetical DoE is shown in Fig. 48.4. Traditionally, a statistical analysis can identify the relative importance of product and process factors on CQAs. The more significant factors in the charts are then further optimized to develop design spaces with interaction and nonlinear terms. In the case of complex bioprocess unit operations, study manageability and cost-benefit is an important consideration. While it would be ideal to study interactions between multiple process parameters, this may not always be feasible from a logistical or process economics standpoint. In these cases, it is permissible to select high risk, CPPs only, or evaluate parameters at extreme end points only (i.e. high impurity feedstocks, rapid processing dwell times, etc.).
4. *Confirmation of the Design Space.* Data from the characterization studies are analyzed to assess the variability in the product CQAs. The analysis can rely on advanced statistical tools to explore interactions between the operating parameters and material attributes when they exist. In the case of a complex bioprocess unit operation, where it may be practical to study only a subset of the variables at a time, simplified approaches may be warranted. For example, if a DoE results in acceptable unit operation performance at each parameter setting, then defining the design space as the DoE may be more feasible than performing complex statistical analyses and further experimentation. Regardless of the strategy, these results establish the final design space that provides the assurance of the product quality and consistency during long-term manufacturing. The simplest confirmation is the traditional three batch validation of a certain point in the design space. More recently, "continuous process verification" is emphasized where the process is continually monitored and evaluated for improvement beyond the initial confirmatory batch results. However, as stated in Q8(R1): "Design space is proposed by the applicant and is subject to regulatory assessment and approval."

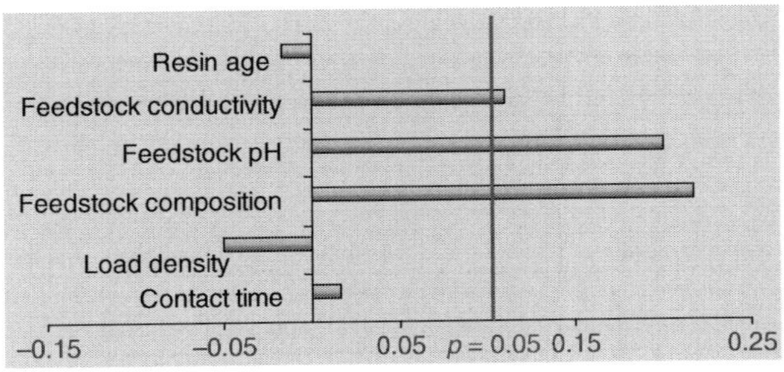

Figure 48.4. Results from the Plackett–Burman screening DoE study showing the effect of operating parameters (RPN > 12) on viral clearance for the flow-through mode AEX. (This figure is available in full color at http://onlinelibrary.wiley.com/book/10.1002/9780470054581.)

48.4.3 Structure of Design Space

Figure 48.5 shows schematic of a design space for viral clearance by AEX for a case study presented later in the chapter. A design space includes (1,7)

- acceptable ranges for material attributes and process parameters;
- scaling factors, if the design space was established at lab scale or if it is intended to span multiple operation scales;
- the scope of the design space, that is, whether it is an independent design space for one unit operation or a single design space that span multiple unit operations or biochemically similar products;
- identification of multivariate/interacting aspects, when feasible.

48.4.4 Generic or Modular Approaches for Design Space

Recently, several groups (9,10) have proposed applying platform approaches to develop a generic downstream purification process for some biotechnology products that share significant biochemical attributes, for example, monoclonal antibodies (mAbs). However, it is not always practical to employ a single complete generic process inflexibly to cover an entire downstream purification process, because mAbs can possess subtle physicochemical differences (e.g. posttranslation, isoelectric point). Also, unit operations can pose a variety of challenges (e.g. subtle impacts of operating parameters or complex raw materials).

Nevertheless, FDA's *Point to Consider Guidance for Monoclonal Antibodies* (11) has defined a "Modular Approach" for viral clearance studies. This is "a study ... that demonstrates virus removal or inactivation of individual steps during the purification process (column chromatography, filtration, pasteurization, solvent/detergent, low pH, etc.). Each module in the purification scheme may be studied independently of the other modules." The PTC-mAb guidance further states that "For mAb manufactured at one site, generic or modular clearance data may be extrapolated to other mAb of the same species and H and L chain class and, generally, subclass, derived from the same source (e.g. ascites or tissue culture) and cell substrate. Generic and/or modular clearance will apply only when the mAb have similar biochemical properties and are purified by identical methods."

In a way, modular and/or generic approaches preceded the concept of design spaces for a specific unit operation(s) of the downstream purification process (e.g. Protein A chromatography, low pH virus inactivation, ion exchange chromatography, or viral filtration) for mAbs. However, current

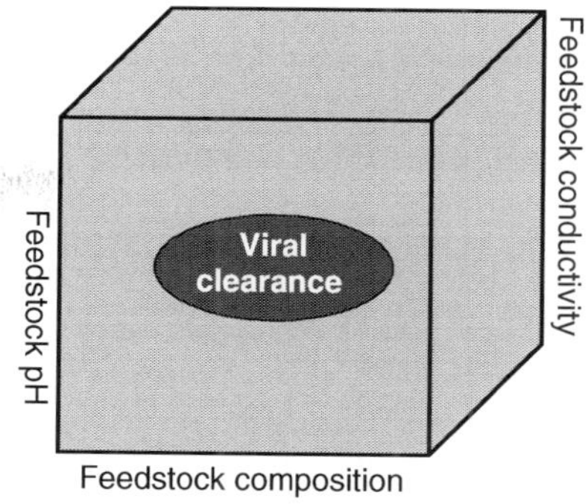

Figure 48.5. Schematic of multidimensional design space for viral removal by AEX. (This figure is available in full color at http://onlinelibrary.wiley.com/book/10.1002/9780470054581.)

thinking is that, applicants should connect the design spaces for sequential unit operations such that the output of one unit operation has the flexibility to be taken as the input of the next unit operations. FDA, in collaboration with Genentech, subsequently published an assessment of critical and key parameters for viral clearance by AEX (12) and low pH inactivation (13). When proposing modular and/or generic approaches for individual biotechnology products and processes, unique issues may need to be addressed (i.e. product specific factors) in order to successfully implement and define the design space of each unit operation.

48.5 PAT IN BIOPHARMACEUTICAL MANUFACTURE

48.5.1 PAT Regulatory Framework

ICH Q8(R1) specifically lists four possible flexible regulatory approaches; (i) risk-based regulatory decisions (section titled "Quality Risk Management for Biopharmaceuticals"), (ii) implementation of design spaces (section titled "Design Space(s) for Biopharmaceuticals"), (iii) reduction of postapproval submissions (this was initially achieved in the United States by implementation of submission categorization in 21 CFR 601.12, but further opportunities for reporting reduction may eventually be pursued), (iv) PAT for process control and real-time release. Thus, PAT can be considered an enabling component of QbD.

In 2004, the US Food and Drug Administration (FDA) published a guidance document entitled "PAT —A Framework for Innovative Pharmaceutical Development, Manufacturing, and Quality Assurance." This document defined "PAT as a system for designing, analyzing, and controlling manufacturing through timely measurements (i.e. during processing) of critical quality and performance attributes of raw and in-process materials and processes, with the goal of ensuring final product quality (14)." The PAT framework document encouraged the development and introduction of new manufacturing technologies that enhance process understanding in order to (i) identify all critical sources of variability; (ii) enable process-managed variability; and, (iii) accurately predict product quality attributes. All of these goals serve to control the product and process variability within the design space.

The core concept of PAT is that science-based, real-time monitoring of material attributes of the raw material or in-process intermediate can facilitate process decisions, improving overall product quality and consistency (15). However, PAT is more than just routine real-time monitoring; it requires process control directed by the above described real-time monitoring at clearly defined points within the process. This strategy allows quality assurance of desired material attributes of the end product of each unit operation to be achieved via adaptive controls during the manufacturing process rather than fixed control batch release quality assurance. Application of PAT to a biotech unit operation includes three phases of implementation: design, analysis, control.

48.5.2 Design and Implementation of PAT

Design of the PAT application should draw from the complete knowledge base applicable for the particular process development goal. The factors to be considered for developing a PAT strategy include QTPP, CPP, material attributes and their relationship to CQA (1) identified as important during QRM (4) for a particular process. Design spaces can also be used to enable a PAT strategy by setting process limits and goals for feedback and feed-forward decisions. The knowledge base can also include process information from other related products (i.e. platform approaches) and/or the scientific literature. For example, mAb processes are often similar (but not identical). Thus, a PAT strategy for one mAb product can often be applied either directly or with modifications to another. Finally, development of PAT should be carefully considered based on a risk and process economic standpoint.

Analysis requires the ability to measure and control the relevant output and input parameters in real-time. The complex nature of biotech process fluids, where the protein API may be only a small percentage of the actual bulk fluid often makes this a difficult undertaking. Thus, opportunities for PAT enabled biotech unit operations may be somewhat limited now, but with improved sensor technology and chemometric approaches (15,16), these are expected to grow in the future. As an example, the changes in bulk fluid that may have a relationship with protein API can be tracked by Near-IR or other PAT sensors and allow control of protein quality.

Control can be defined as the ability to adjust input parameters in response to deviations of measured output parameters. There are two types of real-time PAT-based process control: (i) direct measurement of a product (or raw material) CQAs and (ii) measurement of parameters that directly correlate with a CQA. While the former measurement type is considered to be "classical PAT," the latter is often more achievable and routine in complex biotech unit operations.

Finally, continuous improvement of processes, including PAT, is a goal of the QbD initiative. Often, as multiple batches of a product are manufactured at large scale, opportunities for improvement become apparent. The expansion of the knowledge base that results from PAT is likely to generate such opportunities. This is especially useful when one has established a range of process parameters within which a product or product intermediates can be controlled in real time.

48.6 HYPOTHETICAL QbD CASE STUDY FOR A BIOPHARMACEUTICAL: ANION EXCHANGE CHROMATOGRAPHY

In the case of biopharmaceutical viral safety, the QTPP for the end product is "assurance that an acceptable level of safety in the final product is achieved" (i.e. virtual absence of virus (17,18). It should be understood that "zero virus" in an absolute sense is unattainable because of assay limitations and the logarithmic nature of virus clearance studies (19,20). As illustrated below, QbD approaches including risk assessment, design space, and PAT is applicable to achieve this goal.

48.6.1 Risk Assessment

QRM can be used to assess the potential degree of risk posed by deviations in operating parameters. QRM can be implemented in a systematic manner for all parameters and to prioritize the activities, such as validation exercises and DoE studies, necessary to understand the impact of these parameters on overall process performance. A team consisting of representatives from process development, manufacturing, quality assurance, and other relevant disciplines performs an assessment to list the factors potentially affecting the CQAs of the process, being liberal with regard to likelihood (7); unlikely events tend to be eliminated later in the risk analysis, for example, by low "occurrence" or "detectability" scores in an FMEA.

A hypothetical case study is presented for viral clearance by flow-through mode AEX, an important step in the mAb purification process (Fig. 48.6). Figure 48.2 shows a list of some operating parameters and other factors that can affect the viral clearance from mAb in the form of

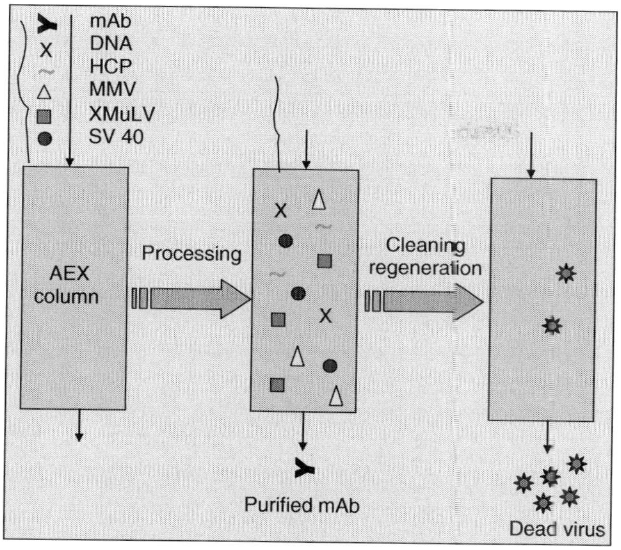

Figure 48.6. Schematic of anion exchange chromatography for purifying mAb. (This figure is available in full color at http://onlinelibrary.wiley.com/book/10.1002/9780470054581.)

fish bone diagram. The operating parameters are typical for flow-through mode AEX and the potential impacts and severity of deviations from validated set points on viral clearance were identified based on the literature studies (12,21,22). It should be noted that the list in our example is nonexhaustive and the present case study depicts the concept and that the factors and scores should not be utilized generically for all mAb products. The hypothetical severity, occurrence, and detection for all these factors (i.e. process deviations from the validated set point or accepted practice) are then listed in a FMEA table as shown in Table 48.1.

The severity score measures the seriousness of a particular failure and is based on an estimate of the severity of the potential failure effect at a local or process level and the potential failure effect at end-product use or patient level. Obviously, failure, in the form of live viruses present in an injectable product, would be a severe outcome. This risk is mitigated by other factors such as other robust viral clearance steps in the purification process and viral testing of upstream intermediates. Because viral safety assurance is understood to be a multilayer approach (17–19), we consider viral clearance of individual unit operations in this case study on a stand alone basis. Thus, the assigned severity values in this exercise are based on failure of this unit operation alone. Occurrence and detection scores are based in our case study on what is known about process complexity and typical biotech manufacturing environments. This knowledge base can be mined to assign possibility scores for particular excursions (manufacturing deviation) outside the operating range and likelihood scores for whether it would be noticed during process. The occurrence is variable between different types of deviations, while in most cases (e.g. use of the wrong buffer) would be very noticeable. Although the occurrence score measures how frequently the failure might occur, the detectability score indicates the timely detection and correction of the excursion or the detection before end-product use. All three scores are multiplied to provide an RPN and the RPN scores are then ranked to identify the parameters with a high enough risk to merit process characterization. It should be noted that, scores from our hypothetical example are likely to differ from real-world examples and these may vary on a firm-to-firm or product-to-product basis.

Figure 48.3 illustrates the FMEA outcome for a viral clearance by AEX step in the mAb purification process. RPN scores are calculated for several process parameters and other potential risks. Process characterization might focus on parameters that have a high impact on the process or product (severity = 5), have high risk of occurrence (occurrence = 5) and are difficult to detect (detection = 5). In contrast, parameters with a low impact on the process (severity = 1), low likelihood of occurrence (occurrence = 1), and a very high ability to detect (detection = 1), can be excluded in further process characterization. Operating parameters with an RPN score > cutoff value can be characterized using a qualified scaled-down model (in our case RPN > 20). For the case study presented here, these would include buffer pH, buffer conductivity, feedstock pH, and feedstock conductivity. Intermediate factors might include contact time, buffer composition, load density, feedstock composition, and resin age (in our case RPN 12–19).

This approach is subjective based on general understandings about bioprocessing, prior knowledge in the firm, and experience about the particular or related processes. It is advantageous to include a diversity of opinions when performing a risk analysis, particularly drawing expertise from several divisions with different functions (i.e. quality, process development, manufacturing, and others). Sometimes, factors with high severity scores, but low occurrence and detectability scores, are considered further in the study, on a case-by-case basis.

48.6.2 Design Space

Rathore and Winkle (7) suggested that a two-phase study is most efficient and effective to define the design space of each specific unit operation for the process of biotech products:

1. screening phase to identify the process parameters that have the most impact on product quality (i.e. viral clearance for this case study);
2. using DoE study to examine and characterize the most significant parameters including their interactions that are decided from the previous screening study.

TABLE 48.1. RPN Scores of Operating Parameters for Viral Clearance by AEX Step in a mAb Purification Process

Operating Parameters	Severity[a]	Occurrence	Detectability	RPN[b]
Resin age[c]	2	5	2	20
Resin lot	2	5	1	10
Contact time[d]	3	2	2	12
Buffer composition	2	2	2	8
Load density	3	2	2	12
Feedstock composition[e]	4	3	4	48
Buffer pH[f]	5	2	1	10
Buffer conductivity[f]	5	2	1	10
Feedstock pH[f]	5	3	2	30
Feedstock conductivity[f]	5	3	2	30
Cleansing buffer[g]	2	2	2	8
Flow rate[d]	3	3	1	9
Sampling frequency[h]	1	1	1	1
Pooling criteria[i]	3	1	2	6
Operator (errors)	4	3	3	36
Plant temperature	2	1	1	2
Plant relative humidity (RH)[j]	1	1	1	1
Plant location	2	2	1	4

[a] In most cases, severity is mitigated by other viral safety measures such as other orthogonal viral clearance steps.
[b] RPN (Risk Priority Number) = Severity × Occurrence × Detectability
[c] Previously demonstrated to have negligible impact on viral clearance when cleaned properly and periodically monitored for band spreading, DNA clearance and/or back pressure (22).
[d] Contact time (bed height ÷ flow rate) generally impacts clearance only under extreme conditions (12,21).
[e] Feedstock are complex mixtures containing host-cell proteins or DNA at variable levels depending on placement in process (i.e. upstream versus downstream).
[f] Buffer and load pH and conductivity previously demonstrated to profoundly impact clearance (12,21).
[g] Improper cleaning can impact clearance over time (22).
[h] Viral safety of the AEX step is achieved by validation, not in process testing. Thus, the occurrence and severity is one.
[i] Product fraction pooling of flow-through chromatography is more simple than for bind and elute.
[j] The bulk intermediate is in a liquid form, thus plant humidity has no impact on the AEX step.

In the above hypothetical Risk Assessment case study, an FMEA was applied to the identified operating parameters of a flow-through mode AEX step for viral clearance. Based on the QRM, process parameters of this unit operation were prioritized for a hypothetical design space analysis. Table 48.1 lists the RPN scores and Fig. 48.3 illustrates the FMEA outcome of the AEX unit operation for viral clearance, which will be used to design the DoE study. In this case study, operating parameters with an RPN score greater than 12 (e.g. feedstock composition, feedstock pH, and feedstock conductivity, resin age, contact time, and load density) are selected for intense study in the DoE (i.e. mostly evaluating ranges). The impact of these operating parameters for viral clearance by the flow-through AEX step is studied using a DoE in a qualified scale-down model.

Statistically designed experiments (e.g. factorial design) are an invaluable tool to identify which process parameters or combination of parameters have significant effects on desired unit operation performance attributes. In our case, six CPPs with RPN scores >12 (feedstock composition, feedstock pH, feedstock conductivity, resin age, contact time, and load density) are selected for testing the effect on the AEX step performance attributes (i.e. viral clearance). High and low parameter set points were studied based on process knowledge and high risk identified in the FMEA analysis. Operator errors, although identified as a significant risk factor, cannot be evaluated in a DoE study, rather, this risk is mitigated by GMPs and training.

Several screening designs can be used to evaluate the relative importance of the variables studied, and a Placket-Burman screening design has been frequently used for small as well as large molecules (23–25). In the current example, the Plackett-Burman method is used to evaluate the effects of each parameter as well as to screen the most important parameters that are statistically significant (e.g. $p < 0.05$). The initial screening experiment consists of 12 AEX runs as listed in Table 48.2. Removal of three validation viruses (Mouse Minute Virus (MMV), SV40, and X-MuLV) was studied in scale-down, multi-spiking studies (Fig. 48.6) (26). Log reduction values (LRV) of each virus were measured by Q-PCR (quantitative polymerase chain reaction) (27).

In the case of MMV and SV40, all 12 hypothetical studies resulted in complete clearance (i.e. LRV > 5.5–6.0 $\log_{10}$, the experimental window). In this case, the design space is defined by the DoE.

In the case of X-MuLV, incomplete clearance was observed in a subset of the studies. In this case, a

TABLE 48.2. Plackett-Burman Screening DoE for Testing the Effects of Operating Parameters for AEX Step of Viral Clearance

Run	Factor Level					
	Contact Time[a]	Load Density[b]	Feedstock Composition[c]	Feedstock pH[d]	Feedstock Conductivity[e]	Resin Age[f]
1	+1	+1	+1	+1	+1	+1
2	+1	−1	+1	+1	+1	−1
3	+1	−1	−1	+1	−1	+1
4	+1	+1	−1	−1	−1	−1
5	−1	+1	+1	+1	−1	−1
6	−1	−1	+1	−1	+1	+1
7	−1	+1	−1	−1	+1	+1
8	+1	−1	−1	−1	+1	−1
9	−1	−1	+1	−1	−1	−1
10	−1	−1	−1	+1	−1	+1
11	−1	+1	−1	+1	+1	−1
12	+1	+1	+1	−1	−1	+1

[a]Contact time (low level = 3 min, high level = 10 min) was studied.
[b]Low (30 mg/mL resin, factor level −1) and high (100 mg/mL resin, factor level +1) load densities were studied.
[c]Low impurity (10 ppm CHO proteins, 10 ng/mL DNA, factor level −1) and high impurity (100 ppm CHO proteins, 100 ng/mL DNA, factor level +1) were studied.
[d]Low pH (7.0, factor level −1) and high pH (8.0, factor level +1) were studied.
[e]Low conductivity (4 mS/cm, factor level −1) and high conductivity (14 mS/cm, factor level +1) were studied.
[f]Resin age (high level = 150 cycles and low level = 10 cycles) was studied.

statistical analysis is used on the preliminary screening data to assess the effects of the six potentially significant operating parameters on the AEX performance response as shown in the chart in Fig. 48.4 These were calculated from Table 48.3 using the statistical analysis function of JMP software. Feedstock composition, feedstock pH, and feedstock conductivity have statistically significant effects ($p < 0.05$) on the viral clearance by the AEX step under the conditions studied, whereas the other parameter values (load density, contact time, and resin age) are less statistically significant. The results of the DoE are consistent with previous observations concerning the impact of load pH on viral clearance (12,21). Because a relatively tight range of conductivity within the known robustness range of AEX was chosen for the DoE, it is not surprising that a conductivity dependence of viral clearance was not found.

In this case, follow-up for further studies and/or setting the design space can consist of one or more of the following three approaches:

- further examination of statistically significant factors at more extended ranges than tested here via additional DoE studies (e.g. Box-Behnken, central composite, or full factorial optimization designs), to more completely understand their interactions;
- more focused experiments to narrow the parameter range until a design space is found where complete clearance is seen;
- claiming the lowest value obtained in the matrix as the clearance value for this step. This is feasible when it can be combined with clearance values from other steps in the purification process to obtain the X-MuLV clearance target for the product.

TABLE 48.3. Hypothetical Results from Screening DoE Case Study Test on the Effects of Operating Parameters for AEX Step of Viral Clearance

Run	Virus LRV[a]		
	X-MuLV[b]	SV40[c]	MMV[c]
1	5.5	>6.0	>5.5
2	5.4	>6.0	>5.5
3	>5.7	>6.0	>5.5
4	5.6	>6.0	>5.5
5	5.3	>6.0	>5.5
6	5.4	>6.0	>5.5
7	5.2	>6.0	>5.5
8	5.5	>6.0	>5.5
9	4.7	>6.0	>5.5
10	>5.7	>6.0	>5.5
11	>5.7	>6.0	>5.5
12	4.5	>6.0	>5.5

[a]Measured by Q-PCR in small-scale multispiking studies.
[b]Variable clearance amenable for statistical study using screening DoE methodology. An alternative approach would be to claim an LRV of 4.5 $\log_{10}$ for entire design space.
[c]Complete clearance seen for both viruses. In this case, the claimed LRVs would be 6.0 $\log_{10}$ (SV40) and 5.5 $\log_{10}$ (MMV).

As described above, manageability and cost-benefits need to be considered when designing DoE studies. For example, the cost of a very large viral clearance study needs to be weighed against the benefit of the processing flexibility that it may ultimately afford (20). In cases where process economics is a constraint, it may be warranted to perform combined validation/DoE studies. In this approach, the process is validated at target conditions as per ICH Q5A (17), using high titer virus spikes and large volume virus testing to obtain a high LRV. This would serve as the regulatory LRV claim for the step. This result is then combined with a DoE establishing the design space, using lower titer spikes or perhaps even bacteriophage to make the DoE study more manageable, The DoE would be designed around the target conditions from the regulatory claim; critical parameters would be varied from the target conditions to assess LRV within parameter ranges. The purpose of the DoE is showing robustness within the space, not to claim the full capacity of viral clearance over the entire design space.

Similar design space can be constructed for other CQAs (e.g. host-cell proteins removal, host-cell DNA removal, or removal of protein A leachables) by the AEX step. The design space for these other performance attributes may differ from the viral clearance design space. Further, the design space for one virus may differ from that of another virus (i.e. some are easier to clear than others; 18,19). Based on all of these, a final comprehensive design space of the flow-through mode AEX step for all CQAs for the product can be created.

48.6.3 PAT Approach

Assurance of viral safety requires assurance of the virtual absence of viruses (i.e. <1 particle/L). At the present time, sensors and/or real-time virus assays with this level of sensitivity do not exist. Because such technology does not currently exist, PAT approaches based on real-time measurements of the absence of virus as a quality attribute are not feasible. However, process parameters that directly correlate with viral clearance, for example, load conductivity and pH as described in the risk analysis above, can be monitored and controlled online. Feedstock conductivity would be monitored by an online conductivity meter attached to the feed line and controlled by spilt stream high salt buffer addition or dilution with Water for Injection (WFI). Feedstock pH would be monitored by an online pH meter attached to the feed line and controlled by split stream acid or base addition (Fig. 48.7). By setting and then maintaining load pH and conductivity by online process control, this approach can be considered to be a form of PAT as they are CPPs that directly correlate to the CQA of interest (15). This will help in minimizing the RPN score of these factors as their detectability score will go down with PAT approach. Thus, continuous real-time monitoring can give insight into the biotech processes.

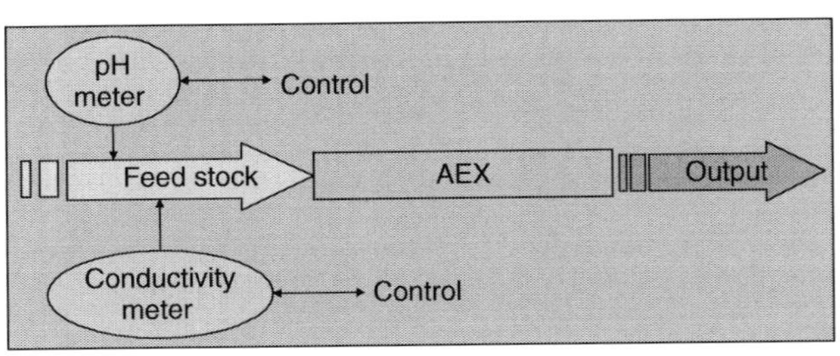

Figure 48.7. Online pH and conductivity meter attached with feedback control to feedstock for AEX. (This figure is available in full color at http://onlinelibrary.wiley.com/book/10.1002/9780470054581.)

48.7 FDA'S QbD PILOT PROGRAM FOR BIOPHARMACEUTICALS

On July 2, 2008, FDA released "Submission of Quality Information for Biotechnology Products in the Office of Biotechnology Products; Notice of Pilot Program" (28,29). This pilot, recently extended, focuses on QbD approaches to the manufacturing of biotechnology products and proposes the use of Expanded Change Protocols. The goals of the pilot include (i) defining clinically relevant attributes for complex products and linking them to the manufacturing process, (ii) providing additional information to FDA for use in facilitating QbD, risk-based approaches for complex molecules, (iii) determining if quality by design is appropriate for an entire original application as well as individual unit operations in supplements, (iv) giving a trial run to the regulatory concept of Expanded Change Protocols.

48.8 SUMMARY

QbD is an umbrella of sometimes overlapping concepts and approaches that include: QTPP, risk assessments, design spaces, PAT, and continuous process improvement among others. It can lead to a future where product quality is assured by flexible, science-based approaches. QbD has been viewed as a paradigm shift applicable to a wide range of medicinal product classes throughout their life cycle, that is, including all phases in the life of a product from the initial development through marketing until the product's discontinuation.

Key enablers for establishing the QbD development are described in other ICH documents; ICH Q8(R1) (pharmaceutical development), ICH Q8R annex, and ICH Q9 (quality risk systems). ICH Q10 is another key enabler which applies to the entire product life cycle for quality systems: pharmaceutical development, technology transfer, commercial manufacturing, and product discontinuation. It identifies four pharmaceutical quality system elements, but allows for regional variance for implementation. The major quality system elements are:

- *Process Performance and Product Quality Monitoring System.* This is essentially a day-to-day system for the monitoring of process performance and product quality.

- *Corrective Action and Preventive Action (CAPA).* System This is a system for implementing corrective actions and preventive actions resulting from multiple inputs (e.g. investigation of complaints, product rejections, nonconformances, recalls, deviations, audits, regulatory inspections, and trending analysis).
- *Change Management System.* A system for process change implementation in response to external scientific innovation, internal continual improvement activities, process and product quality monitoring, CAPAs, consumer demands or regulatory authorities.
- *Management Review of Process Performance and Product Quality.* These periodic review provide assurance that process performance and product quality are managed over the life cycle.

Principles of risk assessment (ICH Q9) should be applied to determine the size, scope and structure of each quality system element specified in ICH Q10, considering the stage of product life cycle, as well as product specific factors (e.g. patient population, complexity of product or manufacturing, etc.). Risk assessment also helps in determining specifications based on risk and science and not merely derived from performance of test batches.

All of the concepts exposed in ICH Q8(R1), Q9, Q10, and the QbD initiative in general can lead to a future where flexible, science-based approaches assure product quality. Design spaces can help gain additional confidence in the process as well as an ability to make change within the design space without triggering any regulatory action. Also it gives an opportunity for continuous process verification, an advanced approach to process validation, in which manufacturing process performance is continuously monitored and evaluated.

QbD also gives an opportunity for real-time release testing leading to a reduction in end-product release testing. Real-time release testing refers to the ability to evaluate and ensure the acceptable quality of in-process and/or final product based on process data. Typically, the PAT component of real-time release includes a valid combination of assessed material attributes and process control. In this, material attributes are measured and controlled along with process parameters. Material attributes can be assessed using direct and/or indirect process analytical methods. As defined in the PAT guidance, real-time release testing builds on parametric release of heat terminally sterilized drug products. It is comparable to alternative analytical procedures for final product release.

Challenges for implementing QbD in biopharmaceutical processes sometimes limit its use within an organization. Introducing new concepts into manufacturing can involve culture change and requires management support, a multidisciplinary approach to development and training, to include a few. However, benefits discussed above might be able to justify its use for biopharmaceutical processes.

It is believed that QbD is likely to not only increase manufacturing efficiency, but also result in higher quality and more consistent medicines for the public. In addition, regulatory flexibility may be possible as stated in ICH Q8 (R1), "The degree of regulatory flexibility is predicated on the level of relevant scientific knowledge provided." Moreover, approaches such as PAT are likely to be useful for real-time release in lieu of end-product testing. These all have the capacity to revolutionize the biopharmaceutical industry.

In conclusion, QbD strategies and goals (identification and justification of target product profiles, and product and process understanding, appropriate control strategies) provides a possibility for flexible bioprocesses which are justified based on risk and science.

48.9 LIST OF ABBREVIATIONS

AEX	anion exchange (chromatography)
API	active pharmaceutical ingredient
CAPA	corrective action and preventive action
CPP	critical process parameters
CQA	critical quality attribute
DoE	design of experiment
FDA	Food and Drug Administration
FMEA	failure mode effect analysis
FMECA	failure mode effect and criticality analysis
HACCP	hazard analysis and critical control points
HAZOP	hazard operability analysis
ICH	International Conference on Harmonization
IR	Infrared (electromagnetic radiation)
LRV	Log_{10} reduction value
MAb	Monoclonal antibody
MMV	Mouse Minute Virus
PAT	process analytical technology
PHA	preliminary hazard analysis
PTC	points to consider
QbD	quality by design
QRM	quality risk management
QTPP	quality target product profile
RPN	risk priority number
SV40	Simian Virus 40
WFI	water for injection
X-MuLV	Xenotropic Murine Leukemia Virus

Acknowledgment

The Authors thank Patrick Swann (CDER/FDA), Christine Moore (CDER/FDA), Qi Chen (Genentech, Inc.) and Greg

Blank (Genentech, Inc.) for reviewing this chapter. This work was supported by the Critical Path Initiative of the Center for Drug Evaluation and Research (FDA), project number 1500 and the Research Participation Program at the Center for Drug Evaluation and Research administered by the Oak Ridge Institute for Science and Education through an interagency agreement between the U.S. Department of Energy and the U.S. Food and Drug Administration.

This scientific contribution is intended to support regulatory policy development. The views presented in this chapter may not necessarily have been adopted as regulatory policies of the Food and Drug Administration at this time.

REFERENCES

1. ICH, Q8(R1): Pharmaceutical Development, International Conference on Harmonisation; 2009; Geneva, Switzerland.
2. Swift R, Nashed-Samuel Y, Liu W, Narhi L, Davis J: Tungsten, Prefilled Syringes and Protein Aggregation, ACS Meeting; 2007; Boston (MA).
3. ICH, Q10: Pharmaceutical Quality Systems, International Conference on Harmonisation. 2009; Geneva, Switzerland.
4. ICH, Q9: Quality Risk Management, International Conference on Harmonisation; 2005; Geneva, Switzerland.
5. PDA, TR42. Process Validation of Protein Manufacturing. Parenteral Drug Association, Bethesda (MD), PDA Journal 59; 2005; supplement 4.
6. ICH, Q6A: Specifications, test procedures and acceptance criteria for new drug substances and new drug products: chemical substances. In: International Conference on Harmonisation; 2000; Geneva, Switzerland.
7. Rathore A, Winkle H. Nat Biotechnol 2008; 27: 26–34.
8. Lepore J, Spavins J. J Pharm Innov 2008; 3: 79–87.
9. Shukla AA, Hubbard B, Tressel T, Guhan S, Low D. J Chromatogr B 2007; 848: 28–39.
10. Sommerfel S, Strube J. Chem Eng Process 2005; 44: 1123–1137.
11. FDA. Points to consider in the manufacture and testing of monoclonal antibody products for human use. Rockville (MD); 1997. Available at http://www.fda.gov/downloads/BiologicsBloodVaccines/GuidanceComplianceRegulatoryInformation/OtherRecommendationsforManufacturers/UCM153182.pdf
12. Curtis S, Lee K, Blank GS, Brorson K, Xu Y. Biotechnol Bioeng 2003; 84: 179–186.
13. Brorson K, Krejci S, Lee K, Hamilton E, Stein K, Xu Y. Biotechnol Bioeng 2003; 82: 321–329.
14. FDA. Guidance for industry, PAT--a framework for innovative pharmaceutical development, manufacturing, and quality assurance. Rockville (MD); 2004.
15. Read E, Shah RB, Park J, Riley B, Brorson K, Rathore A. Biotechnol Bioeng 2009. DOI 10.1002/bit.22528 and 9.
16. Rao G, Moreira A, Brorson K. Biotechnol Bioeng 2009; 102: 348–56.
17. ICH, Q5A(R1). Viral safety evaluation of biotechnology products derived from cell lines of human or animal origin. In: International Conference on Harmonization; 1998; Geneva Switzerland.
18. Farshid M, Taffs R, Scott R, Asher D, Brorson K. Curr Opin Biotechnol 2005; 16: 1–7.
19. Brorson K. Advances in viral clearance. In: Shukla A, Etzel M, Gadam S, editors. Process scale bioseparations for the biopharmaceutical industry. Boca Raton (FL): CRC Press Taylor & Francis; 2007.
20. Sofer G, Carter J. Viral clearance: a strategy for quality by design and the design space. In: R Mhatre, editors. Quality by design for biopharmaceuticals, a Rathore. Hoboken (NJ): John Wiley & Sons, Inc.; 2009.
21. Strauss DM, Gorrell J, Plancarte M, Blank GS, Chen Q, Yang B. Biotechnol Bioeng 2009; 102: 168–175.
22. Norling L, Lute S, Emery R, Khuu W, Voisard M, Xu Y, Chen Q, Blank G, Brorson K. J Chromatogr A 2004; 1069: 79–89.
23. Khan MA, Agarwal V, Vaithiyalingam SR, Nazzal S, Reddy IK, Sastry SV. Control Rel Bioact Mater 1999; 26: 980–981.
24. Vaithiyalingam SR, Reddy IK, Guven N, Khan MA. Particulate Sci Technol 2001; 19: 1–14.
25. Shah RB, Nutan MT, Reddy IK, Khan MA. Clin Res Reg Aff 2004; 21: 231–238.
26. Valera CR, Chen JW, Xu Y. Biotechnol Bioeng 2003; 84: 714–22.
27. Xu Y, Brorson K. Dev Biol 2003; 113: 89–98.
28. FDA. Submission of quality information for biotechnology products in the office of biotechnology products: notice of pilot program. Fed Regist 2008; 73(128): 37972–37974.
29. FDA. Submission of quality information for biotechnology products in the office of biotechnology products: notice of extension of deadlines to request participation in pilot program and submit applications; and notice of increase in number of original applications in the pilot programs. Fed Regist 2009; 74(179): 47806–47807.

49

REGULATORY REQUIREMENTS, EUROPEAN COMMUNITY

Gary Walsh

Associate Professor, Industrial Biochemistry and Materials Surface Sciences Institute, University of Limerick, Limerick City, Ireland

49.1 THE EUROPEAN UNION

The European Union (EU) now consists of 27 member states, geographically covering an area of over 4 million square kilometers, and supporting a population of 490 million people (1). This represents the world's third largest population base, following on from China and India.

Historically, the foundation of what we now know as the EU can be traced back to the early 1950s, particularly, with the establishment in 1951 of the European Coal and Steel Community (ECSC). The aim of the ECSC was to bring about increased cooperation between European countries in the aftermath of World War II. The six participant states (Belgium, Germany, France, Italy, Luxembourg, and the Netherlands) further developed cooperative measures, forming the European Economic community (EEC) in 1957, with the signing of the treaty of Rome. Cooperation, particularly on economic matters, increased between these countries and, in 1973, the original six members were joined by Ireland, Denmark, and the United Kingdom—and subsequently, by an additional 18 member states over the intervening years.

The modern EU now shares very extensive common policies and harmonized laws pertaining to a whole range of economic and political issues. The major organizational institutions underpinning the EU, include its Parliament and Commission, as well as the Council of Europe. These and other major arms of the EU are summarized in Table 49.1.

Earlier treaties between members largely deferred health care–related issues to individual countries. As a consequence, each member state evolved its own unique set of laws governing pharmaceutical activity, and each member state established its own national regulatory authority to enforce national laws (the national regulatory authorities for the 27 member states are listed in Table 49.2).

While the major principles underpinning national pharmaceutical legislation were substantially similar throughout all the European countries, details did differ from country to country. As a consequence, pharmaceutical companies seeking product marketing authorizations were forced to apply separately to each member state. Uniformity of regulatory response was not guaranteed, and each country enforced its own language requirements, scale of fees, processing times, and so on. Moreover, each country's regulatory authority could show subtly different interpretations of various datasets in drug dossier elements submitted. Overall, this approach, therefore, created enormous duplication of effort, for companies and regulators alike.

In response, the European Commission (the European institution responsible for drafting and proposing new laws, including pharmaceutical laws; Table 49.1) began to introduce Europe-wide pharmaceutical legislations in the mid-1980s. In pursuing this objective, the Commission has at its disposal, two legal instruments; "regulations" and "directives". Upon its approval, a regulation must be enforced immediately and without alteration by all EU member states. A directive, in contrast, is a "softer" legal

Downstream Industrial Biotechnology: Recovery and Purification, First Edition. Edited by Michael C. Flickinger.

TABLE 49.1. Major Institutions of the European Union (EU)

Body	Comment
The European Parliament	Elected every 5 years directly by the people of Europe, the present parliament consists of 785 members drawn from all 27 member states. Along with the Council of Europe, it is responsible for passing EU laws, for approving the EU's €100 billion annual budget and for taking policy decisions
The Council of Europe	Formerly known as the *Council of Ministers*, the Council of Europe shares many legal, policy, and budgetary responsibilities with the Parliament. It also bears main responsibility for EU activities relating to common foreign and security policy. The council does not have a static membership, but consists of government ministers drawn from all member states, and specific meetings are attended by the ministers responsible in their home country for the topics on the agenda of a particular meeting
The European Commission	The EU Commission is responsible for drafting proposals for new EU laws—which it then presents to the parliament and council. It is also responsible for ensuring that the member states are abiding by previously enacted treaties/laws and for managing on the ground, the implementation of EU policies and budgetary spending. The commission consists of 27 members—with one commissioner being nominated by the national government of each member state. They are assisted by some 24,000 civil servants, mainly working in offices in Brussels, Belgium
The Court of Justice	Ensures that EU laws are interpreted and applied in the same way throughout all 27 member states
The Court of Auditors	Audits EU budgetary spending
The European Central Bank	Manages the euro, the EU currency

TABLE 49.2. National Bodies of the EU Member States with Regulatory Roles in the Context of Human Medicines

Member State	Regulatory Body	Contact Details
Austria	Bundesamt für Sicherheit im Gesundheitswesen	www.ages.at
Belgium	Directoraat generaal Geneesmiddelen Direction générale Médicaments	www.afigp.fgov.be
Bulgaria	Bulgarian Drug Agency	www.bda.bg
Czech Repbublic	State Institut for Drug Control	www.sukl.cz
Cyprus	Ministry of Health	www.pio.gov.cy
Denmark	Lægemiddelstyrelsen	www.dkma.dk
Estonia	State Agency of Medicines	www.sam.ee
Finland	Lääkelaitos	www.nam.fi
France	Agence Française de Sécurité Sanitaire des Produits de Santé	www.afssaps.sante.fr
Germany	Bundesministerium für Gesundheit und Soziale Sicherung	www.bmgs.bund.de
	Bundesinstitut für Arzneimittel und Medizinprodukte	www.bfarm.de/de/index.php
	Paul-Ehrlich-Institut	www.pei.de
Greece	National Organisation for Medicines	www.eof.gr
Hungary	National Institute of Pharmacy	www.ogyi.hu
Ireland	Irish Medicines Board	www.imb.ie
Italy	Ministero della Salute	www.ministerosalute.it
Latvia	Food and Veterinary Service	zaale.vza.gov.lv
Lithuania	State Medicines Control Agency	www.vvkt.lt
Luxembourg	Ministère de la Santé Division de la Pharmacie et des Médicaments	www.etat.lu/MS
Malta	Medicines Authority	www.medicinesauthority.gov.mt
Netherlands	Staatstoezicht op de volksgezondheid. Inspectie voor de Gezondheidszorg	www.igz.nl
Poland	Office for Medicinal Products	www.urpl.gov.pl
Portugal	Instituto Nacional da Farmácia e do Medicamento	www.infarmed.pt
Romania	National Medicines Agency	www.anm.ro
Slovak Republic	State Institute for Drug Control	www.sukl.sk
	Institute for State Control of Veterinary Biologicals and Medicaments	
Slovenia	Agency of the Republic of Slovenia for Medicinal Products and Medical Devices	www2.gov.si/mz/mz-splet.nsf
Spain	Agencia española del medicamento	www.agemed.es
Sweden	Läkemedelsverket	www.lakemedelsverket.se
United Kingdom	Medicines and Health care products Regulatory Agency	www.mhra.gov.uk

TABLE 49.3. The Volumes Comprising "the Rules Governing Medicinal Products within the European Union"

Volume	Title
1	Pharmaceutical legislation. Medicinal products for human use
2	Notice to applicants. Medicinal products for human use
3	Guidelines. Medicinal products for human use
4	Good manufacturing practices. Medicinal products for human and veterinary use
5	Pharmaceutical legislation. Veterinary medicinal products
6	Notice to applicants. Veterinary medicinal products
7	Guidelines. Veterinary medicinal products
8	Maximum residue limits. Veterinary medicinal products
9	Pharmacovigilance. Medicinal products for human and veterinary use
10	Clinical trials

instrument, requiring member states only to introduce its "essence" or "spirit" into national law.

By the early 1990s, 8 regulations and 18 directives had been introduced, which effectively harmonized pharmaceutical law throughout the EU. In addition to making available the legislative text, the European Commission has also facilitated the preparation and publication of several adjunct documents designed to assist industry and other interested parties, to interpret and conform to the legislative requirements. Collectively, these documents are known as *the rules governing medicinal products in the European Union* and they make compulsory reading for those involved in any aspect of pharmaceutical regulation. The 10-volume (Table 49.3) publication is regularly updated and hard copies may be purchased from the Commission's publication office (2), or may be consulted/downloaded without cost from the relevant EU web site (3).

Volume 1 of "the rules", for example, provides the full text of all 38 directives and regulations introduced by the EU concerning pharmaceutical activity; and collectively, these form the legal basis upon which pharmaceutical activity is regulated and controlled within the EU. Volume 2 of the series focuses upon providing information most immediately relevant to those ready to prepare an actual application to bring a pharmaceutical product onto the market (i.e. those seeking *marketing authorization* in EU terminology). It is actually presented in three parts. Volume 2A overviews the various pathways by which an application can be made (e.g. a centralized application or an application by an alternative route termed *mutual recognition*—as discussed in a later section), which route is permissible/most appropriate for a particular drug type, and what are the actual procedures involved. Volume 2B concerns itself more directly with the presentation and content of the actual application dossier, while Volume 2C presents guidelines intended to assist the applicant in filling out the various sections of the application dossier.

Volume 3 of the rules governing medicinal products in the EU consists of a collection of guidelines in relation to human medicinal products. Unlike the guidelines included in Volume 2C (discussed above), these are not mechanistic guidelines in terms of completing an application form. They are instead scientific/technical guidelines aimed at assisting the pharmaceutical company interpret, and most appropriately, meet European regulatory requirements in terms of exact scientific data requirements for preclinical and clinical scientific studies, development of appropriate manufacturing procedures, and so on. There are over 300 adopted guidelines currently included in Volume 3 of "the rules", grouped under the subheadings of "quality", "biotechnology", " nonclinical", "clinical efficacy and safety", and "multidisciplinary". There are currently some 74 adopted guidelines under the biotechnology heading, with selected examples being presented in Table 49.4.

Volume 4 of the rules concerns itself with principles underpinning the safe and effective manufacture of pharmaceutical products. It consists of nine main chapters (as outlined in Table 49.5), as well as a range of associated guidelines and annexes. In the context of biotech products, the most significant annex is that of "manufacture of biological medicinal products for human use", a document, which considers various elements of good manufacturing practice (GMP) and quality control (QC) specifically in the context of biological/biotechnological-derived pharmaceutical products. Volumes 5–8 inclusive, relate to veterinary products, while Volume 9 concerns itself with pharmacovigilance in the context of both human (Volume 9A) and veterinary (Volume 9B) products. These provide principles and guidelines for both companies and regulatory authorities in the context of postmarketing surveillance of approved products. Finally, Volume 10 of the rules concerns itself with the establishment and conduct of clinical trials for human medicines. Specifically, it deals with issues surrounding applying to undertake a clinical trial, safety monitoring, and reporting adverse events during the trial as well as the manufacture/import of clinical trial material.

49.2 THE EUROPEAN MEDICINES AGENCY (EMEA)

The harmonization of pharmaceutical legislation throughout the 1980s and early 1990s made possible the development and introduction of an EU-wide system for the authorization and subsequent supervision of medicinal products. The European Medicines Agency (originally known as *the European Medicines Evaluation Agency*, and still abbreviated to EMEA) was set up to coordinate and manage this new EU-wide system (4). Based in Canary

TABLE 49.4. Some of the Guidelines (Biologicals) thus Far Adopted by the EMEA in the Context of Human Medicinal Products

Guideline (Year)	Guideline (Year)	Guideline (Year)
Development and manufacture of lentiviral vectors (2005)	Production and quality control of monoclonal antibodies (1995)	Quality of biotechnological products: stability testing of biotechnological/biological products (1996)
Production and quality control of animal immunoglobulins and immunosera for human use (2002)	Gene therapy product quality aspects in the production of vectors and genetically modified somatic cells (1995)	Potency labeling for insulin analog containing products with particular reference to the use of "International Units" or "Units" (2006)
Points to consider on the manufacture and quality control of human somatic cell therapy medicinal products (2001)	Use of transgenic animals in the manufacture of biological medicinal products for human use (1995)	Guidance on the use of bovine serum in the manufacture of human biological medicinal products (2003)
Quality, preclinical, and clinical aspects of gene transfer medicinal products (2001)	Similar biological medicinal product (2005)	Plasma-derived medicinal products (2001)
Quality of biotechnological products: derivation and characterization of cell substrates used for production of biotechnological/biological products (1998)	Minimizing the risk of transmitting animal spongiform encephalopathy agents via human and veterinary medicinal products (2004)	Quality of biotechnological products: viral safety evaluation of biotechnology products derived from cell lines of human or animal origin (1997)
Position statement on DNA and host cell protein (HCP) impurities, routine testing versus validation studies (1997)	Investigation of manufacturing processes for plasma-derived medicinal products with regard to VCJD risk (2004)	Tests on samples of biological origin (1999)
Production and quality control of medicinal products derived by recombinant DNA technology (1995)	Adjuvants in vaccines for human use (2005)	Production and quality control of cytokine products derived by biotechnological process (1990)
Quality of biotechnological products: analysis of the expression construct in cell lines used for production of rDNA-derived protein products (1996)	Submission of marketing authorization applications for pandemic influenza vaccines through the centralized procedure (2004)	Environmental risk assessments for medicinal products containing or consisting of, genetically modified organisms (GMOs) (2007)

VCJD, Variant Creutzfeldt–Jakob disease; rDNA, recombinant DNA.
The year adopted is provided within brackets.

TABLE 49.5. The Main Chapters Included in Volume 4 of the Rules Governing Medicinal Products Within the European Union

Chapter	Title	Comment
1	Quality management	Focuses upon principles of quality assurance, and therefore, good manufacturing practice (GMP) and QC (quality control) in the context of pharmaceutical manufacture
2	Personnel	Focuses upon personnel requirements and organization to ensure safe and consistent manufacture of quality product
3	Premises and equipment	Focuses upon general principles of design, construction, and maintenance of premises and equipment in the context of pharmaceutical manufacture
4	Documentation	Focuses upon documentation requirements in pharmaceutical manufacturing operations, in particular, documents relating to product specifications, manufacturing, operating procedures, and records
5	Production	Focuses upon production procedures appropriate to pharmaceutical manufacture, e.g. procedures to ensure prevention of cross-contamination in production, validation, starting materials and process operating procedures, rejected, recovered, or returned materials, etc.
6	Quality control	Focuses upon QC procedures appropriate to the industry, e.g. good quality control laboratory practice, including testing, sampling, and specifications
7	Contract manufacture and analysis	Focuses upon principles and procedures relating to contract manufacture givers and acceptors
8	Complaints and product recall	Principles of dealing with complaints and recall of defective products
9	Self-inspection	Focuses upon principles underpinning procedures adopted in order to monitor implementation and compliance with GMP and to propose corrective measures

Wharf, London, the agency became operational in 1995. The overall EMEA mission is to protect and promote public and animal health. Its main aims essentially are to:

- protect and promote public heath by ensuring that human (and veterinary) medicines available within the EU are both safe and effective;
- provide patients with speedy access to new therapeutics;
- facilitate the free movement of pharmaceuticals throughout the EU as a single market;
- ensure that appropriate information is available detailing the correct use of approved medicines; and
- protect consumers and harmonize scientific requirements in the context of drug development.

The EMEA pursues these aims in a number of ways, including:

- organizing and coordinating the speedy evaluation of new drug applications for marketing within the EU;
- organizing and coordinating the appropriate surveillance and if necessary the suspension or withdrawal of drugs from the EU market;
- providing national regulatory authorities/additional bodies of member states or the EU itself with scientific advice on drugs;
- providing advice to companies developing drugs; and
- creating databases and other appropriate provisions to facilitate the availability of authoritative and appropriate drug information for the benefit of all interested parties (e.g. patients, health care professionals, industry, etc.).

Day-to-day management of the EMEA is undertaken by its executive director, who operates under the supervision of a management board. The management board consists of two representatives from each member state, as well as two representatives from the European Commission and two from the European Parliament. A simplified structural overview of the organization is provided in Fig. 49.1. A more detailed description is available on the EMEA homepage (4). In the context of biotech drugs, the most significant organizational structures are the unit for preauthorization evaluation of medicines for human use and the unit for postauthorization evaluation of medicines for human use

49.2.1 The CHMP, CVMP, and COMP

At the heart of the functioning of the EMEA are three key scientific committees:

- The Committee for Medicinal Products for Human Use (CHMP)
- The Committee for Medicinal Products for Veterinary Use (CVMP)
- The Committee on Orphan Medicinal Products (COMP).

Each of the three committees are composed of a number of technical experts, the majority of whom are drawn from the national drug regulatory authorities of each EU member state. In addition, the EMEA has at its disposal a bank of some 3500 European technical experts (the majority of whom, again, are drawn from the national regulatory and related authorities). The EMEA draws upon the advice of this expert panel as required.

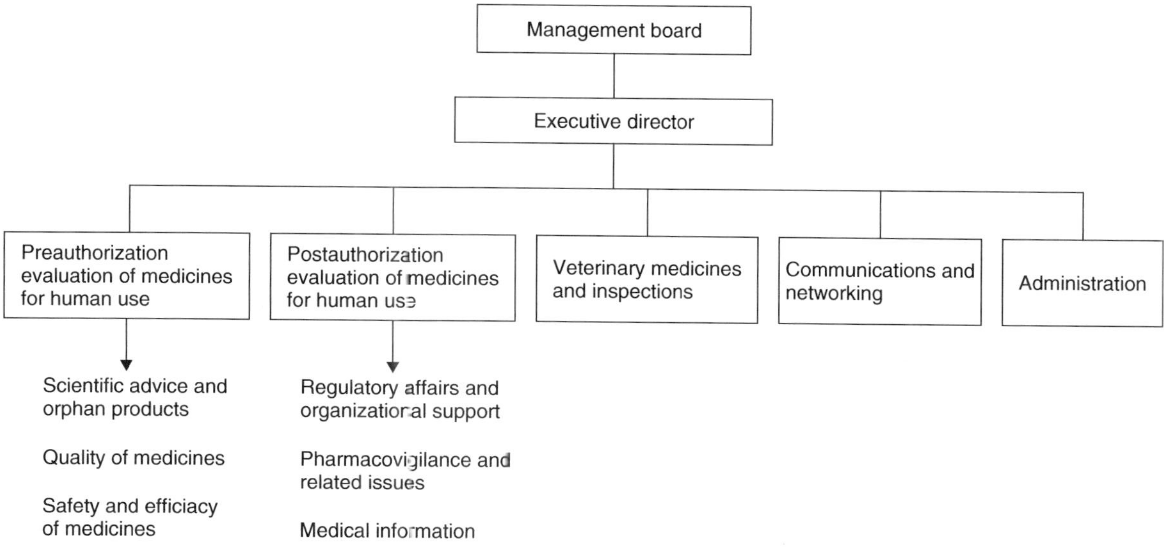

Figure 49.1. Simplified structural overview of the EMEA, with emphasis upon the organizational elements of most relevance to biotechnology drugs.

The CHMP is responsible for considering, formulating, and preparing the EMEA's opinion on all questions relating to drugs for human use. As described in the next section, this committee plays a central role in scientifically assessing new drug application dossiers, including those for biotech drugs. It also plays an important role in drug pharmacovigilance. Ongoing monitoring of the safety of drugs already in the market is largely conducted through a network of member state national drug regulatory authorities. However, the CHMP closely monitors all adverse drug reaction (ADR) reports and, if they consider it necessary, recommend alteration of the conditions of marketing authorization, or indeed full market withdrawal of a problem product. The CHMP (often via subcommittee working parties), also:

- provides assistance to companies developing new drugs
- prepares scientific/regulatory guidelines for the pharmaceutical industry
- cooperates with appropriate international bodies on the harmonization of drug regulatory requirements.

The CHMP, in addition, runs various working parties with expertise in a particular scientific field, which it consults as necessary. The working parties are generally composed of members selected from the list of European experts maintained by the EMEA. A list of the main CHMP working parties is provided in Table 49.6. The main tasks undertaken by such working parties generally

TABLE 49.6. Current Working Parties of the Committee for Medicinal Products for Human Use (CHMP)

Working Party	Comment
Biologics Working Party	Provides recommendations to the EMEA on all matters relating to the quality and safety of biological and biotechnological medicines. Examples of tasks undertaken include the provision of support to the CHMP in evaluating dossiers, helping to ensure consistency in product assessments
Blood Products Working Party	Provides EMEA with recommendations on all issues relating to the safety and efficacy of blood products
Cell-based Products Working Party	Charged with providing recommendations to the EMEA on all issues relating to cell-based products, e.g. preparation of technical requirements for tissue engineering products and supporting dossier evaluation for all new cell-based marketing authorization applications
Efficacy Working Party	Charged with the provision of recommendations to the CHMP on issues relating to clinical aspects of drug development, e.g. preparation and updating guidelines on methodology and interpretation of clinical trials
Gene Therapy Working Party	Provides recommendations to the CHMP on all issues pertaining to gene therapy, e.g. preparing and updating appropriate guidelines and provision to the CHMP of advice on product-specific issues as they arise
Joint CHMP/CVMP Quality Working Party	Provides recommendations to the CHMP and CVMP on issues relating to the quality of human or veterinary medicines, e.g. preparation and update of quality guidelines and provision of advice on product-specific matters
Patients and Consumers Working Party	Works with patients and consumer organizations and provides recommendations to the EMEA on all issues of interest to patients in relation to medicines
Pharmacogenetics Working Party	Provides recommendations on all issues relating to pharmacogenetics, e.g. preparation/update of guidelines, supporting dossier evaluation, provision of advice on pharmacogenetics-related issues, and hosting workshops on the subject
Pharmacovigilance Working Party	Provides recommendations on pharmacovigilance related issues, e.g. provision of advice on product safety and on investigation of adverse events associated with approved drugs
Safety Working Party	Provides recommendations on all issues relating to nonclinical aspects of drug safety, e.g. supporting dossier evaluation and assessing nonclinical safety findings
Scientific Advice Working Party	Provides scientific advice and protocol assistance to companies/sponsors developing human medicines as well as to the World Health Organization and pertinent research institutes
Vaccine Working Party	Provides recommendations on all vaccine-related matters, e.g. preparation and update of appropriate guidelines, supporting dossier evaluation, and provision of scientific advice on both general and product-specific vaccine issues

relates to the provision of scientific advice to the CHMP, undertaking selected tasks associated with the evaluation of a new drug marketing application and drafting/revising scientific guidance documents that fall within their area of expertise. In addition to the main working parties, the CHMP can from time to time form temporary working parties and scientific advisory groups, to assist it with particular technical issues.

The CVMP is organized in a manner, and undertakes tasks substantially similar, to the CHMP, but in the context of veterinary medicines.

The COMP bears responsibility for reviewing applications for drug "orphan" status. An orphan product is one that is intended to diagnose, prevent, or treat serious medical conditions that effect no more that 5 in every 10,000 persons within the EU. As such products will have a small "customer" base, orphan status (if granted) can provide some financial incentive to companies/individuals to develop these products. However, even if orphan status is granted, any such product will still need to undergo standard EMEA evaluation for safety, quality, and efficacy in order to gain an actual marketing licence. Representative biotech drugs holding orphan status within the EU are provided in Table 49.7.

49.3 NEW DRUG APPROVAL ROUTES COORDINATED BY THE EMEA

The rules governing medicinal products in the EU provide for two independent routes by which new potential medicines may be evaluated. These are termed the *centralized* and *decentralized* procedures respectively, and the EMEA plays the major role in both (4). The centralized procedure is compulsory for biotech medicines and may also be used to evaluate new chemical entities.

49.3.1 The Centralized Drug Approvals Procedure

Under the centralized route, marketing authorization applications (dossiers) for the biotech drug are submitted directly to the EMEA. Upon receipt of the application, EMEA staff first validate the dossier, ensuring that all necessary information is present and presented in the

TABLE 49.7. Representative Biotech Drugs Holding Orphan Status within the EU

Drug	Target Indication	Status
Laronidase	Mucopolysaccharidosis type I	Orphan status granted in 2001, approved for general medical use (under trade name Aldurazyme) in 2003
Pegvisomant	Acromegaly	Orphan status granted in 2001, approved for general medical use (under trade name Somavert) in 2002
Human acid alpha-glucosidase	Pompe disease	Orphan status granted in 2001, approved for general medical use (under trade name Myozyme) in 2006
Alpha galactosidase A	Fabry disease	Orphan status granted in 2000, approved for general medical use (under trade name Fabrazyme) in 2001
Pegylated arginine deiminase	Hepatocellular carcinoma	Orphan status granted in 2005
Human acid sphingomyelinase	Niemann–Pick disease	Orphan status granted in 2001
Human alpha mannosidase	Alpha mannosidosis	Orphan status granted in 2005
Human arylsulfatase A	Metachromatic leukodystrophy	Orphan status granted in 2003
Human monoclonal antibody against HLA-DR	Hodgkin's lymphoma	Orphan status granted in 2005
Adeno associated viral vector containing the human calpain 3 gene	Calpinopathy	Orphan status granted in 2006
Adenoviral vector containing the human p53 gene	Li Fraumeni syndrome	Orphan status granted in 2006
Bacterial lipase	Malabsorption due to exocrine pancreatic enzyme insufficiency	Orphan status granted in 2005
Eptacog alpha	Diffuse alveolar hemorrhage	Orphan status granted in 2005
Human monoclonal antibody against CD4	Cutaneous T-cell lymphoma	Orphan status granted in 2001
Adenoviral associated viral vector containing human alpha-glucosidase gene	Pompe disease	Orphan status granted in 2007
Somatropin	AIDS wasting	Orphan status granted in 2000

A full list may be viewed at: http://ec.europa.eu/enterprise/pharmaceuticals/index_en.htm

correct format (as laid down by e.g. Volume 2 of the rules governing medicinal products within the EU; Table 49.3). The validation phase is usually completed within 14 working days. Biotech-based dossiers are termed *Part A applications*, whereas new chemical entities are termed *Part B applications*.

The validated application is subsequently presented at the next monthly meeting of the CHMP. The committee appoints two of its members to act as a "rapporteur" and "corapporteur" for the application. The rapporteurs organize and coordinate technical assessment of the application. Personnel within their home national regulatory agency often play a prominent role in conducting the actual evaluation in conjunction with other appropriate personnel as necessary (e.g. input can be obtained from an appropriate working party or various European experts). Upon completion of the evaluation phase, the rapporteurs draw up a draft report, which they present at the next available meeting to the full CHMP.

The CHMP then issues a scientific opinion on the application, recommending either acceptance or rejection. The EMEA subsequently transmits this scientific opinion to the European Commission. It is the Commission rather than the EMEA that possesses the legal authority to grant marketing authorization. During this final decision-making phase, the commission checks that the marketing authorization complies with all EU laws, and provides a "last chance" for member state submissions regarding any aspect of product evaluation. The commission assigns a community registration number to the product and then issues a final decision on the product (Fig. 49.2), which is binding on all member states. The commission then officially notifies all member states and the marketing authorization applicants of their

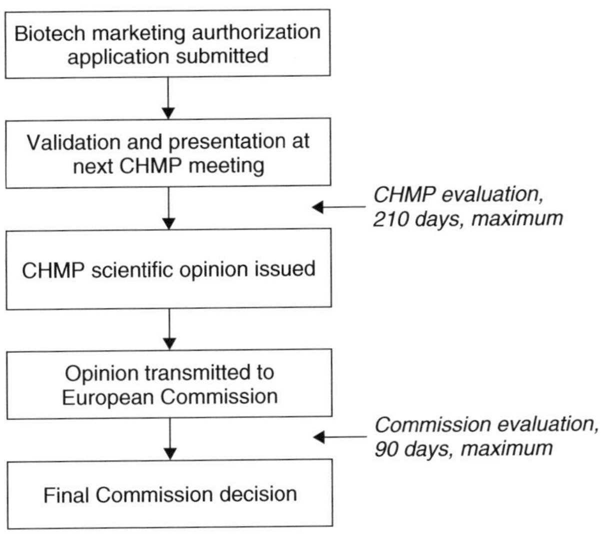

Figure 49.2. Overview of the centralized procedure as applied to biotech (and other appropriate) drug products within the EU. Refer to text for details.

decision, and publishes it in the official journal of the European communities.

Regulatory evaluation of marketing authorization applications must be completed within specific time limits. The EMEA is given 210 days in which to evaluate an application and provide a scientific opinion. If, however, during this process the EMEA requires/seeks further information/clarification on any aspect of the application, the 210-day "clock" stops, only resuming when the sponsoring company provides full and satisfactory responses. The average duration of active EMEA evaluation of biotech-based product applications is in the region of 175 days, comfortably within the 210-day time frame. Clock stop durations will vary from product to product—with durations ranging from 0 days to over 300 days having being recorded thus far. Typically, the "average" application will incur clock stops of the order of 30–80 days.

Upon receipt of the EMEA opinion, the Commission is given a maximum of 90 days in which to translate this opinion into a final decision. Overall therefore, the centralized process is designed to be completed within a maximum of 300 "active" evaluation days, and generally is completed well within this time limit. Biotech products thus far approved for general medical use within the EU under this centralized procedure are summarized in Table 49.8.

Marketing authorizations granted via the centralized procedure are valid throughout the EU for 5 years, and applications for extensions must be made to the EMEA at least 3 months before the authorization is due to expire. Much of the European Commission's activities in the sphere of pharmaceuticals can be viewed on its pharmaceuticals unit website (5), including access to the community register of approved and orphan products, as well as EU pharmaceutical legislation, guidelines and notice to applicants, policy documents, and so on.

49.3.2 Drug Approval within the EU via the Mutual Recognition Procedure

The second route facilitating EU-wide pharmaceutical product authorization is termed *mutual recognition* or the *decentralized procedure*. This is open to nonbiotechnology products and therefore, is of less significance than the centralized procedure, in the context of this specific chapter. In summary, the mutual recognition procedure entails the submission of an authorization application to the national regulatory authority of one member state. This agency then assesses the application and (within a 210-day time frame), formulates an opinion either granting or rejecting the application. If authorization is granted, the company can then apply via "mutual recognition" to extend the market authorization to some or all of the remaining EU states. The extension of market authorization within these remaining countries should follow automatically, as the authorization

TABLE 49.8. Biopharmaceuticals Approved in the European Union via the Centralized Procedure (1995–2006 Inclusive)

Product Trade Name, Nonproprietary Name and Description	Marketing Authorization Holder and (Year of Approval)	Therapeutic Indication
Recombinant blood-related products		
Atryn: antithrombin alpha, produced in the milk of transgenic goats	Leopharma (2006)	Hereditary antithrombin deficiency
Advate: octocog alpha, r human factor VIII produced in CHO cell line	Baxter AG (2004)	Hemophilia
Xigris: drotrecogin alpha, rh activated protein C produced in a mammalian cell line	Eli Lilly (2002)	Severe sepsis
Metalyse: tenecteplase, modified rtPA produced in CHO cells	Boehringer Ingelheim (2001)	Myocardial infarction
Helixate NexGen: octocog alpha, rhFactor VIII produced in BHK cells	Bayer (2000)	Hemophilia A
Kogenate: octocog alpha, rhFactor VIII produced in BHK cells	Bayer (2000)	Hemophilia A
ReFacto: moroctocog alpha, a B-domain deleted rhFactor VIII produced in CHO cells	Genetics Institute (1999)	Hemophilia A
Benefix: nonacog alpha, rhFactor IX produced in CHO cells	Genetics Institute (1997)	Hemophilia B
Refludan: lepirudin, recombinant hirudin produced in *Saccharomyces cerevisiae*	Pharmion (1997)	Anticoagulation therapy for heparin-associated thrombocytopenia
Revasc: desirudin, recombinant hirudin produced in *S. cerevisiae*	Canyon Pharmaceuticals (1997)	Prevention of venous thrombosis
Ecokinase: reteplase, rtPA. Differs from human tPA in that three of its five domains have been deleted. Produced in *Escherichia coli*	Galenus Mannheim (1996)	Acute myocardial infarction
Rapilysin: reteplase, rtPA. See Ecokinase entry above	Boehringer Manheim (1996)	Acute myocardial infarction
NovoSeven: eptacog alpha, rhFactor VIIa produced in BHK cells	Novo-Nordisk (1995)	Some forms of hemophilia
Hormones and related		
Valtropin: somatropin, rhGH produced in *S. cerevisiae*	Biopartners (2006)	Growth failure/GH deficiency
Preotach: parathyroid hormone, rh form of the hormone, produced in *E. coli*	Nycomed (2006)	Osteoporosis
Omnitrope: somatropin, rhGH, produced in *E. coli*	Sandoz (2006)	GH deficinecy/growth failure
Exubera: insulin, human, rhInsulin, produced in *E. coli*	Pfizer (2006)	Diabetes
Kepivance: palifermin, an rKGF produced in *E. coli*	Amgen (2005)	Treatment of severe oral mucositis in selected patients with hematologic malignancies
Apidra: insulin glulisine, rapid acting insulin analogue produced in *E. coli*	Aventis (2004)	Diabetes mellitus
Levemir: insulin detemir, long acting rhInsulin analogue produced in *S. cerevisiae*	Novo Nordisk (2004)	Diabetes mellitus
Forsteo: teriparatide, r shortened form of human parathyroid hormone, produced in *E. coli*	Eli Lilly (2003)	Treatment of osteoporosis in selected postmenopausal women
Somavert: pegvisomant, r engineered hGH analogue (antagonist), produced in *E. Coli*	Pharmacia Enterprises SA (2002)	Treatment of selected patients suffering from acromegaly
Actrapid/Velosulin/Monotard/Insulatard/Protaphane/Mixtard/Actraphane /Ultratard: all contain rhInsulin produced in *S. cerevisiae* formulated as short/intermediate/long acting product	Novo Nordisk (2002)	Diabetes mellitus

(*continued*)

TABLE 49.8. ***(Continued)***

Product Trade Name, Nonproprietary Name and Description	Marketing Authorization Holder and (Year of Approval)	Therapeutic Indication
Nutropin AQ: somatropin, rhGH produced in *E. coli*	IPSEN Ltd (2001)	Growth failure, Turners syndrome
Ovitrelle also termed *Ovidrelle*: choriogonadotropin-alpha, rhCG produced in CHO cells	Serono (2001)	Used in selected assisted reproductive techniques
Novomix 30: contains insulin aspart, short-acting rhInsulin analogue (see NovoRapid entry below) as one ingredient	Novo Nordisk (2000)	Diabetes mellitus
NovoRapid: insulin aspart, short-acting rhInsulin analogue	Novo Nordisk (1999)	Diabetes mellitus
Lantus: insulin glargine, long acting rhInsulin analogue produced in *E. coli*	Aventis Pharma (2000)	Diabetes mellitus
Optisulin: insulin glargine, long acting rhInsulin analogue produced in *E coli*, see also Lantus entry above	Aventis Pharma (2000)	Diabetes mellitus
Thyrogen: thyrotrophin-alpha, rhTSH produced in CHO cells	Genzyme (2000)	Detection/treatment of thyroid cancer
Luveris: lutropin alpha; rhLH produced in CHO cells	Ares-Serono (2000)	Some forms of infertility
Regranex: becaplermin, rhPDGF produced in *S. cerevisiae*	Janssen-Cilag (1999)	Lower extremity diabetic neuropathic ulcers
Forcaltonin: calcitonin, salmon. Recombinant form produced in *E. coli*	Unigene (1999)	Paget's disease
Insuman: rhInsulin produced in *E. Coli*	Hoechst AG (1997)	Diabetes mellitus
Liprolog: bio lysprol, a short-acting insulin analogue produced in *E. coli*	Eli Lilly (1997)	Diabetes mellitus
Humalog: insulin lispro, an insulin analogue produced in *E. coli*	Eli Lilly (1996)	Diabetes mellitus
Puregon: follitropin-beta, rhFSH produced in CHO cells	N.V. Organon (1996)	Anovulation and superovulation
Gonal F : follitropin-alpha, rhFSH produced in CHO cells	Serono (1995)	Anovulation and superovulation
Cytokines and related		
Neulasta: pegfilgrastim, r pegylated CSF. Also marketed as **Neupopeg**	Amgen (2002)	Neutropenia
Nespo: darbepoetin-alpha. See also Aranesp entry below; long-acting rEPO analogue produced in CHO cells	Dompe Biotec (2001)	Treatment of anemia
Aranesp: darbepoetin-alpha, long-acting rEPO analogue produced in CHO cells	Amgen (2001)	Treatment of anemia
Enbrel: etanercept, rTNFR-IgG fragment fusion protein produced in CHO cells	Wyeth Europa (2000)	Rheumatoid arthritis
Intron A: interferon alpha 2b, rIFN-α2b produced in *E. coli*	Schering Plough (2000)	Cancer, genital warts, hepatitis
PegIntron A: interferon alpha 2b, PEGylated rIFN-α-2b produced in *E. coli*	Schering Plough (2000)	Chronic hepatitis C
Viraferon: interferon alpha 2b, rIFN-α-2b produced in *E. coli*	Schering Plough (2000)	Chronic hepatitis B and C
ViraferonPeg: interferon alpha 2b, PEGylated rIFN-α-2b produced in *E. coli*	Schering Plough (2000)	Chronic hepatitis C
Infergen: interferon alfacon 1, rIFN-α, synthetic type I interferon produced in *E. coli*	Yamanouchi Europe (1999)	Chronic hepatitis C
Beromun: tasonermin, rhTNF-α, produced in *E. coli*	Boehringer-Ingelheim (1999)	Adjunct to surgery for subsequent tumor removal, to prevent or delay amputation
Rebif : interferon beta 1a, rh IFN-β-1a, produced in CHO cells	Ares Serono (1998)	Relapsing/remitting multiple sclerosis

(continued)

TABLE 49.8. (*Continued*)

Product Trade Name, Nonproprietary Name and Description	Marketing Authorization Holder and (Year of Approval)	Therapeutic Indication
Avonex: interferon beta 1a, rhIFN-β-1a, produced in CHO cells	Biogen (1997)	Relapsing multiple sclerosis
Neorecormon: epoetin beta, rhEPO produced in CHO cells	Boehringer-Mannheim (1997)	Treatment of anemia
Betaferon: interferon beta 1b, rIFN-β-1b, differs from human protein in that Cys 17 is replaced by Ser. Produced in *E. coli*	Schering AG (1995)	Multiple sclerosis
Vaccines		
Gardasil/Silgard: human papillomavirus vaccine, type 6,11,16,18, recombinant, produced in *S. cerevisiae*	Sanofi Pasteur/ Merch Sharp and Dome (2006)	Vaccine against cervical cancer and related conditions caused by HPV
Dukoral: *Vibrio cholerae* and recombinant cholera toxin B subunit	SBL Vaccin AB (2004)	Active immunization against disease caused by *Vibrio cholerae* subgroup O 1
Ambirix: combination vaccine, containing rHBsAg produced in *S. cerevisiae* as one component)	Glaxo SmithKline (2002)	Immunization against hepatitis A and B
HBVAXPRO: rHBsAg produced in *S. cerevisiae*	Aventis Pharma (2001)	Immunization of children and adolescents against hepatitis B
Infanrix-hexa: combination vaccine, containing rHBsAg produced in *S. cerevisiae* as one component	Smithkline Beecham (2000)	Immunization against diphtheria, tetanus, pertussis, polio, *Haemophilus influenzae* b, and hepatitis B
Infanrix-penta: combination vaccine, containing rHBsAg produced in *S. cerevisiae* as one component)	Smithkline Beecham (2000)	Immunization against diphtheria, tetanus, pertussis, polio, and hepatitis B
Hexavac: combination vaccine, containing rHBsAg produced in *S.cerevisiae* as one component)	Aventis Pasteur (2000)	Immunization against diphtheria, tetanus, pertussis, Hepatitis B, polio, and *Haemophilus influenzae* type b
Hepacare: r S, pre-S, and pre-S2 hepatitis B surface antigens, produced in a mammalian (murine) cell line	Medeva Pharma (2000)	Immunization against hepatitis B
Procomvax: combination vaccine, containing rHBsAg as one component	Pasteur Merieux MSD (1999)	Immunization against *Haemophilus influenzae* type B and hepatitis B
Triacelluvax: combination vaccine containing r(modified) pertussis toxin	Chiron SpA (1999)	Immunization against diphtheria, tetanus, and pertussis
Primavax: combination vaccine, containing rHBsAg produced in *S. cerevisiae* as one component)	Pasteur Merieux MSD (1998)	Immunization against diphtheria, tetanus, and hepatitis B
Infanrix-Hep B, combination vaccine, containing rHBsAg produced in *S. cerevisiae* as one component	Smithkline Beecham (1997)	Immunization against diphtheria, tetanus, pertussis, and hepatitis B
Twinrix: adult and pediatric forms available, combination vaccine containing rHBsAg produced in *S. cerevisiae* as one component	Smithkline Beecham (1996)	Immunization against hepatitis A and B
Tritanrix-HB: combination vaccine, containing rHBsAg produced in *S. cerevisiae* as one component	Smithkline Beecham (1996)	Vaccination against Hepatitis B, diphtheria, tetanus, and pertussis
Monoclonal antibody-based		
Tysabri: natalizumab, a humanized monoclonal antibody raised against selected leukocyte integrins expressed in a murine myeloma cell line	Biogen Idec/Elan (2006)	Treatment of patients with relapsing forms of multiple sclerosis
Xolair: omalizumab, humanized monoclonal which binds immunoglobulin E at the site of high affinity IgE receptor binding	Novartis (2005)	Treatment of adults/adolescents with moderate to severe persistent asthma
Avastin: bevacizumab, humanized monoclonal raised against vascular endothelial growth factor	Roche (2005)	Carcinoma of the colon or rectum

TABLE 49.8. (*Continued*)

Product Trade Name, Nonproprietary Name and Description	Marketing Authorization Holder and (Year of Approval)	Therapeutic Indication
Raptiva: efalizumab, humanized antibody expressed in CHO cell line. Binds to the LFA-1, which is expressed on all leukocytes	Serono (2004)	Treatment of adult patients with chronic moderate to severe plaque psoriasis
Zevalin: ibritumomab tiuxetan, murine monoclonal antibody, produced in a CHO cell line, targeted against the CD20 antigen	Schering AG (2004)	Non-Hodgkin's lymphoma
Erbitux: cetuximab, chimeric antibody raised against human EGF receptor	Merck (2004)	Treatment of EGF receptor- expressing metastatic colorectal cancer
Humira/Trudexa: adalimumab, r anti-TNF human monoclonal antibody created using phage display technology	Abbott (2003)	Rheumatoid arthritis
Mabcampath: alemtuzumab, a humanized monoclonal antibody directed against CD52 surface antigen of B lymphocytes	Millennium and ILEX (2001)	Chronic lymphocytic leukaemia
Herceptin: trastuzumab, humanized antibody directed against HER 2 (human epidermal growth factor receptor 2)	Roche (2000)	Treatment of metastatic breast cancer if tumor over-expresses HER2 protein
Remicade: infliximab, chimaeric Mab directed against TNF-α	Centocor (1999)	Treatment of Crohn's disease
Synagis: palivizumab, humanized Mab directed against an epitope on the surface of respiratory syncytial virus	Abbott (1999)	Prophylaxis of lower respiratory tract disease caused by respiratory syncytial virus in pediatric patients
Zenapax: daclizumab, humanized Mab directed against the α chain of the IL-2 receptor	Hoffman La Roche (1999)	Prevention of acute kidney transplant rejection
Humaspect : votumumab, human Mab directed against cytokeratin tumor associated antigen	Organon Teknika (1998)	Detection of carcinoma of the colon or rectum
Mabthera: rituximab, chimaeric Mab directed against CD20 surface antigen of B lymphocytes	Hoffmann La Roche (1998)	Non-Hodgkin's lymphoma
Simulect: basiliximab, chimaeric Mab directed against the α chain of the IL-2 receptor	Novartis (1998)	Prophylaxis of acute organ rejection in allogeneic renal transplantation
LeukoScan: sulesomab, murine Mab fragment directed against NCA 90, a surface granulocyte nonspecific cross-reacting antigen	Immunomedics (1997)	Diagnostic imaging for infection/inflammation in bone of patients with osteomyelitis
CEA-scan: arcitumomab, murine Mab fragment, directed against human carcinoembryonic antigen, CEA	Immunomedics (1996)	Detection of recurrent/metastatic colorectal cancer
Indimacis 125: igovomab, murine Mab fragment directed against the tumor associated antigen CA 125	CIS Bio (1996)	Diagnosis of ovarian adenocarcimona
Additional products		
Naglazyme: galsulfase, rh *N*-acetylgalactosamine 4 sulfatase produced in a CHO cell line	BioMarin (2006)	Long-term enzyme replacement therapy in patients suffering from Mucopolysaccharidosis VI
Myozyme: algluosidase alpha, rh acid-α-glucosidase, produced in a CHO cell line	Genzyme (2006)	Pompe disease (glycogen storage disease type II)
Aldurazyme: laronidase, rh alpha-L-iduronidase produced in an engineered CHO cell line	Genzyme (2003)	Long tern enzyme replacement therapy in patients suffering from mucopolysaccharidosis I
Inductos: dibotermin alpha, rBone morphogenic protein-2 produced in CHO cells	Genetics institute BV (2002)	Treatment of acute tibia fractures

(*continued*)

TABLE 49.8. (*Continued*)

Product Trade Name, Nonproprietary Name and Description	Marketing Authorization Holder and (Year of Approval)	Therapeutic Indication
Fabrazyme: agalsidase beta, rh α-galactosidase produced in CHO cells	Genzyme (2001)	Fabry disease (α-galactosidase A deficiency)
Replagal: agalsidase alpha, rh α-galactosidase produced in a continuous human cell line	TKT Europe (2001)	Fabry disease (α-galactosidase A deficiency)
Fasturtec: rasburicase, rurate oxidase produced in *S. cerevisiae*	Sanofi-Synthelabo (2001)	Hyperuricaemia
Osteogenic protein 1: osteogenic protein-1, rh form produced in CHO cells	Howmedica (2001)	Treatment of nonunion of tibia
Cerezyme: imiglucerase, rβ-glucocerebrosidase produced in *E. coli*. Differs from native human enzyme by one amino acid, also has modified oligosaccharide component	Genzyme (1997)	Treatment of Gaucher disease

r, recombinant; rh, recombinant human; CHO, Chinese hamster ovary; BHK, baby hamster kidney; Mab, monoclonal antibody; tPA, tissue plasminogen activator; hGH, human growth hormone; LH, luteinizing hormone; hCG, human chorionic gonadotropin; FSH, follicle stimulating hormone; KGF, keratinocyte growth factor; TSH, thyroid stimulating hormone; CSF, colony stimulating factor; EPO, erythropoietin; IFN, interferon; IL, interleukin; HBsAg, hepatitis B surface antigen; PDGF, platelet-derived growth factor; TNFR, tumor necrosis factor receptor; IgG,immunoglobulin G; IgE, immunoglobulin E; TNF, tumor necrosis factor; HPV, human papillomavirus.

requirements are harmonized throughout the EU. If disputes arise, the EMEA then acts as arbitrator and will form a scientific opinion and transmit it to the European Commission. The Commission will then consider the matter and issue a final binding decision.

49.4 THE EMEA IN THE CONTEXT OF DRUG DEVELOPMENT AND MANUFACTURE

As described earlier, the EMEA plays a central role in the drug approval procedure, as well as in pharmacovigilance. Two additional central elements of any drug's life cycle are the development process and routine manufacture. Regulations and guidelines relating to these issues are included in the rules governing medicinal products within the EU (Volumes 1–4 and 10, see Table 49.3). The EMEA plays a coordinating role for some aspects of these activities and helps ensure harmonization relating to these issues throughout the member states. However, much of the day-to-day activities concerning these issues are undertaken within the national regulatory bodies of individual EU member states.

49.5 CONCLUDING REMARKS

Harmonization of pharmaceutical law and the establishment of the EMEA has, within a decade, streamlined the regulatory process for biotech (and other) drugs within Europe. Thus far, well over 80 biotech drugs have gained approval via Europe's centralized process, and the European Commission, in conjunction with the EMEA, is extremely proactive in terms of evolving Europe's pharmaceutical regulatory environment. An example of global leadership in this regard is the development and enactment of a pioneering regulatory framework for biosimiliar medicinal products, and the first two such drugs (Omnitrope and Valtropin; Table 49.8) were approved throughout Europe in 2006. The Commission and the EMEA also continue to play a proactive part in international regulatory harmonization initiatives that should streamline and accelerate the development process and approval of yet more novel biotech drugs.

REFERENCES

1. EU home page. http://www.europa.eu. Accessed 2009.
2. Office for official publications of the European communities. http://www.publications.eu.int. Accessed 2009.
3. EudraLex: the rules governing medicinal products within the European Union. http://ec.europa.eu/enterprise/pharmaceuticals/eudralex/index.htm. Accessed 2009.
4. EMEA homepage. http://www.emea.europa.eu/. Accessed 2009.
5. European Commission's pharmaceuticals unit website. http://ec.europa.eu/enterprise/pharmaceuticals/index_en.htm. Accessed 2009.

INDEX

Downstream Industrial Biotechnology: Recovery and Purification, First Edition. Edited by Michael C. Flickinger.